21. $\int \sqrt{a^2 + x^2}\, dx = \frac{x}{2}\sqrt{a^2 + x^2} + \frac{a^2}{2}\sinh^{-1}\frac{x}{a} + C$

22. $\int x^2\sqrt{a^2 + x^2}\, dx = \frac{x(a^2 + 2x^2)\sqrt{a^2 + x^2}}{8} - \frac{a^4}{8}\sinh^{-1}\frac{x}{a} + C$

23. $\int \frac{\sqrt{a^2 + x^2}}{x}\, dx = \sqrt{a^2 + x^2} - a\sinh^{-1}\left|\frac{a}{x}\right| + C$

24. $\int \frac{\sqrt{a^2 + x^2}}{x^2}\, dx = \sinh^{-1}\frac{x}{a} - \frac{\sqrt{a^2 + x^2}}{x} + C$

25. $\int \frac{x^2}{\sqrt{a^2 + x^2}}\, dx = -\frac{a^2}{2}\sinh^{-1}\frac{x}{a} + \frac{x\sqrt{a^2 + x^2}}{2} + C$

26. $\int \frac{dx}{x\sqrt{a^2 + x^2}} = -\frac{1}{a}\ln\left|\frac{a + \sqrt{a^2 + x^2}}{x}\right| + C$

27. $\int \frac{dx}{x^2\sqrt{a^2 + x^2}} = -\frac{\sqrt{a^2 + x^2}}{a^2 x} + C$
    28. $\int \frac{dx}{\sqrt{a^2 - x^2}} = \sin^{-1}\frac{x}{a} + C$

29. $\int \sqrt{a^2 - x^2}\, dx = \frac{x}{2}\sqrt{a^2 - x^2} + \frac{a^2}{2}\sin^{-1}\frac{x}{a} + C$

30. $\int x^2\sqrt{a^2 - x^2}\, dx = \frac{a^4}{8}\sin^{-1}\frac{x}{a} - \frac{1}{8}x\sqrt{a^2 - x^2}\,(a^2 - 2x^2) + C$

31. $\int \frac{\sqrt{a^2 - x^2}}{x}\, dx = \sqrt{a^2 - x^2} - a\ln\left|\frac{a + \sqrt{a^2 - x^2}}{x}\right| + C$

32. $\int \frac{\sqrt{a^2 - x^2}}{x^2}\, dx = -\sin^{-1}\frac{x}{a} - \frac{\sqrt{a^2 - x^2}}{x} + C$

33. $\int \frac{x^2}{\sqrt{a^2 - x^2}}\, dx = \frac{a^2}{2}\sin^{-1}\frac{x}{a} - \frac{1}{2}x\sqrt{a^2 - x^2} + C$

34. $\int \frac{dx}{x\sqrt{a^2 - x^2}} = -\frac{1}{a}\ln\left|\frac{a + \sqrt{a^2 - x^2}}{x}\right| + C$
    35. $\int \frac{dx}{x^2\sqrt{a^2 - x^2}} = -\frac{\sqrt{a^2 - x^2}}{a^2 x} + C$

36. $\int \frac{dx}{\sqrt{x^2 - a^2}} = \cosh^{-1}\frac{x}{a} + C = \ln\left|x + \sqrt{x^2 - a^2}\right| + C$

37. $\int \sqrt{x^2 - a^2}\, dx = \frac{x}{2}\sqrt{x^2 - a^2} - \frac{a^2}{2}\cosh^{-1}\frac{x}{a} + C$

38. $\int (\sqrt{x^2 - a^2})^n\, dx = \frac{x(\sqrt{x^2 - a^2})^n}{n + 1} - \frac{na^2}{n + 1}\int (\sqrt{x^2 - a^2})^{n-2}\, dx, \quad n \neq -1$

39. $\int \frac{dx}{(\sqrt{x^2 - a^2})^n} = \frac{x(\sqrt{x^2 - a^2})^{2-n}}{(2 - n)a^2} - \frac{n - 3}{(n - 2)a^2}\int \frac{dx}{(\sqrt{x^2 - a^2})^{n-2}}, \quad n \neq 2$

40. $\int x(\sqrt{x^2 - a^2})^n\, dx = \frac{(\sqrt{x^2 - a^2})^{n+2}}{n + 2} + C, \quad n \neq -2$

41. $\int x^2\sqrt{x^2 - a^2}\, dx = \frac{x}{8}(2x^2 - a^2)\sqrt{x^2 - a^2} - \frac{a^4}{8}\cosh^{-1}\frac{x}{a} + C$

42. $\int \frac{\sqrt{x^2 - a^2}}{x}\, dx = \sqrt{x^2 - a^2} - a\sec^{-1}\left|\frac{x}{a}\right| + C$

43. $\int \frac{\sqrt{x^2 - a^2}}{x^2}\, dx = \cosh^{-1}\frac{x}{a} - \frac{\sqrt{x^2 - a^2}}{x} + C$

Continued overleaf.

44. $\displaystyle\int \frac{x^2}{\sqrt{x^2-a^2}}\, dx = \frac{a^2}{2}\cosh^{-1}\frac{x}{a} + \frac{x}{2}\sqrt{x^2-a^2} + C$

45. $\displaystyle\int \frac{dx}{x\sqrt{x^2-a^2}} = \frac{1}{a}\sec^{-1}\left|\frac{x}{a}\right| + C = \frac{1}{a}\cos^{-1}\left|\frac{a}{x}\right| + C$

46. $\displaystyle\int \frac{dx}{x^2\sqrt{x^2-a^2}} = \frac{\sqrt{x^2-a^2}}{a^2 x} + C$ 47. $\displaystyle\int \frac{dx}{\sqrt{2ax-x^2}} = \sin^{-1}\left(\frac{x-a}{a}\right) + C$

48. $\displaystyle\int \sqrt{2ax-x^2}\, dx = \frac{x-a}{2}\sqrt{2ax-x^2} + \frac{a^2}{2}\sin^{-1}\left(\frac{x-a}{a}\right) + C$

49. $\displaystyle\int (\sqrt{2ax-x^2})^n\, dx = \frac{(x-a)(\sqrt{2ax-x^2})^n}{n+1} + \frac{na^2}{n+1}\int (\sqrt{2ax-x^2})^{n-2}\, dx,$

50. $\displaystyle\int \frac{dx}{(\sqrt{2ax-x^2})^n} = \frac{(x-a)(\sqrt{2ax-x^2})^{2-n}}{(n-2)a^2} + \frac{(n-3)}{(n-2)a^2}\int \frac{dx}{(\sqrt{2ax-x^2})^{n-2}}$

51. $\displaystyle\int x\sqrt{2ax-x^2}\, dx = \frac{(x+a)(2x-3a)\sqrt{2ax-x^2}}{6} + \frac{a^3}{2}\sin^{-1}\frac{x-a}{a} + C$

52. $\displaystyle\int \frac{\sqrt{2ax-x^2}}{x}\, dx = \sqrt{2ax-x^2} + a\sin^{-1}\frac{x-a}{a} + C$

53. $\displaystyle\int \frac{\sqrt{2ax-x^2}}{x^2}\, dx = -2\sqrt{\frac{2a-x}{x}} - \sin^{-1}\left(\frac{x-a}{a}\right) + C$

54. $\displaystyle\int \frac{x\, dx}{\sqrt{2ax-x^2}} = a\sin^{-1}\frac{x-a}{a} - \sqrt{2ax-x^2} + C$

55. $\displaystyle\int \frac{dx}{x\sqrt{2ax-x^2}} = -\frac{1}{a}\sqrt{\frac{2a-x}{x}} + C$

56. $\displaystyle\int \sin ax\, dx = -\frac{1}{a}\cos ax + C$ 57. $\displaystyle\int \cos ax\, dx = \frac{1}{a}\sin ax + C$

58. $\displaystyle\int \sin^2 ax\, dx = \frac{x}{2} - \frac{\sin 2ax}{4a} + C$ 59. $\displaystyle\int \cos^2 ax\, dx = \frac{x}{2} + \frac{\sin 2ax}{4a} + C$

60. $\displaystyle\int \sin^n ax\, dx = \frac{-\sin^{n-1} ax \cos ax}{na} + \frac{n-1}{n}\int \sin^{n-2} ax\, dx$

61. $\displaystyle\int \cos^n ax\, dx = \frac{\cos^{n-1} ax \sin ax}{na} + \frac{n-1}{n}\int \cos^{n-2} ax\, dx$

62. (a) $\displaystyle\int \sin ax \cos bx\, dx = -\frac{\cos(a+b)x}{2(a+b)} - \frac{\cos(a-b)x}{2(a-b)} + C, \qquad a^2 \neq b^2$

   (b) $\displaystyle\int \sin ax \sin bx\, dx = \frac{\sin(a-b)x}{2(a-b)} - \frac{\sin(a+b)x}{2(a+b)}, \qquad a^2 \neq b^2$

   (c) $\displaystyle\int \cos ax \cos bx\, dx = \frac{\sin(a-b)x}{2(a-b)} + \frac{\sin(a+b)x}{2(a+b)}, \qquad a^2 \neq b^2$

63. $\displaystyle\int \sin ax \cos ax\, dx = -\frac{\cos 2ax}{4a} + C$

64. $\displaystyle\int \sin^n ax \cos ax\, dx = \frac{\sin^{n+1} ax}{(n+1)a} + C, \qquad n \neq -1$

This table is continued on the endpapers at the back.

7TH EDITION

Calculus and Analytic Geometry

7TH EDITION

Calculus and

GEORGE B. THOMAS, JR.
Massachusetts Institute of Technology

Analytic Geometry

ROSS L. FINNEY

Massachusetts Institute of Technology

ADDISON-WESLEY PUBLISHING COMPANY

Reading, Massachusetts ▪ Menlo Park, California
New York ▪ Don Mills, Ontario ▪ Wokingham, England ▪ Amsterdam ▪ Bonn
Sydney ▪ Singapore ▪ Tokyo ▪ Madrid ▪ Bogotá ▪ Santiago ▪ San Juan

Sponsoring Editor: David Pallai
Developmental Editor: David M. Chelton
Production Supervisor: Marion E. Howe
Copy Editor: Barbara G. Flanagan
Text Designer: Catherine L. Dorin
Layout Artist: Lorraine Hodsdon
Illustrators: C & C Associates
             Dick Morton
Art Consultant: Dick Morton
Manufacturing Supervisor: Roy Logan
Cover: Marshall Henrichs

Library of Congress Cataloging-in-Publication Data

Thomas, George Brinton, 1914-
   Calculus and analytic geometry.

   Includes index.
   1. Calculus.   2. Geometry, Analytic.   I. Finney,
Ross L.   II. Title.
QA303.T42   1987   515'.15   87-14422
ISBN 0-201-16320-9 (set)
ISBN 0-201-16321-7 (v.1)
ISBN 0-201-16322-5 (v.2)

Reprinted with corrections, May 1988

BCDEFGHIJ-DO-898

Preface

In this new edition of *Calculus and Analytic Geometry* we have tried to retain all the qualities and features that users of the previous editions have found most useful. Yet this is one of the most comprehensive revisions of both text and art in the more than thirty-year history of the book. The revision is based on dozens of reviews; numerous conversations with users of earlier editions; letters of advice from friends, students, and instructors from all over the world; and extensive classroom trials of revised lessons. Our overall revision plan has been to make the book more readable and the material more accessible to beginners without compromising the standards or coverage its users want to see.

Audience and Prerequisites   This book provides everything necessary for the standard three-semester or four-quarter calculus sequence in the freshman and sophomore years of college. The prerequisites are the usual exposure to algebra and trigonometry, but to refresh everyone's memory Chapter 1 begins with brief reviews of coordinates, lines, functions, and graphs. There is also a review of trigonometry in Chapter 2. The book is available in a two-volume version: Volume I covers Chapters 1–12 and Volume II covers Chapters 11–20.

Features From Previous Editions

As always, our aim has been to teach the mathematics of calculus and to provide the training readers will need to use calculus effectively in their later academic and professional work. To accomplish this, we have preserved the book's mathematical level, its orientation toward applications, its concentration on worked examples, and the number and variety of the exercises, and continued to show the connections between calculus and some of the numerical methods used in other courses.

Mathematical Level   Although many of the presentations in this new edition are noticeably easier than in earlier editions, the level of rigor is about the same. We try to explain things without belaboring the obvious and without answering questions that readers are not yet ready to ask. For example, we state the max-min theorem for continuous functions on closed intervals and use it to develop the Mean Value Theorem, but we do not prove the max-min theorem or explore the properties of the real number system on which the theorem depends. We give a few simple ϵ-δ proofs about limits in Chapter 1 but put the proofs on the more complicated limit theorems in the appendixes.

Applications Calculus was invented to solve problems in physics and astronomy, and although it has since developed into a far-reaching mathematical discipline in its own right, most of its applications outside of mathematics still involve science and engineering. As in earlier editions, the applications of the present book are directed mainly along these lines. Typical applications include calculating extreme values, centers of mass, work, and hydrostatic force; predicting satellite orbits; and describing fluid flow. (See pp. 192, 345, 356, 363, 827, and 1018.) However, in recent years calculus has become important in many other fields, including economics, business, the life sciences, and even the physics of sports. We have therefore included a variety of applications from these fields as well, on such topics as average daily inventory, birth rates and population growth, and the work required to hit a golf ball or serve a tennis ball, to mention only a few. (See pp. 343, 428, and 361.) Whenever we feel we can do so without burdening the text, we make connections between calculus and real life. We also take more time in this edition with the problem-solving steps in applications that involve mathematical modeling, as in the introduction to related rates (pp. 203–204) and the solution of the hanging cable problem (pp. 577–578).

Worked Examples The exercises are where the readers do the work but the examples are where *we* do the work. We have kept the favorite examples and added a number of fresh ones, often showing solution steps in more detail than before. We have also replaced some of the harder examples with easier ones that make the same points. The topics range from insulating the Trans-Alaska Pipeline to draining and filling a swamp; from household electric power to a method for drawing parabolas; from analyzing the shape of the Gateway Arch to the West in St. Louis to the mysteries of computer calculations and the variation of the temperature below the earth's surface. (See pp. 139, 299, 342, 526, 581, 742, and 864.)

Problem Sets Each problem set contains a mixture that runs from routine mathematical exercises to more challenging problems. In many of the problem sets, the ''ramp up'' is more gradual than before. Nearly every set offers practice in applications and many offer calculator exercises. (See pp. 126–127, 199–203, 299–301.) Each chapter concludes with a section of miscellaneous problems that cover topics in the order in which the mathematics they depend on appears in the chapter. Many of these concluding sections present interesting but seldom-taught applications in which the mathematics is an extension of the material in the chapter. (See pp. 370 and 623.)

Numerical Methods We have retained the discussions of root finding, linear and quadratic approximations of functions, and the numerical approximations of integrals, topics that are becoming increasingly important both in mathematics and in other fields. As in earlier editions, there are occasional calculator exercises and there are references at the ends of many problem sets to microcomputer programs in *The Calculus Toolkit*. However, the text does not require anyone to have experience with or access to either a calculator or a computer.

New Features

In addition to trying to preserve what users have indicated were the best features of earlier editions, we have added a number of new features to meet present classroom needs:

Drawing Lessons Drawing things in three dimensions is often difficult. We have therefore included step-by-step guidelines for drawing planes, cylinders, and quadric surfaces and for making three-dimensional objects look three-dimensional. The surfaces in the drawing lessons and practice exercises are the surfaces readers will work with later in the book when they study multivariable calculus. (See pages 770, 842, and 851.)

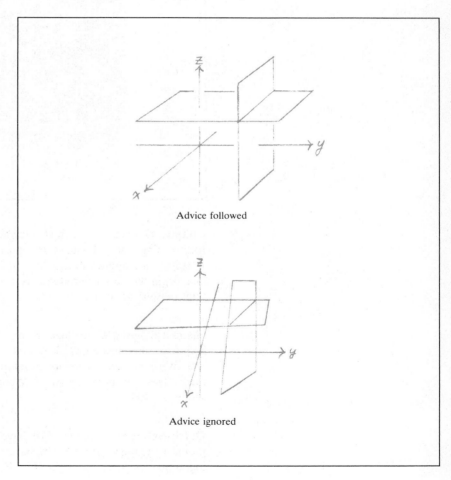

Advice followed

Advice ignored

Enhanced Artwork   We have redrawn many of the older figures and added a number of new ones to make it easier to understand the mathematical arguments and to visualize the curves, surfaces, and solids that arise in the examples and exercises. There are more figures than in previous editions, and the problem sets have more art than before. We have used additional colors to highlight the important parts of many of the three-dimensional figures. (See the figure below, for example.) There are also full-page displays of related figures. (See pp. 261, 265, and 605.)

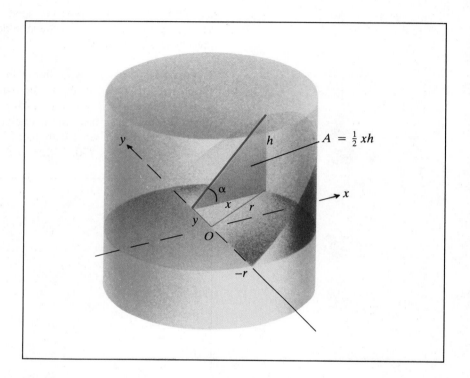

Chapter Overviews   Each chapter begins with an overview that connects the forthcoming material with other topics in the book and describes its importance in theory and applications. (See pp. 236 and 859.) Most articles in the book also begin with brief introductions of their own to set the context for the topics they present. (See pp. 390 and 657.)

Historical Notes   Calculus is the result of centuries of human endeavor, and to help to communicate this idea, we have included marginal notes about people who were involved with its development and about the contributions they made. (See the notes on Maria Agnesi's contributions, p. 556, and Taylor series, p. 704.)

Quick-Reference Charts   We have added quick-reference charts that list groups of related formulas, emphasize procedures, and describe problem-solving strategies. (See pp. 327, 451, 469, and 691.)

Other Changes

There have been significant changes in this edition beyond the new features already mentioned:

Distinction between Core and Optional Topics Calculus books must meet the needs of many different courses simultaneously, with the result that they normally contain more than any individual instructor wants to cover. To help readers distinguish between core topics and topics that may be somewhat tangential, we have marked a number of articles and subsections with open squares (□). The squares identify optional topics not needed for later parts of the book.

Reorganization Differentials are now introduced in Chapter 2 and inverse functions have been moved from Chapter 2 to where they are first needed in Chapter 6. The introduction of the Chain Rule has been separated from the introduction to parametric equations. Long articles have been shortened or divided in two (e.g., the presentation of functions and graphs in Chapter 1 and the presentation of gradients in Chapter 16) and some of the shorter articles have been combined to bring them closer to lesson-length (Rolle's theorem and the Mean Value Theorem now appear in a single lesson). Some of the less frequently covered topics—such as the theorems of Pappus, the substitution $z = \tan (x/2)$, direction cosines, and the division of power series—have been abbreviated, marked as optional, or moved into the problem sets.

Multivariable Calculus The presentation of multivariable calculus has been reorganized and much of it has been rewritten. Quadric surfaces have been removed from the chapter on vectors and placed in a new Chapter 15 with cylinders, cylindrical coordinates, spherical coordinates, and drawing lessons. Chapter 14 on vector functions and motion has been reordered and condensed to simplify the introduction of curvature, torsion, and the **TNB** tangent frame. The old chapter on partial derivatives has been divided into separate chapters, one on theory and the other on applications. The treatment of the vector integral theorems in Chapter 19 is entirely new. It begins with line integrals, vector fields, and Green's theorem in the plane, moves on to surface integrals, the divergence theorem, and Stokes's theorem, and closes with conservative fields and potential functions. We also discuss more of the real-world connections that originally motivated the subject's development.

New Applications We have added dozens of new applications, including the analysis of a truck's motion from its time-distance graph, the calculation of the radial rate of change of a soap bubble, the discussion of the decibel scale of loudness, and the determination of the path of a home run in Boston's Fenway Park. (See pp. 162, 203, 425, and 762.)

Revised Treatments of Standard Topics Among the many topics that now have revised treatments are

Differentials (Article 2.4)

The Chain Rule (Article 2.5)

The Fundamental Theorems of Calculus (Article 4.7)

Volumes of revolution (Articles 5.3–5.4)

Moments and centers of mass (Articles 5.8 and 18.5)

Parametric equations (Articles 2.8 and 8.9)

Sequences, including a subsection on recursion and computer language (Article 11.1)

Convergence tests (Articles 11.5–11.9)

Jacobians, in two optional subsections (Articles 18.3 and 18.6)

Line integrals in vector fields (Article 19.2)

Surface area and surface integrals (Article 19.4)

Path independence and conservative fields (Article 19.7)

Supplements for the Instructor

Solutions Manual   This supplement contains the worked-out solutions for *all* the exercises in the text. Extra care has been taken to ensure the accuracy of this manual.

Complete Answer Book   Contains the answers to all exercises in the text.

Computerized Test Generator (AWTest)   Based on the learning objectives of the text, this easy-to-use algorithm-based system allows the instructor to generate tests or quizzes. Questions are available in open-ended, multiple choice, and true-false formats. AWTest is available for the IBM PC* and is free to adopters. (Questions prepared by Jeffery Cole.)

Printed Test Bank   At least three alternate tests per chapter are included in this valuable supplement. Instructors can use this as a reference for creating tests with or without a computer.

Transparency Masters   Includes a selection of key definitions, theorems, proofs, formulas, tables, and figures that appear in the text.

Computer Supplements for the Instructor and the Student

The Calculus Toolkit   Consisting of twenty-seven programs ranging from functions to vector fields, this software enables the instructor and students to use the microcomputer as an "electronic chalkboard." Three-dimensional graphics are incorporated where appropriate. The Calculus Toolkit is available for both the Apple* and the IBM PC.

Student Edition of MathCAD   This software package is a very powerful free-form scratchpad. When you input equations, MathCAD automatically calculates and displays your results as numbers or graphs. It also allows you to plot the results, annotate your work with text, and print your entire document.

*IBM is a registered trademark. PC is a registered trademark. Apple is a registered trademark.

MathPRO Designed to accompany Chapters 1–12 of the text, this tutorial software package generates drill and practice exercises from four levels of difficulty and provides help screens that are keyed to the text. Available for the IBM PC.

Supplements for the Student

Study Guide By Maurice Weir, Naval Postgraduate School. Organized to correspond with the text, this two-volume workbook in a semi-programmed format increases student proficiency.

Solutions Manual By Alexia Latimer. Available in two volumes, these manuals are designed for the student and contain carefully worked-out solutions to all odd-numbered exercises in the text.

Acknowledgments

We would like to thank and acknowledge the contributions of the participants of the preliminary planning session in New Orleans:

Donald J. Albers
Menlo College

Michael A. Laidacker
Lamar University

Chester I. Palmer
Auburn University

Kirby C. Smith
Texas A&M University

James L. Wayman
Naval Postgraduate School

Maurice D. Weir
Naval Postgraduate School

A great many valuable contributions to the Seventh Edition were made by people who reviewed the manuscript as it developed through its various stages:

M. Kursheed Ali
California State University, Fresno

Robert L. Devaney
Boston University

Sandy Berman
El Cerrito H.S., El Cerrito, CA

John R. Durbin
University of Texas at Austin

Charles H. Bertness
Indiana University of Pennsylvania

David H. Eberly
University of Texas at San Antonio

Michael K. Brozinsky
Queensborough Community College

Bruce H. Edwards
University of Florida

George L. Cain, Jr.
Georgia Institute of Technology

Norbert Ellman
Long Beach City College

Donald R. Cohen
SUNY Agricultural and Technical College at Cobleskill

Elaine P. Genkins
Collegiate School, New York, NY

Larry A. Curnutt
Bellevue Community College

Stuart Goldenberg
California Polytechnic State University, San Luis Obispo

Duane E. Deal
Ball State University

Ralph P. Grimaldi
Rose-Hulman Institute of Technology

F. Lane Hardy
DeKalb Community College

Edward K. Hinson
University of New Hampshire

Dale T. Hoffman
Bellevue Community College

Arnold J. Insel
Illinois State University

Ronald S. Irving
University of Washington

I. Martin Isaacs
University of Wisconsin–Madison

Judy Kasabian
El Camino College

Melvin David Lax
California State University,
Long Beach

David J. Lutzer
Miami University, Oxford

Frank P. May
Evanston Township, H.S.,
Evanston, IL

Mary McCammon
Pennsylvania State University

Charles R. MacCluer
Michigan State University

Maurice L. Monahan
South Dakota State University

Zaven Margosian
Lawrence Institute of Technology

Lois Miller
Golden West College

Kevin T. Phelps
Georgia Institute of Technology

Peter J. Philliou
Northeastern University

Henry C. Pinkham
Columbia University

Monty J. Strauss
Texas Tech University

David F. Ullrich
North Carolina State University

Marjorie Valentine
John Jay H.S., San Antonio, Texas

James K. Washenberger
Virginia Polytechnic and State
University

Peter Westergard
Nickolet H.S., Milwaukee,
Wisconsin

Alan D. Wiederhold
San Jacinto College

George R. Wilkens, Jr.
University of North Carolina

Henry Zatzkis
New Jersey Institute of Technology

We would like to thank the following people for responding (as of this writing) to Addison-Wesley's detailed questionnaire about our revision plans:

Takeo Akasaki
University of California, Irvine

Richard A. Askey
University of Wisconsin–Madison

George L. Baldwin
Texas Tech University

Joyce Becker
Luther College

Jerry Bloomberg
Essex Community College

Megnus V. Braunagle
Regis College

Herb Canage
Alpena Community College

Kirk Cogswell
Northern State College

Leo P. Commerford, Jr.
University of Wisconsin–Parkside

Donald Cook
Albany Junior College

Larry A. Curnutt
Bellevue Community College

Michael B. Dollinger
Pacific Lutheran University

William Dart Dunbar, Jr.
Lansing Community College

Robert Dunkin
Trenton State College

Richard T. Durrett
Cornell University

Elton W. Fors
Northern State College

Thomas A. Fournelle
University of Wisconsin–Parkside

James F. Glazebrook
Rutgers University

Ronald C. Grimmer
Southern Illinois University

William E. Haigh
Northern State College

Earle Hamilton
North Seattle Community College

John O. Herzog
Pacific Lutheran University

Dale T. Hoffman
Bellevue Community College

Kendell H. Hyde
Weber State College

Franklin T. Iha
Leeward Oahu Community College

Geoffrey Jones
El Camino College

Howard J. Jones
Lansing Community College

Leonard F. Klosinski
University of Santa Clara

Roger G. Lautzenheiser
Rose-Hullman Institute of
Technology

Youn W. Lee
University of Wisconsin–Parkside

Kui Li
New Jersey Institute of Technology

Jerry L. Muhasky
University of Washington

A. J. Murray
Carroll College

Michael Ortell
Orange Coast College

Romaiah Devadoss Pandian
North Central College

F. J. Papp
University of Michigan, Dearborn

Charles Pondraza
Northeastern Junior College

Barbara Poole
North Seattle Community College

Kenneth P. Rietz
Asbury College

Judith A. Sanderman
Shoreline Community College

Margaret S. Schoenfelt
County College of Morris

Lawrence E. Schovanec
Texas Tech University

Nedra Shunk
Santa Clara University

Mark D. Smiley
Auburn University at Montgomery

Elaine Sperelove
Brigham Young University

David Stacy
Bellevue Community College

William Stegner
Essex Community College

G. Wayne Sullivan
Volunteer State Community College

Paul B. Venzke
Minot State College

Dwann Veroda
El Camino College

Russell Walker
Carnegie-Mellon University

James S. Wolper
Hamilton College

Douglas Willett
University of Utah

Paul W. Wilson
Lake Superior State College

Paul Wozniak
El Camino College

Arthur Yanushka
Christian Brothers College

Chang Li Yiu
Pacific Lutheran University

Edward Zeidman
Essex Community College

Frank Zizza
University of Washington

We also want to thank the following people for their advice and suggestions in correspondence and conversation:

Richard A. Askey, University of Wisconsin–Madison; Charles W. Austin, California State University, Long Beach; Thomas L. Drucker, University of Wisconsin–Madison, Division of University Outreach; Nathaniel S. Finney, University of Illinois at Urbana-Champaign; Frank R. Giordano, U.S. Military Academy; Adam Grossman, M.I.T.; Richard W. Hamming, Naval Postgraduate School; Edward Hill, Dictionary of Regional English, University of Wisconsin–Madison; Dale T. Hoffman, Bellevue Community College; John P. Hoyt, Lancaster, Pa; John W. Kenelly, Clemson University; Margaret M. LaSalle, University of Southwestern Louisiana; Kathryn Lesh, M.I.T.; Katherine L. Pedersen, Southern Illinois University; Delbert F. Penhall, Long Beach City College; Arthur C. Segal, University of Alabama in Birmingham; Roderick S. Smith, University of Wisconsin–Madison; Neil Stahl, University of Wisconsin–Fox Valley; Norton Starr, Amherst College; Svend Uldall-Hansen, Odense Teknikum, Denmark; Maurice D. Weir, Naval Postgraduate School; Carroll O. Wilde, Naval Postgraduate School; Joseph Wolfson, Georgetown Day High School.

We would like to express our gratitude to Arthur P. Mattuck, whose teaching ideas and supplementary classroom notes for the calculus courses at M.I.T. have influenced the development of this book through its most recent editions. The presentation of partial derivatives of functions of constrained variables in Article 16.5 and the presentation of sensitivity analysis in Article 17.2 are based on his writing.

We would also like to thank our colleagues Francis B. Hildebrand, Frank Morgan, and Alar Toomre for their many thoughtful suggestions.

We want to thank Kenneth R. Manning of M.I.T. for his expertise in researching and writing the new historical notes in this edition.

We would like to express our thanks for many contributions of our colleagues and students in M.I.T.'s Concourse Program.

The computer-generated color images of surfaces in Chapter 16 were produced by Thomas Banchoff and Edward Chang at Brown University Computer Graphics Laboratory. The associated black-and-white images were produced by Thomas Banchoff, Robert Shapire, Cathleen Curry, and Richard Schwartz.

We thank Gary K. Rockswold of Mankato State University for reading the galleys to check the accuracy of the text and examples.

We thank the staff of the Lincoln Public Library, Lincoln, Mass., for helping us track down applications.

It is a pleasure to thank the staff of Addison-Wesley Publishing Company for the expertise and energy they have brought to the development of this new edition.

Dick Morton pioneered the use of air-brushed art in calculus texts with the early editions of this book and it is a special pleasure to be able to call attention to his work and to acknowledge the contribution that it has made to the present edition.

We also wish to make special mention of David Chelton's contribution. He was the developmental editor for the present edition and many of its new features are the direct result of his encouragement, thoughtfulness, and hard work.

The manuscript for this edition was keyed by Maureen Emberley. We appreciate her many cheerful and timely responses to our production schedule.

We also wish to express our thanks to the many others whose names we have not been able to mention. Any errors that may appear are the responsibility of the authors. We will appreciate having these brought to our attention.

Magnolia, Mass.   G.B.T., Jr.
Lincoln, Mass.   R.L.F.

Contents

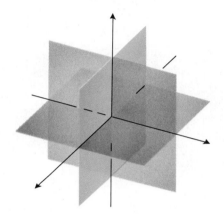

CHAPTER 3

Applications of Derivatives

CHAPTER 4

Integration

CHAPTER 5

Applications of Definite Integrals

CHAPTER 17

Applications of Partial Derivatives

CHAPTER 18

Multiple Integrals

CHAPTER 19

Vector Fields and Integration

CHAPTER 20

Differential Equations

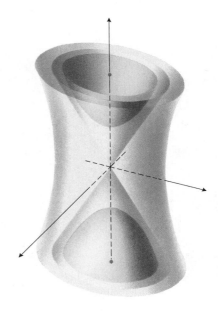

Appendixes

Prologue: What is Calculus?

Calculus is the mathematics of motion and change. Where there is motion or growth, where variable forces are at work producing acceleration, calculus is the right mathematics to apply. This was true in the beginnings of the subject, and it is true today.

Calculus was first created to meet the mathematical needs of the scientists of the seventeenth century. Differential calculus dealt with the problem of calculating rates of change. It enabled people to define slopes of curves, to calculate the velocities and accelerations of moving bodies, to find the firing angle that gave a cannon its greatest range, and to predict the times when planets would be closest together or farthest apart. Integral calculus dealt with the problem of determining a function from information about its rate of change. It enabled people to calculate the future location of a body from its present position and a knowledge of the forces acting on it, to find the areas of irregular regions in the plane, to measure the lengths of curves, and to locate the centers of mass of arbitrary solids.

Before the mathematical developments that culminated in the great discoveries of Sir Isaac Newton (1642–1727) and Baron Gottfried Wilhelm Leibniz (1646–1716), it took the astronomer Johannes Kepler (1571–1630) twenty years of concentration, record-keeping, and arithmetic to discover the three laws of planetary motion that now bear his name:

1. Each planet travels in an ellipse that has one focus at the sun (Fig. P.1).

2. The radius vector from the sun to a planet sweeps out equal areas in equal intervals of time.

3. The squares of the periods of revolution of the planets about the sun are proportional to the cubes of their mean distances from the sun. (If T is the length of a planet's year and D is the mean distance, then the ratio T^2/D^3 has the same constant value for all planets in the solar system.)

With calculus, as you will see if you read Article 14.4, deriving Kepler's laws from Newton's laws of motion is but an afternoon's work.

Today, calculus and its extensions in mathematical analysis are far reaching indeed, and the physicists, mathematicians, and astronomers who first invented the subject would surely be amazed and delighted, as we hope you will be, to see what a profusion of problems it solves and what a wide range of fields now use it in the mathematical models that bring understanding about the universe and the world around us.

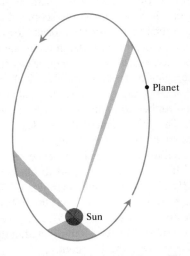

P.1 A planet moving about its sun. The shaded regions shown here have equal areas. According to Kepler's second law, the planet takes the same amount of time to traverse the curved outer boundary of each region. The planet therefore moves faster near the sun than it does farther away.

Calculus helped us predict that Saturn's rings would be ellipses; it also enabled us to send a camera out there to check. The photographs shown here were taken on October 18, 1980 by NASA's Voyager 1 (left) and on July 12, 1981 by Voyager 2 (right). Notice how the features in Saturn's northern hemisphere and the brightness of its rings changed in the nine months between pictures. (Courtesy of NASA.)

Economists use calculus to forecast global trends. Oceanographers use calculus to formulate theories about ocean currents and meteorologists use it to describe the flow of air in the upper atmosphere. Biologists use calculus to forecast population size and to describe the way predators like foxes interact with their prey. Medical researchers use calculus to design ultrasound and x-ray equipment for scanning the internal organs of the body. Space scientists use calculus to design rockets and explore distant planets. Psychologists use calculus to understand optical illusions in visual perception. Physicists use calculus to design inertial navigation systems and to study the nature of time and the universe. Hydraulic engineers use calculus to find safe closure patterns for valves in pipelines. Electrical engineers use it to design stroboscopic flash equipment and to solve the differential equations that describe current flow in computers. Sports equipment manufacturers use calculus to design tennis rackets and baseball bats. Stock market analysts use calculus to predict prices and assess interest rate risk. Physiologists use calculus to describe electrical impulses in neurons in the human nervous system. Drug companies use calculus to determine profitable inventory levels and timber companies use it to decide the most profitable time to harvest trees. The list is practically endless, for almost every professional field today uses calculus in some way.

From an historical point of view, the calculus we use today is an accumulation of the contributions of many people. Its roots can be traced to classical Greek geometry, but its invention is chiefly the work of the scientists of the seventeenth century. Among these were René Descartes (1596–1650), Bonaventura Cavalieri (1598–1647), Pierre de Fermat (1601–1665), John Wallis (1616–1703), and James Gregory (1638–1675). The work culminated in the great creations of Newton and Leibniz. These were the pioneers.

The development of the calculus continued at a furious pace during the next century, and new applications to geometry, mechanics, engineering, and astronomy were found seemingly every day. Among the great contributers were several generations of Bernoullis, chiefly James Bernoulli (1654–1705) and his brother John Bernoulli (1667–1748) (the Bernoulli family was to mathematics what the Bach family was to music); Leonhard Euler (1707–1783)—so inventive and prodigious he was the key mathematical figure of the eighteenth century; Joseph Louis Lagrange (1736–1813); Adrien Marie Legendre (1752–1833); and many, many others.

The perfection of the logical structure behind the procedures of calculus was made by the mathematicians of the nineteenth century, among them Bernhard Bolzano (1781–1848), Augustin Louis Cauchy (1789–1857), and Karl Weierstrass (1815–1897). The nineteenth century also brought another round of spectacular extensions of calculus and great developments in mathematics beyond calculus. You can read about them in Morris Kline's magnificent book *Mathematical Thought from Ancient to Modern Times* (New York: Oxford University Press, 1972).

''The calculus was the first achievement of modern mathematics,'' wrote John von Neumann (1903–1957), one of the great mathematicians of the present century, ''and it is difficult to overestimate its importance. I think it defines more unequivocally than anything else the inception of modern mathematics; and the system of mathematical analysis, which is its logical development, still constitutes the greatest technical advance in exact thinking.''*

World of Mathematics, Vol. 4 (New York: Simon and Schuster, 1960), ''The Mathematician,'' by John von Neumann, pp. 2053–2063.

CHAPTER 1

The Rate of Change of a Function

OVERVIEW

In this chapter we get our first view of the role calculus plays in describing how rapidly things change. We graph the kinds of functions that arise in scientific study, including lines, quadratic functions, square roots, reciprocals, and the basic trigonometric functions. We study the absolute value function and the way it is used to define intervals, and we introduce the greatest integer function as an example of a step function. We also define and calculate the slopes of lines and show how Fermat used the slopes of lines to define the slopes of curves (we still use his method today). Fermat's definition leads to an investigation of the mathematics of limits and to a general process for deriving functions that measure rates of change. The derived functions, called derivatives, are the basic functions of differential calculus, and we shall study them in detail in Chapters 2–4. We look at velocities and other examples of rates of change and present a theorem that enables us to find limits rapidly and with a minimum of fuss. We then see that limits, which we introduced originally for the purpose of defining derivatives, also give us the language we need for describing a special class of functions called *continuous functions*. We conclude the chapter with a description of the properties of continuous functions that account for their importance in scientific work. Our point of departure for all of this is the coordinate plane of Fermat and Descartes.

1.1
Coordinates for the Plane

Analytic Geometry

We do not usually think of geometry today as applied mathematics, but in the minds of the Alexandrian Greeks who first developed the subject it was a tool for studying the physical world. In contrast, the Hindu and Arab mathematicians of the period, who were developing the beginnings of algebra, carried out their proofs and deductions without ever requiring their mathematics to have physical interpretations. For them it was all right to use zero as a number in arithmetic, but not so for the Greeks, to whom zero, with no physical quantity to represent, was only a notational place holder used to denote the absence of a number. To a great extent, geometry and algebra developed separately until only a few hundred years ago.

In the seventeenth century, Fermat and Descartes made the marriage of algebra and geometry that totally changed the face of mathematics. This marriage, which we call analytic geometry today, provided the tool that the seventeenth-century scientists needed for quantifying their work and laid the foundation for astonishing advances in mathematics, physics, astronomy, and biology.

1.1   In Cartesian coordinates, the scaling on each axis is symmetric about the origin.

Cartesian Coordinates

Analytic geometry begins with the assignment of numerical coordinates to all the points in a plane. These coordinates, called **Cartesian coordinates** in honor of Descartes, make it possible to graph algebraic equations in two variables as lines and curves. They also allow us to calculate angles and distances and to write coordinate equations to describe the paths along which objects move. Since most of the theory of calculus can be presented in geometric terms and since the applications of calculus are chiefly about motion and change, the coordinate plane of analytic geometry is the natural setting for learning about calculus and its applications.

To assign coordinates to points in a plane, we start with two number lines that cross at their zero points at right angles. Each line is assumed to represent the real numbers, which are the numbers that can be represented by decimals. Figure 1.1 shows the usual way of drawing the lines, with one line horizontal and the other vertical. The horizontal line is called the **x-axis** and the vertical line the **y-axis.** The point where the lines cross is the **origin.**

On the x-axis, the positive number a lies a units to the right of the origin, and the negative number $-a$ lies a units to the left of the origin. On the y-axis, the positive number b lies b units above the origin while the negative number $-b$ lies b units below the origin.

With the axes in place, we assign a pair (a, b) of real numbers to each point P in the plane. The number a is the number at the foot of the perpendicular from P to the x-axis. The number b is the number at the foot of the perpendicular from P to the y-axis. Figure 1.2 shows the construction. The notation (a, b) is read ''a b.''

1.2   The pair (a, b) is assigned to the point where the perpendicular to the x-axis at a crosses the perpendicular to the y-axis at b.

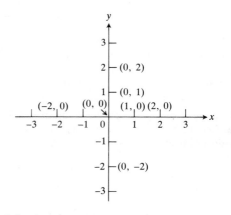

1.3 The points on the axes are now labeled in two ways.

The number a from the x-axis is the **x-coordinate** of P. The number b from the y-axis is the **y-coordinate** of P. The pair (a, b) is the **coordinate pair** of the point P. It is an **ordered pair,** with the x-coordinate first and y-coordinate second. To show that P has the coordinate pair (a, b), we sometimes write the P and (a, b) together: $P(a, b)$.

The construction that assigns an ordered pair of real numbers to each point in the plane can be reversed to assign a point in the plane to each ordered pair of real numbers. The point assigned to the pair (a, b) is the point where the perpendicular to the x-axis at a crosses the perpendicular to the y-axis at b. Thus, the assignment of coordinates is a one-to-one correspondence between the points of the plane and the set of all ordered pairs of real numbers. Every point has a pair and every pair has a point, so to speak.

The points on the coordinate axes now have two kinds of numerical labels: single numbers from the axes and paired numbers from the plane. How do the numbers match up? See Fig. 1.3. Notice that every point on the x-axis has y-coordinate zero and that every point on the y-axis has x-coordinate zero. The coordinates of the origin are $(0, 0)$.

Directions and Quadrants

Motion from left to right along the x-axis is said to be motion in the **positive x-direction.** Motion from right to left is in the **negative x-direction.** Along the y-axis, the positive direction is up, and the negative direction is down.

The origin divides the x-axis into the **positive x-axis** to the right of the origin and the **negative x-axis** to the left of the origin. Similarly, the origin divides the y-axis into the **positive y-axis** and the **negative y-axis.**

The axes divide the plane into four regions called **quadrants,** numbered first, second, third, and fourth. They are labeled I, II, III, and IV in Fig. 1.4.

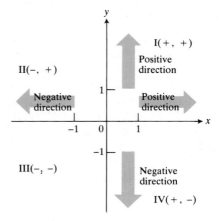

1.4 Directions along the axes: x and y increase in the positive direction and decrease in the negative direction. Roman numerals label the quadrants.

A Word about Scales

When we plot data in the coordinate plane or graph formulas whose variables have different units of measure, the units shown on the two coordinate axes may have very different interpretations. If we plot the size of the national debt at different times, the x-axis might show time in months and the y-axis might show billions of dollars. If we graph steam pressure as a function of boiler temperature, the x-axis might show degrees Fahrenheit and the y-axis pounds per square inch. In cases like these, there is no reason to use the same scale when we draw the two axes. There is no need to place the two 1's on the axes the same number of millimeters or whatever away from the origin.

However, when we graph functions whose variables do not represent physical measurements and when we draw figures in the coordinate plane to learn about their geometry or trigonometry, we shall assume that the scales on the axes we draw are the same. One unit of distance up and down in the plane will then look the same as one unit of distance right and left. As on a surveyor's map or a scale drawing, line segments that are supposed to have the same length will then look as if they do.

PROBLEMS

Figure 1.5 shows four points closely associated with the point $P(5, 2)$:

a) The point Q such that PQ is perpendicular to the x-axis and is bisected by it. Give the coordinates of Q. (P and Q are symmetric with respect to the x-axis.)

b) The point R such that PR is perpendicular to the y-axis and is bisected by it. Give the coordinates of R. (P and R are symmetric with respect to the y-axis.)

c) The point S such that PS is bisected by the origin. Give the coordinates of S. (P and S are symmetric with respect to the origin.)

d) The point T such that PT is perpendicular to and is bisected by the 45° line L through the origin that bisects the first and third quadrants. Give the coordinates of T, assuming equal units on the axes. (P and T are symmetric with respect to L.)

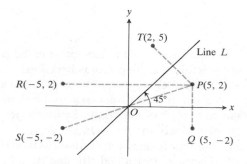

1.5   Symmetry with respect to the x-axis, the y-axis, the origin, and the 45° line L.

Find the coordinates of Q, R, S, and T for the points $P(a, b)$ in Problems 1–12.

1. $(1, -2)$ **2.** $(2, -1)$ **3.** $(-2, 2)$

4. $(-2, 1)$ **5.** $(0, 1)$ **6.** $(1, 0)$

7. $(-2, 0)$ **8.** $(0, -3)$ **9.** $(-1, -3)$

10. $(\sqrt{2}, -\sqrt{2})$ **11.** $(-\pi, -\pi)$ **12.** $(-1.5, 2.3)$

13. If $P = P(x, y)$, then the coordinates of the point Q in (a) above are $(x, -y)$. Give the coordinates of R, S, and T in terms of x and y.

In Problems 14–20, assume that the units of length on the two axes are equal.

14. A line is drawn through the points $(0, 0)$ and $(1, 1)$. What acute angle does it make with the positive x-axis?

15. There are three parallelograms with vertices at $(-1, 1)$, $(2, 0)$, and $(2, 3)$. Sketch them and give the coordinate pairs of the missing vertices.

16. A rectangle with sides parallel to the axes has vertices at $(3, -2)$ and $(-4, -7)$.

a) Find the coordinates of the other two vertices.

b) Find the area of the rectangle.

17. The rectangle in Fig. 1.6 has sides parallel to the axes. It is three times as long as it is wide. Its perimeter is 56 units. Find the coordinates of the vertices A, B, and C.

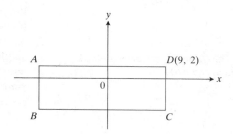

1.6   The rectangle in Problem 17.

18. A circle in quadrant II is tangent to both axes. It touches the y-axis at $(0, 3)$.

a) At what point does it meet the x-axis? Sketch.

b) Find the coordinates of the center of the circle.

19. The line through the points $(1, 1)$ and $(2, 0)$ cuts the y-axis at the point $(0, b)$. Find b by using similar triangles.

20. A 90° rotation counterclockwise about the origin takes $(2, 0)$ to $(0, 2)$ and $(0, 3)$ to $(-3, 0)$, as shown in Fig. 1.7. Where does the rotation take each of the following points?

a) $(4, 1)$ b) $(-2, -3)$ c) $(2, -5)$

d) $(x, 0)$ e) $(0, y)$ f) (x, y)

g) What point is taken to $(10, 3)$?

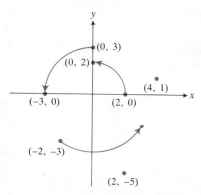

1.7   Points moved to new positions by a counterclockwise 90° rotation about the origin (Problem 20).

1.2
The Slope of a Line

If we keep track of the prices of food or steel or computers, we can watch their progress on graph paper by plotting points and fitting them with a curve. We can extend the curve day by day as new prices appear. To what uses can we then put such a curve? We can see what the price was on any date. We can see by the slope of the curve (whose precise meaning will be given later) the rate at which the prices are rising or falling. If we plot other data on the same sheet of paper, we can perhaps see what relation they have to the rise and fall of prices. The curve may also reveal patterns that would help us to forecast the future with more accuracy than someone who has not graphed the data.

One of the many reasons calculus has proved so useful over the years is that it is the right mathematics for making the connection between rates of change and the slopes of smooth curves. Explaining this connection is one of the goals of this chapter. The basic plan is first to define what we mean by the slope of a line and then to define the slope of a curve at each point on the curve to be the limit of the slopes of selected secant lines through the point. Just how this is done will become clear as the chapter goes on. Our first step is to find a practical way to calculate the slopes of lines.

Civil engineers calculate the slope of a roadbed by calculating the ratio of the distance it rises or falls to the distance it runs horizontally. They call this ratio the **grade** of the roadbed. They usually write grades as percents, as shown in Fig. 1.8. Along the coast, railroad grades are usually less than 2%. In the mountains, they may go as high as 4%. Highway grades are usually less than 5%.

In analytic geometry we calculate slopes the same way, but we do not usually express them as percents.

1.8 Examples of highway grades.

4% grade

100′ 4′ 5′

−5% grade

100′

Increments

When a particle moves from one position in the plane to another, the net changes in the particle's coordinates are calculated by subtracting the coordinates of the starting point from the coordinates of the stopping point.

EXAMPLE 1 Figure 1.9 shows the path of a particle that moved from $A(1, -2)$ to $B(6, 7)$. The net change in the x-coordinate was $\Delta x = 6 - 1 = 5$ units. The y-coordinate increased from $y = -2$ to $y = 7$, for a net change of $\Delta y = 7 - (-2) = 9$ units. ■

The symbols Δx and Δy in Example 1 are read "delta x" and "delta y." They denote net changes or **increments** in the variables x and y. The letter Δ is a capital Greek "dee," for "difference." Neither Δx nor Δy denotes multiplication; Δx is not delta times x nor is Δy delta times y.

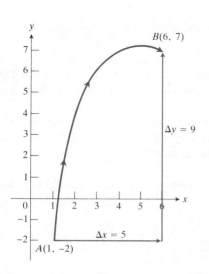

1.9 In going from A to B, $\Delta x = 6 - 1 = 5$ and $\Delta y = 7 - (-2) = 9$.

When a particle moves from (x_1, y_1) to (x_2, y_2), the increments are

$$\Delta x = x_2 - x_1 \quad \text{and} \quad \Delta y = y_2 - y_1. \tag{1}$$

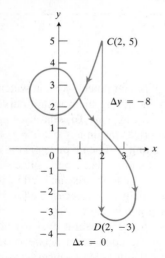

1.10 In going from C to D, $\Delta x = 0$ and $\Delta y = -8$.

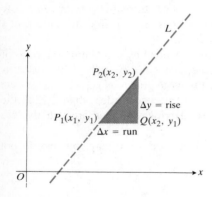

1.11 The slope of the line is

$$m = \frac{\Delta y}{\Delta x} = \frac{\text{rise}}{\text{run}}.$$

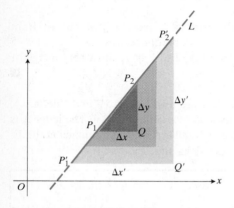

1.12 These triangles are similar. This tells us that $\Delta y / \Delta x = \Delta y' / \Delta x'$ because the ratios of corresponding sides of similar triangles are equal.

EXAMPLE 2   In Fig. 1.10, the net changes in coordinates in moving from $C(2, 5)$ to $D(2, -3)$ along either path are $\Delta x = 2 - 2 = 0$ and $\Delta y = -3 - 5 = -8$. The net change in x is zero, and y decreases for a net change of -8. Notice that the values of Δx and Δy do not depend on the path taken.

■

Slopes of Nonvertical Lines

All lines except vertical lines have slopes. We calculate slopes from changes in coordinates. Once we see how this is done, we shall also see why vertical lines are an exception.

To begin, let L be a nonvertical line in the plane. Let $P_1(x_1, y_1)$ and $P_2(x_2, y_2)$ be two points on L as in Fig. 1.11. We call $\Delta y = y_2 - y_1$ the **rise** from P_1 to P_2 and $\Delta x = x_2 - x_1$ the **run** from P_1 to P_2. Since L is not vertical, $\Delta x \neq 0$ and we may define the **slope** of L to be $\Delta y / \Delta x$, the amount of rise per unit of run. It is conventional to denote the slope by the letter m.

$$\textit{Slope:} \qquad m = \frac{\text{rise}}{\text{run}} = \frac{\Delta y}{\Delta x} = \frac{y_2 - y_1}{x_2 - x_1}. \qquad (2)$$

Suppose that instead of choosing the points P_1 and P_2 to calculate the slope in Eq. (2), we were to choose a different pair of points P_1' and P_2' on L and calculate

$$m' = \frac{y_2' - y_1'}{x_2' - x_1'} = \frac{\Delta y'}{\Delta x'}. \qquad (3)$$

Would we get the same value for the slope? In other words, would $m' = m$? The answer is yes, as we can see from Fig. 1.12. The numbers m and m' are equal because they are ratios of corresponding sides of similar triangles:

$$m' = \frac{\Delta y'}{\Delta x'} = \frac{\Delta y}{\Delta x} = m. \qquad (4)$$

The slope of a line depends only on how steeply the line rises or falls and not on the points we use to calculate it.

A line that goes uphill as x increases, like the line in Fig. 1.13, has a positive slope. A line that goes downhill as x increases, like the line in Fig. 1.14, has a negative slope. A horizontal line has slope zero. The points on it all have the same y-coordinate, so $\Delta y = 0$.

The formula $m = \Delta y / \Delta x$ does not apply to vertical lines because Δx is zero along a vertical line. We express this by saying that vertical lines have no slope or that the slope of a vertical line is undefined.

EXAMPLE 3   Calculate the slope of the line through the points $P_1(1, 2)$ and $P_2(2, 5)$ in Fig. 1.13.

Solution   We use Eq. (2) to get

$$m = \frac{y_2 - y_1}{x_2 - x_1} = \frac{5 - 2}{2 - 1} = \frac{3}{1} = 3.$$

A slope of $m = 3$ means that y increases 3 units every time x increases 1 unit. In other words, the change in y is 3 times the change in x. Along the line, $\Delta y = 3 \Delta x$. The slope $m = 3$ is a proportionality factor.

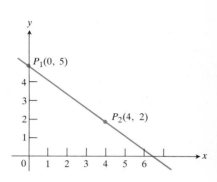

1.13 The slope of this line is

$$m = \frac{(5-2)}{(2-1)} = 3.$$

This means that $\Delta y = 3\,\Delta x$ for every change of position on the line (compare the coordinates of the marked points).

1.14 The slope of this line is

$$m = \frac{\Delta y}{\Delta x} = \frac{(2-5)}{(4-0)} = \frac{-3}{4}.$$

This means that y decreases 3 units every time x increases 4 units.

Suppose we take P_2 as $(1, 2)$ and P_1 as $(2, 5)$ instead of the other way around. Does Eq. (2) still give the answer $m = 3$? Yes, it does, as we can see here:

$$m = \frac{y_2 - y_1}{x_2 - x_1} = \frac{2 - 5}{1 - 2} = \frac{-3}{-1} = 3.$$

Renumbering the points P_1 and P_2 changed the signs of the rise and the run but did not change their ratio.   ■

EXAMPLE 4   Calculate the slope of the line through $P_1(0, 5)$ and $P_2(4, 2)$ in Fig. 1.14.

Solution   Equation (2) gives

$$m = \frac{y_2 - y_1}{x_2 - x_1} = \frac{2 - 5}{4 - 0} = \frac{-3}{4} = -\frac{3}{4}.$$

This tells us that y decreases 3 units whenever x increases 4 units.   ■

Angles of Inclination

The **angle of inclination** of a line that crosses the x-axis is the smallest angle we get when we measure counterclockwise from the x-axis around the point of intersection (Fig. 1.15). The angle of inclination of a horizontal line is taken to be $0°$. Thus, angles of inclination may have any measure from $0°$ up to but not including $180°$.

The slope of a line is the tangent of the line's angle of inclination. The construction in Fig. 1.16 shows why this is true. If m denotes the slope and ϕ the angle, then

1.15   Angles of inclination are measured counterclockwise from the x-axis.

$$m = \tan \phi. \tag{5}$$

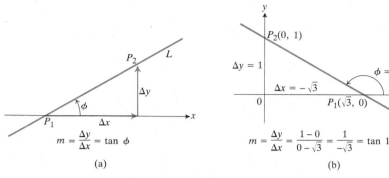

(a)   (b)

1.16   The slope of a line is the tangent of its angle of inclination.

1.17   When $\phi = 90°$, $\tan \phi$ is undefined. Vertical lines have equal angles of inclination but no slope.

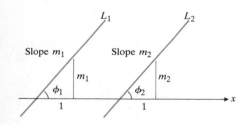

1.18   If $m_1 = m_2$, then $\phi_1 = \phi_2$ and the lines are parallel.

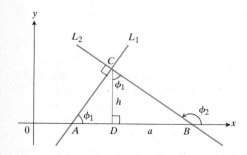

1.19 $\triangle ADC$ is similar to $\triangle CDB$. Hence ϕ_1 is also the upper angle in $\triangle CDB$. From the sides of $\triangle CDB$, we read $\tan \phi_1 = a/h$.

EXAMPLE 5   Let us examine the slopes of lines whose angles of inclination are near 90°, such as

$$\phi_1 = 89°59', \qquad m_1 = \tan \phi_1 \approx 3437.7$$

and

$$\phi_2 = 90°01', \qquad m_2 = \tan \phi_2 \approx -3437.7.$$

The slopes of such lines are numerically very large. By taking ϕ still closer to 90° we can make the slope numerically larger than any number N, no matter how large N may be. This fact can be summarized by saying that the slope of a line "becomes infinite" as its angle of inclination approaches 90° or that a vertical line has "infinite slope." But, strictly speaking, we do not assign any real number to be the slope of a vertical line. Vertical lines have no slope (Fig. 1.17).   ■

Lines That Are Parallel or Perpendicular

Parallel lines have equal angles of inclination. Hence, if they are not vertical, they have the same slope. Conversely, lines with equal slopes have equal angles of inclination and are therefore parallel. See Fig. 1.18.

If neither of two perpendicular lines L_1 and L_2 is vertical, their slopes m_1 and m_2 are related by the equation $m_1 m_2 = -1$. Figure 1.19 shows why:

$$m_1 = \tan \phi_1 = \frac{a}{h}, \qquad \text{while} \quad m_2 = -\frac{h}{a}. \tag{6}$$

Hence,

$$m_1 m_2 = -1 \qquad \text{or} \qquad m_2 = -\frac{1}{m_1}. \tag{7}$$

EXAMPLE 6   A line L is known to be perpendicular to a line whose slope is $-3/4$. Find the slope of L.

Solution   According to Eq. (7), the slope of L is the negative reciprocal of $-3/4$. The slope of L is therefore $4/3$.   ■

Summary

1. The slope of the line through $P_1(x_1, y_1)$ and $P_2(x_2, y_2)$, $x_1 \neq x_2$, is

$$m = \frac{\text{rise}}{\text{run}} = \frac{y_2 - y_1}{x_2 - x_1} = \frac{\Delta y}{\Delta x}.$$

2. $m = \tan \phi$ (ϕ is the angle of inclination).
3. Vertical lines have no slope.
4. Horizontal lines have slope 0.
5. For lines that are neither horizontal nor vertical it is handy to remember:

   a) they are parallel $\Leftrightarrow m_2 = m_1$;

   b) they are perpendicular $\Leftrightarrow m_2 = -1/m_1$.

   (The symbol $\Leftrightarrow$ is read "if and only if.")

PROBLEMS

In Problems 1–6, a particle moves in the plane from A to B. Find Δx and Δy.

1. $A(-1, 1)$, $B(1, 2)$ **2.** $A(1, 2)$, $B(-1, -1)$

3. $A(-3, 2)$, $B(-1, -2)$ **4.** $A(-1, -2)$, $B(-3, 2)$

5. $A(-3, 1)$, $B(-8, 1)$ **6.** $A(0, 4)$, $B(0, -2)$

Problems 7 and 8 refer to Fig. 1.20, which reproduces a multiflash photograph of a ball looping the loop on a circular track. The letters label successive positions, with coordinates rounded to the nearest integer.

7. Find the net horizontal and vertical changes that took place in the coordinates of the center of the ball shown in Fig. 1.20 when the ball moved from
   a) A to G,   b) D to E,   c) C to F.

8. By what increments did the coordinates of the center of the ball in Fig. 1.20 change when the ball later rolled backward from G to F?

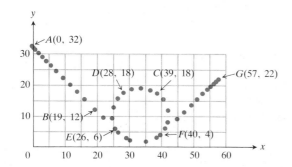

1.20   Figure for Problems 7 and 8.

Suppose that a particle in the plane moves from $P_1(x, y)$ to $P_2(x + \Delta x, y + \Delta y)$ with the increments Δx and Δy given in Problems 9–16. Determine in each problem whether P_2 lies above, below, to the right, or to the left of P_1.

9. $\Delta x = 6$, $\Delta y = 3$ **10.** $\Delta x = 5$, $\Delta y = 0$

11. $\Delta x = -2$, $\Delta y = 0$ **12.** $\Delta x = 0$, $\Delta y = 2$

13. $\Delta x = 3$, $\Delta y = -1$ **14.** $\Delta x = -1$, $\Delta y = -2$

15. $\Delta x = 0$, $\Delta y = -5$ **16.** $\Delta x = -4$, $\Delta y = 0$

In Problems 17–32, plot the points A and B, and find the slope (if any) of the line determined by them. Also find the common slope (if any) of the lines perpendicular to AB.

17. $A(1, -2)$, $B(2, 1)$ **18.** $A(-1, 2)$, $B(-2, -1)$

19. $A(-2, -1)$, $B(1, -2)$ **20.** $A(2, -1)$, $B(-2, 1)$

21. $A(1, 0)$, $B(0, 1)$ **22.** $A(-1, 0)$, $B(1, 0)$

23. $A(2, 3)$, $B(-1, 3)$ **24.** $A(0, 3)$, $B(2, -3)$

25. $A(0, -2)$, $B(-2, 0)$ **26.** $A(1, 2)$, $B(1, -3)$

27. $A(0, 0)$, $B(-2, -4)$ **28.** $A(1/2, 0)$, $B(0, -1/3)$

29. $A(0, 0)$, $B(x, y)$, $(x \neq 0, y \neq 0)$

30. $A(0, 0)$, $B(x, 0)$, $(x \neq 0)$

31. $A(0, 0)$, $B(0, y)$, $(y \neq 0)$

32. $A(a, 0)$, $B(0, b)$, $(a \neq 0, b \neq 0)$

Problems 33–35 are about the stairway shown in Fig. 1.21. The slope of the stairway can be calculated from the riser height R and the tread width T as R/T. The manual from which the drawing was adapted defines a stairway as a stepped footway having a slope not less than 5:16, or 31¼% and not greater than 9:8 or

1.21 A stair profile. R = riser height, T = tread width.

112½%. (The manual goes on to say that below these limits footways become ramps, and above them, stepladders!)

33. What are the minimum and maximum stair angles allowed in Fig. 1.21 by the definition of a stairway?

34. A common angle for household stairs is 40°. If the treads on the stairs are 9 in. wide, about how high are the risers?

35. Find the slope of a 40° stairway.

Solve Problems 36 and 37 by measuring slopes in Fig. 1.22.

1.22 Temperature gradients in a wall. (*Source: Differentiation*, by W. U. Walton et al., Project CALC, Education Development Center, Inc., Newton, Mass. (1975), p. 25.)

36. Find the rate of change of temperature in degrees per inch for
a) gypsum wall board,
b) fiber glass insulation, and
c) wood sheathing.

37. Which of the materials listed in Problem 36 is the best insulator? The poorest? Explain.

38. Plot the points A, B, C, and D. Then determine whether the quadrilateral $ABCD$ is a parallelogram by comparing the slopes of opposite sides.
a) $A(0, 1)$, $B(1, 2)$, $C(2, 1)$, $D(1, 0)$
b) $A(3, 1)$, $B(2, 2)$, $C(0, 1)$, $D(1, 0)$
c) $A(-1, -2)$, $B(2, -1)$, $C(2, 1)$, $D(1, 0)$
d) $A(-2, 2)$, $B(1, 3)$, $C(2, 0)$, $D(-1, -1)$
e) $A(-1, 0)$, $B(0, -1)$, $C(2, 0)$, $D(0, 2)$

39. Use slopes to determine in each case whether the points are collinear (lie on a common straight line).
a) $A(1, 0)$, $B(0, 1)$, $C(2, 1)$
b) $A(-2, 1)$, $B(0, 5)$, $C(-1, 2)$
c) $A(-2, 1)$, $B(-1, 1)$, $C(1, 5)$, $D(2, 7)$
d) $A(-2, 3)$, $B(0, 2)$, $C(2, 0)$
e) $A(-3, -2)$, $B(-2, 0)$, $C(-1, 2)$, $D(1, 6)$

40. A particle moves at a constant velocity v in a straight line past the origin. The coordinate system and the particle's track are shown in Fig. 1.23. The positions of the particle shown on the line are $\Delta t = 1$ sec apart. Why are the areas $A_1, A_2, \ldots, A_5$ in Fig. 1.23 equal? (*Hint:* Drop an altitude from the origin to the line of motion.) As in Kepler's second law (see the Prologue), the line that joins the particle to the origin sweeps out equal areas in equal times.

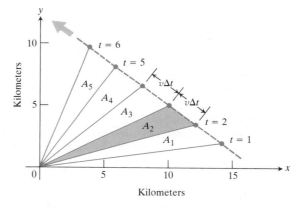

1.23 If a particle moves in a straight line past the origin at a constant velocity v, the line joining the particle to the origin sweeps out equal areas $A_1, A_2, \ldots,$ in equal times (see Problem 40). The axes are scaled in kilometers. Time, denoted by t, is measured in seconds.

41. A particle starts at $A(-2, 3)$ and its coordinates change by increments $\Delta x = 5$, $\Delta y = -6$. Find its new position.

42. A particle starts at $A(6, 0)$ and its coordinates change by increments $\Delta x = -6$, $\Delta y = 0$. Find its new position.

43. The coordinates of a particle change by $\Delta x = 5$ and $\Delta y = 6$ in moving from $A(x, y)$ to $B(3, -3)$. Find x and y.

44. A particle started at $A(1, 0)$, circled the origin once counterclockwise, and returned to $A(1, 0)$. What were the net changes in its coordinates?

1.3
Equations for Lines

An **equation for a line** is an equation that is satisfied by the coordinates of the points that lie on the line and is not satisfied by the coordinates of the points that lie elsewhere.

What do such equations do for us? They tell us when lines are vertical and what their slopes are when they are not vertical. They show us how to calculate the value of y for any value of x on a nonvertical line or the value of x for any value of y on a nonhorizontal line. They also give us a useful way to summarize numerical data and to predict values of unrecorded data. Equations of lines play an important role in estimating roots of equations that are too complicated to solve directly, as we shall see in Chapter 2. In this article we show how to write and interpret equations for lines in the coordinate plane.

Vertical Lines

Every vertical line L has to cross the x-axis at some point $(a, 0)$ (Fig. 1.24). The other points on L lie directly above or below $(a, 0)$. This means that the first coordinate of any point $P(x, y)$ on L must be a, whereas y can be any number. In other words, the coordinates of all the points (x, y) on L satisfy the equation $x = a$.

To be sure that $x = a$ is an equation for L, we must check that the points not on L all have first coordinates different from a. They do, because the perpendiculars from these points to the x-axis do not hit the x-axis at the point $x = a$. (See Fig. 1.24.)

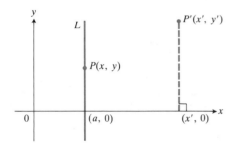

1.24 If $P(x, y)$ lies on L, then $x = a$. Conversely, if $P'(x', y')$ does not lie on L, then $x' \neq a$ because the foot of the perpendicular from P' to the x-axis is different from $(a, 0)$.

The standard equation for the **vertical line** through the point (a, b) is

$$x = a. \tag{1}$$

EXAMPLE 1 The standard equation of the vertical line through the point $(2, -3)$ is $x = 2$. See Fig. 1.25. ∎

Nonvertical Lines

To write an equation for a line L that is not vertical, it is enough to know its slope m and the coordinates of a point $P_1(x_1, y_1)$ on it. If $P(x, y)$ is any other point on L (as in Fig. 1.26), then $x \neq x_1$ and we can write the slope of L as

$$\frac{y - y_1}{x - x_1}. \tag{2}$$

We can then set this expression equal to m to get

$$\frac{y - y_1}{x - x_1} = m. \tag{3}$$

Multiplying both sides of Eq. (3) by $x - x_1$ gives us the more useful equation

$$y - y_1 = m(x - x_1). \tag{4}$$

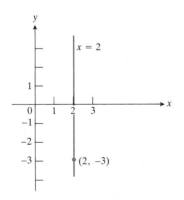

1.25 The standard equation for the vertical line through $(2, -3)$ is $x = 2$.

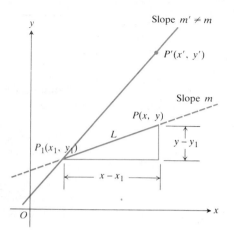

1.26   Suppose L is the line through $P_1(x_1, y_1)$ whose slope is m. Then other points $P(x, y)$ lie on this line if and only if slope $PP_1 = m$. This fact gives us the point-slope equation for L.

Equation (4) is an equation for L, as we can check right away. Every point (x, y) on L satisfies the equation—even the point (x_1, y_1). What about the points not on L? If $P'(x', y')$ is a point not on L, then, as Fig. 1.26 shows, the slope m' of $P'P_1$ is different from m, and the coordinates x' and y' of P' do not satisfy Eqs. (3) and (4).

Equation (4) is called a point-slope equation for L.

DEFINITION

The **point-slope equation** of the line through the point (x_1, y_1) with slope m is

$$y - y_1 = m(x - x_1). \qquad (5)$$

EXAMPLE 2   Write an equation for the line through the point $(1, 2)$ with slope $m = -3/4$.

Solution   We use Eq. (5) with $(x_1, y_1) = (1, 2)$ and $m = -3/4$:

$$y - y_1 = m(x - x_1)$$

$$y - 2 = -\frac{3}{4}(x - 1)$$

$$y = -\frac{3}{4}x + \frac{3}{4} + 2$$

$$y = -\frac{3}{4}x + \frac{11}{4}.$$

EXAMPLE 3   Write an equation for the line through $(-2, -1)$ and $(3, 4)$.

Solution   We first calculate the slope and then use Eq. (5) with one of the given points as (x_1, y_1). The slope is

$$m = \frac{-1 - 4}{-2 - 3} = \frac{-5}{-5} = 1.$$

Either $(-2, -1)$ or $(3, 4)$ can be (x_1, y_1):

With $(x_1, y_1) = (-2, -1)$ With $(x_1, y_1) = (3, 4)$

$$y - (-1) = 1 \cdot (x - (-2))) \qquad y - 4 = 1 \cdot (x - 3)$$

$$y + 1 = x + 2 \qquad\qquad y - 4 = x - 3$$

$$y = x + 1. \qquad\qquad\qquad y = x + 1.$$

Same result

Either way, we find that $y = x + 1$ is an equation for the line.   ∎

Intercepts

The x-coordinate of the point where a line crosses the x-axis is the **x-intercept** of the line. To find it, we set $y = 0$ in an equation for the line and solve for x. The y-coordinate of the point where a line crosses the y-axis is the **y-intercept** of the line. To find it, we set $x = 0$ in an equation for the line and solve for y.

EXAMPLE 4   Find the intercepts of the line $y = 2x - 3$.

Solution   To find the x-intercept, we set $y = 0$ in the equation for the line and solve for x. This gives

$$0 = 2x - 3 \qquad \text{or} \qquad x = \frac{3}{2}.$$

The x-intercept is $3/2$.

To find the y-intercept, we set $x = 0$ in the equation for the line and solve for y. This gives

$$y = 2(0) - 3 \qquad \text{or} \qquad y = -3.$$

The y-intercept is -3.   ∎

Every line that is not vertical must cross the y-axis at some point $(0, b)$. The number b is the y-intercept of the line. If we take $(x_1, y_1) = (0, b)$ in Eq. (5), we find that the line has the point-slope equation

$$y - b = m(x - 0), \tag{6}$$

which is equivalent to

$$y = mx + b. \tag{7}$$

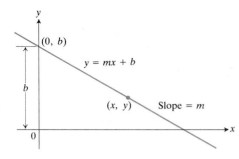

1.27   The line $y = mx + b$ has slope m and y-intercept b.

The equation $y = mx + b$ is called the slope-intercept equation of the line (Fig. 1.27).

DEFINITION

> The **slope-intercept equation** of the line with slope m and y-intercept b is
>
> $$y = mx + b. \tag{8}$$

In Examples 2 and 3, the final form of the equation for each line was its slope-intercept equation.

EXAMPLE 5   The slope-intercept equation of the line whose slope is $m = -3/4$ and whose y-intercept is $b = 5$ is

$$y = -\frac{3}{4}x + 5.$$

This is the line in Fig. 1.14.

EXAMPLE 6   Find the slope and y-intercept of the line $8x + 5y = 20$.

Solution   Solve the equation for y to put the equation in slope-intercept form. Then read the slope and y-intercept from the equation:

$$8x + 5y = 20$$

$$5y = -8x + 20$$

$$y = -\frac{8}{5}x + 4.$$

The slope is $m = -8/5$. The y-intercept is $b = 4$.

Quick Graphing

A quick way to graph a line that crosses both axes is to mark the intercepts and draw a line through the marked points. The method fails only if the line passes through the origin, giving one intercept instead of two, or if the intercepts are hard to plot.

EXAMPLE 7   Graph the line $8x + 5y = 20$.

Solution

1. Find the x-intercept by setting $y = 0$ to obtain $8x = 20$, or $x = 5/2$.
2. Find the y-intercept by setting $x = 0$ to obtain $5y = 20$, or $y = 4$.
3. Plot the intercepts and draw the line (Fig. 1.28).

1.28   To graph $8x + 5y = 20$, mark the intercepts and draw a line through the marked points.

Horizontal Lines

When the line $y = mx + b$ is horizontal, $m = 0$ and the equation reduces to $y = b$.

The standard equation for the **horizontal line** through the point (a, b) is

$$y = b. \tag{9}$$

EXAMPLE 8   The equation of the horizontal line through the point $(1, 2)$ is $y = 2$.

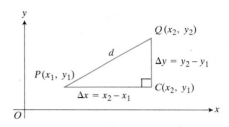

1.29 The distance d between two points $P(x_1, y_1)$ and $Q(x_2, y_2)$ is calculated by applying the Pythagorean theorem to right triangle PCQ. Since $d^2 = (\Delta x)^2 + (\Delta y)^2$, we have $d = \sqrt{(\Delta x)^2 + (\Delta y)^2}$, or $d = \sqrt{(x_2 - x_1)^2 + (y_2 - y_1)^2}$.

The Distance from a Point to a Line

To calculate the distance between the points $P(x_1, y_1)$ and $Q(x_2, y_2)$, we use the following formula:

> **Distance Formula for Points in the Plane**
>
> $$d = \sqrt{(x_2 - x_1)^2 + (y_2 - y_1)^2} \qquad (10)$$

See Fig. 1.29.

To calculate the distance from a point $P(x_1, y_1)$ to a line L, we find the point $Q(x_2, y_2)$ at the foot of the perpendicular from P to L and calculate the distance from P to Q. The next example shows how this is done.

EXAMPLE 9 Find the distance from the point $P(2, 1)$ to the line L: $y = x + 2$.

Solution We solve the problem in three steps: (1) Find an equation for the line L' through P perpendicular to L; (2) find the point Q where L' meets L; and (3) calculate the distance between P and Q. See Fig. 1.30.

STEP 1: We find an equation for the line L' through $P(2, 1)$ perpendicular to L. The slope of L: $y = x + 2$ is $m = 1$. The slope of L' is therefore $m = -1/1 = -1$. We set $(x_1, y_1) = (2, 1)$ and $m = -1$ in Eq. (5) to find L':

$$y - 1 = -1(x - 2)$$
$$y = -x + 2 + 1$$
$$y = -x + 3.$$

STEP 2: We find the point Q by solving the equations for L and L' simultaneously. To find the x-coordinate of Q, we equate the two expressions for y:

$$x + 2 = -x + 3$$
$$2x = 1$$
$$x = \frac{1}{2}.$$

We can now obtain the y-coordinate by substituting $x = 1/2$ in the equation for either line. We choose $y = x + 2$ arbitrarily and find

$$y = \frac{1}{2} + 2 = \frac{5}{2}.$$

The coordinates of Q are $(\frac{1}{2}, \frac{5}{2})$.

STEP 3: We use Eq. (10) to calculate the distance between $P(2, 1)$ and $Q(\frac{1}{2}, \frac{5}{2})$:

$$d = \sqrt{\left(2 - \frac{1}{2}\right)^2 + \left(1 - \frac{5}{2}\right)^2} = \sqrt{\left(\frac{3}{2}\right)^2 + \left(-\frac{3}{2}\right)^2} = \sqrt{\frac{18}{4}} = \frac{3}{2}\sqrt{2}.$$

The distance from P to L is $(3/2)\sqrt{2}$. ∎

1.30 The distance d from $P(2, 1)$ to L is measured along the line L' perpendicular to L. It can be calculated from the coordinates of P and Q.

REMARK   When the coordinate axes are scaled in a common unit, the distance formula $d = \sqrt{(x_2 - x_1)^2 + (y_2 - y_1)^2}$ calculates the distance between points in the plane in that unit. Suppose, for instance, that x and y are in meters. Then $(x_2 - x_1)$ and $(y_2 - y_1)$ are in meters, their squares are in meters squared, the sum of the squares is in meters squared, and the square root of this sum is once again in meters. We do not use the distance formula when the scales on the axes represent different units, for then the formula makes no sense.

The General Linear Equation

The equation

$$Ax + By = C \qquad (A \text{ and } B \text{ not both zero}) \tag{11}$$

is called the **general linear equation** because its graph is always a line and because every line has an equation in this form. We shall not take the time to prove this, but notice that the equations in this article can all be arranged in this form. A few of them are in this form already.

Equations for Lines

$x = a$	Vertical line through (a, b)
$y = b$	Horizontal line through (a, b)
$y = mx + b$	Slope-intercept equation
$y - y_1 = m(x - x_1)$	Point-slope equation
$Ax + By = C$	General linear equation $(A \text{ and } B \text{ not both zero})$

PROBLEMS

In Problems 1–8, write an equation for (a) the vertical line and (b) the horizontal line through the given point.

1. $(2, 3)$ **2.** $(-7, -7)$ **3.** $(0, 0)$

4. $(0, -4)$ **5.** $(-4, 0)$ **6.** $(a, 0)$

7. $(0, b)$ **8.** (x_1, y_1)

In Problems 9–14, write an equation for the line that passes through point P with slope m. Then graph the line.

9. $P(1, 1)$, $m = 1$ **10.** $P(1, -1)$, $m = -1$

11. $P(-1, 1)$, $m = 1$ **12.** $P(-1, 1)$, $m = -1$

13. $P(0, b)$, $m = 2$ **14.** $P(a, 0)$, $m = -2$

In Problems 15–28, find an equation for the line through the two points.

15. $(0, 0)$, $(2, 3)$ **16.** $(1, 1)$, $(2, 1)$

17. $(1, 1)$, $(1, 2)$ **18.** $(-2, 1)$, $(2, -2)$

19. $(-2, 0)$, $(-2, -2)$ **20.** $(1, 3)$, $(3, 1)$

21. $(T, 0)$, $(0, F_0)$ $(T \neq 0, F_0 \neq 0)$

22. $(0, 0)$, $(1, 0)$ **23.** $(0, 0)$, $(0, 1)$

24. $(2, -1)$, $(-2, 3)$ **25.** $(-0.7, 1.5)$, $(1.4, 0.8)$

26. $(\sqrt{2}, \sqrt{2})$, $(\sqrt{5}, \sqrt{5})$ **27.** (x_0, y_0), (x_1, y_1)

28. $(0.5, 10{,}000)$, $(2, 35{,}000)$

In Problems 29–34, write an equation for the line with the given slope and y-intercept. Graph the line.

29. $m = 3$, $b = -2$ **30.** $m = -1$, $b = 2$

31. $m = 1$, $b = \sqrt{2}$ **32.** $m = -1/2$, $b = -3$

33. $m = -5$, $b = 2.5$ **34.** $m = 1/3$, $b = -1$

In Problems 35–48, find the slope and intercepts of the line. Then graph the line.

35. $y = 3x + 5$ **36.** $2y = 3x + 5$

37. $x + y = 2$ **38.** $2x - y = 4$

39. $x - 2y = 4$ **40.** $3x + 4y = 12$

41. $4x - 3y = 12$

42. $x = 2y - 5$

43. $\dfrac{x}{3} + \dfrac{y}{4} = 1$

44. $\dfrac{2x}{5} - \dfrac{y}{3} = 1$

45. $\dfrac{x}{2} - \dfrac{y}{3} = -1$

46. $\dfrac{x}{3} + \dfrac{y}{1} = -1$

47. $1.05x - 0.35y = 7$

48. $0.98x + 1.96y = 9.8$

49. *The intercept equation of a line.* Find the slope of the line $x/a + y/b = 1$ (a, b constants $\neq 0$). What are the x- and y-intercepts of this line? (This is called the intercept equation of the line.)

50. Write an equation for the line in Fig. 1.13.

51. Write an equation for the line in Fig. 1.16(b).

52. *Lines through the origin.*
 a) Write equations for the four lines that pass through the origin and the points $(1, 1/2)$, $(1, 1)$, $(1, 2)$, $(1, 3)$. Graph the lines. Label each line with its slope.
 b) Repeat the instructions in (a) for the points $(-1, 1/2)$, $(-1, 1)$, $(-1, 2)$, and $(-1, 3)$.
 c) Write an equation for the line through the origin with slope m. Find the coordinates of the point of intersection of this line with the line $x = 1$. Illustrate.

In Problems 53–66, find (a) an equation for the line through P parallel to L, (b) an equation for the line L' through P perpendicular to L, and (c) the distance from P to L.

53. $P(2, 1)$, $L: y = x + 2$

54. $P(0, 0)$, $L: y = -x + 2$

55. $P(0, 0)$, $L: x + \sqrt{3}y = 3$

56. $P(1, 2)$, $L: x + 2y = 3$

57. $P(-2, 2)$, $L: 2x + y = 4$

58. $P(3, 6)$, $L: x + y = 3$

59. $P(1, 0)$, $L: 2x - y = -2$

60. $P(-2, 4)$, $L: x = 5$

61. $P(3, 2)$, $L: x = -5$

62. $P(3, 2)$, $L: y = -4$

63. $P(a, b)$, $L: x = -1$

64. $P(3, -h)$, $L: y = 4$ ($h > 0$)

65. $P(4, 6)$, $L: 4x + 3y = 12$

66. $P(2/\sqrt{3}, -1)$, $L: \sqrt{3}x + y = -3$

In Problems 67–74, find the angle of inclination of the given line.

67. $y = x + 2$

68. $y = -x + 2$

69. $x + \sqrt{3}y = 3$

70. $x + 2y = 3$

71. $2x + y = 4$

72. $2x - y = -2$

73. $4x + 3y = 12$

74. $\sqrt{3}x + y = -3$

In Problems 75–78, find the line through the given point with the given angle of inclination ϕ.

75. $(1, 4)$, $\phi = 60°$

76. $(-1, -1)$, $\phi = 135°$

77. $(-2, 3)$, $\phi = 90°$

78. $(3, -2)$, $\phi = 0°$

79. *Pressure under water.* The pressure p experienced by a diver under water is related to the diver's depth d by an equation of the form $p = kd + 1$ (k a constant). When $d = 0$ meters, the pressure is 1 atmosphere. The pressure at 100 meters is about 10.94 atmospheres. Find the pressure at 50 meters.

80. *Reflected light.* A ray of light comes in along the line $x + y = 1$ above the x-axis and reflects off the x-axis. The angle of departure is congruent to the angle of arrival (Fig. 1.31). Write an equation for the line along which the departing light travels.

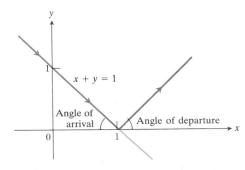

1.31 The path of the light ray in Problem 80.

81. *Expanding railroad track.* The steel in railroad track expands when heated. For the track temperatures encountered in normal outdoor use, the length s of a piece of track is related to its temperature t by a linear equation. An experiment with a piece of track gave the following measurements:

$$t_1 = 65°F, \qquad s_1 = 35 \text{ ft},$$
$$t_2 = 135°F, \qquad s_2 = 35.16 \text{ ft}.$$

Write a linear equation for the relation between s and t.

82. *Temperature scales.* Fahrenheit (F) and Celsius (C) temperature readings are related by a linear equation; that is, the graph of F vs. C is a straight line.
 a) Find an equation that relates F and C, given that C = 0 when F = 32 and C = 100 when F = 212.
 b) Is there a temperature at which C = F? If so, what is it?

83. **CALCULATOR** *The Mt. Washington Cog Railway.* The steepest part of the road on the Mt. Washington Cog Railway in New Hampshire has a staggering 37.1% grade. At this point the passengers in the front of the car are 14 feet above those at the rear. About how far apart are the front and rear rows of seats?

TOOLKIT PROGRAMS

Name That Function   Super * Grapher

1.4
Functions and Graphs

The values of one variable quantity often depend on the values of another. For example:

The pressure in the boiler of a power plant depends on the steam temperature.

The rate at which the water drains from your bathtub when you pull the plug depends on how deep the water is.

The area of a circle depends on its radius.

In each of these examples, the value of one variable quantity, which we might call y, depends on the value of another variable quantity, which we might call x. Since it is also true that the value of y in each case is completely determined by the value of x, we say that y is a function of x.

In mathematics, any rule that assigns to each element in one set some element from another set is called a **function.** The sets may be sets of numbers, sets of number pairs, sets of points, or sets of objects of any kind. The sets do not have to be the same. All the function has to do is assign some element from the second set to each element in the first set. Thus a function is like a machine that assigns an output to every allowable input. The inputs make up the **domain** of the function. The outputs make up the function's **range** (Fig. 1.32).

1.32 A flow diagram for a function f.

DEFINITION

> A **function** from a set D to a set R is a rule that assigns a single element of R to each element in D.

The word *single* in the definition of function does not mean that there must be only one element in the function's range, although this can happen. It means that each input from the domain is assigned exactly one output from the range, no more and no less. In other words, each input appears just once in the list of input-output pairs defined by the function.

Euler invented a symbolic way to say "y is a function of x" by writing

$$y = f(x) \tag{1}$$

which we read "y equals f of x." This notation is shorter than the verbal statements that say the same thing. It also lets us give different functions different names by changing the letters we use. To say that boiler pressure is a function of steam temperature, we can write $p = f(t)$. To say that the area of a circle is a function of its radius, we can write $A = g(r)$. (We use a g here because we just used f for something else.) We have to know what the variables p, t, A, and r mean, of course, for these equations to make sense.

Real-Valued Functions of a Real Variable

In much of our work, the functions will have domains and ranges that are sets of real numbers. Such functions are called **real-valued functions of a real**

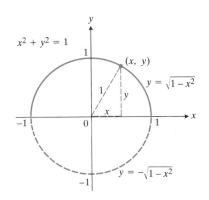

1.33 The circle $x^2 + y^2 = 1$. On the upper half, $y = \sqrt{1 - x^2}$; on the lower half, $y = -\sqrt{1 - x^2}$.

variable. They are often defined by formulas or equations, as in the following examples.

EXAMPLE 1 The formula $A = \pi r^2$ expresses the area A of a circle as a function of its radius r. If $r = 2$, for example, $A = \pi(2)^2 = 4\pi$. In the context of geometry, the domain D of the function is the set of all possible radii—in this case, the set of all positive real numbers. The range is also the set of all positive real numbers. ∎

EXAMPLE 2 The formula $y = x^2$ defines the number y to be the square of the number x. We might call this function the "squaring" function because the output numbers are the squares of the input numbers. If $x = 5$, for example, $y = (5)^2 = 25$. The usual name for the function is "the function $y = x^2$."

The domain of the function $y = x^2$ is the set of allowable values of x—in this case, the set of all real numbers. The range, made up of the resulting values of y, is the set of all nonnegative numbers. ∎

EXAMPLE 3 If a point (x, y) lies on the circle in the plane that is centered at the origin and has radius 1 unit (see Fig. 1.33), then x and y satisfy the equation

$$x^2 + y^2 = 1.$$

This is equivalent to saying that

$$y^2 = 1 - x^2$$

or

$$y = \pm\sqrt{1 - x^2},$$

which gives two possible formulas for y.

If the point (x, y) lies on the upper half of the circle, the semicircle with $y \geq 0$, then $y = +\sqrt{1 - x^2}$. If the point lies on the lower semicircle, then $y = -\sqrt{1 - x^2}$. Each formula defines y as a function of x whose domain is the interval from $x = -1$ to $x = 1$. The range of $y = \sqrt{1 - x^2}$ is $0 \leq y \leq 1$. The range of $y = -\sqrt{1 - x^2}$ is $-1 \leq y \leq 0$. ∎

REMARK The equation $x^2 + y^2 = 1$ does not define y as a single function of x because for each x between -1 and $+1$ there are *two* values of y.

Open interval a b
Symbols: $a < x < b$ or (a, b)

Closed interval a b
Symbols: $a \leq x \leq b$ or $[a, b]$

Half-open interval
Symbols: $a \leq x < b$ or $[a, b)$

Half-open interval
Symbols: $a < x \leq b$ or $(a, b]$

1.34 Finite intervals.

Domains and Ranges Are Often Intervals

The domains and ranges of many functions in mathematics are intervals of real numbers like the ones shown in Fig. 1.34. The set of all real numbers that lie *strictly between* two fixed numbers a and b is an **open interval.** The interval is "open" at each end because it contains neither of its endpoints. Intervals that contain both endpoints are **closed.** Intervals that contain one endpoint but not both are **half-open.** (Half-open intervals could just as well be called half-closed, but no one seems to call them that.)

The domains and ranges of functions can also be **infinite intervals** like the ones in Fig. 1.35.

The set of all real numbers, $(-\infty, \infty)$

The set of numbers greater than a, (a, ∞)

The set of numbers greater than or equal to a, $[a, \infty)$

The set of numbers less than b, $(-\infty, b)$

The set of numbers less than or equal to b, $(-\infty, b]$

1.35 Rays on the number line and the line itself are called *infinite intervals*. The symbol ∞ (infinity) is used in the notation merely for convenience; it is not to be taken as a suggestion that there is a number ∞.

EXAMPLE 4

Function	Domain	Range
$y = x^2$	$-\infty < x < \infty$	$0 \le y$
$y = \sqrt{1 - x^2}$	$-1 \le x \le 1$	$0 \le y \le 1$
$y = \dfrac{1}{x}$	$x \neq 0$	$y \neq 0$
$y = \sqrt{x}$	$0 \le x$	$0 \le y$
$y = \sqrt{4 - x}$	$x \le 4$	$0 \le y$

In each case, the domain of the function is the largest set of real x-values for which the formula gives real y-values.

The formula $y = x^2$ gives a real y-value for any real number x.

The formula $y = \sqrt{1 - x^2}$ gives a real y-value for every value of x in the closed interval from -1 to 1. Beyond this domain, the quantity $1 - x^2$ is negative and its square root is not a real number. (Complex numbers of the form $a + bi$, where $i = \sqrt{-1}$, are excluded from our consideration until Chapter 12.)

The formula $y = 1/x$ gives a real y-value for every x except $x = 0$. We cannot divide 1 (or any other number, for that matter) by 0.

The formula $y = \sqrt{x}$ gives a real y-value only when x is positive or zero. The number $y = \sqrt{x}$ is not a real number when x is negative. The domain of $y = \sqrt{x}$ is therefore the interval $x \ge 0$.

In $y = \sqrt{4 - x}$, the quantity $4 - x$ cannot be negative. That is, $4 - x$ must be greater than or equal to 0. In symbols,

$$0 \le 4 - x \tag{2}$$

or

$$x \le 4. \tag{3}$$

The formula $y = \sqrt{4 - x}$ gives a real y-value for any x less than or equal to 4.   ∎

Independent and Dependent Variables, a Warning about Division by 0, and a Convention about Domains

The variable x in a function $y = f(x)$ is called the **independent variable,** or **argument,** of the function. The variable y, whose value depends on x, is called the **dependent variable** of the function.

We must keep two restrictions in mind when we define functions. First, we *never divide by 0*. When we see $y = 1/x$, we must think $x \neq 0$. Zero is not in the domain of the function. When we see $y = 1/(x - 2)$, we must think "$x \neq 2$." Second, we will deal with real-valued functions only (except for a very short while later in the book). We may therefore have to restrict our domains when we have square roots (or fourth roots, or other even roots). If $y = \sqrt{1 - x^2}$, we should think "x^2 must not be greater than 1. The domain must not extend beyond the interval $-1 \le x \le 1$."

We observe a convention about the domains of functions defined by formulas. If the domain is not stated explicitly, then the domain is automatically the largest set of x-values for which the formula gives real y-values. If we wish to exclude any values from this domain, we must say so. The formula $y = x^2$ gives real y-values for every real x. Therefore, writing

$$y = x^2$$

without restriction tells everyone the intended domain is $-\infty < x < \infty$. To exclude negative values from the domain, we would write

$$y = x^2, \qquad x \geq 0.$$

Graphs and Graphing

The points in the plane whose coordinate pairs are the input-output pairs of a function make up the **graph** of the function. For example, the line L in Fig. 1.30 is the graph of the function $y = x + 2$ because it is the set of points in the plane whose coordinate pairs (x, y) are the function's input-output pairs.

EXAMPLE 5   Graph the function $y = x^2$ over the interval $-2 \leq x \leq 2$.

Solution   To graph the function, we carry out the following steps:

1. Make a table of input-output pairs for the function (Fig. 1.36a).
2. Plot the corresponding points to show the shape of the graph (Fig. 1.36b).
3. Sketch the graph by connecting the points (Fig. 1.37). ∎

In Example 5 we graphed $y = x^2$ over the interval $-2 \leq x \leq 2$. What about the rest of the graph? The domain and range of $y = x^2$ are both infinite, so we cannot hope to draw the entire graph. But we can imagine what the graph looks like by examining the formula $y = x^2$ and by looking at the picture we already have. As x moves away from the interval $-2 \leq x \leq 2$ in either direction, $y = x^2$ increases rapidly. When x is 5, y is 25. When x is 10, y is 100. The graph goes up as shown in Fig. 1.38, continuing the pattern we see in Fig. 1.37.

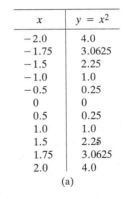

x	$y = x^2$
-2.0	4.0
-1.75	3.0625
-1.5	2.25
-1.0	1.0
-0.5	0.25
0	0
0.5	0.25
1.0	1.0
1.5	2.25
1.75	3.0625
2.0	4.0

(a)

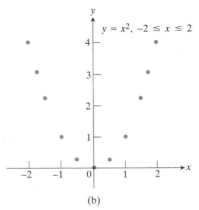

(b)

1.36   (a) Table of input-output pairs for the function $y = x^2$. (b) The points plotted from part (a).

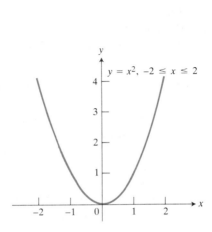

1.37   Figure 1.36 filled in.

1.38   Figure 1.37 extended.

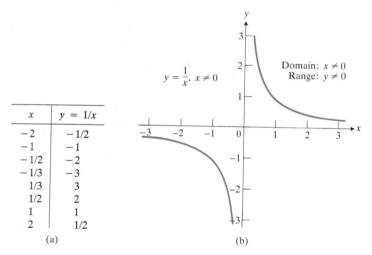

x	$y = 1/x$
-2	$-1/2$
-1	-1
$-1/2$	-2
$-1/3$	-3
$1/3$	3
$1/2$	2
1	1
2	$1/2$

(a)   (b)

1.39   (a) Values of $y = 1/x$ for selected values of x. (b) The graph of $y = 1/x$.

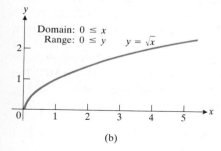

x	$y = \sqrt{x}$
0	0
$1/4$	$1/2$
1	1
2	$\sqrt{2}$
4	2

(a)

(b)

1.40   (a) Values of $y = \sqrt{x}$ for selected values of x. (b) The graph of $y = \sqrt{x}$.

The basic idea for graphing curves that are not straight lines is to plot points until we see the curve's shape and then to use the formula for y to find out how y changes as x moves between or away from the plotted values. But what points do we plot?

Here are some rules about choosing good points to plot. In Chapter 3, we shall have more to say about using the formula $y = f(x)$ to predict how y changes between plotted points.

Choosing Points for Graphing $y = f(x)$

1. Plot any points where the graph crosses or touches the axes. These points are often easy to find by setting $y = 0$ and $x = 0$ in the equation $y = f(x)$.

2. Plot a few points near the origin. When the values of x are small, the values of y are often easy to compute or estimate.

3. Graph the function at or near any endpoints of its domain.

Figures 1.39, 1.40, and 1.41 show the graphs of the functions $y = 1/x$, $y = \sqrt{x}$, and $y = \sqrt{4 - x}$ from Example 4. The graph of the function $y = \sqrt{1 - x^2}$ from Example 4 is the upper semicircle in Fig. 1.33.

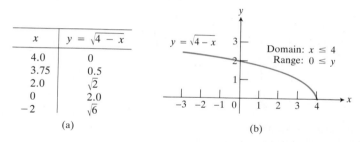

x	$y = \sqrt{4 - x}$
4.0	0
3.75	0.5
2.0	$\sqrt{2}$
0	2.0
-2	$\sqrt{6}$

(a)   (b)

1.41   (a) Values of $y = \sqrt{4 - x}$ for selected values of x. (b) The graph of $y = \sqrt{4 - x}$.

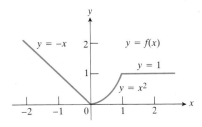

1.42 The function $y = f(x)$ shown here is graphed by applying different formulas to different parts of the function's domain.

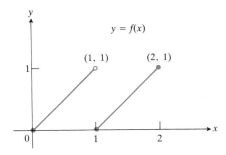

1.43 The graph of the function $y = f(x)$ (Example 7) shown here consists of two line segments. The segment on the left contains the endpoint at the origin (shown by a heavy dot) but does not contain the endpoint (1, 1). The segment on the right contains both endpoints.

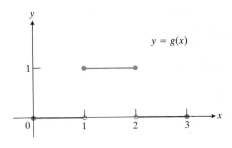

1.44 Functions like the one graphed here are called *step functions.* Example 8 shows how to write a formula for g.

Functions Defined in Pieces

The functions graphed so far have been defined over their domains by single formulas. Some functions, however, are defined by applying different formulas to different parts of their domains. The function in the following example is defined with three different formulas. One applies to the interval $x < 0$, another to the interval $0 \leq x \leq 1$, another to the interval $x > 1$. The function is *just one function,* however. Its domain is $-\infty < x < \infty$.

EXAMPLE 6   The values of the function

$$y = f(x) = \begin{cases} -x & \text{if } x < 0, \\ x^2 & \text{if } 0 \leq x \leq 1, \\ 1 & \text{if } x > 1, \end{cases}$$

are given by the formulas $y = -x$ when $x < 0$, $y = x^2$ when $0 \leq x \leq 1$, and $y = 1$ when $x > 1$. See Fig. 1.42.   ■

EXAMPLE 7   Suppose that the graph of a function $y = f(x)$ consists of the line segments shown in Fig. 1.43. Write a formula for f.

Solution   We find formulas for the segments from (0, 0) to (1, 1) and from (1, 0) to (2, 1) and piece them together in the manner of Example 6.

Segment from (0, 0) to (1, 1).   The line through (0, 0) and (1, 1) has slope $m = (1 - 0)/(1 - 0) = 1$ and y-intercept $b = 0$. Its slope-intercept equation is therefore $y = x$. The segment from (0, 0) to (1, 1) that includes the point (0, 0) but not the point (1, 1) is the graph of the function $y = x$ restricted to the half-open interval $0 \leq x < 1$, namely,

$$y = x, \qquad 0 \leq x < 1.$$

Segment from (1, 0) to (2, 1).   The line through (1, 0) and (2, 1) has slope $m = (1 - 0)/(2 - 1) = 1$ and passes through the point (1, 0). The corresponding point-slope equation for the line is therefore

$$y - 0 = 1(x - 1), \qquad \text{or} \qquad y = x - 1.$$

The segment from (1, 0) to (2, 1) that includes both endpoints is the graph of $y = x - 1$ restricted to the closed interval $1 \leq x \leq 2$, namely,

$$y = x - 1, \qquad 1 \leq x \leq 2.$$

Formula for the function $y = f(x)$ shown in Fig. 1.43.   The values of f on the interval $0 \leq x \leq 2$ are given by combining the formulas we obtained for the two segments of the graph:

$$f(x) = \begin{cases} x & \text{for } 0 \leq x < 1, \\ x - 1 & \text{for } 1 \leq x \leq 2. \end{cases}$$ ■

EXAMPLE 8   The domain of the "step" function $y = g(x)$ graphed in Fig. 1.44 is the closed interval $0 \leq x \leq 3$. Find a formula for $g(x)$.

Solution   The graph consists of three horizontal line segments. The left segment is the half-open interval $0 \leq x < 1$ on the x-axis, which we may think of

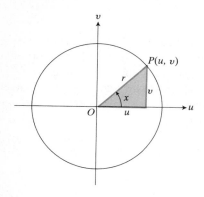

1.45 The angle x in standard position.

as a portion of the line $y = 0$:

$$y = 0, \qquad 0 \le x < 1.$$

The second segment is the portion of the line $y = 1$ that lies over the closed interval $1 \le x \le 2$:

$$y = 1, \qquad 1 \le x \le 2.$$

The third segment is the half-open interval $2 < x \le 3$ on the line $y = 0$:

$$y = 0, \qquad 2 < x \le 3.$$

The values of g are therefore given by the three-piece formula

$$g(x) = \begin{cases} 0 & \text{for } 0 \le x < 1, \\ 1 & \text{for } 1 \le x \le 2, \\ 0 & \text{for } 2 < x \le 3. \end{cases}$$

■

Sines, Cosines, and Tangents

When an angle of measure x (degrees or radians) is placed in standard position in the center of a circle of radius r, as in Fig. 1.45, the values of the sine, cosine, and tangent of the angle are given by the formulas

$$\sin x = \frac{v}{r}, \qquad \cos x = \frac{u}{r}, \qquad \tan x = \frac{v}{u}. \tag{4}$$

The graphs of the functions $y = \sin x$, $y = \cos x$, and $y = \tan x$ in the xy-plane are shown in Fig. 1.46.

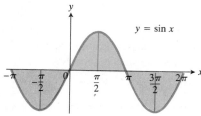

Function: $y = \sin x$
Domain: $-\infty < x < \infty$
Range: $-1 \le y \le 1$

(a)

1.46 Graphs of the (a) sine, (b) cosine, and (c) tangent functions.

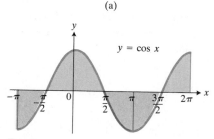

Function: $y = \cos x$
Domain: $-\infty < x < \infty$
Range: $-1 \le y \le 1$

(b)

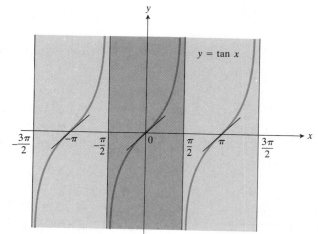

Function: $y = \tan x$
Domain: All real numbers except odd integer multiples of $\pi/2$.
Range: $-\infty < y < \infty$

(c)

TABLE 1.1
Values of sin x, cos x, and tan x for selected values of x.

Degrees	-180	-135	-90	-45	0	45	90	135	180
x (radians)	$-\pi$	$-3\pi/4$	$-\pi/2$	$-\pi/4$	0	$\pi/4$	$\pi/2$	$3\pi/4$	π
sin x	0	$-\sqrt{2}/2$	-1	$-\sqrt{2}/2$	0	$\sqrt{2}/2$	1	$\sqrt{2}/2$	0
cos x	-1	$-\sqrt{2}/2$	0	$\sqrt{2}/2$	1	$\sqrt{2}/2$	0	$-\sqrt{2}/2$	-1
tan x	0	1		-1	0	1		-1	0

The function tan x is not defined for angles for which the denominator u is 0. This means angles of $\pm 90°$, $\pm 270°$, and so on. These angles are excluded from the domain of the tangent function. In radian measure, the excluded angles are $\pm\pi/2$, $\pm 3\pi/2$, ... See Table 1.1 and Fig. 1.46. Notice that

$$\tan x = \frac{\sin x}{\cos x}.$$

This helps to explain why the tangent's graph "blows up" as x nears an odd-integer multiple of $\pi/2$; these are the points where the sine is 1 and the cosine is 0.

Astronomers, navigators, and surveyors measure angles in degrees, but in calculus it is best to measure angles in radians, as we shall see in Chapter 2. For this reason, Table 1.1 shows radians as well as degrees. Do not be alarmed if your grasp of radian measure and trigonometry seems rusty right now. The problems at the end of this article are well within your reach, and we shall review trigonometry in Chapter 2 before we use it in a more active way.

EXAMPLE 9   Find the domain and range of

a) $y = \sin^2 x$,

b) $y = 5\cos 2x$,

c) $y = -\tan x$.

Solution

a) The function $y = \sin^2 x$ is defined wherever sin x is defined, and its values are the squares of the values of sin x. Its domain is the set of all real numbers. Since the values of sin x fill the interval from -1 to 1, the values of $\sin^2 x$ fill the interval from 0 to 1. The range of $y = \sin^2 x$ is the interval $0 \le y \le 1$.

b) To calculate $y = 5\cos 2x$, we multiply x by 2, find the cosine, and multiply by 5. The function is defined for all real values of x. Since $\cos 2x$ ranges from -1 to 1, $y = 5\cos 2x$ ranges from -5 to 5. We conclude: Domain $-\infty < x < \infty$; range $-5 \le y \le 5$.

c) To calculate $y = -\tan x$, we calculate tan x and multiply by -1. The function is defined wherever tan x is defined, which means for all real x except odd-integer multiples of $\pi/2$. The range is the set of all real values, $-\infty < y < \infty$. ∎

Sums, Differences, Products, and Quotients of Functions

If $f(x)$ and $g(x)$ are two functions, with domains D_f and D_g, then the

sum	$f(x) + g(x),$	
differences	$f(x) - g(x),$	$g(x) - f(x),$
product	$f(x) \cdot g(x),$	
quotients	$\dfrac{f(x)}{g(x)}, \quad g(x) \neq 0,$	$\dfrac{g(x)}{f(x)}, \quad f(x) \neq 0,$

are also functions of x, defined for any value of x that lies in both D_f and D_g. The points at which $g(x) = 0$ must be excluded, however, to obtain the domain of the quotient $f(x)/g(x)$. Likewise, any points at which $f(x) = 0$ must be excluded from the domain of the quotient $g(x)/f(x)$.

EXAMPLE 10   Give the domains of

$$f(x) = \sqrt{x}, \qquad g(x) = \sqrt{1 - x},$$

and the corresponding domains of $f + g$, $f - g$, $g - f$, $f \cdot g$, f/g, and g/f.

Solution   See Fig. 1.47. The domains of f and g are

$$D_f = [0, \infty), \qquad D_g = (-\infty, 1].$$

The points common to these domains are the points of the closed interval $[0, 1]$. On $[0, 1]$, we have

sum $f + g$:	$f(x) + g(x) = \sqrt{x} + \sqrt{1 - x},$
differences $f - g$:	$f(x) - g(x) = \sqrt{x} - \sqrt{1 - x},$
$g - f$:	$g(x) - f(x) = \sqrt{1 - x} - \sqrt{x},$
product $f \cdot g$:	$f(x) \cdot g(x) = \sqrt{x(1 - x)},$
quotients f/g:	$\dfrac{f(x)}{g(x)} = \sqrt{\dfrac{x}{1 - x}}, \quad x \neq 1,$
g/f:	$\dfrac{g(x)}{f(x)} = \sqrt{\dfrac{1 - x}{x}}, \quad x \neq 0.$

1.47   The domain of the function $f + g$ is the intersection of the domains of f and g, the interval $[0, 1]$ on the x-axis where these domains overlap. This interval is also the domain of the function $f \cdot g$. See Example 10.

1.48 Two functions can be composed when the range of the first lies in the domain of the second.

The domains of $f + g, f - g, g - f,$ and $f \cdot g$ are all the same closed interval [0, 1].

The number $x = 1$ must be excluded from the domain of f/g because $g(1) = \sqrt{1 - 1} = 0$. The domain of f/g is therefore the half-open interval [0, 1).

Similarly, the number $x = 0$ must be excluded from the domain of g/f because $f(0) = \sqrt{0} = 0$. The domain of g/f is therefore the half-open interval (0, 1]. ■

Composition of Functions

Suppose that the outputs of a function f can be used as inputs of a function g. We can then hook f and g together to form a new function whose inputs are the inputs of f and whose outputs are the numbers $g(f(x))$, as in Fig. 1.48. We say that the function $g(f(x))$ (pronounced "g of f of x") is the **composite** of f and g. It is made by *composing* f and g in the order first f, then g. The usual "stand-alone" notation for this composite is $g \circ f$, which is read as "g of f." Thus, the value of $g \circ f$ at x is $(g \circ f)(x) = g(f(x))$.

EXAMPLE 11 If $f(x) = \sin x$ and $g(x) = -x/2$, write a formula for the composite $g(f(x))$.

Solution As suggested by Fig. 1.48, we can obtain a formula for $g(f(x))$ by substituting $f(x) = \sin x$ for the input variable x in $g(x) = -x/2$:

$$g(x) = -\frac{x}{2},$$

$$g(f(x)) = -\frac{f(x)}{2} = -\frac{\sin x}{2}.$$

The formula we seek is

$$g(f(x)) = -\frac{\sin x}{2}. \qquad ■$$

EXAMPLE 12 Find a formula for $g(f(x))$ if $f(x) = x^2$ and $g(x) = x - 7$. Then find the value of $g(f(2))$.

Solution To find $g(f(x))$, we replace x in the formula for $g(x)$ by the expression given for $f(x)$:

$$g(x) = x - 7,$$
$$g(f(x)) = f(x) - 7 = x^2 - 7.$$

We then find the value of $g(f(2))$ by substituting 2 for x:

$$g(f(2)) = (2)^2 - 7 = 4 - 7 = -3. \qquad ■$$

Changing the order in which functions are composed usually changes the result. In Example 12, we composed $f(x) = x^2$ and $g(x) = x - 7$ in the order first f, then g, obtaining the function $g \circ f$ whose value at x was $g(f(x)) = x^2 - 7$. In the next example, we see what happens when we reverse the order to obtain the function $f \circ g$.

EXAMPLE 13 Find a formula for $f(g(x))$ if $f(x) = x^2$ and $g(x) = x - 7$. Then find $f(g(2))$.

Solution To find $f(g(x))$, we replace x in the formula for $f(x)$ by the expression for $g(x)$:

$$f(x) = x^2,$$

$$f(g(x)) = (g(x))^2 = (x - 7)^2.$$

To find $f(g(2))$, we then substitute 2 for x:

$$f(g(2)) = (2 - 7)^2 = (-5)^2 = 25. \qquad \blacksquare$$

REMARK In the notation for composite functions, the parentheses tell which function comes first:

The notation $g(f(x))$ says "first f, then g." To calculate $g(f(2))$, we calculate $f(2)$ and then apply g.

The notation $f(g(x))$ says "first g, then f." To calculate $f(g(2))$, we calculate $g(2)$ and then apply f.

PROBLEMS

In Problems 1–12, find the domain and range of each function.

1. $y = 2\sqrt{x}$

2. $y = 1 + \sqrt{x}$

3. $y = -\sqrt{x}$

4. $y = \sqrt{-x}$

5. $y = \sqrt{x + 4}$

6. $y = \sqrt{x - 2}$

7. $y = \dfrac{1}{x - 2}$

8. $y = \dfrac{1}{x + 2}$

9. $y = 2 \cos x$

10. $y = -\cos x$

11. $y = -3 \sin x$

12. $y = 2 \sin 4x$

In Problems 13–30, find (a) the domain and (b) the range of the function. Then (c) graph the function.

13. $y = x^2 + 1$

14. $y = x^2 - 2$

15. $y = -x^2$

16. $y = 4 - x^2$

17. $y = \sqrt{x + 1}$

18. $y = \sqrt{4 - x}$

19. $y = 1 + \sqrt{x}$

20. $y = \sqrt{9 - x^2}$

21. $y = (\sqrt{2x})^2$

22. $y = \dfrac{2}{x}$

23. $y = -\dfrac{1}{x}$

24. $y = \dfrac{1}{x^2}$

25. $y = \sin 2x$

26. $y = \cos 2x$

27. $y = \sin^2 x$

28. $y = \cos^2 x$

29. $y = 1 + \sin x$

30. $y = 1 - \cos x$

31. Consider the function $y = 1/\sqrt{x}$.
   a) Can x be negative?
   b) Can $x = 0$?
   c) What is the domain of the function?

32. Consider the function $y = \sqrt{2 - \sqrt{x}}$.
   a) Can x be negative?
   b) Can $\sqrt{x}$ be greater than 2?
   c) What is the domain of the function?

33. Consider the function $y = \sqrt{(1/x) - 1}$.
   a) Can x be negative?
   b) Can $x = 0$?
   c) Can x be greater than 1?
   d) What is the domain of the function?

34. Consider the function $y = \sqrt{(1 + \cos 2x)/2}$.
   a) Can x take on any real value?
   b) How large can $\cos 2x$ become? How small?
   c) How large can $(1 + \cos 2x)/2$ become? How small?
   d) What are the domain and range of $y = \sqrt{(1 + \cos 2x)/2}$?

35. Consider the function $y = \tan(x/2)$.
   a) What values of $x/2$ must be excluded from the domain of $\tan(x/2)$?
   b) What values of x must be excluded from the domain of $\tan(x/2)$?
   c) What values does $y = \tan(x/2)$ assume on the interval $-\pi < x < \pi$?
   d) What are the domain and range of $y = \tan(x/2)$?

36. Which of the graphs in Fig. 1.49 could be the graph of
   a) $y = x^2 - 1$? Why?
   b) $y = (x - 1)^2$? Why?

37. Which of the graphs in Fig. 1.49 could *not* be the graph of $y = 4x^2$? Why?

38. By solving $y^2 = x$ for y, replace the equation by an equivalent system of equations each of which determines y as a function of x. Graph these two equations. (*Hint:* See Fig. 1.40.)

i)

ii)

iii)

iv)

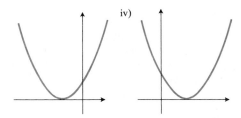

1.49 The graphs for Problems 36 and 37.

e)

f)

In Problems 45–47, give the domains of f and g and the corresponding domains of $f + g$, $f - g$, $f \cdot g$, f/g, and g/f.

45. $f(x) = x$, $g(x) = \sqrt{x - 1}$

46. $f(x) = \dfrac{1}{x - 2}$, $g(x) = \dfrac{1}{\sqrt{x - 1}}$

47. $f(x) = \sqrt{x}$, $g(x) = \sqrt[4]{x + 1}$

48. If $g(x) = 37 - x - 3x^3 + x^4$, then $g(-2)$ is
  a) 31.   b) 75.
  c) 79.   d) None of the above.

49. If $h(x) = 1 + 5/x$, find
  a) $h(-1)$,   b) $h(1/2)$,
  c) $h(5)$,   d) $h(5x)$,
  e) $h(10x)$,   f) $h(1/x)$.

50. If $f(x) = x + 5$ and $g(x) = x^2 - 3$, find
  a) $g(f(0))$,   b) $f(g(0))$,
  c) $g(f(x))$,   d) $f(g(x))$,
  e) $f(f(-5))$,   f) $g(g(2))$,
  g) $f(f(x))$,   h) $g(g(x))$.

51. Let $f(x) = (x - 1)/x$. Show that $f(x) \cdot f(1 - x) = 1$.

52. If $f(x) = 1/x$, find
  a) $f(2)$,   b) $f(x + 2)$,
  c) $(f(x + 2) - f(2))/2$.

53. If $F(t) = 4t - 3$, find $(F(t + h) - F(t))/h$.

54. Copy and complete the following.

$f(x)$	$g(x)$	$(g \circ f)(x)$
a) $x - 7$	$\sqrt{x}$	
b) $x + 2$	$3x$	
c)	$\sqrt{x - 5}$	$\sqrt{x^2 - 5}$
d) $\dfrac{x}{x - 1}$	$\dfrac{x}{x - 1}$	
e)	$1 + \dfrac{1}{x}$	x
f) $\dfrac{1}{x}$		x

39. Make a table of values with $x = 0$, 1, and 2, and graph the function

$$y = \begin{cases} x & \text{when } 0 \le x \le 1, \\ 2 - x & \text{when } 1 \le x \le 2. \end{cases}$$

Graph the functions in Problems 40–43.

40. $y = \begin{cases} 3 - x, & x \le 1, \\ 2x, & 1 < x \end{cases}$
     41. $y = \begin{cases} 1/x, & x < 0, \\ x, & 0 \le x \end{cases}$

42. $y = \begin{cases} 1, & x < 5, \\ 0, & 5 \le x \end{cases}$
     43. $y = \begin{cases} 1, & x < 0, \\ \sqrt{x}, & x \ge 0 \end{cases}$

44. Find formulas for the functions graphed in the following figures.

a)

b)

c)

d)

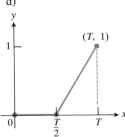

TOOLKIT PROGRAMS

Function Evaluator   Super * Grapher
Name That Function

1.5

Absolute Values

In this article we define the absolute value of a real number and look at the properties of absolute values that make them useful in calculus. We also examine the "greatest integer" function, a function useful in mathematics and computer science.

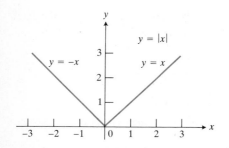

1.50   The absolute value function.

The Absolute Value Function

The absolute value of a number x is the number $\sqrt{x^2}$. If x is positive, its absolute value is simply x. If x is negative, however, its absolute value is $-x$. If x is zero, its absolute value is zero. The notation for the absolute value of x is $|x|$, which is read "the absolute value of x." Thus,

$$|x| = \sqrt{x^2} = \begin{cases} x & \text{if } x \geq 0, \\ -x & \text{if } x \leq 0. \end{cases} \tag{1}$$

The function $y = |x|$ is graphed in Fig. 1.50.

DEFINITION

> The **absolute value** of a number x is the number $|x| = \sqrt{x^2}$.

EXAMPLE 1   The absolute value of 3 is $|3| = 3$. The absolute value of 0 is $|0| = 0$. The absolute value of -5 is $|-5| = -(-5) = 5$.   ■

EXAMPLE 2   Solve the equation $|2x - 3| = 7$.

Solution   From $|2x - 3| = 7$ it follows that

$$2x - 3 = \pm 7$$
$$2x = 3 \pm 7$$
$$2x = 10 \quad \text{or} \quad -4$$
$$x = 5 \quad \text{or} \quad -2.$$

The equation has two solutions, $x = 5$ and $x = -2$.   ■

The absolute value of a product of two numbers is the product of their absolute values. In symbols, we have

$$|ab| = |a||b| \qquad \text{for all numbers } a \text{ and } b. \tag{2}$$

EXAMPLE 3   Examples of $|ab| = |a||b|$:

$$|(-1)(4)| = |-1||4| = (1)(4) = 4,$$
$$|3x| = |3||x| = 3|x|,$$
$$|-2(x + 5)| = |-2||x + 5| = 2|x + 5|.$$ ■

Equation (2) holds because

$$|ab| \;=\; \sqrt{(ab)^2} \;=\; \sqrt{a^2 b^2} \;=\; \sqrt{a^2}\sqrt{b^2} \;=\; |a||b|. \qquad (3)$$

⇧ Definition of $|ab|$ ⇧ Because $(ab)^2 = a^2 b^2$ ⇧ The square root of a product of nonnegative numbers is the product of their square roots.   ⇧ Definitions of $|a|$ and $|b|$

The absolute value of a sum of two numbers is never larger than the sum of their absolute values. When we put this in symbols, we get an inequality known as the triangle inequality.

The Triangle Inequality

$$|a + b| \le |a| + |b| \qquad \text{for all numbers } a \text{ and } b. \qquad (4)$$

EXAMPLE 4   Examples of $|a + b| \le |a| + |b|$:

$$|0 + 5| = 5 \le |0| + |5| = 0 + 5 = 5,$$
$$|-3 + 0| = 3 \le |-3| + |0| = 3 + 0 = 3,$$
$$|3 + 5| = 8 \le |3| + |5| = 3 + 5 = 8,$$
$$|-3 - 5| = 8 \le |-3| + |-5| = 3 + 5 = 8.$$

In all four cases, $|a + b|$ equals $|a| + |b|$.

On the other hand,

$$|-3 + 5| = |2| < |-3| + |5| = 8,$$
$$|3 - 5| = |-2| < |3| + |-5| = 8.$$

In both cases, $|a + b|$ is less than $|a| + |b|$.   ∎

The rule is that $|a + b|$ is less than $|a| + |b|$ when a and b differ in sign. In all other cases, $|a + b|$ equals $|a| + |b|$.

Notice that the absolute value bars in expressions like $|-3 + 5|$ also work like parentheses: We do the addition *before* taking the absolute value.

Absolute Values and Distances

The numbers $|a - b|$ and $|b - a|$ are always equal because

$$|a - b| = |(-1)(b - a)| = |-1||b - a| = |b - a|.$$

They give the distance between a and b on the number line (Fig. 1.51). This is consistent with the square root formula for distance in the plane because

$$\sqrt{(a - b)^2 + (0 - 0)^2} = \sqrt{(a - b)^2} = |a - b|, \qquad (5)$$
$$\sqrt{(0 - 0)^2 + (a - b)^2} = \sqrt{(a - b)^2} = |a - b|. \qquad (6)$$

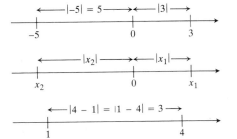

1.51   Absolute values give distances between points on the axes.

$$|a - b| = |b - a| \qquad \text{for all numbers } a \text{ and } b. \qquad (7)$$

This number is the distance between a and b on the number line.

Using Absolute Values to Define Intervals

The connection between absolute values and distance lets us use absolute value inequalities to describe intervals.

An inequality like $|a| < 5$ says that the distance from a to the origin is less than 5 units. This is equivalent to saying that a lies between -5 and 5. In symbols,

$$|a| < 5 \quad \Leftrightarrow \quad -5 < a < 5. \qquad (8)$$

The set of numbers a with $|a| < 5$ is the same as the open interval from -5 to 5 (Fig. 1.52).

In general, if c is any positive number, then the absolute value of a is less than c if and only if a lies in the interval between $-c$ and c.

1.52 $|a| < 5$ means $-5 < a < 5$.

$$|a| < c \quad \Leftrightarrow \quad -c < a < c. \qquad (9)$$

EXAMPLE 5   Find the values of x that satisfy the inequality $|x - 5| < 9$.

Solution   We first use Eq. (9) with $a = x - 5$ and $c = 9$ to change

$$|x - 5| < 9$$

to

$$-9 < x - 5 < 9. \qquad (10)$$

Next, we add 5 to all three quantities in Eq. (10). This isolates x:

$$-9 + 5 < x - 5 + 5 < 9 + 5$$

$$-4 < x < 14.$$

1.53 $|x - 5| < 9$ means $-4 < x < 14$.

These steps show that the values of x that satisfy the inequality $|x - 5| < 9$ are the numbers in the interval $-4 < x < 14$. (See Fig. 1.53.) ∎

EXAMPLE 6   Find the values of x that satisfy the inequality

$$\left| \frac{3x + 1}{2} \right| < 1.$$

Solution   Change

$$\left| \frac{3x + 1}{2} \right| < 1$$

to

$$-1 < \frac{3x + 1}{2} < 1$$

to

$$-2 < 3x + 1 < 2$$

1.54 The inequality

$$\left|\frac{3x+1}{2}\right| < 1$$

holds on the interval $-1 < x < 1/3$.

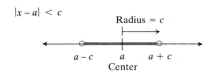

1.55 The open interval $|x - a| < c$ runs from $a - c$ to $a + c$.

to

$$-3 < 3x < 1$$

to

$$-1 < x < \frac{1}{3}.$$

(See Fig. 1.54.) ◼

EXAMPLE 7   Find the endpoints of the interval determined by the inequality $|x - a| < c$.

Solution   To find the endpoints of the interval $|x - a| < c$ (Fig. 1.55), we change

$$|x - a| < c$$

to

$$-c < x - a < c$$

to

$$a - c < x < a + c.$$

The endpoints are $a - c$ and $a + c$. ◼

The Greatest Integer Function

The greatest integer less than or equal to a number x is called the **greatest integer in x.** The symbol for it is $[x]$, which is read "the greatest integer in x." The function $y = [x]$ is graphed in Fig. 1.56.

EXAMPLE 8   *The value of $[x]$ for selected values of x:*

Positive values: $[1.9] = 1$, $[2] = 2$, $[3.4] = 3$
The value zero: $[0.5] = 0$, $[0] = 0$
Negative values: $[-1.2] = -2$, $[-0.5] = -1$

Notice that if x is negative, $[x]$ may have a larger absolute value than x does. ◼

In computer science, the common notation for the greatest integer in x is

$$\lfloor x \rfloor. \qquad \text{(integer floor)} \qquad (11)$$

It suggests an integer floor for x.

The companion notation

$$\lceil x \rceil \qquad \text{(integer ceiling)} \qquad (12)$$

is used for the smallest integer greater than or equal to x. It suggests an integer ceiling for x.

The greatest integer function is a **step function.** Many things around us can be modeled with step functions, such as

The price of priority mail, as a function of weight;

The output of a blinking light, as a function of time;

The number displayed by a machine that gives digital outputs, as a function of time.

Step functions exhibit points of **discontinuity,** where they jump from one value to another without taking on any of the intermediate values. As shown in

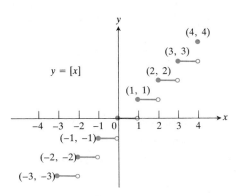

1.56   The graph of $y = [x]$, the greatest integer less than or equal to x. Domain: $-\infty < x < \infty$; range: the integers.

Fig. 1.56, $y = [x]$ jumps from $y = 1$ when $x < 2$ to $y = 2$ at $x = 2$ without taking on any of the values between 1 and 2.

Numerical Properties of Absolute Values

1. $|a| = \sqrt{a^2}$ (the definition of $|a|$)
2. $|ab| = |a||b|$
3. $\left|\dfrac{a}{b}\right| = \dfrac{|a|}{|b|}$
4. $|a - b| = |b - a|$
5. $|a + b| \le |a| + |b|$

Properties 1–5 hold for any numbers a and b.

Intervals and Absolute Values

6. $-c < a < c \iff |a| < c$
7. $a - c < x < a + c \iff |x - a| < c$

PROBLEMS

Find the absolute values in Problems 1–4.

1. $|-3|$ **2.** $|2 - 7|$

3. $|-2 + 7|$ **4.** $|1.1 - 5.2|$

Solve the equations in Problems 5–10.

5. $|x| = 2$ **6.** $|x - 3| = 7$ **7.** $|2x + 5| = 4$

8. $|1 - x| = 1$ **9.** $|8 - 3x| = 9$ **10.** $\left|\dfrac{x}{2} - 1\right| = 1$

In Problems 11–20, match each absolute value inequality with the interval it determines.

11. $|x| < 4$ a. $-2 < x < 1$

12. $|x + 3| < 1$ b. $-1 < x < 3$

13. $|x - 5| < 2$ c. $3 < x < 7$

14. $\left|\dfrac{x}{2}\right| < 1$ d. $-\dfrac{5}{2} < x < -\dfrac{3}{2}$

15. $|1 - x| < 2$ e. $-2 < x < 2$

16. $|2x - 5| \le 1$ f. $-4 < x < 4$

17. $|2x + 4| < 1$ g. $-4 < x < -2$

18. $\left|\dfrac{x - 1}{2}\right| < 1$ h. $2 \le x \le 3$

19. $\left|\dfrac{2x + 1}{3}\right| < 1$ i. $-2 \le x \le 2$

20. $|x^2 - 2| \le 2$

Each absolute value inequality in Problems 21–32 defines one or more intervals. Describe the intervals with inequalities that do not involve absolute values.

21. $|x| < 2$ **22.** $|x| \le 2$ **23.** $|x - 1| \le 2$

24. $|x - 1| < 2$ **25.** $|x + 1| < 3$ **26.** $|x + 2| \le 1$

27. $|2x + 2| < 1$ **28.** $|1 - x| < 1$ **29.** $|1 - 2x| \le 1$

30. $|3x - 6| < 1$ **31.** $\left|\dfrac{1}{x}\right| \le 1$ **32.** $\left|\dfrac{x}{2} - 1\right| \le 1$

In Problems 33–41, use absolute values to describe the given intervals of x- and y-values. It may help to draw a picture of the interval.

33. $-8 < x < 8$ **34.** $-3 < y < 5$

35. $-5 < x < 1$ **36.** $1 < y < 7$

37. $-a < y < a$ **38.** $-1 < x < 2$

39. $L - \epsilon < y < L + \epsilon$ (L and ϵ constant)

40. $1 - \delta < x < 1 + \delta$ (δ constant)

41. $x_0 - 5 < x < x_0 + 5$ (x_0 constant)

42. Do not fall into the trap $|-a| = a$. The equation does not hold for all values of a.
a) Find a value of a for which $|-a| \ne a$.
b) For what values of a does the equation $|-a| = a$ hold?

43. For what values of x does $|1 - x|$ equal $1 - x$? For what values of x does it equal $x - 1$?

i)

ii)

iii)

iv)

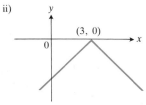

1.57   The graphs for Problem 44.

44. Which of the graphs in Fig. 1.57 is the graph of $y = |x - 3|$?

Graph the functions in Problems 45–48.

45. $y = -|x|$

46. $y = |x - 1|$

47. $y = \dfrac{|x| - x}{2}$

48. $y = \dfrac{|x| + x}{2}$

49. Compare the domains and ranges of the functions $y = \sqrt{x^2}$ and $y = (\sqrt{x})^2$.

50. Find $f(x)$ if $g(x) = \sqrt{x}$ and $(g \circ f)(x) = |x|$.

51. Find $g(x)$ if $f(x) = x^2 + 2x + 1$ and $(g \circ f)(x) = |x + 1|$.

52. Find functions $f(x)$ and $g(x)$ whose composites satisfy the two equations

$$(g \circ f)(x) = |\sin x| \quad \text{and} \quad (f \circ g)(x) = (\sin \sqrt{x})^2.$$

The Greatest Integer Function

53. Graph each function.
a) $y = x - [x], \quad -3 \le x \le 3$
b) $y = \left[\dfrac{x}{2}\right], \quad -3 \le x \le 3$
c) $y = [2x] - 2[x]$
d) $y = \dfrac{1}{2}([x] + x)$

54. For what values of x does $[x] = 0$?

55. When x is positive or zero, $[x]$ is the integer part of the decimal representation of x. What is the corresponding description of $[x]$ when x is negative?

 TOOLKIT PROGRAMS

Super * Grapher

1.6
Tangent Lines and the Slopes of Quadratic and Cubic Curves

We now get our first view of the role calculus plays in describing change. We begin with the average rate with which a quantity changes over a period of time and end with a way to describe the rate at which a quantity is changing at an instant. The transition is accomplished by identifying average rates of change with secant line slopes.

Average Rate of Change

We encounter average rates of change in such forms as average speeds (distance traveled divided by elapsed time, say, in miles per hour), growth rates of populations (in percent per year), and average monthly rainfall (in inches per month). The **average rate of change** in a quantity over a period of time is the amount of change in the quantity divided by the length of the time period.

Experimental biologists are often interested in the rates at which populations grow under controlled laboratory conditions. Figure 1.58 shows data from a fruit fly–growing experiment, the setting for our first example.

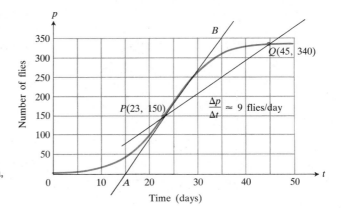

1.58   Growth of a fruit fly population in a controlled experiment. (*Source:* A. J. Lotka, *Elements of Mathematical Biology.* Dover, New York (1956), p. 69.)

EXAMPLE 1   *The average growth rate of a laboratory population.* The graph in Fig. 1.58 shows how the number of fruit flies *(Drosophila)* grew in a controlled 50-day experiment. The graph was made by counting flies at regular intervals, plotting a point for each count, and drawing a smooth curve through the plotted points.

There were 150 flies on day 23 and 340 flies on day 45. This gave an increase of $340 - 150 = 190$ flies in $45 - 23 = 22$ days. The average rate of change in the population from day 23 to day 45 was therefore

Average rate of change: $\dfrac{\Delta p}{\Delta t} = \dfrac{340 - 150}{45 - 23} = \dfrac{190}{22} \approx 9$ flies/day.   (1)

The average change in Eq. (1) is also the slope of the secant line through the two points

$$P(23,\ 150) \qquad \text{and} \qquad Q(45,\ 340)$$

on the population curve. (A line through two points on a curve is called a **secant** to the curve.) The slope of the secant PQ can be calculated from the coordinates of P and Q:

Secant slope: $\dfrac{\Delta p}{\Delta t} = \dfrac{340 - 150}{45 - 23} = \dfrac{190}{22} \approx 9$ flies/day.   (2)

By comparing Eqs. (1) and (2) we can see that the average rate of change in (1) is the same number as the slope in (2), units and all. We can always think of an average rate of change as the slope of a secant line.   ∎

In addition to knowing the average rate at which the population grew from day 23 to day 45, we may also want to know how fast the population was growing on day 23 itself. We can find out by watching the slope of the secant PQ change as we back Q along the curve toward P. The results for four positions of Q are shown in Fig. 1.59.

In terms of geometry, what we see as Q approaches P along the curve is this: The secant PQ approaches the tangent line AB that we drew by eye at P. This means that within the limitations of our drawing, the slopes of the secants approach the slope of the tangent, which we calculate from the coordinates of A and B to be

$$\frac{350 - 0}{35 - 15} = 17 \text{ flies/day.}$$

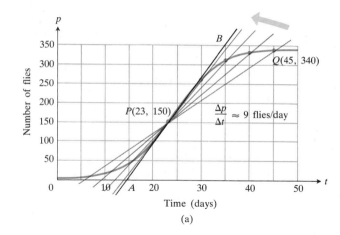

$\dfrac{\Delta p}{\Delta t} \approx 9$ flies/day

Q	Slope of $PQ = \Delta p/\Delta t$ (flies/day)
(45, 340)	$(340 - 150)/(45 - 23) \approx 9$
(40, 330)	$(330 - 150)/(40 - 23) \approx 13$
(35, 310)	$(310 - 150)/(35 - 23) \approx 15$
(30, 265)	$(265 - 150)/(30 - 23) \approx 16.4$

(a)

(b)

In terms of population change, what we see as Q approaches P is this: The average growth rates for increasingly smaller time intervals approach the slope of the tangent to the curve at P (17 flies per day). The slope of the tangent line is therefore the number we take as the rate at which the fly population was changing on day $t = 23$.

Defining Slopes and Tangent Lines

The moral of the fruit fly story would seem to be that we should define the rate at which the value of a function $y = f(x)$ is changing with respect to x at any particular value $x = x_1$ to be the slope of the tangent to the curve $y = f(x)$ at $x = x_1$. But how are we to define the tangent line at an arbitrary point P on the curve and deduce its slope from the formula $y = f(x)$?

It would be hard to overestimate how important it was to the scientists of the early seventeenth century to find an answer to this question. In optics, the angle at which a ray of light strikes the surface of a lens is defined in terms of the tangent to the surface. In physics, the direction of a body's motion at any point of its path is along the tangent to the path. In geometry, the angle between two intersecting curves is the angle between their tangents at the point of intersection.

The answer that Fermat finally found in 1629 proved to be one of that century's major contributions to calculus. We still use his method of defining tangents to produce formulas for slopes of curves and rates of change. It goes like this:

1. We start with what we *can* calculate, namely the slope of a secant through P and a point Q nearby on the curve.

2. We find the limiting value of the secant slope (if it exists) as Q approaches P along the curve.

3. We take this number to be the slope of the curve at P and define the tangent to the curve at P to be the line through P with this slope.

EXAMPLE 2   Find the slope of the parabola $y = x^2$ at the point $P(2, 4)$. Write an equation for the tangent to the parabola at this point.

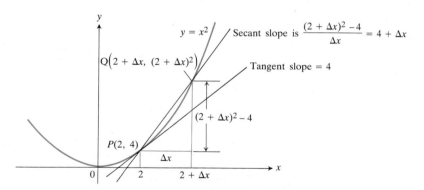

1.60 The slope of the secant PQ approaches 4 as Q approaches P along the curve.

Solution We begin with a secant line that passes through $P(2, 4)$ and a neighboring point $Q(2 + \Delta x, (2 + \Delta x)^2)$ on the curve (Fig. 1.60). We then write an expression for the slope of the secant and watch what happens to the slope as Q approaches P along the curve.

The slope of the secant PQ is

$$m_{\text{sec}} = \frac{\Delta y}{\Delta x} = \frac{(2 + \Delta x)^2 - 2^2}{\Delta x} = \frac{4 + 4\,\Delta x + (\Delta x)^2 - 4}{\Delta x}$$

$$= \frac{4\,\Delta x + (\Delta x)^2}{\Delta x} = 4 + \Delta x. \tag{3}$$

The notation m_{sec} is read ''m secant.''

As Q approaches P along the curve, Δx approaches 0 and $m_{\text{sec}} = 4 + \Delta x$ approaches $4 + 0 = 4$. We describe this behavior by saying that the limit of m_{sec} as Δx approaches 0 is 4. We define this limit to be the slope of the parabola at P. What this means geometrically is that as Q moves toward P along the curve, the secant line PQ approaches the line through P whose slope is $m = 4$. This is the line we define to be the tangent to the parabola $y = x^2$ at the point $P(2, 4)$. Its point-slope equation is obtained in the usual way:

Point:	$(2, 4)$
Slope:	$m = 4$
Equation:	$y - 4 = 4(x - 2)$
	$y - 4 = 4x - 8$
	$y = 4x - 4$ ■

The next example shows how to find a formula for the slope at any point of the parabola $y = x^2$.

EXAMPLE 3 Find a formula for the slope of the parabola $y = x^2$ at any point on the curve.

Solution We carry out the steps of Fermat's method.

Let $P(x_1, x_1^2)$ be any point on the parabola and let Q be a point on the parabola near P (Fig. 1.61). The coordinates of Q can be written as $(x_1 + \Delta x, (x_1 + \Delta x)^2)$, where Δx is the run from P to Q. In terms of these coordi-

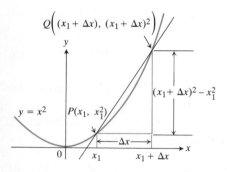

1.61 The slope of the secant PQ is $[(x_1 + \Delta x)^2 - x_1^2]/(\Delta x) = 2x_1 + \Delta x$.

nates, the slope of the secant PQ is

$$m_{\text{sec}} = \frac{\Delta y}{\Delta x} = \frac{(x_1 + \Delta x)^2 - x_1^2}{\Delta x} = \frac{x_1^2 + 2x_1\,\Delta x + (\Delta x)^2 - x_1^2}{\Delta x}$$

$$= \frac{2x_1\,\Delta x + (\Delta x)^2}{\Delta x} = 2x_1 + \Delta x. \tag{4}$$

Now comes an important step. As Q approaches P along the curve, the value of $2x_1$ in the expression $2x_1 + \Delta x$ does not change, but the value of Δx approaches 0. Therefore, the value of $2x_1 + \Delta x$ approaches $2x_1 + 0 = 2x_1$. We describe this by saying that the limit of m_{sec} as Δx approaches 0 is $2x_1$. In accordance with our definition, the slope of the parabola at P is the number $2x_1$.

Since x_1 can be any value of x, we can omit the subscript 1. At any point (x, y) on the parabola, the slope is

$$m = 2x.$$

When $x = 2$, for example, the slope of the parabola is $m = 2 \cdot 2 = 4$, as we found in Example 2. ■

The next example shows how to use the slope formula $m = 2x$ from Example 3 to find equations for tangent lines.

EXAMPLE 4 Find equations for the tangents to the curve $y = x^2$ at the points $(-1/2, 1/4)$ and $(1, 1)$.

Solution We use the slope formula $m = 2x$ from Example 3 to find the point-slope equation for each line.

Tangent at $(-\frac{1}{2}, \frac{1}{4})$ Point: $(-\frac{1}{2}, \frac{1}{4})$

Slope: $m = 2x = 2(-\frac{1}{2}) = -1$

Equation: $y - \frac{1}{4} = -1(x - (-\frac{1}{2}))$

$y - \frac{1}{4} = -x - \frac{1}{2}$

$y = -x - \frac{1}{4}$

Tangent at $(1, 1)$ Point: $(1, 1)$

Slope: $m = 2x = 2(1) = 2$

Equation: $y - 1 = 2(x - 1)$

$y - 1 = 2x - 2$

$y = 2x - 1$

See Fig. 1.62. ■

We summarize our work so far with a definition.

Slope and Tangent

When the slopes of the secants PQ to a curve have a limiting value as Q approaches P along the curve, we define this value to be the **slope** of the curve at P. We then define the **tangent** to the curve at P to be the line through P with this slope.

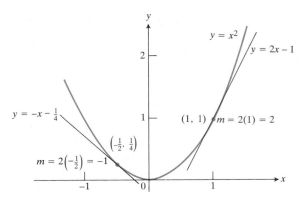

1.62 The slope of the tangent at a point (x, y) on the parabola $y = x^2$ is $m = 2x$.

EXAMPLE 5 Find a formula for the slope of the curve $y = x^3 - 3x + 3$ at an arbitrary point $P(x, y)$ on the curve.

Solution We begin by finding an expression for the slopes of the secants through P. From these we find the slope of the curve at P.

Since $P(x, y)$ lies on the curve, its y-coordinate satisfies the equation $y = x^3 - 3x + 3$. In terms of x, then, the coordinates of P are

$$P(x, x^3 - 3x + 3). \tag{5}$$

The coordinates of any point Q on the curve whose first coordinate differs from x by an increment Δx are

$$Q(x + \Delta x, (x + \Delta x)^3 - 3(x + \Delta x) + 3). \tag{6}$$

The run from P to Q is Δx. The rise from P to Q is

$$\begin{aligned}
\Delta y &= [(x + \Delta x)^3 - 3(x + \Delta x) + 3] - [x^3 - 3x + 3] \\
&= [x^3 + 3x^2\,\Delta x + 3x\,(\Delta x)^2 + (\Delta x)^3 - 3x - 3\,\Delta x + 3] - [x^3 - 3x + 3] \\
&= 3x^2\,\Delta x + 3x\,(\Delta x)^2 + (\Delta x)^3 - 3\,\Delta x. \tag{7}
\end{aligned}$$

The slope of PQ is therefore

$$m_{PQ} = \frac{\Delta y}{\Delta x} = \frac{3x^2\,\Delta x + 3x\,(\Delta x)^2 + (\Delta x)^3 - 3\,\Delta x}{\Delta x}$$

$$= 3x^2 + 3x\,(\Delta x) + (\Delta x)^2 - 3. \tag{8}$$

As Q approaches P along the curve, Δx approaches 0 and m_{PQ} approaches the number

$$m = 3x^2 + 3x(0) + (0)^2 - 3 = 3x^2 - 3. \tag{9}$$

This is the number we call the slope of the curve at P. The formula we seek is

$$m = 3x^2 - 3. \qquad \blacksquare \tag{10}$$

EXAMPLE 6 Find an equation for the line tangent to the curve $y = x^3 - 3x + 3$ at the point $(0, 3)$.

Solution We use the slope formula $m = 3x^2 - 3$ from Example 5 to write a point-slope equation for the line.

Point:	$(0, 3)$
Slope:	$m = 3x^2 - 3 = 3(0)^2 - 3 = -3$
Equation:	$y - 3 = -3(x - 0)$
	$y - 3 = -3x$
	$y = -3x + 3.$

See Fig. 1.63.   ∎

EXAMPLE 7 At what points, if any, does the graph of $y = x^3 - 3x + 3$ have horizontal tangents?

Solution We look for points where the slope is zero. To find them, we set the formula for the slope (found in Example 5) equal to zero and solve:

$$3x^2 - 3 = 0$$
$$3x^2 = 3$$
$$x^2 = 1$$
$$x = \pm 1.$$

The x-coordinates of the points we seek are $x = 1$ and $x = -1$. We find the y-coordinates.

$$\text{When } x = 1, \quad y = (1)^3 - 3(1) + 3 = 1.$$
$$\text{When } x = -1, \quad y = (-1)^3 - 3(-1) + 3 = 5.$$

There are two points where the tangents are horizontal, $(1, 1)$ and $(-1, 5)$. See Fig. 1.63.   ∎

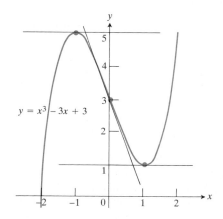

1.63 Three tangents to the curve $y = x^3 - 3x + 3$.

PROBLEMS

1. In Example 1, estimate the increase that took place in the fruit fly population during the tenth day by using a ruler to estimate the slope of the population curve in Fig. 1.58 at $t = 10$.

2. a) Sketch the parabola $y = x^2$, showing the points $(-2, 4)$, $(-1, 1)$, $(0, 0)$, $(1/2, 1/4)$, and $(1, 1)$.
   b) Write an equation for the line tangent to the parabola at $P(1, 1)$.
   c) Calculate the slopes of the secants through $P(1, 1)$ and the four points $(-2, 4)$, $(-1, 1)$, $(0, 0)$, and $(1/2, 1/4)$.
   d) Let Δx be a small increment in x. Express the slope m_{PQ} of the secant through the points $P(1, 1)$ and $Q(1 - \Delta x, (1 - \Delta x)^2)$ as a function of Δx. Find the limit of m_{PQ} as Δx approaches zero. What is the relation between this limit and the slope of the tangent to the parabola at $P(1, 1)$?

3. Let Q be a point on the curve $y = x^3 - 3x + 3$ whose x-coordinate equals $h \neq 0$.
   a) Find the slope of the secant through Q and the point $P(0, 3)$.
   b) Find the limit of the slope of the secant in (a) as Q approaches P.
   c) Write an equation for the tangent to the curve at P.

4. Find equations for the tangents to the curve $y = x^3 - 3x + 3$ of Example 5 at the following points:
   a) $(3, 21)$,
   b) $(-3, -15)$,
   c) $(\sqrt{2}, 3 - \sqrt{2})$.

In Problems 5–20:
   a) Find a formula that gives the slope at any point $P(x, y)$ on the given curve.

b) Use the slope formula from (a) to write an equation for the line tangent to the curve at the given point.

c) Find any points where the curve has a horizontal tangent.

5. $y = x^2 + 1$, $(2, 5)$

6. $y = -x^2$, $(1, -1)$

7. $y = 4 - x^2$, $(-1, 3)$

8. $y = x^2 - 4x$, $(4, 0)$

9. $y = x^2 + 3x + 2$, $(-1, 0)$

10. $y = x^2 - 2x - 3$, $(0, -3)$

11. $y = x^2 + 4x + 4$, $(-2, 0)$

12. $y = 6 + 4x - x^2$, $(2, 10)$

13. $y = x^2 - 4x + 4$, $(1, 1)$

14. $y = 2 - x - x^2$, $(1, 0)$

15. $y = x^3$, $(1, 1)$

16. $y = x^3 - 12x$, $(0, 0)$

17. $y = x^3 - 3x$, $(-1, 2)$

18. $y = 4x^3 + 6x^2 + 1$, $(-1, 3)$

19. $y = x^3 - 3x^2 + 4$, $(1, 2)$

20. $y = 2x^3 + 3x^2 - 12x$, $(2, 4)$

TOOLKIT PROGRAMS

Function Evaluator   Super ∗ Grapher
Secant Lines

1.7
The Slope of the Curve $y = f(x)$: Derivatives

In Article 1.6 we calculated the slopes of quadratic and cubic curves. We now go on to the graphs of other functions. As in Article 1.6, the slope calculations here depend on the notion of limit, a notion that we shall continue to treat informally. After we have presented some applications of derivatives in Article 1.8, we shall return to the notion of limit in Articles 1.9 and 1.10. Our goal there will be to lay a sound mathematical foundation for limits and to prepare for the rapid calculation methods of Chapter 2.

The Derivative of a Function

Let $P(x, y)$ be a point on the graph of the function $y = f(x)$ (Fig. 1.64). If $Q(x + \Delta x, y + \Delta y)$ is another point on the graph, then

$$y + \Delta y = f(x + \Delta x).$$

From this we subtract $y = f(x)$ to obtain

$$\Delta y = f(x + \Delta x) - f(x).$$

The slope of the line PQ is

$$\text{Slope of } PQ = \frac{\Delta y}{\Delta x} = \frac{f(x + \Delta x) - f(x)}{\Delta x}. \tag{1}$$

The division in Eq. (1) can only be *indicated* here because we have no formula for f. For any specific function, like the function $f(x) = x^3 - 3x + 3$ in Example 5 of Article 1.6, this division must be carried out before we do the next operation.

After performing the division, *we hold x fixed* and let Δx approach zero. If the secant slope approaches a value that depends only on x, we call this value

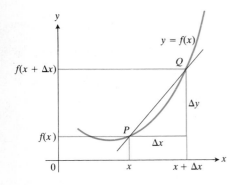

1.64   The slope of the line PQ is

$$\frac{f(x + \Delta x) - f(x)}{\Delta x}.$$

the **slope of the curve** at P:

$$\text{Slope of curve at } P = \lim_{Q \to P} \text{ slope of } PQ$$

$$= \lim_{\Delta x \to 0} \frac{\Delta y}{\Delta x} \qquad (2)$$

$$= \lim_{\Delta x \to 0} \frac{f(x + \Delta x) - f(x)}{\Delta x}.$$

The notation "lim" with $\Delta x \to 0$ beneath is read "the limit, as Δx goes to zero, of."

The fraction

$$\frac{f(x + \Delta x) - f(x)}{\Delta x}$$

in Eq. (2) is called "the difference quotient for f at x."

The slope of the curve at P is itself a function of x, defined at every value of x at which the limit in Eq. (2) exists. We usually denote this slope function by f' ("f prime") and call it the **derivative** of f (in the sense of "function derived from"). Thus the values of the derivative f' of the function f are defined by the rule

$$f'(x) = \lim_{\Delta x \to 0} \frac{f(x + \Delta x) - f(x)}{\Delta x}. \qquad (3)$$

The domain of f', i.e., the set of x-values at which f' is defined, is a subset of the domain of f. For most of the functions in this book, $f'(x)$ will exist at all or at all but a few of the values at which f itself is defined. The exceptions will usually be points where the graph of f has a break, corner, or sharp peak, as in Fig. 1.65.

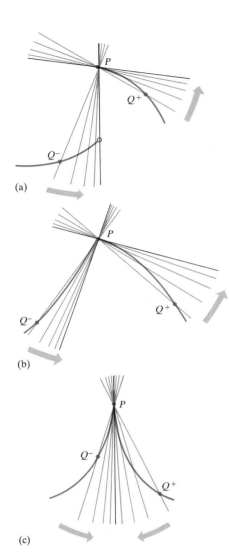

(a)

(b)

(c)

1.65 As the pictures here suggest, the derivative is not defined at a point where the graph of a function has a break, a corner, or a sharp peak. In (a) and (b) there is no single limiting position as $Q \to P$. In (c) the secants become vertical as $Q \to P$; the curve has a vertical tangent at P but no slope.

The Derivative of a Function

The **derivative** of a function f is the function f' whose value at x is defined by the equation

$$f'(x) = \lim_{\Delta x \to 0} \frac{f(x + \Delta x) - f(x)}{\Delta x} \qquad (4)$$

whenever the limit exists.

Differentiable at a Point

A function that has a derivative at a point x is said to be **differentiable at x.**

Differentiable Function

A function that is differentiable at every point of its domain is said to be **differentiable.**

The branch of mathematics that deals with derivatives is called **differential calculus.**

Calculations and Notation

EXAMPLE 1   According to Example 3 of Article 1.6, the derivative of $f(x) = x^2$ is $f'(x) = 2x$. The derivative is defined at every value of x.   ■

EXAMPLE 2   According to Example 5 of Article 1.6, the derivative of $f(x) = x^3 - 3x + 3$ is $f'(x) = 3x^2 - 3$. The derivative is defined at every value of x.   ■

The most common notations for the derivative of $y = f(x)$, besides $f'(x)$, are

$$y' \qquad (\text{``}y \text{ prime''}),$$

$$\frac{dy}{dx} \qquad (\text{``}d\ y\ d\ x\text{''}),$$

$$\frac{df}{dx} \qquad (\text{``}d\ f\ d\ x\text{''}).$$

The d/dx part stands for the operation of taking the derivative with respect to x, and we sometimes write

$$\frac{dy}{dx} = \frac{d}{dx}(y) \qquad \text{or} \qquad \frac{df}{dx} = \frac{d}{dx}(f).$$

Input: function $y = f(x)$ — Operation $\dfrac{d}{dx}$ — Output: derivative $y' = \dfrac{df}{dx}$

1.66   Flow diagram for the operation of taking a derivative with respect to x.

We also read dy/dx as "the derivative of y with respect to x" and df/dx as "the derivative of f with respect to x." See Fig. 1.66.

We now have two definitions for the slope of a line $y = mx + b$: the number m from Article 1.2 and the slope the line has as the graph of the function $f(x) = mx + b$. Whenever we bring in a new definition, it is a good idea to be sure that the new and old definitions agree on objects to which they both apply. We do this in the next example.

EXAMPLE 3   Show that the derivative of $f(x) = mx + b$ is the slope of the line $y = mx + b$. See Fig. 1.67.

Solution   We calculate the limit in Eq. (4) with $f(x) = mx + b$. The calculation takes four steps.

STEP 1:   Write out $f(x + \Delta x)$ and $f(x)$:

$$f(x + \Delta x) = m(x + \Delta x) + b$$
$$= mx + m\,\Delta x + b,$$
$$f(x) = mx + b.$$

STEP 2:   Subtract $f(x)$ from $f(x + \Delta x)$:

$$f(x + \Delta x) - f(x) = m\,\Delta x.$$

STEP 3:   Divide by Δx:

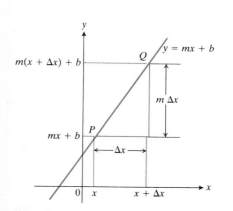

1.67   On the line $y = mx + b$, the slope of every secant is m.

$$\frac{f(x + \Delta x) - f(x)}{\Delta x} = \frac{m\,\Delta x}{\Delta x} = m.$$

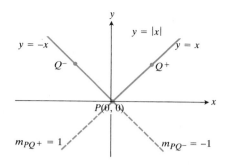

1.68 The lines $y = x$ and $y = -x$ are the only secants to the graph of $y = |x|$ that pass through the origin.

STEP 4: Calculate the limit as $\Delta x \to 0$. Since the difference quotient has the constant value m no matter what Δx is, its limiting value is m as well:

$$f'(x) = \lim_{\Delta x \to 0} \frac{f(x + \Delta x) - f(x)}{\Delta x} = \lim_{\Delta x \to 0} m = m.$$

This shows that at every value of x, the derivative of f is the slope of the line. ∎

EXAMPLE 4   Show that the function $y = |x|$ has a derivative at every point except $x = 0$.

Solution   When x is positive, $y = |x| = x$. From Example 3 with $m = 1$ and $b = 0$, we know that the derivative of $y = x$ is $y' = 1$.

When x is negative, $y = |x| = -x$. The derivative of $y = -x$ (by Example 3 with $m = -1$ and $b = 0$) is $y' = -1$.

Therefore,

$$y' = \begin{cases} 1 & \text{if } x > 0, \\ -1 & \text{if } x < 0. \end{cases} \tag{5}$$

What about $x = 0$? At $x = 0$ there is no derivative. To see why, notice that there are *only two* secants to the curve $y = |x|$ through the point $P(0, 0)$: the line $y = x$ and the line $y = -x$ (Fig. 1.68). If Q is any point on the graph to the right of P, the secant PQ is the line $y = x$, whose slope is $+1$. If Q is any point on the graph to the left of P, the secant PQ is the line $y = -x$, whose slope is -1.

As Q approaches P along the graph, the secants themselves remain stationary. For the limit in Eq. (4) to exist, the slopes of the right- and left-hand secants would have to come together as Q approaches P. They never do. No matter how close Q comes to P, the slope is -1 on the left, $+1$ on the right. ∎

Using h for Δx

To simplify the difference quotient

$$\frac{f(x + \Delta x) - f(x)}{\Delta x},$$

we sometimes replace Δx by the single letter h. The equation that defines the derivative of f at x is then

$$f'(x) = \lim_{h \to 0} \frac{f(x + h) - f(x)}{h}. \tag{6}$$

See Fig. 1.69. This change makes the algebra of calculating derivatives easier.

EXAMPLE 5   Find dy/dx if $y = \sqrt{x}$ and $x > 0$.

Solution   We use Eq. (6) with $f(x + h) = \sqrt{x + h}$ and $f(x) = \sqrt{x}$ to form the quotient

$$\frac{f(x + h) - f(x)}{h} = \frac{\sqrt{x + h} - \sqrt{x}}{h}. \tag{7}$$

Unfortunately, this will involve division by 0 if we replace h by 0. We therefore look for an equivalent expression in which this difficulty does not arise. If we

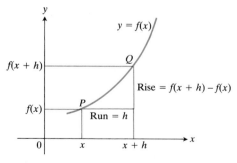

1.69 When the difference between the x-coordinates of P and Q is called h instead of Δx, the defining equation for the derivative of $y = f(x)$ is

$$f'(x) = \lim_{h \to 0} \frac{f(x + h) - f(x)}{h}.$$

rationalize the numerator in Eq. (7), we find

$$\frac{\sqrt{x+h} - \sqrt{x}}{h} = \frac{\sqrt{x+h} - \sqrt{x}}{h} \cdot \frac{\sqrt{x+h} + \sqrt{x}}{\sqrt{x+h} + \sqrt{x}}$$

$$= \frac{(x+h) - x}{h(\sqrt{x+h} + \sqrt{x})} = \frac{1}{\sqrt{x+h} + \sqrt{x}}.$$ (8)

Now as h approaches 0, the denominator in the final form approaches $\sqrt{x} + \sqrt{x} = 2\sqrt{x}$, which is positive because $x > 0$. Therefore,

$$\frac{dy}{dx} = \lim_{h \to 0} \frac{\sqrt{x+h} - \sqrt{x}}{h}$$

$$= \lim_{h \to 0} \frac{1}{\sqrt{x+h} + \sqrt{x}} = \frac{1}{2\sqrt{x}}.$$ (9)

EXAMPLE 6   Find the slope of the curve $y = \sqrt{x}$ at $x = 4$.

Solution   The slope is the value of the derivative of $y = \sqrt{x}$ at $x = 4$. Equation (9) gives

$$\left.\frac{dy}{dx}\right|_{x=4} = \left.\frac{1}{2\sqrt{x}}\right|_{x=4} = \frac{1}{2\sqrt{4}} = \frac{1}{4}.$$

EXAMPLE 7   Find dy/dx if $y = 1/x$.

Solution   We use Eq. (6) with $f(x+h) = 1/(x+h)$ and $f(x) = 1/x$ to form the quotient

$$\frac{f(x+h) - f(x)}{h} = \frac{\dfrac{1}{x+h} - \dfrac{1}{x}}{h}$$

$$= \frac{1}{h} \cdot \frac{x - (x+h)}{x(x+h)}$$

$$= \frac{1}{h} \cdot \frac{-h}{x(x+h)} = \frac{-1}{x(x+h)}.$$ (10)

From this we see that

$$\frac{dy}{dx} = \lim_{h \to 0} \frac{f(x+h) - f(x)}{h} = \lim_{x \to 0} \frac{-1}{x(x+h)}$$

$$= \frac{-1}{x(x+0)} = -\frac{1}{x^2}.$$ (11)

Estimating $f'(x)$ from a Graph of $f(x)$

When we record data in a laboratory or in the field, we are often recording the values of a function $y = f(x)$. We might be recording the pressure in a gas as a function of volume at a given temperature or the size of a population as a function of time. To see what the function looks like we usually plot the data points and fit a curve through them.

Although we may have no formula for the function $y = f(x)$ from which to calculate the derivative $y' = f'(x)$, it is still possible to graph f' by estimating slopes on the graph of f. The following example shows how this can be done and what we can learn from the graph of f'.

EXAMPLE 8 Graph the derivative of the function $y = f(x)$ whose graph is shown in Fig. 1.70(a).

Solution We estimate the slope of the graph of f in y-units per x-unit at frequent intervals. We then plot the estimates in a coordinate plane with the horizontal axis in x-units and the vertical axis in slope units (Fig. 1.70b). We draw a smooth curve through the plotted points.

From the graph of $y' = f'(x)$ we can see at a glance

1. Where f's rate of change is positive, negative, and zero;

2. The rough size of the growth rate at any x and its relation to the size of $f(x)$;

3. Where the rate of change itself is increasing or decreasing. (See Problems 25–28.) ∎

1.70 We made the graph of $y' = f'(x)$ in (b) by plotting slopes from the graph of $y = f(x)$ in (a). The vertical coordinate of B' is the slope at B, and so on. The graph of $y' = f'(x)$ is a visual record of how the slope of f changes with x.

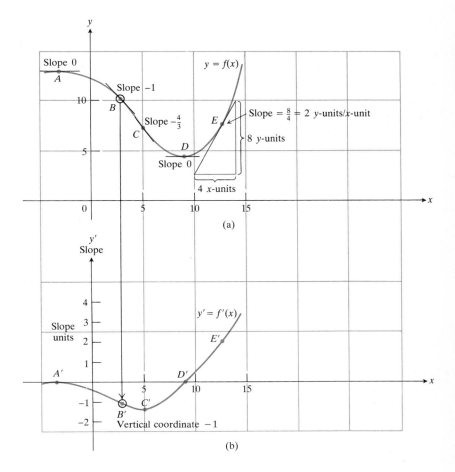

PROBLEMS

In Problems 1–20, use Eq. (4) or (6) to find the derivative f' of the function f. Then find the slope of the curve $y = f(x)$ at $x = 3$ and write an equation for the tangent line.

1. $f(x) = x^2$

2. $f(x) = x^3$

3. $f(x) = 2x + 3$

4. $f(x) = x^2 - x + 1$

5. $f(x) = 1 + \sqrt{x}$

6. $f(x) = \dfrac{1}{x^2}$

7. $f(x) = \dfrac{1}{2x + 1}$

8. $f(x) = \dfrac{x}{x + 1}$

9. $f(x) = 2x^2 - x + 5$

10. $f(x) = x^3 - 12x + 11$

11. $f(x) = x^4$

12. $f(x) = ax^2 + bx + c$
    (a, b, c constants)

13. $f(x) = x - \dfrac{1}{x}$

14. $f(x) = ax + \dfrac{b}{x}$
    (a, b constants)

15. $f(x) = \sqrt{2x}$

16. $f(x) = \sqrt{x + 1}$

17. $f(x) = \sqrt{2x + 3}$

18. $f(x) = \dfrac{1}{\sqrt{x}}$

19. $f(x) = \dfrac{1}{\sqrt{2x + 3}}$

20. $f(x) = \sqrt{x^2 + 1}$

21. Graph the derivative of $f(x) = |x|$ (Example 4). Then graph the function $y = |x|/x$, $x \neq 0$. What can you conclude?

22. Compare the domains and ranges of the function $y = 1/x$ and its derivative $dy/dx = -1/x^2$.

23. The graph of the function $y = f(x)$ in Fig. 1.71 is made of line segments joined end to end.
   a) Graph the derivative of the function. Call the vertical axis the y'-axis. The graph should show a step function.
   b) At what values of x between -3 and 7 is the derivative not defined?

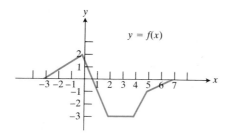

1.71 The graph for Problem 23.

24. Use the following information about a function $y = f(x)$ to graph the function for $-1 \leq x \leq 6$.
   i) The graph of f is made of line segments joined end to end.
   ii) The graph of f starts at the point $(-1, 1)$.
   iii) The derivative of f is the step function shown in Fig. 1.72.

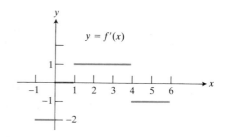

1.72 The graph for Problem 24.

Problems 25–28 are about the graphs shown in Fig. 1.73. The graphs in part (a) show the numbers of rabbits and foxes in a small arctic population. They are plotted as functions of time for 200 days. The number of rabbits increases at first, as the rabbits reproduce. But the foxes prey on the rabbits and as the number of foxes increases, the rabbit population levels off and then drops. Figure 1.73(b) is the graph of the derivative of the rabbit population. It was made by plotting slopes, as in Example 8.

Derivative of the rabbit population

(b)

1.73 Rabbits and foxes in an arctic predator-prey food chain. (*Source: Differentiation,* by W. U. Walton et al., Project CALC, Education Development Center, Inc., Newton, Mass. (1975), p. 86.)

25. What is the value of the derivative of the rabbit population in Fig. 1.73 when the number of rabbits is largest? Smallest?

26. What is the size of the rabbit population in Fig. 1.73 when its derivative is largest? Smallest?

27. In what units should the slope of the fox population curve be measured?

28. a) Use the graphical technique of Example 8 to graph the derivative of the fruit fly population shown in Fig. 1.58.

What units should be used on the horizontal and vertical axes?

b) During what days does the fruit fly population seem to be increasing fastest? Slowest?

TOOLKIT PROGRAMS

Derivative Grapher   Super * Grapher
Secant Lines

1.8
Velocity and Other Rates of Change

Free Fall

When we describe the first few seconds of motion of an object such as a ball bearing or a rock falling from rest, we neglect air resistance and changes in the acceleration of gravity. This simplified motion is called *free fall*.

EXAMPLE 1   Figure 1.74 shows the free fall of a heavy ball bearing released from rest at time $t = 0$. Under these circumstances, the equation that expresses the distance s fallen as a function of the time t is

$$s = \frac{1}{2}gt^2. \tag{1}$$

The number g is the acceleration due to gravity at the surface of the earth. It has been determined experimentally to be about 32 ft/sec^2 ("feet per second squared") if s is measured in feet and t in seconds. Thus,

$$s = \frac{1}{2} \cdot 32t^2 = 16t^2 \text{ ft.} \tag{2}$$

If s is measured in centimeters, then $g = 980$ cm/sec^2 and

$$s = \frac{1}{2} \cdot 980t^2 = 490t^2 \text{ cm.} \tag{3}$$

During the first two seconds, the ball bearing falls

$$s = 16(2)^2 = 64 \text{ ft,}$$

or

$$s = 490(2)^2 = 1960 \text{ cm} = 19.6 \text{ m.} \quad \blacksquare$$

Velocity

Suppose we know the equation of motion of a body along a line to be

$$s = f(t), \tag{4}$$

1.74 Distance fallen by a ball bearing released at $t = 0$ sec.

and we want to find the *velocity* of the body at some instant t. How shall we *define* the instantaneous velocity of a moving body? If we assume that distance and time are the fundamental physical quantities that we can measure, we might reason as follows:

In the interval from time t to time $t + \Delta t$ the body moves from position $s = f(t)$ to position

$$s + \Delta s = f(t + \Delta t), \tag{5}$$

for a net change in position, or **displacement,** of

$$\Delta s = f(t + \Delta t) - f(t). \tag{6}$$

(All this can be measured by clocks and tapelines, say.)

The average velocity of the body over the time interval from t to $t + \Delta t$ is then Δs divided by Δt:

DEFINITION

Average Velocity

The **average velocity** of a body moving along a line is

$$v_{av} = \frac{\text{displacement}}{\text{time traveled}} = \frac{\Delta s}{\Delta t} = \frac{f(t + \Delta t) - f(t)}{\Delta t}. \tag{7}$$

For example, a sprinter running 100 meters in 10 seconds has an average velocity of

$$v_{av} = \frac{\Delta s}{\Delta t} = \frac{100 \text{ m}}{10 \text{ sec}} = 10 \text{ m/sec}.$$

To obtain the instantaneous velocity v of a moving body whose position at time t is $s = f(t)$, we take the limit of the average:

$$v = \lim_{\Delta t \to 0} v_{av} = \lim_{\Delta t \to 0} \frac{\Delta s}{\Delta t}$$

$$= \lim_{\Delta t \to 0} \frac{f(t + \Delta t) - f(t)}{\Delta t}. \tag{8}$$

In other words, we define the instantaneous velocity, as a function of t, to be the derivative of s with respect to t.

DEFINITION

Instantaneous Velocity

The **instantaneous velocity** of a body moving along a line is the derivative of its position $s = f(t)$ with respect to t:

$$v = \frac{ds}{dt} = f'(t). \tag{9}$$

EXAMPLE 2   A body falls freely, as in Example 1, with $s = (1/2)gt^2$. Find its instantaneous velocity as a function of t. How fast is the body falling, in feet per second, 2 seconds after release?

Solution   The instantaneous velocity is the derivative of $s = (1/2)gt^2$ with respect to t. The difference quotient for calculating the derivative is

$$\frac{\Delta s}{\Delta t} = \frac{(1/2)g(t + \Delta t)^2 - (1/2)gt^2}{\Delta t}$$

$$= \frac{1}{2}g\frac{t^2 + 2t\,\Delta t + (\Delta t)^2 - t^2}{\Delta t} = \frac{1}{2}g(2t + \Delta t).$$

The instantaneous velocity at time t is

$$v = \lim_{\Delta t \to 0} \frac{\Delta s}{\Delta t} = \lim_{\Delta t \to 0} \frac{1}{2}g(2t + \Delta t) = \frac{1}{2}g(2t + 0) = gt. \qquad (10)$$

In short,

$$v = gt. \qquad (11)$$

With s in feet and t in seconds, $g = 32$ ft/sec^2 and

$$v = 32t \text{ ft/sec}. \qquad (12)$$

Two seconds after release, the velocity is

$$v = 32(2) = 64 \text{ ft/sec}. \qquad \blacksquare$$

When we graph $s = f(t)$ as a function of t, the average and instantaneous velocities have geometric interpretations (Fig. 1.75). The average velocity is the slope of a secant line and the instantaneous velocity at time t is the slope of the tangent at the point $(t, f(t))$. We can use the interpretation of a tangent slope as a velocity to estimate velocities when our information about a motion is given by a graph rather than an equation. (See Problems 12–14 and 22.)

It is customary to refer to the instantaneous velocity simply as the **velocity,** and we shall do so from now on.

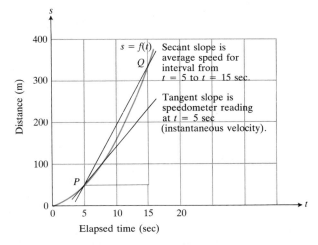

1.75   Time-to-distance graph for a Mercedes 300TD automobile. The slope of the secant is the average speed for the 10-sec interval from $t = 5$ to $t = 15$ sec, in this case 29.5 m/sec or 106.25 km/h. The slope of the tangent at P is the speedometer reading at $t = 5$ sec, about 18 m/sec or 65 km/h. (*Source: Road & Track*, January 1987.)

Other Rates of Change

There are many other applications of average and instantaneous rates.

EXAMPLE 3　The quantity of water Q (gal) in a reservoir at time t (min) is a function of t. Water may flow into or out of the reservoir. As it does so, suppose that Q changes by an amount ΔQ from time t to time $t + \Delta t$. Then the average and instantaneous rates of change of Q with respect to t are

$$\text{Average rate:} \qquad \frac{\Delta Q}{\Delta t} \ (\text{gal/min}),$$

$$\text{Instantaneous rate:} \qquad \frac{dQ}{dt} = \lim_{\Delta t \to 0} \frac{\Delta Q}{\Delta t} \ (\text{gal/min}). \qquad ∎$$

Although it is natural to think of rates of change in terms of motion and time, there is no need to be so restrictive. We can define the average rate of change for any function $y = f(x)$ over any interval in its domain and define the instantaneous rate of change as a limit of average rates of change whenever the limit exists.

DEFINITION

> ### Rates of Change
>
> The **average rate of change** of a function $y = f(x)$ over the interval from x to $x + \Delta x$ is
>
> $$\text{Average rate of change} = \frac{f(x + \Delta x) - f(x)}{\Delta x}.$$
>
> The **instantaneous rate of change** of f at x is the derivative
>
> $$f'(x) = \lim_{\Delta x \to 0} \frac{f(x + \Delta x) - f(x)}{\Delta x} = \lim_{\Delta x \to 0} (\text{Average rate of change}),$$
>
> provided the limit exists.

Thus the instantaneous rate of change of f with respect to x is the value of f' at x. We still use the term "instantaneous" even when x does not represent time.

1.76　Weekly steel production: $y = f(x)$ is the cost of producing x tons. The cost of producing an additional Δx tons is given by $\Delta y = f(x + \Delta x) - f(x)$.

☐ Marginal Cost

Economists often call the derivative of a function the **marginal value** of the function. Suppose, for example, that it costs a company $y = f(x)$ dollars to produce x tons of steel in a week. It costs more to produce $x + \Delta x$ tons a week, say $y + \Delta y$ dollars. The average increase in cost per additional ton is $\Delta y/\Delta x$. The limit of this ratio as Δx approaches zero is the **marginal cost** of producing x tons of steel in a week. (See Fig. 1.76.)

How are we to interpret marginal cost? First of all, it is the slope of the graph of $y = f(x)$ at the point marked P in Fig. 1.76. But there is more.

Figure 1.77 shows an enlarged view of the curve and its tangent at P. We can see that if the company, currently producing x tons, increases production by one ton, then the incremental cost Δy of producing that one ton is approxi-

1.77 As weekly steel production increases from x to $x + 1$ tons, the cost curve rises by the amount Δy. The tangent line rises by the amount Slope · run $= f'(x) \cdot 1 = f'(x)$. Since $\Delta y / \Delta x \approx f'(x)$, we have $\Delta y \approx f'(x)$ when $\Delta x = 1$.

mately $f'(x)$. That is,

$$\Delta y \approx f'(x) \qquad \text{when } \Delta x = 1. \tag{13}$$

Herein lies the economic importance of marginal cost. It gives an estimate or prediction of the cost of producing one more unit beyond the present production level. It is the approximate cost of producing one more car, one more radio, one more washing machine, whatever.

Formulas for Free Fall Near the Earth's Surface

1. $s = \frac{1}{2}gt^2$ $s = $ distance, $t = $ time,
   $v = gt$ $g = $ gravitational constant,
               $v = $ velocity

2. $s = 16t^2$ $s = $ ft, $t = $ sec,
   $v = 32t$ $g = 32$ ft/sec^2, $v = $ ft/sec

3. $s = 490t^2$ $s = $ cm, $t = $ sec,
   $v = 980t$ $g = 980$ cm/sec^2, $v = $ cm/sec

4. $s = 4.9t^2$ $s = $ m, $t = $ sec,
   $v = 9.8t$ $g = 9.8$ m/sec^2, $v = $ m/sec

PROBLEMS

1. If a, b, c are constants and

$$f(t) = at^2 + bt + c,$$

show that

$$f'(t) = \lim_{\Delta t \to 0} \frac{f(t + \Delta t) - f(t)}{\Delta t} = 2at + b.$$

The laws of motion in Problems 2–10 give the position $s = f(t)$ of a moving body as a function of t, with s measured in meters and t in seconds.

a) Find the displacement and average velocity for the time interval from $t = 0$ to $t = 2$ seconds.

b) Use the formula $f'(t) = 2at + b$ from Problem 1 to ex-

press the velocity $v = ds/dt$ as a function of t, by inspection.

c) Use the formula obtained in (b) to find the body's velocity at $t = 2$ seconds.

2. $s = 4.9t^2$ (free fall on Earth)

3. $s = 0.8t^2$ (free fall on the Moon)

4. $s = 1.86t^2$ (free fall on Mars)

5. $s = 2t^2 + 5t - 3$ **6.** $s = t^2 - 3t + 2$

7. $s = 4 - 2t - t^2$ **8.** $s = 3 - 2t^2$

9. $s = 4t + 3$ **10.** $s = (1/2)gt^2 + v_0 t + s_0$
                                     (g, v_0, s_0 constants)

1.78   Two balls falling from rest. (Courtesy of Education Development Center, Inc., Newton, Mass.)

11. Figure 1.78 shows a multiflash photograph of two balls falling after being released from rest. The rulers in the figure are marked in centimeters.
    a) Which of the boxed formulas for free fall give s and v here?
    b) How long did it take the balls to fall the first 160 cm? What was their average velocity for this period?
    c) How long was the time between consecutive flashes?

12. a) Use the time-to-distance graph in Fig. 1.75 to estimate the car's average velocity for the first 10 seconds.
    b) Estimate the car's speedometer reading at $t = 15$ seconds by estimating the slope of the curve with a straight-edge.

13. The following data give the coordinates s of a moving body for various values of t. Plot s versus t on coordinate paper and sketch a smooth curve through the given points. Assuming that this smooth curve represents the motion of the body, estimate the velocity at (a) $t = 1.0$; (b) $t = 2.5$; (c) $t = 2.0$.

s (in ft)	10	38	58	70	74	70	58	38	10
t (in sec)	0	0.5	1.0	1.5	2.0	2.5	3.0	3.5	4.0

14. When a chemical reaction was allowed to run for t minutes, it produced the amounts of substance $A(t)$ shown in the following table.[†]

t (in min)	10	15	20	25	30	35	40
$A(t)$ (in moles)	26.5	36.5	44.8	52.1	57.1	61.3	64.4

    a) Find the average rate of the reaction for the interval from $t = 20$ to $t = 30$.
    b) Plot the data points from the table, draw a smooth curve through them, and estimate the instantaneous rate of the reaction at $t = 25$.

When a model rocket is launched, the propellant burns for a few seconds, accelerating the rocket upward. After burnout, the rocket coasts upward for a while and then begins to fall. A small explosive charge pops out a parachute shortly after the rocket starts down. The parachute slows the rocket to keep it from breaking when it lands.

Figure 1.79 shows velocity data from the flight of a model rocket. Use the data to answer the questions in Problems 15–19.

15. How fast was the rocket climbing when the engine stopped?

16. For how many seconds did the engine burn?

† Data from *Some Mathematical Models in Biology,* Revised Edition, R. M. Thrall, J. A. Mortimer, K. R. Rebman, R. F. Baum, eds., December 1967, PB-202 364, p. 72; distributed by N.T.I.S., U.S. Department of Commerce.

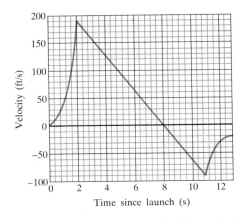

1.79 Velocity of a model rocket (Problems 15–19).

17. When did the rocket reach its highest point? What was its velocity then?

18. When did the parachute pop out? How fast was the rocket falling then?

19. How long did the rocket fall before the parachute opened?

20. When a bactericide was added to a nutrient broth in which bacteria were growing, the bacterium population continued to grow for a while, but then stopped growing and began to decline. The size of the population at time t (hours) was $b(t) = 10^6 + 10^4 t - 10^3 t^2$. Use the result of Problem 1 to find the growth rates at (a) $t = 0$; (b) $t = 5$; and (c) $t = 10$ hours.

21. The number of gallons of water in a tank t minutes after the tank has started to drain is $Q(t) = 200(30 - t)^2$. How fast is the water running out at the end of 10 minutes? What is the *average* rate at which the water flows out during the first 10 minutes?

22. The three graphs in Fig. 1.80 show the distance traveled (miles), velocity (miles per hour), and acceleration (the derivative of the velocity, measured in miles per hour per second) for each second of a two-minute automobile trip. Which graph shows (a) distance; (b) velocity; (c) acceleration? (d) The vertical axis for the velocity graph is marked in units of 5 mph. Estimate the maximum and minimum values of the acceleration.

23. *Marginal cost.* Suppose that the dollar cost of producing x washing machines is $f(x) = 2000 + 100x - 0.1x^2$.
a) Find the average cost of producing 100 washing machines.
b) Find the marginal cost of producing 100 washing machines.
c) Show that the marginal cost for 100 washing machines is approximately the cost of producing one more washing machine after the first 100 have been made, by computing the cost directly.

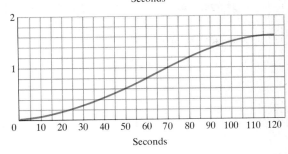

1.80 A two-minute automobile trip. Which of these curves is distance? Velocity? Acceleration?

1.9
Limits

We know that we need to calculate limits to find derivatives. But just what *are* limits and how can we calculate them without doing a lot of work? In this article, we find out.

The Problem

Fermat's method for finding derivatives calls for finding the limiting value of the quotient

$$\frac{f(x + \Delta x) - f(x)}{\Delta x} \tag{1}$$

as Δx approaches zero. We did this in a number of cases by dividing Δx into the numerator to get an expression in which we could find the limit by substituting zero for Δx.

There are, unfortunately, two drawbacks to this algebraic approach: it takes time and it does not always work. There is no way, for instance, to divide Δx into the numerator of the quotient

$$\frac{\sin(x + \Delta x) - \sin x}{\Delta x}, \tag{2}$$

which is the quotient we must use to find the derivative of the sine function. We need a better way to calculate limits.

A Simpler Notation

When we take the limit of

$$\frac{f(x + \Delta x) - f(x)}{\Delta x} \tag{3}$$

as Δx approaches zero, we hold x fixed while Δx varies. In doing so, we treat the entire quotient as a function of a single variable, in this case Δx. As long as we are doing that, we might as well use a simpler notation, F for the function and t for the variable. Our task, then, will be to define what it means for a function F of a single variable t to have a limit as t approaches some predetermined value c. Once we have done so, we can go back to see what this says about the limiting value of $(f(x + \Delta x) - f(x))/\Delta x$ as Δx approaches zero. In the meantime, we don't have to think about derivatives, we can just think about limits.

The Definition of Limit

Suppose we are watching the values of a function $F(t)$ as t moves toward c without actually taking on the value c (just as we will let Δx approach 0 without taking on the value 0). What do we have to know about the values of F to say that they have L as a limit? What observable pattern in their behavior would guarantee their eventual approach to L?

Certainly we want to be able to say that $F(t)$ stays within one-tenth of a unit of L as soon as t stays within a certain radius r_1 of c, as shown here:

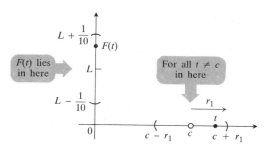

But that in itself is not enough, since as t continues on its course toward c, what is to prevent $F(t)$ from jittering about within the interval $(L - 1/10, L + 1/10)$ instead of tending toward L?

We need to say also that as t continues toward c, $F(t)$ will eventually have to get still closer to L. We might say this by requiring $F(t)$ to lie within $1/100$ of a unit of L for all values of t within some smaller radius r_2 of c:

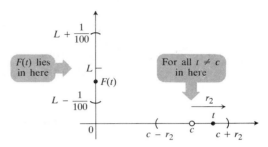

But this is not enough either. What if $F(t)$ skips about within the interval $(L - 1/100, L + 1/100)$ from then on, without heading toward L? We had better require that $F(t)$ also lie within $1/1000$ of a unit of L after a while. That is, for all values of t within some still smaller radius r_3 of c, all the values of $F(t)$ lie in the interval

$$L - \frac{1}{1000} < F(t) < L + \frac{1}{1000},$$

as shown here:

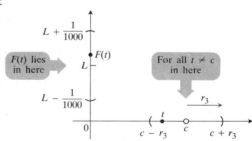

This still does not guarantee that $F(t)$ will now move toward L as t approaches c. Even if $F(t)$ has not skipped about before, it might start now. We need more.

We need to require that for *every* interval about L, no matter how small, we can find an interval of numbers about c all of whose F-values lie within that interval about L. In other words, given *any* positive radius ϵ about L, there exists some positive radius δ about c such that for all t within δ units of c (except $t = c$ itself) the values of $F(t)$ lie within ϵ units of L:

Thus, the closer t stays to c without equaling c, the closer $F(t)$ must stay to L.

DEFINITION

> **Limit**
>
> The **limit** of $F(t)$ as t approaches c is the number L if:
>
> Given any radius $\epsilon > 0$ about L there exists a radius $\delta > 0$ about c such that for all t
>
> $$0 < |t - c| < \delta \qquad \text{implies} \qquad |F(t) - L| < \epsilon. \qquad (4)$$

We can write "the limit of $F(t)$ as t approaches c is L" as

$$\lim_{t \to c} F(t) = L.$$

Roughly speaking, to say that $F(t)$ approaches the limit L as t approaches c means that for any tolerance ϵ there is a small number δ (which depends on ϵ) such that $F(t)$ will stay within ϵ units of L if t is restricted to lie within δ units of c.

We might think of machining something like a generator shaft to a close tolerance. We try for a diameter L, but since nothing is perfect, we must be satisfied to get the diameter $F(t)$ within $L \pm \epsilon$. The δ is how accurate our control setting t must be to guarantee this degree of accuracy in the diameter of the shaft.

Examples—Testing the Definition

Whenever we propose a new definition, it is a good idea to test it against familiar examples to see if it gives results consistent with past experience. For instance, our experience tells us that as t approaches 1, $5t$ approaches $5(1) = 5$ and $5t - 3$ approaches $5(1) - 3 = 2$. If our definition were to tell us that either of these limits were zero, or anything absurd like that, we would throw it out and look for a new definition. The following three examples are included in part to show that the definition in Eq. (4) gives the kinds of results we want.

EXAMPLE 1   Show that $\lim_{t \to 1} (5t - 3) = 2$ according to the definition of limit.

Solution   We apply the definition of limit with $c = 1$, $F(t) = 5t - 3$, and $L = 2$. To satisfy the definition, we need to show that for any $\epsilon > 0$ (radius about $L = 2$) there exists a $\delta > 0$ (radius about $c = 1$) such that for all t,

$$0 < |t - 1| < \delta \quad \Rightarrow \quad |(5t - 3) - 2| < \epsilon.$$

(The symbol $\Rightarrow$ is read "implies.") We change the ϵ-inequality from

$$|(5t - 3) - 2| < \epsilon$$

to

$$|5t - 5| < \epsilon$$

to

$$5|t - 1| < \epsilon$$

to

$$|t - 1| < \epsilon/5.$$

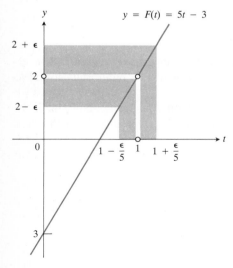

1.81   For the function $F(t) = 5t - 3$, we find that $0 < |t - 1| < \epsilon/5$ will guarantee $|F(t) - 2| < \epsilon$.

These inequalities are all equivalent. Therefore, the original ϵ-inequality will hold if $|t - 1| < \epsilon/5$. We therefore take $\delta = \epsilon/5$. See Fig. 1.81.

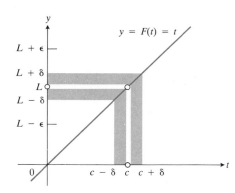

1.82 For the function $F(t) = t$, we find that $0 < |t - c| < \delta$ will guarantee $|F(t) - c| < \epsilon$ whenever $\delta \leq \epsilon$.

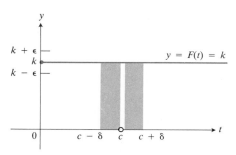

1.83 For the function $F(t) = k$, we find that $|F(t) - k| < \epsilon$ for any positive δ.

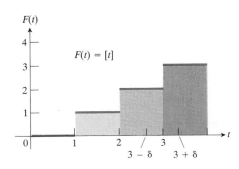

1.84 At each integer, the greatest integer function has different right- and left-hand limits.

The value $\delta = \epsilon/5$ is not the only value of δ that will make the ϵ-inequality hold. Any smaller positive δ will do as well. The definition does not ask for a "best" δ, just one that will work. ∎

EXAMPLE 2 Confirm that $\lim_{t \to c} t = c$ according to the definition of limit.

Solution We apply the definition of limit with $F(t) = t$ and $L = c$. To satisfy the definition, we must show that given any $\epsilon > 0$ (radius about $L = c$) there exists a $\delta > 0$ (radius about c on the t-axis) such that for all t,

$$0 < |t - c| < \delta \quad \Rightarrow \quad |t - c| < \epsilon.$$

The ϵ-inequality will hold if δ is ϵ or any smaller positive number (Fig. 1.82). ∎

When we read

$$F(t) \to L \qquad \text{as} \qquad t \to c$$

as "$F(t)$ approaches L as t approaches c," the verb "approaches" has a connotation of motion that is not always justified. For example, a function that has the constant value $F(t) = k$ for all values of t certainly has the limit k as t approaches any number c. We see this in the following example.

EXAMPLE 3 Let $F(t) = k$ be the function whose value is k for every t. Show that $\lim_{t \to c} F(t) = k$ for any c.

Solution We apply the definition of limit with $F(t) = k$ and $L = k$. We must show that for any $\epsilon > 0$ there exists a $\delta > 0$ such that for all t,

$$0 < |t - c| < \delta \quad \Rightarrow \quad |k - k| < \epsilon.$$

Any positive δ will work because $k - k = 0$ is less than ϵ for all t (Fig. 1.83). ∎

Right-hand Limits and Left-hand Limits

Sometimes the values of a function $F(t)$ tend to different limits as t approaches a number c from different sides. When this happens, we call the limit of F as t approaches c from the right the **right-hand limit** of F at c, and the limit of F as t approaches c from the left the **left-hand limit** of F at c.

EXAMPLE 4 Show that the greatest integer function $F(t) = [t]$ has no limit as t approaches 3 (Fig. 1.84).

Solution Our first guess might be that there is a limit, $L = 3$, since the values of $F(t) = [t]$ are close to 3 when t is equal to or slightly *greater* than 3. But when t is slightly *less* than 3, say $t = 2.9999$, then $[t] = 2$. That is, if δ is any positive number less than 1, then

$$[t] = 2 \quad \text{if} \quad 3 - \delta < t < 3,$$

whereas

$$[t] = 3 \quad \text{if} \quad 3 < t < 3 + \delta.$$

Thus if we are challenged with a small positive ϵ, for example $\epsilon = 0.01$, we cannot find a $\delta > 0$ that makes

$$|[t] - 3| < \epsilon \quad \text{for} \quad 0 < |t - 3| < \delta.$$

In fact, no number L will work as the limit in this case. When t is near 3, *some* of the functional values of $[t]$ are 2 while others are 3. Hence the functional values are not all close to any one number L. Therefore,

$$\lim_{t \to 3} [t] \text{ does not exist.}$$

The right- and left-hand limits do, however, exist at 3. The right-hand limit is

$$L^+ = \lim_{t \to 3^+} [t] = 3,$$

and the left-hand limit is

$$L^- = \lim_{t \to 3^-} [t] = 2.$$

The notation $t \to 3^+$ may be read "t approaches 3 from above" (or "from the right," or "through values larger than 3") with an analogous meaning for $t \to 3^-$.   ∎

The formal definitions of right-hand and left-hand limits go like this:

DEFINITION

Right-hand limit: $\lim\limits_{t \to c^+} F(t) = L$

The limit of the function $F(t)$ as t approaches c from the right equals L if:

Given any $\epsilon > 0$ (radius about L) there exists a $\delta > 0$ (radius to the right of c) such that for all t

$$c < t < c + \delta \quad \Rightarrow \quad |F(t) - L| < \epsilon. \tag{5}$$

DEFINITION

Left-hand limit: $\lim\limits_{t \to c^-} F(t) = L$

The limit of the function $F(t)$ as t approaches c from the left equals L if:

Given any $\epsilon > 0$ (radius about L) there exists a $\delta > 0$ (radius to the left of c) such that for all t

$$c - \delta < t < c \quad \Rightarrow \quad |F(t) - L| < \epsilon. \tag{6}$$

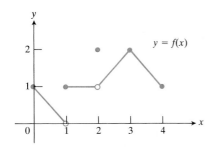

We can see the relation between the right- and left-hand limits of a function at a point and the limit defined earlier by comparing Eqs. (5) and (6) with Eq. (4). If we subtract c from the δ-inequalities in Eqs. (5) and (6), they become

$$0 < t - c < \delta \quad \Rightarrow \quad |F(t) - L| < \epsilon \tag{5'}$$

and

$$-\delta < t - c < 0 \quad \Rightarrow \quad |F(t) - L| < \epsilon. \tag{6'}$$

Together, Eqs. (5') and (6') say the same thing as

$$0 < |t - c| < \delta \quad \Rightarrow \quad |F(t) - L| < \epsilon, \tag{4}$$

which is Eq. (4) in the definition of limit. In short, $F(t)$ has a limit at a point if and only if the right-hand and left-hand limits there exist and are equal.

We sometimes call $\lim_{t \to c} F(t)$ the **two-sided** limit of F at c to distinguish it from the **one-sided** right-hand and left-hand limits of F at c.

Relation Between One-sided and Two-sided Limits

A function $F(t)$ has a limit at point c if and only if the right-hand and left-hand limits at c exist and are equal. In symbols,

$$\lim_{t \to c} F(t) = L \quad \Leftrightarrow \quad \lim_{t \to c^+} F(t) = L \quad \text{and} \quad \lim_{t \to c^-} F(t) = L. \tag{7}$$

EXAMPLE 5   All the following statements about the function $y = f(x)$ graphed in Fig. 1.85 are true.

At $x = 0$: $\quad \lim_{x \to 0^+} f(x) = 1.$

At $x = 1$: $\quad \lim_{x \to 1^-} f(x) = 0$ even though $f(1) = 1$,

$\qquad\qquad \lim_{x \to 1^+} f(x) = 1,$

$f(x)$ has no limit as $x \to 1$. (The right- and left-hand limits at 1 are not equal.)

At $x = 2$: $\quad \lim_{x \to 2^-} f(x) = 1,$

$\qquad\qquad \lim_{x \to 2^+} f(x) = 1,$

$\qquad\qquad \lim_{x \to 2} f(x) = 1$ even though $f(2) = 2.$

1.85   Example 5 discusses the limit properties of the function $y = f(x)$ graphed here.

At $x = 3$: $\lim\limits_{x \to 3^-} f(x) = \lim\limits_{x \to 3^+} f(x) = \lim\limits_{x \to 3} f(x) = f(3) = 2.$

At $x = 4$: $\lim\limits_{x \to 4^-} f(x) = 1.$

At every other point c between 0 and 4, $f(x)$ has a limit as $x \to c$.   ■

EXAMPLE 6   Show that $f(x) = |x|$ has no derivative at $x = 0$.

Solution   We are asked to show that the difference quotient

$$\frac{|0 + \Delta x| - |0|}{\Delta x} = \frac{|\Delta x|}{\Delta x} \tag{8}$$

has no limit as $\Delta x \to 0$. Since $|\Delta x| = \Delta x$ when $\Delta x > 0$, the right-hand limit at 0 is

$$\lim_{\Delta x \to 0^+} \frac{|\Delta x|}{\Delta x} = \lim_{\Delta x \to 0^+} \frac{\Delta x}{\Delta x} = \lim_{\Delta x \to 0^+} 1 = 1. \tag{9}$$

Since $|\Delta x| = -\Delta x$ when $\Delta x < 0$, the left-hand limit is

$$\lim_{\Delta x \to 0^-} \frac{|\Delta x|}{\Delta x} = \lim_{\Delta x \to 0^-} \frac{-\Delta x}{\Delta x} = \lim_{\Delta x \to 0^-} -1 = -1. \tag{10}$$

The right-hand and left-hand limits at 0 are not equal. Therefore, the difference quotient (8) has no limit as $\Delta x \to 0$.   ■

Sometimes a function $f(x)$ has neither a right- nor a left-hand limit at a point. The following example shows how this can happen.

EXAMPLE 7   Show that the function $y = \sin\ (1/x)$ has no limit as $x \to 0$.

Solution   See Fig. 1.86. In every interval about $x = 0$, the function takes on all values between -1 and $+1$. Hence there is no single number L that the function values all stay close to when x is close to 0. This is true even when we restrict x to positive values or to negative values. In other words, this function has neither a right-hand limit nor a left-hand limit as x approaches 0.   ■

1.86   The function $y = \sin(1/x)$ has neither a right-hand nor a left-hand limit as x approaches 0.

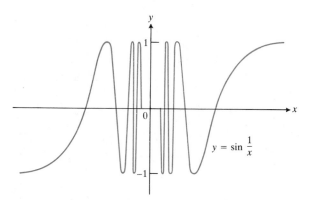

$y = \sin\dfrac{1}{x}$

Calculation Techniques

Now we come to a theorem that tells how to calculate the limits of the sums and products of all the functions whose limits we already know. It also explains how to calculate the limits of many of the ratios of these functions.

THEOREM 1

The Limit Combination Theorem

If $\lim_{t \to c} F_1(t) = L_1$ and $\lim_{t \to c} F_2(t) = L_2$, then

   i) $\lim [F_1(t) + F_2(t)] = L_1 + L_2$,

  ii) $\lim [F_1(t) - F_2(t)] = L_1 - L_2$,

 iii) $\lim [F_1(t)F_2(t)] = L_1 L_2$,

 iv) $\lim [kF_2(t)] = kL_2$ (any number k),

  v) $\lim \dfrac{F_1(t)}{F_2(t)} = \dfrac{L_1}{L_2}$ if $L_2 \neq 0$.

The limits are all to be taken as $t \to c$, and L_1 and L_2 are to be real numbers.

In words, the formulas in Theorem 1 say:

i) The limit of the sum of two functions is the sum of their limits.

ii) The limit of the difference of two functions is the difference of their limits.

iii) The limit of a product of two functions is the product of their limits.

iv) The limit of a constant times a function is the constant times the limit of the function.

v) The limit of a quotient of two functions is the quotient of their limits, provided the denominator does not tend to zero.

Theorem 1 also holds for right-hand limits $(t \to c^+)$ and for left-hand limits $(t \to c^-)$.

In a first course in calculus, it is often thought to be appropriate to use the results of Theorem 1 without proving the theorem first. For those who are interested in a proof we have included one in Appendix 1. Informally, we can paraphrase the theorem in terms that make it highly reasonable: When t is close to c, $F_1(t)$ is close to L_1 and $F_2(t)$ is close to L_2. Then we naturally think that $F_1(t) + F_2(t)$ is close to $L_1 + L_2$; $F_1(t) - F_2(t)$ is close to $L_1 - L_2$; $F_1(t)F_2(t)$ is close to $L_1 L_2$; $kF_2(t)$ is close to kL_2; and $F_1(t)/F_2(t)$ is close to L_1/L_2 if L_2 is not zero.

What keeps this discussion from being a proof is that the word "close" is not precise. The phrases "arbitrarily close to" and "sufficiently close to" might improve the argument a bit, but the full ϵ and δ treatment in Appendix 1 is the clincher.

The importance of Theorem 1 is that it liberates us from having to make ϵ-δ arguments all the time.

■ ■ ■

"GHOSTS OF DEPARTED QUANTITIES"

Newton and Leibniz both used derivatives in their works, even though neither man had a mathematically rigorous conception of the limit, basing their work instead on so-called infinitesimal quantities. This lack of rigor was a major criticism of the philosopher Bishop George Berkeley (1685–1753), who called these infinitesimals "ghosts of departed quantities" that held no higher logical status than questions of religious faith.

Newton's calculus was based on the derivative, or the tangent, whereas Leibniz's was based on the differential, or the secant (Article 2.5). The notation dy/dx is in fact due to Leibniz, and because of its superior notation the Leibnizian form of the calculus dominated in the late eighteenth century. Both forms of the calculus, whatever their original approach, involved a fundamental use of limits. The fact that the limit concept had no rigorous mathematical foundation did not stop mathematicians from using the powerful methods of the calculus to achieve remarkable results in the physical sciences and mathematics.

The rigorous foundation of the calculus is due principally to the nineteenth-century mathematician Augustin Louis Cauchy (1789–1857). In his famous textbook *Cours, d'analyse* (1821) he gave the first rigorous definition of limit and derivative. Cauchy's conception of limit was derived from Lagrange's earlier work on error estimates and his use of the Taylor series remainder in calculating errors. In fact, the Greek letter epsilon in Cauchy's work on limits is an abbreviation of the French *erreur*. [For further reading about Cauchy's accomplishments, see Judith V. Grabiner, *The Origins of Cauchy's Rigorous Calculus* (Cambridge, Mass.: MIT Press, 1981).]

EXAMPLE 8 Since we know from Examples 2 and 3 that $\lim_{t \to c} t = c$ and $\lim_{t \to c} k = k$, we have

$$\lim_{t \to c} t^2 = \lim_{t \to c} (t)(t) = (c)(c) = c^2 \qquad \text{(from iii)},$$

$$\lim_{t \to c} (t^2 - 5) = c^2 - 5 \qquad \text{(from ii)},$$

$$\lim_{t \to c} 4t^2 = 4c^2 \qquad \text{(from iv)},$$

and, if $c \neq 0$,

$$\lim_{t \to c} \frac{t^2 - 5}{4t^2} = \frac{c^2 - 5}{4c^2} \qquad \text{(from v)}. \qquad ■$$

As Example 8 suggests, the limit of any polynomial function $f(t)$ as $t \to c$ is $f(c)$. In other words, the limit can be found by evaluating the polynomial at $t = c$. Similarly, the limit of the ratio $f(t)/g(t)$ of two polynomials as $t \to c$ is $f(c)/g(c)$, provided $g(c) \neq 0$.

Limits of Polynomials

If $f(t) = a_n t^n + a_{n-1} t^{n-1} + \cdots + a_0$ is any polynomial function, then

$$\lim_{t \to c} f(t) = f(c) = a_n c^n + a_{n-1} c^{n-1} + \cdots + a_0.$$

Limits of Quotients of Polynomials

If $f(t)$ and $g(t)$ are polynomials, then

$$\lim_{t \to c} \frac{f(t)}{g(t)} = \frac{f(c)}{g(c)}, \qquad \text{provided } g(c) \neq 0.$$

These are the facts behind the slope and velocity calculations earlier in the chapter. Those who want formal proofs will find them outlined in Appendix 1, Problems 1–5.

EXAMPLE 9 In Example 5 of Article 1.6, we calculated the slope $m = 3x^2 - 3$ of the curve $y = x^3 - 3x + 3$ by evaluating the limit of $3x^2 + 3x \Delta x + (\Delta x)^2 - 3$ as Δx approached zero. With x fixed, this expression is a polynomial function of Δx. The limit is therefore the value of the polynomial at $\Delta x = 0$:

$$\lim_{\Delta x \to 0} [3x^2 + 3x \Delta x + (\Delta x)^2 - 3]$$

$$= 3x^2 + 3x(0) + (0)^2 - 3$$

$$= 3x^2 - 3. \qquad ■$$

EXAMPLE 10 Find

$$\lim_{t \to 2} \frac{t^2 + 2t + 4}{t + 2}.$$

Solution   The function whose limit we are to find is the quotient of two polynomials. The denominator $t + 2$ is not 0 when $t = 2$. Therefore, the limit is the value of the quotient at $t = 2$:

$$\lim_{t \to 2} \frac{t^2 + 2t + 4}{t + 2} = \frac{(2)^2 + 2(2) + 4}{2 + 2} = \frac{12}{4} = 3.$$

■

EXAMPLE 11   Find

$$\lim_{t \to 2} \frac{t^3 - 8}{t^2 - 4}.$$

Solution   The denominator is 0 at $t = 2$ and we cannot calculate the limit by direct substitution. However, if we factor the numerator and denominator we find that

$$\frac{t^3 - 8}{t^2 - 4} = \frac{(t - 2)(t^2 + 2t + 4)}{(t - 2)(t + 2)}.$$

For $t \neq 2$,

$$\frac{t - 2}{t - 2} = 1.$$

Therefore, for all values of t different from 2 (the values that really determine the limit as $t \to 2$),

$$\lim_{t \to 2} \frac{t^3 - 8}{t^2 - 4} = \lim_{t \to 2} \frac{t^2 + 2t + 4}{t + 2} = \frac{(2)^2 + 2(2) + 4}{2 + 2} = \frac{12}{4} = 3.$$

To calculate this limit, we divided the numerator and denominator by a common factor and evaluated the result at $t = 2$.   ■

Example 11 illustrates an important mathematical point about limits: The limit of a function $f(t)$ as $t \to c$ *never* depends on what happens when $t = c$. The limit (if it exists at all) is *entirely determined* by the values f has when $t \neq c$. In Example 11, the quotient $f(t) = (t^3 - 8)/(t^2 - 4)$ is not even defined at $t = 2$. Yet its limit as $t \to 2$ exists and is equal to 3.

The Sandwich Theorem and $(\sin \theta)/\theta$

We conclude this article with a theorem we shall want later on.

THEOREM 2

The Sandwich Theorem

Suppose that

$$f(t) \leq g(t) \leq h(t)$$

for all $t \neq c$ in some interval about c and that $f(t)$ and $h(t)$ approach the same limit L as t approaches c. Then

$$\lim_{t \to c} g(t) = L.$$

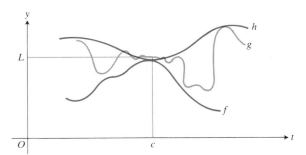

1.87 *The Sandwich Theorem.* If $\lim_{t \to c} f(t)$ and $\lim_{t \to c} h(t)$ both equal L, then $\lim_{t \to c} g(t) = L$ because $f(t) \le g(t) \le h(t)$ for all values of t near c. The graph of g is caught between the graphs of f and h.

See Fig. 1.87. The Sandwich Theorem holds for right-hand and left-hand limits as well as two-sided limits.

The idea behind Theorem 2 is that if $f(t)$ and $h(t)$ head toward L with $g(t)$ "sandwiched" in between, they take $g(t)$ along with them:

$$f(t) \le \quad g(t) \quad \le h(t)$$
$$\downarrow \qquad \downarrow \qquad \downarrow$$
$$L \ \le \lim g(t) \le \ L.$$

We might think of $g(t)$ as a ball between two table tennis paddles that move closer together as $t \to c$. A proof of the theorem can be found at the end of Appendix 1.

EXAMPLE 12   As an application of the Sandwich Theorem, we show that

$$\lim_{\theta \to 0} \sin \theta = 0, \tag{11}$$

$$\lim_{\theta \to 0} \cos \theta = 1, \tag{12}$$

$$\lim_{\theta \to 0} \frac{\sin \theta}{\theta} = 1, \tag{13}$$

provided that θ is measured in radians.

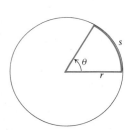

1.88 $\theta = s/r$ radians.

Solution   The radian measure of an angle can be defined by

$$\theta = \frac{s}{r}, \tag{14}$$

where s is the length of the arc the angle intercepts on a circle of radius r when the center of the circle is at the angle's vertex. Figure 1.88 illustrates this definition. (There is more about radian measure in the trigonometry review in Article 2.6.)

In Fig. 1.89, O is the center of a unit circle and θ is the radian measure of an acute angle AOP. Note that $s = \theta$ under these conditions (because $s = r\theta$ and $r = 1$). Now $\triangle APQ$ is a right triangle with legs of length

$$QP = \sin \theta, \qquad AQ = 1 - \cos \theta.$$

From the Pythagorean theorem and the fact that $AP < \theta$, we get

$$\sin^2\theta + (1 - \cos \theta)^2 = (AP)^2 < \theta^2. \tag{15}$$

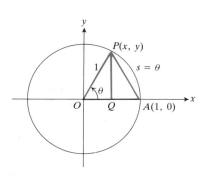

1.89   Radian measure on a unit circle: $r = 1$, $s = \theta$.

Both terms on the left side of Eq. (15) are positive, so each is smaller than their sum and hence is less than θ^2:

$$\sin^2\theta < \theta^2, \tag{16}$$

$$(1 - \cos\theta)^2 < \theta^2, \tag{17}$$

or

$$-\theta < \sin\theta < \theta, \tag{18}$$

$$-\theta < 1 - \cos\theta < \theta. \tag{19}$$

Now we let θ approach 0. Since $\lim_{\theta\to0} -\theta = 0$ and $\lim_{\theta\to0}\theta = 0$, the Sandwich Theorem says that in the limit, Eqs. (18) and (19) become

$$0 \le \lim_{\theta\to0}\sin\theta \le 0, \tag{20}$$

$$0 \le \lim_{\theta\to0}(1 - \cos\theta) \le 0. \tag{21}$$

Therefore, since

$$0 = \lim_{\theta\to0}(1 - \cos\theta) = 1 - \lim_{\theta\to0}\cos\theta,$$

we have

$$\lim_{\theta\to0}\sin\theta = 0, \qquad \lim_{\theta\to0}\cos\theta = 1.$$

To establish Eq. (13), we show that the right-hand and left-hand limits of $(\sin\theta)/\theta$ at 0 are both equal to 1. We will then know that the two-sided limit exists and equals 1 as well.

To show that the right-hand limit is equal to 1, we assume that θ is positive and less than $\pi/2$ in Fig. 1.90. We compare the areas of $\triangle OAP$, sector OAP, and $\triangle OAT$ and note that

$$\text{Area } \triangle OAP < \text{Area sector } OAP < \text{Area } \triangle OAT. \tag{22}$$

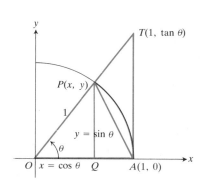

1.90 Area $\triangle OAP <$ area sector $OAP <$ area $\triangle OAT$.

These areas may be expressed in terms of θ as follows:

$$\text{Area } \triangle OAP = \frac{1}{2}\text{ base} \times \text{height} = \frac{1}{2}(1)(\sin\theta) = \frac{1}{2}\sin\theta, \tag{23}$$

$$\text{Area sector } OAP = \frac{1}{2}r^2\theta = \frac{1}{2}(1)^2\theta = \frac{\theta}{2}, \tag{24}$$

$$\text{Area } \triangle OAT = \frac{1}{2}\text{ base} \times \text{height} = \frac{1}{2}(1)(\tan\theta) = \frac{1}{2}\tan\theta, \tag{25}$$

so that

$$\frac{1}{2}\sin\theta < \frac{1}{2}\theta < \frac{1}{2}\tan\theta. \tag{26}$$

The inequality in (26) will go the same way if we divide all three terms by the positive number $(1/2)\sin\theta$:

$$1 < \frac{\theta}{\sin\theta} < \frac{1}{\cos\theta}. \tag{27}$$

We next take reciprocals in (27), which reverses the inequalities:

$$\cos\theta < \frac{\sin\theta}{\theta} < 1. \tag{28}$$

TABLE 1.2

Degrees	θ (radians)	$\sin \theta$	$\dfrac{\sin \theta}{\theta}$
0°	0	0	Undefined
1°	0.017453	0.017452	0.99995
2°	0.034907	0.034899	0.9998
5°	0.08727	0.08716	0.9987
10°	0.17453	0.17365	0.995

Since $\cos \theta$ approaches 1 as θ approaches 0, the Sandwich Theorem tells us that

$$\lim_{\theta \to 0^+} \frac{\sin \theta}{\theta} = 1. \tag{29}$$

The limit in Eq. (29) is a right-hand limit because we have been dealing with values of θ between 0 and $\pi/2$, but the same limit is obtained for $(\sin \theta)/\theta$ as θ approaches 0 from the left. For if $\theta = -\alpha$ and α is positive, then

$$\frac{\sin \theta}{\theta} = \frac{\sin(-\alpha)}{-\alpha} = \frac{-\sin(\alpha)}{-\alpha} = \frac{\sin \alpha}{\alpha}. \tag{30}$$

Therefore,

$$\lim_{\theta \to 0^-} \frac{\sin \theta}{\theta} = \lim_{\alpha \to 0^+} \frac{\sin \alpha}{\alpha} = 1. \tag{31}$$

Together, Eqs. (29) and (31) imply that $\lim_{\theta \to 0} (\sin \theta)/\theta = 1$.

Table 1.2 shows $\sin \theta$ and $(\sin \theta)/\theta$ for selected values of θ near zero. To see $(\sin \theta)/\theta$ approach 1, read from the bottom up. ■

EXAMPLE 13

$$\lim_{x \to 0} \frac{\sin 3x}{x} = \lim_{3x \to 0} \frac{3 \sin 3x}{3x} = 3 \lim_{\theta \to 0} \left(\frac{\sin \theta}{\theta} \right) = 3. \qquad ■$$

EXAMPLE 14

$$\lim_{x \to 0} \frac{\tan x}{x} = \lim_{x \to 0} \left(\frac{\sin x}{x} \cdot \frac{1}{\cos x} \right)$$

$$= \left[\lim_{x \to 0} \left(\frac{\sin x}{x} \right) \right] \left[\lim_{x \to 0} \left(\frac{1}{\cos x} \right) \right] = (1)(1) = 1. \qquad ■$$

PROBLEMS

Find the limits in Problems 1–26.

1. $\lim\limits_{x \to 2} 2x$

2. $\lim\limits_{x \to 0} 2x$

3. $\lim\limits_{x \to 4} 4$

4. $\lim\limits_{x \to -2} 4$

5. $\lim\limits_{x \to 1} (3x - 1)$

6. $\lim\limits_{x \to 1/3} (3x - 1)$

7. $\lim\limits_{x \to 5} x^2$

8. $\lim\limits_{x \to 2} x(2 - x)$

9. $\lim\limits_{x \to 0} (x^2 - 2x + 1)$

10. $\lim\limits_{x \to 5} (x^2 - 3x - 18)$

11. $\lim\limits_{\Delta x \to 0} (2x + \Delta x)$

12. $\lim\limits_{t \to 2} t^2$

13. $\lim\limits_{x \to 1} |x - 1|$

14. $\lim\limits_{x \to -3} |x|$

15. $\lim\limits_{x \to 0} 5(2x - 1)$

16. $\lim\limits_{x \to 1} x(2x - 1)$

17. $\lim\limits_{x \to 2} 3x(2x - 1)$

18. $\lim\limits_{x \to 2} 3(2x - 1)(x + 1)$

19. $\lim\limits_{x \to 2} 3x^2(2x - 1)$

20. $\lim\limits_{x \to 2} 3(2x - 1)(x + 1)^2$

21. $\lim\limits_{x \to -1} (x + 3)$

22. $\lim\limits_{x \to -1} (x + 3)^2$

23. $\lim\limits_{x \to -1} (x^2 + 6x + 9)$

24. $\lim\limits_{x \to -2} (x + 3)^{171}$

25. $\lim\limits_{x \to -4} (x + 3)^{1984}$

26. $\lim\limits_{x \to 1} (x^3 + 3x^2 - 2x - 17)$

27. Suppose $\lim_{x \to c} f(x) = 5$ and $\lim_{x \to c} g(x) = -2$. (a) What is $\lim_{x \to c} f(x)g(x)$? (b) What is $\lim_{x \to c} 2f(x)g(x)$?

28. Suppose $\lim_{x \to b} f(x) = 7$ and $\lim_{x \to b} g(x) = -3$. What is $\lim_{x \to b} [f(x) + g(x)]$?

29. Suppose $\lim_{x \to b} f(x) = 7$ and $\lim_{x \to b} g(x) = -3$. Find
a) $\lim\limits_{x \to b} (f(x) + g(x))$,
b) $\lim\limits_{x \to b} f(x) \cdot g(x)$,
c) $\lim\limits_{x \to b} 4g(x)$,
d) $\lim\limits_{x \to b} f(x)/g(x)$.

30. Suppose $\lim_{x \to -2} p(x) = 4$, $\lim_{x \to -2} r(x) = 0$, and $\lim_{x \to -2} s(x) = -3$. Find (a) $\lim\limits_{x \to -2} (p(x) + r(x) + s(x))$,
(b) $\lim\limits_{x \to -2} p(x) \cdot r(x) \cdot s(x)$.

Find the limits in Problems 31–51.

31. $\lim\limits_{t \to 2} \dfrac{t + 3}{t + 2}$

32. $\lim\limits_{x \to 5} \dfrac{x^2 - 25}{x - 5}$

33. $\lim\limits_{x \to 5} \dfrac{x^2 - 25}{x + 5}$

34. $\lim\limits_{x \to 5} \dfrac{x - 5}{x^2 - 25}$, if it exists

35. $\lim\limits_{x \to 5} \dfrac{x + 5}{x^2 - 25}$, if it exists

36. $\lim\limits_{x \to 1} \dfrac{x^2 - x - 2}{x^2 - 1}$

37. $\lim\limits_{x \to 0} \dfrac{5x^3 + 8x^2}{3x^4 - 16x^2}$

38. $\lim\limits_{y \to 2} \dfrac{y^2 + 5y + 6}{y + 2}$

39. $\lim\limits_{y \to 2} \dfrac{y^2 - 5y + 6}{y - 2}$

40. $\lim\limits_{x \to -5} \dfrac{x^2 + 3x - 10}{x + 5}$

41. $\lim\limits_{x \to 4} \dfrac{x - 4}{x^2 - 5x + 4}$

42. $\lim\limits_{t \to 3} \dfrac{t^2 - 3t + 2}{t^2 - 1}$

43. $\lim\limits_{t \to 1} \dfrac{t^2 - 3t + 2}{t^2 - 1}$

44. $\lim\limits_{x \to 2} \dfrac{x - 2}{x^2 - 6x + 8}$

45. $\lim\limits_{x \to -3} \dfrac{x^2 + 4x + 3}{x - 3}$

46. $\lim\limits_{x \to -3} \dfrac{x + 3}{x^2 + 4x + 3}$

47. $\lim\limits_{x \to -2} \dfrac{x^2 + x - 2}{x^2 - 4}$

48. $\lim\limits_{x \to 1} \dfrac{x^2 + x - 2}{x - 2}$

49. $\lim\limits_{x \to 2} \dfrac{x^2 - 7x + 10}{x - 2}$

50. $\lim\limits_{t \to 1} \dfrac{t^3 - 1}{t - 1}$

51. $\lim\limits_{x \to a} \dfrac{x^3 - a^3}{x^4 - a^4}$

52. Find $\lim_{x \to 1} [(x^n - 1)/(x - 1)]$, n a positive integer. Compare your result with the limits in Problems 50 and 51.

53. Give an example of functions f and g such that $f(x) + g(x)$ approaches a limit as $x \to 0$ even though $f(x)$ and $g(x)$ separately do not approach limits as $x \to 0$.

54. Repeat Problem 53 for $f(x) \cdot g(x)$ in place of $f(x) + g(x)$.

55. Give an example of functions f and g such that $\lim_{x \to 0} f(x)/g(x)$ exists, but at least one of the functions f and g fails to have a limit as $x \to 0$.

Each of the limits in Problems 56 and 57 is the derivative of a function at $x = 0$. Find the function.

56. $\lim\limits_{h \to 0} \dfrac{(1 + h)^2 - 1}{h}$

57. $\lim\limits_{h \to 0} \dfrac{|-1 + h| - |-1|}{h}$

58. Which of the following statements are true of the function f defined for $-1 \le x \le 3$ in Fig. 1.91?
a) $\lim\limits_{x \to -1^+} f(x) = 1$
b) $\lim\limits_{x \to 2} f(x)$ does not exist.
c) $\lim\limits_{x \to 2} f(x) = 2$
d) $\lim\limits_{x \to 1^-} f(x) = 2$
e) $\lim\limits_{x \to 1^+} f(x) = 1$
f) $\lim\limits_{x \to 1} f(x)$ does not exist.
g) $\lim\limits_{x \to 0^+} f(x) = \lim\limits_{x \to 0^-} f(x)$
h) $\lim\limits_{x \to c} f(x)$ exists at every c in $(-1, 1)$.
i) $\lim\limits_{x \to c} f(x)$ exists at every c in $(1, 3)$.

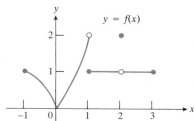

1.91 The function in Problem 58.

Graph the functions in Problems 59 and 60. Then answer the following questions about them:
a) At what points c in the domain of f does $\lim_{x \to c} f(x)$ exist?
b) At what points does only the left-hand limit exist?
c) At what points does only the right-hand limit exist?

59. $f(x) = \begin{cases} \sqrt{1 - x^2} & \text{if } 0 \le x < 1, \\ 1 & \text{if } 1 \le x < 2, \\ 2 & \text{if } x = 2. \end{cases}$

60. $f(x) = \begin{cases} x & \text{if } -1 \le x < 0, \text{ or } 0 < x \le 1, \\ 1 & \text{if } x = 0, \\ 0 & \text{if } x < -1, \text{ or } x > 1. \end{cases}$

Find the limits in Problems 61–68. The square brackets in Problems 61–64 denote the greatest integer function.

61. $\lim\limits_{x \to 0^+} [x]$

62. $\lim\limits_{x \to 0^-} [x]$

63. $\lim\limits_{x \to 0.5} [x]$

64. $\lim\limits_{x \to 2^-} \dfrac{x}{[x]}$

65. $\lim\limits_{x \to 0^+} \dfrac{x}{|x|}$

66. $\lim\limits_{x \to 0^-} \dfrac{x}{|x|}$

67. $\lim\limits_{x \to 3^+} \dfrac{x^2 - 9}{|x - 3|}$

68. $\lim\limits_{x \to 3^-} \dfrac{x^2 - 9}{|x - 3|}$

69. For what values of c does the greatest integer function $f(x) = [x]$ approach a limit as $x \to c$?

70. For what values of c does $f(x) = x/|x|$ approach a limit as $x \to c$?

Find the limits in Problems 71–86.

71. $\lim\limits_{x\to 0} \dfrac{1 + \sin x}{\cos x}$

72. $\lim\limits_{x\to 0^+} \cos x$

73. $\lim\limits_{t\to 0} \dfrac{t}{\sin t}$

74. $\lim\limits_{h\to 0} \dfrac{\sin^2 h}{h}$

75. $\lim\limits_{h\to 0} \dfrac{\sin^2 h}{h^2}$

76. $\lim\limits_{t\to 0} \dfrac{2\sin t \cos t}{t}$

77. $\lim\limits_{x\to 0} \tan x$

78. $\lim\limits_{\theta\to 0} \dfrac{\tan \theta}{\theta}$

79. $\lim\limits_{x\to 0^-} \sin x$

80. $\lim\limits_{x\to 0^+} \dfrac{\sin x}{|x|}$

81. $\lim\limits_{x\to 0^-} \dfrac{\sin x}{|x|}$

82. $\lim\limits_{x\to 0} x \cos x$

83. $\lim\limits_{x\to 0} \dfrac{\sin 2x}{x}$

84. $\lim\limits_{x\to 0} \dfrac{\sin 5x}{\sin 3x}$

85. $\lim\limits_{y\to 0} \dfrac{\tan 2y}{3y}$

86. $\lim\limits_{x\to 0} \dfrac{\sin 2x}{2x^2 + x}$

87. a) Show that $-|x| \le x \sin(1/x) \le |x|$ for all $x \ne 0$.
b) Use the Sandwich Theorem and the inequality in (a) to calculate $\lim_{x\to 0} x \sin(1/x)$.

88. The inequality

$$1 - \frac{x^2}{6} < \frac{\sin x}{x} < 1$$

holds when x is measured in radians and $|x| < 1$. Use this inequality and the Sandwich Theorem to calculate $\lim_{x\to 0} (\sin x)/x$.

89. For each function below, find the limit L of $F(t)$ as $t \to c$. Then show that given $\epsilon > 0$ there exists a $\delta > 0$ such that for all t

$$0 < |t - c| < \delta \;\Rightarrow\; |F(t) - L| < \epsilon.$$

In each case, draw a graph similar to the one in Fig. 1.81.
a) $F(t) = 2t + 3, \quad c = 1$
b) $F(t) = 2t - 3, \quad c = 1$
c) $F(t) = 5 - 3t, \quad c = 2$
d) $F(t) = 7, \quad c = -1$
e) $F(t) = \dfrac{t^2 - 4}{t - 2}, \quad c = 2$
f) $F(t) = \dfrac{t^2 + 6t + 5}{t + 5}, \quad c = 5$
g) $F(t) = \dfrac{3t^2 + 8t - 3}{2t + 6}, \quad c = -3$

h) $F(t) = \dfrac{4}{t}, \quad c = 2$
i) $F(t) = \dfrac{(1/t) - (1/3)}{t - 3}, \quad c = 3$

90. Find a domain $0 < |t - 3| < \delta$ such that, when t is restricted to this domain, the difference between t^2 and 9 will be numerically smaller than (a) $1/10$; (b) $1/100$; and (c) ϵ, where ϵ may be any positive number.

91. Repeat Problem 90 using $t^2 + t$ and 12 in place of t^2 and 9.

92. **CALCULATOR** It is sometimes easy to guess the value of a limit once the limit is known to exist.
a) Ignoring the question of whether the limit below exists (it does and is finite), use a calculator to guess its value. First take $\Delta x = 0.1, 0.01, 0.001, \ldots$, continuing until you are ready to guess the right-hand limit. Then test your guess by using $\Delta x = -0.1, -0.01, \ldots$:

$$\lim_{\Delta x\to 0} \frac{\sqrt{4 + \Delta x} - 2}{\Delta x}.$$

b) Relate the limit in (a) to a derivative.

93. **CALCULATOR** To estimate the value of the derivative of $f(x) = \sqrt{9 - x^2}$ at $x = 0$, write out the appropriate difference quotient and proceed as in Problem 92(a).

94. For each function f and number $\epsilon > 0$, find the x-values for which $4 - \epsilon < f(x) < 4 + \epsilon$.
a) $f(x) = x + 1, \; \epsilon = 0.01$
b) $f(x) = 2x - 2, \; \epsilon = 0.02$
c) $f(x) = x^2 - 5, \; \epsilon = 0.05$
d) $f(x) = \sqrt{19 - x}, \; \epsilon = 0.03$

95. Find $L = \lim_{x\to a} f(x)$ for each function f and number a. Also, find a $\delta > 0$ such that if $0 < |x - a| < \delta$, then $|f(x) - L| < 0.01$.
a) $f(x) = 4x + 3, \; a = 2$
b) $f(x) = \sqrt{x - 7}, \; a = 11$
c) $f(x) = x^2 - 4, \; a = 5$

96. For each function f and $\epsilon > 0$, find the largest $\delta > 0$ such that if $0 < |x - 4| < \delta$, then $|f(x) - 5| < \epsilon$.
a) $f(x) = 9 - x; \; \epsilon = 0.01, 0.001, 0.0001$, general $\epsilon > 0$
b) $f(x) = |3x - 7|; \; \epsilon = 0.003, 0.0003$, general $\epsilon > 0$.

97. Find an interval $I = (5, 5 + \delta), \delta > 0$, such that if x lies in I, then $\sqrt{x - 5} < \epsilon$. What limit is being verified?

TOOLKIT PROGRAMS

| Function Evaluator | Limit Problems |
| Limit Definition | |

1.10
Infinity as a Limit

In this article, we say what it means for the values of a function to approach infinity and what it means for a function $f(x)$ to have a limit as x approaches infinity. While there is no real number "infinity," the word "infinity" pro-

$y = \dfrac{1}{x}$

1.92   The graph of $y = 1/x$.

vides a useful language for describing how some functions behave when their domains or ranges exceed all bounds.

Limits as $x \to \infty$ or $x \to -\infty$

The function

$$ y = \frac{1}{x}, \tag{1}$$

graphed in Fig. 1.92, is defined for all real numbers except $x = 0$. Evidently, the following statements apply.

a) When x is small and positive, $1/x$ is large and positive. For example,

$$\frac{1}{0.001} = 1000.$$

b) When x is small and negative, $1/x$ is large and negative. For example,

$$\frac{1}{-0.001} = -1000.$$

c) When x is large and positive, $1/x$ is small and positive. For example,

$$\frac{1}{10,000} = 0.0001.$$

d) When x is large and negative, $1/x$ is small and negative. For example,

$$\frac{1}{-10,000} = -0.0001.$$

These facts are sometimes abbreviated by saying:

a) As x approaches 0 from the right, $1/x$ tends to ∞.

b) As x approaches 0 from the left, $1/x$ tends to $-\infty$.

c) As x tends to ∞, $1/x$ approaches 0.

d) As x tends to $-\infty$, $1/x$ approaches 0.

The symbol ∞, **infinity,** does not represent any real number. We cannot use ∞ in arithmetic in the usual way, but it is convenient to be able to say things like "the limit of $1/x$ as x approaches infinity is 0," and we can do so according to the following definition.

DEFINITION

Limit as $x \to \infty$ or $x \to -\infty$

1. The limit of the function $f(x)$ as x approaches infinity is the number L,

$$\lim_{x \to \infty} f(x) = L,$$

if: Given any $\epsilon > 0$ there exists a number M such that for all x,

$$M < x \quad \Rightarrow \quad |f(x) - L| < \epsilon. \tag{2}$$

2. The limit of $f(x)$ as x approaches negative infinity is the number L,

$$\lim_{x \to -\infty} f(x) = L,$$

if: Given any $\epsilon > 0$ there exists a number N such that for all x,

$$x < N \quad \Rightarrow \quad |f(x) - L| < \epsilon. \tag{3}$$

In less formal language,

$$\lim_{x \to \infty} f(x) = L$$

means that $f(x)$ can be made as close as desired to L by making x large enough and positive. Similarly,

$$\lim_{x \to -\infty} f(x) = L$$

means that $f(x)$ can be made as close as desired to L by making x large enough and negative (that is, by taking x far enough out on the negative x-axis).

Graphically, the inequalities in statements (2) and (3) mean that the curve $y = f(x)$ stays between the lines $L - \epsilon$ and $L + \epsilon$ for $|x|$ sufficiently large, as in Figs. 1.93 and 1.94, which accompany Examples 1 and 7.

EXAMPLE 1   Show that

$$\lim_{x \to \infty} \frac{1}{x} = 0 \quad \text{and} \quad \lim_{x \to -\infty} \frac{1}{x} = 0 \tag{4}$$

according to the definitions of limit as $x \to \infty$ and $x \to -\infty$.

Solution   See Fig. 1.93. We have

$$\left| \frac{1}{x} - 0 \right| = \left| \frac{1}{x} \right| < \epsilon \tag{5}$$

for any $\epsilon > 0$, provided

$$|x| > \frac{1}{\epsilon}. \tag{6}$$

Thus $1/x$ lies within ϵ of 0 for all $x > M = 1/\epsilon$ and for all $x < N = -1/\epsilon$.

1.93   When $|x| > 1/\epsilon$, the curve $y = 1/x$ lies between the lines $y = \epsilon$ and $y = -\epsilon$.

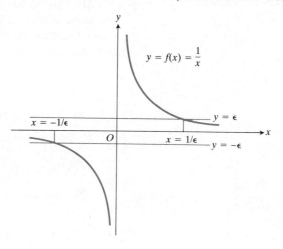

EXAMPLE 2   Let f be the function that has the constant value k for all x. Use the definitions of limit as $x \to \infty$ and $x \to -\infty$ to show that

$$\lim_{x \to \infty} f(x) = k \qquad \text{and} \qquad \lim_{x \to -\infty} f(x) = k. \qquad (7)$$

Solution   When we apply the definitions with $f(x) = k$ and $L = k$, we have

$$|k - k| = 0 < \epsilon \qquad \text{for any } \epsilon > 0. \qquad \blacksquare$$

The following theorem about limits of sums, differences, products, and quotients is analogous to the corresponding theorem for limits when $x \to c$. It tells us how to combine results like those in Examples 1 and 2 to calculate other limits.

THEOREM 3

The Combination Theorem for Limits at Infinity

If

$$\lim_{x \to \infty} f(x) = L_1 \qquad \text{and} \qquad \lim_{x \to \infty} g(x) = L_2,$$

where L_1 and L_2 are (finite) real numbers, then

i) $\lim\limits_{x \to \infty} [f(x) + g(x)] = L_1 + L_2,$

ii) $\lim\limits_{x \to \infty} [f(x) - g(x)] = L_1 - L_2,$

iii) $\lim\limits_{x \to \infty} f(x)g(x) = L_1 L_2,$

iv) $\lim\limits_{x \to \infty} kf(x) = kL_1$ (any number k),

v) $\lim\limits_{x \to \infty} \dfrac{f(x)}{g(x)} = \dfrac{L_1}{L_2}$ if $L_2 \ne 0.$

These results hold for $x \to -\infty$ as well as for $x \to \infty$.

EXAMPLE 3

$$\lim_{x \to \infty} \frac{x}{7x + 4} = \lim_{x \to \infty} \frac{1}{7 + (4/x)} = \frac{1}{7 + 0} = \frac{1}{7} \qquad \blacksquare$$

EXAMPLE 4

$$\lim_{x \to -\infty} \frac{1}{x^2} = \lim_{x \to -\infty} \frac{1}{x} \cdot \frac{1}{x} = 0 \cdot 0 = 0 \qquad \blacksquare$$

EXAMPLE 5

$$\lim_{x \to \infty} \frac{2x^2 - x + 3}{3x^2 + 5} = \lim_{x \to \infty} \frac{2 - (1/x) + (3/x^2)}{3 + (5/x^2)}$$

$$= \frac{2 - 0 + 0}{3 + 0} = \frac{2}{3} \qquad \blacksquare$$

1.94   The graph of $y = 2 + (\sin x)/x$ oscillates about the line $y = 2$. The amplitude of the oscillations decreases toward zero as $x \to \infty$. Because

$$\left| 2 + \frac{\sin x}{x} - 2 \right| = \frac{|\sin x|}{x} \le \frac{1}{x}$$

when $x > 0$, the curve lies between the lines $y = 2 + \epsilon$ and $y = 2 - \epsilon$ when $x > 1/\epsilon$.

EXAMPLE 6

$$\lim_{x \to -\infty} \frac{5x + 3}{2x^2 - 1} = \lim_{x \to -\infty} \frac{(5/x) + (3/x^2)}{2 - (1/x^2)} = \frac{0 + 0}{2 - 0} = 0 \qquad \blacksquare$$

EXAMPLE 7   Find

$$\lim_{x \to \infty} \left(2 + \frac{\sin x}{x} \right).$$

Solution   We have

$$\lim_{x \to \infty} 2 = 2 \qquad \text{and} \qquad \lim_{x \to \infty} \frac{\sin x}{x} = 0$$

because $-1 \le \sin x \le 1$ while $x \to \infty$. Therefore,

$$\lim_{x \to \infty} \left(2 + \frac{\sin x}{x} \right) = \lim_{x \to \infty} 2 + \lim_{x \to \infty} \frac{\sin x}{x} = 2 + 0 = 2.$$

See Fig. 1.94. $\qquad \blacksquare$

Lim $f(x) = \infty$ or lim $f(x) = -\infty$

As suggested by the behavior of $1/x$ as $x \to 0$ and other functions, we sometimes want to say such things as

$$\lim_{x \to c} f(x) = \infty, \tag{8}$$

$$\lim_{x \to c^+} f(x) = \infty, \tag{9}$$

$$\lim_{x \to c^-} f(x) = \infty, \tag{10}$$

$$\lim_{x \to \infty} f(x) = \infty, \tag{11}$$

$$\lim_{x \to -\infty} f(x) = \infty. \tag{12}$$

In every instance, we mean that the value of $f(x)$ eventually exceeds any positive real number B. That is, for any real number B no matter how large, the condition

$$f(x) > B \tag{13}$$

is satisfied for values of x in some restricted set, usually depending on B. The restricted set in (8) has the form

$$0 < |x - c| < \delta.$$

In (9), the set is an interval to the right of c:

$$c < x < c + \delta.$$

In (10), the set is an interval to the left of c:

$$c - \delta < x < c.$$

In (11), the set is an infinite interval of the form

$$M < x < \infty.$$

In (12), the set is an infinite interval of the form

$$-\infty < x < M.$$

By replacing the condition $f(x) > B$ in (13) by the condition

$$f(x) < -B, \qquad (14)$$

where $-B$ is any negative real number, we can similarly define statements like

$$\lim_{x \to c} f(x) = -\infty, \qquad \lim_{x \to c^+} f(x) = -\infty, \qquad \lim_{x \to c^-} f(x) = -\infty,$$

$$\lim_{x \to \infty} f(x) = -\infty, \qquad \lim_{x \to -\infty} f(x) = -\infty.$$

EXAMPLE 8

a) $\lim\limits_{x \to 0^+} \dfrac{1}{x} = \infty$

b) $\lim\limits_{x \to 0^-} \dfrac{1}{x} = -\infty$

c) $\lim\limits_{x \to 0} \dfrac{1}{x^2} = \infty$

d) $\lim\limits_{x \to \infty} \sqrt{x} = \infty$

e) $\lim\limits_{x \to 1^-} \dfrac{1}{x - 1} = -\infty$

f) $\lim\limits_{x \to 2^+} \dfrac{1}{x^2 - 4} = \infty$

g) $\lim\limits_{x \to 2^-} \dfrac{1}{x^2 - 4} = -\infty$

h) $\lim\limits_{x \to -\infty} 2x - \dfrac{3}{x} = -\infty$ ∎

The graphs of $y = 1/x^2$, $y = 1/(x - 1)$, and $y = 1/(x^2 - 4)$ are shown in Fig. 1.95. In Chapter 3 we shall explore the general subject of graphing functions like these.

EXAMPLE 9

$$\lim_{x \to -\infty} \frac{2x^2 - 3}{7x + 4} = \lim_{x \to -\infty} \frac{2x - (3/x)}{7 + (4/x)} = -\infty \qquad ∎$$

The Substitution $x = 1/h$

As many of the preceding examples illustrate, one way to calculate the limit of a quotient of two polynomials as $x \to \infty$ is to divide the numerator and denominator by the largest power of x in the denominator and watch what happens to the new numerator and denominator as $x \to \infty$.

Another way is to let $x = 1/h$ and calculate the limit as $h \to 0^+$.

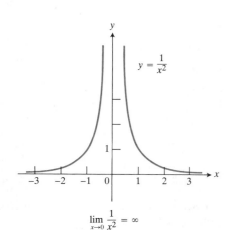

$$\lim_{x \to 0} \frac{1}{x^2} = \infty$$

(a)

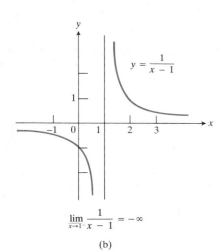

$$\lim_{x \to 1^-} \frac{1}{x - 1} = -\infty$$

(b)

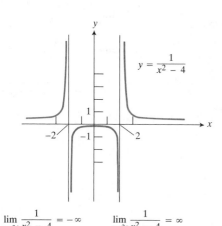

$$\lim_{x \to 2^+} \frac{1}{x^2 - 4} = -\infty \qquad \lim_{x \to 2^-} \frac{1}{x^2 - 4} = \infty$$

(c)

1.95 Graphs of some of the functions in Example 8.

EXAMPLE 10   (Compare with Example 5.)

$$\lim_{x \to \infty} \frac{2x^2 - x + 3}{3x^2 + 5} = \lim_{h \to 0^+} \frac{(2/h^2) - (1/h) + 3}{(3/h^2) + 5}$$

$$= \lim_{h \to 0^+} \frac{2 - h + 3h^2}{3 + 5h^2} = \frac{2 - 0 + 3(0)^2}{3 + 5(0)^2} = \frac{2}{3}$$ ∎

EXAMPLE 11

$$\lim_{x \to \infty} \frac{5x + 3}{2x^2 - 1} = \lim_{h \to 0^+} \frac{(5/h) + 3}{(2/h^2) - 1} = \lim_{h \to 0^+} \frac{5h + 3h^2}{2 - h^2} = \frac{0}{2} = 0$$ ∎

To calculate the limit of a quotient of two polynomials as $x \to -\infty$, we may substitute $h = 1/x$ and calculate the limit as $h \to 0^-$.

EXAMPLE 12   (Compare with Example 9.)

$$\lim_{x \to -\infty} \frac{2x^2 - 3}{7x + 4} = \lim_{h \to 0^-} \frac{(2/h^2) - 3}{(7/h) + 4} = \lim_{h \to 0^-} \frac{2 - 3h^2}{7h + 4h^2} = -\infty$$ ∎

The substitution $x = 1/h$ may help in calculating limits of other functions as well.

EXAMPLE 13

$$\lim_{x \to \infty} x \sin \frac{1}{x} = \lim_{h \to 0^+} \frac{\sin h}{h} = 1$$ ∎

PROBLEMS

Find the limits in Problems 1–32.

1. $\lim_{x \to \infty} \dfrac{2x + 3}{5x + 7}$

2. $\lim_{t \to \infty} \dfrac{t^3 + 7}{t^4}$

3. $\lim_{x \to \infty} \dfrac{x + 1}{x^2 + 3}$

4. $\lim_{x \to \infty} \dfrac{3x^2 - 6x}{4x - 8}$

5. $\lim_{y \to \infty} \dfrac{3y + 7}{y^2 - 2}$

6. $\lim_{x \to \infty} \dfrac{7x - 28}{x^3}$

7. $\lim_{t \to \infty} \dfrac{t^2 - 2t + 3}{2t^2 + 5t - 3}$

8. $\lim_{t \to \infty} \dfrac{t^2 + 1}{t + 1}$

9. $\lim_{x \to \infty} \dfrac{x}{x - 1}$

10. $\lim_{x \to \infty} [x]$

11. $\lim_{x \to -\infty} |x|$

12. $\lim_{x \to -\infty} \dfrac{1}{|x|}$

13. $\lim_{a \to \infty} \dfrac{|a|}{|a| + 1}$

14. $\lim_{t \to -\infty} \dfrac{t}{t + 1}$

15. $\lim_{x \to \infty} \dfrac{3x^3 + 5x^2 - 7}{10x^3 - 11x + 5}$

16. $\lim_{x \to \infty} \left(\dfrac{1}{x} + 1\right)\left(\dfrac{5x^2 - 1}{x^2}\right)$

17. $\lim_{s \to \infty} \left(\dfrac{s}{s + 1}\right)\left(\dfrac{s^2}{5 + s^2}\right)$

18. $\lim_{x \to \infty} \dfrac{8x^{23} - 7x^2 + 5}{2x^{23} + x^{22}}$

19. $\lim_{r \to -\infty} \dfrac{8r^2 + 7r}{4r^2}$

20. $\lim_{x \to \infty} \dfrac{7x^3}{x^3 - 3x^2 + 6x}$

21. $\lim_{y \to \infty} \dfrac{y^4}{y^4 - 7y^3 + 7y^2 + 9}$

22. $\lim_{x \to \infty} \dfrac{5x^3 - 6x + 2}{10x^3 + 5}$

23. $\lim_{x \to \infty} \dfrac{x - 3}{x^2 - 5x + 4}$

24. $\lim_{x \to \infty} \dfrac{9x^4 + x}{2x^4 + 4x^2 - x + 6}$

25. $\lim_{x \to \infty} \dfrac{-2x^3 - 2x + 3}{3x^3 + 3x^2 - 5x}$

26. $\lim_{x \to -\infty} \dfrac{-2x^3 - 2x + 3}{3x^3 + 3x^2 - 5x}$

27. $\lim_{x \to \infty} \dfrac{x + \sin x}{x + \cos x}$

28. $\lim_{x \to -\infty} \dfrac{1 - x^2}{1 + 2x^2}$

29. $\lim_{x \to \infty} \left(\dfrac{1}{x^4} + \dfrac{1}{x}\right)$

30. $\lim_{x \to \infty} \dfrac{\sin 2x}{x}$

31. $\lim_{x \to \infty} \left(1 + \cos \dfrac{1}{x}\right)$

32. $\lim_{y \to \infty} \dfrac{1}{y^2 + 5}$

Find the limits in Problems 33–50.

33. $\lim_{x \to 0^+} \dfrac{1}{3x}$

34. $\lim_{x \to 0^-} \dfrac{2}{x}$

35. $\lim\limits_{x \to 0^+} \dfrac{5}{2x}$

36. $\lim\limits_{t \to 2^+} \dfrac{t^2 + 4}{t - 2}$

37. $\lim\limits_{t \to 2} \dfrac{t^2 - 4}{t - 2}$

38. $\lim\limits_{x \to 2^-} \dfrac{x}{x - 2}$

39. $\lim\limits_{x \to 1^+} \dfrac{x}{x - 1}$

40. $\lim\limits_{x \to 0} \dfrac{|x|}{|x| + 1}$

41. $\lim\limits_{x \to -1^-} \dfrac{1}{x + 1}$

42. $\lim\limits_{x \to 0} \dfrac{1}{|x|}$

43. $\lim\limits_{x \to -2^+} \dfrac{1}{x + 2}$

44. $\lim\limits_{x \to 3^-} \dfrac{x^2}{x - 3}$

45. $\lim\limits_{x \to 3} \dfrac{x - 3}{x^2}$

46. $\lim\limits_{x \to 1^+} \dfrac{2}{x^2 - 1}$

47. $\lim\limits_{x \to 2^-} \dfrac{x^2 + 5}{x - 2}$

48. $\lim\limits_{x \to 2} \dfrac{x - 2}{x^2 + 5}$

49. $\lim\limits_{x \to -5} \dfrac{x^2 + 3x - 10}{x + 5}$

50. $\lim\limits_{x \to 1^+} \dfrac{x + 4}{x^2 + 2x - 3}$

51. Find

$$\lim \frac{x - 1}{2x^2 - 7x + 5}$$

as (a) $x \to 0$, (b) $x \to \infty$, and (c) $x \to 1$.

52. Find the domain and range of the function

$$y = \sqrt{\frac{1}{x} - 1}.$$

Sketch the graphs of the functions in Problems 53–56.

53. $f(x) = \dfrac{1}{x - 2}$

54. $f(x) = \dfrac{1}{x + 1}$

55. $f(x) = 1 + \dfrac{1}{x}$

56. $f(x) = \dfrac{1}{|x|}$

57. Find $\lim_{x \to \infty} f(x)$ if

$$\frac{2x - 3}{x} < f(x) < \frac{2x^2 + 5x}{x^2}.$$

58. Let $f(x) = a_n x^n + a_{n-1} x^{n-1} + \cdots + a_1 x + a_0$ be a polynomial of degree n and $g(x) = b_m x^m + b_{m-1} x^{m-1} + \cdots + b_1 x + b_0$ a polynomial of degree m. Show that $\lim_{x \to \infty} f(x)/g(x)$ is a_n/b_m if $m = n$, 0 if $m > n$, infinite if $m < n$. (*Hint:* Divide the numerator and denominator of the fraction by x^m. What happens to x^n/x^m as $x \to \infty$ if $m = n$? If $m > n$? If $m < n$?)

1.11
Continuity

In this article, we say what it means for a function to be continuous and describe the properties that account for the importance of continuous functions in scientific work.

Continuous Functions

A function $y = f(x)$ that can be graphed over each interval of its domain with one continuous motion of the pen is an example of a **continuous function.** The height of the graph over the interval varies continuously with x. At each interior point of the function's domain, like the point c in Fig. 1.96, the function value $f(c)$ is the limit of the function values on either side; that is,

$$f(c) = \lim_{x \to c} f(x).$$

The function value at each endpoint is also the limit of the nearby function

1.96 Continuity at a, b, and c.

1.97 Discontinuity at $x = 1$ and $x = 2$.

values. At the left endpoint a in Fig. 1.96,

$$f(a) = \lim_{x \to a^+} f(x).$$

At the right endpoint b,

$$f(b) = \lim_{x \to b^-} f(x).$$

To be specific, let us look at the function in Fig. 1.97, whose limits we investigated in Example 5 in Article 1.9.

EXAMPLE 1   The function in Fig. 1.97 is continuous at every point in its domain except $x = 1$ and $x = 2$. At these points there are breaks in the graph. Note the relation between the limit of f and the value of f at each point of the function's domain.

Points of discontinuity:

At $x = 1$: $\lim_{x \to 1} f(x)$ does not exist.

At $x = 2$: $\lim_{x \to 2} f(x) = 1$, but $1 \neq f(2)$.

Points at which f is continuous:

At $x = 0$: $\lim_{x \to 0^+} f(x) = f(0)$.

At $x = 4$: $\lim_{x \to 4^-} f(x) = f(4)$.

At every point $0 < c < 4$ except $x = 1, 2$: $\lim_{x \to c} f(x) = f(c)$.   ∎

We now come to the formal definition of continuity at a point in a function's domain. In the definition we distinguish between continuity at an endpoint (which involves a one-sided limit) and continuity at an interior point (which involves a two-sided limit).

DEFINITIONS

Continuity at an Interior Point

A function $y = f(x)$ is continuous at an interior point c of its domain if

$$\lim_{x \to c} f(x) = f(c). \tag{1}$$

Continuity at an Endpoint

A function $y = f(x)$ is continuous at a left endpoint a of its domain if

$$\lim_{x \to a^+} f(x) = f(a). \tag{2}$$

A function $y = f(x)$ is continuous at a right endpoint b of its domain if

$$\lim_{x \to b^-} f(x) = f(b). \tag{3}$$

Continuous Function

A function is continuous if it is continuous at each point of its domain.

> ### Discontinuity at a Point
>
> If a function f is not continuous at a point c, we say that f is discontinuous at c and call c a point of discontinuity of f.

Functions are usually tested for continuity by applying the following test.

> ### The Continuity Test
>
> The function $y = f(x)$ is continuous at $x = c$ if and only if *all three* of the following statements are true:
>
> 1. $f(c)$ exists   (c is in the domain of f).
> 2. $\lim_{x \to c} f(x)$ exists   (f has a limit as $x \to c$).
> 3. $\lim_{x \to c} f(x) = f(c)$ (the limit equals the function value).

(The limit in the continuity test is to be two-sided if c is an interior point of the domain of f; it is to be the appropriate one-sided limit if c is an endpoint of the domain.)

EXAMPLE 2   When applied to the function $y = f(x)$ of Example 1 at the points $x = 0, 1, 2, 3$, and 4, the continuity test gives the following results. The graph of f is reproduced here as Fig. 1.98.

a) f is continuous at $x = 0$ because
   i) $f(0)$ exists (it equals 1),
   ii) $\lim_{x \to 0^+} f(x) = 1$ (f has a limit as $x \to 0^+$),
   iii) $\lim_{x \to 0^+} = f(0)$ (the limit equals the function value).

b) f is discontinuous at $x = 1$ because $\lim_{x \to 1} f(x)$ does not exist. The function fails part (2) of the test. (The right-hand and left-hand limits exist at $x = 1$, but they are not equal.)

c) f is discontinuous at $x = 2$ because $\lim_{x \to 2} f(x) \neq f(2)$. The function fails part (3) of the test.

d) f is continuous at $x = 3$ because
   i) $f(3)$ exists (it equals 2),
   ii) $\lim_{x \to 3} f(x) = 2$ (f has a limit as $x \to 3$),
   iii) $\lim_{x \to 3} f(x) = f(3)$ (the limit equals the function value).

e) f is continuous at $x = 4$ because
   i) $f(4)$ exists (it equals 1),
   ii) $\lim_{x \to 4^-} f(x) = 1$ (f has a limit as $x \to 4^-$),
   iii) $\lim_{x \to 4^-} f(x) = f(4)$ (the limit equals the function value).   ∎

EXAMPLE 3   The function $y = 1/x$ is continuous at every value of x except $x = 0$. The function is not defined at $x = 0$ and therefore fails part (1) of the continuity test at $x = 0$. See Fig. 1.99.   ∎

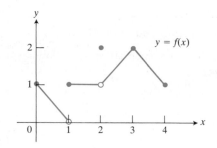

1.98   This function is continuous at $x = 0$, 3, and 4 and discontinuous at $x = 1$ and 2.

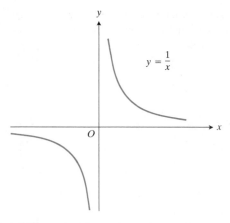

1.99   The function $y = 1/x$ is continuous at every point except $x = 0$.

EXAMPLE 4   The greatest integer function $y = [x]$ is discontinuous at every integer. It does not approach a limit at any integer and therefore fails part (2) of the continuity test at every integer.   ■

EXAMPLE 5   The functions $y = \sin x$ and $y = \cos x$ are continuous at $x = 0$. By Example 12 in Article 1.9,

$$\lim_{x \to 0} \sin x = 0 = \sin 0 \qquad \text{and} \qquad \lim_{x \to 0} \cos x = 1 = \cos 0.$$

We shall see in Chapter 2 that the sine and cosine are continuous at every point.   ■

EXAMPLE 6   *Polynomials and quotients of polynomials.*

a) Every polynomial $f(x) = a_n x^n + \cdots + a_1 x + a_0$ is continuous. We saw in Article 1.9 that $\lim_{x \to c} f(x) = f(c)$ at every point c.

b) Every quotient $f(x)/g(x)$ of polynomials is continuous except where $g(x) = 0$. We saw in Article 1.9 that $\lim_{x \to c} f(x)/g(x) = f(c)/g(c)$ at every point c at which g does not equal zero.   ■

As you may have guessed, algebraic combinations of continuous functions are continuous at every point at which they are defined.

THEOREM 4

> **The Limit Combination Theorem for Continuous Functions**
>
> If the functions f and g are continuous at $x = c$, then all of the following combinations are continuous at $x = c$:
>
> i) $f + g$
>
> ii) $f - g$
>
> iii) fg
>
> iv) kg (any number k)
>
> v) f/g,   provided $g(c) \neq 0$

Proof   Theorem 4 is really a special case of the Limit Combination Theorem in Article 1.9. If the latter were restated for the functions f and g, it would say that if $\lim_{x \to c} f(x) = f(c)$ and $\lim_{x \to c} g(x) = g(c)$, then

i) $\lim_{x \to c} [f(x) + g(x)] = f(c) + g(c),$

ii) $\lim_{x \to c} [f(x) - g(x)] = f(c) - g(c),$

iii) $\lim_{x \to c} f(x)g(x) = f(c)g(c),$

iv) $\lim_{x \to c} kg(x) = kg(c)$ (any number k),

v) $\lim_{x \to c} \dfrac{f(x)}{g(x)} = \dfrac{f(c)}{g(c)},$ provided $g(c) \neq 0.$

In other words, the limits of the functions in (i)–(v) as $x \to c$ exist and equal the function values at $x = c$. Therefore, each function fulfills the three requirements of the continuity test at any interior point $x = c$ of its domain. Similar arguments with right-hand and left-hand limits establish the theorem for continuity at endpoints. ∎

EXAMPLE 7 The functions

$$f(x) = x^{14} + 20x^4, \qquad g(x) = 5x(2 - x) + 1/(x^2 + 1)$$

are continuous at every value of x. The function

$$h(x) = \frac{x + 3}{x^2 - 3x - 10} = \frac{x + 3}{(x - 5)(x + 2)}$$

is continuous at every value of x except $x = 5$ and $x = -2$. ∎

Differentiable Functions Are Continuous

If a function is differentiable at a point c, then it is continuous at c as well.

THEOREM 5

> A function is continuous at every point at which it has a derivative. That is, if $y = f(x)$ has a derivative $f'(c)$ at $x = c$, then f is continuous at $x = c$.

Proof Our task is to show that

$$\lim_{x \to c} f(x) = f(c).$$

If we label the graph of $f(x)$ as in Fig. 1.100, the derivative $f'(c)$ can be calculated from the equation

$$f'(c) = \lim_{x \to c} \frac{f(x) - f(c)}{x - c}. \tag{4}$$

The idea is this: As $x \to c$, the denominator $x - c$ approaches 0. Therefore, if the limit in Eq. (4) is to be finite, the numerator $f(x) - f(c)$ must also approach 0, which means that $f(x)$ must approach $f(c)$.

Formally, we may use the fact that the limit of a product of functions is the product of their limits to show that

$$\lim_{x \to c} [f(x) - f(c)] = \lim_{x \to c} \left[(x - c)\frac{f(x) - f(c)}{x - c} \right]$$

$$= \lim_{x \to c} (x - c) \lim_{x \to c} \frac{f(x) - f(c)}{x - c}$$

$$= 0 \cdot f'(c) = 0. \tag{5}$$

The equation $\lim_{x \to c} [f(x) - f(c)] = 0$ implies that $\lim_{x \to c} f(x) = f(c)$. This is what we set out to show. ∎

EXAMPLE 8 The following functions are continuous:

a) $y = \sqrt{x}$ (differentiable if $x > 0$ by Example 5 in Article 1.7 and continuous at $x = 0$ because $\lim_{x \to 0^+} \sqrt{x} = 0$);

b) $y = x^2$ (differentiable by Example 1 in Article 1.7);

c) $y = |x|$ (differentiable if $x \neq 0$ by Example 4 in Article 1.7 and continuous at $x = 0$ because $\lim_{x \to 0} |x| = 0$). ∎

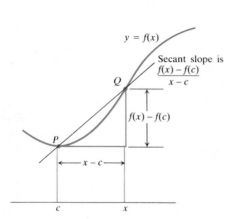

1.100 Figure for the proof that a function is continuous at every point at which it has a derivative.

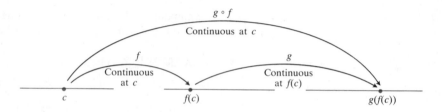

1.101 Composites of continuous functions are continuous.

Although differentiability implies continuity, the converse is not true, as the following example shows.

EXAMPLE 9   The function $y = |x|$ is continuous at $x = 0$ even though it has no derivative at $x = 0$.   ■

Composites of Continuous Functions Are Continuous

All composites of continuous functions are continuous. This means that composites like

$$y = \sin \sqrt{x} \qquad \text{and} \qquad y = |\cos x|$$

are continuous at every point at which they are defined. The idea is that if $f(x)$ is continuous at $x = c$ and $g(x)$ is continuous at $x = f(c)$, then $g \circ f$ is continuous at $x = c$. See Fig. 1.101.

THEOREM 6

> If f is continuous at c and g is continuous at $f(c)$, then the composite $g \circ f$ is continuous at c.

For an outline of the proof of Theorem 6, see Problem 6 in Appendix 1.

EXAMPLE 10   Show that the function

$$y = \left| \frac{x \sin x}{x^2 + 2} \right|$$

is continuous at every value of x.

Solution   The function y is the composite of the continuous functions

$$f(x) = \frac{x \sin x}{x^2 + 2} \qquad \text{and} \qquad g(x) = |x|.$$

The function f is continuous by Theorem 4, the function g by Example 8, and their composite, $g \circ f$ by Theorem 6.   ■

If a composite function $g \circ f$ is continuous at a point $x = c$, its limit as $x \to c$ is $g(f(c))$.

EXAMPLE 11

a) $\displaystyle\lim_{x \to 1} \sin \sqrt{x - 1} = \sin \sqrt{1 - 1} = \sin 0 = 0$

b) $\displaystyle\lim_{x \to 0} |1 + \cos x| = |1 + \cos 0| = |1 + 1| = 2$ ■

Continuous Functions Have Important Properties

We study continuous functions because they are useful in mathematics and applied fields. It turns out that every continuous function is some other function's derivative, as we shall see in Chapter 4. The ability to recover a function from information about its derivative is one of the great powers given to us by calculus. Thus, given a formula $v(t)$ for the velocity of a moving body as a continuous function of time, we shall be able, with the calculus of Chapters 2, 3, and 4, to produce a formula $s(t)$ that tells how far the body has traveled from its starting point at any instant.

In addition, a function that is continuous at every point of a closed interval $[a, b]$ has a maximum value and a minimum value on this interval. We always look for these values when we graph a function, and we shall see the role they play in problem solving (Chapter 3) and in the development of the integral calculus (Chapters 4 and 5).

Finally, a function f that is continuous at every point of a closed interval $[a, b]$ assumes every value between $f(a)$ and $f(b)$. We shall see some consequences of this in a moment.

The proofs of these properties require a detailed knowledge of the real number system and we shall not give them here. They can be found in most texts on advanced calculus.

THEOREM 7

The Max-Min Theorem for Continuous Functions

If f is continuous at every point of the closed interval $[a, b]$, then f takes on a minimum value m and a maximum value M on $[a, b]$. That is, there are numbers α and β in $[a, b]$ such that $f(\alpha) = m$, $f(\beta) = M$, and $m \leq f(x) \leq M$ at all points x in $[a, b]$. (See Fig. 1.102.)

Minimum and maximum at interior points.

Maximum and minimum at endpoints of $[a, b]$.

Minimum and maximum at interior points α and β where the slope is not zero. This function is continuous on $[a, b]$ but is not differentiable at α and β.

Minimum m at interior point α of interval; maximum M at endpoint b.

1.102 A function that is continuous on a closed interval takes on maximum and minimum values somewhere in the interval.

THEOREM 8

The Intermediate Value Theorem

If f is continuous at every point of the closed interval $[a, b]$ and N is any number between $f(a)$ and $f(b)$, then there is at least one point c between a and b where f takes on the value N. (See Fig. 1.103.)

1.103   A function $y = f(x)$ that is continuous on $[a, b]$ takes on every value N between $f(a)$ and $f(b)$.

A Consequence for Graphing: Connectivity   Suppose we want to graph a function $y = f(x)$ that is continuous throughout some interval I on the x-axis. Theorem 8 tells us that the graph of f over I will never move from one y-value to another without taking on the y-values in between. The graph of f over I will be **connected:** it will consist of a single, unbroken curve, like the graph of $y = \sin x$. The graph of f will not have jumps like the graph of $y = [x]$ or separate branches like the graph of $y = \tan x$.

A Consequence for Root Finding   Suppose that $f(x)$ is continuous at every point of a closed interval $[a, b]$ and that $f(a)$ and $f(b)$ differ in sign. Then zero lies between $f(a)$ and $f(b)$, so there is at least one number c between a and b where $f(c) = 0$. In other words, if f is continuous and $f(a)$ and $f(b)$ differ in sign, then the equation $f(x) = 0$ has at least one solution in the open interval (a, b). This helps us locate solutions of equations, as we shall see in Chapter 2. A point at which $f(x) = 0$ is sometimes called a **zero** of f.

EXAMPLE 12   Is there any real number that is 1 less than its cube?

Solution   Any such number must satisfy the equation $x = x^3 - 1$ or $x^3 - x - 1 = 0$. Hence, we are looking for a zero of the function $f(x) = x^3 - x - 1$. By trial we find that $f(1) = -1$ and $f(2) = 5$. We conclude that the equation $f(x) = 0$ has at least one solution $x = c$ between 1 and 2. At this point, $c^3 - c - 1 = 0$, or $c = c^3 - 1$. So, yes, there is a number that is one less than its cube. In Article 2.9, where we study root finding, we shall see that c is about 1.32. See Fig. 1.104. ∎

1.104   The graph of $f(x) = x^3 - x - 1$ crosses the x-axis between $x = 1$ and $x = 2$.

Derivatives Have the Intermediate Value Property

It comes in handy now and then to know that derivatives have the intermediate value property: If f has a derivative at every point of a closed interval $[a, b]$, then f' assumes every value between $f'(a)$ and $f'(b)$. We shall refer to this

A PRIZE ESSAY ON CONTINUOUS FUNCTIONS

Early eighteenth-century mathematicians studied properties of continuous, or "well-behaved," functions to show that if a curve has a point on both sides of a line, then that curve and line in fact intersect. In the second half of the century, however, mathematical problems arose involving more complicated functions that made mathematicians focus on the essential property of continuity. In 1787, the St. Petersburg Academy offered a prize competition for a paper addressing the question *"Whether the arbitrary functions which are achieved by the integration of equations with three or several variables represent any curves or surfaces whatsoever, be they algebraic, transcendental, mechanical, discontinuous, or produced by a voluntary movement of the hand; or whether these functions include only curves represented by an algebraic or transcendental equation."* The prize was won by the little-known mathematician L. F. A. Arbogast, who essentially stated the basic properties of continuous functions. These properties came up anew years later in the works of Bernhard Bolzano and Augustin Louis Cauchy, who did not know of Arbogast's work.

briefly in connection with points of inflection in Article 3.2, but we make no attempt to prove it. There are proofs in more advanced texts.

Continuous Extensions

We can sometimes extend the domain of a function f to include more points where it is continuous. If c is a point where f is not defined but where $\lim_{x \to c} f(x)$ exists, we can define $f(c)$ to be the value of the limit. The extended f is automatically continuous because $f(c)$ exists and equals $\lim_{x \to c} f(x)$.

EXAMPLE 13 Is it possible to define $f(2)$ in a way that extends

$$f(x) = \frac{x^2 + x - 6}{x^2 - 4}$$

to be continuous at $x = 2$? If so, what value should $f(2)$ have?

Solution For f to be continuous at $x = 2$, $f(2)$ must equal $\lim_{x \to 2} f(x)$. Does f have a limit at $x = 2$ and, if so, what is it? To answer this question, we try to factor the numerator and denominator of the expression for $f(x)$ to see if there is a way to rewrite it to avoid division by zero when $x = 2$. We find

$$f(x) = \frac{x^2 + x - 6}{x^2 - 4} = \frac{(x - 2)(x + 3)}{(x - 2)(x + 2)} = \frac{x + 3}{x + 2}.$$

Therefore,

$$\lim_{x \to 2} f(x) = \lim_{x \to 2} \frac{x + 3}{x + 2} = \frac{2 + 3}{2 + 2} = \frac{5}{4},$$

and defining $f(2) = 5/4$ will make

$$f(2) = \lim_{x \to 2} f(x).$$

The extended function

$$f(x) = \begin{cases} \dfrac{x^2 + x - 6}{x^2 - 4} & \text{if } x \neq 2, \\[2mm] \dfrac{5}{4} & \text{if } x = 2, \end{cases} \tag{6}$$

is continuous at $x = 2$ because $\lim_{x \to 2} f(x)$ exists and equals $f(2)$. ■

The function in Eq. (6) is called the **continuous extension** of the original function to the point $x = 2$. Here is another example.

EXAMPLE 14 The function $y = (\sin x)/x$ is not continuous at $x = 0$ but, as we saw in Article 1.9, $\lim_{x \to 0} (\sin x)/x = 1$. It is therefore possible to extend the function to be continuous at $x = 0$. We define

$$f(x) = \begin{cases} \dfrac{\sin x}{x} & \text{if } x \neq 0, \\[2mm] 1 & \text{if } x = 0. \end{cases}$$

1.105 (a) The graph of $f(x) = (\sin x)/x$ for $-\pi/2 \le x \le \pi/2$ does not include the point $(0, 1)$ because the function is not defined at $x = 0$. But we can remove the discontinuity from the graph by defining $f(0) = 1$. When we fill in the missing point this way, we get the continuous curve shown in (b).

(a)

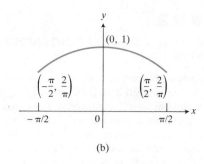

(b)

The function f is continuous at $x = 0$ because $\lim_{x \to 0} f(x) = f(0)$. See Fig. 1.105. ∎

Concluding Remarks

For any function $y = f(x)$ it is important to distinguish between continuity at $x = c$ and having a limit as $x \to c$. The limit, $\lim_{x \to c} f(x)$, is where the function is headed as $x \to c$. Continuity is the property of arriving at the point where $f(x)$ has been heading when x actually gets to c. (Someone is home when you get there, so to speak.) If the limit is what you expect as $x \to c$, and the number $f(c)$ is what you get when $x = c$, then the function is continuous at c if what you expect equals what you get.

Finally, remember the test for continuity:

1. Does $f(c)$ exist?
2. Does $\lim_{x \to c} f(x)$ exist?
3. Does $\lim_{x \to c} f(x) = f(c)$?

For f to be continuous at $x = c$, all three answers must be *yes*.

PROBLEMS

Problems 1–6 are about the function

$$f(x) = \begin{cases} x^2 - 1 & -1 \le x < 0, \\ 2x, & 0 \le x < 1, \\ 1, & x = 1, \\ -2x + 4, & 1 < x < 2, \\ 0, & 2 < x \le 3. \end{cases}$$

This function is graphed in Fig. 1.106.

1. a) Does $f(-1)$ exist?
   b) Does $\lim_{x \to -1^+} f(x)$ exist?
   c) Does $\lim_{x \to -1^+} f(x) = f(-1)$?
   d) Is f continuous at $x = -1$?

2. a) Does $f(1)$ exist?
   b) Does $\lim_{x \to 1} f(x)$ exist?
   c) Does $\lim_{x \to 1} f(x) = f(1)$?
   d) Is f continuous at $x = 1$?

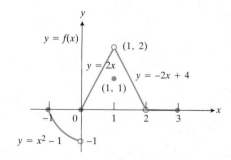

1.106 The function $y = f(x)$ for Problems 1–6.

3. a) Is f defined at $x = 2$? (Look at the definition of f.)
   b) Is f continuous at $x = 2$?

4. At what values of x is f continuous?

5. a) What is the value of $\lim_{x \to 2} f(x)$?

b) What value should be assigned to $f(2)$ to make f continuous at $x = 2$?

6. To what new value should $f(1)$ be changed to make f continuous at $x = 1$?

7. At what points is the function

$$f(x) = \begin{cases} 0, & x < 0, \\ 1, & 0 \leq x \leq 1, \\ 0, & 1 < x, \end{cases}$$

continuous? (*Hint:* Graph the function.)

8. Let $f(x)$ be defined by

$$f(x) = \begin{cases} 1 & \text{for } x < 0, \\ \sqrt{1 - x^2} & \text{for } 0 \leq x \leq 1, \\ x - 1 & \text{for } x > 1. \end{cases}$$

Is f continuous? (*Hint:* Graph the function.)

At what points are the functions in the following problems in Article 1.9 continuous?

9. Problem 58   **10.** Problem 59

11. Problem 60

Find the points (if any) at which the functions in Problems 12–21 are *not* continuous.

12. $y = \dfrac{1}{x - 2}$ **13.** $y = \dfrac{1}{(x + 2)^2}$

14. $y = \dfrac{x}{x + 1}$ **15.** $y = \dfrac{x + 1}{x^2 - 4x + 3}$

16. $y = |x - 1|$ **17.** $y = \dfrac{x + 3}{x^2 - 3x - 10}$

18. $y = \dfrac{x^3 - 1}{x^2 - 1}$ **19.** $y = \dfrac{1}{x^2 + 1}$

20. $y = \dfrac{\cos x}{x}$ **21.** $y = \dfrac{|x|}{x}$

22. The function $f(x)$ is defined by $f(x) = (x^2 - 1)/(x - 1)$ when $x \neq 1$ and by $f(1) = 2$. Is f continuous at $x = 1$? Explain.

23. Define $g(3)$ in a way that extends $g(x) = (x^2 - 9)/(x - 3)$ to be continuous at $x = 3$.

24. Define $h(2)$ in a way that extends $h(x) = (x^2 + 3x - 10)/(x - 2)$ to be continuous at $x = 2$.

25. Define $f(1)$ in a way that extends $f(x) = (x^3 - 1)/(x^2 - 1)$ to be continuous at $x = 1$.

26. Define $g(4)$ in a way that extends $g(x) = (x^2 - 16)/(x^2 - 3x + 4)$ to be continuous at $x = 4$.

27. a) Graph the function

$$f(x) = \begin{cases} x, & 0 \leq x \leq 1, \\ 2 - x, & 1 < x \leq 2. \end{cases}$$

b) Is f continuous at $x = 1$?

c) Does f have a derivative at $x = 1$?

28. How should $f(2)$ be redefined in Fig. 1.97 to make the function continuous at $x = 2$?

29. What value should be assigned to a to make the function

$$f(x) = \begin{cases} x^2 - 1, & x < 3, \\ 2ax, & x \geq 3, \end{cases}$$

continuous at $x = 3$?

30. What value should be assigned to b to make the function

$$g(x) = \begin{cases} x^3, & x < 1/2, \\ bx^2, & x \geq 1/2, \end{cases}$$

continuous at $x = 1/2$?

Find the limits in Problems 31–34.

31. $\lim_{x \to 0} \dfrac{1 + \cos x}{2}$

32. $\lim_{x \to 0} \cos \left(1 - \dfrac{\sin x}{x}\right)$

33. $\lim_{x \to 0} \tan x$

34. $\lim_{x \to 0} \sin \left(\dfrac{\pi}{2} \cos(\tan x)\right)$

35. What is the maximum value of $y = |x|$ for $-1 \leq x \leq 1$? The minimum value?

36. At what values of x does the function in Fig. 1.97 take on its maximum value? Does the function take on a minimum value? Explain.

37. Does the function $y = x^2$ have a maximum value on the open interval $-1 < x < 1$? A minimum value? Explain.

38. On the closed interval $0 \leq x \leq 1$ the greatest integer function $y = [x]$ takes on a minimum value $m = 0$ and a maximum value $M = 1$. It does so even though it is discontinuous at $x = 1$. Does this violate Theorem 7? Why?

39. A continuous function $y = f(x)$ is known to be negative at $x = 0$ and positive at $x = 1$. Why does the equation $f(x) = 0$ have at least one solution between $x = 0$ and $x = 1$? Illustrate with a sketch.

40. Assuming $y = \cos x$ to be continuous, show that the equation $\cos x = x$ has at least one solution. (*Hint:* Show that the function $f(x) = \cos x - x$ has at least one zero.)

41. Show that the function

$$f(x) = \begin{cases} 1, & x \geq 0, \\ -1, & x < 0, \end{cases}$$

is not the derivative of any function. (*Hint:* Does f have the intermediate value property?)

 TOOLKIT PROGRAMS

Continuity at a Point   Super * Grapher
Limit Problems

REVIEW QUESTIONS AND EXERCISES

1. Define the *slope* of a straight line. How would you find the slope of a line from its graph? From an equation for the line?

2. Suppose that m_1, m_2, and m_3 are the slopes of the lines L_1, L_2, and L_3 in Fig. 1.107. List the slopes in order of increasing size.

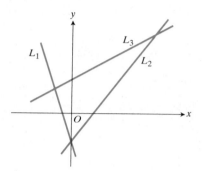

1.107   The lines for Review Question 2.

3. Describe the family of lines $y - y_1 = m(x - x_1)$
   a) if (x_1, y_1) is fixed and lines are drawn for different values of m;
   b) if m and x_1 are fixed and lines are drawn for different values of y_1.

4. Define *function*. What are the *domain* and *range* of a function?

5. Is the circle $x^2 + y^2 = 1$ the graph of a function $y = f(x)$? Explain.

6. If $f(x) = 1/x$ and $g(x) = 1/\sqrt{x}$, what are the domains of f, g, $f + g$, $f - g$, $f \cdot g$, f/g, g/f, $f \circ g$, and $g \circ f$? What is the domain of $h(x) = g(x + 4)$?

7. If $f(x) = \sin x$ and $g(x) = \sqrt{1 - x}$, find a formula for $(g \circ f)(x)$. What are the domain and range of $g \circ f$?

8. In geometry, we define the tangent to a circle to be a line that meets the circle in exactly one point. Is this an appropriate definition for other curves in the plane? Explain. Illustrate your discussion with graphs.

9. Define the *slope* of a curve at a point on the curve.

10. Define average velocity and instantaneous velocity.

11. What general concept includes the notions of slope of a curve and instantaneous velocity?

12. Define the *derivative* of a function at a point in its domain. Illustrate your definition by applying it to the function $f(x) = x^2$ at $x = 2$.

13. Give an example of a continuous function that fails to have a derivative (a) at some point; (b) at several points; (c) at infinitely many points.

14. Suppose that $\lim_{t \to c} F(t) = -7$ and $\lim_{t \to c} G(t) = 0$. Find the limit as t approaches c of each of the following functions.
   a) $3 \cdot F(t)$
   b) $(F(t))^2$
   c) $F(t) \cdot G(t)$
   d) $\dfrac{F(t)}{G(t) + 7}$

15. Define what it means to say that
   a) $\lim_{x \to 2} f(x) = 5$;
   b) $\lim_{x \to 2} g(x) = k$.

16. Evaluate the following limits.
   a) $\lim_{x \to 1} \dfrac{x + 1}{x + 2}$
   b) $\lim_{x \to 1} \dfrac{x - 1}{x + 2}$
   c) $\lim_{x \to 2} \dfrac{x^2 - x - 2}{x^2 - 4}$
   d) $\lim_{x \to 0} \dfrac{\sin 2x}{4x}$

17. Determine which of the following limits exist and evaluate those that do.
   a) $\lim_{x \to \infty} \dfrac{2x + 3}{5x + 7}$
   b) $\lim_{x \to \infty} \dfrac{4z^3 + 3z^2 + 2}{z^4 + 5z^2 + 1}$
   c) $\lim_{x \to \infty} \dfrac{\sin x}{\sqrt{x}}$
   d) $\lim_{x \to \infty} \dfrac{2 + 3x}{1 + 5x}$

18. *Wrong descriptions of limit.* Show by example that the following statements are wrong.
   a) The number L is the limit of $f(x)$ as x approaches c if $f(x)$ gets closer to L as x approaches c.
   b) The number L is the limit of $f(x)$ as x approaches c if, given any $\epsilon > 0$, there is a value of x for which $|f(x) - L| < \epsilon$.

19. Graph and discuss the continuity of
$$f(x) = \begin{cases} x + 1/x & \text{for } x < 0, \\ -x^3 & \text{for } 0 \le x \le 1, \\ -1 & \text{for } 1 < x < 2, \\ 1 & \text{for } x = 2, \\ 0 & \text{for } x > 2. \end{cases}$$

20. Give an example of a continuous extension of a function.

21. Give an example of a function that is defined on $0 \le x \le 1$, continuous in the open interval $0 < x < 1$, and discontinuous at $x = 0$.

22. State and prove a theorem about the relationship between continuity and differentiability of a function at a point in its domain.

23. True, or false? If $y = f(x)$ is continuous, $f(1) = 0$, and $f(2) = 3$, then f takes on the value 2.5 at some point between $x = 1$ and $x = 2$. Explain.

24. The function $y = 1/x$ does not take on either a maximum or a minimum value on the interval $[-1, 1]$. Does this contradict Theorem 7? Why?

25. Read the article "The Lever of Mahomet," by R. Courant and H. Robbins, and the accompanying "Commentary on Continuity," by J. R. Newman, in *World of Mathematics*, Vol. 4 (New York: Simon and Schuster, 1960), pp. 2410–2413.

MISCELLANEOUS PROBLEMS

1. Find the starting position of a particle in the plane whose final position is $B(u, v)$ after its coordinates change by increments $\Delta x = h$ and $\Delta y = k$.

2. A particle moves in the plane from $A(-2, 5)$ to the y-axis in such a way that $\Delta y = 3\,\Delta x$. Find its new coordinates.

3. A particle moves along the parabola $y = x^2$ from the point $A(1, 1)$ to the point $B(a, a^2)$, $a \neq 1$. Sketch the parabola and show that $\Delta y/\Delta x = a + 1$ if $\Delta x \neq 0$.

4. a) Plot the points $A(8, 1)$, $B(2, 10)$, $C(-4, 6)$, $D(2, -3)$, $E(14/3, 6)$.
   b) Find the slopes of the lines AB, BC, CD, DA, CE, BD.
   c) Do four of the five points A, B, C, D, and E form a parallelogram? Why?
   d) Do three of the five points lie on a common straight line? Why?
   e) Does the origin $(0, 0)$ lie on a straight line through two of the five points? Why?
   f) Find equations of the lines AB, CD, AD, CE, BD.
   g) Find the coordinates of the points in which the lines AB, CD, AD, CE, BD intersect the x- and y-axes.

5. a) Find an equation for the line through $P(1, -3)$ perpendicular to the line L: $2y - 3x = 4$.
   b) Find the distance between P and L.

6. Plot the points $A(6, 4)$, $B(4, -3)$, and $C(-2, 3)$.
   a) Is ABC a right triangle? Explain.
   b) Is ABC isosceles? Explain.
   c) Does the origin lie inside, outside, or on the boundary of the triangle? Explain.
   d) The point C is replaced by a point $C'(-2, y)$ that makes angle $C'BA$ a right angle. Find y.

7. Find equations for the lines through the origin that are tangent to the circle of radius 2 centered at $(2, 1)$.

8. Find the coordinates of the midpoint of the line segment joining the points $P_1(x_1, y_1)$ and $P_2(x_2, y_2)$.

9. Find the (a) slope, (b) y-intercept, and (c) x-intercept of the line L: $Ax + By = C$. (d) Find an equation for the line through the origin perpendicular to L.

10. *General formula for the distance between a point and a line in the plane.* Show that the distance between a point $P_1(x_1, y_1)$ and a line $Ax + By = C$ is equal to
$$\frac{|Ax_1 + By_1 - C|}{\sqrt{A^2 + B^2}}.$$
   When $Ax_1 + By_1 = C$: P_1 is on L.
   When $Ax_1 + By_1 > C$: P_1 is on one side of L.
   When $Ax_1 + By_1 < C$: P_1 is on the opposite side of L.
   (There are neat solutions of this problem in *American Mathematical Monthly,* Vol. 59 (1952), pp. 242 and 248.)

11. Suppose that the length of the perpendicular ON from the origin to a line L is p and that ON makes an angle α with the positive x-axis. Show that $x \cos \alpha + y \sin \alpha = p$ is an equation for L. (*Remark:* When $A^2 + B^2 \neq 0$, the general linear equation $Ax + By = C$ can be written $x \cos \alpha + y \sin \alpha = p$, with $\cos \alpha = A/\sqrt{A^2 + B^2}$, $\sin \alpha = B/\sqrt{A^2 + B^2}$, and $p = C/\sqrt{A^2 + B^2}$.)

12. If A, B, C, and C' are constants, and A and B are not both zero, show that (a) the lines
$$Ax + By = C \qquad \text{and} \qquad Ax + By = C'$$
either coincide or are parallel and that (b) the lines
$$Ax + By = C \qquad \text{and} \qquad Bx - Ay = C'$$
are perpendicular.

13. How many circles in the plane are tangent to all three of the following lines?
$$L_1: x + y = 1 \qquad L_2: y = x + 1 \qquad L_3: x - 3y = 1$$
Give the center and radius of one of these circles. You may use the formula in Problem 10.

14. Find the distance between the lines $y = mx + b$ and $y = mx + b'$. Express your answer in terms of b, b', and m.

15. Suppose that L_1 and L_2 are lines with the equations
$$L_1: \quad a_1 x + b_1 y + c_1 = 0, \qquad L_2: \quad a_2 x + b_2 y + c_2 = 0,$$
and that k is a constant. Describe the set of points whose coordinates satisfy the equation
$$(a_1 x + b_1 y + c_1) + k(a_2 x + b_2 y + c_2) = 0.$$

16. Find the coordinates of the point on the line $y = 3x + 1$ that is equidistant from $(0, 0)$ and $(-3, 4)$.

17. Find an equation of a line that is perpendicular to $5x - y = 1$ and is such that the area of the triangle formed by the x-axis, the y-axis, and the straight line is equal to 5. How many lines like this are there?

18. Let $y = (x^2 + 2)/(x^2 - 1)$. Express x in terms of y and find the values of y for which x is real.

19. Express the area A and the circumference C of a circle as functions of the radius r. Then express A as a function of C.

20. If $f(x) = x - (1/x)$, show that $f(1/x) = -f(x) = f(-x)$.

Find the domain and range of each function in Problems 21–24.

21. $y = \dfrac{1}{1 + x}$

22. $y = \dfrac{1}{1 + x^2}$

23. $y = \dfrac{1}{1 + \sqrt{x}}$

24. $y = \dfrac{1}{\sqrt{3 - x}}$

25. Let $f(x) = ax + b$ and $g(x) = cx + d$. What condition must be satisfied by the constants a, b, c, and d to make $f(g(x))$ and $g(f(x))$ identical?

26. Let $f(x) = (ax + b)/(cx + d)$. If $d = -a$, show that $f(f(x)) = x$ for all values of x.

27. If $f(x) = x/(x - 1)$, find
   a) $f(1/x)$,
   b) $f(-x)$,
   c) $f(f(x))$,
   d) $f(1/f(x))$.

28. Without using the absolute value symbol, describe the set of x-values for which $|x + 1| < 4$.

29. Graph the equation $|x| + |y| = 1$. (*Hint:* Work one quadrant at a time, replacing the equation by an equivalent equation without absolute values in each quadrant.)

30. Graph the function $y = |x + 2| + x$ for $-5 \le x \le 2$. Find the function's range.

31. Show that the expression

$$\max(a, b) = \frac{(a + b)}{2} + \frac{|a - b|}{2}$$

is equal to a when $a \ge b$ and is equal to b when $b \ge a$. In other words, $\max(a, b)$ gives the larger of the two numbers a and b. Find a similar expression for $\min(a, b)$, that gives the smaller of the two numbers.

32. For each of the following expressions $f(x)$, sketch first the graph of $y = f(x)$, then the graph of $y = |f(x)|$, and finally the graph of $y = f(x)/2 + |f(x)|/2$.
a) $f(x) = (x - 2)(x + 1)$ b) $f(x) = x^2$
c) $f(x) = -x^2$ d) $f(x) = 4 - x^2$

33. Define $y = [x]$ to be the greatest integer less than x. Graph
a) $y = [x]$,   b) $y = [x] - [x]$.

34. a) Graph the function $y = |4 - x^2|$ for $-3 \le x \le 3$.
b) Find the maximum and minimum values of y on the interval. At what values of x does y take on these values?

35. Using the definition of the derivative, find $f'(x)$ if $f(x)$ is
a) $(x - 1)/(x + 1)$,   b) $x^{3/2}$,   c) $x^{1/3}$.

36. Use the definition of the derivative to find
a) $f'(x)$ if $f(x) = x^2 - 3x - 4$,
b) $\dfrac{dy}{dx}$ if $y = \dfrac{1}{3x} + 2x$,
c) $f'(t)$ if $f(t) = \sqrt{t - 4}$.

37. a) By the Δ-method, find the slope of the curve $y = 2x^3 + 2$ at the point $(1, 4)$.
b) At which point of the curve in (a) is the tangent to the curve parallel to the x-axis? Sketch the curve.

38. If $f(x) = 2x/(x - 1)$, find
a) $f(0),\ f(-1),\ f(1/x)$;   b) $\Delta f(x)/\Delta x$;
c) $f'(x)$, using the result of (b).

39. Using the method of Article 1.6, find the slope of the curve $y = 180x - 16x^2$ at the point (x_1, y_1). Sketch the curve. At what point does the curve have a horizontal tangent?

40. Find the velocity $v = ds/dt$ if a particle's position at time t is $s = 180t - 16t^2$. When does the velocity vanish?

41. If a ball is thrown straight up with a velocity of 32 ft/sec, its height after t seconds is given by the equation $s = 32t - 16t^2$. At what instant will the ball be at its highest point, and how high will it rise?

42. If the pressure P and volume V of a gas are related by the formula $P = 1/V$, find (a) the average rate of change of P with respect to V; (b) the rate of change of P with respect to V at the instant when $V = 2$.

Evaluate the limits in Problems 43–58, or show that they do not exist.

43. $\lim\limits_{x \to \infty} \dfrac{\sin x}{x}$

44. $\lim\limits_{x \to \infty} \dfrac{x + \sin x}{2x + 5}$

45. $\lim\limits_{x \to \infty} \dfrac{1 + \sin x}{x}$

46. $\lim\limits_{x \to 1} \dfrac{x^2 - 4}{x^3 - 8}$

47. $\lim\limits_{x \to 0} \dfrac{x}{\tan 3x}$

48. $\lim\limits_{x \to \infty} \dfrac{x \sin x}{x + \sin x}$

49. $\lim\limits_{x \to a} \dfrac{x^2 - a^2}{x - a}$

50. $\lim\limits_{x \to a} \dfrac{x^2 - a^2}{x + a}$

51. $\lim\limits_{h \to 0} \dfrac{(x + h)^2 - x^2}{h}$

52. $\lim\limits_{h \to 0} \dfrac{\sqrt{x + h} - \sqrt{x}}{h}$

53. $\lim\limits_{\Delta x \to 0} \dfrac{1/(x + \Delta x) - 1/x}{\Delta x}$

54. $\lim\limits_{x \to 0^+} \dfrac{1}{x}$

55. $\lim\limits_{x \to 1} \dfrac{1 - \sqrt{x}}{1 - x}$

56. $\lim\limits_{x \to 1} \dfrac{(2x - 3)(\sqrt{x} - 1)}{2x^2 + x - 3}$

57. $\lim\limits_{x \to \infty} (1 - x \cos x)$

58. $\lim\limits_{x \to 1} \dfrac{\sqrt{x + 1} - \sqrt{2x}}{x^2 - x}$

Find the limits in Problems 59–66.

59. $\lim\limits_{x \to 0^+} \dfrac{|x|}{x}$

60. $\lim\limits_{x \to 0^-} \dfrac{|x|}{x}$

61. $\lim\limits_{x \to 4^-} ([x] - x)$

62. $\lim\limits_{x \to 4^+} ([x] - x)$

63. $\lim\limits_{x \to 3^+} \dfrac{[x]^2 - 9}{x^2 - 9}$

64. $\lim\limits_{x \to 3^-} \dfrac{[x]^2 - 9}{x^2 - 9}$

65. $\lim\limits_{x \to 0} x[x]$

66. $\lim\limits_{x \to 0^+} \dfrac{\sqrt{x}}{\sqrt{4 + \sqrt{x}} - 2}$

67. Given $(x - 1)/(2x^2 - 7x + 5) = f(x)$, find (a) the limit of $f(x)$ as $x \to \infty$; (b) the limit of $f(x)$ as $x \to 1$; (c) $f(-1/x)$, $f(0)$, $1/f(x)$.

68. Find the coordinates of the point of intersection of the straight lines $3x + 5y = 1$ and $(2 + c)x + 5c^2y = 1$ and determine the limiting position of this point as c tends to 1.

69. Find (a) $\lim\limits_{n \to \infty} (\sqrt{n^2 + 1} - n)$, (b) $\lim\limits_{n \to \infty} (\sqrt{n^2 + n} - n)$.

70. Given $\epsilon > 0$, find $\delta > 0$ such that, for all t,
$$0 < |t - 1| < \delta \ \Rightarrow\ \sqrt{t^2 - 1} < \epsilon.$$

71. Given $\epsilon > 0$, find M such that
$$\left| \frac{t^2 + t}{t^2 - 1} - 1 \right| < \epsilon \qquad \text{for all } t > M.$$

72. Suppose that $f(x) = x^3 - 3x^2 - 4x + 12$ and
$$h(x) = \begin{cases} \dfrac{f(x)}{x - 3} & \text{for } x \ne 3, \\ k & \text{for } x = 3. \end{cases}$$

a) Find all zeros of f.
b) Find the value of k that makes h continuous at $x = 3$.
c) Using the value of k found in (b), determine whether h is an even function.

73. Suppose F is a function whose values are all less than or equal to some constant M: $F(t) \le M$. Prove: If $\lim\limits_{t \to c} F(t) = L$, then $L \le M$. (*Suggestion:* An indirect proof may be used to show that $L > M$ is false. If $L > M$, we may take $1/2(L - M)$ as a positive number ϵ, apply the definition of limit, and arrive at a contradiction.)

74. A function f, whose domain is the set of all real numbers, has the property that $f(x + h) = f(x) \cdot f(h)$ for all x and h; and $f(0) \neq 0$.

a) Show that $f(0) = 1$. (*Hint:* Let $h = x = 0$.)

b) If f has a derivative at 0, show that f has a derivative at every real number x and that

$$f'(x) = f(x) \cdot f'(0).$$

75. Can $f(4)$ be defined in a way that extends $f(x) = (x^2 - 16)/|x - 4|$ to be continuous at $x = 4$? If so, what value should $f(4)$ have? If not, why not? (*Hint:* Calculate the right- and left-hand limits of f at $x = 4$.)

76. Can $f(0)$ be defined in a way that extends $f(x) = \sin(1/x)$ to be continuous at $x = 0$? If so, what value should $f(0)$ have? If not, why not? (Figure 1.86 shows the graph of f near the origin.)

77. At what points is the function $y = 1/[x]$ discontinuous?

78. Show that every polynomial of odd degree has at least one real zero.

79. The function $f(x) = |x|$ is continuous at $x = 0$. Given a positive number ϵ, how small must δ be for $|x - 0| < \delta$ to imply $|f(x) - 0| < \epsilon$?

80. a) Graph the function $y = f(x)$ defined by

$$f(x) = \begin{cases} x \sin(1/x) & \text{if } x \neq 0, \\ 0 & \text{if } x = 0. \end{cases}$$

(*Hint:* The graph of f lies in the "bow"-shaped portion of the plane bounded by the lines $y = x$ and $y = -x$ and containing the x-axis. See Fig. 1.86 for a graph of $y = \sin(1/x)$.)

b) Show that the function $y = f(x)$ in (a) is continuous at $x = 0$. (*Hint:* First show that $|x \sin(1/x)| \leq |x|$ for all x. Then answer the question: How small does $|x - 0|$ need to be to make $|x \sin(1/x) - 0|$ less than ϵ?)

81. Let f be a continuous function, and suppose that $f(c)$ is positive. Show that there is some interval about c, say $c - \delta < x < c + \delta$, throughout which $f(x)$ remains positive. Illustrate with a sketch. (*Hint:* Take $\epsilon = f(c)/2$.)

82. *Properties of inequalities.* If a and b are any two real numbers, we say a is less than b and, write $a < b$ if (and only if) $b - a$ is positive. If $a < b$ we also say that b is greater than a $(b > a)$. Prove the following properties of inequalities:

a) If $a < b$, then $a + c < b + c$ and $a - c < b - c$ for any real number c.

b) If $a < b$ and $c < d$, then $a + c < b + d$. Is it also true that $a - c < b - d$? If so, prove it; if not, give a counterexample.

c) If a and b are both positive (or both negative) and $a < b$, then $1/b < 1/a$.

d) If $a < 0 < b$, then $1/a < 0 < 1/b$.

e) If $a < b$ and $c > 0$, then $ac < bc$.

f) If $a < b$ and $c < 0$, then $bc < ac$.

83. *Properties of absolute values.*

a) Prove that $|a| < |b|$ if and only if $a^2 < b^2$.

b) Prove that $|a + b| \leq |a| + |b|$.

c) Prove that $|a - b| \geq ||a| - |b||$.

d) Prove, by mathematical induction, that

$$|a_1 + a_2 + \cdots + a_n| \leq |a_1| + |a_2| + \cdots + |a_n|.$$

(Mathematical induction is reviewed in Appendix 2.)

e) Using the result from (d), prove that

$$|a_1 + a_2 + \cdots + a_n| \geq |a_1| - |a_2| - \cdots - |a_n|.$$

84. *A surprising result.* Suppose that the functions f and g are defined throughout an open interval containing the point x_0, that f is differentiable at x_0, that $f(x_0) = 0$, and that g is continuous at x_0. Show that the product fg is differentiable at x_0. This shows, for example, that while $|x|$ is not differentiable at $x = 0$, the product $x|x|$ *is* differentiable at $x = 0$. Similarly, while $x \sin(1/x)$ is not differentiable at $x = 0$ (Problem 80), the product $x^2 \sin(1/x)$ *is* differentiable at $x = 0$. (*Hint:* Write down the difference quotient for the product fg and see what its limit has to be.)

85. *The Lagrange interpolation formula.* Let (x_1, y_1), (x_2, y_2), $\ldots$, (x_n, y_n) be n points in the plane, no two of them having the same x-coordinate. Find a polynomial $f(x)$ of degree $(n - 1)$ that takes the value y_1 at x_1, y_2 at x_2, $\ldots$, y_n at x_n; that is, $f(x_i) = y_i (i = 1, 2, \ldots, n)$. (*Hint:*

$$f(x) = y_1 \phi_1(x) + y_2 \phi_2(x) + \cdots + y_n \phi_n(x),$$

where $\phi_k(x)$ is a polynomial that is zero at x_i $(i \neq k)$ and $\phi_k(x_k) = 1$.)

86. For what positive values of x does $f(x) = x/(x + 1)$ lie between $y = 1/4$ and $y = 3/4$?

87. Suppose $0 < \epsilon < 4$. Find the largest $\delta > 0$ with the property that $f(x) = x^2$ lies between $y = 4 - \epsilon$ and $y = 4 + \epsilon$ for all x in the interval $I = (2 - \delta, 2 + \delta)$. What happens to δ as ϵ decreases to 0? Graph δ as a function of ϵ.

88. Graph the function

$$f(x) = \begin{cases} 4 - 2x, & x < 1, \\ 6x - 4, & x \geq 1. \end{cases}$$

Then, given $\epsilon > 0$, find the largest δ for which $f(x)$ lies between $y = 2 - \epsilon$ and $y = 2 + \epsilon$ for x in the interval $I = (1 - \delta, 1 + \delta)$.

89. Let $f(x) = |x - 5|/(x - 5)$. Find the set of x-values for which $1 - \epsilon < f(x) < 1 + \epsilon$, for $\epsilon = 4, 2, 1$, and $1/2$. (*Hint:* Graph the function first.)

90. Let $f(x) = 1/(x - 4)$. Find the set of x-values for which $f(x)$ lies between $y = 3 - \epsilon$ and $y = 3 + \epsilon$ if $\epsilon = 4, 2, 1, 1/2$.

CHAPTER 2

Derivatives

OVERVIEW

In Chapter 1 we saw how the slope of a curve is defined as a limit of secant slopes and how this limit, called a *derivative,* enabled the scientists of the seventeenth century to formulate precise and workable definitions of the notions of tangent and instantaneous rate of change.

So far, however, the definition

$$\lim_{\Delta x \to 0} \frac{f(x + \Delta x) - f(x)}{\Delta x}$$

has proved to be workable only in the sense that, with enough time, we can work it out. But that was before we studied limits. We now know enough to be able to calculate derivatives rapidly, and that is the main goal of this chapter—to learn to calculate derivatives rapidly.

The first step is to derive rules for constructing the derivatives of algebraic combinations of functions whose derivatives we already know (sums, products, quotients, and powers). These rules are similar to the limit combination theorems in Article 1.9. As you will see, this is no accident. We then show that the derivative of a composite of two functions is the product of their derivatives (the Chain Rule), and we show how to use derivatives to estimate change (with differentials) and to replace complicated functions by simpler functions that still give the accuracy we want. We then study a shortcut for finding dy/dx when the formula in which y appears cannot be solved directly for y and use this technique to find derivatives of functions raised to fractional powers. The chapter concludes with parametric equations (good for describing motion) and with Newton's method, an amazing technique that uses derivatives to solve equations.

2.1
Polynomial Functions and Their Derivatives

When we calculate the derivative of a function, whether by Fermat's Δ-process or a standard formula, we say that we have *differentiated* the function. In Chapter 1 we differentiated a few polynomials like $y = mx + b$, getting $dy/dx = m$, and $y = x^2$, getting $dy/dx = 2x$. In this article we develop a fast routine for differentiating *any* polynomial.

A single term of the form cx^n, where c is a constant and n is zero or a positive integer, is a **monomial** in x. The sum of a finite number of monomials in x is a **polynomial** in x. Our approach to differentiating polynomials will be to find a formula for differentiating monomials and then to find a rule for constructing the derivative of a polynomial from the derivatives of its monomials.

DEFINITION

Derivative

Let $y = f(x)$ be a function of x. If the limit

$$\frac{dy}{dx} = f'(x) = \lim_{\Delta x \to 0} \frac{f(x + \Delta x) - f(x)}{\Delta x} \tag{1a}$$

exists and is finite, we call this limit the **derivative** of f at x and say that f is **differentiable** at x.

In our work it will save time to write Δy for the increment $f(x + \Delta x) - f(x)$. When we do this, the limit in Eq. (1a) becomes

$$\frac{dy}{dx} = \lim_{\Delta x \to 0} \frac{\Delta y}{\Delta x}. \tag{1b}$$

RULE 1

The derivative of a constant is zero.

2.1 The slope of the graph of $y =$ constant is zero.

Rule 1 says that if $y = f(x)$ has a constant value c, then $dy/dx = 0$. The reason for this rule is the following calculation:

$$\lim_{\Delta x \to 0} \frac{f(x + \Delta x) - f(x)}{\Delta x} = \lim_{\Delta x \to 0} \frac{c - c}{\Delta x} = \lim_{\Delta x \to 0} 0 = 0.$$

See Fig. 2.1.

RULE 2

Power Rule for Positive Integer Powers of x

If n is a positive integer, then

$$\frac{d}{dx}(x^n) = nx^{n-1}. \tag{2}$$

To apply the power rule, we subtract 1 from the original exponent (n) and multiply by n.

EXAMPLE 1

$$\frac{d}{dx}(x) = \frac{d}{dx}(x^1) = 1 \cdot x^0 = 1$$

$$\frac{d}{dx}(x^2) = 2x^1 = 2x$$

$$\frac{d}{dx}(x^3) = 3x^2$$

$$\frac{d}{dx}(x^4) = 4x^3$$

$$\frac{d}{dx}(x^5) = 5x^4$$

$\blacksquare$

Proof of Rule 2   To prove Rule 2, we let $y = f(x) = x^n$. Then

$$\frac{\Delta y}{\Delta x} = \frac{f(x + \Delta x) - f(x)}{\Delta x} = \frac{(x + \Delta x)^n - x^n}{\Delta x}. \tag{3}$$

Since n is a positive integer, we can apply the algebraic formula

$$a^n - b^n = (a - b)(a^{n-1} + a^{n-2}b + \cdots + ab^{n-2} + b^{n-1}),$$

with $a = x + \Delta x$, $b = x$, $a - b = \Delta x$, to the expression $(x + \Delta x)^n - x^n$ on the right-hand side of Eq. (3). This gives

$$\frac{\Delta y}{\Delta x} = \frac{(x + \Delta x)^n - x^n}{\Delta x}$$

$$= \frac{(\Delta x)[(x + \Delta x)^{n-1} + (x + \Delta x)^{n-2}x + \cdots + (x + \Delta x)x^{n-2} + x^{n-1}]}{\Delta x} \tag{4}$$

$$= \underbrace{[(x + \Delta x)^{n-1} + (x + \Delta x)^{n-2}x + \cdots + (x + \Delta x)x^{n-2} + x^{n-1}]}_{n \text{ terms, each with limit } x^{n-1} \text{ as } \Delta x \to 0}.$$

We now let Δx approach zero and find that

$$\frac{dy}{dx} = \lim_{\Delta x \to 0} \frac{\Delta y}{\Delta x}$$

$$= \underbrace{[(x + 0)^{n-1} + (x + 0)^{n-2}x + \cdots + (x + 0)x^{n-2} + x^{n-1}]}_{n \text{ terms}} \tag{5}$$

$$= \underbrace{[x^{n-1} + x^{n-1} + \cdots + x^{n-1} + x^{n-1}]}_{n \text{ copies of } x^{n-1}}$$

$$= nx^{n-1}.$$

In short,

$$\frac{dy}{dx} = nx^{n-1}.$$

$\blacksquare$

The next two rules work for *any* differentiable function.

RULE 3

The (Constant) Multiple Rule

If u is any differentiable function of x, and c is any constant, then

$$\frac{d}{dx}(cu) = c\frac{du}{dx}. \tag{6}$$

Rule 3 says that the derivative of a number times a function is the same number times the derivative of the function.

EXAMPLE 2 The derivative

$$\frac{d}{dx}(7x^5) = 7 \cdot 5x^4 = 35x^4$$

says that if we stretch the graph of $y = x^5$ in the y-direction by multiplying each y-coordinate by 7, then we multiply each slope by 7 as well (Fig. 2.2). ∎

Proof of Rule 3 This follows immediately from the fact that $u = f(x)$ has a derivative:

$$\frac{d}{dx}cu = \lim_{\Delta x \to 0} \frac{cf(x + \Delta x) - cf(x)}{\Delta x}$$

$$= \lim_{\Delta x \to 0} c\frac{f(x + \Delta x) - f(x)}{\Delta x}$$

$$= c \lim_{\Delta x \to 0} \frac{f(x + \Delta x) - f(x)}{\Delta x} \tag{7}$$

$$= c\frac{du}{dx}. \qquad ■$$

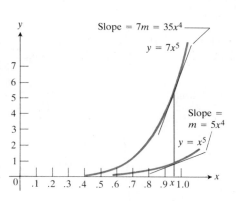

2.2 The graphs of $y = x^5$ and the stretched curve $y = 7x^5$. Multiplying the y-coordinates by 7 multiplies the slopes by 7.

If c is a constant and n a positive integer, then Rules 2 and 3 combine to give

$$\frac{d}{dx}(cx^n) = cnx^{n-1}. \tag{8}$$

EXAMPLE 3 The line $y = 3x + b$ is tangent to the curve $y = 2x^2$. Find b and the point of tangency.

Solution The slope of the line $y = 3x + b$ is 3. The slope of the curve at any point $P(x, y)$ is $dy/dx = 4x$. If P is also the point of tangency, the slope of the curve at P equals the slope of the line. Hence, $4x = 3$ or $x = 3/4$. The y-coordinate of P must then be

$$y = 2\left(\frac{3}{4}\right)^2 = 2\left(\frac{9}{16}\right) = \frac{9}{8}.$$

The point of tangency is therefore $(3/4, 9/8)$.

Since the line $y = 3x + b$ passes through $(3/4, 9/8)$,

$$\frac{9}{8} = 3\left(\frac{3}{4}\right) + b \qquad \text{and} \qquad b = \frac{9}{8} - \frac{9}{4} = -\frac{9}{8}.$$

See Fig. 2.3.

RULE 4

> **The Sum Rule**
>
> If u and v are differentiable functions of x, then their sum $u + v$ is a differential function of x and
>
> $$\frac{d}{dx}(u + v) = \frac{du}{dx} + \frac{dv}{dx} \tag{9}$$
>
> for all values of x at which the derivatives of u and v both exist. Similarly, the derivative of the sum of any finite number of differentiable functions is the sum of their derivatives.

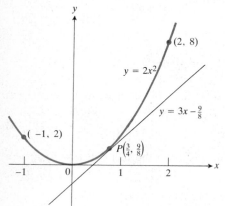

2.3 The line $y = 3x - 9/8$ is tangent to the curve $y = 2x^2$ at $P(3/4, 9/8)$.

The idea is that if u and v both have derivatives at x, then their sum also has a derivative at x that is the sum of the derivatives of u and v at x.

Proof of Rule 4   To prove the first part of Rule 4, let

$$y = u + v$$

be the sum of two differentiable functions of x. If Δu and Δv are the changes in u and v that result from changing x by an amount Δx, the resulting change in y is

$$\Delta y = \Delta u + \Delta v.$$

Hence,

$$\frac{\Delta y}{\Delta x} = \frac{\Delta u}{\Delta x} + \frac{\Delta v}{\Delta x} \qquad \text{and}$$

$$\frac{dy}{dx} = \lim_{\Delta x \to 0} \frac{\Delta y}{\Delta x} = \lim_{\Delta x \to 0} \left(\frac{\Delta u}{\Delta x} + \frac{\Delta v}{\Delta x}\right)$$

$$= \lim_{\Delta x \to 0} \frac{\Delta u}{\Delta x} + \lim_{\Delta x \to 0} \frac{\Delta v}{\Delta x} = \frac{du}{dx} + \frac{dv}{dx}.$$

Therefore,

$$\frac{d(u + v)}{dx} = \frac{du}{dx} + \frac{dv}{dx}.$$

This equation says that the derivative of the sum of two terms is the sum of their derivatives.

We may proceed by mathematical induction (Appendix 2) to establish the result for the sum of any finite number of terms. For example, if

$$y = u_1 + u_2 + u_3$$

is a sum of differentiable functions of x, then we may take

$$u = u_1 + u_2, \qquad v = u_3,$$

and apply the result already established for the sum of two terms, namely,

$$\frac{dy}{dx} = \frac{d(u_1 + u_2)}{dx} + \frac{du_3}{dx}.$$

Since the first term is again a sum of two terms, we have

$$\frac{d(u_1 + u_2)}{dx} = \frac{du_1}{dx} + \frac{du_2}{dx},$$

so

$$\frac{d(u_1 + u_2 + u_3)}{dx} = \frac{du_1}{dx} + \frac{du_2}{dx} + \frac{du_3}{dx}.$$

Finally, if it has been established for some integer n that

$$\frac{d(u_1 + u_2 + \cdots + u_n)}{dx} = \frac{du_1}{dx} + \frac{du_2}{dx} + \cdots + \frac{du_n}{dx},$$

and we let

$$y = u + v,$$

with

$$u = u_1 + u_2 + \cdots + u_n, \qquad v = u_{n+1},$$

then we find in the same way as above that

$$\frac{d(u_1 + u_2 + \cdots u_{n+1})}{dx} = \frac{du_1}{dx} + \frac{du_2}{dx} + \cdots + \frac{du_{n+1}}{dx}.$$

This enables us to conclude that if Rule 4 is true for a sum of n terms it is also true for a sum of $(n + 1)$ terms. Since Rule 4 is already established for the sum of two terms, the mathematical induction principle now guarantees it for the sum of any finite number of terms. ∎

EXAMPLE 4   Find dy/dx if $y = x^3 + 7x^2 - 5x + 4$.

Solution   We find the derivatives of the separate terms and add the results:

$$\frac{dy}{dx} = \frac{d(x^3)}{dx} + \frac{d(7x^2)}{dx} + \frac{d(-5x)}{dx} + \frac{d(4)}{dx}$$

$$= 3x^2 + 14x - 5x^0 + 0$$

$$= 3x^2 + 14x - 5.$$

Second Derivatives

The derivative

$$y' = \frac{dy}{dx}$$

is the **first derivative** of y with respect to x. The first derivative is function of x and may be differentiable. If so, its derivative

$$y'' = \frac{dy'}{dx} = \frac{d}{dx}\left(\frac{dy}{dx}\right)$$

is called the **second derivative** of y with respect to x.

The operation of taking the derivative of a function twice in succession is denoted by

$$\frac{d}{dx}\left(\frac{d}{dx}\cdots\right), \quad \text{or} \quad \frac{d^2}{dx^2}(\cdots).$$

In this notation, we write the second derivative of y with respect to x as

$$\frac{d^2y}{dx^2}.$$

In general terms, the result of differentiating a function $y = f(x)$ n times in succession is denoted by $y^{(n)}$, $f^{(n)}(x)$, or d^ny/dx^n.

EXAMPLE 5   If $y = x^3 - 3x^2 + 2$, then

$$y' = \frac{dy}{dx} = 3x^2 - 6x, \qquad y''' = \frac{d^3y}{dx^3} = 6,$$

$$y'' = \frac{d^2y}{dx^2} = 6x - 6, \qquad y^{(4)} = \frac{d^4y}{dx^4} = 0. \qquad ■$$

Velocity and Acceleration

In studies of motion of a body along a line, we usually assume that the body's position $s = f(t)$ is a twice-differentiable function of time. The first derivative, ds/dt, gives the body's velocity as a function of time; the second derivative, d^2s/dt^2, gives the body's acceleration. Thus, the velocity describes how fast the position is changing, and the acceleration describes how fast the velocity is changing (how quickly the body picks up or loses speed).

EXAMPLE 6   The position of a moving body is given by the equation $s = 160t - 16t^2$, with s in feet and t in seconds. Find the body's velocity and acceleration at time t.

Solution   The velocity is

$$v = \frac{ds}{dt} = \frac{d}{dt}(160t - 16t^2)$$

$$= 160 - (2)(16)t$$

$$= 160 - 32t \text{ ft/sec.}$$

The acceleration is

$$\frac{dv}{dt} = \frac{d}{dt}(160 - 32t)$$

$$= 0 - 32$$

$$= -32 \text{ ft/sec}^2. \qquad ■$$

EXAMPLE 7   A heavy rock blasted vertically upward with a velocity of 160 ft/sec (about 109 mph) reaches a height of $s = 160t - 16t^2$ ft after t sec.

a) How high does the rock go?

b) How fast is the rock traveling when it is 256 ft above the ground on the way up? On the way down?

2.4 The flight of the rock in Example 7.

Solution (a) To find how high the rock goes, we find the value of s when the rock's velocity is zero (Fig. 2.4). The velocity (from Example 6) is

$$v = \frac{ds}{dt} = 160 - 32t \text{ ft/sec.}$$

The velocity is zero when

$$160 - 32t = 0 \quad \text{or} \quad t = 5 \text{ sec.}$$

The rock's height at $t = 5$ sec is

$$s_{max} = s(5) = 160(5) - 16(5)^2 = 800 - 400 = 400 \text{ ft.}$$

(b) To find the rock's velocity at 256 ft on the way up and again on the way down, we find the two values of t for which

$$s(t) = 160t - 16t^2 = 256. \tag{10}$$

To solve Eq. (10) we write

$$16t^2 - 160t + 256 = 0$$
$$16(t^2 - 10t + 16) = 0$$
$$16(t - 2)(t - 8) = 0$$
$$t = 2 \text{ sec,} \quad t = 8 \text{ sec.}$$

The rock is 256 ft above the ground 2 sec after the explosion and again 8 sec after the explosion. The rock's velocities at these times are

$$v(2) = 160 - 32(2) = 160 - 64 = 96 \text{ ft/sec,}$$
$$v(8) = 160 - 32(8) = 160 - 256 = -96 \text{ ft/sec.}$$

The downward velocity is negative because s is decreasing when $t = 8$. ∎

PROBLEMS

In Problems 1–10, find dy/dx and d^2y/dx^2. Try to answer without writing anything down.

1. $y = x$

2. $y = -x$

3. $y = x^2$

4. $y = -10x^2$

5. $y = -x^2 + 3$

6. $y = \frac{x^3}{3} - x$

7. $y = 2x + 1$

8. $y = x^2 + x + 1$

9. $y = \frac{x^3}{3} + \frac{x^2}{2} + x$

10. $y = 1 - x + x^2 - x^3$

In Problems 11–15, s represents the position of a moving body, with s in feet and t in seconds. Find the body's velocity and acceleration.

11. $s = 16t^2 + 3$

12. $s = 832t - 16t^2$

13. $s = 16t^2 - 60t$

14. $s = 6 + 50t - 16t^2$

15. $s = gt^2/2 + v_0t + s_0$ (g, v_0, s_0 constants)

Find $y' = dy/dx$ and $y'' = d^2y/dx^2$ in Problems 16–25.

16. $y = x^4 - 7x^3 + 2x^2 + 15$

17. $y = 5x^3 - 3x^5$

18. $y = 4x^2 - 8x + 1$

19. $y = \frac{x^4}{4} - \frac{x^3}{3} + \frac{x^2}{2} - x + 3$

20. $y = 2x^4 - 4x^2 - 8$

21. $12y = 6x^4 - 18x^2 - 12x$

22. $y = 3x^7 - 7x^3 + 21x^2$

23. $y = x^2(x^3 - 1)$

24. $y = (x - 2)(x + 3)$

25. $y = (3x - 1)(2x + 5)$

26. Find the tangent to each curve at the given point.
   a) $y = x^3$ at $(2, 8)$ b) $y = 2x^2 + 4x - 3$ at $(1, 3)$
   c) $y = x^3 - 6x^2 + 5x$ at the origin

27. Which of the following is the slope of the line tangent to the curve $y = x^2 + 5x$ at $x = 3$?
    a) 24
    b) $-5/2$
    c) 11
    d) 8

28. Which of the following is the slope of the line $3x - 2y + 12 = 0$?
    a) 6
    b) 3
    c) 3/2
    d) 2/3

29. Find the equation of the line perpendicular to the tangent to the curve $y = x^3 - 3x + 1$ at the point $(2, 3)$.

30. The curve $y = x^2 + c$ is to be tangent to the line $y = x$. Find c. (*Hint:* Equate the two slopes.)

31. Find the tangents to the curve $y = x^3 + x$ at the points where the slope is 4. What is the smallest slope on the curve? At what value of x does the curve have this slope?

32. Find the points on the curve $y = 2x^3 - 3x^2 - 12x + 20$ where the tangent is parallel to the x-axis.

33. Find the x- and y-intercepts of the line that is tangent to the curve $y = x^3$ at the point $(-2, -8)$.

34. A line is drawn tangent to the curve $y = x^3 - x$ at the point $(-1, 0)$. Where else does the line intersect the curve?

35. The curve $y = ax^2 + bx + c$ passes through the point $(1, 2)$ and is tangent to the line $y = x$ at the origin. Find a, b, and c.

36. The curves $y = x^2 + ax + b$ and $y = cx - x^2$ are tangent to each other at the point $(1, 0)$. Find a, b, and c.

37. The equations for free fall at the surfaces of Mars and Jupiter (s in meters, t in seconds) are Mars, $s = 1.86t^2$; Jupiter, $s = 11.44t^2$. How long would it take a rock falling from rest to reach a velocity of 16.6 m/sec on each planet? (*Note:* 16.6 m/sec is about 100 km/h.)

38. A rock thrown vertically upward from the surface of the moon at a velocity of 24 m/sec (about 86 km/h) reaches a height of $s = 24t - 0.8t^2$ meters in t seconds.
    a) Find the rock's velocity and acceleration. (The acceleration in this case is the acceleration of gravity on the moon.)
    b) How long did it take the rock to reach its highest point?
    c) How high did the rock go?
    d) How long did it take the rock to reach half its maximum height?
    e) How long was the rock aloft?

39. On the earth, in the absence of air, the rock in Problem 38 would reach a height of $s = 24t - 4.9t^2$ meters in t seconds. How high would the rock go?

40. A 45-caliber bullet fired straight up from the surface of the moon would reach a height of $s = 832t - 2.6t^2$ feet after t seconds. On the earth, in the absence of air, its height would be $s = 832t - 16t^2$ feet after t seconds. How long would it take the bullet to get back down in each case?

41. The position of a body at time t is $s = t^3 - 4t^2 - 3t$. Find the acceleration each time the velocity is zero.

42. Suppose you have plotted the point $P(x, x^2)$ on the graph of $y = x^2$. To construct the tangent to the graph at P, you locate point $T(x/2, 0)$ on the x-axis and draw the line through T and P. Show that this construction is correct.

43. The tangent to the curve $y = x^n$ at $P(x_1, y_1)$ intersects the x-axis at $T(t, 0)$. Express t in terms of x_1 and n. Then show how the result can be used to construct the tangent to an arbitrary point on the curve.

TOOLKIT PROGRAMS

Derivative Grapher

2.2
Products, Powers, and Quotients

When we described in Article 1.9 how to calculate limits, we calculated a few limits directly from the definition and then used limit combination theorems to calculate all the others. We take a similar approach to calculating derivatives. We calculate a few derivatives directly from the definition and then use combination theorems to calculate others. In Article 2.1, we found the rules for calculating the derivatives of constant multiples and sums of differentiable functions. In the present article, we find the formulas for calculating the derivatives of products, integer powers, and quotients of differentiable functions.

Products

The Product Rule

The product of two differentiable functions u and v is differentiable and

$$\frac{d}{dx}(uv) = u\frac{dv}{dx} + v\frac{du}{dx}. \tag{1}$$

As with the Sum Rule in Article 2.1, the Product Rule in Eq. (1) is understood to hold only for values of x at which the derivatives of u and v both exist. At such a value of x, the rule says, the derivative of the product uv is u times the derivative of v, plus v times the derivative of u.

Proof of Rule 5   To prove Rule 5, let $y = uv$, where u and v are differentiable functions of x. Let Δx be an increment in x and let Δu and Δv denote the corresponding changes in u and v. The resulting change in y is then

$$\Delta y = (u + \Delta u)(v + \Delta v) - uv$$
$$= uv + u\Delta v + v\Delta u + \Delta u\Delta v - uv \tag{2}$$
$$= u\Delta v + v\Delta u + \Delta u\Delta v$$

(see Fig. 2.5). Hence,

$$\frac{\Delta y}{\Delta x} = u\frac{\Delta v}{\Delta x} + v\frac{\Delta u}{\Delta x} + \Delta u\frac{\Delta v}{\Delta x}.$$

When Δx approaches zero, so will Δu, since

$$\lim \Delta u = \lim\left(\frac{\Delta u}{\Delta x}\Delta x\right) = \lim\frac{\Delta u}{\Delta x}\lim \Delta x = \frac{du}{dx}\cdot 0 = 0.$$

Thus,

$$\lim\frac{\Delta y}{\Delta x} = \lim\left(u\frac{\Delta v}{\Delta x} + v\frac{\Delta u}{\Delta x} + \Delta u\frac{\Delta v}{\Delta x}\right)$$
$$= \lim u\frac{\Delta v}{\Delta x} + \lim v\frac{\Delta u}{\Delta x} + \lim \Delta u\frac{\Delta v}{\Delta x}$$
$$= \lim u\lim\frac{\Delta v}{\Delta x} + \lim v\lim\frac{\Delta u}{\Delta x} + \lim \Delta u\lim\frac{\Delta v}{\Delta x}$$
$$= u\frac{dv}{dx} + v\frac{du}{dx} + 0\cdot\frac{dv}{dx}.$$

That is,

$$\frac{dy}{dx} = u\frac{dv}{dx} + v\frac{du}{dx},$$

which establishes Eq. (1). ∎

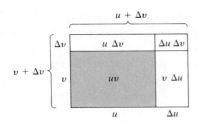

2.5 The area of the shaded rectangle is uv. When u and v change by increments Δu and Δv (shown here as positive), their product $y = uv$ changes by the amount $\Delta y = u\,\Delta v + v\,\Delta u + \Delta u\,\Delta v$.

EXAMPLE 1   Find the derivative of $y = (x^2 + 1)(x^3 + 3)$.

Solution   From the Product Rule with

$$u = x^2 + 1, \qquad v = x^3 + 3,$$

we find

$$\frac{d}{dx}[(x^2 + 1)(x^3 + 3)] = (x^2 + 1)(3x^2) + (2x)(x^3 + 3)$$

$$= 3x^4 + 3x^2 + 2x^4 + 6x$$

$$= 5x^4 + 3x^2 + 6x.$$

This particular example can be done as well (perhaps better) by multiplying out the original expression for y and differentiating the resulting polynomial. We do that now as a check. From

$$y = (x^2 + 1)(x^3 + 3) = x^5 + x^3 + 3x^2 + 3,$$

we obtain

$$\frac{dy}{dx} = 5x^4 + 3x^2 + 6x,$$

in agreement with our first calculation.

There are times, however, when the Product Rule *must* be used, as the next example shows.

EXAMPLE 2   Let $y = uv$ be the product of the functions u and v, and suppose that

$$u(2) = 3, \qquad u'(2) = -4, \qquad v(2) = 1, \qquad \text{and} \qquad v'(2) = 2.$$

Find $y'(2)$.

Solution   From the Product Rule, in the form

$$y' = (uv)' = uv' + vu',$$

we have

$$y'(2) = u(2)v'(2) + v(2)u'(2) = (3)(2) + (1)(-4) = 6 - 4 = 2.$$

Note that the derivative of a product is *not* the product of the derivatives. Instead, we add together two terms $u(dv/dx)$ and $v(du/dx)$. In the first of these we leave u untouched and differentiate v, and in the second we differentiate u and leave v alone. In fact, it is possible to extend the formula, by mathematical induction, to show that the derivative of a product

$$y = u_1 u_2 \cdots u_n$$

of a finite number of differentiable functions is given by

$$\frac{d}{dx}(u_1 u_2 \cdots u_n)$$

$$= \frac{du_1}{dx} \cdot u_2 \cdots u_n + u_1 \frac{du_2}{dx} \cdots u_n + \cdots + u_1 u_2 \cdots u_{n-1} \frac{du_n}{dx}, \tag{3}$$

where the right-hand side of the equation consists of the sum of the n terms obtained by multiplying the derivative of each one of the factors by the other $(n - 1)$ factors undifferentiated.

Positive Integer Powers

RULE 6

Positive Integer Powers of a Differentiable Function

If u is a differentiable function of x, and n is a positive integer, then u^n is differentiable and

$$\frac{d}{dx}(u^n) = nu^{n-1}\frac{du}{dx}. \tag{4}$$

Proof of Rule 6   For $n = 1$ we interpret Eq. (4) as

$$\frac{d}{dx}(u) = u^0\frac{du}{dx} = \frac{du}{dx}, \tag{5}$$

which is certainly true if $u \neq 0$. If $u = 0$, we get 0^0, which is an indeterminate expression but which we interpret here as being 1 to be consistent.

For $n = 2$ we apply the Product Rule to the function $y = u \cdot u$ to obtain

$$\frac{d}{dx}(u^2) = \frac{d}{dx}(u \cdot u) = u\frac{du}{dx} + u\frac{du}{dx} = 2u\frac{du}{dx}. \tag{6}$$

Having established the Power Rule for $n = 2$, we may proceed by mathematical induction to establish it for all positive integer values of n.

Suppose that the rule has been established for some positive integer k, so that

$$\frac{d}{dx}(u^k) = ku^{k-1}\frac{du}{dx}. \tag{7}$$

We may then show that the rule holds for the next integer, $k + 1$, by the following argument. We let $y = u^{k+1}$ and rewrite y as the product

$$y = u \cdot u^k.$$

We then apply the Product Rule,

$$\frac{dy}{dx} = u\frac{dv}{dx} + v\frac{du}{dx},$$

with $v = u^k$, to find the derivative of y:

$$\frac{dy}{dx} = \frac{d}{dx}(u \cdot u^k) = u\frac{d}{dx}(u^k) + u^k\frac{du}{dx}$$

$$= u\left(ku^{k-1}\frac{du}{dx}\right) + u^k\frac{du}{dx} \qquad \text{(by Eq. 7)} \tag{8}$$

$$= ku^k\frac{du}{dx} + u^k\frac{du}{dx} = (k + 1)u^k\frac{du}{dx}.$$

This enables us to conclude that if Rule 6 holds for the exponent $n = k$, it also holds for $n = k + 1$. Since Rule 6 has been established for $n = 1$ and 2, the mathematical induction principle now ensures it for every positive integer n. ∎

Note that Rule 2,

$$\frac{d}{dx}(x^n) = nx^{n-1}, \tag{9}$$

is a special case of the Power Rule,

$$\frac{d}{dx}(u^n) = nu^{n-1}\frac{du}{dx}, \tag{10}$$

obtained by taking $u = x$:

$$\frac{d}{dx}(x^n) = nx^{n-1}\frac{dx}{dx} = nx^{n-1} \cdot 1 = nx^{n-1}. \tag{11}$$

We do not write the dx/dx as part of Eq. (9) because $dx/dx = 1$. To see why du/dx is a factor in Eq. (10), look for its appearance as the Product Rule is used in Eqs. (6) and (8).

EXAMPLE 3 Find the derivative of

$$y = (x^2 - 3x + 1)^5.$$

Solution For the fifth power of a function, the Power Rule says

$$\frac{d}{dx}u^5 = 5u^4\frac{du}{dx}.$$

With $u = x^2 - 3x + 1$, we get

$$\frac{dy}{dx} = \frac{d}{dx}(x^2 - 3x + 1)^5$$

$$= 5(x^2 - 3x + 1)^4 \cdot \frac{d}{dx}(x^2 - 3x + 1)$$

$$= 5(x^2 - 3x + 1)^4 \cdot (2x - 3)$$

$$= 5(2x - 3)(x^2 - 3x + 1)^4. ∎$$

WARNING: Do not forget the du/dx term in Eq. (4). The differentiation is not correct without it. In Example 3, the differentiation would not be correct without the factor

$$\frac{d}{dx}(x^2 - 3x + 1) = (2x - 3).$$

EXAMPLE 4 Find dy/dx if $y = (x^2 + 1)^3(x - 1)^2$.

Solution We could, of course, expand everything here and write y as a polynomial in x, but this is unnecessary. Instead, we first use the Product Rule with

$u = (x^2 + 1)^3$ and $v = (x - 1)^2$:

$$\frac{dy}{dx} = (x^2 + 1)^3 \frac{d}{dx}(x - 1)^2 + (x - 1)^2 \frac{d}{dx}(x^2 + 1)^3.$$

Then we evaluate the remaining derivatives with the Power Rule:

$$\frac{d}{dx}(x - 1)^2 = 2(x - 1)\frac{d}{dx}(x - 1)$$

$$= 2(x - 1)(1)$$

$$= 2(x - 1),$$

$$\frac{d}{dx}(x^2 + 1)^3 = 3(x^2 + 1)^2\frac{d}{dx}(x^2 + 1)$$

$$= 3(x^2 + 1)^2(2x)$$

$$= 6x(x^2 + 1)^2.$$

We substitute these into the equation for dy/dx:

$$\frac{dy}{dx} = (x^2 + 1)^3 2(x - 1) + (x - 1)^2 6x(x^2 + 1)^2$$

$$= 2(x^2 + 1)^2(x - 1)[(x^2 + 1) + 3x(x - 1)]$$

$$= 2(x^2 + 1)^2(x - 1)(4x^2 - 3x + 1). \qquad \blacksquare$$

Quotients

The ratio or quotient u/v of two polynomials in x is usually not a polynomial. Such a ratio is called a **rational function** of x, where the word *ratio* is the key to the use of the word *rational*.

Rational functions play an important role in computation because they are the most elaborate functions a digital computer can evaluate directly. The next differentiation rule goes beyond rational functions to apply to the quotient of any two differentiable functions.

RULE 7

The Quotient Rule

At a point where $v \neq 0$, the quotient $y = u/v$ of two differentiable functions is differentiable, and

$$\frac{d}{dx}\left(\frac{u}{v}\right) = \frac{v\dfrac{du}{dx} - u\dfrac{dv}{dx}}{v^2}. \tag{12}$$

As with the rules for differentiating sums and products of differentiable functions, Eq. (12) in the Quotient Rule is understood to hold only at values of x at which both u and v are differentiable.

Proof of Rule 7   To establish Eq. (12), consider a point x where $v \neq 0$ and where u and v are differentiable. Let x be given an increment Δx and let Δy, Δu, and Δv be the corresponding increments in y, u, and v. Then, as $\Delta x \to 0$,

$$\lim (v + \Delta v) = \lim v + \lim \Delta v$$

while

$$\lim \Delta v = \lim \frac{\Delta v}{\Delta x} \cdot \Delta x = \frac{dv}{dx} \cdot 0 = 0.$$

Therefore, the value of $v + \Delta v$ is close to the value of v when Δx is near zero. In particular, since $v \neq 0$ at x, it follows that $v + \Delta v \neq 0$ when Δx is *near* zero, say when $0 < |\Delta x| < h$. Let Δx be so restricted. Then $v + \Delta v \neq 0$, and

$$y + \Delta y = \frac{u + \Delta u}{v + \Delta v}.$$

From this we subtract $y = u/v$ and obtain

$$\Delta y = \frac{u + \Delta u}{v + \Delta v} - \frac{u}{v} = \frac{(vu + v\,\Delta u) - (uv + u\,\Delta v)}{v(v + \Delta v)} = \frac{v\,\Delta u - u\,\Delta v}{v(v + \Delta v)}.$$

Dividing this by Δx, we have

$$\frac{\Delta y}{\Delta x} = \frac{v\dfrac{\Delta u}{\Delta x} - u\dfrac{\Delta v}{\Delta x}}{v(v + \Delta v)}.$$

When Δx approaches zero,

$$\lim \frac{\Delta u}{\Delta x} = \frac{du}{dx}, \qquad \lim \frac{\Delta v}{\Delta x} = \frac{dv}{dx},$$

and

$$\lim v(v + \Delta v) = \lim v \lim (v + \Delta v) = v^2 \neq 0.$$

Therefore,

$$\frac{dy}{dx} = \lim \frac{\Delta y}{\Delta x} = \frac{\lim \left(v\dfrac{\Delta u}{\Delta x} - u\dfrac{\Delta v}{\Delta x} \right)}{\lim v(v + \Delta v)} = \frac{v\dfrac{du}{dx} - u\dfrac{dv}{dx}}{v^2},$$

which is Eq. (12).   ∎

EXAMPLE 5   Find the derivative of

$$y = \frac{x^2 + 1}{x^2 - 1}.$$

Solution   We apply the Quotient Rule (Eq. 12):

$$\frac{dy}{dx} = \frac{(x^2 - 1) \cdot 2x - (x^2 + 1) \cdot 2x}{(x^2 - 1)^2} = \frac{-4x}{(x^2 - 1)^2}.$$

Negative Integer Powers

By writing u^{-m} as $1/u^m$, we can apply the Quotient Rule to show that the rule for differentiating positive integer powers holds for negative integer powers as well.

RULE 8

> **Negative Integer Powers of a Differentiable Function**
>
> At a point where u is differentiable and not zero, the derivative of
> $$y = u^n$$
> is given by
> $$\frac{d(u^n)}{dx} = nu^{n-1}\frac{du}{dx} \tag{13}$$
> if n is a negative integer.

Proof of Rule 8 To prove Rule 8, we combine the results in Eqs. (12) and (4). Let

$$y = u^{-m} = \frac{1}{u^m},$$

where $-m$ is a negative integer, so that m is a positive integer. Then, using Eq. (12) for the derivative of a quotient, we have

$$\frac{dy}{dx} = \frac{d\left(\dfrac{1}{u^m}\right)}{dx} = \frac{u^m\dfrac{d(1)}{dx} - 1\dfrac{d(u^m)}{dx}}{(u^m)^2} \tag{14}$$

at any point where u is differentiable and not zero. Now the various derivatives on the right-hand side of Eq. (14) can be evaluated by formulas already proved, namely,

$$\frac{d(1)}{dx} = 0,$$

since 1 is a constant, and

$$\frac{d(u^m)}{dx} = mu^{m-1}\frac{du}{dx},$$

since m is a *positive integer*. Therefore,

$$\frac{dy}{dx} = \frac{u^m \cdot 0 - 1 \cdot mu^{m-1}\dfrac{du}{dx}}{u^{2m}} = -mu^{-m-1}\frac{du}{dx}.$$

If $-m$ is replaced by its equivalent value n, this equation reduces to Eq. (13). ∎

EXAMPLE 6 Find the derivative of

$$y = x^2 + \frac{1}{x^2}.$$

Solution We write $y = x^2 + x^{-2}$. Then

$$\frac{dy}{dx} = 2x^{2-1}\frac{dx}{dx} + (-2)x^{-2-1}\frac{dx}{dx} = 2x - 2x^{-3}. \quad ■$$

Practical Matters

As a practical matter in differentiation, it is usually best to treat a reciprocal

$$\frac{1}{[u(x)]^n}$$

as a function raised to a power, not as a quotient.

EXAMPLE 7 The most effective way to find the derivative of

$$y = \frac{1}{(x^2 - 1)^5}$$

is to write $y = (x^2 - 1)^{-5}$ and calculate

$$y' = -5(x^2 - 1)^{-6} \cdot \frac{d}{dx}(x^2 - 1) = \frac{-5}{(x^2 - 1)^6} \cdot (2x) = \frac{-10x}{(x^2 - 1)^6}.$$

If you treat

$$y = \frac{1}{(x^2 - 1)^5}$$

as a quotient, with $u = 1$ and $v = (x^2 - 1)^5$, the first step in calculating y' is

$$y' = \frac{(x^2 - 1)^5 \cdot \dfrac{d}{dx}(1) - 1 \cdot \dfrac{d}{dx}(x^2 - 1)^5}{[(x^2 - 1)^5]^2}.$$

This is correct, but cumbersome.   ∎

As Example 7 suggests, the choice of which rules to use in solving a differentiation problem can make a difference in how much work you have to do. Here are two more examples.

EXAMPLE 8 The easiest way to find y' if

$$y = \left(\frac{2x - 1}{x + 7}\right)^3$$

is to use the Power Rule first and then the Quotient Rule. Start with

$$y' = 3 \cdot \left(\frac{2x - 1}{x + 7}\right)^2 \cdot \frac{d}{dx}\left(\frac{2x - 1}{x + 7}\right).$$

You are in for more work if you begin with

$$y = (2x - 1)^3(x + 7)^{-3},$$
$$y' = (2x - 1)^3 \cdot \frac{d}{dx}(x + 7)^{-3} + (x + 7)^{-3} \cdot \frac{d}{dx}(2x - 1)^3. \qquad ∎$$

EXAMPLE 9 Do not use the Quotient Rule to find the derivative of

$$y = \frac{(x - 1)(x^2 - 2x)}{x^4}.$$

Instead, expand the numerator and divide by x^4:

$$y = \frac{(x - 1)(x^2 - 2x)}{x^4} = \frac{x^3 - 3x^2 + 2x}{x^4}$$
$$= x^{-1} - 3x^{-2} + 2x^{-3}.$$

Then use the Sum and Power Rules:

$$\frac{dy}{dx} = -x^{-2} - 3(-2)x^{-3} + 2(-3)x^{-4} = -\frac{1}{x^2} + \frac{6}{x^3} - \frac{6}{x^4}. \quad \blacksquare$$

PROBLEMS

Find dy/dx in Problems 1–24.

1. $y = \dfrac{x^3}{3} - \dfrac{x^2}{2} + x - 1$

2. $y = (x - 1)^3(x + 2)^4$

3. $y = (x^2 + 1)^5$

4. $y = (x^3 - 3x)^4$

5. $y = (x + 1)^2(x^2 + 1)^{-3}$

6. $y = \dfrac{2x + 1}{x^2 - 1}$

7. $y = \dfrac{2x + 5}{3x - 2}$

8. $y = \left(\dfrac{x + 1}{x - 1}\right)^2$

9. $y = (1 - x)(1 + x^2)^{-1}$

10. $y = (x + 1)^2(x^2 + 2x)^{-2}$

11. $y = \dfrac{5}{(2x - 3)^4}$

12. $y = (x - 1)^3(x + 2)$

13. $y = (5 - x)(4 - 2x)$

14. $y = [(5 - x)(4 - 2x)]^2$

15. $y = (2x - 1)^3(x + 7)^{-3}$

16. $y = \dfrac{x^3 + 7}{x}$

17. $y = (2x^3 - 3x^2 + 6x)^{-5}$

18. $y = \dfrac{x^2}{(x - 1)^2}$

19. $y = \dfrac{(x - 1)^2}{x^2}$

20. $y = \dfrac{-1}{15(5x - 1)^3}$

21. $y = \dfrac{12}{x} - \dfrac{4}{x^3} + \dfrac{3}{x^4}$

22. $y = \dfrac{(x - 1)(x^2 + x + 1)}{x^3}$

23. $y = \dfrac{(x^2 - 1)}{x^2 + x - 2}$

24. $y = \dfrac{(x^2 + x)(x^2 - x + 1)}{x^4}$

Find ds/dt in Problems 25–32.

25. $s = \dfrac{t}{t^2 + 1}$

26. $s = (2t + 3)^3$

27. $s = (t^2 - t)^{-2}$

28. $s = t^2(t + 1)^{-1}$

29. $s = \dfrac{2t}{3t^2 + 1}$

30. $s = (t + t^{-1})^2$

31. $s = (t^2 + 3t)^3$

32. $s = (t^2 - 7t)(5 - 2t^3 + t^4)/t^3$

33. Suppose that u and v are functions of x that are differentiable at $x = 0$ and that

$$u(0) = 5, \quad u'(0) = -3, \quad v(0) = -1, \quad v'(0) = 2.$$

Find the values of the derivatives below at $x = 0$.

a) $\dfrac{d}{dx}(uv)$

b) $\dfrac{d}{dx}\left(\dfrac{u}{v}\right)$

c) $\dfrac{d}{dx}\left(\dfrac{v}{u}\right)$

d) $\dfrac{d}{dx}(7v - 2u)$

e) $\dfrac{d}{dx}(u^3)$

f) $\dfrac{d}{dx}(5v^{-3})$

34. Find an equation for the tangent to the curve $y = x/(x^2 + 1)$ at the origin.

35. Find an equation for the tangent to the curve $y = x + (1/x)$ at $x = 2$.

36. Find the first and second derivatives of

$$f(x) = (x^2 + 3x + 1)^3.$$

37. Find d^2y/dx^2 if $y = (3 - 2x)^{-1}$.

38. Differentiate $x = 5y/(y + 1)$ with respect to y.

To see how Eq. (3) works, calculate the derivatives of the functions in Problems 39–42 in two ways: (a) by applying Eq. (3) directly and (b) by first multiplying the factors to produce a polynomial to differentiate. In these examples, (b) is faster, but that is not always the case.

39. $y = x(x - 1)(x + 1)$

40. $y = (x - 1)(x + 1)(x^2 + 1)$

41. $y = (1 - x)(x + 1)(3 - x^2)$

42. $y = x^2(x - 1)(x^2 + x + 1)$

43. *Industrial production.* Economists often use the expression "rate of growth" in relative rather than absolute terms. For example, in a given industry, let $u = f(t)$ be the number of people in the labor force at time t. (This function will be treated as though it were differentiable even though it is an integer-valued step function. We approximate the step function by a smooth curve.)

Let $v = g(t)$ be the average production per person in the labor force at time t. The total production is then $y = uv$. If the labor force is growing at the rate of 4 percent per year $(du/dt = 0.04u)$ and the production per worker is growing at the rate of 5 percent per year $(dv/dt = 0.05v)$, find the rate of growth of the total production, y.

44. Suppose that the labor force in Problem 43 is decreasing at the rate of 2 percent per year while the production per person is increasing at the rate of 3 percent per year. Is the total production increasing or decreasing, and at what rate?

45. *Rate of a chemical reaction.* When two chemicals, A and B, combine to form an amount p of product, the rate dp/dt at which the product forms is called the **reaction rate.** In many reactions, one molecule of product is formed from one molecule of A and one molecule of B. Suppose that the initial molar masses of A and B are equal, both having value a. Under these conditions, the amount of product at any time t after mixing is given by the function $p(t) = a^2kt/(akt + 1)$. In this equation, k is a positive constant of proportionality from the chemical law of mass action, representing the affinity of the two chemicals.

a) Find dp/dt.

b) Find the time at which the reaction is proceeding at its fastest rate and the value of dp/dt at this time.

46. Let c be a constant and u a differentiable function of x. Show that Rule 3, which says that

$$\frac{d}{dx}(cu) = c\frac{du}{dx},$$

is a special case of the Product Rule.

47. Show that Rule 6, which says that

$$\frac{d}{dx}(u^n) = nu^{n-1}\frac{du}{dx},$$

is a special case of the finite Product Rule in Eq. (3).

TOOLKIT PROGRAMS

Derivative Grapher

2.3
Implicit Differentiation and Fractional Powers

When an equation in x and y defines y as a differentiable function of x, we can often use the rules of differentiation to calculate dy/dx even when we cannot solve the equation for y. In this article, we show how this is done and look briefly at the idea behind the method. We then use this method to show that the Power Rule holds for fractional exponents as well as integers.

EXAMPLE 1 Find dy/dx if $y^2 = x$.

Solution The equation $y^2 = x$ defines two differentiable functions of x, namely $y = \sqrt{x}$ and $y = -\sqrt{x}$ (Fig. 2.6). We know how to find the derivative of each of these, from Example 5 in Article 1.7. But suppose we knew only that the equation $y^2 = x$ defined y as one or more differentiable functions of x without knowing exactly what these functions were. Could we still find dy/dx?

The answer in this case is yes. To find dy/dx we simply differentiate both sides of the equation $y^2 = x$ with respect to x, treating y as a differentiable but otherwise unknown function of x. When we do this, we get

$$y^2 = x$$

$$2y\frac{dy}{dx} = 1$$

$$\frac{dy}{dx} = \frac{1}{2y}.$$

How does this compare with what happens when we solve $y^2 = x$ for y first and then differentiate?

With $y = \sqrt{x}$	With $y = -\sqrt{x}$
$\dfrac{dy}{dx} = \dfrac{1}{2\sqrt{x}}$	$\dfrac{dy}{dx} = -\dfrac{1}{2\sqrt{x}}$
$= \dfrac{1}{2y}$	$= \dfrac{1}{2(-\sqrt{x})} = \dfrac{1}{2y}$

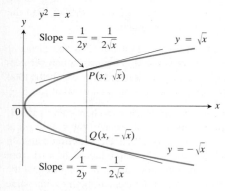

2.6 The slope of the parabola $y^2 = x$, at the points on the curve directly above and below x, is given by the formula $m = 1/(2y)$.

In both cases, the derivative is given by the formula we obtained without solving for y, so the two methods agree. ∎

EXAMPLE 2 Find dy/dx if $x^5 + 4xy^3 - y^5 = 2$.

Solution We assume that the equation defines y as one or more differentiable functions of x. We then differentiate both sides of the equation with respect to x, treating y^3 and y^5 as powers of a differentiable function of x and treating $4xy^3$ as a constant times the product $x \cdot y^3$:

$$\frac{d}{dx}(x^5) + \frac{d}{dx}(4xy^3) - \frac{d}{dx}(y^5) = \frac{d}{dx}(2)$$

$$5x^4 + 4\left(x\frac{d(y^3)}{dx} + y^3\frac{dx}{dx}\right) - 5y^4\frac{dy}{dx} = 0$$

$$5x^4 + 4\left(3xy^2\frac{dy}{dx} + y^3\right) - 5y^4\frac{dy}{dx} = 0$$

$$(12xy^2 - 5y^4)\frac{dy}{dx} = -(5x^4 + 4y^3)$$

$$\frac{dy}{dx} = \frac{5x^4 + 4y^3}{5y^4 - 12xy^2}.$$ ∎

Example 2 differs from Example 1 in that its equation cannot be solved for y to prove that it really defines y as a differentiable function of x. The verification that it does so is a topic we must leave to more advanced texts. We can make the idea seem reasonable, however, and we shall do that next.

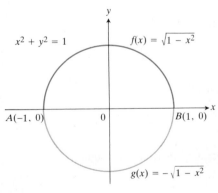

2.7 The graph of an equation in x and y of the form $F(x, y) = 0$ may not be the graph of a function of x. Some vertical lines may intersect it more than once. Portions of the graph, however, such as the arc from A to B, may be considered as the graphs of functions of x.

Implicit Differentiation

Most of the functions we have examined have been given by equations that express y explicitly in terms of x. Quite often, however, we encounter equations like $y^2 - x = 0$ and $x^2 + y^2 - 1 = 0$, which do not give y explicitly in terms of x. Nevertheless, each of these equations defines a relation between x and y. When a definite number from an appropriate domain is substituted for x, the resulting equation determines one or more values for y. The xy-pairs so obtained can be plotted in the plane to graph the equation.

Now, the graph of an arbitrary equation $F(x, y) = 0$ in x and y may fail to be the graph of a function $y = f(x)$ because some vertical lines intersect it more than once. For example, in Fig. 2.7, the numbers y_1, y_2, and y_3 all correspond to the same value $x = x_0$. The pairs (x_0, y_1), (x_0, y_2), and (x_0, y_3) all satisfy the equation $F(x, y) = 0$, and the corresponding points lie on the graph of $F(x, y) = 0$.

However, various *parts* of the curve $F(x, y) = 0$ may be the graphs of functions of x. For example, the segment AB of the curve in Fig. 2.7 is the graph of a function $y = f(x)$ that satisfies $f(x_0) = y_3$ and is defined on an open interval (a, b) containing x_0. If x is any point in (a, b), then the pair $(x, f(x)) = (x, y)$ satisfies the original equation $F(x, y) = 0$. We say that the equation $F(x, y) = 0$ has defined f **implicitly** on (a, b), even though it has not defined f by giving y explicitly in terms of x.

2.8 The graph of the equation $F(x, y) = x^2 + y^2 - 1 = 0$ is the complete circle $x^2 + y^2 = 1$. The upper semicircle AB is the graph of the function $f(x) = \sqrt{1 - x^2}$. The lower semicircle AB is the graph of $g(x) = -\sqrt{1 - x^2}$.

EXAMPLE 3 The graph of $F(x, y) = x^2 + y^2 - 1 = 0$ is the circle $x^2 + y^2 = 1$. Taken as a whole, the circle is not the graph of any function of x (Fig. 2.8). For each x in the interval $(-1, 1)$ there are *two* values of y, namely,

$y = \sqrt{1 - x^2}$ and $y = -\sqrt{1 - x^2}$. However, the upper and lower semicircles are the graphs of the functions $f(x) = \sqrt{1 - x^2}$ and $g(x) = -\sqrt{1 - x^2}$. The pairs $(x, y) = (x, \sqrt{1 - x^2})$ and $(x, y) = (x, -\sqrt{1 - x^2})$ satisfy the equation $x^2 + y^2 - 1 = 0$ whenever $-1 < x < 1$.

As Example 7 will later show, the functions f and g are also differentiable for $-1 < x < 1$. Since their graphs have vertical tangents at $x = \pm 1$ they are not differentiable at these points.   ◼

When may we expect the various functions $y = f(x)$ defined by a relation $F(x, y) = 0$ to be differentiable? The answer is when the graph of the relation is smooth enough to have a tangent line at each point, as it will, for instance, if the formula for F is an algebraic combination of powers of x and y. To calculate the derivatives of these implicitly defined functions, we simply proceed as in Examples 1 and 2—treat y as a differentiable but otherwise unknown function of x and differentiate both sides of the defining equation with respect to x. The method is called **implicit differentiation.** You may wish to review Examples 1 and 2 to see once again how it is done.

Derivatives of Higher Order

Implicit differentiation can also produce derivatives of higher order. We illustrate with an example.

EXAMPLE 4   Find d^2y/dx^2 if $2x^3 - 3y^2 = 7$.

Solution   To start, we differentiate both sides of the equation with respect to x to find $y' = dy/dx$:

$$2x^3 - 3y^2 = 7$$

$$\frac{d}{dx}(2x^3) - \frac{d}{dx}(3y^2) = \frac{d}{dx}(7)$$

$$6x^2 - 6yy' = 0 \tag{1}$$

$$x^2 - yy' = 0$$

$$y' = \frac{x^2}{y} \qquad \text{(when } y \neq 0\text{)}.$$

We now apply the Quotient Rule to find y'':

$$y'' = \frac{d}{dx}\left(\frac{x^2}{y}\right) = \frac{2xy - x^2y'}{y^2} = \frac{2x}{y} - \frac{x^2}{y^2}y'. \tag{2}$$

Finally, we substitute $y' = x^2/y$ to express y'' in terms of x and y:

$$y'' = \frac{2x}{y} - \frac{x^2}{y^2}\left(\frac{x^2}{y}\right) = \frac{2x}{y} - \frac{x^4}{y^3}. \tag{3}$$

The second derivative is not defined at $y = 0$ but is given by Eq. (3) when $y \neq 0$.   ◼

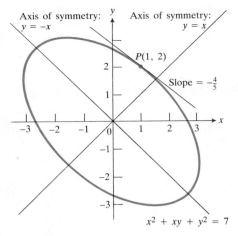

2.9 The graph of $x^2 + xy + y^2 = 7$. The slope of the curve at $P(1, 2)$ is $(dy/dx)_{(1, 2)} = -4/5$.

Tangents

As we have seen, implicit differentiation usually expresses dy/dx in terms of both x and y. To calculate the slope of a curve at a known point (x_1, y_1) in such a case, we must substitute both x_1 and y_1 in the final expression for dy/dx.

EXAMPLE 5 Find the slope of the curve $x^2 + xy + y^2 = 7$ at the point $(1, 2)$. (See Fig. 2.9.)

Solution We differentiate both sides of the equation with respect to x, treating xy as a product of two differentiable functions and y^2 as a differentiable function raised to a power. Thus,

$$\frac{d}{dx}(x^2) + \frac{d}{dx}(xy) + \frac{d}{dx}(y^2) = \frac{d}{dx}(7)$$

$$2x + \left(x\frac{dy}{dx} + y\frac{dx}{dx}\right) + 2y\frac{dy}{dx} = 0$$

$$(x + 2y)\frac{dy}{dx} = -(2x + y)$$

and, whenever $x + 2y \neq 0$,

$$\frac{dy}{dx} = -\frac{2x + y}{x + 2y}.$$

The value of the derivative at the point $(1, 2)$ is

$$\left(\frac{dy}{dx}\right)_{(1, 2)} = -\frac{2(1) + 2}{1 + 2(2)} = -\frac{4}{5}. \qquad (4)$$

The expression on the left side of Eq. (4) is read "the value of dy/dx at the point $(1, 2)$." Another common notation for this value is

$$\frac{dy}{dx}\bigg|_{(1, 2)},$$

which is read the same way.

At points where $2x + y = 0$, the slope of the curve is 0 and the tangent is horizontal. Where $x + 2y = 0$, the tangent is vertical. At these points we cannot solve for dy/dx. ∎

Normal Lines

In the law that describes how light changes direction as it passes through the surface of a lens, the important angles are the angles the light makes with the line perpendicular to the surface at the point of entry (angles A and B in Fig. 2.10). This line is called the *normal* to the surface at the point of entry. In a profile view of a lens like the one in Fig. 2.10, the normal is the line perpendicular to the tangent to the profile curve at the point of entry.

In calculus we define the line **normal** to a differentiable curve at a point P, whether the curve represents the surface of a lens or not, to be the line perpendicular to the tangent to the curve at P.

2.10 The profile or cutaway view of a lens, showing the bending (refraction) of a light ray as it passes through the lens surface.

EXAMPLE 6   Find the lines tangent and normal to the curve

$$y^2 - 6x^2 + 4y + 19 = 0$$

at the point $(2, 1)$.

Solution   We differentiate both sides of the equation with respect to x and solve for dy/dx:

$$2y\frac{dy}{dx} - 12x + 4\frac{dy}{dx} = 0$$

$$\frac{dy}{dx}(2y + 4) = 12x$$

$$\frac{dy}{dx} = \frac{6x}{y + 2}.$$

We then evaluate the derivative at $x = 2$, $y = 1$, to obtain

$$\left.\frac{dy}{dx}\right|_{(2,\,1)} = \left.\frac{6x}{y + 2}\right|_{(2,\,1)} = \frac{12}{3} = 4.$$

Therefore, the tangent to the curve at the point $(2, 1)$ is

$$y - 1 = 4(x - 2).$$

The slope of the normal is $-1/4$, and its equation is

$$y - 1 = -\frac{1}{4}(x - 2). \qquad \blacksquare$$

Fractional Powers

By using implicit differentiation, we can extend the Power Rule to include fractional exponents.

RULE 9

Power Rule for Fractional Exponents

If u is a differentiable function of x, and p and q are integers with $q > 0$, then

$$\frac{d}{dx}u^{p/q} = \frac{p}{q}u^{(p/q)-1}\frac{du}{dx}, \qquad (5)$$

provided $u \neq 0$ if $p/q < 1$.

This is the familiar rule for the derivative of u^n, but extended to the case in which $n = p/q$ is any rational number. As with earlier rules, Eq. (5) is understood to hold only at values of x at which $u^{p/q}$, $u^{(p/q)-1}$, and du/dx are all defined as real numbers. Rule 9 then says that we can obtain the value of the derivative of $u^{p/q}$ at x by evaluating the right-hand side of Eq. (5) at x.

The next example illustrates Eq. (5) and the restrictions that may have to be placed on the domains of x-values for the equation to hold.

EXAMPLE 7

a) $\dfrac{d}{dx}(x^{1/2}) = \dfrac{1}{2}x^{-1/2} = \dfrac{1}{2\sqrt{x}}$ for $x > 0$

b) $\dfrac{d}{dx}(x^{1/5}) = \dfrac{1}{5}x^{-4/5}$ for $x \neq 0$

c) $\dfrac{d}{dx}(x^{-4/3}) = -\dfrac{4}{3}x^{-7/3}$ for $x \neq 0$

d) $\dfrac{d}{dx}(1-x^2)^{1/2} = \dfrac{1}{2}(1-x^2)^{-1/2}(-2x)$

$$= \dfrac{-x}{(1-x^2)^{1/2}}$$ for $|x| < 1$ ■

Proof of Rule 9 To establish Rule 9, let $y = u^{p/q}$, which means that

$$y^q = u^p.$$

Since p and q are integers, we can differentiate both sides of the equation implicitly:

$$qy^{q-1}\dfrac{dy}{dx} = pu^{p-1}\dfrac{du}{dx}.$$

Hence, if $y \neq 0$, we have

$$\dfrac{dy}{dx} = \dfrac{pu^{p-1}}{qy^{q-1}}\dfrac{du}{dx}. \tag{6}$$

But $y^{q-1} = (u^{p/q})^{q-1} = u^{p-(p/q)}$, so

$$\dfrac{dy}{dx} = \dfrac{p}{q}\dfrac{u^{p-1}}{u^{p-(p/q)}}\dfrac{du}{dx}$$

$$= \dfrac{p}{q}u^{(p/q)-1}\dfrac{du}{dx}.$$

This establishes Eq. (5).   ■

The restriction $y \neq 0$ in Eq. (6) is the same as the restriction $u \neq 0$, but it was made without reference to whether p/q was or was not less than 1. The restriction is not needed if $p/q = 1$, since we then have $y = u$ and $dy/dx = du/dx$.

The function $y = x^{3/2}$ is typical of the case $p/q > 1$. In the next example, we investigate its derivative.

EXAMPLE 8 Find the derivative of $y = x^{3/2}$ at $x = 0$.

Solution When the graph of a function stops abruptly at a point, as the graph of $y = x^{3/2}$ does at $x = 0$ (Fig. 2.11), we calculate its derivative as a one-sided limit. The Power Rule still applies, giving in this case

$$\left.\dfrac{dy}{dx}\right|_{x=0} = \left.\dfrac{3}{2}x^{1/2}\right|_{x=0} = 0.$$

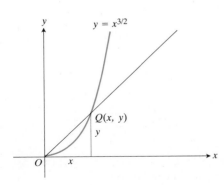

2.11 The graph of $y = x^{3/2}$. The slope of the curve at $x = 0$ is $\lim_{Q \to 0} m_{OQ} = 0$.

We can see why this equation holds by looking at the geometry of the curve. The slope of a typical secant line through the origin and a point $Q(x, y)$ on the curve is

$$m_{OQ} = \frac{y - 0}{x - 0} = \frac{x^{3/2}}{x} = x^{1/2}.$$

As Q approaches the origin from the right, m_{OQ} approaches 0, in agreement with the result from the Power Rule. ∎

Differentiability at an Endpoint

The right-hand limit calculated in Example 8 is a special case of the limit

$$\lim_{\Delta x \to 0^+} \frac{f(x + \Delta x) - f(x)}{\Delta x},$$

which is called the **right-hand derivative** of f at x when it exists. Similarly, the limit

$$\lim_{\Delta x \to 0^-} \frac{f(x + \Delta x) - f(x)}{\Delta x},$$

when it exists, is called the **left-hand derivative** of f at x.

These two definitions extend the notion of differentiability to the endpoints of a function's domain. We say that a function f defined on a closed interval $[a, b]$ is differentiable at a if its right-hand derivative exists at a and is differentiable at b if its left-hand derivative exists at b. For f to be differentiable at any point between a and b, however, the ordinary "two-sided" derivative

$$f'(x) = \lim_{\Delta x \to 0} \frac{f(x + \Delta x) - f(x)}{\Delta x}$$

must exist as usual.

PROBLEMS

In Problems 1–29, find dy/dx.

1. $x^2 + y^2 = 1$

2. $y^2 = \dfrac{x - 1}{x + 1}$

3. $x^2 - xy = 2$

4. $x^2y + xy^2 = 6$

5. $y^2 = x^3$

6. $x^{2/3} + y^{2/3} = 1$

7. $x^{1/2} + y^{1/2} = 1$

8. $x^3 - xy + y^3 = 1$

9. $x^2 = \dfrac{x - y}{x + y}$

10. $y = \dfrac{x}{\sqrt{x^2 + 1}}$

11. $y = x\sqrt{x^2 + 1}$

12. $y^2 = x^2 + \dfrac{1}{x^2}$

13. $2xy + y^2 = x + y$

14. $y = \sqrt{x} + \sqrt[3]{x} + \sqrt[4]{x}$

15. $y^2 = \dfrac{x^2 - 1}{x^2 + 1}$

16. $(x + y)^3 + (x - y)^3 = x^4 + y^4$

17. $(3x + 7)^5 = 2y^3$

18. $y = (x + 5)^4(x^2 - 2)^3$

19. $\dfrac{1}{y} + \dfrac{1}{x} = 1$

20. $y = (x^2 + 5x)^3$

21. $y^2 = x^2 - x$

22. $x^2y^2 = x^2 + y^2$

23. $y = \dfrac{\sqrt[3]{x^2 + 3}}{x}$

24. $y = x^2\sqrt{1 - x^2}$

25. $x^3 + y^3 = 18xy$

26. $y = (3x^2 + 5x + 1)^{3/2}$

27. $y = (2x + 5)^{-1/5}$

28. $y = 3(2x^{-1/2} + 1)^{-1/3}$

29. $y = \sqrt{1 - \sqrt{x}}$

30. Find dT/dL if $T^2 = 4\pi^2 L/g$. (This equation gives the period T of a simple pendulum in terms of its length L and the acceleration of gravity g.)

31. Find the x- and y-intercepts of the line tangent to the curve $y = x^{1/2}$ at $x = 4$.

32. Which of the following may be true if $f''(x) = x^{-1/3}$?

a) $f(x) = \dfrac{3}{2}x^{2/3} - 3$

b) $f(x) = \dfrac{9}{10}x^{5/3} - 7$

c) $f'''(x) = -\dfrac{1}{3}x^{-4/3}$

d) $f'(x) = \dfrac{3}{2}x^{2/3} + 6$

33. Find db/da if $b = a^{2/3}$. What restrictions, if any, should be put on the domain of a for the derivative to exist?

34. a) By differentiating the equation $x^2 - y^2 = 1$ implicitly, show that $dy/dx = x/y$.

b) By differentiating both sides of the equation $dy/dx = x/y$ implicitly, show that

$$\frac{d^2y}{dx^2} = \frac{-1}{y^3}.$$

In Problems 35–42, find dy/dx and then d^2y/dx^2 by implicit differentiation.

35. $x^2 + y^2 = 1$

36. $x^3 + y^3 = 1$

37. $x^{2/3} + y^{2/3} = 1$

38. $y^2 = x^2 + 2x$

39. $y^2 + 2y = 2x + 1$

40. $y^2 = 1 - \dfrac{2}{x}$

41. $y + 2\sqrt{y} = x$

42. $xy + y^2 = 1$

In Problems 43–48, find the lines that are (a) tangent and (b) normal to the curve at the given point.

43. $x^2 + xy - y^2 = 1$, $P(2, 3)$

44. $x^2 + y^2 = 25$, $P(3, -4)$

45. $x^2y^2 = 9$, $P(-1, 3)$

46. $\dfrac{x - y}{x - 2y} = 2$, $P(3, 1)$

47. $y^2 - 2x - 4y - 1 = 0$, $P(-2, 1)$

48. $xy + 2x - 5y = 2$, $P(3, 2)$

49. Find the normals to the curve $xy + 2x - y = 0$ that are parallel to the line $2x + y = 0$.

50. If three normals can be drawn from the point $A(a, 0)$ to the curve $y^2 = x$, show that a must be greater than 1/2. One normal is always the x-axis. Find the value of a for which the other two are perpendicular.

51. The line that is normal to the curve $y = x^2 + 2x - 3$ at $(1, 0)$ intersects the curve at what other point?

52. Show that the normal to the circle $x^2 + y^2 = a^2$ at any point (x_1, y_1) passes through the origin.

53. Find the two points where the curve $x^2 + xy + y^2 = 7$ crosses the x-axis, and show that the tangents to the curve at these points are parallel. What is the common slope of these tangents?

54. Find points on the curve $x^2 + xy + y^2 = 7$ (a) where the tangent is parallel to the x-axis and (b) where the tangent is parallel to the y-axis. (In the latter case, dy/dx is not defined, but dx/dy is. What value does dx/dy have at these points?)

55. The position function of a particle moving on a coordinate line is given by $s(t) = \sqrt{1 + 4t}$, with s in meters and t in seconds. Find the particle's velocity and acceleration when $t = 6$ sec.

56. What horizontal line crosses the curve $y = \sqrt{x}$ at a 45° angle?

57. *Orthogonal curves.* Two curves are said to be *orthogonal* at a point of intersection if their tangents at that point cross at right angles. Show that the curves $2x^2 + 3y^2 = 5$ and $y^2 = x^3$ are orthogonal at $(1, 1)$ and $(1, -1)$.

58. A particle of mass m moves along the x-axis. The velocity $v = dx/dt$ and position x satisfy the equation

$$m(v^2 - v_0^2) = k(x_0^2 - x^2),$$

where k, v_0, and x_0 are constants. Show, by differentiating both sides of this equation with respect to t, that

$$m\frac{dv}{dt} = -kx$$

whenever $v \neq 0$.

TOOLKIT PROGRAMS

Derivative Grapher

2.4
Linear Approximations and Differentials

In science and engineering we can sometimes approximate complicated functions with simpler ones that give the accuracy we want without being so hard to work with. It is important to know how this is done, and in this article we study the simplest of the useful approximations. We also introduce a new symbol, dx, for an increment in the variable x. This symbol, called the *differential* of x, is used more frequently than Δx in the physical sciences. It is also useful in mathematics, as we shall see.

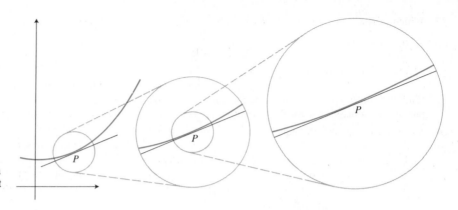

2.12 Successive enlargements show a close fit between a curve and its tangent line.

Linearization

As you can see in Fig. 2.12, the tangent to a curve $y = f(x)$ lies close to the curve near the point of tangency, and, for a brief interval stretching to either side, the y-values along the tangent line give good approximations to the y-values on the curve. What we therefore propose to do is replace the formula for f over this interval by the formula for its tangent line.

In the notation of Fig. 2.13, the tangent passes through the point $P(a, f(a))$ with slope $f'(a)$, so its point-slope equation is

$$y - f(a) = f'(a)(x - a)$$

or

$$y = f(a) + f'(a)(x - a). \tag{1}$$

Thus, the tangent line is the graph of the function

$$L(x) = f(a) + f'(a)(x - a). \tag{2}$$

For as long as the line remains close to the graph of f, $L(x)$ will give a good approximation to $f(x)$. We shall see just how good at the end of Chapter 3.

The function $L(x) = f(a) + f'(a)(x - a)$ is called the linearization of f at a and the approximation $f(x) \approx L(x)$ is called the standard linear approximation of f at a.

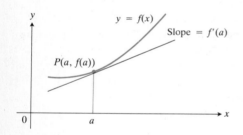

2.13 The equation of the tangent line is $y = f(a) + f'(a)(x - a)$.

DEFINITION

> **Linearization and Standard Linear Approximation**
>
> If $y = f(x)$ is differentiable at $x = a$, then
>
> $$L(x) = f(a) + f'(a)(x - a) \tag{3}$$
>
> is the **linearization** of f at a. The approximation
>
> $$f(x) \approx L(x)$$
>
> is the **standard linear approximation** of f at a.

EXAMPLE 1   Find the linearization of $f(x) = \sqrt{1 + x}$ at $x = 0$. Use it to estimate $\sqrt{1.2}$, $\sqrt{1.05}$, and $\sqrt{1.005}$.

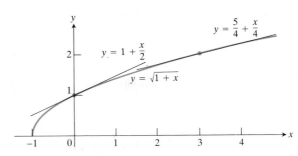

2.14 The graph of $y = \sqrt{1 + x}$ and its linearizations at $x = 0$ and $x = 3$.

Solution We evaluate Eq. (3) for $f(x) = \sqrt{1 + x}$ and $a = 0$. The derivative of f is

$$f'(x) = \frac{1}{2}(1 + x)^{-1/2} = \frac{1}{2\sqrt{1 + x}}.$$

Its value at $x = 0$ is $1/2$. We substitute this along with $a = 0$ and $f(0) = 1$ into Eq. (3):

$$L(x) = f(a) + f'(a)(x - a)$$

$$= 1 + \frac{1}{2}(x - 0)$$

$$= 1 + \frac{x}{2}.$$

The linearization of $\sqrt{1 + x}$ at $x = 0$ is $L(x) = 1 + \frac{x}{2}$. See Fig. 2.14.

How good is the approximation

$$\sqrt{1 + x} \approx 1 + \frac{x}{2}?$$

When $x = 0.2$, 0.05, and 0.005 it gives

$$\sqrt{1.2} \approx 1 + \frac{0.2}{2} = 1.10 \qquad \text{(accurate to 2 decimals)},$$

$$\sqrt{1.05} \approx 1 + \frac{0.05}{2} = 1.025 \qquad \text{(accurate to 3 decimals)},$$

$$\sqrt{1.005} \approx 1 + \frac{0.005}{2} = 1.00250 \quad \text{(accurate to 5 decimals).} \qquad ■$$

EXAMPLE 2 Find the linearization of $f(x) = \sqrt{1 + x}$ at $x = 3$. Use it to estimate $\sqrt{4.2}$. What value does a calculator give for $\sqrt{4.2}$?

Solution We evaluate Eq. (2) for $f(x) = \sqrt{1 + x}$, $f'(x) = 1/(2\sqrt{1 + x})$, and $a = 3$. With

$$f(3) = 2, \qquad f'(3) = \frac{1}{2\sqrt{1 + 3}} = \frac{1}{4},$$

Eq. (2) gives

$$L(x) = 2 + \frac{1}{4}(x - 3) = 2 + \frac{x}{4} - \frac{3}{4} = \frac{5}{4} + \frac{x}{4}.$$

Thus, near $x = 3$,

$$\sqrt{1 + x} \approx \frac{5}{4} + \frac{x}{4}.$$

See Fig. 2.14.

To estimate $\sqrt{4.2}$ we have

$$\sqrt{4.2} = \sqrt{1 + 3.2} \approx \frac{5}{4} + \frac{3.2}{4} = 1.25 + 0.8 = 2.05.$$

A calculator gives $\sqrt{4.2} = 2.04939$ to five places, which differs from 2.05 by less than a thousandth. ∎

The utility of the estimates in Examples 1 and 2 is not in calculating particular square roots. We can do that better with a calculator. The utility lies in finding linear formulas that can replace $\sqrt{1 + x}$ over an *entire interval* with enough accuracy to be useful.

EXAMPLE 3   An approximation used frequently in physics and engineering is

$$(1 + x)^k \approx 1 + kx \qquad \text{for any number } k. \tag{4}$$

(See Problem 54.) The approximation is good for values of x near zero. For example, when x is numerically small,

$$\sqrt{1 + x} \approx 1 + \frac{x}{2}$$

$$\frac{1}{1 - x} = (1 - x)^{-1} \approx 1 + (-1)(-x) = 1 + x$$

$$\tag{5}$$

$$\sqrt[3]{1 + 5x^4} = (1 + 5x^4)^{1/3} \approx 1 + \frac{1}{3}(5x^4) = 1 + \frac{5}{3}x^4$$

$$\frac{1}{\sqrt{1 - x^2}} = (1 - x^2)^{-1/2} \approx 1 + \left(-\frac{1}{2}\right)(-x^2) = 1 + \frac{1}{2}x^2. \qquad ∎$$

Estimating Change

Suppose we know the value of a differentiable function $f(x)$ at a particular point x_0 and want to predict how much this value will change if we move nearby to the point $x_0 + h$. If h is small, f and its linearization L at x_0 will change by nearly the same amount. Since the values of L are always easy to calculate, calculating the change in L gives a practical way to estimate the change in f.

In the notation of Fig. 2.15, the change in f is

$$\Delta f = f(x_0 + h) - f(x_0).$$

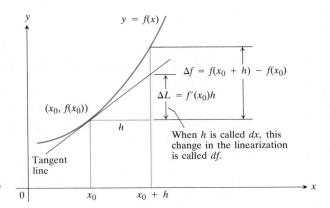

$\Delta f = f(x_0 + h) - f(x_0)$

$\Delta L = f'(x_0)h$

$(x_0, f(x_0))$

When h is called dx, this change in the linearization is called df.

Tangent line

x_0 $x_0 + h$

2.15 If h is small, the change in the linearization of f will be nearly the same as the change in f.

The corresponding change in L is

$$\Delta L = L(x_0 + h) - L(x_0)$$

$$= f(x_0) + f'(x_0)[(x_0 + h) - x_0] - f(x_0) \qquad (6)$$

$$= f'(x_0)h.$$

The formula for Δf will usually be as hard to work with as the formula for f. The formula for ΔL, however, is always easy to work with. As you can see, the change in L is just a constant times h.

The change $\Delta L = f'(x_0)h$ is usually described with the more suggestive notation

$$df = f'(x_0)dx, \qquad (7)$$

in which df denotes the change in the linearization of f that results from the change dx in x. We call dx the **differential** of x and df the corresponding **differential** of f.

If we divide both sides of the equation $df = f'(x)dx$ by dx, we obtain the familiar equation

$$\frac{dy}{dx} = f'(x).$$

This equation now says that we may regard the derivative dy/dx as a quotient of differentials. In many calculations it is convenient to do so, as we shall see in the next article.

From a strictly mathematical point of view, the equation $df = f'(x_0)dx$ is simply an equation that defines a dependent variable, df, as a function of two independent variables, x_0 and dx. When we get around to applying the equation, however, we usually want the domains of the independent variables to be restricted to ensure that their sum $x_0 + dx$ lies in the domain of f. Calculating the change df in the linearization of f does little good if the new point is too far away to lie in the domain of f.

EXAMPLE 4 The radius of a circle is to be increased from an initial value of $r_0 = 10$ by an amount $dr = 0.1$. Estimate the corresponding increase in the circle's area $A = \pi r^2$ by calculating dA. Compare dA with the true change ΔA.

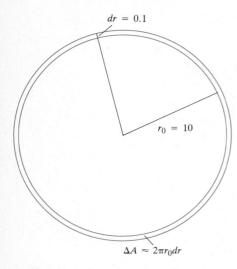

$dr = 0.1$

$r_0 = 10$

$\Delta A \approx 2\pi r_0 dr$

2.16   When dr is small compared to r_0, as it is when $dr = 0.1$ and $r_0 = 10$, the differential $dA = 2\pi r_0\, dr$ gives a good estimate of ΔA. See Example 4.

Solution   To calculate dA, we apply Eq. (7) to the function $A = \pi r^2$:

$$dA = A'(r_0)dr = 2\pi r_0 dr.$$

We then substitute the values $r_0 = 10$ and $dr = 0.1$:

$$dA = 2\pi(10)(0.1) = 2\pi.$$

The estimated change is 2π square units.

A direct calculation of ΔA gives

$$\Delta A = \pi(10.1)^2 - \pi(10)^2$$

$$= (102.01 - 100)\pi$$

$$= \underbrace{2\pi}_{dA} + \underbrace{0.01\pi}_{\text{error}}.$$

The error in the estimate dA is 0.01π square units. As a percentage of the circle's original area, the error is quite small, as we can see from the following calculation:

$$\frac{\text{Error}}{\text{Original area}} = \frac{0.01\pi}{100\pi} = 0.01\%.$$

See Fig. 2.16.   ■

Absolute, Relative, and Percentage Change

When we move from x_0 to a nearby point, we can describe the corresponding change in the value of a function f in three ways:

	True	**Estimate**
Absolute change	Δf	df
Relative change	$\dfrac{\Delta f}{f(x_0)}$	$\dfrac{df}{f(x_0)}$
Percentage change	$\dfrac{\Delta f}{f(x_0)} \times 100$	$\dfrac{df}{f(x_0)} \times 100$

EXAMPLE 5   Estimate the percentage change that will occur in the area of a circle if its radius is increased from $r_0 = 10$ units to 10.1 units.

Solution   From the preceding table we have

$$\text{Estimated percentage change} = \frac{dA}{A(r_0)} \times 100.$$

With $dA = 2\pi$ (from Example 4) and $A(r_0) = 100\pi$, the formula gives

$$\frac{dA}{A(r_0)} \times 100 = \frac{2\pi}{100\pi} \times 100 = 2\%.$$ ■

EXAMPLE 6   An edge of a cube is measured as 6 in. with a possible error of ± 0.05 in. The volume of the cube is to be calculated from this measurement. Estimate the percentage error that would occur in the volume calculation if the absolute value of the error in the edge measurement were as large as 0.05 in.

Solution If x denotes the length of the cube's edge, the volume is $V = x^3$. We have a measured value of 6 in. for x, with a measurement error dx that does not exceed 0.05 in absolute value:

$$|dx| \le 0.05 \text{ in.}$$

We use Eq. (7) to estimate the variation in V caused by the increment dx:

$$dV = \left(\frac{dV}{dx}\right)_{x=6} \cdot dx = (3x^2)_{x=6} \cdot dx = 108 \; dx.$$

Since $|dx| \le 0.05$,

$$|dV| = 108|dx| \le (108)(0.05) = 5.4 \text{ in}^3.$$

The volume calculated with $x = 6$ might differ from the true volume by as much as 5.4 in³ either way. Our estimate of the percentage error in the volume calculation is

$$\frac{dV}{\text{Calculated } V} = \frac{5.4}{6^3} = \frac{5.4}{216} = 0.025 = 2.5\%. \qquad \blacksquare$$

EXAMPLE 7 Suppose the earth were a perfect sphere and that we determined its radius to be 3959 ± 0.1 miles. What effect would the tolerance of ± 0.1 have on our estimate of the earth's surface area?

Solution The surface area of a sphere of radius r is $S = 4\pi r^2$. The uncertainty in the calculation of S that arises from measuring r with a tolerance of dr miles is about

$$dS = \left(\frac{dS}{dr}\right) dr = 8\pi r \; dr.$$

With $r = 3959$ and $dr = 0.1$,

$$dS = 8\pi(3959)(0.1) = 9950 \text{ square miles}$$

to the nearest square mile, which is about the area of the state of Maryland. In absolute terms this may seem like a large error. However, 9950 mi² is a relatively small error when compared to the calculated surface area of the earth:

$$\frac{dS}{\text{Calculated } S} = \frac{9950}{4\pi(3959)^2} \approx \frac{9950}{196,961,284} \approx .005\%. \qquad \blacksquare$$

EXAMPLE 8 About how accurately should we measure the radius r of a sphere to calculate the surface area $S = 4\pi r^2$ within 1% of its true value?

Solution We want any inaccuracy in our measurement to be small enough to make the corresponding increment ΔS in the surface area satisfy the inequality

$$|\Delta S| \le \frac{1}{100} S = \frac{4\pi r^2}{100}. \qquad (8)$$

We replace ΔS in this inequality with

$$dS = \left(\frac{dS}{dr}\right) dr = 8\pi r \; dr.$$

This gives

$$|8\pi r\ dr| \le \frac{4\pi r^2}{100} \qquad \text{and} \qquad |dr| \le \frac{1}{8\pi r} \cdot \frac{4\pi r^2}{100} = \frac{1}{2}\frac{r}{100}.$$

We should measure the radius with an error dr that is no more than 0.5% of the true value.   ■

The Error in the Approximation $\Delta y \approx f'(x)\Delta x$

How well does the quantity $f'(a)\Delta x$ estimate the true increment $\Delta y = f(a + \Delta x) - f(a)$? We measure the error by subtracting one from the other:

$$\text{Approximation error} = \Delta y - f'(a)\Delta x$$

$$= f(a + \Delta x) - f(a) - f'(a)\Delta x$$

$$= \underbrace{\left(\frac{f(a + \Delta x) - f(a)}{\Delta x} - f'(a)\right)}_{\text{Call this part } \epsilon}\Delta x, \tag{9}$$

$$= \epsilon \cdot \Delta x.$$

As $\Delta x \to 0$, the difference quotient

$$\frac{f(a + \Delta x) - f(a)}{\Delta x}$$

approaches $f'(a)$ (remember the definition of $f'(a)$), so the quantity in parentheses becomes a very small number (which is why we called it ϵ). In fact,

$$\epsilon \to 0 \qquad \text{as} \qquad \Delta x \to 0.$$

Thus, when Δx is small, the approximation error $\epsilon \Delta x$ is smaller still.

$$\underset{\text{true change}}{\Delta y} = \underset{\text{estimated change}}{f'(a)\Delta x} + \underset{\text{error}}{\epsilon \Delta x} \tag{10}$$

While we do not know exactly how small the error is and will not be able to make much progress on this front until late in Chapter 3, there is something worth noting here, namely the *form* taken by the equation.

If $y = f(x)$ is differentiable at $x = a$, and x changes from a to $a + \Delta x$, the change Δy in f is given by an equation of the form

$$\Delta y = f'(a)\Delta x + \epsilon \Delta x \tag{11}$$

in which $\epsilon \to 0$ as $\Delta x \to 0$.

Just knowing the form of Eq. (11) can be useful, as we shall see in the next article when we develop a formula for calculating the derivative of the composite of two differentiable functions.

☐ The Conversion of Mass to Energy

Here is an example of how the approximation

$$\frac{1}{\sqrt{1 - x^2}} = 1 + \frac{1}{2}x^2 \tag{12}$$

from Example 3 is used in an applied problem.

Newton's second law,

$$F = \frac{d}{dt}(mv) = m\frac{dv}{dt} = ma,$$

is stated with the assumption that mass is constant, but we know this is not strictly true because the mass of a body increases with velocity. In Einstein's corrected formula, mass has the value

$$m = \frac{m_0}{\sqrt{1 - v^2/c^2}}, \tag{13}$$

where the "rest mass" m_0 represents the mass of a body that is not moving and c is the speed of light, which is about 300,000 kilometers per second. When v is very small compared to c, v^2/c^2 is close to zero and it is safe to use the approximation

$$\frac{1}{\sqrt{1 - v^2/c^2}} = 1 + \frac{1}{2}\left(\frac{v^2}{c^2}\right)$$

(which is Eq. 12 with $x = v/c$) to write

$$m = \frac{m_0}{\sqrt{1 - v^2/c^2}} = m_0\left[1 + \frac{1}{2}\left(\frac{v^2}{c^2}\right)\right] = m_0 + \frac{1}{2}m_0v^2\left(\frac{1}{c^2}\right),$$

or

$$m = m_0 + \frac{1}{2}m_0v^2\left(\frac{1}{c^2}\right). \tag{14}$$

Equation (14) expresses the increase in mass that results from the added velocity v.

In Newtonian physics, $(1/2)m_0v^2$ is the kinetic energy (KE) of the body, and if we rewrite Eq. (14) in the form

$$(m - m_0)c^2 = \frac{1}{2}m_0v^2,$$

we see that

$$(m - m_0)c^2 = \frac{1}{2}m_0v^2 = \frac{1}{2}m_0v^2 - \frac{1}{2}m_0(0)^2 = \Delta(\text{KE}),$$

or

$$(\Delta m)c^2 = \Delta(\text{KE}).$$

In other words, the change in kinetic energy $\Delta(\text{KE})$ in going from velocity 0 to velocity v equals $(\Delta m)c^2$.

Such energy changes ordinarily represent extremely small changes in mass. The energy released by a 20-kiloton atomic bomb, for instance, is the result of converting only one gram of mass into energy. The products of the explosion weigh only one gram less than the material exploded. It may help to put this in perspective if you remember that a penny weighs about 3 grams.

A Frequently Used Linear Approximation

For x near zero: $(1 + x)^k \approx 1 + kx$ for any number k (15)

PROBLEMS

In Problems 1–12, find the linearization $L(x)$ of the given function at the point a. Then use L to estimate the given function value. If you have a calculator, compare your estimate with the calculator's value.

1. $f(x) = x^4$, $a = 1$, $f(1.01)$

2. $f(x) = x^2 + 2x$, $a = 0$, $f(0.1)$

3. $f(x) = x^{-1}$, $a = 2$, $f(2.1)$

4. $f(x) = x^{-1}$, $a = 0.5$, $f(0.6)$

5. $f(x) = x^3 - x$, $a = 1$, $f(1.1)$

6. $f(x) = 2x^2 + 4x - 3$, $a = -1$, $f(-0.9)$

7. $f(x) = x^3 - 2x + 3$, $a = 2$, $f(1.9)$

8. $f(x) = \sqrt{1 + x}$, $a = 8$, $f(8.1)$

9. $f(x) = \sqrt{x}$, $a = 4$, $f(4.1)$

10. $f(x) = \sqrt[3]{x}$, $a = 8$, $f(8.5)$

11. $f(x) = \sqrt{x^2 + 9}$, $a = -4$, $f(-4.2)$

12. $f(x) = \dfrac{x}{x + 1}$, $a = 1$, $f(1.3)$

Use the formula $(1 + x)^k \approx 1 + kx$ to construct linear approximations for the functions in Problems 13–22 for values of x near zero.

13. $(1 + x)^2$

14. $(1 + x)^3$

15. $\dfrac{1}{(1 + x)^5}$

16. $\dfrac{4}{(1 + x)^2}$

17. $\dfrac{2}{(1 - x)^4}$

18. $(1 - x)^6$

19. $2\sqrt{1 + x}$

20. $3(1 + x)^{1/3}$

21. $\dfrac{1}{1 + x}$

22. $\dfrac{1}{\sqrt{1 + x}}$

23. Use the approximation $(1 + x)^k \approx 1 + kx$ to estimate
a) $(1.0002)^{100}$ b) $\sqrt[3]{1.009}$ c) $1/(0.999)$

24. CALCULATOR To illustrate the approximation $\sqrt{1 + x} \approx 1 + x/2$, enter 1.1 into your calculator. Then press the square root key repeatedly, pausing between key presses to read the display. With each key press you will find that the decimal part of the display is approximately halved.

In Problems 25–30, each function $y = f(x)$ changes value when x changes from a to $a + \Delta x$. In each case, find (a) the change $\Delta y = f(a + \Delta x) - f(a)$; (b) the value of the estimate $f'(a)\Delta x$; and (c) the error $\Delta y - f'(a)\Delta x$.

25. $f(x) = x^2 + 2x$, $a = 0$, $\Delta x = 0.1$

26. $f(x) = 2x^2 + 4x - 3$, $a = -1$, $\Delta x = 0.1$

27. $f(x) = x^3 - x$, $a = 1$, $\Delta x = 0.1$

28. $f(x) = x^4$, $a = 1$, $\Delta x = 0.1$

29. $f(x) = x^{-1}$, $a = 0.5$, $\Delta x = 0.1$

30. $f(x) = x^3 - 2x + 3$, $a = 2$, $\Delta x = 0.1$

In Problems 31–38, write a formula that *estimates* the given change in volume or surface area. Formulas for volume and surface area may be found in Appendix 3.

31. The change in the volume of a sphere when the radius changes by an amount dr.

32. The change in the surface area of a sphere when the radius changes by an amount dr.

33. The change in the volume of a cube when the edge lengths all change by an amount dx.

34. The change in the surface area of a cube when the edge lengths all change by an amount dx.

35. The change in volume of a right circular cylinder when the radius changes by an amount dr and the height does not change.

36. The change in the lateral surface area of a right circular cylinder when the height changes by an amount dh and the radius does not change.

37. The change in the volume of a right circular cone when the radius changes by an amount dr and the height does not change.

38. The change in the lateral surface area of a cone when the height changes by an amount dh and the radius does not change.

39. The radius of a circle is increased from 2.00 to 2.02 m.
a) Estimate the resulting change in area.
b) Express the estimate in (a) as a percentage of the circle's original area.

40. The diameter of a tree was 10 in. During the following year, the circumference grew 2 in. About how much did the tree's diameter grow? the tree's cross section area?

41. The edge of a cube is measured as 10 cm with a possible error of 1%. The cube's volume is to be calculated from this measurement. Estimate the percentage error in the volume calculation.

42. The diameter of a sphere is measured as 100 ± 1 cm and the volume is calculated from this measurement. Estimate the percentage error in the volume calculation.

43. About how accurately should you measure the side of a square to be sure of calculating the area within 2% of its true value?

44. a) How accurately should the edge of a cube be measured to be reasonably sure of calculating the cube's surface area with an error of no more than 2%?
b) Suppose the edge is measured with the accuracy required in (a). About how accurately can the cube's volume be calculated from the edge measurement? To find out, estimate the percentage error in the volume calculation that would result from using the edge measurement.

45. The height and radius of a right circular cone are equal, so the volume of the cone is $V = (1/3)\pi h^3$. The volume is to be calculated from a measurement of h and must be calculated with an error of no more than 1% of the true value. Find approximately the greatest error that can be tolerated in the measurement of h, expressed as a percent of h.

46. The circumference of the equator of a sphere is measured as 10 cm, with a possible error of 0.4 cm. The measurement is then used to calculate the radius. The radius is then used to calculate the surface area and volume of the sphere. Estimate the percentage errors in the calculated values of (a) the radius, (b) the surface area, and (c) the volume.

47. Estimate the allowable percentage error in measuring the diameter d of a sphere if the volume is to be calculated correctly to within 3%.

48. a) About how accurately must the interior diameter of a 10-meter-high cylindrical storage tank be measured to calculate the tank's volume to within 1% of its true value?

  b) About how accurately must the tank's exterior diameter be measured to calculate the amount of paint it will take to paint the side of the tank within 5% of the true amount?

49. A manufacturer contracts to mint coins for the federal government. How much variation dr in the radius of the coins can be tolerated in manufacture if the coins are to weigh within 1/1000 of their ideal weight? Assume that the thickness does not vary.

50. *The period of a clock pendulum.* The period T of a clock pendulum (time for one full swing and back) is given by the formula $T^2 = 4\pi^2 L/g$, where T is measured in seconds, $g = 32.2$ ft/sec^2, and L, the length of the pendulum, is measured in feet. Find approximately:

  a) the length of a clock pendulum whose period is $T = 1$ sec;

  b) the change dT in T if the pendulum in (a) is lengthened 0.01 ft; and

  c) the amount the clock gains or loses in a day as a result of the period's changing by the amount dT found in (b).

51. The volume $y = x^3$ of a cube of edge x increases by an amount Δy when x increases by an amount Δx. Show that

Δy can be represented geometrically as the sum of the volumes of

  a) three slabs of dimensions x by x by Δx,

  b) three bars of dimensions x by Δx by Δx,

  c) one cube of dimensions Δx by Δx by Δx.

Illustrate with a sketch.

52. Let $g(x) = \sqrt{x} + \sqrt{1 + x} - 4$.

  a) Find $g(3) < 0$ and $g(4) > 0$ to show (by the Intermediate Value Theorem, Article 1.11) that the equation $g(x) = 0$ has a solution between $x = 3$ and $x = 4$.

  b) To estimate the solution of $g(x) = 0$, replace the square roots by their linearizations at $x = 3$ and solve the resulting linear equation.

  c) **CALCULATOR** Check your estimate in the original equation.

53. Let

$$f(x) = \frac{2}{1 - x} + \sqrt{1 + x} - 3.1.$$

  a) Find $f(0) < 0$ and $f(0.5) > 0$ to show (by the Intermediate Value Theorem, Article 1.11) that the equation $f(x) = 0$ has a solution between $x = 0$ and $x = 0.5$.

  b) To estimate the solution of $f(x) = 0$, replace $2/(1 - x)$ and $\sqrt{1 + x}$ by their linearizations at $x = 0$ and solve the resulting linear equation.

  c) **CALCULATOR** Check your estimate in the original equation.

54. We know that the Power Rule,

$$\frac{d}{dx}(1 + x)^k = k(1 + x)^{k-1},$$

holds for any rational number k. In Chapter 6 we shall show that the rule holds for any irrational number k as well. Assuming this result for now, verify Eq. (4) by showing that the linearization of $f(x) = (1 + x)^k$ at $a = 0$ is $L(x) = 1 + kx$.

55. To what relative speed should a body at rest be accelerated to increase its mass by 1%?

2.5
The Chain Rule

The rule for calculating the derivative of the composite of two differentiable functions is, roughly speaking, that the derivative of their composite is the product of their derivatives. This rule is called the *Chain Rule*. Because most of the functions used in practice are composites of other functions, the Chain Rule is probably the most extensively used rule of differentiation. You have already seen one of its special cases, the Power Rule $du^n/dx = u^{n-1}(du/dx)$, which we present again as Example 8. In this article we shall see why the Chain Rule works and how it applies to calculating derivatives.

EXAMPLE 1   The function $y = 6t - 10 = 2(3t - 5)$ is the composite of $y = 2x$ and $x = 3t - 5$. How are the derivatives of these three functions related?

Solution   We have

$$\frac{dy}{dt} = 6, \qquad \frac{dy}{dx} = 2, \qquad \frac{dx}{dt} = 3.$$

Since $6 = 2 \cdot 3$,

$$\frac{dy}{dt} = \frac{dy}{dx}\frac{dx}{dt}. \qquad \blacksquare$$

EXAMPLE 2   A particle moves along the line $y = 5x - 2$ in such a way that its x-coordinate at time t is $x = 3t$. Find dy/dt.

Solution   As a function of t,

$$y = 5x - 2 = 5(3t) - 2 = 15t - 2.$$

Therefore,

$$\frac{dy}{dt} = \frac{d}{dt}(15t - 2) = 15.$$

Note that $dy/dx = 5$, $dx/dt = 3$, and

$$\frac{dy}{dt} = 3 \cdot 5 = \frac{dy}{dx}\frac{dx}{dt}. \qquad \blacksquare$$

EXAMPLE 3   The function $y = (5t + 1)^2$ is the composite of $y = x^2$ and $x = 5t + 1$. The rule for differentiating powers gives

$$\frac{dy}{dt} = \frac{d}{dt}(5t + 1)^2 = 2(5t + 1) \cdot \frac{d}{dt}(5t + 1) = 2(5t + 1) \cdot 5.$$

The expression on the right is the product of

$$\frac{dy}{dx} = 2x = 2(5t + 1) \qquad \text{and} \qquad \frac{dx}{dt} = \frac{d}{dt}(5t + 1) = 5.$$

Once again,

$$\frac{dy}{dt} = \frac{dy}{dx} \cdot \frac{dx}{dt}. \qquad \blacksquare$$

EXAMPLE 4   In the gear train shown in Fig. 2.17, the ratios of the radii of gears A, B, and C are $3:1:2$. If A turns t times, then B turns $x = 3t$ times and C turns $x/2 = (3/2)t$ times. In terms of derivatives,

$$\frac{dx}{dt} = 3$$

(B turns at three times A's rate, three turns for A's one),

$$\frac{dy}{dx} = \frac{1}{2}$$

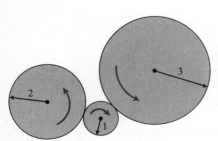

C: *y* turns   B: *x* turns   A: *t* turns

2.17   When wheel A takes t turns, wheel B takes x turns and wheel C takes y turns. By comparing circumferences we see that $dy/dx = 1/2$ and $dx/dt = 3$. What is dy/dt?

(C turns at half B's rate, one half turn for each turn of B), and

$$\frac{dy}{dt} = \frac{3}{2} = \frac{1}{2} \cdot 3 = \frac{dy}{dx}\frac{dx}{dt}$$

(C turns at three-halves A's rate, three halves of a turn for A's one). To calculate dy/dt we can multiply dy/dx by dx/dt. ∎

The Chain Rule

The preceding examples all work because the derivative of a composite $g \circ f$ of two differentiable functions is the product of their derivatives. This is the observation we wish to state formally as the Chain Rule. As in Article 1.4, the notation $g \circ f$ denotes the composite of the functions f and g, with g following f. That is, the value of $g \circ f$ at a point t is $(g \circ f)(t) = g(f(t))$.

The Chain Rule (First Form)

Suppose that $h = g \circ f$ is the composite of the differentiable functions $y = g(x)$ and $x = f(t)$. Then h is a differentiable function of t whose derivative at each value of t is

$$(g \circ f)'_{\text{at } t} = g'_{\text{at } x=f(t)} \cdot f'_{\text{at } t}. \tag{1}$$

In short, at each value of t,

$$h'(t) = g'(f(t)) \cdot f'(t). \tag{2}$$

See Fig. 2.18.

Equations (1) and (2) tell how each derivative is to be evaluated, but once we know that, we can usually get by with writing

$$\frac{dy}{dt} = \frac{dy}{dx}\frac{dx}{dt}. \tag{3}$$

This is usually enough to remind us that if y is a differentiable function of x, and x is a differentiable function of t, then y is a differentiable function of t whose derivative is given by Eq. (1).

2.18 Rates of change multiply: The derivative of $h = g \circ f$ at t is the derivative of f at t times the derivative of g at $f(t)$.

Chain Rule (Short Form)

If y is a differentiable function of x, and x is a differentiable function of t, then y is a differentiable function of t and

$$\frac{dy}{dt} = \frac{dy}{dx}\frac{dx}{dt}. \tag{4}$$

Equation (4) is particularly memorable because, when we think of the derivatives on the right as quotients of differentials, the dx's cancel to produce the fraction on the left.

We now have two formulations of the Chain Rule and, like special tools in a repair shop, they each make some task a little easier. We shall use them both in the examples that follow. But remember, they express the same one rule: the derivative of a composite of differentiable functions is the product of their derivatives.

Proof of the Chain Rule   The idea behind the Chain Rule is this: If $x = f(t)$ is differentiable at t_0, then an increment Δt produces an increment Δx such that

$$\Delta x = f'(t_0)\Delta t + \epsilon_1 \Delta t = [f'(t_0) + \epsilon_1]\Delta t; \tag{5a}$$

and if $y = g(x)$ is differentiable at $x_0 = f(t_0)$, then an increment Δx produces an increment Δy such that

$$\Delta y = g'(x_0)\Delta x + \epsilon_2 \Delta x = [g'(x_0) + \epsilon_2]\Delta x. \tag{5b}$$

Equations (5a) and (5b) are versions of Eq. (11) in Article 2.4, relating the increments Δx and Δy to their tangent line approximations. In these equations, $\epsilon_1 \rightarrow 0$ when $\Delta t \rightarrow 0$, and $\epsilon_2 \rightarrow 0$ when $\Delta x \rightarrow 0$.

Combining Eqs. (5a) and (5b) gives us

$$\Delta y = [g'(x_0) + \epsilon_2]\Delta x = [g'(x_0) + \epsilon_2][f'(t_0) + \epsilon_1]\Delta t. \tag{6}$$

Dividing through by Δt gives

$$\begin{aligned}
\frac{\Delta y}{\Delta t} &= [g'(x_0) + \epsilon_2][f'(t_0) + \epsilon_1] \\
&= g'(x_0)f'(t_0) + \epsilon_2 f'(t_0) + \epsilon_1 g'(x_0) + \epsilon_1 \epsilon_2.
\end{aligned} \tag{7}$$

As Δt approaches zero, so do Δx, ϵ_1, and ϵ_2, and we get

$$\lim \frac{\Delta y}{\Delta t} = g'(x_0)f'(t_0),$$

which is the same as

$$\frac{dy}{dt}(t_0) = g'(f(t_0))f'(t_0). \qquad \blacksquare \tag{8}$$

EXAMPLE 5   Find dy/dt at $t = -1$ if $y = x^3 + 5x - 4$ and $x = t^2 - 1$.

■ ■ ■

EARLY STATUS OF THE CHAIN RULE

Before he first published his version of the calculus in 1684, Leibniz had completed several unpublished papers on the subject. In a manuscript dated 1676, he essentially stated the Chain Rule. In explaining how to differentiate the expression $\sqrt{a + bz + cz^2}$, he said to let $x = a + bz + cz^2$ and then differentiate $\sqrt{x}$ and multiply by dx/dz. The rule itself was employed throughout the eighteenth century with little regard to explicit justification as such. The various versions of the Chain Rule encountered in this book—as in the case of functions of several variables—seemed to eighteenth- and nineteenth-century mathematicians as natural extensions of Leibniz's first and basic observation.

Solution   When $t = -1$, $x = 0$. Therefore

$$\left.\frac{dy}{dt}\right|_{t=-1} = \left.\frac{dy}{dx}\right|_{x=0} \cdot \left.\frac{dx}{dt}\right|_{t=-1}$$

$$= (3x^2 + 5)_{x=0} \cdot (2t)_{t=-1}$$

$$= 5 \cdot -2 = -10. \qquad ■$$

EXAMPLE 6   Express dy/dt in terms of t if $y = x^3 - 3$ and $x = t^2 - 1$.

Solution   The Chain Rule gives

$$\frac{dy}{dt} = \frac{dy}{dx}\frac{dx}{dt}$$

$$= 3x^2 \cdot 2t$$

$$= 3(t^2 - 1)^2 \cdot 2t \qquad (dy/dx \text{ is evaluated at } x = t^2 - 1)$$

$$= 6t(t^2 - 1)^2. \qquad ■$$

EXAMPLE 7   If $g(x) = \sqrt{x + 2}$, and $x = f(t) = t^3 - 1$, find $(d/dt)(g \circ f)$ when $t = 2$.

Solution   By Eq. (1) for the Chain Rule,

$$(g \circ f)'_{\text{at } t=2} = g'_{\text{at } x=f(2)} \cdot f'_{\text{at } t=2}$$

$$= \left.\frac{1}{2\sqrt{x + 2}}\right|_{x=7} \cdot \left. 3t^2 \right|_{t=2}$$

$$= \frac{1}{2\sqrt{9}} \cdot 12$$

$$= 2. \qquad ■$$

EXAMPLE 8   The familiar rule for differentiating powers of functions is a special case of the Chain Rule.

If u is a differentiable function of x, and $y = u^n$, then the Chain Rule in the form

$$\frac{dy}{dx} = \frac{dy}{du} \cdot \frac{du}{dx}$$

gives

$$\frac{dy}{dx} = \frac{d}{du}(u^n) \cdot \frac{du}{dx} = nu^{n-1}\frac{du}{dx}. \qquad ■$$

When using the Chain Rule, it sometimes helps to think in the following way. If $y = g(f(x))$, then

$$y' = g'(f(x)) \cdot f'(x)$$

says: Differentiate g, leaving everything "inside" (that is, $f(x)$) alone; then multiply by the derivative of the "inside." In practice, we then have

$$y = \underbrace{(x^3 + x^2)^{13}}_{\text{"inside" left alone}}$$

$$y' = \overbrace{13(x^3 + x^2)^{12}} \cdot \underbrace{(3x^2 + 2x)}_{\substack{\text{derivative} \\ \text{of the "inside"}}}.$$

☐ **EXAMPLE 9** *Snowball melting.* How long does it take a snowball to melt?

Discussion We start with a mathematical model. Let us assume that the snowball is, approximately, a sphere of radius r and volume $V = (4/3)\pi r^3$. (Of course, the snowball is not a perfect sphere, but we can apply mathematics only to a mathematical model of the situation, and we choose one that seems reasonable and is not too complex.) In the same way, we choose some hypothesis about the rate at which the volume of the snowball is changing. One model is to assume that the volume decreases at a rate that is proportional to the surface area: In mathematical terms,

$$\frac{dV}{dt} = -k(4\pi r^2).$$

We tacitly assume that k, the proportionality factor, is a constant. (It probably depends on several things, like the relative humidity of the surrounding air, the air temperature, the incidence or absence of sunlight, to name a few.) Finally, we need at least one more bit of information: How long has it taken for the snowball to melt some specific percent? We have nothing to guide us unless we make one or more observations, but let us now assume a particular set of conditions in which the snowball melted 1/4 of its volume in two hours. (You could use letters instead of these precise numbers: say n percent in h hours. Then your answer would be in terms of n and h.) Now to work. Mathematically, we have the following problem.

Given: $\qquad V = \dfrac{4}{3}\pi r^3 \qquad$ and $\qquad \dfrac{dV}{dt} = -k(4\pi r^2),$

and $V = V_0$ when $t = 0$, and $V = (3/4)V_0$ when $t = 2$ hr.

To find: The value of t when $V = 0$.

We apply the Chain Rule to differentiate $V = (4/3)\pi r^3$ with respect to t:

$$\frac{dV}{dt} = \frac{4}{3}\pi(3r^2)\frac{dr}{dt} = 4\pi r^2 \frac{dr}{dt}.$$

We set this equal to the given rate, $-k(4\pi r^2)$, and divide by $4\pi r^2$ to get

$$\frac{dr}{dt} = -k.$$

The radius is *decreasing* at the constant rate of k radius units per hour. Thus, in 2 hours the radius decreased by $2k$ centimeters, inches, or whatever. If the length of the radius started at r_0, then 2 hours later its length was $r_2 = r_0 - 2k$.

This equation gives us the value of k:

$$r_2 = r_0 - 2k$$

$$2k = r_0 - r_2$$

$$k = \frac{r_0 - r_2}{2}.$$

The melting time is the value of t that makes $r = 0$, or $kt = r_0$:

$$t_{\text{melt}} = \frac{r_0}{k} = \frac{2r_0}{r_0 - r_2} = \frac{2}{1 - (r_2/r_0)}.$$

But

$$\frac{r_2}{r_0} = \frac{\left(\frac{3}{4\pi}V_2\right)^{1/3}}{\left(\frac{3}{4\pi}V_0\right)^{1/3}} = \left(\frac{V_2}{V_0}\right)^{1/3} = \left(\frac{\frac{3}{4}V_0}{V_0}\right)^{1/3} = \left(\frac{3}{4}\right)^{1/3} \approx 0.91.$$

Therefore,

$$t_{\text{melt}} \approx \frac{2}{1 - 0.91} \approx 22 \text{ hr.}$$

If 1/4 of the snowball melts in 2 hours, it takes nearly 20 hours for the rest of it to melt.

REMARK   If we were natural scientists who were really interested in testing our model, we could collect some data and compare them with the results of the mathematics. One practical application may lie in analyzing the proposal to tow large icebergs from polar waters to offshore locations near southern California to provide fresh water from the melting ice. As a first approximation, we might assume that the iceberg is a large cube, or a pyramid, or a sphere.   ■

PROBLEMS

In Problems 1–10, find dy/dt by the Chain Rule, expressing the results in terms of t.

1. $y = x^2$, $x = 2t - 5$

2. $y = x^4$, $x = \sqrt[3]{t}$

3. $y = 8 - \dfrac{x}{3}$, $x = t^3$

4. $y + 4x^2 = 7$, $x + \dfrac{5}{4}t = 3$

5. $2x - 3y = 9$, $2x + \dfrac{t}{3} = 1$

6. $y = x^{-1}$, $x = t^2 - 3t + 8$

7. $y = \sqrt{x + 2}$, $x = \dfrac{2}{t}$, $t > 0$

8. $y = \dfrac{x^2}{x^2 + 1}$, $x = \sqrt{2t + 1}$

9. $y = x^2 + 3x - 7$, $x = 2t + 1$

10. $y = x^{2/3}$, $x = t^2 + 1$

11. Find dz/dx if $z = w^2 - w^{-1}$, $w = 3x$.

12. Find dy/dx if $y = 2v^3 + 2v^{-3}$, $v = (3x + 2)^{2/3}$.

13. Find dr/dt if $r = (s + 1)^{1/2}$, $s = 16t^2 - 20t$.

14. Find da/db if $a = 7r^3 - 2$, $r = 1 - 1/b$.

15. Find du/dv if $u = t + 1/t$, $t = 1 - 1/v$.

16. Find dy/dx if $y = u^5$ and $u = 3x^2 - 7x + 5$.

In Problems 17–20, use the Chain Rule to express dy/dt in terms of t. Then check your result by expressing y in terms of t and differentiating with respect to t.

17. $y = 3x^{2/3}$, $x = 8t^3$

18. $y = x^2 - 1$, $x = \sqrt{t + 1}$

19. $y = \dfrac{1}{x^2 + 1}$, $x = \sqrt{4t - 1}$

20. $y = 1 - \dfrac{1}{x}$, $x = \dfrac{1}{1 - t}$

In Problems 21–26, find the value of $(d/dt)(g \circ f)$ at the given value of t.

21. $g(x) = x^2 + 1$, $x = f(t) = \sqrt{t + 1}$, $t = 0$

22. $g(x) = \sqrt{x + 5}$, $x = f(t) = 10\sqrt{t}$, $t = 4$

23. $g(x) = \sqrt{1 + x^3}$, $x = f(t) = t^{1/3}$, $t = 1$

24. $g(x) = 1 - \dfrac{1}{x}$, $x = f(t) = \dfrac{1}{1 - t}$, $t = -1$

25. $g(x) = \dfrac{2x}{x^2 + 1}$, $x = f(t) = 10t^2 + t + 1$, $t = 0$

26. $g(x) = \left(\dfrac{x - 1}{x + 1}\right)^2$, $x = f(t) = \dfrac{1}{t^2} - 1$, $t = -1$

27. The velocity of a falling body is $k\sqrt{s}$ meters per second (k constant) at the instant the body has fallen a distance of s meters from its starting point. Show that the acceleration of the body is constant.

28. Let $f(x) = x^2$ and $g(x) = |x|$. Show that the composites $f \circ g$ and $g \circ f$ are both differentiable at $x = 0$ even though g has no derivative at $x = 0$. Does this contradict the Chain Rule? Explain.

29. *Simple pendulum and temperature variation.* For oscillations of small amplitude, the relation between the period T and the length L of a simple pendulum may be approximated by the equation

$$T = 2\pi\sqrt{\dfrac{L}{g}},$$

where $g = 980$ cm/sec^2 is the acceleration due to gravity. When the temperature θ changes, the length L either increases or decreases at a rate that is proportional to L:

$$\dfrac{dL}{d\theta} = kL,$$

where k is a proportionality constant. What is the rate of change of the period with respect to temperature?

30. *Measuring the acceleration of gravity.* When the length L of a clock pendulum is held constant by controlling its temperature, the pendulum's period T depends on the acceleration of gravity g. The period will therefore vary slightly as the clock is moved from place to place on the earth's surface, depending on the change in g. By keeping track of ΔT, one can estimate the variation in g from the equation $T = 2\pi(L/g)^{1/2}$ that relates T, g, and L.

a) With L held constant and g as the independent variable, calculate dT and use it to answer the questions in (b) and (c).

b) If g increases, will T increase, or decrease? Will a pendulum clock speed up, or slow down, if g increases?

c) A clock with a 100-cm pendulum is moved from a location where $g = 980$ cm/sec^2 to a new location. This increases the period by $dT = 0.001$ sec. Find dg and estimate the value of g at the new location.

TOOLKIT PROGRAMS

Parametric Equations   Super * Grapher

2.6
A Brief Review of Trigonometry

Many natural phenomena are periodic; that is, they repeat after definite periods of time. Such phenomena are most readily studied with trigonometric functions, particularly sines and cosines. Our object in this article and the next is to apply the operations of the calculus to these functions, but before we do we shall review some of their properties.

When an angle of measure θ is placed in standard position at the center of a circle of radius r, as in Fig. 2.19, the trigonometric functions of θ are defined by the equations

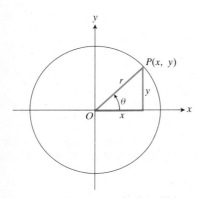

2.19   An angle θ in standard position.

$$\sin \theta = \frac{y}{r}, \qquad \cos \theta = \frac{x}{r}, \qquad \tan \theta = \frac{y}{x},$$

$$\csc \theta = \frac{r}{y}, \qquad \sec \theta = \frac{r}{x}, \qquad \cot \theta = \frac{x}{y}. \tag{1}$$

Observe that $\tan \theta$ and $\sec \theta$ are not defined for values of θ for which $x = 0$. In radian measure, this means that $\pi/2, 3\pi/2, \ldots, -\pi/2, -3\pi/2, \ldots$ are excluded from the domains of the tangent and the secant functions. Similarly, $\cot \theta$ and $\csc \theta$ are not defined for values of θ corresponding to $y = 0$: that is, for $\theta = 0, \pi, 2\pi, \ldots, -\pi, -2\pi, \ldots$. For those values of θ where the functions are defined, it follows from Eqs. (1) that

$$\tan \theta = \frac{\sin \theta}{\cos \theta}, \qquad \csc \theta = \frac{1}{\sin \theta},$$

$$\sec \theta = \frac{1}{\cos \theta}, \qquad \cot \theta = \frac{1}{\tan \theta}.$$

Since $x^2 + y^2 = r^2$ by the Pythagorean theorem, we have

$$\cos^2\theta + \sin^2\theta = \frac{x^2}{r^2} + \frac{y^2}{r^2} = \frac{x^2 + y^2}{r^2} = 1. \tag{2}$$

It is also useful to express the coordinates of $P(x, y)$ in terms of r and θ as follows:

$$x = r \cos \theta, \qquad y = r \sin \theta. \tag{3}$$

When $\theta = 0$ in Fig. 2.19, we have $y = 0$ and $x = r$; hence, from the definitions (1), we obtain

$$\sin 0 = 0, \qquad \cos 0 = 1.$$

Similarly, for a right angle, $\theta = \pi/2$, we have $x = 0$, $y = r$; hence

$$\sin \frac{\pi}{2} = 1, \qquad \cos \frac{\pi}{2} = 0.$$

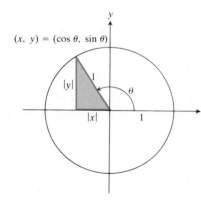

$(x, y) = (\cos \theta, \sin \theta)$

2.20 The acute reference triangle for an angle θ.

Calculating Sines and Cosines

If (x, y) is a point on a circle with radius $r = 1$ unit, then Eqs. (3) become

$$x = \cos \theta, \qquad y = \sin \theta.$$

The cosine and sine of any angle can therefore be calculated from an acute reference triangle made by dropping a perpendicular to the x-axis as shown in Fig. 2.20. The ratios are read from the triangle, and the signs are determined by the quadrant in which the angle lies.

EXAMPLE 1

a) From Fig. 2.21(a):

$$\cos \left(-\frac{\pi}{4}\right) = \frac{\sqrt{2}}{2}, \qquad \sin \left(-\frac{\pi}{4}\right) = -\frac{\sqrt{2}}{2}.$$

b) From Fig. 2.21(b):

$$\cos \left(\frac{2\pi}{3}\right) = -\frac{1}{2}, \qquad \sin \left(\frac{2\pi}{3}\right) = \frac{\sqrt{3}}{2}. \qquad \blacksquare$$

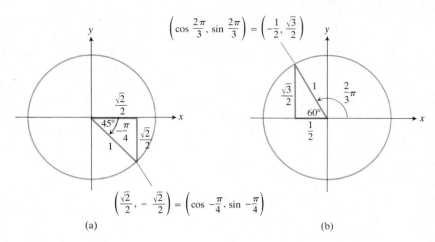

2.21 Calculating the sine and cosine of (a) $-\pi/4$ radians and (b) $2\pi/3$ radians.

(a)   (b)

Radian Measure

The radian measure of angle ABC at the center B of the unit circle in Fig. 2.22 is defined to be the length θ of the circular arc AC. If $A'C'$ is the arc cut by the angle from a second circle centered at B, then the circular sectors $A'B'C'$ and ABC are similar. In particular, their ratios of arc length to radius are equal. In the notation of Fig. 2.22, this means that

$$\frac{\text{Length of arc } A'C'}{r} = \frac{\text{length of arc } AC}{1},$$

or

$$\frac{s}{r} = \frac{\theta}{1} = \theta. \tag{4a}$$

This is true however large or small the radius of the second circle is. Thus for any circle centered at B, the ratio s/r of the length of the intercepted arc to the circle's radius always gives the angle's radian measure.

Equation (4a) is sometimes written in the form

$$s = r\theta, \tag{4b}$$

2.22 The radian measure of the angle centered at B is $s/r = \theta/1 = \theta$.

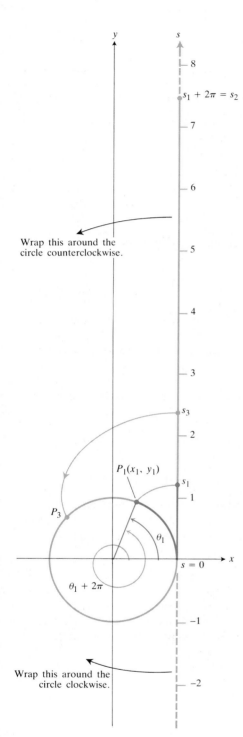

2.23 The s-axis wrapped around the unit circle.

which is handy for calculating the arc length s when the radius r and the angle's radian measure θ are known.

We get a useful interpretation of radian measure by taking $r = 1$ in Eq. (4b). Then the central angle θ, in radians, is just equal to the arc s subtended by θ. We may imagine the circumference of the circle marked off with a scale from which we may read θ. We think of a number scale, like the y-axis shifted one unit to the right, as having been wrapped around the circle. The unit on this number scale is the same as the unit radius. We put the zero of the scale at the place where the initial ray crosses the circle, and then we wrap the positive end of the scale around the circle in the counterclockwise direction and wrap the negative end around in the opposite direction (see Fig. 2.23). Then θ can be read from this curved s-"axis."

Two points on the s-axis that are exactly 2π units apart will map onto the same point on the unit circle when the wrapping is carried out. For example, if $P_1(x_1, y_1)$ is the point to which an arc of length s_1 reaches, then arcs of length $s_1 + 2\pi$, $s_1 + 4\pi$, and so on, will reach exactly the same point after going completely around the circle one or two or more times. Similarly, P_1 will be the image of points on the negative s-axis at $s_1 - 2\pi$, $s_1 - 4\pi$, and so on. Thus, from the wrapped s-axis, we could read

$$\theta_1 = s_1,$$

or

$$\theta_1 + 2\pi, \qquad \theta_1 + 4\pi, \ldots, \qquad \theta_1 - 2\pi, \qquad \theta_1 - 4\pi, \ldots.$$

A unit of arc length $s = 1$ radius subtends a central angle of $57°18'$ (approximately), so

$$1 \text{ radian} \approx 57°18'. \tag{5}$$

We find this and other relations between degree measure and radian measure by using the fact that the full circumference has arc length $s = 2\pi r$ and central angle $360°$. Therefore,

$$360° = 2\pi \text{ radians}, \tag{6a}$$

$$180° = \pi \text{ radians} = 3.14159 \ldots \text{ radians}, \tag{6b}$$

$$\left(\frac{180}{\pi}\right)° = 1 \text{ radian} \approx 57°17'44.8'', \tag{6c}$$

$$1° = \frac{2\pi}{360} = \frac{\pi}{180} \approx 0.01745 \text{ radians}. \tag{6d}$$

It should be emphasized, however, that the radian measure of an angle is dimensionless, since r and s in Eqs. (4a, b) both represent lengths measured in identical units, for instance feet, inches, centimeters, or light-years. Thus $\theta = 2.7$ is to be interpreted as a pure number. The sine and cosine of 2.7 are the x-coordinate and y-coordinate, respectively, of the point $P(x, y)$ on a circle of radius r at the end of an arc of length 2.7 radii. For practical purposes we could convert 2.7 radians to $2.7(180/\pi)$ degrees and say

$$\sin 2.7 = \sin\left[2.7\left(\frac{180}{\pi}\right)°\right] \approx \sin(154°41'55'')$$

$$\approx 0.42738.$$

TABLE 2.1
Values of sin θ, cos θ, and tan θ for selected values of θ

Degrees	-180	-135	-90	-45	0	45	90	135	180
θ (radians)	$-\pi$	$-3\pi/4$	$-\pi/2$	$-\pi/4$	0	$\pi/4$	$\pi/2$	$3\pi/4$	π
sin θ	0	$-\sqrt{2}/2$	-1	$-\sqrt{2}/2$	0	$\sqrt{2}/2$	1	$\sqrt{2}/2$	0
cos θ	-1	$-\sqrt{2}/2$	0	$\sqrt{2}/2$	1	$\sqrt{2}/2$	0	$-\sqrt{2}/2$	-1
tan θ	0	1		-1	0	1		-1	0

Table 2.1 gives the values of the sine, cosine, and tangent functions for selected values of θ.

Periodicity

The mapping from the real numbers s onto points $P(x, y)$ on the unit circle by the wrapping process described above and illustrated in Fig. 2.23 defines the coordinates as functions of s because Eqs. (1) apply, with $\theta = s$ and $r = 1$:

$$x = \cos \theta = \cos s, \qquad y = \sin \theta = \sin s.$$

Because $s + 2\pi$ maps onto the same point that s does, it follows that

$$\cos(\theta + 2\pi) = \cos \theta, \qquad \sin(\theta + 2\pi) = \sin \theta. \tag{7}$$

Equations (7) are **identities;** that is, they are true for all real numbers θ. These identities would be true for $\theta' = \theta + 2\pi$:

$$\cos \theta' = \cos(\theta' - 2\pi) \qquad \text{and} \qquad \sin \theta' = \sin(\theta' - 2\pi). \tag{8}$$

Equations (7) and (8) say that 2π can be added to or subtracted from the domain variable of the sine and cosine functions with no change in the function values. The same process could be repeated any number of times. Consequently,

$$\cos(\theta + 2n\pi) = \cos \theta,$$
$$\sin(\theta + 2n\pi) = \sin \theta, \qquad n = 0, \pm1, \pm2, \ldots . \tag{9}$$

Figure 2.24 shows graphs of the curves $y = \sin x$ and $y = \cos x$. The portion of each curve between 0 and 2π is repeated endlessly to the left and to the right. We also note that the cosine curve is the same as the sine curve shifted to the left an amount $\pi/2$.

2.24 The value of the sine at x is the value of the cosine at $(x - \pi/2)$. That is, $\sin x = \cos(x - \pi/2)$.

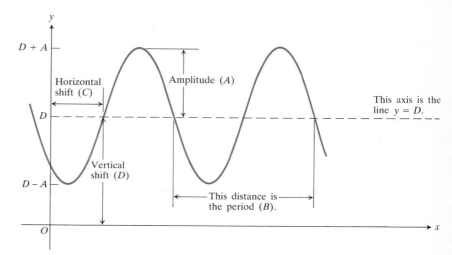

2.25 The general sine curve

$$y = A \sin[(2\pi/B)(x - C)] + D,$$

shown for A, B, C, and D positive.

EXAMPLE 2[†]   The builders of the Trans-Alaska Pipeline used insulated pads to keep the heat from the hot oil in the pipeline from melting the permanently frozen soil beneath. To design the pads, it was necessary to take into account the variation in air temperature throughout the year. The variation was represented in the calculations by a *general sine function* of the form

$$f(x) = A \sin\left[\frac{2\pi}{B}(x - C)\right] + D,$$

where $|A|$ is the *amplitude*, $|B|$ is the *period*, C is the *horizontal shift*, and D is the *vertical shift* (Fig. 2.25).

2.26 Normal mean air temperatures at Fairbanks, Alaska, plotted as data points. The approximating sine function is

$$f(x) = 37 \sin\left[\frac{2\pi}{365}(x - 101)\right] + 25.$$

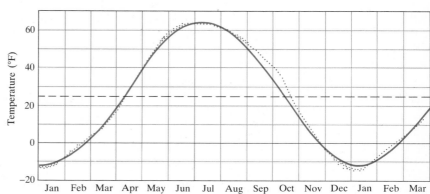

(*Source:* "Is the Curve of Temperature Variation a Sine Curve?" by B. M. Lando and C. A. Lando. *The Mathematics Teacher,* 7:6, Fig. 2, p. 535 (September 1977).)

Figure 2.26 shows how such a function can be used to represent temperature data. The data points in the figure are plots of the mean air temperature for Fairbanks, Alaska, based on records of the National Weather Service from

† From "Is the curve of temperature variation a.sine curve?" by B. M. Lando and C. A. Lando. *The Mathematics Teacher,* September 1977, Vol. 7, No. 6, pp. 534–537.

1941 to 1970. The sine function used to fit the data is

$$f(x) = 37 \sin\left[\frac{2\pi}{365}(x - 101)\right] + 25,$$

where f is temperature in degrees Fahrenheit, and x is the number of the day counting from the beginning of the year.

The fit is remarkably good. ∎

Sum Formulas

Figure 2.27 shows two angles that have equal magnitude but opposite signs. By symmetry, the points where the rays of the angles cross the circle have the same x-coordinate, and their y-coordinates differ only in sign. Hence

$$\sin(-\theta) = -\frac{y}{r} = -\sin\theta, \tag{10a}$$

$$\cos(-\theta) = \frac{x}{r} = \cos\theta. \tag{10b}$$

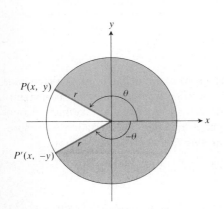

2.27 Angles of opposite sign.

For example,

$$\sin\left(-\frac{\pi}{2}\right) = -\sin\frac{\pi}{2} = -1, \quad \text{and} \quad \cos\left(-\frac{\pi}{2}\right) = \cos\frac{\pi}{2} = 0.$$

It will be helpful, for reasons that will soon be apparent, to review the formulas

$$\sin(A + B) = \sin A \cos B + \cos A \sin B, \tag{11a}$$

$$\cos(A + B) = \cos A \cos B - \sin A \sin B. \tag{11b}$$

These formulas are derived in Appendix 4.

When needed, the formulas for the sine and cosine of the difference of two angles can be obtained from Eqs. (11a) and (11b) by replacing B with $-B$. Since

$$\sin(-B) = -\sin B \quad \text{and} \quad \cos(-B) = \cos B,$$

we get

$$\sin(A - B) = \sin A \cos B - \cos A \sin B, \tag{11c}$$

$$\cos(A - B) = \cos A \cos B + \sin A \sin B. \tag{11d}$$

Double-Angle Formulas

From Eqs. (11a) and (11b) with $A = B = \theta$, we have

$$\sin 2\theta = 2 \sin\theta \cos\theta, \tag{12a}$$

$$\cos 2\theta = \cos^2\theta - \sin^2\theta. \tag{12b}$$

Starting now with the equations

$$\cos^2\theta + \sin^2\theta = 1, \qquad \cos^2\theta - \sin^2\theta = \cos 2\theta,$$

we add to get

$$2\cos^2\theta = 1 + \cos 2\theta,$$

subtract to get

$$2 \sin^2\theta = 1 - \cos 2\theta,$$

and divide by 2 to get

$$\cos^2\theta = \frac{1 + \cos 2\theta}{2} \tag{13a}$$

$$\sin^2\theta = \frac{1 - \cos 2\theta}{2} \tag{13b}$$

A Useful Limit

By turning Eq. (13b) around and dividing through by θ, we get

$$\frac{1 - \cos 2\theta}{2\theta} = \frac{\sin^2\theta}{\theta}, \tag{14}$$

so that

$$\lim_{\theta \to 0} \frac{1 - \cos 2\theta}{2\theta} = \lim_{\theta \to 0} \frac{\sin^2\theta}{\theta} = \lim_{\theta \to 0} \left(\frac{\sin \theta}{\theta} \cdot \sin \theta \right)$$

$$= \lim_{\theta \to 0} \left(\frac{\sin \theta}{\theta} \right) \cdot \lim_{\theta \to 0} \sin \theta = 1 \cdot 0 = 0.$$

The substitution $h = 2\theta$ gives the following simpler expression:

$$\lim_{h \to 0} \frac{1 - \cos h}{h} = 0. \tag{15}$$

We shall use this limit in calculating the derivative of $y = \sin x$ in Article 2.7.

1. *Definitions and Basic Identities*

$$\sin \theta = \frac{y}{r} = \frac{1}{\csc \theta}$$

$$\cos \theta = \frac{x}{r} = \frac{1}{\sec \theta}$$

$$\tan \theta = \frac{y}{x} = \frac{1}{\cot \theta}$$

$$\sin(-\theta) = -\sin \theta, \quad \cos(-\theta) = \cos \theta$$

$$\sin^2\theta + \cos^2\theta = 1$$

$$\sec^2\theta = 1 + \tan^2\theta$$

$$\csc^2\theta = 1 + \cot^2\theta$$

2. *Double-angle formulas*

$$\sin 2\theta = 2 \sin \theta \cos \theta, \quad \cos 2\theta = \cos^2\theta - \sin^2\theta$$

$$\cos^2\theta = \frac{1 + \cos 2\theta}{2}, \quad \sin^2\theta = \frac{1 - \cos 2\theta}{2}$$

3. *Sum formulas*

$$\sin (A + B) = \sin A \cos B + \cos A \sin B$$

$$\cos (A + B) = \cos A \cos B - \sin A \sin B$$

4. *Shift formulas*

$$\sin \left(A - \frac{\pi}{2}\right) = -\cos A, \quad \cos \left(A - \frac{\pi}{2}\right) = \sin A$$

$$\sin \left(A + \frac{\pi}{2}\right) = \cos A, \quad \cos \left(A + \frac{\pi}{2}\right) = -\sin A$$

A more extensive list of trigonometric formulas appears in Appendix 3.

PROBLEMS

Hints for graphing sine and cosine functions:
  i) Find the amplitude, period, and shift.
  ii) Draw the curve roughly.
  iii) Draw and label the axes; then finish the sketch.

Remember: *Angles in radians.*

Graph the equations in Problems 1–20.

1. $y = 2 \sin x$

2. $y = 5 \sin 2x$

3. $y = \sin(-x)$

4. $y = \sin 2\pi x$

5. $y = 2 \cos 3x$

6. $y = \tan(x/3)$

7. $y = \sin(x + (\pi/2))$

8. $y = \cos(x - (\pi/2))$

9. $y = |\cos x|$

10. $y = \frac{1}{2}(|\cos x| + \cos x)$

11. $y = \frac{1}{2}(|\sin x| - \sin x)$

12. $y = \sin^2 x$

13. $y = \cos^2 x$

14. $y = \sin x + \cos x$

15. $y = \sin x - \cos x$

16. $y = \cos 2\pi(x + 1)$

17. $y = 2 \cos(4x - 2\pi), \quad -\pi \le x \le \pi$

18. $y = \sin(x - (\pi/4)), \quad -\pi \le x \le \pi$

19. $y = \sec x$ and $y = \cos x$ together, $\quad -2\pi \le x \le 2\pi$

20. $y = \sin x$, $y = \cos x$, and $y = \tan x$ in a common graph, $-2\pi \le x \le 2\pi$

21. Find the (a) amplitude, (b) period, (c) horizontal shift, and (d) vertical shift of the general sine function

$$f(x) = 37 \sin\left[\frac{2\pi}{365}(x - 101)\right] + 25.$$

22. Use the equation in Problem 21 to approximate the answers to the following questions about the temperatures in Fairbanks, Alaska, shown in Fig. 2.26. Assume that the year has 365 days.
  a) What is the highest mean daily temperature shown?
  b) What is the lowest mean daily temperature shown?
  c) What is the average of the highest and lowest mean daily temperatures shown? Why is this average the vertical shift of the function?

Find the limits in Problems 23–26.

23. $\displaystyle\lim_{h\to 0} \frac{(\sin h)(1 - \cos h)}{h^2}$

24. $\displaystyle\lim_{x\to 0} \frac{1 - \cos x}{x^2}$

25. $\displaystyle\lim_{x\to 0} \frac{1 - \cos x}{\sin x}$
  (*Hint:* Divide numerator and denominator by x.)

26. $\displaystyle\lim_{x\to 0} x \cot x$

27. Take $B = A$ in Eq. (11d). Does the result agree with something you already know?

28. *Even and odd functions.* A function $f(\theta)$ is an *even* function of θ if $f(-\theta) = f(\theta)$ for all θ, and it is an *odd* function of θ if $f(-\theta) = -f(\theta)$ for all θ. Which of the six basic trigonometric functions are even and which are odd?

Use the formulas for $\sin(A + B)$ and $\cos(A + B)$ to derive the identities in Problems 29–34.

29. $\sin(x - (\pi/2)) = -\cos x$

30. $\sin(x + (\pi/2)) = \cos x$

31. $\cos(x - (\pi/2)) = \sin x$

32. $\cos(x + (\pi/2)) = -\sin x$

33. $\sin(\pi - x) = \sin x$

34. $\cos(\pi - x) = -\cos x$

35. *The tangent formulas.* The standard formula for the tangent of the sum of two angles is

$$\tan(A + B) = \frac{\tan A + \tan B}{1 - \tan A \tan B}.$$

a) To derive the formula, write $\tan(A + B)$ as $\sin(A + B)/\cos(A + B)$ and apply formulas (11a) and (11b).

b) Find the analogous formula for $\tan(A - B)$. (*Hint:* $\tan(-B) = -\tan B$.)

TOOLKIT PROGRAMS

| Function Evaluator | Name That Function |
| Limit Problems | Super * Grapher |

2.7
Derivatives of Trigonometric Functions

In this article, we calculate the derivatives of the trigonometric functions. We first differentiate $\sin x$ by a straightforward application of the definition of derivative. Then we use standard differentiation rules and trigonometric identities to obtain the derivatives of the other trigonometric functions.

The Derivative of the Sine

From the definition of derivative, we know that the derivative of $y = \sin x$ with respect to x is the limit

$$\frac{dy}{dx} = \lim_{h \to 0} \frac{\sin(x + h) - \sin x}{h}$$

To calculate this limit, we use three results from our earlier work:

1. $\sin(x + h) = \sin x \cos h + \cos x \sin h$ (Article 2.6, Eq. 11a)

2. $\lim_{h \to 0} \dfrac{\sin h}{h} = 1$ (Article 1.9, Eq. 13)

3. $\lim_{h \to 0} \dfrac{\cos h - 1}{h} = 0$ (Article 2.6, Eq. 15)

Then, taking all limits as $h \to 0$, we have

$$\frac{dy}{dx} = \lim \frac{\sin(x + h) - \sin x}{h}$$

$$= \lim \frac{\sin x \cos h + \cos x \sin h - \sin x}{h}$$

$$= \lim \frac{\sin x(\cos h - 1) + \cos x \sin h}{h}$$

$$= \lim \left(\sin x \frac{\cos h - 1}{h} \right) + \lim \left(\cos x \frac{\sin h}{h} \right)$$

$$= \sin x \lim \frac{\cos h - 1}{h} + \cos x \lim \frac{\sin h}{h}$$

$$= \sin x \cdot 0 + \cos x \cdot 1$$

$$= \cos x.$$

The derivative of $y = \sin x$ with respect to x is

$$\frac{d}{dx}\sin x = \cos x. \tag{1}$$

If u is a differentiable function of x, we can apply the Chain Rule,

$$\frac{dy}{dx} = \frac{dy}{du}\frac{du}{dx},$$

to $y = \sin u$ with the following result:

$$\frac{d}{dx}\sin u = \cos u\frac{du}{dx}. \tag{2}$$

EXAMPLE 1

a) $\dfrac{d}{dx}\sin 2x = \cos 2x\dfrac{d}{dx}(2x)$

$\qquad = 2\cos 2x$

b) $\dfrac{d}{dx}\sin x^5 = \cos x^5\dfrac{d}{dx}(x^5)$

$\qquad = 5x^4\cos x^5$

c) $\dfrac{d}{dx}\sin^5 x = 5\sin^4 x\dfrac{d}{dx}(\sin x)$

$\qquad = 5\sin^4 x\cos x$ ■

The answer to the question "Why do we use radian measure in calculus?" is contained in the argument that the derivative of the sine is the cosine. The argument requires that

$$\lim_{h\to 0}\frac{\sin h}{h} = 1.$$

The limit is 1 *only* if h is measured in radians.

EXAMPLE 2 Find dy/dx by implicit differentiation if

$$xy + \sin y = 0.$$

Solution We differentiate both sides of the equation, treating y as a differentiable function of x:

$$x\frac{dy}{dx} + y + \cos y\frac{dy}{dx} = 0$$

$$(x + \cos y)\frac{dy}{dx} + y = 0$$

$$\frac{dy}{dx} = -\frac{y}{x + \cos y}. \qquad ■$$

The Derivative of the Cosine

To obtain a formula for the derivative of cos u, we use the identities

$$\cos u = \sin\left(\frac{\pi}{2} - u\right), \qquad \sin u = \cos\left(\frac{\pi}{2} - u\right)$$

in the following way:

$$\frac{d}{dx}\cos u = \frac{d}{dx}\sin\left(\frac{\pi}{2} - u\right)$$

$$= \cos\left(\frac{\pi}{2} - u\right)\frac{d}{dx}\left(\frac{\pi}{2} - u\right)$$

$$= \sin u\left(-\frac{du}{dx}\right).$$

Combining these equalities gives

$$\frac{d}{dx}(\cos u) = -\sin u \frac{du}{dx}. \tag{3}$$

The derivative of the cosine of a differentiable function is minus the sine of the function times the derivative of the function.

EXAMPLE 3

a) $\dfrac{d}{dx}\cos 3x = -\sin 3x \dfrac{d}{dx}(3x) = -3 \sin 3x$

b) $\dfrac{d}{dx}\cos^2 3x = 2 \cos 3x \dfrac{d}{dx}\cos 3x$

$\qquad = 2 \cos 3x(-3 \sin 3x)$

$\qquad = -6 \sin 3x \cos 3x$ ∎

NOTE: There is more than one correct way to write the answer in Example 3(b). For example, we can use the identity

$$2 \sin \theta \cos \theta = \sin 2\theta$$

with $\theta = 3x$ to write the answer as

$$-6 \sin 3x \cos 3x = -3 \sin 6x.$$

If you find that your answer to a differentiation problem in trigonometry differs from someone else's, you may both be right.

Derivatives of Other Trigonometric Functions

Since sin x and cos x are differentiable functions of x, the functions

$$\tan x = \frac{\sin x}{\cos x}, \qquad \cot x = \frac{\cos x}{\sin x},$$

$$\sec x = \frac{1}{\cos x}, \qquad \csc x = \frac{1}{\sin x}, \tag{4}$$

are differentiable at every value of x at which they are defined. Their derivatives can be calculated from the Quotient Rule.

EXAMPLE 4

a) Find dy/dx if $y = \tan x$.

b) Find dy/dx if $y = \tan u$ and u is a differentiable function of x.

Solution

a)

$$\frac{d}{dx} \tan x = \frac{d}{dx}\left(\frac{\sin x}{\cos x}\right)$$

$$= \frac{\cos x \dfrac{d}{dx}(\sin x) - \sin x \dfrac{d}{dx}(\cos x)}{\cos^2 x}$$

$$= \frac{\cos x \cos x - \sin x(-\sin x)}{\cos^2 x}$$

$$= \frac{\cos^2 x + \sin^2 x}{\cos^2 x} = \frac{1}{\cos^2 x}$$

$$= \sec^2 x$$

Thus,

$$\frac{d}{dx}\tan x = \sec^2 x. \tag{5}$$

b) If u is a differentiable function of x, then we can apply the Chain Rule,

$$\frac{dy}{dx} = \frac{dy}{du}\frac{du}{dx},$$

to $y = \tan u$ to get

$$\frac{d}{dx}\tan u = \sec^2 u \frac{du}{dx}. \qquad\blacksquare \tag{6}$$

If u is a differentiable function of x, the differentiation rules for the other three functions in Eqs. (4) can be derived in much the same way that the derivative of the tangent was derived in Example 4. The results are

$$\frac{d}{dx}\sec u = \sec u \tan u \frac{du}{dx}, \tag{7}$$

$$\frac{d}{dx}\csc u = -\csc u \cot u \frac{du}{dx}, \tag{8}$$

$$\frac{d}{dx}\cot u = -\csc^2 u \frac{du}{dx}. \tag{9}$$

In differentiation problems you can always get along by converting all the trigonometric functions to sines and cosines before differentiating.

EXAMPLE 5 Find dy/dx if $y = \sec^2 5x = (\cos 5x)^{-2}$.

Solution METHOD 1: We apply the Power Rule,

$$\frac{d(u^n)}{dx} = nu^{n-1}\frac{du}{dx},$$

with $u = \cos 5x$ and $n = -2$:

$$\frac{dy}{dx} = -2(\cos 5x)^{-3}\frac{d(\cos 5x)}{dx} = (-2\sec^3 5x)\left(-\sin 5x\frac{d(5x)}{dx}\right)$$

$$= 10\sec^3 5x \sin 5x. \tag{10}$$

METHOD 2: We use Eq. (7) and the Power Rule to get

$$\frac{d}{dx}\sec^2 5x = 2(\sec 5x)^1 \cdot \frac{d}{dx}(\sec 5x) \tag{11}$$

$$= 2\sec 5x \cdot \sec 5x \tan 5x \cdot \frac{d}{dx}(5x) \tag{12}$$

$$= 10\sec^2 5x \tan 5x. \qquad \blacksquare \tag{13}$$

Why do the answers in Eqs. (10) and (13) differ? Here is another example in which answers that look different are both correct. If we change the $\tan 5x$ in Eq. (13) to

$$\tan 5x = \frac{\sin 5x}{\cos 5x},$$

we find that

$$10\sec^2 5x \tan 5x = 10\sec^2 5x\frac{\sin 5x}{\cos 5x} = 10\sec^2 5x \cdot \frac{1}{\cos 5x} \cdot \sin 5x$$

$$= 10\sec^2 5x \cdot \sec 5x \cdot \sin 5x$$

$$= 10\sec^3 5x \sin 5x,$$

which is the answer in Eq. (10).

EXAMPLE 6 Find dy/dx if $y = \tan\sqrt{3x}$.

Solution

$$\frac{dy}{dx} = \sec^2\sqrt{3x} \cdot \frac{d}{dx}(\sqrt{3x}) \qquad \text{(Chain Rule)}$$

$$= \sec^2\sqrt{3x} \cdot \frac{1}{2\sqrt{3x}}\frac{d}{dx}(3x) \quad \text{(Chain Rule again)}$$

$$= \sec^2\sqrt{3x}\frac{1}{2\sqrt{3x}} \cdot 3$$

$$= \frac{3\sec^2\sqrt{3x}}{2\sqrt{3x}}. \qquad \blacksquare$$

Continuity

Since the six basic trigonometric functions are differentiable, they are automatically continuous by virtue of Theorem 5 in Article 1.11. That is,

$$\lim_{x \to a} f(x) = f(a)$$

for the basic trigonometric functions whenever $f(a)$ is defined. This means that we can calculate the limits of most combinations of trigonometric functions as $x \to a$ by evaluating them at $x = a$.

Linearizations

The standard linearizations of the trigonometric functions come from Eq. (3) in Article 2.4, namely

$$L(x) = f(a) + f'(a)(x - a).$$

EXAMPLE 7   Find the linearization of $f(x) = \tan x$ at $x = 0$.

Solution   We use the equation

$$L(x) = f(a) + f'(a)(x - a)$$

with $f(x) = \tan x$ and $a = 0$. Since

$$f(0) = \tan(0) = 0, \qquad f'(0) = \sec^2(0) = 1,$$

we have $L(x) = 0 + 1(x - 0) = x$. Near $x = 0$,

$$\tan x \approx x.$$ ■

EXAMPLE 8   Find the linearization of $f(x) = \cos x$ at $x = \pi/2$.

Solution   We use the equation

$$L(x) = f(a) + f'(a)(x - a)$$

with $f(x) = \cos x$ and $a = \pi/2$. Since

$$f(\pi/2) = \cos(\pi/2) = 0, \qquad f'(\pi/2) = -\sin(\pi/2) = -1,$$

the linearization is

$$L(x) = 0 - 1\left(x - \frac{\pi}{2}\right) = -x + \frac{\pi}{2}.$$

See Fig. 2.28. ■

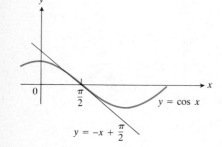

2.28 The graph of $y = \cos x$ and its linearization at $x = \pi/2$. Near $x = \pi/2$, $\cos x \approx -x + (\pi/2)$.

In Problem 54 you will be asked to derive the linearizations of $\sin x$ and $\cos x$ at $x = 0$.

Linearizations at $x = 0$

Function $f(x)$	Linearization $L(x)$
$\sin x$	x
$\cos x$	1
$\tan x$	x

Derivatives of the Basic Trigonometric Functions

If u is a differentiable function of x, then:

1. $\dfrac{d}{dx}\sin u = \cos u \dfrac{du}{dx}$,

2. $\dfrac{d}{dx}\cos u = -\sin u \dfrac{du}{dx}$,

3. $\dfrac{d}{dx}\tan u = \sec^2 u \dfrac{du}{dx}$,

4. $\dfrac{d}{dx}\sec u = \sec u \tan u \dfrac{du}{dx}$,

5. $\dfrac{d}{dx}\csc u = -\csc u \cot u \dfrac{du}{dx}$,

6. $\dfrac{d}{dx}\cot u = -\csc^2 u \dfrac{du}{dx}$.

PROBLEMS

In Problems 1–36, find dy/dx.

1. $y = \sin(x + 1)$
2. $y = -\cos x$
3. $y = \sin(x/2)$
4. $y = \sin(-x)$
5. $y = \cos 5x$
6. $y = \cos(-x)$
7. $y = \cos(-2x)$
8. $y = \sin 7x$
9. $y = \sin(3x + 4)$
10. $y = \cos(2 - x)$
11. $y = x \sin x$
12. $y = \sin 5(x - 1)$
13. $y = x \sin x + \cos x$
14. $y = \dfrac{1}{\sin x}$
15. $y = \dfrac{1}{\cos x}$
16. $y = \dfrac{\sin x}{\cos x}$
17. $y = \sec(x - 1)$
18. $y = \cot(-x)$
19. $y = \sec(1 - x)$
20. $y = \dfrac{2}{\cos 3x}$
21. $y = \tan 2x$
22. $y = \cos(ax + b)$
23. $y = \sin^2 x$
24. $y = \sin^2 x + \cos^2 x$
25. $y = \cos^2 5x$
26. $y = \cot^2 x$
27. $y = \tan(5x - 1)$
28. $y = \sin x - x \cos x$
29. $y = 2 \sin x \cos x$
30. $y = \sec(x^2 + 1)$
31. $y = \sqrt{2 + \cos 2x}$
32. $y = \sin(1 - x^2)$
33. $y = \cos\sqrt{x}$
34. $y = \sec^2 x - \tan^2 x$
35. $y = \sqrt{\dfrac{1 + \cos 2x}{2}}$ (*Hint:* Use a half-angle formula first.)
36. $y = \sin^2 x^2$

Assume that each of the equations in Problems 37–41 defines y as a differentiable function of x. Find dy/dx by implicit differentiation.

37. $x = \tan y$
38. $x = \sin y$
39. $y^2 = \sin^4 2x + \cos^4 2x$
40. $x + \sin y = xy$
41. $x + \tan(xy) = 0$

42. Assume that the equation $2xy + \pi \sin y = 2\pi$ defines y as a differentiable function of x. Find dy/dx when $x = 1$ and $y = \pi/2$.

43. Find an equation for the tangent to the curve $x \sin 2y = y \cos 2x$ at the point $(\pi/4, \pi/2)$.

Find the limits in Problems 44–51.

44. $\lim\limits_{x \to 2} \sin\left(\dfrac{1}{x} - \dfrac{1}{2}\right)$
45. $\lim\limits_{x \to \pi/4} \dfrac{\sin x}{\cos x}$
46. $\lim\limits_{x \to -\pi} \cos^2 x$
47. $\lim\limits_{x \to \pi} \sec(1 + \cos x)$
48. $\lim\limits_{x \to 0} (\sec x + \tan x)$
49. $\lim\limits_{x \to 0} x \csc x$
50. $\lim\limits_{h \to 0} \dfrac{\sin(a + h) - \sin a}{h}$
51. $\lim\limits_{h \to 0} \dfrac{\cos(a + h) + \cos a}{h}$

52. Find an equation for the tangent to the curve $y = \sin mx$ at $x = 0$.

53. Graph $y = \tan x$ and its linearization $y = x$ together for $-\pi/4 \le x \le \pi/4$.

54. Find the linearization $L(x)$ of each function $f(x)$ at the given point. Illustrate with a sketch.

a) $f(x) = \sin x$ at $x = 0$
b) $f(x) = \sin x$ at $x = \pi$
c) $f(x) = \cos x$ at $x = 0$
d) $f(x) = \cos x$ at $x = -\pi/2$
e) $f(x) = \tan x$ at $x = \pi/4$
f) $f(x) = \sec x$ at $x = \pi/4$
g) $f(x) = \tan x$ at $x = -\pi/4$
h) $f(x) = \sec x$ at $x = -\pi/4$

55. Is there a value of b that makes

$$f(x) = \begin{cases} x + b & \text{for } x < 0, \\ \cos x & \text{for } x \geq 0, \end{cases}$$

continuous at $x = 0$? If so, what is it? If not, why not?

56. Figure 2.29 shows a boat 1 km offshore, sweeping the shore with a searchlight. The light turns at a constant rate (angular velocity) $d\theta/dt = -3/5$ radians per second.

a) Express x (see Fig. 2.29) in terms of θ.
b) Differentiate both sides of the equation you obtained in (a) with respect to t. Substitute $d\theta/dt = -3/5$. This will express dx/dt (the rate at which the light moves along the shore) as a function of θ.
c) How fast is the light moving along the shore when it reaches point A?
d) How many revolutions per minute is 0.6 radians per second?

57. Find the linearization of $f(x) = \sqrt{1 + x} + \sin x$ at $x = 0$. How is it related to the individual linearizations of $\sqrt{1 + x}$ and $\sin x$?

58. Carry out the following steps to estimate the solution of

$$2 \cos x = \sqrt{1 + x}.$$

a) Let $f(x) = 2 \cos x - \sqrt{1 + x}$. Find $f(0) > 0$ and $f(\pi/2) < 0$ to show that $f(x)$ has a zero between 0 and $\pi/2$.
b) Find the linearizations of $\cos x$ at $x = \pi/4$ and $\sqrt{1 + x}$ at $x = 0.69$.
c) **CALCULATOR** To estimate the solution of the original equation, replace $\cos x$ and $\sqrt{1 + x}$ by their linearizations from (b) and solve the resulting linear equation for x. Check your estimate in the original equation.

59. Derive Eq. (7) by writing $\sec u = 1/\cos u$ and differentiating with respect to x.

60. Derive Eq. (8) by writing $\csc u = 1/\sin u$ and differentiating with respect to x.

61. Derive Eq. (9) by writing $\cot u = \cos u/\sin u$ and differentiating with respect to x.

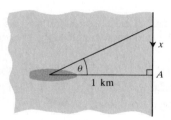

2.29   The boat in Problem 56.

TOOLKIT PROGRAMS

| Derivative Grapher | Super ∗ Grapher |
| Function Evaluator | |

2.8
Parametric Equations

When the path of a particle moving in the plane is not the graph of a function, we cannot describe the path by expressing y directly in terms of x. To avoid this difficulty, we express the particle's coordinates as functions of a third variable by a pair of equations

$$x = f(t), \qquad y = g(t). \tag{1}$$

Equations like these are called **parametric equations** for x and y, and the variable t is called a **parameter.** In many applications, t denotes time, but it might instead denote an angle (as in some of the following examples) or the distance a particle has traveled along the path from its starting point (as it sometimes will when we study motion again in Chapter 14).

In this article we identify a few of the curves described by parametric equations and explore the relationship between the slopes of these curves and the derivatives of the functions that define x and y. There are two practical

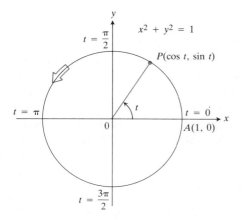

2.30 The equations $x = \cos t$ and $y = \sin t$ describe motion on the unit circle $x^2 + y^2 = 1$. The arrow shows the direction of increasing t.

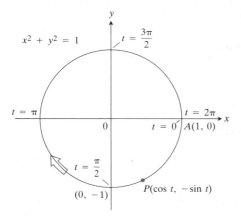

2.31 The point $P(\cos t, -\sin t)$ moves clockwise as t increases.

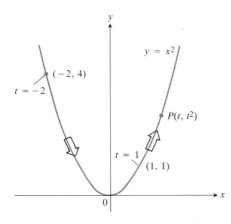

2.32 The arrows show how P moves as t increases.

reasons for introducing parametric equations at this stage. First, the equations used in practice often involve trigonometric functions. Second, we need the Chain Rule to calculate dy/dx and d^2y/dx^2 from the parametric representations of x and y.

EXAMPLE 1 The equations

$$x = \cos t, \qquad y = \sin t, \qquad 0 \le t \le 2\pi,$$

describe the position $P(x, y)$ of a particle that moves counterclockwise around the unit circle $x^2 + y^2 = 1$ as t increases (Fig. 2.30).

Discussion Since $\cos^2 t + \sin^2 t = 1$ for every value of t, the point $P(x, y) = (\cos t, \sin t)$ lies on the circle $x^2 + y^2 = 1$. The parameter t is the radian measure of the angle that radius OP makes with the positive x-axis. The particle starts at $A(1, 0)$, moves up and to the left as t approaches $\pi/2$, and continues around the circle to stop again at $A(1, 0)$ when $t = 2\pi$.

If we replace the interval $0 \le t \le 2\pi$ by an arbitrary interval $t_0 \le t \le t_0 + 2\pi$ of length 2π, the particle will start at $(\cos t_0, \sin t_0)$ and go around once counterclockwise to stop again at $(\cos(t_0 + 2\pi), \sin(t_0 + 2\pi)) = (\cos t_0, \sin t_0)$. ∎

EXAMPLE 2 The equations

$$x = \cos t, \qquad y = \sin t, \qquad 0 \le t \le \pi,$$

describe the position $P(x, y)$ of a particle that traces the upper half of the unit circle counterclockwise from $A(1, 0)$ to $B(-1, 0)$. The motion begins like the motion in Example 1 but stops halfway around. ∎

EXAMPLE 3 The equations

$$x = \cos t, \qquad y = -\sin t, \qquad 0 \le t \le 2\pi,$$

describe the position $P(x, y)$ of a particle moving *clockwise* around the circle $x^2 + y^2 = 1$. The particle starts at $A(1, 0)$, but y initially decreases as t increases. When $t = \pi/2$, for example,

$$P(x, y) = \left(\cos\frac{\pi}{2}, -\sin\frac{\pi}{2}\right) = (0, -1).$$

See Fig. 2.31. ∎

EXAMPLE 4 The equations

$$x = t, \qquad y = t^2, \qquad -\infty < t < \infty,$$

describe the position $P(x, y)$ of a particle on the parabola $y = x^2$. If we eliminate t between the equations for x and y, we find that $y = x^2$. Thus, the coordinates of P satisfy the Cartesian equation $y = x^2$ at every time t. As t increases from negative to positive values, the particle comes down the left side, passes the origin, and then moves up to the right. See Fig. 2.32. ∎

As Example 4 illustrates, the graph of any function $y = f(x)$ has the automatic parameterization $x = x$, $y = f(x)$. This is so simple that we usually don't use it, but the point of view is occasionally helpful.

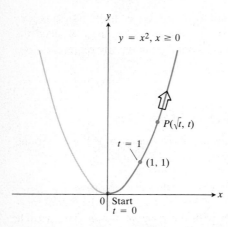

2.33 The equations $x = \sqrt{t}$, $y = t$ describe the motion of a particle that traces the right-hand half of the parabola $y = x^2$.

EXAMPLE 5 The equations

$$x = \sqrt{t}, \qquad y = t, \qquad t \geq 0,$$

describe the motion of a particle along the right-hand half of the parabola $y = x^2$. If we eliminate t by combining the equations for x and y, we see that $y = t = (\sqrt{t})^2 = x^2$. Thus, the point $P(\sqrt{t}, t)$ lies on the parabola $y = x^2$ for every $t \geq 0$. However, the particle's x-coordinate $x = \sqrt{t}$ is never negative, so the motion covers only the right-hand portion of the parabola. The particle starts at the origin and moves into the first quadrant as t increases. See Fig. 2.33. ∎

The Parametric Formula for dy/dx

Suppose that $x = f(t)$ and $y = g(t)$ are differentiable functions of t and that dx/dt is never zero in the interval of t-values with which we are working. Then, for reasons we shall touch on briefly in Article 6.1, the equation $x = f(t)$ that defines x as a differentiable function of t can be reversed to define t as a differentiable function of x. Call this function $t = h(x)$ for the moment. With t a differentiable function of x, and y a differentiable function of t, the composite $y = g \circ h$ gives y as a differentiable function of x, its value at any particular x being $y = g(h(x))$.

How are the derivatives of these functions related? According to the Chain Rule,

$$\frac{dy}{dt} = \frac{dy}{dx}\frac{dx}{dt}. \tag{2}$$

Since $dx/dt \neq 0$, we can divide through by dx/dt to solve for dy/dx:

$$\frac{dy}{dx} = \frac{dy/dt}{dx/dt}. \tag{3}$$

This is the equation we use to calculate dy/dx when x and y are given parametrically. It says: To calculate dy/dx, divide dy/dt by dx/dt. Equation (3) will automatically express dy/dx in terms of t, so we can find the slope of a moving particle's path at every value of t.

EXAMPLE 6 If

$$x = 2t + 3, \qquad y = t^2 - 1,$$

find the value of dy/dx at $t = 6$. Also, find dy/dx as a function of x.

Solution Equation (3) gives dy/dx as a function of t:

$$\frac{dy}{dx} = \frac{dy/dt}{dx/dt} = \frac{2t}{2} = t = \frac{x-3}{2}.$$

When $t = 6$, $dy/dx = 6$. ∎

EXAMPLE 7 If a is a positive constant and $x = a \cos t$, $y = a \sin t$, then

$$\frac{dy}{dx} = \frac{dy/dt}{dx/dt} = \frac{a \cos t}{-a \sin t} = -\frac{x}{y}.$$

We can check the result by combining the equations $x = a \cos t$, $y = a \sin t$ to produce a Cartesian equation satisfied by x and y,

$$x^2 + y^2 = a^2\cos^2 t + a^2\sin^2 t = a^2(\cos^2 t + \sin^2 t) = a^2,$$

and by differentiating with respect to x:

$$x^2 + y^2 = a^2$$

$$2x + 2y\frac{dy}{dx} = 0$$

$$\frac{dy}{dx} = -\frac{x}{y}.$$

∎

The Parametric Formula for d^2y/dx^2

The second derivative of y with respect to x is obtained by differentiating y with respect to x twice:

$$\frac{d^2y}{dx^2} = \frac{d}{dx}\left[\frac{d}{dx}(y)\right].$$

If the parametric equations

$$x = f(t), \qquad y = g(t) = g(h(x))$$

define y as a twice-differentiable function of x, then we may calculate

$$\frac{dy}{dx} = y' = \frac{dy/dt}{dx/dt}$$

from Eq. (3) and calculate d^2y/dx^2 from the equation

$$\frac{d^2y}{dx^2} = \frac{dy'}{dx} = \frac{dy'/dt}{dx/dt}, \tag{4}$$

which we can obtain from Eq. (3) with y' in place of y.

Equation (4) says that to find the second derivative of y with respect to x, we take the following steps:

1. Express $y' = dy/dx$ in terms of t.
2. Differentiate y' with respect to t.
3. Divide the result by dx/dt.

EXAMPLE 8 Find d^2y/dx^2 if $x = t - t^2$ and $y = t - t^3$.

Solution

$$y' = \frac{dy}{dx} = \frac{dy/dt}{dx/dt} = \frac{1 - 3t^2}{1 - 2t},$$

$$\frac{d^2y}{dx^2} = \frac{dy'/dt}{dx/dt} = \frac{\dfrac{d}{dt}\left(\dfrac{1 - 3t^2}{1 - 2t}\right)}{(1 - 2t)}$$

$$= \frac{(1 - 2t) \cdot (-6t) - (1 - 3t^2) \cdot (-2)}{(1 - 2t)^3} = \frac{2 - 6t + 6t^2}{(1 - 2t)^3}$$

∎

PROBLEMS

The parametric equations in Problems 1–8 give the position $P(x, y)$ of a particle moving in the plane. Identify the path traced by the particle. Say where the particle starts and stops and in what direction the particle moves as t increases.

1. $x = \cos t, \quad y = \sin t, \quad 0 \le t \le 2\pi$

2. $x = \cos t, \quad y = \sin t, \quad 0 \le t \le \pi$

3. $x = \cos 2\pi t, \quad y = \sin 2\pi t, \quad 0 \le t \le 1$

4. $x = \cos(\pi - t), \quad y = \sin(\pi - t), \quad 0 \le t \le \pi$

5. $x = 3 \cos t, \quad y = 3 \sin t, \quad 0 \le t \le 2\pi$

6. $x = \cos t, \quad y = -\sin t, \quad 0 \le t \le 2\pi$

7. $x = \cos^2 t, \quad y = \sin^2 t, \quad 0 \le t \le \pi/2$

8. $x = \tan^2 t, \quad y = \sec^2 t, \quad -\pi/3 \le t \le \pi/3$

9. Find parametric equations for the motion of a particle that traces the circle $x^2 + y^2 = 4$ once in the (a) clockwise and (b) counterclockwise direction. Use the parameter interval $0 \le t \le 2\pi$ in each case.

10. Repeat Problem 9 for the parameter domains
    a) $0 \le t \le \pi$,   b) $0 \le t \le 1$.

The parametric equations in Problems 11–20 give the position $P(x, y)$ of a particle moving in the plane. Eliminate t between the two equations to find an equation of the form $y = f(x)$ satisfied by the coordinates of P. Then graph the curve traced by the particle (it may be only a portion of the graph of $y = f(x)$). Indicate the direction in which the particle moves as t increases.

11. $x = 2t - 5, \quad y = 4t - 7$

12. $x = 1 - t, \quad y = 1 + t$

13. $x = 3t, \quad y = 9t^2$

14. $x = t, \quad y = \sqrt{1 - t^2}, \quad -1 \le t \le 1$

15. $x = t, \quad y = \sqrt{1 - t^2}, \quad 0 \le t \le 1$

16. $x = -\sqrt{t}, \quad y = t, \quad t \ge 0$

17. $x = t, \quad y = \sqrt{t}, \quad t \ge 0$

18. $x = t, \quad y = 1 - t, \quad 0 \le t \le 1$

19. $x = 3t, \quad y = 2 - 2t, \quad 0 \le t \le 1$

20. $x = \sqrt{t}, \quad y = \sqrt{t}, \quad t \ge 0$

The parametric equations in Problems 21–28 give the position $P(x, y)$ of a particle in the plane at time t. In each case, eliminate t from the two equations to find a Cartesian coordinate equation satisfied by the coordinates of P. Then calculate dy/dx, dy/dt, and dx/dt and verify that they satisfy the Chain Rule in Eq. (3).

21. $x = 2t, \quad y = 1 + t$ 22. $x = 3t + 1, \quad y = t^2$

23. $x = 5 \cos t, \quad y = 5 \sin t$ 24. $x = t, \quad y = 1/t$

25. $x = t^2 - \pi/2, \quad y = \sin(t^2)$ 26. $x = t^2, \quad y = t^3$

27. $x = \cos t, \quad y = 1 + \sin t$ 28. $x = \cos t, \quad y = 1 - \sin^2 t$

29. If $x = 4t - 5$ and $y = t^2$, which of the following is the value of dy/dx when $t = 2$?
    a) 2   b) 4   c) 1   d) 1/2

30. If $x = 3t^2 + 2$ and $y = 2t^4 - 1$, which of the following is the value of dy/dx when $t = 1$?
    a) 8   b) 4/3   c) 6   d) 3/4

The parametric equations in Problems 31–35 describe curves in the plane. Find (a) the slope of the curve at the point (x, y) at which $t = 2$, and (b) the tangent to the curve at this point.

31. $x = t + 1/t, \quad y = t - 1/t$

32. $x = \sqrt{2t^2 + 1}, \quad y = (2t + 1)^2$

33. $x = t\sqrt{2t + 5}, \quad y = (4t)^{1/3}$

34. $x = \dfrac{t - 1}{t + 1}, \quad y = \dfrac{t + 1}{t - 1}$

35. $x = t^{-2}, \quad y = \sqrt{t^2 + 12}$

36. Find the equation of the tangent to the curve

$$x = \frac{1}{t} + t^2, \quad y = t^2 - t + 1$$

at the point $(2, 1)$.

37. Given $x = 80t$ and $y = 64t - 16t^2$, find the value of t for which $dy/dx = 0$.

38. A particle P moves along the curve $x^2 y^3 = 27$. At the time when P is at $(1, 3)$, $dy/dt = 10$. Find the value of dx/dt at this time.

39. If a point traces the circle $x^2 + y^2 = 25$, and if $dx/dt = 4$ when the point reaches $(3, 4)$, find dy/dt there.

Use Eq. (4) to find d^2y/dx^2 from the parametric equations in Problems 40–49.

40. Problem 7   41. Problem 11

42. Problem 13   43. Problem 17

44. Problem 23   45. Problem 24

46. Problem 25   47. Problem 27

48. Problem 28   49. Problem 37

50. Find d^2y/dx^2 if y is a differentiable function of t, $dy/dx = \sqrt{4 - \sin^2 t}$, and $x = \cos 2t$.

51. Suppose that x and y are differentiable functions of t and that

$$\frac{d^2y}{dx^2} = t^2 + 1, \quad \frac{dy}{dx} = t^3 + 3t.$$

Find dx/dt.

52. **COMPUTER GRAPHER**   If you have access to a parametric equation grapher, you will enjoy graphing

$$x = 6 \cos t + 5 \cos 3t, \quad y = 6 \sin t - 5 \sin 3t$$

for $-\pi \le t \le \pi$.

TOOLKIT PROGRAMS

Parametric Equations   Super ∗ Grapher

2.9
Newton's Method for Approximating Solutions of Equations

When exact formulas are not available for solving an equation $f(x) = 0$, we turn to numerical techniques for approximating the solutions we seek. One of these techniques is Newton's method or, as it is more accurately named, the Newton-Raphson method, discussed in this article. The method is based on the idea of using tangent lines to approximate the graph of $y = f(x)$ near the points where f is zero. Once again we see that linearization is the key to solving a practical problem. If you have access to a computer or a programmable calculator, you can easily write a program to do the arithmetic. If not, you can still see how it can be done. The procedure is as follows.

The Procedure for Newton's Method

1. Guess a first approximation to a root of the equation $f(x) = 0$. A graph of $y = f(x)$ will help.

2. Use the first approximation to get a second, the second to get a third, and so on. To go from the nth approximation x_n to the next approximation x_{n+1}, use the formula

$$x_{n+1} = x_n - \frac{f(x_n)}{f'(x_n)}, \tag{1}$$

where $f'(x_n)$ is the derivative of f at x_n.

We first show how the method works and then go to the theory behind it.
In our first example we find decimal approximations to $\sqrt{2}$ by estimating the positive root of the equation $f(x) = x^2 - 2 = 0$.

EXAMPLE 1 Find the positive root of the equation $f(x) = x^2 - 2 = 0$.

Solution With $f(x) = x^2 - 2$ and $f'(x) = 2x$, Eq. (1) becomes

$$x_{n+1} = x_n - \frac{x_n^2 - 2}{2x_n}. \tag{2}$$

To use our calculator efficiently, we rewrite Eq. (2) in a form that uses fewer arithmetic operations:

$$x_{n+1} = x_n - \frac{x_n^2 - 2}{2x_n} = \frac{x_n^2 + 2}{2x_n} = \frac{x_n}{2} + \frac{1}{x_n}. \tag{3}$$

The equation

$$x_{n+1} = \frac{x_n}{2} + \frac{1}{x_n} \tag{4}$$

2.34 The graph of $f(x) = x^3 - x - 1$ crosses the x-axis just once, at a point between $x = 1$ and $x = 2$.

permits us to go from each approximation to the next by the following steps:

	Operation	Result	Example
x	Enter x	x	1.5
$\frac{1}{x}$ STO	Store the reciprocal	$\frac{1}{x}$	0.66667
$\frac{1}{x}$	Take the reciprocal of the display again	x	1.5
$\div$ 2	Divide by 2	$\frac{x}{2}$	0.75
$+$ RCL $=$	Add memory to display	$\frac{x}{2} + \frac{1}{x}$	1.41667

With the starting value $x_0 = 1$, we get the results in the first column of the following table. (To five decimal places, $\sqrt{2} = 1.41421$.)

	Error	Number of correct figures
$x_0 = 1$	-0.41421	1
$x_1 = 1.5$	$+0.08579$	1
$x_2 = 1.41667$	0.00245	3
$x_3 = 1.41422$	0.00001	5

Newton's method is the method used by most calculators to calculate roots because it converges so fast.† If the arithmetic in the table had been carried to 13 decimal places rather than 5, then going one step further to x_4 would have given $\sqrt{2}$ to more than 10 decimal places. ∎

EXAMPLE 2 Find the x-coordinate of the point where the curve $y = x^3 - x$ crosses the horizontal line $y = 1$.

Solution The curve crosses the line when $x^3 - x = 1$ or $x^3 - x - 1 = 0$. When does $f(x) = x^3 - x - 1$ equal zero? The graph of f (Fig. 2.34) shows a single root, located between $x = 1$ and $x = 2$. We apply Newton's method to f with the starting value $x_0 = 1$. The results are displayed in Table 2.2 and Fig. 2.35.

At $n = 5$ we come to the result $x_5 = x_4 = 1.324717957$. When $x_{n+1} = x_n$, Eq. (1) shows that $f(x_n) = 0$. Hence we have found a solution of $f(x) = 0$ to nine decimals, or so it appears. (Our calculator shows only ten figures, and we

$y = x^3 - x - 1$

$A_1(1.5, 0.875)$

Root sought

$A_0(1, -1)$

2.35 The first three x-values in Table 2.2.

---

†An estimate of the error in Newton's method, calculated from the second derivative of f, may be found in many books on numerical analysis, among them Gerald and Wheatley's *Applied Numerical Analysis,* 3rd ed. (Reading, Mass.: Addison-Wesley, 1984).

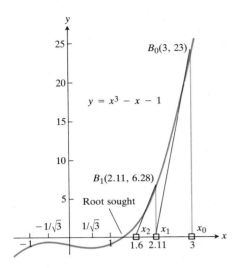

2.36 Any starting value x_0 to the right of $x = 1/\sqrt{3}$ will lead to the root.

TABLE 2.2
The result of applying Newton's method to $f(x) = x^3 - x - 1$ with $x_0 = 1$

n	x_n	$f(x_n)$	$f'(x_n)$	$x_{n+1} = x_n - \dfrac{f(x_n)}{f'(x_n)}$
0	1	-1	2	1.5
1	1.5	0.875	5.75	1.347826087
2	1.347826087	0.100682174	4.449905482	1.325200399
3	1.325200399	0.002058363	4.268468293	1.324718174
4	1.324718174	0.000000925	4.264634722	1.324717957
5	1.324717957	-5×10^{-10}	4.264632997	1.324717957

cannot guarantee the accuracy of the ninth decimal, though we believe it to be correct.) ■

What is the theory behind the method? It is this: We use the tangent to approximate the graph of $y = f(x)$ near the point $P(x_n, y_n)$, where $y_n = f(x_n)$ is small, and we let x_{n+1} be the value of x where that tangent line crosses the x-axis. (We assume that the slope $f'(x_n)$ of the tangent is not zero.) The equation of the tangent is

$$y - y_n = f'(x_n)(x - x_n). \tag{5}$$

We put $y_n = f(x_n)$ and $y = 0$ into Eq. (5) and solve for x to get

$$x = x_n - \frac{f(x_n)}{f'(x_n)}.$$

REMARK 1 The method doesn't work if $f'(x_n) = 0$. In that case, choose a new starting place. Of course, it may happen that $f(x) = 0$ and $f'(x) = 0$ have a common root. To detect whether that is so, we could first find the solutions of $f'(x) = 0$ and then check the value of $f(x)$ at such places.

REMARK 2 In Fig. 2.36 we have indicated that the process might have started at the point $B(3, 23)$ on the curve, with $x_0 = 3$. Point B is quite far from the x-axis, but the tangent at B crosses the x-axis at about $C(2.11, 0)$, so x_1 is still an improvement over x_0. If we use Eq. (1) repeatedly as before, with $f(x) = x^3 - x - 1$ and $f'(x) = 3x^2 - 1$, we confirm the nine-place solution $x_6 = x_5 = 1.324717957$ in six steps.

The curve in Fig. 2.36 has a high turning point at $x = -1/\sqrt{3}$ and a low turning point at $x = +1/\sqrt{3}$. We would not expect good results from Newton's method if we were to start with x_0 between these points, but we can start any place to the right of $x = 1/\sqrt{3}$ and get the answer. It would not be very clever to do so, but we could even begin far to the right of B, for example with $x_0 = 10$. It takes a bit longer, but the process still converges to the same answer as before.

REMARK 3 Newton's method does not always converge. For instance, if

$$f(x) = \begin{cases} \sqrt{x - r} & \text{for } x \geq r, \\ -\sqrt{r - x} & \text{for } x \leq r, \end{cases} \tag{6}$$

the graph will be like that shown in Fig. 2.37. If we begin with $x_0 = r - h$, we get $x_1 = r + h$, and successive approximations go back and forth between these

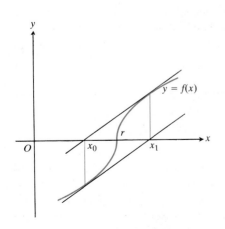

2.37 The graph of a function for which Newton's method fails to converge.

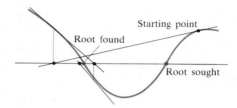

2.38   Newton's method may miss the root you want if you start too far away.

two values. No amount of iteration will bring us any closer to the root r than our first guess.

REMARK 4   If Newton's method does converge, it converges to a root of $f(x)$. However, the method may converge to a root different from the expected one if the starting value is not close enough to the root sought. Figure 2.38 shows two of the ways in which this might happen.

REMARK 5   When will Newton's method converge? It turns out that if the inequality

$$\left| \frac{f(x)f''(x)}{[f'(x)]^2} \right| < 1 \tag{7}$$

holds for all values of x in an interval about a root r of f, then the method will converge to r for any starting value x_0 in that interval. This is a *sufficient*, but not a necessary, condition. The method can (and does) converge in some cases when there is no interval about r in which the inequality (7) holds.

J. Raymond Mouraille (1768) and later Joseph Fourier (1768–1830) independently discovered that Newton's method will always work if the curve $y = f(x)$ is convex ("bulges") toward the axis in the interval between x_0 and the root sought. See Fig. 2.39.

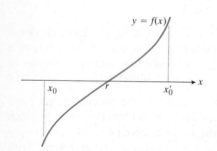

2.39   The curve $y = f(x)$ is convex toward the axis between x_0 and r and between x_0' and r. Newton's method will converge to r from either starting point.

PROBLEMS

Each function in Problems 1–6 has exactly one root in the interval $a \le x \le b$. Sketch the graph of the function over the interval. Then use Newton's method to find the root. Stop when you are sure of the first three decimal places.

1. $f(x) = x^2 + x - 1$, $a = 0$, $b = 1$

2. $f(x) = x^3 + x - 1$, $a = 0$, $b = 1$

3. $f(x) = x^4 + x - 3$, $a = 1$, $b = 2$

4. $f(x) = x^4 - 2$, $a = 1$, $b = 2$

5. $f(x) = 2 - x^4$, $a = -2$, $b = -1$

6. $f(x) = \sqrt{2x + 1} - \sqrt{x + 4}$, $a = 2$, $b = 4$

7. Suppose your first guess is lucky, in the sense that x_0 is a root of $f(x) = 0$. What happens to x_1 and later approximations?

8. You plan to estimate $\pi/2$ to five decimal places by solving the equation $\cos x = 0$ by Newton's method. Does it matter what your starting value is? Explain.

9. **CALCULATOR**   Show that $f(x) = x^3 + 2x - 4$ has a root between $x = 1$ and $x = 2$. Find the root to five decimal places.

10. **CALCULATOR** Show that $f(x) = x^4 - x^3 - 75$ has a root between $x = 3$ and $x = 4$. Find the root to five decimal places.

11. a) Explain why the following four statements ask for the same information:
   i) Find the roots of $f(x) = x^3 - 3x - 1$.
   ii) Find the x-coordinates of the intersections of the curve $y = x^3$ with the line $y = 3x + 1$.
   iii) Find the x-coordinates of the points where the curve $y = x^3 - 3x$ crosses the horizontal line $y = 1$.
   iv) Find the values of x where the derivative of $g(x) = (1/4)x^4 - (3/2)x^2 - x + 5$ equals zero.
   b) Sketch the graph of $f(x) = x^3 - 3x - 1$ over the interval $-2 \leq x \leq 2$.
   c) **CALCULATOR** Find the positive root of $f(x) = x^3 - 3x - 1$ to five decimal places.
   d) **CALCULATOR** Find the two negative roots of $f(x) = x^3 - 3x - 1$ to five decimal places.

12. **CALCULATOR** Estimate π to five decimal places by applying Newton's method to the equation $\tan x = 0$ with $x_0 = 3$. Remember to use radians.

13. **CALCULATOR** Find the point of intersection of the curve $y = \cos x$ with the line $y = x$ to five decimal places.

14. **CALCULATOR** Graphing $f(x) = x - 1 - 0.5 \sin x$ suggests that the function has a root near $x = 1.5$. Use one application of Newton's method to improve this estimate. That is, start with $x_0 = 1.5$ and find x_1. (The value of the root is 1.49870 to five decimal places.) Remember to use radians.

15. **PROGRAMMABLE CALCULATOR** Find two real roots of the equation $x^4 - 2x^3 - x^2 - 2x + 2 = 0$ to six decimal places.

16. **CALCULATOR** (Programmable feature helpful but not necessary.) *The sonobuoy problem.* From C. O. Wilde's *The Contraction Mapping Principle,* UMAP Unit 326 (Arlington, Mass.: COMAP, Inc.). In submarine location problems it is often necessary to find the submarine's closest point of approach (CPA) to a sonobuoy (sound detector) in the water. Suppose that the submarine travels on a parabolic path $y = x^2$ and that the buoy is located at the point $(2, -1/2)$ as shown in Fig. 2.40. As we shall see in Chapter 3 (Article 3.5, Problem 25), the value of x that minimizes the distance between the point (x, x^2) and the point $(2, -1/2)$ is a solution of the equation

$$\frac{1}{x^2 + 1} = x.$$

Solve this equation by Newton's method to find the CPA to five decimal places.

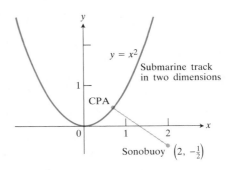

2.40 The diagram for Problem 16. CPA = closest point of approach.

17. Show that Newton's method applied to $f(x)$ in Eq. (6) leads to $x_1 = r + h$ if $x_0 = r - h$, and to $x_1 = r - h$ if $x_0 = r + h$, where $h > 0$. Interpret the result geometrically.

18. (See Remark 3.) Is it possible that successive approximations actually get "worse," in that x_{n+1} is farther away from the root r than x_n is? Can you find such a "pathological" example? (*Hint:* Try cube roots in place of square roots in Eq. (6).)

TOOLKIT PROGRAMS

Function Evaluator   Sequences and Series
Root Finder   Super * Grapher

2.10
Derivative Formulas in Differential Notation

The derivative formulas developed in this chapter are listed below as Formulas 1–12. By multiplying each one by dx we obtain the corresponding differential formulas.

Derivative Formulas	Differential Formulas
1. $\dfrac{dc}{dx} = 0$	1'. $dc = 0$

$$2. \ \frac{d(cu)}{dx} = c\frac{du}{dx} \qquad\qquad\qquad 2'. \ d(cu) = c\,du$$

$$3. \ \frac{d(u+v)}{dx} = \frac{du}{dx} + \frac{dv}{dx} \qquad\qquad 3'. \ d(u+v) = du + dv$$

$$4. \ \frac{d(uv)}{dx} = u\frac{dv}{dx} + v\frac{du}{dx} \qquad\qquad 4'. \ d(uv) = u\,dv + v\,du$$

$$5. \ \frac{d\left(\dfrac{u}{v}\right)}{dx} = \frac{v\dfrac{du}{dx} - u\dfrac{dv}{dx}}{v^2} \qquad 5'. \ d\left(\frac{u}{v}\right) = \frac{v\,du - u\,dv}{v^2}$$

$$6. \ \frac{du^n}{dx} = nu^{n-1}\frac{du}{dx} \qquad\qquad 6'. \ d(u^n) = nu^{n-1}du$$

$$6a. \ \frac{dcx^n}{dx} = cnx^{n-1} \qquad\qquad\quad 6a'. \ d(cx^n) = cnx^{n-1}dx$$

$$7. \ \frac{d\sin u}{dx} = \cos u\frac{du}{dx} \qquad\qquad 7'. \ d(\sin u) = \cos u\,du$$

$$8. \ \frac{d\cos u}{dx} = -\sin u\frac{du}{dx} \qquad\quad 8'. \ d(\cos u) = -\sin u\,du$$

$$9. \ \frac{d\tan u}{dx} = \sec^2 u\frac{du}{dx} \qquad\quad 9'. \ d(\tan u) = \sec^2 u\,du$$

$$10. \ \frac{d\cot u}{dx} = -\csc^2 u\frac{du}{dx} \qquad 10'. \ d(\cot u) = -\csc^2 u\,du$$

$$11. \ \frac{d\sec u}{dx} = \sec u\tan u\frac{du}{dx} \qquad 11'. \ d(\sec u) = \sec u\tan u\,du$$

$$12. \ \frac{d\csc u}{dx} = -\csc u\cot u\frac{du}{dx} \qquad 12'. \ d(\csc u) = -\csc u\cot u\,du$$

To find dy when y is a differentiable function of x, we may either

a) find dy/dx and multiply by dx, or

b) apply one or more of Formulas $1'-12'$.

EXAMPLE 1

a) $d(3x^2 - 6) = 6x\,dx$

b) $d(\cos 3x) = -\sin 3x\,d(3x) = -3\sin 3x\,dx$

c) $d\dfrac{x}{x+1} = \dfrac{(x+1)dx - x\,d(x+1)}{(x+1)^2}$

$$= \frac{x\,dx + dx - x\,dx}{(x+1)^2}$$

$$= \frac{dx}{(x+1)^2} \qquad\blacksquare$$

It should be noted that a differential on one side of an equation also calls for a differential on the other side of the equation. Thus, we never have $dy = 3x^2$ but, instead, $dy = 3x^2\,dx$.

PROBLEMS

In Problems 1–12, find dy.

1. $y = x^3 - 3x$

2. $y = x\sqrt{1 - x^2}$

3. $y = 2x/(1 + x^2)$

4. $y = (3x^2 - 1)^{3/2}$

5. $y + xy - x = 0$

6. $xy^2 + x^2y - 4 = 0$

7. $y = \sin(5x)$

8. $y = \cos(x^2)$

9. $y = 4 \tan(x/2)$

10. $y = \sec(x^2 - 1)$

11. $y = 3 \csc(1 - (x/3))$

12. $y = 2 \cot\sqrt{x}$

In Problems 13–16, express dx and dy in terms of t and dt. Then find dy/dx.

13. $x = t + 1, \quad y = t + t^2/2$

14. $x = 1 + 1/t, \quad y = t - 1/t$

15. $x = \cos t, \quad y = 1 + \sin t$

16. $x = t - \sin t, \quad y = 1 - \cos t$

REVIEW QUESTIONS AND EXERCISES

1. Use the definition of derivative to obtain the formula for the derivative of a product uv of two differentiable functions.

2. Set $v = u$ in the formula for the derivative of uv to get a formula for the derivative of u^2. Repeat the process with $v = u^2$ to get a formula for the derivative of u^3. Extend the result by mathematical induction to obtain the formula for the derivative of u^n for every positive integer n.

3. Explain how the three formulas

   a) $\dfrac{d(x^n)}{dx} = nx^{n-1}$,

   b) $\dfrac{d(cu)}{dx} = c\dfrac{du}{dx}$,

   c) $\dfrac{d(u + v)}{dx} = \dfrac{du}{dx} + \dfrac{dv}{dx}$

   let us differentiate any polynomial.

4. What formula do we need, in addition to the three listed in Exercise 3, to differentiate rational functions?

5. Does the derivative of a polynomial function exist at every point of its domain? What is the largest domain the function can have? Does the derivative of a rational function exist at every point in its domain? What real numbers, if any, must be excluded from the domain of a rational function?

6. What technique in this chapter can be used to find dy/dx if

   $$y^3 - xy^2 + \frac{1}{x^2} - 1 = 0?$$

   What must be assumed about y? Find dy/dx.

7. At what values of x is the function $y = x^{2/3}$ defined? Continuous? Differentiable?

8. Find the derivative of

   $$f(x) = x\sqrt{3x^2 + 1} + \frac{5x^{4/3}}{3x + 2}, \quad x \neq -\frac{2}{3}.$$

   What formulas of this chapter are used in finding derivatives of functions like this one?

9. Suppose that $y = f(x)$ has a derivative at $x = a$ and that we change x by an amount dx. How can we estimate the resulting change in y?

10. What is the linearization of a function $y = f(x)$ at a point where the function has a derivative? Give examples. How are linearizations used?

11. State the Chain Rule for derivatives. Prove it with the book closed.

12. Write expressions for $\sin(A + B)$ and $\cos(A + B)$.

13. Under what assumption is it true that

    $$\lim_{h \to 0} \frac{\sin h}{h} = 1?$$

14. State the double-angle formulas. Use one of them to prove that

    $$\lim_{h \to 0} \frac{1 - \cos h}{h} = 0.$$

    How is this limit used in the proof that

    $$\frac{d}{dx}\sin x = \cos x?$$

15. Let A_n be the area bounded by a regular n-sided polygon inscribed in a circle of radius r. Show that

    $$A_n = \left(\frac{n}{2}\right)r^2\sin\left(\frac{2\pi}{n}\right).$$

    Find $\lim A_n$ as $n \to \infty$. Does the result agree with what you know about the area of a circle?

16. The parametric equations

    $$x = 4\cos t, \quad y = 4\sin t, \quad -\pi \leq t \leq 0,$$

    describe the position $P(x, y)$ of a particle moving in the plane. Where does the particle start and stop? Identify the path along which the particle moves. How is dy/dx related to dy/dt and dx/dt during the motion?

17. If x and y are twice-differentiable functions of t, y is a twice-differentiable function of x, and $dx/dt \neq 0$, what method can you use to find d^2y/dx^2? Give an example.

18. Describe Newton's method for solving equations. Give an example. What is the theory behind the method? What are some of the things to watch out for when you use the method?

MISCELLANEOUS PROBLEMS

Find dy/dx in Problems 1–58.

1. $y = \dfrac{x}{\sqrt{x^2 - 4}}$

2. $x^2 + xy + y^2 - 5x = 2$

3. $xy + y^2 = 1$

4. $x^3 + 4xy - 3y^3 = 2$

5. $x^2y + xy^2 = 10$

6. $y = (x + 1)^2(x^2 + 2x)^{-2}$

7. $y = \cos(1 - 2x)$

8. $y = \dfrac{\cos x}{\sin x}$

9. $y = \dfrac{x}{x + 1}$

10. $y = \sqrt{2x + 1}$

11. $y = x^2\sqrt{x^2 - a^2}$

12. $y = \dfrac{2x + 1}{2x - 1}$

13. $y = \dfrac{x^2}{1 - x^2}$

14. $y = (x^2 + x + 1)^3$

15. $y = \sec^2(5x)$

16. $y^3 = \sin^3 x + \cos^3 x$

17. $y = \dfrac{(2x^2 + 5x)^{3/2}}{3}$

18. $y = \dfrac{3}{(2x^2 + 5x)^{3/2}}$

19. $xy^2 + \sqrt{xy} = 2$

20. $x^2 - y^2 = xy$

21. $x^{2/3} + y^{2/3} = a^{2/3}$

22. $x^{1/2} + y^{1/2} = a^{1/2}$

23. $xy = 1$

24. $\sqrt{xy} = 1$

25. $(x + 2y)^2 + 2xy^2 = 6$

26. $y = \sqrt{\dfrac{1 - x}{1 + x^2}}$

27. $y^2 = \dfrac{x}{x + 1}$

28. $x^2y + xy^2 = 6(x^2 + y^2)$

29. $xy + 2x + 3y = 1$

30. $x^2 + xy + y^2 + x + y + 1 = 0$

31. $x^3 - xy + y^3 = 1$

32. $xy^3 + 3x^2y^2 = 7$

33. $y = \sqrt{\dfrac{1 + x}{1 - x}}$

34. $y = \sqrt{x} + 1 + \dfrac{1}{\sqrt{x}}$

35. $y = (x^3 + 1)^{1/3}$

36. $y = x^2\sin^5 2x$

37. $y = \cot 2x$

38. $y = \sin^2(1 + 3x)$

39. $y = \dfrac{\sin x}{\cos^2 x}$

40. $y = \sin^3 2x$

41. $y = x^2\cos 8x$

42. $y = \sin(\cos^2 x)$

43. $y = \dfrac{\sin x}{1 + \cos x}$

44. $y = \dfrac{\sin^2 x}{\cos x}$

45. $y = \csc x$

46. $y = \cot x^2$

47. $y = \cos(\sin^2 x)$

48. $y = \dfrac{\sin x}{x}$

49. $y = \sec^2 x$

50. $y = \sec x \sin x$

51. $y = \cos(\sin^2 3x)$

52. $y = u^2 - 1, \quad x = u^2 + 1$

53. $y = \sqrt{2t + t^2}, \quad t = 2x + 3$

54. $x = \dfrac{t}{1 + t^2}, \quad y = 1 + t^2$

55. $t = \dfrac{x}{1 + x^2}, \quad y = x^2 + t^2$

56. $x = t^2 - 1, \quad y = 3t^4 - t^2$

57. $x = t^2 + t, \quad y = t^3 - 1$

58. $x = \cos 3t, \quad y = \sin(t^2 + 1)$

59. Find the slope of $y = x/(x^2 + 1)$ at the origin. Write the equation of the tangent line at the origin.

60. Write the equation of the tangent at $(2, 2)$ to the curve
$$x^2 - 2xy + y^2 + 2x + y - 6 = 0.$$

61. What is the slope of the curve $y = 2x^2 - 6x + 3$ at the point on the curve where $x = 2$? Find the tangent to the curve at this point.

62. Find the points on the curve $y = 2x^3 - 3x^2 - 12x + 20$ where the tangent is parallel to the x-axis.

63. If a hemispherical bowl of radius 10 in. is filled with water to a depth of x in., the volume of water is given by $V = \pi[10 - (x/3)]x^2$. Find the rate of increase of the volume per inch increase of depth.

64. A bus will hold 60 people. The number x of people per trip who use the bus is related to the fare charged (p dollars) by the law $p = [3 - (x/40)]^2$. Write an expression for the total revenue $r(x)$ per trip received by the bus company. What number of people per trip will make the marginal revenue dr/dx equal to zero? What is the corresponding fare?

65. A particle projected vertically upward with a speed of a ft/sec reaches an elevation $s = at - 16t^2$ ft at the end of t sec. What must the initial velocity be for the particle to travel 49 ft upward before it starts coming back down?

66. The graph in Fig. 2.41 shows the position $s(t)$ of a truck traveling on a highway. The truck starts at $t = 0$ and returns 15 hours later at $t = 15$.

a) Use the technique described at the end of Article 1.7 to construct a graph of the truck's velocity, $v = ds/dt$.

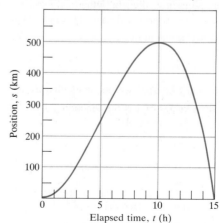

2.41 The time vs. position graph for the truck in Problem 66.

Then repeat the process to construct a graph of the truck's acceleration dv/dt.

b) Suppose $s(t) = 15t^2 - t^3$. Find ds/dt and d^2s/dt^2. Graph them and compare with your results in (a).

67. Find

$$\lim_{\Delta x \to 0} \frac{[2 - 3(x + \Delta x)]^2 - [2 - 3x]^2}{\Delta x}.$$

Of what function $f(x)$ is this the derivative?

68. Find the rate of change of y^2 with respect to x^2 if $y = x - x^2$. Express your answer in terms of x. (*Hint:* Let $u = y^2$ and $v = x^2$. Then relate u to v and find du/dv.)

69. Find the slope of the curve $x^2y + xy^2 = 6$ at the point $(1, 2)$.

70. If $y = x\sqrt{2x - 3}$, find d^2y/dx^2.

71. Find the value of d^2y/dx^2 in the equation $y^3 + y = x$ at the point $(2, 1)$.

72. Find the tangent to the curve $y = 2/\sqrt{x - 1}$ at the point on the curve where $x = 10$.

73. Write an equation for the line through $(2, 1)$ normal to the curve $x^2 = 4y$.

74. Use the definition of the derivative to find dy/dx for $y = \sqrt{2x + 3}$. Check the result with the Power Rule.

75. Find d^3y/dx^3 if

a) $y = \sqrt{2x - 1}$,

b) $y = \dfrac{1}{3x + 2}$,

c) $y = ax^3 + bx^2 + cx + d$.

76. For what value of c is the curve $y = c/(x + 1)$ tangent to the line through the points $(0, 3)$ and $(5, -2)$?

77. Show that the tangent to any point (a, a^3) on the curve $y = x^3$ meets the curve again at a point where the slope is four times the slope at (a, a^3).

78. Find the lines tangent and normal to the curve $(y - x)^2 = 2x + 4$ at the point $(6, 2)$.

79. The circle $(x - h)^2 + (y - k)^2 = a^2$ is tangent to the curve $y = x^2 + 1$ at the point $(1, 2)$.

a) Find the possible locations of the point (h, k).

b) If, in addition, the value of d^2y/dx^2 is the same on both curves at $(1, 2)$, find $h, k,$ and a. Sketch the curve and circle.

80. Which of the following statements could be true if $f''(x) = x^{1/3}$?

I. $f(x) = \dfrac{9}{28}x^{7/3} + 9$ II. $f'(x) = \dfrac{9}{28}x^{7/3} - 2$

III. $f'(x) = \dfrac{3}{4}x^{4/3} + 6$ IV. $f(x) = \dfrac{3}{4}x^{4/3} - 4$

a) I only   b) III only

c) II and IV only   d) I and III only

81. Find the tangents to the curve $x^2y + xy^2 = 6$ at the points where $x = 1$.

82. Find the tangent to the curve at the given point.

a) $x^2 + 2y^2 = 9$ at $(1, 2)$

b) $x^3 + y^2 = 2$ at $(1, 1)$

83. The designer of a 30-ft-diameter spherical hot-air balloon (a cutaway view is shown in Fig. 2.42) wishes to suspend the gondola 8 ft below the bottom of the balloon with suspension cables tangent to the surface of the balloon. Two of the cables are shown running from the top edges of the gondola to their points of tangency, $(-12, -9)$ and $(12, -9)$. How wide must the gondola be?

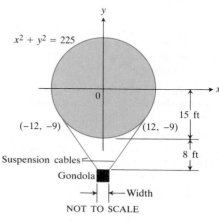

2.42   The balloon in Problem 83.

84. A cylindrical can of height 6 in. and radius r in. has volume $V = 6\pi r^2$ in^3. What is the difference between dV and ΔV as r varies? What is the geometric significance of dV?

85. If $y = 2x^2 - 3x + 5$, find Δy for $x = 3$ and $\Delta x = 0.1$. Approximate Δy by finding dy.

86. To compute the height h of a lamppost, the length a of the shadow of a 6-ft pole is measured (Fig. 2.43). The pole is 20 ft from the lamppost. If $a = 15$ ft, with a possible error of less than 1 in., calculate the height of the lamppost and estimate the error in the result.

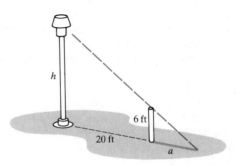

2.43   The lamppost in Problem 86.

87. Suppose $y = f(x)$ is a differentiable function of x, and $x = g(t)$ is a differentiable function of t. Find the value of dy/dt

at $t = 1$ if

$$g(1) = 3, \quad g'(1) = 6, \quad f(3) = 4, \quad f'(3) = 5.$$

88. If $y = x^2 + 1$ and $u = \sqrt{x^2 + 1}$, find dy/du.

89. If $x = y^2 + y$ and $u = (x^2 + x)^{3/2}$, find dy/du.

90. If $f'(x) = \sqrt{3x^2 - 1}$ and $y = f(x^2)$, find dy/dx.

91. If $f'(x) = \sin(x^2)$ and $y = f((2x - 1)/(x + 1))$, find dy/dx.

92. Given $y = 3 \sin 2x$ and $x = u^2 + \pi$, find the value of dy/du when $u = 0$.

93. Find the rate of change of $y = \sqrt{x^2 + 16}$ with respect to $t = x/(x - 1)$ at $x = 3$.

94. Let $f(x) = x^{2/3}$ and $g(x) = x^3$. Show that the composite of these functions in either order is differentiable at $x = 0$ but that f is not differentiable at $x = 0$. Does this contradict the Chain Rule? Explain.

95. Assume that the equations $z = x \sin y - y^2$ and $\cos y = y \sin z$ together define x and y as differentiable functions of z. Find dx/dz.

96. If the identity $\sin(x + a) = \sin x \cos a + \cos x \sin a$ is differentiated with respect to x, is the resulting equation also an identity? Does this principle apply to the equation $x^2 - 2x - 8 = 0$? Explain.

97. A useful linear approximation to

$$\frac{1}{1 + \tan x}$$

can be obtained by combining the approximations

$$\frac{1}{1 + x} \approx 1 - x \quad \text{and} \quad \tan x \approx x$$

to get

$$\frac{1}{1 + \tan x} \approx 1 - x.$$

Show that this is the standard linear approximation of $1/(1 + \tan x)$.

98. Let $f(x) = \sqrt{1 + x} + \sin x - 0.5$.
a) Find $f(-\pi/4) < 0$ and $f(0) > 0$ to show that the equation $f(x) = 0$ has a solution between $-\pi/4$ and 0.
b) To estimate the solution of $f(x) = 0$, replace $\sqrt{1 + x}$ and $\sin x$ by their linearizations at $x = 0$ and solve the resulting linear equation.
c) **CALCULATOR** Check your estimate in the original equation.

99. a) Show that the perimeter P_n of an n-sided regular polygon inscribed in a circle of radius r is $P_n = 2nr \sin(\pi/n)$.
b) Find the limit of P_n as $n \to \infty$. Is the answer consistent with what you know about the circumference of a circle?

100. Find dy/dx at $t = \pi/4$ if
a) $x = \cos^2 t, \quad y = \sin^2 t$;
b) $x = \cos^3 t, \quad y = \sin^3 t$;
c) $x = \tan^2 t, \quad y = \sin 2t$.

101. Find dy/dx and d^2y/dx^2 if $x = \cos 3t$ and $y = \sin^2 3t$.

102. If $x = 3t + 1$ and $y = t^2 + t$, find dy/dt, dx/dt, and dy/dx. Eliminate t to obtain y as a function of x, and then determine dy/dx directly. Do the results check?

103. If $x = t - t^2$, $y = t - t^3$, find the values of dy/dx and d^2y/dx^2 at $t = 1$.

104. *Pythagorean triples.* Suppose that the coordinates of a particle $P(x, y)$ moving in the plane are

$$x(t) = \frac{1 - t^2}{1 + t^2} \quad \text{and} \quad y(t) = \frac{2t}{1 + t^2},$$

for $-\infty < t < \infty$. Show that $x^2 + y^2 = 1$ and hence that the motion takes place on the unit circle. What one point of the circle is not covered by the motion? Sketch the circle and indicate the direction of motion for increasing t. For what values of t does $(x, y) = (0, -1)$? $(1, 0)$? $(0, 1)$?
From $x^2 + y^2 = 1$ we obtain

$$(t^2 - 1)^2 + (2t)^2 = (t^2 + 1)^2,$$

an equation of interest in number theory because it generates *Pythagorean triples* of integers. When t is an integer greater than 1, $a = t^2 - 1$, $b = 2t$, and $c = t^2 + 1$ are positive integers that satisfy the equation $a^2 + b^2 = c^2$.

105. Use mathematical induction (Appendix 2) to prove that if $y = u_1 u_2 \cdots u_n$ is a finite product of differentiable functions, then

$$\frac{dy}{dx} = \frac{du_1}{dx} \cdot u_2 \cdots u_n$$

$$+ u_1 \frac{du_2}{dx} \cdots u_n + \cdots + u_1 u_2 \cdots u_{n-1} \frac{du_n}{dx}.$$

This is Eq. (3) of Article 2.2.

106. If $f(x) = (x - a)^n g(x)$, where $g(x)$ is a polynomial and $g(a) \neq 0$, show that $f(a) = 0 = f'(a) = \cdots = f^{(n-1)}(a)$; but $f^{(n)}(a) = n!\, g(a) \neq 0$.

107. Prove Leibniz's rule:
a) $\dfrac{d^2(uv)}{dx^2} = \dfrac{d^2u}{dx^2} \cdot v + 2\dfrac{du}{dx}\dfrac{dv}{dx} + u\dfrac{d^2v}{dx^2}$,

b) $\dfrac{d^3(uv)}{dx^3} = \dfrac{d^3u}{dx^3} \cdot v + 3\dfrac{d^2u}{dx^2}\dfrac{dv}{dx} + 3\dfrac{du}{dx}\dfrac{d^2v}{dx^2} + u\dfrac{d^3v}{dx^3}$,

c) $\dfrac{d^n(uv)}{dx^n} = \dfrac{d^nu}{dx^n} \cdot v + n\dfrac{d^{n-1}u}{dx^{n-1}}\dfrac{dv}{dx} + \cdots$

$$+ \frac{n(n - 1) \cdots (n - k + 1)}{k!} \frac{d^{n-k}u}{dx^{n-k}}\frac{d^k v}{dx^k} + \cdots + u\frac{d^n v}{dx^n}.$$

The terms on the right-hand side of this equation may be obtained from the terms in the binomial expansion $(a + b)^n$ by replacing $a^{n-k}b^k$ by $(d^{n-k}u/dx^{n-k}) \cdot (d^k v/dx^k)$ for $k = 0, 1, 2, \ldots, n$, and interpreting d^0u/dx^0 as being u itself.

108. Suppose a function f satisfies the following two conditions for all x and y:
i) $f(x + y) = f(x) \cdot f(y)$;
ii) $f(x) = 1 + xg(x)$ where $\lim\limits_{x \to 0} g(x) = 1$.

Prove that (a) the derivative $f'(x)$ exists, and (b) $f'(x) = f(x)$.

109. Find all values of the constants m and b for which the function

$$y = \begin{cases} \sin x & \text{for } x < \pi, \\ mx + b & \text{for } x \geq \pi, \end{cases}$$

is (a) continuous at $x = \pi$; (b) differentiable at $x = \pi$.

110. Does the function

$$f(x) = \begin{cases} \dfrac{1 - \cos x}{x} & \text{for } x \neq 0, \\ 0 & \text{for } x = 0, \end{cases}$$

have a derivative at $x = 0$? Explain.

111. a) Show that the function

$$f(x) = \begin{cases} x^2 \sin \dfrac{1}{x} & \text{for } x \neq 0, \\ 0 & \text{at } x = 0, \end{cases}$$

is differentiable at $x = 0$. (Use the definition of derivative, but also see Chapter 1, Miscellaneous Problem 84.)

b) Find $f'(x)$ for $x \neq 0$.

c) Is f' continuous at $x = 0$? Explain.

112. Use the result of Chapter 1, Miscellaneous Problem 84 to show that the following functions are differentiable at $x = 0$.

a) $|x| \sin x$ b) $x^{2/3} \sin x$ c) $\sqrt[3]{x}(1 - \cos x)$

113. *The linearization gives the best linear approximation.* Suppose that $y = f(x)$ is differentiable at $x = a$ and that $g(x) = m(x - a) + c$ (m and c constants). If the error $e(x) =$ $f(x) - g(x)$ were small enough near $x = a$, we might think of using g as a linear approximation of f instead of the linearization $L(x) = f(a) + f'(a)(x - a)$. Show that if we impose on g the conditions

1. $e(a) = 0$ (the approximation error is zero at $x = a$),

2. $\displaystyle\lim_{x \to a} \frac{e(x)}{x - a} = 0$ $\left(\begin{array}{l}\text{the error is negligible when com-} \\ \text{pared with } (x - a)\end{array}\right)$,

then $g(x) = f(a) + f'(a)(x - a)$. Thus, the linearization gives the only linear approximation whose error is both zero at $x = a$ and negligible in comparison with $(x - a)$.

114. To find $x = \sqrt[q]{a}$, we apply Newton's method to $f(x) = x^q - a$. Here we assume that a is a positive real number and q is a positive integer. Show that x_1 is a "weighted average" of x_0 and a/x_0^{q-1}, and find the coefficients m_0 and m_1 for which

$$x_1 = m_0 x_0 + m_1 \left(\frac{a}{x_0^{q-1}}\right), \qquad \begin{array}{l} m_0 > 0, \; m_1 > 0, \\ m_0 + m_1 = 1. \end{array}$$

What conclusion would you reach if x_0 and a/x_0^{q-1} were equal? What would be the value of x_1 in that case? (You may also wish to read the article by J. P. Ballantine, "An Averaging Method of Extracting Roots," *American Mathematical Monthly*, 63, 1956, pp. 249–252, where more efficient ways of averaging are discussed. Also see J. S. Frame, "The Solution of Equations by Continued Fractions," *American Mathematical Monthly*, 60, 1953, pp. 293–305.)

CHAPTER 3

Applications of Derivatives

OVERVIEW

In this chapter, we see our past work with derivatives pay off in new ways. We use the first and second derivatives of a function to determine the shape of its graph between plotted points. The first derivative tells where the graph rises and falls and the second derivative tells where the graph is concave up and concave down. Many graphs tend to straighten out as x becomes numerically large or approaches selected real values, and we study this phenomenon as well. We then solve the problem of finding the maximum and minimum values of a differentiable function, a problem as current today as when it helped motivate the development of calculus three hundred years ago. We also see how the relation between two variables determines the relation between their rates of change. With relationships like these, we can tell how fast two ships are drawing apart or how fast the radius of a soap bubble will grow as the bubble is inflated. We also explore the Mean Value Theorem, a theorem whose corollaries provide the key to integral calculus. We then use derivatives to calculate limits in an ingenious fashion invented by the Swiss mathematician John Bernoulli but named after a French marquis. Finally, we conclude with a discovery from the late eighteenth century that gives approximations of functions and error estimates in a single easy-to-use formula. With this formula we can tell exactly how accurate a linearization is. We can also see how to add a quadratic term to a linearization to make an approximation that is more accurate still.

3.1
Sketching Curves with the First Derivative

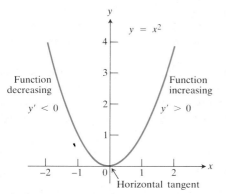

When we know that a function has a derivative at every point of an interval, we also know from the theorems in Article 1.11 that the function is continuous throughout the interval and that its graph over the interval is connected. Thus, the graphs of the differentiable functions $y = \sin x$ and $y = \cos x$ remain unbroken however far extended, as do the graphs of polynomials. The graphs of $y = \tan x$ and $y = 1/x^2$ break only at the points where the functions are undefined. On every interval that avoids these points, the functions are differentiable; therefore they are continuous and have connected graphs.

We can gain information about the shape of a function's graph if we know where the function's derivative is positive, negative, or zero. As we shall see in a moment, this tells us where the graph is rising or falling or has a horizontal tangent. First, some pictures (Figs. 3.1–3.5).

3.1 The function $y = x^2$ decreases on $(-\infty, 0)$, where the derivative $y' = 2x$ is negative, and increases on $(0, \infty)$, where the derivative is positive. In between, $y' = 0$, and the tangent to the curve is horizontal.

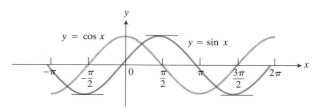

3.2 The graph of $y = \sin x$ rises and falls infinitely many times on its domain. It rises where the derivative $y' = \cos x$ is positive and falls where it is negative. At the points of transition between rise and fall, $y' = 0$, and the tangents are horizontal.

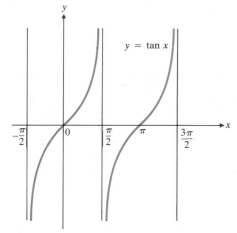

3.3 The graph of $y = \tan x$ has infinitely many separate pieces called "branches." Two of them are shown here. On each branch, $y' = \sec^2 x$ is positive and y is an increasing function of x.

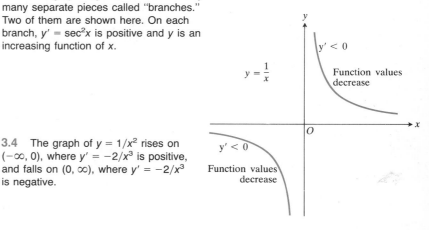

3.4 The graph of $y = 1/x^2$ rises on $(-\infty, 0)$, where $y' = -2/x^3$ is positive, and falls on $(0, \infty)$, where $y' = -2/x^3$ is negative.

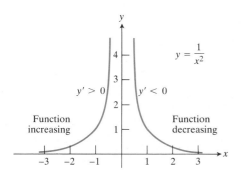

3.5 The graph of $y = 1/x$ falls as x moves from left to right within the intervals $(-\infty, 0)$ and $(0, \infty)$. The derivative $y' = -1/x^2$ is negative throughout each interval but is not defined at $x = 0$.

Increasing and Decreasing Functions

A function $y = f(x)$ is said to increase throughout an interval I if y increases as x increases. That is, whenever $x_2 > x_1$ in I, we find $f(x_2) > f(x_1)$. Similarly, $y = f(x)$ decreases throughout I if y decreases as x increases. Thus, whenever $x_2 > x_1$ in I, we find $f(x_2) < f(x_1)$. The graph of an increasing function rises as x moves from left to right through I; the graph of a decreasing function falls. As we have seen, a function that increases on one interval may well decrease on another interval.

As in Figs. 3.1–3.5, increase is associated with positive derivatives and decrease with negative derivatives. We shall be able to show in Article 3.7 that if f' is positive at every point of an interval I, then f increases on I and that if f' is negative at every point of I, then f decreases on I. We shall assume these facts for now as the first derivative test for rise and fall.

The First Derivative Test for Rise and Fall

Suppose that a function f has a derivative at every point x of an interval I. Then

$$f \text{ increases on } I \text{ if } f'(x) > 0 \text{ for all } x \text{ in } I,$$

$$f \text{ decreases on } I \text{ if } f'(x) < 0 \text{ for all } x \text{ in } I.$$

In geometric terms, the first derivative test says that differentiable functions increase on intervals where their graphs have positive slopes and decrease on intervals where their graphs have negative slopes.

Horizontal Tangents

Since a derivative f' has the intermediate value property on any interval I throughout which it is defined (Article 1.11), it must take on the value zero every time it changes sign on I. Thus, every time f' changes sign on I, the graph of f must have a horizontal tangent.

If f' changes from positive to negative values as x passes from left to right through a point c, then the value of f at c is a **local maximum** value of f, as shown in Fig. 3.6. That is, $f(c)$ is the largest value the function takes on in the

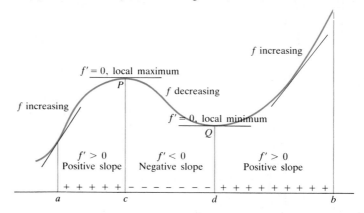

3.6   The function $y = f(x)$ increases on (a, c) where $f' > 0$, decreases on (c, d) where $f' < 0$, and increases again on (d, b). The transitions are marked by horizontal tangents.

x	y	y'
-1	0	9
0	4	0
1	2	-3
2	0	0
3	4	9

(a)

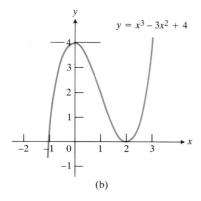

(b)

3.7 (a) Slope table. (b) The graph of $y = x^3 - 3x^2 + 4$ rises on $(-\infty, 0)$ to a local maximum of 4 at $x = 0$, falls to a local minimum of 0 at $x = 2$, and then rises again on $(2, \infty)$.

immediate neighborhood of $x = c$. Similarly, if f' changes from negative to positive values as x passes from left to right through a point d, then the value of f at d is a **local minimum** value of f. That is, $f(d)$ is the smallest value f takes in the immediate neighborhood of d. (We shall give more formal definitions of local maximum and local minimum when we study the theory of maximum and minimum values of functions in Article 3.4.)

Graphing

EXAMPLE 1 Graph the function $y = x^3 - 3x^2 + 4$.

Solution See Fig. 3.7. We first find the intercepts (points where the graph crosses or lies tangent to the axes). The y-intercept is found by setting $x = 0$:

$$y = (0)^3 - 3(0)^2 + 4 = 4.$$

Factoring the polynomial

$$y = x^3 - 3x^2 + 4 = (x + 1)(x - 2)^2$$

shows the x-intercepts to be $x = -1$ and $x = 2$.

We then find where the derivative of the function is positive, negative, and zero. The derivative is

$$y' = 3x^2 - 6x = 3x(x - 2),$$

which is zero at $x = 0$ and $x = 2$. The curve has horizontal tangents at these values.

The derivative y' is positive to the left of $x = 0$, where x and $x - 2$ are both negative, and positive to the right of $x = 2$, where x and $x - 2$ are both positive. The function therefore increases on the intervals $(-\infty, 0)$ and $(2, \infty)$.

Between $x = 0$ and $x = 2$, the derivative $y' = 3x(x - 2)$ is negative because $3x > 0$ while $x - 2 < 0$. The function therefore decreases on the interval $(0, 2)$.

We now construct a short table of function values and slopes (see Fig. 3.7a), which includes the intercepts and the points of transition between the rising and falling portions of the curve.

Finally, we plot the points and use the information about how the curve rises and falls to complete the sketch shown in Fig. 3.7(b). At $x = 0$, the function has a local maximum value of $y = 4$. At $x = 2$, the function has a local minimum value of $y = 0$. ■

EXAMPLE 2 Sketch the curve

$$y = f(x) = \frac{1}{3}x^3 - 2x^2 + 3x + 2.$$

Solution The y-intercept is $y = 2$, but it is not easy to find even one x-intercept because the polynomial does not factor nicely into linear factors. We might notice that there is a root between $x = -1$ and $x = 0$ because $f(-1) < 0$ and $f(0) = 2$, but the root is not easily estimated. Fortunately, we do not need to know the x-intercepts to sketch the curve. The first derivative of the function will tell us all we need to know about where the curve rises, falls, and has horizontal tangents.

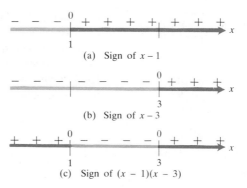

(a) Sign of $x - 1$

(b) Sign of $x - 3$

(c) Sign of $(x - 1)(x - 3)$

3.8   The sign pattern of the product $(x - 1)(x - 3)$.

The first derivative,

$$\frac{dy}{dx} = x^2 - 4x + 3 = (x - 1)(x - 3),$$

is zero when $x = 1$ or $x = 3$. The curve therefore has horizontal tangents at $x = 1$ and $x = 3$. Moreover, these values of x mark the transition between positive and negative slopes, as we shall now see.

The sign of dy/dx depends on the signs of the two factors $x - 1$ and $x - 3$. Since the sign of $x - 1$ is negative when x is to the left of 1 and positive to the right, we have the pattern of signs indicated in Fig. 3.8(a). Similarly, the sign of $x - 3$ is shown in part (b). The sign of the product $dy/dx = (x - 1)(x - 3)$ is shown in Fig. 3.8(c). We can get a rough idea of the shape of the curve just from this pattern of signs if we sketch a curve that is rising when $x < 1$, falling when $1 < x < 3$, and rising again when $x > 3$ (Fig. 3.9). The function has a local maximum at $x = 1$ and a local minimum at $x = 3$.

To get a more accurate curve, we would construct a table of values extending from $x = 0$ to $x = 4$, say, that included the transition points between the rising and falling portions of the curve (Fig. 3.10).   ∎

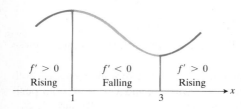

3.9   A function whose derivative has the sign pattern shown in Fig. 3.8(c) must have a graph something like this.

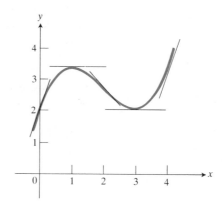

3.10   This graph of

$$y = \frac{1}{3}x^3 - 2x^2 + 3x + 2$$

combines the general information about shape from Fig. 3.9 with a selection of plotted points and slopes.

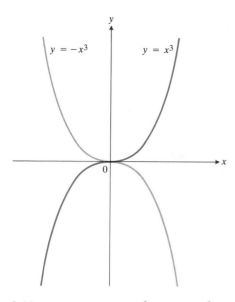

3.11 The graphs of $y = x^3$ and $y = -x^3$ cross their tangents at $x = 0$.

A Horizontal Tangent but No Maximum or Minimum

A derivative need not change sign every time it is zero. The derivative $y' = 3x^2$ of the function $y = x^3$ is zero at the origin but positive on both sides of it. The graph of $y = x^3$ crosses the tangent at $x = 0$ and continues to rise. Similarly, the graph of $y = -x^3$ crosses its horizontal tangent at $x = 0$ and continues to fall. (See Fig. 3.11.)

A Maximum or Minimum without a Horizontal Tangent

We have seen that a function f may take on a local maximum or minimum value at a point where f' is 0. In practice, this happens frequently enough to make the quest for zeros of f' worthwhile. However, you should be aware that a function may take on a local maximum or minimum value at a point where its derivative fails to exist. Thus, the quest for extreme values must go beyond solving the equation $f' = 0$.

EXAMPLE 3 The function $y = |x|$ takes on its minimum value at $x = 0$ where dy/dx does not exist (Fig. 3.12). ∎

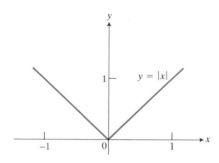

3.12 The function $y = |x|$ takes on its minimum value at a point where it has no derivative.

PROBLEMS

In Problems 1–18, find dy/dx and the intervals of x-values on which $y = f(x)$ is increasing or decreasing. Sketch each curve, showing the points of transition between the rising and falling portions. Find any local maximum and local minimum values that the function has when $y' = 0$.

1. $y = x^2 - x + 1$

2. $y = 12 - 12x + 2x^2$

3. $y = \dfrac{x^3}{3} - \dfrac{x^2}{2} - 2x + \dfrac{1}{3}$

4. $y = 2x^3 - 3x^2 + 3$

5. $y = x^3 - 27x + 36$

6. $y = x^4 - 8x^2 + 16$

7. $y = 3x^2 - 2x^3$

8. $y = (x - 2)(x - 11)(x + 13)$

9. $y = x^4$

10. $y = x^{4/3}$

11. $y = 1/x^3$

12. $y = 1/(x - 1)$

13. $y = 1/(x + 1)^2$

14. $y = 9x - x^3$

15. $y = \cos x,\quad -3\pi/2 < x < 3\pi/2$

16. $y = \sec x,\quad -\pi/2 < x < \pi/2$

17. $y = x|x|$

18. $y = \sin |x|,\quad -2\pi \le x \le 2\pi$

19. Graph $y = \cos|x|,\ -\pi \le x \le \pi$. What are the maximum and minimum values of the function and where are they taken on?

20. Graph $y = (1/2)(|\sin x| + \sin x),\ 0 \le x \le 2\pi$. What are the maximum and minimum values of the function and where are they taken on?

21. Show that the function $y = x/(x + 1)$ increases on every interval in its domain.

22. If $y = f(x)$ has a derivative at $x = c$ and $f'(c) = 0$, must f have a local maximum or local minimum value at $x = c$? Explain.

23. Give an example of a function that is continuous for all x, that decreases on $(-\infty, 0)$ and $(0, \infty)$, but has no derivative at $x = 0$. (*Hint:* Piece two functions together.)

24. Find all values of the constants m and b for which the function

$$f(x) = \begin{cases} mx + b & \text{for } x < 1, \\ x^2 + 1 & \text{for } x \geq 1, \end{cases}$$

is (a) continuous, (b) differentiable.

25. Let $y = x - 2 \sin x$, $0 \leq x \leq \pi$.
a) Find where $y' < 0$, $y' = 0$, and $y' > 0$ on $[0, \pi]$.
b) Plot the endpoints of the curve and any points where $y' = 0$. Then sketch the curve.

c) The equation $x - 2 \sin x = 0$ has two solutions on $[0, \pi]$, one of them being $x = 0$. Estimate the other solution to two decimal places.

26. Suppose that $y = x^n$, and n is a positive integer. Determine the values of x for which y is increasing and decreasing if (a) n is even, (b) n is odd.

TOOLKIT PROGRAMS

Derivative Grapher   Super ∗ Grapher
Function Evaluator

3.2
Concavity and Points of Inflection

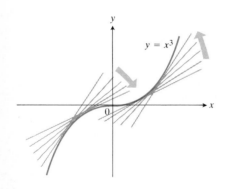

3.13   The graph of $y = x^3$ is concave down on the left, concave up on the right.

This article shows how to make the graphs we draw more accurate by taking the sign of the second derivative into account.

As you can see in Fig. 3.13, the function $y = x^3$ increases as x increases, but the portions of the curve that lie over the intervals $(-\infty, 0)$ and $(0, \infty)$ turn in different ways. If we come in from the left along the curve toward the origin, the curve turns to our right. As we leave the origin, the curve turns to our left. The left portion "bends down." The right portion "bends up."

To describe the turning another way, the tangent to the curve turns clockwise as the point of tangency approaches the origin from the left. The slope of the curve is decreasing. The tangent turns counterclockwise as the point of tangency moves from the origin into the first quadrant. The slope of the curve is increasing.

We say that the curve $y = x^3$ is concave down on the interval $(-\infty, 0)$ where y' decreases and concave up on the interval $(0, \infty)$ where y' increases. Thus we have the following definition of concavity for graphs of differentiable functions.

DEFINITIONS

Concave Down and Concave Up

The graph of a differentiable function $y = f(x)$ is **concave down** on an interval where y' decreases and **concave up** on an interval where y' increases.

If a function $y = f(x)$ has a second derivative as well as a first (as do most of the functions we deal with in this text), we can apply the first derivative test (discussed in Article 3.1) to the function $f' = y'$ to conclude that y' decreases if $y'' < 0$ and increases if $y'' > 0$. We therefore have a test that we can apply to the formula $y = f(x)$ to determine the concavity of its graph. It is called the second derivative test for concavity.

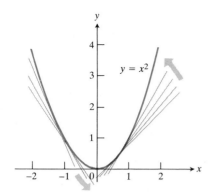

3.14 The graph of $y = x^2$ is concave up. The tangent turns counterclockwise as x increases; y' is increasing.

> **The Second Derivative Test for Concavity**
>
> The graph of $y = f(x)$ is
>
> concave down on any interval where $y'' < 0$,
>
> concave up on any interval where $y'' > 0$.

The idea is that if $y'' < 0$, then y' decreases as x increases and the tangent turns clockwise. Conversely, if $y'' > 0$, then y' increases as x increases and the tangent turns counterclockwise.

EXAMPLE 1 See Fig. 3.14. The curve $y = x^2$ is concave up on the entire x-axis because $y'' = 2 > 0$. ∎

EXAMPLE 2 See Fig. 3.15. The curve $y = \sin x$ is concave down over $0 < x < \pi$ because $y'' = -\sin x < 0$ on this interval. ∎

(a)

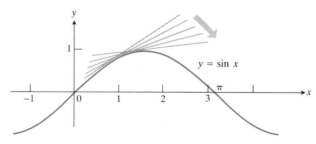

3.15 The graph of $y = \sin x$ is concave down between $x = 0$ and $x = \pi$ because $y'' = -\sin x$ is negative at each point.

Figure 3.16 shows the interplay between a typical function $y = f(x)$ and its first two derivatives. The arc ABC of the y-curve is concave up, CDE is concave down, and EF is again concave up. To focus attention, let us consider a section near A on arc ABC. Here y' is negative and the y-curve slopes downward to the right. But as we travel through A toward B we find that the slope becomes less negative. That is, y' is an increasing function of x. Therefore, the y'-curve slopes upward at A'. Its own slope, which is y'', is positive there. The same kind of argument applies at all points along the arc $A'B'C'$: y' is an increasing function of x, so its derivative y'' is positive. This is indicated by drawing the arc $A''B''C''$ of the y''-curve above the x-axis.

Similarly, where the y-curve is concave down (along CDE) the y'-curve is falling, so its slope y'' is negative. At the point C, where the y-curve changes from concave up to concave down, y'' is zero.

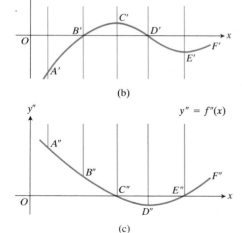

3.16 Compare the graph of the function in (a) with the graphs of its first and second derivatives in (b) and (c).

Points of Inflection

A point on the curve where the concavity changes is called a **point of inflection.** Thus, a point of inflection on a twice-differentiable curve is a point where y'' is positive on one side and negative on the other. At such a point, y'' is zero because derivatives have the intermediate value property.

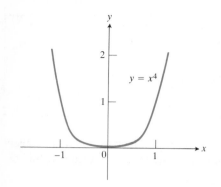

3.17 The graph of $y = x^4$ has no inflection point at the origin even though $y''(0) = 0$.

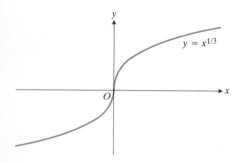

3.18 The graph of $y = x^{1/3}$ shows that a point where y'' fails to exist can be a point of inflection.

It is possible for y'' to be zero at a point that is *not* a point of inflection, and you should watch out for this. We shall see it happen in Example 4. Also, a point of inflection may occur where y'' fails to exist, as we shall see in Example 5.

EXAMPLE 3 See Fig. 3.13. The curve $y = x^3$ has a point of inflection at $x = 0$ where $y'' = 6x$ changes sign. ∎

EXAMPLE 4 See Fig. 3.17. The curve $y = x^4$ has no inflection point at $x = 0$ even though $y'' = 12x^2$ is zero there. The second derivative does not change sign at $x = 0$ (in fact, y'' is never negative). The curve is concave up over the entire x-axis because $y' = 4x^3$ is an increasing function on $(-\infty, \infty)$.

We see in this example that although the condition $y'' > 0$ in the second derivative test for concavity is a sufficient condition, it is not a necessary one. The curve $y = x^4$ is concave up even though y'' fails to be greater than zero at the origin. ∎

EXAMPLE 5 See Fig. 3.18. The curve $y = x^{1/3}$ has a point of inflection at $x = 0$ even though the second derivative does not exist there. To see this, we calculate y'' for $x \neq 0$:

$$y = x^{1/3},$$

$$y' = \frac{1}{3}x^{-2/3},$$

$$y'' = -\frac{2}{9}x^{-5/3}.$$

As $x \to 0$, y'' becomes infinite. Yet the curve is concave up for $x < 0$ (where $y'' > 0$ and y' is increasing) and concave down for $x > 0$ (where $y'' < 0$ and y' is decreasing). ∎

Application to Graphing

We now use what we have learned to graph a polynomial of degree three. The steps 1–5 in the solution give a general procedure for graphing.

EXAMPLE 6 Sketch the curve

$$y = \frac{1}{6}(x^3 - 6x^2 + 9x + 6).$$

Solution

1. We calculate y' and y'':

$$y' = \frac{1}{6}(3x^2 - 12x + 9) = \frac{1}{2}(x^2 - 4x + 3),$$

$$y'' = \frac{1}{2}(2x - 4) = x - 2.$$

2. We find the points where $y' = 0$ and determine where y' is positive and where it is negative. This gives information about where the curve rises

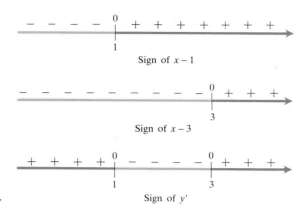

3.19   The sign of $y' = (x - 1)(x - 3)$.

and falls. The points where $y' = 0$ may give local maximum and minimum values of y.

When factored,

$$y' = \frac{1}{2}(x - 1)(x - 3),$$

which is zero when $x = 1$ and $x = 3$. As shown in Fig. 3.19, y' is positive when $x < 1$, negative when $1 < x < 3$, and positive again when $x > 3$. The curve has a local maximum at $x = 1$ (where y' changes from $+$ to $-$) and a local minimum at $x = 3$ (where y' changes from $-$ to $+$).

3. We now find where $y'' = 0$ and determine where y'' is positive and where it is negative. This gives information about concavity. The points where $y'' = 0$ may be points of inflection.

We have

$$y'' = x - 2,$$

which is zero when $x = 2$, negative when $x < 2$, and positive when $x > 2$. The curve is concave down on $(-\infty, 2)$ and concave up on $(2, \infty)$. The point $x = 2$ is a point of inflection because y'' changes sign at $x = 2$.

4. We make a short table of values of y, y', and y'' that includes information we have gathered so far. We also enter the y-intercept and the values of y at a few additional points to get general information about the curve (Table 3.1).

TABLE 3.1

x	y	y'	y''	Conclusions
-1	$-5/3$	$+4$	$-$	Rising, concave down
0	1	$+3/2$	$-$	Rising, concave down
1	$5/3$	0	$-$	Local maximum
2	$4/3$	$-1/2$	0	Falling, point of inflection
3	1	0	$+$	Local minimum
4	$5/3$	$+3/2$	$+$	Rising, concave up

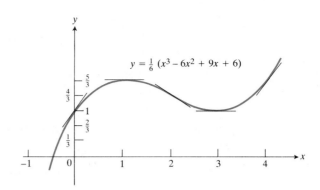

3.20   A graph of

$$y = \frac{1}{6}(x^3 - 6x^2 + 9x + 6)$$

drawn after studying the values of y' and y''.

5. We plot the points from the table, sketch tangents, and use the information about rise, fall, and concavity to draw the graph (Fig. 3.20).   ∎

PROBLEMS

In Problems 1–14, find the intervals of x-values on which the curve is (a) rising, (b) falling, (c) concave up, (d) concave down. Sketch the curve, showing points of inflection and any points where the function assumes a local maximum or minimum value.

1. $y = x^2 - 4x + 3$

2. $y = 20x - x^2$

3. $y = x^{5/3}$

4. $y = x^3 - x$

5. $y = x^3 - 3x + 3$

6. $y = 4 + 3x - x^3$

7. $y = x^3 - 6x^2 + 9x + 1$

8. $y = \dfrac{x^3}{3} - \dfrac{x^2}{2} - 6x$

9. $y = (x - 2)^3 + 1$

10. $y = x^{2/3}$

11. $y = \tan x, \quad -\pi/2 < x < \pi/2$

12. $y = \cos x, \quad 0 \le x \le 2\pi$

13. $y = -x^4$

14. $y = |x^2 - 1|$

15. Sketch a smooth curve $y = f(x)$ with the properties that $f(1) = 0$, $f'(x) < 0$ for $x < 1$, and $f'(x) > 0$ for $x > 1$.

16. Sketch a smooth curve $y = f(x)$ with the properties that $f(1) = 0$, $f''(x) < 0$ for $x < 1$, and $f''(x) > 0$ for $x > 1$.

Sketch the curves in Problems 17–32, indicating inflection points and local maxima and minima.

17. $y = 6 - 2x - x^2$

18. $y = 2x^2 - 4x + 3$

19. $y = x(6 - 2x)^2$

20. $y = (x - 1)^2(x + 2)$

21. $y = 12 - 12x + x^3$

22. $y = x^3 - 3x^2 + 2$

23. $y = x^3 + 3x^2 + 3x + 2$

24. $y = x^3 + 3x^2 - 9x - 11$

25. $y = x^3 - 6x^2 - 135x$

26. $y = x^3 - 33x^2 + 216x$

27. $y = x^4 - 2x^2 + 2$

28. $y = x^4 - 32x + 48$

29. $y = 3x^4 - 4x^3$

30. $y = x + \sin 2x$

31. $y = \sin x + \cos x$

32. $y = |x^3 - x|$

33. Figure 3.21 shows the graph of a function $y = f(x)$. At which of the five points indicated on the graph (a) are $f'(x)$ and $f''(x)$ both negative; (b) is $f'(x)$ negative and $f''(x)$ positive?

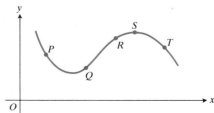

3.21   The curve for Problem 33.

34. Sketch a continuous curve $y = f(x)$ having the following characteristics:

$$f(-2) = 8, \qquad f'(2) = f'(-2) = 0,$$
$$f(0) = 4, \qquad f'(x) < 0 \quad \text{for } |x| < 2,$$
$$f(2) = 0, \qquad f''(x) < 0 \quad \text{for } x < 0,$$
$$f'(x) > 0 \quad \text{for } |x| > 2, \quad f''(x) > 0 \quad \text{for } x > 0.$$

35. Sketch a continuous curve $y = f(x)$ with the following properties. Label coordinates where possible.

x	y	Curve
$x < 2$		Falling, concave up
2	1	Horizontal tangent
$2 < x < 4$		Rising, concave up
4	4	Inflection point
$4 < x < 6$		Rising, concave down
6	7	Horizontal tangent
$x > 6$		Falling, concave down

36. Sketch a continuous curve $y = f(x)$ having

$$f'(x) > 0 \quad \text{for } x < 2, \qquad f'(x) < 0 \quad \text{for } x > 2,$$

a) if $f'(x)$ is continuous at $x = 2$;
b) if $f'(x) \to 1$ as $x \to 2^-$ and $f'(x) \to -1$ as $x \to 2^+$;
c) if $f'(x) = 1$ for all $x < 2$ and $f'(x) = -1$ for all $x > 2$.

37. Sketch a continuous curve $y = f(x)$ for $x > 0$ if

$$f(1) = 0 \quad \text{and} \quad f'(x) = 1/x \quad \text{for all } x > 0.$$

Is such a curve necessarily concave up or concave down?

38. The curve $y = x^3 + bx^2 + cx + d$ (b, c, and d constants) will have a point of inflection when $x = 1$ if b has which of the following values?

a) 2   b) -2
c) 3   d) -3

39. Sketch the curve

$$y = 2x^3 + 2x^2 - 2x - 1$$

after locating its local maximum, minimum, and inflection points. Then answer the following questions from your graph.
a) How many times and approximately where does the curve cross the x-axis?
b) How many times and approximately where would the curve cross the x-axis if $+3$ were added to all the y-values?
c) How many times and approximately where would the curve cross the x-axis if -3 were added to all the y-values?

40. Show that the curve $y = x + \sin x$ has no local maximum or minimum points even though it does have points where dy/dx is zero. Sketch the curve.

41. Find all local maximum points, local minimum points, and points of inflection on the curve $y = x^4 + 8x^3 - 270x^2$.

42. The graph of the function $f(x) = x^4 - 2x^2 - 4$ resembles a molar tooth.
a) Graph the function and label the points of inflection.
b) Estimate the positive root from an initial guess of $x_0 = 2$ with one iteration of Newton's method.

43. The slope of a curve $y = f(x)$ is

$$\frac{dy}{dx} = 3(x - 1)^2(x - 2)^3(x - 3)^4(x - 4).$$

a) For what value or values of x does y have a local maximum?
b) For what value or values of x does y have a local minimum?

TOOLKIT PROGRAMS

Derivative Grapher   Super ∗ Grapher
Function Evaluator

3.3
Asymptotes and Symmetry

In this article, we graph rational functions of x, taking into account their behavior as the denominator nears zero or as x becomes numerically large. The graphs of even and odd functions have symmetries that are useful to know about, so we look into this as well.

Symmetries in the Graphs of Even and Odd Functions

A function $y = f(x)$ is an **even** function of x if $f(-x) = f(x)$ for every x in the function's domain; it is an **odd** function of x if $f(-x) = -f(x)$ for every x in the function's domain.

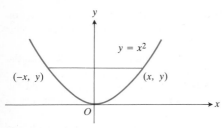

(a) Symmetry about the y-axis

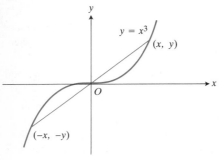

(b) Symmetry about the origin

3.22   Coordinate tests for symmetry.

Saying that a function $y = f(x)$ is even is equivalent to saying that its graph is symmetric about the y-axis. Since $f(-x) = f(x)$, the point (x, y) lies on the curve if and only if the point $(-x, y)$ lies on the curve (Fig. 3.22a).

Saying that a function $y = f(x)$ is odd is equivalent to saying that its graph is symmetric with respect to the origin. Since $f(-x) = -f(x)$, the point (x, y) lies on the curve if and only if the point $(-x, -y)$ lies on the curve (Fig. 3.22b).

The graphs of equations that contain only even powers of x will be symmetric about the y-axis. However, there is no corresponding rule for odd powers and symmetry about the origin.

EXAMPLE 1   The graphs of

$$ y = \frac{1}{x^2} \qquad \text{and} \qquad y = x^2 - \frac{1}{x^2} + 1 $$

are symmetric about the y-axis because the equations contain only even powers of x. In each case, $f(-x) = f(x)$:

$$ \frac{1}{(-x)^2} = \frac{1}{x^2} \qquad \text{and} \qquad (-x)^2 - \frac{1}{(-x)^2} + 1 = x^2 - \frac{1}{x^2} + 1. \qquad ∎ $$

EXAMPLE 2   The graphs of

$$ y = x^3 \qquad \text{and} \qquad y = \frac{x^2 + 1}{x} = x + \frac{1}{x} $$

are symmetric with respect to the origin. In each case, $f(-x) = -f(x)$:

$$ (-x)^3 = -x^3 \qquad \text{and} \qquad -x + \frac{1}{-x} = -\left(x + \frac{1}{x}\right). \qquad ∎ $$

EXAMPLE 3   The functions

$$ y = \frac{1}{x - 1} \qquad \text{and} \qquad y = 1 + \frac{1}{x} $$

are not odd, even though the formulas contain only odd powers of x. In the first case,

$$ f(-x) = \frac{1}{-x - 1} = -\frac{1}{x + 1} \neq -f(x). $$

In the second,

$$ f(-x) = 1 + \frac{1}{-x} = 1 - \frac{1}{x} \neq -f(x). $$

Neither graph is symmetric about the origin.   ∎

Horizontal and Vertical Asymptotes

As a point P on the graph of a function $y = f(x)$ moves increasingly far from the origin, it may happen that the distance between P and some fixed line tends to zero. In other words, the curve "approaches" the line as it gets farther from the origin. In such a case, the line is called an **asymptote** of the graph. For instance, the x-axis and y-axis are asymptotes of the curves $y = 1/x^2$ and $y = 1/x$ (see Figs. 3.4 and 3.5).

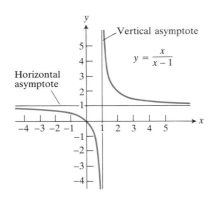

3.23 The graph of $y = x/(x-1)$.

EXAMPLE 4 The asymptotes of

$$y = \frac{x}{x-1}$$

are the lines $y = 1$ and $x = 1$ (Fig. 3.23). This is quickly revealed if we divide by the denominator:

$$
\begin{array}{r}
1 \\
x-1 \overline{\smash{)}x} \\
\underline{x-1} \\
1
\end{array}
$$

This enables us to rewrite the equation as

$$y = 1 + \frac{1}{x-1}.$$

The line $y = 1$ is an asymptote on the right because

$$\lim_{x \to \infty} \left(1 + \frac{1}{x-1} \right) = 1.$$

It is also an asymptote on the left because y again approaches 1 as $x \to -\infty$. For numerically large values of x,

$$y = 1 + \frac{1}{x-1}$$

behaves like $y = 1$.

The line $x = 1$ is a vertical asymptote of the right-hand branch of the graph because

$$\lim_{x \to 1^+} \left(1 + \frac{1}{x-1} \right) = \infty$$

and of the left-hand branch because

$$\lim_{x \to 1^-} \left(1 + \frac{1}{x-1} \right) = -\infty.$$

For x close to 1, the function behaves like $y = 1/(x-1)$.   ∎

In general, we have the following definitions of horizontal and vertical asymptotes.

DEFINITIONS

Horizontal and Vertical Asymptotes

A line $y = b$ is a **horizontal asymptote** of the graph of a function $y = f(x)$ if either

$$\lim_{x \to \infty} f(x) = b \qquad \text{or} \qquad \lim_{x \to -\infty} f(x) = b.$$

A line $x = a$ is a **vertical asymptote** of the graph if either

$$\lim_{x \to a^+} f(x) = \pm\infty \qquad \text{or} \qquad \lim_{x \to a^-} f(x) = \pm\infty.$$

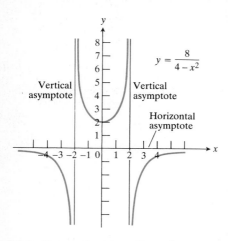

3.24 The graph of $y = 8/(4 - x^2)$. Notice that the curve approaches the x-axis from only one side. Asymptotes do not have to be two-sided.

EXAMPLE 5 Figure 3.24 shows the graph of

$$y = \frac{8}{4 - x^2}.$$

The line $y = 0$ is an asymptote on the right because $y \to 0$ as $x \to \infty$, and it is an asymptote on the left because $y \to 0$ as $x \to -\infty$.

The line $x = 2$ is a vertical asymptote because

$$\lim_{x \to 2^-} \frac{8}{4 - x^2} = \infty$$

and again because

$$\lim_{x \to 2^+} \frac{8}{4 - x^2} = -\infty.$$

Similarly, the line $x = -2$ is a vertical asymptote because $y \to \infty$ as $x \to -2^+$ and again because $y \to -\infty$ as $x \to -2^-$. ∎

Oblique Asymptotes

If a rational function is a quotient of two polynomials that have no common factor, and if the degree of the numerator is one larger than the degree of the denominator, as it is in

$$y = \frac{x^2 - 3}{2x - 4},$$

then the graph will have an **oblique asymptote** (Fig. 3.25). To see why this is so, we divide $x^2 - 3$ by $2x - 4$:

$$
\begin{array}{r}
\frac{x}{2} + 1 \\
2x - 4 \overline{) x^2 - 3 } \\
\underline{x^2 - 2x } \\
2x - 3 \\
\underline{2x - 4} \\
1
\end{array}
$$

3.25 The graph of $y = (x^2 - 3)/(2x - 4)$.

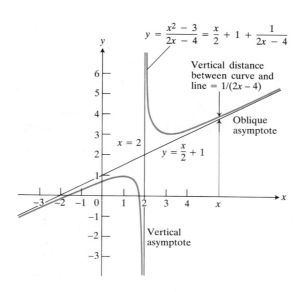

This tells us that

$$y = \frac{x^2 - 3}{2x - 4} = \underbrace{\frac{x}{2} + 1}_{\text{linear}} + \underbrace{\frac{1}{2x - 4}}_{\substack{\text{remainder} \\ \text{goes to } 0 \\ \text{as } x \to \pm\infty}}.$$

(1)

From this representation we see that

$$y - \left(\frac{x}{2} + 1\right) = \frac{1}{2x - 4}$$

and that

$$\lim_{x \to \pm\infty} \left[y - \left(\frac{x}{2} + 1\right)\right] = \lim_{x \to \pm\infty} \frac{1}{2x - 4} = 0.$$

Thus the vertical distance between the curve $y = (x^2 - 3)/(2x - 4)$ and the line $y = x/2 + 1$ approaches zero as $x \to \pm\infty$. Therefore, the line

$$y = \frac{x}{2} + 1$$

is an asymptote of the curve.

Notice too that as $x \to \pm\infty$ the slope

$$y' = \frac{1}{2} - \frac{2}{(2x - 4)^2}$$

of the curve approaches 1/2, which is the slope of the line.

The curve also has a vertical asymptote at $x = 2$.

EXAMPLE 6 Graph the function

$$y = x + \frac{1}{x}.$$

Solution We take into account symmetry, asymptotes, rise and fall, concavity, and dominant terms.

Symmetry. The function is odd (Example 2), so its graph will be symmetric with respect to the origin.

Asymptotes. If we let $x \to 0$ from the right, $y \to \infty$. If we let $x \to 0$ from the left, $y \to -\infty$. The line $x = 0$ is a vertical asymptote.

$$\text{As } x \to \infty, \quad \frac{1}{x} \to 0 \quad \text{and} \quad (y - x) = \frac{1}{x} \to 0.$$

Therefore, the line $y = x$ is also an asymptote of the graph.

Rise and fall. The first derivative,

$$y' = 1 - \frac{1}{x^2} = \frac{x^2 - 1}{x^2} = \frac{(x + 1)(x - 1)}{x^2},$$

equals zero when $x = -1$ and $x = 1$, and it is not defined at $x = 0$. The derivative is positive when $x < -1$, negative when $-1 < x < 0$ and $0 < x < 1$, and positive when $x > 1$ (Fig. 3.26a). Accordingly, the graph rises on $(-\infty, -1)$ and $(1, \infty)$ and falls on $(-1, 0)$ and $(0, 1)$. The function has a local maximum

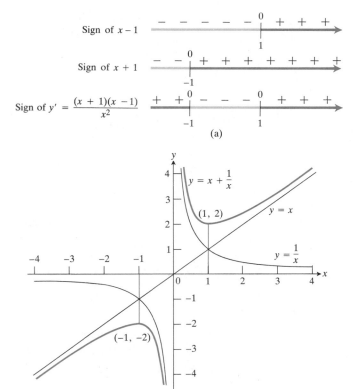

3.26 (a) The sign pattern for $y' = (x + 1)(x - 1)/x^2$. (b) The graph of $y = x + (1/x)$, based on this and other information in Example 6.

value at $x = -1$, where y' changes from $+$ to $-$, and a local minimum value at $x = 1$, where y' changes from $-$ to $+$.

Concavity. The second derivative

$$y'' = \frac{2}{x^3}$$

is negative when $x < 0$ and positive when $x > 0$. The curve is concave down when $x < 0$ and concave up when $x > 0$.

Dominant terms. When x is numerically small, it contributes very little to the sum

$$y = x + \frac{1}{x}$$

in comparison to the contribution of $1/x$. The term $1/x$ is the **dominant term** for values of x close to zero, and we can expect $y = x + 1/x$ to behave like $y = 1/x$ for these values. On the other hand, when x is numerically very large, the term $1/x$ contributes little to the value of y in comparison to the contribution of x. The term x is the dominant term for numerically large values of x, and we can expect $y = x + 1/x$ to behave like $y = x$ for these values. (Hence the asymptote $y = x$.)

Therefore, by looking at the formula $y = x + 1/x$ we can tell two things right away about the behavior of y:

$$y \approx \frac{1}{x} \quad \text{when } x \text{ is small} \qquad \text{and} \qquad y \approx x \quad \text{when } x \text{ is large.} \qquad (2)$$

Of all the observations we can make quickly about y, these are probably the most useful.

The graph. To graph $y = x + 1/x$, we draw the asymptote $y = x$, sketch the curve $y = 1/x$, and plot the local maximum and minimum points. We then sketch the curve that fits these and has the symmetry and other properties we have discovered (Fig. 3.26b). ■

Predicting Behavior from Dominant Terms

We can often use algebra to rewrite a formula in a way that shows what terms dominate at different values of x.

EXAMPLE 7 If

$$y = \frac{2}{x} + 6x^2,$$

then

$$y \approx \frac{2}{x}, \quad x \text{ numerically small,}$$

$$y \approx 6x^2, \quad x \text{ numerically large.} \qquad ■$$

EXAMPLE 8 From Eq. (1) we have

$$y = \frac{x^2 - 3}{2x - 4} = \frac{x}{2} + 1 + \frac{1}{2x - 4},$$

$$y \approx \frac{x}{2} + 1, \quad |x| \text{ large,}$$

$$y \approx \frac{1}{2x - 4}, \quad x \text{ near 2.}$$

(See Fig. 3.25.) ■

EXAMPLE 9 For

$$y = \frac{x + 3}{x + 2} = 1 + \frac{1}{x + 2},$$

we have

$$y \approx 1, \quad |x| \text{ large,}$$

$$y \approx \frac{1}{x + 2}, \quad x \text{ near } -2. \qquad ■$$

EXAMPLE 10   From Example 5 we have

$$y = \frac{8}{4 - x^2} = \frac{8}{(2 - x)(2 + x)},$$

so

$$y \approx \frac{8}{(2 - x)(4)} = \frac{2}{2 - x}, \quad x \text{ near } 2,$$

$$y \approx \frac{8}{4(2 + x)} = \frac{2}{2 + x}, \quad x \text{ near } -2.$$

(See Fig. 3.24.)

PROBLEMS

Say whether the functions in Problems 1–12 are even, odd, or neither. Try to answer without writing anything down (except the answer).

1. $y = x$

2. $y = x^2$

3. $y = x^3$

4. $y = x^4$

5. $y = x + 1$

6. $y = x + x^2$

7. $y = x^2 + 1$

8. $y = x + x^3$

9. $y = \frac{1}{x^2 - 1}$

10. $y = \frac{1}{x - 1}$

11. $y = \frac{x}{x^2 - 1}$

12. $y = \frac{x^2}{x^2 - 1}$

In Problems 13–38, find the domain of the function and investigate the following properties of its graph: (a) symmetry, (b) intercepts, (c) asymptotes, (d) slope at the intercepts, (e) rise and fall, (f) concavity, and (g) dominant terms. Then graph the function.

13. $y = \frac{1}{2x - 3}$

14. $y = \frac{1}{2x + 4}$

15. $y = x - \frac{1}{x}$

16. $y = \frac{x}{2} + \frac{2}{x}$

17. $y = \frac{x + 3}{x + 2}$ (See Example 9.)

18. $y = \frac{x}{x + 1}$ (*Hint:* Divide x by $x + 1$.)

19. $y = \frac{x + 1}{x - 1}$ (*Hint:* Divide $x + 1$ by $x - 1$.)

20. $y = \frac{x - 4}{x - 5}$ (*Hint:* Divide $x - 4$ by $x - 5$.)

21. $y = \frac{1}{x^2 + 1}$

22. $y = \frac{1}{x^2 - 1}$

23. $y = \frac{x}{x^2 - 1}$

24. $y = \frac{x^2}{x - 1}$

25. $y = \frac{x^2}{x^2 - 1}$

26. $y = \frac{x^2 + 1}{x - 1}$

27. $y = \frac{x^2 - 4}{x - 1}$

28. $y = \frac{x^2 - 9}{x - 5}$

29. $y = \frac{x^2 - 1}{2x + 4}$

30. $y = \frac{x^2 + x - 2}{x - 2}$

31. $y = \frac{2}{x} + 6x^2$

32. $y = x^2 + \frac{1}{x^2}$

33. $y = \frac{x^2 - 4}{x^2 - 1}$

34. $y = \frac{x^2 + 8}{x^2 - 4}$

35. $y = \frac{x^2 + 1}{x^2 - 4x + 3}$

36. $y = \frac{1}{x} - \frac{1}{x^2}$

37. $y = \frac{x - 1}{x^2(x - 2)}$

38. $y = \frac{x^2 + 1}{x^3 - 4x}$

39. The graph of the function $f(x) = 2 + (\sin x)/x$, $x > 0$ (see Fig. 1.94) crosses the line $y = 2$ whenever $\sin x = 0$ and therefore crosses the line infinitely often as $x \to \infty$. Show that the slope of the curve nevertheless approaches the slope of the line as $x \to \infty$.

40. Let

$$f(x) = \frac{x^3 - 4x}{x^3 - x}, \quad x \neq 0.$$

a) Find the limits of f as $x \to 0$ and $x \to \pm\infty$.

b) What value should be assigned to $f(0)$ to make f continuous at $x = 0$?

c) Find equations for the vertical and horizontal asymptotes of f.

d) Describe the symmetry of f.

e) Sketch a graph of the continuous extension of f.

Even vs. Odd

Say whether the functions in Problems 41–56 are even, odd, or neither. Do as many as you can without writing anything down but the answer.

41. $\sin x$

42. $\cos x$

43. $\sec x$

44. $\csc x$

45. $\tan x$

46. $\cot x$

47. $\sin 2x$

48. $\cos 3x$

49. $\sin^2 x$

50. $\cos^3 x$

51. $x \cos x$

52. $x^2 \cos x$

53. $x \sin x$

54. $x^2 \sin x$

55. $\sin x + \cos x$

56. $\sin x \cos x$

57. Even, odd, or neither? Say which:

a) $y = |x|$,

b) $y = \begin{cases} x^2 - 9, & x \neq 5, \\ 16, & x = 5. \end{cases}$

58. If u and v are even functions of x, what can be said about $u + v$, $u - v$, u/v, and uv?

59. If u and v are odd functions of x, what can be said about $u + v$, $u - v$, u/v, and uv?

60. If u is an even function of x, and v is an odd function of x, what can be said about uv, u/v, and v/u?

61. a) Suppose that an odd function is known to be increasing on the interval $[-1, 0]$. What can be said about its behavior on $[0, 1]$?

b) Suppose that an even function is known to be increasing on $[-1, 0]$. What can be said about its behavior on $[0, 1]$?

c) Suppose that the graph of an odd function is concave up on $[0, 1]$. What can be said about its concavity on $[-1, 0]$?

d) Suppose that the graph of an even function is concave down on $[0, 1]$. What can be said about its concavity on $[-1, 0]$?

TOOLKIT PROGRAMS

Derivative Grapher   Super * Grapher
Function Evaluator

3.4

Maxima and Minima: Theory

In this article, we show how to use derivatives to locate points where functions take on maximum or minimum values. In the next article, we shall show how the theory described here is put into practice in solving applied problems.

Relative vs. Absolute

Figure 3.27 shows a point $x = c$ where a function $y = f(x)$ has a minimum value. If we move away from c to either side, the curve rises and the function values get larger. When we investigate more of the curve, however, we find that f takes on an even smaller value at $x = d$. We see that $f(c)$ is not the absolute minimum value of f on $[a, b]$ but only a relative or local minimum.

3.27 How maxima and minima are classified.

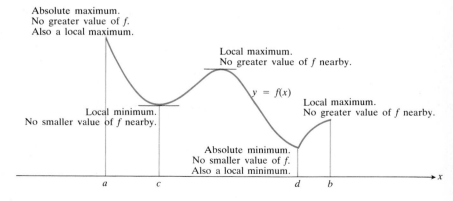

A function f is said to have a **relative** or **local minimum** at $x = c$ if

$$f(c) \le f(x)$$

for all values of x in some open interval about c. (If c is an endpoint of the domain of f, the interval is to be half-open, containing c as an endpoint.) The interval might be small or it might be large, but no value of the function in the interval is less than $f(c)$.

For a **local** or **relative maximum** at $x = c$,

$$f(c) \ge f(x)$$

for all x in some interval about c. No value of the function in this interval is greater than $f(c)$.

The word *relative* or *local* is used to distinguish such a point from an **absolute maximum** or **minimum** that would occur if either of the inequalities

$$f(x) \le f(c) \qquad \text{or} \qquad f(c) \le f(x)$$

held for all x in the domain of f, not just for all x in an appropriate interval about c.

The First Derivative Theorem

In Fig. 3.27, two of the extreme values of the function occur at endpoints of the function's domain, one occurs at a point where f' fails to exist, and two occur at interior points where f' is zero. This is typical for a function defined on a closed interval. As the following theorem says, if a function has a local maximum or minimum value at an interior domain point and also has a derivative there, then the derivative at that point must be zero.

THEOREM 1

The First Derivative Theorem for Local Extreme Values

Suppose that a function f has a local maximum or minimum at an interior point c of an interval on which f is defined. If f' is defined at c, then

$$f'(c) = 0.$$

Proof   Suppose, for instance, that f has a local minimum value at $x = c$, so that $f(x) \ge f(c)$ for all values of x near c (Fig. 3.28). Since c is an interior point of the domain of f, we know that the limit

$$\lim_{x \to c} \frac{f(x) - f(c)}{x - c} \tag{1}$$

that defines $f'(c)$ is a two-sided limit. This means that both the right- and left-hand limits at $x = c$ exist and equal $f'(c)$. When we examine these limits separately, we find that

$$\lim_{x \to c^+} \frac{f(x) - f(c)}{x - c} \ge 0 \tag{2}$$

because $f(x) \ge f(c)$ and $x - c > 0$ when x lies to the right of c and that

$$\lim_{x \to c^-} \frac{f(x) - f(c)}{x - c} \le 0 \tag{3}$$

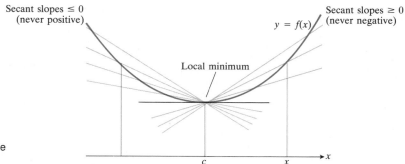

Secant slopes ≤ 0 (never positive)

Secant slopes ≥ 0 (never negative)

$y = f(x)$

Local minimum

3.28   A curve with a local minimum value at $x = c$.

because $f(x) \geq f(c)$ and $x - c < 0$ when x lies to the left of c. The inequality (2) tells us that $f'(c)$ cannot be less than zero. The inequality (3) says that $f'(c)$ also cannot be greater than zero. Therefore, $f'(c)$ equals 0.

The argument that $f'(c) = 0$ when f has a local maximum value at c is so similar to the argument we just gave that we shall omit it.   ■

By ruling out interior points where f' is different from zero, Theorem 1 tells us where to look for a function's local maxima and minima. We can safely confine our investigation, the theorem says, to

1. interior points where f' is zero,
2. interior points where f' does not exist,
3. endpoints of the function's domain.

None of these points is necessarily the location of a local maximum or minimum, but these are the *only* candidates.

The points where $f' = 0$ or fails to exist are commonly called the **critical points** of f. Thus, the only points worth considering in the search for a function's extreme values are critical points and endpoints. Now for some examples.

Continuous Functions on Closed Intervals

Most applications call for finding the absolute maximum or minimum values of a continuous function on a closed interval, so we shall deal with this case first. The number of points where these values can occur is usually so small that we can simply list them and compute the corresponding function values, enabling us to see what the maximum and minimum are and at what point or points they are taken on.

EXAMPLE 1   Find the absolute maximum and minimum values of $y = x^{2/3}$ on the interval $-2 \leq x \leq 3$.

Solution   We find the function values at the critical points and endpoints. The first derivative

$$y' = \frac{2}{3}x^{-1/3} = \frac{2}{3\sqrt[3]{x}}$$

has no zeros but is undefined at $x = 0$. We evaluate the function at this one

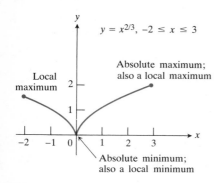

3.29 A function can have a minimum value at a point where its derivative does not exist. One way this can happen is shown here, where the curve has a vertical tangent at $x = 0$. Another way is shown in Fig. 3.12, where $y = |x|$ has no tangent at all at $x = 0$.

critical point and at the domain's two endpoints:

Critical point value: $y(0) = 0$

Endpoint values: $y(-2) = 4^{1/3}$
     $y(3) = 9^{1/3}$

The maximum value is $9^{1/3}$, taken on at $x = 3$. The minimum is 0, taken on at $x = 0$. See Fig. 3.29.   ∎

Example 1 shows that when a maximum or minimum occurs at the *end* of a curve that exists only over a limited interval, the derivative need not vanish at such a point. This does not contradict Theorem 1 in any way. The argument that establishes the theorem does not apply to endpoints because the limit of the difference quotient in Eq. (1) is required to be a two-sided limit at $x = c$ for the subsequent argument to work.

EXAMPLE 2   Find the absolute maximum and minimum values of the function $y = x^3 - 3x + 2$ on the interval $[0, 2]$.

Solution   We find the function values at the critical points and endpoints. The derivative

$$y' = 3x^2 - 3 = 3(x - 1)(x + 1)$$

is defined at every point of the function's domain and is zero at $x = 1$. (It is also zero at $x = -1$, not in the domain.) A comparison of the values

Critical point value: $y(1) = 0$

Endpoint values: $y(0) = 2$
     $y(2) = 4$

reveals an absolute maximum of 4 at $x = 2$ and an absolute minimum of 0 at $x = 1$. See Fig. 3.30.   ∎

Infinite or Open Intervals: The Second Derivative Test

Functions defined on infinite or open intervals may or may not take on maximum or minimum values, and when we list the critical point and endpoint values it may not be immediately clear what to conclude. In such a case, we investigate the function's infinite limits (if any) and, perhaps, the function's second derivative as well.

EXAMPLE 3   Find the maximum and minimum values, if any, of the function

$$y = x + \frac{1}{x}.$$

Solution   We know that the function has no absolute maximum or minimum value because $y \to \infty$ as $x \to \infty$ and $y \to -\infty$ as $x \to -\infty$. The function may still have local maxima and minima, however, at domain endpoints (in this case there aren't any) and at points where the first derivative

$$y' = 1 - \frac{1}{x^2}$$

is zero or fails to exist.

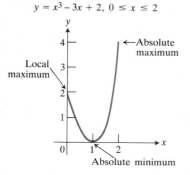

3.30 The graph of $y = x^3 - 3x + 2$ on $[0, 2]$.

3.31 The graph of $y = x + (1/x)$ has only two extreme values: a local maximum at the point $(-1, -2)$ and a local minimum at the point $(1, 2)$.

We see from the formula that y' is zero at $x = \pm 1$ and is undefined at $x = 0$. We disregard $x = 0$ because it does not lie in the domain of the function. The only function values to consider, then, are the critical point values

$$y(-1) = -2 \qquad \text{and} \qquad y(1) = 2.$$

But be careful: Even though -2 is less than 2, it is not safe to conclude that -2 is a local minimum and 2 is a local maximum. In fact, it is just the other way around, as we shall now see.

The second derivative

$$y'' = \frac{2}{x^3}$$

is negative when $x < 0$. Therefore, the curve is concave down at $x = -1$, and $y(-1) = -2$ is a local maximum value. Again, $y'' = 2/x^3$ is positive when $x = 1$ so the curve is concave up at $x = 1$, and $y(1) = 2$ is a local minimum value. Thus, the local minimum value is greater than the local maximum value. How can this be? See Fig. 3.31. ∎

The second derivative test most frequently used to investigate points where the first derivative is zero is the following one.

Second Derivative Test for Local Maxima and Minima

If $f'(c) = 0$ and $f''(c) < 0$, then f has a local maximum at $x = c$.

If $f'(c) = 0$ and $f''(c) > 0$, then f has a local minimum at $x = c$.

Miscellaneous Problem 94 at the end of the chapter indicates why the test works.

The second derivative test does not say what happens when y' and y'' are both zero. At such a point, there may be a maximum, a minimum, or neither (see Problem 38).

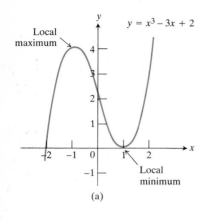

Local maximum

$y = x^3 - 3x + 2$

Local minimum

(a)

+ + + + | – – – – – | + + + + + → x
        −1   1

(b) Sign of $y' = 3(x-1)(x+1)$

3.32 The graph of $y = x^3 - 3x + 2$.

EXAMPLE 4 Find all maxima and minima of the function

$$y = x^3 - 3x + 2, \qquad -\infty < x < \infty.$$

Solution The domain has no endpoints, and the function is differentiable at every point. Therefore, extreme values can occur only where the derivative

$$y' = 3x^2 - 3 = 3(x - 1)(x + 1)$$

is zero, which means at $x = 1$ and $x = -1$. The second derivative $y'' = 6x$ is positive at $x = 1$ and negative at $x = -1$. Hence $y(1) = 0$ is a local minimum value, and $y(-1) = 4$ is a local maximum value. See Fig. 3.32(a).   ■

As a comparison of Examples 2 and 4 suggests, the existence of extreme values can depend as much on a function's domain as on the formula used to compute the function's values. The conclusions we draw about the values of $y = x^3 - 3x + 2$ on [0, 2] in Example 2 are not the same as the conclusions we draw about the values of $y = x^3 - 3x + 2$ on the domain $-\infty < x < \infty$ in Example 4. On [0, 2], the function has absolute maximum and minimum values, as guaranteed by the max-min theorem for continuous functions in Article 1.11 (Theorem 7). On $-\infty < x < \infty$, the maximum and minimum values are relative, not absolute, and the relative maximum is not equal to the absolute maximum found earlier.

Getting Along without y''

In Example 4, we do not need the second derivative test if we watch how $y' = 3(x - 1)(x + 1)$ changes sign as x increases (Fig. 3.32b). The change in y' from + to − at $x = -1$ indicates a local maximum value at $x = -1$. Similarly, the change in y' from − to + at $x = 1$ indicates a local minimum value at $x = 1$.

When the second derivative of a function is hard to find or hard to work with once found, checking y' for a sign change is probably a better route to take. In fact, when the sign changes in y' are easily found, there may be no advantage at all in calculating y''.

Summary

To find the maximum and minimum values of a function $y = f(x)$, locate

1. the points where f' is zero or fails to exist and
2. the endpoints (if any) of the domain of f.

These are the only candidates for the values of x where f takes on an extreme value.

The points where the derivative is zero or fails to exist are called **critical points** of f. By comparing the values of f at the critical points and endpoints with each other and with the values of f at nearby points, decide which of them, if any, are local (or absolute) maxima and minima. Be sure to look for sign changes in y'. Also, the second derivative may help (if it is easy to calculate).

PROBLEMS

Find the critical points for each of the functions in Problems 1–30. For each critical point, determine whether the function has a local maximum, a local minimum, or neither there. If possible, find the absolute maximum and minimum values of the function on the indicated domain.

1. $y = x - x^2$ on $[0, 1]$

2. $y = x - x^3$, $-\infty < x < \infty$

3. $y = x - x^3$ on $[0, 1]$

4. $y = x^3 - 3x^2 + 2$, $-\infty < x < \infty$

5. $y = x^3 - 147x$, $-\infty < x < \infty$

6. $y = x^3 - 2x^2 + x$ on $[-1, 2]$

7. $y = x^2 - 4x + 3$ on $(0, 3)$

8. $y = x^3 - 6x$ on $[0, 2]$

9. $y = x - x^2$ on $(0, 1)$

10. $y = 1/(x - 2)$ on $(1, 3)$

11. $y = 2x$ on $[0, 3]$

12. $y = 1/(3 - x)$ on $[0, 4]$

13. $y = x^2 + (2/x)$, $x > 0$

14. $y = x/(1 + x)$ on $[0, 1]$

15. $y = x^3 + 3x^2 + 3x + 2$, $-\infty < x < \infty$

16. $y = -x^2 + 4x$, $x \geq 0$

17. $y = \sqrt{x} - x$, $x \geq 0$

18. $y = \sqrt{4 - x^2}$, $-2 \leq x \leq 2$

19. $y = x^4 - 4x$ on $[0, 2]$

20. $y = x^4 - x^2$ on $[-1, 1]$

21. $y = \tan x$ on $[0, \pi/2)$

22. $y = \sec x$ on $(-\pi/2, \pi/2)$

23. $y = 2 \sin x + \cos 2x$ on $[0, \pi/2]$

24. $y = x^4 - 2x^2 + 2$ on $[-1, 2]$

25. $y = x^4 - 8x^3 - 270x^2$, $-\infty < x < \infty$

26. $y = x^4 - (x^3/3) - 2x^2 + x - 1$, $-\infty < x < \infty$

27. $y = (x - x^2)^{-1}$ on $(0, 1)$

28. $y = |x^3|$ on $[-2, 3]$. What would occur if the domain were changed to $[-2, 3)$?

29. $y = \begin{cases} -x & \text{for } x \leq 0, \\ 2x - x^2 & \text{for } x > 0 \end{cases}$

30. $y = \begin{cases} 3 - x & \text{on } [0, 2], \\ (1/2)x^2 & \text{on } (2, 3] \end{cases}$

Find the absolute maximum and minimum values (if any) of the functions in Problems 31–36. In each case, start by rewriting the given formula without absolute values by using appropriate versions of the formula on different intervals of the domain. Then

include the boundaries of these intervals among the points you investigate. Unless stated otherwise, the domain is the function's natural (largest possible) domain.

31. $y = \dfrac{x}{1 + |x|}$

32. $y = \dfrac{|x|}{1 + |x|}$

33. $y = \sin|x|$, $-2\pi \leq x \leq 2\pi$

34. $y = \dfrac{|x|}{x}$

35. $y = |x^2 - 1|$, $-1 \leq x \leq 2$

36. $y = |x - x^2|$, $x \geq 0$

37. We do not need calculus to show that $x + 1/x \geq 2$ when $x > 0$. To see why, multiply out the left side of the inequality $(x - 1)^2 \geq 0$ and divide through by x.

38. Calculate the values of the first and second derivatives of $y = x^3$, $y = x^4$, and $y = -x^4$ at the origin to show that y'' has no predictive value when it is zero.

39. Find the critical points, asymptotes, and points of inflection and graph the function

$$y = \frac{x}{x^2 + 1}.$$

40. Test the function

$$y = \frac{x^3}{6} + \frac{x^2}{2} - 1 + \cos x$$

for the existence of a local maximum or minimum at $x = 0$.

41. Suppose that a function $y = f(x)$ is known to be differentiable for all values of x and to have a local maximum at $x = c$. Which of the following must be true of the graph of f'?
   a) It has a point of inflection at $x = c$.
   b) It crosses the x-axis at $x = c$.
   c) It has a local maximum or minimum at $x = c$.

42. Find the maximum height of the curve $y = 4 \sin x - 3 \cos x$ above the x-axis.

43. Find the maximum height of the curve $y = 4 \sin^2 x - 3 \cos^2 x$ above the x-axis.

44. a) Find a value of b that will ensure that the function $y = 2x^3 + bx + c$ has a local minimum value at $x = 1$.
   b) Why will no value of b make $y = 2x^3 + bx + c$ have a local maximum value at $x = 1$?

3.5
Maxima and Minima: Problems

Many problems that arise in science and mathematics call for finding the largest and smallest values that a differentiable function can assume on some particular domain. As we have just seen, calculus is the right mathematical tool for finding these values. In this article, we develop a strategy for solving applied problems.

EXAMPLE 1   Find two positive numbers whose sum is 20 and whose product is as large as possible.

Solution   If one number is x, then the other is $(20 - x)$ and their product is

$$f(x) = x(20 - x) = 20x - x^2.$$

We seek the value or values of x that make the product f as large as possible. In the context of the problem, the domain of f can be restricted to the interval $0 \le x \le 20$. We see from the formula for f that f is continuous on this interval, so it will indeed take on a maximum value. Since f is differentiable as well, we know from our work in Article 3.4 that the maximum is located at either $x = 0$, $x = 20$, or at an interior point where $f' = 0$. The derivative

$$f'(x) = 20 - 2x = 2(10 - x)$$

equals zero only when $x = 10$. The absolute maximum is therefore one of the three numbers

$$f(0) = 0, \qquad f(10) = 10(20 - 10) = 100, \qquad f(20) = 0.$$

The function takes on the largest of these, 100, when $x = 10$ (Fig. 3.33). The two numbers we seek are therefore $x = 10$ and $20 - x = 10$.   ∎

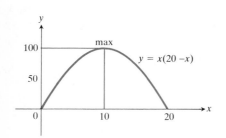

3.33   The product $y = x(20 - x)$ reaches a maximum value of 100 when $x = 10$.

EXAMPLE 2   A rectangle is to be inscribed in a semicircle of radius 2. What is the largest area the rectangle can have and what are its dimensions?

Solution   To describe the dimensions of the rectangle, we place the circle and rectangle in the coordinate plane as shown in Fig. 3.34. The length, height, and area of the rectangle can then be expressed in terms of the position x of the lower right-hand corner:

Length: $2x,$

Height: $\sqrt{4 - x^2},$

Area: $2x \cdot \sqrt{4 - x^2}.$

Our mathematical goal is now to find the absolute maximum value of the continuous function

$$A(x) = 2x\sqrt{4 - x^2}$$

on the interval $0 \le x \le 2$. We do this by examining the values of A at the critical points and endpoints.

The derivative

$$\frac{dA}{dx} = \frac{-2x^2}{\sqrt{4 - x^2}} + 2\sqrt{4 - x^2}$$

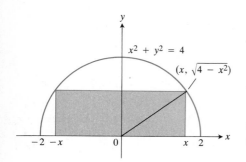

3.34   The rectangle and semicircle in Example 2.

■ ■ ■

THE LEAST AND THE GREATEST

Many problems of the seventeenth century that motivated the development of the calculus were maxima and minima problems. Often these problems came from research in physics, such as finding the maximum range of a cannon. Galileo showed that the maximum range of a cannon is obtained with a firing angle of 45 degrees above the horizontal (see Article 13.2). He also obtained formulas for predicting maximum heights reached by projectiles fired at various angles to the ground. Another common problem in the seventeenth century was that of finding the greatest and least distances of a planet from the sun (see Article 14.4). Pierre de Fermat and René Descartes also worked on other problems of maxima and minima. Fermat's work culminated in his Principle of Least Time (see Example 6), which was generalized by Sir William Hamilton in his Principle of Least Action—one of the most powerful underlying ideas in physics.

is not defined when $x = 2$ and is equal to zero when

$$\frac{-2x^2}{\sqrt{4 - x^2}} + 2\sqrt{4 - x^2} = 0$$

$$-2x^2 + 2(4 - x^2) = 0$$

$$8 - 4x^2 = 0$$

$$x^2 = 2$$

$$x = \pm\sqrt{2}.$$

For $0 \le x \le 2$, we therefore have

Critical point values: $A(\sqrt{2}) = 2\sqrt{2}\sqrt{4 - 2} = 4$
                           $A(2) = 0,$

Endpoint values: $A(0) = 0$
                           $A(2) = 0.$

The area has a maximum value of 4 when the rectangle is $2x = 2\sqrt{2}$ units long by $\sqrt{4 - x^2} = \sqrt{2}$ units high.   ■

EXAMPLE 3   A square sheet of tin a inches on a side is to be used to make an open-top box by cutting a small square of tin from each corner and bending up the sides. How large a square should be cut from each corner to make the box have as large a volume as possible?

Solution   We first draw a figure to illustrate the problem (Fig. 3.35). In the figure, the side of the square cut from each corner is taken to be x inches and the volume of the box in cubic inches is then given by

$$y = x(a - 2x)^2, \qquad 0 \le x \le a/2. \tag{1}$$

The restrictions placed on x in Eq. (1) are imposed by the fact that one can neither cut a negative amount of material from a corner nor cut away more than

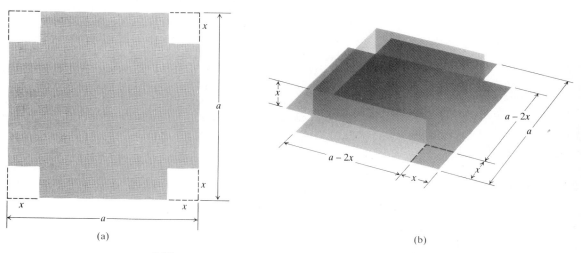

(a)

(b)

3.35   To make an open box, squares are cut from the corners of a square sheet of tin (a) and the sides are bent up (b). What value of x gives the largest volume?

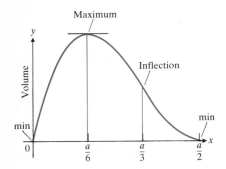

Maximum

Inflection

min

$\dfrac{a}{6}$ $\dfrac{a}{3}$ $\dfrac{a}{2}$

min

3.36 The volume of the box in Fig. 3.35 graphed as a function of x.

the total amount present. It is also evident that $y = 0$ when $x = 0$ or when $x = a/2$, so the maximum volume y must occur at a value of x between 0 and $a/2$. The function in Eq. (1) has a derivative at every such point, and hence the maximum occurs at an interior point of $[0, a/2]$ where $y' = 0$. From Eq. (1) we find

$$y = a^2x - 4ax^2 + 4x^3,$$

$$y' = a^2 - 8ax + 12x^2 = (a - 2x)(a - 6x),$$

so

$$y' = 0 \quad \text{when } x = \frac{a}{6} \quad \text{or} \quad x = \frac{a}{2}.$$

Of these values, only $x = a/6$ lies in the interior of $[0, a/2]$. The maximum therefore occurs at $x = a/6$. Each corner square should have dimensions $a/6$ by $a/6$ to produce a box of maximum volume. The graph of the volume is shown in Fig. 3.36. ∎

EXAMPLE 4 You are about to make a one-quart oil can shaped like a right circular cylinder. What dimensions will use the least material?

Solution Again we start with a figure (Fig. 3.37). The requirement that the can hold a quart of oil is the same as

$$V = \pi r^2 h = a^3, \tag{2a}$$

if the radius r and altitude h are in inches and a^3 is the number of cubic inches in a quart ($a^3 = 57.75$).

How shall we interpret the phrase "least material"? One possibility is to ignore the thickness of the material and the waste in manufacturing. Then we ask for dimensions r and h that make the total surface area

$$A = 2\pi r^2 + 2\pi rh \tag{2b}$$

as small as possible while still satisfying the constraint in Eq. (2a). Problem 40 shows one way we might take the cost of wasted material into account.

In either case, what kind of oil can do we expect? Not a tall, thin one like a six-foot piece of pipe one inch in diameter, nor a short, wide one like a nine-inch pie pan. These both hold about a quart but use more metal than the standard oil can we see in stores and gas stations. We expect something in between.

We are not quite ready to apply the methods used in Examples 1–3 because Eq. (2b) expresses A as a function of *two* variables, r and h, and our methods call for A to be expressed as a function of just *one* variable. However, Eq. (2a) can be used to express either of the variables r or h in terms of the other; in fact, we find

$$h = \frac{a^3}{\pi r^2} \tag{2c}$$

or

$$r = \sqrt{\frac{a^3}{\pi h}}. \tag{2d}$$

$\longleftarrow 2r \longrightarrow$

h

3.37 This one-quart can can be made from the least material when $h = 2r$.

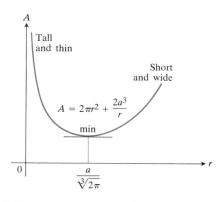

3.38 A graph of $A = 2\pi r^2 + (2a^3/r)$.

The division in Eqs. (2c) and (2d) is legitimate because neither r nor h can be zero, and only the positive square root is used in (2d) because the radius r can never be negative. If we substitute from (2c) into (2b), we have

$$A = 2\pi r^2 + \frac{2a^3}{r}, \qquad 0 < r, \tag{2e}$$

and now we can apply our previous methods. A minimum of A can occur only at a point where

$$\frac{dA}{dr} = 4\pi r - 2a^3 r^{-2} \tag{2f}$$

is zero, that is, where

$$4\pi r = \frac{2a^3}{r^2}, \qquad r^3 = \frac{a^3}{2\pi}, \qquad \text{or} \qquad r = \frac{a}{\sqrt[3]{2\pi}}. \tag{2g}$$

At such a value of r,

$$\frac{d^2A}{dr^2} = 4\pi + 4a^3 r^{-3} = 12\pi > 0,$$

which indicates a local minimum. Since the second derivative is *always positive* for $0 < r$, the curve is concave up and there can be no other local minimum, so we have also found the absolute minimum. From (2g) and (2c), we find

$$r = \frac{a}{\sqrt[3]{2\pi}} = \sqrt[3]{\frac{V}{2\pi}}, \qquad h = \frac{2a}{\sqrt[3]{2\pi}} = 2\sqrt[3]{\frac{V}{2\pi}} = 2r$$

as the dimensions of the can of volume V having minimum surface area. The height of the can equals its diameter. For $V = 1$ qt $= 57.75$ in^3, this comes to

$$r = 2.1 \text{ in.}, \qquad h = 4.2 \text{ in.} \qquad \text{(rounded)}$$

Figure 3.38 shows the graph of A as a function of r. ∎

Some problems about maxima and minima have their solution at an endpoint of the function's domain. The next example discusses one such problem.

EXAMPLE 5 A wire of length L is available for making a circle and a square. How should the wire be divided between the two shapes to maximize the sum of the enclosed areas?

Solution We picture a circle and a square made from a length of wire and label the dimensions involved (Fig. 3.39). In the notation of Fig. 3.39, the total area enclosed by the circle and square is

$$A = \pi r^2 + x^2, \tag{3a}$$

where

$$2\pi r + 4x = L. \tag{3b}$$

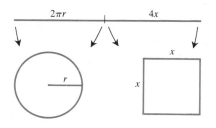

3.39 A circle and a square made from a wire $2\pi r + 4x$ units long. How large an area can they enclose together?

Our mathematical goal is to find the value of r on the interval $0 \le 2\pi r \le L$ that maximizes the differentiable function A. We could solve Eq. (3b) to express x in terms of r and substitute the result in Eq. (3a), but instead we shall

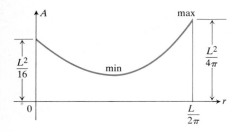

3.40 The sum of the areas in Fig. 3.39 graphed as a function of r.

differentiate (3a) and (3b) with respect to r and combine the resulting expressions for the derivatives:

$$\frac{dA}{dr} = 2\pi r + 2x\frac{dx}{dr}, \tag{3c}$$

$$2\pi + 4\frac{dx}{dr} = \frac{dL}{dr}. \tag{3d}$$

Since L is a constant, $dL/dr = 0$ and we can solve Eq. (3d) for dx/dr,

$$2\pi + 4\frac{dx}{dr} = 0$$

$$\frac{dx}{dr} = -\frac{2\pi}{4} = -\frac{\pi}{2},$$

and rewrite Eq. (3c) as

$$\frac{dA}{dr} = 2\pi r - \pi x. \tag{3e}$$

While (3e) does not express dA/dr in terms of a single variable, it does show us that the second derivative

$$\frac{d^2A}{dr^2} = 2\pi - \pi\frac{dx}{dr} = 2\pi + \frac{\pi^2}{2}$$

is always positive. Thus, the graph of A as a function of r is concave up. The maximum value of A on the interval $0 \le 2\pi r \le L$ therefore occurs at one or both of the endpoints $r = 0$, $r = L/2\pi$. We calculate the endpoint values:

At $r = 0$: $x = \dfrac{L}{4},$ $A = \dfrac{L^2}{16}.$

At $r = \dfrac{L}{2\pi}$: $x = 0,$ $A = \dfrac{L^2}{4\pi}.$

The maximum value of A occurs when $r = L/2\pi$ or $2\pi r = L$. Therefore, to enclose the most area, the wire should not be cut at all but should be bent into a circle.

The graph of A as a function of r looks like the curve in Fig. 3.40, which shows two relative maxima, but we do not need to know this to find the absolute maximum.   ∎

EXAMPLE 6 *Fermat's principle and Snell's law.* The speed of light depends on the medium through which it travels and is generally slower in denser media. In a vacuum, light travels at the famous speed $c = 3 \times 10^8$ m/sec, but in the earth's atmosphere it travels slower than that, and in glass slower still.

Fermat's principle in optics states that light travels from one point to another along a path for which the time of travel is a minimum. Let us find the path that a ray of light will follow in going from a point A in a medium where the speed of light is c_1 to a point B in a second medium where its speed is c_2.

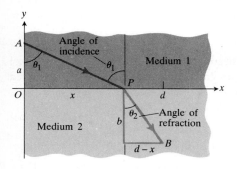

3.41 A light ray is refracted (deflected from its path) when it passes from one medium to another. θ_1 is the angle of incidence, and θ_2 is the angle of refraction.

Solution We assume that the two points lie in the xy-plane and that the x-axis separates the two media as in Fig. 3.41.

dt/dx negative dt/dx zero dt/dx positive

3.42 The sign pattern of dt/dx in Example 6.

In either medium, where the speed of light remains constant, "shortest time" means "shortest path," and the ray of light will follow a straight line. Hence the path from A to B will consist of a line segment from A to a boundary point P, followed by another line segment from P to B. According to the formula distance equals rate times time,

$$\text{time} = \frac{\text{distance}}{\text{rate}}.$$

The time required for light to travel from A to P is therefore

$$t_1 = \frac{AP}{c_1} = \frac{\sqrt{a^2 + x^2}}{c_1}.$$

From P to B the time required is

$$t_2 = \frac{PB}{c_2} = \frac{\sqrt{b^2 + (d - x)^2}}{c_2}.$$

The travel time from A to B is the sum of these:

$$t = t_1 + t_2 = \frac{\sqrt{a^2 + x^2}}{c_1} + \frac{\sqrt{b^2 + (d - x)^2}}{c_2}. \tag{4a}$$

Our mathematical goal is to find any value or values of x in the interval $0 \le x \le d$ at which t assumes its minimum value. We find

$$\frac{dt}{dx} = \frac{x}{c_1\sqrt{a^2 + x^2}} - \frac{(d - x)}{c_2\sqrt{b^2 + (d - x)^2}} \tag{4b}$$

or, if we use the angles θ_1 and θ_2 in the figure,

$$\frac{dt}{dx} = \frac{\sin \theta_1}{c_1} - \frac{\sin \theta_2}{c_2}. \tag{4c}$$

For air and water, the light velocity ratio at room temperature is 1.33 and Snell's law becomes $\sin \theta_1 = 1.33 \sin \theta_2$. In this laboratory photograph, $\theta_1 = 35.5°$, $\theta_2 = 26°$, and $(\sin 35.5°)/(\sin 26°) = 0.581/0.438 \approx 1.33$, as predicted. (Photograph from *PSSC Physics, Second Edition;* 1965; D.C. Heath & Company with Education Development Center, Inc. NCFMF Book of Film Notes, 1974; The MIT Press with Education Development Center, Inc., Newton, Massachusetts. Reprinted by permission.)

We can see from Eq. (4b) that dt/dx is negative at $x = 0$ and positive at $x = d$, so it is zero at some point in between (Fig. 3.42). There is only one such point because dt/dx is an increasing function of x. At this point,

$$\frac{\sin \theta_1}{c_1} = \frac{\sin \theta_2}{c_2}. \tag{4d}$$

This equation is known as **Snell's law** or the **law of refraction.**   ∎

A (Sometimes) Useful Argument

The following argument can be applied to many problems in which we are trying to find an extreme value, say a maximum, of a function. Suppose we know the following:

1. We can restrict our search to a closed interval I.

2. The function is continuous and differentiable everywhere (we might know this from a formula for the function or assume it from physical considerations).

3. The function does not attain a maximum at either endpoint of I.

Then we know that the function has at least one maximum at an interior point of I, at which the derivative must be zero. Therefore, if we find that the derivative of the function is zero at only one interior point of I, then this is where the function takes on its maximum. A similar argument applies to a search for a minimum value.

EXAMPLE 7 A producer can sell x items per week at a price $P = 200 - 0.01x$ cents, and it costs $C = 50x + 20,000$ cents to make x items. What is the most profitable number to make?

Solution The total weekly revenue from x items is

$$xP = 200x - 0.01x^2.$$

The profit T is revenue minus cost:

$$T = xP - C = (200x - 0.01x^2) - (50x + 20,000)$$

$$= 150x - 0.01x^2 - 20,000.$$

For very large values of x, say beyond a million, T is negative. Therefore, T takes on its maximum value somewhere in the interval $0 \le x \le 10^6$. The formula for T shows that T is differentiable at every x, and obviously T does not take on its maximum at either endpoint, 0 or 10^6. The derivative

$$\frac{dT}{dx} = 150 - 0.02x$$

is zero only when

$$x = 7500.$$

Therefore, $x = 7500$ is the production level for maximum profit. ■

To solve the problem in Example 7 it would have been as effective to calculate the second derivative $d^2T/dx^2 = -0.02$ and conclude that T has a local maximum at $x = 7500$. A quick look at the domain of x or the graph of T is all that would then have been needed to show that the maximum was absolute. The virtue of the argument that we gave instead is that it can apply to a function whose second derivative is nonexistent or difficult to compute.

Strategy for Solving Max-Min Problems

1. *Draw a figure*. When possible, draw a figure to illustrate the problem. Label the parts that are important in the problem. Keep track of which letters represent constants and which represent variables.

2. *Write an equation*. Write an equation for the quantity whose maximum or minimum value you wish to find. It is usually desirable to express the quantity as a function of a single variable, say $y = f(x)$. This may require some algebra and the use of information from the statement of the problem. Note the domain in which the values of x are to be found.

3. *Test the critical points and endpoints.* The extreme value of f will be found among the values f takes on at the endpoints of the domain and at the points where f' is zero or fails to exist. List the values of f at these points. If f has an absolute maximum or minimum on its domain, it will appear on the list. You may have to examine the sign pattern of f' or the sign of f'' to decide whether a given value represents a maximum, a minimum, or neither.

PROBLEMS

1. The sum of two nonnegative numbers is 20. Find the numbers (a) if the sum of their squares is to be as large as possible; (b) if the product of the square of one number and the cube of the other is to be as large as possible; (c) if one number plus the square root of the other is as large as possible.

2. What is the largest area possible for a right triangle whose hypotenuse is 5 cm long?

3. What is the smallest perimeter possible for a rectangle of area 16 in^2?

4. Where is the sum $\sin x + \cos x$ smallest on the interval $0 \le x \le \pi/2$?

5. What point on the circle $x = \cos t$, $y = \sin t$, $0 \le t \le 2\pi$ is closest to the point $(1, \sqrt{3})$? (*Hint:* What value of t minimizes the function

$$f(t) = (\cos t - 1)^2 + (\sin t - \sqrt{3})^2$$

that gives the square of the distance between $(\cos t, \sin t)$ and $(1, \sqrt{3})$?)

6. Figure 3.43 shows a rectangle inscribed in an isosceles right triangle whose hypotenuse is 2 units long.
   a) Express the y-coordinate of P in terms of x. You might start by writing an equation for the line AB.
   b) Express the area of the rectangle in terms of x.
   c) What is the largest area possible for a rectangle inscribed in an equilateral right triangle whose hypotenuse is 2 units long?

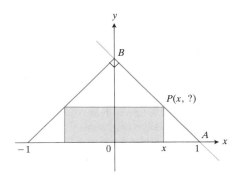

3.43 The rectangle in Problem 6.

7. Find the area of the largest rectangle with lower base on the x-axis and upper vertices on the parabola $y = 12 - x^2$.

8. An open rectangular box is to be made from a piece of cardboard 8 in. wide and 15 in. long by cutting squares from the corners and folding up the sides. Find the dimensions of the box of largest volume.

9. A rectangular plot is to be bounded on one side by a straight river and enclosed on the other three sides by a fence. With 800 m of fence at your disposal, what is the largest area you can enclose?

10. Show that among all rectangles with a given perimeter P, the one with the largest area is a square.

11. An open storage bin with a square base and vertical sides is to be constructed from 108 ft^2 of material. Determine its dimensions if its volume is to be a maximum. Neglect the thickness of the material and waste in construction.

12. A box with a square base and no top is to hold 32 in^3. Find the dimensions that require the least building material. Neglect the thickness of the material and waste in construction.

13. You are planning to close off the corner of the first quadrant with a line segment 20 units long running from $(a, 0)$ to $(0, b)$. Show that the area of the triangle enclosed by the segment is largest when $a = b$.

14. To find the equation of the line $y = mx + b$ through the point $(2, 1)$ that cuts the least area from the first quadrant, follow these steps:
   a) Express m in terms of b.
   b) Find the line's x-intercept.
   c) Find a formula $A(b)$ that expresses the area of the triangle as a function of b.
   d) Find the value of b that minimizes A.

15. Suppose that at time t the position of a particle moving on the x-axis is $x = (t - 1)(t - 4)^4$.
   a) When is the particle at rest?
   b) During what time interval does the particle move to the left?
   c) What is the fastest the particle goes while moving to the left?

16. The equations $x = 1/t$, $y = (t^3/3) - 4t$, $t > 0$, give the position of a particle moving in the plane. At what value of

t does the path traced by the particle have the largest slope? What is the largest slope?

17. You are designing a poster to contain 50 in^2 of printing with margins of 4 in. each at top and bottom and 2 in. at each side. What overall dimensions will minimize the amount of paper used?

18. The height of an object moving vertically is given by

$$s = -16t^2 + 96t + 112,$$

with *s* in feet and *t* in seconds. Find (a) the object's velocity when $t = 0$, (b) its maximum height, and (c) its velocity when $s = 0$.

19. A rectangular plot of land containing 216 m^2 is to be enclosed by a fence and divided into two equal parts by another fence parallel to one of the sides. What dimensions of the outer rectangle require the smallest total length for the two fences? How much fence is needed?

20. Two sides of a triangle are to have lengths *a* cm and *b* cm. What is the largest area the triangle can have? (*Hint:* $A = (1/2)ab \sin \theta$.)

21. What are the dimensions of the lightest open-top right circular cylindrical can that can be made of material weighing 1 gm/cm^2 if the can's volume is to be 1000 cm^3? Compare the result here with the result in Example 4.

22. Find the largest possible value of $2x + y$ if *x* and *y* are the lengths of the sides of a right triangle whose hypotenuse is $\sqrt{5}$ units long.

23. A container with a rectangular base, rectangular sides, and no top is to have a volume of 2 m^3. The width of the base is to be 1 m. When cut to size, the material costs \$10 per square meter for the base and \$5 per square meter for the sides. What is the cost of the least expensive container?

24. The U.S. Postal Service will accept a box for domestic shipment only if the sum of the length and girth (distance around) does not exceed 108 in. Find the dimensions of the largest acceptable box with a square end (Fig. 3.44).

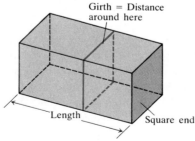

3.44   The box in Problem 24.

25. (Conclusion of the sonobuoy problem, Problem 16 in Article 2.9.) Show that the value of *x* that minimizes the distance between the points (x, x^2) and $(2, -1/2)$ in Fig. 3.45 is a solution of the equation $1/(x^2 + 1) = x$. (*Hint:* Minimize the square of the distance.)

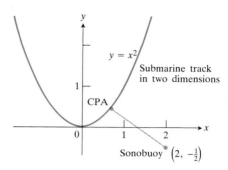

3.45   The submarine track and sonobuoy in Problem 25. CPA = closest point of approach.

26. Compare the answers to the following two construction problems.
   a) A rectangular sheet of dimensions *x* cm by *y* cm is to be rolled into the cylinder shown in Fig. 3.46(a). The perimeter of the sheet is 36 cm. What values of *x* and *y* give the largest cylinder volume? What is this volume?
   b) The rectangular sheet in (a) is revolved about one of the edges of length *y* to sweep out the cylinder in Fig. 3.46(b). What values of *x* and *y* give the largest cylinder volume now? What is this volume?

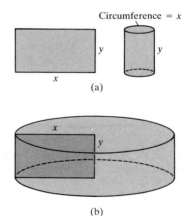

3.46   The rectangular sheet and cylinders in Problem 26.

27. A right triangle of given hypotenuse is rotated about one of its legs to generate a right circular cone. Find the cone of greatest volume.

28. It costs a manufacturer *c* dollars each to manufacture and distribute a certain item. If the items sell at *x* dollars each, the number sold is given by $n = a/(x - c) + b(100 - x)$, where *a* and *b* are certain positive constants. What selling price will bring a maximum profit?

29. What value of *a* makes the function

$$f(x) = x^2 + \frac{a}{x}$$

have
a) a local minimum at $x = 2$?
b) a local minimum at $x = -3$?
c) a point of inflection at $x = 1$?
d) Show that the function cannot have a local maximum for any value of a.

30. What values of a and b make the function
$$f(x) = x^3 + ax^2 + bx + c$$
have
a) a local maximum at $x = -1$ and a local minimum at $x = 3$,
b) a local minimum at $x = 4$ and a point of inflection at $x = 1$?

31. A wire of length L is cut into two pieces, one being bent to form a square and the other to form an equilateral triangle. How should the wire be cut (a) if the sum of the two areas is a minimum; (b) if the sum of the areas is a maximum?

32. Find the point on the curve $y = \sqrt{x}$ nearest the point $(c, 0)$,
a) if $c \geq 1/2$,
b) if $c < 1/2$.

33. Find the volume of the largest right circular cone that can be inscribed in a sphere of radius r.

34. Find the volume of the largest right circular cylinder that can be inscribed in a sphere of radius r.

35. Show that the volume of the largest right circular cylinder that can be inscribed in a given right circular cone is 4/9 the volume of the cone.

36. The strength of a rectangular beam is proportional to the product of its width and the square of its depth. Find the dimensions of the strongest beam that can be cut from a circular cylindrical log of radius r.

37. The stiffness of a rectangular beam is proportional to the product of its breadth and the cube of its depth but is not related to its length. Find the proportions of the stiffest beam that can be cut from a log of given diameter.

38. The intensity of illumination at any point is a constant times the strength of the light source divided by the square of the distance from the source. If two sources of relative strengths a and b are a distance c apart, at what point on the line joining them will the intensity be a minimum? Assume that the intensity at any point is the sum of intensities from the two sources.

39. A window is in the form of a rectangle surmounted by a semicircle. The rectangle is of clear glass while the semicircle is of colored glass that transmits only half as much light per square foot as clear glass does. The total perimeter is fixed. Find the proportions of the window that will admit the most light.

40. Right circular cylindrical tin cans are to be manufactured to contain a given volume. There is no waste involved in cutting the tin for the vertical sides, but the tops of radius r are cut from squares that measure $2r$ units on a side. The total

amount of material consumed by each can is therefore $A = 8r^2 + 2\pi rh$ (rather than the area $A = 2\pi r^2 + 2\pi rh$ in Example 4). Find the ratio of height to diameter for the most economical cans.

41. A closed container is made from a right circular cylinder of radius r and height h with a hemispherical dome on top. Find the relationship between r and h that maximizes the volume for a given surface area.

42. Suppose the L units of wire in Example 5 are to be divided between the circle and square to make the total area enclosed a minimum instead of a maximum. What should the radius of the circle and side length of the square be now? (All the wire is to be used.)

43. Are the following functions ever negative? How do you know?
a) $f(x) = x^2 - x + 1$
b) $f(x) = 3 + 4 \cos x + \cos 2x$

44. The trough shown in Fig. 3.47 is to be constructed from a piece of metal 20 ft long and 3 ft wide. The trough is to be made by turning up strips of width 1 ft to make equal angles θ with the vertical.
a) Express the volume of the trough in terms of the angle θ.
b) Find the maximum possible volume of the trough.

3.47 The trough in Problem 44.

45. Find all maxima and minima of $y = \sin x + \cos x$.

46. A rectangular sheet of $8\frac{1}{2} \times 11$-inch paper is placed on a flat surface and one of the vertices is lifted up and placed on the opposite longer edge, while the other three vertices are held in their original positions. With all four vertices now held fixed, the paper is smoothed flat, as shown in Fig. 3.48.

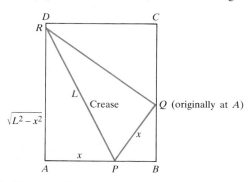

3.48 The paper in Problem 46.

The problem is to find the minimum possible length of the crease.

a) Try it with paper.

b) Show that $L^2 = 2x^3/(2x - 8.5)$, $4.25 < x < 8.5$.

c) Minimize L^2.

47. A silo is to be made in the form of a cylinder surmounted by a hemisphere. The cost of construction per square foot of surface area is twice as great for the hemisphere as for the cylinder. Determine the dimensions to be used if the volume is fixed and the cost of construction is to be a minimum. Neglect the thickness of the silo and waste in construction.

48. If the sum of the surface areas of a cube and a sphere is constant, what is the ratio of an edge of the cube to the diameter of the sphere when (a) the sum of their volumes is a minimum; (b) the sum of their volumes is a maximum?

†49. Two towns, located on the same side of a straight river, agree to construct a pumping station and filtering plant at the river's edge, to be used jointly to supply the towns with water. The distances of the two towns from the river are a and b and the distance between them is c (Fig. 3.49). Show that the sum of the lengths of the pipe lines joining them to the pumping station is at least as great as $\sqrt{c^2 + 4ab}$.

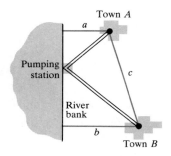

3.49   The towns and pumping station in Problem 49.

†50. Light from a source A is reflected to a point B by a plane mirror. If the time required for the light to travel from A to the mirror and then to B is a minimum, show that the angle of incidence is equal to the angle of reflection. See Fig. 3.50.

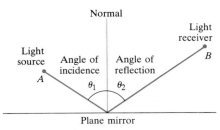

3.50   In studies of light reflection, the angles of incidence and reflection are measured from the line normal to the reflecting surface. Problem 50 asks you to show that if light obeys Fermat's "least-time" principle, then $\theta_1 = \theta_2$.

---

†You may prefer to do this problem without calculus.

51. Let $f(x)$ and $g(x)$ be the differentiable functions on $a \le x \le b$ whose graphs are shown in Fig. 3.51. The point c is the point where the vertical distance D between the curves is the greatest. Show that the tangents to the curves at $x = c$ are parallel.

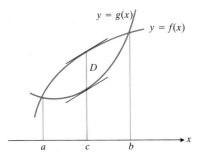

3.51   The graphs for Problem 51.

52. You operate a tour service that offers the following rates:

i) $200 per person if 50 people (the minimum number to book the tour) go on the tour.

ii) For each additional person, up to a maximum of 80 people total, everyone's charge is reduced by $2. It costs you $6000 (a fixed cost) plus $32 per person to conduct the tour. How many people does it take to maximize your profit?

53. At the end of Article 1.8, we defined the *marginal cost* dy/dx of producing x tons of steel per week to be the derivative of the cost y with respect to x. The derivative at any particular x was the approximate cost of producing the next ton of steel.

   Suppose that to sell x tons of steel per week, a company prices its steel at P dollars per ton. The company's *revenue* is then the product xP. Its *marginal revenue* is the derivative of xP with respect to x, or the rate of change of revenue per unit increase in production. The company's profit T is the difference between revenue and cost: $T = xP - y$.

   The company wants to adjust production to achieve maximum profit.

a) Show that if the profit can be maximized then it is maximized at the value of x for which marginal revenue equals marginal cost.

b) What condition involving the second derivatives of P and y would ensure that the point of equality in (a) was one of maximum profit (and not, say, one of minimum profit)?

54. The wall shown in Fig. 3.52 is 8 ft high and stands 27 ft from the building. What is the length of the shortest straight beam that will reach to the side of the building from the ground outside the wall?

55. *Autocatalytic reactions.* A catalyst for a chemical reaction is a substance that controls the rate of the reaction without undergoing any permanent change in itself. An autocatalytic reaction is one whose product is a catalyst for its own

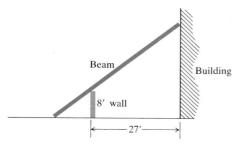

3.52 The diagram for Problem 54.

formation. Such a reaction may proceed slowly at first if the amount of catalyst present is small, and slowly again at the end when most of the original substance is used up. But in between, when both the substance and its product are abundant, the reaction may proceed at a faster rate.

In some cases it is reasonable to assume that the rate $v = dx/dt$ of the reaction is proportional both to the amount of the original substance present and to the amount of product. That is, v may be considered to be a function of x alone, and

$$v = kx(a - x) = kax - kx^2,$$

where

x = the amount of product,

a = the amount of substance at the beginning,

k = a positive constant.

At what value of x does the rate v have a maximum? What is the maximum value of v?

 TOOLKIT PROGRAMS

Derivative Grapher   Super * Grapher

3.6
Related Rates of Change

How fast does the radius change when air is blown into a spherical soap bubble at the rate of 10 cm³/sec? How fast does the water level drop when a cylindrical tank is drained at the rate of 3 liters/sec?

Questions like these ask us to calculate the rate at which one variable changes from the rate at which another variable is known to change. To calculate that rate, we write an equation that relates the two variables and differentiate it to get an equation that relates the rate we seek to the rate we know. This article gives the details.

EXAMPLE 1 *The soap bubble*. How fast does the radius of a spherical soap bubble change when air is blown into it at the rate of 10 cm³/sec?

Solution We are given the rate at which the volume is changing and are asked for the rate at which the radius is changing.

We think of the volume V and radius r as differentiable functions of time t. The derivatives dV/dt and dr/dt give the rates at which V and r are changing. We are told that

$$\frac{dV}{dt} = 10 \qquad \text{(Air is blown in at the rate of 10 cm}^3\text{/sec.)}$$

and we are asked for

$$\frac{dr}{dt}. \qquad \text{(How fast does the radius change?)}$$

To answer the question, we first write an equation that relates V and r:

$$V = \frac{4}{3}\pi r^3. \qquad \text{(The bubble is spherical.)} \qquad (1)$$

We then differentiate both sides of the equation with respect to t to get an equation that relates dr/dt to dV/dt:

$$\frac{dV}{dt} = 4\pi r^2 \frac{dr}{dt}. \tag{2}$$

We substitute the known value $dV/dt = 10$ and solve for dr/dt:

$$\frac{dr}{dt} = \frac{10}{4\pi r^2}. \tag{3}$$

We see from Eq. (3) that the rate at which r changes at any particular time depends on how big r is at the time. When r is small, dr/dt will be large; when r is large, dr/dt will be small:

At $r = 1$ cm: $\dfrac{dr}{dt} = \dfrac{10}{4\pi} \approx 0.8$ cm/sec,

At $r = 10$ cm: $\dfrac{dr}{dt} = \dfrac{10}{400\pi} \approx 0.008$ cm/sec.   ∎

EXAMPLE 2   *The cylindrical tank.* How fast does the water level drop when a cylindrical tank is drained at the rate of 3 liters/sec?

Solution   We draw a picture of a partially filled cylindrical tank, calling its radius r and the height of the water h (Fig. 3.53). We call the volume of water in the tank V.

The radius r is a constant, but V and h change with time. We think of V and h as differentiable functions of time and use t to represent time. The derivatives dV/dt and dh/dt give the rates at which V and h change. We are told that

$$\frac{dV}{dt} = -3 \qquad \text{(The tank is drained at the rate of 3 liters/sec.)}$$

and we are asked for

$$\frac{dh}{dt}. \qquad \text{(How fast does the water level drop?)}$$

To answer the question, we first write an equation that relates V and h:

$$V = \pi r^2 h. \qquad \text{(The tank is cylindrical.)}$$

We then differentiate both sides with respect to t to get an equation that relates dh/dt to dV/dt:

$$\frac{dV}{dt} = \pi r^2 \frac{dh}{dt}.$$

We substitute the known value $dV/dt = -3$ and solve for dh/dt:

$$\frac{dh}{dt} = -\frac{3}{\pi r^2}. \tag{4}$$

The water level is dropping at the constant rate of $3/\pi r^2$ liters/sec.   ∎

3.53   The cylindrical tank in Example 2.

EXAMPLE 3   A ladder 26 ft long rests on horizontal ground and leans against a vertical wall. The foot of the ladder is pulled away from the wall at the

3.54 The ladder in Example 3. When the foot of the ladder slides away from the wall, y decreases and x increases.

rate of 4 ft/sec. How fast is the top sliding down the wall when the foot is 10 ft from the wall?

Solution   We answer the question in stages.

STEP 1: *Draw a picture and name the variables and constants.* We draw a picture of a ladder resting on horizontal ground and leaning against a vertical wall (Fig. 3.54). The variables in the picture are the height y of the top of the ladder above ground and the distance x from the foot of the ladder to the wall. We also show the length of the ladder, 26 ft. We let t denote time and assume x and y to be differentiable functions of t.

STEP 2: *Write down any additional numerical information.* We are told that $dx/dt = 4$ ft/sec.

STEP 3: *Write down what we are asked to find.* We are asked to find the value of dy/dt when $x = 10$ ft.

STEP 4: *Write an equation that relates the variables.* The angle at the base of the wall is a right angle, so x and y are related by the Pythagorean theorem:

$$x^2 + y^2 = 26^2.$$

STEP 5: *Differentiate* to express dy/dt in terms of dx/dt:

$$2x\frac{dx}{dt} + 2y\frac{dy}{dt} = 0$$

$$\frac{dy}{dt} = -\frac{x}{y}\frac{dx}{dt}. \tag{5}$$

When $x = 10$,

$$y = \sqrt{26^2 - 10^2} = 24, \qquad \frac{dx}{dt} = 4, \qquad \text{and} \qquad \frac{dy}{dt} = -\frac{10}{24}(4) = -\frac{5}{3}.$$

The top of the ladder is moving down (y is decreasing) at 5/3 ft/sec when the ladder is 10 ft from the wall. ∎

The steps in Example 3 are the steps of the basic strategy for solving related rate problems.

Strategy for Solving Related Rate Problems

1. *Draw a picture and name the variables and constants.* Use t for time and assume that all variables are differentiable functions of t.

2. *Write down any additional numerical information.*

3. *Write down what you are asked to find.* Usually this is a rate, expressed as a derivative.

4. *Write an equation that relates the variables.* You may have to combine two or more equations to get a single equation that relates the variable whose rate you want to the variable whose rate you know.

5. *Differentiate.*

Keep these steps in mind as you read the next two examples.

3.55 The conical tank in Example 4. To show that the water level is changing in this conical tank, the depth of the water is denoted by a variable, *y*. The rate at which the level changes is then *dy/dt*.

EXAMPLE 4 Water runs into a conical tank at the rate of 2 ft³/min. The tank stands point down and has a height of 10 ft and a base radius of 5 ft. How fast is the water level rising when the water is 6 ft deep?

Solution We draw and label a picture of a partially filled conical tank (Fig. 3.55). The variables in the problem are

V = the volume (ft³) of water in the tank at time t (min),

x = the radius (ft) of the surface of the water at time t,

y = the depth (ft) of water in the tank at time t.

The constants are the dimensions of the tank and the rate

$$\frac{dV}{dt} = 2 \text{ ft}^3/\text{min}$$

at which the tank fills.

We are asked for the value of dy/dt when $y = 6$.

The relation between V and y is expressed by the equation

$$V = \frac{1}{3}\pi x^2 y. \tag{6}$$

This equation involves x as well as V and y, but we can eliminate x in the following way. By similar triangles (Fig. 3.55),

$$\frac{x}{y} = \frac{5}{10} \quad \text{or} \quad x = \frac{1}{2}y.$$

Therefore,

$$V = \frac{1}{3}\pi \left(\frac{1}{2}y\right)^2 y = \frac{1}{12}\pi y^3. \tag{7}$$

We now differentiate both sides of Eq. (7) to express dy/dt in terms of dV/dt:

$$\frac{dV}{dt} = \frac{1}{4}\pi y^2 \frac{dy}{dt} \quad \text{or} \quad \frac{dy}{dt} = \frac{4}{\pi y^2}\frac{dV}{dt}. \tag{8}$$

When $dV/dt = 2$ and $y = 6$,

$$\frac{dy}{dt} = \frac{4 \cdot 2}{\pi \cdot 36} = \frac{2}{9\pi} \approx 0.071 \text{ ft/min}.$$

When the water is 6 ft deep, the water level is rising at 0.071 ft/min.

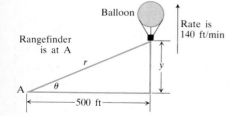

3.56 The rising balloon in Example 5.

EXAMPLE 5 A balloon rising from the ground at 140 ft/min is tracked by a rangefinder at point A, located 500 ft from the point of liftoff (see Fig. 3.56). Find the rate at which the angle at A and the range r are changing when the balloon is 500 ft above the ground.

Solution *The angle at A.* The variable θ is related to y by the equation

$$\tan \theta = \frac{y}{500}.$$

The derivatives of θ and y with respect to time t are therefore related by the equation

$$\sec^2\theta \frac{d\theta}{dt} = \frac{1}{500}\frac{dy}{dt}. \tag{9}$$

When $y = 500$ ft, $\theta = \pi/4$, and $\sec^2\theta = (\sqrt{2})^2 = 2$. Also, $dy/dt = 140$ ft/min. We substitute these values into Eq. (9) to find $d\theta/dt$:

$$2\frac{d\theta}{dt} = \frac{1}{500}(140), \quad \text{or} \quad \frac{d\theta}{dt} = \frac{140}{1000} = 0.14 \text{ radians/min.}$$

When $y = 500$ ft, the angle at A is increasing at the rate of 0.14 radians per minute.

The range r. The variables r and y are related by the equation

$$r^2 = 500^2 + y^2$$

and their time derivatives by the equations

$$2r\frac{dr}{dt} = 2y\frac{dy}{dt}, \quad \text{or} \quad \frac{dr}{dt} = \frac{y}{r}\frac{dy}{dt}.$$

When $y = 500$, $r = \sqrt{500^2 + 500^2} = 500\sqrt{2}$, $dy/dt = 140$ ft/min, and

$$\frac{dr}{dt} = \frac{500}{500\sqrt{2}}140 = \frac{140\sqrt{2}}{2} = 70\sqrt{2} \text{ ft/min.}$$

When $y = 500$ ft, the range is increasing at the rate of $70\sqrt{2}$ feet per minute.

PROBLEMS

1. Let A be the area of a circle of radius r at time t. Write an equation that relates dA/dt to dr/dt.

2. Let S be the surface area of a sphere of radius r at time t. Write an equation that relates dS/dt to dr/dt.

3. Let V be the volume of a cube whose sides have length s at time t. Write an equation that relates dV/dt to ds/dt.

4. When a circular plate of metal is heated in an oven, its radius increases at the rate of 0.01 cm/min. At what rate is the plate's area increasing when the radius is 50 cm?

5. Ohm's law for electrical circuits like the one in Fig. 3.57 states that $V = IR$, where V is the voltage, I is the current in amperes, and R is the resistance in ohms. Suppose that V is

increasing at the rate of 1 volt/sec while I is decreasing at the rate of 1/3 amp/sec. Let t denote time in seconds.
a) What is the value of dV/dt?
b) What is the value of dI/dt?
c) What equation relates dR/dt to dV/dt and dI/dt?
d) Find the rate at which R is changing when $V = 12$ volts and $I = 2$ amp. Is R increasing or decreasing?

6. The length l of a rectangle is decreasing at the rate of 2 cm/sec and the width w is increasing at the rate of 2 cm/sec. When $l = 12$ cm and $w = 5$ cm, find the rates of change of (a) the area, (b) the perimeter, and (c) the lengths of the diagonals of the rectangle. Which of these quantities are decreasing and which are increasing?

7. A baseball diamond is a square 90 ft on a side (Fig. 3.58). A player runs from first base to second base at a rate of 16 ft/sec. At what rate is the player's distance from third base decreasing when the player is 30 ft from first base? Carry out the following steps to answer this question.
a) Assign variables to the distances between the player and second and third bases. What are the values of the variables at the time in question?
b) How are the variables related?

3.57 The current in this circuit obeys Ohm's law (Problem 5).

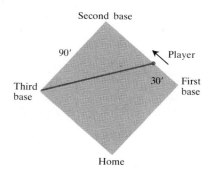

3.58   The baseball diamond in Problem 7.

c) How are the derivatives of the variables related?

d) Calculate the rate at which the player's distance from third base is changing.

8. Let V be the volume and S the total surface area of a solid right circular cylinder that is 5 ft high and has radius r ft. Find dV/dS when $r = 3$.

9. At what rate is the area of the triangle formed by the ladder, ground, and wall in Example 3 changing when $x = 10$? If the motion began with the ladder flat against the wall at $t = 0$ sec, and $x = 4t$, at what time was the area of the triangle largest?

10. Sand falls from a conveyor belt onto a conical pile at the rate of 10 ft³/min. The radius of the base of the pile is always equal to one half its altitude. How fast is the altitude of the pile increasing when the pile is 5 ft high?

11. Suppose that a raindrop is a perfect sphere. Assume that, through condensation, the raindrop accumulates moisture at a rate proportional to its surface area. Show that the radius increases at a constant rate.

12. Point A moves along the x-axis at the constant rate of a ft/sec while point B moves along the y-axis at the constant rate of b ft/sec. Find how fast the distance between them is changing when A is at the point $(x, 0)$ and B is at the point $(0, y)$.

13. A spherical balloon is inflated with gas at the rate of 100 ft³/min. Assuming that the gas pressure remains constant, how fast is the radius of the balloon increasing at the instant when the radius is 3 ft? How fast is the surface area increasing?

14. A boat is pulled in to a dock by a rope with one end attached to the bow of the boat, the other end passing through a ring attached to the dock at a point 4 ft higher than the bow of the boat. If the rope is pulled in at the rate of 2 ft/sec, how fast is the boat approaching the dock when 10 ft of rope are out?

15. A balloon is 200 ft off the ground and rising vertically at the constant rate of 15 ft/sec. An automobile passes beneath it traveling along a straight road at the constant rate of 45 mph = 66 ft/sec. How fast is the distance between them changing one second later?

16. Water is withdrawn from a conical reservoir (vertex down) 8 ft in diameter and 10 ft deep at the constant rate of 5 ft³/min. How fast is the water level falling when the depth of water in the reservoir is 6 ft?

17. A point moves on the curve $3x^2 - y^2 = 12$ so that its y-coordinate increases at the constant rate of 6 m/sec. At what rate is the x-coordinate changing when $x = 4$ m? What is the slope of the curve when $x = 4$ m?

18. A particle moves around the circle $x^2 + y^2 = 1$ with an x-velocity component $dx/dt = y$. Find dy/dt. Does the particle travel clockwise, or counterclockwise, around the circle?

19. A particle in the first quadrant moves along the parabola $y = x^2$ in such a way that the x-coordinate of its position $P(x, x^2)$ increases at a steady 10 m/sec. How fast is the angle of inclination of the line OP that joins P to the origin changing when $x = 3$ m? When $x = 103$ m?

20. a) A particle is moving clockwise around the circle $x^2 + y^2 = 1$ at the rate of one revolution per second. How fast is the particle's x-coordinate increasing the instant the particle passes through the point $(0, 1)$ at the top of the circle? (*Hint:* Let θ be the angle the radius from the origin to P makes with the positive x-axis. What is $d\theta/dt$ in radians per second?)

  b) A wheel of radius 1 ft rolls over level ground at the rate of one revolution per second. How fast is a point on the top of the wheel moving relative to the ground?

21. A man 6 ft tall walks at the rate of 5 ft/sec toward a streetlight that is 16 ft above the ground. At what rate is the tip of his shadow moving? At what rate is the length of his shadow changing when he is 10 ft from the base of the light?

22. A light shines from the top of a pole 50 ft high. A ball is dropped from the same height from a point 30 ft away from the light. How fast is the shadow of the ball moving along the ground 1/2 sec later? (Assume the ball falls a distance $s = 16t^2$ ft in t seconds.)

23. A girl flies a kite at a height of 300 ft, the wind carrying the kite horizontally away from her at a rate of 25 ft/sec. How fast must she let out the string when the kite is 500 ft away from her?

24. A spherical iron ball 8 in. in diameter is coated with a layer of ice of uniform thickness. If the ice melts at the rate of 10 in³/min, how fast is the thickness of the ice decreasing when it is 2 in. thick? How fast is the outer surface area of ice decreasing?

25. A highway patrol plane flies 1 mile above a straight road at a steady ground speed of 120 mph. The pilot sees an oncoming car and, with radar, determines that the line-of-sight distance from plane to car is 1.5 miles, decreasing at the rate of 136 mph. Find the car's speed along the highway.

26. On a day when the sun will pass directly overhead, the shadow of an 80-ft building on level ground is 60 ft long (Fig. 3.59). If the angle θ made by the sun with the ground is increasing at the rate of 0.25 deg/min, at what rate is the

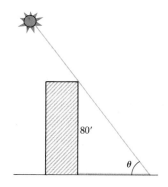

80'

θ

3.59   The building in Problem 26.

shadow length decreasing? (Express your answer in inches per minute, to the nearest tenth.)

27. At noon ship A was 12 nautical miles due north of ship B. Ship A was sailing south at 12 knots (nautical miles per hour) and continued to do so all day. Ship B was sailing east at 8 knots and continued to do so all day.
   a) How rapidly was the distance between the ships changing at noon? One hour later?
   b) The visibility that day was 5 nautical miles. Did the ships ever catch sight of each other?

28. Two ships A and B are sailing straight away from the point O along routes such that the angle $AOB = 120°$. How fast is the distance between them changing if, at a certain instant, $OA = 8$ mi, $OB = 6$ mi, ship A is sailing at the rate of 20 mph, and ship B at the rate of 30 mph? (*Hint:* Use the law of cosines.)

TOOLKIT PROGRAMS

Derivative Grapher   Super * Grapher

3.7
The Mean Value Theorem

Few theorems in calculus are more influential than the Mean Value Theorem and its generalizations. The theorem is so easy to state that one might not at first suspect the importance of its various consequences. Yet it provides the mathematics we need to estimate the amount of error involved in a linear approximation. It explains the first derivative test for rise and fall, and, by showing that constant functions are the only functions whose derivatives are zero, it leads the way to integral calculus. We shall see all of this take place before the present chapter ends.

The key to the Mean Value Theorem is an early version of it, called Rolle's Theorem, with which we now begin.

Rolle's Theorem

There is strong geometric evidence that between any two points where a smooth curve crosses the x-axis there is a point on the curve where the tangent is horizontal (Fig. 3.60). A 300-year-old theorem of Michel Rolle (1652–1719) assures us that this is indeed the case.

THEOREM 2

Rolle's Theorem†

Suppose that $y = f(x)$ is continuous at every point of the closed interval $[a, b]$ and differentiable at every point of its interior (a, b). If

$$f(a) = f(b) = 0,$$

then there is at least one number c between a and b at which

$$f'(c) = 0.$$

---

†Published in 1691 in *Méthode pour Résoudre les Egalités* by the French mathematician Michel Rolle. See ''Rolle's Theorem,'' in *A Source Book in Mathematics* by D. E. Smith (New York: McGraw-Hill, 1929), p. 253.

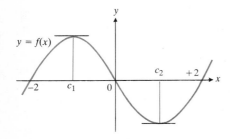

3.60 This smooth curve has horizontal tangents between the points where it crosses the *x*-axis.

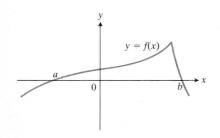

3.61 This curve has no horizontal tangent between the two points where it crosses the *x*-axis.

■ ■ ■

WHO PROVED THE MEAN VALUE THEOREM?

In 1787, Joseph Louis Lagrange (1736–1813) first proved the Mean Value Theorem in his work on the calculus, where he tried to avoid the use of the limit concept altogether. This profound and important theorem can also be found in the works of André Marie Ampère (1775–1836). Although best known for his work on electricity, Ampère focused his early work on the calculus, setting himself the task of challenging and correcting some of Lagrange's ideas on the foundations of the calculus. However, it was A. L. Cauchy who popularized and generalized the Mean Value Theorem in his two famous calculus textbooks, *Cours d'analyse* in 1821 and *Résumé des leçons sur le calcul infinitésimal* in 1823.

Proof of Rolle's Theorem   We know that a continuous function defined on a closed interval assumes maximum and minimum values on the interval. We also know that these values can occur only at endpoints and critical points. For the function we are dealing with, this means endpoints and interior points where f' is zero. Therefore, if f assumes either its maximum or its minimum at an interior point c, then $f'(c) = 0$. If f assumes both its maximum and its minimum at the endpoints, then zero is both the maximum and the minimum value of f. Hence, $f(x) = 0$ for all x in $[a, b]$, and f' is zero throughout (a, b) because f is constant. Either way, we find at least one point c in (a, b) where f' is zero.   ■

EXAMPLE 1   The polynomial

$$y = x^3 - 4x = f(x)$$

is continuous and differentiable for all x, $-\infty < x < +\infty$. If we take

$$a = -2, \qquad b = +2,$$

the hypotheses of Rolle's Theorem are satisfied since

$$f(-2) = f(+2) = 0.$$

Thus

$$f'(x) = 3x^2 - 4$$

must be zero at least once between -2 and $+2$. In fact, we find

$$3x^2 - 4 = 0$$

at

$$x = c_1 = -\frac{2\sqrt{3}}{3} \qquad \text{and} \qquad x = c_2 = +\frac{2\sqrt{3}}{3}.   \blacksquare$$

REMARK   As Fig. 3.61 shows, the differentiability of f on (a, b) is an essential part of Rolle's Theorem. If we allow even one point where f' fails to exist, there may be no horizontal tangent to the curve.

Locating Solutions of Equations

Rolle's original use for his theorem was to show that between every two zeros of a polynomial there may always be found a zero of its derivative. (Rolle distrusted calculus, however, and spent a great deal of time and energy denouncing its use. He used only algebra and geometry in his own work. It is ironic that he is known today only for a contribution he inadvertently made to a subject he tried to suppress.)

The version of Rolle's Theorem we have presented here is not restricted to polynomials. It says that between two zeros of any differentiable function may always be found a zero of its derivative. This observation has a surprising and useful consequence. Suppose that

1. f is continuous on $[a, b]$ and differentiable on (a, b);
2. $f(a)$ and $f(b)$ have opposite signs; and
3. f' is never zero between a and b.

Then f has exactly one zero between a and b. If f had more than one, f' would have to have a zero, too. That f has at least one zero is assured by the Intermediate Value Theorem of Article 1.11.

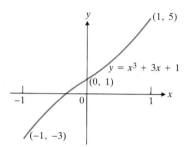

3.62 The only real zero of the polynomial $y = x^3 + 3x + 1$ is the one shown here between -1 and 0.

EXAMPLE 2 Show that the equation $x^3 + 3x + 1 = 0$ has exactly one real solution.

Solution The function $f(x) = x^3 + 3x + 1$ is differentiable at every value of x, and the derivative $f'(x) = 3x^2 + 3$ is never zero. If f had as many as two zeros, f' would have a zero between them. Hence f has at most one zero. On the other hand, f has at least one zero because $f(-1) = -3$ is negative, $f(1) = 5$ is positive, and f is continuous. Therefore, f has exactly one zero (Fig. 3.62).

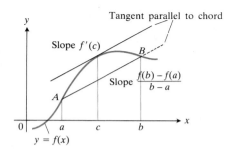

3.63 Geometrically, the Mean Value Theorem says that somewhere between A and B the curve has at least one tangent parallel to chord AB.

The Mean Value Theorem

The Mean Value Theorem is Rolle's Theorem for a chord instead of an interval. You will see what we mean if you look at Fig. 3.63. The figure shows the graph of a differentiable function f defined on an interval $a \leq x \leq b$. There is a point on the curve where the tangent is parallel to the chord AB. In Rolle's Theorem, the line AB is the x-axis and $f'(c) = 0$. Here, the line AB is a chord joining the endpoints of the curve above a and b, and $f'(c) = (f(b) - f(a))/(b - a)$.

THEOREM 3

> **The Mean Value Theorem**
>
> If $y = f(x)$ is continuous at every point of the closed interval $[a, b]$ and differentiable at every point of its interior (a, b), then there is at least one number c between a and b at which
>
> $$\frac{f(b) - f(a)}{b - a} = f'(c). \tag{1}$$

Proof If we graph f over $[a, b]$ and draw the line through the endpoints $A(a, f(a))$ and $B(b, f(b))$, the figure we get, Fig. 3.64(a), resembles the one we drew for Rolle's Theorem. The difference is that the line AB need not be the x-axis because $f(a)$ and $f(b)$ may not be zero. We cannot apply Rolle's Theorem directly to f, but we can apply it to the function that measures the vertical distance between the graph of f and the line AB. This, it turns out, will tell us what we want to know about the derivative of f.

The line AB is the graph of the function

$$g(x) = f(a) + \frac{f(b) - f(a)}{b - a}(x - a) \tag{2}$$

(point-slope equation), and the function we shall use to measure the vertical

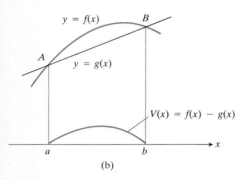

3.64 (a) The graph of a function and its chord *AB*. (b) The function *V* measures the vertical height between the graphs of *f* and *g*.

distance between the graphs of *f* and *g* is $V = f - g$. The formula for $V(x)$ is

$$V(x) = f(x) - g(x) = f(x) - f(a) - \frac{f(b) - f(a)}{b - a}(x - a). \tag{3}$$

Figure 3.64(b) shows the graphs of *f*, *g*, and *V* together.

We can see that *V* satisfies the hypotheses of Rolle's Theorem for the interval [*a*, *b*]. It is continuous on [*a*, *b*] and differentiable in (*a*, *b*) because *f* and *g* are. Both $V(a)$ and $V(b)$ are zero because the graphs of *f* and *g* pass through *A* and *B*. Therefore, V' is zero at some point *c* between *a* and *b*.

To see what this says about the derivative of *f*, we differentiate both sides of the formula for $V(x)$ in Eq. (3) with respect to *x* and set $x = c$. The derivative is

$$V'(x) = f'(x) - \frac{f(b) - f(a)}{b - a}. \tag{4}$$

With *x* set equal to *c*, Eq. (4) becomes

$$0 = f'(c) - \frac{f(b) - f(a)}{b - a},$$

or

$$f'(c) = \frac{f(b) - f(a)}{b - a}, \tag{5}$$

when rearranged. Equation (5) states what we set out to prove: At some point *c* between *a* and *b* the slope of the curve $y = f(x)$ equals the slope of the secant line *AB*.   ∎

If $f'(x)$ is continuous on [*a*, *b*] then by Theorem 7, Article 1.11, it has a maximum value max f' and a minimum value min f' on [*a*, *b*]. Since the number $f'(c)$ cannot exceed max f' or be less than min f', the equation

$$\frac{f(b) - f(a)}{b - a} = f'(c) \tag{6}$$

gives us the inequality

$$\min f' \le \frac{f(b) - f(a)}{b - a} \le \max f', \tag{7}$$

where max and min refer to the values of f' on the interval [*a*, *b*].

If we interpret $y = f(x)$ as distance traveled from time *a* up to time *x*, then Eq. (7) says that the average speed from time *a* to time *b* is no greater than the maximum speed and no less than the minimum speed. Equation (6) says that at some instant *c* during the trip the speed was exactly equal to the average speed for the trip.

The importance of the Mean Value Theorem lies in the numerical estimates that sometimes come from Eq. (7) and in the mathematical conclusions that can be drawn from Eq. (6) (a few of which we shall see in a moment). Estimates based on an extended version of the Mean Value Theorem will be the subject of Article 3.9.

We usually do not know any more about the number *c* than the theorem tells us, which is that *c* exists. In a few exceptional cases one's curiosity about the identity of *c* can be satisfied, as in the next two examples. Keep in mind, however, that our ability to identify *c* is the exception rather than the rule and that the importance of the Mean Value Theorem lies elsewhere.

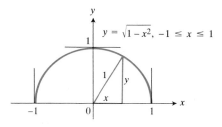

3.65 At $c = \pm\frac{2}{3}\sqrt{3}$, the tangents to the curve are parallel to the chord AB.

EXAMPLE 3 See Fig. 3.65. Let $f(x) = x^3$, $a = -2$, and $b = +2$. Then

$$f'(x) = 3x^2, \qquad f'(c) = 3c^2,$$
$$f(b) = 2^3 = 8, \qquad f(a) = (-2)^3 = -8,$$

so that

$$f'(c) = \frac{f(b) - f(a)}{b - a} = \frac{8 - (-8)}{2 - (-2)} = \frac{16}{4} = 4$$

becomes

$$3c^2 = 4, \qquad c = \pm\frac{2}{3}\sqrt{3}.$$

There are thus two values of c between $a = -2$ and $b = +2$ where the tangent to the curve $y = x^3$ is parallel to the chord through $A(-2, -8)$ and $B(2, 8)$. ∎

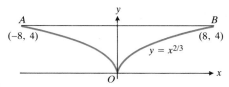

3.66 The vertical tangents at $x = -1$ and $x = 1$ do not keep $y = \sqrt{1 - x^2}$ from satisfying the hypotheses (and conclusion) of the Mean Value Theorem on the interval $[-1, 1]$.

EXAMPLE 4 See Fig. 3.66. The function $y = \sqrt{1 - x^2}$ satisfies the hypotheses (and conclusion) of the Mean Value Theorem on the interval $[-1, 1]$. It is continuous at each point of the closed interval, and its derivative

$$y' = \frac{-x}{\sqrt{1 - x^2}}$$

is defined at each point of the interior $(-1, 1)$. The graph has a horizontal tangent at $x = 0$. Notice that the function is not differentiable at $x = -1$ and $x = 1$. It does not need to be for the theorem to apply. ∎

EXAMPLE 5 The Mean Value Theorem does not apply to the function $f(x) = x^{2/3}$ on the interval $[-8, 8]$ (Fig. 3.67). The derivative of f is

$$f'(x) = \frac{2}{3}x^{-1/3} = \frac{2}{3\sqrt[3]{x}}.$$

We find that

$$\frac{f(b) - f(a)}{b - a} = \frac{8^{2/3} - (-8)^{2/3}}{8 - (-8)} = \frac{4 - 4}{16} = 0,$$

but that

$$f'(c) = \frac{2}{3\sqrt[3]{c}}$$

3.67 Having a tangent at each point does not mean having a derivative at each point and does not guarantee the conclusion of the Mean Value Theorem. The graph of $y = x^{2/3}$ has a tangent at every point (the tangent at O is vertical), but none of the tangents is parallel to the chord AB.

is not zero for any value of c in the interval $(-8, 8)$. The difficulty can be traced to the failure of f' to exist at $x = 0$. The Mean Value Theorem does not apply on a closed interval unless the function is differentiable at every point of the interval's interior. ∎

In the next example, we show how Eq. (7) is sometimes used to estimate the value of a function f at a particular value of x when a and f' are known.

EXAMPLE 6 Estimate $f(1)$ if

$$f'(x) = \frac{1}{5 - x^2} \qquad \text{and} \qquad f(0) = 2.$$

Solution   With $a = 0$, $b = 1$, and $f'(x) = 1/(5 - x^2)$, Eq. (7) becomes

$$\min \frac{1}{5 - x^2} \leq \frac{f(1) - f(0)}{1 - 0} \leq \max \frac{1}{5 - x^2},$$

where min and max refer to [0, 1]. On [0, 1] the value of $1/(5 - x^2)$ is smallest when $x = 0$ and largest when $x = 1$, so the inequalities in the preceding equation give

$$\frac{1}{5 - 0} \leq f(1) - 2 \leq \frac{1}{5 - 1},$$

$$2 + \frac{1}{5} \leq f(1) \leq 2 + \frac{1}{4},$$

$$2.2 \leq f(1) \leq 2.25. \qquad \blacksquare$$

Three Corollaries

We now come to three corollaries that account, in part, for the importance of the Mean Value Theorem in the development of calculus. The first corollary gives us the first derivative test for rise and fall and a little more. The second says that only constant functions can have zero derivatives. The third says that functions with identical derivatives must differ by a constant value. We shall prove the first corollary and leave the others as exercises.

COROLLARY 1

> ***f* increases when $f' > 0$ and decreases when $f' < 0$.**
>
> Suppose that f is continuous at each point of the closed interval [a, b] and differentiable at each point of its interior (a, b). If $f' > 0$ at each point of (a, b), then f increases throughout [a, b]. If $f' < 0$ at each point of (a, b), then f decreases throughout [a, b]. In either case, f is one-to-one.

Proof   Let x_1 and x_2 be any two numbers in [a, b] with $x_1 < x_2$. Apply the Mean Value Theorem to f on $[x_1, x_2]$:

$$f(x_2) - f(x_1) = f'(c)(x_2 - x_1) \qquad (8)$$

for some c between x_1 and x_2. The sign of the right-hand side of Eq. (8) is the same as the sign of $f'(c)$ because $x_2 - x_1$ is positive. Therefore

$$f(x_2) > f(x_1) \qquad \text{if } f'(x) \text{ is positive on } (a, b)$$

(f is increasing) and

$$f(x_2) < f(x_1) \qquad \text{if } f'(x) \text{ is negative on } (a, b)$$

(f is decreasing). In either case, $x_1 \neq x_2$ implies that $f(x_1) \neq f(x_2)$, so f is one-to-one. $\qquad \blacksquare$

COROLLARY 2

> **If $F' = 0$, then F is a constant.**
>
> If $F'(x) = 0$ for all x in (a, b), then F has a constant value throughout (a, b). In other words, there is a constant C such that $F(x) = C$ for all x in (a, b).

Corollary 2 is the converse of the rule that says the derivative of a constant is zero. While the derivatives of nonconstant functions may be zero from time to time, the only functions whose derivatives are zero throughout an entire interval are the functions that are constant on the interval.

COROLLARY 3

Functions with identical derivatives differ by a constant.

If $F_1'(x) = F_2'(x)$ at each point x of an open interval (a, b), then there is a constant C such that

$$F_1(x) - F_2(x) = C$$

for all x in (a, b).

Corollary 3 says that the only way two functions can have identical rates of change throughout an interval is for their values on the interval to differ by a constant. For example, we know that the derivative of the function x^2 is $2x$. Therefore, every other function whose derivative is $2x$ is given by the formula $F(x) = x^2 + C$ for some value of C. No other functions have $2x$ as their derivative.

In determining that $F(x) = x^2 + C$, we say that we have *determined F up to a constant*. Techniques for determining functions from their rates of change are extremely important in science and engineering, and Corollary 3 gives us just the mathematics we need to develop them, as we shall see in Chapter 4.

PROBLEMS

Rolle's Theorem

In Problems 1–3, show that the equation has exactly one solution on the given interval.

1. $x^4 + 3x + 1 = 0$, $[-2, -1]$

2. $x^4 + 2x^3 - 2 = 0$, $[0, 1]$

3. $2x^3 - 3x^2 - 12x - 6 = 0$, $[-1, 0]$

4. Let f and its first two derivatives f' and f'' be continuous on the interval $a \leq x \leq b$. Suppose the graph of f intersects the x-axis in at least three places in the interval. Show that f'' has at least one zero between a and b. Generalize this result.

5. a) Plot the zeros of each polynomial on a line together with the zeros of its first derivative:
   i) $y = x^2 - 4$
   ii) $y = x^2 + 8x + 15$
   iii) $y = x^3 - 3x^2 + 4 = (x + 1)(x - 2)^2$
   iv) $y = x^3 - 33x^2 + 216x = x(x - 9)(x - 24)$
   What pattern do you see?
   b) Use Rolle's Theorem to prove that between every two zeros of the polynomial $x^n + a_{n-1}x^{n-1} + \cdots + a_1x + a_0$ there lies a zero of the polynomial
   $$nx^{n-1} + (n - 1)a_{n-1}x^{n-2} + \cdots + a_1.$$

6. The function
$$y = f(x) = \begin{cases} x & \text{if } 0 \leq x < 1, \\ 0 & \text{if } x = 1, \end{cases}$$
is zero at $x = 0$ and at $x = 1$. Its derivative, $y' = 1$, is different from zero at every point between 0 and 1. Why doesn't that contradict Rolle's Theorem?

The Mean Value Theorem

In Problems 7–10, a, b, and c refer to the equation
$$f(b) - f(a) = (b - a)f'(c),$$
which expresses the Mean Value Theorem. Given $f(x)$, a, and b, find c.

7. $f(x) = x^2 + 2x - 1$, $a = 0$, $b = 1$

8. $f(x) = x^{2/3}$, $a = 0$, $b = 1$

9. $f(x) = x + \dfrac{1}{x}$, $a = 1/2$, $b = 2$

10. $f(x) = \sqrt{x - 1}$, $a = 1$, $b = 3$

11. Suppose you know that $f(x)$ is differentiable and that $f'(x)$ always has a value between -1 and $+1$. Show that

$$|f(x) - f(a)| \le |x - a|.$$

12. Suppose that a function $y = f(x)$ is continuous on $[a, b]$ and differentiable on (a, b). Show that if f' is never zero on (a, b), then $f(a) \ne f(b)$.

13. Use the Mean Value Theorem to show that

$$|\sin b - \sin a| \le |b - a|.$$

14. Does the function $y = x$, $0 \le x \le 1$, satisfy the hypotheses of the Mean Value Theorem? If not, why not? If so, what value or values could c have?

15. The function $y = |1 - x^2|$, $-2 \le x \le 2$, has a horizontal tangent at $x = 0$ even though the function is not differentiable at $x = -1$ and $x = 1$. Does this contradict the Mean Value Theorem? Explain.

16. Show that $f(x) = \tan x$ increases throughout any interval on which $\cos x \ne 0$.

17. The greatest integer function $y = [x]$ is defined at every point of the interval $[0, 1]$. Yet it fails to satisfy the conclusion of the Mean Value Theorem for the interval $[0, 1]$. What is going on?

18. Suppose that $y = f(x)$ is continuous on $[0, 1]$ and differentiable on $(0, 1)$ and that $f(0) = 0$, $f(1) = 1$. Show that the derivative of f must equal 1 at some point between $x = 0$ and $x = 1$.

19. A motorist drove 30 miles during a one-hour trip. Show that the car's speed was equal to 30 mph at least once during the trip.

20. Suppose that $f'(x) = 1/(1 + \cos x)$ for $0 \le x \le \pi/2$ and that $f(0) = 3$. Estimate $f(\pi/2)$.

21. Suppose that f is differentiable for all values of x, $f(-3) = -3$, $f(3) = 3$, and $|f'(x)| \le 1$. Show that $f(0) = 0$.

22. Suppose that f is continuous at every point of $[a, b]$ and differentiable at every point of (a, b) and that $f(a) < f(b)$. Show that f' is positive at some point between a and b.

23. The formula $f(x) = 3x + b$ gives a different function for every value of b. All of these functions, however, have the same derivative with respect to x, namely, $f'(x) = 3$. Are these the only differentiable functions of x whose derivative is 3? Could there be any others? Explain.

24. Show that

$$\frac{d}{dx}\left(\frac{x}{x + 1}\right) = \frac{d}{dx}\left(-\frac{1}{x + 1}\right),$$

even though

$$\frac{x}{x + 1} \ne -\frac{1}{x + 1}.$$

Doesn't this contradict Corollary 3 of the Mean Value Theorem? Explain.

25. Prove Corollary 2.

26. Prove Corollary 3.

 TOOLKIT PROGRAMS

Derivative Grapher   Super * Grapher
Function Evaluator

3.8
Indeterminate Forms and L'Hôpital's Rule

In the late seventeenth century, John Bernoulli discovered a rule for calculating limits of fractions whose numerators and denominators both approach zero. The rule is known today as **l'Hôpital's rule**, named after Guillaume François Antoine de l'Hôpital (1661–1704), Marquis de St. Mesme, a French nobleman who wrote the first introductory differential calculus text, in which the rule first appeared.

L'Hôpital's rule is easy to apply and gives fast results even when other methods are slow or unavailable.

The Indeterminate Form 0/0

If f and g are continuous at $x = a$ but $f(a) = g(a) = 0$, the limit

$$\lim_{x \to a} \frac{f(x)}{g(x)} \tag{1}$$

cannot be evaluated by substituting $x = a$, since this produces $0/0$, a meaningless expression known as an **indeterminate form.**

As we have seen, the value of the limit in Eq. (1) is hard to predict:

$$\lim_{x \to 2} \frac{x^2 - 4}{x - 2} = \lim_{x \to 2} (x + 2) = 4$$

$$\lim_{x \to 0} \frac{\sin x}{x} = 1$$

$$\lim_{x \to 0} \frac{1 - \cos x}{x} = 0.$$

The limit

$$f'(a) = \lim_{x \to a} \frac{f(x) - f(a)}{x - a}$$

from which we calculate derivatives always produces the indeterminate form $0/0$. Our success in calculating derivatives suggests that we might turn things around and use derivatives to calculate limits that lead to indeterminate forms. For example,

$$\lim_{x \to 0} \frac{\sin x}{x} = \lim_{x \to 0} \frac{\sin x - \sin 0}{x - 0}$$

$$= \frac{d}{dx}(\sin x)\bigg|_{x=0}$$

$$= \cos 0 = 1.$$

L'Hôpital's rule gives an explicit connection between derivatives and limits that lead to the indeterminate form $0/0$.

THEOREM 4

L'Hôpital's Rule (First Form)

Suppose that $f(a) = g(a) = 0$, that $f'(a)$ and $g'(a)$ exist, and that $g'(a) \neq 0$. Then

$$\lim_{x \to a} \frac{f(x)}{g(x)} = \frac{f'(a)}{g'(a)}.$$

Proof Working backward from $f'(a)$ and $g'(a)$, which are themselves limits, we have

$$\frac{f'(a)}{g'(a)} = \frac{\displaystyle\lim_{x \to a} \frac{f(x) - f(a)}{x - a}}{\displaystyle\lim_{x \to a} \frac{g(x) - g(a)}{x - a}} = \lim_{x \to a} \frac{\dfrac{f(x) - f(a)}{x - a}}{\dfrac{g(x) - g(a)}{x - a}}$$

$$= \lim_{x \to a} \frac{f(x) - f(a)}{g(x) - g(a)} = \lim_{x \to a} \frac{f(x) - 0}{g(x) - 0} = \lim_{x \to a} \frac{f(x)}{g(x)}. \qquad \blacksquare$$

EXAMPLE 1

a) $\lim\limits_{x\to 0}\dfrac{3x - \sin x}{x} = \dfrac{3 - \cos x}{1}\bigg|_{x=0} = 2$

b) $\lim\limits_{x\to 0}\dfrac{\sqrt{1+x} - 1}{x} = \dfrac{\frac{1}{2\sqrt{1+x}}}{1}\bigg|_{x=0} = \dfrac{1}{2}$

c) $\lim\limits_{x\to 0}\dfrac{x - \sin x}{x^3} = \dfrac{1 - \cos x}{3x^2}\bigg|_{x=0} = ?$ ■

What can we do about the limit in Example 1(c)? The first form of l'Hôpital's rule does not tell us what the limit is because the derivative of $g(x) = x^3$ is zero at $x = 0$. However, there is a stronger form of l'Hôpital's rule that says that whenever the rule gives $0/0$ we can apply it again, repeating the process until we get a different result. With this stronger rule we can finish the work begun in Example 1(c):

$$\lim\limits_{x\to 0}\frac{x - \sin x}{x^3} = \lim\limits_{x\to 0}\frac{1 - \cos x}{3x^2} \qquad \left(\text{Still } \frac{0}{0}; \text{ apply the rule again.}\right)$$

$$= \lim\limits_{x\to 0}\frac{\sin x}{6x} \qquad \left(\text{Still } \frac{0}{0}; \text{ apply the rule again.}\right)$$

$$= \lim\limits_{x\to 0}\frac{\cos x}{6} = \frac{1}{6}. \qquad \text{(A different result. Stop.)}$$

Notice that to apply l'Hôpital's rule to f/g we divide the derivative of f by the derivative of g. Do not fall into the trap of taking the derivative of f/g. The quotient to use is f'/g', not $(f/g)'$.

THEOREM 5

L'Hôpital's Rule (Stronger Form)

Suppose that

$$f(x_0) = g(x_0) = 0$$

and that the functions f and g are both differentiable on an open interval (a, b) that contains the point x_0. Suppose also that $g' \neq 0$ at every point in (a, b) except possibly x_0. Then

$$\lim\limits_{x\to x_0}\frac{f(x)}{g(x)} = \lim\limits_{x\to x_0}\frac{f'(x)}{g'(x)}, \qquad (2)$$

provided the limit on the right exists.

The proof of the stronger form of l'Hôpital's rule is based on a special version of the Mean Value Theorem. It can be found in Appendix 5.

A MISNAMED RULE AND THE FIRST CALCULUS TEXTBOOK

L'Hôpital's calculus textbook, published in 1696, made known many of the results of Leibniz and the Bernoulli brothers and offered the rule for calculating limits that lead to expressions of the form 0/0. Although l'Hôpital himself acknowledged a general indebtedness to his predecessors, he did not give any specific credit for the rule. In fact, l'Hôpital paid the younger Bernoulli brother, John, to tutor him in the new calculus and to keep him up to date on new developments in the calculus. John did this, and, in particular, he wrote l'Hôpital a letter on 22 July 1694 in which he revealed what we now call l'Hôpital's rule. [For further reading, see D. J. Struik, *Mathematics Teacher,* 56, 257–260 (1963); see also C. B. Boyer, "The First Calculus Textbooks," *The Mathematics Teacher,* 39, 159–167 (1946).]

EXAMPLE 2

$$\lim_{x \to 0} \frac{\sqrt{1 + x} - 1 - (x/2)}{x^2} \qquad \left(\frac{0}{0}\right)$$

$$= \lim_{x \to 0} \frac{(1/2)(1 + x)^{-1/2} - (1/2)}{2x} \qquad \left(\text{still } \frac{0}{0}\right)$$

$$= \lim_{x \to 0} \frac{-(1/4)(1 + x)^{-3/2}}{2} = -\frac{1}{8} \qquad ■$$

As you apply l'Hôpital's rule, look for a change from 0/0 to something else. This is where the limit's value is revealed.

EXAMPLE 3

$$\lim_{x \to 0} \frac{1 - \cos x}{x + x^2} \qquad \left(\frac{0}{0}\right)$$

$$= \lim_{x \to 0} \frac{\sin x}{1 + 2x} = \frac{0}{1} = 0$$

If we continue to differentiate in an attempt to apply l'Hôpital's rule once more, we get

$$\lim_{x \to 0} \frac{1 - \cos x}{x + x^2} = \lim_{x \to 0} \frac{\sin x}{1 + 2x} = \lim_{x \to 0} \frac{\cos x}{2} = \frac{1}{2},$$

which is wrong. ■

If we reach a point where one of the derivatives is zero and the other is not, then the limit in question is either zero, as in Example 3, or infinity, as in the next example.

EXAMPLE 4

$$\lim_{x \to 0^+} \frac{\sin x}{x^2} \qquad \left(\frac{0}{0}\right)$$

$$= \lim_{x \to 0^+} \frac{\cos x}{2x} = \infty \qquad ■$$

Other Indeterminate Forms

In more advanced books it is proved that l'Hôpital's rule applies to the indeterminate form ∞/∞ as well as 0/0. If $f(x)$ and $g(x)$ both approach infinity as x approaches a, then

$$\lim_{x \to a} \frac{f(x)}{g(x)} = \lim_{x \to a} \frac{f'(x)}{g'(x)},$$

provided the limit on the right exists.† In the notation $x \to a$, a may be either finite or infinite.

---

†There is also a proof of l'Hôpital's rule for both forms 0/0 and ∞/∞ in the article "L'Hôpital's Rule," by A. E. Taylor, *American Mathematical Monthly,* 59, 20–24 (1952).

EXAMPLE 5

$$\lim_{x \to \pi/2} \frac{\tan x}{1 + \tan x} = \lim_{x \to \pi/2} \frac{\sec^2 x}{\sec^2 x} = 1$$

$$\lim_{x \to \infty} \frac{x - 2x^2}{3x^2 + 5x} = \lim_{x \to \infty} \frac{1 - 4x}{6x + 5}$$

$$= \lim_{x \to \infty} \frac{-4}{6} = -\frac{2}{3}. \qquad \blacksquare$$

The forms $\infty \cdot 0$ and $\infty - \infty$ can sometimes be handled by changing the expressions algebraically to get $0/0$ or ∞/∞ instead. (Here again we do not mean to suggest that there is a number $\infty \cdot 0$ or a number $\infty - \infty$, any more than we meant to suggest that there is a number $0/0$ or ∞/∞. These forms are not numbers but descriptions of limits.)

EXAMPLE 6 The limit

$$\lim_{x \to \infty} x \sin \frac{1}{x}$$

leads to the form $\infty \cdot 0$, but we can change it to the form $0/0$ by writing $x = 1/t$ and letting $t \to 0^+$:

$$\lim_{x \to \infty} x \sin \frac{1}{x} = \lim_{t \to 0^+} \frac{1}{t} \cdot \sin t$$

$$= \lim_{t \to 0^+} \frac{\sin t}{t} = 1. \qquad \blacksquare$$

EXAMPLE 7 Find

$$\lim_{x \to 0} \left(\frac{1}{\sin x} - \frac{1}{x} \right).$$

Solution If $x \to 0^+$, then $\sin x \to 0^+$ and $1/\sin x \to +\infty$, while $1/x \to +\infty$. The expression $(1/\sin x) - (1/x)$ formally becomes $+\infty - (+\infty)$, which is indeterminate. On the other hand, if $x \to 0^-$, then $1/\sin x \to -\infty$ and $1/x \to -\infty$, so that $(1/\sin x) - (1/x)$ becomes $-\infty + \infty$, which is also indeterminate. But we may also write

$$\frac{1}{\sin x} - \frac{1}{x} = \frac{x - \sin x}{x \sin x}$$

and apply l'Hôpital's rule to the expression on the right. Thus,

$$\lim_{x \to 0} \left(\frac{1}{\sin x} - \frac{1}{x} \right) = \lim_{x \to 0} \frac{x - \sin x}{x \sin x} \qquad \left(\frac{0}{0} \right)$$

$$= \lim_{x \to 0} \frac{1 - \cos x}{\sin x + x \cos x} \qquad \left(\text{still } \frac{0}{0} \right)$$

$$= \lim_{x \to 0} \frac{\sin x}{2 \cos x - x \sin x} = 0. \qquad \blacksquare$$

> **When to Use L'Hôpital's Rule and When to Stop**
>
> To find
>
> $$\lim_{x \to a} \frac{f(x)}{g(x)}$$
>
> by l'Hôpital's rule, proceed to differentiate f and g as long as you still get the form $0/0$ or ∞/∞ at $x = a$. Stop differentiating as soon as you get something else. L'Hôpital's rule does not apply when either the numerator or denominator has a finite nonzero limit.

PROBLEMS

Find the limits in Problems 1–23.

1. $\lim\limits_{x \to 2} \dfrac{x - 2}{x^2 - 4}$

2. $\lim\limits_{t \to \infty} \dfrac{6t + 5}{3t - 8}$

3. $\lim\limits_{x \to \infty} \dfrac{5x^2 - 3x}{7x^2 + 1}$

4. $\lim\limits_{x \to 1} \dfrac{x^3 - 1}{4x^3 - x - 3}$

5. $\lim\limits_{t \to 0} \dfrac{\sin t^2}{t}$

6. $\lim\limits_{x \to \pi/2} \dfrac{2x - \pi}{\cos x}$

7. $\lim\limits_{x \to 0} \dfrac{\sin 5x}{x}$

8. $\lim\limits_{t \to 0} \dfrac{\cos t - 1}{t^2}$

9. $\lim\limits_{\theta \to \pi} \dfrac{\sin \theta}{\pi - \theta}$

10. $\lim\limits_{x \to \pi/2} \dfrac{1 - \sin x}{1 + \cos 2x}$

11. $\lim\limits_{x \to \pi/4} \dfrac{\sin x - \cos x}{x - \pi/4}$

12. $\lim\limits_{x \to \pi/3} \dfrac{\cos x - 0.5}{x - \pi/3}$

13. $\lim\limits_{x \to (\pi/2)} -\left(x - \dfrac{\pi}{2}\right) \tan x$

14. $\lim\limits_{x \to 0} \dfrac{2x}{x + 7\sqrt{x}}$

15. $\lim\limits_{x \to 1} \dfrac{2x^2 - (3x + 1)\sqrt{x} + 2}{x - 1}$

16. $\lim\limits_{x \to 2} \dfrac{\sqrt{x^2 + 5} - 3}{x^2 - 4}$

17. $\lim\limits_{x \to 0} \dfrac{\sqrt{a(a + x)} - a}{x}, \quad a > 0$

18. $\lim\limits_{t \to 0} \dfrac{10\,(\sin t - t)}{t^3}$

19. $\lim\limits_{x \to 0} \dfrac{x(\cos x - 1)}{\sin x - x}$

20. $\lim\limits_{h \to 0} \dfrac{\sin(a + h) - \sin a}{h}$

21. $\lim\limits_{r \to 1} \dfrac{a(r^n - 1)}{r - 1}, \quad n$ a positive integer

22. $\lim\limits_{x \to 0^+} \left(\dfrac{1}{x} - \dfrac{1}{\sqrt{x}}\right)$

23. $\lim\limits_{x \to \infty} (x - \sqrt{x^2 + x})$

24. Which is correct, (a) or (b)? Explain.

a) $\lim\limits_{x \to 3} \dfrac{x - 3}{x^2 - 3} = \lim\limits_{x \to 3} \dfrac{1}{2x} = \dfrac{1}{6}$ b) $\lim\limits_{x \to 3} \dfrac{x - 3}{x^2 - 3} = \dfrac{0}{6} = 0$

25. L'Hôpital's rule does not help with $\lim\limits_{x \to \infty} \dfrac{\sqrt{10x + 1}}{\sqrt{x + 1}}$. Find the limit some other way.

26. L'Hôpital's rule does not work with

$$\lim_{x \to \pi/2} \frac{\sec x}{\tan x}.$$

Try it—you just keep on going. Find the limit some other way.

27. Let $y = \sec x + \tan x$, $-\pi/2 < x < \pi/2$.

a) Show that y, y', and y'' are positive.

b) Find

$$\lim_{x \to -(\pi/2)^+} (\sec x + \tan x)$$

c) Graph y for $-\pi/2 < x < \pi/2$.

28. In Fig. 3.68, the circle has radius OA equal to 1, and AB is tangent to the circle at A. The arc AC has radian measure θ and the segment AB also has length θ. The line through B and C crosses the x-axis at $P(x, 0)$.

a) Show that the length of PA is

$$1 - x = \frac{\theta(1 - \cos \theta)}{\theta - \sin \theta}.$$

b) Find $\lim\limits_{\theta \to 0}(1 - x)$.

c) Show that $\lim\limits_{\theta \to \infty}[(1 - x) - (1 - \cos \theta)] = 0$.

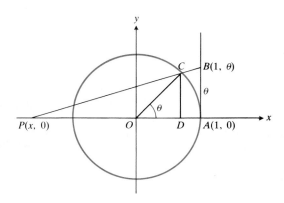

3.68 The diagram for Problem 28.

29. In Fig. 3.69, the right triangle has one leg of length 1, another of length y, and a hypotenuse of length r. The angle opposite y has radian measure θ. Find the limits as $\theta \rightarrow \pi/2$ of

a) $r - y$,   b) $r^2 - y^2$,   c) $r^3 - y^3$.

3.69 The right triangle in Problem 29.

30. Consider the functions

$$f(x) = \begin{cases} x + 2 & \text{for } x \neq 0, \\ 0 & \text{for } x = 0, \end{cases}$$

$$g(x) = \begin{cases} x + 1 & \text{for } x \neq 0, \\ 0 & \text{for } x = 0, \end{cases}$$

and let $x_0 = 0$. Show that

$$\lim_{x \to x_0} \frac{f'(x)}{g'(x)} = 1 \quad \text{but} \quad \lim_{x \to x_0} \frac{f(x)}{g(x)} = 2.$$

Does this contradict l'Hôpital's rule?

TOOLKIT PROGRAMS

Limit Problems   Super ∗ Grapher

3.9
Quadratic Approximations and Approximation Errors: Extending the Mean Value Theorem

In this article, we produce a version of the Mean Value Theorem that gives the linearization of a function and a formula for its error in a single equation. We also see how adding one more term to the equation produces the quadratic approximations commonly used in science and engineering. We shall also see how to control the errors in these approximations.

The Extended Mean Value Theorem (First Form)

In a slightly altered form, the Mean Value Theorem of Article 3.7 says that if a function f is continuous at every point of a closed interval $[a, b]$ and differentiable at every point of its interior (a, b), then there is at least one number c between a and b at which

$$f(b) = f(a) + f'(c)(b - a). \tag{1a}$$

If we think of b as an independent variable, we can rewrite Eq. (1a) in the form

$$f(x) = f(a) + f'(c)(x - a), \tag{1b}$$

which is valid for the interval joining a and x, with c in between.

The right-hand side of Eq. (1b) looks like the linearization

$$L(x) = f(a) + f'(a)(x - a).$$

The resemblance is more than coincidental, as the next theorem shows.

THEOREM 6

> **The Extended Mean Value Theorem (First Form)**
>
> If f and f' are continuous at every point of $[a, b]$, and f' is differentiable at every point in (a, b), then there is at least one number c_2 between a and b at which
>
> $$f(b) = f(a) + f'(a)(b - a) + \frac{f''(c_2)}{2}(b - a)^2. \tag{2}$$

The significance of Eq. (2) is not that some number k satisfies the equation

$$f(b) = f(a) + f'(a)(b - a) + k(b - a)^2. \tag{3}$$

It is hardly news that such a k exists, for we can always solve Eq. (3) for it. The importance of Eq. (2) is that the value of k is half the value of f'' at some point between a and b. This tells us that the size of k is controlled by f''. In a moment we shall see how useful this information can be.

Proof Equation (3) says that when $x = b$, the function $f(x)$ has the same value as the function

$$f(a) + f'(a)(x - a) + k(x - a)^2.$$

These two functions also have the same value when $x = a$ (namely, $f(a)$), but generally their difference,

$$F(x) = f(x) - f(a) - f'(a)(x - a) - k(x - a)^2, \tag{4}$$

is not zero.

The function $F(x)$ satisfies all the hypotheses of Rolle's Theorem on the interval $[a, b]$: $F(a) = 0$, $F(b) = 0$, F is continuous on $[a, b]$ because f, $(x - a)$, and $(x - a)^2$ are, and F is differentiable on (a, b) for the same reason. Therefore, $F' = 0$ at some point c_1 between a and b:

$$F'(c_1) = 0, \qquad a < c_1 < b.$$

Because $F'(c_1) = 0$, the derivative

$$F'(x) = f'(x) - f'(a) - 2k(x - a) \tag{5}$$

satisfies all the hypotheses of Rolle's Theorem on the interval $[a, c_1]$: $F'(c_1) = 0$, and substituting $x = a$ in Eq. (5) shows that $F'(a) = 0$. Also, F' is continuous on $[a, c_1]$ and differentiable on (a, c_1) because both f' and $(x - a)$ are. Therefore, the derivative

$$F''(x) = f''(x) - 2k$$

is zero at some point c_2 between a and c_1, which means that

$$0 = f''(c_2) - 2k \qquad \text{or} \qquad k = \frac{f''(c_2)}{2}.$$

Substituting this value of k in Eq. (3) gives Eq. (2), which is what we set out to prove.   ∎

REMARK The proof and theorem remain valid if $b < a$, provided we refer to "the interval with endpoints a and b" rather than explicitly to $[a, b]$ or (a, b). If we do this and take similar care with the other intervals that appear in the proof, then the arguments go through as before.

Measuring the Error in a Linear Approximation

We are now in a position to calculate the error in the linear approximation

$$f(x) \approx f(a) + f'(a)(x - a)$$

that we used in Article 2.4. We begin by regarding b as an independent variable and rewriting Eq. (2) in the form

$$f(x) = f(a) + f'(a)(x - a) + \frac{f''(c_2)}{2}(x - a)^2, \qquad (6)$$

with the understanding that c_2 lies between a and x. This equation holds for $x < a$ as well as $x > a$, by the remark at the end of the proof of Theorem 6. From Eq. (6) we get the linear approximation

$$f(x) \approx f(a) + f'(a)(x - a), \qquad x \approx a, \qquad (7)$$

which is valid on the interval from a to x with an error of

$$e_1(x) = \frac{f''(c_2)}{2}(x - a)^2. \qquad (8)$$

If f'' is continuous on the closed interval from a to x, then it has a maximum value on the interval and $e_1(x)$ satisfies the inequality

$$|e_1(x)| \leq \frac{1}{2}\max|f''|(x - a)^2, \qquad (9)$$

where max refers to the interval joining a and x. When we use this inequality to estimate the error, however, we usually cannot find the exact value of $\max|f''|$ and have to replace it with an upper bound or ''worst case'' value instead. If M is *any* upper bound for $\max|f''|$, then

$$|e_1(x)| \leq \frac{1}{2}M(x - a)^2. \qquad (10)$$

This is the inequality we normally use in estimating $|e_1(x)|$. We find the best M we can and go on from there. To make $|e_1(x)|$ small for a given M, we just make $(x - a)^2$ small.

The Error in the Linear Approximation of $f(x)$ near $x = a$

If f, f', and f'' are continuous at every point of the closed interval joining x and a, then the error $e_1(x)$ incurred in replacing $f(x)$ by its linearization

$$L(x) = f(a) + f'(a)(x - a)$$

on this interval satisfies the inequality

$$|e_1(x)| \leq \frac{1}{2}M(x - a)^2, \qquad (11)$$

where M is any upper bound for the values of $|f''|$ on the interval joining a and x.

EXAMPLE 1   The linearization of $f(x) = 1/(1 - x)$ at $x = 0$ is $L(x) = 1 + x$. How good is the approximation

$$\frac{1}{1 - x} \approx 1 + x$$

if $|x| \leq 0.1$?

Solution We use inequality (11) to find an upper bound on $|e_1(x)|$ for the given values of x. We start by finding the second derivative of $f(x) = (1 - x)^{-1}$:

$$f(x) = (1 - x)^{-1}, \qquad f'(x) = (1 - x)^{-2}, \qquad f''(x) = 2(1 - x)^{-3}.$$

We then look for an upper bound M for the values of $|f''(x)|$ on the interval $-0.1 \le x \le 0.1$. On this interval,

$$|f''(x)| = \frac{2}{|1 - x|^3} \le \frac{2}{|1 - 0.1|^3} = \frac{2}{(0.9)^3} < 2.8.$$

We may safely take $M = 2.8$. With this value of M and with $a = 0$, inequality (11) gives

$$|e_1(x)| \le \frac{1}{2}(2.8)(x - 0)^2$$

$$\le 1.4x^2$$

$$\le 1.4(0.1)^2 \qquad \text{(because } |x| \le 0.1)$$

$$\le 0.014.$$

The error in the approximation $(1 - x)^{-1} \approx 1 + x$ is no more than 0.014 if $|x| \le 0.1$. ∎

Quadratic Approximations

To get more accuracy in a linear approximation, we add a quadratic term. Typical quadratic approximations near $x = 0$ are

$$\frac{1}{1 - x} \approx 1 + x + x^2$$

$$\sqrt{1 + x} \approx 1 + \frac{x}{2} - \frac{x^2}{8}$$

$$\cos x \approx 1 - \frac{x^2}{2}$$

$$\sin x \approx x.$$

Others can be obtained from these with algebra. For $x \approx 0$ we have

$$\frac{1}{2 - x} = \frac{1}{2}\left(\frac{1}{1 - x/2}\right) \approx \frac{1}{2}\left[1 + \frac{x}{2} + \left(\frac{x}{2}\right)^2\right] = \frac{1}{2} + \frac{x}{4} + \frac{x^2}{8}$$

$$\frac{1 + x}{1 - x} \approx (1 + x)(1 + x + x^2) \approx 1 + 2x + 2x^2 \qquad \text{(ignoring the } x^3 \text{ term).}$$

To combine approximations correctly, each approximation must be valid at the time it is applied:

Right: $\qquad \dfrac{1}{1 - \sin x} \approx \dfrac{1}{1 - x} \approx 1 + x + x^2, \qquad x \approx 0$

Wrong: $\qquad \sin\left(\dfrac{1}{1 - x}\right)_{(1)} \approx \dfrac{1}{1 - x}_{(2)} \approx 1 + x + x^2.$

The second combination is wrong: approximation (1) requires $1/(1 - x) \approx 0$, and hence $|x|$ to be large, while approximation (2) requires $|x|$ to be small.

We can get a good estimate for the error in quadratic approximations by extending the Mean Value Theorem one step further in the following way.

THEOREM 7

The Extended Mean Value Theorem (Second Form)

If f, f', and f'' are continuous on $[a, b]$, and f'' is also differentiable on (a, b), then there exists a number c_3 between a and b for which

$$f(b) = f(a) + f'(a)(b - a) + \frac{f''(a)}{2}(b - a)^2$$
$$+ \frac{f'''(c_3)}{6}(b - a)^3. \tag{12}$$

The proof of this second extended mean value theorem, which resembles the proof of the first, is a special case of the proof outlined in Miscellaneous Problem 95 at the end of the chapter.

In applications we usually write Eq. (12) with x in place of b,

$$f(x) = f(a) + f'(a)(x - a) + \frac{f''(a)}{2}(x - a)^2 + \frac{f'''(c_3)}{6}(x - a)^3, \tag{13}$$

with the understanding that c_3 lies between a and x. From Eq. (13) we get the quadratic approximation

$$f(x) \approx f(a) + f'(a)(x - a) + \frac{f''(a)}{2}(x - a)^2, \tag{14}$$

which is valid on the interval between a and x with an error of

$$e_2(x) = \frac{f'''(c_3)}{6}(x - a)^3. \tag{15}$$

Note that the first two terms on the right-hand side of approximation (14) give the standard linear approximation of f. To get the quadratic approximation we have only to add the quadratic term without changing the linear part.

If $f'''(x)$ is continuous on the closed interval from a to x, then it has a maximum value on the interval, and Eq. (15) tells us that

$$|e_2(x)| \le \frac{1}{6}\max\left|f'''(x)\right||x - a|^3. \tag{16}$$

We usually stand no more chance of finding $\max|f'''|$ than we do of finding $\max|f''|$. We therefore replace it by an upper bound M and estimate $|e_2(x)|$ with the inequality

$$|e_2(x)| \le \frac{1}{6}M|x - a|^3. \tag{17}$$

The Quadratic Approximation of $f(x)$ near $x = a$

If f, f', f'', and f''' are continuous at every point of the closed interval joining x and a, then the error $e_2(x)$ incurred in replacing $f(x)$ by its quadratic approximation

$$Q(x) = f(a) + f'(a)(x - a) + \frac{f''(a)}{2}(x - a)^2 \qquad (18)$$

on this interval satisfies the inequality

$$|e_2(x)| \leq \frac{1}{6}M|x - a|^3, \qquad (19)$$

where M is any upper bound for the values of $|f'''|$ on the interval joining a and x.

EXAMPLE 2 Find the quadratic approximation of $f(x) = 1/(1 - x)$ near $x = 0$. How accurate is the approximation if $|x| \leq 0.1$?

Solution To find the quadratic approximation, we use Eq. (18) with $a = 0$:

$$f(x) = (1 - x)^{-1} \qquad f(0) = 1$$
$$f'(x) = (1 - x)^{-2} \qquad f'(0) = 1$$
$$f''(x) = 2(1 - x)^{-3} \qquad f''(0) = 2$$
$$f'''(x) = 6(1 - x)^{-4},$$

$$Q(x) = f(a) + f'(a)(x - a) + \frac{f''(a)}{2}(x - a)^2$$

$$= 1 + (1)(x - 0) + \frac{2}{2}(x - 0)^2$$

$$= 1 + x + x^2.$$

The quadratic approximation of $1/(1 - x)$ near $x = 0$ is

$$\frac{1}{1 - x} \approx 1 + x + x^2.$$

To estimate the approximation error, we first find an upper bound M for the values of $|f'''(x)|$ on the interval $-0.1 \leq x \leq 0.1$. On this interval,

$$|f'''(x)| = \frac{6}{|1 - x|^4} \leq \frac{6}{(0.9)^4} < 9.15,$$

so we may safely take $M = 9.15$. We then use inequality (19) with $a = 0$ and $M = 9.15$:

$$|e_2(x)| \leq \frac{1}{6}(9.15)|x - 0|^3$$

$$\leq \frac{1}{6}(9.15)(0.1)^3 \qquad \text{(because } |x| \leq 0.1)$$

$$\leq 0.0016.$$

The error in the approximation is no more than 0.0016 if $|x| \le 0.1$. This is an improvement over the upper bound of 0.014 found for the linear approximation of $1/(1 - x)$ in Example 1. ∎

EXAMPLE 3 Find the quadratic approximation of $\sin x$ near $x = 0$. Give an upper bound for the error in the approximation on the interval $|x| \le 0.3$.

Solution To find the quadratic approximation, we use Eq. (18) with $f(x) = \sin x$ and $a = 0$:

$$f(x) = \sin x \qquad f(0) = 0$$
$$f'(x) = \cos x \qquad f'(0) = 1$$
$$f''(x) = -\sin x \qquad f''(0) = 0$$
$$f'''(x) = -\cos x,$$

$$Q(x) = f(a) + f'(a)(x - a) + f''(a)(x - a)^2$$
$$= 0 + (1)(x - 0) + 0$$
$$= x.$$

The quadratic approximation of $\sin x$ near $x = 0$ is $\sin x \approx x$.

How can the approximation be quadratic if there is no x^2 term? There *is* a quadratic term, but its coefficient is zero. The important fact is that no quadratic term is missing and we can estimate the approximation error with $e_2(x)$ instead of $e_1(x)$.

To find an upper bound for $|e_2(x)|$ on the interval $-0.3 \le x \le 0.3$, we first find an upper bound for $|f'''(x)|$. Since $|\cos x|$ never exceeds 1,

$$|f'''(x)| = |-\cos x| \le 1,$$

and we may safely take $M = 1$. With $M = 1$ and $a = 0$, inequality (19) gives

$$|e_2(x)| \le \frac{1}{6}(1)|x - 0|^3$$

$$\le \frac{1}{6}|x|^3$$

$$\le \frac{1}{6}(0.3)^3 \qquad \text{(because } |x| \le 0.3\text{)}$$

$$\le 0.0045.$$

The error on the interval $-0.3 \le x \le 0.3$ will never exceed 0.0045. ∎

PROBLEMS

Use formulas from Table 3.2 to find quadratic approximations for the functions in Problems 1–5 for the given values of x.

1. $x \sin x, \quad x \approx 0$

2. $\sqrt{1 + \sin x}, \quad x \approx 0$

3. $\cos \sqrt{1 + x}, \quad x \approx -1$

4. $\dfrac{\sec x}{1 - x}, \quad x \approx 0$

5. $\sqrt{x}, \quad x \approx 1$ (*Hint*: $x = 1 + (x - 1)$, with $(x - 1) \approx 0$.)

6. Find the quadratic approximation of $f(x) = \sec x + \tan x$ near $x = 0$.

7. Find upper bounds for the error $|e_1(x)|$ in the following linearizations.

a) $\sqrt{1 + x} \approx 1 + (x/2), \quad |x| \le 0.1$
b) $\cos x \approx 1, \quad |x| \le 0.1$

TABLE 3.2
Approximation Formulas

Linear	Quadratic

Approximation

$$L(x) = f(a) + f'(a)(x - a)$$

$$Q(x) = f(a) + f'(a)(x - a) + \frac{f''(a)}{2}(x - a)^2$$

Error bound

$$|e_1(x)| \le \frac{1}{2}M(x - a)^2$$

M is any upper bound for $\max|f''|$ on the interval joining x and a.

$$|e_2(x)| \le \frac{1}{6}M|x - a|^3$$

M is any upper bound for $\max|f'''|$ on the interval joining x and a.

Common approximations for $x \approx 0$

$$\frac{1}{1 - x} \approx 1 + x$$

$$\sqrt{1 + x} \approx 1 + \frac{x}{2}$$

$$\sin x \approx x$$

$$\cos x \approx 1$$

$$(1 + x)^k \approx 1 + kx$$

for any number k

$$\frac{1}{1 - x} \approx 1 + x + x^2$$

$$\sqrt{1 + x} \approx 1 + \frac{x}{2} - \frac{x^2}{8}$$

$$\sin x \approx x$$

$$\cos x \approx 1 - \frac{x^2}{2}$$

$$(1 + x)^k \approx 1 + kx + \frac{k(k - 1)}{2}x^2$$

for any number k

In Problems 8–10, (a) derive the quadratic approximation for the function near $x = 0$; (b) find a numerical upper bound for the error $|e_2(x)|$ of each approximation for the interval $|x| \le 0.1$.

8. $\sqrt{1 + x}$ **9.** $\tan x$ **10.** $\cos x$

In Problems 11–14, (a) find the quadratic approximation of the function near the given point a; (b) use Eq. (19) to find a numerical upper bound for the error of each approximation for the interval $|x - a| \le 0.1$.

11. $\sin x$, $a = \pi/2$ **12.** $\sin x$, $a = \pi$

13. $\cos x$, $a = \pi/2$ **14.** $\cos x$, $a = \pi$

15. According to Eq. (19), for what values of x will the error $|e_2(x)|$ in the approximation $\sin x \approx x$ be (a) less than $1/100$; (b) less than 1% of $|x|$?

16. According to Eq. (19), for what values of x will the error $|e_2(x)|$ in the approximation

$$\cos x \approx 1 - \frac{x^2}{2}$$

be (a) less than $1/100$; (b) less than 1% of $|x|$?

17. Assuming that the differentiation formula

$$\frac{d}{dx}u^k = ku^{k-1}\frac{du}{dx}$$

holds for any number k, show that the quadratic approximation of $(1 + x)^k$ near $x = 0$ is

$$(1 + x)^k \approx 1 + kx + \frac{k(k - 1)}{2}x^2.$$

18. Verify Eq. (6) for $f(x) = 3x^2 + 2x + 4$, $a = 1$.

19. Verify Eq. (13) for $f(x) = x^3 + 5x - 7$, $a = 1$.

20. Use Eq. (6) to prove the following: Let f be continuous and have continuous first and second derivatives. Suppose that $f'(a) = 0$. Then
a) f has a local maximum at a if $f'' \le 0$ throughout an interval whose interior contains a;
b) f has a local minimum at a if $f'' \ge 0$ throughout an interval whose interior contains a.

 TOOLKIT PROGRAMS

Derivative Grapher   Super ∗ Grapher

REVIEW QUESTIONS AND EXERCISES

1. Discuss the significance of the signs of the first and second derivatives. Sketch a small portion of a curve on which
   a) both y' and y'' are positive;
   b) $y' > 0$, $y'' < 0$;
   c) $y' < 0$, $y'' > 0$;
   d) $y' < 0$, $y'' < 0$.

2. Define *point of inflection*. How do you find points of inflection from an equation of a curve?

3. Give an example of a function whose graph has an oblique asymptote; a vertical asymptote; a horizontal asymptote. Sketch the graphs and asymptotes.

4. Define *even function* and *odd function*. What symmetries do the graphs of such functions have?

5. How do you locate the local maximum and minimum values of a function? Include critical points and endpoints in your discussion. Illustrate with a graph.

6. How do you locate the absolute maximum and minimum values of a continuous function on a closed interval? Give an example.

7. Let n be a positive integer. For which values of n does the curve $y = x^n$ have (a) a local minimum at the origin; (b) a point of inflection at the origin?

8. Outline a general method for solving max-min problems.

9. Outline a general method for solving related rate problems.

10. What are the hypotheses of Rolle's Theorem? What is the conclusion?

11. With the book closed, state and prove the Mean Value Theorem. What is its geometric interpretation?

12. We know that if $F(x) = x^2$, then $F'(x) = 2x$. If someone knows a function G such that $G'(x) = 2x$ but $G(x) \neq x^2$, what can be said about the difference $G(2) - G(1)$? Explain.

13. Describe l'Hôpital's rule. How do you know how many times to use it in a given problem? Give an example.

14. Give the linear and quadratic approximations of three different functions. What are the general formulas for the linear and quadratic approximations of a function? How can we find upper bounds for the errors in these approximations?

15. Read the article ''Mathematics in Warfare'' by F. W. Lanchester, *World of Mathematics*, Vol. 4 (New York: Simon and Schuster, 1960), pp. 2138–2157, as a discussion of a practical problem in related rates.

MISCELLANEOUS PROBLEMS

In Problems 1–18, find y' and y''. Determine in each case the sets of values of x for which
   a) y is increasing (as x increases),
   b) y is decreasing (as x increases),
   c) the graph is concave up,
   d) the graph is concave down.
Also sketch the graph in each case, indicating maxima, minima, asymptotes, and points of inflection.

1. $y = 9x - x^2$
2. $y = x^3 - 5x^2 + 3x$
3. $y = 4x^3 - x^4$
4. $y = 4x + x^{-1}$
5. $y = x^2 + 4x^{-1}$
6. $y = x + 4x^{-2}$
7. $y = 5 - x^{2/3}$
8. $y = \dfrac{x - 1}{x + 1}$
9. $y = x - \dfrac{4}{x}$
10. $y = x^4 - 2x^2$
11. $y = \dfrac{x^2}{ax + b}$, $a > 0$, $b > 0$
12. $y = 2x^3 - 9x^2 + 12x$
13. $y = (x - 1)(x + 1)^2$
14. $y = x^2 - (1/6)x^3$
15. $y = 2x^3 - 9x^2 + 12x - 4$
16. $y = 3x(x - 1)(x + 1)^2$

17. $xy^2 = 3(1 - x)$
18. $y = 2 \cos x + \cos^2 x$

In Problems 19–26, find the domain of the function and investigate the following properties of its graph: (a) symmetry, (b) intercepts, (c) asymptotes, (d) slope at the intercepts, (e) rise and fall, (f) concavity, and (g) dominant terms. Then graph the function.

19. a) $y = \dfrac{x + 1}{x^2 + 1}$ b) $y = \dfrac{x^2 + 1}{x + 1}$

20. a) $y = x + \dfrac{1}{x^2}$ b) $y = x^2 + \dfrac{1}{x^2}$

21. $y = x(x + 1)(x - 2)$
22. $y = \dfrac{8}{4 + x^2}$
23. $y = \dfrac{8}{4 - x^2}$
24. $y = \dfrac{8x}{4 + x^2}$
25. $x^2y - y = 4(x - 2)$
26. $y = \dfrac{x^2 + 1}{x^2 - 1}$

27. A certain graph has an equation of the form
$$y = \frac{ax + b}{cx^2 + dx + e},$$
where the constants a, b, c, d, and e are either 0 or 1. Use the following information to determine the constants. Give

a reason for your choice in each case.

1. The graph has no y-intercept.
2. The x-axis is an asymptote.
3. The x-intercept is $(-1, 0)$.

Sketch the graph.

28. *Symmetry with respect to the x-axis.* The test for symmetry of a curve with respect to the x-axis is this: The point $(x, -y)$ lies on the curve whenever the point (x, y) lies on the curve. Suppose that the graph of $F(x, y) = 0$ is symmetric with respect to both the y-axis and the origin. Must it be symmetric with respect to the x-axis as well? If so, why? If not, find a curve that has the first two symmetries but not the third.

29. The slope of a curve at any point (x, y) is given by the equation

$$\frac{dy}{dx} = 6(x - 1)(x - 2)^2(x - 3)^3(x - 4)^4.$$

a) For what value (or values) of x does y have a local maximum? Why?
b) For what value (or values) of x does y have a local minimum? Why?

30. Divide the number 20 into two parts (not necessarily integers) in a way that makes the product of one part with the square of the other as large as possible.

31. Find the largest value of $f(x) = 4x^3 - 8x^2 + 5x$ for $0 \le x \le 2$. Give reasons for your answer.

32. Find two *positive* numbers whose sum is 36 and whose product is as large as possible. Can the problem be solved if the product is to be as small as possible?

33. For what values of a, b, c, and d does $y = ax^3 + bx^2 + cx + d$ have a local maximum at $(-1, 10)$ and an inflection point at $(1, -6)$?

34. Find the number that most exceeds its square.

35. The perimeter p and area A of a circular sector ("piece of pie") of radius r and arc length s are given by the formulas $p = 2r + s$ and $A = (1/2)rs$. If the perimeter is 100 ft, what value of r will produce a maximum area?

36. If a ball is thrown vertically upward with a velocity of 32 ft/sec, its height after t sec is given by the equation $s = 32t - 16t^2$. At what instant will the ball be at its highest point, and how high will it rise?

37. A right circular cone has an altitude of 12 ft and a base radius of 6 ft. A second cone is inscribed in the first, with its vertex at the center of the base of the first cone and its base parallel to the base of the first cone. Find the dimensions of the cone of maximum volume that can be so inscribed.

38. Find all maxima and minima of $f(x) = \sin x + \cos x$.

39. An isosceles triangle is drawn with its vertex at the origin, its base parallel to and above the x-axis, and the vertices of its base on the curve $12y = 36 - x^2$. Find the area of the largest such triangle.

40. A tire manufacturer can make x (hundred) grade A tires and y (hundred) grade B tires per day, where

$$y = \frac{(40 - 10x)}{(5 - x)},$$

with $0 \le x \le 4$. If the profit on each grade A tire is twice the profit on a grade B tire, what is the most profitable number of grade A tires to make per day?

41. Find the points on the curve $x^2 - y^2 = 1$ nearest the point $P(a, 0)$ if
a) $a = 4$,   b) $a = 2$,   c) $a = \sqrt{2}$.

42. A motorist in a desert 5 mi from point A, which is the nearest point on a long, straight road, wishes to get to point B on the road. If the car can travel at 15 mph on the desert and 39 mph on the road, find the point where it must meet the road to get to B in the shortest possible time if
a) B is 5 mi from A,
b) B is 10 mi from A,
c) B is 1 mi from A.

43. A drilling rig, a km offshore, is to be connected by pipe to a refinery on shore, b km down the coast from the rig, as shown in Fig. 3.70. If underwater pipe costs w dollars/km and land-based pipe costs $l < w$ dollars/km, what values of x and y give the least expensive connection?

3.70   The pipe diagram for Problem 43.

44. Points A and B are ends of a diameter of a circle and C is a point on the circumference. Which of the following statements about triangle ABC is (or are) true?
a) The area is largest when the triangle is isosceles.
b) The area is smallest when the triangle is isosceles.
c) The perimeter is largest when the triangle is isosceles.
d) The perimeter is smallest when the triangle is isosceles.

45. The base and the perimeter of a triangle are fixed. Determine the remaining two sides if the area is to be a maximum. The area of a triangle with sides a, b, and c and semiperimeter s is

$$A = [s(s - a)(s - b)(s - c)]^{1/2}.$$

46. The base b and the area A of a triangle are fixed. Determine the base angles if the angle at the vertex opposite b is to be as large as possible.

47. A line is drawn through a fixed point (a, b) to meet the axes Ox, Oy in P and Q. Show that the minimum values of PQ, $OP + OQ$, and $OP \cdot OQ$ are, respectively,

$$(a^{2/3} + b^{2/3})^{3/2}, \quad (\sqrt{a} + \sqrt{b})^2, \quad \text{and} \quad 4ab.$$

48. Find the smallest value of the constant m that will make $mx - 1 + (1/x)$ greater than or equal to zero for all positive values of x.

49. Let s be the distance between the fixed point $P_1(x_1, y_1)$ and a point $P(x, y)$ on the line

$$L: \quad ax + by = c.$$

Using calculus methods,

a) show that s^2 is a minimum when P_1P is perpendicular to L;

b) show that the minimum distance is

$$|ax_1 + by_1 - c|/\sqrt{a^2 + b^2}.$$

50. A playing field is to be built in the shape of a rectangle plus a semicircular region at each end. A 400-m racetrack is to form the perimeter of the field. What dimensions will give the rectangular part the largest area?

51. If $ax + (b/x) \geq c$ for all positive values of x, where a, b, and c are positive constants, show that $ab \geq c^2/4$.

52. Prove that if $ax^2 + (b/x) \geq c$ for all positive values of x, where a, b, and c are positive constants, then $27ab^2 \geq 4c^3$.

53. Show that if $a > 0$, then $f(x) = ax^2 + 2bx + c \geq 0$ for all real x if and only if $b^2 - ac \leq 0$.

54. *Schwarz's inequality.* In Problem 53, take

$$f(x) = (a_1x + b_1)^2 + (a_2x + b_2)^2 + \cdots + (a_nx + b_n)^2,$$

and deduce Schwarz's inequality:

$$(a_1b_1 + a_2b_2 + \cdots + a_nb_n)^2$$
$$\leq (a_1^2 + a_2^2 + \cdots + a_n^2)(b_1^2 + b_2^2 + \cdots + b_n^2).$$

55. In Problem 54, prove that equality can hold only if there is a real number x such that $b_i = -a_i x$ for every $i = 1$, $2, \ldots, n$.

56. If x is positive and m is greater than 1, prove that $x^m - 1 - m(x - 1)$ is not negative.

57. What are the dimensions of the rectangular plot of greatest area that can be laid out within a triangle of base 36 ft and altitude 12 ft? Assume that one side of the rectangle lies on the base of the triangle.

58. Find the width across the top of an isosceles trapezoid of base 12 in. and slant sides 6 in. if its area is as large as possible.

59. The positions of two particles on the x-axis are $x_1 = \sin t$ and $x_2 = \sin[t + (\pi/3)]$. Find the greatest distance between them.

60. Assuming that the cost per hour of running the *Queen Elizabeth II* is $a + bv^n$, where a, b, and n are positive constants, $n > 1$, and v is the velocity through the water, find the speed for making the run from Liverpool to New York at minimum cost.

61. A flower bed is to be in the shape of a circular sector ("piece of pie") of radius r and central angle θ. Find r and θ if the area is fixed and the perimeter is a minimum.

62. A light hangs above the center of a table of radius r ft. The illumination at any point on the table is directly proportional to the cosine of the angle of incidence (the angle a ray of light makes with the normal) and is inversely proportional to the square of the distance from the light. How far should the light be above the table to give the strongest illumination at the edge of the table?

63. Let A be the area of the region R between two concentric circles of radius r_1 and r_2 (Fig. 3.71).

a) How fast is A increasing (or decreasing) when $r_1 = 4$ cm and is increasing at the rate of 0.02 cm/sec while $r_2 = 6$ cm and is increasing at the rate of 0.01 cm/sec?

b) Suppose that at time $t = 0$, r_1 is 3 cm and r_2 is 5 cm, and that for $t > 0$, r_1 increases at the constant rate of a cm/sec and r_2 increases at the constant rate of b cm/sec. If $(3/5)a < b < a$, find when the area A will be largest.

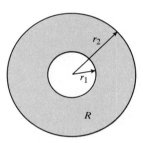

3.71 The region in Problem 63.

64. Let V be the volume of the region D between two concentric spheres of radius r_1 and r_2, with $r_1 < r_2$. Suppose that at time $t = 0$, $r_1 = r$ in. and $r_2 = R$ in., and that for $t > 0$, r_1 increases at the constant rate of a in/sec and r_2 increases at the constant rate of b in/sec. If $a > b > ar^2/R^2$, find when V will be largest.

65. Suppose that the position of a particle on a straight line at time t is $s(t) = at - (1 + a^4)t^2$. Show that the particle moves forward initially when a is positive but ultimately retreats. Show also that for different values of a the maximum possible distance that the particle can move forward is $1/8$.

66. Let $h = fg$, with f and g differentiable functions of x.

a) If f and g are positive, with local minima at $x = a$, and if f' and g' change sign at $x = a$, does h have a local maximum at $x = a$?

b) If f and g both have a point of inflection at $x = a$, does h have a point of inflection at $x = a$?

For both (a) and (b), either give a proof or construct a numerical example showing that the statement is false.

67. The numbers $c_1, c_2, \ldots, c_n$ are recorded in an experiment. What value of x makes the sum

$$(c_1 - x)^2 + (c_2 - x)^2 + (c_3 - x)^2 + \cdots + (c_n - x)^2$$

as small as possible?

68. The four points

$$\left(-2, -\frac{1}{2}\right), \quad (0, 1), \quad (1, 2), \quad \text{and} \quad (3, 3)$$

are observed to lie more or less close to a line $y = mx + 1$. Find m if the sum

$$(y_1 - mx_1 - 1)^2 + (y_2 - mx_2 - 1)^2$$
$$+ (y_3 - mx_3 - 1)^2 + (y_4 - mx_4 - 1)^2$$

is to be a minimum, where $(x_1, y_1), \ldots, (x_4, y_4)$ are the coordinates of the given points.

69. *Geometric and arithmetic means.* The geometric mean of the n positive numbers $a_1, a_2, \ldots, a_n$ is the nth root of $a_1 a_2 \cdots a_n$, and the arithmetic mean is $(a_1 + a_2 + \cdots + a_n)/n$. Show that if $a_1, a_2, \ldots, a_{n-1}$ are fixed and $a_n = x$ is permitted to vary over the set of positive real numbers, the ratio of the arithmetic mean to the geometric mean is a minimum when x is the arithmetic mean of $a_1, a_2, \ldots, a_{n-1}$.

70. Suppose that it costs a company $y = a + bx$ dollars to produce x units per week. It can sell x units per week at a price of $P = c - ex$ dollars per unit.
   a) What production level (units per week) maximizes the profit?
   b) What is the corresponding price?
   c) What is the weekly profit at this level of production?
   d) At what price should each item be sold to maximize profits if the government imposes a tax of t dollars per item sold? Comment on the difference between this price and the price before tax.

71. A particle moves along the x-axis with velocity $v = dx/dt = f(x)$. Show that its acceleration is $f(x)f'(x)$.

72. A meteorite entering the earth's atmosphere has velocity inversely proportional to $\sqrt{s}$ when it is s km from the center of the earth. Show that its acceleration is inversely proportional to s^2.

73. The volume of a cube is increasing at a rate of 300 cm^3/min at the instant when the edge is 20 cm. Find the rate at which the edge is changing.

74. Sand falling at the rate of 3 ft^3/min forms a conical pile whose radius always equals twice its height. Find the rate at which the height is changing at the instant the height is 10 ft.

75. The volume of a sphere is decreasing at the rate of 12π m^3/min. Find the rates at which the radius and the surface area are changing at the instant when the radius is 20 m. Also find approximately how much the radius and surface area may be expected to change in the following 6 sec.

76. At a certain instant, airplane A is flying a level course at 500 mph. At the same time, airplane B is straight above airplane A and flying at the rate of 700 mph on a course that intercepts A's course at a point C that is 4 mi from B and 2 mi from A.
   a) At the instant in question, how fast is the distance between the airplanes decreasing?
   b) How near to each other will the airplanes come if they continue at their present course and speed?

77. A point moves along the curve $y^2 = x^3$ in such a way that its distance from the origin increases at the constant rate of 2 units per second. Find dx/dt at $(2, 2\sqrt{2})$.

78. How fast is the area of the triangle formed by the ladder, wall, and ground in Example 3 of Article 3.6 changing when $x = 17\sqrt{2}$?

79. Suppose the cone in Fig. 3.72 has a small opening at the vertex through which the water escapes at the rate of $0.08\sqrt{y}$ ft^3/min when its depth is y. Water is also running into the cone at a constant rate of c ft^3/min. When the depth is 6.25 ft, the depth of the water is observed to be increasing at the rate of 0.02 ft/min. Will the tank fill? Give a reason for your answer.

Water in (c ft^3/min)

Water out, $0.08\sqrt{y}$ ft^3/min

3.72   The conical tank in Problem 79.

80. A particle projected vertically upward from the surface of the earth with initial velocity v_0 has velocity $\sqrt{v_0^2 - 2gR[1 - (R/s)]}$ when it reaches a distance $s \geq R$ from the *center* of the earth. Here R is the radius of the earth. Show that the acceleration is inversely proportional to s^2.

81. The segment DE in Fig. 3.73 is parallel to the base BC of triangle ABC and lies x units above the base. Show that the derivative with respect to x of the area of $BCED$ is equal to the length of DE.

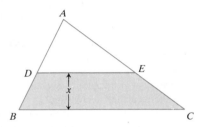

3.73   The triangle in Problem 81.

82. Points A and B move along the x- and y-axes, respectively, in such a way that the distance r (in inches) along the perpendicular from the origin to AB remains constant. How fast is OA changing, and is it increasing, or decreasing, when $OB = 2r$ and B is moving toward O at the rate of 0.3r in./sec?

83. Ships A and B start from O at the same time. Ship A travels due east at a rate of 15 mph. Ship B travels in a straight course making an angle of $60°$ with the path of ship A at a rate of 20 mph. How fast are they separating at the end of 2 hr?

84. Water is being poured into a conical tank (vertex down) at the rate of 2 ft³/min. How fast is the water level rising when the depth of the water is 5 ft? The radius of the base of the cone is 3 ft and the altitude is 10 ft.

85. The curve $(y + 1)^3 = x^2$ passes through the points $(1, 0)$ and $(-1, 0)$. Does Rolle's Theorem justify the conclusion that $dy/dx = 0$ for some value of x in the interval $-1 \le x \le 1$? Give reasons for your answer.

86. a) If $a < 0 < b$ and $f(x) = x^{-1/3}$, show that there is no c that satisfies Eq. (1) in Article 3.7. Illustrate with a sketch of the graph.
   b) If $a < 0 < b$ and $f(x) = x^{1/3}$, show that there is a value of c that satisfies Eq. (1) in Article 3.7 even though the function fails to have a derivative at $x = 0$. Illustrate with a sketch of the graph.

87. Graph $y = \sin x \sin(x + 2) - \sin^2(x + 1)$ over the interval $-\pi \le x \le \pi$. (*Hint:* Find y' first.)

88. CALCULATOR   Show that the equation $f(x) = 2x^3 - 3x^2 + 6x + 6 = 0$ has exactly one real root and find its value as accurately as you can. (*Hint:* $f(-1) = -5, f(0) = +6$, and $f'(x) > 0$ for all real x.)

89. If $f'(x) \le 2$ for all x, what is the most f can increase on the interval $[0, 6]$?

90. Suppose that f is continuous and differentiable on $[a, b]$. Show that if $f'(x) \le 0$ on $[a, c]$ and $f'(x) \ge 0$ on $(c, b]$, $a < c < b$, then $f(x)$ is never less than $f(c)$ on $[a, b]$.

91. a) Show that
$$-\frac{1}{2} \le \frac{x}{1 + x^2} \le \frac{1}{2}$$
   for every value of x.

   b) Let f be a function whose derivative f' is defined by
$$f'(x) = \frac{x}{1 + x^2}.$$
   Use the result in (a) to show that
$$|f(b) - f(a)| \le \frac{1}{2}|b - a|$$
   for any a and b with $a \ne b$.

92. Evaluate the following limits.
   a) $\lim\limits_{x \to 0} \dfrac{2 \sin 5x}{3x}$
   b) $\lim\limits_{x \to 0} \sin 5x \cot 3x$
   c) $\lim\limits_{x \to 0} x \csc^2 \sqrt{2x}$
   d) $\lim\limits_{x \to \pi/2} (\sec x - \tan x)$
   e) $\lim\limits_{x \to 0} \dfrac{x - \sin x}{x - \tan x}$
   f) $\lim\limits_{x \to 0} \dfrac{\sin x^2}{x \sin x}$
   g) $\lim\limits_{x \to 0} \dfrac{\sec x - 1}{x^2}$
   h) $\lim\limits_{x \to 2} \dfrac{x^3 - 8}{x^2 - 4}$

93. L'Hôpital's rule does not help with the following limits. Find them some other way.
   a) $\lim\limits_{x \to \infty} \dfrac{\sqrt{x + 5}}{\sqrt{x} + 5}$
   b) $\lim\limits_{x \to \infty} \dfrac{2x}{x + 7\sqrt{x}}$

94. *The second derivative test.* The second derivative test for local maxima and minima (Article 3.4) says:
   If $f'(c) = 0$ and $f''(c) < 0$, then f has a local maximum at $x = c$.
   If $f'(c) = 0$ and $f''(c) > 0$, then f has a local minimum at $x = c$.
   To prove the test, let $\epsilon = (1/2)|f''(c)|$ and use the fact that
$$f''(c) = \lim_{h \to 0} \frac{f'(c + h) - f'(c)}{h} = \lim_{h \to 0} \frac{f'(c + h)}{h}$$
   to conclude that, for some $\delta > 0$,
$$0 < |h| < \delta \implies \frac{f'(c + h)}{h} < f''(c) + \epsilon < 0.$$
   Thus $f'(c + h)$ is positive for $-\delta < h < 0$ and is negative for $0 < h < \delta$.

95. *Taylor's formula.* The extended mean value theorems of Article 3.9 are special cases of a formula published by Brook Taylor in 1712. The formula was known to James Gregory (1638–1675) in 1670, who did not publish it; was discovered independently a few years later by Leibniz, who did not publish it; and was discovered again by John Bernoulli, who did publish it, in 1694. Nevertheless, it is known today as Taylor's formula. It goes like this:

   Suppose $f(x)$ and its derivatives $f'(x), f''(x), \ldots, f^{(n)}(x)$ of order 1 through n are continuous on $a \le x \le b$, and $f^{(n+1)}(x)$ exists for $a < x < b$. If
$$F(x) = f(x) - f(a) - (x - a)f'(a) - (x - a)^2 f''(a)/2!$$
$$- \cdots - \frac{(x - a)^n f^{(n)}(a)}{n!} - k(x - a)^{n+1},$$

where k is chosen so that $F(b) = 0$, show that

a) $F(a) = F(b) = 0$,

b) $F'(a) = F''(a) = \cdots = F^{(n)}(a) = 0$,

c) there exist numbers $c_1, c_2, c_3, \ldots, c_{n+1}$ such that

$$a < c_{n+1} < c_n < \cdots < c_2 < c_1 < b$$

and such that

$$F'(c_1) = 0 = F''(c_2) = F'''(c_3) = \cdots$$
$$= F^{(n)}(c_n) = F^{(n+1)}(c_{n+1}).$$

d) Hence, deduce that $k = [f^{(n+1)}(c_{n+1})]/(n + 1)!$ for c_{n+1} as in (c); or, in other words, since $F(b) = 0$,

$$f(b) = f(a) + f'(a)(b - a) + \frac{f''(a)}{2!}(b - a)^2 + \cdots$$
$$+ \frac{f^{(n)}(a)}{n!}(b - a)^n + \frac{f^{(n+1)}(c_{n+1})}{(n + 1)!}(b - a)^{(n+1)}$$

for some c_{n+1}, $a < c_{n+1} < b$.†

96. In Problem 95, rewrite Taylor's formula for $f(b)$ with x in place of b. Then use the formula with $f(x) = 1/(1 - x)$, $a = 0$, and $n = 3$ to find a cubic approximation to $1/(1 - x)$. Give an upper bound for the magnitude of the error in this approximation when $|x| < 0.1$.

97. *Estimating reciprocals of numbers.* Newton's method (Article 2.9) can be used to estimate the reciprocal of a positive number a without ever dividing by a, by taking $f(x) = (1/x) - a$.

a) Show that the formula for the method gives

$$x_{n+1} = x_n(2 - ax_n),$$

where x_n and x_{n+1} are two successive approximations to $1/a$.

b) To determine the interval of x-values for which a positive initial guess x_0 will lead to $1/a$, carry out the following steps:

i) Graph f to see what is going on for $x > 0$. What is the x-intercept of the graph?

ii) Show that if $x_0 > 2/a$, then $x_1 < 0$, but that if $0 < x_0 < 2/a$, then $x_1 > 0$. What happens if $x_0 = 2/a$?

iii) What is the appropriate interval for choosing x_0?

---

†James Wolfe, *American Mathematical Monthly*, 60 (1953), p. 415.

CHAPTER 4

Integration

OVERVIEW

One of the early accomplishments of calculus was predicting the future location of a moving body by using the body's location at an earlier time and the body's velocity function. Today, we view this as one of a number of occasions on which we recover functions from information about their derivatives. We calculate the velocities of bodies whose accelerations we know. We calculate the future sizes of populations by using their present sizes and the rates at which they are changing. We calculate from the decay rates of radioactive wastes how long it will take for the wastes to become harmless.

The theory of finding functions whose derivatives are known is part of **integral calculus.** To integrate a function is to find all the functions that have it as a derivative—to find all of the given function's "antiderivatives," so to speak. There is also a second meaning of the word *integrate* that is nearly the same as the nontechnical meaning, "to give the sum or total of." When used in the sense of summation, integration is the mathematical process that enables us to calculate the area of a region that has curved boundaries, the volume of a solid in space, the force of water against a dam, or the amount of energy it takes to put a satellite into orbit. As we shall see, the two kinds of integration are closely related.

This chapter works out a systematic method of integration and presents the fundamental theorems of Leibniz and Newton that reveal the connection between antiderivatives and summation. This will complement our earlier work with derivatives and set the stage for the applications of integration in Chapter 5.

4.1
Indefinite Integrals

We begin this article by finding a function that gives the velocity $v(t)$ of a body that falls from rest with a constant acceleration of 9.8 m/sec². We do this by solving the equation $dv/dt = 9.8$ for v subject to the condition that $v(0)$ be zero. This is one of a number of differential equations (equations with derivatives in them) that can be solved by reversing known derivative formulas. We solve some of these equations and practice finding antiderivatives. The set of all antiderivatives of a function is called the **indefinite integral** of the function.

Finding a Velocity

The acceleration due to gravity near the surface of the earth is 9.8 m/sec². This means that the velocity v of a body falling freely in a vacuum near the earth's surface changes at the rate of

$$\frac{dv}{dt} = 9.8 \text{ m/sec}^2. \tag{1}$$

If the body is dropped from rest, what will its velocity be t seconds after it is released?

If we are to determine v, it must be from the only two facts we know about v as a function of t, which are

1. $\dfrac{dv}{dt} = 9.8$ (the acceleration is 9.8 m/sec²);

2. $v(0) = 0$ (the initial velocity of the body is zero).

We start with the equation for the acceleration and ask, "What functions of t have derivatives exactly equal to 9.8?" Our experience tells us that one answer is

$$v = 9.8t,$$

but there are other answers as well, since

$$v = 9.8t + C \tag{2}$$

is an answer for any value of the constant C.

Does Eq. (2) account for all functions whose derivatives are 9.8? The answer is yes, because Corollary 3 of the Mean Value Theorem (Article 3.7) says that any function whose derivative is 9.8 can differ from the function $9.8t$ only by a constant.

Having established that the velocity of the falling body is

$$v(t) = 9.8t + C \tag{3}$$

for some value of C, we ask, "What is the right value of C for this particular problem?" To find out, we substitute $t = 0$ into Eq. (3) and use the fact that $v(0) = 0$ to find that

$$0 = 9.8(0) + C, \quad \text{or} \quad C = 0.$$

The velocity of the falling body t seconds after release is therefore

$$v(t) = 9.8t \text{ m/sec.} \tag{4}$$

Differential Equations

From a mathematical point of view, the problem of determining the velocity of a falling body from its acceleration is a special case of finding a function $y = F(x)$ whose derivative is given by an equation

$$\frac{dy}{dx} = f(x) \tag{5}$$

over some interval of x-values. An equation like (5) that has a derivative in it is called a **differential equation.**

Equation (5) gives dy/dx as a function of x. A more complicated differential equation might express dy/dx in terms of y as well:

$$\frac{dy}{dx} = 2xy^2.$$

A differential equation may also involve higher-order derivatives:

$$\frac{d^2y}{dx^2} + 6xy\frac{dy}{dx} + 3x^2y^3 = 0.$$

For the time being, however, we shall restrict our attention to equations that contain a single derivative. Differential equations of a more general nature will be considered in Chapter 20.

A function $F(x)$ is called a **solution** of the differential equation $dy/dx = f(x)$ over the interval I if F is differentiable at every point of I and if

$$\frac{d}{dx}F(x) = f(x)$$

at every point of I. We also say that $F(x)$ is an **antiderivative** of $f(x)$.

To **solve** the equation $dy/dx = f(x)$ on the interval I means to find all functions defined on I that are antiderivatives of f. The interval may be finite or infinite. In the falling body example, we solved the equation $dv/dt = 9.8$ on the interval $t \geq 0$ by finding the **general solution** $v(t) = 9.8t + C$. We then solved the falling body problem by using the **initial condition** $v(0) = 0$ to select the **particular solution** $v(t) = 9.8t$ from the general solution.

Indefinite Integrals

If $F(x)$ is an antiderivative of $f(x)$, then $F(x) + C$ is an antiderivative of $f(x)$ for every value of the constant C. For, if $dF/dx = f$, then

$$\frac{d}{dx}(F + C) = \frac{dF}{dx} + \frac{dC}{dx} = f(x) + 0 = f(x). \tag{6}$$

Are there any other antiderivatives of f besides those given by the formula $F(x) + C$? Corollary 3 of the Mean Value Theorem says, "No. Every antiderivative of f is given by this formula for some value of C." Therefore, if $F(x)$ is any solution whatever of $dy/dx = F(x)$, the formula $y = F(x) + C$ gives all other solutions.

The set of all antiderivatives of a function $f(x)$ is called the **indefinite integral of f with respect to x.** The notation for the indefinite integral is

$$\int f(x)\, dx.$$

When a formula $F(x) + C$ gives all the antiderivatives of f, we indicate this by writing

$$\int f(x)\, dx = F(x) + C.$$

(7)

This equation can be read in two ways: "The integral of $f(x)$ with respect to x is $F(x)$ plus C" or "The integral of $f(x)\, dx$ is $F(x)$ plus C." The symbol $\int$ is an **integral sign.** The function f is the **integrand** of the integral, and C is the **constant of integration.** The dx tells us that the **variable of integration** is x.

In the falling body example, we found that the integral of 9.8 with respect to t is $9.8t + C$:

$$\int 9.8\, dt = 9.8t + C.$$

The integrand here is the constant function 9.8, and the variable of integration is t.

EXAMPLE 1 Solve the differential equation

$$\frac{dy}{dx} = 3x^2.$$

Solution We know from experience that

$$\frac{d}{dx}(x^3) = 3x^2.$$

Therefore,

$$y = \int 3x^2\, dx = x^3 + C. \qquad \blacksquare$$

Separation of Variables

If both x and y occur in the expression for dy/dx in a differential equation, we have an equation that does not give one variable explicitly in terms of the other. Solving the differential equation

$$\frac{dy}{dx} = x^2\sqrt{y}$$

(8)

means finding a function $y = f(x)$ whose derivative dy/dx at any particular x equals x^2 times the value of $\sqrt{y}$ at x. While equations like (8) usually do have solutions, as we shall see in Chapter 20, the solutions are not always easy to find. A method that sometimes works involves combining all the y terms with dy/dx on one side of the equation and putting all the x terms on the other side. This is called **separating the variables.** If we can separate the variables, we may then be able to find y by integration, as in the following example.

EXAMPLE 2   Solve the differential equation

$$\frac{dy}{dx} = x^2\sqrt{y}, \qquad y > 0. \tag{8}$$

Solution   We divide both sides of the equation by $\sqrt{y}$ to separate the variables and obtain

$$y^{-1/2}\frac{dy}{dx} = x^2. \tag{9}$$

Under the assumption that the original equation defines y as a differentiable function of x, we see that the left side of Eq. (9) is the derivative of the function $2y^{1/2}$ with respect to x. The right side is the derivative of $x^3/3$ with respect to x. Therefore, when we integrate both sides of Eq. (9) with respect to x we get

$$\int \left(y^{-1/2}\frac{dy}{dx}\right) dx = \int x^2 \, dx$$

$$2y^{1/2} + C_1 = \frac{x^3}{3} + C_2$$

$$2y^{1/2} = \frac{x^3}{3} + C,$$

where we have combined C_1 and C_2 into a single constant $C = C_2 - C_1$. There is no need to describe the family of solutions with two constants when one will do, nor is any greater generality achieved by doing so.   ∎

Integration Formulas

Integration, as described above, requires the ability to guess the answer. But the following formulas help to reduce the amount of guesswork in many cases. In these formulas, u and v denote differentiable functions of x, and a, n, and C are constants.

Integration Formulas

INTEGRAL	MATCHING DERIVATIVE
1. $\displaystyle\int \frac{du}{dx}\, dx = u(x) + C$	$\dfrac{d}{dx}(u(x) + C) = \dfrac{du}{dx}$
2. $\displaystyle\int au(x)\, dx = a \int u(x)\, dx$ (if a is a constant)	$\dfrac{d}{dx}(au) = a\dfrac{du}{dx}$
3. $\displaystyle\int (u(x) + v(x))\, dx = \int u(x)\, dx$ $\displaystyle\qquad\qquad + \int v(x)\, dx$	$\dfrac{d}{dx}(u + v) = \dfrac{du}{dx} + \dfrac{dv}{dx}$
4. $\displaystyle\int u^n \frac{du}{dx}\, dx = \frac{u^{n+1}}{n+1} + C$ $(n \neq -1)$	$\dfrac{d}{dx}\left(\dfrac{u^{n+1}}{n+1}\right) = u^n\dfrac{du}{dx}$

In words, these formulas say

1. The integral of the derivative of a differentiable function u is u plus an arbitrary constant.

2. A constant may be moved across the integral sign. (*Caution:* Functions of the variable of integration must not be moved across the integral sign.)

3. The integral of the sum of two functions is the sum of their integrals. This extends to the sum of any finite number of functions:

$$\int (u_1 + u_2 + \cdots + u_n)\, dx = \int u_1\, dx + \int u_2\, dx + \cdots + \int u_n\, dx.$$

4. If $n \neq -1$, the integral of $u^n\, du/dx$ is obtained by adding 1 to the exponent, dividing by the new exponent, and adding an arbitrary constant.

Formula (1) is a restatement of the definition of the indefinite integral as the set of all functions with a given derivative. It says that any function whose derivative is du/dx must be given by the formula $u(x) + C$ for some value of C. The remaining formulas come from "reversing" derivative formulas from Chapter 2. We shall not derive these integration formulas but instead shall show by example how they are used to evaluate indefinite integrals.

EXAMPLE 3

$$\int (5x - x^2 + 2)\, dx = 5 \int x\, dx - \int x^2\, dx + \int 2\, dx$$

$$= \frac{5}{2}x^2 - \frac{x^3}{3} + 2x + C.$$

$$\int x^{1/2}\, dx = \frac{x^{3/2}}{3/2} + C = \frac{2}{3}x^{3/2} + C.$$

$$\int (x^2 + 5)^2\, dx = \int (x^4 + 10x^2 + 25)\, dx = \frac{x^5}{5} + \frac{10}{3}x^3 + 25x + C.$$

$$\int (x^2 + 5)^2\, 2x\, dx = \frac{(x^2 + 5)^3}{3} + C.$$

In the last equation we applied the formula

$$\int u^n \frac{du}{dx}\, dx = \frac{u^{n+1}}{n+1} + C, \qquad n \neq -1,$$

with $u = x^2 + 5$, $n = 2$, and $du/dx = 2x$. ∎

EXAMPLE 4 *Adjusting the integrand by a constant.* Evaluate the integral

$$\int x^3\, dx.$$

Solution We know that

$$\frac{d}{dx}(x^4) = 4x^3,$$

so

$$\int 4x^3\, dx = x^4 + C.$$

We can therefore evaluate the original integral by rewriting the integrand as

$$x^3 = \frac{1}{4} \cdot 4x^3$$

and moving the 1/4 outside the integral:

$$\int x^3 \, dx = \int \frac{1}{4} \cdot 4x^3 \, dx = \frac{1}{4} \int 4x^3 \, dx$$

$$= \frac{1}{4}(x^4 + C)$$

$$= \frac{1}{4}x^4 + C',$$

where $C' = C/4$.   ■

The (Educated) Guess and Check Strategy

The examples we have seen so far succeeded because our experience with derivatives allowed us to guess the answers. But what if we don't know what to guess?

There is a procedure to try if we don't see how to integrate a given function right away. The procedure assumes that we have enough experience to make a reasonable guess at the answer, but it does not require us to be right the first time. The steps are

1. Write down a reasonable candidate for the answer.
2. Compare its derivative with the integrand.
3. Modify the trial answer accordingly.
4. Check the result and make improvements as necessary.
5. Add C.

We use this procedure in the next example. We shall introduce a more systematic method of integration, integration by substitution, in Article 4.3 after we have spent more time with antiderivatives.

EXAMPLE 5 Evaluate the integral

$$\int \sqrt{2x + 1} \, dx.$$

Solution We seek a function whose derivative is $(2x + 1)^{1/2}$. We add 1 to the exponent and try $(2x + 1)^{3/2}$, whose derivative is

$$\frac{3}{2}(2x + 1)^{1/2} \cdot 2 = 3(2x + 1)^{1/2}.$$

This differs from the integrand $(2x + 1)^{1/2}$ by a factor of 3. In other words, our trial function is 3 times too large. We divide it by 3 to obtain the next candidate,

$$\frac{1}{3}(2x + 1)^{3/2}.$$

The derivative of this new function is

$$\frac{3}{2} \cdot \frac{1}{3}(2x + 1)^{1/2} \cdot 2 = (2x + 1)^{1/2},$$

the function whose integral we were asked to find. We conclude that

$$\int \sqrt{2x + 1} \, dx = \frac{1}{3}(2x + 1)^{3/2} + C.$$ ■

PROBLEMS

Evaluate the integrals in Problems 1–12. Do as much as you can without writing anything down.

1. a) $\int 2x \, dx$ b) $\int 3 \, dx$

  c) $\int (2x + 3) \, dx$

2. a) $\int 6x \, dx$ b) $\int -2 \, dx$

  c) $\int (6x - 2) \, dx$

3. a) $\int 3x^2 \, dx$ b) $\int x^2 \, dx$

  c) $\int (x^2 + 2x) \, dx$

4. a) $\int 8x^7 \, dx$ b) $\int x^7 \, dx$

  c) $\int (x^7 - 6x) \, dx$

5. a) $\int -3x^{-4} \, dx$ b) $\int x^{-4} \, dx$

  c) $\int (x^{-4} + 2x + 3) \, dx$

6. a) $\int \frac{1}{x^2} \, dx$ b) $\int \frac{-5}{x^2} \, dx$

  c) $\int \left(2 - \frac{5}{x^2}\right) dx$

7. a) $\int \frac{3}{2}\sqrt{x} \, dx$ b) $\int 4\sqrt{x} \, dx$

  c) $\int (x^2 - 4\sqrt{x}) \, dx$

8. a) $\int \frac{3}{2}\sqrt{x + 1} \, dx$ b) $\int \sqrt{x + 1} \, dx$

  c) $\int \sqrt{5x + 1} \, dx$

9. a) $\int (2x - 1) \, dx$ b) $\int (2x - 1)^2 \, dx$

  c) $\int (2x - 1)^3 \, dx$

10. a) $\int 5(x - 2)^4 \, dx$ b) $\int (x - 2)^4 \, dx$

  c) $\int 2(x - 2)^4 \, dx$

11. a) $\int 2(x^2 - 3)2x \, dx$ b) $\int (x^2 - 3)x \, dx$

12. a) $\int 2(2x^3 + 1)6x^2 \, dx$ b) $\int (2x^3 + 1)x^2 \, dx$

Solve the differential equations in Problems 13–36.

13. $\frac{dy}{dx} = 2x - 7$ **14.** $\frac{dy}{dx} = 7 - 2x$

15. $\frac{dy}{dx} = x^2 + 1$ **16.** $\frac{dy}{dx} = \frac{1}{x^2}, \quad x > 0$

17. $\frac{dy}{dx} = \frac{-5}{x^2}, \quad x > 0$ **18.** $\frac{dy}{dx} = 3x^2 - 2x + 5$

19. $\frac{dy}{dx} = (x - 2)^4$ **20.** $\frac{dy}{dx} = (5x - 2)^4$

21. $\frac{dy}{dx} = \frac{1}{x^2} + x, \quad x > 0$ **22.** $\frac{dy}{dx} = x + \sqrt{2x}$

23. $\frac{dy}{dx} = \frac{x}{y}, \quad y > 0$ **24.** $\frac{dy}{dx} = \sqrt{\frac{x}{y}}, \quad y > 0$

25. $\frac{dy}{dx} = \frac{x + 1}{y - 1}, \quad y > 1$ **26.** $\frac{dy}{dx} = \frac{\sqrt{x + 1}}{\sqrt{y - 1}}, \quad y > 1$

27. $\frac{dy}{dx} = \sqrt{xy}, \quad x > 0, \quad y > 0$

28. $\frac{dy}{dx} = x\sqrt{x^2 + 1}$

29. $\frac{dy}{dx} = 2xy^2, \quad y > 0$

30. $\frac{dy}{dx} = \sqrt[3]{y/x}, \quad x > 0, \quad y > 0$

31. $x^3\frac{dy}{dx} = -2, \quad x > 0$

32. $x^2 \dfrac{dy}{dx} = \dfrac{1}{y^2 + \sqrt{y}}, \quad x > 0, \quad y > 0$

33. $\dfrac{ds}{dt} = 3t^2 + 4t - 6$

34. $\dfrac{dx}{dt} = 8\sqrt{x}, \quad x > 0$

35. $\dfrac{dy}{dt} = (2t + t^{-1})^2, \quad t > 0$

36. $\dfrac{dy}{dz} = \sqrt{(z^2 - z^{-2})^2 + 4}, \quad z > 0$

37. On the moon, the acceleration due to gravity is 1.6 m/sec². If a rock is dropped into a crevasse, how fast will it be going just before it hits the bottom 30 seconds later?

38. A rocket lifts off the surface of Earth with a constant acceleration of 20 m/sec². How fast will the rocket be going 1 minute later?

4.2
Selecting a Value for the Constant of Integration

When we solve a differential equation $dy/dx = f(x)$, we usually want to find a particular solution that satisfies numerical conditions set down in advance. To do so, we first find the general solution $y = F(x) + C$ that gives all possible solutions and then find the value of C that gives us the particular solution we want.

The graphs of the solution curves $y = F(x) + C$ are all obtained by shifting the solution curve $y = F(x)$ through a vertical displacement C. Thus, the graphs form a family of "parallel" curves that fit together to fill up the plane above and below the domain of F. You can see what we mean if you look at Fig. 4.1, where we have graphed a few of the curves that make up the general solution $y = x^3 + C$ of the equation $dy/dx = 3x^2$.

If we pick a point x_0 in the domain of F and select an arbitrary value y_0, we can find the solution whose graph passes through the point (x_0, y_0) by substituting $x = x_0$ and $y = y_0$ in the equation $y = F(x) + C$ and solving for C. This gives

$$y_0 = F(x_0) + C \qquad \text{or} \qquad C = y_0 - F(x_0).$$

The curve $y = F(x) + (y_0 - F(x_0))$ is the one that passes through (x_0, y_0).

The condition that y equal y_0 when x equals x_0 is called an **initial condition.** The name comes from problems in which time is the independent variable and y_0 is the velocity or the position of a moving body at an initial time x_0. What we have just seen is that we can always select from the general solution $y = F(x) + C$ the particular solution that satisfies a given initial condition, provided x_0 lies in the domain of F.

EXAMPLE 1   Find the curve whose slope at the point (x, y) is $3x^2$ if the curve is also required to pass through the point $(1, -1)$. See Fig. 4.1.

Solution   In mathematical language, we want to solve the problem that consists of

$$\text{The differential equation:} \qquad \frac{dy}{dx} = 3x^2, \tag{1}$$

$$\text{The initial condition:} \qquad y = -1 \quad \text{when } x = 1. \tag{2}$$

First we find the general solution of the differential equation by integrating both sides with respect to x:

$$\frac{dy}{dx} = 3x^2$$

$$y = \int 3x^2 \, dx = x^3 + C.$$

Then, to identify the particular solution whose graph passes through the point $(1, -1)$, we find the value of C for which $y = -1$ when $x = 1$:

$$y = x^3 + C$$

$$-1 = (1)^3 + C$$

$$C = -2.$$

The solution we seek is therefore $y = x^3 - 2$. The slope of this curve is $y' = 3x^2$, and the curve passes through the point $(1, -1)$. ■

EXAMPLE 2   At time t, the velocity of a moving body is given by

$$\frac{ds}{dt} = at,$$

where a is a constant and s is the body's position at time t. If $s = s_0$ when $t = 0$, find s as a function of t.

Solution   We want to solve the problem that consists of

The differential equation: $\dfrac{ds}{dt} = at,$

The initial condition: $s = s_0$ when $t = 0$.

We find the general solution of the equation by integrating both sides with respect to t:

$$\frac{ds}{dt} = at$$

$$s = \int at \, dt = a\frac{t^2}{2} + C.$$

We then find the value of C that gives the particular solution that satisfies the condition that s equal s_0 when t equals 0. To do so, we substitute $s = s_0$ and $t = 0$ in the formula for the general solution and solve for C:

$$s_0 = a\frac{(0)^2}{2} + C$$

$$C = s_0.$$

The solution we seek is

$$s = a\frac{t^2}{2} + s_0.$$ ■

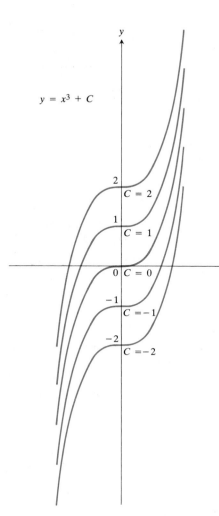

$y = x^3 + C$

4.1   Curves from the family $y = x^3 + C$. The entire family, which constitutes the general solution of $dy/dx = 3x^2$, would fill up the plane.

EXAMPLE 3   A projectile is fired straight up from a platform 10 ft above the ground, with an initial velocity of 160 ft/sec. Assume that the only force affecting the motion of the projectile during its flight is from gravity, which produces a downward acceleration of 32 ft/sec². Find an equation for the height of the projectile above the ground as a function of time t if $t = 0$ when the projectile is fired.

Solution   If s denotes the height of the projectile above the ground, then the velocity and acceleration of the projectile are

$$v = \frac{ds}{dt} \quad \text{and} \quad a = \frac{dv}{dt} = \frac{d^2s}{dt^2}.$$

Since the projectile is fired upward, against the force of gravity, the velocity will be a decreasing function of t. Therefore, the equation to solve is

$$\frac{dv}{dt} = -32 \text{ ft/sec}^2,$$

subject to the initial conditions

$$v(0) = 160 \text{ ft/sec} \quad \text{and} \quad s(0) = 10 \text{ ft}.$$

We integrate once to find

$$v = \frac{ds}{dt} = -32t + C_1$$

and a second time to find

$$s = -16t^2 + C_1 t + C_2.$$

We determine appropriate values for C_1 and C_2 from the initial conditions:

$$C_1 = v(0) = 160, \qquad C_2 = s(0) = 10.$$

The equation of motion is

$$s = -16t^2 + 160t + 10. \qquad ■$$

PROBLEMS

In Problems 1–6, find the position s as a function of time t from the given velocity $v = ds/dt$. Then evaluate the constant of integration so that $s = s_0$ when $t = 0$.

1. $v = 3t^2, \quad s_0 = 4$ **2.** $v = 2t + 1, \quad s_0 = 0$

3. $v = (t + 1)^2, \quad s_0 = 0$ **4.** $v = t^2 + 1, \quad s_0 = 1$

5. $v = (t + 1)^{-2}, \quad s_0 = -5$ **6.** $v = 8\sqrt{s}, \quad s_0 = 9$

In Problems 7–12, find the velocity $v = ds/dt$ and position s as functions of t from the given acceleration $a = dv/dt$. Then evaluate the constants of integration so that $v = v_0$ and $s = s_0$ when $t = 0$.

7. $a = 9.8, \quad v_0 = 20, \quad s_0 = 0$

8. $a = 9.8, \quad v_0 = 0, \quad s_0 = 20$

9. $a = 2t, \quad v_0 = 1, \quad s_0 = 1$

10. $a = 6t, \quad v_0 = 0, \quad s_0 = 5$

11. $a = 2t + 2, \quad v_0 = 1, \quad s_0 = 0$

12. $a = 4/v, \quad v_0 = 0, \quad s_0 = 25$

In Problems 13–24, solve the differential equations subject to the given initial conditions.

13. $\dfrac{dy}{dx} = 3x^2 + 2x + 1, \quad y = 0$ when $x = 1$

14. $\dfrac{dy}{dx} = 9x^2 - 4x + 5, \quad y = 0$ when $x = -1$

15. $\dfrac{dy}{dx} = 4(x - 7)^3, \quad y = 10$ when $x = 8$

16. $\dfrac{dy}{dx} = x^{1/2} + x^{1/4}, \quad y = -2$ when $x = 0$

17. $\dfrac{dy}{dx} = x\sqrt{y}, \quad y = 1$ when $x = 0$

18. $\dfrac{dy}{dx} = \dfrac{x^2 + 1}{x^2}$, $y = 1$ when $x = 1$

19. $\dfrac{dy}{dx} = 2xy^2$, $y = 1$ when $x = 1$

20. $\dfrac{dy}{dx} = (x + x^{-1})^2$, $y = 1$ when $x = 1$

21. $\dfrac{d^2y}{dx^2} = 2 - 6x$, $y = 1$ and $\dfrac{dy}{dx} = 4$ when $x = 0$

22. $\dfrac{d^3y}{dx^3} = 6$, $y = 5$, $\dfrac{dy}{dx} = 0$, and $\dfrac{d^2y}{dx^2} = -8$ when $x = 0$

23. $\dfrac{d^2y}{dx^2} = \dfrac{3x}{8}$,   the graph of y passes through the point (4, 4) with slope 3.

24. $\dfrac{d^2y}{dx^2} = 2x - 3x^2 + 1$,   the graph of y passes through the origin and the point (1, 1).

25. The graph of $y = f(x)$ passes through the point (9, 4). Also, the line tangent to the graph at any point (x, y) has the slope $3\sqrt{x}$. Find f.

26. If $y = 3$ when $x = 3$, and $dy/dx = 2x/y^2$, find the value of y when $x = 1$.

27. With approximately what velocity does a diver enter the water after diving from a 30-m platform? (Use $g = 9.8$ m/sec^2.)

28. The acceleration of gravity near the surface of Mars is 3.72 m/sec^2. A rock is thrown straight up from the surface with an initial velocity of 23 m/sec. How high does it go? (*Hint:* When is the velocity zero?)

29. Figure 4.2 gives a graphical representation of the function that solves one of the following differential equations with the given initial condition. Which one? How do you know?
    a) $dy/dx = 2x$, $y(1) = 0$
    b) $dy/dx = x^2$, $y(1) = 1$
    c) $dy/dx = 2x + 2$, $y(1) = 1$
    d) $dy/dx = 2x$, $y(1) = 1$.

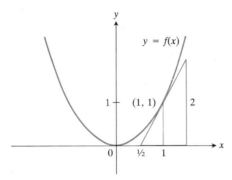

$y = f(x)$

(1, 1)

1

2

0   ½   1

4.2   Figure for Problem 29.

30. If we open a valve to drain the water from a cylindrical tank (Fig. 4.3), the water will flow fast when the tank is full but slow down as the tank drains. It turns out that the rate at which the water level drops is proportional to the square

4.3   The draining tank in Problem 30.

root of the water's depth. In the notation of Fig. 4.3, this means that

$$\frac{dy}{dt} = -k\sqrt{y}. \qquad (3)$$

The number k depends on the acceleration of gravity, the cross-section area of the tank, and the cross-section area of the drain hole. Equation (3) has a minus sign because y decreases with time.
a) Solve Eq. (3) for y as a function of t.
b) Suppose t is measured in minutes and $k = 1/10$. Find y as a function of t if $y = 9$ ft when $t = 0$.
c) How long does it take to drain the tank if the water is 9 ft deep to start with?

31. *Escape velocity.* The gravitational attraction exerted by Earth on a particle of mass m at distance s from Earth's center is given by $F = -mgR^2s^{-2}$, where R is the radius of Earth. The force F is negative because the force acts in opposition to increasing s (Fig. 4.4). If a particle is pro-

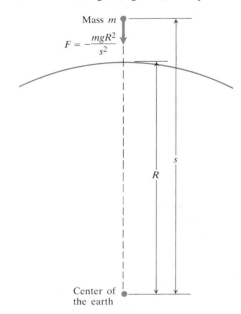

Mass m

$F = -\dfrac{mgR^2}{s^2}$

s

R

Center of the earth

4.4   A mass m that is s km from the earth's center. See Problem 31.

jected vertically upward from the surface of the earth with initial velocity $v_0 = \sqrt{2gR}$, apply Newton's second law $F = ma$ with $a = dv/dt = (dv/ds)(ds/dt) = v(dv/ds)$ to show that $v = v_0\sqrt{R/s}$ and that $s^{3/2} = R^{3/2}[1 + (3v_0t/2R)]$.

The initial velocity $v_0 = \sqrt{2gR}$ (11.2 km/sec) is known as the "escape velocity" since the displacement s tends to infinity with increasing t if the initial velocity is this large or greater. Actually, a somewhat larger initial velocity

is required to escape Earth's gravitational attraction because of the retardation effect of air resistance, which we neglected here for the sake of simplicity.

TOOLKIT PROGRAMS

Antiderivatives and Direction Fields

4.3
The Substitution Method of Integration

A change in variable can often change an unfamiliar integral into one we know how to evaluate. The method for doing this is called the **substitution method of integration.** It is the principal method by which integrals are evaluated. We first show how the method works and then show why it works.

EXAMPLE 1   To evaluate

$$\int (x^4 - 1)^2 \cdot 4x^3 \, dx,$$

we carry out these steps:

$$\int (x^4 - 1)^2 \cdot 4x^3 \, dx = \int u^2 \, du = \frac{u^3}{3} + C = \frac{(x^4 - 1)^3}{3} + C.$$

1. Substitute   2. Integrate.   3. Substitute
   $u = x^4 - 1$,   back.
   $du = 4x^3 \, dx$.

This works because

$$4x^3 = \frac{d}{dx}(x^4 - 1) = \frac{du}{dx}$$

is part of the integrand. By the Chain Rule (run backward) we have

$$\int (x^4 - 1)^2 4x^3 \, dx = \int (x^4 - 1)^2 \frac{d}{dx}(x^4 - 1) \, dx$$

$$= \int \frac{d}{dx}\left[\frac{(x^4 - 1)^3}{3}\right] dx = \frac{(x^4 - 1)^3}{3} + C.$$

⇧

Chain Rule
(run backward)

EXAMPLE 2   To evaluate

$$\int \sqrt{1 + x^2} \cdot 2x \, dx,$$

we carry out these steps:

$$\int \sqrt{1 + x^2} \cdot 2x \, dx = \int \sqrt{u} \, du = \frac{2}{3} u^{3/2} + C = \frac{2}{3}(1 + x^2)^{3/2} + C$$

                1. Substitute   2. Integrate.   3. Substitute
                   $u = 1 + x^2$,   back.
                   $du = 2x \, dx$.

This works because

$$2x = \frac{d}{dx}(1 + x^2)$$

is part of the integrand. By the Chain Rule, run backward, we have

$$\int \sqrt{1 + x^2} \cdot 2x \, dx = \int \sqrt{1 + x^2} \cdot \frac{d}{dx}(1 + x^2) \, dx$$

$$= \int \frac{d}{dx}\left[\frac{2}{3}(1 + x^2)^{3/2}\right] dx = \frac{2}{3}(1 + x^2)^{3/2} + C.$$

⇧

Chain Rule
(run backward)

The rule illustrated by Examples 1 and 2 is this:

$$\int f(g(x)) \cdot g'(x) \, dx = \int f(u) \, du$$ 1. Substitute $u = g(x)$,
                                         $du = g'(x) \, dx$.

$$= F(u) + C$$ 2. Evaluate by finding an
⇧   antiderivative of $f(u)$.

any
antiderivative
of $f(u)$

$$= F(g(x)) + C$$ 3. Substitute back.

These three steps are the steps of the substitution method of integration.

The Substitution Method of Integration

To evaluate the integral

$$\int f(g(x))g'(x) \, dx$$

when f and g' are continuous functions, carry out the following steps:

1. Substitute $u = g(x)$ and $du = g'(x) \, dx$ to obtain the integral

$$\int f(u) \, du.$$

2. Integrate with respect to u.
3. Replace u by $g(x)$ in the result.

EXAMPLE 3 Evaluate

$$\int \sqrt{4x - 1} \; dx.$$

Solution We substitute

$$u = 4x - 1, \qquad du = 4 \; dx, \qquad \frac{1}{4} \; du = dx.$$

Then

$$\int \sqrt{4x - 1} \; dx = \int \sqrt{u} \cdot \frac{1}{4} \; du$$

$$= \frac{1}{4} \int \sqrt{u} \; du$$

$$= \frac{1}{4} \left(\frac{2}{3} u^{3/2} \right) + C = \frac{1}{6} u^{3/2} + C$$

$$= \frac{1}{6} (4x - 1)^{3/2} + C.$$ ∎

EXAMPLE 4 Evaluate

$$\int \frac{x \; dx}{\sqrt{4 - x^2}}.$$

Solution We substitute

$$u = 4 - x^2, \qquad du = -2x \; dx, \qquad -\frac{1}{2} \; du = x \; dx.$$

Then

$$\int \frac{x \; dx}{\sqrt{4 - x^2}} = \int \frac{(-1/2) \; du}{\sqrt{u}} = -\frac{1}{2} \int \frac{du}{\sqrt{u}}$$

$$= -\frac{1}{2} (2u^{1/2}) + C = -u^{1/2} + C$$

$$= -\sqrt{4 - x^2} + C.$$ ∎

There is often more than one way to make a successful substitution, as the next example shows.

EXAMPLE 5 Evaluate

$$\int \frac{(z + 1) \; dz}{\sqrt[3]{3z^2 + 6z + 5}}.$$

Solution The substitution method can be used as an exploratory tool: We take the most troublesome part of the integrand, substitute for it, and see how things work out. For the integral here, we might try $u = 3z^2 + 6z + 5$, or we might even press our luck and try $u = \sqrt[3]{3z^2 + 6z + 5}$. Here is what happens when we do this.

We substitute

$$u = 3z^2 + 6z + 5, \qquad du = (6z + 6)\ dz = 6(z + 1)\ dz,$$

$$\frac{1}{6}\ du = (z + 1)\ dz.$$

Then

$$\int \frac{(z + 1)\ dz}{\sqrt[3]{3z^2 + 6z + 5}} = \int \frac{(1/6)\ du}{u^{1/3}}$$

$$= \frac{1}{6} \int u^{-1/3}\ du = \frac{1}{6} \cdot \frac{3}{2} u^{2/3} + C$$

$$= \frac{1}{4}(3z^2 + 6z + 5)^{2/3} + C.$$

We substitute

$$u = \sqrt[3]{3z^2 + 6z + 5}, \qquad 3u^2\ du = 6z\ dz + 6\ dz = 6(z + 1)\ dz,$$

$$u^3 = 3z^2 + 6z + 5, \qquad \frac{1}{2}u^2\ du = (z + 1)\ dz.$$

Then

$$\int \frac{(z + 1)\ dz}{\sqrt[3]{3z^2 + 6z + 5}} = \int \frac{(1/2)u^2\ du}{u} = \frac{1}{2} \int u\ du$$

$$= \frac{1}{2}\left(\frac{u^2}{2}\right) + C = \frac{1}{4}u^2 + C$$

$$= \frac{1}{4}(3z^2 + 6z + 5)^{2/3} + C. \qquad \blacksquare$$

Why Substitution Works

If F is an antiderivative of f, then, by the Chain Rule,

$$\frac{d}{dx}F(g(x)) = F'(g(x))g'(x) = f(g(x))g'(x). \tag{1}$$

Therefore,

$$\int f(g(x))g'(x)\ dx = \int F'(g(x))g'(x)\ dx$$

$$= \int \frac{d}{dx}(F(g(x)))\ dx \qquad \text{(By the Chain Rule)} \tag{2}$$

$$= F(g(x)) + C.$$

Setting $u = g(x)$ and $du = g'(x)\ dx$ in the integral gives the same result:

$$\int f(g(x))g'(x)\ dx = \int f(u)\ du = F(u) + C = F(g(x)) + C. \tag{3}$$

Integral Formulas in Differential Notation

Differential notation enables us to express integral formulas in a useful shorthand. We shall not do much of it here, but this is a good place to mention it for future reference. For example,

$$\int \frac{du}{dx} dx = u + C \quad \text{becomes} \quad \int du = u + C,$$

and

$$\int u^n \frac{du}{dx} dx = \frac{u^{n+1}}{n+1} + C \quad \text{becomes} \quad \int u^n \, du = \frac{u^{n+1}}{n+1} + C, \qquad n \neq -1.$$

Similarly,

$$\int \frac{d(uv)}{dx} dx = \int \left(u \frac{dv}{dx} + v \frac{du}{dx} \right) dx = \int u \frac{dv}{dx} dx + \int v \frac{du}{dx} dx \qquad (4)$$

becomes

$$\int d(uv) = \int u \, dv + \int v \, du. \qquad (5)$$

Abbreviations like these can come in handy as a way of remembering more complicated formulas, as we shall see in Chapters 5 and 7.

PROBLEMS

Evaluate the integrals in Problems 1–30.

1. $\int (x-1)^{243} \, dx$

2. $\int \sqrt{1-x} \, dx$

3. $\int \frac{1}{\sqrt{1-x}} \, dx$

4. $\int 2x\sqrt{x^2-1} \, dx$

5. $\int x\sqrt{2x^2-1} \, dx$

6. $\int (3x-1)^5 \, dx$

7. $\int (2-t)^{2/3} \, dt$

8. $\int x^2\sqrt{1+x^3} \, dx$

9. $\int (1+x^3)^2 \, dx$

10. $\int (1+x^3)^2 3x^2 \, dx$

11. $\int x(x^2+1)^{10} \, dx$

12. $\int \frac{dt}{2\sqrt{1+t}}$

13. $\int \frac{x^2}{\sqrt{1+x^3}} \, dx$

14. $\int \sqrt{2+5y} \, dy$

15. $\int \frac{dx}{(3x+2)^2}$

16. $\int 5r\sqrt{1-r^2} \, dr$

17. $\int \frac{3r \, dr}{\sqrt{1-r^2}}$

18. $\int \frac{y \, dy}{\sqrt{2y^2+1}}$

19. $\int x^4(7-x^5)^3 \, dx$

20. $\int \frac{x^3 \, dx}{\sqrt[4]{1+x^4}}$

21. $\int \frac{ds}{(s+1)^3}$

22. $\int \frac{s \, ds}{(s^2+1)^2}$

23. $\int \frac{1}{x^2+4x+4} \, dx$

24. $\int \frac{1}{y^2-2y+1} \, dy$

25. $\int \frac{x+1}{2\sqrt{x+1}} \, dx$

26. $\int x\sqrt{a^2-x^2} \, dx$

27. $\int (y^3+6y^2+12y+8)(y^2+4y+4) \, dy$

28. $\int \frac{(z+1) \, dz}{\sqrt[3]{z^2+2z+2}}$

29. $\int \frac{1}{\sqrt{x}(1+\sqrt{x})^2} \, dx$

30. $\int (y^4+4y^2+1)^2(y^3+2y) \, dy$

Solve the differential equations in Problems 31–36 subject to the given initial conditions.

31. $\dfrac{dy}{dx} = x\sqrt{1+x^2}, \quad y = 0$ when $x = 0$

32. $\dfrac{dy}{dx} = 3x^2\sqrt{1+x^3}, \quad y = 1$ when $x = 0$

33. $\dfrac{dr}{dz} = 24z(3z^2-1)^3, \quad r = -3$ when $z = 0$

34. $\dfrac{dy}{dx} = 4x(x^2 - 8)^{-1/3}, \quad y = 0$ when $x = 0$

35. $2y\dfrac{dy}{dx} = 3x\sqrt{x^2 + 1}\sqrt{y^2 + 1}, \quad y = 0$ when $x = 0$

36. $\dfrac{dy}{dx} = \dfrac{4\sqrt{(1 + y^2)^3}}{y}, \quad y = 0$ when $x = 0$

37. Which of the following methods could be used to evaluate

$$\int 3x^2(x^3 - 1)^5 \, dx?$$

a) Expand $(x^3 - 1)^5$ and then multiply by $3x^2$ to get a polynomial to integrate term by term.
b) Factor x^2 out to get an integral of the form $3x^2 \int u^n \, du$.
c) Use the substitution $u = x^3 - 1$ to get an integral of the form $\int u^n \, du$.

4.4
Integrals of Trigonometric Functions

We devote this article to integrating trigonometric functions.
From the derivative formulas

$$\cos x = \frac{d}{dx}(\sin x), \qquad \sin x = \frac{d}{dx}(-\cos x)$$

we get the integration formulas

$$\int \cos x \, dx = \sin x + C, \tag{1a}$$

$$\int \sin x \, dx = -\cos x + C. \tag{1b}$$

If u is a differentiable function of x, the Chain Rule applied to $\sin u$ gives

$$\frac{d}{dx}\sin u = \cos u \frac{du}{dx}.$$

When we turn this equation around, to get

$$\cos u \frac{du}{dx} = \frac{d}{dx}\sin u, \qquad \text{or} \qquad \cos u \, du = d(\sin u),$$

we see that

$$\int \cos u \, du = \sin u + C. \tag{2a}$$

Similarly,

$$\frac{d}{dx}(-\cos u) = \sin u \frac{du}{dx}$$

$$d(-\cos u) = \sin u \, du,$$

and

$$\int \sin u \, du = -\cos u + C. \tag{2b}$$

Equations (1) and (2) enable us to evaluate a variety of trigonometric integrals.

EXAMPLE 1   Evaluate

$$\int \cos 2x \, dx.$$

Solution   We substitute

$$u = 2x, \quad du = 2 \, dx, \quad \frac{1}{2} \, du = dx.$$

Then

$$\int \cos 2x \, dx = \int \cos u \cdot \frac{1}{2} \, du = \frac{1}{2} \int \cos u \, du$$

$$= \frac{1}{2} \sin u + C = \frac{1}{2} \sin 2x + C. \qquad \blacksquare$$

EXAMPLE 2   Evaluate

$$\int \sin(7x + 5) \, dx.$$

Solution   We substitute

$$u = 7x + 5, \quad du = 7 \, dx, \quad \frac{1}{7} \, du = dx.$$

Then

$$\int \sin (7x + 5) \, dx = \int \sin u \cdot \frac{1}{7} \, du = \frac{1}{7} \int \sin u \, du$$

$$= \frac{1}{7} (-\cos u) + C = -\frac{1}{7} \cos(7x + 5) + C. \qquad \blacksquare$$

EXAMPLE 3   Evaluate

$$\int \frac{\cos 2x}{\sin^3 2x} \, dx.$$

Solution   We substitute

$$u = \sin 2x, \quad du = 2 \cos 2x \, dx, \quad \frac{1}{2} \, du = \cos 2x \, dx.$$

Then

$$\int \frac{\cos 2x}{\sin^3 2x}\, dx = \int \frac{(1/2)\, du}{u^3} = \frac{1}{2}\int u^{-3}\, du$$

$$= \frac{1}{2}\left(-\frac{1}{2}u^{-2}\right) + C = -\frac{1}{4}(\sin 2x)^{-2} + C$$

$$= -\frac{1}{4\sin^2 2x} + C.$$ ■

As we showed in the preceding article, there can be more than one way to make a successful substitution. In the next example, there are two ways to substitute for the "troublesome" part of the integral, and they both work well.

EXAMPLE 4 Evaluate

$$\int 16x\, \sin^3(2x^2 + 1)\, \cos(2x^2 + 1)\, dx.$$

Solution with u = 2x² + 1 We substitute

$$u = 2x^2 + 1, \qquad du = 4x\, dx.$$

Then

$$\int 16x\, \sin^3(2x^2 + 1)\, \cos(2x^2 + 1)\, dx = \int 4\, \sin^3 u \cos u\, du$$

$$= \sin^4 u + C = \sin^4(2x^2 + 1) + C.$$

Solution with u = sin(2x² + 1) We substitute

$$u = \sin(2x^2 + 1), \qquad du = \cos(2x^2 + 1)\cdot 4x\, dx$$

$$4\, du = 16x\, \cos(2x^2 + 1)\, dx.$$

Then

$$\int 16x\, \sin^3(2x^2 + 1)\, \cos(2x^2 + 1)\, dx = \int 4u^3\, du$$

$$= u^4 + C = \sin^4(2x^2 + 1) + C.$$ ■

EXAMPLE 5 Evaluate

$$\int \sin^n x \cos x\, dx, \qquad n \neq -1.$$

Solution We substitute

$$u = \sin x, \qquad du = \cos x\, dx.$$

Then

$$\int \sin^n x \cos x\, dx = \int u^n\, du = \frac{u^{n+1}}{n+1} + C = \frac{\sin^{n+1} x}{n+1} + C.$$ ■

EXAMPLE 6   Evaluate

$$\int \cos^n x \, \sin x \, dx, \qquad n \neq -1.$$

Solution   We substitute

$$u = \cos x, \qquad du = -\sin x \, dx.$$

Then,

$$\int \cos^n x \, \sin x \, dx = \int u^n \cdot -du$$

$$= -\frac{u^{n+1}}{n+1} + C$$

$$= -\frac{\cos^{n+1} x}{n+1} + C. \qquad \blacksquare$$

What do we do with the integrals in Examples 5 and 6 if $n = -1$? When $n = -1$, the integrals are

$$\int (\sin x)^{-1} \cos x \, dx = \int \cot x \, dx, \qquad \int (\cos x)^{-1} \sin x \, dx = \int \tan x \, dx.$$

While $\cot x$ and $\tan x$ have antiderivatives, they cannot be expressed in terms of the functions we have worked with so far. We shall have to wait for the introduction of the natural logarithm function in Chapter 6 to see what the integrals are.

Integrals of Other Trigonometric Functions

The formulas for the derivatives of the tangent, cotangent, secant, and cosecant functions lead to the following integral formulas.

$$\int \sec^2 u \, du = \tan u + C$$

$$\int \csc^2 u \, du = -\cot u + C$$

$$\int \sec u \, \tan u \, du = \sec u + C$$

$$\int \csc u \, \cot u \, du = -\csc u + C$$

In each formula, u is assumed to be a differentiable function of x. Each formula can be checked by differentiating the right-hand side with respect to x. In each case, the Chain Rule applies to produce the integrand on the left.

EXAMPLE 7

a) $\displaystyle\int \frac{1}{\cos^2 2x}\,dx = \int \sec^2 2x\,dx = \frac{1}{2}\int \sec^2 2x\,d(2x) = \frac{1}{2}\tan 2x + C$

b) $\displaystyle\int \frac{\sin x}{\cos^2 x}\,dx = \int \frac{1}{\cos x}\cdot\frac{\sin x}{\cos x}\,dx = \int \sec x \tan x\,dx = \sec x + C$

c) $\displaystyle\int \sec^2 x \tan x\,dx = \int \sec x\,(\sec x \tan x\,dx) = \int \sec x\,d(\sec x)$

$$= \frac{1}{2}\sec^2 x + C.$$ ∎

EXAMPLE 8   Evaluate

$$\int \tan^3 x \sec^2 x\,dx.$$

Solution   We substitute

$$u = \tan x, \qquad du = \sec^2 x\,dx.$$

Then

$$\int \tan^3 x \sec^2 x\,dx = \int u^3\,du$$

$$= \frac{1}{4}u^4 + C = \frac{1}{4}\tan^4 x + C.$$ ∎

Using Trigonometric Identities

We can often use trigonometric identities to change an unfamiliar integral into one we can integrate. Here are some examples.

EXAMPLE 9   Evaluate

$$\int \tan^2 x\,dx.$$

Solution   We use the identity $\tan^2 x = \sec^2 x - 1$:

$$\int \tan^2 x\,dx = \int (\sec^2 x - 1)\,dx$$

$$= \int \sec^2 x\,dx - \int dx = \tan x - x + C.$$ ∎

EXAMPLE 10   Evaluate

$$\int \cos^2 x\,dx.$$

Solution   Using the double-angle formula

$$\cos^2 x = \frac{1 + \cos 2x}{2},$$

we have

$$\int \cos^2x \, dx = \int \frac{1 + \cos 2x}{2} \, dx = \int \frac{1}{2} \, dx + \int \frac{\cos 2x}{2} \, dx$$

$$= \frac{x}{2} + \frac{1}{2} \frac{\sin 2x}{2} + C = \frac{x}{2} + \frac{\sin 2x}{4} + C. \quad \blacksquare$$

EXAMPLE 11 Evaluate

$$\int \sin^2x \, dx.$$

Solution Using the double-angle formula

$$\sin^2x = \frac{1 - \cos 2x}{2},$$

we have

$$\int \sin^2x \, dx = \int \frac{1 - \cos 2x}{2} \, dx = \int \frac{1}{2} \, dx - \int \frac{\cos 2x}{2} \, dx$$

$$= \frac{x}{2} - \frac{1}{2} \frac{\sin 2x}{2} + C = \frac{x}{2} - \frac{\sin 2x}{4} + C. \quad \blacksquare$$

EXAMPLE 12 Evaluate

$$\int \sin^3x \, dx.$$

Solution We write $\sin^3x$ as

$$\sin^3x = \sin^2x \cdot \sin x = (1 - \cos^2x) \cdot \sin x.$$

Then

$$\int \sin^3x \, dx = \int (1 - \cos^2x) \sin x \, dx$$

$$= \int \sin x \, dx - \int \cos^2x \sin x \, dx$$

$$= -\cos x + \frac{1}{3} \cos^3x + C.$$

In evaluating the integral of $\cos^2x \sin x$ with respect to x, we used the result of Example 6. $\quad \blacksquare$

PROBLEMS

In Problems 1–54, evaluate the integrals.

1. $\int \sin 3x \, dx$

2. $\int \cos(2x + 4) \, dx$

5. $\int \csc(x + \pi/2) \cot(x + \pi/2) \, dx$

3. $\int \sec^2(x + 2) \, dx$

4. $\int \sec 2x \tan 2x \, dx$

6. $\int \csc^2(2x - 3) \, dx$

7. $\int x \sin(2x^2) \, dx$

8. $\int (\cos\sqrt{x}) \dfrac{dx}{\sqrt{x}}$

44. $\int x^2 \cos(x^3 + 1) \, dx$

9. $\int \sin 2t \, dt$

10. $\int \cos(3\theta - 1) \, d\theta$

45. $\int \cos^3 x \, dx$

11. $\int 4 \cos 3y \, dy$

12. $\int 2 \sin z \cos z \, dz$

46. $\int \tan^2 4x \, dx$

13. $\int \sin^2 x \cos x \, dx$

14. $\int \cos^2 2y \sin 2y \, dy$

47. $\int \cos^5 x \, dx$ (Hint: $\cos^5 x = \cos^4 x \cos x$.)

15. $\int \sec^2 2\theta \, d\theta$

16. $\int \sec^3 x \tan x \, dx$

48. $\int \sin^{-3} 5x \cos 5x \, dx$

17. $\int \sec \dfrac{x}{2} \tan \dfrac{x}{2} \, dx$

18. $\int \dfrac{d\theta}{\cos^2\theta}$

49. $\int \cos^{-4} 2x \sin 2x \, dx$

19. $\int \dfrac{d\theta}{\sin^2\theta}$

20. $\int \csc^2 5\theta \cot 5\theta \, d\theta$

50. $\int \dfrac{\cos \sqrt{x}}{\sqrt{x} \sin^2\sqrt{x}} \, dx$

21. $\int \cos^2 y \, dy$

22. $\int \sin^2(x/2) \, dx$

51. $\int \dfrac{\tan^2\sqrt{x}}{\sqrt{x}} \, dx$

52. $\int \dfrac{1}{x^2} \sin \dfrac{1}{x} \cos \dfrac{1}{x} \, dx$

23. $\int (1 - \sin^2 3t) \cos 3t \, dt$

24. $\int \dfrac{\sin x \, dx}{\cos^2 x}$

53. $\int \dfrac{x \cos \sqrt{3x^2 - 6}}{\sqrt{3x^2 - 6}} \, dx$

25. $\int \dfrac{\cos x \, dx}{\sin^2 x}$

26. $\int \sqrt{2 + \sin 3t} \cos 3t \, dt$

54. $\int \dfrac{\sin((z - 1)/3)}{\cos^2((z - 1)/3)} \, dz$

27. $\int \dfrac{\sin 2t \, dt}{\sqrt{2 - \cos 2t}}$

28. $\int \sin^3 \dfrac{y}{2} \cos \dfrac{y}{2} \, dy$

55. Which of the following are antiderivatives of $f(x) = \sec x \tan x$?
a) $-\sec x + \pi/6$
b) $-\tan x + \pi/3$
c) $\sec x \, (\sec^2 x) + \tan x \, (\sec x \tan x)$
d) $\sec x - \pi/4$

29. $\int \cos^2 \dfrac{2x}{3} \sin \dfrac{2x}{3} \, dx$

30. $\int \dfrac{\sec^2 u}{\tan^2 u} \, du$

31. $\int \sec \theta \, (\sec \theta + \tan \theta) \, d\theta$ **32.** $\int (1 + \tan^2\theta) \, d\theta$

56. Which of the following are antiderivatives of $f(x) = \csc^2 x$?
a) $-2 \csc x \, (\csc x \cot x) + C$
b) $-\cot x + \pi/6$
c) $\cot x - \pi/3$
d) $-(\cot x + C)$

33. $\int (\sec^2 y + \csc^2 y) \, dy$

34. $\int (1 + \sin 2t)^{3/2} \cos 2t \, dt$

35. $\int (3 \sin 2x + 4 \cos 3x) \, dx$

In Problems 57–60, solve the differential equations subject to the given initial conditions.

36. $\int \sin t \cos t \, (\sin t + \cos t) \, dt$

57. $2y \dfrac{dy}{dx} = 5x - 3 \sin x, \quad y = 0$ when $x = 0$

37. $\int \tan^2 x \sec^2 x \, dx$

38. $\int \tan^3 5x \sec^2 5x \, dx$

58. $\dfrac{dy}{dx} = \dfrac{\sqrt{y^2 + 1}}{y} \cos x, \quad y = \sqrt{3}$ when $x = \pi$

39. $\int \cot^3 x \csc^2 x \, dx$

59. $\dfrac{dy}{dx} = \dfrac{\pi \cos \pi x}{\sqrt{y}}, \quad y = 1$ when $x = 1/2$

40. $\int \sin^4 x \cos^3 x \, dx$ (Hint: $\cos^3 x = \cos^2 x \cos x$.)

60. $y^{(4)} = \cos x, \quad y = 3, \ y' = 2, \ y'' = 1, \ \text{and} \ y''' = 0$ when $x = 0$

41. $\int \sqrt{\tan x} \sec^2 x \, dx$

61. The velocity of a particle moving back and forth on a line is $v = ds/dt = 6 \sin 2t$ m/sec for all t. If $s = 0$ when $t = 0$, find the value of s when $t = \pi/2$ sec.

42. $\int (\sec x)^{3/2} \tan x \, dx$

62. The acceleration of a particle moving back and forth on a line is $a = d^2 s/dt^2 = \pi^2 \cos \pi t$ m/sec^2 for all t. If $s = 0$ and $v = 8$ m/sec when $t = 0$, find s when $t = 1$ sec.

43. $\int (\sin x)^{3/2} \cos x \, dx$

63. It looks as if we can integrate $2 \sin x \cos x$ with respect to x in three different ways:

a) $\displaystyle \int 2 \sin x \cos x \, dx = \int 2 \sin x \frac{d}{dx} (\sin x) \, dx$

$$= \sin^2 x + C_1$$

b) $\displaystyle \int 2 \sin x \cos x \, dx = \int -2 \cos x \frac{d}{dx} (\cos x) \, dx$

$$= -\cos^2 x + C_2$$

c) $\displaystyle \int 2 \sin x \cos x \, dx = \int \sin 2x \, dx$

$$= -\frac{1}{2} \cos 2x + C_3$$

Can all three integrations be correct? Explain.

64. The substitution $u = \tan x$ gives

$$\int \sec^2 x \tan x \, dx = \int \tan x \cdot \sec^2 x \, dx$$

$$= \int u \, du = \frac{u^2}{2} + C = \frac{\tan^2 x}{2} + C.$$

The substitution $u = \sec x$ gives

$$\int \sec^2 x \tan x \, dx = \int \sec x \cdot \sec x \tan x \, dx$$

$$= \int u \, du = \frac{u^2}{2} + C = \frac{\sec^2 x}{2} + C.$$

Can both integrations be correct? Explain.

65. To evaluate

$$\int \sqrt{1 + \sin^2(x - 1)} \, \sin(x - 1) \cos(x - 1) \, dx,$$

try these sequences of substitutions:
a) first $u = x - 1$, then $v = \sin u$, then $w = 1 + v^2$;
b) first $v = \sin(x - 1)$, then $w = 1 + v^2$.

TOOLKIT PROGRAMS

Antiderivatives and   Function Evaluator
Direction Fields

4.5
Definite Integrals: The Area under a Curve

We now turn our attention to the second kind of integration, the integration associated with summation. The integrals we define are called definite integrals, to distinguish them from the indefinite integrals we have calculated up to now. Definite integrals are numerical limits, not families of antiderivatives, and you may well wonder why two such different mathematical entities are both called integrals and what connection they could possibly have. The connection is a surprising one, as we shall see in Article 4.7, and its discovery and formulation by Leibniz and Newton ranks as one of the outstanding intellectual achievements of our civilization.

The approach we take is first to define what is meant by the area of the region between the graph of a nonnegative continuous function $y = f(x)$ and an interval $a \le x \le b$ of the x-axis. We start by filling as much of the region as we can with vertical inscribed rectangles in the manner suggested by Fig. 4.5. The sum of the rectangle areas approximates the area of the region. The more rectangles we use, the better the approximation becomes. We define the area of the region to be the limit of the rectangle area sums as the rectangles become smaller and smaller and the number of rectangles we use approaches infinity. By the time we have said all this in precise mathematical terms, two additional features of our construction will become apparent. First, we get exactly the same limit if we use circumscribed rectangles (Fig. 4.6) or any other kinds of rectangles whose bases lie along the x-axis and whose tops touch the curve. Second, the limit of the sums of the rectangle areas exists for any continuous function whatever, not just for the nonnegative continuous functions with which we now begin.

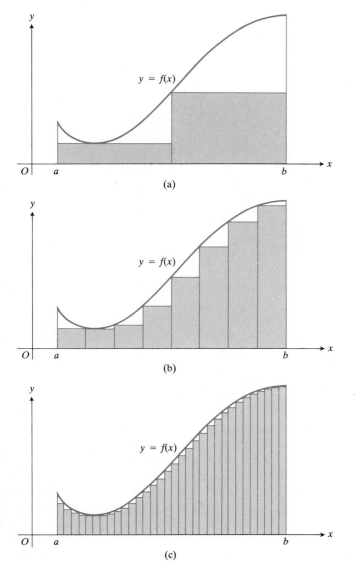

(a)

(b)

(c)

4.5 To define the area of the region beneath the graph of *f* from *a* to *b*, we approximate the region with inscribed rectangles and add the areas of the rectangles. As (a), (b), and (c) suggest, the approximation improves as the rectangles become narrower and the number of rectangles increases.

4.6 Circumscribed rectangles, like the ones shown here, would do as well as the inscribed rectangles in Fig. 4.5 for estimating the area of a region under a curve.

(a)

(b)

(c)

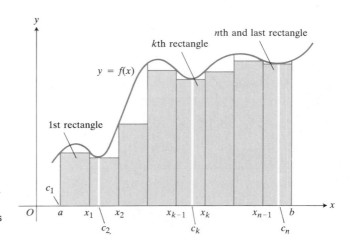

4.7 When we use inscribed rectangles to estimate the area under the graph of a continuous nonnegative function, the height of each rectangle is the minimum value f takes on the rectangle's base.

Approximating Areas with Rectangles

Let $y = f(x)$ be a function that is continuous and nonnegative throughout a closed interval $[a, b]$. We want to define the area of the region enclosed by the graph of f, the lines $x = a$ and $x = b$, and the x-axis. We shall call this the area under the curve $y = f(x)$ from $x = a$ to $x = b$.

We start with an estimate. To get it, we divide the region into n thin strips of uniform width $\Delta x = (b - a)/n$ by lines perpendicular to the x-axis. These lines pass through the endpoints $x = a$ and $x = b$ and many intermediate points, which we label $x_1, x_2, \ldots, x_{k-1}, x_k, \ldots, x_{n-1}$ (Fig. 4.7). We approximate each strip with an inscribed rectangle, which reaches from the strip's base on the x-axis to the lowest point of the curve above the base. If c_k is a point at which f takes on its minimum value in the interval from x_{k-1} to x_k (we know there is such a point because f is continuous), the height of the rectangle there is $f(c_k)$. The area of the rectangle, its height times its base length, is

$$f(c_k)(x_k - x_{k-1}) = f(c_k)\, \Delta x. \tag{1}$$

Thus, the area of the first inscribed rectangle in Fig. 4.7 is $f(c_1)\, \Delta x$, the area of the second is $f(c_2)\, \Delta x$, and so on down to the nth and last rectangle, whose area is $f(c_n)\, \Delta x$. The sum of these areas,

$$S_n = f(c_1)\, \Delta x + f(c_2)\, \Delta x + \cdots + f(c_n)\, \Delta x, \tag{2}$$

provides our estimate of the area under the curve $y = f(x)$ from $x = a$ to $x = b$.

EXAMPLE 1   Estimate the area under the curve $y = x^2 + 1$ from $a = 0$ to $b = 1$ with $n = 4$ inscribed rectangles.

Solution   We sketch the curve over the interval $0 \le x \le 1$ (Fig. 4.8). We then divide the interval into four parts with the points

$$x_1 = \frac{1}{4}, \qquad x_2 = \frac{1}{2}, \qquad \text{and} \qquad x_3 = \frac{3}{4}.$$

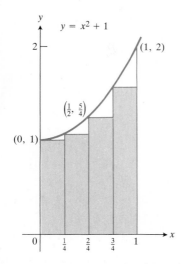

4.8 Rectangles for estimating the area under the graph of $y = x^2 + 1$ from $x = 0$ to $x = 1$.

The length of each subinterval is $\Delta x = 1/4$. We sketch the inscribed rectangle for each subinterval. The height of each rectangle in this case is the length of its

■ ■ ■

SUMMING IT UP

Since antiquity, the problem of finding areas and volumes has captured the imagination of mathematicians. Archimedes, in particular, verified calculations of areas and volumes through an ingenious method later called ''exhaustion''—where infinitesimal quantities were avoided altogether and formal logic alone was used. The medieval mathematician Nicole Oresme (c. 1320) calculated areas under curves in discussions of motion. A resurgence of interest in area and volume calculations emerged with the scientific revolution of the seventeenth century, as in the works of the Italian mathematicians Cavalieri and Torricelli, as well as the astronomer Kepler, whose second law treats the notion of areas explicitly. Johannes Kepler (1571–1630) approached the area swept out by a planet as a collection of infinitely small triangles with one vertex at the sun and the other two vertices at points infinitely close together along the planet's orbit. Then he used a type of crude calculus to find a sum. Kepler was also concerned with calculating the volumes of commercial wine casks. But throughout the sixteenth and early seventeenth centuries, area and volume calculations remained idiosyncratic, with most scientists and mathematicians hitting on an interesting method here and there. Newton and Leibniz were the first to apply the integral calculus in a systematic way for the calculation of areas and volumes.

left-hand edge. The sum of the areas of the rectangles is therefore

$$S_4 = f(c_1)\ \Delta x + f(c_2)\ \Delta x + f(c_3)\ \Delta x + f(c_4)\ \Delta x$$

$$= f(0)\ \Delta x + f\left(\frac{1}{4}\right)\Delta x + f\left(\frac{1}{2}\right)\Delta x + f\left(\frac{3}{4}\right)\Delta x$$

$$= ((0)^2 + 1)\left(\frac{1}{4}\right) + \left(\left(\frac{1}{4}\right)^2 + 1\right)\left(\frac{1}{4}\right) + \left(\left(\frac{1}{2}\right)^2 + 1\right)\left(\frac{1}{4}\right)$$

$$+ \left(\left(\frac{3}{4}\right)^2 + 1\right)\left(\frac{1}{4}\right)$$

$$= \frac{16}{64} + \frac{17}{64} + \frac{20}{64} + \frac{25}{64} = \frac{78}{64} = 1.21875.$$

The sum, $S_4 = 1.21875$, gives an estimate of the area under the curve $y = x^2 + 1$ from $x = 0$ to $x = 1$. Since the rectangles do not cover the entire region under the curve, we expect 1.21875 to underestimate the area. As we shall see in Example 7 of Article 4.7, the area is exactly 4/3, so the estimate is about 8% too small. ■

Summation Notation

We often abbreviate sums like

$$S_n = f(c_1)\ \Delta x + f(c_2)\ \Delta x + \cdots + f(c_n)\ \Delta x \tag{3}$$

by using the capital Greek letter Σ (sigma) to denote the word ''sum'':

$$S_n = \sum_{k=1}^{n} f(c_k)\ \Delta x \tag{4}$$

(pronounced ''S_n is the sum from $k = 1$ to n of f of c_k delta x''). This is called writing the sum in sigma notation. Notice that each term of the sum in Eq. (3) has the form $f(c_k)\ \Delta x$, with only the subscript on c changing from one term to another. We have indicated the subscript by k, but we could have used i or j or any other symbol not currently in use for something else. In the first term, the subscript is $k = 1$; in the second, $k = 2$; and so on to the last, or nth term, in which $k = n$. We indicate this by writing $k = 1$ below the Σ, to say that the sum is to *start* with the term we get by replacing k by 1 in the expression that follows. The n above the sigma tells us where to *stop*. For instance, if $n = 4$, we have

$$\sum_{k=1}^{4} f(c_k)\ \Delta x = f(c_1)\ \Delta x + f(c_2)\ \Delta x + f(c_3)\ \Delta x + f(c_4)\ \Delta x.$$

The only thing that changes from one term to the next is the number k. We replace k by 1, then by 2, then 3, then 4. Then we add.

EXAMPLE 2

a) $\displaystyle\sum_{k=1}^{5} k^2 = 1^2 + 2^2 + 3^2 + 4^2 + 5^2$

b) $\displaystyle\sum_{k=1}^{3} \frac{k}{k+1} = \frac{1}{1+1} + \frac{2}{2+1} + \frac{3}{3+1} = \frac{1}{2} + \frac{2}{3} + \frac{3}{4}$

c) $\displaystyle\sum_{j=0}^{2} \frac{j+1}{j+2} = \frac{0+1}{0+2} + \frac{1+1}{1+2} + \frac{2+1}{2+2} = \frac{1}{2} + \frac{2}{3} + \frac{3}{4}$

d) $\displaystyle\sum_{i=1}^{4} x_i = x_1 + x_2 + x_3 + x_4$ e) $\displaystyle\sum_{k=1}^{4} x^k = x + x^2 + x^3 + x^4$

The Area under a Curve

We turn our attention once more to the area under a curve. We know general formulas for areas of regions enclosed by triangles, trapezoids, and circles, which are all shapes from classical Greek geometry. But there are no general formulas from precalculus times for the arbitrary shapes under the graphs of continuous nonnegative functions. We have reached the point where we have to *define* these areas, and we have just discussed the mechanism for doing so. We shall define them as limits of the sums of areas of inscribed rectangles. That these limits always exist is a consequence of a theorem we shall state later in this article.

DEFINITION

> **Area under a Curve**
>
> The *area* under the graph of a nonnegative continuous function f over an interval $[a, b]$ is the limit of the sums of the areas of inscribed rectangles of equal base length as their number n increases without bound. In symbols,
>
> $$A = \lim_{n \to \infty} (f(c_1)\,\Delta x + f(c_2)\,\Delta x + \cdots + f(c_n)\,\Delta x)$$
>
> (5)
>
> $$= \lim_{n \to \infty} \sum_{k=1}^{n} f(c_k)\,\Delta x,$$
>
> where $f(c_k)$ is the smallest value of f on the interval $[x_{k-1}, x_k]$.

The limit in Eq. (5) always exists, as we shall explain shortly, and we shall develop a number of techniques for computing it later in the chapter.

Riemann Integrals

The existence of the limit in Eq. (5) is a consequence of a more general theorem that applies to any continuous function on the interval $[a, b]$. In the more general theorem, the function may have negative values as well as positive ones. We shall introduce the theorem first, and then discuss why it works, but we shall not go through the rigors of a complete proof. For convenience, the pictures we draw will show positive functions, but the general processes illustrated by the pictures are valid for arbitrary continuous functions.

Given a function f, continuous on $[a, b]$, we begin by inserting points

$$x_1, \quad x_2, \quad \ldots, \quad x_{k-1}, \quad x_k, \quad \ldots, \quad x_{n-1}$$

between a and b as shown in Fig. 4.9. These points subdivide $[a, b]$ into n

(a)

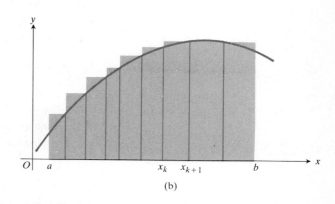

(b)

The sum of the areas of rectangles inscribed under the curve gives us a lower sum L that is less than . . .

. . . the upper sum U obtained by adding the areas of rectangles circumscribed about the curve.

(c)

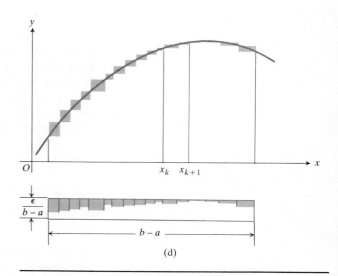

(d)

The difference between the upper sum and the lower sum can be made very small—less than $\epsilon \cdot (b - a)$.

As we subdivide $[a, b]$ more finely, the difference $U - L$ becomes even smaller— smaller than any given ϵ if we make the division fine enough.

subintervals of lengths

$$\Delta x_1 = x_1 - a, \quad \Delta x_2 = x_2 - x_1, \quad \ldots, \quad \Delta x_n = b - x_{n-1},$$

which need not be equal.

Since f is continuous, it has a minimum value $\min_k$ and a maximum value $\max_k$ on each subinterval. The areas of the shaded rectangles in Fig. 4.9(a) add up to what we call the **lower sum:**

$$L = \min_1 \Delta x_1 + \min_2 \Delta x_2 + \cdots + \min_n \Delta x_n. \qquad (6)$$

The areas of the shaded rectangles in Fig. 4.9(b) add up to the **upper sum:**

$$U = \max_1 \Delta x_1 + \max_2 \Delta x_2 + \cdots + \max_n \Delta x_n. \qquad (7)$$

The difference $U - L$ between the upper and lower sums is the sum of the areas of the shaded blocks in Fig. 4.9(c).

The idea conveyed by Fig. 4.9(c), and the idea we want to pursue, is that the more finely we subdivide $[a, b]$, the less area there will be in $U - L$. To make this important idea useful, we have to say more precisely what we mean by subdividing $[a, b]$ more finely. Our goal is to improve the way the tops of the rectangles approximate the curve and thereby decrease the difference between U and L. At the very least, we want the number of x's to increase in a way that makes the bases of the rectangles smaller. In other words, we want "subdivide more finely" to mean "subdivide $[a, b]$ in a way that makes the longest of the subintervals smaller than before." We therefore define the **norm** of the subdivision

$$a, \ x_1, \ x_2, \ \ldots, \ x_{n-1}, \ b$$

to be the longest subinterval length. Then, as norm $\to 0$, the subintervals become more numerous and shorter at the same time. In Fig. 4.9(c) this means that as norm $\to 0$, the blocks increase in number and get less wide (although their total length remains $b - a$). As they get less wide they also get less tall. As Fig. 4.9(d) suggests, we can make the difference $U - L$ less than any prescribed positive ϵ by taking the norm of the subdivision of $[a, b]$ close enough to zero. In other words

$$\lim_{\text{norm} \to 0} (U - L) = 0, \qquad (8)$$

and, as shown in more advanced texts,

$$\lim_{\text{norm} \to 0} L = \lim_{\text{norm} \to 0} U. \qquad (9)$$

The fact that Eqs. (8) and (9) hold for any continuous function, and not just for the one shown in Fig. 4.9, is the consequence of a special property continuous functions have on a bounded closed interval, called **uniform continuity.** This property guarantees that as the norm approaches zero, the blocks shown in Fig. 4.9(c) that make up the difference between U and L get less tall as well as less wide, and that we can make them as short as we like by making them narrow enough. The fact that we are passing over the $\epsilon - \delta$ arguments associated with uniform continuity here is one of the things that keeps the derivation of Eq. (9) from being a proof. But the argument is right in spirit and gives a faithful portrait of the real proof.

Now, assuming that we know that (9) holds for any continuous function on $[a, b]$, suppose that in each interval $[x_{k-1}, x_k]$ of a subdivision of $[a, b]$ we choose a point c_k to form the sum

$$S = \sum f(c_k)\, \Delta x_k. \tag{10}$$

Like the sums

$$L = \sum \min_k \Delta x_k \qquad \text{and} \qquad U = \sum \max_k \Delta x_k,$$

S is a sum of function values times interval lengths. But each point c_k is randomly chosen and all we know about the number $f(c_k)$ is that

$$\min_k \leq f(c_k) \leq \max_k.$$

This is enough, however, to tell us that

$$L \leq S \leq U, \tag{11}$$

and therefore that

$$\lim_{\text{norm}\to 0} L = \lim_{\text{norm}\to 0} S = \lim_{\text{norm}\to 0} U \tag{12}$$

(by a slightly modified version of the Sandwich Theorem in Article 1.9). In other words, S has the same limit that L and U have.

Pause for a moment to see how remarkable the conclusion in (12) really is. It says that *no matter how* we choose the points c_k to form the sum

$$S = \sum f(c_k)\, \Delta x_k$$

for a continuous function on an interval $[a, b]$, we always get the same limit as the norm of the subdivision approaches zero. We can choose every c_k so that $f(c_k)$ is the maximum value of f on $[x_{k-1}, x_k]$. The limit is the same. We can choose c_k so that $f(c_k)$ is the minimum value of f on $[x_{k-1}, x_k]$. The limit is the same. We can choose every c_k at random. No matter how we choose the c_k's, the limit is the same.

This fact was first discovered (without uniform continuity) by Cauchy in 1823 and later put on a solid logical foundation (with uniform continuity) by other mathematicians of the nineteenth century. The limit, called the **Riemann integral** of f over the interval $[a, b]$ and denoted by

$$\int_a^b f(x)\, dx,$$

bears the name of Georg Friedrich Bernhard Riemann (1826–1866). It was his idea to trap the limit between upper and lower sums. The sum

$$S = \sum f(c_k)\, \Delta x_k$$

is called an **approximating sum** or **Riemann sum** for the integral. The numbers a and b are called the **limits of integration,** a being the **lower limit** and b the **upper limit.**

THEOREM 1

The Integral Existence Theorem

EXISTENCE OF THE RIEMANN INTEGRAL

If f is continuous on $[a, b]$, then

$$\int_a^b f(x)\, dx = \lim_{\text{norm}\to 0} \sum f(c_k)\, \Delta x_k$$

exists and is the same number for any choice of the numbers c_k.

The full proof of this theorem can be found in most texts on mathematical analysis.

The area we defined earlier in Eq. (5) for the region under the graph of a nonnegative function f over an interval $[a, b]$ was the Riemann integral of f:

$$\text{Area} = \int_a^b f(x)\, dx, \quad \text{when } f(x) \geq 0.$$

As we shall see in Article 4.7 and again in Chapter 5, Riemann integrals can have many other interpretations as well.

Definite Integrals

To distinguish it from the indefinite integral, the Riemann integral

$$\int_a^b f(x)\, dx$$

is called the **definite integral** of f over $[a, b]$. It may also be denoted by

$$\int_a^b f(t)\, dt, \qquad \int_a^b f(u)\, du,$$

and so on. The variable of integration can be any letter like x, t, or u that is not currently in use for something else.

The point to remember is that $\int_a^b f(x)\, dx$ is a number defined as a limit of approximating sums over the interval from a to b on the x-axis. If we use another name for the axis, say the t-axis, then the appropriate symbol for the integral is $\int_a^b f(t)\, dt$, *but the integral is still the same number.*

Algebraic Properties of Definite Integrals

We often want to add and subtract definite integrals, multiply them by constants, and compare them with other integrals. The properties listed in Table 4.1 make this easy to do. All of the properties except the first two follow immediately from the way integrals are defined as limits of finite sums: the finite sums have these properties, so their limits do, too. For example, Property 3,

$$\int_a^b kf(x)\, dx = k \int_a^b f(x)\, dx \qquad \text{(any number } k\text{)},$$

says that the integral of k times a function is k times the integral of the function. This is true because

$$\int_a^b kf(x)\, dx = \lim \sum kf(c_i)\, \Delta x_i = \lim k \sum f(c_i)\, \Delta x_i$$

$$= k \lim \sum f(c_i)\, \Delta x_i = k \int_a^b f(x)\, dx. \tag{13}$$

Property 1,

$$\int_a^a f(x)\, dx = 0,$$

is a definition. We want the integral of a function over an interval of zero length to be zero.

Table 4.1 Algebraic Properties of Definite Integrals

1. $\displaystyle\int_a^a f(x)\,dx = 0$ (a definition)

2. $\displaystyle\int_b^a f(x)\,dx = -\int_a^b f(x)\,dx$ (a definition)

3. $\displaystyle\int_a^b kf(x)\,dx = k\int_a^b f(x)\,dx$ (any number k)

4. $\displaystyle\int_a^b [f(x) + g(x)]\,dx = \int_a^b f(x)\,dx + \int_a^b g(x)\,dx$

5. $\displaystyle\int_a^b [f(x) - g(x)]\,dx = \int_a^b f(x)\,dx - \int_a^b g(x)\,dx$

6. $\displaystyle\int_a^b f(x)\,dx \geq 0$ if $f(x) \geq 0$ on $[a, b]$

7. $\displaystyle\int_a^b f(x)\,dx \leq \int_a^b g(x)\,dx$ if $f(x) \leq g(x)$ on $[a, b]$

8. $\min f \cdot (b - a) \leq \displaystyle\int_a^b f(x)\,dx \leq \max f \cdot (b - a)$, where min and max refer to the minimum and maximum value of f on $[a, b]$

9. $\displaystyle\int_a^b f(x)\,dx + \int_b^c f(x)\,dx = \int_a^c f(x)\,dx$

The equality in Property 9 holds for any a, b, and c provided f is continuous on the intervals joining them. Figure 4.10 illustrates Property 9 for areas.

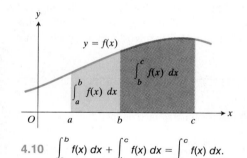

4.10 $\displaystyle\int_a^b f(x)\,dx + \int_b^c f(x)\,dx = \int_a^c f(x)\,dx.$

Property 2,

$$\int_b^a f(x)\,dx = -\int_a^b f(x)\,dx,$$

is a definition, too. It says that reversing the order of integration changes the sign of the integral. It gives us a way to change the sign of an integral. It also gives us a way to combine integrals. For example,

$$\int_1^2 f(x)\,dx + \int_2^1 g(x)\,dx = \int_1^2 f(x)\,dx - \int_1^2 g(x)\,dx \qquad \text{(Property 2)}$$

$$= \int_1^2 [f(x) - g(x)]\,dx. \qquad \text{(Property 5)}$$

EXAMPLE 3 Suppose that f, g, and h are continuous functions, that $h(x) \geq f(x)$ on $[-1, 1]$, and that

$$\int_{-1}^1 f(x)\,dx = 5, \qquad \int_1^4 f(x)\,dx = -2, \qquad \int_{-1}^1 g(x)\,dx = 7.$$

Then

1. $\displaystyle\int_{-1}^{1} 3f(x)\, dx = 3\int_{-1}^{1} f(x)\, dx = 15,$

2. $\displaystyle\int_{-1}^{1} [2f(x) + 3g(x)]\, dx = 2(5) + 3(7) = 31,$

3. $\displaystyle\int_{-1}^{1} [f(x) - g(x)]\, dx = 5 - 7 = -2,$

4. $\displaystyle\int_{-1}^{4} f(x)\, dx = \int_{-1}^{1} f(x)\, dx + \int_{1}^{4} f(x)\, dx = 5 - 2 = 3,$

5. $\displaystyle\int_{-1}^{1} h(x)\, dx \geq \int_{-1}^{1} f(x)\, dx = 5.$ ∎

The notation for the definite integral suggests the indefinite integrals we worked with in Articles 4.1–4.4. The connection between the two kinds of integrals, one of the most important relationships in all of calculus, will be developed in Article 4.7. But first, in Article 4.6, we shall calculate a few integrals without the benefit of the more modern mathematical tools of Article 4.7. The contrast between the methods of Articles 4.6 and 4.7 is striking and typifies the difference between the mathematics of the world into which Leibniz and Newton were born and the mathematics of the world that Leibniz and Newton left behind.

PROBLEMS

In Problems 1–5, sketch the graph of the given equation over the interval $a \leq x \leq b$. Divide the interval into $n = 4$ subintervals each of length $\Delta x = (b - a)/4$. (a) Sketch the inscribed rectangles and compute the sum of their areas. (b) Do the same using the circumscribed in place of the inscribed rectangle in each subinterval (Fig. 4.11).

1. $y = 2x + 1, \quad a = 0, \quad b = 1$
2. $y = x^2, \quad a = -1, \quad b = 1$
3. $y = \sin x, \quad a = 0, \quad b = \pi$
4. $y = 1/x, \quad a = 1, \quad b = 2$
5. $y = \sqrt{x}, \quad a = 0, \quad b = 4$

Write out the following sums, as in Example 2.

6. $\displaystyle\sum_{k=1}^{5} \frac{1}{k}$ 7. $\displaystyle\sum_{i=-1}^{3} 2^i$ 8. $\displaystyle\sum_{n=1}^{4} \cos n\pi x$

Find the value of each sum.

9. $\displaystyle\sum_{n=0}^{4} \frac{n}{4}$ 10. $\displaystyle\sum_{k=1}^{3} \frac{k-1}{k}$ 11. $\displaystyle\sum_{m=0}^{5} \sin \frac{m\pi}{2}$

12. $\displaystyle\sum_{i=1}^{4} (i^2 - 1)$ 13. $\displaystyle\sum_{i=0}^{3} (i^2 + 5)$ 14. $\displaystyle\sum_{k=0}^{5} \frac{1}{2^k}$

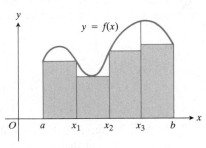

Inscribed rectangles.

The approximating sum is $S_4 = \displaystyle\sum_{k=1}^{4} \min_k \Delta x.$

(a)

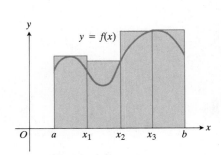

Circumscribed rectangles.

The approximating sum is $S_4 = \displaystyle\sum_{k=1}^{4} \max_k \Delta x.$

(b)

4.11 Problems 1–5 use inscribed and circumscribed rectangles like the ones shown here.

15. Which of the following express $1 + 2 + 4 + 8 + 16 + 32$ in summation notation?

a) $\displaystyle\sum_{j=2}^{7} 2^{j-2}$ b) $\displaystyle\sum_{k=0}^{5} 2^k$

c) $\displaystyle\sum_{j=0}^{5} 2^j$ d) $\displaystyle\sum_{j=1}^{6} 2^{j-1}$

16. Suppose f and g are continuous and that

$$\int_1^2 f(x)\,dx = -4, \quad \int_2^5 f(x)\,dx = 6, \quad \int_1^5 g(x)\,dx = 8.$$

Find

a) $\displaystyle\int_1^5 f(x)\,dx,$ b) $\displaystyle\int_5^1 -4f(x)\,dx,$

c) $\displaystyle\int_1^5 [4f(x) - 2g(x)]\,dx.$

17. Suppose f and h are continuous and that

$$\int_1^7 f(x)\,dx = -1, \quad \int_7^9 f(x)\,dx = 5, \quad \int_7^9 h(x) = 4.$$

Find

a) $\displaystyle\int_1^9 -2f(x)\,dx,$ b) $\displaystyle\int_7^9 [2f(x) - h(x)]\,dx,$

c) $\displaystyle\int_9^7 f(x)\,dx,$ d) $\displaystyle\int_7^9 [f(x) + h(x)]\,dx.$

18. Suppose f is continuous on the interval $[0, 4]$ and that

$$\int_0^3 f(x)\,dx = 3, \quad \int_0^4 f(z)\,dz = 7.$$

Find

$$\int_3^4 f(y)\,dy.$$

19. *Increasing functions.* Suppose that the graph of f is always rising toward the right between $x = a$ and $x = b$, as in Fig.

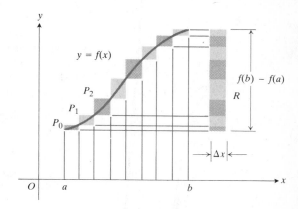

4.12 If f is an increasing function, the blocks in $U - L$ fill the rectangle on the right without overlapping. See Problems 19–21.

4.12. Take $\Delta x = (b - a)/n$. Show, by reference to Fig. 4.12 that the difference between the upper and lower sums is representable graphically as the area $[f(b) - f(a)]\,\Delta x$ of the rectangle R. (*Hint:* The difference $U - L$ is the sum of areas of rectangles with diagonals P_0P_1, P_1P_2, and so forth along the curve, and there is no overlapping when these are all displaced horizontally into the rectangle R.)

20. *Decreasing functions.* Draw a figure to represent a continuous curve $y = f(x)$ that is always falling to the right between $x = a$ and $x = b$. Suppose the subdivisions Δx_k are again equal, with $\Delta x_k = \Delta x = (b - a)/n$. Obtain an expression for the difference $U - L$ analogous to the expression in Problem 19.

21. In Problem 19 or 20, if the Δx_k's are not all equal, show that

$$U - L \le |f(b) - f(a)|(\Delta x_{\max}),$$

where $\Delta x_{\max}$ is the largest of the Δx_k's for

$$k = 1, 2, \ldots, n.$$

 TOOLKIT PROGRAMS

Integration

4.6
Calculating Definite Integrals by Summation

In Article 4.5 we defined the area under the graph of a continuous nonnegative function $y = f(x)$ from $x = a$ to $x = b$ to be a limit of sums of areas of inscribed rectangles. We also saw that this limit was a special case of a limit called the definite integral of f, which could be defined for any continuous function.

In this article we shall compute definite integrals with the help of algebraic formulas, which we first develop. Our purpose is to show how definite integrals might be calculated with the mathematics of the late Renaissance, before the seventeenth-century development of the fundamental theorems of calculus that are the subject of the next article. As you will see, each calculation requires its own bit of ingenuity, and the formulas developed to solve one problem do not apply readily to any other. This lack of generality is one of the features that distinguished the early calculus of the Renaissance from the later methods of Leibniz and Newton.

Formulas for $\sum_{k=1}^{n} k$ and $\sum_{k=1}^{n} k^2$

To begin, we need the formulas

$$\sum_{k=1}^{n} k = 1 + 2 + 3 + \cdots + n = \frac{n(n+1)}{2},$$

$$\sum_{k=1}^{n} k^2 = 1^2 + 2^2 + 3^2 + \cdots + n^2 = \frac{n(n+1)(2n+1)}{6},$$

which we shall prove by mathematical induction. This method consists of showing that each formula is true when $n = 1$ and that if the formula is true for any integer n, then it is also true for the next integer, $n + 1$. We shall also show how such formulas might be discovered.

First, consider the sum of the first n integers,

$$F(n) = 1 + 2 + 3 + \cdots + n.$$

Table 4.2 shows briefly how $F(n)$ increases with n. The last column exhibits $F(n)/n$, the ratio of $F(n)$ to n.

The last column suggests that $F(n)/n$ is equal to $(n + 1)/2$. At least this is the case for all the entries in the table. In other words, the formula

$$\frac{F(n)}{n} = \frac{n+1}{2},$$

or

$$1 + 2 + 3 + \cdots + n = \frac{n(n+1)}{2}, \tag{1}$$

is true for $n = 1, 2, 3, 4, 5, 6$. Suppose now that n is any integer for which

---

TABLE 4.2

n	$F(n) = 1 + 2 + 3 + \cdots + n$	$F(n)/n$
1	1	$1 = 2/2$
2	$1 + 2 = 3$	$3/2 = 3/2$
3	$1 + 2 + 3 = 6$	$6/3 = 4/2$
4	$1 + 2 + 3 + 4 = 10$	$10/4 = 5/2$
5	$1 + 2 + 3 + 4 + 5 = 15$	$15/5 = 6/2$
6	$1 + 2 + 3 + 4 + 5 + 6 = 21$	$21/6 = 7/2$

Eq. (1) is known to be true (at the moment, n could be any integer from 1 through 6). Then if $(n + 1)$ were added to both sides of the equation, the new equation

$$1 + 2 + 3 + \cdots + n + (n + 1) = \frac{n(n + 1)}{2} + (n + 1) \tag{2}$$

would also be true for that same n. But the right side of Eq. (2) is

$$\frac{n(n + 1)}{2} + (n + 1) = \frac{(n + 1)}{2}(n + 2) = \frac{(n + 1)(n + 2)}{2},$$

so Eq. (2) becomes

$$1 + 2 + 3 + \cdots + n + (n + 1) = \frac{(n + 1)[(n + 1) + 1]}{2},$$

which is just like Eq. (1) except that n is replaced by $n + 1$. Thus if Eq. (1) is true for an integer n, it is also true for the next integer, $n + 1$. Hence we now know that it is true for $n + 1 = 7$, since it was true for $n = 6$. Then we can say it is true for $n + 1 = 8$, since it is true for $n = 7$. By the principle of mathematical induction, it is true for every positive integer n. (For an overview of mathematical induction, see Appendix 2.)

Now let's consider

$$Q(n) = 1^2 + 2^2 + 3^2 + \cdots + n^2,$$

the sum of the squares of the first n positive integers. Obviously, $Q(n)$ grows faster than $F(n)$, the sum of first powers, but let us look at the ratio of $Q(n)$ to $F(n)$ to compare them. (See Table 4.3.)

Notice how regular the last column is: 3/3, 5/3, 7/3, and so on. In fact it is just $(2n + 1)/3$ for $n = 1, 2, 3, 4, 5, 6$; that is,

$$Q(n) = F(n) \cdot \frac{2n + 1}{3}.$$

But from Eq. (1), $F(n) = n(n + 1)/2$, and hence

$$Q(n) = 1^2 + 2^2 + 3^2 + \cdots + n^2 = \frac{n(n + 1)(2n + 1)}{6} \tag{3}$$

TABLE 4.3

n	$F(n)$	$Q(n) = 1^2 + 2^2 + 3^2 + \cdots + n^2$	$Q(n)/F(n)$
1	1	$1^2 = 1$	$1/1 = 3/3$
2	3	$1^2 + 2^2 = 5$	$5/3 = 5/3$
3	6	$1^2 + 2^2 + 3^2 = 14$	$14/6 = 7/3$
4	10	$1^2 + 2^2 + 3^2 + 4^2 = 30$	$30/10 = 9/3$
5	15	$1^2 + 2^2 + 3^2 + 4^2 + 5^2 = 55$	$55/15 = 11/3$
6	21	$1^2 + 2^2 + 3^2 + 4^2 + 5^2 + 6^2 = 91$	$91/21 = 13/3$

is true for the integers n from 1 through 6. To establish it for all other positive integers, we proceed as before. Start with any n for which Eq. (3) is true and add $(n + 1)^2$. Then

$$1^2 + 2^2 + 3^2 + \cdots + n^2 + (n + 1)^2 = \frac{n(n + 1)(2n + 1)}{6} + (n + 1)^2$$

$$= \frac{(n + 1)}{6}[n(2n + 1) + 6(n + 1)] \tag{4}$$

$$= \frac{(n + 1)}{6}(2n^2 + 7n + 6)$$

$$= \frac{(n + 1)(n + 2)(2n + 3)}{6}.$$

The last expression in Eq. (4) is the same as the last expression in (3) with n replaced by $n + 1$. In other words, if the formula in (3) is true for any integer n, we have just shown that it is true for $n + 1$. Since we know it is true for $n = 6$, it is also true for $n + 1 = 7$. And now that we know it is true for $n = 7$, it follows that it is true for $n + 1 = 8$, and so on. It is true for every positive integer n by the principle of mathematical induction.

The Areas under the Curves $y = mx$ and $y = x^2$

We now apply formulas (1) and (3) to find areas under two curves.

EXAMPLE 1   Let a, b, and m be positive numbers, with $a < b$. Find the area under the curve $y = mx$ from $x = a$ to $x = b$ (Fig. 4.13).

Solution   We calculate

$$\int_a^b mx\,dx$$

as a limit of sums. Let n be a positive integer and divide the interval $[a, b]$ into n subintervals, each of length $\Delta x = (b - a)/n$, by inserting the points

$$x_1 = a + \Delta x,$$

$$x_2 = a + 2\,\Delta x,$$

$$x_3 = a + 3\,\Delta x,$$

$$\vdots \qquad \vdots$$

$$x_{n-1} = a + (n - 1)\,\Delta x.$$

The inscribed rectangles have areas

$$f(a)\,\Delta x = ma \cdot \Delta x,$$

$$f(x_1)\,\Delta x = m(a + \Delta x) \cdot \Delta x,$$

$$f(x_2)\,\Delta x = m(a + 2\,\Delta x) \cdot \Delta x,$$

$$\vdots \qquad \vdots$$

$$f(x_{n-1})\,\Delta x = m(a + (n - 1)\,\Delta x) \cdot \Delta x,$$

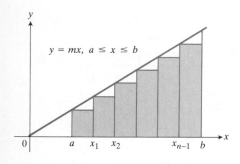

4.13   Rectangles for approximating the area under the curve $y = mx$ from a to b.

whose sum is

$$S_n = m(a + (a + \Delta x) + (a + 2\,\Delta x) + \cdots + (a + (n-1)\,\Delta x)) \cdot \Delta x$$

$$= m(na + (1 + 2 + \cdots + (n-1))\,\Delta x)\,\Delta x$$

$$= m\left(na + \frac{(n-1)n}{2}\,\Delta x\right)\Delta x$$

$$= m\left(a + \frac{n-1}{2}\,\Delta x\right)n\,\Delta x$$

$$= m\left(a + \frac{b-a}{2}\cdot\frac{n-1}{n}\right)\cdot(b-a). \qquad \left(\Delta x = \frac{b-a}{n}\right)$$

The area under the curve is defined to be the limit of S_n as $n \to \infty$. In the final form, the only place n appears is in the fraction

$$\frac{n-1}{n} = 1 - \frac{1}{n},$$

and $1/n \to 0$ as $n \to \infty$, so

$$\lim \frac{n-1}{n} = 1 - 0 = 1.$$

Therefore,

$$\int_a^b mx\,dx = \lim S_n = m\left(a + \frac{b-a}{2}\right)\cdot(b-a)$$

$$= m\left(\frac{b+a}{2}\right)\cdot(b-a) = m\left(\frac{b^2}{2} - \frac{a^2}{2}\right).$$

This is the area of a trapezoid with bases ma and mb (in this case vertical) and with altitude $b - a$.   ∎

EXAMPLE 2 Find the area under the curve $y = x^2$ from $x = 0$ to $x = b$ (Fig. 4.14).

Solution We calculate

$$\int_0^b x^2\,dx$$

as a limit of sums. Divide the interval $0 \le x \le b$ into n subintervals, each of length $\Delta x = b/n$, by inserting the points

$$x_1 = \Delta x, \quad x_2 = 2\,\Delta x, \quad x_3 = 3\,\Delta x, \quad \ldots, \quad x_{n-1} = (n-1)\,\Delta x.$$

The inscribed rectangles have areas

$$f(0)\,\Delta x = 0,$$

$$f(x_1)\,\Delta x = (\Delta x)^2\,\Delta x,$$

$$f(x_2)\,\Delta x = (2\,\Delta x)^2\,\Delta x,$$

$$f(x_3)\,\Delta x = (3\,\Delta x)^2\,\Delta x,$$

$$\vdots \qquad\qquad \vdots$$

$$f(x_{n-1})\,\Delta x = [(n-1)\,\Delta x]^2\,\Delta x.$$

$y = x^2, \ 0 \le x \le b$

4.14 Rectangles for approximating the area under the curve $y = x^2$ from 0 to b.

The sum of these areas is

$$S_n = [1^2 + 2^2 + 3^2 + \cdots + (n-1)^2](\Delta x)^3$$

$$= \frac{(n-1)n(2n-1)}{6} \cdot \left(\frac{b}{n}\right)^3$$

$$= \frac{b^3}{6} \cdot \frac{n-1}{n} \cdot \frac{n}{n} \cdot \frac{2n-1}{n}$$

$$= \frac{b^3}{6} \cdot \left(1 - \frac{1}{n}\right) \cdot \left(2 - \frac{1}{n}\right).$$

To find the area under the curve, we let n increase without bound and get

$$A = \int_0^b x^2 \, dx = \lim S_n = \frac{b^3}{6} \cdot 1 \cdot 2 = \frac{b^3}{3}.$$

Therefore the area is $1/3$ the base b times the "altitude" b^2. The triangle OBP in Fig. 4.14 has area $(1/2)b \cdot b^2 = b^3/2$, and the area under the curve turns out to be somewhat smaller, as we would expect.

The formula

$$\int_0^b x^2 \, dx = \frac{b^3}{3}$$

gives

$$\int_0^1 x^2 \, dx = \frac{1}{3}, \qquad \int_0^2 x^2 \, dx = \frac{8}{3}, \qquad \int_0^3 x^2 \, dx = \frac{27}{3} = 9,$$

and so on. ∎

PROBLEMS

1. Verify the formula

$$\sum_{k=1}^{n} k^3 = 1^3 + 2^3 + \cdots + n^3 = \left(\frac{n(n+1)}{2}\right)^2$$

for $n = 1, 2, 3$. Then add $(n+1)^3$ and thereby prove by mathematical induction (as in the text) that the formula is true for all positive integers n.

2. Use the result of Problem 1 and the method of Example 2 in the text to show that the area under the graph of $y = x^3$ over the interval $0 \le x \le b$ is $b^4/4$.

3. Find the area under the graph of $y = mx$ from $x = a$ to $x = b$ by using circumscribed rectangles in place of the inscribed rectangles of Example 1.

4. Find the area under the curve $y = x^2$ from $x = 0$ to $x = b$ by using circumscribed rectangles in place of the inscribed rectangles of Example 2.

5. Do Problem 2 with circumscribed rectangles instead of inscribed rectangles.

Establish the formulas in Problems 6 and 7 for every positive integer n, by showing (a) that the formula is correct for $n = 1$, and (b) if true for n, the formula is also true for $n + 1$.

6. $\displaystyle\sum_{k=1}^{n} (2k - 1) = 1 + 3 + 5 + \cdots + (2n - 1) = n^2$

7. $\displaystyle\sum_{k=1}^{n} \frac{1}{k(k+1)} = \frac{1}{1 \cdot 2} + \frac{1}{2 \cdot 3} + \cdots + \frac{1}{n \cdot (n+1)}$

$$= \frac{n}{n+1}$$

8. Use the results of Examples 1 and 2 and the properties of definite integrals from Table 4.1 in Article 4.5 to calculate the following integrals.

a) $\displaystyle\int_0^2 3x \, dx$

b) $\displaystyle\int_2^3 4x \, dx$

c) $\displaystyle\int_0^2 x^2 \, dx$

d) $\displaystyle\int_0^2 (x^2 - 5x) \, dx$

e) $\displaystyle\int_0^1 x^2 \, dx$

f) $\displaystyle\int_1^2 x^2 \, dx$

g) $\displaystyle\int_1^3 x^2\,dx$ h) $\displaystyle\int_2^3 x^2\,dx$

9. Interpret

$$\int_0^1 (x^2 + 1)\,dx$$

as an area and calculate the integral from the result of Example 2. (The graph of $y = x^2 + 1$ is shown in Fig. 4.8.)

10. Let

$$S_n = \frac{1}{n}\left[\frac{1}{n} + \frac{2}{n} + \frac{3}{n} + \cdots + \frac{n-1}{n}\right].$$

Calculate $\lim_{n\to\infty} S_n$ by showing that S_n is an approximating sum of the integral

$$\int_0^1 x\,dx,$$

whose value we know from Example 1. (*Hint:* Subdivide [0, 1] into n intervals of equal length and write out the approximating sum for inscribed rectangles.)

11. Let

$$S_n = \frac{1^2}{n^3} + \frac{2^2}{n^3} + \cdots + \frac{(n-1)^2}{n^3}.$$

To calculate $\lim_{n\to\infty} S_n$, show that

$$S_n = \frac{1}{n}\left[\left(\frac{1}{n}\right)^2 + \left(\frac{2}{n}\right)^2 + \cdots + \left(\frac{n-1}{n}\right)^2\right]$$

and interpret S_n as an approximating sum of the integral

$$\int_0^1 x^2\,dx,$$

whose value we know from Example 2. (*Hint:* Subdivide [0, 1] into n intervals of equal length and write out the approximating sum for inscribed rectangles.)

12. Use the formula

$$\sin h + \sin 2h + \sin 3h + \cdots + \sin mh$$
$$= \frac{\cos(h/2) - \cos(m + \tfrac{1}{2})h}{2\sin(h/2)}$$

to find the area under the curve $y = \sin x$ from $x = 0$ to $x = \pi/2$, in two steps:
a) Divide the interval $[0, \pi/2]$ into n equal subintervals and calculate the corresponding upper sum U; then
b) find the limit of U as $n \to \infty$ and

$$\Delta x = (b - a)/n \to 0.$$

TOOLKIT PROGRAMS

Integration   Sequences and Series

4.7
The Fundamental Theorems of Integral Calculus

In Article 4.5 we defined the area under a curve as a definite integral and showed how to estimate it by adding areas of inscribed rectangles. Nothing more than arithmetic was involved in these calculations, but we got only an estimate of the true area. In Article 4.6 we used the definition of the definite integral as a limit to compute areas exactly, but this required extensive algebra. In this article we follow the trail blazed by Leibniz and Newton to show how definite integrals can be computed with calculus. As you will see, everything follows from one key observation: when f is continuous, the integral

$$\int_a^x f(t)\,dt$$

is a differentiable function of x. This is the theorem that makes the all-important connection between definite integrals and antiderivatives.

Functions Defined by Integrals: The Fundamental Theorems of Calculus

The definite integral of any continuous function $f(t)$ from $t = a$ to $t = x$ defines a number

$$F(x) = \int_a^x f(t)\,dt \qquad (1)$$

■ ■ ■

WHO INVENTED THE CALCULUS?

As we have seen, some aspects of the calculus can be traced back to the works of many different mathematicians. Fermat and Descartes had principally worked out many aspects of finding tangents, or the differential calculus, while Cavalieri and Huygens, for example, had made considerable headway in calculating areas, or the integral calculus. In what way, then, do we consider Newton and Leibniz the inventors of the calculus? These men were the first to understand that the processes of finding tangents and of finding areas are mutually inverse. This profound result had escaped their predecessors.

But the question "Who invented the calculus?" was asked in another and more pointed way at the time. The question of priority between Newton and Leibniz over the invention of the calculus led to one of the most bitter controversies in the history of mathematics. Leibniz was accused of plagiarizing Newton's work, and a committee of the Royal Society of London, convened to investigate the matter, did not exonerate Leibniz from this charge. As a result, there was a pervasive split in the mathematical world for more than a century and a half. The disciples of Newton, mostly British, followed his approach and methods religiously; followers of Leibniz, mostly Continentals, pursued his methods. Due to the superior Leibnizian notation of the differential over the awkward Newtonian one of the fluxion, Leibniz's Continental followers pushed the calculus and mathematical physics much further than did their British counterparts. Not until the early nineteenth century did the British start participating in the exciting mathematical work of the French, Germans, and Swiss.

Newton did, in fact, arrive at his conception of the calculus several years before Leibniz arrived at his, but Leibniz published his version first. It is generally held by historians of mathematics today that the two inventions were simultaneous but independent. Both Newton and Leibniz are credited with the achievement.

that we can treat as a function of x. For every value of x in the domain of f, the integral gives the output $F(x)$.

This gives an important way of defining new functions. For example, in Chapter 6 we shall define the natural logarithm of a number $x > 0$ by the formula

$$\ln x = \int_1^x \frac{1}{t}\, dt, \qquad x > 0. \tag{2}$$

The **error function**

$$\text{erf}(x) = \frac{2}{\sqrt{\pi}} \int_0^x e^{-t^2}\, dt \tag{3}$$

appears in probability as well as in the theories of heat flow and signal transmission. The **sine integral**

$$\text{si}(x) = \int_0^x \frac{\sin t}{t}\, dt \tag{4}$$

and others like it come up in engineering.

People who first encounter functions defined by integrals sometimes feel that the integral definitions serve only to give complicated descriptions of functions that must have simpler definitions somewhere else. But there are no simpler formulas for $\ln x$, $\text{erf}(x)$, and $\text{si}(x)$ than Eqs. (2)–(4). Nor do we need simpler descriptions. However unfamiliar, the integral formulas enable us to calculate the values of the functions they define as accurately as we please by any of a large number of numerical methods for estimating integrals. We shall practice with two of the easier methods in Article 4.9.

Writing

$$\ln 2 = \int_1^2 \frac{1}{t}\, dt$$

for the value of the natural logarithm of 2 is really no different from writing π for the ratio of the circumference of a circle to its diameter. There is no way to write the exact value of this ratio except by writing π. But we can calculate a numerical approximation of the ratio to as many decimals as we please whenever we want.

The formula

$$F(x) = \int_a^x f(t)\, dt$$

provides the connection between antiderivatives and definite integrals: If f is any continuous function, then F is a differentiable function of x, and $dF/dx = f(x)$. If you were being sent to a desert island and could take only one formula with you, the equation

$$\frac{d}{dx} \int_a^x f(t)\, dt = f(x) \tag{5}$$

might well be your choice. It says that the differential equation $dF/dx = f(x)$ has a solution for any continuous function f. It says that every continuous function (f) is the derivative of some other function $(\int_a^x f(t)\, dt)$. It says that every continuous function has an antiderivative. (This was how we knew in Article 4.4 that $\tan x$ and $\cot x$ had antiderivatives even when we could not find them.) Equation (5) is so important that we call it the First Fundamental Theorem of Calculus.

THEOREM 2

> **The First Fundamental Theorem of Calculus**
>
> If f is continuous on $[a, b]$, then
>
> $$F(x) = \int_a^x f(t)\, dt$$
>
> is differentiable at every point x in $[a, b]$, and
>
> $$\frac{dF}{dx} = \frac{d}{dx}\int_a^x f(t)\, dt = f(x). \tag{6}$$

COROLLARY

> **The Existence of Antiderivatives of Continuous Functions**
>
> If $y = f(x)$ is continuous on $[a, b]$, then there exists a function $F(x)$ whose derivative on $[a, b]$ is f.

Proof of the Corollary Take $F(x) = \int_a^x f(t)\, dt$. The integral exists by the Integral Existence Theorem in Article 4.5, and $dF/dx = f(x)$ by the First Fundamental Theorem. ∎

In a moment we shall prove the First Fundamental Theorem by calculating dF/dx from its definition as the limit as $\Delta x \to 0$ of the quotient

$$\frac{F(x + \Delta x) - F(x)}{\Delta x} = \frac{\displaystyle\int_a^{x+\Delta x} f(t)\, dt - \int_a^x f(t)\, dt}{\Delta x} = \frac{\displaystyle\int_x^{x+\Delta x} f(t)\, dt}{\Delta x}.$$

That is, we shall calculate

$$\lim_{\Delta x \to 0} \frac{\displaystyle\int_x^{x+\Delta x} f(t)\, dt}{\Delta x} \tag{7}$$

and show that its value is $f(x)$. But first, let us look at some of the geometry behind the theorem.

If f is positive, its integral from x to $x + \Delta x$ is the area of the strip under the graph of f from x to $x + \Delta x$ (Fig. 4.15). The strip is about $f(x)$ high by Δx wide, so

$$\text{Area} = \int_x^{x+\Delta x} f(t)\, dt \approx f(x)\, \Delta x.$$

Dividing through by Δx gives

$$\frac{\text{Area}}{\Delta x} = \frac{\displaystyle\int_x^{x+\Delta x} f(t)\, dt}{\Delta x} \approx f(x).$$

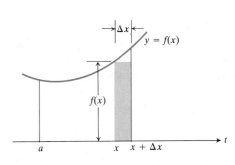

4.15 When Δx is small, the integral of f from x to $x + \Delta x$ is nearly equal to the area of the shaded rectangle. In symbols,

$$\int_x^{x+\Delta x} f(t)\, dt \approx f(x)\, \Delta x.$$

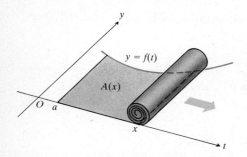

4.16 The rate at which the carpet covers the floor at the point x is the width of the carpet's leading edge as it rolls past x. In symbols, $dA/dx = f(x)$.

The approximation improves as $\Delta x \to 0$, and we expect (and find) equality in the limit.

We can also think of Eq. (6) dynamically. Imagine covering the region under the curve $y = f(t)$ from left to right by unrolling a carpet of variable width $f(t)$ (Fig. 4.16). The rate at which the floor is being covered as the carpet rolls past x is $f(x)$.

Now to the proof.

Proof of the First Fundamental Theorem   We prove the theorem for arbitrary continuous functions by showing that

$$\lim \frac{\displaystyle\int_{x}^{x+\Delta x} f(t)\, dt}{\Delta x} = f(x)$$

holds for $\Delta x \to 0^{+}$ and for $\Delta x \to 0^{-}$. This will show that the two-sided limit as $\Delta x \to 0$ exists and equals $f(x)$.

For positive values of Δx and $a \le x < b$, the calculation goes as follows: Since $a \le x < b$, we can start with values of Δx small enough to make $x + \Delta x$ lie between a and b. Since f is continuous on the closed interval from x to $x + \Delta x$, it has a minimum value min f and a maximum value max f on the interval. The integral inequality in Property 8 in Table 4.1 in Article 4.5, applied to the interval $[x, x + \Delta x]$, therefore gives

$$(\min f)\, \Delta x \le \int_{x}^{x+\Delta x} f(t)\, dt \le (\max f)\, \Delta x,$$

$$\min f \le \frac{\displaystyle\int_{x}^{x+\Delta x} f(t)\, dt}{\Delta x} \le \max f. \tag{8}$$

Now, f, being continuous, takes on the value max f at some point c in $[x, x + \Delta x]$ and the value min f at some point c' in $[x, x + \Delta x]$. That is,

$$\min f = f(c') \qquad \text{and} \qquad \max f = f(c).$$

With these substitutions, Eq. (8) becomes

$$f(c') \le \frac{\displaystyle\int_{x}^{x+\Delta x} f(t)\, dt}{\Delta x} \le f(c). \tag{9}$$

As $\Delta x \to 0^{+}$, c' and c both approach x because they are trapped between x and $x + \Delta x$, and $f(c')$ and $f(c)$ both approach $f(x)$ because f is continuous at x. Therefore, the right-hand and left-hand sides of Eq. (9) both approach $f(x)$, and

$$\lim_{\Delta x \to 0^{+}} \frac{\displaystyle\int_{x}^{x+\Delta x} f(t)\, dt}{\Delta x} = f(x) \tag{10}$$

by the Sandwich Theorem.

A similar argument, with Δx negative and $a < x \le b$, shows that

$$\lim_{\Delta x \to 0^{-}} \frac{\displaystyle\int_{x}^{x+\Delta x} f(t)\, dt}{\Delta x} = f(x). \tag{11}$$

Together Eqs. (10) and (11) prove that

$$\frac{d}{dx}\int_a^x f(t)\, dt = f(x) \tag{12}$$

at every point x in $[a, b]$.

This concludes the proof of the First Fundamental Theorem. ■

EXAMPLE 1

$$\frac{d}{dx}\int_{-\pi}^x \cos t\, dt = \cos x$$

$$\frac{d}{dx}\int_0^x \frac{\sin t}{t^2 + 1}\, dt = \frac{\sin x}{x^2 + 1}$$

■

EXAMPLE 2 Find dy/dx if $y = \displaystyle\int_0^{x^2} \cos t\, dt$.

Solution The Chain Rule is the key to doing this problem. We treat y as the composite of

$$f(u) = \int_0^u \cos t\, dt \qquad \text{and} \qquad u = x^2.$$

From the Chain Rule, in the form

$$\frac{dy}{dx} = \frac{df}{du}\frac{du}{dx},$$

we get

$$\frac{dy}{dx} = \cos u \cdot \frac{d}{dx}(x^2) = \cos x^2 \cdot 2x = 2x \cos x^2.$$

Therefore,

$$\frac{d}{dx}\int_0^{x^2} \cos t\, dt = 2x \cos x^2.$$

■

EXAMPLE 3 Find dy/dx if

$$y = \int_{x^2}^0 \cos t\, dt.$$

Solution Here the variable limit of integration is the lower limit, not the upper limit, so we must reverse the order of integration before applying the First Fundamental Theorem. We change the integral by writing

$$y = \int_{x^2}^0 \cos t\, dt = -\int_0^{x^2} \cos t\, dt.$$

We can then apply the result of Example 2:

$$\frac{dy}{dx} = \frac{d}{dx}\int_{x^2}^0 \cos t\, dt = -\frac{d}{dx}\int_0^{x^2} \cos t\, dt = -2x \cos x^2 \,.$$

■

The Integral Evaluation Theorem

We now come to a remarkable theorem that tells us how to use antiderivatives to evaluate definite integrals. With this theorem in hand, we shall never again be obliged to calculate definite integrals as limits. The theorem is so useful that it is often called the Second Fundamental Theorem of Calculus.

THEOREM 2

> **The Integral Evaluation Theorem (Second Fundamental Theorem of Calculus)**
>
> If f is continuous at every point of $[a, b]$, and F is any antiderivative of f on $[a, b]$, then
>
> $$\int_a^b f(x)\,dx = F(b) - F(a). \tag{13}$$

The Integral Evaluation Theorem says that to calculate the integral of f over $[a, b]$, all we need to do is

1. find an antiderivative F of f, and
2. calculate the number $F(b) - F(a)$.

This number will be $\int_a^b f(x)\,dx$. The existence of the antiderivative F is assured by the First Fundamental Theorem. To carry out the calculation, all we need to do is find F and evaluate it.

Proof   To prove the theorem, we use Corollary 3 of the Mean Value Theorem in Article 3.7, which says that any two functions that have f as a derivative on $[a, b]$ must differ by some fixed constant throughout $[a, b]$. We already know one function whose derivative equals f, namely,

$$G(x) = \int_a^x f(t)\,dt.$$

Therefore, if F is any other such function, then

$$F(x) = G(x) + C \tag{14}$$

throughout $[a, b]$ for some constant C. When we use Eq. (14) to calculate $F(b) - F(a)$, we find that

$$F(b) - F(a) = [G(b) + C] - [G(a) + C]$$

$$= G(b) - G(a)$$

$$= \int_a^b f(t)\,dt - \int_a^a f(t)\,dt$$

$$= \int_a^b f(t)\,dt - 0$$

$$= \int_a^b f(t)\,dt,$$

which establishes Eq. (13).   ∎

It is conventional to use the notation $F(x)]_a^b$ for the number $F(b) - F(a)$, and we shall do so in the following examples.

EXAMPLE 4   The area under the line $y = mx$ in Example 1 in Article 4.6 is

$$\int_a^b mx \, dx = \frac{mx^2}{2}\bigg]_a^b = \frac{mb^2}{2} - \frac{ma^2}{2}. \qquad \blacksquare$$

EXAMPLE 5   The area under the curve $y = x^2$ in Example 2 in Article 4.6 is

$$\int_0^b x^2 \, dx = \frac{x^3}{3}\bigg]_0^b = \frac{b^3}{3} - 0 = \frac{b^3}{3}. \qquad \blacksquare$$

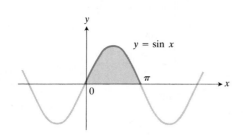

EXAMPLE 6   Calculate the area of the region enclosed by the x-axis and one arch of the curve $y = \sin x$.

Solution   We choose the arch that extends from $x = 0$ to $x = \pi$ (Fig. 4.17). The area enclosed by this arch and the x-axis is

$$\int_0^\pi \sin x \, dx = -\cos x\bigg]_0^\pi = -\cos \pi - (-\cos 0) = -(-1) + 1 = 2. \qquad \blacksquare$$

4.17   Example 6 calculates the area of the shaded region shown here.

EXAMPLE 7   This is an illustration of some of the algebraic properties of definite integrals in Table 4.1 in Article 4.5.

a) $\displaystyle\int_0^\pi 5 \sin x \, dx = 5 \int_0^\pi \sin x \, dx = 5 \cdot 2 = 10 \qquad \left(\begin{array}{l}\text{using the result}\\ \text{of Example 6}\end{array}\right)$

b) $\displaystyle\int_0^1 (x^2 + 1) \, dx = \int_0^1 x^2 \, dx + \int_0^1 dx$

$$= \frac{x^3}{3}\bigg]_0^1 + x\bigg]_0^1$$

$$= \left(\frac{1}{3} - 0\right) + (1 - 0) = \frac{4}{3}$$

c) $\displaystyle\int_1^2 \left(3 - \frac{6}{x^2}\right) dx = \int_1^2 3 \, dx - \int_1^2 \frac{6}{x^2} \, dx$

$$= \left[3x\right]_1^2 - \left[-\frac{6}{x}\right]_1^2 = 6 - 3 + 3 - 6 = 0$$

The integral in (b) gives the exact value of the area we approximated in Example 1 of Article 4.5. $\qquad \blacksquare$

If a continuous function $f(x)$ has no negative values on an interval $[a, b]$, then its integral from a to b gives the area between its graph and the x-axis. If f is negative or nonpositive over the interval, its integral gives the negative of the area, as in the next example.

EXAMPLE 8   Find

a) the area between the curve $y = x^2 - 4$ and the x-axis from $x = -2$ to $x = 2$;

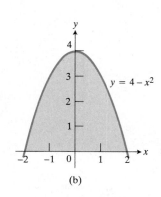

4.18 The graphs in (a) and (b) enclose the same amount of area with the x-axis but the definite integrals of the functions from -2 to 2 differ in sign.

(a)   (b)

b) the area between the curve $y = 4 - x^2$ and the x-axis from $x = -2$ to $x = 2$.

See Fig. 4.18.

Solution

a) If we integrate $f(x) = x^2 - 4$ over the interval from -2 to 2, we find

$$\int_{-2}^{2} (x^2 - 4)\, dx = \frac{x^3}{3} - 4x \Big]_{-2}^{2} = \left(\frac{8}{3} - 8\right) - \left(-\frac{8}{3} + 8\right) = -\frac{32}{3}.$$

The curve and the x-axis from $x = -2$ to $x = 2$ enclose $32/3$ units of area.

b) The graph of

$$y = g(x) = -f(x) = 4 - x^2, \qquad -2 \le x \le 2$$

in Fig. 4.18(b) is the mirror image of the graph of f across the x-axis. The area between the graph of g and the x-axis is

$$\int_{-2}^{2} (4 - x^2)\, dx = 4x - \frac{x^3}{3} \Big]_{-2}^{2} = \frac{32}{3}. \qquad \blacksquare$$

If the graph of a function f over the interval $a \le x \le b$ lies partly above and partly below the x-axis, as in Fig. 4.19, the integral of f over the interval will be the algebraic sum of signed areas. Areas above the axis will be counted as positive and areas below the axis will be counted as negative. The value of the

4.19 The integral of f from a to b is the algebraic sum of signed areas.

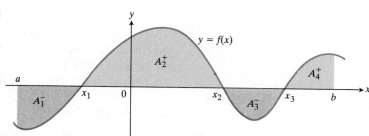

integral in such a case will be less than the total area between the curve and the *x*-axis.

For example, if the absolute values of the areas between the curve and the *x*-axis in Fig. 4.19 are A_1, A_2, A_3, and A_4, then

$$\int_a^b f(x)\, dx = -A_1 + A_2 - A_3 + A_4,$$

which is less than the total area $A_1 + A_2 + A_3 + A_4$. To calculate the total area, we find the points x_1, x_2, and x_3 where the curve crosses the axis and compute a separate integral for each part:

$$\int_a^{x_1} f(x)\, dx = -A_1, \qquad \int_{x_1}^{x_2} f(x)\, dx = A_2,$$

$$\int_{x_2}^{x_3} f(x)\, dx = -A_3, \qquad \int_{x_3}^b f(x)\, dx = A_4.$$

We then add the absolute values of these integrals.

EXAMPLE 9   Find the area of the region between the curve $y = x^3 - 4x$ and the *x*-axis.

Solution   We graph $y = x^3 - 4x = x(x - 2)(x + 2)$ to determine its shape and the limits of integration (Fig. 4.20). The curve lies above the *x*-axis from $x = -2$ to $x = 0$ and below the *x*-axis from $x = 0$ to $x = 2$. We evaluate the integrals:

$$\int_{-2}^0 (x^3 - 4x)\, dx = \frac{x^4}{4} - 2x^2 \Big]_{-2}^0 = 0 - (4 - 8) = +4 = A_1$$

$$\int_0^2 (x^3 - 4x)\, dx = \frac{x^4}{4} - 2x^2 \Big]_0^2 = (4 - 8) - 0 = -4 = -A_2.$$

The area of the region is $A = 4 + |-4| = 8$. ∎

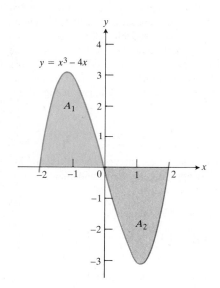

$y = x^3 - 4x$

4.20   The graph of $y = x^3 - 4x$ from $x = -2$ to $x = 2$.

PROBLEMS

In Problems 1–10, find the area bounded by the *x*-axis, the given curve $y = f(x)$, and the given vertical lines.

1. $y = x^2 + 1$, $x = 0$, $x = 3$

2. $y = 2x + 3$, $x = 0$, $x = 1$

3. $y = \sqrt{2x + 1}$, $x = 0$, $x = 4$

4. $y = \dfrac{1}{\sqrt{2x + 1}}$, $x = 0$, $x = 4$

5. $y = \dfrac{1}{(2x + 1)^2}$, $x = 1$, $x = 2$

6. $y = (2x + 1)^2$, $x = -1$, $x = 3$

7. $y = x^3 + 2x + 1$, $x = 0$, $x = 2$

8. $y = x\sqrt{2x^2 + 1}$, $x = 0$, $x = 2$

9. $y = \dfrac{x}{\sqrt{2x^2 + 1}}$, $x = 0$, $x = 2$

10. $y = \dfrac{x}{(2x^2 + 1)^2}$, $x = 0$, $x = 2$

11. Find the area bounded by the coordinate axes and the line $x + y = 1$.

12. Find the area between the curve $y = 1/\sqrt{x}$, the *x*-axis, and the lines $x = 1$ and $x = 4$.

13. Find the area between the curve $y = \sqrt{1 - x}$ and the coordinate axes.

14. Find the area contained between the *x*-axis and one arch of the curve $y = \cos 3x$.

15. Find the area between the curve $x = 1 - y^2$ and the *y*-axis. (*Hint:* Integrate with respect to *y*.)

16. Show that the area between the x-axis and one arch of the curve $y = \sin ax$ is $2/a$, $a > 0$.

17. Find the area between the x-axis and one arch of the curve $y = \sin^2 3x$.

18. a) Find the area under the parabolic arch $y = 6 - x - x^2$, $-3 \le x \le 2$.
   b) Find the height of the arch above the x-axis.
   c) Show that the area is two-thirds the base times the height.

19. Find the area of the triangular region in the plane that is bounded below by the positive x-axis and above by the curves $y = x^2$ and $y = 2 - x$.

20. The graph of $y = \sqrt{a^2 - x^2}$ over $-a \le x \le a$ is a semicircle of radius a.
   a) Using this fact, explain why
   $$\int_{-a}^{a} \sqrt{a^2 - x^2}\, dx = \frac{1}{2}\pi a^2.$$
   b) Evaluate
   $$\int_{0}^{a} \sqrt{a^2 - x^2}\, dx.$$

Evaluate the integrals in Problems 21–42.

21. $\int_{1}^{2} (2x + 5)\, dx$

22. $\int_{0}^{1} (x^2 - 2x + 3)\, dx$

23. $\int_{-1}^{1} (x + 1)^2\, dx$

24. $\int_{0}^{2} \sqrt{4x + 1}\, dx$

25. $\int_{0}^{\pi} \sin x\, dx$

26. $\int_{0}^{\pi} \cos x\, dx$

27. $\int_{\pi/4}^{\pi/2} \frac{\cos x\, dx}{\sin^2 x}$

28. $\int_{0}^{\pi/6} \frac{\sin 2x}{\cos^2 2x}\, dx$

29. $\int_{0}^{\pi} \sin^2 x\, dx$

30. $\int_{0}^{2\pi/\omega} \cos^2(\omega t)\, dt \quad (\omega \text{ constant})$

31. $\int_{0}^{1} \frac{dx}{(2x + 1)^3}$

32. $\int_{-1}^{0} x\sqrt{1 - x^2}\, dx$

33. $\int_{0}^{1} \sqrt{5x + 4}\, dx$

34. $\int_{-2}^{0} (4 - w)^2\, dw$

35. $\int_{-1}^{0} \left(\frac{x^7}{2} - x^{15}\right) dx$

36. $\int_{0}^{2} (t + 1)(t^2 + 4)\, dt$

37. $\int_{1}^{2} \frac{x^2 + 1}{x^2}\, dx$

38. $\int_{9}^{4} \frac{1 - \sqrt{u}}{\sqrt{u}}\, du$

39. $\int_{\pi/6}^{\pi/2} \cos^2\theta \sin\theta\, d\theta$

40. $\int_{0}^{\pi} x \cos\left(2x - \frac{\pi}{2}\right) dx$

41. $\int_{-4}^{4} |x|\, dx$

42. $\int_{0}^{\pi} \frac{1}{2}(\cos x + |\cos x|)\, dx$

43. In Article 4.5 we introduced a definition that extended the notion of area from the class of triangles and polygons to a larger class of regions bounded by continuous curves. Whenever we bring in a new definition like that it is a good idea to be sure that the new and old definitions agree on objects to which they both apply. For example, does the integral definition of area give $A = (1/2)bh$ for the area of a right triangle with base b and height h? To find out, use an integral to calculate the area of the triangle in Fig. 4.21.

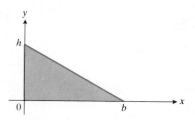

4.21 Is the area of this triangle still $(1/2)bh$? See Problem 43.

44. Calculate
$$\frac{d}{dx}\int_{0}^{x} \cos t\, dt$$
two ways: (a) by evaluating the integral and differentiating the result and (b) by applying the First Fundamental Theorem directly to the integral.

Find dF/dx in Problems 45–54.

45. $F(x) = \int_{0}^{x} \sqrt{1 + t^2}\, dt$

46. $F(x) = \int_{1}^{x} \frac{dt}{t}$

47. $F(x) = \int_{x}^{1} \sqrt{1 - t^2}\, dt$

48. $F(x) = \int_{0}^{x} \frac{dt}{1 + t^2}$

49. $F(x) = \int_{1}^{2x} \cos(t^2)\, dt$

50. $F(x) = \int_{1}^{x^2} \frac{dt}{1 + \sqrt{1 - t}}$

51. $F(x) = \int_{\sin x}^{0} \frac{dt}{2 + t}$

52. $F(x) = \int_{1/x}^{1} \frac{1}{t}\, dt$

53. $F(x) = \int_{\cos x}^{0} \frac{1}{1 - t^2}\, dt$

54. $F(x) = \int_{\sqrt{x}}^{10} \sin t^2\, dt$

55. Suppose that the function f is continuous and positive at every point of the interval $0 \le x \le 1$ and that the area between its graph and the interval $[0, x]$ is $\sin x$. Find $f(x)$.

56. Use l'Hôpital's rule to find
$$\lim_{x \to 0} \frac{1}{x^3}\int_{0}^{x} \frac{t^2}{t^4 + 1}\, dt.$$

57. Find (a) the linearization and (b) the quadratic approximation of the function
$$f(x) = 2 + \int_{0}^{x} \frac{10}{1 + t}\, dt$$
near $x = 0$.

58. Suppose that f has a positive derivative for all values of x and that $f(1) = 0$. Which of the following statements about

the function

$$y = \int_0^x f(t)\ dt$$

must be true?

a) y is a differentiable function of x.

b) y is a continuous function of x.

c) The graph of y has a horizontal tangent at $x = 1$.

d) y has a local maximum at $x = 1$.

e) y has a local minimum at $x = 1$.

f) The graph of y has an inflection point at $x = 1$.

g) The graph of dy/dx crosses the x-axis at $x = 1$.

59. Find $f(4)$ if

a) $\displaystyle\int_0^x f(t)\ dt = x \cos \pi x,$

b) $\displaystyle\int_0^{x^2} f(t)\ dt = x \cos \pi x,$

c) $\displaystyle\int_0^{f(x)} t^2\ dt = x \cos \pi x.$ (*Hint:* Integrate.)

60. Find $f(\pi/2)$ from the following two pieces of information: (i) $f(x)$ is continuous, and (ii) the area under the graph of f and over the interval $[0, a]$ is

$$\frac{a^2}{2} + \frac{a}{2}\sin a + \frac{\pi}{2}\cos a.$$

TOOLKIT PROGRAMS

Integration   Derivative Grapher

4.8

Substitution in Definite Integrals

When we evaluate a definite integral by substitution, we have a choice of evaluating the transformed integral with transformed limits or of going on to evaluate the original integral with the original limits. The examples in this article will show you how it is done.

The formula we use for evaluating definite integrals by substitution first appeared in a book written by Isaac Barrow (1630–1677), Newton's teacher and predecessor at Cambridge University. It looks like this:

The Substitution Formula for Definite Integrals

$$\int_a^b f(g(x))g'(x)\ dx = \int_{u=g(a)}^{u=g(b)} f(u)\ du. \qquad \left(\begin{matrix} u = g(x), \\ du = g'(x)\ dx \end{matrix}\right) \qquad (1)$$

The formula will hold if g' is continuous on the interval joining a and b and if f is continuous on the set of values taken on by g.

To prove Eq. (1), let F be an antiderivative of f, so that $F' = f$. Then

$$\int_a^b f(g(x))g'(x)\ dx = \int_a^b F'(g(x))g'(x)\ dx$$

$$= F(g(x))\ \Big]_{x=a}^{x=b} = F(g(b)) - F(g(a)) = \int_{g(a)}^{g(b)} f(u)\ du.$$

EXAMPLE 1   Use Eq. (1) to evaluate

$$\int_0^{\pi/2} \sin^2 x \cos x\ dx.$$

Solution   We use the substitutions

$$u = g(x) = \sin x, \qquad\qquad g(0) = \sin(0) = 0,$$
$$du = g'(x)\,dx = \cos x\,dx, \qquad g(\pi/2) = \sin(\pi/2) = 1,$$

to get

$$\int_0^{\pi/2} \sin^2 x \cos x\,dx = \int_{u=g(0)}^{u=g(\pi/2)} u^2\,du = \int_0^1 u^2\,du = \frac{u^3}{3}\Big]_0^1 = \frac{1}{3}. \qquad \blacksquare$$

EXAMPLE 2   Evaluate

$$\int_0^1 15x^2\sqrt{5x^3 + 4}\,dx.$$

Solution   We have two choices here:

Choice 1: Transform the integral and evaluate the transformed integral with transformed limits, as in Eq. (1). We substitute

$$u = g(x) = 5x^3 + 4, \qquad du = g'(x)\,dx = 15x^2\,dx.$$

Then

$$\int_0^1 15x^2\sqrt{5x^3 + 4}\,dx = \int_{g(0)}^{g(1)} \sqrt{u}\,du$$

$$= \int_4^9 u^{1/2}\,du$$

$$= \frac{2}{3}u^{3/2}\Big]_4^9 = \frac{2}{3}(27 - 8) = \frac{38}{3}.$$

Choice 2: Transform the integral as before, integrate, change back to x, and use the original x-limits. With

$$u = 5x^3 + 4, \qquad du = 15x^2\,dx,$$

we get

$$\int 15x^2\sqrt{5x^3 + 4}\,dx = \int \sqrt{u}\,du = \frac{2}{3}u^{3/2} + C = \frac{2}{3}(5x^3 + 4)^{3/2} + C.$$

Therefore,

$$\int_0^1 15x^2\sqrt{5x^3 + 4}\,dx = \frac{2}{3}(5x^3 + 4)^{3/2}\Big]_0^1$$

$$= \frac{2}{3}(9)^{3/2} - \frac{2}{3}(4)^{3/2} = \frac{2}{3}(27 - 8) = \frac{38}{3}.$$

Which method is better—evaluating the transformed integral with transformed limits or evaluating the original integral with original limits? Here they seem equally good, but sometimes one is easier than the other. As a general rule, it is best to know both methods and use whichever seems better at the time. $\blacksquare$

PROBLEMS

Evaluate the integrals in Problems 1–18.

1. a) $\displaystyle\int_0^3 \sqrt{y+1}\,dy$ b) $\displaystyle\int_{-1}^0 \sqrt{y+1}\,dy$

2. a) $\displaystyle\int_0^1 r\sqrt{1-r^2}\,dr$ b) $\displaystyle\int_{-1}^1 r\sqrt{1-r^2}\,dr$

3. a) $\displaystyle\int_0^{\pi/4} \tan x \sec^2 x\,dx$ b) $\displaystyle\int_{-\pi/4}^0 \tan x \sec^2 x\,dx$

4. a) $\displaystyle\int_0^1 x^3(1+x^4)^3\,dx$ b) $\displaystyle\int_{-1}^1 x^3(1+x^4)^3\,dx$

5. a) $\displaystyle\int_0^1 \frac{x^3}{\sqrt{x^4+9}}\,dx$ b) $\displaystyle\int_{-1}^0 \frac{x^3}{\sqrt{x^4+9}}\,dx$

6. a) $\displaystyle\int_{-1}^1 \frac{x}{(1+x^2)^2}\,dx$ b) $\displaystyle\int_0^1 \frac{x}{(1+x^2)^2}\,dx$

7. a) $\displaystyle\int_0^{\sqrt{7}} x(x^2+1)^{1/3}\,dx$ b) $\displaystyle\int_{-\sqrt{7}}^0 x(x^2+1)^{1/3}\,dx$

8. a) $\displaystyle\int_0^{\pi} 3\cos^2 x \sin x\,dx$ b) $\displaystyle\int_{2\pi}^{3\pi} 3\cos^2 x \sin x\,dx$

9. a) $\displaystyle\int_0^{\pi/6} (1-\cos 3x)\sin 3x\,dx$ b) $\displaystyle\int_{\pi/6}^{\pi/3} (1-\cos 3x)\sin 3x\,dx$

10. a) $\displaystyle\int_0^{\sqrt{3}} \frac{4x}{\sqrt{x^2+1}}\,dx$ b) $\displaystyle\int_{-\sqrt{3}}^{\sqrt{3}} \frac{4x}{\sqrt{x^2+1}}\,dx$

11. a) $\displaystyle\int_0^{2\pi} \frac{\cos x}{\sqrt{2+\sin x}}\,dx$ b) $\displaystyle\int_{-\pi}^{\pi} \frac{\cos x}{\sqrt{2+\sin x}}\,dx$

12. a) $\displaystyle\int_{-\pi/2}^0 \frac{\sin x}{(3+\cos x)^2}\,dx$ b) $\displaystyle\int_0^{\pi/2} \frac{\sin x}{(3+\cos x)^2}\,dx$

13. a) $\displaystyle\int_{-\pi}^{\pi} x\cos(2x^2)\,dx$ b) $\displaystyle\int_{-\pi}^0 x\cos(2x^2)\,dx$

14. $\displaystyle\int_0^{\pi^2/4} \frac{\cos\sqrt{x}}{\sqrt{x}}\,dx$

15. $\displaystyle\int_0^1 \sqrt{t^5+2t}\,(5t^4+2)\,dt$

16. $\displaystyle\int_1^4 \frac{dy}{2\sqrt{y}\,(1+\sqrt{y})^2}$

17. $\displaystyle\int_0^{\pi/2} \cos^3 2x \sin 2x\,dx$

18. $\displaystyle\int_{-\pi/4}^{\pi/4} \tan^2 x \sec^2 x\,dx$

19. a) Graph the curve $y = x\sqrt{3-x^2}$.
  b) Find the area between the curve and the x-axis.

20. Evaluate
$$\int_0^{2\pi} |\cos x|\,dx.$$

21. Suppose $F(x)$ is an antiderivative of
$$f(x) = \frac{\sin x}{x}, \qquad x \neq 0.$$
Express
$$\int_1^3 \frac{\sin 2x}{x}\,dx$$
in terms of F.

22. Suppose that
$$\int_0^1 f(x)\,dx = 3.$$
Find
$$\int_{-1}^0 f(x)\,dx$$
if (a) f is odd, (b) f is even.

23. Suppose that the function $h(x)$ is even and continuous for all x.
  a) Show that the product $h(x)\sin x$ is odd.
  b) Show that, for any number a,
  $$\int_{-a}^0 h(x)\sin x\,dx = -\int_0^a h(x)\sin x\,dx.$$
  (*Hint:* Use the substitution $u = -x$.)
  c) Use the result in (b) to show that
  $$\int_{-a}^a h(x)\sin x\,dx = 0.$$
  d) Show that
  $$\int_{-\pi/4}^{\pi/4} \sec x \sin x\,dx = 0.$$

24. Show that
$$\int_{-a}^a h(x)\,dx = \begin{cases} 0 & \text{if } h \text{ is odd} \\ 2\displaystyle\int_0^a h(x)\,dx & \text{if } h \text{ is even.} \end{cases}$$

The shift property for definite integrals. A basic property of the definite integral is its invariance under translation, as expressed by the equation
$$\int_a^b f(x)\,dx = \int_{a-c}^{b-c} f(x+c)\,dx. \qquad (2)$$
This equation will hold whenever f is continuous and defined for the necessary values of x. For example,
$$\int_0^1 x^3\,dx = \int_{-2}^{-1} (x+2)^3\,dx = \int_2^3 (x-2)^3\,dx.$$
See Fig. 4.22.

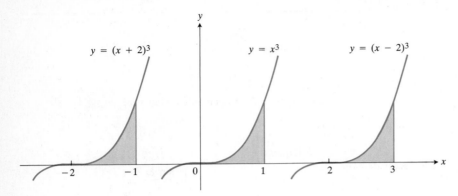

4.22 The shaded regions, being congruent, have the same area. See the introduction, to Problems 25–28.

For each of the functions in Problems 25–27, graph $f(x)$ and $f(x + c)$ together and convince yourself that Eq. (2) is reasonable. Then evaluate both sides of Eq. (2).

25. $f(x) = x^2$, $a = 0$, $b = 1$, $c = 1$

26. $f(x) = \sin x$, $a = 0$, $b = \pi$, $c = \pi/2$

27. $f(x) = \sqrt{x - 4}$, $a = 4$, $b = 8$, $c = 5$

28. Use the formula for changing limits of integration (Eq. (1)) to verify Eq. (2).

4.9
Rules for Approximating Definite Integrals

To evaluate definite integrals of functions like

$$\frac{\sin x}{x}, \qquad \sin x^2,$$

and

$$\sqrt{1 + x^4}$$

whose antiderivatives have no elementary formula, we turn to numerical methods like the trapezoidal rule and Simpson's rule. These rules also enable us to estimate the integral of a function from a table of values even if we do not know a formula for the function. This comes in handy when the only information we have is a set of specific values measured in the laboratory or in the field.

The Trapezoidal Rule

The trapezoidal rule for the value of a definite integral is based on approximating the region between a curve and the x-axis with trapezoids instead of rectangles, as in Fig. 4.23. It is not necessary for the subdivision points $x_1, x_2, \ldots, x_{n-1}$ in the figure to be evenly spaced, but the resulting formula is simpler if they are. We therefore suppose that the length of each subinterval is

$$h = \Delta x = \frac{b - a}{n}.$$

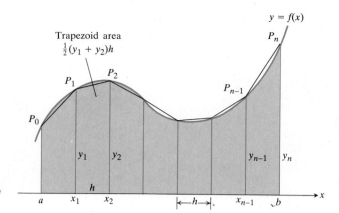

4.23 The trapezoidal rule approximates short stretches of curve with line segments. To estimate the shaded area, we add the areas of the trapezoids made by joining the ends of these segments to the x-axis.

The sum of the areas of the trapezoids is then

$$T = \frac{1}{2}(y_0 + y_1)h + \frac{1}{2}(y_1 + y_2)h + \cdots$$

$$+ \frac{1}{2}(y_{n-2} + y_{n-1})h + \frac{1}{2}(y_{n-1} + y_n)h$$

$$= h\left(\frac{1}{2}y_0 + y_1 + y_2 + \cdots + y_{n-1} + \frac{1}{2}y_n\right) \tag{1}$$

$$= \frac{h}{2}(y_0 + 2y_1 + 2y_2 + \cdots + 2y_{n-1} + y_n),$$

where

$$y_0 = f(a), \quad y_1 = f(x_1), \quad \ldots, \quad y_{n-1} = f(x_{n-1}), \quad y_n = f(b).$$

The trapezoidal rule says: Use T to estimate the integral of f from a to b.

The Trapezoidal Rule

To approximate

$$\int_a^b f(x)\, dx,$$

use

$$T = \frac{h}{2}(y_0 + 2y_1 + 2y_2 + \cdots + 2y_{n-1} + y_n) \tag{2}$$

(n subintervals of length $h = (b - a)/n$).

EXAMPLE 1 Use the trapezoidal rule with $n = 4$ to estimate

$$\int_1^2 x^2\, dx.$$

Compare the estimate with the exact value of the integral.

TABLE 4.4

x	$y = x^2$
1	1
$\dfrac{5}{4}$	$\dfrac{25}{16}$
$\dfrac{6}{4}$	$\dfrac{36}{16}$
$\dfrac{7}{4}$	$\dfrac{49}{16}$
2	4

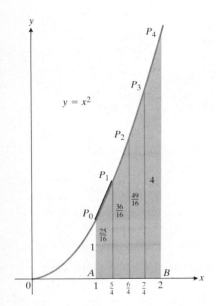

4.24 The trapezoidal approximation of the area under the graph of $y = x^2$ from $x = 1$ to $x = 2$ is a slight overestimate.

Solution   The exact value of the integral is

$$\int_1^2 x^2 \, dx = \frac{x^3}{3}\bigg]_1^2 = \frac{7}{3}.$$

To find the trapezoidal approximation, we divide the interval of integration into four subintervals of equal length and list the values of $y = x^2$ at the endpoints and subdivision points (see Table 4.4).

We then evaluate Eq. (2) with $n = 4$ and $h = 1/4$:

$$T = \frac{h}{2}(y_0 + 2y_1 + 2y_2 + 2y_3 + y_4)$$

$$= \frac{1}{8}\left(1 + 2\left(\frac{25}{16}\right) + 2\left(\frac{36}{16}\right) + 2\left(\frac{49}{16}\right) + 4\right)$$

$$= \frac{75}{32} = 2.34375.$$

The approximation overestimates the area by about half a percent of its true value. Each trapezoid contains slightly more than the corresponding strip under the curve (Fig. 4.24).   ∎

Error Estimates for the Trapezoidal Rule

As n increases and the step size $h = \Delta x$ approaches zero, T approaches the exact value of $\int_a^b f(x) \, dx$. To see why, write

$$T = h\left(\frac{1}{2}y_0 + y_1 + y_2 + \cdots + y_{n-1} + \frac{1}{2}y_n\right)$$

$$= (y_1 + y_2 + \cdots + y_n)\,\Delta x + \frac{1}{2}(y_0 - y_n)\,\Delta x \qquad (3)$$

$$= \sum f(x_k)\,\Delta x + \frac{1}{2}[f(a) - f(b)]\,\Delta x.$$

As $n \to \infty$ and $\Delta x \to 0$,

$$\sum f(x_k) \, \Delta x \to \int_a^b f(x) \, dx \qquad \text{and} \qquad \frac{1}{2}[f(a) - f(b)] \, \Delta x \to 0.$$

Therefore,

$$\lim_{n \to \infty} T = \int_a^b f(x) \, dx + 0 = \int_a^b f(x) \, dx.$$

This means that in theory we can make the difference between T and the integral as small as we want by taking n large enough. In practice, though, how do we tell how large n should be for a given tolerance?

We answer this question with a result from advanced calculus, which says that if f'' is continuous on $[a, b]$, then

$$\int_a^b f(x) \, dx = T - \frac{b - a}{12} h^2 f''(c) \tag{4}$$

for some number c between a and b. Thus, as h approaches zero, the error,

$$E_T = -\frac{b - a}{12} h^2 f''(c), \tag{5}$$

approaches zero as the *square* of h.

The inequality

$$|E_T| \le \frac{b - a}{12} h^2 \, \max|f''(x)|, \tag{6}$$

where max refers to the interval $[a, b]$, gives an upper bound for the magnitude of the error. In practice, we usually cannot find the exact value of $\max|f''(x)|$ and have to estimate an upper bound or "worst case" value for it instead. If M is any upper bound for the values of $|f''(x)|$ on $[a, b]$, then

$$|E_T| \le \frac{b - a}{12} h^2 M. \tag{7}$$

This is the inequality we normally use in estimating $|E_T|$. We find the best M we can and go on to estimate $|E_T|$ from there. This may sound careless, but it works. To make $|E_T|$ small for a given M, we just make h small.

The Error Estimate for the Trapezoidal Rule

If f'' is continuous and M is any upper bound for the values of $|f''|$ on $[a, b]$, then the error E_T in the trapezoidal approximation of the integral of f from a to b satisfies the inequality

$$|E_T| \le \frac{b - a}{12} h^2 M. \tag{8}$$

EXAMPLE 2 Find an upper bound for the error in the approximation found in Example 1 for the integral

$$\int_1^2 x^2 \, dx.$$

Solution We first find an upper bound M for the magnitude of the second derivative of $f(x) = x^2$ on the interval $1 \le x \le 2$. Since $f''(x) = 2$ for all x, we may safely take $M = 2$. With $b - a = 1$ and $h = 1/4$, Eq. (8) gives

$$|E_T| \le \frac{b - a}{12} h^2 M = \frac{1}{12}\left(\frac{1}{4}\right)^2 (2) = \frac{1}{96}.$$

This is precisely what we find when we subtract $T = 75/32$ from $\int_1^2 x^2 \, dx = 7/3$, since $7/3 - 75/32 = -1/96$. Here we are able to give the error *exactly* because the second derivative of $f(x) = x^2$ is a constant and we have no uncertainty in the term $f''(c)$ in Eq. (4). We are not always this lucky; in most cases the best we can do is to *estimate* the difference between the integral and T.

EXAMPLE 3 The trapezoidal rule is to be used to estimate the value of

$$\int_0^1 x \sin x \, dx$$

with $n = 10$ steps. Find an upper bound for the error in the estimate.

Solution We use the formula

$$|E_T| \le \frac{b - a}{12} h^2 M$$

with $b = 1$, $a = 0$, and $h = 1/n = 1/10$. This gives

$$|E_T| \le \frac{1}{12}\left(\frac{1}{10}\right)^2 M = \frac{1}{1200} M.$$

The number M can be any upper bound for the values of $|f''|$ on $[0, 1]$. To choose a value for M, we calculate f'' to see how big it might be. A straightforward differentiation gives

$$f'' = \cos x + (1 - x) \sin x.$$

Hence,

$$|f''| \le |\cos x| + |1 - x||\sin x| \le 1 + (1)(1) = 2$$

because $0 \le x \le 1$ and $|\cos x|$ and $|\sin x|$ never exceed 1. We can safely take $M = 2$. Therefore,

$$|E_T| \le \frac{1}{1200}(2) = \frac{1}{600} < 0.00167.$$

The error is no greater than 1.67×10^{-3}.

For greater accuracy we would not try to improve M but would take more steps. With $n = 100$ steps, for example, $h = 1/100$ and

$$|E_T| \le \frac{1}{12}\left(\frac{1}{100}\right)^2 (2) < 1.67 \times 10^{-5}.$$

EXAMPLE 4 How many subdivisions should be used in the trapezoidal rule to approximate

$$\ln 2 = \int_1^2 \frac{1}{x} dx$$

with an error whose absolute value is less than 10^{-4}?

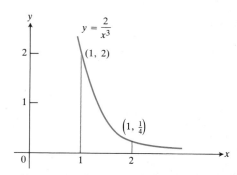

4.25 The continuous function $y = 2/x^3$ has its maximum value on $[1, 2]$ at $x = 1$.

Solution To determine n, the number of subdivisions, we use Eq. (8) with

$$b - a = 2 - 1 = 1, \qquad h = \frac{b-a}{n} = \frac{1}{n},$$

$$f''(x) = \frac{d^2}{dx^2}(x^{-1}) = 2x^{-3} = \frac{2}{x^3}.$$

Then,

$$|E_T| \le \frac{b-a}{12} h^2 \max|f''(x)| = \frac{1}{12}\left(\frac{1}{n}\right)^2 \max\left|\frac{2}{x^3}\right|,$$

where max refers to $[1, 2]$.

This is one of the rare cases in which we actually can find $\max|f''|$ rather than having to settle for an upper bound. On $[1, 2]$, $y = 2/x^3$ decreases steadily from a maximum of $y = 2$ to a minimum of $y = 1/4$ (Fig. 4.25). Therefore,

$$|E_T| \le \frac{1}{12}\left(\frac{1}{n}\right)^2 \cdot 2 = \frac{1}{6n^2}.$$

The error's absolute value will therefore be less than 10^{-4} if

$$\frac{1}{6n^2} < 10^{-4}, \qquad \frac{10^4}{6} < n^2, \qquad \frac{100}{\sqrt{6}} < n, \qquad \text{or} \qquad 40.83 < n.$$

The first integer beyond 40.83 is $n = 41$. With $n = 41$ subdivisions we can guarantee calculating $\ln 2$ with an error of magnitude less than 10^{-4}. Any larger n will work, too. ■

Simpson's Rule

Any three noncollinear points in the plane can be fitted with a parabola, and Simpson's rule is based on approximating curves with parabolas instead of trapezoids. The shaded area under the parabola in Fig. 4.26 is

$$A_p = \frac{h}{3}(y_0 + 4y_1 + y_2),$$

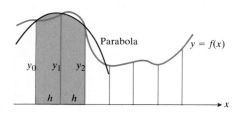

4.26 Simpson's rule approximates short stretches of curve with parabolas.

and applying this formula successively along a continuous curve $y = f(x)$ from $x = a$ to $x = b$ leads to an estimate of $\int_a^b f(x)\,dx$ that is generally more accurate than T for a given step size h.

We can derive the formula for A_p in the following way. To simplify the algebra, we use the coordinate system shown in Fig. 4.27. The area under the parabola will be the same no matter where the y-axis is, as long as we preserve the vertical scale. The parabola has an equation of the form

$$y = Ax^2 + Bx + C,$$

so the area under it from $x = -h$ to $x = h$ is

$$A_p = \int_{-h}^{h} (Ax^2 + Bx + C)\,dx$$

$$= \frac{Ax^3}{3} + \frac{Bx^2}{2} + Cx \Big]_{-h}^{h}$$

$$= \frac{2Ah^3}{3} + 2Ch = \frac{h}{3}(2Ah^2 + 6C).$$

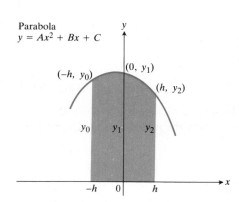

4.27 By integrating from $-h$ to h, the shaded area is found to be

$$A_p = \frac{h}{3}(y_0 + 4y_1 + y_2).$$

Since the curve passes through the three points $(-h, y_0)$, $(0, y_1)$, and (h, y_2), we also have

$$y_0 = Ah^2 - Bh + C, \qquad y_1 = C, \qquad y_2 = Ah^2 + Bh + C,$$

from which we obtain

$$C = y_1,$$
$$Ah^2 - Bh = y_0 - y_1,$$
$$Ah^2 + Bh = y_2 - y_1,$$
$$2Ah^2 = y_0 + y_2 - 2y_1.$$

Hence, expressing the area A_p in terms of the ordinates y_0, y_1, and y_2, we have

$$A_p = \frac{h}{3}(2Ah^2 + 6C) = \frac{h}{3}((y_0 + y_2 - 2y_1) + 6y_1),$$

or

$$A_p = \frac{h}{3}(y_0 + 4y_1 + y_2).$$

Simpson's rule follows from applying the formula for A_p to successive pieces of the curve $y = f(x)$ between $x = a$ and $x = b$. Each separate piece of the curve, covering an x-subinterval of width $2h$, is approximated by an arc of a parabola through its ends and midpoint. The areas under the parabolic arcs are then added to give Simpson's rule.

Simpson's Rule

To approximate

$$\int_a^b f(x)\, dx,$$

use

$$S = \frac{h}{3}(y_0 + 4y_1 + 2y_2 + 4y_3 + \cdots + 2y_{n-2} + 4y_{n-1} + y_n) \qquad (9)$$

(n even, $h = (b - a)/n$).

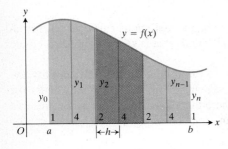

4.28 The y's in Eq. (9) are the values of f at the points of subdivision.

The y's in Eq. (9) are the values of $y = f(x)$ at the points

$$a, \quad x_1 = a + h, \quad x_2 = a + 2h, \quad \ldots, \quad x_{n-1} = a + (n-1)h, \quad b$$

that subdivide $[a, b]$ into n equal subintervals of length $h = (b - a)/n$. See Fig. 4.28. The number n must be even to apply the rule because each parabolic arc uses two subintervals.

Error Estimates for Simpson's Rule

To estimate the error in Simpson's rule, we start with a result from advanced calculus that says that if the fourth derivative $f^{(4)}$ is continuous, then

$$\int_a^b f(x)\, dx = S - \frac{b - a}{180} h^4 f^{(4)}(c) \qquad (10)$$

for some point c between a and b. Thus, as h approaches zero, the error,

$$E_S = -\frac{b-a}{180}h^4 f^{(4)}(c),$$ (11)

approaches zero as the *fourth power* of h. (This helps to explain why Simpson's rule is likely to give better results than the trapezoidal rule.)

The inequality

$$|E_S| \le \frac{b-a}{180}h^4 \max|f^{(4)}(x)|,$$ (12)

where max refers to the interval $[a, b]$, gives an upper bound for the magnitude of the error. As with $\max|f''|$ in the error formula for the trapezoidal rule, we usually cannot find the exact value of $\max|f^{(4)}(x)|$ and have to replace it with an upper bound. If M is any upper bound for the values of $|f^{(4)}|$ on $[a, b]$, then

$$|E_S| \le \frac{b-a}{180}h^4 M.$$ (13)

This is the formula we usually use in estimating the error in Simpson's rule. We find a reasonable value for M and go on to estimate $|E_S|$ from there.

The Error Estimate for Simpson's Rule

If $f^{(4)}$ is continuous and M is any upper bound for the values of $|f^{(4)}|$ on $[a, b]$, then the error E_S in the Simpson's rule approximation of the integral of f from a to b satisfies the inequality

$$|E_S| \le \frac{b-a}{180}h^4 M.$$ (14)

EXAMPLE 5 Use Simpson's rule with $n = 4$ to approximate

$$\int_0^1 5x^4 \, dx.$$

What estimate does Eq. (14) give for the error in the approximation?

Solution Again we have chosen an integral whose exact value we can calculate directly:

$$\int_0^1 5x^4 \, dx = x^5 \Big]_0^1 = 1.$$

To find the Simpson approximation, we divide the interval of integration into four subintervals and list the values of $f(x) = 5x^4$ at the endpoints and subdivision points (see Table 4.5).

We then evaluate Eq. (9) with $n = 4$ and $h = 1/4$:

$$S = \frac{h}{3}(y_0 + 4y_1 + 2y_2 + 4y_3 + y_4)$$

$$= \frac{1}{12}\left(0 + 4\left(\frac{5}{256}\right) + 2\left(\frac{80}{256}\right) + 4\left(\frac{405}{256}\right) + 5\right)$$

$$= 1.00260 \quad \text{(rounded)}.$$

TABLE 4.5

x	$y = 5x^4$
0	0
$\frac{1}{4}$	$\frac{5}{256}$
$\frac{2}{4}$	$\frac{80}{256}$
$\frac{3}{4}$	$\frac{405}{256}$
1	5

To estimate the error with Eq. (14), we first find an upper bound M for the magnitude of the fourth derivative of $f(x) = 5x^4$ on the interval $0 \le x \le 1$. Since the fourth derivative has the constant value $f^{(4)}(x) = 120$, we may safely take $M = 120$. With $(b - a) = 1$ and $h = 1/4$, Eq. (14) then gives

$$|E_S| \le \frac{b - a}{180} h^4 M = \frac{1}{180} \left(\frac{1}{4}\right)^4 (120) = \frac{1}{384} < 0.00261. \qquad \blacksquare$$

Polynomials of Low Degree

If $f(x)$ is a polynomial of degree less than four, then its fourth derivative is zero, and

$$E_S = -\frac{b - a}{180} h^4 f^{(4)}(c) = -\frac{b - a}{180} h^4 (0) = 0.$$

Thus, there will be no error in the Simpson approximation of any integral of f. In other words, if f is a constant, a linear function, or a quadratic or cubic polynomial, Simpson's rule will give the value of any integral of f exactly, whatever the number of subdivisions.

Similarly, if f is a constant or a linear function, then its second derivative is zero and

$$E_T = -\frac{b - a}{12} h^2 f''(c) = -\frac{b - a}{12} h^2 (0) = 0.$$

The trapezoidal rule will therefore give the exact value of any integral of f. This is no surprise, for the trapezoids fit the graph perfectly.

EXAMPLE 6 Estimate

$$\int_0^2 x^3 \, dx$$

with Simpson's rule.

Solution The fourth derivative of $f(x) = x^3$ is zero, so we expect Simpson's rule to give the integral's exact value with any (even) number of steps. Indeed, with $n = 2$ and $h = (2 - 0)/2 = 1$,

$$S = \frac{h}{3}(y_0 + 4y_1 + y_2)$$

$$= \frac{1}{3}((0)^3 + 4(1)^3 + (2)^3)$$

$$= \frac{12}{3} = 4,$$

while

$$\int_0^2 x^3 \, dx = \frac{x^4}{4}\Big]_0^2 = \frac{16}{4} - 0 = 4. \qquad \blacksquare$$

146 ft 122 ft 76 ft 54 ft 40 ft 30 ft 13 ft Ignored

Horizontal spacing = 20 ft

4.29 The swamp in Example 7.

Working with Numerical Data

The next example shows how to use Simpson's rule to estimate the integral of a function from values measured in the laboratory or in the field. The trapezoidal rule could be used the same way.

EXAMPLE 7 A town wants to drain and fill the small swamp shown in Fig. 4.29. The swamp averages 5 ft deep. About how many cubic yards of dirt will be needed to fill the area after the swamp is drained?

Solution To calculate the volume of the swamp, we estimate the surface area and multiply by 5. To estimate the area, we use Simpson's rule with $h = 20$ ft and the y's equal to the distances measured across the swamp shown in Fig. 4.29:

$$S = \frac{h}{3}(y_0 + 4y_1 + 2y_2 + 4y_3 + 2y_4 + 4y_5 + y_6)$$

$$= \frac{20}{3}(146 + 488 + 152 + 216 + 80 + 120 + 13)$$

$$= \frac{20}{3}(1215) = 8100.$$

The volume is about $(8100)(5) = 40{,}500$ ft^3 or 1500 yd^3. ∎

Round-off Errors

Although decreasing the step size h reduces the error in the Simpson and trapezoidal approximations in theory, it may not always do so in practice. When h is very small, say $h = 10^{-5}$, the round-off errors in the arithmetic required to evaluate S and T may accumulate to such an extent that the error formulas no longer describe what is going on. Shrinking h below a certain size can actually make things worse rather than better. While this will not be an issue in the present book, you should consult a text on numerical analysis for alternative methods if you are having problems with round-off.

PROBLEMS

Approximate each of the integrals in Problems 1–6 with $n = 4$ by (a) the trapezoidal rule and (b) Simpson's rule. Compare your answers with (c) the integral's exact value.

1. $\int_0^2 x\, dx$ **2.** $\int_0^2 x^2\, dx$ **3.** $\int_0^2 x^3\, dx$

4. $\int_1^2 \frac{1}{x^2}\, dx$ **5.** $\int_0^4 \sqrt{x}\, dx$ **6.** $\int_0^\pi \sin x\, dx$

7. Estimate the error in using
  a) the trapezoidal rule and
  b) Simpson's rule to estimate the value of

$$\ln 2 = \int_1^2 \frac{1}{t}\, dt$$

  with $n = 10$ steps.

  c) What accuracy can you expect from each rule if you use $n = 4$ steps instead?

8. Repeat Example 4 with Simpson's rule in place of the trapezoidal rule.

In Problems 9–14, estimate the minimum number of subdivisions needed to approximate the integral with an error of absolute value less than 10^{-4} by (a) the trapezoidal rule, (b) Simpson's rule.

9. $\int_0^2 x\, dx$ **10.** $\int_0^2 x^2\, dx$ **11.** $\int_0^2 x^3\, dx$

12. $\int_1^2 \frac{1}{x^3}\, dx$ **13.** $\int_1^4 \sqrt{x}\, dx$ **14.** $\int_0^\pi \sin x\, dx$

15. The value of the integral

$$\int_1^2 f(x)\,dx$$

is to be estimated by Simpson's rule with an error whose magnitude is less than 10^{-5}. You have determined that $|f^{(4)}(x)| \le 3$ throughout the interval of integration. According to the Simpson's rule error formula, how many subdivisions are needed to guarantee the required accuracy?

16. As the fish and game warden of your township, you are responsible for stocking the town pond with fish before fishing season. The average depth of the pond is 20 ft. You plan to start the season with one fish per 1000 cubic feet. You intend to have at least 25% of the opening day's fish population left at the end of the season. What is the maximum number of licenses the town can sell if the average seasonal catch is 20 fish per license? The pond is sketched to scale in Fig. 4.30.

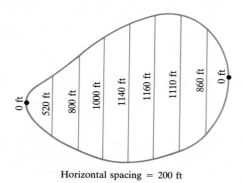

Horizontal spacing = 200 ft

4.30   The fish pond in Problem 16.

17. CALCULATOR The design of a new airplane requires a gasoline tank of constant cross-section area in each wing. A scale drawing of a cross section is shown in Fig. 4.31. The tank must hold 5000 lb of gasoline that has a density of 42 lb/ft³. Estimate the length of the tank.

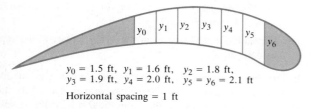

$y_0 = 1.5$ ft, $y_1 = 1.6$ ft, $y_2 = 1.8$ ft,
$y_3 = 1.9$ ft, $y_4 = 2.0$ ft, $y_5 = y_6 = 2.1$ ft

Horizontal spacing = 1 ft

4.31   The airplane wing and tank cross section for Problem 17.

18. CALCULATOR The rate at which flashbulbs give off light varies during the flash. For some bulbs, the rate at which light is produced, measured in lumens, reaches a peak and

(a)

(b)

4.32   Flashbulb output data from Tables 4.6 and 4.7 plotted and connected by smooth curves.

fades quickly, as shown in Fig. 4.32(a). For other bulbs, the light, instead of reaching a peak, stays at a moderate level for a relatively longer period of time, as shown in Fig. 4.32(b). To calculate how much light reaches the film in a camera, we must know when the shutter opens and closes. A typical focal-plane shutter opens 20 milliseconds and closes 70 milliseconds after the button is pressed. The amount A of light emitted by the flashbulb in this interval is

$$A = \int_{20}^{70} L(t)\,dt \text{ lumen-milliseconds,}$$

where $L(t)$ is the lumen output of the bulb as a function of time. Use the trapezoidal rule and the numerical data from Tables 4.6 and 4.7 to estimate A for each of the given bulbs and find out which bulb gets more light to the film.†

19. CALCULATOR As we shall see in Chapter 12, the function

$$f(x) = \begin{cases} \dfrac{\sin x}{x}, & x \ne 0, \\ 1, & x = 0, \end{cases}$$

has derivatives of all orders at every value of x. In particu-

†From *Integration,* by W. U. Walton, et al., Project CALC, Education Development Center, Inc., Newton, MA (1975), p. 83.

lar, its graph is smooth and we can expect good results from Simpson's rule.

a) Use the fact that $|f^{(4)}(x)| \le 1$ on $[-\pi/2,\ \pi/2]$ to give an upper bound for the error that would occur if

$$\int_{-\pi/2}^{\pi/2} f(x)\ dx$$

were estimated by Simpson's rule with $n = 4$.

b) Estimate the integral in part (a) by Simpson's rule with $n = 4$.

c) Express the upper bound in (a) as a percent of the estimate in (b).

TABLE 4.6
Light output (in millions of lumens) vs. time (in milliseconds) for No. 22 flashbulb

Time after ignition	Light output	Time after ignition	Light output
0	0	30	1.7
5	0.2	35	0.7
10	0.5	40	0.35
15	2.6	45	0.2
20	4.2	50	0
25	3.0		

Data from *Photographic Lamp and Equipment Guide*, P4-15P, General Electric Company, Cleveland, Ohio.

TABLE 4.7
Light output (in millions of lumens) vs. time (in milliseconds) for No. 31 flashbulb

Time after ignition	Light output	Time after ignition	Light output
0	0	50	1.3
5	0.1	55	1.4
10	0.3	60	1.3
15	0.7	65	1.0
20	1.0	70	0.8
25	1.2	75	0.6
30	1.0	80	0.3
35	0.9	85	0.2
40	1.0	90	0
45	1.1		

Data from *Photographic Lamp and Equipment Guide*, P4-15P, General Electric Company, Cleveland, Ohio.

TOOLKIT PROGRAMS

Integration   Integral Evaluator

REVIEW QUESTIONS AND EXERCISES

1. Can a function have more than one antiderivative? If so, how are the antiderivatives related? What theorem in Chapter 3 is the key to the relationship?

2. If the acceleration of a body moving in a straight line is given as a function of time t, what more do you need to know to find the body's position $s = f(t)$? How is s found?

3. Give an example of the substitution method of integration. What do you do about the limits in definite integrals?

4. How can trigonometric identities help you to evaluate integrals of trigonometric functions? Give some examples.

5. How are definite integrals defined?

6. Write a formula for the area $A(x)$ bounded above by the semicircle $y = \sqrt{4 - t^2}$, below by the t-axis, on the left by the y-axis, and on the right by the vertical line $t = x$. Find dA/dx.

7. State the First Fundamental Theorem of Calculus. How does the Integral Evaluation Theorem follow from the First Fundamental Theorem?

8. Must every continuous function be the derivative of some function? Explain.

9. True, or false?

a) If $\int_a^b f(x)\ dx$ exists, then f is differentiable.

b) If f is differentiable, then $\int_a^b f(x)\ dx$ exists.

c) If f is continuous on $[a,\ b]$, then $\int_a^b f(x)\ dx$ exists.

d) If f is continuous on $[a,\ b]$, then $F(x) = \int_a^x f(t)\ dt$ is continuous on $[a,\ b]$.

e) If f is continuous on $[a,\ b]$ then $F(x) = \int_a^x f(t)\ dt$ is differentiable on $[a,\ b]$.

10. What numerical methods do you know for evaluating definite integrals? How accurate are they? How can the accuracy sometimes be improved?

11. Look up the Euler-Maclaurin summation formula in a book on numerical analysis or in *Handbook of Mathematical Functions* (Dover Publications). How does it relate to the trapezoidal rule?

MISCELLANEOUS PROBLEMS

Solve the differential equations in Problems 1–6.

1. $\dfrac{dy}{dx} = xy^2$

2. $\dfrac{dy}{dx} = \sqrt{1 + x + y + xy}$

3. $\dfrac{dy}{dx} = \dfrac{x^2 - 1}{y^2 + 1}$

4. $\dfrac{dy}{dx} = \dfrac{x - \sqrt{x}}{y\sqrt{y}}$

5. $\dfrac{dr}{ds} = \left(\dfrac{2 + r}{3 - s}\right)^2$

6. $\dfrac{dr}{ds} = \dfrac{r^2}{s^2} + r^2$

7. Solve each of the following differential equations subject to the prescribed initial conditions.

a) $\dfrac{dy}{dx} = x\sqrt{x^2 - 4}, \quad y = 3$ when $x = 2$

b) $\dfrac{dy}{dx} = xy^3, \quad y = 1$ when $x = 0$

c) $\dfrac{dy}{dx} = x^3 y^2, \quad y = 4$ when $x = 1$

d) $\sqrt{y + 1}\,\dfrac{dy}{dx} = \dfrac{1}{x^2}, \quad y = 3$ when $x = -3$

8. Does any function $y = f(x)$ satisfy all of the following conditions?

i) $d^2y/dx^2 = 0$ for all x

ii) $dy/dx = 1$ when $x = 0$

iii) $y = 0$ when $x = 0$

Explain.

9. Find an equation for the curve whose slope at the point (x, y) is $3x^2 + 2$, if the curve is required to pass through the point $(1, -1)$.

10. A particle moves along the x-axis with acceleration $d^2x/dt^2 = -t^2$. At $t = 0$, the particle is at the origin. In the course of its motion, it reaches the point $x = b$, where $b > 0$, but no point beyond b. Determine its velocity at $t = 0$.

11. A particle moves with acceleration $a = \sqrt{t} - (1/\sqrt{t})$. Assuming that the velocity $v = 4/3$ and the position $s = -4/15$ when $t = 0$, find

a) the velocity v in terms of t,

b) the position s in terms of t.

12. A particle's acceleration is $3 + 2t$ at time t. At $t = 0$, the velocity is 4. Find the velocity as a function of time and find the distance between the positions of the particle at $t = 0$ and $t = 4$.

13. The acceleration of a particle moving along the x-axis is $d^2x/dt^2 = -4x$. If the particle starts from rest at $x = 5$, find its velocity when it first reaches $x = 3$.

14. A shovelful of dirt is thrown up from the bottom of a 16-ft hole with an initial velocity of 48 ft/sec. How fast is the dirt moving as it emerges from the hole?

Solve the equations in Problems 15–20 subject to the given initial conditions.

15. $dy/dx = x\sqrt{1 + x^2}, \quad y = -2$ when $x = 0$

16. $dy/dx = 1/(y\sqrt{x}) + (\sec x \tan x)/y, \quad y = \sqrt{6}$ when $x = 0$

17. $\sqrt{x}\, y\, (dy/dx) = x + 1, \quad y = 2$ when $x = 1$

18. $du/dv = 2u^2(4v^3 + 4v^{-3}), \quad u > 0, \quad v > 0; \quad u = 1$ when $v = 1$

19. $dy/dx = x\sqrt{9y + x^2y}, \quad y = 36$ when $x = 0$

20. $dy/dx = 27 \csc^2 2y \sqrt{9x + 16}, \quad y = 0$ when $x = 1$

21. The acceleration of gravity near the earth's surface is 32 ft/sec^2. A stone is thrown upward from the ground with a speed of 96 ft/sec. How high will the stone rise in t seconds? How high will the stone go? (Neglect air resistance.)

22. Suppose the brakes of an automobile produce a constant deceleration of k ft/sec^2.

a) What must k be to bring an automobile traveling 60 mph (88 ft/sec) to rest in a distance of 100 ft from the point where the brakes are applied?

b) With the same k, how far would a car traveling 30 mph travel before being brought to a stop?

23. A body is moving with a velocity 16 ft/sec when it is suddenly subjected to a deceleration. The deceleration is proportional to the square root of the velocity, and the body comes to rest in 4 sec.

a) How fast is the body moving 2 sec after it begins decelerating?

b) How far does the body travel before coming to rest?

24. Let $f(x)$ and $g(x)$ be continuously differentiable functions satisfying the relationships $f'(x) = g(x)$ and $f''(x) = -f(x)$. Let $h(x) = f^2(x) + g^2(x)$. If $h(0) = 5$, find $h(10)$.

Approximating Finite Sums with Integrals

In many applications of calculus, integrals are used to approximate finite sums—the reverse of the usual procedure of using finite sums to approximate integrals. Here is an example.

EXAMPLE Estimate the sum of the square roots of the first n positive integers, $\sqrt{1} + \sqrt{2} + \cdots + \sqrt{n}$.

Solution See Fig. 4.33. The integral

$$\int_0^1 \sqrt{x}\, dx = \frac{2}{3} x^{3/2} \Big]_0^1 = \frac{2}{3}$$

is the limit of the sums

$$S_n = \sqrt{\frac{1}{n}} \cdot \frac{1}{n} + \sqrt{\frac{2}{n}} \cdot \frac{1}{n} + \cdots + \sqrt{\frac{n}{n}} \cdot \frac{1}{n}$$

$$= \frac{\sqrt{1} + \sqrt{2} + \cdots + \sqrt{n}}{n^{3/2}}.$$

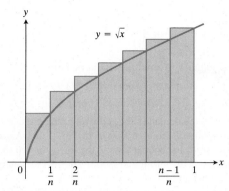

4.33 The relation of the circumscribed rectangles to the integral $\int_0^1 \sqrt{x}\,dx$ leads to an estimate of the sum $1 + \sqrt{2} + \sqrt{3} + \cdots + \sqrt{n}$. See the introduction to Problems 25–28.

Therefore, when n is large, S_n will be close to $2/3$ and we will have

$$\text{Root sum} = \sqrt{1} + \sqrt{2} + \cdots + \sqrt{n} = S_n \cdot n^{3/2} \approx \frac{2}{3} n^{3/2}.$$

The following table shows how good the approximation can be.

n	Root sum	$(2/3)n^{3/2}$	Relative error
10	22.468	21.082	$1.386/22.468 \approx 6\%$
50	239.04	235.70	1.4%
100	671.46	666.67	0.7%
1000	21,097	21,081	0.07%

25. Evaluate

$$\lim_{n \to \infty} \frac{1^5 + 2^5 + 3^5 + \cdots + n^5}{n^6}$$

by showing that the limit is

$$\int_0^1 x^5\,dx$$

and evaluating the integral.

26. (See Problem 25.) Evaluate

$$\lim_{n \to \infty} \frac{1}{n^4}[1^3 + 2^3 + 3^3 + \cdots + n^3].$$

27. Let $f(x)$ be a continuous function. Express

$$\lim_{n \to \infty} \frac{1}{n}\left[f\left(\frac{1}{n}\right) + f\left(\frac{2}{n}\right) + \cdots + f\left(\frac{n}{n}\right)\right]$$

as a definite integral.

28. Use the result of Problem 27 to evaluate

a) $\displaystyle \lim_{n \to \infty} \frac{1}{n^{16}}[1^{15} + 2^{15} + 3^{15} + \cdots + n^{15}],$

b) $\displaystyle \lim_{n \to \infty} \frac{\sqrt{1} + \sqrt{2} + \sqrt{3} + \cdots + \sqrt{n}}{n^{3/2}},$

c) $\displaystyle \lim_{n \to \infty} \frac{1}{n}\left[\sin\frac{\pi}{n} + \sin\frac{2\pi}{n} + \sin\frac{3\pi}{n} + \cdots + \sin\frac{n\pi}{n}\right].$

What can be said about the following limits?

d) $\displaystyle \lim_{n \to \infty} \frac{1}{n^{17}}[1^{15} + 2^{15} + 3^{15} + \cdots + n^{15}]$

e) $\displaystyle \lim_{n \to \infty} \frac{1}{n^{15}}[1^{15} + 2^{15} + 3^{15} + \cdots + n^{15}]$

29. a) Show that the area A_n of an n-sided regular polygon in a circle of radius r is

$$A_n = \frac{nr^2}{2}\sin\frac{2\pi}{n}.$$

b) Find the limit of A_n as $n \to \infty$. Is this answer consistent with what you know about the area of a circle?

30. Let

$$S_n = 1 + 2 + \cdots + n = \frac{n(n + 1)}{2}$$

be the sum of the first n integers.

a) Use mathematical induction to show that

$$S_n^{(2)} = S_1 + S_2 + \cdots + S_n$$
$$= \frac{n(n + 1)(n + 2)}{2 \cdot 3}.$$

b) Use mathematical induction to show that

$$S_n^{(3)} = S_1^{(2)} + S_2^{(2)} + \cdots + S_n^{(2)}$$
$$= \frac{n(n + 1)(n + 2)(n + 3)}{2 \cdot 3 \cdot 4}.$$

c) For any $k > 1$, let

$$S_n^{(k)} = S_1^{(k-1)} + \cdots + S_n^{(k-1)}.$$

Guess a formula for $S_n^{(k)}$.

Evaluate the integrals in Problems 31–42.

31. $\displaystyle \int_0^\pi \cos 4x \sin 4x \, dx$

32. $\displaystyle \int_1^4 \frac{dt}{t\sqrt{2t}}$

33. $\displaystyle \int_{-1}^0 \frac{12\,dx}{(2 - 3x)^2}$

34. $\displaystyle \int_{-1}^1 x \cos(1 - x^2)\,dx$

35. $\displaystyle \int_0^{\pi/2} \frac{\sin x \cos x}{\sqrt{1 + 3\sin^2 x}}\,dx$

36. $\displaystyle \int_1^2 \frac{x^3 + 1}{x^2}\,dx$

37. $\displaystyle \int_{-\pi/2}^{\pi/2} 15 \sin^4 3x \cos 3x \, dx$

38. $\displaystyle\int_1^4 \frac{(1 + \sqrt{u})^{1/2}}{\sqrt{u}}\, du$

39. $\displaystyle\int_0^1 \frac{dr}{\sqrt[3]{(7 - 5r)^2}}$

40. $\displaystyle\int_0^1 t^{1/3}(1 + t^{4/3})^{-7}\, dt$

41. $\displaystyle\int_0^1 \pi x^2 \sec^2(\pi x^3/3)\, dx$

42. $\displaystyle\int_0^{\pi/4} \cot^2\!\left(x + \frac{\pi}{4}\right) dx$

43. Find

$$\int_{-1}^3 f(x)\, dx \quad \text{if } f(x) = \begin{cases} 3 - x, & x \le 2, \\ x/2, & x > 2. \end{cases}$$

44. The area bounded by the x-axis, the curve $y = f(x)$, and the lines $x = 1$ and $x = b$, is equal to $\sqrt{b^2 + 1} - \sqrt{2}$ for all $b > 1$. Find $f(x)$.

45. Find

a) $\displaystyle\lim_{h \to 0} \frac{1}{h}\int_x^{x+h} \frac{du}{u + \sqrt{u^2 + 1}}$,

b) $\displaystyle\lim_{x \to 2}\left[\frac{x}{x - 2}\int_2^x f(t)\, dt\right]$.

46. Suppose x and y are related by the equation

$$x = \int_0^y \frac{1}{\sqrt{1 + 4t^2}}\, dt.$$

Show that d^2y/dx^2 is proportional to y and find the constant of proportionality.

47. Prove that

$$\int_0^x \left(\int_0^u f(t)\, dt\right) du = \int_0^x f(u)(x - u)\, du.$$

(*Hint:* Express the integral on the right-hand side as the difference of two integrals. Then show that both sides of the original equation have the same derivative with respect to x.)

48. Show that

$$y(x) = \frac{1}{a}\int_0^x f(t)\, \sin a(x - t)\, dt$$

is a solution of the differential equation

$$y'' + a^2 y = f(x), \quad y(0) = y'(0) = 0.$$

(*Hint:* $\sin(ax - at) = \sin ax \cos at - \cos ax \sin at$.)

49. To meet the demand for additional parking space, your town has allocated the area shown in Fig. 4.34. As town engineer, you have been asked by the town council to find out if the lot can be built for $11,000. The cost to clear the land is $0.10 per square foot and the lot will cost $2.00 per square foot to pave. Can the job be done for $11,000?

Horizontal spacing = 15 ft

4.34   The parking lot in Problem 49.

50. CALCULATOR   The error function

$$\text{erf}(x) = \frac{2}{\sqrt{\pi}}\int_0^x e^{-t^2}\, dt$$

must be evaluated numerically because there is no elementary expression for the antiderivative of e^{-t^2}.

a) Use Simpson's rule with $n = 10$ to estimate erf(1).

b) On [0, 1],

$$\left|\frac{d^4}{dt^4} e^{-t^2}\right| \le 12.$$

Give an upper bound for the absolute value of error of the estimate in (a).

CHAPTER 5

Applications of Definite Integrals

OVERVIEW

The importance of integral calculus stems from the fact that so many of the things we want to know are calculated with integrals. For instance, the areas enclosed by curves, the volumes of three-dimensional shapes, the lengths of curves, the areas of curved surfaces, moments of inertia, root-mean-square voltages, and the forces against dams can all be defined in natural ways as limits of finite sums. In every case, the sums are Riemann sums of a continuous function whose identity is clear from the context in which the calculation takes place. Thus, the limits of these sums exist and can be calculated as definite integrals with the Integral Evaluation Theorem. In this chapter, we look at these and other calculations made possible by the theorems of integral calculus.

In many cases, the structures and mechanical systems studied in engineering and physics behave as if their entire mass were concentrated at a single point called the *center of mass,* and it is important to know how to locate this point. It turns out that the problem of locating the center of mass is a purely mathematical one and one that can be solved by integration. In Article 5.8, we derive the integral formulas for finding the centers of mass of wires, thin rods, and thin flat plates. The presentation is entirely self-contained and assumes no previous acquaintance with either engineering or physics. Locating the centers of mass of three-dimensional solids usually requires repeated integrations and we shall have to wait until Chapter 18 to see how these particular calculations go.

5.1
The Net Change in Position and Distance Traveled by a Moving Body

To find the net change in the position $s(t)$ of a body moving along a line from time $t = a$ to time $t = b$, we integrate the body's velocity function $v(t)$ from a to b. This is the reverse of what we did in Articles 1.8 and 2.1, where we differentiated the body's position function to find its velocity.

To find the total distance traveled by a body moving back and forth along a line, we integrate the body's speed $|v(t)|$ instead of the body's velocity $v(t)$. This article shows why all this is true and how the calculations go.

Net Change in Position

If a body moves along a line and its velocity $v(t)$ is a continuous function of time, we can determine the body's position function up to a constant by integration:

$$s(t) = \int v(t)\, dt = F(t) + C, \tag{1}$$

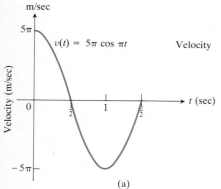

where F is any antiderivative of v. To find the net change in the body's position during any particular time interval from $t = a$ to $t = b$, we then subtract $s(a)$ from $s(b)$. When we do this formally with the expression for s in Eq. (1), we find that

$$s(b) - s(a) = (F(b) + C) - (F(a) + C)$$
$$= F(b) - F(a)$$
$$= \int_a^b v(t)\, dt.$$

Thus, the net change in the body's position is the integral of the body's velocity from a to b.

$$\text{Net change in position} = \int_a^b v(t)\, dt \tag{2}$$

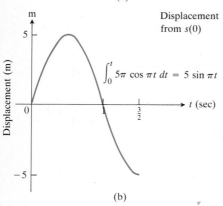

In physics, a net change in position is called a **displacement.**

EXAMPLE 1 The velocity of a body moving along a line is

$$v(t) = 5\pi \cos \pi t \text{ m/sec}.$$

Find the net change in the body's position from time $t = 0$ to time $t = 3/2$.

5.1 (a) The velocity and (b) the displacement of the body in Examples 1 and 2. The velocity is positive at first and the corresponding displacement is positive. But the particle stops at $t = 1/2$ and begins to move to the left. By time $t = 3/2$, the particle lies 5 m to the left of where it started.

Solution According to Eq. (2), the net change is

$$\int_0^{3/2} 5\pi \cos \pi t\, dt = 5 \sin \pi t \Big]_0^{3/2} = 5(-1 - 0) = -5.$$

Thus, the net effect of the motion from $t = 0$ to $t = 3/2$ is to shift the body 5 meters to the left. See Fig. 5.1.

Notice that we did not need to find $s(t)$ to make the calculation in Example 1, which is a good thing because all we can tell from the information given is that $s(t) = 5 \sin \pi t + C$ for some value of C. To determine this value, we would have to know the value of s for some value of t. This does not stop us from calculating the net change in s from $t = 0$ to $t = 3/2$, however, because the value of C drops out of the calculation:

$$s\left(\frac{3}{2}\right) - s(0) = \left(5 \sin \frac{3}{2}\pi + C\right) - (5 \sin 0 + C) = 5 \sin \frac{3}{2}\pi = -5.$$

Thus, the integral formula in Eq. (2) gives the net change in the particle's position even when we do not know exactly where the particle is.

Distance Traveled

If a body moving along the line reverses direction while it travels, the magnitude of the net change in the body's position will be less than the total distance traveled by the body. For example, if the body moves 5 meters forward and then 5 meters backward to its original position, the net change in the body's position is zero, but the total distance traveled is 10 meters. To calculate the total distance traveled with an integral, we have to keep the contributions of the forward motion and backward motion from canceling each other out. We accomplish this by integrating the absolute value of the velocity from a to b. The absolute value of the velocity is called the body's **speed.**

$$\text{Total distance traveled along a line} = \int_a^b |v(t)|\, dt \qquad (3)$$

EXAMPLE 2   The velocity of a body moving along a line is

$$v(t) = 5\pi \cos \pi t \text{ m/sec.}$$

Find the total distance traveled by the body from $t = 0$ to $t = 3/2$.

Solution   We graph v to see where it changes sign (Fig. 5.1a) and apply Eq. (3):

$$\text{Total distance traveled} = \int_0^{3/2} |5\pi \cos \pi t|\, dt$$

$$= \int_0^{1/2} 5\pi \cos \pi t\, dt + \int_{1/2}^{3/2} -5\pi \cos \pi t\, dt$$

$$= 5 \sin \pi t \Big]_0^{1/2} - 5 \sin \pi t \Big]_{1/2}^{3/2}$$

$$= 5(1 - 0) - 5(-1 - 1)$$

$$= 5 + 10$$

$$= 15.$$

During the motion, the body travels 5 meters forward and 10 meters backward for a total distance of 15 meters. The resulting change in the body's position is a 5-meter shift to the left, as we saw in Example 1.   ∎

PROBLEMS

In Problems 1–8, v is the velocity in meters per second of a body moving along a line. (a) Graph v to find where it is positive and where it is negative. Then find (b) the total distance traveled by the body during the given time interval and (c) the net change in the body's position.

1. $v = 5\pi \cos \pi t, \quad 0 \le t \le 2$

2. $v = \sin \pi t, \quad 0 \le t \le 2$

3. $v = 49 - 9.8t, \quad 0 \le t \le 10$

4. $v = 8 - 1.6t, \quad 0 \le t \le 10$

5. $v = 6(t - 1)(t - 2), \quad 0 \le t \le 2$

6. $v = 6(t - 1)(t - 2), \quad 0 \le t \le 3$

7. $v = 6 \sin 3t, \quad 0 \le t \le \pi/2$

8. $v = 4 \cos 2t, \quad 0 \le t \le \pi$

Problems 9–12 give the velocity $v = f(t)$ m/sec of a body for a time interval $a \le t \le b$ and the body's initial position $s(a)$. Find
   a) the position $s(t)$ as a function of t,
   b) the total distance traveled by the body from $t = a$ to $t = b$, and
   c) the net change in the body's position.
Calculate the net change in part (c) both as an integral and directly from the formula for $s(t)$ obtained in part (a).

9. $v = \cos t, \quad 0 \le t \le 2\pi, \quad s(0) = 0$

10. $v = -\sin t, \quad 0 \le t \le 2\pi, \quad s(0) = 1$

11. $v = 5\pi \cos \pi t, \quad 0 \le t \le 3/2, \quad s(0) = 5$

12. $v = \sin \pi t, \quad 0 \le t \le 3/2, \quad s(0) = 0$

In Problems 13–16, a is the acceleration in meters per second per second (m/sec²) of a body moving along a line from time $t = 0$ to time $t = 2$. The body's velocity at $t = 0$ is $v(0)$. Find the net change in the body's position over the time interval. (*Hint:* First find a formula for $v(t)$.)

13. $a = -4\pi^2 \cos 2\pi t, \quad v(0) = 2$

14. $a = 9.8 + \sin \pi t, \quad v(0) = 0$

15. $a = g \text{ (constant)}, \quad v(0) = 0$

16. $a = \sqrt{4t + 1}, \quad v(0) = -13/3$

TABLE 5.1
Selected Speeds of a Car on a 2-minute Trip

Time (sec)	Speed (mph)	Time (sec)	Speed (mph)
0	0	70	66
10	32	80	66
20	51	90	58
30	57	100	40
40	54	110	6
50	64	120	0
60	66		

(a)

(b)

(c)

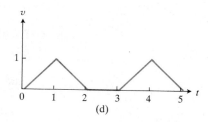
(d)

5.2 The velocity graphs for Problem 17. Time is measured in seconds and velocity in meters per second.

17. Figure 5.2 shows the velocity graphs of four bodies moving along straight lines for various intervals of time. Find the total distance traveled by each body and the net change in the body's position.

18. **CALCULATOR** Table 5.1 shows the speed of an automobile every 10 seconds from start to stop on a 2-minute trip. Use Simpson's Rule to estimate how far the car went. Compare your results with the graphs in Fig. 1.80.

TOOLKIT PROGRAMS

Derivative Grapher   Super ∗ Grapher
Integral Evaluator

5.2
Areas between Curves

In this article, we show how to calculate the areas of regions that lie between curves in the plane.

Suppose that $y = f_1(x)$ and $y = f_2(x)$ are continuous and that $f_1(x) \geq f_2(x)$ for $a \leq x \leq b$. Then the graph of f_1 lies above the graph of f_2 from a to b, as in Fig. 5.3. We define the area of the region enclosed by the two graphs and the vertical lines $x = a$ and $x = b$ in the following way.

We first subdivide the interval $a \leq x \leq b$ into n subintervals of length $\Delta x = (b - a)/n$. To do this, we mark off points $a = x_0, x_1, x_2, \ldots, x_{n-1}$, $x_n = b$ the way we did when we defined Riemann sums in Chapter 4. We then approximate the region between the curves with vertical rectangles that stretch from curve to curve, one for each subinterval, like the one shown in Fig. 5.3. The area of a typical rectangle is

$$(f_1(x_k) - f_2(x_k)) \, \Delta x.$$

The sum of the areas of the n rectangles is

$$S_n = \sum_{k=1}^{n} (f_1(x_k) - f_2(x_k)) \, \Delta x.$$

As $n \to \infty$, the sums S_n tend to the limit

$$\lim_{n \to \infty} S_n = \int_a^b (f_1(x) - f_2(x)) \, dx$$

by virtue of the Integral Existence Theorem in Article 4.5. This limit is the number we define to be the area between the curves from a to b.

DEFINITION

The Area between Two Curves

If $f_1(x) \geq f_2(x)$ throughout the interval $a \leq x \leq b$, the area between the graphs of f_1 and f_2 from a to b is the integral of $(f_1(x) - f_2(x))$ from a to b.

$$\text{Area} = \int_a^b (f_1(x) - f_2(x)) \, dx \tag{1}$$

5.3 The area between two curves can be approximated by adding the areas of rectangular strips that reach from one curve to the other. The area of the strip shown here is $(f_1(x_k) - f_2(x_k)) \, \Delta x$.

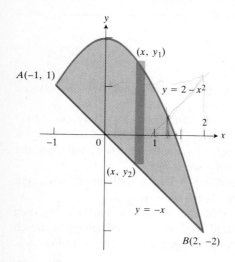

5.4 We can calculate the area between $y = 2 - x^2$ and $y = -x$ by integrating a vertical strip over the interval from $x = -1$ to $x = 2$.

One way to remember the formula in Eq. (1) is to think of adding, by an integral, vertical strips of "area" $(f_1(x) - f_2(x))\,dx$ along the x-axis from $x = a$ to $x = b$. The differential dx plays a dual role in the notation here, representing both the width of the rectangular strip and the variable of integration in the integral.

EXAMPLE 1 Find the area of the region bounded above by the parabola $y = 2 - x^2$ and below by the line $y = -x$.

Solution To find the area, we need to know where the region begins and ends. We therefore sketch the parabola and line together (Fig. 5.4). The limits of integration for the area integral in Eq. (1) will be the x-coordinates of the points where the parabola and line intersect. We find these by solving the equations $y = 2 - x^2$ and $y = -x$ simultaneously for x. This gives

$$2 - x^2 = -x$$
$$x^2 - x - 2 = 0$$
$$(x - 2)(x + 1) = 0$$
$$x = -1, \qquad x = 2.$$

The limits of integration are -1 and 2.

The parabola $y = 2 - x^2$ lies above the line $y = -x$ for all values of x between -1 and 2. This tells us that in Eq. (1) we should take $f_1(x) = 2 - x^2$ and $f_2(x) = -x$:

$$\text{Area} = \int_{-1}^{2} ((2 - x^2) - (-x))\,dx = \int_{-1}^{2} (2 - x^2 + x)\,dx$$

$$= \left[2x - \frac{x^3}{3} + \frac{x^2}{2} \right]_{-1}^{2} = \left[4 - \frac{8}{3} + \frac{4}{2} \right] - \left[-2 + \frac{1}{3} + \frac{1}{2} \right]$$

$$= 6 - \frac{9}{3} + \frac{3}{2} = 3 + \frac{3}{2} = \frac{9}{2}. \qquad \blacksquare$$

Instead of finding an area by integrating vertical strips over an interval on the x-axis, it is sometimes easier to find the area by integrating horizontal strips over an interval on the y-axis, as shown in the first part of the next example.

EXAMPLE 2 Find the area of the region bounded on the right by the line $y = x - 2$, on the left by the parabola $x = y^2$, and below by the x-axis.

Solution METHOD 1: *Integration with respect to y.* We sketch the region (Fig. 5.5). The y-coordinates of the points of intersection of the parabola and line can be found by solving the equations $y = x - 2$ and $x = y^2$ simultaneously for y. We substitute $x = y^2$ into $y = x - 2$ to find

$$y = y^2 - 2,$$
$$y^2 - y - 2 = 0,$$
$$(y - 2)(y + 1) = 0,$$
$$y = 2, \qquad y = -1.$$

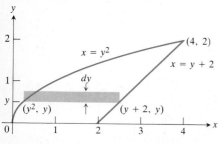

5.5 The area of the horizontal strip is length × width = $(y + 2 - y^2)\,dy$.

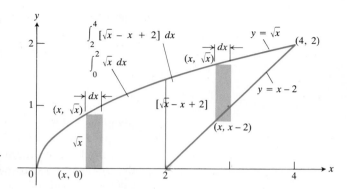

5.6 The area of this region can be expressed as the sum of the two integrals shown.

Of the two possibilities, only $y = 2$ gives a point of intersection in the first quadrant where the region lies, namely the point $(4, 2)$.

We now imagine a thin horizontal strip reaching across the region from the parabola on the left to the line on the right. It runs from the point (y^2, y) to the point $(y + 2, y)$. The length of the strip is $y + 2 - y^2$ and its width is dy. The strip's area is therefore $(y + 2 - y^2)\, dy$. We find the area between the curves by integrating $(y + 2 - y^2)\, dy$ from $y = 0$ to $y = 2$:

$$\text{Area} = \int_0^2 (y + 2 - y^2)\, dy = \left[\frac{y^2}{2} + 2y - \frac{y^3}{3}\right]_0^2 = 2 + 4 - \frac{8}{3} = \frac{10}{3}.$$

METHOD 2: *Integration with respect to x* (Fig. 5.6). Integrating with respect to x is not as easy in this case as integrating with respect to y because we end up with two integrals instead of one. When we move a vertical strip across the region from $x = 0$ to $x = 4$, the formula for the length of the strip changes at $x = 2$. To the left of $x = 2$, the length is $\sqrt{x} - 0 = \sqrt{x}$. To the right of $x = 2$, the length is $\sqrt{x} - (x - 2) = \sqrt{x} - x + 2$. To find the area on the left, we integrate $\sqrt{x}$ from $x = 0$ to $x = 2$. To find the area on the right, we integrate $(\sqrt{x} - x + 2)$ from $x = 2$ to $x = 4$. We then add the results to find the total area of the region:

$$\text{Area} = \int_0^2 \sqrt{x}\, dx + \int_2^4 (\sqrt{x} - x + 2)\, dx$$

$$= \left[\frac{2}{3}x^{3/2}\right]_0^2 + \left[\frac{2}{3}x^{3/2} - \frac{x^2}{2} + 2x\right]_2^4$$

$$= \frac{2}{3}(2)^{3/2} + \left(\frac{2}{3}(4)^{3/2} - \frac{16}{2} + 8\right) - \left(\frac{2}{3}(2)^{3/2} - \frac{4}{2} + 4\right)$$

$$= \frac{2}{3}(8) - 2$$

$$= \frac{10}{3}.$$

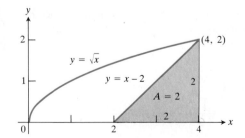

5.7 The area of the unshaded region is also the area under the curve $y = \sqrt{x}$, $0 \le x \le 4$, minus the area of the triangle.

METHOD 3: *Subtracting areas* (Fig. 5.7). (This is the fastest method because of the geometry of this particular problem.) The area we wish to calculate is the area between the x-axis and the curve $y = \sqrt{x}$, $0 \le x \le 4$, *minus* the area of a triangle with base 2 and height 2:

$$\text{Area} = \int_0^4 \sqrt{x}\, dx - \frac{1}{2}(2)(2) = \frac{2}{3}x^{3/2}\Big]_0^4 - 2$$

$$= \frac{2}{3}(8) - 2 = \frac{10}{3}.$$ ∎

Example 2 shows that it is sometimes easier to find the area of a region by integrating with respect to y instead of x. You should therefore examine each region beforehand to determine which method, if either, will be easier. Sketching the region may also reveal how to use geometry to simplify the calculation, as it did in method 3.

PROBLEMS

In Problems 1–24, find the areas of the regions enclosed by the given curves and lines.

1. The curve $y = x^2 - 2$ and the line $y = 2$

2. The x-axis and the curve $y = 2x - x^2$

3. The y-axis and the curve $x = y^2 - y^3$

4. The curve $y^2 = x$ and the line $x = 4$

5. The curve $y = 2x - x^2$ and the line $y = -3$

6. The curve $y = x^2$ and the line $y = x$

7. The curve $x = 3y - y^2$ and the line $x + y = 3$

8. The curves $y = x^4 - 2x^2$ and $y = 2x^2$

9. The curve $x = y^2$ and the line $x = y + 2$

10. The curve $y = x^4$ and the line $y = 8x$

11. The curves $x = y^3,\ x = y^2$

12. The curve $y^3 = x$ and the line $y = x$

13. The curve $y = x^2 - 2x$ and the line $y = x$

14. The curve $x = 10 - y^2$ and the line $x = 1$

15. The curves $x = y^2,\quad x = -2y^2 + 3$

16. The curves $y = x^2,\quad y = -x^2 + 4x$

17. The line $y = x$ and the curve $y = 2 - (x - 2)^2$

18. The curves $y = 7 - 2x^2$ and $y = x^2 + 4$

19. The curves $y = 4 - 4x^2$ and $y = x^4 - 1$

20. The curves $y = x^2/4$ and $y = x^4/16$

21. The curve $y = \cos x$ and the line $y = -1$, for $-\pi \le x \le \pi$

22. The curves $y = \cos(\pi x/2),\quad y = 1 - x^2$

23. The curve $y = \sin(\pi x/2)$ and the line $y = x$

24. The curves $y = \sec^2 x,\ y = \tan^2 x$ and the lines $x = -\pi/4$, $x = \pi/4$

25. Find the area of the "triangular" region in the first quadrant bounded by the y-axis and the curves $y = \sin x,\ y = \cos x$.

26. Find the area of the region in the first quadrant bounded by $y = \sqrt{4 - x},\ x = 0,\ y = 0$.

27. Find the area of the region bounded on the right by $x + y = 2$, on the left by $y = x^2$, and below by the x-axis.

28. Find the area of the region bounded on the right by $y = 6 - x$, on the left by $y = \sqrt{x}$, and below by $y = 1$.

29. Find the area of the region between $y = 3 - x^2$ and $y = -1$ (a) by integration with respect to x and (b) by integration with respect to y.

30. Find the area of the region enclosed by the curve $y = \sin|x|$ and the x-axis for $-\pi \le x \le \pi$.

31. Find the area of the region enclosed by $y = \cos x$ and $y = \sin x$ for $\pi/4 \le x \le 5\pi/4$.

32. The area of the region between the curve $y = x^2$ and the line $y = 4$ is divided into two equal portions by the line $y = c$.
    a) Find c by integrating with respect to y. (This puts c into the limits of integration.)
    b) Find c by integrating with respect to x. (This puts c into the integrand.)

33. Figure 5.8 shows triangle AOC inscribed in the shaded region cut from the parabola $y = x^2$ by the line $y = a^2$. Find

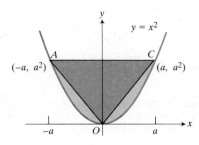

5.8 The figure for Problem 33.

the limit of the ratio of the area of the triangle to the area of the parabolic region as a approaches zero.

34. Show that the largest triangle that can be inscribed in the parabolic region shown in Fig. 5.9, with base on the x-axis and vertex on the parabola, has an area equal to three-fourths of the area of the region.

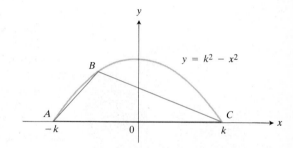

5.9 The inscribed triangle of Problem 34. What position of B makes the area of $\triangle ABC$ largest? What fraction of the parabolic region does the triangle occupy when B is in this position?

TOOLKIT PROGRAMS

Integral Evaluator   Super * Grapher

5.3

Calculating Volumes by Slicing. Volumes of Revolution

Now that we can calculate the areas of a wide variety of plane regions, we can extend our method of forming Reimann sums to find the volumes of solids that have these regions as cross sections. In this article, we show how to define and calculate the volumes of such solids. We also show how to determine the volume of a related type of solid, called a solid of revolution.

The Method of Slicing

Our first step is to define the volume of any cylinder of base area A and height h to be Ah. This extends the solid geometry formula for circular cylinders

$$\text{Volume} = \text{base} \times \text{height}$$

to cylinders with arbitrary bases like the one in Fig. 5.10.

Now suppose we want to find the volume of a solid like the one in Fig. 5.11. The solid is bounded by planes perpendicular to the x-axis at a and b and

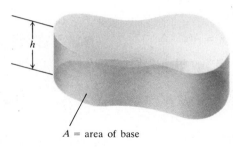

$A = \text{area of base}$

5.10 The volume of a cylinder like this is defined in the usual way as Ah.

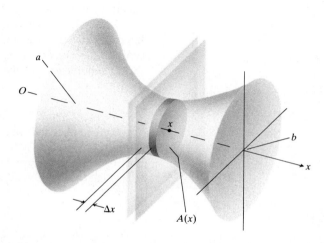

5.11 The volume of the solid is the sum of the volumes of slices of thickness Δx.

Volume $A_{max}\ \Delta x$

ΔV

Volume $A_{min}\ \Delta x$

Δx

5.12 The trumpet-shaped solid slice lies between two cylinders whose volumes we can calculate.

varies in shape between these two ends. Common sense tells us that the solid must have a volume, but how are we to find it or even define it?

We imagine the solid to be cut into thin slices of thickness Δx by planes perpendicular to the x-axis. The volume V of the solid is then the sum of the volumes of the slices. But how are we to find the volume ΔV of a typical slice?

However we define ΔV, we want it to be at least as big as $A_{min}\ \Delta x$, which is the volume of a cylinder based on the smallest cross-section area in the slice. In Fig. 5.12, this is the small cylinder lying inside the slice. Likewise, we want ΔV to be no larger than $A_{max}\ \Delta x$, which is the volume of a cylinder based on the largest cross-section area in the slice. In Fig. 5.12, this is the large cylinder surrounding the slice. In symbols, we want

$$A_{min}\ \Delta x \le \Delta V \le A_{max}\ \Delta x. \tag{1}$$

Smallest cylinder volume   Largest cylinder volume

If the cross-section area of the solid perpendicular to the x-axis is a continuous function $A(x)$, it takes on its minimum value for the interval $[x, x + \Delta x]$ at some point c in the interval. It takes on its maximum value at some point c'. That is,

$$A_{min}\ \Delta x = A(c)\ \Delta x, \qquad A_{max}\ \Delta x = A(c')\ \Delta x. \tag{2}$$

With these substitutions, (1) becomes

$$A(c)\ \Delta x \le \Delta V \le A(c')\ \Delta x. \tag{3}$$

Adding the volumes of all the slices from a to b gives

$$\sum_a^b A(c)\ \Delta x \le \sum_a^b \Delta V \le \sum_a^b A(c')\ \Delta x, \tag{4}$$

or, since $\Sigma_a^b\ \Delta V = V$,

$$\sum_a^b A(c)\ \Delta x \le V \le \sum_a^b A(c')\ \Delta x. \tag{5}$$

Thus, however we define it, we want V to be a number that satisfies the inequalities in (5) for any subdivision of the interval $a \le x \le b$.

This requirement tells us exactly what V has to be. We know from the Integral Existence Theorem that the sums on the left and right sides of (5) approach $\int_a^b A(x)\ dx$ as Δx approaches zero. Therefore, if we want V to satisfy (5) for all subdivisions of the interval $a \le x \le b$, we must define it to be

$$V = \int_a^b A(x)\ dx. \tag{6}$$

No other number will do.

DEFINITION

> The **volume** of a solid of known cross-section area $A(x)$ from $x = a$ to $x = b$ is
>
> $$\text{Volume} = \int_a^b A(x)\ dx. \tag{6}$$

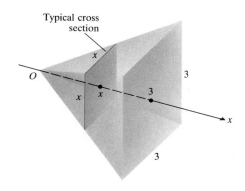

Typical cross section

O

5.13 The cross sections of the pyramid in Example 1 are squares.

To use this definition we do not need $A(x)$ to be continuous as long as the integral exists.

The practical steps involved in using Eq. (6) to calculate volumes of solids are as follows:

STEP 1: Draw a figure and sketch a typical cross section.

STEP 2: Find $A(x)$.

STEP 3: Find the limits of integration.

STEP 4: Integrate.

EXAMPLE 1   A pyramid 3 meters high has a square base that is 3 meters on a side. The cross section of the pyramid perpendicular to the altitude x units down from the vertex is a square x units on a side. Find the volume of the pyramid.

Solution   We draw the pyramid with its altitude along the x-axis and its vertex at the origin. We then sketch a typical cross section (Fig. 5.13). Since the cross section is a square x meters on a side, its area is $A(x) = x^2$. The volume of the pyramid is the integral of $A(x)$ from $x = 0$ to $x = 3$:

$$\text{Volume} = \int_0^3 A(x)\, dx = \int_0^3 x^2\, dx = \frac{x^3}{3} \Big]_0^3 = 9.$$

The volume is 9 m³. The result agrees with the value

$$V = \frac{1}{3}(\text{base area})(\text{height}) = \frac{1}{3}(9)(3) = 9,$$

given by the formula from solid geometry. ∎

EXAMPLE 2   A curved wedge is cut from a right circular cylinder of radius r by two planes. One plane is perpendicular to the axis of the cylinder. The second plane makes an acute angle α with the first and intersects it at the center of the cylinder. Find the volume of the wedge.

Solution 1   *Cross sections perpendicular to the y-axis.* We draw the wedge and sketch a typical cross section perpendicular to the y-axis (Fig. 5.14). The cross section is a triangle of area

$$A = \frac{1}{2}xh.$$

To express A in terms of y, we observe that

$$h = x \tan \alpha \qquad \text{(by trigonometry)}$$

and

$$x^2 = r^2 - y^2. \qquad \text{(by the Pythagorean Theorem)}$$

Hence,

$$A(y) = \frac{1}{2}xh = \frac{1}{2}x(x \tan \alpha) = \frac{1}{2}(r^2 - y^2)\tan \alpha.$$

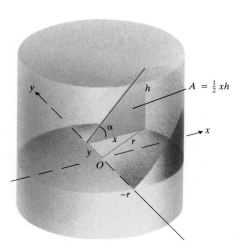

5.14 The curved wedge can be cut into triangular slices.

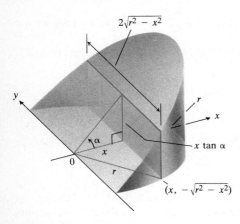

5.15 When the wedge of Example 2 is sliced perpendicular to the x-axis, the cross sections are rectangles.

The volume is the integral of $A(y)$ from $-r$ to r:

$$V = \int_{-r}^{r} A(y) \, dy = \frac{1}{2} \tan \alpha \int_{-r}^{r} (r^2 - y^2) \, dy$$

$$= \frac{1}{2} \tan \alpha \left[r^2 y - \frac{y^3}{3} \right]_{y=-r}^{y=r}$$

$$= \frac{1}{2} \tan \alpha \left[\left(r^3 - \frac{r^3}{3} \right) - \left(-r^3 + \frac{r^3}{3} \right) \right]$$

$$= \frac{1}{2} \tan \alpha \left[\frac{2}{3} r^3 + \frac{2}{3} r^3 \right] = \frac{2}{3} r^3 \tan \alpha.$$

Solution 2 *Cross sections perpendicular to the x-axis.* In Solution 1, the use of cross sections perpendicular to the y-axis was arbitrary. We get equally good results with cross sections perpendicular to the x-axis (Fig. 5.15). These cross sections are rectangles, a typical area being

$$A(x) = (\text{height})(\text{width}) = (x \tan \alpha)(2\sqrt{r^2 - x^2}).$$

The volume of the wedge is the integral of $A(x)$ from $x = 0$ to $x = r$:

$$\text{Volume} = \int_{0}^{r} A(x) \, dx = 2 \tan \alpha \int_{0}^{r} x\sqrt{r^2 - x^2} \, dx. \qquad (7)$$

To evaluate the integral, we substitute

$$u^2 = r^2 - x^2, \qquad -u \, du = x \, dx.$$

Then

$$\text{Volume} = 2 \tan \alpha \int_{r}^{0} -u\sqrt{u^2} \, du$$

$$= 2 \tan \alpha \int_{0}^{r} u^2 \, du \qquad (\sqrt{u^2} = u \text{ because } u \geq 0)$$

$$= 2 \tan \alpha \left. \frac{u^3}{3} \right]_{0}^{r}$$

$$= \frac{2}{3} r^3 \tan \alpha. \qquad \blacksquare$$

The Volume of a Solid of Revolution

A solid of revolution is a solid made by revolving a region in the plane about an axis in the plane. If the region is the region between the graph of $y = f(x)$, $a \leq x \leq b$, and the x-axis, and the axis of revolution happens to be the x-axis, then the cross sections of the solid perpendicular to the x-axis are disks (Fig. 5.16). The area $A(x)$ of a typical disk is π times its radius squared. Since the radius at x is $f(x)$,

$$A(x) = \pi(\text{radius})^2 = \pi(f(x))^2.$$

To calculate the volume of the solid, we substitute this expression for $A(x)$ in Eq. (6). The result of the substitution is the following formula.

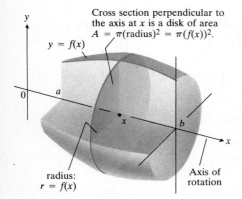

5.16 The solid generated by revolving the region bounded by $y = f(x)$, the x-axis, and the lines $x = a$ and $x = b$ about the x-axis.

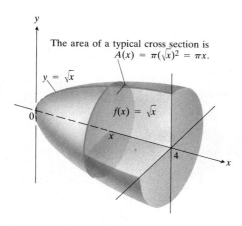

The area of a typical cross section is
$$A(x) = \pi(\sqrt{x})^2 = \pi x.$$

$y = \sqrt{x}$

$f(x) = \sqrt{x}$

5.17 A slice perpendicular to the axis of the solid of revolution in Example 3.

> **Volume of the Solid of Revolution (Rotation about the x-axis)**
>
> $$\text{Volume} = \int_a^b \pi(\text{radius})^2 \, dx = \int_a^b \pi(f(x))^2 \, dx \qquad (8)$$

EXAMPLE 3 The curve $y = \sqrt{x}$, $0 \le x \le 4$, is revolved about the x-axis to generate the shape in Fig. 5.17. Find its volume.

Solution The radius of the cross-section disk at x is $\sqrt{x}$. The volume of the solid is therefore

$$\text{Volume} = \int_a^b \pi(\text{radius})^2 \, dx = \int_0^4 \pi(\sqrt{x})^2 \, dx$$

$$= \pi \int_0^4 x \, dx$$

$$= \pi \frac{x^2}{2}\Big]_0^4 = 8\pi.$$

EXAMPLE 4 The semicircle $y = \sqrt{a^2 - x^2}$ is revolved about the x-axis to generate a sphere. Find the volume of the sphere.

Solution We sketch the sphere and a typical cross-section disk (Fig. 5.18). The radius of the disk is $y = \sqrt{a^2 - x^2}$. The volume of the sphere is therefore

$$\text{Volume} = \int_{-a}^a \pi(\text{radius})^2 \, dx$$

$$= \int_{-a}^a \pi(\sqrt{a^2 - x^2})^2 \, dx$$

$$= \pi \int_{-a}^a (a^2 - x^2) \, dx = \pi\left[a^2 x - \frac{x^3}{3}\right]_{-a}^a = \frac{4}{3}\pi a^3.$$

As in Example 1, this result agrees with the formula from solid geometry.

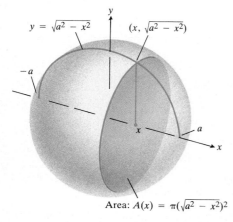

$y = \sqrt{a^2 - x^2}$ $(x, \sqrt{a^2 - x^2})$

$-a$

a

Area: $A(x) = \pi(\sqrt{a^2 - x^2})^2$

5.18 The sphere generated by rotating the semicircle $y = \sqrt{a^2 - x^2}$ about the x-axis.

The axis of revolution in the next example is not the x-axis, but the rule for calculating the volume is the same: integrate $\pi(\text{radius})^2$ between appropriate limits.

EXAMPLE 5 Find the volume generated by revolving the region bounded by $y = \sqrt{x}$ and the lines $y = 1$ and $x = 4$ about the line $y = 1$.

Solution We draw a figure and sketch a typical cross section (Fig. 5.19). The area of the cross section at x is

$$A(x) = \pi(\text{radius})^2 = \pi(\sqrt{x} - 1)^2.$$

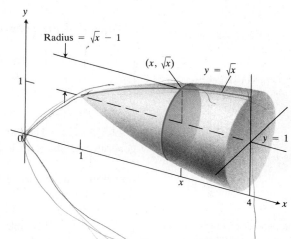

5.19 The solid swept out by revolving the region bounded by $y = \sqrt{x}$, $x = 4$, and $y = 1$ about the line $y = 1$.

The volume is the integral of A from $x = 1$ to $x = 4$:

$$\text{Volume} = \int_1^4 \pi(\text{radius})^2 \, dx = \int_1^4 \pi(\sqrt{x} - 1)^2 \, dx$$

$$= \pi \int_1^4 (x - 2\sqrt{x} + 1) \, dx$$

$$= \pi \left[\frac{x^2}{2} - \frac{4}{3}x^{3/2} + x \right]_1^4$$

$$= \frac{7\pi}{6}. \qquad \blacksquare$$

PROBLEMS

Volumes of Revolution

In Problems 1–10, find the volumes generated by revolving the regions bounded by the given lines and curves about the *x*-axis.

1. $x + y = 2$, $\quad x = 0$, $\quad y = 0$

2. $y = \sin x$, $\quad y = 0$, $\quad 0 \le x \le \pi$

3. $y = x - x^2$, $\quad y = 0$

4. $y = -3x - x^2$, $\quad y = 0$

5. $y = x^2 - 2x$, $\quad y = 0$

6. $y = x^3$, $\quad x = 2$, $\quad y = 0$

7. $y = x^4$, $\quad x = 1$, $\quad y = 0$

8. $y = \sqrt{\cos x}$, $\quad 0 \le x \le \pi/2$, $\quad x = 0$, $\quad y = 0$

9. $y = \sec x$, $\quad x = -\pi/4$, $\quad x = \pi/4$, $\quad y = 0$

10. $y = x^3 + 1$, $\quad x = 2$, $\quad y = 0$

In Problems 11–16, find the volumes generated by revolving the regions bounded by the given lines and curves about the *y*-axis.

11. $y = x/2$, $\quad x = 0$, $\quad y = 2$

12. $x = \sqrt{4 - y}$, $\quad x = 0$, $\quad y = 0$

13. $x = 1 - y^2$, $\quad x = 0$

14. $x = y^{3/2}$, $\quad x = 0$, $\quad y = 3$

15. $xy = 1$, $\quad x = 0$, $\quad y = 1$, $\quad y = 2$

16. $x = 2/(y + 1)$, $\quad x = 0$, $\quad y = 0$, $\quad y = 1$

17. Find the volume enclosed by revolving the arch $y = 2 \sin 2x$, $0 \le x \le \pi/2$, about the *x*-axis.

18. Find the volume generated by revolving the region in the first quadrant bounded by $y = \tan x$ and the line $x = \pi/3$ about the *x*-axis.

19. Find the volume generated by revolving the region bounded by $y = \sqrt{x}$ and the lines $y = 2$ and $x = 0$
a) about the *x*-axis,
b) about the line $y = 2$.

20. Find the volume generated by revolving the region bounded above by the parabola $y = 3 - x^2$ and below by the line $y = -1$ about the line $y = -1$.

21. Find the volume generated by revolving the region bounded by the curve $y = x^2$, the line $y = 1$, and the y-axis about the line $y = 1$.

22. Find the volume generated by revolving the region bounded by $y = x^{3/2}$, the x-axis, and the line $x = 1$ about the line $x = 1$.

23. Find the volume generated by revolving the region bounded by $y = \cos x$, $-\pi \le x \le \pi$, and the line $y = -1$ about the line $y = -1$.

24. Find the volume generated by revolving the region bounded by $y = x^{2/3}$, the x-axis, and the line $x = 1$ about the line $x = 1$.

25. Find the volume generated by revolving the region bounded below by the parabola $y = 3x^2 + 1$ and above by the line $y = 4$ about the line $y = 4$.

26. Find the volume generated by revolving the region bounded by $y = \sin x$ and the lines $x = 0$, $x = \pi$, and $y = 2$ about the line $y = 2$.

27. The arch $y = \sin x$, $0 \le x \le \pi$, is revolved about the line $y = c$ to generate the solid shown in Fig. 5.20. Find the value of c that minimizes the volume.

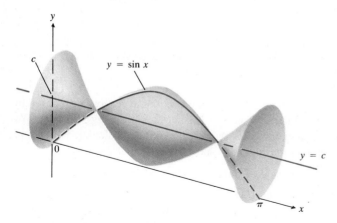

5.20 Problem 27 asks for the value of c that minimizes this volume.

28. By integration, find the volume generated by revolving the triangle with vertices at $(0, 0)$, $(h, 0)$, (h, r)
a) about the x-axis,
b) about the y-axis.

29. a) A hemispherical bowl of radius a contains water to a depth h. Find the volume of water in the bowl.
b) *(Related rates)* Water runs into a hemispherical bowl of radius 5 ft at the rate of 0.2 ft³/sec. How fast is the water level in the bowl rising when the water is 4 ft deep?

30. A football has a volume that is approximately the same as the volume generated by revolving the region inside the ellipse $b^2x^2 + a^2y^2 = a^2b^2$ (where a and b are constants) about the x-axis. Find the volume so generated.

Other Volumes

31. The vertex of a pyramid lies at the origin, and the base is perpendicular to the x-axis at $x = 4$. The cross sections of the pyramid perpendicular to the x-axis are squares whose diagonals run from the curve $y = -5x^2$ to the curve $y = 5x^2$. Find the volume of the pyramid.

32. The cross sections of a solid cut by planes perpendicular to the x-axis are circles with diameters extending from the curve $y = x^2$ to the curve $y = 8 - x^2$. The solid lies between the points of intersection of these two curves. Find its volume.

33. The base of a solid is the circle $x^2 + y^2 = a^2$. Each section of the solid cut by a plane perpendicular to the x-axis is a square with one edge in the base of the solid. Find the volume of the solid.

34. Two great circles, lying in planes that are perpendicular to each other, are marked on a sphere of radius a. A portion of the sphere is then shaved off in such a manner that any plane section of the remaining solid, perpendicular to the common diameter of the two great circles, is a square with vertices on these circles. Find the volume of the solid that remains.

35. The base of a solid is the circle $x^2 + y^2 = a^2$. Each section of the solid cut by a plane perpendicular to the y-axis is an isosceles right triangle with one leg in the base of the solid. Find the volume.

36. The base of a solid is the region between the x-axis and the curve $y = \sin x$ between $x = 0$ and $x = \pi/2$. Each plane section of the solid perpendicular to the x-axis is an equilateral triangle with one side in the base of the solid. Find the volume.

37. CALCULATOR A rectangular swimming pool is 30 ft wide and 50 ft long. The depth of water h ft at distance x ft from one end of the pool is measured at 5-ft intervals and is found to be as follows:

x (ft)	h (ft)	x (ft)	h (ft)
0	6.0	30	11.5
5	8.2	35	11.9
10	9.1	40	12.3
15	9.9	45	12.7
20	10.5	50	13.0
25	11.0		

Use the trapezoidal rule to estimate the volume of water in the pool.

5.4
Volumes Modeled with Washers and Cylindrical Shells

(a)

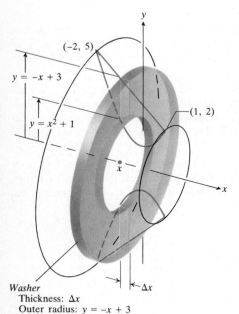

Washer
Thickness: Δx
Outer radius: $y = -x + 3$
Inner radius: $y = x^2 + 1$

(b)

5.21 When the region bounded by the line $y = -x + 3$ and the parabola $y = x^2 + 1$ in (a) is revolved about the x-axis to generate the solid in (b), the shaded strip sweeps out a washer. The outer radius of the washer is $y = -x + 3$. The inner radius is $y = x^2 + 1$.

In the preceding article, we found the volumes of solids of revolution that could be divided into disks perpendicular to the axis of revolution. However, not all solids of revolution can be divided this way, and when they cannot we turn to the two methods described in the present article. The methods start with covering the region being revolved with thin rectangular strips and adding the volumes of the figures swept out by the strips during the revolution. If the strips are perpendicular to the axis of revolution, the figures they sweep out are washers. If the strips lie parallel to the axis of revolution, the figures they sweep out are cylindrical shells. In either case, the sums of the volumes swept out by the strips are Riemann sums for an integral whose value gives the volume of the solid of revolution. We now derive the formulas for these integrals and show how they are applied.

Washers

Let us begin with an example.

EXAMPLE 1 The region bounded by the curve $y = x^2 + 1$ and the line $y = -x + 3$ is revolved about the x-axis to generate a solid (Fig. 5.21). Find the volume of the solid.

Solution When the region is revolved about the axis, the vertical strip of width Δx shown in Fig. 5.21(a) sweeps out the washer shown in Fig. 5.21(b). We think of the washer as a thickened disk with a hole in it and calculate its volume as its face area $A(x)$ times its thickness Δx:

Area of disk: $\quad \pi(\text{outer radius})^2 = \pi(-x + 3)^2$

Area of hole: $\quad \pi(\text{inner radius})^2 = \pi(x^2 + 1)^2$

Area of washer: $\quad A(x) = \pi(-x + 3)^2 - \pi(x^2 + 1)^2$
$$= \pi(8 - 6x - x^2 - x^4).$$

Volume of washer = face area × thickness = $A(x)\,\Delta x$.

We approximate the volume swept out by the region by adding the volumes of all the washers from $x = -2$ to $x = 1$. We then take the limit of the washer sums as Δx approaches zero to find the volume of the solid:

$$\text{Volume} = \lim \sum_{-2}^{1} \pi(8 - 6x - x^2 - x^4)\,\Delta x$$

$$= \int_{-2}^{1} \pi(8 - 6x - x^2 - x^4)\,dx$$

$$= \pi\left[8x - 3x^2 - \frac{x^3}{3} - \frac{x^5}{5}\right]_{-2}^{1} = \frac{117\pi}{5}. \qquad \blacksquare$$

We can get a general formula for the washer method by thinking of the typical washer as a difference of two disks of thickness dx, a large disk of radius R minus a small disk of radius r. The washer's volume is then

$$\text{Washer volume} = \pi R^2\, dx - \pi r^2\, dx = \pi(R^2 - r^2)\, dx.$$

The volume of the solid the washers make up is the integral of $\pi(R^2 - r^2)\, dx$ between appropriate limits:

Formula for Calculating Volumes by Washers

$$\text{Volume} = \int_a^b \pi(R^2(x) - r^2(x))\, dx, \tag{1}$$

$$R = \text{outer radius}, \qquad r = \text{inner radius}.$$

Notice that the function being integrated here is $\pi(R^2 - r^2)$, not $\pi(R - r)^2$. In Example 1, we had $R(x) = (3 - x)$, $r(x) = (x^2 + 1)$, and

$$\text{Volume} = \int_{-2}^{1} \pi((3 - x)^2 - (x^2 + 1)^2)\, dx.$$

To calculate the volume of a solid generated by revolving a region about the y-axis, we may use Eq. (1) with y in place of x.

EXAMPLE 2 The region bounded by the parabola $y = x^2$ and the line $y = 2x$ in the first quadrant is revolved about the y-axis. Find the volume swept out.

Solution We draw a picture to identify the dimensions of a typical washer and the limits of integration (Fig. 5.22). We find:

Outer radius: $R(y) = \sqrt{y}$

Inner radius: $r(y) = \dfrac{y}{2}$

Volume of solid: $\displaystyle\int_0^4 \pi(R^2 - r^2)\, dy = \int_0^4 \pi\left((\sqrt{y})^2 - \left(\frac{y}{2}\right)^2\right) dy$

$$= \pi \int_0^4 \left(y - \frac{y^2}{4}\right) dy$$

$$= \pi \left[\frac{y^2}{2} - \frac{y^3}{12}\right]_0^4 = \frac{8}{3}\pi. \qquad \blacksquare$$

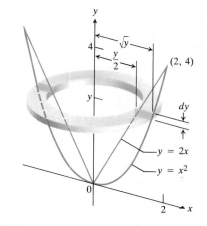

5.22 The washer volume is
$\pi(R^2 - r^2)\, dy = \pi((\sqrt{y})^2 - (y/2)^2)\, dy$.

When the axis of a solid of revolution is not one of the coordinate axes, the rule for calculating the volume is still the same: find formulas for the outer radius R and the inner radius r and integrate $\pi(R^2 - r^2)$ between appropriate limits.

EXAMPLE 3 The region between the parabola $y = x^2$ and the line $y = 2x$ is revolved about the line $x = 2$. Find the volume swept out.

5.23 The inner and outer radii of the washer are measured (as always) from the axis of rotation.

Solution We draw a picture to identify the radii of a typical washer and the limits of integration (Fig. 5.23). We find:

Outer radius: $R(y) = \left(2 - \dfrac{y}{2}\right)$

Inner radius: $r(y) = 2 - \sqrt{y}$

Volume of solid: $\displaystyle\int_0^4 \pi(R^2 - r^2)\, dy$

$$= \int_0^4 \pi\left(\left(2 - \frac{y}{2}\right)^2 - (2 - \sqrt{y})^2\right) dy$$

$$= \int_0^4 \pi\left(\frac{y^2}{4} - 3y + 4\sqrt{y}\right) dy = \frac{8}{3}\pi. \quad\blacksquare$$

Cylindrical Shells

If the rectangular strips that approximate a region being revolved about an axis lie parallel to the axis, they sweep out cylindrical shells instead of washers. Cylindrical shells are sometimes easier to work with than washers because the formula they lead to does not involve squaring. We arrive at the formula in the following way.

Suppose the tinted region $PQRS$ in Fig. 5.24 is revolved about the y-axis to generate a solid. We approximate the region with rectangular strips parallel to the y-axis from $x = a$ to $x = b$ and then approximate the volume of the solid by adding the volumes of the cylindrical shells generated by the strips.

Figure 5.24 shows a typical strip of width Δx. The base of the strip is centered at the point x, and the height of the strip is $f(x)$. The cylindrical shell generated by revolving the strip about the y-axis has inner radius $x - (\Delta x/2)$, outer radius $x + (\Delta x/2)$, and height $f(x)$.

The base of the shell is a ring bounded by two concentric circles. The dimensions and area of the ring are

Inner radius: $r_1 = x - \dfrac{\Delta x}{2}$

Outer radius: $r_2 = x + \dfrac{\Delta x}{2}$

Ring area: $\Delta A = \pi r_2^2 - \pi r_1^2 = \pi(r_2^2 - r_1^2)$

$$= \pi(r_2 + r_1)(r_2 - r_1)$$
$$= \pi(2x)(\Delta x) \tag{2}$$
$$= 2\pi x\, \Delta x.$$

5.24 The solid swept out by revolving $PQRS$ about the y-axis is approximated by a union of cylindrical shells like the one shown here.

The volume ΔV of the cylindrical shell is its base area times its height, which is $2\pi x\, \Delta x$ times $f(x)$:

Shell volume: $\Delta V = 2\pi x f(x)\, \Delta x. \tag{3}$

The sum of the volumes of the cylindrical shells generated by the strips that cover region $PQRS$ from a to b is

$$S_n = \sum_a^b \Delta V = \sum_a^b 2\pi x f(x)\, \Delta x. \tag{4}$$

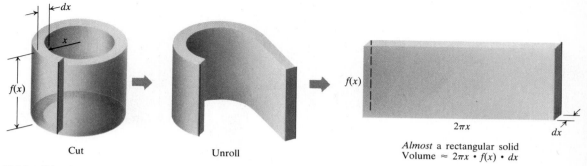

Height: $f(x)$ Inner circumference: $2\pi x$ Thickness: dx

5.25   How to remember the integral formula for cylindrical shells.

The volume of the solid swept out by revolving $PQRS$ about the y-axis is the limit approached by the sums S_n as Δx goes to zero, which is the integral of $2\pi x f(x)$ with respect to x from a to b.

Formula for Calculating Volume with Thin Cylindrical Shells of Inner Radius x and Height $f(x)$

$$\text{Volume} = \int_a^b 2\pi(\text{radius})(\text{height})\,dx = \int_a^b 2\pi x f(x)\,dx. \qquad (5)$$

An easy way to remember Eq. (5) is to imagine that a cylindrical shell of average circumference $2\pi x$, height $f(x)$, and thickness dx has been cut along a generator of the cylinder and rolled flat like a sheet of tin (Fig. 5.25). The sheet is almost a rectangular solid of dimensions $2\pi x$ by $f(x)$ by dx. Hence, the shell's volume is about $2\pi x f(x)\,dx$. Equation (5) says that the volume of the solid is the integral of $2\pi x f(x)\,dx$ from $x = a$ to $x = b$.

EXAMPLE 4   The region bounded by $y = \sqrt{x}$, $y = 0$, and $x = 4$ is revolved about the y-axis. Find the volume swept out.

Solution   We draw a picture to identify the radius and height of a typical cylinder and the limits of integration (Fig. 5.26). We find:

Radius: x

Height: $f(x) = \sqrt{x}$

Volume: $\displaystyle\int_0^4 2\pi(\text{radius})(\text{height})\,dx$

$$= \int_0^4 2\pi x \sqrt{x}\,dx$$

$$= 2\pi \int_0^4 x^{3/2}\,dx = 2\pi\left[\frac{2}{5}x^{5/2}\right]_0^4 = \frac{128\pi}{5}. \qquad \blacksquare$$

To calculate the volume of the solid generated by revolving a region about the x-axis (instead of the y-axis), we use Eq. (5) with y in place of x.

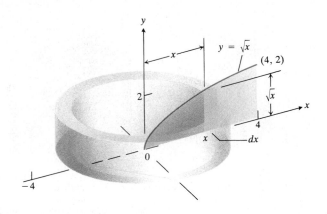

5.26 The cylindrical shell and integration limits in Example 4.

EXAMPLE 5 The region bounded by $y = \sqrt{x}$, $y = 0$, and $x = 4$ is revolved about the x-axis. Find the volume swept out.

Solution We draw a picture to show the dimensions of a typical cylinder and the limits of integration (Fig. 5.27). We find

Radius: y

Height: $4 - y^2$

Volume: $\displaystyle\int_0^2 2\pi y(4 - y^2)\, dy = 2\pi \left[2y^2 - \frac{y^4}{4} \right]_0^2 = 8\pi,$

in agreement with the disk calculation in Example 3, Article 5.3.   ◼

One thing we have not proved, and shall not, is that the three different ways we have developed for calculating the volumes of solids of revolution always agree. We use all three methods in the next example, however, and, as you will see, they produce the same result.

EXAMPLE 6 The disk bounded by the circle $x^2 + y^2 = a^2$ is revolved about the y-axis to generate a solid sphere. A hole of diameter a is then bored through the sphere along the y-axis. Find the volume of the "cored" sphere.

Solution The cored sphere could have been generated by revolving about the y-axis the part of the disk that lies to the right of the line $x = a/2$ (Fig. 5.28a). Thus, there are three methods we might use to find the volume: disks, washers, and cylindrical shells. In this example, they can all be used effectively.

METHOD 1: *Disks and subtraction.* Figure 5.28(b) shows the solid sphere with the core pulled out. The core is a circular cylinder with a spherical cap at each end. Our plan is to subtract the volume of the core from the volume of the solid sphere. We shall simplify matters by imagining that the two caps have been sliced off first, by planes perpendicular to the y-axis at $y = a\sqrt{3}/2$ and $y = -a\sqrt{3}/2$. When these two caps have been removed, the truncated sphere has volume V_1, say. From this truncated sphere we remove a right circular cylinder of radius $a/2$ and altitude $2(a\sqrt{3}/2) = a\sqrt{3}$. The volume of this flat-ended

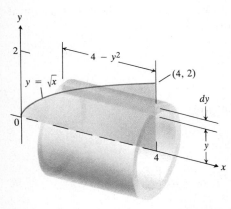

5.27 The cylindrical shell and integration limits in Example 5.

(b)

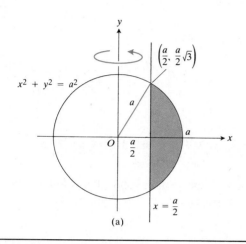

(a)

The volume of the cored sphere is the volume swept out by the shaded region as it revolves about the y-axis.

With the method of disks, we can calculate the volume of the core and subtract it from the volume of the sphere.

(c)

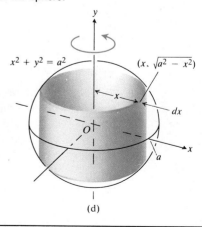

(d)

With the method of washers, we can calculate the volume of the cored sphere directly by modeling it as a stack of washers perpendicular to the y-axis.

With the method of cylinders, we can calculate the volume of the cored sphere directly by modeling it as a union of cylindrical shells parallel to the y-axis.

cylinder is

$$V_2 = \pi \left(\frac{a}{2}\right)^2 (a\sqrt{3}) = \frac{\pi\sqrt{3}}{4} a^3.$$

The volume we wish to calculate is $V = V_1 - V_2$.

To find V_1, we observe that the truncated sphere is a solid of revolution whose cross sections perpendicular to the y-axis are disks. The radius of a typical disk is $\sqrt{a^2 - y^2}$. Therefore,

$$V_1 = \int_{-a\sqrt{3}/2}^{a\sqrt{3}/2} \pi(\text{radius})^2 \, dy$$

$$= \int_{-a\sqrt{3}/2}^{a\sqrt{3}/2} \pi(a^2 - y^2) \, dy$$

$$= \pi \left[a^2 y - \frac{y^3}{3} \right]_{-a\sqrt{3}/2}^{a\sqrt{3}/2} = \frac{3\pi\sqrt{3}}{4} a^3.$$

Hence,

$$V = V_1 - V_2 = \frac{3\pi\sqrt{3}}{4} a^3 - \frac{\pi\sqrt{3}}{4} a^3 = \frac{\pi\sqrt{3}}{2} a^3.$$

METHOD 2: *Washers.* The cored sphere is a solid of revolution whose cross sections perpendicular to the y-axis are washers (Fig. 5.28c). The radii of a typical washer are

Outer radius: $R = \sqrt{a^2 - y^2}$,

Inner radius: $r = \dfrac{a}{2}$.

The volume of the solid is therefore

$$V = \int_{-a\sqrt{3}/2}^{a\sqrt{3}/2} \pi(R^2 - r^2) \, dy = \int_{-a\sqrt{3}/2}^{a\sqrt{3}/2} \pi\left(a^2 - y^2 - \frac{a^2}{4}\right) dy$$

$$= \pi \left[\frac{3a^2 y}{4} - \frac{y^3}{3} \right]_{-a\sqrt{3}/2}^{a\sqrt{3}/2} = \frac{\pi\sqrt{3}}{2} a^3.$$

METHOD 3: *Cylindrical shells.* We model the volume of the cored sphere with cylindrical shells like the one shown in Fig. 5.28(d). The typical shell has radius x and height $2\sqrt{a^2 - x^2}$. The volume of the solid is therefore

$$V = \int_{x=a/2}^{a} 2\pi(\text{radius})(\text{height}) \, dx$$

$$= \int_{a/2}^{a} 4\pi x\sqrt{a^2 - x^2} \, dx = 4\pi \left[-\frac{1}{3}(a^2 - x^2)^{3/2} \right]_{a/2}^{a}$$

$$= \frac{4\pi}{3}\left(a^2 - \frac{a^2}{4}\right)^{3/2} = \frac{4\pi}{3}\left(\frac{3a^2}{4}\right)^{3/2} = \frac{\pi\sqrt{3}}{2} a^3. \qquad \blacksquare$$

Table 5.1 summarizes the methods of finding volumes with washers and shells.

Table 5.1 WASHERS VS. SHELLS

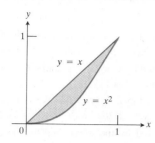

The region bounded by

$$y = x, \qquad y = x^2 \quad (x = \sqrt{y})$$

REVOLVED ABOUT X-AXIS

Washer: $\pi(R^2 - r^2)\, dx$

$$V = \int_{x=0}^{x=1} \pi((x)^2 - (x^2)^2)\, dx = \frac{2\pi}{15}$$

REVOLVED ABOUT Y-AXIS

Washer: $\pi(R^2 - r^2)\, dy$

RECTANGLE PERPENDICULAR TO AXIS: USE WASHERS.

$$V = \int_{y=0}^{y=1} \pi((\sqrt{y})^2 - (y)^2)\, dy = \frac{\pi}{6}$$

Shell: 2π(radius)(height) dy

$$V = \int_{y=0}^{y=1} 2\pi(y)(\sqrt{y} - y)\, dy = \frac{2\pi}{15}$$

Shell: 2π(radius)(height) dx

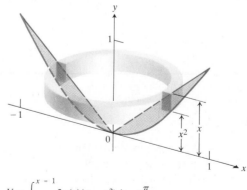

RECTANGLE PARALLEL TO AXIS: USE SHELLS.

$$V = \int_{x=0}^{x=1} 2\pi(x)(x - x^2)\, dx = \frac{\pi}{6}$$

PROBLEMS

In Problems 1–8, find the volume generated when the region bounded by the given curves and lines is revolved about the x-axis. (*Note:* $x = 0$ is the y-axis and $y = 0$ is the x-axis.)

1. $x + y = 2$, $x = 0$, $y = 0$
2. $x = 2y - y^2$, $x = 0$
3. $y = 3x - x^2$, $y = x$
4. $y = x$, $y = 1$, $x = 0$
5. $y = x^2$, $y = 4$
6. $y = 3 + x^2$, $y = 4$
7. $y = x^2 + 1$, $y = x + 3$
8. $y = 4 - x^2$, $y = 2 - x$

In Problems 9–22, find the volume generated by revolving the given region about the given axis.

9. The region bounded by $y = x^4$, $x = 1$, and $y = 0$ about the y-axis.
10. The region bounded by $y = x^3$, $x = 2$, and $y = 0$ about the y-axis.
11. The triangle with vertices $(1, 1)$, $(1, 2)$, and $(2, 2)$ about (a) the x-axis, (b) the y-axis.
12. The region in the first quadrant bounded by the curve $x = y - y^3$ and the y-axis about (a) the x-axis, (b) the y-axis.
13. The region in the first quadrant bounded by $x = y - y^3$, $x = 1$, and $y = 1$ about (a) the x-axis, (b) the y-axis, (c) the line $x = 1$, (d) the line $y = 1$.
14. The triangular region bounded by the lines $2y = x + 4$, $y = x$, and $x = 0$ about (a) the x-axis, (b) the y-axis, (c) the line $x = 4$, (d) the line $y = 8$.
15. The region in the first quadrant bounded by $y = x^3$ and $y = 4x$ about (a) the x-axis, (b) the line $y = 8$.

16. The region bounded by $y = \sqrt{x}$ and $y = x^2/8$ about (a) the x-axis, (b) the y-axis.
17. The region bounded by $y = 2x - x^2$ and $y = x$ about (a) the y-axis, (b) the line $x = 1$.
18. The region bounded by $y = \sqrt{x}$, $y = 2$, $x = 0$ about (a) the x-axis, (b) the y-axis, (c) the line $x = 4$, (d) the line $y = 2$.
19. The region bounded by the y-axis and by the curves $y = \cos x$ and $y = \sin x$, $0 \le x \le \pi/4$, about the x-axis.
20. The region bounded by $y = 0$ and the curve $y = 8x^2 - 8x^3$, $0 \le x \le 1$, about the y-axis.
21. The region between the curves $y = 2x^2$ and $y = x^4 - 2x^2$ about the y-axis.
22. The region in the first quadrant bounded by $y = x^2$, $x + y = 2$, and the x-axis, about the x-axis.
23. Use cylindrical shells and the formula $\int x \sin x \, dx = \sin x - x \cos x + C$ to find the volume generated by revolving about the y-axis the region bounded by $y = 0$ and the curve $y = \sin x$, $0 \le x \le \pi$.
24. The region bounded by the curve $y = x^2$ and the line $y = 4$ generates various solids when revolved

    a) about the y-axis,
    b) about the line $y = 4$,
    c) about the x-axis,
    d) about the line $y = -1$,
    e) about the line $x = 2$.

    Find the volume of each one.
25. The disk $x^2 + y^2 \le a^2$ is revolved about the line $x = b$ ($b > a$) to generate a torus (a solid shaped like a doughnut). Find the volume of the torus. (*Hint:* $\int_{-a}^{a} \sqrt{a^2 - x^2} \, dx = \pi a^2/2$, since it is the area of a semicircle of radius a.)

5.5
Lengths of Plane Curves

We approximate the length of a curved path in the plane the way we use a ruler to estimate the length of a curved road on a map: by adding up the lengths of straight segments whose ends touch the curve. There is always a limit to the accuracy of such an estimate, imposed by how accurately we measure and the number of segments we are willing to use. With calculus we can do better because we can imagine using straight line segments as short as we please, each set of segments making a polygon that fits the curve more tightly than before. When we proceed this way, the lengths of the polygons approach a number we can calculate with an integral. In this article, we find out what that integral is.

The Basic Cartesian Formula

Suppose that the curve whose length we wish to find is the graph of $y = f(x)$ from $x = a$ to $x = b$ (Fig. 5.29). We subdivide the curve into n pieces and connect the successive endpoints with line segments. The length of a typical segment PQ is

$$\sqrt{(\Delta x_k)^2 + (\Delta y_k)^2}.$$

The length of the curve from $x = a$ to $x = b$ is approximated by the sum

$$\sum_{k=1}^{n} \sqrt{(\Delta x_k)^2 + (\Delta y_k)^2}. \tag{1}$$

We expect the approximation to improve as the number of subdivisions is increased indefinitely and the lengths of the individual segments tend to zero, and we would like to show that the sum in (1) approaches a calculable limit. To see that it does, we rewrite it in a form to which we can apply the Integral Existence Theorem.

Suppose that f has a derivative that is continuous at every point of $[a, b]$. Then, by the Mean Value Theorem, there is a point (c_k, d_k) on the curve between P and Q where the tangent to the curve is parallel to the chord PQ. That is,

$$f'(c_k) = \frac{\Delta y_k}{\Delta x_k} \qquad \text{or} \qquad \Delta y_k = f'(c_k)\,\Delta x_k.$$

With this substitution for Δy_k, the sum in (1) takes the form

$$\sum_{k=1}^{n} \sqrt{(\Delta x_k)^2 + (f'(c_k)\,\Delta x_k)^2} = \sum_{k=1}^{n} \sqrt{1 + (f'(c_k))^2}\,\Delta x_k. \tag{2}$$

The sum can now be seen to be an approximating sum for the integral

$$\int_{a}^{b} \sqrt{1 + (f'(x))^2}\,dx. \tag{3}$$

Therefore, the sum will have this integral as a limit as the subdivision of the curve becomes finer and finer. For this reason, we define the integral of $\sqrt{1 + (f'(x))^2}$ from a to b to be the length of the curve from a to b. We usually simplify the final integral formula by writing y' in place of $f'(x)$.

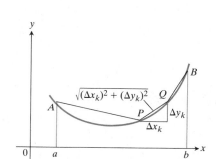

5.29 The arc AB is approximated by the polygonal path $APQB$. The length of the arc is defined to be the limit (when it exists) of the lengths of successively finer polygonal approximations.

DEFINITION

> **Length of the Curve $y = f(x)$ from $x = a$ to $x = b$**
>
> $$L = \int_{a}^{b} \sqrt{1 + (y')^2}\,dx \tag{4}$$

EXAMPLE 1 Find the length of the curve

$$y = \frac{4\sqrt{2}}{3}x^{3/2} - 1$$

from $x = 0$ to $x = 1$.

Solution   We calculate the length with Eq. (4):

$$y = \frac{4\sqrt{2}}{3}x^{3/2} - 1$$

$$y' = \frac{4\sqrt{2}}{3} \cdot \frac{3}{2}x^{1/2} = 2\sqrt{2}x^{1/2}$$

$$1 + (y')^2 = 1 + (2\sqrt{2}x^{1/2})^2 = 1 + 8x$$

$$L = \int_0^1 \sqrt{1 + (y')^2}\,dx = \int_0^1 \sqrt{1 + 8x}\,dx = \frac{2}{3} \cdot \frac{1}{8}(1 + 8x)^{3/2}\Big]_0^1 = \frac{13}{6}.$$

■

The Parametric Formula

There is a particularly useful formula for calculating the length of a curve given by parametric equations. Suppose the equations are

$$x = g(t), \qquad y = h(t), \qquad a \le t \le b,$$

and that the point $P(x(t), y(t))$ traces the curve exactly once as t runs from a to b. Instead of subdividing the x-axis to subdivide the curve, we subdivide the interval $a \le t \le b$. This induces a subdivision of the curve like the one shown in Fig. 5.30, in which consecutive points have coordinates $P(g(t_k), h(t_k))$ and $Q(g(t_{k+1}), h(t_{k+1}))$. The length of the line segment PQ can be calculated with the Pythagorean Theorem to be

$$\sqrt{(g(t_{k+1}) - g(t_k))^2 + (h(t_{k+1}) - h(t_k))^2}$$

(a formula soon to be simplified).

If g and h have first derivatives that are continuous for $a \le t \le b$, the Mean Value Theorem may be applied to give

$$g(t_{k+1}) - g(t_k) = g'(t_k')\,\Delta t_k, \qquad h(t_{k+1}) - h(t_k) = h'(t_k'')\,\Delta t_k, \qquad (5)$$

where t_k' and t_k'' are suitably chosen values between t_k and t_{k+1}. The sum of the lengths of the line segments approximating the curve therefore takes the form

$$\sum_{k=1}^{n} \sqrt{(g'(t_k'))^2 + (h'(t_k''))^2}\,\Delta t_k. \qquad (6)$$

This sum is not the Riemann sum of any function because the points t_k' and t_k'' need not be the same. However, a theorem called Bliss's Theorem (proved in more advanced texts) assures us that the sums converge to the integral we would like them to, namely the following integral.

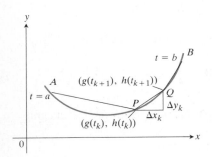

5.30   If the curve is given parametrically by the equations $x = g(t)$, $y = h(t)$, $a \le t \le b$, then

$$\Delta x_k = g(t_{k+1}) - g(t_k), \quad \Delta y_k = h(t_{k+1}) - h(t_k),$$

and the length of PQ is

$$\sqrt{(\Delta x_k)^2 + (\Delta y_k)^2}$$
$$= \sqrt{(g(t_{k+1}) - g(t_k))^2 + (h(t_{k+1}) - h(t_k))^2}.$$

Length of the Parametric Curve $x = x(t)$, $y = y(t)$, $a \le t \le b$

$$L = \int_a^b \sqrt{\left(\frac{dx}{dt}\right)^2 + \left(\frac{dy}{dt}\right)^2}\,dt \qquad (7)$$

EXAMPLE 2   Find the distance traveled between $t = 0$ and $t = \pi/2$ by a particle $P(x, y)$ whose position at time t is given by $x = \sin^2 t$, $y = \cos^2 t$.

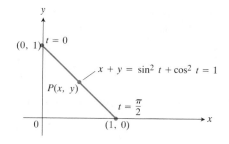

5.31 The path traced by the particle $P(x, y)$ whose position at time t is given by $x = \sin^2 t$, $y = \cos^2 t$, $0 \le t \le \pi/2$.

Solution We calculate the length of the path traced by the particle from $t = 0$ to $t = \pi/2$ from Eq. (7) as follows:

$$x = \sin^2 t \qquad\qquad y = \cos^2 t$$

$$\frac{dx}{dt} = 2 \sin t \cos t \qquad \frac{dy}{dt} = -2 \cos t \sin t$$

$$\int_0^{\pi/2} \sqrt{\left(\frac{dx}{dt}\right)^2 + \left(\frac{dy}{dt}\right)^2} \, dt = \int_0^{\pi/2} \sqrt{8 \sin^2 t \cos^2 t} \, dt$$

$$= \int_0^{\pi/2} \sqrt{2} \cdot 2 \sin t \cos t \, dt$$

$$= \sqrt{2} \cdot \sin^2 t \Big]_0^{\pi/2}$$

$$= \sqrt{2}.$$

In this case we can check our result geometrically, since

$$x + y = \sin^2 t + \cos^2 t = 1$$

for all values of t. The path of the particle is the segment of the line $x + y = 1$ that runs from $(0, 1)$, where $t = 0$, to $(1, 0)$, where $t = \pi/2$, as shown in Fig. 5.31. The length of this segment is

$$\sqrt{(0 - 1)^2 + (1 - 0)^2} = \sqrt{2}. \qquad \blacksquare$$

The Short Differential Formula

The equation

$$L = \int_a^b \sqrt{\left(\frac{dx}{dt}\right)^2 + \left(\frac{dy}{dt}\right)^2} \, dt$$

is usually written with differentials instead of derivatives. This is done formally by thinking of the derivatives as quotients of differentials and bringing the dt inside the radical as dt^2 to cancel the denominators:

$$\sqrt{\left(\frac{dx}{dt}\right)^2 + \left(\frac{dy}{dt}\right)^2} \, dt = \sqrt{\frac{dx^2}{dt^2} + \frac{dy^2}{dt^2}} \, dt$$

$$= \sqrt{\frac{dx^2}{dt^2} \, dt^2 + \frac{dy^2}{dt^2} \, dt^2}$$

$$= \sqrt{dx^2 + dy^2}.$$

The resulting formula for arc length is

$$L = \int \sqrt{dx^2 + dy^2}. \tag{8}$$

Of course, dx and dy must be expressed in terms of one and the same variable, and appropriate limits must be supplied before the integration in (8) is performed.

Equation (8) can be shortened still further. We think of dx and dy as two sides of a small triangle whose hypotenuse

$$ds = \sqrt{dx^2 + dy^2} \tag{9}$$

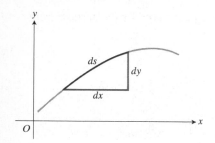

5.32 Diagram for remembering the equation $ds = \sqrt{dx^2 + dy^2}$.

is a differential of arc length that can be integrated between appropriate limits to give the length of the curve (Fig. 5.32). Then Eq. (8) becomes

$$L = \int ds. \tag{10}$$

EXAMPLE 3 Show that the formula $L = \int ds$ gives the correct result for the circumference of a circle of radius r.

Solution We parameterize the circle with the equations

$$x = r \cos t, \qquad y = r \sin t, \qquad 0 \le t \le 2\pi$$

(Fig. 5.33). Then

$$dx = -r \sin t \, dt, \qquad dy = r \cos t \, dt,$$

$$ds^2 = dx^2 + dy^2 = r^2(\sin^2 t + \cos^2 t) \, dt^2 = r^2 \, dt^2$$

and

$$L = \int ds = \int_{t=0}^{t=2\pi} r \, dt = rt \Big]_0^{2\pi} = 2\pi r. \qquad \blacksquare$$

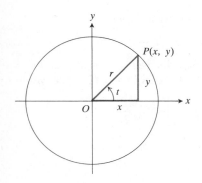

5.33 Example 3 calculates the circumference of this circle as the length of the path traced by the point P as t runs from 0 to 2π.

Dealing with Discontinuities in dy/dx

At a point on a curve where dy/dx fails to exist, dx/dy may exist and we may be able to find the curve's length with one or more applications of the formula

$$L = \int_c^d \sqrt{1 + \left(\frac{dx}{dy}\right)^2} \, dy, \tag{11}$$

which is Eq. (4) with x and y interchanged.

EXAMPLE 4 Find the length of the curve $y = x^{2/3}$ between $x = -1$ and $x = 8$.

Solution We sketch the curve (Fig. 5.34) and examine the derivative,

$$\frac{dy}{dx} = \frac{2}{3} x^{-1/3}.$$

Since this becomes infinite at the origin, we calculate the length of the curve with Eq. (11) instead of Eq. (4). We solve $y = x^{2/3}$ for x to find

$$x = \pm y^{3/2}$$

$$\frac{dx}{dy} = \pm \frac{3}{2} y^{1/2}$$

$$1 + \left(\frac{dx}{dy}\right)^2 = 1 + \left(\pm \frac{3}{2} y^{1/2}\right)^2 = 1 + \frac{9}{4} y.$$

The length of the portion of the curve between $A(-1, 1)$ and the origin is

$$L_1 = \int_0^1 \sqrt{1 + \frac{9}{4} y} \, dy.$$

5.34 To compute the length of $y = x^{2/3}$ between A and B, write $x = -y^{3/2}$ for the part from A to O, and $x = y^{3/2}$ for the part from O to B, and use Eq. (11) twice. The discussion in Example 4 explains why.

The length of the portion of the curve between the origin and $B(8, 4)$ is

$$L_2 = \int_0^4 \sqrt{1 + \frac{9}{4}y}\; dy.$$

The total length of the curve is $L = L_1 + L_2$.

We have to calculate L with two separate integrals because the equation $y = x^{2/3}$ defines x as two separate functions of y. The arc AO is given by the formula $x = -y^{3/2},\ 0 \le y \le 1$, while the arc OB is given by $x = +y^{3/2},\ 0 \le y \le 4$.

To evaluate the integrals for L_1 and L_2, we substitute

$$u = 1 + \frac{9}{4}y, \qquad du = \frac{9}{4}\, dy, \qquad dy = \frac{4}{9}\, du$$

to get

$$\int \left(1 + \frac{9}{4}y\right)^{1/2} dy = \frac{4}{9} \int u^{1/2}\, du = \frac{8}{27} u^{3/2} + C$$

and

$$L = \frac{8}{27}\left(1 + \frac{9}{4}y\right)^{3/2}\Big]_0^1 + \frac{8}{27}\left(1 + \frac{9}{4}y\right)^{3/2}\Big]_0^4$$

$$= \frac{1}{27}(13\sqrt{13} + 80\sqrt{10} - 16) \approx 10.5.$$

To check against gross errors, we can calculate the sum of the lengths of the chords AO and OB:

$$AO + OB = \sqrt{1 + 1} + \sqrt{64 + 16} = \sqrt{2} + \sqrt{80} \approx 10.4,$$

so our answer seems reasonable. ∎

The curve in Fig. 5.34 has a cusp at the origin where the slope becomes infinite. Were we to attempt to derive Eq. (4) for this particular curve, we would see that the step requiring the Mean Value Theorem could not have been taken for any chord PQ that bridged the cusp. We *can* apply the Mean Value Theorem to a chord PQ that ends or starts at the cusp, however, because the theorem does not require differentiability at an endpoint. Hence, the formula we would derive in place of Eq. (4) would have two integrals, one giving the length from A up to O, the other giving the length from O to B. This is the usual rule: when one or more cusps occur in a curve, calculate the length of the curve by integrating from cusp to cusp and adding the results. (This is how to handle the curve in Problem 7, for example.)

PROBLEMS

Find the lengths of the curves in Problems 1–6.

1. $y = (1/3)(x^2 + 2)^{3/2}$ from $x = 0$ to $x = 3$
2. $y = x^{3/2}$ from $(0, 0)$ to $(4, 8)$
3. $9x^2 = 4y^3$ from $(0, 0)$ to $(2\sqrt{3}, 3)$
4. $y = (x^3/3) + 1/(4x)$ from $x = 1$ to $x = 3$
5. $x = (y^4/4) + 1/(8y^2)$ from $y = 1$ to $y = 2$

6. $(y + 1)^2 = 4x^3$ from $x = 0$ to $x = 1$

7. The curve $x = a \cos^3 t$, $y = a \sin^3 t$, $0 \le t \le 2\pi$, in Fig. 5.35 consists of four congruent pieces that meet on the coordinate axes in cusps. Because of its starlike shape, the curve is sometimes called an astroid. Find the curve's total length.

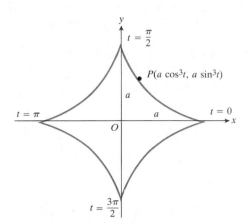

5.35 The astroid $x = a \cos^3 t$, $y = a \sin^3 t$, $0 \le t \le 2\pi$. The point P traces the curve counterclockwise as t goes from 0 to 2π. See Problem 7.

8. Find the length of the curve

$$y = \int_0^x \sqrt{\cos 2t} \ dt$$

from $x = 0$ to $x = \pi/4$.

9. Find the distance traveled between $t = 0$ and $t = \pi$ by the particle $P(x, y)$ whose position at time t is

$$x = \cos t, \qquad y = t + \sin t.$$

(*Hint:* $\sqrt{2 + 2 \cos t} = 2\sqrt{(1 + \cos t)/2}$.)

10. Find the length of the curve

$$x = t - \sin t, \qquad y = 1 - \cos t, \qquad 0 \le t \le 2\pi.$$

(*Hint:* $\sqrt{2 - 2 \cos t} = 2\sqrt{(1 - \cos t)/2}$.)

11. Find the distance traveled between $t = 0$ and $t = \pi/2$ by a particle $P(x, y)$ whose position at time t is given by

$$x = a \cos t + at \sin t, \qquad y = a \sin t - at \cos t,$$

where a is a positive constant.

12. Find the distance traveled by the particle $P(x, y)$ between $t = 0$ and $t = 4$ if its position at time t is given by

$$x = \frac{t^2}{2}, \qquad y = \frac{1}{3}(2t + 1)^{3/2}.$$

13. The position of a particle $P(x, y)$ at time t is given by

$$x = \frac{1}{3}(2t + 3)^{3/2}, \qquad y = \frac{t^2}{2} + t.$$

Find the distance it travels between $t = 0$ and $t = 3$.

14. Find the length of the loop in the curve $x = t^2$, $y = (t^3/3) - t$ (Fig. 5.36).

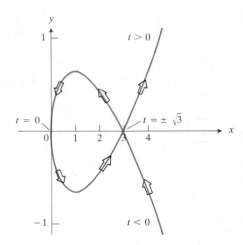

5.36 The curve $x = t^2$, $y = (t^3/3) - t$ in Problem 14. The arrows show the direction of increasing t.

15. Find the length of the curve $5y^3 = x^2$ that lies inside the circle $x^2 + y^2 = 6$.

16. Find the length of the curve $y = mx + b$ between $x = 0$ and $x = c$.

17. The graph of any function $y = f(x)$, $a \le x \le b$, has the automatic parameterization $x = x$, $y = f(x)$, $a \le x \le b$. Show that Eqs. (4) and (7) give the same result for such a curve.

18. **CALCULATOR** A company wants to make sheets of corrugated iron roofing like the one shown in Fig. 5.37. The cross sections of the corrugated sheets conform to the curve

$$y = \sin \frac{3\pi}{20}x, \qquad 0 \le x \le 20 \text{ in.}$$

If the roofing is to be stamped from flat sheets by a process that does not stretch the material, how wide should the original material be? To find out, use Simpson's Rule with $n = 10$ to approximate the length of the sine curve.

5.37 How wide does the original sheet have to be? See Problem 18.

19. CALCULATOR Your engineering firm is bidding for the contract to construct the tunnel shown in Fig. 5.38. The tunnel is 300 ft long and 50 ft wide at the base. The cross section has the shape of one arch of the curve $y = 25 \cos(\pi x/50)$.

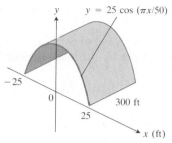

$y = 25 \cos(\pi x/50)$

5.38 The tunnel in Problem 19 (not to scale).

Upon completion, the inside surface of the tunnel (not counting the roadway) will be waterproofed with a sealer that costs $1.75 per ft^2 to apply. How much will it cost to apply the sealer? (*Hint:* Use numerical integration to find the length of the cosine curve.)

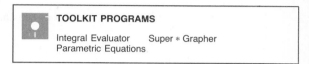

TOOLKIT PROGRAMS

Integral Evaluator   Super ∗ Grapher
Parametric Equations

5.6
The Area of a Surface of Revolution

When you jump rope, the rope sweeps out a surface in the space around you. As you can imagine, the area of the surface depends on the length of the rope and on how far away each part of the rope swings. In this article, we explore the relation between the area of a surface of revolution and the length and reach of the curve that generates it.

The Definition of Surface Area and the Cartesian Formula

Suppose that a curve AB like the one in Fig. 5.39(a) is revolved about the x-axis to generate a surface. If AB is approximated by an inscribed polygon like the ones used to define arc length in Article 5.5, then each segment PQ of the polygon sweeps out part of a cone whose axis lies along the x-axis (magnified view in Fig. 5.39b). This part is called a **frustum** of the cone (*frustum* is Latin for "piece"). If the base radii of the frustum are r_1 and r_2, as in Fig. 5.39(c),

(a)

(b)

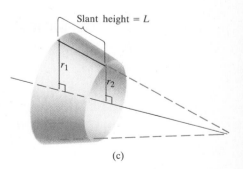

Slant height = L

r_1

r_2

(c)

5.39 (a) The surface swept out by revolving the curve AB about the x-axis is a union of bands like the one swept out by the arc PQ. (b) The line segment joining P to Q sweeps out a frustum of a cone. (c) The important dimensions of the frustum.

and its slant height is L, then its lateral surface area A is

$$A = \pi(r_1 + r_2)L. \tag{1}$$

The sum of the frustum areas swept out by the segments of the inscribed polygon from A to B gives an approximation to the area S of the surface swept out by the curve AB. This approximation leads to an integral formula for S in the following way.

If we let the coordinates of P be (x, y) and of Q be $(x + \Delta x, y + \Delta y)$, then the dimensions of the frustum swept out by the line segment PQ are

$$r_1 = y, \qquad r_2 = y + \Delta y, \qquad L = \sqrt{(\Delta x)^2 + (\Delta y)^2}. \tag{2}$$

The lateral surface area of the frustum, from Eq. (1), is

$$\begin{aligned} \text{Frustum area} &= \pi(r_1 + r_2)L \\ &= \pi(2y + \Delta y)\sqrt{(\Delta x)^2 + (\Delta y)^2}. \end{aligned} \tag{3}$$

Adding the individual frustum areas over the interval $[a, b]$ from left to right gives

$$\text{Cone frustum sum} = \sum_{x=a}^{b} \pi(2y + \Delta y)\sqrt{(\Delta x)^2 + (\Delta y)^2} \tag{4}$$

$$= \sum_{a}^{b} 2\pi\left(y + \frac{1}{2}\Delta y\right)\sqrt{1 + \left(\frac{\Delta y}{\Delta x}\right)^2}\,\Delta x.$$

If y and dy/dx are continuous functions of x, it can be shown (although we shall not do so here) that these sums approach the limit

$$S = \int_a^b 2\pi y\sqrt{1 + \left(\frac{dy}{dx}\right)^2}\,dx, \tag{5}$$

which we obtain by ignoring the $(1/2)\Delta y$ in Eq. (4). We therefore define the area of the surface to be the value of this integral.

DEFINITION

Surface Area

The area of the surface swept out by revolving the curve $y = f(x)$, $a \leq x \leq b$, about the x-axis is

$$S = \int_a^b 2\pi y\sqrt{1 + (y')^2}\,dx. \tag{6}$$

EXAMPLE 1 Find the area of the surface obtained by revolving the curve

$$y = \sqrt{x}, \qquad 0 \leq x \leq 2,$$

about the x-axis (Fig. 5.40).

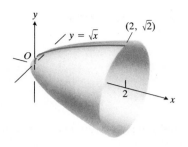

5.40 The area of this surface is calculated in Example 1.

Solution To find the surface area we use Eq. (6) with $y = \sqrt{x}$, $a = 0$, and $b = 2$:

$$y = \sqrt{x}, \qquad y' = \frac{1}{2\sqrt{x}},$$

$$\sqrt{1 + (y')^2} = \sqrt{1 + \frac{1}{4x}} = \sqrt{\frac{4x + 1}{4x}} = \frac{\sqrt{4x + 1}}{2\sqrt{x}},$$

$$\int_0^2 2\pi y \sqrt{1 + (y')^2}\, dx = \int_0^2 2\pi \sqrt{x}\, \frac{\sqrt{4x + 1}}{2\sqrt{x}}\, dx = \pi \int_0^2 \sqrt{4x + 1}\, dx$$

$$= \pi \cdot \frac{2}{3} \cdot \frac{1}{4} (4x + 1)^{3/2} \Big]_0^2 = \frac{\pi}{6}(27 - 1) = \frac{13\pi}{3}.$$

■

Other Forms of the Integral

If the axis of revolution is the y-axis, the formula that replaces (6) is

$$S = \int_c^d 2\pi x \sqrt{1 + \left(\frac{dx}{dy}\right)^2}\, dy. \tag{7}$$

If the curve that sweeps out the surface is given in parametric form, with x and y as functions of a third variable t that varies from a to b, then we may compute S from the formula

$$S = \int_a^b 2\pi\rho \sqrt{\left(\frac{dx}{dt}\right)^2 + \left(\frac{dy}{dt}\right)^2}\, dt, \tag{8}$$

where ρ is the distance from the axis of revolution to the element of arc length and is expressed as a function of t.

The formulas for surface area all have the form

$$S = \int 2\pi(\text{radius})(\text{band width}) = \int 2\pi \rho\, ds, \tag{9}$$

where ρ is the radius from the axis of revolution to the element of arc length ds (Fig. 5.41). If you wish to remember only one formula for surface area, you might make it this one. In any particular problem, you would then express the

5.41 The area of the surface swept out by revolving arc AB about the axis shown here is $\int_a^b 2\pi\rho\, ds$. The exact expression depends on the formulas for ρ and ds.

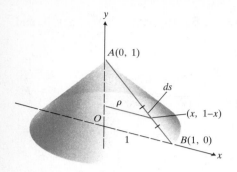

5.42 Revolving the line segment *AB* about the *y*-axis generates a cone.

radius function ρ and arc length differential ds in terms of some one variable and supply limits of integration.

EXAMPLE 2 The line segment

$$x = \sin^2 t, \qquad y = \cos^2 t, \qquad 0 \le t \le \frac{\pi}{2}$$

of Example 2, Article 5.5, is revolved about the *y*-axis to generate a cone. Find its surface area.

Solution See Fig. 5.42. We use Eq. (8). The distance from the axis of revolution to a typical element of arc ds is

$$\rho = x = \sin^2 t.$$

With this expression in place of ρ and with

$$\frac{dx}{dt} = 2\sin t \cos t, \qquad \frac{dy}{dt} = -2\cos t \sin t,$$

Equation 8 gives

$$\int_a^b 2\pi\rho \sqrt{\left(\frac{dx}{dt}\right)^2 + \left(\frac{dy}{dt}\right)^2}\, dt = \int_0^{\pi/2} 2\pi \sin^2 t \sqrt{8 \sin^2 t \cos^2 t}\, dt$$

$$= 4\pi\sqrt{2} \int_0^{\pi/2} \sin^3 t \cos t\, dt$$

$$= 4\pi\sqrt{2}\,\frac{1}{4}\,\sin^4 t \Big]_0^{\pi/2}$$

$$= \pi\sqrt{2}(1 - 0)$$

$$= \pi\sqrt{2}.$$

In this case we can also calculate the surface area from the formula for the lateral surface area of a cone:

$$\text{Lateral surface area} = \frac{\text{base circumference}}{2} \times \text{slant height} = \pi\sqrt{2}. \quad \blacksquare$$

EXAMPLE 3 Find the area of the sphere generated by revolving the circle $x^2 + y^2 = a^2$ about the *x*-axis (Fig. 5.43).

Solution We represent the top half of the circle, which generates the entire sphere, by

$$x = a \cos\theta, \qquad y = a \sin\theta, \qquad 0 \le \theta \le \pi.$$

Then

$$dx = -a \sin\theta\, d\theta, \qquad dy = a \cos\theta\, d\theta$$

and

$$ds = \sqrt{dx^2 + dy^2} = \sqrt{a^2 \sin^2\theta + a^2 \cos^2\theta}\, d\theta = a\, d\theta.$$

With

$$\rho = y = a \sin\theta,$$

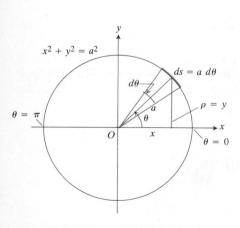

5.43 To find the surface area of the sphere obtained by revolving the circle $x^2 + y^2 = a^2$ about the *x*-axis, we let θ vary from 0 to π. The top half of the circle generates the entire sphere.

Eq. (9) now gives

$$S = \int_{\theta=0}^{\theta=\pi} 2\pi\rho \, ds = \int_0^\pi 2\pi \, a \sin \theta \, a \, d\theta$$

$$= 2\pi a^2 \int_0^\pi \sin \theta \, d\theta$$

$$= 2\pi a^2 \left[-\cos \theta \right]_0^\pi = 2\pi a^2 \, (1 + 1)$$

$$= 4\pi a^2.$$

EXAMPLE 4 The circle $x^2 + y^2 = a^2$ is revolved about the line $y = -a$, which is tangent to the circle at the point $(0, -a)$. Find the area of the surface generated. See Fig. 5.44.

Solution Here it takes the whole circle to generate the surface. This means that we must let θ vary from 0 to 2π. The radius of rotation is

$$\rho = y + a,$$

and we have

$$2\pi\rho \, ds = 2\pi(y + a)a \, d\theta = 2\pi(a \sin \theta + a)a \, d\theta = 2\pi a^2(\sin \theta + 1) \, d\theta.$$

Hence, Eq. (9) gives

$$S = \int_{\theta=0}^{\theta=2\pi} 2\pi\rho \, ds = \int_0^{2\pi} 2\pi a^2 \, (\sin \theta + 1) \, d\theta$$

$$= 2\pi a^2 \left[-\cos \theta + \theta \right]_0^{2\pi}$$

$$= 2\pi a^2((-1 + 2\pi) - (-1 + 0))$$

$$= 4\pi^2 a^2.$$

5.44 Revolving the circle $x^2 + y^2 = a^2$ about the line $y = -a$ generates a doughnut-like surface. Its area is found in Example 4.

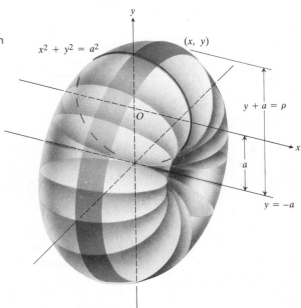

PROBLEMS

In Problems 1–8, find the area of the surface generated by revolving the given curve about the given axis.

1. The curve $y = x^3$, $0 \leq x \leq 1$, about the x-axis

2. The curve $y = x^2$, $0 \leq x \leq 2$, about the y-axis

3. The curve $y = (x^3/3) + 1/(4x)$, $1 \leq x \leq 3$, about the line $y = -1$

4. The curve $y = (x^3/6) + 1/(2x)$, $1 \leq x \leq 3$, about the x-axis

5. The curve $x = (y^4/4) + 1/(8y^2)$, $1 \leq y \leq 2$, about the x-axis

6. The curve $y = (x^2/2) + 1/2$, $0 \leq x \leq 1$, about the y-axis

7. The curve $y = (1/3)(x^2 + 2)^{3/2}$, $0 \leq x \leq 3$, about the y-axis

8. The line segment joining the points $(0, 4)$ and $(4, 1)$ about the x-axis

9. Use Eq. (6) to calculate the lateral surface area of the cone frustum in Fig. 5.39(c).

10. The region bounded by $y = 4x$ and $y = x^3$ in the first quadrant is revolved about the x-axis to generate a solid. Find the total surface area of the solid.

11. a) Find the area of the surface generated by revolving the curve $x = \cos t$, $y = 1 + \sin t$, $0 \leq t \leq 2\pi$, about the x-axis.

    b) Graph the curve and compare the result in (a) with the result in Example 4.

12. Find the area of the surface generated by revolving the curve $x = t^2$, $y = t$, $0 \leq t \leq 1$, about the x-axis.

13. Find the area of the surface generated by revolving the curve $x = a \cos^3 t$, $y = a \sin^3 t$, $0 \leq t \leq \pi$, shown in Fig. 5.45 about the x-axis.

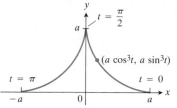

5.45 The curve $x = a \cos^3 t$, $y = a \sin^3 t$, $0 \leq t \leq \pi$ in Problem 13.

14. Find the area of the surface generated by revolving the curve
$$x = t + 1, \qquad y = \frac{t^2}{2} + t, \qquad 0 \leq t \leq 4,$$
about the y-axis.

15. Did you know that if you cut a spherical loaf of bread into slices of equal width, each slice will have the same amount of crust? To see why, suppose that the semicircle $y = \sqrt{r^2 - x^2}$ in Fig. 5.46 is revolved about the x-axis to generate a sphere. Let AB be an arc of the semicircle that stands above an interval of length h on the x-axis. Show that the area swept out by AB does not depend on the location of the interval. (It does, however, depend on the length of the interval.)

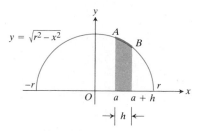

5.46 The semicircle in Problem 15.

5.7
The Average Value of a Function

The process of finding the average value of a finite number of data is used in situations familiar to all of us. For example, if $y_1, y_2, \ldots, y_n$ are the grades of a class of n students on a calculus quiz, then the class average on the quiz is

$$y_{\text{av}} = \frac{y_1 + y_2 + \cdots + y_n}{n}. \tag{1}$$

When the number of data are infinite, as they are when y is given by a continuous function $y = f(x)$ over an interval $a \leq x \leq b$, it is not feasible to use Eq. (1) because it takes on the meaningless form ∞/∞. We get around this problem by defining the average value of y with respect to x by an integral in the following way.

DEFINITION

> **Average Value of a Function**
>
> The average value of $y = f(x)$ with respect to x is
>
> $$y_{av} = \frac{1}{b-a} \int_a^b f(x)\, dx. \tag{2}$$

Equation (2) is a *definition* and hence is not subject to proof. Nevertheless, some discussion may help to explain why this particular formula is used to define the average. One might arrive at it as follows. From the total "population" of x-values, $a \le x \le b$, we select a representative "sample," $x_1, x_2, \ldots, x_n$, uniformly distributed, with $a < x_1 < \cdots < x_n = b$. Then, using Eq. (1), we calculate the average of the functional values

$$y_1 = f(x_1), \qquad y_2 = f(x_2), \qquad \ldots, \qquad y_n = f(x_n)$$

associated with these representative x's. This gives us

$$\frac{y_1 + y_2 + \cdots + y_n}{n} = \frac{f(x_1) + f(x_2) + \cdots + f(x_n)}{n}. \tag{3}$$

Since we require that the x's be uniform, let us take the spacing to be Δx, with

$$x_2 - x_1 = x_3 - x_2 = \cdots = x_n - x_{n-1} = \Delta x$$

and

$$\Delta x = \frac{b-a}{n}.$$

Then, in Eq. (3), let us replace the n in the denominator by $(b-a)/\Delta x$, thus obtaining

$$\frac{f(x_1) + f(x_2) + \cdots + f(x_n)}{(b-a)/\Delta x} = \frac{f(x_1)\,\Delta x + f(x_2)\,\Delta x + \cdots + f(x_n)\,\Delta x}{b-a}.$$

Now, if n is large and Δx is small, the expression

$$f(x_1)\,\Delta x + f(x_2)\,\Delta x + \cdots + f(x_n)\,\Delta x = \sum_{k=1}^{n} f(x_k)\,\Delta x$$

is nearly equal to $\int_a^b f(x)\, dx$. In fact, if we take limits, letting $n \to \infty$, we obtain precisely

$$\lim_{n \to \infty} \frac{f(x_1)\,\Delta x + f(x_2)\,\Delta x + \cdots + f(x_n)\,\Delta x}{b-a} = \frac{1}{b-a} \int_a^b f(x)\, dx.$$

This is the expression that defines the average value of y in Eq. (2).

EXAMPLE 1 The average value of the function $y = \sqrt{x}$ with respect to x from $x = 0$ to $x = 4$ is

$$y_{av} = \frac{1}{4} \int_0^4 \sqrt{x}\, dx = \frac{1}{4} \cdot \frac{2}{3} x^{3/2} \Big]_0^4$$

$$= \frac{1}{6} \cdot 8 = \frac{4}{3}.$$

∎

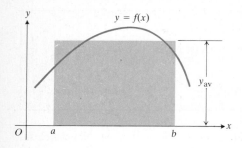

5.47 One interpretation of

$$y_{av} = \left[\frac{1}{(b-a)}\right]\int_a^b f(x)\,dx$$

comes from interpreting the integral as an area.

EXAMPLE 2 (This is a geometric interpretation of Eq. (2) for $f \geq 0$.)
If both sides of Eq. (2) are multiplied by $b - a$, we have

$$y_{av} \cdot (b - a) = \int_a^b f(x)\,dx. \tag{4}$$

The right-hand side of Eq. (4) is the area between the curve $y = f(x)$ and the x-axis, from $x = a$ to $x = b$. The left-hand side of the equation can be interpreted as the area of a rectangle of altitude y_{av} and of base $b - a$. Hence, Eq. (4) provides a geometric interpretation of y_{av} as the height on the curve $y = f(x)$ that can be used to construct a rectangle whose base is the interval $[a, b]$ and whose area is the area under the curve (Fig. 5.47).   ■

We use average values in calculating the effective voltage and current in an electric circuit.

EXAMPLE 3 The electric current in our household power supply is an alternating current whose flow can be modeled by a sine function

$$i = I \sin \omega t \tag{5}$$

like the one graphed in Fig. 5.48. Equation (5) expresses the current i in amperes as a function of time t in seconds. The amplitude I is the current's peak value, and $2\pi/\omega$ is the period.

The average value of i over a half-cycle is

$$
\begin{aligned}
i_{av} &= \frac{1}{\pi/\omega - 0}\int_0^{\pi/\omega} I \sin \omega t\,dt \\
&= \frac{I\omega}{\pi}\int_0^{\pi/\omega} \sin \omega t\,dt = \frac{I\omega}{\pi}\left[-\frac{\cos \omega t}{\omega}\right]_0^{\pi/\omega} \\
&= \frac{I}{\pi}(-\cos \pi + \cos 0) = \frac{2}{\pi}I.
\end{aligned}
$$

The average value of i over an entire cycle is

$$i_{av} = \frac{\omega}{2\pi}\int_0^{2\pi/\omega} I \sin \omega t\,dt = 0.$$

If the current were measured with a standard moving coil galvanometer, the meter would read zero.

To measure the current effectively, we use an instrument that measures the square root of the average value of the square of the current, namely

$$I_{rms} = \sqrt{(i^2)_{av}}. \tag{6}$$

The subscript rms stands for "root mean square." Since the average value of $i^2 = I^2 \sin^2 \omega t$ over an entire cycle is

$$(i^2)_{av} = \frac{\omega}{2\pi}\int_0^{2\pi/\omega} I^2 \sin^2 \omega t\,dt = \frac{I^2}{2}, \tag{7}$$

the rms current is

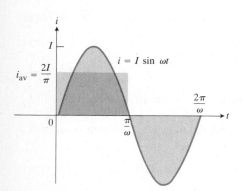

5.48 The average value of a sinusoidal current over a half-cycle is $2I/\pi$. The average over a complete cycle is zero.

$$I_{rms} = \sqrt{\frac{I^2}{2}} = \frac{I}{\sqrt{2}}. \tag{8}$$

In the same way, the root-mean-square value of a sinusoidal voltage $v = V \sin \omega t$ is

$$V_{\text{rms}} = \frac{V}{\sqrt{2}}. \tag{9}$$

The values given for household voltages and currents are always rms values. Thus, "115 volts ac" means that the rms voltage is 115 volts. The **peak value** of the voltage, obtained from Eq. (9) as

$$V = \sqrt{2} \, V_{\text{rms}} = \sqrt{2} \cdot 115 \approx 163 \text{ volts,}$$

is considerably higher. ■

EXAMPLE 4 The distance traveled by a body moving with speed $|v(t)|$ along a straight line from time $t = a$ to time $t = b$ is

$$\text{Distance traveled} = \int_a^b |v(t)| \, dt.$$

The average speed for the trip is therefore

$$\text{Average speed} = \frac{\text{distance traveled}}{b - a} = \frac{1}{b - a} \int_a^b |v(t)| \, dt.$$

Thus, the average speed is the average value of $|v(t)|$ over $[a, b]$. ■

☐ Average Daily Inventory

The notion of average value is used in economic theory to study average daily inventory. If $I(x)$ is the number of radios or shoes or whatever a firm has on hand on day x, then

$$I_{\text{av}} = \frac{1}{b - a} \int_a^b I(x) \, dx$$

is called the **average daily inventory** of that item for the period $a \leq x \leq b$. The costs of warehouse space, utilities, insurance, and security can be a large part of the expense of doing business, and the firm's daily inventory can play a significant role in determining these costs.

EXAMPLE 5 Suppose a wholesaler receives a shipment of 1200 cases of chocolate bars every 30 days. The chocolate is sold to retailers at a steady rate; and x days after the shipment arrives, the inventory of cases still on hand is $I(x) = 1200 - 40x$. Find the average daily inventory. Also find the average daily holding cost for the chocolate if the cost of holding one case is 3 cents a day.

Solution   The average daily inventory is

$$I_{\text{av}} = \frac{1}{30 - 0} \int_0^{30} (1200 - 40x) \, dx = \frac{1}{30} \left[1200x - 20x^2 \right]_0^{30} = 600.$$

The average daily holding cost for the chocolate is the dollar cost of holding one case times the average daily inventory. This works out to

$$\text{Average daily holding cost} = (0.03)(600) = 18$$

or \$18 a day. ■

PROBLEMS

In Problems 1–4, find the average value with respect to x, over the given domain, of the given function $f(x)$. In each case, graph $y = f(x)$ and sketch a rectangle whose altitude is the average y-coordinate.

1. a) $\sin x, \quad 0 \le x \le \pi/2$ b) $\sin x, \quad 0 \le x \le 2\pi$

2. a) $\sin^2 x, \quad 0 \le x \le \pi/2$ b) $\sin^2 x, \quad \pi \le x \le 2\pi$

3. $\sqrt{2x + 1}, \quad 4 \le x \le 12$

4. $1/2 + (1/2) \cos 2x, \quad 0 \le x \le \pi$

5. Find the average values of the functions graphed in Fig. 5.49.

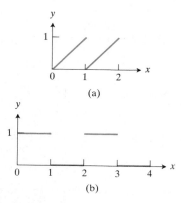

5.49   The graphs for Problem 5.

6. Carry out the integration in Eq. (7), Example 3.

7. Show that the root-mean-square value of the voltage $v = V \sin \omega t$ in Example 3 is $V_{\text{rms}} = V/\sqrt{2}$.

8. If a household electric circuit is rated 20 amps (rms), what is the peak value (amplitude) of the allowable current?

9. Solon Container receives 450 drums of plastic pellets every 30 days. The inventory function (drums on hand as a function of days) is $I(x) = 450 - x^2/2$. Find the average daily inventory. If the holding cost for one drum is 2¢ per day, find the average daily holding cost.

10. Mitchell Mailorder receives a shipment of 600 cases of athletic socks every 60 days. The number of cases on hand x days after the shipment arrives is $I(x) = 600 - 20\sqrt{15x}$. Find the average daily inventory. If the holding cost for one case is $1/2$¢ per day, find the average daily holding cost.

11. CALCULATOR   Compute the average value of the temperature function

$$f(x) = 37 \sin \left[\frac{2\pi}{365}(x - 101) \right] + 25$$

for a 365-day year (see Example 2 in Article 2.6). This is one way to estimate the annual mean air temperature in Fairbanks, Alaska. The National Weather Service's official

figure, a numerical average of the daily normal mean air temperatures for the year, is 25.7°F, which is slightly higher than the average value of $f(x)$. Figure 2.26 shows why.

12. CALCULATOR

a) Use the trapezoidal rule to estimate the average lumen output of the No. 22 flashbulb of Problem 18, Article 4.9, for the time interval from $t = 0$ to $t = 60$ milliseconds. Take the data from Table 4.6.

b) How many seconds would a 60-watt incandescent lightbulb rated at 765 lumens have to burn to put out as many lumen-milliseconds as the No. 22 flashbulb did in (a)?

13. *Fall from rest on the earth.* When distance is measured in feet and time in seconds, the motion of a body falling from rest in a vacuum near the surface of the earth satisfies the equations

$$s = 16t^2, \quad v = 32t, \quad \text{and} \quad v = 8\sqrt{s}.$$

During the first 2 sec, the body will fall 64 ft.

a) Graph $v = 32t$ for $0 \le t \le 2$ and $v = 8\sqrt{s}$ for $0 \le s \le 64$.

b) Find the average velocity of the body with respect to time for the first 2 sec of fall.

c) Find the average velocity of the body with respect to distance for the first 2 sec of fall.

14. *Fall from rest on the moon.* For a body falling from rest near the surface of the moon, where the acceleration of gravity is 1.62 m/sec^2,

$$s = 0.81t^2, \quad v = 1.62t, \quad \text{and} \quad v = \sqrt{3.24s}.$$

Here, s is in meters, t in seconds, v in meters per second. During the first two seconds after being released from rest a body will fall $s = 0.81(2)^2 = 3.24$ m.

a) Graph $v = 1.62t$ for $0 \le t \le 2$, and $v = \sqrt{3.24s}$ for $0 \le s \le 3.24$.

b) Find the average velocity of the body with respect to time for the first 2 sec of fall.

c) Find the average velocity of the body with respect to distance for the first 2 sec of fall.

15. Let f be a function that is differentiable on $[a, b]$. In Chapter 1 we defined the average rate of change of f on $[a, b]$ to be

$$\frac{f(b) - f(a)}{b - a}$$

and the instantaneous rate of change of f at x to be $f'(x)$. In this article, we defined the average value of a function. For the new definition of average to be consistent with the old one, we should have

$$\frac{f(b) - f(a)}{b - a} = \text{average value of } f' \text{ on } [a, b].$$

Show that this is the case.

5.8
Moments and Centers of Mass

5.50 A multiflash photograph of a falling wrench whose center of mass has been marked with a cross. Although the wrench turns in the air as it falls, its center of mass falls in a straight line, as you can verify with a straightedge. (Photograph from *PSSC Physics, Second Edition*; 1965; D.C. Heath & Company with Education Development Center, Inc.)

In many cases, the structures and mechanical systems studied in engineering and physics behave as if their entire mass were concentrated at a single point, called the center of mass. The behavior of a planet revolving about its sun is the behavior of one point mass revolving about another. When we balance a flat plate on the tip of a finger, our fingertip is at the plate's center of mass. When we balance a long, thin rod on a knife edge, the knife is right below the rod's center of mass. A seesaw balances at its center of mass. A wrench may rotate about its center of mass as it falls through the air, but the center of mass itself falls in a straight line (Fig. 5.50).

It is important to know how to locate a center of mass, and it turns out that this is basically a mathematical problem. In this article, we show how the problem is solved and find the centers of mass of a number of standard shapes. We shall concentrate on one- and two-dimensional shapes here. Three-dimensional shapes, which normally require more complicated integrations, will be treated in Chapter 18.

Masses along a Line

If we place three masses m_1, m_2, and m_3 on the x-axis and place a support or fulcrum at the origin as in Fig. 5.51, the system might or might not balance. Each mass exerts a downward force $m_i g$ equal to the magnitude of the mass times the acceleration of gravity, and each of these forces has a tendency to turn the axis about the origin. This turning effect is called a **torque** and is defined to be the product of the force $m_i g$ with the signed distance x_i from the mass to the origin. For masses to the left of the origin, x_i is negative. For masses to the right, x_i is positive. Thus, the masses on the left and right of the origin exert opposing torques.

The total torque, or tendency of the system to rotate about the origin, is the sum $m_1 g x_1 + m_2 g x_2 + m_3 g x_3$. The system will balance if this sum is zero and will not balance if it is different from zero.

$$\text{Torque} = m_1 g x_1 + m_2 g x_2 + m_3 g x_3. \tag{1}$$

If we factor out g, we see that the torque is

$$g\underbrace{(m_1 x_1 + m_2 x_2 + m_3 x_3)}_{\substack{\text{A feature of} \\ \text{the system}}} = g \sum m_i x_i. \tag{2}$$

A feature of
the environment

Thus, the torque is the product of the gravitational acceleration g, which is a feature of the environment in which the system happens to reside, and the

5.51 Masses on the x-axis.

number $(m_1x_1 + m_2x_2 + m_3x_3)$, which is a feature of the system itself, a constant that stays the same no matter where the system lies.

The number $(m_1x_1 + m_2x_2 + m_3x_3)$, which we shall now write as $\sum m_i x_i$, is called the **moment of the system about the origin:**

$$M_O = \text{Moment about origin} = \sum m_i x_i.$$

The moment is the same wherever the system is put—on the moon, on Mars, or on Earth.

At what point $\bar{x}$ on the axis could we locate the total mass $\sum m_i$ to produce the same torque? To find out, we equate the two torques and solve for $\bar{x}$:

$$\bar{x} \sum m_i g = \sum x_i m_i g, \qquad (3)$$

or

$$\bar{x} = \frac{\sum x_i m_i g}{\sum m_i g}$$

$$= \frac{\sum x_i m_i}{\sum m_i} \qquad \text{(the } g\text{'s factor out and cancel)}$$

$$= \frac{\sum x_i m_i}{M} \qquad \left(M = \sum m_i \right)$$

$$= \frac{\text{system moment}}{\text{system mass}}.$$

There are two things to notice here. First, $\bar{x}$ is computed by dividing the system's moment about the origin by the system's mass. Second, the absence of g tells us that $\bar{x}$ is an intrinsic feature of the system, unaffected by environment.

The significance of $\bar{x}$ is that $\bar{x}$ is the system's **center of mass.** The system will balance when the fulcrum is placed at $\bar{x}$. The torque of each mass about $\bar{x}$ is $m_i g(x_i - \bar{x})$, the product of the weight $m_i g$ and the signed distance $(x_i - \bar{x})$ of m_i from $\bar{x}$. The sum of these torques is zero, as the following calculation shows:

$$\text{Torque about } \bar{x} = \sum m_i g(x_i - \bar{x})$$

$$= g\left(\sum m_i x_i - \bar{x} \sum m_i \right)$$

$$= g\left(\sum m_i x_i - \bar{x}M \right)$$

$$= g\left(\sum m_i x_i - \frac{\sum m_i x_i}{M} M \right)$$

$$= g(0) = 0.$$

Wires and Thin Rods

In many applications, we want to know the center of mass of a straight wire or rod or thin strip of metal. In cases like this, where the distribution of mass can be assumed to be continuous, the summation signs in our formulas become integrals, in a manner we shall now describe.

5.52 To derive a formula for the moment of a thin strip about the origin of the *x*-axis, we first picture the strip as a system of small masses.

Imagine a long thin strip cut up into small pieces of mass Δm_i, as in Fig. 5.52. Each piece is Δx units long and lies approximately x_i units from the origin. Then observe three things. First, the strip's center of mass $\bar{x}$ will be nearly the same as the center of mass of the system of masses Δm_i:

$$\bar{x} \approx \frac{\text{system moment}}{\text{system mass}}. \tag{4}$$

Second, the moment of each piece of the strip about the origin is approximately $x_i \, \Delta m_i$. Thus, the moment of the entire system of masses about the origin is approximately the sum of the $x_i \, \Delta m_i$:

$$\text{System moment} \approx \sum x_i \, \Delta m_i. \tag{5}$$

Third, if the density of the strip at x_i is $\delta(x_i)$, expressed in terms of mass per unit length, then Δm_i is approximately equal to $\delta(x_i) \, \Delta x$ (mass per unit length times length):

$$\Delta m_i \approx \delta(x_i) \, \Delta x. \tag{6}$$

Combining these three observations gives

$$\bar{x} \approx \frac{\text{system moment}}{\text{system mass}} \approx \frac{\sum x_i \, \Delta m_i}{\sum \Delta m_i} \approx \frac{\sum x_i \, \delta(x_i) \, \Delta x}{\sum \delta(x_i) \, \Delta x}. \tag{7}$$

The sums in the last quotient in (7) are approximating sums for integrals. As the subdivision of the strip becomes finer and the length Δx approaches zero, the approximations improve and lead to the equation

$$\bar{x} = \frac{\displaystyle\int_a^b x\delta \, dx}{\displaystyle\int_a^b \delta \, dx}. \tag{8}$$

This is the formula we use to calculate $\bar{x}$.

Moment, Mass, and Center of Mass of a Thin Rod, Wire, or Strip along the *x*-axis

$$\text{Moment about the origin:} \quad M_O = \int_a^b x\delta \, dx$$

$$\text{Mass:} \quad M = \int_a^b \delta \, dx \tag{9}$$

$$\text{Center of mass:} \quad \bar{x} = \frac{M_O}{M}$$

If the density of the material is not constant, but varies with the position x, then the δ in Eqs. (9) is a function of x that must be taken into account when the integrals are evaluated. If, however, the density δ is a constant function, as it would be if the strip had a uniform width and thickness and were made of the same material throughout, then the constant δ could be brought outside the integrals to cancel in the calculation of $\bar{x}$.

EXAMPLE 1   A thin strip of uniform (i.e., constant) density stretches along the x-axis from $x = a$ to $x = b$ (Fig. 5.53). Show that the center of mass is located halfway between the two ends.

Solution   Since the density δ is constant, Eqs. (9) give

$$\bar{x} = \frac{\displaystyle\int_a^b x\delta \, dx}{\displaystyle\int_a^b \delta \, dx} = \frac{\delta \displaystyle\int_a^b x \, dx}{\delta \displaystyle\int_a^b dx} = \frac{\displaystyle\int_a^b x \, dx}{\displaystyle\int_a^b dx}$$

$$= \frac{\left. \dfrac{x^2}{2} \right]_a^b}{\left. x \right]_a^b} = \frac{\dfrac{1}{2}(b^2 - a^2)}{(b - a)} = \frac{a + b}{2}.$$

5.53   The center of mass of a uniform strip or rod lies at the midpoint.

Thus, a uniform strip (or rod or wire) has its center of mass at the midpoint between the two ends.   ■

EXAMPLE 2   A metal rod with one end at the origin and the other end at $x = 10$ thickens from left to right so that its density, instead of being a constant mass per unit length, is

$$\delta(x) = 1 + \frac{x}{10} \text{ kg/m.}$$

Find the rod's center of mass.

Solution   The moment of the rod about the origin is

$$M_O = \int_0^{10} x\delta \, dx = \int_0^{10} x\left(1 + \frac{x}{10}\right) dx$$

$$= \frac{x^2}{2} + \frac{x^3}{30} \bigg]_0^{10} = 50 + \frac{100}{3} = \frac{250}{3} \text{ kg} \cdot \text{m.}$$

The mass of the rod is

$$M = \int_0^{10} \delta \, dx = \int_0^{10} \left(1 + \frac{x}{10}\right) dx = x + \frac{x^2}{20} \bigg]_0^{10} = 10 + 5 = 15 \text{ kg.}$$

The center of mass is located at the point

$$\bar{x} = \frac{M_O}{M} = \frac{250}{3} \cdot \frac{1}{15} = \frac{50}{9} \approx 5.56 \text{ m.} \qquad ■$$

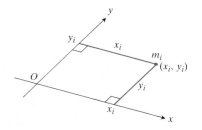

5.54 The mass m_i has a moment about each axis.

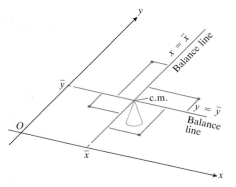

5.55 A two-dimensional array of masses balances on its center of mass.

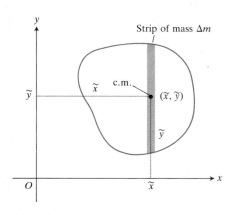

5.56 A plate cut into thin strips parallel to the y-axis. The moment exerted by a typical strip about each axis is the moment its mass Δm would exert if concentrated at the strip's center of mass $(\tilde{x}, \tilde{y})$.

Masses Distributed over a Plane Region

Suppose we have a finite collection of masses located in the coordinate plane, the mass m_i being located at the point (x_i, y_i) (see Fig. 5.54). The total mass of the system is

System mass: $\qquad M = \sum m_i.$

Each mass m_i has a moment about each axis. Its moment about the x-axis is $m_i y_i$, and its moment about the y-axis is $m_i x_i$. The moments of the entire system about the two axes are

Moment about x-axis: $\qquad M_x = \sum m_i y_i,$

Moment about y-axis: $\qquad M_y = \sum m_i x_i.$

The x-coordinate of the system's center of mass is defined to be

$$\bar{x} = \frac{M_y}{M} = \frac{\sum m_i x_i}{\sum m_i}. \tag{10}$$

With this choice of $\bar{x}$, as in the one-dimensional case, the system balances about the line $x = \bar{x}$ (Fig. 5.55).

The y-coordinate of the system's center of mass is defined to be

$$\bar{y} = \frac{M_x}{M} = \frac{\sum m_i y_i}{\sum m_i}. \tag{11}$$

With this choice of $\bar{y}$, the system balances about the line $y = \bar{y}$ as well. The torques exerted by the masses about the line $y = \bar{y}$ cancel out. Thus, as far as balance is concerned, the system behaves as if all its mass were at the single point $(\bar{x}, \bar{y})$ that we call the system's center of mass.

Thin Plates

In many applications, we need to find the center of mass of a thin flat plate: a disk of aluminum, say, or a triangular sheet of steel. In such cases, we assume the distribution of mass to be continuous, and the formulas we use to calculate $\bar{x}$ and $\bar{y}$ have integrals in them instead of finite sums. The integrals arise in the following way.

Imagine the plate occupying a region in the xy-plane, cut into thin strips parallel to one of the axes (in Fig. 5.56, the y-axis). The center of mass of a typical strip is $(\tilde{x}, \tilde{y})$. (The symbol $\sim$ over the x and y is a *tilde*, pronounced to rhyme with "Hilda." Thus, $\tilde{x}$ is read "x tilde," and so on.) We treat the strip's mass Δm as if it were concentrated at $(\tilde{x}, \tilde{y})$. The moment of the strip about the y-axis is then $\tilde{x}\,\Delta m$, and the moment of the strip about the x-axis is $\tilde{y}\,\Delta m$. Equations (10) and (11) then become

$$\bar{x} = \frac{M_y}{M} = \frac{\sum \tilde{x}\,\Delta m}{\sum \Delta m}, \qquad \bar{y} = \frac{M_x}{M} = \frac{\sum \tilde{y}\,\Delta m}{\sum \Delta m}. \tag{12}$$

As in the one-dimensional case, the sums in the numerator and denominator are approximating sums for integrals and approach these integrals as limiting values as the strips into which the plate is cut become narrower and narrower. We

write these integrals symbolically as

$$\bar{x} = \frac{\displaystyle\int \tilde{x} \, dm}{\displaystyle\int dm} \quad \text{and} \quad \bar{y} = \frac{\displaystyle\int \tilde{y} \, dm}{\displaystyle\int dm} \tag{13}$$

The examples that follow show how these integrals are evaluated.

Moments, Mass, and Center of Mass of a Thin Plate Covering a Region in the xy-plane

Moment about the x-axis: $M_x = \displaystyle\int \tilde{y} \, dm$

Moment about the y-axis: $M_y = \displaystyle\int \tilde{x} \, dm$ (14)

Mass: $M = \displaystyle\int dm$

Center of mass: $\bar{x} = \dfrac{M_y}{M}, \quad \bar{y} = \dfrac{M_x}{M}$

To evaluate these integrals, we picture the plate in the coordinate plane and sketch a strip of mass parallel to one of the coordinate axes. We then express the strip's mass dm and the coordinates $(\tilde{x}, \tilde{y})$ of the strip's center of mass in terms of x or y. Finally, we integrate $\tilde{y} \, dm$, $\tilde{x} \, dm$, and dm between limits of integration determined by the plate's location in the plane.

EXAMPLE 3 The triangular plate shown in Fig. 5.57, bounded by the lines $y = 0$, $y = 2x$, and $x = 1$, has a uniform density of $\delta = 3$ gm/cm². Find (a) the plate's moment M_y about the y-axis, (b) the plate's mass M, and (c) the x-coordinate of the plate's center of mass.

Solution METHOD 1: *Vertical strips* (Fig. 5.57a).

a) The moment M_y: The typical vertical strip has

center of mass (c.m.): $(\tilde{x}, \tilde{y}) = (x, x)$,

length: $2x$,

width: dx,

area: $dA = 2x \, dx$,

mass: $dm = \delta \, dA = 3 \cdot 2x \, dx = 6x \, dx$,

distance of c.m. from y-axis: $\tilde{x} = x$.

The moment of the strip about the y-axis is

$$\tilde{x} \, dm = x \cdot 6x \, dx = 6x^2 \, dx.$$

The moment of the plate about the y-axis is therefore

$$M_y = \int \tilde{x} \, dm = \int_0^1 6x^2 \, dx = 2x^3 \Big]_0^1 = 2 \text{ gm} \cdot \text{cm}.$$

(a)

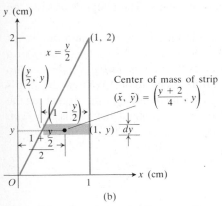

(b)

5.57 Two ways to model the calculation of the moment M_y of the triangular plate in Example 3.

b) The plate's mass:

$$M = \int dm = \int_0^1 6x \, dx = 3x^2 \Big]_0^1 = 3 \text{ gm.}$$

c) The x-coordinate of the plate's center of mass:

$$\bar{x} = \frac{M_y}{M} = \frac{2 \text{ gm} \cdot \text{cm}}{3 \text{ gm}} = \frac{2}{3} \text{ cm.}$$

By a similar computation, we could find M_x and $\bar{y} = M_x/M$.

METHOD 2. *Horizontal strips* (Fig. 5.57b).

a) The moment M_y: The typical horizontal strip has

$$\text{c.m.:} \quad (\tilde{x}, \tilde{y}) = \left(\frac{1}{2}\left(1 + \frac{y}{2}\right), y \right) = \left(\frac{y + 2}{4}, y \right),$$

$$\text{length:} \quad 1 - \frac{y}{2} = \frac{2 - y}{2},$$

$$\text{width:} \quad dy,$$

$$\text{area:} \quad dA = \frac{2 - y}{2} \, dy,$$

$$\text{mass:} \quad dm = \delta \, dA = 3 \cdot \frac{2 - y}{2} \, dy,$$

$$\text{distance of c.m. to } y\text{-axis:} \quad \tilde{x} = \frac{y + 2}{4}.$$

The moment of the strip about the y-axis is

$$\tilde{x} \, dm = \frac{y + 2}{4} \cdot 3 \cdot \frac{2 - y}{2} \, dy = \frac{3}{8}(4 - y^2) \, dy.$$

The moment of the plate about the y-axis is

$$M_y = \int \tilde{x} \, dm = \int_0^2 \frac{3}{8}(4 - y^2) \, dy = \frac{3}{8}\left[4y - \frac{y^3}{3} \right]_0^2$$

$$= \frac{3}{8}\left(\frac{16}{3} \right) = 2 \text{ gm} \cdot \text{cm.}$$

b) The plate's mass:

$$M = \int dm = \int_0^2 \frac{3}{2}(2 - y) \, dy = \frac{3}{2}\left[2y - \frac{y^2}{2} \right]_0^2 = \frac{3}{2}[4 - 2] = 3 \text{ gm.}$$

c) The x-coordinate of the plate's center of mass:

$$\bar{x} = \frac{M_y}{M} = \frac{2 \text{ gm} \cdot \text{cm}}{3 \text{ gm}} = \frac{2}{3} \text{ cm.}$$

By a similar computation, we could find M_x and $\bar{y}$.   ■

EXAMPLE 4   Find the center of mass of a thin homogeneous (constant density δ) plate covering the region bounded above by the parabola $y = 4 - x^2$ and below by the x-axis. See Fig. 5.58.

Solution   Since the plate is symmetric about the y-axis and its density is constant, the center of mass lies on the y-axis. This means that $\bar{x} = 0$. It remains to find $\bar{y} = M_x/M$.

A trial calculation with horizontal strips (Fig. 5.58a) leads to a difficult integration:

$$M_x = \int_0^4 2\delta y \sqrt{4 - y}\, dy.$$

We therefore model the distribution of mass with vertical strips instead (Fig. 5.58b). The typical vertical strip has

center of mass (c.m.): $(\tilde{x}, \tilde{y}) = \left(x, \dfrac{4 - x^2}{2}\right),$

length: $4 - x^2,$

width: $dx,$

area: $dA = (4 - x^2)\, dx,$

mass: $dm = \delta\, dA = \delta(4 - x^2)\, dx,$

distance from c.m. to x-axis: $\tilde{y} = \dfrac{4 - x^2}{2}.$

The moment of the strip about the x-axis is

$$\tilde{y}\, dm = \frac{4 - x^2}{2} \cdot \delta(4 - x^2)\, dx = \frac{\delta}{2}(4 - x^2)^2\, dx.$$

The moment of the plate about the x-axis is

$$M_x = \int \tilde{y}\, dm = \int_{-2}^{2} \frac{\delta}{2}(4 - x^2)^2\, dx = \frac{256}{15}\delta. \tag{15}$$

The mass of the plate is

$$M = \int dm = \int_{-2}^{2} \delta(4 - x^2)\, dx = \frac{32}{3}\delta. \tag{16}$$

Therefore,

$$\bar{y} = \frac{M_x}{M} = \frac{\dfrac{256}{15}\delta}{\dfrac{32}{3}\delta} = \frac{8}{5}.$$

The plate's center of mass is the point

$$(\bar{x}, \bar{y}) = \left(0, \frac{8}{5}\right). \qquad \blacksquare$$

(a)

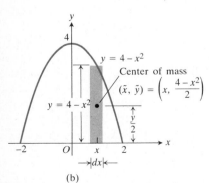

(b)

5.58   Modeling the problem in Example 4 with horizontal strips (a) leads to a difficult integration, so we model with vertical strips (b) instead.

EXAMPLE 5 *(Variable density)* Find the center of mass of the plate in Example 4 if the density at any point (x, y) is proportional to x^2, the square of the distance from the point to the y-axis.

Solution The density function is

$$\delta = kx^2,$$

for some constant k. The mass distribution is still symmetric about the y-axis, so

$$\bar{x} = 0.$$

With $\delta = kx^2$, Eqs. (15) and (16) become

$$M_x = \int \tilde{y} \, dm = \int_{-2}^{2} \frac{\delta}{2}(4 - x^2)^2 \, dx$$

$$= \int_{-2}^{2} \frac{kx^2}{2}(4 - x^2)^2 \, dx = \frac{1024}{105}k$$

$$M = \int dm = \int_{-2}^{2} \delta(4 - x^2) \, dx$$

$$= \int_{-2}^{2} kx^2(4 - x^2) \, dx = \frac{128}{15}k.$$

Therefore,

$$\bar{y} = \frac{M_x}{M} = \frac{1024}{105} \cdot \frac{15}{128} = \frac{8}{7}.$$

The plate's new center of mass is

$$(\bar{x}, \bar{y}) = \left(0, \frac{8}{7}\right).$$

■

EXAMPLE 6 Show that the center of mass of a thin homogeneous triangular plate of base b and altitude h lies at the intersection of the medians.

Solution You may recall that the medians of a triangle meet at the one point inside the triangle that is one-third of the way from each side to the opposite vertex. Our plan is to show that the center of mass is this point by showing that it, too, lies one-third of the way from each side to the opposite vertex.

We base one side of the triangle on the x-axis, with the opposite vertex on the positive y-axis (Fig. 5.59). The mass of a typical horizontal strip is

$$dm = \delta \, dA = \delta L \, dy,$$

where δ is the density and L is the width of the triangle at distance y above its base. By similar triangles,

$$\frac{L}{b} = \frac{h - y}{h} \qquad \text{or} \qquad L = \frac{b}{h}(h - y),$$

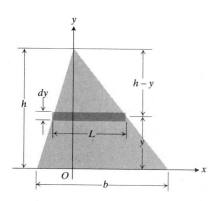

5.59 The dimensions and variables used in calculating the mass of a thin triangular plate.

so that

$$dm = \delta \frac{b}{h}(h - y)\, dy.$$

For the y-coordinate of the center of mass of the strip, we have $\tilde{y} = y$. For the entire plate,

$$\bar{y} = \frac{\int y\, dm}{\int dm} = \frac{\int_0^h \delta \frac{b}{h} y(h - y)\, dy}{\int_0^h \delta \frac{b}{h}(h - y)\, dy} = \frac{1}{3}h.$$

Thus the center of mass lies above the base of the triangle one-third of the way toward the opposite vertex. By considering each side in turn as a base of the triangle, this result shows that the center of mass lies at the intersection of the medians. ∎

EXAMPLE 7   A thin homogeneous wire is bent to form a semicircle of radius a. Find its center of mass.

Solution   We model the wire with the semicircle $y = \sqrt{a^2 - x^2}$ (Fig. 5.60). The mass of a typical small segment of wire is then

$$dm = \delta\, ds,$$

where ds is an element of arc length of the semicircle and

$$\delta = \frac{M}{L} = \frac{M}{\pi a}$$

is the mass per unit length of the wire. In terms of the central angle θ measured in radians (as usual), we have

$$ds = a\, d\theta$$

and

$$\tilde{x} = a \cos\theta, \qquad \tilde{y} = a \sin\theta.$$

Hence

$$\bar{x} = \frac{\int \tilde{x}\, dm}{\int dm} = \frac{\int_0^\pi a \cos\theta\, \delta\, a\, d\theta}{\int_0^\pi \delta\, a\, d\theta} = \frac{\delta\, a^2 \Big[\sin\theta\Big]_0^\pi}{\delta\, a \Big[\theta\Big]_0^\pi} = 0,$$

$$\bar{y} = \frac{\int \tilde{y}\, dm}{\int dm} = \frac{\int_0^\pi a \sin\theta\, \delta\, a\, d\theta}{\int_0^\pi \delta\, a\, d\theta} = \frac{\delta\, a^2 \Big[-\cos\theta\Big]_0^\pi}{\delta\, a \Big[\theta\Big]_0^\pi} = \frac{2}{\pi}a.$$

The center of mass is therefore on the y-axis at a distance $2/\pi$ (roughly two-thirds) of the way up from the origin toward the intercept $(0, a)$. Notice that the center of mass does not lie on the wire. ∎

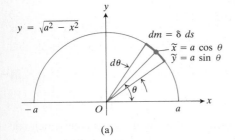

$y = \sqrt{a^2 - x^2}$

$dm = \delta\, ds$

$\tilde{x} = a \cos\theta$
$\tilde{y} = a \sin\theta$

$d\theta$

θ

$-a \qquad O \qquad a$

(a)

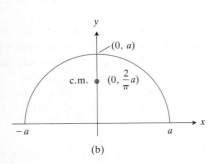

$(0, a)$

c.m. $\left(0, \frac{2}{\pi}a\right)$

$-a \qquad \qquad a$

(b)

5.60   (a) The dimensions and variables used in calculating the mass of a semicircular wire. (b) The wire and its center of mass.

Centers of Gravity, Centroids, Homogeneity, and Uniformity

As you read elsewhere, you will find some variety in the vocabulary used in connection with centers of mass.

When physicists discuss the effects of gravity on a system of masses, they may call the center of mass the **center of gravity.**

Material that has a constant density δ is also said to be **homogeneous,** or to be **uniform,** or to have **uniform density.**

When the density function is constant, it cancels out of the numerator and denominator of the formulas for $\bar{x}$ and $\bar{y}$. This happened in nearly every example in this article. As far as $\bar{x}$ and $\bar{y}$ were concerned, δ might as well have been 1. Thus, when the density is constant, the location of the center of mass is a feature of the geometry of the object and not of the material from which it is made. In such cases, engineers may call the center of mass the **centroid** of the shape, as in "Find the centroid of a triangle or a solid cone." To do so, we just set δ equal to 1 and proceed to find $\bar{x}$ and $\bar{y}$ as before, by dividing moments by masses.

PROBLEMS

1. Two children are balancing on a seesaw. The 80-lb child is 5 ft from the fulcrum. How far from the fulcrum is the 100-lb child?

2. The ends of two thin uniform rods of equal length are welded together to make a right-angled frame (Fig. 5.61). Use the result of Example 1 to help locate the frame's center of mass. Does the size of the angle really matter?

Right-angled weld

5.61 The welded rods in Problem 2.

Problems 3–6 give density functions of thin rods of length L that lie along the positive x-axis with one end at the origin. Find each rod's moment about the origin, mass, and center of mass.

3. $\delta(x) = 1, \quad 0 \le x \le L$

4. $\delta(x) = 1 + \dfrac{x}{L}, \quad 0 \le x \le L$

5. $\delta(x) = \left(1 + \dfrac{x}{L}\right)^2, \quad 0 \le x \le L$

6. $\delta(x) = \begin{cases} 2, & 0 \le x \le \dfrac{L}{2} \\ 1, & \dfrac{L}{2} \le x \le L \end{cases}$

7. Find the y-coordinate of the center of mass of the plate in Example 3.

8. Find the center of mass of the parabolic plate in Example 4 if, instead of being constant, $\delta(x) = |3x|$.

In Problems 9–20, find the center of mass of a thin homogeneous plate covering the given region.

9. The region bounded by the y-axis and the curve $x = y - y^3$, $0 \le y \le 1$

10. The region bounded by the curve $y = x^2$ and the line $y = 4$

11. The region bounded by the curve $y = x - x^2$ and the line $x + y = 0$

12. The region bounded by the curve $x = y^2 - y$ and the line $y = x$

13. The first quadrant of the circle $x^2 + y^2 = a^2$

14. The region bounded by the parabola $y = h^2 - x^2$ and the x-axis

15. The "triangular" region in the first quadrant between the circle $x^2 + y^2 = a^2$ and the lines $x = a, y = a$

16. The region between the x-axis and the curve $y = \sin x$ from $x = 0$ to $x = \pi$. (*Hint:* Take $dA = y\, dx$ and $\tilde{y} = (1/2)y$.)

17. The region between the y-axis and the curve $x = 2y - y^2$

18. The region bounded by the x-axis and the semicircle $y = \sqrt{a^2 - x^2}$. Compare the answer here with the result in Example 7.

19. The region bounded by the curves $y = 2x^2 - 4x$ and $y = 2x - x^2$

20. The region bounded above by $y = 1 - x^n$, n even, and below by the x-axis. What is the limiting position of the centroid as $n \to \infty$?

In Problems 21–24, use the result of Example 6 to locate the centroids of the triangles whose vertices are given.

21. $(-1, 0)$, $(1, 0)$, $(0, 3)$

22. $(0, 0)$, $(0, 1)$, $(1, 0)$

23. $(0, 0)$, $(3, 0)$, $(0, 3)$

24. $(a, 0)$, $(b, 0)$, $(0, c)$

25. Find the moment about the x-axis of a wire of constant density that lies along the curve $y = \sqrt{x}$ from $(0, 0)$ to $(2, \sqrt{2})$. Compare your calculation with the surface area calculation in Example 1, Article 5.6.

26. Find the moment about the x-axis of a wire of constant density that lies along the curve $y = x^3$ from $(0, 0)$ to $(1, 1)$. Compare the calculation here with the calculation required by Problem 1, Article 5.6.

27. The density of a triangular plate with base b cm and height h cm is proportional to the square root of the distance from the base. How far from the base is the center of mass?

28. Suppose that the density of the plate in Problem 27 is proportional to the square of the distance from the base. How far from the base is the center of mass now?

29. The density of a thin plate bounded by the curves $y = x^2$ and the line $y = x$ in the first quadrant is $\delta(x) = 12x$. Find the plate's center of mass.

30. A uniform tapered rod 1 m long is shaped like a frustum of an elongated cone. Its diameter is 1 cm at one end and 2 cm at the other. How far from the heavier end is the balance point?

31. Suppose the density of the wire in Example 7 is $\delta = k \sin \theta$, k being a constant. Find the center of mass.

5.9

Work

In everyday life we use the term *work* to describe any activity that takes muscular or mental effort. In science, however, the term is used in a narrower sense that involves the application of a force to a body and the body's subsequent displacement.

When a body moves a distance d along a straight line while acted on by a force of constant magnitude F in the direction of motion, the **work** done is defined to be

$$W = Fd, \tag{1}$$

measured in foot-pounds, newton-meters, or whatever. (It takes a force of about 1 newton to lift an apple from a table.) We can see right away that there may be quite a difference between what we are used to calling work and what this formula says work is. If we push a car down the street we are doing work, both by our own reckoning and according to Eq. (1). However, if we push against the car and the car doesn't move, Eq. (1) tells us we are doing no work at all, no matter how hard or how long we push.

When the force F varies along the way, as it will when it is compressing a spring, the equation $W = Fd$ cannot be used directly to calculate the work done. It can, however, be used to approximate the work over a short interval, which suggests a way to calculate the work done as an integral. In this article, we find out what that integral is and how it is used in applications. We also see how to formulate other integrals that calculate work.

The Force-Distance Integral for Work

In our first calculation of work, we assume that the force performing the work varies continuously along a line that we take to be the x-axis. We are interested in the work done along the interval from $x = a$ to $x = b$. We imagine the interval subdivided in the usual way into subintervals of length Δx, choose a point c_i in each subinterval, and form the sum

$$\sum_{a}^{b} F(c_i)\, \Delta x.$$

Since $F(c_i)\, \Delta x$ approximates the work done by F in the ith subinterval, the sum approximates the work done by F from a to b. If F is continuous, the sum approaches the integral of F from a to b as Δx approaches zero. We therefore define the work F does from a to b to be the value of its integral from a to b.

DEFINITION

Work Done by a Force $F(x)$ Acting along the x-axis from $x = a$ to $x = b$

$$W = \int_{a}^{b} F(x)\, dx \qquad (2)$$

Hooke's Law for Springs: $F = kx$

For most springs there is a range over which the amount of force F required to stretch or compress the spring from its natural length can be approximated by the linear equation

$$F = kx, \qquad (3)$$

where x is the amount the spring has been displaced from its natural or unstressed length, and k is a constant characteristic of the spring, called the **spring constant** (Fig. 5.62a). Beyond this range, the metal of the spring becomes distorted and Eq. (3) is no longer a reliable description. We assume that the springs in this article are never stretched or compressed to such an extent. Equation (3) is known as **Hooke's Law.**

EXAMPLE 1   Find the work in foot-pounds required to compress a spring from its natural length of 1 ft to a length of 0.75 ft if the spring constant is $k = 16$.

Solution   We picture an uncompressed spring laid out along the x-axis with its movable end at the origin and the fixed end at $x = 1$ ft (Fig. 5.62). This enables us to describe the force required to compress the spring from 0 to x with the formula $F = 16x$. As the spring is compressed from 0 to 0.25 ft, the force varies from

$$F(0) = 16(0) = 0 \text{ lb}$$

to

$$F(0.25) = 16(0.25) = 4 \text{ lb}.$$

The work done by F over this interval is

$$\text{Work} = \int_{0}^{0.25} F(x)\, dx = \int_{0}^{0.25} 16x\, dx = 4x^2 \Big]_{0}^{0.25} = 0.25 \text{ ft} \cdot \text{lb}.$$

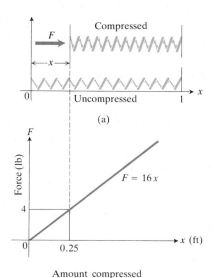

5.62   The force F required to hold a spring under compression increases linearly as the spring is compressed.

5.63   A 24-newton weight stretches this spring 0.8 m beyond its unstressed length.

EXAMPLE 2   A spring has a natural length of $L = 1$ m. A force of 24 newtons stretches the spring to a length of 1.8 m. Find the spring constant k. Then calculate how much work it takes to stretch the spring from its natural length to a length of 3 m. How much work is done in stretching the spring from a length of 2 m to a length of 3 m? How far beyond its natural length will a 45-newton force stretch the spring?

Solution   We find the spring constant from Eq. (3). A force of 24 N stretches the spring 0.8 m, so

$$24 = k(0.8) \quad \text{or} \quad k = 30.$$

To find how much work it takes to stretch the spring 2 m beyond its unstressed length, we imagine the spring hanging parallel to the x-axis as in Fig. 5.63. Then the force required to pull the spring x meters from the origin is

$$F(x) = kx = 30x$$

and the work to stretch the spring 2 m beyond its unstressed length is

$$W = \int_0^2 30x \, dx = 15x^2 \Big]_0^2 = 60 \text{ N} \cdot \text{m}.$$

The work done in stretching the spring from a length of 2 m to a length of 3 m is the work done in applying the stretching force $F(x) = 30x$ from $x = 1$ to $x = 2$ (not $x = 2$ to $x = 3$):

$$W = \int_{x=1}^{x=2} 30x \, dx = 15x^2 \Big]_1^2 = 45 \text{ N} \cdot \text{m}.$$

How far will a 45-newton force stretch the spring beyond its natural length? We substitute $F = 45$ in the equation $F = 30x$ to find $45 = 30x$, or $x = 1.5$ m. No calculus is involved in this last calculation.   ■

Pumping Liquids from Containers

We now consider another instance in which we calculate work as an integral obtained by applying the formula $W = Fd$ to small pieces.

EXAMPLE 3   Calculate the amount of work required to pump all the water from a full hemispherical bowl of radius r feet to a height h feet above the top of the bowl (Fig. 5.64).

Solution   We introduce coordinate axes as shown in Fig. 5.64 and imagine the water divided into thin slices by planes perpendicular to the x-axis between $x = 0$ and $x = r$. The representative slice between the planes at x and $x + \Delta x$ has a volume ΔV that is approximately

$$\Delta V \approx \pi y^2 \, \Delta x = \pi (r^2 - x^2) \, \Delta x \text{ ft}^3.$$

The force F required to lift *this slice* is equal to its weight:

$$w \, \Delta V \approx \pi w (r^2 - x^2) \, \Delta x \text{ lb},$$

where w is the weight of a cubic foot of water. Finally, the *distance* through which this force must act is approximately $(x + h)$ feet, so the work ΔW done in lifting this one slice is approximately

$$\Delta W \approx \pi w (r^2 - x^2)(x + h) \, \Delta x \text{ foot-pounds}.$$

5.64   To calculate the work required to pump the water from a bowl, think of lifting the water out one slab at a time.

The total work is

$$W = \lim_{\Delta x \to 0} \sum_0^r \pi w(r^2 - x^2)(x + h)\Delta x$$

$$= \int_0^r \pi w(x + h)(r^2 - x^2)dx$$

$$= \int_0^r x \cdot \pi w(r^2 - x^2) \, dx + hw \int_0^r \pi(r^2 - x^2) \, dx$$

$$= M_y + hwV$$

$$= \bar{x}wV + hwV$$

$$= wV(\bar{x} + h) \text{ ft} \cdot \text{lb}.$$

Here wV is the weight of the whole bowlful of water of volume V. The first integral gives the work required to lift that much water from the depth of the center of mass of the water to the level $x = 0$. The second integral gives the work done in lifting the water h feet beyond that. The actual evaluation of the integrals leads to the formula

$$W = \frac{2}{3} \pi r^3 w \left(\frac{3}{8} r + h\right). \qquad \blacksquare$$

It turns out (see Problem 20) that the formula $W = wV(\bar{x} + h)$ discovered in Example 3 does not depend on the shape of the container or the choice of liquid. It is a useful equation to know if you have to solve a lot of pumping problems.

The Work Required to Empty a Container

The work required to pump a volume V of liquid of weight-density w from a container to a height h above the original surface of the liquid is given by the equation

$$\text{Work} = wV(\bar{x} + h). \qquad (4)$$

The quantity $wV\bar{x}$ is the work required to lift the weight of the liquid a distance equal to the distance from the liquid's center of mass to the liquid's surface. The quantity wVh is the work required to lift the weight of the liquid the additional h units of distance.

EXAMPLE 4 A cylindrical storage tank 10 ft in diameter and 20 ft long is buried horizontally with its top edge 6 ft belowground. The tank is full of gasoline, whose weight-density is 42 lb/ft³. How much work does it take to pump the gasoline to 2 ft above ground level?

Solution Before pumping, the gasoline's center of mass lies on the axis of the tank, $6 + 5 = 11$ ft belowground. The gasoline weighs

$$wV = 42\pi(5)^2(20) = 21{,}000 \, \pi \text{ lb}.$$

The amount of work required to lift it the 5 ft from its center of mass to the top of the tank is

$$wV\bar{x} = (21{,}000\ \pi)(5) = 105{,}000\ \pi\ \text{ft} \cdot \text{lb}.$$

The amount of work to lift the gasoline from there to a point 2 ft aboveground, a lift of $6 + 2 = 8$ ft, is

$$wVh = (21{,}000\ \pi)(8) = 168{,}000\ \pi\ \text{ft} \cdot \text{lb}.$$

The total work required is the sum of these:

$$W = 105{,}000\ \pi + 168{,}000\ \pi = 273{,}000\ \pi\ \text{ft} \cdot \text{lb}. \qquad \blacksquare$$

Work and Kinetic Energy

Equation (2) leads to the Work–Kinetic Energy Theorem of mechanics.

THEOREM

> **The Work–Kinetic Energy Theorem**
>
> The work done by a force acting on a body equals the change in the body's kinetic energy.

To prove the theorem, we assume that the force is applied along the x-axis from x_1 to x_2 and bring together three ideas:

1. The equation for work: $W = \displaystyle\int_{x_1}^{x_2} F(x)\ dx$

2. The definition of kinetic energy: $K = \dfrac{1}{2}mv^2$

3. Newton's second law of motion: $F = m\dfrac{dv}{dt}$.

We apply the Chain Rule in the form

$$\frac{dv}{dt} = \frac{dv}{dx}\frac{dx}{dt} = v\frac{dv}{dx}$$

to rewrite the work integral in the form

$$
\begin{aligned}
W = \int_{x_1}^{x_2} F(x)\ dx &= \int_{x_1}^{x_2} m\frac{dv}{dt}\ dx \\
&= \int_{x_1}^{x_2} mv\frac{dv}{dx}\ dx \\
&= \int_{v_1}^{v_2} mv\ dv \qquad \left(\begin{array}{l} v_1 \text{ and } v_2 \text{ are the body's} \\ \text{velocities at } x_1 \text{ and } x_2 \end{array}\right) \\
&= \frac{1}{2}mv^2 \Big]_{v_1}^{v_2} \\
&= \frac{1}{2}mv_2^2 - \frac{1}{2}mv_1^2.
\end{aligned}
\qquad (5)
$$

Thus, the work done by F is the kinetic energy at x_2 minus the kinetic energy at x_1 or, more simply,

$$\text{Work} = \Delta K.$$

The virtue of this result is that if we know the body's mass and the velocities v_1 and v_2, then we can find the work without any knowledge of the force F and displacement $x_2 - x_1$. Thus, we can find the work it takes to pitch a baseball at 90 mph without knowing how hard the ball was thrown or how long it was held. The same idea applies to tennis. If we know how fast a ball is moving right after it is served, we can calculate how much work was put into accelerating the ball without knowing how much pressure the racket exerted on the ball or how long the racket was in contact with the ball.

EXAMPLE 5   A 2-oz tennis ball was served at 160 ft/sec (about 109 mph). How many foot-pounds of work were done on the ball to get it to this speed?

Solution   We use Eq. (5) with $v_1 = 0$, $v_2 = 160$, and $m =$ the mass of the ball. To find the mass from the weight of the ball, we express the weight in pounds and divide by 32, the acceleration of gravity. The resulting engineering unit is called a *slug*:

$$\text{Weight} = 2 \text{ oz} = \frac{1}{8} \text{ lb,} \qquad \text{Mass} = \frac{1}{8} \cdot \frac{1}{32} = \frac{1}{256} \text{ slug.}$$

Equation (5) then gives

$$\text{Work} = \frac{1}{2}mv_2^2 - \frac{1}{2}mv_1^2$$

$$= \frac{1}{2}\left(\frac{1}{256}\right)(160)^2 - 0$$

$$= 50.$$

It took 50 ft · lb of work to accelerate the ball.   ∎

PROBLEMS

1. If a force of 6 newtons is required to extend a spring 0.4 m beyond its natural length, how much work is required to extend the spring 0.2 m beyond its natural length?

2. A force of 90 newtons stretches a spring 1 m beyond its natural length. How much work does it take to stretch the spring 5 m beyond its natural length?

3. A spring has a natural length of 10 in. An 800-lb force stretches the spring to 14 in.
   a) Find the spring constant.
   b) How much work is done in stretching the spring from 10 in. to 12 in.?
   c) How far beyond its natural length will a 1600-lb force stretch the spring?

4. A 10,000-lb force compresses a spring from its natural length of 12 in. to a length of 11 in. How much work is done in compressing the spring
   a) from 12 to 11.5 in.?   b) from 11.5 in. to 11 in.?

5. A spring has a natural length of 2 ft. A 1-lb force stretched the spring 5 ft (from 2 ft to 7 ft). How much work did the 1-lb force do? If the spring is stretched by a 2-lb force, what will its total length be?

6. A bathroom scale is depressed 1/16 in. when a 150-lb person stands on it. Assuming that the scale behaves like a spring, find how much work is required to depress the scale 1/8 in. from its natural height. How much weight is required to compress the scale this much?

7. Two electrons repel each other with a force inversely proportional to the square of the distance between them.

   a) Suppose one electron is held fixed at the point (1, 0) on
      the x-axis. Find the work required to move a second
      electron along the x-axis from the point (−1, 0) to the
      origin.
   b) Suppose two electrons are held stationary at the points
      (−1, 0) and (1, 0) on the x-axis. Find the work done in
      moving a third electron from (5, 0) to (3, 0) along the
      x-axis.

8. If a straight hole could be bored through the center of the
   earth, a particle of mass m falling in this hole would be
   attracted toward the center of the earth with a force $mg(r/R)$
   when it is at distance r from the center. (R is the radius of
   the earth; g is the acceleration due to gravity at the surface
   of the earth.) How much work is done on the particle as it
   falls from the surface to the center of the earth?

9. A bag of sand originally weighing 144 lb is lifted at a constant rate of 3 ft/min. The sand leaks out uniformly at such
   a rate that half of the sand is lost when the bag has been
   lifted 18 ft. Find the work done in lifting the bag this far.

10. Find the work done in pumping all the gasoline from the
    tank in Example 4 if the cylinder is vertical instead of hori-
    zontal, with its top 1 ft belowground. Calculate the work
    two ways: (a) by the slab method of Example 3 and (b) with
    Eq. (4). Compare your answer with the result in Example 4.

11. About how much work is done in sucking up a milk shake
    from a conical container (initially full, wide end up) if the
    top of the straw is always π inches above the top of the
    container and the weight-density of the milk shake is
    48 lb/ft³? Express your answer in terms of the radius and
    altitude of the cone.

12. A vertical cylindrical tank 30 ft high and 20 ft in diameter is
    filled to a depth of 20 ft with kerosene weighing
    51.2 lb/ft³. How much work is done in pumping the kero-
    sene to the level of the top of the tank?

13. The curve $y = x^2$, $0 \le x \le 4$, is revolved about the y-axis to
    generate a parabolic bowl. A steel container in the shape of
    this bowl (measurements in meters) is full of water. How
    much work is done in pumping the water to the level of the
    top of the bowl? (A cubic meter of water weighs about 9800
    newtons.)

14. Suppose the tank in Example 4 is only half full to start with.
    How much work is done in emptying it?

15. Find the work done in pumping all the water out of a conical
    reservoir of radius 10 ft at the top, altitude 8 ft, to a height
    of 6 ft above the top of the reservoir. (A cubic foot of water
    weighs about 62.5 lb.)

16. Find the work done in Problem 15 if, at the beginning, the
    reservoir is filled to a depth of 5 ft and the water is pumped
    just to the top of the reservoir.

17. The driver of a leaky 800-gal tank truck carrying water from
    the base to the summit of Mt. Washington discovered upon

arrival that the tank was only half full. The truck started out
full, climbed at a steady rate, and took 50 minutes to accomplish an elevation change of 4750 ft. The water leaked
out at a steady rate. How much work was done in carrying
the water to the top? (Do not include the work done on the
truck and driver. Water weighs 8 lb/U.S. gal.)

18. You are in charge of the evacuation and repair of the storage
    tank shown in Fig. 5.65. The tank is a hemisphere of radius
    10 ft and is full of benzene with a density of 56 lb/ft³. A
    firm states that it can empty the tank at a cost of 1/2 cent per
    foot-pound of work. Find the work required to empty the
    tank by pumping the liquid to an outlet 2 ft above the top of
    the tank. If you have \$5000 budgeted for the pumping job,
    can you afford to hire the firm?

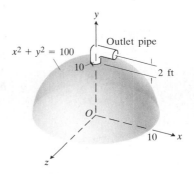

5.65   The tank in Problem 18.

19. Your town has decided to drill a well to increase its water
    supply. As the town engineer, you have determined that a
    water tower will be necessary to provide the pressure
    needed for distribution, and you have designed the system
    shown in Fig. 5.66. The well is to be 300 ft deep. The water
    will be lifted through a 4-in. pipe into the base of a cylindri-
    cal tank 20 ft in diameter and 25 ft high. The bottom of the
    tank will be 60 ft above the ground. The pump will be sub-
    merged in the well beneath the surface of the water. The

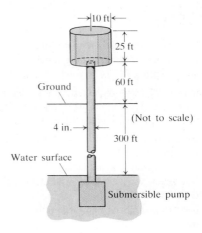

5.66   The water tower in Problem 19.

pump performs work at the rate of 1650 foot-pounds per second. How long will it take to fill the tank the first time? (Include the work to fill the pipe.)

20. Prove that, no matter what the shape of the container in Fig. 5.64, the total work is the sum of two terms, one of which is

$$W_1 = hw \int dV$$

and represents the total work done in lifting a bowlful of liquid of weight density w a distance h, while the other is

$$W_2 = w \int x\, dV$$

and represents the work done in lifting a bowlful of liquid of weight density w a distance equal to the original depth of the center of gravity of the bowl.

21. *Work and kinetic energy.* How many foot-pounds of work does it take to throw a baseball 90 mph? (A baseball weighs 5 oz = 0.3125 lb, mass = weight/g, and g = 32 ft/sec^2 if time is measured in seconds.)

22. *Work and kinetic energy.* A 1.6-oz golf ball is hit off the tee at a speed of 120 ft/sec. How many foot-pounds of work are done getting the ball into the air?

5.10
Hydrostatic Force

We make dams thicker at the bottom than at the top because the water pressure against a dam increases with depth. The deeper the water, the stronger a dam has to be. It is a remarkable fact that the water pressure at any spot on the surface of a dam depends only on how far below the surface the spot is and not at all on the volume of water the dam is holding back. The water pressure p in pounds per square foot at any point h feet below the surface on the dam's face may always be calculated with the simple formula

$$p = 62.5h,$$

where 62.5 is the weight-density of water in pounds per cubic foot. It doesn't matter whether we are talking about Lake Tahoe, Lake Michigan, or a bathtub.

The formula $p = 62.5h$ is an instance of the more general formula

$$p = wh$$

for the pressure p exerted against the walls of a container at depth h by a fluid whose weight-density is w. Typical weight-densities (lb/ft^3) are

gasoline	42
mercury	849
milk	64.5
olive oil	57
seawater	64
water	62.5

In this article, we see how to use the formula $p = wh$ to calculate the total force exerted by a standing liquid against a containing wall. Such a force is called a **hydrostatic force.**

The Force of a Liquid against a Wall

In a flat-bottomed container filled with water to a depth h, the force against the bottom of the container due to the weight of the liquid is

$$F = whA, \tag{1}$$

5.67 The hydrostatic force on the bottoms of these containers is the same.

where w is the weight-density of the water and A is the area of the bottom of the container. Obviously, the units in Eq. (1) must be compatible, say h in feet, A in square feet, w in pounds per cubic foot, giving F in pounds. This force does not depend on the shape of the sides of the container. The force on the bottoms of the containers in Fig. 5.67 will be the same as long as the containers have the same base area A and are filled to the same depth h. The **pressure, or force per unit area,** at the bottom of each container is therefore

$$p = F/A = whA/A = wh. \tag{2}$$

Next consider the forces exerted by any body of water, such as the water in a reservoir or behind a dam. According to Pascal's Principle, the pressure $p = wh$ at depth h in such a body of water is the same *in all directions*. For a flat plate submerged *horizontally*, the downward *force* acting on its upper face due to this liquid pressure is the same as that given by Eq. (1). If the plate is submerged *vertically*, however, then the pressure against it will be different at different depths and Eq. (1) is no longer usable in that form because we would have different h factors for points at different depths.

We get around this difficulty by approximating the plate with narrow rectangular strips parallel to the surface of the liquid (Fig. 5.68). The typical strip is L units long by Δh units wide. Its area is $L\,\Delta h$, and its top edge lies h units below the surface. The pressure on the strip varies from wh to $w(h + \Delta h)$. Equation (1) therefore suggests that the force ΔF on the strip will lie between $wh \cdot L\,\Delta h$ and $w(h + \Delta h) \cdot L\,\Delta h$:

$$whL\,\Delta h \le \Delta F \le w(h + \Delta h)L\,\Delta h. \tag{3}$$

When we add the forces on all the strips, we get

$$\sum whL\,\Delta h \le \sum \Delta F \le \sum w(h + \Delta h)L\,\Delta h. \tag{4}$$

If we then let Δh approach zero, we are led to the following equation for the force F on one side of the plate:

$$F = \int_a^b whL\,dh = w\int_a^b hL\,dh. \tag{5}$$

The depth of the plate's center of mass is

$$\overline{h} = \frac{\displaystyle\int_a^b hL\,dh}{\displaystyle\int_a^b L\,dh}.$$

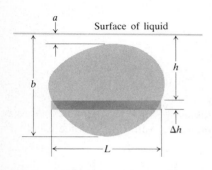

5.68 A plate submerged vertically beneath the surface of a liquid. The force exerted by the liquid on one side of the rectangular strip is approximately $whL\,\Delta h$, where w is the weight-density (weight per unit volume) of the liquid.

Hence,

$$\int_a^b hL\,dh = \overline{h}\int_a^b L\,dh = \overline{h}(\text{plate area}) = \overline{h}A, \tag{6}$$

and Eq. (5) becomes

$$F = w\overline{h}A. \tag{7}$$

Equation (7) states that the force on one side of a submerged vertical plate is the same as it would be if the plate's entire area lay $\overline{h}$ units beneath the surface. For many shapes, $\overline{h}$ can be found in a table, and Eq. (7) gives a practical way to find F. Of course, the location of the center of mass given in the table was calculated by someone who performed an integration equivalent to evaluating the integral in Eq. (5). We recommend that, for now, you find F

by thinking through the steps that lead to Eq. (5) by integration and then checking your results, when you can conveniently do so, by Eq. (7).

EXAMPLE 1 An isosceles trapezoid is submerged vertically in water with its upper edge 4 ft below the surface and its lower edge 10 ft below the surface. The upper and lower edges are respectively 6 ft and 8 ft long. Find the total force on one face of the trapezoid.

Solution In Fig. 5.69(a) we have, by similar triangles,

$$\frac{L-6}{2} = \frac{h-4}{6}.$$

Routine algebra gives

$$L = \frac{h+14}{3}.$$

Then the force is

$$F = \int_4^{10} whL\ dh = \int_4^{10} wh\left(\frac{h+14}{3}\right) dh = \frac{w}{3}\left[\frac{h^3}{3} + 7h^2\right]_4^{10} = 300w.$$

Since $w = 62.5$ lb/ft^3 for water,

$$F = (300)(62.5) = 18{,}750\ \text{lb}.$$

To check this result with Eq. (7), we divide the trapezoid into a parallelogram and a triangle (Fig. 5.69b). For the parallelogram,

$$\overline{h}_1 = 7, \qquad A_1 = 36, \qquad \text{and} \qquad F_1 = wh_1A_1 = 252w.$$

For the triangle,

$$\overline{h}_2 = 8, \qquad A_2 = 6, \qquad \text{and} \qquad F_2 = wh_2A_2 = 48w.$$

For the trapezoid,

$$F = F_1 + F_2 = 300w.$$

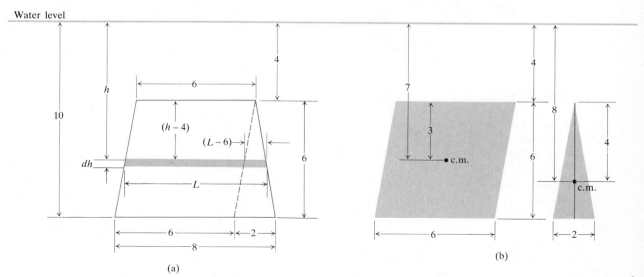

5.69 (a) The trapezoid in Example 1 submerged vertically beneath the surface of the water. (b) The trapezoid divided into a parallelogram and an isosceles triangle. The resulting centroids are marked "c.m." (center of mass).

PROBLEMS

1. The vertical ends of a water trough are inverted isosceles triangles of base 4 ft and altitude 3 ft. Find the force on one end if the trough is full of water weighing 62.5 lb/ft³.

2. Find the force in the previous problem if the water level in the trough is lowered 1 ft.

3. A triangular plate ABC is submerged in water with its plane vertical. The side AB, 4 ft long, is 1 ft below the surface, while C is 5 ft below AB. Find the total force on one face of the plate.

4. Find the force on one face of the triangle ABC of Problem 3 is AB is 1 ft below the surface as before, but the triangle is revolved 180° about AB to bring the vertex C 4 ft *above* the surface.

5. A semicircular plate is submerged in water with its plane vertical and its diameter in the surface. Find the force on one face of the plate if its diameter is 2 ft.

6. The face of a dam is a rectangle, $ABCD$, of dimensions $AB = CD = 100$ ft, $AD = BC = 26$ ft. Instead of being vertical, the plane $ABCD$ is inclined as indicated in Fig. 5.70, so that the top of the dam is 24 ft higher than the bottom. Find the force due to the water pressure on the dam when the surface of the water is level with the top of the dam.

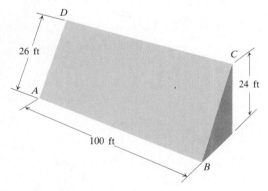

5.70   The dam in Problem 6.

7. Each end plate of the trough shown in Fig. 5.71 was designed to withstand a force of 6667 pounds. How many cubic feet of water can the tank hold without exceeding design limitations? Use $w = 62.5$ lb/ft³.

8. The cubical metal container shown in Fig. 5.72 is used to store liquids. The container has a parabolic gate (enlarged in Fig. 5.72b) held in place by bolts. The gate was designed to withstand a force of 160 lb without rupturing. The liquid you plan to store in the tank has a density of 50 lb/ft³.
   a) What will the force on the gate be when the liquid is 2 ft deep?
   b) What is the maximum height to which the container can be filled without exceeding the design limitation on the gate?

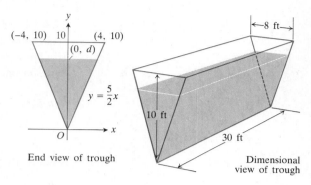

5.71   The trough in Problem 7.

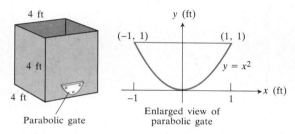

5.72   The storage tank in Problem 8.

9. Water pours into the tank shown in Fig. 5.73 at a rate of 4 ft³ per minute. The cross sections of the tank are semicircles 4 ft in diameter. One end of the tank is movable, but moving it to increase the volume of the tank compresses a spring. The spring constant is $k = 100$ lb/ft. If the end of the tank moves 5 ft against the spring, water will flow out of a hole in the bottom at a rate of 5 ft³/min. Will the tank overflow?

5.73   The tank in Problem 9.

REVIEW QUESTIONS AND EXERCISES

1. List the applications of integration in this chapter.

2. How do you define and calculate the area of a region that is bounded above by a curve $y = f_1(x)$, below by the curve $y = f_2(x)$, and on the sides by $x = a$ and $x = b$? Give an example.

3. How do you define and calculate the volumes of solids of revolution? Give examples.

4. How do you define and calculate the length of a plane curve? Give examples.

5. How do you define and calculate the area of a surface of revolution? Give examples.

6. What is the average value of a function over an interval? Give an example.

7. How are moments and centers of mass defined and calculated? Give examples.

8. Define the work done by a variable force acting along a straight line. Give an example.

9. How do you calculate the amount of work involved in pumping out a tank? Give an example.

10. How do you define the hydrostatic force on one face of a vertical plate submerged in a liquid? Give an example.

MISCELLANEOUS PROBLEMS

1. The function $v = 3t^2 - 15t + 18$ gives the velocity (ft/sec) of a body moving along a line as a function of time for $0 \le t \le 3$. Find the total distance traveled by the body and the net change in the body's position.

2. The function $v = t^3 - 3t^2 + 2t$ gives the velocity (m/sec) of a body moving along a line as a function of time for $0 \le t \le 2$. Find the total distance traveled by the body and the net change in the body's position.

In Problems 3–18, find the area of the region bounded by the given curves and lines.

3. $y = 2 - x^2$, $y = -x$

4. $y = x$, $y = 1/x^2$, $x = 2$

5. $y = x$, $y = 1/\sqrt{x}$, $x = 2$

6. $y = x + 1$, $y = 3 - x^2$

7. $y = 2x^2$, $y = x^2 + 2x + 3$

8. $x = 2y^2$, $x = 0$, $y = 3$

9. $4x = y^2 - 4$, $4x = y + 16$

10. $y = x$, $y = x^3$, $-1 \le x \le 1$

11. $y = \sin x$, $y = \sqrt{2}x/2$

12. $y = \sin x$, $y = x$, $0 \le x \le \pi/4$

13. $y^2 = 9x$, $y = 3x^2/8$

14. $y = x\sqrt{2x^2 + 1}$, $x = 0$, $x = 2$

15. $y^2 = 4x$ and $y = 4x - 2$

16. $y = 2 - x^2$ and $y = x^2 - 6$

17. $y = |\cos x|$, $y = 1$, $0 \le x \le \pi$

18. $y = 2 \sin x$ and $y = \sin 2x$, $0 \le x \le 2\pi$

19. Find the maximum and minimum points of the curve $y = x^3 - 3x^2$ and find the total area of the region that lies between the curve and the x-axis.

20. Find the area of the region cut from the first quadrant by the curve $x^{1/2} + y^{1/2} = a^{1/2}$.

21. Find the volume of the solid generated by revolving about the y-axis the region bounded above by the x-axis and below by the curve $y = x^2 - 2x$.

22. Find the volume of the solid generated by revolving about the x-axis the region bounded by the curve $y = 2 \tan x$, the x-axis, and the lines $x = -\pi/4$ and $x = \pi/4$.

23. A solid is generated by revolving about the x-axis the region bounded by the curve $y = f(x)$, the x-axis, and the lines $x = a$, $x = b$. Its volume, for all $b > a$, is $b^2 - ab$. Find $f(x)$.

24. Find the volume of the solid generated by revolving the region bounded by the given curves and lines about the line indicated.
   a) $y = x^2$, $y = 0$, $x = 3$, about the x-axis
   b) $y = x^2$, $y = 0$, $x = 3$, about the line $x = -3$
   c) $y = x^2$, $y = 0$, $x = 3$, about the y-axis, first integrating with respect to x and then integrating with respect to y
   d) $x = 4y - y^2$, $x = 0$, about the y-axis
   e) $x = 4y - y^2$, $x = 0$, about the x-axis.

25. The region bounded by the curve $y^2 = 4x$ and the straight line $y = x$ is revolved about the x-axis to generate a solid. Find the volume of the solid.

26. Sketch the region bounded by the curve $y^2 = 4ax$, the line $x = a$, and the x-axis. Find the volumes of the solids generated by revolving this region in the following ways: (a) about the x-axis, (b) about the line $x = a$, (c) about the y-axis.

27. The region bounded by the curve $y = x/\sqrt{x^3 + 8}$, x-axis, and the line $x = 2$ is revolved about the y-axis to generate a solid. Find the volume of the solid.

28. Find the volume of the solid generated by revolving the larger region bounded by $y^2 = x - 1$, $x = 3$, and $y = 1$ about the y-axis.

29. The region bounded by the curve $y^2 = 4ax$ and the line $x = a$ is revolved about the line $x = 2a$ to generate a solid. Find the volume of the solid.

30. A twisted solid is generated as follows: We are given a fixed line L in space and a square of side s in a plane perpendicular to L. One vertex of the square is on L. As this vertex moves a distance h along L, the square turns through a full revolution, with L as the axis. Find the volume of the solid generated by this motion. What would the volume be if the square had turned through two full revolutions in moving the same distance along L?

31. Find the volume of the solid generated by revolving about the x-axis the region bounded by the x-axis and one arch of the curve $y = \sin 2x$.

32. A round hole of radius $\sqrt{3}$ ft is bored through the center of a solid sphere of radius 2 ft. Find the volume cut out.

33. The cross section of a solid in any plane perpendicular to the x-axis is a circle having diameter AB with A on the curve $y^2 = 4x$ and B on the curve $x^2 = 4y$. Find the volume of the solid lying between the points of intersection of the curves.

34. The base of a solid is the region bounded by $y^2 = 4ax$ and $x = a$. Each cross section perpendicular to the x-axis is an equilateral triangle. Find the volume of the solid.

35. A solid is generated by revolving the continuous curve $y = f(x)$, $0 \le x \le a$, about the x-axis. Its volume for all a is $a^2 + a$. Find $f(x)$.

36. Suppose that f is a continuous function with the property that, for every number $a > 0$, the volume swept out by revolving the region enclosed by the x-axis and the graph of f from $x = 0$ to $x = a$ is πa^3. Find $f(x)$.

37. Find the lengths of the following curves.

a) $y = 2\sqrt{7}x^{3/2} - 1$, $0 \le x \le 1$
b) $y = (2/3)x^{3/2} - (1/2)x^{1/2}$, $0 \le x \le 4$
c) $x = (3/5)y^{5/3} - (3/4)y^{1/3}$, $0 \le y \le 1$

38. Suppose that f is a differentiable function of x for $x \ge 0$ and that $f(0) = a$. Let $s(x)$ denote the length of the curve $y = f(x)$ from $(0, a)$ to $(x, f(x))$.

a) Find $f(x)$ if $s(x) = Cx$ for some constant C. What are the allowable values for C?
b) Could $s(x)$ equal x^n for any $n > 1$? Explain.

39. Find the area of the surface swept out by revolving
a) the curve in Problem 37(b) about the y-axis;
b) the curve in Problem 37(c) about the line $y = -1$.

40. Find the area of the surface generated by revolving the curve

$$x = t^{2/3}, \quad y = t^2/2, \quad 0 \le t \le 2$$

about the x-axis.

41. *The Mean Value Theorem for Integrals.* Prove the Mean Value Theorem for Integrals, which says that if a function f is continuous at every point of an interval $a \le x \le b$, then there is at least one point c in the interval at which

$$f(c) = \frac{1}{b - a} \int_a^b f(x)\, dx.$$

(*Hint:* Show that

$$\min f \le \frac{1}{b - a} \int_a^b f(x)\, dx \le \max f$$

and apply the Intermediate Value Theorem.)

42. Find the average value of $y = \sqrt{ax}$ over the interval from $x = 0$ to $x = a$.

43. Suppose AB is the diameter of a circle of radius a. Chords are drawn perpendicular to AB, cutting AB into segments of equal length. Find the limit of the average lengths of these chords as the number of chords approaches infinity. (*Hint:*

$$\int_{-a}^{a} \sqrt{a^2 - x^2}\, dx = \frac{\pi a^2}{2}$$

because the integral gives the area of a semicircle of radius a.)

44. Solve Problem 43 under the assumption that the chords divide the circle into arcs of equal length.

45. Repeat Problem 43 using the *squares* of the lengths of the chords in place of the lengths of the chords.

46. Repeat Problem 44 using the *squares* of the lengths of the chords in place of the lengths of the chords.

47. A point moves in a straight line during the time from $t = 0$ to $t = 3$ according to the law $s = 120t - 16t^2$.
a) Find the average value of the velocity, with respect to time, for these three seconds. (Compare this with the average velocity $\Delta s/\Delta t$ for the interval.)
b) Find the average value of the velocity with respect to the distance s during the three seconds.

48. Suppose that the area from 0 to x under the curve $y = f(x)$ is
$$A = (1 + 3x)^{1/2} - 1, \quad x \ge 0.$$
a) Find the *average* rate of change of A with respect to x as x increases from 1 to 8.
b) Find the *instantaneous* rate of change of A with respect to x at $x = 5$.
c) Find $f(x)$.
d) Find the average value of f with respect to x as x increases from 1 to 8.

49. Find the center of mass of a thin homogeneous plate covering the region enclosed by the curves $y^2 = 8x$ and $y = x^2$.

50. Find the center of mass of a homogeneous plate covering the region in the first quadrant bounded by the curve $4y = x^2$, the y-axis, and the line $y = 4$.

51. Find the center of mass of a thin homogeneous plate bounded by the curves $y^2 = x$ and $x = 2y$.

52. Find the center of mass of a thin plate of constant density δ covering the region bounded by the curve $y = 4x - x^2$ and the line $2x - y = 0$.

53. Find the center of mass of a thin homogeneous plate covering the portion of the xy-plane in the first quadrant bounded by the curve $y = x^2$, the x-axis, and the line $x = 1$.

54. Suppose that a thin metal plate of area A has a constant density δ. Show that, if its first moment about the y-axis is

M, its moment about the line $x = b$ is $M - b\ \delta A$. Explain why this result shows that the center of mass is a physical property of the body, independent of the coordinate system used for finding its location.

55. Suppose *a* is a positive constant. Find the center of mass of a thin plate covering the region bounded by the curve $y^2 = 4ax$ and the line $x = a$ if the density at (x, y) is directly proportional to (a) *x*, (b) $|y|$.

56. Find the centroid of the region in the first quadrant bounded by two concentric circles and the coordinate axes, if the circles have radii *a* and *b*, $b > a > 0$, and their centers are at the origin. Find the limits of the coordinates of the centroid as *a* approaches *b*. Discuss the meaning of the result.

57. Find the centroid of the arc of the curve $x = a\ \cos^3 t$, $y = a\ \sin^3 t$ that lies in the first quadrant.

58. A triangular corner is cut from a square 12 in. on a side. The area of the cutoff triangle is 36 in^2. If the centroid of the remaining region is 7 in. from one side of the original square, how far is it from the remaining sides?

59. A right circular conical tank, point down, with top radius 5 ft and height 10 ft, is filled with a liquid of density 60 lb/ft^3. Find the work done in pumping the liquid to a trough located 2 ft above the tank. If the pump is run by a 1/2-hp motor, how long will it take to empty the tank? (One horsepower is 550 foot-pounds per second.)

60. A particle of mass *M* starts from rest at time $t = 0$ and is moved with constant acceleration *a* from $x = 0$ to $x = h$ against a variable force $F(t) = t^2$. Find the work done.

61. When a particle of mass *M* is at $(x, 0)$, it is attracted toward the origin with a force whose magnitude is k/x^2. If the particle starts from rest at $x = b$ and is acted on by no other forces, find the work done on it by the time it reaches $x = a$, $0 < a < b$.

62. Below the surface of the earth, the force of its gravitational attraction is directly proportional to the distance from its center. Find the work done in lifting an object, whose weight at the surface is *w* lb, from a distance *r* ft below the earth's surface to the surface.

63. A storage tank is a right circular cylinder 20 ft long and 8 ft in diameter with its axis horizontal. If the tank is half full of olive oil weighing 57 lb/ft^3, find the work done in emptying it through a pipe that runs from the bottom of the tank to an outlet that is 6 ft above the top of the tank.

64. Suppose that the gas in a circular cylinder of cross-section area *A* is being compressed by a piston.
a) If *p* is the pressure of the gas in pounds per square inch and *V* is the volume in cubic inches, show that the work done in compressing the gas from state (p_1, V_1) to state (p_2, V_2) is given by the equation

$$\text{Work} = \int_{(p_1,\ V_1)}^{(p_2,\ V_2)} p\ dV.$$

(*Hint:* In the coordinates suggested in Fig. 5.74, $dV = A\ dx$. The force against the piston is pA.)

5.74   The cylinder and piston in Problem 64.

b) Use the integral in part (a) to find the work done in compressing the gas from $V_1 = 243$ in^3 to $V_2 = 32$ in^3 if $p_1 = 50$ lb/in^3 and *p* and *V* obey the law $pV^{1.4} = $ constant.

65. A flat vertical metal gate in the face of a dam is shaped like the parabolic region between the curve $y = 4x^2$ and the line $y = 4$, with measurements in feet. The top of the gate lies 5 ft below the surface of the water. Find the hydrostatic force against the gate. (Use $w = 62.5$ lb/ft^3.)

66. Water is running at the rate of 1000 ft^3/hr into the swimming pool shown in Fig. 5.75.
a) Find the force against the triangular drain plate after 9 hours of filling.
b) The plate is designed to withstand a force of 520 lb. How high can the pool be filled without exceeding this design limitation?

Triangular drain plate

Enlarged view of drain plate

5.75   The swimming pool in Problem 66.

67. A triangular plate *ABC* is submerged in water with its plane vertical. The side *AB*, 4 ft long, lies 6 ft below the surface of the water, while the vertex *C* lies 2 ft below the surface. Find the force exerted by the water on one face of the plate.

68. A dam is in the form of a vertical trapezoid, with its two horizontal sides 200 and 100 ft, respectively, the longer side being at the top; the height is 20 ft. Find the hydrostatic force on the dam when the water is level with the top of the dam.

69. The **center of pressure** on a submerged plane region is defined to be the point at which the total force could be applied without changing its total moment about any axis in the plane. Find the depth of the center of pressure
  a) on a vertical rectangle of height h and width b if its upper edge is in the surface of the water,
  b) on a vertical triangle of height h and base b if the vertex opposite b is a ft below the surface and the base b is $(a + h)$ ft below the surface.

70. A container is filled with two nonmixing liquids with weight-densities w_1 and w_2, $w_1 < w_2$. Find the force on one face of a square plate $ABCD$, $6\sqrt{2}$ ft on a side, immersed in the liquids with the diagonal AC normal to the free surface, if the highest point A of the square is 2 ft below the free surface and BD lies on the surface separating the two liquids. See Fig. 5.76.

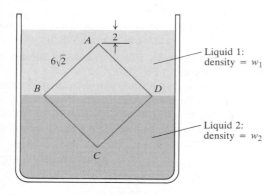

5.76 The plate in Problem 70.

The Theorems of Pappus

In the third century, an Alexandrian Greek named Pappus discovered two formulas that relate centers of mass to surfaces and volumes of revolution. These formulas, easy to remember, provide useful shortcuts to a number of otherwise lengthy calculations.

THEOREM 1 If a plane region is revolved once about an axis in the plane that does not pass through the region's interior, then the volume of the solid swept out by the region is equal to the region's area times the distance traveled by the region's center of mass. In symbols,

$$V = 2\pi \bar{y} A.$$

THEOREM 2 If an arc of a plane curve is revolved once about a line in the plane that does not cut through the interior of the arc, then the area of the surface swept out by the arc is equal to the length of the arc times the distance traveled by the arc's center of mass. In symbols,

$$S = 2\pi \bar{y} L.$$

EXAMPLE 1 The volume of the torus (doughnut) generated by revolving a circle of radius a about an axis in its plane at a distance $b \geq a$ from its center (see Fig. 5.77) is

$$V = (2\pi b)(\pi a^2) = 2\pi^2 ba^2.$$ ■

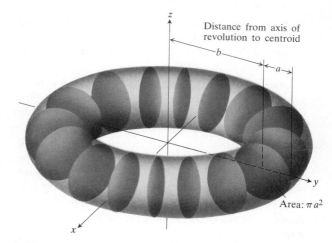

5.77 The volume swept out by the revolving disk is $(2\pi b)(\pi a^2)$.

EXAMPLE 2 The surface area of the torus in Example 1 is

$$S = (2\pi b)(2\pi a) = 4\pi^2 ba.$$ ■

71. The square region with vertices $(0, 2)$, $(2, 0)$, $(4, 2)$ and $(2, 4)$ is revolved about the x-axis to generate a solid. Find the volume and surface area of the solid.

72. Use a Theorem of Pappus to find the volume generated by revolving about the line $x = 5$ the triangular region bounded by the coordinate axes and the line $2x + y = 6$. (Do you remember how to locate the center of mass of a homogeneous triangular plate?)

73. Find the volume of the torus generated by revolving the circle $(x - 2)^2 + y^2 = 1$ about the y-axis.

74. Use the Theorems of Pappus to find the lateral surface area and the volume of a right circular cone.

75. Use the second Theorem of Pappus and the fact that the surface area of a sphere of radius a is $4\pi a^2$ to find the center of mass of the semicircle $y = \sqrt{a^2 - x^2}$.

76. As found in Problem 75, the center of mass of the semicircle $y = \sqrt{a^2 - x^2}$ lies at the point $(0, 2a/\pi)$. Find the area of the surface swept out by revolving the semicircle about the line $y = a$.

77. Use the first Theorem of Pappus and the fact that the volume of a sphere of radius a is $V = (4/3)\pi a^3$ to find the center of mass of the region enclosed by the x-axis and the semicircle $y = \sqrt{a^2 - x^2}$.

78. As found in Problem 77, the center of mass of the region enclosed by the x-axis and the semicircle $y = \sqrt{a^2 - x^2}$ lies at the point $(0, 4a/3\pi)$. Find the volume of the solid generated by revolving this region about the line $y = -a$.

79. The region of Problem 78 is revolved about the line $y = x - a$ to generate a solid. Find the volume of the solid.

80. As found in Problem 75, the center of mass of the semicircle $y = \sqrt{a^2 - x^2}$ lies at the point $(0, 2a/\pi)$. Find the area of the surface generated by revolving the semicircle about the line $y = x - a$.

CHAPTER 6

Transcendental Functions

OVERVIEW

If $y = f(t)$ measures a quantity that varies with time, then the equation $dy/dt = ky$ or

$$y' = ky \qquad (k \text{ constant}) \qquad (1)$$

says that at any time t, the rate at which y changes is proportional to the amount of y present. Depending on the function f, this might describe the loss of heat from a body immersed in a cooling medium (hot silver plunged into water); the variation of current in a battery-driven electric circuit; or the decay of a radioactive element like carbon-14 (the number of atoms disintegrating per unit time is proportional to the number of radioactive atoms that are left).

In the present chapter we shall see that one solution of $y' = ky$ is the exponential function

$$y = e^{kt} \qquad (e \approx 2.71828), \qquad (2)$$

which is one of a number of so-called transcendental functions. Euler coined the word "transcendental" to describe numbers that are not roots of polynomial equations. Such numbers "transcend the power of algebraic methods," as he put it. Today, we call a function $y = f(x)$ *transcendental* if it satisfies no equation of the form

$$P_n(x)y^n + \cdots + P_1(x)y + P_0(x) = 0 \qquad (3)$$

in which the coefficients $P_0(x)$, $P_1(x)$, . . . , $P_n(x)$ are polynomials in x with rational coefficients.

Functions that do satisfy an equation like (3) are called *algebraic*. For instance, $y = 1/\sqrt{(x + 1)}$ is algebraic because it satisfies the equation $(x + 1)y^2 - 1 = 0$. Here the polynomial coefficients are $P_2(x) = x + 1$, $P_1(x) = 0$, and $P_0(x) = -1$. Polynomials and rational functions are algebraic, and all sums, products, quotients, powers, and roots of algebraic functions are algebraic.

The six basic trigonometric functions are transcendental, as are the inverse trigonometric functions and the exponential and logarithmic functions that are the main subject of the present chapter. Transcendental functions are important in engineering and physics, as well as mathematics, and we shall see a number of their applications as the chapter goes on.

6.1

Inverse Functions

6.1 The inverse of a function f sends every output of f back to the input from which it came.

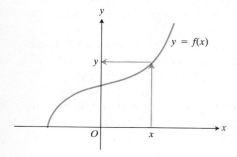

6.2 If $y = f(x)$ is a one-to-one function, then $f^{-1}(f(x)) = x$ and $f(f^{-1}(y)) = y$. Each of the composites $f^{-1} \circ f$ and $f \circ f^{-1}$ is the identity function on its domain.

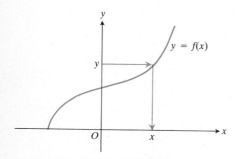

6.3 To find the value of f at x, we go up to the curve and over to the y-axis.

To proceed with our development of calculus we need to define functions that are best introduced as inverses of functions we have already worked with. In this article, we discuss what it means for two functions to be inverses of one another and what this tells us about their formulas, graphs, and derivatives.

One-to-One Functions Have Inverses

As you know, a function is a rule that assigns a number in its range to each number in its domain. Some functions, like

$$y = \sin x, \qquad y = x^2, \qquad \text{and} \qquad y = 3,$$

can give the same output for different inputs. But other functions, like

$$y = \sqrt{x}, \qquad y = x^3, \qquad \text{and} \qquad y = 4x - 4,$$

always give different outputs for different inputs. Functions that always give different outputs for different inputs are called **one-to-one** functions.

Since each output of a one-to-one function comes from just one input, any one-to-one function can be reversed to turn the outputs back into the inputs from which they came (Fig. 6.1). The function defined by reversing a one-to-one function f is called the **inverse** of f. The symbol for the inverse of f is f^{-1}, read "f inverse." The symbol -1 in f^{-1} is *not* an exponent: $f^{-1}(x)$ does not mean $1/f(x)$.

As you can see in Fig. 6.2, the result of composing f and f^{-1} in either order is the **identity function,** the function that assigns each number to itself. This is the test for whether two functions f and g are inverses of one another: Compute $f \circ g$ and $g \circ f$. If both composites are the identity function, then f and g are inverses of one another; otherwise they are not.

What functions have inverses? Increasing functions do and decreasing functions do, because they are all one-to-one (Problem 22). Among continuous functions, these are the only functions with inverses, although we shall not prove this fact.

The Graphs of Inverses

How is the graph of the inverse of a function related to the graph of the function? If the function is increasing, say, its graph will rise from left to right like the graph in Fig. 6.3. To read the graph, we start at the point x on the x-axis, go up to the graph and then over to the y-axis to read the value of y. If we start with y and want to find the x from which it came, we reverse the process (Fig. 6.4).

The graph of f is already the graph of f^{-1}, although it is not drawn in the normal way with the domain axis horizontal and the range axis vertical. To draw the graph of f^{-1} the way we are used to seeing graphs, we have to reconstruct it from the graph of f in the following way. We rotate the graph of f (Fig. 6.5a) counterclockwise to make the y-axis horizontal and the x-axis vertical (Fig. 6.5b). Then we reflect the graph across the vertical axis as if the axis were a mirror (Fig. 6.5c) to make the y-axis point to the right. Finally, we write x for y and y for x (Fig. 6.5d). We now have a normal-looking graph of f^{-1} as a function of x.

6.4 To find the x that gives y, we go over to the curve and down to the x-axis.

(a) y independent and x dependent.

(b) Rotate.

(c) Reflect.

(d) Switch letters: x independent and y dependent.

6.5 The steps in constructing the graph of f^{-1} as the graph of a function of x.

Now that we see how the graph of $y = f^{-1}(x)$ is related to the graph of $y = f(x)$, we can find a quicker way to draw it. What we did in going from Fig. 6.5(a) to 6.5(d) amounted to solving for x in terms of y and interchanging the letters x and y. This has exactly the same effect as reflecting the graph of $y = f(x)$ in the line $y = x$. An example will show you what we mean.

EXAMPLE 1 Find the inverse of the function $y = \sqrt{x}$, expressed as a function of x. Then graph $y = \sqrt{x}$ and its inverse together.

Solution We solve the equation $y = \sqrt{x}$ for x in terms of y and interchange the letters x and y:

$$y = \sqrt{x}, \qquad x = y^2, \qquad y = x^2.$$

The inverse of the function $y = \sqrt{x}$ is the function $y = x^2$, $x \geq 0$. The restriction $x \geq 0$, implicit in the range of $y = \sqrt{x}$, must be stated for the domain of its inverse $y = x^2$ because the unrestricted domain of $y = x^2$ is $-\infty < x < \infty$, which is too large.

As a check, we can write

$$f(x) = \sqrt{x}, \qquad g(x) = x^2, \qquad x \geq 0,$$

and verify that the composite in either order is the identity function:

$$g(f(x)) = (\sqrt{x})^2 = x, \qquad f(g(x)) = \sqrt{x^2} = |x| = x. \tag{1}$$

We can drop the absolute value bars in the last equation because $x \geq 0$.

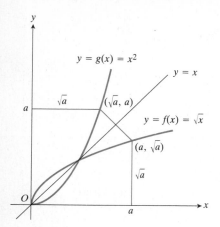

6.6 The graph of a function and its inverse are symmetric with respect to the line $y = x$.

When we graph $y = x^2$, $x \geq 0$, and $y = \sqrt{x}$ together (Fig. 6.6), we see the symmetry along the line $y = x$. The graph of $y = \sqrt{x}$ is made up of the points $(a, \sqrt{a})$, $a \geq 0$. The graph of $y = x^2$, $x \geq 0$, is made up of the points $(\sqrt{a}, a)$, $a \geq 0$. ∎

REMARK   The equations in (1) show that the function $y = \sqrt{x}$ is also the inverse of the function $y = x^2$, $x \geq 0$. Every one-to-one function is the inverse of its inverse:

$$(f^{-1})^{-1} = f.$$

EXAMPLE 2   Find the inverse of

$$y = \frac{1}{4}x + 3.$$

Solution   We solve the given equation for x in terms of y:

$$x = 4y - 12.$$

We then interchange x and y in the formula $x = 4y - 12$ to get

$$y = 4x - 12.$$

The inverse of $y = (1/4)x + 3$ is $y = 4x - 12$.
    To check, we let

$$f(x) = \frac{1}{4}x + 3, \qquad g(x) = 4x - 12.$$

Then,

$$g(f(x)) = 4\left(\frac{1}{4}x + 3\right) - 12 = x + 12 - 12 = x,$$

$$f(g(x)) = \frac{1}{4}(4x - 12) + 3 = x - 3 + 3 = x. \quad ∎$$

EXAMPLE 3   Find the inverse of the function $y = 8x^3$.

Solution   We solve the equation $y = 8x^3$ for x in terms of y, getting

$$x = \frac{\sqrt[3]{y}}{2}.$$

We interchange x and y to get

$$y = \frac{\sqrt[3]{x}}{2}.$$

The inverse of $y = 8x^3$ is $y = \sqrt[3]{x}/2$.
    To check the calculation, we form the composites of the two functions to be sure that each composite is the identity function:

$$y = 8\left(\frac{\sqrt[3]{x}}{2}\right)^3 = 8\left(\frac{x}{8}\right) = x, \qquad y = \frac{1}{2}\sqrt[3]{8x^3} = \frac{1}{2}(2x) = x. \quad ∎$$

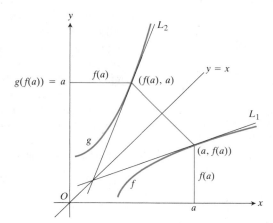

6.7 When the graphs of f and g are mirror images of one another across the line $y = x$, the tangent to the graph of f at the point $(a, f(a))$ reflects into the tangent to the graph of g at the point $(f(a), a)$. Thus, the slope of the graph of g at $x = f(a)$ is the reciprocal of the slope of the graph of f at $x = a$, provided $f'(a) \neq 0$.

Derivatives of Inverses of Differentiable Functions

If f and g are inverses of one another, their graphs are mirror images of one another across the line $y = x$, as in Fig. 6.7. Thus if L_1 is a line tangent to the graph of f at $(a, f(a))$ and L_2 is the mirror image of L_1 across the line $y = x$, it is reasonable to expect L_2 to be tangent to the graph of g at $(f(a), a)$. Since rise/run on L_2 corresponds to run/rise on L_1, we see that the slope m_2 of L_2 is the reciprocal of the slope m_1 of L_1:

$$m_2 = \frac{1}{m_1} = \frac{1}{f'(a)}. \tag{2}$$

If L_2 really is tangent to the graph of g at $(f(a), a)$, then the slope of L_2 is $g'(f(a))$, and we have

$$g'(f(a)) = \frac{1}{f'(a)}. \tag{3}$$

The rule for calculating derivatives of inverses of differentiable functions (proved in more advanced texts) is the following.

RULE 11

> **The Derivative Rule for Inverse Functions**
>
> If f is differentiable at every point of an interval I, and f' is never zero on I, then $g = f^{-1}$ is differentiable at every point of the interior of the interval $f(I)$ and its value at the point $f(x)$ is
>
> $$g'(f(x)) = \frac{1}{f'(x)}. \tag{4}$$

We showed as a corollary of the Mean Value Theorem in Article 3.7 that if f is differentiable at every point of an open interval I, and if f' is never zero on I, then f is one-to-one on I. Such a function automatically has an inverse. Rule 11 says that f^{-1} is also differentiable and that its derivative is the reciprocal of the derivative of f, as in Eq. (4).

In discussing the Chain Rule for parametric equations in Article 2.8, we said that if the derivative of $x = f(t)$ were never zero on an interval I, then the equation $x = f(t)$ defined t implicitly as a differentiable function of x. We can now see why this is so: f has an inverse because its derivative is never zero, and f^{-1} is differentiable by Rule 11. In Article 2.8 we called the inverse $t = h(x)$.

EXAMPLE 4 If $f(x) = \sqrt{x}$ and $g(x) = x^2$, then $f(9) = \sqrt{9} = 3$. Show that $g'(3) = 1/f'(9)$, as Eq. (4) predicts.

Solution We have

$$g'(x) = 2x, \qquad f'(x) = \frac{1}{2\sqrt{x}},$$

$$g'(3) = 6, \qquad f'(9) = \frac{1}{2 \cdot 3} = \frac{1}{6}.$$

Therefore,

$$g'(3) = \frac{1}{f'(9)}. \qquad \blacksquare$$

EXAMPLE 5 Verify Eq. (4) for the function $f(x) = 4x - 12$ and its inverse

$$g(x) = \frac{1}{4}x + 3.$$

Solution We have

$$f'(x) = 4, \qquad g'(x) = \frac{1}{4} = \frac{1}{f'(x)}$$

for all values of x. $\qquad \blacksquare$

EXAMPLE 6 Verify Eq. (4) for the function $f(x) = x^3$ and its inverse $g(x) = \sqrt[3]{x}$ at the point $x = 2$. That is, show that

$$g'(f(2)) = \frac{1}{f'(2)}.$$

Solution We have

$$f(2) = 2^3 = 8, \qquad g'(x) = \frac{1}{3}x^{-2/3},$$

$$f'(x) = 3x^2, \qquad g'(8) = \frac{1}{3}(8)^{-2/3} = \frac{1}{3 \cdot 8^{2/3}} = \frac{1}{3 \cdot 4} = \frac{1}{12},$$

$$f'(2) = 3 \cdot 2^2 = 12.$$

Thus, $g'(8) = 1/f'(2)$. $\qquad \blacksquare$

The virtue of Eq. (4) is that it tells exactly how the derivative of $g = f^{-1}$ is to be calculated at $f(x)$: Take the reciprocal of the value of f' at x. As with the Chain Rule, there is a shorter formulation that gives less information but may be easier to remember: If $y = f(x)$ and its inverse $x = g(y)$ are differentiable,

then

$$\frac{dx}{dy} = \frac{1}{\dfrac{dy}{dx}}. \tag{5}$$

EXAMPLE 7 Verify Eq. (5) for the inverse pair $y = \sqrt{x}$, $x = y^2$.

Solution The derivatives of the functions are

$$y = \sqrt{x}, \qquad x = y^2,$$

$$\frac{dy}{dx} = \frac{1}{2\sqrt{x}}, \qquad \frac{dx}{dy} = 2y.$$

Thus,

$$\frac{1}{\dfrac{dy}{dx}} = 2\sqrt{x} = 2y = \frac{dx}{dy}. \qquad \blacksquare$$

☐ Other Ways to Look at Rule 11

Equation (4) can be written in the form

$$g'(f(x)) \cdot f'(x) = 1, \tag{6}$$

which may remind you of the Chain Rule. Indeed, there is a connection. If f and g are differentiable functions that are inverses of one another, then

$$(g \circ f)(x) = x,$$

$$(g \circ f)'(x) = 1,$$

and the Chain Rule gives

$$(g \circ f)'(x) = g'(f(x)) \cdot f'(x),$$

or

$$1 = g'(f(x)) \cdot f'(x). \tag{7}$$

If $f'(x) \neq 0$, then Eq. (7) can be solved for $g'(f(x))$ to obtain

$$g'(f(x)) = \frac{1}{f'(x)},$$

which is Eq. (4). (Our derivation of Eq. 4 from the Chain Rule does not prove Rule 11, however, because it assumes that $g = f^{-1}$ is differentiable.)

Still another way to look at Rule 11 is this: If $y = f(x)$ is differentiable at $x = a$, then

$$dy = f'(a) \, dx.$$

This means that y is changing about $f'(a)$ times as fast as x, and x is changing about $1/f'(a)$ times as fast as y.

PROBLEMS

In Problems 1–8,

    a) find the inverse $g(x)$ of the function $f(x)$,

    b) graph f and g together,

    c) verify that Rule 11 applies to f and g at the points c and $f(c)$.

1. $f(x) = 2x + 3, \quad c = -1$

2. $f(x) = 5 - 4x, \quad c = 1/2$

3. $f(x) = (1/5)x + 7, \quad c = -1$

4. $f(x) = 2x^2, \quad x \geq 0, \quad c = 1$

5. $f(x) = x^2 + 1, \quad x \geq 0, \quad c = 5$

6. $f(x) = (x - 1)^{1/3}, \quad c = 9$

7. $f(x) = x^3 - 1, \quad c = 2$

8. $f(x) = x^2 - 2x + 1, \quad x \geq 1, \quad c = 4$

In Problems 9–16, find the inverse $f^{-1}(x)$ of each function and verify that $f(f^{-1}(x)) = f^{-1}(f(x)) = x$.

9. $f(x) = x^5$

10. $f(x) = x^4, \quad x \geq 0$

11. $f(x) = x^{2/3}, \quad x \geq 0$

12. $f(x) = (1/2)x - 7/2$

13. $f(x) = (x - 1)^2, \quad x \geq 1$

14. $f(x) = x^3 + 1$

15. $f(x) = x^{-2}, \quad x > 0$

16. $f(x) = x^{-3}, \quad x \neq 0$

17. a) Sketch the graphs of $y = x^3$ and $y = x^{1/3}$ for $-2 \leq x \leq 2$, and sketch the lines tangent to them at $(1, 1)$ and $(-1, -1)$.

    b) Which of these functions fails to have a derivative at one value of x, and what is that x? What is the slope of the other curve at x? What lines are tangent to the curves at that x?

18. a) Sketch the curve $y = 1/x$ and observe that it is symmetric about the line $y = x$.

    b) What is the slope of the curve at $P(a, 1/a)$? at $P'(1/a, a)$? What is the product of these slopes?

    c) Find the inverse of the function $f(x) = 1/x$.

19. Let $f(x) = x^2 - 4x - 3, x > 2$, and let g be the inverse of f. Find the value of g' when $f(x) = 2$.

20. Find the inverse $g(x)$ of the function $f(x) = 1 + 1/x, x \neq 0$. Then show that $f(g(x)) = g(f(x)) = x$ and that $g'(f(x)) = 1/f'(x)$ whenever f and g are both defined.

21. The volume of a spherical flask of radius 10 cm is to be increased by 1 cm^3 by eating away the interior with hydrofluoric acid. Assuming the acid eats into the surface at a uniform 0.1 mm/hr, about how long should the flask be kept full of acid?

22. *Increasing and decreasing functions.* Recall from Article 3.1 that f is an *increasing function* if, for all x_1 and x_2 in the domain of f, $x_1 < x_2 \Rightarrow f(x_1) < f(x_2)$. Similarly, f is a *decreasing function* if, for all x_1 and x_2 in its domain, $x_1 < x_2 \Rightarrow f(x_1) > f(x_2)$. Show that increasing functions and decreasing functions are one-to-one.

TOOLKIT PROGRAMS

Picard's Fixed Point Method (Among other things, this program graphs functions and their inverses together.)
Super * Grapher

6.2
The Inverse Trigonometric Functions

The inverse trigonometric functions arise in problems that require finding angles from side measurements in triangles. They also provide antiderivatives for a wide variety of functions and hence appear in solutions to a number of differential equations that arise in mathematics, engineering, and physics. In this article we show how the functions are defined, graphed, and evaluated. In Article 6.3, we shall look at their derivatives and companion integrals.

The Arc Sine

The function $y = \sin x$ is not one-to-one; it runs through its full range of values from -1 to 1 twice on every interval of length 2π. However, if we restrict the domain of the sine to the interval from $-\pi/2$ to $\pi/2$, we find that the restricted function

$$y = \sin x, \qquad -\pi/2 \leq x \leq \pi/2, \tag{1}$$

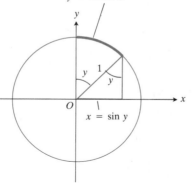

If $x = \sin y$, the length of this arc is y. That is, y is the length of arc whose sine is x. In symbols, $y = \arcsin x$.

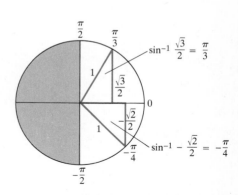

6.8 The geometry of $y = \arcsin x$ when y is positive.

6.9 The angle whose measure is $y = \sin^{-1}x$ ranges from $-\pi/2$ to $\pi/2$.

is one-to-one. It therefore has an inverse (Article 6.1), which we denote by

$$y = \sin^{-1}x. \tag{2}$$

This equation is read "y equals the **arc sine** of x" or "y equals arc sine x" and is often written as

$$y = \arcsin x. \tag{3}$$

In case you are wondering what the "arc" is doing there, look at Fig. 6.8, which gives the geometric interpretation of $y = \sin^{-1}x$ when y is positive. If $x = \sin y$, then y is the *arc* on the unit circle whose sine is x. For every value of x in the interval $[-1, 1]$, $y = \sin^{-1}x$ is the number in the interval $[-\pi/2, \pi/2]$ whose sine is x. For instance (Fig. 6.9),

$$\sin^{-1}0 = 0 \qquad \text{because} \quad \sin 0 = 0,$$
$$\sin^{-1}\sqrt{3}/2 = \pi/3 \qquad \text{because} \quad \sin \pi/3 = \sqrt{3}/2,$$
$$\sin^{-1}1 = \pi/2 \qquad \text{because} \quad \sin \pi/2 = 1,$$
$$\sin^{-1}(-\sqrt{2}/2) = -\pi/4 \qquad \text{because} \quad \sin(-\pi/4) = -\sqrt{2}/2.$$

The graph of $y = \sin^{-1}x$ is shown in Fig. 6.10. The gray curve in the figure is the reflection of the graph of $y = \sin x$ across the line $y = x$, and so it is the graph of $x = \sin y$. The graph of $y = \sin^{-1}x$ is the portion of this curve between $y = -\pi/2$ and $y = \pi/2$.

The -1 in $y = \sin^{-1}x$ is not an exponent; it means "inverse," not "reciprocal." The *reciprocal* of $\sin x$ is

$$(\sin x)^{-1} = \frac{1}{\sin x} = \csc x.$$

The graph of the arc sine in Fig. 6.10 is symmetric about the origin because the graph of $x = \sin y$ is symmetric about the origin. Algebraically, this means that

$$\sin^{-1}(-x) = -\sin^{-1}x \tag{4}$$

for every x in the domain of the arc sine, which is another way to say that the function $y = \sin^{-1}x$ is odd.

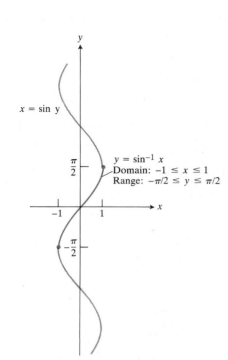

$x = \sin y$

$y = \sin^{-1} x$
Domain: $-1 \le x \le 1$
Range: $-\pi/2 \le y \le \pi/2$

6.10 The graph of $y = \sin^{-1}x$.

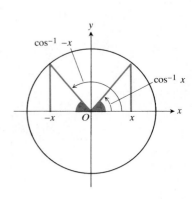

6.11   The graph of

$$y = \cos^{-1}x = \frac{\pi}{2} - \sin^{-1}x.$$

6.12 $\cos^{-1}x + \cos^{-1}(-x) = \pi.$

The Arc Cosine

Like the sine function, the cosine function $y = \cos x$ is not one-to-one, but its restriction to the interval $[0, \pi]$,

$$y = \cos x, \qquad 0 \le x \le \pi, \tag{5}$$

is one-to-one. The restricted function therefore has an inverse,

$$y = \cos^{-1}x, \tag{6}$$

which we call the **arc cosine** of x. For each value of x in the interval $[-1, 1]$, $y = \cos^{-1}x$ is the number in the interval $[0, \pi]$ whose cosine is x. The graph of $y = \cos^{-1}x$ is shown in Fig. 6.11.

As we can see from Fig. 6.12, the arc cosine of x satisfies the identity

$$\cos^{-1}x + \cos^{-1}(-x) = \pi, \tag{7}$$

or

$$\cos^{-1}(-x) = \pi - \cos^{-1}x. \tag{8}$$

We can also see from the triangle in Fig. 6.13 that for $x > 0$,

$$\sin^{-1}x + \cos^{-1}x = \pi/2 \tag{9}$$

because $\sin^{-1}x$ and $\cos^{-1}x$ are then complementary angles in a right triangle whose hypotenuse is 1 unit long and one of whose legs is x units long. Equation (9) holds for the other values of x in $[-1, 1]$ as well, but we cannot draw this conclusion from the geometry of the triangle in Fig. 6.13. It is, however, a consequence of Eqs. (4) and (8) (Problem 48).

6.13   In this figure,

$$\sin^{-1}x + \cos^{-1}x = \pi/2.$$

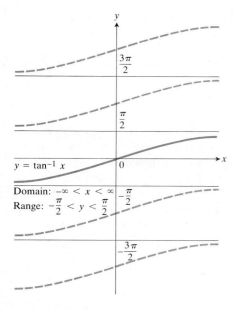

$y = \tan^{-1} x$

Domain: $-\infty < x < \infty$

Range: $-\dfrac{\pi}{2} < y < \dfrac{\pi}{2}$

6.14 The branch chosen for $y = \tan^{-1}x$ is the one through the origin.

The Inverses of tan x, sec x, csc x, cot x

The other four basic trigonometric functions, $y = \tan x$, $y = \sec x$, $y = \csc x$, and $y = \cot x$, also have inverses when suitably restricted. The inverse of

$$y = \tan x, \qquad -\pi/2 < x < \pi/2, \qquad (10)$$

is denoted by

$$y = \tan^{-1}x. \qquad (11)$$

The domain of the arc tangent is the entire real line, and the range is the open interval $(-\pi/2, \pi/2)$. For every value of x, $y = \tan^{-1}x$ is the angle between $-\pi/2$ and $\pi/2$ whose tangent is x. The graph of $y = \tan^{-1}x$ is shown in Fig. 6.14.

The graph of $y = \tan^{-1}x$ is symmetric about the origin because it is a branch of the graph of $x = \tan y$ that is symmetric about the origin. Algebraically, this means that

$$\tan^{-1}(-x) = -\tan^{-1}x. \qquad (12)$$

Like the arc sine, the arc tangent is an odd function of x.

The inverses of the (restricted) functions

$$y = \cot x, \qquad 0 < x < \pi, \qquad (13)$$

$$y = \sec x, \qquad 0 \le x \le \pi, \quad x \ne \pi/2, \qquad (14)$$

$$y = \csc x, \qquad -\pi/2 \le x \le \pi/2, \quad x \ne 0, \qquad (15)$$

are chosen to be the functions graphed in Figs. 6.15, 6.16, and 6.17. They are chosen this way to satisfy the relationships

$$\cot^{-1}x = \pi/2 - \tan^{-1}x, \qquad (16)$$

$$\sec^{-1}x = \cos^{-1}(1/x), \qquad (17)$$

$$\csc^{-1}x = \sin^{-1}(1/x). \qquad (18)$$

$y = \cot^{-1} x$

Domain: $-\infty < x < \infty$

Range: $0 < y < \pi$

6.15 The graph of

$$y = \cot^{-1}x = \pi/2 - \tan^{-1}x.$$

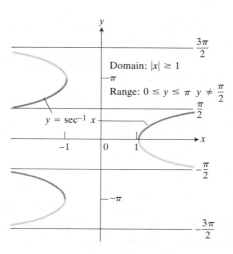

Domain: $|x| \ge 1$

Range: $0 \le y \le \pi$ $y \ne \dfrac{\pi}{2}$

$y = \sec^{-1} x$

6.16 $y = \sec^{-1}x = \cos^{-1}(1/x)$ is defined for $|x| \ge 1$.

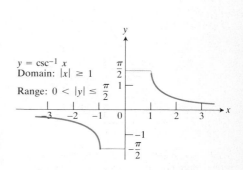

$y = \csc^{-1} x$

Domain: $|x| \ge 1$

Range: $0 < |y| \le \dfrac{\pi}{2}$

6.17 The graph of

$$y = \csc^{-1}x = \sin^{-1}(1/x).$$

We shall not dwell on these relationships, but they are handy for finding the values of $\cot^{-1}x$, $\sec^{-1}x$, and $\csc^{-1}x$ on a calculator that gives only $\tan^{-1}x$, $\cos^{-1}x$, and $\sin^{-1}x$.

In Problem 47 you will be asked to establish one further identity:

$$\sec^{-1}(-x) = \pi - \sec^{-1}x. \qquad (19)$$

It follows from Eqs. (8) and (17).

WARNING ABOUT THE ARC SECANT: Some writers choose $\sec^{-1}x$ to lie between 0 and $\pi/2$ when x is positive and between $-\pi$ and $-\pi/2$ when x is negative (hence as a negative angle in the third quadrant, as shown by the gray curve in Fig. 6.16). This has the advantage of simplifying the formula for the derivative of $\sec^{-1}x$ but the disadvantage of failing to satisfy Eq. (17) when x is negative. Also, some mathematical tables give third-quadrant values for $\sec^{-1}x$ instead of the second-quadrant values used in this book. Watch out for this when you use tables.

Right-Triangle Interpretations

The right-triangle interpretations of the inverse trigonometric functions in Fig. 6.18 can be useful in integration problems that require substitutions. We shall use some of them in Chapter 7.

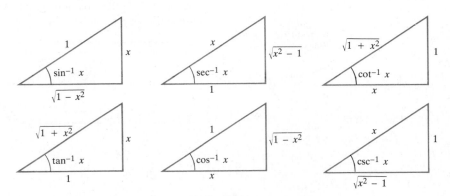

6.18   Right-triangle interpretations of the inverse trigonometric function angles (first-quadrant values).

6.19 If $\alpha = \sin^{-1}(\sqrt{3}/2)$, then the values of the trigonometric functions of α can be read from this triangle.

EXAMPLE 1   Given that

$$\alpha = \sin^{-1}\frac{\sqrt{3}}{2},$$

find $\csc \alpha$, $\cos \alpha$, $\sec \alpha$, $\tan \alpha$, and $\cot \alpha$.

Solution   We draw a reference triangle like the one shown in Fig. 6.19, with hypotenuse 2 and vertical side $\sqrt{3}$. We then calculate the length of the remaining side to be $\sqrt{(2)^2 - (\sqrt{3})^2} = 1$. The values of the trigonometric functions

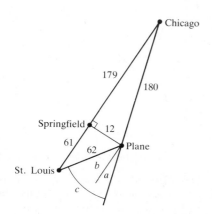

6.20 Diagram for drift correction (Example 2), with distances in miles (drawing not to scale).

of α are then read from the triangle as ratios of side lengths:

$$\csc \alpha = \frac{2}{\sqrt{3}} = \frac{2\sqrt{3}}{3}, \qquad \cos \alpha = \frac{1}{2}, \qquad \sec \alpha = 2,$$

$$\tan \alpha = \frac{\sqrt{3}}{1} = \sqrt{3}, \qquad \cot \alpha = \frac{1}{\sqrt{3}} = \frac{\sqrt{3}}{3}. \qquad \blacksquare$$

EXAMPLE 2 *Drift correction.* During an airplane flight from Chicago to St. Louis the navigator determines that the plane is 12 miles off course, as shown in Fig. 6.20. Find the angle a for a course parallel to the original, correct course, the angle b, and the correction angle $c = a + b$.

Solution

$$a = \sin^{-1}\frac{12}{180} \approx 0.067 \text{ radians} \approx 3.8°,$$

$$b = \sin^{-1}\frac{12}{62} \approx 0.195 \text{ radians} \approx 11.2°,$$

$$c = a + b \approx 15°. \qquad \blacksquare$$

Domains and Ranges, for Quick Reference

Function	Domain	Range
$y = \sin^{-1}x$	$-1 \leq x \leq 1$	$-\dfrac{\pi}{2} \leq y \leq \dfrac{\pi}{2}$
$y = \cos^{-1}x$	$-1 \leq x \leq 1$	$0 \leq y \leq \pi$
$y = \tan^{-1}x$	$-\infty < x < \infty$	$-\dfrac{\pi}{2} < y < \dfrac{\pi}{2}$
$y = \cot^{-1}x$	$-\infty < x < \infty$	$0 < y < \pi$
$y = \sec^{-1}x$	$\lvert x \rvert \geq 1$	$0 \leq y \leq \pi, \quad y \neq \dfrac{\pi}{2}$
$y = \csc^{-1}x$	$\lvert x \rvert \geq 1$	$-\dfrac{\pi}{2} \leq y \leq \dfrac{\pi}{2}, \quad y \neq 0$

PROBLEMS

Use reference triangles like the ones in Figs. 6.18 and 6.19 to find the function values in Problems 1–12.

1. a) $\tan^{-1}1$ b) $\tan^{-1}\sqrt{3}$ c) $\tan^{-1}\left(\dfrac{1}{\sqrt{3}}\right)$

2. a) $\tan^{-1}(-1)$ b) $\tan^{-1}(-\sqrt{3})$ c) $\tan^{-1}\left(\dfrac{-1}{\sqrt{3}}\right)$

3. a) $\sin^{-1}\left(\dfrac{1}{2}\right)$ b) $\sin^{-1}\left(\dfrac{1}{\sqrt{2}}\right)$ c) $\sin^{-1}\left(\dfrac{\sqrt{3}}{2}\right)$

4. a) $\sin^{-1}\left(\dfrac{-1}{2}\right)$ b) $\sin^{-1}\left(\dfrac{-1}{\sqrt{2}}\right)$ c) $\sin^{-1}\left(\dfrac{-\sqrt{3}}{2}\right)$

5. a) $\cos^{-1}\left(\dfrac{1}{2}\right)$ b) $\cos^{-1}\left(\dfrac{1}{\sqrt{2}}\right)$ c) $\cos^{-1}\left(\dfrac{\sqrt{3}}{2}\right)$

6. a) $\cos^{-1}\left(\dfrac{-1}{2}\right)$ b) $\cos^{-1}\left(\dfrac{-1}{\sqrt{2}}\right)$ c) $\cos^{-1}\left(\dfrac{-\sqrt{3}}{2}\right)$

7. a) $\sec^{-1}\sqrt{2}$ b) $\sec^{-1}\left(\dfrac{2}{\sqrt{3}}\right)$ c) $\sec^{-1}2$

8. a) $\sec^{-1}(-\sqrt{2})$ b) $\sec^{-1}\left(\dfrac{-2}{\sqrt{3}}\right)$ c) $\sec^{-1}(-2)$

9. a) $\csc^{-1}\sqrt{2}$ b) $\csc^{-1}\left(\dfrac{2}{\sqrt{3}}\right)$ c) $\csc^{-1}2$

10. a) $\csc^{-1}(-\sqrt{2})$ b) $\csc^{-1}\left(\dfrac{-2}{\sqrt{3}}\right)$ c) $\csc^{-1}(-2)$

11. a) $\cot^{-1}1$ b) $\cot^{-1}\sqrt{3}$ c) $\cot^{-1}\left(\dfrac{1}{\sqrt{3}}\right)$

12. a) $\cot^{-1}(-1)$ b) $\cot^{-1}(-\sqrt{3})$ c) $\cot^{-1}\left(\dfrac{-1}{\sqrt{3}}\right)$

13. Given that $\alpha = \sin^{-1}(1/2)$, find $\cos\alpha$, $\tan\alpha$, $\sec\alpha$, $\csc\alpha$.

14. Given that $\alpha = \cos^{-1}(-1/2)$, find $\sin\alpha$, $\tan\alpha$, $\sec\alpha$, $\csc\alpha$.

Evaluate the expressions in Problems 15–34.

15. $\sin\left(\cos^{-1}\dfrac{\sqrt{2}}{2}\right)$

16. $\tan\left(\sin^{-1}\left(-\dfrac{1}{2}\right)\right)$

17. $\sec\left(\cos^{-1}\dfrac{1}{2}\right)$

18. $\cot\left(\sin^{-1}\left(-\dfrac{1}{2}\right)\right)$

19. $\csc(\sec^{-1}2)$

20. $\cos(\tan^{-1}(-\sqrt{3}))$

21. $\cos(\cot^{-1}1)$

22. $\csc\left(\sin^{-1}\left(-\dfrac{\sqrt{2}}{2}\right)\right)$

23. $\cot(\cos^{-1}0)$

24. $\sec\left(\tan^{-1}\left(-\dfrac{1}{2}\right)\right)$

25. $\tan(\sec^{-1}1)$

26. $\sin(\csc^{-1}(-1))$

27. $\sin^{-1}(1) - \sin^{-1}(-1)$

28. $\tan^{-1}(1) - \tan^{-1}(-1)$

29. $\sec^{-1}(2) - \sec^{-1}(-2)$

30. $\sin(\sin^{-1}0.735)$

31. $\cos(\sin^{-1}0.8)$

32. $\tan^{-1}(\tan \pi/3)$

33. $\cos^{-1}(-\sin \pi/6)$

34. $\sec^{-1}(\sec(-30°))$ (The answer is *not* $-30°$.)

Find the limits in Problems 35–42. (*Hint:* If in doubt, look at the graph of the function.)

35. $\lim\limits_{x\to1^-} \sin^{-1}x$

36. $\lim\limits_{x\to-1^+} \cos^{-1}x$

37. $\lim\limits_{x\to\infty} \tan^{-1}x$

38. $\lim\limits_{x\to-\infty} \tan^{-1}x$

39. $\lim\limits_{x\to\infty} \sec^{-1}x$

40. $\lim\limits_{x\to-\infty} \sec^{-1}x$

41. $\lim\limits_{x\to\infty} \csc^{-1}x$

42. $\lim\limits_{x\to-\infty} \csc^{-1}x$

43. CALCULATOR Find the angle α in Fig. 6.21.

6.21   Find α. (*Hint:* $\alpha + \beta = 65°$.)

44. A picture a feet high is placed on a vertical wall with its base b feet above the level of an observer's eye. If the observer stands x feet from the wall, show that the angle of vision α subtended by the picture can be calculated from the equation

$$\alpha = \cot^{-1}\dfrac{x}{a+b} - \cot^{-1}\dfrac{x}{b}.$$

(*Hint:* Begin by drawing a side view of the picture and a perpendicular from the observer's eye to the wall.)

45. For what vertex angle (Fig. 6.22) does a cone whose slant height is 3 m enclose the largest possible volume?

6.22   The cone in Problem 45.

46. Find the volume generated by revolving the region in the first quadrant bounded by $y = \tan^{-1}x$, $y = 0$, and $x = 1$ about the y-axis. (*Hint:* Use washers.)

47. Combine Eqs. (8) and (17) to show that

$$\sec^{-1}(-x) = \pi - \sec^{-1}x.$$

48. Figure 6.13 shows that the equation

$$\sin^{-1}x + \cos^{-1}x = \pi/2$$

holds for $0 < x < 1$. To show that the equation holds for all x in $[-1, 1]$, carry out the following steps:
a) Show by direct calculation that

$$\sin^{-1}(1) + \cos^{-1}(1) = \pi/2,$$
$$\sin^{-1}(0) + \cos^{-1}(0) = \pi/2,$$
$$\sin^{-1}(-1) + \cos^{-1}(-1) = \pi/2.$$

b) Then, for values of x in $(-1, 0)$, let $x = -a$, $a > 0$, and apply Eqs. (4) and (8) to the sum $\sin^{-1}(-a) + \cos^{-1}(-a)$.

49. Draw a figure similar to Fig. 6.8 that gives a geometric interpretation of $y = \cos^{-1}x$.

TOOLKIT PROGRAMS

Function Evaluator   Super * Grapher

6.3
The Derivatives of the Inverse Trigonometric Functions: Related Integrals

In this article, we list the standard formulas for the derivatives of the inverse trigonometric functions (Table 6.1), show how they are derived, and discuss their companion integral formulas. As we shall see, the restrictions on the domains of the inverse trigonometric functions show up in natural ways as restrictions on the domains of the derivatives.

TABLE 6.1

Derivatives	**Differentials**
1. $\dfrac{d(\sin^{-1}u)}{dx} = \dfrac{du/dx}{\sqrt{1-u^2}}, \quad -1 < u < 1$	1'. $d(\sin^{-1}u) = \dfrac{du}{\sqrt{1-u^2}}, \quad -1 < u < 1$
2. $\dfrac{d(\cos^{-1}u)}{dx} = -\dfrac{du/dx}{\sqrt{1-u^2}}, \quad -1 < u < 1$	2'. $d(\cos^{-1}u) = -\dfrac{du}{\sqrt{1-u^2}}, \quad -1 < u < 1$
3. $\dfrac{d(\tan^{-1}u)}{dx} = \dfrac{du/dx}{1+u^2}$	3'. $d(\tan^{-1}u) = \dfrac{du}{1+u^2}$
4. $\dfrac{d(\cot^{-1}u)}{dx} = -\dfrac{du/dx}{1+u^2}$	4'. $d(\cot^{-1}u) = -\dfrac{du}{1+u^2}$
5. $\dfrac{d(\sec^{-1}u)}{dx} = \dfrac{du/dx}{|u|\sqrt{u^2-1}}, \quad |u| > 1$	5'. $d(\sec^{-1}u) = \dfrac{du}{|u|\sqrt{u^2-1}}, \quad |u| > 1$
6. $\dfrac{d(\csc^{-1}u)}{dx} = \dfrac{-du/dx}{|u|\sqrt{u^2-1}}, \quad |u| > 1$	6'. $d(\csc^{-1}u) = \dfrac{-du}{|u|\sqrt{u^2-1}}, \quad |u| > 1$

EXAMPLE 1

a) $\dfrac{d}{dx}\sin^{-1}x^2 = \dfrac{1}{\sqrt{1-(x^2)^2}} \cdot \dfrac{d}{dx}(x^2) = \dfrac{2x}{\sqrt{1-x^4}}$

b) $\dfrac{d}{dx}\tan^{-1}\sqrt{x+1} = \dfrac{1}{1+(\sqrt{x+1})^2} \cdot \dfrac{d}{dx}(\sqrt{x+1})$

$$= \dfrac{1}{x+2} \cdot \dfrac{1}{2\sqrt{x+1}}$$

$$= \dfrac{1}{2\sqrt{x+1}(x+2)}$$

c) $\dfrac{d}{dx}\sec^{-1}(3x) = \dfrac{1}{|3x|\sqrt{(3x)^2-1}} \cdot \dfrac{d}{dx}(3x)$

$$= \dfrac{3}{|3x|\sqrt{9x^2-1}} = \dfrac{1}{|x|\sqrt{9x^2-1}}. \qquad \blacksquare$$

To show how the derivative formulas listed above may be derived, we shall prove Formulas 1 and 5.

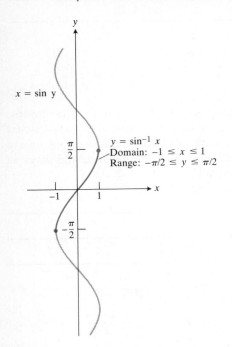

6.23   The graph of $y = \sin^{-1}x$.

The Derivative of $y = \sin^{-1}u$

We know that the function $x = \sin y$ is differentiable in the open interval $-\pi/2 < y < \pi/2$ and that its derivative, the cosine, is positive there. Rule 11 in Article 6.1 therefore assures us that the inverse function $y = \sin^{-1}x$ is differentiable throughout the interval $-1 < x < 1$. We cannot expect it to be differentiable at $x = 1$ or $x = -1$, however, because the tangents to the graph are vertical at these points (see Fig. 6.23).

To calculate the derivative of $y = \sin^{-1}x$, we differentiate both sides of the equation $\sin y = x$ with respect to x:

$$\sin y = x,$$

$$\frac{d}{dx}\sin y = 1, \tag{1}$$

$$\cos y \frac{dy}{dx} = 1.$$

We then divide through by $\cos y$ (> 0 for $-\pi/2 < y < \pi/2$) to get

$$\frac{dy}{dx} = \frac{1}{\cos y} = \frac{1}{\sqrt{1 - \sin^2 y}} = \frac{1}{\sqrt{1 - x^2}}. \tag{2}$$

The derivative of $y = \sin^{-1}x$ with respect to x is

$$\frac{d}{dx}\sin^{-1}x = \frac{1}{\sqrt{1 - x^2}}. \tag{3}$$

If u is a differentiable function of x, we apply the Chain Rule in the form

$$\frac{dy}{dx} = \frac{dy}{du}\frac{du}{dx}$$

to $y = \sin^{-1}u$ to obtain

$$\frac{d}{dx}\sin^{-1}u = \frac{1}{\sqrt{1 - u^2}}\frac{du}{dx}. \tag{4}$$

The Derivative of $y = \sec^{-1}u$

We begin by differentiating both sides of the equation $\sec y = x$ with respect to x:

$$\sec y = x,$$

$$\frac{d}{dx}\sec y = 1,$$

$$\sec y \tan y \frac{dy}{dx} = 1, \tag{5}$$

$$\frac{dy}{dx} = \frac{1}{\sec y \tan y}.$$

To express the result in terms of x, we use the relations

$$\sec y = x \quad \text{and} \quad \tan y = \pm\sqrt{\sec^2 y - 1} = \pm\sqrt{x^2 - 1}.$$

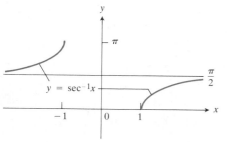

6.24 The graph of $y = \sec^{-1}x$.

Hence,

$$\frac{dy}{dx} = \pm \frac{1}{x\sqrt{x^2 - 1}}.$$

What do we do about the sign? A glance at Fig. 6.24 shows that the slope of the graph of $y = \sec^{-1}x$ is always positive. Therefore,

$$\frac{d}{dx}\sec^{-1}x = \begin{cases} \dfrac{1}{x\sqrt{x^2 - 1}} & \text{if } x > 1 \\[3mm] -\dfrac{1}{x\sqrt{x^2 - 1}} & \text{if } x < -1. \end{cases} \tag{6}$$

With absolute values, we can write Eq. (6) as a single formula:

$$\frac{d}{dx}\sec^{-1}x = \frac{1}{|x|\sqrt{x^2 - 1}}, \qquad |x| > 1. \tag{7}$$

We can then apply the Chain Rule to obtain

$$\frac{d}{dx}\sec^{-1}u = \frac{1}{|u|\sqrt{u^2 - 1}}\frac{du}{dx}, \qquad |u| > 1, \tag{8}$$

where u is a differentiable function of x.

Integration Formulas

We might expect the six derivative formulas in Table 6.1 to lead to six new integration formulas, but there are only three that matter:

Integrals Leading to Inverse Trigonometric Functions

1. $\displaystyle \int \frac{du}{\sqrt{1 - u^2}} = \sin^{-1}u + C$ (valid for $u^2 < 1$)

3. $\displaystyle \int \frac{du}{1 + u^2} = \tan^{-1}u + C$ (valid for all u)

5. $\displaystyle \int \frac{du}{u\sqrt{u^2 - 1}} = \int \frac{d(-u)}{(-u)\sqrt{u^2 - 1}} = \sec^{-1}|u| + C = \cos^{-1}\left|\frac{1}{u}\right| + C$

    (valid for $u^2 > 1$)

The others add nothing new:

2. $\displaystyle \int \frac{du}{\sqrt{1 - u^2}} = -\cos^{-1}u + C,$

4. $\displaystyle \int \frac{du}{1 + u^2} = -\cot^{-1}u + C,$

6. $\displaystyle \int \frac{du}{u\sqrt{u^2 - 1}} = -\csc^{-1}|u| + C.$

An isosceles right triangle:

$\sec^{-1}\sqrt{2} = \dfrac{\pi}{4}$

A 30-60-90 triangle:

$\sec^{-1}\dfrac{2}{\sqrt{3}} = \dfrac{\pi}{6}$

6.25 Triangles for finding $\sec^{-1}\sqrt{2}$ and $\sec^{-1}(2/\sqrt{3})$.

EXAMPLE 2   Evaluate

a) $\displaystyle\int_{0}^{1} \frac{dx}{1 + x^2}$ and   b) $\displaystyle\int_{2/\sqrt{3}}^{\sqrt{2}} \frac{dx}{x\sqrt{x^2 - 1}}$.

Solution

a) $\displaystyle\int_{0}^{1} \frac{dx}{1 + x^2} = \tan^{-1}x\Big]_{0}^{1} = \tan^{-1}1 - \tan^{-1}0 = \frac{\pi}{4} - 0 = \frac{\pi}{4}$

b) $\displaystyle\int_{2/\sqrt{3}}^{\sqrt{2}} \frac{dx}{x\sqrt{x^2 - 1}} = \sec^{-1}x\Big]_{2/\sqrt{3}}^{\sqrt{2}} = \frac{\pi}{4} - \frac{\pi}{6} = \frac{\pi}{12}$

If you do not see right away how to evaluate $\sec^{-1}x$ at the limits of integration, draw triangles like the ones in Fig. 6.25.

EXAMPLE 3   Evaluate

$$\int \frac{x^2\, dx}{\sqrt{1 - x^6}}.$$

Solution   The resemblance between the given integral and the standard form

$$\int \frac{du}{\sqrt{1 - u^2}} = \sin^{-1}u + C$$

suggests the substitution

$$u = x^3, \qquad du = 3x^2\, dx.$$

Indeed,

$$\int \frac{x^2\, dx}{\sqrt{1 - x^6}} = \frac{1}{3} \int \frac{3x^2\, dx}{\sqrt{1 - (x^3)^2}}$$

$$= \frac{1}{3} \int \frac{du}{\sqrt{1 - u^2}}$$

$$= \frac{1}{3} \sin^{-1}u + C$$

$$= \frac{1}{3} \sin^{-1}(x^3) + C.$$

EXAMPLE 4   Evaluate

$$\int \frac{dx}{\sqrt{9 - x^2}}.$$

Solution   This is the integral for the arc sine of x with a 9 in place of the 1. To replace the 9 with a 1, we factor a 9 from $9 - x^2$ and bring it outside the radical as a 3:

$$\sqrt{9 - x^2} = \sqrt{9\left(1 - \frac{x^2}{9}\right)} = 3\sqrt{1 - \left(\frac{x}{3}\right)^2}.$$

Notice that we also write $x^2/9$ as $(x/3)^2$. Then

$$\int \frac{dx}{\sqrt{9-x^2}} = \int \frac{dx}{3\sqrt{1-(x/3)^2}}.$$

We now substitute

$$u = \frac{x}{3} \quad \text{and} \quad du = \frac{dx}{3}, \quad \text{or} \quad dx = 3 \, du.$$

This gives

$$\int \frac{dx}{\sqrt{9-x^2}} = \int \frac{dx}{3\sqrt{1-(x/3)^2}}$$

$$= \int \frac{3 \, du}{3\sqrt{1-u^2}} = \int \frac{du}{\sqrt{1-u^2}}$$

$$= \sin^{-1}u + C = \sin^{-1}\left(\frac{x}{3}\right) + C. \qquad \blacksquare$$

PROBLEMS

In Problems 1–20, find dy/dx.

1. $y = \cos^{-1}x^2$ 2. $y = \cos^{-1}(1/x)$

3. $y = 5 \tan^{-1}3x$ 4. $y = \cot^{-1}\sqrt{x}$

5. $y = \sin^{-1}(x/2)$ 6. $y = \sin^{-1}(1-x)$

7. $y = \sec^{-1}5x$ 8. $y = (1/3) \tan^{-1}(x/3)$

9. $y = \csc^{-1}(x^2+1)$ 10. $y = \cos^{-1}2x$

11. $y = \csc^{-1}\sqrt{x} + \sec^{-1}\sqrt{x}$ 12. $y = \csc^{-1}\sqrt{x+1}$

13. $y = \cot^{-1}\sqrt{x-1}$ 14. $y = x\sqrt{1-x^2} - \cos^{-1}x$

15. $y = \sqrt{x^2-4} - 2 \sec^{-1}(x/2)$

16. $y = \cot^{-1}\dfrac{2}{x} + \tan^{-1}\dfrac{x}{2}$

17. $y = \tan^{-1}\dfrac{x-1}{x+1}$ 18. $y = x \sin^{-1}x + \sqrt{1-x^2}$

19. $y = x(\sin^{-1}x)^2 - 2x + 2\sqrt{1-x^2} \sin^{-1}x$

20. $y = x \cos^{-1}2x - (1/2)\sqrt{1-4x^2}$

Evaluate the integrals in Problems 21–40.

21. $\displaystyle\int_0^{1/2} \frac{dx}{\sqrt{1-x^2}}$ 22. $\displaystyle\int_{-1}^1 \frac{dx}{1+x^2}$

23. $\displaystyle\int_{\sqrt{2}}^2 \frac{dx}{x\sqrt{x^2-1}}$ 24. $\displaystyle\int_{-2}^{-\sqrt{2}} \frac{dx}{x\sqrt{x^2-1}}$

25. $\displaystyle\int_{-1}^0 \frac{4 \, dx}{1+x^2}$ 26. $\displaystyle\int_{\sqrt{3}/3}^{\sqrt{3}} \frac{6 \, dx}{1+x^2}$

27. $\displaystyle\int_0^{\sqrt{2}/2} \frac{x \, dx}{\sqrt{1-x^4}}$ 28. $\displaystyle\int_0^{1/4} \frac{dx}{\sqrt{1-4x^2}}$

29. $\displaystyle\int_{1/\sqrt{3}}^1 \frac{dx}{x\sqrt{4x^2-1}}$ 30. $\displaystyle\int_0^1 \frac{x}{1+x^4} \, dx$

31. $\displaystyle\int_0^{\sqrt{2}} \frac{4x \, dx}{\sqrt{4-x^4}}$ 32. $\displaystyle\int_0^1 \frac{dx}{\sqrt{4-x^2}}$

33. $\displaystyle\int_{\sqrt{2}}^{\sqrt[4]{2}} \frac{x \, dx}{x^2\sqrt{x^4-1}}$

34. $\displaystyle\int_2^4 \frac{dx}{2x\sqrt{x-1}}$ (Hint: Let $x = u^2$.)

35. $\displaystyle\int_0^2 \frac{dx}{1+(x-1)^2}$ 36. $\displaystyle\int_1^3 \frac{2 \, dx}{\sqrt{x}(1+x)}$

37. $\displaystyle\int_{1/2}^{3/4} \frac{dx}{\sqrt{x} \sqrt{1-x}}$ 38. $\displaystyle\int_{3/2}^{1+\sqrt{2}/2} \frac{dx}{\sqrt{1-(x-1)^2}}$

39. $\displaystyle\int_{-2/3}^{-\sqrt{2}/3} \frac{dx}{x\sqrt{9x^2-1}}$ 40. $\displaystyle\int_{-\pi/2}^{\pi/2} \frac{2 \cos x \, dx}{1+\sin^2x}$

Use l'Hôpital's rule (Article 3.8) to evaluate the limits in Problems 41–44.

41. $\displaystyle\lim_{x\to0} \frac{\sin^{-1}2x}{x}$ 42. $\displaystyle\lim_{x\to0} \frac{2 \tan^{-1}3x}{5x}$

43. $\lim\limits_{x \to 0} x^{-3} (\sin^{-1}x - x)$ **44.** $\lim\limits_{x \to 0} x^{-3} (\tan^{-1}x - x)$

45. Let $f(x) = \sin^{-1}x + \cos^{-1}x$. Find
  a) $f'(x)$,   b) $f(0.32)$.   _Look for an identity_

46. Find a curve whose slope at the point (x, y) is $(1 - x^2)^{-1/2}$ and that passes through the point $(0, 1)$.

47. A beachcomber, walking 4 km/hr along a straight shore, is tracked by a rotating light 1 km offshore. Find the rate (radians per hour) at which the light rotates when the beachcomber is 2 km from the point on the shore closest to the light. (The beachcomber is walking toward this point.)

48. Find the volume of the solid generated by revolving about the x-axis the "triangular" region in the first quadrant bounded by the curve $y = 1/\sqrt{1 + x^2}$ and the lines $x = 0$ and $y = x/\sqrt{2}$.

49. Can the integrations in (a) and (b) both be correct? Explain.

  a) $\displaystyle\int \frac{dx}{\sqrt{1 - x^2}} = \sin^{-1}x + C$

  b) $\displaystyle\int \frac{dx}{\sqrt{1 - x^2}} = -\int -\frac{dx}{\sqrt{1 - x^2}} = -\cos^{-1}x + C$

50. a) Show that the functions

$$f(x) = \sin^{-1}\frac{x - 1}{x + 1} \quad \text{and} \quad g(x) = 2 \tan^{-1}\sqrt{x},$$

  both defined for $x \geq 0$, have the same derivative and therefore that

$$f(x) = g(x) + C. \tag{9}$$

  b) Find C. (*Hint:* Evaluate both sides of Eq. 9 for a particular value of x.)

51. The integral

$$\int \frac{x \, dx}{\sqrt{1 - x^2}}$$

does not involve inverse trigonometric functions. Evaluate it some other way.

52. Let $f(x) = \displaystyle\int_0^x \frac{dt}{1 + t^2}$.
  a) Show that $f(x) + f(1/x) = $ constant. (*Hint:* Find the derivative of the sum.)
  b) Find the value of the constant in (a).

In Problems 53–56, solve the differential equations subject to the given initial conditions.

53. $(x^2 + 1) \dfrac{dy}{dx} = -y^2, \quad y = 1$ when $x = 0$

54. $\sqrt{1 - x^2} \dfrac{dy}{dx} = \sqrt{1 - y^2}, \quad y = 0$ when $x = 0$

55. $x\sqrt{x^2 - 1} \dfrac{dy}{dx} = \sqrt{1 - y^2}, \quad y = -1/2$ when $x = 2$

56. $\dfrac{dy}{dx} = -\dfrac{\sqrt{1 - y^2}}{x^2 + 1}, \quad y = 1/2$ when $x = 1$

57. Verify the following derivative formulas.

  a) $\dfrac{d(\cos^{-1}u)}{dx} = -\dfrac{du/dx}{\sqrt{1 - u^2}}, \quad -1 < u < 1$

  b) $\dfrac{d(\tan^{-1}u)}{dx} = \dfrac{du/dx}{1 + u^2}$

  c) $\dfrac{d(\cot^{-1}u)}{dx} = -\dfrac{du/dx}{1 + u^2}$

  d) $\dfrac{d(\csc^{-1}u)}{dx} = \dfrac{-du/dx}{|u|\sqrt{u^2 - 1}}, \quad |u| > 1$

TOOLKIT PROGRAMS

Derivative Grapher   Super ∗ Grapher

6.4
The Natural Logarithm and Its Derivative

Toward the end of the sixteenth century, a Scottish baron, John Napier (1550–1617), invented a device called a *logarithm* that simplified arithmetic by replacing multiplication by addition. It used the equation

$$\text{logarithm of } ax = \text{logarithm of } a + \text{logarithm of } x. \tag{1}$$

To multiply two positive numbers a and x, you looked up the logarithms of a and x in a table. You then added the logarithms on a piece of paper, found the sum in the body of the table, and read the table backward to find the product ax.

Having the table was the key, of course, and Napier spent the last two decades of his life working on a table he never finished (while the astronomer Tycho Brahe waited in vain for the information he needed to speed his calculations). The table was finished after Napier's death by Henry Briggs, a friend of Napier's in London, and subsequently publicized throughout Europe by Johannes Kepler, Tycho Brahe having long been dead.

Napier's discovery, the biggest single improvement in arithmetic before the advent of computers, made possible the decimal calculations of astronomy, navigation, and trigonometry—all because of the equation $\log(ax) = \log a + \log x$. Today we use calculators, not tables, to find the logarithms in this equation, but the equation is no less important. In the next few articles we shall see how logarithms are defined and why they have the algebraic properties they do. As we shall also see, there are many different kinds of logarithms. We begin with the one most useful in calculus, the "natural" logarithm, defined as an integral.

The Natural Logarithm and Its Derivative

The **natural logarithm** of a positive number x, denoted by $\ln x$, is defined to be the value of the integral of the function $1/t$ from $t = 1$ to $t = x$.

DEFINITION

> **The Natural Logarithm Function $y = \ln x$**
>
> $$\ln x = \int_1^x \frac{1}{t}\, dt, \qquad x > 0 \tag{2}$$

For any x greater than 1, this integral represents the area of the region that is bounded above by the curve $y = 1/t$, below by the t-axis, on the left by the line $t = 1$, and on the right by the line $t = x$. (See Fig. 6.26.)

If $x = 1$, the left and right boundaries of the region are identical and the area is zero:

$$\ln 1 = \int_1^1 \frac{1}{t}\, dt = 0. \tag{3}$$

If x is less than 1, then the left boundary is the line $t = x$ and the right boundary is $t = 1$. In this case,

$$\ln x = \int_1^x \frac{1}{t}\, dt = -\int_x^1 \frac{1}{t}\, dt \tag{4}$$

is the negative of the area under the curve between x and 1.

The value of the definite integral in (2) can be calculated by Simpson's rule to as many decimal places as we please. (For another method, see Article 12.4.) We shall study the range of $y = \ln x$ in Article 6.5.

Since the function $y = \ln x$ is defined by the integral

$$\ln x = \int_1^x \frac{1}{t}\, dt, \qquad x > 0,$$

it follows at once, from the First Fundamental Theorem of Calculus (Article 4.7), that

$$\frac{d}{dx} \ln x = \frac{1}{x}. \tag{5}$$

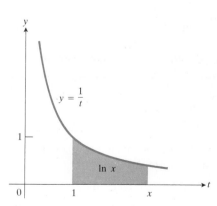

6.26 $\ln x = \displaystyle\int_1^x \frac{1}{t}\, dt, \quad \dfrac{d(\ln x)}{dx} = \dfrac{1}{x}.$

If u is a differentiable function of x, then the Chain Rule,

$$\frac{dy}{dx} = \frac{dy}{du}\frac{du}{dx},$$

gives the more general formula

$$\frac{d}{dx}\ln u = \frac{1}{u}\frac{du}{dx}. \tag{6}$$

EXAMPLE 1   Find dy/dx if $y = \ln(3x^2 + 4)$.

Solution

$$\frac{dy}{dx} = \frac{1}{3x^2 + 4}\frac{d(3x^2 + 4)}{dx}$$

$$= \frac{6x}{3x^2 + 4}.$$

The Integral $\int \frac{1}{u}\, du$

The integral formula

$$\int u^n\, du = \frac{u^{n+1}}{n+1} + C, \qquad n \neq -1,$$

that we derived in Chapter 4 failed to cover the case $n = -1$ because there is no power of u whose derivative is $1/u$.

We are now in a position to treat this exceptional case because Eq. (6) leads at once to the equation

$$\int u^{-1}\, du = \int \frac{1}{u}\, du = \int \frac{1}{u}\frac{du}{dx}\, dx = \ln u + C, \tag{7a}$$

provided u is positive. But if u is negative, then $-u$ is positive and

$$\int \frac{du}{u} = \int \frac{d(-u)}{-u} = \ln(-u) + C'. \tag{7b}$$

The two results (7a, b) can be combined into a single result, namely,

$$\int \frac{du}{u} = \begin{cases} \ln u + C & \text{if } u > 0, \\ \ln(-u) + C' & \text{if } u < 0. \end{cases} \tag{8}$$

If u does not change sign on the domain given for it, then a single constant of integration is sufficient and we may use the formula

$$\int \frac{du}{u} = \ln|u| + C. \tag{9}$$

In applications, it is important to remember that the function u here can be any differentiable function $u = f(x)$. Equation (9) says that integrals of a certain *form* lead to logarithms. That is,

$$\int \frac{f'(x)}{f(x)}\, dx = \ln|f(x)| + C \tag{10}$$

whenever $f(x)$ is a differentiable function that does not change sign on the domain given for it.

EXAMPLE 2 Evaluate

$$\int \frac{6x}{3x^2 + 4}\, dx.$$

Solution This integral has the form

$$\int \frac{f'(x)}{f(x)}\, dx$$

with $f(x) = 3x^2 + 4$ and $f'(x) = 6x$. The function f does not change sign (it is always positive). Therefore Eq. (10) applies, and

$$\int \frac{6x}{3x^2 + 4}\, dx = \ln|3x^2 + 4| + C = \ln(3x^2 + 4) + C.$$

We can do without the absolute value bars because $3x^2 + 4$ is positive. ■

EXAMPLE 3 Find a function $y = f(x)$ such that $dy/dx = 1/x$, $f(1) = 1$, and $f(-1) = 2$. Sketch the solution.

Solution There are constants C and C' such that

$$y = \begin{cases} \ln x + C & \text{if } x \text{ is positive,} \\ \ln(-x) + C' & \text{if } x \text{ is negative.} \end{cases}$$

Substituting $y = 1$ when $x = 1$, we get

$$1 = \ln 1 + C = 0 + C, \qquad \text{so} \qquad C = 1.$$

Likewise, putting $y = 2$ when $x = -1$, we get

$$2 = \ln(-(-1)) + C' = \ln 1 + C' = 0 + C', \qquad \text{so} \qquad C' = 2.$$

The complete solution is therefore

$$y = f(x) = \begin{cases} \ln x + 1 & \text{if } x \text{ is positive,} \\ \ln(-x) + 2 & \text{if } x \text{ is negative.} \end{cases}$$

A graph of the solution is shown in Fig. 6.27. (For a general analysis of the graph of $y = \ln x$, see Article 6.5.)

REMARK It is unusual to have two initial conditions for the solution of a first order differential equation. We have two here because the general solution of $dy/dx = 1/x$ consists of two families of curves, one for $x > 0$ and one for $x < 0$. To select a complete particular solution, we need to choose a curve from each family. We chose the curve for $x > 0$ with the condition $f(1) = 1$ and the curve for $x < 0$ with the condition $f(-1) = 2$. ■

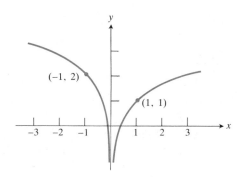

6.27 Graph of a function $f(x)$ whose derivative is $1/x$ with $f(-1) = 2$ and $f(1) = 1$:

$$f(x) = \begin{cases} \ln x + 1 & \text{when } x \text{ is positive,} \\ \ln(-x) + 2 & \text{when } x \text{ is negative.} \end{cases}$$

The two constants of integration C and C' in Example 3 are not equal.

EXAMPLE 4   Evaluate

$$\int_{-\pi/2}^{\pi/2} \frac{\cos \theta \, d\theta}{2 + \sin \theta}.$$

Solution   To evaluate the associated indefinite integral, we substitute

$$u = 2 + \sin \theta, \qquad du = \cos \theta \, d\theta.$$

Then

$$\int \frac{\cos \theta \, d\theta}{2 + \sin \theta} = \int \frac{du}{u} = \ln|u| + C = \ln(2 + \sin \theta) + C.$$

From this, we find

$$\int_{-\pi/2}^{\pi/2} \frac{\cos \theta \, d\theta}{2 + \sin \theta} = \ln(2 + \sin \theta) \Big]_{-\pi/2}^{\pi/2}$$

$$= \ln(2 + 1) - \ln(2 + (-1))$$

$$= \ln 3 - \ln 1$$

$$= \ln 3.$$

EXAMPLE 5   Evaluate

$$\int \frac{\ln x}{x} \, dx.$$

Solution   We substitute

$$u = \ln x, \qquad du = \frac{1}{x} \, dx.$$

Then

$$\int \frac{\ln x}{x} \, dx = \int u \, du = \frac{u^2}{2} + C = \frac{(\ln x)^2}{2} + C.$$

EXAMPLE 6   Evaluate

$$\int_{1}^{4} \frac{dx}{\sqrt{x}(1 + \sqrt{x})}.$$

Solution   To evaluate the associated indefinite integral, we substitute

$$u = 1 + \sqrt{x}, \qquad du = \frac{1}{2\sqrt{x}} \, dx, \qquad 2 \, du = \frac{1}{\sqrt{x}} \, dx.$$

Then

$$\int \frac{dx}{\sqrt{x}(1 + \sqrt{x})} = 2 \int \frac{du}{u} = 2 \ln|u| + C = 2 \ln(1 + \sqrt{x}) + C.$$

From this we find

$$\int_{1}^{4} \frac{dx}{\sqrt{x}(1 + \sqrt{x})} = 2 \ln(1 + \sqrt{x}) \Big]_{1}^{4} = 2(\ln 3 - \ln 2).$$

The Integrals of $y = \tan x$ and $y = \cot x$

The formula

$$\int \frac{du}{u} = \ln|u| + C$$

enables us at last to integrate the tangent and cotangent functions. For the tangent, we have

$$\int \tan x \, dx = \int \frac{\sin x \, dx}{\cos x} = -\int \frac{-\sin x \, dx}{\cos x}$$

$$= -\int \frac{du}{u} \qquad \left(\begin{array}{l} u = \cos x, \\ du = -\sin x \, dx \end{array} \right) \qquad (11)$$

$$= -\ln|u| + C$$

$$= -\ln|\cos x| + C$$

$$= \ln|\sec x| + C.$$

(The identity $-\ln(a) = \ln(1/a)$, used in the last step, will be justified in Article 6.5.)

For the cotangent we have

$$\int \cot x \, dx = \int \frac{\cos x \, dx}{\sin x}$$

$$= \int \frac{du}{u} \qquad \left(\begin{array}{l} u = \sin x, \\ du = \cos x \, dx \end{array} \right) \qquad (12)$$

$$= \ln|u| + C$$

$$= \ln|\sin x| + C.$$

The general formulas are

$$\int \tan u \, du = -\ln|\cos u| + C = \ln|\sec u| + C \qquad (13)$$

$$\int \cot u \, du = \ln|\sin u| + C \qquad (14)$$

EXAMPLE 7 Evaluate

$$\int 2x \tan(5x^2 - 1) \, dx.$$

Solution   We substitute

$$u = 5x^2 - 1, \qquad du = 10x \, dx, \qquad 2x \, dx = \frac{1}{5} \, du.$$

Then

$$\int 2x \tan(5x^2 - 1) \, dx = \int \frac{1}{5} \tan u \, du = \frac{1}{5} \int \tan u \, du$$

$$= \frac{1}{5} \ln|\sec u| + C = \frac{1}{5} \ln|\sec(5x^2 - 1)| + C. \quad \blacksquare$$

Important Formulas

1. $\ln x = \displaystyle\int_1^x \frac{1}{t}\, dt, \qquad x > 0 \qquad$ (definition of $\ln x$)

2. $\dfrac{d}{dx} \ln x = \dfrac{1}{x}$

3. $\dfrac{d}{dx} \ln u = \dfrac{1}{u}\dfrac{du}{dx}$

4. $\displaystyle\int \frac{du}{u} = \begin{cases} \ln u + C & \text{if } u > 0, \\ \ln(-u) + C' & \text{if } u < 0 \end{cases}$

5. $\displaystyle\int \frac{du}{u} = \ln|u| + C \qquad \left(\begin{array}{l}\text{if } u \text{ does not change sign on the} \\ \text{domain of integration}\end{array}\right)$

PROBLEMS

In Problems 1–20, find dy/dx.

1. $y = \ln 2x$

2. $y = \ln 5x$

3. $y = \ln kx$ (k constant)

4. $y = (\ln x)^2$

5. $y = \ln(10/x)$

6. $y = \ln(x^2 + 2x)$

7. $y = (\ln x)^3$

8. $y = \ln(\cos x)$

9. $y = \ln(\sec x + \tan x)$

10. $y = x \ln x - x$

11. $y = x^3 \ln(2x)$

12. $y = \ln(\csc x)$

13. $y = \tan^{-1}(\ln x)$

14. $y = \ln(\ln x)$

15. $y = x^2 \ln(x^2)$

16. $y = \ln(x^2 + 4) - x \tan^{-1}\dfrac{x}{2}$

17. $y = \ln x - \dfrac{1}{2}\ln(1 + x^2) - \dfrac{\tan^{-1}x}{x}$

18. $y = x(\ln x)^3$

19. $y = x[\sin(\ln x) + \cos(\ln x)]$

20. $y = x \ln(a^2 + x^2) - 2x + 2a \tan^{-1}\dfrac{x}{a}$

Evaluate the integrals in Problems 21–40.

21. $\displaystyle\int \frac{dx}{x}$

22. $\displaystyle\int \frac{2\, dx}{x}$

23. $\displaystyle\int \frac{dx}{2x}$

24. $\displaystyle\int \frac{dx}{x + 2}$

25. $\displaystyle\int_0^1 \frac{dx}{x + 1}$

26. $\displaystyle\int_{-1}^0 \frac{dx}{1 - x}$

27. $\displaystyle\int_{-1}^0 \frac{dx}{2x + 3}$

28. $\displaystyle\int_{-1}^0 \frac{3\, dx}{2 - 3x}$

29. $\displaystyle\int_0^1 \frac{x\, dx}{4x^2 + 1}$

30. $\displaystyle\int_0^\pi \frac{\sin x\, dx}{2 - \cos x}$

31. $\displaystyle\int \tan 3x\, dx$

32. $\displaystyle\int \cot 5x\, dx$

33. $\displaystyle\int \frac{x^2\, dx}{4 - x^3}$

34. $\displaystyle\int \frac{\sec^2 2x\, dx}{1 + \tan 2x}$

35. $\displaystyle\int \frac{dx}{x \ln x}$

36. $\displaystyle\int \frac{dx}{x\,(\ln x)^2}$

37. $\displaystyle\int_1^2 \frac{(\ln x)^2\, dx}{x}$

38. $\displaystyle\int_1^3 \frac{\cos(\ln x)\, dx}{x}$

39. $\displaystyle\int \frac{\sec^2 x + \sec x \tan x}{\sec x + \tan x}\, dx$

40. $\displaystyle\int \frac{dx}{(1 + x^2) \tan^{-1}x}$

Evaluate the limits in Problems 41–44.

41. $\displaystyle\lim_{x \to \infty} \frac{\ln x}{x}$

42. $\displaystyle\lim_{x \to \infty} \frac{\ln(\ln x)}{\ln x}$

43. $\displaystyle\lim_{t \to 0} \frac{\ln(1 + 2t) - 2t}{t^2}$

44. $\displaystyle\lim_{\theta \to 0^+} \frac{\ln(\sin \theta)}{\cot \theta}$

45. Find the area of the "triangular" region in the first quadrant bounded by the lines $x = 1$ and $y = 1$ and the hyperbola $xy = 2$.

46. Find the center of mass of a thin homogeneous plate bounded by the curve $y = 1/x$, the x-axis, and the lines $x = 1$ and $x = 2$.

47. Find the center of mass of a thin plate of constant density $\delta = 1$ bounded by the curves $y = 1/(1 + x^2)$ and $y = -1/(1 + x^2)$ and by the lines $x = 0$ and $x = 1$.

48. Solve the differential equation

$$y'' = \sec^2 x$$

subject to the condition that $y = 0$ and $y' = 1$ when $x = 0$.

49. Find the length of the curve

$$x = \ln(\sec t + \tan t) - \sin t, \quad y = \cos t, \quad 0 \le t \le \pi/3.$$

See Fig. 6.28.

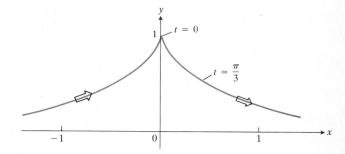

6.28 The curve in Problem 49 is a portion of a curve that covers the entire x-axis as t runs between $-\pi/2$ and $\pi/2$. The arrows show the direction of increasing t.

6.5

Properties of Natural Logarithms. The Graph of $y = \ln x$

In this article we establish the following important properties of natural logarithms and use them to graph the function $y = \ln x$:

$$\ln ax = \ln a + \ln x \tag{1}$$

$$\ln \frac{x}{a} = \ln x - \ln a \tag{2}$$

$$\ln x^n = n \ln x \tag{3}$$

These properties hold provided x and a are positive. For the moment there is also an added restriction that the exponent n be a rational number, but we shall remove this restriction in Article 6.7.

EXAMPLE 1

$$\ln\left(\frac{1}{8}\right) = \ln 1 - \ln 8 = 0 - \ln 2^3 = -3 \ln 2$$

$$\ln 4 - \ln 5 = \ln\left(\frac{4}{5}\right) = \ln 0.8$$

$$\ln\sqrt[3]{25} = \ln(25)^{1/3} = \frac{1}{3} \ln 25 = \frac{1}{3} \ln 5^2 = \frac{2}{3} \ln 5 \qquad \blacksquare$$

The proofs of Eqs. (1)–(3) are based on the fact that $y = \ln x$ satisfies the differential equation

$$\frac{dy}{dx} = \frac{1}{x} \qquad \text{for all} \qquad x > 0,$$

plus the fact that if two functions have the same derivative, then the two functions can differ only by a constant.

The Identity ln ax = ln a + ln x

To prove Eq. (1), notice that ln ax and ln x have the same derivative for all positive values of x:

$$\frac{d}{dx} \ln ax = \frac{1}{ax} \frac{d}{dx}(ax) = \frac{a}{ax} = \frac{1}{x} = \frac{d}{dx} \ln x.$$

Therefore,

$$\ln ax = \ln x + C$$

for some constant C. We can find the value of C by setting x equal to 1:

$$\ln a = \ln 1 + C = 0 + C,$$

$$C = \ln a.$$

Hence,

$$\ln ax = \ln x + \ln a,$$

which is Eq. (1).

The Identity ln(x/a) = ln x − ln a

To prove Eq. (2), we first put $x = 1/a$ in Eq. (1) and recall that ln $1 = 0$:

$$0 = \ln 1 = \ln\left(a \cdot \frac{1}{a}\right) = \ln a + \ln\left(\frac{1}{a}\right)$$

so that

$$\ln \frac{1}{a} = -\ln a. \tag{4}$$

We now apply Eq. (1) with a replaced by $1/a$ and ln a replaced by $\ln(1/a) = -\ln a$:

$$\ln \frac{x}{a} = \ln\left(x \cdot \frac{1}{a}\right) = \ln x + \ln \frac{1}{a} = \ln x - \ln a.$$

The result is Eq. (2).

The Identity ln x^n = n ln x

To prove Eq. (3), notice that ln x^n and n ln x have the same derivative for all positive values of x:

$$\frac{d}{dx} \ln x^n = \frac{1}{x^n} \cdot \frac{d}{dx}(x^n) = \frac{1}{x^n} \cdot nx^{n-1} = \frac{n}{x} = \frac{d}{dx}(n \ln x).$$

Therefore,

$$\ln x^n = n \ln x + C$$

for some constant C. By taking $x = 1$, we find $C = 0$, which gives Eq. (3).
If $n = 1/m$, where m is a positive integer, Eq. (3) gives

$$\ln \sqrt[m]{x} = \ln x^{1/m} = \frac{1}{m} \ln x. \tag{5}$$

Simplifying Derivative Calculations

We can use the arithmetic properties in Eqs. (1)–(3) to simplify our work when we calculate the derivatives of logarithms of products, quotients, and powers of functions.

EXAMPLE 2   Find dy/dx if

$$y = \ln \frac{x\sqrt{x + 5}}{(x - 1)^3}.$$

Solution   By applying Eqs. (1)–(3) we find that

$$y = \ln \frac{x\sqrt{x + 5}}{(x - 1)^3}$$

$$= \ln x\sqrt{x + 5} - \ln(x - 1)^3 \qquad \text{(from Eq. 2)}$$

$$= \ln x + \ln\sqrt{x + 5} - \ln(x - 1)^3 \qquad \text{(from Eq. 1)}$$

$$= \ln x + \frac{1}{2}\ln(x + 5) - 3\ln(x - 1). \qquad \text{(from Eq. 3)}$$

Therefore,

$$\frac{dy}{dx} = \frac{1}{x} + \frac{1}{2(x + 5)} - \frac{3}{x - 1}.$$

Logarithmic Differentiation

The derivative of a function given by a complicated equation can sometimes be calculated more quickly if we take the logarithm of both sides of the equation before differentiating. The process, called **logarithmic differentiation,** is illustrated in the next example.

EXAMPLE 3   Find dy/dx if

$$y = \frac{\sqrt{\cos x}}{x^2 \sin x}, \qquad 0 < x < \frac{\pi}{2}. \tag{6}$$

Solution   We take the logarithm of both sides of the equation and use the arithmetic properties in Eqs. (1)–(3) to simplify the right-hand side:

$$\ln y = \ln\left(\frac{\sqrt{\cos x}}{x^2 \sin x}\right)$$

$$= \ln\sqrt{\cos x} - \ln(x^2 \sin x) \tag{7}$$

$$= \frac{1}{2}\ln \cos x - 2\ln x - \ln \sin x.$$

We then take the derivative of both sides, using implicit differentiation on the left:

$$\frac{1}{y}\frac{dy}{dx} = \frac{1}{2}\frac{-\sin x}{\cos x} - \frac{2}{x} - \frac{\cos x}{\sin x}. \tag{8}$$

Finally, we solve for dy/dx:

$$\frac{dy}{dx} = y\left[-\frac{1}{2}\tan x - \frac{2}{x} - \cot x\right]. \qquad \blacksquare$$

EXAMPLE 4   Find dy/dx if

$$y^{2/3} = \frac{(x^2 + 1)(3x + 4)^{1/2}}{\sqrt[5]{2x - 4}}. \qquad (9)$$

Solution   The steps are the same as in Example 3. Take the logarithm of both sides of the equation:

$$\frac{2}{3}\ln y = \ln(x^2 + 1) + \frac{1}{2}\ln(3x + 4) - \frac{1}{5}\ln(2x - 4). \qquad (10)$$

Take the derivative of both sides, using implicit differentiation on the left:

$$\frac{2}{3} \cdot \frac{1}{y}\frac{dy}{dx} = \frac{2x}{x^2 + 1} + \frac{1}{2} \cdot \frac{3}{3x + 4} - \frac{1}{5} \cdot \frac{2}{2x - 4}.$$

Solve for dy/dx:

$$\frac{dy}{dx} = \frac{3y}{2}\left[\frac{2x}{x^2 + 1} + \frac{3}{6x + 8} - \frac{1}{5x - 10}\right]. \qquad \blacksquare$$

The Graph of $y = \ln x$

The slope of the curve

$$y = \ln x \qquad (11)$$

is given by

$$\frac{dy}{dx} = \frac{1}{x}, \qquad (12)$$

which is positive for all $x > 0$. Hence the graph of $y = \ln x$ steadily rises from left to right. Since the derivative is continuous, the function $\ln x$ is itself continuous, and the curve has a continuously turning tangent.

The second derivative,

$$\frac{d^2y}{dx^2} = -\frac{1}{x^2}, \qquad (13)$$

is always negative, so the graph is always concave down.

The curve passes through the point $(1, 0)$, since $\ln 1 = 0$. At this point its slope is $+1$, so the tangent line at this point makes an angle of $45°$ with the x-axis (if we use equal units on the x- and y-axes).

If we refer to the definition of $\ln 2$ as an integral,

$$\ln 2 = \int_1^2 \frac{1}{t}\, dt,$$

we see that it may be interpreted as the area in Fig. 6.26 with $x = 2$. By considering the areas of rectangles of base 1 and altitudes 1 or 1/2, respectively circumscribed over and inscribed under the given area, we see that

$$0.5 < \ln 2 < 1.0.$$

In fact, by more extensive calculations, the value of $\ln 2$ is found to be

$$\ln 2 \approx 0.69315,$$

to five decimal places. By Eq. (3) we then have

$$\ln 4 = \ln 2^2 = 2 \ln 2 \approx 1.38630,$$

$$\ln 8 = \ln 2^3 = 3 \ln 2 \approx 2.07944,$$

$$\ln \frac{1}{2} = \ln 2^{-1} = -\ln 2 \approx -0.69315,$$

$$\ln \frac{1}{4} = \ln 2^{-2} = -2 \ln 2 \approx -1.38630,$$

and so on.

We now plot the points that correspond to $x = 1/4, 1/2, 1, 2, 4, 8$ on the curve $y = \ln x$ and connect them with a smooth curve. The curve we draw should have slope $1/x$ at the point $(x, \ln x)$ and should be concave down. The curve is shown in Fig. 6.29.

Since $\ln 2$ is greater than 0.5, and $\ln 2^n = n \ln 2$, we have

$$\ln 2 > 0.5$$

$$\ln 4 > 2(0.5) = 1$$

$$\ln 8 > 3(0.5) = 1.5$$

$$\ln 16 > 4(0.5) = 2$$

$$\vdots$$

$$\ln 2^n > n(0.5) = \frac{n}{2},$$

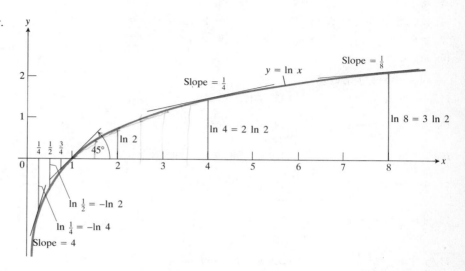

6.29   Graph of $y = \ln x$.

and hence $\ln x$ increases without limit as x does. That is,

$$\ln x \to +\infty \qquad \text{as} \qquad x \to +\infty. \tag{14}$$

On the other hand, as x approaches zero through positive values, $1/x$ tends to plus infinity. Hence, from Eq. (4) we have

$$\ln x = -\ln \frac{1}{x} \to -\infty \qquad \text{as} \qquad x \to 0^+. \tag{15}$$

The y-axis is a vertical asymptote of the graph of $y = \ln x$.

Properties of $y = \ln x$

1. Domain: The set of all positive real numbers, $x > 0$.
2. Range: The set of all real numbers $-\infty < y < \infty$.
3. It is a continuous, increasing function everywhere on its domain. If $x_1 > x_2 > 0$, then $\ln x_1 > \ln x_2$. It is a one-to-one function from its domain to its range. (It therefore has an inverse, which will be the subject of the next article.)
4. Products, quotients, and powers: If a and x are any two positive numbers, then

$$\ln ax = \ln a + \ln x, \tag{1}$$

$$\ln \frac{x}{a} = \ln x - \ln a, \tag{2}$$

$$\ln x^n = n \ln x. \tag{3}$$

PROBLEMS

Express the logarithms in Problems 1–10 in terms of $\ln 2$ and $\ln 3$. For example, $\ln 1.5 = \ln(3/2) = \ln 3 - \ln 2$.

1. $\ln 16$
2. $\ln\sqrt[3]{9}$
3. $\ln 2\sqrt{2}$
4. $\ln 0.25$
5. $\ln 4/9$
6. $\ln 12$
7. $\ln 9/8$
8. $\ln 36$
9. $\ln 4.5$
10. $\ln\sqrt{13.5}$

In Problems 11–22, find dy/dx.

11. $y = \ln\sqrt{x^2 + 5}$
12. $y = \ln x^{3/2}$
13. $y = \ln \dfrac{1}{x\sqrt{x + 1}}$
14. $y = \ln\sqrt[3]{\cos x}$
15. $y = \ln(\sin x \sin 2x)$
16. $y = \ln(x\sqrt{x^2 + 1})$
17. $y = \ln(3x\sqrt{x + 2})$
18. $y = \dfrac{1}{2}\ln\dfrac{1 + x}{1 - x}$
19. $y = \dfrac{1}{3}\ln\dfrac{x^3}{1 + x^3}$
20. $y = \ln\dfrac{x}{2 + 3x}$
21. $y = \ln\dfrac{(x^2 + 1)^5}{\sqrt{1 - x}}$
22. $y = \displaystyle\int_{\sqrt{x}}^{\sqrt[3]{x}} \ln t \, dt$

In Problems 23–31, find dy/dx by logarithmic differentiation.

23. $y^2 = x(x + 1), \quad x > 0$
24. $y = \sqrt[3]{\dfrac{x + 1}{x - 1}}, \quad x > 1$
25. $y = \sqrt{x + 3} \sin x \cos x, \quad 0 < x < \pi/2$
26. $y = \dfrac{x\sqrt{x^2 + 1}}{(x + 1)^{2/3}}, \quad x > 0$
27. $y = \sqrt[3]{\dfrac{x(x - 2)}{x^2 + 1}}, \quad x > 2$
28. $y^5 = \sqrt{\dfrac{(x + 1)^5}{(x + 2)^{10}}}$
29. $y = \sqrt[3]{\dfrac{x(x + 1)(x - 2)}{(x^2 + 1)(2x + 3),}}, \quad x > 2$
30. $y^{4/5} = \dfrac{\sqrt{\sin x \cos x}}{1 + 2\ln x}$
31. $\sqrt{y} = \dfrac{x^5 \tan^{-1}x}{(3 - 2x)\sqrt[3]{x}}$

32. a) What is the largest possible domain of

$$y = \sqrt{\frac{(x + 1)(x + 2)}{(3 - x)(4 - x)}} \,?$$

  b) Find dy/dx.

Evaluate the integrals in Problems 33–40.

33. $\displaystyle\int_{-1}^{1} \frac{dx}{x + 3}$

34. $\displaystyle\int_{0}^{6} \frac{dx}{x + 2}$

35. $\displaystyle\int_{\pi/4}^{\pi/2} \cot x \, dx$

36. a) $\displaystyle\int_{0}^{\sqrt{3}} \frac{dx}{1 + x^2}$ b) $\displaystyle\int_{0}^{\sqrt{3}} \frac{x \, dx}{1 + x^2}$

37. $\displaystyle\int_{2}^{4} \frac{2x - 5}{x} \, dx$

38. $\displaystyle\int_{0}^{\pi/3} \frac{\sec x \tan x \, dx}{2 + \sec x}$

39. a) $\displaystyle\int_{0}^{3/5} \frac{x \, dx}{1 - x^2}$ b) $\displaystyle\int_{0}^{3/5} \frac{dx}{\sqrt{1 - x^2}}$

  c) $\displaystyle\int_{0}^{3/5} \frac{x \, dx}{\sqrt{1 - x^2}}$

40. a) $\displaystyle\int_{-1}^{3} \frac{dx}{\sqrt{2x + 3}}$ b) $\displaystyle\int_{-1}^{3} \frac{dx}{2x + 3}$

  c) $\displaystyle\int_{-1}^{3} \frac{dx}{(2x + 3)^2}$

41. Graph (a) $y = \ln|x|$, (b) $y = |\ln x|$.

42. Graph $y = \ln x$ and $y = \ln 2x$ together. (*Hint:* Before you start, apply Eq. 1 to $\ln 2x$.)

43. Find

$$\lim_{x \to \infty} \int_{x}^{2x} (1/t) \, dt.$$

44. a) Find the area of the region in the first quadrant bounded by the x-axis, the curve $y = \tan x$, and the line $x = \pi/3$.

  b) Find the volume of the solid generated by revolving the region in (a) about the x-axis.

45. Repeat Problem 44 for the region bounded by the curves $y = 4/x$ and $y = (x - 3)^2$.

46. The region bounded by the curve $y = 1/\sqrt{x}$, the x-axis, and the lines $x = 1/2$ and $x = 4$ is revolved about the x-axis to generate a solid. Find the volume of the solid.

47. *Linear and quadratic approximations of* $\ln(1 + x)$. The standard approximations of $\ln(1 + x)$ near $x = 0$ are

  Linear: $\ln(1 + x) \approx x$ (16)

  Quadratic: $\ln(1 + x) \approx x - \dfrac{x^2}{2}.$ (17)

  a) Use the formulas in Table 3.2 (Article 3.9), to verify these approximations.

  b) Estimate the errors involved in replacing $\ln(1 + x)$ by its linear and quadratic approximations for the interval $0 \le x \le 0.1$.

48. a) Graph $y = x$ and $y = \ln(1 + x)$ together to show that the maximum error in the approximation $\ln(1 + x) \approx x$ on the interval $0 \le x \le 0.1$ occurs at $x = 0.1$. (*Hint:* What is the slope of the curve $y = \ln(1 + x)$ at $x = 0$?)

  b) **CALCULATOR** Find the difference $\ln(1.1) - 0.1$ with a calculator.

49. *Calculating values of* $\ln x$ *with Simpson's rule.* While Eqs. (16) and (17) in Problem 47 are good for replacing $\ln(1 + x)$ by x and $(x - x^2/2)$ over short intervals, when it comes to estimating the value of a *particular* logarithm, Simpson's rule gives better results.

As a case in point, the values of $\ln(1.2)$ and $\ln(0.8)$ to five decimal places are

$$\ln(1.2) = 0.18232, \qquad \ln(0.8) = -0.22314.$$

Calculate $\ln(1.2)$ and $\ln(0.8)$ first with Eqs. (16) and (17) and then with Simpson's rule with $n = 2$. The accuracy of Simpson's rule is impressive, isn't it?

50. **CALCULATOR** Use Simpson's rule with $n = 8$ to estimate $\ln 5$. Compare the result with the value your calculator gives for $\ln 5$.

TOOLKIT PROGRAM

Super ∗ Grapher

6.6
The Exponential Function e^x

We now come to the inverse of the function $y = \ln x$, a function that appears with amazing frequency in mathematics and its applications.

The Number e

Since $\ln 2$ is less than 1, and $\ln 4$ is greater than 1, the Intermediate Value Theorem tells us there is a number between 2 and 4 whose logarithm is equal to 1. Because $\ln x$ is one-to-one, there is only one such number. We denote it by the letter e, which stands for Euler, who named the number after himself when he wrote about it in the early eighteenth century. Thus,

$$e = \ln^{-1}1 \quad \text{and} \quad \ln e = 1. \tag{1}$$

If you hold a ruler along the line $y = 1$ in Fig. 6.29, you will see that e lies between 2.5 and 3. In Chapter 12 we shall see how to compute the value of e to any desired number of decimal places. Its value to 15 places is

$$e = 2.7 \; 1828 \; 1828 \; 45 \; 90 \; 45 \ldots \tag{2}$$

The Function $y = e^x$

When x is a rational number, we can define e^x the way we define rational powers of any other positive number. For these powers of e we have

$$\ln e^x = x \ln e = x \cdot 1 = x. \tag{3}$$

That is, when x is rational, e^x is the number whose natural logarithm is x. In symbols,

$$e^x = \ln^{-1}x, \qquad x \text{ rational}, \tag{4}$$

Equation (4) says that the functions e^x and $\ln^{-1}x$ are the same function when restricted to rational values of x. This is the key to defining e^x for other values of x.

Although e^x has been defined only for rational values of x so far, the function $\ln^{-1}x$ has been defined for all values of x, irrational as well as rational. We can therefore use the formula $e^x = \ln^{-1}x$ to define e^x for the values of x for which e^x has no previous definition. In fact, there is no other way to define e^x if we want the resulting function to be continuous. By fiat, then, $e^x = \ln^{-1}x$ for all x. Thus, $y = e^x$ if and only if $x = \ln y$.

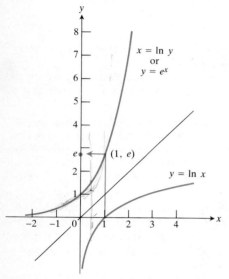

6.30 The graph of $y = \ln x$ and its inverse $y = e^x$.

DEFINITION

The Function $y = e^x$

For every real number x,

$$e^x = \ln^{-1}x. \tag{5}$$

The function $y = e^x$ is often called the **exponential function** with **base e** and **exponent x.** An alternate notation for e^x is **exp(x)**, pronounced "exponential of x." The "exp" notation is preferred when the expression for x is complicated. Appendix 12 gives a table of values of e^x for small values of x.

The graph of $y = e^x$ may be obtained by reflecting the graph of $y = \ln x$ across the line $y = x$ (Fig. 6.30). The graph of $y = e^x$ is the same as the graph of $x = \ln y$. Notice that the graph crosses the y-axis at $y = e^0 = 1$, that

$$\lim_{x \to -\infty} e^x = 0, \tag{6}$$

and that

$$\lim_{x \to \infty} e^x = \infty. \tag{7}$$

A TRANSCENDENTAL NUMBER AGAINST A MATHEMATICAL DELUSION

Historically, to actually know whether a number is transcendental, or to prove that fact, was very difficult. Even the easier task of proving that a number is irrational was itself difficult. One of the "three famous problems of antiquity," that of squaring the circle with straightedge and compass alone, called the quadrature problem, was thought to rest on the question of the irrationality of π. (The other two famous problems were the duplication of the cube and the trisection of an angle.)

The numbers π and e were proved to be irrational in the mid-eighteenth century by Johann Heinrich Lambert (1728–1777). But the proof of the irrationality of π does *not* dispel the question of circle-squaring; that question is, however, dispelled by the transcendence of π—a proof beyond the scope of this book. The mathematician Charles Hermite proved in 1873 that e is transcendental. In 1882, C. L. F. Lindemann, using Hermite's method, proved that π, too, is transcendental and thus disposed of the last point in this famous construction problem of geometry. Still, purported and self-deluded circle-squarers exist today, sending their erroneous solutions to university mathematics departments throughout the world.

Equations Involving ln x and e^x

Because $y = e^x$ and $y = \ln x$ are inverses of one another,

$$e^{\ln x} = x \qquad \text{for all } x > 0 \tag{8}$$

and

$$\ln e^x = x \qquad \text{for all } x. \tag{9}$$

EXAMPLE 1

a) $\ln e^2 = 2$

b) $\ln e^{-1} = -1$

c) $\ln\sqrt{e} = \dfrac{1}{2}$

d) $e^{\ln 2} = 2$

e) $e^{\ln(x^2+1)} = x^2 + 1$

f) $\ln e^{\sin x} = \sin x$

g) $\ln \dfrac{e^{2x}}{5} = \ln e^{2x} - \ln 5 = 2x - \ln 5.$ ■

EXAMPLE 2   Solve for y.

a) $\ln y = x^2$

b) $e^{3y} = 2 + \cos x$

c) $\ln(y - 1) - \ln y = 3x$

Solution

a) $\ln y = x^2$

   Exponentiate: $e^{\ln y} = e^{(x^2)}$

$$y = e^{(x^2)}.$$

b) $e^{3y} = 2 + \cos x$

   Take the logarithm of both sides: $\ln e^{3y} = \ln(2 + \cos x)$

$$3y = \ln(2 + \cos x)$$

$$y = \frac{1}{3}\ln(2 + \cos x).$$

c) $\ln(y - 1) - \ln y = 3x$

   Combine the logarithms: $\ln \dfrac{y - 1}{y} = 3x,$

   Exponentiate: $\dfrac{y - 1}{y} = e^{3x},$

   Solve as usual: $y - 1 = ye^{3x}$

$$y - ye^{3x} = 1$$

$$y(1 - e^{3x}) = 1$$

$$y = \frac{1}{1 - e^{3x}}.$$ ■

> **Two Useful Operating Rules**
>
> To remove logarithms from an equation, exponentiate both sides.
> To remove exponentials, take the logarithm of both sides.

The Laws of Exponents

Even though e^x was defined in a seemingly roundabout way as an inverse logarithm, it obeys the familiar rules of exponents from algebra:

$$e^{x_1} \cdot e^{x_2} = e^{x_1 + x_2}, \tag{10}$$

$$e^{-x} = \frac{1}{e^x}. \tag{11}$$

These equations hold for all real numbers and can be derived from the relationship that e^x has with $\ln x$. To establish Eq. (10), let

$$y_1 = e^{x_1} \qquad \text{and} \qquad y_2 = e^{x_2}.$$

Then, by definition,

$$x_1 = \ln y_1 \qquad \text{and} \qquad x_2 = \ln y_2,$$

so that

$$x_1 + x_2 = \ln y_1 + \ln y_2 = \ln y_1 y_2$$

by Eq. (1), Article 6.5. Therefore, by definition again,

$$y_1 y_2 = e^{x_1 + x_2},$$

which establishes Eq. (10).

To establish Eq. (11), let $y = e^{-x}$. Then, by definition, $-x = \ln y$ and, from Eq. (4), Article 6.5,

$$x = -\ln y = \ln \left(\frac{1}{y} \right).$$

Therefore

$$\frac{1}{y} = e^x, \qquad \text{or} \qquad y = \frac{1}{e^x},$$

which establishes Eq. (11).

EXAMPLE 3   Simplify.

a) $e^{\ln 2 + 3 \ln x}$

b) $e^{2x - \ln x}$

Solution

a) $e^{\ln 2 + 3 \ln x} = e^{\ln 2} \cdot e^{3 \ln x}$ (from Eq. 10)

$\qquad\qquad = 2 \cdot e^{\ln x^3}$ (definition of e^x and a property of logarithms)

$\qquad\qquad = 2x^3$

b) $e^{2x - \ln x} = e^{2x} \cdot e^{-\ln x}$ (from Eq. 10)

$\qquad = \dfrac{e^{2x}}{e^{\ln x}}$ (from Eq. 11)

$\qquad = \dfrac{e^{2x}}{x}$ (definition of e^x)   ∎

The Derivative and Integral of $y = e^x$

The function $y = e^x$ is differentiable because it is the inverse of a differentiable function whose derivative is never zero. To find the derivative of

$$y = e^x$$

we take the logarithm of both sides,

$$\ln y = x,$$

and differentiate implicitly with respect to x. Then

$$\frac{1}{y}\frac{dy}{dx} = 1 \qquad \text{or} \qquad \frac{dy}{dx} = y.$$

Since $y = e^x$, this gives

$$\frac{de^x}{dx} = e^x.$$

Here is a function that is not changed by differentiation! It is indestructible. It can be differentiated again and again without changing. In this, the exponential function is like the story of a student who asked a guru what was holding up the earth. The answer was that the earth was held up by an elephant, and the student naturally wanted to know what held up that elephant. The guru paused a moment, then replied "It's elephants all the way down."

If we didn't know it already, we could determine that $y = e^x$ is an increasing function of x from the fact that its derivative is positive.

We obtain a formula for the derivative of e^u, where u is a differentiable function of x, by applying the Chain Rule:

$$\frac{de^u}{dx} = \frac{de^u}{du}\frac{du}{dx} = e^u\frac{du}{dx}. \tag{12}$$

This in turn leads to the integration formula

$$\int e^u \, du = e^u + C. \tag{13}$$

EXAMPLE 4

$$\frac{d}{dx}e^{\sin x} = e^{\sin x}\frac{d}{dx}(\sin x) = e^{\sin x}\cos x \qquad ∎$$

EXAMPLE 5   Evaluate

$$\int_0^1 \frac{e^{\tan^{-1}x}\,dx}{1+x^2}.$$

Solution   We substitute

$$u = \tan^{-1}x, \qquad du = \frac{1}{1+x^2}\,dx, \qquad 0 = \tan^{-1}0, \qquad \frac{\pi}{4} = \tan^{-1}1.$$

Then

$$\int_0^1 \frac{e^{\tan^{-1}x}\,dx}{1+x^2} = \int_0^{\pi/4} e^u\,du = e^u\Bigg]_0^{\pi/4} = e^{\pi/4} - e^0 = e^{\pi/4} - 1. \qquad \blacksquare$$

EXAMPLE 6   Solve the differential equation

$$\frac{dy}{dx} = 2xe^{-y}, \quad x > \sqrt{3} \tag{14}$$

subject to the condition that $y = 0$ when $x = 2$.

Solution   We separate the variables in Eq. (14) by multiplying both sides of the equation by e^y:

$$e^y \cdot \frac{dy}{dx} = e^y \cdot 2xe^{-y} = 2xe^{y-y} = 2xe^0 = 2x,$$

or

$$e^y \frac{dy}{dx} = 2x.$$

We integrate both sides with respect to x to obtain

$$e^y = x^2 + C.$$

We use the condition $y = 0$ when $x = 2$ to find the right value for C:

$$C = e^0 - 4 = 1 - 4 = -3.$$

Therefore,

$$e^y = x^2 - 3. \tag{15}$$

To solve this equation for y, we take the logarithm of both sides:

$$\ln e^y = \ln(x^2 - 3),$$

$$y = \ln(x^2 - 3). \tag{16}$$

Notice that the solution is valid for $x > \sqrt{3}$.

It is always a good idea to check the solution of a differential equation in the original equation. From Eq. (16) and then Eq. (15), we have

$$\frac{dy}{dx} = \frac{d}{dx}\ln(x^2 - 3) = \frac{2x}{x^2 - 3} = \frac{2x}{e^y} = 2xe^{-y}.$$

$$\Uparrow \qquad\qquad\qquad \Uparrow$$
$$\text{Eq. (16)} \qquad\qquad \text{Eq. (15)}$$

The function y and its derivative dy/dx therefore satisfy Eq. (14). $\qquad \blacksquare$

> **Properties of $y = e^x$**
>
> 1. The exponential function $y = e^x$ is the inverse of the natural logarithm function $y = \ln x$; that is, $e^x = \ln^{-1} x$.
>
>   Domain: The set of all real numbers, $-\infty < x < \infty$.
>   Range: The set of all positive numbers, $y > 0$.
>
> 2. Its derivative is
>
> $$\frac{d}{dx}(e^x) = e^x.$$
>
> 3. It is continuous (because it is differentiable) and is an increasing function of x.
>
> 4. If u is any differentiable function of x, then
>
> $$\frac{d}{dx}e^u = e^u\frac{du}{dx} \quad \text{and} \quad \int e^u\, du = e^u + C.$$
>
> 5. $e^{x_1} \cdot e^{x_2} = e^{x_1 + x_2} \quad$ and $\quad e^{-x} = 1/e^x$

PROBLEMS

Simplify the expressions in Problems 1–12.

1. $e^{\ln x}$ **2.** $\ln(e^x)$ **3.** $e^{-\ln(x^2)}$

4. $\ln(e^{-x^2})$ **5.** $\ln(e^{1/x})$ **6.** $\ln(1/e^x)$

7. $e^{\ln 2 + \ln x}$ **8.** $e^{2\ln x}$ **9.** $\ln(e^{x-x^2})$

10. $\ln(x^2 e^{-2x})$ **11.** $e^{x + \ln x}$ **12.** $e^{\ln x - 2\ln y}$

In Problems 13–18, solve for y.

13. $e^{\sqrt{y}} = x^2$

14. $e^{2y} = x^2$

15. $e^{(x^2)} \cdot e^{(2x+1)} = e^y$

16. $\ln(y - 1) = x + \ln x$

17. $\ln(y - 2) = \ln(\sin x) - x$

18. $\ln(y^2 - 1) - \ln(y + 1) = \sin x$

Find dy/dx in Problems 19–42.

19. $y = e^{3x}$ **20.** $y = e^{(x+1)}$

21. $y = e^{5-7x}$ **22.** $y = \cos e^x$

23. $y = x^2 e^x$ **24.** $y = \sin e^{-x}$

25. $y = e^{\sin x}$ **26.** $y = e^{(x^2)} \cdot e^{-x}$

27. $y = \ln(3xe^{-x})$ **28.** $y = \ln\dfrac{e^x}{1 + e^x}$

29. $y = e^{\sin^{-1} x}$ **30.** $y = (1 + 2x)e^{-2x}$

31. $y = (9x^2 - 6x + 2)e^{3x}$ **32.** $y = \dfrac{ax - 1}{a^2}e^{ax}$

33. $y = x^2 e^{-x^2}$ **34.** $y = e^x \ln x$

35. $y = \tan^{-1}(e^x)$ **36.** $y = \sec^{-1}(e^{2x})$

37. $y = x^3 e^{-2x} \cos 5x$ (*Hint:* Use logarithmic differentiation.)

38. $y = \displaystyle\int_0^{\ln x} \sin e^t\, dt, \quad x > 0$

39. $\ln y = x \sin x$ **40.** $\ln xy = e^{x+y}$

41. $e^{2x} = \sin(x + 3y)$ **42.** $\tan y = e^x + \ln x$

Evaluate the integrals in Problems 43–56.

43. $\displaystyle\int_{\ln 3}^{\ln 5} e^{2x}\, dx$ **44.** $\displaystyle\int_{-1}^{1} xe^{x^2}\, dx$

45. $\displaystyle\int_0^{\pi} e^{\sin x} \cos x\, dx$ **46.** $\displaystyle\int_0^{\ln 8} e^{x/3}\, dx$

47. $\displaystyle\int_{-\ln(a+1)}^{0} e^{-x}\, dx$ **48.** $\displaystyle\int_0^{2} e^{x/2}\, dx$

49. $\displaystyle\int_0^{1} e^{\ln\sqrt{x}}\, dx$ **50.** $\displaystyle\int_0^{1} \frac{dx}{e^x}$

51. $\displaystyle\int_0^{\ln 2} \frac{24\, dx}{e^{3x}}$ **52.** $\displaystyle\int_0^{1} \frac{e^x\, dx}{1 + e^x}$

53. $\displaystyle\int_0^{\ln 13} \frac{e^x\, dx}{1 + 2e^x}$ **54.** $\displaystyle\int_e^{e^2} \frac{dx}{x \ln x}$

55. $\displaystyle\int_0^{\ln 2} \frac{e^x\, dx}{1 + e^{2x}}$ (*Hint:* Let $u = e^x$.)

56. $\displaystyle\int_1^{4} \frac{e^{\sqrt{x}}\, dx}{\sqrt{x}}$ (*Hint:* Let $u = \sqrt{x}$.)

Use l'Hôpital's rule (Article 3.8) to evaluate the limits in Problems 57–60.

57. $\lim\limits_{h \to 0} \dfrac{e^h - (1 + h)}{h^2}$

58. $\lim\limits_{x \to 0} \dfrac{\sin x}{e^x - 1}$

59. $\lim\limits_{x \to \infty} \dfrac{x^2 + e^x}{x + e^x}$

60. $\lim\limits_{x \to \infty} xe^{-x}$

61. a) Find all maxima and minima (relative and absolute) of the function

$$y = e^{\sin x}, \quad -\pi \le x \le 2\pi.$$

b) Graph the function. (You may use the tables in Appendix 12.)

62. Find the absolute maximum and minimum values of

$$f(x) = e^x - 2x$$

on the interval [0, 1].

63. Find the maximum value of $f(x) = x^2 \ln(1/x)$.

64. The graph of $y = (x - 3)^2 e^x$ has a horizontal tangent at the point $P(3, 0)$. Does y have a relative extreme value at $x = 3$, or is P a point of inflection?

65. *Linear and quadratic approximations of e^x.* The standard linear and quadratic approximations of e^x near $x = 0$ are

Linear: $e^x \approx 1 + x$

Quadratic: $e^x \approx 1 + x + \dfrac{x^2}{2}$.

a) Use the formulas in Table 3.2 (Article 3.9) to verify these approximations.

b) Estimate the errors involved in replacing e^x by its linear and quadratic approximations for $0 \le x \le 0.1$.

c) Graph e^x, $L(x) = 1 + x$, and $Q(x) = 1 + x + x^2/2$ together over the interval $-1 \le x \le 1$.

66. The region between the curve $y = e^{-x}$ and the x-axis from $x = 0$ to $x = \ln 10$ is revolved about the x-axis to generate a solid (Fig. 6.31). Find the volume of the solid.

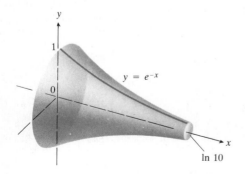

6.31 The solid in Problem 66 runs from the origin to $x =$ ln 10, which is about 2.3.

67. Let $A(t)$ be the area of the region in the first quadrant enclosed by the coordinate axes, the curve $y = e^{-x}$, and the line $x = t > 0$ (Fig. 6.32). Let $V(t)$ be the volume of the solid generated by revolving the region about the x-axis. Find the following limits.

a) $\lim\limits_{t \to \infty} A(t)$

b) $\lim\limits_{t \to \infty} V(t)/A(t)$

c) $\lim\limits_{t \to 0^+} V(t)/A(t)$

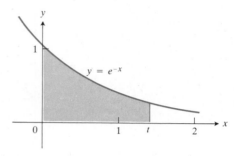

6.32 The graph of $y = e^{-x}$ and the region in Problem 67.

68. Find the length of the curve $x = e^t \sin t$, $y = e^t \cos t$, $0 \le t \le \pi$.

69. a) Show that $y = Ce^{ax}$ is a solution of the differential equation $dy/dx = ay$ for any choice of the constant C.

b) Using the result of (a), find a solution of the differential equation $dy/dt = -2y$ satisfying the initial condition $y = 3$ when $t = 0$.

70. Find the value of the constant r for which $y = e^{rx}$ is a solution of the differential equation

$$y'' - 4y' + 4y = 0.$$

In Problems 71–74, solve the differential equations subject to the given initial conditions.

71. $\dfrac{dy}{dx} = e^{-x}, \quad y = 0$ when $x = 4$

72. $e^y \dfrac{dy}{dx} = 2xe^{x^2-1}, \quad y = 0$ when $x = 1$

73. $\dfrac{1}{y+1} \dfrac{dy}{dx} = \dfrac{1}{2x}, \quad x > 0, \quad y = 1$ when $x = 2$

74. $\dfrac{1}{y+1} \dfrac{dy}{dx} = \dfrac{1}{x^2}, \quad x > 0, \quad y = 0$ when $x = 1$

Hyperbolic Functions

Problems 75–78 introduce the hyperbolic sine and hyperbolic cosine functions, treated in detail in Chapter 9. These new functions resemble the trigonometric sine and cosine in many ways.

Their definitions are

Hyperbolic sine: $\quad \sinh x = \dfrac{1}{2}(e^x - e^{-x})$

Hyperbolic cosine: $\quad \cosh x = \dfrac{1}{2}(e^x + e^{-x}).$

Figure 6.33 shows their graphs together with the graphs of $e^x/2$ and $e^{-x}/2$. Besides being of mathematical interest, these functions appear in the solutions of many differential equations and play a prominent role in Einstein's theory of relativity.

Using these definitions, prove the following.

75. $\dfrac{d}{dx}(\cosh x) = \sinh x, \quad \dfrac{d}{dx}(\sinh x) = \cosh x$

76. $\cosh^2 x - \sinh^2 x = 1, \quad \cosh^2 x + \sinh^2 x = \cosh(2x)$

77. $\cosh(-x) = \cosh x, \quad \sinh(-x) = -\sinh x$

78. Show that $y = \cosh x$ satisfies the differential equation $y'' = \sqrt{1 + (y')^2}$. (This is one of the "hanging cable" equations discussed in Article 9.3.)

TOOLKIT PROGRAMS

Integral Evaluator	Root Finder
Name That Function	Super * Grapher
Parametric Equations	

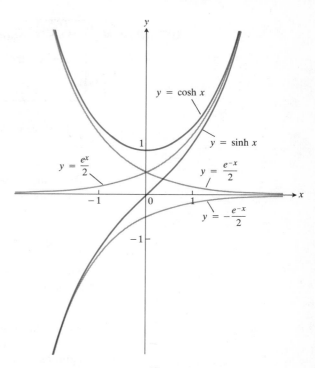

6.33 The graphs of the hyperbolic sine and cosine.

6.7
The Functions a^x and a^u

While we have not yet devised a way to raise positive numbers to any but rational powers, we have an exception in the number e. The definition $e^x = \ln^{-1} x$ defines e^x for *every* real number x, irrational as well as rational. In this article, we show how this good fortune enables us to raise any other positive number to an arbitrary power. We also prove the power rule for differentiation in its final form (good for all real exponents) and define functions like x^x and $(\sin x)^{\tan x}$ that involve raising the values of one function to powers given by another function.

The Function a^x

If a is a positive number and x is any number whatsoever, we define the function a^x ("a to the x") by the equation

$$a^x = (e^{\ln a})^x = e^{x \ln a}. \tag{1}$$

This works because $\ln x$ is defined at $x = a$, and e can be raised to any power.

DEFINITION

> **The Function $y = a^x$**
>
> If a is positive and x is any number, then
> $$a^x = e^{x \ln a}. \tag{2}$$

We can now remove the restriction that n be a rational number in the equation

$$\ln a^n = n \ln a. \tag{3}$$

To do so, we take the logarithm of both sides of Eq. (2), which holds for any real number x, to get

$$\ln a^x = \ln e^{x \ln a} = x \ln a. \tag{4}$$

Equation (2) is the basis for the algorithm used by some small calculators to compute y^x. Humans have no trouble computing $(-2)^3$, because we know that it is just $-(2^3)$, or -8. But suppose we thought we should calculate the result using Eq. (2):

$$(-2)^3 = e^{3 \ln(-2)}, \tag{5}$$

and we have learned that -2 is not in the domain of the function $\ln x$. We would do something similar to what these calculators do: flash a signal that signifies "error." The number a in Eq. (2) *must* be positive.

The Law of Exponents

From Eqs. (2) and (4), we obtain the law of exponents for arbitrary powers of positive numbers.

> **Law of Exponents**
>
> For any positive number a,
> $$a^{xy} = (a^x)^y = (a^y)^x. \tag{6}$$

Thus,

$$e^{3 \ln 2} = (e^3)^{\ln 2} = (e^{\ln 2})^3 = (2)^3 = 8. \tag{7}$$

This law generalizes the integer exponent law

$$a^{mn} = (a^m)^n,$$

which says, for instance, that

$$x^6 = (x^2)^3 = (x^3)^2.$$

The steps in the derivation of Eq. (6) are laid out in Problem 43.

EXAMPLE 1 From Eq. (6) we have

$$e^{3 \ln x} = (e^{\ln x})^3 = x^3,$$

in agreement with the calculation

$$e^{3 \ln x} = e^{\ln x^3} = x^3,$$

which was part of Example 3(a) in Article 6.6. ■

The Derivatives of a^x and a^u

To find a formula for the derivative of a^x with respect to x when a is a positive number, we differentiate both sides of Eq. (2):

$$\frac{d}{dx}a^x = \frac{d}{dx}e^{x \ln a}$$

$$= e^{x \ln a} \cdot \frac{d}{dx}x \ln a \qquad \text{(Chain Rule)}$$

$$= e^{x \ln a} \cdot \ln a$$

$$= a^x \ln a.$$

Thus, we have the following formula.

Derivative of a^x

If $a > 0$, then

$$\frac{d}{dx}a^x = a^x \ln a. \qquad\qquad (8)$$

Equation (8) shows why e is the most desirable base to use in calculus. When $a = e$, $\ln a = \ln e = 1$, and Eq. (8) reduces to

$$\frac{d}{dx}e^x = e^x.$$

EXAMPLE 2 From Eq. (8) we have

$$\frac{d}{dx}3^x = 3^x \ln 3. \qquad\qquad ■$$

From the formula

$$\frac{d}{dx}a^x = a^x \ln a, \qquad a > 0,$$

we can see that the derivative of $y = a^x$ is positive if $a > 1$ and negative if $0 < a < 1$. Thus, $y = a^x$ is an increasing function of x if $a > 1$ and a decreasing function of x if $0 < a < 1$. In either case, $y = a^x$ is one-to-one. It therefore has an inverse (which will be the subject of the next article). Figure 6.34 shows the graphs of $y = 2^x$ (increasing and one-to-one) and $y = (1/2)^x$ (decreasing and one-to-one).

Combining Eq. (8) with the Chain Rule,

$$\frac{dy}{dx} = \frac{dy}{du}\frac{du}{dx},$$

gives a formula for the derivative of $y = a^u$ when u is any differentiable function of x:

$$\frac{d}{dx}a^u = \frac{d}{du}a^u \cdot \frac{du}{dx} = a^u \ln a \frac{du}{dx}.$$

Thus we have the following formula for the derivative of a^u.

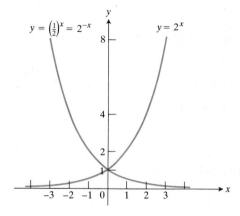

6.34 The functions $y = 2^x$ and $y = (1/2)^x$ are both one-to-one.

Derivative of a^u

If $a > 0$ and u is any differentiable function of x, then

$$\frac{d}{dx}a^u = a^u \ln a \frac{du}{dx}. \tag{9}$$

EXAMPLE 3   From Eq. (9) we have

$$\frac{d}{dx}3^{\sin x} = 3^{\sin x} \ln 3 \cdot \frac{d}{dx}\sin x = 3^{\sin x} \ln 3 \cos x.$$

In practice, there is little reason to memorize Eq. (9) because the derivative of a^u can always be calculated by the method of logarithmic differentiation. The main reason for producing the formula here is to derive from it the companion integration formula. We shall come to that shortly.

EXAMPLE 4   Calculate the derivative of

$$y = 3^{\sin x} \tag{10}$$

by logarithmic differentiation.

Solution   We take the natural logarithm of both sides of Eq. (10), differentiate, and solve for dy/dx:

$$y = 3^{\sin x}$$

$$\ln y = \ln 3^{\sin x} = \sin x \ln 3$$

$$\frac{1}{y}\frac{dy}{dx} = \ln 3 \cos x$$

$$\frac{dy}{dx} = y \ln 3 \cos x = 3^{\sin x} \ln 3 \cos x.$$

This agrees with the result in Example 3.

The Integral of a^u

When $a \neq 1$, so that $\ln a \neq 0$, the formula in (9) can be rewritten as

$$a^u \frac{du}{dx} = \frac{1}{\ln a}\frac{d}{dx}a^u. \tag{11}$$

Integrating both sides of this equation with respect to x then yields

$$\int a^u \frac{du}{dx}dx = \int \frac{1}{\ln a}\left(\frac{d}{dx}a^u\right)dx = \frac{1}{\ln a}\int \left(\frac{d}{dx}a^u\right)dx = \frac{1}{\ln a}a^u + C. \tag{12}$$

The integral on the left in Eq. (12) is usually written as

$$\int a^u \frac{du}{dx}dx = \int a^u \, du,$$

so that (12) takes the following form:

$$\int a^u \, du = \frac{1}{\ln a} a^u + C, \qquad a \neq 1, \quad a > 0. \tag{13}$$

EXAMPLE 5   From Eq. (13) we have

a) $\displaystyle\int 2^x \, dx = \frac{1}{\ln 2} 2^x + C,$

b) $\displaystyle\int 2^{\sin x} \cos x \, dx = \frac{1}{\ln 2} 2^{\sin x} + C.$

To evaluate the first integral, we apply Eq. (13) with

$$a = 2, \qquad u = x, \qquad du = dx.$$

To evaluate the second integral, we apply Eq. (13) with

$$a = 2, \qquad u = \sin x, \qquad du = \cos x \, dx.$$

This gives

$$\int 2^{\sin x} \cos x \, dx = \int 2^u \, du = \frac{1}{\ln 2} 2^u + C$$

$$= \frac{1}{\ln 2} 2^{\sin x} + C. \qquad \blacksquare$$

The Power Rule for Differentiation (Final Form)

Now that we can raise all positive numbers to arbitrary powers, we can extend the power rule for differentiation to hold for all real numbers, irrational as well as rational.

RULE 12

Power Rule (Final Form)

If n is any real constant and u is a positive differentiable function of x, then

$$\frac{d}{dx} u^n = n u^{n-1} \frac{du}{dx} \qquad \text{for any real number } n. \tag{14}$$

Proof of the Power Rule.   Let $y = u^n$. Then

$$\ln y = \ln u^n = n \ln u, \qquad (\text{Eq. 3})$$

$$\frac{1}{y} \frac{dy}{dx} = n \frac{1}{u} \frac{du}{dx},$$

and

$$\frac{dy}{dx} = n \frac{y}{u} \frac{du}{dx} = n \frac{u^n}{u} \frac{du}{dx} = n u^{n-1} \frac{du}{dx}. \qquad \blacksquare$$

The ability to raise positive numbers to arbitrary real powers makes it possible to define functions like

$$x^x, \qquad x^{\sin x}, \qquad \text{and} \qquad (1 + x)^{1/x}$$

for $x > 0$. The derivatives of functions like these are found by logarithmic differentiation.

EXAMPLE 6 Find dy/dx if $y = x^x$, $x > 0$.

Solution

$$\ln y = \ln x^x = x \ln x$$

$$\frac{1}{y}\frac{dy}{dx} = x \cdot \frac{1}{x} + \ln x = 1 + \ln x$$

$$\frac{dy}{dx} = y(1 + \ln x) = x^x(1 + \ln x)$$

Using Logarithms and Exponentials to Find Limits

Logarithms can be used to calculate limits like the one in the next example. The idea is this: To calculate the limit of a positive function, we calculate the limit of the logarithm of the function and then exponentiate that limit.

EXAMPLE 7 Show that

$$\lim_{n \to \infty} \left(1 + \frac{1}{n}\right)^n = \lim_{x \to 0^+} (1 + x)^{1/x} = e. \tag{15}$$

Solution We let $f(x) = (1 + x)^{1/x}$ and work with the expression

$$\ln f(x) = \ln(1 + x)^{1/x} = \frac{1}{x}\ln(1 + x) = \frac{\ln(1 + x)}{x}.$$

We find from l'Hôpital's Rule that

$$\lim_{x \to 0^+} \ln f(x) = \lim_{x \to 0^+} \frac{\ln(1 + x)}{x} = \lim_{x \to 0^+} \frac{\dfrac{1}{1 + x}}{1} = 1.$$

Therefore,

$$\ln f(x) \to 1 \qquad \text{as} \qquad x \to 0^+.$$

We can now exponentiate to see that as $x \to 0^+$,

$$(1 + x)^{1/x} = f(x) = e^{\ln f(x)} \to e^1 = e$$

(because the exponential function is continuous). That is,

$$\lim_{x \to 0^+} (1 + x)^{1/x} = e,$$

which is Eq. (15).

REMARK The equation

$$\lim_{n \to \infty} \left(1 + \frac{1}{n}\right)^n = e$$

gives a way to define e independently of the definition of the natural logarithm. However, the proof that this limit exists would then have to be different from the proof here which uses the logarithm. ■

Properties of $y = a^x$, $a > 0$, $a \neq 1$

If a is a positive real number and $a \neq 1$, then the function $y = a^x$ has the following properties.

1. It is defined by the equation

$$a^x = e^{x \ln a}.$$

   Domain: the set of all real numbers, $-\infty < x < \infty$.

   Range: the set of all positive real numbers, $y > 0$.

2. Its derivative is

$$\frac{d}{dx} a^x = a^x \ln a.$$

3. It is continuous (because it is differentiable), increasing if $a > 1$, decreasing if $0 < a < 1$, and one-to-one in either case.

4. If u is any differentiable function of x, then

$$\frac{d}{dx} a^u = a^u \ln a \frac{du}{dx} \quad \text{and} \quad \int a^u \, du = \frac{1}{\ln a} a^u + C.$$

PROBLEMS

In Problems 1–16, find dy/dx.

1. $y = 2^x$ **2.** $y = 2^{3x}$ **3.** $y = 8^x$

4. $y = 3^{2x}$ **5.** $y = 9^x$ **6.** $y = 2^x 3^x$

7. $y = (2^x)^2$

8. $y = x^{\sin x}, \quad x > 0$

9. $y = (\sin x)^{\tan x}, \quad \sin x > 0$

10. $y = 2^{\sec x}$

11. $y = x^{\ln x}, \quad x > 0$

12. $y = (\cos x)^x, \quad \cos x > 0$

13. $y = (1 - x)^x, \quad x < 1$

14. $y = x 2^{(x^2)}$

15. $y = 2^x \ln x$

16. $y = (\cos x)^{\sqrt{x}}, \quad x > 0, \quad \cos x > 0$

Evaluate the integrals in Problems 17–28.

17. $\displaystyle\int_0^1 5^x \, dx$ **18.** $\displaystyle\int_{-1}^0 2^x \, dx$

19. $\displaystyle\int_0^1 \frac{1}{2^x} \, dx$ **20.** $\displaystyle\int_{-1}^1 \left(\frac{1}{10}\right)^x \, dx$

21. $\displaystyle\int_0^1 3^{2x} \, dx$ **22.** $\displaystyle\int_{-1}^1 2^{(x+1)} \, dx$

23. $\displaystyle\int_{-1}^0 4^{-x} \ln 2 \, dx$ **24.** $\displaystyle\int_{-2}^0 5^{-x} \, dx$

25. $\displaystyle\int_1^2 5^{(2x-2)} \, dx$ **26.** $\displaystyle\int_1^{\sqrt{2}} x 2^{-x^2} \, dx$

27. $\displaystyle\int_0^{\pi/2} 2^{\cos x} \sin x \, dx$ **28.** $\displaystyle\int_0^{\pi/3} 2^{\sec x} \sec x \tan x \, dx$

29. Which integral has the larger value: a, or b?

  a) $\displaystyle\int_0^1 2^{(3x)} \, dx$ b) $\displaystyle\int_0^1 3^{(2x)} \, dx$

30. Find the derivative with respect to x of the following functions.

  a) $y = 2^{\ln x}$ b) $y = \ln 2^x$

  c) $y = \ln x^2$ d) $y = (\ln x)^2$

Find the limits in Problems 31–36.

31. $\displaystyle\lim_{x \to \infty} 2^{-x}$ **32.** $\displaystyle\lim_{x \to -\infty} 3^x$

33. $\displaystyle\lim_{x\to 0}\frac{3^{\sin x}-1}{x}$

34. $\displaystyle\lim_{x\to 1^+} x^{1/(x-1)}$

35. $\displaystyle\lim_{x\to 0}(e^x+x)^{1/x}$

36. $\displaystyle\lim_{x\to\infty} x^{1/x}$

37. Logarithms are no help in finding the following limits. Find them some other way.

a) $\displaystyle\lim_{x\to\infty}\frac{3^x-5}{4(3^x+2)}$

b) $\displaystyle\lim_{x\to-\infty}\frac{3^x-5}{4(3^x+2)}$

38. CALCULATOR Find out how close you can come to the limit

$$e=\lim_{n\to\infty}(1+(1/n))^n\approx 2.718281828459045$$

with your calculator by taking $n=10,\ 10^2,\ 10^3,\ \dots\ .$ You can expect the approximations to approach e at first, but on some calculators they will move away after a while as round-off errors take their toll.

39. CALCULATOR The curves $y=x^2$ and $y=2^x$ intersect at $x=2$ and $x=4$. There is also a third intersection between -1 and 0 (Fig. 6.35). Find its coordinates, as accurately as your calculator will allow, by applying Newton's method with $f(x)=2^x-x^2$.

40. Find the maximum value of

a) $x^{1/x}$ for $x>0$,

b) x^{1/x^2} for $x>0$,

c) x^{1/x^n} for $x>0$ and n a positive integer.

41. Show that

$$\lim_{x\to\infty} x^{1/x^n}=1$$

if n is a positive integer.

42. For what positive values of x does $x^{(x^x)}=(x^x)^x$?

43. The steps in the derivation of Eq. (6) are as follows:

$$a^{xy}=e^{xy\ln a} \qquad\qquad\text{(a)}$$
$$=e^{y\cdot x\ln a} \qquad\qquad\text{(b)}$$
$$=e^{y\cdot\ln(a^x)} \qquad\qquad\text{(c)}$$
$$=(a^x)^y. \qquad\qquad\text{(d)}$$

Explain the roles that Eqs. (2) and (4) play in these steps.

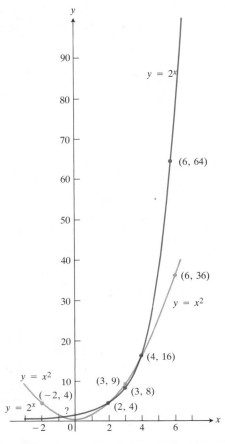

6.35 Where do the curves $y=x^2$ and $y=2^x$ intersect? At $x=2$, $x=4$, and $x=?$ See Problem 39.

44. Show that $(ab)^u=a^ub^u$ if a and b are any positive numbers and u is any real number.

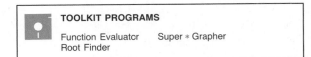

TOOLKIT PROGRAMS

Function Evaluator   Super ∗ Grapher
Root Finder

6.8
The Functions $y=\log_a u$: Rates of Growth

In Article 6.4 we mentioned that the natural logarithm was but one of many logarithmic functions. What are the others? They are the inverses of the exponential functions a^x, just as the natural logarithm is the inverse of e^x. These new inverses are surprisingly easy to evaluate, as we shall see in a moment. They have important applications in science and engineering and we shall mention a few of these at the end of the article. In passing, we shall also look at how rapidly logarithms and exponentials grow as x becomes large.

This article completes the development of our ideas about logarithms and exponentials. We began with

$$\ln x = \int_1^x \frac{1}{t}\,dt, \qquad x > 0.$$

This enabled us to define e^x by the equation

$$e^x = \text{inverse of } \ln x.$$

Then we defined a^x for any positive number a by the rule

$$a^x = e^{(\ln a)x}.$$

Now, as the final step, we close the circle by defining the base a logarithm of x to be

$$\log_a x = \text{inverse of } a^x.$$

Base a Logarithms

From the previous article we know that if a is any positive number other than 1, the function $y = a^x$ is differentiable and one-to-one. Furthermore, its derivative, $a^x \ln a$, is never zero. The function therefore has a differentiable inverse, which we call the **logarithm of x to the base a** and denote by

$$y = \log_a x. \tag{1}$$

Since $y = a^x$ and $y = \log_a x$ are inverses of one another, their composite in either order is the identity function. Thus we obtain the equations

$$\log_a(a^x) = x \qquad \text{for all } x, \tag{2a}$$

and

$$a^{(\log_a x)} = x \qquad \text{for each positive } x. \tag{2b}$$

In words, Eq. (2b) says that the logarithm of x to the base a is the exponent to which we raise a to get x. For example,

$$\log_a(1) = 0 \qquad \text{since} \qquad a^0 = 1,$$
$$\log_a(a) = 1 \qquad \text{since} \qquad a^1 = a,$$
$$\log_5 25 = 2 \qquad \text{since} \qquad 5^2 = 25,$$
$$\log_2\left(\frac{1}{4}\right) = -2 \qquad \text{since} \qquad 2^{-2} = \frac{1}{4}.$$

In every case, the logarithm of x is the exponent to which the base is raised to give x. In the equation

$$2^{-2} = \frac{1}{4},$$

2 is the base, and -2, the logarithm of $1/4$ to the base 2, is the power to which we raise 2 to get $1/4$.

Figure 6.36 shows the graphs of the functions $y = 2^x$ and $y = \log_2 x$.

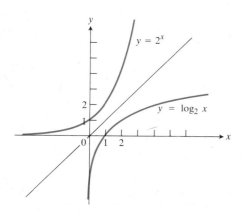

6.36 The graph of $y = 2^x$ and its inverse $y = \log_2 x$.

Calculation of $\log_a x$

The number $y = \log_a x$ can always be calculated from the natural logarithms of a and x by the following formula:

$$\log_a x = \frac{\ln x}{\ln a}. \tag{3}$$

This formula can be derived in the following way. If

$$y = \log_a x,$$

then

$$a^y = x.$$

Therefore,

$$\ln a^y = \ln x, \quad y \ln a = \ln x,$$

and

$$y = \frac{\ln x}{\ln a},$$

which is Eq. (3).

EXAMPLE 1 From Eq. (3) we find

$$\log_2(10) = \frac{\ln 10}{\ln 2} \approx \frac{2.30259}{0.69315} \approx 3.32193. \qquad \blacksquare$$

EXAMPLE 2 Calculate $y = \log_4 8$.

Solution From Eq. (3) we have

$$y = \log_4 8 = \frac{\ln 8}{\ln 4} = \frac{\ln 2^3}{\ln 2^2} = \frac{3 \ln 2}{2 \ln 2} = \frac{3}{2}. \qquad \blacksquare$$

Rules of Arithmetic

If we combine Eq. (3) with the three rules

$$\ln uv = \ln u + \ln v, \tag{4a}$$

$$\ln \frac{u}{v} = \ln u - \ln v, \tag{4b}$$

$$\ln u^v = v \ln u, \tag{4c}$$

established in Article 6.5 for any positive real values (called a, x, and n there instead of u and v), we get

$$\log_a uv = \log_a u + \log_a v \tag{5a}$$

$$\log_a \frac{u}{v} = \log_a u - \log_a v \tag{5b}$$

$$\log_a u^v = v \log_a u \tag{5c}$$

Equation (5c), for example, is derived as follows:

$$\log_a u^v = \frac{\ln u^v}{\ln a} = \frac{v \ln u}{\ln a} = v \log_a u,$$

where we use Eq. (3), then Eq. (4) of Article 6.7, then Eq. (3) again.

The Derivative of $\log_a u$

If u is a differentiable positive function of x, we may differentiate both sides of

$$\log_a u = \frac{\ln u}{\ln a}$$

to obtain

$$\frac{d}{dx} \log_a u = \frac{d}{dx} \frac{\ln u}{\ln a} = \frac{1}{\ln a} \frac{d}{dx} \ln u = \frac{1}{\ln a} \frac{1}{u} \frac{du}{dx} = \frac{1}{u \ln a} \frac{du}{dx}.$$

In short,

$$\frac{d}{dx} \log_a u = \frac{1}{u \ln a} \frac{du}{dx}. \qquad (6)$$

EXAMPLE 3 Calculate the derivative of $y = \log_{10}(3x + 1)$.

Solution From Eq. (6) with $a = 10$ and $u = 3x + 1$, we have

$$\frac{d}{dx} \log_{10}(3x + 1) = \frac{1}{(3x + 1)\ln 10} \frac{d}{dx}(3x + 1) = \frac{3}{(3x + 1)\ln 10}. \qquad \blacksquare$$

Integrals involving $\log_a x$ can always be evaluated as integrals involving $\ln x$.

EXAMPLE 4 Evaluate

$$\int \frac{\log_2 x}{x} dx.$$

Solution We express $\log_2 x$ in terms of $\ln x$:

$$\log_2 x = \frac{\ln x}{\ln 2}.$$

Then

$$\int \frac{\log_2 x}{x} dx = \frac{1}{\ln 2} \int \frac{\ln x}{x} dx$$

$$= \frac{1}{\ln 2} \int u \, du \qquad \left(u = \ln x, \ du = \frac{1}{x} dx \right)$$

$$= \frac{1}{\ln 2} \frac{u^2}{2} + C$$

$$= \frac{(\ln x)^2}{2 \ln 2} + C. \qquad \blacksquare$$

Relative Rates of Growth

You may have noticed that exponential functions like

$$y = 2^x \quad \text{and} \quad y = e^x$$

seem to grow much more rapidly as x gets large than the polynomials and rational functions we graphed in Chapter 3. These exponentials certainly grow more rapidly than the function $y = x$, as Figs. 6.30 and 6.34 show, and you can see $y = 2^x$ outgrowing $y = x^2$ as x increases in Fig. 6.35. In fact, as $x \to \infty$, the functions $y = 2^x$ and $y = e^x$ grow faster than any positive power of x, even $x^{1,000,000}$ (Problem 44).

To get a feeling for how rapidly the values of $y = e^x$ grow with increasing x, think of graphing the function on a large blackboard, with the axes scaled in centimeters. At $x = 1$ cm, the graph is $e^1 \approx 3$ cm above the x-axis. At $x = 6$ cm, the graph is $e^6 \approx 403$ cm ≈ 4 m high (it is about to go through the ceiling if it hasn't done so already). At $x = 10$ cm, the graph is $e^{10} \approx 22{,}026$ cm ≈ 220 m high, higher than most buildings. At $x = 24$ cm, the graph is more than halfway to the moon, and at $x = 43$ cm from the origin, the graph is high enough to reach past the nearest neighboring star, Proxima Centauri:

$$e^{43} \approx 4.7 \times 10^{18} \text{ cm}$$

$$= 4.7 \times 10^{13} \text{ km}$$

$$\approx 1.57 \times 10^8 \text{ light seconds} \tag{7}$$
$$\text{(light travels at 300,000 km/sec in a vacuum)}$$

$$\approx 5.0 \text{ light years}$$

The distance to Proxima Centauri is about 4.3 light years. Yet, with $x = 43$ cm from the origin, the graph is still less than 2 feet to the right of the y-axis.

In contrast, logarithmic functions like

$$y = \log_2 x \quad \text{and} \quad y = \ln x$$

grow more slowly as $x \to \infty$ than any positive power of x (see Problem 46). With axes scaled in centimeters, you have to go more than 4 light-years out on the x-axis to find a point where the graph of $y = \ln x$ is even $y = 43$ cm high.

These important comparisons of exponential, polynomial, and logarithmic functions can be made precise by defining what it means for a function $y = f(x)$ to grow faster than another function $y = g(x)$ as $x \to \infty$.

DEFINITION

Rates of Growth

$f(x)$ grows faster than $g(x)$ as $x \to \infty$ if

$$\lim_{x \to \infty} \frac{f(x)}{g(x)} = \infty, \tag{8}$$

and f grows at the same rate as g as $x \to \infty$ if

$$\lim_{x \to \infty} \frac{f(x)}{g(x)} = L \neq 0. \quad \text{(L finite and not zero)} \tag{9}$$

According to these definitions, $y = 2x$ does not grow faster than $y = x$. The two functions grow at the same rate because

$$\lim_{x \to \infty} \frac{2x}{x} = \lim_{x \to \infty} 2 = 2,$$

which is a finite nonzero limit. The reason for this apparent disregard of common sense is that we want "f grows faster than g" to mean that for large x-values, g is negligible when compared to f.

EXAMPLE 5 $y = e^x$ grows faster than $y = x^2$ because

$$\lim_{x \to \infty} \frac{e^x}{x^2} = \infty,$$

as we can see by two applications of l'Hôpital's rule:

$$\lim_{x \to \infty} \frac{e^x}{x^2} = \lim_{x \to \infty} \frac{e^x}{2x} = \lim_{x \to \infty} \frac{e^x}{2} = \infty. \qquad ■$$

EXAMPLE 6 $y = 3^x$ grows faster than $y = 2^x$ as $x \to \infty$ because

$$\lim_{x \to \infty} \frac{3^x}{2^x} = \lim_{x \to \infty} \left(\frac{3}{2}\right)^x = \infty. \qquad ■$$

EXAMPLE 7 $y = x^2$ grows faster than $y = \ln x$ as $x \to \infty$ because

$$\lim_{x \to \infty} \frac{x^2}{\ln x} = \lim_{x \to \infty} \frac{2x}{1/x} = \lim_{x \to \infty} 2x^2 = \infty. \qquad ■$$

⇧
l'Hôpital's
rule

EXAMPLE 8 Compare the rates of growth of

$$y = \log_2 x$$

and

$$y = \ln x$$

as $x \to \infty$.

Solution We take the ratio of the functions in either order (it does not matter which) and calculate the limit of the ratio as $x \to \infty$:

$$\lim_{x \to \infty} \frac{\log_2 x}{\ln x} = \lim_{x \to \infty} \frac{\ln x / \ln 2}{\ln x} = \frac{1}{\ln 2}.$$

⇧
Eq. (3)

The limit is finite and not zero. The logarithms are therefore seen to grow at the same rate even though their bases are different. ■

As Example 8 suggests, any two logarithmic functions $y = \log_a x$ and $y = \log_b x$ grow at the same rate as $x \to \infty$. To see why, we calculate the limit

$$\lim_{x \to \infty} \frac{\log_a x}{\log_b x} = \lim_{x \to \infty} \frac{\ln x / \ln a}{\ln x / \ln b} = \frac{\ln b}{\ln a}.$$

This limit is always finite and different from zero.

In contrast to the way logarithmic functions behave, two different exponential functions

$$y = a^x \qquad \text{and} \qquad y = b^x$$

grow at different rates as $x \to \infty$, as we can see from the calculation

$$\lim_{x \to \infty} \frac{a^x}{b^x} = \lim_{x \to \infty} \left(\frac{a}{b} \right)^x = \begin{cases} \infty & \text{if } a > b, \\ 0 & \text{if } a < b. \end{cases}$$

If $a > b$, then a^x grows faster than b^x.

If f grows at the same rate as g and g grows at the same rate as h as $x \to \infty$, then f grows at the same rate as h. The reason is that

$$\lim \frac{f}{g} = L_1 \qquad \text{and} \qquad \lim \frac{g}{h} = L_2$$

imply

$$\lim \frac{f}{h} = \lim \frac{f}{g} \cdot \frac{g}{h} = L_1 L_2.$$

EXAMPLE 9   Show that $\sqrt{x^2 + 5}$ and $(2\sqrt{x} - 1)^2$ grow at the same rate as $x \to \infty$.

Solution   We show that the functions grow at the same rate by showing that they both grow at the same rate as x:

$$\lim_{x \to \infty} \frac{\sqrt{x^2 + 5}}{x} = \lim_{x \to \infty} \sqrt{1 + \frac{5}{x^2}} = 1$$

$$\lim_{x \to \infty} \frac{(2\sqrt{x} - 1)^2}{x} = \lim_{x \to \infty} \left(\frac{2\sqrt{x} - 1}{\sqrt{x}} \right)^2$$

$$= \lim_{x \to \infty} \left(2 - \frac{1}{\sqrt{x}} \right)^2 = 4. \qquad \blacksquare$$

☐ Base 10 Logarithms

Base 10 logarithms, often called **common logarithms,** appear in many scientific formulas. For example, earthquake intensity is reported on the Richter scale. Here the formula is

$$\text{Magnitude } R = \log_{10} \left(\frac{a}{T} \right) + B, \tag{10}$$

where a is the amplitude of the ground motion in microns at the receiving station, T is the period of the seismic wave in seconds, and B is an empirical factor that allows for weakening of the seismic wave with increasing distance from the epicenter of the earthquake. For an earthquake 10,000 km from the receiving station, $B = 6.8$. If the recorded vertical ground motion is $a = 10$

microns and the period is $T = 1$ second, then the magnitude of the earthquake is

$$R = \log_{10}\left(\frac{10}{1}\right) + 6.8 = 7.8 \qquad (11)$$

on the Richter scale. An earthquake of this magnitude does great damage near its epicenter. Damage begins at a magnitude of 5, and destruction is almost total at a magnitude of 8. The 1964 earthquake in Anchorage, Alaska, registered 8.4 on the Richter scale. The San Fernando, California, earthquake of 1971 (with a damage bill of billions of dollars) registered 6.8.

The pH scale for measuring the acidity of a solution is a logarithmic scale. The pH (pronounced ''p, h'') value (hydrogen potential) of the solution is the common logarithm of the reciprocal of the solution's hydronium ion concentration, $[H_3O^+]$:

$$pH = \log_{10}\frac{1}{[H_3O^+]} = -\log_{10}[H_3O^+]. \qquad (12)$$

The hydronium ion concentration is measured in moles per liter. Vinegar has a pH of 3, distilled water a pH of 7, seawater a pH of 8.15, and household ammonia a pH of 12. The total scale ranges from about 0.1 for normal hydrochloric acid to 14 for a normal solution of sodium hydroxide. Most foods are acidic. Some typical pH values are

Food	pH value
bananas	4.5–4.7
grapefruit	3.0–3.3
oranges	3.0–4.0
limes	1.8–2.0
milk	6.3–6.6
soft drinks	2.0–4.0
spinach	5.1–5.7

In astronomy, the relation between a star's absolute magnitude M, apparent magnitude m, and distance d in parsecs (a parsec is 3.262 light years) is given by the equation

$$M = m + 5 - 5\log_{10}d.$$

This equation can be used to find how far away the star is when m and M are known.

Another example of the use of common logarithms is the db (''d, b'') scale for measuring loudness in decibels. If I is the intensity of the sound in watts per square meter, then the decibel level of the sound is

$$\text{Sound level} = 10\log(I \times 10^{12}) \text{ db.} \qquad (13)$$

Typical sound levels are

Threshold of hearing	0 db
Rustle of leaves	10 db
Average whisper	20 db
Quiet automobile	50 db
Ordinary conversation	65 db
Pneumatic drill 10 feet away	90 db
Threshold of pain	120 db

If you ever wondered why doubling the power of your audio amplifier increased the sound level by only a few decibels, Eq. (13) provides the answer. As the following calculation shows, doubling I adds only about 3 db:

$$10 \log(2I \times 10^{12}) = 10 \log(I \times 10^{12}) + 10 \log 2$$

$$\approx 10 \log(I \times 10^{12}) + 3.$$

REMARK ON NOTATION Many advanced texts and research publications in mathematics use log x, with no base specified, to represent the natural logarithm ln x. Most texts in the physical sciences use log x to represent $\log_{10}x$. Most calculators use ln x for the natural logarithm and log x for base 10 logarithms. Computers, however, may use LOG(X) for the natural logarithm. One then evaluates $\log_{10}x$ as (LOG(X))/(LOG(10)).

Brief Summary of the Definitions in This Chapter

$$\ln x = \int_1^x \frac{1}{t}\,dt, \qquad x > 0$$

$$e^x = \text{inverse of } \ln x$$

$$a^x = e^{(\ln a)x} \qquad \text{for any positive number } a$$

$$\log_a x = \text{inverse of } a^x$$

PROBLEMS

In Problems 1–8, express each logarithm as a rational number.

1. $\log_4 16$

2. $\log_8 32$

3. $\log_5 0.04$

4. $\log_{0.5} 4$

5. $\log_2 4$

6. $\log_4 2$

7. $\log_8 16$

8. $\log_{32} 4$

In Problems 9 and 10, solve for x.

9. $3^{\log_3 7} + 2^{\log_2 5} = 5^{\log_5 x}$

10. $8^{\log_8 3} - e^{\ln 5} = x^2 - 7^{\log_7 3x}$

Find the limits in Problems 11–14.

11. $\lim\limits_{x \to \infty} \dfrac{\log_2 x}{\log_3 x}$

12. $\lim\limits_{x \to \infty} \dfrac{\log_2 x}{\log_8 x}$

13. $\lim\limits_{x \to \infty} \dfrac{\log_9 x}{\log_3 x}$

14. $\lim\limits_{x \to \infty} \dfrac{\log_{\sqrt{10}} x}{\log_{\sqrt{2}} x}$

In Problems 15–28, find dy/dx.

15. $y = \log_4 x$

16. $y = \log_4 x^2$

17. $y = \log_{10} e^x$

18. $y = \log_5 \sqrt{x}$

19. $y = \ln 2 \cdot \log_2 x$

20. $y = \log_2(1/x)$

21. $y = \log_{10} \sqrt{x + 1}$

22. $y = \log_2(3x + 1)$

23. $y = 1/\log_2 x$

24. $y = \ln 10^x$

25. $y = \log_5(x + 1)^2$

26. $y = \log_2(\ln x)$

27. $y = \log_7(\sin x)$

28. $y = e^{\log_{10} x}$

29. Which function is changing faster at $x = 10$,

$$y = \ln x \quad \text{or} \quad y = \log_2 x?$$

30. Show that $\log_b u = \log_a u \cdot \log_b a$ if a, b, and u are positive numbers and neither a nor b equals 1.

Evaluate the integrals in Problems 31–38.

31. $\displaystyle\int_1^{10} \frac{\log_{10} x}{x}\,dx$

32. $\displaystyle\int_1^4 \frac{\log_2 x}{x}\,dx$

33. $\displaystyle\int_1^8 \frac{\log_4(x^2)}{x}\,dx$

34. $\displaystyle\int_0^1 \frac{\log_2(3x + 1)}{3x + 1}\,dx$

35. $\displaystyle\int_{1}^{125} \frac{(\log_5 x)^2}{x}\,dx$

36. $\displaystyle\int_{e}^{e^2} \frac{dx}{x\,\log_2 x}$

37. $\displaystyle\int_{\sqrt{e}}^{e} \frac{dx}{x\,\log_{10} x}$

38. $\displaystyle\int_{2}^{8} \frac{dx}{x(\log_8 x)^2}$

39. Which of the following functions grow slower than $y = e^x$ as $x \to \infty$?

a) $y = x + 3$
b) $y = x^3 - 3x + 1$
c) $y = \sqrt{x}$
d) $y = 4^x$
e) $y = (5/2)^x$
f) $y = \ln x$
g) $y = \log_{10} x$
h) $y = e^{-x}$
i) $y = e^{x+1}$
j) $y = (1/2)e^x$

40. Which of the following functions grow faster than $y = x^2 - 1$ as $x \to \infty$? Which grow at the same rate as $y = x^2 - 1$? Which grow slower?

a) $y = x^2 + 4x$
b) $y = x^3 + 3$
c) $y = x^5$
d) $y = 15x + 3$
e) $y = \sqrt{x^4 + 5x}$
f) $y = (x + 1)^2$
g) $y = \ln x$
h) $y = \ln(x^2)$
i) $y = \ln(10^x)$
j) $y = 2^x$

41. Which of the following functions grow at the same rate as $y = \ln x$ as $x \to \infty$?

a) $y = \log_3 x$
b) $y = \log_2 x^2$
c) $y = \log_{10} \sqrt{x}$
d) $y = 1/x$
e) $y = 1/\sqrt{x}$
f) $y = e^{-x}$
g) $y = x$
h) $y = 5 \ln x$
i) $y = 2$
j) $y = \sin x$

42. Order the following functions from fastest growing to slowest growing as $x \to \infty$.

a) e^x
b) x^x
c) $(\ln x)^x$
d) $e^{x/2}$

43. Compare the growth rates as $x \to \infty$ of $y = \ln x$ and $y = \ln(\ln x)$.

44. Show that $y = e^x$ grows faster as $x \to \infty$ than $y = x^n$ for any positive integer n (even $x^{1,000,000}$). (*Hint:* What is the nth derivative of x^n?)

45. What do the conclusions about the limits in Article 1.10, Problem 58, imply about the relative growth rates of polynomials as $x \to \infty$?

46. Show that $y = \ln x$ grows slower as $x \to \infty$ than $y = x^{1/n}$ for any positive number n.

47. In any solution, the product of the concentration $[H_3O^+]$ of hydronium ions and the concentration $[OH^-]$ of hydroxyl ions is about 10^{-14}.

a) What value of $[H_3O^+]$ minimizes the sum of the concentrations, $S = [H_3O^+] + [OH^-]$? (*Hint:* Change notation. Let $x = [H_3O^+]$.)

b) What is the pH of a solution in which S has this minimum value?

c) What ratio of $[H_3O^+]$ to $[OH^-]$ minimizes S?

48. *Stirling's formula for estimating $n!$* Stirling's formula says that, if n is large, then

$$n! \sim \sqrt{2\pi}\; n^{n+(1/2)} e^{-n}.$$

For what value of the constant m can we replace e^{-n} by 10^{-mn} in the formula?

49. a) Verify Eq. (5a).   b) Verify Eq. (5b).

6.9
Applications of Exponential and Logarithmic Functions

In this article, we present some of the applications of logarithmic and exponential functions that account for their importance in engineering and science.

The Law of Exponential Change

In many physical, biological, ecological, and economic phenomena, some quantity y grows or declines at a rate that at any given time t is proportional to the amount that is present. This leads to the equation

$$\frac{dy}{dt} = ky, \tag{1}$$

where k is a constant that is positive if y is increasing and negative if y is decreasing.

To solve Eq. (1), we divide through by y to get

$$\frac{1}{y}\frac{dy}{dt} = k$$

and then integrate both sides with respect to t to get

$$\ln y = kt + C. \qquad (2)$$

We can omit the usual absolute value bars on the left because y is positive.

It follows from Eq. (2) that

$$y = e^{kt+C} = e^{kt} \cdot e^{C} = Ae^{kt},$$

where $A = e^{C}$. If y_0 denotes the value of y when $t = 0$, then $A = y_0$ and $y = y_0 e^{kt}$. This equation is called the **law of exponential change.**

The Law of Exponential Change

$$y = y_0 e^{kt} \qquad (3)$$

EXAMPLE 1 *The growth of a cell.* In an ideal environment, the mass m of a cell will grow exponentially, at least early on. Chemicals pass quickly through the cell wall, and growth is limited only by the metabolism within the cell, which in turn depends on the mass of participating molecules. If we make the reasonable assumption that, at each instant of time, the cell's growth rate dm/dt is proportional to the mass that has already been accumulated, then

$$\frac{dm}{dt} = km$$

and

$$m = m_0 e^{kt}.$$

There are limitations, of course, and in any particular case we would expect this equation to provide reliable information only for values of m below a certain size. ∎

EXAMPLE 2 *Birth rates and population growth.** Strictly speaking, the number of individuals in a population (of people, plants, foxes, or whatever) is a discontinuous function of time because it takes on discrete values. However, as soon as the number of individuals becomes large enough, it may safely be described with a continuous and even differentiable function. If we assume that the proportion of reproducing individuals remains constant and assume a constant fertility, then at any instant t, the birth rate is proportional to the number $y(t)$ of individuals present. If, further, we neglect departures, arrivals, and deaths, the growth rate dy/dt will be the same as the birth rate ky. In other words,

$$\frac{dy}{dt} = ky.$$

Once again, we find that $y = y_0 e^{kt}$. ∎

---

*For more information about mathematical applications to the life sciences, see Edward Baschelet, *Introduction to Mathematics for Life Scientists,* 2nd ed. (New York: Springer-Verlag, 1976).

Continuously Compounded Interest

If you invest an amount A_0 of money at a fixed annual interest rate r and interest is added to your account k times a year, it turns out that the amount of money you will have at the end of t years is

$$A_t = A_0\left(1 + \frac{r}{k}\right)^{kt}. \qquad (4)$$

The money might be added ("compounded," bankers say) monthly ($k = 12$), weekly ($k = 52$), daily ($k = 365$), or even more frequently, say by the hour or even by the minute. But there is still a limit to how much you will earn that way, and the limit is

$$\lim_{k \to \infty} A_t = \lim_{k \to \infty} A_0\left(1 + \frac{r}{k}\right)^{kt} = A_0\, e^{rt}. \qquad (5)$$

(See Problem 10.)

The resulting formula for the amount of money in your account after t years is

$$A(t) = A_0\, e^{rt}. \qquad (6)$$

This formula is called the **continuous compound interest formula.** Interest paid according to this formula is said to be **compounded continuously.**

- **EXAMPLE 3** Suppose you deposit $621 in a bank account that pays 6% compounded continuously. How much money will you have in the account 8 years later?

Solution We use Eq. (6) with $A_0 = 621$, $r = 0.06$, and $t = 8$:

$$A(8) = 621\, e^{(0.06)(8)} = 621\, e^{0.48} = 1003.58$$

(to the nearest cent).

Had the bank paid interest quarterly ($k = 4$ in Eq. 4), the amount in your account would have been an even $1000. Thus, the effect of continuous compounding, as compared with quarterly compounding, has been an addition of $3.58. A bank might decide it would be worth this additional amount to be able to advertise "we compound your money every second, night and day—better than that, we compound the interest continuously."   ∎

Radioactivity

When a radioactive atom emits some of its mass as radiation, the remainder of the atom re-forms to make an atom of some new substance. This process of radiation and change is called **radioactive decay,** and an element whose atoms go through this process spontaneously is called **radioactive.** Thus, radioactive carbon-14 decays into nitrogen; radium, through a number of intervening radioactive steps, decays into lead.

Experiments have shown that at any given time the rate at which a radioactive element decays (as measured by the number of nuclei that change per unit time) is approximately proportional to the number of radioactive nuclei present.

Thus, the decay of a radioactive element is described by the equation $dy/dt = ky$ and the number of radioactive nuclei present at time t will be

$$y = y_0\, e^{kt}, \tag{7}$$

where y_0 is the number that were present at time zero.

In Eq. (7), the **decay constant** k is a negative number whose value is a characteristic of the element that is decaying. For example, when time is measured in years, k is -1.2×10^{-4} for carbon-14 and -4.3×10^{-4} for radium-226.

EXAMPLE 4 *The half-life of a radioactive element.* The **half-life** of a radioactive element is the time required for half of the radioactive nuclei present in a sample to decay. It is a remarkable fact that the half-life is a constant that does not depend on the number of radioactive nuclei initially present in the sample.

To see why, let y_0 be the number of radioactive nuclei initially present in the sample. Then the number y present at any later time t will be

$$y = y_0\, e^{kt}.$$

We seek the value of t at which

$$y_0\, e^{kt} = \frac{1}{2} y_0,$$

for this will be the time when the number of radioactive nuclei present equals half the original number. The y_0's cancel in this equation to give

$$e^{kt} = \frac{1}{2}$$

$$kt = \ln\frac{1}{2} = -\ln 2$$

$$t = -\frac{\ln 2}{k}. \tag{8}$$

This value of t is the half-life of the element. As Eq. (8) shows, it depends only on the value of k and not on the number y_0 of nuclei originally present. ∎

Heat Transfer: Newton's Law of Cooling

When hot chocolate is left standing in a tin cup, it cools until its temperature drops to that of the surrounding air. When a hot silver ingot is put into water to cool, its temperature drops until it equals that of the surrounding water. In situations like these, the rate at which the temperature of the object changes is roughly proportional to the difference between the temperature of the object and the temperature of the surrounding medium. This rule is called **Newton's law of cooling,** although it applies to warming as well. It can be written in an equation in the following way.

If $T(t)$ is the temperature of the object at time t, and T_s is the surrounding temperature, then

$$\frac{dT}{dt} = k(T - T_s). \tag{9}$$

Since T_s is constant, this is the same as $dy/dt = ky$ with $y = (T - T_s)$. Hence, the solution of Eq. (9) is

$$y = y_0 \, e^{kt}$$

or

$$T - T_s = (T_0 - T_s)e^{kt}, \tag{10}$$

where T_0 is the value of T at time zero.

EXAMPLE 5 A hard-boiled egg at 98°C is put in a sink of 18°C water to cool. After 5 minutes, the egg's temperature is found to be 38°C. Assuming that the water has not warmed appreciably, how much longer will it take the egg to reach 20°C?

Solution We find how long it would take the egg to cool from 98°C to 20°C and subtract the 5 minutes that have already elapsed.

According to Eq. (10), the egg's temperature t minutes after it is put in the sink is

$$T - 18 = (98 - 18)e^{kt}, \quad \text{or} \quad T = 18 + 80e^{kt}.$$

To find k, we use the information that $T = 38$ when $t = 5$. This gives

$$38 = 18 + 80e^{5k}$$

$$e^{5k} = \frac{1}{4}$$

$$5k = \ln\frac{1}{4} = -\ln 4$$

$$k = -\frac{1}{5}\ln 4 = -0.28$$

(to two decimal places). The egg's temperature at time t is

$$T = 18 + 80e^{-0.28t}.$$

When will $T = 20$? When

$$20 = 18 + 80e^{-0.28t}$$

$$80e^{-0.28t} = 2$$

$$e^{-0.28t} = \frac{1}{40}$$

$$-0.28t = \ln\frac{1}{40} = -\ln 40$$

$$t = \frac{\ln 40}{0.28} = 13 \text{ min.} \quad \text{(to two digits)}$$

The egg's temperature will reach 20°C 13 minutes after it is put in to cool. Since it took 5 minutes to reach 38°C, it will take 8 more to reach 20°C.

6.37 The *R–L* circuit.

The *R–L* Circuit

The diagram in Fig. 6.37 represents an electrical circuit whose total resistance is *R* ohms and whose self-inductance, shown schematically as a coil, is *L* henries (hence the name "*R–L* circuit"). There is also a switch whose terminals at *a* and *b* can be closed to connect a constant electrical source of *V* volts.

Ohm's law, $V = RI$, has to be modified for such a circuit. The modified form is

$$L\frac{di}{dt} + Ri = V, \tag{11}$$

where *i* is the current in amperes and *t* is the time in seconds. (See Sears, Zemansky, and Young, *University Physics*, 7th ed., 1987, Chapter 33.) From this equation it is possible to predict how the current will flow after the switch is closed.

EXAMPLE 6 The switch in the *R–L* circuit in Fig. 6.37 is closed at time $t = 0$. How will the current flow as a function of time?

Solution We solve Eq. (11) for *i* under the assumption that $i = 0$ when $t = 0$. The calculations go like this:

$$L\frac{di}{dt} + Ri = V$$

$$\frac{di}{V - Ri} = \frac{1}{L}dt \qquad \text{(variables separated)}$$

$$\frac{-\frac{1}{R}du}{u} = \frac{1}{L}dt \qquad \left(\begin{array}{c}\text{substitute } u = V - Ri, \\ du = -R\,di, \quad di = -(1/R)\,du\end{array}\right)$$

$$\frac{du}{u} = -\frac{R}{L}dt \qquad \text{(variables separated)}$$

$$\ln|u| = -\frac{R}{L}t + C \qquad \text{(integrate)}$$

$$u = \pm e^{-(Rt/L)+C} \qquad \text{(exponentiate)}$$

$$= Ae^{-Rt/L} \qquad \text{(write } A \text{ for } \pm e^{C}\text{)}$$

$$V - Ri = Ae^{-Rt/L} \qquad \text{(change from } u \text{ back to } i\text{)}$$

To find *A*, we use the condition that $i = 0$ when $t = 0$:

$$V - R(0) = Ae^{-R(0)/L}$$

$$V = A.$$

The equation for the current is

$$V - Ri = Ve^{-Rt/L}$$

or

$$i = \frac{V}{R}(1 - e^{-Rt/L}). \tag{12}$$

6.38 (a) R–L circuit, solution of Example 6. Growth of current in a circuit containing inductance and resistance. I is the current's steady-state value. (b) Problem 20. Decay of current in a circuit containing inductance and resistance.

We see from this that the current is always less than V/R but that it approaches V/R as a **steady-state value:**

$$\lim_{t \to \infty} \frac{V}{R}(1 - e^{-Rt/L}) = \frac{V}{R}(1 - 0) = \frac{V}{R}. \tag{13}$$

The current $I = V/R$ is the current that will flow in the circuit if either $L = 0$ (no inductance) or $di/dt = 0$ (steady current, $i =$ constant). The graph of the current-versus-time relation (12) is shown in Fig. 6.38(a).   ∎

Law of exponential change:	$y = y_0\, e^{kt}$
Continuous compound interest:	$A(t) = A_0\, e^{rt}$
Newton's law of cooling:	$T - T_s = (T_0 - T_s)e^{kt}$
R-L circuit equations:	$L\dfrac{di}{dt} + Ri = V$
	$i = \dfrac{V}{R}(1 - e^{-Rt/L})$

PROBLEMS

1. *Growth of bacteria.* Suppose that the bacteria in a colony can grow unchecked, by the law of exponential change. The colony starts with one bacterium and doubles every half hour. How many bacteria will the colony contain at the end of 24 hours? (Under favorable laboratory conditions, the number of cholera bacteria can double every 30 minutes. Of course, in an infected person, many of the bacteria are destroyed, but this example helps to explain why a person who feels well in the morning may be dangerously ill by evening.)

2. *Growth of bacteria.* A colony of bacteria is grown under ideal conditions in a laboratory so that the population increases exponentially with time. At the end of 3 hours there are 10,000 bacteria. At the end of 5 hours there are 40,000. How many bacteria were present initially?

3. *The incidence of a disease.* In the course of any given year, the number y of cases of a disease is reduced by 10%. If there are 10,000 cases today, about how many years will it take to reduce the number of cases to less than 1000?

4. *Atmospheric pressure.* The earth's atmospheric pressure p is often modeled by assuming that the rate dp/dh at which p changes with the altitude h above sea level is proportional to p. Suppose that the pressure at sea level is 1013 millibars

(about 14.7 pounds per square inch) and that the pressure at an altitude of 20 km is 50 millibars.

a) Solve the equation $dp/dh = kp$ (k a constant) to express p in terms of h. Determine the values of k and the constant of integration from the given initial conditions.

b) What is the atmospheric pressure at $h = 50$ km?

c) At what altitude is the pressure equal to 900 millibars?

5. *Discharging capacitor.* As a result of leakage, an electrical capacitor discharges at a rate proportional to the charge. If the charge Q has the value Q_0 at time $t = 0$, find Q as a function of t.

6. *First order chemical reactions.* In some chemical reactions, the rate at which the amount of a substance changes with time is proportional to the amount present. For the change of δ-glucono lactone into gluconic acid, for example,

$$\frac{dy}{dt} = -0.6y$$

when t is measured in hours. If there are 100 grams of δ-glucono lactone present when $t = 0$, how many grams will be left after the first hour?

7. *Benjamin Franklin's will.* The Franklin Technical Institute of Boston owes its existence to a provision in a codicil to the will of Benjamin Franklin. In part, it reads:

I was born in Boston, New England and owe my first instruction in Literature to the free Grammar Schools established there: I have therefore already considered those schools in my Will. . . . I have considered that among Artisans good Apprentices are most likely to make good citizens . . . I wish to be useful even after my Death, if possible, in forming and advancing other young men that may be serviceable to their Country in both Boston and Philadelphia. To this end I devote Two thousand Pounds Sterling, which I give, one thousand thereof to the Inhabitants of the Town of Boston in Massachusetts, and the other thousand to the inhabitants of the City of Philadelphia, in Trust and for the Uses, Interests and Purposes hereinafter mentioned and declared.

Franklin's plan was to lend money to young apprentices at 5% interest with the provision that each borrower should pay each year

. . . with the yearly Interest, one tenth part of the Principal, which sums of Principal and Interest shall be again let to fresh Borrowers. . . . If this plan is executed and succeeds as projected without interruption for one hundred Years, the Sum will then be one hundred and thirty-one thousand Pounds of which I would have the Managers of the Donation to the Inhabitants of the Town of Boston, then lay out at their discretion one hundred thousand Pounds in Public Works. . . . The remaining thirty-one thousand Pounds, I would

have continued to be let out on Interest in the manner above directed for another hundred Years. . . . At the end of this second term if no unfortunate accident has prevented the operation the sum will be Four Millions and Sixty-one Thousand Pounds.

It was not always possible to find as many borrowers as Franklin had planned, but the managers of the trust did the best they could; they lent money to medical students as well as others. At the end of 100 years from the reception of the Franklin gift, in January 1894, the fund had grown from 1000 pounds to almost exactly 90,000 pounds. In 100 years the original capital had multiplied about 90 times instead of the 131 times Franklin had imagined.

What rate of interest, compounded continuously for 100 years, would have multiplied Benjamin Franklin's original capital by 90?

8. In Benjamin Franklin's estimate that the original 1000 pounds would grow to 131,000 in 100 years, he was using an annual rate of 5% and compounding once each year. What rate of interest per year when compounded continuously for 100 years would multiply the original amount by 131?

9. *The Rule of 70.* If you use the approximation ln 2 ≈ 0.70 (in place of 0.69315 . . .), you can derive a rule of thumb that says, "To find how many years it will take an amount of money to double when invested at r percent compounded continuously, divide r into 70." For instance, an amount of money invested at 5% will double in about $70/5 = 14$ years. If you want it to double in 10 years instead, you have to invest it at $70/10 = 7\%$. Show how the Rule of 70 is derived. (A similar "Rule of 72" uses 72 instead of 70, probably because it has more integer factors.)

10. *John Napier's question.* John Napier, who invented natural logarithms, was the first person to answer the question "What happens if you invest an amount of money at 100% interest, compounded continuously?"

a) What does happen?

b) How long does it take to triple your money?

c) How much can you earn in a year?

11. *Carbon-14 dating.* The half-lives of radioactive elements can sometimes be used to date events from the Earth's past. The ages of rocks more than 2 billion years old have been measured by the extent of the radioactive decay of uranium (half-life 4.5 billion years!). In a living organism, the ratio of radioactive carbon, carbon-14, to ordinary carbon stays fairly constant during the lifetime of the organism, being approximately equal to the ratio in the organism's surroundings at the time. After the organism's death, however, no new carbon is ingested, and the proportion of carbon-14 in the organism's remains decreases as the carbon-14 decays. Since the half-life of carbon-14 is known to be about 5700 years, it is possible to estimate the age of organic remains by comparing the proportion of carbon-14 they contain with

the proportion assumed to have been in the organism's environment at the time it lived. Archeologists have dated shells (which contain $CaCO_3$), seeds, and wooden artifacts this way. The estimate of 15,500 years for the age of the cave paintings at Lascaux, France, is based on carbon-14 dating.
a) Find k in Eq. (3) for carbon-14.
b) What is the age of a sample of charcoal in which 90% of the carbon-14 has decayed?
c) The charcoal from a tree killed in the volcanic eruption that formed Crater Lake in Oregon contained $44\frac{1}{2}\%$ of the carbon-14 found in living matter. About how old is Crater Lake?

12. To see the effect of a relatively small error in the estimate of the amount of carbon-14 in a sample being dated, consider this hypothetical situation:
a) A fossilized bone found in central Illinois in the year 2000 AD is found to contain 17% of its original carbon-14 content. Estimate the year the animal died.
b) Repeat (a) assuming 18% instead of 17%.
c) Repeat (a) assuming 16% instead of 17%.

13. *Polonium.* The half-life of polonium is 140 days, but your sample will not be useful to you after 90% of the radioactive nuclei originally present have disintegrated. About how many days can you use the polonium?

14. *Cooling cocoa.* Suppose that a cup of cocoa cooled from 90°C to 60°C after 10 minutes in a room whose temperature was 20°C. Use Newton's law of cooling to answer the following questions.
a) How much longer would it take the cocoa to cool to 35°C?
b) Instead of being left to stand in the room, the cup of 90°C cocoa is put in a freezer whose temperature is −15°C. How long will it take the cocoa to cool from 90°C to 35°C?

15. *Body of unknown temperature.* A body of unknown temperature was placed in a room that was held at 30°F. After 10 minutes, the body's temperature was 0°F, and 20 minutes after the body was placed in the room the body's temperature was 15°F. Use Newton's law of cooling to estimate the body's initial temperature.

16. *Surrounding medium of unknown temperature.* A pan of warm water (46°C) was put in a refrigerator. Ten minutes later, the water's temperature was 39°C; 10 minutes after that, it was 33°C. Use Newton's law of cooling to estimate how cold the refrigerator was.

17. *Silver cooling in air.* The temperature of an ingot of silver is 60°C above room temperature right now. Twenty minutes ago, it was 70°C above room temperature. How far above room temperature will the silver be 15 minutes from now? Two hours from now? When will the silver be 10°C above room temperature?

18. *The R–L circuit.* How many seconds after the switch in the circuit in Example 6 is closed will it take the current to reach half of its steady-state value? Notice that this time does not depend on V. Compare your answer with Eq. (8).

19. What will be the current in the R–L circuit when $t = L/R$ seconds? (The number L/R is called the *time constant* of the circuit.)

20. If there is a steady current in an R–L circuit and the switch in Fig. 6.37 is thrown open, the decaying current obeys the equation

$$L\frac{di}{dt} + Ri = 0$$

(Eq. 11 with $V = 0$).
a) Solve this equation for i.
b) How long after the switch is thrown will it take the current to decay to half its original value?
c) What is the value of the current when $t = L/R$? (This value of t is called the *time constant* for the open circuit.) See Fig. 6.38(b).

21. *Resistance proportional to velocity.* Suppose that a body of mass m moving in a straight line with velocity v encounters a resistance proportional to the velocity and that this is the only force acting on the body. If the body starts with velocity v_0, how far does it travel in time t? Assume $F = d(mv)/dt$.

22. *Blood sugar.* If glucose is fed intravenously at a constant rate, the change in the overall concentration $c(t)$ of glucose in the blood with respect to time may be described by the differential equation

$$\frac{dc}{dt} = \frac{G}{100V} - kc.$$

In this equation, G, V, and k are positive constants, G being the rate at which glucose is admitted, in milligrams per minute, and V the volume of blood in the body, in liters (around 5 liters for an adult). The concentration $c(t)$ is measured in milligrams per centiliter. The term $-kc$ is included because the glucose is assumed to be changing continually into other molecules at a rate proportional to its concentration.
a) Solve the equation for $c(t)$, using c_0 to denote $c(0)$.
b) Find the steady-state concentration, $\lim_{t \to \infty} c(t)$.

23. *The inversion of sugar.* The processing of raw sugar has a step called "inversion" that changes the sugar's molecular structure. Once the process has begun, the rate of change of the amount of raw sugar is proportional to the amount of raw sugar remaining. If 1000 kg of raw sugar reduces to 800 kg of raw sugar during the first 10 hours, how much raw sugar will remain after another 14 hours?

TOOLKIT PROGRAMS

Antiderivatives and Direction Fields
Sequences and Series
Super * Grapher

REVIEW QUESTIONS AND EXERCISES

1. Define the terms "algebraic function" and "transcendental function." To which class (algebraic or transcendental) do you think the greatest integer function (Fig. 6.39) belongs?

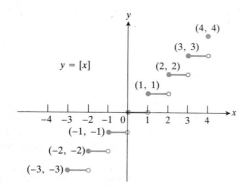

6.39 The greatest integer function, $y = [x]$. See Exercise 1.

[It is often difficult to prove that a particular function belongs to one class or the other. Numbers are also classified as algebraic and transcendental. *References:* W. J. Le Veque, *Topics in Number Theory* (Reading, Mass.: Addison-Wesley, 1956), Vol. II, Chap. 5, pp. 161–200; I. M. Niven, *Irrational Numbers*, Carus monograph number 11 of the Mathematical Association of America (1956).]

2. Give an example of a function that is (a) one-to-one; (b) not one-to-one.

3. Suppose that the domain of a one-to-one function f is $[a, b]$ and that the range is $[f(a), f(b)]$. Describe the inverse of f. What are its domain and range?

4. *Continuation of Exercise 3*. If $g(y) = x$ whenever $f(x) = y$, then

$$g(f(x)) = x$$

for all $a \le x \le b$. Suppose that f is differentiable. Apply the Chain Rule to the composite $h = g \circ f$ to deduce a general formula for the derivative of an inverse function. What restriction must be placed on the derivative of f?

5. Define the arc sine, arc cosine, arc tangent, and arc secant functions. What are their domains? Ranges? Derivatives? Sketch their graphs.

6. Define the natural logarithm function. What is its domain? Range? Derivative? Sketch its graph.

7. Prove that $\lim_{x \to \infty} (\ln x)/x = 0$, starting with the definition of $\ln x$ as an integral.

8. What is the inverse of the natural logarithm function? Its domain and range? Its derivative? Sketch its graph.

9. How would you calculate $(8.73)^{2.75}$? (If you have a calculator with a y^x key, of course, you just put $y = 8.73$ and $x = 2.75$.) How would you calculate $(-8)^{1/3}$? (When this was tried on an old calculator, the "error" symbol came on. Why?)

10. What logarithmic functions are there besides $\ln x$? How are they defined and evaluated? What are their derivatives?

11. What does it mean for a function $f(x)$ to grow faster than a function $g(x)$ as x approaches infinity? What does it mean for f to grow at the same rate as g? Give examples.

12. What is the physical meaning of the differential equation $dy/dt = ky$? What solution of this equation satisfies the initial condition $y = y_0$ when $t = 0$? Give examples of applications of this equation.

MISCELLANEOUS PROBLEMS

1. Solve for x: $\tan^{-1}x - \cot^{-1}x = \pi/4$.

2. A particle starts at the origin and moves along the x-axis in such a way that its velocity at x is $dx/dt = \cos^2 \pi x$. How long will it take the particle to reach $x = 1/4$? Will it ever reach $x = 1/2$? Why?

Find the limits in Problems 3 and 4.

3. $\lim\limits_{b \to 1^-} \int_0^b \dfrac{dx}{\sqrt{1 - x^2}}$

4. $\lim\limits_{x \to \infty} \dfrac{\displaystyle\int_0^x \tan^{-1} t \, dt}{x}$

5. Find the volume generated by revolving the region bounded by $x = \sec y$, $x = 0$, $y = 0$, and $y = \pi/3$ about the y-axis.

6. You are under contract to build a solar station at ground level on the east-west line between two tall buildings, one of height h, the other of height $2h$. The buildings are a distance d apart. How far from the taller building should you place the station to maximize the number of hours it will be in the sun on a day when the sun passes directly overhead?

7. *Optimal branching angle for pipes.* When a smaller pipe branches off from a larger one in a flow system, we may want it to run off at an angle that is "best" from some energy-saving point of view. We might require, for instance, that energy loss due to friction be minimized along the section AOB shown in Fig. 6.40. In this diagram, B is a given point to be reached by the smaller pipe, A is a point in the larger pipe upstream from B, and O is the point where the branching occurs. A law due to Poiseuille states that the loss of energy due to friction in nonturbulent flow is proportional to the length of the path and inversely proportional to the fourth power of the radius. Thus, the loss along AO is

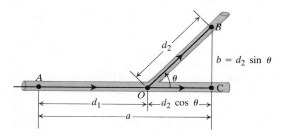

6.40 The smaller pipe *OB* branches away from the larger *AOC* at an angle θ that minimizes the friction loss along *AO* and *OB*. The optimum angle is found to be $\theta_c = \cos^{-1}(r^4/R^4)$, where r is the radius of the smaller pipe and R is the radius of the larger (Problem 7).

$(kd_1)/R^4$ and along *OB* is $(kd_2)/r^4$, where k is a constant, d_1 is the length of *AO*, d_2 is the length of *OB*, R is the radius of the larger pipe, and r is the radius of the smaller pipe. The angle θ is to be such as to minimize the sum of these two losses:

$$L = k\frac{d_1}{R^4} + k\frac{d_2}{r^4}.$$

In our model, we assume that $AC = a$ and $BC = b$ are fixed. Thus we have the relations

$$d_1 + d_2 \cos\theta = a, \quad d_2 \sin\theta = b,$$

so that

$$d_2 = b\csc\theta$$

and

$$d_1 = a - d_2\cos\theta = a - b\cot\theta.$$

We can express the total loss L as a function of θ:

$$L = k\left(\frac{a - b\cot\theta}{R^4} + \frac{b\csc\theta}{r^4}\right). \tag{11}$$

a) Show that the critical value of θ for which $dL/d\theta$ equals zero is

$$\theta_c = \cos^{-1}\frac{r^4}{R^4}.$$

b) **CALCULATOR OR TABLES** If the ratio of the pipe radii is $r/R = 5/6$, estimate to the nearest degree the optimal branching angle given in part (a).

The mathematical analysis described here is also used to explain the angles at which arteries branch in an animal's body. See Edward Batschelet's *Introduction to Mathematics for Life Scientists*, 2nd ed. (New York: Springer-Verlag, 1976).

8. When a ray of light passes from one medium (say, air) where it travels with a velocity c_1 into a second medium (for example, water) where it travels with velocity c_2, the angle of incidence θ_1 and angle of refraction θ_2 are related by Snell's Law:

$$\frac{\sin\theta_1}{c_1} = \frac{\sin\theta_2}{c_2}.$$

(Fig. 6.41). The quotient $c_1/c_2 = n_{12}$ is called the *index of refraction* of medium 2 with respect to medium 1.
a) Express θ_2 as a function of θ_1.
b) Find the largest value of θ_1 for which the expression for θ_2 in part (a) is defined. (For values of θ_1 larger than this, the incoming light will be reflected.)

Angle of incidence
Angle of refraction

6.41 A ray refracted as it passes from one medium to another (Problem 8).

9. Show that

$$\int_0^1 \frac{x^4(1-x)^4}{1+x^2}dx = \frac{22}{7} - \pi$$

(*Hint:* Expand the numerator and divide by the denominator before attempting to integrate.)

10. Show that the functions $\ln 5x$ and $\ln 3x$ differ by a constant. What is the constant?

In Problems 11–28, find dy/dx.

11. $y = x\ln x$ **12.** $y = \sqrt{x}\ln x$

13. $y = (\ln x)/x$ **14.** $y = (\ln x)/\sqrt{x}$

15. $y = x/(\ln x)$ **16.** $y = x(\ln x)^3$

17. $y = x^3\ln x$ **18.** $y = \ln(\ln x)$

19. $y = \ln(3x^2)$

20. $y = \ln(ax^b), \quad a > 0, b > 0$

21. $y = (1/2)\ln((1+x)/(1-x))$

22. $y = \ln\sqrt{1+x^2} - \tan^{-1}x$

23. $y = \ln(x/\sqrt{x^2+1})$

24. $y = \ln(x - \sqrt{x^2+1})$

25. $y = x\sec^{-1}x - \ln(x + \sqrt{x^2-1}), \quad x > 1$

26. $y = (2x/\sqrt{x^2-1})^{-2}$

27. $y = (x(x-2)/(x^2+1))^{1/3}$

28. $y = (2x-5)(8x^2+1)^{1/2}/(x^3+2)^2$

Evaluate the integrals in Problems 29–40.

29. $\displaystyle\int_0^{-8/3}\frac{dx}{4-3x}$ **30.** $\displaystyle\int_0^5\frac{x\,dx}{x^2+1}$

31. $\displaystyle\int_0^2\frac{x\,dx}{x^2+2}$ **32.** $\displaystyle\int_0^2\frac{x\,dx}{(x^2+2)^2}$

33. $\displaystyle\int_1^2\frac{5\,dx}{x-3}$

34. $\displaystyle\int_0^7 \frac{x}{2x+2}\,dx$ (*Hint*: First divide x by $2x+2$.)

35. $\displaystyle\int \frac{x^3}{x^4+1}\,dx$

36. $\displaystyle\int \frac{\tan\sqrt{x}}{\sqrt{x}}\,dx$

37. $\displaystyle\int x^2 \cot(2+x^3)\,dx$

38. $\displaystyle\int \frac{\sin 3x}{5-2\cos 3x}\,dx$

39. $\displaystyle\int \frac{dx}{\sqrt{1-x^2}(3+\sin^{-1}x)}$

40. $\displaystyle\int \frac{dx}{x\sec^{-1}x\sqrt{x^2-1}}$

41. Evaluate the following limits.
  a) $\displaystyle\lim_{h\to0^+} h\ln h$
  b) $\displaystyle\lim_{x\to0^+} x^p\ln x,\quad p>0$

42. Prove by l'Hôpital's rule and mathematical induction that
$$\lim_{x\to\infty}\frac{(\ln x)^n}{x}=0$$
for any positive integer n.

43. Let p be a positive integer greater than or equal to 2. Show that
$$\lim_{n\to\infty}\left(\frac{1}{n+1}+\frac{1}{n+2}+\cdots+\frac{1}{p\cdot n}\right)=\ln p.$$

44. Show that $(x^2/4)<x-\ln(1+x)<(x^2/2)$ if
$$0<x<1.$$
(*Hint:* Let $f(x)=x-\ln(1+x)$ and show that $(x/2)<f'(x)<x$.)

45. Find the quadratic approximation of
  a) $f(x)=\ln(\sec x+\tan x)$ near $x=0$;
  b) $f(x)=\ln x$ near $x=1$.

46. Find the length of the curve
$$\frac{x}{a}=\left(\frac{y}{b}\right)^2-\frac{1}{8}\left(\frac{b^2}{a^2}\right)\ln\left(\frac{y}{b}\right)$$
from $y=b$ to $y=3b$, assuming a and b to be positive constants.

47. Find the length of the arc of the curve
$$x=a\left(\cos t+\ln\tan\frac{t}{2}\right),\quad y=a\sin t,\quad \frac{\pi}{2}\le t<\pi$$
that extends from the point $(0,a)$ to the point (x_1,y_1).

48. A particle moves in a straight line with acceleration $a=4/(4-t)^2$. If when $t=0$ the velocity is equal to 2, find how far the particle moves between $t=1$ and $t=2$.

49. Solve the differential equation
$$\frac{dy}{dx}=\frac{1+(1/x)}{1+(1/y)},\quad x>0,\quad y>0,$$
subject to the condition that $x=1$ when $y=1$.

50. Prove that the area under the graph of $y=1/x$ over the interval $a\le x\le b$ $(a>0)$ is the same as the area over the interval $ka\le x\le kb$ for any $k>0$.

51. A curve $y=f(x)$ goes through the points $(0,0)$ and (a,b). It divides the rectangle $0\le x\le a,\ 0\le y\le b$ into two regions: A above the curve and B below. Find the curve if the area of A is twice the area of B for all choices of $a>0$ and $b>0$.

52. $P(x_1,y_1)$ and $Q(x_2,y_2)$ are any two points in the first quadrant lying on the hyperbola $xy=R$ (R positive). Show that the area bounded by the arc PQ, the lines $x=x_1,\ x=x_2$, and the x-axis is equal to the area bounded by the arc PQ, the lines $y=y_1,\ y=y_2$, and the y-axis.

53. The portion of a tangent to a curve included between the x-axis and the point of tangency is bisected by the y-axis. If the curve passes through $(1,2)$, find its equation.

54. Use mathematical induction (Appendix 2) and the rule $\ln ax=\ln a+\ln x$ to show that at any value of x where the functions $u_1(x),\ \ldots,\ u_n(x)$ are all positive,
$$\ln(u_1u_2\cdots u_n)=\ln u_1+\ln u_2+\cdots+\ln u_n.$$

In Problems 55–74, find dy/dx.

55. $y=e^{1/x}$

56. $y=\ln e^x$

57. $y=xe^{-x}$

58. $y=x^2e^x$

59. $y=(\ln x)/e^x$

60. $x=\ln y$

61. $y=\ln(2xe^{2x})$

62. $y=e^{\sec x}$

63. $y=e^{\sin^2x}$

64. $y=e^{\ln(\sin e^x)}$

65. $y=\ln(e^x/(1+e^x))$

66. $y=e^{-x}\sin 2x$

67. a) $y=2x-\dfrac{1}{2}e^{2x}$

  b) $y=e^{(2x-(1/2)e^{2x})}$

68. a) $y=x-e^x$

  b) $y=e^{(x-e^x)}$

69. $y=\dfrac{e^{2x}-e^{-2x}}{e^{2x}+e^{-2x}}$

70. $y=x^2e^{2x}\sin 3x$

71. $y=\sin^{-1}(x^2)-xe^{(x^2)}$

72. $y=e^x(\sin 2x-2\cos 2x)$

73. $y=e^{2x}(\sin 3x+\cos 3x)$

74. $\ln(x-y)=e^{xy},\quad x>y$

Evaluate the integrals in Problems 75–80.

75. $\displaystyle\int_1^e \frac{x+1}{x}\,dx$

76. $\displaystyle\int_0^1 (e^x+1)\,dx$

77. $\displaystyle\int_0^{\ln 2} e^{-2x}\,dx$

78. $\displaystyle\int_1^{e^2} \frac{2\ln x}{x}\,dx$

79. $\displaystyle\int_{-1}^1 \frac{e^x-e^{-x}}{e^x+e^{-x}}\,dx$

80. $\displaystyle\int_0^{\pi/6} \frac{\sec^3x+e^{\sin x}}{\sec x}\,dx$

Find the limits in Problems 81–84.

81. $\displaystyle\lim_{x\to\infty}\frac{x^5}{e^x}$

82. $\displaystyle\lim_{x\to0}\frac{xe^x}{4-4e^x}$

83. $\displaystyle\lim_{x\to4}\frac{e^{x-4}+4-x}{\cos^2(\pi x)}$

84. $\displaystyle\lim_{x\to0^+}(e^x+x)^{1/x}$

85. Find the linearization of $f(x)=x+e^{4x}$ at $x=0$.

86. Find the quadratic approximation of
$$\text{erf}(x)=\frac{2}{\sqrt{\pi}}\int_0^x e^{-t^2}\,dt$$
near $x=0$.

87. Graph each of the following functions, taking into account extreme values, inflection points, and concavity.
   a) $y = (\ln x)/\sqrt{x}, \quad x > 0$
   b) $y = e^{-x^2}$
   c) $y = (1 + x)e^{-x}$

88. Calculate $f'(2)$ if $f(x) = e^{g(x)}$ and

$$g(x) = \int_2^x \frac{t}{1 + t^4}\, dt.$$

89. Find a curve passing through the origin and such that the length s of the curve between the origin and any point (x, y) of the curve is given by

$$s = e^x + y - 1.$$

90. If $y = (e^{2x} - 1)/(e^{2x} + 1)$, show that $dy/dx = 1 - y^2$.

91. Find the area of the surface generated by revolving the curve

$$y = \frac{1}{2}(e^x + e^{-x}), \quad 0 \le x \le \ln\sqrt{2},$$

about the x-axis.

92. Find the volume generated by revolving about the x-axis the region bounded by $y = e^x$, $y = 0$, $x = 0$, $x = 2$.

93. Find the area of the region bounded by the curve

$$y = \left(\frac{a}{2}\right)(e^{x/a} + e^{-x/a}),$$

the x-axis, and the lines $x = -a$ and $x = +a$.

94. Find the area under the curve $y = t \sin(t - 1)$, $x = \ln t$ from $x = 0$ to $x = \ln(\pi + 1)$.

95. A triangle is bounded by the y-axis, the horizontal line through the highest point on the curve $y = e^{-2x^2}$, and the line that passes through the origin and the point of inflection of the curve $y = e^{-2x^2}$ that lies in the first quadrant. Find the area of the triangle.

96. a) Show that

$$\int_1^a \ln x\, dx + \int_0^{\ln a} e^y dy = a \ln a$$

for any number greater than 1. (*Hint:* Interpret the integrals as areas.)

   b) Solve the equation in (a) to evaluate $\int_1^a \ln x\, dx$.

97. Show that the tangent to the curve $y = e^x$ at the point $P(x, y)$ and the perpendicular from P to the x-axis always intersect the axis at points that are 1 unit apart (Fig. 6.42). This is one way to show how rapidly the graph of $y = e^x$ climbs.

98. If $dy/dx = 2/e^y$ and $y = 0$ when $x = 5$, find y as a function of x.

99. Solve the equation $dy/dx = y^2 e^{-x}$ subject to the condition that $y = 2$ when $x = 0$.

100. A point $P(x, y)$ moves in the plane in such a way that

$$\frac{dx}{dt} = \frac{1}{t + 2} \quad \text{and} \quad \frac{dy}{dt} = 2t$$

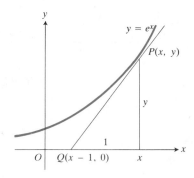

6.42 Line QP is tangent to the curve at P (Problem 97). Figure not to scale.

for $t \ge 0$.
   a) Express x and y as functions of t if $x = \ln 2$ and $y = 1$ when $t = 0$.
   b) Express y in terms of x.
   c) Express x in terms of y.
   d) Find the average rate of change in y with respect to x as t varies from 0 to 2.
   e) Find dy/dx when $t = 1$.

101. If a particle moves along the x-axis so that its position at time t is given by $x = ae^{\omega t} + be^{-\omega t}$, where a, b, and ω are constants, show that it is repelled from the origin with a force proportional to the displacement. (Assume that force = mass × acceleration.)

102. Prove that
   a) $\lim_{x \to \infty} (\ln x)/x^h = 0$ if h is any positive constant,
   b) $\lim_{x \to +\infty} x^n/e^x = 0$ if n is any constant.

103. Prove by mathematical induction (Appendix 2) that

$$\frac{d^n}{dx^n}(xe^x) = (x + n)e^x$$

for any positive integer n.

104. a) If $(\ln x)/x = (\ln 2)/2$, does it necessarily follow that $x = 2$?
   b) If $(\ln x)/x = -2 \ln 2$, does it necessarily follow that $x = 1/2$?

105. **CALCULATOR** *Evaluating ln x with square roots.*
   a) Show that

$$\lim_{h \to 0} \frac{e^h - 1}{h} = 1.$$

   b) Show that if x is any positive number, then

$$\lim_{n \to \infty} n(\sqrt[n]{x} - 1) = \ln x.$$

(*Hint:* Take $x = e^{nh}$ and apply the result in (a).)

   The result in (b) provides a method for finding $\ln x$ to any desired finite number of decimal places using nothing fancier than repeated use of the operation of extracting square roots. For we may take $n = 2^k$, and then $\sqrt[n]{x}$ is obtained from x by taking k successive square roots.

106. Find the limit, as $n \to \infty$, of

$$\frac{e^{1/n} + e^{2/n} + \cdots + e^{(n-1)/n} + e^{n/n}}{n}.$$

107. Find dy/dx.

a) $y = x^{\tan 3x}$ b) $x^{\ln y} = 2$

c) $y = (x^2 + 2)^{2-x}$ d) $x^{(1/x)}$

108. Solve for x.

a) $4^x = 2^x$ b) $x^x = 2^x, \quad x > 0$

c) $3^x = 2^{x+1}$ d) $4^{-x} = 3^{x+2}$

109. Evaluate the following integrals.

a) $\displaystyle\int_0^{\pi/6} (\cos x)4^{-\sin x}\, dx$ b) $\displaystyle\int_{\ln \log_4 \ln 4}^{\ln \log_4 \ln 16} e^x 4^{(e^x)}\, dx$

110. Let a be a number greater than 1. Prove that the graph of $y = a^x$ has the following characteristics.

a) If $x_1 > x_2$, then $a^{x_1} > a^{x_2}$.

b) The graph is concave up.

c) The graph lies entirely above the x-axis.

d) The slope at any point is proportional to the y-coordinate there, and the proportionality factor is the slope at the y-intercept of the graph.

e) The curve approaches the x-axis as $x \to -\infty$.

111. Suppose that a and b are positive numbers such that $a^b = b^a$. (For example, $2^4 = 4^2$.)

a) Show that $(\ln a)/a = (\ln b)/b$.

b) Show that the graph of the function $f(x) = (\ln x)/x$ has a maximum at $x = e$, a point of inflection at $x = e^{3/2}$, approaches minus infinity as x approaches zero through positive values, and approaches zero as x approaches plus infinity. Sketch its graph.

c) Use the results of parts (a) and (b) to show that
  i) if $0 < a \le 1$ and $a^b = b^a$, then $b = a$, and
  ii) if $1 < a < e$, or if $a > e$, then there is exactly one number $b \ne a$ such that $a^b = b^a$.

112. *Blood tests.* During World War II it was necessary to administer blood tests to large numbers of recruits. There are two standard ways to administer a blood test to N people: In method 1, each person is tested separately. In method 2, the blood samples of x people are pooled and tested as one large sample. If the test is negative, this one test is enough for all x people. If the test is positive, then each of the x people is tested separately, requiring a total of $x + 1$ tests. Using the second method and some probability theory it can be shown that, on the average, the total number of tests y will be

$$y = N\left(1 - q^x + \frac{1}{x}\right).$$

With $q = 0.99$ and $N = 1000$, show that the integer value of x that minimizes y is $x = 11$ and that the integer value of x that maximizes y is $x = 895$. (The second result is not important to the real-life situation.) The group testing method was used in World War II with a saving of 80% over the individual testing method, but not with the given q.

113. *Population growth.* It is estimated that the population of a certain country is now increasing at a rate of 2% per year. Assuming that this (instantaneous) rate will continue indefinitely, estimate what the population N will be t years from now, if the population now is N_0. How many years will it take for the population to double?

114. *Voltage in a discharging capacitor.* In a capacitor discharging electricity, the rate of change of the voltage in volts per second is proportional to the voltage, being numerically equal to minus one-fortieth of the voltage. Express the voltage as a function of the time. In how many seconds will the voltage decrease to 10% of its original value?

115. *Velocity proportional to displacement.* The velocity of a particle moving along the x-axis is proportional to x. At time $t = 0$ the particle is located at $x = 2$ and at time $t = 10$ it is at $x = 4$. Find its position at $t = 5$.

116. *Transport through a cell membrane.** Under some conditions the result of the movement of a dissolved substance across a cell's membrane is described by the equation

$$\frac{dy}{dt} = k\frac{A}{V}(c - y).$$

In this equation, y is the concentration of the substance inside the cell and dy/dt is the rate with which y changes over time. The letters k, A, V, and c stand for constants, k being the *permeability coefficient* (a property of the membrane), A the surface area of the membrane, V the cell's volume, and c the concentration of the substance outside the cell. The equation says that the rate at which the concentration changes within the cell is proportional to the difference between it and the outside concentration.

a) Solve the equation for $y(t)$, using y_0 to denote $y(0)$.

b) Find the steady-state concentration, $\lim_{t \to \infty} y(t)$.

---

*Based on *Some Mathematical Models in Biology,* Revised Edition, R. M. Thrall, J. A. Mortimer, K. R. Rebman, R. F. Baum, eds., December 1967, PB–202 364, pp. 101–103; distributed by N.T.I.S., U.S. Department of Commerce.

CHAPTER 7

Methods of Integration

OVERVIEW

The goal of every method of integration is to change unfamiliar integrals into integrals we recognize or can find in a table. The chief method is the method of substitution, the one we have been using all along. There are three other methods, however, somewhat more specialized but just right when they do apply. They are integration by parts, the use of trigonometric substitutions and identities to combine squares and eliminate square roots, and integration by partial fractions.

Integration by partial fractions is the basic technique for preparing rational functions for integration. It enables us to write an arbitrary quotient of polynomials as a sum of simpler quotients that can then be integrated by substitution.

As we go along, we shall look at the integrals of a number of products and powers of trigonometric functions (important because of the roles they play in both pure and applied mathematics) and practice with the integration tables on the endpapers of the book. We also describe a technique for calculating integrals over open intervals and infinite intervals. We close the chapter with a brief discussion of reduction formulas, formulas that enable us to express integrals of high powers of functions in terms of integrals of lower powers that can be evaluated directly.

We begin with a review of progress so far and a few words of advice.

7.1

Basic Integration Formulas

Since indefinite integration is the inverse of differentiation, evaluating an integral

$$\int f(x) \, dx$$

is equivalent to finding a function $F(x)$ whose derivative is $f(x)$ or whose differential is $f(x) \, dx$. To reduce the amount of effort involved in finding F, we build up a table of standard integral forms by reversing formulas for differentials, as we have done in previous chapters. We then try to match any integral we are confronted with against one of the standard forms. Table 7.1 displays the standard forms we have developed so far. A more extensive list can be found inside the covers of this book.

When the integral we want to evaluate does not match any of the forms in the tables available to us, we try to change it into one of these forms with substitutions and algebra. The examples and problems in this chapter are designed to develop your skill with the substitutions and algebraic transformations that have been found to be most helpful. To concentrate on the techniques involved without becoming entangled in the algebra, these problems and examples have been kept fairly simple. In fact, *these particular problems* can frequently be solved immediately by consulting an integral table. But solving them this way right now would defeat the purpose of developing the skill likely to be needed for later work. And it is this skill that is important, rather than the specific answer to any given problem.

Since integration is based on differentiation, we have included in Table 7.1 the differential formulas from earlier chapters along with their integral counterparts. At present, just consider this list a handy reference and *not* a challenge to your memory! A bit of examination will probably convince you that you are already familiar with the first fifteen formulas for differentials. And, so far as integration is concerned, your training in technique will enable you to get the integrals in the last three cases without memorizing the formulas for the differentials of the inverse trigonometric functions.

Just what types of functions can you integrate directly with Table 7.1?

Powers

$$\int u^n \, du, \qquad \int \frac{du}{u}$$

Exponentials

$$\int e^u \, du, \qquad \int a^u \, du$$

Trigonometric Functions

$$\int \sin u \, du, \qquad \int \cos u \, du, \qquad \int \tan u \, du, \qquad \int \cot u \, du$$

Algebraic Functions

$$\int \frac{du}{\sqrt{1 - u^2}}, \qquad \int \frac{du}{1 + u^2}, \qquad \int \frac{du}{u\sqrt{u^2 - 1}}$$

Additional combinations of trigonometric functions, such as $\int \sec^2 u \, du$, etc., are integrable accidentally, so to speak.

TABLE 7.1
The integral formulas developed so far (standard forms)

Differentials	Integrals
1. $du = \dfrac{du}{dx} \, dx$	1. $\displaystyle\int du = u + C$
2. $d(au) = a \, du$	2. $\displaystyle\int a \, du = a \int du$
3. $d(u + v) = du + dv$	3. $\displaystyle\int (du + dv) = \int du + \int dv$
4. $d(u^n) = nu^{n-1} \, du$	4. $\displaystyle\int u^n \, du = \dfrac{u^{n+1}}{n + 1} + C, \quad n \neq -1$
5. $d(\ln u) = \dfrac{du}{u}$	5. $\displaystyle\int \dfrac{du}{u} = \ln|u| + C$
6. $d(e^u) = e^u \, du$	6. $\displaystyle\int e^u \, du = e^u + C$
7. $d(a^u) = a^u \ln a \, du$	7. $\displaystyle\int a^u \, du = \dfrac{a^u}{\ln a} + C$
8. $d(\sin u) = \cos u \, du$	8. $\displaystyle\int \cos u \, du = \sin u + C$
9. $d(\cos u) = -\sin u \, du$	9. $\displaystyle\int \sin u \, du = -\cos u + C$
10. $d(\ln|\sec u|) = \tan u$	10. $\displaystyle\int \tan u \, du = \ln|\sec u| + C$
11. $d(\ln|\sin u|) = \cot u$	11. $\displaystyle\int \cot u \, du = \ln|\sin u| + C$
12. $d(\tan u) = \sec^2 u \, du$	12. $\displaystyle\int \sec^2 u \, du = \tan u + C$
13. $d(\cot u) = -\csc^2 u \, du$	13. $\displaystyle\int \csc^2 u \, du = -\cot u + C$
14. $d(\sec u) = \sec u \tan u \, du$	14. $\displaystyle\int \sec u \tan u \, du = \sec u + C$
15. $d(\csc u) = -\csc u \cot u \, du$	15. $\displaystyle\int \csc u \cot u \, du = -\csc u + C$
16. $d(\sin^{-1} u) = \dfrac{du}{\sqrt{1 - u^2}}$	16. $\displaystyle\int \dfrac{du}{\sqrt{1 - u^2}} = \sin^{-1} u + C$
17. $d(\tan^{-1} u) = \dfrac{du}{1 + u^2}$	17. $\displaystyle\int \dfrac{du}{1 + u^2} = \tan^{-1} u + C$
18. $d(\sec^{-1} u) = \dfrac{du}{|u|\sqrt{u^2 - 1}}$	18. $\displaystyle\int \dfrac{du}{u\sqrt{u^2 - 1}} = \sec^{-1}|u| + C$

What common types of functions are *not* included in the table?

Logarithms

$$\int \ln u \, du, \qquad \int \log_a u \, du$$

Trigonometric Functions

$$\int \sec u \, du, \qquad \int \csc u \, du$$

Algebraic Functions

$$\int \frac{du}{a^2 \pm u^2} \quad \text{with } a^2 \neq 1, \qquad \int \frac{du}{au^2 + bu + c},$$

$$\int \sqrt{a^2 \pm u^2} \, du, \qquad \int \sqrt{u^2 - a^2} \, du, \qquad \int \sqrt{au^2 + bu + c} \, du, \quad \text{etc.}$$

Inverse Functions

$$\int \sin^{-1} u \, du, \qquad \int \tan^{-1} u \, du, \quad \text{etc.}$$

We shall eventually see how to evaluate all of the integrals in the table and some others. However, no method will solve *all* integration problems in terms of elementary functions, and some problems lie beyond the reach of any method.

The examples that follow review the substitution method of integration. Our goal is to reduce integrals to ones we recognize or can find in a table, in this case Table 7.1.

EXAMPLE 1 Evaluate

$$\int \frac{dx}{1 + 4x^2}.$$

Solution The given integral resembles the standard form

$$\int \frac{du}{1 + u^2} = \tan^{-1} u + C$$

(Formula 17 in the list) with $2x$ in place of u. We substitute

$$u = 2x, \qquad du = 2 \, dx \qquad \text{and} \qquad dx = \frac{1}{2} \, du.$$

Then

$$\int \frac{dx}{1 + 4x^2} = \int \frac{(1/2) \, du}{1 + u^2} = \frac{1}{2} \int \frac{du}{1 + u^2}$$

$$= \frac{1}{2} \tan^{-1} u + C = \frac{1}{2} \tan^{-1} 2x + C. \qquad \blacksquare$$

EXAMPLE 2 Evaluate

$$\int \sqrt{1 - 5x}\ dx.$$

Solution We substitute

$$u = 1 - 5x, \qquad du = -5\ dx, \qquad dx = -\frac{1}{5}\ du,$$

and use Formula 4 with $n = 1/2$ to obtain

$$\int \sqrt{1 - 5x}\ dx = \int \sqrt{u} \cdot -\frac{1}{5}\ du = -\frac{1}{5} \int \sqrt{u}\ du$$

$$= -\frac{1}{5} \cdot \frac{2}{3} u^{3/2} + C = -\frac{2}{15}(1 - 5x)^{3/2} + C. \qquad \blacksquare$$

EXAMPLE 3 Evaluate

$$\int \frac{\sin(\ln x)}{x}\ dx.$$

Solution We substitute

$$u = \ln x, \qquad du = \frac{1}{x}\ dx.$$

Then

$$\int \frac{\sin(\ln x)}{x}\ dx = \int \sin u\ du = -\cos u + C = -\cos(\ln x) + C. \qquad \blacksquare$$

EXAMPLE 4 Evaluate

$$\int_0^{\pi/2} \sqrt{1 + \sin x}\ \cos x\ dx.$$

Solution We substitute

$$u = 1 + \sin x, \qquad du = \cos x\ dx.$$

Then

$$\int \sqrt{1 + \sin x}\ \cos x\ dx = \int \sqrt{u}\ du = \int u^{1/2}\ du$$

$$= \frac{2}{3} u^{3/2} + C = \frac{2}{3}(1 + \sin x)^{3/2} + C.$$

What about the limits of integration? Our options (Article 4.8) are (1) to evaluate the transformed integral with transformed limits or (2) to evaluate the original integral with the original limits.

METHOD 1: *With transformed limits.* When $x = 0$,

$$u = 1 + \sin(0) = 1,$$

and when $x = \pi/2$,

$$u = 1 + \sin(\pi/2) = 2.$$

Therefore,

$$\int_0^{\pi/2} \sqrt{1 + \sin x}\, \cos x\, dx = \int_1^2 \sqrt{u}\, du = \frac{2}{3}u^{3/2}\Big]_1^2 = \frac{2}{3}(2\sqrt{2} - 1).$$

METHOD 2: *With the original limits.*

$$\int_0^{\pi/2} \sqrt{1 + \sin x}\, \cos x\, dx = \frac{2}{3}(1 + \sin x)^{3/2}\Big]_0^{\pi/2}$$

$$= \frac{2}{3}(1 + 1)^{3/2} - \frac{2}{3}(1 + 0)^{3/2}$$

$$= \frac{2}{3}(2\sqrt{2} - 1).$$

EXAMPLE 5 Evaluate

$$\int_0^{\pi/2} \cos^4 x\, \sin^3 x\, dx.$$

Solution We use the trigonometric identity $\sin^2 x = 1 - \cos^2 x$ to change the integral into one we can integrate:

$$\int_0^{\pi/2} \cos^4 x\, \sin^3 x\, dx = \int_0^{\pi/2} \cos^4 x(1 - \cos^2 x)\sin x\, dx$$

$$= \int_0^{\pi/2} (\cos^4 x - \cos^6 x)\sin x\, dx = -\frac{\cos^5 x}{5} + \frac{\cos^7 x}{7}\Big]_0^{\pi/2}$$

$$= (-0 + 0) - \left(-\frac{1}{5} + \frac{1}{7}\right)$$

$$= \frac{1}{5} - \frac{1}{7} = \frac{2}{35}.$$

If you do not see right away how to integrate $(\cos^4 x - \cos^6 x)\sin x$, substitute

$$u = \cos x, \qquad du = -\sin x\, dx.$$

Then

$$\int_0^{\pi/2} (\cos^4 x - \cos^6 x)\sin x\, dx = \int_{u=1}^{u=0} (u^4 - u^6)(-du)$$

$$= \int_0^1 (u^4 - u^6)du = \frac{u^5}{5} - \frac{u^7}{7}\Big]_0^1$$

$$= \frac{1}{5} - \frac{1}{7} = \frac{2}{35}.$$

☐ Nonelementary Integrals

The work of Joel Moses and others at MIT in developing the computer program MACSYMA, which evaluates indefinite integrals by symbolic manipulation, has led to a renewed interest in determining which integrals can be expressed as finite algebraic combinations of elementary functions (the functions we have been studying) and which require infinite series (Chapter 12) or numerical methods for their evaluation. Examples of the latter include the error function

$$\text{erf}(x) = \frac{2}{\sqrt{\pi}} \int_0^x e^{-t^2} \, dt$$

and integrals like

$$\int \sin x^2 \, dx \qquad \text{and} \qquad \int \sqrt{1 + x^4} \, dx$$

that arise in engineering and physics. These, and a number of others, like

$$\int \frac{e^x}{x} \, dx, \qquad \int e^{(e^x)} \, dx, \qquad \int \frac{1}{\ln x} \, dx,$$

$$\int \ln(\ln x) \, dx, \qquad \int \frac{\sin x}{x} \, dx,$$

$$\int \sqrt{1 - k^2 \sin^2 x} \, dx, \qquad 0 < k < 1,$$

look so easy they tempt us to try them just to see how they turn out. It can be proved, however, that there are no simple expressions for evaluating these integrals. That is, there is no way to express these integrals as finite combinations of elementary functions. The same applies to integrals that can be changed into these by substitutions. None of the integrals you will be asked to evaluate in the present chapter falls into this category, but you may encounter nonelementary integrals from time to time in your other work.

If you would like to know more about this topic, you might start with D. G. Mead's article "Integration," *American Mathematical Monthly*, 152–156 (February 1961). An outline of the history of the search for antiderivatives, through 1979, may be found in Tom Kasper's article, "Integration in Finite Terms: The Liouville Theory," *Mathematics Magazine*, 53:4, 195–201 (September 1980).

PROBLEMS

Evaluate the integrals in Problems 1–16 by using the given substitution to reduce the integral to a standard form.

1. $\int 6x\sqrt{3x^2 + 5} \, dx, \quad u = 3x^2 + 5$

2. $\int \frac{16x \, dx}{8x^2 + 2}, \quad u = 8x^2 + 2$

3. $\int_0^{\sqrt{\ln 2}} xe^{x^2} \, dx, \quad u = x^2$

4. $\int \frac{\cos x \, dx}{\sqrt{1 + \sin x}}, \quad u = 1 + \sin x$

5. $\int_0^1 \frac{x \, dx}{\sqrt{8x^2 + 1}}, \quad u = 8x^2 + 1$

6. $\int \frac{\sin x}{3 + 4 \cos x} \, dx, \quad u = 3 + 4 \cos x$

7. $\int e^x \sec^2(e^x) \, dx, \quad u = e^x$

8. $\displaystyle\int_1^3 \frac{dy}{\sqrt{y}(1+y)}, \quad u = \sqrt{y}$

9. $\displaystyle\int e^x\sqrt{3+4e^x}\, dx, \quad u = 3+4e^x$

10. $\displaystyle\int_4^9 \frac{dx}{x-\sqrt{x}}, \quad u = \sqrt{x}$

11. $\displaystyle\int \frac{dx}{x\ln x}, \quad u = \ln x$

12. $\displaystyle\int_0^2 \frac{e^x\, dx}{1+e^x}, \quad u = 1+e^x$

13. $\displaystyle\int_0^1 e^{\sqrt{x}}\, dx, \quad u = \sqrt{x}$

14. $\displaystyle\int \frac{\tan\sqrt{x}}{\sqrt{x}}\, dx, \quad u = \sqrt{x}$

15. $\displaystyle\int \tan x\, \sec^2 x\, dx, \quad u = \tan x$

16. $\displaystyle\int \tan x\, \sec^2 x\, dx, \quad u = \sec x$

In Problems 17–20, use the given identities and substitutions to change the integral into a standard form. Then evaluate the integral.

17. $\displaystyle\int \cos^3 x\, dx, \quad \cos^2 x = 1 - \sin^2 x, \quad u = \sin x$

18. $\displaystyle\int \sin^4 x\, \cos^3 x\, dx, \quad \cos^2 x = 1 - \sin^2 x, \quad u = \sin x,$

19. $\displaystyle\int \tan^3 x\, \sec x\, dx, \quad \tan^2 x = \sec^2 x - 1, \quad u = \sec x,$

20. $\displaystyle\int \tan^3 x\, \sec^3 x\, dx, \quad \tan^2 x = \sec^2 x - 1, \quad u = \sec x$

Evaluate the integrals in Problems 21–80 by finding substitutions that reduce them to standard forms.

21. $\displaystyle\int_{1/2}^3 \sqrt{2x+3}\, dx$

22. $\displaystyle\int_1^{40} \frac{dx}{3x+5}$

23. $\displaystyle\int \frac{dx}{(2x-7)^2}$

24. $\displaystyle\int \frac{(x+1)\, dx}{x^2+2x+3}$

25. $\displaystyle\int_{-\pi}^0 \frac{\sin x\, dx}{2+\cos x}$

26. $\displaystyle\int \tan^3 2x\, \sec^2 2x\, dx$

27. $\displaystyle\int \sqrt{\sin x}\, \cos x\, dx$

28. $\displaystyle\int \frac{\sec^2 x}{2+\tan x}\, dx$

29. $\displaystyle\int \frac{\cos x}{(1+\sin x)^2}\, dx$

30. $\displaystyle\int_0^{\pi/4} 8\cos^3 2x\, \sin 2x\, dx$

31. $\displaystyle\int_{-\pi/2}^{\pi/2} \sin^2 2x\, \cos^3 2x\, dx$
(*Hint:* $\cos^3 2x = \cos 2x(1 - \sin^2 2x)$.)

32. $\displaystyle\int \csc^4 x\, dx$ (*Hint:* $\csc^4 x = \csc^2 x(1 + \cot^2 x)$.)

33. $\displaystyle\int \frac{x\, dx}{\sqrt{1-4x^2}}$

34. $\displaystyle\int x^{1/3}\sqrt{x^{4/3}-1}\, dx$

35. $\displaystyle\int_0^{\sqrt{2}/2} \frac{dx}{\sqrt{1-x^2}}$

36. $\displaystyle\int_0^{\pi/2} \sin 2x\, dx$

37. $\displaystyle\int_0^1 \frac{3x}{1+x^2}\, dx$

38. $\displaystyle\int_0^\pi x\sin(x^2)\, dx$

39. $\displaystyle\int_0^{\pi/4} \frac{\sin^2 2x}{1+\cos 2x}\, dx$ (*Hint:* $\sin^2 2x = 1 - \cos^2 2x$.)

40. $\displaystyle\int_{\sqrt{2}}^2 \frac{dy}{y\sqrt{y^2-1}}$

41. $\displaystyle\int_0^{1/2} \frac{2\, dx}{\sqrt{1-4x^2}}$

42. $\displaystyle\int_0^{1/\sqrt{2}} \frac{2v\, dv}{\sqrt{1-v^4}}$

43. $\displaystyle\int \frac{x\, dx}{(3x^2+4)^3}$

44. $\displaystyle\int x^2\sqrt{x^3+5}\, dx$

45. $\displaystyle\int \frac{x^2\, dx}{\sqrt{x^3+5}}$

46. $\displaystyle\int \frac{x\, dx}{4x^2+1}$

47. $\displaystyle\int_0^{\ln 2} e^{2x}\, dx$

48. $\displaystyle\int e^{\cos x}\sin x\, dx$

49. $\displaystyle\int \frac{dx}{e^{3x}}$

50. $\displaystyle\int \frac{e^{\sqrt{x+1}}}{\sqrt{x+1}}\, dx$

51. $\displaystyle\int \frac{e^x\, dx}{1+e^{2x}}$

52. $\displaystyle\int_0^{\sqrt{3}/3} \frac{dt}{1+9t^2}$

53. $\displaystyle\int \cos^2 x\, \sin x\, dx$

54. $\displaystyle\int \frac{\cos x\, dx}{\sin^3 x}$

55. $\displaystyle\int_{\pi/6}^{\pi/2} \cot^3 x\, \csc^2 x\, dx$

56. $\displaystyle\int_0^{\pi/3} \tan 3x\, \sec^2 3x\, dx$

57. $\displaystyle\int \frac{e^{2x}+e^{-2x}}{e^{2x}-e^{-2x}}\, dx$

58. $\displaystyle\int_0^\pi \sin 2x\, \cos^2 2x\, dx$

59. $\displaystyle\int_{-\pi/2}^{\pi/2} (1+\cos\theta)^3\sin\theta\, d\theta$

60. $\displaystyle\int te^{-t^2}\, dt$

61. $\displaystyle\int \frac{dt}{t\sqrt{4t^2-1}}$

62. $\displaystyle\int \frac{dx}{\sqrt{e^{2x}-1}}$

(*Hint:* Multiply numerator and denominator by e^x.)

63. $\displaystyle\int \frac{\cos x\, dx}{\sin x}$

64. $\displaystyle\int_0^{\pi/2} \frac{\cos x\, dx}{1+\sin x}$

65. $\displaystyle\int \sec^3 x\, \tan x\, dx$

66. $\displaystyle\int \frac{\sin\theta\, d\theta}{\sqrt{1+\cos\theta}}$

67. $\displaystyle\int e^{\tan 3x}\sec^2 3x\, dx$

68. $\displaystyle\int \cos 2t\sqrt{4-\sin 2t}\, dt$

69. $\displaystyle\int_{\pi/6}^{\pi/4} \frac{1 + \cos 2x}{\sin^2 2x}\, dx$

70. $\displaystyle\int_{-\pi/4}^{\pi/4} \frac{\sin^2 2x}{1 + \cos 2x}\, dx$

79. $\displaystyle\int_{1}^{9} \frac{dx}{\sqrt{x}\,(1 + \sqrt{x})}$

71. $\displaystyle\int \frac{\csc^2 2t}{\sqrt{1 + \cot 2t}}\, dt$

72. $\displaystyle\int_{0}^{\ln 3} e^{3x}\, dx$

80. $\displaystyle\int_{0}^{\sqrt{e-1}} \ln(x^2 + 1) \cdot 2x(x^2 + 1)^{-1}\, dx$

73. $\displaystyle\int_{0}^{1/2} \frac{e^{\tan^{-1} 2t}}{1 + 4t^2}\, dt$

74. $\displaystyle\int xe^{-x^2}\, dx$

81. Evaluate

$$\int xe^{(x^2)}\, dx$$

with the substitutions (a) $u = x^2$ and (b) $u = e^{(x^2)}$.

75. $\displaystyle\int_{1}^{4} \frac{2^{\sqrt{x}}}{2\sqrt{x}}\, dx$

76. $\displaystyle\int 10^{2x}\, dx$

77. $\displaystyle\int \frac{\ln x}{x}\, dx$

78. $\displaystyle\int \frac{\cos(\ln x)}{x}\, dx$

7.2
Integration by Parts

Integration by parts is a technique for replacing difficult integrals by ones that are usually easier to integrate. It applies mainly to integrals of the form

$$\int f(x)g(x)\, dx \tag{1}$$

in which f can be differentiated repeatedly to become zero and g can be integrated repeatedly without difficulty. The integral

$$\int xe^x\, dx$$

is such an integral because $f(x) = x$ can be differentiated twice to become zero and $g(x) = e^x$ can be integrated repeatedly without difficulty. Integration by parts also applies to important integrals like

$$\int e^x \sin x\, dx,$$

which are not of the form described in integral (1).

In this article, we describe integration by parts and show how it is applied. In addition to being a powerful technique for evaluating indefinite integrals, integration by parts plays an important role in preparing definite integrals for computer evaluation, but we will not take time for this topic here.

The Formula

The formula for integration by parts comes from the product rule,

$$\frac{d}{dx}(uv) = u\frac{dv}{dx} + v\frac{du}{dx}.$$

In its differential form, the rule becomes

$$d(uv) = u\, dv + v\, du,$$

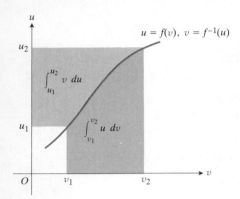

7.1 The area of the blue region, $\int_{v_1}^{v_2} u\,dv$, is equal to the area of the large rectangle, u_2v_2, minus the areas of the small rectangle, u_1v_1, and the gray region, $\int_{u_1}^{u_2} v\,du$. In symbols,

$$\int_{v_1}^{v_2} u\,dv = (u_2v_2 - u_1v_1) - \int_{u_1}^{u_2} v\,du.$$

which is then written as

$$u\,dv = d(uv) - v\,du$$

and integrated to give

The Integration-by-Parts Formula

$$\int u\,dv = uv - \int v\,du. \tag{2}$$

The integration-by-parts formula expresses one integral, $\int u\,dv$, in terms of a second integral, $\int v\,du$. With a proper choice of u and v, the second integral may be easier to evaluate than the first. This is the reason for the importance of the formula. When faced with an integral we cannot handle, we replace it by one with which we might have more success.

The equivalent formula for definite integrals is

$$\int_{v_1}^{v_2} u\,dv = (u_2v_2 - u_1v_1) - \int_{u_1}^{u_2} v\,du. \tag{3}$$

Figure 7.1 shows how the different parts of the formula may be interpreted as areas.

EXAMPLE 1 Evaluate

$$\int x \cos x\,dx.$$

Solution We use the formula

$$\int u\,dv = uv - \int v\,du$$

with

$$u = x, \qquad dv = \cos x\,dx,$$
$$du = dx, \qquad v = \sin x \qquad \text{(simplest function with } dv = \cos x\,dx\text{)}.$$

Then

$$\int x \cos x\,dx = x \sin x - \int \sin x\,dx$$

$$= x \sin x + \cos x + C.$$

Notice that we chose v to be the simplest function whose differential is $\cos x\,dx$. A more complicated function like $\sin x + 5$ or $\sin x + C_1$ would only lengthen the calculation; no greater generality would be achieved in the answer. (See Problem 51, however.) ∎

EXAMPLE 2 Find the moment about the y-axis of a thin homogeneous plate of density δ covering the region in the first quadrant bounded by the curve $y = e^x$ and the line $x = 1$. See Fig. 7.2.

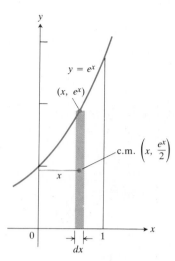

7.2 The moment of the strip about the *y*-axis is $x \, \delta \, dA = \delta \, xe^x \, dx$.

Solution   A typical vertical strip has

$$\text{center of mass (c.m.):}\quad (\tilde{x}, \tilde{y}) = \left(x, \frac{e^x}{2}\right),$$

length: e^x,

width: dx,

area: $dA = e^x \, dx$,

mass: $dm = \delta \, dA = \delta \, e^x \, dx$.

The moment of the strip about the *y*-axis is therefore

$$\tilde{x} \, dm = x \cdot \delta \, e^x \, dx = \delta \, x \, e^x \, dx.$$

The moment of the plate about the *y*-axis is the integral

$$M_y = \int \tilde{x} \, dm = \delta \int_0^1 x \, e^x \, dx.$$

To evaluate this integral we use the formula

$$\int u \, dv = uv - \int v \, du$$

with

$$u = x, \qquad dv = e^x \, dx,$$

$$du = dx, \qquad v = e^x \quad \text{(simplest function with } dv = e^x \, dx\text{)}.$$

Then

$$\int xe^x \, dx = xe^x - \int e^x \, dx$$

and

$$\int_0^1 xe^x \, dx = xe^x \bigg]_0^1 - \int_0^1 e^x \, dx = e - \left[e^x\right]_0^1 = e - [e - 1] = 1.$$

The moment of the plate about the *y*-axis is therefore

$$M_y = \delta \int_0^1 xe^x \, dx = \delta \cdot 1 = \delta.$$

Choosing *u* and *dv*

The integration in Example 2 illustrates a rule for choosing the *u* and the *dv* for integration by parts.

> **Rule for Choosing *u* and *dv***
>
> For *u*: Choose something that becomes simpler when differentiated.
>
> For *dv*: Choose something whose integral is simple.

It is not always possible to follow this rule, but when we can, the integral on the right side of the equation

$$\int u \, dv = uv - \int v \, du$$

will be simpler than the one on the left.

Right: With

$$u = x, \qquad v = \int dv = \int e^x \, dx = e^x,$$

we obtain

$$\int xe^x \, dx = xe^x - \int e^x \, dx = xe^x - e^x + C. \qquad (4)$$

Wrong: With

$$u = e^x, \qquad v = \int dv = \int x \, dx = \frac{x^2}{2},$$

we obtain

$$\int xe^x \, dx = e^x \cdot \frac{x^2}{2} - \int \frac{x^2}{2} e^x \, dx. \qquad (5)$$

Equation (5) is correct, but the integral on the right is more complicated than the original integral. It leaves us worse off than before.

EXAMPLE 3 Evaluate

$$\int \ln x \, dx.$$

Solution We use the formula

$$\int u \, dv = uv - \int v \, du$$

with

$$u = \ln x, \qquad \text{(simplifies when differentiated)}$$
$$du = \frac{1}{x} \, dx,$$
$$dv = dx, \qquad \text{(easy to integrate)}$$
$$v = x \qquad \text{(simplest function with } dv = dx\text{)}.$$

Then

$$\int \ln x \, dx = x \ln x - \int x \cdot \frac{1}{x} \, dx = x \ln x - \int dx = x \ln x - x + C. \quad \blacksquare$$

Repeated Use

Sometimes we have to use integration by parts more than once to obtain an answer. Here is an example.

EXAMPLE 4 Evaluate

$$\int x^2 e^x \, dx.$$

Solution We first use the formula

$$\int u \, dv = uv - \int v \, du$$

with

$$u = x^2, \qquad dv = e^x \, dx,$$
$$du = 2x \, dx \qquad v = e^x.$$

This gives

$$\int x^2 e^x \, dx = x^2 e^x - 2 \int x e^x \, dx.$$

The integral on the right now requires another integration by parts. As in Example 2, its value is

$$\int x e^x \, dx = x e^x - e^x + C'.$$

Hence,

$$\int x^2 e^x \, dx = x^2 e^x - 2x e^x + 2e^x + C.$$ ∎

Solving for the Unknown Integral

Integrals like the one in the next example occur in electrical engineering problems. Their evaluation requires two integrations by parts, followed by solving for the unknown integral.

EXAMPLE 5 Evaluate

$$\int e^{2x} \cos 3x \, dx.$$

Solution We first use the formula

$$\int u \, dv = uv - \int v \, du$$

with

$$u = e^{2x}, \qquad dv = \cos 3x \, dx,$$
$$du = 2e^{2x}, \qquad v = \frac{1}{3} \sin 3x.$$

Then

$$\int e^{2x} \cos 3x \, dx = \frac{e^{2x} \sin 3x}{3} - \frac{2}{3} \int e^{2x} \sin 3x \, dx.$$

The second integral is like the first, except it has $\sin 3x$ in place of $\cos 3x$. To evaluate it, we use integration by parts with

$$U = e^{2x}, \qquad dV = \sin 3x \, dx,$$
$$dU = 2e^{2x} \, dx, \qquad V = -\frac{1}{3} \cos 3x.$$

Then

$$\int e^{2x} \cos 3x \, dx = \frac{e^{2x} \sin 3x}{3} - \frac{2}{3}\left(-\frac{e^{2x} \cos 3x}{3} + \frac{2}{3}\int e^{2x} \cos 3x \, dx\right).$$

Now the unknown integral appears on the right with coefficient $-4/9$, as well as on the left. Combining the two gives

$$\frac{13}{9}\int e^{2x} \cos 3x \, dx = \frac{e^{2x} \sin 3x}{3} + \frac{2e^{2x} \cos 3x}{9}$$

$$= \frac{3e^{2x} \sin 3x + 2e^{2x} \cos 3x}{9}.$$

Dividing by $13/9$ then gives

$$\int e^{2x} \cos 3x \, dx = \frac{3e^{2x} \sin 3x + 2e^{2x} \cos 3x}{13} + C.$$

General formulas for the integrals of $e^{ax} \cos bx$ and the closely related $e^{ax} \sin bx$ can be found on the endpapers of this book.   ■

☐ Tabular Integration

As we have seen, integrals of the form

$$\int f(x)g(x) \, dx$$

in which f can be differentiated repeatedly to become zero and g can be integrated repeatedly without difficulty are natural candidates for integration by parts. If many repetitions are required, however, the calculations can be cumbersome. In situations like this, there is a way to organize the calculations that saves a great deal of work. It is called **tabular integration** and is illustrated in the following examples.

EXAMPLE 6 Evaluate

$$\int x^2 e^x \, dx$$

by tabular integration.

Solution With $f(x) = x^2$ and $g(x) = e^x$, we have

$f(x)$ and its derivatives		$g(x)$ and its integrals
x^2	$(+)$	e^x
$2x$	$(-)$	e^x
2	$(+)$	e^x
0		$e^x.$

The products of the functions connected by the arrows are added, with the middle sign changed, to obtain

$$\int x^2 e^x \, dx = x^2 e^x - 2xe^x + 2e^x + C.$$

Notice the agreement with Example 4. ∎

EXAMPLE 7 Evaluate

$$\int x^3 \sin x \, dx$$

by tabular integration.

Solution With $f(x) = x^3$ and $g(x) = \sin x$, we have

$f(x)$ and its derivatives		$g(x)$ and its integrals
x^3	$(+)$	$\sin x$
$3x^2$	$(-)$	$-\cos x$
$6x$	$(+)$	$-\sin x$
6	$(-)$	$\cos x$
0		$\sin x.$

The products of the functions connected by the arrows are added, with every other sign changed, to obtain

$$\int x^3 \sin x \, dx = -x^3 \cos x + 3x^2 \sin x + 6x \cos x - 6 \sin x + C. \quad ∎$$

PROBLEMS

In Problems 1–38, evaluate the integrals.

1. $\displaystyle\int x \sin x \, dx$

2. $\displaystyle\int x \cos 2x \, dx$

3. $\displaystyle\int x^2 \sin x \, dx$

4. $\displaystyle\int x^2 \cos x \, dx$

5. $\displaystyle\int_1^2 x \ln x \, dx$

6. $\displaystyle\int x^2 \ln x \, dx$

7. $\displaystyle\int x^3 \ln x \, dx$

8. $\displaystyle\int_0^1 \ln(x + 1) \, dx$

9. $\displaystyle\int \tan^{-1} x \, dx$

10. $\displaystyle\int \tan^{-1} ax \, dx$

11. $\displaystyle\int \sin^{-1} x \, dx$

12. $\displaystyle\int \sin^{-1} ax \, dx$

13. $\displaystyle\int x \sec^2 x \, dx$

14. $\displaystyle\int 4x \sec^2 2x \, dx$

15. $\displaystyle\int x^3 e^x \, dx$

16. $\displaystyle\int x^4 e^{-x} \, dx$

17. $\displaystyle\int (x^2 - 5x) e^x \, dx$

18. $\displaystyle\int (x^2 + x + 1) e^x \, dx$

19. $\displaystyle\int x^5 e^x \, dx$

20. $\displaystyle\int x^2 e^{4x} \, dx$

21. $\displaystyle\int_0^{\pi/2} x^2 \sin 2x \, dx$

22. $\displaystyle\int_0^{\pi/2} x^3 \cos 2x \, dx$

23. $\displaystyle\int x^4 \cos x \, dx$

24. $\displaystyle\int x^5 \sin x \, dx$

25. $\displaystyle\int x^2 \cos ax \, dx$

26. $\displaystyle\int x \cos(2x + 1) \, dx$

27. $\displaystyle\int_1^2 x \sec^{-1} x \, dx$

28. $\displaystyle\int_1^4 \sec^{-1} \sqrt{x} \, dx$

29. $\displaystyle\int_{1}^{e} \frac{\ln x}{x}\, dx$

30. $\displaystyle\int_{0}^{1} x\sqrt{1-x}\, dx$

31. $\displaystyle\int x \sin^{-1}\!\left(\frac{1}{x}\right) dx$

32. $\displaystyle\int x \sin^{-1}\!\left(\frac{a}{x}\right) dx$

33. $\displaystyle\int e^{x}\sin x\, dx$

34. $\displaystyle\int e^{-x}\cos x\, dx$

35. $\displaystyle\int e^{2x}\cos 3x\, dx$

36. $\displaystyle\int e^{-2x}\sin 2x\, dx$

37. $\displaystyle\int \sin(\ln x)\, dx$

38. $\displaystyle\int x(\ln x)^{2}dx$

39. Find the area of the region enclosed by the x-axis and the curve $y = x \sin x$ for (a) $0 \le x \le \pi$, (b) $\pi \le x \le 2\pi$.

40. Use cylindrical shells to find the volume swept out by revolving the region bounded by $x = 0$, $y = 0$, and $y = \cos x$, $0 \le x \le \pi/2$, about the y-axis.

41. Find the volume swept out by revolving about the y-axis the region in the first quadrant bounded by the coordinate axes, the curve $y = e^{-x}$, and the line $x = 1$.

42. Find the center of mass of a thin homogeneous plate bounded by the curve $y = \sin x$, $0 \le x \le \pi$, and the x-axis.

43. Find the center of mass of a thin homogeneous plate bounded by the curve $y = x^{2}e^{x}$, the x-axis, and the line $x = 1$.

44. Find the center of mass of a thin homogeneous plate bounded by the curve $y = \ln x$ and the lines $x = 1$ and $y = 1$.

45. Find the volume swept out when the region in the first quadrant bounded by the x-axis and the curve $y = x \sin x$, $0 \le x \le \pi$, is revolved (a) about the x-axis, (b) about the line $x = \pi$.

46. Find the center of mass of a thin homogeneous plate bounded by the curve $y = \ln x$, the x-axis, and the line $x = 2$.

47. Find the moment about the y-axis of a thin plate of density $\delta = (1 + x)$ covering the region bounded by the x-axis and the curve $y = \sin x$, $0 \le x \le \pi$.

48. Find the volume of the solid generated by revolving about the y-axis the region bounded by the x-axis and the curve $y = x\sqrt{1-x}$, $0 \le x \le 1$.

49. Find the area of the surface generated by revolving the curve

$$x = e^{t}\sin t, \quad y = e^{t}\cos t, \quad 0 \le t \le \frac{\pi}{2}$$

about the x-axis (Fig. 7.3).

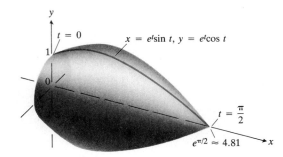

7.3   The surface in Problem 49.

50. Evaluate $\int_{0}^{1} e^{\sqrt{x}}\, dx$ by integration by parts. (*Hint:* First let $u^{2} = x$.)

51. Although we usually drop the constant of integration in determining v as $\int dv$ in integration by parts, choosing the constant to be different from zero can occasionally be helpful. As a case in point, evaluate

$$\int x \tan^{-1}x\, dx$$

with $u = \tan^{-1}x$ and $v = (x^{2}/2) + 1/2$.

7.3

Products and Powers of Trigonometric Functions (Other Than Even Powers of Sines and Cosines)

Integrals of combinations of trigonometric functions arise frequently in mathematics and science and it is important to know how to deal with them when they do. You have seen many of these combinations already, so this article will concentrate on the others. The integrals we need to study are the integrals of odd powers of sines and cosines (even powers, too, but they come in the next article); the integrals of the secant and its cube; the integrals of even powers of secants and tangents; and the integrals of products of sines and cosines like $\cos 3x \cos 5x$, $\sin 2x \cos 2x$, and $\sin 3x \sin 4x$. After looking at all these, we shall spend a few moments with definite integrals of odd functions and show how the integral of the secant is used to scale the maps we use in navigation.

Positive Odd Powers of Sines and Cosines

To integrate an odd positive power of $\sin x$, say $\sin^{2n+1} x$, we split off a factor of $\sin x$ and rewrite the remaining even power in terms of the cosine. The integration is then accomplished by setting $u = \cos x$ and $-du = \sin x\,dx$. This turns the integrand into a polynomial in u that is then easy to integrate.

EXAMPLE 1 Evaluate

$$\int \sin^3 x\,dx.$$

Solution We write

$$\int \sin^3 x\,dx = \int \sin^2 x\,\sin x\,dx$$

$$= \int (1 - \cos^2 x)\sin x\,dx$$

$$= \int (1 - u^2)(-du) \qquad \left(\begin{array}{l} u = \cos x \\ -du = \sin x\,dx \end{array}\right)$$

$$= \int (u^2 - 1)\,du$$

$$= \frac{u^3}{3} - u + C = \frac{\cos^3 x}{3} - \cos x + C. \qquad \blacksquare$$

The integrals of other positive odd powers of $\sin x$ are handled the same way. We write

$$\sin^{2n+1} x = \sin^{2n} x\,\sin x = (\sin^2 x)^n \sin x = (1 - \cos^2 x)^n \sin x.$$

We then set $u = \cos x$, $-du = \sin x\,dx$ and evaluate the integral of $\sin^{2n+1} x$ as

$$\int \sin^{2n+1} x\,dx = \int (1 - \cos^2 x)^n \sin x\,dx$$

$$= -\int (1 - u^2)^n\,du. \tag{1}$$

The expression $(1 - u^2)^n$ can now be expanded by the binomial theorem and the result integrated as a polynomial in u.

To integrate an odd positive power of the cosine, we proceed as we did with the sine. We write

$$\cos^{2n+1} x = \cos^{2n} x\,\cos x = (\cos^2 x)^n \cos x = (1 - \sin^2 x)^n \cos x$$

and set $u = \sin x$, $du = \cos x\,dx$ to get

$$\int \cos^{2n+1} x\,dx = \int (1 - \sin^2 x)^n \cos x\,dx = \int (1 - u^2)^n\,du. \tag{2}$$

EXAMPLE 2 Evaluate

$$\int \cos^5 x\,dx.$$

Solution   We write

$$\int \cos^5 x \, dx = \int \cos^4 x \cos x \, dx$$

$$= \int (1 - \sin^2 x)^2 \cos x \, dx$$

$$= \int (1 - u^2)^2 \, du \quad \left(\begin{matrix} u = \sin x \\ du = \cos x \, dx \end{matrix}\right)$$

$$= \int (1 - 2u^2 + u^4) \, du = u - \frac{2}{3}u^3 + \frac{1}{5}u^5 + C$$

$$= \sin x - \frac{2}{3}\sin^3 x + \frac{1}{5}\sin^5 x + C. \qquad \blacksquare$$

The Integrals of the Secant and Cosecant

Isaac Barrow, Isaac Newton's teacher at Cambridge University, gave the first intelligible calculation of the integral of the secant in his book *Geometrical Lectures*.* In modern notation, it goes like this:

$$\int \sec x \, dx = \int \frac{1}{\cos x} \, dx$$

$$= \int \frac{\cos x}{\cos^2 x} \, dx$$

$$= \int \frac{\cos x}{1 - \sin^2 x} \, dx$$

$$= \int \frac{\cos x}{(1 - \sin x)(1 + \sin x)} \, dx$$

$$= \frac{1}{2} \int \frac{\cos x}{1 - \sin x} + \frac{\cos x}{1 + \sin x} \, dx$$

$$= \frac{1}{2}[-\ln|1 - \sin x| + \ln|1 + \sin x|] + C$$

$$= \frac{1}{2} \ln \left| \frac{1 + \sin x}{1 - \sin x} \right| + C \tag{3}$$

$$= \frac{1}{2} \ln \left| \frac{1 + \sin x}{1 - \sin x} \cdot \frac{1 + \sin x}{1 + \sin x} \right| + C$$

$$= \frac{1}{2} \ln \left| \frac{(1 + \sin x)^2}{1 - \sin^2 x} \right| + C$$

$$= \frac{1}{2} \ln \left| \frac{(1 + \sin x)^2}{(\cos x)^2} \right| + C$$

$$= \ln \left| \frac{1 + \sin x}{\cos x} \right| + C = \ln|\sec x + \tan x| + C.$$

---

*For a delightful historical account of the integral of the secant, read Philip M. Tuchinsky and V. Frederick Rickey's article "An Application of Geography to Mathematics: History of the Integral of the Secant," *Mathematics Magazine,* 53:3 (May 1980).

If you happen to think of it, there is a trick that makes the integration much shorter. Write

$$\sec x = \frac{\sec x(\tan x + \sec x)}{\sec x + \tan x} = \frac{\sec x \tan x + \sec^2 x}{\sec x + \tan x}.$$

In this form, the numerator is the derivative of the denominator. Therefore,

$$\int \sec x \, dx = \int \frac{\sec x \tan x + \sec^2 x}{\sec x + \tan x} \, dx$$

$$= \int \frac{du}{u} \qquad (u = \sec x + \tan x)$$

$$= \ln|u| + C.$$

Once again,

$$\int \sec x \, dx = \ln|\sec x + \tan x| + C. \tag{4}$$

A similar integration shows that the companion formula for the integral of the cosecant is

$$\int \csc x \, dx = -\ln|\csc x + \cot x| + C. \tag{5}$$

See Problem 23.

Now that we have formulas for integrating the secant and cosecant, evaluating the integrals of their cubes is straightforward.

EXAMPLE 3 Evaluate

$$\int \sec^3 x \, dx.$$

Solution We use integration by parts. With

$$u = \sec x, \; du = \sec x \tan x \, dx, \; dv = \sec^2 x \, dx, \; v = \tan x$$

the integral becomes

$$\int \sec^3 x \, dx = \sec x \tan x - \int \tan x \sec x \tan x \, dx$$

$$= \sec x \tan x - \int (\sec^2 x - 1)\sec x \, dx$$

$$= \sec x \tan x + \int \sec x \, dx - \int \sec^3 x \, dx.$$

Hence,

$$2 \int \sec^3 x \, dx = \sec x \tan x + \int \sec x \, dx$$

and

$$\int \sec^3 x \, dx = \frac{1}{2} \sec x \tan x + \frac{1}{2} \ln|\sec x + \tan x| + C. \qquad \blacksquare$$

See Problem 24 for the integral of $\csc^3 x$.

Positive Even Powers of Secants and Tangents

For these integrands, we use the identities

$$\tan^2 x = \sec^2 x - 1 \qquad \text{and} \qquad \sec^2 x = 1 + \tan^2 x.$$

EXAMPLE 4 Evaluate

$$\int \tan^4 x \, dx.$$

Solution

$$\int \tan^4 x \, dx = \int \tan^2 x \cdot \tan^2 x \, dx = \int \tan^2 x \cdot (\sec^2 x - 1) \, dx$$

$$= \int \tan^2 x \sec^2 x \, dx - \int \tan^2 x \, dx$$

$$= \int \tan^2 x \sec^2 x \, dx - \int (\sec^2 x - 1) \, dx$$

$$= \int \tan^2 x \sec^2 x \, dx - \int \sec^2 x \, dx + \int dx.$$

In the first integral, we let

$$u = \tan x, \qquad du = \sec^2 x \, dx$$

and have

$$\int u^2 \, du = \frac{1}{3} u^3 + C'.$$

The remaining integrals are standard forms, so

$$\int \tan^4 x \, dx = \frac{1}{3} \tan^3 x - \tan x + x + C.$$ ∎

Products of Sines and Cosines

The integrals

$$\int \sin mx \sin nx \, dx,$$

$$\int \sin mx \cos nx \, dx,$$

$$\int \cos mx \cos nx \, dx$$

arise in alternating-current theory, heat transfer problems, bending of beams, cable stress analysis in suspension bridges, and many other places where trigonometric series are applied to problems in mathematics, science, and engineering. They can be evaluated by integration by parts, but two such integrations are required in each case. A simpler way to evaluate them is to exploit the identities

$$\sin mx \sin nx = \frac{1}{2}[\cos(m - n)x - \cos(m + n)x], \tag{6}$$

$$\sin mx \cos nx = \frac{1}{2}[\sin(m - n)x + \sin(m + n)x], \tag{7}$$

$$\cos mx \cos nx = \frac{1}{2}[\cos(m - n)x + \cos(m + n)x]. \tag{8}$$

These identities follow at once from

$$\cos(A + B) = \cos A \cos B - \sin A \sin B,$$
$$\cos(A - B) = \cos A \cos B + \sin A \sin B, \tag{9}$$

and

$$\sin(A + B) = \sin A \cos B + \cos A \sin B,$$
$$\sin(A - B) = \sin A \cos B - \cos A \sin B. \tag{10}$$

For example, if we add the two equations in (9) and then divide by 2, we obtain (8) by taking $A = mx$ and $B = nx$. The identity in (6) is obtained in a similar way by subtracting the first equation in (9) from the second equation. Finally, if we add the two equations in (10) we are led to the identity in (7).

EXAMPLE 5   From Eq. (7) with $m = 3$ and $n = 5$, we get

$$\int \sin 3x \cos 5x \, dx = \frac{1}{2} \int [\sin(-2x) + \sin 8x] \, dx$$

$$= \frac{1}{2} \int (\sin 8x - \sin 2x) \, dx$$

$$= -\frac{\cos 8x}{16} + \frac{\cos 2x}{4} + C. \qquad ∎$$

EXAMPLE 6   Show that if m and n are integers, and $m^2 \neq n^2$, then

$$\int_0^{2\pi} \cos mx \cos nx \, dx = 0.$$

Solution   From Eq. (8) we have

$$\int_0^{2\pi} \cos mx \cos nx \, dx = \int_0^{2\pi} \frac{1}{2}[\cos(m - n)x + \cos(m + n)x] \, dx$$

$$= \frac{1}{2}\left[\frac{\sin(m - n)x}{m - n} + \frac{\sin(m + n)x}{m + n} \right]_0^{2\pi}. \tag{11}$$

(The integration requires $m - n \neq 0$ and $m + n \neq 0$, both of which are true if $m^2 \neq n^2$.) Now, $(m - n)$ and $(m + n)$ are integers because m and n are integers. Furthermore, the sine of any integer multiple of 2π is zero. The terms in brackets in Eq. (11) are therefore zero at both the upper and lower limits of integration, and

$$\int_0^{2\pi} \cos mx \cos nx \, dx = 0. \qquad ∎$$

Integrals of Odd Functions

In many applications, we integrate functions over intervals like $-a \le x \le a$ that are symmetric about the origin. In an amazing number of cases, these integrals are zero. For example,

$$\int_{-1}^{1} 2x \, dx = x^2 \Big]_{-1}^{1} = 1 - 1 = 0,$$

$$\int_{-\pi}^{\pi} \sin x \, dx = -\cos x \Big]_{-\pi}^{\pi}$$

$$= -\cos \pi + \cos(-\pi) = -(-1) + (-1) = 0,$$

$$\int_{-a}^{a} \frac{\sin x}{\cos^2 x} \, dx = \frac{1}{\cos x} \Big]_{-a}^{a} = \frac{1}{\cos a} - \frac{1}{\cos(-a)} = 0.$$

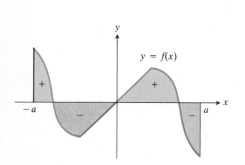

7.4 The graph of an odd function over an interval that is symmetric about the origin encloses as much area above the axis as below. The integral of the function from $-a$ to a is therefore zero.

These three integrals are zero because they are integrals of odd functions over intervals that are symmetric about the origin. For each function, $f(-x) = -f(x)$. Hence, the amount of area enclosed above the x-axis to the left of the origin equals the amount of area enclosed below the x-axis to the right of the origin and cancels it out in integration. Similarly, the area enclosed beneath the x-axis to the left of the origin cancels the area enclosed above the x-axis to the right of the origin (see Fig. 7.4).

What functions are odd?

Odd powers of x: $x, \quad x^{-1}, \quad x^3, \quad x^{-3}$

Sines: $\sin ax, \quad a \ne 0$

Odd integer powers of odd functions: $\sin^3 x, \quad \dfrac{1}{\sin^5 x}$

The product of an odd function with an even function: $\cos x \sin 3x,$ $x^2 \sin x$

The quotient of an odd function and an even function: $\tan x = \dfrac{\sin x}{\cos x}$

☐ Mercator's World Map

7.5 A flight with a constant bearing of 45° E of N from the Galápagos Islands in the Pacific to Franz Josef Land in the Arctic Ocean.

The integral of the secant plays an important role in making maps for compass navigation. The easiest course for a sailor to steer is a course whose compass heading is constant. This might be a course of 45° (northeast), for example, or a course of 225° (southwest), or whatever. Such a course will lie along a spiral that winds around the globe toward one of the poles (Fig. 7.5), unless the course runs due north or south or lies parallel to the equator.

In 1569, Gerhard Krämer, a Flemish surveyor and geographer known to us by his Latinized last name, Mercator, made a world map on which all spirals of constant compass heading appeared as straight lines (Fig. 7.6). This fantastic achievement met what must have been one of the most pressing navigational needs of all time. A sailor could then read the compass heading for a voyage between any two points from the direction of a straight line connecting them on Mercator's map (Fig. 7.7).

If you look closely at Fig. 7.7, you will see that the vertical lines of longitude that meet at the poles on a globe have been spread apart to lie parallel on the map. The horizontal lines of latitude that are shown every 10° are paral-

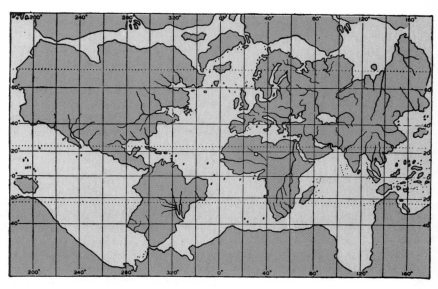

7.6 Sketch of Mercator's world map of 1569.

7.7 The flight of Fig. 7.5 traced on a Mercator map.

lel also, as they are on the globe, but they are not evenly spaced. The spacing between them increases toward the poles.

The secant function plays a role in determining the correct spacing of all these lines. The scaling factor by which horizontal distances are increased at a fixed latitude $\theta°$ to spread the lines of longitude apart is precisely sec θ. There is no spread at the equator, where sec $0° = 1$. At latitude $30°$ north or south, the spreading is accomplished by multiplying all horizontal distances by the factor sec $30° = 2/\sqrt{3} \approx 1.15$. At $60°$, the factor is sec $60° = 2$. The closer you move toward the poles, the more the longitudes have to be spread to be parallel. The lines of latitude are spread apart toward the poles to match the spreading of the longitudes, but the formula for the spreading is complicated by the fact that the scaling factor sec θ increases with the latitude θ. Thus, the factor to be used for stretching an interval of latitude is not a constant on the interval. This difficulty is overcome by integration. If R is the radius of the globe being modeled (Fig. 7.8), then the distance D between the lines that are drawn on the map to show the equator and the latitude $\alpha°$ is R times the integral of the secant

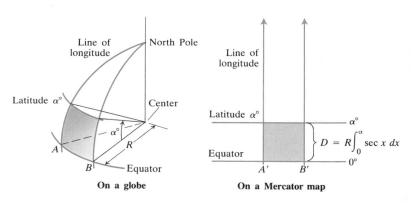

7.8 Lines of latitude and longitude.

from zero to α:

$$D = R \int_0^\alpha \sec x \, dx. \qquad (12)$$

Therefore, the map distance between two lines of north latitude, say, at $\alpha°$ and $\beta°$ ($\alpha < \beta$) is

$$R \int_0^\beta \sec x \, dx - R \int_0^\alpha \sec x \, dx = R \int_\alpha^\beta \sec x \, dx$$

$$= R \ln|\sec x + \tan x| \Big]_\alpha^\beta. \qquad (13)$$

EXAMPLE 7   Suppose that the equatorial length of a Mercator map just matches the equator of a globe of radius 25 cm. Then Eq. (13) gives the spacing on the map between the equator and the latitude 20° north as

$$25 \int_0^{20°} \sec x \, dx = 25 \ln|\sec x + \tan x| \Big]_0^{20°} \approx 9 \text{ cm}.$$

The spacing between 60° and 80° north is given by Eq. (13) as

$$25 \int_{60°}^{80°} \sec x \, dx = 25 \ln|\sec x + \tan x| \Big]_{60°}^{80°} \approx 28 \text{ cm}.$$

The navigational properties of a Mercator map are achieved at the expense of a considerable distortion of distance. (For a very readable derivation and discussion of the mathematics of Mercator maps, see Philip M. Tuchinsky's *Mercator's World Map and the Calculus*, UMAP Unit No. 206, COMAP, Inc., Arlington, MA, 1979.) ∎

TABLE 7.2
How to integrate odd positive powers of sines and cosines

To evaluate	Write	Substitute
$\int \sin^{2n+1} x \, dx$	$(\sin^2 x)^n \sin x = (1 - \cos^2 x)^n \sin x \, dx$	$u = \cos x$
		$du = -\sin x \, dx$
$\int \cos^{2n+1} x \, dx$	$(\cos^2 x)^n \cos x = (1 - \sin^2 x)^n \cos x \, dx$	$u = \sin x$
		$du = \cos x \, dx$

PROBLEMS

Evaluate the integrals in Problems 1–22. In evaluating the definite integrals, you may wish to refer to the triangles in Fig. 7.9.

1. $\int \sin^5 x \, dx$

2. $\int_0^\pi \sin^5 \dfrac{x}{2} \, dx$

3. $\int_{-\pi/2}^{\pi/2} \cos^3 x \, dx$

4. $\int \cos^5 3x \, dx$

5. $\int \sin^7 x \, dx$

6. $\int \cos^7 x \, dx$

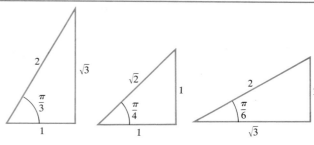

7.9   Reference triangles for evaluating trigonometric functions at $\pi/6$, $\pi/4$, and $\pi/3$ radians.

7. $\displaystyle\int \cos^{2/3}x \, \sin^5x \, dx$

(*Hint:* Put one factor of sin x with dx and express the rest in terms of the cosine.)

8. $\displaystyle\int \sin^{3/2}x \, \cos^3x \, dx$

(*Hint:* Put one factor of cos x with dx and express the rest in terms of the sine.)

9. $\displaystyle\int_0^{\pi/3} \sec x \, dx$

10. $\displaystyle\int \sec 4t \, dt$

11. $\displaystyle\int_{-\pi/3}^{0} \sec^3x \, dx$

12. $\displaystyle\int_0^{\pi/6} \frac{2 \sin^2x}{\cos x} \, dx$

13. $\displaystyle\int e^x\sec^3e^x \, dx$

14. $\displaystyle\int_0^{\pi/4} \sec^4x \, dx$

15. $\displaystyle\int_{\pi/4}^{\pi/2} \frac{dx}{\sin^4x}$

16. $\displaystyle\int \sec^43x \, dx$

17. $\displaystyle\int_0^{\pi/4} \sec^8x \, dx$

18. $\displaystyle\int \frac{2x \, dx}{\cos^3(x^2)}$

19. $\displaystyle\int_0^{\pi/4} \tan^3x \, dx$

20. $\displaystyle\int_{-\pi/4}^{\pi/4} \tan^4x \, dx$

21. $\displaystyle\int \tan^6x \, dx$

22. a) $\displaystyle\int \cot^2x \, dx$ b) $\displaystyle\int \cot^4x \, dx$

23. Show that

$$\int \csc x \, dx = -\ln|\csc x + \cot x| + C.$$

(*Hint:* Repeat the short derivation of the integral of the secant with cofunctions.)

24. Use the result in Problem 23 to show that

$$\int \csc^3x \, dx = -\frac{1}{2} \csc x \cot x - \frac{1}{2} \ln|\csc x + \cot x| + C.$$

25. Evaluate

a) $\displaystyle\int_0^{\pi/4} \frac{dx}{\sqrt{1 - \sin^2x}}$,   b) $\displaystyle\int_{\pi/3}^{\pi/2} \frac{dx}{\sqrt{1 - \cos^2x}}$.

26. Evaluate

$$\int_0^{\pi/3} \frac{dx}{1 + \sin x}.$$

(*Hint:* Multiply numerator and denominator by $1 - \sin x$.)

Evaluate the integrals in Problems 27–32.

27. $\displaystyle\int_{-\pi}^{0} \sin 3x \cos 2x \, dx$

28. $\displaystyle\int_0^{\pi/2} \cos 3x \sin 2x \, dx$

29. $\displaystyle\int_{-\pi}^{\pi} \sin^23x \, dx$

30. $\displaystyle\int_0^{\pi/2} \sin x \cos x \, dx$

31. $\displaystyle\int_0^{\pi} \cos 3x \cos 4x \, dx$

32. $\displaystyle\int_{-\pi/2}^{\pi/2} \cos x \cos 7x \, dx$

33. Use the result in Problem 23 and the identity $2 \sin A \cos A = \sin 2A$ to evaluate

$$\int \frac{\sec 2x \csc 2x}{2} \, dx.$$

34. Which integrals are zero and which are not? (You can do most of these without writing anything down.)

a) $\displaystyle\int_{-\pi}^{\pi} \sin x \cos^2x \, dx$ b) $\displaystyle\int_{-1}^{1} \sin 3x \cos 5x \, dx$

c) $\displaystyle\int_{-L}^{L} \sqrt[3]{\sin x} \, dx$ d) $\displaystyle\int_{-a}^{a} x\sqrt{a^2 - x^2} \, dx$

e) $\displaystyle\int_{-\pi/4}^{\pi/4} x \sec x \, dx$ f) $\displaystyle\int_{-\pi/4}^{\pi/4} \tan^3x \, dx$

g) $\displaystyle\int_{-\pi/2}^{\pi/2} x \sin x \, dx$ h) $\displaystyle\int_{-\pi/2}^{\pi/2} x \cos x \, dx$

i) $\displaystyle\int_{-a}^{a} \sin mx \cos mx \, dx, \ m \neq 0$ j) $\displaystyle\int_{-\pi}^{\pi} \sin^5x \, dx$

k) $\displaystyle\int_{-\pi/2}^{\pi/2} \cos^3x \, dx$ l) $\displaystyle\int_{-\pi}^{\pi} \cos^5x \, dx$

m) $\displaystyle\int_{-\ln 2}^{\ln 2} x(e^x + e^{-x}) \, dx$ n) $\displaystyle\int_{-\pi/2}^{\pi/2} \sin^2x \cos x \, dx$

o) $\displaystyle\int_{-\pi/2}^{\pi/2} \sin x \sin 2x \, dx$ p) $\displaystyle\int_{-\pi/4}^{\pi/4} \sec x \tan x \, dx$

q) $\displaystyle\int_{-a}^{a} (e^x\sin x + e^{-x}\sin x) \, dx$ r) $\displaystyle\int_{-1}^{1} \frac{\sin x \, dx}{e^x + e^{-x}}$

35. Find the area of the region bounded above by $y = 2 \cos x$ and below by $y = \sec x$, $-\pi/4 \leq x \leq \pi/4$.

36. Find the length of the curve

$$y = \ln(\cos x), \quad 0 \leq x \leq \pi/3.$$

37. Find the length of the curve

$$y = \ln(\sec x), \quad 0 \leq x \leq \pi/4.$$

38. Find the centroid of the region bounded by the x-axis, the curve $y = \sec x$, and the lines $x = -\pi/4$, $x = \pi/4$.

39. **CALCULATOR OR TABLES** How far apart should the following lines of latitude be on the Mercator map of Example 7?
a) 30° and 45° north (about the latitudes of New Orleans, La., and Minneapolis, Minn.);
b) 45° and 60° north (about the latitudes of Salem, Ore., and Seward, Al.).

40. Show that

a) $\displaystyle\int_0^{2\pi} \sin mx \sin nx \, dx = 0,$

b) $\displaystyle\int_0^{2\pi} \sin px \cos qx \, dx = 0$,

when m, n, p, and q are integers and $m^2 \neq n^2$.

41. Two functions f and g are said to be *orthogonal* on an interval $a \leq x \leq b$ if $\int_a^b f(x)g(x) \, dx = 0$.

a) Prove that $\sin mx$ and $\sin nx$ are orthogonal on any interval of length 2π provided m and n are integers and $m^2 \neq n^2$.

b) Prove the same for $\cos mx$ and $\cos nx$.

c) Prove the same for $\sin mx$ and $\cos nx$ even if $m = n$.

(Example 6 and Problem 40 prove these statements for the interval $[0, 2\pi]$.)

42. *A proof that the integral of an odd continuous function $f(x)$ from $x = -a$ to $x = a$ is zero.*

a) First, show that

$$\int_{-a}^0 f(x) \, dx = -\int_0^a f(x) \, dx.$$

You can do this by substituting $u = -x$, $du = -dx$ in the integral on the left.

b) Then use the result in (a) to show that

$$\int_{-a}^a f(x) \, dx = \int_{-a}^0 f(x) \, dx + \int_0^a f(x) \, dx = 0.$$

7.4
Even Powers of Sines and Cosines

In this article, we see how identities like

$$\sin^2\theta = \frac{1 - \cos 2\theta}{2}, \tag{1}$$

$$\cos^2\theta = \frac{1 + \cos 2\theta}{2}, \tag{2}$$

and a number of others with which we have been working enable us to evaluate integrals by reducing even powers and eliminating square roots.

Reducing Even Powers

EXAMPLE 1

$$\int \cos^4 x \, dx = \int (\cos^2 x)^2 \, dx = \int \frac{1}{4}(1 + \cos 2x)^2 \, dx \qquad \text{(Eq. 2 with } \theta = x)$$

$$= \frac{1}{4} \int (1 + 2\cos 2x + \cos^2 2x) \, dx$$

$$= \frac{1}{4} \int \left[1 + 2\cos 2x + \frac{1}{2}(1 + \cos 4x)\right] dx \qquad \left(\begin{array}{c}\text{Eq. 2 with} \\ \theta = 2x\end{array}\right)$$

$$= \frac{3}{8}x + \frac{1}{4}\sin 2x + \frac{1}{32}\sin 4x + C. \qquad \blacksquare$$

An integral like

$$\int \sin^2 x \cos^4 x \, dx,$$

which involves even powers of both $\sin x$ and $\cos x$, can be changed to a sum of integrals that involve powers of only one of them. Then these integrals may be handled by the method of Example 1.

EXAMPLE 2

$$\int \sin^2 x \cos^4 x \, dx = \int (1 - \cos^2 x) \cos^4 x \, dx = \int \cos^4 x \, dx - \int \cos^6 x \, dx$$

We evaluated $\int \cos^4 x \, dx$ above, and

$$\int \cos^6 x \, dx = \int (\cos^2 x)^3 \, dx = \frac{1}{8} \int (1 + \cos 2x)^3 \, dx$$

$$= \frac{1}{8} \int (1 + 3 \cos 2x + 3 \cos^2 2x + \cos^3 2x) \, dx.$$

We now know how to handle each term of the integrand. The result is

$$\int \cos^6 x \, dx = \frac{5}{16} x + \frac{1}{4} \sin 2x + \frac{3}{64} \sin 4x - \frac{1}{48} \sin^3 2x + C.$$

Combining this with the result of Example 1 gives

$$\int \sin^2 x \cos^4 x \, dx = \int \cos^4 x \, dx - \int \cos^6 x \, dx$$

$$= \frac{1}{16} x - \frac{1}{64} \sin 4x + \frac{1}{48} \sin^3 2x + C. \qquad \blacksquare$$

Eliminating Square Roots

EXAMPLE 3 Evaluate

$$\int_0^{\pi/4} \sqrt{1 + \cos 4x} \, dx.$$

Solution To eliminate the square root we use the identity

$$\cos^2 \theta = \frac{1 + \cos 2\theta}{2}, \qquad \text{or} \qquad 1 + \cos 2\theta = 2 \cos^2 \theta,$$

from Eq. (2). With $\theta = 2x$, this becomes

$$1 + \cos 4x = 2 \cos^2 2x.$$

Therefore,

$$\int_0^{\pi/4} \sqrt{1 + \cos 4x} \, dx = \int_0^{\pi/4} \sqrt{2 \cos^2 2x} \, dx = \int_0^{\pi/4} \sqrt{2} \sqrt{\cos^2 2x} \, dx$$

$$= \sqrt{2} \int_0^{\pi/4} |\cos 2x| \, dx$$

$$= \sqrt{2} \int_0^{\pi/4} \cos 2x \, dx \qquad \left(\begin{array}{l} \text{because } \cos 2x \geq 0 \\ \text{on } [0, \pi/4] \end{array} \right)$$

$$= \sqrt{2} \left[\frac{\sin 2x}{2} \right]_0^{\pi/4}$$

$$= \frac{\sqrt{2}}{2} [1 - 0] = \frac{\sqrt{2}}{2}. \qquad \blacksquare$$

EXAMPLE 4 Evaluate

$$\int_{-\pi/2}^{\pi/2} \sqrt{1 - \cos^2 t} \, dt.$$

Solution 1 We can use the substitution

$$\sin^2 t = 1 - \cos^2 t$$

to write

$$\int_{-\pi/2}^{\pi/2} \sqrt{1 - \cos^2 t} \, dt = \int_{-\pi/2}^{\pi/2} \sqrt{\sin^2 t} \, dt = \int_{-\pi/2}^{\pi/2} |\sin t| \, dt,$$

but we must not remove the absolute value signs without first checking the sign of $\sin t$ on the domain $-\pi/2 \le t \le \pi/2$. Indeed, $\sin t$ is negative to the left of 0 on this domain so that

$$|\sin t| = \begin{cases} -\sin t, & -\pi/2 \le t \le 0, \\ \sin t, & 0 \le t \le \pi/2. \end{cases}$$

To integrate correctly we must therefore write

$$\int_{-\pi/2}^{\pi/2} |\sin t| \, dt = \int_{-\pi/2}^{0} -\sin t \, dt + \int_{0}^{\pi/2} \sin t \, dt$$

$$= \cos t \Big]_{-\pi/2}^{0} - \cos t \Big]_{0}^{\pi/2}$$

$$= (1 - 0) - (0 - 1)$$

$$= 2.$$

If we disregard the absolute value bars and mistakenly write $\sqrt{1 - \cos^2 t} = \sin t$, we get

$$\int_{-\pi/2}^{\pi/2} \sqrt{1 - \cos^2 t} \, dt = \int_{-\pi/2}^{\pi/2} \sin t \, dt = -\cos t \Big]_{-\pi/2}^{\pi/2} = -0 + 0 = 0,$$

which is wrong.

Solution 2 Another way to evaluate this integral is to begin as in Solution 1, observe that the function $|\sin t|$ is even, and conclude that

$$\int_{-\pi/2}^{\pi/2} |\sin t| \, dt = 2 \int_{0}^{\pi/2} |\sin t| \, dt = 2 \int_{0}^{\pi/2} \sin t \, dt$$

$$= -2 \cos t \Big]_{0}^{\pi/2} = 2.$$

EXAMPLE 5 Evaluate

$$\int_{\pi/4}^{3\pi/4} \sqrt{\csc^2 x - 1} \, dx.$$

Solution The identity $\csc^2 x - 1 = \cot^2 x$ enables us to eliminate the square root and write

$$\int_{\pi/4}^{3\pi/4} \sqrt{\csc^2 x - 1} \, dx = \int_{\pi/4}^{3\pi/4} \sqrt{\cot^2 x} \, dx = \int_{\pi/4}^{3\pi/4} |\cot x| \, dx.$$

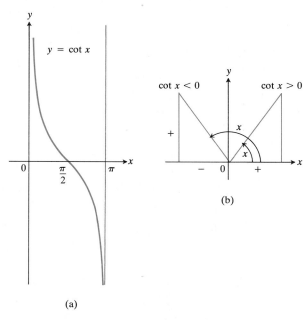

7.10 Diagrams for determining the sign of cot x on the interval $\pi/4 \le x \le 3\pi/4$.

To see how to remove the absolute value signs, we check the sign of cot x on the domain of integration. A glance at the graph in Fig. 7.10(a) or the triangles in Fig. 7.10(b) shows that the cotangent is positive between $\pi/4$ and $\pi/2$ and negative between $\pi/2$ and $3\pi/4$. We therefore split the interval of integration at $x = \pi/2$ and continue the calculation with

$$\int_{\pi/4}^{3\pi/4} |\cot x|\, dx = \int_{\pi/4}^{\pi/2} |\cot x|\, dx + \int_{\pi/2}^{3\pi/4} |\cot x|\, dx$$

$$= \int_{\pi/4}^{\pi/2} \cot x\, dx + \int_{\pi/2}^{3\pi/4} -\cot x\, dx$$

$$= \ln|\sin x|\Big]_{\pi/4}^{\pi/2} - \ln|\sin x|\Big]_{\pi/2}^{3\pi/4}$$

$$= \ln|1| - \ln\left|\frac{\sqrt{2}}{2}\right| - \ln\left|-\frac{\sqrt{2}}{2}\right| + \ln|1|$$

$$= -2\ln\left|\frac{\sqrt{2}}{2}\right| = \ln\left(\frac{2}{\sqrt{2}}\right)^2 = \ln 2. \qquad \blacksquare$$

TABLE 7.3

How to integrate a product $\sin^{2m}x\, \cos^{2n}x$ of even, nonnegative powers of sin x and cos x

---

1. Change the product into a power of sin x or cos x alone, whichever is easier to do. (Use $\cos^2 x = 1 - \sin^2 x$ or $\sin^2 x = 1 - \cos^2 x$.)

2. Replace the resulting even power with lower powers by one or more applications of $\sin^2 x = (1 - \cos 2x)/2$ or $\cos^2 x = (1 + \cos 2x)/2$.

PROBLEMS

Evaluate the integrals in Problems 1–26.

1. $\displaystyle\int_{-\pi}^{\pi} \sin^2x \, dx$

2. $\displaystyle\int_{-\pi}^{\pi} \cos^2t \, dt$

3. $\displaystyle\int \sin^2 2t \, dt$

4. $\displaystyle\int \cos^2 3\theta \, d\theta$

5. $\displaystyle\int_{-\pi/4}^{\pi/4} \sin^2x \cos^2x \, dx$

6. $\displaystyle\int_{0}^{\pi} \sin^4x \, dx$

7. $\displaystyle\int_{0}^{\pi/a} \sin^4 ax \, dx$

8. $\displaystyle\int_{0}^{1} \cos^4 2\pi t \, dt$

9. $\displaystyle\int \frac{\sin^4x}{\cos^2x} \, dx$

10. $\displaystyle\int \frac{\cos^6x}{\sin^2x} \, dx$

11. $\displaystyle\int_{0}^{\pi} \sin^4y \cos^2y \, dy$

12. $\displaystyle\int_{-\pi/4}^{\pi/4} 4 \sin^4t \cos^4t \, dt$ (*Hint:* $2 \sin \theta \cos \theta = \sin 2\theta$.)

13. $\displaystyle\int \frac{\sin^6\theta}{\cos^2\theta} \, d\theta$

14. $\displaystyle\int_{-\pi}^{0} \sin^6x \, dx$

15. $\displaystyle\int_{0}^{2\pi} \sqrt{\frac{1 - \cos t}{2}} \, dt$

16. $\displaystyle\int_{0}^{\pi} \sqrt{1 - \cos 2x} \, dx$

17. $\displaystyle\int_{0}^{\pi/10} \sqrt{1 + \cos 5\theta} \, d\theta$

18. $\displaystyle\int_{0}^{2\pi} \sqrt{1 + \cos(y/4)} \, dy$

19. $\displaystyle\int_{0}^{\pi/2} \theta\sqrt{1 - \cos \theta} \, d\theta$

(*Hint:* Substitute $\sqrt{1 - \cos \theta} = \sqrt{2} \sin(\theta/2)$.)

20. $\displaystyle\int_{0}^{\pi} \sqrt{1 - \sin^2t} \, dt$

21. $\displaystyle\int_{-\pi/4}^{\pi/4} \sqrt{1 + \tan^2x} \, dx$

22. $\displaystyle\int_{-\pi/4}^{\pi/4} \sqrt{\sec^2x - 1} \, dx$

23. $\displaystyle\int_{0}^{\pi} \sqrt{1 - \cos^2\theta} \, d\theta$

24. $\displaystyle\int_{0}^{\pi} \sqrt{1 - \cos^2 2x} \, dx$

25. $\displaystyle\int_{\pi/4}^{3\pi/4} \sqrt{\cot^2\theta + 1} \, d\theta$

26. $\displaystyle\int_{-\pi}^{\pi} (1 - \cos^2t)^{3/2} \, dt$

Evaluate the integrals in Problems 27–34. They look as if they might involve reducing powers or eliminating square roots, but they are more quickly done some other way.

27. $\displaystyle\int_{-\pi}^{\pi} \sqrt{1 - \cos^2x} \sin x \, dx$

28. $\displaystyle\int \frac{1}{\cos^2t} \, dt$

29. $\displaystyle\int \frac{1}{\sin^4x} \, dx$

30. $\displaystyle\int \frac{\sin^3\theta}{\cos^2\theta} \, d\theta$

31. $\displaystyle\int_{-\pi/4}^{\pi/4} \frac{\sin^2x}{\cos^2x} \, dx$

32. $\displaystyle\int \frac{\cos 2t}{\sin^4 2t} \, dt$

33. $\displaystyle\int \frac{\sin^2t}{\cos t} \, dt$

34. $\displaystyle\int \sin 2x(1 - \cos 2x)^{3/2} \, dx$

35. Find the area between the x-axis and the curve $y = \sqrt{1 + \cos 4x}$, $0 \le x \le \pi$.

36. Find the volume generated by revolving the arch $y = \sin x$, $0 \le x \le \pi$, about the x-axis.

37. The region bounded by the curve $y = x \sin x$ and the interval $0 \le x \le \pi$ is revolved about the x-axis to generate a solid (Fig. 7.11). Find the volume of the solid.

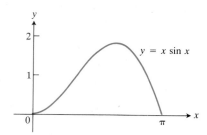

7.11   The curve in Problem 37.

38. The graph of $x = t - \sin t$, $y = 1 - \cos t$, $0 \le t \le 2\pi$, is an arch standing on the x-axis (Fig. 7.12). Find the surface area generated by revolving the arch about the x-axis.

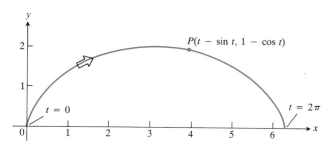

7.12   The curve in Problem 38, called a cycloid, is the kind of curve traced by a pebble in the tread of a rolling tire. Article 8.9 explores some of the interesting properties of this curve.

7.5
Trigonometric Substitutions that Replace $a^2 - u^2$, $a^2 + u^2$, and $u^2 - a^2$ by Single Squared Terms

We now embark on a three-step program whose completion will enable us to integrate all rational functions of x. The first step, taken in the present article, is to study substitutions that change binomials like

$$a^2 - u^2, \qquad a^2 + u^2, \qquad \text{and} \qquad u^2 - a^2$$

into single squared terms. The second step, taken in Article 7.6, will be to simplify integrals involving $ax^2 + bx + c$ by completing the square and then replacing the resulting sums and differences of squares by single squared terms. The third and last step, taken in Article 7.7, will be to express rational functions of x as sums of polynomials (which we already know how to integrate), fractions with linear-factored denominators (which become logarithms or fractions when integrated), and fractions with quadratic denominators (which we will be able to integrate by the techniques of the present article and the next).

Trigonometric Substitutions for Combining Squares

Trigonometric substitutions enable us to replace the binomials

$$a^2 - u^2, \qquad a^2 + u^2, \qquad \text{and} \qquad u^2 - a^2$$

by single squared terms, thereby transforming a number of important integrals into integrals we recognize or can find in a table. The substitutions most commonly used are $u = a \sin \theta$, $u = a \tan \theta$, and $u = a \sec \theta$. With $u = a \sin \theta$,

$$a^2 - u^2 = a^2 - a^2\sin^2\theta = a^2(1 - \sin^2\theta) = a^2\cos^2\theta. \tag{1}$$

With $u = a \tan \theta$,

$$a^2 + u^2 = a^2 + a^2\tan^2\theta = a^2(1 + \tan^2\theta) = a^2\sec^2\theta. \tag{2}$$

With $u = a \sec \theta$,

$$u^2 - a^2 = a^2\sec^2\theta - a^2 = a^2(\sec^2\theta - 1) = a^2\tan^2\theta. \tag{3}$$

In short,

$$
\begin{array}{llllr}
u = a \sin \theta & \text{replaces} & a^2 - u^2 & \text{by} & a^2\cos^2\theta, \quad (4) \\
u = a \tan \theta & \text{replaces} & a^2 + u^2 & \text{by} & a^2\sec^2\theta, \quad (5) \\
u = a \sec \theta & \text{replaces} & u^2 - a^2 & \text{by} & a^2\tan^2\theta. \quad (6)
\end{array}
$$

When we make substitutions, we always want them to be reversible so that we can change the integrals back to express the final results in terms of the original variables. For example, if $u = a \sin \theta$, we want to be able to set

$$\theta = \sin^{-1}\frac{u}{a}$$

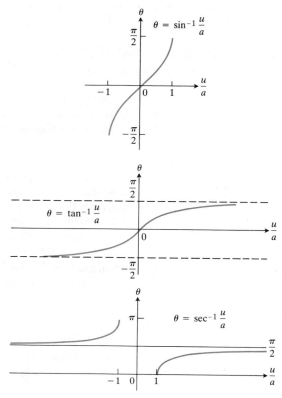

7.13   The arc sine, arc tangent, and arc secant of u/a, graphed as functions of u/a.

after the integration has taken place. If $u = a \tan \theta$, we want to be able to set

$$\theta = \tan^{-1} \frac{u}{a}$$

when we're done, and similarly for $u = a \sec \theta$.

As we know from Article 6.2, these substitutions have inverses only for selected values of θ. For reversibility,

$$u = a \sin \theta \text{ requires } \theta = \sin^{-1} \frac{u}{a} \text{ with } -\frac{\pi}{2} \le \theta \le \frac{\pi}{2},$$

$$u = a \tan \theta \text{ requires } \theta = \tan^{-1} \frac{u}{a} \text{ with } -\frac{\pi}{2} < \theta < \frac{\pi}{2},$$

$$u = a \sec \theta \text{ requires } \theta = \sec^{-1} \frac{u}{a} \text{ with } \begin{cases} 0 \le \theta < \dfrac{\pi}{2} & \text{if } \dfrac{u}{a} \ge 1 \\[2ex] \dfrac{\pi}{2} < \theta \le \pi & \text{if } \dfrac{u}{a} \le -1. \end{cases}$$

See Fig. 7.13.

EXAMPLE 1   Evaluate

$$\int \frac{du}{a^2 + u^2}.$$

Solution We set

$$u = a \tan \theta, \qquad du = a \sec^2\theta \; d\theta,$$

$$a^2 + u^2 = a^2 + a^2\tan^2\theta = a^2(1 + \tan^2\theta) = a^2\sec^2\theta$$

to obtain

$$\int \frac{du}{a^2 + u^2} = \int \frac{a \sec^2\theta \; d\theta}{a^2\sec^2\theta} = \frac{1}{a} \int d\theta$$

$$= \frac{1}{a} \cdot \theta + C = \frac{1}{a} \tan^{-1}\frac{u}{a} + C. \qquad \blacksquare$$

(7)

EXAMPLE 2 Evaluate

$$\int \frac{du}{\sqrt{a^2 - u^2}}, \qquad a > 0.$$

Solution To replace $a^2 - u^2$ by a single squared term (and thus eliminate the square root), we set

$$u = a \sin \theta, \qquad \theta = \sin^{-1}\frac{u}{a}, \qquad -\frac{\pi}{2} < \theta < \frac{\pi}{2}$$

$$du = a \cos \theta \; d\theta,$$

$$a^2 - u^2 = a^2(1 - \sin^2\theta) = a^2\cos^2\theta.$$

Then

$$\int \frac{du}{\sqrt{a^2 - u^2}} = \int \frac{a \cos \theta \; d\theta}{\sqrt{a^2\cos^2\theta}}$$

$$= \int \frac{a \cos \theta \; d\theta}{|a \cos \theta|}$$

$$= \int d\theta \qquad (a \cos \theta > 0 \text{ for } -\pi/2 < \theta < \pi/2) \qquad (8)$$

$$= \theta + C$$

$$= \sin^{-1}\frac{u}{a} + C. \qquad \blacksquare$$

EXAMPLE 3 Evaluate

$$\int \frac{du}{\sqrt{a^2 + u^2}}, \qquad a > 0.$$

Solution To replace $a^2 + u^2$ by a single squared term, we set

$$u = a \tan \theta, \qquad -\frac{\pi}{2} < \theta < \frac{\pi}{2}$$

$$du = a \sec^2\theta \; d\theta,$$

$$a^2 + u^2 = a^2(1 + \tan^2\theta) = a^2\sec^2\theta.$$

Then

$$\int \frac{du}{\sqrt{a^2 + u^2}} = \int \frac{a \sec^2\theta \, d\theta}{\sqrt{a^2\sec^2\theta}}$$

$$= \int \frac{\sec^2\theta \, d\theta}{|\sec \theta|} \qquad (a > 0)$$

$$= \int \sec \theta \, d\theta \qquad \left(\sec \theta > 0 \text{ when } -\frac{\pi}{2} < \theta < \frac{\pi}{2}\right) \tag{9}$$

$$= \ln|\sec \theta + \tan \theta| + C$$

$$= \ln\left|\frac{\sqrt{a^2 + u^2}}{a} + \frac{u}{a}\right| + C \qquad \text{(from Fig. 7.14)}$$

$$= \ln\left|\sqrt{a^2 + u^2} + u\right| + C',$$

7.14 Trigonometric functions of θ can be expressed in terms of u by reference to this triangle, in which $u = a \tan \theta$. For example, $\sec \theta = (\sqrt{a^2 + u^2})/a$.

where $C' = C - \ln a$. Notice how we expressed $\ln|\sec \theta + \tan \theta|$ in terms of u: We drew a reference triangle that described the original substitution (Fig. 7.14) and read the ratios from the triangle. ■

EXAMPLE 4 Evaluate

$$\int \frac{du}{\sqrt{u^2 - a^2}}, \qquad \left|\frac{u}{a}\right| > 1.$$

Solution To replace $u^2 - a^2$ by a single squared term, we set

$$u = a \sec \theta, \qquad \theta = \sec^{-1}\frac{u}{a}$$

$$du = a \sec \theta \tan \theta,$$

$$u^2 - a^2 = a^2(\sec^2\theta - 1) = a^2\tan^2\theta.$$

With this substitution, we have

$$0 < \theta < \frac{\pi}{2} \quad \text{for} \quad \frac{u}{a} > 1 \qquad \text{and} \qquad \frac{\pi}{2} < \theta < \pi \quad \text{for} \quad \frac{u}{a} < -1$$

(see Fig. 7.13). Then

$$\int \frac{du}{\sqrt{u^2 - a^2}} = \int \frac{a \sec \theta \tan \theta \, d\theta}{\sqrt{a^2\tan^2\theta}}$$

$$= \int \frac{\sec \theta \tan \theta \, d\theta}{|\tan \theta|} \qquad (a > 0)$$

$$= \pm\int \sec \theta \, d\theta$$

$$= \pm\ln|\sec \theta + \tan \theta| + C$$

$$= \pm\ln\left|\frac{u}{a} \pm \frac{\sqrt{u^2 - a^2}}{a}\right| + C. \qquad \text{(Fig. 7.15)}$$

(a)

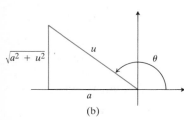

(b)

7.15 (a) Reference triangle for

$u = a \sec \theta, \ 0 < \theta < \pi/2;$

(b) reference triangle for

$u = a \sec \theta, \ \pi/2 < \theta < \pi.$

What do we do about the signs? When $0 < \theta < \pi/2$, the tangent and secant are positive and both signs are $+$. When $\pi/2 < \theta < \pi$, the secant and tangent are both negative. Therefore,

$$\int \frac{du}{\sqrt{u^2 - a^2}} = \begin{cases} \ln\left|\frac{u}{a} + \frac{\sqrt{u^2 - a^2}}{a}\right| + C \\ \text{or} \\ -\ln\left|\frac{u}{a} - \frac{\sqrt{u^2 - a^2}}{a}\right| + C. \end{cases} \qquad (10)$$

Fortunately, we do not have to settle for this two-line formula because the two forms on the right are actually equal:

$$-\ln\left|\frac{u}{a} - \frac{\sqrt{u^2 - a^2}}{a}\right| = \ln\left|\frac{a}{u - \sqrt{u^2 - a^2}}\right|$$

$$= \ln\left|\frac{a(u + \sqrt{u^2 - a^2})}{(u - \sqrt{u^2 - a^2})(u + \sqrt{u^2 - a^2})}\right|$$

$$= \ln\left|\frac{a(u + \sqrt{u^2 - a^2})}{a^2}\right| = \ln\left|\frac{u + \sqrt{u^2 - a^2}}{a}\right|$$

$$= \ln|u + \sqrt{u^2 - a^2}| - \ln a.$$

Therefore,

$$\int \frac{du}{\sqrt{u^2 - a^2}} = \ln|u + \sqrt{u^2 - a^2}| + C'. \qquad ■ \quad (11)$$

EXAMPLE 5 Evaluate

$$\int \frac{x^2 \, dx}{\sqrt{9 - x^2}}.$$

Solution To replace $9 - x^2$ by a single squared term, we set

$$x = 3 \sin \theta, \qquad -\frac{\pi}{2} < \theta < \frac{\pi}{2}$$

$$dx = 3 \cos \theta \, d\theta$$

$$9 - x^2 = 9(1 - \sin^2\theta) = 9 \cos^2\theta.$$

Then

$$\int \frac{x^2 \, dx}{\sqrt{9 - x^2}} = \int \frac{9 \sin^2\theta \cdot 3 \cos \theta \, d\theta}{|3 \cos \theta|}$$

$$= 9 \int \sin^2\theta \, d\theta \qquad \left(\cos \theta > 0 \text{ for } -\frac{\pi}{2} < \theta < \frac{\pi}{2}\right)$$

$$= 9 \int \frac{1 - \cos 2\theta}{2} \, d\theta = \frac{9}{2}\left(\theta - \frac{\sin 2\theta}{2}\right) + C$$

$$= \frac{9}{2}(\theta - \sin \theta \cos \theta) + C$$

$$= \frac{9}{2}\left(\sin^{-1}\frac{x}{3} - \frac{x}{3} \cdot \frac{\sqrt{9 - x^2}}{3}\right) + C \qquad \text{(Fig. 7.16)}$$

$$= \frac{9}{2} \sin^{-1}\frac{x}{3} - \frac{x}{2}\sqrt{9 - x^2} + C. \qquad ■$$

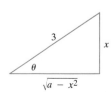

7.16 Reference triangle for $x = 3 \sin \theta$, $0 < \theta < \pi/2$.

PROBLEMS

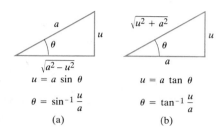

$u = a \sin \theta$

$\theta = \sin^{-1} \dfrac{u}{a}$

(a)

$u = a \tan \theta$

$\theta = \tan^{-1} \dfrac{u}{a}$

(b)

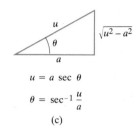

$u = a \sec \theta$

$\theta = \sec^{-1} \dfrac{u}{a}$

(c)

Reference triangles.

In Problems 1–15, evaluate the integrals.

1. $\displaystyle\int_{-2}^{2} \frac{dx}{4 + x^2}$

2. $\displaystyle\int_{0}^{2} \frac{dx}{8 + 2x^2}$

3. $\displaystyle\int \frac{dx}{1 + 4x^2}$

4. $\displaystyle\int_{0}^{3/2} \frac{dx}{\sqrt{9 - x^2}}$

5. $\displaystyle\int_{0}^{1/2\sqrt{2}} \frac{2\, dx}{\sqrt{1 - 4x^2}}$

6. $\displaystyle\int_{0}^{3\sqrt{2}/4} \frac{dx}{\sqrt{9 - 4x^2}}$

7. $\displaystyle\int \frac{dy}{\sqrt{25 + y^2}}$

8. $\displaystyle\int \frac{3\, dy}{\sqrt{1 + 9y^2}}$

9. $\displaystyle\int \frac{dy}{\sqrt{25 + 9y^2}}$

10. $\displaystyle\int \frac{dz}{\sqrt{z^2 - 4}}$

11. $\displaystyle\int \frac{3\, dz}{\sqrt{9z^2 - 1}}$

12. $\displaystyle\int \frac{dz}{\sqrt{25z^2 - 9}}$

13. $\displaystyle\int_{8/\sqrt{3}}^{8} \frac{dx}{x\sqrt{x^2 - 16}}$

14. $\displaystyle\int_{2}^{\sqrt{6}} \frac{dx}{x\sqrt{x^2 - 3}}$

15. $\displaystyle\int_{\sqrt{2}/2}^{1} \frac{dx}{2x\sqrt{4x^2 - 1}}$

In Problems 16–22, evaluate the integrals with the substitutions given.

16. $\displaystyle\int \frac{dy}{y^2\sqrt{y^2 - 16}}, \quad y = 4 \sec u$

17. $\displaystyle\int_{2}^{4} \sqrt{x^2 - 4}\, dx, \quad x = 2 \sec u$

18. $\displaystyle\int_{0}^{4/5} \frac{x^3\, dx}{\sqrt{1 - x^2}}, \quad x = \cos u$

19. $\displaystyle\int \frac{dx}{x^2\sqrt{9 - x^2}}, \quad x = 3 \sin u$

20. $\displaystyle\int_{0}^{1/2} \frac{dx}{\sqrt{1 + x^2}}, \quad x = \tan u$

21. $\displaystyle\int_{5/4}^{5/3} \frac{dx}{x^2\sqrt{x^2 - 1}}, \quad x = \csc u$

22. $\displaystyle\int_{1/2}^{1} \frac{\sqrt{1 - x}}{x}\, dx, \quad x = \cos^2 u$

Evaluate the integrals in Problems 23–38.

23. $\displaystyle\int_{0}^{5} \sqrt{25 - x^2}\, dx$

24. $\displaystyle\int \frac{dx}{\sqrt{1 - 4x^2}}$

25. $\displaystyle\int_{1}^{2} \frac{dx}{\sqrt{4 - (x - 1)^2}}$

26. $\displaystyle\int_{0}^{2} \frac{dx}{\sqrt{4 + x^2}}$

27. $\displaystyle\int_{0}^{1} \frac{12\, dx}{\sqrt{4 - x^2}}$

28. $\displaystyle\int_{0}^{3/2} \frac{x\, dx}{\sqrt{4 + x^2}}$

29. $\displaystyle\int_{0}^{1} \frac{x^3\, dx}{\sqrt{x^2 + 1}}$

30. $\displaystyle\int \frac{x + 1}{\sqrt{4 - x^2}}\, dx$

31. $\displaystyle\int \frac{dx}{x\sqrt{x^2 - 1/4}}$

32. $\displaystyle\int \frac{dx}{\sqrt{2 - 5x^2}}$

33. $\displaystyle\int \frac{\sqrt{1 - x^2}}{x^2}\, dx$

34. $\displaystyle\int_{0}^{(1/2)\ln 3} \frac{e^x\, dx}{1 + e^{2x}}$

35. $\displaystyle\int_{3/4}^{4/5} \frac{dx}{x^2\sqrt{1 - x^2}}$

36. $\displaystyle\int \frac{4x^2\, dx}{(1 - x^2)^{3/2}}$

37. $\displaystyle\int \frac{dx}{(a^2 - x^2)^{3/2}}$

38. $\displaystyle\int \frac{\sin \theta\, d\theta}{\sqrt{2 - \cos^2\theta}}$

39. Evaluate

$$\int \frac{y\, dy}{\sqrt{16 - y^2}}$$

a) without a trigonometric substitution,

b) with a trigonometric substitution.

40. Evaluate

a) $\displaystyle\int_0^2 \frac{x \, dx}{4 + x^2}$,

b) $\displaystyle\int_0^2 \frac{dx}{4 + x^2}$.

41. Evaluate

$$\int \frac{x \, dx}{(x^2 - 1)^{3/2}}$$

a) without a trigonometric substitution,
b) with a trigonometric substitution.

42. Evaluate

a) $\displaystyle\int_0^{\sqrt{3}/2} \frac{x \, dx}{\sqrt{1 - x^2}}$,

b) $\displaystyle\int_0^{\sqrt{3}/2} \frac{dx}{\sqrt{1 - x^2}}$.

43. Use the substitution $u = az$ to evaluate the following integrals.

a) $\displaystyle\int \frac{du}{u^2 + a^2}$

b) $\displaystyle\int \frac{du}{\sqrt{a^2 - u^2}}$

44. Suppose $\theta = \sin^{-1}(u/2)$. Express $\cos \theta$ and $\tan \theta$ in terms of u.

45. Express in terms of u and a:

a) $\sin\left(\tan^{-1}\dfrac{u}{a}\right)$

b) $\cos\left(\sec^{-1}\dfrac{u}{a}\right)$

46. Find the area of the region bounded by the curve $y = \sqrt{1 - (x^2/9)}$ in the first quadrant.

47. Find the area of the surface generated by revolving the curve

$$x = 2 \cos t, \quad y = \sin t, \quad 0 \le t \le \pi,$$

about the x-axis. See Fig. 7.17.

48. Find the length of the segment of the parabola $y = x^2$ that extends from $x = 0$ to $x = \sqrt{3}/2$.

7.17 The curve $x = 2 \cos t$, $y = \sin t$, $0 \le t \le \pi$ in Problem 47 is the top half of an ellipse. The arrow shows the direction of increasing t.

49. Find the solution of the differential equation

$$\frac{dy}{dx} = y^2\left(1 - \frac{1}{\sqrt{4 - x^2}}\right)$$

that passes through the point $(\sqrt{2}, 4/\pi)$.

50. Find the area of the surface generated by revolving the curve $y = \sqrt{x^2 + 2}$, $-1 \le x \le 1$ about the x-axis. See Fig. 7.18.

51. Find the moment about the x-axis of a thin homogeneous wire of density δ that lies along the curve $y = e^x$, $0 \le x \le \ln \sqrt{3}$.

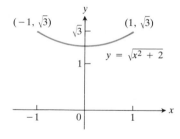

7.18 The curve in Problem 50.

7.6
Integrals Involving $ax^2 + bx + c$

By using the algebraic process called completing the square, we can convert any quadratic

$$ax^2 + bx + c, \qquad a \ne 0$$

to the form

$$a(u^2 \pm A^2).$$

We can then use one of the trigonometric substitutions from the previous article to write the expression as a times a single squared term. As the examples below show, this combination of completing the square and trigonometric substitution enables us to evaluate a number of new integrals.

EXAMPLE 1 *Completing the square.*

$$4x^2 + 4x + 2 = 4(x^2 + x) + 2$$

$$= 4\left(x^2 + x + \frac{1}{4} - \frac{1}{4}\right) + 2$$

$$= 4\left(x^2 + x + \frac{1}{4}\right) + 2 - 1$$

$$= 4\left(x + \frac{1}{2}\right)^2 + 1$$

$$= 4\left(\left(x + \frac{1}{2}\right)^2 + \frac{1}{4}\right) = 4(u^2 + A^2).$$ ■

EXAMPLE 2 *Completing the square.*

$$2x^2 - 6x + 4 = 2(x^2 - 3x) + 4$$

$$= 2\left(x^2 - 3x + \frac{9}{4} - \frac{9}{4}\right) + 4$$

$$= 2\left(x^2 - 3x + \frac{9}{4}\right) + 4 - \frac{9}{2}$$

$$= 2\left(x - \frac{3}{2}\right)^2 - \frac{1}{2}$$

$$= 2\left(\left(x - \frac{3}{2}\right)^2 - \left(\frac{1}{2}\right)^2\right) = 2(u^2 - A^2).$$ ■

The general pattern is this:

$$ax^2 + bx + c = a\left(x^2 + \frac{b}{a}x\right) + c = a\left(x^2 + \frac{b}{a}x + \frac{b^2}{4a^2}\right) + c - \frac{b^2}{4a}$$

$$= a\left(x + \frac{b}{2a}\right)^2 + \frac{4ac - b^2}{4a}$$

$$= a\left(\left(x + \frac{b}{2a}\right)^2 + \frac{4ac - b^2}{4a^2}\right)$$

$$= a(u^2 \pm A^2), \tag{1}$$

where

$$x + \frac{b}{2a} = u \qquad \text{and} \qquad \frac{4ac - b^2}{4a^2} = \pm A^2.$$

EXAMPLE 3 *Completing the square.*

$$2x - x^2 = -(x^2 - 2x) = -(x^2 - 2x + 1) + 1$$

$$= -(x - 1)^2 + 1$$

$$= -((x - 1)^2 - 1) \qquad \text{(the form in Eq. 1)}$$

$$= -(u^2 - 1)$$

$$= 1 - u^2. \qquad \text{(a more convenient form)}$$ ■

The next three examples show how the computations in Examples 1–3 would take place in the context of integration.

EXAMPLE 4 Evaluate

$$\int \frac{dx}{\sqrt{2x - x^2}}.$$

Solution From Example 3 we have

$$2x - x^2 = 1 - (x - 1)^2 = 1 - u^2$$

and the integral becomes

$$\int \frac{dx}{\sqrt{2x - x^2}} = \int \frac{du}{\sqrt{1 - u^2}}$$

$$= \sin^{-1} u + C$$

$$= \sin^{-1}(x - 1) + C.$$

EXAMPLE 5 Evaluate

$$\int \frac{dx}{4x^2 + 4x + 2}.$$

Solution With

$$4x^2 + 4x + 2 = 4(x^2 + x) + 2$$

$$= 4\left(x^2 + x + \frac{1}{4}\right) + 1$$

$$= 4\left(\left(x + \frac{1}{2}\right)^2 + \frac{1}{4}\right)$$

$$= 4\left(u^2 + \left(\frac{1}{2}\right)^2\right),$$

the integral becomes

$$\int \frac{dx}{4x^2 + 4x + 2} = \frac{1}{4}\int \frac{du}{u^2 + (1/2)^2}$$

$$= \frac{1}{4}\int \frac{du}{u^2 + a^2} \qquad \left(a = \frac{1}{2}\right)$$

$$= \frac{1}{4} \cdot \frac{1}{a} \tan^{-1}\frac{u}{a} + C \qquad \text{(Article 7.5, Example 1)}$$

$$= \frac{1}{2} \tan^{-1}(2x + 1) + C.$$

EXAMPLE 6 Evaluate

$$\int \frac{(x + 1)\, dx}{\sqrt{2x^2 - 6x + 4}}.$$

Solution   We complete the square to get

$$2x^2 - 6x + 4 = 2(x^2 - 3x) + 4$$

$$= 2\left(x^2 - 3x + \frac{9}{4}\right) + 4 - \frac{9}{2}$$

$$= 2\left(x - \frac{3}{2}\right)^2 - \frac{1}{2}$$

$$= 2\left(\left(x - \frac{3}{2}\right)^2 - \left(\frac{1}{2}\right)^2\right).$$

With

$$u = x - \frac{3}{2}, \qquad x + 1 = u + \frac{5}{2}, \qquad du = dx, \qquad \text{and} \qquad a = \frac{1}{2},$$

the integral becomes

$$\int \frac{(x + 1)\, dx}{\sqrt{2x^2 - 6x + 4}} = \int \frac{(u + 5/2)\, du}{\sqrt{2(u^2 - a^2)}}$$

$$= \frac{1}{\sqrt{2}} \int \frac{u\, du}{\sqrt{u^2 - a^2}} + \frac{5}{2\sqrt{2}} \int \frac{du}{\sqrt{u^2 - a^2}}.$$

In the first integral we let

$$z = u^2 - a^2, \qquad dz = 2u\, du, \qquad u\, du = \frac{1}{2}\, dz,$$

and to the second we apply Eq. (11) from Article 7.5. This gives

$$\int \frac{u\, du}{\sqrt{u^2 - a^2}} = \int \frac{dz}{2\sqrt{z}}$$

$$= \sqrt{z} + C_1$$

$$= \sqrt{u^2 - a^2} + C_1$$

and

$$\int \frac{du}{\sqrt{u^2 - a^2}} = \ln\left|u + \sqrt{u^2 - a^2}\right| + C_2.$$

Hence,

$$\int \frac{(x + 1)\, dx}{\sqrt{2x^2 - 6x + 4}}$$

$$= \sqrt{\frac{u^2 - a^2}{2}} + \frac{5}{2\sqrt{2}} \ln\left|u + \sqrt{u^2 - a^2}\right| + C$$

$$= \sqrt{\frac{x^2 - 3x + 2}{2}} + \frac{5}{2\sqrt{2}} \ln\left|x - \frac{3}{2} + \sqrt{x^2 - 3x + 2}\right| + C. \quad \blacksquare$$

PROBLEMS

In Problems 1–24, evaluate the integrals.

1. $\displaystyle\int_1^3 \frac{dx}{x^2 - 2x + 5}$

2. $\displaystyle\int \frac{dx}{\sqrt{x^2 - 2x + 5}}$

3. $\displaystyle\int_1^3 \frac{x\,dx}{x^2 - 2x + 5}$

4. $\displaystyle\int_1^{5/2} \frac{x\,dx}{\sqrt{x^2 - 2x + 5}}$

5. $\displaystyle\int_1^2 \frac{dx}{x^2 - 2x + 4}$

6. $\displaystyle\int \frac{3\,dx}{9x^2 - 6x + 5}$

7. $\displaystyle\int \frac{dx}{\sqrt{9x^2 - 6x + 5}}$

8. $\displaystyle\int \frac{3x\,dx}{9x^2 - 6x + 5}$

9. $\displaystyle\int \frac{x\,dx}{\sqrt{9x^2 - 6x + 5}}$

10. $\displaystyle\int \frac{dx}{\sqrt{x^2 - 2x}}$

11. $\displaystyle\int \frac{dx}{\sqrt{x^2 + 2x}}$

12. $\displaystyle\int_{-2}^1 \frac{dx}{\sqrt{x^2 + 4x + 13}}$

13. $\displaystyle\int_{-2}^2 \frac{x\,dx}{\sqrt{x^2 + 4x + 13}}$

14. $\displaystyle\int_{-1}^0 \frac{dx}{\sqrt{3 - 2x - x^2}}$

15. $\displaystyle\int \frac{dx}{\sqrt{x^2 - 2x - 3}}$

16. $\displaystyle\int \frac{dx}{(x + 1)\sqrt{x^2 + 2x}}$

17. $\displaystyle\int \frac{(x + 1)\,dx}{\sqrt{2x - x^2}}$

18. $\displaystyle\int \frac{(x - 1)\,dx}{\sqrt{x^2 - 4x + 3}}$

19. $\displaystyle\int \frac{x\,dx}{\sqrt{5 + 4x - x^2}}$

20. $\displaystyle\int_5^6 \frac{dx}{\sqrt{x^2 - 2x - 8}}$

21. $\displaystyle\int_0^1 \frac{(1 - x)\,dx}{\sqrt{8 + 2x - x^2}}$

22. $\displaystyle\int \frac{x\,dx}{\sqrt{x^2 + 4x + 5}}$

23. $\displaystyle\int_{-2}^{-1} \frac{x\,dx}{x^2 + 4x + 5}$

24. $\displaystyle\int \frac{(2x + 3)\,dx}{4x^2 + 4x + 5}$

25. Find the volume of the solid generated by revolving the region bounded by the x-axis, the curve $y =$

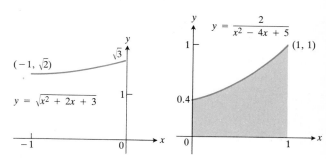

7.19   The solid swept out by the region in Problem 25.

$20/\sqrt{x^2 - 2x + 17}$, and the lines $x = -2$ and $x = 11$ about the x-axis. See Fig. 7.19.

26. Find the average value of the function
$$f(x) = \frac{4}{(x^2 - 4x + 8)}$$
over the interval from $x = 2$ to $x = 4$.

27. Find the area of the surface generated by revolving the arc $y = \sqrt{x^2 + 2x + 3},\ -1 \le x \le 0$ about the x-axis. See Fig. 7.20.

28. a) Find the area of the region bounded by the curve $y = 2/(x^2 - 4x + 5)$, the coordinate axes, and the line $x = 1$ (Fig. 7.21).
b) Find the moment about the y-axis of a thin homogeneous plate of density δ that occupies the region in (a).

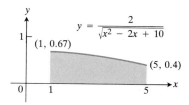

7.20   The arc in Problem 27. 7.21   The region in Problem 28.

29. Find the center of mass of a thin homogeneous plate bounded by the x-axis, the curve $y = 2/\sqrt{x^2 - 2x + 10}$, and the lines $x = 1$ and $x = 5$ (Fig. 7.22).

7.22   The plate in Problem 29.

Find the limits in Problems 30–31.

30. $\displaystyle\lim_{a \to 1^+} \int_a^3 \frac{dx}{(x + 1)\sqrt{x^2 + 2x - 3}}$

31. $\displaystyle\lim_{a \to -5^+} \int_a^{-4} \frac{dx}{\sqrt{-x^2 - 8x - 15}}$

TOOLKIT PROGRAM

Super * Grapher

7.7

The Integration of Rational Functions—Partial Fractions

We now come to the method of partial fractions, the basic technique for preparing rational functions for integration. This technique enables us to write an arbitrary quotient of polynomials (for example, a rational function we want to integrate) as a sum of simpler quotients that can be integrated by techniques already at our command. For instance, the equation

$$\frac{4x^5 - 15x^3 + 4x^2 + 9x + 8}{4x^4 - 4x^3 - 3x^2 + 2x + 1} = \underbrace{x + 1}_{(1)} - \underbrace{\frac{2}{x - 1} + \frac{1}{(x - 1)^2}}_{(2)} + \underbrace{\frac{1}{4x^2 + 4x + 2}}_{(3)}$$

expresses the rational function on the left as the sum of (1) a polynomial, (2) fractions with linear-factored denominators that integrate to logarithms or fractions, and (3) a fraction with a quadratic denominator that can be integrated with the substitutions of Articles 7.5 and 7.6.

Do not be alarmed at the size of this example—you will not be asked to work with anything so complex in the problems at the end of this article. We include it here only to show how the techniques you have been learning can now be brought to bear, in theory at least, to integrate any rational function of x. We shall now look at some simpler examples, present the method of partial fractions in general terms, and then describe a shortcut called the **Heaviside "cover-up method"** (named after Oliver Heaviside, 1850–1925, a pioneer in electrical engineering and vector analysis).

EXAMPLE 1 *An example of the kind of addition we want to reverse.*

$$\frac{2}{x + 1} + \frac{3}{x - 3} = \frac{2(x - 3) + 3(x + 1)}{(x + 1)(x - 3)} = \frac{5x - 3}{(x + 1)(x - 3)} \tag{1}$$

Adding the fractions on the left produces the fraction on the right. The reverse process consists in finding constants A and B such that

$$\frac{5x - 3}{(x + 1)(x - 3)} = \frac{A}{x + 1} + \frac{B}{x - 3}.$$

(Pretend, for a moment, we don't know that $A = 2$, $B = 3$ will work.) We call A and B **undetermined coefficients.** Clearing of fractions, we have

$$5x - 3 = A(x - 3) + B(x + 1)$$

$$= (A + B)x - 3A + B.$$

This will be an identity in x if and only if coefficients of like powers of x on the two sides of the equation are equal:

$$A + B = 5, \qquad -3A + B = -3.$$

These two equations in two unknowns determine

$$A = 2, \qquad B = 3.$$

EXAMPLE 2 *Two linear factors in the denominator.* Evaluate

$$\int \frac{5x - 3}{(x + 1)(x - 3)}\ dx.$$

Solution From Example 1,

$$\int \frac{5x - 3}{(x + 1)(x - 3)}\ dx = \int \frac{2}{x + 1}\ dx + \int \frac{3}{x - 3}\ dx$$

$$= 2 \ln |x + 1| + 3 \ln |x - 3| + C. \qquad \blacksquare$$

EXAMPLE 3 *A repeated linear factor in the denominator.* Express

$$\frac{6x + 7}{(x + 2)^2}$$

as a sum of partial fractions.

Solution Since the denominator has a repeated linear factor, $(x + 2)^2$, we must express the fraction in the form

$$\frac{6x + 7}{(x + 2)^2} = \frac{A}{x + 2} + \frac{B}{(x + 2)^2}. \qquad (2)$$

Clearing Eq. (2) of fractions gives

$$6x + 7 = A(x + 2) + B = Ax + (2A + B).$$

Matching coefficients of like terms gives $A = 6$, $B = -5$. Hence,

$$\frac{6x + 7}{(x + 2)^2} = \frac{6}{x + 2} - \frac{5}{(x + 2)^2}. \qquad \blacksquare \ (3)$$

EXAMPLE 4 *A quadratic factor in the denominator.* Express

$$\frac{-2x + 4}{(x^2 + 1)(x - 1)^2}$$

as a sum of partial fractions.

Solution Since the denominator has a quadratic factor as well as a repeated linear factor, we write

$$\frac{-2x + 4}{(x^2 + 1)(x - 1)^2} = \frac{Ax + B}{x^2 + 1} + \frac{C}{x - 1} + \frac{D}{(x - 1)^2}. \qquad (4)$$

Note that the numerator over $x^2 + 1$ is linear and not constant. Then

$$-2x + 4 = (Ax + B)(x - 1)^2 + C(x - 1)(x^2 + 1) + D(x^2 + 1)$$

$$= (A + C)x^3 + (-2A + B - C + D)x^2$$

$$+ (A - 2B + C)x + (B - C + D).$$

For this to be an identity in x, it is both necessary and sufficient that the coefficient of each power of x be the same on the left side of the equation as it is on the right side. Equating these coefficients leads to the following equations:

Coefficient of x^3:	$0 = A + C,$
Coefficient of x^2:	$0 = -2A + B - C + D$
Coefficient of x^1:	$-2 = A - 2B + C,$
Coefficient of x^0:	$4 = B - C + D.$

If we subtract the fourth equation from the second, we obtain

$$-4 = -2A, \qquad A = 2.$$

Then from the first equation, we have

$$C = -A = -2.$$

Knowing A and C, we find B from the third equation,

$$B = 1.$$

Finally, from the fourth equation, we have

$$D = 4 - B + C = 1.$$

Hence

$$\frac{-2x + 4}{(x^2 + 1)(x - 1)^2} = \frac{2x + 1}{x^2 + 1} - \frac{2}{x - 1} + \frac{1}{(x - 1)^2}.$$ ∎

EXAMPLE 5   Evaluate

$$\int \frac{-2x + 4}{(x^2 + 1)(x - 1)^2}\, dx.$$

Solution   We expand the integrand by partial fractions, as in Example 4, and integrate the terms of the expansion:

$$\int \frac{-2x + 4}{(x^2 + 1)(x - 1)^2}\, dx$$

$$= \int \left(\frac{2x + 1}{x^2 + 1} - \frac{2}{x - 1} + \frac{1}{(x - 1)^2} \right) dx$$

$$= \int \left(\frac{2x}{x^2 + 1} + \frac{1}{x^2 + 1} - \frac{2}{x - 1} + \frac{1}{(x - 1)^2} \right) dx$$

$$= \ln(x^2 + 1) + \tan^{-1}x - 2\ln|x - 1| - \frac{1}{x - 1} + C.$$ ∎

We now integrate the rational function with which we began this article, just to show you how it works out. Again, do not be alarmed by the complexity of the integrand. There will be nothing like it in the problems that follow.

EXAMPLE 6   Evaluate

$$\int \frac{4x^5 - 15x^3 + 4x^2 + 9x + 8}{4x^4 - 4x^3 - 3x^2 + 2x + 1}\, dx.$$

Solution   Although it takes considerable effort to obtain the partial fraction expansion here, the steps we take are the same as in every other case. The degree of the numerator is greater than the denominator. Hence we divide first, obtaining

$$\frac{4x^5 - 15x^3 + 4x^2 + 9x + 8}{4x^4 - 4x^3 - 3x^2 + 2x + 1} = x + 1 - \frac{8x^3 - 5x^2 - 6x - 7}{4x^4 - 4x^3 - 3x^2 + 2x + 1}. \quad (5)$$

We then factor the denominator on the right,

$$4x^4 - 4x^3 - 3x^2 + 2x + 1 = (x - 1)^2(4x^2 + 4x + 2),$$

and, after some more work, find that

$$\frac{8x^3 - 5x^2 - 6x - 7}{4x^4 - 4x^3 - 3x^2 + 2x + 1} = \frac{2}{x - 1} - \frac{1}{(x - 1)^2} - \frac{1}{4x^2 + 4x + 2}. \quad (6)$$

Finally, we substitute from Eq. (6) into Eq. (5), multiply by dx, and integrate. The end result is

$$\int \frac{4x^5 - 15x^3 + 4x^2 + 9x + 8}{4x^4 - 4x^3 - 3x^2 + 2x + 1}\, dx$$

$$= \int \left(x + 1 - \frac{2}{x - 1} + \frac{1}{(x - 1)^2} + \underbrace{\frac{1}{4x^2 + 4x + 2}} \right) dx$$

$$\left(\begin{array}{l} \text{Article 7.6,} \\ \text{Example 5} \end{array} \right)$$

$$= \frac{x^2}{2} + x - 2\,\ln|x - 1| - \frac{1}{x - 1} + \frac{1}{2}\,\tan^{-1}(2x + 1) + C. \quad \blacksquare$$

General Description of the Method

Success in separating

$$\frac{f(x)}{g(x)}$$

into a sum of partial fractions hinges on two things:

1. *The degree of $f(x)$ must be less than the degree of $g(x)$.* (If this is not the case, we first perform a long division, then work with the remainder term. The remainder can always be put into the required form.)

2. *The factors of $g(x)$ must be known.* (Theoretically, any polynomial with real coefficients can be expressed as a product of real linear and quadratic factors. In practice, it may be hard to factor.)

If these two conditions are met we can carry out the following steps.

STEP 1: Let $x - r$ be a linear factor of $g(x)$. Suppose $(x - r)^m$ is the highest power of $x - r$ that divides $g(x)$. Then assign the sum of m partial fractions to this factor, as follows:

$$\frac{A_1}{x - r} + \frac{A_2}{(x - r)^2} + \cdots + \frac{A_m}{(x - r)^m}.$$

Do this for each distinct linear factor of $g(x)$.

STEP 2: Let $x^2 + px + q$ be an irreducible quadratic factor of $g(x)$. Suppose

$$(x^2 + px + q)^n$$

is the highest power of this factor that divides $g(x)$. Then, to this factor, assign the sum of the n partial fractions:

$$\frac{B_1 x + C_1}{x^2 + px + q} + \frac{B_2 x + C_2}{(x^2 + px + q)^2} + \cdots + \frac{B_n x + C_n}{(x^2 + px + q)^n}.$$

Do this for each distinct quadratic factor of $g(x)$.

STEP 3: Set the original fraction $f(x)/g(x)$ equal to the sum of all these partial fractions. Clear the resulting equation of fractions and arrange the terms in decreasing powers of x.

STEP 4: Equate the coefficients of corresponding powers of x and solve the resulting equations for the undetermined coefficients.

Proofs that $f(x)/g(x)$ can be written as a sum of partial fractions as described here are given in advanced algebra texts.

☐ The Heaviside "Cover-up" Method

When the degree of the polynomial $f(x)$ is less than the degree of $g(x)$, and

$$g(x) = (x - r_1)(x - r_2) \cdots (x - r_n)$$

is a product of n distinct linear factors, each to the first power, there is a quick way to expand $f(x)/g(x)$ by partial fractions.

EXAMPLE 7 Find A, B, and C in the partial-fraction expansion

$$\frac{x^2 + 1}{(x - 1)(x - 2)(x - 3)} = \frac{A}{x - 1} + \frac{B}{x - 2} + \frac{C}{x - 3}. \tag{7}$$

Solution If we multiply both sides of Eq. (7) by $(x - 1)$ to get

$$\frac{x^2 + 1}{(x - 2)(x - 3)} = A + \frac{B(x - 1)}{x - 2} + \frac{C(x - 1)}{x - 3} \tag{8}$$

and set $x = 1$, the resulting equation gives the value of A:

$$\frac{(1)^2 + 1}{(1 - 2)(1 - 3)} = A + 0 + 0,$$
$$A = 1.$$

Thus, the value of A is the number we would have obtained if we had covered the factor $(x - 1)$ in the denominator of the original fraction

$$\frac{x^2 + 1}{(x - 1)(x - 2)(x - 3)} \tag{9}$$

and evaluated the rest at $x = 1$:

$$A = \frac{(1)^2 + 1}{\boxed{(x - 1)} \; (1 - 2)(1 - 3)} = \frac{2}{(-1)(-2)} = 1. \tag{10}$$

⇧
Cover

Similarly, we find the value of B in Eq. (7) by covering the factor $(x - 2)$ in (9) and evaluating the rest at $x = 2$:

$$B = \frac{(2)^2 + 1}{(2 - 1)\ \boxed{(x - 2)}\ (2 - 3)} = \frac{5}{(1)(-1)} = -5. \tag{11}$$

$$\Uparrow$$

Cover

Finally, C is found by covering the $(x - 3)$ in (9) and evaluating the rest at $x = 3$:

$$C = \frac{(3)^2 + 1}{(3 - 1)(3 - 2)\ \boxed{(x - 3)}} = \frac{10}{(2)(1)} = 5. \qquad \blacksquare \tag{12}$$

$$\Uparrow$$

Cover

The steps in the cover-up method are these:

STEP 1: Write the quotient with $g(x)$ factored:

$$\frac{f(x)}{g(x)} = \frac{f(x)}{(x - r_1)(x - r_2) \cdots (x - r_n)}. \tag{13}$$

STEP 2: Cover the factors $(x - r_i)$ of $g(x)$ in (13) one at a time, each time replacing all the uncovered x's by the number r_i. This gives a number A_i for each root r_i:

$$A_1 = \frac{f(r_1)}{(r_1 - r_2) \cdots (r_1 - r_n)},$$

$$A_2 = \frac{f(r_2)}{(r_2 - r_1)(r_2 - r_3) \cdots (r_2 - r_n)}, \tag{14}$$

$$\vdots$$

$$A_n = \frac{f(r_n)}{(r_n - r_1)(r_n - r_2) \cdots (r_n - r_{n-1})}.$$

STEP 3: Write the partial-fraction expansion of $f(x)/g(x)$ as

$$\frac{f(x)}{g(x)} = \frac{A_1}{(x - r_1)} + \frac{A_2}{(x - r_2)} + \cdots + \frac{A_n}{(x - r_n)}. \tag{15}$$

EXAMPLE 8 Evaluate

$$\int \frac{x + 4}{x^3 + 3x^2 - 10x}\, dx.$$

Solution The degree of $f(x) = x + 4$ is less than the degree of $g(x) = x^3 + 3x^2 - 10x$, and, with $g(x)$ factored,

$$\frac{x + 4}{x^3 + 3x^2 - 10x} = \frac{x + 4}{x(x - 2)(x + 5)}.$$

The roots of $g(x)$ are $r_1 = 0$, $r_2 = 2$, and $r_3 = -5$. We find

$$A_1 = \frac{0 + 4}{\boxed{x}\, (0 - 2)(0 + 5)} = \frac{4}{(-2)(5)} = -\frac{2}{5},$$

⇧ Cover

$$A_2 = \frac{2 + 4}{2\, \boxed{(x - 2)}\, (2 + 5)} = \frac{6}{(2)(7)} = \frac{3}{7},$$

⇧ Cover

$$A_3 = \frac{-5 + 4}{(-5)(-5 - 2)\, \boxed{(x + 5)}} = \frac{-1}{(-5)(-7)} = -\frac{1}{35}.$$

⇧ Cover

Therefore,

$$\frac{x + 4}{x(x - 2)(x + 5)} = -\frac{2}{5x} + \frac{3}{7(x - 2)} - \frac{1}{35(x + 5)},$$

and

$$\int \frac{x + 4}{x(x - 2)(x + 5)}\, dx = -\frac{2}{5} \ln|x| + \frac{3}{7} \ln|x - 2| - \frac{1}{35} \ln|x + 5| + C. \qquad \blacksquare$$

Other Ways to Determine the Constants

Another way to determine the constants that appear in partial fractions is to differentiate, as in the next example.

EXAMPLE 9 Find A, B, and C in the equation

$$\frac{x - 1}{(x + 1)^3} = \frac{A}{x + 1} + \frac{B}{(x + 1)^2} + \frac{C}{(x + 1)^3}.$$

Solution We first clear of fractions to get

$$x - 1 = A(x + 1)^2 + B(x + 1) + C.$$

Substituting $x = -1$ shows $C = -2$. We then differentiate both sides with respect to x to get

$$1 = 2A(x + 1) + B.$$

Substituting $x = -1$ shows $B = 1$. We differentiate again to get $0 = 2A$, which shows $A = 0$. Hence

$$\frac{x - 1}{(x + 1)^3} = \frac{1}{(x + 1)^2} - \frac{2}{(x + 1)^3}. \qquad \blacksquare$$

In many problems, assigning small values to x, like $x = 0$, ± 1, ± 2, to get equations in A, B, C, D, and so on, provides a fast alternative to the Heaviside method.

EXAMPLE 10 Find A, B, and C in

$$\frac{x^2 + 1}{(x - 1)(x - 2)(x - 3)} = \frac{A}{x - 1} + \frac{B}{x - 2} + \frac{C}{x - 3}.$$

Solution Clear of fractions to get

$$x^2 + 1 = A(x - 2)(x - 3) + B(x - 1)(x - 3) + C(x - 1)(x - 2).$$

Then let $x = 1$, 2, 3 successively to find A, B, and C:

$$x = 1: \quad (1)^2 + 1 = A(-1)(-2) + B(0) + C(0)$$
$$2 = 2A$$
$$A = 1,$$

$$x = 2: \quad (2)^2 + 1 = A(0) + B(1)(-1) + C(0)$$
$$5 = -B$$
$$B = -5,$$

$$x = 3: \quad (3)^2 + 1 = A(0) + B(0) + C(2)(1)$$
$$10 = 2C$$
$$C = 5.$$

Conclusion:

$$\frac{x^2 + 1}{(x - 1)(x - 2)(x - 3)} = \frac{1}{x - 1} - \frac{5}{x - 2} + \frac{5}{x - 3}.$$

PROBLEMS

In Problems 1–10, expand by partial fractions.

1. $\dfrac{5x - 13}{(x - 3)(x - 2)}$

2. $\dfrac{5x - 7}{x^2 - 3x + 2}$

3. $\dfrac{x + 4}{(x + 1)^2}$

4. $\dfrac{2x + 2}{x^2 - 2x + 1}$

5. $\dfrac{x + 1}{x^2(x - 1)}$

6. $\dfrac{z}{z^3 - z^2 - 6z}$

7. $\dfrac{x^2 + 8}{x^2 - 5x + 6}$ (Remember to divide first.)

8. $\dfrac{4}{x^3 + 4x}$

9. $\dfrac{3}{x^2(x^2 + 9)}$

10. $\dfrac{x^3 - 1}{(x^2 + x + 1)^2}$

In Problems 11–49, evaluate the integrals.

11. $\displaystyle\int_0^{1/2} \frac{dx}{1 - x^2}$

12. $\displaystyle\int_1^2 \frac{dx}{x^2 + 2x}$

13. $\displaystyle\int_0^{2\sqrt{2}} \frac{x^3}{x^2 + 1} \, dx$

14. $\displaystyle\int_1^2 \frac{dx}{x^3 + x}$

15. $\displaystyle\int_{1/4}^{3/4} \frac{dx}{x - x^2}$

16. $\displaystyle\int_{-1}^0 \frac{x \, dx}{x^2 - 3x + 2}$

17. $\displaystyle\int \frac{x + 4}{x^2 + 5x - 6} \, dx$

18. $\displaystyle\int \frac{2x + 1}{x^2 - 7x + 12} \, dx$

19. $\displaystyle\int_0^1 \frac{3x^2}{x^2 + 2x + 1} \, dx$

20. $\displaystyle\int \frac{d\theta}{\theta^3 + \theta^2 - 2\theta}$

21. $\displaystyle\int \frac{x \, dx}{x^2 + 4x - 5}$

22. $\displaystyle\int_4^8 \frac{x \, dx}{x^2 - 2x - 3}$

23. $\displaystyle\int \frac{(x + 1) \, dx}{x^2 + 4x - 5}$

24. $\displaystyle\int \frac{x^3 \, dx}{x^2 + 2x + 1}$

25. $\displaystyle\int_1^3 \frac{dx}{x(x + 1)^2}$

26. $\displaystyle\int_0^1 \frac{dx}{(x + 1)(x^2 + 1)}$

27. $\displaystyle\int \frac{(x + 3) \, dx}{2x^3 - 8x}$

28. $\displaystyle\int \frac{\cos x \, dx}{\sin^2 x + \sin x - 6}$

29. $\displaystyle\int_0^{\sqrt{3}} \frac{5x^2 \, dx}{x^2 + 1}$

30. $\displaystyle\int_2^6 \frac{x^3 \, dx}{x^2 - 2x + 1}$

31. $\displaystyle\int_{-1}^1 \frac{x^3 + x}{x^2 + 1} \, dx$

32. $\displaystyle\int_{\pi/3}^{\pi/2} \frac{\sin \theta \, d\theta}{\cos^2 \theta + \cos \theta - 2}$

33. $\displaystyle\int \frac{3x^2 + x + 4}{x^3 + x} \, dx$

34. $\displaystyle\int \frac{dx}{(x^2 - 1)^2}$

35. $\displaystyle\int \frac{x^3 + 4x^2}{x^2 + 4x + 3} \, dx$

36. $\displaystyle\int \frac{4x + 4}{x^2(x^2 + 2)} \, dx$

37. $\displaystyle\int_0^1 \frac{x^2 + 2x + 1}{(x^2 + 1)^2} \, dx$

38. $\displaystyle\int_{-1}^0 \frac{x^3 - x}{(x^2 + 1)(x - 1)^2} \, dx$

39. $\displaystyle\int_{-1}^{0} \frac{2x}{(x^2+1)(x-1)^2}\,dx$

40. $\displaystyle\int \frac{x^2}{(x-1)(x^2+2x+1)}\,dx$

41. $\displaystyle\int_{0}^{\ln 2} \frac{e^t\,dt}{e^{2t}+3e^t+2}$ **42.** $\displaystyle\int_{0}^{1} \frac{dx}{(x^2+1)^2}$

43. $\displaystyle\int_{0}^{1} \frac{x^4\,dx}{(x^2+1)^2}$ **44.** $\displaystyle\int \frac{(4x^3-20x)\,dx}{x^4-10x^2+9}$

45. $\displaystyle\int \frac{1-\sqrt{x}}{1+\sqrt{x}}\,dx$ (*Hint:* Substitute $u=\sqrt{x}$.)

46. $\displaystyle\int \frac{dx}{1+\sqrt{x}}$ (*Hint:* Substitute $u=\sqrt{x}$.)

47. $\displaystyle\int \frac{dx}{\sqrt{x}+\sqrt[3]{x}}$ (*Hint:* Substitute $x=u^6$.)

48. $\displaystyle\int x\ln(x+5)\,dx$

49. $\displaystyle\int_{0}^{1} \ln(x^2+1)\,dx$

50. To integrate $\int x^2\,dx/(x^2-1)$ by partial fractions, we would first divide x^2 by x^2-1 to get

$$\frac{x^2}{x^2-1}=1+\frac{1}{x^2-1}.$$

Suppose that we ignore the fact that x^2 and x^2-1 have the same degree and try to write

$$\frac{x^2}{x^2-1}=\frac{x^2}{(x-1)(x+1)}=\frac{A}{x-1}+\frac{B}{x+1}.$$

What goes wrong? Find out by trying to solve for A and B.

51. Find the volume swept out by revolving about the x-axis the region bounded by the x-axis, the curve $y=3/\sqrt{3x-x^2}$, and the lines $x=1/2$ and $x=5/2$. See Fig. 7.23.

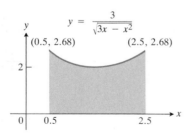

$y=\dfrac{3}{\sqrt{3x-x^2}}$

$(0.5, 2.68)$ $(2.5, 2.68)$

7.23   The region in Problem 51.

52. Find the x-coordinate of the center of mass of a thin homogeneous plate in the first quadrant bounded by the x-axis, the curve $y=\tan^{-1}x$, and the line $x=\sqrt{3}$.

53. Find the x-coordinate of the center of mass of a thin homogeneous plate bounded by the x-axis, the lines $x=3$ and

$x=5$, and the curve

$$y=\frac{4x^2+13x-9}{x^3+2x^2-3x}$$

(Fig. 7.24).

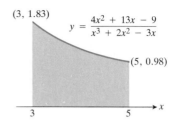

$(3, 1.83)$ $y=\dfrac{4x^2+13x-9}{x^3+2x^2-3x}$

$(5, 0.98)$

7.24   The plate in Problem 53.

54. Find the length of the arc

$$y=\ln(1-x^2),\ 0\le x\le\frac{1}{2}.$$

55. *Social diffusion.* Sociologists sometimes use the phrase "social diffusion" to describe the way information spreads through a population. The information might be a rumor, a cultural fad, or news about a technical innovation. In a sufficiently large population, the number of people x who have the information is treated as a differentiable function of time t and the rate of diffusion, dx/dt, is assumed to be proportional to the number of people who have the information times the number of people who do not. This leads to the equation

$$\frac{dx}{dt}=kx(N-x),$$

where N is the number of people in the population.
Suppose t is measured in days, $k=1/250$, and two people start a rumor at time $t=0$ in a population of $N=1000$ people.
a) Find x as a function of t.
b) When will half the population have heard the rumor? (This is when the rumor will be spreading the fastest.)

56. *Second-order chemical reactions.* Many chemical reactions are the result of the interaction of two molecules that undergo a change to produce a new product. The rate of the reaction typically depends on the concentrations of the two kinds of molecules. If a is the amount of substance A, and b is the amount of substance B at time $t=0$, and if x is the amount of product at time t, then the rate of formation of x may be given by the differential equation

$$\frac{dx}{dt}=k(a-x)(b-x),$$

or

$$\frac{1}{(a-x)(b-x)}\frac{dx}{dt}=k.$$

(k a constant for the reaction). Integrate both sides of this equation with respect to t to obtain a relation between x and *t* (a) if $a = b$, and (b) if $a \neq b$. Assume in each case that $x = 0$ when $t = 0$.

57. *Autocatalytic reactions.* The equation that describes the autocatalytic reaction of Article 3.5, Problem 55, can be written as

$$\frac{dx}{dt} = kx(a - x).$$

Read Problem 55 for background (there is no need to solve Problem 55). Then solve the equation above to find *x* as a function of *t*. Assume that $x = x_0$ when $t = 0$.

TOOLKIT PROGRAMS

Partial Fractions

7.8
Improper Integrals

$$\int_a^c f(x)\,dx = \lim_{b \to c^-} \int_a^b f(x)\,dx$$

7.25 To integrate a function *f* over the half-open interval [*a*, *c*), we (1) integrate *f* over a closed interval [*a*, *b*] inside [*a*, *c*) and (2) take the limit as [*a*, *b*] expands to fill [*a*, *c*).

In many applications we want to integrate a function over an interval that is not closed. To do so, we first integrate the function over a closed interval inside the nonclosed interval; then we take the limit of this integral as the closed interval expands to fill the nonclosed interval. Figure 7.25 shows how this is done for a half-open interval [*a*, *c*). It is conventional to call the resulting limit, whether it exists or not, the **improper integral** of the function over the nonclosed interval.

How can we tell if the improper integral exists? That is the basic question and the one we address in this article. Once the integral is known to exist, its value, if not immediately apparent, can be found by numerical methods.

Convergence and Divergence

In lifting-line theory in aerodynamics, the function

$$f(x) = \sqrt{\frac{1 + x}{1 - x}} \tag{1}$$

needs to be integrated over the interval from $x = 0$ to $x = 1$. The function is not defined at $x = 1$, although it is defined and continuous everywhere else in [0, 1]. To integrate *f* from zero to one, we integrate *f* from zero to a positive number *b* less than 1 and take the limit of the resulting definite integral as *b* approaches 1. If the limit exists, we define the integral of *f* from zero to 1 to be this value and write

$$\int_0^1 \sqrt{\frac{1 + x}{1 - x}}\,dx = \lim_{b \to 1^-} \int_0^b \sqrt{\frac{1 + x}{1 - x}}\,dx. \tag{2}$$

In this case, we also say that the integral

$$\int_0^1 \sqrt{\frac{1 + x}{1 - x}}\,dx$$

converges and say that the area under the curve $y = \sqrt{(1 + x)/(1 - x)}$ from 0 to 1 is the value of the integral. If the limit in (2) does not exist, we say that the integral **diverges.**

EXAMPLE 1 Determine whether

$$\int_0^1 \sqrt{\frac{1 + x}{1 - x}}\,dx$$

converges or diverges.

7.26 (a) To evaluate

$$\int_0^1 \sqrt{(1 + x)/(1 - x)}\ dx,$$

we evaluate

$$\int_0^b \sqrt{(1 + x)/(1 - x)}\ dx$$

and let b approach 1 from below. (b) If we treat the curve as the graph of the function $x = (y^2 - 1)/(y^2 + 1)$, we can make an equivalent calculation by integrating from 1 to c with respect to y and letting c approach infinity. We then add 1 to the result to account for the area of the gray square.

(a)

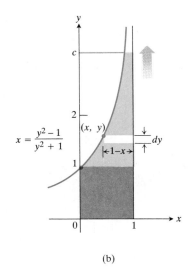

(b)

Solution See Fig. 7.26(a). Multiplying numerator and denominator by $\sqrt{1 + x}$ gives

$$\int \sqrt{\frac{1 + x}{1 - x}}\ dx = \int \frac{1 + x}{\sqrt{1 - x^2}}\ dx = \int \frac{1}{\sqrt{1 - x^2}}\ dx + \int \frac{x}{\sqrt{1 - x^2}}\ dx$$

$$= \sin^{-1}x - \sqrt{1 - x^2} + C.$$

Therefore,

$$\lim_{b \to 1^-} \int_0^b \sqrt{\frac{1 + x}{1 - x}}\ dx = \lim_{b \to 1^-} [\sin^{-1}x - \sqrt{1 - x^2}]_0^b$$

$$= \lim_{b \to 1^-} [\sin^{-1}b - \sqrt{1 - b^2} + 1]$$

$$= \sin^{-1}1 - 0 + 1 = (\pi/2) + 1.$$

The integral converges to $(\pi/2) + 1$.

The computation above integrates vertical strips with respect to x. We can get the same result by integrating horizontal strips with respect to y (Fig. 7.26b). In this case,

$$dA = (1 - x)\ dy = \frac{2}{y^2 + 1}\ dy$$

and the portion of the area above the shaded square is

$$\int_1^\infty \frac{2}{y^2 + 1}\ dy = \lim_{c \to \infty} \int_1^c \frac{2}{y^2 + 1}\ dy$$

$$= \lim_{c \to \infty} [2 \tan^{-1}y]_1^c$$

$$= \lim_{c \to \infty} 2 \tan^{-1}c - 2 \cdot \frac{\pi}{4} = 2 \cdot \frac{\pi}{2} - \frac{\pi}{2} = \frac{\pi}{2}.$$

Including the shaded square, the area is $(\pi/2) + 1$, in agreement with our first calculation.

The notation

$$\int_a^b f(x)\ dx$$

for improper integrals is the same as the notation for definite integrals. In any given case it is usually a simple matter to tell whether a particular integral is to be calculated as an ordinary definite integral or as a limit. If a and b are finite and f is continuous at every point of $[a,\ b]$, the integral is an ordinary definite integral. If f becomes infinite at one or more points in the interval of integration, or one or both of the limits of integration is infinite, then the designated integral is improper and is to be calculated as a limit.

DEFINITION

Improper Integrals

$\displaystyle\int_a^b f(x)\ dx$ denotes an **improper integral** if

1. f becomes infinite at one or more points of the interval of integration, or
2. one or both of the limits of integration is infinite, or
3. both (1) and (2) hold.

EXAMPLE 2 In the integral

$$\int_0^1 \frac{dx}{x},$$

the function

$$f(x) = \frac{1}{x}$$

becomes infinite at $x = 0$. We cut off the point $x = 0$ and start our integration at some positive number $b < 1$. (See Fig. 7.27.) That is, we consider the integral

$$\int_b^1 \frac{dx}{x} = \ln x\ \Big]_b^1 = \ln 1 - \ln b = \ln \frac{1}{b}$$

and investigate its behavior as b approaches zero from the right. Since

$$\lim_{b \to 0^+} \int_b^1 \frac{dx}{x} = \lim_{b \to 0^+} \left(\ln \frac{1}{b} \right) = +\infty,$$

we say that the integral from $x = 0$ to $x = 1$ *diverges*. ∎

The method to be used when the function f becomes infinite at an interior point of the interval of integration is illustrated in the following example.

EXAMPLE 3 In the integral

$$\int_0^3 \frac{dx}{(x - 1)^{2/3}},$$

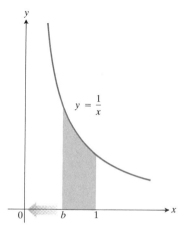

7.27 $\displaystyle\int_0^1 \frac{dx}{x} = \lim_{b \to 0^+} \int_b^1 \frac{dx}{x}.$

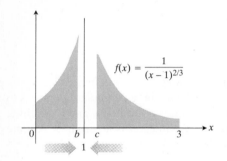

the function

$$f(x) = \frac{1}{(x - 1)^{2/3}}$$

becomes infinite at $x = 1$, which lies between the limits of integration 0 and 3. In such a case, we again cut out the point where $f(x)$ becomes infinite. This time we integrate from 0 to b, where b is slightly less than 1, and start again on the other side of 1 at c and integrate from c to 3 (Fig. 7.28). This gives two integrals to investigate:

$$\int_0^b \frac{dx}{(x - 1)^{2/3}} \quad \text{and} \quad \int_c^3 \frac{dx}{(x - 1)^{2/3}}.$$

If the first of these has a definite limit as $b \to 1^-$ and if the second also has a definite limit as $c \to 1^+$, then we say that the integral of f from 0 to 3 *converges* and that its value is

$$\int_0^3 \frac{dx}{(x - 1)^{2/3}} = \lim_{b \to 1^-} \int_0^b \frac{dx}{(x - 1)^{2/3}} + \lim_{c \to 1^+} \int_c^3 \frac{dx}{(x - 1)^{2/3}}.$$

If either limit fails to exist, we say that the integral of f from 0 to 3 *diverges*. For this example,

$$\lim_{b \to 1^-} \int_0^b (x - 1)^{-2/3} \, dx = \lim_{b \to 1^-} [3(b - 1)^{1/3} - 3(0 - 1)^{1/3}] = +3$$

and

$$\lim_{c \to 1^+} \int_c^3 (x - 1)^{-2/3} \, dx = \lim_{c \to 1^+} [3(3 - 1)^{1/3} - 3(c - 1)^{1/3}] = 3\sqrt[3]{2}.$$

Since both limits exist and are finite, the integral of f converges and its value is $3 + 3\sqrt[3]{2}$.   ∎

7.28 $\int_0^3 f(x)\, dx$

$= \lim_{b \to 1^-} \int_0^b f(x)\, dx + \lim_{c \to 1^+} \int_c^3 f(x)\, dx.$

The Integral $\int_1^\infty dx/x^p$

The convergence of the integral

$$\int_1^\infty \frac{dx}{x^p}$$

depends on the value of the exponent p. The next example illustrates this with $p = 1$ and $p = 2$.

EXAMPLE 4   Do the integrals

$$\int_1^\infty \frac{dx}{x} \quad \text{and} \quad \int_1^\infty \frac{dx}{x^2}$$

converge, or diverge?

Solution   The curves

$$y = \frac{1}{x} \quad \text{and} \quad y = \frac{1}{x^2}$$

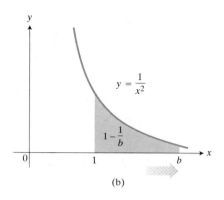

7.29 (a) The shaded area, ln b, does not approach a finite limit as $b \to \infty$. (b) The shaded area, $1 - (1/b)$, approaches 1 as $b \to \infty$.

both approach the x-axis as $x \to \infty$ (Fig. 7.29). In the first case,

$$\int_1^b \frac{dx}{x} = \ln x \Big]_1^b = \ln b,$$

so that

$$\lim_{b \to \infty} \int_1^b \frac{dx}{x} = \lim_{b \to \infty} \ln b = \infty.$$

Therefore,

$$\int_1^\infty \frac{dx}{x} = \infty$$

and the integral *diverges*.

In the second case,

$$\int_1^b \frac{dx}{x^2} = -\frac{1}{x} \Big]_1^b = 1 - \frac{1}{b},$$

so that

$$\int_1^\infty \frac{dx}{x^2} = \lim_{b \to \infty} \int_1^b \frac{dx}{x^2} = \lim_{b \to \infty} \left(1 - \frac{1}{b}\right) = 1.$$

The integral converges and its value is 1. ∎

Generally, the integral

$$\int_1^\infty \frac{dx}{x^p}$$

converges when $p > 1$ but diverges when $p \le 1$ (Problem 43).

The Domination Test for Convergence and Divergence

Sometimes we can determine whether an improper integral converges without having to evaluate it. Instead, we compare it to an integral whose convergence or divergence we already know. This is the case with the next example, an integral important in probability theory.

EXAMPLE 5 Determine whether the improper integral

$$\int_1^\infty e^{-x^2} \, dx$$

converges or diverges.

Solution Even though we cannot find any simpler expression for

$$I(b) = \int_1^b e^{-x^2} \, dx,$$

we can show that $I(b)$ has a finite limit as $b \to \infty$.

The function $I(b)$ represents the area between the x-axis and the curve $y = e^{-x^2}$ from $x = 1$ to $x = b$. It is an increasing function of b. Therefore, it either becomes infinite as $b \to \infty$ or has a finite limit as $b \to \infty$.

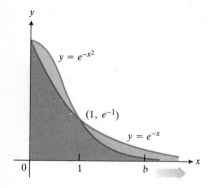

7.30   The graphs of $y = e^{-x^2}$ and $y = e^{-x}$.

We show that the first alternative cannot be the true one. We do this by comparing the area under the given curve $y = e^{-x^2}$ with the area under the curve $y = e^{-x}$ (Fig. 7.30). The latter area, from $x = 1$ to $x = b$, is given by the integral

$$\int_1^b e^{-x}\, dx = -e^{-x}\Big]_1^b = e^{-1} - e^{-b},$$

which approaches the finite limit e^{-1} as $b \to \infty$. Since e^{-x^2} is less than e^{-x} for all x greater than 1, the area under the given curve is certainly no greater than e^{-1} no matter how large b is.

The discussion above is summarized in the following inequalities:

$$I(b) = \int_1^b e^{-x^2}\, dx \le \int_1^b e^{-x}\, dx = e^{-1} - e^{-b} \le e^{-1};$$

that is, $I(b) \le e^{-1} < 0.37$. Therefore $I(b)$ does not become infinite as $b \to \infty$. The first alternative is ruled out and the second alternative must hold; that is,

$$\int_1^\infty e^{-x^2}\, dx = \lim_{b \to \infty} \int_1^b e^{-x^2}\, dx$$

converges to a definite finite value. We have not calculated what the limit is, but we know that it exists and is less than 0.37. ∎

We say that a positive function f **dominates** a positive function g as $x \to \infty$ if

$$g(x) \le f(x)$$

for all values of x beyond some point a. For instance, $f(x) = e^{-x}$ dominates $g(x) = e^{-x^2}$ as $x \to \infty$ because $e^{-x^2} \le e^{-x}$ for all $x > a = 1$.

If f dominates g as $x \to \infty$, then

$$\int_a^b g(x)\, dx \le \int_a^b f(x)\, dx, \qquad b > a,$$

and from this it can be argued as in Example 5 that

$$\int_a^\infty g(x)\, dx \text{ converges if } \int_a^\infty f(x)\, dx \text{ converges.}$$

Turning this around says that

$$\int_a^\infty f(x)\, dx \text{ diverges if } \int_a^\infty g(x)\, dx \text{ diverges.}$$

We state these results as a theorem and then give examples.

THEOREM 1

Domination Test for Convergence and Divergence of Improper Integrals

If $0 \le g(x) \le f(x)$ for all $x > a$, then

1. $\displaystyle\int_a^\infty g(x)\, dx$ converges if $\displaystyle\int_a^\infty f(x)\, dx$ converges,

2. $\displaystyle\int_a^\infty f(x)\, dx$ diverges if $\displaystyle\int_a^\infty g(x)\, dx$ diverges.

Theorem 1 assumes that f and g are integrable over every finite interval $[a, b]$, which will be true, for instance, if the functions are continuous.

EXAMPLE 6

a) $\displaystyle\int_1^\infty \frac{1}{x^2 + 1} \, dx$ converges because $\dfrac{1}{x^2 + 1} < \dfrac{1}{x^2}$ and $\displaystyle\int_1^\infty \frac{1}{x^2} \, dx$ converges.

b) $\displaystyle\int_1^\infty \frac{1}{e^{2x}} \, dx$ converges because $\dfrac{1}{e^{2x}} < \dfrac{1}{e^x}$ and $\displaystyle\int_1^\infty \frac{1}{e^x} \, dx$ converges.

c) $\displaystyle\int_1^\infty \frac{1}{\sqrt{x}} \, dx$ diverges because $\dfrac{1}{\sqrt{x}} > \dfrac{1}{x}$ for $x > 1$ and $\displaystyle\int_1^\infty \frac{1}{x} \, dx$ diverges.

d) $\displaystyle\int_1^\infty \left(\frac{1}{x} + \frac{1}{x^2} \right) dx$ diverges because $\dfrac{1}{x} + \dfrac{1}{x^2} > \dfrac{1}{x}$ and $\displaystyle\int_1^\infty \frac{1}{x} \, dx$ diverges.

∎

The Limit Comparison Test

Another useful result, which we shall not prove, is the limit comparison test for the convergence and divergence of improper integrals. (You will find a similar result for infinite series in Chapter 11.) It goes like this:

THEOREM 2

> **Limit Comparison Test for Convergence and Divergence of Improper Integrals**
>
> Suppose that $f(x)$ and $g(x)$ are positive functions and that
>
> $$\lim_{x \to \infty} \frac{f(x)}{g(x)} = L, \qquad 0 < L < \infty.$$
>
> Then, $\displaystyle\int_a^\infty f(x) \, dx$ and $\displaystyle\int_a^\infty g(x) \, dx$ both converge or both diverge.

Like Theorem 1, Theorem 2 assumes that f and g are integrable over every finite interval $[a, b]$.

In the language of Article 6.8, Theorem 2 says that *if two positive functions grow at the same rate as $x \to \infty$, then their integrals from a to ∞ behave alike: they both converge or both diverge.* This does not mean, however, that their integrals must have the same value, as the next example shows.

EXAMPLE 7   Compare

$$\int_1^\infty \frac{dx}{x^2} \qquad \text{and} \qquad \int_1^\infty \frac{dx}{1 + x^2}$$

with the limit comparison test.

Solution   With

$$f(x) = \frac{1}{x^2}, \qquad g(x) = \frac{1}{1 + x^2},$$

we have

$$\lim_{x \to \infty} \frac{1}{x^2} \frac{1 + x^2}{1} = \lim_{x \to \infty} \frac{1 + x^2}{x^2} = \lim_{x \to \infty} \left(\frac{1}{x^2} + 1 \right) = 0 + 1 = 1,$$

as a positive finite limit. Therefore,

$$\int_1^\infty \frac{dx}{1 + x^2} \quad \text{converges because} \quad \int_1^\infty \frac{dx}{x^2} \quad \text{converges.}$$

The integrals converge to different values, however:

$$\int_1^\infty \frac{dx}{x^2} = 1,$$

from Example 4, and

$$\int_1^\infty \frac{dx}{1 + x^2} = \lim_{b \to \infty} \int_1^b \frac{dx}{1 + x^2} \qquad \text{(the definition)}$$

$$= \lim_{b \to \infty} [\tan^{-1} b - \tan^{-1} 1]$$

$$= \frac{\pi}{2} - \frac{\pi}{4} = \frac{\pi}{4}. \qquad \blacksquare$$

EXAMPLE 8

$$\int_1^\infty \frac{3}{e^x + 5} \, dx \text{ converges because } \int_1^\infty \frac{1}{e^x} \, dx \text{ converges}$$

and

$$\lim_{x \to \infty} \frac{1}{e^x} \cdot \frac{e^x + 5}{3} = \lim_{x \to \infty} \frac{e^x + 5}{3e^x} = \lim_{x \to \infty} \left(\frac{1}{3} + \frac{5}{3e^x} \right) = \frac{1}{3} + 0 = \frac{1}{3},$$

a positive finite limit. As far as the convergence of the improper integral is concerned, $3/(e^x + 5)$ behaves like $1/e^x$. $\blacksquare$

EXAMPLE 9

$$\int_3^\infty \frac{1}{e^{2x} - 10e^x} \, dx \text{ converges because } \int_3^\infty \frac{1}{e^{2x}} \, dx \text{ converges}$$

and

$$\lim_{x \to \infty} \frac{1}{e^{2x}} \cdot \frac{e^{2x} - 10e^x}{1} = \lim_{x \to \infty} \left(1 - \frac{10}{e^x} \right) = 1 + 0 = 1,$$

a positive finite limit. As far as the convergence of the improper integral is concerned, $1/(e^{2x} - 10e^x)$ behaves like $1/e^{2x}$. $\blacksquare$

Concluding Remarks

We know that

$$\int_1^\infty \frac{1}{x^2} \, dx$$

converges, but what about integrals like

$$\int_2^\infty \frac{1}{x^2}\, dx \qquad \text{and} \qquad \int_{100}^\infty \frac{1}{x^2}\, dx?$$

The answer is that they converge, too. The existence of the limits

$$\lim_{b\to\infty} \int_2^b \frac{1}{x^2}\, dx \qquad \text{and} \qquad \lim_{b\to\infty} \int_{100}^b \frac{1}{x^2}\, dx$$

does not depend on the starting points $a = 2$ and $a = 100$, but only on the values of $1/x^2$ as x approaches infinity. Indeed, we find for any positive a that

$$\lim_{b\to\infty} \int_a^b \frac{1}{x^2}\, dx = \lim_{b\to\infty}\left(-\frac{1}{b} + \frac{1}{a}\right) = \frac{1}{a}.$$

The *value* of the limit depends on the value of a, but the *existence* of the limit does not. The limit exists for any positive a.

Similarly, the integrals

$$\int_2^\infty \frac{1}{x}\, dx \qquad \text{and} \qquad \int_{100}^\infty \frac{1}{x}\, dx$$

both diverge. We find for any positive a that

$$\lim_{b\to\infty} \int_a^b \frac{1}{x}\, dx = \lim_{b\to\infty}(\ln b - \ln a) = \infty.$$

The convergence and divergence of integrals that are improper only because the upper limit is infinite never depend on the lower limit of integration

In our final example, we show that an improper integral may diverge without becoming infinite.

EXAMPLE 10 The integral

$$\int_0^b \cos x\, dx = \sin b$$

takes all values between -1 and $+1$ as b varies between $2n\pi - \pi/2$ and $2n\pi + \pi/2$, where n is any integer. Hence,

$$\lim_{b\to\infty} \int_0^b \cos x\, dx$$

does not exist. We might say that this integral "diverges by oscillation." ∎

PROBLEMS

In Problems 1–10, evaluate the integrals.

1. $\displaystyle\int_0^\infty \frac{dx}{x^2 + 1}$

2. $\displaystyle\int_0^1 \frac{dx}{\sqrt{x}}$

5. $\displaystyle\int_0^4 \frac{dx}{\sqrt{4 - x}}$

6. $\displaystyle\int_0^1 \frac{dx}{\sqrt{1 - x^2}}$

3. $\displaystyle\int_{-1}^1 \frac{dx}{x^{2/3}}$

4. $\displaystyle\int_1^\infty \frac{dx}{x^{1.001}}$

7. $\displaystyle\int_0^1 \frac{dx}{x^{0.999}}$

8. $\displaystyle\int_{-\infty}^2 \frac{dx}{4 - x}$

9. $\displaystyle\int_{2}^{\infty} \frac{1}{x^2 - x} \, dx$

10. $\displaystyle\int_{0}^{\infty} \frac{dx}{(1 + x)\sqrt{x}}$

In Problems 11–40, determine whether the integrals converge or diverge. (In some cases, you may not need to evaluate the integral to decide. Name any tests you use.)

11. $\displaystyle\int_{1}^{\infty} \frac{dx}{\sqrt{x}}$

12. $\displaystyle\int_{1}^{\infty} \frac{dx}{x^3}$

13. $\displaystyle\int_{1}^{\infty} \frac{dx}{x^3 + 1}$

14. $\displaystyle\int_{0}^{\infty} \frac{dx}{x^3}$

15. $\displaystyle\int_{0}^{\infty} \frac{dx}{x^{3/2} + 1}$

16. $\displaystyle\int_{0}^{\infty} \frac{dx}{1 + e^x}$

17. $\displaystyle\int_{0}^{\pi/2} \tan x \, dx$

18. $\displaystyle\int_{-1}^{1} \frac{dx}{x^2}$

19. $\displaystyle\int_{-1}^{1} \frac{dx}{x^{2/5}}$

20. $\displaystyle\int_{0}^{\infty} \frac{dx}{\sqrt{x}}$

21. $\displaystyle\int_{2}^{\infty} \frac{dx}{\sqrt{x - 1}}$

22. $\displaystyle\int_{1}^{\infty} \frac{5}{x} \, dx$

23. $\displaystyle\int_{0}^{2} \frac{dx}{1 - x^2}$

24. $\displaystyle\int_{2}^{\infty} \frac{dx}{(x + 1)^2}$

25. $\displaystyle\int_{0}^{\infty} \frac{dx}{\sqrt{x^6 + 1}}$

26. $\displaystyle\int_{-1}^{1} \frac{dx}{\sqrt[3]{x}}$

27. $\displaystyle\int_{0}^{\infty} x^2 e^{-x} \, dx$

28. $\displaystyle\int_{1}^{\infty} \frac{\sqrt{x + 1}}{x^2} \, dx$

29. $\displaystyle\int_{\pi}^{\infty} \frac{2 + \cos x}{x} \, dx$

30. $\displaystyle\int_{1}^{\infty} \frac{\ln x}{x} \, dx$

31. $\displaystyle\int_{6}^{\infty} \frac{1}{\sqrt{x + 5}} \, dx$

32. $\displaystyle\int_{1}^{\infty} \frac{dx}{\sqrt{2x + 10}}$

33. $\displaystyle\int_{2}^{\infty} \frac{2}{x^2 - 1} \, dx$

34. $\displaystyle\int_{1}^{\infty} \frac{1}{e^{\ln x}} \, dx$

35. $\displaystyle\int_{2}^{\infty} \frac{1}{\ln x} \, dx$

36. $\displaystyle\int_{1}^{\infty} \frac{1}{\sqrt{e^x - x}} \, dx$

37. $\displaystyle\int_{1}^{\infty} \frac{1}{e^x - 2^x} \, dx$

38. $\displaystyle\int_{2}^{\infty} \frac{1}{x^3 - 5} \, dx$

39. $\displaystyle\int_{0}^{\infty} \frac{dx}{\sqrt{x + x^4}}$

(*Hint*: Compare the integral with $\int dx/\sqrt{x}$ for x near zero and with $\int dx/x^2$ for large x.)

40. $\displaystyle\int_{0}^{\infty} e^{-x}\cos x \, dx$

41. *Estimating the value of a convergent improper integral whose domain is infinite.* Show that

$$\int_{3}^{\infty} e^{-3x} \, dx = \frac{1}{3}e^{-9} < 0.000042,$$

and hence that $\int_{3}^{\infty} e^{-x^2} \, dx < 0.000042$. Therefore, $\int_{0}^{\infty} e^{-x^2} \, dx$ can be replaced by $\int_{0}^{3} e^{-x^2} \, dx$ without introducing more than this much error. Evaluate this last integral by Simpson's Rule with $n = 6$. (This illustrates one method by which a convergent improper integral may be approximated numerically.)

42. *The infinite paint can.* As Example 4 shows, the integral $\int_{1}^{\infty} (dx/x)$ diverges. This means that the integral

$$\int_{1}^{\infty} 2\pi \frac{1}{x} \sqrt{1 + \frac{1}{x^4}} \, dx,$$

which measures the *surface area* of the solid of revolution traced out by revolving the curve $y = 1/x$, $1 \le x$, about the x-axis, diverges also. For, by comparing the two integrals, we see that, for every finite value $b > 1$,

$$\int_{1}^{b} 2\pi \frac{1}{x} \sqrt{1 + \frac{1}{x^4}} \, dx > \int_{1}^{b} \frac{dx}{x}.$$

However, the integral

$$\int_{1}^{\infty} \pi \left(\frac{1}{x}\right)^2 \, dx$$

for the *volume* of the solid converges. Calculate it. This solid of revolution is sometimes described as a can that does not hold enough paint to cover its outside surface. (See Fig. 7.31.)

7.31   If the solid shown here is extended to the right indefinitely, it produces an "infinite" solid with a finite volume but an infinite surface area. To learn why, read Problem 42.

43. Show that

$$\int_{1}^{\infty} \frac{dx}{x^p} = \frac{1}{p - 1} \quad \text{when} \quad p > 1,$$

but that the integral is infinite when $p < 1$. Example 4 shows what happens when $p = 1$.

44. Find the values of p for which each integral converges:

a) $\displaystyle\int_{1}^{2} \frac{dx}{x(\ln x)^p}$,

b) $\displaystyle\int_{2}^{\infty} \frac{dx}{x(\ln x)^p}$.

Problems 45–48 are about the region in the first quadrant between the curve $y = e^{-x}$ and the x-axis.

45. Find the area of the region.

46. Find the centroid of the region.

47. Find the volume swept out by revolving the region about the y-axis.

48. Find the volume swept out by revolving the region about the x-axis.

49. Find the center of mass of the region bounded by the curves $y = \pm(1 - x^2)^{-1/2}$ and the lines $x = 0$, $x = 1$.

50. Find the area of the region that lies between the curves $y = \sec x$ and $y = \tan x$ for $0 \le x \le \pi/2$.

51. Show that the area of the region between the curve $y = 1/(1 + x^2)$ and the entire x-axis is the same as the area of the unit disk, $x^2 + y^2 \le 1$.

7.9
Using Integral Tables

The numbered integration formulas on the endpapers at the front and back of this book are stated in terms of constants a, b, c, m, n, and so on. These constants can usually assume any real value and need not be integers. Occasional limitations on their values are stated along with the formulas. Formula 5 requires $n \ne -1$, for example, and Formula 11 requires $n \ne -2$.

The formulas also assume that the constants do not take on values that require dividing by zero or taking even roots of negative numbers. For example, Formula 8 assumes $a \ne 0$, and Formula 13(a) cannot be used unless b is negative.

The following examples show how the integration formulas on the endpapers of this book are commonly used. More extensive tables of integrals can be found in the books listed at the end of the present article.

EXAMPLE 1 Evaluate

$$\int x(2x + 5)^{-1} \, dx.$$

Solution We use Formula 8 (not 7, which requires $n \ne -1$),

$$\int x(ax + b)^{-1} \, dx = \frac{x}{a} - \frac{b}{a^2} \ln|ax + b| + C,$$

with $a = 2$ and $b = 5$:

$$\int x(2x + 5)^{-1} \, dx = \frac{x}{2} - \frac{5}{4} \ln|2x + 5| + C. \quad \blacksquare$$

EXAMPLE 2 Evaluate

$$\int \frac{dx}{x\sqrt{2x + 4}}.$$

Solution We use Formula 13(b),

$$\int \frac{dx}{x\sqrt{ax + b}} = \frac{1}{\sqrt{b}} \ln\left|\frac{\sqrt{ax + b} - \sqrt{b}}{\sqrt{ax + b} + \sqrt{b}}\right| + C, \qquad \text{if } b > 0,$$

with $a = 2$ and $b = 4$:

$$\int \frac{dx}{x\sqrt{2x+4}} = \frac{1}{\sqrt{4}} \ln \left| \frac{\sqrt{2x+4}-\sqrt{4}}{\sqrt{2x+4}+\sqrt{4}} \right| + C$$

$$= \frac{1}{2} \ln \left| \frac{\sqrt{2x+4}-2}{\sqrt{2x+4}+2} \right| + C.$$

Formula 13(a), which requires $b < 0$, would not have been appropriate here. It *is* appropriate, however, in the next example. ∎

EXAMPLE 3 Evaluate

$$\int \frac{dx}{x\sqrt{2x-4}}.$$

Solution We use Formula 13(a),

$$\int \frac{dx}{x\sqrt{ax+b}} = \frac{2}{\sqrt{-b}} \tan^{-1} \sqrt{\frac{ax+b}{-b}} + C, \qquad \text{if } b < 0,$$

with $a = 2$ and $b = -4$:

$$\int \frac{dx}{x\sqrt{2x-4}} = \frac{2}{\sqrt{-(-4)}} \tan^{-1} \sqrt{\frac{2x-4}{-(-4)}} + C = \tan^{-1} \sqrt{\frac{x-2}{2}} + C. \quad ∎$$

EXAMPLE 4 Evaluate

$$\int \frac{dx}{x^2\sqrt{2x-4}}.$$

Solution We begin with Formula 15,

$$\int \frac{dx}{x^2\sqrt{ax+b}} = -\frac{\sqrt{ax+b}}{bx} - \frac{a}{2b} \int \frac{dx}{x\sqrt{ax+b}} + C,$$

with $a = 2$ and $b = -4$:

$$\int \frac{dx}{x^2\sqrt{2x-4}} = -\frac{\sqrt{2x-4}}{-4x} + \frac{2}{2 \cdot 4} \int \frac{dx}{x\sqrt{2x-4}} + C.$$

We then use Formula 13(a) to evaluate the integral on the right (Example 3) to obtain

$$\int \frac{dx}{x^2\sqrt{2x+4}} = \frac{\sqrt{2x-4}}{4x} + \frac{1}{4} \tan^{-1} \sqrt{\frac{x-2}{2}} + C. \quad ∎$$

EXAMPLE 5 Evaluate

$$\int x \sin^{-1} x \, dx.$$

Solution We use Formula 99,

$$\int x^n \sin^{-1} ax \, dx = \frac{x^{n+1}}{n+1} \sin^{-1} ax - \frac{a}{n+1} \int \frac{x^{n+1} \, dx}{\sqrt{1-a^2x^2}}, \qquad n \neq -1,$$

with $n = 1$ and $a = 1$:

$$\int x \sin^{-1}x \, dx = \frac{x^2}{2} \sin^{-1}x - \frac{1}{2} \int \frac{x^2 \, dx}{\sqrt{1 - x^2}}.$$

The integral on the right is found in the table as Formula 33,

$$\int \frac{x^2}{\sqrt{a^2 - x^2}} \, dx = \frac{a^2}{2} \sin^{-1}\frac{x}{a} - \frac{1}{2}x\sqrt{a^2 - x^2} + C,$$

with $a = 1$:

$$\int \frac{x^2 \, dx}{\sqrt{1 - x^2}} = \frac{1}{2} \sin^{-1}x - \frac{1}{2}x\sqrt{1 - x^2} + C.$$

The combined result is

$$\int x \sin^{-1}x \, dx = \frac{x^2}{2} \sin^{-1}x - \frac{1}{2}\left(\frac{1}{2} \sin^{-1}x - \frac{1}{2}x\sqrt{1 - x^2}\right) + C$$

$$= \left(\frac{x^2}{2} - \frac{1}{4}\right) \sin^{-1}x + \frac{1}{4}x\sqrt{1 - x^2} + C. \qquad \blacksquare$$

EXAMPLE 6 The region in the first quadrant enclosed by the x-axis, the line $x = 1$, and the curve $y = \sin^{-1}x$ is revolved about the y-axis to generate a solid. Find the volume of the solid.

Solution We sketch the region (Fig. 7.32) and decide on the method of cylindrical shells. The volume is

$$V = \int_0^1 2\pi(\text{radius})(\text{height}) \, dx$$

$$= \int_0^1 2\pi x \sin^{-1}x \, dx$$

$$= 2\pi\left[\left(\frac{x^2}{2} - \frac{1}{4}\right) \sin^{-1}x + \frac{1}{4}x\sqrt{1 - x^2}\right]_0^1 \qquad \text{(Example 5)}$$

$$= 2\pi\left[\left(\frac{1}{4}\right)\left(\frac{\pi}{2}\right) + 0\right] - 2\pi \, [0 + 0]$$

$$= \frac{\pi^2}{4}. \qquad \blacksquare$$

References More extensive tables of integrals can be found in the following books.

Handbook of Mathematical Functions with Formulas, Graphs, and Mathematical Tables, edited by Milton Abramowitz and Irene A. Stegun, Dover Publications, Inc.

Standard Mathematical Tables, 27th Edition, edited by William H. Beyer, CRC Press, Inc., Boca Raton, Florida, 1984.

Tables of Indefinite Integrals, G. Petit Bois, Dover Publications, Inc.

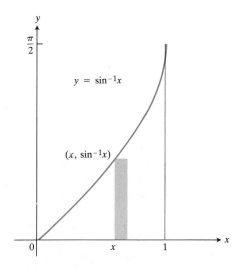

7.32 The region in Example 6.

PROBLEMS

Use the integral formulas on the endpapers of the book to evaluate the integrals in Problems 1–16.

1. $\displaystyle\int_0^\infty e^{-x^2}\,dx$

2. $\displaystyle\int x\cos^{-1}x\,dx$

3. $\displaystyle\int_6^9 \frac{dx}{x\sqrt{x-3}}$

4. $\displaystyle\int_0^{1/2} x\tan^{-1}2x\,dx$

5. $\displaystyle\int \frac{dx}{(9-x^2)^2}$

6. $\displaystyle\int_4^{10} \frac{\sqrt{4x+9}}{x^2}\,dx$

7. $\displaystyle\int_3^{11} \frac{dx}{x^2\sqrt{7+x^2}}$

8. $\displaystyle\int \frac{dx}{x^2\sqrt{7-x^2}}$

9. $\displaystyle\int_{-2}^{-\sqrt{2}} \frac{\sqrt{x^2-2}}{x}\,dx$

10. $\displaystyle\int_{-\pi/12}^{\pi/4} \frac{dx}{5+4\sin 2x}$

11. $\displaystyle\int \frac{dx}{4+5\sin 2x}$

12. $\displaystyle\int_3^6 \frac{x}{\sqrt{x-2}}\,dx$

(*Hint*: The n in Formula 7 need not be an integer.)

13. $\displaystyle\int x\sqrt{2x-3}\,dx$

14. $\displaystyle\int \frac{\sqrt{3x-4}}{x}\,dx$

15. $\displaystyle\int_0^\infty x^{10}e^{-x}\,dx$

16. $\displaystyle\int_0^1 x^2\tan^{-1}x\,dx$

In Problems 17–20, use the given substitutions to change the integral into one that can be found in the tables on the endpapers of this book. Then evaluate the integral.

17. $\displaystyle\int_0^1 \sin^{-1}\sqrt{x}\,dx, \quad u=\sqrt{x}$

18. $\displaystyle\int_{3/4}^1 \frac{\cos^{-1}\sqrt{x}}{\sqrt{x}}\,dx, \quad u=\sqrt{x}$

19. $\displaystyle\int_0^{1/2} \frac{\sqrt{x}}{\sqrt{1-x}}\,dx, \quad u=\sqrt{x}$

20. $\displaystyle\int_{\pi/4}^{\pi/2} \cot x\sqrt{1-\sin^2 x}\,dx, \quad u=\sin x$

21. Find the centroid of the region cut from the first quadrant by the curve $y=1/\sqrt{x+1}$ and the line $x=3$.

22. A thin plate of constant density $\delta=1$ occupies the region enclosed by the curve $y=36/(2x+3)$ and the line $x=3$ in the first quadrant. Find the moment of the plate about the y-axis.

Problems 23–26 refer to the formulas on the endpapers of this book by number.

23. Verify Formula 55 by differentiating the right side.

24. Verify Formula 76 by differentiating the right side.

25. Verify Formula 9 by integrating

$$\int \frac{x}{(ax+b)^2}\,dx$$

with the substitution $u=ax+b$.

26. Verify Formula 46 by integrating

$$\int \frac{dx}{x^2\sqrt{x^2-a^2}}$$

with the substitution $x=a\sec u$.

7.10
Reduction Formulas

A reduction formula is a formula that expresses the integral of a power of a function in terms of the integral of a lower power of the function, thus "reducing" the size of the exponent. By applying such a formula repeatedly, we may eventually express the integral of a high power of a function in terms of the integral of a power low enough to be evaluated directly. In this article, we derive a few of the common reduction formulas and show how they are used. As you will see, most of the derivations use integration by parts.

Do not memorize the formulas that appear in this article. In practice, you can always look them up when you need them. The point here is to learn that formulas like these are available and to develop some skill in applying them.

EXAMPLE 1   Express

$$\int \cos^n x \ dx$$

in terms of an integral of a lower power of cos x.

Solution   We use integration by parts with

$$u = \cos^{n-1}x, \qquad\qquad\qquad dv = \cos x \ dx,$$
$$du = (n - 1)\cos^{n-2}x(-\sin x \ dx) \qquad v = \sin x.$$
$$= -(n - 1) \cos^{n-2}x \sin x \ dx.$$

The formula

$$\int u \ dv = uv - \int v \ du$$

then gives

$$\int \cos^n x \ dx = \int \cos^{n-1}x \cos x \ dx$$

$$= \cos^{n-1}x \sin x + (n - 1) \int \cos^{n-2}x \ \sin^2 x \ dx$$

$$= \cos^{n-1}x \sin x + (n - 1) \int \cos^{n-2}x(1 - \cos^2 x) \ dx,$$

$$= \cos^{n-1}x \sin x + (n - 1) \int \cos^{n-2}x \ dx - (n - 1) \int \cos^n x \ dx.$$

If we add

$$(n - 1) \int \cos^n x \ dx$$

to both sides of this equation, we obtain

$$n \int \cos^n x \ dx = \cos^{n-1}x \sin x + (n - 1) \int \cos^{n-2}x \ dx.$$

We then divide through by n, and the final result is

$$\int \cos^n x \ dx = \frac{\cos^{n-1}x \sin x}{n} + \frac{n - 1}{n} \int \cos^{n-2}x \ dx. \qquad (1)$$

This allows us to reduce the exponent on cos x by 2 and is a very useful formula. When n is a positive integer, we may apply the formula repeatedly until the remaining integral is either

$$\int \cos x \ dx = \sin x + C \qquad \text{or} \qquad \int \cos^0 x \ dx = \int dx = x + C. \quad \blacksquare$$

The companion formula for sines (Problem 29) is

$$\int \sin^n x \; dx = -\frac{\sin^{n-1}x \cos x}{n} + \frac{n-1}{n}\int \sin^{n-2}x \; dx. \qquad (2)$$

EXAMPLE 2 Evaluate

$$\int \cos^4 x \; dx.$$

Solution With $n = 4$ in Eq. (1), we get

$$\int \cos^4 x \; dx = \frac{\cos^3 x \sin x}{4} + \frac{3}{4}\int \cos^2 x \; dx,$$

and with $n = 2$,

$$\int \cos^2 x \; dx = \frac{\cos x \sin x}{2} + \frac{1}{2}\int dx.$$

Therefore,

$$\int \cos^4 x \; dx = \frac{\cos^3 x \sin x}{4} + \frac{3}{4}\left(\frac{\cos x \sin x}{2} + \frac{1}{2}x\right) + C. \qquad ■$$

EXAMPLE 3 Find a reduction formula for

$$\int \tan^n x \; dx, \quad n > 1.$$

Solution Instead of using integration by parts, we replace a factor of $\tan^2 x$ by $(\sec^2 x - 1)$:

$$\int \tan^n x \; dx = \int \tan^{n-2}x \, \tan^2 x \; dx = \int \tan^{n-2}x(\sec^2 x - 1)dx$$

$$= \int \tan^{n-2}x \, \sec^2 x \; dx - \int \tan^{n-2}x \; dx$$

$$= \frac{\tan^{n-1}x}{n-1} - \int \tan^{n-2}x \; dx.$$

The final result is

$$\int \tan^n x \; dx = \frac{\tan^{n-1}x}{n-1} - \int \tan^{n-2}x \; dx. \qquad ■ \quad (3)$$

The reduction formula for cotangents when $n > 1$ is

$$\int \cot^n x \; dx = -\frac{\cot^{n-1}x}{n-1} - \int \cot^{n-2}x \; dx. \qquad (4)$$

EXAMPLE 4   Evaluate

$$\int \cot^5 x \, dx.$$

Solution   We apply Eq. (4) with $n = 5$ to get

$$\int \cot^5 x \, dx = -\frac{\cot^4 x}{4} - \int \cot^3 x \, dx.$$

We then apply Eq. (4) again, with $n = 3$, to evaluate the remaining integral:

$$\int \cot^3 x \, dx = -\frac{\cot^2 x}{2} - \int \cot x \, dx$$

$$= -\frac{\cot^2 x}{2} - \ln|\sin x| + C.$$

The combined result is

$$\int \cot^5 x \, dx = -\frac{\cot^4 x}{4} + \frac{\cot^2 x}{2} + \ln|\sin x| + C. \qquad \blacksquare$$

EXAMPLE 5   Find a reduction formula for

$$\int \sec^n x \, dx, \quad n > 1.$$

Solution   We split off a factor of $\sec^2 x$ to go with dx and use integration by parts with

$$u = \sec^{n-2} x, \qquad\qquad dv = \sec^2 x \, dx,$$
$$du = (n-2)\sec^{n-3} x \, d(\sec x) \qquad v = \tan x.$$
$$= (n-2)\sec^{n-2} x \tan x \, dx,$$

With these substitutions, the formula

$$\int u \, dv = uv - \int v \, du$$

gives

$$\int \sec^n x \, dx = \int \sec^{n-2} x \sec^2 x \, dx$$

$$= \sec^{n-2} x \tan x - \int \tan x \, (n-2) \sec^{n-2} x \tan x \, dx$$

$$= \sec^{n-2} x \tan x - (n-2) \int \sec^{n-2} x \tan^2 x \, dx$$

$$= \sec^{n-2} x \tan x - (n-2) \int \sec^{n-2} x \, (\sec^2 x - 1) \, dx$$

$$= \sec^{n-2} x \tan x + (n-2) \int \sec^{n-2} x \, dx - (n-2) \int \sec^n x \, dx.$$

Solving for $\int \sec^n x \, dx$ then gives

$$\int \sec^n x \, dx = \frac{\sec^{n-2}x \tan x}{n-1} + \frac{n-2}{n-1}\int \sec^{n-2}x \, dx. \quad \blacksquare \quad (5)$$

The companion formula for the cosecant is

$$\int \csc^n x \, dx = -\frac{\csc^{n-2}x \cot x}{n-1} + \frac{n-2}{n-1}\int \csc^{n-2}x \, dx. \quad (6)$$

To test Eq. (5), we use it to evaluate the integral of $\sec^3 x$, a result we already know (Article 7.3, Example 3).

EXAMPLE 6   Evaluate

$$\int \sec^3 x \, dx.$$

Solution   We use Eq. (5) with $n = 3$:

$$\int \sec^3 x \, dx = \frac{\sec x \tan x}{2} + \frac{1}{2}\int \sec x \, dx$$

$$= \frac{\sec x \tan x}{2} + \frac{1}{2}\ln|\sec x + \tan x| + C. \quad \blacksquare$$

PROBLEMS

Use the reduction formulas in this article to evaluate the integrals in Problems 1–24.

1. $\displaystyle\int_{-\pi}^{\pi} \cos^4 x \, dx$

2. $\displaystyle\int \cos^4 2x \, dx$

3. $\displaystyle\int \cos^6 x \, dx$

4. $\displaystyle\int_{\pi/6}^{\pi/3} \cos^6 3x \, dx$

5. $\displaystyle\int_0^{\pi} \sin^4 x \, dx$

6. $\displaystyle\int \sin^4\left(\frac{x}{2}\right) dx$

7. $\displaystyle\int_0^{\pi/2} \sin^5 x \, dx$

8. $\displaystyle\int_0^{\pi/4} \sin^5 2x \, dx$

9. $\displaystyle\int \tan^3 2x \, dx$

10. $\displaystyle\int_0^{\pi/2} \tan^4\left(\frac{x}{2}\right) dx$

11. $\displaystyle\int \tan^5 x \, dx$

12. $\displaystyle\int \tan^5 2x \, dx$

13. $\displaystyle\int \cot^3 x \, dx$

14. $\displaystyle\int \cot^3\left(\frac{x}{3}\right) dx$

15. $\displaystyle\int_{\pi/4}^{3\pi/4} \cot^4 x \, dx$

16. $\displaystyle\int \cot^4 2x \, dx$

17. $\displaystyle\int_{-\pi/3}^{\pi/3} \sec^4 x \, dx$

18. $\displaystyle\int_0^{\pi/16} \sec^4 4x \, dx$

19. $\displaystyle\int \sec^5 x \, dx$

20. $\displaystyle\int \sec^5\left(\frac{x}{2}\right) dx$

21. $\displaystyle\int_{\pi/4}^{3\pi/4} \csc^3 x \, dx$

22. $\displaystyle\int_{\pi}^{2\pi} \csc^3\left(\frac{x}{3}\right) dx$

23. $\displaystyle\int \csc^4 x \, dx$

24. $\displaystyle\int \csc^5 x \, dx$

To evaluate the integrals in Problems 25–28, make a trigonometric substitution and apply a reduction formula.

25. $\displaystyle\int_0^1 (x^2 + 1)^{-3/2} \, dx$

26. $\displaystyle\int_0^1 (x^2 + 1)^{3/2} dx$

27. $\displaystyle\int_0^{3/5} \frac{dx}{(1 - x^2)^3}$

28. $\displaystyle\int_1^2 \frac{(x^2 - 1)^{3/2}}{x} dx$

29. Verify the formula

$$\int \sin^n x \, dx = -\frac{\sin^{n-1}x \cos x}{n} + \frac{n-1}{n}\int \sin^{n-2}x \, dx$$

by differentiating the right-hand side and combining the results to get $\sin^n x$.

30. Verify the formula

$$\int \csc^n x \, dx = -\frac{\csc^{n-2}x \cot x}{n-1} + \frac{n-2}{n-1}\int \csc^{n-2}x \, dx$$

by differentiating the right-hand side and combining the results to get $\csc^n x$.

31. a) Derive the formula

$$\int x^n \sin x \, dx = -x^n \cos x + n \int x^{n-1} \cos x \, dx.$$

b) What is the corresponding formula for $x^n \sin ax$?

32. a) Derive the formula

$$\int x^n \cos x \, dx = x^n \sin x - n \int x^{n-1} \sin x \, dx.$$

b) What is the corresponding formula for $x^n \cos ax$?

33. Derive the formula

$$\int x^m(\ln x)^n \, dx = \frac{x^{m+1}(\ln x)^n}{m+1} - \frac{n}{m+1}\int x^m(\ln x)^{n-1} \, dx$$

and use it to evaluate

$$\int_1^3 x^3 (\ln x)^2 \, dx.$$

34. Derive the formula

$$\int x^n e^x \, dx = x^n e^x - n \int x^{n-1}e^x \, dx$$

and use it to evaluate

$$\int_0^1 x^3 e^x \, dx.$$

REVIEW QUESTIONS AND EXERCISES

1. What are the general methods for finding indefinite integrals?

2. What substitution(s) would you consider if the integrand contained the following terms?

a) $\sqrt{x^2 + 9}$
b) $\sqrt{x^2 - 9}$
c) $\sqrt{9 - x^2}$
d) $\sin^3 x \cos^2 x$
e) $\sin^2 x \cos^2 x$

3. What method(s) would you try if the integrand contained the following terms?

a) $\sin^{-1} x$
b) $\ln x$
c) $\sqrt{1 + 2x - x^2}$
d) $x \sin x$
e) $\dfrac{2x + 3}{x^2 - 5x + 6}$
f) $\sin 5x \cos 3x$

g) $\dfrac{1 - \sqrt{x}}{1 + \sqrt[4]{x}}$
h) $x\sqrt{2x + 3}$

4. Discuss two types of improper integral. Define convergence and divergence of each type. Give examples of convergent and divergent integrals of each type. What tests are there for convergence and divergence of improper integrals?

5. What is a reduction formula? How are reduction formulas used? Illustrate by using the formula

$$\int \tan^n ax = \frac{\tan^{n-1} ax}{a(n-1)} - \int \tan^{n-2} ax \, dx$$

to evaluate

$$\int_0^{\pi/2} \tan^3\left(\frac{x}{2}\right) dx.$$

MISCELLANEOUS PROBLEMS

Evaluate the following integrals.

1. $\displaystyle\int \frac{\cos x \, dx}{\sqrt{1 + \sin x}}$

2. $\displaystyle\int_0^{\sqrt{2}/2} \frac{\sin^{-1}x \, dx}{\sqrt{1 - x^2}}$

3. $\displaystyle\int \frac{\tan x \, dx}{\cos^2 x}$

4. $\displaystyle\int \frac{y}{y^4 + 1} \, dy$

5. $\displaystyle\int_0^9 e^{\ln \sqrt{x}}dx$

6. $\displaystyle\int \frac{\cos \sqrt{x}}{\sqrt{x}} \, dx$

7. $\displaystyle\int \frac{dx}{\sqrt{x^2 + 2x + 2}}$

8. $\displaystyle\int_4^5 \frac{(3x - 7) \, dx}{(x - 1)(x - 2)(x - 3)}$

9. $\displaystyle\int x^2 e^x \, dx$

10. $\displaystyle\int \sqrt{x^2 + 1} \, dx$

11. $\displaystyle\int \frac{e^t \, dt}{1 + e^{2t}}$

12. $\displaystyle\int \frac{dx}{e^x + e^{-x}}$

13. $\displaystyle\int \frac{dx}{(x + 1)\sqrt{x}}$

14. $\displaystyle\int_0^{64} \frac{dx}{\sqrt{1 + \sqrt{x}}}$

15. $\displaystyle\int t^{2/3}(t^{5/3} + 1)^{2/3} \, dt$

16. $\displaystyle\int_{\pi/6}^{\pi/2} \frac{\cot x \, dx}{\ln(e \sin x)}$

17. $\displaystyle\int \frac{dt}{\sqrt{e^t + 1}}$

18. $\displaystyle\int \frac{\sin xe^{\sec x}}{\cos^2 x} \, dx$

19. $\displaystyle\int_0^{\pi/2} \frac{\cos x \, dx}{1 + \sin^2 x}$

20. $\displaystyle\int_{1/2}^{(1+\sqrt{2})/2} \frac{dx}{\sqrt{2x - x^2}}$

21. $\displaystyle\int_0^{\pi/2} \frac{\sin x \, dx}{1 + \cos^2 x}$

22. $\displaystyle\int_0^{\pi/4} \frac{\cos 2t}{1 + \sin 2t} \, dt$

23. $\displaystyle\int_{\pi/4}^{\pi/3} \frac{dx}{\sin x \cos x}$

24. $\displaystyle\int \sqrt{1 + \sin x}\, dx$

25. $\displaystyle\int_{-\pi/2}^{0} \sqrt{1 - \sin x}\, dx$

26. $\displaystyle\int \frac{dx}{\sqrt{(a^2 + x^2)^3}}$

27. $\displaystyle\int \frac{e^{2x}\, dx}{\sqrt[3]{1 + e^x}}$

28. $\displaystyle\int_0^1 \ln \sqrt{1 + x^2}\, dx$

29. $\displaystyle\int_{5/4}^{5/3} \frac{dx}{(x^2 - 1)^{3/2}}$

30. $\displaystyle\int_0^{3/5} \frac{x^3\, dx}{\sqrt{1 - x^2}}$

31. $\displaystyle\int \frac{dx}{x(2 + \ln x)}$

32. $\displaystyle\int \frac{\cos 2x - 1}{\cos 2x + 1}\, dx$

33. $\displaystyle\int_0^{\ln 2} \frac{e^{2x}\, dx}{\sqrt[4]{e^x + 1}}$

34. $\displaystyle\int \frac{2 \sin \sqrt{x}\, dx}{\sqrt{x} \sec \sqrt{x}}$

35. $\displaystyle\int_0^3 (16 + x^2)^{-3/2}\, dx$

36. $\displaystyle\int_0^{(e-1)^2} \frac{dx}{\sqrt{x + 1}}$

37. $\displaystyle\int \sin \sqrt{x + 1}\, dx$

38. $\displaystyle\int \cos \sqrt{1 - x}\, dx$

39. $\displaystyle\int_0^1 \frac{dx}{4 - x^2}$

40. $\displaystyle\int_0^1 \frac{dx}{x^3 + 1}$

41. $\displaystyle\int \frac{dy}{y(2y^3 + 1)^2}$

42. $\displaystyle\int \frac{x\, dx}{1 + \sqrt{x}}$

43. $\displaystyle\int \frac{dx}{x(x^2 + 1)^2}$

44. $\displaystyle\int \ln \sqrt{x - 1}\, dx$

45. $\displaystyle\int \frac{dx}{e^x - 1}$

46. $\displaystyle\int \frac{(x + 1)\, dx}{x^2(x - 1)}$

47. $\displaystyle\int \frac{x\, dx}{x^2 + 4x + 3}$

48. $\displaystyle\int_{\ln 2}^{\ln 3} \frac{15\, du}{(e^u - e^{-u})^2}$

49. $\displaystyle\int \frac{4\, dx}{x^3 + 4x}$

50. $\displaystyle\int \frac{dx}{5x^2 + 8x + 5}$

51. $\displaystyle\int \frac{\sqrt{x^2 - 1}}{x}\, dx$

52. $\displaystyle\int e^x \cos 2x\, dx$

53. $\displaystyle\int \frac{dx}{x(3\sqrt{x} + 1)}$

54. $\displaystyle\int \frac{dx}{x(1 + \sqrt[3]{x})}$

55. $\displaystyle\int_{\pi/6}^{\pi/2} \frac{\cot \theta\, d\theta}{1 + \sin^2\theta}$

56. $\displaystyle\int \frac{z^5\, dz}{\sqrt{1 + z^2}}$

57. $\displaystyle\int \frac{e^{4t}\, dt}{(1 + e^{2t})^{2/3}}$

58. $\displaystyle\int \frac{dx}{x^{1/5}\sqrt{1 + x^{4/5}}}$

59. $\displaystyle\int \frac{(x^3 + x^2)\, dx}{x^2 + x - 2}$

60. $\displaystyle\int_2^3 \frac{x^3 + 1}{x^3 - x}\, dx$

61. $\displaystyle\int \frac{x\, dx}{(x - 1)^2}$

62. $\displaystyle\int \frac{(x + 1)\, dx}{(x^2 + 2x - 3)^{2/3}}$

63. $\displaystyle\int \frac{dy}{(2y + 1)\sqrt{y^2 + y}}$

64. $\displaystyle\int \frac{dx}{x^2\sqrt{a^2 - x^2}}$

65. $\displaystyle\int (1 - x^2)^{3/2}\, dx$

66. $\displaystyle\int \ln (x + \sqrt{1 + x^2})\, dx$

67. $\displaystyle\int x \tan^2 x\, dx$

68. $\displaystyle\int x \cos^2 x\, dx$

69. $\displaystyle\int_0^{\pi} x^2 \sin x\, dx$

70. $\displaystyle\int x \sin^2 x\, dx$

71. $\displaystyle\int_0^1 \frac{dt}{t^4 + 4t^2 + 3}$

72. $\displaystyle\int \frac{du}{e^{4u} + 4e^{2u} + 3}$

73. $\displaystyle\int_0^2 x \ln \sqrt{x + 2}\, dx$

74. $\displaystyle\int (x + 1)^2 e^x\, dx$

75. $\displaystyle\int \sec^{-1}x\, dx$

76. $\displaystyle\int \frac{8\, dx}{x^4 + 2x^3}$

77. $\displaystyle\int \frac{x\, dx}{x^4 - 16}$

78. $\displaystyle\int_0^{\pi/2} \frac{\cos x\, dx}{\sqrt{1 + \cos x}}$

79. $\displaystyle\int \frac{\cos x\, dx}{\sin^3 x - \sin x}$

80. $\displaystyle\int_0^{\ln 2} \frac{du}{(e^u + e^{-u})^2}$

81. $\displaystyle\int_0^1 \frac{x\, dx}{1 + \sqrt{x} + x}$

82. $\displaystyle\int \frac{\sec^2 t\, dt}{\sec^2 t - 3 \tan t + 1}$

83. $\displaystyle\int \frac{dt}{\sec^2 t + \tan^2 t}$

84. $\displaystyle\int_0^{\tan^{-1}\sqrt{2}} \frac{dx}{1 + \cos^2 x}$

85. $\displaystyle\int e^{2t} \cos (e^t)\, dt$

86. $\displaystyle\int_0^1 \ln \sqrt{x^2 + 1}\, dx$

87. $\displaystyle\int x \ln(x^3 + x)\, dx$

88. $\displaystyle\int_0^1 x^3 e^{x^2}\, dx$

89. $\displaystyle\int \frac{\cos x\, dx}{\sqrt{4 - \cos^2 x}}$

90. $\displaystyle\int_0^{\pi/4} \frac{\sec^2 x\, dx}{\sqrt{4 - \sec^2 x}}$

91. $\displaystyle\int x^2 \sin (1 - x)\, dx$

92. $\displaystyle\int_0^1 \frac{dx}{(x^2 + 1)(2 + \tan^{-1}x)}$

93. $\displaystyle\int \frac{dx}{\cot^3 x}$

94. $\displaystyle\int_0^{1/3} x \ln \sqrt[3]{3x + 1}\, dx$

95. $\displaystyle\int \frac{x^3\, dx}{(x^2 + 1)^2}$

96. $\displaystyle\int_0^{3/4} \frac{x\, dx}{\sqrt{1 - x}}$

97. $\displaystyle\int_0^4 x\sqrt{2x + 1}\, dx$

98. $\displaystyle\int \ln (x + \sqrt{x^2 - 1})\, dx$

99. $\displaystyle\int \ln(x - \sqrt{x^2 - 1})\, dx$

100. $\displaystyle\int_0^{\ln \sqrt{3}} e^{-x} \tan^{-1} (e^x)\, dx$

101. $\displaystyle\int \ln(x + \sqrt{x})\, dx$

102. $\displaystyle\int \tan^{-1} \sqrt{x}\, dx$

103. $\displaystyle\int_1^2 \ln(x^2 + x)\, dx$

104. $\displaystyle\int_0^{\pi^2} \cos \sqrt{x}\, dx$

105. $\displaystyle\int_0^{\pi^2/4} \sin \sqrt{x}\, dx$

106. $\int_0^2 \tan^{-1} \sqrt{x+1}\, dx$ **107.** $\int \sqrt{1-x^2}\, \sin^{-1} x\, dx$

108. $\int_0^\pi x \sin^2 (2x)\, dx$ **109.** $\int \dfrac{\tan x\, dx}{\tan x + \sec x}$

110. $\int \dfrac{dt}{\sqrt{e^{2t} + 1}}$ **111.** $\int \dfrac{dx}{(\cos^2 x + 4 \sin x - 5)\cos x}$

112. $\int \dfrac{dt}{a + be^{ct}}, \quad a, b, c \neq 0$

113. $\int \sqrt{\dfrac{1 - \cos x}{\cos \alpha - \cos x}}\, dx, \quad \alpha$ constant, $0 < \alpha < x < \pi$

114. $\int \dfrac{dx}{9x^4 + x^2}$ **115.** $\int_0^1 \ln(2x^2 + 4)\, dx$

116. $\int \dfrac{\sin x\, dx}{\cos^2 x - 5 \cos x + 4}$ **117.** $\int \dfrac{dt}{\sqrt{1 - e^{-t}}}$

118. $\int \dfrac{\tan^{-1} x}{x^2}\, dx$

We dare you to evaluate the integrals in Problems 119–128.

119. $\int (\sin^{-1} x)^2\, dx$ **120.** $\int \dfrac{dx}{x(x+1)(x+2)\cdots(x+m)}$

121. $\int x \sin^{-1} x\, dx$ **122.** $\int \sin^{-1} \sqrt{x}\, dx$

123. $\int \dfrac{d\theta}{1 - \tan^2 \theta}$ **124.** $\int \ln(\sqrt{x} + \sqrt{1+x})\, dx$

125. $\int \dfrac{dt}{t - \sqrt{1 - t^2}}$ **126.** $\int \dfrac{(2e^{2x} - e^x)\, dx}{\sqrt{3e^{2x} - 6e^x - 1}}$

127. $\int \dfrac{dx}{x^4 + 4}$ **128.** $\int \dfrac{dx}{x^6 - 1}$

In Problems 129–131, evaluate each of the limits by identifying it with an appropriate definite integral and evaluating the latter.

129. $\lim\limits_{n \to \infty} \left(\dfrac{n}{n^2 + 0^2} + \dfrac{n}{n^2 + 1^2} + \dfrac{n}{n^2 + 2^2} + \cdots \right.$

$$\left. + \dfrac{n}{n^2 + (n+1)^2} \right)$$

130. $\lim\limits_{n \to \infty} \sum\limits_{k=1}^{n} \ln \sqrt[n]{1 + \dfrac{k}{n}}$ **131.** $\lim\limits_{n \to \infty} \sum\limits_{k=0}^{n-1} \dfrac{1}{\sqrt{n^2 - k^2}}$

132. Show that $\int_0^\infty x^3 e^{-x^2}\, dx$ converges and find its value.

133. Show that $\int_0^1 \ln x\, dx$ converges and find its value.

134. At points of the curve $y = 2\sqrt{x}$, lines of length $h = y$ are drawn perpendicular to the coordinate plane (Fig. 7.33).

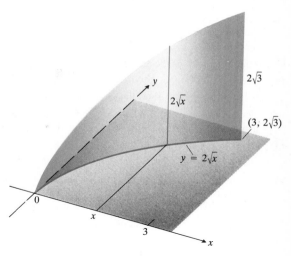

7.33   The surface in Problem 134.

Find the area of the surface formed by these lines from $(0, 0)$ to $(3, 2\sqrt{3})$.

135. A plane figure is bounded by a straight line and a 90° arc of a circle of radius a. Find its area and its centroid.

136. Find the coordinates of the center of mass of the region bounded by the curves $y = e^x$, $x = 1$, and $y = 1$.

137. At points on a circle of radius a, perpendiculars to its plane are erected, the perpendicular at each point P being of length ks, where s is the length of the arc of the circle from $(a, 0)$ to P and k is a positive constant. Find the area of the surface formed by the perpendiculars along the arc beginning at $(a, 0)$ and extending once around the circle. See Fig. 7.34.

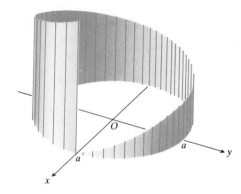

7.34   The surface in Problem 137.

138. A plate in the first quadrant, bounded by the curves $y = e^x$, $y = 1$, and $x = 1$, is submerged vertically in water with its upper corner on the surface. The surface of the water is given by the line $y = e$. Find the total force on one side of the plate if water weighs 62.5 lb/ft^3 and x and y are measured in feet.

139. Find the length of the arc of $y = \ln x$ from $x = 1$ to $x = e$.

140. The arc in Problem 139 is revolved about the y-axis to generate a surface. Find the area of the surface.

141. The curve $y = e^x$, $0 \le x \le 2$, is revolved about the x-axis to generate a surface. Find the area of the surface.

142. Find the area of the surface generated by revolving the curve $y = \cos x$, $-\pi/2 \le x \le \pi/2$ about the x-axis.

143. The region in the first quadrant enclosed by the curve $y = \ln(1/x)$ and the coordinate axes is revolved about the x-axis to generate a solid. Find the volume of the solid.

144. Find the area of the region that lies between the x-axis and the curve $x = a \ln(\sec t + \tan t) - a \sin t$, $y = a \cos t$, $-\pi/2 < t < \pi/2$. See Fig. 7.35.

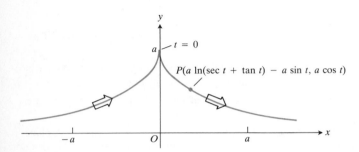

$P(a \ln(\sec t + \tan t) - a \sin t, a \cos t)$

7.35   The curve in Problem 144 covers the entire x-axis. The arrows show the direction of increasing t.

145. The region in Problem 144 is revolved about the x-axis to generate a solid. Find the volume of the solid.

146. The curve in Problem 144 is revolved about the x-axis to generate a surface. Find the area of the surface.

147. Does the integral

$$\int_1^\infty \frac{\ln x \, dx}{x^2}$$

converge, or diverge?

148. Sketch the following curves and show their behavior for large $|x|$.
a) $y = x - e^x$
b) $y = e^{(x - e^x)}$

Show that the following integrals converge and compute their values:

c) $\int_{-\infty}^b e^{(x - e^x)} \, dx$,   d) $\int_{-\infty}^\infty e^{(x - e^x)} \, dx$.

149. Find

$$\lim_{n \to \infty} \int_0^1 \frac{n y^{n-1}}{1 + y} \, dy.$$

150. Show that if $p(x)$ is a polynomial of degree n then

$$\int e^x p(x) \, dx = e^x [p(x) - p'(x) + p''(x) - \cdots + (-1)^n p^{(n)}(x)].$$

The Substitution $z = \tan(x/2)$

The substitution

$$z = \tan \frac{x}{2} \tag{1}$$

reduces the problem of integrating any rational function of $\sin x$ and $\cos x$ to a problem involving a rational function of z. This in turn can be integrated by partial fractions. Thus the substitution (1) is a very powerful tool. It is cumbersome, however, and used only when simpler methods fail.

Figure 7.36 shows how $\tan(x/2)$ expresses a rational function of $\sin x$ and $\cos x$. To see the effect of the substitution, we calculate

$$\cos x = 2 \cos^2 \frac{x}{2} - 1 = \frac{2}{\sec^2(x/2)} - 1$$

$$= \frac{2}{1 + \tan^2(x/2)} - 1 = \frac{2}{1 + z^2} - 1$$

or

$$\cos x = \frac{1 - z^2}{1 + z^2}, \tag{2}$$

and

$$\sin x = 2 \sin \frac{x}{2} \cos \frac{x}{2} = 2 \frac{\sin(x/2)}{\cos(x/2)} \cdot \cos^2 \frac{x}{2}$$

$$= 2 \tan \frac{x}{2} \cdot \frac{1}{\sec^2(x/2)} = \frac{2 \tan(x/2)}{1 + \tan^2(x/2)}$$

or

$$\sin x = \frac{2z}{1 + z^2}. \tag{3}$$

Finally, $x = 2 \tan^{-1} z$, so that

$$dx = \frac{2 \, dz}{1 + z^2}. \tag{4}$$

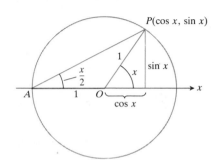

7.36   From this figure, we can read the relation

$$\tan(x/2) = \frac{(\sin x)}{(1 + \cos x)}.$$

See the discussion of the substitution $z = \tan(x/2)$.

EXAMPLE 1

$$\int \frac{dx}{1 + \cos x} = \int \frac{1 + z^2}{2} \frac{2\, dz}{1 + z^2}$$

$$= \int dz = z + C = \tan \frac{x}{2} + C \quad \blacksquare$$

EXAMPLE 2

$$\int \frac{dx}{2 + \sin x} = \int \left(\frac{1 + z^2}{2 + 2z + 2z^2} \right) \frac{2\, dz}{1 + z^2}$$

$$= \int \frac{dz}{z^2 + z + 1} = \int \frac{dz}{(z + 1/2)^2 + 3/4}$$

$$= \int \frac{du}{u^2 + a^2} \quad \left(u = z + \frac{1}{2}, \quad a = \frac{\sqrt{3}}{2} \right)$$

$$= \frac{1}{a} \tan^{-1} \frac{u}{a} + C = \frac{2}{\sqrt{3}} \tan^{-1} \frac{2z + 1}{\sqrt{3}} + C$$

$$= \frac{2}{\sqrt{3}} \tan^{-1} \frac{1 + 2\tan(x/2)}{\sqrt{3}} + C. \quad \blacksquare$$

Use the substitutions (1)–(4) to evaluate the integrals in Problems 151–158. Integrals like the ones in Problems 151 and 153 arise in calculating the average angular velocity of the output shaft of a universal joint when the input and output shafts are not aligned.

151. $\displaystyle\int_0^\pi \frac{dx}{1 + \sin x}$

152. $\displaystyle\int_{\pi/2}^\pi \frac{dx}{1 - \cos x}$

153. $\displaystyle\int \frac{dx}{1 - \sin x}$

154. $\displaystyle\int_0^{\pi/2} \frac{dx}{2 + \cos x}$

155. $\displaystyle\int \frac{\cos x\, dx}{1 - \cos x}$

156. $\displaystyle\int \frac{dx}{1 + \sin x + \cos x}$

157. $\displaystyle\int \frac{dx}{\sin x - \cos x}$

158. $\displaystyle\int_{\pi/2}^{2\pi/3} \frac{dx}{\sin x + \tan x}$

CHAPTER 8

Conic Sections and Other Plane Curves

OVERVIEW

This chapter is mostly about how the conic sections that originated in Greek geometry are described today as the graphs of quadratic equations in the coordinate plane. The Greeks of Plato's time described these curves as the curves formed by cutting through a double cone with a plane; hence the name "conic section." There are a number of possibilities, as you can see in Fig. 8.1. Toward the end of the chapter we shall look at parametric equations for the conic sections and introduce cycloids and some other special plane curves.

Kepler's use of conic sections in his *Commentaries on the Motions of Mars* in 1609, the work in which he announced his first two laws (elliptical orbits, equal areas in equal times), led to an energetic reexamination of conic sections for properties that might be useful in astronomy. Optics, an interest of mathematicians since Greek times (Claudius Ptolemy experimented with refraction, but when Snell's law eluded him he fitted his data with a parabola), received greatly increased attention after the invention of the telescope and microscope in the beginning of the seventeenth century. The resulting study of lenses and mirrors led to investigations of their surface shapes and, since these surfaces are surfaces of revolution, to an examination of their generating curves. For many of the mirrors used in reflecting telescopes, these curves are conic sections.

The acceptance of the idea that planets revolve around the sun (instead of around the earth) called for new principles of mechanics to account for their orbits and, more generally, for the path of any object moving under the action of a force. The resulting studies led to an examination of plane curves, from the paths of the planets to the paths followed by projectiles. By the seventeenth century, cannons could hit targets thousands of meters away and accurate predictions of the paths of projectiles became increasingly important. (John Napier, who invented logarithms, also invented an artillery piece that could hit a cow a mile away and was so horrified by the weapon's accuracy that he stopped its development.) To a first approximation, the path of a projectile is a conic section, as we shall see in Chapter 13.

As the mathematicians of the sixteenth and seventeenth centuries studied Greek works, they began to realize that the Greek methods of proof lacked generality. Nearly every theorem required its own special arguments and calculations, as we saw when we calculated areas in Article 4.6. The approach to conic sections was eventually changed from a purely geometric one (sections of a cone) to one that used the notions of coordinates and distance. For example, by 1579, Guidobaldo del Monte was defining an ellipse as the set of points in a plane the sum of whose distances from two fixed points is a constant. (We shall use this definition in Article 8.4.)

From an algebraic point of view, the present chapter sets the stage for the amazing result that the graph (if any) of a quadratic equation in x and y in the plane is either a conic section or two parallel lines. No other configuration can be the graph of a quadratic equation in x and y, ever.

From the point of view of applications (and here is where the calculus comes in), the mathematics of conic sections is just what we need to describe the paths of planets, comets, moons, asteroids, satellites, or anything else that is moved through space by gravitational forces. Once we know that the path of a moving body is a conic section, we know volumes of information about its velocity, its acceleration, and the force that drives it, as we shall see in Chapter 14.

If you enjoy history, you will find a great deal to interest you about conics in Morris Kline's *Mathematical Thought from Ancient to Modern Times* (Oxford University Press, 1972). Another fine reference is Morris Kline's *Mathematics in Western Culture* (Oxford University Press, 1964).

(a)

Circle: plane perpendicular to cone axis

Ellipse

Parabola: plane parallel to side of cone

Hyperbola: plane parallel to cone axis

(b)

Point: plane through cone vertex only

Single line: plane tangent to cone

Pair of intersecting lines

8.1 (a) The "standard" conic sections are the curves in which a plane cuts a double cone. In the case of a hyperbola, the intersection consists of two curves, not one, each called a "branch" of the hyperbola. The "degenerate" conic sections in (b) are obtained by passing the plane through the cone's vertex.

8.1
Equations from the Distance Formula

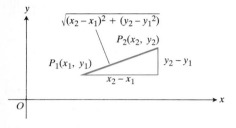

In this article we use the distance formula to generate equations for curves in the Cartesian plane.

As you know, the distance d between two points (x_1, y_1) and (x_2, y_2) in the plane is calculated from their coordinates by the formula

$$d = \sqrt{(x_2 - x_1)^2 + (y_2 - y_1)^2}. \tag{1}$$

8.2 The distance between two points in the plane is calculated from the points' coordinates with the formula from the Pythagorean theorem.

This formula comes from applying the Pythagorean theorem to the triangle in Fig. 8.2. The formula is particularly useful in finding equations for curves whose geometric character depends on one or more distances, as in the following example.

EXAMPLE 1   Find an equation for the set of points $P(x, y)$ that are equidistant from the origin O and the line L: $x = 4$.

Solution   The distance between P and L is the perpendicular distance PQ between $P(x, y)$ and the point $Q(4, y)$ on L that has the same y-coordinate as P. (See Fig. 8.3.) Thus,

$$PQ = \sqrt{(4 - x)^2 + (y - y)^2} = |4 - x|.$$

The distance OP is

$$OP = \sqrt{x^2 + y^2}.$$

The condition to be satisfied by P is

$$OP = PQ, \quad \text{or} \quad \sqrt{x^2 + y^2} = |4 - x|. \tag{2}$$

If (2) holds, so does the equation we get by squaring:

$$x^2 + y^2 = 16 - 8x + x^2$$
$$y^2 = 16 - 8x. \tag{3}$$

That is, if a point is equidistant from O and L, then its coordinates must satisfy Eq. (3).

It is also true that any point whose coordinates satisfy Eq. (3) lies equidistant from O and L. For, if

$$y^2 = 16 - 8x,$$

then

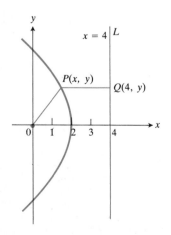

8.3   The curve traced by a point P that stays equidistant from the origin and the line $x = 4$.

$$\sqrt{x^2 + y^2} = \sqrt{x^2 + (16 - 8x)} = \sqrt{(x - 4)^2} = |x - 4| = |4 - x|,$$

and hence

$$OP = PQ.$$

Therefore Eq. (3) expresses both the necessary and sufficient condition for $P(x, y)$ to be equidistant from O and L. In other words, Eq. (3) is an equation for these equidistant points. ∎

PROBLEMS

Use the distance formula to find equations for the sets of points $P(x, y)$ that satisfy the conditions in Problems 1–11.

1. P is equidistant from the origin and the line $y = -4$.

2. P is equidistant from the point $(0, 1)$ and the line $y = -1$.

3. P is equidistant from the two points $A(-2, 1)$ and $B(2, -3)$.

4. The distance from P to $F_1(-1, 0)$ is twice its distance to $F_2(2, 0)$.

5. The product of the distances from P to $F_1(-2, 0)$ and $F_2(2, 0)$ is 4.

6. The sum of the distances from P to $F_1(1, 0)$ and $F_2(0, 1)$ is constant, and the curve passes through the origin.

7. The distance of P from the line $x = -2$ is 2 times its distance from the point $(2, 0)$.

8. The distance from P to $(-\sqrt{2}, -\sqrt{2})$ is 2 plus the distance from P to $(\sqrt{2}, \sqrt{2})$.

9. The distance of P from the point $(-3, 0)$ is 4 more than its distance from the point $(3, 0)$.

10. The distance of P from the line $y = 1$ is 3 less than its distance from the origin.

11. P is 3 units from the point $(2, 3)$.

12. Find the points on the line $x - y = 1$ that are 2 units from the point $(3, 0)$.

13. Find a point that is equidistant from the three points $A(0, 1)$, $B(1, 0)$, $C(4, 3)$. What is the radius of the circle through A, B, and C?

8.2

Circles

In this article, we derive equations for circles in the coordinate plane and show how the circles' radii and centers can be found directly from these equations.

DEFINITION

A **circle** is the set of points in a plane whose distance from a given fixed point in the plane is a constant.

Equations for Circles

Let $C(h, k)$ be the given fixed point, the **center** of the circle. Let a be the constant distance, the **radius** of the circle. Let $P(x, y)$ be a point on the circle (Fig. 8.4a). Then

$$CP = a, \tag{1}$$

$$\sqrt{(x - h)^2 + (y - k)^2} = a,$$

$$(x - h)^2 + (y - k)^2 = a^2. \tag{2}$$

If (1) is satisfied, so is (2), and if (2) is satisfied, so is (1). Therefore (2) is an equation for the points of the circle.

The standard equation for the circle of radius a centered at the point (h, k) is

$$(x - h)^2 + (y - k)^2 = a^2. \tag{3}$$

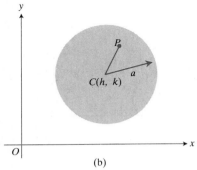

8.4 (a) $(x - h)^2 + (y - k)^2 = a^2$. (b) The region $(x - h)^2 + (y - k)^2 < a^2$ is the interior of the circle with center $C(h, k)$ and radius a.

EXAMPLE 1 Find an equation for the circle with center at the origin and with radius a.

Solution If $h = k = 0$, Eq. (3) becomes

$$x^2 + y^2 = a^2. \qquad \blacksquare \quad (4)$$

EXAMPLE 2 Find the circle through the origin with center at $C(2, -1)$.

Solution With $(h, k) = (2, -1)$, Eq. (3) takes the form

$$(x - 2)^2 + (y + 1)^2 = a^2.$$

Since the circle goes through the origin, $x = y = 0$ must satisfy the equation. Hence

$$(0 - 2)^2 + (0 + 1)^2 = a^2 \quad \text{or} \quad a^2 = 5.$$

The equation is

$$(x - 2)^2 + (y + 1)^2 = 5. \qquad \blacksquare$$

EXAMPLE 3 What points $P(x, y)$ satisfy the inequality

$$(x - h)^2 + (y - k)^2 < a^2? \qquad (5)$$

Solution The left side of (5) is the square of the distance CP from $C(h, k)$ to $P(x, y)$. The inequality is satisfied if and only if

$$CP < a,$$

that is, if and only if P lies inside the circle of radius a with center at $C(h, k)$ (Fig. 8.4b). $\qquad \blacksquare$

EXAMPLE 4 Find the center and radius of the circle

$$x^2 + y^2 + 4x - 6y = 12.$$

Solution We complete the squares in the x terms and y terms and get

$$(x^2 + 4x + 4) + (y^2 - 6y + 9) = 12 + 4 + 9,$$

or

$$(x + 2)^2 + (y - 3)^2 = 25.$$

This is Eq. (3) with $(h, k) = (-2, 3)$ and $a^2 = 25$. It therefore represents a circle with

Center: $C(-2, 3)$,

Radius: $a = 5$.

See Fig. 8.5. $\qquad \blacksquare$

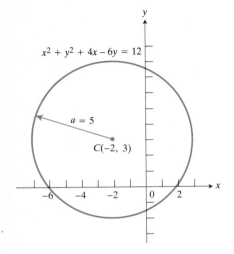

8.5 The graph of $x^2 + y^2 + 4x - 6y = 12$.

The Equation $Ax^2 + Ay^2 + Dx + Ey + F = 0$, $A \neq 0$

An equation of the form

$$Ax^2 + Ay^2 + Dx + Ey + F = 0, \qquad A \neq 0, \qquad (6)$$

can often be reduced to the form of Eq. (3) by completing the squares as we did in Example 4. We first divide (6) by A and write

$$\left(x^2 + \frac{D}{A}x\right) + \left(y^2 + \frac{E}{A}y\right) = -\frac{F}{A}. \tag{7}$$

We then complete the squares, adding $(D/2A)^2 = D^2/(4A^2)$ for x and $(E/2A)^2 = E^2/(4A^2)$ for y. Of course we must add to both sides of Eq. (7), thus obtaining

$$\left(x + \frac{D}{2A}\right)^2 + \left(y + \frac{E}{2A}\right)^2 = -\frac{F}{A} + \frac{D^2}{4A^2} + \frac{E^2}{4A^2} = \frac{D^2 + E^2 - 4AF}{4A^2}. \tag{8}$$

Equation (8) is like Eq. (3), with

$$a^2 = \frac{D^2 + E^2 - 4AF}{4A^2}, \tag{9}$$

provided the expression on the right is positive. Then Eq. (6) represents a circle with

Center: $\left(-\dfrac{D}{2A}, -\dfrac{E}{2A}\right)$,

Radius: $a = \sqrt{(D^2 + E^2 - 4AF)/4A^2}$. $\qquad$ (10)

If the right-hand side of Eq. (9) is zero, the curve reduces to a single point. If the right-hand side of Eq. (9) is negative, no points in the plane satisfy either (8) or (6).

In analyzing the equation

$$Ax^2 + Ay^2 + Dx + Ey + F = 0, \qquad A \neq 0, \tag{11}$$

we recommend that you use the method of Example 4 instead of memorizing the formulas in (10). It is enough to remember that *a quadratic equation like (6), in which x^2 and y^2 have equal coefficients, and with no xy term, represents a circle* (or a single point, or no point at all).

Dividing Eq. (6) by A gives an equation of the form

$$x^2 + y^2 + ax + by + c = 0, \tag{12}$$

where a, b, and c are constants. The constants a, b, and c can often be determined from geometric conditions. For instance, the circle might be known to pass through three specific (noncollinear) points, or to be tangent to three specific nonconcurrent lines, or to be tangent to two lines and to pass through a point not on either line.

EXAMPLE 5 Find an equation for the circle through the points (8, 0), (0, 6), and (0, 0).

Solution We substitute the coordinates of the three points into Eq. (12) to get three equations that can be solved simultaneously for a, b, and c:

$$x^2 + y^2 + ax + by + c = 0$$

(8, 0): $64 + 0 + 8a + 0 + c = 0$ $\qquad$ (13)

(0, 6): $0 + 36 + 0 + 6b + c = 0$ $\qquad$ (14)

(0, 0): $0 + 0 + 0 + 0 + c = 0.$ $\qquad$ (15)

Equation (15) tells us that $c = 0$. Setting $c = 0$ in Eqs. (13) and (14) gives

$$64 + 8a = 0, \qquad 36 + 6b = 0,$$
$$a = -8, \qquad b = -6.$$

The equation we want is Eq. (12) with $a = -8$, $b = -6$, and $c = 0$:

$$x^2 + y^2 - 8x - 6y = 0. \qquad \blacksquare$$

PROBLEMS

In Problems 1–4, find the circle with center $C(h, k)$ and radius a.

1. $C(0, 2), \quad a = 2$ **2.** $C(-2, 0), \quad a = 3$

3. $C(3, -4), \quad a = 5$ **4.** $C(1, 1), \quad a = \sqrt{2}$

5. Write an inequality that describes the points that lie inside the circle with center $C(-2, -1)$ and radius $a = \sqrt{6}$.

6. Write an inequality that describes the points that lie outside the circle with center $C(-4, 2)$ and radius $a = 4$.

In Problems 7–16, find the center and radius of the given circle. Then draw the circle. (*Hint for drawing circles:* Draw the circle first and mark the center. Then draw the coordinate axes, using the center and radius to decide where they should go.)

7. $x^2 + y^2 = 16$ **8.** $x^2 + y^2 + 6y = 0$

9. $x^2 + y^2 - 2y = 3$ **10.** $x^2 + y^2 + 2x = 8$

11. $x^2 + 4x + y^2 = 12$ **12.** $3x^2 + 3y^2 + 6x = 1$

13. $x^2 + y^2 + 2x + 2y = -1$

14. $x^2 + y^2 - 6x + 2y + 1 = 0$

15. $2x^2 + 2y^2 + x + y = 0$

16. $2x^2 + 2y^2 - 28x + 12y + 114 = 0$

What points satisfy the inequalities in Problems 17 and 18?

17. $x^2 + y^2 + 2x - 4y + 5 \le 0$

18. $x^2 + y^2 + 4x + 4y + 9 \ge 0$

19. Find an equation for the circle through the point $(4, 5)$ whose center is at the point $(2, 2)$.

20. Find an equation for the circle through the points $(0, 0)$ and $(6, 0)$ that is tangent to the line $y = -1$.

21. Find an equation for the circle centered at $(-1, 1)$ that is tangent to the line $x + 2y = 4$.

22. Find an equation for the circle through the points $(0, 0)$ and $(17, 7)$ whose center lies on the line $12x - 5y = 0$.

23. Find an equation for the circle through the points $(2, 3)$, $(3, 2)$, and $(-4, 3)$.

24. Find an equation for the circle through the points $(1, 0)$, $(0, 1)$, and $(2, 2)$.

25. Find an equation for the circle through the points $(7, 1)$, $(0, 0)$, and $(-1, 7)$. Find the center and radius of the circle.

26. Is the point $(0.1, 3.1)$ inside, outside, or on the circle

$$x^2 + y^2 - 2x - 4y + 3 = 0?$$

Why?

27. If the distance from $P(x, y)$ to the point $(6, 0)$ is twice its distance from the point $(0, 3)$, show that P lies on a circle and find the center and radius.

28. The sum of the squares of the distances from point $P(x, y)$ to the two points $A(-5, 2)$ and $B(1, 4)$ is 52. Show that the set of possible locations for P make up a circle whose center is the midpoint of the line segment AB. Do A and B lie inside the circle, outside the circle, or on the circle?

29. Find the dimensions of the rectangle of largest area that will fit inside the smaller segment cut from the circle $x^2 + y^2 = 36$ by the line $x = 1$. Assume that one side of the rectangle lies along the line.

30. Find the length of the upper base of the trapezoid of largest area that can be inscribed in a semicircle of radius a if its lower base lies along the diameter. What is the area of the trapezoid?

31. Show geometrically that the length s of either tangent to the circle $(x - h)^2 + (y - k)^2 = a^2$ from the exterior point (x_1, y_1) satisfies the equation $s^2 = (x_1 - h)^2 + (y_1 - k)^2 - a^2$.

32. Show that the line perpendicular to the circle $x^2 + y^2 = a^2$ at any point (x_1, y_1) passes through the origin.

33. In Fig. 8.6, the line PT is tangent to the circle at T, and the line through P and the circle's center C intersects the circle at M and N. Prove that $(PM)(PN) = (PT)^2$.

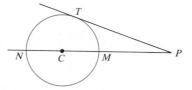

8.6 The circle and lines in Problem 33.

34. As you know, any angle inscribed in a semicircle is a right angle. Prove the converse: If angle OPA is a right angle for every choice of the point $P(x, y)$ on a curve C joining O and A, then the curve lies along a circle with diameter OA.

TOOLKIT PROGRAMS

Conic Sections   Super * Grapher

8.3
Parabolas

In this article, we derive equations for parabolas whose axes are parallel to one of the coordinate axes and show how to find the parabolas' foci and directrices directly from these equations. We also explain why parabolic mirrors make good reflectors and show how to draw parabolas with string the way Kepler once did.

DEFINITION

> A **parabola** is the set of points in a plane that are equidistant from a given fixed point and fixed line in the plane.

The fixed point is called the **focus** of the parabola and the fixed line the **directrix.**

If the focus F lies on the directrix L, then the parabola is nothing more than the line through F perpendicular to L. We consider this to be a degenerate case.

Parabolas That Open Upward

If F does not lie on L, then we may choose a coordinate system that results in a simple equation for the parabola by taking the positive y-axis through F perpendicular to L and taking the origin halfway between F and L. If the distance between F and L is $2p$, then F is the point $(0, p)$ and L is the line $y = -p$, as in Fig. 8.7.

In the notation of Fig. 8.7, a point $P(x, y)$ lies on the parabola if and only if the distances PF and PQ are equal:

$$PF = PQ. \tag{1}$$

From the distance formula,

$$PF = \sqrt{x^2 + (y - p)^2} \quad \text{and} \quad PQ = \sqrt{(y + p)^2}.$$

When we equate these two expressions, square, and simplify, we get

$$x^2 = 4py. \quad \text{(Opens upward)} \tag{2}$$

This equation is satisfied by any point on the parabola. Conversely, if (2) is satisfied then

$$PF = \sqrt{x^2 + (y - p)^2} = \sqrt{4py + (y^2 - 2py + p^2)}$$
$$= \sqrt{(y + p)^2} = PQ$$

and $P(x, y)$ lies on the parabola. In short, the parabola is the graph of the equation $x^2 = 4py$.

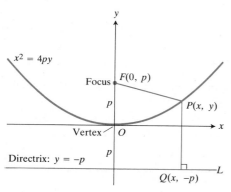

8.7 The parabola $x^2 = 4py$. Notice that a parabola always passes between its directrix and focus.

Axis and Vertex

Since p is positive in Eq. (2), y cannot be negative (if x is to be real) and the curve lies on or above the x-axis.

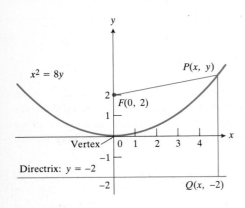

$x^2 = 8y$

8.8   The parabola $x^2 = 8y$.

The curve is symmetric about the y-axis since x appears only to an even power.

The axis of symmetry of the parabola is called the **axis** of the parabola. The point on this axis midway between the focus and the directrix lies on the parabola, since it is equidistant from the focus and the directrix. It is called the **vertex** of the parabola. The origin is the vertex of the parabola in Fig. 8.7. The tangent to a parabola at its vertex is parallel to the directrix. From Eq. (2) we find that the slope of the tangent at any point is $dy/dx = x/2p$, which is zero at the origin.

EXAMPLE 1   Find the focus and directrix of the parabola

$$x^2 = 8y. \tag{3}$$

Solution   See Fig. 8.8. Equation (3) is Eq. (2) with

$$4p = 8, \qquad p = 2.$$

The focus lies on the y-axis $p = 2$ units from the vertex, that is, at

   Focus: $F(0, 2)$.

The directrix $y = -p$ is the line $y = -2$:

   Directrix: $y = -2$.   ∎

Parabolas That Open Downward

If the parabola opens downward, as in Fig. 8.9, with its focus at $F(0, -p)$ and its directrix the line $y = p$, then Eq. (2) becomes

$$x^2 = -4py. \qquad \text{(Opens downward)} \tag{4}$$

Parabolas That Open to the Right or Left

If we interchange the roles of x and y in Eqs. (2) and (4), the resulting equations

$$y^2 = 4px \qquad \text{(Opens to the right)} \tag{5a}$$

and

$$y^2 = -4px \qquad \text{(Opens to the left)} \tag{5b}$$

also represent parabolas (Fig. 8.10). These parabolas are symmetric about the x-axis because y appears only to an even power. The vertex is still at the origin. The directrix is perpendicular to the axis of symmetry, p units from the vertex. The focus lies on the axis of symmetry, also p units from the vertex, and "inside" the curve. The parabola in (5a) opens toward the right because x must be greater than or equal to zero, while the parabola in (5b) opens toward the left.

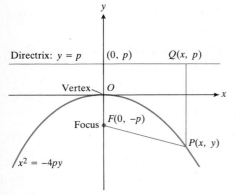

$x^2 = -4py$

8.9   The parabola $x^2 = -4py$.

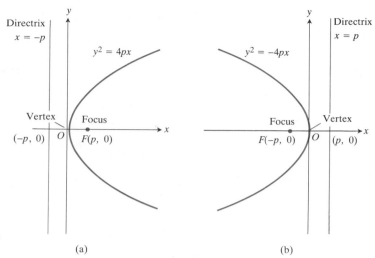

8.10 (a) The parabola $y^2 = 4px$. (b) The parabola $y^2 = -4px$.

(a)

(b)

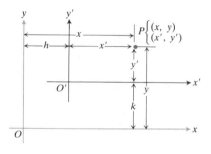

8.11 A translation of the coordinate axes. Note that $h + x' = x$ and $k + y' = y$. Solving for x' and y' gives $x' = x - h$, $y' = y - k$.

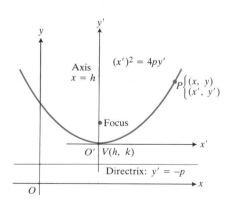

8.12 A parabola with vertex at $V(h, k)$ opening upward.

Translating the Axes

If the vertex of the parabola is at a point $V(h, k)$ different from the origin, Eqs. (2), (4), and (5) no longer apply. However, we can find the appropriate equation by introducing a new coordinate system, with its origin O' at V and its axes parallel to the original axes. Every point P in the plane then has two sets of coordinates, say (x, y) in the original system and (x', y') in the new system. To move from O to P, we make a horizontal displacement x and a vertical displacement y. The value of x is obtained from two horizontal displacements: h from O to O' and x' from O' to P (Fig. 8.11). Similarly, y is obtained from two vertical displacements: k from O to O' and y' from O' to P. Thus the two sets of coordinates are related by the equations

$$\begin{cases} x = x' + h \\ y = y' + k \end{cases} \tag{6a}$$

or

$$\begin{cases} x' = x - h \\ y' = y - k. \end{cases} \tag{6b}$$

Equations (6) are called the equations for **translation of axes** because the new coordinate axes may be obtained by translating the old ones.

Suppose, now, we have a parabola, with vertex $V(h, k)$ and opening upward as in Fig. 8.12. In terms of $x'y'$-coordinates, Eq.(2) provides an equation of the parabola in the form

$$(x')^2 = 4py'. \tag{7}$$

By using Eqs. (6a), we may express this in xy-coordinates by the equation

$$(x - h)^2 = 4p(y - k). \tag{8a}$$

The axis of the parabola in (8a) is the line $x = h$. The equation of this line is obtained by setting the quadratic term $(x - h)^2$ in (8a) equal to zero. The focus lies on the axis of symmetry, p units above the vertex at $x = h$, $y = k + p$. The directrix lies p units below the vertex and perpendicular to the axis of symmetry, thus having equation $y = k - p$.

Other forms of equations of parabolas are included with Eq. (8a) in the following list.

Standard Equations for Parabolas with Vertex (h, k) and Axis Parallel to a Coordinate Axis

$$(x - h)^2 = 4p(y - k) \qquad \text{(Opens upward)} \qquad\qquad \text{(8a)}$$

$$(x - h)^2 = -4p(y - k) \qquad \text{(Opens downward)} \qquad\quad \text{(8b)}$$

$$(y - k)^2 = 4p(x - h) \qquad \text{(Opens to the right)} \qquad\quad \text{(8c)}$$

$$(y - k)^2 = -4p(x - h) \qquad \text{(Opens to the left)} \qquad\qquad \text{(8d)}$$

$$p = \text{focus–vertex and vertex–directrix distance}$$

Parabolas (8a) and (8b) are symmetric about the line $x = h$; (8c) and (8d) are symmetric about the line $y = k$.

EXAMPLE 2   Find an equation for the parabola with vertex $V(-1, 2)$ and focus $F(3, 2)$. What is the directrix?

Solution   Since the parabola opens to the right, we use Eq. (8c) with $h = -1$ and $k = 2$:

$$(y - 2)^2 = 4p(x + 1).$$

The number p is the distance between V and F:

$$p = \sqrt{(-1 - 3)^2 + (2 - 2)^2} = \sqrt{(-4)^2 + (0)^2} = 4.$$

The equation we seek is

$$(y - 2)^2 = 4(4)(x + 1) \qquad \text{or} \qquad (y - 2)^2 = 16(x + 1).$$

The directrix is the vertical line p units to the left of the vertex $(-1, 2)$. Since $p = 4$, the directrix is the line $x = -5$. ∎

EXAMPLE 3   Find an equation for the parabola with vertex $V(1, 2)$ and directrix $y = 3$. What are the coordinates of the focus?

Solution   Since the parabola opens downward, we use Eq. (8b) with $h = 1$ and $k = 2$:

$$(x - 1)^2 = -4p(y - 2).$$

The number p, which is the distance between the vertex $(1, 2)$ and the directrix $y = 3$, is

$$p = 3 - 2 = 1.$$

$$y^2 - 2y - 12x - 23 = 0,$$
or
$$(y - 1)^2 = 12(x + 2)$$

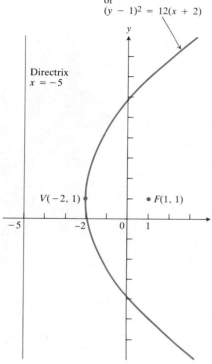

Directrix
$x = -5$

$V(-2, 1)$ • • $F(1, 1)$

-5 -2 0 1

8.13 The parabola
$$y^2 - 2y - 12x - 23 = 0$$
in Example 4.

The equation we seek is

$$(x - 1)^2 = -4(y - 2).$$

The focus lies $p = 1$ unit below the vertex $(1, 2)$, at the point $F(1, 1)$. ∎

Reducing Equations for Parabolas to Standard Form

The key feature of an equation that represents a parabola in the xy-plane with its axis parallel to one of the coordinate axes is that it is quadratic in that coordinate and linear in the other. Whenever we have such an equation, we may reduce it to one of the four standard forms in Eq. (8) by taking the following steps:

1. Complete the square in the coordinate that appears quadratically.
2. Put the remaining terms in the form $\pm 4p(x - h)$ or $\pm 4p(y - k)$, with p positive.

The parabola's vertex, distance from vertex to focus, axis of symmetry, and direction of opening can then be found in the usual way from the standard form.

EXAMPLE 4   Find the vertex, axis, focus, and directrix of the parabola

$$y^2 - 2y - 12x - 23 = 0.$$

Solution   The equation is quadratic in y and linear in x. We rearrange the equation as

$$y^2 - 2y = 12x + 23$$

and complete the square by adding $(-2/2)^2 = 1$ to both sides. This gives

$$y^2 - 2y + 1 = 12x + 24 \qquad \text{or} \qquad (y - 1)^2 = 12(x + 2).$$

The latter equation has the form

$$(y - k)^2 = 4p(x - h)$$

with $h = -2, k = 1, p = 3$. Hence, the vertex is $V(-2, 1)$. The parabola's axis is the line $y = k$ or $y = 1$. The parabola opens to the right because x must satisfy the inequality $(x + 2) \geq 0$ or $x \geq -2$ for y to be real. The focus lies $p = 3$ units to the right of the vertex $(-2, 1)$, at the point $F(1, 1)$. The directrix is the line parallel to the y-axis p units to the left of the vertex, $(-2, 1)$. Since $p = 3$, the directrix is the line $x = -5$. See Fig. 8.13. ∎

The Reflective Property of Parabolas

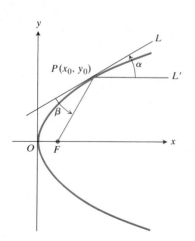

8.14 A parabolic reflector sends all light from the focus out parallel to the parabola's axis.

The chief application of parabolas involves their use as reflectors of light and radio waves. Rays originating at a parabola's focus are reflected out of the parabola parallel to the axis, and rays coming into the parabola parallel to the axis are reflected to the focus. Figure 8.14 shows why. The line L is a tangent to the parabola, the point F is the focus, and the ray PL' is parallel to the axis of the parabola. This makes angles α and β equal (Problem 34) so that any ray from F to P will be reflected out along PL'. Similarly, any ray coming in along PL' will be reflected toward F.

(Photograph from *PSSC Physics, Second Edition*; 1965; D.C. Heath & Company with Education Development Center, Inc. *NCFMF Book of Film Notes*, 1974; the MIT Press with Education Development Center, Inc., Newton, Massachusetts.)

This property is used in the parabolic mirrors of telescopes, in automobile headlamps, in spotlights of all kinds, in parabolic radar and microwave antennas, and in television "dish" receivers. The ripple tank photograph presented above shows waves entering a parabolic dam to be reflected toward the parabola's focus.

Kepler's Construction

Kepler's method for drawing a parabola (with more modern tools) requires a string the length of a T square and a table whose edge can serve as the parabola's directrix. Pin one end of the string to the point where you want the focus to be and the other end to the T square, as shown in Fig. 8.15. Then, holding the string taut against the T square with a pencil, slide the T square along the table's edge. As the T square moves, the pencil will trace a parabola: *PF* will always equal *PB* because the string has the constant length *AB*.

8.15   How Kepler used string to draw a parabola.

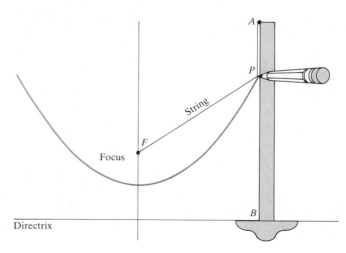

Applications

For a very readable account of the many roles played by parabolas in astronomy, radio communications, radar, wind tunnel photography, submarine tracking, and bridge construction, see Lee Whitt's "The Standup Conic Presents: The Parabola and Applications," *The UMAP Journal*, 3:3, 285–313 (1982). The shape of the surface of a liquid in a vertical cylinder rotating at a constant rate is parabolic, but attempts to make telescopes with revolving mercury have failed because of the difficulties in keeping the surface quiet enough to be really smooth.

PROBLEMS

Each of Problems 1–8 gives the vertex V and focus F of a parabola. Find equations for the parabola and its directrix. Then sketch the parabola, showing the focus, vertex, and directrix.

1. $V(0, 0)$, $F(0, 2)$
2. $V(0, 0)$, $F(-2, 0)$
3. $V(0, 0)$, $F(0, -4)$
4. $V(0, 0)$, $F(4, 0)$
5. $V(-2, 3)$, $F(-2, 4)$
6. $V(0, 3)$, $F(-1, 3)$
7. $V(-3, 1)$, $F(0, 1)$
8. $V(1, 3)$, $F(1, 0)$

Each of Problems 9–14 gives the vertex V and the directrix of a parabola. Find an equation for the parabola and the coordinates of its focus. Then sketch the parabola, showing the focus, vertex, and directrix.

9. $V(2,0)$,   the y-axis
10. $V(1, -2)$,   the x-axis
11. $V(-3, 1)$ the line $x = 1$
12. $V(-2, -2)$,   the line $y = -3$
13. $V(0, 1)$,   the line $x = -1$
14. $V(0, 1)$,   the line $y = 2$

In Problems 15–26, find the vertex, axis, focus, and directrix of the given parabola. Then sketch the parabola, showing these features.

15. $y^2 = 8x$
16. $y^2 = -36x$
17. $x^2 = 100y$
18. $x^2 = -9y$
19. $x^2 - 2x + 8y - 7 = 0$
20. $x^2 + 8x - 4y + 4 = 0$
21. $y^2 + 4x - 8 = 0$
22. $x^2 + 8y - 4 = 0$
23. $x^2 + 2x - 4y - 3 = 0$
24. $x^2 + 2x + 4y - 3 = 0$
25. $y^2 - 4y - 8x - 12 = 0$
26. $y^2 + 6y + 2x + 5 = 0$

27. What regions in the plane satisfy the inequalities $y^2 < x$ and $y^2 > x$? Sketch.

28. What regions in the plane satisfy the inequalities $x^2 < 8y$ and $x^2 > 8y$? Sketch.

29. Show that the tangents to the curve $y^2 = 4px$ from any point on the line $x = -p$ are perpendicular.

30. *Archimedes' formula for the area enclosed by a parabolic arch.*
  a) Find an equation for the parabolic arch of base b and altitude h in Fig. 8.16.

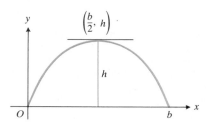

8.16   The parabolic arch in Problem 30.

  b) Show that the area enclosed by the arch and the x-axis is $(2/3)bh$.

31. *Archimedes' formula for the volume of a parabolic solid.* The region enclosed by the parabola $y = (4h/b^2)x^2$ and the line $y = h$ is revolved about the y-axis to generate a solid. Show that the volume of the solid is $3/2$ the volume of the corresponding inscribed cone. See Fig. 8.17.

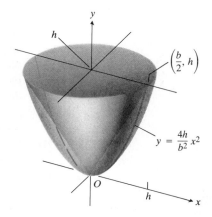

8.17   The cone and parabolic solid in Problem 31.

32. *Suspension bridge cables.* The condition for equilibrium in a section OP of a cable that supports a uniform weight of w kilograms per horizontal meter (Fig. 8.18) is

$$\frac{dy}{dx} = \tan \phi = \frac{\sin \phi}{\cos \phi} = \frac{T \sin \phi}{T \cos \phi} = \frac{wx}{H}$$

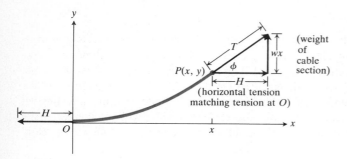

8.18 The cable section in Problem 32.

or

$$\frac{dy}{dx} + \left(\frac{w}{H}\right)x,$$

where the origin O is taken at the low point of the cable and H is the horizontal tension at O. Show that the curve in which the cable hangs is a parabola. (For this reason, parabolas are used to model the main cables in suspension bridges. Cables and chains that hang "freely" when suspended from their two ends, with no additional load, hang in curves called *catenaries*, studied in Article 9.3. The word catenary comes from *catena*, the Latin word for chain.)

33. *Constructing tangents to parabolas.* Show that the tangent to the parabola $y^2 = 4px$ at $P_1(x_1, y_1)$ meets the axis of symmetry x_1 units to the left of the vertex. This provides a simple way to construct the tangent to the parabola at any point other than the vertex (where you already have the y-axis): mark the point P on the parabola where you want the tangent, drop a perpendicular to the x-axis, measure $2x_1$ units to the left, mark that point, and draw a line from there through P.

34. *The reflective property of parabolas.*
   a) Show that the angles α and β in Fig. 8.14 are equal. (*Hint:* Find where the tangent line L crosses the x-axis. What angle does L make with the x-axis?)
   b) Assume that when a ray of light is reflected by a mirror, the angle of incidence is equal to the angle of reflection. If a mirror is formed by rotating a parabola about its axis and silvering the resulting surface, show that a ray of light coming from the focus of the parabola is reflected parallel to the axis.

> **TOOLKIT PROGRAMS**
>
> Conic Sections   Super * Grapher

8.4
Ellipses

In this article, we derive equations for ellipses in the coordinate plane and show how to find the ellipses' dimensions and foci directly from these equations. We also introduce a number called the *eccentricity* of an ellipse, which measures the extent to which the ellipse is out of round. This number is used to describe the shapes of the elliptical orbits of the comets and planets that revolve around the sun and of the satellites that orbit the earth.

DEFINITION

An **ellipse** is the set of points in a plane whose distances from two fixed points in the plane have a constant sum.

Equations for Ellipses

If the two fixed points, called **foci,** are taken at $F_1(-c, 0)$ and $F_2(c, 0)$ as in Fig. 8.19, and the sum of the distances $PF_1 + PF_2$ is denoted by $2a$, then the coordinates of a point $P(x, y)$ on the ellipse satisfy the equation

$$\sqrt{(x + c)^2 + y^2} + \sqrt{(x - c)^2 + y^2} = 2a.$$

To simplify this equation, we move the second radical to the right side of the

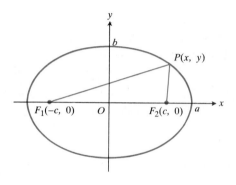

8.19 The ellipse shown here is defined by the equation

$$PF_1 + PF_2 = 2a.$$

As we see in the text, this leads to the co-ordinate equation

$$\frac{x^2}{a^2} + \frac{y^2}{b^2} = 1.$$

equation, and square and simplify twice to obtain

$$\frac{x^2}{a^2} + \frac{y^2}{a^2 - c^2} = 1. \tag{1}$$

Since the sum $PF_1 + PF_2 = 2a$ of two sides of the triangle F_1F_2P is greater than the third side $F_1F_2 = 2c$, the term $(a^2 - c^2)$ in (1) is positive and has a real positive square root, which we denote by b:

$$b = \sqrt{a^2 - c^2}. \tag{2}$$

Then (1) takes the more compact form

$$\frac{x^2}{a^2} + \frac{y^2}{b^2} = 1. \tag{3}$$

Equation (3) reveals that the curve is symmetric about both axes and lies inside the rectangle bounded by the lines $x = a$, $x = -a$, $y = b$, $y = -b$. The intercepts of the curve are $(\pm a, 0)$ and $(0, \pm b)$. The curve intersects each axis at an angle of 90°, since the slope

$$\frac{dy}{dx} = \frac{-b^2x}{a^2y}$$

is zero at $x = 0$, $y = \pm b$ and is infinite at $y = 0$, $x = \pm a$.

We have shown that the coordinates of P satisfy (1) if P satisfies the geometric condition $PF_1 + PF_2 = 2a$. We now prove the converse. Suppose that x and y satisfy (1) with $0 < c < a$. Then

$$y^2 = (a^2 - c^2)\frac{a^2 - x^2}{a^2}.$$

Substituting this in the radicals in Eqs. (4), we find that

$$PF_1 = \sqrt{(x + c)^2 + y^2} = \left| a + \frac{c}{a}x \right| \tag{4a}$$

and

$$PF_2 = \sqrt{(x - c)^2 + y^2} = \left| a - \frac{c}{a}x \right|. \tag{4b}$$

Since x is restricted to the interval $-a \le x \le a$, the value of $(c/a)x$ lies between $-c$ and c, so both $a + (c/a)x$ and $a - (c/a)x$ are positive, both being between $a + c$ and $a - c$. Hence the absolute values in (4a) and (4b) yield

$$PF_1 = a + \frac{c}{a}x, \qquad PF_2 = a - \frac{c}{a}x. \tag{5}$$

Adding these, we see that $PF_1 + PF_2$ has the value $2a$ for every position of P on the curve. Thus the *geometric property* and *algebraic equation* are equivalent.

Axes

In Eq. (3), $b^2 = a^2 - c^2$ is less than a^2. The **major axis** of the ellipse is the segment of length $2a$ between the x-intercepts $(\pm a, 0)$. The **minor axis** is the segment of length $2b$ between the y-intercepts $(0, \pm b)$. The number a is called the **semimajor axis,** and the number b the **semiminor axis.** If $a = 4$, $b = 3$, then Eq. (3) is

$$\frac{x^2}{16} + \frac{y^2}{9} = 1. \tag{6a}$$

If we interchange x and y in (6a), we get the equation

$$\frac{x^2}{9} + \frac{y^2}{16} = 1, \tag{6b}$$

which represents an ellipse with its major axis vertical rather than horizontal. See Fig. 8.20.

There is never any cause for confusion in analyzing equations like (6a) and (6b). We simply find the intercepts on the axes of symmetry; then we know which way the major axis runs because it is the longer of the two axes. The foci always lie on the major axis.

If we use the letters a, b, and c to represent the semimajor axis, semiminor axis, and **half-distance** between foci, Eq. (2) tells us that

$$b^2 = a^2 - c^2 \qquad \text{or} \qquad a^2 = b^2 + c^2. \tag{7}$$

Hence a is the hypotenuse of a right triangle of sides b and c, as in Fig. 8.20. When we start with an equation like (6a) or (6b), we can read off a^2 and b^2 from it at once. Then Eq. (7) determines c^2 as their difference. So in either of Eqs. (6a) and (6b) we have

$$c^2 = a^2 - b^2 = 16 - 9 = 7.$$

Therefore the foci are $c = \sqrt{7}$ units from the center of the ellipse.

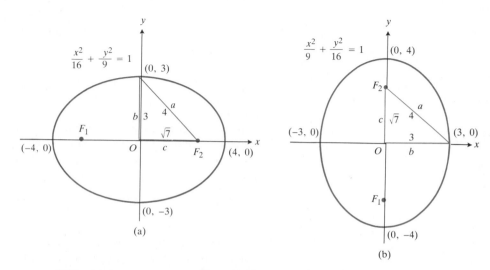

8.20   (a) The major axis of $(x^2/16) + (y^2/9) = 1$ is horizontal. (b) The major axis of $(x^2/9) + (y^2/16) = 1$ is vertical.

Center Not at the Origin

The **center** of an ellipse is the point of intersection of its axes. The standard equations for the ellipses whose axes are parallel to the coordinate axes and whose centers lie at the point (h, k) are listed in the following box.

Standard Equations for Ellipses Centered at (h, k) with Axes Parallel to the Coordinate Axes

$$\frac{(x - h)^2}{a^2} + \frac{(y - k)^2}{b^2} = 1 \qquad \text{(Major axis horizontal)} \qquad (8)$$

Vertices: $(h \pm a, k)$ Foci: $(h \pm \sqrt{a^2 - b^2}, k)$

$$\frac{(x - h)^2}{b^2} + \frac{(y - k)^2}{a^2} = 1 \qquad \text{(Major axis vertical)} \qquad (9)$$

Vertices: $(h, k \pm a)$ Foci: $(h, k \pm \sqrt{a^2 - b^2})$

In each case, a is the semimajor axis and b is the semiminor axis.

Equations (8) and (9) are derived by applying the translation $x' = x - h$, $y' = y - k$ and observing that the resulting equations in primed coordinates are

$$\frac{x'^2}{a^2} + \frac{y'^2}{b^2} = 1 \qquad \text{and} \qquad \frac{x'^2}{b^2} + \frac{y'^2}{a^2} = 1.$$

EXAMPLE 1   Find the center, vertices, and foci of the ellipse

$$9x^2 + 4y^2 + 36x - 8y + 4 = 0.$$

Solution   We collect the x and y terms separately to get

$$9(x^2 + 4x) + 4(y^2 - 2y) = -4$$

and complete the square in each set of parentheses to obtain

$$9(x^2 + 4x + 4) + 4(y^2 - 2y + 1) = -4 + 36 + 4.$$

We then divide both sides by 36 and write the equation in standard form:

$$\frac{(x + 2)^2}{4} + \frac{(y - 1)^2}{9} = 1.$$

Since $9 > 4$, this is Eq. (9) with $a = 3$, $b = 2$, $h = -2$, and $k = 1$. The center of the ellipse is the point $(h, k) = (-2, 1)$:

Center: $(-2, 1)$.

The vertices are the points $(h, k \pm a) = (-2, 1 \pm 3)$:

Vertices: $(-2, 4)$ and $(-2, -2)$.

The foci are the points $(h, k \pm \sqrt{a^2 - b^2}) = (-2, 1 \pm \sqrt{9 - 4}) = (-2, 1 \pm \sqrt{5})$:

Foci: $(-2, 1 + \sqrt{5})$ and $(-2, 1 - \sqrt{5})$.

See Fig. 8.21. ∎

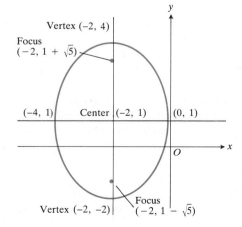

8.21 The ellipse

$$9x^2 + 4y^2 + 36x - 8y + 4 = 0.$$

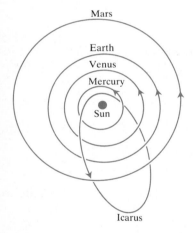

8.22 The eccentricity of an ellipse describes its shape.

8.23 The orbit of the asteroid Icarus.

Eccentricity

To discuss the properties of the ellipse in more detail, we shall assume that its equation has been reduced to the form

$$\frac{x^2}{a^2} + \frac{y^2}{b^2} = 1, \qquad a > b > 0. \tag{10}$$

Although the distance c from the center of the ellipse to either focus does not appear in its equation, we may still determine c from the equation

$$c = \sqrt{a^2 - b^2}. \tag{11}$$

If we keep a fixed and vary the focal distance c over the interval $0 \le c \le a$, the resulting ellipses will vary in shape. They are circular when $c = 0$ (so that $a = b$) and become flatter as c increases, until in the extreme case $c = a$, the "ellipse" becomes the line segment $F_1 F_2$ joining the two foci (Fig. 8.22). The ratio

$$e = \frac{c}{a} = \frac{\sqrt{a^2 - b^2}}{a}, \tag{12}$$

called the **eccentricity** of the ellipse, varies from 0 to 1 and indicates the degree of departure from circularity.

The planets in the solar system revolve around the sun in elliptical orbits with the sun at one focus. Most of the planets, including the earth, have orbits that are nearly circular, as can be seen from the eccentricities in Table 8.1. Pluto, however, has a fairly eccentric orbit, with $e = 0.25$, as does Mercury, with $e = 0.21$. Other members of the solar system have orbits that are even more eccentric. Icarus, an asteroid about 1 mile wide that revolves around the sun every 409 Earth days, has an orbital eccentricity of 0.83 (Fig. 8.23).

TABLE 8.1
Eccentricities of planetary orbits

Mercury	0.21	Saturn	0.06
Venus	0.01	Uranus	0.05
Earth	0.02	Neptune	0.01
Mars	0.09	Pluto	0.25
Jupiter	0.05		

EXAMPLE 2　The orbit of the comet Kohoutek (Fig. 8.24) is about 44 astronomical units wide by 3600 astronomical units long. (One *astronomical unit* (AU) is the semimajor axis of the Earth's orbit, about 92,600,000 miles.) Find the eccentricity of the orbit.

Solution

$$e = \frac{\sqrt{a^2 - b^2}}{a} = \frac{\sqrt{(1800)^2 - (22)^2}}{1800} \approx 0.9999.$$

8.24 The orbit of comet Kohoutek, shown approximately to scale. The small circle shows the outer limit of the orbit of Pluto.

Halley's Comet

Edmund Halley (1656–1742), the British biologist, geologist, sea captain, spy, Antarctic voyager, astronomer, adviser on fortifications, company founder and director, and classicist, was also the mathematician who financed the publication of Newton's *Principia* and used Newton's theory to calculate the orbit of the great comet of 1682:

> "wherefore if according to what we have already said [the comet] should return again about the year 1758, candid posterity will not refuse to acknowledge that this was first discovered by an Englishman."

Indeed, candid posterity did not refuse—ever since the comet's return in 1758, it has been known as Halley's comet.

Last seen rounding the sun during the winter and spring of 1985–86, the comet is due to return again in the year 2062. Records of its passing go back 30 orbit cycles to 240 BC. A recent study* indicates that the comet has made about 2000 cycles so far with about the same number to go before the sun erodes it away completely.

The orbit of Halley's comet is an ellipse 36.18 AU long by 9.12 AU wide. Its eccentricity is

$$e = \frac{\sqrt{a^2 - b^2}}{a} = \frac{\sqrt{(18.09)^2 - (4.56)^2}}{18.09} \approx 0.97.$$

Classifying Conic Sections by Eccentricity

While a parabola has one focus and one directrix, each ellipse has two foci and two directrices. The directrices are the lines perpendicular to the major axis of the ellipse at distances $\pm a/e$ from its center. The *parabola* has the property that

$$PF = 1 \cdot PD \tag{13}$$

for any point P on it, where F is the focus and D is the point nearest P on the directrix. For an *ellipse,* it is not difficult to show that the equations that replace (13) are

$$PF_1 = e \cdot PD_1, \qquad PF_2 = e \cdot PD_2. \tag{14}$$

Here e is the eccentricity, P is any point on the ellipse, F_1 and F_2 are the foci, and D_1 and D_2 are the points nearest P on the two directrices. In Eq. (14), the directrix and focus must correspond; that is, if we use the distance from P to the focus F_1, we must also use the distance from P to the directrix at the same end of the ellipse (see Fig. 8.25). We thus associate the directrix $x = -a/e$ with the focus $F_1(-c, 0)$ and the directrix $x = a/e$ with the focus $F_2(c, 0)$. In terms of the semimajor axis a and eccentricity $e < 1$, as we move away from the center along the major axis, we find successively

- a **focus** at distance ae from the center,
- a **vertex** at distance a from the center,
- a **directrix** at distance a/e from the center.

---

*D. W. Hughes, *Monthly Notices Roy. Ast. Soc.,* 213:103 (1985).

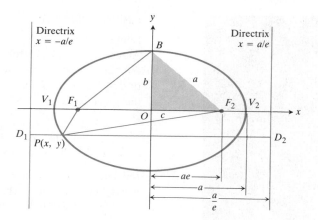

8.25 The ellipse $(x^2/a^2) + (y^2/b^2) = 1$.

The "focus–directrix" property unites the parabola, ellipse, and hyperbola (next article) in the following way. Suppose that the distance PF of a point P from a fixed point F (the focus) is a constant multiple of its distance from a fixed line (the directrix). That is, suppose

$$PF = e \cdot PD, \tag{15}$$

where e is the constant of proportionality. Then the path traced by P is

a) a *parabola* if $e = 1$,

b) an *ellipse* of eccentricity e if $e < 1$, and

c) a *hyperbola* of eccentricity e if $e > 1$.

As you can see, the parabola is a very special case.

Construction

The quickest way to construct an ellipse uses the definition directly. A loop of string around two tacks F_1 and F_2 is pulled tight with a pencil point P and held taut as the pencil traces a curve (Fig. 8.26). The curve is an ellipse because $PF_1 + PF_2$ is constant (being equal to the length of the loop minus the distance between the tacks).

Applications

Ellipses appear in airplane wings (British Spitfire) and sometimes in gears designed for racing bicycles. Stereo systems often have elliptical styli, and water pipes are sometimes designed with elliptical cross sections to allow the pipe to expand without breaking when the water freezes. Ellipses also appear in instruments used to study aircraft noise in wind tunnels (sound at one focus can be received at the other with relatively little noise from other sources). The triggering mechanisms in some lasers are elliptical, and stones on a beach become more and more elliptical as they are ground down by waves. There are

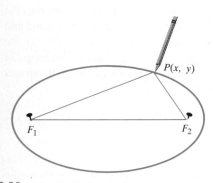

8.26 How to draw an ellipse.

also applications of ellipses to fossil formation. The ellipsolith, once inter-preted as a separate species, is now known to be an elliptically deformed nauti-lus. For details of these applications and many more, see Lee Whitt's ''The Standup Conic Presents: The Ellipse and Applications,'' *The UMAP Journal*, 4:2 (1983).

PROBLEMS

In Problems 1–4, find an equation for the ellipse that has given center C, focus F, and semimajor axis a. Then calculate the eccentricity and sketch the ellipse.

1. $C(0, 0)$, $F(0, 2)$, $a = 4$

2. $C(0, 0)$, $F(-3, 0)$, $a = 5$

3. $C(0, 2)$, $F(0, 0)$, $a = 3$

4. $C(-3, 0)$, $F(-3, -2)$, $a = 4$

Find the centers, vertices, and foci of the ellipses in Problems 5–18.

5. $\dfrac{(x - 7)^2}{4} + \dfrac{(y - 5)^2}{25} = 1$

6. $\dfrac{(x + 3)^2}{16} + \dfrac{(y - 1)^2}{4} = 1$

7. $\dfrac{(x + 1)^2}{9} + \dfrac{(y + 4)^2}{25} = 1$

8. $\dfrac{(x - 8)^2}{25} + \dfrac{y^2}{81} = 1$

9. $25(x - 3)^2 + 4(y - 1)^2 = 100$

10. $9(x - 4)^2 + 16(y - 3)^2 = 144$

11. $x^2 + 10x + 25y^2 = 0$

12. $x^2 + 16y^2 + 96y + 128 = 0$

13. $x^2 + 9y^2 - 4x + 18y + 4 = 0$

14. $4x^2 + y^2 - 32x + 16y + 124 = 0$

15. $4x^2 + y^2 - 16x + 4y + 16 = 0$

16. $x^2 + 4y^2 + 2x + 8y + 1 = 0$

17. $9x^2 + 16y^2 + 18x - 96y + 9 = 0$

18. $25x^2 + 9y^2 - 100x + 54y - 44 = 0$

19. Sketch the following ellipses.
    a) $9x^2 + 4y^2 = 36$
    b) $4x^2 + 9y^2 = 144$
    c) $\dfrac{(x - 1)^2}{16} + \dfrac{(y + 2)^2}{4} = 1$
    d) $4x^2 + y^2 = 1$
    e) $16(x - 2)^2 + 9(y + 3)^2 = 144$

20. Find an equation for the ellipse that passes through the origin and has foci at $(-1, 1)$ and $(1, 1)$.

21. The endpoints of the major and minor axes of an ellipse are $(1, 1)$, $(3, 4)$, $(1, 7)$, and $(-1, 4)$. Sketch the ellipse, give an equation for it, and find its foci.

22. Find an equation for the ellipse of eccentricity 2/3 that has the line $x = 9$ as a directrix and the point $(4, 0)$ as the corresponding focus.

Sketch the regions whose points satisfy the inequalities or sets of inequalities in Problems 23–26.

23. $9x^2 + 16y^2 \leq 144$

24. $x^2 + y^2 \geq 1$ and $4x^2 + y^2 \leq 4$

25. $x^2 + 4y^2 \geq 4$ and $4x^2 + 9y^2 \leq 36$

26. $(x^2 + y^2 - 4)(x^2 + 9y^2 - 9) \leq 0$

27. Draw an ellipse of eccentricity 4/5.

28. Draw the orbit of Pluto to scale.

29. *Halley's comet.*
    a) Write an equation for the orbit of Halley's comet in a coordinate system in which the sun lies at the origin and the other focus lies on the positive x-axis, scaled in as-tronomical units.
    b) About how close does the comet come to the sun in astronomical units? In miles?
    c) What is the farthest the comet gets from the sun in astro-nomical units? In miles?

30. Find the dimensions of the rectangle of largest area that can be inscribed in the ellipse $x^2 + 4y^2 = 4$ with its sides paral-lel to the coordinate axes. What is the area of the rectangle?

31. Find the center of mass of a thin homogeneous plate that is bounded below by the x-axis and above by the ellipse $(x^2/9) + (y^2/16) = 1$.

32. Find the volume of the solid generated by revolving the region enclosed by the ellipse $(x^2/4) + (y^2/9) = 1$ about (a) the x-axis, (b) the y-axis.

33. What values of the constants a, b, and c make the ellipse
$$4x^2 + y^2 + ax + by + c = 0$$
tangent to the x-axis at the origin and pass through the point $(-1, 2)$?

34. *The reflective property of ellipses.* An ellipsoid is generated by rotating an ellipse about its major axis. The inside sur-

face of the ellipsoid is silvered to produce a mirror. Show that a ray of light emanating from one focus will be reflected to the other focus. Sound waves also follow such paths, and this property of ellipsoids is used to construct "whispering galleries." (*Hint:* Put the ellipse in standard position in the plane, as in Fig. 8.27, and show that the lines from a point P on the ellipse to the two foci make equal angles with the tangent at P.)

8.27 The ellipse in Problem 34.

TOOLKIT PROGRAMS		
Conic Sections		Super ∗ Grapher

8.5
Hyperbolas

In this article, we derive equations for hyperbolas in the coordinate plane and show how to find the foci and asymptotes of hyperbolas directly from their equations.

DEFINITION

> A **hyperbola** is the set of points in a plane whose distances from two fixed points in the plane have a constant difference.

The Equation of a Hyperbola

If we take the two fixed points, called **foci,** at $F_1(-c, 0)$ and $F_2(c, 0)$ and the constant equal to $2a$ (see Fig. 8.28), then a point $P(x, y)$ lies on the hyperbola if and only if

$$\sqrt{(x + c)^2 + y^2} - \sqrt{(x - c)^2 + y^2} = 2a$$

or

$$\sqrt{(x - c)^2 + y^2} - \sqrt{(x + c)^2 + y^2} = 2a.$$

The second equation is like the first, with $2a$ replaced by $-2a$. Hence we write the first one with $\pm 2a$, move one radical to the right side of the equation, square, and simplify. One radical still remains. We isolate it and square again. We then obtain the equation

$$\frac{x^2}{a^2} + \frac{y^2}{a^2 - c^2} = 1. \tag{1}$$

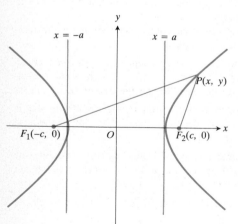

8.28 A hyperbola has two branches. For points on the right-hand branch of the hyperbola shown here, $PF_1 - PF_2 = 2a$. For points on the left-hand branch, $PF_2 - PF_1 = 2a$.

So far, this is just like the equation for an ellipse. But now $a^2 - c^2$ is negative, because the *difference* in two sides of the triangle F_1F_2P is less than the third side:

$$2a < 2c.$$

So in this case $c^2 - a^2$ is positive and has a real, positive square root, which we call b:

$$b = \sqrt{c^2 - a^2}, \qquad \text{or} \qquad a^2 - c^2 = -b^2. \tag{2}$$

The equation of the hyperbola now becomes

$$\frac{x^2}{a^2} - \frac{y^2}{b^2} = 1, \tag{3}$$

which is analogous to the equation of an ellipse. The only differences are the minus sign in the equation of the hyperbola and the new relation among a, b, and c given by Eq. (2).

The hyperbola, like the ellipse, is symmetric with respect to both axes and the origin, but it has no real y-intercepts. In fact, no portion of the curve lies between the lines $x = a$ and $x = -a$.

If we start with a point $P(x, y)$ whose coordinates satisfy Eq. (3), the distances PF_1 and PF_2 will be given by

$$PF_1 = \sqrt{(x + c)^2 + y^2} = \left| a + \frac{c}{a}x \right|, \tag{4a}$$

$$PF_2 = \sqrt{(x - c)^2 + y^2} = \left| a - \frac{c}{a}x \right|, \tag{4b}$$

as for the ellipse. But now c is greater than a, and P lies either to the right of the line $x = a$, that is, $x > a$, or else P lies to the left of the line $x = -a$, and then $x < -a$. The absolute values in Eqs.(4) work out to be

$$\left. \begin{aligned} PF_1 &= a + \frac{c}{a}x \\[2mm] PF_2 &= \frac{c}{a}x - a \end{aligned} \right\} \quad \text{if } x > a \tag{5a}$$

and

$$\left. \begin{aligned} PF_1 &= -\left(a + \frac{c}{a}x \right) \\[2mm] PF_2 &= a - \frac{c}{a}x \end{aligned} \right\} \quad \text{if } x < -a. \tag{5b}$$

Thus, when P lies to the right of the line $x = a$, $PF_1 - PF_2 = 2a$, while if P lies to the left of $x = -a$, $PF_2 - PF_1 = 2a$ (Fig. 8.28). In either case, any point P that satisfies the geometric conditions satisfies Eq. (3). Conversely, any point that satisfies Eq. (3) satisfies the geometric conditions.

Asymptotes

As we saw in Chapter 3, it may happen that as a point P on a curve moves farther and farther away from the origin, the distance between P and some fixed line tends to zero. We call such a line an **asymptote** of the curve. As we shall now see, the hyperbola

$$\frac{x^2}{a^2} - \frac{y^2}{b^2} = 1 \qquad (3)$$

has two asymptotes, the lines

$$y = \pm \frac{b}{a}x.$$

The left-hand side of Eq. (3) can be factored and the equation written in the form

$$\left(\frac{x}{a} - \frac{y}{b}\right)\left(\frac{x}{a} + \frac{y}{b}\right) = 1,$$

or

$$\frac{x}{a} - \frac{y}{b} = \frac{ab}{bx + ay}. \qquad (6a)$$

An analysis of Eq. (3) shows that one branch of the curve lies in the first quadrant and has infinite extent. If the point P moves steadily away from the origin on this branch, x and y become infinite, and the right-hand side of (6a) tends to zero. Hence, the left-hand side must do likewise. That is,

$$\lim_{\substack{x \to \infty \\ y \to \infty}} \left(\frac{x}{a} - \frac{y}{b}\right) = 0, \qquad (6b)$$

which leads us to think that the line

$$\frac{x}{a} - \frac{y}{b} = 0, \qquad \text{or} \qquad y = \frac{b}{a}x \qquad (7)$$

may be an asymptote of the curve.

To see that it is, we investigate the vertical distance between the line and the curve

$$y = \frac{b}{a}\sqrt{x^2 - a^2}$$

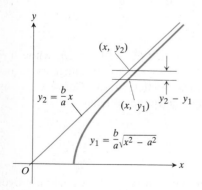

$y_2 = \frac{b}{a}x$

(x, y_2)

(x, y_1) $y_2 - y_1$

$y_1 = \frac{b}{a}\sqrt{x^2 - a^2}$

8.29   As $x \to \infty$, the point (x, y_1) rises on the hyperbola toward the asymptote $y_2 = (b/a)x$.

(Fig. 8.29). As $x \to \infty$, we find

$$\lim_{x \to \infty} \left(\frac{b}{a}x - \frac{b}{a}\sqrt{x^2 - a^2}\right) = b \lim_{x \to \infty} \left(\frac{x}{a} - \frac{\sqrt{x^2 - a^2}}{a}\right)$$

$$= b \lim_{\substack{x \to \infty \\ y \to \infty}} \left(\frac{x}{a} - \frac{y}{b}\right) \qquad (8)$$

$$= b \cdot 0 \qquad \text{(by Eq. 6b)}$$

$$= 0.$$

Since the vertical distance between the line and the hyperbola approaches zero as $x \to 0$, the perpendicular distance from points on the hyperbola to the line $y = (b/a)x$ also approaches zero. Therefore, the line $y = (b/a)x$ is an asymptote of the hyperbola.

By symmetry, the line

$$\frac{x}{a} + \frac{y}{b} = 0, \qquad \text{or} \qquad y = -\frac{b}{a}x \tag{9}$$

is also an asymptote of the hyperbola.

Both asymptotes may be obtained by replacing the 1 on the right-hand side of Eq. (3) by 0 and then factoring.

REMARK Some definitions of asymptote also require the slope of the curve to approach the slope of the asymptote as $x \to \infty$. To show that this is the case here, we calculate the slope of the curve:

$$y = \frac{b}{a}\sqrt{x^2 - a^2}, \qquad \frac{dy}{dx} = \frac{b}{a}\frac{x}{\sqrt{x^2 - a^2}},$$

from which we see that

$$\lim_{x \to \infty} \frac{dy}{dx} = \frac{b}{a} \lim_{x \to \infty} \frac{x}{\sqrt{x^2 - a^2}} = \frac{b}{a} \cdot 1 = \frac{b}{a}, \tag{10}$$

which is the slope of the asymptote $y = (b/a)x$.

Graphing Hyperbolas

In graphing a hyperbola, it helps to mark the points $\pm a$ on the x-axis and $\pm b$ on the y-axis and to draw a rectangle whose sides pass through these points parallel to the coordinate axes (Fig. 8.30). The diagonals of the rectangle, extended, are the asymptotes of the hyperbola. The semidiagonal $c = \sqrt{a^2 + b^2}$ can also be used as the radius of a circle that will cut the x-axis at the foci, $F_1(-c, 0)$ and $F_2(c, 0)$.

If we interchange x and y in Eq. (3), the new equation,

$$\frac{y^2}{a^2} - \frac{x^2}{b^2} = 1, \tag{11}$$

represents a hyperbola with foci on the y-axis (Fig. 8.31).

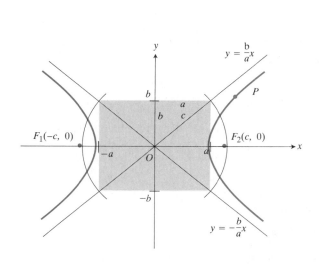

8.30 The hyperbola $(x^2/a^2) - (y^2/b^2) = 1$.

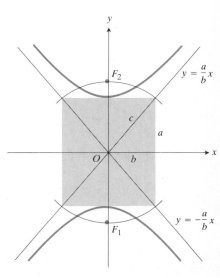

8.31 The hyperbola $(y^2/a^2) - (x^2/b^2) = 1$.

Center Not at the Origin

The **center** of a hyperbola is the point of intersection of its axes of symmetry. The standard equations for hyperbolas whose axes are parallel to the coordinate axes and whose centers lie at the point (h, k) are listed in the following box.

Standard Equations for Hyperbolas Centered at (h, k) with Axes Parallel to the Coordinate Axes

$$\frac{(x - h)^2}{a^2} - \frac{(y - k)^2}{b^2} = 1 \qquad \text{(Line of foci parallel to } x\text{-axis)} \quad (12)$$

Vertices: $(h \pm a, k)$

Foci: $(h \pm \sqrt{a^2 + b^2}, k)$

Asymptotes: $(y - k) = \pm(b/a)(x - h)$

$$\frac{(y - k)^2}{a^2} - \frac{(x - h)^2}{b^2} = 1 \qquad \text{(Line of foci parallel to } y\text{-axis)} \quad (13)$$

Vertices: $(h, k \pm a)$

Foci: $(h, k \pm \sqrt{a^2 + b^2})$

Asymptotes: $(y - k) = \pm(a/b)(x - h)$

Equations (12) and (13) are derived by applying the translation $x' = x - h$, $y' = y - k$ and observing that the resulting equations in primed coordinates are

$$\frac{x'^2}{a^2} - \frac{y'^2}{b^2} = 1 \qquad \text{and} \qquad \frac{y'^2}{a^2} - \frac{x'^2}{b^2} = 1. \tag{14}$$

Notice that we have b over a in the asymptote equation for (12), but a over b in the asymptote equation for (13).

EXAMPLE 1   Find the center, vertices, foci, and asymptotes of the hyperbola

$$4x^2 - y^2 + 8x + 2y - 1 = 0.$$

Solution   We complete the squares in the x and y terms separately and write the equation in standard form:

$$4x^2 - y^2 + 8x + 2y - 1 = 0$$

$$4(x^2 + 2x + 1) - (y^2 - 2y + 1) = 1 + 4 - 1$$

$$4(x + 1)^2 - (y - 1)^2 = 4$$

$$(x + 1)^2 - \frac{(y - 1)^2}{4} = 1.$$

This is Eq. (12) with $a = 1, b = 2, h = -1, k = 1$. The center of the hyperbola is the point $(h, k) = (-1, 1)$:

   Center: $(-1, 1)$.

The vertices are the points $(h \pm a, k) = (-1 \pm 1, 1)$:

   Vertices: $(0, 1)$ and $(-2, 1)$.

$$4x^2 - y^2 + 8x + 2y - 1 = 0,$$
or
$$(x + 1)^2 - \frac{(y - 1)^2}{4} = 1$$

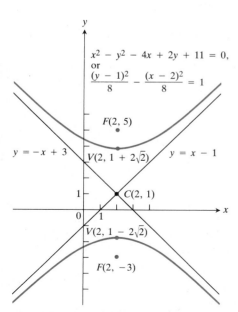

8.32 The hyperbola in Example 1.

The foci are the points

$$(h \pm \sqrt{a^2 + b^2}, k) = (-1 \pm \sqrt{1 + 4}, 1) = (-1 \pm \sqrt{5}, 1):$$

Foci: $(-1 + \sqrt{5}, 1)$ and $(-1 - \sqrt{5}, 0).$

The asymptotes are the lines $(y - k) = \pm(b/a)(x - h)$ or

$$(y - 1) = \pm\frac{2}{1}(x + 1):$$

Asymptotes: $y = 2x + 3$ and $y = -2x - 1.$

See Fig. 8.32.   ◼

EXAMPLE 2 Find the center, vertices, foci, and asymptotes of the hyperbola

$$x^2 - y^2 - 4x + 2y + 11 = 0.$$

Solution We rearrange the equation, complete the squares, and write the resulting equation in standard form:

$$x^2 - y^2 - 4x + 2y + 11 = 0$$
$$x^2 - 4x + 4 - (y^2 - 2y + 1) = -11 + 4 - 1$$
$$(x - 2)^2 - (y - 1)^2 = -8$$
$$\frac{(y - 1)^2}{8} - \frac{(x - 2)^2}{8} = 1.$$

This is Eq. (13) with $a = 2\sqrt{2}$, $b = 2\sqrt{2}$, $h = 2$, $k = 1$. The center of the hyperbola is $(h, k) = (2, 1)$:

Center: $(2, 1).$

The vertices are the points $(h, k \pm a) = (2, 1 \pm 2\sqrt{2})$:

Vertices: $(2, 1 + 2\sqrt{2})$ and $(2, 1 - 2\sqrt{2}).$

The foci are the points $(h, k \pm \sqrt{a^2 + b^2}) = (2, 1 \pm \sqrt{8 + 8}) = (2, 1 \pm 4)$:

Foci: $(2, 5)$ and $(2, -3).$

The asymptotes are the lines $(y - k) = \pm(a/b)(x - h)$ or

$$(y - 1) = \pm\frac{2\sqrt{2}}{2\sqrt{2}}(x - 2) = \pm(x - 2):$$

Asymptotes: $y = x - 1$ and $y = -x + 3.$

See Fig. 8.33.   ◼

Eccentricity

There is no restriction $a > b$ for the hyperbola as there is for the ellipse. The direction in which the hyperbola opens is controlled by the *signs* rather than by the relative *sizes* of the coefficients of the quadratic terms.

8.33 The hyperbola in Example 2.

In our further discussion of the hyperbola, we shall assume that it has been referred to axes through its center and that its equation has the form

$$\frac{x^2}{a^2} - \frac{y^2}{b^2} = 1. \tag{15}$$

Then

$$b = \sqrt{c^2 - a^2}$$

and

$$c^2 = a^2 + b^2. \tag{16}$$

As for the ellipse, we define the **eccentricity** e of the hyperbola to be

$$e = \frac{c}{a}.$$

Since $c > a$, the eccentricity of a hyperbola is greater than 1. The lines

$$x = \frac{a}{e}, \qquad x = -\frac{a}{e}$$

are the **directrices** of the hyperbola.

The Focus–Directrix Property

We shall now verify that a point $P(x, y)$ whose coordinates satisfy Eq. (15) also has the property that

$$PF_1 = e \cdot PD_1 \tag{17a}$$

and

$$PF_2 = e \cdot PD_2, \tag{17b}$$

where $F_1(-c, 0)$ and $F_2(c, 0)$ are the foci and $D_1(-a/e, y)$ and $D_2(a/e, y)$ are the points nearest P on the directrices.

We shall content ourselves with establishing (17a and b) for the right branch of the hyperbola; the method is the same for the left branch. Reference to Eqs. (5a) then shows that

$$PF_1 = \frac{c}{a}x + a = e\left(x + \frac{a}{e}\right), \qquad PF_2 = \frac{c}{a}x - a = e\left(x - \frac{a}{e}\right), \tag{18a}$$

while we see from Fig. 8.34 that

$$PD_1 = x + \frac{a}{e}, \qquad PD_2 = x - \frac{a}{e}. \tag{18b}$$

These results combine to establish the ''focus–directrix'' properties of the hyperbola expressed in Eqs. (17a and b). Conversely, if Eqs. (18a) are satisfied, it is also true that

$$PF_1 - PF_2 = 2a.$$

That is, P satisfies the requirement that the difference of its distances from the two foci is constant.

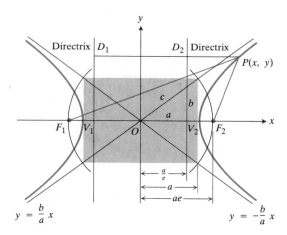

8.34 $PF_1 = e \cdot PD_1$ and $PF_2 = e \cdot PD_2$.

Applications

Hyperbolic paths arise in Einstein's theory of relativity and form the basis for the LORAN radio navigation system (LORAN is short for Long Range Navigation). A comet that does not return to its sun follows a hyperbolic path (the probability of its being parabolic is zero). Reflecting telescopes like the 200-inch Hale telescope on Mount Palomar in California, and the space telescope NASA plans to launch in 1988, use small hyperbolic mirrors in conjunction with their larger parabolic ones (schematic drawing in Fig. 8.35). For more about applications of hyperbolas, see Lee Whitt's "The Standup Conic Presents: The Hyperbola and Applications," *The UMAP Journal,* 5:1 (1984).

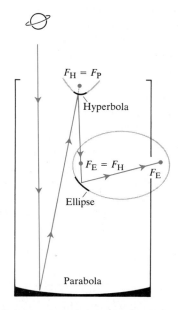

8.35 In this schematic drawing of a reflecting telescope, light from a planet reflects off a primary parabolic mirror toward the mirror's focus F_P. It is then reflected by a small hyperbolic mirror, whose focus is $F_H = F_P$, toward the second focus of the hyperbola, $F_E = F_H$. Since this focus is shared by an ellipse, the light is reflected by the elliptical mirror to the ellipse's second focus to be seen by an observer. See Problem 24 for more about the reflective property of the hyperbola used here.

PROBLEMS

In Problems 1–4, sketch the hyperbolas and their asymptotes.

1. $\dfrac{x^2}{9} - \dfrac{y^2}{16} = 1,$ **2.** $\dfrac{x^2}{16} - \dfrac{y^2}{9} = 1,$

3. $\dfrac{y^2}{9} - \dfrac{x^2}{16} = 1,$ **4.** $\dfrac{x^2}{9} - \dfrac{y^2}{16} = -1.$

In Problems 5–14, find the center, vertices, foci, eccentricity, and asymptotes of the given hyperbola. Then sketch the hyperbola, showing these features.

5. $9(x - 2)^2 - 4(y + 3)^2 = 36$

6. $x^2 - 9(y - 1)^2 = 9$

7. $4(x - 2)^2 - 9(y + 3)^2 = 36$

8. $(y + 3)^2 - 9(x - 2)^2 = 9$

9. $16(y - 3)^2 - 9(x - 4)^2 = 144$

10. $y^2 - (x + 2)^2 = 8$

11. $5x^2 - 4y^2 + 20x + 8y = 4$

12. $4x^2 = y^2 - 4y + 8$

13. $4y^2 = x^2 - 4x$

14. $4x^2 - 5y^2 - 16x + 10y + 31 = 0$

15. Find an equation for the hyperbola with foci at $(0, 0)$ and $(0, 4)$ that passes through the point $(12, 9)$.

16. One focus of a hyperbola is located at the point $(1, -3)$ and the corresponding directrix is the line $y = 2$. Find an equation for the hyperbola if its eccentricity is $3/2$.

17. *LORAN.* A radio signal was sent simultaneously from towers A and B located several hundred miles apart on the California coast. A ship offshore received the signal from A 1400 microseconds before it received the signal from B.
a) Assume that the radio signals traveled at 980 ft per microsecond. What can be said about the approximate location of the ship relative to the two towers?
b) Find out what you can about LORAN and other hyperbolic radio navigation systems. (See, for example, Nathaniel Bowditch's *American Practical Navigator,* Vol. I, U.S. Defense Mapping Agency Hydrographic Center, Publication No. 9, 1977, Chapter 43.)

18. Find out what you can about the DECCA system of air navigation. (See *Time,* Feb. 23, 1959, p. 87, or the reference in Problem 17.)

19. Find the volume of the solid generated when the region bounded by the line $x = 5$ and the right-hand branch of the hyperbola $x^2 - y^2 = 9$ is revolved (a) about the x-axis, (b) about the y-axis.

20. Find the volume of the solid generated by revolving the region enclosed by the hyperbola $x^2 - y^2 = 1$ and the lines $y = -3$ and $y = 3$ (a) about the x-axis, (b) about the y-axis.

21. The region bounded by the right-hand branch of the hyperbola $3x^2 - 4y^2 = 12$ and the line $x = 5$ is revolved about the y-axis to generate a solid. Find the volume of the solid.

22. Find the center of mass of a thin homogeneous plate bounded on the left and right by the hyperbola $x^2 - y^2 = 9$, below by the x-axis, and above by the line $y = 4$.

23. The circular waves in the photograph below were made by touching the surface of a ripple tank first at A and shortly afterward at B. As the waves expanded, their point of intersection traced a hyperbola whose foci lay at A and B. Why? (*Hint:* The waves traveled with the same constant velocity.) (Photograph from *PSSC Physics, Second Edition;* 1965; D.C. Heath & Company with Education Development Center, Inc. *NCFMF Book of Film Notes,* 1974; The MIT Press with Education Development Center, Inc., Newton, Massachusetts.)

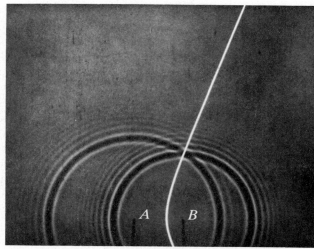

24. *The reflective property of hyperbolas.* Show that a ray of light directed toward one focus of a hyperbolic mirror, as in Fig. 8.36, is reflected toward the other focus. (*Hint:* Show that the tangent to the hyperbola at the point P in Fig. 8.36 bisects the angle made by the segments PF_1 and PF_2.)

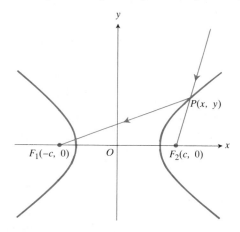

8.36 The reflective property of hyperbolas described in Problem 24.

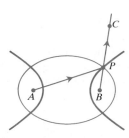

8.37 The confocal ellipse and hyperbola in Problem 25.

25. Show that an ellipse and a hyperbola that have the same foci, *A* and *B*, as in Fig. 8.37, cross at right angles at their points of intersection. (*Hint:* A ray of light emanating from the focus *A* that met the hyperbola at *P* would be reflected from the hyperbola as if it came directly from *B* (Problem 24). The same light ray would be reflected off the ellipse to pass through *B*. Thus, *BPC* is a straight line.)

	TOOLKIT PROGRAMS
	Conic Sections Super * Grapher

8.6
The Graphs of Quadratic Equations

In this article, we establish one of the most amazing results in analytic geometry, which is that the Cartesian graph of any equation of the form

$$Ax^2 + Bxy + Cy^2 + Dx + Ey + F = 0, \qquad (1)$$

in which *A*, *B*, and *C* are not all zero is nearly always a conic section. The only exceptions are the case in which the graph consists of two parallel lines and the case in which there is no graph at all. It is conventional to call all graphs of Eq. (1) **quadratic curves,** curved or not, and we shall use that convention here.

TABLE 8.2
Examples of quadratic curves

$Ax^2 + Bxy + Cy^2 + Dx + Ey + F = 0$

	A	*B*	*C*	*D*	*E*	*F*	*Equation*	*Remarks*
Circle	1		1			−4	$x^2 + y^2 = 4$	$A = C$
Parabola			1	−9			$y^2 = 9x$	Quadratic in y, linear in x
Ellipse	4		9			−36	$4x^2 + 9y^2 = 36$	A, C have same sign, $A \neq C$
Hyperbola	1		−1			−1	$x^2 - y^2 = 1$	A, C have opposite signs
One line (still a conic section)	1						$x^2 = 0$	y-axis
Intersecting lines (still a conic section)		1		1	−1	−1	$xy + x - y - 1 = 0$	Factors to $(x - 1)(y + 1) = 0$, so $x = 1$, $y = -1$.
Parallel lines (not a conic section)	1			−3		2	$x^2 - 3x + 2 = 0$	Factors to $(x - 1)(x - 2) = 0$, so $x = 1$, $x = 2$.
Point	1		1				$x^2 + y^2 = 0$	The origin
No graph	1					1	$x^2 = -1$	

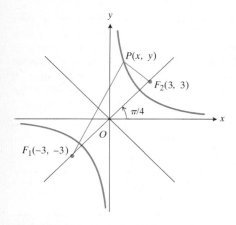

8.38 The focal axis (axis of symmetry through the foci) of the hyperbola $2xy = 9$ makes an angle of $\pi/4$ radians with the positive x-axis.

The Cross-Product Term

You may have noticed that the term Bxy did not appear in the equations we used for conic sections earlier in this chapter. This happened because the coordinate axes always ran parallel to the axes of the conic sections. To see what happens when the parallelism is absent, let us write an equation for the hyperbola in Fig. 8.38, whose foci lie at $F_1(-3, -3)$ and $F_2(3, 3)$. For a point P on the hyperbola, the equation $|PF_1 - PF_2| = 2a$ then becomes $|PF_1 - PF_2| = 2(3) = 6$ and

$$\sqrt{(x + 3)^2 + (y + 3)^2} - \sqrt{(x - 3)^2 + (y - 3)^2} = \pm 6.$$

When we transpose one radical, square, solve for the radical that still appears, and square again, this reduces to

$$2xy = 9, \tag{2}$$

which is a special case of Eq. (1) in which the cross-product term is present. The asymptotes of the hyperbola in Eq. (2) are the x- and y-axes, and the axis of symmetry on which the foci lie makes an angle of $\pi/4$ radians with the positive x-axis. As in this example, the cross-product term is present in Eq. (1) only when the axes of the conic are tilted.

Rotating the Coordinate Axes to Eliminate the Cross-Product Term

To eliminate the xy term from the equation of a conic, we rotate the coordinate axes to eliminate the "tilt" in the axes of the conic. The equations for the rotations we use are derived in the following way. In the notation of Fig. 8.39, which shows a counterclockwise rotation about the origin through an angle α,

$$x = OM = OP \cos(\theta + \alpha) = OP \cos \theta \cos \alpha - OP \sin \theta \sin \alpha,$$
$$y = MP = OP \sin(\theta + \alpha) = OP \cos \theta \sin \alpha + OP \sin \theta \cos \alpha. \tag{3a}$$

Since

$$OP \cos \theta = OM' = x' \quad \text{and} \quad OP \sin \theta = M'P = y', \tag{3b}$$

the equations in (3a) reduce to

Equations for Rotating the Coordinate Axes

$$x = x' \cos \alpha - y' \sin \alpha.$$
$$y = x' \sin \alpha + y' \cos \alpha. \tag{4}$$

EXAMPLE 1 The x- and y-axes are rotated through an angle of $\pi/4$ radians about the origin. Find an equation for the hyperbola $2xy = 9$ in the new coordinates.

Solution Since $\cos \pi/4 = \sin \pi/4 = 1\sqrt{2}$, we substitute

$$x = \frac{x' - y'}{\sqrt{2}}, \qquad y = \frac{x' + y'}{\sqrt{2}}$$

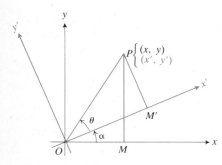

8.39 A counterclockwise rotation through angle α about the origin.

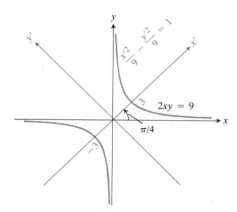

8.40 The hyperbola in Example 1.

from Eqs. (4) into the equation $2xy = 9$ and obtain

$$2\left(\frac{x' - y'}{\sqrt{2}}\right)\left(\frac{x' + y'}{\sqrt{2}}\right) = 9$$

$$x'^2 - y'^2 = 9$$

$$\frac{x'^2}{9} - \frac{y'^2}{9} = 1.$$

See Fig. 8.40. ■

If we apply the rotation equations in (4) to the general quadratic equation (1), we obtain a new quadratic equation

$$A'x'^2 + B'x'y' + C'y'^2 + D'x' + E'y' + F' = 0. \qquad (5)$$

The new coefficients are related to the old ones by the equations

$$A' = A \cos^2\alpha + B \cos \alpha \sin \alpha + C \sin^2\alpha,$$

$$B' = B \cos 2\alpha + (C - A) \sin 2\alpha,$$

$$C' = A \sin^2\alpha - B \sin \alpha \cos \alpha + C \cos^2\alpha,$$

$$D' = D \cos \alpha + E \sin \alpha, \qquad (6)$$

$$E' = -D \sin \alpha + E \cos \alpha,$$

$$F' = F.$$

These equations show, among other things, that if we start with an equation for a curve in which the cross-product term is present ($B \neq 0$), we can find a rotation angle α that produces an equation in which no cross-product term appears ($B' = 0$). To find α, we put $B' = 0$ in the second equation in (6) and solve the resulting equation,

$$B \cos 2\alpha + (C - A)\sin 2\alpha = 0,$$

for α. In practice, this means finding α from one of the two equations

$$\cot 2\alpha = \frac{A - C}{B} \qquad \text{or} \qquad \tan 2\alpha = \frac{B}{A - C}. \qquad (7)$$

EXAMPLE 2 The coordinate axes are to be rotated through an angle α to produce an equation for the curve

$$x^2 + xy + y^2 - 6 = 0 \qquad (8)$$

that has no cross-product term. Find α and the new equation.

Solution 1 Using Eqs. (7), (6), and (5).

Equation (8) has $A = B = C = 1$. We substitute these values into Eq. (7) to find α:

$$\cot 2\alpha = \frac{A - C}{B} = \frac{1 - 1}{1} = 0, \qquad 2\alpha = \frac{\pi}{2}, \qquad \alpha = \frac{\pi}{4}.$$

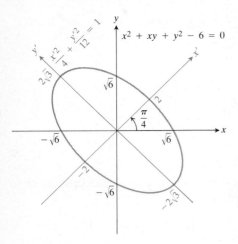

8.41   The ellipse in Example 2.

Substituting $\alpha = \pi/4$, $A = B = C = 1$, $D = E = 0$, and $F = -6$ into Eqs. (6) gives

$$A' = \frac{3}{2}, \qquad B' = 0, \qquad C' = \frac{1}{2}, \qquad D' = E' = 0, \qquad F' = -6.$$

Equation (5) then gives

$$\frac{3}{2}x'^2 + \frac{1}{2}y'^2 - 6 = 0, \qquad \text{or} \qquad \frac{x'^2}{4} + \frac{y'^2}{12} = 1.$$

This is the equation of an ellipse with foci on the new y'-axis. See Fig. 8.41.

Solution 2   Using the rotation equations directly.

This method brings in fewer formulas but takes more arithmetic. We begin by finding α as in Solution 1 and substitute the value found, $\alpha = \pi/4$, into the rotation equations to get

$$x = x' \cos \alpha - y' \sin \alpha = \frac{\sqrt{2}}{2}x' - \frac{\sqrt{2}}{2}y',$$

$$y = x' \sin \alpha + y' \cos \alpha = \frac{\sqrt{2}}{2}x' + \frac{\sqrt{2}}{2}y'.$$

We then substitute the primed expressions for x and y in the original equation, $x^2 + xy + y^2 - 6 = 0$. This gives

$$\left(\frac{\sqrt{2}}{2}x' - \frac{\sqrt{2}}{2}y'\right)^2 + \left(\frac{\sqrt{2}}{2}x' - \frac{\sqrt{2}}{2}y'\right)\left(\frac{\sqrt{2}}{2}x' + \frac{\sqrt{2}}{2}y'\right)$$

$$+ \left(\frac{\sqrt{2}}{2}x' + \frac{\sqrt{2}}{2}y'\right)^2 - 6 = 0,$$

or, skipping over the arithmetic,

$$\frac{x'^2}{4} + \frac{y'^2}{12} = 1$$

when rearranged.   ◼

The Graphs of Quadratic Equations

We now return to our analysis of the graph of the general quadratic equation.

Since axes may always be rotated to eliminate the cross-product term, there is no loss of generality in assuming that this has been done, and our equation has the form

$$Ax^2 + Cy^2 + Dx + Ey + F = 0. \tag{9}$$

Equation 9 represents

a) a *circle* if $A = C \neq 0$ (special cases: the graph is a point or there is no graph at all);

b) a *parabola* if Eq. (9) is quadratic in one variable and linear in the other;

c) an *ellipse* if A and C are both positive or both negative (special cases: a single point or no graph at all);

d) a *hyperbola* if A and C have opposite signs (special case: a pair of intersecting lines);

e) a *straight line* if A and C are zero and at least one of D and E is different from zero;

f) *one or two straight lines* if the left-hand side of Eq. (9) can be factored into the product of two linear factors.

☐ How Calculators Use Rotations

Some calculators use rotations to calculate sines and cosines of arbitrary angles. The procedure goes something like this: The calculator has, stored,

1. ten angles or so, say

$$\alpha_1 = \sin^{-1}(10^{-1}), \quad \alpha_2 = \sin^{-1}(10^{-2}), \ldots, \quad \alpha_{10} = \sin^{-1}(10^{-10}),$$

and

2. twenty numbers, the sines and cosines of the angles $\alpha_1, \alpha_2, \ldots, \alpha_{10}$.

To calculate the sine and cosine of an arbitrary angle θ, one enters θ (in radian measure) into the calculator. The calculator subtracts or adds multiples of 2π to θ to replace θ by the angle between 0 and 2π that has the same sine and cosine as θ (we shall continue to call the angle θ). The calculator then "writes" θ as a sum of multiples of α_1 (as many as possible, without overshooting) plus multiples of α_2 (again, as many as possible), and so on, working its way to α_{10}. This gives

$$\theta \approx m_1\alpha_1 + m_2\alpha_2 + \cdots + m_{10}\alpha_{10}.$$

The calculator then rotates the point $(1, 0)$ through m_1 copies of α_1 (through α_1, m_1 times in succession), plus m_2 copies of α_2, and so on, finishing off with m_{10} copies of α_{10}. See Fig. 8.42. The coordinates of the final position of $(1, 0)$ on the unit circle are the values the calculator gives for $(\cos \theta, \sin \theta)$.

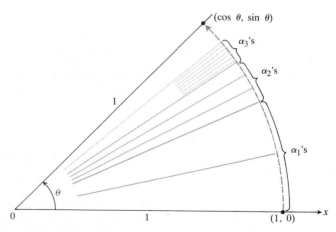

8.42 To calculate the sine and cosine of an angle θ between 0 and 2π, the calculator rotates the point $(1, 0)$ to an appropriate location on the unit circle and displays the resulting coordinates.

PROBLEMS

In Problems 1–10, rotate the coordinate axes to change the given equation into an equation that has no cross-product term. Then identify the graph of the equation. (The new equations will vary with the size and sense of the rotation you choose.)

1. $xy = 2$
2. $x^2 + xy + y^2 = 1$
3. $3x^2 + 2\sqrt{3}xy + y^2 - 8x + 8\sqrt{3}y = 0$
4. $x^2 - \sqrt{3}xy + 2y^2 = 1$
5. $x^2 - 2xy + y^2 = 2$
6. $x^2 - 3xy + y^2 = 5$
7. $\sqrt{2}x^2 + 2\sqrt{2}xy + \sqrt{2}y^2 - 8x + 8y + 4\sqrt{2} = 0$
8. $xy - y - x + 1 = 0$
9. $3x^2 + 2xy + 3y^2 = 19$
10. $3x^2 + 4\sqrt{3}xy - y^2 = 7$

11. Find the sine and cosine of an angle through which the coordinate axes may be rotated to eliminate the cross-product term from the equation

$$14x^2 + 16xy + 2y^2 - 10x + 26{,}370y - 17 = 0.$$

Do not carry out the rotation.

12. An ellipse with foci $F_1(-1, 0)$ and $F_2(0, \sqrt{3})$ passes through the point $(1, 0)$. Write an equation for the ellipse (use the definition). Then find an angle α through which to rotate the coordinate axes to eliminate the cross-product term in your equation. Do not carry out the rotation, however.

13. Show that the equation $x^2 + y^2 = a^2$ becomes $x'^2 + y'^2 = a^2$ for every choice of the angle α in the rotation equations.

14. Show that rotating the axes through an angle of $\pi/4$ radians will eliminate the cross-product term from Eq. (1) whenever $A = C$.

15. Find the equation of the curve $x^2 + 2xy + y^2 = 1$ after a rotation of axes that makes $A' = 0$ in Eq. (6).

TOOLKIT PROGRAMS

Conic Sections   Super * Grapher

8.7

Parabola, Ellipse, or Hyperbola? The Discriminant Tells

In this article, we present a quick way to tell whether the graph of

$$Ax^2 + Bxy + Cy^2 + Dx + Ey + F = 0 \tag{1}$$

is a parabola, ellipse, or hyperbola. The test does not require eliminating the xy term.

The Discriminant

As we know from the previous article, if B is not zero then a rotation of axes through the angle α determined by the equation

$$\cot 2\alpha = \frac{A - C}{B} \tag{2}$$

will change Eq. (1) into the equivalent form

$$A'x'^2 + C'y'^2 + D'x' + E'y' + F' = 0 \tag{3}$$

without a cross-product term.

Now, the graph of Eq. (3) is a (real or degenerate)

a) *parabola* if A' or $C' = 0$; that is, if $A'C' = 0$;

b) *ellipse* if A' and C' have the same sign; that is, if $A'C' > 0$;

c) *hyperbola* if A' and C' have opposite signs; that is, if $A'C' < 0$.

It can also be verified, by using Eqs. (6) in the previous article, that for any rotation of axes,

$$B^2 - 4AC = B'^2 - 4A'C'. \tag{4}$$

This means that the quantity $B^2 - 4AC$ is not changed by a rotation. But when we rotate through the angle α given by Eq. (2), B' becomes zero, so that

$$B^2 - 4AC = -4A'C'.$$

Since the curve is a parabola if $A'C' = 0$, an ellipse if $A'C' > 0$, and a hyperbola if $A'C' < 0$, the curve must be

a) a *parabola* if $B^2 - 4AC = 0$,

b) an *ellipse* if $B^2 - 4AC < 0$, $\qquad\qquad$ (5)

c) a *hyperbola* if $B^2 - 4AC > 0$.

The number $B^2 - 4AC$ is called the **discriminant** of Eq. (1). What we have just seen is that the graph of Eq. (1) is a parabola if the discriminant is zero, an ellipse if the discriminant is negative, and a hyperbola if the discriminant is positive (with the understanding that occasional degenerate cases may arise).

EXAMPLE 1

a) $3x^2 - 6xy + 3y^2 + 2x - 7 = 0$ represents a parabola because

$$B^2 - 4AC = (-6)^2 - 4 \cdot 3 \cdot 3 = 36 - 36 = 0.$$

b) $x^2 + xy + y^2 - 1 = 0$ represents an ellipse because

$$B^2 - 4AC = (1)^2 - 4 \cdot 1 \cdot 1 = -3 < 0.$$

c) $xy - y^2 - 5y + 1 = 0$ represents a hyperbola because

$$B^2 - 4AC = (1)^2 - 4(0)(-1) = 1 > 0. \qquad\blacksquare$$

The Invariant $A + C$

Another invariant associated with Eqs. (1) and (3) is the sum of the coefficients of the squared terms. Since $\sin^2\alpha + \cos^2\alpha = 1$ for any angle α, Eqs. (6) in Article 8.6 give

$$A' + C' = A(\cos^2\alpha + \sin^2\alpha) + C(\sin^2\alpha + \cos^2\alpha) = A + C. \tag{6}$$

We can use the invariance of $B^2 - 4AC$ and $A + C$ to check against numerical errors when we rotate axes. We can also use them to find the coefficients A' and C' in Eq. (3) when the axes are rotated to make $B' = 0$.

EXAMPLE 2   Find the equation to which

$$x^2 + xy + y^2 = 1$$

reduces when the axes are rotated to eliminate the cross-product term.

Solution   From the original equation we find

$$B^2 - 4AC = -3, \qquad A + C = 2.$$

Then, taking $B' = 0$, we have from Eqs. (4) and (6),

$$-4A'C' = -3, \qquad A' + C' = 2,$$

from which we know the curve to be an ellipse. Substituting $C' = 2 - A'$ from the second of these into the first, we obtain

$$-4A'(2 - A') = -3,$$

$$4A'^2 - 8A' + 3 = 0,$$

$$(2A' - 3)(2A' - 1) = 0,$$

and

$$A' = \frac{3}{2} \quad \text{or} \quad A' = \frac{1}{2}.$$

The corresponding values of C' are

$$C' = \frac{1}{2} \quad \text{or} \quad C' = \frac{3}{2}.$$

The equation in the new coordinates is therefore

$$\frac{3}{2}x'^2 + \frac{1}{2}y'^2 = 1 \tag{7a}$$

or

$$\frac{1}{2}x'^2 + \frac{3}{2}y'^2 = 1. \tag{7b}$$

The advantage of Eqs. (7a and b) over the original equation is that they give ready information about the shape of the ellipse: the lengths of the axes, the distance from center to foci, and the eccentricity.   ■

PROBLEMS

Use the discriminant to decide whether the equations in Problems 1–17 represent circles, parabolas, ellipses, or hyperbolas.

1. $x^2 - y^2 - 1 = 0$

2. $25x^2 + 9y^2 - 225 = 0$

3. $y^2 - 4x - 4 = 0$

4. $x^2 + y^2 - 10 = 0$

5. $x^2 + 4y^2 - 4x - 8y + 4 = 0$

6. $x^2 + y^2 + xy + x - y = 3$

7. $2x^2 - y^2 + 4xy - 2x + 3y = 6$

8. $x^2 + 4xy + 4y^2 - 3x = 6$

9. $x^2 + y^2 + 3x - 2y = 10$

10. $xy + y^2 - 3x = 5$

11. $3x^2 + 6xy + 3y^2 - 4x + 5y = 12$

12. $x^2 - y^2 = 1$

13. $2x^2 + 3y^2 - 4x = 7$

14. $x^2 - 3xy + 3y^2 + 6y = 7$

15. $25x^2 - 4y^2 - 350x = 0$

16. $6x^2 + 3xy + 2y^2 + 17y + 2 = 0$

17. $3x^2 + 12xy + 12y^2 + 435x - 9y + 72 = 0$

18. *A nice area formula for ellipses.* When $B^2 - 4AC$ is negative, the equation

$$Ax^2 + Bxy + Cy^2 = 1$$

represents an ellipse. If the semiaxes are a and b, the area of the ellipse is πab. Show that the area of the ellipse is also given by the formula $2\pi / \sqrt{4AC - B^2}$. (*Hint:* Rotate the coordinate axes to eliminate the cross-product term and apply Eq. 4 to the new equation.)

19. Use Eqs. (6) in Article 8.6 to show that

$$D'^2 + E'^2 = D^2 + E^2$$

for every angle of rotation α.

20. If $C = -A$ in Eq. (1), show that there is a rotation of axes for which $A' = C' = 0$ in the resulting Eq. (3). Find an angle α that makes $A' = C' = 0$ in this case. (*Hint:* Since $A' + C' = 0$, you need only to make the further requirement that $A' = 0$ in Eq. 3.)

21. *Proof of Eq.* (4). Use Eqs. (6) in Article 8.6 to verify that $B'^2 - 4A'C' = B^2 - 4AC$ for any rotation of axes. (The calculation works out nicely but requires patience.)

8.8
Sections of a Cone

In this article, we show that every curve obtained by cutting through a double cone with a plane satisfies the equation $PF = e \cdot PD$ for a suitably chosen F and D. Thus, every "geometric" conic section is also an "algebraic" conic section.

Suppose the cutting plane makes an acute angle α with the axis of the cone, and let the acute angle between the side and axis of the cone be β (see Fig. 8.43). Then the section is

i) a circle if $\alpha = 90°$;

ii) an ellipse if $\beta < \alpha < 90°$;

iii) a parabola if $\alpha = \beta$;

iv) a hyperbola if $0 \le \alpha < \beta$.

The connection between these curves as we have defined them and the sections of a cone can be seen in Fig. 8.44. The figure shows an ellipse, but the argument works for the other cases as well. The general construction goes as follows:

A sphere is inscribed tangent to the cone along a circle C and tangent to the cutting plane at a point F. Point P is any point on the conic section. We shall see that F is a focus and that the line L, in which the cutting plane and the plane

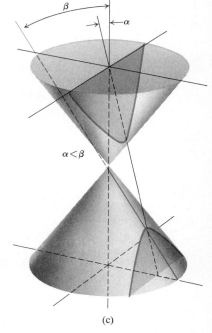

(a) (b) (c)

8.43 The intersection of a plane and double cone in (a) an ellipse, (b) a parabola, (c) a hyperbola. The angle β is the angle between the side and axis of the cone. The angle α is the acute angle between the plane and the cone's axis.

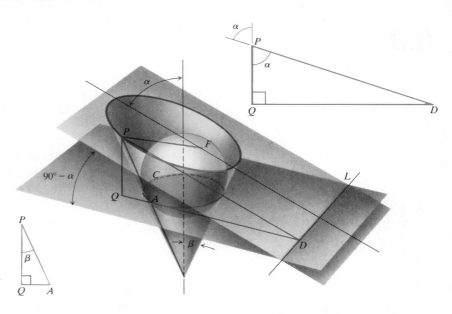

8.44 The line L is the directrix that corresponds to the focus F of the ellipse.

of the circle C intersect, is a directrix of the curve. To this end, let Q be the point where the line through P parallel to the axis of the cone intersects the plane of C, let A be the point where the line joining P to the vertex of the cone touches C, and let PD be perpendicular to line L at D. Then PA and PF are two lines tangent to the same sphere from a common point P and hence have the same length:

$$PA = PF.$$

Also, from the right triangle PQA, we have

$$PQ = PA \cos \beta;$$

and from the right triangle PQD, we find that

$$PQ = PD \cos \alpha.$$

Hence

$$PA \cos \beta = PD \cos \alpha, \qquad \text{or}$$

$$\frac{PA}{PD} = \frac{\cos \alpha}{\cos \beta}. \tag{1}$$

But since $PA = PF$, this means that

$$\frac{PF}{PD} = \frac{\cos \alpha}{\cos \beta}, \qquad \text{or} \tag{2}$$

$$PF = \frac{\cos \alpha}{\cos \beta} PD. \tag{3}$$

Since α and β are constant for a given cone and cutting plane, Eq. (3) has the form

$$PF = e \cdot PD. \tag{4}$$

This characterizes P as belonging to a parabola, an ellipse, or a hyperbola, with focus at F and directrix L, depending on whether $e = 1$, $e < 1$, or $e > 1$, where

$$e = \cdot \frac{\cos \alpha}{\cos \beta}$$

is thus identified with the eccentricity.

PROBLEMS

1. Sketch a figure similar to Fig. 8.44 when the conic section is a parabola, and carry through the argument of this article on the basis of your figure.

2. Sketch a figure similar to Fig. 8.44 when the conic section is a hyperbola, and carry through the argument of this article on the basis of your figure.

3. Which parts of the construction described in this article become impossible when the conic section is a circle?

4. Let one directrix be the line $x = -p$ and take the corresponding focus at the origin. Use the equation $PF = e \cdot PD$ to derive an equation for the general conic section of eccentricity e. If e is neither 0 nor 1, show that the center of the conic section is the point

$$\left(\frac{pe^2}{1 - e^2}, 0 \right).$$

8.9

Parametric Equations for Conics and Other Curves

In this article we examine parametric equations that describe the motions of particles along conic sections and other curves in the plane. Among the other curves are trochoids and cycloids, which are the paths traced through the air by particles on rolling wheels. You may wish to review Article 2.8 briefly before you begin.

Parametric Equations for Conic Sections

Parametric equations for the position of a particle moving in the plane are sometimes called parametric equations for the path traced by the particle. Thus, equations that describe the motion of a particle around a circle may be called parametric equations for the circle, and equations that describe the motion of a particle that traces out a parabola may be called parametric equations for the parabola, as in the examples that follow.

EXAMPLE 1 *Parametric equations for the circle $x^2 + y^2 = a^2$.* Describe the motion of the particle whose position $P(x, y)$ at time t is given by

$$x = a \cos t, \qquad y = a \sin t, \qquad 0 \le t \le 2\pi. \tag{1}$$

Solution Since

$$x^2 + y^2 = a^2\cos^2 t + a^2\sin^2 t = a^2,$$

the particle moves on the circle of radius a centered at the origin. The particle begins at the point $(a, 0)$ when t equals zero and moves once counterclockwise around the circle as t increases from 0 to 2π. See Fig. 8.45. ∎

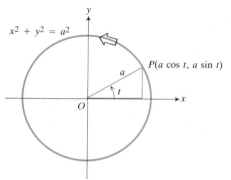

8.45 The equations $x = a \cos t$, $y = a \sin t$, $0 \le t \le 2\pi$ are parametric equations for the circle $x^2 + y^2 = a^2$.

EXAMPLE 2   *Parametric equations for the parabola $y^2 = x$.* Describe the motion of a particle whose position $P(x, y)$ at time t is given by

$$x = t^2, \qquad y = t, \qquad -\infty < t < \infty.$$

Solution   We find a Cartesian equation for the coordinates of the particle by eliminating t between the equations for x and y. Since

$$y^2 = t^2 = x,$$

we see that the motion takes place on the parabola

$$y^2 = x$$

(Fig. 8.46). As t increases between $-\infty$ and ∞, the particle comes in on the lower half of the parabola, reaches the origin when t is zero, and moves out into the first quadrant as t continues to increase.

As we saw with circles in Article 2.8, a given curve can be traversed in either direction and have many different parameterizations. Among the other parameterizations of the parabola here are $y = (\tan^{-1}t)^2$, $x = (\tan^{-1}t)$, $-\pi/2 < t < \pi/2$ (same direction, finite interval) and $x = t^2$, $y = -t$, $-\infty < t < \infty$ (direction reversed). ■

EXAMPLE 3   *Parametric equations for the ellipse $x^2/a^2 + y^2/b^2 = 1$.* Describe the motion of a particle whose position $P(x, y)$ at time t is given by the equations

$$x = a \cos t, \qquad y = b \sin t, \qquad 0 \le t \le 2\pi.$$

Solution   We find a Cartesian equation for the coordinates of the particle by eliminating t between the equations for x and y. Since

$$\frac{x^2}{a^2} + \frac{y^2}{b^2} = \frac{a^2\cos^2 t}{a^2} + \frac{b^2\sin^2 t}{b^2} = \cos^2 t + \sin^2 t = 1,$$

the motion takes place on the ellipse $x^2/a^2 + y^2/b^2 = 1$. The particle begins at $(a, 0)$ when t equals zero and moves counterclockwise around the ellipse, traversing it exactly once as t moves from 0 to 2π. See Fig. 8.47. ■

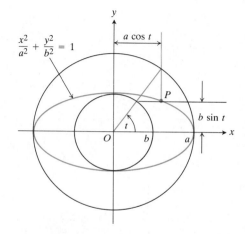

8.46 The equations $x = t^2$, $y = t$, $-\infty < t < \infty$, are parametric equations for the parabola $y^2 = x$.

8.47 The coordinates of P are $x = a \cos t$, $y = b \sin t$.

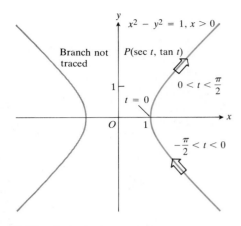

8.48 The equations $x = \sec t$, $y = \tan t$, $-\pi/2 < t < \pi/2$, are parametric equations for the right-hand branch of the hyperbola $x^2 - y^2 = 1$.

EXAMPLE 4 *Parametric equations for the right-hand branch of the hyperbola $x^2 - y^2 = 1$.* Describe the motion of the particle whose position $P(x, y)$ at time t is given by the equations

$$x = \sec t, \qquad y = \tan t, \qquad -\frac{\pi}{2} < t < \frac{\pi}{2}.$$

Solution We find a Cartesian equation for the coordinates of P by eliminating t between the equation for x and y. Since

$$x^2 - y^2 = \sec^2 t - \tan^2 t = 1,$$

we see that the motion takes place somewhere on the hyperbola $x^2 - y^2 = 1$. Since $x = \sec t$ is always positive for the parameter values $-\pi/2 < t < \pi/2$, the motion takes place on the hyperbola's right-hand branch. As t moves from $-\pi/2$ to $\pi/2$, the particle comes in along the lower half of the right-hand branch, reaching the origin at $t = 0$. It then moves into the first quadrant to complete the coverage of the right-hand branch as t approaches $\pi/2$. See Fig. 8.48. ∎

EXAMPLE 5 *A parabolic arch.* Sketch the curve traced by the point $P(x, y)$ whose coordinates satisfy the equations

$$x = \cos t, \qquad y = 1 - \cos 2t, \qquad -\infty < t < \infty. \qquad (2)$$

Solution We find a Cartesian equation for the curve by eliminating t:

$$y = 1 - \cos 2t = 1 - 2\cos^2 t + 1 = 2 - 2x^2.$$

Thus every point of the graph of (2) lies on the parabola

$$y = 2 - 2x^2. \qquad (3)$$

The parametric equations in (2), however, describe only the portion of the parabola (Fig. 8.49) for which

$$-1 \le x = \cos t \le 1 \qquad \text{and} \qquad 0 \le y = 1 - \cos 2t \le 2.$$

From (2) we see that the point $P(x, y)$ starts at $A(1, 0)$ when $t = 0$. It then moves up and to the left as t increases, arriving at $B(0, 2)$ when $t = \pi/2$. It continues on to $C(-1, 0)$ as t increases to π. As t varies from π to 2π, the point retraces the arch CBA back to A. Since x and y are periodic, x with period 2π and y with period π, any further variation in t results in retracing a portion of the arch. ∎

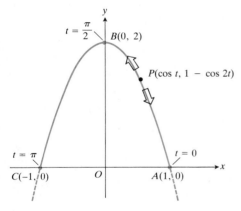

8.49 As t varies from $-\infty$ to ∞, the point P traces and retraces the parabolic arch.

EXAMPLE 6 *Trochoids and cycloids.* A wheel of radius a rolls along a horizontal straight line without slipping. Find the curve traced by a point P on a spoke of the wheel b units from its center. Such a curve is called a **trochoid** (one Greek word for wheel is *trochos*). When $b = a$, P is on the circumference, and the curve is called a **cycloid.** This is like the path traveled by a pebble in the tread of a rolling tire.

Solution In Fig. 8.50, we take the x-axis to be the line the wheel rolls along, with the y-axis through a low point of the trochoid. It is customary to use the angle t through which CP has rotated as the parameter. Since the circle rolls without slipping, the distance OM that the wheel has moved horizontally is just equal to the circular arc $MN = at$. (Roll the wheel back. Then N will fall at the

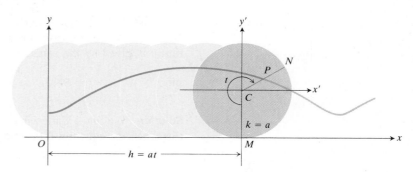

8.50 The trochoid $x = at - b \sin t$, $y = a - b \cos t$, shown for $t \leq 0$.

origin O.) The xy-coordinates of C are therefore

$$h = at, \qquad k = a. \tag{4}$$

We now introduce $x'y'$-axes parallel to the xy-axes and having their origin at C (Fig. 8.51). The xy- and $x'y'$-coordinates of P are related by the equations

$$x = h + x', \qquad y = k + y'. \tag{5}$$

From Fig. 8.51 we may immediately read

$$x' = b \cos \theta, \qquad y' = b \sin \theta,$$

or, since

$$\theta = \frac{3\pi}{2} - t,$$

$$x' = -b \sin t, \qquad y' = -b \cos t. \tag{6}$$

We substitute these results and Eqs. (5) into (4) and obtain

$$x = at - b \sin t, \qquad y = a - b \cos t \tag{7}$$

as parametric equations of the trochoid.

The cycloid (Fig. 8.52),

$$x = a(t - \sin t), \qquad y = a(1 - \cos t), \tag{8}$$

obtained from (7) by taking $b = a$, is the most important special case. ■

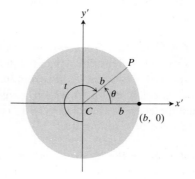

8.51 The $x'y'$-coordinates of P are $x' = b \cos \theta$, $y' = b \sin \theta$.

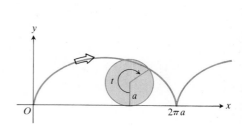

8.52 The cycloid $x = a(t - \sin t)$, $y = a(1 - \cos t)$, shown for $t \geq 0$.

Brachistochrones and Tautochrones

If we turn Fig. 8.52 upside down, Eqs. (8) still apply, and the resulting curve (Fig. 8.53) has several interesting properties, one of which we shall now discuss without proof. The proofs belong to a branch of mathematics known as the calculus of variations. Much of the fundamental theory of this subject is attributed to the Bernoulli brothers, John and James, who were friendly rivals and challenged each other with mathematical problems. One of these, the brachistochrone (''shortest time'') problem, was: Among all smooth curves joining two given points, find that one along which a bead, subject only to the force of gravity, might slide *in the shortest time*.

The two points, labeled P_0 and P_1 in Fig. 8.54, may be taken to lie in a vertical plane at the origin and at (x_1, y_1), respectively. We can formulate the problem in mathematical terms as follows. The kinetic energy of the bead at the start is zero, since its velocity is zero. The work done by gravity in moving the bead from $(0, 0)$ to any point (x, y) is mgy and this must be equal to the change in kinetic energy; that is,

$$mgy = \frac{1}{2}mv^2 - \frac{1}{2}m(0)^2.$$

Thus the velocity

$$v = ds/dt$$

that the bead has when it reaches $P(x, y)$ is

$$v = \sqrt{2gy}.$$

That is,

$$\frac{ds}{dt} = \sqrt{2gy} \qquad \text{or} \qquad dt = \frac{ds}{\sqrt{2gy}} = \frac{\sqrt{1 + \left(\dfrac{dy}{dx}\right)^2}\, dx}{\sqrt{2gy}}. \tag{9}$$

The time T_1 required for the bead to slide from P_0 to P_1 depends on the particular curve $y = f(x)$ along which it moves and is given by

$$T_1 = \int_0^{x_1} dt = \int_0^{x_1} \sqrt{\frac{1 + (dy/dx)^2}{2gy}}\, dx. \tag{10}$$

The problem is *to find the curve* $y = f(x)$ (if there is one) that passes through the points $P_0(0, 0)$ and $P_1(x_1, y_1)$ and minimizes the value of the integral in (10).

At first sight, we might guess that the straight line joining P_0 and P_1 would also yield the shortest time, but a moment's reflection will cast some doubt on this conjecture. There may be some gain in time by having the bead start to fall vertically at first, thereby building up its velocity more quickly than if it were to slide along an inclined path. With this increased velocity, the bead might travel over a longer path and still reach P_1 in a shorter time. The solution of the problem is beyond the present book, but the brachistochrone curve, when one exists, is actually an arc of a cycloid through P_0 and P_1, having a cusp at the origin and a horizontal tangent at P_1.

If we write Eq. (9) in the equivalent form

$$T_1 = \int_0^{x_1} \sqrt{\frac{dx^2 + dy^2}{2gy}}$$

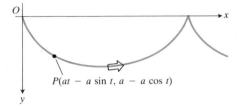

8.53 To study motion along an upside-down cycloid under the influence of gravity, we turn Fig. 8.52 upside down. This points the *y*-axis in the direction of the gravitational force and makes all the downward *y*-coordinates positive. The equations for the cycloid are still $x = a(t - \sin t)$, $y = a(1 - \cos t)$.

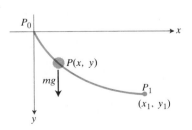

8.54 A bead sliding down a frictionless cycloid under the force of gravity.

8.55 Beads released on the cycloid at O, A, and B will all take the same amount of time to reach C.

■ ■ ■

PENDULUM CLOCKS

One trouble with pendulum clocks is that the rate at which the pendulum swings changes with the size of the swing. The wider the swing, the longer it takes the pendulum to return. This would create no problem if clocks never ran down, but they do. As the spring unwinds, the force exerted on the pendulum decreases, the pendulum swings through increasingly shorter arcs, and the clock speeds up. The ticks come faster as the clock winds down.

Christiaan Huygens (1629–1695), the Dutch mathematician, physicist, and astronomer, needed an accurate clock to make careful astronomical measurements. In 1673 he published the first description of an ideal pendulum clock, whose bob swings in a cycloid. The period of the swing of Huygens's clock did not depend on the amplitude and would not change as the clock wound down.

How does one make a pendulum bob swing in a cycloid? Hang it from a fine wire constrained by "cheeks," which cause the bob to draw up as it swings to the side. And what is the shape of the cheeks? They are cycloids.

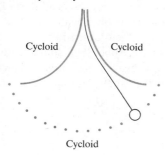

Cycloid   Cycloid

Cycloid

and then substitute Eqs. (8) into this, we obtain

$$T_1 = \int_0^{t_1} \sqrt{\frac{a^2(2 - 2\cos t)}{2ga(1 - \cos t)}}\, dt = t_1 \sqrt{\frac{a}{g}}$$

as the time required for the bead to slide from P_0 to P_1. The time required to reach the bottom of the arc is obtained by taking $t_1 = \pi$. Now it is a remarkable fact, which we shall demonstrate, that the time required for the bead to slide along the cycloid from $(0, 0)$ to the lowest point $(a\pi, 2a)$ is the same as the time required for the bead, starting from rest, to slide from *any intermediate point* of the arc, say (x_0, y_0), to $(a\pi, 2a)$. For the latter case, the bead's velocity is

$$v = \sqrt{2g(y - y_0)}$$

and the time required is

$$T = \int_{t_0}^{\pi} \sqrt{\frac{a^2(2 - 2\cos t)}{2ag(\cos t_0 - \cos t)}}\, dt = \sqrt{\frac{a}{g}} \int_{t_0}^{\pi} \sqrt{\frac{1 - \cos t}{\cos t_0 - \cos t}}\, dt$$

$$= \sqrt{\frac{a}{g}} \int_{t_0}^{\pi} \sqrt{\frac{2\sin^2(t/2)}{[2\cos^2(t_0/2) - 1] - [2\cos^2(t/2) - 1]}}\, dt$$

$$= \sqrt{\frac{a}{g}} \int_{t_0}^{\pi} \frac{\sin(t/2)\, dt}{\sqrt{\cos^2(t_0/2) - \cos^2(t/2)}}$$

$$= \sqrt{\frac{a}{g}} \int_{t=t_0}^{t=\pi} \frac{-2\, du}{\sqrt{a^2 - u^2}} \quad \left(\begin{aligned} u &= \cos(t/2) \\ -2du &= \sin(t/2)\, dt \\ a &= \cos(t_0/2) \end{aligned} \right)$$

$$= 2\sqrt{\frac{a}{g}} \left[-\sin^{-1}\frac{u}{a} \right]_{t=t_0}^{t=\pi}$$

$$= 2\sqrt{\frac{a}{g}} \left[-\sin^{-1}\frac{\cos(t/2)}{\cos(t_0/2)} \right]_{t_0}^{\pi} = 2\sqrt{\frac{a}{g}}(-\sin^{-1}0 + \sin^{-1}1) = \pi\sqrt{\frac{a}{g}}.$$

Since this answer is independent of the value of t_0, it follows that the same length of time is required to reach the lowest point on the cycloid no matter where on the arc the particle is released from rest. Thus, in Fig. 8.55, three particles that start at the same time from O, A, and B will reach C simultaneously. In this sense, the cycloid is a **tautochrone** (meaning "the same time") as well as a brachistochrone.

Standard Parametric Equations

Circle: $x^2 + y^2 = a^2$ $\begin{cases} x = a\cos t, \\ y = a\sin t, \end{cases}$ $0 \le t \le 2\pi$

Ellipse: $\dfrac{x^2}{a^2} + \dfrac{y^2}{b^2} = 1$ $\begin{cases} x = a\cos t, \\ y = b\sin t, \end{cases}$ $0 \le t \le 2\pi$

Cycloid: $x = a(t - \sin t), \quad y = a(1 - \cos t)$

PROBLEMS

In Problems 1–16, sketch the curve traced by the point $P(x, y)$ as the parameter t varies over the given domain. Also find a Cartesian equation for each curve.

1. $x = \cos t, \quad y = \sin t, \quad 0 \le t \le 2\pi$

2. $x = \cos 2t, \quad y = \sin 2t, \quad 0 \le t \le \pi$

3. $x = 4 \cos t, \quad y = 2 \sin t, \quad 0 \le t \le 2\pi$

4. $x = 4 \cos t, \quad y = 5 \sin t, \quad 0 \le t \le 2\pi$

5. $x = \cos 2t, \quad y = \sin t, \quad 0 \le t \le 2\pi$

6. $x = \cos t, \quad y = \sin 2t, \quad 0 \le t \le 2\pi$

7. $x = -\sec t, \quad y = \tan t, \quad -\pi/2 < t < \pi/2$

8. $x = \csc t, \quad y = \cot t, \quad 0 < t < \pi$

9. $x = t - \sin t, \quad y = 1 - \cos t, \quad 0 \le t \le 2\pi$

10. $x = 2 + 4 \sin t, \quad y = 3 - 2 \cos t, \quad 0 \le t \le 2\pi$

11. $x = t^3, \quad y = t^2, \quad -\infty < t < \infty$

12. $x = 2t + 3, \quad y = 4t^2 - 9, \quad -\infty < t < \infty$

13. $x = \sec^2 t - 1, \quad y = \tan t, \quad -\pi/2 < t < \pi/2$

14. $x = 2 + 1/t, \quad y = 2 - t, \quad 0 < t < \infty$

15. $x = t + 1, \quad y = t^2 + 4, \quad 0 \le t < \infty$

16. $x = t^2 + t, \quad y = t^2 - t, \quad -\infty < t < \infty$

17. Find parametric equations for the semicircle

$$x^2 + y^2 = a^2, \qquad y > 0,$$

using as parameter the slope $t = dy/dx$ of the tangent to the curve at (x, y).

18. Find parametric equations for the circle

$$x^2 + y^2 = a^2,$$

using as parameter the arc length s measured counterclockwise from the point $(a, 0)$ to the point (x, y).

19. *Involute of a circle.* If a string wound around a fixed circle is unwound while held taut in the plane of the circle, its end traces an *involute* of the circle (Fig. 8.56). Let the fixed circle be located with its center at the origin O and have radius a. Let the initial position of the tracing point P be $A(a, 0)$ and let the unwound portion of the string PT be tangent to the circle at T. Derive parametric equations for the involute, using the angle AOT as the parameter t.

20. *Epicycloids.* When a circle rolls externally along the circumference of a second, fixed circle, any point P on the circumference of the rolling circle describes an *epicycloid* (Fig. 8.57). Let the fixed circle have its center at the origin O and have radius a. Let the radius of the rolling circle be b and let the initial position of the tracing point P be $A(a, 0)$. Determine parametric equations for the epicycloid, using as parameter the angle θ from the positive x-axis to the line of centers.

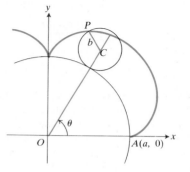

8.57 An epicycloid with $b = a/4$ (Problem 20).

21. *Hypocycloids.* When a circle rolls on the inside of a fixed circle, any point P on the circumference of the rolling circle describes a *hypocycloid*. Let the fixed circle be $x^2 + y^2 = a^2$, let the radius of the rolling circle be b, and let the initial position of the tracing point P be $A(a, 0)$. Use the angle θ from the positive x-axis to the line of centers as parameter and determine parametric equations for the hypocycloid. In particular, if $b = a/4$, as in Fig. 8.58, show that the hypocycloid is the astroid

$$x = a \cos^3 \theta, \qquad y = a \sin^3 \theta.$$

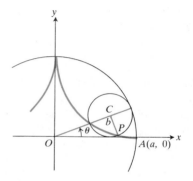

8.58 A hypocycloid with $b = a/4$ (Problem 21).

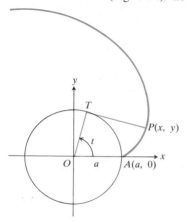

8.56 The involute of a circle (Problem 19).

22. *The witch of Maria Agnesi.* The witch of Maria Agnesi is a bell-shaped curve that may be constructed as follows: Let C be a circle of radius a having its center at $(0, a)$ on the y-axis (Fig. 8.59). The variable line OA through the origin O intersects the line $y = 2a$ in the point A and intersects the circle in the point B. A point P on the witch is now located by taking the intersection of lines through A and B parallel to the y- and x-axes, respectively.

a) Find parametric equations for the witch, using as parameter the angle θ from the x-axis to the line OA.

b) Also find a Cartesian equation for the witch.

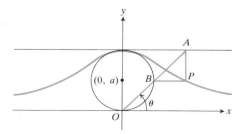

8.59   The witch of Maria Agnesi (Problem 22).

Historical note: Maria Gaetana Agnesi (1718–1799), the daughter of a mathematics professor at the University of Bologna, wrote the first comprehensive calculus text. In four books, the text treated algebra and geometry, differential calculus, integral calculus, and differential equations. (L'Hôpital's earlier book treated only analytic geometry and differential calculus.) The text was translated into French and English, and it is a mistranslation that is responsible for our calling Agnesi's bell-shaped curve "the witch" today. This name, in fact, is found only in texts written in English. Agnesi's own name for the curve was "versiera," from the Latin verb *vertere,* to turn. The translator, a Cambridge scholar who had learned Italian expressly for the purpose of translating Agnesi's text, probably confused the Latin *versiera* with the Italian *avversiera,* "wife of the devil," carefully translating the latter as "the witch."

23. *Parametric equations for lines in the plane.*

a) Show that the equations

$$x = x_0 + (x_1 - x_0)t, \qquad y = y_0 + (y_1 - y_0)t,$$

$-\infty < t < \infty$ are parametric equations for the line through the points (x_0, y_0) and (x_1, y_1).

b) Write parametric equations for the line through a point (x_0, y_0) and the origin.

c) Write parametric equations for the line through $(-1, 0)$ and $(0, 1)$.

24. The question presented in Fig. 8.60 appeared in a college entrance examination some years ago. All the answers offered with the question were wrong. What is the correct answer? What would the correct answer be if the smaller circle were inside the larger circle instead of outside?

25. Find the point on the parabola $x = t$, $y = t^2$ closest to the point $(2, 1/2)$.

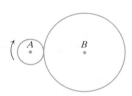

In the figure above, the radius of circle A is 1/3 the radius of circle B. Starting from the position shown in the figure, circle A rolls around circle B. At the end of how many revolutions of circle A will the center of circle A first reach its starting point?

(A) $\dfrac{3}{2}$ (B) 3   (C) 6   (D) $\dfrac{9}{2}$ (E) 9

Figure 8.60

26. Find the point on the ellipse $x = 2 \cos t$, $y = \sin t$, $0 \le t \le 2\pi$ closest to the point $(3/4, 0)$.

27. Find the length of one arch of the cycloid

$$x = a(t - \sin t), \quad y = a(1 - \cos t).$$

28. Find the area of the surface generated by revolving one arch of the cycloid $x = t - \sin t$, $y = 1 - \cos t$ about the x-axis.

29. Find the volume swept out by revolving the region bounded by the x-axis and one arch of the cycloid $x = t - \sin t$, $y = 1 - \cos t$ about the x-axis. (*Hint:* $dV = \pi y^2 dx = \pi y^2 (dx/dt) dt$.)

30. The region bounded by the ellipse $x = 2 \cos t$, $y = \sin t$, $0 \le t \le 2\pi$, is revolved about the y-axis to generate a solid. Find the volume of the solid.

31. CALCULATOR   An automobile tire of radius 1 ft has a pebble stuck in the tread. Estimate to the nearest foot the length of the arched path traced by the pebble when the car goes 1 mile. Start by finding the ratio of the length of one arch of a cycloid to its base length.

32. COMPUTER GRAPHER   If you have access to a parametric equation grapher, try the following equations. The screen displays are gorgeous.

a) *Deltoid:* $x = 2 \cos t + \cos 2t$
$ y = 2 \sin t - \sin 2t$
$ 0 \le t \le 2\pi$

b) *Epicycloid:* $x = 9 \cos t - \cos 9t$
$ y = 9 \sin t - \sin 9t$
$ 0 \le t \le 2\pi$

c) *Hypocycloid:* $x = 8 \cos t + 2 \cos 4t$
$ y = 8 \sin t - 2 \sin 4t$
$ 0 \le t \le 2\pi$

d) *Hypotrochoid:* $x = \cos t + 5 \cos 3t$
$ y = 6 \cos t - 5 \sin 3t$
$ 0 \le t \le 2\pi$

TOOLKIT PROGRAMS

Parametric Equations   Super * Grapher

REVIEW QUESTIONS AND EXERCISES

1. Name the conic sections. Where does the name "conic" come from?

2. How are parabolas defined in terms of distance? Give typical equations for parabolas. Graph one of the equations and include the parabola's vertex, focus, axis, and directrix. What is the eccentricity of a parabola?

3. What reflective property do parabolas have?

4. How are ellipses defined in terms of distance? Give typical equations for ellipses. Graph one of the equations and include the ellipse's vertices and foci. How is the eccentricity of an ellipse defined? Sketch an ellipse whose eccentricity is close to 1 and another whose eccentricity is close to 0.

5. How are hyperbolas defined in terms of distance? Give typical equations for hyperbolas. Graph one of the equations and include the hyperbola's vertices, foci, axes, and asymptotes. What values can the eccentricities of hyperbolas have?

6. Explain the equation $PF = e \cdot PD$.

7. What can be said about the graph of the equation

$$Ax^2 + Bxy + Cy^2 + Dx + Ey + F = 0$$

if A, B, and C are not all zero? What can you tell from the number $B^2 - 4AC$?

8. Give equations for translating and rotating axes. Illustrate each set of equations with a diagram. How are translations and rotations used in the present chapter?

9. Give parametric equations for a circle, parabola, and ellipse. In each case, give a parameter domain that covers the conic exactly once.

10. Give parametric equations for one branch of a hyperbola. What is the appropriate domain for the parameter if the curve is to be traced out exactly once by your equations?

11. What is a cycloid? What are typical parametric equations for a cycloid?

MISCELLANEOUS PROBLEMS

1. Let $P(x, y)$ be a point on the curve

$$x^2 + xy + y^2 = 3$$

and $P'(kx, ky)$ be a point on the line OP from the origin to P. If k is held constant, find an equation for the curve traced by P' as P traces the curve C.

2. *Symmetry with respect to a circle.* Two points P and Q are called symmetric with respect to a circle if they lie on the same ray from the center and if the product of their distances from the center is equal to the square of the circle's radius. If Q traverses the line $x + 2y - 5 = 0$, find the path of the point P that is symmetric to Q with respect to the circle $x^2 + y^2 = 4$.

3. A point $P(x, y)$ moves so that the ratio of its distances from two fixed points is a constant k. Show that the point traces a circle if $k \neq 1$ and a straight line if $k = 1$.

4. Find the center and radius of the circle that passes through the points $A(2, 0)$ and $B(6, 0)$ and is tangent to the curve $y = x^2$.

5. Find the center of the circle that passes through the point $(0, 1)$ and is tangent to the curve $y = x^2$ at $(2, 4)$.

6. **CALCULATOR** Graph the equation $x^{2n} + y^{2n} = a^{2n}$ for the following values of n: (a) 1, (b) 2, (c) 100. In each instance, find where the curve cuts the line $y = x$.

7. Find an equation for the parabola with focus $(4, 0)$ and directrix $x = 3$. Sketch the parabola together with its vertex, focus, and directrix.

8. Find the vertex, focus, and directrix of the parabola

$$x^2 - 6x - 12y + 9 = 0.$$

9. Show that if a line is drawn tangent to the parabola $y^2 = kx$ at a point P not at the origin, then the portion of the tangent

that lies between the x-axis and P is bisected by the y-axis.

10. If lines are drawn parallel to the coordinate axes through a point P on the parabola $y^2 = kx$, the parabola divides the rectangular region bounded by these lines and the axes into two smaller regions.

a) If these two smaller regions are revolved about the y-axis, show that they generate two solids whose volumes are in the ratio $4:1$.

b) What is the ratio of the volumes of the solids generated by revolving the regions about the x-axis?

11. Show that the centers of all chords of the parabola $x^2 = 4py$ with slope m lie on a straight line, and find its equation.

12. The line through the focus F and the point $P(x_1, y_1)$ on the parabola $y^2 = 4px$ intersects the parabola in a second point $Q(x_2, y_2)$. Find the coordinates of Q in terms of y_1 and p. If the line through P and the origin cuts the directrix at R, prove that QR is parallel to the parabola's axis.

13. *A semicubical parabola.* Find the point (or points) on the curve $x^2 = y^3$ nearest the point $P(0, 4)$. Sketch the curve and the shortest line segment from P to the curve.

14. A comet moves in a parabolic orbit with the sun at the focus. When the comet is 4×10^7 miles from the sun, the line from the sun to the comet makes an angle of $60°$ with the axis of the orbit (drawn in the direction in which the orbit opens). How near does the comet come to the sun?

15. Suppose that the line tangent to the parabola $y^2 = 4px$ at a point Q crosses the parabola's axis at A and that the line through Q parallel to the parabola's axis crosses the directrix at D. Show that Q, A, D, and the parabola's focus F are the four vertices of a rhombus.

16. Find an equation for the curve traced by point $P(x, y)$ if the

distance from P to the vertex of the parabola $x^2 = 8y$ is twice the distance from P to the focus. Identify the curve.

17. Prove that the tangent to a parabola at a point P cuts the axis of the parabola at a point whose distance from the vertex equals the distance from P to the tangent at the vertex.

18. Find an equation of an ellipse with foci at $(1, 0)$ and $(5, 0)$ and one vertex at the origin.

19. Let $F_1 = (3, 0)$, $F_2 = (0, 5)$, $P = (-1, 3)$.
a) Find the distances F_1P and F_2P.
b) Does the origin lie inside or outside the ellipse through P that has F_1 and F_2 as its foci? Why?

20. Find the eccentricity and center of the ellipse
$$x^2 + 12y^2 - 6x - 48y + 9 = 0.$$

21. Find the eccentricity, center, foci, and vertices of the ellipse
$$x^2 + 9y^2 - 6x - 36y - 99 = 0.$$

22. Show that the line $y = mx + c$ is tangent to the curve $Ax^2 + y^2 - 1 = 0$ if and only if the constants A, m, and c satisfy the equation $A(c^2 - 1) = m^2$.

23. Find an equation for the ellipse of eccentricity 2/3 that has one vertex at $(3, 1)$ and the nearer focus at $(1, 1)$.

24. A line segment of length $a + b$ runs from the x-axis to the y-axis in the first quadrant. The point P on the segment lies a units from one end and b units from the other end. Show that P traces an elliptical path as the ends of the segment slide along the axes.

25. An ellipse in standard position was translated to a new location in the plane, where its equation became
$$x^2 + 4y^2 - 4x - 8y + 4 = 0.$$
What translation was used? What are the new coordinates of the ellipse's center?

26. A ripple tank is made by bending a strip of tin around the perimeter of an ellipse for the wall of the tank and soldering a flat bottom onto this. An inch or two of water is put in the tank and the experimenter pokes a finger into it, right at one focus of the ellipse. Ripples radiate outward through the water, reflect from the strip around the edge of the tank, and in a short time a drop of water spurts up at the second focus. Why?

27. Set up the integrals that give (a) the area of a quadrant of the circle $x^2 + y^2 = a^2$, (b) the area of a quadrant of the ellipse $b^2x^2 + a^2y^2 = a^2b^2$. Show that the integral in (b) is b/a times the integral in (a), and deduce the area of the ellipse from the known area of the circle.

28. *The weighted rope.* A rope with a ring in one end is looped over two pegs in a horizontal line. The free end, after being passed through the ring, has a weight suspended from it to make the rope hang taut. If the rope slips freely over the pegs and through the ring, the weight will descend as far as possible. Assume that the length of the rope is at least four times as great as the distance between the pegs and that the configuration of the rope is symmetric with respect to the line of the vertical part of the rope.

8.61 Problem 28 asks how big angle A will be when the rope is pulled tight by the weight.

a) Find the angle formed at the bottom of the loop (Fig. 8.61).
b) Show that for each fixed position of the ring on the rope, the possible locations of the ring in space lie on an ellipse with foci at the pegs.
c) Justify the original symmetry assumption by combining the result in (b) with the assumption that the rope and weight will take a rest position of minimal potential energy.

29. Two radar stations lie 20 km apart along an east-west line. A low-flying plane traveling from west to east is known to have a speed of v_0 km/sec. At $t = 0$ a signal is sent from the station at $(-10, 0)$, bounces off the plane, and is received at $(10, 0)$ $30/c$ seconds later (c is the velocity of the signal). When $t = 10/v_0$, another signal is sent out from the station at $(-10, 0)$, reflects off the plane, and is once again received $30/c$ seconds later by the other station. Find the position of the plane when it reflects the second signal under the assumption that v_0 is much less than c.

30. Find the center, vertices, foci, and asymptotes of the hyperbola
$$3x^2 - y^2 + 12x - 6y = 0.$$

31. Find the center, vertices, foci, and asymptotes of the hyperbola
$$9x^2 - 4y^2 - 18x - 16y + 29 = 0.$$

32. Find an equation for the hyperbola with eccentricity $\sqrt{2}$ and with vertices at $(2, 0)$ and $(-2, 0)$.

33. If c is a fixed positive constant, then
$$\frac{x^2}{t^2} + \frac{y^2}{t^2 - c^2} = 1, \qquad c^2 < t^2,$$
defines a family of ellipses, any member of which is characterized by a particular value of t. Show that every member of the family
$$\frac{x^2}{t^2} - \frac{y^2}{c^2 - t^2} = 1, \qquad t^2 < c^2,$$
intersects any member of the first family at right angles.

34. Show that if the tangent to a curve at a point $P(x, y)$ passes through the origin, then $dy/dx = y/x$ at P. Hence, show that no tangent can be drawn from the origin to the hyperbola $x^2 - y^2 = 1$.

35. Two vertices A, B of a triangle are fixed. The third vertex $C(x, y)$ moves in such a way that $\angle A = 2(\angle B)$. Find the path traced by C.

36. Let p and q be positive numbers with $q < p$. If r is a third number, prove that the equation

$$\frac{x^2}{p - r} + \frac{y^2}{q - r} = 1$$

represents (a) an ellipse if $r < q$, (b) a hyperbola if $q < r < p$, and (c) nothing if $p < r$. Prove that all these ellipses and hyperbolas have the same foci, and find these foci.

37. On a level plane the sound of a rifle and that of the bullet striking the target are heard at the same instant. What is the location of the hearer?

38. Show that any tangent to the hyperbola $xy = a^2$ determines with its asymptotes a triangle of area $2a^2$.

39. Find the eccentricity of the hyperbola $xy = 1$.

40. Find all points on the curve $x^2 - 2xy - 3y^2 + 3 = 0$ at which the tangent line is perpendicular to the line $x + y = 1$. Identify the curve.

41. A line PT is drawn tangent to the curve $xy = x + y$ at the point $P(-2, 2/3)$. Find the equations of two lines that are normal to the curve and perpendicular to PT.

42. Find an equation for the tangent to the curve

$$x^2 - 2xy + y^2 + 2x + y - 6 = 0$$

at the point $(2, 2)$. Identify the curve.

43. Starting from the equation

$$Ax^2 + Bxy + Cy^2 + Dx + Ey + F = 0,$$

find an equation for the conic section that has the following properties:
a) it is symmetric with respect to the origin;
b) it passes through the point $(1, 0)$;
c) the line $y = 1$ is tangent to it at the point $(-2, 1)$.

44. Show that the equation $xy - x - y - 1 = 0$ represents a hyperbola. Find the hyperbola's center, vertices, foci, axes, and asymptotes. Show them all in a sketch.

45. Show that the equation $\sqrt{2}y - 2xy = 2$ represents a hyperbola. Find the hyperbola's center, vertices, foci, axes, and asymptotes.

46. a) Show that the line

$$b^2xx_1 + a^2yy_1 - a^2b^2 = 0$$

is tangent to the ellipse $b^2x^2 + a^2y^2 - a^2b^2 = 0$ at the point (x_1, y_1) on the ellipse.
b) Show that the line

$$b^2xx_1 - a^2yy_1 - a^2b^2 = 0$$

is tangent to the hyperbola $b^2x^2 - a^2y^2 - a^2b^2 = 0$ at the point (x_1, y_1) on the hyperbola.

c) Show that the tangent to the conic section

$$Ax^2 + Bxy + Cy^2 + Dx + Ey + F = 0$$

at a point (x_1, y_1) on it has an equation that may be written in the form

$$Axx_1 + B\left(\frac{x_1y + xy_1}{2}\right) + Cyy_1$$

$$+ D\left(\frac{x + x_1}{2}\right) + E\left(\frac{y + y_1}{2}\right) + F = 0.$$

47. Find the equation into which $x^{1/2} + y^{1/2} = a^{1/2}$ is transformed by rotating the axes counterclockwise through an angle of $\pi/4$ radians. Eliminate the radicals from the new equation, and identify the curve's shape.

48. If

$$x = x' \cos \alpha - y' \sin \alpha,$$
$$y = x' \sin \alpha + y' \cos \alpha,$$

then

$$dx = dx' \cos \alpha - dy' \sin \alpha,$$
$$dy = dx' \sin \alpha + dy' \cos \alpha.$$

Show that $dx^2 + dy^2 = dx'^2 + dy'^2$ and that

$$x\, dy - y\, dx = x'\, dy' - y'\, dx'.$$

Thus, the quantities $dx^2 + dy^2$ and $x\, dy - y\, dx$ are not changed by rotating the axes.

What points satisfy the equations and inequalities in Problems 49–58? Draw a figure for each problem.

49. $(2x + y - 2)(x^2 + y^2 - 4)(x^2 - y) = 0$

50. $(x^2 + 4y)(x^2 - y^2 - 1)(x^2 + y^2 - 25)(x^2 + 4y^2 - 4) = 0$

51. $(x + y)(x^2 + y^2 - 1) = 0$

52. $(y - x + 2)(2y + x - 4) = 0$

53. $(x^2/a^2) + (y^2/b^2) \leq 1$ **54.** $(x^2/a^2) - (y^2/b^2) \leq 1$

55. $(9x^2 + 4y^2 - 36)(4x^2 + 9y^2 - 36) \leq 0$

56. $(9x^2 + 4y^2 - 36)(4x^2 + 9y^2 - 36) > 0$

57. $x^4 - (y^2 - 9)^2 = 0$ **58.** $x^2 + xy + y^2 < 3$

59. As the point N moves along the line $y = a$ in Fig. 8.62, P moves in such a way that $OP = MN$. Find parametric equations for the coordinates of P as functions of the angle t that the line ON makes with the positive y-axis.

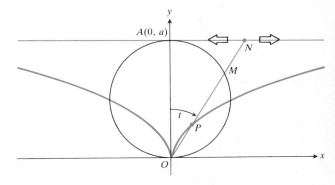

8.62 The construction in Problem 59.

60. Figure 8.63 shows a circle of radius a inside a circle of radius $2a$. The point P, shown as a point of tangency in the figure, is attached to the smaller circle. Find the path traced by P as the smaller circle rolls around the inside of the larger circle.

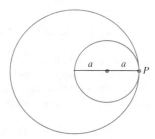

8.63   The circles in Problem 60.

61. A wheel of radius 4 in. rolls along the x-axis with angular velocity 2 radians/sec. Find parametric equations for the curve described by a point on a spoke and 2 in. from the center of the wheel if it starts from the point $(0, 2)$ at time $t = 0$.

62. Show that the slope of the cycloid
$$x = a(t - \sin t), \qquad y = a(1 - \cos t)$$
is $dy/dx = \cot (t/2)$. In particular, the tangent to the cycloid is vertical when t is 0 or 2π.

63. Show that the slope of the trochoid
$$x = at - b \sin t, \qquad y = a - b \cos t$$
is always finite if $b < a$.

64. Find parametric equations and a Cartesian equation for the figure traced by the point $P(x, y)$ if
$$\frac{dx}{dt} = -2y, \qquad \frac{dy}{dt} = \cos t,$$
and $x = 3$, $y = 0$ when $t = 0$. Identify the figure.

65. **CALCULATOR**   *An elliptic integral.*
   a) Show that the perimeter of the ellipse
$$x = a \cos t, \quad y = b \sin t, \quad 0 \le t \le 2\pi$$
   is given by the integral
$$4a \int_0^{\pi/2} \sqrt{1 - e^2 \cos^2 t} \; dt,$$
   where e is the eccentricity of the ellipse.
   b) The integral in (a) is called an *elliptic integral*. It has no elementary antiderivative. Estimate the length of the ellipse when $a = 1$ and $e = 1/2$ by the trapezoidal rule with $n = 10$.
   c) The absolute value of the second derivative of $f(t) = \sqrt{1 - e^2 \cos^2 t}$ is less than 1. Based on this, what estimate does Eq. (8), Article 4.9, give of the error in the approximation in (b)?

66. Find the center of mass of a thin plate of constant density $\delta = 1$ bounded by the x-axis and one arch of the cycloid
$$x = a(t - \sin t), \qquad y = a(1 - \cos t).$$
(*Hint:* $dA = y \, dx = y(dx/dt) \, dt$.)

Orthogonal Curves

Two curves are said to be *orthogonal* if their tangents meet at right angles at every point where the curves intersect. Problems 67–70 are about orthogonal conic sections.

67. Sketch the curves $xy = 2$ and $x^2 - y^2 = 3$ in one diagram and show that they intersect at right angles.

68. Sketch the curves $y^2 = 4x + 4$ and $y^2 = 64 - 16x$ together and show that they meet at right angles.

69. Show that the curves $2x^2 + 3y^2 = a^2$ and $ky^2 = x^3$ are orthogonal for all values of the constants a and k ($a \ne 0$, $k \ne 0$). Sketch the four curves corresponding to $a = 2$, $a = 4$, $k = 1/2$, $k = -2$ in one diagram.

70. Show that the parabolas
$$y^2 = 4a(a - x)$$
and
$$y^2 = 4b(b + x),$$
$a > 0$ and $b > 0$, have a common focus, the same for any a and b. Show that the parabolas intersect at the points $(a - b, \pm 2\sqrt{ab})$ and that each a-parabola is orthogonal to every b-parabola. (By varying a and b, we obtain two families of confocal parabolas. Each family is said to be a set of *orthogonal trajectories* of the other family. See Fig. 8.64.)

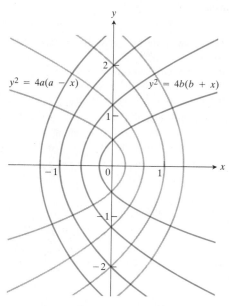

8.64   Some of the parabolas in Problem 70.

CHAPTER 9

Hyperbolic Functions

OVERVIEW

Some combinations of e^x and e^{-x} arise so frequently in formulating the solutions of differential equations in engineering and physics that we give them special names and study them as functions in their own right. For reasons that will become clear in Article 9.1, these combinations are called **hyperbolic functions.** Hyperbolic functions are used to describe the motions of waves in elastic solids; the shapes of electric power lines; temperature distributions in the metal fins that cool hot pipes; pursuit curves (more about this later); and the geometry of the general theory of relativity. The designers of the Gateway Arch to the West in St. Louis used hyperbolic functions to predict the arch's internal forces, and the arch itself follows a hyperbolic cosine curve.

After introducing the hyperbolic functions in Article 9.1 and examining their derivatives and associated integrals in Article 9.2, we shall study the shapes of hanging cables in Article 9.3. The calculations associated with hanging cables lead in a natural way to inverse hyperbolic functions. These, in turn, provide a number of useful new integral formulas, as we shall see in Article 9.4. These formulas, and others like them, appear frequently in integral tables.

9.1

Definitions and Identities

In this article, we define the hyperbolic sine, cosine, tangent, cotangent, secant, and cosecant functions and compare them with the six trigonometric or "circular" functions that bear the same names.

The hyperbolic sine and hyperbolic cosine are defined by the following equations:

Hyperbolic sine of u: $\sinh u = \dfrac{1}{2}(e^u - e^{-u})$

Hyperbolic cosine of u: $\cosh u = \dfrac{1}{2}(e^u + e^{-u})$

(1)

The notation $\cosh u$ is often read "kosh u," which rhymes with "gosh you," and $\sinh u$ is pronounced as if spelled "cinch u" or "shine u." (They all sound a bit strange.)

The hyperbolic sine and cosine are related to each other by rules very much like the rules that relate the functions $\cos u$ and $\sin u$. And just as $\cos u$ and $\sin u$ may be identified with the point (x, y) on the unit circle $x^2 + y^2 = 1$, the functions $\cosh u$ and $\sinh u$ may be identified with the coordinates of a point (x, y) on the **unit hyperbola** $x^2 - y^2 = 1$.

To check that the point with coordinates $x = \cosh u$ and $y = \sinh u$ lies on the unit hyperbola, we substitute the defining relations (1) into the equation of the hyperbola:

$$x^2 - y^2 = 1,$$

$$\cosh^2 u - \sinh^2 u \overset{?}{=} 1,$$

$$\frac{1}{4}(e^{2u} + 2 + e^{-2u}) - \frac{1}{4}(e^{2u} - 2 + e^{-2u}) \overset{?}{=} 1,$$

$$\frac{1}{4}(e^{2u} + 2 + e^{-2u} - e^{2u} + 2 - e^{-2u}) \overset{?}{=} 1,$$

$$\frac{1}{4}(4) \overset{?}{=} 1.   \text{(Yes!)}$$

Actually, if we let

$$x = \cosh u = \frac{1}{2}(e^u + e^{-u}),$$

$$y = \sinh u = \frac{1}{2}(e^u - e^{-u}),$$

(2)

then when u varies from $-\infty$ to $+\infty$, the point $P(x, y)$ describes the right-hand branch of the hyperbola $x^2 - y^2 = 1$. The direction in which the curve is traced is shown by the arrows in Fig. 9.1. Since e^u is always positive and $e^{-u} = 1/e^u$ is also positive, $x = \cosh u = \frac{1}{2}(e^u + e^{-u})$ is positive for all real values of u, $-\infty < u < +\infty$. Hence the point (x, y) lies to the right of the y-axis.

TRIGONOMETRY BEYOND THE HORIZON

Roger Cotes (1682–1716), Abraham de Moivre (1667–1754), and Leonhard Euler (1707–1783) each had developed, in one form or another, certain aspects of hyperbolic functions. However, it was Johann Heinrich Lambert, a Swiss-German mathematician and friend of Euler, who first studied hyperbolic functions in a systematic way. Lambert juxtaposed the functions of an angle whose reference is the unit circle (circular functions) to analogous functions of an angle whose reference is the equilateral hyperbola (hyperbolic functions). He was motivated by his work in astronomy, which required data on celestial bodies below the horizon. He needed to know the sine of an arc whose measure is an imaginary number, and in general the trigonometric functions of such arcs. Lambert used hyperbolic trigonometry to make these calculations.

The first bit of hyperbolic trigonometry that we have just established is the basic identity

$$\cosh^2 u - \sinh^2 u = 1. \tag{3}$$

This is analogous to, but not the same as, the ordinary trigonometric identity, $\cos^2 u + \sin^2 u = 1$.

The remaining hyperbolic functions are defined in terms of sinh u and cosh u as follows.

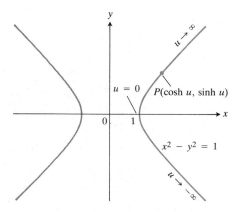

Hyperbolic tangent:	$\tanh u = \dfrac{\sinh u}{\cosh u} = \dfrac{e^u - e^{-u}}{e^u + e^{-u}}$	
Hyperbolic cotangent:	$\coth u = \dfrac{\cosh u}{\sinh u} = \dfrac{e^u + e^{-u}}{e^u - e^{-u}}$	(4)
Hyperbolic secant:	$\operatorname{sech} u = \dfrac{1}{\cosh u} = \dfrac{2}{e^u + e^{-u}}$	
Hyperbolic cosecant:	$\operatorname{csch} u = \dfrac{1}{\sinh u} = \dfrac{2}{e^u - e^{-u}}$	

If we divide the identity $\cosh^2 u - \sinh^2 u = 1$ by $\cosh^2 u$, we get

$$1 - \tanh^2 u = \operatorname{sech}^2 u. \tag{5}$$

If we divide $\cosh^2 u - \sinh^2 u = 1$ by $\sinh^2 u$, we get

$$\coth^2 u - 1 = \operatorname{csch}^2 u. \tag{6}$$

From the definitions of cosh u and sinh u in (1) we also find that

$$\cosh u + \sinh u = e^u \tag{7}$$

and

$$\cosh u - \sinh u = e^{-u}. \tag{8}$$

Thus, any combination of e^u and e^{-u} can be replaced by a combination of sinh u and cosh u, and conversely.

Since e^{-u} is positive, the equation $\cosh u - \sinh u = e^{-u}$ shows that cosh u is always greater than sinh u. For large positive values of u, however, e^{-u} is small and cosh u is approximately equal to sinh u.

The graphs of the hyperbolic functions are shown in Fig. 9.2.

Comparisons with the Trigonometric Functions

Since cosh $0 = 1$ and sinh $0 = 0$, the hyperbolic functions (all six of them) have the same values at zero that the corresponding trigonometric functions have.

The hyperbolic cosine is an even function because

$$\cosh(-x) = \frac{e^{-x} + e^{-(-x)}}{2} = \frac{e^{-x} + e^x}{2} = \cosh x$$

9.1 The equations $x = \cosh u$, $y = \sinh u$, $-\infty < u < \infty$, are parametric equations for the right-hand branch of the unit hyperbola $x^2 - y^2 = 1$. (For the left-hand branch, you would take $x = -\cosh u$, $y = \sinh u$.)

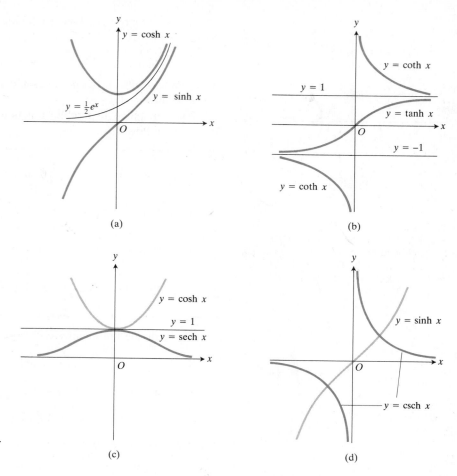

9.2 The graphs of the six hyperbolic functions.

for all x. The hyperbolic sine is odd. For all x,

$$\sinh(-x) = \frac{e^{-x} - e^{-(-x)}}{2} = \frac{e^{-x} - e^{x}}{2} = -\sinh x.$$

Thus, the graph of $y = \cosh x$ is symmetric about the y-axis and the graph of $y = \sinh x$ is symmetric about the origin. Here again, the hyperbolic functions behave like their trigonometric counterparts.

There are nevertheless major differences between the hyperbolic and trigonometric functions. The trigonometric functions are periodic while the hyperbolic functions are not. The ranges of the hyperbolic and trigonometric functions are also very different:

$\sin x$ oscillates between -1 and $+1$;
$\sinh x$ increases steadily from $-\infty$ to $+\infty$.

$\cos x$ oscillates between -1 and $+1$;
$\cosh x$ varies from $+\infty$ to $+1$ to $+\infty$.

$|\sec x|$ is never less than 1;
$\operatorname{sech} x$ is never greater than 1 and is always positive.

$\tan x$ varies from $-\infty$ to $+\infty$;
$\tanh x$ varies from -1 to $+1$.

Another difference lies in the behavior of the functions as $x \to \pm\infty$. We can say nothing very specific about the behavior of the circular functions $\sin x$, $\cos x$, $\tan x$, etc., for large values of x. But the hyperbolic functions behave very much like $e^x/2$, $e^{-x}/2$, or the constant functions zero and one:

For x large and positive	For x negative, $|x|$ large	
$\cosh x \approx \sinh x \approx \dfrac{1}{2}e^x$	$\cosh x \approx -\sinh x \approx \dfrac{1}{2}e^{-x}$	
$\tanh x \approx \coth x \approx 1$	$\tanh x \approx \coth x \approx -1$	(9)
$\operatorname{sech} x \approx \operatorname{csch} x \approx 2e^{-x} \approx 0$	$\operatorname{sech} x \approx -\operatorname{csch} x \approx 2e^{x} \approx 0$	

Identities

It requires only a bit of routine algebra to produce the identities

$$\sinh(x + y) = \sinh x \cosh y + \cosh x \sinh y$$
$$\cosh(x + y) = \cosh x \cosh y + \sinh x \sinh y \tag{10}$$

from the definitions in (1). These in turn give

$$\sinh 2x = 2 \sinh x \cosh x, \tag{11a}$$

$$\cosh 2x = \cosh^2 x + \sinh^2 x, \tag{11b}$$

when we take $y = x$. The second of these leads to useful "double-angle" formulas when we combine it with the identity

$$1 = \cosh^2 x - \sinh^2 x. \tag{3}$$

If we add (11b) and (3), we have

$$\cosh 2x + 1 = 2 \cosh^2 x. \tag{12a}$$

If we subtract (3) from (11b), we get

$$\cosh 2x - 1 = 2 \sinh^2 x. \tag{12b}$$

Dividing (12a) and (12b) by 2 gives

$$\cosh^2 u = \frac{\cosh 2u + 1}{2} \tag{12c}$$

$$\sinh^2 u = \frac{\cosh 2u - 1}{2}. \tag{12d}$$

PROBLEMS

Each of Problems 1–6 gives the value of one of the six hyperbolic functions of u. Use the definitions and identities given in the text to determine the values of the remaining five hyperbolic functions.

1. $\sinh u = -\dfrac{3}{4}$

2. $\cosh u = \dfrac{17}{15}, \quad u > 0$

3. $\tanh u = -\dfrac{7}{25}$

4. $\coth u = \dfrac{13}{12}$

5. $\operatorname{sech} u = \dfrac{3}{5}$

6. $\operatorname{csch} u = \dfrac{5}{12}$

Rewrite the expressions in Problems 7–14 in terms of exponentials. Write the final results as simply as you can.

7. $2 \cosh(\ln x)$

8. $\sinh(2 \ln x)$

9. $\tanh(\ln x)$

10. $\dfrac{1}{\cosh x - \sinh x}$

11. $\cosh 5x + \sinh 5x$

12. $(\sinh x + \cosh x)^4$

13. $\cosh 3x - \sinh 3x$

14. $\ln(\cosh x + \sinh x) + \ln(\cosh x - \sinh x)$

Say whether the functions in Problems 15–24 are even, odd, or neither.

15. $\tanh x$

16. $\coth x$

17. $\operatorname{sech} x$

18. $\operatorname{csch} x$

19. $\cosh 3x$

20. $\sinh 2x$

21. $\sinh x \cosh x$

22. $(\sinh x)^2$

23. $\operatorname{sech} x + \cosh x$

24. $\sinh x + \cosh x$

In Problems 25 and 26, solve the given equation for x.

25. $\cosh x = \sinh x + \dfrac{1}{2}$

26. $\tanh x = \dfrac{3}{5}$

Verify the identities in Problems 27 and 28.

27. $\sinh(u + v) = \sinh u \cosh v + \cosh u \sinh v$

28. $\cosh(u + v) = \cosh u \cosh v + \sinh u \sinh v$

Use the identities in Problems 27 and 28 to derive the identities in Problems 29 and 30.

29. $\sinh(u - v) = \sinh u \cosh v - \cosh u \sinh v$

30. $\cosh(u - v) = \cosh u \cosh v - \sinh u \sinh v$

Use the identities in Problems 27–30 to verify the identities in Problems 31–36.

31. $\sinh u \cosh v = \dfrac{1}{2} \sinh(u + v) + \dfrac{1}{2} \sinh(u - v)$

32. $\cosh u \sinh v = \dfrac{1}{2} \sinh(u + v) - \dfrac{1}{2} \sinh(u - v)$

33. $\cosh u \cosh v = \dfrac{1}{2} \cosh(u + v) + \dfrac{1}{2} \cosh(u - v)$

34. $\sinh u \sinh v = \dfrac{1}{2} \cosh(u + v) - \dfrac{1}{2} \cosh(u - v)$

35. $\sinh 3u = \sinh^3 u + 3 \cosh^2 u \sinh u = 3 \sinh u + 4 \sinh^3 u$

36. $\cosh 3u = \cosh u + 4 \sinh^2 u \cosh u = 4 \cosh^3 u - 3 \cosh u$

37. Show that
$$\sinh^2 u - \sinh^2 v = \cosh^2 u - \cosh^2 v.$$

38. Show that
$$(\cosh x + \sinh x)^n = \cosh nx + \sinh nx.$$

39. Show that the equations
$$x = -\cosh u, \qquad y = \sinh u, \qquad -\infty < u < \infty$$
are parametric equations for the left-hand branch of the hyperbola $x^2 - y^2 = 1$.

40. Show that the tangent to the hyperbola $x^2 - y^2 = 1$ at the point $P(\cosh u, \sinh u)$ cuts the x-axis at the point $(\operatorname{sech} u, 0)$ and, except when vertical, cuts the y-axis at the point $(0, -\operatorname{csch} u)$.

41. Show that the distance r from the origin O to the point $P(\cosh u, \sinh u)$ on the hyperbola $x^2 - y^2 = 1$ is $r = \sqrt{\cosh 2u}$.

42. Show that the line tangent to the hyperbola at its vertex in Fig. 9.1 intersects the line OP in the point $(1, \tanh u)$. This gives a geometric representation of $\tanh u$.

43. If θ lies in the interval $-\pi/2 < \theta < \pi/2$ and $\sinh x = \tan \theta$, show that $\cosh x = \sec \theta$, $\tanh x = \sin \theta$, $\coth x = \csc \theta$, $\operatorname{csch} x = \cot \theta$, and $\operatorname{sech} x = \cos \theta$.

TOOLKIT PROGRAMS

Function Evaluator   Super ∗ Grapher

9.2
Derivatives and Integrals

The six hyperbolic functions are rational combinations of the differentiable function e^u and so have derivatives at every point at which they are defined. In this article, we find out what these derivatives are and look at integrals that involve hyperbolic functions.

Let u be a differentiable function of x and differentiate

$$\sinh u = \frac{1}{2}(e^u - e^{-u}), \qquad \cosh u = \frac{1}{2}(e^u + e^{-u}) \tag{1}$$

with respect to x. Applying the formulas

$$\frac{de^u}{dx} = e^u \frac{du}{dx}, \qquad \frac{de^{-u}}{dx} = e^{-u}\frac{d(-u)}{dx} = -e^{-u}\frac{du}{dx},$$

we get

$$\frac{d(\sinh u)}{dx} = \cosh u \frac{du}{dx}, \qquad \frac{d(\cosh u)}{dx} = \sinh u \frac{du}{dx}. \qquad (2)$$

Then, if we differentiate

$$y = \tanh u = \frac{\sinh u}{\cosh u}$$

as a fraction, we get

$$\frac{d(\tanh u)}{dx} = \frac{\cosh u(d(\sinh u)/dx) - \sinh u(d(\cosh u)/dx)}{\cosh^2 u}$$

$$= \frac{\cosh^2 u(du/dx) - \sinh^2 u(du/dx)}{\cosh^2 u}$$

$$= \frac{(\cosh^2 u - \sinh^2 u)(du/dx)}{\cosh^2 u} = \frac{1}{\cosh^2 u} \frac{du}{dx} = \operatorname{sech}^2 u \frac{du}{dx}.$$

In a similar manner, we may establish the other formulas in the following list.

$$\frac{d(\tanh u)}{dx} = \operatorname{sech}^2 u \frac{du}{dx}$$

$$\frac{d(\coth u)}{dx} = -\operatorname{csch}^2 u \frac{du}{dx}$$

$$\frac{d(\operatorname{sech} u)}{dx} = -\operatorname{sech} u \tanh u \frac{du}{dx} \qquad (3)$$

$$\frac{d(\operatorname{csch} u)}{dx} = -\operatorname{csch} u \coth u \frac{du}{dx}$$

These derivative formulas produce the following integral formulas.

$$\int \sinh u \, du = \cosh u + C$$

$$\int \cosh u \, du = \sinh u + C$$

$$\int \operatorname{sech}^2 u \, du = \tanh u + C$$

$$\int \operatorname{csch}^2 u \, du = -\coth u + C \qquad (4)$$

$$\int \operatorname{sech} u \tanh u \, du = -\operatorname{sech} u + C$$

$$\int \operatorname{csch} u \coth u \, du = -\operatorname{csch} u + C$$

EXAMPLE 1 Evaluate

$$\int \coth 5x \, dx.$$

Solution

$$\int \coth 5x \, dx = \int \frac{\cosh 5x}{\sinh 5x} \, dx$$

$$= \frac{1}{5} \int \frac{du}{u} \qquad (u = \sinh 5x)$$

$$= \frac{1}{5} \ln|u| + C = \frac{1}{5} \ln|\sinh 5x| + C.$$ ∎

EXAMPLE 2 Evaluate

$$\int_0^1 \sinh^2 x \, dx.$$

Solution From the identity (12d) in Article 9.1,

$$\sinh^2 x = \frac{\cosh 2x - 1}{2}.$$

Therefore,

$$\int_0^1 \sinh^2 x \, dx = \int_0^1 \frac{\cosh 2x - 1}{2} \, dx$$

$$= \frac{\sinh 2x}{4} - \frac{x}{2} \bigg]_0^1 = \frac{\sinh 2}{4} - \frac{1}{2}$$

$$\approx 0.40672.$$ ∎

EXAMPLE 3 Evaluate

$$\int x \sinh x \, dx.$$

Solution *Integration by parts.* We let

$$u = x, \qquad dv = \sinh x \, dx, \qquad v = \cosh x,$$

so that

$$\int x \sinh x \, dx = x \cosh x - \int \cosh x \, dx$$

$$= x \cosh x - \sinh x + C.$$ ∎

EXAMPLE 4 Evaluate

$$\int_0^{\ln 4} e^x \sinh x \, dx.$$

Solution From the definition of sinh x we have

$$\int_0^{\ln 4} e^x \sinh x \, dx = \int_0^{\ln 4} e^x \cdot \frac{e^x - e^{-x}}{2} \, dx$$

$$= \int_0^{\ln 4} \frac{e^{2x} - 1}{2} \, dx = \frac{e^{2x}}{4} - \frac{x}{2} \Big]_0^{\ln 4}$$

$$= \frac{e^{2 \ln 4}}{4} - \frac{\ln 4}{2} - \frac{1}{4} = \frac{e^{\ln 16}}{4} - \frac{2 \ln 2}{2} - \frac{1}{4}$$

$$= 4 - \ln 2 - \frac{1}{4} = \frac{15}{4} - \ln 2.$$

EXAMPLE 5 Find the area of the infinite region in the first quadrant that lies between the curve $y = \tanh x$ and the line $y = 1$.

Solution We sketch the region (Fig. 9.3) and note that the area is given by the improper integral

$$\int_0^{\infty} (1 - \tanh x) \, dx = \lim_{b \to \infty} \int_0^b (1 - \tanh x) \, dx.$$

In terms of b, the value of the integral on the right is

$$\int_0^b (1 - \tanh x) \, dx = \int_0^b \left(1 - \frac{\sinh x}{\cosh x} \right) dx = x - \ln \cosh x \Big]_0^b$$

$$= b - \ln \cosh b = \ln e^b - \ln \cosh b$$

$$= \ln \frac{e^b}{\cosh b} = \ln \frac{2e^b}{e^b + e^{-b}}$$

$$= \ln \frac{2}{1 + e^{-2b}}.$$

The value of the improper integral is therefore

$$\int_0^{\infty} (1 - \tanh x) \, dx = \lim_{b \to \infty} \ln \frac{2}{1 + e^{-2b}}$$

$$= \ln \frac{2}{1 + 0} = \ln 2.$$

The area is $\ln 2$.

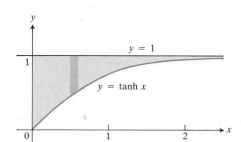

9.3 The area of the region between $y =$ tanh x and $y = 1$ in the first quadrant can be calculated with an improper integral. As shown in Example 5, the area is ln 2.

PROBLEMS

Find dy/dx in Problems 1–16.

1. $y = \sinh 3x$

2. $y = \cosh^2 5x$

3. $y = \cosh^2 5x - \sinh^2 5x$

4. $y = \tanh 2x$

5. $y = \coth(\tan x)$

6. $y = \operatorname{sech}^3 x$

7. $y = 4 \operatorname{csch}(x/4)$

8. $\sinh y = \tan x$

9. $y = \operatorname{sech}^2 x + \tanh^2 x$

10. $y = \operatorname{csch}^2 x$

11. $y = \sin^{-1}(\tanh x)$

12. $y = x - (1/4) \coth 4x$

13. $y = \ln|\tanh(x/2)|$

14. $y = x^4 \sinh x$

15. $y = x \sinh 2x - (1/2) \cosh 2x$

16. $y = x \sinh x - \cosh x$

17. Show that

$$\int \text{sech } x \, dx = \sin^{-1}(\tanh x) + C.$$

18. Show that

$$\int \text{csch } x \, dx = \ln\left|\tanh \frac{x}{2}\right| + C.$$

Evaluate the integrals in Problems 19–28.

19. $\displaystyle\int_{-1}^{1} \cosh 5x \, dx$

20. $\displaystyle\int_{-1}^{0} \cosh(2x + 1) \, dx$

21. $\displaystyle\int_{-3}^{3} \sinh x \, dx$

22. $\displaystyle\int_{-\pi}^{\pi} \tanh 2x \, dx$

23. $\displaystyle\int_{0}^{1} x \cosh x \, dx$

24. $\displaystyle\int_{\ln 2}^{\ln 4} \coth x \, dx$

25. $\displaystyle\int_{0}^{1/2} \frac{\sinh x}{e^x} \, dx$

26. $\displaystyle\int_{0}^{\ln 2} 4e^x \cosh x \, dx$

27. $\displaystyle\int_{0}^{1} \sinh \sqrt{x} \, dx$

28. $\displaystyle\int_{1}^{2} \frac{\cosh(\ln x) \, dx}{x}$

Evaluate the integrals in Problems 29–46.

29. $\displaystyle\int \cosh(2x + 1) \, dx$

30. $\displaystyle\int \tanh x \, dx$

31. $\displaystyle\int \frac{\sinh x}{\cosh^4 x} \, dx$

32. $\displaystyle\int \frac{4 \, dx}{(e^x + e^{-x})^2}$

33. $\displaystyle\int \frac{e^x - e^{-x}}{e^x + e^{-x}} \, dx$

34. $\displaystyle\int \tanh^2 x \, dx$

35. $\displaystyle\int \frac{\sinh \sqrt{x}}{\sqrt{x}} \, dx$

36. $\displaystyle\int \cosh^2 3x \, dx$

37. $\displaystyle\int \sqrt{\cosh x - 1} \, dx$

38. $\displaystyle\int \cosh^2 5x \, dx$

39. $\displaystyle\int 2 \cosh x \sinh x \, dx$

40. $\displaystyle\int \cosh^3 x \, dx$

41. $\displaystyle\int \frac{\sinh x}{1 + \cosh x} \, dx$

42. $\displaystyle\int x^2 \sinh x \, dx$

43. $\displaystyle\int x^2 \cosh x \, dx$

44. $\displaystyle\int_{0}^{\ln 2} \frac{1 - e^{-2x}}{1 + e^{-2x}} \, dx$

(*Hint:* Multiply the numerator and denominator of the integrand by e^x.)

45. $\displaystyle\int \text{sech}^3 5x \tanh 5x \, dx$

46. $\displaystyle\int \text{csch}^2 x \coth x \, dx$

Use tables to evaluate the integrals in Problems 47–50.

47. $\displaystyle\int \sinh^4 3x \, dx$

48. $\displaystyle\int \text{sech } 7x \, dx$

49. $\displaystyle\int e^{3x} \cosh 2x \, dx$

50. $\displaystyle\int \tanh^3 x \, dx$

51. a) Evaluate $\int \text{csch } x \, dx$ by multiplying the integrand by

$$\frac{\text{csch } x + \coth x}{\text{csch } x + \coth x}$$

and integrating the result.

b) Show that the answer you obtained in (a) is equivalent to the answer in Problem 18.

52. Find the area of the infinite region in the first quadrant that lies between the curves $y = \sinh x$ and $y = \cosh x$.

53. The region between the curves $y = \sinh x$ and $y = \cosh x$ and the lines $x = 0$ and $x = 3$ is revolved about the x-axis to generate a solid. Find the volume of the solid.

54. Find the length of the curve $y = \cosh x$, $0 \le x \le 1$.

55. Find the surface area swept out by revolving the arc $y = \cosh x$, $0 \le x \le \ln 2$ about the x-axis.

56. Find the volume enclosed by revolving the entire curve $y = \text{sech } x$, $-\infty < x < \infty$, about the x-axis.

57. Find the volume swept out by revolving about the x-axis the region in the first quadrant bounded below by the x-axis, on the left by the line $x = \ln 2$, and above by the curve $y = \text{csch } x$.

58. Let R denote the region in the first quadrant that lies between the curves $y = \sinh x$ and $y = \cosh x$.
a) *A finite volume:* Find the volume swept out by revolving R about the y-axis.
b) *An infinite volume:* Show that the volume swept out by revolving R about the x-axis is infinite.

59. Use the integral formula in Problem 17 to find the center of mass of a thin homogeneous plate covering the infinite region bounded by the x-axis and the curve $y = \text{sech } x$.

60. Find the center of mass of a thin uniform wire that lies along the curve $y = \cosh x$, $0 \le x \le \ln 2$.

61. The region in the first quadrant between the curve $y = \tanh x$ and the line $y = 1$ is revolved around the x-axis to generate an infinite solid. Find the volume of the solid.

62. One of the analogies between hyperbolic and circular functions is that the variable u in the parametric equations

$$x = \cosh u, \quad y = \sinh u, \quad -\infty < u < \infty$$

for the right-hand branch of the hyperbola $x^2 - y^2 = 1$ is twice the area of the sector AOP pictured in Fig. 9.4. To see why this is so, carry out the following steps.
a) Show that the area $A(u)$ of sector AOP is given by the formula

$$A(u) = \frac{1}{2} \cosh u \sinh u - \int_{1}^{\cosh u} \sqrt{x^2 - 1} \, dx.$$

b) Differentiate both sides of the equation in (a) to show that

$$A'(u) = \frac{1}{2}.$$

c) Solve this last equation for $A(u)$. What is the value of $A(0)$?

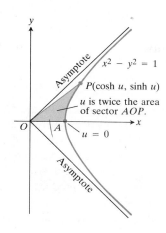

$x^2 - y^2 = 1$

$P(\cosh u, \sinh u)$

u is twice the area of sector AOP.

$u = 0$

Asymptote

Asymptote

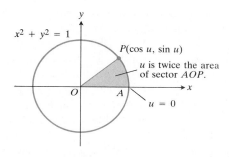

$x^2 + y^2 = 1$

$P(\cos u, \sin u)$

u is twice the area of sector AOP.

$u = 0$

9.4 One of the analogies between hyperbolic and circular functions is revealed by these two diagrams. See Problem 62.

63. a) Show that the function
$$y = A \cosh x + B \sinh x + C \cos x + D \sin x$$
is a solution of the differential equation
$$y^{(4)} - y = 0.$$
(In the formula for y, A, B, C, and D are constants.)

b) For what values of A, B, C, and D does y satisfy the initial conditions
$$y(0) = y'(0) = 0, \quad y''(0) = y'''(0) = 2?$$

64. NUMERICAL ROOT FINDER OR PROGRAMMABLE CALCULATOR. To find the natural frequencies of vibration of a beam that is held fast at both ends, we have to solve equations of the form
$$\cosh x \cos x = 1.$$

a) Sketch $y = 1/\cosh x$ and $y = \cos x$ in a common graph. How many solutions does the equation $\cosh x \cos x = 1$ have?

b) Use Newton's method or some other root finder to estimate the value of the smallest positive solution of the equation
$$\cosh x \cos x = 1.$$

65. Show that the functions
$$x = -\frac{2}{\sqrt{3}} \sinh \frac{t}{\sqrt{3}},$$
$$y = \frac{1}{\sqrt{3}} \sinh \frac{t}{\sqrt{3}} + \cosh \frac{t}{\sqrt{3}},$$
taken together, satisfy the differential equations
$$\frac{dx}{dt} + 2 \frac{dy}{dt} + x = 0,$$
$$\frac{dx}{dt} - \frac{dy}{dt} + y = 0$$
and the initial conditions $x(0) = 0$, $y(0) = 1$. Systems of differential equations like these often have hyperbolic function solutions.

TOOLKIT PROGRAMS

Root Finder   Super * Grapher

9.3
Hanging Cables

In this article, we show that clotheslines, chains, telephone lines, and electric power cables that are strung from one support to another always hang in the shape of a hyperbolic cosine curve. In contrast, the cables of suspension bridges, which do not hang freely but support a uniform load per horizontal foot, hang in parabolas (Article 8.3, Problem 32).

The starting point for the mathematical description of a hanging cable's shape is the observation that the forces acting on any section of a cable hanging at rest must cancel each other out. We know this because the cable is not moving. If we think about the forces acting on a section that runs from the cable's lowest point A to a point P partway up one of the sides, we find that

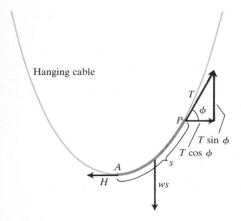

9.5 The forces on a typical section *AP* of a hanging cable. Each arrow in the diagram is labeled with its length.

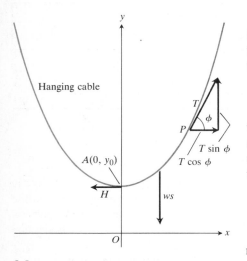

9.6 To find the shape of a hanging cable, we match it to a curve in a coordinate plane.

they can be described as

1. the horizontal tension at *A*,

2. the tangential tension at *P*, and

3. the downward force exerted on the cable section by gravity.

Figure 9.5 shows a typical section of cable with the forces acting on it. The horizontal tension at *A* is shown by drawing an arrow of length *H* pointing to the left. The tangential tension at *P* is shown by drawing an arrow of length *T* pointing up and to the right along the tangent to the cable at *P*. The force of gravity is shown by an arrow pointing straight down. Its length is *ws*, the weight of the cable section, expressed as a product of the cable's weight per unit length, *w*, and the section's length, *s*. The tension at *P* has a horizontal component of magnitude $T \cos \phi$ and a vertical component of magnitude $T \sin \phi$.

To make full use of the information that the forces acting on section *AP* are in balance, we introduce a coordinate system in the plane in which the cable hangs (Fig. 9.6) and assume that the cable lies along a differentiable curve in that plane. This enables us to write equations for the relationships among *H*, *T*, and *ws* and, from these equations, to derive a formula for the curve. We take the *x*-axis perpendicular to the force of gravity and the *y*-axis vertical with the positive *y*-axis passing through *A*. For the moment, we let the *y*-coordinate of *A* be y_0. Later we shall find a value for y_0 that makes the cable's equation in the coordinate system particularly simple.

The angle ϕ that the tangent to the cable at *P* makes with the horizontal is also the angle of inclination of the tangent line. The slope dy/dx of the curve along which the cable lies therefore has the value $\tan \phi$ at *P*. At *A*, the curve's lowest point, $dy/dx = 0$.

Since the cable section *AP* is moving neither left nor right, the magnitude of the horizontal component of the tension at *P* just equals the horizontal tension at *A*:

$$T \cos \phi = H. \tag{1}$$

Similarly, the fact that section *AP* is moving neither up nor down tells us that the magnitude of the vertical component of the tension at *P* equals the weight of the cable section:

$$T \sin \phi = ws. \tag{2}$$

These are the only two facts we have about the hanging cable and, if we are to determine its shape by finding *y* as a function of *x*, this must be where we start. The only other information we have, an artifact of the coordinate system and our assumption that the curve along which the cable lies is differentiable, is that

$$\frac{dy}{dx} = \tan \phi. \tag{3}$$

The importance of Eq. (3) is that it gets *x* and *y* into the picture, for we can now write

$$\frac{dy}{dx} = \tan \phi = \frac{T \sin \phi}{T \cos \phi} = \frac{ws}{H}$$

or, simply,

$$\frac{dy}{dx} = \frac{ws}{H}. \tag{4}$$

While the equations we have so far do not give a way to express s directly in terms of x and y, we do know from our work with arc length that

$$\frac{ds}{dx} = \sqrt{1 + \left(\frac{dy}{dx}\right)^2}. \tag{5}$$

We can therefore eliminate s from Eq. (4) by differentiating both sides with respect to x and replacing the resulting ds/dx by $\sqrt{1 + (dy/dx)^2}$. This gives

$$\frac{d^2y}{dx^2} = \frac{w}{H}\sqrt{1 + \left(\frac{dy}{dx}\right)^2}. \tag{6}$$

We now have an equation that shows how the coordinates of the points on the hanging cable are related.

To find y as a function of x, we solve Eq. (6) subject to the conditions that

$$\frac{dy}{dx}(0) = 0 \qquad \text{(The cable is horizontal at } x = 0.)$$

$$y(0) = y_0 \qquad \text{(The cable crosses the } y\text{-axis at } y = y_0.)$$

To make the work easier we can replace w/H temporarily with a single letter, say a. Equation (6) then becomes

$$\frac{d^2y}{dx^2} = a\sqrt{1 + \left(\frac{dy}{dx}\right)^2}. \tag{7}$$

While we have had little experience with second order differential equations, we can draw on our experience with first order equations by observing that Eq. (7) is a first order equation in the variable dy/dx. If we represent dy/dx by the single letter p, writing

$$\frac{dy}{dx} = p,$$

then the second derivative becomes

$$\frac{d^2y}{dx^2} = \frac{dp}{dx}$$

and Eq. (7) becomes

$$\frac{dp}{dx} = a\sqrt{1 + p^2}.$$

From this we get

$$\frac{dp}{\sqrt{1 + p^2}} = a\,dx$$

and

$$\int \frac{dp}{\sqrt{1 + p^2}} = \int a \, dx = ax + C. \tag{8}$$

We evaluate the integral on the left by substituting

$$p = \sinh u, \qquad dp = \cosh u \, du,$$

$$1 + p^2 = 1 + \sinh^2 u = \cosh^2 u.$$

Then

$$\int \frac{dp}{\sqrt{1 + p^2}} = \int \frac{\cosh u \, du}{\sqrt{\cosh^2 u}}$$

$$= \int \frac{\cosh u \, du}{\cosh u} \qquad (\cosh u > 0) \tag{9}$$

$$= \int du = u + C'.$$

The function $p = \sinh u$, an increasing function of u, has an inverse $u = \sinh^{-1} p$. We may therefore continue Eq. (9) by writing

$$\int \frac{dp}{\sqrt{1 + p^2}} = \sinh^{-1} p + C'. \tag{10}$$

This last equation combines with Eq. (8) to give

$$\sinh^{-1} p = ax + C_1, \tag{11}$$

where $C_1 = C - C'$.

The initial condition

$$p(0) = \frac{dy}{dx}(0) = 0$$

determines the value of C_1:

$$C_1 = \sinh^{-1}(0) - a \cdot 0 = 0.$$

Therefore,

$$\sinh^{-1} p = ax$$

and

$$p = \sinh ax. \tag{12}$$

When restated in terms of x and y, Eq. (12) becomes

$$\frac{dy}{dx} = \sinh ax,$$

an equation readily solved for y to give

$$y = \frac{1}{a} \cosh ax + C_2.$$

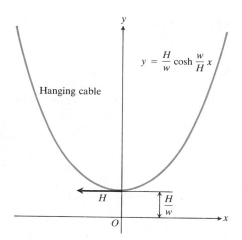

9.7 In a coordinate system chosen to match H and w in the manner shown, a hanging cable lies along the hyperbolic cosine $y = (H/w)\cosh(wx/H)$.

The remaining initial condition, $y(0) = y_0$, determines C_2:

$$y(0) = \frac{1}{a}\cosh 0 + C_2$$

$$y_0 = \frac{1}{a} + C_2$$

$$C_2 = \frac{1}{a} - y_0.$$

We now go back and adjust the coordinate system for the cable to set the value of y_0 at

$$y_0 = \frac{1}{a} = \frac{H}{w}.$$

This makes C_2 equal to zero, and the final equation for the curve of the hanging cable is

$$y = \frac{1}{a}\cosh ax = \frac{H}{w}\cosh\frac{w}{H}x. \tag{13}$$

To summarize, here is what we have learned about the shape of a hanging cable: If we choose a coordinate system for the plane of the cable in which the x-axis is horizontal, the force of gravity is straight down, the positive y-axis points straight up, and the lowest point of the cable lies at the point $y = H/w$ on the y-axis, then the cable lies along the graph of the hyperbolic cosine

$$y = \frac{H}{w}\cosh\frac{w}{H}x. \tag{14}$$

See Fig. 9.7.

REMARK 1   In case you are wondering why we substituted $p = \sinh u$ instead of $p = \tan u$ to evaluate the integral in Eq. (8), the answer is that the tangent substitution leads to a logarithmic expression for p that is hard to integrate.

REMARK 2   Our result for hanging cables also explains the shape of the Gateway Arch in St. Louis. When a cable hangs freely, all its internal forces are in equilibrium and there are no transverse forces working to push the cable out of shape. If an arch were built hanging down in the shape of a hyperbolic cosine curve, there would be no transverse forces in the arch. If the arch were then turned right side up, all the forces, reversed, would still be in equilibrium, with no transverse forces present to encourage collapse. The inherent stability of the hyperbolic cosine shape was what led the architects to choose it for the Gateway Arch.

EXAMPLE 1   Show that, in the coordinate system of Fig. 9.7, the magnitude T of the tension at any point $P(x, y)$ along a hanging cable is given by the equation

$$T = wy.$$

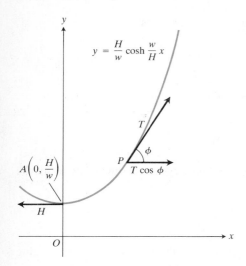

$$y = \frac{H}{w} \cosh \frac{w}{H} x$$

$A\left(0, \frac{H}{w}\right)$

T

$T \cos \phi$

ϕ

P

H

O

9.8 As explained in Example 1, $T = wy$ in this coordinate system.

Solution We work in a coordinate system in which the cable lies along the curve

$$y = \frac{H}{w} \cosh \frac{w}{H} x.$$

We sketch the section of cable from the point $A(0, H/w)$ up to an arbitrary point $P(x, y)$ and show the tensions at A and P (Fig. 9.8). The cable's slope at P is

$$\tan \phi = \frac{dy}{dx} = \sinh \frac{w}{H} x.$$

The tension at P satisfies the equation

$$T \cos \phi = H,$$

from which we find

$$T = H \sec \phi = H \sqrt{1 + \tan^2 \phi}$$

$$= H \sqrt{1 + \sinh^2 \frac{w}{H} x} = H \sqrt{\cosh^2 \frac{w}{H} x}$$

$$= H \cosh \frac{w}{H} x = w \frac{H}{w} \cosh \frac{w}{H} x = wy.$$

What does it mean for T to equal wy at every point along the hanging cable in the chosen coordinate system? It means that the magnitude of the tension at $P(x, y)$ is exactly equal to the weight of y units of cable. Thus, if the cable to the right of P were allowed to hang straight down from P over a smooth peg and were cut off at the point Q where it crossed the x-axis (Fig. 9.9), the weight of the cable length PQ would be just enough to keep the cable from slipping at P. Thus, the cable could be draped over two smooth pegs without slipping, provided the ends just reached the x-axis. ∎

The curve $y = a \cosh(x/a)$ is sometimes called a **chain curve** or **catenary,** the latter deriving from the Latin word *catena,* meaning "chain." The x-axis is called the **directrix** of the catenary.

A typical electric power line of 4-gauge copper wire might span 350 ft between poles and have a tension of 690 lb at the poles (achieved by winching the wire over a pulley). The wire would then make an angle of about 2° with the horizontal. As we can see from the equation

$$H = T \cos \phi = T \cos 2°$$

$$= (690)(0.9994) = 689.6 \quad \text{(rounded)},$$

such a flat span would exhibit very little difference between T and H. The difference between T and H is more noticeable in wires that sag a lot. If the wire meets the pole at an angle of 60° with the horizontal, then the tension at the wire's lowest point is only half the tension at the pole, as we can see from the equation

$$H = T \cos 60° = \frac{T}{2}.$$

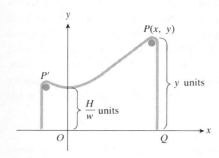

$P(x, y)$

P'

y units

$\frac{H}{w}$ units

O

Q

9.9 If the axes are positioned so that the lowest point of the uniform cable is H/w units above the x-axis, then the tension at each point $P(x, y)$ on the cable is exactly equal to the weight of a piece of cable y units long.

For a fine article about how construction crews determine the tensions in the cables they string, read Thomas O'Neil's "Constructing Power Lines," *The UMAP Journal,* 4:3, 259–264 (1983).

PROBLEMS

1. Find the length of the segment of the catenary $y = a \cosh(x/a)$ from $A(0, a)$ to $P(x_1, y_1)$, $x_1 > 0$.

2. If a segment of arc length s of the catenary $y = a \cosh(x/a)$ is measured from the lowest point, show that $dy/dx = s/a$.

3. Show that the area of the region bounded by the x-axis, the catenary $y = a \cosh x/a$, the y-axis, and the vertical line through $P_1(x_1, y_1)$, $x_1 > 0$, is the same as the area of a rectangle of altitude a and base s, where s is the length of the arc from $A(0, a)$ to P_1.

4. The catenary $y = a \cosh x/a$ is revolved about the x-axis. Find the surface area generated by the portion of the curve between the points $A(0, a)$ and $P_1(x_1, y_1)$, $x_1 > 0$. (Incidentally, of all continuously differentiable curves $y = f(x)$, $f(x) > 0$, from $A(0, a)$ to $P(x_1, y_1)$, the catenary generates the surface of revolution of least area.)

5. Find the center of mass of the arc of the catenary $y = a \cosh x/a$ between two symmetrically located points $P_0(-x_1, y_1)$ and $P_1(x_1, y_1)$.

6. Find the volume generated when the region of Problem 3 is revolved about the x-axis.

7. The length of the arc AP (Fig. 9.8) is $s = a \sinh x/a$.
   a) Show that the coordinates of $P(x, y)$ may be expressed as functions of the arc length s, as follows:
   $$x = a \sinh^{-1} \frac{s}{a},$$
   $$y = \sqrt{s^2 + a^2}.$$
   b) Calculate dx/ds and dy/ds from part (a) above and verify that $(dx/ds)^2 + (dy/ds)^2 = 1$.

8. **CALCULATOR OR TABLES** A cable 32 ft long and weighing 2 lb/ft has its ends fastened at the same level to two posts 30 ft apart.

a) Show that the constant a in Eq. (13) must satisfy the equation
$$\sinh u = \frac{16}{15}u, \quad u = \frac{15}{a}.$$
   (See Problem 7a.)
b) Sketch graphs of the curves $y_1 = \sinh u$, $y_2 = (16/15)u$ and show (with the aid of a calculator or tables) that they intersect at $u = 0$ and $u = \pm 0.6$ (approximately).
c) Using the results of part (b), find the sag in the cable at its center.
d) Using the results of part (b), find the tension in the cable at its lowest point.

9. **CALCULATOR** a) The line $y = (x/2) + 1$ crosses the catenary $y = \cosh x$ at the point $(0, 1)$. Experiment with a calculator to collect evidence for the fact that the line also crosses the catenary approximately at the point $(0.9308, 1.4654)$.
b) Two successive poles supporting an electric power line are 100 ft apart, the supporting members being at the same level. If the wire dips 25 ft at the center, estimate (i) the length of the wire between supports, and (ii) the tension in the wire at its lowest point if its weight is $w = 0.3$ lb/ft. (*Hint:* First approximate a from the equation $(25/a) + 1 = \cosh(50/a)$, which is the equation
$$\frac{x}{2} + 1 = \cosh x$$
with $x = 50/a$.

TOOLKIT PROGRAMS

Root Finder   Super ∗ Grapher

9.4
Inverse Hyperbolic Functions

As we saw in Article 9.3, the substitutions
$$p = \sinh u, \quad u = \sinh^{-1} p$$
enable us to evaluate the integral
$$\int \frac{dp}{\sqrt{1 + p^2}}$$
and express the final result once again in terms of p. The substitutions work because the function
$$y = \sinh x$$

is an increasing function of x and therefore has an inverse. Although we did not use this fact in Article 9.3, the inverse is also a differentiable function because the hyperbolic sine is differentiable and its derivative is never zero.

In this article, we shall see that the hyperbolic sine is not alone in this respect. All hyperbolic functions have inverses that are useful in integration and interesting as differentiable functions in their own right.

The Inverses of the Hyperbolic Functions

Since

$$\frac{d}{dx} \sinh x = \cosh x \tag{1}$$

is positive for every value of x, the hyperbolic sine is an increasing function of x and therefore has an inverse, which we denote by

$$y = \sinh^{-1} x. \tag{2}$$

This equation is read "y equals the arc hyperbolic sine of x" or "y equals the inverse hyperbolic sine of x." For every value of x in the interval $-\infty < x < \infty$, the value of $y = \sinh^{-1} x$ is the number whose hyperbolic sine is x. The graphs of $y = \sinh x$ and $y = \sinh^{-1} x$ are shown in Fig. 9.10.

The function $y = \cosh x$ is not one-to-one, as we can see quickly enough from its graph in Fig. 9.2(a). But the restricted function

$$y = \cosh x, \qquad x \ge 0, \tag{3}$$

is one-to-one and therefore has an inverse, denoted by

$$y = \cosh^{-1} x. \tag{4}$$

For every value of $x \ge 1$, $y = \cosh^{-1} x$ is the number in the interval $0 \le y < \infty$ whose hyperbolic cosine is x. The graphs of $y = \cosh x$, $x \ge 0$, and $y = \cosh^{-1} x$ are shown in Fig. 9.10(b).

(a)

(b)

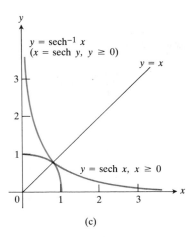

(c)

9.10   The graphs of $y = \sinh^{-1} x$, $y = \cosh^{-1} x$, and $y = \sech^{-1} x$. Note the symmetries about the line $y = x$.

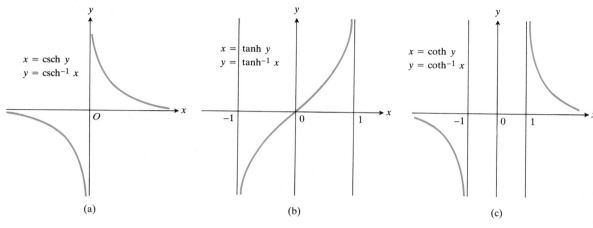

9.11 The graphs of $y = \text{csch}^{-1}x$, $y = \tanh^{-1}x$, and $y = \coth^{-1}x$.

Like $y = \cosh x$, the function

$$y = \text{sech } x = \frac{1}{\cosh x}$$

fails to be one-to-one, but its restriction

$$y = \text{sech } x, \qquad x \ge 0, \tag{5}$$

to nonnegative values of x does have an inverse, denoted by

$$y = \text{sech}^{-1}x. \tag{6}$$

For every value of x in the interval $(0, 1]$, $y = \text{sech}^{-1}x$ is the nonnegative number whose hyperbolic secant is x. The graphs of $y = \text{sech } x$, $x \ge 0$, and $y = \text{sech}^{-1}x$ are shown in Fig. 9.10(c).

The hyperbolic tangent, cotangent, and cosecant are one-to-one on their domains and therefore have inverses, denoted by

$$y = \tanh^{-1}x, \qquad y = \coth^{-1}x, \qquad y = \text{csch}^{-1}x. \tag{7}$$

These are graphed in Fig. 9.11.

Useful Identities

The inverse hyperbolic secant, cosecant, and cotangent satisfy the identities

$$\text{sech}^{-1}x = \cosh^{-1}\frac{1}{x}, \tag{8}$$

$$\text{csch}^{-1}x = \sinh^{-1}\frac{1}{x}, \tag{9}$$

$$\coth^{-1}x = \tanh^{-1}\frac{1}{x}. \tag{10}$$

These identities come in handy when we have to calculate the values of $\text{sech}^{-1}x$, $\text{csch}^{-1}x$, and $\coth^{-1}x$ on a calculator that gives only $\cosh^{-1}x$, $\sinh^{-1}x$, and $\tanh^{-1}x$.

Numerical values of inverse hyperbolic functions are given in tables such as the *Handbook of Mathematical Functions* (formerly published by the National Bureau of Standards, now printed by Dover Publications), which gives their values to ten decimal places. They are also given by many scientific calculators to eight or ten places.

When hyperbolic function keys are not available on a calculator, it is still possible to evaluate the inverse hyperbolic functions by expressing them first in terms of logarithms. The conversion formulas are listed in the following table.

Logarithm Formulas for Evaluating Inverse Hyperbolic Functions

$$\sinh^{-1}x = \ln(x + \sqrt{x^2 + 1}), \qquad -\infty < x < \infty$$

$$\cosh^{-1}x = \ln(x + \sqrt{x^2 - 1}), \qquad x \geq 1$$

$$\tanh^{-1}x = \frac{1}{2} \ln \frac{1 + x}{1 - x}, \qquad |x| < 1$$

$$\text{sech}^{-1}x = \ln\left(\frac{1 + \sqrt{1 - x^2}}{x}\right) = \cosh^{-1}\left(\frac{1}{x}\right), \qquad 0 < x \leq 1,$$

$$\text{csch}^{-1}x = \ln\left(\frac{1}{x} + \frac{\sqrt{1 + x^2}}{|x|}\right) = \sinh^{-1}\left(\frac{1}{x}\right), \qquad x \neq 0,$$

$$\coth^{-1}x = \frac{1}{2} \ln \frac{x + 1}{x - 1} = \tanh^{-1}\left(\frac{1}{x}\right), \qquad |x| > 1.$$

(11)

To show where the logarithms in (11) come from, we derive the formula for $\tanh^{-1}x$. The other formulas may be derived in a similar way.

EXAMPLE 1   Derive the formula

$$\tanh^{-1}x = \frac{1}{2} \ln \frac{1 + x}{1 - x}, \qquad |x| < 1.$$

Solution   Let $y = \tanh^{-1}x$. Then

$$x = \tanh y = \frac{\sinh y}{\cosh y} = \frac{(e^y - e^{-y})/2}{(e^y + e^{-y})/2} = \frac{e^y - (1/e^y)}{e^y + (1/e^y)} = \frac{e^{2y} - 1}{e^{2y} + 1}.$$

We now solve this equation for e^{2y}:

$$xe^{2y} + x = e^{2y} - 1$$

$$1 + x = e^{2y}(1 - x)$$

$$e^{2y} = \frac{1 + x}{1 - x}.$$

Taking logarithms of both sides now gives

$$2y = \ln\left(\frac{1 + x}{1 - x}\right)$$

$$y = \frac{1}{2} \ln\left(\frac{1 + x}{1 - x}\right).$$

Hence,

$$y = \tanh^{-1}x = \frac{1}{2} \ln \frac{1 + x}{1 - x}, \qquad |x| < 1. \tag{12}$$

The restriction $|x| < 1$ comes from the fact that $\tanh y$ lies between -1 and 1 for every value of y. ∎

EXAMPLE 2

$$\tanh^{-1} 0.25 = \frac{1}{2} \ln \frac{1.25}{0.75} = \frac{1}{2} \ln \frac{5}{3}$$

$$= \frac{1}{2}(\ln 5 - \ln 3) \approx 0.25541 \qquad ∎$$

Derivatives and Integrals

The chief merit of the inverse hyperbolic functions lies in their usefulness in integration. You will see why after we have derived the following formulas for their derivatives:

$$\frac{d(\sinh^{-1}u)}{dx} = \frac{1}{\sqrt{1 + u^2}} \frac{du}{dx}, \tag{13}$$

$$\frac{d(\cosh^{-1}u)}{dx} = \frac{1}{\sqrt{u^2 - 1}} \frac{du}{dx}, \qquad u > 1 \tag{14}$$

$$\frac{d(\tanh^{-1}u)}{dx} = \frac{1}{1 - u^2} \frac{du}{dx}, \qquad |u| < 1, \tag{15}$$

$$\frac{d(\coth^{-1}u)}{dx} = \frac{1}{1 - u^2} \frac{du}{dx}, \qquad |u| > 1, \tag{16}$$

$$\frac{d(\operatorname{sech}^{-1}u)}{dx} = \frac{-du/dx}{u\sqrt{1 - u^2}}, \qquad 0 < u < 1, \tag{17}$$

$$\frac{d(\operatorname{csch}^{-1}u)}{dx} = \frac{-du/dx}{|u|\sqrt{1 + u^2}}, \qquad u \neq 0. \tag{18}$$

To show how these formulas may be obtained, we derive the formula for the derivative of $\cosh^{-1}u$ for $u > 1$. (Although $\cosh^{-1}u$ is defined at $u = 1$, we do not expect it to be differentiable there because the tangent to the graph is vertical.)

EXAMPLE 3 Show that if u is a differentiable function of x whose values are greater than 1, then

$$\frac{d}{dx} \cosh^{-1}u = \frac{1}{\sqrt{u^2 - 1}} \frac{du}{dx}.$$

Solution   We first find the derivative of $y = \cosh^{-1}x$ for x greater than 1. To do so, we differentiate both sides of the equation $\cosh y = x$ with respect to x:

$$\cosh y = x, \qquad \frac{d}{dx}\cosh y = 1, \qquad \sinh y\,\frac{dy}{dx} = 1.$$

We then divide through by $\sinh y$ to get

$$\frac{dy}{dx} = \frac{1}{\sinh y}$$

$$= \frac{1}{\sqrt{\cosh^2 y - 1}} \qquad \begin{array}{l}(\sinh y \text{ is positive}\\ \text{because } y = \cosh^{-1}x\\ \text{is positive})\end{array}$$

$$= \frac{1}{\sqrt{x^2 - 1}}.$$

Finally, we apply the Chain Rule in the form

$$\frac{dy}{dx} = \frac{dy}{du}\frac{du}{dx}$$

to $y = \cosh^{-1}u$ to obtain

$$\frac{d}{dx}\cosh^{-1}u = \frac{1}{\sqrt{u^2 - 1}}\frac{du}{dx}. \qquad\blacksquare$$

The restrictions $|u| < 1$ and $|u| > 1$ on the derivative formulas in (15) and (16) arise from the natural restrictions on the values of the hyperbolic tangent and cotangent. (See Figs. 9.11a and b.) The distinction between $|u| < 1$ and $|u| > 1$ becomes important when we reverse the derivative formulas to get integration formulas, since otherwise we would be unable to tell whether to write $\tanh^{-1}u + C$ or $\coth^{-1}u + C$ for the integral of $1/(1 - u^2)$.

The integration formulas below follow at once from Eqs. (13)–(18). The integrals in Formulas 1, 2, 4, and 5 can be evaluated by trigonometric substitutions as well, and the integral in Formula 3 can be evaluated by partial fractions. (See Problems 47–50.)

1. $\displaystyle\int \frac{du}{\sqrt{1 + u^2}} = \sinh^{-1}u + C$

2. $\displaystyle\int \frac{du}{\sqrt{u^2 - 1}} = \cosh^{-1}u + C, \qquad u > 1$

3. $\displaystyle\int \frac{du}{1 - u^2} = \begin{cases}\tanh^{-1}u + C & \text{if } |u| < 1\\ \coth^{-1}u + C & \text{if } |u| > 1\end{cases} = \frac{1}{2}\ln\left|\frac{1 + u}{1 - u}\right| + C$

4. $\displaystyle\int \frac{du}{u\sqrt{1 - u^2}} = -\operatorname{sech}^{-1}|u| + C = -\cosh^{-1}\left(\frac{1}{|u|}\right) + C$

5. $\displaystyle\int \frac{du}{u\sqrt{1 + u^2}} = -\operatorname{csch}^{-1}|u| + C = -\sinh^{-1}\left(\frac{1}{|u|}\right) + C$

EXAMPLE 4 Evaluate the integral

$$\int_0^{1/2} 2x \tanh^{-1}x \, dx.$$

Solution We begin with integration by parts. With

$$u = \tanh^{-1}x,$$

$$dv = 2x \, dx,$$

$$du = \frac{1}{1 - x^2} \, dx,$$

$$v = x^2,$$

the indefinite integral becomes

$$\int 2x \tanh^{-1}x \, dx = x^2 \tanh^{-1}x - \int x^2 \frac{1}{1 - x^2} \, dx$$

$$= x^2 \tanh^{-1}x + \int \frac{x^2}{x^2 - 1} \, dx. \tag{19}$$

The integrand on the right can be expanded by partial fractions:

$$\frac{x^2}{x^2 - 1} = 1 + \frac{1}{x^2 - 1} \qquad \text{(by division)}$$

$$= 1 + \frac{1}{2} \frac{1}{x - 1} - \frac{1}{2} \frac{1}{x + 1}. \tag{20}$$

Together, Eqs. (19) and (20) give

$$\int 2x \tanh^{-1}x \, dx = x^2\tanh^{-1}x + x + \frac{1}{2} \ln|x - 1| - \frac{1}{2} \ln|x + 1| + C$$

$$= x^2\tanh^{-1}x + x + \frac{1}{2} \ln\left|\frac{x - 1}{x + 1}\right| + C. \tag{21}$$

To evaluate $\tanh^{-1}x$ at the given limits of integration, we express it as a logarithm. Then

$$\int_0^{1/2} 2x \tanh^{-1}x \, dx = \frac{x^2}{2} \ln\frac{1 + x}{1 - x} + x + \frac{1}{2} \ln\left|\frac{x - 1}{x + 1}\right|\Bigg]_0^{1/2}$$

$$= \frac{1}{8} \ln\frac{3/2}{1/2} + \frac{1}{2} + \frac{1}{2} \ln\frac{1/2}{3/2}$$

$$= \frac{1}{8} \ln 3 + \frac{1}{2} + \frac{1}{2} \ln\frac{1}{3}$$

$$= \frac{1}{8} \ln 3 + \frac{1}{2} - \frac{1}{2} \ln 3$$

$$= \frac{1}{2} - \frac{3}{8} \ln 3.$$

PROBLEMS

Use the logarithm formulas given in the text to find the values in Problems 1–16.

1. $\sinh^{-1}(0)$

2. $\sinh^{-1}(3/4)$

3. $\sinh^{-1}(-4/3)$

4. $\sinh^{-1}(-5/12)$

5. $\cosh^{-1}(5/4)$

6. $\cosh^{-1}(5/3)$

7. $\cosh^{-1}(2/\sqrt{3})$

8. $\cosh^{-1}(13/12)$

9. $\tanh^{-1}(1/2)$

10. $\coth^{-1}(5/4)$

11. $\coth^{-1}(-2)$

12. $\tanh^{-1}(-3/5)$

13. $\text{sech}^{-1}(3/5)$

14. $\text{sech}^{-1}(4/5)$

15. $\text{csch}^{-1}(5/12)$

16. $\text{csch}^{-1}(-1/\sqrt{3})$

In Problems 17–32, find dy/dx.

17. $y = \sinh^{-1}(2x)$

18. $y = \tanh^{-1}(\cos x)$

19. $y = \cosh^{-1}(\sec x)$

20. $y = \coth^{-1}(\sec x)$

21. $y = \text{sech}^{-1}(\sin 2x)$

22. $y = \cosh^{-1} x^2$

23. $y = \sinh^{-1}\sqrt{x-1}$

24. $y = \text{csch}^{-1}(\tan x)$

25. $y = \sinh^{-1}(1/x)$

26. $y = \cosh^{-1}\sqrt{x+1}$

27. $y = \sinh^{-1}(\tan x)$

28. $y = \coth^{-1}(\csc x)$

29. $y = \sqrt{1+x^2} - \sinh^{-1}\dfrac{1}{x}$

30. $y = \dfrac{x}{2}\sqrt{x^2-1} - \dfrac{1}{2}\cosh^{-1}x$

31. $y = 2\cosh^{-1}\left(\dfrac{x}{2}\right) + \dfrac{x}{2}\sqrt{x^2-4}$

32. $y = x^2\text{sech}^{-1}x - \sqrt{1-x^2}$

Evaluate the integrals in Problems 33–46. Use the logarithm formulas in Eqs. (11) to complete the evaluation of the definite integrals.

33. $\displaystyle\int_0^{4/3} \frac{dx}{\sqrt{1+4x^2}}$

34. $\displaystyle\int_0^{2\sqrt{3}} \frac{dx}{\sqrt{4+x^2}}$

35. $\displaystyle\int_0^{0.5} \frac{dx}{1-x^2}$

36. $\displaystyle\int_{5/4}^{2} \frac{dx}{1-x^2}$

37. $\displaystyle\int_1^2 \frac{dx}{x\sqrt{4+x^2}}$

38. $\displaystyle\int_0^{\pi} \frac{\cos x}{\sqrt{1+\sin^2 x}}\,dx$

39. $\displaystyle\int_{-1}^{1} \sinh^{-1}x\,dx$

40. $\displaystyle\int_0^{1/2} \tanh^{-1}x\,dx$

41. $\displaystyle\int \sqrt{x^2+1}\,dx$

42. $\displaystyle\int (x^2+1)^{3/2}\,dx$

43. $\displaystyle\int_2^5 4x\coth^{-1}x\,dx$

44. $\displaystyle\int_{3/5}^{4/5} 2x\,\text{sech}^{-1}x\,dx$

45. $\displaystyle\int \frac{dx}{\sqrt{x^2-4x+3}}$

46. $\displaystyle\int \frac{x-1}{\sqrt{x^2-4x+3}}\,dx$

Use trigonometric substitutions to evaluate the integrals in Problems 47–49.

47. $\displaystyle\int \frac{du}{\sqrt{u^2-1}},\quad u \geq 1$

48. $\displaystyle\int \frac{du}{u\sqrt{u^2-1}}$

49. $\displaystyle\int \frac{du}{u\sqrt{u^2+1}}$

50. Evaluate the following integral by partial fractions:

$$\int \frac{du}{1-u^2}.$$

Use tables to evaluate the integrals in Problems 51–56.

51. $\displaystyle\int \frac{\sqrt{x^2-25}}{x}\,dx$

52. $\displaystyle\int \frac{\sqrt{x^2-25}}{x^2}\,dx$

53. $\displaystyle\int \frac{x^2}{\sqrt{x^2-25}}\,dx$

54. $\displaystyle\int x^2\sqrt{x^2+3}\,dx$

55. $\displaystyle\int \sqrt{x^2-4}\,dx$

56. $\displaystyle\int x^2\sqrt{x^2-4}\,dx$

57. Find the length of the curve $y = x^2$, $0 \le x \le 1$. (Use tables to evaluate the integral.)

58. Find the area of the surface generated by revolving the curve $y = x^2$, $0 \le x \le 1$ about the *x*-axis. (Use tables to evaluate the integral.)

59. Find the length of the curve $y = \ln x$, $3/4 \le x \le 1$. (Use tables to evaluate the integral.)

60. *A tractrix.* A point *P* at the origin of the *xy*-plane is attached by a string of length *a* to a mass *M* located at the point $(a, 0)$. The point *P* then moves up the *y*-axis, dragging the mass *M* by the string. As it moves, the mass traces a differentiable curve $y = f(x)$, shown in Fig. 9.12. The curve is called a *tractrix,* after the Latin word *tractum,* for "drag." The tractrix plays an important role in the study of non-Euclidean geometry because the surface generated by revolving it about the *y*-axis is a model for the Lobachevsky plane.

a) Show that the function $y = f(x)$ whose graph is traced by *M* satisfies the differential equation

$$\frac{dy}{dx} = -\frac{\sqrt{a^2 - x^2}}{x}.$$

b) Solve the differential equation in (a), subject to the condition that $y = 0$ when $x = a$.

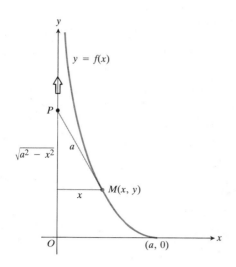

9.12 As *P* moves up the *y*-axis, it pulls *M* after it with a string *a* units long. The curve $y = f(x)$ traced by *M* is called a *tractrix.* The goal of Problem 60 is to find a formula for this curve.

61. *A pursuit curve.* A rabbit starts at the origin and runs at a constant speed up the *y*-axis. At the same time, a dog, running at the same speed as the rabbit, starts at the point $(1, 0)$ and pursues the rabbit. It can be shown that the dog's path (Fig. 9.13) is the graph of a function $y = f(x)$ that satisfies

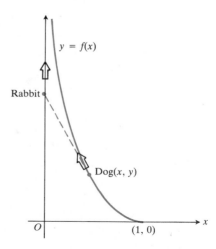

9.13 *A pursuit curve.* The dog runs straight for the rabbit, but the rabbit moves too, so that the dog's path is curved. Problem 61 gives a differential equation for the dog's path.

the differential equation

$$x\frac{d^2y}{dx^2} = \sqrt{1 + \left(\frac{dy}{dx}\right)^2}.$$

Solve this equation subject to the condition that *y* and dy/dx equal zero when $x = 1$.

62. *Retarded free fall.* If a body of mass *m* falling from rest under the action of gravity encounters an air resistance proportional to the square of the velocity, then the velocity *v* at time *t* satisfies the differential equation

$$m\left(\frac{dv}{dt}\right) = mg - kv^2,$$

where *k* is a constant of proportionality and $v = 0$ when $t = 0$. Show that

$$v = \sqrt{\frac{mg}{k}}\, \tanh\left(\sqrt{\frac{gk}{m}}\, t\right),$$

and hence deduce that the body approaches a limiting velocity equal to $\sqrt{mg/k}$ as $t \to \infty$.

63. Solve the equation $x = \sinh y = (e^y - e^{-y})/2$ for e^y in terms of *x*, and thus show that $y = \ln(x + \sqrt{x^2 + 1})$. (This equation expresses $\sinh^{-1} x$ as a logarithm.)

64. Express $\cosh^{-1} x$ as a logarithm by using the method of Problem 63.

65. Establish Eq. (13).

66. Establish Eq. (15).

67. Establish Eq. (17).

TOOLKIT PROGRAMS

First Order Initial Value Problem
Second Order Initial Value Problem
Super * Grapher

REVIEW QUESTIONS AND EXERCISES

1. Define the six hyperbolic functions. Sketch their graphs. Give their domains and ranges.

2. What do hyperbolic functions have to do with hyperbolas?

3. State four trigonometric identities, such as the formulas for $\sin(A + B)$, $\cos(A - B)$, $\cos^2 A + \sin^2 A$, and $2 \sin^2 A$. What are the corresponding hyperbolic identities? Verify them.

4. State some of the differences between the trigonometric functions and their hyperbolic counterparts.

5. Derive formulas for the derivatives of the six hyperbolic functions.

6. If $y = C_1 \sin at + C_2 \cos at$, then $y'' = -a^2 y$. What corresponding differential equation is satisfied by $y = C_1 \sinh at + C_2 \cosh at$?

7. What shapes are assumed by hanging cables? How do the relations among the magnitudes of the forces acting on a section of a hanging cable lead to a differential equation for the cable's shape? What is the differential equation? How is it solved?

8. Define the six inverse hyperbolic functions. Sketch their graphs. Give their domains and ranges. How can their values be calculated?

9. Give formulas for the derivatives of the inverse hyperbolic functions. Derive one of the formulas.

10. What integrals lead naturally to inverse hyperbolic functions?

MISCELLANEOUS PROBLEMS

1. Prove the hyperbolic identity $\cosh 2x = \cosh^2 x + \sinh^2 x$.

2. Verify that $\tanh x = \sinh 2x / (1 + \cosh 2x)$.

3. Find $\lim_{x \to \infty} (\cosh x - \sinh x)$.

4. If $\cosh x = 5/4$, find $\sinh x$ and $\tanh x$.

5. If $\operatorname{csch} x = -9/40$, find $\cosh x$ and $\tanh x$.

6. If $\tanh x > 5/13$, show that $\sinh x > 0.4$ and $\operatorname{sech} x < 0.95$.

7. Let P be a point on the curve $y = \tanh x$ (Fig. 9.14). Let AB be the vertical line segment through P with A and B on the asymptotes of the curve. Let C be a semicircle with AB as diameter. Let the line through P perpendicular to AB cut C in the point Q. Show that $PQ = \operatorname{sech} x$.

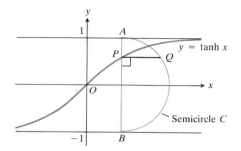

9.14   The construction in Problem 7.

8. Find parametric equations for the semicircle
$$x^2 + y^2 = a^2, \qquad y > 0,$$
using as parameter the variable t defined by the equation $x = a \tanh t$.

9. Find an equation for the asymptote of the hyperbolic curve $y = \tanh((1/2) \ln |x|)$.

10. *Accelerations whose magnitudes are proportional to displacement.* A particle moves along the x-axis according to one of the following laws:
    a) $x = a \cos kt + b \sin kt$,
    b) $x = a \cosh kt + b \sinh kt$.
    In both cases, show that the acceleration is proportional to x but that in the first case it is always directed toward the origin while in the second case it is directed away from the origin.

11. Show that $y = \cosh 2x$, $\sinh 2x$, $\cos 2x$, and $\sin 2x$ all satisfy the relationship $d^4 y/dx^4 = 16y$.

12. Graph the function
$$y = \frac{1}{2} \ln \frac{1 + \tanh x}{1 - \tanh x}.$$

In Problems 13–22, find dy/dx.

13. $y = \sinh^2 3x$

14. $\tan x = \tanh^2 y$

15. $\sin^{-1} x = \operatorname{sech} y$

16. $\sinh y = \sec x$

17. $\tan^{-1} y = \tanh^{-1} x$

18. $y = \tanh(\ln x)$

19. $x = \cosh(\ln y)$

20. $y = \sinh(\tan^{-1} e^{3x})$

21. $y = \sinh^{-1}(\tan x)$

22. $y^2 + x \cosh y + \sinh^2 x = 50$

Evaluate the integrals in Problems 23–32.

23. $\displaystyle\int_{-\ln 2}^{0} \frac{d\theta}{\sinh \theta + \cosh \theta}$

24. $\displaystyle\int_{-\ln 2}^{\ln 2} \frac{\cosh \theta \, d\theta}{\sinh \theta + \cosh \theta}$

25. $\displaystyle\int_{0}^{(\ln 3)/2} \sinh^3 x \, dx$

26. $\displaystyle\int_{-\ln 2}^{0} e^x \sinh 2x \, dx$

27. $\displaystyle\int_{1}^{\sqrt{2}} \frac{e^{2x} - 1}{e^{2x} + 1}\, dx$

(*Hint:* Multiply numerator and denominator by e^{-x}.)

28. $\displaystyle\int_{0}^{1} \frac{dx}{4 - x^2}$

29. $\displaystyle\int_{0}^{1} \frac{dx}{\sqrt{x} - x\sqrt{x}}$ (*Hint:* Let $u = \sqrt{x}$.)

30. $\displaystyle\int_{0}^{\ln(3/4)} \frac{e^t\, dt}{\sqrt{2 + e^{2t}}}$

31. $\displaystyle\int_{\pi/3}^{\pi/2} \frac{\sin x\, dx}{1 - \cos^2 x}$

32. $\displaystyle\int_{\pi/4}^{\pi/3} \frac{\sec^2\theta\, d\theta}{\sqrt{\tan^2\theta - 1}}$

33. Find the limit, as $x \to \infty$, of $\cosh^{-1}x - \ln x$.

34. Evaluate

$$\lim_{x \to \infty} \int_{1}^{x} \left(\frac{1}{\sqrt{1 + t^2}} - \frac{1}{t} \right) dt.$$

CHAPTER 10

Polar Coordinates

OVERVIEW

One of the original ideas in Newton's *The Method of Fluxions and Infinite Series* was the use of new coordinate systems for the plane, among them the polar coordinate system that locates points by their distance and compass direction from the origin. Because Newton's book, written in Latin around 1671, was not published in English until 1736, credit for the invention of polar coordinates is usually given to James Bernoulli, who invented them independently and described them in an article that appeared in 1691.

Polar coordinates are important in physics and astronomy because the equations of ellipses, parabolas, and hyperbolas all have the same form in this system. We do not need different forms for different conic sections as we do in Cartesian coordinates. This means we can describe the orbits of planets, moons, comets, and satellites with a single equation. Once we know an orbit's eccentricity and the distance from one focus to its associated directrix, we automatically know the orbit's equation in a polar coordinate system centered at the focus. In Article 10.3, we shall see what this equation is.

Polar coordinate equations generate other interesting curves, as we shall see in Article 10.2, where we study graphing, and again in Article 10.3, where we look at particular curves. In Article 10.4, we shall see how to calculate lengths and areas associated with these curves. The equations we use could all be expressed in Cartesian coordinates as well, but the use of polar coordinates makes the calculations easier.

We begin our presentation with an introduction to polar coordinates.

10.1
The Polar Coordinate System

In this article, we define polar coordinates and study their relation to Cartesian coordinates. One of the distinctions between polar and Cartesian coordinates is that while a point in the plane has just one pair of Cartesian coordinates, it has infinitely many pairs of polar coordinates. This has interesting consequences for the graphs of polar equations, as we shall see in the next article.

How Polar Coordinates Are Defined

To define polar coordinates, we first fix an **origin** O and an **initial ray** from O, as shown in Fig. 10.1. Then each point P can be located by assigning to it a **polar coordinate pair** (r, θ), in which the first number, r, gives the directed distance from O to P and the second number, θ, gives the directed angle from the initial ray to the segment OP:

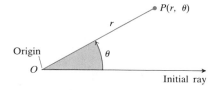

10.1 To define polar coordinates for the plane, we start with an origin and an initial ray.

Polar Coordinates

$$P(r, \theta) \qquad (1)$$

Directed distance from O to P Directed angle from initial ray to OP

As in trigonometry, the angle θ is *positive* when measured counterclockwise and negative when measured clockwise (Fig. 10.1). But the angle associated with a given point is not unique (Fig. 10.2). For instance, the point 2 units from the origin along the ray $\theta = 30°$ has polar coordinates $r = 2$, $\theta = 30°$. It also has coordinates $r = 2$, $\theta = -330°$, and $r = 2$, $\theta = 390°$.

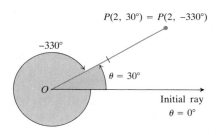

10.2 The ray $\theta = 30°$ is the same as the ray $\theta = -330°$.

Negative Values of r; Changing to Radian Measure

There are occasions when we wish to allow r to be negative. That is why we say "directed distance" in (1). The ray $\theta = 30°$ and the ray $\theta = 210°$ together make a complete line through O (Fig. 10.3). The point $P(2, 210°)$ 2 units from

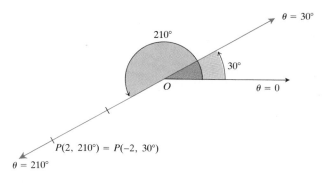

10.3 The rays $\theta = 30°$ and $\theta = 210°$ make a line.

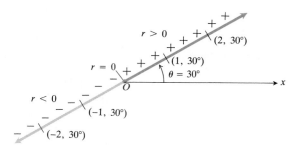

10.4   The ray $\theta = 30°$ and its opposite.

O on the ray $\theta = 210°$ has polar coordinates $r = 2$, $\theta = 210°$. It can be reached by turning $210°$ counterclockwise from the initial ray and going forward 2 units. It can also be reached by turning $30°$ counterclockwise from the initial ray and going *backward* two units. So we say that the point also has polar coordinates $r = -2$, $\theta = 30°$.

Whenever the angle between two rays is $180°$, the rays make a straight line. We then say that each ray is the **opposite** of the other. Points on the ray $\theta = \alpha$ have polar coordinates (r, α) with $r \geq 0$. Points on the opposite ray, $\theta = \alpha + 180°$, have coordinates (r, α) with $r \leq 0$. See Fig. 10.4.

EXAMPLE 1   Find all polar coordinates of the point $(2, 30°)$. Express the angles in radians as well as in degrees.

Solution   We sketch the initial ray of the coordinate system, draw the line through the origin that makes a $30°$ angle with the initial ray, and mark the point $(2, 30°)$ (Fig. 10.5). We then find formulas for the coordinate pairs in which $r = 2$ and $r = -2$ and convert the formulas to radian measure.

For $r = 2$: The angles

$$30° + 1 \cdot 360° = 390° \qquad 30° - 1 \cdot 360° = -330°$$

$$30° + 2 \cdot 360° = 750° \qquad 30° - 2 \cdot 360° = -690°$$

$$30° + 3 \cdot 360° = 1110° \qquad 30° - 3 \cdot 360° = -1050°$$

$$\tag{2}$$

$$\vdots \qquad\qquad\qquad \vdots$$

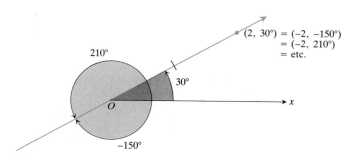

10.5   The point P $(2, 30°)$ has many different polar coordinates.

all end in the same ray as the angle 30°. Thus, the polar coordinates

$$(2, 30° + n \cdot 360°), \qquad n = 0, \pm1, \pm2, \ldots \tag{3}$$

all identify the point (2, 30°).

For r = −2: Numerous as they are, the coordinates in (3) are not the only polar coordinates of the point (2, 30°). The angles

$$
\begin{array}{ll}
-150° & -150° \\
-150° + 360° = 210° & -150° - 360° = -510° \\
-150° + 720° = 570° & -150° - 720° = -870°
\end{array}
$$

$$\vdots \qquad\qquad \vdots \tag{4}$$

all define the ray opposite the ray $\theta = 30°$. Hence, the polar coordinates

$$(-2, -150° + n \cdot 360°), \qquad n = 0, \pm1, \pm2, \ldots \tag{5}$$

represent the point (2, 30°) as well.

Radian measure: If we measure angles in radians, the formulas that correspond to (3) and (5) are

$$\left(2, \frac{\pi}{6} + 2n\pi\right), \qquad n = 0, \pm1, \pm2, \ldots \tag{6}$$

and

$$\left(-2, -\frac{5\pi}{6} + 2n\pi\right), \qquad n = 0, \pm1, \pm2, \ldots . \tag{7}$$

When $n = 0$, these formulas give

$$(2, \pi/6) \qquad \text{and} \qquad (-2, -5\pi/6).$$

When $n = 1$, they give

$$(2, 13\pi/6) \qquad \text{and} \qquad (-2, 7\pi/6),$$

and so on. ∎

Although nothing in the definition of polar coordinates requires the use of radian measure, we shall need to have all angles in radians when we differentiate and integrate trigonometric functions of θ. We shall therefore use radian measure exclusively from now on.

Elementary Coordinate Equations and Inequalities

If we hold r fixed at a constant nonzero value $r = a$, then the point $P(r, \theta)$ lies $|a|$ units from the origin. As θ varies over any interval of length 2π radians, P traces a circle of radius $|a|$ centered at the origin. The equation

$$r = a$$

is thus an equation for this circle (Fig. 10.6). The equation $r = 1$ is an equation for the circle of radius 1 centered at the origin. The equation $r = -1$ is an equation for the same circle.

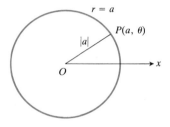

10.6 The polar equation for this circle is $r = a$.

(a)

(b)

(c)

(d)

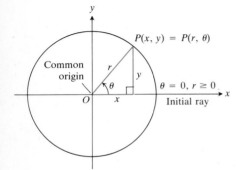

10.7   The graphs of typical inequalities in r and θ.

If we hold θ fixed at a constant value $\theta = \theta_0$ and let r take on all real values, the point $P(r, \theta)$ traces a line through the origin that makes an angle of θ_0 radians with the coordinate system's initial ray. The equation

$$\theta = \theta_0$$

is thus an equation for this line. The equation $\theta = \pi/6$ is an equation for the line in Fig. 10.5. The equations $\theta = 7\pi/6$ and $\theta = -5\pi/6$ are also equations for this line.

Equations of the form $r = a$ and $\theta = \theta_0$ can be combined in various ways to define regions, segments, and rays in the coordinate plane, as in the next example.

EXAMPLE 2   Graph the sets of points whose polar coordinates satisfy the following conditions.

a) $1 \le r \le 2$ and $0 \le \theta \le \dfrac{\pi}{2}$

b) $-3 \le r \le 2$ and $\theta = \dfrac{\pi}{4}$

c) $r \le 0$ and $\theta = \dfrac{\pi}{4}$

d) $\dfrac{2\pi}{3} \le \theta \le \dfrac{3\pi}{4}$ (no restriction on r)

Solution   The graphs are shown in Fig. 10.7.   ■

Cartesian vs. Polar Equations

When we use both polar and Cartesian coordinates in a plane, we usually place the polar origin at the Cartesian origin and take the initial ray of the polar coordinate system to be the positive x-axis. The ray $\theta = \pi/2$, $r \ge 0$, is the nonnegative y-axis. As Fig. 10.8 shows, the two sets of coordinates are then related by the equations

$$x = r \cos \theta, \quad y = r \sin \theta, \quad \text{or} \quad x^2 + y^2 = r^2, \quad \frac{y}{x} = \tan \theta. \quad (8)$$

These equations define $\sin \theta$ and $\cos \theta$ when r is positive. They are also valid if r is negative because $\cos(\theta + \pi) = -\cos \theta$, $\sin(\theta + \pi) = -\sin \theta$, so positive r's on the ray $\theta + \pi$ correspond to negative r's on the ray θ. If $r = 0$, then $x = y = 0$, and P is the origin.

These equations may be used to rewrite polar equations in Cartesian form and vice versa, as in the next example.

10.8   The usual way to relate polar and Cartesian coordinates.

EXAMPLE 3

Polar equation	Cartesian equivalent
$r \cos \theta = 2$	$x = 2$
$r^2 \cos \theta \sin \theta = 4$	$xy = 4$
$r^2 \cos^2\theta - r^2 \sin^2\theta = 1$	$x^2 - y^2 = 1$
$r = 1 + 2r \cos \theta$	$y^2 - 3x^2 - 4x - 1 = 0$
$r = 1 - \cos \theta$	$x^4 + y^4 + 2x^2y^2 + 2x^3 + 2xy^2 - y^2 = 0$

With some curves, we are better off with polar coordinates; with others, we aren't. ■

EXAMPLE 4 Find a Cartesian equation for the curve

$$r \cos \left(\theta - \frac{\pi}{3}\right) = 3.$$

Solution We use the identity

$$\cos(A - B) = \cos A \cos B + \sin A \sin B$$

with $A = \theta$ and $B = \pi/3$:

$$r \cos \left(\theta - \frac{\pi}{3}\right) = 3,$$

$$r\left(\cos \theta \cos \frac{\pi}{3} + \sin \theta \sin \frac{\pi}{3}\right) = 3,$$

$$r \cos \theta \cdot \frac{1}{2} + r \sin \theta \cdot \frac{\sqrt{3}}{2} = 3,$$

$$\frac{1}{2}x + \frac{\sqrt{3}}{2}y = 3,$$

$$x + \sqrt{3}y = 6. \qquad ■$$

EXAMPLE 5 Replace the following polar equations by equivalent Cartesian equations, and identify their graphs.

a) $r \cos \theta = -4$

b) $r^2 = 4r \cos \theta$

c) $r = \dfrac{4}{2 \cos \theta - \sin \theta}$

Solution We use the substitutions $r \cos \theta = x$, $r \sin \theta = y$, $r^2 = x^2 + y^2$.

a) $r \cos \theta = -4$

The Cartesian equation: $r \cos \theta = -4$
$$x = -4$$

The graph: Vertical line through $x = -4$ on the x-axis

b) $r^2 = 4r \cos \theta$

The Cartesian equation:
$$r^2 = 4r \cos \theta$$
$$x^2 + y^2 = 4x$$
$$x^2 - 4x + y^2 = 0$$
$$x^2 - 4x + 4 + y^2 = 4 \qquad \binom{\text{completing}}{\text{the square}}$$
$$(x - 2)^2 + y^2 = 4$$

The graph: Circle, radius 2, center $(h, k) = (2, 0)$

c) $r = \dfrac{4}{2 \cos \theta - \sin \theta}$

The Cartesian equation:
$$r(2 \cos \theta - \sin \theta) = 4$$
$$2r \cos \theta - r \sin \theta = 4$$
$$2x - y = 4$$
$$y = 2x - 4$$

The graph: Line, slope $m = 2$, y-intercept $b = -4$ ■

PROBLEMS

All angles are given in radians.

1. Pick out the polar coordinate pairs that label the same point.
a) $(3, 0)$
b) $(-3, 0)$
c) $(-3, \pi)$
d) $(-3, 2\pi)$
e) $(2, 2\pi/3)$
f) $(2, -\pi/3)$
g) $(2, 7\pi/3)$
h) $(-2, \pi/3)$
i) $(2, -\pi/3)$
j) $(2, -4\pi/3)$
k) $(-2, -\pi/3)$
l) $(-2, 2\pi/3)$
m) (r, θ)
n) $(r, \theta + \pi)$
o) $(-r, \theta + \pi)$
p) $(-r, \theta)$

2. Find the Cartesian coordinates of the points whose polar coordinates are given in parts (a)–(l) of Problem 1.

3. Plot the following points (given in polar coordinates). Then find all the polar coordinates of each point.
a) $(2, \pi/2)$
b) $(2, 0)$
c) $(-2, \pi/2)$
d) $(-2, 0)$

4. Plot the following points (given in polar coordinates). Then find all the polar coordinates of each point.
a) $(3, \pi/4)$
b) $(-3, \pi/4)$
c) $(3, -\pi/4)$
d) $(-3, -\pi/4)$

5. Find the Cartesian coordinates of the following points (given in polar coordinates).
a) $(\sqrt{2}, \pi/4)$
b) $(1, 0)$
c) $(0, \pi/2)$
d) $(-\sqrt{2}, \pi/4)$
e) $(-3, 5\pi/6)$
f) $(5, \tan^{-1}(4/3))$
g) $(-1, 7\pi)$
h) $(2\sqrt{3}, 2\pi/3)$

6. Find all polar coordinates of the origin.

Graph the sets of points whose polar coordinates satisfy the equations and inequalities in Problems 7–22.

7. $r = 2$
8. $0 \le r \le 2$
9. $r \ge 1$
10. $1 \le r \le 2$
11. $0 \le \theta \le \pi/6, \quad r \ge 0$
12. $\theta = 2\pi/3, \quad r \le -2$
13. $\theta = \pi/3, \quad -1 \le r \le 3$
14. $\theta = 11\pi/4, \quad r \ge -1$
15. $\theta = \pi/2, \quad r \ge 0$
16. $\theta = \pi/2, \quad r \le 0$
17. $0 \le \theta \le \pi, \quad r = 1$
18. $0 \le \theta \le \pi, \quad r = -1$
19. $\pi/4 \le \theta \le 3\pi/4, \quad 0 \le r \le 1$
20. $-\pi/4 \le \theta \le \pi/4, \quad -1 \le r \le 1$
21. $-\pi/2 \le \theta \le \pi/2, \quad 1 \le r \le 2$
22. $0 \le \theta \le \pi/2, \quad 1 \le |r| \le 2$

Replace the polar equations in Problems 23–40 by equivalent Cartesian equations. Then graph the equations.

23. $r \cos \theta = 2$
24. $r \sin \theta = -1$
25. $r \sin \theta = 4$
26. $r \cos \theta = 0$
27. $r \sin \theta = 0$
28. $r \cos \theta = -3$
29. $r \cos \theta + r \sin \theta = 1$
30. $r \sin \theta = r \cos \theta$
31. $r^2 = 1$
32. $r^2 = 4r \sin \theta$
33. $r \sin \theta = e^{r \cos \theta}$
34. $r^2 + 2r^2 \cos \theta \sin \theta = 1$
35. $r = \dfrac{5}{\sin \theta - 2 \cos \theta}$
36. $r = 4 \cos \theta$
37. $r = 4 \sin \theta$
38. $r = 2 \cos \theta + 2 \sin \theta$
39. $r = 4 \tan \theta \sec \theta$
40. $r + \sin \theta = 2 \cos \theta$

Replace the Cartesian equations in Problems 41–50 by equivalent polar equations.

41. $x = 7$
42. $y = 1$
43. $x = y$
44. $x - y = 3$

45. $x^2 + y^2 = 4$ **46.** $x^2 - y^2 = 1$

47. $\dfrac{x^2}{9} + \dfrac{y^2}{4} = 1$ **48.** $xy = 2$

49. $y^2 = 4x$

50. $x^2 - y^2 = 25\sqrt{x^2 + y^2}$

In Problems 51–54, use the trigonometric identities for $\sin(A \pm B)$ and $\cos(A \pm B)$ to replace the polar equations by equivalent Cartesian equations. Then graph the equations.

51. $r \cos\left(\theta - \dfrac{\pi}{3}\right) = 3$ **52.** $r \sin\left(\theta + \dfrac{\pi}{4}\right) = 4$

53. $r \sin\left(\dfrac{\pi}{4} - \theta\right) = \sqrt{2}$ **54.** $r \cos\left(\dfrac{\pi}{6} - \theta\right) = 0$

10.2
Graphing in Polar Coordinates

The graph of the equation

$$F(r, \theta) = 0$$

consists of the points whose polar coordinates in some form satisfy the equation. We say "in some form" because it is a sad fact, but true, that some coordinate pairs of a point on the graph may not satisfy the equation even though others do.

To speed our work, we look for symmetries, for values of θ at which the curve passes through the origin, and for points at which r takes on extreme values. When the curve passes through the origin, we also try to calculate its slope there.

About the origin
(a)

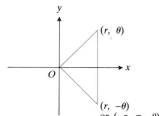

About the x-axis
(b)

Symmetry and Slope at the Origin

Three kinds of symmetry are easy to detect in the graph of an equation $F(r, \theta) = 0$. As Fig. 10.9 shows, the graph is

a) symmetric about the origin if the equation is unchanged when r is replaced by $-r$, or when θ is replaced by $\theta + \pi$;

b) symmetric about the x-axis if the equation is unchanged when θ is replaced by $-\theta$, or the pair (r, θ) by the pair $(-r, \pi - \theta)$;

c) symmetric about the y-axis if the equation is unchanged when θ is replaced by $\pi - \theta$, or the pair (r, θ) by the pair $(-r, -\theta)$.

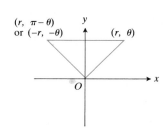

About the y-axis
(c)

10.9 Three polar coordinate tests for symmetry.

EXAMPLE 1 *A cardioid.* Graph the curve $r = a(1 - \cos \theta)$, $a > 0$.

Solution Since $\cos(-\theta) = \cos \theta$, the equation is unaltered when θ is replaced by $-\theta$; hence the curve is symmetric about the x-axis (Fig. 10.9b). Also, since

$$-1 \le \cos \theta \le 1,$$

the values of r vary between 0 and $2a$. The minimum value, $r = 0$, occurs at $\theta = 0$, and the maximum value, $r = 2a$, occurs at $\theta = \pi$. Moreover, as θ varies from 0 to π, $\cos \theta$ decreases from 1 to -1; hence $1 - \cos \theta$ increases from 0 to 2. That is, r increases from 0 to $2a$ as the radius vector OP swings

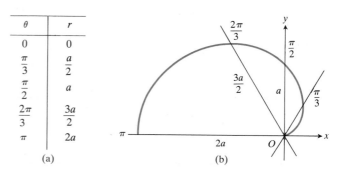

θ	r
0	0
$\dfrac{\pi}{3}$	$\dfrac{a}{2}$
$\dfrac{\pi}{2}$	a
$\dfrac{2\pi}{3}$	$\dfrac{3a}{2}$
π	$2a$

(a)

(b)

10.10   (a) Values of $r = a(1 - \cos \theta)$ for selected values of θ. (b) A smooth curve sketched through the points from part (a).

from $\theta = 0$ to $\theta = \pi$. We make a table of values (Fig. 10.10a) and plot the corresponding points.

Before sketching the curve, we investigate the slope at the origin. We imagine a point P approaching the origin along the curve in the first quadrant (Fig. 10.11). The slope of the curve at the origin is the limit of the slope of secant OP as $P \to 0$, which is

$$\lim_{\theta \to 0} (\text{slope of } OP) = \lim_{\theta \to 0} \tan \theta = 0.$$

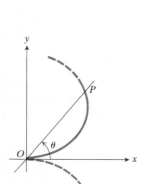

10.11   As $P \to O$ along the cardioid, $\theta \to 0$. At the origin, the cardioid has a horizontal tangent.

We now sketch a curve through the plotted points, making sure that the tangent at the origin is horizontal (Fig. 10.10b). We expect r to increase steadily as θ increases from 0 to π because

$$\frac{dr}{d\theta} = a \sin \theta$$

is positive for $0 < \theta < \pi$.

Finally, we exploit the symmetry of the curve and reflect the part we have just drawn across the x-axis to complete the sketch (Fig. 10.12). The completed graph is called a **cardioid** because of its heart-shaped appearance.   ∎

The process by which we found the tangent to the cardioid at the origin in Example 1 works for any smooth curve through the origin. If the curve passes through the origin when $\theta = \theta_0$, then the discussion in Example 1 would be modified only to the extent of saying that $P \to O$ along the curve as $\theta \to \theta_0$, and hence

$$\left(\frac{dy}{dx}\right)_{\theta = \theta_0} = \lim_{\theta \to \theta_0} (\tan \theta) = \tan \theta_0.$$

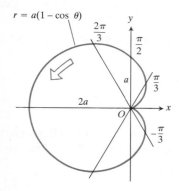

10.12   The completed graph of the cardioid $r = a(1 - \cos \theta)$. The arrow shows the direction of increasing θ.

But $(dy/dx)_{\theta=\theta_0}$ is also the tangent of the angle between the x-axis and the curve at this point. Hence the line $\theta = \theta_0$ is tangent to the curve at the origin. In other words, whenever a curve passes through the origin for a value θ_0 of θ, it does so *tangent* to the line $\theta = \theta_0$ provided the derivative $dr/d\theta$ exists at that point. See Fig. 10.13.

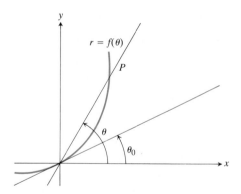

10.13   If the curve $r = f(\theta)$ passes through the origin at $\theta = \theta_0$, and if f has a derivative at $\theta = \theta_0$, then the line $\theta = \theta_0$ is tangent to the curve at the origin.

EXAMPLE 2   Graph the curve $r^2 = 4a^2 \cos \theta$.

Solution   This curve is symmetric about the origin. Two values of r,

$$r = \pm 2a\sqrt{\cos \theta},$$

correspond to each value of θ for which $\cos \theta > 0$, namely,

$$-\frac{\pi}{2} < \theta < \frac{\pi}{2}.$$

Furthermore, the curve is symmetric about the x-axis because $\cos(-\theta) = \cos \theta$ for all θ. The curve passes through the origin at $\theta = \pi/2$ and is tangent to the y-axis at this point. Since $\cos \theta \le 1$, the maximum value of r is $2a$, which occurs at $\theta = 0$. As θ increases from 0 to $\pi/2$, $|r|$ decreases from $2a$ to 0. See Fig. 10.14.   ■

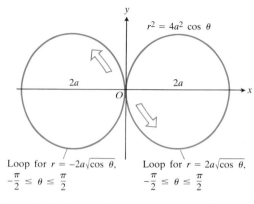

10.14   The graph of $r^2 = 4a^2 \cos \theta$. The arrows show the direction of increasing θ.

A Technique for Graphing

One way to graph a polar equation $r = f(\theta)$ is to make a table of (r, θ) values, plot the corresponding points, and connect them in order of increasing θ. This can work well if enough points have been plotted to reveal all the loops and dimples in the graph. In this section we describe another method of graphing that is usually quicker and more reliable. The steps in the new method are these:

1. First graph $r = f(\theta)$ in the *Cartesian $r\theta$-plane* (that is, plot the values of θ on a horizontal axis and the corresponding values of r along a vertical axis).

2. Then use the Cartesian graph as a "table" and guide to sketch the *polar coordinate* graph.

This method is better than simple point plotting because the Cartesian graph, even when hastily drawn, shows at a glance where r is positive, negative, and nonexistent, as well as where r is increasing and decreasing. As examples, we graph $r = 1 + \cos(\theta/2)$ and $r^2 = \sin 2\theta$.

EXAMPLE 3   Graph the curve

$$r = 1 + \cos\frac{\theta}{2}.$$

Solution   We first graph r as a function of θ in the Cartesian $r\theta$-plane. Since the cosine has period 2π, we must let θ run from 0 to 4π to produce the entire graph. See Fig. 10.15. The arrows from the θ-axis to the curve give the radius vectors for graphing $r = 1 + \cos(\theta/2)$ in the polar plane, as shown in Fig. 10.16.   ■

EXAMPLE 4   *A lemniscate.* Graph the curve $r^2 = \sin 2\theta$.

Solution   Here we begin by plotting r^2 (not r) as a function of θ in the Cartesian $r^2\theta$-plane. See Fig. 10.17. We pass from there to the graph of $r = \pm\sqrt{\sin 2\theta}$ in the $r\theta$-plane (Fig. 10.18), and then draw the polar graph (Fig. 10.19). The graph in Fig. 10.18 "covers" the final polar graph in Fig. 10.19 twice. We could have managed with either loop alone, with the two upper halves, or with the two lower halves. The double covering does no harm, however, and we actually learn a little more about the behavior of the function this way.   ■

Finding the Points Where Curves Intersect

The fact that a point may be represented in different ways in polar coordinates makes extra care necessary in deciding when a point lies on the graph of a polar equation and in determining the points at which the graphs of polar equations intersect. The problem is that a point of intersection may appear in the equation of one curve with different polar coordinates than it has in the equation of another curve. Thus, solving the equations of two curves simultaneously may not identify all their points of intersection. The only sure way to identify all the points of intersection is to graph the equations.

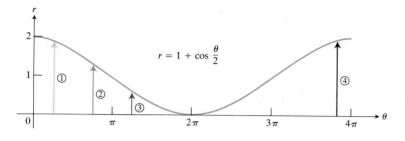

10.15   The graph of $r = 1 + \cos(\theta/2)$ in the Cartesian $r\theta$-plane gives us any number of radius vectors from the θ-axis to the curve.

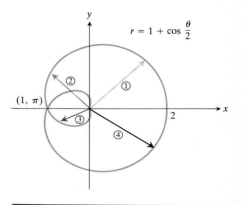

10.16   The radius vectors from the previous sketch help us draw the graph in the polar $r\theta$-plane.

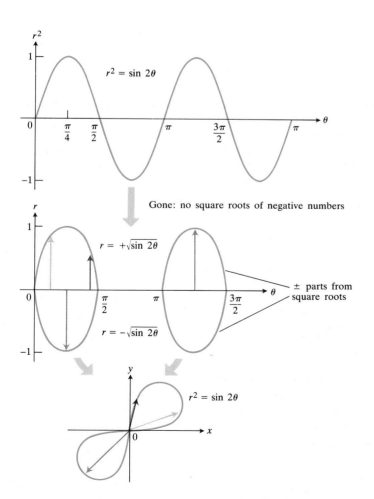

10.17   The graph of $r^2 = \sin 2\theta$ in the Cartesian $r^2\theta$-plane includes negative values of the dependent variable r^2 as well as positive values.

10.18   When we graph r vs θ in the Cartesian $r\theta$-plane we ignore the points where r^2 is negative but plot $+$ and $-$ parts from the points where r^2 is positive.

10.19   In the polar $r\theta$-plane, the radius vectors from the previous sketch cover the final graph twice.

605

EXAMPLE 5   Show that the point $(2, \pi/2)$ lies on the curve $r = 2 \cos 2\theta$.

Solution   It may seem at first that the point $(2, \pi/2)$ does not lie on the curve because substituting the given coordinates into the equation gives

$$2 = 2 \cos 2\left(\frac{\pi}{2}\right) = 2 \cos \pi = -2,$$

which is not a true equality. The magnitude is right, but the sign is wrong. This suggests looking for a pair of coordinates for the given point in which r is negative, for example,

$$\left(-2, -\frac{\pi}{2}\right).$$

When we try these in the equation $r = 2 \cos 2\theta$, we find

$$-2 = 2 \cos 2\left(-\frac{\pi}{2}\right) = 2(-1) = -2,$$

and the equation is satisfied. The point $(2, \pi/2)$ does lie on the curve after all.   ◼

EXAMPLE 6   Find the points of intersection of the curves

$$r^2 = 4a^2 \cos \theta \qquad \text{and} \qquad r = a(1 - \cos \theta), \qquad a > 0.$$

Solution   Our experience with equations in Cartesian coordinates might suggest finding the points of intersection by solving the equations $r^2 = 4a^2 \cos \theta$ and $r = a(1 - \cos \theta)$ simultaneously. However, as we shall see, simultaneous solution reveals only two of the four points in which the curves intersect. The others must be found by graphing.

If we substitute $\cos \theta = r^2/4a^2$ in the equation $r = a(1 - \cos \theta)$, we get

$$r = a(1 - \cos \theta)$$
$$r = a\left(1 - \frac{r^2}{4a^2}\right)$$
$$r^2 + 4ra - 4a^2 = 0$$
$$r = -2a \pm 2a\sqrt{2}.$$

The value $-2a - 2a\sqrt{2}$ has too large an absolute value to represent a point on either curve, but the value $r = -2a + 2a\sqrt{2} = 2a(\sqrt{2} - 1)$ is acceptable.

When we graph the original equations (Fig. 10.20), as we can easily do by combining the graphs in Figs. 10.12 and 10.14, we find that the curves also intersect at the origin, where $r = 0$, and at the point $(2a, \pi)$, where $r = 2a$. Why weren't these values of r revealed by simultaneous solution? The answer is simple enough: The points $(0, 0)$ and $(2a, \pi)$ are not on the curves "simultaneously" in the sense of being reached at the same value of θ on each curve. The point $(2a, \pi)$ is reached on the curve $r = a(1 - \cos \theta)$ when $\theta = \pi$. It is reached on the curve $r^2 = 4a^2 \cos \theta$ when $\theta = 0$, where it is identified not by the coordinates $(2a, \pi)$, which do not satisfy the equation, but by the coordinates $(-2a, 0)$, which do. Similarly, the origin is reached on the cardioid when $\theta = 0$ but on the curve $r^2 = 4a^2 \cos \theta$ when $\theta = \pi/2$.

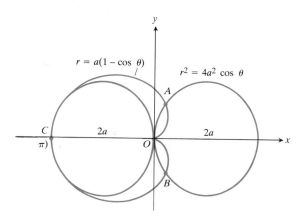

10.20 The four points of intersection of the curves $r = a(1 - \cos\theta)$ and $r^2 = 4a^2 \cos\theta$. Only two of the four (A and B) were found by simultaneous solution. The other two (O and C) were disclosed by graphing.

We conclude that the curves intersect at the four points

$$(0, 0), \qquad (2a, \pi), \qquad (r_1, \theta_1), \qquad (r_1, -\theta_1), \tag{1}$$

where

$$r_1 = (-2 + 2\sqrt{2})a, \qquad \cos\theta_1 = 1 - \frac{r_1}{a} = 3 - 2\sqrt{2}. \tag{2}$$

See Problem 26. ∎

PROBLEMS

Discuss and sketch the curves in Problems 1–8.

1. $r = a(1 + \cos\theta)$
2. $r = a(1 - \sin\theta)$
3. $r = a \sin 2\theta$
4. $r^2 = 2a^2 \cos 2\theta$
5. $r = a(2 + \sin\theta)$
6. $r = a(1 + 2 \sin\theta)$
7. $r = \theta$
8. $r = a \sin(\theta/2)$

Graph the limaçons in Problems 9–12. *Limaçon* (pronounced "leemasahn") is an Old French word for "snail." You will see why the word is appropriate when you graph the limaçons in Problem 9. Equations for limaçons have the form $r = a \pm b \cos\theta$ or $r = a \pm b \sin\theta$. There are four basic shapes.

9. *Limaçon with an inner loop*

a) $r = \dfrac{1}{2} + \cos\theta$
b) $r = \dfrac{1}{2} + \sin\theta$

10. *Cardioid*

a) $r = 1 - \cos\theta$
b) $r = -1 + \sin\theta$

11. *Dimpled limaçon*

a) $r = \dfrac{3}{2} + \cos\theta$
b) $r = \dfrac{3}{2} - \sin\theta$

12. *Convex limaçon*

a) $r = 2 + \cos\theta$
b) $r = -2 + \sin\theta$

13. Sketch the region defined by the inequality $0 \le r \le 2 - 2\cos\theta$.

14. Sketch the region defined by the inequality $0 \le r^2 \le \cos\theta$.

15. Show that the point $(2, 3\pi/4)$ lies on the curve $r = 2 \sin 2\theta$.

16. Show that the point $(1/2, 3\pi/2)$ lies on the curve $r = -\sin(\theta/3)$.

Find the points of intersection of the pairs of curves in Problems 17–24.

17. $r = a(1 + \cos\theta), \quad r = a(1 - \sin\theta)$
18. $r^2 = \sin 2\theta, \quad r^2 = \cos 2\theta, \quad a > 0$
19. $r = 1 - \cos\theta, \quad r^2 = \cos\theta$
20. $r^2 = \sin\theta, \quad r^2 = \cos\theta$
21. $r = 1, \quad r = 2 \sin 2\theta$
22. $r = a \cos 2\theta, \quad r = a \sin 2\theta, \quad a > 0$

23. $r = 1 + \cos \dfrac{\theta}{2}, \quad r = 1 - \sin \dfrac{\theta}{2}$

24. $r^2 = 2a^2 \sin 2\theta, \quad r = a, \quad a > 0$

25. Show that the equations

$$r = 1 + \cos \theta \qquad \text{and} \qquad r = -1 + \cos \theta$$

represent the same curve.

26. The simultaneous solution of the equations

$$r^2 = 4a^2 \cos \theta, \tag{3}$$

$$r = a(1 - \cos \theta), \tag{4}$$

in the text did not reveal the points $(0, 0)$ and $(2a, \pi)$ in which their graphs intersected.

a) We could have found the point $(2a, \pi)$, however, by replacing the (r, θ) in Eq. (3) by the equivalent $(-r, \theta + \pi)$, to obtain

$$r^2 = 4a^2 \cos \theta$$
$$(-r)^2 = 4a^2 \cos(\theta + \pi) \tag{5}$$
$$r^2 = -4a^2 \cos \theta.$$

Solve Eqs. (4) and (5) simultaneously to show that $(2a, \pi)$ is a common solution. (This will still not reveal that the graphs intersect at $(0, 0)$.)

b) The origin is still a special case. (It often is.) One way to handle it is the following: Set $r = 0$ in Eqs. (3) and (4) and solve each equation for a corresponding value of θ. Since $(0, \theta)$ is the origin for *any* θ, this will show that both curves pass through the origin even if they do so for different θ-values.

27. *Generating a cardioid with circles.* Show that if you roll a circle of radius a about another circle of radius a, as shown in Fig. 10.21, the original point of contact P will trace a

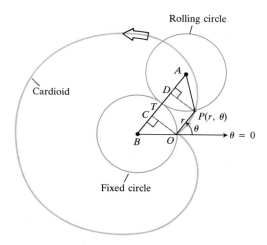

10.21 The circles in Problem 27. As the circle with center A rolls around the circle with center B, the point P traces a cardioid.

cardioid in the plane. (*Hint:* Show that angles OBC and PAD both have measure θ.)

28. COMPUTER GRAPHER If you have access to a computer grapher, you will enjoy graphing the following curves.

a) *The nephroid of Freeth:* $r = 1 + 2 \sin \dfrac{\theta}{2}$

b) *Roses:* $r = \cos m\theta$, for $m = \dfrac{1}{3}$, 2, 3, and 7

10.3
Polar Equations for Conic Sections and Other Curves

In this article, we encounter the remarkable fact that ellipses, parabolas, and hyperbolas can all be described with a single polar equation. We also find polar equations for lines and circles. Still other curves are mentioned in the problems at the end of the article.

EXAMPLE 1 *Lines.* Suppose that the perpendicular from the origin to the line L meets L at the point $N(p, \beta)$. Find a polar equation for L.

Solution We let $P(r, \theta)$ be a typical point on L (Fig. 10.22) and read the result

$$\cos(\theta - \beta) = \frac{p}{r} \tag{1}$$

or

$$r \cos(\theta - \beta) = p \tag{2}$$

from the right triangle ONP.

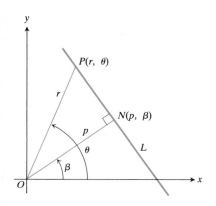

10.22 The relation $p/r = \cos(\theta - \beta)$, or $r \cos(\theta - \beta) = p$, can be read from triangle ONP in this picture.

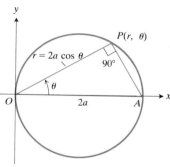

10.23 To find an equation for the circle shown above, apply the law of cosines to triangle OCP.

10.24 The circle $r = 2a \cos \theta$.

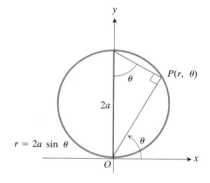

10.25 $r = 2a \sin \theta$ is the equation of a circle through the origin with its center on the positive y-axis.

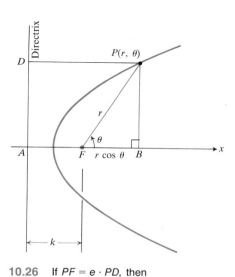

10.26 If $PF = e \cdot PD$, then

$$r = e(k + r \cos \theta).$$

When this equation is solved for r, we get $r = ke/(1 - e \cos \theta)$.

If L were perpendicular to the x-axis, β would be 0 and Eq. (2) would reduce to

$$r \cos \theta = p \qquad \text{or} \qquad x = p. \qquad \blacksquare \quad (3)$$

EXAMPLE 2 *Circles.* Find a polar equation for the circle of radius a with center at (b, β).

Solution We let $P(r, \theta)$ be a representative point on the circle and apply the law of cosines to the triangle OCP (Fig. 10.23) to obtain

$$a^2 = b^2 + r^2 - 2br \cos(\theta - \beta). \qquad (4)$$

If the circle passes through the origin, then $b = a$ and the equation takes the simpler form

$$r(r - 2a \cos(\theta - \beta)) = 0 \qquad (5)$$

or

$$r = 2a \cos(\theta - \beta). \qquad (6)$$

If $\beta = 0$, Eq. (6) reduces to

$$r = 2a \cos \theta \qquad (7)$$

(Fig. 10.24). If $\beta = 90°$, so the center of the circle lies on the y-axis (Fig. 10.25), Eq. (6) reduces to

$$r = 2a \sin \theta. \qquad \blacksquare \quad (8)$$

EXAMPLE 3 *Ellipses, parabolas, and hyperbolas.* Find a polar equation for the conic section of eccentricity e if the focus lies at the origin and the associated directrix is the line $x = -k$.

Solution We adopt the notation of Fig. 10.26 and use the focus–directrix property

$$PF = e \cdot PD \qquad (9)$$

from Article 8.4. This enables us to handle the ellipse, parabola, and hyperbola all at the same time.

By taking the origin at the focus F, we have

$$PF = r,$$

while

$$PD = AB = AF + FB = k + r \cos \theta.$$

Then Eq. (9) is the same as

$$r = e(k + r \cos \theta). \tag{10}$$

We solve this equation for r to put it in standard form:

The Polar Equation for Conic Sections

$$r = \frac{ke}{1 - e \cos \theta}. \tag{11}$$

One focus at the origin

Directrix vertical, to the left of origin

k = distance from origin to directrix

e = eccentricity   ■

Equation (11) is the standard equation for conic sections in polar coordinates. Typical cases are obtained by taking $e = 1/2$, 1, and 2:

$$e = \frac{1}{2}: \quad \text{ellipse} \qquad r = \frac{k}{2 - \cos \theta} \tag{12}$$

$$e = 1: \quad \text{parabola} \qquad r = \frac{k}{1 - \cos \theta} \tag{13}$$

$$e = 2: \quad \text{hyperbola} \qquad r = \frac{2k}{1 - 2 \cos \theta} \tag{14}$$

The denominator in Eq. (12) for the ellipse is never less than 1, so r is never greater than k. But r becomes infinite as θ approaches 0 in Eq. (13) and as θ approaches $\pi/3$ in Eq. (14).

From the diagram for the ellipse in Fig. 10.27, we see that k is related to the eccentricity e and semimajor axis a by the equation

$$k = \frac{a}{e} - ea. \tag{15}$$

From this, we find that $ke = a(1 - e^2)$. The equation for an ellipse with semimajor axis a and eccentricity e is therefore

$$r = \frac{a(1 - e^2)}{1 - e \cos \theta}. \tag{16}$$

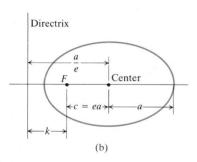

10.27 In an ellipse with semimajor axis a, the distance from focus to directrix is $k = (a/e) - ea$, so $ke = a(1 - e^2)$.

EXAMPLE 4 Find a polar equation for the ellipse with semimajor axis 39.44 astronomical units (AU) and eccentricity 0.25. This is the approximate size and shape of the orbit of Pluto around the sun. One astronomical unit is the length of the semimajor axis of the Earth's orbit.

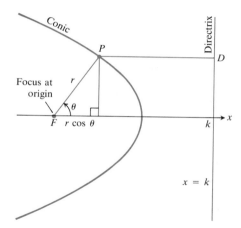

10.28 The diagram for finding the standard polar equation for a conic of eccentricity e whose focus lies at the origin and whose directrix is a vertical line k units to the right of the origin. The distance PD is $(k - r \cos \theta)$.

Solution We use Eq. (16) with $a = 39.44$ and $e = 0.25$ to find

$$r = \frac{39.44(1 - (0.25)^2)}{1 - 0.25 \cos \theta}$$

$$= \frac{147.9}{4 - \cos \theta}.$$

At its point of closest approach (perihelion), Pluto is

$$r = \frac{147.9}{4 + 1} = 29.58 \text{ AU}$$

from the sun. At its most distant point (aphelion), Pluto is

$$r = \frac{147.9}{4 - 1} = 49.3 \text{ AU}$$

from the sun.   ∎

EXAMPLE 5 Find the distance from one focus of the ellipse in Example 4 to the associated directrix.

Solution We use Eq. (15) with $a = 39.44$ and $e = 0.25$ to find

$$k = 39.44\left(\frac{1}{0.25} - 0.25\right) = 147.9 \text{ AU}.$$ ∎

If the directrix of a conic is taken to be the line $x = k$ instead of the line $x = -k$ (Fig. 10.28), the equation $PF = e \cdot PD$ takes the form

$$r = e(k - r \cos \theta), \tag{17}$$

which, upon solving for r, becomes

$$r = \frac{ke}{1 + e \cos \theta}. \tag{18}$$

Except for the sign in the denominator, this is the same as Eq. (11).

EXAMPLE 6 Find an equation for the hyperbola with eccentricity 3/2 and directrix $x = 2$.

Solution We use Eq. (18) with $k = 2$ and $e = 3/2$ to get

$$r = \frac{2(3/2)}{1 + (3/2) \cos \theta} \quad \text{or} \quad r = \frac{6}{2 + 3 \cos \theta}.$$ ∎

EXAMPLE 7 Find the directrix of the parabola

$$r = \frac{25}{10 + 10 \cos \theta}.$$

Solution We divide the numerator and denominator by 10 to put the equation in standard form:

$$r = \frac{5/2}{1 + \cos \theta}.$$

This is the equation

$$r = \frac{ke}{1 + e \cos \theta}$$

with $k = 5/2$ and $e = 1$. The equation of the directrix is $x = 5/2$. ∎

PROBLEMS

Sketch the curves in Problems 1–18. Find a Cartesian equation for each curve.

1. $r = 4 \cos \theta$ **2.** $r = 6 \sin \theta$

3. $r = -2 \cos \theta$ **4.** $r = -2 \sin \theta$

5. $r = \sin 2\theta$ **6.** $r = \sin 3\theta$

7. $r^2 = 8 \cos 2\theta$ **8.** $r^2 = 4 \sin 2\theta$

9. $r = 8(1 - 2 \cos \theta)$ **10.** $r = 1/(2 - \cos \theta)$

11. $r = 4(1 - \cos \theta)$ **12.** $r(2 - 2 \cos \theta) = 4$

13. $r(3 - 6 \cos \theta) = 12$ **14.** $r(10 - 5 \cos \theta) = 25$

15. $r(2 + \cos \theta) = 4$ **16.** $r(2 + 3 \cos \theta) = 1$

17. $r(3 + 3 \cos \theta) = 2$ **18.** $r(16 + 8 \cos \theta) = 416$

Graph the equations in Problems 19–23.

19. $r = 2 \cos\left(\theta + \dfrac{\pi}{4}\right)$ **20.** $r = 4 \csc\left(\theta - \dfrac{\pi}{6}\right)$

21. $r = 5 \sec\left(\dfrac{\pi}{3} - \theta\right)$ **22.** $r = 3 \sin\left(\theta + \dfrac{\pi}{6}\right)$

23. $r = a + a \cos\left(\theta - \dfrac{\pi}{6}\right)$

24. Sketch the region defined by the inequality $0 \le r \le 2 \cos \theta$.

Each of the graphs in Problems 25–32 is the graph of exactly one of the equations (a)–(l) in the list below. Find the equation for each graph.

a) $r = \cos 2\theta$ b) $r \cos \theta = 1$

c) $r = \dfrac{6}{1 - 2 \cos \theta}$ d) $r = \sin 2\theta$

e) $r = \theta$ f) $r^2 = \cos 2\theta$

g) $r = 1 + \cos \theta$ h) $r = 1 - \sin \theta$

i) $r = \dfrac{2}{1 - \cos \theta}$ j) $r^2 = \sin 2\theta$

k) $r = -\sin \theta$ l) $r = 2 \cos \theta + 1$

25. Four-leaved rose

26. Spiral

27. Limaçon

28. Lemniscate

29. Circle

30. Cardioid

31. Parabola

32. Lemniscate

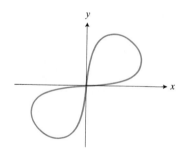

33. *Archimedes' spiral.* The graph of an equation of the form $r = a\theta$, where a is a nonzero constant, is called an Archimedes spiral. Show that such a spiral cuts any ray from the origin into congruent segments. In other words, show that the width between successive turns of the spiral remains the same.

34. *A rose within a rose.* Graph the equation $r = 1 - 2 \sin 3\theta$.

35. *Planetary orbits.* In Example 4, we found a polar equation for the orbit of Pluto. Use the data in Table 1 to find polar equations for the orbits of the other planets.

TABLE 1
Semimajor Axes and Eccentricities of the Planets in Our Solar System

Planet	Semimajor axis (astronomical units)	Eccentricity
Mercury	0.3871	0.2056
Venus	0.7233	0.0068
Earth	1.000	0.0167
Mars	1.524	0.0934
Jupiter	5.203	0.0484
Saturn	9.539	0.0543
Uranus	19.18	0.0460
Neptune	30.06	0.0082
Pluto	39.44	0.2481

36. *Perihelion and aphelion.* (See Fig. 10.29.) A planet travels about its sun in an ellipse whose semimajor axis has length a.
a) Show that $r = a(1 - e)$ when the planet is closest to the sun and that $r = a(1 + e)$ when the planet is farthest from the sun.
b) Use the data in Table 1 to find how close each planet in our solar system comes to the sun and how far away each planet gets from the sun.

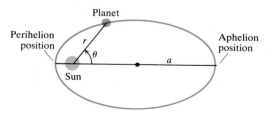

10.29 The positions of a planet closest to and farthest from its sun are called *perihelion* and *aphelion*.

37. a) Find Cartesian equations for the curves $r = 2 \sin \theta$ and $r = \csc \theta$.
b) Graph the curves in part (a) together, and label the points of intersection in both Cartesian and polar coordinates.

38. Repeat Problem 37 for the curves $r = 2 \cos \theta$ and $r = \sec \theta$.

39. The focus of a parabola lies at the origin, and its directrix is the line $r \cos \theta = -4$. Find a polar equation for the parabola.

40. Sketch the ellipse $r = 2/(2 - \cos \theta)$ and locate its center.

41. One focus of a hyperbola of eccentricity $5/4$ lies at the origin, and the corresponding directrix is the line $r \cos \theta = 9$. Find the polar coordinates of the second focus. Also find a polar equation for the hyperbola.

42. Write a polar equation for the parabola with focus at the origin and directrix the line $r \cos(\theta - \pi/2) = 2$.

	TOOLKIT PROGRAMS
	Parametric Equations Super ∗ Grapher

10.4
Integrals in Polar Coordinates

In this article, we calculate areas and lengths of curves in polar coordinates. Although the general methods for setting up the integrals are the same as for Cartesian coordinates, the resulting formulas are different.

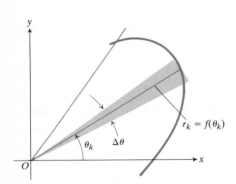

10.30 To derive a formula for the area swept out by the radius OP as P moves from A to B along the curve, we divide the area into sectors.

Area in the Plane

The region AOB in Fig. 10.30 is bounded by the rays $\theta = \alpha$, $\theta = \beta$, and the curve $r = f(\theta)$. We divide angle AOB into n parts and approximate a typical sector POQ by a *circular* sector of radius r and central angle $\Delta\theta$ (Fig. 10.31). Then

$$\text{Area of } POQ \approx \frac{1}{2}r^2\,\Delta\theta,$$

and

$$A = \text{Area } AOB \approx \sum_{\theta=\alpha}^{\beta} \frac{1}{2}r^2\,\Delta\theta.$$

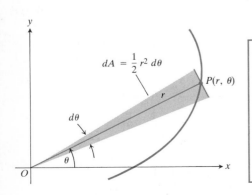

10.31 For some θ_k between θ and $\theta + \Delta\theta$, the area of the shaded circular sector just equals the area of the sector POQ bounded by the curve shown in Fig. 10.30.

If the function $r = f(\theta)$, which represents the polar curve, is a continuous function of θ for $\alpha \le \theta \le \beta$, then there is a θ_k between θ and $\theta + \Delta\theta$ such that the circular sector of radius

$$r_k = f(\theta_k)$$

and central angle $\Delta\theta$ gives the *exact* area of POQ (Fig. 10.31). Then the entire area is given exactly by

$$A = \sum \frac{1}{2}r_k^2\,\Delta\theta = \sum \frac{1}{2}(f(\theta_k))^2\,\Delta\theta.$$

If we let $\Delta\theta \to 0$, we see that

$$A = \lim_{\Delta\theta \to 0} \sum \frac{1}{2}(f(\theta_k))^2\,\Delta\theta = \int_\alpha^\beta \frac{1}{2}f^2(\theta)\,d\theta = \int_\alpha^\beta \frac{1}{2}r^2\,d\theta.$$

Area Between the Origin and $r = f(\theta)$, $\quad \alpha \le \theta \le \beta$

$$A = \int_\alpha^\beta \frac{1}{2}r^2\,d\theta. \qquad (1)$$

This is the integral of the **area differential**

$$dA = \frac{1}{2}r^2\,d\theta. \qquad (2)$$

10.32 The area differential dA.

See Fig. 10.32.

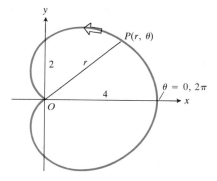

10.33   The cardioid $r = 2(1 + \cos \theta)$.

EXAMPLE 1   Find the area of the region enclosed by the cardioid $r = 2(1 + \cos \theta)$.

Solution   We graph the cardioid (Fig. 10.33) and determine that the radius OP will sweep out the region (exactly once) if we let θ run from 0 to 2π. The area is therefore

$$\int_{\theta=0}^{\theta=2\pi} \frac{1}{2} r^2 \, d\theta = \int_0^{2\pi} \frac{1}{2} \cdot 4(1 + \cos \theta)^2 \, d\theta$$

$$= \int_0^{2\pi} 2(1 + 2 \cos \theta + \cos^2 \theta) \, d\theta$$

$$= \int_0^{2\pi} \left(2 + 4 \cos \theta + 2 \frac{1 + \cos 2\theta}{2} \right) d\theta$$

$$= \int_0^{2\pi} (3 + 4 \cos \theta + \cos 2\theta) \, d\theta$$

$$= \left[3\theta + 4 \sin \theta + \frac{\sin 2\theta}{2} \right]_0^{2\pi}$$

$$= 6\pi - 0$$

$$= 6\pi. \qquad \blacksquare$$

EXAMPLE 2   Find the area inside the smaller loop of the limaçon

$$r = 2 \cos \theta + 1.$$

Solution   After sketching the curve (Fig. 10.34), we see that the smaller loop is traced out by the point (r, θ) as θ increases from $\theta = 2\pi/3$ to $\theta = 4\pi/3$. Since the curve is symmetric about the x-axis (the equation is unaltered when we replace θ by $-\theta$), we may calculate the area of the shaded half of the inner loop by integrating from $\theta = 2\pi/3$ to $\theta = \pi$. The area A we seek will be twice the value of the resulting integral:

$$A = 2 \int_{2\pi/3}^{\pi} \frac{1}{2} r^2 \, d\theta = \int_{2\pi/3}^{\pi} r^2 \, d\theta.$$

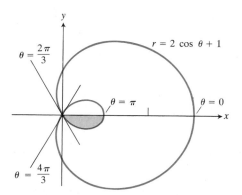

10.34   The limacon in Example 2.

Since

$$r^2 = (2 \cos \theta + 1)^2$$

$$= 4 \cos^2\theta + 4 \cos \theta + 1 = 4 \cdot \frac{1 + \cos 2\theta}{2} + 4 \cos \theta + 1$$

$$= 2 + 2 \cos 2\theta + 4 \cos \theta + 1 = 3 + 2 \cos 2\theta + 4 \cos \theta,$$

we have

$$A = \int_{2\pi/3}^{\pi} (3 + 2 \cos 2\theta + 4 \cos \theta)\, d\theta$$

$$= \left[3\theta + \sin 2\theta + 4 \sin \theta \right]_{2\pi/3}^{\pi}$$

$$= (3\pi) - \left(2\pi - \frac{\sqrt{3}}{2} + 4 \cdot \frac{\sqrt{3}}{2} \right)$$

$$= \pi - \frac{3\sqrt{3}}{2}.$$

Areas Between Curves

To find the area of a region like the one in Fig. 10.35, which lies between two polar curves from $\theta = \alpha$ to $\theta = \beta$, we subtract the integral of $(1/2)r_1^2\, d\theta$ from the integral of $(1/2)r_2^2\, d\theta$. This leads to the following formula.

Area of the Region $r_1(\theta) \leq r \leq r_2(\theta)$, $\alpha \leq \theta \leq \beta$

$$A = \int_{\alpha}^{\beta} \frac{1}{2} r_2^2\, d\theta - \int_{\alpha}^{\beta} \frac{1}{2} r_1^2\, d\theta = \int_{\alpha}^{\beta} \frac{1}{2} (r_2^2 - r_1^2)\, d\theta \qquad (3)$$

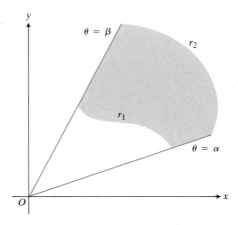

10.35 The area of the shaded region is calculated by subtracting the area of the region between r_1 and the origin from the area of the region between r_2 and the origin.

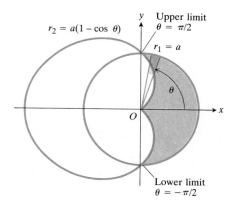

y
Upper limit
$\theta = \pi/2$

$r_2 = a(1 - \cos\ \theta)$

$r_1 = a$

θ

O

x

Lower limit
$\theta = -\pi/2$

10.36 The region and limits of integration in Example 3.

EXAMPLE 3 Find the area of the region that lies inside the circle $r = 1$ and outside the cardioid $r = 1 - \cos \theta$.

Solution We sketch the region to determine its boundaries and find the limits of integration (Fig. 10.36). The outer curve is $r_2 = 1$, the inner curve is $r_1 = 1 - \cos \theta$, and θ runs from $-\pi/2$ to $\pi/2$. The area, from Eq. (3), is

$$A = \int_{-\pi/2}^{\pi/2} \frac{1}{2}(r_2^2 - r_1^2)\ d\theta$$

$$= \frac{1}{2}\int_{-\pi/2}^{\pi/2} ((1)^2 - (1 - \cos\ \theta)^2)\ d\theta$$

$$= \frac{1}{2}\int_{-\pi/2}^{\pi/2} (1 - (1 - 2 \cos\ \theta + \cos^2\theta))\ d\theta$$

$$= \frac{1}{2}\int_{-\pi/2}^{\pi/2} (2 \cos\ \theta - \cos^2\theta)\ d\theta$$

$$= \frac{1}{2}\int_{-\pi/2}^{\pi/2} \left(2 \cos\ \theta - \frac{\cos 2\theta + 1}{2}\right) d\theta$$

$$= \frac{1}{2}\left[2 \sin\ \theta - \frac{\sin 2\theta}{4} - \frac{\theta}{2}\right]_{-\pi/2}^{\pi/2}$$

$$= \frac{1}{2}\left(2 - \frac{\pi}{4}\right) - \frac{1}{2}\left(-2 + \frac{\pi}{4}\right)$$

$$= 1 - \frac{\pi}{8} + 1 - \frac{\pi}{8}$$

$$= 2 - \frac{\pi}{4}. \qquad \blacksquare$$

Arc Length

We obtain an expression for the arc length differential ds by squaring and adding the differentials

$$dx = d(r \cos\ \theta) = -r \sin\ \theta\ d\theta + \cos\ \theta\ dr,$$
$$dy = d(r \sin\ \theta) = r \cos\ \theta\ d\theta + \sin\ \theta\ dr.$$

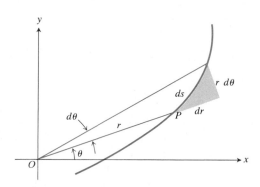

10.37   For arc length, $ds^2 = r^2\, d\theta^2 + dr^2$.

After the arithmetic is done we find that

$$ds^2 = dx^2 + dy^2 = r^2\, d\theta^2 + dr^2. \tag{4}$$

Think of $r\, d\theta$, dr, and ds as the sides and hypotenuse of the ''right triangle'' in Fig. 10.37.

To Find the Length of a Curve in Polar Coordinates:

1. Find ds from the equation

$$ds^2 = r^2\, d\theta^2 + dr^2. \tag{5}$$

2. Integrate between appropriate limits on θ.

EXAMPLE 4   Find the length of the cardioid $r = a(1 - \cos\theta)$.

Solution   We graph the cardioid (Fig. 10.38) to determine the limits of integration. As θ goes from 0 to 2π, we start at the origin, trace the cardioid once counterclockwise, and return to the origin.

To calculate the length of the curve we substitute

$$r = a(1 - \cos\theta), \qquad dr = a\sin\theta\, d\theta,$$

into Eq. (5) to obtain

$$\begin{aligned}
ds^2 &= r^2\, d\theta^2 + dr^2 \\
&= a^2(1 - \cos\theta)^2\, d\theta^2 + a^2\sin^2\theta\, d\theta^2 \\
&= 2a^2\, d\theta^2(1 - \cos\theta)
\end{aligned}$$

and

$$ds = a\sqrt{2}\sqrt{1 - \cos\theta}\; d\theta.$$

To integrate the expression on the right, we substitute

$$1 - \cos\theta = 2\sin^2\frac{\theta}{2}$$

so that

$$ds = 2a\left|\sin\frac{\theta}{2}\right|\, d\theta.$$

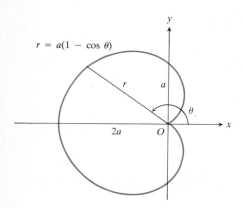

$r = a(1 - \cos\theta)$

10.38   Example 4 calculates the length of this cardioid.

Since $\sin \theta/2 \geq 0$ for $0 \leq \theta \leq 2\pi$,

$$\text{Length of cardioid} = \int_0^{2\pi} 2a \sin \frac{\theta}{2} \, d\theta = -4a \cos \frac{\theta}{2} \Big]_0^{2\pi} = 8a.$$

If we wish to take advantage of the symmetry of the cardioid, we can calculate the length of the upper portion from $\theta = 0$ to $\theta = \pi$ and double the result:

Length of upper half: $\quad \int_0^{\pi} 2a \sin \frac{\theta}{2} \, d\theta = -4a \cos \frac{\theta}{2} \Big]_0^{\pi} = 4a,$

Length of cardioid: $\quad 2(4a) = 8a.$ ∎

Surface Area

The formula for the area of a surface of revolution is

$$S = \int 2\pi \rho \, ds, \tag{6}$$

just as it is in rectangular coordinates, but now we express the integral in terms of r and θ instead of x and y. The next example shows how this is done.

EXAMPLE 5 Find the area of the surface generated by revolving the right-hand loop of the lemniscate $r^2 = 2a^2 \cos 2\theta$ about the y-axis.

Solution We sketch the curve to determine the limits of integration and the radius ρ of a typical arc of length ds (Fig. 10.39). The curve runs from $\theta = -\pi/4$ to $\theta = \pi/4$, and $\rho = x = r \cos \theta$. Hence,

$$2\pi \rho \, ds = 2\pi r \cos \theta \sqrt{dr^2 + r^2 \, d\theta^2} = 2\pi \cos \theta \sqrt{r^2 \, dr^2 + r^4 \, d\theta^2}.$$

From the equation for the curve we get

$$r \, dr = -2a^2 \sin 2\theta \, d\theta,$$

so

$$r^2 \, dr^2 + r^4 \, d\theta^2 = (2a^2 \, d\theta)^2 (\sin^2 2\theta + \cos^2 2\theta)$$

and

$$2\pi \rho \, ds = 4\pi a^2 \cos \theta \, d\theta.$$

The area of the surface is

$$\int_{-\pi/4}^{\pi/4} 4\pi a^2 \cos \theta \, d\theta = 4\pi a^2 \sin \theta \Big]_{-\pi/4}^{\pi/4}$$

$$= 4\pi a^2 \left(\frac{\sqrt{2}}{2} + \frac{\sqrt{2}}{2} \right) = 4\pi a^2 \sqrt{2}. \quad ∎$$

(a)

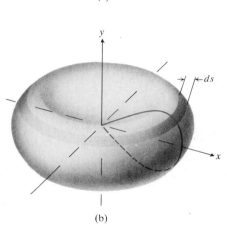

(b)

10.39 The right-hand half of a lemniscate (a) is revolved about the y-axis to generate a surface (b), whose area is calculated in Example 5.

PROBLEMS

Find the areas of the regions described in Problems 1–20. The letter a, when it appears, stands for a positive constant.

1. Inside the circle $r = \cos \theta$ between the rays $\theta = 0$ and $\theta = \pi/4$ in the first quadrant

2. Shared by the circle $r = a$ and the cardioid $r = a(1 - \cos \theta)$

3. Inside the limaçon $r = 4 + 2 \cos \theta$

4. Inside the cardioid $r = a(1 + \cos \theta)$

5. Inside the circle $r = 2a \sin \theta$

6. Inside the lemniscate $r^2 = 2a^2 \cos 2\theta$

7. Inside the lemniscate $r^2 = 2a^2 \cos 2\theta$ but not included in the circle $r = a$

8. Inside the curve $r = a(2 + \cos \theta)$

9. Common to the circles $r = 2a \cos \theta$ and $r = 2a \sin \theta$

10. Inside the circle $r = 3a \cos \theta$ but outside the cardioid $r = a(1 + \cos \theta)$

11. Inside the circle $r = -2 \cos \theta$ and outside the circle $r = 1$

12. Shared by the circles $r = a$ and $r = 2a \sin \theta$

13. Shared by the cardioids $r = a(1 + \cos \theta)$ and $r = a(1 - \cos \theta)$

14. Inside one leaf of the rose $r = \cos 2\theta$

15. Inside one loop of the lemniscate $r^2 = 4 \sin 2\theta$

16. Inside the curve $r = -4 + 2 \cos \theta$

17. Inside the six-leaved rose $r^2 = 2a^2 \sin 3\theta$

18. a) Inside the large loop of the limaçon of Example 2
    b) Inside the large loop of the limaçon but outside the small loop

19. Enclosed by the rays $\theta = 0$ and $\theta = \pi$ and the curve $r = \sqrt{\theta} e^\theta$, $0 \le \theta \le \pi$

20. Inside the circle $r = 6$ and above the line $r \sin \theta = 3$

21. As usual, when we have a new formula, it is a good idea to try it out on familiar objects to be sure it gives the results we want. Use the formula $ds^2 = r^2 \, d\theta^2 + dr^2$ from Eq. (5) to calculate the circumferences of the following circles:
    a) $r = a$,   b) $r = a \cos \theta$,   c) $r = a \sin \theta$.

22. Find the length of the cardioid $r = a(1 + \cos \theta)$. (*Hint:* $\int \sqrt{1 + \cos \theta} \, d\theta = \int \sqrt{2} |\cos(\theta/2)| \, d\theta$.)

23. Find the length of the curve $r = a \sin^2(\theta/2)$ from $\theta = 0$ to $\theta = \pi$.

24. Find the length of the parabolic spiral $r = a\theta^2$ between $\theta = 0$ and $\theta = \pi$.

25. Find the length of the curve $r = a \sin^3(\theta/3)$ between $\theta = 0$ and $\theta = 3\pi$.

26. The equations $r = e^{2t}$, $\theta = 3t$, $0 \le t \le \pi/6$ define a curve in the polar coordinate plane.
    a) Find the area of the region bounded by the curve in the first quadrant.
    b) Find the length of the curve.

27. Find the area of the surface generated by revolving the lemniscate $r^2 = 2a^2 \cos 2\theta$ about the x-axis.

28. Find the area of the surface generated by revolving the circle $r = 2a \cos \theta$ about the y-axis.

29. Find the area of the surface generated by revolving the first-quadrant portion of the cardioid $r = 1 + \cos \theta$ about the x-axis. (*Hint:* Use the identities $1 + \cos \theta = 2 \cos^2(\theta/2)$ and $\sin \theta = 2 \sin(\theta/2)\cos(\theta/2)$ to simplify the integral.)

30. *Average value.* The average value of r over the curve $r = f(\theta)$, $\alpha \le \theta \le \beta$, with respect to θ is the value of the integral

$$r_{av} = \frac{1}{\beta - \alpha} \int_\alpha^\beta f(\theta) \, d\theta.$$

Find the average value of r with respect to θ over
    a) the cardioid $r = a(1 - \cos \theta)$,
    b) the circle $r = a$,
    c) the circle $r = a \cos \theta$, $-\pi/2 \le \theta \le \pi/2$.

TOOLKIT PROGRAMS

Parametric Equations   Super ∗ Grapher

REVIEW QUESTIONS AND EXERCISES

1. Make a diagram to show the standard relations between Cartesian coordinates (x, y) and polar coordinates (r, θ). Express each set of coordinates in terms of the other kind.

2. If a point has polar coordinates (r_1, θ_1), what other polar coordinates does the point have?

3. How do you test the graph of the equation $F(r, \theta) = 0$ for symmetry about the origin? About the x-axis? About the y-axis? Give examples.

4. Describe a technique for graphing the equation $r = f(\theta)$ that involves the Cartesian $r\theta$-plane. Give an example.

5. What are the standard polar coordinate equations for lines and circles?

6. Discuss the equation $r = ke/(1 - e \cos \theta)$.

7. *A satellite orbit.* A satellite is in an orbit that passes over the North and South Poles of the earth. When it is over the North Pole it is at the highest point of its orbit, 1000 miles above the earth's surface. Above the South Pole it is at the lowest point of its orbit, 300 miles above the earth's surface.

   a) Assuming that the orbit (with reference to the earth) is an ellipse with one focus at the center of the earth, find its

eccentricity. (Take the diameter of the earth to be 8000 miles.)

   b) Using the north-south axis of the earth as the *x*-axis and the center of the earth as origin, find a polar equation for the orbit.

8. How are area and arc length calculated in polar coordinates? Give examples.

MISCELLANEOUS PROBLEMS

Sketch the curves in Problems 1–14 (where a is a positive constant). When you can, identify the curves.

1. $r = a\theta$

2. $r = a(1 + \cos 2\theta)$

3. a) $r = a \sec \theta$
   b) $r = a \csc \theta$
   c) $r = a \sec \theta + a \csc \theta$

4. $r = a \sin\left(\theta + \dfrac{\pi}{3}\right)$

5. $r^2 + 2r(\cos \theta + \sin \theta) = 7$

6. $r = a \cos \theta - a \sin \theta$

7. $r \cos(\theta/2) = a$

8. $r^2 = a^2 \sin \theta$

9. $r^2 = 2a^2 \sin 2\theta$

10. $r = a(1 - 2 \sin 3\theta)$

11. a) $r = \cos 2\theta$
    b) $r^2 = \cos 2\theta$

12. a) $r = 1 + \cos \theta$
    b) $r = \dfrac{1}{1 + \cos \theta}$

13. a) $r = \dfrac{2}{1 - \cos \theta}$
    b) $r = \dfrac{2}{1 + \sin \theta}$

14. a) $r = \dfrac{1}{2 + \cos \theta}$
    b) $r = \dfrac{1}{2 + \sin \theta}$

Find the points of intersection of the curves in Problems 15–22. In each case, a is a positive constant.

15. $r = a \cos \theta, \quad r = a \sin \theta$

16. $r = a, \quad r = 2a \sin \theta$

17. $r = a, \quad r = a(1 - \sin \theta)$

18. $r = a \sec \theta, \quad r = 2a \sin \theta$

19. $r = a \cos \theta, \quad r = a(1 + \cos \theta)$

20. $r = a(1 + \sin \theta), \quad r = 2a \sin \theta$

21. $r = a(1 + \cos 2\theta), \quad r = a \cos 2\theta$

22. $r^2 = 4 \cos 2\theta, \quad r^2 = \sec 2\theta$

Find Cartesian equations for the conic sections in Problems 23–26.

23. a) $r = \dfrac{1}{1 - \cos \theta}$

    b) $r = \dfrac{1}{1 + \cos \theta}$

24. a) $r = \dfrac{1}{1 - \sin \theta}$

    b) $r = \dfrac{1}{1 + \sin \theta}$

25. a) $r = \dfrac{1}{1 - 2 \cos \theta}$

    b) $r = \dfrac{1}{1 + 2 \cos \theta}$

26. a) $r = \dfrac{2}{2 - \cos \theta}$

    b) $r = \dfrac{2}{2 + \cos \theta}$

27. Find a polar equation for the parabola with focus at the origin and vertex at the point $(r, \theta) = (1, 0)$.

28. Find a polar equation for the line with intercepts a and b on the rays $\theta = 0$, $\theta = \pi/2$.

29. Find a polar equation for the circle of radius a with center on the ray $\theta = \pi$ and passing through the origin.

30. Find a polar equation for the parabola with focus at the origin and vertex at $(a, \pi/4)$.

31. Find a polar equation for the ellipse with foci at the origin and $(2, 0)$ and one vertex at $(4, 0)$.

32. Find a polar equation for the hyperbola with one focus at the origin, center at $(2, \pi/2)$, and vertex at $(1, \pi/2)$.

Find the total area enclosed by the curves in Problems 33–38. In each case, a is a positive constant.

33. $r^2 = a^2 \cos 2\theta$

34. $r = a(2 - \cos \theta)$

35. $r = a(1 + \cos 2\theta)$

36. $r = 2a \sin 3\theta$

37. $r^2 = 2a^2 \sin 3\theta$

38. $r^2 = 2a^2 \cos^2(\theta/2)$

39. Find the area of the region that lies inside the cardioid $r = a(1 + \sin \theta)$ and outside the circle $r = a \sin \theta$.

40. Find the area of the region that lies inside the curve $r = 2a \cos 2\theta$ and outside the curve $r = a\sqrt{2}$.

41. Sketch the regions bounded by the curves
$$r = 2a \cos^2(\theta/2)$$
and
$$r = 2a \sin^2(\theta/2)$$
and find the area of the portion of the plane they have in common.

42. Find the area of the region that lies inside the lemniscate $r^2 = 2a^2 \cos 2\theta$ and outside the circle $r = a$.

43. If $r = a \cos^3(\theta/3)$, show that $ds = a \cos^2(\theta/3) \, d\theta$ and find the perimeter of the curve.

44. Find the area of the surface generated by revolving the cardioid $r = a(1 - \cos \theta)$ about the x-axis. (*Hint:* Use the identities $\sin \theta = 2 \sin(\theta/2)\cos(\theta/2)$ and $1 - \cos \theta = 2 \sin^2(\theta/2)$ to simplify the integral.)

The Angle Between the Radius Vector and the Tangent Line

In Cartesian coordinates, when we want to discuss the direction of a curve at a point, we use the angle ϕ measured counterclockwise from the positive x-axis to the tangent line. In polar coordinates, it is more convenient to calculate the angle ψ from the *radius vector* to the tangent line (Fig. 10.40). The angle ϕ can

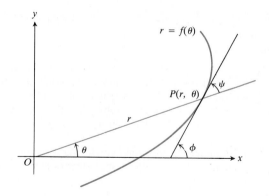

10.40 The angle ψ between the tangent line and the radius vector.

then be calculated from the relation

$$\phi = \theta + \psi, \tag{1}$$

which comes from applying the exterior angle theorem to the triangle in Fig. 10.40.

Suppose the equation of the curve is given in the form $r = f(\theta)$, where $f(\theta)$ is a differentiable function of θ. Then

$$x = r \cos \theta \quad \text{and} \quad y = r \sin \theta \tag{2}$$

are differentiable functions of θ with

$$\frac{dx}{d\theta} = -r \sin \theta + \cos \theta \, \frac{dr}{d\theta},$$

$$\frac{dy}{d\theta} = r \cos \theta + \sin \theta \, \frac{dr}{d\theta}. \tag{3}$$

Since $\psi = \phi - \theta$ from (1),

$$\tan \psi = \tan(\phi - \theta) = \frac{\tan \phi - \tan \theta}{1 + \tan \phi \tan \theta}.$$

Furthermore,

$$\tan \phi = \frac{dy}{dx} = \frac{dy/d\theta}{dx/d\theta}$$

because $\tan \phi$ is the slope of the curve at P. Also,

$$\tan \theta = \frac{y}{x}.$$

Hence

$$\tan \psi = \frac{\dfrac{dy/d\theta}{dx/d\theta} - \dfrac{y}{x}}{1 + \dfrac{y}{x}\dfrac{dy/d\theta}{dx/d\theta}}$$

$$= \frac{x\dfrac{dy}{d\theta} - y\dfrac{dx}{d\theta}}{x\dfrac{dx}{d\theta} + y\dfrac{dy}{d\theta}}. \tag{4}$$

The numerator in the last expression in Eq. (4) is found from Eqs. (2) and (3) to be

$$x\frac{dy}{d\theta} - y\frac{dx}{d\theta} = r^2.$$

Similarly, the denominator is

$$x\frac{dx}{d\theta} + y\frac{dy}{d\theta} = r\frac{dr}{d\theta}.$$

When we substitute these into Eq. (4), we obtain

$$\tan\psi = \frac{r}{dr/d\theta}. \qquad (5)$$

This is the equation we use for finding ψ.

45. Show, by reference to a figure, that the angle β between the tangents to two curves at a point of intersection may be found from the formula

$$\tan\beta = \frac{\tan\psi_2 - \tan\psi_1}{1 + \tan\psi_2 \tan\psi_1}. \qquad (6)$$

When will the two curves intersect at right angles?

46. Find the value of $\tan\psi$ for the curve $r = \sin^4(\theta/4)$.

47. Find the angle between the curve $r = 2a\sin 3\theta$ and its tangent when $\theta = \pi/3$.

48. For the hyperbolic spiral $r\theta = a$ show that $\psi = 3\pi/4$ when $\theta = 1$ radian, and that $\psi \to \pi/2$ as the spiral winds around the origin. Sketch the curve and indicate ψ for $\theta = 1$ radian.

49. The circles $r = \sqrt{3}\cos\theta$ and $r = \sin\theta$ intersect at the point $(\sqrt{3}/2, \pi/3)$. Show that their tangents are perpendicular there.

50. Sketch the cardioid $r = a(1 + \cos\theta)$ and circle $r = 3a\cos\theta$ in one diagram and find the angle between their tangents at the point of intersection that lies in the first quadrant.

51. Find the points of intersection of the parabolas

$$r = \frac{1}{1 - \cos\theta}$$

and

$$r = \frac{3}{1 + \cos\theta}$$

and the angles between their tangents at these points.

52. Find points on the cardioid $r = a(1 + \cos\theta)$ where the tangent line is (a) horizontal, (b) vertical.

53. Show that the parabolas $r = a/(1 + \cos\theta)$ and $r = b/(1 - \cos\theta)$ are orthogonal at each point of intersection $(ab \neq 0)$.

54. Find the angle at which the cardioid $r = a(1 - \cos\theta)$ crosses the ray $\theta = \pi/2$.

55. Find the angle between the line $r = 3\sec\theta$ and the cardioid $r = 4(1 + \cos\theta)$ at one of their intersections.

56. Find the slope of the tangent line to the curve $r = a\tan(\theta/2)$ at $\theta = \pi/2$.

57. Find the angle at which the parabolas $r = 1/(1 - \cos\theta)$ and $r = 1/(1 - \sin\theta)$ intersect in the first quadrant.

58. The equation $r^2 = 2\csc 2\theta$ represents a curve in polar coordinates.
a) Sketch the curve.
b) Find an equation for the curve in rectangular coordinates.
c) Find the angle at which the curve intersects the ray $\theta = \pi/4$.

59. Suppose that the angle ψ from the radius vector to the tangent line of the curve $r = f(\theta)$ has the constant value α.
a) Show that the area bounded by the curve and two rays $\theta = \theta_1$, $\theta = \theta_2$, is proportional to $r_2^2 - r_1^2$, where (r_1, θ_1) and (r_2, θ_2) are polar coordinates of the ends of the arc of the curve between these rays. Find the factor of proportionality.
b) Show that the length of the arc of the curve in part (a) is proportional to $r_2 - r_1$ and find the proportionality constant.

60. Let P be a point on the hyperbola $r^2\sin 2\theta = 2a^2$. Show that the triangle formed by OP, the tangent at P, and the initial line is isosceles.

Centers of Mass

Since the center of mass of a triangle is located on each median, two-thirds of the way from the vertex to the opposite base, the lever arm for the moment about the x-axis of the thin triangular region in Fig. 10.41 is about $(2/3)r\sin\theta$. Similarly, the lever

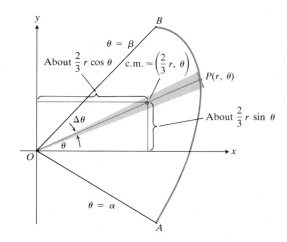

10.41 The moment of the thin triangular sector about the x-axis is approximately

$$\frac{2}{3}r\sin\theta\, dA = \frac{2}{3}r\sin\theta \cdot \frac{1}{2}r^2 d\theta = \frac{1}{3}r^3\sin\theta\, d\theta.$$

arm for the moment of the region about the y-axis is about $(2/3)r \cos \theta$. These approximations improve as $\Delta\theta \to 0$ and lead to the following formulas for the coordinates of the center of mass of the region AOB:

$$\bar{x} = \frac{\displaystyle\int \frac{2}{3}r \cos \theta \cdot \frac{1}{2}r^2 \, d\theta}{\displaystyle\int \frac{1}{2}r^2 \, d\theta} = \frac{\displaystyle\frac{2}{3}\int r^3 \cos \theta \, d\theta}{\displaystyle\int r^2 \, d\theta},$$

$$\bar{y} = \frac{\displaystyle\int \frac{2}{3}r \sin \theta \cdot \frac{1}{2}r^2 \, d\theta}{\displaystyle\int \frac{1}{2}r^2 \, d\theta} = \frac{\displaystyle\frac{2}{3}\int r^3 \sin \theta \, d\theta}{\displaystyle\int r^2 \, d\theta},$$

with limits $\theta = \alpha$ to $\theta = \beta$ on all integrals.

61. Find the center of mass of the region enclosed by the cardioid $r = a(1 + \cos \theta)$.

62. Find the center of mass of the region enclosed by a semicircle of radius a.

63. Find the center of mass of a thin uniform wire bent into the shape of the cardioid $r = a(1 + \cos \theta)$. (*Hint:* You can evaluate $\int \cos \theta \cos(\theta/2) \, d\theta$ by substituting $\cos \theta = 1 - 2 \sin^2(\theta/2)$ and then letting $u = \sin(\theta/2)$.)

CHAPTER 11

Infinite Sequences and Infinite Series

OVERVIEW

In this chapter and the next we deal with two related topics: infinite sequences and infinite series. Because series involve sequences, we begin with the study of sequences; and since the simplest sequences are sequences of constants, we begin with them. In Chapter 12 we shall examine sequences and series in which each term is a function.

Here and in more advanced mathematics, the word *series* always implies an infinite number of terms to be combined by adding in a definite order. The fact that we cannot really *add* an infinite number of terms leads to the idea of convergence to a limit—do the finite sums we *can* compute approach a limit? If so, what is that limit? A repertory of ways to handle these questions is useful, and we use both differentiation and integration to develop appropriate tests for convergence (or its opposite, divergence).

Sequences and series arise in many areas of physical science, computer science, and higher mathematics. Because series can be used to approximate the values of many functions to any desired accuracy, they are commonly used to prepare tables of values of trigonometric, exponential, and logarithmic functions. Solutions of many differential equations are best approximated by using a finite number of terms of the infinite series that give the exact solutions. And sometimes we have no choice but to use some method of successive approximations to find the ''solution'' of an equation like Kepler's equation for determining the position of a planet or a satellite in its orbit. Often, these successive approximations are the early terms of a sequence of numbers generated by an appropriate computer program.

Many applications of sequences and series are beyond the scope of this book, but the material in these chapters can prepare you for later study as well as provide techniques that you can use as needed.

11.1
Sequences of Numbers

Informally, a sequence is a collection of numbers in a particular order. How do sequences arise? What can we do with them? We have already seen some applications: for example, in using Newton's method for finding solutions of equations $f(x) = 0$ (Article 2.9), in approximating a definite integral by use of the Trapezoidal Rule or Simpson's rule (Article 4.9), and in earlier courses in mathematics. In geometry, the area enclosed by a circle of radius R is defined as the limit of areas of inscribed or circumscribed regular polygons having n sides, as n tends to infinity. Those areas form sequences of numbers $A_3, A_4, A_5,$ . . . , where A_n is the area of a regular polygon of n sides.

We are also familiar with many sequences from arithmetic, such as the sequence of positive integers $1, 2, 3, 4,$ . . . ; the sequence of even integers $2, 4, 6, 8,$ . . . , the sequence of prime numbers $2, 3, 5, 7, 11, 13,$ . . . ; and the sequence of squares $1, 4, 9, 16,$ . . . . Those particular sequences, however, do not approach finite limits—and sequences that do approach finite limits are of primary interest in this chapter. For instance, the sequence of reciprocals $1, 1/2, 1/3, 1/4, 1/5,$ . . . , $1/n,$ . . . , is a sequence that approaches a limit as $n \to \infty$; that limit is 0. However, before investigating further the properties of sequences that approach finite limits, let us proceed with a more precise definition of sequence.

DEFINITION 1

A **sequence** of numbers is a function whose domain is the set of positive integers.

Sequences are defined by rules the way other functions are, some typical rules being

$$a(n) = n - 1, \qquad a(n) = 1 - \frac{1}{n}, \qquad a(n) = \frac{\ln n}{n^2}. \tag{1}$$

To signal the fact that the domains are restricted to the set of positive integers, it is conventional to use a letter like n from the middle of the alphabet for the independent variable, instead of the x, y, z, and t used so widely in other contexts. The formulas in the defining rules, however, like the ones above, are often valid for domains much larger than the set of positive integers. This can prove to be an advantage, as we shall see later.

The number $a(n)$ is called the nth term of the sequence, or the term with **index** n. For example, if $a(n) = (n - 1)/n$, then the terms are

First term	Second term	Third term		nth term
$a(1) = 0,$	$a(2) = \dfrac{1}{2},$	$a(3) = \dfrac{2}{3},$	. . . ,	$a(n) = \dfrac{n-1}{n}.$

$$(2)$$

When we use the simpler notation a_n for $a(n)$, the sequence in (2) becomes

$$a_1 = 0, \qquad a_2 = \frac{1}{2}, \qquad a_3 = \frac{2}{3}, \qquad . . . , \qquad a_n = \frac{n-1}{n}. \tag{3}$$

To describe sequences, we often write the first few terms as well as a formula for the nth term.

EXAMPLE 1

We write	For the sequence whose defining rule is
$0, 1, 2, \ldots, n - 1, \ldots$	$a_n = n - 1$
$1, \dfrac{1}{2}, \dfrac{1}{3}, \ldots, \dfrac{1}{n}, \ldots$	$a_n = \dfrac{1}{n}$
$1, -\dfrac{1}{2}, \dfrac{1}{3}, -\dfrac{1}{4}, \ldots, (-1)^{n+1}\dfrac{1}{n}, \ldots$	$a_n = (-1)^{n+1}\dfrac{1}{n}$
$0, \dfrac{1}{2}, \dfrac{2}{3}, \dfrac{3}{4}, \ldots, \dfrac{n-1}{n}, \ldots$	$a_n = \dfrac{n-1}{n}$
$0, -\dfrac{1}{2}, \dfrac{2}{3}, -\dfrac{3}{4}, \ldots, (-1)^{n+1}\left(\dfrac{n-1}{n}\right), \ldots$	$a_n = (-1)^{n+1}\left(\dfrac{n-1}{n}\right)$
$3, 3, 3, \ldots, 3, \ldots$	$a_n = 3$ ∎

Notation We refer to the sequence whose nth term is a_n as "the sequence $\{a_n\}$." Here, the curly braces $\{\ \}$ indicate that we have in mind all the terms of the sequence, not just a single term. Thus, the first two sequences in Example 1 would be $\{n - 1\}$ and $\{1/n\}$ while the last would be $\{3\}$.

Graphs See Figure 11.1. Strictly speaking, the graph of a sequence $\{a_n\}$ consists of those points in the xy-plane for which $x = n$ and $y = a_n$, for $n = 1$, 2, 3, ... A less precise, but often useful, way to visualize a sequence is to think of the numbers a_n as displayed on a number line, say the x-axis. If we plot these points in order, one after the other, it may be obvious that they are approaching a limit. That would also be evident if the graph of the points (n, a_n) approaches an asymptote $y = L$ as n increases.

Convergence and Divergence

As Fig. 11.1 shows, the sequences of Example 1 exhibit different kinds of behavior. The sequences $\{1/n\}$, $\{(-1)^{n+1}(1/n)\}$, and $\{(n - 1)/n\}$ seem to approach single limiting values as n increases, and the sequence $\{3\}$ is already at a limiting value from the very first. On the other hand, terms of the sequence $\{(-1)^{n+1}(n - 1)/n\}$ seem to accumulate near two different values, -1 and 1, while the terms of $\{n - 1\}$ get larger and larger and do not accumulate anywhere.

To distinguish sequences that approach a unique limiting value L, as n increases, from those that do not, we say that they *converge*, according to the following definition.

DEFINITION 2

The sequence $\{a_n\}$ **converges** to the number L if to every positive number ϵ there corresponds an index N such that

$$|a_n - L| < \epsilon \qquad \text{for all } n > N. \tag{4}$$

If no such limit exists, we say that $\{a_n\}$ **diverges**.

The sequences of Example 1 are graphed here in two different ways: by plotting the numbers a_n on a horizontal axis and by plotting the points (n, a_n) in the coordinate plane.

The terms $a_n = n - 1$ eventually surpass every integer, so the sequence $\{a_n\}$ diverges, . . .

$$a_n = n - 1$$

. . . however, the terms $a_n = 1/n$ decrease steadily and get arbitrarily close to 0 as n increases, so the sequence $\{a_n\}$ converges to 0.

$$a_n = \frac{1}{n}$$

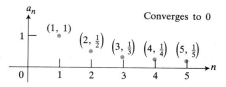

The terms $a_n = (-1)^{n+1} (1/n)$ alternate in sign but still converge to 0.

$$a_n = (-1)^{n+1} \frac{1}{n}$$

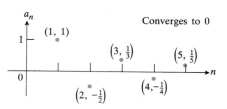

The terms $a_n = (n - 1)/n$ approach 1 steadily and get arbitrarily close as n increases, so the sequence $\{a_n\}$ converges to 1.

$$a_n = \frac{n-1}{n}$$

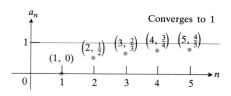

The terms $a_n = (-1)^{n+1} (n - 1)/n$ alternate in sign. The positive terms approach 1 and the negative terms approach -1 as n increases, so the sequence $\{a_n\}$ diverges.

$$a_n = (-1)^{n+1} \left(\frac{n-1}{n} \right)$$

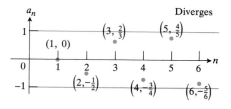

The terms in the sequence of constants $a_n = 3$ have the same value regardless of n; so we say the sequence $\{a_n\}$ converges to 3.

$$a_n = 3$$

11.2 $a_n \to L$ if L is a horizontal asymptote of $\{(n, a_n)\}$. In this figure, all the a_n's after a_N lie within ϵ of L.

In other words, $\{a_n\}$ converges to L if, for every positive ϵ, there is an index N such that all terms after the Nth lie between $L - \epsilon$ and $L + \epsilon$ or such that all but finitely many (namely the first N) terms of the sequence lie within a radius ϵ of L. (See Fig. 11.2 and look once more at the sequences in Fig. 11.1.) We indicate the fact that $\{a_n\}$ converges to L by writing

$$\lim_{n \to \infty} a_n = L \qquad \text{or} \qquad a_n \to L \qquad \text{as } n \to \infty,$$

and we call L the **limit** of the sequence $\{a_n\}$.

EXAMPLE 2 Common sense tells us that the sequence $\{1/n\}$ converges to zero. Take $L = 0$ in Definition 2 and find how large N must be to satisfy condition (4).

Solution Let $\epsilon > 0$ be given. We begin by writing down the inequality (4), with $a_n = 1/n$ and $L = 0$. This gives

$$|a_n - L| = \left| \frac{1}{n} - 0 \right| = \frac{1}{n} < \epsilon, \tag{5}$$

and therefore we seek an integer N such that

$$\frac{1}{n} < \epsilon \qquad \text{for all } n > N. \tag{6}$$

Certainly

$$\frac{1}{n} < \epsilon \qquad \text{for all } n > \frac{1}{\epsilon}, \tag{7}$$

but there is no reason to expect $1/\epsilon$ to be an integer. This minor difficulty is easily overcome: We just choose any integer $N > 1/\epsilon$. Then every index n greater than N will automatically be greater than $1/\epsilon$. In short, for this choice of N we can guarantee (6). The criterion set forth in Definition 2 for convergence to 0 is satisfied. ∎

EXAMPLE 3 If k is any number, then the constant sequence $\{k\}$ is defined by $a_n = k$ for all n. What is its limit?

Solution The limit must be k. To show this, let $\epsilon > 0$ be given. When we take both $a_n = k$ and $L = k$ on the left of the inequality in (4), we get

$$|a_n - L| = |k - k| = 0, \tag{8}$$

which is less than any positive ϵ for every $n \geq 1$. Hence, $N = 1$ will work. ∎

$$a_n = (-1)^{n+1} \left(\frac{n-1}{n} \right)$$

Neither the ϵ-interval about 1 nor the ϵ-interval about -1 contains a complete tail of the sequence.

Neither of the ϵ-bands shown here contains all the points (n, a_n) from some index onward.

11.3 The sequence $\{(-1)^{n+1}[(n-1)/n]\}$ diverges.

EXAMPLE 4 Show that the sequence $\{(-1)^{n+1}(n-1)/n\}$ diverges.

Solution Take a positive ϵ smaller than 1 so that the bands shown in Fig. 11.3 about the lines $y = 1$ and $y = -1$ do not overlap. Any $\epsilon < 1$ will do. Convergence to 1 would require every point of the graph beyond a certain index N to lie inside the upper band, but this will never happen. As soon as a point (n, a_n) lies within the upper band, every alternate point starting with $(n + 1, a_{n+1})$ will lie within the lower band. Hence the sequence cannot converge to 1. Likewise, it cannot converge to -1. On the other hand, because the terms of the sequence get increasingly close to 1 and -1 alternately, they never accumulate near any other value. Therefore, the sequence diverges.   ■

Uniqueness of a Limit

Can a sequence converge to two different limits? This question is answered by the following uniqueness theorem.

THEOREM 1

The Uniqueness Theorem

If a sequence $\{a_n\}$ converges, then its limit is unique.

Proof   If $\{a_n\}$ converged to both L and L', $L \neq L'$, we could take $\epsilon = \frac{1}{2}|L - L'|$, which would be a positive number, and then apply the definition of limit to both L and L' as follows:

$$|a_n - L| < \epsilon \qquad \text{when } n > N \tag{9a}$$

and

$$|a_n - L'| < \epsilon \qquad \text{when } n > N'. \tag{9b}$$

But there are infinitely many values of n greater than both N and N', and for any of these values of n, both (9a) and (9b) hold. Combining these equations, we get

$$
\begin{aligned}
|L - L'| &= |L - a_n + a_n - L'| \\
&\leq |L - a_n| + |a_n - L'| \\
&< 2\epsilon \\
&= |L - L'|.
\end{aligned}
$$

But this is absurd; no number is less than itself. Hence, if a sequence converges, its limit is unique.   ■

A **tail** of a sequence $\{a_n\}$ is the collection of all the terms whose indices are greater than some index N; in other words, one of the sets $\{a_n \mid n > N\}$. Another way to say $a_n \to L$ is to say that every ϵ-interval about L contains a tail of $\{a_n\}$. The convergence (or divergence) of a sequence, as well as the limit of a convergent sequence, depends only on the tail behavior of the sequence.

The behavior of the sequence $\{(-1)^{n+1}(n - 1)/n\}$ is qualitatively different from that of $\{n - 1\}$, which diverges because it outgrows every real number L. We describe the behavior of $\{n - 1\}$ by writing

$$\lim_{n \to \infty} (n - 1) = \infty.$$

In speaking of infinity as a limit of a sequence $\{a_n\}$, we do not mean that the difference between a_n and infinity becomes small as n increases. We mean that a_n becomes numerically large as n increases.

Sequences Defined Recursively

In the foregoing examples we had explicit formulas to give a_n directly in terms of n. But sequences are often produced another way—recursively. By this we mean that the terms are generated by a process that has these two features:

1. the first term is given (or first few terms are given), and

2. a rule is given by which any later term can be calculated from those terms that precede it.

We say that the sequence is defined **recursively**, and the rule number 2 is called a **recursion formula.**

The simplest example might be to say

$$a_1 = 1 \qquad \text{and} \qquad a_{n+1} = a_n + 1.$$

This recursion process would produce the sequence of positive integers 1, 2, 3, ... Another simple example might be

$$x_1 = 1 \qquad \text{and} \qquad x_{n+1} = x_n + (2n + 1),$$

which produces the sequence of squares: 1, 4, 9, 16, 25, ... In both examples, we have two ways to describe the sequences: by the recursion formulas or by the explicit formulas $\{a_n\} = \{n\}$ and $\{x_n\} = \{n^2\}$, respectively. However, we often do not have anything except the recursion process to define a particular sequence, as, for example, when we use Newton's method to find solutions of an equation $f(x) = 0$. (See Article 2.9.) When both an explicit formula and a recursion formula are available for a given sequence, the actual calculations of successive terms may still be easier if done recursively. The sequence of factorials 1, 2, 6, 24, . . . , $n!$, . . . , is one such example. Here we would probably rely on the formula $(n + 1)! = (n + 1)n!$ to go from one term to the next, rather than always beginning at 1 and multiplying all the numbers from 1 to n to get $n!$. (If we wanted just one particular factorial, say 5!, we would naturally multiply $1 \times 2 \times 3 \times 4 \times 5$, but if we wanted several terms in the *sequence* $\{n!\}$, then we would more likely use the recursion process with $x_1 = 1$ and $x_{n+1} = (n + 1)x_n$.)

Sequences and Computer Language: Computer Loops You may already be familiar with BASIC as a computer programming language, but if not, you will still find the following examples easy to follow. For convenience, the steps in a program are usually labeled 10, 20, 30, and so on, leaving the numbers in between for corrections if we discover a flaw in our first attempt to write the program. The next example shows a possible program for listing the paired numbers $(n, 2n)$ for $n = 1, 2, 3, . . . , 10$. Our program starts with $x_1 = 2$ and uses the recursion formula $x_{n+1} = x_n + 2$.

EXAMPLE 5 Write and discuss a BASIC program for listing the ordered pairs $(n, 2n)$ for $n = 1, 2, 3, . . . , 10$, using the recursion formula $x_{n+1} = x_n + 2$.

Solution   One possible program looks like this:

PROGRAM	Comments
10 LET x = 2	We start with $x_1 = 2$.
20 FOR n = 1 TO 10	We have chosen these values of n.
30 PRINT n;x	We want a list of (n, x_n) values.
40 LET x = x + 2	The next x will be $x_{n+1} = x_n + 2$.
50 NEXT n	This step creates the loop and sends the program back to line 30, with n replaced by $n + 1$ and x_n replaced by x_{n+1}.
60 END	Tells the computer to stop. ■

In statements like $x = x + 2$ in line 40, we are suppressing the subscripts on the x's in the equation $x_{n+1} = x_n + 2$. Remember that the *new* value of x is on the *left* of the equation and the *old* value is on the *right*. (In machine language, this is like a command that says "Load the contents of x into a register. Add 2, and store the contents of the register back into x.")

It was not necessary to use a recursion formula to compute x_n in Example 5: ordinary multiplication would easily work to give $x_n = 2n$ directly. In the next example, however, that kind of alternative is not available.

EXAMPLE 6   Write and discuss a BASIC program for computing and listing n, x_n, and s_n for $n = 1$ to 12, where

$$x_1 = 1, \quad x_2 = 1, \quad x_3 = \frac{1}{2!}, \quad \ldots, \quad x_n = \frac{1}{(n-1)!}$$

and

$$s_1 = x_1, \quad s_2 = x_1 + x_2, \quad \ldots, \quad s_n = x_1 + x_2 + \cdots + x_n.$$

Solution   The equations that define x_n and s_n can be expressed in ways that are convenient for computing recursively, as follows:

$$x_1 = 1 \qquad \text{and} \qquad x_{n+1} = \frac{x_n}{n}$$

and

$$s_1 = x_1 \qquad \text{and} \qquad s_{n+1} = s_n + x_{n+1}.$$

With these equations to guide us, we write the following program.

PROGRAM	Comments
10 LET x = 1	The first term in the x-sequence.
20 LET s = x	The first term in the s-sequence.
30 FOR n = 1 to 12	We want 12 terms in each sequence.
40 PRINT n;x;s	This gives the list we want.

50 LET x = x/n   The new value of x is the old
                         value divided by n.

60 LET s = s + x   The new value of s is the old value
                         plus the new value of x.

70 NEXT n   Sends the program back to line 40
                         with new values of n, x, and s.

80 END   Computer stops after $n = 12$.

Table 11.1 shows the results we got on a small computer. The value of x_{10} is shown in scientific notation as 2.75573 E–06, which means 2.75573×10^{-6}, with similar notations for x_{11} and x_{12}.

You have probably guessed that the sequence of numbers $\{s_n\}$ is approaching the number e as a limit. We shall have more to say about that later. ■

As a final example, we illustrate how we might use a computer to apply Newton's method to estimate where the graphs of $y = \sin x$ and $y = x^2$ intersect. A first estimate is $x_1 = 1$. Some new notation will be introduced and explained.

EXAMPLE 7 See Fig. 11.4. Write and run a BASIC program based on Newton's method for estimating a root of the equation $\sin x = x^2$.

Solution The calculations to be made are as follows.

1. Start with $x_1 = 1$.

2. We want to make the function $f(x) = \sin x - x^2$ small, zero if possible, and its derivative is $f'(x) = \cos x - 2x$.

3. Given any x_n, the next value of x is

$$x_{n+1} = x_n - \frac{f(x_n)}{f'(x_n)}.$$

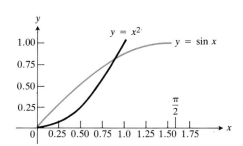

11.4 The graphs of $y = x^2$ and $y = \sin x$ intersect near $x = 1$. A better approximation is $x = 0.8767262154$.

TABLE 11.1

n	x_n	s_n
1	1	1
2	1	2
3	0.5	2.5
4	0.16666 66667	2.66666 6667
5	0.04166 66667	2.70833 3333
6	0.00833 33333	2.71666 6667
7	0.00138 88889	2.71805 5556
8	0.00019 84127	2.71825 3968
9	0.00002 48016	2.71827 8770
10	2.75573 E–06	2.71828 1526
11	2.75573 E–07	2.71828 1801
12	2.50521 E–08	2.71828 1826

(See Article 2.9.) The computer does not recognize x^2 unless we write it as $x \wedge 2$, as shown in line 20 in the program that follows. Likewise, $10 \wedge (-10)$ in line 40 is the way we write 10^{-10} in the program. The asterisk sign in line 50, $2*x$, indicates multiplication—the machine does not recognize $2x$ as 2 times x.

Now, let's look again at line 40. We don't know how many iterations may be needed to find a "sufficiently accurate" estimate of the root of $f(x) = 0$, so we arbitrarily decided to stop when a value of x is reached for which the absolute value of $f(x)$ is less than 10^{-10}. The entire command in line 40 is known as an "IF . . . THEN" conditional. If the condition *is* satisfied, the machine goes to line 80, which stops the computing; but if it is *not* satisfied, the machine proceeds to the instructions in lines 50, 60, and 70. Line 70 creates the loop that gives an additional iteration. Here is the program.

PROGRAM	Comment
10 LET x = 1	The starting value of x is 1.
20 LET f = SIN(x) − x $\wedge$ 2	The function is $f(x) = \sin x - x^2$.
30 PRINT x;f	So we can see what is going on.
40 IF ABS(f) < 10 $\wedge$ (−10) THEN 80	Conditional command discussed in text.
50 LET g = COS(x) − 2*x	We let g stand for the derivative f'.
60 LET x = x − f/g	This is the heart of the Newton method: $x_{n+1} = x_n - f(x_n)/f'(x_n)$.
70 GO TO 20	If the condition ABS(f) < 10 $\wedge$ (−10) is not yet satisfied, do it again.
80 END	The program stops only after ABS(f) < 10 $\wedge$ (−10).

In Table 11.2, we include the index n along with x_n and $f(x_n)$.

REMARK   In theory, the calculations in the table could have been continued indefinitely, thus generating a complete sequence; but in practice, in such an application, we are quite willing to stop when the degree of accuracy we asked for has been reached. Are these results an example of a *sequence*? No, if we

---

TABLE 11.2
Estimates of the root of $\sin x - x^2 = 0$.

n	x_n	$f(x_n)$
1	1	− 0.15852 90152
2	0.89139 59953	− 0.01663 71742
3	0.87698 48448	− 0.00028 81492
4	0.87672 62985	− 9.25402 E−08
5	0.87672 62154	− 4.2 E−13*

*An unexpected result: more accuracy than we asked for.

insist on a sequence going on indefinitely. We could convert our results into such a sequence by defining the early terms x_1, x_2, x_3, x_4, x_5 to be the numbers shown in the table and then saying $x_n = x_5$ for $n > 5$ (no longer requiring the x's to be those given by Newton's method).   ■

If you have ready access to a computer and want to write short programs to explore some of the topics later in this chapter and the next, fine. But if you don't, there is no reason to worry—we will not pursue the topic of computers any further. Courses in computer science are widely available and give opportunity for hands-on experience. We have included this small discussion about computer language to show some of the differences and some of the similarities in the symbols used to represent sequences—especially sequences generated by recursion. One of the most noticeable differences in notation is the prolific use of subscripts in talking about sequences in this book and the disappearance of the subscripts in a computer program. In the computer, each reference to a variable, say x, on the right side of an equation, refers to its value just before that line of the program goes into effect. If the same variable also appears on the left side of the equation, it refers to the value that the variable will have just after that line has been completed. The corresponding mathematical notation would be x_n on the right and x_{n+1} on the left.

PROBLEMS

In Problems 1–6, write out in explicit form the first four terms of the sequence $\{a_n\}$. If the sequence converges, what is its limit?

1. $a_n = \dfrac{n+1}{2n+1}$

2. $a_n = \dfrac{2n-1}{n+1}$

3. $a_n = \dfrac{1-2n}{1+2n}$

4. $a_n = \dfrac{2^n-1}{2^n}$

5. $a_n = \dfrac{2^n}{2^{n+1}}$

6. $a_n = 1 + \dfrac{(-1)^n}{2^n}$

In Problems 7–11, write out in explicit form each of the terms x_1, x_2, x_3, x_4, x_5, x_6.

7. $x_1 = 1$, $x_{n+1} = x_n + \left(\dfrac{1}{2}\right)^n$

8. $x_1 = 1$, $x_{n+1} = \dfrac{x_n}{n+1}$

9. $x_1 = 2$, $x_{n+1} = \dfrac{x_n}{2}$

10. $x_1 = -2$, $x_{n+1} = \dfrac{n}{n+1} x_n$

11. $x_1 = x_2 = 1$, $x_{n+2} = x_{n+1} + x_n$ (This is called the Fibonacci sequence. The ratio $r_n = x_n/x_{n+1}$ gives a related sequence r_1, r_2, ...)

12. The first term of a sequence is $x_1 = 1$. Each succeeding term is the sum of all those that come before it:
$$x_{n+1} = x_1 + x_2 + \cdots + x_n.$$

Write out enough of the early terms of the sequence to deduce a general formula for x_n that holds for all $n \geq 2$.

13. A sequence of rational numbers is described as follows:
$$\frac{1}{1}, \frac{3}{2}, \frac{7}{5}, \frac{17}{12}, \ldots, \frac{a}{b}, \frac{a+2b}{a+b}, \ldots$$

Here the numerators form one sequence, the denominators form a second sequence, and their ratios form a third sequence. Let x_n and y_n be, respectively, the numerator and the denominator of the nth fraction $r_n = x_n/y_n$.

a) Verify that $x_1^2 - 2y_1^2 = -1$, $x_2^2 - 2y_2^2 = +1$ and, more generally, that if $a^2 - 2b^2 = -1$ or $+1$, then
$$(a + 2b)^2 - 2(a + b)^2 = +1 \quad \text{or} \quad -1,$$
respectively.

b) The fractions $r_n = x_n/y_n$ approach a limit as n increases. What is that limit? (*Hint:* Use part (a) to show that $r_n^2 - 2 = \pm(1/y_n)^2$; and that y_n is not less than n.)

Calculator or Computer Problems

14. Verify (or correct) the results given in Example 7 for approximations to the solution, near $x = 1$, of $\sin x = x^2$.

15. Use Newton's method to find approximations to the solution of $\cos x = x$.

16. Start with $x_0 = 1$ and let $x_{n+1} = \cos(x_n)$. Does this process seem to lead to a solution of $\cos x = x$? If so, for what value of x?

17. Start with $x_1 = 1$ and let $x_{n+1} = \sqrt{\sin(x_n)}$. Does this process seem to lead to a limiting value? If so, what equation does the limit satisfy (that is, in relation to the iterative formula)?

18. CALCULATOR *Kepler's equation.* To determine the position of a planet (or satellite) in its orbit at any time t, it becomes important to solve Kepler's equation: $E - e \sin E = M$, where e is the eccentricity of the orbit, M is the mean anomaly (known in terms of t), and E (the eccentric anomaly) is to be found. Once E is known, the position of the planet in its orbit can be determined. (See, for example, S. W. McCuskey, *Introduction to Celestial Mechanics,* Reading, Mass., Addison-Wesley Publishing Co., 1963, pp. 45–50.) Using Newton's method, find approximations to the solution of Kepler's equation for E, given $M = \pi/3$ and $e = 0.1$. (For small values of e, the iteration usually begins with $E_1 = M$.)

19. CALCULATOR Sometimes it is important to know that certain inequalities are valid over some definite interval. For example, the graph of $y = \tan x$ is concave upward for $0 < x < \pi/2$, and the line $y = x$ is tangent to the graph at the origin. This fact shows that $\tan x > x$ for $0 < x < \pi/2$. If we draw a line through the origin with slope $m > 1$, that line will be *above* the graph of $y = \tan x$ until a root of the equation $\tan x = mx$ is reached.

a) Use Newton's method to find where the line $y = 2x$ intersects $y = \tan x$ near $x = \pi/3$.

b) With this information, is it true, or not true, that $x < \tan x < 2x$ for $0 < x < \pi/3$?

c) For x in an interval $0 < x < b$, for which it is true that $x < \tan x < 2x$, prove that it is also true that $x^2/2 < \ln \sec x < x^2$. (*Hint:* Integrate the first set of inequalities from 0 to h and then replace h by x.)

In Problems 20–23, write a BASIC program that would generate x_n and print n and x_n for $n = 1$ to 10.

20. For the sequence of Problem 7.

21. For the sequence of Problem 8.

22. For the sequence of Problem 9.

23. For the sequence of Problem 10.

24. The Fibonacci sequence of Problem 11 is defined recursively as follows: $x_1 = 1$, $x_2 = 1$, and $x_{n+2} = x_n + x_{n+1}$. It can also be described in another way that amounts to "zippering" two sequences together. (Although this may seem unnecessary, we found it easier to write a BASIC program doing it this way.) Let $a_n = x_{2n-1}$ and $b_n = x_{2n}$. Then the Fibonacci sequence can be written as

$$a_1, b_1, a_2, b_2, a_3, b_3, \ldots, a_k, b_k, a_{k+1}, b_{k+1}, \ldots$$

The recursion formulas now become $a_1 = b_1 = 1$ and $a_{n+1} = a_n + b_n$, $b_{n+1} = a_{n+1} + b_n$. Verify that the following BASIC program will thus give the first twelve terms of the Fibonacci sequence, and fill in the rest of the table. (You don't need a computer to do the calculations.)

PROGRAM		n	a_n	b_n
10	LET a = 1	1	1	1
20	LET b = 1	2	2	
30	FOR n = 1 to 6	3		
40	PRINT n;a;b	4		
50	LET a = a + b	5		
60	LET b = a + b	6		
70	NEXT n			
80	END			

Note: The a on the right in line 60 comes from the a on the left in line 50.

11.2
Limit Theorems

The study of limits would be a cumbersome business if every question about convergence had to be answered by applying Definition 2 directly, as we have done so far. Fortunately, there are three theorems that will make this process largely unnecessary from now on. The first two are practically the same as Theorems 1 and 2 in Article 1.9.

THEOREM 2

If $A = \lim_{n \to \infty} a_n$ and $B = \lim_{n \to \infty} b_n$ both exist and are finite, then

i) $\lim \{a_n + b_n\} = A + B$,

ii) $\lim \{ka_n\} = kA$ (k any number),

iii) $\lim \{a_n \cdot b_n\} = A \cdot B$,

> iv) $\lim \left\{ \dfrac{a_n}{b_n} \right\} = \dfrac{A}{B}$,   provided $B \neq 0$ and b_n is never 0,
>
> it being understood that all of the limits are taken as $n \to \infty$.

By combining Theorem 2 with Examples 2 and 3 of Article 11.1, we can proceed immediately to

$$\lim_{n \to \infty} \left(-\frac{1}{n} \right) = -1 \cdot \lim_{n \to \infty} \frac{1}{n} = -1 \cdot 0 = 0,$$

$$\lim_{n \to \infty} \left(\frac{n-1}{n} \right) = \lim_{n \to \infty} \left(1 - \frac{1}{n} \right) = \lim_{n \to \infty} 1 - \lim_{n \to \infty} \frac{1}{n} = 1 - 0 = 1,$$

$$\lim_{n \to \infty} \frac{5}{n^2} = 5 \cdot \lim_{n \to \infty} \frac{1}{n} \cdot \lim_{n \to \infty} \frac{1}{n} = 5 \cdot 0 \cdot 0 = 0,$$

$$\lim_{n \to \infty} \frac{4 - 7n^6}{n^6 + 3} = \lim_{n \to \infty} \frac{(4/n^6) - 7}{1 + (3/n^6)} = \frac{0 - 7}{1 + 0} = -7.$$

A corollary of Theorem 2 that will be useful later is that every nonzero multiple of a divergent sequence is divergent.

COROLLARY

> If the sequence $\{a_n\}$ diverges, and if c is any number different from 0, then the sequence $\{ca_n\}$ diverges.

Proof of the Corollary   Suppose, on the contrary, that $\{ca_n\}$ converges. Then, by taking $k = 1/c$ in part (ii) of Theorem 2, we see that the sequence

$$\left\{ \frac{1}{c} \cdot ca_n \right\} = \{a_n\}$$

converges. Thus $\{ca_n\}$ cannot converge unless $\{a_n\}$ converges. If $\{a_n\}$ does not converge, then $\{ca_n\}$ does not converge.   ∎

The next theorem is the sequence version of the Sandwich Theorem of Article 1.9.

THEOREM 3

> If $a_n \leq b_n \leq c_n$ for all n beyond some index N, and if $\lim a_n = \lim c_n = L$, then $\lim b_n = L$ also.

An immediate consequence of Theorem 3 is that, if $|b_n| \leq c_n$ and $c_n \to 0$, then $b_n \to 0$ because $-c_n \leq b_n \leq c_n$. We use this fact in the next example.

EXAMPLE 1

$$\frac{\cos n}{n} \to 0 \qquad \text{because} \qquad 0 \leq \left| \frac{\cos n}{n} \right| = \frac{|\cos n|}{n} \leq \frac{1}{n}$$

and $1/n \to 0$.   ∎

EXAMPLE 2

$$\frac{1}{2^n} \to 0 \qquad \text{because} \qquad 0 \le \frac{1}{2^n} \le \frac{1}{n}$$

and $1/n \to 0$.

EXAMPLE 3

$$(-1)^n \frac{1}{n} \to 0 \qquad \text{because} \qquad 0 \le \left| (-1)^n \frac{1}{n} \right| \le \frac{1}{n}$$

and $1/n \to 0$.

The application of Theorems 2 and 3 is broadened by a theorem stating that the result of applying a continuous function to a convergent sequence is again a convergent sequence. We state the theorem without proof. (See Problem 65 for an outline of the proof.)

THEOREM 4

If $a_n \to L$ and if f is a function that is continuous at L and defined at all the a_n's, then $f(a_n) \to f(L)$.

EXAMPLE 4 Find $\lim_{n \to \infty} \sqrt{(n + 1)/n}$.

Solution Let $f(x) = \sqrt{x}$ and $a_n = (n + 1)/n$. Then

$$a_n \to 1 \qquad \text{and} \qquad f(a_n) = \sqrt{a_n} \to f(1) = \sqrt{1} = 1$$

because $f(x)$ is continuous at $x = 1$.

EXAMPLE 5 See Fig. 11.5. Find $\lim_{n \to \infty} 2^{1/n}$.

Solution Let $f(x) = 2^x$ and $a_n = 1/n$. Then

$$a_n \to 0 \qquad \text{and} \qquad f(a_n) = 2^{1/n} \to f(0) = 2^0 = 1$$

because 2^x is continuous at $x = 0$.

THEOREM 5

Suppose that $f(x)$ is a function defined for all $x \ge n_0$ and $\{a_n\}$ is a sequence such that $a_n = f(n)$ when $n \ge n_0$. If

$$\lim_{x \to \infty} f(x) = L, \qquad \text{then} \qquad \lim_{n \to \infty} a_n = L.$$

Proof Let $\epsilon > 0$. Suppose that L is a finite limit such that

$$\lim_{x \to \infty} f(x) = L.$$

Then there is a number M such that

$$x > M \Rightarrow |f(x) - L| < \epsilon.$$

n	$2^{1/n}$
1	2
2	1.414213562
4	1.189207115
10	1.071773463
100	1.006955555
1000	1.000693387
10000	1.000069317

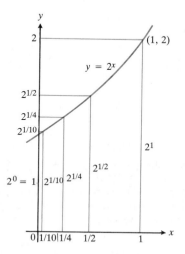

11.5 Since 2^x is continuous at $x = 0$ as $n \to \infty$, $x = 1/n \to 0$ and $y = 2^{1/n} \to 2^0 = 1$.

Let N be an integer such that

$$N \geq n_0 \qquad \text{and} \qquad N > M.$$

Then

$$n > N \Rightarrow a_n = f(n) \qquad \text{and} \qquad |a_n - L| = |f(n) - L| < \epsilon. \qquad \blacksquare$$

The proof would require modification in case $f(x) \to +\infty$ or $f(x) \to -\infty$, but in either of these cases the sequence $\{a_n\}$ would diverge. Theorem 5 allows us to use l'Hôpital's rule to determine the limits of some sequences. The next example shows how.

EXAMPLE 6   Find $\lim_{n \to \infty} (\ln n)/n$.

Solution   The function $(\ln x)/x$ is defined for all $x \geq 1$ and agrees with the given sequence on the positive integers. Therefore $\lim_{n \to \infty} (\ln n)/n$ will equal $\lim_{x \to \infty} (\ln x)/x$ if the latter exists. A single application of l'Hôpital's rule shows that

$$\lim_{x \to \infty} \frac{\ln x}{x} = \lim_{x \to \infty} \frac{1/x}{1} = \frac{0}{1} = 0.$$

We conclude that $\lim_{n \to \infty} (\ln n)/n = 0$. $\qquad \blacksquare$

When we use l'Hôpital's rule to find the limit of a sequence, we often treat n as a continuous real variable and differentiate directly with respect to n. This saves us from having to rewrite the formula for a_n as we did in Example 6.

EXAMPLE 7   Find $\lim_{n \to \infty} (2^n/5n)$.

Solution   By l'Hôpital's rule,

$$\lim_{n \to \infty} \frac{2^n}{5n} = \lim_{n \to \infty} \frac{2^n \cdot \ln 2}{5} = \infty. \qquad \blacksquare$$

When the terms of a sequence are ratios of polynomials in n, we have a choice of methods for finding the limit: either use l'Hôpital's rule or divide the numerator and denominator by an appropriate power of n. (For example, we could divide both numerator and denominator by the highest power of n in one or the other.) The following example shows what happens when the degree of the numerator is equal to, less than, or greater than the degree of the denominator.

EXAMPLE 8   Find the following limits as $n \to \infty$.

$$\lim \frac{n^2 - 2n + 1}{2n^2 + 5} = \lim \frac{1 - (2/n) + (1/n^2)}{2 + (5/n^2)} = \frac{1}{2}$$

$$\lim \frac{n^3 + 5n}{n^4 - 6} = \lim \frac{(1/n) + (5/n^3)}{1 - (6/n^4)} = 0$$

$$\lim \frac{n^2 - 5}{n + 1} = \lim \frac{2n}{1} = \infty \qquad \text{(l'Hôpital's rule)} \qquad \blacksquare$$

PROBLEMS

In Problems 1–10, write a_1, a_2, a_3, and a_4 for each sequence $\{a_n\}$. Determine which of the sequences converge and which diverge. Find the limit of each sequence that converges.

1. $a_n = \dfrac{1 - n}{n^2}$

2. $a_n = \dfrac{n}{2^n}$

3. $a_n = \left(\dfrac{1}{3}\right)^n$

4. $a_n = \dfrac{1}{n!}$

5. $a_n = \dfrac{(-1)^{n+1}}{2n - 1}$

6. $a_n = 2 + (-1)^n$

7. $a_n = \cos \dfrac{n\pi}{2}$

8. $a_n = 8^{1/n}$

9. $a_n = \dfrac{(-1)^{n-1}}{\sqrt{n}}$

10. $a_n = \sin^2 \dfrac{1}{n} + \cos^2 \dfrac{1}{n}$

In Problems 11–54, determine which of the sequences $\{a_n\}$ converge and which diverge. Find the limit of each sequence that converges.

11. $a_n = \dfrac{1}{10n}$

12. $a_n = \dfrac{n}{10}$

13. $a_n = 1 + \dfrac{(-1)^n}{n}$

14. $a_n = \dfrac{1 + (-1)^n}{n}$

15. $a_n = (-1)^n \left(1 - \dfrac{1}{n}\right)$

16. $a_n = 1 + (-1)^n$

17. $a_n = \dfrac{2n + 1}{1 - 3n}$

18. $a_n = \dfrac{n^2 - n}{2n^2 + n}$

19. $a_n = \sqrt{\dfrac{2n}{n + 1}}$

20. $a_n = \dfrac{\sin n}{n}$

21. $a_n = \sin \pi n$

22. $a_n = \sin\left(\dfrac{\pi}{2} + \dfrac{1}{n}\right)$

23. $a_n = n\pi \cos n\pi$

24. $a_n = \dfrac{\sin^2 n}{2^n}$

25. $a_n = \dfrac{n^2}{(n + 1)^2}$

26. $a_n = \dfrac{\sqrt{n - 1}}{\sqrt{n}}$

27. $a_n = \dfrac{1 - 5n^4}{n^4 + 8n^3}$

28. $a_n = \sqrt[n]{3^{2n+1}}$

29. $a_n = \tanh n$

30. $a_n = \dfrac{\ln n}{\sqrt{n}}$

31. $a_n = \dfrac{2(n + 1) + 1}{2n + 1}$

32. $a_n = \dfrac{(n + 1)!}{n!}$

33. $a_n = 5$

34. $a_n = 5^n$

35. $a_n = (0.5)^n$

36. $a_n = \dfrac{10^{n+1}}{10^n}$

37. $a_n = \dfrac{n^n}{(n + 1)^{n+1}}$

38. $a_n = (0.03)^{1/n}$

39. $a_n = \sqrt{2 - \dfrac{1}{n}}$

40. $a_n = 2 + (0.1)^n$

41. $a_n = \dfrac{3^n}{n^3}$

42. $a_n = \dfrac{\ln(n + 1)}{n + 1}$

43. $a_n = \ln n - \ln(n + 1)$

44. $a_n = \dfrac{1 - 2^n}{2^n}$

45. $a_n = \dfrac{n^2 - 2n + 1}{n - 1}$

46. $a_n = \dfrac{n + (-1)^n}{n}$

47. $a_n = \left(-\dfrac{1}{2}\right)^n$

48. $a_n = \dfrac{\ln n}{\ln 2n}$

49. $a_n = \tan^{-1} n$

50. $a_n = \sinh(\ln n)$

51. $a_n = n \sin \dfrac{1}{n}$

52. $a_n = \dfrac{2n + \sin n}{n + \cos 5n}$

53. $a_n = \dfrac{n^2}{2n - 1} \sin \dfrac{1}{n}$

54. $a_n = n\left(1 - \cos \dfrac{1}{n}\right)$

55. Show that $\lim_{n \to \infty} (n!/n^n) = 0$. (*Hint:* Expand the numerator and denominator and compare the quotient with $1/n$.)

56. CALCULATOR The formula $x_{n+1} = (x_n + a/x_n)/2$ is the one produced by Newton's method to generate a sequence of approximations to the positive solution of $x^2 - a = 0$, $a > 0$. Starting with $x_1 = 1$ and $a = 3$, use the formula to calculate successive terms of the sequence until you have approximated $\sqrt{3}$ as accurately as your calculator permits.

57. CALCULATOR If your calculator has a square root key, enter $x = 10$ and take successive square roots to approximate the terms of the sequence $10^{1/2}$, $10^{1/4}$, $10^{1/8}$, ..., continuing as far as your calculator permits. Repeat, with $x = 0.1$. Try other positive numbers above and below 1. When you have enough evidence, guess the answers to these questions: Does $\lim_{n \to \infty} x^{1/n}$ exist when $x > 0$? Does it matter what x is?

58. CALCULATOR If you start with a reasonable value of x_1, then the rule $x_{n+1} = x_n + \cos x_n$ will generate a sequence that converges to $\pi/2$. Figure 11.6 shows why. The convergence is rapid. With $x_1 = 1$, calculate x_2, x_3, and x_4. Find out what happens when you start with $x_1 = 5$. Remember to use radians.

11.6 The length $\pi/2$ of the circular arc is approximated by $x_n + \cos x_n$.

59. Suppose that $f(x)$ is defined for all $0 \le x \le 1$, that f is differentiable at $x = 0$, and that $f(0) = 0$. Define a sequence $\{a_n\}$ by the rule $a_n = nf(1/n)$. Show that $\lim a_n = f'(0)$.

Use the result of Problem 59 to find the limits of the sequences in Problems 60–62.

60. $a_n = n \tan^{-1} \dfrac{1}{n}$

61. $a_n = n(e^{1/n} - 1)$

62. $a_n = n \ln\left(1 + \dfrac{2}{n}\right)$

63. Under the hypotheses of Theorem 5, prove that if $\lim_{x \to \infty} f(x) = +\infty$, then the sequence $\{a_n\}$ diverges.

64. Prove Theorem 3.

65. Prove Theorem 4. (*Outline of proof:* Assume the hypotheses of the theorem and let ϵ be any positive number. For this ϵ there is a $\delta > 0$ such that

$$|f(x) - f(L)| < \epsilon \quad \text{when} \quad |x - L| < \delta.$$

For such a $\delta > 0$, there is an index N such that

$$|a_n - L| < \delta \quad \text{when} \quad n > N.$$

What is the conclusion?)

66. Prove the "zipper" theorem for sequences: If $\{a_n\}$ and $\{b_n\}$ both converge to L, then the sequence

$$a_1, b_1, a_2, b_2, \ldots, a_n, b_n, \ldots$$

also converges to L.

TOOLKIT PROGRAMS

Sequences and Series

11.3
Limits That Arise Frequently

Some limits arise so frequently that they are worth special attention. In this article we investigate these limits and look at examples in which they occur.

---

1. $\displaystyle \lim_{n \to \infty} \frac{\ln n}{n} = 0$

2. $\displaystyle \lim_{n \to \infty} \sqrt[n]{n} = 1$

3. $\displaystyle \lim_{n \to \infty} x^{1/n} = 1 \qquad (x > 0)$

4. $\displaystyle \lim_{n \to \infty} x^n = 0 \qquad (|x| < 1)$

5. $\displaystyle \lim_{n \to \infty} \left(1 + \frac{x}{n}\right)^n = e^x \qquad \text{(any } x)$

6. $\displaystyle \lim_{n \to \infty} \frac{x^n}{n!} = 0 \qquad \text{(any } x)$

It is important to note that x is fixed and that only n varies in Formulas 3 through 6.

1. $\displaystyle \lim_{n \to \infty} \frac{\ln n}{n} = 0$

This limit was calculated in Example 6 of Article 11.2.

2. $\displaystyle \lim_{n \to \infty} \sqrt[n]{n} = 1$

Let $a_n = n^{1/n}$. Then

$$\ln a_n = \ln n^{1/n} = \frac{1}{n} \ln n \to 0, \tag{1}$$

so that, by applying Theorem 4 of Article 11.2 to $f(x) = e^x$, we have

$$n^{1/n} = a_n = e^{\ln a_n} = f(\ln a_n) \to f(0) = e^0 = 1. \tag{2}$$

3. $\lim\limits_{n\to\infty} x^{1/n} = 1$, if $x > 0$

Let $a_n = x^{1/n}$. Then

$$\ln a_n = \ln x^{1/n} = \frac{1}{n}\ln x \to 0 \tag{3}$$

because x remains fixed while n gets large. Thus, again by Theorem 4, with $f(x) = e^x$,

$$x^{1/n} = a_n = e^{\ln a_n} \to e^0 = 1. \tag{4}$$

4. $\lim\limits_{n\to\infty} x^n = 0$, if $|x| < 1$

Our scheme here is to show that the criteria of Definition 2 of Article 11.1 are satisfied with $L = 0$. That is, we show that to each $\epsilon > 0$ there corresponds an index N so large that

$$|x^n| < \epsilon \qquad \text{for } n > N. \tag{5}$$

Since $\epsilon^{1/n} \to 1$, while $|x| < 1$, there is an index N for which

$$|x| < \epsilon^{1/N}. \tag{6}$$

In other words,

$$|x^N| = |x|^N < \epsilon. \tag{7}$$

This is the index we seek because

$$|x^n| < |x^N| \qquad \text{for } n > N. \tag{8}$$

Combining (7) and (8) produces

$$|x^n| < |x^N| < \epsilon \qquad \text{for } n > N, \tag{9}$$

which is just what we needed to show.

5. $\lim\limits_{n\to\infty} \left(1 + \dfrac{x}{n}\right)^n = e^x$ (any x)

Let

$$a_n = \left(1 + \frac{x}{n}\right)^n.$$

Then

$$\ln a_n = \ln\left(1 + \frac{x}{n}\right)^n = n\ln\left(1 + \frac{x}{n}\right) \to x,$$

as we can see by the following application of l'Hôpital's rule, in which we differentiate with respect to n:

$$\lim_{n\to\infty} n\ln\left(1 + \frac{x}{n}\right) = \lim_{n\to\infty} \frac{\ln(1 + x/n)}{1/n}$$

$$= \lim_{n\to\infty} \frac{\left(\dfrac{1}{1 + x/n}\right)\cdot\left(-\dfrac{x}{n^2}\right)}{-1/n^2} = \lim_{n\to\infty} \frac{x}{1 + x/n} = x.$$

Thus, by Theorem 4 of Article 11.2, with $f(x) = e^x$,

$$\left(1 + \frac{x}{n}\right)^n = a_n = e^{\ln a_n} \to e^x.$$

6. $\displaystyle \lim_{n \to \infty} \frac{x^n}{n!} = 0$ **(any x)**

Since

$$-\frac{|x|^n}{n!} \le \frac{x^n}{n!} \le \frac{|x|^n}{n!},$$

all we really need to show is that $|x|^n/n! \to 0$. The first step is to choose an integer $M > |x|$, so that

$$\frac{|x|}{M} < 1 \qquad \text{and} \qquad \left(\frac{|x|}{M}\right)^n \to 0.$$

We then restrict our attention to values of $n > M$. For these values of n, we can write

$$\frac{|x|^n}{n!} = \frac{|x|^n}{1 \cdot 2 \cdots M \cdot \underbrace{(M+1)(M+2) \cdots n}_{(n-M) \text{ factors}}}$$

$$\le \frac{|x|^n}{M! \, M^{n-M}} = \frac{|x|^n M^M}{M! \, M^n} = \frac{M^M}{M!}\left(\frac{|x|}{M}\right)^n.$$

Thus,

$$0 \le \frac{|x|^n}{n!} \le \frac{M^M}{M!}\left(\frac{|x|}{M}\right)^n.$$

Now, the constant $M^M/M!$ does not change with n. Thus the Sandwich Theorem tells us that

$$\frac{|x|^n}{n!} \to 0 \qquad \text{because} \qquad \left(\frac{|x|}{M}\right)^n \to 0.$$

A large number of limits can be found directly from the six limits we have just calculated.

EXAMPLES Calculate the following limits as $n \to \infty$.

1. If $|x| < 1$, then $x^{n+4} = x^4 \cdot x^n \to x^4 \cdot 0 = 0$.

2. $\sqrt[n]{2n} = \sqrt[n]{2}\sqrt[n]{n} \to 1 \cdot 1 = 1$

3. $\left(1 + \dfrac{1}{n}\right)^{2n} = \left[\left(1 + \dfrac{1}{n}\right)^n\right]^2 \to e^2$

4. $\dfrac{100^n}{n!} \to 0$

5. $\dfrac{x^{n+1}}{(n+1)!} = \dfrac{x}{(n+1)} \cdot \dfrac{x^n}{n!} \to 0 \cdot 0 = 0$ ∎

Still other limits can be calculated by using logarithms or l'Hôpital's rule, as in the calculations of limits 2, 3, and 5 at the beginning of this article.

EXAMPLE 6   Find $\lim_{n \to \infty} (\ln(3n + 5))/n$.

Solution   By l'Hôpital's rule,

$$\lim_{n \to \infty} \frac{\ln(3n + 5)}{n} = \lim_{n \to \infty} \frac{3/(3n + 5)}{1} = 0.$$

EXAMPLE 7   Find $\lim_{n \to \infty} \sqrt[n]{3n + 5}$.

Solution   Let

$$a_n = \sqrt[n]{3n + 5} = (3n + 5)^{1/n}.$$

Then

$$\ln a_n = \ln(3n + 5)^{1/n} = \frac{\ln(3n + 5)}{n} \to 0,$$

as in Example 6. Therefore,

$$a_n = e^{\ln a_n} \to e^0 = 1,$$

by Theorem 4 of Article 11.2.

We know that $\ln n$ increases more slowly than n does as $n \to \infty$, because $(\ln n)/n \to 0$. But we can say much more: $\ln n$ increases more slowly than $\sqrt{n}$, $\sqrt[3]{n}$, or even $n^{0.00001}$. In fact, if c is any positive constant, $(\ln n)/n^c \to 0$ as $n \to \infty$. The next example establishes this fact.

EXAMPLE 8   Show that

$$\lim_{n \to \infty} \frac{\ln n}{n^c} = 0$$

if c is any positive constant.

Solution   Apply l'Hôpital's rule and get

$$\lim \frac{\ln n}{n^c} = \lim \frac{1/n}{cn^{c-1}} = \lim \frac{1}{cn^c} = 0.$$

PROBLEMS

In Problems 1–33, determine which of the sequences $\{a_n\}$ converge and which diverge. Find the limit of each sequence that converges.

1. $a_n = \dfrac{1 + \ln n}{n}$

2. $a_n = \dfrac{\ln n}{3n}$

3. $a_n = \dfrac{(-4)^n}{n!}$

4. $a_n = \sqrt[n]{10n}$

5. $a_n = (0.5)^n$

6. $a_n = \dfrac{1}{(0.9)^n}$

7. $a_n = \left(1 + \dfrac{7}{n}\right)^n$

8. $a_n = \left(\dfrac{n + 5}{n}\right)^n$

9. $a_n = \dfrac{\ln(n + 1)}{n}$

10. $a_n = \sqrt[n]{n + 1}$

11. $a_n = \dfrac{n!}{10^{6n}}$

12. $a_n = \dfrac{1}{\sqrt{2^n}}$

13. $a_n = \sqrt[2n]{n}$

14. $a_n = (n + 4)^{1/(n+4)}$

15. $a_n = \dfrac{1}{3^{2n-1}}$

16. $a_n = \ln\left(1 + \dfrac{1}{n}\right)^n$

17. $a_n = \left(\dfrac{n}{n+1}\right)^n$

18. $a_n = \left(1 + \dfrac{1}{n}\right)^{-n}$

19. $a_n = \dfrac{\ln(2n+1)}{n}$

20. $a_n = \sqrt[n]{2n+1}$

21. $a_n = \sqrt[n]{\dfrac{x^n}{2n+1}}, \quad x > 0$

22. $a_n = \sqrt[n]{n^2}$

23. $a_n = \sqrt[n]{n^2 + n}$

24. $a_n = \dfrac{3^n \cdot 6^n}{2^{-n} \cdot n!}$

25. $a_n = \left(\dfrac{3}{n}\right)^{1/n}$

26. $a_n = \sqrt[n]{4^n n}$

27. $a_n = \left(1 - \dfrac{1}{n}\right)^n$

28. $a_n = \left(1 - \dfrac{1}{n^2}\right)^n$

29. $a_n = \dfrac{\ln(n^2)}{n}$

30. $a_n = \dfrac{(\ln n)^{200}}{n}$

31. $a_n = \dfrac{\ln n}{n^{1/n}}$

32. $a_n = \dfrac{1}{n}\displaystyle\int_1^n \dfrac{1}{x}\,dx$

33. $a_n = \displaystyle\int_1^n \dfrac{1}{x^p}\,dx, \quad p > 1$

CALCULATOR In Problems 34–36, use a calculator to find a value of N such that the given inequality is satisfied for $n \geq N$.

34. $|\sqrt[n]{0.5} - 1| < 10^{-3}$

35. $|\sqrt[n]{n} - 1| < 10^{-3}$

36. $\dfrac{2^n}{n!} < 10^{-9}$

(*Hint:* If you do not have a factorial key, then write
$$\dfrac{2^n}{n!} = \left(\dfrac{2}{1}\right)\left(\dfrac{2}{2}\right)\cdots\left(\dfrac{2}{n}\right).$$
That is, calculate successive terms by multiplying by 2 and dividing by the next value of n.)

37. In Example 8, we assumed that obviously if $c > 0$, then $1/n^c \to 0$ as $n \to \infty$. Write out a formal proof of this fact. (*Hint:* If $\epsilon = 0.001$ and $c = 0.04$, how large should N be in order to be sure that $|1/n^c - 0| < \epsilon$ when $n > N$?)

TOOLKIT PROGRAMS

Sequences and Series

11.4

Infinite Series

Before giving a definition of the term *infinite series,* let us consider the following question: What significance, if any, can be given to such an expression as

$$1 + \dfrac{1}{2} + \dfrac{1}{4} + \dfrac{1}{8} + \dfrac{1}{16} + \dfrac{1}{32} + \cdots ?$$

Can we actually add an infinite number of terms? The answer we give is the one usually given: Start at the beginning and add a finite number of terms, in order, one at a time (Fig. 11.7). This process yields the related **sequence** of numbers $\{s_n\}$

$$\left\{1, 1 + \dfrac{1}{2}, 1 + \dfrac{1}{2} + \dfrac{1}{4}, \cdots\right\} = \left\{1, \dfrac{3}{2}, \dfrac{7}{4}, \dfrac{15}{8}, \dfrac{31}{16}, \dfrac{63}{32}, \cdots\right\},$$

$$1 + \dfrac{1}{2} + \dfrac{1}{4} + \dfrac{1}{8} + \cdots \to 2$$

11.7 If lengths 1, ½, ¼, ⅛, . . . , are added, the sum approaches 2.

■ ■ ■

THE LONG HISTORY OF INFINITE SERIES

Although infinite series were used extensively in the early history of the calculus, especially by Newton, they have appeared in the history of mathematics since antiquity. In the Middle Ages the mathematician Nicole Oresme, in his work *Questiones Super Geometriam Euclidis* (c. 1360), showed that the harmonic series is divergent by using a grouping similar to that in Example 9. Still, throughout the eighteenth century, mathematicians did not understand general notions of divergence and convergence of series, even though they could find the sum of specific series, developed limited tests of convergence for certain ones, and knew that others diverged. Many mathematicians, including the great Euler, treated series rather loosely, as if a series were just another long addition problem or a long polynomial in the case of a power series. A. L. Cauchy, in the nineteenth century, was the first mathematician to give systematic, clear, and general criteria for convergence. He did so in terms of his rigorous definition of the limit, on which these criteria fundamentally rest.

which appears to converge to the limit 2. Indeed, if we write

$$s_n = 1 + \frac{1}{2} + \frac{1}{4} + \frac{1}{8} + \cdots + \frac{1}{2^{n-1}},$$

we can show that

$$s_n = 2 - \frac{1}{2^{n-1}}.$$

As n increases without bound, $(1/2)^{n-1}$ approaches zero and we come to the conclusion that

$$\lim s_n = 2.$$

In this sense we say

"the sum of the series $1 + \frac{1}{2} + \frac{1}{4} + \frac{1}{8} + \cdots$"

is 2. Is the sum of any finite number of terms in this series 2? No. Can we actually *add* an infinite number of terms? No. It is only in the sense that the **sequence of partial sums s_n converges to the limit** 2 that we say "the sum of the series" is 2.

In the foregoing example, the successive summands form one sequence

$$a_1 = 1, \quad a_2 = \frac{1}{2}, \quad a_3 = \frac{1}{4}, \quad \ldots, \quad a_n = \frac{1}{2^{n-1}}.$$

By adding successive terms from this sequence, we form a new sequence

$$s_1 = a_1$$
$$s_2 = a_1 + a_2$$
$$\vdots$$
$$s_n = a_1 + a_2 + \cdots + a_n$$

because it was the *sum* of the terms in the first sequence that we wanted to study.

There are many examples in mathematics where such a sum is important. For example, the mathematical constants π and e as well as tables of such functions as $\sin x$, $\cos x$, e^x, $\ln x$, $\tan^{-1}(x)$, and so on can be computed from series. These and other applications will be presented in Chapters 12 and 20.

The word *series* as used in mathematics is different from the way it is used in everyday speech—for example, when we speak of a TV series, a concert series, or a world championship series. *Webster's Third New International Dictionary* includes the following two distinct meanings:

1: a group of usually three or more things or events standing or succeeding in order and having a like relationship to each other; a spatial or temporal succession of persons or things.

2: the expression obtained from a mathematical sequence by connecting its terms with plus signs.

For our purposes in this chapter, the second of these definitions is incorporated in the following definition.

DEFINITION

Given a sequence of numbers $\{a_n\}$, an expression of the form

$$a_1 + a_2 + a_3 + \cdots + a_n + \cdots \tag{1a}$$

is called an **infinite series.** The number a_n is called the **nth term** of the series. With such an expression as (1a), we associate a second sequence $\{s_n\}$ defined by

$$s_1 = a_1$$

$$s_2 = a_1 + a_2$$

$$\vdots \tag{1b}$$

$$s_n = a_1 + a_2 + \cdots + a_n = \sum_{k=1}^{n} a_k.$$

The sequence $\{s_n\}$ is called the **sequence of partial sums** of the series. If this sequence of partial sums converges to a limit L, we say that the **series converges** and that its **sum** is L. In this case, we also write

$$a_1 + a_2 + \cdots + a_n + \cdots = \sum_{n=1}^{\infty} a_n = L. \tag{1c}$$

If the sequence of partial sums of the series does not converge, we say that the **series diverges.**

Notations When we first begin to study a given infinite series, we may not know whether it converges or diverges. In either case, it is convenient to use the summation sigma, Σ, to denote the series and to write

$$\sum_{n=1}^{\infty} a_n \qquad \text{or} \qquad \sum_{k=1}^{\infty} a_k \qquad \text{or} \qquad \sum a_n.$$

The first of these is read "summation, from $n = 1$ to infinity, of terms a_n"; the second as "summation, from $k = 1$ to infinity, of terms a_k"; and the third as "summation a_n." All three notations are simply shorthand ways of writing $a_1 + a_2 + a_3 + \cdots + a_n + \cdots$.

Some Convergent Series

We illustrate one method of finding the sum of an infinite series with the repeating decimal

$$0.3333 \ldots = \frac{3}{10} + \frac{3}{100} + \frac{3}{1000} + \frac{3}{10,000} + \cdots$$

$$s_1 = \frac{3}{10},$$

$$s_2 = \frac{3}{10} + \frac{3}{10^2},$$

$$s_n = \frac{3}{10} + \frac{3}{10^2} + \cdots + \frac{3}{10^n}.$$

We can obtain a simple expression for s_n in closed form as follows: We multiply both sides of the equation for s_n by $1/10$, obtaining

$$\frac{1}{10}s_n = \frac{3}{10^2} + \frac{3}{10^3} + \cdots + \frac{3}{10^n} + \frac{3}{10^{n+1}}.$$

When we subtract this from s_n, we have

$$s_n - \frac{1}{10}s_n = \frac{3}{10} - \frac{3}{10^{n+1}} = \frac{3}{10}\left(1 - \frac{1}{10^n}\right).$$

Therefore,

$$\frac{9}{10}s_n = \frac{3}{10}\left(1 - \frac{1}{10^n}\right) \quad \text{and} \quad s_n = \frac{1}{3}\left(1 - \frac{1}{10^n}\right).$$

As $n \to \infty$, $(1/10)^n \to 0$ and

$$\lim_{n \to \infty} s_n = \frac{1}{3}.$$

We therefore say that the sum of the infinite series

$$\frac{3}{10^1} + \frac{3}{10^2} + \frac{3}{10^3} + \cdots + \frac{3}{10^n} + \cdots$$

is $1/3$, and we write

$$\sum_{n=1}^{\infty} \frac{3}{10^n} = \frac{1}{3}.$$

The decimal $0.333 \ldots$ is a special kind of *geometric series*.

DEFINITION

> **Geometric Series**
>
> A series of the form
>
> $$a + ar + ar^2 + ar^3 + \cdots + ar^{n-1} + \cdots \qquad (2)$$
>
> is called a **geometric series.**

The ratio r can be positive, as in

$$1 + \frac{1}{2} + \frac{1}{4} + \cdots + \frac{1}{2^{n-1}} + \cdots, \qquad (3a)$$

or negative, as in

$$1 - \frac{1}{3} + \frac{1}{9} - \cdots + (-1)^{n-1}\frac{1}{3^{n-1}} + \cdots. \qquad (3b)$$

Partial Sums of a Geometric Series

The sum of the first n terms of (2) is

$$s_n = a + ar + ar^2 + \cdots + ar^{n-1}. \qquad (4)$$

Multiplying both sides of (4) by r gives

$$rs_n = ar + ar^2 + \cdots + ar^{n-1} + ar^n. \qquad (5)$$

When we subtract (5) from (4), nearly all the terms cancel on the right side, leaving

$$s_n - rs_n = a - ar^n,$$

or

$$(1 - r)s_n = a(1 - r^n). \qquad (6)$$

If $r \neq 1$, we may divide (6) by $(1 - r)$ to obtain

$$s_n = \frac{a(1 - r^n)}{1 - r}, \qquad r \neq 1. \qquad (7a)$$

On the other hand, if $r = 1$ in (4), we get

$$s_n = na, \qquad r = 1. \qquad (7b)$$

We are interested in the limit as $n \to \infty$ in Eqs. (7a) and (7b). Clearly, (7b) has no finite limit if $a \neq 0$. If $a = 0$, the series (2) is just

$$0 + 0 + 0 + \cdots,$$

which converges to the sum zero.

If $r \neq 1$, we use (7a). In the right side of (7a), n appears only in the expression r^n. This approaches zero as $n \to \infty$ if $|r| < 1$. Therefore,

$$\lim_{n \to \infty} s_n = \lim_{n \to \infty} \frac{a(1 - r^n)}{1 - r} = \frac{a}{1 - r} \qquad \text{if } |r| < 1. \qquad (8)$$

If we recall that $r^0 = 1$ when $r \neq 0$ we can summarize by writing

$$a + ar + ar^2 + \cdots + ar^{n-1} + \cdots = a \sum_{n=1}^{\infty} r^{n-1} = \frac{a}{1 - r}, \qquad (9)$$

or

$$a + ar + ar^2 + \cdots + ar^{n-1} + \cdots = a \sum_{n=0}^{\infty} r^n = \frac{a}{1 - r} \qquad (10)$$

if $0 < |r| < 1$. If $r = 0$, the series still converges, to $a/(1 - r) = a$. If $|r| > 1$, then $|r^n| \to \infty$, and (2) diverges.

The remaining case is where $r = -1$. Then $s_1 = a$, $s_2 = a - a = 0$, $s_3 = a$, $s_4 = 0$, and so on. If $a \neq 0$, this sequence of partial sums has no limit as $n \to \infty$, and the series (2) diverges.

We have thus proved the following theorem.

THEOREM 6

Sum of a Geometric Series

If $|r| < 1$, the geometric series

$$a + ar + ar^2 + \cdots + ar^{n-1} + \cdots$$

converges to $a/(1 - r)$. If $|r| \geq 1$, the series diverges unless $a = 0$. If $a = 0$, the series converges to 0.

EXAMPLE 1   Geometric series with $a = 1/9$ and $r = 1/3$:

$$\frac{1}{9} + \frac{1}{27} + \frac{1}{81} + \cdots = \frac{1}{9}\left(1 + \frac{1}{3} + \frac{1}{3^2} + \cdots\right) = \frac{1/9}{1 - 1/3} = \frac{1}{6}. \quad \blacksquare$$

EXAMPLE 2   Geometric series with $a = 4$ and $r = -1/2$:

$$4 - 2 + 1 - \frac{1}{2} + \frac{1}{4} - \cdots = 4\left(1 - \frac{1}{2} + \frac{1}{4} - \frac{1}{8} + \frac{1}{16} - \cdots\right)$$

$$= \frac{4}{1 + (1/2)} = \frac{8}{3}. \quad \blacksquare$$

EXAMPLE 3   A ball is dropped from a meters above a flat surface. Each time the ball hits the surface after falling a distance h, it rebounds a distance rh, where r is a positive number less than 1. Find the total distance the ball travels up and down. See Fig. 11.8.

Solution   The distance is given by the series

$$s = a + 2ar + 2ar^2 + 2ar^3 + \cdots.$$

The terms following the first term form a geometric series of sum $2ar/(1 - r)$. Hence the distance is

$$s = a + \frac{2ar}{1 - r} = a\frac{1 + r}{1 - r}.$$

For instance, if a is 6 meters and $r = 2/3$, the distance is

$$s = 6\frac{1 + (2/3)}{1 - (2/3)} = 6\frac{5/3}{1/3} = 30 \text{ meters.} \quad \blacksquare$$

REMARK 1   We were fortunate in the case of the geometric series to have found the closed-form expressions

$$s_n = \begin{cases} a\dfrac{1 - r^n}{1 - r} & \text{when } r \neq 1, \\[2ex] na & \text{when } r = 1, \end{cases}$$

from which we could get the precise results given by Theorem 6. There are not many other types of series where such closed-form expressions are available. (The next example is one of those rare cases.) Most of the remainder of this chapter is devoted to tests that we can apply to the individual terms a_n of the series Σa_n to determine whether the series converges or diverges, without having to calculate the partial sums s_n. It turns out that we can do so for a great many series. If a series does converge, we still have the question of determining its sum. Chapter 12 will help to some extent in doing that, but for a great many series we will still be left with no alternative but to compute numerical values of the partial sums and use those to estimate the true sum.

As noted above, the next example is another series whose sum can be found exactly by first finding a closed expression, or formula, for the kth partial sum s_k.

(a)

(b)

11.8   (a) The height of each rebound is reduced by the factor r. (b) A stroboscopic photo of a bouncing ball. (Photograph from *PSSC Physics, Second Edition;* 1965; D.C. Heath & Company with Education Development Center, Inc.)

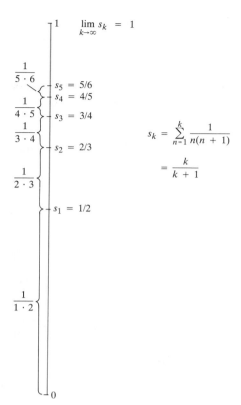

$$\lim_{k \to \infty} s_k = 1$$

$$s_5 = 5/6$$
$$s_4 = 4/5$$
$$s_3 = 3/4$$

$$s_k = \sum_{n=1}^{k} \frac{1}{n(n+1)}$$

$$s_2 = 2/3$$

$$= \frac{k}{k+1}$$

$$s_1 = 1/2$$

11.9 The sum of the first k terms of the series

$$\sum \frac{1}{n(n+1)}$$

is $k/(k+1)$ and the sum of the series is

$$\lim_{k \to \infty} \frac{k}{k+1} = 1.$$

EXAMPLE 4 Determine whether $\sum_{n=1}^{\infty} [1/n(n+1)]$ converges. If it does, find the sum.

Solution See Fig. 11.9. We begin by looking for a pattern in the sequence of partial sums that might lead us to a closed expression for s_k. The key to success here, as in the integration

$$\int \frac{dx}{x(x+1)} = \int \frac{dx}{x} - \int \frac{dx}{x+1},$$

is the use of partial fractions:

$$\frac{1}{k(k+1)} = \frac{1}{k} - \frac{1}{k+1}. \tag{11a}$$

This permits us to write the partial sum

$$\sum_{n=1}^{k} \frac{1}{n(n+1)} = \frac{1}{1 \cdot 2} + \frac{1}{2 \cdot 3} + \cdots + \frac{1}{k \cdot (k+1)}$$

as

$$s_k = \left(\frac{1}{1} - \frac{1}{2} \right) + \left(\frac{1}{2} - \frac{1}{3} \right) + \cdots + \left(\frac{1}{k} - \frac{1}{k+1} \right). \tag{11b}$$

By removing parentheses on the right and combining terms, we find that the sum telescopes into

$$s_k = 1 - \frac{1}{k+1} = \frac{k}{k+1}. \tag{12}$$

From this expression for s_k, we see immediately that $s_k \to 1$. Therefore the series does converge, and

$$\sum_{n=1}^{\infty} \frac{1}{n(n+1)} = 1. \tag{13}$$

Some Divergent Series

There are other series that diverge besides geometric series with $|r| \ge 1$.

EXAMPLE 5 The series

$$\sum_{n=1}^{\infty} n^2 = 1 + 4 + 9 + \cdots + n^2 + \cdots$$

diverges because the partial sums grow beyond every number L. The number $s_n = 1 + 4 + 9 + \cdots + n^2$ is greater than or equal to n^2 at each stage. ▪

EXAMPLE 6 The series

$$\sum_{n=1}^{\infty} \frac{n+1}{n} = \frac{2}{1} + \frac{3}{2} + \frac{4}{3} + \cdots + \frac{n+1}{n} + \cdots$$

diverges because the sequence of partial sums eventually outgrows every preassigned number. Each term is greater than 1, so the sum of n terms is greater than n. ■

A series can diverge without having its partial sums become large. For instance, the partial sums may oscillate between two extremes, as they do in the next example.

EXAMPLE 7 The series $\sum_{n=1}^{\infty} (-1)^{n+1}$ diverges because its partial sums alternate between 1 and 0:

$$s_1 = (-1)^2 = 1,$$

$$s_2 = (-1)^2 + (-1)^3 = 1 - 1 = 0,$$

$$s_3 = (-1)^2 + (-1)^3 + (-1)^4 = 1 - 1 + 1 = 1,$$

and so on. ■

The next theorem provides a quick way to detect the kind of divergence that occurred in Examples 5, 6, and 7.

THEOREM 7

The nth-Term Test for Divergence

If $\lim_{n \to \infty} a_n \neq 0$, or if $\lim_{n \to \infty} a_n$ fails to exist, then $\sum_{n=1}^{\infty} a_n$ diverges.

When we apply Theorem 7 to the series in Examples 5, 6, and 7, we find that

$$\sum_{n=1}^{\infty} n^2 \quad \text{diverges because } n^2 \to \infty,$$

$$\sum_{n=1}^{\infty} \frac{n+1}{n} \quad \text{diverges because } \frac{n+1}{n} \to 1 \neq 0,$$

$$\sum_{n=1}^{\infty} (-1)^{n+1} \quad \text{diverges because } \lim_{n \to \infty} (-1)^{n+1} \text{ does not exist.}$$

Proof of the Theorem We prove Theorem 7 by showing that if Σa_n converges, then $\lim_{n \to \infty} a_n = 0$. Let

$$s_n = a_1 + a_2 + \cdots + a_n,$$

and suppose that Σa_n converges to S; that is,

$$s_n \to S.$$

When n is large, so is $n - 1$, and both s_n and s_{n-1} are close to S. Their difference, a_n, must then be close to zero. More formally,

$$a_n = s_n - s_{n-1} \to S - S = 0.$$ ■

In the next example, we use both Theorems 6 and 7.

EXAMPLE 8 Determine whether each series converges or diverges. If it converges, find its sum.

a) $\displaystyle\sum_{n=1}^{\infty} 2\left(\cos\frac{\pi}{3}\right)^n$

b) $\displaystyle\sum_{n=0}^{\infty} \left(\tan\frac{\pi}{4}\right)^n$

c) $\displaystyle\sum_{n=1}^{\infty} \frac{n}{2n+5}$

d) $\displaystyle\sum_{n=1}^{\infty} \frac{5(-1)^n}{4^n}$

Solution

a) $\cos(\pi/3) = 1/2$. This is a geometric series with first term $a_1 = 2(\cos\pi/3) = 1$ and ratio $r = \cos(\pi/3) = 1/2$; so the series *converges* and its *sum* is $a_1/(1-r) = 1/(1-1/2) = 2$.

b) $\tan(\pi/4) = 1$. The nth term does not have zero as its limit, so the series *diverges*.

c) $a_n = \dfrac{n}{2n+5}$ and $\displaystyle\lim_{n\to\infty}\frac{n}{2n+5} = \frac{1}{2} \neq 0$. The series *diverges*.

d) This is a geometric series with first term $a_1 = -5/4$ and ratio $r = -1/4$. The series *converges* and its *sum* is

$$\frac{a_1}{1-r} = \frac{-5/4}{1+(1/4)} = -1.$$ ∎

A Necessary Condition for Convergence of a Series

Because of the way in which Theorem 7 is proved, it is often stated in the following, shorter way.

THEOREM 8

> If $\sum_{n=1}^{\infty} a_n$ converges, then $a_n \to 0$.

A WORD OF CAUTION Theorem 8 does *not* say that if $a_n \to 0$ then $\sum a_n$ converges. A series $\sum a_n$ may diverge even though $a_n \to 0$. Thus, $\lim a_n = 0$ is a *necessary* but *not a sufficient* condition for the series $\sum a_n$ to converge.

EXAMPLE 9 The series

$$1 + \underbrace{\frac{1}{2} + \frac{1}{2}}_{2\text{ terms}} + \underbrace{\frac{1}{4} + \frac{1}{4} + \frac{1}{4} + \frac{1}{4}}_{4\text{ terms}} + \cdots + \underbrace{\frac{1}{2^n} + \frac{1}{2^n} + \cdots + \frac{1}{2^n}}_{2^n\text{-terms}} + \cdots$$

diverges even though its terms form a sequence that converges to 0. ∎

Whenever we have two convergent series we can add them, subtract them, and multiply them by constants to make other convergent series. The next theorem gives the details.

THEOREM 9

> If $\Sigma\, a_n$ converges to A and $\Sigma\, b_n$ converges to B, then
>
> i) $\Sigma\, (a_n + b_n)$ converges to $A + B$,
>
> ii) $\Sigma\, ka_n$ converges to kA (k any number).

Proof of the Theorem   Let

$$A_n = a_1 + a_2 + \cdots + a_n, \qquad B_n = b_1 + b_2 + \cdots + b_n.$$

Then the partial sums of $\Sigma\, (a_n + b_n)$ are

$$S_n = (a_1 + b_1) + (a_2 + b_2) + \cdots + (a_n + b_n)$$

$$= (a_1 + \cdots + a_n) + (b_1 + \cdots + b_n)$$

$$= A_n + B_n.$$

Since $A_n \to A$ and $B_n \to B$, we have $S_n \to A + B$. The partial sums of $\Sigma\, (ka_n)$ are

$$S_n = ka_1 + ka_2 + \cdots + ka_n = k(a_1 + a_2 + \cdots + a_n) = kA_n,$$

which converge to kA.   ■

REMARK 2   If you think there should be two more parts of Theorem 9 to match those of Theorem 2 in Article 11.2, see Problems 44–46. Also see Problem 30 in Article 11.5, and Problem 39, Article 11.8.

Part (ii) of Theorem 9 says that every multiple of a convergent series converges. A companion to this is the next corollary, which says that every *nonzero* multiple of a divergent series diverges.

COROLLARY

> If $\Sigma\, a_n$ diverges, and if c is any number different from 0, then the series of multiples $\Sigma\, ca_n$ diverges.

REMARK 3   An immediate consequence of Theorem 9 is that if $A = \Sigma\, a_n$ and $B = \Sigma\, b_n$, then

$$\Sigma\, (a_n - b_n) = \Sigma\, a_n + \Sigma\, (-1)b_n = \Sigma\, a_n - \Sigma\, b_n = A - B. \tag{14}$$

The series $\Sigma\, (a_n - b_n)$ is called the **difference** of $\Sigma\, a_n$ and $\Sigma\, b_n$, whereas $\Sigma\, (a_n + b_n)$ is called their **sum.**

EXAMPLE 10

a) $\displaystyle\sum_{n=1}^{\infty} \frac{4}{2^{n-1}} = 4 \sum_{n=1}^{\infty} \frac{1}{2^{n-1}} = 4\,\frac{1}{1 - \frac{1}{2}} = 8$

b) $\displaystyle\sum_{n=0}^{\infty} \frac{3^n - 2^n}{6^n} = \sum_{n=0}^{\infty} \left(\frac{1}{2^n} - \frac{1}{3^n} \right) = \sum_{n=0}^{\infty} \frac{1}{2^n} - \sum_{n=0}^{\infty} \frac{1}{3^n}$

$$= \frac{1}{1 - (1/2)} - \frac{1}{1 - (1/3)} = 2 - \frac{3}{2} = \frac{1}{2}$$ ▪

REMARK 4 A finite number of terms can always be deleted from or added to a series without altering its convergence or divergence. If $\sum_{n=1}^{\infty} a_n$ converges and k is an index greater than 1, then $\sum_{n=k}^{\infty} a_n$ converges, and

$$\sum_{n=1}^{\infty} a_n = a_1 + a_2 + \cdots + a_{k-1} + \sum_{n=k}^{\infty} a_n. \tag{15}$$

Conversely, if $\sum_{n=k}^{\infty} a_n$ converges for any $k > 1$, then $\sum_{n=1}^{\infty} a_n$ converges and the sums continue to be related as in Eq. (15). Thus, for example,

$$\sum_{n=1}^{\infty} \frac{1}{5^n} = \frac{1}{5} + \frac{1}{25} + \frac{1}{125} + \sum_{n=4}^{\infty} \frac{1}{5^n} \tag{16}$$

and

$$\sum_{n=4}^{\infty} \frac{1}{5^n} = \sum_{n=1}^{\infty} \frac{1}{5^n} - \frac{1}{5} - \frac{1}{25} - \frac{1}{125}. \tag{17}$$

Note that while the addition or removal of a finite number of terms from a series has no effect on the convergence or divergence of the series, these operations can change the *sum* of a convergent series.

REMARK 5 The indexing of the terms of a series can be changed without altering the convergence of the series. For example, the geometric series that starts with

$$1 + \frac{1}{2} + \frac{1}{4} + \cdots$$

can be described as

$$\sum_{n=0}^{\infty} \frac{1}{2^n} \quad \text{or} \quad \sum_{n=-4}^{\infty} \frac{1}{2^{n+4}} \quad \text{or} \quad \sum_{n=5}^{\infty} \frac{1}{2^{n-5}}. \tag{18}$$

The partial sums remain the same no matter what indexing is chosen, so we are free to start indexing with whatever integer we want. Preference is usually given to an indexing that leads to a simple expression. In Example 11(b) we chose to start with $n = 0$ instead of $n = 1$ because this allowed us to describe the series we had in mind as

$$\sum_{n=0}^{\infty} \frac{3^n - 2^n}{6^n} \quad \text{instead of} \quad \sum_{n=1}^{\infty} \frac{3^{n-1} - 2^{n-1}}{6^{n-1}}. \tag{19}$$

PROBLEMS

In Problems 1–8, find a closed expression for the sum s_n of the first n terms of each series. Then compute the sum of the series if the series converges.

1. $2 + \dfrac{2}{3} + \dfrac{2}{9} + \dfrac{2}{27} + \cdots + \dfrac{2}{3^{n-1}} + \cdots$

2. $\dfrac{9}{100} + \dfrac{9}{100^2} + \dfrac{9}{100^3} + \cdots + \dfrac{9}{100^n} + \cdots$

3. $1 + e^{-1} + e^{-2} + \cdots + e^{-(n-1)} + \cdots$

4. $1 - \dfrac{1}{2} + \dfrac{1}{4} - \dfrac{1}{8} + \cdots + (-1)^{n-1}\dfrac{1}{2^{n-1}} + \cdots$

5. $1 - 2 + 4 - 8 + \cdots + (-1)^{n-1}2^{n-1} + \cdots$

6. $\dfrac{1}{2 \cdot 3} + \dfrac{1}{3 \cdot 4} + \dfrac{1}{4 \cdot 5} + \cdots + \dfrac{1}{(n+1)(n+2)} + \cdots$

7. $\ln \dfrac{1}{2} + \ln \dfrac{2}{3} + \ln \dfrac{3}{4} + \cdots + \ln \dfrac{n}{n+1} + \cdots$

8. $1 + 2 + 3 + \cdots + n + \cdots$

9. The series in Problem 6 can be described as

$$\sum_{n=1}^{\infty} \frac{1}{(n+1)(n+2)}.$$

It can also be described as a summation beginning with $n = -1$:

$$\sum_{n=-1}^{\infty} \frac{1}{(n+3)(n+4)}.$$

Describe the series as a summation beginning with
a) $n = -2$,   b) $n = 0$,   c) $n = 5$.

10. a) A ball is dropped from a height of 4 meters. Each time it strikes the pavement after falling from a height of h meters, it rebounds to a height of $0.75h$ meters. Find the total distance traveled up and down by the ball.
   b) Calculate the total time the ball is traveling. (*Hint:* The formula $s = 4.9t^2$ gives $t = \sqrt{s/4.9}$, with s in meters and t in seconds.)

In Problems 11–18, find the sum of the series.

11. $\displaystyle\sum_{n=0}^{\infty} \frac{1}{4^n}$

12. $\displaystyle\sum_{n=2}^{\infty} \frac{1}{4^n}$

13. $\displaystyle\sum_{n=1}^{\infty} \frac{7}{4^n}$

14. $\displaystyle\sum_{n=0}^{\infty} (-1)^n \frac{5}{4^n}$

15. $\displaystyle\sum_{n=0}^{\infty} \left(\frac{5}{2^n} + \frac{1}{3^n} \right)$

16. $\displaystyle\sum_{n=0}^{\infty} \left(\frac{5}{2^n} - \frac{1}{3^n} \right)$

17. $\displaystyle\sum_{n=0}^{\infty} \left(\frac{2^n}{5^n} \right)$

18. $\displaystyle\sum_{n=0}^{\infty} \left(\frac{2^{n+1}}{5^n} \right)$

Use partial fractions to find the sum of the series in Problems 19–22.

19. $\displaystyle\sum_{n=1}^{\infty} \frac{4}{(4n-3)(4n+1)}$

20. $\displaystyle\sum_{n=1}^{\infty} \frac{1}{(4n-3)(4n+1)}$

21. $\displaystyle\sum_{n=3}^{\infty} \frac{4}{(4n-3)(4n+1)}$

22. $\displaystyle\sum_{n=1}^{\infty} \frac{2n+1}{n^2(n+1)^2}$

23. a) Express the repeating decimal

$$0.234\ 234\ 234 \ldots$$

as an infinite series and give the sum as a ratio p/q of two integers.
   b) Is it true that *every* repeating decimal is a rational number p/q? Give a reason for your answer.

24. Express the decimal number

$$1.24\ 123\ 123\ 123 \ldots,$$

which begins to repeat after the first three figures, as a rational number p/q.

In Problems 25–38, determine whether each series converges or diverges. If it converges, find the sum.

25. $\displaystyle\sum_{n=0}^{\infty} \left(\frac{1}{\sqrt{2}} \right)^n$

26. $\displaystyle\sum_{n=1}^{\infty} \ln \frac{1}{n}$

27. $\displaystyle\sum_{n=1}^{\infty} (-1)^{n+1} \frac{3}{2^n}$

28. $\displaystyle\sum_{n=1}^{\infty} (\sqrt{2})^n$

29. $\displaystyle\sum_{n=0}^{\infty} \cos n\pi$

30. $\displaystyle\sum_{n=0}^{\infty} \frac{\cos n\pi}{5^n}$

31. $\displaystyle\sum_{n=0}^{\infty} e^{-2n}$

32. $\displaystyle\sum_{n=1}^{\infty} \frac{n^2 + 1}{n}$

33. $\displaystyle\sum_{n=1}^{\infty} (-1)^{n+1} n$

34. $\displaystyle\sum_{n=1}^{\infty} \frac{2}{10^n}$

35. $\displaystyle\sum_{n=0}^{\infty} \frac{2^n - 1}{3^n}$

36. $\displaystyle\sum_{n=1}^{\infty} \left(1 - \frac{1}{n} \right)^n$

37. $\displaystyle\sum_{n=0}^{\infty} \frac{n!}{1000^n}$

38. $\displaystyle\sum_{n=0}^{\infty} \frac{1}{x^n}, \quad |x| > 1$

In Problems 39 and 40, the equalities are instances of Theorem 6. Give the value of a and of r in each case.

39. $\dfrac{1}{1+x} = \displaystyle\sum_{n=0}^{\infty} (-1)^n x^n, \quad |x| < 1$

40. $\dfrac{1}{1+x^2} = \displaystyle\sum_{n=0}^{\infty} (-1)^n x^{2n}, \quad |x| < 1$

41. Figure 11.10 shows the first five of an infinite sequence of squares. The outermost square has an area of 4, and each of the other squares is obtained by joining the midpoints of the sides of the square before it. Find the sum of the areas of all the squares.

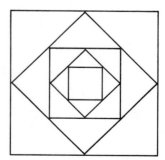

Figure 11.10

42. Find a closed-form expression for the nth partial sum of the series $\sum (-1)^{n+1}$.

43. Show by example that the term-by-term sum of two divergent series may converge.

44. Find convergent geometric series $A = \Sigma\, a_n$ and $B = \Sigma\, b_n$ that illustrate the fact that $\Sigma\, a_n b_n$ may converge without being equal to $A \cdot B$.

45. Show by example that $\Sigma\, (a_n/b_n)$ may diverge even though $\Sigma\, a_n$ and $\Sigma\, b_n$ converge and no $b_n = 0$.

46. Show by example that $\Sigma\, (a_n/b_n)$ may converge to something other than A/B even when $A = \Sigma\, a_n$, $B = \Sigma\, b_n \neq 0$, and no $b_n = 0$.

47. Show that if $\Sigma\, a_n$ converges, and $a_n \neq 0$ for all n, then $\Sigma\, (1/a_n)$ diverges.

48. a) Verify by long division that

$$\frac{1}{1 + t} = 1 - t + t^2 - t^3 + \cdots$$

$$+ (-1)^n t^n + \frac{(-1)^{n+1} t^{n+1}}{1 + t}.$$

b) By integrating both sides of the equation in part (a) with respect to t, from 0 to x, show that

$$\ln(1 + x) = x - \frac{x^2}{2} + \frac{x^3}{3} - \frac{x^4}{4} + \cdots$$

$$+ (-1)^n \frac{x^{n+1}}{n + 1} + R,$$

where

$$R = (-1)^{n+1} \int_0^x \frac{t^{n+1}}{1 + t}\, dt.$$

c) If $x > 0$, show that

$$|R| \leq \int_0^x t^{n+1}\, dt = \frac{x^{n+2}}{n + 2}.$$

(*Hint:* As t varies from 0 to x, $1 + t \geq 1$.)

d) If $x = 1/2$, how large should n be in part (c) if we want to be able to guarantee that $|R| < 0.001$? Write a polynomial that approximates $\ln(1 + x)$ to this degree of accuracy for $0 \leq x \leq 1/2$.

e) If $x = 1$, how large should n be in part (c) if we want to be able to guarantee that $|R| < 0.001$?

TOOLKIT PROGRAMS

Sequences and Series

11.5

Series with Nonnegative Terms: Comparison and Integral Tests

Given a series $\Sigma\, a_n$, we have two questions:

1. Does the series converge?

2. If it converges, what is its sum?

Most of the rest of this chapter is devoted to the first question. But as a practical matter, the second question is just as important for a scientist or engineer, and we come back to that question later.

In this article and the next we shall study series that do not have negative terms. The reason for this restriction is that the partial sums of these series always form nondecreasing sequences, and nondecreasing sequences *that are bounded from above* always converge, as we shall see. Thus, to show that a series of nonnegative terms converges, we need only show that there is some number beyond which the partial sums never go.

It may at first seem to be a drawback that this approach establishes the fact of convergence without actually producing the sum of the series in question. Surely it would be better to compute sums of series directly from nice formulas for their partial sums. But in most cases such formulas are not available, and in their absence we have to turn instead to a two-step procedure of first establishing convergence and then approximating the sum.

Surprisingly enough, beginning our study of convergence with the temporary exclusion of series that have one or more negative terms is not a severe restriction. As we shall see in Article 11.7, a series $\Sigma\ a_n$ will converge whenever the corresponding series of absolute values $\Sigma\ |a_n|$ converges. (The converse is not true. Articles 11.7 and 11.8 will deal with these matters more fully.) Thus, once we know that

$$\sum_{n=1}^{\infty} \frac{1}{n^2} = 1 + \frac{1}{4} + \frac{1}{9} + \frac{1}{16} + \frac{1}{25} + \cdots \tag{1}$$

converges, we know that *all* of the series like

$$1 - \frac{1}{4} + \frac{1}{9} - \frac{1}{16} + \frac{1}{25} + \cdots \tag{2}$$

and

$$-1 - \frac{1}{4} + \frac{1}{9} + \frac{1}{16} - \frac{1}{25} - \cdots \tag{3}$$

that can be obtained from (1) by changing the sign of one or more terms also converge! We might not know at first what they converge to, but at least we know they converge, and that is a first and necessary step toward estimating their sums.

Nondecreasing Sequences

Suppose that $\Sigma\ a_n$ is an infinite series without negative terms. That is, $a_n \geq 0$ for every n. Then, when we calculate the partial sums s_1, s_2, s_3, and so on, we see that each one is greater than or equal to its predecessor because $s_{n+1} = s_n + a_n$. That is,

$$s_1 \leq s_2 \leq s_3 \leq \cdots \leq s_n \leq s_{n+1} \leq \cdots. \tag{4}$$

A sequence $\{s_n\}$ like the one in (4), with the property that $s_n \leq s_{n+1}$ for every n, is called a nondecreasing sequence.

There are two types of nondecreasing sequences—those that increase beyond any finite bound and those that don't. The former diverge to infinity. We turn our attention to the other kind: those that do not grow beyond all bounds. Such a sequence is said to be **bounded from above,** and any number M such that

$$s_n \leq M \qquad \text{for all } n$$

is called an **upper bound** of the sequence.

EXAMPLE 1 If $s_n = n/(n + 1)$, then 1 is an upper bound and so is any larger number, like $\sqrt{2}$, 5, or 17. No number smaller than 1 is an upper bound, so for this sequence 1 is the **least upper bound.** ∎

When a nondecreasing sequence is bounded from above, we may ask, "Must it have a *least* upper bound?" The answer is yes, but we shall not prove this fact. We shall prove that if L is the *least upper bound*, then the sequence *converges* to L. The following argument shows why L is the limit.

When we plot the points $(1, s_1), (2, s_2), \ldots, (n, s_n)$ in the xy-plane, if M is an upper bound of the sequence, all these plotted points will lie on or below the line $y = M$ (Fig. 11.11). It seems clear that there ought to be a lowest such

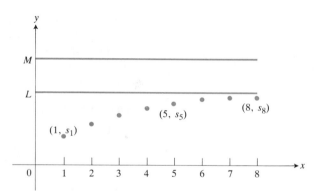

11.11 When the terms of a nondecreasing sequence have an upper bound M, they have a limit $L \leq M$.

line $y = L$. That would mean that none of the points (n, s_n) is above $y = L$ but that some do lie above any lower line $y = L - \epsilon$, if ϵ is a positive number. Then the sequence converges to L as limit, because the number L has the properties

a) $s_n \leq L$ for *all* values of n and

b) given any $\epsilon > 0$, there exists at least one integer N such that

$$s_N > L - \epsilon.$$

Then the fact that $\{s_n\}$ is a nondecreasing sequence tells us further that

$$s_n \geq s_N > L - \epsilon \qquad \text{for all } n \geq N.$$

This means that *all* the numbers s_n, beyond the Nth one in the sequence, lie within ϵ distance of L. This is precisely the condition for L to be the limit of the sequence s_n,

$$L = \lim_{n \to \infty} s_n.$$

The facts for nondecreasing sequences are summarized in the following theorem.

THEOREM 10

The Nondecreasing Sequence Theorem

Let $\{s_n\}$ be a nondecreasing sequence of real numbers. Then one of the following alternatives must hold:

a) The terms of the sequence are all less than or equal to some finite constant M. In this case, the sequence has a finite limit L that is also less than or equal to M.

b) The sequence diverges to positive infinity; that is, the numbers in the sequence ultimately exceed every preassigned number.

Alternative (b) of the theorem occurs when there are points (n, s_n) above any given line $y = M$, no matter how large M may be.

Let us now apply Theorem 10 to the convergence of infinite series of nonnegative numbers. If Σa_n is such a series, its sequence of partial sums $\{s_n\}$ is a nondecreasing sequence. Therefore, $\{s_n\}$, and hence Σa_n, will converge if and only if the numbers s_n have an upper bound. The question is how to find out in any particular instance whether the s_n's do have an upper bound.

Sometimes we can show that the s_n's are bounded above by showing that each one is less than or equal to the corresponding partial sum of a series that is already known to converge. The next example shows how this can happen.

EXAMPLE 2   The series

$$\sum_{n=0}^{\infty} \frac{1}{n!} = 1 + \frac{1}{1!} + \frac{1}{2!} + \frac{1}{3!} + \cdots \tag{5}$$

converges because its terms are all positive and less than or equal to the corresponding terms of

$$1 + \sum_{n=0}^{\infty} \frac{1}{2^n} = 1 + 1 + \frac{1}{2} + \frac{1}{2^2} + \cdots . \tag{6}$$

To see how this relationship between these two series leads to an upper bound for the partial sums of $\sum_{n=0}^{\infty} (1/n!)$, let

$$s_n = 1 + \frac{1}{1!} + \frac{1}{2!} + \cdots + \frac{1}{n!},$$

and observe that, for each n,

$$s_n \le 1 + 1 + \frac{1}{2} + \frac{1}{2^2} + \cdots + \frac{1}{2^n} < 1 + \sum_{n=0}^{\infty} \frac{1}{2^n} = 1 + \frac{1}{1 - (1/2)} = 3.$$

Thus the partial sums of $\sum_{n=0}^{\infty} (1/n!)$ are all less than 3. Therefore, $\sum_{n=0}^{\infty} (1/n!)$ converges.

Just because 3 is an upper bound for the partial sums of $\sum_{n=0}^{\infty} (1/n!)$ we cannot conclude that the series converges to 3. The series actually converges to $e = 2.71828 \ldots$. ∎

Comparison Test for Convergence

We established the convergence of the series in Example 2 by comparing it with a series that was already known to converge. This kind of comparison is typical of a procedure called the *Comparison Test* for convergence of series of nonnegative terms.

Comparison Test for Series of Nonnegative Terms

Let $\Sigma\, a_n$ be a series with no negative terms.

a) Test for **convergence** of $\Sigma\, a_n$. The series $\Sigma\, a_n$ converges if there is a convergent series of nonnegative terms $\Sigma\, c_n$ with $a_n \le c_n$ for all $n > n_0$, for some positive integer n_0.

b) Test for **divergence** of $\Sigma\, a_n$. The series $\Sigma\, a_n$ diverges if there is a divergent series of nonnegative terms $\Sigma\, d_n$ with $a_n \ge d_n$ for all $n > n_0$.

In part (a), the partial sums of the series $\Sigma\ a_n$ are bounded from above by

$$M = a_1 + a_2 + \cdots + a_{n_0} + \sum_{n=n_0+1}^{\infty} c_n.$$

Therefore, they form a nondecreasing sequence with a limit L that is less than or equal to M.

In part (b), the partial sums for $\Sigma\ a_n$ are not bounded from above. If they were, the partial sums for $\Sigma\ d_n$ would be bounded by

$$M' = d_1 + d_2 + \cdots + d_{n_0} + \sum_{n=n_0+1}^{\infty} a_n.$$

This would imply convergence of $\Sigma\ d_n$. Therefore, divergence of $\Sigma\ d_n$ implies divergence of $\Sigma\ a_n$.

To apply the Comparison Test to a series, we do not have to include the early terms of the series. We can start the test with any index N, provided we include all the terms of the series being tested from there on.

EXAMPLE 3 The convergence of the series

$$5 + \frac{2}{3} + 1 + \frac{1}{7} + \frac{1}{2} + \frac{1}{3!} + \frac{1}{4!} + \cdots + \frac{1}{k!} + \cdots$$

can be established by ignoring the first four terms and comparing the remainder of the series from the fifth term on (the fifth term is $1/2$) with the convergent geometric series

$$\sum_{n=1}^{\infty} \frac{1}{2^n} = \frac{1}{2} + \frac{1}{4} + \frac{1}{8} + \cdots. \qquad \blacksquare$$

To apply the Comparison Test we need to have on hand a list of series that are known to converge and a list of series that are known to diverge. Our next example adds a divergent series to the list. Example 4 also serves as an introduction to the *Integral Test,* which will, in turn, add more series, both convergent and divergent, to our lists. After the Integral Test, we shall return to the Comparison Test in a stronger form known as the Limit Comparison Test.

EXAMPLE 4 The **harmonic series**

$$\sum_{n=1}^{\infty} \frac{1}{n} = 1 + \frac{1}{2} + \frac{1}{3} + \frac{1}{4} + \cdots$$

diverges because its sequence of partial sums is not bounded. To see that this is so, see Fig. 11.12, which shows some of the shaded rectangles whose combined areas represent

$$s_n = 1 + \frac{1}{2} + \frac{1}{3} + \cdots + \frac{1}{n}.$$

These rectangles cover an area that is greater than the area under the graph of

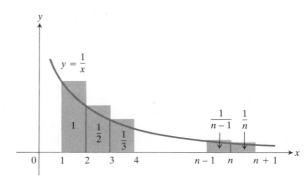

11.12 $1 + \dfrac{1}{2} + \dfrac{1}{3} + \cdots + \dfrac{1}{n} > \displaystyle\int_1^{n+1} \dfrac{1}{x}\, dx = \ln(n+1).$

$y = 1/x$, over the interval from $x = 1$ to $x = n + 1$:

$$s_n > \int_1^{n+1} \frac{dx}{x} = \ln(n+1).$$

Therefore $s_n \to +\infty$ because $\ln(n+1) \to +\infty$. The series

$$1 + \frac{1}{2} + \frac{1}{3} + \cdots + \frac{1}{n} + \cdots$$

diverges to positive infinity. ∎

The harmonic series $\sum_{n=1}^{\infty} (1/n)$ is another series whose divergence cannot be detected by the nth-Term Test for divergence. The series diverges in spite of the fact that $1/n \to 0$.

We know that every nonzero multiple of a divergent series diverges (corollary of Theorem 9 in the preceding article). Therefore, the divergence of the harmonic series implies the divergence of series like

$$\sum_{n=1}^{\infty} \frac{1}{2n} = \frac{1}{2} + \frac{1}{4} + \frac{1}{6} + \frac{1}{8} + \cdots$$

and

$$\sum_{n=1}^{\infty} \frac{1}{100n} = \frac{1}{100} + \frac{1}{200} + \frac{1}{300} + \frac{1}{400} + \cdots.$$

The Integral Test

In Example 4 we deduced the divergence of the harmonic series by comparing its sequence of partial sums with a divergent sequence of integrals. This comparison is a special case of a general comparison process called the *Integral Test*, a test that gives criteria for convergence as well as for divergence of series whose terms are positive.

Integral Test

Let $a_n = f(n)$ where $f(x)$ is a continuous, positive, decreasing function of x for all $x \geq 1$. Then the series $\Sigma\, a_n$ and the integral

$$\int_1^\infty f(x)\ dx$$

both converge or both diverge.

Proof We start with the assumption that f is a decreasing function with $f(n) = a_n$ for every n. This leads us to observe that the rectangles in Fig. 11.13(a), which have areas $a_1, a_2, \ldots, a_n$, collectively enclose more area than that under the curve $y = f(x)$ from $x = 1$ to $x = n + 1$. That is,

$$\int_1^{n+1} f(x)\ dx \leq a_1 + a_2 + \cdots + a_n.$$

In Fig. 11.13(b) the rectangles have been faced to the left instead of to the right. If we momentarily disregard the first rectangle, of area a_1, we see that

$$a_2 + a_3 + \cdots + a_n \leq \int_1^n f(x)\ dx.$$

If we include a_1, we have

$$a_1 + a_2 + \cdots + a_n \leq a_1 + \int_1^n f(x)\ dx.$$

Combining these results gives

$$\int_1^{n+1} f(x)\ dx \leq a_1 + a_2 + \cdots + a_n \leq a_1 + \int_1^n f(x)\ dx. \qquad (7)$$

If the integral $\int_1^\infty f(x)\ dx$ is finite, the right-hand inequality shows that $\Sigma\, a_n$ is also finite. But if $\int_1^\infty f(x)\ dx$ is infinite, then the left-hand inequality shows that the series is also infinite.

Hence the series and the integral are both finite or both infinite. ∎

11.13 $\int_1^{n+1} f(x)\ dx \leq a_1 + a_2 + \cdots + a_n$

$$\leq a_1 + \int_1^n f(x)\ dx.$$

EXAMPLE 5 *The p-series*. If p is a real constant, the series

$$\sum_{n=1}^\infty \frac{1}{n^p} = \frac{1}{1^p} + \frac{1}{2^p} + \frac{1}{3^p} + \cdots + \frac{1}{n^p} + \cdots \qquad (8)$$

converges if $p > 1$ and diverges if $p \leq 1$. To prove this, let

$$f(x) = \frac{1}{x^p}.$$

Then, if $p > 1$, we have $-p + 1 < 0$ and

$$\int_1^\infty x^{-p}\ dx = \lim_{b\to\infty} \frac{x^{-p+1}}{-p+1}\Big|_1^b = \frac{1}{1-p} \lim_{b\to\infty} (b^{-p+1} - 1) = \frac{1}{p-1},$$

which is finite. Hence the p-series converges if p is greater than 1.

If $p = 1$, we have

$$1 + \frac{1}{2} + \frac{1}{3} + \cdots + \frac{1}{n} + \cdots,$$

which we already know diverges. Or, by the Integral Test,

$$\int_1^\infty x^{-1}\, dx = \lim_{b \to \infty} \ln x \Big|_1^b = +\infty,$$

and, since the integral diverges, the series does likewise.

Finally, if $p < 1$, then the terms of the p-series are greater than the corresponding terms of the divergent harmonic series. Hence the p-series diverges, by the Comparison Test.

Thus, we have convergence for $p > 1$ but divergence for every other value of p. ∎

Observe that we can combine the Comparison Tests with the information we have about geometric series and the various p-series.

The Limit Comparison Test

We now present a more powerful form of the Comparison Test, known as the *Limit Comparison Test*. It is particularly handy in dealing with series in which a_n is a rational function of n. The next example will illustrate what we mean.

EXAMPLE 6 Do you think $\Sigma\, a_n$ converges? Diverges? Why?

a) $a_n = \dfrac{2n}{n^2 - n + 1}$ \qquad\qquad b) $a_n = \dfrac{2n^3 + 100n^2 + 1000}{(1/8)n^6 - n + 2}$

Discussion In determining convergence or divergence, only the tails count. And, when n is very large, the highest powers of n in numerator and denominator are what count the most. So, in (a), we reason this way:

$$a_n = \frac{2n}{n^2 - n + 1}$$

behaves about like $2n/n^2 = 2/n$, and, by comparison with $\Sigma\, 1/n$, we guess that $\Sigma\, a_n$ diverges. In (b), we reason that a_n will behave about like $2n^3/(1/8)n^6 = 16/n^3$ and, by comparison with $\Sigma\, 1/n^3$, a p-series with $p = 3$, we guess that the series will converge.

To be more precise, in part (a) we can take

$$a_n = \frac{2n}{n^2 - n + 1} \qquad \text{and} \qquad d_n = \frac{1}{n}$$

and look at the ratio

$$\frac{a_n}{d_n} = \frac{2n^2}{n^2 - n + 1} = \frac{2}{1 - \left(\dfrac{1}{n}\right) + \left(\dfrac{1}{n^2}\right)}.$$

Clearly, as $n \to \infty$ the limit is 2:

$$\lim \frac{a_n}{d_n} = 2.$$

This means that, in particular, if we take $\epsilon = 1$ in the definition of limit, we know there is an index N such that a_n/d_n is within 1 unit of this limit for all $n \geq N$:

$$2 - 1 \leq a_n/d_n \leq 2 + 1 \qquad \text{for} \qquad n \geq N.$$

Thus $a_n \geq d_n$ for $n \geq N$. Therefore, by the Comparison Test, $\Sigma \, a_n$ diverges because $\Sigma \, d_n$ diverges.

In part (b), if we let $c_n = 1/n^3$, we can easily show that

$$\lim \frac{a_n}{c_n} = 16.$$

Taking $\epsilon = 1$ in the definition of limit, we can conclude that there is an index N' such that a_n/c_n is between 15 and 17 when $n \geq N'$. Since $\Sigma \, c_n$ converges, so also does $\Sigma \, 17c_n$ and thus $\Sigma \, a_n$. ∎

Our rather rough guesswork paved the way for successful choices of comparison series. We make all of this more precise in the following *Limit Comparison Test*.

Limit Comparison Test

a) *Test for convergence.* If $a_n \geq 0$ for $n \geq n_0$ and there is a convergent series $\Sigma \, c_n$ such that $c_n > 0$ and

$$\lim \frac{a_n}{c_n} < \infty, \tag{9}$$

then $\Sigma \, a_n$ converges.

b) *Test for divergence.* If $a_n \geq 0$ for $n \geq n_0$ and there is a divergent series $\Sigma \, d_n$ such that $d_n > 0$ and

$$\lim \frac{a_n}{d_n} > 0, \tag{10}$$

then $\Sigma \, a_n$ diverges.

We shall not formally prove these results. Part (a) follows easily from the fact that if (9) holds, then there is an index $N \geq n_0$ and a constant M such that $a_n < Mc_n$ when $n > N$, and $\Sigma \, Mc_n$ converges. Similarly, if (10) holds, there is an index $N' \geq n_0$ and a constant $k > 0$ such that $a_n > kd_n$ for $n \geq N'$, and $\Sigma \, kd_n$ diverges.

A simpler version of the Limit Comparison Test combines parts (a) and (b) as follows.

Simplified Limit Comparison Test

If all the terms of the two series $\Sigma \, a_n$ and $\Sigma \, b_n$ are positive for $n \geq n_0$, and the limit of a_n/b_n is finite and positive, then both series converge or both diverge.

In practice, this Simplified Limit Comparison Test is the one we use most often.

EXAMPLE 7 Do the following series converge, or diverge?

a) $\dfrac{3}{4} + \dfrac{5}{9} + \dfrac{7}{16} + \dfrac{9}{25} + \cdots = \displaystyle\sum_{n=1}^{\infty} \dfrac{2n+1}{(n+1)^2}$

b) $\dfrac{1}{2} + \dfrac{2}{3} + \dfrac{3}{4} + \dfrac{4}{5} + \dfrac{5}{6} + \cdots = \displaystyle\sum_{n=1}^{\infty} \dfrac{n}{n+1}$

c) $\dfrac{101}{3} + \dfrac{102}{10} + \dfrac{103}{29} + \cdots = \displaystyle\sum_{n=1}^{\infty} \dfrac{100+n}{n^3+2}$

d) $\dfrac{1}{1} + \dfrac{1}{3} + \dfrac{1}{7} + \cdots = \displaystyle\sum_{n=1}^{\infty} \dfrac{1}{2^n-1}$

Solution

a) Let $a_n = (2n+1)/(n^2+2n+1)$ and $d_n = 1/n$. Then

$$\sum d_n \text{ diverges} \qquad \text{and} \qquad \lim \frac{a_n}{d_n} = \lim \frac{2n^2+n}{n^2+2n+1} = 2,$$

so $\sum a_n$ diverges.

b) Let $b_n = n/(n+1)$. Then $\lim b_n = 1 \neq 0$, so $\sum b_n$ diverges, by the nth-Term Test.

c) Let $a_n = (100+n)/(n^3+2)$. When n is large, this ought to compare with $n/n^3 = 1/n^2$, so we let $c_n = 1/n^2$. Then we apply the Limit Comparison Test:

$$\sum c_n \text{ converges} \qquad \text{and} \qquad \lim \frac{a_n}{c_n} = \lim \frac{n^3+100n^2}{n^3+2} = 1,$$

so $\sum a_n$ converges.

d) Let $a_n = 1/(2^n-1)$ and $c_n = 1/2^n$. (We reason that 2^n-1 behaves somewhat like 2^n when n is large.) Then

$$\frac{a_n}{c_n} = \frac{2^n}{2^n-1} = \frac{1}{1-(1/2)^n} \to 1 \qquad \text{as } n \to \infty.$$

Because $\sum c_n$ converges, we conclude that $\sum a_n$ does also.   ∎

When we use the p-series for comparison, it is essential that p be a constant, as shown by the next example.

EXAMPLE 8 Does the series

$$1^{-2} + 2^{-3/2} + 3^{-4/3} + 4^{-5/4} + \cdots + n^{-(n+1)/n} + \cdots$$

converge, or diverge?

Solution Let

$$a_n = n^{-(n+1)/n} = \frac{1}{n^{1+(1/n)}}.$$

This looks a bit like $1/n^p$ with $p = 1 + (1/n)$, which is greater than 1. But $1 + (1/n)$ isn't constant, so we need to be careful about drawing a conclusion. In fact

$$n^{1+(1/n)} = n(n)^{1/n} \quad \text{and} \quad (n)^{1/n} \to 1 \quad \text{as } n \to \infty.$$

Let's use the Limit Comparison Test with $d_n = 1/n$. Then

$$\frac{a_n}{d_n} = \frac{n}{n(n)^{1/n}}$$

$$= \frac{1}{(n)^{1/n}} \to 1 \quad \text{as } n \to \infty.$$

We know that $\Sigma\, d_n$ diverges, and when we apply the Limit Comparison Test we conclude that $\Sigma\, a_n$ also diverges. ∎

EXAMPLE 9 Does the series $\Sigma_{n=1}^{\infty} (\ln n)/n^{3/2}$ converge?

Solution Let $a_n = (\ln n)/n^{3/2}$ and $b_n = n^c/n^{3/2}$, where c is a positive constant that is less than $1/2$. For example, we might choose $c = 1/4$. Then $b_n = 1/n^{3/2-c} = 1/n^p$ with $p = 3/2 - c > 1$. (When $c = 1/4$, $p = 5/4$.) Hence $\Sigma\, b_n$ converges. Because $\ln n$ goes to infinity more slowly than n^c, for any positive constant c (Article 11.3, Example 8), we have reason to believe that $\Sigma\, a_n$ also converges. To verify this hunch, we apply the Limit Comparison Test:

$$\lim \frac{a_n}{b_n} = \lim \frac{\ln n}{n^c} = \lim \frac{1/n}{cn^{c-1}} \quad \text{(by l'Hôpital's rule)}$$

$$= \lim \frac{1}{cn^c} = 0.$$

The given series converges. ∎

Notice in Example 9 that we do not apply the *Simplified* Limit Comparison Test because the limit of a_n/b_n is 0, not positive.

In the next three articles, we introduce additional tests for convergence, but a partial summary of what we know so far may be helpful. A more complete list of tests, in the form of a flow chart, will be given at the end of Article 11.8. The matter of estimating the sum of a convergent series will be dealt with in Article 11.9.

Partial Summary of Tests for Convergence

1. *Definition:* The series $\Sigma\, a_n$ converges if and only if the sequence of partial sums $\{s_n\}$ converges, where

$$s_n = a_1 + a_2 + \cdots + a_n.$$

2. *nth-Term Test for Divergence:* If $a_n \nrightarrow 0$ as $n \to \infty$, the series diverges.

3. *Geometric Series Test:* If $\Sigma\, a_n = a + ar + ar^2 + \cdots$ is a geometric series, then the series converges to $a/(1 - r)$ if $|r| < 1$, and the series diverges if $|r| \geq 1$ unless $a = 0$.

4. *Boundedness Test for Series with Nonnegative Terms:* If $a_n \geq 0$ for all $n \geq 1$, then the series $\Sigma\, a_n$ converges if and only if the sequence of partial sums $\{s_n\}$ is bounded from above.

5. a) *Comparison Test for Convergence:* If $\Sigma\, a_n$ is a series of nonnegative terms and there exists a convergent series $\Sigma\, b_n$ such that $a_n \leq b_n$ for all n, then $\Sigma\, a_n$ converges.

   b) *Limit Comparison Test for Convergence:* If $\Sigma\, a_n$ is a series of nonnegative terms and there exists a convergent series $\Sigma\, b_n$ of positive terms such that $\lim(a_n/b_n)$ is finite, then $\Sigma\, a_n$ converges.

6. a) *Comparison Test for Divergence:* If $\Sigma\, a_n$ is a series of nonnegative terms and there exists a divergent series $\Sigma\, d_n$ of nonnegative terms such that $a_n \geq d_n$ for all n, then $\Sigma\, a_n$ diverges.

   b) *Limit Comparison Test for Divergence:* If $\Sigma\, a_n$ is a series of nonnegative terms and there exists a divergent series $\Sigma\, d_n$ of positive terms such that $\lim(a_n/d_n)$ is positive, then $\Sigma\, a_n$ diverges.

7. *Simplified Limit Comparison Test:* If $\Sigma\, a_n$ and $\Sigma\, b_n$ are series of positive numbers such that $\lim(a_n/b_n)$ is positive and finite, then $\Sigma\, a_n$ and $\Sigma\, b_n$ both converge or both diverge.

8. *Integral Test:* Let $a_n = f(n)$ where $f(x)$ is a continuous, positive, decreasing function of x for all $x \geq 1$. Then the series $\Sigma\, a_n$ and the integral

$$\int_1^\infty f(x)\, dx$$

both converge or both diverge.

PROBLEMS

In Problems 1–24, determine whether the given series converges or diverges. In each case, give a reason for your answer.

1. $\displaystyle\sum_{n=1}^{\infty} \frac{1}{10^n}$

2. $\displaystyle\sum_{n=1}^{\infty} \frac{n}{n+2}$

3. $\displaystyle\sum_{n=1}^{\infty} \frac{\sin^2 n}{2^n}$

4. $\displaystyle\sum_{n=1}^{\infty} \frac{5}{n}$

5. $\displaystyle\sum_{n=1}^{\infty} \frac{1 + \cos n}{n^2}$

6. $\displaystyle\sum_{n=1}^{\infty} -\frac{1}{8^n}$

7. $\displaystyle\sum_{n=2}^{\infty} \frac{\ln n}{n}$

8. $\displaystyle\sum_{n=1}^{\infty} \frac{1}{n\sqrt{n}}$

9. $\displaystyle\sum_{n=1}^{\infty} \frac{2^n}{3^n}$

10. $\displaystyle\sum_{n=0}^{\infty} \frac{-2}{n+1}$

11. $\displaystyle\sum_{n=1}^{\infty} \frac{1}{1 + \ln n}$

12. $\displaystyle\sum_{n=1}^{\infty} \frac{\ln n}{n\sqrt{n+1}}$

13. $\displaystyle\sum_{n=1}^{\infty} \frac{2^n}{n+1}$

14. $\displaystyle\sum_{n=1}^{\infty} \left(\frac{n}{3n+1}\right)^n$

15. $\displaystyle\sum_{n=1}^{\infty} \frac{1}{\sqrt{n^3 + 2}}$

16. $\displaystyle\sum_{n=1}^{\infty} \frac{1}{\sqrt{n}}\frac{(\ln n)^{10}}{n^{2/3}}$

17. $\displaystyle\sum_{n=1}^{\infty} \frac{n}{n^2 + 1}$

18. $\displaystyle\sum_{n=1}^{\infty} \frac{1}{\sqrt[n]{2}}$

19. $\displaystyle\sum_{n=1}^{\infty} \left(1 + \frac{1}{n}\right)^n$

20. $\displaystyle\sum_{n=1}^{\infty} \frac{\sqrt{n}}{n^2 + 1}$

21. $\displaystyle\sum_{n=1}^{\infty} \frac{1 - n}{n \cdot 2^n}$

22. $\displaystyle\sum_{n=1}^{\infty} \frac{1}{(\ln 2)^n}$

23. $\displaystyle\sum_{n=1}^{\infty} \frac{1}{3^{n-1} + 1}$

24. $\displaystyle\sum_{n=1}^{\infty} \frac{10n + 1}{n(n+1)(n+2)}$

25. Show that the series

$$\sum_{n=1}^{\infty} \frac{1}{2n - 1} = 1 + \frac{1}{3} + \frac{1}{5} + \cdots$$

diverges. (*Hint:* Use the Integral Test.)

26. *Logarithmic p-series*. Let p be a positive constant. Show that

$$\int_{2}^{\infty} \frac{dx}{x(\ln x)^p}$$

converges if and only if $p > 1$. (The integral does not start at 1, but at 2, because $\ln 1 = 0$.) What can you deduce about convergence or divergence of the following series?

a) $\sum_{n=2}^{\infty} \frac{1}{n \ln n}$

b) $\sum_{n=2}^{\infty} \frac{1}{n(\ln n)^{1.01}}$

c) $\sum_{n=5}^{\infty} \frac{n^{1/2}}{(\ln n)^3}$

d) $\sum_{n=3}^{\infty} \frac{1}{n \ln(n^3)}$

e) $\sum_{n=2}^{\infty} \frac{1}{n(\ln n)^{(n+1)/n}}$

27. **CALCULATOR** To estimate partial sums of the divergent harmonic series, Inequality (7) with $f(x) = 1/x$ tells us that

$$\ln n < 1 + \frac{1}{2} + \cdots + \frac{1}{n} < 1 + \ln n.$$

Suppose that the summation started with $s_1 = 1$ thirteen billion years ago (one estimate of the age of the universe) and that a new term has been added every *second* since then. How large would you expect s_n to be today?

28. There are no values of x for which $\sum_{n=1}^{\infty} (1/nx)$ converges. Why?

29. Show that if $\sum_{n=1}^{\infty} a_n$ is a convergent series of nonnegative numbers, then the series $\sum_{n=1}^{\infty} (a_n/n)$ converges.

30. Show that if $\sum a_n$ and $\sum b_n$ are convergent series with $a_n \geq 0$ and $b_n \geq 0$, then $\sum a_n b_n$ converges. (*Hint:* From some index on, $a_n b_n \leq a_n + b_n$.)

31. A sequence of numbers $\{s_n\}$ in which $s_n \geq s_{n+1}$ for every n is called a **nonincreasing sequence**. A sequence $\{s_n\}$ is bounded from below if there is a finite constant M with $M \leq s_n$ for every n. Such a number M is called a lower bound for the sequence. Deduce from Theorem 10 that a nonincreasing sequence that is bounded from below converges, and that a nonincreasing sequence that is not bounded from below diverges.

32. *The Cauchy condensation test*. The Cauchy condensation test says: Let $\{a_n\}$ be a nonincreasing sequence ($a_n \geq a_{n+1}$ for all n) of positive terms that converges to 0. Then $\sum a_n$ converges if and only if $\sum 2^n a_{2^n}$ converges. For example,

$\sum (1/n)$ diverges because $\sum 2^n \cdot (1/2^n) = \sum 1$ diverges. Show why the test works.

33. Use the Cauchy condensation test of Problem 32 to show that

a) $\sum_{n=2}^{\infty} \frac{1}{n \ln n}$ diverges,

b) $\sum_{n=1}^{\infty} \frac{1}{n^p}$ converges if $p > 1$ and diverges if $p \leq 1$.

34. Pictures like the one in Fig. 11.12 suggest that, as n increases, there is very little change in the difference between the sum

$$1 + \frac{1}{2} + \cdots + \frac{1}{n}$$

and the integral

$$\ln n = \int_{1}^{n} \frac{1}{x} \, dx.$$

To explore this idea, carry out the following steps.
a) By taking $f(x) = 1/x$ in Inequality (7), show that

$$\ln(n + 1) \leq 1 + \frac{1}{2} + \cdots + \frac{1}{n} \leq 1 + \ln n$$

or

$$0 < \ln(n + 1) - \ln n \leq 1 + \frac{1}{2} + \cdots + \frac{1}{n} - \ln n \leq 1.$$

Thus, the sequence

$$a_n = 1 + \frac{1}{2} + \cdots + \frac{1}{n} - \ln n$$

is bounded from below and from above.
b) Show that

$$\frac{1}{n + 1} < \int_{n}^{n+1} \frac{1}{x} \, dx = \ln(n + 1) - \ln n,$$

and use this result to show that the sequence $\{a_n\}$ in part (a) is decreasing.

Since a decreasing sequence that is bounded from below converges (Problem 31), the numbers a_n defined in (a) converge:

$$1 + \frac{1}{2} + \cdots + \frac{1}{n} - \ln n \to \gamma.$$

The number γ, whose value is $0.5772\ldots$, is called *Euler's constant*. In contrast to other special numbers like π and e, no other expression with a simple law of formulation has ever been found for γ.

35. The prime numbers form a sequence $\{p_n\} = \{2, 3, 5, 7, 11, 13, 17, 19, \ldots\}$. It is known that $\lim_{n \to \infty}[(n \ln n)/p_n] = 1$. Using this fact, show that

$$\sum_{n=1}^{\infty} \frac{1}{p_n} = \frac{1}{2} + \frac{1}{3} + \frac{1}{5} + \frac{1}{7} + \frac{1}{11} + \cdots + \frac{1}{p_n} + \cdots$$

diverges. (See Problem 26.)

11.6

Series with Nonnegative Terms: Ratio and Root Tests

Convergence tests that depend on comparing one series with another series or with an integral are called **extrinsic** tests. They are very useful, but there are reasons to look for tests that do not involve such comparisons. As a practical matter, we may not readily find either another series or an integrable function to compare. And, in principle, all the information about a given series should be contained in its own terms. We therefore turn our attention to **intrinsic** tests—those that depend only on the series at hand.

The Ratio Test

The next two tests are intrinsic and are easy to apply. The first of these, the *Ratio Test,* measures the rate of growth (or decline) of a series by examining the ratio a_{n+1}/a_n. For a geometric series, this rate of growth is a constant, and the series converges if and only if its ratio is less than 1 in absolute value. But even if the ratio is not constant, we may be able to find a geometric series for comparison, as illustrated in the next example.

EXAMPLE 1   Let $a_1 = 1$ and define a_{n+1} to be

$$a_{n+1} = \frac{n}{2n+1} a_n.$$

Does the series $\Sigma\, a_n$ converge, or diverge?

Solution   We begin by writing out a few terms of the series:

$$a_1 = 1, \quad a_2 = \frac{1}{3} a_1 = \frac{1}{3}, \quad a_3 = \frac{2}{5} a_2 = \frac{1 \cdot 2}{3 \cdot 5}, \quad a_4 = \frac{3}{7} a_3 = \frac{1 \cdot 2 \cdot 3}{3 \cdot 5 \cdot 7}.$$

We observe that each term is somewhat less than $1/2$ the term before it, because $n/(2n+1)$ is less than $1/2$. Therefore the terms of the given series are less than or equal to the terms of the geometric series

$$1 + \left(\frac{1}{2}\right) + \left(\frac{1}{4}\right) + \cdots + \left(\frac{1}{2}\right)^{n-1} + \cdots$$

that converges to 2. So our series also converges, and its sum is less than 2.   ■

In proving the Ratio Test, we shall make a comparison with appropriate geometric series as in the example above, but when we *apply* it we do not actually make a direct comparison.

The Ratio Test

Let $\Sigma\, a_n$ be a series with positive terms, and suppose that

$$\lim_{n \to \infty} \frac{a_{n+1}}{a_n} = \rho \qquad \text{(Greek letter rho).}$$

Then

a) the series *converges* if $\rho < 1$,

b) the series *diverges* if $\rho > 1$,

c) the series *may converge or it may diverge* if $\rho = 1$. (The test provides no information.)

Proof

a) Assume first that $\rho < 1$ and let r be a number between ρ and 1. Then the number

$$\epsilon = r - \rho$$

is positive. Since

$$\frac{a_{n+1}}{a_n} \to \rho,$$

a_{n+1}/a_n must lie within ϵ of ρ when n is large enough, say for all $n \geq N$. In particular,

$$\frac{a_{n+1}}{a_n} < \rho + \epsilon = r, \qquad \text{when } n > N.$$

That is,

$$a_{N+1} < ra_N,$$
$$a_{N+2} < ra_{N+1} < r^2 a_N,$$
$$a_{N+3} < ra_{N+2} < r^3 a_N,$$

$$\vdots$$

$$a_{N+m} < ra_{N+m-1} < r^m a_N.$$

These inequalities show that the terms of our series, after the Nth term, approach zero more rapidly than the terms in a geometric series with ratio $r < 1$. More precisely, consider the series $\Sigma\ c_n$, where $c_n = a_n$ for $n = 1$, 2, . . . , N and $c_{N+1} = ra_N$, $c_{N+2} = r^2 a_N$, . . . , $c_{N+m} = r^m a_N$, ... Now $a_n \leq c_n$ for all n, and

$$\sum_{n=1}^{\infty} c_n = a_1 + a_2 + \cdots + a_{N-1} + a_N + ra_N + r^2 a_N + \cdots$$

$$= a_1 + a_2 + \cdots + a_{N-1} + a_N(1 + r + r^2 + \cdots).$$

Because $|r| < 1$, the geometric series $1 + r + r^2 + \cdots$ converges, and hence so does $\Sigma\ c_n$. By the Comparison Test, $\Sigma\ a_n$ also converges.

b) Next, suppose $\rho > 1$. Then, from some index M on, we have

$$\frac{a_{n+1}}{a_n} > 1 \qquad \text{or} \qquad a_M < a_{M+1} < a_{M+2} < \cdots.$$

Hence, the terms of the series do not approach zero as n becomes infinite, and the series diverges, by the nth-Term Test.

c) Finally, the two series

$$\sum_{n=1}^{\infty} \frac{1}{n} \qquad \text{and} \qquad \sum_{n=1}^{\infty} \frac{1}{n^2}$$

show that, when $\rho = 1$, some other test for convergence must be used.

$$\text{For } \sum_{n=1}^{\infty} \frac{1}{n}: \quad \frac{a_{n+1}}{a_n} = \frac{1/(n+1)}{1/n} = \frac{n}{n+1} \to 1.$$

$$\text{For } \sum_{n=1}^{\infty} \frac{1}{n^2}: \quad \frac{a_{n+1}}{a_n} = \frac{1/(n+1)^2}{1/n^2} = \left(\frac{n}{n+1}\right)^2 \to 1^2 = 1.$$

In both cases $\rho = 1$, yet the first series diverges while the second converges. ∎

The Ratio Test is often effective when the terms of the series contain factorials of expressions involving n or expressions raised to the nth power or combinations, as in the next example. Recall that

$$n! = 1 \cdot 2 \cdot 3 \cdot \,\cdots\, \cdot n$$

implies that

$$(n+1)! = (n+1)n!$$

and that $(2(n+1))! = (2n+2)! = (2n+2)(2n+1)(2n)!$. These facts are used in parts (a) and (b) of the example. Part (b) also uses the fact that $4^{n+1}/4^n = 4$. Making use of appropriate cancellation properties of factorials and powers leads to simplified expressions for the ratio a_{n+1}/a_n.

EXAMPLE 2   Test the following series for convergence or divergence, using the Ratio Test.

a) $\displaystyle\sum_{n=1}^{\infty} \frac{n!n!}{(2n)!}$ b) $\displaystyle\sum_{n=1}^{\infty} \frac{4^n n!n!}{(2n)!}$ c) $\displaystyle\sum_{n=0}^{\infty} \frac{2^n + 5}{3^n}$

Solution

a) If $a_n = n!n!/(2n)!$, then $a_{n+1} = (n+1)!(n+1)!/(2n+2)!$, and

$$\frac{a_{n+1}}{a_n} = \frac{(n+1)!(n+1)!(2n)!}{n!n!(2n+2)(2n+1)(2n)!}$$

$$= \frac{(n+1)(n+1)}{(2n+2)(2n+1)} = \frac{n+1}{4n+2} \to \frac{1}{4}.$$

The series converges because $\rho = 1/4$ is less than 1.

b) If $a_n = 4^n n!n!/(2n)!$, then

$$\frac{a_{n+1}}{a_n} = \frac{4^{n+1}(n+1)!(n+1)!}{(2n+2)(2n+1)(2n)!} \times \frac{(2n)!}{4^n n!n!}$$

$$= \frac{4(n+1)(n+1)}{(2n+2)(2n+1)} = \frac{2(n+1)}{2n+1} \to 1.$$

Because the limit is $\rho = 1$, we cannot decide on the basis of the Ratio Test alone whether the series converges or diverges. However, when we note that $a_{n+1}/a_n = (2n+2)(2n+1)$, we conclude that a_{n+1} is always greater than a_n because $(2n + 2)/(2n + 1)$ is always greater than 1. Therefore, all terms are greater than or equal to $a_1 = 2$, and the nth term does not go to zero as n tends to infinity. Hence, by the nth-Term Test, the series diverges.

c) For the series $\sum_{n=0}^{\infty} (2^n + 5)/3^n$,

$$\frac{a_{n+1}}{a_n} = \frac{(2^{n+1} + 5)/3^{n+1}}{(2^n + 5)/3^n} = \frac{1}{3} \cdot \frac{2^{n+1} + 5}{2^n + 5}$$

$$= \frac{1}{3} \cdot \left(\frac{2 + 5 \cdot 2^{-n}}{1 + 5 \cdot 2^{-n}} \right) \to \frac{1}{3} \cdot \frac{2}{1} = \frac{2}{3}.$$

The series converges because $\rho = 2/3$ is less than 1.

This does *not* mean that 2/3 is the sum of the series. In fact,

$$\sum_{n=0}^{\infty} \frac{2^n + 5}{3^n} = \sum_{n=0}^{\infty} \left(\frac{2}{3} \right)^n + \sum_{n=0}^{\infty} \frac{5}{3^n} = \frac{1}{1 - (2/3)} + \frac{5}{1 - (1/3)} = \frac{21}{2}. \quad \blacksquare$$

REMARK   While the Ratio Test is useful in the types of series just discussed, it is not very useful for series like p-series.

In the next example, the series is expressed in terms of powers of x. By applying the Ratio Test, we can learn for what values of x the series converges. For those values, the series defines a function of x. (In Chapter 12, we discuss such series in more detail.)

EXAMPLE 3   For what values of x does the series

$$x + \frac{x^3}{3} + \frac{x^5}{5} + \frac{x^7}{7} + \cdots + \frac{x^{2n-1}}{2n - 1} + \cdots$$

converge?

Solution   The nth term of the series is

$$a_n = \frac{x^{2n-1}}{2n - 1}.$$

We consider first the case where x is positive. Then the series is a positive series and

$$\frac{a_{n+1}}{a_n} = \frac{x^{2(n+1)-1}}{2(n + 1) - 1} \cdot \frac{2n - 1}{x^{2n-1}} = \frac{(2n - 1)x^2}{2n + 1} \to x^2.$$

The Ratio Test therefore tells us that the series converges if x is positive and less than 1 and diverges if x is greater than 1.

Since only odd powers of x occur in the series, we see that the series simply changes sign when x is replaced by $-x$. Therefore the series also converges for $-1 < x \leq 0$ and diverges for $x < -1$. The series converges to zero when $x = 0$.

We know, thus far, that the series

$$\text{converges for } |x| < 1, \qquad \text{diverges for } |x| > 1,$$

but we don't know what happens when $|x| = 1$. To test at $x = 1$, we apply the

Integral Test to the series

$$1 + \frac{1}{3} + \frac{1}{5} + \frac{1}{7} + \cdots + \frac{1}{2n - 1} + \cdots,$$

which we get by taking $x = 1$ in the original series. The companion integral is

$$\int_1^\infty \frac{dx}{2x - 1} = \lim_{b \to \infty} \left(\frac{1}{2} \ln(2x - 1) \Big|_1^b \right) = \infty.$$

Hence the series diverges to $+\infty$ when $x = 1$. It diverges to $-\infty$ when $x = -1$. Therefore the only values of x for which the given series converges are $-1 < x < 1$. ∎

The *n*th-Root Test

We return to the question "Does $\Sigma\, a_n$ converge?" When there is a simple formula for a_n, we can try one of the tests we already have. But consider the following example.

EXAMPLE 4 Let $a_n = f(n)/2^n$, where

$$f(n) = \begin{cases} n & \text{if } n \text{ is a prime number,} \\ 1 & \text{otherwise.} \end{cases}$$

Does the series $\Sigma\, a_n$ converge?

Solution We write out several terms of the series:

$$\Sigma\, a_n = \frac{1}{2} + \frac{2}{4} + \frac{3}{8} + \frac{1}{16} + \frac{5}{32} + \frac{1}{64} + \frac{7}{128} + \cdots + \frac{f(n)}{2^n} + \cdots.$$

Clearly, this is not a geometric series. The nth term approaches zero as $n \to \infty$, so we don't know that the series diverges. The Integral Test doesn't look promising. The Ratio Test produces

$$\frac{a_{n+1}}{a_n} = \frac{1}{2} \frac{f(n + 1)}{f(n)} = \begin{cases} \dfrac{1}{2} & \text{if neither } n \text{ nor } n + 1 \text{ is a prime,} \\[2mm] \dfrac{1}{2n} & \text{if } n \text{ is a prime} \geq 3, \\[2mm] \dfrac{n + 1}{2} & \text{if } n + 1 \text{ is a prime} \geq 5. \end{cases}$$

The ratio is sometimes close to zero, sometimes is very large, and sometimes is $1/2$. It has no limit because there are infinitely many primes. A test that will answer the question (affirmatively—yes, the series does converge) is the nth-Root Test. To apply it, we consider the following:

$$\sqrt[n]{a_n} = \frac{\sqrt[n]{f(n)}}{2} = \begin{cases} \dfrac{\sqrt[n]{n}}{2} & \text{if } n \text{ is a prime,} \\[2mm] \dfrac{1}{2} & \text{otherwise.} \end{cases}$$

In this example, therefore, it will always be true that

$$\frac{1}{2} \leq \sqrt[n]{a_n} \leq \frac{\sqrt[n]{n}}{2}$$

and $\lim \sqrt[n]{a_n} = 1/2$ (by the Sandwich Limit Theorem). Because this limit is less than 1, the *nth-Root Test* tells us that the given series converges, as we shall now see. ■

The *n*th-Root Test

Let $\Sigma\, a_n$ be a series with $a_n \geq 0$ for $n \geq n_0$, and suppose that

$$\sqrt[n]{a_n} \to \rho.$$

Then

    a) the series *converges* if $\rho < 1$,

    b) the series *diverges* if $\rho > 1$,

    c) the test is *not conclusive* if $\rho = 1$.

Proof

a) Suppose that $\rho < 1$, and choose an $\epsilon > 0$ so small that $\rho + \epsilon < 1$ also. Since $\sqrt[n]{a_n} \to \rho$, the terms $\sqrt[n]{a_n}$ eventually get closer than ϵ to ρ. In other words, there exists an index $N \geq n_0$ such that

$$\sqrt[n]{a_n} < \rho + \epsilon \qquad \text{when } n \geq N.$$

Then it is also true that

$$a_n < (\rho + \epsilon)^n \qquad \text{for } n \geq N.$$

Now, $\Sigma_{n=N}^{\infty} (\rho + \epsilon)^n$, a geometric series with ratio $(\rho + \epsilon) < 1$, converges. By comparison, $\Sigma_{n=N}^{\infty} a_n$ converges, from which it follows that

$$\sum_{n=1}^{\infty} a_n = a_1 + \cdots + a_{N-1} + \sum_{n=N}^{\infty} a_n$$

converges.

b) Suppose that $\rho > 1$. Then, for all indices beyond some index M, we have $\sqrt[n]{a_n} > 1$, so that $a_n > 1$ for $n > M$, and the terms of the series do not converge to zero. The series therefore diverges by the *n*th-Term Test.

c) The series $\Sigma_{n=1}^{\infty} (1/n)$ and $\Sigma_{n=1}^{\infty} (1/n^2)$ show that the test is not conclusive when $\rho = 1$. The first series diverges and the second converges, but in both cases $\sqrt[n]{a_n} \to 1$. ■

EXAMPLE 5 For the series $\Sigma_{n=1}^{\infty} (1/n^n)$,

$$\sqrt[n]{\frac{1}{n^n}} = \frac{1}{n} \to 0.$$

The series converges. ■

EXAMPLE 6   For the series $\sum_{n=1}^{\infty} (2^n/n^2)$,

$$\sqrt[n]{\frac{2^n}{n^2}} = \frac{2}{\sqrt[n]{n^2}} = \frac{2}{(\sqrt[n]{n})^2} \to \frac{2}{1^2} = 2.$$

The series diverges.   ■

EXAMPLE 7   For the series $\sum_{n=1}^{\infty} (1 - 1/n)^n = 0 + (1/4) + (8/27) + \cdots$,

$$\sqrt[n]{\left(1 - \frac{1}{n}\right)^n} = \left(1 - \frac{1}{n}\right) \to 1.$$

Because $\rho = 1$, the Root Test is not conclusive. However, if we apply the nth-Term Test for divergence, we find that

$$\left(1 - \frac{1}{n}\right)^n = \left(1 + \frac{-1}{n}\right)^n \to e^{-1} = \frac{1}{e}.$$

The series diverges.   ■

List of Tests

We now have nine tests for divergence and convergence of infinite series:

1. The Geometric Series Test.
2. The nth-Term Test for divergence (applies to all series).
3. The "bounded from above" test (applies to partial sums of nonnegative series).
4. A. The Comparison Test for convergence (nonnegative series for $n \geq n_0$).
   B. The Comparison Test for divergence (nonnegative series for $n \geq n_0$).
5. A. The Limit Comparison Test for convergence (as for 4A).
   B. The Limit Comparison Test for divergence (as for 4B).
6. The Simplified Limit Comparison Test (positive series).
7. The Integral Test (positive decreasing series).
8. The Ratio Test (positive series).
9. The nth-Root Test (nonnegative series for $n \geq n_0$).

NOTE: These tests can also be applied to settle questions about the convergence or divergence of series of nonpositive or negative terms. Just factor -1 from the series in question and test the resulting series of nonnegative or positive terms.

EXAMPLE 8

$$\sum_{n=1}^{\infty} -\frac{1}{n} = -1 \cdot \sum_{n=1}^{\infty} \frac{1}{n} \text{ diverges.}$$ ■

EXAMPLE 9

$$\sum_{n=0}^{\infty} -\frac{1}{2^n} = -1 \cdot \sum_{n=0}^{\infty} \frac{1}{2^n} = -1 \cdot 2 = -2$$ ■

PROBLEMS

In Problems 1–26, determine whether the given series converges or diverges. Give reasons for your answers.

1. $\displaystyle\sum_{n=1}^{\infty} \frac{n^2}{2^n}$

2. $\displaystyle\sum_{n=1}^{\infty} \frac{n!}{10^n}$

3. $\displaystyle\sum_{n=1}^{\infty} \frac{n^{10}}{10^n}$

4. $\displaystyle\sum_{n=1}^{\infty} n^2 e^{-n}$

5. $\displaystyle\sum_{n=1}^{\infty} \left(\frac{n-2}{n}\right)^n$

6. $\displaystyle\sum_{n=1}^{\infty} \frac{2 + (-1)^n}{1.25^n}$

7. $\displaystyle\sum_{n=1}^{\infty} n! e^{-n}$

8. $\displaystyle\sum_{n=1}^{\infty} \frac{(-2)^n}{3^n}$

9. $\displaystyle\sum_{n=1}^{\infty} \left(1 - \frac{3}{n}\right)^n$

10. $\displaystyle\sum_{n=1}^{\infty} \left(1 - \frac{1}{n^2}\right)^n$

11. $\displaystyle\sum_{n=1}^{\infty} \sin\left(\frac{1}{n}\right)$

12. $\displaystyle\sum_{n=1}^{\infty} \sin\left(\frac{2}{n^2}\right)$

13. $\displaystyle\sum_{n=1}^{\infty} \left[1 - \cos\left(\frac{1}{n}\right)\right]$

14. $\displaystyle\sum_{n=1}^{\infty} \sin^2\left(\frac{1}{n}\right)$

15. $\displaystyle\sum_{n=1}^{\infty} \tan\left(\frac{\ln n}{n}\right)$

16. $\displaystyle\sum_{n=1}^{\infty} \tan(2^{-n} \ln n)$

17. $\displaystyle\sum_{n=1}^{\infty} \ln\left(\frac{n+2}{n+1}\right)$

18. $\displaystyle\sum_{n=1}^{\infty} \ln\left(\frac{n^2+1}{n^2}\right)$

19. $\displaystyle\sum_{n=2}^{\infty} n \sin\left(\frac{1}{n^2}\right)$

20. $\displaystyle\sum_{n=1}^{\infty} n^2 \tan\left(\frac{2+n}{n^2+5}\right)$

21. $\displaystyle\sum_{n=1}^{\infty} \frac{(n+1)(n+2)}{n!}$

22. $\displaystyle\sum_{n=1}^{\infty} e^{-n}(n^3)$

23. $\displaystyle\sum_{n=1}^{\infty} \frac{(n+3)!}{3! n! 3^n}$

24. $\displaystyle\sum_{n=1}^{\infty} -\frac{n^2}{2^n}$

25. $\displaystyle\sum_{n=1}^{\infty} \frac{1}{(2n+1)!}$

26. $\displaystyle\sum_{n=1}^{\infty} \frac{n!}{n^n}$

In Problems 27–32, find all values of x for which the given series converges. *Suggestion:* Look first for geometric series, then try the Ratio Test or nth-Root Test, and apply other tests if needed.

27. $\displaystyle\sum_{n=1}^{\infty} \frac{|nx^n|}{2^n}$

28. $\displaystyle\sum_{n=1}^{\infty} \frac{(n+1)x^{2n-1}}{4^n}$

29. $\displaystyle\sum_{n=1}^{\infty} \left(\frac{x^2+1}{3}\right)^n$

30. $\displaystyle\sum_{n=1}^{\infty} \left(\frac{1}{|x|}\right)^n$

31. $\displaystyle\sum_{n=1}^{\infty} \frac{x^{2n+1}}{n^2}$

32. $\displaystyle\sum_{n=1}^{\infty} \frac{n! n! |2x|^n}{(2n)!}$

In Problems 33–41, the terms of a series are defined by the formulas given. Does the series $\sum_{n=1}^{\infty} a_n$ converge, or diverge? Give reasons for your answers.

33. $a_1 = 2, \quad a_{n+1} = \dfrac{1 + \sin n}{n} a_n$

34. $a_1 = \dfrac{1}{3}, \quad a_{n+1} = \dfrac{3n-1}{2n+5} a_n$

35. $a_1 = 3, \quad a_{n+1} = \dfrac{n}{n+1} a_n$

36. $a_1 = 2, \quad a_{n+1} = \dfrac{2}{n} a_n$

37. $a_1 = -1, \quad a_{n+1} = \dfrac{1 + \ln n}{n} a_n$

38. $a_1 = \dfrac{1}{2}, \quad a_{n+1} = \dfrac{n + \ln n}{n + 10} a_n$

39. $a_n = \dfrac{2^n n! n!}{(2n)!}$

40. $a_n = \dfrac{(3n)!}{n!(n+1)!(n+2)!}$

41. $a_1 = 1, \quad a_{n+1} = \dfrac{n(n+1)}{(n+2)(n+3)} a_n$ (*Hint:* Write out several terms, see what factors cancel, and then generalize.)

11.7

Absolute Convergence

We now extend to series that have both positive and negative terms the techniques we have developed for answering questions about the convergence of series of nonnegative numbers. The extension is made possible by a theorem that says that, if a series converges after all its negative terms have been made positive, then the unaltered series converges also.

THEOREM 11

> **The Absolute Convergence Theorem**
>
> If $\sum_{n=1}^{\infty} |a_n|$ converges, then $\sum_{n=1}^{\infty} a_n$ converges.

Proof of the Theorem   For each n,

$$-|a_n| \leq a_n \leq |a_n|,$$

so

$$0 \leq a_n + |a_n| \leq 2|a_n|.$$

If $\sum_{n=1}^{\infty} |a_n|$ converges, then $\sum_{n=1}^{\infty} 2|a_n|$ converges and, by the Comparison Test, the nonnegative series

$$\sum_{n=1}^{\infty} (a_n + |a_n|)$$

converges. The equality $a_n = (a_n + |a_n|) - |a_n|$ now lets us express $\sum_{n=1}^{\infty} a_n$ as the difference of two convergent series:

$$\sum_{n=1}^{\infty} a_n = \sum_{n=1}^{\infty} (a_n + |a_n| - |a_n|) = \sum_{n=1}^{\infty} (a_n + |a_n|) - \sum_{n=1}^{\infty} |a_n|.$$

Therefore, $\sum_{n=1}^{\infty} a_n$ converges.   ■

An obvious corollary of Theorem 11 (which is also called the *contrapositive* form of the theorem) is as follows.

COROLLARY

> If $\sum a_n$ diverges, then $\sum |a_n|$ diverges.

(See Problem 19.)

DEFINITION

> A series $\sum_{n=1}^{\infty} a_n$ is said to **converge absolutely** if $\sum_{n=1}^{\infty} |a_n|$ converges.

Our theorem can now be rephrased to say that *every absolutely convergent series converges.* We shall see in the next article, however, that the converse of this statement is false. Many convergent series do not converge absolutely. That is, there are many series whose convergence depends on the presence of infinitely many positive and negative terms arranged in a particular order.

Here are some examples of how the theorem can and cannot be used to determine convergence.

EXAMPLE 1   For

$$\sum_{n=1}^{\infty} (-1)^{n+1} \frac{1}{n^2} = 1 - \frac{1}{4} + \frac{1}{9} - \frac{1}{16} + \cdots,$$

the corresponding series of absolute values is

$$\sum_{n=1}^{\infty} \frac{1}{n^2} = 1 + \frac{1}{4} + \frac{1}{9} + \frac{1}{16} + \cdots,$$

which converges because it is a *p*-series with $p = 2 > 1$ (Article 11.5). Therefore

$$\sum_{n=1}^{\infty} (-1)^{n+1} \frac{1}{n^2}$$

converges absolutely. Therefore

$$\sum_{n=1}^{\infty} (-1)^{n+1} \frac{1}{n^2}$$

converges.

EXAMPLE 2   For

$$\sum_{n=1}^{\infty} \frac{\sin n}{n^2} = \frac{\sin 1}{1} + \frac{\sin 2}{4} + \frac{\sin 3}{9} + \cdots,$$

the corresponding series of absolute values is

$$\sum_{n=1}^{\infty} \left| \frac{\sin n}{n^2} \right| = \frac{|\sin 1|}{1} + \frac{|\sin 2|}{4} + \cdots,$$

which converges by comparison with $\sum_{n=1}^{\infty}(1/n^2)$ because $|\sin n| \leq 1$ for every *n*. The original series converges absolutely; therefore it converges.

EXAMPLE 3   For

$$\sum_{n=1}^{\infty} (-1)^{n+1} \frac{1}{n} = 1 - \frac{1}{2} + \frac{1}{3} - \frac{1}{4} + \cdots,$$

the corresponding series of absolute values is

$$\sum_{n=1}^{\infty} \frac{1}{n} = 1 + \frac{1}{2} + \frac{1}{3} + \frac{1}{4} + \cdots,$$

which diverges. *We can draw no conclusion from this about the convergence or divergence of the original series.* Some other test must be found. In fact, the original series converges, as we shall see in the next article.

EXAMPLE 4   The series

$$\sum_{n=1}^{\infty} (-1)^n \frac{n}{5n + 1} = -\frac{1}{6} + \frac{2}{11} - \frac{3}{16} + \frac{4}{21} - \cdots$$

does not converge, by the *n*th-Term Test. Therefore, by the corollary of Theorem 11, the series does not converge absolutely.

REMARK   We know that $\Sigma \, a_n$ converges if $\Sigma \, |a_n|$ converges, but the two series will generally not converge to the same sum. For example,

$$\sum_{n=0}^{\infty} \left| \frac{(-1)^n}{2^n} \right| = \sum_{n=0}^{\infty} \frac{1}{2^n} = \frac{1}{1 - (1/2)} = 2,$$

while

$$\sum_{n=0}^{\infty} \frac{(-1)^n}{2^n} = \frac{1}{1 + (1/2)} = \frac{2}{3}.$$

In fact, when a series $\Sigma \, a_n$ converges absolutely, we can expect $\Sigma \, a_n$ to equal $\Sigma \, |a_n|$ only if none of the numbers a_n is negative.

One other important fact about absolutely convergent series is the following theorem.

THEOREM 12

> **The Rearrangement Theorem for Absolutely Convergent Series**
>
> If $\Sigma \, a_n$ converges absolutely, and $b_1, b_2, \ldots, b_n, \ldots$ is any arrangement of the sequence $\{a_n\}$, then $\Sigma \, b_n$ converges and
>
> $$\sum_{n=1}^{\infty} b_n = \sum_{n=1}^{\infty} a_n.$$

(For an outline of the proof, see Problem 22.)

EXAMPLE 5   As we saw in Example 1, the series

$$1 - \left(\frac{1}{4} \right) + \left(\frac{1}{9} \right) - \left(\frac{1}{16} \right) + \cdots + (-1)^{n-1} \left(\frac{1}{n^2} \right) + \cdots$$

converges absolutely. A possible rearrangement of the terms of the series might start with a positive term, then 2 negative terms, then 3 positive terms, then 4 negative terms, and so on: After k terms of one sign, take $k + 1$ terms of the opposite sign. The first ten terms of this series look like this:

$$1 - \frac{1}{4} - \frac{1}{16} + \frac{1}{9} + \frac{1}{25} + \frac{1}{49} - \frac{1}{36} - \frac{1}{64} - \frac{1}{100} - \frac{1}{144} + \cdots.$$

The Rearrangement Theorem says that both series converge to the same value. In this example, if we had the second series to begin with, we would probably be glad to exchange it for the first, if we knew that we could. We can do even better: The sum of either series is also equal to

$$\sum_{n=1}^{\infty} \frac{1}{(2n-1)^2} - \sum_{n=1}^{\infty} \frac{1}{(2n)^2}.$$

(See Problem 23.)   ∎

Multiplication of Series

Don't panic: It is possible and often useful to multiply two series, by applying the distributive law. For convenience of notation, we shall write the first series as

$$a_0 + a_1 + a_2 + \cdots + a_n + \cdots \tag{1}$$

and the second as

$$b_0 + b_1 + b_2 + \cdots + b_n + \cdots \tag{2}$$

and the product as

$$a_0b_0 + (a_0b_1 + a_1b_0) + (a_0b_2 + a_1b_1 + a_2b_0) + \cdots$$
$$+ (a_0b_n + a_1b_{n-1} + \cdots + a_kb_{n-k} + \cdots + a_nb_0) + \cdots. \tag{3}$$

Thus every term in the first series appears as a factor with every term in the second series, in the form a_kb_{n-k}.

EXAMPLE 6 Let

$$a_0 = 1, \qquad a_1 = \frac{1}{2}, \qquad \cdots, \qquad a_n = \frac{1}{2^n} \tag{4}$$

and

$$b_0 = 1, \qquad b_1 = \frac{1}{3}, \qquad \cdots, \qquad b_n = \frac{1}{3^n}. \tag{5}$$

The products in (3) are

$$a_0b_0 = 1, \quad a_0b_1 = \frac{1}{3}, \quad a_1b_0 = \frac{1}{2}, \quad a_0b_2 = \frac{1}{9}, \quad a_1b_1 = \frac{1}{6}, \quad a_2b_0 = \frac{1}{4}$$

and, in general, all terms are of the form

$$a_kb_{n-k} = \frac{1}{2^k3^{n-k}}, \quad n = 0, 1, 2, \ldots, \quad k = 0, 1, 2, \ldots, n. \qquad \blacksquare$$

The following theorem tells us that the sum of all these products is the sum of the first series, which is 2, times the sum of the second series, which is 3/2 (both are geometric series: $r_1 = 1/2$ and $r_2 = 1/3$). By the way in which the terms of the product series are formed, we see that every number that is not divisible by a prime greater than 3 appears as a denominator, precisely once. By rearranging the terms (using Theorem 12) we can indicate the sum as

$$\frac{1}{1} + \frac{1}{2} + \frac{1}{3} + \frac{1}{4} + \frac{1}{6} + \frac{1}{8} + \frac{1}{9} + \frac{1}{12} + \frac{1}{16} + \frac{1}{18} + \frac{1}{24} + \frac{1}{27} + \cdots. \tag{6}$$

The sum of the series in (6) is 3.

We now state Theorem 13 (without proof).

THEOREM 13

The Series Multiplication Theorem

Let $\Sigma\, a_n$ and $\Sigma\, b_n$ be absolutely convergent series. Define $\Sigma\, c_n$ by the equations

$$c_0 = a_0b_0, \qquad c_1 = a_0b_1 + a_1b_0, \ldots, c_n = \sum_{k=0}^{n} a_kb_{n-k}. \tag{7}$$

Then $\Sigma\, c_n$ converges absolutely and

$$\sum_{n=0}^{\infty} c_n = \left(\sum_{n=0}^{\infty} a_n\right) \times \left(\sum_{n=0}^{\infty} b_n\right). \tag{8}$$

EXAMPLE 7   The sum of the reciprocals of those integers that can be expressed as products of the form $3^a 7^b$, where a and b independently take all nonnegative integer values, is

$$\left(1 + \frac{1}{3} + \frac{1}{9} + \cdots + \frac{1}{3^n} + \cdots\right) \times \left(1 + \frac{1}{7} + \frac{1}{49} + \cdots + \frac{1}{7^n} + \cdots\right)$$

$$= \frac{1}{1-(1/3)} \times \frac{1}{1-(1/7)} = \frac{3}{2} \times \frac{7}{6} = \frac{7}{4}. \qquad \blacksquare$$

We shall do more with multiplication of series in Chapter 12.

PROBLEMS

In Problems 1–18, determine whether the series converge absolutely. In each case give a reason for the convergence or divergence of the corresponding series of absolute values.

1. $\displaystyle\sum_{n=1}^{\infty} \frac{1}{n^2}$

2. $\displaystyle\sum_{n=1}^{\infty} \frac{1}{(-n)^3}$

3. $\displaystyle\sum_{n=1}^{\infty} \frac{1-n}{n^2}$

4. $\displaystyle\sum_{n=1}^{\infty} \left(-\frac{1}{5}\right)^n$

5. $\displaystyle\sum_{n=1}^{\infty} \frac{-1}{n^2 + 2n + 1}$

6. $\displaystyle\sum_{n=1}^{\infty} \frac{(-1)^n}{2n}$

7. $\displaystyle\sum_{n=1}^{\infty} \frac{\cos n\pi}{n\sqrt{n}}$

8. $\displaystyle\sum_{n=1}^{\infty} \frac{-10}{n}$

9. $\displaystyle\sum_{n=0}^{\infty} \frac{(-1)^n}{(2n)!}$

10. $\displaystyle\sum_{n=0}^{\infty} \frac{(-1)^n}{(2n+1)!}$

11. $\displaystyle\sum_{n=2}^{\infty} (-1)^n \frac{n}{n+1}$

12. $\displaystyle\sum_{n=1}^{\infty} \frac{-n}{2^n}$

13. $\displaystyle\sum_{n=1}^{\infty} (5)^{-n}$

14. $\displaystyle\sum_{n=1}^{\infty} \left(\frac{1}{2^n} - 1\right)$

15. $\displaystyle\sum_{n=1}^{\infty} \frac{(-100)^n}{n!}$

16. $\displaystyle\sum_{n=2}^{\infty} (-1)^n \frac{\ln n}{\ln n^2}$

17. $\displaystyle\sum_{n=1}^{\infty} \frac{2-n}{n^3}$

18. $\displaystyle\sum_{n=1}^{\infty} \left(\frac{1}{2^n} - \frac{1}{3^n}\right)$

19. Show that if $\Sigma_{n=1}^{\infty}\, a_n$ diverges, then $\Sigma_{n=1}^{\infty}\, |a_n|$ diverges.

20. Show that if $\Sigma_{n=1}^{\infty}\, a_n$ converges absolutely, then

$$\left|\sum_{n=1}^{\infty} a_n\right| \le \sum_{n=1}^{\infty} |a_n|.$$

21. Show that if $\Sigma_{n=1}^{\infty}\, a_n$ and $\Sigma_{n=1}^{\infty}\, b_n$ both converge absolutely, then so does

a) $\Sigma\, (a_n + b_n)$ b) $\Sigma\, (a_n - b_n)$
c) $\Sigma\, ka_n$ (k any number)

22. Prove Theorem 12. *Outline of proof:* Assume the hypotheses. Let $\epsilon > 0$. Show there is an index N_1 such that

$$\sum_{n=N_1}^{\infty} |a_n| < \frac{\epsilon}{2}$$

and an index $N_2 \ge N_1$ such that $|s_{N_2} - L| < \epsilon/2$, where $s_n = \Sigma_{k=1}^n\, a_k$. Because all the terms $a_1, a_2, \ldots, a_{N_2}$ appear somewhere in the sequence $\{b_n\}$, there is an index $N_3 \ge N_2$ such that if $n \ge N_3$, then $\Sigma_{k=1}^n\, b_k - s_{N_2}$ is, at most, a sum of terms of the form a_m with $m \ge N_1$. Therefore, if $n \ge N_3$,

$$\left|\sum_{k=1}^{n} b_k - L\right| \le \left|\sum_{k=1}^{n} b_k - s_{N_2}\right| + |s_{N_2} - L|$$

$$\le \sum_{k=N_1}^{\infty} |a_k| + |s_{N_2} - L| < \epsilon.$$

23. Establish the following: If $\Sigma\, |a_n|$ converges and $b_n = a_n$ when $a_n \ge 0$, while $b_n = 0$ when $a_n < 0$, then $\Sigma\, b_n$ converges. Likewise $\Sigma\, c_n$ converges, where $c_n = 0$ when $a_n \ge 0$ and $c_n = -a_n$ when $a_n < 0$. In other words, when the original series converges absolutely, its positive terms by themselves form a convergent series, and so do the negative terms. And

$$\sum_{n=1}^{\infty} a_n = \sum_{n=1}^{\infty} b_n - \sum_{n=1}^{\infty} c_n$$

because $b_n = (a_n + |a_n|)/2$ and $c_n = (-a_n + |a_n|)/2$.

24. In Example 6, where $a_n = 1/2^n$ and $b_n = 1/3^n$, let $c_n = a_0 b_n + a_1 b_{n-1} + \cdots + a_n b_0$, as in Eq. (3). Since the number $a_k b_{n-k}$ is less than or equal to $(1/2)^n$, show that $c_n \leq (n + 1)/2^n$ and then prove that $\Sigma\, c_n$ converges. Does it converge absolutely?

TOOLKIT PROGRAMS

Sequences and Series

11.8
Alternating Series and Conditional Convergence

A series in which the terms are alternately positive and negative is called an **alternating series.** Here are three examples:

$$1 - \frac{1}{2} + \frac{1}{3} - \frac{1}{4} + \frac{1}{5} - \cdots + \frac{(-1)^{n+1}}{n} + \cdots \tag{1}$$

$$-2 + 1 - \frac{1}{2} + \frac{1}{4} - \frac{1}{8} + \cdots + \frac{(-1)^n 4}{2^n} + \cdots \tag{2}$$

$$1 - 2 + 3 - 4 + 5 - 6 + \cdots + (-1)^{n-1} n + \cdots \tag{3}$$

Do these series converge, or diverge?

Series (1), called the **alternating harmonic series,** converges, as we shall see in a moment. However, it does not converge absolutely because the corresponding series of absolute values is the harmonic series

$$1 + \frac{1}{2} + \frac{1}{3} + \frac{1}{4} + \cdots + \frac{1}{n} + \cdots,$$

which we know diverges. A series like this one, which converges but does not converge absolutely, is said to **converge conditionally.**

Series (2) is a geometric series with ratio $r = -1/2$. We know that it converges and its sum is $-2/(1 + 1/2) = -4/3$. It also converges absolutely because the corresponding series of absolute values is

$$2 + 1 + \frac{1}{2} + \frac{1}{4} + \frac{1}{8} + \cdots + \frac{4}{2^n} + \cdots$$

Series (3) diverges because the nth term does not approach zero.

We can prove the convergence of the alternating harmonic series by applying a more general result, known as **Leibniz's theorem.**

THEOREM 14

Leibniz's Theorem

The series

$$\sum_{n=1}^{\infty} (-1)^{n+1} a_n = a_1 - a_2 + a_3 - a_4 + \cdots \tag{4}$$

converges if all three of the following conditions are satisfied:

1. The a_n's are all positive.
2. $a_n \geq a_{n+1}$ for all n.
3. $a_n \to 0$.

Proof If n is an even integer, say $n = 2m$, then the sum of the first n terms is

$$S_{2m} = (a_1 - a_2) + (a_3 - a_4) + \cdots + (a_{2m-1} - a_{2m})$$
$$= a_1 - (a_2 - a_3) - (a_4 - a_5) - \cdots - (a_{2m-2} - a_{2m-1}) - a_{2m}. \tag{5}$$

The first equality shows that S_{2m} is the sum of m nonnegative terms, since each term in parentheses is positive or zero. Hence $S_{2m+2} \geq S_{2m}$, and the sequence $\{S_{2m}\}$ is nondecreasing. The second equality shows that $S_{2m} \leq a_1$. Since $\{S_{2m}\}$ is nondecreasing and is bounded from above, it has a limit, say

$$\lim_{m \to \infty} S_{2m} = L. \tag{6}$$

If n is an odd integer, say $n = 2m + 1$, then the sum of the first n terms is

$$S_{2m+1} = S_{2m} + a_{2m+1}.$$

Since $a_n \to 0$,

$$\lim_{m \to \infty} a_{2m+1} = 0$$

and, as $m \to \infty$,

$$S_{2m+1} = S_{2m} + a_{2m+1} \to L + 0 = L. \tag{7}$$

When we combine the results of (6) and (7), we get

$$\lim_{n \to \infty} S_n = L. \qquad \blacksquare$$

Here are some examples of what Theorem 14 can do.

EXAMPLE 1 The *alternating harmonic series*

$$\sum_{n=1}^{\infty} (-1)^{n+1} \frac{1}{n} = 1 - \frac{1}{2} + \frac{1}{3} - \frac{1}{4} + \cdots$$

satisfies the three requirements of the theorem; therefore it converges. It converges conditionally because the corresponding series of absolute values is the harmonic series, which diverges. ▪

EXAMPLE 2 The series

$$\sum_{n=1}^{\infty} (-1)^{n+1}\sqrt{n} = 1 - \sqrt{2} + \sqrt{3} - \sqrt{4} + \cdots$$

diverges by the nth-Term Test. It fails to satisfy conditions 2 and 3 of Leibniz's theorem. ∎

EXAMPLE 3 Theorem 14 gives no information about

$$\frac{2}{1} - \frac{1}{1} + \frac{2}{2} - \frac{1}{2} + \frac{2}{3} - \frac{1}{3} + \cdots + \frac{2}{n} - \frac{1}{n} + \cdots.$$

The sequence 2/1, 1/1, 2/2, 1/2, 2/3, 1/3, . . . is not a monotone sequence. Some other test must be found. When we group the terms of the series in consecutive pairs

$$\left(\frac{2}{1} - \frac{1}{1}\right) + \left(\frac{2}{2} - \frac{1}{2}\right) + \left(\frac{2}{3} - \frac{1}{3}\right) + \cdots + \left(\frac{2}{n} - \frac{1}{n}\right) + \cdots,$$

we see that the $2n$th partial sum of the given series is the same number as the nth partial sum of the harmonic series. Thus the sequence of partial sums, and hence the series, diverges. ∎

EXAMPLE 4 Does the series $\sum_{n=2}^{\infty} (-1)^n (\ln n)/(n+1)$ converge? While it is clear that this is an alternating series and $a_n = (\ln n)/(n+1)$ approaches 0 as $n \to \infty$, it isn't clear at first glance that $a_n \geq a_{n+1}$. We consider the corresponding function

$$f(x) = \frac{\ln x}{x+1}$$

whose derivative is

$$f'(x) = \frac{(x+1)/x - \ln x}{(x+1)^2}.$$

The derivative is negative, and the function f is decreasing, if $\ln x$ is greater than $(x+1)/x = 1 + 1/x$. For $x = 1, 2, \ldots$, we have $1 + (1/x) \leq 2$, so the sequence decreases when $\ln n > 2$. Since $e^2 \approx 7.4$, we conclude that the original sequence decreases for $n \geq 8$. That is, we can apply Leibniz's theorem to conclude that $\sum_{n=8}^{\infty} (-1)^n (\ln n)/(n+1)$ converges. Therefore the original series also converges. ∎

We use the following graphical interpretation of the partial sums to gain added insight into the way in which an alternating series converges to its limit L when the three conditions of the theorem are satisfied. Starting from the origin O on a scale of real numbers (Fig. 11.14), we lay off the positive distance

$$s_1 = a_1.$$

To find the point corresponding to

$$s_2 = a_1 - a_2,$$

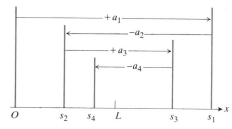

11.14 The partial sums of an alternating series that satisfies the hypotheses of Leibniz's theorem straddle their limit.

we must back up a distance equal to a_2. Since $a_2 \leq a_1$, we do not back up any farther than O at most. Next we go forward a distance a_3 and mark the point corresponding to

$$s_3 = a_1 - a_2 + a_3.$$

Since $a_3 \leq a_2$, we go forward by an amount that is no greater than the previous backward step; that is, $s_3 \leq s_1$. We continue in this seesaw fashion, backing up or going forward as the signs in the series demand. But each forward or backward step is shorter than (or at most the same size as) the preceding step, because $a_{n+1} \leq a_n$. And since the nth term approaches zero as n increases, the size of step we take forward or backward gets smaller and smaller. We thus oscillate across the limit L, but the amplitude of oscillation continually decreases and approaches zero as its limit. The even-numbered partial sums s_2, s_4, s_6, . . . , s_{2m} continually increase toward L, while the odd-numbered sums s_1, s_3, s_5, . . . , s_{2m+1} continually decrease toward L. The limit L is between any two successive sums s_n and s_{n+1} and hence differs from s_n by an amount less than a_{n+1}.

It is because

$$|L - s_n| < a_{n+1} \qquad \text{for every } n \tag{8}$$

that we can make useful estimates of the sums of convergent alternating series.

THEOREM 15

The Alternating Series Estimation Theorem

If the series

$$\sum_{n=1}^{\infty} (-1)^{n+1} a_n$$

is an alternating series that satisfies the three conditions of Leibniz's theorem, then

$$s_n = a_1 - a_2 + \cdots + (-1)^{n+1} a_n$$

approximates the sum L of the series with an error whose absolute value is less than a_{n+1}, the numerical value of the first unused term. Furthermore, the remainder, $L - s_n$, has the same sign as the first unused term.

We will leave the determination of the sign of the remainder as an exercise (see Problem 37).

EXAMPLE 5 Let us first try the Estimation Theorem on an alternating series whose sum we already know, namely, the geometric series

$$\sum_{n=0}^{\infty} (-1)^n \frac{1}{2^n} = 1 - \frac{1}{2} + \frac{1}{4} - \frac{1}{8} + \frac{1}{16} - \frac{1}{32} + \frac{1}{64} - \frac{1}{128} \; \vdots \; + \frac{1}{256} - \cdots.$$

Theorem 15 says that when we truncate the series after the eighth term, we throw away a total that is positive and less than $1/256$. A rapid calculation shows that the sum of the first eight terms is

$$0.6640625.$$

The sum of the series is

$$\frac{1}{1-(-1/2)} = \frac{1}{3/2} = \frac{2}{3}. \tag{9}$$

The difference,

$$\frac{2}{3} - 0.6640625 = 0.0026041666 \ldots ,$$

is positive and less than

$$\frac{1}{256} = 0.00390625.$$ ■

Series for $\ln(1 + x)$

We shall show in Chapter 12 that a series for computing $\ln(1 + x)$ when $|x| < 1$ is

$$\ln(1 + x) = x - \frac{x^2}{2} + \frac{x^3}{3} - \cdots + (-1)^{n+1}\frac{x^n}{n}\cdots. \tag{10}$$

For $0 < x < 1$, this series satisfies all three conditions of Theorem 14, and we may use the Estimation Theorem to see how good an approximation of $\ln(1 + x)$ we get from the first few terms of the series.

EXAMPLE 6   Calculate $\ln(1.1)$ with the approximation

$$\ln(1 + x) \approx x - \frac{x^2}{2}, \tag{11}$$

and estimate the error involved. Is $x - (x^2/2)$ too large or too small in this case?

Solution

$$\ln(1.1) \approx (0.1) - \frac{(0.1)^2}{2} = 0.095$$

This approximation differs from the exact value of $\ln 1.1$ by less than

$$\frac{(0.1)^3}{3} = 0.000333 \ldots .$$

Since the sign of this, the first unused term, is positive, the remainder is positive. That is, $0.095 < \ln 1.1 < 0.09533\overline{3}$.   ■

EXAMPLE 7   How many terms of the series (10) do we need to use in order to be sure of calculating $\ln(1.2)$ with an error of less than 10^{-6}?

Solution

$$\ln(1.2) = (0.2) - \frac{(0.2)^2}{2} + \frac{(0.2)^3}{3} - \cdots$$

We find by trial that the eighth term

$$-\frac{(0.2)^8}{8} = -3.2 \times 10^{-7}$$

is the first term in the series whose absolute value is less than 10^{-6}. Therefore the sum of the first seven terms will give $\ln(1.2)$ with an error of less than 10^{-6}. The use of more terms would give an approximation that is even better, but seven terms are enough to guarantee the accuracy we wanted. Note also that we have not shown that six terms would *not* provide that accuracy. ∎

CAUTION ABOUT REARRANGEMENTS If we rearrange an infinite number of terms of a conditionally convergent series, we can get results that are far different from the sum of the original series. The next example illustrates some of the things that can happen.

EXAMPLE 8 Consider the alternating harmonic series

$$\frac{1}{1} - \frac{1}{2} + \frac{1}{3} - \frac{1}{4} + \frac{1}{5} - \frac{1}{6} + \frac{1}{7} - \frac{1}{8} + \frac{1}{9} - \frac{1}{10} + \frac{1}{11} - \cdots.$$

Here, the series of terms $\Sigma\, 1/(2n-1)$ diverges to $+\infty$ and the series of terms $\Sigma\, -1/2n$ diverges to $-\infty$. We can always add enough positive terms (no matter how far out in the sequence of odd-numbered terms we begin) to get an arbitrarily large sum. Similarly, with the negative terms, no matter how far out we start, we can add enough consecutive even-numbered terms to get a negative sum of arbitrarily large absolute value. If we wished to do so, we could start adding odd numbered terms until we had a sum equal to $+3$, say and then follow that with enough consecutive negative terms to make the new total less than -4. We could then add enough positive terms to make the total greater than $+5$ and follow with consecutive unused negative terms to make a new total less than -6, and so on. In this way, we could make the swings arbitrarily large in both the positive and negative directions.

Another possibility, with the same series, is to focus on a particular limit. Suppose we try to get sums that converge to 1. We start with the first term, $1/1$, and then subtract $1/2$. Next we add $1/3$ and $1/5$, which brings the total back to 1 or above. Then we add consecutive negative terms until the total is less than 1. Continue in this manner: When the sum is below 1, add positive terms until the total is 1 or more; then subtract (i.e., add negative) terms until the total is once more less than 1. This process can be continued indefinitely. Because both the odd-numbered terms and the even-numbered terms of the original series approach zero as $n \to \infty$, the amount by which our partial sums exceed 1 or fall below it approaches zero. So the new series converges to 1. The rearranged series starts like this:

$$\frac{1}{1} - \frac{1}{2} + \frac{1}{3} + \frac{1}{5} - \frac{1}{4} + \frac{1}{7} + \frac{1}{9} - \frac{1}{6} + \frac{1}{11} + \frac{1}{13} - \frac{1}{8} + \frac{1}{15} + \frac{1}{17} - \frac{1}{10}$$

$$+ \frac{1}{19} + \frac{1}{21} - \frac{1}{12} + \frac{1}{23} + \frac{1}{25} - \frac{1}{14} + \frac{1}{27} - \frac{1}{16} + \cdots. \quad ∎$$

The kind of behavior illustrated by this example is typical of what can happen with any conditionally convergent series. Moral: Add the terms of such a series in the order given.

PROBLEMS

In Problems 1–10, determine which of the alternating series converge and which diverge.

1. $\displaystyle\sum_{n=1}^{\infty} (-1)^{n+1}\frac{1}{n^2}$

2. $\displaystyle\sum_{n=2}^{\infty} (-1)^{n+1}\frac{1}{\ln n}$

3. $\displaystyle\sum_{n=1}^{\infty} (-1)^{n+1}$

4. $\displaystyle\sum_{n=1}^{\infty} (-1)^{n+1}\frac{10^n}{n^{10}}$

5. $\displaystyle\sum_{n=1}^{\infty} (-1)^{n+1}\frac{\sqrt{n}+1}{n+1}$

6. $\displaystyle\sum_{n=1}^{\infty} (-1)^{n+1}\frac{\ln n}{n}$

7. $\displaystyle\sum_{n=1}^{\infty} (-1)^{n+1}\frac{1}{n^{3/2}}$

8. $\displaystyle\sum_{n=1}^{\infty} (-1)^{n+1}\frac{\ln n}{\ln n^2}$

9. $\displaystyle\sum_{n=1}^{\infty} (-1)^{n} \ln\left(1 + \frac{1}{n}\right)$

10. $\displaystyle\sum_{n=1}^{\infty} (-1)^{n+1}\frac{3\sqrt{n}+1}{\sqrt{n}+1}$

In Problems 11–28, determine whether the series are absolutely convergent, conditionally convergent, or divergent.

11. $\displaystyle\sum_{n=1}^{\infty} (-1)^{n+1}(0.1)^n$

12. $\displaystyle\sum_{n=1}^{\infty} (-1)^{n+1}\frac{1}{\sqrt{n}}$

13. $\displaystyle\sum_{n=1}^{\infty} (-1)^{n+1}\frac{n}{n^3+1}$

14. $\displaystyle\sum_{n=1}^{\infty} \frac{n!}{2^n}$

15. $\displaystyle\sum_{n=1}^{\infty} (-1)^n\frac{1}{n+3}$

16. $\displaystyle\sum_{n=1}^{\infty} (-1)^n\frac{\sin n}{n^2}$

17. $\displaystyle\sum_{n=1}^{\infty} (-1)^{n+1}\frac{3+n}{5+n}$

18. $\displaystyle\sum_{n=2}^{\infty} (-1)^n\frac{1}{\ln n^3}$

19. $\displaystyle\sum_{n=1}^{\infty} (-1)^{n+1}\frac{1+n}{n^2}$

20. $\displaystyle\sum_{n=1}^{\infty} \frac{(-2)^{n+1}}{n+5^n}$

21. $\displaystyle\sum_{n=1}^{\infty} n^2(2/3)^n$

22. $\displaystyle\sum_{n=1}^{\infty} (-1)^{n+1}(\sqrt[n]{10})$

23. $\displaystyle\sum_{n=1}^{\infty} (-1)^n\frac{\tan^{-1}n}{n^2+1}$

24. $\displaystyle\sum_{n=2}^{\infty} (-1)^{n+1}\frac{1}{n \ln n}$

25. $\displaystyle\sum_{n=1}^{\infty} \left(\frac{1}{n} - \frac{1}{2n}\right)$

26. $\displaystyle\sum_{n=1}^{\infty} (-1)^{n+1}\frac{(0.1)^n}{n}$

27. $\displaystyle\sum_{n=1}^{\infty} (-1)^{n+1}(\sqrt{n+1} - \sqrt{n})$

28. $\displaystyle\sum_{n=1}^{\infty} \frac{(-1)^{n+1}(n!)^2}{(2n)!}$

In Problems 29–32, estimate the magnitude of the error if the first four terms are used to approximate the series.

29. $\displaystyle\sum_{n=1}^{\infty} (-1)^{n+1}\frac{1}{n}$

30. $\displaystyle\sum_{n=1}^{\infty} (-1)^{n+1}\frac{1}{10^n}$

31. $\ln(1.01) = \displaystyle\sum_{n=1}^{\infty} (-1)^{n+1}\frac{(0.01)^n}{n}$

32. $\dfrac{1}{1+t} = \displaystyle\sum_{n=0}^{\infty} (-1)^n t^n, \quad 0 < t < 1$

Approximate the sums in Problems 33 and 34 to five decimal places (magnitude of the error less than 5×10^{-6}).

33. $\displaystyle\sum_{n=0}^{\infty} (-1)^n\frac{1}{(2n)!}$ $\left(\begin{array}{l}\text{This is cos 1, the cosine}\\\text{of one radian.}\end{array}\right)$

34. $\displaystyle\sum_{n=0}^{\infty} (-1)^n\frac{1}{n!}$ (This is $1/e$.)

35. a) The series

$$\frac{1}{3} - \frac{1}{2} + \frac{1}{9} - \frac{1}{4} + \frac{1}{27} - \frac{1}{8} + \cdots + \frac{1}{3^n} - \frac{1}{2^n} + \cdots$$

   does not meet one of the conditions of Theorem 14. Which one?

   b) Find the sum of the series in (a).

36. The limit L of an alternating series that satisfies the conditions of Theorem 14 lies between the values of any two consecutive partial sums. This suggests using the average

$$\frac{s_n + s_{n+1}}{2} = s_n + \frac{1}{2}a_{n+1}$$

to estimate L. Compute

$$s_{20} + \frac{1}{2} \cdot \frac{1}{21}$$

as an approximation to the sum of the alternating harmonic series. The exact sum is $\ln 2 = 0.6931 \ldots$

37. Show that whenever an alternating series is approximated by one of its partial sums, if the three conditions of Leibniz's theorem are satisfied, then the *remainder* (sum of the unused terms) has the same sign as the first unused term. (*Hint:* Group the terms of the remainder in consecutive pairs.)

38. Show that the sum of the first $2n$ terms of the series

$$1 - \frac{1}{2} + \frac{1}{2} - \frac{1}{3} + \frac{1}{3} - \frac{1}{4} + \frac{1}{4} - \frac{1}{5} + \frac{1}{5} - \frac{1}{6} + \cdots$$

is the same as the sum of the first n terms of the series

$$\frac{1}{1 \cdot 2} + \frac{1}{2 \cdot 3} + \frac{1}{3 \cdot 4} + \frac{1}{4 \cdot 5} + \frac{1}{5 \cdot 6} + \cdots$$

Do these series converge? What is the sum of the first $2n + 1$ terms of the first series? If the series converge, what is their sum?

39. Show by example that $\sum_{n=1}^{\infty} a_n b_n$ may diverge even though $\sum_{n=1}^{\infty} a_n$ and $\sum_{n=1}^{\infty} b_n$ both converge.

40. Show that the alternating series

$$\sum_{n=2}^{\infty} \frac{(-1)^n}{\sqrt{n} + (-1)^n}$$

diverges. Which of the three conditions of Leibniz's theorem does it not satisfy? (*Hint:* First show that

$$\frac{(-1)^n}{\sqrt{n} + (-1)^n} = \frac{(-1)^n}{\sqrt{n}} - \frac{1}{n + (-1)^n \sqrt{n}}$$

and write the given series as the difference of two series, one of which converges and the other of which diverges. Give reasons for convergence or divergence for each.)

41. **CALCULATOR** In Example 8, suppose the goal is to arrange the terms to get a new series that converges to $-1/2$. Start the new arrangement with the first negative term, which is $-1/2$. Whenever you have a sum that is less than or equal to $-1/2$, start introducing positive terms, taken in order, until the new total is greater than $-1/2$. Then add negative terms until the total is less than or equal to $-1/2$ again. Continue this process until your partial sums have been above the target at least three times and finish at or below it. If s_n is the sum of the first n terms of your new series, plot the points (n, s_n) to illustrate how the sums are behaving.

 TOOLKIT PROGRAMS

Sequences and Series

11.9
Recapitulation

We now have several tests for convergence or divergence of a series, and we introduce a flow chart (Fig. 11.15) to show how the tests might be applied. But such a chart is not to be used as a substitute for good judgment. For example, if you have an alternating series that satisfies the conditions of Leibniz's theorem, the matter is settled right there—you need not start at the top of the chart and work your way down. Similarly, if the series is a geometric series or a p-series, you will probably recognize that fact and know whether the series converges, or diverges.

The word *maybe* is inserted in the chart at places where the answer to the question posed may not be apparent. For example, if $a_n = 3^n n! n!/(2n)!$, can you tell quickly whether $\lim a_n = 0$? With exponents and factorials in the expression for a_n, it is usually a good idea to go to the Ratio Test right away. In this instance, some algebra will show that $a_{n+1}/a_n = 3(n + 1)/(4n + 2)$, and the limit of this ratio is $3/4$. Therefore the series $\sum a_n$ converges and now we also know that $\lim a_n = 0$.

In applying the Ratio Test to $\sum |a_n|$, if $|a_{n+1}/a_n| \geq 1$ for all n, then $|a_{n+1}| \geq |a_n|$ and the terms do not approach zero as $n \to \infty$: the series diverges.

The Ratio Test and nth-Root Test can sometimes be used with equal effectiveness. For practice, it may be a good idea to vary your strategy from time to time to see which of these two tests you prefer. For example, if $a_n = n/2^n$, then

$$\sqrt[n]{a_n} = \frac{1}{2}\sqrt[n]{n} \to \frac{1}{2} \qquad \text{as } n \to \infty$$

and

$$\frac{a_{n+1}}{a_n} = \frac{1}{2}\frac{n + 1}{n} \to \frac{1}{2} \qquad \text{as } n \to \infty.$$

By either test, we see that $\sum a_n$ converges. If you ask, "Which test should I

11.15   Does Σa_n converge?

use?'' we reply, ''Whichever comes to mind first, but don't rely on one to the exclusion of the other.''

When the terms of a series are defined by a recursion formula, it may be possible to determine the matter of convergence directly by applying the Ratio Test. For example, if $a_1 = 1$ and $a_{n+1} = (3/4)a_n$ for all $n \geq 1$, then $a_{n+1}/a_n = 3/4$ is a constant. The series is a geometric series with ratio $|r| < 1$, so it converges. If the ratio a_{n+1}/a_n varies with n, is always less than 1, but approaches 1 as a limit as $n \to \infty$, then it may be best to write out several terms of the series.

EXAMPLE   Suppose that $a_1 = 4$ and $a_{n+1} = (n/(n + 2))a_n$. Does $\Sigma\, a_n$ converge?

Solution   Because the ratio $a_{n+1}/a_n = n/(n + 2)$ is less than 1 but approaches 1 as $n \to \infty$, we write out some terms of the series, using the recursion formula $a_{n+1} = (n/(n + 2))a_n$:

$$n = 1: \quad a_2 = \frac{1}{3}a_1 = \frac{4}{3}$$

$$n = 2: \quad a_3 = \frac{2}{4}a_2 = \frac{4 \cdot 2}{3 \cdot 4}$$

$$n = 3: \quad a_4 = \frac{3}{5}a_3 = \frac{4 \cdot 2 \cdot 3}{3 \cdot 4 \cdot 5} = \frac{4 \cdot 2}{4 \cdot 5}$$

$$n = 4: \quad a_5 = \frac{4}{6}a_4 = \frac{4 \cdot 2 \cdot 3 \cdot 4}{3 \cdot 4 \cdot 5 \cdot 6} = \frac{4 \cdot 2}{5 \cdot 6}.$$

We observe that, for $n = 3, 4,$ and 5, the terms satisfy the formula

$$a_n = \frac{4 \cdot 2}{n(n + 1)}.$$

Indeed, this formula also holds for $n = 1$ and 2, and we can show by mathematical induction that it holds for all n. So the series we are concerned with is $\Sigma\ (4 \times 2)/(n(n + 1))$, and this series converges because its terms are all less than the corresponding terms of the series $\Sigma\ 8/n^2$, which converges. (In fact, we showed in an earlier example that $\Sigma\ 1/(n(n + 1)) = 1$, so our series converges to 8.) ∎

Finally, a word about the last box on the flow chart. Most textbooks on advanced calculus and analysis contain material on infinite sequences and series. One book that is devoted exclusively to those topics is K. Knopp, *Infinite Sequences and Infinite Series,* translated from German by Frederick Bagemihl, New York, Dover Publications, 1956.

PROBLEMS

In Problems 1–20, determine whether $\Sigma\ a_n$ converges.

1. $a_n = \dfrac{(-1)^{n-1}}{\ln(n + 1)}$

2. $a_n = \dfrac{4^n n! n!}{(2n)!}$

3. $a_1 = 1,\quad a_{n+1} = -\dfrac{n}{n + 1} a_n$

4. $a_n = n!/n^n$

5. $a_n = n^n/n!$

6. $a_1 = 4,\quad a_{n+1} = (a_n)^{1/n}$

7. $a_n = n^2/2^n$

8. $a_n = 2^n/n^2$

9. $a_1 = 2,\quad a_{n+1} = \dfrac{1}{2} a_n$

10. $a_1 = 1,\quad a_{n+1} = 2a_n$

11. $a_1 = 2,\quad a_{n+1} = \dfrac{n}{n + 1} a_n$

12. $a_1 = \dfrac{1}{2},\quad a_{n+1} = \dfrac{n + 1}{n} a_n$

13. $a_1 = 1,\quad a_{n+1} = \dfrac{1}{(1 + a_n)}$

14. $a_1 = 0,\quad a_{n+1} = na_n$

15. $a_1 = 1,\quad a_{n+1} = na_n$

16. $a_1 = 10,\quad a_{n+1} = \dfrac{a_n}{n}$

17. $a_n = 1$ if n is odd, $\quad a_n = -1$ if n is even

18. $a_n = 1/3^n$ if n is odd, $\quad a_n = -n/3^n$ if n is even

19. $a_n = \dfrac{\sin n}{n^2}$

20. $a_n = \dfrac{2n + 3}{n^3 + 2}$

21. a) Prove the following theorem: If $\{c_n\}$ is a sequence of numbers such that every sum $t_n = \Sigma_{k=1}^{n} c_k$ is bounded, then the series $\Sigma\ c_n/n$ converges and is equal to $\Sigma\ t_n/(n(n + 1))$.

(*Outline of proof:* Replace c_1 by t_1 and c_n by $t_n - t_{n-1}$ for $n \geq 2$. If $s_{2n+1} = \Sigma_{k=1}^{2n+1} c_k/k$, show that

$$s_{2n+1} = t_1\left(1 - \frac{1}{2}\right) + t_2\left(\frac{1}{2} - \frac{1}{3}\right)$$

$$+ \cdots + t_{2n}\left(\frac{1}{2n} - \frac{1}{2n + 1}\right) + \frac{t_{2n+1}}{2n + 1}$$

$$= \sum_{k=1}^{2n} \frac{t_k}{k(k + 1)} + \frac{t_{2n+1}}{2n + 1}.$$

Because $|t_k| < M$ for some constant M, the series $\Sigma_{k=1}^{\infty}\ t_k/k(k + 1))$ converges absolutely and s_{2n+1} has a limit as $n \to \infty$. Finally, if $s_{2n} = \Sigma_{k=1}^{2n} c_k/k$, then $s_{2n+1} - s_{2n} = c_{2n+1}/(2n + 1)$ approaches zero as $n \to \infty$ because $|c_{2n+1}| = |t_{2n+1} - t_{2n}| < 2M$. Hence the sequence of partial sums of the series $\Sigma\ c_k/k$ converges and the limit is $\Sigma_{k=1}^{\infty}\ t_k/(k(k + 1))$.

b) Show how the foregoing theorem applies to the alternating harmonic series

$$1 - \frac{1}{2} + \frac{1}{3} - \frac{1}{4} + \frac{1}{5} - \frac{1}{6} + \cdots.$$

c) Show that the series

$$1 - \frac{1}{2} - \frac{1}{3} + \frac{1}{4} + \frac{1}{5} - \frac{1}{6} - \frac{1}{7} + \cdots$$

converges. (After the first term, the signs are two negative, two positive, two negative, two positive, and so on in that pattern.)

□ **11.10**

Estimating the Sum of a Series of Constant Terms

We now turn our attention to a question that was posed earlier: If a series converges, what is its sum? Usually, we can determine the sum only to a specified number of decimals, depending on the capacity of the computing device we use. There are two possible sources of error in any estimate we get: (1) truncation errors and (2) round-off errors.

Truncation Errors If the sum of an infinite series is S, the partial sums approach S as a limit:

$$\lim s_n = S,$$

where s_n is the sum of the first n terms of the series. In theory, this means that, given any $\epsilon > 0$, there is an index N such that

$$|s_n - S| < \epsilon \qquad \text{for all } n > N.$$

If there were no round-off errors, we could specify the degree of accuracy we want by specifying ϵ and then use s_n as an approximation to S, choosing any $n > N$. The difference, $R_n = S - s_n$, between the sum of the series and its nth partial sum is called a **remainder** or **truncation error.**

Round-off Errors Because most of the arithmetic calculations in evaluating s_n are done only to a limited number of decimal (or binary) places, there is the potential for a round-off error in evaluating each individual term of the series and therefore in the partial sums. We may expect some balance between positive and negative round-off errors, so that a partial sum should be more accurate than a "worst case scenario" might indicate.

Estimation of Remainders by Integrals

As noted above, the difference $R_n = S - s_n$ between the sum of a convergent series and its nth partial sum is called a *remainder* or a *truncation error*. Since R_n itself is given as an infinite series, which, in principle, is as difficult to evaluate as the original series, you might think there would be no advantage in singling out R_n for attention. But sometimes even a crude estimate for R_n can lead to an estimate of S that is closer to S than s_n is.

For example, suppose we are interested in learning the numerical value of the series

$$\sum_{k=1}^{\infty} \frac{1}{k^2} = \frac{1}{1^2} + \frac{1}{2^2} + \frac{1}{3^2} + \cdots.$$

This is a p-series with $p = 2$, and hence it converges. This means that the sequence of partial sums

$$s_n = \frac{1}{1^2} + \frac{1}{2^2} + \cdots + \frac{1}{n^2}$$

has a limit S. If we want to know S to a couple of decimal places, we might try

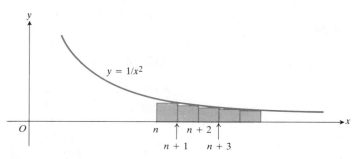

11.16   The rectilinear area R_n is

$$\frac{1}{(n+1)^2} + \frac{1}{(n+2)^2} + \frac{1}{(n+3)^2} + \cdots < \int_n^\infty \frac{dx}{x^2}.$$

to find an integer n such that the corresponding *finite* sum s_n differs from S by less than, say, 0.005. Then we would use this s_n in place of S, to two decimal places. If we write

$$S = \sum_{k=1}^\infty \frac{1}{k^2} = \frac{1}{1^2} + \frac{1}{2^2} + \cdots + \frac{1}{n^2} + \frac{1}{(n+1)^2} + \cdots,$$

we see that

$$R_n = S - s_n = \frac{1}{(n+1)^2} + \cdots.$$

We can estimate the error R_n by comparing it with the area under the curve

$$y = \frac{1}{x^2}$$

from $x = n$ to ∞. From Fig. 11.16, we see that

$$R_n < \int_n^\infty \frac{1}{x^2}\, dx = \frac{1}{n}.$$

This tells us that, by taking 200 terms of the series, we can be sure that the difference between the sum S of the entire series and the sum s_{200} of these 200 terms will be less than $1/200 = 0.005$.

A somewhat closer estimate of R_n results from using the Trapezoidal Rule to approximate the area under the curve in Fig. 11.16. Let us write u_k for $1/k^2$ and consider the trapezoidal approximation

$$T_n = \sum_{k=n}^\infty \frac{1}{2}(u_k + u_{k+1}) = \frac{1}{2}(u_n + u_{n+1}) + \frac{1}{2}(u_{n+1} + u_{n+2}) + \cdots$$

$$= \frac{1}{2}u_n + u_{n+1} + u_{n+2} + \cdots = \frac{1}{2}u_n + R_n.$$

Now since the curve $y = 1/x^2$ is concave upward,

$$T_n > \int_n^\infty \frac{1}{x^2}\, dx = \frac{1}{n},$$

and we have

$$R_n = T_n - \frac{1}{2}u_n > \frac{1}{n} - \frac{1}{2n^2}.$$

We now know that

$$\frac{1}{n} > R_n > \frac{1}{n} - \frac{1}{2n^2},$$

and $S = s_n + R_n$ may be estimated as follows:

$$s_n + \frac{1}{n} > S > s_n + \frac{1}{n} - \frac{1}{2n^2}. \tag{1}$$

Thus, by using $s_n + 1/n$ in place of s_n to estimate S, we shall be making an error which is numerically less than $1/(2n^2)$. By taking $n \geq 10$, this error is then made less than 0.005. The difference in time required to compute the sum of 10 terms versus 200 terms may be sufficiently great to make this sharper analysis of practical importance.

What we have done in the case of this specific example can be done in any case where the graph of the function $y = f(x)$ is concave upward as in Fig. 11.16. We find that when $\int_n^\infty f(x)\,dx$ exists,

$$u_1 + u_2 + \cdots + u_n + \int_n^\infty f(x)\,dx \tag{2}$$

tends to overestimate the value of the series, but by an amount that is less than $u_n/2$.

Using the Ratio Test to Estimate Remainders We consider a series $\Sigma\, a_n$ of positive terms and suppose that the rate of growth ratio a_{n+1}/a_n has a limit $p < 1$. In particular, suppose that for some index N it is true that

$$r_1 \leq \frac{a_{n+1}}{a_n} \leq r_2 \qquad \text{for } n \geq N,$$

where r_1 and r_2 are constants that are both less than 1. Then the inequalities

$$r_1 a_n \leq a_{n+1} \leq r_2 a_n \qquad (n = N, N+1, N+2, \ldots)$$

enable us to deduce that

$$a_N(r_1 + r_1^2 + r_1^3 + \cdots) \leq \sum_{n=N+1}^{\infty} a_n \leq a_N(r_2 + r_2^2 + r_2^3 + \cdots). \tag{3}$$

The two geometric series have sums

$$r_1 + r_1^2 + r_1^3 + \cdots = \frac{r_1}{1 - r_1}, \qquad r_2 + r_2^2 + r_2^3 + \cdots = \frac{r_2}{1 - r_2}.$$

Hence, the truncation error

$$R_N = \sum_{n=N+1}^{\infty} a_n$$

lies between

$$a_N \frac{r_1}{1 - r_1} \quad \text{and} \quad a_N \frac{r_2}{1 - r_2}.$$

That is,

$$a_N \frac{r_1}{1 - r_1} \leq S - s_N \leq a_N \frac{r_2}{1 - r_2} \tag{4}$$

if

$$0 \leq r_1 \leq \frac{a_{n+1}}{a_n} \leq r_2 < 1 \quad \text{for } n \geq N.$$

EXAMPLE   Estimate the remainder for the series

$$1 + \frac{1}{3} + \frac{1 \cdot 2}{3 \cdot 5} + \frac{1 \cdot 2 \cdot 3}{3 \cdot 5 \cdot 7} + \cdots = a_1 + a_2 + a_3 + \cdots$$

where

$$a_1 = 1 \quad \text{and} \quad a_{n+1} = \frac{n}{2n + 1} a_n \quad \text{for } n \geq 1.$$

Solution   We see that the ratio

$$\frac{a_{n+1}}{a_n} = \frac{n}{2n + 1}$$

is less than 1/2 for all n. So we can take $r_2 = 1/2$ in (3) and (4). We also observe that the sequence $\{n/(2n + 1)\}$ is an increasing sequence, so we can take $r_1 = N/(2N + 1)$. When we substitute these values of r_1 and r_2 in (4), we get

$$a_N \frac{N}{N + 1} \leq S - s_N \leq a_N.$$

For $N = 10$, for example, the difference between the sum of all terms of the series and the sum of the first ten terms is between $(10/11)a_{10}$ and a_{10}. For $N = 20$, the difference is between $(20/21)a_{20}$ and a_{20}. We have calculated s_{20} and a_{20} to be

$$s_{20} = 1.570795962, \quad a_{20} = 3.80 \times 10^{-7}.$$

Therefore, R_{20} lies between 3.80×10^{-7} and $(20/21)a_{20}$, which is 3.62×10^{-7}. We estimate the sum S for the entire series by $S = s_{20} + R_{20}$ and get

$$1.570796324 \leq S \leq 1.570796342. \qquad \blacksquare$$

CAUTION   These calculations involve round-off errors that can easily affect the last decimal place or more. It seems safe to say that the first six decimals are correct. Scientists and engineers who need to be certain of accuracy beyond that obtained here should consult textbooks on numerical analysis.

For excellent articles on the subject of estimating remainders, see "Estimating Remainders," by R. P. Boas, *Mathematics Magazine,* 51:2, 83–89 (March 1978), and "Partial Sums of Infinite Series and How They Grow," by R. P. Boas, *American Mathematical Monthly,* 84:4, 237–258 (April 1977). For a treatment of round-off errors and computer arithmetic see *Numerical Analysis,* 3rd ed., by Richard L. Burden and J. Douglas Faires (Boston: Prindle, Weber & Schmidt), 1985, pp. 8ff.

PROBLEMS

1. COMPUTER Let

$$s_n = \sum_{k=1}^{n} \frac{1}{k^2}, \qquad r_n = \sum_{k=n+1}^{\infty} \frac{1}{k^2}.$$

We have found that r_n is remarkably well approximated by $x + x^2/2 + x^3/6$ with $x = 1/(n + 1)$. Write, and run, a BASIC program for computing s_n and $s_n + x + x^2/2 + x^3/6$ for $n = 1$ to 40. Comment on the results. (The series $\sum 1/n^2$ converges to $\pi^2/6 = 1.644934067$ to nine decimals.) We found the approximation to r_n by considering the trapezoidal approximation to the area under the curve $y = 1/x^2$ from $x = n + 1$ to ∞ and estimating the sum of the areas of the tiny pieces that make up the difference between the curve and the trapezoids. See Fig. 11.17. From $x = k$ to $k + 1$, the difference is

$$\frac{1}{2}\left(\frac{1}{k^2} + \frac{1}{(k+1)^2}\right) - \left(\frac{1}{k} - \frac{1}{k+1}\right) = \frac{1}{2}\frac{1}{k^2(k+1)^2}$$

and we used

$$\int_{n+1}^{\infty} \frac{1}{2}\frac{dx}{x^4} = \frac{1}{6}\frac{1}{(n+1)^3}$$

as an approximation to

$$\sum_{k=n+1}^{\infty} \frac{1}{2}\frac{1}{k^2(k+1)^2}.$$

2. COMPUTER *A midpoint estimation formula.* Let $f(x) = 1/x^2$ and $u_k = 1/k^2$. Suppose that we have computed a partial sum $s_N = u_1 + u_2 + \cdots + u_N$ and want to estimate the remainder $R_N = u_{N+1} + u_{N+2} + \cdots$. This time, we compare the area of a rectangle of height u_k with base extending from $k - \frac{1}{2}$ to $k + \frac{1}{2}$ and the corresponding area under the curve $y = 1/x^2$. See Fig. 11.18. Call the latter area A_k:

$$A_k = \int_{k-1/2}^{k+1/2} \frac{1}{x^2}\,dx = -1/x \Big]_{k-1/2}^{k+1/2} = 1/(k^2 - 1/4).$$

Now $R_N = \sum_{k=N+1}^{\infty} u_k$ is approximately equal to the area under $y = 1/x^2$ from $x = N + 1/2$ to ∞, which is $1/(N + 1/2)$. We can improve this estimate by looking at the amount by which the area under the curve from $x = k$ to $k + 1$ exceeds the area of the rectangle. This excess is

$$\frac{1}{k^2 - 1/4} - \frac{1}{k^2} = \frac{1/4}{k^2(k^2 - 1/4)}.$$

To correct the previous estimate of R_N we would need to subtract the sum, from $k = N + 1$ to ∞, of these excess amounts. To simplify matters, we now replace that sum by

$$\int_{N+1/2}^{\infty} \frac{1}{4}\frac{1}{x^4}\,dx = \frac{1}{12}\frac{1}{(N + 1/2)^3}.$$

This leads to the estimate

$$R_N \approx w - w^3/12, \qquad \text{where } w = 1/(N + 1/2)$$

and $S = s_N + R_N \approx s_N + w - w^3/12$. The correct sum is $(\pi^2)/6 \approx 1.644934067$. The estimates we got on a com-

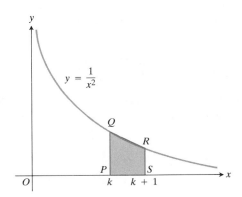

11.17 The area of the trapezoid $PQRS$ is

$$\frac{1}{2}\left(\frac{1}{k^2} + \frac{1}{(k+1)^2}\right).$$

The corresponding area under the curve is

$$\int_{k}^{k+1} \frac{1}{x^2}\,dx = \frac{1}{k} - \frac{1}{k+1}.$$

The area between the curve and the chord QR is

$$\frac{1}{2}\left(\frac{1}{k^2} + \frac{1}{(k+1)^2}\right) - \left(\frac{1}{k} - \frac{1}{k+1}\right) = \frac{1}{2k^2(k+1)^2}.$$

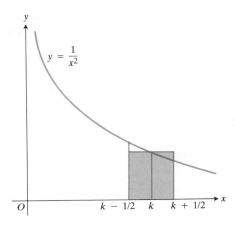

11.18 The rectangle has area $u_k = 1/k^2$ because the base from $k - \frac{1}{2}$ to $k + \frac{1}{2}$ has length equal to 1. The area under the corresponding part of the curve $y = 1/x^2$ is $1/(k^2 - \frac{1}{4})$. The area under the curve exceeds the area of the rectangle by $(\frac{1}{4})/[k^2(k^2 - \frac{1}{4})]$. The remainder $R_N = u_{N+1} + u_{N+2} + \cdots$ is approximately equal to

$$\int_{N+1/2}^{\infty} \left(\frac{1}{x^2}\right)dx - \frac{1}{4}\int_{N+1/2}^{\infty} \left(\frac{1}{x^4}\right)dx.$$

puter, for selected values of N, are as follows:

N	Estimated sum
10	1.64493 3923
34	1.64493 4066
≥ 38	1.64493 4067

Write, and run, a BASIC program for computing s_N and $s_N + w - w^3/12$ for $N = 1$ to 40.

3. Let $a_n = n/3^n$. Find numbers r_1 and r_2, both less than 1, such that $r_1 \leq a_{n+1}/a_n \leq r_2$ when (a) $n \geq 10$, (b) $n \geq 100$. Use the results to estimate the remainders $R_{10} = \Sigma_{n=11}^{\infty} a_n$ and $R_{100} = \Sigma_{n=101}^{\infty} a_n$.

REVIEW QUESTIONS AND EXERCISES

1. Define "sequence," "series," and "sequence of partial sums of a series."

2. Define "convergence" (a) of a sequence, (b) of an infinite series.

3. Which of the following statements are true, and which are false?
   a) If a sequence does not converge, then it diverges.
   b) If a sequence $\{f(n)\}$ does not converge, then $f(n)$ tends to infinity as n does.
   c) If a series does not converge, then its nth term does not approach zero as n tends to infinity.
   d) If the nth term of a series does not approach zero as n tends to infinity, then the series diverges.
   e) If a sequence $\{f(n)\}$ converges, then there is a number L such that $f(n)$ lies within 1 unit of L (i) for all values of n, (ii) for all but a finite number of values of n.
   f) If all partial sums of a series are less than some constant L, then the series converges.
   g) If a series converges, then its partial sums s_n are bounded (that is, $m \leq s_n \leq M$ for some constants m and M).

4. List three tests for convergence (or divergence) of an infinite series.

5. Under what circumstances do you know that a bounded sequence converges?

6. Define "absolute convergence" and "conditional convergence." Give examples of series that are (a) absolutely convergent, (b) conditionally convergent.

7. What test is commonly used to decide whether a given alternating series converges? Give examples of convergent and divergent alternating series.

MISCELLANEOUS PROBLEMS

1. Find explicitly the nth partial sum of the series $\Sigma_{n=2}^{\infty} \ln(1 - 1/n^2)$, and thereby determine whether the series converges.

2. Evaluate $\Sigma_{k=2}^{\infty} 1/(k^2 - 1)$ by finding the nth partial sum and taking the limit as n becomes infinite.

3. Prove that the sequence $\{x_n\}$ and the series $\Sigma_{k=1}^{\infty} (x_{k+1} - x_k)$ both converge or both diverge.

4. Does the series

$$\sum_{n=1}^{\infty} \frac{(-1)^{n+1}}{\sqrt{n}}$$

   a) converge,
   b) converge absolutely,
   c) converge conditionally,
   d) diverge?

5. Assuming $|x| > 1$, show that

$$\frac{1}{1-x} = -\frac{1}{x} - \frac{1}{x^2} - \frac{1}{x^3} - \cdots.$$

6. Does the series $\Sigma_{n=1}^{\infty}$ sech n converge? Why?

7. Does $\Sigma_{n=1}^{\infty} (-1)^n$ tanh n converge? Why?

Establish the convergence or divergence of the series whose nth terms are given in Problems 8–19.

8. $\dfrac{1}{\ln(n + 1)}$

9. $\dfrac{n}{2(n + 1)(n + 2)}$

10. $\dfrac{\sqrt{n + 1} - \sqrt{n}}{\sqrt{n}}$

11. $\dfrac{1}{n(\ln n)^2}, \quad n \geq 2$

12. $\dfrac{1 + (-2)^{n-1}}{2^n}$

13. $\dfrac{n}{1000n^2 + 1}$

14. $e^n/n!$

15. $\dfrac{1}{n\sqrt{n^2 + 1}}$

16. $\dfrac{1}{n^{1+1/n}}$

17. $\dfrac{1 \cdot 3 \cdot 5 \cdots (2n - 1)}{2 \cdot 4 \cdot 6 \cdots (2n)}$

18. $\dfrac{n^2}{n^3 + 1}$

19. $\dfrac{n + 1}{n!}$

20. If the following series converges, find the sum:

$$\sum_{n=1}^{\infty} \frac{1}{(n+1)(n+2)}.$$

21. a) Suppose $a_1, a_2, a_3, \ldots, a_n$ are positive numbers satisfying the following conditions:

  i) $a_1 \geq a_2 \geq a_3 \geq \cdots$,

  ii) the series $a_2 + a_4 + a_8 + a_{16} + \cdots$ diverges.

  Show that the series

$$\frac{a_1}{1} + \frac{a_2}{2} + \frac{a_3}{3} + \cdots$$

  diverges.

b) Use the result above to show that

$$1 + \sum_{n=2}^{\infty} \frac{1}{n \ln n}$$

  diverges.

22. Given $a_n \neq 1$, $a_n > 0$, Σa_n converges.

a) Show that Σa_n^2 converges.

b) Does $\Sigma a_n/(1 - a_n)$ converge? Explain.

23. If p is a constant, show that the series

$$\sum_{n=3}^{\infty} \frac{1}{n \ln n (\ln(\ln n))^p}$$

a) converges if $p > 1$,

b) diverges if $p \leq 1$.

(*Remark:* If $f_1(x) = x$; $f_{n+1}(x) = \ln(f_n(x))$; and we let n take on values $1, 2, 3, \ldots$, we find $f_2(x) = \ln x$, $f_3(x) = \ln(\ln x)$, and so on. Provided the lower limit on the integral is such that $f_n(a) > 1$, then

$$\int_a^{\infty} \frac{dx}{f_1(x)f_2(x) \cdots f_n(x)(f_{n+1}(x))^p}$$

(i) converges when $p > 1$ and (ii) diverges when $p \leq 1$.)

24. a) Prove the following theorem: Let $\{c_n\}$ be a sequence of numbers such that the sums $t_n = \Sigma_{k=1}^{n} c_k$ satisfy the condition $|t_n| \leq Mn^h$ for some constants M and h with $h < 1$. Then the series $\Sigma c_n/n$ converges and is equal to $\Sigma t_n/n(n + 1)$. (See Problem 21, Art. 11.9.)

b) Use the theorem of part (a) to prove convergence of the series

$$1 - \frac{1}{2} - \frac{1}{3} + \frac{1}{4} + \frac{1}{5} + \frac{1}{6} - \frac{1}{7} - \frac{1}{8} - \frac{1}{9} - \frac{1}{10} + \cdots$$

where k positive terms are followed by $k + 1$ negative terms, which in turn are followed by $k + 2$ positive terms, for $k = 1, 3, 5, 7, \ldots$.

c) COMPUTER Estimate the sum of the series in part (b). (Our estimate to seven decimals is 0.5171004.)

CHAPTER **12**

Power Series

OVERVIEW

In Chapter 11 we dealt mainly with sequences and series of constants. All of the tests for convergence or divergence that we used there are useful in this chapter as well, but here we shall examine sequences and series where the nth term is a function $u_n(x)$. In particular, we shall study the general power series

$$a_0 + a_1x + a_2x^2 + a_3x^3 + \cdots + a_nx^n + \cdots.$$

Where do power series come from? And what good are they? Power series and other types of infinite series are useful in solving the differential equations of mathematical physics, including mathematical descriptions of vibrating strings, heat flow, transmission of electrical current, motion of a simple pendulum, and many other phenomena. In the branch of mathematics called numerical analysis, power series can be used to determine how many decimal places may be required in a computation to guarantee a specified accuracy. We can also use power series to extend the domains of definition of e^x, $\sin x$, $\cos x$, and other elementary functions from real numbers to complex numbers. (Such functions are useful, for example, in physics and electrical engineering.) Most of these applications are dealt with in those specific subjects (the theory of functions of complex variables is an advanced subject in mathematics) and will not be discussed in this book.

12.1

Introduction

In this chapter we shall examine sequences and series where the nth term is a function $u_n(x)$. In most instances, $u_n(x)$ will be a constant times x^n or $(x - a)^n$. For example, if

$$u_0(x) = 1, \qquad u_1(x) = x, \qquad u_2(x) = x^2, \qquad \ldots, \qquad u_n(x) = x^n, \quad (1)$$

then the **sequence** $\{u_n(x)\} = \{x^n\}$ converges to zero if $-1 < x < 1$; converges to 1 if $x = 1$; and diverges elsewhere.

When we connect the terms of the sequence $\{x^n\}$ with plus signs, we get the geometric **series**

$$\sum_{n=0}^{\infty} x^n = 1 + x + x^2 + x^3 + \cdots + x^n + \cdots, \qquad (2)$$

which we know converges to $1/(1 - x)$ if $|x| < 1$ and diverges otherwise. We say that $\Sigma \, x^n$ defines a function on $(-1, 1)$, whether we know a closed formula for $\Sigma \, x^n$ or not.

Formal Power Series

If we have a more general sequence of terms

$$u_0 = a_0, \quad u_1 = a_1 x, \quad u_2 = a_2 x^2, \quad u_3 = a_3 x^3, \quad \ldots, \quad u_n = a_n x^n, \quad \ldots, \quad (3)$$

and we connect the terms of this sequence with plus signs, we get what is called a **formal power series**

$$\sum_{n=0}^{\infty} a_n x^n = a_0 + a_1 x + a_2 x^2 + a_3 x^3 + \cdots + a_n x^n + \cdots. \qquad (4)$$

In the expressions (3) and (4), the a_n's are constants (independent of x) and x is a variable whose domain at the moment may be any set of real numbers. By applying such tests as the Ratio Test or nth-Root Test to the series in (4), we often can determine the values of x for which the given series converges. For such values of x, the series represents some function of x. For instance, it is appropriate to write

$$\frac{1}{1 - x} = 1 + x + x^2 + x^3 + \cdots + x^n + \cdots \qquad \text{for } -1 < x < 1, \quad (5)$$

because the series on the right side of Eq. (5) converges to the same number as the functional value on the left side of the equation for every value of x between -1 and $+1$.

The series whose terms are $u_n = n! x^n$ is rather bizarre:

$$0! + 1! x + 2! x^2 + 3! x^3 + \cdots + n! x^n + \cdots. \qquad (6)$$

If $x = 0$, this series reduces to its first term, $0!$, and has the value 1. For all other values of x, the nth term $u_n = n! x^n$ does not go to zero as $n \to \infty$, so the series diverges. This series has no practical use, except as an example of a series that converges at just one point. Most of the power series that we shall study will converge on some interval of values of x. Most of the power series we shall study will be those generated by the process of looking for polynomial approximations to the standard elementary functions like e^x, $\sin x$, $\cos x$, $\ln(1 + x)$, and $\tan^{-1} x$. Linear and quadratic approximations of such functions, for small values of x, have been presented in earlier chapters. For some applications these approximations are good enough; for others, they are not. So we now turn our attention to the study of Maclaurin series and Taylor series: What are they? How do we get them? Where do they converge? When they converge, do they give actual values of the functions that generate them?

12.2
Taylor Polynomials

Not every function generates (or has associated with it) a power series with infinitely many terms. However, every function f that is defined in a neighborhood of $x = 0$ and has finite derivatives f', f'', . . . , $f^{(n)}$ at 0, generates polynomials $p_0(x)$, $p_1(x)$, $p_2(x)$, . . . , $p_n(x)$ that approximate the given function $f(x)$ for values of x near 0. The approximations usually get better as the degree of the polynomial increases.

What do we mean by "polynomials generated by the function f"? For any integer k from 0 to n, let the polynomial p_k be

$$p_k(x) = a_0 + a_1 x + a_2 x^2 + \cdots + a_k x^k. \tag{1}$$

The coefficients a_0, a_1, a_2, . . . , a_k are to be determined. We focus our attention on a portion of the curve $y = f(x)$ near the point $A(0, f(0))$, as shown in Fig. 12.1.

1. The graph of the polynomial $p_0(x) = a_0$ of degree zero will pass through $(0, f(0))$ if we take

$$a_0 = f(0).$$

2. The graph of the polynomial $p_1(x) = a_0 + a_1 x$ will pass through $(0, f(0))$ and have the same slope as the given curve at that point if we choose

$$a_0 = f(0) \qquad \text{and} \qquad a_1 = f'(0).$$

3. The graph of the polynomial $p_2(x) = a_0 + a_1 x + a_2 x^2$ will pass through $(0, f(0))$ and have the same first and second derivatives as the given curve at that point if

$$a_0 = f(0), \qquad a_1 = f'(0),$$

and

$$a_2 = \frac{f''(0)}{2}.$$

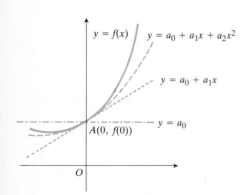

12.1 The function $f(x)$ is approximated near $x = 0$ by polynomials whose derivatives at $x = 0$ match the derivatives of f.

4. In general, the polynomial $p_k(x) = a_0 + a_1 x + a_2 x^2 + \cdots + a_k x^k$, which we choose to approximate the graph of $y = f(x)$ near $x = 0$, is the polynomial whose graph passes through $(0, f(0))$ and whose first k derivatives match the corresponding derivatives of f at $x = 0$. We make them match by

putting $x = 0$ in each of the following equations:

$$p_k(x) = a_0 + a_1 x + a_2 x^2 + a_3 x^3 + \cdots + a_k x^k$$

$$p_k'(x) = a_1 + 2a_2 x + 3a_3 x^2 + \cdots + ka_k x^{k-1}$$

$$p_k''(x) = 2a_2 + 3 \cdot 2a_3 x + \cdots + k(k-1)a_k x^{k-2}$$

$$\vdots$$

$$p_k^{(k)}(x) = (k!)a_k.$$

Putting $x = 0$, $p_k(0) = f(0)$, $p_k'(0) = f'(0)$, $p_k''(0) = f''(0)$, . . . , $p_k^{(k)}(0) = f^{(k)}(0)$, and solving for $a_0, a_1, a_2, \ldots, a_k$, we obtain

$$a_0 = f(0), \quad a_1 = f'(0), \quad a_2 = \frac{f''(0)}{2!}, \quad a_3 = \frac{f'''(0)}{3!}, \quad \ldots, \quad a_k = \frac{f^{(k)}(0)}{k!}.$$

Thus,

$$p_k(x) = f(0) + f'(0)\,x + \frac{f''(0)}{2!}x^2 + \frac{f'''(0)}{3!}x^3 + \cdots + \frac{f^{(k)}(0)}{k!}x^k \qquad (2)$$

is the polynomial that matches f and its first k derivatives at $x = 0$. It is called the kth order **Taylor polynomial** generated by f at $x = 0$. Such a Taylor polynomial exists for every k from 0 to n.

When the function f has derivatives of *all* orders at $x = 0$, as in the next example, there is a Taylor polynomial $p_n(x)$ generated by f for *every* nonnegative integer n.

EXAMPLE 1 Find the Taylor polynomials $p_n(x)$ generated by $f(x) = e^x$ at $x = 0$.

Solution Expressed in terms of x, the given function and its derivatives are

$$f(x) = e^x, \qquad f'(x) = e^x, \qquad \ldots, \qquad f^{(n)}(x) = e^x,$$

so

$$f(0) = e^0 = 1, \qquad f'(0) = 1, \qquad \ldots, \qquad f^{(n)}(0) = 1,$$

and

$$p_n(x) = 1 + x + \frac{x^2}{2!} + \frac{x^3}{3!} + \cdots + \frac{x^n}{n!}. \qquad (3)$$

See Fig. 12.2. ∎

EXAMPLE 2 Find the Taylor polynomials $p_n(x)$ generated by $f(x) = \cos x$ at $x = 0$.

Solution The cosine and its derivatives are

$$f(x) = \cos x, \qquad f'(x) = -\sin x,$$

$$f''(x) = -\cos x, \qquad f^{(3)}(x) = \sin x,$$

$$\vdots \qquad\qquad \vdots$$

$$f^{(2k)}(x) = (-1)^k \cos x, \qquad f^{(2k+1)}(x) = (-1)^{k+1}\sin x.$$

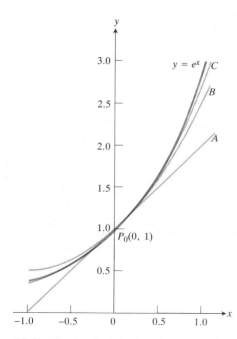

12.2 The graph of the function $y = e^x$ and graphs of three approximating polynomials: (A) a straight line, (B) a parabola, and (C) a cubic curve.

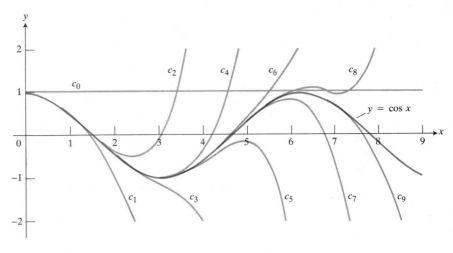

12.3 The polynomials

$$c_n(x) = \sum_{k=0}^{n} [(-1)^k x^{2k}/(2k)!]$$

converge to $\cos x$ as $n \to \infty$. (Adapted from Helen M. Kammerer, *American Mathematical Monthly*, 43 (1936), pp. 293–294.)

■ ■ ■

WHAT'S IN A NAME?

Taylor series were not first described by Taylor, and Maclaurin series are not due to Maclaurin. In 1688, when Taylor was three years old and ten years before Maclaurin was born, the English mathematician James Gregory published Taylor series for functions, and he also knew the Maclaurin series for $\tan x$, $\sec x$, $\tan^{-1} x$, and $\sec^{-1} x$. At about the same time, the mathematician Nicolaus Mercator discovered the Maclaurin series for $\ln(1 + x)$.

Brook Taylor (1685–1731) was apparently unaware of Gregory's work when he published his book *Methodus incrementorum directa et inversa* in 1715, which contained what we now call Taylor series. Taylor's work was quoted by Scotsman Colin Maclaurin (1698–1746) in his *Treatise of Fluxions*, published in 1742. This influential work made series representations of functions widely known to contemporaries, and although Maclaurin made no claim to having discovered these results, Taylor series expansions about the value $a = 0$ became known as Maclaurin series. There is a certain element of justice in this usage—Maclaurin, a brilliant mathematician, was the original discoverer of the method for solving simultaneous linear equations that we now call Cramer's rule.

When $x = 0$, the cosines are 1 and the sines are 0, so

$$f^{(2k)}(0) = (-1)^k, \qquad f^{(2k+1)}(0) = 0.$$

In this case the Taylor polynomials of order $2k$ and of order $2k + 1$ are identical:

$$p_{2k}(x) = p_{2k+1}(x) = 1 - \frac{x^2}{2!} + \frac{x^4}{4!} - \cdots + (-1)^k \frac{x^{2k}}{(2k)!}. \tag{4}$$

Figure 12.3 shows how well these polynomials can be expected to approximate $y = \cos x$ near $x = 0$. Only the right-hand portions of the graphs are shown because the graphs are symmetric about the y-axis.   ■

Maclaurin and Taylor Series

If a function f has derivatives of *all* orders at $x = 0$, it generates Taylor polynomials approximating the function near $x = 0$ with no bound on the degrees of those polynomials. Also, the coefficients for terms up to and including x^n do not change when more terms are added. All of these polynomials are partial sums of the formal power series known as the **Maclaurin series** generated by the function f.

The Maclaurin Series Generated by f

$$f(0) + \frac{f'(0)}{1!}x + \frac{f''(0)}{2!}x^2 + \frac{f'''(0)}{3!}x^3 + \cdots + \frac{f^{(n)}(0)}{n!}x^n + \cdots \tag{5}$$

Although the Maclaurin series generated by f has partial sums whose values and derivatives match those of f, at $x = 0$, we do not know that the series actually converges to $f(x)$ for any x other than $x = 0$. Therefore, we do not have an equation in (5)—it is just an expression into which values of x can be substituted. Most of this chapter will deal with the questions of convergence of the Maclaurin series (and the similar Taylor series, to which we soon shall turn our attention).

Question 1: Does the formal power series in (5) converge for values of x different from $x = 0$?

Question 2: When the series does converge, does it converge to the proper value, namely $f(x)$?

For some functions, the answer to both questions is yes. The graphs in Figs. 12.2 and 12.3 are encouraging, and the next few articles will confirm that we can normally expect a Maclaurin series to converge to its generating function in an interval about the origin. For many functions, this interval is the entire x-axis.

If, instead of approximating the values of f near zero, we are concerned with values of x near some other point a, we write our approximating polynomials in powers of $(x - a)$:

$$p_n(x) = a_0 + a_1(x - a) + a_2(x - a)^2 + \cdots + a_n(x - a)^n. \qquad (6)$$

When we now determine the coefficients $a_0, a_1, \ldots, a_n$, so that the polynomial and its first n derivatives agree with the given function and its derivatives at $x = a$, we are led to a series called the **Taylor series** generated by f at $x = a$.

The Taylor Series Generated by f at $x = a$

$$f(a) + f'(a)(x - a) + \frac{f''(a)}{2!}(x - a)^2 + \cdots + \frac{f^{(n)}(a)}{n!}(x - a)^n + \cdots. \qquad (7)$$

There are two things to notice here. First, Maclaurin series are Taylor series with $a = 0$. Second, a function cannot generate a Taylor series expansion about $x = a$ unless it has finite derivatives of all orders at $x = a$. For instance, $f(x) = \ln x$ does not generate a Maclaurin series expansion because the function itself, to say nothing of its derivatives, does not have a finite value at $x = 0$. On the other hand, it does generate a Taylor series expansion in powers of $(x - 1)$, since $\ln x$ and all its derivatives are finite at $x = 1$.

Here are some examples of Taylor series.

EXAMPLE 3 From the formula derived for the Taylor polynomials of $\cos x$ in Example 2, it follows immediately that

$$\sum_{k=0}^{\infty} (-1)^k \frac{x^{2k}}{(2k)!} = 1 - \frac{x^2}{2!} + \frac{x^4}{4!} - \frac{x^6}{6!} + \cdots$$

is the Maclaurin series generated by $\cos x$. ∎

EXAMPLE 4 Find the Taylor series generated by $\cos x$ at the point $a = 2\pi$.

Solution The values of $\cos x$ and its derivatives at $a = 2\pi$ are the same as their values at $a = 0$. Therefore,

$$f^{(2k)}(2\pi) = f^{(2k)}(0) = (-1)^k \qquad \text{and} \qquad f^{(2k+1)}(2\pi) = f^{(2k+1)}(0) = 0,$$

as in Example 2. The required series is

$$\sum_{k=0}^{\infty} (-1)^k \frac{(x - 2\pi)^{2k}}{(2k)!} = 1 - \frac{(x - 2\pi)^2}{2!} + \frac{(x - 2\pi)^4}{4!} - \cdots. \qquad \blacksquare$$

REMARK There is a convention about how formulas like

$$\frac{x^{2k}}{(2k)!} \qquad \text{and} \qquad \frac{(x - 2\pi)^{2k}}{(2k)!},$$

which arise in the power series of Examples 3 and 4, are to be evaluated when $k = 0$. Besides the usual agreement that $0! = 1$, we also assume that

$$\frac{x^0}{0!} = \frac{1}{1} = 1 \qquad \text{and} \qquad \frac{(x - 2\pi)^0}{0!} = \frac{1}{1} = 1,$$

even when $x = 0$ or 2π.

EXAMPLE 5 What are the first five terms of the Taylor series generated by $f(x) = 1/x$ at $x = 2$?

Solution We need to compute $f(2)$, $f'(2)$, $f''(2)$, $f'''(2)$ and $f^{(4)}(2)$. Taking derivatives, we have

$$f(x) = x^{-1}, \qquad\qquad f(2) = 2^{-1} = \frac{1}{2},$$

$$f'(x) = -x^{-2}, \qquad\qquad f'(2) = -\frac{1}{2^2},$$

$$f''(x) = 2x^{-3}, \qquad\qquad f''(2) = \frac{2!}{2^3},$$

$$f'''(x) = -6x^{-4}, \qquad\qquad f'''(2) = -\frac{(3!)}{2^4},$$

$$f^{(4)}(x) = 24x^{-5}, \qquad\qquad f^{(4)}(2) = \frac{4!}{2^5}.$$

The first five terms of the Taylor series are

$$\frac{1}{2} - \frac{(x - 2)}{2^2} + \frac{(x - 2)^2}{2^3} - \frac{(x - 2)^3}{2^4} + \frac{(x - 2)^4}{2^5}.$$

These appear to be the terms of a geometric series with first term $1/2$ and ratio $r = -(x - 2)/2$. If the rest of the series follows the same pattern, then the series converges to

$$\frac{1/2}{1 - r} = \frac{1/2}{1 + (x - 2)/2}$$

$$= \frac{1}{2 + (x - 2)}$$

$$= \frac{1}{x},$$

provided

$$\left|\frac{x-2}{2}\right| < 1 \qquad \text{or} \qquad |x - 2| < 2 \qquad \text{or} \qquad 0 < x < 4.$$

So, for this example, it would be correct to write

$$\frac{1}{x} = \frac{1}{2} - \frac{(x-2)}{2^2} + \frac{(x-2)^2}{2^3} - \frac{(x-2)^3}{2^4} + \cdots$$
$$+ (-1)^{n+1}\frac{(x-2)^{n-1}}{2^n} + \cdots$$

for $0 < x < 4$ because we know that the geometric series on the right converges to $1/x$ for $0 < x < 4$. ∎

EXAMPLE 6   What is the Maclaurin series generated by $f(x) = (1 + x)^3$?

Solution   We use expression (5) and calculate the derivatives and their values at $x = 0$, as follows:

$$f(x) = (1 + x)^3, \qquad f(0) = 1,$$
$$f'(x) = 3(1 + x)^2, \qquad f'(0) = 3,$$
$$f''(x) = 6(1 + x), \qquad f''(0) = 6,$$
$$f'''(x) = 6 \qquad f'''(0) = 6.$$

All higher order derivatives are identically zero, so the Maclaurin series generated by f is

$$f(0) + \frac{f'(0)}{1!}x + \frac{f''(0)}{2!}x^2 + \frac{f'''(0)}{3!}x^3 + \cdots$$
$$= 1 + 3x + \frac{6}{2}x^2 + \frac{6}{6}x^3 + \frac{0}{24}x^4 + \cdots + \frac{0}{n!}x^n + \cdots$$
$$= 1 + 3x + 3x^2 + x^3.$$

In this case, the infinite Maclaurin series has only four nonzero terms, and so it converges for all values of x. We recognize that $1 + 3x + 3x^2 + x^3$ is the binomial expansion

$$(1 + x)^3 = 1 + 3x + 3x^2 + x^3.$$

The two questions about convergence have affirmative answers:

1. The series converges for all values of x.
2. When the series converges, it converges to the proper value, namely $f(x) = (1 + x)^3$. ∎

See Review Question 2 at the end of this chapter for an example of a function that generates a Maclaurin series that converges for all values of x but does *not* converge to the value of the function that generates it (except at $x = 0$).

PROBLEMS

In Problems 1–9, use Eq. (2) to write the Taylor polynomials $p_3(x)$ and $p_4(x)$ for each of the following functions $f(x)$ at $x = 0$. In each case, your first step should be to complete a table like the one shown below.

n	$f^{(n)}(x)$	$f^{(n)}(0)$
0		
1		
2		
3		
4		

1. e^{-x}

2. $\sin x$

3. $\cos x$

4. $\sin\left(x + \dfrac{\pi}{2}\right)$

5. $\sinh x$

6. $\cosh x$

7. $x^4 - 2x + 1$

8. $x^3 - 2x + 1$

9. $x^2 - 2x + 1$

In Problems 10–13, find the Maclaurin series generated by each function.

10. $\dfrac{1}{1 + x}$

11. x^2

12. $(1 + x)^2$

13. $(1 + x)^{3/2}$

14. Find the Maclaurin series for $f(x) = 1/(1 - x)$. Show that the series diverges when $|x| \geq 1$ and converges when $|x| < 1$.

In Problems 15–20, use Formula (7) to write the Taylor series generated by the given function about the given point a.

15. $f(x) = e^x, \quad a = 10$

16. $f(x) = x^2, \quad a = \dfrac{1}{2}$

17. $f(x) = \ln x, \quad a = 1$

18. $f(x) = \sqrt{x}, \quad a = 4$

19. $f(x) = \dfrac{1}{x}, \quad a = -1$

20. $f(x) = \cos x, \quad a = -\dfrac{\pi}{4}$

In Problems 21 and 22, write the sum of the first three terms of the Taylor series for the given function about the given point a.

21. $f(x) = \tan x, \quad a = \dfrac{\pi}{4}$

22. $f(x) = \ln \cos x, \quad a = \dfrac{\pi}{3}$

12.3
Taylor's Theorem with Remainder: Sines, Cosines, and e^x

In the previous article, we asked when a Taylor series generated by a function can be expected to converge to the function. In this article, we answer the question with a theorem named after the English mathematician Brook Taylor (1685–1731).

THEOREM 1

Taylor's Theorem

If f and its first n derivatives $f', f'', \ldots, f^{(n)}$ are continuous on $[a, b]$ or on $[b, a]$, and $f^{(n)}$ is differentiable on (a, b) or on (b, a), then there exists a number c between a and b such that

$$f(b) = f(a) + f'(a)(b - a) + \frac{f''(a)}{2!}(b - a)^2 + \cdots$$

$$+ \frac{f^{(n)}(a)}{n!}(b - a)^n + \frac{f^{(n+1)}(c)}{(n + 1)!}(b - a)^{n+1}.$$

Proof   We assume that f satisfies the hypotheses of the theorem and define the Taylor polynomial about a of order n:

$$p_n(x) = f(a) + f'(a)(x - a) + \frac{f''(a)}{2!}(x - a)^2 + \cdots + \frac{f^{(n)}(a)}{n!}(x - a)^n. \quad (1)$$

This polynomial and its first n derivatives match the function f and its first n derivatives at $x = a$. We do not disturb that matching by adding another term of the form $K(x - a)^{n+1}$, where K is any constant, because such a function and its first n derivatives are all equal to zero at $x = a$. Therefore, the new function

$$\phi_n(x) = p_n(x) + K(x - a)^{n+1} \tag{2}$$

and its first n derivatives still agree with f and its first n derivatives at $x = a$.

We now choose the particular value of K that makes the curve $y = \phi_n(x)$ agree with the original curve $y = f(x)$ at $x = b$. This can be done: We need only satisfy

$$f(b) = p_n(b) + K(b - a)^{n+1} \tag{3a}$$

or

$$K = \frac{f(b) - p_n(b)}{(b - a)^{n+1}}. \tag{3b}$$

With K defined by Eq. (3b), let $F(x) = f(x) - \phi_n(x)$, so that $F(x)$ measures the difference between the original function f and the approximating function ϕ_n, for each x in $[a, b]$ or in $[b, a]$ if $b < a$. To simplify the notation, we assume $a < b$, so a is the left endpoint of all intervals mentioned. The same proof is valid if a is the right endpoint instead of the left endpoint (for example, $[b, a]$, (b, a), (c_1, a), . . . , (c_n, a)).

The remainder of the proof makes repeated use of Rolle's theorem. First, because $F(a) = F(b) = 0$ and both F and F' are continuous on $[a, b]$, we know that

$$F'(c_1) = 0 \qquad \text{for some } c_1 \text{ in } (a, b).$$

Next, because $F'(a) = F'(c_1) = 0$ and both F' and F'' are continuous on $[a, c_1]$, we know that

$$F''(c_2) = 0 \qquad \text{for some } c_2 \text{ in } (a, c_1).$$

Rolle's theorem, applied successively to F'', F''', . . . , $F^{(n-1)}$ implies the existence of

$$c_3 \text{ in } (a, c_2) \qquad \text{such that } F'''(c_3) = 0,$$
$$c_4 \text{ in } (a, c_3) \qquad \text{such that } F^{(4)}(c_4) = 0,$$
$$\vdots$$
$$c_n \text{ in } (a, c_{n-1}) \qquad \text{such that } F^{(n)}(c_n) = 0.$$

Finally, because $F^{(n)}$ is continuous on $[a, c_n]$ and differentiable on (a, c_n) and $F^{(n)}(a) = F^{(n)}(c_n) = 0$, Rolle's theorem implies that there is a number c_{n+1} in (a, c_n) such that

$$F^{(n+1)}(c_{n+1}) = 0. \tag{4}$$

When we differentiate

$$F(x) = f(x) - p_n(x) - K(x - a)^{n+1}$$

$n + 1$ times, we get

$$F^{(n+1)}(x) = f^{(n+1)}(x) - 0 - (n + 1)! \, K. \tag{5}$$

Equations (4) and (5) together lead to the result

$$K = \frac{f^{(n+1)}(c)}{(n+1)!} \qquad \text{for some number } c = c_{n+1} \text{ in } (a, b). \tag{6}$$

Combining Eqs. (3b) and (6), we have

$$\frac{f(b) - p_n(b)}{(b-a)^{n+1}} = \frac{f^{(n+1)}(c)}{(n+1)!}$$

or

$$f(b) = p_n(b) + \frac{f^{(n+1)}(c)}{(n+1)!}(b-a)^{n+1} \qquad \text{for some } c \text{ between } a \text{ and } b. \quad \blacksquare$$

COROLLARY

If f has derivatives of all orders in an open interval I containing a, then for each positive integer n and for each x in I,

$$f(x) = f(a) + f'(a)(x-a) + \frac{f''(a)}{2!}(x-a)^2 + \cdots \tag{7a}$$

$$+ \frac{f^{(n)}(a)}{n!}(x-a)^n + R_n(x),$$

where

$$R_n(x) = \frac{f^{(n+1)}(c)}{(n+1)!}(x-a)^{n+1} \qquad \text{for some } c \text{ between } a \text{ and } x. \tag{7b}$$

The corollary follows at once from Taylor's theorem because the existence of derivatives of all orders in an interval I implies the continuity of those derivatives and we have merely replaced b by x in the final formula.

The function $R_n(x)$ is called the **remainder of order** n: It's the difference $f(x) - p_n(x)$ where $p_n(x)$ is the Taylor polynomial of order n used to approximate $f(x)$ near $x = a$. This difference, also called the "error" in the approximation $p_n(x)$, can often be estimated by using Eq. (7b) as in the next example.

When $R_n(x) \to 0$ as $n \to \infty$, for all x in some interval around $x = a$, we say that the Taylor series expansion for $f(x)$ converges to $f(x)$ on that interval and write

$$f(x) = \sum_{k=0}^{\infty} \frac{f^{(k)}(a)}{k!}(x-a)^k. \tag{8}$$

EXAMPLE 1 *Series for e^x.* Show that the Taylor series generated by $f(x) = e^x$ at $a = 0$ converges to $f(x)$ for every real value of x.

Solution Let $f(x) = e^x$. This function and all its derivatives are continuous at every point, so Taylor's theorem may be applied with any convenient value of a. We take $a = 0$, since the values of f and its derivatives are easy to compute there. Taylor's theorem leads to

$$e^x = 1 + x + \frac{x^2}{2!} + \frac{x^3}{3!} + \cdots + \frac{x^n}{n!} + R_n(x), \tag{9a}$$

where

$$R_n(x) = \frac{e^c}{(n+1)!}x^{n+1} \qquad \text{for some } c \text{ between } 0 \text{ and } x. \qquad (9b)$$

Because e^x is an increasing function of x, and c is between 0 and x, the value of e^c is between 1 and e^x. Therefore, if x is negative, so is c, and $e^c < 1$; if x is positive, so is c, and $e^c < e^x$. Thus we can write

$$|R_n(x)| < \frac{|x|^{n+1}}{(n+1)!} \qquad \text{when } x < 0, \qquad (9c)$$

and

$$R_n(x) < e^x\frac{x^{n+1}}{(n+1)!} \qquad \text{when } x > 0. \qquad (9d)$$

When $x = 0$, the first term of the series in (9a) is $1 = e^0$, so the "error" is zero. Finally, because

$$\lim_{n\to\infty} \frac{x^{n+1}}{(n+1)!} = 0 \qquad \text{for every } x,$$

it is also true that

$$\lim_{n\to\infty} R_n(x) = 0 \qquad \text{for every value of } x:$$

$$e^x = \sum_{k=0}^{\infty} \frac{x^k}{k!} = 1 + x + \frac{x^2}{2!} + \cdots + \frac{x^k}{k!} + \cdots. \qquad (10)$$

Estimating the Remainder

It is often possible to estimate $R_n(x)$ as we did in Example 1. This method of estimation is so convenient that we state it as a theorem for future reference.

THEOREM 2

The Remainder Estimation Theorem

If there are positive constants M and r such that $|f^{(n+1)}(t)| \le Mr^{n+1}$ for all t between a and x, inclusive, then the remainder term $R_n(x)$ in Taylor's theorem satisfies the inequality

$$|R_n(x)| \le M\frac{r^{n+1}|x-a|^{n+1}}{(n+1)!}.$$

Furthermore, if these conditions hold for every n and all the other conditions of Taylor's theorem are satisfied by $f(x)$, then the series converges to $f(x)$.

In the simplest examples, we can take $r = 1$ provided f and all its derivatives are bounded by some constant M. But if $f(x) = 2\cos(3x)$, each time we differentiate we get a factor of 3, so we could take $r = 3$ and $M = 2$.

712   Chapter 12: Power Series

We are now ready to look at some examples of how the Remainder Estimation Theorem and Taylor's theorem can be used together to settle questions of convergence. As you will see, they can also be used to determine the accuracy with which a function is approximated by one of its Taylor polynomials.

EXAMPLE 2 *Series for* $\sin x$. Show that the Maclaurin series for $\sin x$ converges to $\sin x$ for all x.

Solution Expressed in terms of x, the function and its derivatives are

$$f(x) \quad = \quad \sin x, \qquad f'(x) \quad = \quad \cos x,$$

$$f''(x) \quad = \quad -\sin x, \qquad f'''(x) \quad = \quad -\cos x,$$

$$\vdots$$

$$f^{(2k)}(x) = (-1)^k \sin x, \qquad f^{(2k+1)}(x) = (-1)^k \cos x,$$

so

$$f^{(2k)}(0) = 0 \quad \text{and} \quad f^{(2k+1)}(0) = (-1)^k.$$

The series has only odd-powered terms and, for $n = 2k + 1$, Taylor's theorem gives

$$\sin x = x - \frac{x^3}{3!} + \frac{x^5}{5!} - \cdots + \frac{(-1)^k x^{2k+1}}{(2k+1)!} + R_{2k+1}(x).$$

Now, since all the derivatives of $\sin x$ have absolute values less than or equal to 1, we can apply the Remainder Estimation Theorem with $M = 1$ and $r = 1$ to obtain

$$|R_{2k+1}(x)| \le 1 \cdot \frac{|x|^{2k+2}}{(2k+2)!}.$$

Since $[|x|^{2k+2}/(2k+2)!] \to 0$ as $k \to \infty$, whatever the value of x,

$$R_{2k+1}(x) \to 0,$$

and the Maclaurin series for $\sin x$ converges to $\sin x$ for every x:

$$\sin x = \sum_{k=0}^{\infty} \frac{(-1)^k x^{2k+1}}{(2k+1)!} = x - \frac{x^3}{3!} + \frac{x^5}{5!} - \frac{x^7}{7!} + \cdots. \qquad (11)$$

EXAMPLE 3 *Series for* $\cos x$. Show that the Maclaurin series for $\cos x$ converges to $\cos x$ for every value of x.

Solution We begin by adding the remainder term to the Taylor polynomial for $\cos x$ in Eq. (4) of the previous article, to obtain Taylor's formula for $\cos x$ with $n = 2k$:

$$\cos x = 1 - \frac{x^2}{2!} + \frac{x^4}{4!} - \cdots + (-1)^k \frac{x^{2k}}{(2k)!} + R_{2k}(x).$$

Since the derivatives of the cosine have absolute value less than or equal to 1, we apply the Remainder Estimation Theorem with $M = 1$ and $r = 1$ to obtain

$$|R_{2k}(x)| \leq 1 \cdot \frac{|x|^{2k+1}}{(2k+1)!}.$$

For every value of x, $R_{2k} \to 0$ as $k \to \infty$. Therefore, the series converges to $\cos x$ for every value of x:

$$\cos x = \sum_{k=0}^{\infty} \frac{(-1)^k x^{2k}}{(2k)!} = 1 - \frac{x^2}{2!} + \frac{x^4}{4!} - \frac{x^6}{6!} + \cdots. \qquad (12)$$

EXAMPLE 4 Find the Maclaurin series for $\cos 2x$ and show that it converges to $\cos 2x$ for every value of x.

Solution The Maclaurin series for $\cos x$ converges to $\cos x$ for every value of x and therefore converges for every value of $2x$:

$$\cos 2x = \sum_{k=0}^{\infty} \frac{(-1)^k (2x)^{2k}}{(2k)!} = 1 - \frac{(2x)^2}{2!} + \frac{(2x)^4}{4!} - \frac{(2x)^6}{6!} + \cdots. \quad \blacksquare$$

Taylor series can be added, subtracted, and multiplied by constants, just as other series can, and the results are once again Taylor series. The Taylor series for $f(x) + g(x)$ is the sum of the Taylor series for $f(x)$ and $g(x)$ because the nth derivative of $f(x) + g(x)$ is $f^{(n)}(x) + g^{(n)}(x)$, and so on. In the next example, we add the series for e^x and e^{-x} and divide by 2, to obtain the Taylor series for $\cosh x$.

EXAMPLE 5 *Series for* $\cosh x$. Find the Maclaurin series for $\cosh x$.

Solution

$$e^x = 1 + x + \frac{x^2}{2!} + \frac{x^3}{3!} + \frac{x^4}{4!} + \frac{x^5}{5!} + \cdots,$$

$$e^{-x} = 1 - x + \frac{x^2}{2!} - \frac{x^3}{3!} + \frac{x^4}{4!} - \frac{x^5}{5!} + \cdots,$$

$$\cosh x = \frac{e^x + e^{-x}}{2} = 1 \quad + \frac{x^2}{2!} \quad + \frac{x^4}{4!} \quad + \cdots = \sum_{k=0}^{\infty} \frac{x^{2k}}{(2k)!}. \quad \blacksquare$$

The Identity $e^{i\theta} = \cos \theta + i \sin \theta$

Up to this point we have not used imaginary numbers in our study of series. We recall that complex numbers occur in solving quadratic equations. The formula for the roots of the quadratic equation

$$ax^2 + bx + c = 0$$

is

$$x = \frac{-b \pm \sqrt{b^2 - 4ac}}{2a},$$

when a, b, and c are real numbers and $a \neq 0$. When the discriminant $D = b^2 - 4ac$ is negative, the two roots are complex numbers

$$u + iv \quad \text{and} \quad u - iv,$$

where

$$u = -\frac{b}{2a},$$

$$v = \frac{\sqrt{4ac - b^2}}{2a}.$$

We review these facts mainly to recall the symbol i,

$$i = \sqrt{-1},$$

and to remind ourselves that

$$i^2 = -1, \quad i^3 = i^2 i = -i, \quad i^4 = i^2 i^2 = 1, \quad i^5 = i^4 i = i,$$

and so on.

With these facts in mind, we replace x by $i\theta$ in the Maclaurin series for e^x and simplify to get

$$e^{i\theta} = 1 + \frac{i\theta}{1!} + \frac{i^2\theta^2}{2!} + \frac{i^3\theta^3}{3!} + \frac{i^4\theta^4}{4!} + \frac{i^5\theta^5}{5!} + \frac{i^6\theta^6}{6!} + \cdots$$

$$= \left(1 - \frac{\theta^2}{2!} + \frac{\theta^4}{4!} - \frac{\theta^6}{6!} + \cdots \right) + i\left(\theta - \frac{\theta^3}{3!} + \frac{\theta^5}{5!} - \cdots \right)$$

$$= \cos \theta + i \sin \theta.$$

It would not be accurate to say that the calculations just completed have proved that

$$e^{i\theta} = \cos \theta + i \sin \theta. \tag{13}$$

Rather, we shall adopt the point of view that Eq. (13) is the *definition* of $e^{i\theta}$. This definition, which is standard, is motivated by the series expansions for $\cos \theta$, $\sin \theta$, and e^x with $x = i\theta$.

Once we have accepted Eq. (13) as the definition of $e^{i\theta}$, we quickly verify the following law.

The Law of Addition of Imaginary Exponents

If θ_1 and θ_2 are any real numbers, then

$$e^{i\theta_1} e^{i\theta_2} = e^{i(\theta_1 + \theta_2)}. \tag{14}$$

Proof   By definition,

$$e^{i\theta_1} = \cos \theta_1 + i \sin \theta_1, \quad e^{i\theta_2} = \cos \theta_2 + i \sin \theta_2.$$

Multiplying and simplifying, we get

$$e^{i\theta_1}e^{i\theta_2} = (\cos \theta_1 \cos \theta_2 - \sin \theta_1 \sin \theta_2)$$
$$+ i(\sin \theta_1 \cos \theta_2 + \cos \theta_1 \sin \theta_2)$$
$$= \cos(\theta_1 + \theta_2) + i \sin(\theta_1 + \theta_2)$$
$$= e^{i(\theta_1+\theta_2)}.$$ ∎

Notice also that when $\theta = 0$, $i\theta = 0$ and Eq. (13) yields

$$e^0 = \cos 0 + i \sin 0 = 1,$$

which is consistent with $e^x = 1$ when $x = 0$.

If $\theta_1 = \theta$ and $\theta_2 = -\theta$, then $\theta_1 + \theta_2 = 0$, and Eq. (14) yields the result

$$e^{i\theta}e^{-i\theta} = e^0 = 1,$$

so

$$e^{-i\theta} = \frac{1}{e^{i\theta}}. \tag{15}$$

Thus the usual laws for the exponential function continue to apply to the function $e^{i\theta}$ defined by Eq. (13).

Truncation Error

Here are some examples of how to use the Remainder Estimation Theorem to estimate truncation error.

EXAMPLE 6 Calculate e with an error of less than 10^{-6}.

Solution We can use the result of Example 1, Eq. (9a), with $x = 1$ to write

$$e = 1 + 1 + \frac{1}{2!} + \cdots + \frac{1}{n!} + R_n(1),$$

with

$$R_n(1) = e^c \frac{1}{(n + 1)!} \qquad \text{for some } c \text{ between 0 and 1.}$$

For the purposes of this example, we do not assume that we already know that $e = 2.71828 \ldots$, but we have earlier shown that $e < 3$. Hence, we are certain that

$$\frac{1}{(n + 1)!} < R_n(1) < \frac{3}{(n + 1)!}$$

because $1 < e^c < 3$.

By experiment we find that $1/9! > 10^{-6}$, while $3/10! < 10^{-6}$. Thus we should take $(n + 1)$ to be at least 10, or n to be at least 9. With an error of less than 10^{-6},

$$e = 1 + 1 + \frac{1}{2} + \frac{1}{3!} + \cdots + \frac{1}{9!} \approx 2.718282.$$ ∎

EXAMPLE 7   For what values of x can $\sin x$ be replaced by $x - (x^3/3!)$ with an error of magnitude no greater than 3×10^{-4}?

Solution   Here we can take advantage of the fact that the Maclaurin series for $\sin x$ is an alternating series for every nonzero value of x. According to the Alternating Series Estimation Theorem in Article 11.8, the error in truncating

$$\sin x = x - \frac{x^3}{3!} + \frac{x^5}{5!} - \cdots$$

after $(x^3/3!)$ is no greater than

$$\left|\frac{x^5}{5!}\right| = \frac{|x|^5}{120}.$$

Therefore the error will be less than or equal to 3×10^{-4} if

$$\frac{|x|^5}{120} < 3 \times 10^{-4} \qquad \text{or} \qquad |x| < \sqrt[5]{360 \times 10^{-4}} \approx 0.514.$$

The Alternating Series Estimation Theorem tells us something that the Remainder Estimation Theorem does not: namely, that the estimate $x - (x^3/3!)$ for $\sin x$ is an underestimate when x is positive because then $x^5/120$ is positive.

Figure 12.4 shows the graph of $\sin x$, along with the graphs of a number of its approximating Taylor polynomials. Note that the graph of $s_1 = x - (x^3/3!)$ is almost indistinguishable from the sine curve when $-1 \le x \le 1$. However, it crosses the x-axis at $\pm\sqrt{6} \approx \pm 2.45$, whereas the sine curve crosses the axis at $\pm\pi \approx \pm 3.14$.

You might wonder how the estimate given by the Remainder Estimation Theorem compares with the one just obtained from the Alternating Series Estimation Theorem. If we write

$$\sin x = x - \frac{x^3}{3!} + R_3,$$

then the Remainder Estimation Theorem gives

$$|R_3| \le 1 \cdot \frac{|x|^4}{4!} = \frac{|x|^4}{24},$$

12.4   The polynomials

$$s_n(x) = \sum_{k=0}^{n} [(-1)^k x^{2k+1}/(2k+1)!]$$

converge to $\sin x$ as $n \to \infty$. (Adapted from Helen M. Kammerer, *American Mathematical Monthly*, 43 (1936), pp. 293–294.)

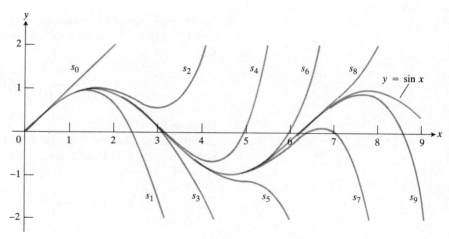

which is not very good. But when we recognize that $x - x^3/3! = 0 + x + 0x^2 - x^3/3! + 0x^4$ is the Taylor polynomial of order 4 as well as of order 3, then we have

$$\sin x = x - \frac{x^3}{3!} + 0 + R_4,$$

and the Remainder Estimation Theorem with $M = r = 1$ gives

$$|R_4| \le 1 \cdot \frac{|x|^5}{5!} = \frac{|x|^5}{120}.$$

This is what we had from the Alternating Series Estimation Theorem. ∎

In the preceding examples of application of the Remainder Estimation Theorem, we have been able to take $r = 1$. In the next example, $f(x) = \sin 2x$, a factor of 2 enters each time we differentiate, so that we have $r = 2$.

EXAMPLE 8 Let $f(x) = \sin 2x$. For what values of x is the approximation $\sin 2x \approx (2x) - (2x)^3/3! + (2x)^5/5!$ not in error by more than 5×10^{-6}?

Solution Because the Taylor polynomials of order 5 and of order 6 for $\sin 2x$ are identical, in that they differ only in a term that has a zero coefficient, we are certain that the error is no greater than $2^7|x|^7/7!$ (by comparison with Example 7). We therefore solve the inequality

$$\frac{|2x|^7}{7!} < 5 \times 10^{-6}.$$

The result is

$$|2x| < \sqrt[7]{7! \times 5 \times 10^{-6}} = \sqrt[7]{0.025200} \approx 0.59106$$

or $|x| < 0.29553$ (radians). ∎

PROBLEMS

In Problems 1–6, write the Maclaurin series for each function.

1. $e^{x/2}$

2. $\sin 3x$

3. $5 \cos \dfrac{x}{\pi}$

4. $\sinh x$

5. $\dfrac{x^2}{2} - 1 + \cos x$

6. $\cos^2 x = \dfrac{1 + \cos 2x}{2}$

7. Use series to verify that
   a) $\cos(-x) = \cos x$,
   b) $\sin(-x) = -\sin x$

8. Show that

$$e^x = e^a \left[1 + (x - a) + \frac{(x - a)^2}{2!} + \cdots \right].$$

In Problems 9–11, write Taylor's formula (Eq. 7a), with $n = 2$ and $a = 0$, for the given function.

9. $\dfrac{1}{1 + x}$

10. $\ln(1 + x)$

11. $\sqrt{1 + x}$

12. Find the Taylor series for e^x at $a = 1$. Compare your series with the result in Problem 8.

13. For approximately what values of x can one replace $\sin x$ by $x - (x^3/6)$ with an error of magnitude no greater than 5×10^{-4}?

14. If $\cos x$ is replaced by $1 - (x^2/2)$ and $|x| < 0.5$, what estimate can be made of the error? Does $1 - (x^2/2)$ tend to be too large, or too small?

15. How close is the approximation $\sin x = x$ when $|x| < 10^{-3}$? For which of these values of x is $x < \sin x$?

16. The estimate $\sqrt{1 + x} = 1 + (x/2)$ is used when $|x|$ is small. Estimate the error when $|x| < 0.01$.

17. The approximation $e^x = 1 + x + (x^2/2)$ is used when x is small. Use the Remainder Estimation Theorem to estimate the error when $|x| < 0.1$.

18. When $x < 0$, the series for e^x is an alternating series. Use the Alternating Series Estimation Theorem to estimate the error that results from replacing e^x by $1 + x + (x^2/2)$ when $-0.1 < x < 0$. Compare with Problem 17.

19. Estimate the error in the approximation $\sinh x = x + (x^3/3!)$ when $|x| < 0.5$. (*Hint:* Use R_4, not R_3.)

20. When $0 \le h \le 0.01$, show that e^h may be replaced by $1 + h$ with an error of magnitude no greater than 0.6% of h. Use $e^{0.1} = 1.105$.

21. Let $f(x)$ and $g(x)$ have derivatives of all orders at $a = 0$. Show that the Maclaurin series for $f + g$ is

$$\sum_{n=0}^{\infty} \frac{f^{(n)}(0) + g^{(n)}(0)}{n!} x^n.$$

22. Each of the following sums is the value of an elementary function at some point. Find the function and the point.

a) $(0.1) - \dfrac{(0.1)^3}{3!} + \dfrac{(0.1)^5}{5!} - \cdots + \dfrac{(-1)^k(0.1)^{2k+1}}{(2k+1)!} + \cdots$

b) $1 - \dfrac{\pi^2}{4^2 \cdot 2!} + \dfrac{\pi^4}{4^4 \cdot 4!} - \cdots + \dfrac{(-1)^k(\pi)^{2k}}{4^{2k} \cdot (2k)!} + \cdots$

c) $1 + \dfrac{1}{2!} + \dfrac{1}{4!} + \cdots + \dfrac{1}{(2k)!} + \cdots$

23. Express each of the following in the form $u + iv$, with u and v real.

a) $e^{i\pi}$

b) $e^{i\pi/4}$

c) $e^{-i\pi/2}$

d) $e^{i\pi} \cdot e^{-i\pi/2}$

24. Using Eq. (13), show that

$$\cos\theta = \frac{e^{i\theta} + e^{-i\theta}}{2} \quad \text{and} \quad \sin\theta = \frac{e^{i\theta} - e^{-i\theta}}{2i}.$$

These are sometimes called Euler's identities.

25. Use the results of Problem 24 to show that

$$\cos^3\theta = \frac{1}{4}\cos 3\theta + \frac{3}{4}\cos\theta$$

and

$$\sin^3\theta = -\frac{1}{4}\sin 3\theta + \frac{3}{4}\sin\theta.$$

26. When a and b are real, we define $e^{(a+ib)x}$ to be $e^{ax}(\cos bx + i \sin bx)$. From this definition, show that

$$\frac{d}{dx} e^{(a+ib)x} = (a + ib)e^{(a+ib)x}.$$

27. Two complex numbers, $a + ib$ and $c + id$, are equal if and only if $a = c$ and $b = d$. Use this fact to evaluate

$$\int e^{ax} \cos bx \, dx \quad \text{and} \quad \int e^{ax} \sin bx \, dx$$

from

$$\int e^{(a+ib)x} \, dx = \frac{a - ib}{a^2 + b^2} e^{(a+ib)x} + C,$$

where

$$C = C_1 + iC_2$$

is a complex constant of integration.

TOOLKIT PROGRAMS

Taylor Series

12.4

Expansion Points, the Binomial Theorem, Arctangents, and π

In this article we continue examining useful series expansions for common elementary functions and use them for approximating functional values. However, before we do that, we first consider what effect our choice of the expansion point, a, has on the efficiency of our series calculations.

Choosing an Expansion Point

The Taylor series expansion

$$f(x) = f(a) + f'(a)(x - a) + \frac{f''(a)}{2!}(x - a)^2 + \cdots$$

$$+ \frac{f^{(n)}(a)}{n!}(x - a)^n + R_n(x) \tag{1}$$

expresses the value of the function at x in terms of its value and the values of its derivatives at a, plus a remainder term, which we hope is so small that it may safely be omitted. In applying series to numerical computations, it is therefore necessary that a be chosen so that $f(a), f'(a), f''(a), \ldots$, are known. In dealing with the trigonometric functions, for example, we might take $a = 0$, $\pm \pi/6$, $\pm \pi/4$, $\pm \pi/3$, $\pm \pi/2$, and so on. It is also clear that it is *desirable* to choose the value of a near the value of x for which the function is to be computed, in order to make $(x - a)$ small so that the terms of the series decrease rapidly as n increases.

EXAMPLE 1 What value of a might one choose in the Taylor series (1) to compute sin 35°? Also discuss truncation and round-off errors.

Solution The radian measure for $35°$ is $35\pi/180$. We could choose $a = 0$ and use the series

$$\sin x = x - \frac{x^3}{3!} + \frac{x^5}{5!} - \cdots$$

$$+ (-1)^n \frac{x^{2n+1}}{(2n+1)!} + 0 \cdot x^{2n+2} + R_{2n+2}(x), \tag{2}$$

or we could choose $a = \pi/6$ (which corresponds to 30°) and use the series

$$\sin x = \sin \frac{\pi}{6} + \cos \frac{\pi}{6}\left(x - \frac{\pi}{6}\right) - \sin \frac{\pi}{6} \frac{(x - \pi/6)^2}{2!} - \cos \frac{\pi}{6} \frac{(x - \pi/6)^3}{3!}$$

$$+ \cdots + \sin\left(\frac{\pi}{6} + n\frac{\pi}{2}\right) \frac{(x - \pi/6)^n}{n!} + R_n(x).$$

The remainder in the series (2) satisfies the inequality

$$|R_{2n+2}(x)| \le \frac{|x|^{2n+3}}{(2n+3)!},$$

which tends to zero as $n \to \infty$, no matter how large $|x|$ may be. We could therefore calculate sin 35° by placing

$$x = \frac{35\pi}{180} \approx 0.6108652$$

in the approximation

$$\sin x \approx x - \frac{x^3}{6} + \frac{x^5}{120} - \frac{x^7}{5040},$$

with a truncation error of magnitude no greater than 3.3×10^{-8}, since

$$\left| R_8\left(\frac{35\pi}{180}\right) \right| < \frac{(0.611)^9}{9!} < 3.3 \times 10^{-8}.$$

By using the series with $a = \pi/6$, we could obtain equal accuracy with a smaller exponent n, but at the expense of introducing $\cos \pi/6 = \sqrt{3}/2$ as one of the coefficients. In this series, with $a = \pi/6$, we would take

$$x = \frac{35\pi}{180},$$

and the quantity that appears raised to the various powers is

$$x - \frac{\pi}{6} = \frac{5\pi}{180} \approx 0.0872665,$$

which decreases rapidly when raised to high powers.

Round-off Error   What additional error is caused by using 0.6108652 in place of $35\pi/180$ for x in the polynomial

$$f(x) = x - \frac{x^3}{3!} + \frac{x^5}{5!} - \frac{x^7}{7!}?$$

There are two ways to answer this question:

1. by using a computer, and

2. by using differentials.

Using a computer, we calculated $f(x)$ with the computer's 14-decimal approximation to $35\pi/180 \approx 0.61086\ 52381\ 9833$ and found the value to be

$$f(x) = 0.57357\ 64038 \qquad \text{(to 10 decimals)}.$$

From a table of sines we found

$$\sin 35° = 0.57357\ 64363\ 51046 \qquad \text{(to 15 decimals)}.$$

These results indicate a *truncation* error for $\sin x - f(x)$ approximately equal to 3.26×10^{-8}.

Continuing to use the computer, we calculated $f(x_0)$ with $x_0 = 0.6108652$ (which differs from $35\pi/180$ by about 3.82×10^{-8}). The computed value was

$$f(x_0) = 0.57357\ 63725 \qquad \text{(to 10 decimals)},$$

with an indicated error (*truncation* error plus *round-off* error) of

$$\sin x - f(x_0) \approx 6.38 \times 10^{-8}.$$

Hence the *round-off* error alone is approximately

$$(6.38 - 3.26) \times 10^{-8} = 3.12 \times 10^{-8}.$$

Using differentials, we let

$$dx = \frac{35\pi}{180} - x_0 \approx 3.82 \times 10^{-8}.$$

The corresponding difference,

$$f(x_0 + dx) - f(x_0) \approx f'(x_0)\, dx,$$

is nearly equal to

$$\cos x_0\, dx \approx 0.82\, dx \approx 3.13 \times 10^{-8}.$$

Both methods of estimating the round-off error lead to nearly identical results and show that the round-off error in this example is nearly equal to the truncation error.   ■

Comparable results are obtained if we use the Taylor series with $a = \pi/6$, and we omit the details.

The Binomial Series

One of the most useful series of all time, the **binomial series,** is the Maclaurin series for the function $f(x) = (1 + x)^m$. Newton used it to estimate integrals (we will, too, in Article 12.5), and it can be used to give accurate estimates of roots. To derive the series, we first list the function and its derivatives:

$$f(x) = (1 + x)^m,$$
$$f'(x) = m(1 + x)^{m-1},$$
$$f''(x) = m(m - 1)(1 + x)^{m-2},$$
$$f'''(x) = m(m - 1)(m - 2)(1 + x)^{m-3},$$
$$\vdots$$
$$f^{(k)}(x) = m(m - 1)(m - 2) \cdots (m - k + 1)(1 + x)^{m-k}.$$

We then evaluate these at $x = 0$ and substitute in the Maclaurin series formula to obtain

$$1 + mx + \frac{m(m - 1)}{2!}x^2 + \cdots$$

$$+ \frac{m(m - 1)(m - 2) \cdots (m - k + 1)}{k!}x^k + \cdots. \tag{3}$$

If m is an integer greater than or equal to zero, the series stops after $(m + 1)$ terms because the coefficients from $k = m + 1$ on are zero. But when m is not an integer, the series is infinite. The Maclaurin series extends the binomial theorem beyond positive integral exponents. (For a proof that the series converges to $(1 + x)^m$ when $|x| < 1$, see Courant and John's *Introduction to Calculus and Analysis,* Wiley-Interscience, 1974.)

EXAMPLE 2 Use the binomial series to estimate $\sqrt{1.25}$ with an error of less than 0.001.

Solution We take $x = 1/4$ and $m = 1/2$ in (3) to obtain

$$\left(1 + \frac{1}{4}\right)^{1/2} = 1 + \frac{1}{2}\left(\frac{1}{4}\right) + \frac{(1/2)(-1/2)}{2!}\left(\frac{1}{4}\right)^2 + \frac{(1/2)(-1/2)(-3/2)}{3!}\left(\frac{1}{4}\right)^3 + \cdots$$

$$= 1 + \frac{1}{8} - \frac{1}{128} + \frac{1}{1024} - \frac{5}{32768} + \cdots.$$

The series alternates after the first term, so the approximation

$$\sqrt{1.25} \approx 1 + \frac{1}{8} - \frac{1}{128} = 1.1171875$$

is within $1/1024$ of the exact value and thus has the required accuracy. Rounded to three decimals, $\sqrt{1.25} \approx 1.117$.   ■

Computation of Logarithms

Natural logarithms may be computed from series. The starting point is the series for $\ln(1 + x)$ in powers of x:

$$\ln(1 + x) = x - \frac{x^2}{2} + \frac{x^3}{3} - \cdots + (-1)^{n-1}\frac{x^n}{n} + \cdots. \tag{4}$$

This Maclaurin series (or Taylor series with $a = 0$) can be found directly by applying Taylor's theorem with remainder and showing that the remainder approaches zero as $n \to \infty$ if $-1 < x \le 1$.

In particular, when $x = 1$, the series on the right becomes the alternating harmonic series, which we know converges, and now we know that it converges to $\ln 2$.

Computation of π

The numbers π, $\sqrt{2}$, and e are probably the best-known irrational numbers. No doubt the most famous of the three is π. Over the centuries, many people have calculated rational approximations, either as fractions like 22/7, and 355/113 or as decimals like 3.14159 26535 89793 23846. A French mathematician, Viéta (1540–1603), gave the formula

$$\frac{2}{\pi} = \sqrt{\tfrac{1}{2}} \times \sqrt{(\tfrac{1}{2} + \tfrac{1}{2}\sqrt{\tfrac{1}{2}})} \times \sqrt{(\tfrac{1}{2} + \tfrac{1}{2}\sqrt{(\tfrac{1}{2} + \tfrac{1}{2}\sqrt{\tfrac{1}{2}})})} \times \cdots, \tag{5}$$

which Turnbull[†] calls "the first actual formula for the time-honoured number π." Other formulas include

$$\frac{\pi}{4} = \frac{2 \times 4 \times 4 \times 6 \times 6 \times 8 \times \cdots}{3 \times 3 \times 5 \times 5 \times 7 \times 7 \times \cdots}, \tag{6}$$

discovered by the English mathematician Wallis, and

$$\frac{\pi}{4} = 1 - \frac{1}{3} + \frac{1}{5} - \frac{1}{7} + \cdots,$$

known as Leibniz's formula. John Machin discovered the formula

$$\frac{\pi}{4} = 4 \tan^{-1}\frac{1}{5} - \tan^{-1}\frac{1}{239}$$

and used it, together with the series

$$\tan^{-1}x = x - \frac{x^3}{3} + \frac{x^5}{5} - \frac{x^7}{7} + \cdots,$$

which was discovered by James Gregory in 1671. Machin's computation of π, correct to 100 decimals, was published in 1706. (For many more details about the history of computations of decimal approximations to π, including a bibliography of 55 items, see "The Evolution of Extended Decimal Approximations to π," by J. W. Wrench, Jr., in *The Mathematics Teacher* (Dec. 1960), pp. 644–650.)

The first approximations with more than 1000 decimals were evidently made by June 1949, by Wrench and L. B. Smith, to "about 1120 decimal

---

[†] *World of Mathematics,* Vol. 1 (New York: Simon and Schuster, 1956), p. 121.

places.'' In September 1949, George W. Reitwiesner and his associates at the Ballistic Research Laboratories, Aberdeen Proving Ground, evaluated π to ''about 2037 places (2040 working decimals)'' using the ENIAC (electronic numerical integrator and computer), one of the earliest computers. Machin's formula was used in these computations, and the ENIAC used a total time, ''including card handling,'' of 70 hours.

The most recent information we have (1986) is that the value of π has been computed to 29,360,128 digits, using a newly developed Cray-2 supercomputer. The program was written (in his spare time!) by David H. Bailey of the NASA Ames Research Center, California, using an algorithm discovered by Jonathan M. Borwein and Peter B. Borwein of Dalhousie University in Halifax, Nova Scotia. Bailey's program took up 28 hours of processing time and 138 million words of main memory and required 12 trillion arithmetic operations. To check the result, Bailey ran a second program based on an older algorithm, also developed by the Borweins. Only the last 17 digits of π were different.

Why does anyone continue calculating π to this degree? From a theoretical point of view, π continues to generate mathematical questions, such as the statistical problem of the randomness of π's digits. Practically speaking, calculations of π continue to be of interest as a way of checking the reliability, accuracy, and efficiency of computer equipment.

We now turn our attention to the series for $\tan^{-1}x$, since, as noted above, it leads to the Leibniz formula and others from which π has been computed to many decimal places.

Since

$$\tan^{-1}x = \int_0^x \frac{dt}{1 + t^2},$$

we integrate the geometric series, with remainder,

$$\frac{1}{1 + t^2} = 1 - t^2 + t^4 - t^6 + \cdots + (-1)^n t^{2n} + \frac{(-1)^{n+1} t^{2n+2}}{1 + t^2}. \qquad (7)$$

Thus

$$\tan^{-1}x = x - \frac{x^3}{3} + \frac{x^5}{5} - \frac{x^7}{7} + \cdots + (-1)^n \frac{x^{2n+1}}{2n + 1} + R,$$

where

$$R = \int_0^x \frac{(-1)^{n+1} t^{2n+2}}{1 + t^2}\, dt.$$

The denominator of the integrand is greater than or equal to 1; hence

$$|R| \leq \int_0^{|x|} t^{2n+2} dt = \frac{|x|^{2n+3}}{2n + 3}.$$

If $|x| \leq 1$, the right side of this inequality approaches zero as $n \to \infty$. Therefore R also approaches zero and we have

$$\tan^{-1}x = \sum_{n=0}^{\infty} \frac{(-1)^n x^{2n+1}}{2n + 1}.$$

Frequently Used Maclaurin Series

$$\frac{1}{1 - x} = 1 + x + x^2 + \cdots + x^n + \cdots = \sum_{n=0}^{\infty} x^n, \quad |x| < 1$$

$$\frac{1}{1 + x} = 1 - x + x^2 - \cdots + (-x)^n + \cdots = \sum_{n=0}^{\infty} (-1)^n x^n, \quad |x| < 1$$

$$e^x = 1 + x + \frac{x^2}{2!} + \cdots + \frac{x^n}{n!} + \cdots = \sum_{n=0}^{\infty} \frac{x^n}{n!}, \quad |x| < \infty$$

$$\sin x = x - \frac{x^3}{3!} + \frac{x^5}{5!} - \cdots + (-1)^n \frac{x^{2n+1}}{(2n + 1)!} + \cdots = \sum_{n=0}^{\infty} \frac{(-1)^n x^{2n+1}}{(2n + 1)!}, \quad |x| < \infty$$

$$\cos x = 1 - \frac{x^2}{2!} + \frac{x^4}{4!} - \cdots + (-1)^n \frac{x^{2n}}{(2n)!} + \cdots = \sum_{n=0}^{\infty} \frac{(-1)^n x^{2n}}{(2n)!}, \quad |x| < \infty$$

$$\ln(1 + x) = x - \frac{x^2}{2} + \frac{x^3}{3} - \cdots + (-1)^{n-1} \frac{x^n}{n} + \cdots = \sum_{n=1}^{\infty} \frac{(-1)^{n-1} x^n}{n}, \quad -1 < x \le 1$$

$$\ln\frac{1 + x}{1 - x} = 2 \tanh^{-1} x = 2\left(x + \frac{x^3}{3} + \frac{x^5}{5} + \cdots + \frac{x^{2n+1}}{2n + 1} + \cdots\right) = 2 \sum_{n=0}^{\infty} \frac{x^{2n+1}}{2n + 1}, \quad |x| < 1$$

$$\tan^{-1} x = x - \frac{x^3}{3} + \frac{x^5}{5} - \cdots + (-1)^{n-1} \frac{x^{2n-1}}{2n - 1} + \cdots = \sum_{n=1}^{\infty} \frac{(-1)^{n-1} x^{2n-1}}{2n - 1}, \quad |x| \le 1$$

Binomial Series

$$(1 + x)^m = 1 + mx + \frac{m(m - 1)x^2}{2!} + \frac{m(m - 1)(m - 2)x^3}{3!} + \cdots + \frac{m(m - 1)(m - 2)\cdots(m - k + 1)x^k}{k!} + \cdots$$

$$= 1 + \sum_{k=1}^{\infty} \binom{m}{k} x^k, \quad |x| < 1$$

where

$$\binom{m}{1} = m, \qquad \binom{m}{2} = \frac{m(m - 1)}{2!}, \qquad \binom{m}{k} = \frac{m(m - 1)\cdots(m - k + 1)}{k!} \quad \text{for } k \ge 3.$$

NOTE: It is customary to define $\binom{m}{0}$ to be 1 and to take $x^0 = 1$ (even in the usually excluded case where $x = 0$) in order to write the binomial series compactly as

$$(1 + x)^m = \sum_{k=0}^{\infty} \binom{m}{k} x^k, \quad |x| < 1.$$

If m is a *positive integer*, the series terminates at x^m, and the result converges for all x.

$$\tan^{-1}x = x - \frac{x^3}{3} + \frac{x^5}{5} - \frac{x^7}{7} + \cdots, \qquad |x| \le 1. \tag{8}$$

When we put $x = 1$, $\tan^{-1} 1 = \pi/4$, in Eq. (8), we get Leibniz's formula:

$$\frac{\pi}{4} = 1 - \frac{1}{3} + \frac{1}{5} - \frac{1}{7} + \frac{1}{9} - \cdots + \frac{(-1)^{n-1}}{2n-1} + \cdots.$$

Because this series converges very slowly, it is not used in approximating π to many decimal places. The series for $\tan^{-1}x$ converges most rapidly when x is near zero. For that reason, people who use the series for $\tan^{-1}x$ to compute π use various trigonometric identities. For example, if

$$\alpha = \tan^{-1}\frac{1}{2} \qquad \text{and} \qquad \beta = \tan^{-1}\frac{1}{3},$$

then

$$\begin{aligned}
\tan(\alpha + \beta) &= \frac{\tan \alpha + \tan \beta}{1 - \tan \alpha \tan \beta} \\
&= \frac{(1/2) + (1/3)}{1 - (1/6)} \\
&= 1 \\
&= \tan\frac{\pi}{4}
\end{aligned}$$

and

$$\frac{\pi}{4} = \alpha + \beta = \tan^{-1}\frac{1}{2} + \tan^{-1}\frac{1}{3}. \tag{9}$$

Now Eq. (8) may be used with $x = 1/2$ to evaluate $\tan^{-1}1/2$ and with $x = 1/3$ to give $\tan^{-1}1/3$. The sum of these results, multiplied by 4, gives π†.

---

† For a delightful account of attempts to compute, and even to *legislate* (!) the value of π, see Chapter 12 of David A. Smith's *Interface: Calculus and the Computer*, 2d ed. (Philadelphia: Saunders College Publishing Company, 1984).

PROBLEMS

In Problems 1–6, use a suitable series to calculate the indicated quantity to three decimal places. In each case, show that the remainder term does not exceed 5×10^{-4}.

1. $\cos 31°$ **2.** $\tan 46°$ **3.** $\sin 6.3$

4. $\cos 69$ **5.** $\ln 1.25$ **6.** $\tan^{-1}1.02$

7. Find the Maclaurin series for $\ln (1 + 2x)$. For what values of x does the series converge?

8. For what values of x can one replace $\ln (1 + x)$ by x with an error of magnitude no greater than 1% of the absolute value of x?

Use series to evaluate the integrals in Problems 9 and 10 to three decimals.

9. $\displaystyle\int_0^{0.1} \frac{\sin x}{x}\, dx$ **10.** $\displaystyle\int_0^{0.1} e^{-x^2}\, dx$

11. Show that the ordinate of the catenary $y = a \cosh x/a$ deviates from the ordinate of the parabola $x^2 = 2a(y - a)$ by less than $0.003|a|$ over the range $|x/a| \leq 1/3$.

12. a) Replace x by $-x$ in the series for $\ln (1 + x)$ to obtain a series for $\ln (1 - x)$. Combine this with the series for $\ln (1 + x)$ to show that

$$\ln\frac{1 + x}{1 - x} = 2\left(x + \frac{x^3}{3} + \frac{x^5}{5} + \cdots\right) \quad \text{for } |x| < 1.$$

b) For what value of x is $(1 + x)/(1 - x) = 2$? Use that value of x in the series of part (a) to estimate $\ln 2$ to three decimals.

13. Find the sum of the series

$$\frac{1}{2} - \frac{1}{2}\left(\frac{1}{2}\right)^2 + \frac{1}{3}\left(\frac{1}{2}\right)^3 - \frac{1}{4}\left(\frac{1}{2}\right)^4 + \cdots.$$

14. How many terms of the series for $\tan^{-1} 1$ would you have to add for the Alternating Series Estimation Theorem to guarantee a calculation of $\pi/4$ to two decimals?

15. **CALCULATOR** Equation (8) and the equation

$$\pi = 48 \tan^{-1}\frac{1}{18} + 32 \tan^{-1}\frac{1}{57} - 20 \tan^{-1}\frac{1}{239}$$

yield a series that converges to $\pi/4$ fairly rapidly. Estimate π to three decimal places with this series. In contrast, the convergence of $\Sigma_{n=1}^{\infty} (1/n^2)$ to $\pi^2/6$ is so slow that even 50 terms will not yield two-place accuracy.

16. **CALCULATOR**
a) Find π to two decimals with the formula of Wallis.
b) If your calculator is programmable, use Viéta's formula to calculate π to five decimal places.

17. **CALCULATOR** A special case of Salamin's algorithm for estimating π begins with defining sequences $\{a_n\}$ and $\{b_n\}$ by the rules

$$a_0 = 1, \qquad\qquad b_0 = \frac{1}{\sqrt{2}}.$$

$$a_{n+1} = \frac{(a_n + b_n)}{2}, \qquad b_{n+1} = \sqrt{a_n b_n}.$$

Then the sequence $\{c_n\}$ defined for $n \geq 1$ by

$$c_n = \frac{4a_n b_n}{1 - \Sigma_{j=1}^{n} 2^{j+1}(a_j^2 - b_j^2)}$$

converges to π. Calculate c_3. (E. Salamin, "Computation of π using arithmetic-geometric mean," *Mathematics of Computation,* vol. 30 (July 1976), pp. 565–570.)

18. Show that the series in Eq. (8) for $\tan^{-1}x$ diverges for $|x| > 1$.

19. Show that

$$\int_0^x \frac{dt}{1 - t^2} = \int_0^x \left(1 + t^2 + t^4 + \cdots + t^{2n} + \frac{t^{2n+2}}{1 - t^2}\right) dt$$

or, in other words, that

$$\tanh^{-1}x = x + \frac{x^3}{3} + \frac{x^5}{5} + \cdots + \frac{x^{2n+1}}{2n + 1} + R,$$

where

$$R = \int_0^x \frac{t^{2n+2}}{1 - t^2}\, dt.$$

20. Show that R in Problem 19 is no greater than

$$\frac{1}{1 - x^2} \cdot \frac{|x|^{2n+3}}{2n + 3}, \quad \text{if } x^2 < 1.$$

21. a) Differentiate the identity

$$\frac{1}{1 - x} = 1 + x + x^2 + \cdots + x^n + \frac{x^{n+1}}{1 - x}$$

to obtain the expansion

$$\frac{1}{(1 - x)^2} = 1 + 2x + 3x^2 + \cdots + nx^{n-1} + R.$$

b) Prove that, if $|x| < 1$, then $R \to 0$ as $n \to \infty$.

c) In one throw of two dice, the probability of getting a roll of 7 is $p = 1/6$. If the dice are thrown repeatedly, the probability that a 7 will appear for the first time at the nth throw is $q^{n-1}p$, where $q = 1 - p = 5/6$. The expected number of throws until a 7 first appears is $\Sigma_{n=1}^{\infty} nq^{n-1}p$. Evaluate this series numerically.

d) In applying statistical quality control to an industrial operation, an engineer inspects items taken at random from the assembly line. Each item sampled is classified as "good" or "bad." If the probability of a good item is p and of a bad item is $q = 1 - p$, the probability that the first bad item found is the nth inspected is $p^{n-1}q$. The average number inspected up to and including the first bad item found is $\Sigma_{n=1}^{\infty} np^{n-1}q$. Evaluate this series, assuming $0 < p < 1$.

22. In probability theory, a random variable X may assume the values 1, 2, 3, . . . , with probabilities $p_1, p_2, p_3, \ldots,$ where p_k is the probability that X is equal to k ($k = 1, 2, \ldots$). It is customary to assume $p_k \geq 0$ and $\Sigma_{k=1}^{\infty} p_k = 1$. The *expected value of X,* denoted by $E(X)$, is defined as $\Sigma_{k=1}^{\infty} kp_k$, provided this series converges. In each of the following cases, show that $\Sigma p_k = 1$ and find $E(X)$, if it exists. (*Hint:* See Problem 21.)

a) $p_k = 2^{-k}$

b) $p_k = \frac{5^{k-1}}{6^k}$

c) $p_k = \frac{1}{k(k + 1)} = \frac{1}{k} - \frac{1}{k + 1}$

TOOLKIT PROGRAMS

Sequences and Series   Taylor Series

12.5

Convergence of Power Series; Differentiation, Integration, Multiplication, and Division

Now that we have studied power series for the common elementary functions, we shall deal with questions about power series that may be generated some other way—for instance, in solving a differential equation. (Examples occur in mathematical physics, and some simpler examples are given in Chapter 20.) We also discuss such operations as multiplication, division, differentiation, and integration of power series. A word of caution: Other kinds of series, including trigonometric (Fourier) series are also important in engineering and physics, and some operations that are valid for power series do not apply to some of those series. For example, a theorem that is valid for power series says, in effect, that the derivative of a *power* series $\sum a_n x^n$ is $\sum n a_n x^{n-1}$. However, it is not true that the derivative of the trigonometric series $\sum (1/n)\sin nx$ is $\sum \cos nx$. Moreover, we omit proofs of some of the most important theorems because the proofs are more difficult than those usually included in a first course in calculus. (You can find proofs in more advanced textbooks.) But the results are useful and easy to apply.

Convergence of Power Series

We now know that some power series, like the series for $\sin x$, $\cos x$, and e^x, converge for all values of x, while others, like the series for $\ln(1 + x)$ and $\tan^{-1}x$, converge only on finite intervals. But we learned all this by analyzing remainder formulas, and we have yet to face the question of how to investigate the convergence of a power series when there is no remainder formula to analyze. Moreover, all of the power series we have worked with have been Taylor series generated by functions we already know. What about other power series? Are they Taylor series, too, of functions otherwise unknown?

The first step in answering these questions is to note that a power series $\sum_{n=0}^{\infty} a_n x^n$ defines a function whenever it converges, namely, the function f whose value at each x is the number

$$f(x) = \sum_{n=0}^{\infty} a_n x^n. \tag{1}$$

We can then ask what kind of domain f has, how f is to be differentiated and integrated (if at all), whether f has a Taylor series, and, if it has, how its Taylor series is related to the defining series $\sum_{n=0}^{\infty} a_n x^n$.

The questions of what domain f has and for what values the series (1) may be expected to converge are answered by Theorem 3 and the discussion that follows it. We will prove Theorem 3 and then, after looking at examples, will proceed to Theorems 4 and 5, which answer the questions of whether f can be differentiated and integrated and *how* to do so when it can be. Theorem 5 also solves a problem that arose many chapters ago but that has remained unsolved until now: that of finding convenient expressions for evaluating integrals like

$$\int_0^1 \sin x^2 \, dx \quad \text{and} \quad \int_0^{0.5} \sqrt{1 + x^4} \, dx,$$

which frequently arise in applications. (See Examples 8, 9, and 10 later in this article.) Finally, we will see that, in the interior of its domain of definition, the function f does have a Maclaurin series and that this is none other than the defining series $\sum_{n=0}^{\infty} a_n x^n$.

THEOREM 3

The Convergence Theorem for Power Series

If a power series

$$\sum_{n=0}^{\infty} a_n x^n = a_0 + a_1 x + a_2 x^2 + \cdots \tag{2}$$

converges for $x = c \ (c \neq 0)$, then it converges absolutely for all $|x| < |c|$. If the series diverges for $x = d$, then it diverges for all $|x| > |d|$.

Proof   Suppose the series

$$\sum_{n=0}^{\infty} a_n c^n \tag{3}$$

converges. Then

$$\lim_{n \to \infty} a_n c^n = 0.$$

Hence, there is an index N such that

$$|a_n c^n| < 1 \qquad \text{for all } n \geq N.$$

That is,

$$|a_n| < \frac{1}{|c|^n} \qquad \text{for } n \geq N. \tag{4}$$

Now take any x such that $|x| < |c|$ and consider

$$|a_0| + |a_1 x| + \cdots + |a_{N-1} x^{N-1}| + |a_N x^N| + |a_{N+1} x^{N+1}| + \cdots.$$

There is only a finite number of terms prior to $|a_N x^N|$, and their sum is finite. Starting with $|a_N x^N|$ and beyond, the terms are less than

$$\left|\frac{x}{c}\right|^N + \left|\frac{x}{c}\right|^{N+1} + \left|\frac{x}{c}\right|^{N+2} + \cdots \tag{5}$$

by virtue of the inequality (4). But the series in (5) is a geometric series with ratio $r = |x/c|$, which is less than 1, since $|x| < |c|$. Hence the series (5) converges, so that the original series (3) converges absolutely. This proves the first half of the theorem.

The second half of the theorem involves nothing new. If the series diverges at $x = d$ and converges at a value x_0 with $|x_0| > |d|$, we may take $c = x_0$ in the first half of the theorem and conclude that the series converges absolutely at d. But the series cannot converge absolutely and diverge at one and the same time. Hence, if it diverges at d, it diverges for all $|x| > |d|$. ∎

Figure 12.5

The Radius and Interval of Convergence

The significance of Theorem 3 is that a power series always behaves in exactly *one* of the following three ways (Fig. 12.5).

Possible Behavior of a Power Series

1. It converges at $x = 0$ and diverges everywhere else.

2. There is a positive number c such that the series diverges for $|x| > c$ but converges absolutely for $|x| < c$. It may or may not converge at either of the endpoints $x = c$ and $x = -c$.

3. It converges absolutely for every x.

In case 2, the set of points at which the series converges is a finite interval. We know from past examples that this interval may be open, half-open, or closed, depending on the series in question. But no matter which kind of interval it is, c is called the **radius of convergence** of the series, and the convergence is absolute at every point in the interior of the interval. The interval is called the **interval of convergence.** If a power series converges absolutely for all values of x, we say that its radius of convergence is infinite. If it converges only at $x = 0$, we say that the radius of convergence is zero.

As examples of power series whose radii of convergence are infinite, we have the Taylor series of $\sin x$, $\cos x$, and e^x. These series converge for every value of x.

As examples of series whose radii of convergence are finite we have the following:

Series	Interval of convergence
$\dfrac{1}{1 - x} = 1 + x + x^2 + \cdots$	$-1 < x < 1$
$\ln(1 + x) = x - \dfrac{x^2}{2} + \dfrac{x^3}{3} - \dfrac{x^4}{4} + \cdots$	$-1 < x \leq 1$
$\tan^{-1}x = x - \dfrac{x^3}{3} + \dfrac{x^5}{5} - \cdots$	$-1 \leq x \leq 1$

Finding the Interval of Convergence

The interval of convergence of a power series $\sum_{n=0}^{\infty} a_n x^n$ can often be found by applying the Ratio Test or the Root Test to the series of absolute values,

$$\sum_{n=0}^{\infty} |a_n x^n|.$$

Thus, if

$$\rho = \lim_{n \to \infty} \left| \frac{a_{n+1}x^{n+1}}{a_n x^n} \right| \qquad \text{or if} \qquad \rho = \lim_{n \to \infty} \sqrt[n]{|a_n x^n|},$$

then,

a) $\Sigma |a_n x^n|$ converges at all values of x for which $\rho < 1$,

b) $\Sigma |a_n x^n|$ diverges at all values of x for which $\rho > 1$,

c) $\Sigma |a_n x^n|$ may either converge or diverge at a value of x for which $\rho = 1$.

How do these three alternatives translate into statements about the series $\Sigma a_n x^n$? Case (a) says that $\Sigma a_n x^n$ converges absolutely at all values of x for which $\rho < 1$. Case (c) does not tell us anything more about the series $\Sigma a_n x^n$ than it does about the series $\Sigma |a_n x^n|$. Either series might converge or diverge at a value of x for which $\rho = 1$. In case (b), we can actually conclude that $\Sigma a_n x^n$ diverges at all values of x for which $\rho > 1$. The argument goes like this: As you may recall from the discussions in Article 11.6, the fact that ρ is greater than 1 means that either

$$0 < |a_n x^n| < |a_{n+1} x^{n+1}| < |a_{n+2} x^{n+2}| < \cdots$$

or

$$\sqrt[n]{|a_n x^n|} > 1$$

for n sufficiently large. Thus the terms of the series do not approach zero as $n \to \infty$, and the series diverges *with or without absolute values,* by the nth-Term Test.

Therefore, the Ratio and Root Tests, when successfully applied to $\Sigma |a_n x^n|$, lead us to the following conclusions about $\Sigma a_n x^n$:

a) $\Sigma a_n x^n$ converges absolutely for all values of x for which $\rho < 1$,

b) $\Sigma a_n x^n$ diverges at all values of x for which $\rho > 1$,

c) $\Sigma a_n x^n$ may either converge or diverge at a value of x for which $\rho = 1$.

In case (c), another test is needed. The Comparison Test, Limit Comparison Test, or Alternating Series Test often works. (Also see Problem 46, which describes Raabe's test.)

EXAMPLE 1 Find the interval of convergence of

$$\sum_{n=1}^{\infty} \frac{x^n}{n}. \tag{6}$$

Solution We apply the Ratio Test to the series of absolute values and find

$$\rho = \lim_{n \to \infty} \left| \frac{x^{n+1}}{n+1} \cdot \frac{n}{x^n} \right| = |x|.$$

Therefore the original series converges absolutely if $|x| < 1$ and diverges if $|x| > 1$. When $x = +1$, the series becomes

$$1 + \frac{1}{2} + \frac{1}{3} + \frac{1}{4} + \cdots,$$

which diverges. When $x = -1$, the series becomes

$$-\left(1 - \frac{1}{2} + \frac{1}{3} - \frac{1}{4} + \cdots \right),$$

which converges, by Leibniz's theorem. Therefore the series (6) converges for $-1 \le x < 1$ and diverges for all other values of x. ∎

EXAMPLE 2   For what values of x does the series

$$\sum_{n=1}^{\infty} \frac{(2x-5)^n}{n^2}$$

converge?

Solution   We treat the series as a power series in the variable $2x - 5$. An application of the Root Test to the series of absolute values yields

$$\rho = \lim_{n \to \infty} \sqrt[n]{\left| \frac{(2x-5)^n}{n^2} \right|} = \lim_{n \to \infty} \frac{|2x-5|}{\sqrt[n]{n^2}} = \frac{|2x-5|}{1} = |2x-5|.$$

(The Ratio Test leads to the same conclusion.) The series converges absolutely for

$$|2x - 5| < 1 \qquad \text{or} \qquad -1 < 2x - 5 < 1$$

or

$$4 < 2x < 6 \qquad \text{or} \qquad 2 < x < 3.$$

When $x = 2$, the series is $\sum_{n=1}^{\infty} [(-1)^n / n^2]$, which converges.
When $x = 3$, the series is $\sum_{n=1}^{\infty} [(1)^n / n^2]$, which converges. Therefore, the interval of convergence is $2 \le x \le 3$.   ∎

Sometimes the Comparison Test does as well as any.

EXAMPLE 3   For what values of x does

$$\sum_{n=1}^{\infty} \frac{\cos^n x}{n!}$$

converge?

Solution   For every value of x,

$$\left| \frac{\cos^n x}{n!} \right| \le \frac{1}{n!}.$$

The series converges for every value of x.   ∎

Differentiation of Power Series

The next theorem says that a function defined by a power series has derivatives of all orders at every point in the interior of its interval of convergence. The derivatives can be obtained as power series by differentiating the terms of the original series. The first derivative is obtained by differentiating the terms of the original series once:

$$\frac{d}{dx} \sum_{n=0}^{\infty} (a_n x^n) = \sum_{n=0}^{\infty} \frac{d}{dx}(a_n x^n) = \sum_{n=0}^{\infty} n a_n x^{n-1}.$$

For the second derivative, the terms are differentiated again, and so on. We state the theorem without proof and go directly to the examples.

THEOREM 4

> **The Term-by-Term Differentiation Theorem**
>
> If $f(x) = \sum_{n=0}^{\infty} a_n x^n$ has radius of convergence c, then
>
> 1. $\sum_{n=0}^{\infty} n a_n x^{n-1}$ also has radius of convergence c,
>
> 2. $f(x)$ is differentiable on $(-c, c)$, and
>
> 3. $f'(x) = \sum_{n=0}^{\infty} n a_n x^{n-1}$ on $(-c, c)$.

Ostensibly, Theorem 4 mentions only f and f'. But because f' has the same radius of convergence that f has, the theorem applies equally well to f', saying that it has a derivative f'' on $(-c, c)$. This in turn implies that f'' is differentiable on $(-c, c)$, and so on. Thus, if $f(x) = \sum_{n=0}^{\infty} a_n x^n$ converges on $(-c, c)$, it has derivatives of all orders at every point of $(-c, c)$.

EXAMPLE 4 The relation $(d/dx)(\sin x) = \cos x$ is easily checked by differentiating the series for $\sin x$ term by term:

$$\sin x = x - \frac{x^3}{3!} + \frac{x^5}{5!} - \frac{x^7}{7!} + \cdots$$

$$\frac{d}{dx}(\sin x) = 1 - \frac{x^2}{2!} + \frac{x^4}{4!} - \frac{x^6}{6!} + \cdots$$

$$= \cos x. \qquad \blacksquare$$

Convergence at one or both endpoints of the interval of convergence of a power series may be lost in the process of differentiation. That is why Theorem 4 mentions only the *open* interval $(-c, c)$.

EXAMPLE 5 The series $f(x) = \sum_{n=1}^{\infty} (x^n/n)$ of Example 1 converges for $-1 \leq x < 1$. The series of derivatives

$$f'(x) = \sum_{n=1}^{\infty} x^{n-1} = 1 + x + x^2 + x^3 + \cdots$$

is a geometric series that converges only for $-1 < x < 1$. The series diverges at the endpoint $x = -1$ as well as at the endpoint $x = 1$. $\qquad \blacksquare$

Integration of Power Series

Just as a power series may be differentiated term by term, it may also be integrated term by term. The new series will surely converge in the open interval where the original series converges, and it may converge at one or both endpoints as well. The justification for term-by-term integration of a series is the following theorem, which we also state without proof.

THEOREM 5

The Term-by-Term Integration Theorem

If $f(x) = \sum_{n=0}^{\infty} a_n x^n$ has radius of convergence c, then

1. $\displaystyle\sum_{n=0}^{\infty} \frac{a_n x^{n+1}}{n+1}$ also has radius of convergence c,

2. $\displaystyle\int f(x)\,dx$ exists for x in $(-c, c)$,

3. $\displaystyle\int f(x)\,dx = \sum_{n=0}^{\infty} \frac{a_n x^{n+1}}{n+1} + C$ on $(-c, c)$.

EXAMPLE 6 The series

$$\frac{1}{1+t} = 1 - t + t^2 - t^3 \cdots$$

converges on the open interval $-1 < t < 1$. Therefore,

$$\ln(1+x) = \int_0^x \frac{1}{1+t}\,dt$$

$$= t - \frac{t^2}{2} + \frac{t^3}{3} - \frac{t^4}{4} + \cdots \bigg]_0^x,$$

$$= x - \frac{x^2}{2} + \frac{x^3}{3} - \frac{x^4}{4} + \cdots, \qquad -1 < x < 1.$$

As you know, the latter series also converges at $x = 1$, but that was not guaranteed by the theorem. ∎

EXAMPLE 7 By replacing t by t^2 in the series of Example 6, we obtain

$$\frac{1}{1+t^2} = 1 - t^2 + t^4 - t^6 + \cdots, \qquad -1 < t < 1.$$

Therefore

$$\tan^{-1}x = \int_0^x \frac{1}{1+t^2}\,dt = t - \frac{t^3}{3} + \frac{t^5}{5} - \frac{t^7}{7} + \cdots \bigg]_0^x$$

$$= x - \frac{x^3}{3} + \frac{x^5}{5} - \frac{x^7}{7} + \cdots, \qquad -1 < x < 1.$$

This is not as refined a result as the one we obtained in Article 12.4, where we were able to show that the interval of convergence was $-1 \le x \le 1$ by analyzing a remainder. But the result here is obtained more quickly. ∎

EXAMPLE 8 Express

$$\int \sin x^2\,dx$$

as a power series.

Solution   From the series for $\sin x$ we obtain

$$\sin x^2 = x^2 - \frac{x^6}{3!} + \frac{x^{10}}{5!} - \frac{x^{14}}{7!} + \cdots, \qquad -\infty < x < \infty.$$

Therefore,

$$\int \sin x^2 \, dx = C + \frac{x^3}{3} - \frac{x^7}{7 \cdot 3!} + \frac{x^{11}}{11 \cdot 5!} - \frac{x^{15}}{15 \cdot 7!} + \cdots,$$

$$-\infty < x < \infty. \qquad \blacksquare$$

EXAMPLE 9   Estimate $\int_0^1 \sin x^2 \, dx$ with an error of less than 0.001.

Solution   From the indefinite integral in Example 8,

$$\int_0^1 \sin x^2 \, dx = \frac{1}{3} - \frac{1}{7 \cdot 3!} + \frac{1}{11 \cdot 5!} - \frac{1}{15 \cdot 7!} + \frac{1}{19 \cdot 9!} - \cdots.$$

The series alternates, and we find by experiment that

$$\frac{1}{11 \cdot 5!} \approx 0.00076$$

is the first term to be numerically less than 0.001. The sum of the preceding two terms gives

$$\int_0^1 \sin x^2 \, dx \approx \frac{1}{3} - \frac{1}{42} \approx 0.310.$$

With two more terms we could estimate

$$\int_0^1 \sin x^2 \, dx \approx 0.310268$$

with an error of less than 10^{-6}, and with only one term beyond that we have

$$\int_0^1 \sin x^2 \, dx \approx \frac{1}{3} - \frac{1}{42} + \frac{1}{1320} - \frac{1}{75600} + \frac{1}{6894720} \approx 0.310268303,$$

with an error of less than 10^{-9}. To guarantee this accuracy with the error formula for the Trapezoidal Rule would require using about 13,000 subintervals. $\qquad \blacksquare$

EXAMPLE 10   Estimate $\int_0^{0.5} \sqrt{1 + x^4} \, dx$ with an error of less than 10^{-4}.

Solution   The binomial expansion of $(1 + x^4)^{1/2}$ is

$$(1 + x^4)^{1/2} = 1 + \frac{1}{2}x^4 - \frac{1}{8}x^8 + \cdots,$$

a series whose terms alternate in sign after the second term. Therefore,

$$\int_0^{0.5} \sqrt{1 + x^4} \, dx = x + \frac{1}{2 \cdot 5}x^5 - \frac{1}{8 \cdot 9}x^9 + \cdots \Bigg]_0^{0.5}$$

$$= 0.5 + 0.0031 - 0.00003 + \cdots \approx 0.5031,$$

with an error of magnitude less than 0.00003. $\qquad \blacksquare$

The Maclaurin Series Generated by $\sum\limits_{n=0}^{\infty} a_n x^n$

At the beginning of this article we asked whether a function

$$f(x) = \sum_{n=0}^{\infty} a_n x^n$$

defined by a convergent power series has a Taylor series. We can now answer that a function defined by a power series with a radius of convergence $c > 0$ has a Maclaurin series that converges to the function at every point of $(-c, c)$. Why? Because the Maclaurin series generated by the function $f(x) = \sum_{n=0}^{\infty} a_n x^n$ is the series $\sum_{n=0}^{\infty} a_n x^n$ itself. To see this, we differentiate

$$f(x) = a_0 + a_1 x + a_2 x^2 + \cdots + a_n x^n + \cdots$$

term by term and substitute $x = 0$ in each derivative $f^{(n)}(x)$. This produces

$$f^{(n)}(0) = n!\, a_n \quad \text{or} \quad a_n = \frac{f^{(n)}(0)}{n!}$$

for every n. Thus,

$$f(x) = \sum_{n=0}^{\infty} a_n x^n$$

$$= \sum_{n=0}^{\infty} \frac{f^{(n)}(0)}{n!} x^n, \qquad -c < x < c. \tag{7}$$

An immediate consequence of this is that series like

$$x \sin x = x^2 - \frac{x^4}{3!} + \frac{x^6}{5!} - \frac{x^8}{7!} + \cdots$$

and

$$x^2 e^x = x^2 + x^3 + \frac{x^4}{2!} + \frac{x^5}{3!} + \cdots,$$

which are obtained by multiplying Maclaurin series by powers of x, as well as series obtained by integration and differentiation of power series are themselves the Maclaurin series generated by the functions they represent.

Equality of Power Series

Another consequence of (7) is that if two power series $\sum_{n=0}^{\infty} a_n x^n$ and $\sum_{n=0}^{\infty} b_n x^n$ are equal for all values of x in an open interval that contains the origin $x = 0$, then $a_n = b_n$ for every n. For if

$$f(x) = \sum_{n=0}^{\infty} a_n x^n$$

$$= \sum_{n=0}^{\infty} b_n x^n, \qquad -c < x < c,$$

then a_n and b_n are both equal to $f^{(n)}(0)/n!$.

Multiplication of Power Series

We illustrate with an example.

EXAMPLE 11   Find the terms through x^4 in the Maclaurin series expansion for $e^x \cos x$ by multiplying together the series for e^x and for $\cos x$.

Solution   From Article 12.3,

$$e^x = 1 + x + \frac{x^2}{2!} + \frac{x^3}{3!} + \frac{x^4}{4!} + \cdots,$$

$$\cos x = 1 - \frac{x^2}{2!} + \frac{x^4}{4!} - \cdots.$$

Obviously, if we need only terms involving x^n for $n \leq 4$, we can truncate both series at their x^4 terms and multiply the resulting polynomials, discarding everything involving higher powers like $x^5, \ldots, x^8$. The result is

$$e^x \cos x = 1 + x + x^2\left(\frac{1}{2!} - \frac{1}{2!}\right) + x^3\left(\frac{1}{3!} - \frac{1}{2!}\right)$$

$$+ x^4\left(\frac{1}{4!} - \frac{1}{2!2!} + \frac{1}{4!}\right) + \cdots$$

$$= 1 + x - \frac{x^3}{3} - \frac{x^4}{6} + \cdots. \qquad \blacksquare \quad (8)$$

We shall not prove it here, but using $\cos x = (e^{ix} + e^{-ix})/2$ we could establish the result

$$e^x \cos x = \frac{1}{2} \sum_{n=0}^{\infty} \frac{(1+i)^n + (1-i)^n}{n!} x^n = \sum_{n=0}^{\infty} \frac{(\sqrt{2})^n \cos (n\pi/4)}{n!} x^n, \qquad (9)$$

because

$$1 + i = \sqrt{2}\, e^{i\pi/4}, \qquad 1 - i = \sqrt{2}\, e^{-i\pi/4}.$$

From Eq. (9), it would be easy to show that the series for $e^x \cos x$ converges absolutely for all real values of x. This is also guaranteed by the following theorem, which we shall not prove.

THEOREM 6

> **The Series Multiplication Theorem for Power Series**
>
> If both $\Sigma\, a_n x^n$ and $\Sigma\, b_n x^n$ converge absolutely for $|x| < R$, and
>
> $$c_n = a_0 b_n + a_1 b_{n-1} + a_2 b_{n-2} + \cdots + a_{n-1} b_1 + a_n b_0$$
>
> $$= \sum_{k=0}^{n} a_k b_{n-k}, \qquad (10a)$$
>
> then the series $\Sigma\, c_n x^n$ also converges absolutely for $|x| < R$, and
>
> $$(a_0 + a_1 x + a_2 x^2 + \cdots) \cdot (b_0 + b_1 x + b_2 x^2 + \cdots)$$
>
> $$= c_0 + c_1 x + c_2 x^2 + \cdots. \qquad (10b)$$

☐ Division of Power Series

Again, we illustrate with an example.

EXAMPLE 12   Find some of the terms in the Maclaurin series for $\tan x$ by dividing the series for $\sin x$ by the series for $\cos x$.

Solution   We proceed as in ordinary algebraic long division, keeping track of terms up to x^5 and disregarding all higher powers of x:

$$
\cos x = 1 - \frac{x^2}{2!} + \frac{x^4}{4!} - \cdots \,\overline{\smash{\big)}\, x - \frac{x^3}{3!} + \frac{x^5}{5!} - \cdots = \sin x}
$$

$$x + \frac{x^3}{3} + \frac{2}{15}x^5 + \cdots = \tan x$$

$$
\begin{array}{l}
x - \dfrac{x^3}{2!} + \dfrac{x^5}{4!}\cdots \\[2mm]
\hline
\dfrac{x^3}{3} - \dfrac{x^5}{30}\cdots \\[2mm]
\dfrac{x^3}{3} - \dfrac{x^5}{6}\cdots \\[2mm]
\hline
\dfrac{2}{15}x^5\cdots \\[2mm]
\dfrac{2}{15}x^5 \\[2mm]
\hline
\end{array}
$$

To terms in x^5, we thus have

$$
\tan x = \frac{\sin x}{\cos x} = x + \frac{x^3}{3} + \frac{2}{15}x^5 + \cdots. \tag{11}
$$

We shall use these first few terms of the series for $\tan x$ in Article 12.6, Example 2. ■

REMARK   Because $\cos(\pi/2) = 0$, we certainly cannot expect the Maclaurin series for $\tan x$ to converge outside the interval $|x| < \pi/2$. For the full Maclaurin series, which is not easy to obtain either by long division or by direct application of Taylor's formula, the reader is referred to the *Handbook of Mathematical Functions*, Applied Mathematics Series, U.S. Department of Commerce, National Bureau of Standards, AMS 55, edited by M. Abramowitz and I. A. Stegun (New York: Dover Publications), p. 75, Eq. 4.3.67.

Another way to calculate the coefficients in the Maclaurin series for the quotient $f(x)/g(x)$ is readily adaptable to machine computation. The facts are as follows:

1. If $f(x) = \Sigma\, a_n x^n$ for $|x| < R_1$, $g(x) = \Sigma\, b_n x^n$ for $|x| < R_2$, and $b_0 = g(0) \neq 0$, then $f(x)/g(x)$ has a power series representation $\Sigma\, c_n x^n$ on some interval $(-h, h)$.
2. Within that interval,

$$
\sum a_n x^n = f(x) = g(x) \cdot \frac{f(x)}{g(x)} = \sum b_n x^n \cdot \sum c_n x^n = \sum_n \left(\sum_{k=0}^{n} b_k c_{n-k} \right) x^n,
$$

and hence

$$a_n = \sum_{k=0}^{n} b_k c_{n-k}.$$

In other words,

$$a_0 = b_0 c_0, \qquad a_1 = b_0 c_1 + b_1 c_0, \qquad \text{etc.},$$

so that the coefficients c_n can be found one after the other in this way:

$$c_0 = \frac{a_0}{b_0}, \qquad c_1 = \frac{a_1 - b_1 c_0}{b_0}, \tag{12a}$$

and, for all $n \geq 1$,

$$c_n = \frac{a_n - (b_1 c_{n-1} + b_2 c_{n-2} + \cdots + b_n c_0)}{b_0}$$

$$\tag{12b}$$

$$= \frac{a_n - \sum_{k=1}^{n} b_k c_{n-k}}{b_0}.$$

EXAMPLE 13 Let

$$f(x) = \sin x = x - \frac{x^3}{3!} + \frac{x^5}{5!} - \cdots$$

so that

$$a_0 = 0, \quad a_1 = 1, \quad a_2 = 0, \quad a_3 = -\frac{1}{3!}, \quad a_4 = 0, \ldots$$

and let

$$g(x) = \cos x = 1 - \frac{x^2}{2!} + \frac{x^4}{4!} - \frac{x^6}{6!} + \cdots$$

so that

$$b_0 = 1, \quad b_1 = 0, \quad b_2 = -\frac{1}{2!}, \quad b_3 = 0, \quad b_4 = \frac{1}{4!}, \text{ etc.}$$

Then

$$\tan x = \frac{\sin x}{\cos x} = c_0 + c_1 x + c_2 x^2 + \cdots$$

with

$$c_0 = \frac{a_0}{b_0} = 0,$$

$$c_1 = \frac{a_1 - b_1 c_0}{b_0} = \frac{1 - 0}{1} = 1,$$

and so on. When we know the values of $c_0, c_1, \ldots, c_{n-1}$, the value of c_n is given by Eq. (12b) in terms of known coefficients. ■

PROBLEMS

In Problems 1–20, find the interval of absolute convergence. If the interval is finite, determine whether the series converges at each endpoint.

1. $\sum_{n=0}^{\infty} x^n$

2. $\sum_{n=0}^{\infty} n^2 x^n$

3. $\sum_{n=1}^{\infty} \frac{n x^n}{2^n}$

4. $\sum_{n=0}^{\infty} \frac{(2x)^n}{n!}$

5. $\sum_{n=0}^{\infty} \frac{(-1)^n x^{2n+1}}{(2n+1)!}$

6. $\sum_{n=1}^{\infty} (-1)^{n-1} \frac{(x-1)^n}{n}$

7. $\sum_{n=0}^{\infty} \frac{n^2}{2^n} (x+2)^n$

8. $\sum_{n=0}^{\infty} \frac{x^{2n+1}}{2n+1}$

9. $\sum_{n=0}^{\infty} (-1)^n \frac{x^{2n+1}}{2n+1}$

10. $\sum_{n=1}^{\infty} \frac{(x-2)^n}{n^2}$

11. $\sum_{n=0}^{\infty} \frac{\cos nx}{2^n}$

12. $\sum_{n=1}^{\infty} \frac{2^n x^n}{n^5}$

13. $\sum_{n=0}^{\infty} \frac{x^n e^n}{n+1}$

14. $\sum_{n=1}^{\infty} \frac{(\cos x)^n}{n^n}$

15. $\sum_{n=0}^{\infty} n^n x^n$

16. $\sum_{n=0}^{\infty} \frac{(3x+6)^n}{n!}$

17. $\sum_{n=1}^{\infty} (-2)^n (n+1)(x-1)^n$

18. $\sum_{n=1}^{\infty} \frac{(-1)^{n+1}(x-2)^n}{n \cdot 2^n}$

19. $\sum_{n=0}^{\infty} \left(\frac{x^2 - 1}{2} \right)^n$

20. $\sum_{n=1}^{\infty} \frac{(x+3)^{n-1}}{n}$

21. Find the sum of the series in Problem 16.

22. When the series of Problem 19 converges, to what does it converge?

23. Use series to verify that

a) $\frac{d}{dx}(\cos x) = -\sin x$,   b) $\int_0^x \cos t \, dt = \sin x$,

c) $y = e^x$ is a solution of the equation $y' = y$.

24. Obtain the Maclaurin series for $1/(1+x)^2$ from the series for $-1/(1+x)$.

25. Use the Maclaurin series for $1/(1-x^2)$ to obtain a series for $2x/(1-x^2)^2$.

26. Use the identity $\sin^2 x = (1 - \cos 2x)/2$ to obtain a series for $\sin^2 x$. Then differentiate this series to obtain a series for $2 \sin x \cos x$. Check that this is the series for $\sin 2x$.

CALCULATOR In Problems 27–34, use series and a calculator to estimate each integral with an error of magnitude less than 0.001.

27. $\int_0^{0.2} \sin x^2 \, dx$

28. $\int_0^{0.1} \tan^{-1} x \, dx$

29. $\int_0^{0.1} x^2 e^{-x^2} \, dx$

30. $\int_0^{0.1} \frac{\tan^{-1} x}{x} \, dx$

31. $\int_0^{0.4} \frac{1 - e^{-x}}{x} \, dx$

32. $\int_0^{0.1} \frac{\ln(1+x)}{x} \, dx$

33. $\int_0^{0.1} \frac{1}{\sqrt{1+x^4}} \, dx$

34. $\int_0^{0.25} \sqrt[3]{1+x^2} \, dx$

35. CALCULATOR a) Obtain a power series for

$$\sinh^{-1} x = \int_0^x \frac{dt}{\sqrt{1+t^2}}.$$

b) Use the result of (a) to estimate $\sinh^{-1} 0.25$ to three decimal places.

36. CALCULATOR Estimate $\int_0^1 \cos x^2 \, dx$ with an error of less than one millionth.

37. Show by example that there are power series that converge only at $x = 0$.

38. Show by examples that the convergence of a series at an endpoint of its interval of convergence may be either conditional or absolute.

39. Let r be any positive number. Use Theorem 3 to show that if $\sum_{n=0}^{\infty} a_n x^n$ converges for $-r < x < r$, then it converges absolutely for $-r < x < r$.

40. Use the Ratio Test to show that the binomial series converges for $|x| < 1$. (This still does not show that the series converges to $(1+x)^m$.)

41. Find terms through x^5 of the Maclaurin series for $e^x \sin x$ by appropriate multiplication. The series is the imaginary part of the series for

$$e^x \cdot e^{ix} = e^{(1+i)x}.$$

Use this fact to check your answer. For what values of x should the series for $e^x \sin x$ converge? Why?

42. Divide 1 by a sufficient number of terms of the Maclaurin series for $\cos x$ to obtain terms through x^4 in the Maclaurin series for $\sec x$. Where do you think the resulting complete Maclaurin series should converge?

43. Integrate the first three nonzero terms of the Maclaurin series for $\tan t$ from 0 to x to get the first three nonzero terms in the Maclaurin series for $\ln \sec x$.

44. *Continuation of Problem 43.* Another way to get some of the terms in the Maclaurin series for $\ln \sec x$ is as follows: Let $\sec x = 1 + y$, so that $y = x^2/2 + 5x^4/24 + \cdots$ (from

Problem 42). Then, for $|y| < 1$,

$$\ln \sec x = \ln (1 + y) = y - \frac{y^2}{2} + \cdots.$$

Neglecting powers of x higher than x^4, show that this also leads to

$$\ln \sec x = \frac{x^2}{2} + \frac{x^4}{12} + \cdots.$$

45. A circle of radius $r_1 = 1$ is inscribed in an equilateral triangle. A circle of radius r_2 passes through the vertices of that triangle and is inscribed in a square. A circle of radius r_3 passes through the vertices of that square and is inscribed in a regular pentagon. Continue in this fashion: a circle of radius r_{n-1} passes through the vertices of a regular polygon of n sides and is inscribed in a regular polygon of $n + 1$ sides.

a) Show that $r_2 = r_1 \sec(\pi/3)$, $r_3 = r_2 \sec(\pi/4)$, and, in general,

$$r_n = r_{n-1} \sec \frac{\pi}{n + 1}.$$

b) Next, show that

$$\ln r_n = \ln r_1 + \ln \sec \frac{\pi}{3} + \ln \sec \frac{\pi}{4} + \cdots$$

$$+ \ln \sec \frac{\pi}{n + 1}.$$

c) Does r_n have a finite limit as n tends to infinity? (*Suggestion:* Compare the series $\Sigma_{n=3}^{\infty} \ln \sec (\pi/n)$ with the series $\Sigma_{n=3}^{\infty} 1/n^2$ using the Limit Comparison Test.) You may wish to use the answer to Problem 44.

46. *Raabe's (or Gauss's) test.* This can be thought of as an extension of the Ratio Test. In effect, Raabe's test compares

the rate of growth of our series $\Sigma\ u_n$ with the corresponding rate of growth of the p-series $\Sigma\ 1/n^p$. (If our series grows no faster than a convergent p-series, then our series converges. If it grows at least as fast as a divergent p-series, then our series diverges.) We state the test without proof. (You can find proofs in F. B. Hildebrand, *Advanced Calculus for Applications,* Prentice-Hall, 1976, and John M. H. Olmsted, *Advanced Calculus,* Prentice-Hall, 1961.)

Raabe's test: If $\Sigma\ u_n$ is a series of positive constants and there exist constants C, K, and N such that

$$\frac{u_{n+1}}{u_n} = 1 - \frac{C}{n} + \frac{f(n)}{n^2}, \tag{i}$$

where $|f(n)| < K$ for $n \geq N$, then $\Sigma\ u_n$

$$\text{converges if } C > 1 \tag{iia}$$

and

$$\text{diverges if } C \leq 1. \tag{iib}$$

a) Show that the results of Raabe's test agree with what you already know about convergence and divergence of the series $\Sigma\ 1/n^2$ and $\Sigma\ 1/n$.

b) The terms of the series $\Sigma\ u_n$ are defined recursively by the formulas $u_1 = 1$ and

$$u_{n+1} = \frac{(2n - 1)(2n - 1)}{2n(2n + 1)} u_n.$$

Apply Raabe's test to determine whether the series converges or diverges.

TOOLKIT PROGRAMS

Taylor Series

□ 12.6
Indeterminate Forms

In considering the ratio of two functions $f(x)$ and $g(x)$ in the neighborhood of a point a where $f(a)$ and $g(a)$ are zero, we may quickly determine the limit

$$\lim_{x \to a} \frac{f(x)}{g(x)} \tag{1}$$

if we have known Taylor series that converge to these functions in a neighborhood of $x = a$. The following examples show how this is done.

EXAMPLE 1   Evaluate $\lim_{x \to 1} [(\ln x)/(x - 1)]$.

Solution Let $f(x) = \ln x$, $g(x) = x - 1$. The Taylor series for $f(x)$, with $a = 1$, is found as follows:

$$f(x) = \ln x, \qquad f(1) = \ln 1 = 0,$$
$$f'(x) = 1/x, \qquad f'(1) = 1,$$
$$f''(x) = -1/x^2, \qquad f''(1) = -1,$$

so

$$\ln x = 0 + (x - 1) - \frac{1}{2}(x - 1)^2 + \cdots,$$
$$\frac{\ln x}{x - 1} = 1 - \frac{1}{2}(x - 1) + \cdots,$$

and

$$\lim_{x \to 1} \frac{\ln x}{x - 1} = \lim_{x \to 1} \left[1 - \frac{1}{2}(x - 1) + \cdots \right] = 1. \qquad \blacksquare$$

EXAMPLE 2 Evaluate $\lim_{x \to 0} [(\sin x - \tan x)/x^3]$.

Solution The Maclaurin series for $\sin x$ and $\tan x$, to terms in x^5, are

$$\sin x = x - \frac{x^3}{3!} + \frac{x^5}{5!} - \cdots, \qquad \tan x = x + \frac{x^3}{3} + \frac{2x^5}{15} + \cdots.$$

Hence

$$\sin x - \tan x = -\frac{x^3}{2} - \frac{x^5}{8} - \cdots = x^3 \left(-\frac{1}{2} - \frac{x^2}{8} - \cdots \right)$$

and

$$\lim_{x \to 0} \frac{\sin x - \tan x}{x^3} = \lim_{x \to 0} \left(-\frac{1}{2} - \frac{x^2}{8} - \cdots \right) = -\frac{1}{2}. \qquad \blacksquare$$

When we apply series to compute the limit $\lim_{x \to 0} (1/\sin x - 1/x)$ of Example 7, Article 3.8, we not only compute the limit successfully but also discover a nice approximation formula for $\csc x$.

EXAMPLE 3 Find

$$\lim_{x \to 0} \left(\frac{1}{\sin x} - \frac{1}{x} \right).$$

Solution

$$\frac{1}{\sin x} - \frac{1}{x} = \frac{x - \sin x}{x \sin x} = \frac{x - \left(x - \dfrac{x^3}{3!} + \dfrac{x^5}{5!} - \cdots \right)}{x \cdot \left(x - \dfrac{x^3}{3!} + \dfrac{x^5}{5!} - \cdots \right)}$$

$$= \frac{x^3 \left(\dfrac{1}{3!} - \dfrac{x^2}{5!} + \cdots \right)}{x^2 \left(1 - \dfrac{x^2}{3!} + \cdots \right)} = x \, \frac{\dfrac{1}{3!} - \dfrac{x^2}{5!} + \cdots}{1 - \dfrac{x^2}{3!} + \cdots}.$$

Therefore,

$$\lim_{x \to 0} \left(\frac{1}{\sin x} - \frac{1}{x} \right) = \lim_{x \to 0} \left(x \cdot \dfrac{\dfrac{1}{3!} - \dfrac{x^2}{5!} + \cdots}{1 - \dfrac{x^2}{3!} + \cdots} \right) = 0.$$

In fact, from the series expressions above we can see that if $|x|$ is small, then

$$\frac{1}{\sin x} - \frac{1}{x} \approx x \cdot \frac{1}{3!} = \frac{x}{6} \qquad \text{or} \qquad \csc x \approx \frac{1}{x} + \frac{x}{6}. \qquad \blacksquare$$

PROBLEMS

Use series to evaluate the limits in Problems 1–20.

1. $\lim\limits_{h \to 0} \dfrac{\sin h}{h}$

2. $\lim\limits_{x \to 0} \dfrac{e^x - (1 + x)}{x^2}$

3. $\lim\limits_{t \to 0} \dfrac{1 - \cos t - \dfrac{1}{2}t^2}{t^4}$

4. $\lim\limits_{x \to \infty} x \sin \dfrac{1}{x}$

5. $\lim\limits_{x \to 0} \dfrac{x^2}{1 - \cosh x}$

6. $\lim\limits_{h \to 0} \dfrac{(\sin h)/h - \cos h}{h^2}$

7. $\lim\limits_{x \to 0} \dfrac{1 - \cos x}{\sin x}$

8. $\lim\limits_{x \to 0} \dfrac{\sin x}{e^x - 1}$

9. $\lim\limits_{z \to 0} \dfrac{\sin (z^2) - \sinh (z^2)}{z^6}$

10. $\lim\limits_{t \to 0} \dfrac{\cos t - \cosh t}{t^2}$

11. $\lim\limits_{x \to 0} \dfrac{\sin x - x + \dfrac{x^3}{6}}{x^5}$

12. $\lim\limits_{x \to 0} \dfrac{e^x - e^{-x} - 2x}{x - \sin x}$

13. $\lim\limits_{x \to 0} \dfrac{x - \tan^{-1} x}{x^3}$

14. $\lim\limits_{x \to 0} \dfrac{\tan x - \sin x}{x^3 \cos x}$

15. $\lim\limits_{x \to \infty} x^2 (e^{-1/x^2} - 1)$

16. $\lim\limits_{x \to 0} \dfrac{\ln(1 + x^2)}{1 - \cos x}$

17. $\lim\limits_{x \to 0} \dfrac{\tan 3x}{x}$

18. $\lim\limits_{x \to 1} \dfrac{\ln x^2}{x - 1}$

19. $\lim\limits_{x \to \infty} \dfrac{x^{100}}{e^x}$

20. $\lim\limits_{x \to 0} \left(\dfrac{1}{2 - 2\cos x} - \dfrac{1}{x^2} \right)$

21. a) Prove that $\int_0^x e^{t^2}\, dt \to +\infty$ as $x \to +\infty$.
    b) Find $\lim\limits_{x \to \infty} x \int_0^x e^{t^2 - x^2}\, dt$.

22. Find values of r and s such that
$$\lim_{x \to 0} (x^{-3} \sin 3x + rx^{-2} + s) = 0.$$

23. **CALCULATOR** The approximation for $\csc x \approx 1/x + x/6$ in Example 3 leads to the approximation $\sin x \approx 6x/(6 + x^2)$. Evaluate both sides of this latter approximation for $x = \pm 1.0, \pm 0.1$, and ± 0.01 radians. Try these values of x in the approximation $\sin x \approx x$. Which approximation for $\sin x$ appears to give better results?

TOOLKIT PROGRAMS

Taylor Series

☐ **12.7**
A Computer Mystery

You may never have to actually use a series to compute the value of π, $\ln 2$, e, $\sin 35°$, or the like, because if you need any of these numbers you can either look them up in a table or (more likely) enter x in your computer and ask for $\ln x$, or whatever. But it was an exciting new idea in mathematics (and still can be an exciting idea) to realize that we can get *all* the information about e^x, for example, from just knowing that $e^0 = 1$ and that the derivative of e^x is e^x. That is enough to generate the Maclaurin series for e^x, which converges to e^x for all x. By multiplying the series for e^x by the series for e^y, we can also show that $e^x \cdot e^y = e^{x+y}$; by differentiation, we can show that the derivative of the series for e^x is that same series, and so on. (Some years ago, an eminent American mathematician, Tomlinson Fort, wrote a calculus book that started with the subject of series and developed the rest from there.)

Those of you who are interested in computer science may find the following "detective story" challenging. We wanted to compare, for purposes of illustration, three different ways of calculating e.

1. e is the root of the equation $\ln x = 1$. (Newton's method produced 2.718281828 on the third iteration, starting with $x_1 = 3$.)

2. e is the sum of the series

$$1 + \frac{1}{1} + \frac{1}{2!} + \frac{1}{3!} + \frac{1}{4!} + \frac{1}{5!} + \cdots.$$

(This produced the same degree of precision when we included 14 or more terms.)

3. $e = \lim_{n \to \infty} \left(1 + \frac{1}{n}\right)^n$

This is often taken as the definition of e. With a computer, one can just pick any large value of n and tell the computer to print the value of $s_n = (1 + 1/n)^n$. Of course, it is also instructive to see what happens for various values of n along the way. When we did that, we got a surprise. Here is a portion of the table of values we got, starting at $n = 10^5$.

n	$\left(1 + \dfrac{1}{n}\right)^n$	
10^5	2.71826 8237	
2×10^5	2.71827 5305	
3×10^5	2.71827 6664	
4×10^5	2.71827 7479	
5×10^5	2.71827 7751	
6×10^5	2.71827 9110	
7×10^5	2.71827 9110	(What's this? A repetition?)
8×10^5	2.71828 1828	(Hooray!)
9×10^5	2.71827 9110	(The machine likes this number?)
10^6	2.71828 1828	(Well, that's better.)
10^{13}	1	(What's going on?)
$\vdots$		
10^{19}	1	

We offer this as a puzzle. What is the machine doing? (Don't read on until you give it your best try!) Want a clue? Recall the definition of a^b.

A Computer Mystery Solved

We have just presented some data for $(1 + 1/n)^n$ for selected values of n between 10^5 and 10^{19}. How can we account for the obvious errors in some of these answers?

Let's start with the definition of a^b as $\exp(b \ln a)$:

$$\left(1 + \frac{1}{n}\right)^n = \exp\left(n \ln\left(1 + \frac{1}{n}\right)\right).\tag{1}$$

For large values of n, $1/n$ is small, and we look at the series for $\ln(1 + x)$ with $x = 1/n$. The result is

$$\ln\left(1 + \frac{1}{n}\right) = \frac{1}{n} - \frac{1}{2n^2} + \frac{1}{3n^3} - \cdots.\tag{2}$$

Clearly, this series approaches zero as $n \to \infty$. But Eq. (1) requires that we multiply this series by n. Doing so, we get

$$n \ln\left(1 + \frac{1}{n}\right) = 1 - \frac{1}{2n} + \frac{1}{3n^2} - \cdots.\tag{3}$$

For large values of n, the terms from $1/3n^2$ and beyond are very small compared to the first two terms. So, for a first approximation, we have

$$\left(1 + \frac{1}{n}\right)^n \approx \exp\left(1 - \frac{1}{2n}\right).\tag{4}$$

How about the irregular data that we got from the computer? We guessed that it might come from inaccurate values for $\ln(1 + 1/n)^n$, so we programmed the computer to give those data. At the same time, we had the computer give the values of $1 - (1/2n)$ and $1 - (1/2n) + (1/3n^2)$. These latter values were the same for all values of n shown in the table. And they are certainly more accurate than the machine's values of $n \ln(1 + 1/n)$.

n	$n \ln(1 + 1/n)$	$1 - (1/2n) + (1/3n^2)$
10^5	0.99999 5	0.99999 5
2×10^5	0.99999 76	0.99999 75
3×10^5	0.99999 81	0.99999 8333
4×10^5	0.99999 84	0.99999 875
5×10^5	0.99999 85	0.99999 90
6×10^5	0.99999 90	0.99999 91667
7×10^5	0.99999 90	0.99999 92857
8×10^5	1	0.99999 94444
9×10^5	0.99999 9	0.99999 95
10^6	1	0.99999 95455
1.1×10^6	1.00000 01	0.99999 95833
1.2×10^6	0.99999 6	0.99999 96154
1.3×10^6	0.99999 0	0.99999 96429
1.4×10^6	1.00000 04	0.99999 96667
1.5×10^6	1.00000 05	0.99999 96875
2×10^6	1	0.99999 97500

Naturally, if the computer has wrong values for $n \ln(1 + 1/n)$, it will give wrong answers for $\exp(n \ln(1 + 1/n))$. So, we have tracked down the source of the trouble; the values of $\ln(1 + 1/n)$ are not accurate enough to give right answers when multiplied by n, if $n \geq 3 \times 10^5$. For $n = 10^{13}$, the computer says $\ln(1.0000000000001) = 0$. When it multiplies this by 10^{13} it still says "zero" and $\exp(0) = 1$. We have gone beyond the machine's realm of reliability. It isn't the computer's fault—nor ours. But we have gained some insight by solving the mystery.

One more remark: If we use the approximation (4) and the fact that

$$\exp\left(1 - \frac{1}{2n}\right) = \exp(1) \cdot \exp\left(-\frac{1}{2n}\right), \tag{5}$$

we can go a step further. Remember that we are talking about $n \geq 3 \times 10^5$, so $(1/2n) \leq (1/6) \times 10^{-5}$. If we put $h = -1/2n$ in the Maclaurin series for $\exp(h) = e^h$, we get

$$\exp\left(-\frac{1}{2n}\right) = 1 - \frac{1}{2n} \tag{6}$$

with an error less than $(1/2)(1/2n)^2$ by the Alternating Series Estimation Theorem. This error is less than 0.3×10^{-11} for $n \geq 3 \times 10^5$. Therefore, for large values of n, the combined results of Eqs. (4), (5), and (6) yield

$$\left(1 + \frac{1}{n}\right)^n \approx e \cdot \left(1 - \frac{1}{2n}\right) \approx e - \frac{e}{2n}.$$

For $n = 10^6$, the machine should *not* give 2.718281828 (which seems so very accurate), but 2.718280469 to nine decimals.

REVIEW QUESTIONS AND EXERCISES

1. State Taylor's theorem with remainder.

2. It can be shown (though not very simply) that at every value of x the function defined by

$$f(x) = \begin{cases} 0 & \text{when } x = 0, \\ e^{-1/x^2} & \text{when } x \neq 0, \end{cases}$$

is continuous and has derivatives of all orders. At zero, the derivatives are all zero.
   a) Write the Maclaurin series generated by this function.
   b) What is the remainder $R_n(x)$ for this function? Does the series converge for values of x different from zero? Does it converge to $f(x)$ at some $x \neq 0$? Give reasons for your answers.

3. If a Taylor series in powers of $x - a$ is to be used for the numerical evaluation of a function, what is necessary or desirable in the choice of a?

4. Describe a method that may be useful in finding $\lim_{x \to a} f(x)/g(x)$ if $f(a) = g(a) = 0$. Illustrate.

5. What tests may be used to find the interval of convergence of a power series? Do they also work at the endpoints of the interval? Illustrate with examples.

MISCELLANEOUS PROBLEMS

1. a) Find the Maclaurin series for the function $x^2/(1 + x)$.
   b) Does the series converge when $x = 2$? (Give a brief reason.)

2. Obtain the Maclaurin series expansion for $\sin^{-1}x$ by integrating the series for $(1 - t^2)^{-1/2}$ from 0 to x. Find the intervals of convergence of these series.

3. Obtain the first four terms in the Maclaurin series for $e^{\sin x}$ by substituting the series for $y = \sin x$ in the series for e^y.

4. Assuming $|x| > 1$, obtain the expansions

$$\tan^{-1}x = \frac{\pi}{2} - \frac{1}{x} + \frac{1}{3x^3} - \frac{1}{5x^5} + \cdots, \quad x > 1,$$

$$\tan^{-1}x = -\frac{\pi}{2} - \frac{1}{x} + \frac{1}{3x^3} - \frac{1}{5x^5} + \cdots, \quad x < -1,$$

by integrating the series

$$\frac{1}{1 + t^2} = \frac{1}{t^2} \cdot \frac{1}{1 + (1/t^2)} = \frac{1}{t^2} - \frac{1}{t^4} + \frac{1}{t^6} - \frac{1}{t^8} + \cdots$$

from $x \, (>1)$ to $+\infty$ or from $-\infty$ to $x \, (<-1)$.

5. a) Obtain the Maclaurin series, through the term in x^6, for $\ln (\cos x)$ by substituting the series for $y = 1 - \cos x$ in the series for $\ln (1 - y)$.

b) Use the result of part (a) to estimate

$$\int_0^{0.1} \ln (\cos x) \, dx$$

to five decimal places.

6. Compute $\int_0^1 [(\sin x)/x] \, dx$ to three decimal places.

7. Compute $\int_0^1 e^{-x^2} \, dx$ to three decimal places.

8. Expand the function $f(x) = \sqrt{1 + x^2}$ in powers of $(x - 1)$, obtaining three nonzero terms.

9. Expand the function $f(x) = 1/(1 - x)$ in powers of $(x - 2)$, and find the interval of convergence.

10. Expand $f(x) = 1/(x + 1)$ in powers of $(x - 3)$.

11. Expand $\cos x$ in powers of $(x - \pi/3)$.

12. Find the first three terms of the Taylor series expansion of the function $1/x$ about the point π.

13. Let f and g be functions satisfying the following conditions: (a) $f(0) = 1$, (b) $f'(x) = g(x)$, $g'(x) = f(x)$, (c) $g(0) = 0$. Estimate $f(1)$ to three decimal places.

14. Suppose $f(x) = \sum_{n=0}^{\infty} a_n x^n$. Prove that
a) if $f(x)$ is an even function, then $a_1 = a_3 = a_5 = \cdots = 0$;
b) if $f(x)$ is an odd function, then $a_0 = a_2 = a_4 = \cdots = 0$.

15. Find the first four terms (up to x^3) of the Maclaurin series of $f(x) = e^{(e^x)}$.

16. a) Use Taylor's theorem with remainder to give an upper bound to the error in the approximation $e^h = 1 + h$, if h is between 0 and 0.001.

b) **CALCULATOR/COMPUTER** Compute and list values of $n^{1/n}$ for $n = 10^m$, $m = 1, 2, 3, 4, \ldots, 10$.

c) Given $\ln 10 = 2.30258 \ 50929 \ 9404$ and $n^{1/n} = \exp(1/n \ln n)$, show that

$$n^{1/n} \approx 1 + 10^{-m}(m \ln 10)$$
$$\approx 1.00000 \ 00023 \ 02585$$

is correct to 15 decimals when $m = 10$.

d) If $n = 10^{10}$, show that there is an error in the last few places in the following estimate of $n^{1/n}$:

$$n^{1/n} = 1.00000 \ 00023 \ 02585 \ 09299 \ 404,$$

and find the approximation that is good to 21 decimals.

17. If $(1 + x)^{1/3}$ is replaced by $1 + x/3$ and $0 \le x \le 1/10$, what estimate can be given for the error?

18. Use series to find

$$\lim_{x \to 0} \frac{\ln (1 - x) - \sin x}{1 - \cos^2 x}.$$

19. Find $\lim_{x \to 0} [(\sin x)/x]^{1/x^2}$.

20. Suppose that $\Sigma \, a_n$ is a convergent series of positive numbers. Does $\Sigma \ln (1 + a_n)$ converge? Explain.

In Problems 21–28, find the interval of convergence of each series. Test for convergence at the endpoints if the interval is finite.

21. $1 + \dfrac{x + 2}{3 \cdot 1} + \dfrac{(x + 2)^2}{3^2 \cdot 2} + \cdots + \dfrac{(x + 2)^n}{3^n \cdot n} + \cdots$

22. $1 + \dfrac{(x - 1)^2}{2!} + \dfrac{(x - 1)^4}{4!} + \cdots + \dfrac{(x - 1)^{2n-2}}{(2n - 2)!} + \cdots$

23. $\displaystyle\sum_{n=1}^{\infty} \frac{x^n}{n^n}$

24. $\displaystyle\sum_{n=1}^{\infty} \frac{n! x^n}{n^n}$

25. $\displaystyle\sum_{n=0}^{\infty} \frac{n + 1}{2n + 1} \frac{(x - 3)^n}{2^n}$

26. $\displaystyle\sum_{n=0}^{\infty} \frac{n + 1}{2n + 1} \frac{(x - 2)^n}{3^n}$

27. $\displaystyle\sum_{n=1}^{\infty} \frac{(-1)^{n-1}(x - 1)^n}{n^2}$

28. $\displaystyle\sum_{n=1}^{\infty} \frac{x^n}{\sqrt{n}}$

In Problems 29–31, find *all* the values of x for which the series converge.

29. $\displaystyle\sum_{n=1}^{\infty} \frac{(x - 2)^{3n}}{n!}$

30. $\displaystyle\sum_{n=1}^{\infty} \frac{2^n (\sin x)^n}{n^2}$

31. $\displaystyle\sum_{n=1}^{\infty} \frac{1}{n} \left(\frac{x - 1}{x} \right)^n$

32. A function is defined by the power series

$$y = 1 + \frac{1}{6}x^3 + \frac{1}{180}x^6 + \cdots$$
$$+ \frac{1 \cdot 4 \cdot 7 \cdots (3n - 2)}{(3n)!} x^{3n} + \cdots.$$

a) Find the interval of convergence of the series.
b) Show that the function defined by the series satisfies a differential equation of the form $y'' = x^a y + b$, and find the constants a and b.

33. If $a_n > 0$ and the series $\sum_{n=1}^{\infty} a_n$ converges, prove that $\sum_{n=1}^{\infty} a_n/(1 + a_n)$ converges.

34. If $1 > a_n > 0$ and $\sum_{n=1}^{\infty} a_n$ converges, prove that $\sum_{n=1}^{\infty} \ln (1 - a_n)$ converges. (*Hint:* First show that $|\ln(1 - a_n)| \le a_n/(1 - a_n)$.)

35. An infinite product, indicated by $\prod_{n=1}^{\infty} (1 + a_n)$, is said to converge if the series $\sum_{n=1}^{\infty} \ln (1 + a_n)$ converges. (The series is the natural logarithm of the product.) Prove that the product converges if every $a_n > -1$ and $\sum_{n=1}^{\infty} |a_n|$ converges. (*Hint:* Show that

$$|\ln(1 + a_n)| \le \frac{|a_n|}{1 - |a_n|} < 2|a_n|$$

when $|a_n| < 1/2$.)

36. By multiplying the appropriate terms of the Maclaurin series for $\tan^{-1}x$ and $\ln(1+x)$, find the terms through x^5 in the Maclaurin series for the product $(\tan^{-1}x)\cdot\ln(1+x)$.

37. By appropriate division or multiplication of series, find the terms through x^5 in the Maclaurin series for $\tan^{-1}x/(1-x)$.

38. *Continuous extension of quotients.* Suppose that a function f is represented by a power series

$$f(x) = \Sigma\, a_nx^n = a_0 + a_1x + a_2x^2 + \cdots$$

and that $f(0) = 0$. Then $a_0 = 0$ and the series may be divided through by x. Now, if a function g is represented by a series $\Sigma\, b_nx^n$ and $g(0) = 0$, but $g'(0) \neq 0$, then $b_0 = 0$, but $b_1 \neq 0$. We may then cancel an x from the numerator and denominator of the quotient $f(x)/g(x) = \Sigma\, a_nx^n/\Sigma\, b_nx^n$ to represent f/g as the quotient of two new series. In this new representation, the denominator does not vanish at $x = 0$ because $b_1 \neq 0$ is the constant term. Thus the new quotient extends the domain of the function f/g to include the point $x = 0$.

For example, if $f(x) = x$ and $g(x) = e^x - 1$, the formula

$$\frac{x}{e^x-1} = \frac{1}{1+\dfrac{x}{2}+\dfrac{x^2}{3!}+\dfrac{x^3}{4!}+\cdots+\dfrac{x^{n-1}}{n!}+\cdots}$$

extends the quotient f/g to $x = 0$. The extension, being the quotient of the two continuous functions 1 and $1 + (x/2) + (x^2/3!) + \cdots$, is also continuous. Divide the numerator by the denominator on the right to obtain the first three coefficients c_0, c_1, and c_2 for the series $\Sigma\, c_nx^n$ that represents the continuous extension of f/g.

39. a) Show by long division or otherwise that

$$\frac{1}{1+t} = 1 - t + t^2 - t^3 + \cdots + (-1)^nt^n$$
$$+ \frac{(-1)^{n+1}t^{n+1}}{1+t}.$$

b) By integrating the equation of part (a) with respect to t from 0 to x, show that

$$\ln(1+x) = x - \frac{x^2}{2} + \frac{x^3}{3} - \frac{x^4}{4} + \cdots$$
$$+ (-1)^n\frac{x^{n+1}}{n+1} + R_{n+1}$$

where

$$R_{n+1} = (-1)^{n+1}\int_0^x \frac{t^{n+1}}{1+t}\,dt.$$

c) If $x \geq 0$, show that

$$|R_{n+1}| \leq \int_0^x t^{n+1}\,dt = \frac{x^{n+2}}{n+2}.$$

$\left(\textit{Hint: }\text{As } t \text{ varies from } 0 \text{ to } x,\right.$

$$1 + t \geq 1 \qquad \text{and} \qquad t^{n+1}/(1+t) \leq t^{n+1},$$

and $\left|\displaystyle\int_0^x f(t)\,dt\right| \leq \displaystyle\int_0^x \left|f(t)\right|dt.\Big)$

d) If $-1 < x < 0$, show that

$$\left|R_{n+1}\right| \leq \left|\int_0^x \frac{t^{n+1}}{1-|x|}\,dt\right| = \frac{|x|^{n+2}}{(n+2)(1-|x|)}.$$

$\left(\textit{Hint: }\text{ If } x < t \leq 0, \text{ then } |1+t| \geq 1-|x| \text{ and}\right.$

$$\left|\frac{t^{n+1}}{1+t}\right| \leq \frac{|t|^{n+1}}{1-|x|}.\Big)$$

e) Use the foregoing results to prove that the series

$$x - \frac{x^2}{2} + \frac{x^3}{3} - \frac{x^4}{4} + \cdots + \frac{(-1)^nx^{n+1}}{n+1} + \cdots$$

converges to $\ln(1+x)$ for $-1 < x \leq 1$.

CHAPTER 13

Vectors

OVERVIEW

Some of the things we measure are completely determined by their magnitudes. To record mass, length, or time, for example, we need only write down a number and name an appropriate unit of measure. But we need more than that to describe a force, displacement, or velocity, for these quantities have direction as well as magnitude. To describe a force, we need to record the direction in which it acts as well as how large it is. To describe a body's displacement, we have to say in what direction it moved as well as how far it moved. To describe a body's velocity at any given time, we have to know where the body is headed as well as how fast it is going.

Quantities that have direction as well as magnitude are usually represented by arrows that point in the direction of the action and whose lengths give the magnitude of the action in terms of a suitably chosen unit. When we describe these arrows abstractly, as directed line segments in the plane or in space, we call them *vectors*.

We begin this chapter by describing vectors in the coordinate plane and showing how they are used to write equations that model the behavior of projectiles near the surface of the earth. We then define vectors in space, study two ways to multiply two vectors together, and use equations involving vector multiplication to describe lines in the plane and lines and planes in space. Products involving three vectors also have a

place in engineering and mathematics, and we shall study these at the end of the chapter.

All this sets the stage for the more detailed study of motion in Chapter 14. That is where the calculus will come in. The vectors we work with there will be differentiable functions of time, and we shall combine our knowledge of calculus with our knowledge of vectors to describe the positions, velocities, and accelerations of bodies moving in space.

13.1
Vectors in the Plane

In this article we study the arithmetic of vectors in the plane and look briefly at how vectors are used to prove theorems in geometry.

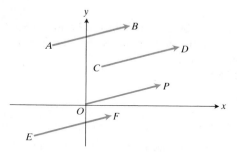

Vectors

A **vector** in the plane is a directed line segment. Since the essential ingredients of a vector are length and direction, we shall call two vectors "equal" or even "the same" if they have the same length and direction. This means that the arrows we use when we draw vectors are understood to represent the same vector if they have the same length, are parallel, and point the same way. Figure 13.1 shows four arrows representing the same vector.

13.1   The four arrows shown here have the same length and direction. They therefore represent the same vector, and we write $\overrightarrow{AB} = \overrightarrow{CD} = \overrightarrow{OP} = \overrightarrow{EF}$.

Geometric Addition: The Parallelogram Law

Two vectors $\mathbf{v}_1$ and $\mathbf{v}_2$ may be added geometrically by drawing a representative of $\mathbf{v}_1$, say from A to B as in Fig. 13.2, and then a vector equal to $\mathbf{v}_2$ starting from the terminal point of $\mathbf{v}_1$.† In Fig. 13.2, $\mathbf{v}_2 = \overrightarrow{BC}$. The sum $\mathbf{v}_1 + \mathbf{v}_2$ is then the vector from the initial point A of $\mathbf{v}_1$ to the terminal point C of $\mathbf{v}_2$. That is, if

$$\mathbf{v}_1 = \overrightarrow{AB} \qquad \text{and} \qquad \mathbf{v}_2 = \overrightarrow{BC},$$

then

$$\mathbf{v}_1 + \mathbf{v}_2 = \overrightarrow{AB} + \overrightarrow{BC} = \overrightarrow{AC}.$$

This description of addition is sometimes called the **parallelogram law** of addition because $\mathbf{v}_1 + \mathbf{v}_2$ is given by the diagonal of the parallelogram determined by $\mathbf{v}_1$ and $\mathbf{v}_2$, as shown in Fig. 13.3.

13.2   The sum of $\mathbf{v}_1$ and $\mathbf{v}_2$.

Components

Whenever a vector $\mathbf{v}$ can be written as a sum

$$\mathbf{v} = \mathbf{v}_1 + \mathbf{v}_2,$$

the vectors $\mathbf{v}_1$ and $\mathbf{v}_2$ are said to be **components** of $\mathbf{v}$.

The most common algebra of vectors is based on a representation of each vector in terms of components parallel to the axes of a Cartesian coordinate system. This is accomplished by choosing the same unit of length on the two axes, with vectors of unit length along the axes used as **basic vectors** in terms of which other vectors in the plane may be expressed.

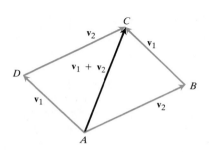

13.3   The parallelogram law of addition. Quadrilateral $ABCD$ is a parallelogram because its opposite sides have equal lengths. The law was used by Aristotle to describe the combined action of two forces.

---

† In print, vectors are usually represented by single boldface roman letters. The vector defined by the directed line segment from point A to point B, however, would appear as $\overline{AB}$. In handwritten work, it is customary to draw small arrows above all letters representing vectors. Thus, the equation $\mathbf{v} = \overline{AB}$ in print would appear as $\vec{v} = \overrightarrow{AB}$ when written by hand.

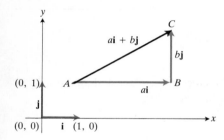

13.4 Any vector $\overrightarrow{AC}$ can be expressed as a multiple of **i** plus a multiple of **j**.

Along the x-axis we choose the vector **i** from $(0, 0)$ to $(1, 0)$ as one basic vector, as in Fig. 13.4. Along the y-axis we choose the vector **j** from $(0, 0)$ to $(0, 1)$ as our second basic vector. Then $a\mathbf{i}$, a being a scalar, represents a vector parallel to the x-axis having magnitude $|a|$ and pointing to the right if a is positive, to the left if a is negative. Similarly, $b\mathbf{j}$ is a vector parallel to the y-axis and having the same direction as **j** if b is positive or the opposite direction if b is negative. Figure 13.4 shows how the vector $\mathbf{v} = \overrightarrow{AC}$ is "resolved" into its **i**- and **j**-components as the sum

$$\mathbf{v} = a\mathbf{i} + b\mathbf{j}.$$

We call $a\mathbf{i}$ and $b\mathbf{j}$ the **vector components of v in the directions of i and j**. The numbers a and b are called the **scalar components of v in the directions of i and j**. (We shall say more about direction later.)

Components give us a way to define the equality of vectors algebraically:

DEFINITION

Equality of Vectors (Algebraic Definition)

$$a\mathbf{i} + b\mathbf{j} = a'\mathbf{i} + b'\mathbf{j} \Leftrightarrow a = a' \text{ and } b = b' \tag{1}$$

That is, two vectors are equal if and only if their scalar components in the directions of **i** and **j** are equal.

Algebraic Addition

Two vectors may be added algebraically by adding their corresponding scalar components, as shown in Fig. 13.5. If

$$\mathbf{v}_1 = a_1\mathbf{i} + b_1\mathbf{j}, \qquad \mathbf{v}_2 = a_2\mathbf{i} + b_2\mathbf{j},$$

then

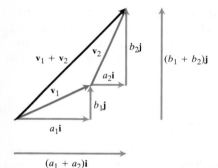

13.5 If $\mathbf{v}_1 = a_1\mathbf{i} + b_1\mathbf{j}$ and $\mathbf{v}_2 = a_2\mathbf{i} + b_2\mathbf{j}$, then $\mathbf{v}_1 + \mathbf{v}_2 = (a_1 + a_2)\mathbf{i} + (b_1 + b_2)\mathbf{j}$.

$$\mathbf{v}_1 + \mathbf{v}_2 = (a_1 + a_2)\mathbf{i} + (b_1 + b_2)\mathbf{j}. \tag{2}$$

EXAMPLE 1

$$(2\mathbf{i} - 4\mathbf{j}) + (5\mathbf{i} + 3\mathbf{j}) = (2 + 5)\mathbf{i} + (-4 + 3)\mathbf{j} = 7\mathbf{i} - \mathbf{j}. \qquad \blacksquare$$

Subtraction

The negative of a vector **v** is the vector $-\mathbf{v}$ that has the same length as **v** but points in the opposite direction. To subtract a vector $\mathbf{v}_2$ from a vector $\mathbf{v}_1$, we add $-\mathbf{v}_2$ to $\mathbf{v}_1$. This may be done geometrically by drawing $-\mathbf{v}_2$ from the tip of $\mathbf{v}_1$ and then drawing the vector from the initial point of $\mathbf{v}_1$ to the tip of $-\mathbf{v}_2$, as shown in Fig. 13.6(a), where

$$\overrightarrow{AD} = \overrightarrow{AB} + \overrightarrow{BD} = \mathbf{v}_1 + (-\mathbf{v}_2) = \mathbf{v}_1 - \mathbf{v}_2.$$

Another way to subtract $\mathbf{v}_2$ from $\mathbf{v}_1$ is to draw them both with a common initial point and then draw the vector from the tip of $\mathbf{v}_2$ to the tip of $\mathbf{v}_1$. This is illustrated in Fig. 13.6(b), where

$$\overrightarrow{CB} = \overrightarrow{CA} + \overrightarrow{AB} = -\mathbf{v}_2 + \mathbf{v}_1 = \mathbf{v}_1 - \mathbf{v}_2.$$

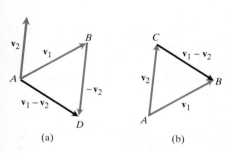

13.6 Two ways to draw $\mathbf{v}_1 - \mathbf{v}_2$: (a) as $\mathbf{v}_1 + (-\mathbf{v}_2)$ and (b) as the vector from the tip of $\mathbf{v}_2$ to the tip of $\mathbf{v}_1$.

THE CONTROVERSY OVER VECTORS AND A STONE ETCHING

Modern-day vectors had their forerunners in quaternions, a kind of extension of the couple $a + bi$ to the ordered quadruple $a + bi + cj + dk$. Quaternion algebra was developed by the Irishman William R. Hamilton (1805–1865). One day as he strolled along Dublin's Royal Canal with his wife, the insight for the formulation of his thoughts over the previous fifteen years came to him. Immediately Hamilton etched with a knife the fundamental properties of quaternions on a stone on Brougham Bridge: $i^2 = j^2 = k^2 = ijk = -1$. He had created a new algebra, but in doing so he had discarded the commutative law for multiplication.

At the same time, the German mathematician Hermann Günther Grassmann (1809–1877) developed a hypernumber with n components. As is the case with quaternions, multiplication is noncommutative. The result is a general calculus of vectors in any number of dimensions, which Grassmann published in 1844, one year after Hamilton published his work on quaternions.

The American Josiah Willard Gibbs (1839–1903), a mathematical physicist at Yale University, promoted the use of vectors, having been influenced by Grassmann's work and less so by Hamilton's. Mathematical friends of Hamilton, particularly Peter Guthrie Tait, urged physicists to adopt quaternions; engineers, especially Oliver Heaviside, promoted vector analysis. Some physicists, most notably James Clerk Maxwell, used both notions of quaternion and vector (separating the scalar and vector parts). Gibbs's and Heaviside's vector analysis won out, with engineers being the first adherents, physicists the first converts, and mathematicians the last, well near the turn of the century.

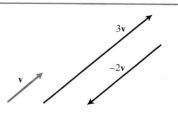

13.7 Scalar multiples of **v**.

Thus, $\overrightarrow{CB}$ is the vector that when added to $\mathbf{v}_2$ gives $\mathbf{v}_1$:

$$\overrightarrow{CB} + \mathbf{v}_2 = (\mathbf{v}_1 - \mathbf{v}_2) + \mathbf{v}_2 = \mathbf{v}_1.$$

In terms of components, vector subtraction follows the simple algebraic law

$$\mathbf{v}_1 - \mathbf{v}_2 = (a_1 - a_2)\mathbf{i} + (b_1 - b_2)\mathbf{j}, \tag{3}$$

which says that corresponding scalar components are subtracted.

EXAMPLE 2

$$(6\mathbf{i} + 2\mathbf{j}) - (3\mathbf{i} - 5\mathbf{j}) = (6 - 3)\mathbf{i} + (2 - (-5))\mathbf{j} = 3\mathbf{i} + 7\mathbf{j}. \qquad ■$$

Length of a Vector

The length of the vector $\mathbf{v} = a\mathbf{i} + b\mathbf{j}$ is usually denoted by $|\mathbf{v}|$, which may be read "the magnitude of **v**." Figure 13.4 shows that **v** is the hypotenuse of a right triangle whose legs have lengths $|a|$ and $|b|$. Hence we may apply the theorem of Pythagoras to obtain

$$|\mathbf{v}| = |a\mathbf{i} + b\mathbf{j}| = \sqrt{a^2 + b^2}. \tag{4}$$

EXAMPLE 3

$$|3\mathbf{i} - 5\mathbf{j}| = \sqrt{(3)^2 + (-5)^2} = \sqrt{9 + 25} = \sqrt{34}. \qquad ■$$

Multiplication by Scalars

The algebraic operation of multiplying a vector $\mathbf{v} = a\mathbf{i} + b\mathbf{j}$ by a scalar c is also simple, namely,

$$c(a\mathbf{i} + b\mathbf{j}) = (ca)\mathbf{i} + (cb)\mathbf{j}. \tag{5}$$

Notice how the vectors **i** and **j** allow us to keep the components separated from one another when we operate on vectors algebraically. Geometrically, $c\mathbf{v}$ is a vector whose length is $|c|$ times the length of **v**:

$$|c\mathbf{v}| = |(ca)\mathbf{i} + (cb)\mathbf{j}| = \sqrt{(ca)^2 + (cb)^2} = |c|\sqrt{a^2 + b^2} = |c|\,|\mathbf{v}|. \tag{6}$$

The direction of $c\mathbf{v}$ agrees with that of **v** if c is positive and is opposite to that of **v** if c is negative (Fig. 13.7). If $c = 0$, the vector $c\mathbf{v}$ has no direction.

EXAMPLE 4   Let $c = 2$ and $\mathbf{v} = -3\mathbf{i} + 4\mathbf{j}$. Then

$$|\mathbf{v}| = |-3\mathbf{i} + 4\mathbf{j}| = \sqrt{(-3)^2 + (4)^2} = \sqrt{9 + 16} = \sqrt{25} = 5,$$

and

$$|2\mathbf{v}| = |2(-3\mathbf{i} + 4\mathbf{j})| = |-6\mathbf{i} + 8\mathbf{j}| = \sqrt{(-6)^2 + (8)^2} = \sqrt{36 + 64}$$
$$= \sqrt{100} = 10 = 2|\mathbf{v}|.$$

If c had been -2 instead of 2, we would have found
$$|-2\mathbf{v}| = |-2(-3\mathbf{i} + 4\mathbf{j})| = |6\mathbf{i} - 8\mathbf{j}| = \sqrt{(6)^2 + (-8)^2} = \sqrt{100}$$
$$= 10 = |-2| \, |\mathbf{v}|.$$ ∎

Zero Vector

The vector
$$\mathbf{0} = 0\mathbf{i} + 0\mathbf{j} \qquad (7)$$

is called the **zero vector.** It is the only vector whose length is zero, as we can see from the fact that

$$|a\mathbf{i} + b\mathbf{j}| = \sqrt{a^2 + b^2} = 0 \qquad \Leftrightarrow \qquad a = b = 0.$$

Unit Vector

Any vector $\mathbf{u}$ whose length is equal to the unit of length used along the coordinate axes is called a **unit vector.** The vectors $\mathbf{i}$ and $\mathbf{j}$ are unit vectors:

$$|\mathbf{i}| = |1\,\mathbf{i} + 0\,\mathbf{j}| = \sqrt{1^2 + 0^2} = 1, \qquad\qquad |\mathbf{j}| = |0\,\mathbf{i} + 1\,\mathbf{j}| = \sqrt{0^2 + 1^2} = 1.$$

If $\mathbf{u}$ is the unit vector obtained by rotating $\mathbf{i}$ through an angle θ in the positive direction, then (Fig. 13.8) $\mathbf{u}$ has a horizontal component

$$u_x = \cos \theta$$

and a vertical component

$$u_y = \sin \theta,$$

so that

$$\mathbf{u} = (\cos \theta)\mathbf{i} + (\sin \theta)\mathbf{j}. \qquad (8)$$

If we allow the angle θ in Eq. (8) to vary from 0 to 2π, then the point P in Fig. 13.8 traces the unit circle $x^2 + y^2 = 1$ once in the counterclockwise direction. Since this takes in all possible directions, every unit vector in the plane is given by Eq. (8) for some value of θ.

In handwritten work it is common to denote unit vectors with small "hats," as in $\hat{\mathbf{u}}$ (pronounced "u hat"). In hat notation, $\mathbf{i}$ and $\mathbf{j}$ become $\hat{\mathbf{i}}$ and $\hat{\mathbf{j}}$.

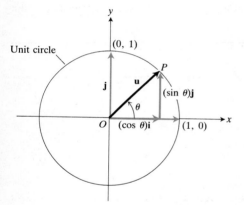

13.8 Every unit vector $\mathbf{u}$ has the form

$$\mathbf{u} = (\cos \theta)\mathbf{i} + (\sin \theta)\mathbf{j}$$

for some angle θ.

Direction

It is common in subjects like classical electricity and magnetism, which use vectors a great deal, to define the **direction** of a nonzero vector $\mathbf{A}$ to be the unit vector obtained by dividing $\mathbf{A}$ by its own length:

$$\text{Direction of } \mathbf{A} \neq \frac{\mathbf{A}}{|\mathbf{A}|}. \qquad (9)$$

Instead of just saying that the unit vector $\mathbf{A}/|\mathbf{A}|$ *represents* the direction of $\mathbf{A}$, we say that it *is* the direction of $\mathbf{A}$.

To see that $\mathbf{A}/|\mathbf{A}|$ is indeed a unit vector, we can calculate its length directly:

$$\text{Length of } \frac{\mathbf{A}}{|\mathbf{A}|} = \left|\frac{\mathbf{A}}{|\mathbf{A}|}\right| = \left|\frac{1}{|\mathbf{A}|}\mathbf{A}\right| = \frac{1}{|\mathbf{A}|}|\mathbf{A}| = 1. \tag{10}$$

The zero vector $\mathbf{0}$ has no defined direction.

EXAMPLE 5 The direction of $\mathbf{A} = 3\mathbf{i} - 4\mathbf{j}$ is

$$\text{Direction of } \mathbf{A} = \frac{\mathbf{A}}{|\mathbf{A}|} = \frac{3\mathbf{i} - 4\mathbf{j}}{\sqrt{(3)^2 + (-4)^2}} = \frac{3\mathbf{i} - 4\mathbf{j}}{\sqrt{25}} = \frac{3}{5}\mathbf{i} - \frac{4}{5}\mathbf{j}.$$

To check the length, we can calculate

$$\left|\frac{3}{5}\mathbf{i} - \frac{4}{5}\mathbf{j}\right| = \sqrt{\left(\frac{3}{5}\right)^2 + \left(\frac{4}{5}\right)^2} = \sqrt{\frac{9}{25} + \frac{16}{25}} = \sqrt{\frac{25}{25}} = 1. \quad \blacksquare$$

It follows from the definition of direction that two nonzero vectors $\mathbf{A}$ and $\mathbf{B}$ have the same direction if and only if

$$\frac{\mathbf{A}}{|\mathbf{A}|} = \frac{\mathbf{B}}{|\mathbf{B}|}, \tag{11}$$

or

$$\mathbf{A} = \frac{|\mathbf{A}|}{|\mathbf{B}|}\mathbf{B}. \tag{12}$$

Thus, if $\mathbf{A}$ and $\mathbf{B}$ have the same direction, then $\mathbf{A}$ is a positive scalar multiple of $\mathbf{B}$. Conversely, if

$$\mathbf{A} = k\mathbf{B}, \qquad k > 0,$$

then

$$\frac{\mathbf{A}}{|\mathbf{A}|} = \frac{k\mathbf{B}}{|k\mathbf{B}|} = \frac{k}{|k|}\frac{\mathbf{B}}{|\mathbf{B}|} = \frac{k}{k}\frac{\mathbf{B}}{|\mathbf{B}|} = \frac{\mathbf{B}}{|\mathbf{B}|}. \tag{13}$$

Therefore, two nonzero vectors $\mathbf{A}$ and $\mathbf{B}$ point in the same direction if and only if $\mathbf{A}$ is a positive scalar multiple of $\mathbf{B}$.

We say that two nonzero vectors $\mathbf{A}$ and $\mathbf{B}$ point in *opposite* directions if their directions are opposite in sign:

$$\frac{\mathbf{A}}{|\mathbf{A}|} = -\frac{\mathbf{B}}{|\mathbf{B}|}. \tag{14}$$

From this it follows (see Problem 23) that $\mathbf{A}$ and $\mathbf{B}$ have opposite directions if and only if $\mathbf{A}$ is a negative scalar multiple of $\mathbf{B}$.

EXAMPLE 6

a) Same direction: $\mathbf{A} = 3\mathbf{i} - 4\mathbf{j}$ and $\mathbf{B} = \frac{3}{2}\mathbf{i} - 2\mathbf{j} = \frac{1}{2}\mathbf{A}$

($\mathbf{B}$ is a positive scalar multiple of $\mathbf{A}$.)

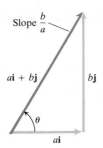

13.9 If $a \neq 0$, the vector $a\mathbf{i} + b\mathbf{j}$ has slope $b/a = \tan \theta$.

b) Opposite directions: $\mathbf{A} = 3\mathbf{i} - 4\mathbf{j}$ and $\mathbf{B} = -9\mathbf{i} + 12\mathbf{j} = -3\mathbf{A}$

($\mathbf{B}$ is a negative scalar multiple of $\mathbf{A}$.)   ∎

Slopes, Tangents, and Normals

Two vectors are said to be **parallel** if they are either positive or negative scalar multiples of one another or, equivalently, if the line segments representing them are parallel. Similarly, a vector is parallel to a line if the segments that represent the vector are parallel to the line. The **slope** of a vector that is not parallel to the y-axis is the slope shared by the lines parallel to the vector. Thus, when $a \neq 0$, the vector $\mathbf{v} = a\mathbf{i} + b\mathbf{j}$ has a well-defined slope, which can be calculated from the components of $\mathbf{v}$ as the number b/a. (See Fig. 13.9.)

When we say that a vector is **tangent** or **normal** to a curve at a point, we mean that the vector is parallel or normal to the line that is tangent to the curve at the point. The next example shows how such vectors may be found.

EXAMPLE 7 Find unit vectors tangent and normal to the curve

$$y = \frac{x^3}{2} + \frac{1}{2}$$

at the point $(1, 1)$.

Solution We find the unit vectors that are parallel and normal to the curve's tangent line at the point $(1, 1)$, shown in Fig. 13.10.

The slope of the line tangent to the curve at $(1, 1)$ is

$$y' = \frac{3x^2}{2}\bigg|_{x=1} = \frac{3}{2}.$$

We find a unit vector with this slope. The vector $\mathbf{v} = 2\mathbf{i} + 3\mathbf{j}$ has slope $3/2$, as does every nonzero multiple of $\mathbf{v}$. To find a multiple of $\mathbf{v}$ that is a unit vector, we divide $\mathbf{v}$ by its length,

$$|\mathbf{v}| = \sqrt{2^2 + 3^2} = \sqrt{13}.$$

This produces the unit vector

$$\mathbf{u} = \frac{\mathbf{v}}{|\mathbf{v}|} = \frac{2}{\sqrt{13}}\mathbf{i} + \frac{3}{\sqrt{13}}\mathbf{j}.$$

The vector $\mathbf{u}$ is tangent to the curve at $(1, 1)$ because it has the same direction as $\mathbf{v}$. Of course, the vector

$$-\mathbf{u} = -\frac{2}{\sqrt{13}}\mathbf{i} - \frac{3}{\sqrt{13}}\mathbf{j},$$

which points in the opposite direction, is also tangent to the curve at $(1, 1)$. Without some additional requirement, there is no reason to prefer one of these vectors to the other.

To find unit vectors normal to the curve at $(1, 1)$, we look for unit vectors whose slopes are the negative reciprocal of the slope of $\mathbf{u}$. This is quickly done by interchanging the scalar components of $\mathbf{u}$ and changing the sign of one of

13.10 The unit tangent and normal vectors at the point $(1, 1)$ on the curve $y = (x^3/2) + 1/2$.

them. We obtain

$$\mathbf{n} = -\frac{3}{\sqrt{13}}\mathbf{i} + \frac{2}{\sqrt{13}}\mathbf{j} \quad \text{and} \quad -\mathbf{n} = \frac{3}{\sqrt{13}}\mathbf{i} - \frac{2}{\sqrt{13}}\mathbf{j}.$$

Again, either one will do. The vectors have opposite directions but both are normal to the curve at the point (1, 1). ∎

Geometric Proofs with Vectors

Vectors provide a convenient and efficient language for proving theorems in geometry. Here is an example of a vector proof.

EXAMPLE 8 Prove that the line segment joining the midpoints of two sides of a triangle is parallel to and half as long as the third side.

Solution We draw a triangle, mark the midpoints of two sides, and sketch the relevant vectors (Fig. 13.11). In the notation of Fig. 13.11, our goal is to show that

$$\overrightarrow{HK} = \frac{1}{2}\overrightarrow{AC},$$

for this one vector equation tells us both that $\overrightarrow{HK}$ is parallel to $\overrightarrow{AC}$ and that the length of $\overrightarrow{HK}$ is half the length of $\overrightarrow{AC}$. Notice how efficient vector notation can be. The steps in the proof are straightforward:

$$\overrightarrow{HK} = \overrightarrow{AK} - \overrightarrow{AH} \quad \text{(because } \overrightarrow{HK} + \overrightarrow{AH} = \overrightarrow{AK}\text{)}$$

$$= (\overrightarrow{AB} + \overrightarrow{BK}) - \overrightarrow{AH}$$

$$= \left(\overrightarrow{AB} + \frac{1}{2}\overrightarrow{BC}\right) - \frac{1}{2}\overrightarrow{AB}$$

$$= \frac{1}{2}(\overrightarrow{AB} + \overrightarrow{BC})$$

$$= \frac{1}{2}\overrightarrow{AC}. \quad ∎$$

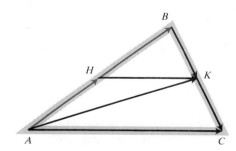

13.11 A vector diagram for proving that the segment joining the midpoints of two sides of a triangle is parallel to and half as long as the third side.

PROBLEMS

1. The three vectors **A**, **B**, and **C** shown below lie in a plane. Copy them on a sheet of paper.

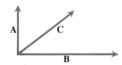

Then, by arranging the vectors in head-to-tail fashion, sketch

a) **A** + **B**, to which then add **C**

b) **A** + **C**, to which then add **B**

c) **B** + **C**, to which then add **A**

d) **A** − **B**, to which then add **C**

e) **A** + **C**, from which then subtract **B**

In Problems 2–11 express each of the vectors in the form $a\mathbf{i} + b\mathbf{j}$. Indicate all quantities graphically.

2. $\overrightarrow{P_1P_2}$ if P_1 is the point (1, 3) and P_2 is the point (2, −1)

3. $\overrightarrow{OP_3}$ if O is the origin and P_3 is the midpoint of the vector $\overrightarrow{P_1P_2}$ joining $P_1(2, -1)$ and $P_2(-4, 3)$

4. The vector from the point $A(2, 3)$ to the origin

5. The sum of the vectors $\overrightarrow{AB}$ and $\overrightarrow{CD}$, given the four points $A(1, -1)$, $B(2, 0)$, $C(-1, 3)$, and $D(-2, 2)$

6. A unit vector making an angle of 30° with the positive *x*-axis

7. The unit vector obtained by rotating **j** 120° in the clockwise direction

8. A unit vector having the same direction as the vector $3\mathbf{i} - 4\mathbf{j}$

9. A unit vector tangent to the curve $y = x^2$ at the point (2, 4)

10. A unit vector normal to the curve $y = x^2$ at the point $P(2, 4)$ and pointing from P toward the inside of the curve (that is, an "inner" normal)

11. a) The unit vector tangent to the curve $y = x^2 + 2x$ at the point (1, 3) that points toward the *x*-axis
   b) The unit vector normal to the curve at the point (1, 3) obtained by rotating the unit tangent vector found in part (a) 90° clockwise

12. Show the unit vectors given by Eq. (8) for $\theta = 0$, $\pi/4$, $\pi/2$, $2\pi/3$, $5\pi/4$, and $5\pi/3$, together with a graph of the circle $x^2 + y^2 = 1$.

In Problems 13–18, find the length and direction of each vector and the angle it makes with the positive *x*-axis.

13. $\mathbf{i} + \mathbf{j}$

14. $2\mathbf{i} - 3\mathbf{j}$

15. $\sqrt{3}\mathbf{i} + \mathbf{j}$

16. $-2\mathbf{i} + 3\mathbf{j}$

17. $5\mathbf{i} + 12\mathbf{j}$

18. $-5\mathbf{i} - 12\mathbf{j}$

19. The *speed* of a particle moving with velocity **v** is defined to be $|\mathbf{v}|$, the length (or magnitude) of **v**. Find the speed of a particle whose velocity is $\mathbf{v} = -4\mathbf{i} + 2\mathbf{j}$.

20. Show that $\mathbf{A} = 3\mathbf{i} + 6\mathbf{j}$ and $\mathbf{B} = -\mathbf{i} - 2\mathbf{j}$ have opposite directions. Sketch them together.

21. Show that $\mathbf{C} = 3\mathbf{i} + 6\mathbf{j}$ and $\mathbf{D} = \frac{1}{2}\mathbf{i} + \mathbf{j}$ have the same direction.

22. Let **v** be a vector in the plane not parallel to the *y*-axis. Will the slope of $-\mathbf{v}$ be the same as the slope of **v** or the negative of the slope of **v**? Explain.

23. Show that any two nonzero vectors **A** and **B** have opposite directions if and only if **A** is a negative scalar multiple of **B**. (*Hint:* See Eq. 14.)

24. Let A', B', C', D' be the midpoints of the sides AB, BC, CD, and DA of a quadrilateral $ABCD$. Prove that $A'B'C'D'$ is a parallelogram. (*Hint:* First show that $\overrightarrow{A'B'} = \overrightarrow{D'C'} = \frac{1}{2}\overrightarrow{AC}$.)

25. Show that the diagonals of a parallelogram bisect each other. (*Hint:* Let A be one vertex and let M and N be the midpoints of the diagonals. Then show that $\overrightarrow{AM} = \overrightarrow{AN}$.)

13.2
Modeling Projectile Motion

When we shoot a projectile into the air we usually want to know beforehand how far it will go (will it reach the target?), how high it will rise (will it clear the hill?), and when it will land (when do we get results?). We get this information from equations that calculate the answers we want from the direction and magnitude of the projectile's initial velocity vector, described in terms of the angle and speed at which the projectile is fired. The equations come from combining calculus and Newton's second law of motion in vector form. In this article, we derive these equations and show how to use them to get the information we want about projectile motion.

The Parametric Equations for Ideal Projectile Motion

To derive equations for projectile motion, we assume that the projectile behaves like a particle moving in a vertical coordinate plane and that the only force acting on the projectile during its flight is the constant force of gravity, which always points straight down. In practice, none of these assumptions really holds. The ground moves beneath the projectile as the earth turns, the air creates a frictional force that varies with the projectile's speed and altitude, and the force of gravity changes as the projectile moves along. All this must be taken into account by applying corrections to the predictions of the ideal equations we are about to derive. The corrections, however, are not the subject of this article.

We begin with Newton's second law,

$$\mathbf{F} = m\mathbf{a} = m\frac{d\mathbf{v}}{dt}. \tag{1}$$

The component form,

$$F_x\,\mathbf{i} + F_y\,\mathbf{j} = m\frac{dv_x}{dt}\mathbf{i} + m\frac{dv_y}{dt}\mathbf{j} \tag{2}$$

(see Fig. 13.12), is equivalent to writing

$$F_x = m\frac{dv_x}{dt} \quad\text{and}\quad F_y = m\frac{dv_y}{dt}. \tag{3}$$

Since we know the position and velocity of the projectile at the time it is fired, we should be able to find the position of the projectile at any later time by integrating the equations in (3) with respect to time and determining the constants of integration from the initial conditions.

To describe the motion in terms of a specific coordinate system, we suppose that our projectile is fired at time $t = 0$ from the origin of the xy-plane. We further suppose the projectile moves into the first quadrant with an initial velocity vector of magnitude v_0 that makes an angle α with the positive x-axis. See Fig. 13.13. At any time $t \geq 0$, the projectile's position is given by a coordinate pair $x(t)$, $y(t)$. We assume x and y to have continuous derivatives with respect to t.

The initial conditions for the projectile's motion are

$$x(0) = 0, \qquad y(0) = 0 \qquad \text{(The projectile starts at the origin.)}$$

$$v_x(0) = \frac{dx}{dt}\bigg|_{t=0} = v_0 \cos \alpha,$$

$$\left(\begin{array}{c}\text{The initial speed is } v_0 \text{ and} \\ \text{the firing angle is } \alpha.\end{array}\right) \tag{4}$$

$$v_y(0) = \frac{dy}{dt}\bigg|_{t=0} = v_0 \sin \alpha.$$

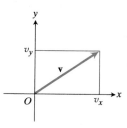

The force and velocity vectors of the motion of a particle in a plane might look like this at a particular time t.

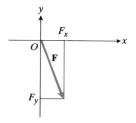

The scalar components v_x and v_y of **v**.

The scalar components F_x and F_y of **F**.

13.12 When we describe the motion of a particle in a plane, we draw the force and velocity vectors as arrows starting at the particle's position.

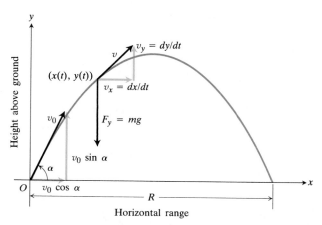

13.13 Ideal projectiles move along parabolas. The diagram here shows the velocity vector and its components, first at time $t = 0$ and then later in the flight. Each vector is labeled with its magnitude.

The scalar components of the force **F** exerted on the projectile by gravity are

$F_x = 0,$ (The force acts parallel to the y-axis.)

$F_y = -mg.$ (m is the projectile's mass, g is the acceleration of gravity, and the force acts downward.)

With these values for F_x and F_y, the equations in (3) become

$$0 = m\frac{d^2x}{dt^2}, \qquad -mg = m\frac{d^2y}{dt^2}. \tag{5}$$

To solve these equations, we integrate each one twice. This introduces four constants of integration, which we evaluate by using the initial conditions in (4). The first equation in (5) gives

$$\frac{d^2x}{dt^2} = 0, \qquad \frac{dx}{dt} = C_1, \qquad x = C_1 t + C_2. \tag{6}$$

The second equation in (5) gives

$$\frac{d^2y}{dt^2} = -g, \qquad \frac{dy}{dt} = -gt + C_3, \qquad y = -\frac{1}{2}gt^2 + C_3 t + C_4. \tag{7}$$

From the initial conditions in (4), we find

$$C_1 = v_0 \cos \alpha, \qquad C_2 = 0, \qquad C_3 = v_0 \sin \alpha, \qquad C_4 = 0. \tag{8}$$

The position of the projectile t seconds after firing is therefore

$$x = (v_0 \cos \alpha)t, \qquad y = (v_0 \sin \alpha)t - \frac{1}{2}gt^2. \tag{9}$$

The parametric equations in (9) express the coordinates of the projectile's position in terms of the **initial speed** v_0 and **angle of elevation** or **firing angle** α. If time is measured in seconds and distance in meters, g is 9.8 m/sec^2 and the equations give x and y in meters. If time is measured in seconds and distance in feet, g is 32 ft/sec^2 and the equations give x and y in feet.

EXAMPLE 1   A projectile is fired over horizontal ground at an initial speed of 500 m/sec at an angle of elevation of 60°. Where will the projectile be 10 sec later?

Solution   We use Eqs. (9) with $v_0 = 500$, $\alpha = 60°$, $g = 9.8$, and $t = 10$ to find the coordinates of the projectile to the nearest meter t seconds after firing:

$$x = (v_0 \cos \alpha)t = 500 \cdot \frac{1}{2} \cdot 10 = 2500 \text{ m}$$

$$y = (v_0 \sin \alpha)t - \frac{1}{2}gt^2$$

$$= 500 \cdot \frac{\sqrt{3}}{2} \cdot 10 - \frac{1}{2} \cdot 9.8(10)^2$$

$$= 2500\sqrt{3} - 490 = 3840 \text{ m.} \qquad \text{(rounded)}$$

Ten seconds after firing, the projectile is 3840 m in the air and 2500 m downrange.   ■

Height, Range, and Flight Time

The parametric equations in (9) enable us to answer most questions about the flight of an ideal projectile.

EXAMPLE 2 A projectile is fired at an angle of elevation α and at an initial speed v_0. When does the projectile reach its highest point? How high does the projectile rise?

Solution The projectile will reach its highest point when its vertical velocity component is zero, that is, when

$$\frac{dy}{dt} = v_0 \sin \alpha - gt = 0, \qquad \text{or} \qquad t = \frac{v_0 \sin \alpha}{g}.$$

For this value of t, the value of y is

$$y_{\max} = (v_0 \sin \alpha)\left(\frac{v_0 \sin \alpha}{g}\right) - \frac{1}{2}g\left(\frac{v_0 \sin \alpha}{g}\right)^2 = \frac{(v_0 \sin \alpha)^2}{2g}. \quad ■ \quad (10)$$

EXAMPLE 3 A projectile is fired over horizontal ground. When and where does it land?

Solution To find when the projectile lands, we set y equal to zero in Eq. (9) and solve for t:

$$y = t\left(v_0 \sin \alpha - \frac{1}{2}gt\right) = 0$$

when

$$t = 0 \qquad \text{or} \qquad t = \frac{2v_0 \sin \alpha}{g}. \tag{11}$$

Since $t = 0$ is the instant the projectile is fired, $t = (2v_0 \sin \alpha)/g$ must be when the projectile strikes the ground. The corresponding value for the projectile's horizontal range R is obtained by evaluating x at this value of t:

$$R = v_0 \cos \alpha \, \frac{2v_0 \sin \alpha}{g} = \frac{v_0^2}{g}(2 \sin \alpha \cos \alpha) = \frac{v_0^2}{g}\sin 2\alpha. \tag{12}$$

The projectile will strike $(v_0^2 \sin 2\alpha)/g$ distance units downrange.   ■

EXAMPLE 4 What angle of elevation gives the maximum horizontal range?

Solution The value of

$$R = \frac{v_0^2}{g}\sin 2\alpha$$

from Eq. (12) is greatest when $\sin 2\alpha = 1$, which, for the angle in question, means $2\alpha = 90°$ or $\alpha = 45°$. The maximum horizontal range is achieved with an elevation of 45°.   ■

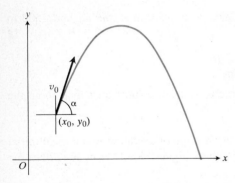

13.14 The path of a projectile fired from (x_0, y_0) with an initial speed v_0 at an angle of α degrees with the horizontal.

EXAMPLE 5 Find the equations for the motion of a projectile fired into the first quadrant from an arbitrary point (x_0, y_0), as shown in Fig. 13.14.

Solution The initial conditions

$$x(0) = x_0, \qquad \frac{dx}{dt}(0) = v_0 \cos \alpha,$$

$$y(0) = y_0, \qquad \frac{dy}{dt}(0) = v_0 \sin \alpha,$$

determine the constants in Eqs. (6) and (7) to be

$$C_1 = v_0 \cos \alpha, \qquad C_2 = x_0, \qquad C_3 = v_0 \sin \alpha, \qquad C_4 = y_0.$$

Hence,

$$x = x_0 + (v_0 \cos \alpha)t,$$

$$y = y_0 + (v_0 \sin \alpha)t - \frac{1}{2}gt^2. \tag{13}$$

EXAMPLE 6 We are about to hit a golf ball horizontally off the top edge of a 100-ft building that stands on level ground. How fast will we have to launch the ball to land it 450 ft away?

Solution We use a coordinate system in which the x-axis lies along the ground and the ball's position coordinates at $t = 0$ are

$$x_0 = 0, \qquad y_0 = 100.$$

See Fig. 13.15.

We then solve Eqs. (13) for v_0, given that $\alpha = 0$ and that $x = 450$ ft when $y = 0$. These equations tell us that at the time the ball lands,

$$450 = 0 + (v_0 \cos 0)t, \qquad \text{and} \qquad 0 = 100 + (v_0 \sin 0)t - \frac{1}{2}gt^2.$$

If we measure time in seconds, g equals 32 ft/sec^2 and the equations reduce to

$$450 = v_0 t, \qquad 0 = 100 - 16t^2,$$

which give

$$t = 2.5 \qquad \text{and} \qquad v_0 = 180$$

when solved simultaneously. To land the ball 450 ft away, we must give it an initial horizontal speed of 180 ft/sec.

13.15 What initial horizontal speed gets the golf ball 450 ft downrange? To find out, see Example 6.

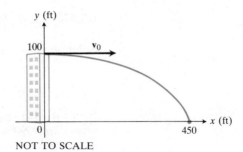

NOT TO SCALE

Ideal Trajectories Are Parabolic

It is often claimed that the water from a hose traces a parabola in the air, but anyone who looks closely enough will see that this isn't so. The air slows the water down, and its forward progress is too slow at the end to match the rate at which it falls. What is really being claimed is that ideal projectiles move along parabolas, and this we can see from the equations in (9). If we substitute $t = x/(v_0 \cos \alpha)$ from the first equation in (9) into the second, we obtain the Cartesian equation

$$y = -\left(\frac{g}{2v_0^2 \cos^2\alpha}\right)x^2 + (\tan \alpha)x. \qquad (14)$$

This equation is quadratic in x and linear in y. Its graph is therefore a parabola.

PROBLEMS

The projectiles in the following problems are to be treated as ideal projectiles whose behavior is faithfully portrayed by the equations derived in the text. Most of the arithmetic, however, is realistic and is best done with a calculator. *All firing angles are assumed to be measured from the horizontal.*

1. A projectile is fired at a speed of 840 m/sec at an angle of 60°. How long will it take to get 21 km downrange?

2. Find the muzzle speed of a gun whose maximum range is 24.5 km.

3. A projectile is fired over level ground with an initial speed of 500 m/sec at an angle of elevation of 45°.
   a) When and how far away will the projectile strike?
   b) How high overhead will the projectile be when it is 5 km downrange?
   c) What is the highest the projectile will go?

4. A baseball is thrown from the stands 32 ft above the field at an angle of 30°. When and how far away will the ball strike the ground if its initial speed is 32 ft/sec?

5. Show that a projectile fired at an angle of α degrees, $0 < \alpha < 90$, has the same range as a projectile fired at the same speed at an angle of $(90 - \alpha)$ degrees.

6. What two angles of elevation will enable a projectile to reach a target 16 km downrange on the same level as the gun if the projectile's initial speed is 400 m/sec?

7. A spring gun at ground level fires a golf ball at an angle of 45°. The ball lands 10 m away. What was the ball's initial speed? For the same initial speed, find the two firing angles that make the range 6 m.

8. An electron in a TV tube is beamed horizontally at a speed of 5×10^6 m/sec toward the face of the tube 40 cm away. About how far will the electron drop before it hits?

9. Laboratory tests designed to find how far golf balls of different hardness go when hit with a driver showed that a 100-compression, Surlyn-covered, two-piece ball hit with a club-head speed of 100 mph at a launch angle of 9° carried 248.8 yd. What was the launch speed of the ball? (It was not 100 mph. The data are from "Does Compression Really Matter?" by Lew Fishman, *Golf Digest,* August 1986, pp. 35–37.)

10. Show that doubling the initial speed of a projectile multiplies its maximum height and range by 4. By about what percentage should you increase the initial speed to double the height and range?

11. Show that a projectile attains three-quarters of its maximum height in half the time it takes to reach the maximum height.

12. A human cannonball is to be fired with an initial speed of $v_0 = 80\sqrt{10}/3$ ft/sec. The circus performer (of the right caliber, naturally) hopes to land on a special cushion located 200 ft downrange. The circus is being held in a large room with a flat ceiling 75 ft high. Can the performer be fired to the cushion without striking the ceiling? If so, what should the cannon's angle of elevation be?

13. A golf ball leaves the ground at a 30° angle at a speed of 90 ft/sec. Will it clear the top of a 30-ft tree 135 ft away?

14. A golf ball is hit from the tee to an elevated green with an initial speed of 116 ft/sec at an angle of elevation of 45°, as shown in Fig. 13.16. Will the ball reach the pin?

13.16 The tee and green in Problem 14.

13.17   The Green Monster, the left-field wall at Fenway Park in Boston.

15. A baseball hit by a Boston Red Sox player at a 20° angle from 3 ft above the ground just cleared the left end of the "Green Monster," the left-field wall in Fenway Park (Fig. 13.17). The wall there is 37 ft high and 315 ft from home plate. About how fast was the ball going? How long did it take the ball to reach the wall?

16. The multiflash photograph in Figure 13.18 shows a model train engine moving at a constant speed on a straight track. As the engine moved along, a marble was fired into the air by a spring in the engine's smokestack. The marble, which continued to move with the same forward speed as the engine, rejoined the engine 1 second after it was fired. Measure the angle the marble's path makes with the horizontal

and use the information to find out how high the marble went and how fast the engine was moving.

17. Figure 13.19 shows an experiment with two marbles. Marble A is launched toward marble B at an initial speed v_0. The launch angle is α. At the same instant, marble B, suspended directly over a spot R units downrange from A, is released to fall from a height of $R \tan \alpha$. It is found that the marbles collide regardless of the value of v_0. Must this happen? Why?

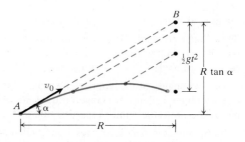

13.19   The marbles in Problem 17.

13.18   The train in Problem 16. (Photograph from *PSSC Physics, Second Edition;* 1965; D.C. Heath & Company with Education Development Center, Inc.)

18. An ideal projectile is fired down an inclined plane as shown in profile in Fig. 13.20. Show that the maximum downhill range is achieved when the initial velocity vector bisects the angle *AOR*. If the projectile were fired uphill instead of down, what firing angle would maximize its range?

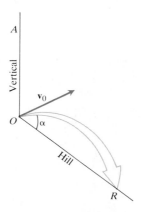

13.20 Maximum downhill range occurs when the velocity vector bisects angle *AOR*. See Problem 18.

19. Suppose a target moving at the constant rate of a ft/sec is level with and b ft away from a gun at the instant the gun is fired. If the target moves in a horizontal line directly away from the gun, show that the muzzle velocity v_0 and angle of elevation α must satisfy the equation

$$v_0^2 \sin 2\alpha - 2\, av_0 \sin \alpha - bg = 0$$

if the projectile is to strike the target.

TOOLKIT PROGRAMS

Parametric Equations   Super * Grapher

13.3
Coordinates and Vectors in Space

Our goal in this article is to describe the three-dimensional Cartesian coordinate system and learn our way around in space. This means defining distance, practicing with the arithmetic of vectors (same rules as in the plane but with an extra term), and making connections between sets of points in space and equations and inequalities. Everyone we know finds it harder to draw in three dimensions than in two, so we have included a few tips for drawing as well. You will find them on pages 770–771.

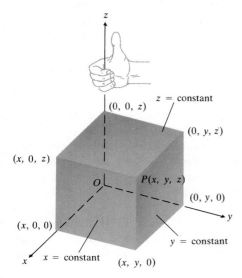

13.21 The Cartesian coordinate system is right-handed.

Cartesian Coordinates

To locate points in space, we use three mutually perpendicular coordinate axes, arranged as in Fig. 13.21. The axes Ox, Oy, and Oz shown there make what is known as a *right-handed* coordinate frame. When you hold your right hand so that the fingers curl from the positive x-axis toward the positive y-axis, your thumb points along the positive z-axis.

The Cartesian coordinates (x, y, z) of a point P in space are the numbers where the planes through P perpendicular to the three axes cut the axes.

Points that lie on the x-axis have their y- and z-coordinates equal to zero. That is, they have coordinates of the form $(x, 0, 0)$. Similarly, points on the y-axis have coordinates of the form $(0, y, 0)$, and points on the z-axis have coordinates of the form $(0, 0, z)$.

The points in a plane perpendicular to the x-axis all have the same x-coordinate, the number at which that plane cuts the x-axis. Similarly, the points in a plane perpendicular to the y-axis have a common y-coordinate, and the points

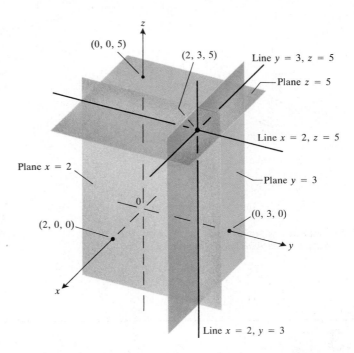

13.22 The planes $x = 2$, $y = 3$, and $z = 5$ determine three lines through the point $(2, 3, 5)$.

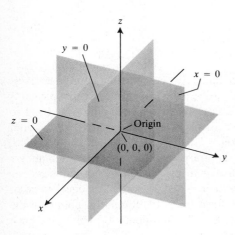

13.23 The planes $x = 0$, $y = 0$, and $z = 0$ are the planes determined by the coordinate axes. They divide space into eight cells called octants.

in a plane perpendicular to the z-axis have a common z-coordinate. It is therefore easy to write equations for these planes—we just name the common coordinate's value. The equation $x = 2$ is an equation for the plane perpendicular to the x-axis at $x = 2$. The equation $y = 3$ is an equation for the plane perpendicular to the y-axis at $y = 3$. The equation $z = 5$ is an equation for the plane perpendicular to the z-axis at $z = 5$. Figure 13.22 shows the planes $x = 2$, $y = 3$, and $z = 5$, together with their intersection point $(2, 3, 5)$.

The planes $x = 2$ and $y = 3$ in Fig. 13.22 intersect in a line that runs parallel to the z-axis. This line is described by the *pair* of equations $x = 2$, $y = 3$. A point (x, y, z) lies on the line if and only if x equals 2 and y equals 3. Similarly, the line of intersection of the planes $y = 3$ and $z = 5$ is described by the equation pair $y = 3$, $z = 5$. This line runs parallel to the x-axis. The line of intersection of the planes $x = 2$ and $z = 5$, parallel to the y-axis, is described by the equation pair $x = 2$, $z = 5$.

The planes determined by the three coordinate axes are the **xy-plane**, whose standard equation is $z = 0$; the **yz-plane**, whose standard equation is $x = 0$; and the **xz-plane**, whose standard equation is $y = 0$. They meet in the point $(0, 0, 0)$, which is called the **origin** of the coordinate system. See Fig. 13.23.

The three **coordinate planes** $x = 0$, $y = 0$, and $z = 0$ divide space into eight cells called **octants**. The octant in which the point coordinates are all positive is called the **first octant**, but there is no conventional numbering for the remaining seven octants.

In the following examples, we match a number of coordinate equations and inequalities with the sets of points they define in space.

EXAMPLE 1

Defining equations and inequalities	Verbal description
$z \geq 0$	The half-space consisting of the points on and above the xy-plane.
$x = -3$	The plane perpendicular to the x-axis at $x = -3$. This plane lies parallel to the yz-plane and 3 units behind it.
$z = 0, \; x \leq 0, \; y \geq 0$	The second quadrant of the xy-plane.
$x \geq 0, \; y \geq 0, \; z \geq 0$	The first octant.
$-1 \leq y \leq 1$	The slab between the planes $y = -1$ and $y = 1$ (planes included).
$y = -2, \; z = 2$	The line in which the planes $y = -2$ and $z = 2$ intersect. Alternatively, the line through the point $(0, -2, 2)$ parallel to the x-axis. ∎

EXAMPLE 2 Identify the set of points $P(x, y, z)$ whose coordinates satisfy the two equations

$$x^2 + y^2 = 4$$

and

$$z = 3.$$

Solution The points lie in the horizontal plane $z = 3$ and, in this plane, make up the circle $x^2 + y^2 = 4$. We call this set of points "the circle $x^2 + y^2 = 4$ in the plane $z = 3$" or, more simply, "the circle $x^2 + y^2 = 4, z = 3$." See Fig. 13.24. ∎

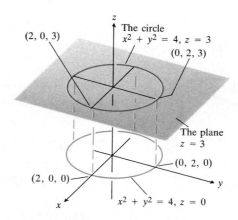

13.24 The circle $x^2 + y^2 = 4$, $z = 3$.

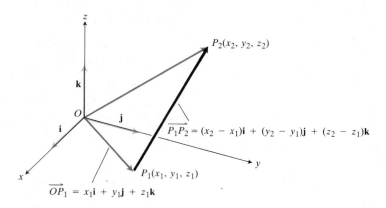

13.25 The vector from P_1 to P_2 is

$$\overrightarrow{P_1P_2} = (x_2 - x_1)\mathbf{i} + (y_2 - y_1)\mathbf{j} + (z_2 - z_1)\mathbf{k}.$$

Vectors in Space

The directed line segments we use to represent forces, displacements, and velocities in space are called vectors just as they are in the plane. The same rules of addition, subtraction, and scalar multiplication apply.

The vectors from the origin to the points $(1, 0, 0)$, $(0, 1, 0)$, and $(0, 0, 1)$ are the basic vectors. We denote them by $\mathbf{i}$, $\mathbf{j}$, and $\mathbf{k}$. The position vector $\mathbf{R}$ from the origin O to the typical point $P(x, y, z)$ is

$$\mathbf{R} = x\,\mathbf{i} + y\,\mathbf{j} + z\,\mathbf{k}. \tag{1}$$

The Vector between Two Points

Often we want to express the vector $\overrightarrow{P_1P_2}$ from the point $P_1(x_1, y_1, z_1)$ to the point $P_2(x_2, y_2, z_2)$ in terms of the coordinates of P_1 and P_2. To do so, we first observe that $\overrightarrow{P_1P_2} = \overrightarrow{P_1O} + \overrightarrow{OP_2}$, as in Fig. 13.25. We then write $\overrightarrow{P_1O}$ as $-\overrightarrow{OP_1}$ and express the results in terms of $\mathbf{i}$, $\mathbf{j}$, and $\mathbf{k}$:

$$\overrightarrow{P_1P_2} = \overrightarrow{P_1O} + \overrightarrow{OP_2}$$
$$= \overrightarrow{OP_2} - \overrightarrow{OP_1}$$
$$= x_2\mathbf{i} + y_2\mathbf{j} + z_2\mathbf{k} - x_1\mathbf{i} - y_1\mathbf{j} - z_1\mathbf{k}$$
$$= (x_2 - x_1)\mathbf{i} + (y_2 - y_1)\mathbf{j} + (z_2 - z_1)\mathbf{k}.$$

> The vector from $P_1(x_1, y_1, z_1)$ to $P_2(x_2, y_2, z_2)$ is
> $$\overrightarrow{P_1P_2}. = (x_2 - x_1)\mathbf{i} + (y_2 - y_1)\mathbf{j} + (z_2 - z_1)\mathbf{k}. \tag{2}$$

Length and Direction

As in the plane, the important features of a vector are its length and direction. The length of a vector $a\mathbf{i} + b\mathbf{j} + c\mathbf{k}$ is calculated by applying the Pythagorean theorem twice. In the notation of Fig. 13.26, we have

$$|\overrightarrow{AC}| = |a\mathbf{i} + b\mathbf{j}| = \sqrt{a^2 + b^2},$$

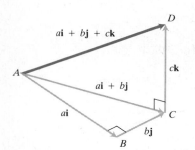

13.26 The length of the vector $\overrightarrow{AD}$ can be determined by applying the Pythagorean theorem to the right triangles ABC and ACD.

from triangle ABC, and then

$$|a\mathbf{i} + b\mathbf{j} + c\mathbf{k}| = |\overrightarrow{AD}| = \sqrt{|\overrightarrow{AC}|^2 + |\overrightarrow{CD}|^2}$$

$$= \sqrt{a^2 + b^2 + c^2},$$

from triangle ACD.

The **length** of the vector $\mathbf{A} = a\mathbf{i} + b\mathbf{j} + c\mathbf{k}$ is

$$|\mathbf{A}| = |a\mathbf{i} + b\mathbf{j} + c\mathbf{k}| = \sqrt{a^2 + b^2 + c^2}. \qquad (3)$$

EXAMPLE 3 The length of $\mathbf{A} = \mathbf{i} - 2\mathbf{j} + 3\mathbf{k}$ is

$$|\mathbf{A}| = \sqrt{(1)^2 + (-2)^2 + (3)^2} = \sqrt{1 + 4 + 9} = \sqrt{14}. \qquad \blacksquare$$

Again as with vectors in the plane, vectors of unit length are called **unit vectors.** The vectors $\mathbf{i}$, $\mathbf{j}$, and $\mathbf{k}$ are unit vectors because

$$|\mathbf{i}| = |1\mathbf{i} + 0\mathbf{j} + 0\mathbf{k}| = \sqrt{1^2 + 0^2 + 0^2} = 1,$$

$$|\mathbf{j}| = |0\mathbf{i} + 1\mathbf{j} + 0\mathbf{k}| = \sqrt{0^2 + 1^2 + 0^2} = 1,$$

$$|\mathbf{k}| = |0\mathbf{i} + 0\mathbf{j} + 1\mathbf{k}| = \sqrt{0^2 + 0^2 + 1^2} = 1.$$

The direction of a nonzero vector $\mathbf{A}$ is the unit vector obtained by dividing $\mathbf{A}$ by its length $|\mathbf{A}|$.

The **direction** of $\mathbf{A}$ is $\dfrac{\mathbf{A}}{|\mathbf{A}|}$. $\qquad (4)$

EXAMPLE 4 In Example 3 we found the length of $\mathbf{A} = \mathbf{i} - 2\mathbf{j} + 3\mathbf{k}$ to be $\sqrt{14}$. The direction of $\mathbf{A}$ is therefore

$$\frac{\mathbf{A}}{|\mathbf{A}|} = \frac{\mathbf{i} - 2\mathbf{j} + 3\mathbf{k}}{\sqrt{14}} = \frac{1}{\sqrt{14}}(\mathbf{i} - 2\mathbf{j} + 3\mathbf{k}). \qquad \blacksquare$$

EXAMPLE 5 Find a unit vector $\mathbf{u}$ in the direction of the vector from $P_1(1, 0, 1)$ to $P_2(3, 2, 0)$.

Solution The vector we want is the direction of $\overrightarrow{P_1P_2}$. To find it, we divide $\overrightarrow{P_1P_2}$ by its own length. This gives

$$\overrightarrow{P_1P_2} = (3 - 1)\mathbf{i} + (2 - 0)\mathbf{j} + (0 - 1)\mathbf{k} = 2\mathbf{i} + 2\mathbf{j} - \mathbf{k},$$

$$|\overrightarrow{P_1P_2}| = \sqrt{(2)^2 + (2)^2 + (-1)^2} = \sqrt{4 + 4 + 1} = \sqrt{9} = 3,$$

$$\mathbf{u} = \frac{\overrightarrow{P_1P_2}}{|\overrightarrow{P_1P_2}|} = \frac{2\mathbf{i} + 2\mathbf{j} - \mathbf{k}}{3} = \frac{2}{3}\mathbf{i} + \frac{2}{3}\mathbf{j} - \frac{1}{3}\mathbf{k}. \qquad \blacksquare$$

If we multiply a vector $\mathbf{A} = a_1\mathbf{i} + a_2\mathbf{j} + a_3\mathbf{k}$ by a scalar c, the length of $c\mathbf{A}$ is $|c|$ times the length of $\mathbf{A}$. The reason is that

$$c\mathbf{A} = ca_1\mathbf{i} + ca_2\mathbf{j} + ca_3\mathbf{k},$$

so

$$|c\mathbf{A}| = \sqrt{(ca_1)^2 + (ca_2)^2 + (ca_3)^2}$$
$$= \sqrt{c^2 a_1^2 + c^2 a_2^2 + c^2 a_3^2}$$
$$= |c|\sqrt{a_1^2 + a_2^2 + a_3^2}$$
$$= |c|\|\mathbf{A}\|.$$

EXAMPLE 6   Find a vector 6 units long in the direction of $\mathbf{A} = 2\mathbf{i} + 2\mathbf{j} - \mathbf{k}$.

Solution   The vector we want is

$$6\frac{\mathbf{A}}{|\mathbf{A}|} = 6\frac{2\mathbf{i} + 2\mathbf{j} - \mathbf{k}}{\sqrt{2^2 + 2^2 + (-1)^2}} = 6\frac{2\mathbf{i} + 2\mathbf{j} - \mathbf{k}}{3} = 4\mathbf{i} + 4\mathbf{j} - 2\mathbf{k}. \quad \blacksquare$$

Distance in Space

To find the distance between two points P_1 and P_2 in space, we find the length of $\overrightarrow{P_1P_2}$. Equation (5) is the resulting formula:

> **The Distance between $P_1(x_1, y_1, z_1)$ and $P_2(x_2, y_2, z_2)$**
> $$|\overrightarrow{P_1P_2}| = \sqrt{(x_2 - x_1)^2 + (y_2 - y_1)^2 + (z_2 - z_1)^2} \qquad (5)$$

EXAMPLE 7   The distance between

$$P_1(2, 1, 5) \qquad \text{and} \qquad P_2(-2, 3, 0)$$

is

$$|\overrightarrow{P_1P_2}| = \sqrt{(-2 - 2)^2 + (3 - 1)^2 + (0 - 5)^2} = \sqrt{16 + 4 + 25}$$
$$= \sqrt{45} = 3\sqrt{5}. \quad \blacksquare$$

Spheres

We can use Eq. (5) to write equations for spheres. Since a point $P(x, y, z)$ lies on the sphere of radius a centered at $P_0(x_0, y_0, z_0)$ if and only if it lies a units from P_0, it lies on the sphere if and only if

$$|\overrightarrow{P_0P}| = a$$

or

$$(x - x_0)^2 + (y - y_0)^2 + (z - z_0)^2 = a^2.$$

This equation is the *standard equation* for the sphere. See Fig. 13.27.

> **The Standard Equation for the Sphere of Radius a and Center (x_0, y_0, z_0)**
> $$(x - x_0)^2 + (y - y_0)^2 + (z - z_0)^2 = a^2 \qquad (6)$$

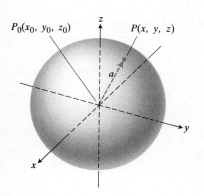

13.27   The standard equation of the sphere of radius a centered at the point (x_0, y_0, z_0) is

$$(x - x_0)^2 + (y - y_0)^2 + (z - z_0)^2 = a^2.$$

EXAMPLE 8   Find the center and radius of the sphere

$$x^2 + y^2 + z^2 + 2x - 4y = 0.$$

Solution   Complete the squares in the given equation to obtain

$$x^2 + 2x + 1 + y^2 - 4y + 4 + z^2 = 0 + 1 + 4$$

$$(x + 1)^2 + (y - 2)^2 + z^2 = 5.$$

This is Eq. (6) with $x_0 = -1$, $y_0 = 2$, $z_0 = 0$, and $a = \sqrt{5}$. The center is $(-1, 2, 0)$ and the radius is $\sqrt{5}$.   ■

EXAMPLE 9   *Sets bounded by spheres or portions of spheres.*

Defining equations and inequalities	Description
a) $x^2 + y^2 + z^2 < 4$	The interior of the sphere $x^2 + y^2 + z^2 = 4$.
b) $x^2 + y^2 + z^2 \le 4$	The solid ball bounded by the sphere $x^2 + y^2 + z^2 = 4$. Alternatively, the sphere $x^2 + y^2 + z^2 = 4$ together with its interior.
c) $x^2 + y^2 + z^2 > 4$	The exterior of the sphere $x^2 + y^2 + z^2 = 4$.
d) $x^2 + y^2 + z^2 = 4$, $z \le 0$	The lower hemisphere cut from the sphere $x^2 + y^2 + z^2 = 4$ by the xy-plane (the plane $z = 0$). ■

Midpoints of Line Segments

The coordinates of the midpoint M of the line segment joining two points $P_1(x_1, y_1, z_1)$ and $P_2(x_2, y_2, z_2)$ are found by averaging the coordinates of P_1 and P_2. That is,

$$M = \left(\frac{x_1 + x_2}{2}, \frac{y_1 + y_2}{2}, \frac{z_1 + z_2}{2} \right). \tag{7}$$

To see why, we have only to observe that these coordinates are the scalar components of the position vector $\overrightarrow{OM}$ (see Fig. 13.28) and that

$$\overrightarrow{OM} = \overrightarrow{OP_1} + \frac{1}{2}(\overrightarrow{P_1P_2}) = \overrightarrow{OP_1} + \frac{1}{2}(\overrightarrow{OP_2} - \overrightarrow{OP_1}) = \frac{1}{2}(\overrightarrow{OP_1} + \overrightarrow{OP_2})$$

$$= \frac{x_1 + x_2}{2}\mathbf{i} + \frac{y_1 + y_2}{2}\mathbf{j} + \frac{z_1 + z_2}{2}\mathbf{k}.$$

EXAMPLE 10   The midpoint of the segment joining $P_1(3, -2, 0)$ and $P_2(7, 4, 4)$ is

$$\left(\frac{3 + 7}{2}, \frac{-2 + 4}{2}, \frac{0 + 4}{2} \right) = (5, 1, 2).$$ ■

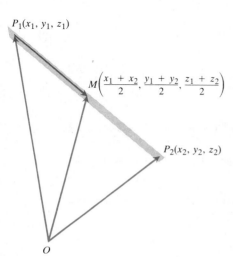

13.28   The coordinates of the point half-way between P_1 and P_2 are found by averaging the coordinates of P_1 and P_2.

Tips for Drawing

Here are some tips for making pictures of three-dimensional objects look three-dimensional. Sketching with a pencil is safer than sketching with a pen because you can erase, but the advice below applies no matter what medium you choose.

HOW TO DRAW THREE-DIMENSIONAL OBJECTS TO LOOK THREE-DIMENSIONAL

1. Break lines. When one line passes behind another, break it to show that it doesn't touch and that part of it is hidden.

Crossing

CD behind AB

AB behind CD

2. Make the angle between the positive x-axis and the positive y-axis large enough.

This

Not this

3. Draw planes parallel to the coordinate planes as if they were rectangles with sides parallel to the coordinate axes.

4. Dot hidden portions of lines. Don't let the line touch the boundary of the parallelogram that represents the plane, unless the line lies in the plane.

Line below plane

Line above plane

Line *in* plane

5. Spheres: Draw the sphere first (outline and equator); draw axes, if any, later. Use line breaks and dotted lines.

Hidden part dotted

Sphere first

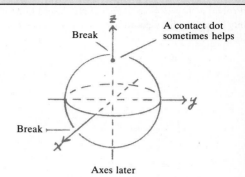

Break

A contact dot sometimes helps

Break

Axes later

6. A general rule for perspective: Draw the object as if it lies some distance away, below, and to the left.

Advice ignored

Advice followed

7. To draw a plane that intersects all three coordinate axes, follow the steps shown here: (a) Sketch the axes and mark the intercepts. (b) Connect the intercepts to form a triangle. (c) Expand the triangle into a parallelogram. Then make the parallelogram larger by drawing lines parallel to its sides. (d) Finally, darken the exposed parts, break hidden lines, and, if desired, dash hidden portions of the axes.

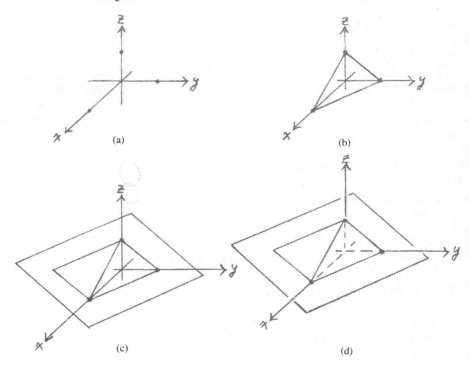

(a)

(b)

(c)

(d)

PROBLEMS

In Problems 1–12, give a geometric description of the set of points in space whose coordinates satisfy the given pairs of equations.

1. $x = 2, \quad y = 3$

2. $x = -1, \quad z = 0$

3. $y = 0, \quad z = 0$

4. $x = 1, \quad y = 0$

5. $x^2 + y^2 = 4, \quad z = 0$

6. $x^2 + y^2 = 4, \quad z = -2$

7. $x^2 + z^2 = 4, \quad y = 0$

8. $y^2 + z^2 = 1, \quad x = 0$

9. $x^2 + y^2 + z^2 = 1, \quad x = 0$

10. $x^2 + y^2 + z^2 = 25, \quad y = -4$

11. $x^2 + y^2 + (z + 3)^2 = 25, \quad z = 0$

12. $x^2 + (y - 1)^2 + z^2 = 4, \quad y = 0$

In Problems 13–18, describe the sets of points in space whose coordinates satisfy the given inequalities or combinations of equations and inequalities.

13. a) $x \geq 0, \quad y \geq 0, \quad z = 0$
   b) $x \geq 0, \quad y \leq 0, \quad z = 0$

14. a) $0 \leq x \leq 1$
   b) $0 \leq x \leq 1, \quad 0 \leq y \leq 1$
   c) $0 \leq x \leq 1, \quad 0 \leq y \leq 1, \quad 0 \leq z \leq 1$

15. a) $x^2 + y^2 + z^2 \leq 1$
   b) $x^2 + y^2 + z^2 > 1$

16. a) $x^2 + y^2 \leq 1, \quad z = 0$
   b) $x^2 + y^2 \leq 1, \quad z = 3$
   c) $x^2 + y^2 \leq 1, \quad$ no restriction on z

17. a) $x^2 + y^2 + z^2 = 1, \quad z \geq 0$
   b) $x^2 + y^2 + z^2 \leq 1, \quad z \geq 0$

18. a) $x = y, \quad z = 0$
   b) $x = y, \quad$ no restriction on z

In Problems 19–28, describe the given set with a single equation or with a pair of equations.

19. The plane perpendicular to the
   a) x-axis at $(3, 0, 0)$,
   b) y-axis at $(0, -1, 0)$,
   c) z-axis at $(0, 0, -2)$

20. The plane through the point $(3, -1, 2)$ perpendicular to the
   a) x-axis,   b) y-axis,   c) z-axis

21. The plane through the point $(3, -1, 1)$ parallel to the
   a) xy-plane,   b) yz-plane,   c) xz-plane

22. The circle of radius 2 centered at $(0, 0, 0)$ and lying in the
   a) xy-plane,   b) yz-plane,   c) xz-plane

23. The circle of radius 2 centered at $(0, 2, 0)$ and lying in the
   a) xy-plane,   b) yz-plane,   c) plane $y = 2$

24. The circle of radius 1 centered at $(-3, 4, 1)$ and lying in a plane parallel to the
   a) xy-plane,   b) yz-plane,   c) xz-plane

25. The line through the point $(1, 3, -1)$ parallel to the
   a) x-axis,   b) y-axis,   c) z-axis

26. The set of points in space equidistant from the origin and the point $(0, 2, 0)$

27. The circle in which the plane through the point $(1, 1, 3)$ perpendicular to the z-axis meets the sphere of radius 5 centered at the origin

28. The set of points in space that lie 2 units from the point $(0, 0, 1)$ and, at the same time, 2 units from the point $(0, 0, -1)$

Write inequalities to describe the sets in Problems 29–34.

29. The slab bounded by the planes $z = 0$ and $z = 1$ (planes included)

30. The solid cube in the first octant bounded by the planes $x = 2$, $y = 2$, and $z = 2$

31. The half-space consisting of the points on and below the xy-plane

32. The upper hemisphere of the sphere of radius 1 centered at the origin

33. The (a) interior and (b) exterior of the sphere of radius 1 centered at the point $(1, 1, 1)$

34. The closed region bounded by the spheres of radius 1 and radius 2 centered at the origin. (*Closed* means the spheres are to be included. Had we wanted the spheres left out, we would have asked for the *open* region bounded by the spheres. This is analogous to the way we use "closed" and "open" to describe intervals: "closed" means endpoints included, "open" means endpoints left out. Closed sets include boundaries; open sets leave them out.)

Find the lengths and directions of the vectors in Problems 35–46.

35. $2\mathbf{i} + \mathbf{j} - 2\mathbf{k}$

36. $3\mathbf{i} - 6\mathbf{j} + 2\mathbf{k}$

37. $\mathbf{i} + 4\mathbf{j} - 8\mathbf{k}$

38. $9\mathbf{i} - 2\mathbf{j} + 6\mathbf{k}$

39. $5\mathbf{k}$

40. $-\dfrac{1}{3}\mathbf{j} + \dfrac{1}{4}\mathbf{k}$

41. $\dfrac{1}{\sqrt{2}}\mathbf{i} - \dfrac{1}{\sqrt{2}}\mathbf{k}$

42. $\dfrac{1}{\sqrt{6}}\mathbf{i} - \dfrac{1}{\sqrt{6}}\mathbf{j} - \dfrac{1}{\sqrt{6}}\mathbf{k}$

43. $\dfrac{\mathbf{i}}{\sqrt{3}} + \dfrac{\mathbf{j}}{\sqrt{3}} + \dfrac{\mathbf{k}}{\sqrt{3}}$

44. $\dfrac{3}{5}\mathbf{i} + \dfrac{4}{5}\mathbf{k}$

45. $6\mathbf{i}$

46. $-4\mathbf{j}$

47. Find the vectors whose lengths and directions are given. Try to answer without writing anything down.

Length	Direction
a) 2	$\mathbf{i}$
b) -4	$\mathbf{j}$
c) $\sqrt{3}$	$\mathbf{k}$
d) $\frac{1}{2}$	$\frac{3}{5}\mathbf{j} + \frac{4}{5}\mathbf{k}$
e) 7	$\frac{6}{7}\mathbf{i} - \frac{2}{7}\mathbf{j} + \frac{3}{7}\mathbf{k}$
f) a	$u_1\mathbf{i} + u_2\mathbf{j} + u_3\mathbf{k}$

48. Find a unit vector in the direction of $\mathbf{A} = 12\mathbf{i} - 5\mathbf{k}$.

49. Find a vector 5 units long in the direction of $\mathbf{A} = \mathbf{i} + \mathbf{j} + \mathbf{k}$.

50. Find a vector $\frac{1}{2}$ unit long in the direction of $\mathbf{A} = \mathbf{i} - \mathbf{j}$.

Spheres and Distance

Find the centers and radii of the spheres in Problems 51–54.

51. $x^2 + y^2 + z^2 + 4x - 4z = 0$

52. $2x^2 + 2y^2 + 2z^2 + x + y + z = 9$

53. $x^2 + y^2 + z^2 - 2az = 0$

54. $3x^2 + 3y^2 + 3z^2 + 2y - 2z = 9$

55. Find a formula for the distance between the point $P(x, y, z)$ and the (a) x-axis, (b) y-axis, (c) z-axis.

56. Find a formula for the distance between the point $P(x, y, z)$ and the (a) xy-plane, (b) yz-plane, (c) xz-plane.

Geometry with Vectors

57. Suppose that the points
$$A(4, 2, 0), \quad B(1, 3, 0), \quad C(1, 1, 3)$$
are the corners of a thin homogeneous triangular plate in space (Fig. 13.29).
a) Find the vector from C to the midpoint M of side AB.
b) Find the vector from C to the point that lies two thirds of the way from C to M on the median CM.
c) Find the coordinates of the point in which the medians of $\triangle ABC$ intersect. According to Example 6, Article 5.8, this point is the plate's center of mass (c. m.).

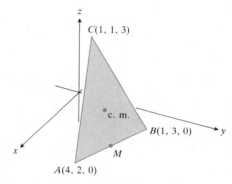

13.29 The triangular plate in Problem 57.

58. Let $ABCD$ be a general (not necessarily planar) quadrilateral in space. Show that the two segments joining the midpoints of opposite sides of $ABCD$ bisect each other. (*Hint:* Show that the segments have the same midpoint.)

13.4
Dot Products

We now introduce the dot product of two vectors, the first of the two methods we shall learn for multiplying vectors together. Our motivation is the need to calculate the work done by a constant force in displacing a mass. When the force and displacement are represented as vectors, the dot product of the two vectors gives the work done by the force during the displacement.

Dot products, also called *scalar products* because the resulting products are numbers and not vectors, have applications in mathematics as well as in engineering and physics. In this article we present the algebraic and geometric properties on which many of these applications depend. The second kind of vector product, the cross product, will be described in the next article.

13.30 The work done by a constant force **F** during a displacement $\overrightarrow{PQ}$ is $(|\mathbf{F}| \cos \theta)|\overrightarrow{PQ}|$.

The Definition

In mechanics, the work done by a constant force **F** when the point of application of **F** undergoes a displacement $\overrightarrow{PQ}$ is defined by the equation

$$\text{Work} = (|\mathbf{F}| \cos \theta)|\overrightarrow{PQ}| = |\mathbf{F}||\overrightarrow{PQ}| \cos \theta. \tag{1}$$

The angle θ is the angle between the direction of **F** and the direction of $\overrightarrow{PQ}$. The work is the product of the magnitude of the displacement with the scalar component of the force in the direction of the displacement. See Fig. 13.30. For instance, if $|\mathbf{F}| = 40$ newtons (about 9 pounds), $|\overrightarrow{PQ}| = 3$ m, and $\theta = 60°$ so that $\cos \theta = 1/2$, then the work done by **F** as its point of application moves from P to Q is

$$\text{Work} = (40)(3)\left(\frac{1}{2}\right) = 60 \text{ newton-meters.}$$

In mathematics, the quantity $|\mathbf{F}||\overrightarrow{PQ}| \cos \theta$ is called the *scalar product* or *dot product* of **F** and $\overrightarrow{PQ}$.

DEFINITION

> **Scalar or Dot Product**
>
> The **scalar product** $\mathbf{A} \cdot \mathbf{B}$ ("**A** dot **B**") or **dot product** of two vectors **A** and **B** is the number
>
> $$\mathbf{A} \cdot \mathbf{B} = |\mathbf{A}||\mathbf{B}| \cos \theta, \tag{2}$$
>
> where θ measures the smaller angle made by **A** and **B** when their initial points coincide (as in Fig. 13.31).

In words, the scalar product of **A** and **B** is the length of **A** times the length of **B** times the cosine of the angle between **A** and **B**. The product is a scalar, not a vector. It is often called the dot product because of the dot in the notation $\mathbf{A} \cdot \mathbf{B}$.

From Eq. (2) we see that the scalar product of two vectors is positive when the angle between them is acute and negative when the angle is obtuse.

Since the angle a vector **A** makes with itself is zero, and the cosine of the angle is 1,

$$\mathbf{A} \cdot \mathbf{A} = |\mathbf{A}||\mathbf{A}| \cos 0 = |\mathbf{A}||\mathbf{A}|(1) = |\mathbf{A}|^2 \quad \text{or} \quad |\mathbf{A}| = \sqrt{\mathbf{A} \cdot \mathbf{A}}. \tag{3}$$

This gives us a handy way to calculate a vector's length, as we shall see in a moment.

13.31 $\mathbf{A} \cdot \mathbf{B}$ is the number $|\mathbf{A}||\mathbf{B}| \cos \theta$.

Calculation

To calculate $\mathbf{A} \cdot \mathbf{B}$ from the components of **A** and **B**, we let

$$\mathbf{A} = a_1\mathbf{i} + a_2\mathbf{j} + a_3\mathbf{k},$$
$$\mathbf{B} = b_1\mathbf{i} + b_2\mathbf{j} + b_3\mathbf{k},$$

and

$$\mathbf{C} = \mathbf{B} - \mathbf{A} = (b_1 - a_1)\mathbf{i} + (b_2 - a_2)\mathbf{j} + (b_3 - a_3)\mathbf{k}.$$

Then we apply the law of cosines to a triangle whose sides represent the vectors

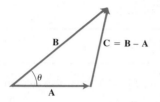

13.32 Equation (5) is obtained by applying the law of cosines to a triangle whose sides represent **A**, **B**, and **C** = **B** − **A**.

A, **B**, and **C** (Fig. 13.32) and obtain

$$|\mathbf{C}|^2 = |\mathbf{A}|^2 + |\mathbf{B}|^2 - 2|\mathbf{A}||\mathbf{B}| \cos \theta,$$

$$|\mathbf{A}||\mathbf{B}| \cos \theta = \frac{|\mathbf{A}|^2 + |\mathbf{B}|^2 - |\mathbf{C}|^2}{2}. \tag{4}$$

The left side of this equation is **A** · **B**, and we may evaluate the right side of (4) by applying Eq. (3) of Article 13.3 to find the lengths of **A**, **B**, and **C**. The result of this algebra is the formula

$$\mathbf{A} \cdot \mathbf{B} = a_1 b_1 + a_2 b_2 + a_3 b_3. \tag{5}$$

Thus, to find the scalar product of two given vectors we multiply their corresponding **i**, **j**, and **k** components and add the results. In particular, from Eq. (3) we have

$$|\mathbf{A}| = \sqrt{\mathbf{A} \cdot \mathbf{A}} = \sqrt{a_1^2 + a_2^2 + a_3^2}. \tag{6}$$

EXAMPLE 1 Find the angle θ between **A** = **i** − 2**j** − 2**k** and **B** = 6**i** + 3**j** + 2**k**.

Solution

$$\mathbf{A} \cdot \mathbf{B} = (1)(6) + (-2)(3) + (-2)(2) = 6 - 6 - 4 = -4$$

from Eq. (5), while

$$\mathbf{A} \cdot \mathbf{B} = |\mathbf{A}||\mathbf{B}| \cos \theta$$

from Eq. (2). Since $|\mathbf{A}| = \sqrt{1 + 4 + 4} = 3$ and $|\mathbf{B}| = \sqrt{36 + 9 + 4} = 7$, we have

$$\cos \theta = \frac{\mathbf{A} \cdot \mathbf{B}}{|\mathbf{A}||\mathbf{B}|} = \frac{-4}{21}, \quad \text{or} \quad \theta = \cos^{-1} \frac{-4}{21} \approx 101°. \quad \blacksquare$$

Laws of Multiplication

From the equation $\mathbf{A} \cdot \mathbf{B} = a_1 b_1 + a_2 b_2 + a_3 b_3$, we can see right away that

$$\mathbf{A} \cdot \mathbf{B} = \mathbf{B} \cdot \mathbf{A}. \tag{7}$$

In other words, scalar multiplication is commutative. We can also see that if c is any number, then

$$(c\mathbf{A}) \cdot \mathbf{B} = \mathbf{A} \cdot (c\mathbf{B}) = c(\mathbf{A} \cdot \mathbf{B}) \tag{8}$$

(see Problem 32). If

$$\mathbf{C} = c_1 \mathbf{i} + c_2 \mathbf{j} + c_3 \mathbf{k}$$

is any third vector, then

$$\mathbf{A} \cdot (\mathbf{B} + \mathbf{C}) = a_1(b_1 + c_1) + a_2(b_2 + c_2) + a_3(b_3 + c_3)$$
$$= (a_1 b_1 + a_2 b_2 + a_3 b_3) + (a_1 c_1 + a_2 c_2 + a_3 c_3)$$
$$= \mathbf{A} \cdot \mathbf{B} + \mathbf{A} \cdot \mathbf{C}.$$

Hence scalar multiplication obeys the distributive law:

$$\mathbf{A} \cdot (\mathbf{B} + \mathbf{C}) = \mathbf{A} \cdot \mathbf{B} + \mathbf{A} \cdot \mathbf{C}. \qquad (9)$$

If we combine this with the commutative law, Eq. (7), it is also evident that

$$(\mathbf{A} + \mathbf{B}) \cdot \mathbf{C} = \mathbf{A} \cdot \mathbf{C} + \mathbf{B} \cdot \mathbf{C}. \qquad (10)$$

Equations (9) and (10) together permit us to multiply sums of vectors by the familiar laws of algebra. For example,

$$(\mathbf{A} + \mathbf{B}) \cdot (\mathbf{C} + \mathbf{D}) = \mathbf{A} \cdot \mathbf{C} + \mathbf{A} \cdot \mathbf{D} + \mathbf{B} \cdot \mathbf{C} + \mathbf{B} \cdot \mathbf{D}. \qquad (11)$$

Orthogonal Vectors

Two vectors whose scalar product is zero are said to be **orthogonal.** From the equation $\mathbf{A} \cdot \mathbf{B} = |\mathbf{A}||\mathbf{B}| \cos \theta$, we can see that when neither $|\mathbf{A}|$ nor $|\mathbf{B}|$ is zero, $\mathbf{A} \cdot \mathbf{B}$ is zero if and only if $\cos \theta$ is zero, that is, when θ equals $90°$. Hence, for the vectors we are dealing with, "orthogonal" means the same as "perpendicular." (In other scientific contexts in which orthogonal is used, there may be no such geometric interpretation.)

The zero vector $\mathbf{0} = 0\mathbf{i} + 0\mathbf{j} + 0\mathbf{k}$ is orthogonal to every vector because its dot product with every vector is zero.

13.33 Pulling on the box has the effect of moving the box in the direction of **A.** The force in this direction is represented by the vector projection of **B** onto **A.**

EXAMPLE 2 The vectors $\mathbf{A} = 3\mathbf{i} - 2\mathbf{j} + \mathbf{k}$ and $\mathbf{B} = 2\mathbf{j} + 4\mathbf{k}$ are orthogonal because

$$\mathbf{A} \cdot \mathbf{B} = (3)(0) - (2)(2) + (1)(4) = 0. \qquad \blacksquare$$

Vector Projections and Scalar Components

The vector we get by projecting a vector **B** onto the line through a vector **A** is called the **vector projection of B onto A,** sometimes denoted

$$\text{proj}_\mathbf{A}\mathbf{B} \qquad \text{("the vector projection of B onto A").}$$

If **B** represents a force, then the vector projection of **B** onto **A** represents the effective force in the direction of **A.** See Fig. 13.33.

If the angle between **B** and **A** is acute, the length of the vector projection of **B** onto **A** is $|\mathbf{B}| \cos \theta$. If the angle is obtuse, its cosine is negative and the length of the vector projection of **B** onto **A** is $-|\mathbf{B}| \cos \theta$. In either case, the number $|\mathbf{B}| \cos \theta$ is called the **scalar component of B in the direction of A.** See Fig. 13.34.

The scalar component of **B** in the direction of **A** can be found by dividing both sides of the equation $\mathbf{A} \cdot \mathbf{B} = |\mathbf{A}||\mathbf{B}| \cos \theta$ by $|\mathbf{A}|$. This gives

(a)

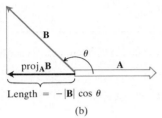

(b)

13.34 (a) When θ is acute, the length of the vector projection of **B** onto **A** is $|\mathbf{B}| \cos \theta$, the scalar component of **B** in the direction of **A.** (b) When θ is obtuse, the scalar component of **B** in the direction of **A** is negative and the length of the vector projection of **B** onto **A** is $-|\mathbf{B}| \cos \theta$.

$$|\mathbf{B}| \cos \theta = \frac{\mathbf{A} \cdot \mathbf{B}}{|\mathbf{A}|} = \mathbf{B} \cdot \frac{\mathbf{A}}{|\mathbf{A}|}. \qquad (12)$$

Equation (12) says that the scalar component of **B** in the direction of **A** can be obtained by "dotting" **B** with the direction of **A.**

The vector projection of **B** onto **A** is the scalar component of **B** in the direction of **A** times the direction of **A.** If the angle between **A** and **B** is acute, the vector projection has length $|\mathbf{B}| \cos \theta$ and direction $\mathbf{A}/|\mathbf{A}|$. If the angle is obtuse, the vector projection has length $-|\mathbf{B}|\cos\theta$ and direction $-\mathbf{A}/|\mathbf{A}|$. In either case,

$$\text{proj}_{\mathbf{A}}\mathbf{B} = (|\mathbf{B}|\cos\theta)\frac{\mathbf{A}}{|\mathbf{A}|}. \tag{13}$$

Equations (12) and (13) together give a useful way to find $\text{proj}_{\mathbf{A}}\mathbf{B}$:

$$\text{proj}_{\mathbf{A}}\mathbf{B} = (|\mathbf{B}|\cos\theta)\frac{\mathbf{A}}{|\mathbf{A}|} \qquad \text{(Eq. 13)}$$

$$= \left(\frac{\mathbf{A}\cdot\mathbf{B}}{|\mathbf{A}|}\right)\frac{\mathbf{A}}{|\mathbf{A}|} \qquad \text{(Eq. 12)}$$

$$= \frac{\mathbf{A}\cdot\mathbf{B}}{\mathbf{A}\cdot\mathbf{A}}\mathbf{A}. \qquad (|\mathbf{A}|^2 = \mathbf{A}\cdot\mathbf{A}) \tag{14}$$

EXAMPLE 3 Find the vector projection of $\mathbf{B} = 6\mathbf{i} + 3\mathbf{j} + 2\mathbf{k}$ onto $\mathbf{A} = \mathbf{i} - 2\mathbf{j} - 2\mathbf{k}$ and the scalar component of $\mathbf{B}$ in the direction of $\mathbf{A}$.

Solution We find $\text{proj}_{\mathbf{A}}\mathbf{B}$ from Eq. (14):

$$\text{proj}_{\mathbf{A}}\mathbf{B} = \frac{\mathbf{A}\cdot\mathbf{B}}{\mathbf{A}\cdot\mathbf{A}}\mathbf{A}$$

$$= \frac{6 - 6 - 4}{1 + 4 + 4}(\mathbf{i} - 2\mathbf{j} - 2\mathbf{k})$$

$$= -\frac{4}{9}(\mathbf{i} - 2\mathbf{j} - 2\mathbf{k})$$

$$= -\frac{4}{9}\mathbf{i} + \frac{8}{9}\mathbf{j} + \frac{8}{9}\mathbf{k}.$$

We find the scalar component of $\mathbf{B}$ in the direction of $\mathbf{A}$ from Eq. (12):

$$|\mathbf{B}|\cos\theta = \mathbf{B}\cdot\frac{\mathbf{A}}{|\mathbf{A}|}$$

$$= (6\mathbf{i} + 3\mathbf{j} + 2\mathbf{k})\cdot\left(\frac{1}{3}\mathbf{i} - \frac{2}{3}\mathbf{j} - \frac{2}{3}\mathbf{k}\right)$$

$$= 2 - 2 - \frac{4}{3} = -\frac{4}{3}.$$

In mechanics, we often want to express a vector $\mathbf{B}$ as the sum of a vector $\mathbf{B}_1$ parallel to a particular vector $\mathbf{A}$ and a vector $\mathbf{B}_2$ orthogonal to $\mathbf{A}$. (See Fig. 13.35.) This can be done in a straightforward way by writing $\mathbf{B}$ as the sum of its vector projection onto $\mathbf{A}$ plus whatever is left over, because the leftover part will automatically be orthogonal to $\mathbf{A}$. To see why this is so, write $\mathbf{B}$ as the sum

$$\mathbf{B} = \underbrace{\frac{\mathbf{A}\cdot\mathbf{B}}{\mathbf{A}\cdot\mathbf{A}}\mathbf{A}}_{\mathbf{B}_1} + \underbrace{\left(\mathbf{B} - \frac{\mathbf{A}\cdot\mathbf{B}}{\mathbf{A}\cdot\mathbf{A}}\mathbf{A}\right)}_{\mathbf{B}_2}. \tag{15}$$

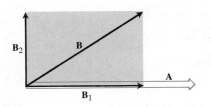

13.35 The vector $\mathbf{B}$ as the sum of vectors parallel and orthogonal to $\mathbf{A}$.

Then $\mathbf{B}_1$, the vector projection of $\mathbf{B}$ onto $\mathbf{A}$, is parallel to $\mathbf{A}$, while $\mathbf{B}_2$ can be seen to be orthogonal to $\mathbf{A}$ because $\mathbf{A} \cdot \mathbf{B}_2$ is zero:

$$\mathbf{A} \cdot \mathbf{B}_2 = \mathbf{A} \cdot \left(\mathbf{B} - \frac{\mathbf{A} \cdot \mathbf{B}}{\mathbf{A} \cdot \mathbf{A}}\mathbf{A}\right)$$

$$= \mathbf{A} \cdot \mathbf{B} - \frac{\mathbf{A} \cdot \mathbf{B}}{\mathbf{A} \cdot \mathbf{A}}\mathbf{A} \cdot \mathbf{A} \tag{16}$$

$$= \mathbf{A} \cdot \mathbf{B} - \mathbf{A} \cdot \mathbf{B}$$

$$= 0.$$

EXAMPLE 4   Express $\mathbf{B} = 2\mathbf{i} + \mathbf{j} - 3\mathbf{k}$ as the sum of a vector parallel to $\mathbf{A} = 3\mathbf{i} - \mathbf{j}$ and a vector orthogonal to $\mathbf{A}$.

Solution   We use Eq. (15). With

$$\mathbf{A} \cdot \mathbf{B} = 6 - 1 = 5 \quad \text{and} \quad \mathbf{A} \cdot \mathbf{A} = 9 + 1 = 10,$$

Eq. (15) gives

$$\mathbf{B} = \frac{\mathbf{A} \cdot \mathbf{B}}{\mathbf{A} \cdot \mathbf{A}}\mathbf{A} + \left(\mathbf{B} - \frac{\mathbf{A} \cdot \mathbf{B}}{\mathbf{A} \cdot \mathbf{A}}\mathbf{A}\right)$$

$$= \frac{5}{10}(3\mathbf{i} - \mathbf{j}) + \left(2\mathbf{i} + \mathbf{j} - 3\mathbf{k} - \frac{5}{10}(3\mathbf{i} - \mathbf{j})\right)$$

$$= \left(\frac{3}{2}\mathbf{i} - \frac{1}{2}\mathbf{j}\right) + \left(\frac{1}{2}\mathbf{i} + \frac{3}{2}\mathbf{j} - 3\mathbf{k}\right).$$

Check: The first vector in the sum is parallel to $\mathbf{A}$ because it is $(1/2)\mathbf{A}$. The second vector in the sum is orthogonal to $\mathbf{A}$ because

$$\left(\frac{1}{2}\mathbf{i} + \frac{3}{2}\mathbf{j} - 3\mathbf{k}\right) \cdot (3\mathbf{i} - \mathbf{j}) = \frac{3}{2} - \frac{3}{2} = 0. \qquad \blacksquare$$

Lines in the Plane and Distances from Points to Lines

Dot products give a new understanding of the equations we write for lines in the plane and a quick way to calculate distances from points to lines.

Suppose L is a line through a point $P_0(x_0, y_0)$ perpendicular to a vector $\mathbf{N} = a\mathbf{i} + b\mathbf{j}$ (Fig. 13.36). Then a point $P(x, y)$ lies on L if and only if

$$\mathbf{N} \cdot \overrightarrow{P_0P} = 0 \tag{17}$$

or

$$a(x - x_0) + b(y - y_0) = 0.$$

The second equation may be rearranged as

$$ax + by = ax_0 + by_0$$

or, more simply,

$$ax + by = c, \quad \text{where} \quad c = ax_0 + by_0. \tag{18}$$

Notice that the coefficients of x and y in Eq. (18) are the x- and y-components of a vector perpendicular to the line.

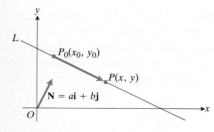

13.36   A point $P(x, y)$ lies on the line through P_0 perpendicular to $\mathbf{N}$ if and only if $\mathbf{N} \cdot \overrightarrow{P_0P} = 0$.

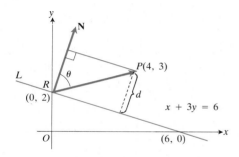

13.37 The distance d from P to line L is the length of the vector projection of $\overrightarrow{RP}$ onto **N**.

EXAMPLE 5 Find an equation for the line through $P_0(3, 5)$ perpendicular to **N** = **i** + 2**j**.

Solution We use Eq. (18) to get

$$c = 1(3) + 2(5) = 13 \quad \text{and} \quad x + 2y = 13.$$ ∎

EXAMPLE 6 Find the distance from the point $P(4, 3)$ to the line $x + 3y = 6$.

Solution We find a point R on the line (Fig. 13.37) and calculate the distance as the length of the vector projection of $\overrightarrow{RP}$ onto a vector **N** normal to the line. Any point on the line will do for R and we can find **N** directly from the equation for the line: **N** = **i** + 3**j**. With $R = (0, 2)$, say, we then have

$$\overrightarrow{RP} = (4 - 0)\mathbf{i} + (3 - 2)\mathbf{j} = 4\mathbf{i} + \mathbf{j}$$

and

$$\text{Distance} = \left| \text{proj}_{\mathbf{N}} \overrightarrow{RP} \right|$$

$$= \left| \overrightarrow{RP} \cdot \frac{\mathbf{N}}{|\mathbf{N}|} \right|$$

$$= \left| (4\mathbf{i} + \mathbf{j}) \cdot \frac{\mathbf{i} + 3\mathbf{j}}{\sqrt{10}} \right|$$

$$= \frac{4 + 3}{\sqrt{10}}$$

$$= \frac{7\sqrt{10}}{10}.$$ ∎

PROBLEMS

In Problems 1–12, find $\mathbf{A} \cdot \mathbf{B}$, $|\mathbf{A}|$, $|\mathbf{B}|$, the cosine of the angle between **A** and **B**, the scalar component of **B** in the direction of **A**, and the vector projection of **B** onto **A**.

1. $\mathbf{A} = 3\mathbf{i} + 2\mathbf{j}, \quad \mathbf{B} = 5\mathbf{j} + \mathbf{k}$

2. $\mathbf{A} = \mathbf{i}, \quad \mathbf{B} = 5\mathbf{j} - 3\mathbf{k}$

3. $\mathbf{A} = 3\mathbf{i} - 2\mathbf{j} - \mathbf{k}, \quad \mathbf{B} = -2\mathbf{j}$

4. $\mathbf{A} = -2\mathbf{i} + 7\mathbf{j}, \quad \mathbf{B} = \mathbf{k}$

5. $\mathbf{A} = 5\mathbf{j} - 3\mathbf{k}, \quad \mathbf{B} = \mathbf{i} + \mathbf{j} + \mathbf{k}$

6. $\mathbf{A} = \dfrac{1}{\sqrt{2}}\mathbf{i} + \dfrac{1}{\sqrt{3}}\mathbf{j} + \dfrac{1}{\sqrt{6}}\mathbf{k}, \quad \mathbf{B} = \dfrac{1}{\sqrt{2}}\mathbf{j} - \mathbf{k}$

7. $\mathbf{A} = -\mathbf{i} + \mathbf{j}, \quad \mathbf{B} = \sqrt{2}\mathbf{i} + \sqrt{3}\mathbf{j} + 2\mathbf{k}$

8. $\mathbf{A} = \mathbf{i} + \mathbf{k}, \quad \mathbf{B} = \mathbf{i} + \mathbf{j} + \mathbf{k}$

9. $\mathbf{A} = 2\mathbf{i} - 4\mathbf{j} + \sqrt{5}\mathbf{k}, \quad \mathbf{B} = -2\mathbf{i} + 4\mathbf{j} - \sqrt{5}\mathbf{k}$

10. $\mathbf{A} = -5\mathbf{i} + \mathbf{j}, \quad \mathbf{B} = 2\mathbf{i} + \sqrt{17}\mathbf{j} + 10\mathbf{k}$

11. $\mathbf{A} = 10\mathbf{i} + 11\mathbf{j} - 2\mathbf{k}, \quad \mathbf{B} = 3\mathbf{j} + 4\mathbf{k}$

12. $\mathbf{A} = 2\mathbf{i} + 10\mathbf{j} - 11\mathbf{k}, \quad \mathbf{B} = 2\mathbf{i} + 2\mathbf{j} + \mathbf{k}$

13. Find the interior angles of the triangle ABC whose vertices are $A(-1, 0, 2)$, $B(2, 1, -1)$, and $C(1, -2, 2)$.

14. Find the angle between $\mathbf{A} = 2\mathbf{i} + 2\mathbf{j} + \mathbf{k}$ and $\mathbf{B} = 2\mathbf{i} + 10\mathbf{j} - 11\mathbf{k}$.

15. Find the angle between the diagonal of a cube and the diagonal of one of its faces. (*Hint:* Use a cube whose edges represent **i**, **j**, and **k**.)

16. Find the angle between the diagonal of a cube and one of the edges it meets at a vertex.

17. Find the point $A(a, a, 0)$ at the foot of the perpendicular from the point $B(2, 4, -3)$ to the line $y = x$ in the xy-plane.

18. How many lines through the origin make angles of $60°$ with both the y- and z-axes? What angles do these lines make with the positive x-axis?

19. Write the vector $\mathbf{B} = 3\mathbf{j} + 4\mathbf{k}$ as the sum of a vector $\mathbf{B}_1$ parallel to $\mathbf{A} = \mathbf{i} + \mathbf{j}$ and a vector $\mathbf{B}_2$ perpendicular to **A**.

20. Repeat Problem 19 for $\mathbf{B} = \mathbf{i} + \mathbf{j} + \mathbf{k}$.

21. Find the distance in the xy-plane between the line $x + 3y = 6$ and
a) the point $(2, 8)$,   b) the origin.

22. Find the distance between the lines $2x + y = 2$ and $2x + y = 4$.

23. Show that the vectors

$$\mathbf{A} = \frac{1}{\sqrt{3}}(\mathbf{i} - \mathbf{j} + \mathbf{k}), \quad \mathbf{B} = \frac{1}{\sqrt{2}}(\mathbf{j} + \mathbf{k}),$$

$$\mathbf{C} = \frac{1}{\sqrt{6}}(-2\mathbf{i} - \mathbf{j} + \mathbf{k})$$

are orthogonal to one another.

24. Find the vector projections of $\mathbf{D} = \mathbf{i} + \mathbf{j} + \mathbf{k}$ on the vectors $\mathbf{A}, \mathbf{B},$ and $\mathbf{C}$ of Problem 23. Then show that $\mathbf{D}$ is the sum of these vector projections.

25. Find the work done by a force $\mathbf{F} = -5\mathbf{k}$ (magnitude 5 newtons) in moving an object along the line from the origin to the point $(1, 1, 1)$.

26. A locomotive exerted a constant force of 60,000 newtons on a freight train while drawing it for 1 km along a straight track. How much work did the locomotive do?

27. How much work does it take to slide a crate 20 m along a loading dock by pulling on it with a constant force of 200 newtons at an angle of 30° from the horizontal?

28. Suppose you know that $\mathbf{A} \cdot \mathbf{B}_1 = \mathbf{A} \cdot \mathbf{B}_2$ and that $\mathbf{A}$ is not zero, but you know nothing more about the vectors $\mathbf{B}_1$ and $\mathbf{B}_2$. Is it all right to cancel $\mathbf{A}$ from both sides of the equation? Give a reason for your answer.

29. In Fig. 13.38 it looks as if $\mathbf{v}_1 + \mathbf{v}_2$ and $\mathbf{v}_1 - \mathbf{v}_2$ are orthogonal. Is this mere coincidence, or are there circumstances under which we may expect the sum of two vectors to be orthogonal to the difference of the same two vectors? Find

13.38   Problem 29 is about the vectors in this figure.

out by expanding the left-hand side of the equation

$$(\mathbf{v}_1 + \mathbf{v}_2) \cdot (\mathbf{v}_1 - \mathbf{v}_2) = 0.$$

30. Show that the diagonals of a rhombus are perpendicular.

31. Show that squares are the only rectangles with perpendicular diagonals.

32. Show that if c is any number, and $\mathbf{A}$ and $\mathbf{B}$ any vectors, then

$$(c\mathbf{A}) \cdot \mathbf{B} = c(\mathbf{A} \cdot \mathbf{B}) = \mathbf{A} \cdot (c\mathbf{B}).$$

Angles between Curves

The angles between two differentiable curves at a point of intersection are the angles between their tangents there. In Problems 33–37, find the angles between the curves at each intersection point.

33. $3x + y = 5, \quad 2x - y = 4$

34. $y = \sqrt{3}x - 1, \quad y = -\sqrt{3}x + 2$

35. $y = \sqrt{(3/4)} + x, \quad y = \sqrt{(3/4)} - x$

36. $y = x^3, \quad y = \sqrt{x}$ (two points of intersection)

37. $y = x^2, \quad y = \sqrt[3]{x}$ (two points of intersection)

13.5
Cross Products

When we turn a bolt by applying a force to a wrench, as in Fig. 13.39, the torque we produce acts along the axis of the bolt to drive the bolt forward. The magnitude of the torque depends on how far along the wrench the force is applied and on how much of the force is actually perpendicular to the wrench at that point. The number we use to measure the torque's magnitude is a product made up of the length of the vector arm $\mathbf{A}$ and the scalar component of $\mathbf{B}$ perpendicular to $\mathbf{A}$. In the notation of Fig. 13.39,

Magnitude of the torque vector $= |\mathbf{A}||\mathbf{B}|\sin \theta$.

If we let $\mathbf{n}$ be a unit vector along the axis of the bolt in the direction of the

Torque

n

A

Component of **B**
perpendicular to **A**

θ

B

13.39   The torque vector describes the
tendency of the force **B** to drive the bolt forward.

A × B

n

θ

A

B

13.40   The construction of **A** × **B**.

B

θ

−n

A

B × A

13.41   The construction of **B** × **A**.

torque, then the complete description of the torque vector is

$$\text{Torque vector} = \mathbf{n}|\mathbf{A}||\mathbf{B}|\sin\theta.$$

We call the torque the **vector product** of **A** and **B**.

Vector products are widely used to describe the effects of forces in studies of electricity, magnetism, fluid mechanics, and planetary motion. The goal of this article is to acquaint you with the mathematical properties of vector products that account for their use in these fields. We shall see a number of these applications ourselves in Chapters 14 and 19, where we study motion in space and the integrals associated with fluid flow. In the next article we shall also see how vector products are combined with scalar products to produce equations for lines and planes in space.

The Vector Product of Two Vectors in Space

When we define vector products in mathematics, we start with two nonzero vectors **A** and **B** in space without requiring them to have any particular physical interpretation. If **A** and **B** are not parallel, they determine a plane. We select a unit vector **n** perpendicular to the plane by the **right-hand rule.** This means we choose **n** to be the unit (normal) vector that points the way your right thumb points when your fingers curl through the angle θ from **A** to **B** (Fig. 13.40). We then define the vector product **A** × **B** (pronounced ''A cross B'') of **A** and **B** to be the vector

$$\mathbf{A} \times \mathbf{B} = \mathbf{n}|\mathbf{A}||\mathbf{B}|\sin\theta. \tag{1}$$

Since **A** × **B** is a scalar multiple of **n**, it is perpendicular to both **A** and **B**.

If θ approaches 0° or 180° in Eq. (1), the length of **A** × **B** approaches zero. We therefore define **A** × **B** to be **0** if **A** and **B** are parallel (and fail to determine a plane). This is consistent with our torque interpretation as well. If the force **B** in Fig. 13.39 is parallel to the wrench, meaning that we are trying to turn the bolt by pushing or pulling straight along the handle of the wrench, the torque produced is **0**.

If one or both of **A** and **B** is zero, we define **A** × **B** to be zero as well. Thus, the cross product of two vectors **A** and **B** is zero if and only if **A** and **B** are parallel or one or both of them is the zero vector.

The vector product of **A** and **B** is often called the **cross product** of **A** and **B** because of the cross in the notation **A** × **B**. In contrast to the dot product **A** · **B**, which is a scalar, the cross product of **A** and **B** is a vector.

A × B vs. B × A

Reversing the order of the factors in a nonzero vector product reverses the direction of the resulting vector. When the fingers of our right hand curl through the angle θ from **B** to **A**, our thumb points the opposite way and the unit vector we choose in forming **B** × **A** is the negative of the one we choose in forming **A** × **B** (Fig. 13.41). Thus,

$$\mathbf{B} \times \mathbf{A} = -(\mathbf{A} \times \mathbf{B}) \tag{2}$$

for all vectors **A** and **B**. Unlike the dot product, the cross product is not commutative.

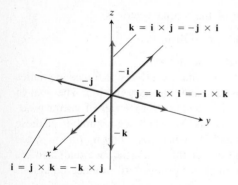

13.42 The pairwise cross products of **i, j,** and **k** can be read from this diagram.

When we apply the definition of the cross product to the unit vectors, **i, j,** and **k,** we find

$$i \times j = -(j \times i) = k,$$
$$j \times k = -(k \times j) = i, \tag{3}$$
$$k \times i = -(i \times k) = j,$$

and

$$i \times i = j \times j = k \times k = 0.$$

See Fig. 13.42.

$|A \times B|$ Is the Area of a Parallelogram

Because **n** is a unit vector, the magnitude of $A \times B$ is

$$|A \times B| = |n||A||B||\sin \theta| = |A||B||\sin \theta|.$$

This is the area of the parallelogram determined by **A** and **B** (Fig. 13.43), $|A|$ being the base of the parallelogram and $|B||\sin \theta|$ the height.

The Associative and Distributive Laws

As a rule, cross-product multiplication is not associative because $(A \times B) \times C$ lies in the plane of **A** and **B** whereas $A \times (B \times C)$ lies in the plane of **B** and **C**, as we shall see in Article 13.7. However, the **scalar distributive law**

$$(rA) \times (sB) = (rs)A \times B \tag{4}$$

does hold, as do the **vector distributive laws**

$$A \times (B + C) = A \times B + A \times C \tag{5}$$

and

$$(B + C) \times A = B \times A + C \times A. \tag{6}$$

As a special case of (4) we have

$$(-A) \times B = A \times (-B) = -(A \times B).$$

The scalar distributive law can be verified by applying Eq. (1) to the products on both sides of Eq. (4) and comparing the results. The vector distributive law in Eq. (5) is not easy to prove, however. We shall assume it here and leave the proof to Appendix 6. Equation (6) follows from Eq. (5): just multiply both sides of Eq. (5) by -1 and reverse the orders of the products.

The Determinant Formula for $A \times B$

Our next objective is to show how to calculate the components of $A \times B$ from the components of **A** and **B**.

Suppose

$$A = a_1 i + a_2 j + a_3 k, \qquad B = b_1 i + b_2 j + b_3 k.$$

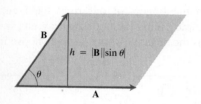

13.43 The parallelogram determined by **A** and **B**. Its area is $|A \times B|$.

Then the distributive laws and the rules for multiplying $\mathbf{i}$, $\mathbf{j}$, and $\mathbf{k}$ tell us that

$$\mathbf{A} \times \mathbf{B} = (a_1\mathbf{i} + a_2\mathbf{j} + a_3\mathbf{k}) \times (b_1\mathbf{i} + b_2\mathbf{j} + b_3\mathbf{k})$$

$$= a_1b_1\mathbf{i} \times \mathbf{i} + a_1b_2\mathbf{i} \times \mathbf{j} + a_1b_3\mathbf{i} \times \mathbf{k}$$

$$+ a_2b_1\mathbf{j} \times \mathbf{i} + a_2b_2\mathbf{j} \times \mathbf{j} + a_2b_3\mathbf{j} \times \mathbf{k} \tag{7}$$

$$+ a_3b_1\mathbf{k} \times \mathbf{i} + a_3b_2\mathbf{k} \times \mathbf{j} + a_3b_3\mathbf{k} \times \mathbf{k}$$

$$= (a_2b_3 - a_3b_2)\mathbf{i} + (a_3b_1 - a_1b_3)\mathbf{j} + (a_1b_2 - a_2b_1)\mathbf{k}.$$

The terms at the end of Eq. (7) are the same as the terms in the expansion of the symbolic determinant

$$\begin{vmatrix} \mathbf{i} & \mathbf{j} & \mathbf{k} \\ a_1 & a_2 & a_3 \\ b_1 & b_2 & b_3 \end{vmatrix}$$

We may therefore use the following rule to calculate $\mathbf{A} \times \mathbf{B}$.

If $\mathbf{A} = a_1\mathbf{i} + a_2\mathbf{j} + a_3\mathbf{k}$ and $\mathbf{B} = b_1\mathbf{i} + b_2\mathbf{j} + b_3\mathbf{k}$, then

$$\mathbf{A} \times \mathbf{B} = \begin{vmatrix} \mathbf{i} & \mathbf{j} & \mathbf{k} \\ a_1 & a_2 & a_3 \\ b_1 & b_2 & b_3 \end{vmatrix}. \tag{8}$$

Equation (8) is remarkable, given that neither the associative nor commutative law holds for cross-product multiplication.

EXAMPLE 1 Find $\mathbf{A} \times \mathbf{B}$ and $\mathbf{B} \times \mathbf{A}$ if

$$\mathbf{A} = 2\mathbf{i} + \mathbf{j} + \mathbf{k}, \qquad \mathbf{B} = -4\mathbf{i} + 3\mathbf{j} + \mathbf{k}.$$

Solution We use Eq. (8) to find $\mathbf{A} \times \mathbf{B}$:

$$\mathbf{A} \times \mathbf{B} = \begin{vmatrix} \mathbf{i} & \mathbf{j} & \mathbf{k} \\ 2 & 1 & 1 \\ -4 & 3 & 1 \end{vmatrix}$$

$$= \begin{vmatrix} 1 & 1 \\ 3 & 1 \end{vmatrix} \mathbf{i} - \begin{vmatrix} 2 & 1 \\ -4 & 1 \end{vmatrix} \mathbf{j} + \begin{vmatrix} 2 & 1 \\ -4 & 3 \end{vmatrix} \mathbf{k}$$

$$= -2\mathbf{i} - 6\mathbf{j} + 10\mathbf{k}.$$

Equation (2) then gives $\mathbf{B} \times \mathbf{A}$:

$$\mathbf{B} \times \mathbf{A} = -\mathbf{A} \times \mathbf{B} = 2\mathbf{i} + 6\mathbf{j} - 10\mathbf{k}. \qquad \blacksquare$$

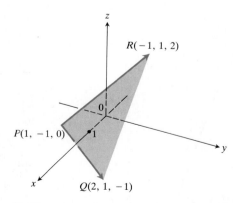

13.44 The area of triangle PQR is half of $|\overrightarrow{PQ} \times \overrightarrow{PR}|$.

EXAMPLE 2 Find the area of the triangle whose vertices are $P(1, -1, 0)$, $Q(2, 1, -1)$, and $R(-1, 1, 2)$ (Fig. 13.44).

Solution   The area of the triangle is $(1/2)|\overrightarrow{PQ} \times \overrightarrow{PR}|$. In terms of components,

$$\overrightarrow{PQ} = (2 - 1)\mathbf{i} + (1 + 1)\mathbf{j} + (-1 - 0)\mathbf{k} = \mathbf{i} + 2\mathbf{j} - \mathbf{k},$$

$$\overrightarrow{PR} = (-1 - 1)\mathbf{i} + (1 + 1)\mathbf{j} + (2 - 0)\mathbf{k} = -2\mathbf{i} + 2\mathbf{j} + 2\mathbf{k},$$

$$\overrightarrow{PQ} \times \overrightarrow{PR} = \begin{vmatrix} \mathbf{i} & \mathbf{j} & \mathbf{k} \\ 1 & 2 & -1 \\ -2 & 2 & 2 \end{vmatrix}$$

$$= \begin{vmatrix} 2 & -1 \\ 2 & 2 \end{vmatrix} \mathbf{i} - \begin{vmatrix} 1 & -1 \\ -2 & 2 \end{vmatrix} \mathbf{j} + \begin{vmatrix} 1 & 2 \\ -2 & 2 \end{vmatrix} \mathbf{k}$$

$$= 6\mathbf{i} + 6\mathbf{k}.$$

Hence, the triangle's area is

$$\frac{1}{2}|6\mathbf{i} + 6\mathbf{k}| = \frac{1}{2}\sqrt{72} = 3\sqrt{2}. \qquad\blacksquare$$

EXAMPLE 3   Find a unit vector perpendicular to both $\mathbf{A} = \mathbf{i} + 2\mathbf{k}$ and $\mathbf{B} = 2\mathbf{i} - 2\mathbf{j}$.

Solution   We find the direction $\mathbf{n}$ of $\mathbf{A} \times \mathbf{B}$:

$$\mathbf{A} \times \mathbf{B} = \begin{vmatrix} \mathbf{i} & \mathbf{j} & \mathbf{k} \\ 1 & 0 & 2 \\ 2 & -2 & 0 \end{vmatrix} = 4\mathbf{i} + 4\mathbf{j} - 2\mathbf{k},$$

$$|\mathbf{A} \times \mathbf{B}| = \sqrt{16 + 16 + 4} = \sqrt{36} = 6,$$

$$\mathbf{n} = \frac{\mathbf{A} \times \mathbf{B}}{|\mathbf{A} \times \mathbf{B}|} = \frac{4\mathbf{i} + 4\mathbf{j} - 2\mathbf{k}}{6} = \frac{2}{3}\mathbf{i} + \frac{2}{3}\mathbf{j} - \frac{1}{3}\mathbf{k}.$$

The unit vector $\mathbf{n}$ is perpendicular to both $\mathbf{A}$ and $\mathbf{B}$. See Fig. 13.45. The vector $-\mathbf{n}$ would do as well. $\blacksquare$

13.45   To find a unit vector $\mathbf{n}$ perpendicular to $\mathbf{A}$ and $\mathbf{B}$, we divide $\mathbf{A} \times \mathbf{B}$ by $|\mathbf{A} \times \mathbf{B}|$, in this case 6.

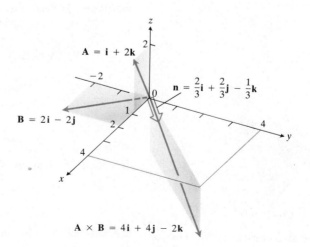

PROBLEMS

In Problems 1–8, find the length and direction of $\mathbf{A} \times \mathbf{B}$ and $\mathbf{B} \times \mathbf{A}$.

1. $\mathbf{A} = 2\mathbf{i} - 2\mathbf{j} - \mathbf{k}, \quad \mathbf{B} = \mathbf{i} + \mathbf{j} + \mathbf{k}$
2. $\mathbf{A} = 2\mathbf{i} + 3\mathbf{j}, \quad \mathbf{B} = -\mathbf{i} + \mathbf{j}$
3. $\mathbf{A} = 2\mathbf{i} - 2\mathbf{j} + 4\mathbf{k}, \quad \mathbf{B} = -\mathbf{i} + \mathbf{j} - 2\mathbf{k}$
4. $\mathbf{A} = \mathbf{i} + \mathbf{j} - \mathbf{k}, \quad \mathbf{B} = \mathbf{0}$
5. $\mathbf{A} = 2\mathbf{i}, \quad \mathbf{B} = -3\mathbf{j}$
6. $\mathbf{A} = \mathbf{i} \times \mathbf{j}, \quad \mathbf{B} = \mathbf{j} \times \mathbf{k}$
7. $\mathbf{A} = -8\mathbf{i} - 2\mathbf{j} - 4\mathbf{k}, \quad \mathbf{B} = 2\mathbf{i} + 2\mathbf{j} + \mathbf{k}$
8. $\mathbf{A} = \mathbf{i} - \dfrac{1}{2}\mathbf{j} + \mathbf{k}, \quad \mathbf{B} = -\dfrac{3}{2}\mathbf{i} - \mathbf{j} - 3\mathbf{k}$

In Problems 9–14, sketch $\mathbf{A}$, $\mathbf{B}$, and $\mathbf{A} \times \mathbf{B}$ together with the coordinate axes.

9. $\mathbf{A} = \mathbf{i}, \quad \mathbf{B} = \mathbf{j}$
10. $\mathbf{A} = \mathbf{i} + \mathbf{k}, \quad \mathbf{B} = \mathbf{j}$
11. $\mathbf{A} = \mathbf{i} - \mathbf{k}, \quad \mathbf{B} = \mathbf{j} + \mathbf{k}$
12. $\mathbf{A} = 2\mathbf{i} - \mathbf{j}, \quad \mathbf{B} = \mathbf{i} + 2\mathbf{j}$
13. $\mathbf{A} = \mathbf{i} + 3\mathbf{j} + 2\mathbf{k}, \quad \mathbf{B} = \mathbf{k}$
14. $\mathbf{A} = \mathbf{i} + 2\mathbf{j}, \quad \mathbf{B} = 2\mathbf{j} + \mathbf{k}$

15. Find a vector $\mathbf{N}$ perpendicular to the plane determined by the points $A(1, -1, 2)$, $B(2, 0, -1)$, and $C(0, 2, 1)$.

16. Find a vector that is perpendicular to both $\mathbf{A} = \mathbf{i} + \mathbf{j} + \mathbf{k}$ and $\mathbf{B} = \mathbf{i} + \mathbf{j}$.

17. Find the distance from the origin to the plane ABC of Problem 15 by projecting $\overrightarrow{OA}$ onto the normal vector $\mathbf{N}$.

18. Find the area of triangle ABC in Problem 15.

19. If $\overrightarrow{AB} \times \overrightarrow{AC} = 2\mathbf{i} - 4\mathbf{j} + 4\mathbf{k}$, find the area of triangle ABC.

20. a) Find a vector normal to the plane through the points $A(0, 2, 1)$, $B(2, 1, 2)$, and $C(1, 1, 3)$.
    b) Find the area of the triangle formed by the three points in part (a).

21. If $\mathbf{A} = 2\mathbf{i} - \mathbf{j}$ and $\mathbf{B} = \mathbf{i} + 3\mathbf{j} - 2\mathbf{k}$, find $\mathbf{A} \times \mathbf{B}$. Then calculate $(\mathbf{A} \times \mathbf{B}) \cdot \mathbf{A}$ and $(\mathbf{A} \times \mathbf{B}) \cdot \mathbf{B}$.

22. Is $(\mathbf{A} \times \mathbf{B}) \cdot \mathbf{A}$ always zero? Explain. What about $(\mathbf{A} \times \mathbf{B}) \cdot \mathbf{B}$?

23. Let $\mathbf{A} = 5\mathbf{i} - \mathbf{j} + \mathbf{k}$, $\mathbf{B} = \mathbf{j} - 5\mathbf{k}$, $\mathbf{C} = -15\mathbf{i} + 3\mathbf{j} - 3\mathbf{k}$. Which pairs of vectors are (a) perpendicular, (b) parallel?

24. Given vectors $\mathbf{A}$, $\mathbf{B}$, and $\mathbf{C}$, use dot-product and cross-product notation to describe the following vectors.
    a) The vector projection of $\mathbf{A}$ onto $\mathbf{B}$
    b) A vector orthogonal to $\mathbf{A}$ and $\mathbf{B}$
    c) A vector with the length of $\mathbf{A}$ and the direction of $\mathbf{B}$
    d) A vector orthogonal to $\mathbf{A} \times \mathbf{B}$ and $\mathbf{C}$
    e) A vector in the plane of $\mathbf{B}$ and $\mathbf{C}$ perpendicular to $\mathbf{A}$

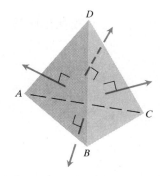

13.46 The tetrahedron in Problem 25.

25. Suppose we associate with each face of the tetrahedron $ABCD$ in Fig. 13.46 an outward normal vector whose length is equal to the area of the face. Show that the sum of these four normal vectors is $\mathbf{0}$.

26. Let $\mathbf{A}$ be a nonzero vector. Show that (a) $\mathbf{A} \times \mathbf{B} = \mathbf{A} \times \mathbf{C}$ does not guarantee $\mathbf{B} = \mathbf{C}$ (see Problem 28, Article 13.4); (b) $\mathbf{A} \cdot \mathbf{B} = \mathbf{A} \cdot \mathbf{C}$ and $\mathbf{A} \times \mathbf{B} = \mathbf{A} \times \mathbf{C}$ together imply $\mathbf{B} = \mathbf{C}$.

Areas in the Plane

To find the area of a parallelogram in a plane, picture the plane as the xy-plane in space. Then write vectors for a pair of adjacent sides (their $\mathbf{k}$-components will be zero) and calculate the magnitude of their cross product.

Find the areas of the parallelograms whose vertices are given in Problems 27–30.

27. $A(1, 0), \quad B(0, 1), \quad C(-1, 0), \quad D(0, -1)$
28. $A(0, 0), \quad B(7, 3), \quad C(9, 8), \quad D(2, 5)$
29. $A(-1, 2), \quad B(2, 0), \quad C(7, 1), \quad D(4, 3)$
30. $A(-6, 0), \quad B(1, -4), \quad C(3, 1), \quad D(-4, 5)$

Find the areas of the triangles whose vertices are given in Problems 31–34.

31. $A(0, 0), \quad B(-2, 3), \quad C(3, 1)$
32. $A(-1, -1), \quad B(3, 3), \quad C(2, 1)$
33. $A(-5, 3), \quad B(1, -2), \quad C(6, -2)$
34. $A(-6, 0), \quad B(10, -5), \quad C(-2, 4)$

35. Show that the area of a triangle in the xy-plane with one vertex at the origin and the other two vertices at (a_1, a_2) and (b_1, b_2) is

$$\pm \frac{1}{2} \begin{vmatrix} a_1 & a_2 \\ b_1 & b_2 \end{vmatrix}.$$

What controls the sign?

13.6
Equations for Lines, Line Segments, and Planes in Space

In this article we show how dot products and cross products lead to equations for lines, line segments, and planes in space. The equations we use for lines and line segments are parametric equations, and you may wish to review the discussion of parametric equations for plane curves in Article 8.9 before you begin.

Equations for Lines and Line Segments

Suppose L is a line in space that passes through a point $P_0(x_0, y_0, z_0)$ and lies parallel to a vector $\mathbf{v} = A\mathbf{i} + B\mathbf{j} + C\mathbf{k}$. Then L is the set of all points $P(x, y, z)$ for which the vector $\overrightarrow{P_0P}$ is parallel to $\mathbf{v}$ (Fig. 13.47). That is, P lies on L if and only if

$$\overrightarrow{P_0P} = t\mathbf{v} \tag{1}$$

for some number t.

When we write Eq. (1) in terms of components,

$$(x - x_0)\mathbf{i} + (y - y_0)\mathbf{j} + (z - z_0)\mathbf{k} = t(A\mathbf{i} + B\mathbf{j} + C\mathbf{k}), \tag{2}$$

and equate the corresponding components on the two sides, we get three equations in the parameter t:

$$x - x_0 = tA, \qquad y - y_0 = tB, \qquad z - z_0 = tC. \tag{3}$$

When rearranged, these become the standard parametric equations of line L:

$$x = x_0 + tA, \qquad y = y_0 + tB, \qquad z = z_0 + tC. \tag{4}$$

As the parameter t increases from $-\infty$ to ∞, the point $P(x, y, z)$ traces the line exactly once. As t traverses the closed interval $a \le t \le b$, P traces the segment from the point where $t = a$ to the point where $t = b$.

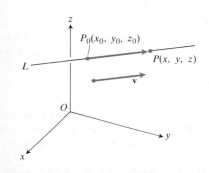

13.47   A point P lies on the line through P_0 parallel to $\mathbf{v}$ if and only if $\overrightarrow{P_0P}$ is a scalar multiple of $\mathbf{v}$.

EXAMPLE 1   Find parametric equations for the line through the point $(-2, 0, 4)$ parallel to the vector $\mathbf{v} = 2\mathbf{i} + 4\mathbf{j} - 2\mathbf{k}.$

Solution   With $P_0(x_0, y_0, z_0) = (-2, 0, 4)$ and

$$A\mathbf{i} + B\mathbf{j} + C\mathbf{k} = 2\mathbf{i} + 4\mathbf{j} - 2\mathbf{k},$$

Eqs. (4) become

$$x = -2 + 2t, \qquad y = 4t, \qquad z = 4 - 2t.$$

See Fig. 13.48.   ∎

EXAMPLE 2   Find parametric equations for the line through the points $P(-3, 2, -3)$ and $Q(1, -1, 4)$.

Solution   The vector

$$\overrightarrow{PQ} = 4\mathbf{i} - 3\mathbf{j} + 7\mathbf{k}$$

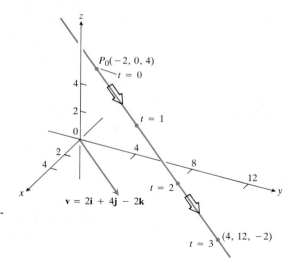

13.48 Selected points and parameter values on the line $x = -2 + 2t$, $y = 4t$, $z = 4 - 2t$. The arrows show the direction of increasing t.

is parallel to the line, and Eqs. (4) with $(x_0, y_0, z_0) = (-3, 2, -3)$ give

$$x = -3 + 4t, \qquad y = 2 - 3t, \qquad z = -3 + 7t. \tag{5}$$

We could equally well have chosen $Q(1, -1, 4)$ as the "base point" and written

$$x = 1 + 4t, \qquad y = -1 - 3t, \qquad z = 4 + 7t \tag{6}$$

as equations for the line. The parameterizations in (5) and (6) place you at different points for a given value of t, but each set of equations covers the line completely as t runs from $-\infty$ to ∞. ∎

To find equations for the line segment joining two points, we first find equations for the line through the points. We then find the t-values for which the line passes through the points and restrict t to lie within the closed interval bounded by these values. The line equations, together with this added restriction, are called equations for the segment.

EXAMPLE 3 Find parametric equations for the line segment joining the points $P(-3, 2, -3)$ and $Q(1, -1, 4)$.

Solution We begin with equations for the line through P and Q, taking them, in this case, from Example 2:

$$x = -3 + 4t, \qquad y = 2 - 3t, \qquad z = -3 + 7t. \tag{7}$$

We notice that the point

$$(x, y, z) = (-3 + 4t, 2 - 3t, -3 + 7t)$$

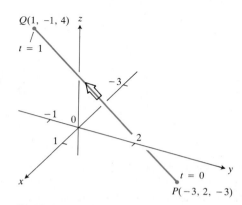

13.49 Example 3 derives parametric equations for the line segment joining P and Q. The arrow shows the direction of increasing t.

passes through $P(-3, 2, -3)$ at $t = 0$ and $Q(1, -1, 4)$ at $t = 1$. We add the restriction $0 \leq t \leq 1$ to Eqs. (7) to get equations for the segment:

$$x = -3 + 4t, \qquad y = 2 - 3t, \qquad z = -3 + 7t, \qquad 0 \leq t \leq 1.$$

See Fig. 13.49. ∎

The Distance from a Point to a Line

To find the distance from a point P to a line L, we carry out the following steps:

STEP 1: Find the point Q on L closest to P.

STEP 2: Calculate the distance from P to Q.

The next example shows how this is done.

EXAMPLE 4   Find the distance from the point $P(1, 1, 5)$ to the line

$$x = 1 + t, \qquad y = 3 - t, \qquad z = 2t.$$

Solution STEP 1: We find the point on the line closest to P. The coordinates of a typical point Q on the line are $Q(1 + t, 3 - t, 2t)$.

We want the value of t that minimizes the distance from P to Q. To avoid working with square roots, we find this value of t by minimizing the *square* of the distance instead. The formula for the square of the distance from P to Q is

$$
\begin{aligned}
f(t) &= (1 + t - 1)^2 + (3 - t - 1)^2 + (2t - 5)^2 \\
&= (t)^2 + (2 - t)^2 + (2t - 5)^2 \\
&= t^2 + 4 - 4t + t^2 + 4t^2 - 20t + 25 \\
&= 6t^2 - 24t + 29.
\end{aligned}
$$

The derivative,

$$\frac{df}{dt} = 12t - 24,$$

equals 0 when $t = 2$. This value of t gives a minimum value for f because $d^2f/dt^2 = 12$ is positive. The point Q closest to P is $Q(1 + t, 3 - t, 2t)_{t=2} = Q(3, 1, 4)$.

STEP 2: We find the distance from $P(1, 1, 5)$ to $Q(3, 1, 4)$:

$$
\begin{aligned}
d(P, Q) &= \sqrt{(1 - 3)^2 + (1 - 1)^2 + (5 - 4)^2} \\
&= \sqrt{(-2)^2 + (0)^2 + (1)^2} \\
&= \sqrt{4 + 1} \\
&= \sqrt{5}.
\end{aligned}
$$

The distance from P to the line is $\sqrt{5}$.   ∎

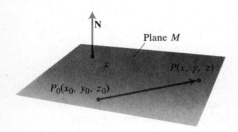

13.50   The standard equation for a plane in space is defined in terms of a vector normal to the plane: a point P lies in the plane through P_0 normal to $\mathbf{N}$ if and only if $\mathbf{N} \cdot \overrightarrow{P_0P} = 0$.

Equations for Planes

Suppose M is a plane in space that passes through a point $P_0(x_0, y_0, z_0)$ and is perpendicular (normal) to the nonzero vector $\mathbf{N} = A\mathbf{i} + B\mathbf{j} + C\mathbf{k}$. Then M consists of all points $P(x, y, z)$ for which the vector $\overrightarrow{P_0P}$ is orthogonal to $\mathbf{N}$. See Fig. 13.50. That is, P lies on M if and only if

$$\mathbf{N} \cdot \overrightarrow{P_0P} = 0 \tag{8}$$

or

$$A(x - x_0) + B(y - y_0) + C(z - z_0) = 0.$$

This becomes

$$Ax + By + Cz = Ax_0 + By_0 + Cz_0$$

when rearranged or, more simply,

$$Ax + By + Cz = D, \quad \text{where} \quad D = Ax_0 + By_0 + Cz_0. \tag{9}$$

Notice that the coefficients of x, y, and z in Eq. (9) are the **i**-, **j**-, and **k**-components of a vector normal to the plane.

EXAMPLE 5 Find an equation for the plane through $P_0(-3, 0, 7)$ perpendicular to $\mathbf{N} = 5\mathbf{i} + 2\mathbf{j} - \mathbf{k}.$

Solution We use Eq. (9) to get

$$D = 5(-3) + 2(0) - 1(7) = -15 - 7 = -22$$

and

$$5x + 2y - z = -22. \qquad \blacksquare$$

EXAMPLE 6 Find an equation for the plane through the points $A(0, 0, 1)$, $B(2, 0, 0)$, and $C(0, 3, 0)$.

Solution We find a vector normal to the plane and use it with one of the points to write an equation for the plane.

The cross product

$$\overrightarrow{AB} \times \overrightarrow{AC} = \begin{vmatrix} \mathbf{i} & \mathbf{j} & \mathbf{k} \\ 2 & 0 & -1 \\ 0 & 3 & -1 \end{vmatrix} = 3\mathbf{i} + 2\mathbf{j} + 6\mathbf{k}$$

is normal to the plane. We substitute the components of this vector and the coordinates of the point $(0, 0, 1)$ into Eq. (9) to get $D = 3(0) + 2(0) + 6(1) = 6$ and

$$3x + 2y + 6z = 6$$

as an equation for the plane. $\qquad \blacksquare$

EXAMPLE 7 Find the point in which the line

$$x = \frac{8}{3} + 2t, \qquad y = -2t, \qquad z = 1 + t$$

meets the plane $3x + 2y + 6z = 6$.

Solution The point

$$\left(\frac{8}{3} + 2t, \ -2t, \ 1 + t \right)$$

will lie in the plane if its coordinates satisfy the equation of the plane; that is, if

$$3\left(\frac{8}{3} + 2t \right) + 2(-2t) + 6(1 + t) = 6$$

$$8 + 6t - 4t + 6 + 6t = 6$$

$$8t = -8$$

$$t = -1.$$

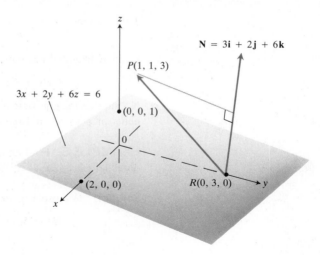

13.51 The distance from P to the plane is the length of the vector projection of $\overrightarrow{RP}$ onto a vector **N** normal to the plane.

The point of intersection is

$$(x,\, y,\, z)_{t=-1} = \left(\frac{8}{3} - 2,\, 2,\, 1 - 1\right) = \left(\frac{2}{3},\, 2,\, 0\right). \qquad \blacksquare$$

EXAMPLE 8   Find the distance from the point $P(1,\, 1,\, 3)$ to the plane $3x + 2y + 6z = 6$.

Solution   The approach we take is similar to the one we took to find the distance from a point P to a line L in the coordinate plane. There, we found a point R on the line and found the distance from P to L by calculating the length of the vector projection of $\overrightarrow{RP}$ onto a vector **N** normal to L. Here, we find a point R in the plane (it does not matter which one) and calculate the length of the vector projection of $\overrightarrow{RP}$ onto a vector **N** normal to the plane. See Fig. 13.51.

We find that the point $R(0,\, 3,\, 0)$ lies on the plane (put $x = z = 0$ and solve for y). We also know from the plane's equation that $\mathbf{N} = 3\mathbf{i} + 2\mathbf{j} + 6\mathbf{k}$ is normal to the plane. The length of **N** is $|\mathbf{N}| = \sqrt{9 + 4 + 36} = 7$. The vector $\overrightarrow{RP}$ is found from the coordinates of R and P:

$$\overrightarrow{RP} = (1 - 0)\mathbf{i} + (1 - 3)\mathbf{j} + (3 - 0)\mathbf{k} = \mathbf{i} - 2\mathbf{j} + 3\mathbf{k}.$$

The length of the vector projection of $\overrightarrow{RP}$ onto **N** (from Article 13.4, Eq. 12) is the absolute value of

$$\overrightarrow{RP} \cdot \frac{\mathbf{N}}{|\mathbf{N}|} = (\mathbf{i} - 2\mathbf{j} + 3\mathbf{k}) \cdot \left(\frac{3}{7}\mathbf{i} + \frac{2}{7}\mathbf{j} + \frac{6}{7}\mathbf{k}\right) = \frac{3}{7} - \frac{4}{7} + \frac{18}{7} = \frac{17}{7}.$$

The distance from P to the plane is therefore $17/7$. $\qquad \blacksquare$

Angles between Planes; Lines of Intersection

The angle between two intersecting planes is defined to be the (acute) angle made by their normal vectors (Fig. 13.52).

EXAMPLE 9   Find the angle between the planes $3x - 6y - 2z = 15$ and $2x + y - 2z = 5$.

13.52 The angle between two planes is obtained from the angle between their normals.

Solution   The vectors

$$\mathbf{N}_1 = 3\mathbf{i} - 6\mathbf{j} - 2\mathbf{k}, \qquad \mathbf{N}_2 = 2\mathbf{i} + \mathbf{j} - 2\mathbf{k}$$

are normals to the planes. The angle between them is found from their dot product:

$$\cos \theta = \frac{\mathbf{N}_1 \cdot \mathbf{N}_2}{|\mathbf{N}_1||\mathbf{N}_2|} = \frac{4}{21}, \qquad \theta = \cos^{-1}\frac{4}{21} \approx 79°.$$ ■

EXAMPLE 10 Find a vector parallel to the line of intersection of the planes $3x - 6y - 2z = 15$ and $2x + y - 2z = 5$.

Solution Any vector parallel to the line of intersection will be parallel to both planes and therefore perpendicular to their normals. Conversely, any vector perpendicular to the planes' normals will be parallel to both planes and hence parallel to their line of intersection. These requirements are met by the vector

$$\mathbf{v} = \mathbf{N}_1 \times \mathbf{N}_2 = \begin{vmatrix} \mathbf{i} & \mathbf{j} & \mathbf{k} \\ 3 & -6 & -2 \\ 2 & 1 & -2 \end{vmatrix} = 14\mathbf{i} + 2\mathbf{j} + 15\mathbf{k}.$$

Any nonzero scalar multiple of $\mathbf{v}$ will do as well. ■

EXAMPLE 11 Find parametric equations for the line in which the planes $3x - 6y - 2z = 15$ and $2x + y - 2z = 5$ intersect.

Solution We find a vector parallel to the line and a point on the line and use Eqs. (4).
   Example 10 gives a vector parallel to the line, namely, $\mathbf{v} = 14\mathbf{i} + 2\mathbf{j} + 15\mathbf{k}$. To find a point on the line, we find a point common to the two planes. Substituting $z = 0$ in the plane equations and solving for x and y simultaneously gives the point $(3, -1, 0)$. The line is

$$x = 3 + 14t, \qquad y = -1 + 2t, \qquad z = 15t.$$ ■

Equations for Lines and Planes in Space

Line through $P(x_0, y_0, z_0)$ parallel to $A\mathbf{i} + B\mathbf{j} + C\mathbf{k}$:

$$x = x_0 + tA, \qquad y = y_0 + tB, \qquad z = z_0 + tC$$

Plane through $P(x_0, y_0, z_0)$ perpendicular to $A\mathbf{i} + B\mathbf{j} + C\mathbf{k}$:

$$Ax + By + Cz = D, \qquad D = Ax_0 + By_0 + Cz_0$$

PROBLEMS

Find parametric equations for the lines in Problems 1–12.

1. The line through $P(3, -4, -1)$ parallel to the vector $\mathbf{i} + \mathbf{j} + \mathbf{k}$

2. The line through $P(1, 2, -1)$ and $Q(-1, 0, 1)$

3. The line through $P(-2, 0, 3)$ and $Q(3, 5, -2)$

4. The line through $P(1, 2, 0)$ and $Q(1, 1, -1)$

5. The line through the origin parallel to the vector $2\mathbf{j} + \mathbf{k}$

6. The line through the point $(3, -2, 1)$ parallel to the line $x = 1 + 2t$, $y = 2 - t$, $z = 3t$

7. The line through $(1, 1, 1)$ parallel to the z-axis

8. The line through $(2, 4, 5)$ perpendicular to the plane $3x + 7y - 5z = 21$

9. The line through $(0, -7, 0)$ perpendicular to the plane $x + 2y + 2z = 13$

10. The line through $(2, 3, 0)$ perpendicular to the vectors $\mathbf{A} = \mathbf{i} + 2\mathbf{j} + 3\mathbf{k}$ and $\mathbf{B} = 3\mathbf{i} + 4\mathbf{j} + 5\mathbf{k}$

11. The x-axis

12. The z-axis

Find parametric equations for the line segments joining the points in Problems 13–20. Sketch each segment together with the coordinate axes and indicate the direction of motion for increasing t in your parameterization.

13. $(0, 0, 0)$, $(1, 1, 1)$

14. $(0, 0, 0)$, $(1, 0, 0)$

15. $(1, 0, 0)$, $(1, 1, 0)$

16. $(1, 1, 0)$, $(1, 1, 1)$

17. $(0, -1, 1)$, $(0, 1, 1)$

18. $(3, 0, 0)$, $(0, 2, 0)$

19. $(2, 2, 0)$, $(1, 2, -2)$

20. $(1, -1, -2)$, $(0, 2, 1)$

Find equations for the planes in Problems 21–30.

21. The plane through $P_0(0, 2, -1)$ perpendicular to $\mathbf{N} = 3\mathbf{i} - 2\mathbf{j} - \mathbf{k}$

22. The plane through $(1, -1, 3)$ parallel to the plane $3x + y + z = 7$

23. The plane through $(1, 1, -1)$, $(2, 0, 2)$, and $(0, -2, 1)$

24. The plane through $(2, 4, 5)$, $(1, 5, 7)$, and $(-1, 6, 8)$

25. The plane through $(1, 0, 0)$, $(0, 1, 0)$, and $(0, 0, 1)$

26. The plane through $P_0(2, 4, 5)$ perpendicular to the line
$$x = 5 + t, \quad y = 1 + 3t, \quad z = 4t$$

27. The plane through the origin that contains the line
$$x = 1 + 2t, \quad y = -1 + 3t, \quad z = 4t$$

28. The plane through $A(1, -2, 1)$ perpendicular to the vector from the origin to A

29. The plane through $P_0(2, 1, -1)$ perpendicular to the line of intersection of the planes $2x + y - z = 3$, $x + 2y + z = 2$

30. The plane through the points $P_0(1, 2, 3)$ and $P_2(3, 2, 1)$ perpendicular to the plane $4x - y + 2z = 7$

In Problems 31–36, find the distance from the point to the line.

31. $x = 4t$, $y = -2t$, $z = 2t$; $P(0, 0, 12)$

32. $x = 5 + 3t$, $y = 5 + 4t$, $z = -3 - 5t$; $P(0, 0, 0)$

33. $x = 2 + 2t$, $y = 1 + 6t$, $z = 3$; $P(2, 1, 3)$

34. $x = 2t$, $y = 1 + 2t$, $z = 2t$; $P(2, 1, -1)$

35. $x = 4$, $y = 4 + 2t$, $z = 4t$; $P(2, -1, -10)$

36. $x = 10 + 4t$, $y = -3$, $z = 4t$; $P(-1, 4, 3)$

In Problems 37–42, find the distance from the point to the plane.

37. $(2, -3, 4)$, $x + 2y + 2z = 13$

38. $(0, 0, 0)$, $3x + 2y + 6z = 6$

39. $(0, 1, 1)$, $4y + 3z = -12$

40. $(2, 2, 3)$, $2x + y + 2z = 4$

41. $(0, -1, 0)$, $2x + y + 2z = 4$

42. $(1, 0, -1)$, $-4x + y + z = 4$

In Problems 43–46, find the point in which the line meets the plane.

43. $x = 1 - t$, $y = 3t$, $z = 1 + t$; $2x - y + 3z = 6$

44. $x = 2$, $y = 3 + 2t$, $z = -2 - 2t$; $6x + 3y - 4z = -12$

45. $x = 1 + 2t$, $y = 1 + 5t$, $z = 3t$; $x + y + z = 2$

46. $x = -1 + 3t$, $y = -2$, $z = 5t$; $2x - 3z = 7$

Find the angles between the planes in Problems 47–50.

47. $2x + 2y + 2z = 3$, $2x - 2y - z = 5$

48. $x + y = 1$, $2x + y - 2z = 2$

49. $5x + y - z = 10$, $x - 2y + 3z = -1$

50. $x + y - 3z = 1$, $-2x - 2y + 6z = 7$

Find parametric equations for the lines of intersection of the planes in Problems 51–54.

51. $x + y + z = 1$, $x + y = 2$

52. $3x - 6y - 2z = 3$, $2x + y - 2z = 2$

53. $x - 2y + 4z = 2$, $x + y - 2z = 5$

54. $5x - 2y = 11$, $4y - 5z = -17$

55. Find the points in which the line $x = 1 + 2t$, $y = -1 - t$, $z = 3t$ meets the three coordinate planes.

56. a) Find an equation for the plane through $P_0(4, 2, 1)$ perpendicular to $\mathbf{N} = 6\mathbf{i} - 2\mathbf{j} + 3\mathbf{k}.$
b) Find the distance from the origin to the plane in part (a).
c) Find the distance from the point $(6, 6, 8)$ to the plane in part (a).
d) Find equations for the line through $P_0(4, 2, -1)$ perpendicular to the plane in part (a).

13.7
Products of Three or More Vectors

The products $(\mathbf{A} \times \mathbf{B}) \cdot \mathbf{C}$ and $(\mathbf{A} \times \mathbf{B}) \times \mathbf{C}$ are often used to write formulas and work out calculations in studies of electricity, magnetism, and fluid flow. In this article, we describe the geometry of $(\mathbf{A} \times \mathbf{B}) \cdot \mathbf{C}$ (we shall call upon it when we derive the formula for surface area in Article 19.4) and derive the identities people usually use when they calculate $(\mathbf{A} \times \mathbf{B}) \cdot \mathbf{C}$ and $(\mathbf{A} \times \mathbf{B}) \times \mathbf{C}$ from the components of $\mathbf{A}$, $\mathbf{B}$, and $\mathbf{C}$.

The Triple Scalar or Box Product

The product $(\mathbf{A} \times \mathbf{B}) \cdot \mathbf{C}$ is called the **triple scalar product** of $\mathbf{A}$, $\mathbf{B}$, and $\mathbf{C}$ (in that order). As you can see from the formula

$$(\mathbf{A} \times \mathbf{B}) \cdot \mathbf{C} = |\mathbf{A} \times \mathbf{B}| \, |\mathbf{C}| \cos \theta,$$

the product is the volume of the parallelogram-sided box (parallelepiped) determined by $\mathbf{A}$, $\mathbf{B}$, and $\mathbf{C}$ in Fig. 13.53. The number $|\mathbf{A} \times \mathbf{B}|$ is the area of the base parallelogram, and $|\mathbf{C}| \cos \theta$ is the height of the tip of $\mathbf{C}$ above the plane of $\mathbf{A}$ and $\mathbf{B}$. If θ is greater than $90°$, then $\cos \theta$ is negative and we must take the absolute value of $(\mathbf{A} \times \mathbf{B}) \cdot \mathbf{C}$ to get the volume. Because of the geometry here, $(\mathbf{A} \times \mathbf{B}) \cdot \mathbf{C}$ is often called the **box product** of $\mathbf{A}$, $\mathbf{B}$, and $\mathbf{C}$.

By treating the planes of $\mathbf{B}$ and $\mathbf{C}$ and of $\mathbf{C}$ and $\mathbf{A}$ as the base planes of the parallelepiped determined by $\mathbf{A}$, $\mathbf{B}$, and $\mathbf{C}$, we can see that

$$(\mathbf{A} \times \mathbf{B}) \cdot \mathbf{C} = (\mathbf{B} \times \mathbf{C}) \cdot \mathbf{A} = (\mathbf{C} \times \mathbf{A}) \cdot \mathbf{B}. \tag{1}$$

Since the dot product is commutative, Eq. (1) also gives

$$(\mathbf{A} \times \mathbf{B}) \cdot \mathbf{C} = \mathbf{A} \cdot (\mathbf{B} \times \mathbf{C}). \tag{2}$$

In other words, the dot and cross may be interchanged in a triple scalar product without altering its value.

The triple scalar product can be evaluated as a determinant in the following way:

$$\mathbf{A} \cdot (\mathbf{B} \times \mathbf{C}) = \mathbf{A} \cdot \left[\begin{vmatrix} b_2 & b_3 \\ c_2 & c_3 \end{vmatrix} \mathbf{i} - \begin{vmatrix} b_1 & b_3 \\ c_1 & c_3 \end{vmatrix} \mathbf{j} + \begin{vmatrix} b_1 & b_2 \\ c_1 & c_2 \end{vmatrix} \mathbf{k} \right]$$

$$= a_1 \begin{vmatrix} b_2 & b_3 \\ c_2 & c_3 \end{vmatrix} - a_2 \begin{vmatrix} b_1 & b_3 \\ c_1 & c_3 \end{vmatrix} + a_3 \begin{vmatrix} b_1 & b_2 \\ c_1 & c_2 \end{vmatrix}$$

$$= \begin{vmatrix} a_1 & a_2 & a_3 \\ b_1 & b_2 & b_3 \\ c_1 & c_2 & c_3 \end{vmatrix}.$$

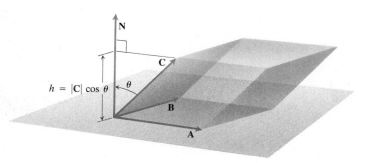

13.53 Except perhaps for sign, the number $(\mathbf{A} \times \mathbf{B}) \cdot \mathbf{C}$ is the volume of the parallelepiped (parallelogram-sided box) shown here.

The number $\mathbf{A} \cdot (\mathbf{B} \times \mathbf{C})$ can be calculated with the following formula:

$$\mathbf{A} \cdot (\mathbf{B} \times \mathbf{C}) = \begin{vmatrix} a_1 & a_2 & a_3 \\ b_1 & b_2 & b_3 \\ c_1 & c_2 & c_3 \end{vmatrix}. \qquad (3)$$

EXAMPLE 1 Find the volume of the box (parallelepiped) determined by $\mathbf{A} = \mathbf{i} + 2\mathbf{j} - \mathbf{k}$, $\mathbf{B} = -2\mathbf{i} + 3\mathbf{k}$, and $\mathbf{C} = 7\mathbf{j} - 4\mathbf{k}$.

Solution The volume is the absolute value of

$$\mathbf{A} \cdot (\mathbf{B} \times \mathbf{C}) = \begin{vmatrix} 1 & 2 & -1 \\ -2 & 0 & 3 \\ 0 & 7 & -4 \end{vmatrix}$$

$$= \begin{vmatrix} 0 & 3 \\ 7 & -4 \end{vmatrix} - 2 \begin{vmatrix} -2 & 3 \\ 0 & -4 \end{vmatrix} - \begin{vmatrix} -2 & 0 \\ 0 & 7 \end{vmatrix}$$

$$= -21 - 16 + 14 = -23,$$

or 23.   ■

Projecting Parallelograms onto Planes

Our goal here is to establish a result we shall need when we derive the formula for calculating the areas of curved surfaces in Article 19.4. The result is that if $\mathbf{A}$ and $\mathbf{B}$ are adjacent sides of a parallelogram and $\mathbf{n}$ is a unit vector, then $(\mathbf{A} \times \mathbf{B}) \cdot \mathbf{n}$ is the area of the orthogonal projection of the parallelogram onto any plane perpendicular to $\mathbf{n}$.

The way we establish this result is somewhat indirect: we show that if parallelograms $PQRS$ and $P'Q'R'S'$ are any two plane sections of a prism with sides parallel to a unit vector $\mathbf{n}$ (Fig. 13.54), then the box determined by $PQRS$

13.54 If $|\mathbf{n}| = 1$, then $(\mathbf{A} \times \mathbf{B}) \cdot \mathbf{n}$ is the area of the projection of the parallelogram determined by $\mathbf{A}$ and $\mathbf{B}$ on a plane perpendicular to $\mathbf{n}$.

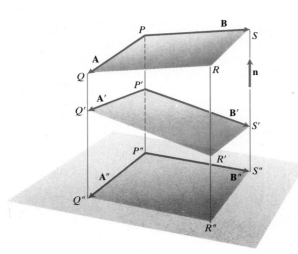

and **n** and the box determined by $P'Q'R'S'$ and **n** have the same volume. In particular, if $P''Q''R''S''$ is a right section of the prism, then the height of the box it makes with **n** is the length of **n**, one unit, and the volume is the area of the parallelogram itself. We can see why all this is true if we change over to vector notation.

Let

$$\mathbf{A} = \overrightarrow{PQ}, \qquad \mathbf{A}' = \overrightarrow{P'Q'}, \qquad \mathbf{A}'' = \overrightarrow{P''Q''}$$

$$\mathbf{B} = \overrightarrow{PS}, \qquad \mathbf{B}' = \overrightarrow{P'S'}, \qquad \mathbf{B}'' = \overrightarrow{P''S''}.$$

Then

$$\mathbf{A} = \overrightarrow{PQ} = \overrightarrow{PP'} + \overrightarrow{P'Q'} + \overrightarrow{Q'Q}$$

$$= \overrightarrow{P'Q'} + (\overrightarrow{PP'} - \overrightarrow{QQ'})$$

$$= \mathbf{A}' + s\mathbf{n}$$

for some scalar s, since both $\overrightarrow{PP'}$ and $\overrightarrow{QQ'}$ are parallel to **n.** Similarly,

$$\mathbf{B} = \mathbf{B}' + t\mathbf{n}$$

for some scalar t. Hence

$$\mathbf{A} \times \mathbf{B} = (\mathbf{A}' + s\mathbf{n}) \times (\mathbf{B}' + t\mathbf{n})$$

$$= \mathbf{A}' \times \mathbf{B}' + t(\mathbf{A}' \times \mathbf{n}) + s(\mathbf{n} \times \mathbf{B}') + st(\mathbf{n} \times \mathbf{n}). \tag{4}$$

But $\mathbf{n} \times \mathbf{n} = \mathbf{0}$, while $\mathbf{A}' \times \mathbf{n}$ and $\mathbf{n} \times \mathbf{B}'$ are both perpendicular to **n.** Therefore when we dot both sides of (4) with **n** we get

$$(\mathbf{A} \times \mathbf{B}) \cdot \mathbf{n} = (\mathbf{A}' \times \mathbf{B}') \cdot \mathbf{n}. \tag{5}$$

Thus the volumes of the boxes made by any two cross sections of the prism with **n** are equal. In particular, they are both equal to $(\mathbf{A}'' \times \mathbf{B}'') \cdot \mathbf{n}$, which is not only the volume of the box made by $P''Q''R''S''$ and **n** but is also the area of $P''Q''R''S''$.

Since $P''Q''R''S''$ could have been cut from the prism by any plane orthogonal to **n,** we have the result we need for Article 19.4:

$(\mathbf{A} \times \mathbf{B}) \cdot \mathbf{n}$ is the area of the orthogonal projection of the parallelogram determined by **A** and **B** onto any plane with unit normal **n.** $\qquad$ (6)

The statement in (6) assumes $\mathbf{A} \times \mathbf{B}$ and **n** lie on the same side of the plane determined by **A** and **B.** If they lie on opposite sides, $(\mathbf{A} \times \mathbf{B}) \cdot \mathbf{n}$ is negative, and its absolute value gives the area. Except possibly for sign, then,

$$(\mathbf{A} \times \mathbf{B}) \cdot \mathbf{i} = \text{Area of projection on the } yz\text{-plane,}$$

$$(\mathbf{A} \times \mathbf{B}) \cdot \mathbf{j} = \text{Area of projection on the } xz\text{-plane,} \tag{7}$$

$$(\mathbf{A} \times \mathbf{B}) \cdot \mathbf{k} = \text{Area of projection on the } xy\text{-plane.}$$

EXAMPLE 2 Calculate the area of the projection on the xy-plane of the parallelogram determined by the points $P(0, 0, 3)$, $Q(2, -1, 2)$, $R(3, 2, 1)$, and $S(1, 3, 2)$.

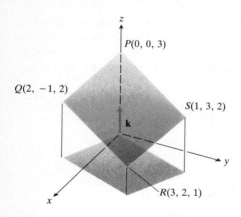

$P(0, 0, 3)$

$Q(2, -1, 2)$

$S(1, 3, 2)$

$\mathbf{k}$

$R(3, 2, 1)$

13.55   Example 2 calculates the area of the orthogonal projection of parallelogram $PQRS$ on the xy-plane.

Solution   We have

$$(\overrightarrow{PQ} \times \overrightarrow{PS}) \cdot \mathbf{k} = \begin{vmatrix} 2 & -1 & -1 \\ 1 & 3 & -1 \\ 0 & 0 & 1 \end{vmatrix} = \begin{vmatrix} 2 & -1 \\ 1 & 3 \end{vmatrix} = 7.$$

The area is 7. See Fig. 13.55.   ∎

Triple Vector Products

The **triple vector products** $(\mathbf{A} \times \mathbf{B}) \times \mathbf{C}$ and $\mathbf{A} \times (\mathbf{B} \times \mathbf{C})$ are usually not equal, but they can both be evaluated rather simply from the formulas we are about to derive.

We start by showing that the vector product $(\mathbf{A} \times \mathbf{B}) \times \mathbf{C}$ is given by

$$(\mathbf{A} \times \mathbf{B}) \times \mathbf{C} = (\mathbf{A} \cdot \mathbf{C})\mathbf{B} - (\mathbf{B} \cdot \mathbf{C})\mathbf{A}. \tag{8}$$

In this formula, $\mathbf{A} \cdot \mathbf{C}$ is a scalar multiplying $\mathbf{B}$, and $\mathbf{B} \cdot \mathbf{C}$ is a scalar multiplying $\mathbf{A}$.

If one of the vectors is the zero vector, Eq. (8) is true because both sides are zero.

If none of the vectors is zero but $\mathbf{B} = s\mathbf{A}$ for some scalar s, then both sides of Eq. (8) are zero again.

Suppose now that none of the vectors is zero and that $\mathbf{A}$ and $\mathbf{B}$ are not parallel. The vector on the left in Eq. (8) is perpendicular to $\mathbf{A} \times \mathbf{B}$ and consequently is parallel to the plane determined by $\mathbf{A}$ and $\mathbf{B}$. Hence it is possible to find scalars m and n such that

$$(\mathbf{A} \times \mathbf{B}) \times \mathbf{C} = m\mathbf{A} + n\mathbf{B}. \tag{9}$$

To calculate m and n, we introduce orthogonal unit vectors $\mathbf{I}$ and $\mathbf{J}$ in the plane of $\mathbf{A}$ and $\mathbf{B}$ with $\mathbf{I} = \mathbf{A}/|\mathbf{A}|$ (Fig. 13.56). We also introduce a third unit vector $\mathbf{K} = \mathbf{I} \times \mathbf{J}$ and write all our vectors in terms of these unit vectors $\mathbf{I}$, $\mathbf{J}$, and $\mathbf{K}$:

$$\mathbf{A} = a_1\mathbf{I},$$
$$\mathbf{B} = b_1\mathbf{I} + b_2\mathbf{J}, \tag{10}$$
$$\mathbf{C} = c_1\mathbf{I} + c_2\mathbf{J} + c_3\mathbf{K}.$$

Then

$$\mathbf{A} \times \mathbf{B} = a_1b_2\mathbf{K}$$

and

$$(\mathbf{A} \times \mathbf{B}) \times \mathbf{C} = a_1b_2c_1\mathbf{J} - a_1b_2c_2\mathbf{I}. \tag{11}$$

Combining this with Eq. (9) gives

$$m(a_1\mathbf{I}) + n(b_1\mathbf{I} + b_2\mathbf{J}) = a_1b_2c_1\mathbf{J} - a_1b_2c_2\mathbf{I}.$$

This is equivalent to the pair of scalar equations

$$ma_1 + nb_1 = -a_1b_2c_2,$$
$$nb_2 = a_1b_2c_1.$$

$\mathbf{J}$

$\mathbf{B}$

θ

$\mathbf{I}$

$\mathbf{A}$

13.56   Orthogonal vectors $\mathbf{I}$ and $\mathbf{J}$ in the plane of $\mathbf{A}$ and $\mathbf{B}$.

If b_2 were equal to zero, **A** and **B** would be parallel, contrary to hypothesis. Hence b_2 is not zero and we may solve the last equation for n. We find

$$n = a_1c_1 = \mathbf{A} \cdot \mathbf{C}.$$

Then, by substitution,

$$ma_1 = -nb_1 - a_1b_2c_2$$
$$= -a_1c_1b_1 - a_1b_2c_2,$$

and since $|\mathbf{A}| = a_1 \neq 0$, we may divide by a_1 and have

$$m = -(b_1c_1 + b_2c_2) = -(\mathbf{B} \cdot \mathbf{C}).$$

When these values are substituted for m and n in Eq. (9), we obtain Eq. (8).

The identity

$$(\mathbf{B} \times \mathbf{C}) \times \mathbf{A} = (\mathbf{B} \cdot \mathbf{A})\mathbf{C} - (\mathbf{C} \cdot \mathbf{A})\mathbf{B} \qquad (12)$$

follows from Eq. (8) by interchanging the letters **A**, **B**, and **C**. If we now interchange the factors **B** × **C** and **A** we must change the sign on the right-hand side of the equation. This gives the following identity, which is a companion of Eq. (8):

$$\boxed{\mathbf{A} \times (\mathbf{B} \times \mathbf{C}) = (\mathbf{A} \cdot \mathbf{C})\mathbf{B} - (\mathbf{A} \cdot \mathbf{B})\mathbf{C}.} \qquad (13)$$

EXAMPLE 3   Verify Eq. (8) for

$$\mathbf{A} = \mathbf{i} - \mathbf{j} + 2\mathbf{k}, \qquad \mathbf{B} = 2\mathbf{i} + \mathbf{j} + \mathbf{k}, \qquad \mathbf{C} = \mathbf{i} + 2\mathbf{j} - \mathbf{k}.$$

Solution   Since

$$\mathbf{A} \cdot \mathbf{C} = -3 \quad \text{and} \quad \mathbf{B} \cdot \mathbf{C} = 3,$$

the right-hand side of Eq. (8) is

$$(\mathbf{A} \cdot \mathbf{C})\mathbf{B} - (\mathbf{B} \cdot \mathbf{C})\mathbf{A} = -3\mathbf{B} - 3\mathbf{A} = -3(3\mathbf{i} + 3\mathbf{k}) = -9\mathbf{i} - 9\mathbf{k}.$$

To calculate the left-hand side of Eq. (8) we have

$$\mathbf{A} \times \mathbf{B} = \begin{vmatrix} \mathbf{i} & \mathbf{j} & \mathbf{k} \\ 1 & -1 & 2 \\ 2 & 1 & 1 \end{vmatrix} = -3\mathbf{i} + 3\mathbf{j} + 3\mathbf{k},$$

so that

$$(\mathbf{A} \times \mathbf{B}) \times \mathbf{C} = \begin{vmatrix} \mathbf{i} & \mathbf{j} & \mathbf{k} \\ -3 & 3 & 3 \\ 1 & 2 & -1 \end{vmatrix} = -9\mathbf{i} - 9\mathbf{k}. \qquad \blacksquare$$

EXAMPLE 4   Express $(\mathbf{A} \times \mathbf{B}) \times (\mathbf{C} \times \mathbf{D})$ in terms of scalar multiplication and cross products involving no more than two factors.

Solution   For convenience, write

$$\mathbf{C} \times \mathbf{D} = \mathbf{V}.$$

Then use Eq. (8) to evaluate

$$(A \times B) \times V = (A \cdot V)B - (B \cdot V)A$$

or

$$(A \times B) \times (C \times D) = (A \cdot C \times D)B - (B \cdot C \times D)A.$$

This expresses the answer as a scalar times **B** minus a scalar times **A**. We could use Eq. (13) instead of Eq. (8) to express the answer as a scalar times **C** minus a scalar times **D**. See Problem 16. ■

PROBLEMS

In Problems 1–3, take

$$A = 4i - 8j + k, \quad B = 2i + j - 2k,$$
$$C = 3i - 4j + 12k.$$

1. Find the volume of the box determined by **A**, **B**, and **C**.

2. Find $(A \cdot B)C$ and $A(B \cdot C)$.

3. a) Find $A \times B$ and then $(A \times B) \times C$.
   b) Find $(A \times B) \times C$ another way.

4. Find the volume of the parallelepiped whose edges are determined by $A = 3j$, $B = -2i + j$, and $C = -i + j + k$.

In Problems 5–8, find the volume of the tetrahedron whose vertices are the given points.

5. $(0, 0, 0)$, $(2, 0, 0)$, $(0, 2, 0)$, $(0, 0, 2)$

6. $(0, 0, 0)$, $(1, -1, 1)$, $(2, 1, -2)$, $(-1, 2, -1)$

7. $(0, 0, 0)$, $(2, 1, 0)$, $(2, -1, 1)$, $(1, 0, 2)$

8. $(1, 0, 3)$, $(2, 1, 1)$, $(0, 0, 2)$, $(3, 4, 1)$

9. Let $P(1, 2, -1)$, $Q(3, -1, 4)$, and $R(2, 6, 2)$ be three vertices of a parallelogram $PQRS$.
   a) Find the coordinates of S.
   b) Find the area of $PQRS$.
   c) Find the area of the projection of $PQRS$ in the xy-plane; in the yz-plane; in the xz-plane.

10. Find the line in the plane of $P(0, 0, 0)$, $Q(2, 2, 0)$, and $R(0, 1, -2)$ that passes through P perpendicular to the line
$$x = 3t, \quad y = 2t, \quad z = t/2.$$

11. Which of the following are *not always* true?
    a) $A \times B = B \times A$
    b) $A \times (B + C) = A \times B + A \times C$

c) $A \times (-A) = 0$
d) $(A \times B) \cdot B = 0$
e) $A \cdot (B \times C) = (A \cdot B) \times C$
f) $(A \times B) \cdot A = B \cdot (A \times B)$
g) $(A \times A) \cdot A = 0$
h) $(A \times B) \cdot C = (B \times C) \cdot A$
i) $(A \times B) \times C = A \times (B \times C)$

12. Suppose that
$$A \cdot A = 4, \quad B \cdot B = 4, \quad A \cdot B = 0,$$
$$(A \times B) \times C = 0, \quad (A \times B) \cdot C = 8.$$

Find
a) $A \cdot C$,   b) $|C|$,   c) $B \times C$.

(*Hint:* Picture the vectors, and think geometrically. Use basic, coordinate-free definitions. Avoid long calculations.)

13. Prove that any vector **A** satisfies the identity
$$A = \frac{1}{2}[i \times (A \times i) + j \times (A \times j) + k \times (A \times k)].$$

14. Express the product $R = (A \times B) \times (C \times D)$ in the form $aC + bD$ with scalars a and b.

15. Use Eq. (3) to show that
    a) $A \cdot (C \times B) = -A \cdot (B \times C)$,
    b) $A \cdot (A \times B) = 0$,
    c) $(A + D) \cdot (B \times C) = A \cdot (B \times C) + D \cdot (B \times C)$.

Interpret the results geometrically.

16. Show that
$$(A \times B) \times (C \times D) = (A \times B \cdot D)C - (A \times B \cdot C)D.$$
(*Hint:* Set $V = A \times B$ and apply Eq. 13.)

REVIEW QUESTIONS AND EXERCISES

1. When are two vectors equal?

2. How are vectors added and subtracted?

3. How are the length and direction of a vector calculated?

4. If a vector is multiplied by a scalar, how is the result related to the original vector? What if the scalar is zero? Negative?

5. Give an example of a geometric proof using vectors.

6. What are the equations for ideal projectile motion? How do you determine from the projectile's initial position and initial velocity vector the projectile's range, flight time, and maximum height?

7. Define the *scalar product* of two vectors. Which algebraic laws (commutative, associative, distributive) are satisfied by the operations of addition and scalar multiplication of vectors? Which of these laws is (are) not satisfied? Explain. When is the scalar product of two vectors equal to zero?

8. What is the vector projection of a vector **B** onto a vector **A**? How do you write **B** as the sum of a vector parallel to **A** and a vector orthogonal to **A**?

9. Define the *vector product* of two vectors. Which algebraic laws (commutative, associative, distributive) are satisfied by the vector product operation (combined with addition) and which are not? Explain. When is the vector product of two vectors equal to zero?

10. Derive the formula for expressing the vector product of two vectors as a determinant. What is the effect of interchanging the order of the two vectors and the corresponding rows of the determinant?

11. How are vector and scalar products used to find equations for lines, line segments, and planes? Give examples.

12. How can vectors be used to calculate the distance between a point and a line? A point and a plane? Give examples.

13. What is the geometric interpretation of the product $(\mathbf{A} \times \mathbf{B}) \cdot \mathbf{C}$ as a volume? An area? When is the product positive? Zero? Negative? How may the product be calculated from the components of **A**, **B**, and **C**?

14. What are the formulas for expressing $(\mathbf{A} \times \mathbf{B}) \times \mathbf{C}$ and $\mathbf{A} \times (\mathbf{B} \times \mathbf{C})$ in terms of scalar multiples of **A**, **B**, and **C**?

MISCELLANEOUS PROBLEMS

1. In Fig. 13.57, D is the midpoint of side AB, and E is one-third of the way between C and B. *Using vectors*, prove that F is midpoint of the line CD.

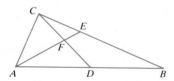

13.57   The triangle in Problem 1.

2. The vectors $2\mathbf{i} + 3\mathbf{j}$, $4\mathbf{i} + \mathbf{j}$, and $5\mathbf{i} + y\mathbf{j}$ have their initial points at the origin. Find the value of y that will make the vectors terminate on a common straight line.

3. **A** and **B** are vectors from the origin to the two points A and B. The point P is determined by the vector $\overrightarrow{OP} = x\mathbf{A} + y\mathbf{B}$, where x and y are positive numbers whose sum is equal to 1. Prove that P lies on the line segment AB.

4. Let ABC be a triangle and let M be the midpoint of AB. Let P be the point on CM that is two-thirds of the way from C to M. Let O be any point in space.
   a) Show that
   $$\overrightarrow{OP} = \left(\frac{\overrightarrow{OA} + \overrightarrow{OB} + \overrightarrow{OC}}{3} \right).$$
   b) Show how the result in part (a) leads to the conclusion that the medians of a triangle meet in a point.

5. Suppose A, B, and C are the vertices of a triangle and a, b, and c are the midpoints of the opposite sides. Show that
   $$\overrightarrow{Aa} + \overrightarrow{Bb} + \overrightarrow{Cc} = \mathbf{0}.$$
   Interpret the result geometrically.

6. Vectors are drawn from the center of a regular polygon to its vertices. Show that their sum is zero.

7. *Using vectors*, prove that for any four numbers a, b, c, and d we have the inequality
   $$(a^2 + b^2)(c^2 + d^2) \geq (ac + bd)^2.$$
   (*Hint:* Let $\mathbf{A} = a\mathbf{i} + b\mathbf{j}$ and $\mathbf{B} = c\mathbf{i} + d\mathbf{j}$.)

8. A howitzer at the point $(1720, 0)$ in the xy-plane fires a practice round whose position t seconds after firing is
   $$\mathbf{R}(t) = (256t + 1720)\mathbf{i} + (185t - 4.9t^2)\mathbf{j},$$
   with distances measured in meters. Downrange, 7860 m from the howitzer, is a hill that rises 1100 m above the x-axis. Assuming no other obstructions, will the round clear the hill? If so, by how many meters?

9. A golf ball is hit with an initial speed v at an angle α to the horizontal from a point that is at the foot of a hill inclined at an angle ϕ to the horizontal, where
   $$0 < \phi < \alpha < \frac{\pi}{2}.$$
   Show that the ball reaches the ground at a distance
   $$\frac{2v^2 \cos \alpha}{g \cos^2 \phi} \sin (\alpha - \phi)$$
   measured up the face of the hill. Hence show that the greatest range achieved for a given v is when $\alpha = (\phi/2) + (\pi/4)$.

10. Let **A**, **B**, **C** be vectors from a common point O to points A, B, C.
    a) If A, B, C are collinear, show that three constants x, y, z (not all zero) exist such that
    $$x + y + z = 0 \quad \text{and} \quad x\mathbf{A} + y\mathbf{B} + z\mathbf{C} = \mathbf{0}.$$
    b) Conversely, if three constants x, y, z (not all zero) exist such that
    $$x + y + z = 0 \quad \text{and} \quad x\mathbf{A} + y\mathbf{B} + z\mathbf{C} = \mathbf{0},$$
    show that A, B, C are collinear.

11. Find the vector projection of **B** onto **A** if
$$\mathbf{A} = 3\mathbf{i} - \mathbf{j} + \mathbf{k}, \quad \mathbf{B} = 2\mathbf{i} + \mathbf{j} - 2\mathbf{k}.$$

12. Show that $|\mathbf{A} + \mathbf{B}| \le |\mathbf{A}| + |\mathbf{B}|$ for any vectors **A** and **B**.

13. Suppose that vectors **A** and **B** are not parallel and that $\mathbf{A} = \mathbf{C} + \mathbf{D}$, where **C** is parallel to **B** and **D** is orthogonal to **B**. Express **C** and **D** in terms of **A** and **B**.

14. If $a = |\mathbf{A}|$ and $b = |\mathbf{B}|$, show that
$$\mathbf{C} = a\mathbf{B} + b\mathbf{A}$$
bisects the angle between **A** and **B**.

15. With the same notation as in Problem 14, show that $a\mathbf{B} + b\mathbf{A}$ and $b\mathbf{A} - a\mathbf{B}$ are orthogonal.

16. Show that scalar multiplication is *positive definite;* that is, $\mathbf{A} \cdot \mathbf{A} \ge 0$ for every vector **A**, and $\mathbf{A} \cdot \mathbf{A} = 0$ if and only if **A** is the zero vector.

17. Using vector methods, show that the distance d between the point (x_1, y_1) and the line $ax + by = c$ is
$$d = \frac{|ax_1 + by_1 - c|}{\sqrt{a^2 + b^2}}.$$

18. Show that the points $(0, -1, 0)$, $(1/2, 0, 1/2)$, and $(1, 1, 1)$ are collinear.

19. Find the vector from the origin to the point of intersection of the medians of the triangle whose vertices are the three points
$$A(1, -1, 2), \quad B(2, 1, 3), \quad C(-1, 2, -1).$$

20. By forming the cross product of two appropriate vectors, derive the trigonometric identity
$$\sin(A - B) = \sin A \cos B - \cos A \sin B.$$

21. Show that
$$\mathbf{i} \times (\mathbf{j} \times (\mathbf{i} + \mathbf{j})) \ne (\mathbf{i} \times \mathbf{j}) \times (\mathbf{i} + \mathbf{j}).$$

22. Find a vector normal to the plane determined by the points
$$A(1, 0, -1), \quad B(2, -1, 1), \quad C(-1, 1, 2).$$

23. The three vectors
$$\mathbf{A} = 3\mathbf{i} - \mathbf{j} + \mathbf{k}, \quad \mathbf{B} = \mathbf{i} + 2\mathbf{j} - \mathbf{k},$$
$$\mathbf{C} = \mathbf{i} + \mathbf{j} + \mathbf{k}$$
are all drawn from the origin. Find an equation for the plane through their endpoints.

24. Show that the line of intersection of the planes
$$x + 2y - 2z = 5 \quad \text{and} \quad 5x - 2y - z = 0$$
is parallel to the line
$$x = -3 + 2t, \quad y = 3t, \quad z = 1 + 4t.$$

25. Find the point in which the line through the origin normal to the plane $2x - y - z = 4$ meets the plane $x + y - 2z = 2$.

26. Find the point in which the line through $P(3, 2, 1)$ normal to the plane $2x - y + 2z = -2$ meets that plane.

27. Find a vector parallel to the plane $2x - y - z = 4$ and perpendicular to the vector $\mathbf{i} + \mathbf{j} + \mathbf{k}$.

28. Find a vector of *length 2* parallel to the line of intersection of the planes
$$x + 2y + z - 1 = 0 \quad \text{and} \quad x - y + 2z + 7 = 0.$$

29. What angle does the line of intersection of the planes
$$2x + y - z = 0 \quad \text{and} \quad x + y + 2z = 0$$
make with the nonnegative x-axis?

30. Find equations for the line in the plane $z = 3$ that makes an angle of $30°$ with **i** and an angle of $60°$ with **j**.

31. The line L given by the equations
$$x = 3 + 2t, \quad y = 2t, \quad z = t$$
intersects the plane
$$x + 3y - z + 4 = 0.$$
Find the point of intersection P and find equations for the line in this plane that passes through P perpendicular to L.

32. Which of the following are equations for the plane through the points $P(1, 1, -1)$, $Q(3, 0, 2)$, and $R(-2, 1, 0)$?
a) $(2\mathbf{i} - 3\mathbf{j} + 3\mathbf{k}) \cdot ((x + 2)\mathbf{i} + (y - 1)\mathbf{j} + z\mathbf{k}) = 0$
b) $x = 3 - t, \quad y = -11t, \quad z = 2 - 3t$
c) $(x + 2) + 11(y - 1) = -3z$
d) $(2\mathbf{i} - 3\mathbf{j} + 3\mathbf{k}) \times ((x + 2)\mathbf{i} + (y - 1)\mathbf{j} + z\mathbf{k}) = 0$
e) $(2\mathbf{i} - \mathbf{j} + 3\mathbf{k}) \times (-3\mathbf{i} + \mathbf{k}) \cdot ((x + 2)\mathbf{i} + (y - 1)\mathbf{j} + z\mathbf{k}) = 0$

33. Show that for every real number k the plane
$$x - 2y + z + 3 + k(2x - y - z + 1) = 0$$
contains the line of intersection of the planes $x - 2y + z + 3 = 0$ and $2x - y - z + 1 = 0$.

34. The equation $\mathbf{N} \cdot \overrightarrow{P_0 P} = 0$ represents a plane through P_0 perpendicular to **N**. What set does the inequality $\mathbf{N} \cdot \overrightarrow{P_0 P} > 0$ represent? Explain.

35. Show that
$$\begin{vmatrix} x_1 - x & y_1 - y & z_1 - z \\ x_2 - x & y_2 - y & z_2 - z \\ x_3 - x & y_3 - y & z_3 - z \end{vmatrix} = 0$$
is an equation for the plane through the three noncollinear points
$$(x_1, y_1, z_1), \quad (x_2, y_2, z_2), \quad \text{and} \quad (x_3, y_3, z_3).$$

36. Let $P_i(x_i, y_i, z_i)$, $i = 1, 2, 3$, be three noncollinear points. What set is described by the equation
$$\begin{vmatrix} x & y & z & 1 \\ x_1 & y_1 & z_1 & 1 \\ x_2 & y_2 & z_2 & 1 \\ x_3 & y_3 & z_3 & 1 \end{vmatrix} = 0?$$

37. Given the two lines
$$x = a_1 s + b_1, \quad y = a_2 s + b_2, \quad z = a_3 s + b_3,$$
$$x = c_1 t + d_1, \quad y = c_2 t + d_2, \quad z = c_3 t + d_3,$$
where s and t are parameters. Show that a necessary and

sufficient condition that the lines either intersect or are parallel is

$$\begin{vmatrix} a_1 & c_1 & b_1 - d_1 \\ a_2 & c_2 & b_2 - d_2 \\ a_3 & c_3 & b_3 - d_3 \end{vmatrix} = 0.$$

38. Find an equation for the sphere that has the planes

$$x + y + z - 3 = 0, \quad x + y + z - 9 = 0$$

as tangent planes, if the planes

$$2x - y = 0, \quad 3x - z = 0$$

pass through the center of the sphere. (*Hint:* Find equations for the line in which the planes $2x - y = 0$ and $3x - z = 0$ intersect. Then use the formula from Problem 50.)

39. If

$$\mathbf{A} = 2\mathbf{i} - \mathbf{j} + \mathbf{k}, \quad \mathbf{B} = \mathbf{i} + 2\mathbf{j} - \mathbf{k},$$
$$\mathbf{C} = \mathbf{i} + \mathbf{j} - 2\mathbf{k},$$

find a *unit vector* orthogonal to $\mathbf{A}$ in the plane of $\mathbf{B}$ and $\mathbf{C}$.

40. Find an equation for the plane through $A(-2, 0, -3)$ and $B(1, -2, 1)$ that lies parallel to the line through $C(-2, -13/5, 26/5)$ and $D(16/5, -13/5, 0)$.

41. Find the angle between the planes $x = 7$ and $x + y + \sqrt{2}z = -3$.

42. A parallelogram in Fig. 13.58 has vertices at $A(2, -1, 4)$, $B(1, 0, -1)$, $C(1, 2, 3)$, and D.
a) Find the coordinates of D.
b) Find the cosine of the interior angle at B.
c) Find the vector projection of $\overrightarrow{BA}$ onto $\overrightarrow{BC}$.
d) Find the area of the parallelogram.
e) Find an equation for the plane in which the parallelogram lies.

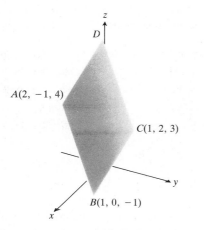

13.58 The parallelogram in Problem 42.

43. Two surface ships on maneuvers are trying to determine a submarine's course and speed to prepare for an aircraft in-

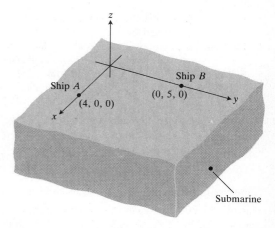

13.59 The ships and submarine in Problem 43. The picture is not to scale.

tercept. As shown in Fig. 13.59, ship A is located at $(4, 0, 0)$ while ship B is located at $(0, 5, 0)$. All coordinates are given in thousands of feet. Ship A locates the submarine in the direction of the vector $2\mathbf{i} + 3\mathbf{j} - (1/3)\mathbf{k}$, and ship B locates it in the direction of the vector $18\mathbf{i} - 6\mathbf{j} - \mathbf{k}$. Four minutes ago, the submarine was located at $(2, -1, -1/3)$. The aircraft is due in 20 minutes. Assuming the submarine moves in a straight line at a constant speed, to what position should the surface ships direct the aircraft?

44. Two helicopters, H_1 and H_2, are traveling together. At time $t = 0$ hours, they separate and follow different straight line paths given by

$$H_1: \quad x = 6 + 40t, \quad y = -3 + 10t, \quad z = -3 + 2t$$
$$H_2: \quad x = 6 + 110t, \quad y = -3 + 4t, \quad z = -3 + t,$$

all coordinates measured in miles. Due to system malfunctions, H_2 stops its flight at $(446, 13, 1)$ and, in a negligible amount of time, lands at $(446, 13, 0)$. Two hours later, H_1 is advised of this fact and heads toward H_2 at 150 mph. How long will it take H_1 to reach H_2?

45. Find the distance from the point $P(2, 2, 0)$ to the line $x = -t, y = t, z = -1 + t$.

46. Find the distance from the point $P(0, 4, 1)$ to the line $x = 2 + t, y = 2 + t, z = t$.

47. Find the distance from the point $P(1, 4, 0)$ to the plane through the points $A(0, 0, 0)$, $B(2, 0, -1)$, and $C(2, -1, 0)$.

48. Find the distance from the point $P(2, 2, 3)$ to the plane $2x + 3y + 5z = 0$.

49. a) Show that the distance between the parallel planes

$$Ax + By + Cz = D_1, \quad Ax + By + Cz = D_2$$

is given by the formula

$$\text{Distance between planes} = \frac{|D_1 - D_2|}{|A\mathbf{i} + B\mathbf{j} + C\mathbf{k}|}. \quad (1)$$

b) Use Eq. (1) to find the distance between the planes $2x + 3y - z = 6$ and $2x + 3y - z = 12$.

c) Find an equation for the plane parallel to the plane $2x - y + 2z = -4$ if the point $(3, 2, -1)$ is equidistant from both planes.

d) Write equations for the planes that lie parallel to and 5 units away from the plane $x - 2y + z = 3$.

50. Show, by vector methods, that the distance from the point $P_1(x_1, y_1, z_1)$ to the plane $Ax + By + Cz - D = 0$ is

$$\frac{|Ax_1 + By_1 + Cz_1 - D|}{\sqrt{A^2 + B^2 + C^2}}.$$

(Compare this formula with the one in Problem 17.)

51. Find the distance between the line L_1 through the two points $A(1, 0, -1)$, $B(-1, 1, 0)$ and the line L_2 through the two points $C(3, 1, -1)$, $D(4, 5, -2)$. (The distance is to be measured along a line perpendicular to L_1 and L_2. First find a vector $\mathbf{N}$ perpendicular to both lines, and then project $\overrightarrow{AC}$ onto $\mathbf{N}$.)

52. Repeat Problem 51 for the points $A(4, 0, 2)$, $B(2, 4, 1)$, $C(1, 3, 2)$, $D(2, 2, 4)$.

53. If

$$\mathbf{A} = \mathbf{i} + \mathbf{j} - \mathbf{k}, \quad \mathbf{B} = 2\mathbf{i} + \mathbf{j} + \mathbf{k},$$

$$\mathbf{C} = -\mathbf{i} - 2\mathbf{j} + 3\mathbf{k},$$

find

a) $\mathbf{A} \cdot (\mathbf{B} \times \mathbf{C})$,   b) $\mathbf{A} \times (\mathbf{B} \times \mathbf{C})$.

54. Prove that four points A, B, C, D are coplanar (lie in a common plane) if and only if $\overrightarrow{AD} \cdot (\overrightarrow{AB} \times \overrightarrow{BC}) = 0$.

55. Prove or disprove the formula

$$\mathbf{A} \times [\mathbf{A} \times (\mathbf{A} \times \mathbf{B})] \cdot \mathbf{C} = -|\mathbf{A}|^2 \mathbf{A} \cdot \mathbf{B} \times \mathbf{C}.$$

56. If the four vectors $\mathbf{A}, \mathbf{B}, \mathbf{C}, \mathbf{D}$ are coplanar, show that

$$(\mathbf{A} \times \mathbf{B}) \times (\mathbf{C} \times \mathbf{D}) = \mathbf{0}.$$

57. Show that if $\mathbf{A}, \mathbf{B},$ and $\mathbf{C}$ are any vectors, then

a) $\mathbf{A} \times (\mathbf{B} \times \mathbf{C}) + \mathbf{B} \times (\mathbf{C} \times \mathbf{A}) + \mathbf{C} \times (\mathbf{A} \times \mathbf{B}) = \mathbf{0}$.

b) $\mathbf{A} \times \mathbf{B} = [\mathbf{A} \cdot (\mathbf{B} \times \mathbf{i})]\mathbf{i}$
$\qquad + [\mathbf{A} \cdot (\mathbf{B} \times \mathbf{j})]\mathbf{j} + [\mathbf{A} \cdot (\mathbf{B} \times \mathbf{k})]\mathbf{k}$.

58. Show that if $\mathbf{A}, \mathbf{B}, \mathbf{C},$ and $\mathbf{D}$ are any vectors, then

$$(\mathbf{A} \times \mathbf{B}) \cdot (\mathbf{C} \times \mathbf{D}) = \begin{vmatrix} \mathbf{A} \cdot \mathbf{C} & \mathbf{B} \cdot \mathbf{C} \\ \mathbf{A} \cdot \mathbf{D} & \mathbf{B} \cdot \mathbf{D} \end{vmatrix}.$$

Unit Vectors and Direction Cosines

Suppose that $\mathbf{u}$ is a unit vector that makes angles $\alpha, \beta,$ and γ with the coordinate axes when we represent it by an arrow with its initial point at the origin. Figure 13.60 shows the case in which the three angles are acute. Then

$$\mathbf{u} \cdot \mathbf{i} = |\mathbf{u}||\mathbf{i}| \cos \alpha = \cos \alpha,$$
$$\mathbf{u} \cdot \mathbf{j} = |\mathbf{u}||\mathbf{j}| \cos \beta = \cos \beta, \qquad (2)$$
$$\mathbf{u} \cdot \mathbf{k} = |\mathbf{u}||\mathbf{k}| \cos \gamma = \cos \gamma.$$

If we write

$$\mathbf{u} = u_1\mathbf{i} + u_2\mathbf{j} + u_3\mathbf{k},$$

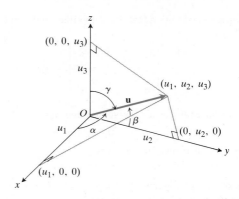

13.60 If $\mathbf{u} = u_1\mathbf{i} + u_2\mathbf{j} + u_3\mathbf{k}$ is a unit vector, then $u_1 = \cos \alpha$, $u_2 = \cos \beta$, $u_3 = \cos \gamma$. We get this information from the three right triangles shown here.

then

$$\mathbf{u} \cdot \mathbf{i} = u_1, \quad \mathbf{u} \cdot \mathbf{j} = u_2, \quad \mathbf{u} \cdot \mathbf{k} = u_3. \qquad (3)$$

Thus,

$$u_1 = \cos \alpha, \quad u_2 = \cos \beta, \quad u_3 = \cos \gamma. \qquad (4)$$

That is, the components of any unit vector $\mathbf{u}$ are the cosines of the angles it makes with the coordinate axes:

$$\mathbf{u} = u_1\mathbf{i} + u_2\mathbf{j} + u_3\mathbf{k} = \mathbf{i} \cos \alpha + \mathbf{j} \cos \beta + \mathbf{k} \cos \gamma. \qquad (5)$$

If $\mathbf{A}$ is any nonzero vector, then $\mathbf{A}$ and its direction $\mathbf{u} = \mathbf{A}/|\mathbf{A}|$ make the same angles with the coordinate axes. To find the cosines of these angles we need only calculate the components of the unit vector

$$\frac{\mathbf{A}}{|\mathbf{A}|} = \mathbf{i} \cos \alpha + \mathbf{j} \cos \beta + \mathbf{k} \cos \gamma.$$

These cosines are called the *direction cosines* of $\mathbf{A}$. In other words, the direction cosines of $\mathbf{A}$ are the components of the direction of $\mathbf{A}$.

Find the direction cosines of the vectors in Problems 59–68.

59. $10\mathbf{i} + 11\mathbf{j} - 2\mathbf{k}$ **60.** $3\mathbf{i} + 4\mathbf{k}$

61. $\mathbf{j} + \sqrt{3}\mathbf{k}$ **62.** $6\mathbf{i} - 2\mathbf{j} + 3\mathbf{k}$

63. $\mathbf{i}$ **64.** $\mathbf{i} + \mathbf{j}$

65. $\mathbf{i} + \mathbf{j} + \mathbf{k}$ **66.** $8\mathbf{i} - 11\mathbf{j} - 2\mathbf{k}$

67. $\dfrac{2}{7}\mathbf{i} - \dfrac{3}{7}\mathbf{j} + \dfrac{6}{7}\mathbf{k}$

68. $-\dfrac{1}{\sqrt{2}}\mathbf{i} + \dfrac{1}{\sqrt{3}}\mathbf{j} + \dfrac{1}{\sqrt{6}}\mathbf{k}$

69. Let $\alpha, \beta,$ and γ be the angles that a line through the origin makes with the positive x-, y-, and z-axes. Show that $\cos^2\alpha + \cos^2\beta + \cos^2\gamma = 1$.

TOOLKIT PROGRAMS

3D Grapher

CHAPTER **14**

Vector Functions and Motion

OVERVIEW

In this chapter, we combine calculus and vectors to study the motion of a body in space. We do this by treating the scalar components of the radius vector $\mathbf{R}(t)$ from the origin to the body as differentiable functions of time (Fig. 14.1). The vectors $\mathbf{v}(t)$ and $\mathbf{a}(t)$ that describe the body's velocity and acceleration are then calculated by differentiating the components of $\mathbf{R}$:

$$\mathbf{R}(t) = x(t)\mathbf{i} + y(t)\mathbf{j} + z(t)\mathbf{k},$$

$$\mathbf{v}(t) = \frac{d\mathbf{R}}{dt} = \frac{dx}{dt}\mathbf{i} + \frac{dy}{dt}\mathbf{j} + \frac{dz}{dt}\mathbf{k},$$

$$\mathbf{a}(t) = \frac{d^2\mathbf{R}}{dt^2} = \frac{d^2x}{dt^2}\mathbf{i} + \frac{d^2y}{dt^2}\mathbf{j} + \frac{d^2z}{dt^2}\mathbf{k}.$$

The vectors $\mathbf{R}$, $\mathbf{v}$, and $\mathbf{a}$ are vector functions of t and, as the preceding formulas suggest, the rules for differentiating vector functions are natural extensions of the rules for differentiating real-valued functions, or

scalar functions as they are sometimes called. As we shall see, the rules for differentiating dot products and cross products of vectors have the same form as the product rule for scalar functions.

After introducing derivatives, we turn our attention to the tangent and normal vectors associated with travel in space. One of the important parameters here is arc length, and we shall study that along with the notions of curvature and torsion, which are measures of how much a body's path turns and twists as it moves along. Finally, we shall apply our vector calculus to satellite motion and show how Kepler's laws for planetary and satellite orbits may be derived from Newton's laws of motion and gravitation.

Throughout our discussion, we shall assume that any motion that takes place in a plane takes place in the xy-plane in space. The formulas for planar motion will then be special cases of the formulas for motion in space, obtained by setting the $\mathbf{k}$-components of the vectors involved equal to zero.

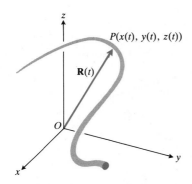

14.1   We use a variable radius vector, $\mathbf{R}(t) = x(t)\mathbf{i} + y(t)\mathbf{j} + z(t)\mathbf{k}$, to describe the different positions of a body moving in space.

14.1
Vector Functions and Motion

In this article, we define derivatives of vector functions and develop the rules for differentiating dot products and cross products. As we shall see, these rules have the same form as the rule for differentiating products of real-valued functions.

The most important vector function is the radius vector

$$\mathbf{R}(t) = x(t)\mathbf{i} + y(t)\mathbf{j} + z(t)\mathbf{k}$$

from the origin to the point $P(x(t), y(t), z(t))$ that gives the position at time t of a particle moving through space. This is the particle's **position vector.** Its first and second derivatives with respect to t are the particle's velocity and acceleration.

We begin with an example in which we find a vector function that describes the velocity of an ideal projectile at any time t during its flight.

The Velocity of an Ideal Projectile

Suppose a projectile is launched at time $t = 0$ from a point (x_0, y_0) into the first quadrant of a vertical xy-plane. The projectile has an initial speed v_0 and is launched at an angle α. What will the projectile's speed and direction be t_1 seconds later?

To answer this question, we let

$$\mathbf{R}(t) = x(t)\mathbf{i} + y(t)\mathbf{j}$$

be the projectile's position vector at time t and assume that x and y are differentiable functions of t. The vector

$$\Delta\mathbf{R} = \mathbf{R}(t_1 + \Delta t) - \mathbf{R}(t_1)$$

then gives the projectile's displacement for the time interval from t_1 to $t_1 + \Delta t$ (Fig. 14.2). As Δt approaches zero, the secant line PQ approaches the tangent

14.2 A projectile's displacement, described as an increment in the position vector.

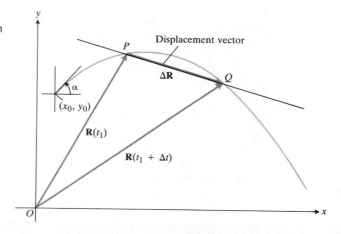

to the projectile's path at P, and $\Delta \mathbf{R}$ contracts to zero. The quotient

$$\frac{\Delta \mathbf{R}}{\Delta t} = \frac{\Delta x}{\Delta t}\mathbf{i} + \frac{\Delta y}{\Delta t}\mathbf{j}$$

does not approach zero, however, but approaches the vector

$$\mathbf{v}(t_1) = \left(\frac{dx}{dt}\bigg|_{t=t_1}\right)\mathbf{i} + \left(\frac{dy}{dt}\bigg|_{t=t_1}\right)\mathbf{j}.$$

This vector is tangent to the curve at P and points in the direction of the projectile's motion. Furthermore, its length is

$$\sqrt{\left(\frac{dx}{dt}\right)^2 + \left(\frac{dy}{dt}\right)^2} = \frac{ds}{dt},$$

the speed of the projectile at t_1.

To answer our original question, the projectile's speed t_1 seconds into the flight is $|\mathbf{v}(t_1)|$ and its direction is $\mathbf{v}_1(t)/|\mathbf{v}(t_1)|$. In short, the projectile's velocity at $t = t_1$ is the vector $\mathbf{v}(t_1)$.

EXAMPLE 1 A projectile is fired into the first quadrant from the origin at an initial speed of 500 m/sec at an angle of 60°. What will its velocity, speed, and direction be 10 sec later?

Solution According to Article 13.2, Example 5, the parametric equations for a projectile fired into the first quadrant from the point (x_0, y_0) with an initial speed v_0 at an angle α are

$$x(t) = x_0 + (v_0 \cos \alpha)t,$$

$$y(t) = y_0 + (v_0 \sin \alpha)t - \frac{1}{2}gt^2.$$

With $x_0 = 0$, $y_0 = 0$, $v_0 = 500$, $\alpha = 60°$, and $g = 9.8$, these equations become

$$x(t) = 250t, \qquad y = 250\sqrt{3}t - 4.9t^2.$$

The projectile's position vector is therefore

$$\mathbf{R}(t) = (250t)\mathbf{i} + (250\sqrt{3}t - 4.9t^2)\mathbf{j}$$

and its velocity vector at time t is

$$\mathbf{v}(t) = \frac{d}{dt}(250t)\mathbf{i} + \frac{d}{dt}(250\sqrt{3}t - 4.9t^2)\mathbf{j}$$

$$= 250\mathbf{i} + (250\sqrt{3} - 9.8t)\mathbf{j}.$$

Ten seconds into the flight, when $t = 10$, the projectile's velocity, speed, and direction are

Velocity: $\mathbf{v}(10) = 250\mathbf{i} + (250\sqrt{3} - 98)\mathbf{j}$

                    $= 250\mathbf{i} + 335\mathbf{j},$ (rounded)

Speed: $|\mathbf{v}(10)| = \sqrt{(250)^2 + (335)^2} = 418$ m/sec,   (rounded)

Direction: $\mathbf{v}(10)/|\mathbf{v}(10)| = 0.6\mathbf{i} + 0.8\mathbf{j}.$ (rounded)

The Derivative of a Vector Function

If $x(t)$, $y(t)$, and $z(t)$ are real-valued functions of a real variable t, the vector

$$\mathbf{F}(t) = x(t)\mathbf{i} + y(t)\mathbf{j} + z(t)\mathbf{k}$$

is a **vector-valued function** of t or, for short, a **vector function** of t. The domain of $\mathbf{F}$ is the set of t-values common to the domains of x, y, and z. In our previous discussions, $\mathbf{F}(t)$ was the position vector of a moving body, but it need not have this or any other interpretation to be a vector function of t.

If $x(t)$, $y(t)$, and $z(t)$ are differentiable functions of t, the vector

$$\mathbf{F}'(t) = \frac{d\mathbf{F}}{dt} = \frac{dx}{dt}\mathbf{i} + \frac{dy}{dt}\mathbf{j} + \frac{dz}{dt}\mathbf{k} \tag{1}$$

is called the **derivative** of $\mathbf{F}$ with respect to t. The domain of $\mathbf{F}'$ is the set of t-values common to the domains of the derivatives of x, y, and z. A vector function $\mathbf{F}(t)$ is said to be **differentiable at a value** $t = t_1$ if x, y, and z are differentiable at t_1. A vector function $\mathbf{F}(t)$ is said to be **differentiable** if it is differentiable at every t-value in its domain.

EXAMPLE 2   Find $d\mathbf{F}/dt$ at $t = \pi/3$ if

$$\mathbf{F}(t) = (\sin t)\mathbf{i} + (\ln t)\mathbf{j} + (\tan t)\mathbf{k}.$$

Solution   We find $d\mathbf{F}/dt$ by differentiating the components of $\mathbf{F}$ with respect to t:

$$\mathbf{F}'(t) = (\cos t)\mathbf{i} + \frac{1}{t}\mathbf{j} + (\sec^2 t)\mathbf{k}.$$

The derivative's value at $t = \pi/3$ is the vector

$$\mathbf{F}'\left(\frac{\pi}{3}\right) = \frac{1}{2}\mathbf{i} + \frac{3}{\pi}\mathbf{j} + 4\mathbf{k}. \qquad \blacksquare$$

EXAMPLE 3   Find $d\mathbf{R}/dt$ if

$$\mathbf{R}(t) = (250t)\mathbf{i} + (250\sqrt{3}t - 4.9t^2)\mathbf{j}.$$

Solution

$$\frac{d\mathbf{R}}{dt} = 250\mathbf{i} + (250\sqrt{3} - 9.8t)\mathbf{j}.$$

This derivative is the velocity vector of the projectile in Example 1. $\blacksquare$

Position, Velocity, Speed, and Acceleration

If $\mathbf{R}(t) = x(t)\mathbf{i} + y(t)\mathbf{j} + z(t)\mathbf{k}$ is the position vector of a body in space, we call $d\mathbf{R}/dt$ and $d^2\mathbf{R}/dt^2$, when they exist, the body's **velocity** and **acceleration.** As in the plane, the length of the velocity vector is the **speed** with which the body moves along its path, and the direction of the velocity vector is the direction of the motion. We shall say more about this in the next article.

DEFINITIONS

Position: $\mathbf{R}(t) = x(t)\mathbf{i} + y(t)\mathbf{j} + z(t)\mathbf{k}$

Velocity: $\mathbf{v} = \dfrac{d\mathbf{R}}{dt}$

Speed: $|\mathbf{v}|$

Direction: $\dfrac{\mathbf{v}}{|\mathbf{v}|}$

Acceleration: $\dfrac{d^2\mathbf{R}}{dt^2}$

EXAMPLE 4   The vector

$$\mathbf{R}(t) = (3 \cos t)\mathbf{i} + (3 \sin t)\mathbf{j} + t^2\mathbf{k}$$

gives the position of a moving body at time t. Find the body's speed and direction when $t = 2$. At what times, if any, are the body's velocity and acceleration orthogonal?

Solution   We have

$$\mathbf{R}(t) = (3 \cos t)\mathbf{i} + (3 \sin t)\mathbf{j} + t^2\mathbf{k},$$

$$\mathbf{v}(t) = \frac{d\mathbf{R}}{dt} = -(3 \sin t)\mathbf{i} + (3 \cos t)\mathbf{j} + 2t\mathbf{k},$$

$$\mathbf{a}(t) = \frac{d^2\mathbf{R}}{dt^2} = -(3 \cos t)\mathbf{i} - (3 \sin t)\mathbf{j} + 2\mathbf{k}.$$

When $t = 2$, the body's speed and direction are

Speed: $|\mathbf{v}(2)| = \sqrt{(-3 \sin 2)^2 + (3 \cos 2)^2 + (4)^2} = 5$

Direction: $\mathbf{v}(2)/|\mathbf{v}(2)| = -\left(\dfrac{3}{5} \sin 2\right)\mathbf{i} + \left(\dfrac{3}{5} \cos 2\right)\mathbf{j} + \dfrac{4}{5}\mathbf{k}.$

To find when $\mathbf{v}$ and $\mathbf{a}$ are orthogonal, we look for values of t for which

$$\mathbf{v}(t) \cdot \mathbf{a}(t) = 9 \sin t \cos t - 9 \cos t \sin t + 4t = 4t = 0.$$

The only value is $t = 0$.   ∎

The Chain Rule

If $\mathbf{R}(t)$ gives the position of a particle moving along a path and we change variables by substituting a function of s for t, the particle's position becomes a function of s. The particle moves along the same path as s varies, but with a different velocity (unless s happens to equal t). The derivatives $d\mathbf{R}/dt$ and $d\mathbf{R}/ds$ are related by the rule

$$\frac{d\mathbf{R}}{ds} = \frac{d\mathbf{R}}{dt} \frac{dt}{ds}. \tag{2}$$

At each point, the derivative dt/ds is a scalar by which $d\mathbf{R}/dt$ may be multiplied to give $d\mathbf{R}/ds$.

To see why this is so, suppose that

$$\mathbf{F}(t) = x(t)\mathbf{i} + y(t)\mathbf{j} + z(t)\mathbf{k}$$

is a differentiable vector function of t and that $t = g(s)$ is a differentiable scalar function of s. Then x, y, and z, being differentiable functions of t, are also differentiable functions of s and the Chain Rule for differentiable real-valued functions gives

$$
\begin{aligned}
\frac{d\mathbf{F}}{ds} &= \frac{dx}{ds}\mathbf{i} + \frac{dy}{ds}\mathbf{j} + \frac{dx}{ds}\mathbf{k} \\
&= \frac{dx}{dt}\frac{dt}{ds}\mathbf{i} + \frac{dy}{dt}\frac{dt}{ds}\mathbf{j} + \frac{dz}{dt}\frac{dt}{ds}\mathbf{k} \\
&= \left(\frac{dx}{dt}\mathbf{i} + \frac{dy}{dt}\mathbf{j} + \frac{dz}{dt}\mathbf{k}\right)\frac{dt}{ds} \\
&= \frac{d\mathbf{F}}{dt}\frac{dt}{ds}.
\end{aligned}
\tag{3}
$$

In short,

$$\frac{d\mathbf{F}}{ds} = \frac{d\mathbf{F}}{dt}\frac{dt}{ds}, \tag{4}$$

a result we call the **Chain Rule** for vector-valued functions of a real variable.

EXAMPLE 5   Find $d\mathbf{F}/ds$ as a function of s if

$$\mathbf{F}(t) = \mathbf{i} + \sin(t + 1)\mathbf{j} + \sqrt{t}\,\mathbf{k} \quad \text{and} \quad t = s^2 - 1.$$

Solution   We apply the Chain Rule, Eq. (4), and use the substitution $t = s^2 - 1$ to express the result in terms of s:

$$\frac{d\mathbf{F}}{ds} = \frac{d\mathbf{F}}{dt}\frac{dt}{ds} = \left(\cos(t + 1)\mathbf{j} + \frac{1}{2\sqrt{t}}\mathbf{k}\right)(2s)$$

$$= (2s \cos s^2)\mathbf{j} + \frac{s}{\sqrt{s^2 - 1}}\mathbf{k}. \qquad \blacksquare$$

Derivatives of Dot Products and Cross Products

If $\mathbf{U}(t)$ and $\mathbf{V}(t)$ are differentiable vector functions of t, then the products $\mathbf{U} \cdot \mathbf{V}$ and $\mathbf{U} \times \mathbf{V}$ are differentiable functions of t and their derivatives may be calculated from the rules

$$\frac{d}{dt}(\mathbf{U} \cdot \mathbf{V}) = \frac{d\mathbf{U}}{dt} \cdot \mathbf{V} + \mathbf{U} \cdot \frac{d\mathbf{V}}{dt}, \tag{5}$$

$$\frac{d}{dt}(\mathbf{U} \times \mathbf{V}) = \frac{d\mathbf{U}}{dt} \times \mathbf{V} + \mathbf{U} \times \frac{d\mathbf{V}}{dt}. \tag{6}$$

Notice the similarity between Eqs. (5) and (6) and the product rule for real-valued functions. Also, remember to preserve the order of the factors in

Eq. (6): first **U,** then **V.** If **U** comes first on the left, it must also come first on the right or the signs will be wrong.

One way to derive Eqs. (5) and (6) is to express **U** and **V** in terms of components, multiply the products out, and apply the ordinary rules for differentiating products of scalar functions. The problem with this method, however, is that it is cumbersome. It is better to apply the Δ-process directly to the vectors themselves.

To verify Eq. (6), for example, let

$$\mathbf{W} = \mathbf{U} \times \mathbf{V},$$

where **U** and **V** are differentiable functions of t. Let Δt be an increment in t and let $\Delta\mathbf{W}$, $\Delta\mathbf{U}$, and $\Delta\mathbf{V}$ be the corresponding increments in **W, U,** and **V.** Then

$$\mathbf{W} + \Delta\mathbf{W} = (\mathbf{U} + \Delta\mathbf{U}) \times (\mathbf{V} + \Delta\mathbf{V})$$

$$= \mathbf{U} \times \mathbf{V} + \mathbf{U} \times \Delta\mathbf{V} + \Delta\mathbf{U} \times \mathbf{V} + \Delta\mathbf{U} \times \Delta\mathbf{V},$$

and

$$\frac{\Delta\mathbf{W}}{\Delta t} = \mathbf{U} \times \frac{\Delta\mathbf{V}}{\Delta t} + \frac{\Delta\mathbf{U}}{\Delta t} \times \mathbf{V} + \frac{\Delta\mathbf{U}}{\Delta t} \times \Delta\mathbf{V}.$$

Now take limits as $\Delta t \to 0$:

$$\lim\frac{\Delta\mathbf{W}}{\Delta t} = \lim\left(\frac{\Delta w_1}{\Delta t}\mathbf{i} + \frac{\Delta w_2}{\Delta t}\mathbf{j} + \frac{\Delta w_3}{\Delta t}\mathbf{k}\right)$$

$$= \frac{dw_1}{dt}\mathbf{i} + \frac{dw_2}{dt}\mathbf{j} + \frac{dw_3}{dt}\mathbf{k}$$

$$= \frac{d\mathbf{W}}{dt}.$$

Similarly,

$$\lim\frac{\Delta\mathbf{U}}{\Delta t} = \frac{d\mathbf{U}}{dt}, \qquad \lim\frac{\Delta\mathbf{V}}{\Delta t} = \frac{d\mathbf{V}}{dt},$$

and

$$\lim \Delta\mathbf{V} = \lim\frac{\Delta\mathbf{V}}{\Delta t} \cdot \lim \Delta t = \frac{d\mathbf{V}}{dt} \cdot 0 = 0.$$

The net result is

$$\frac{d\mathbf{W}}{dt} = \mathbf{U} \times \frac{d\mathbf{V}}{dt} + \frac{d\mathbf{U}}{dt} \times \mathbf{V}, \tag{7}$$

which is equivalent to Eq. (6).

Derivatives of Vectors of Constant Length

If **V** is a differentiable function of t whose length is a constant, then

$$\mathbf{V} \cdot \mathbf{V} = |\mathbf{V}|^2$$

is a constant and we may differentiate both sides of this equation to get

$$\mathbf{V} \cdot \frac{d\mathbf{V}}{dt} + \frac{d\mathbf{V}}{dt} \cdot \mathbf{V} = 0, \qquad 2\mathbf{V} \cdot \frac{d\mathbf{V}}{dt} = 0,$$

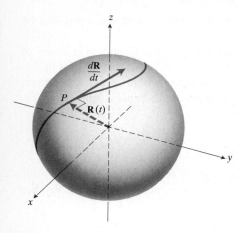

14.3 The velocity vector of a particle P that moves on the surface of a sphere is tangent to the sphere.

or, simply,

$$\mathbf{V} \cdot \frac{d\mathbf{V}}{dt} = 0. \tag{8}$$

Equation (8) says that $\mathbf{V}$ and its derivative $d\mathbf{V}/dt$ are orthogonal. In other words, the derivative of a vector of constant length is always a vector orthogonal to it.

Conversely, if $\mathbf{V}$ is a differentiable function of t and $\mathbf{V} \cdot d\mathbf{V}/dt = 0$, then the length of $\mathbf{V}$ is a constant (Problem 31).

EXAMPLE 6   Show that

$$V(t) = (\sin t)\mathbf{i} + (\cos t)\mathbf{j} + \sqrt{3}\mathbf{k}$$

has a constant length and is orthogonal to its derivative.

Solution   We have

$$\mathbf{V}(t) = (\sin t)\mathbf{i} + (\cos t)\mathbf{j} + \sqrt{3}\mathbf{k}$$
$$|\mathbf{V}(t)| = \sqrt{\sin^2 t + \cos^2 t + 3} = 2$$
$$\mathbf{V}'(t) = (\cos t)\mathbf{i} - (\sin t)\mathbf{j}$$
$$\mathbf{V}(t) \cdot \mathbf{V}'(t) = \sin t \cos t - \cos t \sin t = 0. \quad \blacksquare$$

EXAMPLE 7   Suppose a particle on a sphere centered at the origin moves in such a way that its position vector $\mathbf{R}(t)$ is differentiable. Show that the velocity $d\mathbf{R}/dt$ is always orthogonal to $\mathbf{R}$ (Fig. 14.3).

Solution   The length of $\mathbf{R}$ is constant because it always equals the radius of the sphere. Therefore, $\mathbf{R} \cdot d\mathbf{R}/dt = 0.$ $\quad \blacksquare$

Finding $\mathbf{R}(t)$ by Integration

The next example shows how we can sometimes find the path of a particle's motion by integrating its acceleration vector when the particle's initial velocity and position are given.

EXAMPLE 8   The acceleration of a particle in the plane is

$$\frac{d^2\mathbf{R}}{dt^2} = (\cos t)\mathbf{i} + (\sin t)\mathbf{j}. \tag{9}$$

Find the particle's position as a function of t if $\mathbf{v}(0) = \mathbf{j}$ and $\mathbf{R}(0) = 2\mathbf{i}$.

Solution   We integrate Eq. (9) twice, applying the initial conditions to determine the constants of integration as we go along. The first integration gives

$$\frac{d\mathbf{R}}{dt} = (\sin t)\mathbf{i} - (\cos t)\mathbf{j} + \mathbf{C}_1,$$

where $\mathbf{C}_1$ is a *vector*. The condition $\mathbf{v}(0) = \mathbf{j}$ gives

$$\mathbf{j} = (0)\mathbf{i} - (1)\mathbf{j} + \mathbf{C}_1, \qquad \mathbf{C}_1 = 2\mathbf{j}.$$

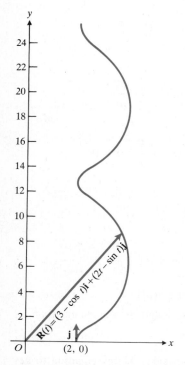

14.4 The path traced by the particle in Example 8.

Hence,

$$\frac{d\mathbf{R}}{dt} = (\sin t)\mathbf{i} + (2 - \cos t)\mathbf{j}.$$

A second integration gives

$$\mathbf{R}(t) = -(\cos t)\mathbf{i} + (2t - \sin t)\mathbf{j} + \mathbf{C}_2,$$

and the condition $\mathbf{R}(0) = 2\mathbf{i}$ gives

$$2\mathbf{i} = -\mathbf{i} + \mathbf{C}_2, \qquad \mathbf{C}_2 = 3\mathbf{i}.$$

The particle's position vector is therefore

$$\mathbf{R}(t) = (3 - \cos t)\mathbf{i} + (2t - \sin t)\mathbf{j}.$$

The path traced by the particle is shown in Fig. 14.4. ∎

PROBLEMS

In Problems 1–12, $\mathbf{R}(t)$ gives the position of a particle at time t. Find the particle's velocity and acceleration vectors. Then find the particle's speed at the instant given.

1. $\mathbf{R}(t) = (2 \cos t)\mathbf{i} + (3 \sin t)\mathbf{j}, \quad t = \pi/4$

2. $\mathbf{R}(t) = (t + 1)\mathbf{i} + (t^2 - 1)\mathbf{j}, \quad t = 2$

3. $\mathbf{R}(t) = (\cos 2t)\mathbf{i} + (2 \sin t)\mathbf{j}, \quad t = 0$

4. $\mathbf{R}(t) = e^t\mathbf{i} + \frac{2}{9}e^{2t}\mathbf{j}, \quad t = \ln 3$

5. $\mathbf{R}(t) = (\sec t)\mathbf{i} + (\tan t)\mathbf{j}, \quad t = \pi/6$

6. $\mathbf{R}(t) = [\ln(t + 1)]\mathbf{i} + t^2\mathbf{j}, \quad t = 1$

7. $\mathbf{R}(t) = (a \cos \omega t)\mathbf{i} + (a \sin \omega t)\mathbf{j}, \quad t = \pi/(3\omega)$ (a and ω are positive constants)

8. $\mathbf{R}(t) = (e^{-t})\mathbf{i} + (2 \cos 3t)\mathbf{j} + (2 \sin 3t)\mathbf{k}, \quad t = 0$

9. $\mathbf{R}(t) = (1 + t)\mathbf{i} + (t^2/\sqrt{2})\mathbf{j} + (t^3/3)\mathbf{k}, \quad t = 1$

10. $\mathbf{R}(t) = (4 \cos t)\mathbf{i} - (3 \sin t)\mathbf{j} + 2t\mathbf{k}, \quad t = \pi/3$

11. $\mathbf{R}(t) = (\sqrt{2}/(t + 1))\mathbf{i} + (t^3/3)\mathbf{j} + (\sqrt{2} \ln t)\mathbf{k}, \quad t = 1$

12. $\mathbf{R}(t) = (\cosh 3t)\mathbf{i} + (2 \sinh t)\mathbf{j}, \quad t = 0$

In Problems 13–16, $\mathbf{R}(t)$ is the position vector of a body at time t. Find the time or times in the given time interval when the velocity and acceleration vectors are orthogonal.

13. $\mathbf{R}(t) = (t - \sin t)\mathbf{i} + (1 - \cos t)\mathbf{j}, \quad 0 \le t \le 2\pi$

14. $\mathbf{R}(t) = (\sin t)\mathbf{i} + t\mathbf{j} + (\cos t)\mathbf{k}, \quad 0 \le t$

15. $\mathbf{R}(t) = \mathbf{i} + (5 \cos t)\mathbf{j} + (3 \sin t)\mathbf{k}, \quad 0 \le t \le \pi$

16. $\mathbf{R}(t) = (v_0\cos \alpha)t\mathbf{i} + \left((v_0\sin \alpha)t - \frac{1}{2}gt^2\right)\mathbf{j}, \quad 0 \le t$

In Problems 17–20, $\mathbf{R}(t)$ is the position vector of a particle at time t. Find the angle between the particle's velocity and acceleration vectors when $t = 0$.

17. $\mathbf{R}(t) = (e^t)\mathbf{i} + (e^t \sin t)\mathbf{j} + (e^t \cos t)\mathbf{k}$

18. $\mathbf{R}(t) = (\tan t)\mathbf{i} + (\sinh 2t)\mathbf{j} + (\text{sech } 3t)\mathbf{k}$

19. $\mathbf{R}(t) = (\ln(t^2 + 1))\mathbf{i} + (\tan^{-1}t)\mathbf{j} + (\sqrt{t^2 + 1})\mathbf{k}$

20. $\mathbf{R}(t) = (v_0\cos \alpha)t\mathbf{i} + \left((v_0\sin \alpha)t - \frac{1}{2}gt^2\right)\mathbf{j}$

In Problems 21–24, find $\mathbf{R}(t)$ from the given acceleration and initial conditions.

21. $\mathbf{a}(t) = -3t\mathbf{i}, \quad \mathbf{v}(0) = 2\mathbf{j}, \quad \mathbf{R}(0) = 4\mathbf{i}$

22. $\mathbf{a}(t) = 3t\mathbf{i} + 4\mathbf{j} + \mathbf{k}, \quad \mathbf{v}(0) = 4\mathbf{i}, \quad \mathbf{R}(0) = 5\mathbf{j}$

23. $\mathbf{a}(t) = (1 + t)^{-1/2}\mathbf{i} - e^{-t}\mathbf{j}, \quad \mathbf{v}(0) = -\mathbf{i} + \mathbf{j}, \quad \mathbf{R}(0) = (1/3)\mathbf{i} - \mathbf{j} + \mathbf{k}$

24. $\mathbf{a}(t) = -9.8\mathbf{j}, \quad \mathbf{v}(0) = (v_0\cos \alpha)\mathbf{i} + (v_0\sin \alpha)\mathbf{j}, \quad \mathbf{R}(0) = x_0\mathbf{i} + y_0\mathbf{j}$

25. Let $\mathbf{R}(t) = t\mathbf{i} + t^2\mathbf{j}$ be the position vector of a particle moving in the xy-plane.
   a) Find $\mathbf{v}$ and $\mathbf{a}$.
   b) Sketch the path of the motion, and draw $\mathbf{v}$ and $\mathbf{a}$ for $t = 2$ as vectors starting at the point $(2, 4)$.

26. *Resistance proportional to velocity.* If a projectile encounters a resistance proportional to velocity, the total force $m\, d^2\mathbf{R}/dt^2$ on the projectile satisfies the equation

$$m\frac{d^2\mathbf{R}}{dt^2} = -mg\mathbf{j} - k\frac{d\mathbf{R}}{dt}.$$

Show that one integration of this equation leads to the equation

$$\frac{d\mathbf{R}}{dt} + \frac{k}{m}\mathbf{R} = \mathbf{v}(0) - gt\mathbf{j}.$$

Solve this equation. (*Hint:* Multiply both sides by $e^{(k/m)t}$. Then the left side will be the derivative of the product $\mathbf{R}e^{(k/m)t}$ and both sides of the equation can be integrated.)

27. CALCULATOR A ball is thrown horizontally from the top edge of a 100-m building at a speed of 20 m/sec. What will the ball's velocity and speed be 2 sec later?

28. A particle is fired from the origin into the first octant at time $t = 0$ with velocity $\mathbf{v}(0)$ (Fig. 14.5). The acceleration is the constant vector $\mathbf{a} = -g\mathbf{k}$. Find the particle's velocity and position as functions of t.

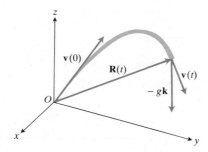

14.5 If a projectile is fired over the xy-plane from the origin with initial velocity $\mathbf{v}(0)$, its velocity and position at any later time can be found by integration (Problem 28).

29. A particle moves around the ellipse $(x/3)^2 + (y/2)^2 = 1$ in such a way that its position at time t is $\mathbf{R}(t) = (3 \cos t)\mathbf{i} + (2 \sin t)\mathbf{j}$. Find the maximum and minimum values of the magnitude of the acceleration.

30. Suppose the position vectors of particles P_1 and P_2 moving in space are

$$\mathbf{R}_1(s) = 2s\mathbf{i} + (s - 5)\mathbf{j} + s\mathbf{k},$$
$$\mathbf{R}_2(t) = (3 + t)\mathbf{i} - (2 + t)\mathbf{j} + (1 + t)\mathbf{k},$$

with s and t both representing time.
a) Find the point of intersection of the lines traced by P_1 and P_2. At what time does each particle reach the intersection point?
b) Since s and t both represent time, set s equal to t and find the distance between P_1 and P_2 as a function of t. At what time are the particles closest?

31. Let $\mathbf{V}(t)$ be a differentiable function of t. Show that if $\mathbf{V} \cdot \mathbf{V}' = 0$ for all t, then the length of $\mathbf{V}$ is constant.

32. Derive Eq. (5) by the Δ-process.

33. *Derivatives of triple scalar products.*
a) Use Eqs. (5) and (6) to show that

$$\frac{d}{dt}(\mathbf{U} \cdot \mathbf{V} \times \mathbf{W}) =$$

$$\frac{d\mathbf{U}}{dt} \cdot \mathbf{V} \times \mathbf{W} + \mathbf{U} \cdot \frac{d\mathbf{V}}{dt} \times \mathbf{W} + \mathbf{U} \cdot \mathbf{V} \times \frac{d\mathbf{W}}{dt}. \quad (10)$$

b) Show that Eq. (10) is equivalent to

$$\frac{d}{dt}\begin{vmatrix} u_1 & u_2 & u_3 \\ v_1 & v_2 & v_3 \\ w_1 & w_2 & w_3 \end{vmatrix} = \begin{vmatrix} \dfrac{du_1}{dt} & \dfrac{du_2}{dt} & \dfrac{du_3}{dt} \\ v_1 & v_2 & v_3 \\ w_1 & w_2 & w_3 \end{vmatrix} + \begin{vmatrix} u_1 & u_2 & u_3 \\ \dfrac{dv_1}{dt} & \dfrac{dv_2}{dt} & \dfrac{dv_3}{dt} \\ w_1 & w_2 & w_3 \end{vmatrix}$$

$$+ \begin{vmatrix} u_1 & u_2 & u_3 \\ v_1 & v_2 & v_3 \\ \dfrac{dw_1}{dt} & \dfrac{dw_2}{dt} & \dfrac{dw_3}{dt} \end{vmatrix} \quad (11)$$

(Equation 11 says that the derivative of a determinant of order three is the sum of the three determinants obtained from the original by differentiating one row at a time. The result may be extended to determinants of any order.)

34. Suppose that $\mathbf{F}(t) = x(t)\mathbf{i} + y(t)\mathbf{j} + z(t)\mathbf{k}$ and that x, y, and z have derivatives of order one, two, and three. Use Eq. (10) of Problem 33 or, equivalently, Eq. (11) to show that

$$\frac{d}{dt}\left[F \cdot \left(\frac{d\mathbf{F}}{dt} \times \frac{d^2\mathbf{F}}{dt^2} \right) \right] = F \cdot \left(\frac{d\mathbf{F}}{dt} \times \frac{d^3\mathbf{F}}{dt^3} \right).$$

(*Hint:* Carry out the differentiation with respect to t on the left and look for vectors whose products are zero.)

35. The derivation of Eq. (7) tacitly assumed that

$$\lim_{\Delta t \to 0}\left(\mathbf{U} \times \frac{\Delta \mathbf{V}}{\Delta t} \right) = \mathbf{U} \times \lim_{\Delta t \to 0}\frac{\Delta \mathbf{V}}{\Delta t}. \quad (12)$$

Use the determinant formula for cross products (Article 13.5, Eq. 8) to show that this is indeed the case.

 TOOLKIT PROGRAMS

Parametric Equations   Super * Grapher

14.2

Directed Distance and the Unit Tangent Vector **T**

The curve traced by a body whose position vector $\mathbf{R}(t)$ is a differentiable function of t is called a **differentiable curve**. As you can imagine, differentiable curves, especially those with continuous first and second derivatives, have been subjects of intense study, for their mathematical interest as well as their applications. In this article and the next, we describe some of the features that account for the popularity of these curves.

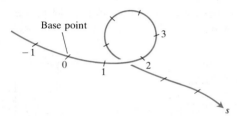

14.6 Continuously differentiable curves can all be scaled like coordinate axes. The coordinate given to each point is its directed distance from a preselected base point.

One of the special features of continuously differentiable curves is that they are smooth enough to have a measurable length. This enables us to locate points on these curves by giving their directed distance s along the curve from some chosen base point, the way we locate points on coordinate axes by giving their directed distance from the origin. Thus, every continuously differentiable curve has the potential to be a curved coordinate axis (Fig. 14.6). While time is the natural parameter for describing a moving body's velocity and acceleration, the directed distance coordinate s is the natural parameter for studying a curve's geometry.

To simplify the presentation, we shall refer to the curve traced by the tip of a position vector $\mathbf{R}(t) = x(t)\mathbf{i} + y(t)\mathbf{j} + z(t)\mathbf{k}$ as "the curve $\mathbf{R}(t) = x(t)\mathbf{i} + y(t)\mathbf{j} + z(t)\mathbf{k}$."

Directed Distance along a Curve

As Fig. 14.7 suggests, we define the length of the curve $\mathbf{R}(t) = x(t)\mathbf{i} + y(t)\mathbf{j} + z(t)\mathbf{k}$ as a limit of lengths

$$\sum_{a}^{b} \sqrt{(x_{k+1} - x_k)^2 + (y_{k+1} - y_k)^2 + (z_{k+1} - z_k)^2} \tag{1}$$

of approximating polygonal paths in much the same way that we defined the length of a plane curve in Article 5.5. The limit exists when x, y, and z have continuous first derivatives and is given by the formula

$$L = \int_{a}^{b} \sqrt{x'(t)^2 + y'(t)^2 + z'(t)^2}\, dt. \tag{2}$$

Thus, the length of the curve from $t = a$ to $t = b$ is found by integrating the length of the velocity vector of the motion from a to b. Dropping the z-term from Eq. (2) gives the usual arc length formula for curves in the plane.

DEFINITION	**Length of a Curve**

The length of the curve $\mathbf{R}(t) = x(t)\mathbf{i} + y(t)\mathbf{j} + z(t)\mathbf{k}$ from $t = a$ to $t = b$ is

$$L = \int_{a}^{b} \sqrt{x'(t)^2 + y'(t)^2 + z'(t)^2}\, dt = \int_{a}^{b} |\mathbf{v}(t)|\, dt. \tag{3}$$

14.7 The length of the curve from $t = a$ to $t = b$ is a limit of lengths of polygonal paths.

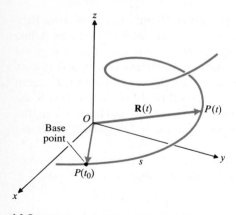

14.8 The directed distance along the curve from $P(t_0)$ to any point $P(t)$ is

$$s = \int_{t_0}^{t} |\mathbf{v}| \, d\tau.$$

If we choose a base point $P(t_0)$ on the curve, as shown in Fig. 14.8, the integral of $|\mathbf{v}|$ from t_0 to t gives the directed distance s along the curve from $P(t_0)$ to $P(t)$.

Directed Distance along a Curve from t_0 to t

$$s = \int_{t_0}^{t} \sqrt{x'(\tau)^2 + y'(\tau)^2 + z'(\tau)^2} \, d\tau = \int_{t_0}^{t} |\mathbf{v}| d\tau \qquad (4)$$

The value of s is positive if t is greater than t_0 and negative if t is less than t_0. If the derivatives beneath the radical are continuous, the First Fundamental Theorem of Calculus tells us that s is a differentiable function of t whose derivative is

$$\frac{ds}{dt} = \sqrt{x'(t)^2 + y'(t)^2 + z'(t)^2} = |\mathbf{v}(t)|, \qquad (5)$$

the magnitude of the velocity vector of $\mathbf{R}(t)$. Once again we find that the speed with which a particle moves along its path is the magnitude of $\mathbf{v}$:

$$\text{Speed} = \frac{ds}{dt} = |\mathbf{v}|. \qquad (6)$$

Notice that while the base point $P(t_0)$ plays a role in defining s, it plays no role in Eq. (5). The rate at which s changes somewhere does not depend on how far away the base point is.

Notice also that as long as $|\mathbf{v}|$ is different from zero, *as we shall assume it to be from now on,* ds/dt is positive and s is an increasing function of t.

EXAMPLE 1 Find the length of one turn of the helix

$$\mathbf{R}(t) = (\cos t)\mathbf{i} + (\sin t)\mathbf{j} + t\mathbf{k}.$$

Solution This is the helix in Fig. 14.9. Since $\cos t$ and $\sin t$ have period 2π, the helix makes one full turn as t runs from 0 to 2π. According to Eq. (3), therefore, the length we seek is

$$\int_{0}^{2\pi} |\mathbf{v}| \, dt = \int_{0}^{2\pi} \sqrt{(-\sin t)^2 + (\cos t)^2 + (1)^2} \, dt = \int_{0}^{2\pi} \sqrt{2} \, dt = 2\pi\sqrt{2}.$$

This is $\sqrt{2}$ times the length of the unit circle in the xy-plane over which the helix stands. (Also, see Problem 15.) ∎

EXAMPLE 2 Show that if $\mathbf{u} = u_1\mathbf{i} + u_2\mathbf{j} + u_3\mathbf{k}$ is a unit vector, then the directed distance along the line

$$x = x_0 + tu_1, \qquad y = y_0 + tu_2, \qquad z = z_0 + tu_3$$

from the base point $P_0(x_0, y_0, z_0)$ to the point

$$P_t(x_0 + tu_1, y_0 + tu_2, z_0 + tu_3)$$

is $s = t$.

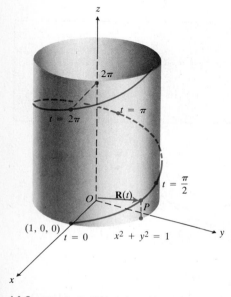

14.9 The helix $\mathbf{R}(t) = (\cos t)\mathbf{i} + (\sin t)\mathbf{j} + t\mathbf{k}$ spirals up from the xy-plane as t increases from zero.

Solution The vector equation for the line is

$$\mathbf{R}(t) = (x_0 + tu_1)\mathbf{i} + (y_0 + tu_2)\mathbf{j} + (z_0 + tu_3)\mathbf{k}.$$

Hence,

$$\mathbf{v}(t) = u_1\mathbf{i} + u_2\mathbf{j} + u_3\mathbf{k} = \mathbf{u}$$

and the directed distance from P_0 to P_t is

$$s = \int_0^t |\mathbf{v}|d\tau = \int_0^t |\mathbf{u}|d\tau = \int_0^t 1 \, d\tau = \tau\Big]_0^t = t. \qquad\blacksquare$$

The Unit Tangent Vector **T**

Suppose that the component functions of

$$\mathbf{R}(t) = x(t)\mathbf{i} + y(t)\mathbf{j} + z(t)\mathbf{k}$$

have continuous first derivatives and that $s(t)$ gives the directed distance along the curve traced by the tip of $\mathbf{R}$, measured from some preselected base point $P(t_0)$. Since ds/dt is required to be nonzero, s is one-to-one and has an inverse that gives t as a differentiable function of s (Article 6.1). The derivative of the inverse function is

$$\frac{dt}{ds} = \frac{1}{ds/dt} = \frac{1}{|\mathbf{v}|}. \qquad (7)$$

This makes $\mathbf{R}$ a differentiable function of s whose derivative can be calculated from the Chain Rule to be

$$\frac{d\mathbf{R}}{ds} = \frac{d\mathbf{R}}{dt}\frac{dt}{ds} = \frac{\mathbf{v}}{|\mathbf{v}|}. \qquad (8)$$

From this equation we see that

$$\left|\frac{d\mathbf{R}}{ds}\right| = \frac{|\mathbf{v}|}{|\mathbf{v}|} = 1. \qquad (9)$$

Together, Eqs. (8) and (9) say that $d\mathbf{R}/ds$ is a unit vector that points in the direction of $\mathbf{v}$. We call $d\mathbf{R}/ds$ the *unit tangent vector* of the curve traced by the tip of $\mathbf{R}$ and denote it by $\mathbf{T}$. See Fig. 14.10.

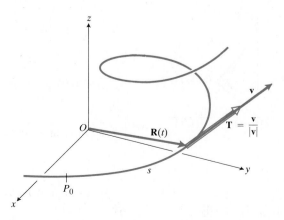

14.10   We divide **v** by $|\mathbf{v}| = ds/dt$ to get the unit tangent vector **T**.

DEFINITION

Unit Tangent Vector

$$\mathbf{T} = \frac{d\mathbf{R}}{ds} = \frac{d\mathbf{R}/dt}{ds/dt} = \frac{\mathbf{v}}{|\mathbf{v}|}.$$

The vector $\mathbf{T}$ is a differentiable function of the parameter t whenever $\mathbf{v}$ is a differentiable function of t. As we shall see in the next article, $\mathbf{T}$ is one of three unit vectors in a traveling reference frame that describes the motion of space vehicles and other bodies moving in three dimensions.

EXAMPLE 3   Find $\mathbf{T}$ for the helix

$$\mathbf{R}(t) = (\cos t)\mathbf{i} + (\sin t)\mathbf{j} + t\mathbf{k}.$$

Solution   With

$$\mathbf{v}(t) = -(\sin t)\mathbf{i} + (\cos t)\mathbf{j} + \mathbf{k},$$
$$|\mathbf{v}(t)| = \sqrt{(-\sin t)^2 + (\cos t)^2 + (1)^2}$$
$$= \sqrt{2},$$

we have

$$\mathbf{T} = \frac{\mathbf{v}}{|\mathbf{v}|} = -\frac{\sin t}{\sqrt{2}}\mathbf{i} + \frac{\cos t}{\sqrt{2}}\mathbf{j} + \frac{1}{\sqrt{2}}\mathbf{k}.$$

EXAMPLE 4   Find $\mathbf{T}$ for the plane curve

$$\mathbf{R}(t) = (\cos t + t \sin t)\mathbf{i} + (\sin t - t \cos t)\mathbf{j}, \qquad t > 0.$$

Solution

$$\mathbf{v} = \frac{d\mathbf{R}}{dt} = (-\sin t + t \cos t + \sin t)\mathbf{i} + (\cos t + t \sin t - \cos t)\mathbf{j}$$

$$= (t \cos t)\mathbf{i} + (t \sin t)\mathbf{j},$$
$$|\mathbf{v}| = \sqrt{t^2\cos^2 t + t^2\sin^2 t} = |t| = t,$$

$$\mathbf{T} = \frac{\mathbf{v}}{|\mathbf{v}|} = (\cos t)\mathbf{i} + (\sin t)\mathbf{j}.$$

See Fig. 14.20.

EXAMPLE 5   For the counterclockwise motion

$$\mathbf{R}(t) = (\cos t)\mathbf{i} + (\sin t)\mathbf{j}$$

around the unit circle,

$$\mathbf{v} = (-\sin t)\mathbf{i} + (\cos t)\mathbf{j}$$

is already a unit vector, so $\mathbf{v} = \mathbf{T}$. In fact, $\mathbf{T}$ is $\mathbf{R}$ rotated 90° counterclockwise (Fig. 14.11).

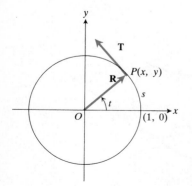

14.11 $\mathbf{R}(t) = (\cos t)\mathbf{i} + (\sin t)\mathbf{j}.$

PROBLEMS

In Problems 1–10, the vector $\mathbf{R}(t)$ is the position vector of a particle moving in the plane or in space. Find the unit tangent vector $\mathbf{T}$ for the curve traced by the particle. Also, find the length of the indicated section of the curve.

1. $\mathbf{R}(t) = (2 \cos t)\mathbf{i} + (2 \sin t)\mathbf{j} + \sqrt{5}t\mathbf{k}$, section from $t = 0$ to $t = \pi$

2. $\mathbf{R}(t) = (\cos^3 t)\mathbf{i} + (\sin^3 t)\mathbf{j}$, section from $t = 0$ to $t = \pi/2$

3. $\mathbf{R}(t) = t\mathbf{i} + (2/3)t^{3/2}\mathbf{j}$, section from $t = 0$ to $t = 8$

4. $\mathbf{R}(t) = (2 + t)\mathbf{i} - (t + 1)\mathbf{j} + t\mathbf{k}$, section from $t = 0$ to $t = 3$

5. $\mathbf{R}(t) = t^3(6\mathbf{i} - 2\mathbf{j} - 3\mathbf{k})$, section from $t = -1$ to $t = 1$

6. $\mathbf{R}(t) = (6 \sin 2t)\mathbf{i} + (6 \cos 2t)\mathbf{j} + 5t\mathbf{k}$, section from $t = 0$ to $t = \pi$

7. $\mathbf{R}(t) = (t \cos t)\mathbf{i} + (t \sin t)\mathbf{j} + (2\sqrt{2}/3)t^{3/2}\mathbf{k}$, section from $t = 0$ to $t = \pi$

8. $\mathbf{R}(t) = (t \sin t + \cos t)\mathbf{i} + (t \cos t - \sin t)\mathbf{j}$, section from $t = 0$ to $t = a$

9. $\mathbf{R}(t) = (a \cos \omega t)\mathbf{i} + (a \sin \omega t)\mathbf{j} + b\mathbf{k}$, $a > 0$, $\omega > 0$, section from $t = 0$ to $t = 2\pi/\omega$

10. $\mathbf{R}(t) = (3 \cosh 2t)\mathbf{i} + (3 \sinh 2t)\mathbf{j} + 6t\mathbf{k}$, section from $t = 0$ to $t = \ln 2$

In Problems 11 and 12, assume that the directed distance along the given curve is measured from the point where $t = 0$. Find the directed distance coordinate s as a function of t by evaluating the integral

$$s = \int_0^t |\mathbf{v}(\tau)|\,d\tau.$$

Also, find the length of the indicated section of the curve.

11. $\mathbf{R}(t) = (e^t \cos t)\mathbf{i} + (e^t \sin t)\mathbf{j} + e^t\mathbf{k}$, section from $t = -\pi$ to $t = \pi$

12. $\mathbf{R}(t) = (\sinh t)\mathbf{i} + (\cosh t)\mathbf{j} + t\mathbf{k}$, section from $t = 0$ to $t = \ln 2$

13. Find $\mathbf{v}$, $|\mathbf{v}|$, $\mathbf{T}$, and $\mathbf{a}$ at $t = 0$ if $\mathbf{R}(t) = 3t^2\mathbf{i} + (6t + 7)\mathbf{j} + 2t^4\mathbf{k}$.

14. Find the length of the curve traced by

$$\mathbf{R}(t) = (\sqrt{2}t)\mathbf{i} + (\sqrt{2}t)\mathbf{j} + (1 - t^2)\mathbf{k}$$

from $(0, 0, 1)$ to $(\sqrt{2}, \sqrt{2}, 0)$.

15. The length $2\pi\sqrt{2}$ of one turn of the helix in Example 1 is also the length of the diagonal of a square 2π units on a side. Show how to obtain this square by cutting away and flattening a portion of the cylinder around which the helix winds.

TOOLKIT PROGRAMS

Parametric Equations   Super ∗ Grapher

14.3
Curvature, Torsion, and the **TNB** Frame

In this article, we define a frame of mutually orthogonal unit vectors that always travels along with a body moving in space (Fig. 14.12). There are three vectors in the frame. One of them is the unit tangent vector $\mathbf{T}$. Another is the vector $\mathbf{N}$ that gives the direction of $d\mathbf{T}/ds$. The third is $\mathbf{B} = \mathbf{T} \times \mathbf{N}$. These vectors and their derivatives, when available, give useful information about

14.12 Every moving body travels with a **TNB** frame that describes how the body moves.

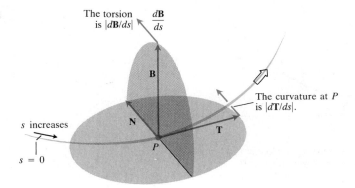

moving vehicles and the paths they trace. For example, the magnitude of the derivative $d\mathbf{T}/ds$ tells how much a vehicle's path turns to the left or right as it moves along; it is called the *curvature* of the vehicle's path. The number $|d\mathbf{B}/ds|$ tells how much a vehicle's path rotates or twists as the vehicle moves along; it is called the *torsion* of the vehicle's path. Look at Fig. 14.12 again. If P were a train climbing up a curved track, the rate at which the headlight turned from side to side per unit distance would be the curvature of the track. The rate at which the engine rotated about its longitudinal axis (the line of $\mathbf{T}$) would be the torsion.

We begin with curves in the plane.

The Curvature of a Plane Curve

As we move along a differentiable curve in the plane, the unit tangent vector $\mathbf{T}$ turns as the curve bends. We measure the rate at which $\mathbf{T}$ turns by measuring the change in the angle ϕ that $\mathbf{T}$ makes with $\mathbf{i}$ (Fig. 14.13). At each point P, the absolute value of $d\phi/ds$, stated in radians per unit of length along the curve, is called the **curvature** at P. If $|d\phi/ds|$ is large, $\mathbf{T}$ turns sharply as we pass through P and the curvature at P is large. If $|d\phi/ds|$ is close to zero, $\mathbf{T}$ turns more slowly and the curvature at P is small. On circles and lines, the curvature is constant, as we shall see in Examples 1 and 2. On other curves, the curvature can vary from place to place. The traditional symbol for the curvature function is the Greek letter κ (kappa).

If $\mathbf{R}(t) = x(t)\mathbf{i} + y(t)\mathbf{j}$ is a vector equation for a curve in the plane, we can derive a formula for the curvature κ in the following way. In Newton's dot notation, in which $\dot{y}$ means dy/dt, $\ddot{y}$ means d^2y/dt^2, and so on, we have

$$\tan \phi = \frac{dy}{dx} = \frac{dy/dt}{dx/dt} = \frac{\dot{y}(t)}{\dot{x}(t)} \qquad (1)$$

and

$$\phi = \tan^{-1}\left(\frac{\dot{y}}{\dot{x}}\right).$$

Hence,

$$\frac{d\phi}{ds} = \frac{d\phi}{dt}\frac{dt}{ds}$$

$$= \frac{1}{1 + (\dot{y}/\dot{x})^2} \frac{d}{dt}\left(\frac{\dot{y}}{\dot{x}}\right) \frac{1}{(\dot{x}^2 + \dot{y}^2)^{1/2}}$$

$$= \frac{\dot{x}^2}{(\dot{x}^2 + \dot{y}^2)^{3/2}} \frac{\dot{x}\ddot{y} - \dot{y}\ddot{x}}{\dot{x}^2} \qquad (2)$$

$$= \frac{\dot{x}\ddot{y} - \dot{y}\ddot{x}}{|\mathbf{v}|^3}.$$

The curvature, therefore, is

$$\left|\frac{d\phi}{ds}\right| = \frac{|\dot{x}\ddot{y} - \dot{y}\ddot{x}|}{|\mathbf{v}|^3}. \qquad (3)$$

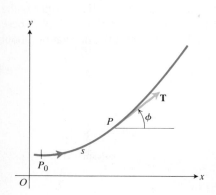

14.13 The value of $|d\phi/ds|$ at P is called the curvature of the curve at P.

The observation that $|\ddot{x}\ddot{y} - \dot{y}\ddot{x}|$ is the magnitude of the vector

$$\mathbf{v} \times \mathbf{a} = \begin{vmatrix} \mathbf{i} & \mathbf{j} & \mathbf{k} \\ \dot{x} & \dot{y} & 0 \\ \ddot{x} & \ddot{y} & 0 \end{vmatrix} \qquad (4)$$

enables us to write Eq. (3) in the following compact vector form.

Curvature

$$\kappa = \frac{|\mathbf{v} \times \mathbf{a}|}{|\mathbf{v}|^3} \qquad (5)$$

Equation (5) calculates the curvature, a geometric property of the curve, from the velocity and acceleration of any vector representation of the curve in which $|\mathbf{v}|$ is different from zero. Take a moment to think about how remarkable this really is: From any formula for motion along a curve, no matter how variable the motion may be (as long as $\mathbf{v}$ is never zero), we can calculate a physical property of the curve that seems to have nothing to do with the way the curve is traversed.

EXAMPLE 1 Show that the curvature of a straight line is zero.

Solution On a straight line, ϕ is constant (Fig. 14.14). Therefore, $d\phi/ds$ and $\kappa = |d\phi/ds|$ are both zero. ■

EXAMPLE 2 Show that the curvature of a circle of radius a is $1/a$.

Solution We start with

$$\mathbf{R}(t) = (a \cos t)\mathbf{i} + (a \sin t)\mathbf{j} + (0)\mathbf{k},$$

$$\mathbf{v}(t) = (-a \sin t)\mathbf{i} + (a \cos t)\mathbf{j},$$

$$\mathbf{a}(t) = (-a \cos t)\mathbf{i} - (a \sin t)\mathbf{j}.$$

14.14 Along a line, the angle ϕ stays the same from point to point and the curvature, $d\phi/ds$, is zero.

Then

$$\mathbf{v} \times \mathbf{a} = \begin{vmatrix} \mathbf{i} & \mathbf{j} & \mathbf{k} \\ -a\sin t & a\cos t & 0 \\ -a\cos t & -a\sin t & 0 \end{vmatrix} = (a^2\sin^2 t + a^2\cos^2 t)\mathbf{k} = a^2\mathbf{k},$$

$$|\mathbf{v}|^3 = \left[\sqrt{(-a\sin t)^2 + (a\cos t)^2} \right]^3 = a^3,$$

and

$$\kappa = \frac{|\mathbf{v} \times \mathbf{a}|}{|\mathbf{v}|^3} = \frac{|a^2\mathbf{k}|}{a^3} = \frac{a^2|\mathbf{k}|}{a^3} = \frac{1}{a}.$$

■

Circle of Curvature and Radius of Curvature

The **circle of curvature** or **osculating circle** at a point P on a plane curve where $\kappa \neq 0$ is the circle in the plane of the curve that

1. is tangent to the curve at P (has the same tangent that the curve has);
2. has the same curvature the curve has at P; and
3. lies toward the concave or inner side of the curve (as in Fig. 14.15).

The **radius of curvature** of the curve at P is the radius of the circle of curvature, which, according to Example 2, is

$$\text{Radius of curvature} = \rho = \frac{1}{\kappa}. \tag{6}$$

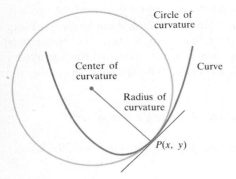

14.15 The osculating circle or circle of curvature at $P(x, y)$ lies toward the inner side of the curve.

To calculate ρ, we calculate κ and take its reciprocal.

The **center of curvature** of the curve at P is the center of the circle of curvature.

If we write a Cartesian equation for the circle of curvature at a point P on the curve $y = f(x)$ and differentiate the circle equation implicitly to find dy/dx and d^2y/dx^2, the values of these derivatives will agree with the values of f' and f'' at P. (We shall not prove this, but see Problem 40.)

The Principal Unit Normal Vector for Curves in the Plane

The vectors $d\mathbf{T}/ds$ and $d\mathbf{T}/d\phi$ are related by the Chain Rule equation

$$\frac{d\mathbf{T}}{ds} = \frac{d\mathbf{T}}{d\phi}\frac{d\phi}{ds}. \tag{7}$$

Furthermore, since $\mathbf{T}$ is a vector of constant length, $d\mathbf{T}/ds$ and $d\mathbf{T}/d\phi$ are both orthogonal to $\mathbf{T}$. They have the same direction if $d\phi/ds$ is positive and opposite directions if $d\phi/ds$ is negative. Now,

$$\mathbf{T} = (\cos\phi)\mathbf{i} + (\sin\phi)\mathbf{j}, \tag{8}$$

and

$$\frac{d\mathbf{T}}{d\phi} = -(\sin\phi)\mathbf{i} + (\cos\phi)\mathbf{j} = \cos\left(\phi + \frac{\pi}{2}\right)\mathbf{i} + \sin\left(\phi + \frac{\pi}{2}\right)\mathbf{j} \tag{9}$$

is the unit vector obtained by rotating $\mathbf{T}$ counterclockwise through $\pi/2$ radians. Therefore, if we stand at a point on the curve facing in the direction of $\mathbf{T}$, the vector $d\mathbf{T}/ds = (d\mathbf{T}/d\phi)(d\phi/ds)$ will point toward the left if $d\phi/ds$ is positive

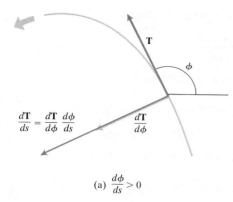

$$\frac{d\mathbf{T}}{ds} = \frac{d\mathbf{T}}{d\phi}\frac{d\phi}{ds}$$

(a) $\dfrac{d\phi}{ds} > 0$

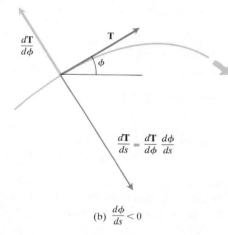

$$\frac{d\mathbf{T}}{ds} = \frac{d\mathbf{T}}{d\phi}\frac{d\phi}{ds}$$

(b) $\dfrac{d\phi}{ds} < 0$

14.16 The vector $d\mathbf{T}/ds$ always points inward.

and toward the right if $d\phi/ds$ is negative. In other words, $d\mathbf{T}/ds$ will point toward the concave side of the curve (Fig. 14.16).

Since $d\mathbf{T}/d\phi$ is a unit vector, the magnitude of $d\mathbf{T}/ds$ at any point on the curve is the curvature at that point, as we can see from the equation

$$\left|\frac{d\mathbf{T}}{ds}\right| = \left|\frac{d\mathbf{T}}{d\phi}\right|\left|\frac{d\phi}{ds}\right| = (1)(\kappa) = \kappa. \tag{10}$$

When $d\mathbf{T}/ds \neq \mathbf{0}$, its direction is given by the unit vector

$$\mathbf{N} = \frac{d\mathbf{T}/ds}{|d\mathbf{T}/ds|} = \frac{1}{\kappa}\frac{d\mathbf{T}}{ds}. \tag{11}$$

Since $\mathbf{N}$ points the same way $d\mathbf{T}/ds$ does, $\mathbf{N}$ is always orthogonal to $\mathbf{T}$ and directed toward the concave side of the curve. The vector $\mathbf{N}$ is called the **principal unit normal vector** of the curve.

Because the directed distance on a curve $\mathbf{R}(t) = x(t)\mathbf{i} + y(t)\mathbf{j}$ is defined with ds/dt positive, the Chain Rule gives

$$\mathbf{N} = \frac{d\mathbf{T}/ds}{|d\mathbf{T}/ds|} = \frac{(d\mathbf{T}/dt)(dt/ds)}{|d\mathbf{T}/dt|\,|dt/ds|} = \frac{d\mathbf{T}/dt}{|d\mathbf{T}/dt|}. \tag{12}$$

This formula enables us to find $\mathbf{N}$ without having to find ϕ, κ, or s first.

EXAMPLE 3 Find $\mathbf{T}$ and $\mathbf{N}$ for the circular motion

$$\mathbf{R}(t) = (\cos 2t)\mathbf{i} + (\sin 2t)\mathbf{j}.$$

Solution We first find $\mathbf{T}$:

$$\mathbf{v} = -(2\sin 2t)\mathbf{i} + (2\cos 2t)\mathbf{j},$$

$$|\mathbf{v}| = \sqrt{4\sin^2 2t + 4\cos^2 2t} = 2,$$

$$\mathbf{T} = \frac{\mathbf{v}}{|\mathbf{v}|} = -(\sin 2t)\mathbf{i} + (\cos 2t)\mathbf{j}.$$

From this we find

$$\frac{d\mathbf{T}}{dt} = -(2\cos 2t)\mathbf{i} - (2\sin 2t)\mathbf{j},$$

$$\left|\frac{d\mathbf{T}}{dt}\right| = \sqrt{4\cos^2 2t + 4\sin^2 2t} = 2,$$

and

$$\mathbf{N} = \frac{d\mathbf{T}/dt}{|d\mathbf{T}/dt|} = -(\cos 2t)\mathbf{i} - (\sin 2t)\mathbf{j}. \quad \blacksquare$$

Curvature for Curves in Space

In space there is no natural way to find an angle like ϕ with which to measure the change in $\mathbf{T}$ along a differentiable curve. But we still have s, the directed distance along the curve, and can define the curvature to be

$$\kappa = \left|\frac{d\mathbf{T}}{ds}\right|,$$

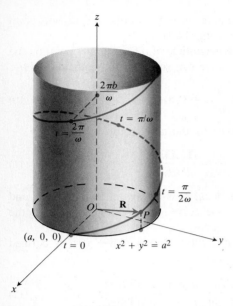

14.17   The helix

$\mathbf{R}(t) = (a \cos \omega t)\mathbf{i} + (a \sin \omega t)\mathbf{j} + bt\mathbf{k}$,

drawn here with a and b positive and $t \geq 0$.

as it worked out to be for curves in the plane. As before,

$$\kappa = \frac{|\mathbf{v} \times \mathbf{a}|}{|\mathbf{v}|^3}, \tag{13}$$

as you will see if you do the calculation in Problem 43.

EXAMPLE 4   *The general helix.* Find the curvature of the helix

$$\mathbf{R}(t) = (a \cos \omega t)\mathbf{i} + (a \sin \omega t)\mathbf{j} + bt\mathbf{k}, \qquad a, b, \text{ and } \omega > 0,$$

shown in Fig. 14.17. How do the values of a and b affect κ?

Solution   We use Eq. (13) to calculate κ:

$$\mathbf{v}(t) = -(a\omega \sin \omega t)\mathbf{i} + (a\omega \cos \omega t)\mathbf{j} + b\mathbf{k},$$

$$\mathbf{a}(t) = -(a\omega^2 \cos \omega t)\mathbf{i} - (a\omega^2 \sin \omega t)\mathbf{j},$$

$$\mathbf{v} \times \mathbf{a} = \begin{vmatrix} \mathbf{i} & \mathbf{j} & \mathbf{k} \\ -a\omega \sin \omega t & a\omega \cos \omega t & b \\ -a\omega^2 \cos \omega t & -a\omega^2 \sin \omega t & 0 \end{vmatrix}$$

$$= (ab\omega^2 \sin \omega t)\mathbf{i} - (ab\omega^2 \cos \omega t)\mathbf{j} + a^2\omega^3\mathbf{k},$$

and

$$\kappa = \frac{|\mathbf{v} \times \mathbf{a}|}{|\mathbf{v}|^3} = \frac{\sqrt{a^2 b^2 \omega^4 + a^4 \omega^6}}{(a^2\omega^2 + b^2)^{3/2}} = \frac{a\omega^2\sqrt{a^2\omega^2 + b^2}}{(a^2\omega^2 + b^2)^{3/2}} = \frac{a\omega^2}{a^2\omega^2 + b^2}. \tag{14}$$

From Eq. (14) we can see that increasing b for a fixed a and ω decreases the curvature. Decreasing a for a fixed ω and b eventually decreases the curvature as well. In other words, stretching a spring tends to straighten it.

If $b = 0$, the helix reduces to a circle and Eq. (14) reduces to $\kappa = 1/a$, as it should. If $a = 0$, the helix reduces to the z-axis and Eq. (14) reduces to $\kappa = 0$, again as it should.

In the general case, the curvature of the helix is a constant that is less than $1/a$, the curvature of the circle in the xy-plane over which the helix stands.

■

N for Curves in Space

To define the principal unit normal vector of a curve in space, we use the same definition we use for curves in the plane:

$$\mathbf{N} = \frac{d\mathbf{T}/ds}{|d\mathbf{T}/ds|} = \frac{1}{\kappa} \frac{d\mathbf{T}}{ds} = \frac{d\mathbf{T}/dt}{|d\mathbf{T}/dt|}, \tag{15}$$

whenever $d\mathbf{T}/ds \neq \mathbf{0}$. As before, $d\mathbf{T}/ds$ and its unit vector direction $\mathbf{N}$ are orthogonal to $\mathbf{T}$ because $\mathbf{T}$ has constant length—in this case 1.

EXAMPLE 5   Find $\mathbf{N}$ for the helix in Example 4.

Solution   The formula for $\mathbf{v}$ in Example 4 gives

$$\mathbf{T} = \frac{\mathbf{v}}{|\mathbf{v}|} = \frac{-(a\omega \sin \omega t)\mathbf{i} + (a\omega \cos \omega t)\mathbf{j} + b\mathbf{k}}{\sqrt{a^2\omega^2 + b^2}}.$$

Therefore,

$$\frac{d\mathbf{T}}{dt} = -\frac{a\omega^2}{\sqrt{a^2\,\omega^2 + b^2}}((\cos \omega t)\mathbf{i} + (\sin \omega t)\mathbf{j})$$

and

$$\mathbf{N} = \frac{d\mathbf{T}/dt}{|d\mathbf{T}/dt|} = -(\cos \omega t)\mathbf{i} - (\sin \omega t)\mathbf{j}. \qquad \blacksquare$$

Torsion and the Binormal Vector

The vector $\mathbf{B} = \mathbf{T} \times \mathbf{N}$, a unit vector orthogonal to both $\mathbf{T}$ and $\mathbf{N}$, is the **binormal vector** of the curve $\mathbf{R}(t) = x(t)\mathbf{i} + y(t)\mathbf{j} + z(t)\mathbf{k}$. The three unit vectors $\mathbf{T}$, $\mathbf{N}$, and $\mathbf{B}$ define a moving right-handed vector frame that plays a significant role in calculating the flight paths of space vehicles.

The three planes determined by $\mathbf{T}$, $\mathbf{N}$, and $\mathbf{B}$ are shown in Fig. 14.18. The curvature $\kappa = |d\mathbf{T}/ds|$ can be thought of as the rate at which the normal plane turns as the point P moves along the curve. Similarly, the **torsion**, $\tau = |d\mathbf{B}/ds|$, is the rate at which the osculating plane lifts as P moves along the curve. It is a measure of how much the curve twists.

The most widely used formula for the torsion is

$$\tau = \pm \frac{\begin{vmatrix} \dot{x} & \dot{y} & \dot{z} \\ \ddot{x} & \ddot{y} & \ddot{z} \\ \dddot{x} & \dddot{y} & \dddot{z} \end{vmatrix}}{|\mathbf{v} \times \mathbf{a}|^2}, \qquad (16)$$

derived in more advanced texts. It calculates the torsion directly from $\mathbf{v}$, $\mathbf{a}$, and $\dot{\mathbf{a}}$. The sign is chosen to make $\tau \geq 0$. Problem 44 gives another formula for τ.

EXAMPLE 6   Find the torsion of the helix

$$\mathbf{R}(t) = (\cos t)\mathbf{i} + (\sin t)\mathbf{j} + t\mathbf{k}.$$

Solution   We use Eq. (16):

$$\mathbf{v} = -(\sin t)\mathbf{i} + (\cos t)\mathbf{j} + \mathbf{k},$$

$$\mathbf{a} = -(\cos t)\mathbf{i} - (\sin t)\mathbf{j},$$

$$\dot{\mathbf{a}} = (\sin t)\mathbf{i} - (\cos t)\mathbf{j},$$

$$\begin{vmatrix} -\sin t & \cos t & 1 \\ -\cos t & -\sin t & 0 \\ \sin t & -\cos t & 0 \end{vmatrix} = \cos^2 t + \sin^2 t = 1$$

$$\mathbf{v} \times \mathbf{a} = \begin{vmatrix} \mathbf{i} & \mathbf{j} & \mathbf{k} \\ -\sin t & \cos t & 1 \\ -\cos t & -\sin t & 0 \end{vmatrix}$$

$$= (\sin t)\mathbf{i} - (\cos t)\mathbf{j} + \mathbf{k}$$

$$|\mathbf{v} \times \mathbf{a}|^2 = 2$$

$$\tau = \frac{1}{2}. \qquad \blacksquare$$

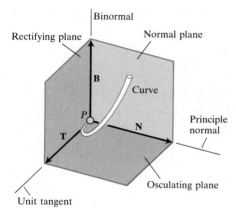

14.18   The three planes determined by $\mathbf{T}$, $\mathbf{N}$, and $\mathbf{B}$.

A curve whose torsion is zero lies in a plane. The binormal vector is a constant and **T** and **N** remain perpendicular to this constant direction. See Problem 47.

The Tangential and Normal Components of Acceleration

As we can see from the equations

$$\mathbf{v} = \frac{d\mathbf{R}}{dt} = \frac{d\mathbf{R}}{ds}\frac{ds}{dt} = \mathbf{T}\frac{ds}{dt} \tag{17}$$

and

$$\begin{aligned}
\mathbf{a} &= \frac{d\mathbf{v}}{dt} = \frac{d}{dt}\left(\mathbf{T}\frac{ds}{dt}\right) \\
&= \frac{d^2s}{dt^2}\mathbf{T} + \frac{ds}{dt}\frac{d\mathbf{T}}{dt} \\
&= \frac{d^2s}{dt^2}\mathbf{T} + \frac{ds}{dt}\left(\frac{d\mathbf{T}}{ds}\frac{ds}{dt}\right) \\
&= \frac{d^2s}{dt^2}\mathbf{T} + \frac{ds}{dt}\left(\kappa\mathbf{N}\frac{ds}{dt}\right) \\
&= \frac{d^2s}{dt^2}\mathbf{T} + \kappa\left(\frac{ds}{dt}\right)^2\mathbf{N},
\end{aligned} \tag{18}$$

the acceleration of a moving body always lies in the osculating plane spanned by **T** and **N**. This is one reason we use the **TNB** frame in studying motion.

The equalities in (18) combine to give

$$\mathbf{a} = \frac{d^2s}{dt^2}\mathbf{T} + \kappa\left(\frac{ds}{dt}\right)^2\mathbf{N}. \tag{19}$$

This equation expresses a in terms of its tangential and normal components. The **tangential scalar component,**

$$a_{\mathrm{T}} = \frac{d^2s}{dt^2} = \frac{d}{dt}|\mathbf{v}|, \tag{20}$$

is the derivative of the particle's speed along the path. The **normal scalar component,**

$$a_{\mathrm{N}} = \kappa\left(\frac{ds}{dt}\right)^2 = \kappa|\mathbf{v}|^2, \tag{21}$$

is the curvature times the square of the speed. This explains the large normal force that must be supplied by tire friction to hold a car on the road during a sharp (large κ) or high-speed (large $|\mathbf{v}|^2$) turn.

If a particle moves around a circle at a constant speed, d^2s/dt^2 is zero and all the acceleration points along **N** toward the center of the circle. If the particle's speed is not constant, **a** has a nonzero tangential component (Fig. 14.19).

To calculate a_{T} we use Eq. (20), but instead of using Eq. (21) to calculate a_{N}, we usually use the equation

$$a_{\mathrm{N}} = \sqrt{|\mathbf{a}|^2 - a_{\mathrm{T}}^2}, \tag{22}$$

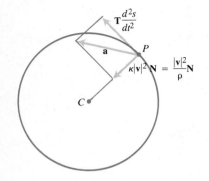

14.19 The tangential and normal components of the acceleration vector of a particle P moving counterclockwise around a circle with a nonconstant velocity.

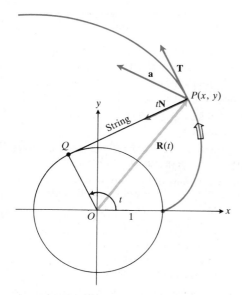

14.20 The *involute* of a circle is the path traced by the endpoint P of a string unwinding from a circle. The string shown here unwinds counterclockwise, tracing the path

$$\mathbf{R}(t) = (\cos t + t \sin t)\mathbf{i} + (\sin t - t \cos t)\mathbf{j}.$$

The tangential and normal vector components of the acceleration $\mathbf{a}$ in this case are $\mathbf{T}$ and $t\mathbf{N}$, as shown in Example 7.

which comes from solving the equation $|\mathbf{a}|^2 = \mathbf{a} \cdot \mathbf{a} = a_\mathrm{T}^2 + a_\mathrm{N}^2$ for a_N. This way we find a_N without having to calculate κ.

EXAMPLE 7   The position of a particle is

$$\mathbf{R}(t) = (\cos t + t \sin t)\mathbf{i} + (\sin t - t \cos t)\mathbf{j}$$

for $t \geq 0$. Without finding $\mathbf{N}$ and $\mathbf{T}$, write $\mathbf{a}$ in the form

$$\mathbf{a} = a_\mathrm{T}\mathbf{T} + a_\mathrm{N}\mathbf{N} \qquad \text{for } t > 0.$$

Solution   We use Eqs. (20) and (22):

$$\mathbf{v} = (t \cos t)\mathbf{i} + (t \sin t)\mathbf{j},$$

$$\mathbf{a} = (\cos t - t \sin t)\mathbf{i} + (\sin t + t \cos t)\mathbf{j},$$

$$|\mathbf{v}| = \sqrt{t^2\cos^2 t + t^2\sin^2 t} = t,$$

$$a_\mathrm{T} = \frac{d}{dt}|\mathbf{v}| = 1,$$

$$|\mathbf{a}|^2 = (\cos t - t \sin t)^2 + (\sin t + t \cos t)^2 = t^2 + 1$$

$$a_\mathrm{N} = \sqrt{|\mathbf{a}|^2 - a_\mathrm{T}^2} = \sqrt{t^2} = t$$

$$\mathbf{a} = a_\mathrm{T}\mathbf{T} + a_\mathrm{N}\mathbf{N} = \mathbf{T} + t\mathbf{N}.$$

See Fig. 14.20.   ∎

Selected Formulas for Motion ($|\mathbf{v}| \neq 0$)

Unit tangent vector

$$\mathbf{T} = \frac{\mathbf{v}}{|\mathbf{v}|}$$

Principal unit normal vector

$$\mathbf{N} = \frac{d\mathbf{T}/dt}{|d\mathbf{T}/dt|}$$

Binormal vector

$$\mathbf{B} = \mathbf{T} \times \mathbf{N}$$

Curvature

$$\kappa = \left|\frac{d\mathbf{T}}{ds}\right| = \frac{|\mathbf{v} \times \mathbf{a}|}{|\mathbf{v}|^3}$$

Torsion

$$\tau = \left|\frac{d\mathbf{B}}{ds}\right| = \pm \frac{\begin{vmatrix} \dot{x} & \dot{y} & \dot{z} \\ \ddot{x} & \ddot{y} & \ddot{z} \\ \dddot{x} & \dddot{y} & \dddot{z} \end{vmatrix}}{|\mathbf{v} \times \mathbf{a}|^2}$$

Tangential and normal scalar components of acceleration

$$\mathbf{a} = a_\mathrm{T}\mathbf{T} + a_\mathrm{N}\mathbf{N}$$

$$a_\mathrm{T} = \frac{d}{dt}|\mathbf{v}|$$

$$a_\mathrm{N} = \sqrt{|\mathbf{a}|^2 - a_\mathrm{T}^2}$$

PROBLEMS

Find **T**, **N**, and κ for the plane curves in Problems 1–8.

1. $\mathbf{R}(t) = t\mathbf{i} + (\ln \cos t)\mathbf{j}, \ -\pi/2 < t < \pi/2$

2. $\mathbf{R}(t) = t\mathbf{i} + e^{2t}\mathbf{j}$

3. $\mathbf{R}(t) = (\ln \sec t)\mathbf{i} + t\mathbf{j}, \ -\pi/2 < t < \pi/2$

4. $\mathbf{R}(t) = (2t + 3)\mathbf{i} + (5 - t^2)\mathbf{j}$

5. $\mathbf{R}(t) = (t^3/3)\mathbf{i} + (t^2/2)\mathbf{j}, \ t > 0$

6. $\mathbf{R}(t) = (\cos t + t \sin t)\mathbf{i} + (\sin t - t \cos t)\mathbf{j}, \ t \geq 0$

7. $\mathbf{R}(t) = (\cos^3 t)\mathbf{i} + (\sin^3 t)\mathbf{j}, \ 0 < t < \pi/2$

8. $\mathbf{R}(t) = t\mathbf{i} + (a \cosh(t/a))\mathbf{j}$

Find **T**, **N**, κ, **B**, and τ for the curves in Problems 9–14.

9. $\mathbf{R}(t) = (3 \sin t)\mathbf{i} + (3 \cos t)\mathbf{j} + 4t\mathbf{k}$

10. $\mathbf{R}(t) = (\cos t + t \sin t)\mathbf{i} + (\sin t - t \cos t)\mathbf{j} + 3\mathbf{k}$

11. $\mathbf{R}(t) = (e^t\cos t)\mathbf{i} + (e^t\sin t)\mathbf{j} + 2\mathbf{k}$

12. $\mathbf{R}(t) = (6 \sin 2t)\mathbf{i} + (6 \cos 2t)\mathbf{j} + 5t\mathbf{k}$

13. $\mathbf{R}(t) = (4/9)(1 + t)^{3/2}\mathbf{i} + (4/9)(1 - t)^{3/2}\mathbf{j} + (1/3)t\mathbf{k}$

   (**T**, **N**, and κ only)

14. $\mathbf{R}(t) = (\cosh t)\mathbf{i} + (\sinh t)\mathbf{j} + t\mathbf{k}$

In Problems 15–20, write **a** in the form $\mathbf{a} = a_T\mathbf{T} + a_N\mathbf{N}$ (without finding **N** and **T**).

15. $\mathbf{R}(t) = (2t + 3)\mathbf{i} + (t^2 - 1)\mathbf{j}$

16. $\mathbf{R}(t) = \ln(t^2 + 1)\mathbf{i} + (t - 2 \tan^{-1}t)\mathbf{j}$

17. $\mathbf{R}(t) = (2 \cos t)\mathbf{i} + (2 \sin t)\mathbf{j}$

18. $\mathbf{R}(t) = (e^t\cos t)\mathbf{i} + (e^t\sin t)\mathbf{j}$

19. $\mathbf{R}(t) = (a \cos \omega t)\mathbf{i} + (a \sin \omega t)\mathbf{j} + bt\mathbf{k}$

20. $\mathbf{R}(t) = (\cosh t)\mathbf{i} + (\sinh t)\mathbf{j} + t\mathbf{k}$

In Problems 21–26, write **a** in the form $\mathbf{a} = a_T\mathbf{T} + a_N\mathbf{N}$ at the given value of t without finding **N** and **T**.

21. $\mathbf{R}(t) = (t + 1)\mathbf{i} + 2t\mathbf{j} + t^2\mathbf{k}, \quad t = 1$

22. $\mathbf{R}(t) = (t \cos t)\mathbf{i} + (t \sin t)\mathbf{j} + t^2\mathbf{k}, \quad t = 0$

23. $\mathbf{R}(t) = t^2\mathbf{i} + \left(t + \frac{1}{3}t^3\right)\mathbf{j} + \left(t - \frac{1}{3}t^3\right)\mathbf{k}, \quad t = 0$

24. $\mathbf{R}(t) = (e^t\cos t)\mathbf{i} + (e^t\sin t)\mathbf{j} + \sqrt{2}e^t\mathbf{k}, \quad t = 0$

25. $\mathbf{R}(t) = (2 + 3t + 3t^2)\mathbf{i} + (4t + 4t^2)\mathbf{j} - (6 \cos t)\mathbf{k}, \quad t = 0$

26. $\mathbf{R}(t) = (2 + t)\mathbf{i} + (t + 2t^2)\mathbf{j} + (1 + t^2)\mathbf{k}, \quad t = 0$

In Problems 27–30, find equations for the osculating, normal, and rectifying planes of the curves at the given values of t. (See Fig. 14.18.)

27. $\mathbf{R}(t) = (\cos t)\mathbf{i} + (\sin t)\mathbf{j}, \quad t = \pi/4$

28. $\mathbf{R}(t) = (\cos t)\mathbf{i} + (\sin t)\mathbf{j} + \mathbf{k}, \quad t = 0$

29. $\mathbf{R}(t) = (2 \cos t)\mathbf{i} + (2 \sin t)\mathbf{j} + t\mathbf{k}, \quad t = 0$

30. $\mathbf{R}(t) = (\cosh t)\mathbf{i} + (\sinh t)\mathbf{j} + t\mathbf{k}, \quad t = 0$

31. Show that κ and τ are both zero for the line

$$\mathbf{R}(t) = (x_0 + At)\mathbf{i} + (y_0 + Bt)\mathbf{j} + (z_0 + Ct)\mathbf{k}.$$

32. Show that the torsion of any (sufficiently differentiable) plane curve is zero, when defined.

33. Show that a moving particle will move in a straight line if the normal component of its acceleration is zero.

34. Show that if a particle's speed is constant, then the acceleration is directed along **N**.

35. Show that a particle's speed will be constant if the force acting on it is always perpendicular to the velocity.

36. a) Show that the curvature of the graph of a twice-differentiable function $y = f(x)$ is given by the formula

$$\kappa = \frac{|y''|}{(1 + (y')^2)^{3/2}}.$$

   (*Hint:* Let $\mathbf{R}(t) = x\mathbf{i} + f(x)\mathbf{j}$.)

   b) Apply the formula in (a) to $y = \ln(\cos x)$ and compare your result with the one in Problem 1.

37. Equation (3) says that the curvature of the plane curve $x = x(t)$, $y = y(t)$ is

$$\kappa = \frac{|\dot{x}\ddot{y} - \dot{y}\ddot{x}|}{(\dot{x}^2 + \dot{y}^2)^{3/2}}.$$

Apply this formula to the curve $x = \ln(\sec t)$, $y = t$, $-\pi/2 < t < \pi/2$, and compare your result with the one in Problem 3.

38. Find an equation for the circle of curvature at $t = 1$ of the plane curve

$$\mathbf{R}(t) = (2 \ln t)\mathbf{i} - (t + (1/t))\mathbf{j}.$$

39. Find an equation for the circle of curvature of the curve $y = \sin x$ at the point $(\pi/2, 1)$.

40. Find an equation for the osculating circle of the curve $y = e^x$ at the point $(0, 1)$. Calculate dy/dx and d^2y/dx^2 at the point $(0, 1)$ from the equation of this circle. Show that these

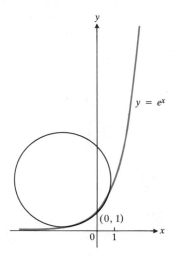

14.21   The curve $y = e^x$ and its osculating circle at $(0, 1)$. See Problem 40.

derivatives have the same values as the corresponding derivatives of $y = e^x$. See Fig. 14.21.

41. Suppose that the functions $x(t)$ and $y(t)$ have continuous first derivatives with respect to t and that the derivative of $\phi = \tan^{-1}(\dot{y}/\dot{x})$ with respect to s is zero. Show that $y = C_1 x + C_2$. In other words, show that the path traced by $\mathbf{R}(t) = x(t)\mathbf{i} + y(t)\mathbf{j}$ is a straight line if $d\phi/ds = 0$.

42. When a_N and $|\mathbf{v}|$ are known, the equation $a_N = \kappa|\mathbf{v}|^2$ gives a convenient way to find the curvature. Use it to find the curvature and radius of curvature of the curve

$$\mathbf{R}(t) = (\cos t + t \sin t)\mathbf{i} + (\sin t - t \cos t)\mathbf{j}$$

in Example 7 for $t > 0$.

43. *How to derive the curvature formula:* Use Eqs. (17) and (19) to show that

$$|\mathbf{v} \times \mathbf{a}| = \left|\mathbf{B}\kappa\left(\frac{ds}{dt}\right)^3\right| = \kappa|\mathbf{v}|^3.$$

44. *Another formula for torsion.* Use the definition $\tau = d\mathbf{B}/ds$ to show that the torsion is given by the formula

$$\tau = \frac{|d\mathbf{B}/dt|}{|\mathbf{v}|}.$$

Use this formula to recalculate the torsion of the helix

$$\mathbf{R}(t) = (\cos t)\mathbf{i} + (\sin t)\mathbf{j} + t\mathbf{k}$$

in Example 6.

45. In Example 4 we found the curvature of the helix

$$\mathbf{R}(t) = (a \cos \omega t)\mathbf{i} + (a \sin \omega t)\mathbf{j} + bt\mathbf{k},$$

where a, b, and $\omega > 0$, to be $\kappa = a\omega^2/(a^2\omega^2 + b^2)$. For a given ω and b, what is the largest value κ can have?

46. Find the torsion of the helix in Problem 45. What is the value of the torsion when the curvature has the maximum value found in Problem 45? For a given a and ω, what is the largest value the torsion can have?

47. *Differentiable curves with zero torsion lie in a plane.* That a differentiable curve with zero torsion lies in a plane is a special case of the fact that a particle whose acceleration and velocity remain normal to a fixed constant vector $\mathbf{C}$ moves in a plane perpendicular to $\mathbf{C}$. This, in turn, can be viewed as the solution of the following problem in calculus:

Suppose that $\mathbf{R}(t) = x(t)\mathbf{i} + y(t)\mathbf{j} + z(t)\mathbf{k}$ is twice differentiable, that $\mathbf{R}(0) = 0$, and that $\mathbf{v} \cdot \mathbf{k} = 0$ and $\mathbf{a} \cdot \mathbf{k} = 0$ for all t. Then $z(t) = 0$ for all t.

Solve this problem. (*Hint:* Start with $\mathbf{a} = d^2\mathbf{R}/dt^2$ and apply the initial conditions in reverse order.)

TOOLKIT PROGRAMS

Parametric Equations   Super ∗ Grapher

14.4
Planetary Motion and Satellites

In this article we introduce the standard vector equations for motion in the polar coordinate plane and extend these to space by adding a z-axis. We then show that any body that obeys Newton's laws of gravitation and motion moves in a plane. We introduce polar coordinates in this plane and use polar coordinate equations to derive Kepler's three laws of planetary motion. The derivation, while lengthy in comparison to what we have seen so far, is one of the triumphs of vector calculus and the calculations in polar coordinates are typical of work done in orbital mechanics today.

Equations for Motion in Polar Coordinates

When a particle moves along a curve in the polar coordinate plane, we express its position, velocity, and acceleration in terms of the moving unit vectors

$$\mathbf{u}_r = (\cos \theta)\mathbf{i} + (\sin \theta)\mathbf{j},$$
$$\mathbf{u}_\theta = -(\sin \theta)\mathbf{i} + (\cos \theta)\mathbf{j} \tag{1}$$

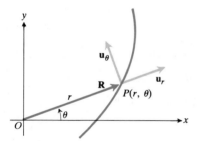

14.22 The length of $\mathbf{R}$ is the polar coordinate r of the point P. Thus, $\mathbf{u}_r$, which is $\mathbf{R}/|\mathbf{R}|$, is also $\mathbf{R}/r$.

shown in Fig. 14.22. The vector $\mathbf{u}_r$ points along the radius vector $\mathbf{R} = \overrightarrow{OP}$, so $\mathbf{R} = r\mathbf{u}_r$. The vector $\mathbf{u}_\theta$, orthogonal to $\mathbf{u}_r$, points in the direction of increasing θ.

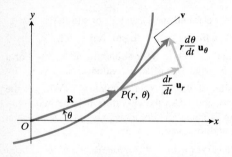

14.23 In polar coordinates, the velocity vector is

$$\mathbf{v} = (dr/dt)\mathbf{u}_r + r(d\theta/dt)\mathbf{u}_\theta.$$

We find from (1) that

$$\frac{d\mathbf{u}_r}{d\theta} = -(\sin\,\theta)\mathbf{i} + (\cos\,\theta)\mathbf{j} = \mathbf{u}_\theta,$$

$$\frac{d\mathbf{u}_\theta}{d\theta} = -(\cos\,\theta)\mathbf{i} - (\sin\,\theta)\mathbf{j} = -\mathbf{u}_r. \tag{2}$$

When we differentiate $\mathbf{u}_r$ and $\mathbf{u}_\theta$ with respect to t to find how they change with time, the Chain Rule gives

$$\frac{d\mathbf{u}_r}{dt} = \frac{d\mathbf{u}_r}{d\theta}\frac{d\theta}{dt} = \frac{d\theta}{dt}\mathbf{u}_\theta,$$

$$\frac{d\mathbf{u}_\theta}{dt} = \frac{d\mathbf{u}_\theta}{d\theta}\frac{d\theta}{dt} = -\frac{d\theta}{dt}\mathbf{u}_r. \tag{3}$$

Hence,

$$\mathbf{v} = \frac{d\mathbf{R}}{dt} = \frac{d}{dt}(r\mathbf{u}_r)$$

$$= \frac{dr}{dt}\mathbf{u}_r + r\frac{d\mathbf{u}_r}{dt} \tag{4}$$

$$= \frac{dr}{dt}\mathbf{u}_r + r\frac{d\theta}{dt}\mathbf{u}_\theta.$$

See Fig. 14.23.

The acceleration is

$$\mathbf{a} = \frac{d\mathbf{v}}{dt} = \left(\frac{d^2r}{dt^2}\mathbf{u}_r + \frac{dr}{dt}\frac{d\mathbf{u}_r}{dt}\right) + \left(r\frac{d^2\theta}{dt^2}\mathbf{u}_\theta + \frac{dr}{dt}\frac{d\theta}{dt}\mathbf{u}_\theta + r\frac{d\theta}{dt}\frac{d\mathbf{u}_\theta}{dt}\right). \tag{5}$$

When Eqs. (3) are used to evaluate $d\mathbf{u}_r/dt$ and $d\mathbf{u}_\theta/dt$ and the components are separated, the equation for acceleration becomes

$$\mathbf{a} = \left(\frac{d^2r}{dt^2} - r\left(\frac{d\theta}{dt}\right)^2\right)\mathbf{u}_r + \left(r\frac{d^2\theta}{dt^2} + 2\frac{dr}{dt}\frac{d\theta}{dt}\right)\mathbf{u}_\theta. \tag{6}$$

To extend the equations of motion to space, we add $z(t)\mathbf{k}$ to the right-hand side of the equation $\mathbf{R} = r\mathbf{u}_r$. Then

$$\mathbf{R} = r\mathbf{u}_r + z\mathbf{k}, \tag{7a}$$

$$\mathbf{v} = \frac{dr}{dt}\mathbf{u}_r + r\frac{d\theta}{dt}\mathbf{u}_\theta + \frac{dz}{dt}\mathbf{k}, \tag{7b}$$

$$\mathbf{a} = \left(\frac{d^2r}{dt^2} - r\left(\frac{d\theta}{dt}\right)^2\right)\mathbf{u}_r + \left(r\frac{d^2\theta}{dt^2} + 2\frac{dr}{dt}\frac{d\theta}{dt}\right)\mathbf{u}_\theta + \frac{d^2z}{dt^2}\mathbf{k}. \tag{7c}$$

The vectors $\mathbf{u}_r$, $\mathbf{u}_\theta$, and $\mathbf{k}$ make a right-handed frame in which

$$\begin{aligned}\mathbf{u}_r \times \mathbf{u}_\theta &= \mathbf{k},\\ \mathbf{k} \times \mathbf{u}_r &= \mathbf{u}_\theta,\\ \mathbf{u}_\theta \times \mathbf{k} &= \mathbf{u}_r.\end{aligned} \tag{8}$$

See Fig. 14.24.

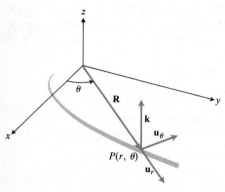

14.24 Motion in space, with polar coordinates in the xy-plane and the z-axis as usual.

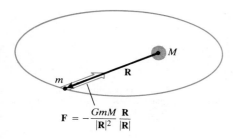

$$\mathbf{F} = -\frac{GmM}{|\mathbf{R}|^2}\frac{\mathbf{R}}{|\mathbf{R}|}$$

14.25 The force of gravity is directed along the line joining the centers of mass.

Coordinates for Planetary Motion; Initial Conditions

Newton's law of gravitation says that if $\mathbf{R}$ is the radius vector from the center of mass of a sun of mass M to the center of mass of a planet of mass m, then the force $\mathbf{F}$ of the gravitational attraction between the planet and the sun is given by the equation

$$\mathbf{F} = -\frac{GmM}{|\mathbf{R}|^2}\frac{\mathbf{R}}{|\mathbf{R}|} \qquad (9)$$

(Fig. 14.25). The constant G in this equation is called the (universal) **gravitational constant.** If the masses are measured in kilograms, time in seconds, and distance in meters, G is about 6.6720×10^{-11} Nm^{2}kg^{-2}.

Combining Eq. (9) with Newton's second law of motion, $\mathbf{F} = m\, d^2\mathbf{R}/dt^2$, gives

$$m\frac{d^2\mathbf{R}}{dt^2} = -\frac{GmM}{|\mathbf{R}|^2}\frac{\mathbf{R}}{|\mathbf{R}|},$$

or

$$\frac{d^2\mathbf{R}}{dt^2} = -\frac{GM}{|\mathbf{R}|^2}\frac{\mathbf{R}}{|\mathbf{R}|}. \qquad (10)$$

This equation is typical of a **central force,** one that points toward a fixed center at all times—in this case the center of mass of the sun.

Equation (10) tells us that $d^2\mathbf{R}/dt^2$ is proportional to $\mathbf{R}$ and hence that

$$\mathbf{R} \times \frac{d^2\mathbf{R}}{dt^2} = \mathbf{0}. \qquad (11)$$

The left-hand side of Eq. (11) is the derivative of $\mathbf{R} \times d\mathbf{R}/dt$:

$$\frac{d}{dt}\left(\mathbf{R} \times \frac{d\mathbf{R}}{dt}\right) = \frac{d\mathbf{R}}{dt} \times \frac{d\mathbf{R}}{dt} + \mathbf{R} \times \frac{d^2\mathbf{R}}{dt^2} = \mathbf{0} + \mathbf{R} \times \frac{d^2\mathbf{R}}{dt^2}. \qquad (12)$$

Thus, Eq. (11) is equivalent to

$$\frac{d}{dt}\left(\mathbf{R} \times \frac{d\mathbf{R}}{dt}\right) = \mathbf{0},$$

which says that

$$\mathbf{R} \times \frac{d\mathbf{R}}{dt} = \mathbf{C} \qquad (13)$$

for some constant vector $\mathbf{C}$.

Equation (13) tells us that $\mathbf{R}$ and $\mathbf{v} = d\mathbf{R}/dt$ always lie in a plane perpendicular to $\mathbf{C}$ and hence that the planet moves in a fixed plane through the center of mass of its sun.

We now adjust the coordinate system for space, if necessary, to place the origin at the sun's center of mass and make the xy-plane the plane of the planet's motion. We also position the z-axis in a way that makes $\mathbf{k}$ the direction of $\mathbf{C}$. We then introduce polar coordinates in the xy-plane, choosing the direction of the initial ray to be the direction $\mathbf{R}$ has when the planet is closest to the sun. This runs the initial ray through the **perihelion** position of the planet (Fig. 14.26).

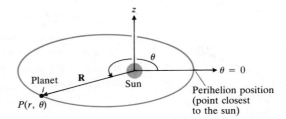

14.26 The coordinate system for planetary motion.

If we also measure time so that t is zero when the planet is at its perihelion, then

$$r(0) = r_0 \text{ is the minimum value of } r,$$

$$\left(\frac{dr}{dt}\right)_{t=0} = 0, \qquad \binom{\text{because } r \text{ has a minimum}}{\text{value at } t = 0} \tag{14}$$

$$\theta(0) = 0, \qquad \text{(the planet lies on the initial ray when } t = 0\text{)}$$

$$|\mathbf{v}(0)| = v_0 = \left|\frac{dr}{dt}\mathbf{u}_r + r\frac{d\theta}{dt}\mathbf{u}_\theta\right|_{t=0} = \left|r\frac{d\theta}{dt}\right|_{t=0}.$$

Kepler's Second Law (the Equal Area Law)

We start with Kepler's second law because we need it to prove Kepler's first law. Kepler's second law says that the radius vector from a sun to a planet sweeps out equal areas in equal times (Fig. 14.27). To derive this law, we use Eq. (4) to evaluate the cross product in (13):

$$\mathbf{C} = \mathbf{R} \times \mathbf{v}$$

$$= r\mathbf{u}_r \times \left(\frac{dr}{dt}\mathbf{u}_r + r\frac{d\theta}{dt}\mathbf{u}_\theta\right)$$

$$= r\mathbf{u}_r \times r\frac{d\theta}{dt}\mathbf{u}_\theta \tag{15}$$

$$= r\left(r\frac{d\theta}{dt}\right)\mathbf{k}.$$

Setting t equal to zero then shows that $\mathbf{C} = r_0 v_0 \mathbf{k}$. Equation (15) can therefore be rewritten as

$$r_0 v_0 \mathbf{k} = r^2 \frac{d\theta}{dt}\mathbf{k},$$

14.27 The line joining a planet to its sun sweeps over equal areas in equal times.

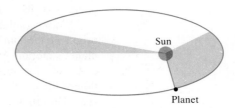

or

$$r^2 \frac{d\theta}{dt} = r_0 v_0. \tag{16}$$

Here is where the area comes in. From Article 10.4, the area element in polar coordinates is

$$dA = \frac{1}{2} r^2 d\theta.$$

Accordingly,

$$\frac{dA}{dt} = \frac{1}{2} r^2 \frac{d\theta}{dt} = \frac{1}{2} r_0 v_0. \tag{17}$$

In other words, dA/dt is constant, which is Kepler's second law.

For the Earth, r_0 is about 150,000,000 km, v_0 about 30 km/sec, and dA/dt about 2,250,000,000 km^2/sec.

Kepler's First Law (the Conic Section Law)

Kepler's first law says that a planet's path is a conic section whose equation is

$$r = \frac{\dfrac{r_0^2 v_0^2}{GM}}{1 + \left(\dfrac{r_0 v_0^2}{GM} - 1 \right) \cos \theta}. \tag{18}$$

The path is a circle if the eccentricity $e = (r_0 v_0{}^2/GM) - 1$ is zero, an ellipse if e lies between zero and 1, a parabola if e equals 1, and a hyperbola if e is greater than 1. Perihelion occurs when θ is zero, and the distance at perihelion from the planet to the focus occupied by the sun is indeed r_0, as we can check by setting $\cos \theta$ equal to 1 in Eq. (18).

To derive Eq. (18) we need to express r in terms of θ. This requires a long sequence of calculations and some substitutions that are not altogether obvious.

By equating the coefficients of $\mathbf{u}_r$ in Eqs. (6) and (10) we get

$$\frac{d^2 r}{dt^2} - r \left(\frac{d\theta}{dt} \right)^2 = -\frac{GM}{r^2}. \tag{19}$$

We cannot integrate this with respect to t because we do not know how r varies with t. We therefore eliminate $d\theta/dt$ by substituting $r_0 v_0/r^2$ for it, from Eq. (16):

$$\frac{d^2 r}{dt^2} - \frac{r_0^2 v_0^2}{r^3} = -\frac{GM}{r^2}. \tag{20}$$

We then replace dr/dt by a single letter p and use the Chain Rule to express $d^2 r/dt^2$ in terms of p and r:

$$\frac{dr}{dt} = p, \qquad \frac{d^2 r}{dt^2} = \frac{dp}{dt} = \frac{dp}{dr} \frac{dr}{dt} = p \frac{dp}{dr}.$$

When $d^2 r/dt^2$ is replaced by $p(dp/dr)$ and the variables are separated, Eq. (20)

becomes

$$pdp = \left(\frac{r_0^2 v_0^2}{r^3} - \frac{GM}{r^2} \right) dr. \tag{21}$$

Integration then gives

$$p^2 = \left(\frac{dr}{dt} \right)^2 = -\frac{r_0^2 v_0^2}{r^2} + \frac{2GM}{r} + C_1. \tag{22}$$

We use the initial conditions

$$r = r_0 \quad \text{and} \quad \frac{dr}{dt} = 0 \quad \text{when } t = 0$$

to evaluate C_1:

$$C_1 = v_0^2 - \frac{2GM}{r_0}.$$

When the value of C_1 is substituted in Eq. (22), the result is

$$\left(\frac{dr}{dt} \right)^2 = v_0^2 \left(1 - \frac{r_0^2}{r^2} \right) + 2GM \left(\frac{1}{r} - \frac{1}{r_0} \right). \tag{23}$$

If we square both sides of Eq. (16) and divide the results into the corresponding sides of Eq. (23), we get

$$\frac{1}{r^4} \left(\frac{dr}{d\theta} \right)^2 = \frac{1}{r_0^2} - \frac{1}{r^2} + \frac{2GM}{r_0^2 v_0^2} \left(\frac{1}{r} - \frac{1}{r_0} \right). \tag{24}$$

For simplicity, we take

$$h = \frac{GM}{r_0^2 v_0^2}$$

and rewrite Eq. (24) as

$$\frac{1}{r^4} \left(\frac{dr}{d\theta} \right)^2 = \frac{1}{r_0^2} - \frac{1}{r^2} + 2h \left(\frac{1}{r} - \frac{1}{r_0} \right). \tag{25}$$

To simplify the equation further, we substitute

$$u = \frac{1}{r}, \qquad u_0 = \frac{1}{r_0},$$

$$\frac{du}{d\theta} = -\frac{1}{r^2} \frac{dr}{d\theta}, \qquad \left(\frac{du}{d\theta} \right)^2 = \frac{1}{r^4} \left(\frac{dr}{d\theta} \right)^2,$$

obtaining

$$\left(\frac{du}{d\theta} \right)^2 = u_0^2 - u^2 + 2hu - 2hu_0$$

$$= (u_0 - h)^2 - (u - h)^2. \tag{26}$$

When we take square roots in Eq. (26) to solve for $du/d\theta$, we get

$$\frac{du}{d\theta} = \pm\sqrt{(u_0 - h)^2 - (u - h)^2}. \tag{27}$$

What do we do about the sign? We know that

$$\frac{d\theta}{dt} = \frac{r_0 v_0}{r^2}$$

is positive. Also, because r starts from a *minimum* value at $t = 0$, it should not decrease, and so we find $dr/dt \geq 0$, at least for early positive values of t. Therefore

$$\frac{dr}{d\theta} = \frac{dr/dt}{d\theta/dt} \geq 0 \qquad \text{and} \qquad \frac{du}{d\theta} = -\frac{1}{r^2}\frac{dr}{d\theta} \leq 0,$$

and so we select the negative sign in Eq. (27) and get

$$\frac{-du}{\sqrt{(u_0 - h)^2 - (u - h)^2}} = d\theta. \qquad (28)$$

The integral of Eq. (28) is

$$\cos^{-1}\left(\frac{u - h}{u_0 - h}\right) = \theta + C_2,$$

and the constant of integration C_2 is zero because

$$u = u_0 \qquad \text{when} \qquad \theta = 0$$

and

$$\cos^{-1}(1) = 0 + C_2 = 0.$$

Therefore

$$\frac{u - h}{u_0 - h} = \cos\theta,$$

or

$$u = \frac{1}{r} = h + (u_0 - h)\cos\theta.$$

A few more algebraic maneuvers produce

$$r = \frac{(1/h)}{1 + e\cos\theta}, \qquad (29)$$

with

$$\frac{1}{h} = \frac{r_0^2 v_0^2}{GM} \qquad \text{and} \qquad e = \frac{1}{r_0 h} - 1 = \frac{r_0 v_0^2}{GM} - 1. \qquad (30)$$

This is the modern formulation of Kepler's first law.

Kepler's Third Law (the Mean Distance-Time Law)

The time T it takes a planet to go once around its sun is the planet's **orbital period**. The semimajor axis a of the planet's orbit is the planet's **mean distance** from the sun. Kepler's third law says that T and a are related by the equation

$$\frac{T^2}{a^3} = \frac{4\pi^2}{GM}. \qquad (31)$$

Since all the quantities on the right side of this equation are constant within a given solar system, the ratio T^2/a^3 is the same for every planet in the system.

Kepler's third law is the starting point for working out the size of our solar system. It allows the semimajor axis of each planetary orbit to be expressed in astronomical units, Earth's semimajor axis being one unit. The distance between any two planets at any time can then be predicted in astronomical units and all that remains is to find one of these distances in kilometers. This can be done by bouncing radar waves off Venus, for example. The astronomical unit is now known, after a series of such measurements, to be 149,597,870 km.

To derive Kepler's third law, we combine two formulas for the area enclosed by a planet's ellipse:

$$\text{Area} = \pi ab \qquad (b \text{ is the semiminor axis})$$

and, from Eq. (17),

$$\begin{aligned}
\text{Area} &= \int_0^T dA \\
&= \int_0^T \frac{1}{2} r_0 v_0 \, dt \\
&= \frac{1}{2} T r_0 v_0.
\end{aligned}$$

Equating these gives

$$T = \frac{2\pi ab}{r_0 v_0}.$$

Now, for any ellipse, $b = a\sqrt{1 - e^2}$, so

$$T = \frac{2\pi a^2}{r_0 v_0}\sqrt{1 - e^2}. \tag{32}$$

It remains only to express a and e in terms of r_0, v_0, G, and M. Equation (30) does this for e. For a, we observe that setting θ equal to π in Eq. (29) gives

$$r_{\max} = \frac{(1/h)}{1 - e} = \frac{r_0}{2r_0 h - 1}.$$

Hence,

$$2a = r_0 + r_{\max} = \frac{2r_0^2 h}{2r_0 h - 1} = \frac{2r_0 GM}{2GM - r_0 v_0^2}. \tag{33}$$

Squaring both sides of (32) and substituting the results of Eqs. (30) and (33) leads to Kepler's third law, as we ask you to verify in Problem 12.

14.28 This multiflash photograph shows a body moving along a hyperbola when deflected by an inverse square law force. (Photograph from *PSSC Physics, Second Edition*; 1965; D.C. Heath & Company with Education Development Center, Inc.)

Astronomical Data

Although Kepler discovered his laws empirically and stated them only for the six planets known at the time, we can see from our derivations that they apply to any body driven by an inverse square law force. They apply to Halley's

TABLE 14.1
Values of *a*, *e*, and *T* for the major planets

Planet	Semimajor axis $a^{\dagger}$	Eccentricity e	Period T
Mercury	57.95	0.2056	87.967 days
Venus	108.11	0.0068	224.701 days
Earth	149.57	0.0167	365.256 days
Mars	227.84	0.0934	1.8808 years
Jupiter	778.14	0.0484	11.8613 years
Saturn	1427.0	0.0543	29.4568 years
Uranus	2870.3	0.0460	84.0081 years
Neptune	4499.9	0.0082	164.784 years
Pluto	5909	0.2481	248.35 years

†Millions of kilometers

comet and the asteroid Icarus. They apply to the moon's orbit about Earth, and they applied to the orbit of the spacecraft Apollo 8 about the moon. They also applied to the air puck shown in Fig. 14.28 being deflected by an inverse square law force—its path is a hyperbola. Even charged particles fired at the nuclei of atoms scatter along hyperbolic paths.

Tables 14.1–14.3 give additional data for planetary orbits and for the orbits of eight of Earth's artificial satellites (Fig. 14.29). Vanguard 1 was nicknamed "Grapefruit" because of its size and shape. The data it sent back revealed differences between the levels of Earth's oceans and provided the first determination of the precise locations of some of the more isolated Pacific islands. The data also verified that the gravitation of the sun and moon would affect the orbits of Earth's satellites and that solar light could exert enough pressure to deform an orbit. Syncom 3 is one of a series of U.S. Department of Defense telecommunications satellites. Tiros 11 (for "television infrared observation satellite") is one of a series of weather satellites. GOES 4 (for "geostationary

TABLE 14.2
Data on Earth's satellites

Name	Launch date	Time or expected time aloft	Weight at launch (kg)	Period (min)	Perigee height (km)	Apogee height (km)	Semimajor axis a (km)	Eccentricity
Sputnik 1	Oct. 1957	57.6 days	83.6	96.2	215	939	6,955	0.052
Vanguard 1	March 1958	300 years	1.47	138.5	649	4,340	8,872	0.208
Syncom 3	Aug. 1964	$> 10^6$ years	39	1436.2	35,718	35,903	42,189	0.002
Skylab 4	Nov. 1973	84.06 days	13,980	93.11	422	437	6,808	0.001
Tiros 11	Oct. 1978	500 years	734	102.12	850	866	7,236	0.001
GOES 4	Sept. 1980	$> 10^6$ years	627	1436.2	35,776	35,800	42,166	0.0003
Intelsat 5	Dec. 1980	$> 10^6$ years	1,928	1417.67	35,143	35,707	41,803	0.007

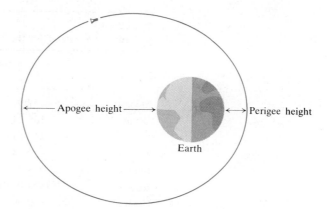

14.29 The orbit of an Earth satellite:
$2a$ = Diameter of earth + perigee height + apogee height.

operational environmental satellite'') is one of a series of satellites designed to gather information about Earth's atmosphere. Its orbital period, 1436.2 minutes, is nearly the same as Earth's rotational period of 1436.1 minutes, and its orbit is nearly circular ($e = 0.0003$). Intelsat 5 is a heavy-capacity commercial telecommunications satellite.

For circular orbits, e is zero and the orbit equation (29) becomes

$$r = \frac{1}{h} = \frac{r_0^2 v_0^2}{GM},$$

which reduces to

$$r = \frac{GM}{v^2} \tag{34}$$

since r and v are constant (see Problem 13). Also, Kepler's third law becomes

$$T^2 = \frac{4\pi^2 r^3}{GM} \tag{35}$$

for circular orbits, because $a = r$.

TABLE 14.3
Numerical data

Gravitational constant: $G = 6.6720 \times 10^{-11}$ Nm^{2}kg^{-2}
(When you use this value of G in a calculation, remember to express distance in meters, mass in kilograms, and time in seconds.)

Sun's mass: 1.99×10^{30} kg

Earth's mass: 5.975×10^{24} kg

Equatorial radius of Earth: 6378.533 km

Polar radius of Earth: 6356.912 km

Earth's rotational period: 1436.1 min

Earth's orbital period: 1 year = 365.256 days

PROBLEMS

Reminder: When a calculation involves the gravitational constant G, express distance in meters, mass in kilograms, and time in seconds.

1. Since the orbit of Skylab 4 had a semimajor axis of $a = 6808$ km, Kepler's third law with M equal to the earth's mass should give the period. Calculate it. Compare your result with the value in Table 14.2.

2. Earth's distance from the sun at perihelion is approximately 149,577,000 km, and the eccentricity of the earth's orbit about the sun is 0.0167. Compute the velocity v_0 of the earth in its orbit at perihelion. (Use Eq. 30.)

3. In July 1965, the USSR launched Proton I, weighing 12,200 kg (at launch), with a perigee height of 183 km, an apogee height of 589 km, and a period of 92.25 minutes. Using the relevant data for the mass of the earth and the gravitational constant G, compute the semimajor axis a of the orbit from Eq. (31). Compare your answer with the number you get by adding the perigee and apogee heights to the diameter of the earth.

4. a) The Viking 1 orbiter, which surveyed Mars from August 1975 to June 1976, had a period of 1639 min. Use this and the fact that the mass of Mars is 6.418×10^{23} kg to find the semimajor axis of the Viking 1 orbit.
   b) The Viking 1 orbiter was 1499 km from the surface of Mars at its closest point and 35,800 km from the surface at its farthest point. Use this information together with the value you obtained in part (a) to estimate the average diameter of Mars.

5. The Viking 2 orbiter, which surveyed Mars from September 1975 to August 1976, moved in an ellipse whose semimajor axis was 22,030 km. What was the orbital period? (Express your answer in minutes.)

6. If a satellite is to hold a geostationary orbit, what must the semimajor axis of its orbit be? Compare your result with the semimajor axes of the satellites in Table 14.2.

7. The mass of Mars is 6.418×10^{23} kg. If a satellite revolving about Mars is to hold a stationary orbit (have the same period as the period of Mars's rotation, which is 1477.4 min), what must the semimajor axis of its orbit be?

8. The period of the moon's rotation about the earth is 2.36055×10^6 sec. About how far away is the moon?

9. For what values of v_0 is the orbit in Eq. (29) a parabola? A circle? An ellipse? A hyperbola?

10. A satellite moves around the earth in a circular orbit. Express the satellite's speed as a function of the orbit's radius.

11. If T is measured in seconds and a in meters, what is the value of T^2/a^3 for planets in our solar system? For satellites orbiting the earth? For satellites orbiting the moon? (The moon's mass is 7.354×10^{22} kg.)

12. Complete the derivation of Kepler's third law (the part following Eq. 33).

13. Show that a planet in a circular orbit moves with a constant speed. (*Hint:* This is a consequence of one of Kepler's laws.)

14. Without introducing coordinates, give a geometric argument for the validity of the equation

$$\frac{dA}{dt} = \frac{1}{2}\left|\mathbf{R} \times \frac{d\mathbf{R}}{dt}\right|,$$

where $\mathbf{R}$ is the position vector of a particle moving along a plane curve and dA/dt is the rate at which that vector sweeps out area.

In Problems 15–20, r and θ are the polar coordinates of a particle moving in the plane. For each motion, express $\mathbf{v}$ and $\mathbf{a}$ in terms of $\mathbf{u}_r$ and $\mathbf{u}_\theta$.

15. $r = a(1 - \cos \theta)$ and $\dfrac{d\theta}{dt} = 3$

16. $r = a \sin 2\theta$ and $\dfrac{d\theta}{dt} = 2t$

17. $r = e^{a\theta}$ and $\dfrac{d\theta}{dt} = 2$

18. $r = a(1 + \sin t)$ and $\theta = 1 - e^{-t}$

19. $r = 2 \cos 4t$ and $\theta = 2t$

20. a) $r = \sin \theta$ and $\theta = \pi t$
    b) Find $\mathbf{v}$ and $\mathbf{a}$ when $t = 1/4$.

REVIEW QUESTIONS AND EXERCISES

1. State the rules for differentiating dot products and cross products of vector functions. Give examples.

2. What do you know about differentiating vectors of constant length? Give an example.

3. Define curvature, radius of curvature, circle of curvature, and center of curvature for plane curves. Give sample calculations. What plane curves have zero curvature?

4. Define $\mathbf{N}$ for plane curves. Which way does $\mathbf{N}$ point?

5. How is length defined for space curves? Calculate the length of one turn of a helix.

6. Define **T**, κ, **N**, **B**, and τ for space curves. Give sample calculations. What curves have zero torsion? What curves have zero torsion as well as zero curvature?

7. What are the formulas for the tangential and normal components of the acceleration of a moving body? What if the body moves at a constant speed? At a constant speed around a circle?

8. State Kepler's laws. To what do they apply?

MISCELLANEOUS PROBLEMS

1. The coordinates of a particle moving in the plane are

$$x = 1/\sqrt{1 + t^2}, \quad y = t/\sqrt{1 + t^2}.$$

Find the particle's highest speed.

2. A circular wheel rolls toward the right along the x-axis at a half-turn per second (Fig. 14.30). The position of the point P on the circumference is $\mathbf{R}(t) = (\pi t - \sin \pi t)\mathbf{i} + (1 - \cos \pi t)\mathbf{j}$. The center of the circle is C. The point that is topmost at any given time is called Q.
a) Find $\mathbf{v}(t)$ and $\mathbf{a}(t)$.
b) Find the slopes of the lines PC and PQ as functions of t.
c) How are $\mathbf{v}$ and $\mathbf{a}$ related to lines PC and PQ?

14.30 The wheel in Problem 2.

3. Find parametric equations for the line tangent to the curve

$$\mathbf{R}(t) = e^t\mathbf{i} + (\sin t)\mathbf{j} + \ln(1 - t)\mathbf{k}$$

at $t = 0$.

4. Find the length of the curve

$$\mathbf{R}(t) = (3 \cos t)\mathbf{i} + (3 \sin t)\mathbf{j} + t^2\mathbf{k}$$

from $t = 0$ to $t = 2$.

5. The line through the origin and the point $A(1, 1, 1)$ is the axis of rotation of a rigid body rotating with a constant angular speed of 6 rad/sec. The rotation appears to be clockwise when we look toward the origin from A. Find the velocity of the point of the body that is at the position $(1, 3, 2)$.

6. A particle moves around the unit circle in the xy-plane. At time t, its position is

$$\mathbf{R}(t) = x(t)\mathbf{i} + y(t)\mathbf{j}.$$

The scalar component of the velocity in the direction of $\mathbf{i}$ is $dx/dt = y$. Find dy/dt. Does the particle move clockwise, or counterclockwise?

7. A message is shot through a pneumatic tube that follows the curve $9y = x^3$ (distance in meters). At the point $(3, 3)$, $\mathbf{v} \cdot \mathbf{i} = 4$ and $\mathbf{a} \cdot \mathbf{i} = -2$. Find the values of $\mathbf{v} \cdot \mathbf{j}$ and $\mathbf{a} \cdot \mathbf{j}$ at the point $(3, 3)$.

8. At time $t \geq 0$, the position of a particle moving in space is

$$\mathbf{R}(t) = 2\mathbf{i} + \left(4 \sin \frac{t}{2}\right)\mathbf{j} + \left(3 - \frac{t}{\pi}\right)\mathbf{k}.$$

Find the first time $\mathbf{R}(t)$ is perpendicular to the vector $\mathbf{B} = \mathbf{i} - \mathbf{j}$.

9. Suppose $\mathbf{R}(t) = (e^t\cos t)\mathbf{i} + (e^t\sin t)\mathbf{j}$.
a) Show that $\mathbf{a} = 2\mathbf{v} - 2\mathbf{R}$.
b) Show that the angle between $\mathbf{R}$ and $\mathbf{a}$ is constant.
c) Find the angle.

10. The motion of a particle in the xy-plane is given by

$$\mathbf{R}(t) = (t \cos t)\mathbf{i} + (t \sin t)\mathbf{j}.$$

Find the speed and the tangential and normal components of the acceleration when $t = \sqrt{3}$.

11. A particle moves so that the derivative of its position vector is always perpendicular to the position vector. Show that the particle moves on a circle with center at the origin.

12. The velocity and acceleration of a particle moving in the plane are $\mathbf{v} = 3\mathbf{i} + 4\mathbf{j}$ and $\mathbf{a} = 5\mathbf{i} + 15\mathbf{j}$ at point P. Find the curvature of the particle's path at P.

13. Find the point on the curve $y = e^x$ where the radius of curvature is smallest.

14. Express the curvature of the path

$$\mathbf{R}(\theta) = \left(\int_0^\theta \cos \left(\frac{1}{2}\pi t^2\right) dt\right)\mathbf{i} + \left(\int_0^\theta \sin \left(\frac{1}{2}\pi t^2\right) dt\right)\mathbf{j}$$

as a function of the directed distance s measured from $(0, 0)$.

15. Show that the radius of curvature of a plane curve $\mathbf{R}(t) = x(t)\mathbf{i} + y(t)\mathbf{j}$ is given by

$$\rho = \frac{(\dot{x}^2 + \dot{y}^2)}{\sqrt{\ddot{x}^2 + \ddot{y}^2 - \ddot{s}^2}},$$

where the dots mean derivatives with respect to t ($\dot{x} =$

dx/dt, $\ddot{x} = d^2x/dt^2$, and so on) and

$$\ddot{s} = \frac{d}{dt}\sqrt{\dot{x}^2 + \dot{y}^2}.$$

16. Find the normal plane of the curve $\mathbf{R}(t) = t\mathbf{i} + t\mathbf{j} + (2/3)t^{3/2}\mathbf{k}$ at the point $(1, 1, 2/3)$.

17. Find $\mathbf{T}$, $\mathbf{N}$, κ, $\mathbf{B}$, and τ at $t = 0$ for the curve

$$\mathbf{R}(t) = (e^t\sin 2t)\mathbf{i} + (e^t\cos 2t)\mathbf{j} + (2e^t)\mathbf{k}.$$

18. Find equations for the osculating, normal, and rectifying planes of the curve

$$\mathbf{R}(t) = t\mathbf{i} + t^2\mathbf{j} + t^3\mathbf{k}$$

at the point $(1, 1, 1)$.

19. Figure 14.31 shows directed distance s on the circle $x^2 + y^2 = a^2$, measured counterclockwise from the positive x-axis to a point P. It also shows the angle ϕ that the tangent to the circle at P makes with the x-axis. Use the equations $s = a\theta$ and $\phi = \theta + \pi/2$ to calculate the circle's curvature directly from the definition $\kappa = |d\phi/ds|$.

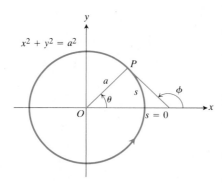

14.31   The circle in Problem 19.

20. A particle moves so that its position at time t has the polar coordinates $r = t$, $\theta = t$. Find the velocity $\mathbf{v}$, the acceleration $\mathbf{a}$, and the curvature κ at $t = 0$.

21. Find an expression for the curvature of the plane curve whose equation in polar coordinates is $r = f(\theta)$.

22. The polar coordinates of a particle at time t are

$$r = e^{\omega t} + e^{-\omega t}, \quad \theta = t,$$

where ω is a constant. Find the acceleration vector when $t = 0$.

23. A slender rod, passing through the fixed point O, is rotating about O in a plane at the constant rate of 3 rad/min. An insect is crawling along the rod toward O at the constant rate of 1 in./min. Use polar coordinates in the plane, with point O as the origin, and assume that the insect starts at the point $r = 2$, $\theta = 0$.

a) Find, in polar form, the velocity and acceleration of the insect when it is halfway to the origin.

b) What will be the length of the path that the insect has traveled when it reaches the origin?

24. *A Kepler equation.* The problem of locating a planet in its orbit at a given time and date leads to solving "Kepler" equations of the form

$$f(x) = x - 1 - \frac{1}{2}\sin x = 0.$$

a) Show that $f(x)$ has a zero between $x = 0$ and $x = 2$.

b) Estimate the solution to the equation $f(x) = 0$ by replacing $\sin x$ by its linearization at $x = \pi/2$ and solving the resulting linear equation.

c) **CALCULATOR**   Check your solution in the original equation.

25. In Article 14.4, we found the velocity vector of a particle moving in a plane to be

$$\mathbf{v} = \frac{d\mathbf{R}}{dt} = \frac{dx}{dt}\mathbf{i} + \frac{dy}{dt}\mathbf{j} = \frac{dr}{dt}\mathbf{u}_r + \frac{r\,d\theta}{dt}\mathbf{u}_\theta.$$

a) Express dx/dt and dy/dt in terms of dr/dt and $r\,d\theta/dt$ by computing $\mathbf{v} \cdot \mathbf{i}$ and $\mathbf{v} \cdot \mathbf{j}$.

b) Express dr/dt and $r\,d\theta/dt$ in terms of dx/dt and dy/dt by computing $\mathbf{v} \cdot \mathbf{u}_r$ and $\mathbf{v} \cdot \mathbf{u}_\theta$.

26. If a particle moves in an ellipse whose polar equation is $r = c/(1 - e\cos\theta)$ and the acceleration is directed toward the origin, show that the magnitude of the acceleration is proportional to $1/r^2$.

27. What percentage of Earth's surface could the astronauts see when Skylab 4 was at its apogee height, 437 km above the surface?

CHAPTER 15

Surfaces, Coordinate Systems, and Drawing

OVERVIEW

Surfaces form the common geometric thread of this and the next three chapters of the book. Surfaces are the graphs of functions of two variables and play the same role for these functions that curves play for functions of a single variable. They give information about rates of change, extreme values, and the existence of roots. They also form the boundaries of regions in space, and their formulas appear in the integrals that give the volumes, surface areas, and centers of mass of these regions. They model the membranes across which fluids flow, the plates over which electrical charges are distributed, and the walls of containers that are subjected to pressures of various kinds. We shall see all this and more in the chapters to come.

In the present chapter, our goal is to become acquainted with the surfaces most commonly used in the theory and applications of the calculus of functions of more than one variable. This means finding out what the surfaces look like, what their equations are, and how to draw them.

We shall also look at two new coordinate systems for space: cylindrical coordinates and spherical coordinates. A surface may have an equation in one of these systems that is simpler than its Cartesian coordinate equation. When it does, we can save time by switching over.

15.1
Cylinders

The simplest surfaces to draw and write equations for, besides planes, are cylinders. A cylinder is a surface composed of all the lines that pass through a plane curve and are parallel to a fixed line. Figure 15.1, for example, shows the cylinder made by the lines that pass through the curve $y = x^2$ in the xy-plane and are parallel to the z-axis. In this article, we show how to draw cylinders and write equations for them.

In solid geometry, the word "cylinder" sometimes means "circular cylinder," but that is not the case here. The cross sections of the cylinder in Fig. 15.1 are parabolas. We will allow our cylinders to have cross sections of any kind.

EXAMPLE 1   Find an equation for the parabolic cylinder in Fig. 15.1.

Solution   Suppose that the point $P_0(x_0, x_0^2, 0)$ lies on the parabola $y = x^2$ in the xy-plane. Then, for any value of z, the point $Q(x_0, x_0^2, z)$ will lie on the cylinder because it lies on the line $x = x_0$, $y = x_0^2$ through P_0 parallel to the z-axis. Conversely, any point $Q(x_0, x_0^2, z)$ whose y-coordinate is the square of its x-coordinate lies on the cylinder because it lies on the line $x = x_0$, $y = x_0^2$ through P_0 parallel to the z-axis. See Fig. 15.2.

Regardless of the value of z, therefore, the points on the surface are the points whose coordinates satisfy the equation $y = x^2$. In other words, $y = x^2$ is an equation for the cylinder.   ∎

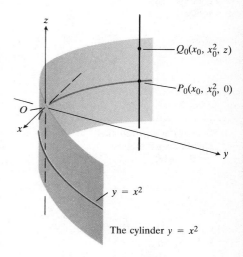

15.1   The cylinder generated by lines parallel to the z-axis and passing through the parabola $y = x^2$ in the xy-plane.

15.2   Every point of the cylinder in Fig. 15.1 has coordinates of the form (x_0, x_0^2, z).

As Example 1 suggests, any curve $f(x, y) = C$ in the xy-plane defines a cylinder parallel to the z-axis whose equation is also $f(x, y) = C$. The equation $x^2 + y^2 = 1$ defines the circular cylinder made by the lines parallel to the z-axis and passing through the circle $x^2 + y^2 = 1$ in the xy-plane. The equation $x^2 + 4y^2 = 9$ defines the elliptical cylinder made by the lines parallel to the z-axis and passing through the ellipse $x^2 + 4y^2 = 9$ in the xy-plane, and so on.

In a similar manner, any curve $g(x, z) = C$ in the xz-plane defines a cylinder parallel to the y-axis whose equation is also $g(x, z) = C$, and any curve $h(y, z) = C$ defines a cylinder parallel to the x-axis whose equation is also $h(y, z) = C$.

HOW TO DRAW CYLINDERS PARALLEL TO THE COORDINATE AXES

1. Sketch all three coordinate axes *very lightly.*

2. Sketch the trace of the cylinder in the coordinate plane of the two variables that appear in the cylinder's equation. Sketch *very lightly.*

3. Sketch traces in parallel planes on either side (again, lightly).

In short, an equation in any two of the three Cartesian coordinates defines a cylinder made of lines parallel to the axis of the third coordinate. See Figures 15.3 and 15.4.

Drawing Cylinders Parallel to the Coordinate Axes

Here is some advice about drawing cylinders. As always, pencil is safer than pen because you can erase, but the advice applies no matter what medium you choose. Just determine which axis the cylinder is parallel to and carry out the steps shown.

4. Add parallel outer edges to give the shape definition.

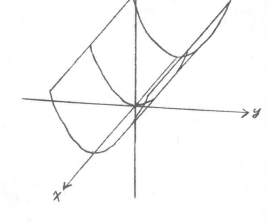

5. If more definition is required, darken the parts of the lines that are exposed to view. Leave the hidden parts light. Use line breaks when you can.

The elliptical cylinder $x^2 + 4z^2 = 4$

15.3   The elliptic cylinder $x^2 + 4z^2 = 4$ is made of lines parallel to the y-axis and passing through the ellipse $x^2 + 4z^2 = 4$ in the xz-plane. The cross sections or "traces" of the cylinder in planes perpendicular to the y-axis are ellipses congruent to the generating ellipse. The cylinder extends along the entire y-axis but we can draw only a finite portion of it.

The hyperbolic cylinder $y^2 - z^2 = 1$

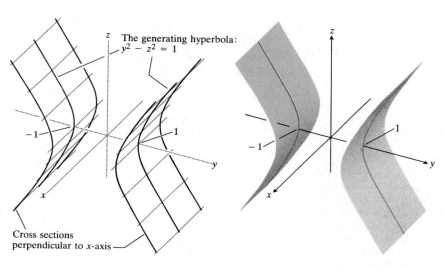

15.4   The hyperbolic cylinder $y^2 - z^2 = 1$ is made of lines parallel to the x-axis and passing through the hyperbola $y^2 - z^2 = 1$ in the yz-plane. The cross sections or "traces" of the cylinder in planes perpendicular to the x-axis are hyperbolas congruent to the generating hyperbola.

PROBLEMS

Sketch the cylinders in Problems 1–18.

1. $x^2 + y^2 = 4$
2. $x^2 + z^2 = 4$
3. $y^2 + z^2 = 1$
4. $z = y^2/4$
5. $z = y^2 - 1$
6. $x = y^2$
7. $z = 4 - x^2$
8. $x = 4 - y^2$
9. $y = x^2$
10. $y = x^2 - 2$
11. $y = -x$
12. $z = -y$

13. $y^2 + 4z^2 = 16$
14. $4x^2 + y^2 = 36$
15. $z^2 + 4y^2 = 9$
16. $y^2 - z^2 = 4$
17. $z^2 - y^2 = 1$
18. $yz = 1$

19. True, or false? Every plane is a cylinder. Explain.

20. The helix $\mathbf{R}(t) = (\cos t)\mathbf{i} + t\mathbf{j} + (\sin t)\mathbf{k}$ lies on a cylinder. Sketch the cylinder and, on it, the portion of the helix from $t = -2\pi$ to $t = 2\pi$.

21. a) Show that the curve

$$\mathbf{R}(t) = (\cos t)\mathbf{i} + (\sin t)\mathbf{j} + (1 - \cos t)\mathbf{k}$$

is the intersection of a plane with a cylinder. What cylinder?
What plane?

b) Sketch the plane and cylinder together.

c) Write an integral for the length of the curve but don't try
to evaluate the integral—it's nonelementary.

TOOLKIT PROGRAM

3D Grapher

15.2
Quadric Surfaces

Most of the cylinders in Article 15.1 were **quadric surfaces,** surfaces whose
equations combine quadratic terms with linear terms and constants. In this
article, we describe some other quadric surfaces and show how to draw them.
The constants a, b, and c in the examples are all positive.

EXAMPLE 1   The **ellipsoid** (Fig. 15.5)

$$\frac{x^2}{a^2} + \frac{y^2}{b^2} + \frac{z^2}{c^2} = 1 \tag{1}$$

cuts the coordinate axes at $(\pm a, 0, 0)$, $(0, \pm b, 0)$, and $(0, 0, \pm c)$. It lies inside
the rectangular box

$$|x| \le a, \qquad |y| \le b, \qquad |z| \le c.$$

Since only even powers of x, y, and z occur in the equation, this surface is
symmetric with respect to each coordinate plane. The sections cut out by the
coordinate planes are ellipses. For example,

$$\frac{x^2}{a^2} + \frac{y^2}{b^2} = 1 \qquad \text{when} \quad z = 0.$$

15.5   A typical ellipsoid.

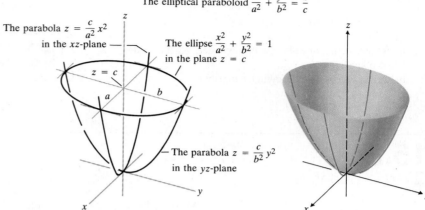

15.6 The surface $(x^2/a^2) + (y^2/b^2) = z/c$ is an elliptic paraboloid. The cross sections perpendicular to the z-axis above the xy-plane are ellipses. The cross sections in the planes that contain the z-axis are parabolas.

Each section cut out by a plane

$$z = z_1, \qquad |z_1| < c,$$

is an ellipse

$$\frac{x^2}{a^2[1 - (z_1^2/c^2)]} + \frac{y^2}{b^2[1 - (z_1^2/c^2)]} = 1. \tag{2}$$

When two of the three semiaxes a, b, and c are equal, the surface is an ellipsoid of revolution, and when all three are equal, it is a sphere. ∎

EXAMPLE 2   The **elliptic paraboloid** (Fig. 15.6)

$$\frac{x^2}{a^2} + \frac{y^2}{b^2} = \frac{z}{c} \tag{3}$$

is symmetric with respect to the planes $x = 0$ and $y = 0$. The only intercept on the axes is at the origin. Except for this point, the surface lies above the xy-plane because z is positive whenever either x or y is different from zero. The sections cut by the coordinate planes are

$$x = 0: \quad \text{the parabola } z = \frac{c}{b^2}y^2,$$

$$y = 0: \quad \text{the parabola } z = \frac{c}{a^2}x^2, \tag{4}$$

$$z = 0: \quad \text{the point } (0, 0, 0).$$

Each plane $z = z_1$ above the xy-plane cuts the surface in the ellipse

$$\frac{x^2}{a^2} + \frac{y^2}{b^2} = \frac{z_1}{c}. \qquad\qquad ∎$$

EXAMPLE 3   The **circular paraboloid** or **paraboloid of revolution**

$$\frac{x^2}{a^2} + \frac{y^2}{a^2} = \frac{z}{c} \tag{5}$$

is obtained by taking $b = a$ in Eq. (3) for the elliptic paraboloid. The cross sections of the surface by planes perpendicular to the z-axis are circles centered

■ ■ ■

THREE-DIMENSIONAL ANALYTIC GEOMETRY

Seventeenth-century mathematicians such as Fermat, Descartes, and La Hire suggested three-dimensional analytic geometry. But John Bernoulli, in a letter to Leibniz in 1715, formulated the coordinate system we use today. In the eighteenth century, Alexis Clairaut (1713–1765) and Leonhard Euler (1707–1783) were the most notable mathematicians to develop three-dimensional geometry. Clairaut, in particular, specified that a surface can be represented by an equation in three coordinates and that two such equations are required to describe a curve in space. He presented his ideas in his book *Recherche sur les courbes à double courbure* in 1731, in which he gave the equations of several quadric surfaces, such as the sphere, cylinder, hyperboloid, and ellipsoid. His ultimate concern was with the shape of the earth, which he believed to be an ellipsoid of some kind. Gaspard Monge, the preeminent geometer of the eighteenth century, wrote extensively on three-dimensional analytic geometry as well.

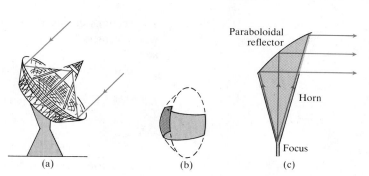

15.7 Many antennas are shaped like pieces of paraboloids of revolution. (a) Radio telescopes use the same principles as optical telescopes. (b) A "rectangular-cut" radar reflector. (c) The profile of a horn antenna in a microwave radio link.

on the z-axis. The cross sections by planes containing the z-axis are congruent parabolas with a common focus at the point $(0, 0, a^2/4c)$.

Shapes cut from circular paraboloids are used for antennas in radio telescopes, satellite trackers, and microwave radio links (Fig. 15.7). ∎

EXAMPLE 4 The **elliptic cone**

$$\frac{x^2}{a^2} + \frac{y^2}{b^2} = \frac{z^2}{c^2} \tag{6}$$

is symmetric with respect to the three coordinate planes (Fig. 15.8). The sections cut by the coordinate planes are

$$x = 0: \quad \text{the lines } z = \pm\frac{c}{b}y, \tag{7}$$

$$y = 0: \quad \text{the lines } z = \pm\frac{c}{a}x, \tag{8}$$

$$z = 0: \quad \text{the point } (0, 0, 0).$$

15.8 The surface $(x^2/a^2) + (y^2/b^2) = z^2/c^2$ is an elliptic cone. Planes perpendicular to the z-axis cut the cone in ellipses above and below the xy-plane. Vertical planes that contain the z-axis cut it in pairs of intersecting lines.

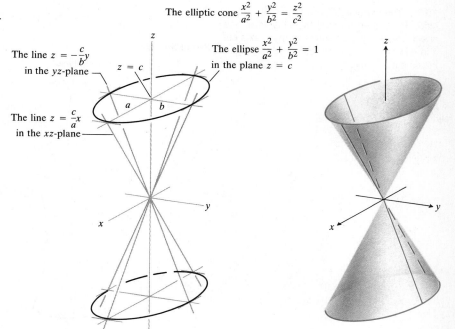

The line $z = -\dfrac{c}{b}y$ in the yz-plane

The line $z = \dfrac{c}{a}x$ in the xz-plane

$z = c$

The elliptic cone $\dfrac{x^2}{a^2} + \dfrac{y^2}{b^2} = \dfrac{z^2}{c^2}$

The ellipse $\dfrac{x^2}{a^2} + \dfrac{y^2}{b^2} = 1$ in the plane $z = c$

The sections cut by planes $z = z_1$ above and below the xy-plane are ellipses whose centers lie on the z-axis and whose vertices lie on the lines in Eqs. (7) and (8).

If $a = b$, the cone is a right circular cone. ∎

EXAMPLE 5   The **hyperboloid of one sheet** (Fig. 15.9)

$$\frac{x^2}{a^2} + \frac{y^2}{b^2} - \frac{z^2}{c^2} = 1 \tag{9}$$

is symmetric with respect to each of the three coordinate planes. The sections cut out by the coordinate planes are

$$x = 0: \quad \text{the hyperbola } \frac{y^2}{b^2} - \frac{z^2}{c^2} = 1,$$

$$y = 0: \quad \text{the hyperbola } \frac{x^2}{a^2} - \frac{z^2}{c^2} = 1, \tag{10}$$

$$z = 0: \quad \text{the ellipse } \frac{x^2}{a^2} + \frac{y^2}{b^2} = 1.$$

The plane $z = z_1$ cuts the surface in an ellipse with center on the z-axis and vertices on one of the hyperbolas in (10).

The surface is connected, meaning that it is possible to travel from any point on it to any other point on it without leaving the surface. For this reason, it is said to have *one* sheet, in contrast to the hyperboloid in the next example, which has two sheets.

If $a = b$, the hyperboloid is a surface of revolution. ∎

15.9   The surface $(x^2/a^2) + (y^2/b^2) - (z^2/c^2) = 1$ is an (elliptic) hyperboloid of one sheet. Planes perpendicular to the z-axis cut it in ellipses. Vertical planes containing the z-axis cut it in hyperbolas.

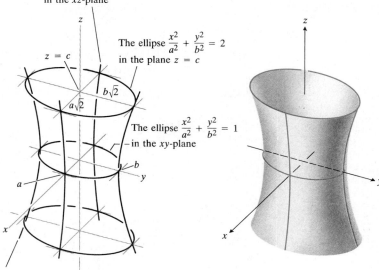

The hyperboloid $\dfrac{x^2}{a^2} + \dfrac{y^2}{b^2} - \dfrac{z^2}{c^2} = 1$

Part of the hyperbola $\dfrac{x^2}{a^2} - \dfrac{z^2}{c^2} = 1$ in the xz-plane

The ellipse $\dfrac{x^2}{a^2} + \dfrac{y^2}{b^2} = 2$ in the plane $z = c$

The ellipse $\dfrac{x^2}{a^2} + \dfrac{y^2}{b^2} = 1$ in the xy-plane

The hyperboloid $\dfrac{z^2}{c^2} - \dfrac{x^2}{a^2} - \dfrac{y^2}{c^2} = 1$

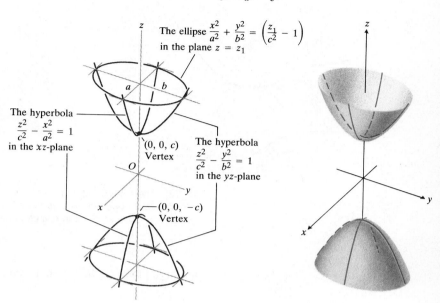

The ellipse $\dfrac{x^2}{a^2} + \dfrac{y^2}{b^2} = \left(\dfrac{z_1}{c^2} - 1\right)$ in the plane $z = z_1$

The hyperbola $\dfrac{z^2}{c^2} - \dfrac{x^2}{a^2} = 1$ in the xz-plane

The hyperbola $\dfrac{z^2}{c^2} - \dfrac{y^2}{b^2} = 1$ in the yz-plane

$(0, 0, c)$ Vertex

$(0, 0, -c)$ Vertex

15.10 The surface $(z^2/c^2) - (x^2/a^2) - (y^2/b^2) = 1$ is an (elliptic) hyperboloid of two sheets. Planes perpendicular to the z-axis above and below the vertices cut it in ellipses. Vertical planes containing the z-axis cut it in hyperbolas.

EXAMPLE 6 The **hyperboloid of two sheets** (Fig. 15.10)

$$\frac{z^2}{c^2} - \frac{x^2}{a^2} - \frac{y^2}{b^2} = 1 \tag{11}$$

is symmetric with respect to the three coordinate planes. The plane $z = 0$ does not intersect the surface; in fact, for the plane to intersect the surface, we must have

$$|z| \geq c$$

for real values of x and y in Eq. (11). The hyperbolic sections

$$x = 0: \quad \frac{z^2}{c^2} - \frac{y^2}{b^2} = 1,$$

$$y = 0: \quad \frac{z^2}{c^2} - \frac{x^2}{a^2} = 1,$$

have their vertices and foci on the z-axis. The surface is separated into two portions, one above the plane $z = c$ and the other below the plane $z = -c$. This accounts for its name.

Equations (9) and (11) differ in the number of negative terms that each contains on the left side when the right side is $+1$. The number of negative signs is the same as the number of sheets of the hyperboloid. If we compare with Eq. (6), we see that replacing the one on the right side of either Eq. (9) or (11) by zero gives the equation of a cone. This cone (Fig. 15.11) is, in fact, asymptotic to both of the hyperboloids (9) and (11) in the same way that the lines

$$\frac{x^2}{a^2} - \frac{y^2}{b^2} = 0$$

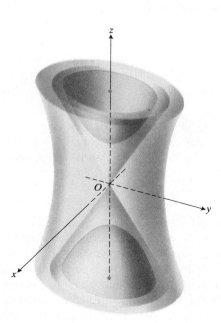

15.11 The cone is asymptotic to both hyperboloids.

15.12   The surface $(y^2/b^2) - (x^2/a^2) = z/c$ is a hyperbolic paraboloid. The cross sections in planes perpendicular to the z-axis above and below the xy-plane are hyperbolas. The cross sections in planes perpendicular to the other axes are parabolas.

are asymptotic to the two hyperbolas

$$\frac{x^2}{a^2} - \frac{y^2}{b^2} = \pm 1$$

in the xy-plane.   ∎

EXAMPLE 7   The **hyperbolic paraboloid** (Fig. 15.12)

$$\frac{y^2}{b^2} - \frac{x^2}{a^2} = \frac{z}{c} \tag{12}$$

has symmetry with respect to the planes $x = 0$ and $y = 0$. The sections in these planes are

$$x = 0: \quad \text{the parabola } z = \frac{c}{b^2}y^2, \tag{13}$$

$$y = 0: \quad \text{the parabola } z = -\frac{c}{a^2}x^2. \tag{14}$$

In the plane $x = 0$, the parabola opens upward and has vertex at the origin. The parabola in the plane $y = 0$ has the same vertex, but it opens downward.

If we cut the surface by a plane $z = z_1 > 0$, the section is a hyperbola,

$$\frac{y^2}{b^2} - \frac{x^2}{a^2} = \frac{z_1}{c}, \tag{15}$$

with its focal axis parallel to the y-axis and its vertices on the parabola in (13). If, on the other hand, z_1 is negative in Eq. (15), then the focal axis of the hyperbola is parallel to the x-axis, and its vertices lie on the parabola in (14).

Near the origin, the surface is shaped like a saddle. To a person traveling along the surface in the yz-plane, the origin looks like a minimum. To a person traveling in the xz-plane, on the other hand, the origin looks like a maximum. Such a point is called a **minimax** or **saddle point** of a surface (Fig. 15.12). We shall discuss maximum and minimum points on surfaces in Chapter 16.   ∎

HOW TO DRAW QUADRIC SURFACES

$$x^2 + \frac{y^2}{4} + z^2 = 1$$

$$z = 4 - x^2 - y^2$$

1. Lightly sketch the three co-ordinate axes.

2. Decide on a scale and mark the intercepts on the axes.

3. Sketch cross sections in the coordinate planes and in a few parallel planes, but don't clutter the picture. Use tangent lines as guides.

4. If more is required, darken the parts exposed to view. Leave the rest light. Use line breaks when you can.

PROBLEMS

Sketch the surfaces in Problems 1–34. If the surface is listed as a paraboloid, hyperboloid, or cone, say what kind it is.

Ellipsoids

1. $9x^2 + y^2 + z^2 = 9$
2. $4x^2 + 4y^2 + z^2 = 16$
3. $x^2 + y^2 + z^2 = 4$
4. $9x^2 + 4y^2 + z^2 = 36$
5. $4x^2 + 9y^2 + 4z^2 = 36$
6. $9x^2 + 4y^2 + 36z^2 = 36$

Paraboloids

7. $x^2 + y^2 = z$
8. $x^2 + z^2 = y$
9. $x^2 + 4y^2 = z$
10. $z = x^2 + 9y^2$
11. $z = 8 - x^2 - y^2$
12. $z = 18 - x^2 - 9y^2$
13. $x = 4 - 4y^2 - z^2$
14. $y = 1 - x^2 - z^2$
15. $z = x^2 + y^2 + 1$
16. $z = 4x^2 + y^2 - 4$

Cones

17. $x^2 + y^2 = z^2$
18. $y^2 + z^2 = x^2$
19. $x^2 + z^2 = y^2$
20. $4x^2 + 9y^2 = z^2$
21. $9x^2 + 4y^2 = 36z^2$
22. $4x^2 + 9z^2 = 9y^2$

Hyperboloids

23. $x^2 + y^2 - z^2 = 1$
24. $y^2 + z^2 - x^2 = 1$
25. $(y^2/4) + (z^2/9) - (x^2/4) = 1$
26. $(x^2/4) + (y^2/4) - (z^2/9) = 1$
27. $(x^2/9) + (y^2/4) - z^2 = 1$
28. $z^2 - x^2 - y^2 = 1$
29. $z^2 - (x^2/4) - y^2 = 1$
30. $(y^2/4) - (x^2/4) - z^2 = 1$
31. $(x^2/4) - (y^2/9) - z^2 = 1$
32. $(x^2/4) - (z^2/4) - y^2 = 1$

Hyperbolic paraboloids

33. $y^2 - x^2 = z$
34. $x^2 - y^2 = z$

The surfaces in Problems 35–38 are slanted parabolic cylinders. Sketch them as you would other quadric surfaces.

35. $z = x^2 + y$
36. $x = y^2 - z$
37. $y = x^2 + z$
38. $z = 4 - x^2 - y$

39. The curve
$$\mathbf{R}(t) = (2\sqrt{t}\ \cos t)\mathbf{i} + (3\sqrt{t}\ \sin t)\mathbf{j} + \sqrt{1 - t}\,\mathbf{k},$$
$0 \le t \le 1$, lies on a quadric surface. Find an equation for the surface and identify it.

40. a) Express the area $A(z_1)$ of the cross section cut from the ellipsoid
$$x^2 + \frac{y^2}{4} + \frac{z^2}{9} = 1$$
by the plane $z = z_1$ as a function of z_1. (The area of an ellipse with semiaxes A and B is πAB.)
   b) By integration, find the volume of the ellipsoid in part (a). Use slices perpendicular to the z-axis.
   c) Find the volume of the ellipsoid
$$\frac{x^2}{a^2} + \frac{y^2}{b^2} + \frac{z^2}{c^2} = 1.$$
   Does your answer give the volume of a sphere of radius a if $a = b = c$?

  TOOLKIT PROGRAM

3D Grapher

15.3

Cylindrical Coordinates and Spherical Coordinates

In this article we introduce two new systems of coordinates for space, the cylindrical coordinate system and the spherical coordinate system. In the cylindrical coordinate system, cylinders whose axes lie along the z-axis and planes that contain the z-axis have especially simple equations. In the spherical coordinate system, spheres centered at the origin and single cones at the origin whose axes lie along the z-axis have especially simple equations. When your work involves these shapes and you need a coordinate system, these may be the best systems to use.

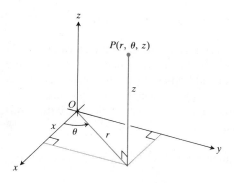

15.13 The cylindrical coordinates of a point in space are r, θ, and z.

Cylindrical Coordinates

We obtain cylindrical coordinates for space by combining polar coordinates in the xy-plane with the usual z-axis. This assigns to every point in space one or more coordinate triples of the form (r, θ, z), as shown in Fig. 15.13.

The values of x, y, r, and θ in cylindrical coordinates are related by the usual equations:

$$x = r \cos \theta, \quad r^2 = x^2 + y^2, \quad y = r \sin \theta, \quad \tan \theta = y/x. \quad (1)$$

Cylindrical coordinates were the coordinates we used when we first studied planetary motion in Article 14.4.

In cylindrical coordinates, the equation $r = a$ describes not just a circle in the xy-plane but an entire cylinder about the z-axis (Fig. 15.14). The z-axis itself is given by the equation $r = 0$. The equation $\theta = \theta_0$ describes the plane that contains the z-axis and makes an angle of θ_0 radians with the positive x-axis.

EXAMPLE 1 Describe the points in space whose cylindrical coordinates satisfy the equations

$$r = 2, \qquad \theta = \frac{\pi}{4}.$$

Solution These points make up the line in which the cylinder $r = 2$ cuts the portion of the plane $\theta = \pi/4$ in which r is positive (Fig. 15.15). This is the line through the point $(2, \pi/4, 0)$ parallel to the z-axis. ∎

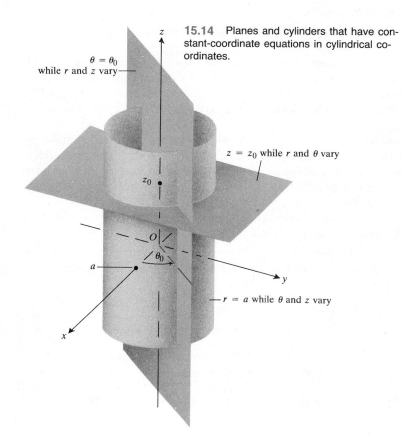

15.14 Planes and cylinders that have constant-coordinate equations in cylindrical coordinates.

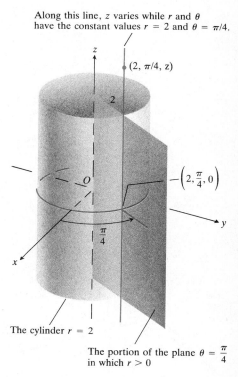

15.15 The points whose first two cylindrical coordinates are $r = 2$ and $\theta = \pi/4$ form a line parallel to the z-axis.

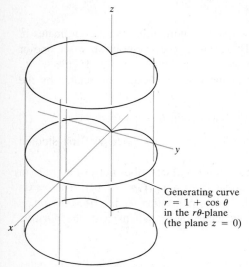

15.16 The cylindrical-coordinate equation $r = 1 + \cos\theta$ defines a cylinder in space whose cross sections perpendicular to the z-axis are cardioids.

EXAMPLE 2   Sketch the surface $r = 1 + \cos\theta$.

Solution   The equation involves only r and θ; the coordinate variable z is missing. Therefore, the surface is a cylinder of lines that pass through the cardioid $r = 1 + \cos\theta$ in the $r\theta$-plane and lie parallel to the z-axis. The rules for sketching the cylinder are the same as always: sketch the x-, y-, and z-axes, draw a few perpendicular cross sections, connect the cross sections with parallel lines, and darken the exposed parts. See Fig. 15.16.   ■

EXAMPLE 3   Find a Cartesian equation for the surface $z = r^2$ and identify the surface.

Solution   From Eqs. (1) we have $z = r^2 = x^2 + y^2$. The surface is the circular paraboloid $x^2 + y^2 = z$.   ■

EXAMPLE 4   Find an equation for the circular cylinder $4x^2 + 4y^2 = 9$ in cylindrical coordinates.

Solution   The cylinder consists of the points whose distance from the z-axis is $\sqrt{x^2 + y^2} = 3/2$. The corresponding equation in cylindrical coordinates is $r = 3/2$.   ■

Spherical Coordinates

Spherical coordinates locate points in space with two angles and a distance, as shown in Fig. 15.17.

The first coordinate, $\rho = |\overrightarrow{OP}|$, is the point's distance from the origin. Unlike r, the variable ρ is never negative. The second coordinate, ϕ, is the angle the vector $\overrightarrow{OP}$ makes with the positive z-axis. It is required to lie in the interval from 0 to π. The third coordinate is the angle θ from cylindrical coordinates.

The equation $\rho = a$ describes the sphere of radius a centered at the origin (Fig. 15.18). The equation $\phi = \phi_0$ describes a single cone whose vertex lies at the origin and whose axis lies along the z-axis. (We have to broaden our interpretation here to include the xy-plane as the cone $\phi = \pi/2$.) If ϕ_0 is greater than $\pi/2$, the cone $\phi = \phi_0$ opens downward.

A few books give spherical coordinates in the order (ρ, θ, ϕ), with the θ and ϕ reversed. Watch out for this when you read elsewhere.

15.17   The spherical coordinates ρ, ϕ, and θ and their relation to x, y, z, and r.

Selected Equations Relating Cartesian, Cylindrical, and Spherical Coordinates

$$r = \rho \sin\phi, \qquad x = r\cos\theta = \rho\sin\phi\cos\theta,$$

$$z = \rho\cos\phi, \qquad y = r\sin\theta = \rho\sin\phi\sin\theta,$$

$$\rho = \sqrt{x^2 + y^2 + z^2} = \sqrt{r^2 + z^2}$$

(2)

EXAMPLE 5   Find a spherical coordinate equation for the sphere

$$x^2 + y^2 + (z - 1)^2 = 1.$$

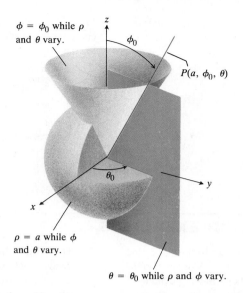

$\phi = \phi_0$ while ρ and θ vary.

$\rho = a$ while ϕ and θ vary.

$\theta = \theta_0$ while ρ and ϕ vary.

15.18 Spheres whose centers are at the origin, and single cones at the origin whose axes lie along the z-axis, have simple equations in spherical coordinates.

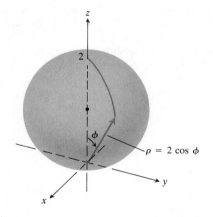

$\rho = 2\cos\phi$

15.19 The sphere $\rho = 2\cos\phi$. Notice that the equation restricts ϕ to lie in the interval $0 \leq \phi \leq \pi/2$ because the spherical coordinate ρ is not allowed to be negative.

Solution From Eqs. (2) we find that the left side of the equation is

$$\rho^2\sin^2\phi(\cos^2\theta + \sin^2\theta) + \rho^2\cos^2\phi - 2\rho\cos\phi + 1 = \rho^2 - 2\rho\cos\phi + 1.$$

Hence the original equation transforms into

$$\rho^2 - 2\rho\cos\phi + 1 = 1,$$
$$\rho^2 = 2\rho\cos\phi,$$
$$\rho = 2\cos\phi.$$

See Fig. 15.19. ∎

PROBLEMS

The following table gives the coordinates of specific points in space in one of three coordinate systems. In Problems 1–10, find coordinates for each point in the other two systems. There may be more than one right answer because points in cylindrical and spherical coordinates may have more than one coordinate triple.

Cartesian (x, y, z)	Cylindrical (r, θ, z)	Spherical (ρ, ϕ, θ)
1. (0, 0, 0)		
2. (1, 0, 0)		
3. (0, 1, 0)		
4. (0, 0, 1)		
5.	(1, 0, 0)	
6.	($\sqrt{2}$, 0, 1)	
7.		(1, $\pi/2$, 1)
8.		($\sqrt{3}$, $\pi/3$, $-\pi/2$)
9.		($2\sqrt{2}$, $\pi/2$, $3\pi/2$)
10.		($\sqrt{2}$, π, $3\pi/2$)

In Problems 11–20, translate the equations from the given coordinate system (Cartesian, cylindrical, spherical) into equations in the other two systems. Also, identify the set of points defined by the equation.

11. $r = 0$

12. $x^2 + y^2 = 5$

13. $z = 0$

14. $z = -2$

15. $\rho\cos\phi = 3$

16. $\sqrt{x^2 + y^2} = z$

17. $\rho\sin\phi\cos\theta = 0$

18. $\tan^2\phi = 1$

19. $x^2 + y^2 + z^2 = 4$

20. $\rho = 6\cos\phi$

In Problems 21–28, describe the sets of points whose cylindrical coordinates satisfy the given equations or pairs of equations. Sketch.

21. $r = 4$

22. $r^2 + z^2 = 1$

23. $r = 1 - \cos\theta$

24. $r = 2\cos\theta$

25. $r = 2, \quad z = 3$

26. $\theta = \pi/6, \quad z = r$

27. $r = 3, \quad z = \theta/2$

28. $r^2 - z^2 = 1$

In Problems 29–34, describe the sets of points whose spherical coordinates satisfy the given equations or pairs of equations. Sketch.

29. $\phi = \pi/6$

30. $\rho = 6, \quad \phi = \pi/6$

31. $\rho = 5, \quad \theta = \pi/4$

32. $\theta = \pi/4, \quad \phi = \pi/4$

33. $\theta = \pi/2, \quad \rho = 4\cos\phi$

34. $\rho = 1 - \cos\phi$ (*Hint:* The absence of θ indicates symmetry with respect to the z-axis. What is the trace of the surface in the yz-plane?)

REVIEW QUESTIONS AND EXERCISES

1. What is a cylinder? Give examples of equations that define cylinders in Cartesian coordinates; in cylindrical coordinates. What advice can you give about drawing cylinders?

2. Give examples of ellipsoids, paraboloids, cones, and hyperboloids (equations and sketches). What advice can you give about sketching these surfaces?

3. How are cylindrical and spherical coordinates defined? Draw diagrams that show how cylindrical and spherical coordinates are related to Cartesian space coordinates. What sets have constant-coordinate equations (like $x = 1, r = 1,$ or $\phi = \pi/3$) in the three coordinate systems?

MISCELLANEOUS PROBLEMS

The equations in Problems 1–12 define sets both in the plane and in three-dimensional space. Identify both sets for each equation.

1. $x + y = 1$

2. $x = 0$

3. $x^2 + y^2 = 4$

4. $x^2 + 4y^2 = 16$

5. $y^2 = -x$

6. $r = 1 + \cos\theta$

7. $r = \cos 2\theta$

8. $r = \dfrac{2}{(2 - \cos\theta)}$

9. $r = -\sin\theta$

10. $r\cos\theta = 1$

11. $r = \sin 2\theta$

12. $r = \theta$

Describe the sets in space defined by the equations and inequalities in Problems 13–18.

13. $3 \le \rho \le 5$

14. $r \ge 2, \quad \rho \le 5$

15. $0 \le \theta \le \dfrac{\pi}{4}, \quad 0 \le \phi \le \dfrac{\pi}{4}$

16. $4x^2 + 9y^2 \le 36$

17. $\rho = 1, \quad \theta = \phi, \quad 0 \le \theta \le \dfrac{\pi}{2}$

18. $x^2 + y^2 = f(z)$

19. What symmetry will you find in a surface whose spherical coordinate equation has the form $\rho = f(\phi)$?

20. Suppose PO is the distance from $P(x, y, z)$ to the origin and PA is the distance from P to the point $A(0, 0, 3)$. Write Cartesian equations for the surfaces defined by the following equations:

a) $PO = 2\,PA$, b) $PO + PA = 6$, c) $|PO - PA| = 2$.

21. The equation

$$r^2 + z^2 = 4r\cos\theta + 6r\sin\theta + 2z$$

is the equation of a sphere in cylindrical coordinates. Find the rectangular coordinates of the sphere's center.

22. Find parametric equations $x = x(\theta), y = y(\theta),$ and $z = z(\theta)$ for the curve of intersection of the sphere $\rho = a$ and the plane $y + z = 0$. Then find the length of the curve.

23. A barrel has the shape of an ellipsoid with equal pieces cut from the ends by planes perpendicular to the z-axis (Fig. 15.20). The height of the barrel is $2h$, the midsection radius

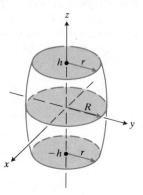

15.20 The barrel in Problem 23.

is R, and the end radii are both r. Find a formula for the volume of the barrel. Then check two things. First, suppose the sides of the barrel are straightened to turn it into a cylinder of radius R and height $2h$. Then $r = R$. Does your formula give the cylinder's volume? Second, suppose $r = 0$ and $h = R$ so the barrel is a sphere. Does your formula give the sphere's volume?

24. Show that the volume of the segment cut from the elliptic paraboloid

$$\frac{x^2}{a^2} + \frac{y^2}{b^2} = \frac{z}{c}$$

by the plane $z = h$ equals half the area of the segment's base times its altitude.

25. a) Find the volume of the solid bounded by the hyperboloid

$$\frac{x^2}{a^2} + \frac{y^2}{b^2} - \frac{z^2}{c^2} = 1$$

and the planes $z = 0$ and $z = h$, with $h > 0$.

b) Express your answer to part (a) in terms of h and the areas A_0 and A_h of the regions cut from the planes $z = 0$ and $z = h$ by the hyperboloid.

c) Show that the volume in (a) is also given by the formula

$$V = \frac{h}{6}(A_0 + 4A_m + A_h),$$

where A_m is the area of the region cut by the hyperboloid from the plane $z = h/2$.

26. A particle P slides without friction along a coil spring having the form of a right circular helix (Fig. 15.21). If the positive z-axis is taken downward, the cylindrical coordinates of P at time t satisfy the equations $r = a$, $z = b\theta$, where a and b are positive constants. If the particle starts at $r = a$, $\theta = 0$, with zero velocity and falls under gravity, the law of conservation of energy tells us that its speed after it has fallen a vertical distance z is $\sqrt{2gz}$.

a) Find the angular velocity $d\theta/dt$ when $\theta = 2\pi$.

b) Express θ and z as functions of t.

c) Determine the tangential and normal components of the velocity $d\mathbf{R}/dt$ and acceleration $d^2\mathbf{R}/dt^2$ as functions of t. Is there any component of acceleration in the direction of the binormal vector $\mathbf{B}$?

27. Suppose the curve in Problem 26 is replaced by the conical helix $r = a\theta$, $z = b\theta$ shown in Fig. 15.22.

a) Express the angular velocity $d\theta/dt$ as a function of θ.

b) Express the distance that the particle travels along this helix as a function of θ.

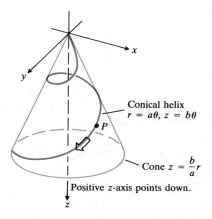

15.22 A particle sliding down a conical helix. See Problem 27.

28. When the position of a particle moving in space is given in cylindrical coordinates, the unit vectors we use to describe its position and motion are $\mathbf{u}_r = (\cos\theta)\mathbf{i} + (\sin\theta)\mathbf{j}$, $\mathbf{u}_\theta = -(\sin\theta)\mathbf{i} + (\cos\theta)\mathbf{j}$, and $\mathbf{k}$. See Fig. 15.23. Thus, the particle's position vector is $\mathbf{R} = r\mathbf{u}_r + z\mathbf{k}$.

a) Show that $\mathbf{u}_r$, $\mathbf{u}_\theta$, and $\mathbf{k}$, in this order, form a right-handed frame of unit vectors.

b) Show that

$$\frac{d\mathbf{u}_r}{d\theta} = \mathbf{u}_\theta \qquad \text{and} \qquad \frac{d\mathbf{u}_\theta}{d\theta} = -\mathbf{u}_r.$$

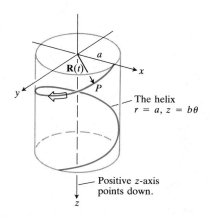

15.21 A particle sliding down a cylindrical helix. See Problem 26.

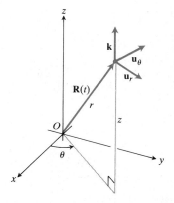

15.23 The cylindrical-coordinate unit vectors for describing motion. See Problem 28.

c) Assuming that the appropriate derivatives of r, θ, and z with respect to t exist, express $\mathbf{v}(t) = d\mathbf{R}/dt$ and $\mathbf{a}(t) = d^2\mathbf{R}/dt^2$ in terms of $\mathbf{u}_r$, $\mathbf{u}_\theta$, $\mathbf{k}$, and the derivatives of r and θ with respect to t. (Article 14.4 shows how the vectors mentioned here are used in describing planetary motion.)

29. Hold two of the three spherical coordinates ρ, ϕ, θ of point P in Fig. 15.24 constant while letting the other coordinate

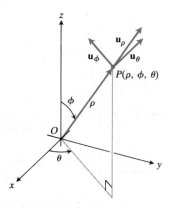

15.24   The spherical-coordinate unit vectors $\mathbf{u}_\rho$, $\mathbf{u}_\phi$, and $\mathbf{u}_\theta$. See Problem 29.

increase. Let $\mathbf{u}$, with subscript corresponding to the coordinate that is permitted to vary, denote the unit vector that points in the direction in which P starts to move under these conditions.

a) Express the three unit vectors $\mathbf{u}_\rho$, $\mathbf{u}_\phi$, and $\mathbf{u}_\theta$ obtained in this manner in terms of ρ, ϕ, θ and the unit vectors $\mathbf{i}$, $\mathbf{j}$, $\mathbf{k}$.

b) Show that $\mathbf{u}_\rho \cdot \mathbf{u}_\phi = 0$.

c) Show that $\mathbf{u}_\theta = \mathbf{u}_\rho \times \mathbf{u}_\phi$.

d) Are $\mathbf{u}_\rho$, $\mathbf{u}_\phi$, $\mathbf{u}_\theta$ mutually orthogonal? Is the system they form in the order given right-handed, or left-handed?

30. If the spherical coordinates ρ, ϕ, θ of a moving point P are differentiable functions of the time t, and $\mathbf{R} = \overrightarrow{OP}$ is the vector from the origin to P, express $\mathbf{R}$ and $d\mathbf{R}/dt$ in terms of ρ, ϕ, θ and their derivatives and the unit vectors $\mathbf{u}_\rho$, $\mathbf{u}_\phi$, $\mathbf{u}_\theta$ of Problem 29.

31. Express $ds^2 = dx^2 + dy^2 + dz^2$ in terms of (a) cylindrical coordinates r, θ, z, and (b) spherical coordinates ρ, ϕ, θ (see Problem 30). Interpret your results geometrically in terms of the sides and a diagonal of a rectangular box. Sketch.

32. Using the results of Problem 31, find the lengths of the following curves between $\theta = 0$ and $\theta = \ln 8$.

a) $z = r = ae^\theta$

b) $\phi = \pi/6$, $\rho = 2e^\theta$

CHAPTER 16

Functions of Two or More Variables and Their Derivatives

OVERVIEW

In this chapter we begin our study of functions that have more than one independent variable. Functions with two or more variables appear more often in science than functions of a single variable, and their calculus is even richer. Their derivatives are more varied and more interesting because of the different ways in which the variables can interact. Their integrals lead to a greater variety of applications. The studies of probability, statistics, fluid dynamics, and electricity, to mention only a few, all lead in natural ways to functions of more than one variable and the mathematical analysis associated with these functions is one of the finest achievements in science. These are glowing words, as you may have noticed; they are also true. We shall get more than a glimpse of this achievement in the chapters ahead.

In the present chapter, we introduce functions of two or more variables, define and calculate their derivatives, learn how to write the Chain Rule, and find out how to calculate the rates at which functions change as we move in different directions through their domains.

We shall also spend a few moments practicing our newly acquired skills on some of the differential equations that arise in physics and chemistry.

In Chapter 17, which is mainly about applications of derivatives, we learn how to estimate the values of functions of two or more variables and how to calculate their maxima and minima on a given domain. Chapter 18 deals with integrals of functions of two and three variables. These are the integrals with which we calculate the volumes, moments, and centers of mass of three-dimensional solids. In Chapter 19 we define surface integrals, the integrals with which we calculate surface area and the rates at which fluids flow through membranes. We also derive the integral theorems of vector calculus that arise in the studies of fluid flow and electricity in physics and engineering. Finally, in Chapter 20, a brief introduction to differential equations, we examine techniques for solving equations of the form $dy/dx = f(x, y)$, equations in which the derivative dy/dx of an unknown function y of x is given as a function of the two variables x and y.

16.1
Functions of Two or More Independent Variables

The values of many functions are determined by more than one independent variable. For example, the function $V = \pi r^2 h$ calculates the volume of a circular cylinder from its radius and height. The function $f(x, y) = x^2 + y^2$ calculates the height of the paraboloid $z = x^2 + y^2$ above the point $P(x, y)$ from the two coordinates of P. The function

$$w(x, t) = \cos(1.7 \times 10^{-2}t - 0.2x)e^{-0.2x},$$

in Example 5 below, calculates the temperature x feet below ground on the tth day of the year as a fraction of the surface temperature on that day.

In this article, we define functions of more than one independent variable and discuss two ways to display their values graphically.

Definitions

Suppose D is a collection of n-tuples of real numbers

$$(x_1, x_2, \ldots, x_n).$$

A **real valued function** f with domain D is a rule that assigns a real number

$$w = f(x_1, x_2, \ldots, x_n)$$

to each n-tuple of numbers in D. The function's range is the set of values the function takes on. The symbol w is the **dependent variable** of f and f is said to be a function of the n **independent variables** x_1 to x_n.

In the function $V = \pi r^2 h$, the dependent variable is V. The independent variables are r and h.

Functions given by formulas are evaluated in the usual way by substituting values of the independent variables and calculating the corresponding values of the dependent variable. Thus the value of the function $w = \sqrt{x^2 + y^2 + z^2}$ at the point $(3, 0, 4)$ is $\sqrt{(3)^2 + (0)^2 + (4)^2} = \sqrt{25} = 5$.

Domains

In defining functions of more than one variable, we follow the usual practice of excluding inputs that lead to complex numbers or division by zero. Thus, if $f(x, y) = \sqrt{y - x^2}$, we require y to be greater than or equal to x^2, and if $f(x, y) = xy/(x^2 - y^2)$, we do not allow x^2 and y^2 to be equal.

These are the only exclusions, however, and the domains of functions are otherwise assumed to be the largest sets for which their defining rules generate real numbers. These sets are sometimes called the "natural" domains of the functions.

EXAMPLE 1   Sketch the domain of

$$f(x, y) = \sqrt{y - x^2}.$$

What is the function's range?

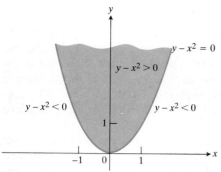

16.1 The domain of $f(x, y) = \sqrt{y - x^2}$ consists of the shaded region and its bounding parabola $y = x^2$. Elsewhere in the plane, $y - x^2 < 0$ and f is not defined.

Solution According to our convention, the domain D is the set of all number pairs (x, y) in the plane for which $\sqrt{y - x^2}$ is real (Fig. 16.1). This means the points for which

$$y - x^2 \geq 0 \qquad \text{or} \qquad y \geq x^2.$$

Therefore, D is the set of points that lie above and on the parabola $y = x^2$. The range of f is the set of all nonnegative numbers, $z \geq 0$. ∎

EXAMPLE 2 What are the domain and range of

$$f(x, y) = \frac{xy}{x^2 - y^2}?$$

Solution By our convention, x and y may take on any values except the ones for which $x^2 - y^2 = 0$, or $x^2 = y^2$. The domain therefore consists of all points in the plane except the points on the lines $y = x$ and $y = -x$. The range of f is the entire set of real numbers $-\infty < z < \infty$, as we may see by letting the point $P(x, y)$ traverse the right-hand branch of the hyperbola $x^2 - y^2 = 1$. Along this curve, $x = \sqrt{1 + y^2}$ and the values of f are given by the formula

$$f(x, y) = \frac{xy}{x^2 - y^2} = \frac{(\sqrt{1 + y^2})y}{1} = y\sqrt{1 + y^2}.$$

As y varies from $-\infty$ to ∞, so does $y\sqrt{1 + y^2}$. ∎

Graphs and Level Curves

There are two standard ways to picture the values of a function $f(x, y)$. One is to draw some of its **level curves,** the curves in the domain along which the function has a constant value $f(x, y) = c$. The other is to sketch the surface $z = f(x, y)$ in space. The surface, consisting of the points $(x, y, f(x, y))$, is the **graph** of f.

EXAMPLE 3 Graph the function

$$f(x, y) = 100 - x^2 - y^2$$

in space and plot the level curves $f(x, y) = 0, f(x, y) = 51$, and $f(x, y) = 75$ in the domain of f in the plane.

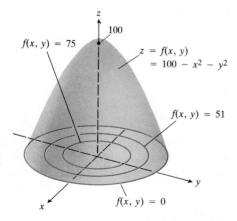

16.2 The graph and selected level curves of the function $f(x, y) = 100 - x^2 - y^2$.

Solution The domain of f is the entire xy-plane and the range of f is the set of real numbers less than or equal to 100. The graph is the paraboloid $z = 100 - x^2 - y^2$, a portion of which is shown in Fig. 16.2.

The level curve $f(x, y) = 0$ is the set of points in the xy-plane at which

$$f(x, y) = 100 - x^2 - y^2 = 0, \qquad \text{or} \qquad x^2 + y^2 = 100,$$

which is the circle of radius 10 centered at the origin (Fig. 16.2). Similarly, the level curves $f(x, y) = 51$ and $f(x, y) = 75$ are the circles

$$f(x, y) = 100 - x^2 - y^2 = 75, \qquad \text{or} \qquad x^2 + y^2 = 25,$$
$$f(x, y) = 100 - x^2 - y^2 = 51, \qquad \text{or} \qquad x^2 + y^2 = 49.$$

∎

The contour line $f(x, y) = 100 - y^2 - x^2 = 75$ is the circle $x^2 + y^2 = 25$ in the plane $z = 75$.

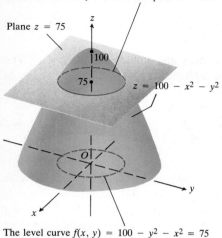

The level curve $f(x, y) = 100 - y^2 - x^2 = 75$ is the circle $x^2 + y^2 = 25$ in the xy-plane.

16.3 The graph of $f(x, y) = 100 - x^2 - y^2$ and its intersection with the plane $z = 75$.

Contour Lines

The curve in space in which the plane $z = c$ cuts a surface $z = f(x, y)$ is made up of the points that represent the function value $f(x, y) = c$. It is called the **contour line** $f(x, y) = c$ to distinguish it from the level curve $f(x, y) = c$ in the domain of f. Figure 16.3 shows the contour line $f(x, y) = 75$ on the surface $z = 100 - x^2 - y^2$ defined by the function $f(x, y) = 100 - x^2 - y^2$. The contour line lies directly above the circle $x^2 + y^2 = 25$, which is the level curve $f(x, y) = 75$ in the function's domain.

Not everyone makes this distinction, however, and you may wish to call both kinds of curves by a single name and rely on context to convey which one you have in mind. On most maps, for example, the curves that represent constant elevation (height above sea level) are called contours, not level curves (Fig. 16.4).

16.4 Contours on Mt. Washington in north-central New Hampshire. The streams, which follow paths of steepest descent, run perpendicular to the contours. So does the Cog Railway. (Based on a map by Louis F. Cutter, © 1987 Appalachian Mountain Club, Boston, Mass., by permission.)

Level Surfaces

In the plane, the points at which a function of two independent variables takes on a constant value $f(x, y) = c$ is called a level curve of the function. In space, the points at which a function of three independent variables takes on a constant value $f(x, y, z) = c$ make up a surface called a **level surface** of the function.

EXAMPLE 4   Describe the level surfaces of the function

$$f(x, y, z) = \sqrt{x^2 + y^2 + z^2}.$$

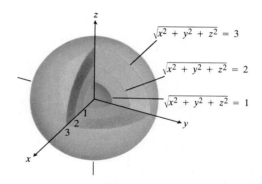

16.5 The level surfaces of the function $f(x, y, z) = \sqrt{x^2 + y^2 + z^2}$ are concentric spheres.

Solution The value of f is the distance from the origin to the point (x, y, z). Each level surface

$$\sqrt{x^2 + y^2 + z^2} = c, \qquad c > 0,$$

is a sphere of radius c centered at the origin. Figure 16.5 shows a cutaway view of three of these spheres. The level surface $\sqrt{x^2 + y^2 + z^2} = 0$ is not really a surface but consists of the origin alone.

Notice that we are not graphing the function here. The graph of the function, made up of the points $(x, y, z, \sqrt{x^2 + y^2 + z^2})$, lies in a four-variable space. Instead, we are looking at surfaces in the function's domain.

The function's level surfaces show how the function's values change as we move about in its domain. If we remain on a sphere of radius c centered at the origin, the function maintains a constant value, namely c. If we move from one sphere to another, the function's value changes. It increases if we move away from the origin and decreases if we move toward the origin. Thus, the way the function's values change depends on the direction we take. The dependence of change on direction is important and we shall return to it in Article 16.6.

Computer Graphing

The three-dimensional graphing programs now available for computers make it possible to graph functions of two variables with only a few keystrokes. We can often get information about a function more quickly from one of these graphs than we can from a function's formula.

EXAMPLE 5 Figure 16.6 shows a computer-generated graph of the function

$$w(x, t) = \cos(1.7 \times 10^{-2}t - 0.2x)e^{-0.2x},$$

where t is in days and x is in feet. The graph shows how the temperature beneath the earth's surface varies with time. The variation is given as a fraction of the variation at the surface. At a depth of 15 ft, the variation (vertical amplitude in the figure) is about 5 percent of the surface variation. At 30 ft, there is almost no variation in temperature during the year.

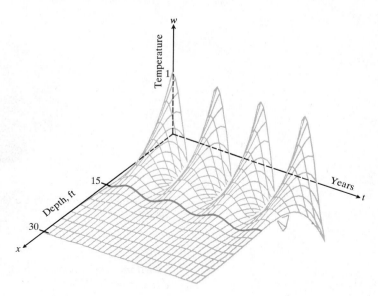

16.6   This computer-generated graph of

$$w(x, t) = \cos(1.7 \times 10^{-2}t - 0.2x)e^{-0.2x}$$

shows the seasonal variation of the temperature below ground as a fraction of surface temperature. $\Delta x = 0.375$ ft, $\Delta t = 15.625$ days. At $x = 15$ ft the variation is only 5% of the variation at the surface. At $x = 30$ ft the variation is less than 0.25% of the surface variation. (Adapted by John G. Aspinall from art provided by Norton Starr for G. C. Berresford's "Differential Equations and Root Cellars," *The UMAP Journal*, Volume 2, Number 3 (1981), pp. 53–75.)

Another thing the graph shows that is not immediately apparent from the equation for f is that the temperature 15 ft below the surface is about half a year out of phase with the surface temperature. When the temperature is lowest on the surface (late January, say) it is at its highest 15 ft below. Fifteen feet below, the seasons are reversed. ∎

PROBLEMS

In Problems 1–8, give the function's domain and range and describe the level curves in the domain.

1. $f(x, y) = x + y$

2. $f(x, y) = \sqrt{x + y}$

3. $f(x, y) = \ln(x + y)$

4. $f(x, y) = \sqrt{4 - x^2 - y^2}$

5. $f(x, y) = \tan^{-1}(y/x)$

6. $f(x, y) = \cos xy$

7. $f(x, y) = \sqrt{y}/x$

8. $f(x, y) = y/x^2$

Describe the level surfaces in the domains of the functions in Problems 9–18.

9. $f(x, y, z) = x^2 + y^2 + z^2$

10. $f(x, y, z) = \ln(x^2 + y^2 + z^2)$

11. $f(x, y, z) = 1/(x^2 + y^2 + z^2)$

12. $f(x, y, z) = (x^2/25) + (y^2/16) + (z^2/9)$

13. $f(x, y, z) = x$

14. $f(x, y, z) = x + y$

15. $f(x, y, z) = x + y + z$

16. $f(x, y, z) = x^2 + y^2$

17. $f(x, y, z) = y^2 + z^2$

18. $f(x, y, z) = x^2 - y$

Display the values of the functions in Problems 19–24 in two ways: (a) by sketching the surface $z = f(x, y)$, and (b) by drawing level curves in the function's domain. Label each level curve you draw with the appropriate equation $f(x, y) = c$.

19. $f(x, y) = 4 - y^2$

20. $f(x, y) = \sqrt{1 - x^2}$

21. $f(x, y) = 4x^2 + y^2$

22. $f(x, y) = \sqrt{x^2 + y^2}$

23. $f(x, y) = \sqrt{4 - x^2 - y^2}$

24. $f(x, y) = \sqrt{y - x^2}$ (This is the function in Example 1.)

25. Sketch the level curves of

$$f(x, y) = -(x - 1)^2 - y^2 + 1$$

for $f(x, y) = 1, 0, -3, -8$.

26. Sketch the level curves of $f(x, y) = 2x^2 - y + 1$ for $f(x, y) = 0, 1, 2$.

27. Find an equation for the level curve of $f(x, y) = 16 - x^2 - y^2$ that passes through the point $(2\sqrt{2}, \sqrt{2})$.

28. Find an equation for the level curve of the function $f(x, y) = \sqrt{x^2 - 1}$ that passes through the point $(1, 0)$.

29. *The Concorde's sonic booms.** The width w of the region in which people on the ground hear the Concorde's sonic boom directly, and not reflected from a layer in the atmosphere, is a function of

T = air temperature at ground level (in degrees Kelvin),

h = the Concorde's altitude (in km),

d = the vertical temperature gradient
       (temperature drop in degrees Kelvin per km).

The formula for w is

$$w = 4(Th/d)^{1/2}.$$

See Fig. 16.7.

The Washington-bound Concorde approaches the United States from Europe on a course that takes it south of Nantucket Island at an altitude of 16.8 km. If the surface temperature is 290K and the vertical temperature gradient is 5K/km, how far south of Nantucket must the plane be flown to keep its sonic boom carpet away from the island?

30. Find the maximum value of the function $f(x, y, z) = xyz$ on the line $x = 20 - t$, $y = t$, $z = 20$. At what point or points

---

*From N. K. Balachandra, W. L. Donn, and D. H. Rind, "Concorde Sonic Booms as an Atmospheric Probe," *Science,* 1 July 1977, Vol. 197, p. 47.

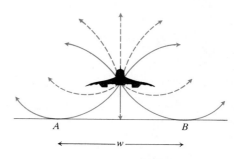

Sonic boom carpet

16.7 Sound waves from the Concorde bend as the temperature changes above and below the altitude at which the plane flies. The sonic boom carpet is the region on the ground that receives shock waves directly from the plane, not reflected from the atmosphere or diffracted along the ground. The carpet is determined by the grazing rays striking the ground from the point directly under the plane.

on the line does the maximum occur? (*Hint:* Along the line, f can be expressed as a differentiable function of the single variable t.)

16.2
Limits and Continuity

In this article, we define limits and continuity for functions of two or more variables. The definitions are the same as they were for functions of a single variable, except that there are now more variables to watch. You will find more going on in the examples, however, because the direction of approach may turn out to be an issue, as we shall soon explain.

Limits

If the values of a function $z = f(x, y)$ can be made as close as we like to a fixed number L by taking the point (x, y) close to the point (x_0, y_0), but not equal to (x_0, y_0), we say that L is the limit of f as (x, y) approaches (x_0, y_0). In symbols, we write

$$\lim_{(x, y) \to (x_0, y_0)} f(x, y) = L, \tag{1}$$

and we say "the limit of f as (x, y) approaches (x_0, y_0) equals L." This is like the limit of a function of one variable, except that there are two independent variables involved instead of one.

For (x, y) to be "close" to (x_0, y_0) means that the Cartesian distance $\sqrt{(x - x_0)^2 + (y - y_0)^2}$s is small in some sense. Since

$$|x - x_0| = \sqrt{(x - x_0)^2} \le \sqrt{(x - x_0)^2 + (y - y_0)^2} \tag{2}$$

and

$$|y - y_0| = \sqrt{(y - y_0)^2} \le \sqrt{(x - x_0)^2 + (y - y_0)^2}, \tag{3}$$

the inequality

$$\sqrt{(x - x_0)^2 + (y - y_0)^2} < \delta$$

implies

$$|x - x_0| < \delta \qquad \text{and} \qquad |y - y_0| < \delta.$$

Conversely, if for some $\delta > 0$ both $|x - x_0| < \delta$ and $|y - y_0| < \delta$, then

$$\sqrt{(x - x_0)^2 + (y - y_0)^2} < \sqrt{\delta^2 + \delta^2} = \sqrt{2}\delta,$$

16.8   (a) The open square $|x - x_0| < \delta$, $|y - y_0| < \delta$ lies inside the open disk $\sqrt{(x - x_0)^2 + (y - y_0)^2} < \sqrt{2}\delta$ and contains the open disk $\sqrt{(x - x_0)^2 + (y - y_0)^2} < \delta$. Thus, we can make both $|x - x_0|$ and $|y - y_0|$ small by making $\sqrt{(x - x_0)^2 + (y - y_0)^2}$ small, and conversely. (b) For every point (x, y) within δ of (x_0, y_0), the value of $f(x, y)$ lies within ϵ of L.

(a)

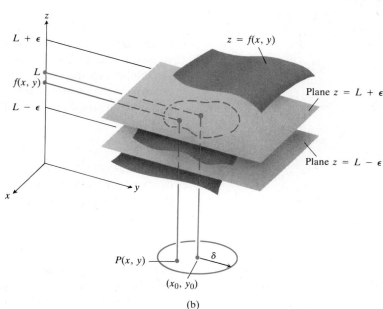

(b)

which is small if δ is small. See Fig. 16.8. Thus, in calculating limits we may think either in terms of the distance in the plane or in terms of differences in individual coordinates. We therefore have two equivalent definitions of limit.

Two Equivalent Definitions of Limit

The limit of $f(x, y)$ as $(x, y) \to (x_0, y_0)$ is the number L if for any $\epsilon > 0$ there exists a $\delta > 0$ such that for all points $(x, y) \neq (x_0, y_0)$ in the domain of f, either

1. $\sqrt{(x - x_0)^2 + (y - y_0)^2} < \delta \;\; \Rightarrow \;\; |f(x, y) - L| < \epsilon$ or

2. $|x - x_0| < \delta$ and $|y - y_0| < \delta \;\; \Rightarrow \;\; |f(x, y) - L| < \epsilon$.

As we did with functions of a single variable in Article 1.9, we may use the definition of limit to prove directly that

$$\lim_{(x, y) \to (x_0, y_0)} x = x_0, \qquad \lim_{(x, y) \to (x_0, y_0)} y = y_0, \qquad \text{and} \qquad \lim_{(x, y) \to (x_0, y_0)} k = k. \quad (4)$$

The proofs are similar and we shall not repeat them here.

We can also state a theorem like Theorem 1 of Article 1.9 that says that the limit of a sum $f(x, y) + g(x, y)$ is the sum of the limits of f and g (when they exist), with similar results for differences, products, constant multiples, and quotients.

THEOREM 1

The Limit Combination Theorem

If

$$\lim_{(x, y) \to (x_0, y_0)} f(x, y) = L_1 \qquad \text{and} \qquad \lim_{(x, y) \to (x_0, y_0)} g(x, y) = L_2,$$

then

i) $\lim[f(x, y) + g(x, y)] = L_1 + L_2,$

ii) $\lim[f(x, y) - g(x, y)] = L_1 - L_2,$

iii) $\lim[f(x, y)g(x, y)] = L_1 L_2,$

iv) $\lim[kf(x, y)] = kL_1$ (any number k),

v) $\lim \dfrac{f(x, y)}{g(x, y)} = \dfrac{L_1}{L_2}$ if $L_2 \neq 0$.

The limits are all to be taken as $(x, y) \to (x_0, y_0)$.

When we apply Theorem 1 to the functions and limits in (4) we obtain the useful result that the limits of polynomials and rational functions like

$$x^2 + y^2, \qquad \frac{x - xy + 3}{x^2 y + 5xy - y^3}$$

as $(x, y) \to (x_0, y_0)$ may be calculated by evaluating the functions at the point (x_0, y_0). The only requirement is that the functions be defined there. In other words, polynomials and rational functions in two variables are *continuous* wherever they are defined, as we shall see in a moment.

EXAMPLE 1

a) $\lim\limits_{(x,\,y)\to(3,\,-4)} (x^2 + y^2) = (3)^2 + (-4)^2 = 9 + 16 = 25.$

b) $\lim\limits_{(x,\,y)\to(0,\,1)} \dfrac{x - xy + 3}{x^2 y + 5xy - y^3} = \dfrac{0 - 0(1) + 3}{(0)^2(1) + 5(0)(1) - (1)^3} = -3.$ ∎

Continuity

DEFINITION

A function $f(x, y)$ is said to be **continuous** at the point (x_0, y_0) if

1. f is defined at (x_0, y_0),

2. $\lim\limits_{(x,\,y)\to(x_0,\,y_0)} f(x, y)$ exists, and   (5)

3. $\lim\limits_{(x,\,y)\to(x_0,\,y_0)} f(x, y) = f(x_0, y_0).$

A consequence of Theorem 1, which we shall not prove, is that if $f(x, y)$ and $g(x, y)$ are both continuous at a point, then their sum $f(x, y) + g(x, y)$ is continuous at that point. Similar results hold for differences, products, and multiples of continuous functions. Also, the quotient of two continuous functions is continuous wherever it is defined. It follows that polynomials and rational functions in two variables are continuous at any point at which they are defined.

If $z = f(x, y)$ is a continuous function of x and y, and $w = g(z)$ is a continuous function of z, then the composite

$$w = g(f(x, y))$$

is continuous. Thus, composites like

$$e^{x-y}, \qquad \cos \frac{xy}{x^2 + 1}, \qquad \ln(1 + x^2 y^2)$$

are continuous at every point (x, y).

As with functions of a single variable, the general rule is that composites of continuous functions are continuous. The only requirement is that each function be continuous where it is applied.

EXAMPLE 2   Given $\epsilon > 0$, how close to $(0, 0)$ should we take (x, y) to make $|f(x, y) - f(0, 0)| < \epsilon$ if

$$f(x, y) = \frac{x + y}{x^2 + y^2 + 1}?$$

Solution   Since $f(0, 0) = 0$,

$$|f(x, y) - f(0, 0)| = \left| \frac{x + y}{x^2 + y^2 + 1} \right|,$$

and we are being asked how close to $(0, 0)$ we should take (x, y) to make

$$\left| \frac{x + y}{x^2 + y^2 + 1} \right| < \epsilon.$$

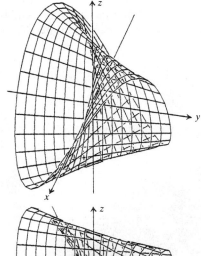

Since the denominator of the fraction is never less than 1, we have

$$\left| \frac{x + y}{x^2 + y^2 + 1} \right| \le |x + y| \le |x| + |y|$$

$$\le \sqrt{x^2 + y^2} + \sqrt{x^2 + y^2} = 2\sqrt{x^2 + y^2},$$

which will be less than ϵ if

$$\sqrt{x^2 + y^2} < \frac{\epsilon}{2}.$$

Therefore, the original ϵ inequality will hold if the distance from (x, y) to $(0, 0)$ is less than $\epsilon/2$. That is, we may safely choose our δ in the definition of limit to be the number $\epsilon/2$. ∎

EXAMPLE 3 Show that the function

$$f(x, y) = \begin{cases} \dfrac{xy}{x^2 + y^2}, & (x, y) \ne (0, 0), \\ 0, & (x, y) = (0, 0), \end{cases}$$

is continuous at every point except the origin. See Fig. 16.9.

Solution The function f is continuous at any point $(x, y) \ne (0, 0)$ because its values are then given by a rational function of x and y.

At $(0, 0)$ the value of f is defined, but f, we claim, has no limit as $(x, y) \to (0, 0)$. The reason is that different paths of approach to the origin can lead to different results, as we shall now see.

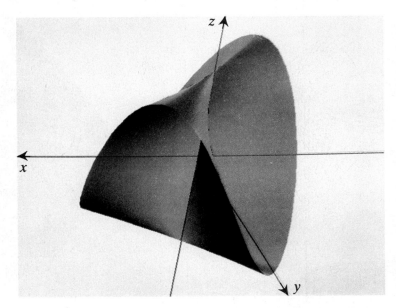

16.9 The graph of

$$f(x, y) = \begin{cases} \dfrac{xy}{x^2 + y^2}, & (x, y) \ne (0, 0) \\ 0, & (x, y) = (0, 0). \end{cases}$$

The only point of the z-axis that belongs to this surface is the point $(0, 0, 0)$. Images created by Thomas Banchoff and student assistants at Brown University Computer Graphics Laboratory.

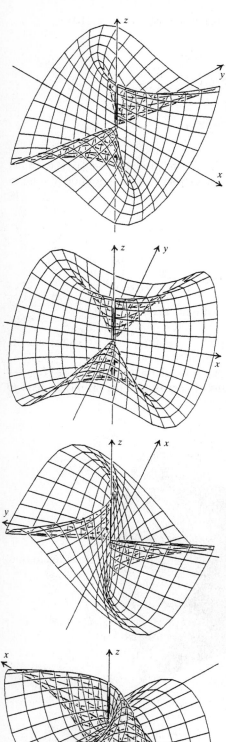

If $(x, y) \to (0, 0)$ along the line $y = mx$, but $(x, y) \neq (0, 0)$, then

$$f(x, y) = \frac{xy}{x^2 + y^2} = \frac{x(mx)}{x^2 + m^2x^2} = \frac{m}{1 + m^2},$$

$$\lim_{\substack{(x, y) \to (0, 0) \\ \text{along } y = mx}} f(x, y) = \frac{m}{1 + m^2}.$$

This limit changes as m changes. In fact, the substitution $m = \tan \theta$ produces

$$f(x, y) = \frac{m}{1 + m^2} = \frac{\tan \theta}{\sec^2\theta} = \frac{1}{2}\sin 2\theta,$$

which shows that the limit varies from $-1/2$ to $1/2$ depending on the angle of approach. There is no single number L that we may call the limit of f as $(x, y) \to (0, 0)$. Therefore, the limit fails to exist and the function cannot be continuous at $(0, 0)$. Problem 33 describes another way to reveal the discontinuity. ∎

EXAMPLE 4   Examine the limits of

$$f(x, y) = \frac{x^2y}{x^4 + y^2}, \qquad (x, y) \neq (0, 0),$$

as $(x, y) \to (0, 0)$ along the line $y = mx$ and along the parabola $y = x^2$. Does f have a limit as $(x, y) \to (0, 0)$?

16.10   The graph of

$$f(x, y) = \begin{cases} \dfrac{x^2y}{x^4 + y^2}, & (x, y) \neq (0, 0) \\ 0, & (x, y) = (0, 0). \end{cases}$$

The only point of the z-axis that belongs to this surface is the point (0, 0, 0). Images created by Thomas Banchoff and student assistants at Brown University Computer Graphics Laboratory.

Solution   See Fig. 16.10. Along the line $y = mx$,

$$f(x, y) = f(x, mx) = \frac{x^2(mx)}{x^4 + m^2x^2} = \frac{mx}{x^2 + m^2}.$$

Therefore,

$$\lim_{\substack{(x, y) \to (0, 0) \\ \text{along } y = mx}} f(x, y) = \lim_{x \to 0} \frac{mx}{x^2 + m^2} = 0.$$

The fact that the limit, taken along all *linear* paths to $(0, 0)$, exists and equals zero may lead you to suspect that $\lim_{(x, y) \to (0, 0)} f(x, y)$ exists. However, along the parabola $y = x^2$,

$$f(x, y) = f(x, x^2) = \frac{x^2(x^2)}{x^4 + x^4} = \frac{1}{2}.$$

Therefore,

$$\lim_{\substack{(x, y) \to (0, 0) \\ \text{along } y = x^2}} f(x, y) = \frac{1}{2}.$$

For f to have a limit as $(x, y) \to (0, 0)$, the limits along all paths of approach would have to agree. Since we have reached different limits on different paths, we conclude that f has no limit as $(x, y) \to (0, 0)$. ∎

Functions of Three or More Variables

The definitions of limit and continuity for functions of two variables and the conclusions about limits and continuity for sums, products, quotients, and composites all extend to functions of three or more variables. Thus, functions like

$$\ln \sqrt{x^2 + y^2 + z^2} \qquad \text{and} \qquad \frac{e^{x-y+z}}{z^2 + \cos\sqrt{xy}}$$

are continuous at every point at which they are defined, and limits like

$$\lim_{P \to (1, 0, 1)} \frac{e^{x-y+z}}{z^2 + \cos(\sqrt{xy})} = \frac{e^{1-0+1}}{(1)^2 + \cos(0)} = \frac{e^2}{2},$$

where P denotes the point (x, y, z), may be evaluated by direct substitution.

Continuous Functions Defined on Closed, Bounded Regions

As we know, a function of a single variable that is continuous throughout a closed, bounded interval $[a, b]$ takes on an absolute maximum value and an absolute minimum value at least once in $[a, b]$. The same is true of a function $z = f(x, y)$ that is continuous on a closed, bounded set R in the plane (like a line segment, disc, or filled-in triangle). The function takes on an absolute maximum value at some point in R and an absolute minimum value at some point in R.

Similar theorems hold for functions of three or more variables. A continuous function $w = f(x, y, z)$, for example, must take on absolute maximum and minimum values on any closed, bounded set (solid ball or cube, spherical shell, rectangular plate) on which it is defined.

Techniques for finding these maximum and minimum values will be the subject of Articles 17.1 and 17.3.

PROBLEMS

Find the limits in Problems 1–16.

1. $\displaystyle\lim_{(x,\,y)\to(0,\,0)} \frac{3x^2 - y^2 + 5}{x^2 + y^2 + 2}$

2. $\displaystyle\lim_{(x,\,y)\to(1,\,1)} \ln|1 + x^2 y^2|$

3. $\displaystyle\lim_{(x,\,y)\to(0,\,\ln 2)} e^{x-y}$

4. $\displaystyle\lim_{(x,\,y)\to(0,\,4)} \frac{x}{\sqrt{y}}$

5. $\displaystyle\lim_{P\to(1,\,3,\,4)} \sqrt{x^2 + y^2 + z^2 - 1}$

6. $\displaystyle\lim_{P\to(1,\,2,\,6)} \left(\frac{1}{x} + \frac{1}{y} + \frac{1}{z}\right)$

7. $\displaystyle\lim_{(x,\,y)\to(0,\,0)} \frac{e^y \sin x}{x}$

8. $\displaystyle\lim_{(x,\,y)\to(0,\,\pi/4)} \sec x \tan y$

9. $\displaystyle\lim_{(x,\,y)\to(-1,\,2)} \frac{xy + y - 2x - 2}{x + 1}$

10. $\displaystyle\lim_{(x,\,y)\to(0,\,0)} \cos \frac{x^2 + y^2}{x + y + 1}$

11. $\displaystyle\lim_{(x,\,y)\to(1,\,1)} \cos \sqrt[3]{|xy| - 1}$

12. $\displaystyle\lim_{(x,\,y)\to(1,\,0)} \frac{x \sin y}{x^2 + 1}$

13. $\displaystyle\lim_{(x,\,y)\to(1,\,1)} \frac{x^2 - 2xy + y^2}{x - y}, \quad x \neq y$

14. $\displaystyle\lim_{(x,\,y)\to(1,\,1)} \frac{x^2 - y^2}{x - y}, \quad x \neq y$

15. $\displaystyle\lim_{(x,\,y)\to(0,\,0)} \tan^{-1}(1/\sqrt{x^2 + y^2})$

16. $\displaystyle\lim_{(x,\,y)\to(0,\,2)} \left(\frac{\cos x - 1}{x^2}\right)\left(\frac{y - 2}{y^2 - 4}\right)$

By considering different lines of approach, show that the functions in Problems 17–22 have no limit as $(x, y) \to (0, 0)$.

17. $\dfrac{x + y}{x - y}$

18. $\dfrac{x - y}{x + y}$

19. $\dfrac{x^2 - y^2}{x^2 + y^2}$

20. $\dfrac{x}{\sqrt{x^2 + y^2}}$

21. $\dfrac{x^2 + y^2}{xy}$

22. $\dfrac{xy}{|xy|}$

In Problems 23–26, show that you can make $|f(x, y) - f(0, 0)| < 0.01$ by taking the point (x, y) close enough to the origin.

23. $f(x, y) = x^2 + y^2$

24. $f(x, y) = y/(x^2 + 1)$

25. $f(x, y) = (x + y)/(x^2 + 1)$

26. $f(x, y) = (x + y)/(2 + \cos x)$

In Problems 27–30, show that for the given value of ϵ you can make $|f(x, y, z) - f(0, 0, 0)| < \epsilon$ by taking the point (x, y, z) close enough to the origin.

27. $f(x, y, z) = x^2 + y^2 + z^2, \quad \epsilon = 0.01$

28. $f(x, y, z) = xyz, \quad \epsilon = 0.008$

29. $f(x, y, z) = \dfrac{x + y + z}{x^2 + y^2 + z^2 + 1}, \quad \epsilon = 0.015$

30. $f(x, y, z) = \tan^2 x + \tan^2 y + \tan^2 z, \quad \epsilon = 0.03$

31. Is $f(x, y, z) = \sqrt{x^2 + y^2 + z^2}$ continuous at $(0, 0, 0)$? Explain.

32. Show that the function $f(x, y) = x^2/(x^2 + y)$ has no limit as $(x, y) \to (0, 0)$ by examining the limits of f as $(x, y) \to (0, 0)$ along the parabola $y = kx^2$ for selected values of k.

33. Sketch the level curves $f(x, y) = 0$, $f(x, y) = 1/2$, and $f(x, y) = -1/2$ for the function

$$f(x, y) = \begin{cases} \dfrac{xy}{x^2 + y^2}, & (x, y) \neq (0, 0), \\ 0, & (x, y) = (0, 0), \end{cases}$$

in Example 3. This is another way to reveal the lack of continuity at the origin.

TOOLKIT PROGRAM

3D Grapher

16.3
Partial Derivatives

Partial derivatives are the derivatives we get when we hold constant all but one of the independent variables in a function and differentiate with respect to that one. In this article, we show how partial derivatives arise geometrically and how they may be calculated by applying the rules we already know for differentiating functions of a single variable. We begin with functions of two independent variables.

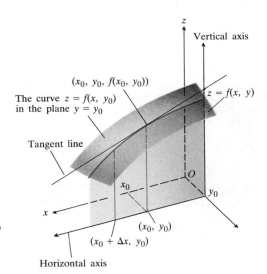

$(x_0, y_0, f(x_0, y_0))$

The curve $z = f(x, y_0)$
in the plane $y = y_0$

$z = f(x, y)$

Vertical axis

Tangent line

x_0

O

y_0

x

(x_0, y_0)

$(x_0 + \Delta x, y_0)$

Horizontal axis

16.11 The intersection of the plane $y = y_0$ with the surface $z = f(x, y)$, viewed from a point above the fourth quadrant of the xy-plane.

The Definition

If (x_0, y_0) is a point in the domain of a function $z = f(x, y)$, the plane $y = y_0$ will cut the surface $z = f(x, y)$ in the curve $z = f(x, y_0)$, as shown in Fig. 16.11. This curve is the graph of the function $z = f(x, y_0)$ in the plane $y = y_0$. The vertical coordinate in this plane is z, the height above (below) the xy-plane. The horizontal coordinate is x.

The derivative of $z = f(x, y_0)$ with respect to x at $x = x_0$ is defined in the usual way as the limit

$$\frac{\partial f}{\partial x}(x_0, y_0) = \frac{d}{dx}f(x, y_0) \Big|_{\text{at } x = x_0} = \lim_{\Delta x \to 0} \frac{f(x_0 + \Delta x, y_0) - f(x_0, y_0)}{\Delta x}, \qquad (1)$$

provided that the limit exists. The limit is called the **partial derivative of f with respect to x** at the point (x_0, y_0). The slope of the curve $z = f(x, y_0)$ in the plane $y = y_0$ at the point $(x_0, y_0, f(x_0, y_0))$ is the value of the partial derivative with respect to x at (x_0, y_0). The tangent to the curve $z = f(x, y_0)$ at $x = x_0$ is the line in the plane $y = y_0$ that passes through the point $(x_0, y_0, f(x_0, y_0))$ with this slope.

The usual notations for the partial derivative of $z = f(x, y)$ with respect to x at (x_0, y_0) are as follows:

$\dfrac{\partial f}{\partial x}(x_0, y_0)$ or $f_x(x_0, y_0)$ "Partial derivative of f with respect to x at (x_0, y_0)" or "f sub x at (x_0, y_0)." Convenient for stressing the point (x_0, y_0).

$\dfrac{\partial z}{\partial x}\Big|_{(x_0, y_0)}$ "Partial derivative of z with respect to x at (x_0, y_0)." Common in science and engineering when you are dealing with variables and do not mention the function explicitly.

$f_x,\ \dfrac{\partial f}{\partial x},\ z_x,$ or $\dfrac{\partial z}{\partial x}$ "Partial derivative of f (or z) with respect to x." Convenient when you regard the partial derivative as a function in its own right.

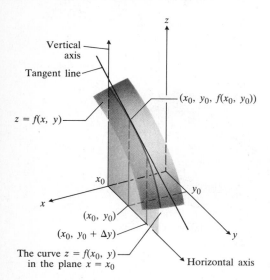

16.12 The intersection of the plane $x = x_0$ with the surface $z = f(x, y)$, viewed from the first quadrant of the xy-plane.

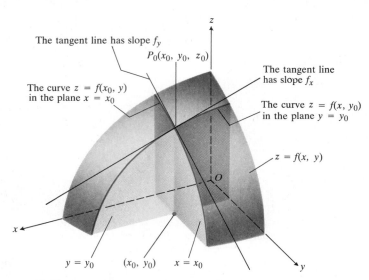

16.13 The plane $y = y_0$ cuts the surface $z = f(x, y)$ in the curve $z = f(x, y_0)$. At each x, the slope of this curve is $f_x(x, y_0)$. Similarly, the plane $x = x_0$ cuts the surface in a curve whose slope is $f_y(x_0, y)$.

Calculations

As Eq. (1) shows, $\partial f / \partial x$ is the ordinary derivative of f with respect to x with y held fixed. Thus, to calculate $\partial f / \partial x$ we differentiate f with respect to x in the usual way while treating y as a constant.

EXAMPLE 1 Find the value of $\partial f / \partial x$ at the point $(4, 5)$ if

$$f(x, y) = x^2 + 3xy + y - 1.$$

Solution We regard y as a constant and differentiate with respect to x:

$$\frac{\partial f}{\partial x} = \frac{\partial}{\partial x}(x^2 + 3xy + y - 1) = 2x + 3y + 0 - 0 = 2x + 3y.$$

The value of $\partial f / \partial x$ at $(4, 5)$ is $2(4) + 3(5) = 23$. ∎

The definition of the partial derivative $\partial f / \partial y$ of $z = f(x, y)$ with respect to y at (x_0, y_0) is like the definition of $\partial f / \partial x$. We regard $z = f(x_0, y)$ as a function of the single independent variable y (Fig. 16.12) and define

$$\frac{\partial f}{\partial y}(x_0, y_0) = \frac{d}{dy} f(x_0, y) \bigg|_{y=y_0} = \lim_{\Delta y \to 0} \frac{f(x_0, y_0 + \Delta y) - f(x_0, y_0)}{\Delta y}, \quad (2)$$

provided the limit exists. This limit is also taken to be the slope of the curve $z = f(x_0, y)$ in the plane $x = x_0$. The tangent to the curve is the line in the plane $x = x_0$ that passes through $(x_0, y_0, f(x_0, y_0))$ with this slope.

We now have two tangent lines associated with $z = f(x, y)$ at (x_0, y_0). Figure 16.13 shows them together. If $\partial f / \partial x$ and $\partial f / \partial y$ are both continuous at (x_0, y_0), the plane determined by these two lines will be tangent to the surface

$z = f(x, y)$ at (x_0, y_0). Tangency is more complicated for surfaces than it is for curves, however, and we shall have to wait until Article 16.7 to deal with it.

EXAMPLE 2   Find the value of $\partial f/\partial y$ at the point $(4, 5)$ if

$$f(x, y) = x^2 + 3xy + y - 1.$$

Solution   We regard x as a constant and differentiate with respect to y:

$$\frac{\partial f}{\partial y} = \frac{\partial}{\partial y}(x^2 + 3xy + y - 1) = 0 + 3x + 1 - 0 = 3x + 1.$$

The value of $\partial f/\partial y$ at $(4, 5)$ is $3(4) + 1 = 13.$ ∎

EXAMPLE 3   Find $\partial f/\partial y$ if

$$f(x, y) = y \sin xy.$$

Solution   We treat x as a constant and f as a product of y and $\sin xy$:

$$\frac{\partial f}{\partial y} = \frac{\partial}{\partial y}(y \sin xy) = y\frac{\partial}{\partial y}\sin xy + \sin xy\frac{\partial}{\partial y}(y)$$

$$= y \cos xy\frac{\partial}{\partial y}(xy) + \sin xy = xy \cos xy + \sin xy.$$ ∎

EXAMPLE 4   Find f_x and f_y if

$$f(x, y) = \frac{2y}{y + \cos x}.$$

Solution   We treat f as the quotient of $2y$ divided by $(y + \cos x)$. With y held constant, we get

$$f_x = \frac{\partial}{\partial x}\left(\frac{2y}{y + \cos x}\right) = \frac{(y + \cos x)\dfrac{\partial}{\partial x}(2y) - 2y\dfrac{\partial}{\partial x}(y + \cos x)}{(y + \cos x)^2}$$

$$= \frac{(y + \cos x)(0) - 2y(-\sin x)}{(y + \cos x)^2} = \frac{2y \sin x}{(y + \cos x)^2}.$$

With x held constant, we get

$$f_y = \frac{\partial}{\partial y}\left(\frac{2y}{y + \cos x}\right) = \frac{(y + \cos x)\dfrac{\partial}{\partial y}(2y) - 2y\dfrac{\partial}{\partial y}(y + \cos x)}{(y + \cos x)^2}$$

$$= \frac{(y + \cos x)(2) - 2y(1)}{(y + \cos x)^2} = \frac{2 \cos x}{(y + \cos x)^2}.$$ ∎

Functions of More Than Two Variables

The definitions of the partial derivatives of functions of more than two independent variables are like the definitions for functions of two variables. They are ordinary derivatives with respect to one variable, taken while all the other independent variables are regarded as constants.

EXAMPLE 5   Find

$$\frac{\partial}{\partial z}(x \sin(y + 3z)).$$

Solution

$$\frac{\partial}{\partial z}(x \sin(y + 3z)) = x\frac{\partial}{\partial z} \sin(y + 3z)$$

$$= x \cos(y + 3z)\frac{\partial}{\partial z}(y + 3z)$$

$$= 3x \cos(y + 3z). \qquad \blacksquare$$

EXAMPLE 6   Let $\mathbf{R}(x, y, z) = x\mathbf{i} + y\mathbf{j} + z\mathbf{k}$ be the position vector of the Cartesian coordinate point (x, y, z) and let $x = \rho \sin \phi \cos \theta$, $y = \rho \sin \phi \sin \theta$, and $z = \rho \cos \phi$. Find $\partial \mathbf{R}/\partial \theta$.

Solution

$$\frac{\partial \mathbf{R}}{\partial \theta} = \frac{\partial x}{\partial \theta}\mathbf{i} + \frac{\partial y}{\partial \theta}\mathbf{j} + \frac{\partial z}{\partial \theta}\mathbf{k}$$

$$= -(\rho \sin \phi \sin \theta)\mathbf{i} + (\rho \sin \phi \cos \theta)\mathbf{j}. \qquad \blacksquare$$

EXAMPLE 7   Three resistors of resistances R_1, R_2, and R_3 connected in parallel produce a resistance R given by

$$\frac{1}{R} = \frac{1}{R_1} + \frac{1}{R_2} + \frac{1}{R_3}.$$

Find $\partial R/\partial R_2$.

Solution   Treat R_1 and R_3 as constants and differentiate both sides of the equation with respect to R_2. Then

$$-\frac{1}{R^2}\frac{\partial R}{\partial R_2} = -\frac{1}{R_2^2}, \qquad \text{or} \qquad \frac{\partial R}{\partial R_2} = \left(\frac{R}{R_2}\right)^2. \qquad \blacksquare$$

PROBLEMS

In Problems 1–20, find $\partial f/\partial x$ and $\partial f/\partial y$.

1. $f(x, y) = 2x$

2. $f(x, y) = -3y$

3. $f(x, y) = -4$

4. $f(x, y) = 2x - 3y - 4$

5. $f(x, y) = x(y - 1)$

6. $f(x, y) = x^2 + y^2$

7. $f(x, y) = x^2 - xy + y^2$

8. $f(x, y) = (x + 2)(y + 3)$

9. $f(x, y) = 5xy - 7x^2 - y^2 + 3x - 6y + 2$

10. $f(x, y) = 1/(x + y)$

11. $f(x, y) = \sqrt{x^2 + y^2}$

12. $f(x, y) = x/(x^2 + y^2)$

13. $f(x, y) = (x + y)/(xy - 1)$

14. $f(x, y) = (y - x)e^x$

15. $f(x, y) = e^x\ln y$

16. $f(x, y) = \sin(x + y)$

17. $f(x, y) = e^x\sin(y + 1)$

18. $f(x, y) = \tan^{-1}(y/x)$

19. $f(x, y) = \tanh(2x + 5y)$

20. $f(x, y) = \cosh(y/x)$

In Problems 21–28, find the partial derivative of f with respect to each variable.

21. $f(x, y, z) = xy + yz + zx$

22. $f(x, y, z) = (x^2 + y^2 + z^2)^{-1/2}$

23. $f(x, y, z) = 1 + y^2 + 2z^2$

24. $f(x, y, z) = x - \sqrt{y^2 + z^2}$

25. $f(P, Q, R) = 1/PQR$

26. $f(u, v, w, x) = \ln(uvwx)$

27. $f(\rho, \phi, \theta) = \rho \sin \phi \cos \theta$

28. $f(r, \theta, z) = r(1 - \cos \theta) - z$

29. Assume that the equation

$$xy + z^3 x - 2yz = 0$$

defines z as a differentiable function of the two independent variables x and y. Find the value of $\partial z / \partial x$ at the point $(x, y, z) = (1, 1, 1)$.

30. Assume that the equation

$$xz + y \ln x + x^2 + 4 = 0$$

defines x as a differentiable function of the two independent variables y and z. Find the value of $\partial x / \partial z$ at the point $(x, y, z) = (1, -1, -3)$.

31. Assume that the equation

$$x \sin(y + z) - xy + z^2 = 0$$

defines x as a differentiable function of the two independent variables y and z. Find the values of $\partial x / \partial y$ and $\partial x / \partial z$ at the point $(1, -1, 1)$.

32. Let $\mathbf{R}(x, y, z) = x\mathbf{i} + y\mathbf{j} + z\mathbf{k}$ be the position vector of the point (x, y, z) in Cartesian coordinates and let r, θ, and z be cylindrical coordinates. Express $\partial \mathbf{R} / \partial r$, $\partial \mathbf{R} / \partial \theta$, and $\partial \mathbf{R} / \partial z$ as vector functions of r, θ, and z.

Problems 33 and 34 are about the angles and sides of the triangle in Fig. 16.14.

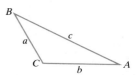

16.14 The triangle in Problems 33 and 34.

33. Express A implicitly as a function of a, b, and c and calculate $\partial A / \partial a$ and $\partial A / \partial b$.

34. Express a implicitly as a function of A, b, and B and calculate $\partial a / \partial A$ and $\partial a / \partial B$.

In Problems 35–40, express the spherical coordinates ρ, ϕ, and θ as functions of the Cartesian coordinates x, y, and z and calculate the indicated derivatives.

35. $\dfrac{\partial \rho}{\partial x}$

36. $\dfrac{\partial \phi}{\partial z}$

37. $\dfrac{\partial \theta}{\partial y}$

38. $\dfrac{\partial \theta}{\partial z}$

39. $\dfrac{\partial \phi}{\partial x}$

40. $\dfrac{\partial \theta}{\partial x}$

41. Let $\mathbf{R}(x, y, z) = x\mathbf{i} + y\mathbf{j} + z\mathbf{k}$ be the position vector of the point (x, y, z) in Cartesian coordinates. Express x, y, and z as functions of the spherical coordinates ρ, ϕ, and θ and calculate $\partial \mathbf{R} / \partial \rho$, $\partial \mathbf{R} / \partial \phi$, and $\partial \mathbf{R} / \partial \theta$. Then express these derivatives in terms of the unit vectors

$$\mathbf{u}_\rho = (\sin \phi \cos \theta)\mathbf{i} + (\sin \phi \sin \theta)\mathbf{j} + (\cos \phi)\mathbf{k},$$
$$\mathbf{u}_\phi = (\cos \phi \cos \theta)\mathbf{i} + (\cos \phi \sin \theta)\mathbf{j} - (\sin \phi)\mathbf{k},$$
$$\mathbf{u}_\theta = -(\sin \theta)\mathbf{i} + (\cos \theta)\mathbf{j}.$$

16.4
The Chain Rule

When we are interested in the temperature $f(x, y, z)$ at points on a path

$$x = x(t), \qquad y = y(t), \qquad z = z(t)$$

in space or in the pressure or density at points along a path through a gas or fluid, we may think of f as a function of the single variable t. At each t the temperature at the point $(x(t), y(t), z(t))$ on the path is the value of the composite $f(x(t), y(t), z(t))$. If we then wish to know the rate at which f changes with respect to t along the path, we have only to differentiate this composite with respect to t (provided, of course, df/dt exists).

Sometimes df/dt can be found by substituting $x(t)$, $y(t)$, and $z(t)$ into the formula for f and differentiating directly with respect to t. But often we work with functions whose formulas are too complicated for convenient substitution or for which formulas are not readily available. To find a function's derivatives in circumstances like these, we use the Chain Rule. The form the Chain Rule takes depends on the number of variables involved, but, except for the presence of additional variables, it works just like the Chain Rule from Chapter 2.

The Chain Rule for Functions of Two Variables

In Chapter 2, when we had a differentiable function $u = f(x)$ and x was a differentiable function of t, we found that u was a differentiable function of t with

$$\frac{du}{dt} = \frac{df}{dx}\frac{dx}{dt}. \tag{1}$$

The analogous formula for a function $w = f(x, y)$, when x and y are differentiable functions of t, is

$$\frac{dw}{dt} = \frac{\partial f}{\partial x}\frac{dx}{dt} + \frac{\partial f}{\partial y}\frac{dy}{dt}. \tag{2}$$

This formula holds when f, $\partial f/\partial x$, and $\partial f/\partial y$ are continuous.

The derivation of Eq. (2) is much the same as in the single-variable case. A theorem similar to Eq. (11) of Article 2.4 says that if Δx, Δy, and Δw are the increments that result from changing t from t_0 to $t_0 + \Delta t$, then

$$\Delta w = \left(\frac{\partial f}{\partial x}\right)_{t_0}\Delta x + \left(\frac{\partial f}{\partial y}\right)_{t_0}\Delta y + \epsilon_1\Delta x + \epsilon_2\Delta y. \tag{3}$$

The ϵ_1 and ϵ_2 in this formula are like the ϵ in the old Eq. (11) in that they both go to zero as Δt goes to zero. To find the value of dw/dt at t_0, we divide Eq. (3) through by Δt and let Δt approach zero. The division gives

$$\frac{\Delta w}{\Delta t} = \left(\frac{\partial f}{\partial x}\right)_{t_0}\frac{\Delta x}{\Delta t} + \left(\frac{\partial f}{\partial y}\right)_{t_0}\frac{\Delta y}{\Delta t} + \epsilon_1\frac{\Delta x}{\Delta t} + \epsilon_2\frac{\Delta y}{\Delta t} \tag{4}$$

and letting Δt approach zero gives the value of dw/dt at t_0:

$$\left(\frac{dw}{dt}\right)_{t_0} = \left(\frac{\partial f}{\partial x}\right)_{t_0}\left(\frac{dx}{dt}\right)_{t_0} + \left(\frac{\partial f}{\partial y}\right)_{t_0}\left(\frac{dx}{dt}\right)_{t_0} + 0\cdot\left(\frac{dx}{dt}\right)_{t_0} + 0\cdot\left(\frac{dy}{dt}\right)_{t_0}. \tag{5}$$

Since the last two expressions on the right are zero, the formula boils down to

$$\frac{dw}{dt} = \frac{\partial f}{\partial x}\frac{dx}{dt} + \frac{\partial f}{\partial y}\frac{dy}{dt} \tag{6}$$

when we leave off the subscript t_0 that tells where the derivatives are evaluated.

For those who would like to know more about Eq. (3) and why w, $\partial f/\partial x$, and $\partial f/\partial y$ are assumed continuous, we have included a more detailed derivation in Appendix 9.

Equation (6) is also written as

$$\frac{df}{dt} = \frac{\partial f}{\partial x}\frac{dx}{dt} + \frac{\partial f}{\partial y}\frac{dy}{dt}, \tag{7}$$

in which the letter f replaces the letter w. Both formulations are common. Using dw/dt emphasizes the independent variable; using df/dt emphasizes the function. In either case, the way to remember the Chain Rule is to picture the diagram in Fig. 16.15.

Chain Rule

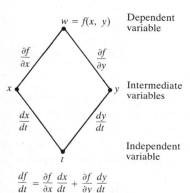

Dependent variable

Intermediate variables

Independent variable

$$\frac{df}{dt} = \frac{\partial f}{\partial x}\frac{dx}{dt} + \frac{\partial f}{\partial y}\frac{dy}{dt}$$

16.15 To find df/dt, start at w and read down each route to t, multiplying derivatives along the way; then add the products.

EXAMPLE 1 Use the Chain Rule to find the derivative of

$$f(x, y) = xy$$

with respect to t along the path $x = \cos t$, $y = \sin t$. What is the value of the derivative at $t = \pi/2$?

Solution We evaluate the right-hand side of Eq. (7) with $f(x, y) = xy$, $x = \cos t$, and $y = \sin t$:

$$\frac{\partial f}{\partial x} = y = \sin t, \quad \frac{\partial f}{\partial y} = x = \cos t,$$

$$\frac{dx}{dt} = -\sin t, \quad \frac{dy}{dt} = \cos t$$

$$\frac{df}{dt} = \frac{\partial f}{\partial x}\frac{dx}{dt} + \frac{\partial f}{\partial y}\frac{dy}{dt}$$

$$= (\sin t)(-\sin t) + (\cos t)(\cos t)$$

$$= -\sin^2 t + \cos^2 t$$

$$= \cos 2t.$$

In this case we can check our result by a more direct calculation. As a function of t,

$$f(x, y) = xy = \cos t \sin t = \frac{1}{2}\sin 2t.$$

Therefore,

$$\frac{df}{dt} = \frac{d}{dt}\left(\frac{1}{2}\sin 2t\right) = \cos 2t.$$

In either case, the value of df/dt at $t = \pi/2$ is

$$\left(\frac{df}{dt}\right)_{\pi/2} = \cos \pi = -1.$$

∎

The Chain Rule for Functions of Three Variables

If $w = f(x, y, z)$ and its partial derivatives are continuous and $x = x(t)$, $y = y(t)$, and $z = z(t)$ are differentiable functions of t, then the appropriate formula for df/dt is

$$\frac{df}{dt} = \frac{\partial f}{\partial x}\frac{dx}{dt} + \frac{\partial f}{\partial y}\frac{dy}{dt} + \frac{\partial f}{\partial z}\frac{dz}{dt}. \tag{8}$$

The derivation is identical with the derivation of Eq. (3) except that there are three intermediate variables (x, y, and z) instead of two. The diagram for remembering the equation is similar as well (Fig. 16.16).

EXAMPLE 2 Find the value of df/dt at $t = 0$ if $f(x, y, z) = xy + z$ and $x = \cos t$, $y = \sin t$, $z = t$.

Chain Rule

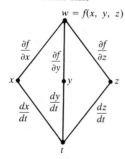

$$\frac{df}{dt} = \frac{\partial f}{\partial x}\frac{dx}{dt} + \frac{\partial f}{\partial y}\frac{dy}{dt} + \frac{\partial f}{\partial z}\frac{dz}{dt}$$

16.16

Solution The derivative df/dt shows how the values of f vary along a helix in its domain. From Eq. (8) we have

$$\frac{df}{dt} = \frac{\partial f}{\partial x}\frac{dx}{dt} + \frac{\partial f}{\partial y}\frac{dy}{dt} + \frac{\partial f}{\partial z}\frac{dz}{dt}$$

$$= (y)(-\sin t) + (x)(\cos t) + 1(1)$$

$$= -\sin^2 t + \cos^2 t + 1$$

$$= 1 + \cos 2t.$$

Therefore,

$$\left(\frac{df}{dt}\right)_0 = 1 + \cos(0) = 2.$$ ∎

The Chain Rule for Functions Defined on Surfaces

If we are interested in the temperature $f(x, y, z)$ at points on a globe in space, we might prefer to think of x, y, and z as functions of the variables r and s that give the points' latitude and longitude. We would then express the temperature as a composite

$$f(x(r, s), y(r, s), z(r, s)).$$

If f and its partial derivatives with respect to x, y, and z are continuous, and if x, y, and z and their partial derivatives with respect to r and s are continuous, then f has continuous partial derivatives with respect to r and s. Furthermore,

$$\frac{\partial f}{\partial r} = \frac{\partial f}{\partial x}\frac{\partial x}{\partial r} + \frac{\partial f}{\partial y}\frac{\partial y}{\partial r} + \frac{\partial f}{\partial z}\frac{\partial z}{\partial r} \tag{9}$$

and

$$\frac{\partial f}{\partial s} = \frac{\partial f}{\partial x}\frac{\partial x}{\partial s} + \frac{\partial f}{\partial y}\frac{\partial y}{\partial s} + \frac{\partial f}{\partial z}\frac{\partial z}{\partial s}. \tag{10}$$

Equation (9) can be derived from Eq. (8) by holding s fixed and setting t equal to r. Similarly, Eq. (10) can be derived by holding r fixed and setting t equal to s.

The tree diagrams for Eqs. (9) and (10) are shown in Fig. 16.17.

Chain Rule

$$\frac{\partial f}{\partial r} = \frac{\partial f}{\partial x}\frac{\partial x}{\partial r} + \frac{\partial f}{\partial y}\frac{\partial y}{\partial r} + \frac{\partial f}{\partial z}\frac{\partial z}{\partial r}$$

(a)

$$\frac{\partial f}{\partial s} = \frac{\partial f}{\partial x}\frac{\partial x}{\partial s} + \frac{\partial f}{\partial y}\frac{\partial y}{\partial s} + \frac{\partial f}{\partial z}\frac{\partial z}{\partial s}$$

(b)

16.17

Chain Rule

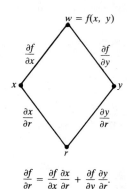

$w = f(x, y)$

$$\frac{\partial f}{\partial x} \qquad \frac{\partial f}{\partial y}$$

$$x \qquad y$$

$$\frac{\partial x}{\partial r} \qquad \frac{\partial y}{\partial r}$$

$$r$$

$$\frac{\partial f}{\partial r} = \frac{\partial f}{\partial x}\frac{\partial x}{\partial r} + \frac{\partial f}{\partial y}\frac{\partial y}{\partial r}.$$

16.18

EXAMPLE 3   Find $\partial f/\partial r$ and $\partial f/\partial s$ as functions of r and s if

$$f(x, y, z) = x + 2y + z^2, \qquad x = \frac{r}{s}, \qquad y = r^2 + \ln s, \qquad \text{and} \qquad z = 2r.$$

Solution   From Eq. (9),

$$\frac{\partial f}{\partial r} = \frac{\partial f}{\partial x}\frac{\partial x}{\partial r} + \frac{\partial f}{\partial y}\frac{\partial y}{\partial r} + \frac{\partial f}{\partial z}\frac{\partial z}{\partial r}$$

$$= (1)\left(\frac{1}{s}\right) + (2)(2r) + (2z)(2) = \frac{1}{s} + 12r.$$

From Eq. (10),

$$\frac{\partial f}{\partial s} = \frac{\partial f}{\partial x}\frac{\partial x}{\partial s} + \frac{\partial f}{\partial y}\frac{\partial y}{\partial s} + \frac{\partial f}{\partial z}\frac{\partial z}{\partial s}$$

$$= (1)\left(-\frac{r}{s^2}\right) + (2)\left(\frac{1}{s}\right) + (2z)(0)$$

$$= -\frac{r}{s^2} + \frac{2}{s}. \qquad \blacksquare$$

If f is a function of only two variables, x and y, the z term drops out of Eq. (9) to give

$$\frac{\partial f}{\partial r} = \frac{\partial f}{\partial x}\frac{\partial x}{\partial r} + \frac{\partial f}{\partial y}\frac{\partial y}{\partial r} \qquad (11)$$

(Fig. 16.18). There is a similar result for $\partial f/\partial s$.

EXAMPLE 4   Find $\partial f/\partial r$ and $\partial f/\partial s$ as functions of r and s if

$$f(x, y) = x^2 + y^2, \qquad x = r - s, \qquad \text{and} \qquad y = r + s.$$

Solution

$$\frac{\partial f}{\partial r} = \frac{\partial f}{\partial x}\frac{\partial x}{\partial r} + \frac{\partial f}{\partial y}\frac{\partial y}{\partial r}$$

$$= (2x)(1) + (2y)(1)$$

$$= 2(r - s) + 2(r + s) = 4r.$$

$$\frac{\partial f}{\partial s} = \frac{\partial f}{\partial x}\frac{\partial x}{\partial s} + \frac{\partial f}{\partial y}\frac{\partial y}{\partial s}$$

$$= (2x)(-1) + (2y)(1)$$

$$= -2(r - s) + 2(r + s) = 4s. \qquad \blacksquare$$

Chain Rule

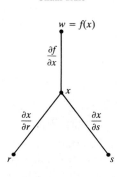

$$w = f(x)$$

$$\frac{\partial f}{\partial x}$$

$$x$$

$$\frac{\partial x}{\partial r} \qquad \frac{\partial x}{\partial s}$$

$$r \qquad\qquad s$$

$$\frac{\partial f}{\partial r} = \frac{df}{dx}\frac{\partial x}{\partial r}.$$

$$\frac{\partial f}{\partial s} = \frac{df}{dx}\frac{\partial x}{\partial s}.$$

16.19

If f is a function of x alone, then Eqs. (9) and (10) reduce to

$$\frac{\partial f}{\partial r} = \frac{df}{dx}\frac{\partial x}{\partial r} \qquad \text{and} \qquad \frac{\partial f}{\partial s} = \frac{df}{dx}\frac{\partial x}{\partial s}. \qquad (12)$$

Here, df/dx is the ordinary (single-variable) derivative. See Fig. 16.19.

EXAMPLE 5 Show that any differentiable function of the form $w = f(s)$, where $s = y + ax$, is a solution of the partial differential equation

$$\frac{\partial w}{\partial x} - a\frac{\partial w}{\partial y} = 0 \qquad (a \neq 0 \text{ constant}).$$

Solution From the Chain Rule we find

$$\frac{\partial w}{\partial x} = \frac{df}{ds}\frac{\partial s}{\partial x} = a\frac{df}{ds}, \qquad \frac{\partial w}{\partial y} = \frac{df}{ds}\frac{\partial s}{\partial y} = \frac{df}{ds}.$$

Therefore,

$$\frac{\partial w}{\partial x} - a\frac{\partial w}{\partial y} = a\frac{df}{ds} - a\frac{df}{ds} = 0. \qquad \blacksquare$$

EXAMPLE 6 Suppose $f(u, v, w)$ and its partial derivatives with respect to u, v, and w are continuous and that $u = x - y$, $v = y - z$, and $w = z - x$. Show that

$$\frac{\partial f}{\partial x} + \frac{\partial f}{\partial y} + \frac{\partial f}{\partial z} = 0.$$

Solution In this problem, x, y, and z are the independent variables and u, v, and w the intermediate variables. We therefore have

$$\frac{\partial f}{\partial x} = \frac{\partial f}{\partial u}\frac{\partial u}{\partial x} + \frac{\partial f}{\partial v}\frac{\partial v}{\partial x} + \frac{\partial f}{\partial w}\frac{\partial w}{\partial x}$$

$$= \frac{\partial f}{\partial u}(1) + \frac{\partial f}{\partial v}(0) + \frac{\partial f}{\partial w}(-1)$$

$$= \frac{\partial f}{\partial u} - \frac{\partial f}{\partial w}.$$

Similar calculations give

$$\frac{\partial f}{\partial y} = -\frac{\partial f}{\partial u} + \frac{\partial f}{\partial v} \qquad \text{and} \qquad \frac{\partial f}{\partial z} = -\frac{\partial f}{\partial v} + \frac{\partial f}{\partial w}.$$

Hence,

$$\frac{\partial f}{\partial x} + \frac{\partial f}{\partial y} + \frac{\partial f}{\partial z} = 0. \qquad \blacksquare$$

Remembering the Different Forms of the Chain Rule

How are we to remember all the different forms of the Chain Rule? The answer is that there is no need to remember them all. The best thing to do is to remember a few key forms, like Eqs. (7), (8), and (9). The others, when needed, can be constructed from tree diagrams. The appropriate diagram can always be drawn by placing the dependent variable at the top, the intermediate variables in the middle, and the selected independent variable at the bottom.

The Chain Rule for Functions of Many Variables

Suppose $f(x, y, z, u, \ldots, v)$ is a differentiable function of the variables x, y, z, u, $\ldots$, v (a finite set) and that these variables are themselves differentiable functions of the variables p, q, r, s, $\ldots$, t (another finite set). Suppose also that all the functions involved are continuous. Then f is a differentiable function of the variables p through t and the partial derivatives of f with respect to these variables are given by equations of the form

$$\frac{\partial f}{\partial p} = \frac{\partial f}{\partial x}\frac{\partial x}{\partial p} + \frac{\partial f}{\partial y}\frac{\partial y}{\partial p} + \cdots + \frac{\partial f}{\partial v}\frac{\partial v}{\partial p}. \tag{13}$$

The other equations are obtained by replacing p by q, r, s, and so on, one at a time.

A quick way to remember Eq. (13) is to think of the right-hand side as the dot product of two vectors whose components are

$$\left(\frac{\partial f}{\partial x}, \frac{\partial f}{\partial y}, \ldots, \frac{\partial f}{\partial v}\right) \quad \text{and} \quad \left(\frac{\partial x}{\partial p}, \frac{\partial y}{\partial p}, \ldots, \frac{\partial v}{\partial p}\right). \tag{14}$$

The first vector contains the derivatives of f with respect to the intermediate variables. The second contains the derivatives of the intermediate variables with respect to the chosen independent variable.

PROBLEMS

In Problems 1–8, use the Chain Rule to express df/dt in terms of t. Then evaluate df/dt at the given value of t.

1. $f(x, y, z) = \ln(x + y + z)$, $\quad x = \cos^2 t$, $\quad y = \sin^2 t$, $\quad z = t$; $\quad t = \pi$

2. $f(x, y, z) = (x + y)/z$, $\quad x = \cos t$, $\quad y = \sin t$, $\quad z = 1/t$; $\quad t = 3$

3. $f(x, y) = x^2 + y^2$, $\quad x = \cos t$, $\quad y = \sin t$; $\quad t = \pi$

4. $f(x, y, z) = x^2 + y^2 + z^2$, $\quad x = \cos t$, $\quad y = \sin t$, $\quad z = t$; $\quad t = \sqrt{2}$

5. $f(x, y, z) = \sqrt{x^2 + y^2 + z^2}$, $\quad x = 2 + t$, $\quad y = -t - 1$, $\quad z = t$; $\quad t = -2$

6. $f(x, y, z) = x^2 + y^2 + z^2$, $\quad x = \cos t + \sin t$, $\quad y = \cos t - \sin t$, $\quad z = \sin t$; $\quad t = 0$

7. $f(x, y, z) = 1/\sqrt{x^2 + y^2 + z^2}$; $\quad x = y = z = t$; $\quad t = 1$

8. $f(x, y) = xy/(x^2 - y^2)$, $\quad x = e^t + e^{-t}$, $\quad y = e^t - e^{-t}$; $\quad t = 0$

In Problems 9–16, draw a tree diagram and write a formula for each derivative.

9. $\dfrac{dz}{dt}$ for $z = f(x, y)$, $\quad x = x(t)$, $\quad y = y(t)$

10. $\dfrac{dz}{dt}$ for $z = f(u, v, w)$, $\quad u = u(t)$, $\quad v = v(t)$, $\quad w = w(t)$

11. $\dfrac{\partial w}{\partial y}$ for $w = g(u, v)$, $\quad u = u(x, y)$, $\quad v = v(x, y)$

12. $\dfrac{\partial z}{\partial s}$ for $z = f(x, y)$, $\quad x = x(t, s)$, $\quad z = z(t, s)$

13. $\dfrac{\partial y}{\partial r}$ for $y = f(u)$, $\quad u = u(r, s)$

14. $\dfrac{\partial w}{\partial u}$ for $w = f(x, y)$, $\quad x = x(u)$, $\quad y = y(v)$

15. $\dfrac{\partial w}{\partial p}$ for $w = f(x, y, z, t)$, $\quad x = x(p, q)$, $\quad y = y(p, q)$, $\quad z = z(p, q)$, $\quad t = t(p, q)$

16. $\dfrac{\partial w}{\partial s}$ for $w = g(x, y)$, $\quad x = x(r, s, t)$, $\quad y = y(r, s, t)$

17. If $u = x^2 + e^{y^2}$, $x = \sin 2t$, and $y = \cos t^2$, which of the following give du/dt?

a) $\dfrac{\partial u}{\partial x}\dfrac{dx}{dt} + \dfrac{\partial u}{\partial y}\dfrac{dy}{dt}$ ~~yes~~

b) $4x \cos 2t - 4ye^{y^2} t \sin t^2$ ~~yes~~

c) $2x \cos 2t - 2ye^{y^2} \sin t^2$ ~~No~~

18. Find $\partial w/\partial v$ when $u = 0$, $v = 0$ if
$w = (x^2 + y - 2)^4 + (x - y + 2)^3$, $\quad x = u - 2v + 1$, and $\quad y = 2u + v - 2$.

19. Suppose $f(u, v, w)$ and its partial derivatives with respect to u, v, and w are continuous and that $u = x - y$, $v = y - z$, and $w = z - x$. Show that

$$\frac{\partial f}{\partial y} = -\frac{\partial f}{\partial u} + \frac{\partial f}{\partial v} \quad \text{and} \quad \frac{\partial f}{\partial z} = -\frac{\partial f}{\partial v} + \frac{\partial f}{\partial w}.$$

This will supply the remaining calculations in Example 6.

20. Find $\partial z / \partial v$ if $z = x^2 + 2xy$, $x = u \cos v$, $y = u \sin v$. (Express your answer in terms of x and y.)

21. Find $\partial w / \partial x$ if $w = uv + \ln v$, $u = x + y^2$, $v = e^x \cos y$. (Express your answer in terms of u and v.)

22. Find $\partial z / \partial u$ if $z = e^x + xy^2$, $x = u + v$, $y = e^{u+v}$. (Express your answer in terms of x and y.)

23. Find $\partial w / \partial u$ if $w = x^2 + y^2$, $x = u - v$, $y = ve^{2u}$. (Express your answer in terms of x and y.)

24. Find $\partial z / \partial u$ when $u = 0$, $v = 1$ if

$$z = \sin xy + x \sin y, \quad x = u^2 + v^2, \quad y = uv.$$

25. Find $\partial w / \partial v$ when $u = 0$, $v = 0$ if

$$w = (x^2 + y - 2)^4 + (x - y + 2)^3, \quad x = u - 2v + 1,$$
$$y = 2u + v - 2.$$

26. Find $\partial w / \partial x$ at the point $(x, y, z) = (1, 1, 1)$ if

$$w = \cos uv, \quad u = xyz, \quad v = \pi / (4(x^2 + y^2)).$$

27. Find $\partial F / \partial x$ at $(x, y) = (\pi, 1)$ if

$$F(x, y) = \int_0^{x^2 y} \cos \sqrt{t} \; dt.$$

28. If $z = f(t)$ and $t = (x + y)/xy$, show that

$$x^2 \frac{\partial z}{\partial x} = y^2 \frac{\partial z}{\partial y}.$$

29. If a and b are constants, $w = f(u)$, and $u = ax + by$, show that

$$a \frac{\partial w}{\partial y} = b \frac{\partial w}{\partial x}.$$

30. If $w = f(u)$ and $u = xy/(x^2 + y^2)$, show that

$$x \frac{\partial w}{\partial x} + y \frac{\partial w}{\partial y} = 0.$$

31. If $w = f(u, v)$, $u = x + y$, and $v = x - y$, show that

$$\frac{\partial w}{\partial x} \frac{\partial w}{\partial y} = \left(\frac{\partial f}{\partial u} \right)^2 - \left(\frac{\partial f}{\partial v} \right)^2.$$

32. The dimensions of a rectangular solid at a given instant t_0 are $a(t_0) = 13$ cm, $b(t_0) = 9$ cm, and $c(t_0) = 5$ cm. If a and c are increasing at 2 cm/sec and b is decreasing at 4 cm/sec, find the rates at which the volume V and surface area S are changing. Are V and S increasing, decreasing, or neither?

33. Suppose that the partial derivatives of $f(x, y, z)$ at points of the helix $x = \cos t$, $y = \sin t$, $z = t$ are

$$\frac{\partial f}{\partial x} = \cos t, \quad \frac{\partial f}{\partial y} = \sin t, \quad \frac{\partial f}{\partial z} = t^2 + t - 2.$$

At what points on the curve, if any, can f take on extreme values?

34. Let $f(x, y, z) = x^2 e^{2y} \cos 3z$. Find the value of df/dt at the point $(1, \ln 2, 0)$ on the curve $x = \cos t$, $y = \ln(t + 2)$, $z = t$.

35. a) Let $f(x, y)$ be the temperature at the point (x, y) on the circle $x = \cos t$, $y = \sin t$ and suppose that

$$\frac{\partial f}{\partial x} = 8x - 4y \quad \text{and} \quad \frac{\partial f}{\partial y} = 8y - 4x.$$

Find where the maximum and minimum temperatures on the circle are taken on. (*Hint:* Find df/dt and $d^2 f/dt^2$ as functions of t.)

b) Suppose that the values of the function f in (a) are given by the formula

$$f(x, y) = 4x^2 - 4xy + 4y^2.$$

What are the maximum and minimum values of f on the circle?

36. If we substitute polar coordinates $x = r \cos \theta$ and $y = r \sin \theta$ in a continuous function $w = f(x, y)$ that has continuous partial derivatives, show that

$$\frac{\partial w}{\partial r} = f_x \cos \theta + f_y \sin \theta,$$

and

$$\frac{1}{r} \frac{\partial w}{\partial \theta} = -f_x \sin \theta + f_y \cos \theta.$$

37. Use determinants to solve the equations in Problem 36 for f_x and f_y in terms of $(\partial w / \partial r)$ and $(\partial w / \partial \theta)$.

38. In connection with Problem 36, show that

$$\left(\frac{\partial w}{\partial r} \right)^2 + \frac{1}{r^2} \left(\frac{\partial w}{\partial \theta} \right)^2 = f_x^2 + f_y^2.$$

16.5

Partial Derivatives with Constrained Variables*

In finding partial derivatives of functions like $w = f(x, y)$, so far we have assumed x and y to be independent. But in many applications this is not the case. For example, the internal energy U of a gas may be expressed in terms of

---

*This article is based on notes written for MIT by Arthur P. Mattuck.

its pressure p, volume v, and temperature T:

$$U = f(p, v, T).$$

If the gas is an ideal gas, however, then p, v, and T obey the ideal gas law

$$pv = nRT \qquad (n, R \text{ constant})$$

and are therefore not independent. Finding partial derivatives in situations like these can be complicated. But it is better to face the complication now than to meet it for the first time while you are also trying to learn economics, engineering, or physics.

Decide Which Variables Are Dependent and Which Are Independent

If the variables in a function $w = f(x, y, z)$ are constrained by a relation like the one imposed on x, y, and z by the equation $z = x^2 + y^2$, the geometric meanings and the numerical values of the partial derivatives of f will depend on which variables are chosen to be dependent and which are chosen to be independent. To see how this choice can affect the outcome, we consider the calculation of $\partial w / \partial x$ when $w = x^2 + y^2 + z^2$ and $z = x^2 + y^2$.

EXAMPLE 1   Find $\partial w / \partial x$ if

$$w = x^2 + y^2 + z^2 \qquad \text{and} \qquad z = x^2 + y^2.$$

Solution   We are given two equations in the four unknowns x, y, z, and w. Like many such systems, the one here can be solved for two of the unknowns (the dependent variables) in terms of the others (the independent variables). In being asked for $\partial w / \partial x$, we are told that w is to be a dependent variable and x an independent variable. Thus, the possible choices are

Dependent	Independent
w, z	x, y
w, y	x, z

In either case, we can express w explicitly in terms of the selected independent variables. We do this by using the second equation to eliminate the other dependent variable in the first equation.

In the first case, the other dependent variable is z. We eliminate it from the first equation by replacing it by $x^2 + y^2$. The resulting expression for w is

$$w = x^2 + y^2 + z^2 = x^2 + y^2 + (x^2 + y^2)^2$$
$$= x^2 + y^2 + x^4 + 2x^2y^2 + y^4$$

and

$$\frac{\partial w}{\partial x} = 2x + 4x^3 + 4xy^2. \tag{1}$$

This is the formula for $\partial w / \partial x$ when x and y are the independent variables.

In the second case, where the independent variables are x and z and the other dependent variable is y, we eliminate the dependent variable y in the expression for w by replacing y^2 by $z - x^2$. This gives

$$w = x^2 + y^2 + z^2 = x^2 + (z - x^2) + z^2 = z + z^2$$

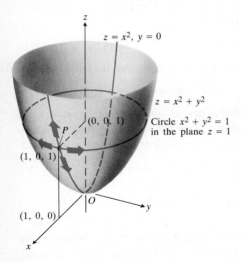

16.20 (a) As *x* changes, with *y* = 0, *P* moves up or down the surface on the parabola $z = x^2$ in the *xz*-plane. (b) As *x* changes, with *z* = 1, *P* moves on the circle $x^2 + y^2 = 1$, *z* = 1.

and

$$\frac{\partial w}{\partial x} = 0. \tag{2}$$

This is the formula for $\partial w/\partial x$ when *x* and *z* are the independent variables.

The formulas for $\partial w/\partial x$ in Eqs. (1) and (2) are genuinely different. We cannot change either into the other by using the relation $z = x^2 + y^2$. There is not just one $\partial w/\partial x$, there are two, and we see that the original instruction to find $\partial w/\partial x$ was incomplete. *Which* $\partial w/\partial x$? we ask.

The answers in Eqs. (1) and (2) have geometric interpretations that help to explain why they differ. The function $w = x^2 + y^2 + z^2$ measures the square of the distance from the point (x, y, z) to the origin. The condition $z = x^2 + y^2$ says that the point (x, y, z) lies on the surface of the paraboloid of revolution shown in Fig. 16.20. What does it mean to calculate $\partial w/\partial x$ at a point $P(x, y, z)$ that can move only on this surface? What is the value of $\partial w/\partial x$ when the coordinates of *P* are, say, (1, 0, 1)?

If we take *x* and *y* to be independent, then we find $\partial w/\partial x$ by holding *y* fixed (at *y* = 0 in this case) and letting *x* vary. This means *P* moves along the parabola $z = x^2$ in the *xz*-plane. As *P* moves on this parabola, *w*, which is the square of the distance from *P* to the origin, changes. We calculate $\partial w/\partial x$ in this case (our first solution above) to be

$$\frac{\partial w}{\partial x} = 2x + 4x^3 + 4xy^2.$$

At the point $P(1, 0, 1)$, the value of this derivative is

$$\frac{\partial w}{\partial x} = 2 + 4 + 0 = 6.$$

If we take *x* and *z* to be independent, then we find $\partial w/\partial x$ by holding *z* fixed while *x* varies. Since the *z*-coordinate of *P* is 1, varying *x* moves *P* along a circle in the plane *z* = 1. As *P* moves along this circle, its distance from the origin remains constant, and *w*, being the square of this distance, does not change. That is,

$$\frac{\partial w}{\partial x} = 0,$$

as we found in our second solution. ∎

As we saw in Example 1, a typical routine for finding $\partial w/\partial x$ when the variables in the function $w = f(x, y, z)$ are related by a second equation consists of three steps.

STEP 1: Decide which variables are to be dependent and which are to be independent. (In practice, the decision is based on the physical or theoretical context of our work. The problems at the end of this article will always say which variables are which.)

STEP 2: Eliminate the other dependent variable(s) in the expression for *w*.

STEP 3: Differentiate as usual.

If we cannot carry out Step 2 after deciding which variables are dependent, we differentiate the equations as they are and try to solve for $\partial w/\partial x$ afterward. The next example shows how this is done.

EXAMPLE 2   Find $\partial w/\partial x$ at the point $(x, y, z) = (2, -1, 1)$ if

$$w = x^2 + y^2 + z^2, \qquad z^3 - xy + yz + y^3 = 0,$$

and x and y are the independent variables.

Solution   It is not convenient to eliminate z in the expression for w. We therefore differentiate both equations implicitly with respect to x, treating x and y as independent variables and w and z as dependent variables. This gives

$$\frac{\partial w}{\partial x} = 2x + 2z\frac{\partial z}{\partial x} \tag{3}$$

and

$$3z^2\frac{\partial z}{\partial x} - y + y\frac{\partial z}{\partial x} + 0 = 0. \tag{4}$$

These equations may now be combined to express $\partial w/\partial x$ in terms of x, y, and z. We solve Eq. (4) for $\partial z/\partial x$ to get

$$\frac{\partial z}{\partial x} = \frac{y}{y + 3z^2}$$

and substitute into Eq. (3) to get

$$\frac{\partial w}{\partial x} = 2x + \frac{2yz}{y + 3z^2}.$$

The value of this derivative at $(x, y, z) = (2, -1, 1)$ is

$$\left(\frac{\partial w}{\partial x}\right)_{(2, -1, 1)} = 2(2) + \frac{2(-1)(1)}{-1 + 3(1)^2}$$

$$= 4 + \frac{-2}{2}$$

$$= 3. \qquad\blacksquare$$

To show what variables are assumed to be independent in calculating a derivative, we can use the following notation:

$$\left(\frac{\partial w}{\partial x}\right)_y \qquad \partial w/\partial x \text{ with } x \text{ and } y \text{ independent,} \tag{5}$$

$$\left(\frac{\partial w}{\partial x}\right)_z \qquad \partial w/\partial x \text{ with } x \text{ and } z \text{ independent,} \tag{6}$$

$$\left(\frac{\partial f}{\partial y}\right)_{x, t} \qquad \partial f/\partial y \text{ with } y, x, \text{ and } t \text{ independent.} \tag{7}$$

EXAMPLE 3   Find

a) $\left(\dfrac{\partial w}{\partial x}\right)_{y, z}$ and   b) $\left(\dfrac{\partial w}{\partial x}\right)_{t, z}$

if $w = x^2 + y - z + \sin t$ and $x + y = t$.

Solution

a) With x, y, z independent, we have

$$t = x + y, \qquad w = x^2 + y - z + \sin(x + y),$$

$$\left(\frac{\partial w}{\partial x}\right)_{y, z} = 2x + 0 - 0 + \cos(x + y)\frac{\partial}{\partial x}(x + y)$$

$$= 2x + \cos(x + y).$$

b) With x, t, z independent, we have

$$y = t - x, \qquad w = x^2 + (t - x) - z + \sin t,$$

$$\left(\frac{\partial w}{\partial x}\right)_{t, z} = 2x - 1 - 0 + 0 = 2x - 1.$$ ∎

Arrow Diagrams

In solving problems like the one in Example 3, it often helps to start with an arrow diagram that shows how the variables and functions are related. If

$$w = x^2 + y - z + \sin t \qquad \text{and} \qquad x + y = t$$

and we are asked to find $\partial w / \partial x$ when x, y, and z are independent, the appropriate diagram is one like this:

$$\begin{pmatrix} x \\ y \\ z \end{pmatrix} \quad \longrightarrow \quad \begin{pmatrix} x \\ y \\ z \\ t \end{pmatrix} \quad \longrightarrow \quad w \tag{8}$$

Independent   Intermediate   Dependent
variables   variables and   variable
                 relations
                 $x = x$
                 $y = y$
                 $z = z$
                 $t = x + y$

The diagram shows the independent variables on the left, the intermediate variables and their relation to the independent variables in the middle, and the dependent variable on the right.

To find $\partial w / \partial x$, we first apply the four-variable form of the Chain Rule to w, getting

$$\frac{\partial w}{\partial x} = \frac{\partial w}{\partial x}\frac{\partial x}{\partial x} + \frac{\partial w}{\partial y}\frac{\partial y}{\partial x} + \frac{\partial w}{\partial z}\frac{\partial z}{\partial x} + \frac{\partial w}{\partial t}\frac{\partial t}{\partial x}. \tag{9}$$

We then use the formula for w,

$$w = x^2 + y - z + \sin t,$$

to evaluate the partial derivatives of w that appear on the right-hand side of Eq. (9). This gives

$$\frac{\partial w}{\partial x} = 2x\frac{\partial x}{\partial x} + (1)\frac{\partial y}{\partial x} + (-1)\frac{\partial z}{\partial x} + \cos t\frac{\partial t}{\partial x}$$

$$= 2x\frac{\partial x}{\partial x} + \frac{\partial y}{\partial x} - \frac{\partial z}{\partial x} + \cos t\frac{\partial t}{\partial x}. \tag{10}$$

To calculate the remaining partial derivatives, we apply what we know about the dependence and independence of the variables involved. As shown in the diagram (8), the variables x, y, and z are independent and $t = x + y$. Hence,

$$\frac{\partial x}{\partial x} = 1, \quad \frac{\partial y}{\partial x} = 0, \quad \frac{\partial z}{\partial x} = 0, \quad \frac{\partial t}{\partial x} = \frac{\partial}{\partial x}(x + y) = (1 + 0) = 1. \quad (11)$$

We substitute these values into Eq. (10) to find $\partial w / \partial x$:

$$\left(\frac{\partial w}{\partial x}\right)_{y,\,z} = 2x(1) + 0 - 0 + (\cos t)(1)$$

$$= 2x + \cos t$$

$$= 2x + \cos(x + y). \qquad \text{(in terms of the independent variables)}$$

Working without Specific Formulas

In applications, we often have to find an expression for the derivative of a function when neither the function nor the relation among its variables is given explicitly. Two observations can help us here.

The first observation is that if $F(r, s)$ has a constant value throughout some region R of the rs-plane, where F, $\partial F / \partial r$, and $\partial F / \partial s$ are continuous and r and s are independent variables, then

$$\frac{\partial F}{\partial r} = \frac{\partial F}{\partial s} = 0 \qquad (12)$$

throughout R. In particular, Eq. (12) holds if $F(r, s) = 0$ throughout R.

To make the second observation, suppose that

$$f(x, y, z) = f(x(r, s), y(r, s), z(r, s)) = F(r, s) \qquad (13)$$

throughout some region R of the rs-plane, where all the functions and appropriate partial derivatives are continuous and r and s are independent variables. To display this information, we draw an arrow diagram:

$$\qquad (14)$$

Independent variables	Intermediate variables and relations	Dependent variable
	$x = x(r, s)$	$w = f(x, y, z)$
	$y = y(r, s)$	$= F(r, s)$
	$z = z(r, s)$	

Suppose further that $F(r, s) = 0$ throughout R. Then

$$\frac{\partial f}{\partial x}\frac{\partial x}{\partial r} + \frac{\partial f}{\partial y}\frac{\partial y}{\partial r} + \frac{\partial f}{\partial z}\frac{\partial z}{\partial r} = 0 \qquad (15)$$

and

$$\frac{\partial f}{\partial x}\frac{\partial x}{\partial s} + \frac{\partial f}{\partial y}\frac{\partial y}{\partial s} + \frac{\partial f}{\partial z}\frac{\partial z}{\partial s} = 0. \qquad (16)$$

The left-hand sides of these equations are $\partial F/\partial r$ and $\partial F/\partial s$, and these derivatives are zero because $F(r, s)$ has a constant value throughout R.

EXAMPLE 4   Suppose that the equation $f(x, y, z) = 0$ determines z as a differentiable function of the independent variables x and y and that $\partial f/\partial z \neq 0$. Show that

$$\left(\frac{\partial z}{\partial x}\right)_y = -\frac{\partial f/\partial x}{\partial f/\partial z}.$$

Solution   We regard all three of x, y, and z to be functions of the two independent variables x and y as in the diagram

$$\begin{pmatrix} x \\ y \end{pmatrix} \rightarrow \begin{pmatrix} x \\ y \\ z \end{pmatrix} \xrightarrow{f} w \tag{17}$$

$$x = x$$
$$y = y$$
$$z = z(x, y)$$

This is the diagram in (14) with $r = x$, $s = y$, and a few details left out. From Eq. (15) with $r = x$ we have

$$\frac{\partial f}{\partial x}\frac{\partial x}{\partial x} + \frac{\partial f}{\partial y}\frac{\partial y}{\partial x} + \frac{\partial f}{\partial z}\frac{\partial z}{\partial x} = 0. \tag{18}$$

⇧   ⇧

Equals 1   Equals 0 because
                y is independent of x

Hence,

$$\frac{\partial f}{\partial x} + \frac{\partial f}{\partial z}\frac{\partial z}{\partial x} = 0 \quad \text{and} \quad \frac{\partial z}{\partial x} = -\frac{\partial f/\partial x}{\partial f/\partial z}. \qquad ∎$$

PROBLEMS

In Problems 1–3, in each case, begin by drawing a diagram that shows the relations among the variables.

1. If $w = x^2 + y^2 + z^2$ and $z = x^2 + y^2$, find

a) $\left(\dfrac{\partial w}{\partial y}\right)_z$ b) $\left(\dfrac{\partial w}{\partial z}\right)_x$ c) $\left(\dfrac{\partial w}{\partial z}\right)_y$.

2. If $w = x^2 + y - z + \sin t$ and $x + y = t$, find

a) $\left(\dfrac{\partial w}{\partial y}\right)_{x,z}$ b) $\left(\dfrac{\partial w}{\partial y}\right)_{z,t}$ c) $\left(\dfrac{\partial w}{\partial z}\right)_{x,y}$

d) $\left(\dfrac{\partial w}{\partial z}\right)_{y,t}$ e) $\left(\dfrac{\partial w}{\partial t}\right)_{x,z}$ f) $\left(\dfrac{\partial w}{\partial t}\right)_{y,z}$.

3. Let $U = f(p, v, T)$ be the internal energy of a gas that obeys the ideal gas law $pv = nRT$ (n and R constant). Find

a) $\left(\dfrac{\partial U}{\partial p}\right)_v$ b) $\left(\dfrac{\partial U}{\partial T}\right)_v$.

4. Find

a) $\left(\dfrac{\partial w}{\partial x}\right)_y$ b) $\left(\dfrac{\partial w}{\partial z}\right)_y$

at the point $(x, y, z) = (0, 1, \pi)$ if

$$w = x^2 + y^2 + z^2 \quad \text{and} \quad y \sin z + z \sin x = 1.$$

5. Find

a) $\left(\dfrac{\partial w}{\partial y}\right)_x$ b) $\left(\dfrac{\partial w}{\partial y}\right)_z$

at the point $(w, x, y, z) = (4, 2, 1, -1)$ if

$$w = x^2y^2 + yz - z^3 \quad \text{and} \quad x^2 + y^2 + z^2 = 6.$$

6. Find $\left(\dfrac{\partial u}{\partial y}\right)_x$ at the point $(u, v) = (\sqrt{2}, 1)$ if $x = u^2 + v^2$ and $y = uv$.

7. Suppose that $x^2 + y^2 = r^2$ and $x = r \cos \theta$, as in polar coordinates. Find

$$\left(\frac{\partial x}{\partial r}\right)_\theta \quad \text{and} \quad \left(\frac{\partial r}{\partial x}\right)_y.$$

8. Suppose that the equation $g(x, y, z) = 0$ determines z as a differentiable function of the independent variables x and y and that $g_z \neq 0$. Show that

$$\left(\frac{\partial z}{\partial y}\right)_x = -\frac{\partial g/\partial y}{\partial g/\partial z}.$$

9. Establish the fact, widely used in hydrodynamics, that if $f(x, y, z) = 0$, then

$$\left(\frac{\partial x}{\partial y}\right)_z \left(\frac{\partial y}{\partial z}\right)_x \left(\frac{\partial z}{\partial x}\right)_y = -1.$$

(*Hint:* Express all the derivatives in terms of the formal partial derivatives $\partial f/\partial x$, $\partial f/\partial y$, and $\partial f/\partial z$.)

10. If $z = x + f(u)$, where $u = xy$, show that

$$x\frac{\partial z}{\partial x} - y\frac{\partial z}{\partial y} = x.$$

11. Suppose

$$w = x^2 - y^2 + 4z + t \quad \text{and} \quad x + 2z + t = 25.$$

Show that the equations

$$\frac{\partial w}{\partial x} = 2x - 1 \quad \text{and} \quad \frac{\partial w}{\partial x} = 2x - 2$$

each give $\partial w/\partial x$, depending on which variables are chosen to be dependent and which are chosen to be independent. Identify the independent variables in each case.

16.6
Gradients and Directional Derivatives

We know from Article 16.4 that if $f(x, y, z)$ and its partial derivatives are continuous, then the rate at which the values of f change with respect to t along a differentiable curve

$$x = x(t), \qquad y = y(t), \qquad z = z(t)$$

may be calculated from the formula

$$\frac{df}{dt} = \frac{\partial f}{\partial x}\frac{dx}{dt} + \frac{\partial f}{\partial y}\frac{dy}{dt} + \frac{\partial f}{\partial z}\frac{dz}{dt}. \tag{1}$$

To use this formula at any particular point

$$P_0(x_0, y_0, z_0) = P_0(x(t_0), y(t_0), z(t_0)),$$

we evaluate the partial derivatives of f at P_0 and the derivatives of x, y, and z at t_0.

The derivative df/dt in Eq. (1) is the rate of change of f at P_0 with respect to increasing t and therefore depends, among other things, on the direction of motion along the curve. This observation is particularly important when the curve is a straight line and the parameter t is the distance along the line measured from P_0 in the direction of a given unit vector **u**. For then df/dt is the rate of change of f with respect to distance in its domain in the direction of **u**. By varying **u**, we can find the rates at which f changes with respect to distance as we move through P_0 in different directions. These "directional derivatives" are extremely useful in science and engineering (as well as in mathematics). The goal of this article is to develop a simple formula for calculating them.

Calculating Directional Derivatives

Suppose that the partial derivatives of a function $f(x, y, z)$ are continuous throughout some region R in space, that $P_0(x_0, y_0, z_0)$ is a point in R, and that

Line $x = x_0 + su_1,\ y = y_0 + su_2,\ z = z_0 + su_3$

$\mathbf{u} = u_1\mathbf{i} + u_2\mathbf{j} + u_3\mathbf{k}$

$P_0(x_0, y_0, z_0)$ Domain of $f(x, y, z)$

16.21   The derivative of f in the direction of the unit vector $\mathbf{u}$ at a point P_0 in its domain is the value of df/ds at P_0.

$\mathbf{u} = u_1\mathbf{i} + u_2\mathbf{j} + u_3\mathbf{k}$ is a unit vector (Fig. 16.21). Then

$$x = x_0 + su_1, \qquad y = y_0 + su_2, \qquad z = z_0 + su_3 \qquad (2)$$

are parametric equations for the line through P_0 parallel to $\mathbf{u}$, and the parameter s measures the directed distance along this line from P_0 to the point $P_s(x_0 + su_1, y_0 + su_2, z_0 + su_3)$, as we saw in Example 2 in Article 14.2. To calculate df/ds, the rate at which f changes with respect to distance in the direction of $\mathbf{u}$ at P_0, we substitute the derivatives

$$\frac{dx}{ds} = u_1, \qquad \frac{dy}{ds} = u_2, \qquad \frac{dz}{ds} = u_3$$

into the equation

$$\frac{df}{ds} = \frac{\partial f}{\partial x}\frac{dx}{ds} + \frac{\partial f}{\partial y}\frac{dy}{ds} + \frac{\partial f}{\partial z}\frac{dz}{ds}$$

and evaluate the partial derivatives of f at P_0. This gives

$$\frac{df}{ds} = \left(\frac{\partial f}{\partial x}\right)_{P_0} u_1 + \left(\frac{\partial f}{\partial y}\right)_{P_0} u_2 + \left(\frac{\partial f}{\partial z}\right)_{P_0} u_3. \qquad (3)$$

The expression on the right-hand side of Eq. (3) is the dot product of $\mathbf{u}$ and the vector

$$\nabla f = \left(\frac{\partial f}{\partial x}\right)_{P_0}\mathbf{i} + \left(\frac{\partial f}{\partial y}\right)_{P_0}\mathbf{j} + \left(\frac{\partial f}{\partial z}\right)_{P_0}\mathbf{k}.$$

This vector is called the *gradient* of f at the point P_0. It is customary to picture the gradient as a vector in the domain of f. Its components are calculated by evaluating the three partial derivatives of f at (x_0, y_0, z_0). The derivative on the left-hand side of Eq. (3) is called *the derivative of f at the point P_0 in the direction of $\mathbf{u}$.* It is often denoted by

$$(D_\mathbf{u}f)_{P_0}.$$

DEFINITION

Definitions of Gradient and Directional Derivative

1. If the partial derivatives of $f(x, y, z)$ are defined at $P_0(x_0, y_0, z_0)$, then the **gradient** of f at P_0 is the vector

$$\nabla f = \frac{\partial f}{\partial x}\mathbf{i} + \frac{\partial f}{\partial y}\mathbf{j} + \frac{\partial f}{\partial z}\mathbf{k} \qquad (4)$$

   obtained by evaluating the partial derivatives of f at P_0.

2. If $f(x, y, z)$ has continuous partial derivatives at $P_0(x_0, y_0, z_0)$ and $\mathbf{u}$ is a unit vector, then the **derivative of f at P_0 in the direction of $\mathbf{u}$** is the number

$$(D_\mathbf{u}f)_{P_0} = (\nabla f)_{P_0} \cdot \mathbf{u}, \qquad (5)$$

   which is the scalar product of the gradient of f at P_0 and $\mathbf{u}$.

Another notation for the gradient of f is

$$\mathrm{grad}\, f,$$

read the way it is written. The symbol ∇f is read "grad f" as well as "gradient of f" and "del f."

Another common notation for the directional derivative is

$$\left(\frac{df}{ds}\right)_{\mathbf{u},P_0}$$

EXAMPLE 1 Find the derivative of

$$f(x, y, z) = x^3 - xy^2 - z$$

at $P_0(1, 1, 0)$ in the direction of the vector $\mathbf{A} = 2\mathbf{i} - 3\mathbf{j} + 6\mathbf{k}$.

Solution The direction of $\mathbf{A}$ is obtained by dividing $\mathbf{A}$ by its length:

$$|\mathbf{A}| = \sqrt{(2)^2 + (-3)^2 + (6)^2} = \sqrt{49} = 7,$$

$$\mathbf{u} = \frac{\mathbf{A}}{|\mathbf{A}|} = \frac{2}{7}\mathbf{i} - \frac{3}{7}\mathbf{j} + \frac{6}{7}\mathbf{k}.$$

The partial derivatives of f at P_0 are

$$f_x = 3x^2 - y^2|_{(1,\,1,\,0)} = 2,$$
$$f_y = -2xy|_{(1,\,1,\,0)} = -2,$$
$$f_z = -1|_{(1,\,1,\,0)} = -1.$$

The gradient of f at P_0 is

$$\nabla f|_{(1,\,1,\,0)} = 2\mathbf{i} - 2\mathbf{j} - \mathbf{k}.$$

The derivative of f at P_0 in the direction $\mathbf{A}$ is therefore

$$(D_{\mathbf{u}}f)|_{(1,\,1,\,0)} = \nabla f|_{(1,\,1,\,0)} \cdot \mathbf{u}$$

$$= (2\mathbf{i} - 2\mathbf{j} - \mathbf{k}) \cdot \left(\frac{2}{7}\mathbf{i} - \frac{3}{7}\mathbf{j} + \frac{6}{7}\mathbf{k}\right)$$

$$= \frac{4}{7} + \frac{6}{7} - \frac{6}{7}$$

$$= \frac{4}{7}.$$

The directional derivative plays the role of an ordinary derivative when we want to estimate how much a function f changes if we move a small distance ds from a point P_0 to a nearby point. If f were a function of a single variable, we would have

$$df = f'(P_0)ds. \qquad \text{(ordinary derivative} \times \text{increment)}$$

For a function of two or more variables, we use the formula

$$df = \nabla f|_{P_0} \cdot \mathbf{u}\, ds, \qquad \text{(directional derivative} \times \text{increment)} \qquad (6)$$

where $\mathbf{u}$ is the direction of the motion away from P_0.

EXAMPLE 2 Estimate how much

$$f(x, y, z) = xe^y + yz$$

will change if the point $P(x, y, z)$ is moved from $P_0(2, 0, 0)$ straight toward $P_1(4, 1, -2)$ a distance of $ds = 0.1$ units.

Solution We first find the derivative of f at P_0 in the direction of the vector

$$\overrightarrow{P_0P_1} = 2\mathbf{i} + \mathbf{j} - 2\mathbf{k}.$$

The direction of this vector is

$$\mathbf{u} = \frac{\overrightarrow{P_0P_1}}{|\overrightarrow{P_0P_1}|} = \frac{\overrightarrow{P_0P_1}}{3} = \frac{2}{3}\mathbf{i} + \frac{1}{3}\mathbf{j} - \frac{2}{3}\mathbf{k}.$$

The gradient of f at P_0 is

$$\nabla f|_{(2, 0, 0)} = (e^y\mathbf{i} + (xe^y + z)\mathbf{j} + y\mathbf{k})|_{(2, 0, 0)}$$

$$= \mathbf{i} + 2\mathbf{j}.$$

Therefore,

$$\nabla f|_{P_0} \cdot \mathbf{u} = (\mathbf{i} + 2\mathbf{j}) \cdot \left(\frac{2}{3}\mathbf{i} + \frac{1}{3}\mathbf{j} - \frac{2}{3}\mathbf{k}\right) = \frac{2}{3} + \frac{2}{3} = \frac{4}{3}.$$

The change df in f that results from moving $ds = 0.1$ units away from P_0 in the direction of $\mathbf{u}$ is approximately

$$df = \nabla f|_{P_0} \cdot \mathbf{u}\; ds = \left(\frac{4}{3}\right)(0.1) \approx 0.13.$$

We will say more about estimating change in Article 17.2.

If we evaluate the dot product in the formula for the directional derivative, to obtain

$$D_\mathbf{u}f = \nabla f \cdot \mathbf{u} = |\nabla f|\,|\mathbf{u}|\cos\theta = |\nabla f|\cos\theta, \tag{7}$$

the following facts come to light:

Properties of the Directional Derivative $D_\mathbf{u}f = \nabla f \cdot \mathbf{u} = |\nabla f|\cos\theta$

1. The directional derivative has its largest positive value when $\cos\theta = 1$, or when $\mathbf{u}$ is the direction of the gradient. That is, f increases most rapidly at any point in its domain in the direction of ∇f. The derivative in this direction is

$$D_\mathbf{u}f = |\nabla f|\cos(0) = |\nabla f|.$$

2. Similarly, f decreases most rapidly in the direction of $-\nabla f$. The derivative in this direction is

$$D_\mathbf{u}f = |\nabla f|\cos(\pi) = -|\nabla f|.$$

3. $D_{-\mathbf{u}}f = \nabla f \cdot (-\mathbf{u}) = -\nabla f \cdot \mathbf{u} = -D_\mathbf{u}f$. That is, the derivative of f in the direction of $-\mathbf{u}$ is the negative of the derivative of f in the direction of $\mathbf{u}$.

4. The relationships of the partial derivatives of f to the directional derivative are

$$D_\mathbf{i}f = \nabla f \cdot \mathbf{i} = f_x, \qquad D_\mathbf{j}f = \nabla f \cdot \mathbf{j} = f_y, \qquad D_\mathbf{k}f = \nabla f \cdot \mathbf{k} = f_z.$$

Thus,

$$f_x = \text{derivative of } f \text{ in the } \mathbf{i} \text{ direction,}$$

$$f_y = \text{derivative of } f \text{ in the } \mathbf{j} \text{ direction,}$$

$$f_z = \text{derivative of } f \text{ in the } \mathbf{k} \text{ direction.}$$

Combining these results with Step 3 gives

$$D_{-\mathbf{i}}f = -f_x, \qquad D_{-\mathbf{j}}f = -f_y, \qquad D_{-\mathbf{k}}f = -f_z.$$

5. Any direction $\mathbf{u}$ normal (perpendicular) to the gradient is a direction of zero change in f because

$$D_{\mathbf{u}}f = |\nabla f| \cos(\pi/2) = |\nabla f| \cdot 0 = 0.$$

EXAMPLE 3   Find the directions in which $f(x, y, z) = x^2 + y^2 + z^2$ increases and decreases most rapidly at the point $(1, 1, 1)$ in its domain. At what rates does f change in these directions?

Solution   The most rapid increase is in the direction of

$$\nabla f = 2x\,\mathbf{i} + 2y\,\mathbf{j} + 2z\,\mathbf{k}.$$

At $(1, 1, 1)$ we have

$$\nabla f = 2\mathbf{i} + 2\mathbf{j} + 2\mathbf{k},$$

$$|\nabla f| = \sqrt{4 + 4 + 4} = 2\sqrt{3}.$$

The direction of ∇f is

$$\mathbf{u} = \frac{\nabla f}{|\nabla f|} = \frac{1}{\sqrt{3}}(\mathbf{i} + \mathbf{j} + \mathbf{k}).$$

The derivative of f in this direction is

$$|\nabla f| = 2\sqrt{3}.$$

The direction of most rapid decrease at $(1, 1, 1)$ is

$$-\mathbf{u} = -\frac{1}{\sqrt{3}}(\mathbf{i} + \mathbf{j} + \mathbf{k})$$

and the derivative of f in this direction is

$$-|\nabla f| = -2\sqrt{3}.$$

Results for Functions of Two Variables

For functions of two variables, we get results much like the ones for functions of three variables. The two-variable formulas are obtained by dropping the z-terms from the three-variable formulas. Thus, for a function $f(x, y)$ and a unit vector $\mathbf{u} = u_1\mathbf{i} + u_2\mathbf{j}$,

$$\nabla f = \frac{\partial f}{\partial x}\mathbf{i} + \frac{\partial f}{\partial y}\,\mathbf{j}, \tag{8}$$

and

$$D_{\mathbf{u}}f = \nabla f \cdot \mathbf{u} = \frac{\partial f}{\partial x}u_1 + \frac{\partial f}{\partial y}u_2. \qquad (9)$$

EXAMPLE 4

a) Find the derivative of

$$f(x, y) = x^2 + y^2$$

at the point $P_0(1, 1)$ in the direction of the unit vector $\mathbf{u} = u_1\mathbf{i} + u_2\mathbf{j}$.

b) In what direction in its domain (the xy-plane) does f increase most rapidly at $(1, 1)$? What is the derivative of f in this direction?

c) Find the directions in which the derivative of f at $(1, 1)$ is zero.

Solution

a) We have

$$f(x, y) = x^2 + y^2$$

$$f_x(1, 1) = 2x\big|_{(1, 1)} = 2,$$

$$f_y(1, 1) = 2y\big|_{(1, 1)} = 2,$$

$$(\nabla f)_{(1, 1)} = f_x(1, 1)\mathbf{i} + f_y(1, 1)\mathbf{j} = 2\mathbf{i} + 2\mathbf{j},$$

and

$$(D_{\mathbf{u}}f)_{(1, 1)} = (\nabla f)_{(1, 1)} \cdot \mathbf{u} = (2\mathbf{i} + 2\mathbf{j}) \cdot \mathbf{u} = 2u_1 + 2u_2.$$

The derivative of f at $(1, 1)$ in the direction of $\mathbf{u}$ is $2u_1 + 2u_2$.

b) The function increases most rapidly at $(1, 1)$ in the direction of ∇f at $(1, 1)$, which is

$$\left(\frac{\nabla f}{|\nabla f|}\right)_{(1, 1)} = \frac{2\mathbf{i} + 2\mathbf{j}}{\sqrt{2^2 + 2^2}} = \frac{2}{\sqrt{8}}\mathbf{i} + \frac{2}{\sqrt{8}}\mathbf{j} = \frac{1}{\sqrt{2}}\mathbf{i} + \frac{1}{\sqrt{2}}\mathbf{j}.$$

The derivative of f in this direction at $(1, 1)$ is

$$|\nabla f|_{(1, 1)} = 2\sqrt{2}.$$

See Fig. 16.22.

c) The derivative of f is zero in the directions perpendicular to ∇f. We can obtain one of these directions by interchanging the components of $(\nabla f/|\nabla f|)_{(1, 1)}$ and changing the sign of the new first component (see Example 7, Article 13.1). This gives

$$\mathbf{n} = -\frac{1}{\sqrt{2}}\mathbf{i} + \frac{1}{\sqrt{2}}\mathbf{j}.$$

As a check, we calculate $\nabla f \cdot \mathbf{n}$ to see that it is zero:

$$\nabla f \cdot \mathbf{n} = (2\mathbf{i} + 2\mathbf{j}) \cdot \left(-\frac{1}{\sqrt{2}}\mathbf{i} + \frac{1}{\sqrt{2}}\mathbf{j}\right) = -\frac{2}{\sqrt{2}} + \frac{2}{\sqrt{2}} = 0.$$

The other direction of zero change in f at $(1, 1)$ is

$$-\mathbf{n} = \frac{1}{\sqrt{2}}\mathbf{i} - \frac{1}{\sqrt{2}}\mathbf{j}. \qquad \blacksquare$$

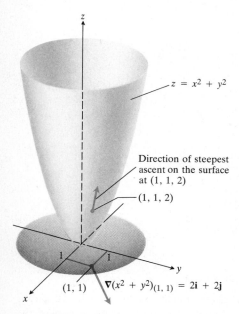

16.22 The gradient of $f(x, y) = x^2 + y^2$ at the point $(1, 1)$ is the vector $\nabla f = 2\mathbf{i} + 2\mathbf{j}$ in the xy-plane. This vector points in the direction in which f increases most rapidly at $(1, 1)$. This corresponds to the direction of steepest ascent on the surface $z = x^2 + y^2$ at the point $(1, 1, 2)$.

Algebraic Properties of Gradient Vectors

If $f(x, y, z)$ and $g(x, y, z)$ have partial derivatives, then the constant multiple kf, the sum $f + g$, and the product fg have gradient vectors, and

1. $\nabla(kf) = k\nabla f$ (any number k),
2. $\nabla(f + g) = \nabla f + \nabla g$, (10)
3. $\nabla(fg) = f\nabla g + g\nabla f$.

The proofs are left as exercises.

EXAMPLE 5 If

$$f(x, y, z) = e^x, \qquad g(x, y, z) = y - z,$$

then

$$\nabla f = e^x \mathbf{i}, \qquad \nabla g = \mathbf{j} - \mathbf{k}.$$

Therefore,

1. $\nabla(2f) = \nabla(2e^x) = 2e^x \mathbf{i} = 2\nabla f$,
2. $\nabla(f + g) = \nabla(e^x + y - z) = e^x \mathbf{i} + \mathbf{j} - \mathbf{k} = \nabla f + \nabla g$,
3. $\nabla(fg) = \nabla(ye^x - ze^x)$
   $= (ye^x - ze^x)\mathbf{i} + e^x\mathbf{j} - e^x\mathbf{k}$
   $= (y - z)e^x\mathbf{i} + e^x(\mathbf{j} - \mathbf{k})$
   $= g\nabla f + f\nabla g.$ ∎

PROBLEMS

In Problems 1–8, find ∇f at the given point.

1. $f(x, y, z) = x^2 + y^2 - 2z^2$, $(1, 1, 1)$
2. $f(x, y, z) = 2z^3 - 3(x^2 + y^2)z$, $(1, 1, 1)$
3. $f(x, y, z) = e^{-2y}\cos 2x$, $(\pi/4, 0, 0)$
4. $f(x, y, z) = \ln \sqrt{x^2 + y^2}$, $(3, 4, 0)$
5. $f(x, y, z) = (x^2 + y^2 + z^2)^{-1/2}$, $(1, 2, -2)$
6. $f(x, y, z) = e^{x+y}\cos z$, $(0, 0, \pi/6)$
7. $f(x, y, z) = \cos x \cos y \sinh z$, $(0, \pi/4, 0)$
8. $f(x, y, z) = \cosh xyz$, $(0, 0, 0)$

In Problems 9–18, find the derivative of f at P_0 in the direction of $\mathbf{A}$.

9. $f(x, y) = x^2 + y^2$, $P_0(1, 0)$, $\mathbf{A} = \mathbf{i} - \mathbf{j}$
10. $f(x, y) = e^x \sin \pi y$, $P_0(0, 1)$, $\mathbf{A} = 4\mathbf{i} + 4\mathbf{j}$
11. $f(x, y) = \cos xy$, $P_0(2, \pi/4)$, $\mathbf{A} = 4\mathbf{i} - \mathbf{j}$
12. $f(x, y) = \dfrac{x - y^2}{x}$, $P_0(1, 1)$, $\mathbf{A} = 12\mathbf{i} + 5\mathbf{j}$
13. $f(x, y) = x^2 + 2xy - 3y^2$, $P_0(1/2, 1/2)$, $\mathbf{A} = \sqrt{3}\mathbf{i} + \mathbf{j}$
14. $f(x, y) = x \tan^{-1}(y/x)$, $P_0(1, 1)$, $\mathbf{A} = 2\mathbf{i} - \mathbf{j}$
15. $f(x, y, z) = xy + yz + zx$, $P_0(1, -1, 2)$, $\mathbf{A} = 10\mathbf{i} + 11\mathbf{j} - 2\mathbf{k}$
16. $f(x, y, z) = x^2 + 2y^2 + 3z^2$, $P_0(1, 1, 1)$, $\mathbf{A} = \mathbf{i} + \mathbf{j} + \mathbf{k}$

17. $f(x, y, z) = e^x \cos yz$, $P_0(0, 0, 0)$, $\mathbf{A} = 2\mathbf{i} + \mathbf{j} - 2\mathbf{k}$
18. $f(x, y, z) = \cos xy + e^{yz} + \ln zx$, $P_0(1, 0, 1/2)$, $\mathbf{A} = \mathbf{i} + 2\mathbf{j} + 2\mathbf{k}$

In Problems 19–24, find the directions in which f increases and decreases most rapidly at P_0 and the rates at which f changes in these directions.

19. $f(x, y) = x^2 + \cos xy$, $P_0(1, 0)$
20. $f(x, y, z) = e^{xy} + z^2$, $P_0(0, 2, 3)$
21. $f(x, y, z) = (x + y - 2)^2 + (3x - y - 6)^2$, $P_0(1, 1, 0)$
22. $f(x, y) = x^2 + xy + y^2$, $P_0(-1, 1)$
23. $f(x, y, z) = (x + y)^2 + (y + z)^2 + (z + x)^2$, $P_0(2, -1, 2)$
24. $f(x, y, z) = z \ln(x^2 + y^2 - 1)$, $P_0(1, 1, 1)$
25. Estimate how much the value of $f(x, y) = \cos \pi xy + xy^2$ will change if (x, y) is moved from $(-1, -1)$ a distance of $ds = 0.1$ units in the direction of $\mathbf{i} + \mathbf{j}$.
26. By about how much will

    $$f(x, y, z) = \ln \sqrt{x^2 + y^2 + z^2}$$

    change if the point (x, y, z) is moved from $(3, 4, 12)$ a distance of $ds = 0.1$ units in the direction of $3\mathbf{i} + 6\mathbf{j} - 2\mathbf{k}$?
27. By about how much will

    $$f(x, y, z) = e^x \cos yz$$

change as the point $P(x, y, z)$ moves from the origin at a distance of $ds = 0.1$ units in the direction of $2\mathbf{i} + \mathbf{j} - 2\mathbf{k}$?

28. By about how much will

$$f(x, y, z) = x + x \cos z - y \sin z + y$$

change if the point $P(x, y, z)$ moves from $P_0(2, -1, 0)$ a distance of $ds = 0.2$ units toward the point $P_1(0, 1, 2)$?

29. In which two directions is the derivative of $f(x, y) = xy + y^2$ at the point $P_0(2, 5)$ equal to zero?

30. In which two directions is the derivative of $f(x, y) = (x^2 - y^2)/(x^2 + y^2)$ at $P_0(1, 1)$ equal to zero?

31. The derivative of $f(x, y)$ at $P_0(1, 2)$ in the direction of the vector $\mathbf{i} + \mathbf{j}$ is $2\sqrt{2}$ and in the direction of the vector $-2\mathbf{j}$ is -3. What is the derivative of f in the direction of $-\mathbf{i} - 2\mathbf{j}$?

32. The derivative of $f(x, y, z)$ at a given point P is greatest in the direction of the vector $\mathbf{A} = \mathbf{i} + \mathbf{j} - \mathbf{k}$. In this direction the value of the derivative is $2\sqrt{3}$.
a) Find ∇f at P.
b) Find the derivative of f at P in the direction of $\mathbf{i} + \mathbf{j}$.

33. Find the derivative of $f(x, y, z) = xyz$ in the direction of the velocity vector of the helix

$$\mathbf{R}(t) = (\cos 3t)\mathbf{i} + (\sin 3t)\mathbf{j} + 3t\mathbf{k}$$

at time $t = \pi/3$.

34. For the function $f(x, y) = x^2y + 2y^2x$ at the point $P_0(1, 3)$, find
a) the direction of greatest increase in f,
b) the derivative of f in the direction of greatest increase in f,
c) the direction of greatest decrease in f,
d) the directions in which the derivative of f is zero.

35. Suppose cylindrical coordinates r, θ, z are introduced into a function $w = f(x, y, z)$ to yield $w = F(r, \theta, z)$ and let $\mathbf{u}_r$ and $\mathbf{u}_\theta$ be the unit vectors

$$\mathbf{u}_r = (\cos \theta)\mathbf{i} + (\sin \theta)\mathbf{j}$$
$$\mathbf{u}_\theta = -(\sin \theta)\mathbf{i} + (\cos \theta)\mathbf{j}.$$

Show that

$$\nabla w = \frac{\partial w}{\partial r}\mathbf{u}_r + \frac{1}{r}\frac{\partial w}{\partial \theta}\mathbf{u}_\theta + \frac{\partial w}{\partial z}\mathbf{k}.$$

(*Hint:* The scalar component of ∇w in the direction of $\mathbf{u}_r$ is equal to the derivative of w in that direction. But this is precisely $\partial w/\partial r$. Reason similarly for the scalar components of ∇w in the directions of $\mathbf{u}_\theta$ and $\mathbf{k}$.)

36. Verify the result in Problem 35 by transforming the right-hand side of the equation for ∇w into $\mathbf{i}$, $\mathbf{j}$, $\mathbf{k}$ components, replacing the cylindrical coordinates r, θ by Cartesian coordinates x, y, and using the Chain Rule.

37. Suppose spherical coordinates ρ, ϕ, θ are introduced into a function $w = f(x, y, z)$ to produce $w = F(\rho, \phi, \theta)$. Express ∇w in terms of spherical coordinates and the unit vectors

$$\mathbf{u}_\rho = (\sin \phi \sin \theta)\mathbf{i} + (\sin \phi \sin \theta)\mathbf{j} + (\cos \phi)\mathbf{k},$$
$$\mathbf{u}_\phi = (\cos \phi \cos \theta)\mathbf{i} + (\cos \phi \sin \theta)\mathbf{j} - (\sin \phi)\mathbf{k},$$
$$\mathbf{u}_\theta = -(\sin \theta)\mathbf{i} + (\cos \theta)\mathbf{j}.$$

Use a geometrical argument to determine the scalar component of ∇w in the directions of each of these vectors. See the hint for Problem 35.

38. Verify the expression you obtained for ∇w in Problem 37 by transforming it back into Cartesian coordinates. Use the Chain Rule.

39. Verify the gradient formulas in Eq. (10).

16.7
Tangent Planes and Normal Lines

Gradients provide a natural and efficient way to define tangent planes and normal lines at points on surfaces. In this article, we show how this is done.

Tangent Planes and Normal Lines for the Surface $f(x, y, z) = c$

Suppose that $f(x, y, z)$ and its partial derivatives are continuous, that

$$x = x(t), \qquad y = y(t), \qquad z = z(t)$$

is a differentiable curve that lies on the surface S defined by the equation $f(x, y, z) = c$, and that this curve passes through the point P_0 on the surface. Then

$$f(x(t), y(t), z(t)) = c$$

for every value of t. Taking the derivative of both sides of the equation for S

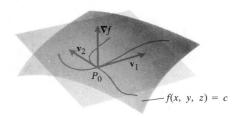

16.23 ∇f is perpendicular to the velocity vector of every differentiable curve in the surface through P_0. The velocity vectors at P_0 therefore lie in a common plane, which we call the tangent plane at P_0.

with respect to t gives

$$\frac{\partial f}{\partial x}\frac{dx}{dt} + \frac{\partial f}{\partial y}\frac{dy}{dt} + \frac{\partial f}{\partial z}\frac{dz}{dt} = 0 \tag{1}$$

or

$$\nabla f \cdot \mathbf{v} = 0, \tag{2}$$

where $\mathbf{v}$ is the velocity vector of the curve.

We conclude from Eq. (2) that the gradient of f is orthogonal to the velocity vector of every differentiable curve on S that passes through P_0. The gradient is therefore perpendicular to the tangent line of every differentiable curve on S that passes through P_0. These tangent lines therefore lie in a common plane, namely the plane through P_0 normal to ∇f. We define this plane to be the **tangent plane of the surface** $f(x, y, z) = c$ at the point P_0 (Fig. 16.23). Its equation is

$$f_x(P_0)(x - x_0) + f_y(P_0)(y - y_0) + f_z(P_0)(z - z_0) = 0. \tag{3}$$

The line normal to the surface at P_0 is defined to be the line normal to the tangent plane at P_0. Its parametric equations in terms of ∇f may be taken as

$$x = x_0 + f_x(P_0)t, \qquad y = y_0 + f_y(P_0)t, \qquad z = z_0 + f_z(P_0)t. \tag{4}$$

EXAMPLE 1 Find equations for the tangent plane and normal line of the surface

$$f(x, y, z) = x^2 + y^2 + z - 9 = 0 \qquad \text{(a circular paraboloid)}$$

at the point $P_0(1, 2, 4)$.

Solution The surface is shown in Fig. 16.24.

The tangent plane is the plane through P_0 perpendicular to the gradient of f at P_0. The gradient is

$$\nabla f_{P_0} = (2x\mathbf{i} + 2y\mathbf{j} + \mathbf{k})_{(1, 2, 4)} = 2\mathbf{i} + 4\mathbf{j} + \mathbf{k}.$$

The plane is therefore the plane

$$2(x - 1) + 4(y - 2) + (z - 4) = 0$$

or

$$2x + 4y + z = 14.$$

The line normal to the surface at P_0 is

$$x = 1 + 2t, \qquad y = 2 + 4t, \qquad z = 4 + t.$$

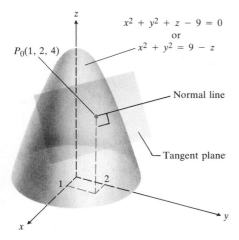

$x^2 + y^2 + z - 9 = 0$
or
$x^2 + y^2 = 9 - z$

$P_0(1, 2, 4)$

Normal line

Tangent plane

16.24 The tangent plane and normal line to the surface $x^2 + y^2 + z - 9 = 0$ at $P_0(1, 2, 4)$.

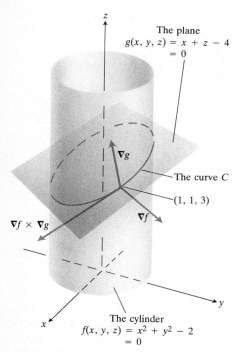

16.25 The cylinder $f(x, y, z) = x^2 + y^2 - 2 = 0$ and the plane $g(x, y, z) = x + z - 4 = 0$ intersect in a curve C. Any vector tangent to C will automatically be orthogonal to ∇f and ∇g and hence parallel to $\nabla f \times \nabla g$. This information enables us to write an equation for the tangent line in Example 2.

EXAMPLE 2   The surfaces

$$f(x, y, z) = x^2 + y^2 - 2 = 0 \qquad \text{(a cylinder)}$$

and

$$g(x, y, z) = x + z - 4 = 0 \qquad \text{(a plane)}$$

meet in a curve C (Fig. 16.25). Find parametric equations for the line tangent to C at the point $P_0(1, 1, 3)$.

Solution   The tangent line is perpendicular to both ∇f and ∇g at P_0. The vector $\mathbf{v} = \nabla f \times \nabla g$ will therefore be parallel to the line. We use the components of $\mathbf{v}$ and the coordinates of P_0 to write equations for the line. We have

$$\nabla f_{(1, 1, 3)} = (2x\mathbf{i} + 2y\mathbf{j})_{(1, 1, 3)} = 2\mathbf{i} + 2\mathbf{j}$$

and

$$\nabla g_{(1, 1, 3)} = (\mathbf{i} + \mathbf{k})_{(1, 1, 3)} = \mathbf{i} + \mathbf{k}.$$

Hence,

$$\mathbf{v} = (2\mathbf{i} + 2\mathbf{j}) \times (\mathbf{i} + \mathbf{k}) = \begin{vmatrix} \mathbf{i} & \mathbf{j} & \mathbf{k} \\ 2 & 2 & 0 \\ 1 & 0 & 1 \end{vmatrix} = 2\mathbf{i} - 2\mathbf{j} - 2\mathbf{k}.$$

The line is

$$x = 1 + 2t, \qquad y = 1 - 2t, \qquad z = 3 - 2t. \qquad \blacksquare$$

Analogous Results for Functions of Two Variables

Just as the gradient of $f(x, y, z)$ is perpendicular to the level surfaces $f(x, y, z) = c$, the gradient of a function $f(x, y)$ of two variables is perpendicular to the level curves $f(x, y) = c$. (See Miscellaneous Problem 57.)

EXAMPLE 3   Find a vector normal to the level curve

$$\frac{x^2}{4} + y^2 = 2$$

of the function $f(x, y) = (x^2/4) + y^2$ at the point $P_0(-2, 1)$. Write an equation for the tangent to the curve at P_0. (See Fig. 16.26.)

Solution   The gradient vector

$$\nabla f_{(-2, 1)} = \left(\frac{x}{2}\mathbf{i} + 2y\mathbf{j} \right)_{(-2, 1)} = -\mathbf{i} + 2\mathbf{j}$$

is normal to the curve at the point $(-2, 1)$. The tangent at this point is the line

$$-1(x + 2) + 2(y - 1) = 0, \qquad \text{or} \qquad x - 2y = -4. \qquad \blacksquare$$

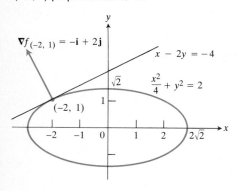

16.26 The tangent to the level curve $f(x, y) = 2$ of the function $f(x, y) = (x^2/4) + y^2$ at the point $(-2, 1)$ is the line through $(-2, 1)$ perpendicular to ∇f.

PROBLEMS

Each of Problems 1–18 gives a point P_0 on a surface $f(x, y, z) = c$. Find equations for (a) the tangent plane at P_0 and (b) the line normal to the surface at P_0.

1. $x^2 + y^2 + z^2 = 3$, $P_0(1, 1, 1)$
2. $x^2 + y^2 - z^2 = 18$, $P_0(3, 5, -4)$
3. $z^2 - x^2 - y^2 = 0$, $P_0(3, 4, -5)$
4. $2z - x^2 = 0$, $P_0(2, 0, 2)$
5. $z - \ln(x^2 + y^2) = 0$, $P_0(1, 0, 0)$
6. $(x + y)^2 + z^2 = 25$, $P_0(1, 2, 4)$
7. $x^2 + 2xy - y^2 + z^2 = 7$, $P_0(1, -1, 3)$
8. $\cos \pi x - x^2 y + e^{xz} + yz = 4$, $P_0(0, 1, 2)$
9. $x^2 - xy - y^2 - z = 0$, $P_0(1, 1, -1)$
10. $\tan^{-1}(y/x) - z = 0$, $P_0(1, 1, \pi/4)$
11. $\cos(\pi x/2) - z = 0$, $P_0(1, 0, 0)$
12. $x + y + z = 1$, $P_0(0, 1, 0)$
13. $x^2 + y^2 - 2xy - x + 3y - z = -4$, $P_0(2, -3, 18)$
14. $x - y^3 + zy - 2z^2 = 0$, $P_0(-4, -2, 1)$
15. $x + y^2 + z^2 = 1$, $P_0(0, 0, 1)$
16. $x^2 + y + 4z^2 = 4$, $P_0(0, 0, 1)$
17. $\sin x - y = 0$, $P_0(0, 0, 0)$
18. $(x + y)/(xy - 1) - z = 0$, $P_0(1, 2, 3)$

In Problems 19–28, sketch the surface together with its gradient vector at the given point.

19. $x^2 + y^2 - z = 0$, $(1, 1, 2)$
20. $x^2 + y^2 - z^2 = 0$, $(2, -1, -\sqrt{5})$
21. $x^2 + y^2 + z = 4$, $(-1, -1, 2)$
22. $x^2 + y^2 + z^2 = 9$, $(1, 2, 2)$
23. $x^2 + y^2 = 4$, $(-1, \sqrt{3}, 1)$
24. $x^2 + z^2 = 2$, $(1, -1, 1)$
25. $4y^2 + 9z^2 = 36$, $(0, 3/2, \sqrt{3})$
26. $3/\sqrt{x^2 + y^2 + z^2} = 1$, $(1, 2, -2)$
27. $x^2 + y^2 - z^2 = 1$, $(1, 1, -1)$
28. $2x + 3y + 6z = 18$, $(3, 2, 1)$

In Problems 29–34, find parametric equations for the line that is tangent to the curve of intersection of the surfaces at the given point.

29. Surfaces: $x + y^2 + 2z = 4$, $x = 1$
    Point: $(1, 1, 1)$

30. Surfaces: $xyz = 1$, $x^2 + 2y^2 + 3z^2 = 6$
    Point: $(1, 1, 1)$

31. Surfaces: $x^2 + 2y + 2z = 4$, $y = 1$
    Point: $(1, 1, 1/2)$

32. Surfaces: $x + y^2 + z = 2$, $y = 1$
    Point: $(1/2, 1, 1/2)$

33. Surfaces: $x^3 + 3x^2 y^2 + y^3 + 4xy - z^2 = 0$,
        $x^2 + y^2 + z^2 = 11$
    Point: $(1, 1, 3)$

34. Surfaces: $x^2 + y^2 = 4$, $x^2 + y^2 - z = 0$
    Point: $(\sqrt{2}, \sqrt{2}, 4)$

In Problems 35–40, sketch the level curve $f(x, y) = c$. Then find a vector normal to the curve at P_0 and an equation for the line tangent to the curve at P_0.

35. $x^2 + y^2 = 4$, $P_0(\sqrt{2}, \sqrt{2})$
36. $x^2 - y^2 = 3$, $P_0(2, -1)$
37. $4x^2 + 9y^2 = 36$, $P_0(2, 2\sqrt{5}/3)$
38. $x^2 - y = 1$, $P_0(\sqrt{2}, 1)$
39. $xy = -4$, $P_0(2, -2)$
40. $x^2 + xy + y^2 = 7$, $P_0(1, 2)$

(This was Example 5 in Article 2.3.)

41. Find the points on the surface
$$(y + z)^2 + (z - x)^2 = 16$$
where the normal line is parallel to the yz-plane.

42. Find the points on the surface
$$xy + yz + zx - x - z^2 = 0$$
where the tangent plane is parallel to the xy-plane.

43. Show that the curve
$$\mathbf{R}(t) = \sqrt{t}\,\mathbf{i} + \sqrt{t}\,\mathbf{j} - \frac{1}{4}(t + 3)\mathbf{k}$$
is normal to the surface $x^2 + y^2 - z = 1$ when $t = 1$. That is, show that the curve's velocity vector is a scalar multiple of the gradient of the function $f(x, y, z) = x^2 + y^2 - z$ at that point.

44. Show that the curve
$$\mathbf{R}(t) = \sqrt{t}\,\mathbf{i} + \sqrt{t}\,\mathbf{j} + (2t - 1)\mathbf{k}$$
is tangent to the surface $f(x, y, z) = x^2 + y^2 - z = 1$ when $t = 1$. That is, show that $d\mathbf{R}/dt$ is orthogonal to ∇f at the point of contact.

16.8
Higher Order Derivatives. Partial Differential Equations from Physics

The derivatives we obtain by repeated partial differentiation are called higher order derivatives (meaning higher than first order). These derivatives appear in the equations that express important physical laws for wave motion, heat flow, and gravitation. They also appear in tests that help to identify the maxima and minima of functions of more than one variable, as we shall see in the next chapter.

Differential equations that have partial derivatives in them are usually hard to solve, in part because their solutions can take on so many different forms. You will not be asked to solve any of the partial differential equations you are about to see, but only to verify that given functions are solutions of particular equations. There are still benefits to be gained from this: a chance to become acquainted with important equations and an opportunity to practice the Chain Rule in another important professional context.

Second Order Derivatives

When we differentiate a function $f(x, y)$ twice, we produce its second order derivatives. These derivatives are usually denoted by

$$\frac{\partial^2 f}{\partial x^2}, \qquad \frac{\partial^2 y}{\partial y^2}, \qquad \frac{\partial^2 f}{\partial x\, \partial y}, \qquad \frac{\partial^2 f}{\partial y\, \partial x}$$

or

$$f_{xx}, \qquad f_{yy}, \qquad f_{yx}, \qquad f_{xy}$$

and defined by the equations

$$\frac{\partial^2 f}{\partial x^2} = \frac{\partial}{\partial x}\left(\frac{\partial f}{\partial x}\right), \qquad \frac{\partial^2 f}{\partial x\, \partial y} = \frac{\partial}{\partial x}\left(\frac{\partial f}{\partial y}\right),$$

and so on. Notice the order in which the derivatives are taken:

$$\frac{\partial^2 f}{\partial x\, \partial y} \qquad \text{Differentiate first with respect to } y, \text{ then with respect to } x.$$

$$f_{yx} \qquad \text{Means the same thing.}$$

EXAMPLE 1   If

$$f(x, y) = x \cos y + y e^x,$$

then

$$\frac{\partial f}{\partial x} = \cos y + y e^x,$$

$$\frac{\partial^2 f}{\partial y\, \partial x} = \frac{\partial}{\partial y}\left(\frac{\partial f}{\partial x}\right) = -\sin y + e^x,$$

$$\frac{\partial^2 f}{\partial x^2} = \frac{\partial}{\partial x}\left(\frac{\partial f}{\partial x}\right) = y e^x,$$

$$\frac{\partial^3 f}{\partial x^3} = \frac{\partial}{\partial x}\left(\frac{\partial^2 f}{\partial x^2}\right) = ye^x,$$

$$\frac{\partial^3 f}{\partial y\,\partial x^2} = \frac{\partial}{\partial y}\left(\frac{\partial^2 f}{\partial x^2}\right) = e^x,$$

while

$$\frac{\partial f}{\partial y} = -x\,\sin y + e^x,$$

$$\frac{\partial^2 f}{\partial x\,\partial y} = \frac{\partial}{\partial x}\left(\frac{\partial f}{\partial y}\right) = -\sin y + e^x,$$

$$\frac{\partial^2 f}{\partial y^2} = \frac{\partial}{\partial y}\left(\frac{\partial f}{\partial y}\right) = -x\,\cos y,$$

$$\frac{\partial^3 f}{\partial^2 x\,\partial y} = \frac{\partial}{\partial x}\left(\frac{\partial^2 f}{\partial x\,\partial y}\right) = e^x. \qquad\blacksquare$$

The Mixed Derivative Theorem

You may have noticed in Example 1 that the "mixed" second order derivatives

$$\frac{\partial^2 f}{\partial y\,\partial x} \qquad \text{and} \qquad \frac{\partial^2 f}{\partial x\,\partial y}$$

were equal. This was no mere coincidence. The mixed partial derivatives f_{xy} and f_{yx} have to be equal at any point at which f, f_x, f_y, f_{xy}, and f_{yx} are continuous. We call this fact the mixed derivative theorem.

THEOREM 2

> **The Mixed Derivative Theorem**
>
> If $f(x, y)$ and its partial derivatives f_x, f_y, f_{xy}, f_{yx} are defined in a region containing a point (a, b) and are all continuous at (a, b), then
>
> $$f_{xy}(a, b) = f_{yx}(a, b). \tag{1}$$

A proof of Theorem 2 is given in Appendix 9.

EXAMPLE 2 Find z_{xy} if

$$z = f(u, v), \qquad u = x + y, \qquad v = xy,$$

and f and its partial derivatives are continuous.

Solution We first find z_x with the Chain Rule:

$$z_x = \frac{\partial}{\partial x}f(u, v) = f_u u_x + f_v v_x$$

$$= f_u(1) + f_v(y) \tag{2}$$

$$= f_u + yf_v.$$

We then take the partial derivative of both sides of this equation with respect to

y, treating yf_v as the product of y and the function f_v:

$$z_{xy} = \frac{\partial}{\partial y}(f_u + yf_v)$$

$$= \frac{\partial}{\partial y}(f_u) + y\frac{\partial}{\partial y}(f_v) + f_v\frac{\partial}{\partial y}(y) \qquad (3)$$

$$= \frac{\partial}{\partial y}(f_u) + y\frac{\partial}{\partial y}(f_v) + f_v.$$

We use the Chain Rule again to calculate the two remaining derivatives:

$$\frac{\partial}{\partial y}(f_u) = f_{uu}\frac{\partial u}{\partial y} + f_{uv}\frac{\partial v}{\partial y} = f_{uu} + xf_{uv},$$

$$\frac{\partial}{\partial y}(f_v) = f_{vu}\frac{\partial u}{\partial y} + f_{vv}\frac{\partial v}{\partial y} = f_{vu} + xf_{vv}.$$

With these equalities, we continue the work in Eq. (3):

$$z_{xy} = \frac{\partial}{\partial y}(f_u) + y\frac{\partial}{\partial y}(f_v) + f_v$$

$$= f_{uu} + xf_{uv} + y(f_{vu} + xf_{vv}) + f_v$$

$$= f_{uu} + xf_{uv} + yf_{vu} + xyf_{vv} + f_v \qquad (4)$$

$$= f_{uu} + (x + y)f_{uv} + xyf_{vv} + f_v \qquad (5)$$

$$= f_{uu} + uf_{uv} + vf_{vv} + f_v.$$

We used the equality $f_{vu} = f_{uv}$ to get from (4) to (5).   ∎

EXAMPLE 3   Find d^2w/dt^2 if

$$w = f(x, y), \qquad x = e^t, \qquad y = 2t - 1,$$

and w and its partial derivatives are all continuous.

Solution   We find dw/dt, and then differentiate it with respect to t. From the Chain Rule,

$$\frac{dw}{dt} = f_x\frac{dx}{dt} + f_y\frac{dy}{dt} = f_x \cdot e^t + f_y \cdot 2 = e^tf_x + 2f_y.$$

Therefore,

$$\frac{d^2w}{dt^2} = \frac{d}{dt}(e^tf_x) + \frac{d}{dt}(2f_y)$$

$$= e^tf_x + e^t\frac{d}{dt}(f_x) + 2\frac{d}{dt}(f_y). \qquad (6)$$

We use the Chain Rule again to calculate the two remaining derivatives:

$$\frac{d}{dt}(f_x) = f_{xx}\frac{dx}{dt} + f_{xy}\frac{dy}{dt} = f_{xx}\,e^t + f_{xy} \cdot 2 = e^tf_{xx} + 2f_{xy},$$

$$\frac{d}{dt}(f_y) = f_{yx}\frac{dx}{dt} + f_{yy}\frac{dy}{dt} = f_{yx}\,e^t + f_{yy} \cdot 2 = e^tf_{yx} + 2f_{yy}.$$

Substituting these in Eq. (6) gives

$$\frac{d^2w}{dt^2} = e^t f_x + e^t(e^t f_{xx} + 2f_{xy}) + 2(e^t f_{yx} + 2f_{yy})$$

$$= e^t f_x + e^{2t} f_{xx} + 2e^t f_{xy} + 2e^t f_{yx} + 4f_{yy}$$

$$= e^t f_x + e^{2t} f_{xx} + 4e^t f_{xy} + 4f_{yy}.$$

We used the equality $f_{yx} = f_{xy}$ to complete the last step in this calculation.

∎

Partial Derivatives of Still Higher Order

If we refer once more to Example 1, we note not only that

$$\frac{\partial^2 f}{\partial x \, \partial y} = \frac{\partial^2 f}{\partial y \, \partial x}$$

but also that

$$\frac{\partial^3 f}{\partial x^2 \, \partial y} = \frac{\partial^3 f}{\partial y \, \partial x^2}.$$

This equality may be derived from Theorem 2 as follows:

$$\frac{\partial^3 f}{\partial x^2 \, \partial y} = \frac{\partial}{\partial x}\left(\frac{\partial^2 f}{\partial x \, \partial y}\right) = \frac{\partial}{\partial x}\left(\frac{\partial^2 f}{\partial y \, \partial x}\right)$$

$$= \frac{\partial}{\partial x}\left(\frac{\partial}{\partial y}f_x\right) = \frac{\partial}{\partial y}\left(\frac{\partial}{\partial x}f_x\right)$$

$$= \frac{\partial}{\partial y}\left(\frac{\partial^2 f}{\partial x^2}\right) = \frac{\partial^3 f}{\partial y \, \partial x^2}.$$

In fact, if all the partial derivatives that appear are continuous, the notation

$$\frac{\partial^{m+n} f}{\partial x^m \, \partial y^n}$$

may be used to denote the result of differentiating the function $f(x, y)$ m times with respect to x and n times with respect to y in any order. The final result will be the same whatever the order of differentiation.

Being free to control the order of differentiation can work to our advantage, as the next example shows.

EXAMPLE 4   Calculate

$$\frac{\partial^5}{\partial x^2 \, \partial y^3}(x \sin y + e^y).$$

Solution   We postpone the differentiation with respect to y and differentiate first with respect to x to show without further work that the final result is zero:

$$\frac{\partial^5}{\partial x^2 \, \partial y^3}(x \sin y + e^y) = \frac{\partial^3}{\partial y^3}\frac{\partial^2}{\partial x^2}(x \sin y + e^y)$$

$$= \frac{\partial^3}{\partial y^3}\frac{\partial}{\partial x}(\sin y) = \frac{\partial^3}{\partial y^3}(0) = 0.$$

It takes longer to reach this conclusion if we differentiate first with respect to y.

∎

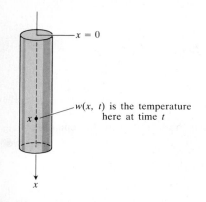

16.27 The temperature distribution in a heat-conducting rod satisfies the equation

$$w_{xx} = \frac{1}{c^2} w_t.$$

The One-dimensional Heat Equation (= Diffusion Equation = Telegraph Equation)

If $w(x, t)$ represents the temperature at position x at time t in a uniform conducting rod with perfectly insulated sides (no heat flow through the sides—see Fig. 16.27), then the partial derivatives w_{xx} and w_t satisfy a differential equation of the following form:

$$w_{xx} = \frac{1}{c^2} w_t. \tag{7}$$

This equation is called the **one-dimensional heat equation.** The value of c^2 is determined by the material from which the rod is made. It has been determined experimentally for a broad range of materials, and for a given application one finds the appropriate value in a table.

Equation (7) can be used effectively to model the seasonal temperature variation beneath the earth's surface, $w(x, t)$ then being the temperature on the tth day of the year x feet below the surface. (One thinks of a vertical core of the earth's surface as an insulated rod.) For dry soil,

$$c^2 = 0.19 \text{ ft}^2/\text{day},$$

and Eq. (7) becomes

$$w_{xx} = \frac{1}{0.19} w_t, \quad \text{or} \quad w_t = 0.19 w_{xx}. \tag{8}$$

The solution that matches the seasonal temperature variation at the earth's surface is

$$w(x, t) = \cos(1.7 \times 10^{-2} t - 0.2x)e^{-0.2x}. \tag{9}$$

This can be verified by calculating w_{xx} and w_t and seeing that $w_t = 0.19 w_{xx}$. The calculation is tedious, however, and we shall omit it. The solution in (9) was graphed in Fig. 16.6.

In chemistry and biochemistry, the heat equation is known as the **diffusion equation.** In this context, $w(x, t)$ represents the concentration of a dissolved substance, a salt for instance, diffusing along a tube filled with liquid. The value of $w(x, t)$ is the concentration at point x at time t. In other applications, $w(x, t)$ represents the diffusion of a gas down a long, thin pipe.

In electrical engineering, the heat equation appears in the forms

$$v_{xx} = RCv_t \tag{10}$$

and

$$i_{xx} = RCi_t, \tag{11}$$

which are known as the **telegraph equations.** These equations describe the voltage v and the flow of current i in a coaxial cable or in any other cable in which leakage and inductance are negligible. The functions and constants in these equations are

$$v(x, t) = \text{voltage at point } x \text{ at time } t,$$

$$R = \text{resistance per unit length},$$

$$C = \text{capacitance to ground per unit of cable length},$$

$$i(x, t) = \text{current at point } x \text{ at time } t.$$

PROBLEMS

In Problems 1–6, find $\partial^2 f/\partial x^2$, $\partial^2 f/\partial y^2$, and $\partial^2 f/\partial y\,\partial x$.

1. $f(x, y) = \ln(x^2 + y^2)$

2. $f(x, y) = e^x \ln(3 - y^2)$

3. $f(x, y) = x^2 y + \cos y + y \sin x$

4. $f(x, y) = (x - y)/xy$

5. $f(x, y, z) = xyz$

6. $f(x, y, z) = xy + yz + zx$

In Problems 7–10, verify that $w_{xy} = w_{yx}$.

7. $w = \ln(2x + 3y)$

8. $w = \tan^{-1}(y/x)$

9. $w = xy^2 + x^2 y^3 + x^3 y^4$

10. $w = e^x \sinh y + \cos(2x - 3y)$

11. Which order of differentiation will calculate f_{xy} faster: x first or y first? Try to answer without writing anything down.

a) $f(x, y) = x \sin y + e^y$

b) $f(x, y) = 1/x$

c) $f(x, y) = y + (x/y)$

d) $f(x, y) = y + x^2 y + 4y^3 - \ln(y^2 + 1)$

e) $f(x, y) = x^2 + 5xy + \sin x + 7e^x$

f) $f(x, y) = x \ln xy$

12. The derivative $\partial^5 f/\partial x^2\,\partial y^3$ is zero for each of the following functions. To show this as quickly as possible, which variable should one differentiate with respect to first: x or y? Try to answer without writing anything down.

a) $f(x, y) = y^2 x^4 e^x + 2$

b) $f(x, y) = y + y^2[\ln(x^2 + 5) + \sin x - 7x^3]$

c) $f(x, y) = xe^y + x \sin y - x \cos y$

d)
$$f(x, y) = x \int_1^y e^{t^2/2}\, dt$$

In Problems 13–18, assume that the necessary derivatives exist and that the functions and derivatives are all continuous.

13. Find w_{xy} if $w = f(u, v)$, $u = x + y$, and $v = xy$. Express the answer in terms of u, v, f_u, and f_v.

14. Find z_{xx} if $z = f(u, v)$, $u = x^2 - y^2$, and $v = 2xy$. Express the answer in terms of x, y, f_u, and f_v.

15. Find u_{ss} if $u = f(x, y)$, $x = r^2 + s^2$, and $y = 2rs$. Express the answer in terms of r, s, f_x, and f_y.

16. Find $\partial^2 w/\partial x\,\partial y$ if $w = f(u, v)$, $u = x + y$, and $v = y^2$. Express the answer in terms of x, y, f_u, and f_v.

17. Find $d^2 w/dt^2$ if $w = f(x, y)$, $x = \sin t$, and $y = t^2$.

18. Let $w = f(u)$, where $u = xg(y)$. Show that
$$w_{xx} = f''(xg(y)) \cdot g^2(y).$$

19. Find the value of $\partial^2 w/\partial\theta^2$ at $r = 2$, $\theta = \pi/2$, if $w = f(x, y)$, $x = r \cos \theta$, $y = r \sin \theta$, and $f_y = f_{xx} = 1$ when $r = 2$, $\theta = \pi/2$. (Be sure to work out all the derivatives before substituting numerical values.)

Laplace Equations

The *three-dimensional Laplace equation*
$$\frac{\partial^2 f}{\partial x^2} + \frac{\partial^2 f}{\partial y^2} + \frac{\partial^2 f}{\partial z^2} = 0 \tag{12}$$

is satisfied by steady-state temperature distributions $T = f(x, y, z)$ in space, by gravitational potentials, and by electrostatic potentials. The *two-dimensional Laplace equation*
$$\frac{\partial^2 f}{\partial x^2} + \frac{\partial^2 f}{\partial y^2} = 0, \tag{13}$$

obtained by dropping the $\partial^2 f/\partial z^2$ term from (12), describes potentials and steady-state temperature distributions in a plane. See Fig. 16.28.

(a)

(b)

Boundary temperatures controlled

16.28 Steady-state temperature distributions in planes and solids satisfy Laplace equations. The plane (a) may be treated as a thin slice of the solid (b) perpendicular to the z-axis.

Show that the functions in Problems 20–27 satisfy a Laplace equation.

20. $f(x, y, z) = x^2 + y^2 - 2z^2$

21. $f(x, y, z) = 2z^3 - 3(x^2 + y^2)z$

22. $f(x, y) = e^{-2y}\cos 2x$

23. $f(x, y) = \ln \sqrt{x^2 + y^2}$

24. $f(x, y, z) = (x^2 + y^2 + z^2)^{-1/2}$

25. $f(x, y, z) = e^{3x+4y} \cos 5z$

26. $f(x, y) = \tan^{-1}(y/x)$

27. $f(x, y, z) = \cos 3x \cos 4y \sinh 5z$

28. Show that if $w = f(u, v)$ satisfies the Laplace equation $f_{uu} + f_{vv} = 0$, and if $u = (x^2 - y^2)/2$, $v = xy$, then $w_{xx} + w_{yy} = 0$.

29. Show that $u = f(x - iy) + g(x + iy)$ is a solution of the equation

$$\frac{\partial^2 u}{\partial x^2} + \frac{\partial^2 u}{\partial y^2} = 0,$$

if all the necessary partial derivatives exist. (Here, $i = \sqrt{-1}$. Let $x - iy = v$ and $x + iy = w$.)

30. For what values of n does

$$f(x, y, z) = (x^2 + y^2 + z^2)^n$$

satisfy the three-dimensional Laplace equation (12)?

31. Find all solutions of the two-dimensional Laplace equation (13) of the form
a) $f(x, y) = ax^2 + bxy + cy^2$,
b) $f(x, y) = ax^3 + bx^2y + cxy^2 + dy^3$.

The Wave Equation

If we stand on an ocean shore and take a snapshot of the waves, the picture shows a regular pattern of peaks and valleys at an instant in time (Fig. 16.29). We see periodic vertical motion in

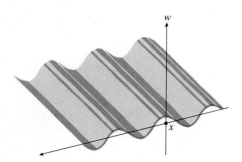

16.29 Waves in water at an instant in time. As time passes,

$$\frac{\partial^2 w}{\partial t^2} = c^2 \frac{\partial^2 w}{\partial x^2}.$$

space, with respect to distance. If we stand in the water, we can feel the rise and fall of the water as the waves go by. We see periodic vertical motion in time. In physics, this beautiful symmetry is expressed by the *one-dimensional wave equation*

$$\frac{\partial^2 w}{\partial t^2} = c^2 \frac{\partial^2 w}{\partial x^2}, \qquad (14)$$

where w is the wave height, x is the distance variable, t is the time variable, and c is the velocity with which the waves are propagated.

In our example, x is the distance across the ocean's surface, but in other applications x might be the distance along a vibrating string, distance through air (sound waves), or distance through space (light waves). The number c varies with the medium and type of wave.

Show that the functions in Problems 32–37 are solutions of the wave equation, Eq. (14).

32. $w = \sin(x + ct)$

33. $w = \cos(2x + 2ct)$

34. $w = \sin(x + ct) + \cos(2x + 2ct)$

35. $w = \ln(2x + 2ct)$

36. $w = \tan(2x - 2ct)$

37. $w = 5 \cos(3x + 3ct) - 7 \sinh(4x - 4ct)$

38. If c is a constant and $w = f(u) + g(v)$, where $u = x + ct$ and $v = x - ct$, show that

$$\frac{\partial^2 w}{\partial t^2} = c^2 \frac{\partial^2 w}{\partial x^2} = c^2(f''(u) + g''(v)),$$

assuming that all the necessary derivatives exist.

The Heat Equation

39. Find all solutions of the one-dimensional heat equation $w_t = c^2 w_{xx}$ of the form $w = e^{rt} \sin \pi x$, r a constant.

40. Find all solutions of the one-dimensional heat equation $w_t = c^2 w_{xx}$ that have the form $w = e^{rt} \sin kx$ and that satisfy the conditions that $w(0, t) = 0$ and $w(L, t) = 0$. What happens to the solution as $t \to \infty$?

41. Let $u = f(y)$ be a differentiable function of y, and let $y = x - tg(u)$, where g is any function of u. Show that u satisfies the equation

$$\frac{\partial u}{\partial t} + g(u)\frac{\partial u}{\partial x} = 0.$$

REVIEW QUESTIONS AND EXERCISES

1. Let $f(x, y)$ be a function of two independent variables defined on a region R in the xy-plane. Describe two ways to display the values of f graphically.

2. Define level surface and level curve.

3. Give two equivalent definitions of

$$\lim_{(x, y) \to (x_0, y_0)} f(x, y) = L.$$

What is the basic theorem for calculating limits of sums, differences, products, and quotients of functions of two variables?

4. When is a function of two variables continuous at a point of its domain? Give an example of a function that is discontinuous at some point(s) of its domain.

5. When is a function of three variables continuous at a point of its domain? Give an example of a function of three independent variables that is continuous at some points of its domain but discontinuous at others.

6. Let $f(x, y)$ be a function of the two independent variables x and y. How are the derivatives $\partial f/\partial x$ and $\partial f/\partial y$ defined? What is their geometric meaning? How are they calculated? Give sample calculations.

7. Give examples of the Chain Rule for partial derivatives of functions of two and three variables. Draw tree diagrams for your examples.

8. Define the gradient of a function $f(x, y, z)$ and describe the role the gradient plays in defining directional derivatives, tangent planes, and normal lines. What is the relation between ∇f and the directions in which f changes most rapidly? What are the analogous results for functions of two variables? Give examples.

9. What is the basic theorem about mixed second order partial derivatives of functions of two variables? What about mixed partial derivatives of higher order? Give examples.

10. Give three important partial differential equations from physics, along with sample solutions.

MISCELLANEOUS PROBLEMS

1. Sketch the hyperbolas that bound the domain of the function
$$f(x, y) = \sqrt{\sin(\pi xy)}, \quad xy \ge 0.$$
Then complete the picture by shading the function's domain.

2. Calculate the following limits.
a) $\displaystyle\lim_{(x, y)\to(0, 0)} y \sin(1/x)$

b) $\displaystyle\lim_{(x, y)\to(1, 1)} \frac{x^3y^3 - 1}{xy - 1}$

3. Let $f(x, y) = (x^2 - y^2)/(x^2 + y^2)$ for $x^2 + y^2 \ne 0$. Is it possible to define $f(0, 0)$ in a way that makes f continuous at $(0, 0)$? Why?

4. Let the function $f(x, y)$ be defined by the equations
$$f(x, y) = \frac{\sin^2(x - y)}{|x| + |y|} \quad \text{for} \quad |x| + |y| \ne 0, \quad f(0, 0) = 0.$$
Is f continuous at the origin?

5. Let $f(x, y)$ be defined and continuous for all x, y (differentiability not assumed). Show that it is always possible to find arbitrarily many points $(x_1, y_1), (x_2, y_2), \ldots, (x_n, y_n)$ where f has the same value.

6. Prove that if $f(x, y)$ is defined for all x, y by
$$f(x, y) = \begin{cases} \dfrac{2xy}{x^2 + y^2} & \text{if } (x, y) \ne (0, 0), \\ 0 & \text{if } (x, y) = (0, 0), \end{cases}$$
then
a) for any fixed x, $f(x, y)$ is a continuous function of y;
b) for any fixed y, $f(x, y)$ is a continuous function of x;
c) $f(x, y)$ is not continuous at $(0, 0)$;
d) $\partial f/\partial x$ and $\partial f/\partial y$ exist at $(0, 0)$ but are not continuous there.
(This example shows that a function may have partial derivatives at all points of a region yet not be continuous in the region. This is very different from what we find for functions of one variable, where the existence of a derivative implies continuity.)

7. Find the first partial derivatives of the following functions:
a) $(\sin xy)^2$ b) $\sin((xy)^2)$.

8. Let α, β, γ be the angles that a line passing through the origin into the first octant makes with the positive coordinate axes. Consider γ as a function of α and β. Find the value of $\partial\gamma/\partial\alpha$ when $\alpha = \pi/4$, $\beta = \pi/3$, $\gamma = \pi/3$.

9. Let (r, θ) and (x, y) be polar coordinates and Cartesian coordinates in the plane. Show geometrically why $\partial r/\partial x$ does not equal $(\partial x/\partial r)^{-1}$ by appealing to the definitions of these derivatives.

10. If $z = xe^{xy}$, which of the following statements are true about the value of the derivative $\partial z/\partial x$ at the point $(1, 1)$?
i) The value of the derivative is $2e$.
ii) Geometrically, the value of the derivative represents the slope of the line tangent to the surface $z = xe^{xy}$ at the point $(x, y, z) = (1, 1, e)$ and lying in the plane $y = 1$.
iii) This derivative is evaluated by holding the variable x constant, taking an ordinary derivative with respect to the variable y, and evaluating the resulting function at $(x, y) = (1, 1)$.

11. Let
$$f(r, \theta) = \begin{cases} \dfrac{\sin 6r}{6r}, & r \ne 0 \\ 1, & r = 0. \end{cases}$$
(See Fig. 16.30.) Find (a) $\lim_{r\to 0} f(r, \theta)$, (b) $f_r(0, 0)$, (c) $f_\theta(r, \theta)$, $r \ne 0$.

16.30 The surface of Problem 11.

12. Let $z = f(u, v)$, $u = x$, and $v = \ln(x^2 + 1)$. Express $\partial z/\partial x$ in terms of x, f_x, and f_y.

13. Let $w = f(r, \theta)$, $r = \sqrt{x^2 + y^2}$, and $\theta = \tan^{-1}(y/x)$. Find $\partial w/\partial x$ and $\partial w/\partial y$ and express the answers in terms of r and θ.

14. Let $z = f(u, v)$, $u = ax + by$, and $v = ax - by$. Find formulas for $\partial z/\partial x$ and $\partial z/\partial y$.

15. If a and b are constants, $w = u^3 + \tanh u + \cos u$, and $u = ax + by$, show that

$$a\frac{\partial w}{\partial y} = b\frac{\partial w}{\partial x}.$$

16. If $w = \sqrt{x^2 + y^2 + z^2}$, $x = e^r \cos s$, $y = e^r \sin s$, and $z = e^s$, find $\partial w/\partial r$ and $\partial w/\partial s$ by the Chain Rule. Check your answer another way.

17. If $w = \ln(x^2 + y^2 + 2z)$, $x = r + s$, $y = r - s$, and $z = 2rs$, find $\partial w/\partial r$ and $\partial w/\partial s$ by the Chain Rule. Check your answer another way.

18. The equations

$$e^u \cos v - x = 0 \quad \text{and} \quad e^u \sin v - y = 0$$

define u and v as functions of x and y, say $u = u(x, y)$ and $v = v(x, y)$. Show that the angle between the vectors

$$\left(\frac{\partial u}{\partial x}\right)\mathbf{i} + \left(\frac{\partial u}{\partial y}\right)\mathbf{j}$$

and

$$\left(\frac{\partial v}{\partial x}\right)\mathbf{i} + \left(\frac{\partial v}{\partial y}\right)\mathbf{j}$$

is constant.

19. If the substitution $u = (x - y)/2$, $v = (x + y)/2$, changes $f(u, v)$ into $F(x, y)$, express $\partial F/\partial x$ and $\partial F/\partial y$ in terms of the derivatives of $f(u, v)$ with respect to u and v.

20. Suppose that n is a fixed integer, that $f(x, y, z)$ is continuous and has continuous partial derivatives, and that f satisfies the equation

$$f(tx, ty, tz) = t^n f(x, y, z)$$

for all x, y, z, and t. Show that

$$\frac{\partial f}{\partial x}x + \frac{\partial f}{\partial y}y + \frac{\partial f}{\partial z}z = nf.$$

(*Hint:* Differentiate the first equation with respect to t and set $t = 1$.)

Leibniz's Rule

In applications we sometimes encounter functions defined by integrals that have variable upper limits and variable lower limits at the same time:

$$y = \int_{\sin x}^{x^2} (1 + t)\, dt, \quad y = \int_{\sqrt{x}}^{2\sqrt{x}} \sin t^2\, dt.$$

The first integral can be evaluated directly, but the second cannot. We may calculate the derivative of either integral, however,

by a formula called Leibniz's rule:

> **LEIBNIZ'S RULE**
>
> If f is continuous on $[a, b]$, and $u(x)$ and $v(x)$ are differentiable functions of x whose values lie in $[a, b]$, then
>
> $$\frac{d}{dx}\int_{u(x)}^{v(x)} f(t)\, dt = f(v(x))\frac{dv}{dx} - f(u(x))\frac{du}{dx}.$$

Figure 16.31 gives a geometric interpretation of Leibniz's rule. It shows a carpet of variable height $f(t)$ that is being rolled up at the left at the same time x as it is being unrolled at the right. (In this interpretation time is x, not t.) At time x, the floor is covered from $u(x)$ to $v(x)$. The rate du/dx at which carpet is being rolled up need not be the same as the rate dv/dx at which the carpet is being laid down. At any given time x, the area covered by carpet is

$$A(x) = \int_{u(x)}^{v(x)} f(t)\, dt.$$

Now, at what rate is the covered area changing? At the instant x, $A(x)$ is increasing by the width $f(v(x))$ of the unrolling carpet times the rate dv/dx at which the carpet is being unrolled. That is, $A(x)$ is being increased at the rate

$$f(v(x))\frac{dv}{dx}.$$

At the same time, A is being decreased at the rate

$$f(u(x))\frac{du}{dx},$$

the width at the end that is being rolled up times the rate du/dx. The net change in A is therefore

$$\frac{dA}{dx} = f(v(x))\frac{dv}{dx} - f(u(x))\frac{du}{dx},$$

which is precisely Leibniz's rule.

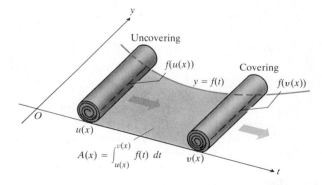

16.31 Maya's carpet: a geometric interpretation of Leibniz's rule.

$$\frac{dA}{dx} = f(v(x))\frac{dv}{dx} - f(u(x))\frac{du}{dx}.$$

21. Prove Leibniz's rule by setting

$$g(u, v) = \int_u^v f(t)dt, \quad u = u(x), \quad v = v(x),$$

and using a Chain Rule to calculate dg/dx.

22. Use Leibniz's rule to calculate dF/dx if

a) $F(x) = \displaystyle\int_{1/x}^{x} \frac{1}{t}\, dt$

b) $F(x) = \displaystyle\int_{\cos x}^{\sin x} \frac{1}{1 - t^2}\, dt$

c) $F(x) = \displaystyle\int_{\sqrt{x}}^{2\sqrt{x}} \sin t^2 dt$

23. Find the value of x that maximizes the integral

$$\int_x^{x+3} t(5 - t)\, dt.$$

(Problems like this arise in the mathematical theory of political elections. See the article "The Entry Problem in a Political Race," by Steven J. Brams and Philip D. Straffin, Jr., in *Political Equilibrium,* Peter Ordeshook and Kenneth Shepfle, Editors, Kluwer-Nijhoff, Boston, 1982, pp. 181–195.)

24. Let $u = xyz$. Show that if x and y are the independent variables (so u and z are functions of x and y), then

$$\partial u/\partial x = xy(\partial z/\partial x) + yz,$$

but that if x, y, and z are the independent variables,

$$\partial u/\partial x = yz.$$

25. When y is eliminated from the two equations $z = f(x, y)$ and $g(x, y) = 0$, the result is expressible in the form $z = h(x)$. Express the derivative $h'(x)$ in terms of $\partial f/\partial x$, $\partial f/\partial y$, $\partial g/\partial x$, $\partial g/\partial y$. Check your formula by computing $h(x)$ and $h'(x)$ explicitly for $f(x, y) = x^2 + y^2$ and $g(x, y) = x^3 + y^2 - x$.

26. Suppose the equation $F(x, y, z) = 0$ defines z as a function of x and y, say $z = f(x, y)$, with derivatives $\partial f/\partial x$ and $\partial f/\partial y$. Suppose that the equation $F(x, y, z) = 0$ also defines x as a function of y and z, say $x = g(y, z)$, with derivatives $\partial g/\partial y$ and $\partial g/\partial z$. Prove that

$$\frac{\partial g}{\partial y} = -\frac{\partial f/\partial y}{\partial f/\partial x},$$

and express $\partial g/\partial z$ in terms of $\partial f/\partial x$ and $\partial f/\partial y$.

27. If $z = f(u)$ and $u = (x - y)/y$, show that $x(\partial z/\partial x) + y(\partial z/\partial y) = 0$.

28. If $f(x, y, z) = 0$ and $z = x + y$, find dz/dx.

29. Find the derivative of

$$f(x, y, z) = (x - 1)^2 + 2(y + 1)^2 + 3(z - 2)^2 - 6$$

in the direction of $i - j + 2k$ at the point $(2, 0, 1)$.

30. Find the derivative of the function

$$f(x, y, z) = x^2 - 2y^2 + z^2$$

at the point $(3, 3, 1)$ in the direction of the vector $2i + j - k$.

31. What is the maximum value of the directional derivative of

$$f(x, y, z) = x^2 + y^2 - 3z$$

at the point $(1, 3, 5)$?

32. Find the direction of greatest change of the function

$$f(x, y, z) = x^2 - y^2 + z^2$$

at the point $(1, 1, \sqrt{14})$.

33. Find the derivative of $f(x, y, z) = xyz$ in the direction of the vector $i + j + k$ at the point $(1, 1, 1)$. What is the largest value any directional derivative of f at $(1, 1, 1)$ can have?

34. At the point $(1, 2)$, the function $f(x, y)$ has a derivative of $+2$ in the direction toward $(2, 2)$ and a derivative of -2 in the direction toward $(1, 1)$. What is the derivative of f at $(1, 2)$ in the direction toward $(4, 6)$?

35. In a smoothly flowing fluid, the density $\rho(x, y, z, t)$ depends on position and time. If

$$V = V(x, y, z, t)$$

is the velocity of the fluid particle at the point (x, y, z) at time t, then

$$\frac{d\rho}{dt} = V \cdot \nabla\rho + \frac{\partial\rho}{\partial t} = V_1\frac{\partial\rho}{\partial x} + V_2\frac{\partial\rho}{\partial y} + V_3\frac{\partial\rho}{\partial z} + \frac{\partial\rho}{\partial t},$$

where $V = V_1 i + V_2 j + V_3 k$. Explain the physical and geometrical meaning of this relation.

36. If the gradient of a function $f(x, y, z)$ is always parallel to the vector $xi + yj + zk$, show that the function assumes the same value at the points $(0, 0, a)$ and $(0, 0, -a)$.

37. Show that the directional derivative of

$$r = \sqrt{x^2 + y^2 + z^2}$$

equals 1 in any direction at the origin but that r does not have a gradient vector at the origin.

38. Let $R = xi + yj + zk$ and $r = |R|$.
a) From its geometrical interpretation, show that $\nabla r = R/r$.
b) Show that $\nabla(r^n) = nr^{n-2}R$.
c) Find a function whose gradient is equal to R.
d) Show that $R \cdot dR = r\, dr$.
e) If A is a constant vector, show that $\nabla(A \cdot R) = A$.

39. Let $\theta(x, y)$ be the polar coordinate θ as a function of x and y in the xy-plane. Find the direction and magnitude of $\nabla\theta$ in both rectangular and polar coordinates.

40. If A, B are fixed points and θ is the angle at $P(x, y, z)$ subtended by the line segment AB, show that $\nabla\theta$ is normal to the circle through A, B, P.

41. a) Show that $f(x, y) = ay - bx + C$ and $f(x, y) = e^{ay-bx} + C$ (a and b constant) both satisfy the equation $af_x + bf_y = 0$.

b) Show that any differentiable function $f(x, y)$ whose gradient has the form $-bk(x, y)\,\mathbf{i} + ak(x, y)\,\mathbf{j}$ satisfies the equation in part (a).

42. Let $w = f(u, v)$ be a function of u, v with continuous partial derivatives, where u, v in turn are functions of independent variables x, y, z with continuous partial derivatives. Show that if w is regarded as a function of x, y, z, its gradient at any point (x_0, y_0, z_0) lies in a common plane with the gradients of $u = u(x, y, z)$ and $v = v(x, y, z)$.

43. Find a unit vector normal to the surface $x^2 + y^2 = 3z$ at the point $(1, 3, 10/3)$.

44. Find an equation for the plane tangent to the surface $x^2y + 2y^2x - z = 0$ at the point $(1, 3, 21)$.

45. Show that the line normal to the surface $xy + z = 2$ at the point $(1, 1, 1)$ passes through the origin.

46. a) Find the gradient of $f(x, y) = x^5 + 4xy^3 - 3y^5$.

b) Find an equation for the line in the xy-plane that is tangent to the curve
$$x^5 + 4xy^3 - 3y^5 = 2 \text{ at the point } (1, 1).$$

47. a) Find a vector $\mathbf{N}(x, y, z)$ normal to the surface $z = \sqrt{x^2 + y^2} + (x^2 + y^2)^{3/2}$ at the point (x, y, z) of the surface.

b) Find the cosine of the angle γ between $\mathbf{N}(x, y, z)$ and the z-axis and the limit of $\cos \gamma$ as $(x, y, z) \to (0, 0, 0)$.

48. Find an equation for the plane tangent to the surface
$$x^3z + y^2x^2 + \sin(yz) + 54 = 0$$
at the point $(3, 0, -2)$. Also find equations for the line normal to the surface at $(3, 0, -2)$.

49. a) Sketch the surface $x^2 - y^2 + z^2 = 4$.

b) Find a vector normal to this surface at $(2, -3, 3)$.

c) Find equations for the tangent plane and normal line at $(2, -3, 3)$.

50. a) Find an equation for the plane tangent to the surface
$$x^3 + xy^2 + y^3 + z^3 + 1 = 0$$
at the point $(-2, 1, 2)$.

b) Find equations for the line normal to the surface at the point $(-2, 1, 2)$.

51. Find all points (a, b, c) in space for which the spheres
$$(x - a)^2 + (y - b)^2 + (z - c)^2 = 1$$
and
$$x^2 + y^2 + z^2 = 1$$
intersect orthogonally (their tangent planes are perpendicular at each point of intersection).

52. a) Find the gradient of
$$f(x, y, z) = x^2 + 2xy - y^2 + z^2$$
at the point $P_0(1, -1, 3)$.

b) Find an equation for the plane that is tangent to the surface
$$x^2 + 2xy - y^2 + z^2 = 7$$
at $P_0(1, -1, 3)$.

53. For what value of the constant a will the planes tangent to the spheres
$$(x - a)^2 + y^2 + z^2 = 3 \quad \text{and} \quad x^2 + (y - 1)^2 + z^2 = 1$$
be perpendicular to each other at each point where the spheres intersect?

54. A curve is said to be tangent to a surface at a point P if the line tangent to the curve lies in the plane that is tangent to the surface at P.

a) Show that the curve
$$\mathbf{R}(t) = (\ln t)\mathbf{i} + (t \ln t)\mathbf{j} + t\mathbf{k}$$
is tangent to the surface
$$xz^2 - yz + \cos xy = 1$$
at $P(0, 0, 1)$.

b) Show that the curve
$$\mathbf{R}(t) = \left(\frac{t^3}{4} - 2\right)\mathbf{i} + \left(\frac{4}{t} - 3\right)\mathbf{j} + \cos(t - 2)\mathbf{k}$$
is tangent to the surface
$$x^3 + y^3 + z^3 - xyz = 0$$
at $t = 2$.

55. Let C be any curve that lies on the surface $f(x, y, z) = 0$ and passes through the point $P_0(x_0, y_0, z_0)$. Prove that the line tangent to C at P_0 lies in the plane determined by the tangent lines to the curves C_x and C_y at P_0, where C_x is the curve in which the surface meets the plane $y = y_0$ and C_y is the curve in which the surface meets the plane $x = x_0$. (Assume that f and its first partial derivatives are continuous.)

56. If $f(x, y)$ and its first partial derivatives are continuous at the point $P_0(x_0, y_0)$, which of the following statements are true?

a) If $\mathbf{u}$ is a unit vector, then the derivative of f in the direction of $\mathbf{u}$ is $(f_x(x_0, y_0)\mathbf{i} + f_y(x_0, y_0)\mathbf{j}) \cdot \mathbf{u}$.

b) The derivative of f in the direction of $\mathbf{u}$ is a vector.

c) The directional derivative of f at P_0 has its greatest value in the direction of ∇f.

d) At (x_0, y_0), ∇f is normal to the level curve $f(x, y) = f(x_0, y_0)$.

57. Suppose that $f(x, y)$ and its first partial derivatives are continuous. Suppose also that f has the constant value c along the differentiable curve $x = x(t)$, $y = y(t)$; that is,
$$f(x(t), y(t)) = c$$
for all values of t. Differentiate both sides of this equation with respect to t to show that ∇f is normal to the tangent vector at every point of the curve.

58. Suppose $w = f(x, y)$, $x = u + v$, and $y = u - v$. Show that
$$\frac{\partial^2 w}{\partial u \, \partial v} = \frac{\partial^2 w}{\partial x^2} - \frac{\partial^2 w}{\partial y^2}.$$

59. The substitution $u = x + y$, $v = xy^2$ changes the function $f(u, v)$ into $F(x, y)$. Express the partial derivative $\partial^2 F/\partial x\, \partial y$ in terms of x, y, and the partial derivatives of $f(u, v)$ with respect to u and v.

60. Suppose that a and b are constants, that $z = u(x, y) \cdot e^{ax+by}$, that $u(x, y)$ is a function of x and y, and that $\partial^2 u/\partial x\, \partial y = 0$. Find values of a and b that will make the expression

$$\frac{\partial^2 z}{\partial x\, \partial y} - \frac{\partial z}{\partial x} - \frac{\partial z}{\partial y} + z$$

equal zero for all values of x and y.

61. Introducing polar coordinates $x = r \cos\theta$ and $y = r \sin\theta$ changes $f(x, y)$ into $g(r, \theta)$. Find the value of $\partial^2 g/\partial\theta^2$ at the point where $r = 2$ and $\theta = \pi/2$, given that

$$\frac{\partial f}{\partial x} = \frac{\partial f}{\partial y} = \frac{\partial^2 f}{\partial x^2} = \frac{\partial^2 f}{\partial y^2} = 1$$

at that point.

62. Assuming that the equation $f(x, y) = 0$ defines y as a twice-differentiable function of x and that f and all the necessary partial derivatives are continuous, find d^2y/dx^2.

63. Suppose that f is a function of r, that $r = \sqrt{x^2 + y^2 + z^2}$, and that $f_{xx} + f_{yy} + f_{zz} = 0$. Show that

$$f(r) = \frac{a}{r} + b$$

for some constants a and b.

64. A function $f(x, y)$, defined and differentiable for all x and y, is said to be homogeneous of degree n (a nonnegative integer) if $f(tx, ty) = t^n f(x, y)$ for all t, x, and y. For such a function prove:

a) $x(\partial f/\partial x) + y(\partial f/\partial y) = nf(x, y)$ and express this in vector form;

b) $x^2\left(\dfrac{\partial^2 f}{\partial x^2}\right) + 2xy\left(\dfrac{\partial^2 f}{\partial x\, \partial y}\right) + y^2\left(\dfrac{\partial^2 f}{\partial y^2}\right) = n(n-1)$

if f has continuous second partial derivatives;

c) a homogeneous function of degree zero is a constant.

65. Let z be defined implicitly as a function of x and y by the equation $\sin(x + y) + \sin(y + z) = 1$. Compute $\partial^2 z/\partial x\, \partial y$ in terms of x, y, and z.

66. If $z = xy^2 - y \sin x$, find $y(\partial^2 z/\partial y\, \partial x) - \partial z/\partial x$.

67. Let $w = z \tan^{-1}(x/y)$. Find

$$\frac{\partial^2 w}{\partial x^2} + \frac{\partial^2 w}{\partial y^2} + \frac{\partial^2 w}{\partial z^2}.$$

68. Show that the function satisfies the differential equation.

a) $f(x, t) = \displaystyle\int_0^{x/(2\sqrt{kt})} e^{-\sigma^2}\, d\sigma, \quad kf_{xx} - f_t = 0$ (k constant)

b) $f(x, t) = \phi(x + at) + \psi(x - at), \quad f_{tt} = a^2 f_{xx}$

69. Let

$$f(x, y) = \begin{cases} xy\dfrac{x^2 - y^2}{x^2 + y^2}, & (x, y) \ne (0, 0), \\ 0, & (x, y) = (0, 0). \end{cases}$$

Find $f_{yx}(0, 0)$ and $f_{xy}(0, 0)$. See Fig. 16.32.

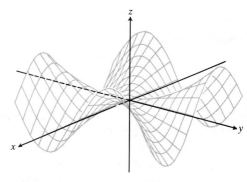

16.32 The surface of Problem 69.

70. Find a function $w = f(x, y)$ such that $\partial w/\partial x = 1 + e^x \cos y$ and $\partial w/\partial y = 2y - e^x \sin y$, or else explain why no such function exists.

71. In thermodynamics the five quantities S, T, u, p, and v are such that any two of them may be considered independent variables, the others then being determined. They are connected by the differential relation $T\, dS = du + p\, dv$. Show that

$$\left(\frac{\partial S}{\partial v}\right)_T = \left(\frac{\partial p}{\partial T}\right)_v \quad \text{and} \quad \left(\frac{\partial v}{\partial S}\right)_p = \left(\frac{\partial T}{\partial p}\right)_s.$$

CHAPTER 17

Applications of Partial Derivatives

OVERVIEW

As we mentioned at the end of Article 16.2, continuous functions defined on closed and bounded domains in the xy-plane take on maximum and minimum values on these domains. It is important to be able to find these values and at what locations they occur. For example, what is the greatest temperature encountered on the edge of a heated metal plate? Where does a given surface attain its highest point above a given patch of the xy-plane? As we shall see in Article 17.1 and again in Article 17.3, questions like these can often be answered by examining a function's partial derivatives.

Partial derivatives play an equally important role in constructing linear approximations, calculating increments, and estimating approximation errors (Article 17.2). They also play a role in applying the method of Lagrange multipliers, a method for finding the extreme values of differentiable functions on restricted subsets of their domains. With this method, for example, we can find the greatest and smallest values of the function $T = 400xyz^2$ on the sphere $x^2 + y^2 + z^2 = 1$.

Partial derivatives also form the basis for a clever and powerful two-variable version of Taylor's formula. We introduce the formula in (optional) Article 17.4 and use it to establish some of the mathematical results on which the work on extreme values and estimation earlier in the chapter depends. We also use the formula to generate quadratic approximations of functions of two variables and to estimate the associated approximation errors.

17.1

Maxima, Minima, and Saddle Points

Functions of two independent variables can have local maximum and minimum values just the way functions of a single variable can (Figs. 17.1 and 17.2). In this article, we show how to use partial derivatives to find the local maximum and minimum values of a continuous function $f(x, y)$ on a region R in the xy-plane. The first step is to use the function's first partial derivatives to make a (usually) short and comprehensive list of points where f can assume its local extreme values. What we do next then depends on whether R is closed and bounded. If it is, we appeal to a theorem from advanced calculus that a continuous function on a closed, bounded region R assumes an absolute maximum value and an absolute minimum value on R, and we look through the list to find

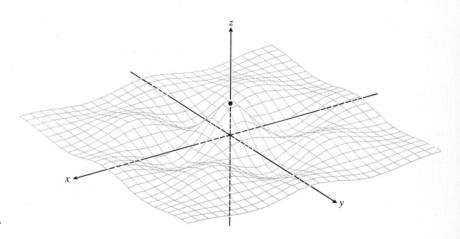

17.1 The function

$$z = (\cos x)(\cos y)e^{-\sqrt{x^2+y^2}}$$

has many local maxima and minima.

17.2 The "ridge" surface

$$z = \frac{1}{2}(\|x| - |y\| - |x| - |y|)$$

viewed from the point (10, 15, 20).

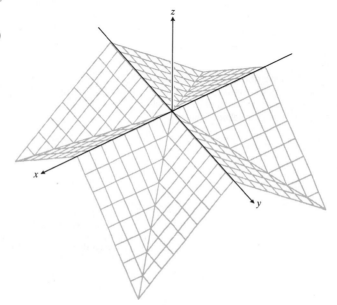

what these values are. If R is not closed and bounded, the function may not have absolute maximum and minimum values on R. However, we can still try to use the function's second partial derivatives to tell which points on the list, if any, give local maximum and minimum values. As you will see, the routine is much the same as the routine for functions of a single variable. The main difference is that the first and second derivative tests now involve more derivatives.

The Derivative Tests

For a function of a single variable, we look for a local maximum, a local minimum, or a point of inflection by examining points where the graph has a horizontal tangent line. We also examine the values of the function at the boundary points of its domain and at any points where the first derivative does not exist.

For a function $f(x, y)$ of two independent variables, we look for points where the surface $z = f(x, y)$ has a horizontal tangent plane. At such points we then look for a local maximum, a local minimum, or a **saddle point** (Fig. 17.3). We also examine the values of the function at the boundary points in its domain and at any points where the first partial derivatives f_x or f_y do not exist.

We organize the search for the extreme values assumed by a continuous function $f(x, y)$ on a region R into three steps:

STEP 1: *Make a list that includes the points where f has its local maxima and minima and evaluate f at all the points on the list.*

As we shall see later, the local maxima and minima of f can occur only at

i) boundary points of R,
ii) interior points of R where $f_x = f_y = 0$ and points where f_x or f_y fail to exist. (We call these the **critical points** of f.)

Therefore, to make the list, we find the maximum and minimum values of f on the boundary of R and find the values of f at the critical points.

17.3 How values of $f(x, y)$ are classified. Saddle points are the three-dimensional analogues of points of inflection. Local values are often called "relative" values.

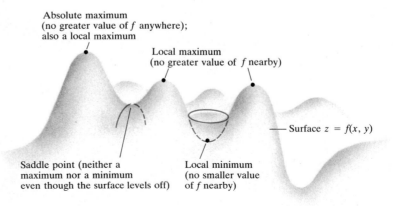

Absolute maximum
(no greater value of f anywhere);
also a local maximum

Local maximum
(no greater value of f nearby)

Surface $z = f(x, y)$

Saddle point (neither a
maximum nor a minimum
even though the surface levels off)

Local minimum
(no smaller value
of f nearby)

In theory, such a list could be very long and might even include infinitely many points, but that is usually not the case in practice and definitely not the case in this book. The lists in the examples and problems will be short.

STEP 2: *If R is closed and bounded, look through the list for the maximum and minimum values of f. These will be the absolute maximum and minimum values of f on R.* As we mentioned in the introduction, a function that is continuous on a closed bounded region of the plane has an absolute maximum value on the region and an absolute minimum value on the region. Since absolute maxima and minima are also local maxima and minima, the absolute maximum and minimum values of f already appear somewhere in the list we made in Step 1. We have only to glance at the list to see what they are. If we then wish to learn which of the remaining values, if any, are local maxima and minima, we can go on to Step 3.

STEP 3: *If R is not closed and bounded, try the following second derivative test.*

The fact that $f_x = f_y = 0$ at an interior point (a, b) of R does not in itself guarantee that f will have an extreme value there. However, if f and its first and second partial derivatives are continuous on R, there is a second derivative test that may identify the behavior of f at (a, b). The test goes like this:

If $f_x(a, b) = f_y(a, b) = 0$, then

i) f has a **local maximum** at (a, b) if $f_{xx} < 0$ and $f_{xx}f_{yy} - f_{xy}^2 > 0$ at (a, b);

ii) f has a **local minimum** at (a, b) if $f_{xx} > 0$ and $f_{xx}f_{yy} - f_{xy}^2 > 0$ at (a, b);

iii) f has a **saddle point** at (a, b) if $f_{xx}f_{yy} - f_{xy}^2 < 0$ at (a, b).

iv) The test is *inconclusive* at (a, b) if $f_{xx}f_{yy} - f_{xy}^2 = 0$ at (a, b). We must find some other way to determine the behavior of f at (a, b).

The expression $f_{xx}f_{yy} - f_{xy}^2$ is called the **discriminant** of f. It is sometimes easier to remember in the determinant form

$$f_{xx}f_{yy} - f_{xy}^2 = \begin{vmatrix} f_{xx} & f_{xy} \\ f_{yx} & f_{yy} \end{vmatrix}.$$

We shall now look at examples that show these tests at work. After that, we shall show why the condition $f_x = f_y = 0$ is a necessary condition for having an extreme value at an interior point of the domain of a differentiable function. The mathematics behind the second derivative test will be discussed in Article 17.4.

The Tests at Work

In the first example we look at the function $f(x, y) = x^2 + y^2$, whose behavior we already know from looking at the formula: Its value is zero at the origin and increases steadily as (x, y) moves away from the origin. The point of Example 1 is to show how the derivative tests reveal this.

EXAMPLE 1 Find the extreme values of

$$f(x, y) = x^2 + y^2.$$

17.4 The graph of the function $f(x, y) = x^2 + y^2$ is the paraboloid $z = x^2 + y^2$. The function has only one extreme value, an absolute minimum value of 0 at the origin.

Solution   The domain of f has no boundary points, for it is the entire plane (Fig. 17.4). The derivatives $f_x = 2x$ and $f_y = 2y$ exist everywhere. Therefore, local maxima and minima can occur only where

$$f_x = 2x = 0 \qquad \text{and} \qquad f_y = 2y = 0.$$

The only possibility is the origin, where the value of f is zero. Since f is never negative, we see that zero is an absolute minimum.

We have not needed the second derivative test at all. Had we used it, we would have found

$$f_{xx} = 2, \qquad f_{yy} = 2, \qquad f_{xy} = 0$$

and

$$f_{xx}f_{yy} - f_{xy}^2 = (2)(2) - (0)^2 = 4 > 0,$$

identifying $(0, 0)$ as a local minimum. This in itself does not identify $(0, 0)$ as an absolute minimum. It takes more information to do that.   ■

One virtue of the procedure we have described for finding extreme values is that it applies even to functions whose graphs are too complicated to draw. The next example illustrates this point.

EXAMPLE 2   Find the extreme values of the function

$$f(x, y) = xy - x^2 - y^2 - 2x - 2y + 4.$$

Solution   The function is defined and differentiable for all x and y and its domain has no boundary points. The function therefore has extreme values only at the points where f_x and f_y are simultaneously zero. This leads to

$$f_x = y - 2x - 2 = 0, \qquad f_y = x - 2y - 2 = 0,$$

or

$$x = y = -2.$$

Therefore, the point $(-2, -2)$ is the only point where f may take on an extreme value. To see if it does so, we calculate

$$f_{xx} = -2, \qquad f_{yy} = -2, \qquad f_{xy} = 1.$$

The discriminant of f at $(a, b) = (-2, -2)$ is

$$f_{xx}f_{yy} - f_{xy}^2 = (-2)(-2) - (1)^2 = 4 - 1 = 3.$$

The combination

$$f_{xx} < 0 \qquad \text{and} \qquad f_{xx}f_{yy} - f_{xy}^2 > 0$$

tells us that f has a local maximum at $(-2, -2)$. The value of f at this point is $f(-2, -2) = 8$.   ■

EXAMPLE 3   Find the extreme values of

$$f(x, y) = xy.$$

Solution   Since the function is differentiable everywhere and its domain has no boundary points (Fig. 17.5), the function can assume extreme values only

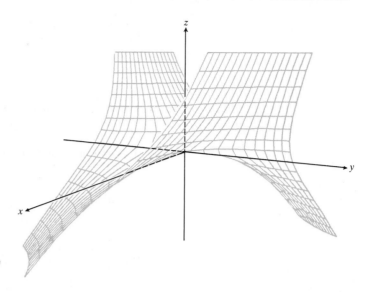

17.5 The surface $z = xy$ has a saddle point at the origin.

where

$$f_x = y = 0 \qquad \text{and} \qquad f_y = x = 0.$$

Thus, the origin is the only point where f might have an extreme value. To see what happens there, we calculate

$$f_{xx} = 0, \qquad f_{yy} = 0, \qquad f_{xy} = 1.$$

The discriminant,

$$f_{xx}f_{yy} - f_{xy}^2 = -1,$$

is negative. Therefore the function has a saddle point at $(0, 0)$. We conclude that $f(x, y) = xy$ assumes no extreme values at all. ■

EXAMPLE 4 Find the absolute maximum and minimum values of

$$f(x, y) = 2 + 2x + 2y - x^2 - y^2$$

on the triangular plate in the first quadrant bounded by the lines $x = 0$, $y = 0$, $y = 9 - x$.

Solution The only places where f can assume these values are points on the boundary of the triangle (Fig. 17.6) and points inside the triangle where $f_x = f_y = 0$.

Boundary points. We take the triangle one side at a time:

1. On the segment OA, $y = 0$. The function

$$f(x, y) = f(x, 0) = 2 + 2x - x^2$$

may now be regarded as a function of x defined on the closed interval $0 \le x \le 9$.

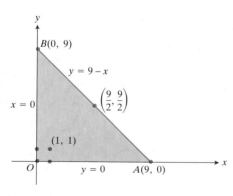

17.6 This triangular plate is the domain of the function in Example 4.

Its extreme values (we know from Chapter 3) may occur at the end-points

$$x = 0 \quad \text{where } f(0, 0) = 2,$$

$$x = 9 \quad \text{where } f(9, 0) = 2 + 18 - 81 = -61,$$

and at the interior points where

$$f'(x, 0) = 2 - 2x = 0.$$

The only interior point where $f'(x, 0) = 0$ is $x = 1$, where

$$f(x, 0) = f(1, 0) = 3.$$

2. On the segment OB, $x = 0$ and

$$f(x, y) = f(0, y) = 2 + 2y - y^2.$$

We know from the symmetry of f in x and y and from the analysis we just carried out that the candidates on this segment are

$$f(0, 0) = 2, \quad f(0, 9) = -61, \quad f(0, 1) = 3.$$

3. We have already accounted for the values of f at the endpoints of AB, so we have only to look at the interior points of AB. With

$$y = 9 - x,$$

we have

$$f(x, y) = 2 + 2x + 2(9 - x) - x^2 - (9 - x)^2 = -61 + 18x - 2x^2.$$

Setting

$$f'(x, 9 - x) = 18 - 4x = 0$$

gives

$$x = \frac{18}{4} = \frac{9}{2}.$$

At this value of x,

$$y = 9 - \frac{9}{2} = \frac{9}{2},$$

and

$$f(x, y) = f\left(\frac{9}{2}, \frac{9}{2}\right) = -\frac{41}{2}.$$

Interior points. For these we have

$$f_x = 2 - 2x = 0, \quad f_y = 2 - 2y = 0,$$

or

$$(x, y) = (1, 1),$$

where

$$f(1, 1) = 4.$$

Summary. We look at all the candidates:

$$4, \quad 2, \quad -61, \quad 3, \quad -\frac{41}{2}.$$

The maximum is 4, which f assumes at $(1, 1)$. The minimum is -61, assumed at $(0, 9)$ and $(9, 0)$. ■

The Condition $f_x(a, b) = f_y(a, b) = 0$

Suppose that a local maximum value of the function $f(x, y)$ occurs at an interior point (a, b) of the function's domain and that $\partial f/\partial x$ and $\partial f/\partial y$ both exist at (a, b). (See Fig. 17.7). Then

1. $x = a$ is an interior point of the domain of the curve $z = f(x, b)$ in which the plane $y = b$ cuts the surface $z = f(x, y)$.
2. The function $z = f(x, b)$ has a relative maximum value at $x = a$.
3. The value of the derivative of $z = f(x, b)$ at $x = a$ is therefore zero.

Since this derivative is precisely $\partial f/\partial x|_{(a, b)}$, we conclude that

$$\left.\frac{\partial f}{\partial x}\right|_{(a, b)} = 0.$$

A similar argument with the function $z = f(a, y)$ shows that

$$\left.\frac{\partial f}{\partial y}\right|_{(a, b)} = 0.$$

Thus,

$$f_x(a, b) = f_y(a, b) = 0$$

is a necessary condition for f to have an extreme value at an interior point (a, b) where both f_x and f_y exist.

Remember, though, that these derivatives are also zero at a saddle point and that a function can assume an extreme value at a point where one or both of the derivatives fail to exist.

17.7 The maximum of f occurs at $x = a$, $y = b$.

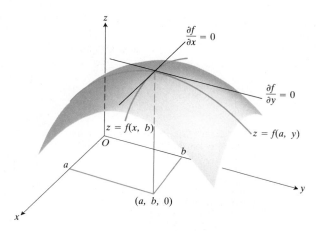

Extreme Values on Parameterized Curves

To find the extreme values of a function $f(x, y)$ on a curve $x = x(t)$, $y = y(t)$, we treat f as a function of the single variable t and use the Chain Rule to find where df/dt is zero. As in any other single-variable case, the extreme values of f are then found among the values at the

a) critical points (points where df/dt is zero or fails to exist), and

b) endpoints of the parameter domain.

EXAMPLE 5   Find the absolute maximum and minimum values of $f(x, y) = xy$ on the curve

$$x = 3 \cos t, \qquad y = 2 \sin t, \qquad 0 \le t \le \pi.$$

Solution   The curve is shown in Fig. 17.8. The derivative of f with respect to t is

$$\frac{df}{dt} = \frac{\partial f}{\partial x}\frac{dx}{dt} + \frac{\partial f}{\partial y}\frac{dy}{dt} = y\frac{dx}{dt} + x\frac{dy}{dt}$$

$$= (2 \sin t)(-3 \sin t) + (3 \cos t)(2 \cos t)$$

$$= -6 \sin^2 t + 6 \cos^2 t = 6 \cos 2t.$$

The derivative is defined at every value of t and equals zero when $\cos 2t = 0$. On the domain $0 \le t \le \pi$, this means $t = \pi/4$ and $t = 3\pi/4$. The candidates for the extreme values of f on the curve are therefore

Critical point values: $\quad f\left(\dfrac{\pi}{4}\right) = 6 \cos\dfrac{\pi}{4} \sin\dfrac{\pi}{4} = 3 \sin\left(\dfrac{\pi}{2}\right) = 3$

$$f\left(\frac{3\pi}{4}\right) = 6 \cos\frac{3\pi}{4} \sin\frac{3\pi}{4} = 3 \sin\left(\frac{3\pi}{2}\right) = -3$$

Endpoint values: $\qquad f(0) = 3 \sin (0) = 0$

$$f(\pi) = 3 \sin (2\pi) = 0.$$

A comparison of these values shows that, on the curve in question, f has an absolute maximum value of 3 at the point

$$\left(3 \cos\frac{\pi}{4}, 2 \sin\frac{\pi}{4}\right) = \left(\frac{3\sqrt{2}}{2}, \sqrt{2}\right)$$

and an absolute minimum value of -3 at the point

$$\left(3 \cos\frac{3\pi}{4}, 2 \sin\frac{3\pi}{4}\right) = \left(-\frac{3\sqrt{2}}{2}, \sqrt{2}\right). \qquad \blacksquare$$

17.8   Example 5 deals with the values of $f(x, y) = xy$ on the upper half of the ellipse $(x^2/9) + (y^2/4) = 1$.

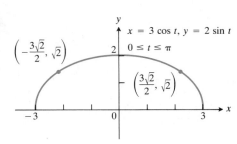

Summary of Max-Min Tests

The extreme values of $f(x, y)$ can occur only at

i) boundary points of the domain of f,

ii) interior points where $f_x = f_y = 0$ or points where f_x or f_y fail to exist. (These are the *critical points* of f.)

If f and its first and second order partial derivatives are continuous throughout some open disk containing (a, b), and if $f_x(a, b) = f_y(a, b) = 0$, then

i) $f_{xx} < 0$ and $f_{xx}f_{yy} - f_{xy}^2 > 0$ at (a, b) $\Rightarrow$ *local maximum,*

ii) $f_{xx} > 0$ and $f_{xx}f_{yy} - f_{xy}^2 > 0$ at (a, b) $\Rightarrow$ *local minimum,*

iii) $f_{xx}f_{yy} - f_{xy}^2 < 0$ at (a, b) $\Rightarrow$ *saddle point,*

iv) $f_{xx}f_{yy} - f_{xy}^2 = 0$ at (a, b) $\Rightarrow$ test is *inconclusive.*

On a curve $x = x(t)$, $y = y(t)$, we treat f as a function of t and find the extreme values of f as we would those of any other function of a single variable.

PROBLEMS

Test the following functions for maxima, minima, and saddle points. Find the function values at these points.

1. $f(x, y) = x^2 + xy + y^2 + 3x - 3y + 4$
2. $f(x, y) = x^2 + 3xy + 3y^2 - 6x + 3y - 6$
3. $f(x, y) = 5xy - 7x^2 + 3x - 6y + 2$
4. $f(x, y) = 2xy - 5x^2 - 2y^2 + 4x + 4y - 4$
5. $f(x, y) = x^2 + xy + 3x + 2y + 5$
6. $f(x, y) = y^2 + xy - 2x - 2y + 2$
7. $f(x, y) = 2xy - 5x^2 - 2y^2 + 4x - 4$
8. $f(x, y) = 2xy - x^2 - 2y^2 + 3x + 4$
9. $f(x, y) = x^2 + xy + y^2 + 3y + 3$
10. $f(x, y) = 3x^2 + 6xy + 7y^2 - 2x + 4y$
11. $f(x, y) = 2x^2 + 3xy + 4y^2 - 5x + 2y$
12. $f(x, y) = 4x^2 - 6xy + 5y^2 - 20x + 26y$
13. $f(x, y) = x^2 - 4xy + y^2 + 5x - 2y$
14. $f(x, y) = x^2 + y^2 - 2x + 4y + 6$
15. $f(x, y) = x^2 - y^2 - 2x + 4y + 6$
16. $f(x, y) = x^2 - 2xy + 2y^2 - 2x + 2y + 1$
17. $f(x, y) = x^2 + 2xy$
18. $f(x, y) = 3 + 2x + 2y - 2x^2 - 2xy - y^2$
19. $f(x, y) = x^2 + xy + y^2 + x - 4y + 5$
20. $f(x, y) = x^2 - xy + y^2 + 2x + 2y - 4$
21. $f(x, y) = 3x^2 - xy + 2y^2 - 8x + 9y + 10$
22. $f(x, y) = 5x^2 - 4xy + 2y^2 + 4x - 4y + 10$

23. $f(x, y) = x^3 - y^3 - 2xy + 6$
24. $f(x, y) = x^3 + y^3 + 3x^2 - 3y^2 - 8$
25. $f(x, y) = 6x^2 - 2x^3 + 3y^2 + 6xy$
26. $f(x, y) = 9x^3 + y^3/3 - 4xy$
27. $f(x, y) = x^3 + y^3 - 3xy + 15$
28. $f(x, y) = x^3 + 3xy + y^3$
29. $f(x, y) = 4xy - x^4 - y^4$
30. $f(x, y) = x \sin y$

In Problems 31–38, find the absolute maxima and minima of the functions on the given domains.

31. $f(x, y) = 2x^2 - 4x + y^2 - 4y + 1$ on the closed triangular plate bounded by the lines $x = 0$, $y = 2$, $y = 2x$ in the first quadrant

32. $D(x, y) = x^2 - xy + y^2 + 1$ on the closed triangular plate in the first quadrant bounded by the lines $x = 0$, $y = 4$, $y = x$

33. $f(x, y) = x^2 + y^2$ on the closed triangular plate bounded by the lines $x = 0$, $y = 0$, $y + 2x = 2$ in the first quadrant

34. $T(x, y) = x^2 + xy + y^2 - 6x$ on the rectangular plate $0 \le x \le 5$, $-3 \le y \le 3$

35. $T(x, y) = x^2 + xy + y^2 - 6x + 2$ on the rectangular plate $0 \le x \le 5$, $-3 \le y \le 0$

36. $f(x, y) = 48xy - 32x^3 - 24y^2$ on the rectangular plate $0 \le x \le 1$, $0 \le y \le 1$

37. $f(x, y) = (x^2 - 4x) \cos y$ over the region $1 \le x \le 3$, $-\pi/4 \le y \le \pi/4$

38. $f(x, y) = x - (x - 1/4)(y + 1/2)$ on the closed triangular region bounded by the lines $x = 0$, $y = 0$, $x + y = 1$ in the first quadrant

39. A flat circular plate has the shape of the region $x^2 + y^2 \le 1$. The plate, including the boundary where $x^2 + y^2 = 1$, is heated so that the temperature at any point (x, y) is

$$T(x, y) = x^2 + 2y^2 - x.$$

Find the hottest and coldest points on the plate and the temperature at each of these points.

40. Find the critical point of

$$f(x, y) = xy + 2x - \ln x^2 y$$

in the open first quadrant $(x > 0, y > 0)$ and show that f takes on a minimum there.

41. Find all local and absolute maxima and minima of

$$f(x, y) = -x^2 - y^2 + 2x + 2y + 2$$

on the closed first quadrant $x \ge 0$, $y \ge 0$.

42. Find the maxima, minima, and saddle points of $f(x, y)$, if any, given that
a) $f_x = 2x - 4y$ and $f_y = 2y - 4x$,
b) $f_x = 2(x - 1)$ and $f_y = 2(y - 2)$,
c) $f_x = 9x^2 - 9$ and $f_y = 2y + 4$.

43. Sketch the surface

$$z = \sqrt{x^2 + y^2}$$

over the region R: $|x| \le 1$, $|y| \le 1$. Find the high and low points of the surface over R. Discuss the existence and the values of $\partial z/\partial x$ and $\partial z/\partial y$ at these points.

44. The discriminant $f_{xx}f_{yy} - f_{xy}^2$ of each of the following functions is zero at the origin. Determine whether the function has a maximum, a minimum, or neither at the origin by imagining in each case what the surface $z = f(x, y)$ looks like.
a) $f(x, y) = x^2 y^2$
b) $f(x, y) = 1 - x^2 y^2$

c) $f(x, y) = xy^2$
d) $f(x, y) = x^3 y^2$
e) $f(x, y) = x^3 y^3$
f) $f(x, y) = x^4 y^4$

Extreme Values on Curves

45. Find the absolute maximum and minimum values of the following functions on
   i) the quarter-circle $x^2 + y^2 = 4$ in the first quadrant,
   ii) the half-circle $x^2 + y^2 = 4$, $y \ge 0$, and
   iii) the full circle $x^2 + y^2 = 4$.

a) $f(x, y) = xy$
b) $f(x, y) = x + y$
c) $f(x, y) = 2x^2 + y^2$
Use the parameterization $x = 2 \cos t$, $y = 2 \sin t$.

46. Find the absolute maximum and minimum values, if any, of the function $f(x, y) = xy$ on the following.
a) The line $x = 2t$, $y = t + 1$
b) The line segment $x = 2t$, $y = t + 1$, $-1 \le t \le 0$
c) The line segment $x = 2t$, $y = t + 1$, $0 \le t \le 1$

47. Find the absolute maximum and minimum values of the following functions on
   i) the quarter-ellipse $(x^2/9) + (y^2/4) = 1$ in the first quadrant,
   ii) the half-ellipse $(x^2/9) + (y^2/4) = 1$, $y \ge 0$, and
   iii) the full ellipse $(x^2/9) + (y^2/4) = 1$.

a) $f(x, y) = x^2 + 3y^2$
b) $f(x, y) = 2x + 3y$

48. Find the absolute maximum and minimum values of $f(x, y) = xy$ on the ellipse $x^2 + 4y^2 = 8$.

TOOLKIT PROGRAM

3D Grapher

17.2

Linear Approximations and Increment Estimates

In science and mathematics we can often replace complicated functions of two variables by simpler functions that give the accuracy we want without being so hard to work with. What we do is similar to what we did in Article 2.4 for functions of a single variable, except that now the approximation comes from a tangent plane instead of a tangent line. In this article, we show how these tangent plane approximations are defined and used.

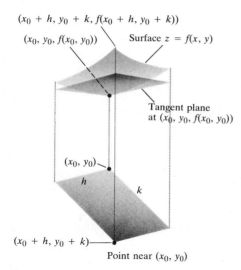

$(x_0 + h, y_0 + k, f(x_0 + h, y_0 + k))$

$(x_0, y_0, f(x_0, y_0))$ Surface $z = f(x, y)$

Tangent plane
at $(x_0, y_0, f(x_0, y_0))$

(x_0, y_0)

h

k

$(x_0 + h, y_0 + k)$

Point near (x_0, y_0)

17.9 A portion of the surface $z = f(x, y)$ and its tangent plane at $(x_0, y_0, f(x_0, y_0))$. The plane stays close to the surface near the point of tangency.

The Linearization of a Function $f(x, y)$

As you can see in Fig. 17.9, the plane tangent to the surface $z = f(x, y)$ at the point $(x_0, y_0, f(x_0, y_0))$ lies close to the surface near the point of tangency. For points $(x_0 + h, y_0 + k)$ near (x_0, y_0), therefore, the z-values along the tangent plane give good approximations to the z-values on the surface. Hence, to approximate the values of f near (x_0, y_0) we can replace the formula for f by the formula for its tangent plane.

If we set $F(x, y, z)$ equal to $f(x, y) - z$, the surface $z = f(x, y)$ in Fig. 17.9 also has the equation $F(x, y, z) = 0$. The plane tangent to the surface at $P_0(x_0, y_0, f(x_0, y_0))$ is therefore the plane

$$F_x(P_0)(x - x_0) + F_y(P_0)(y - y_0) + F_z(P_0)(z - f(x_0, y_0)) = 0.$$

This becomes

$$f_x(x_0, y_0)(x - x_0) + f_y(x_0, y_0)(x - x_0) - (z - f(x_0, y_0)) = 0,$$

or

$$z = f(x_0, y_0) + f_x(x_0, y_0)(x - x_0) + f_y(x_0, y_0)(y - y_0) \tag{1}$$

when expressed in terms of f. The function

$$L(x, y) = f(x_0, y_0) + f_x(x_0, y_0)(x - x_0) + f_y(x_0, y_0)(y - y_0) \tag{2}$$

on the right-hand side of Eq. (1) is called the *linearization* of f at (x_0, y_0). For as long as the tangent plane lies close to the surface, the values of L will give good approximations to the values of f.

DEFINITION

> **The Linearization and Standard Linear Approximation of $f(x, y)$ at (x_0, y_0)**
>
> The **linearization** of the function $f(x, y)$ at (x_0, y_0) is the function
>
> $$L(x, y) = f(x_0, y_0) + f_x(x_0, y_0)(x - x_0) + f_y(x_0, y_0)(y - y_0). \tag{3}$$
>
> The approximation
>
> $$L(x, y) \approx f(x, y) \tag{4}$$
>
> is the **standard linear approximation** of f near (x_0, y_0).

In many instances, we can use the formula for L as a replacement for the formula for f. In using the formula for L, we assume, as usual, that f and its first partial derivatives are continuous in the neighborhood of (x_0, y_0).

EXAMPLE 1 Show that

$$f(x, y) = \frac{1}{1 + x - y} \approx 1 - x + y \tag{5}$$

near $(0, 0)$.

Solution We show that the linearization of f is

$$L(x, y) = 1 - x + y.$$

The function f is continuous for $|x - y| < 1$, as are its partial derivatives

$$f_x(x, y) = \frac{-1}{(1 + x - y)^2}, \qquad f_y(x, y) = \frac{1}{(1 + x - y)^2}. \qquad (6)$$

At $(x_0, y_0) = (0, 0)$, the linearization

$$L(x, y) = f(x_0, y_0) + f_x(x_0, y_0)(x - x_0) + f_y(x_0, y_0)(y - y_0)$$

is

$$L(x, y) = f(0, 0) + f_x(0, 0)x + f_y(0, 0)y$$
$$= 1 + (-1)x + (1)y$$
$$= 1 - x + y.$$

Hence, $f(x, y) \approx 1 - x + y$ near $(0, 0)$, as claimed in Eq. (5). ∎

EXAMPLE 2   Find the linearization of

$$f(x, y) = \frac{1}{1 + x - y}$$

at the point $(2, 1)$.

Solution   According to Eq. (2), the linearization of f at the point $(2, 1)$ is

$$L(x, y) = f(2, 1) + f_x(2, 1)(x - 2) + f_y(2, 1)(y - 1).$$

We use the formulas in Eq. (6) to evaluate f_x and f_y at the point $(2, 1)$. The result is

$$L(x, y) = \frac{1}{2} + \left(-\frac{1}{4}\right)(x - 2) + \left(\frac{1}{4}\right)(y - 1)$$

$$= \frac{3}{4} - \frac{x}{4} + \frac{y}{4}. \qquad ∎$$

How Accurate Is the Standard Linear Approximation?

To see how accurate the standard linear approximation is, we need to consider the second order partial derivatives of f.

The result we are about to quote assumes that f and its first and second order partial derivatives are continuous throughout a rectangular region

$$R: \qquad |x - x_0| \leq h, \qquad |y - y_0| \leq k$$

centered at (x_0, y_0). See Fig. 17.10. Since R is closed and bounded and f_{xx}, f_{yy}, and f_{xy} are continuous on R, the absolute values of these second derivatives have maximum values on R. If we let

$$B = \text{max on } R \text{ of } |f_{xx}|, \quad |f_{yy}|, \quad |f_{xy}|, \qquad (7)$$

it turns out, for reasons explained in Article 17.4, that

$$|f(x, y) - L(x, y)| \leq \frac{1}{2} B(|x - x_0| + |y - y_0|)^2 \qquad (8)$$

throughout R.

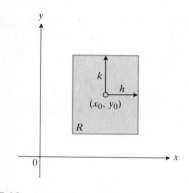

17.10   The rectangular region

$R: |x - x_0| \leq h, \quad |y - y_0| \leq k$

in the xy-plane.

The inequality in (8) shows us how the error

$$E(x, y) = f(x, y) - L(x, y)$$

in the linear approximation of f on the rectangle R is bounded:

$$|E(x, y)| \leq \frac{1}{2}B(|x - x_0| + |y - y_0|)^2. \qquad (9)$$

When we use this inequality to estimate E, we usually cannot find the values of f_{xx}, f_{yy}, and f_{xy} that determine B and we have to replace B itself with an upper-bound or "worst-case" value instead. If M is any common upper bound for $|f_{xx}|$, $|f_{yy}|$, and $|f_{xy}|$ on R, then B will be less than or equal to M and we will know that

$$|E(x, y)| \leq \frac{1}{2}M(|x - x_0| + |y - y_0|)^2. \qquad (10)$$

This is the inequality we normally use in estimating E. When we need to make $|E(x, y)|$ small for a given M, we just make $|x - x_0|$ and $|y - y_0|$ small.

The Error in the Standard Linear Approximation of $f(x, y)$ near (x_0, y_0)

If f and its first and second partial derivatives are continuous throughout a rectangle R centered at (x_0, y_0), then the error $E(x, y)$ incurred in replacing $f(x, y)$ on R by its linearization

$$L(x, y) = f(x_0, y_0) + f_x(x_0, y_0)(x - x_0) + f_y(x_0, y_0)(y - y_0) \qquad (11)$$

satisfies the inequality

$$|E(x, y)| \leq \frac{1}{2}M(|x - x_0| + |y - y_0|)^2, \qquad (12)$$

where M is any upper bound for the values of $|f_{xx}|$, $|f_{yy}|$, and $|f_{xy}|$ on R.

EXAMPLE 3 Find the linearization $L(x, y)$ of

$$f(x, y) = x^2 - xy + \frac{1}{2}y^2 + 3$$

near the point $(3, 2)$. Find an upper bound for the error in the approximation

$$f(x, y) \approx L(x, y)$$

over the rectangle

$$R: \qquad |x - 3| \leq 0.1, \qquad |y - 2| \leq 0.1.$$

Express the upper bound for the error as a percentage of $f(3, 2)$, the value of f at the center of the rectangle.

Solution With

$$f(3, 2) = 8,$$

$$f_x(3, 2) = 2x - y|_{(3, 2)} = 4,$$

$$f_y(3, 2) = -x + y|_{(3, 2)} = -1,$$

Eq. (11) gives

$$L(x, y) = f(3, 2) + f_x(3, 2)(x - 3) + f_y(3, 2)(y - 2)$$
$$= 8 + 4(x - 3) - (y - 2)$$
$$= 4x - y - 2.$$

To get an upper bound for the error in the approximation of f by L on R, we use Eq. (12):

$$|E(x, y)| \leq \frac{1}{2}M(|x - x_0| + |y - y_0|)^2.$$

To find a suitable value for M, we calculate

$$|f_{xx}| = |2| = 2,$$
$$|f_{yy}| = |1| = 1,$$
$$|f_{xy}| = |-1| = 1.$$

The largest of these is 2, so we may safely take $M = 2$. With $(x_0, y_0) = (3, 2)$, we then know that

$$|E(x, y)| \leq \frac{1}{2}(2)(|x - 3| + |y - 2|)^2 = (|x - 3| + |y - 2|)^2$$

throughout R.

Finally, since

$$|x - 3| \leq 0.1 \qquad \text{and} \qquad |y - 2| \leq 0.1$$

on R, we have

$$|E(x, y)| \leq (0.1 + 0.1)^2 = (0.2)^2 = 0.04.$$

As a percentage of $f(3, 2) = 8$, the error is no greater than

$$\frac{0.04}{8} \times 100 = 0.5\%.$$

As long as (x, y) stays in R, the approximation $f(x, y) \approx L(x, y)$ will be in error by no more than 0.04, which is $1/2\%$ of the value of f at the center of R. ∎

Predicting Change. Differentials

Suppose we know the value of a differentiable function $f(x, y)$ at a particular point (x_0, y_0) and we want to predict how much this value will change if we move nearby to $(x_0 + h, y_0 + k)$. If h and k are small, f and its linearization at (x_0, y_0) will change by approximately the same amount. Since the values of L are always easy to calculate, calculating the change in L gives a practical way to estimate the corresponding change in f.

In the notation of Fig. 17.11, the change in f is

$$\Delta f = f(x_0 + h, y_0 + k) - f(x_0, y_0).$$

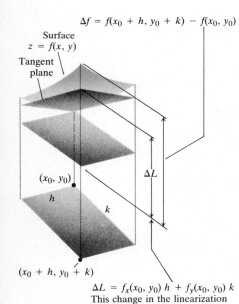

$\Delta f = f(x_0 + h, y_0 + k) - f(x_0, y_0)$

Surface
$z = f(x, y)$

Tangent plane

(x_0, y_0)

h

k

ΔL

$(x_0 + h, y_0 + k)$

$\Delta L = f_x(x_0, y_0)\, h + f_y(x_0, y_0)\, k$
This change in the linearization is also called df.

17.11 If h and k are small, the change in the linearization of f will be nearly the same as the change in f.

The corresponding change in L is

$$\Delta L = L(x_0 + h, y_0 + k) - L(x_0, y_0)$$

$$= f(x_0, y_0) + f_x(x_0, y_0)((x_0 + h) - x_0)$$

$$+ f_y(x_0, y_0)((y_0 + k) - y_0) - f(x_0, y_0) \tag{13}$$

$$= f_x(x_0, y_0)h + f_y(x_0, y_0)k.$$

The formula for Δf will usually be as hard to work with as the formula for f. The formula for ΔL, however, is always easy to work with because it is linear. The change in L is just a constant times h plus a constant times k.
   The change

$$\Delta L = f_x(x_0, y_0)h + f_y(x_0, y_0)k$$

is usually described with the more suggestive notation

$$df = f_x(x_0, y_0)dx + f_y(x_0, y_0)dy, \tag{14}$$

in which df denotes the change in the linearization of f resulting from changes dx and dy in x and y. We call dx and dy the **differentials** in x and y and df the corresponding **differential** in f.
   From a strictly mathematical point of view, Eq. (14) is simply an equation that defines a dependent variable df as a function of the four independent variables x_0, y_0, dx, and dy. When we get around to applying the equation, however, we usually want the domains of the independent variables to be restricted to ensure that the point $(x_0 + dx, y_0 + dy)$ lies within the domain of f. Calculating the change in a linearization does little good if the new point is too far away to lie in the domain of the function.

EXAMPLE 4 Suppose that a right circular cylinder initially has a radius of 1 unit and a height of 5 units. How sensitive will the cylinder's volume be to small changes in height and radius?

Solution As a function of the radius r and height h, the cylinder's volume is

$$V = \pi r^2 h.$$

The change in volume caused by small changes dr and dh in r and h is approximately

$$dV = V_r(1, 5)dr + V_h(1, 5)dh$$

$$= (2\pi r h)_{(1, 5)} \, dr + (\pi r^2)_{(1, 5)} \, dh \tag{15}$$

$$= 10\pi dr + \pi dh.$$

Thus, a 1-unit change in r will change V by about 10π units. A 1-unit change in h will change V by about π units. The volume of a cylinder of radius $r = 1$ and height $h = 5$ is nearly 10 times as sensitive to small changes in r as it is to small changes in h.
   In contrast, if the values of r and h are reversed to make $r = 5$ and $h = 1$, then

$$dV = (2\pi r h)_{(5, 1)} \, dr + (\pi r^2)_{(5, 1)} \, dh = 10\pi dr + 25\pi dh.$$

(a)   (b)

17.12 The volume of cylinder (a) is more sensitive to a small change in r than it is to an equally small change in h. The volume of cylinder (b) is more sensitive to small changes in h than it is to small changes in r.

Now the volume is more sensitive to small changes in h than it is to small changes in r. See Fig. 17.12.

The general rule to be learned from this example is that functions are most sensitive to small changes in the variables that give the largest partial derivatives. ∎

Absolute, Relative, and Percentage Change

When we move from (x_0, y_0) to a nearby point, we can describe the corresponding change in the value of a function $f(x, y)$ in three different ways.

	True	Estimate	
Absolute change:	Δf	df	
Relative change:	$\dfrac{\Delta f}{f(x_0, y_0)}$	$\dfrac{df}{f(x_0, y_0)}$	(16)
Percentage change:	$\dfrac{\Delta f}{f(x_0, y_0)} \times 100$	$\dfrac{df}{f(x_0, y_0)} \times 100$	

EXAMPLE 5   Suppose that the variables r and h change from the initial values of $(r_0, h_0) = (1, 5)$ by the amounts $dr = 0.03$ and $dh = -0.1$. Estimate the resulting absolute, relative, and percentage changes in the values of the function $V = \pi r^2 h$.

Solution   To estimate the absolute change in V we evaluate

$$dV = V_r(r_0, h_0)dr + V_h(r_0, h_0)dh \tag{17}$$

to get

$$dV = 2\pi r_0 h_0 dr + \pi r_0^2 dh$$

$$= 2\pi(1)(5)(0.03) + \pi(1)^2(-0.1)$$

$$= 0.3\pi - 0.1\pi = 0.2\pi.$$

We divide this by $V(r_0, h_0)$ to estimate the relative change:

$$\frac{dV}{V(r_0, h_0)} = \frac{0.2\pi}{\pi r_0^2 h_0} = \frac{0.2\pi}{\pi(1)^2(5)} = 0.04.$$

We multiply this by 100 to estimate the percentage change:

$$\frac{dV}{V(r_0, h_0)} \times 100 = 0.04 \times 100 = 4\%. \qquad ∎$$

EXAMPLE 6   The volume $V = \pi r^2 h$ of a right circular cylinder is to be calculated from measured values of r and h. Suppose that r is measured with an error of no more than 2% and h with an error of no more than 0.5%. Estimate the resulting possible percentage error in the calculation of V.

Solution   We are told that

$$\left| \frac{dr}{r} \times 100 \right| \leq 2 \quad \text{and} \quad \left| \frac{dh}{h} \times 100 \right| \leq 0.5.$$

Since

$$\frac{dV}{V} = \frac{2\pi r h dr + \pi r^2 dh}{\pi r^2 h} = \frac{2dr}{r} + \frac{dh}{h},$$

we have

$$\left| \frac{dV}{V} \times 100 \right| = \left| 2\frac{dr}{r} \times 100 + \frac{dh}{h} \times 100 \right|$$

$$\leq 2 \left| \frac{dr}{r} \times 100 \right| + \left| \frac{dh}{h} \times 100 \right|$$

$$\leq 2(2) + 0.5 = 4.5.$$

We estimate the error in the volume calculation to be at most 4.5%.   ■

How accurately do we have to measure r and h to have a reasonable chance of calculating $V = \pi r^2 h$ with an error, say, of less than 2%? Questions like this are hard to answer for functions of more than one variable because there is usually no single right answer. Since

$$\frac{dV}{V} = 2\frac{dr}{r} + \frac{dh}{h},$$

we see that dV/V is controlled by a combination of dr/r and dh/h. If we can measure h with great accuracy, we might come out all right even if we are sloppy about measuring r. On the other hand, our measurement of h might have so large a dh that the resulting dV/V would be too crude an estimate of $\Delta V/V$ to be useful even if dr were zero.

What we do in such cases is look for a "reasonable" square about the measured values (r_0, h_0) in which V will not vary by more than the allowed amount from $V_0 = \pi r_0^2 h_0$. The next example shows how this is done.

EXAMPLE 7   Find a reasonable square about the point $(r_0, h_0) = (5, 12)$ in which the value of $V = \pi r^2 h$ will not vary by more than ± 0.1.

Solution   We approximate the variation ΔV by the differential

$$dV = 2\pi r_0 h_0 dr + \pi r_0^2 dh$$

$$= 2\pi(5)(12)dr + \pi(5)^2 dh$$

$$= 120\pi dr + 25\pi dh.$$

Since the region to which we are restricting our attention is a square (Fig. 17.13), we may set $dh = dr$ to get

$$dV = 120\pi dr + 25\pi dr = 145\pi dr.$$

We then ask, How small must we take dr to be sure that $|dV|$ is no larger than 0.1? To answer, we start with the inequality

$$|dV| \leq 0.1,$$

express dV in terms of dr,

$$|145\pi dr| \leq 0.1,$$

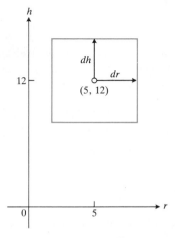

17.13 A small square about the point (5, 12) in the *rh*-plane.

and find a corresponding upper bound for dr:

$$|dr| \leq \frac{0.1}{145\pi} \approx 2.1 \times 10^{-4} \qquad \left(\begin{array}{l} \text{We round down to make sure} \\ dr \text{ won't accidentally be too big.} \end{array} \right)$$

With $dh = dr$, then, the square we want is described by the inequalities

$$|r - 5| \leq 2.1 \times 10^{-4}, \qquad |h - 12| \leq 2.1 \times 10^{-4}.$$

As long as (r, h) stays in this square, we may expect $|dV|$, and hence $|\Delta V|$, to be less than or equal to 0.1. ∎

Results for Functions of More Than Two Variables

Analogous results, which we state without proof, hold for functions of any finite number of variables.

Linearization and Increment Estimation Formulas for $f(x, y, z)$

1. *Linearization:* The linearization of f at the point (x_0, y_0, z_0) is the function

$$L(x, y, z) = f(x_0, y_0, z_0) + f_x(x_0, y_0, z_0)(x - x_0)$$
$$+ f_y(x_0, y_0, z_0)(y - y_0) + f_z(x_0, y_0, z_0)(z - z_0). \tag{18}$$

When (x, y, z) lies close to (x_0, y_0, z_0), the values of L give a good approximation to the values of f.

2. *Error bound:* Suppose that $f(x, y, z)$ and its first and second partial derivatives are continuous throughout the three-dimensional rectangular region R centered at (x_0, y_0, z_0) and defined by the inequalities

$$|x - x_0| \leq h, \qquad |y - y_0| \leq k, \qquad |z - z_0| \leq m.$$

Suppose also that $|f_{xx}|, |f_{yy}|, |f_{zz}|, |f_{xy}|, |f_{xz}|$, and $|f_{yz}|$ are all less than or equal to some number M throughout R. Then, throughout R, the magnitude of the error $E(x, y, z) = f(x, y, z) - L(x, y, z)$ in the approximation of f by L is bounded by the inequality

$$|E| \leq \frac{1}{2} M(|x - x_0| + |y - y_0| + |z - z_0|)^2. \tag{19}$$

3. *Increment estimate:* Suppose that f and its first partial derivatives are continuous and that x, y, and z are changed by small amounts dx, dy, and dz. Then the differential df defined by the equation

$$df = f_x(x, y, z)dx + f_y(x, y, z)dy + f_z(x, y, z)dz$$

is a good approximation to the resulting change in f.

EXAMPLE 8   Find the linearization $L(x, y, z)$ of

$$f(x, y, z) = x^2 - xy + 3 \sin z$$

at the point $(x_0, y_0, z_0) = (2, 1, 0)$. Give an upper bound for the error in the approximation $f(x, y, z) \approx L(x, y, z)$ over the rectangle

$$R: \qquad |x - 2| \leq 0.01, \qquad |y - 1| \leq 0.02, \qquad |z| \leq 0.01.$$

Solution We get the linearization from Eq. (18). With

$$f(2, 1, 0) = (2)^2 - (2)(1) + 3 \sin(0) = 2,$$
$$f_x(2, 1, 0) = (2x - y)_{(2, 1, 0)} = 3,$$
$$f_y(2, 1, 0) = (-x)_{(2, 1, 0)} = -2,$$
$$f_z(2, 1, 0) = (3 \cos z)_{(2, 1, 0)} = 3,$$

Eq. (18) gives

$$L(x, y, z) = 2 + 3(x - 2) + (-2)(y - 1) + 3(z - 0)$$
$$= 3x - 2y + 3z - 2.$$

We get the upper bound for the error from Eq. (19). Since

$$f_{xx} = 2, \qquad f_{yy} = 0, \qquad f_{zz} = -3 \sin z,$$
$$f_{xy} = -1, \qquad f_{xz} = 0, \qquad f_{yz} = 0,$$

we can safely take

$$M = \max(2, |-3 \sin z|) = 3$$

as an upper bound for the absolute values of the second derivatives on R. Hence,

$$|E| \leq \frac{1}{2}(3)(0.01 + 0.02 + 0.01)^2 = 0.0024.$$

The error in the approximation

$$x^2 - xy + 3 \sin z \approx 3x - 2y + 3z - 2$$

will be no greater than 0.0024 on R. ■

EXAMPLE 9 *Deflection of loaded beams.* A rectangular beam, supported at its two ends, will sag when subjected to a uniform load (Fig. 17.14). The amount of sag S, called the deflection of the beam, is calculated from the formula

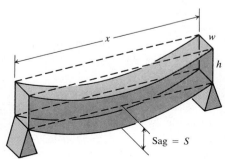

$$S = C \frac{px^4}{wh^3},$$

where

p = the load (newtons per meter of beam length),
x = the length between supports (m),
w = the width of the beam (m),
h = the height of the beam (m),
C = a constant that depends on the units of measurement and on the material from which the beam is made.

17.14 A beam supported at its two ends before and after loading.

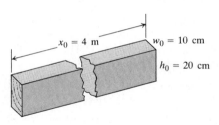

17.15 The dimensions of the beam in Example 9.

Find dS for a beam 4 m long, 10 cm wide, and 20 cm high that is subjected to a load of 100 N/m (Fig. 17.15). What conclusions can be drawn about the beam from the expression for dS?

Solution Since S is a function of the four independent variables p, x, w, and h, its differential dS is given by the equation

$$dS = S_p \, dp + S_x \, dx + S_w \, dw + S_h \, dh.$$

When we write this out for a particular set of values p_0, x_0, w_0, and h_0 and simplify the result, we find that

$$dS = S_0 \left(\frac{dp}{p_0} + \frac{4dx}{x_0} - \frac{dw}{w_0} - \frac{3dh}{h_0} \right), \tag{20}$$

where $S_0 = S(p_0, x_0, w_0, h_0) = C p_0 x_0^4 / w_0 h_0^3$.

If $p_0 = 100$ N/m, $x_0 = 4$ m, $w_0 = 0.1$ m, and $h_0 = 0.2$ m, then

$$dS = S_0 \left(\frac{dp}{100} + dx - 10dw - 15dh \right). \tag{21}$$

Here is what we can learn from Eq. (21). Since dp and dx appear with positive coefficients, increases in p and x will increase the sag. But dw and dh appear with negative coefficients, so increases in w and h will decrease the sag (make the beam stiffer). The sag is not very sensitive to changes in load because the coefficient of dp is $1/100$. The magnitude of the coefficient of dh is greater than the magnitude of the coefficient of dw. Making the beam 1 cm higher will therefore decrease the sag more than making the beam 1 cm wider.   ∎

PROBLEMS

In Problems 1–6, find the linearization $L(x, y)$ of the given function at each point.

1. $f(x, y) = x^2 + y^2 + 1$ at   (a) (0, 0),   (b) (1, 1)

2. $f(x, y) = x^3 y^4$ at   (a) (1, 1),   (b) (0, 0)

3. $f(x, y) = e^x \cos y$ at   (a) (0, 0),   (b) (0, $\pi/2$)

4. $f(x, y) = (x + y + 2)^2$ at   (a) (0, 0),   (b) (1, 2)

5. $f(x, y) = 3x - 4y + 5$ at   (a) (0, 0),   (b) (1, 1)

6. $f(x, y) = e^{2y-x}$ at   (a) (0, 0),   (b) (1, 2)

In Problems 7–12, find the linearization $L(x, y)$ of the function $f(x, y)$ at P_0. Then use Eq. (12) to find an upper bound for the magnitude of the error E in the approximation $f(x, y) \approx L(x, y)$ over the rectangle R.

7. $f(x, y) = x^2 - 3xy + 5$ at $P_0(2, 1)$,
   R: $|x - 2| \le 0.1$, $|y - 1| \le 0.1$

8. $f(x, y) = (1/2)x^2 + xy + (1/4)y^2 + 3x - 3y + 4$ at P_0 (2, 2),
   R: $|x - 2| \le 0.1$, $|y - 2| \le 0.1$

9. $f(x, y) = 1 + y + x \cos y$ at $P_0(0, 0)$,
   R: $|x| \le 0.2$, $|y| \le 0.2$
   (Use $|\cos y| \le 1$ and $|\sin y| \le 1$ in estimating E.)

10. $f(x, y) = \ln x + \ln y$ at $P_0(1, 1)$,
   R: $|x - 1| \le 0.2$, $|y - 1| \le 0.2$

11. $f(x, y) = e^x \cos y$ at $P_0(0, 0)$,
   R: $|x| \le 0.1$, $|y| \le 0.1$
   (Use $e^x \le 1.11$ and $|\cos y| \le 1$ in estimating E.)

12. $f(x, y) = xy^2 + y \cos(x - 1)$ at $P_0(1, 2)$,
   R: $|x - 1| \le 0.1$, $|y - 2| \le 0.1$

13. What relationship must hold between r_0 and h_0 for the function $V = \pi r^2 h$ to be equally sensitive to small changes in r and h at the point (r_0, h_0)? See Example 4.

14. You plan to calculate the area of a long, thin rectangle from measurements of its length and width. Which measurement should you be more careful with? Why?

15. a) Around the point (1, 0), is $f(x, y) = x^2(y + 1)$ more sensitive to changes in x, or to changes in y?
   b) What ratio of dx to dy will make df equal zero at (1, 0)?

16. Suppose T is to be found from the formula $T = x \cosh y$, where x and y are found to be 2 and $\ln 2$ with maximum possible errors of $|dx| = 0.1$ and $|dy| = 0.02$. Estimate the maximum possible error in the computed value of T.

17. About how accurately may $V = \pi r^2 h$ be calculated from measurements of r and h that are in error by 1%?

18. If $r = 5.0$ cm and $h = 12.0$ cm to the nearest millimeter, what should we expect the maximum percentage error in calculating $V = \pi r^2 h$ to be?

19. To estimate the volume of a cylinder of radius about 2 m and height about 3 m, about how accurately should the radius and height be measured so that the error in the volume estimate will not exceed 0.1 m^3? Assume that the possible error dr in measuring r is equal to the possible error dh in measuring h.

20. Give a reasonable square centered at $(1, 1)$ over which the value of $f(x, y) = x^3y^4$ will not vary by more than ± 0.1.

21. When an x-ohm and a y-ohm resistor are in parallel, the resistance R they produce may be calculated from the formula $1/R = 1/x + 1/y$. By about what percentage will R change if x increases from 20 to 20.1 ohms and y decreases from 25 to 24.9 ohms?

22. a) If $x = 3 \pm 0.01$ and $y = 4 \pm 0.01$, with approximately what accuracy can the polar coordinates r and θ of the point (x, y) be calculated from the formulas $r^2 = x^2 + y^2$ and $\theta = \tan^{-1}(y/x)$? Express your estimates as percentages of the values of r and θ at the point $(x_0, y_0) = (3, 4)$.

    b) At the point $(x_0, y_0) = (3, 4)$, are r and θ more sensitive to changes in x, or to changes in y? Draw a figure showing x, y, r, and θ and confirm your results by appealing to the geometry of the figure.

In Problems 23–28, find the linearization $L(x, y, z)$ of the given function at each point.

23. $f(x, y, z) = xy + yz + xz$ at
    a) $(1, 1, 1)$ b) $(1, 0, 0)$ c) $(0, 0, 0)$

24. $f(x, y, z) = x^2 + y^2 + z^2$ at
    a) $(1, 1, 1)$ b) $(0, 1, 0)$ c) $(1, 0, 0)$

25. $f(x, y, z) = \sqrt{x^2 + y^2 + z^2}$ at
    a) $(1, 0, 0)$ b) $(1, 1, 0)$ c) $(1, 2, 2)$

26. $f(x, y, z) = (\sin xy)/z$ at
    a) $(\pi/2, 1, 1)$ b) $(2, 0, 1)$

27. $f(x, y, z) = e^x + \cos(y + z)$ at
    a) $(0, 0, 0)$ b) $\left(0, \dfrac{\pi}{2}, 0\right)$ c) $\left(0, \dfrac{\pi}{4}, \dfrac{\pi}{4}\right)$

28. $f(x, y, z) = \tan^{-1}(xyz)$ at
    a) $(1, 0, 0)$ b) $(1, 1, 0)$ c) $(1, 1, 1)$

In Problems 29–32, find the linearization $L(x, y, z)$ of the function $f(x, y, z)$ at P_0. Then use Eq. (19) to find an upper bound for the magnitude of the error E in the approximation $f(x, y, z) \approx L(x, y, z)$ over the region R.

29. $f(x, y, z) = xz - 3yz + 2$ at $P_0(1, 1, 2)$
    R: $|x - 1| \le 0.01$, $|y - 1| \le 0.01$, $|z - 2| \le 0.02$

30. $f(x, y, z) = x^2 + xy + yz + (1/4)z^2$ at $P_0(1, 1, 2)$
    R: $|x - 1| \le 0.01$, $|y - 1| \le 0.01$, $|z - 2| \le 0.08$

31. $f(x, y, z) = xy + 2yz - 3xz$ at $P_0(1, 1, 0)$
    R: $|x - 1| \le 0.01$, $|y - 1| \le 0.01$, $|z| \le 0.01$

32. $f(x, y, z) = \sqrt{2} \cos x \sin(y + z)$ at $P_0(0, 0, \pi/4)$
    R: $|x| \le 0.01$, $|y| \le 0.01$, $|z - \pi/4| \le 0.01$

33. The beam in Example 9 is tipped on its side so that $h = 0.1$ m and $w = 0.2$ m.
    a) What is the value of dS now?

b) Compare the sensitivity of the newly positioned beam to a small change in height with its sensitivity to a change of the same size in width.

34. If $|a|$ is much greater than $|b|$, $|c|$, and $|d|$, to which entry is the value of the determinant

$$\begin{vmatrix} a & b \\ c & d \end{vmatrix}$$

most sensitive?

35. Estimate how much simultaneous errors of 2% in a, b, and c affect the calculation of the product abc.

36. Estimate the amount of material in a hollow rectangular wooden box whose inside measurements are 5 ft long by 3 ft wide by 2 ft deep if the box is made of lumber 1/2-in. thick and the box has no top.

37. The area of a triangle is $(1/2)ab \sin C$, where a and b are the lengths of two sides of the triangle and C is the measure of the included angle. In surveying a triangular plot, you have measured a, b, and C to be 150 ft, 200 ft, and 60°, respectively. By about how much will your area calculation be in error if your values of a and b are off by half a foot each and your measurement of C is off by 2°? See Fig. 17.16.

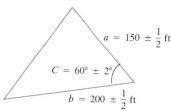

17.16 How accurately can the triangle's area be calculated from the measurements shown? See Problem 37.

38. Suppose that $u = xe^y + y \sin z$ and that x, y, and z can be measured with maximum possible errors of ± 0.2, ± 0.6, and $\pm \pi/180$, respectively. Estimate the maximum possible error in calculating u from the measured values $x = 2$, $y = \ln 3$, $z = \pi/2$.

39. *The Wilson lot size formula.* The Wilson lot size formula in economics says that the most economical quantity Q of goods (radios, shoes, brooms, whatever) for a store to order is given by the formula $Q = \sqrt{2KM/h}$, where K is the cost of placing the order, M is the number of items sold per week, and h is the weekly holding cost for each item (cost of space, utilities, security, and so on). To which of the variables K, M, and h is Q most sensitive near the point $(K_0, M_0, h_0) = (2, 20, 0.05)$?

40. Let $u = u(x, y)$ and $v = v(x, y)$. Assuming that all the necessary partial derivatives exist, establish the following differential formulas:
    a) $d(cu) = c\, du$ (any constant c)
    b) $d(u^n) = n\, u^{n-1}\, du$ (any number n)
    c) $d(uv) = u\, dv + v\, du$
    d) $d\left(\dfrac{u}{v}\right) = \dfrac{v\, du - u\, dv}{v^2}$

17.3

Lagrange Multipliers

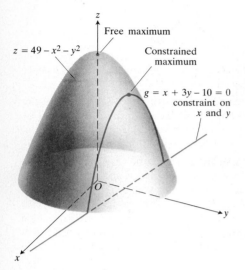

$z = 49 - x^2 - y^2$

Free maximum

Constrained maximum

$g = x + 3y - 10 = 0$
constraint on x and y

17.17 $f(x, y) = 49 - x^2 - y^2$, subject to the constraint $g(x, y) = x + 3y - 10 = 0$.

As we saw in Article 17.1, we sometimes need to find the maximum and minimum values of functions whose domains are confined to lie within some particular subset of the plane—a disk, for example, or a triangular region. But, as Fig. 17.17 suggests, functions may be subject to other kinds of constraints as well.

In this article, we explore a powerful method for finding the maxima and minima of constrained functions: the method of **Lagrange multipliers.** Lagrange developed the method in 1755 to solve sophisticated max-min problems in geometry, and the method has since become important in economics, in engineering (where it is used in designing multistage rockets, for example), and in mathematics itself.

We begin with two examples, then describe the method in general terms, and look at more examples.

Constrained Maxima and Minima

EXAMPLE 1   Find the point $P(x, y, z)$ on the plane

$$2x + y - z - 5 = 0$$

that lies closest to the origin.

Solution   The problem asks us to find the minimum value of the function

$$|\overrightarrow{OP}| = \sqrt{(x - 0)^2 + (y - 0)^2 + (z - 0)^2} = \sqrt{x^2 + y^2 + z^2}$$

subject to the constraint that

$$2x + y - z - 5 = 0.$$

Since $|\overrightarrow{OP}|$ has a minimum value wherever the function

$$f(x, y, z) = x^2 + y^2 + z^2$$

has a minimum value, we may solve the problem by finding the minimum value of $f(x, y, z)$ subject to the constraint $2x + y - z - 5 = 0$. If we regard x and y as the independent variables in this equation and write z as

$$z = 2x + y - 5,$$

our problem reduces to one of finding the points (x, y) at which the function

$$h(x, y) = f(x, y, 2x + y - 5) = x^2 + y^2 + (2x + y - 5)^2$$

has its minimum value or values. Since the domain of h is the entire xy-plane, the first derivative test of Article 17.1 tells us that any minima that h might have must occur at points where

$$h_x = 2x + 2(2x + y - 5)(2) = 0, \qquad h_y = 2y + 2(2x + y - 5) = 0.$$

This leads to

$$10x + 4y = 20, \qquad 4x + 4y = 10,$$

LAGRANGE MULTIPLIERS

The calculus of variations, invented by John Bernoulli (1667–1748), involves problems like that of finding a path $y = f(x)$ from a point $P_1(x_1, y_1)$ to a point $P_2(x_2, y_2)$ that minimizes the value of an integral

$$\int_{x_1}^{x_2} g(x, y) \, dx.$$

One such problem is that of finding a path from P_1 to P_2 that minimizes the time it takes for an object to slide from one point to the other under the force of gravity (see Article 8.9). Another is that of finding a path from P_1 to P_2 that sweeps out the minimum surface area when revolved about the x-axis. (It will be a catenary like the one in Article 9.3.)

Newton, Leibniz, and Euler all worked on variational problems as part of calculus, but the contributions of Joseph Louis Lagrange (1736–1813), beginning about 1760, forged the calculus of variations into a self-standing branch of mathematics. Euler considered Lagrange's work in this new field to be superior to his own and delayed publishing his own work so that Lagrange could receive full acclaim. While working on the calculus of variations Lagrange developed the elegant procedure that we know today as the method of Lagrange multipliers.

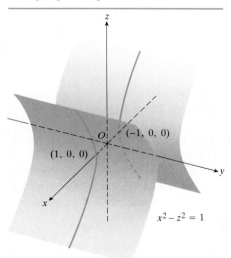

17.18 The hyperbolic cylinder $x^2 - z^2 - 1 = 0$ in Example 2.

and the solution

$$x = \frac{5}{3}, \qquad y = \frac{5}{6}.$$

We may apply either geometric arguments or the second derivative test to show that these values minimize h. The z-coordinate of the corresponding point on the plane $z = 2x + y - 5$ is

$$z = 2\left(\frac{5}{3}\right) + \frac{5}{6} - 5 = -\frac{5}{6}.$$

Therefore, the point we seek is

Closest point: $P\left(\dfrac{5}{3}, \dfrac{5}{6}, -\dfrac{5}{6}\right).$ ■

Attempts to solve a constrained maximum or minimum problem by substitution, as we might call the method of Example 1, do not always go smoothly, as the next example shows. This is one of the reasons for learning the new method of this article, which does not require us to decide in advance which of the constrained variables to regard as independent.

EXAMPLE 2 Find the points on the surface

$$x^2 - z^2 - 1 = 0$$

closest to the origin.

Solution 1 The surface is shown in Fig. 17.18. We seek the points on the surface closest to the origin. These are the points whose coordinates minimize the value of the function

$$f(x, y, z) = x^2 + y^2 + z^2 \qquad \text{(square of the distance)}$$

subject to the constraint that $x^2 - z^2 - 1 = 0$. If we regard x and y as independent variables in the constraint equation, then

$$z^2 = x^2 - 1$$

and the values of $f(x, y, z) = x^2 + y^2 + z^2$ on the surface are given by the function

$$h(x, y) = x^2 + y^2 + (x^2 - 1) = 2x^2 + y^2 - 1.$$

To find the points on the surface whose coordinates minimize f, we look for the points in the xy-plane whose coordinates minimize h. The only extreme value of h occurs where

$$h_x = 4x = 0 \qquad \text{and} \qquad h_y = 2y = 0,$$

i.e., at the point $(0, 0)$. But now we're in trouble—there are no points on the surface where both x and y are zero. What went wrong?

What happened was that the first derivative test found (as it should have) the point *in the domain of h* where h has a minimum value. We, on the other hand, want the points *on the surface* where h has a minimum value. While the domain of h is the entire xy-plane, the domain from which we can select the first two coordinates of the points (x, y, z) on the surface is restricted to the

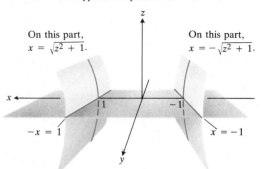

The hyperbolic cylinder $x^2 - z^2 = 1$

On this part,
$x = \sqrt{z^2 + 1}.$

On this part,
$x = -\sqrt{z^2 + 1}.$

17.19 The region in the xy-plane from which the first two coordinates of the points (x, y, z) on the surface $x^2 - z^2 = 1$ are selected excludes the band $-1 < x < 1$ in the xy-plane.

"shadow" of the surface on the xy-plane; it does not include the band between the lines $x = -1$ and $x = 1$ (Fig. 17.19).

We can avoid this problem if we treat y and z as independent variables (instead of x and y) and express x in terms of y and z as

$$x^2 = z^2 + 1.$$

With this substitution, $f(x, y, z) = x^2 + y^2 + z^2$ becomes

$$k(y, z) = (z^2 + 1) + y^2 + z^2 = 1 + y^2 + 2z^2$$

and we look for the points where k takes on its smallest value. The domain of k in the yz-plane now matches the domain from which we select the y- and z-coordinates of the points (x, y, z) on the surface. Hence, the points that minimize k in the plane will have corresponding points on the hyperboloid. The smallest values of k occur where

$$k_y = 2y = 0 \qquad \text{and} \qquad k_z = 4z = 0,$$

or where $y = z = 0$. This leads to

$$x^2 = z^2 + 1 = 1, \qquad x = \pm 1.$$

The corresponding points on the surface are the points $(\pm 1, 0, 0)$. We can see from the inequality

$$k(y, z) = 1 + y^2 + 2z^2 \geq 1$$

that the points $(\pm 1, 0, 0)$ give a minimum value for k. We can also see that the minimum distance from the origin to a point on the surface is 1 unit.

Solution 2   Another way to find the points on the surface closest to the origin is to imagine a small sphere centered at the origin expanding like a soap bubble until it just touches the surface (Fig. 17.20). At each point of contact, the surface and sphere have the same tangent plane and normal line. Therefore, if the sphere and surface are represented as level surfaces of the functions

$$f(x, y, z) = x^2 + y^2 + z^2 - a^2 = 0, \qquad g(x, y, z) = x^2 - z^2 - 1 = 0,$$

then the gradients ∇f and ∇g will be parallel where the surfaces touch. At any point of contact we will therefore be able to find a scalar λ such that

$$\nabla f = \lambda \nabla g,$$

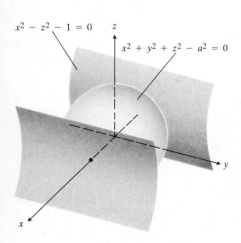

$x^2 - z^2 - 1 = 0$

$x^2 + y^2 + z^2 - a^2 = 0$

17.20   The sphere obtained by expanding a soap bubble centered at the origin until it just touches the hyperbolic cylinder $x^2 - z^2 - 1 = 0$. See Solution 2 of Example 2.

or

$$2x\mathbf{i} + 2y\mathbf{j} + 2z\mathbf{k} = \lambda(2x\mathbf{i} - 2z\mathbf{k}).$$

Thus, the coordinates x, y, and z of any point of tangency will have to satisfy the three scalar equations

$$2x = 2\lambda x, \qquad 2y = 0, \qquad 2z = -2\lambda z. \qquad (1)$$

For what values of λ will a point (x, y, z) whose coordinates satisfy the equations in (1) also lie on the surface $x^2 - z^2 - 1 = 0$? To answer this question, we use the fact that no point on the surface has a zero x-coordinate to conclude that $x \neq 0$ in the first equation in (1). This means that $2x = 2\lambda x$ only if

$$2 = 2\lambda \qquad \text{or} \qquad \lambda = 1.$$

For $\lambda = 1$, the equation $2z = -2\lambda z$ becomes $2z = -2z$. If this equation is to be satisfied as well, z must be zero. Since $y = 0$ also (from the equation $2y = 0$), we conclude that the points we seek all have coordinates of the form

$$(x, 0, 0).$$

Finally, what points on the surface $x^2 - z^2 = 1$ have coordinates of this form? The points $(x, 0, 0)$ for which

$$x^2 - (0)^2 = 1, \qquad x^2 = 1, \qquad \text{or} \qquad x = \pm 1.$$

The points on the surface closest to the origin are the points $(\pm 1, 0, 0)$. ∎

The Method of Lagrange Multipliers

In Solution 2 above we solved the problem presented in Example 2 by the **method of Lagrange multipliers.** In general terms, the method says that the extreme values of a function $f(x, y, z)$ whose variables are subject to a constraint of the form $g(x, y, z) = 0$ are to be found on the surface $g = 0$ at the points where

$$\nabla f = \lambda \nabla g$$

for some scalar λ (called a **Lagrange multiplier**).

To explore the method further and see why it works, we first make the following observation, which we state as a theorem.

THEOREM 1

> Suppose that $f(x, y, z)$ and its first partial derivatives are continuous in a region that contains the differentiable curve
>
> $$C: \quad \mathbf{R}(t) = x(t)\mathbf{i} + y(t)\mathbf{j} + z(t)\mathbf{k}.$$
>
> If P_0 is a point on C where f has a local maximum or minimum relative to its values on C, then ∇f is perpendicular to C at P_0.

Proof We show that $\nabla f \cdot \mathbf{v} = 0$ at the points in question, where $\mathbf{v} = d\mathbf{R}/dt$ is the tangent vector to C. The values of f on C are given by the composite function

$$f(x(t), y(t), z(t)),$$

whose derivative with respect to t is

$$\frac{df}{dt} = \frac{\partial f}{\partial x}\frac{dx}{dt} + \frac{\partial f}{\partial y}\frac{dy}{dt} + \frac{\partial f}{\partial z}\frac{dz}{dt} = \nabla f \cdot \mathbf{v}.$$

At any point P_0 where f has a local maximum or minimum relative to its values on the curve, $df/dt = 0$, so that

$$\nabla f \cdot \mathbf{v} = 0. \qquad \blacksquare$$

By dropping the z-terms in the statement and proof of Theorem 1 we obtain a similar result for functions of two variables: The points on a differentiable curve

$$\mathbf{R}(t) = x(t)\mathbf{i} + y(t)\mathbf{j}$$

where a function $f(x, y)$ takes on its local maxima and minima relative to its values on the curve are the points where $\nabla f \cdot d\mathbf{R}/dt = 0$.

Theorem 1 is the key to why the method of Lagrange multipliers works, as we shall now see. Suppose that $f(x, y, z)$ and $g(x, y, z)$ have continuous first partial derivatives and that P_0 is a point on the surface $g(x, y, z) = 0$ where f has a local maximum or minimum value relative to its other values on the surface. Then f takes on a local maximum or minimum at P_0 relative to its values on every differentiable curve through P_0 on the surface $g(x, y, z) = 0$. Therefore, ∇f is perpendicular to the velocity vector of every such differentiable curve through P_0. But so is ∇g (because ∇g is perpendicular to the level surface $g = 0$, as we saw in Article 16.7). Therefore, at P_0, ∇f is some scalar multiple λ of ∇g.

The argument for the two-variable case is left as an exercise (Miscellaneous Problem 17).

The Method of Lagrange Multipliers

Suppose that $f(x, y, z)$ and $g(x, y, z)$ and their first partial derivatives are continuous. To find the local maximum and minimum values of f subject to the constraint $g(x, y, z) = 0$, find the values of x, y, z, and λ that simultaneously satisfy the equations

$$\nabla f = \lambda \nabla g \qquad \text{and} \qquad g(x, y, z) = 0. \qquad (2)$$

Comment on Notation   Some books describe the method of Lagrange multipliers without vector notation, in the following equivalent way:

To maximize or minimize a function $f(x, y, z)$ subject to the constraint $g(x, y, z) = 0$, construct the auxiliary function

$$H(x, y, z, \lambda) = f(x, y, z) - \lambda g(x, y, z). \qquad (3)$$

Then find the values of x, y, z, and λ for which the partial derivatives of H are all zero:

$$H_x = 0, \qquad H_y = 0, \qquad H_z = 0, \qquad H_\lambda = 0. \qquad (4)$$

These requirements are equivalent to the requirements in (2), as we can see by calculating the partial derivatives of H. From (4) we have

$$H_x = f_x - \lambda g_x = 0 \quad \text{or} \quad f_x = \lambda g_x,$$
$$H_y = f_y - \lambda g_y = 0 \quad \text{or} \quad f_y = \lambda g_y,$$
$$H_z = f_z - \lambda g_z = 0 \quad \text{or} \quad f_z = \lambda g_z,$$
$$H_\lambda = -g(x, y, z) = 0 \quad \text{or} \quad g(x, y, z) = 0.$$

(5)

The first three equations give $\nabla f = \lambda \nabla g$, and the last is $g(x, y, z) = 0$.

EXAMPLE 3 Find the greatest and smallest values that the function

$$f(x, y) = xy$$

takes on the ellipse

$$\frac{x^2}{8} + \frac{y^2}{2} = 1.$$

Solution We are asked to find the extreme values of $f(x, y) = xy$ subject to the constraint

$$g(x, y) = \frac{x^2}{8} + \frac{y^2}{2} - 1 = 0.$$

To do so, we first find the values of x, y, and λ for which

$$\nabla f = \lambda \nabla g \quad \text{and} \quad g(x, y) = 0.$$

The gradient equation gives

$$y\mathbf{i} + x\mathbf{j} = \frac{\lambda}{4}x\mathbf{i} + \lambda y\mathbf{j},$$

from which we find

$$y = \frac{\lambda}{4}x, \qquad x = \lambda y,$$

and

$$y = \frac{\lambda}{4}(\lambda y) = \frac{\lambda^2}{4}y, \quad \text{so that} \quad y = 0 \quad \text{or} \quad \lambda = \pm 2.$$

We now consider these two cases.

CASE 1: If $y = 0$, then $x = y = 0$, but the point $(0, 0)$ is not on the ellipse. Hence, $y \neq 0$.

CASE 2: If $y \neq 0$, then $\lambda = \pm 2$ and

$$x = \pm 2y.$$

Substituting this in the equation $g(x, y) = 0$ gives

$$\frac{(\pm 2y)^2}{8} + \frac{y^2}{2} = 1,$$
$$4y^2 + 4y^2 = 8,$$
$$y = \pm 1.$$

The function $f(x, y) = xy$ therefore takes on its extreme values on the ellipse at the four points $(\pm 2, 1)$, $(\pm 2, -1)$. The extreme values are $xy = 2$ and $xy = -2$.

The Geometry of the Solution   (Fig. 17.21) The level curves of the function $f(x, y) = xy$ are the hyperbolas $xy = c$. The farther the hyperbolas lie from the origin, the larger the absolute value of f. We want to find the extreme values of $f(x, y)$, given that the point (x, y) also lies on the ellipse $x^2 + 4y^2 = 8$. Which hyperbolas intersecting the ellipse lie farthest from the origin? The hyperbolas that just graze the ellipse, the ones that are tangent to it. At these points, any vector normal to the hyperbola is normal to the ellipse, so the gradient $\nabla f = y\mathbf{i} + x\mathbf{j}$ is a multiple ($\lambda = \pm 2$) of the gradient $\nabla g = (x/4)\mathbf{i} + y\mathbf{j}$. At the point $(2, 1)$, for example,

$$\nabla f = \mathbf{i} + 2\mathbf{j}, \qquad \nabla g = \frac{1}{2}\mathbf{i} + \mathbf{j}, \qquad \text{and} \qquad \nabla f = 2\nabla g.$$

At the point $(-2, 1)$,

$$\nabla f = \mathbf{i} - 2\mathbf{j}, \qquad \nabla g = -\frac{1}{2}\mathbf{i} + \mathbf{j}, \qquad \text{and} \qquad \nabla f = -2\nabla g. \qquad \blacksquare$$

EXAMPLE 4   Find the maximum and minimum values of the function

$$3x + 4y$$

on the circle

$$x^2 + y^2 = 1.$$

17.21   When subjected to the constraint $g(x, y) = x^2/8 + y^2/2 - 1 = 0$, the function $f(x, y) = xy$ takes on extreme values at the four points $(\pm 2, \pm 1)$. These are the points on the ellipse where ∇f (black) is a multiple of ∇g (blue).

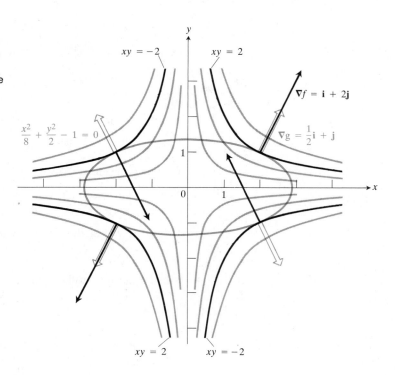

Solution We model this as a Lagrange multiplier problem with

$$f(x, y) = 3x + 4y, \qquad g(x, y) = x^2 + y^2 - 1$$

and look for the values of x, y, and λ that satisfy the equations

$$\nabla f = \lambda \nabla g: \qquad 3\mathbf{i} + 4\mathbf{j} = 2x\lambda\mathbf{i} + 2y\lambda\mathbf{j},$$

$$g(x, y) = 0: \qquad x^2 + y^2 - 1 = 0. \tag{6}$$

The gradient equation implies that $\lambda \neq 0$ and gives

$$x = \frac{3}{2\lambda}, \qquad y = \frac{2}{\lambda}.$$

These equations tell us, among other things, that x and y have the same sign. With these values for x and y, the equation $g(x, y) = 0$ gives

$$\left(\frac{3}{2\lambda}\right)^2 + \left(\frac{2}{\lambda}\right)^2 - 1 = 0,$$

$$\frac{9}{4\lambda^2} + \frac{4}{\lambda^2} = 1,$$

$$9 + 16 = 4\lambda^2,$$

$$4\lambda^2 = 25,$$

$$\lambda = \pm\frac{5}{2}.$$

Thus,

$$x = \frac{3}{2\lambda} = \pm\frac{3}{5}, \qquad y = \frac{2}{\lambda} = \pm\frac{4}{5},$$

and $f(x, y) = 3x + 4y$ has extreme values at the points

$$(x, y) = \pm\left(\frac{3}{5}, \frac{4}{5}\right).$$

(There are only two points instead of four because x and y have the same sign.)

By calculating the value of $3x + 4y$ at the points $\pm(3/5, 4/5)$, we see that its maximum and minimum values on the circle $x^2 + y^2 = 1$ are

$$3\left(\frac{3}{5}\right) + 4\left(\frac{4}{5}\right) = \frac{25}{5} = 5$$

and

$$3\left(-\frac{3}{5}\right) + 4\left(-\frac{4}{5}\right) = -\frac{25}{5} = -5.$$

The Geometry of the Solution (Fig. 17.22) The level curves of $f(x, y) = 3x + 4y$ are the lines $3x + 4y = c$. The farther the lines lie from the origin, the larger the absolute value of f. We want to find the extreme values of $f(x, y)$ given that the point (x, y) also lies on the circle $x^2 + y^2 = 1$. Which lines intersecting the circle lie farthest from the origin? The lines tangent to the

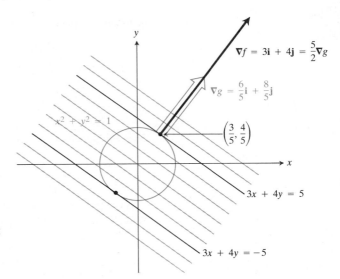

17.22 The function $f(x, y) = 3x + 4y$ takes on its largest value on the unit circle $g(x, y) = x^2 + y^2 - 1 = 0$ at the point $(3/5, 4/5)$ and its smallest value at the point $(-3/5, -4/5)$. At each of these points, $\nabla f = (5/2)\, \nabla g$. The figure shows the gradients at the first point but not the second.

circle. At the points of tangency, any vector normal to the line is normal to the circle, so the gradient $\nabla f = 3\mathbf{i} + 4\mathbf{j}$ is a multiple ($\lambda = \pm 5/2$) of the gradient $\nabla g = 2x\mathbf{i} + 2y\mathbf{j}$. At the point $(3/5, 4/5)$, for example,

$$\nabla f = 3\mathbf{i} + 4\mathbf{j}, \qquad \nabla g = \frac{6}{5}\mathbf{i} + \frac{8}{5}\mathbf{j}, \qquad \text{and} \qquad \nabla f = \frac{5}{2}\nabla g. \qquad \blacksquare$$

Lagrange Multipliers with Two Constraints

If there are two constraints on the variables of $f(x, y, z)$, say

$$g_1(x, y, z) = 0 \qquad \text{and} \qquad g_2(x, y, z) = 0,$$

we may find the constrained local maxima and minima of f by introducing two Lagrange multipliers λ and μ. That is, we locate the points $P(x, y, z)$ where f takes on its constrained extreme values by finding the values of x, y, z, λ, and μ that simultaneously satisfy the equations

$$\nabla f = \lambda \nabla g_1 + \mu \nabla g_2, \tag{7a}$$

$$g_1(x, y, z) = 0, \tag{7b}$$

$$g_2(x, y, z) = 0. \tag{7c}$$

The functions f, g_1, and g_2 are required to be continuous and to have continuous first partial derivatives.

The equations in (7) have a nice geometric interpretation. The surfaces $g_1 = 0$ and $g_2 = 0$ (usually) intersect in a differentiable curve, say C (Fig. 17.23), and along this curve we seek the points where f has local maximum and minimum values relative to its other values on the curve. These are the points where ∇f is normal to C, as we saw in Theorem 1. But ∇g_1 and ∇g_2 are also normal to C at these points because C lies in the surfaces $g_1 = 0$ and $g_2 = 0$. Therefore ∇f lies in the plane determined by ∇g_1 and ∇g_2, which means that

$$\nabla f = \lambda \nabla g_1 + \mu \nabla g_2$$

for some λ and μ. Since the points we seek also lie in both surfaces, their

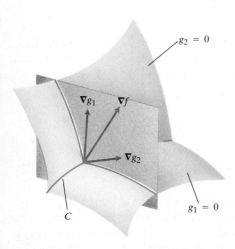

17.23 The vectors ∇g_1 and ∇g_2 lie in a plane perpendicular to the curve C because ∇g_1 is normal to the surface $g_1 = 0$ and ∇g_2 is normal to the surface $g_2 = 0$.

coordinates must satisfy the equations

$$g_1(x, y, z) = 0 \quad \text{and} \quad g_2(x, y, z) = 0,$$

which are the remaining requirements in (7).

EXAMPLE 5 The cone $z^2 = x^2 + y^2$ is cut by the plane $z = 1 + x + y$ in a conic section C. Find the points on C that are nearest to and farthest from the origin.

Solution We model this as a Lagrange multiplier problem in which we find the extreme values of

$$f(x, y, z) = x^2 + y^2 + z^2$$

(the square of the function that measures distance from the origin) subject to the two constraints

$$g_1(x, y, z) = x^2 + y^2 - z^2 = 0, \tag{8a}$$

$$g_2(x, y, z) = 1 + x + y - z = 0. \tag{8b}$$

The gradient equation (7a) gives

$$2x\mathbf{i} + 2y\mathbf{j} + 2z\mathbf{k} = \lambda(2x\mathbf{i} + 2y\mathbf{j} - 2z\mathbf{k}) + \mu(\mathbf{i} + \mathbf{j} - \mathbf{k}),$$

which leads to the scalar equations

$$\left.\begin{array}{r} 2x = 2x\lambda + \mu \\ 2y = 2y\lambda + \mu \\ 2z = -2z\lambda - \mu \end{array}\right\} \quad \begin{array}{l} \Rightarrow \quad x - y = (x - y)\lambda, \quad (9a) \\ \Rightarrow \quad y + z = (y - z)\lambda. \quad (9b) \end{array}$$

The equation $x - y = (x - y)\lambda$ is satisfied if $x = y$ or if $x \neq y$ and $\lambda = 1$.

The value $\lambda = 1$ does not lead to a point on the cutting plane (that is, to a point whose coordinates satisfy Eq. 8b). For $\lambda = 1$ implies

$$y + z = y - z, \quad \text{(from Eq. 9b)}$$

$$2z = 0,$$

$$z = 0,$$

and Eq. (8a) then gives

$$x^2 + y^2 = 0, \quad x = 0, \quad y = 0.$$

The point $(0, 0, 0)$ does not satisfy the constraint (8b).

Therefore $\lambda \neq 1$ and we have $x = y$. The plane $x = y$ meets the plane $z = 1 + x + y$ in a line that cuts the cone in just two points, as we can see by substituting $y = x$ and $z = 1 + 2x$ into the cone equation

$$z^2 = x^2 + y^2,$$

$$(1 + 2x)^2 = x^2 + x^2,$$

$$2x^2 + 4x + 1 = 0,$$

$$x = -1 \pm \frac{\sqrt{2}}{2}.$$

The points are

$$A = \left(-1 - \sqrt{\frac{1}{2}}, \; -1 - \sqrt{\frac{1}{2}}, \; -1 - \sqrt{2}\right),$$

$$B = \left(-1 + \sqrt{\frac{1}{2}}, \; -1 + \sqrt{\frac{1}{2}}, \; -1 + \sqrt{2}\right).$$

(10)

Now, we know that C is either an ellipse or a hyperbola. If it is an ellipse, we would conclude that B is the point on it nearest the origin and A the point farthest from the origin. But if it is a hyperbola, then there is no point on it that is farthest from the origin, and the points A and B are the points on the two branches that are nearest the origin. Problem 36 asks you to think about these possibilities and decide between them. It seems obvious that the critical points should satisfy the condition $x = y$ because all three of the functions f, g_1, and g_2 treat x and y alike.   ∎

PROBLEMS

1. Find the points on the ellipse $x^2 + 2y^2 = 1$ where $f(x, y) = xy$ has its extreme values.

2. Find the extreme values of $f(x, y) = xy$ subject to the constraint $g(x, y) = x^2 + y^2 - 10 = 0$.

3. Find the maximum value of $f(x, y) = 49 - x^2 - y^2$ on the line $x + 3y = 10$ (Fig. 17.17).

4. How close does the line $y = x + 1$ come to the parabola $y^2 = x$?

5. Find the extreme values of $f(x, y) = x^2 y$ on the line $x + y = 3$.

6. Find the points on the curve $x^2 y = 2$ nearest the origin.

7. Use the method of Lagrange multipliers to find
   a) the minimum value of $x + y$, subject to the constraints $xy = 16$, $x > 0$, $y > 0$;
   b) the maximum value of xy, subject to the constraint $x + y = 16$.

   Comment on the geometry of each solution.

8. Find the points on the curve $x^2 + xy + y^2 = 1$ that are nearest to and farthest from the origin.

9. Find the dimensions of the closed circular can of smallest surface area whose volume is 16π cm^3.

10. Use the method of Lagrange multipliers to find the dimensions of the rectangle of greatest area that can be inscribed in the ellipse $x^2/16 + y^2/9 = 1$ with sides parallel to the coordinate axes.

11. The temperature at a point (x, y) on a metal plate is $T(x, y) = 4x^2 - 4xy + y^2$. An ant on the plate walks around the circle of radius 5 centered at the origin. What are the highest and lowest temperatures encountered by the ant?

12. Your firm has been asked to design a storage tank for liquid petroleum gas. The customer's specifications call for a cylindrical tank with hemispherical ends, and the tank is to hold 8000 m^3 of gas. The customer wants to use the smallest amount of material possible in building the tank. What radius and height do you recommend for the cylindrical portion of the tank?

13. Find the maximum and minimum values of $x^2 + y^2$ subject to the constraint $x^2 - 2x + y^2 - 4y = 0$.

14. A pentagon is made by mounting an isosceles triangle on top of a rectangle. What dimensions minimize the perimeter for a given area A?

15. Find the point on the plane $x + 2y + 3z = 13$ closest to the point $(1, 1, 1)$.

16. Find the maximum and minimum values of

$$f(x, y, z) = x - 2y + 5z$$

on the sphere

$$x^2 + y^2 + z^2 = 30.$$

17. Find the minimum distance from the surface $x^2 + y^2 - z^2 = 1$ to the origin.

18. Find the point on the surface $z = xy + 1$ nearest the origin.

19. Find the points on the surface $z^2 = xy + 4$ closest to the origin.

20. Find the points on the sphere $x^2 + y^2 + z^2 = 25$ where $f(x, y, z) = x + 2y + 3z$ has its maximum and minimum values.

21. The temperature T at any point (x, y, z) in space is $T = 400xyz^2$. Find the highest temperature on the unit sphere $x^2 + y^2 + z^2 = 1$.

22. Find three real numbers whose sum is 9 and the sum of whose squares is as small as possible.

23. Find the largest product the positive numbers x, y, and z can have if $x + y + z^2 = 16$.

24. If a, b, and c are positive numbers, find the maximum value that $f(x, y, z) = ax + by + cz$ can take on the sphere $x^2 + y^2 + z^2 = 1$.

25. Find the maximum and minimum values of $f(x, y) = x^2 + 3y^2 + 2y$ on the unit disk $x^2 + y^2 \leq 1$.

26. Find the extreme values of $f(x, y, z) = x^2yz + 1$ on the intersection of the plane $z = 1$ with the sphere $x^2 + y^2 + z^2 = 2$.

27. A space probe in the shape of the ellipsoid

$$4x^2 + y^2 + 4z^2 = 16$$

enters the earth's atmosphere and its surface begins to heat. After one hour, the temperature at the point (x, y, z) on the probe's surface is

$$T(x, y, z) = 8x^2 + 4yz - 16z + 600.$$

Find the hottest point on the probe's surface.

28. You are in charge of erecting a radio telescope on a newly discovered planet. To minimize interference, you want to place it where the magnetic field of the planet is weakest. The planet is spherical with a radius of 6 units. Based on a coordinate system whose origin is at the center of the planet, the strength of the magnetic field is given by $M(x, y, z) = 6x - y^2 + xz + 60$. Where should you locate the radio telescope?

29. *An example from economics.* In economics, the usefulness or *utility* of amounts x and y of two capital goods G_1 and G_2 is sometimes measured by a function $U(x, y)$. For example, G_1 and G_2 might be two chemicals a pharmaceutical company needs to have on hand and $U(x, y)$ the gain from manufacturing a product whose synthesis requires different amounts of the chemicals depending on the process used. If G_1 costs a dollars per kilogram, G_2 costs b dollars per kilogram, and the total amount allocated for the purchase of G_1 an G_2 together is c dollars, then the company wants to maximize $U(x, y)$ given that $ax + by = c$. Thus, they need to solve a typical Lagrange multiplier problem.

  Suppose

$$U(x, y) = xy + 2x$$

and that the equation $ax + by = c$ becomes

$$2x + y = 30$$

when reduced to lowest terms. Find the maximum value of U and the corresponding values of x and y subject to this latter constraint.

30. *The condition $\nabla f = \lambda \nabla g$ is not sufficient.* While $\nabla f = \lambda \nabla g$ is a necessary condition for the occurrence of an extreme value of $f(x, y)$ subject to the condition $g(x, y) = 0$, it does not in itself guarantee that one exists. As a case in point, try using the method of Lagrange multipliers to find a maximum value of $f(x, y) = x + y$ subject to the constraint that $xy = 16$. The method will identify the two points $(4, 4)$ and $(-4, -4)$ as candidates for the location of extreme values. Yet the sum $(x + y)$ has no maximum value on the hyperbola $xy = 16$. The farther you go from the origin on this

hyperbola in the first quadrant, the larger the sum $f(x, y) = x + y$ becomes.

31. Given n positive numbers $a_1, a_2, \ldots, a_n$, find the maximum value of the expression

$$w = a_1x_1 + a_2x_2 + \cdots + a_nx_n$$

if the variables $x_1, x_2, \ldots, x_n$ are restricted so that the sum of their squares is 1.

32. A plane of the form

$$z = Ax + By + C$$

is to be "fitted" to the following points (x_i, y_i, z_i):

$$(0, 0, 0), \quad (0, 1, 1), \quad (1, 1, 1), \quad (1, 0, -1).$$

Find the plane that minimizes the sum of the squares of the deviations,

$$\sum_{i=1}^{4} (Ax_i + By_i + C - z_i)^2.$$

Lagrange Multipliers with Two Constraints

33. Find the point closest to the origin on the intersection of the planes $y + 2z = 12$ and $x + z = 6$.

34. Find the maximum value of $f(x, y, z) = x^2 + 2y - z^2$ subject to the constraints $2x - y = 0$ and $y + z = 0$.

35. a) Find the maximum value of $w = xyz$ among all points lying on the intersection of the two planes $x + y + z = 40$ and $z = x + y$.

    b) Give a geometric argument supporting the fact that you have found a maximum (and not a minimum) value of xyz subject to the constraints.

36. *Continuation of Example 5.* In the solution of Example 5, two points A and B were located as candidates for maximum or minimum distances from the origin. Use cylindrical coordinates r, θ, z to express the equations of the cone, the plane $z = 1 + x + y$, and the cylinder that contains their curve of intersection and has elements parallel to the z-axis. Is this cylinder circular, elliptical, parabolic, or hyperbolic? (Consider its intersection with the xy-plane.) Express the distance from the origin to a point on the curve of intersection of the cone and the plane as a function of θ. Does it have a minimum? A maximum? What can you now say about the points A and B of Eqs. (10) as solutions in Example 5?

37. In Example 5, the extreme values of $f(x, y, z)$ on the cone $z^2 = x^2 + y^2$ and the plane $z = 1 + x + y$ were found to satisfy the equation $y = x$ as well. Thus $z = 1 + 2x$, and the function to be maximized or minimized is $x^2 + y^2 + z^2 = x^2 + x^2 + (1 + 2x)^2$, which can also be written as $6(x + 1/3)^2 + 1/3$. This is obviously a minimum when $x = -1/3$. But the point we get with

$$y = x, \quad z = 1 + 2x, \quad x = -\frac{1}{3},$$

is $(-1/3, -1/3, 1/3)$, which is not on the cone. What's

wrong? (A sketch of the situation in the plane $y = x$ may throw some light on the question.)

38. In Example 5, we can determine that for all points on the cone $z^2 = x^2 + y^2$, the square of the distance from the origin to $P(x, y, z)$ is $w = 2(x^2 + y^2)$, which is a function of two independent variables, x and y. But if P is also to be on the plane

$$z = 1 + x + y$$

as well as the cone, then

$$(1 + x + y)^2 = x^2 + y^2.$$

Show that these points have coordinates that satisfy the equation

$$2xy + 2x + 2y + 1 = 0.$$

Interpret this equation in two ways:
a) as a curve in the xy-plane and
b) as a set of points on a cylinder in 3-space.

Sketch the curve of (a) and find the point or points on it for which w is a minimum. Are there points on this curve for which w is a maximum? Use the information you now have to complete the discussion of Example 5.

☐ 17.4

Taylor's Formula, the Second Derivative Test, and Error Estimates

In this article, we explore the theory behind the second derivative test in Article 17.1 and the formulas we used to estimate the errors in the linear approximations in Article 17.2. The underlying mathematics in each case is the mathematics of Taylor's formula from Article 12.3. Our use of the "single-variable" Taylor's formula in this context leads us to extend Taylor's formula to functions of two variables. This extended formula provides quadratic approximations of functions of two variables along with estimates of the associated errors.

The Derivation of the Second Derivative Test

The second derivative test from Article 17.1 says that if $f(x, y)$ and its first and second partial derivatives are continuous throughout some region R whose interior contains a point (a, b) at which f_x and f_y are zero, then

1. f has a *local maximum* at (a, b) if $f_{xx} < 0$ and $f_{xx}f_{yy} - f_{xy}^2 > 0$ at (a, b);
2. f has a *local minimum* at (a, b) if $f_{xx} > 0$ and $f_{xx}f_{yy} - f_{xy}^2 > 0$ at (a, b);
3. f has a *saddle point* at (a, b) if $f_{xx}f_{yy} - f_{xy}^2 < 0$ at (a, b);
4. the test is *inconclusive* at (a, b) if $f_{xx}f_{yy} - f_{xy}^2 = 0$ at (a, b). We must find some other way to determine the behavior of f at (a, b).

The expression $f_{xx}f_{yy} - f_{xy}^2 = 0$ is called the *discriminant* of f. The derivation of the test involves a clever use of Taylor's formula from Article 12.3. The argument goes like this:

We assume that f and its first and second order partial derivatives are continuous in some region R about $P(a, b)$, as shown in Fig. 17.24. We let $S(a + h, b + k)$ be a point close enough to P for the segment PS to lie in R and we describe PS with the parametric equations

$$x = a + th, \qquad y = b + tk, \qquad 0 \le t \le 1. \tag{1}$$

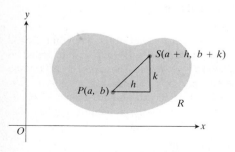

17.24 A region R whose interior contains the point $P(a, b)$.

We study the values of $f(x, y)$ along PS by considering the function

$$F(t) = f(a + ht, b + kt).$$

First, F is a differentiable function of the single variable t because the first partial derivatives of f are continuous and $x = a + th$, $y = b + tk$ are differentiable functions of t. The Chain Rule gives

$$F'(t) = \frac{\partial f}{\partial x}\frac{dx}{dt} + \frac{\partial f}{\partial y}\frac{dy}{dt} = h\frac{\partial f}{\partial x} + k\frac{\partial f}{\partial y}. \tag{2}$$

Also, F' is continuous on the closed interval $0 \le t \le 1$ because $\partial f/\partial x$ and $\partial f/\partial y$ are continuous throughout R.

Next, $F'(t)$ is a differentiable function of t because $\partial f/\partial x$ and $\partial f/\partial y$ are continuous and have continuous partial derivatives. The Chain Rule gives

$$F''(t) = h\frac{\partial F'(t)}{\partial x} + k\frac{\partial F'(t)}{\partial y} = h\left(h\frac{\partial^2 f}{\partial x^2} + k\frac{\partial^2 f}{\partial x\,\partial y}\right) + k\left(h\frac{\partial^2 f}{\partial y\,\partial x} + k\frac{\partial^2 f}{\partial y^2}\right) \tag{3}$$

$$= h^2\frac{\partial^2 f}{\partial x^2} + 2hk\frac{\partial^2 f}{\partial x\,\partial y} + k^2\frac{\partial^2 f}{\partial y^2}.$$

Thus, $F(t)$ satisfies the hypotheses of Taylor's theorem for $n = 2$ on the interval $[0, 1]$. Namely, F and F' are continuous on $[0, 1]$ and F' is differentiable on $(0, 1)$. Therefore,

$$F(1) = F(0) + F'(0)(1 - 0) + F''(c)\frac{(1 - 0)^2}{2},$$

or

$$F(1) = F(0) + F'(0) + \frac{1}{2}F''(c), \tag{4}$$

for some number c between 0 and 1.

If we express F and its derivatives in terms of f, we find

$$F(1) = f(a + h, b + k),$$

$$F(0) = f(a, b),$$

$$F'(0) = hf_x(a, b) + kf_y(a, b), \tag{5}$$

$$F''(c) = (h^2f_{xx} + 2hkf_{xy} + k^2f_{yy})|_{(a+ch, b+ck)}.$$

With these substitutions, Eq. (4) becomes

$$f(a + h, b + k) = f(a, b) + hf_x(a, b) + kf_y(a, b)$$

$$+ \frac{1}{2}(h^2f_{xx} + 2hkf_{xy} + k^2f_{yy})|_{(a+ch, b+ck)}. \tag{6}$$

This is the equation from which we now derive the second derivative test.

Suppose f_x and f_y are both zero at (a, b) and we wish to find out if f has a maximum or minimum at (a, b). With $f_x = f_y = 0$, Eq. (6) becomes

$$f(a + h, b + k) - f(a, b) = \frac{1}{2}(h^2f_{xx} + 2hkf_{xy} + k^2f_{yy})|_{(a+ch, b+ck)}. \tag{7}$$

Since a maximum or minimum value of f at (a, b) depends on the sign of $f(a + h, b + k) - f(a, b)$, we are led to considering the sign of

$$Q(c) = (h^2f_{xx} + 2hkf_{xy} + k^2f_{yy})|_{(a+ch, b+ck)}, \tag{8}$$

the quadratic expression on the right-hand side of Eq. (7) (without the 1/2).

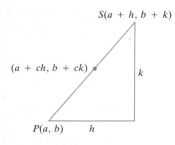

17.25 The geometry of the quantities in Eq. (6).

The second order derivatives in $Q(c)$ are to be evaluated at some point $(a + ch, b + ck)$ on the segment PS in Fig. 17.25. Since f_{xx}, f_{xy}, and f_{yy} are continuous throughout R, the values of these derivatives at $(a + ch, b + ck)$ will be nearly the same as their values at (a, b) when h and k are small. In particular, if $Q(0) \neq 0$, the sign of $Q(c)$ will be the same as the sign of $Q(0)$ for sufficiently small values of h and k.

We can predict the sign of

$$Q(0) = h^2 f_{xx}(a, b) + 2hk f_{xy}(a, b) + k^2 f_{yy}(a, b) \tag{9}$$

(and hence the sign of $Q(c)$) from the signs of f_{xx} and $f_{xx}f_{yy} - f_{xy}^2$ at (a, b). To do so, we multiply both sides of Eq. (9) by f_{xx} and rearrange the right-hand side to get

$$f_{xx}Q(0) = (hf_{xx} + kf_{xy})^2 + (f_{xx}f_{yy} - f_{xy}^2)k^2. \tag{10}$$

The sign of $Q(0)$ can be determined from this equation, and we are led to the following criteria for the behavior of $f(x, y)$ at (a, b).

1. If $f_{xx} < 0$ and $f_{xx}f_{yy} - f_{xy}^2 > 0$ at (a, b), then $Q(0) < 0$ for all sufficiently small nonzero values of h and k, and f has a *local maximum* value at (a, b).

2. If $f_{xx} > 0$ and $f_{xx}f_{yy} - f_{xy}^2 > 0$ at (a, b), then $Q(0) > 0$ for all sufficiently small nonzero values of h and k, and f has a *local minimum* value at (a, b).

3. If $f_{xx}f_{yy} - f_{xy}^2 < 0$ at (a, b), then it can be shown that there are combinations of arbitrarily small nonzero values of h and k for which $Q(0) > 0$, and others for which $Q(0) < 0$. Thus, arbitrarily close to the point $P_0(a, b, f(a, b))$ on the surface $z = f(x, y)$ there are points above P_0 and points below P_0. The function f therefore has a *saddle point* at (a, b).

4. Finally, if $f_{xx}f_{yy} - f_{xy}^2 = 0$, we can draw *no conclusion* about the sign of $Q(c)$, and some other test is needed. It follows from Eq. (10) that $f_{xx}Q(0) \geq 0$ for all choices of h and k, but if $Q(0) = 0$ no conclusion can be drawn about the sign of $Q(c)$.

This completes the derivation of the second derivative test.

The Derivation of the Error Estimate for the Standard Linear Approximations of $f(x, y)$

In Article 17.2 we stated that if $f(x, y)$ and its first and second order partial derivatives are continuous throughout a rectangle R centered at a point (x_0, y_0), then the absolute value $|E|$ of the difference between f and its linearization at a point (x_0, y_0) in R satisfies the inequality

$$|E| \leq \frac{1}{2}B(|x - x_0| + |y - y_0|)^2. \tag{11}$$

The source of this inequality becomes clear if we change the notation in Eq. (6) of this article. With

$$(a, b) = (x_0, y_0), \qquad a + h = x, \qquad b + k = y,$$

$$h = (x - x_0), \qquad k = (y - y_0), \tag{12}$$

$$(a + ch, b + ck) = (x_0 + c(x - x_0), y_0 + c(y - y_0)),$$

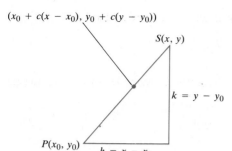

$(x_0 + c(x - x_0), y_0 + c(y - y_0))$

$S(x, y)$

$k = y - y_0$

$P(x_0, y_0)$

$h = x - x_0$

17.26 The geometry of the quantities in Eqs. (12) and (13). (Figure 17.25 in new notation.)

Fig. 17.25 becomes Fig. 17.26, and Eq. (6) becomes

$$f(x, y) = \underbrace{f(x_0, y_0) + f_x(x_0, y_0)(x - x_0) + f_y(x_0, y_0)(y - y_0)}_{\text{Linearization } L(x, y)} \tag{13}$$

$$\underbrace{+ \frac{1}{2}((x - x_0)^2 f_{xx} + 2(x - x_0)(y - y_0)f_{xy} + (y - y_0)^2 f_{yy})_{(x_0 + c(x - x_0),\, y_0 + c(y - y_0))}.}_{\text{Error } E(x, y)}$$

We now have an explicit formula for the error $E(x, y)$ in the linearization of f. From this formula we can get a useful upper bound for $|E|$ in the following way. First,

$$|E| \le \frac{1}{2}((x - x_0)^2|f_{xx}| + 2|x - x_0||y - y_0||f_{xy}| + (y - y_0)^2|f_{yy}|). \tag{14}$$

Hence, if B is an upper bound for the values of $|f_{xx}|$, $|f_{xy}|$, and $|f_{yy}|$ on R,

$$|E| \le \frac{1}{2}((x - x_0)^2 B + 2|x - x_0||y - y_0|B + (y - y_0)^2 B)$$

$$\le \frac{1}{2}B((x - x_0)^2 + 2|x - x_0||y - y_0| + (y - y_0)^2) \tag{15}$$

$$\le \frac{1}{2}B(|x - x_0| + |y - y_0|)^2.$$

Combining these last inequalities gives the inequality we set out to derive, namely,

$$|E| \le \frac{1}{2}B(|x - x_0| + |y - y_0|)^2. \tag{16}$$

Taylor's Formula for Functions of Two Variables

The formula

$$F'(t) = \left(h\frac{\partial f}{\partial x} + k\frac{\partial f}{\partial y}\right)$$

from Eq. (2) may be interpreted to say that applying the operator d/dt to $F(t) = f(a + ht, b + kt)$ gives the same result as applying the operator

$$\left(h\frac{\partial}{\partial x} + k\frac{\partial}{\partial y}\right)$$

to $f(x, y)$. Similarly, the formula

$$F''(t) = h^2\frac{\partial^2 f}{\partial x^2} + 2hk\frac{\partial^2 f}{\partial x\,\partial y} + k^2\frac{\partial^2 f}{\partial y^2}$$

from Eq. (3) says that applying d^2/dt^2 to $F(t)$ gives the same result as applying

$$\left(h\frac{\partial}{\partial x} + k\frac{\partial}{\partial y}\right)^2 = h^2\frac{\partial^2}{\partial x^2} + 2hk\frac{\partial^2}{\partial x\,\partial y} + k^2\frac{\partial^2}{\partial y^2}$$

to $f(x, y)$. These are the first two instances of a more general formula,

$$F^{(n)}(t) = \frac{d^n}{dt^n} F(t) = \left(h\frac{\partial}{\partial x} + k\frac{\partial}{\partial y} \right)^n f(x, y), \tag{17}$$

which says that applying d^n/dt^n to $F(t)$ gives the same result as applying the operator

$$\left(h\frac{\partial}{\partial x} + k\frac{\partial}{\partial y} \right)^n,$$

after expanding it by the binomial theorem, to $f(x, y)$.

Suppose now that $f(x, y)$ and its partial derivatives through order $n + 1$ are continuous throughout a rectangular region centered at the point (a, b). Then we may extend the Taylor formula for $F(t)$ to more terms,

$$F(t) = F(0) + F'(0) \cdot t + \frac{F''(0)}{2!}t^2 + \cdots + \frac{F^{(n)}(0)}{n!}t^n + \text{Remainder},$$

and take $t = 1$ to obtain

$$F(1) = F(0) + F'(0) + \frac{F''(0)}{2!} + \cdots + \frac{F^{(n)}(0)}{n!} + \text{Remainder}.$$

When we replace the first n derivatives on the right of this last series by their equivalent expressions from Eq. (17) evaluated at $t = 0$ and add the appropriate remainder term, we arrive at the following formula.

Taylor's Formula for $f(x, y)$ at the Point (a, b)

Suppose $f(x, y)$ and its partial derivatives through order $n + 1$ are continuous throughout a rectangular region R centered at a point (a, b). Then, throughout R,

$$f(a + h, b + k) = f(a, b) + (hf_x + kf_y)_{(a, b)}$$

$$+ \frac{1}{2!}(h^2 f_{xx} + 2hk f_{xy} + k^2 f_{yy})_{(a, b)}$$

$$+ \frac{1}{3!}(h^3 f_{xxx} + 3h^2 k f_{xxy} + 3hk^2 f_{xyy} + k^3 f_{yyy})_{(a, b)} \tag{18}$$

$$+ \cdots + \frac{1}{n!}\left(h\frac{\partial}{\partial x} + k\frac{\partial}{\partial y} \right)^n f \bigg|_{(a, b)}$$

$$+ \frac{1}{(n + 1)!}\left(h\frac{\partial}{\partial x} + k\frac{\partial}{\partial y} \right)^{n+1} f \bigg|_{(a+ch, b+ck)}$$

The derivatives in the first n derivative terms are evaluated at (a, b). The derivative in the last term is to be evaluated at some point $(a + ch, b + ck)$ on the line segment joining (a, b) and $(a + h, b + k)$.

If $(a, b) = (0, 0)$ and we treat h and k as independent variables, Eq. (18) assumes the following simpler form.

Taylor's Formula for $f(x, y)$ at the Origin

$$f(x, y) = f(0, 0) + xf_x + yf_y$$

$$+ \frac{1}{2!}(x^2f_{xx} + 2xyf_{xy} + y^2f_{yy})$$

$$+ \frac{1}{3!}(x^3f_{xxx} + 3x^2yf_{xxy} + 3xy^2f_{xyy} + y^3f_{yyy}) \qquad (19)$$

$$+ \cdots + \frac{1}{n!}\left(x\frac{\partial}{\partial x} + y\frac{\partial}{\partial y}\right)^n f$$

$$+ \frac{1}{(n+1)!}\left(x\frac{\partial}{\partial x} + y\frac{\partial}{\partial y}\right)^{n+1} f \bigg|_{(cx, cy)}.$$

The partial derivatives in the first n derivative terms are to be evaluated at $(0, 0)$. The partial derivatives in the final term are to be evaluated at (cy, cx), a point on the line segment joining the origin and (x, y).

EXAMPLE 1 Use Taylor's formula to find a quadratic polynomial that approximates

$$f(x, y) = \sin x \sin y$$

near the origin. How accurate is the approximation if $|x| \le 0.1$ and $|y| \le 0.1$?

Solution We take $n = 2$ in Eq. (19):

$$f(x, y) = f(0, 0) + (xf_x + yf_y)$$

$$+ \frac{1}{2}(x^2f_{xx} + 2xyf_{xy} + y^2f_{yy})$$

$$+ \frac{1}{6}(x^3f_{xxx} + 3x^2yf_{xxy} + 3xy^2f_{xyy} + y^3f_{yyy})_{(cx, cy)},$$

where all the first and second partial derivatives are to be evaluated at the origin and the third partial derivatives are to be evaluated at (cx, cy). With

$$f(x, y) = \sin x \sin y,$$

$$f(0, 0) = \sin x \sin y|_{(0, 0)} = 0,$$

$$f_x(0, 0) = \cos x \sin y|_{(0, 0)} = 0,$$

$$f_y(0, 0) = \sin x \cos y|_{(0, 0)} = 0,$$

$$f_{xx}(0, 0) = -\sin x \sin y|_{(0, 0)} = 0,$$

$$f_{xy}(0, 0) = \cos x \cos y|_{(0, 0)} = 1,$$

$$f_{yy}(0, 0) = -\sin x \sin y|_{(0, 0)} = 0,$$

we have

$$\sin x \sin y \approx 0 + 0 + 0 + \frac{1}{2}(x^2(0) + 2xy(1) + y^2(0))$$

or

$$\sin x \sin y \approx xy.$$

This is the standard quadratic approximation of $\sin x \sin y$ at the origin. The error in the approximation is

$$E(x, y) = \frac{1}{6}(x^3 f_{xxx} + 3x^2 y f_{xxy} + 3xy^2 f_{xyy} + y^3 f_{yyy})_{(cx, cy)}.$$

The third derivatives never exceed 1 in absolute value because they are products of sines and cosines. Also, $|x| \leq 0.1$ and $|y| \leq 0.1$. Hence

$$|E(x, y)| \leq \frac{1}{6}((0.1)^3 + 3(0.1)^3 + 3(0.1)^3 + (0.1)^3)$$

$$\leq \frac{8}{6}(0.1)^3$$

$$\leq 0.00134.$$

The error in the approximation will not exceed 0.00134 as long as $|x| \leq 0.1$ and $|y| \leq 0.1$.   ∎

PROBLEMS

In Problems 1–6, use Taylor's formula to find a quadratic polynomial that approximates $f(x, y)$ near the origin.

1. $f(x, y) = e^x \cos y$

2. $f(x, y) = e^x \ln(1 + y)$

3. $f(x, y) = \sin(x^2 + y^2)$

4. $f(x, y) = \cos(x^2 + y^2)$

5. $f(x, y) = \dfrac{1}{1 - x - y}$

6. $f(x, y) = \dfrac{1}{1 - x - y + xy}$

7. Use Taylor's formula to find a quadratic approximation of $f(x, y) = \cos x \cos y$ at the origin. Estimate the error in the approximation if $|x| \leq 0.1$ and $|y| \leq 0.1$.

8. Use Taylor's formula to find a quadratic approximation of $e^x \sin y$ at the origin. Estimate the error in the approximation if $|x| \leq 0.1$ and $|y| \leq 0.1$.

9. *The error bound for linear approximations of functions of three variables (Article 17.2, Eq. 19)*
   a) Suppose that $f(x, y, z)$ and its first and second order partial derivatives are continuous throughout a three-dimensional rectangular region R centered at the point $P(a, b, d)$. Suppose also that the increments h, k, and m are small enough for the line segment PS joining P to the point $S(a + h, b + k, d + m)$ to lie in the interior of R.

Parameterize PS by the equations

$$x = a + th, \quad y = b + tk, \quad z = d + tm, \quad 0 \leq t \leq 1.$$

Apply Taylor's theorem to the function

$$F(t) = f(a + th, b + tk, d + tm)$$

to show that

$$f(a + h, b + k, d + m)$$
$$= f(a, b, d) + h f_x(a, b, d)$$
$$\quad + k f_y(a, b, d) + m f_z(a, b, d)$$
$$\quad + \frac{1}{2}\left| h^2 f_{xx} + k^2 f_{yy} + m^2 f_{zz} + 2hk f_{xy} \right.$$
$$\quad + 2hm f_{xz} + 2km f_{yz}\left.\right|_{(a+ch,\, b+ck,\, d+cm)} \quad (6')$$

for some c between 0 and 1. This is the three-variable version of Eq. (6) of the present article.

   b) Use Eq. (6') from part (a) to derive the inequality in Article 17.2, Eq. (19).

10. Prove the Mean Value Theorem for functions of two variables,

$$f(x + h, y + k) - f(x, y) = f_x(x + \theta h, y + \theta k)h$$
$$+ f_y(x + \theta h, y + \theta k)k,$$
$$0 < \theta < 1,$$

with suitable assumptions about f. What are the assumptions? (Apply the Mean Value Theorem for functions of one variable to $F(t) = f(x + ht, y + kt)$.)

REVIEW QUESTIONS AND EXERCISES

1. Describe a way to find the extreme values of a function $f(x, y)$ that is continuous and has continuous first and second order partial derivatives (a) at an interior point of its domain and (b) at a boundary point. Give examples.

2. How does one look for the extreme values of a function $f(x, y)$ along a differentiable curve $x = x(t)$, $y = y(t)$ in the function's domain? Give an example.

3. What is the linearization $L(x, y)$ of a function $f(x, y)$ at a point (x_0, y_0)? Give an upper bound for the error in the approximation of f by L. What formula may be used to estimate increments in f? Give examples.

4. Suppose $\mathbf{R}(t) = x(t)\mathbf{i} + y(t)\mathbf{j} + z(t)\mathbf{k}$ is a curve in the domain of a function $f(x, y, z)$. Describe the relation between df/dt, ∇f, and $\mathbf{v} = d\mathbf{R}/dt$. What can be said about ∇f and $\mathbf{v}$ at points of the curve where f has a local maximum or minimum relative to its values on the curve?

5. Describe the method of Lagrange multipliers and its geometric interpretation as it applies to the problem of maxi-

mizing or minimizing a function $f(x, y)$ subject to the constraint $g(x, y) = 0$.

6. If you studied Lagrange multipliers with two constraints, describe the general method of finding the extreme values of a function $f(x, y, z)$ subject to the simultaneous constraints $g_1(x, y, z) = 0$ and $g_2(x, y, z) = 0$.

7. Sketch some of the level curves of $f(x, y) = 2x + 3y$ and find points on the curve $g(x, y) = x^2 + xy + y^2 - 3 = 0$ that (a) maximize f, (b) minimize f. Does it seem reasonable that the level curve of f at each of these points should be tangent to the curve $g(x, y) = 0$ at each of these points? Is it true?

8. If you studied Taylor's formula for a function $f(x, y)$ (Article 17.4), quote the formula and use it to generate a quadratic approximation for $f(x, y) = e^{x+y}$ near the origin. Show how to estimate the error in your approximation if $|x| \leq 0.1$ and $|y| \leq 0.1$.

MISCELLANEOUS PROBLEMS

In Problems 1–8, find the absolute maximum and minimum values of $f(x, y)$ on the region R.

1. $f(x, y) = x^2 + xy + y^2 - 3x + 3y$
   R: The triangular region cut from the first quadrant by the line $x + y = 4$

2. $f(x, y) = x^2 - y^2 - 2x + 4y + 1$
   R: The rectangular region in the first quadrant bounded by the coordinate axes and the lines $x = 4$ and $y = 2$

3. $f(x, y) = y^2 - xy - 3y + 2x$
   R: The square region enclosed by the lines $x = \pm 2$ and $y = \pm 2$.

4. $f(x, y) = 2xy - x^2 - 2y^2 + 3x + 4$
   R: The rectangular region in the first quadrant bounded by the coordinate axes and the lines $x = 10$ and $y = 4$

5. $f(x, y) = x^2 - y^2 - 2x + 4y$
   R: The triangular region bounded below by the x-axis, above by the line $y = x + 2$, and on the right by the line $x = 2$

6. $f(x, y) = 4xy - x^4 - y^4 + 16$
   R: The triangular region bounded below by the line $y = -2$, above by the line $y = x$, and on the right by the line $x = 2$

7. $f(x, y) = x^3 + y^3 + 3x^2 - 3y^2$
   R: The square region enclosed by the lines $x = \pm 1$ and $y = \pm 1$

8. $f(x, y) = x^3 + 3xy + y^3 + 1$
   R: The square region enclosed by the lines $x = \pm 1$ and $y = \pm 1$

9. Find the maximum and minimum values of $f(x, y) = x^2 - y^2$ on the disk $x^2 + y^2 \leq 4$.

10. Find the largest and smallest values (if any) of z on each of the following surfaces.
    a) $x^2 + y^2 + z^2 = 3$
    b) $x^2 + y^2 = 2z$
    c) $x^2 - y^2 = 2z$

11. Prove that the only possible maxima and minima of z on the surface $z = x^3 + y^3 - 9xy + 27$ occur at $(0, 0)$ and $(3, 3)$. Prove that $(0, 0)$ is neither a maximum nor a minimum. Determine whether z has a maximum or a minimum at $(3, 3)$.

12. Find the maximum value of the function $xye^{-(2x+3y)}$ in the first quadrant.

13. Find the point(s) on the surface $xyz = 1$ closest to the origin.

14. A closed rectangular box is to have a volume of V in^3. The cost of the material used in the box is a cents/in^2 for top and bottom, b cents/in^2 for front and back, and c cents/in^2 for the remaining two sides. What dimensions minimize the total cost of materials?

15. Find the minimum volume for a region bounded by the

planes $x = 0$, $y = 0$, $z = 0$, and a plane tangent to the ellipsoid

$$\frac{x^2}{a^2} + \frac{y^2}{b^2} + \frac{z^2}{c^2} = 1$$

at a point in the first octant.

16. Prove the following theorem: If $f(x, y)$ is defined in a region R, and f_x and f_y exist and are bounded in R, then $f(x, y)$ is continuous in R. (The assumption of boundedness is essential.)

17. *An argument for the two-variable version of the method of Lagrange multipliers.* Suppose that $f(x, y)$ and $g(x, y)$ have continuous first partial derivatives and that we wish to find the relative maximum and minimum values of f subject to the constraint that $g(x, y) = 0$. To find these values of f we find the values of x, y, and λ that satisfy the equations

$$\nabla f = \lambda \nabla g \quad \text{and} \quad g(x, y) = 0$$

simultaneously.

To prove this, let P_0 be a point on the curve $g(x, y) = 0$ at which f has a local maximum or minimum value relative to its values on the curve. Then carry out the following steps:

STEP 1: Let $x = x(t)$, $y = y(t)$ be a parameterization of the curve $g(x, y) = 0$ in an interval about P_0. Show that $df/dt = 0$ at P_0 and therefore that ∇f is perpendicular to the velocity vector of the curve at P_0. (Assume that neither vector vanishes at P_0.)

STEP 2: Combine the result in Step 1 above with the result of Miscellaneous Problem 57 in Chapter 16 to show that ∇f and ∇g are parallel at P_0 and therefore that $\nabla f = \lambda \nabla g$ at P_0 for some number λ.

CHAPTER 18

Multiple Integrals

OVERVIEW

In this chapter, we show how to calculate multiple integrals—the integrals of functions of two or more variables. Multiple integrals are defined in much the same way as are integrals of functions of single variables, and they have even more far-reaching interpretations.

We first introduce double integrals over regions in the plane. These are the integrals with which we calculate the areas of regions with curved boundaries and the moments, masses, centers of mass, and radii of gyration of thin flat plates of irregular shape. We show how to carry out these calculations in polar coordinates as well as in rectangular coordinates and then discuss other ways to change variables. After that, we move on to triple integrals, the integrals with which we calculate the volumes, moments, masses, centers of mass, and radii of gyration of regular and irregular solids in space. We show how to carry out these calculations in cylindrical and spherical coordinates as well as in rectangular coordinates and then discuss substitutions in triple integrals in a more general setting.

In the next chapter, we shall combine multiple integrals with vectors to calculate the areas of curved surfaces and the rates at which fluids circulate and flow across two- and three-dimensional boundaries. Multiple integrals play important roles in statistics, electrical engineering, and the physical theory of electromagnetism, but we shall not go into these applications here.

18.1
Double Integrals

In this article, we show how to calculate the integral of a function $f(x, y)$ over a bounded region in the xy-plane. We call this integral the double integral of f over the region. We begin with rectangular regions and then generalize to bounded regions of a more general nature.

You will find many similarities here with the theory of "single" integrals, as we shall now call the integrals we defined in Chapter 4 for functions of a single variable. Indeed, the connection is very strong. The basic theorem for evaluating double integrals says that every double integral can be evaluated in stages, using the single-integration methods already at our command.

Double Integrals Over Rectangles

Suppose that $f(x, y)$ is defined on a rectangular region R defined by

$$R: \quad a \le x \le b, \quad c \le y \le d.$$

We imagine R to be covered by a network of lines parallel to the x- and y-axes, as shown in Fig. 18.1. These lines divide R into small pieces of area

$$\Delta A = \Delta x \, \Delta y.$$

We number these in some order

$$\Delta A_1, \quad \Delta A_2, \quad \ldots, \quad \Delta A_n,$$

choose a point (x_k, y_k) in each piece ΔA_k, and form the sum

$$S_n = \sum_{k=1}^{n} f(x_k, y_k) \, \Delta A_k. \tag{1}$$

If f is continuous throughout R, then, as we refine the mesh width to make both Δx and Δy go to zero, the sums in (1) approach a limit called the **double integral** of f over R. The notation for it is

$$\iint_R f(x, y) \, dA \qquad \text{or} \qquad \iint_R f(x, y) \, dx \, dy.$$

Thus,

$$\iint_R f(x, y) \, dA = \lim_{\Delta A \to 0} \sum_{k=1}^{n} f(x_k, y_k) \, \Delta A_k. \tag{2}$$

As with functions of a single variable, the sums approach this limit no matter how the intervals $[a, b]$ and $[c, d]$ that determine R are subdivided, as long as the norms of the subdivisions both go to zero. The limit in (2) is also independent of the order in which the areas ΔA_k are numbered and independent of the choice of the point (x_k, y_k) within each ΔA_k. The values of the individual approximating sums S_n depend on these choices, but the sums approach the same limit in the end. The proof of the existence and uniqueness of this limit for a continuous function f is given in more advanced texts. The continuity of f is a sufficient condition for the existence of the double integral, but not a necessary one, and the limit in question exists for many discontinuous functions as well.

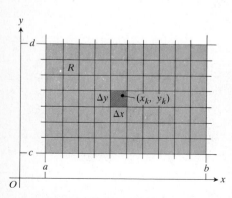

18.1 Rectangular grid subdividing the region R into small rectangles of area $\Delta A_k = \Delta x_k \Delta y_k$.

Properties of Double Integrals

Like single integrals, double integrals of continuous functions have algebraic properties that are useful in computations and applications. These properties hold because they hold for the sums with which the integrals are defined. Among these properties are the following.

1. $\displaystyle\iint_R kf(x, y)\, dA = k \iint_R f(x, y)\, dA$ (any number k)

2. $\displaystyle\iint_R (f(x, y) + g(x, y))\, dA = \iint_R f(x, y)\, dA + \iint_R g(x, y)\, dA$

3. $\displaystyle\iint_R (f(x, y) - g(x, y))\, dA = \iint_R f(x, y)\, dA - \iint_R g(x, y)\, dA$

4. $\displaystyle\iint_R f(x, y)\, dA \geq 0$ if $f(x, y) \geq 0$ on R

5. $\displaystyle\iint_R f(x, y)\, dA \geq \iint_R g(x, y)\, dA$ if $f(x, y) \geq g(x, y)$ on R

These are like properties 1–5 in Article 4.5, and the proofs, which we omit, are similar.

There is also a "domain additivity" property:

6. $\displaystyle\iint_R f(x, y)\, dA = \iint_{R_1} f(x, y)\, dA + \iint_{R_2} f(x, y)\, dA,$

which holds when R is the union of two nonoverlapping rectangles R_1 and R_2, as shown in Fig. 18.2. Again, we shall omit the proof.

Interpreting the Double Integral as a Volume

When $f(x, y)$ is positive, we may interpret the double integral of f over a rectangular region R as the volume of the solid prism bounded below by R and above by the surface $z = f(x, y)$ (Fig. 18.3). Each term $f(x_k, y_k)\, \Delta A_k$ in the sum

$$S_n = \sum f(x_k, y_k)\, \Delta A_k$$

is the volume of a vertical rectangular prism that approximates the volume of the portion of the solid that stands directly above the base ΔA_k. The sum S_n thus approximates what we want to call the total volume of the solid, and we *define* this volume to be

$$\text{Volume} = \lim S_n = \iint_R f(x, y)\, dA. \tag{3}$$

As you might expect, this more general method of calculating volume agrees with the methods in Chapter 5 (it would be useless if it did not), but we shall not prove this here.

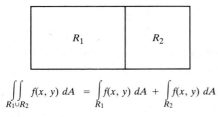

$$\iint_{R_1 \cup R_2} f(x, y)\, dA = \int_{R_1} f(x, y)\, dA + \int_{R_2} f(x, y)\, dA$$

18.2 Double integrals have the same kind of domain additivity property that single integrals have.

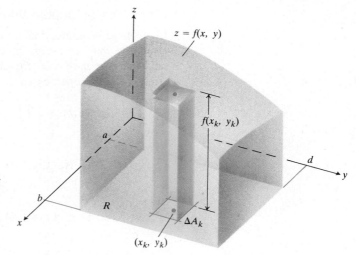

18.3 Approximating solids with rectangular prisms leads us to define the volumes of more general prisms as double integrals. The volume of the prism shown here is the double integral of $f(x, y)$ over the base region R.

Fubini's Theorem for Calculating Double Integrals

We are now ready to calculate our first double integral.

Suppose we wish to calculate the volume under the plane $z = 4 - x - y$ over the rectangular region R: $0 \le x \le 2$, $0 \le y \le 1$ in the xy-plane. If we apply the method of slicing from Article 5.3, with slices perpendicular to the x-axis (Fig. 18.4), then the volume is

$$\int_{x=0}^{x=2} A(x) \, dx, \tag{4}$$

where $A(x)$ is the cross-section area at x. For each value of x we may calculate $A(x)$ as the integral

$$A(x) = \int_{y=0}^{y=1} (4 - x - y) \, dy, \tag{5}$$

which is the area under the curve $z = 4 - x - y$ in the plane of the cross section at x. In calculating $A(x)$, x is held fixed and the integration takes place with respect to y. Combining (4) and (5), we see that the volume of the entire solid is

$$
\begin{aligned}
\text{Volume} &= \int_{x=0}^{x=2} A(x) \, dx = \int_{x=0}^{x=2} \left(\int_{y=0}^{y=1} (4 - x - y) \, dy \right) dx \\
&= \int_{x=0}^{x=2} \left[4y - xy - \frac{y^2}{2} \right]_{y=0}^{y=1} dx \tag{6} \\
&= \int_{x=0}^{x=2} \left(\frac{7}{2} - x \right) dx = \left[\frac{7}{2}x - \frac{x^2}{2} \right]_{0}^{2} = 5.
\end{aligned}
$$

If we had just wanted to write instructions for calculating the volume, without carrying out any of the integrations, we could have written

$$\text{Volume} = \int_{0}^{2} \int_{0}^{1} (4 - x - y) \, dy \, dx.$$

The expression on the right, called an **iterated** or **repeated integral,** says that the volume is to be obtained by integrating $4 - x - y$ with respect to y from

18.4 The cross-section area $A(x)$ is obtained by holding x fixed and integrating with respect to y.

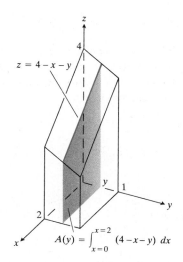

$z = 4 - x - y$

$$A(y) = \int_{x=0}^{x=2} (4-x-y)\, dx$$

18.5 The cross-section area $A(y)$ is obtained by holding y fixed and integrating with respect to x.

$y = 0$ to $y = 1$ holding x fixed and then by integrating the resulting expression in x with respect to x from $x = 0$ to $x = 2$.

What would have happened if we had calculated the volume by slicing with planes perpendicular to the y-axis, as shown in Fig. 18.5? As a function of y, the typical cross-section area is

$$A(y) = \int_{x=0}^{x=2} (4 - x - y)\, dx = \left[4x - \frac{x^2}{2} - xy \right]_{x=0}^{x=2} = 6 - 2y. \qquad (7)$$

The volume of the entire solid is therefore

$$\text{Volume} = \int_{y=0}^{y=1} A(y)\, dy = \int_{y=0}^{y=1} (6 - 2y)\, dy = \left[6y - y^2 \right]_0^1 = 5,$$

in agreement with our earlier calculation.

Again, we may give instructions for calculating the volume as an iterated integral by writing

$$\text{Volume} = \int_0^1 \int_0^2 (4 - x - y)\, dx\, dy.$$

The expression on the right says that the volume may be obtained by integrating $4 - x - y$ with respect to x from $x = 0$ to $x = 2$ (as we did in Eq. 7) and by integrating the result with respect to y from $y = 0$ to $y = 1$. In this iterated integral the order of integration is first x and then y, the reverse of the order we used in (6).

What do these two volume calculations with iterated integrals have to do with the double integral

$$\iint\limits_R (4 - x - y)\, dA$$

over the rectangle $R: 0 \le x \le 2, 0 \le y \le 1$? The answer is that they both give the value of the double integral. A theorem proved by Guido Fubini (1879–1943) and published in 1907 says that the double integral of any continuous function over a rectangle can always be calculated as an iterated integral in either order of integration. (Fubini proved his theorem in greater generality, but this is how it translates into what we're doing now.)

THEOREM 1

Fubini's Theorem (First Form)

If $f(x, y)$ is continuous on the rectangular region $R: a \le x \le b, c \le y \le d$, then

$$\iint\limits_R f(x, y)\, dA = \int_c^d \int_a^b f(x, y)\, dx\, dy = \int_a^b \int_c^d f(x, y)\, dy\, dx.$$

Fubini's theorem says that double integrals over rectangles can always be calculated as iterated integrals. This means that we can evaluate a double integral by integrating one variable at a time, using the integration techniques we already know for functions of a single variable.

Fubini's theorem also says that we may calculate the double integral by integrating in *either* order, a genuine convenience, as we shall see in Example 3. In particular, when we calculate a volume by slicing, we may use either planes perpendicular to the x-axis or planes perpendicular to the y-axis. We get the same answer either way.

Even more important is the fact that Fubini's theorem holds for *any* continuous function $f(x, y)$. In particular, f may have negative values as well as positive values on R, and the integrals we calculate with Fubini's theorem may represent other things besides volumes (as we shall see later).

EXAMPLE 1   Calculate $\iint_R f(x, y)\, dA$ for

$$f(x, y) = 1 - 6x^2y \qquad \text{and} \qquad R: \ 0 \le x \le 2,\ -1 \le y \le 1.$$

Solution   By Fubini's theorem,

$$\iint_R f(x, y)\, dA = \int_{-1}^{1} \int_{0}^{2} (1 - 6x^2y)\, dx\, dy = \int_{-1}^{1} \left[x - 2x^3y \right]_{x=0}^{x=2} dy$$

$$= \int_{-1}^{1} (2 - 16y)\, dy = \left[2y - 8y^2 \right]_{-1}^{1}$$

$$= (2 - 8) - (-2 - 8) = 4.$$

Reversing the order of integration gives the same answer:

$$\int_{0}^{2} \int_{-1}^{1} (1 - 6x^2y)\, dy\, dx = \int_{0}^{2} \left[y - 3x^2y^2 \right]_{y=-1}^{y=1} dx$$

$$= \int_{0}^{2} \left[(1 - 3x^2) - (-1 - 3x^2) \right] dx$$

$$= \int_{0}^{2} 2\, dx = 4. \qquad \blacksquare$$

Double Integrals Over Bounded Nonrectangular Regions

To define the double integral of a function $f(x, y)$ over a bounded nonrectangular region like the one shown in Fig. 18.6, we again imagine R to be covered by a rectangular grid, but we include in the partial sum only the small pieces of area $\Delta A = \Delta x\, \Delta y$ that lie entirely within the region (shaded in the figure). We number the pieces in some order, choose an arbitrary point (x_k, y_k) in each ΔA_k, and form the sum

$$S_n = \sum_{k=1}^{n} f(x_k, y_k)\, \Delta A_k.$$

The only difference between this sum and the one in Eq. (1) for rectangular regions is that now the areas ΔA_k may not cover all of R. But as the mesh becomes increasingly fine and the number of terms in S_n increases, more and more of R is included. If f is continuous and the boundary of R is made up of a finite number of line segments or smooth curves pieced together end to end, then the sums S_n will have a limit as Δx and Δy approach zero. We call the limit the *double integral* of f over R:

$$\iint_R f(x, y)\, dA = \lim_{\Delta A \to 0} \sum f(x_k, y_k)\, \Delta A_k.$$

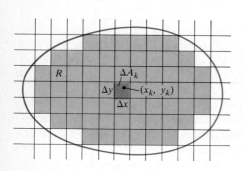

18.6   A rectangular grid subdividing a bounded nonrectangular region into cells.

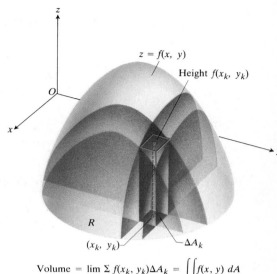

$$\iint\limits_{R} f(x, y)\, dA = \iint\limits_{R_1} f(x, y)\, dA + \iint\limits_{R_2} f(x, y)\, dA$$

$$\text{Volume} = \lim \Sigma\, f(x_k, y_k)\Delta A_k = \iint\limits_{R} f(x, y)\, dA$$

18.7 The domain additivity property stated earlier for rectangular regions holds for other regions as well.

18.8 We define the volumes of solids with curved bases the same way we define the volumes of solids with rectangular bases.

This limit may also exist under less restrictive circumstances, but we shall not pursue the point here.

Double integrals of continuous functions over nonrectangular regions have all the algebraic properties listed earlier for integrals over rectangular regions. The domain additivity property corresponding to property 6 says that if R is decomposed into nonoverlapping regions R_1 and R_2 with boundaries that are again made of line segments or smooth curves (see Fig. 18.7 for an example), then

$$6'. \quad \iint\limits_{R} f(x, y)\, dA = \iint\limits_{R_1} f(x, y)\, dA + \iint\limits_{R_2} f(x, y)\, dA.$$

If $f(x, y)$ is positive and continuous over R (Fig. 18.8), we define the volume of the solid region between R and the surface $z = f(x, y)$ to be $\iint_R f(x, y)\, dA$, as before.

If R is a region like the one shown in the xy-plane in Fig. 18.9, bounded "above" and "below" by the curves $y = f_2(x)$ and $y = f_1(x)$ and on the sides by the lines $x = a$, $x = b$, we may again calculate the volume by the method of slicing. We first calculate the cross-section area

$$A(x) = \int_{y=f_1(x)}^{y=f_2(x)} f(x, y)\, dy$$

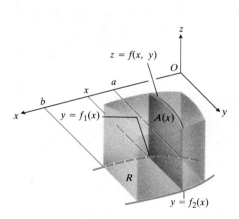

18.9 The area of the vertical slice shown here is

$$A(x) = \int_{f_1(x)}^{f_2(x)} f(x, y)\, dy.$$

The volume of the solid is calculated by integrating this area from $x = a$ to $x = b$.

and then integrate $A(x)$ from $x = a$ to $x = b$ to get the volume as an iterated integral:

$$V = \int_{a}^{b} A(x)\, dx = \int_{a}^{b} \int_{f_1(x)}^{f_2(x)} f(x, y)\, dy\, dx. \tag{8}$$

Similarly, if R is a region like the one shown in Fig. 18.10, bounded by the curves $x = g_2(y)$ and $x = g_1(y)$ and the lines $y = c$ and $y = d$, then the volume

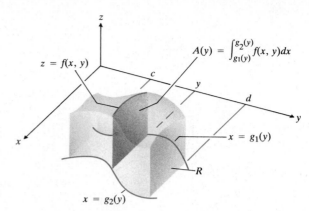

18.10 The volume of the solid shown here is

$$\int_c^d A(y)\,dy = \int_c^d \int_{g_1(y)}^{g_2(y)} f(x, y)\,dx\,dy.$$

calculated by slicing is given by the iterated integral

$$\text{Volume} = \int_c^d \int_{g_1(y)}^{g_2(y)} f(x, y)\,dx\,dy. \tag{9}$$

The fact that the iterated integrals in Eqs. (8) and (9) both give the volume that we defined to be the double integral of f over R is a consequence of the following stronger form of Fubini's theorem.

THEOREM 2

Fubini's Theorem (Stronger Form)

Let $f(x, y)$ be continuous on a region R.

1. If R is defined by $a \le x \le b$, $f_1(x) \le y \le f_2(x)$, with f_1 and f_2 continuous on $[a, b]$, then

$$\iint_R f(x, y)\,dA = \int_a^b \int_{f_1(x)}^{f_2(x)} f(x, y)\,dy\,dx.$$

2. If R is defined by $c \le y \le d$, $g_1(y) \le x \le g_2(y)$, with g_1 and g_2 continuous on $[c, d]$, then

$$\iint_R f(x, y)\,dA = \int_c^d \int_{g_1(y)}^{g_2(y)} f(x, y)\,dx\,dy.$$

EXAMPLE 2 Find the volume of the prism whose base is the triangle in the xy-plane bounded by the x-axis and the lines $y = x$ and $x = 1$ and whose top lies in the plane

$$z = f(x, y) = 3 - x - y.$$

Solution See Fig. 18.11(a). For any x between 0 and 1, y may vary from $y = 0$ to $y = x$ (Fig. 18.11b). Hence,

$$V = \int_0^1 \int_0^x (3 - x - y)\,dy\,dx = \int_0^1 \left[3y - xy - \frac{y^2}{2} \right]_{y=0}^{y=x} dx$$

$$= \int_0^1 \left(3x - \frac{3x^2}{2} \right) dx = \left[\frac{3x^2}{2} - \frac{x^3}{2} \right]_{x=0}^{x=1} = 1.$$

Figure 18.11

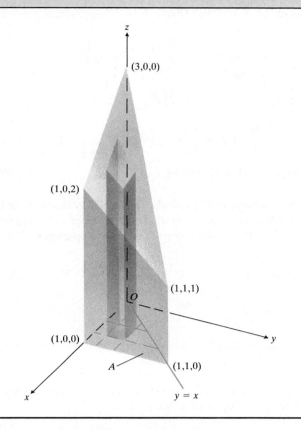

(a) Prism with a triangular base in the *xy*-plane. The volume of this prism is defined as a double integral over *A*. To evaluate it as an iterated integral, we may integrate first with respect to *y* and then with respect to *x*, or the other way around.

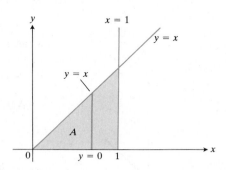

(b) Integration limits of

$$\int_{x=0}^{x=1} \int_{y=0}^{y=x} F(x, y)\, dy\, dx.$$

If we integrate first with respect to *y*, we integrate along a vertical line through *A* and then integrate to include all the vertical lines in *A*.

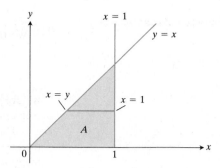

(c) Integration limits of

$$\int_{y=0}^{y=1} \int_{x=y}^{x=1} F(x, y)\, dx\, dy.$$

If we integrate first with respect to *x*, we integrate along a horizontal line through *A* and then integrate to include all the horizontal lines in *A*.

When the order of integration is reversed (Fig. 18.11c), the integral for the volume is

$$V = \int_0^1 \int_y^1 (3 - x - y) \, dx \, dy = \int_0^1 \left[3x - \frac{x^2}{2} - xy \right]_{x=y}^{x=1} dy$$

$$= \int_0^1 \left(3 - \frac{1}{2} - y - 3y + \frac{y^2}{2} + y^2 \right) dy$$

$$= \int_0^1 \left(\frac{5}{2} - 4y + \frac{3}{2}y^2 \right) dy = \left[\frac{5}{2}y - 2y^2 + \frac{y^3}{2} \right]_{y=0}^{y=1} = 1.$$

The two integrals are equal, as they should be.   ∎

While Fubini's theorem assures us that a double integral may be calculated as an iterated integral in either order of integration, the value of one integral may be easier to find than the value of the other. The next example shows how this can happen.

EXAMPLE 3   Calculate

$$\iint_A \frac{\sin x}{x} \, dA,$$

where A is the triangle in the xy-plane bounded by the x-axis, the line $y = x$, and the line $x = 1$.

Solution   The region of integration is shown in Fig. 18.12. If we integrate first with respect to y and then with respect to x, we find

$$\int_0^1 \left(\int_0^x \frac{\sin x}{x} \, dy \right) dx = \int_0^1 \left(y \, \frac{\sin x}{x} \right)_{y=0}^{y=x} dx$$

$$= \int_0^1 \sin x \, dx = -\cos (1) + 1 \approx 0.46.$$

If we reverse the order of integration and attempt to calculate

$$\int_0^1 \int_y^1 \frac{\sin x}{x} \, dx \, dy,$$

we are stopped by the fact that $\int (\sin x / x) \, dx$ cannot be expressed in terms of elementary functions.

There is no general rule for predicting which order of integration will be the good one in circumstances like these, so don't worry about how to start your integrations. Just forge ahead and if the order you choose first doesn't work, try the other.   ∎

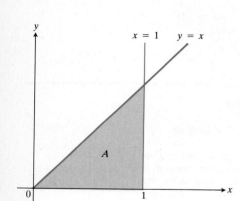

18.12   The region of integration in Example 3.

Determining the Limits of Integration

The hardest part of evaluating a double integral can be finding the limits of integration. Fortunately, there is a good procedure to follow.

If we want to evaluate

$$\iint_R f(x, y) \, dA$$

(a)

(b)

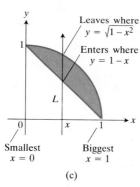

Smallest
$x = 0$

Biggest
$x = 1$

(c)

18.13 This sequence of pictures shows how to find the limits of integration in a double integral if the first integration takes place with respect to y.

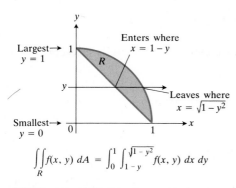

$$\iint_R f(x, y) \, dA = \int_0^1 \int_{1-y}^{\sqrt{1-y^2}} f(x, y) \, dx \, dy$$

18.14 How to find the limits of integration in a double integal if the first integration takes place with respect to x.

over the region R shown in Fig. 18.13(a), integrating first with respect to y and then with respect to x, we take the following steps:

STEP 1: We imagine a vertical line L cutting through R in the direction of increasing y (Fig. 18.13b).

STEP 2: We integrate from the y-value where L enters R to the y-value where L leaves R (Fig. 18.13c).

STEP 3: We choose x-limits that include all the vertical lines that pass through R (Fig. 18.13c). The integral is

$$\int_{x=0}^{x=1} \int_{y=1-x}^{y=\sqrt{1-x^2}} f(x, y) \, dy \, dx.$$

To calculate the same double integral as an iterated integral with the order of integration reversed, the procedure uses horizontal lines (Fig. 18.14) to give

$$\int_{y=0}^{y=1} \int_{x=1-y}^{x=\sqrt{1-y^2}} f(x, y) \, dx \, dy.$$

EXAMPLE 4 Sketch the region over which the integration

$$\int_0^2 \int_{x^2}^{2x} (4x + 2) \, dy \, dx$$

takes place and write an equivalent integral with the order of integration reversed. Evaluate both integrals.

Solution The region of integration is given by the inequalities $x^2 \le y \le 2x$ and $0 \le x \le 2$. It is therefore the region bounded by the curves $y = x^2$ and $y = 2x$ between $x = 0$ and $x = 2$, as shown in Fig. 18.15.

To find the limits for integrating in the reverse order, we imagine a horizontal line passing from left to right through the region. It enters at $x = y/2$ and leaves at $x = \sqrt{y}$. To include all such lines we let y run from $y = 0$ to $y = 4$. The integral is

$$\int_0^4 \int_{y/2}^{\sqrt{y}} (4x + 2) \, dx \, dy.$$

The value of the first integral is

$$\int_0^2 \int_{x^2}^{2x} (4x + 2) \, dy \, dx = \int_0^2 \left[4xy + 2y \right]_{y=x^2}^{y=2x} dx$$

$$= \int_0^2 (8x^2 + 4x - 4x^3 - 2x^2) \, dx$$

$$= \int_0^2 (6x^2 + 4x - 4x^3) \, dx$$

$$= \left[2x^3 + 2x^2 - x^4 \right]_{x=0}^{x=2}$$

$$= 16 + 8 - 16 = 8.$$

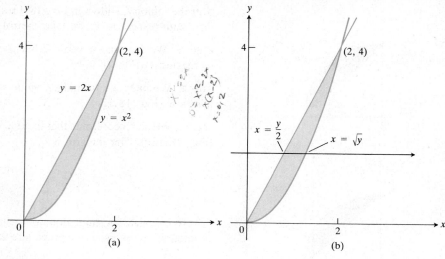

(a)   (b)

18.15   To write

$$\int_0^2 \int_{x^2}^{2x} (4x + 2)\, dy\, dx$$

as a double integral with the order of integration reversed, we (a) sketch the region of integration and (b) work out what the new limits of integration have to be. Example 4 gives the details.

The value of the second integral is

$$\int_0^4 \int_{y/2}^{\sqrt{y}} (4x + 2)\, dx\, dy = \int_0^4 \left[2x^2 + 2x \right]_{x=y/2}^{x=\sqrt{y}} dy$$

$$= \int_0^4 \left(2y + 2\sqrt{y} - \frac{y^2}{2} - y \right) dy$$

$$= \int_0^4 \left(y + 2\sqrt{y} - \frac{y^2}{2} \right) dy$$

$$= \left[\frac{y^2}{2} + \frac{4}{3} y^{3/2} - \frac{y^3}{6} \right]_0^4$$

$$= 8 + \frac{4}{3}(8) - \frac{32}{3} = 8.$$

The integrals have the same value, as we expect from Fubini's theorem. ■

PROBLEMS

Evaluate the following integrals and sketch the region over which each integration takes place.

1. $\int_0^3 \int_0^2 (4 - y^2)\, dy\, dx$

2. $\int_{-1}^0 \int_{-1}^1 (x + y + 1)\, dx\, dy$

3. $\int_0^3 \int_{-2}^0 (x^2 y - 2xy)\, dy\, dx$

4. $\displaystyle\int_{\pi}^{2\pi}\int_{0}^{\pi}(\sin x + \cos y)\,dx\,dy$

5. $\displaystyle\int_{0}^{\pi}\int_{0}^{x}x\sin y\,dy\,dx$ **6.** $\displaystyle\int_{1}^{\ln 8}\int_{0}^{\ln y}e^{x+y}\,dx\,dy$

7. $\displaystyle\int_{0}^{\pi}\int_{0}^{\sin x}y\,dy\,dx$ **8.** $\displaystyle\int_{1}^{2}\int_{y}^{y^2}dx\,dy$

Evaluate the following integrals.

9. $\displaystyle\int_{10}^{1}\int_{0}^{1/y}ye^{xy}\,dx\,dy$ **10.** $\displaystyle\int_{0}^{1}\int_{0}^{x^3}e^{y/x}\,dy\,dx$

In Problems 11–16, integrate the function $f(x, y)$ over the given region.

11. $f(x, y) = x/y$ over the region in the first quadrant bounded by the lines $y = x$, $y = 2x$, $x = 1$, $x = 2$

12. $f(x, y) = x^2 + y^2$ over the triangular region whose vertices are $(0, 0)$, $(1, 0)$ and $(0, 1)$

13. $f(x, y) = y - \sqrt{x}$ over the triangular region cut from the first quadrant by the line $x + y = 1$

14. $f(x, y) = x^2 + 3xy$ over the rectangle R: $0 \le x \le 1$, $0 \le y \le 1$

15. $f(x, y) = 1/xy$ over the rectangle R: $1 \le x \le 2$, $1 \le y \le 2$

16. $f(x, y) = y \cos xy$ over the rectangle R: $0 \le x \le \pi$, $0 \le y \le 1$

In Problems 17–24, sketch the region over which the integration takes place and write an equivalent integral with the order of integration reversed. Evaluate both integrals.

17. $\displaystyle\int_{0}^{2}\int_{0}^{4-2x}dy\,dx$ **18.** $\displaystyle\int_{0}^{1}\int_{2}^{4-2x}dy\,dx$

19. $\displaystyle\int_{0}^{1}\int_{y}^{\sqrt{y}}dx\,dy$ **20.** $\displaystyle\int_{0}^{4}\int_{\sqrt{y}}^{2}dx\,dy$

21. $\displaystyle\int_{1}^{2}\int_{1}^{e^x}dy\,dx$ **22.** $\displaystyle\int_{0}^{1}\int_{\sqrt{y}}^{1}dx\,dy$

23. $\displaystyle\int_{0}^{\sqrt{2}}\int_{-\sqrt{4-2y^2}}^{\sqrt{4-2y^2}}y\,dx\,dy$ **24.** $\displaystyle\int_{0}^{1}\int_{0}^{4-x^2}dy\,dx$

In Problems 25–30, write an equivalent iterated integral with the order of integration reversed. *Do not integrate.* It will help to sketch the region over which the integration takes place.

25. $\displaystyle\int_{0}^{1}\int_{x^2}^{x}f(x, y)\,dy\,dx$ **26.** $\displaystyle\int_{0}^{1}\int_{x}^{2x}f(x, y)\,dy\,dx$

27. $\displaystyle\int_{0}^{1}\int_{1}^{e^x}dy\,dx$ **28.** $\displaystyle\int_{0}^{1}\int_{\sqrt{x}}^{1}\cos(x + y)\,dy\,dx$

29. $\displaystyle\int_{0}^{2}\int_{0}^{x^3}f(x, y)\,dy\,dx$ **30.** $\displaystyle\int_{0}^{1}\int_{-\sqrt{y}}^{\sqrt{y}}f(x, y)\,dx\,dy$

Evaluate the integrals in Problems 31–36 by integrating the equivalent integral obtained by reversing the order of integration.

31. $\displaystyle\int_{0}^{\pi}\int_{x}^{\pi}\frac{\sin y}{y}\,dy\,dx$ **32.** $\displaystyle\int_{0}^{1}\int_{2y}^{2}\cos(x^2)\,dx\,dy$

33. $\displaystyle\int_{0}^{1}\int_{y}^{1}x^2e^{xy}\,dx\,dy$ **34.** $\displaystyle\int_{0}^{2}\int_{x}^{2}y^2\sin xy\,dy\,dx$

35. $\displaystyle\int_{0}^{8}\int_{\sqrt[3]{x}}^{2}\frac{dy\,dx}{y^4 + 1}$

36. $\displaystyle\int_{0}^{2}\int_{0}^{4-x^2}\frac{xe^{2y}}{4 - y}\,dy\,dx$

37. Find the volume of the solid whose base is the region in the xy-plane that is bounded by the parabola $y = 4 - x^2$ and the line $y = 3x$, while the top of the solid is bounded by the plane $z = x + 4$.

38. Find the volume in the first octant bounded by the coordinate planes, the cylinder $x^2 + y^2 = 4$, and the plane $z + y = 3$.

39. Find the volume of the solid in the first octant bounded by the coordinate planes, the plane $x = 3$, and the parabolic cylinder $z = 4 - y^2$.

40. Find the volume of the solid cut from the first octant by the surface $z = 4 - x^2 - y$.

41. Evaluate the integral

$$\int_{0}^{2}\int_{y/2}^{1}e^{x^2}\,dx\,dy.$$

42. The volume under the paraboloid $z = x^2 + y^2$ and above a certain region R in the xy-plane is

$$V = \int_{0}^{1}\int_{0}^{y}(x^2 + y^2)\,dx\,dy + \int_{1}^{2}\int_{0}^{2-y}(x^2 + y^2)\,dx\,dy.$$

Sketch the region and express the volume as an iterated integral with the order of integration reversed. Evaluate the integral to find the volume.

43. The base of a solid is the region in the xy-plane that is bounded by the circle $x^2 + y^2 = a^2$, while the top of the solid is bounded by the paraboloid $az = x^2 + y^2$. Find the volume.

44. Evaluate the integral

$$\int_{0}^{2}(\tan^{-1}\pi x - \tan^{-1}x)\,dx.$$

(*Hint:* Write the integrand as an integral.)

TOOLKIT PROGRAM

Double Integral

18.2

Areas, Moments, and Centers of Mass

In this article we show how to use double integrals to calculate the areas of bounded regions in the plane and the masses, moments, centers of mass, and radii of gyration of thin plates covering these regions. These applications are similar to the ones in Chapter 5, but with double integrals we can now handle a much wider variety of shapes.

Areas of Bounded Regions in the Plane

If we take $f(x, y) = 1$ in the definition of the double integral over a region R in the preceding article, the partial sums reduce to

$$S_n = \sum_{k=1}^{n} f(x_k, y_k)\, \Delta A_k = \sum_{k=1}^{n} \Delta A_k \tag{1}$$

and approximate what we would like to call the area of R. As Δx and Δy approach zero, the coverage of R by the ΔA_k's (Fig. 18.16) becomes increasingly complete, and we define the area of R to be the limit

$$\text{Area} = \lim \sum \Delta A_k = \iint_R dA. \tag{2}$$

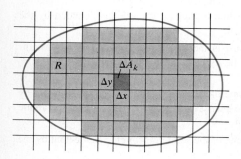

18.16   The first step in defining the area of a region is to divide the interior of the region into cells.

The area of a closed bounded plane region R is given by the formula

$$\text{Area} = \iint_R dA. \tag{3}$$

As with the other definitions in this chapter, the definition here applies to a greater variety of regions than does the earlier single-variable definition of area, but it agrees with the earlier definition on regions to which they both apply.

To evaluate the area integral in (3), we integrate the constant function $f(x, y) = 1$ over R.

EXAMPLE 1   Find the area of the region R bounded by $y = x$ and $y = x^2$ in the first quadrant.

Solution   We sketch the region (Fig. 18.17) and calculate the area as

$$A = \int_0^1 \int_{x^2}^{x} dy\, dx = \int_0^1 (x - x^2)\, dx = \left[\frac{x^2}{2} - \frac{x^3}{3} \right]_0^1 = \frac{1}{6}. \qquad \blacksquare$$

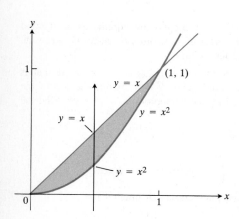

18.17   The area of the region between the parabola and the line is given by the double integral

$$\int_0^1 \int_{x^2}^{x} dy\, dx.$$

EXAMPLE 2   Find the area of the region R enclosed by the parabola $y = x^2$ and the line $y = x + 2$.

Solution   If we divide R into the regions R_1 and R_2 shown in Fig. 18.18, we

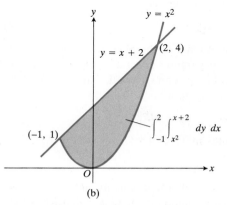

18.18 Calculating the area shown here takes (a) two integrals if the first integration is with respect to x, but (b) only one if the first integration is with respect to y.

may calculate the area as

$$A = \iint_{R_1} dA + \iint_{R_2} dA = \int_0^1 \int_{-\sqrt{y}}^{\sqrt{y}} dx \, dy + \int_1^4 \int_{y-2}^{\sqrt{y}} dx \, dy.$$

On the other hand, reversing the order of integration gives

$$A = \int_{-1}^2 \int_{x^2}^{x+2} dy \, dx.$$

Clearly, this result is simpler and is the only one we would bother to write down in practice. Evaluation of this integral leads to the result

$$A = \int_{-1}^2 \Big[y\Big]_{x^2}^{x+2} dx = \int_{-1}^2 (x + 2 - x^2) \, dx = \left[\frac{x^2}{2} + 2x - \frac{x^3}{3}\right]_{-1}^2 = \frac{9}{2}. \quad \blacksquare$$

First and Second Moments and Centers of Mass

To find the moments and centers of mass of thin sheets and plates, we use formulas similar to those in Chapter 5. The main difference is that now, with double integrals, we can accommodate a greater variety of shapes and density functions. The formulas are given in Table 18.1.

TABLE 18.1
Mass and moment formulas for thin plates covering regions in the xy-plane

Density: $\delta(x, y)$

Mass: $M = \iint \delta(x, y) \, dA$

First moments: $M_x = \iint y\delta(x, y) \, dA$

$M_y = \iint x\delta(x, y) \, dA$

Center of mass: $\bar{x} = \dfrac{M_y}{M}, \quad \bar{y} = \dfrac{M_x}{M}$

Moments of inertia (second moments):

About the x-axis: $I_x = \iint y^2\delta(x, y) \, dA$

About the y-axis: $I_y = \iint x^2\delta(x, y) \, dA$

About the origin: $I_0 = \iint (x^2 + y^2)\delta(x, y) \, dA = I_x + I_y$

Radii of gyration:

About the x-axis: $R_x = \sqrt{I_x/M}$
About the y-axis: $R_y = \sqrt{I_y/M}$
About the origin: $R_0 = \sqrt{I_0/M}$

The examples that follow show how these formulas are used.

Moments of inertia are important in calculating the kinetic energies of rotating bodies. They also appear in the formulas that are used to calculate the stiffness of beams.

The difference between the **first moments** M_x and M_y and the **moments of inertia,** or **second moments,** I_x and I_y is that the second moments use the *squares* of the "lever-arm" distances x and y.

The moment I_0 is also called the **polar moment** of inertia about the origin. It is calculated by integrating the density $\delta(x, y)$ times $r^2 = x^2 + y^2$, the square of the distance from a representative point (x, y) to the origin. Notice that $I_0 = I_x + I_y$; once we find two, we get the third free. (The moment I_0 is sometimes called I_z, for moment of inertia about the z-axis. The observation that $I_z = I_x + I_y$ is then called the **Perpendicular Axis Theorem.**)

The **radius of gyration** R_x is defined by the equation

$$I_x = MR_x^2.$$

It tells how far from the x-axis the entire mass of the plate might be concentrated to give the same I_x. The radius of gyration gives a convenient way to express the moment of inertia in terms of a mass and a length. The radii R_y and R_0 are defined in a similar way, by the equations

$$I_y = MR_y^2 \quad \text{and} \quad I_0 = MR_0^2.$$

We take square roots to get the formulas in Table 18.1.

EXAMPLE 3   A thin plate covers the triangular region bounded by the x-axis and the lines $x = 1$ and $y = 2x$ in the first quadrant. The plate's density at the point (x, y) is $\delta(x, y) = 6x + 6y + 6$. Find the plate's mass, first moments, center of mass, moments of inertia, and radii of gyration about the coordinate axes.

Solution   We sketch the plate and put in enough detail to determine the limits of integration for the integrals we will have to evaluate (Fig. 18.19).

The plate's mass is

$$M = \int_0^1 \int_0^{2x} \delta(x, y)\, dy\, dx = \int_0^1 \int_0^{2x} (6x + 6y + 6)\, dy\, dx$$

$$= \int_0^1 \Big[6xy + 3y^2 + 6y \Big]_{y=0}^{y=2x} dx = \int_0^1 (24x^2 + 12x)\, dx$$

$$= \Big[8x^3 + 6x \Big]_0^1 = 14.$$

The first moment about the x-axis is

$$M_x = \int_0^1 \int_0^{2x} y\delta(x, y)\, dy\, dx = \int_0^1 \int_0^{2x} (6xy + 6y^2 + 6y)\, dy\, dx$$

$$= \int_0^1 \Big[3xy^2 + 2y^3 + 3y^2 \Big]_{y=0}^{y=2x} dx = \int_0^1 (28x^3 + 12x^2)\, dx$$

$$= \Big[7x^4 + 4x^3 \Big]_0^1 = 11.$$

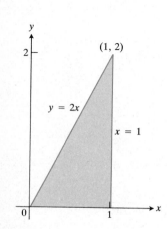

18.19 The triangular region covered by the plate in Example 3.

A similar calculation gives

$$M_y = \int_0^1 \int_0^{2x} x\delta(x, y) \, dy \, dx = 10.$$

The coordinates of the center of mass are therefore

$$\bar{x} = \frac{M_y}{M} = \frac{10}{14} = \frac{5}{7}, \qquad \bar{y} = \frac{M_x}{M} = \frac{11}{14}.$$

The moment of inertia about the x-axis is

$$I_x = \int_0^1 \int_0^{2x} y^2 \delta(x, y) \, dy \, dx$$

$$= \int_0^1 \int_0^{2x} (6xy^2 + 6y^3 + 6y^2) \, dy \, dx$$

$$= \int_0^1 \left[2xy^3 + \frac{3}{2}y^4 + 2y^3 \right]_{y=0}^{y=2x} dx$$

$$= \int_0^1 (40x^4 + 16x^3) \, dx$$

$$= \left[8x^5 + 4x^4 \right]_0^1 = 12.$$

Similarly, the moment of inertia about the y-axis is

$$I_y = \int_0^1 \int_0^{2x} x^2 \delta(x, y) \, dy \, dx = \frac{39}{5}.$$

Since we now know I_x and I_y, we do not need to evaluate an integral to calculate I_0 but can use the equation $I_0 = I_x + I_y$:

$$I_0 = 12 + \frac{39}{5} = \frac{60 + 39}{5} = \frac{99}{5}.$$

The three radii of gyration are

$$R_x = \sqrt{I_x/M} = \sqrt{12/14} = \sqrt{6/7},$$

$$R_y = \sqrt{I_y/M} = \sqrt{\left(\frac{39}{5}\right)/14} = \sqrt{39/70},$$

$$R_0 = \sqrt{I_0/M} = \sqrt{\left(\frac{99}{5}\right)/14} = \sqrt{99/70}. \qquad \blacksquare$$

Centroids of Geometric Figures

When the density of an object is constant, it cancels out of the numerator and denominator of the formulas for $\bar{x}$ and $\bar{y}$. As far as $\bar{x}$ and $\bar{y}$ are concerned, δ might as well be 1. Thus, when δ is constant, the location of the center of mass becomes a feature of the object's shape and not of the material of which it is made. In such cases, engineers may call the center of mass the **centroid** of the shape. To find a centroid, we just set δ equal to 1 and proceed to find $\bar{x}$ and $\bar{y}$ as before, by dividing first moments by masses.

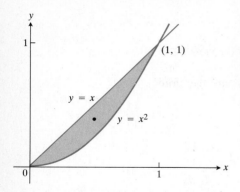

18.20 Example 4 calculates the coordinates of the centroid of the region shown here.

EXAMPLE 4 Find the centroid of the region in the first quadrant that is bounded above by the line $y = x$ and below by the parabola $y = x^2$.

Solution We sketch the region and include enough detail to determine the limits of integration (Fig. 18.20). We then set δ equal to 1 and evaluate the appropriate formulas from Table 18.1:

$$M = \int_0^1 \int_{x^2}^x 1 \, dy \, dx = \int_0^1 \left[y \right]_{y=x^2}^{y=x} dx = \int_0^1 (x - x^2) \, dx$$

$$= \left[\frac{x^2}{2} - \frac{x^3}{3} \right]_0^1 = \frac{1}{6},$$

$$M_x = \int_0^1 \int_{x^2}^x y \, dy \, dx = \int_0^1 \left[\frac{y^2}{2} \right]_{y=x^2}^{y=x} dx$$

$$= \int_0^1 \left(\frac{x^2}{2} - \frac{x^4}{2} \right) dx = \left[\frac{x^3}{6} - \frac{x^5}{10} \right]_0^1 = \frac{1}{15},$$

$$M_y = \int_0^1 \int_{x^2}^x x \, dy \, dx = \int_0^1 \left[xy \right]_{y=x^2}^{y=x} dx$$

$$= \int_0^1 (x^2 - x^3) \, dx = \left[\frac{x^3}{3} - \frac{x^4}{4} \right]_0^1 = \frac{1}{12}.$$

From these values of M, M_x, and M_y, we find

$$\bar{x} = \frac{M_y}{M} = \frac{1/12}{1/6} = \frac{1}{2} \quad \text{and} \quad \bar{y} = \frac{M_x}{M} = \frac{1/15}{1/6} = \frac{2}{5}.$$

The centroid is the point $\left(\frac{1}{2}, \frac{2}{5} \right)$. ∎

PROBLEMS

In Problems 1–8, find the area of the region bounded by the given curves and lines by means of double integration.

1. The coordinate axes and the line $x + y = a$
2. The x-axis, the curve $y = e^x$, and the lines $x = 0$, $x = 1$
3. The y-axis, the line $y = 2x$, and the line $y = 4$
4. The curve $y^2 + x = 0$ and the line $y = x + 2$
5. The curves $x = y^2$, $x = 2y - y^2$
6. The parabola $x = y - y^2$ and the line $x + y = 0$
7. The semicircle $y = \sqrt{1 - x^2}$, the lines $x = \pm 1$, and the line $y = -1$
8. Above by $y = x^2$, below by $y = -1$, on the left by $x = -2$, and on the right by $y = 2x - 1$

The integrals in Problems 9–14 give areas of regions in the xy-plane. Sketch each region. Label each bounding curve with its equation, and give the coordinates of the boundary points where the curves intersect. Then find the area of the region.

9. $\int_0^1 \int_y^{\sqrt{y}} dx \, dy$

10. $\int_0^3 \int_{-x}^{x(2-x)} dy \, dx$

11. $\int_0^{\pi/4} \int_{\sin x}^{\cos x} dy \, dx$

12. $\int_{-1}^2 \int_{y^2}^{y+2} dx \, dy$

13. $\int_{-1}^0 \int_{-2x}^{1-x} dy \, dx + \int_0^2 \int_{-x/2}^{1-x} dy \, dx$

14. $\int_0^2 \int_{x^2-4}^0 dy \, dx + \int_0^4 \int_0^{\sqrt{x}} dy \, dx$

Constant Density

15. Find the center of mass of a thin plate of density $\delta = 3$ bounded by the lines $x = 0$, $y = x$, and the parabola $y = 2 - x^2$ in the first quadrant.

16. Find the moments of inertia and radii of gyration about the coordinate axes of a thin rectangular plate of constant den-

sity δ bounded by the lines $x = a$ and $y = b$ in the first quadrant.

17. Find the centroid of the region in the first quadrant bounded by the x-axis, the parabola $y^2 = 2x$, and the line $x + y = 4$.

18. Find the centroid of the triangular region cut from the first quadrant by the line $x + y = a$, $a > 0$.

19. Find the center of mass and the moments of inertia and radii of gyration about the coordinate axes of a thin plate of constant density $\delta = 60$ bounded on the right by the parabola $y^2 + x = 0$ and on the left by the line $y = x + 2$.

20. Find the centroid of the semicircular region bounded by the x-axis and the curve $y = \sqrt{1 - x^2}$.

21. Find the moment of inertia and radius of gyration about the x-axis of a triangular region of constant density δ whose base is the interval $0 \le x \le b$ on the x-axis and whose vertex lies on the line $y = h$ above the x-axis. (It does not matter where the vertex lies on this line. As you will see, all such triangles have the same moment of inertia and radius of gyration about the x-axis.)

22. The area of the region in the first quadrant bounded by the parabola $y = 6x - x^2$ and the line $y = x$ is $125/6$ square units. Find the centroid.

23. Find the centroid of the region cut from the first quadrant by the circle $x^2 + y^2 = a^2$.

24. Find the moment of inertia about the x-axis of a thin plate of density $\delta = 1$ bounded by the circle $x^2 + y^2 = a^2$. Then use your result to find I_y and I_0 for the plate.

25. Find the centroid of the region between the x-axis and the arch $y = \sin x$, $0 \le x \le \pi$.

26. Find the moment of inertia with respect to the y-axis of a thin sheet of constant density $\delta = 1$ bounded by the curve $y = (\sin^2 x)/x^2$ and the interval $\pi \le x \le 2\pi$ of the x-axis.

27. Find the polar moment of inertia about the center of a thin rectangular sheet of constant density $\delta = 1$ bounded by the lines $x = \pm a$, $y = \pm b$. (*Hint:* Find I_x. Then use the formula for I_x to find I_y and add the two together to find I_0.)

28. Find the moment of inertia and radius of gyration about the x-axis of a thin sheet of constant density $\delta = 1$ bounded by the ellipse $(x^2/a^2) + (y^2/b^2) = 1$.

Variable Density

29. Find the center of mass, the moments of inertia and radii of gyration about the coordinate axes, and the polar moment of inertia and radius of gyration of a thin triangular plate bounded by the lines $y = x$, $y = -x$, and $y = 1$ if $\delta(x, y) = y + 1$.

30. Repeat Problem 29 for $\delta(x, y) = 3x^2 + 1$.

31. Find the center of mass of a thin triangular plate bounded by the y-axis and the lines $y = x$ and $y = 2 - x$ if $\delta(x, y) = 6x + 3y + 3$.

32. Find the mass of a thin plate occupying the smaller region cut from the ellipse $x^2 + 4y^2 = 12$ by the parabola $x = 4y^2$ if $\delta(x, y) = 5x$.

33. Find the center of mass and moment of inertia about the x-axis of a thin plate bounded by the curves $x = y^2$ and $x = 2y - y^2$ if the density at the point (x, y) is $\delta(x, y) = y + 1$.

34. Find the moment of inertia and radius of gyration about the x-axis of a thin plate bounded by the parabola $x = y - y^2$ and the line $x + y = 0$ if $\delta(x, y) = x + y$.

35. Find the center of mass and the moment of inertia and radius of gyration about the y-axis of a thin rectangular plate cut from the first quadrant by the lines $x = 6$ and $y = 1$ if $\delta(x, y) = x + y + 1$.

36. Find the center of mass and the moment of inertia and radius of gyration about the y-axis of a thin plate bounded by the line $y = 1$ and the parabola $y = x^2$ if the density is $\delta(x, y) = y + 1$.

37. Find the center of mass and the moment of inertia and radius of gyration about the y-axis of a thin plate bounded by the x-axis, the lines $x = \pm 1$, and the parabola $y = x^2$ if $\delta(x, y) = 7y + 1$.

38. Find the center of mass and the polar moment of inertia of a thin square plate bounded by the lines $x = \pm 1$ and $y = \pm 1$ if the density is $\delta(x, y) = 3x^2 + 3y^2 + 1/5$.

39. Find the center of mass and moment of inertia and radius of gyration about the x-axis of a thin rectangular plate bounded by the lines $x = 0$, $x = 20$, $y = -1$, and $y = 1$ if $\delta(x, y) = 1 + (x/20)$.

40. Find the center of mass and the moments of inertia and radii of gyration about the coordinate axes of a thin plate bounded by $y = x$ and $y = x^2$ if the density is $\delta(x, y) = x + 1$.

Additional Problems

41. A rectangular plate of constant density $\delta(x, y) = 1$ occupies the region bounded by the lines $x = 4$ and $y = 2$ in the first quadrant. The moment of inertia I_a of the rectangle about the line $y = a$ is given by the integral

$$I_a = \int_0^4 \int_0^2 (y - a)^2 \, dy \, dx.$$

Find the value of a that minimizes I_a.

42. Find the centroid of the region in the xy-plane bounded by the curves $y = 1/\sqrt{1 - x^2}$, $y = -1/\sqrt{1 - x^2}$, and the lines $x = 0$, $x = 1$. (*Hint:* Note that $\bar{y} = 0$, by symmetry.)

43. The average value of a function $f(x, y)$ over a region R is defined to be

$$\frac{1}{\text{Area } R} \iint\limits_R f(x, y) \, dA.$$

a) Calculate the average value of the derivative of the function $w = (x^2 + y^2)/2$ in the direction of the unit vector $\mathbf{u} = u_1\mathbf{i} + u_2\mathbf{j}$ over the region enclosed by the triangle whose vertices are $(0, 0)$, $(0, 1)$ and $(1, 0)$.

b) Show in general that if $w = (x^2 + y^2)/2$, then the average value of $D_{\mathbf{u}}w$ over a region R is the value of $D_{\mathbf{u}}w$ at the centroid of R.

44. Find the radius of gyration of a uniform slender rod with constant density δ and length L with respect to an axis
a) perpendicular to the axis of the rod through the rod's center of mass,
b) perpendicular to the axis of the rod at one end,

c) parallel to the rod at a distance d from the axis of the rod. Assume that d is very large compared with the radius of the rod.

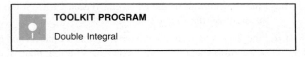

TOOLKIT PROGRAM

Double Integral

18.3
Double Integrals in Polar Form

Integrals are sometimes easier to evaluate if we change from rectangular coordinates to polar coordinates. In this article, we show how to accomplish this change and how to evaluate integrals over domains whose boundaries are given in polar form. We also look briefly at Jacobians, the determinants that we use to change variables in multiple integrals.

Integrals in Polar Coordinates

When we defined the double integral of a function over a region R in the xy-plane, we began by cutting R into rectangles whose sides were parallel to the coordinate axes. These were the natural shapes to use because their sides have either constant x-values or constant y-values. In polar coordinates, the natural shape is a "polar rectangle" of the kind we are about to describe.

Suppose that a function $F(r, \theta)$ is defined over a region R that is bounded by the rays $\theta = \alpha$ and $\theta = \beta$ and by the continuous curves $r = f_1(\theta)$ and $r = f_2(\theta)$. Suppose also that $0 \le f_1(\theta) \le f_2(\theta) \le a$ for every value of θ between α and β. Then R lies in a fan-shaped region Q defined by the inequalities $0 \le r \le a$ and $\alpha \le \theta \le \beta$. See Fig. 18.21.

We cover Q by a grid of circular arcs and rays. The arcs are cut from circles centered at the origin, with radii

$$\Delta r, \quad 2\Delta r, \quad \ldots, \quad m\Delta r,$$

18.21 The region R: $f_1(\theta) \le r \le f_2(\theta)$, $\alpha \le \theta \le \beta$ is contained in the fan-shaped region Q: $0 \le r \le a$, $\alpha \le \theta \le \beta$. The subdivision of Q by circular arcs and rays induces a subdivision of R.

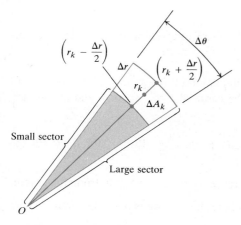

18.22 The observation that

$$\Delta A_k = \left(\begin{array}{c}\text{area of}\\\text{large sector}\end{array}\right) - \left(\begin{array}{c}\text{area of}\\\text{small sector}\end{array}\right)$$

leads to the formula $\Delta A_k = r_k \Delta r\, \Delta\theta$. The text explains why.

where $\Delta r = a/m$. The rays are given by

$$\theta = \alpha, \quad \theta = \alpha + \Delta\theta, \quad \theta = \alpha + 2\Delta\theta, \quad \ldots, \quad \theta = \alpha + n\Delta\theta = \beta.$$

The arcs and rays divide Q into small patches called "polar rectangles."

We number the polar rectangles that lie inside R (the order doesn't matter), calling their areas

$$\Delta A_1, \quad \Delta A_2, \quad \ldots, \quad \Delta A_{n'}.$$

We let (r_k, θ_k) be the center of the polar rectangle whose area is ΔA_k. By "center" we mean the point that lies halfway between the circular arcs on the ray that bisects the arcs. We then form the sum

$$S_{n'} = \sum_{k=1}^{n'} F(r_k, \theta_k)\, \Delta A_k. \tag{1}$$

If F is continuous throughout R, this sum will approach a limit as we refine the grid to make Δr and $\Delta\theta$ go to zero. The limit is called the double integral of F over R. In symbols,

$$\lim S_n = \iint\limits_R F(r, \theta)\, dA.$$

To see how to evaluate this limit, we first have to write the sum S_n in a way that expresses ΔA_k in terms of Δr and $\Delta\theta$. Here is how we do that. The radius of the inner arc bounding ΔA_k is $r_k - (\Delta r/2)$ (Fig. 18.22). The area of the circular sector subtended by this arc at the origin is therefore

$$\frac{1}{2}\left(r_k - \frac{\Delta r}{2}\right)^2 \Delta\theta. \tag{2}$$

Similarly, the radius of the outer boundary of ΔA_k is $r_k + (\Delta r/2)$. The area of the sector it subtends is

$$\frac{1}{2}\left(r_k + \frac{\Delta r}{2}\right)^2 \Delta\theta. \tag{3}$$

Therefore,

$$\Delta A_k = \text{area of large sector} - \text{area of small sector}$$

$$= \frac{\Delta\theta}{2}\left[\left(r_k + \frac{\Delta r}{2}\right)^2 - \left(r_k - \frac{\Delta r}{2}\right)^2\right]$$

$$= \frac{\Delta\theta}{2}[2r_k\, \Delta r]$$

$$= r_k\, \Delta r\, \Delta\theta.$$

Combining this result with Eq. (1) gives

$$S_n = \sum_{k=1}^{\infty} F(r_k, \theta_k)\, r_k\, \Delta r\, \Delta\theta. \tag{4}$$

A version of Fubini's theorem now says that the limit approached by these sums can be evaluated by repeated single integrations with respect to r and θ as

$$\iint\limits_R F(r, \theta)\, dA = \int_{\theta=\alpha}^{\theta=\beta} \int_{r=f_1(\theta)}^{r=f_2(\theta)} F(r, \theta)\, r\, dr\, d\theta. \tag{5}$$

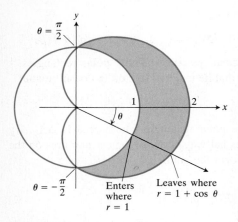

18.23 Example 1 describes the process by which we find limits of integration for calculating the area of the shaded region between the circle $r = 1$ and the cardioid $r = 1 + \cos \theta$.

Limits of Integration

The procedure we used for finding limits of integration for integrals in rectangular coordinates also works for polar coordinates.

EXAMPLE 1 Find the limits of integration for integrating a function $F(r, \theta)$ over the region R that lies inside the cardioid $r = 1 + \cos \theta$ and outside the circle $r = 1$.

Solution We graph the cardioid and circle (Fig. 18.23) and carry out the following steps:

STEP 1: Holding θ fixed, let r increase to trace a ray out from the origin.

STEP 2: Integrate from the r-value where the ray enters R to the r-value where the ray leaves R.

STEP 3: Choose θ-limits to include all the rays from the origin that intersect R.

The result is the integral

$$\int_{-\pi/2}^{\pi/2} \int_{r=1}^{r=1+\cos \theta} F(r, \theta) \, r \, dr \, d\theta.$$ ∎

If $F(r, \theta)$ is the constant function whose value is 1, then the value of the integral of F over a region R is the area of R:

The area of a closed and bounded region R in the polar coordinate plane is given by the formula

$$\text{Area of } R = \iint_R r \, dr \, d\theta. \tag{6}$$

As you might expect, this formula for area is consistent with all earlier formulas, although we shall not prove the fact.

EXAMPLE 2 Find the area enclosed by the lemniscate $r^2 = 2a^2 \cos 2\theta$.

Solution We graph the lemniscate (Fig. 18.24) and calculate the area of the right-hand half to be

$$\int_{-\pi/4}^{\pi/4} \int_0^{\sqrt{2a^2 \cos 2\theta}} r \, dr \, d\theta = \int_{-\pi/4}^{\pi/4} \left[\frac{r^2}{2} \right]_{r=0}^{r=\sqrt{2a^2 \cos 2\theta}} d\theta$$

$$= \int_{-\pi/4}^{\pi/4} a^2 \cos 2\theta \, d\theta = \frac{a^2}{2} \sin 2\theta \Big]_{-\pi/4}^{\pi/4}$$

$$= \frac{a^2}{2}[1 - (-1)] = a^2.$$

The total area is therefore $2a^2$. ∎

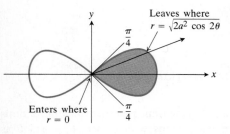

18.24 To integrate over the shaded region bounded by the lemniscate $r^2 = 2a^2 \cos 2\theta$, we let r run from 0 to $\sqrt{2a^2 \cos 2\theta}$ and then let θ run from $-\pi/4$ to $\pi/4$.

Changing Cartesian Integrals into Polar Integrals

The procedure for changing a Cartesian integral

$$\iint_R F(x, y) \, dx \, dy$$

into a polar integral has two steps:

STEP 1: Substitute

$$x = r \cos \theta, \qquad y = r \sin \theta, \qquad \text{and} \quad dx \, dy = r \, dr \, d\theta$$

into the Cartesian integral.

STEP 2: Supply polar limits of integration for the boundary of R.

The Cartesian integral then becomes

$$\iint_R F(x, y) \, dx \, dy = \iint_G F(r \cos \theta, r \sin \theta) \, r \, dr \, d\theta, \qquad (7)$$

where G denotes the region of integration in polar coordinates. This is like the substitution method in Chapter 4 except that there are now two variables to substitute for instead of one. Notice that $dx \, dy$ does not become $dr \, d\theta$; it becomes $r \, dr \, d\theta$. We shall go into the reasons for that briefly at the end of the article.

EXAMPLE 3 Find the polar moment of inertia about the origin of a thin plate of density $\delta(x, y) = 1$ bounded by the quarter-circle $x^2 + y^2 = 1$ in the first quadrant.

Solution We sketch the region of integration to determine the limits of integration (Fig. 18.25).

In Cartesian coordinates, the polar moment is the value of the integral

$$\int_0^1 \int_0^{\sqrt{1-x^2}} (x^2 + y^2) \, dy \, dx.$$

Integration with respect to y gives

$$\int_0^1 \left(x^2 \sqrt{1 - x^2} + \frac{(1 - x^2)^{3/2}}{3} \right) dx,$$

an integral difficult to evaluate without tables.

Things go better if we change the original integral to polar coordinates. Substituting

$$x = r \cos \theta, \qquad y = r \sin \theta, \qquad dx \, dy = r \, dr \, d\theta,$$

we get

$$\int_0^1 \int_0^{\sqrt{1-x^2}} (x^2 + y^2) \, dy \, dx = \int_0^{\pi/2} \int_0^1 (r^2) \, r \, dr \, d\theta$$

$$= \int_0^{\pi/2} \left[\frac{r^4}{4} \right]_{r=0}^{r=1} d\theta = \int_0^{\pi/2} \frac{1}{4} \, d\theta = \frac{\pi}{8}.$$

Why was the polar coordinate transformation so effective? One reason is that $x^2 + y^2$ was simplified to r^2. Another is that the limits of integration became constants. ∎

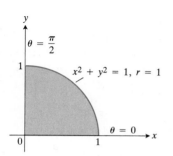

18.25 Example 3 evaluates the integral of a function over the region bounded by the quarter-circle $x^2 + y^2 = 1$ in the first quadrant. In polar coordinates, this region is described by simple inequalities:

$$0 \le r \le 1 \quad \text{and} \quad 0 \le \theta \le \pi/2.$$

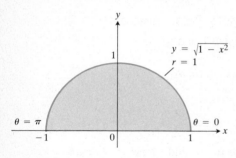

18.26 The semicircular region in Example 4 is described by the polar coordinate inequalities

$$0 \le r \le 1 \quad \text{and} \quad 0 \le \theta \le \pi.$$

EXAMPLE 4 Evaluate

$$\iint\limits_{R} e^{x^2+y^2} \, dy \, dx,$$

where R is the semicircular region bounded by the x-axis and the curve $y = \sqrt{1 - x^2}$ (Fig. 18.26).

Solution In Cartesian coordinates, the integral in question is a nonelementary integral and there is no direct way to integrate $e^{x^2+y^2}$ with respect to either x or y. Yet this integral and others like it are important in mathematics—in statistics, for example—and we must find a way to evaluate it. Polar coordinates save the day. The substitutions

$$x = r \cos \theta, \qquad y = r \sin \theta, \qquad dy \, dx = r \, dr \, d\theta$$

enable us to evaluate the integral as

$$\iint\limits_{R} e^{x^2+y^2} \, dy \, dx = \int_{0}^{\pi} \int_{0}^{1} e^{r^2} r \, dr \, d\theta$$

$$= \int_{0}^{\pi} \left[\frac{1}{2} e^{r^2} \right]_{0}^{1} d\theta$$

$$= \int_{0}^{\pi} \frac{1}{2} (e - 1) \, d\theta$$

$$= \frac{\pi}{2} (e - 1).$$

The r in the $r \, dr \, d\theta$ was just what we needed to integrate e^{r^2}. Without it we would have been stuck, as we were at the beginning. ∎

Polar substitutions are often good for evaluating Cartesian integrals over regions whose descriptions are simpler when given in polar coordinates. The next example is a case in point.

EXAMPLE 5 Find the moment of inertia about the y-axis of the region enclosed by the cardioid

$$r = 1 - \cos \theta.$$

Solution We take $\delta(x, y) = 1$ when working with a geometric figure. Thus

$$I_y = \iint x^2 \delta(x, y) \, dA = \iint x^2 \, dA.$$

We substitute

$$x = r \cos \theta, \qquad dA = r \, dr \, d\theta,$$

and sketch the region to determine the limits of integration (Fig. 18.27):

$$I_y = \int_{0}^{2\pi} \int_{0}^{1-\cos \theta} r^3 \cos^2 \theta \, dr \, d\theta$$

$$= \int_{0}^{2\pi} \frac{1}{4} \cos^2 \theta \, (1 - \cos \theta)^4 \, d\theta.$$

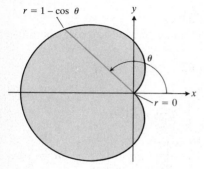

18.27 To integrate over the region enclosed by the cardioid, we let r run from 0 to $1 - \cos \theta$ and then let θ run from 0 to 2π.

The powers of $\cos \theta$ that now arise may be integrated by using the reduction formula

$$\int_0^{2\pi} \cos^n\theta \, d\theta = \frac{\cos^{n-1}\theta \sin \theta}{n}\bigg]_0^{2\pi} + \frac{n-1}{n} \int_0^{2\pi} \cos^{n-2}\theta \, d\theta$$

from Article 7.9. Since $\sin \theta$ vanishes at both limits,

$$\int_0^{2\pi} \cos^n\theta \, d\theta = \frac{n-1}{n} \int_0^{2\pi} \cos^{n-2}\theta \, d\theta.$$

Thus

$$\int_0^{2\pi} \cos^2\theta \, d\theta = \frac{1}{2} \int_0^{2\pi} d\theta = \pi,$$

$$\int_0^{2\pi} \cos^3\theta \, d\theta = \frac{2}{3} \int_0^{2\pi} \cos \theta \, d\theta = \frac{2}{3} \sin \theta \bigg]_0^{2\pi} = 0,$$

$$\int_0^{2\pi} \cos^4\theta \, d\theta = \frac{3}{4} \int_0^{2\pi} \cos^2\theta \, d\theta = \frac{3\pi}{4},$$

$$\int_0^{2\pi} \cos^5\theta \, d\theta = \frac{4}{5} \int_0^{2\pi} \cos^3\theta \, d\theta = 0,$$

$$\int_0^{2\pi} \cos^6\theta \, d\theta = \frac{5}{6} \int_0^{2\pi} \cos^4\theta \, d\theta = \frac{5}{6} \frac{3\pi}{4} = \frac{5\pi}{8}.$$

Therefore

$$I_y = \frac{1}{4} \int_0^{2\pi} (\cos^2\theta - 4 \cos^3\theta + 6 \cos^4\theta - 4 \cos^5\theta + \cos^6\theta) \, d\theta$$

$$= \frac{1}{4}\left[1 + \frac{18}{4} + \frac{5}{8}\right]\pi$$

$$= \frac{49\pi}{32}.$$

Substitutions in Double Integrals

The polar coordinate substitution is a special case of a more general substitution method for double integrals, a method that pictures changes in variables as transformations of regions.

Suppose that a region G in the uv-plane is transformed into the region R in the xy-plane by equations of the form

$$x = g(u, v), \qquad y = h(u, v),$$

as suggested in Fig. 18.28. Then any function $F(x, y)$ defined on R can be thought of as a function $F(g(u, v), h(u, v))$ defined on G as well. How is the integral of $F(x, y)$ over R related to the integral of $F(g(u, v), h(u, v))$ over G?

The answer is: If g, h, and F are continuous and have continuous partial derivatives, then the integrals are related by the equation

$$\iint_R F(x, y) \, dx \, dy = \iint_G F(g(u, v), h(u, v)) \, |J(u, v)| \, du \, dv. \qquad (8)$$

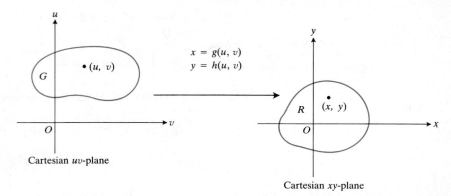

18.28 The equations $x = g(u, v)$ and $y = h(u, v)$ allow us to write an integral over a region R in the xy-plane as an integral over a region G in the uv-plane.

The factor $J(u, v)$, whose absolute value appears in this formula, is the determinant

$$J(u, v) = \begin{vmatrix} \dfrac{\partial x}{\partial u} & \dfrac{\partial x}{\partial v} \\[2mm] \dfrac{\partial y}{\partial u} & \dfrac{\partial y}{\partial v} \end{vmatrix} = \frac{\partial(x, y)}{\partial(u, v)}. \tag{9}$$

It is called the **Jacobian determinant** or **Jacobian** of the coordinate transformation $x = g(u, v)$, $y = h(u, v)$, named after the mathematician Carl Gustav Jacob Jacobi (1804–1851). The alternative notation $\partial(x, y)/\partial(u, v)$ may help you to remember how the determinant is constructed from the partial derivatives of x and y. The derivation of Eq. (8) is intricate and we shall not give it here.

For polar coordinates, we have r and θ in place of u and v. With $x = r \cos \theta$ and $y = r \sin \theta$, the Jacobian is

$$J(r, \theta) = \begin{vmatrix} \dfrac{\partial x}{\partial r} & \dfrac{\partial x}{\partial \theta} \\[2mm] \dfrac{\partial y}{\partial r} & \dfrac{\partial y}{\partial \theta} \end{vmatrix} = \begin{vmatrix} \cos \theta & -r \sin \theta \\[1mm] \sin \theta & r \cos \theta \end{vmatrix} = r(\cos^2\theta + \sin^2\theta) = r.$$

Hence, Eq. (8) becomes

$$\iint\limits_{R} F(x, y) \, dx \, dy = \iint\limits_{G} F(r \cos \theta, r \sin \theta) \, |r| \, dr \, d\theta$$

$$= \iint\limits_{G} F(r \cos \theta, r \sin \theta) \, r \, dr \, d\theta, \tag{10}$$

which is Eq. (7).

Figure 18.29 shows how the equations $x = r \cos \theta$, $y = r \sin \theta$ transform the rectangle G: $0 \le r \le 1$, $0 \le \theta \le \pi/2$ into the quarter-circle R bounded by $x^2 + y^2 = 1$ in the first quadrant of the xy-plane.

Notice that the integral on the right-hand side of Eq. (10) is not the integral of $F(r \cos \theta, r \sin \theta)$ over a region in the polar coordinate plane. It is the integral of the product of $F(r \cos \theta, r \sin \theta)$ and r over a region G in the *Cartesian $r\theta$-plane*.

Here is an example of another substitution.

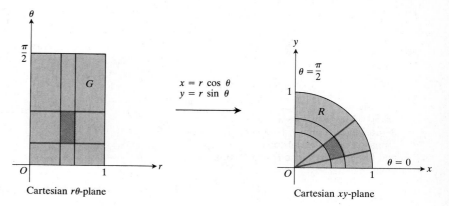

18.29 The equations $x = r \cos \theta$, $y = r \sin \theta$ transform G into R.

EXAMPLE 6 Evaluate

$$\int_0^4 \int_{x=y/2}^{x=(y/2)+1} \frac{2x - y}{2} \, dx \, dy$$

by applying the transformation

$$u = \frac{2x - y}{2}, \qquad v = \frac{y}{2} \tag{11}$$

and integrating over an appropriate region in the uv-plane.

Solution We sketch the region R of integration in the xy-plane and identify its boundaries (Fig. 18.30).

To apply Eq. (8), we need to find the corresponding uv-region G and the Jacobian of the transformation. To find them, we first solve Eqs. (11) for x and y in terms of u and v. Routine algebra gives

$$x = u + v, \qquad y = 2v. \tag{12}$$

We then find the boundaries of G by substituting these expressions into the equations for the boundaries of R:

Boundary of R	Corresponding boundary of G	Simplified equation
$x = y/2$	$u + v = 2v/2 = v$	$u = 0$
$x = (y/2) + 1$	$u + v = (2v/2) + 1 = v + 1$	$u = 1$
$y = 0$	$2v = 0$	$v = 0$
$y = 4$	$2v = 4$	$v = 2$

See Fig. 18.30.

The Jacobian of the transformation (again from Eqs. 12) is

$$J(u, v) = \begin{vmatrix} \dfrac{\partial x}{\partial u} & \dfrac{\partial x}{\partial v} \\ \dfrac{\partial y}{\partial u} & \dfrac{\partial y}{\partial v} \end{vmatrix} = \begin{vmatrix} \dfrac{\partial}{\partial u}(u + v) & \dfrac{\partial}{\partial v}(u + v) \\ \dfrac{\partial}{\partial u}(2v) & \dfrac{\partial}{\partial v}(2v) \end{vmatrix} = \begin{vmatrix} 1 & 1 \\ 0 & 2 \end{vmatrix} = 2.$$

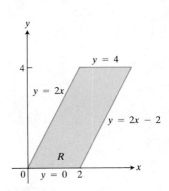

18.30 The equations $x = u + v$ and $y = 2v$ transform G into R. Reversing the transformation by the equations $u = (2x - y)/2$ and $v = y/2$ transforms R into G. See Example 6.

We now have everything we need to apply Eq. (8):

$$\int_0^4 \int_{x=y/2}^{x=(y/2)+1} \frac{2x - y}{2} \, dx \, dy = \int_{v=0}^{v=2} \int_{u=0}^{u=1} u \, |J(u, v)| \, du \, dv$$

$$= \int_0^2 \int_0^1 (u)(2) \, du \, dv = \int_0^2 \left[u^2 \right]_0^1 dv$$

$$= \int_0^2 dv = 2.$$

Problem 31 asks you to evaluate the original xy-integral directly to show that its value is also 2. ∎

PROBLEMS

In Problems 1–14, change the Cartesian integral into an equivalent polar integral and evaluate the polar integral.

1. $\int_{-1}^{1} \int_0^{\sqrt{1-x^2}} dy \, dx$

2. $\int_{-1}^{1} \int_{-\sqrt{1-x^2}}^{\sqrt{1-x^2}} dy \, dx$

3. $\int_0^1 \int_0^{\sqrt{1-y^2}} (x^2 + y^2) \, dx \, dy$

4. $\int_{-1}^{1} \int_{-\sqrt{1-y^2}}^{\sqrt{1-y^2}} (x^2 + y^2) \, dx \, dy$

5. $\int_{-a}^{a} \int_{-\sqrt{a^2-x^2}}^{\sqrt{a^2-x^2}} dy \, dx$

6. $\int_0^a \int_0^{\sqrt{a^2-y^2}} (x^2 + y^2) \, dx \, dy$

7. $\int_0^{a/\sqrt{2}} \int_y^{\sqrt{a^2-y^2}} x \, dx \, dy$

8. $\int_0^2 \int_0^x y \, dy \, dx$

9. $\int_0^3 \int_0^{\sqrt{3}x} \frac{dy \, dx}{\sqrt{x^2 + y^2}}$

10. $\int_0^2 \int_0^{\sqrt{4-x^2}} \frac{xy}{\sqrt{x^2 + y^2}} \, dy \, dx$

11. $\int_0^1 \int_0^{\sqrt{1-x^2}} 5\sqrt{x^2 + y^2} \, dy \, dx$

12. $\int_0^1 \int_0^{\sqrt{1-x^2}} e^{-(x^2+y^2)} \, dy \, dx$

13. $\int_0^2 \int_0^{\sqrt{1-(x-1)^2}} \frac{x + y}{x^2 + y^2} \, dy \, dx$

14. $\int_0^{\infty} \int_0^{\infty} e^{-(x^2+y^2)} \, dx \, dy$

15. Find the area of the region cut from the first quadrant by the curve $r = (2 - \sin 2\theta)^{1/2}$.

16. Find the area of the region that lies inside the cardioid $r = 1 + \cos \theta$ and outside the circle $r = 1$.

17. Find the area enclosed by one leaf of the rose $r = \cos 3\theta$.

18. Find the area of the region enclosed by the positive x-axis and the spiral $r = 4\theta/3$, $0 \le \theta \le 2\pi$.

19. Find the area of the region cut from the first quadrant by the cardioid $r = 1 + \sin \theta$.

20. Find the area of the region common to the interiors of the cardioids $r = 1 + \cos \theta$ and $r = 1 - \cos \theta$.

21. Integrate the function $f(x, y) = 1/(1 - x^2 - y^2)$ over the disk $x^2 + y^2 \le 3/4$.

22. Integrate the function $f(x, y) = (\ln (x^2 + y^2))/(x^2 + y^2)$ over the region between the circles $x^2 + y^2 = 1$ and $x^2 + y^2 = e^2$.

23. Find the moment about the x-axis of a thin plate of constant density $\delta(x, y) = 3$, bounded below by the x-axis and above by the cardioid $r = 1 - \cos \theta$.

24. Find the moment of inertia about the x-axis and the polar moment of inertia about the origin of a thin disk bounded by the circle $x^2 + y^2 = a^2$ if the density is $\delta(x, y) = k(x^2 + y^2)$, k a constant.

25. Find the centroid of the region enclosed by the cardioid $r = 1 + \cos \theta$.

26. Find the centroid of the region that lies inside the cardioid $r = 1 + \cos \theta$ and outside the circle $r = 1$.

27. Find the moment of inertia about the origin of a thin plate bounded by the cardioid $r = 1 + \cos \theta$ if the density is $\delta(x, y) = 1$.

28. Find the moment of inertia about the origin of a thin plate covering the region that lies inside the cardioid $r = 1 + \cos \theta$ and outside the circle $r = 1$ if the density is $\delta(x, y) = 1$.

29. The region that lies inside the cardioid $r = 1 + \cos \theta$ and outside the circle $r = 1$ is the base of a solid right cylinder. The top of the cylinder lies in the plane $z = x$. Find the volume of the cylinder.

30. The region enclosed by the lemniscate $r^2 = 2 \cos 2\theta$ is the base of a solid right cylinder whose top is bounded by the sphere $z = \sqrt{2 - r^2}$. Find the volume of the cylinder.

Substitutions

31. Evaluate the integral

$$\int_0^4 \int_{x=y/2}^{x=(y/2)+1} \frac{2x - y}{2} \, dx \, dy$$

from Example 6 directly by integration with respect to x and y to confirm that its value is 2.

32. Find the Jacobian $\partial(x, y)/\partial(u, v)$ for the transformation
   a) $x = u \cos v, \quad y = u \sin v,$
   b) $x = u \sin v, \quad y = u \cos v.$

33. a) Solve the system

$$u = x - y, \quad v = 2x + y$$

for x and y in terms of u and v. Then find the value of the Jacobian $\partial(x, y)/\partial(u, v)$.
   b) Let R be the region in the first quadrant bounded by the lines $y = -2x + 4$, $y = -2x + 7$, $y = x - 2$, and $y = x + 1$. Evaluate

$$\iint_R (2x^2 - xy - y^2) \, dx \, dy$$

by changing variables with the equations in (a) and integrating over a region G in the uv-plane.

34. a) Find the Jacobian of the transformation $x = u$, $y = uv$, and sketch the region G: $1 \le u \le 2$, $1 \le uv \le 2$ in the uv-plane.
   b) Then use Eq. (8) to transform the integral

$$\int_1^2 \int_1^2 \frac{y}{x} \, dy \, dx$$

into an integral over G, and evaluate both integrals.

35. Let R be the region in the first quadrant of the xy-plane bounded by the hyperbolas $xy = 1$, $xy = 9$ and the lines $y = x$, $y = 4x$ (Fig. 18.31). Use the transformation $x = u/v$, $y = uv$ with $u > 0$ and $v > 0$ to rewrite

$$\iint_R \left(\sqrt{\frac{y}{x}} + \sqrt{xy} \right) dx \, dy$$

as an integral over an appropriate region G in the uv-plane. Then evaluate the uv-integral over G.

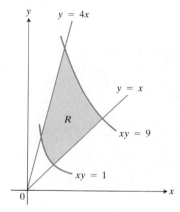

18.31 The xy-region in Problem 35.

36. The area πab of the ellipse $x^2/a^2 + y^2/b^2 = 1$ can be found by integrating the function $f(x, y) = 1$ over the region bounded by the ellipse in the xy-plane. Evaluating the integral directly requires a trigonometric substitution. An easier way to evaluate the integral is to use the transformation $x = au$, $y = bv$ and evaluate the transformed integral over the disk G: $u^2 + v^2 \le 1$ in the uv-plane. Find the area this way.

37. A thin plate of uniform thickness and density covers the region bounded by the ellipse $x^2/a^2 + y^2/b^2 = 1$ in the xy-plane. Find the moment of the plate about the origin. (*Hint:* Use the transformation $x = ar \cos \theta$, $y = br \sin \theta$.)

38. Use the transformation $x = u + 1/2v$, $y = v$ to evaluate the integral

$$\int_0^2 \int_{y/2}^{(y+4)/2} y^3(2x - y)e^{(2x-y)^2} \, dx \, dy$$

by first writing it as an integral over a region G in the uv-plane.

18.4
Triple Integrals in Rectangular Coordinates. Volumes and Average Values

Triple integrals are used to calculate the volumes of irregular three-dimensional shapes and the average values of functions over three-dimensional regions. They are also the integrals we use to calculate masses and moments of three-dimensional solids. When combined with vectors, as they will be in Chapter 19, they are the integrals we use to describe many of the phenomena associated with fluid flow and electromagnetism.

In the present article, we define triple integrals and use them to calculate volumes and average values. Masses and moments are the subject of the next article, Article 18.5. Triple integrals in cylindrical and spherical coordinates are the subject of the chapter's last article, Article 18.6.

Triple Integrals

If $F(x, y, z)$ is a function defined on a closed bounded region D in space—the region occupied by a solid ball, for example, or a lump of clay—then the integral of F over D may be defined in the following way. We partition a rectangular region about D into rectangular cells by planes parallel to the coordinate planes, as shown in Fig. 18.32. The cells have dimensions Δx by Δy by Δz. We number the cells that lie inside D in some order

$$\Delta V_1, \quad \Delta V_2, \quad \ldots, \quad \Delta V_n,$$

choose a point (x_k, y_k, z_k) in each ΔV_k, and form the sum

$$S_n = \sum_{k=1}^{n} F(x_k, y_k, z_k)\, \Delta V_k. \tag{1}$$

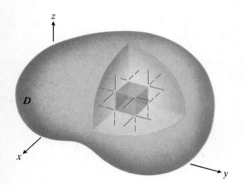

18.32 Partitioning a solid with rectangular cells of volume ΔV.

If F is continuous and the bounding surface of D is made of smooth surfaces joined along continuous curves, then as Δx, Δy, and Δz all approach zero the sums S_n approach a limit

$$\lim_{n \to \infty} S_n = \iiint_D F(x, y, z)\, dV. \tag{2}$$

We call this limit the **triple integral of F over D.** The limit also exists for some discontinuous functions.

Properties of Triple Integrals

Triple integrals have the same algebraic properties as double integrals and single integrals. If $F = F(x, y, z)$ and $G = G(x, y, z)$ are both integrable, then

1. $\displaystyle\iiint_D kF\, dV = k \iiint_D F\, dV$ (any number k),

2. $\displaystyle\iiint_D [F + G]\, dV = \iiint_D F\, dV + \iiint_D G\, dV,$

3. $\iiint\limits_{D} [F - G]\, dV = \iiint\limits_{D} F\, dV - \iiint\limits_{D} G\, dV,$

4. $\iiint\limits_{D} F\, dV \geq 0 \quad \text{if } F \geq 0 \text{ on } D,$

5. $\iiint\limits_{D} F\, dV \geq \iiint\limits_{D} G\, dV \quad \text{if } F \geq G \text{ on } D.$

The integrals have these properties because the sums that approximate them have these properties.

Triple integrals also have a domain additivity property that proves useful in physics and engineering as well as in mathematics. If the domain D of a continuous function F is partitioned by smooth surfaces into a finite number of cells $D_1, D_2, \ldots, D_n$, then

6. $\iiint\limits_{D} F\, dV = \iiint\limits_{D_1} F\, dV + \iiint\limits_{D_2} F\, dV + \cdots + \iiint\limits_{D_n} F\, dV.$

Volume of a Region in Space

If F is the constant function whose value is 1, then the sums in Eq. (1) reduce to

$$S_n = \sum F(x_k, y_k, z_k)\, \Delta V_k = \sum 1 \cdot \Delta V_k = \sum \Delta V_k. \qquad (3)$$

As Δx, Δy, and Δz all approach zero, the cells ΔV_k become smaller and more numerous and fill up more and more of D. We therefore define the volume of D to be the triple integral

$$\lim \sum_{k=1}^{n} \Delta V_k = \iiint\limits_{D} dV.$$

The volume of a closed bounded region D in space is given by the formula

$$\text{Volume of } D = \iiint\limits_{D} dV. \qquad (4)$$

As we shall see in a moment, this formula enables us to calculate the volumes of solids enclosed by curved surfaces.

Evaluation of Triple Integrals

The triple integral is seldom evaluated directly from its definition as a limit. Instead, we apply a three-dimensional version of Fubini's theorem to evaluate the integral by repeated single integrations.

For example, suppose we want to integrate a continuous function $F(x, y, z)$ over a region D that is bounded below by a surface $z = f_1(x, y)$,

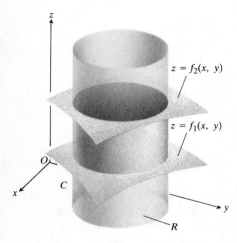

18.33 The enclosed volume can be found by evaluating

$$V = \iint_R \int_{z=f_1(x,y)}^{z=f_2(x,y)} dz\, dy\, dx.$$

The curve in which the upper and lower bounding surfaces intersect. Along this curve, $f_1(x, y) = f_2(x, y)$.

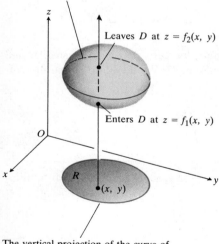

The vertical projection of the curve of intersection onto the xy-plane. Along this curve, $f_1(x, y) = f_2(x, y)$ as well, and we can use this equation as an equation for the boundary of R.

18.34 A schematic diagram for finding the limits of integration for a triple integral of a function F over a three-dimensional region D enclosed by two surfaces. For the region here,

$$\iiint_D F\, dV = \iint_R \int_{z=f_1(x,y)}^{z=f_2(x,y)} F\, dz\, dy\, dx.$$

above by the surface $z = f_2(x, y)$, and on the side by a cylinder C parallel to the z-axis (Fig. 18.33). Let R denote the vertical projection of D onto the xy-plane, which is the region in the xy-plane enclosed by C. The integral of F over D is then evaluated as

$$\iiint_D F(x, y, z)\, dV = \iint_R \left(\int_{z=f_1(x,y)}^{z=f_2(x,y)} F(x, y, z)\, dz \right) dy\, dx,$$

or

$$\iiint_D F(x, y, z)\, dV = \iint_R \int_{z=f_1(x,y)}^{z=f_2(x,y)} F(x, y, z)\, dz\, dy\, dx, \qquad (5)$$

if we omit the large parentheses.

The z-limits of integration indicate that for every (x, y) in the region R, z may extend from the lower surface $z = f_1(x, y)$ to the upper surface $z = f_2(x, y)$. The y- and x-limits of integration have not been given explicitly in Eq. (5) but are to be determined in the usual way from the boundaries of R.

In case the lateral surface of the cylinder reduces to zero, as in Fig. 18.34 and the example that follows, we may find the equation of the boundary of R by eliminating z between the two equations $z = f_1(x, y)$ and $z = f_2(x, y)$. This gives

$$f_1(x, y) = f_2(x, y),$$

an equation that contains no z and that defines the boundary of R in the xy-plane.

To supply the z-limits of integration in any particular instance we may use a procedure like the one for double integrals. We imagine a line L through a point (x, y) in R and parallel to the z-axis. As z increases, the line enters D at $z = f_1(x, y)$ and leaves D at $z = f_2(x, y)$. These give the lower and upper limits of the integration with respect to z. The result of this integration is now a function of x and y alone, which we integrate over R, supplying limits in the usual way.

EXAMPLE 1 Find the volume of the three-dimensional region enclosed by the surfaces

$$z = x^2 + 3y^2$$

and

$$z = 8 - x^2 - y^2.$$

Solution The two surfaces (Fig. 18.35) intersect on the elliptic cylinder

$$x^2 + 3y^2 = 8 - x^2 - y^2,$$

or

$$x^2 + 2y^2 = 4.$$

The three-dimensional region projects into the two-dimensional region R in the xy-plane that is enclosed by the ellipse having this same equation. In the double integral with respect to y and x over R, if we integrate first with respect to y, holding x fixed, y varies from $-\sqrt{(4 - x^2)/2}$ to $+\sqrt{(4 - x^2)/2}$. Then x var-

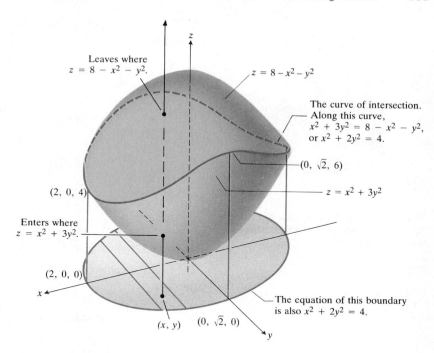

Leaves where
$z = 8 - x^2 - y^2$.

$z = 8 - x^2 - y^2$

The curve of intersection.
Along this curve,
$x^2 + 3y^2 = 8 - x^2 - y^2$,
or $x^2 + 2y^2 = 4$.

$(0, \sqrt{2}, 6)$

$(2, 0, 4)$

$z = x^2 + 3y^2$

Enters where
$z = x^2 + 3y^2$.

$(2, 0, 0)$

The equation of this boundary
is also $x^2 + 2y^2 = 4$.

(x, y) $(0, \sqrt{2}, 0)$

18.35 The volume of the region enclosed by these two paraboloids is calculated in Example 1.

ies from -2 to $+2$. Thus we have

$$V = \int_{-2}^{2} \int_{-\sqrt{(4-x^2)/2}}^{\sqrt{(4-x^2)/2}} \int_{x^2+3y^2}^{8-x^2-y^2} dz\,dy\,dx$$

$$= \int_{-2}^{2} \int_{-\sqrt{(4-x^2)/2}}^{\sqrt{(4-x^2)/2}} (8 - 2x^2 - 4y^2)\,dy\,dx$$

$$= \int_{-2}^{2} \left[(8 - 2x^2)y - \frac{4}{3}y^3 \right]_{y=-\sqrt{(4-x^2)/2}}^{y=\sqrt{(4-x^2)/2}} dx$$

$$= \int_{-2}^{2} \left(2(8 - 2x^2)\sqrt{\frac{4-x^2}{2}} - \frac{8}{3}\left(\frac{4-x^2}{2}\right)^{3/2} \right) dx$$

$$= \int_{-2}^{2} \left(8\left(\frac{4-x^2}{2}\right)^{3/2} - \frac{8}{3}\left(\frac{4-x^2}{2}\right)^{3/2} \right) dx$$

$$= \frac{4\sqrt{2}}{3} \int_{-2}^{2} (4 - x^2)^{3/2}\,dx$$

$$= 8\pi\sqrt{2}. \qquad \text{(After integration with the substitution } x = 2\sin u)$$

As we know, there are sometimes (but not always) two different orders in which the single integrations that evaluate a double integral may be worked. For triple integrals, there are sometimes (but not always) as many as *six* workable orders of integration. The next example shows an extreme case in which all six are workable.

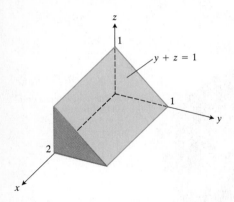

18.36 Example 2 shows how to calculate the volume of this prism with six different iterated triple integrals.

EXAMPLE 2 Each of the following integrals gives the volume of the solid shown in Fig. 18.36.

a) $\displaystyle\int_0^1 \int_0^{1-z} \int_0^2 dx\, dy\, dz$

b) $\displaystyle\int_0^1 \int_0^{1-y} \int_0^2 dx\, dz\, dy$

c) $\displaystyle\int_0^1 \int_0^2 \int_0^{1-z} dy\, dx\, dz$

d) $\displaystyle\int_0^2 \int_0^1 \int_0^{1-z} dy\, dz\, dx$

e) $\displaystyle\int_0^1 \int_0^2 \int_0^{1-y} dz\, dx\, dy$

f) $\displaystyle\int_0^2 \int_0^1 \int_0^{1-y} dz\, dy\, dx$ ∎

Average Value of a Function in Space

The average value of a function F over a region D in space is defined by the formula

$$\text{Average value of } F \text{ over } D = \frac{1}{\text{volume of } D} \iiint_D F\, dV. \tag{6}$$

If $F(x, y, z) = \sqrt{x^2 + y^2 + z^2}$, then the average value of F over D is the average distance of points in D from the origin. If $F(x, y, z)$ is the density of a solid that occupies a region D in space, then the average value of F over D is the average density of the solid in units of mass per unit volume.

EXAMPLE 3 Find the average value of $F(x, y, z) = xyz$ over the cube bounded by the planes $x = 2$, $y = 2$, and $z = 2$ in the first octant.

Solution We sketch the cube with enough detail to show the limits of integration (Fig. 18.37). We then use Eq. (6) to calculate the average value of F over the cube.

The volume of the cube is $(2)(2)(2) = 8$.

The value of the integral of F over the cube is

$$\int_0^2 \int_0^2 \int_0^2 xyz\, dx\, dy\, dz = \int_0^2 \int_0^2 \left[\frac{x^2}{2}\, yz\right]_{x=0}^{x=2} dy\, dz$$

$$= \int_0^2 \int_0^2 2yz\, dy\, dz = \int_0^2 \left[y^2 z\right]_{y=0}^{y=2} dz$$

$$= \int_0^2 4z\, dz = \left[2z^2\right]_0^2 = 8.$$

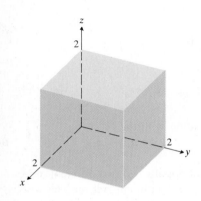

18.37 The solid cube bounded by the planes $x = 2$, $y = 2$, and $z = 2$ is the region of integration in Example 3.

With these values, Eq. (6) gives

$$\begin{array}{c}\text{Average value of}\\ xyz \text{ over the cube}\end{array} = \frac{1}{\text{volume}} \iiint_{\text{cube}} xyz\, dV = \left(\frac{1}{8}\right)(8) = 1.$$

In evaluating the integral, we chose the order dx, dy, dz, but any of the other five possible orders would have done as well.   ∎

PROBLEMS

1. What is the common value of the integrals in Example 2?

2. Write six different iterated triple integrals for the volume of the rectangular solid in the first octant bounded by the coordinate planes and the planes $x = 1$, $y = 2$, and $z = 3$. Evaluate one of the integrals.

3. Write six different iterated triple integrals for the volume of the tetrahedron cut from the first octant by the plane $6x + 3y + 2z = 6$. Evaluate one of the integrals.

4. Write six different iterated triple integrals for the volume of the region in the first octant enclosed by the cylinder $x^2 + z^2 = 4$ and the plane $y = 3$. Evaluate one of the integrals.

Evaluate the integrals in Problems 5–18.

5. $\displaystyle\int_0^1 \int_0^1 \int_0^1 (x^2 + y^2 + z^2)\, dz\, dy\, dx$

6. $\displaystyle\int_0^{\sqrt{2}} \int_0^{3y} \int_{x^2+3y^2}^{8-x^2-y^2} dz\, dx\, dy$

7. $\displaystyle\int_1^e \int_1^e \int_1^e \frac{1}{xyz}\, dx\, dy\, dz$

8. $\displaystyle\int_0^1 \int_0^{3-3x} \int_0^{3-3x-y} dz\, dy\, dx$

9. $\displaystyle\int_0^1 \int_0^{\pi} \int_0^{\pi} y \sin z\, dx\, dy\, dz$

10. $\displaystyle\int_{-1}^1 \int_{-1}^1 \int_{-1}^1 (x + y + z)\, dy\, dx\, dz$

11. $\displaystyle\int_{-1}^1 \int_{-\sqrt{1-z^2}}^{\sqrt{1-z^2}} \int_{-\sqrt{1-y^2-z^2}}^{\sqrt{1-y^2-z^2}} dx\, dy\, dz$

12. $\displaystyle\int_0^3 \int_0^{\sqrt{9-x^2}} \int_0^{\sqrt{9-x^2}} dz\, dy\, dx$

13. $\displaystyle\int_0^2 \int_{-\sqrt{4-y^2}}^{\sqrt{4-y^2}} \int_0^{2x+y} dz\, dx\, dy$

14. $\displaystyle\int_0^1 \int_0^{2-x} \int_0^{2-x-y} dz\, dy\, dx$

15. $\displaystyle\int_{-1}^1 \int_0^{\sqrt{1-x^2}} \int_0^1 (2 - 2x^2 - 2y^2)\, dz\, dy\, dx$

16. $\displaystyle\int_0^1 \int_0^{1-x^2} \int_3^{4-x^2-y} x\, dz\, dy\, dx$

17. $\displaystyle\int_0^{\pi} \int_0^{\pi} \int_0^{\pi} \cos(x + y + z)\, dx\, dy\, dz$

18. $\displaystyle\int_0^1 \int_0^{1-x} \int_{4x^2}^{5-x^2} (x - y + 1)\, dz\, dy\, dx$

In Problems 19–22, find the average value of $F(x, y, z)$ over the given region.

19. $F(x, y, z) = x^2 + 9$ over the cube in the first octant bounded by the coordinate planes and the planes $x = 2$, $y = 2$, and $z = 2$

20. $F(x, y, z) = x + y - z$ over the rectangular solid in the first octant bounded by the coordinate planes and the planes $x = 1$, $y = 1$, and $z = 2$

21. $F(x, y, z) = x^2 + y^2 + z^2$ over the cube in the first octant bounded by the coordinate planes and the planes $x = 1$, $y = 1$, and $z = 1$

22. $F(x, y, z) = xyz$ over the cube in the first octant bounded by the coordinate planes and the planes $x = a$, $y = a$, and $z = a$

Find the volumes of the figures and regions in Problems 23–38.

23. The region bounded above by the paraboloid $z = 2 - x^2 - y^2$ and below by the paraboloid $z = x^2 + y^2$

24. The region between the cylinder $z = y^2$ and the xy-plane that is bounded by the four vertical planes $x = 0$, $x = 1$, $y = -1$, $y = 1$

25. The region in the first octant bounded by the planes $x + z = 1$, $y + 2z = 2$

26. The region in the first octant bounded by the cylinder $x = 4 - y^2$ and the planes $z = y$, $x = 0$, $z = 0$

27. The wedge cut from the cylinder $x^2 + y^2 = 1$ by the plane $z = y$ above the plane $z = 0$ below

28. The region in the first octant bounded by the coordinate planes, the cylinder $x^2 + y^2 = 4$, and the plane $x + z = 3$

29. The region in the first octant bounded by the coordinate planes, above by the cylinder $x^2 + z = 1$, and on the right by the paraboloid $y = x^2 + z^2$. (*Hint:* Integrate first with respect to y.)

30. The region enclosed by the cylinders $z = 5 - x^2$, $z = 4x^2$, and the planes $y = 0$, $x + y = 1$

31. The region enclosed by the cylinder $y^2 + 4z^2 = 16$ and the planes $x = 0$, $x + y = 4$

32. The region bounded below by the plane $z = 0$, laterally by the elliptic cylinder $x^2 + 4y^2 = 4$, and above by the plane $z = x + 2$.

33. The tetrahedron in the first octant bounded by the coordinate planes and the plane $x/a + y/b + z/c = 1$

34. The region in the first octant bound by the coordinate planes, the plane $y = 1 - x$, and the surface $z = \cos(\pi x/2)$

35. The region common to the interiors of the cylinders $x^2 + y^2 = a^2$ and $x^2 + z^2 = a^2$ (Fig. 18.38)

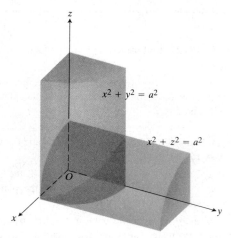

18.38 One-eighth of the volume common to the cylinders $x^2 + y^2 = a^2$, $x^2 + z^2 = a^2$ in Problem 35.

36. The region bounded by the elliptic paraboloids $z = x^2 + 9y^2$ and $z = 18 - x^2 - 9y^2$

37. The region enclosed by the ellipsoid $(x/a)^2 + (y/b)^2 + (z/c)^2 = 1$.

38. The region in the first octant that is bounded above by the surface $z = 4 - x^2 - y$ and below by the plane $z = 3$

39. Sketch the domain of integration of the integral

$$\int_{-1}^{1} \int_{x^2}^{1} \int_{0}^{1-y} dz \, dy \, dx.$$

Then rewrite the integral as an equivalent iterated integral in the order

a) $dy \, dz \, dx$,
b) $dy \, dx \, dz$,
c) $dx \, dy \, dz$,
d) $dx \, dz \, dy$,
e) $dz \, dx \, dy$.

40. Repeat Problem 39 for the integral

$$\int_{0}^{1} \int_{0}^{1} \int_{0}^{y^2} dz \, dy \, dx.$$

TOOLKIT PROGRAM

 Double Integral

18.5
Masses and Moments in Three Dimensions

In this article, we show how to calculate the masses and moments of three-dimensional objects in Cartesian coordinates. For calculations in spherical and cylindrical coordinates, see Article 18.6.

Masses and Moments

If $\delta(x, y, z)$ is the density of an object occupying a region D in space, then the integral of δ over D gives the mass of the object. To see why, imagine subdividing D as in Fig. 18.39. The object's mass is the limit

$$M = \lim \sum_{k} \Delta m_k = \lim \sum_{k} \delta(x_k, y_k, z_k)\Delta V_k = \iiint_{D} \delta(x, y, z) \, dV. \quad (1)$$

If $r(x, y, z)$ is the distance from the point (x, y, z) in D to a line L, then the moment of inertia of the mass

$$\Delta m_k = \delta(x_k, y_k, z_k)\Delta V_k$$

(shown in Fig. 18.39) about the line L is approximately

$$\Delta I_k = r^2(x_k, y_k, z_k)\Delta m_k.$$

The moment of inertia of the entire object about L is

$$I_L = \lim \sum_{k} \Delta I_k = \lim \sum_{k} r^2(x_k, y_k, z_k)\delta(x_k, y_k, z_k)\Delta V_k = \iiint_{D} r^2\delta \, dV.$$

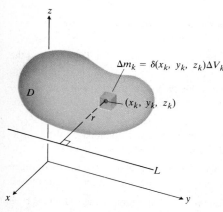

18.39 To define I_L we first imagine D to be subdivided into a finite number of mass elements Δm_k.

If L is the x-axis, then $r^2 = y^2 + z^2$ (Fig. 18.40) and

$$I_x = \iiint_D (y^2 + z^2)\delta\, dV.$$

Similarly,

$$I_y = \iiint_D (x^2 + z^2)\delta\, dV \quad \text{and} \quad I_z = \iiint_D (x^2 + y^2)\delta\, dV.$$

These and other useful formulas are summarized in the following list.

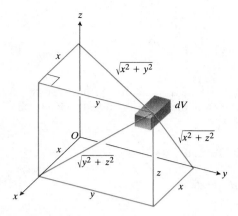

18.40 Distances from dV to the coordinate planes and axes.

Mass and Moment Formulas for Objects in Space

Mass: $\quad M = \iiint_D \delta\, dV \qquad (\delta = \text{density})$

First moments about the coordinate planes:

$$M_{yz} = \iiint_D x\,\delta\, dV, \qquad M_{xz} = \iiint_D y\,\delta\, dV, \qquad M_{xy} = \iiint_D z\,\delta\, dV$$

Center of mass:

$$\bar{x} = \frac{\iiint x\,\delta\, dV}{M}, \qquad \bar{y} = \frac{\iiint y\,\delta\, dV}{M}, \qquad \bar{z} = \frac{\iiint z\,\delta\, dV}{M}$$

Moments of inertia (second moments):

$$I_x = \iiint (y^2 + z^2)\delta\, dV, \qquad I_y = \iiint (x^2 + z^2)\delta\, dV,$$

$$I_z = \iiint (x^2 + y^2)\delta\, dV, \qquad I_L = \iiint r^2\delta\, dV,$$

$$r(x, y, z) = \text{distance from point } (x, y, z) \text{ to line } L$$

Radius of gyration about a line L: $\quad R = \sqrt{I_L/M}$

EXAMPLE 1 Find I_x, I_y, I_z for the rectangular solid of constant density δ shown in Fig. 18.41.

Solution The preceding formula for I_x gives

$$I_x = \int_{-c/2}^{c/2} \int_{-b/2}^{b/2} \int_{-a/2}^{a/2} (y^2 + z^2)\delta\, dx\, dy\, dz. \tag{2}$$

We can avoid some of the work of integration by observing that $(y^2 + z^2)\delta$ is an even function of x, y, and z and therefore

$$I_x = 8 \int_0^{c/2} \int_0^{b/2} \int_0^{a/2} (y^2 + z^2)\delta\, dx\, dy\, dz = 4a\delta \int_0^{c/2} \int_0^{b/2} (y^2 + z^2)\, dy\, dz$$

$$= 4a\delta \int_0^{c/2} \left[\frac{y^3}{3} + z^2 y \right]_{y=0}^{y=b/2} dz = 4a\delta \int_0^{c/2} \left(\frac{b^3}{24} + \frac{z^2 b}{2} \right) dz$$

$$= 4a\delta \left(\frac{b^3 c}{48} + \frac{c^3 b}{48} \right) = \frac{abc\delta}{12}(b^2 + c^2) = \frac{M}{12}(b^2 + c^2).$$

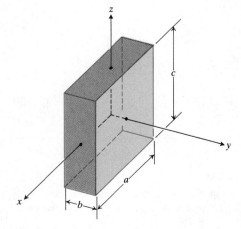

18.41 Example 1 calculates I_x, I_y, and I_z for the block shown here. The origin lies at the center of the block.

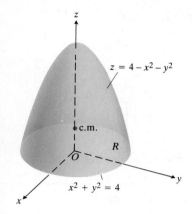

18.42 Example 2 calculates the coordinates of the center of mass of this solid.

Similarly,

$$I_y = \frac{M}{12}(a^2 + c^2) \quad \text{and} \quad I_z = \frac{M}{12}(a^2 + b^2).$$ ■

EXAMPLE 2 Find the center of mass of a solid of constant density δ bounded below by the disk $R: x^2 + y^2 \le 4$ in the plane $z = 0$ and above by the paraboloid $z = 4 - x^2 - y^2$ (Fig. 18.42).

Solution By symmetry, $\bar{x} = \bar{y} = 0$. To find $\bar{z}$ we first calculate

$$M_{xy} = \iint_R \int_{z=0}^{z=4-x^2-y^2} z\, \delta \, dz \, dy \, dx$$

$$= \iint_R \left[\frac{z^2}{2}\right]_{z=0}^{z=4-x^2-y^2} \delta \, dy \, dx$$

$$= \frac{\delta}{2} \iint_R (4 - x^2 - y^2)^2 \, dy \, dx$$

$$= \frac{\delta}{2} \int_0^{2\pi} \int_0^2 (4 - r^2)^2 \, r \, dr \, d\theta \quad \text{(polar coordinates)}$$

$$= \frac{\delta}{2} \int_0^{2\pi} \left[-\frac{1}{6}(4 - r^2)^3\right]_{r=0}^{r=2} d\theta$$

$$= \frac{16\delta}{3} \int_0^{2\pi} d\theta$$

$$= \frac{32\pi\delta}{3}.$$

A similar calculation gives

$$M = \iint_R \int_0^{4-x^2-y^2} \delta \, dz \, dy \, dx = 8\pi\delta.$$

Therefore,

$$\bar{z} = \frac{M_{xy}}{M} = \frac{4}{3},$$

and the center of mass is $(\bar{x}, \bar{y}, \bar{z}) = (0, 0, 4/3)$. ■

PROBLEMS

Constant Density

In Problems 1–12, whenever the density is not mentioned it is assumed to be $\delta = 1$.

1. Evaluate the integral for I_x in Eq. (2) directly to show that the shortcut in Example 1 gives the same answer. Use the results in Example 1 to find the radius of gyration of the rectangular solid about each coordinate axis.

2. The axes shown in Fig. 18.43 run through the centroid of the solid wedge parallel to its edges. Using the dimensions shown, find I_x, I_y, and I_z.

3. Find the moments of inertia of the rectangular solid shown in Fig. 18.44 with respect to its edges by calculating I_x, I_y, and I_z.

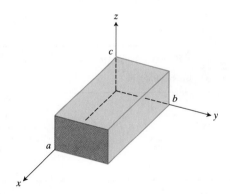

18.43 Figure for Problem 2.

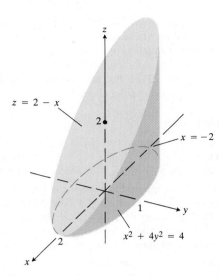

18.45 The solid in Problem 6.

18.44 The rectangular solid in Problem 3.

4. a) Find the centroid and the moments of inertia I_x, I_y, and I_z of the tetrahedron whose vertices are the points $(0, 0, 0)$, $(1, 0, 0)$, $(0, 1, 0)$ and $(0, 0, 1)$.

b) Find the radius of gyration of the tetrahedron about the x-axis. Compare it with the distance from the centroid to the x-axis.

5. A solid "trough" of constant density δ is bounded below by the surface $z = 4y^2$, above by the plane $z = 4$, and on the ends by the planes $x = 1$ and $x = -1$. Find the center of mass and the moments of inertia with respect to the three axes.

6. A solid of constant density $\delta = 1$ is bounded below by the plane $z = 0$, on the sides by the elliptic cylinder $x^2 + 4y^2 = 4$, and above by the plane $z = 2 - x$ (Fig. 18.45).

a) Find $\bar{x}$ and $\bar{y}$.

b) Evaluate the integral

$$M_{xy} = \int_{-2}^{2} \int_{-(1/2)\sqrt{4-x^2}}^{(1/2)\sqrt{4-x^2}} \int_{0}^{2-x} z \, dz \, dy \, dx,$$

using integral tables to carry out the final integration with respect to x. Then divide M_{xy} by M to verify that $\bar{z} = 5/2$.

7. a) Find the center of mass of a homogeneous solid bounded below by the paraboloid $z = x^2 + y^2$ and above by the plane $z = 4$.

b) Find the plane $z = c$ that divides the solid into two parts of equal volume. This plane does not pass through the center of mass.

8. A solid cube, 2 units on a side, is bounded by the planes $x = \pm 1$, $z = \pm 1$, $y = 3$, and $y = 5$. Find the center of mass and the moments of inertia and radii of gyration about the coordinate axes. Suppose that the cube's mass is concentrated at the center of mass. How, if at all, does that change the moments of inertia and radii of gyration about the coordinate axes?

9. A wedge shaped like the one in Fig. 18.43 has $a = 4$, $b = 6$, and $c = 3$. Make a quick sketch to check for yourself that the square of the distance from a typical point (x, y, z) of the wedge to the line $L: z = 0$, $y = 6$ is $r^2 = (y - 6)^2 + z^2$. Then calculate the moment of inertia and radius of gyration of the wedge about L.

10. A wedge shaped like the one in Fig. 18.43 has $a = 4$, $b = 6$, and $c = 3$. Make a quick sketch to check for yourself that the square of the distance from a typical point (x, y, z) of the wedge to the line $L: x = 4$, $y = 0$ is $r^2 = (x - 4)^2 + y^2$. Then calculate the moment of inertia and radius of gyration of the wedge about L.

11. A rectangular solid like the one in Fig. 18.44 has $a = 4$, $b = 2$, and $c = 1$. Make a quick sketch to check for yourself that the square of the distance between a typical point (x, y, z) of the solid and the line $L: y = 2$, $z = 0$ is $r^2 = (y - 2)^2 + z^2$. Then find the moment of inertia and radius of gyration of the solid about L.

12. A rectangular solid like the one in Fig. 18.44 has $a = 4$, $b = 2$, and $c = 1$. Make a quick sketch to check for yourself that the square of the distance between a typical point (x, y, z) of the solid and the line L: $x = 4$, $y = 0$ is $r^2 = (x - 4)^2 + y^2$. Then find the moment of inertia and radius of gyration of the solid about L.

Variable Density

13. A solid region in the first octant is bounded by the coordinate planes and the plane $x + y + z = 2$. The density of the solid is $\delta(x, y, z) = 2x$. Find the center of mass.

14. A solid wedge shaped like the one in Fig. 18.43 has dimensions $a = 4$, $b = 6$, and $c = 3$. The density is $\delta(x, y, z) = x$. Find the center of mass and the moments of inertia and radii of gyration about the coordinate axes. Notice that if the density is constant, the center of mass will be $(2, 2, 1)$.

15. A solid cube in the first octant is bounded by the coordinate planes and by the planes $x = 1$, $y = 1$, and $z = 1$. Find its center of mass and the moments of inertia and radii of gyration about the coordinate axes if

$$\delta(x, y, z) = x + y + z + 1$$

16. A solid in the first octant is bounded by the planes $y = 0$ and $z = 0$ and by the surfaces $z = 4 - x^2$ and $x = y^2$ (Fig. 18.46). Its density function is $\delta(x, y, z) = kxy$.
    a) Find the solid's mass.   b) Find $\bar{x}$.
    c) Find $\bar{y}$.   d) Find $\bar{z}$.

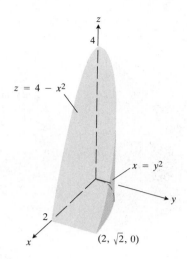

18.46   The solid in Problem 16.

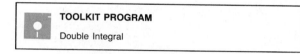

TOOLKIT PROGRAM

Double Integral

18.6
Triple Integrals in Cylindrical and Spherical Coordinates

If we are working with a solid like a cone or a cylindrical shell that has an axis of symmetry, we can often simplify our calculations by taking the axis to be the z-axis in a cylindrical coordinate system. In like manner, if we have to work with a shape like a ball or spherical shell that has a center of symmetry, we can usually save time by choosing the center to be the original in a spherical coordinate system.

In this article, we show how to calculate masses, moments, and volumes in cylindrical and spherical coordinates.

Cylindrical Coordinates

Cylindrical coordinates (Fig. 18.47) are good for describing cylinders whose axes run along the z-axis and planes that either contain the z-axis or lie perpendicular to the z-axis. As we saw in Article 15.3, surfaces like these have simple

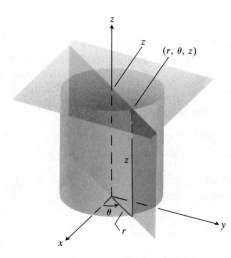

18.47 Cylindrical coordinates and typical surfaces of constant coordinate value.

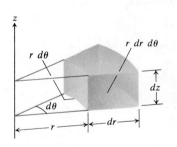

18.48 The volume element in cylindrical coordinates is $dV = dz \, r \, dr \, d\theta$.

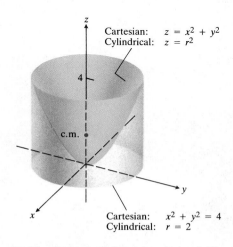

18.49 Example 1 shows how to locate the center of mass of this solid.

equations of constant coordinate value:

$$r = 4 \qquad \text{(cylinder, radius 4, axis the } z\text{-axis),}$$

$$\theta = \frac{\pi}{3} \qquad \text{(plane containing the } z\text{-axis),}$$

$$z = 2 \qquad \text{(plane perpendicular to the } z\text{-axis).}$$

The volume element for subdividing a region in space with cylindrical coordinates is taken to be

$$dV = dz \, r \, dr \, d\theta, \tag{1}$$

as shown in Fig. 18.48. Triple integrals in cylindrical coordinates are then evaluated as iterated integrals, as in the following example.

EXAMPLE 1 Find the centroid of the solid that is enclosed by the cylinder $x^2 + y^2 = 4$ and is also bounded above by the paraboloid $z = x^2 + y^2$ and below by the xy-plane.

Solution We sketch the figure to show the surfaces and determine the limits of integration (Fig. 18.49). In cylindrical coordinates, the surfaces are

The cylinder: $r = 2$
The paraboloid: $z = r^2$
The xy-plane: $z = 0$.

From the figure we see that r runs from 0 to 2, θ runs from 0 to 2π, and z runs from 0 to r^2.

The solid's center of mass lies on its axis of symmetry, in this case the z-axis, so $\bar{x} = \bar{y} = 0$. To find $\bar{z}$, we divide the moment M_{xy} by the mass M. The value of M_{xy} is

$$M_{xy} = \int_0^{2\pi} \int_0^2 \int_0^{r^2} z \, dz \, r \, dr \, d\theta = \int_0^{2\pi} \int_0^2 \left[\frac{z^2}{2} \right]_0^{r^2} r \, dr \, d\theta$$

$$= \int_0^{2\pi} \int_0^2 \frac{r^5}{2} \, dr \, d\theta = \int_0^{2\pi} \left[\frac{r^6}{12} \right]_0^2 d\theta = \int_0^{2\pi} \frac{16}{3} \, d\theta = \frac{32\pi}{3}.$$

The value of M is

$$M = \int_0^{2\pi} \int_0^2 \int_0^{r^2} dz \, r \, dr \, d\theta = \int_0^{2\pi} \int_0^2 \left[z \right]_0^{r^2} r \, dr \, d\theta$$

$$= \int_0^{2\pi} \int_0^2 r^3 \, dr \, d\theta = \int_0^{2\pi} \left[\frac{r^4}{4} \right]_0^2 d\theta$$

$$= \int_0^{2\pi} 4 \, d\theta = 8\pi.$$

Therefore,

$$\bar{z} = \frac{M_{xy}}{M} = \frac{32\pi}{3} \frac{1}{8\pi} = \frac{4}{3},$$

and the solid's center of mass is the point $(0, 0, 4/3)$.

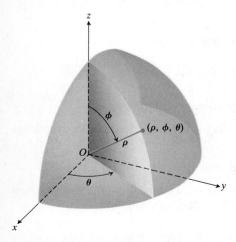

18.50 Spherical coordinates are measured with a distance and two angles.

Spherical Coordinates

Spherical coordinates (Fig. 18.50) are good for describing spheres centered at the origin, half-planes hinged along the z-axis, and single-napped cones whose vertices lie at the origin and whose axes lie along the z-axis. Surfaces like these have equations of constant coordinate value:

$$\rho = 4 \qquad \text{(sphere, radius 4, center at origin),}$$

$$\phi = \frac{\pi}{3} \qquad \begin{array}{l}\text{(cone opening up from the origin, making an}\\ \text{angle of } \pi/3 \text{ radians with the positive } z\text{-axis),}\end{array}$$

$$\theta = \frac{\pi}{3} \qquad \begin{array}{l}\text{(half-plane, hinged along the } z\text{-axis, making}\\ \text{an angle of } \pi/3 \text{ radians with the positive } x\text{-axis).}\end{array}$$

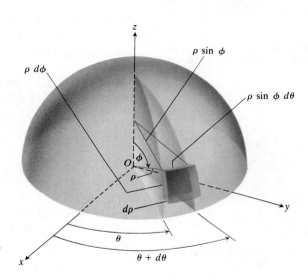

18.51 The volume element in spherical coordinates is $dV = d\rho \cdot \rho\, d\phi \cdot \rho \sin\phi\, d\theta = \rho^2 \sin\phi\, d\rho\, d\phi\, d\theta$.

18.52 The region cut from the solid sphere $\rho \leq 1$ by the cone $\phi = \pi/3$.

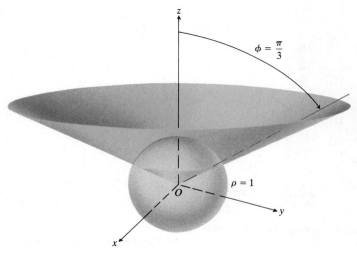

The volume element in spherical coordinates is

$$dV = \rho^2 \sin \phi \; d\rho \; d\phi \; d\theta \tag{2}$$

(as shown in Fig. 18.51), and triple integrals take the form

$$\iiint F(\rho, \phi, \theta) \; dV = \iiint F(\rho, \phi, \theta) \; \rho^2 \sin \phi \; d\rho \; d\phi \; d\theta. \tag{3}$$

To evaluate these integrals we usually integrate first with respect to ρ. The procedure for finding the limits of integration for a region D in space is therefore the following:

STEP 1: Hold ϕ and θ fixed and let ρ increase. This gives a ray out from the origin.

STEP 2: Integrate from the ρ-value where the ray first enters D to the ρ-value where the ray leaves D. This gives the limits for ρ.

STEP 3: Hold θ fixed and let ϕ increase. (This gives a family of rays that make a vertical "fan.") Integrate over the ϕ-values for which the rays pass through D.

STEP 4: Choose θ-limits that include all the fans that intersect D.

EXAMPLE 2 Find the volume of the region cut from the solid sphere $\rho \leq 1$ by the cone $\phi = \pi/3$.

Solution We sketch the sphere and cone to determine the limits of integration (Fig. 18.52). From the figure, we see that ρ runs from 0 to 1, ϕ runs from 0 to $\pi/3$, and θ runs from 0 to 2π. The volume is therefore

$$V = \int_0^{2\pi} \int_0^{\pi/3} \int_0^1 \rho^2 \sin \phi \; d\rho \; d\phi \; d\theta = \int_0^{2\pi} \int_0^{\pi/3} \left[\frac{\rho^3}{3} \right]_0^1 \sin \phi \; d\phi \; d\theta$$

$$= \int_0^{2\pi} \int_0^{\pi/3} \frac{1}{3} \sin \phi \; d\phi \; d\theta = \int_0^{2\pi} \left[-\frac{1}{3} \cos \phi \right]_0^{\pi/3} d\theta$$

$$= \int_0^{2\pi} \left(-\frac{1}{6} + \frac{1}{3} \right) d\theta = \frac{1}{6}(2\pi) = \frac{\pi}{3}. \qquad \blacksquare$$

EXAMPLE 3 Find the moment of inertia about the z-axis of the region in Example 2.

Solution In rectangular coordinates, the moment is

$$I_z = \iiint (x^2 + y^2) \; dV.$$

In spherical coordinates, $x^2 + y^2 = (\rho \sin \phi \cos \theta)^2 + (\rho \sin \phi \sin \theta)^2 = \rho^2 \sin^2 \phi$. Hence,

$$I_z = \iiint (\rho^2 \sin^2 \phi) \; \rho^2 \sin \phi \; d\rho \; d\phi \; d\theta$$

$$= \iiint \rho^4 \sin^3 \phi \; d\rho \; d\phi \; d\theta.$$

For the region in Example 2, this becomes

$$I_z = \int_0^{2\pi} \int_0^{\pi/3} \int_0^1 \rho^4 \sin^3 \phi \, d\rho \, d\phi \, d\theta$$

$$= \int_0^{2\pi} \int_0^{\pi/3} \left[\frac{\rho^5}{5} \right]_0^1 \sin^3 \phi \, d\phi \, d\theta$$

$$= \frac{1}{5} \int_0^{2\pi} \int_0^{\pi/3} (1 - \cos^2 \phi) \sin \phi \, d\phi \, d\theta$$

$$= \frac{1}{5} \int_0^{2\pi} \left[-\cos \phi + \frac{\cos^3 \phi}{3} \right]_0^{\pi/3} d\theta$$

$$= \frac{1}{5} \int_0^{2\pi} \left(-\frac{1}{2} + 1 + \frac{1}{24} - \frac{1}{3} \right) d\theta$$

$$= \frac{1}{5} \int_0^{2\pi} \frac{5}{24} \, d\theta = \frac{1}{24}(2\pi) = \frac{\pi}{12}.$$

■

Substitutions in Triple Integrals

The cylindrical and spherical coordinate substitutions are special cases of a substitution method that pictures changes of variables in triple integrals as transformations of three-dimensional regions. The method is just like the method we described at the end of Article 18.3 for double integrals, except that now we work in three dimensions instead of two.

Suppose that a region G in uvw-space is transformed into the region R in xyz-space by differentiable equations of the form

$$x = g(u, v, w), \qquad y = h(u, v, w), \qquad z = k(u, v, w),$$

as suggested in Fig. 18.53. Then any function $F(x, y, z)$ defined on R can be thought of as a function

$$F(g(u, v, w), h(u, v, w), k(u, v, w)) = H(u, v, w)$$

defined on G. The integral of $F(x, y, z)$ over R is related to the integral of $H(u, v, w)$ over G by the equation

$$\iiint_R F(x, y, z) \, dx \, dy \, dz = \iiint_G H(u, v, w) \, |J(u, v, w)| \, du \, dv \, dw. \quad (4)$$

18.53 The equations $x = g(u, v, w)$, $y = h(u, v, w)$, and $z = k(u, v, w)$ allow us to write an integral over a region R in Cartesian xyz-space as an integral over a region G in Cartesian uvw-space.

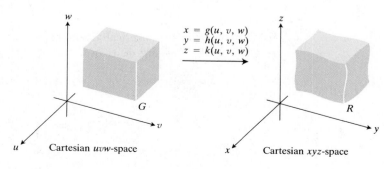

$$x = g(u, v, w)$$
$$y = h(u, v, w)$$
$$z = k(u, v, w)$$

Cartesian uvw-space   Cartesian xyz-space

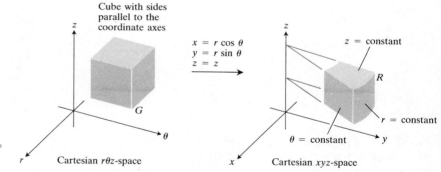

18.54 The equations $x = r \cos \theta$, $y = r \sin \theta$, and $z = z$ transform G into R.

The factor $J(u, v, w)$, whose absolute value appears in this equation, is the **Jacobian determinant**

$$J(u, v, w) = \begin{vmatrix} \dfrac{\partial x}{\partial u} & \dfrac{\partial x}{\partial v} & \dfrac{\partial x}{\partial w} \\[2mm] \dfrac{\partial y}{\partial u} & \dfrac{\partial y}{\partial v} & \dfrac{\partial y}{\partial w} \\[2mm] \dfrac{\partial z}{\partial u} & \dfrac{\partial z}{\partial v} & \dfrac{\partial z}{\partial w} \end{vmatrix} = \dfrac{\partial(x, y, z)}{\partial(u, v, w)}. \tag{5}$$

As in the two-dimensional case, the derivation of the change-of-variable formula in Eq. (4) is complicated and we shall not go into it here.

For cylindrical coordinates, we have r, θ, and z in place of u, v, and w. The transformation from Cartesian $r\theta z$-space to Cartesian xyz-space is given by the equations

$$x = r \cos \theta, \qquad y = r \sin \theta, \qquad z = z$$

(Fig. 18.54). The Jacobian of the transformation is

$$J(r, \theta, z) = \begin{vmatrix} \dfrac{\partial x}{\partial r} & \dfrac{\partial x}{\partial \theta} & \dfrac{\partial x}{\partial z} \\[2mm] \dfrac{\partial y}{\partial r} & \dfrac{\partial y}{\partial \theta} & \dfrac{\partial y}{\partial z} \\[2mm] \dfrac{\partial z}{\partial r} & \dfrac{\partial z}{\partial \theta} & \dfrac{\partial z}{\partial z} \end{vmatrix}$$

$$= \begin{vmatrix} \cos \theta & -r \sin \theta & 0 \\ \sin \theta & r \cos \theta & 0 \\ 0 & 0 & 1 \end{vmatrix}$$

$$= r \cos^2 \theta + r \sin^2 \theta = r.$$

The corresponding version of Eq. (4) is

$$\iiint_R F(x, y, z) \, dx \, dy \, dz = \iiint_G H(r, \theta, z) \, |r| \, dr \, d\theta \, dz, \tag{6}$$

and we can drop the absolute value signs whenever $r \geq 0$.

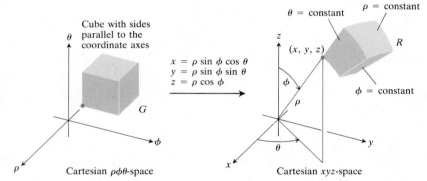

18.55   The equations $x = \rho \sin \phi \cos \theta$, $y = \rho \sin \phi \sin \theta$, and $z = \rho \cos \phi$ transform G into R.

For spherical coordinates, we have ρ, ϕ, and θ in place of u, v, and w. The transformation from Cartesian $\rho\phi\theta$-space to Cartesian xyz-space is given by the equations

$$x = \rho \sin \phi \cos \theta,$$
$$y = \rho \sin \phi \sin \theta,$$
$$z = \rho \cos \phi$$

(Fig. 18.55). The Jacobian of the transformation is

$$J(\rho, \phi, \theta) = \begin{vmatrix} \dfrac{\partial x}{\partial \rho} & \dfrac{\partial x}{\partial \phi} & \dfrac{\partial x}{\partial \theta} \\[2mm] \dfrac{\partial y}{\partial \rho} & \dfrac{\partial y}{\partial \phi} & \dfrac{\partial y}{\partial \theta} \\[2mm] \dfrac{\partial z}{\partial \rho} & \dfrac{\partial z}{\partial \phi} & \dfrac{\partial z}{\partial \theta} \end{vmatrix} = \rho^2 \sin \phi, \tag{7}$$

as you will see if you do Problem 57. The corresponding version of Eq. (4) is

$$\iiint_R F(x, y, z)\, dx\, dy\, dz = \iiint_G H(\rho, \phi, \theta)\, |\rho^2 \sin \phi|\, d\rho\, d\phi\, d\theta, \tag{8}$$

and we can drop the absolute value signs because $\sin \phi$ is never negative. Here is an example of another substitution.

EXAMPLE 4   Evaluate

$$\int_0^3 \int_0^4 \int_{x=y/2}^{x=(y/2)+1} \left(\frac{2x - y}{2} + \frac{z}{3} \right) dx\, dy\, dz$$

by applying the transformation

$$u = \frac{2x - y}{2}, \qquad v = \frac{y}{2}, \qquad w = \frac{z}{3} \tag{9}$$

and integrating over an appropriate region in uvw-space.

Solution   We sketch the region R of integration in xyz-space and identify its boundaries (Fig. 18.56). In this case, the surfaces that bound R are all planes.

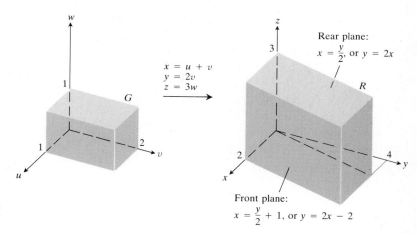

18.56 The equations $x = u + v$, $y = 2v$, and $z = 3w$ transform G into R. Reversing the transformation by the equations $u = (2x - y)/2$, $v = y/2$, and $w = z/3$ transforms R into G. See Example 4.

To apply Eq. (4), we need to find the corresponding uvw-region G and the Jacobian of the transformation. To find them, we first solve Eqs. (9) for x, y, and z in terms of u, v, and w. Routine algebra gives

$$\begin{aligned} x &= u + v, \\ y &= 2v, \\ z &= 3w. \end{aligned} \tag{10}$$

We then find the boundaries of G by substituting these expressions into the equations for the boundaries of R:

Boundary of R	Corresponding boundary of G	Simplified equation
$x = y/2$	$u + v = 2v/2 = v$	$u = 0$
$x = (y/2) + 1$	$u + v = (2v/2) + 1 = v + 1$	$u = 1$
$y = 0$	$2v = 0$	$v = 0$
$y = 4$	$2v = 4$	$v = 2$
$z = 0$	$3w = 0$	$w = 0$
$z = 3$	$3w = 3$	$w = 1$

See Fig. 18.56.

The Jacobian of the transformation, again from Eqs. (10), is

$$J(u, v, w) = \begin{vmatrix} \dfrac{\partial x}{\partial u} & \dfrac{\partial x}{\partial v} & \dfrac{\partial x}{\partial w} \\[2mm] \dfrac{\partial y}{\partial u} & \dfrac{\partial y}{\partial v} & \dfrac{\partial y}{\partial w} \\[2mm] \dfrac{\partial z}{\partial u} & \dfrac{\partial z}{\partial v} & \dfrac{\partial z}{\partial w} \end{vmatrix} = \begin{vmatrix} 1 & 1 & 0 \\ 0 & 2 & 0 \\ 0 & 0 & 3 \end{vmatrix} = 6.$$

We now have everything we need to apply Eq. (4):

$$\int_0^3 \int_0^4 \int_{x=y/2}^{x=(y/2)+1} \left(\frac{2x-y}{2} + \frac{z}{3}\right) dx\,dy\,dz$$

$$= \int_0^1 \int_0^2 \int_0^1 (u+w)\,|J(u,\,v,\,w)|\,du\,dv\,dw$$

$$= \int_0^1 \int_0^2 \int_0^1 (u+w)(6)\,du\,dv\,dw$$

$$= 6 \int_0^1 \int_0^2 \left[\frac{u^2}{2} + uw\right]_0^1 dv\,dw$$

$$= 6 \int_0^1 \int_0^2 \left(\frac{1}{2} + w\right) dv\,dw$$

$$= 6 \int_0^1 (1 + 2w)\,dw$$

$$= 6\left[w + w^2\right]_0^1 = 6(2) = 12.$$

Problem 58 asks you to evaluate the original integral directly to show that its value is also 12.   ∎

Coordinate Conversion Formulas

Polar to rectangular	Spherical to cylindrical	Spherical to rectangular
$x = r \cos \theta$	$r = \rho \sin \phi$	$x = \rho \sin \phi \cos \theta$
$y = r \sin \theta$	$z = \rho \cos \phi$	$y = \rho \sin \phi \sin \theta$
$z = z$	$\theta = \theta$	$z = \rho \cos \phi$

(Details in Article 15.3)

Volume: $dV = dx\,dy\,dz = dz\,r\,dr\,d\theta = \rho^2 \sin \phi\,d\rho\,d\phi\,d\theta$

PROBLEMS

Evaluate the cylindrical-coordinate integrals in Problems 1–6.

1. $\int_0^{2\pi} \int_0^1 \int_r^{\sqrt{2-r^2}} dz\,r\,dr\,d\theta$

2. $\int_0^{2\pi} \int_0^3 \int_{r^2/3}^{\sqrt{18-r^2}} dz\,r\,dr\,d\theta$

3. $\int_0^{2\pi} \int_0^{\theta/2\pi} \int_0^{3+24r^2} dz\,r\,dr\,d\theta$

4. $\int_0^{\pi} \int_0^{\theta/\pi} \int_{-\sqrt{4-r^2}}^{3\sqrt{4-r^2}} z\,dz\,r\,dr\,d\theta$

5. $\int_0^{2\pi} \int_0^1 \int_r^{1/\sqrt{2-r^2}} 3dz\,r\,dr\,d\theta$

6. $\int_0^{2\pi} \int_0^1 \int_{-1/2}^{1/2} (r^2\sin^2\theta + z^2)\,dz\,r\,dr\,d\theta$

Evaluate the spherical-coordinate integrals in Problems 7–12.

7. $\int_0^{\pi} \int_0^{\pi} \int_0^{2\sin\phi} \rho^2\sin\phi\,d\rho\,d\phi\,d\theta$

8. $\int_0^{2\pi} \int_0^{\pi/4} \int_0^2 (\rho\cos\phi)\rho^2\sin\phi\,d\rho\,d\phi\,d\theta$

9. $\displaystyle\int_0^{2\pi}\int_0^{\pi}\int_0^{(1-\cos\phi)/2}\rho^2\sin\phi\;d\rho\;d\phi\;d\theta$

10. $\displaystyle\int_0^{3\pi/2}\int_0^{\pi}\int_0^{1}5\rho^3\sin^3\phi\;d\rho\;d\phi\;d\theta$

11. $\displaystyle\int_0^{2\pi}\int_0^{\pi/3}\int_{\sec\phi}^{2}3\rho^2\sin\phi\;d\rho\;d\phi\;d\theta$

12. $\displaystyle\int_0^{2\pi}\int_0^{\pi/2}\int_0^{\sec\phi}(\rho\cos\phi)\rho^2\sin\phi\;d\rho\;d\phi\;d\theta$

13. Set up an iterated triple integral for the volume of the sphere $x^2 + y^2 + z^2 = 4$ in (a) spherical, (b) cylindrical, and (c) rectangular coordinates.

14. Let D be the smaller spherical cap cut from a solid ball of radius 2 units by a plane 1 unit from the center of the sphere. Express the volume of D as an iterated triple integral in (a) spherical, (b) cylindrical, and (c) rectangular coordinates.

15. Convert the following integral (a) to rectangular coordinates, (b) to spherical coordinates:

$$\int_0^{2\pi}\int_0^{1}\int_0^{\sqrt{4-r^2}} r\;dz\;r\;dr\;d\theta.$$

16. Let D denote the region in the first octant enclosed below by the cone $z^2 = x^2 + y^2$ and above by the sphere $x^2 + y^2 + z^2 = 8$. Express the volume V of D as an iterated triple integral in (a) cylindrical, and (b) spherical coordinates. Then (c) find V.

17. Give the limits of integration for evaluating the iterated integral

$$\iiint f(r, \theta, z)\;dz\;r\;dr\;d\theta$$

over the region D that is bounded below by the plane $z = 0$, laterally by the circular cylinder $r = \cos\theta$, and above by the paraboloid $z = 3r^2$.

18. Express $I_z = \iiint (x^2 + y^2)\delta dV$ as an iterated triple integral over an arbitrary region D in (a) cylindrical and (b) spherical coordinates.

19. Express the moment of inertia I_z of the solid bounded below by the plane $z = 0$ and above by the sphere $x^2 + y^2 + z^2 = 1$ as an iterated triple integral in (a) cylindrical and (b) spherical coordinates.

20. Convert the integral

$$\int_{-1}^{1}\int_0^{\sqrt{1-y^2}}\int_0^{x}(x^2 + y^2)\;dz\;dx\;dy$$

to an equivalent integral in cylindrical coordinates and evaluate the result.

Cylindrical Coordinates

In problems that ask for masses and moments, assume that the density is $\delta = 1$ unless stated otherwise.

21. Find the volume of the region bounded below by the plane $z = 0$, laterally by the cylinder $x^2 + y^2 = 1$, and above by the paraboloid $z = x^2 + y^2$.

22. Find the volume of the region bounded below by the paraboloid $z = x^2 + y^2$, laterally by the cylinder $x^2 + y^2 = 1$, and above by the paraboloid $z = x^2 + y^2 + 1$.

23. Find the volume of the region bounded below by the plane $z = 0$ and above by the paraboloid $z = 4 - x^2 - y^2$.

24. Find the volume of the region enclosed by the cylinder $x^2 + y^2 = 4$ and the planes $z = 0$ and $y + z = 4$. (*Hint:* In cylindrical coordinates, $z = 4 - y$ becomes $z = 4 - r\sin\theta$.)

25. Find the volume of the region bounded above by the paraboloid $z = 5 - x^2 - y^2$ and below by the paraboloid $z = 4x^2 + 4y^2$.

26. Find the volume of the region bounded above by the paraboloid $z = 9 - x^2 - y^2$, below by the xy-plane, and that lies *outside* the cylinder $x^2 + y^2 = 1$.

27. Find the volume of the region cut from the solid sphere $x^2 + y^2 + z^2 \leq 4a^2$ by the cylinder $x^2 + y^2 = a^2$.

28. Find the volume of the region bounded above by the sphere $x^2 + y^2 + z^2 = 2a^2$ and below by the paraboloid $az = x^2 + y^2$.

29. A hemispherical bowl of radius 5 cm is filled with water to within 3 cm of the top. Find the volume of water in the bowl.

30. Find the volume of the region in the first octant that lies between the cylinders $r = 1$ and $r = 2$ and that is bounded below by the xy-plane and above by the surface $z = xy$.

31. Find the average value of the function $F(r, \theta, z) = r$ over the region bounded by the cylinder $r = 1$ between the planes $z = -1$ and $z = 1$.

32. Find the average value of the function $F(r, \theta, z) = r$ over the solid ball bounded by the sphere $r^2 + z^2 = 1$. (This is the sphere $x^2 + y^2 + z^2 = 1$.)

33. A solid is bounded below by the plane $z = 0$, above by the cone $z = r$, $r \geq 0$, and on the sides by the cylinder $r = 1$. Find the center of mass.

34. Find the centroid of the region that is bounded below by the surface $z = \sqrt{r}$ and above by the plane $z = 2$.

35. Find the moment of inertia and radius of gyration about the z-axis of a thick-walled right circular cylinder bounded on the inside by the cylinder $r = 1$, on the outside by the cylinder $r = 2$, and on the top and bottom by the planes $z = 4$ and $z = 0$.

36. A solid of constant density δ in the first octant is bounded above by the cone $z^2 = x^2 + y^2$, below by the plane $z = 0$, and on the sides by the cylinder $x^2 + y^2 = 4$ and the planes $x = 0$ and $y = 0$. Find the center of gravity of the solid. (*Hint:* $\bar{x} = \bar{y}$.)

37. Find the centroid of the region bounded above by the sphere $x^2 + y^2 + z^2 = 2a^2$ and below by the paraboloid $az = x^2 + y^2$.

38. Find the moment of inertia of a solid circular cylinder of radius 1 and height 2 (a) about the axis of the cylinder, (b) about a line through the centroid perpendicular to the axis of the cylinder.

39. Find the moment of inertia of a right circular cone of base radius 1 and height 1 about an axis through the vertex parallel to the base.

40. Find the moment of inertia of a solid sphere of radius a about a diameter.

41. Find the centroid of the portion of the ball enclosed by the sphere $r^2 + z^2 = 1$ that lies between the half-planes $\theta = -\pi/3$, $r \geq 0$ and $\theta = \pi/3$, $r \geq 0$.

42. A solid is bounded on the top by the paraboloid $z = r^2$, on the bottom by the plane $z = 0$, and on the sides by the cylinder $r = 1$. Find the center of mass and the moment of inertia and radius of gyration about the z-axis if the density is (a) $\delta(r, \theta, z) = z$, (b) $\delta(r, \theta, z) = r$.

43. A solid is bounded below by the cone $z = \sqrt{x^2 + y^2}$ and above by the plane $z = 1$. Find the center of mass and the moment of inertia and radius of gyration about the z-axis if the density is (a) $\delta(r, \theta, z) = z$, (b) $\delta(r, \theta, z) = z^2$.

44. A solid right circular cylinder is bounded by the cylinder $r = a$ and the planes $z = 0$ and $z = h$, $h > 0$. Find the center of mass and the moment of inertia and radius of gyration about the z-axis if the density is $\delta(r, \theta, z) = z + 1$.

Spherical Coordinates

In problems that ask for masses and moments, assume that the density is $\delta = 1$ unless stated otherwise.

45. Find the volume inside the sphere $\rho = a$ that lies between the cones $\phi = \pi/3$ and $\phi = 2\pi/3$.

46. Find the volume cut from the sphere $\rho = a$ by the half-planes $\theta = 0$ and $\theta = \pi/6$ in the first octant.

47. Find the smaller volume cut from the sphere $\rho = 2$ by the plane $z = 1$.

48. Find the volume of the region bounded inside by the surface $\rho = 1 + \cos \phi$ and outside by the sphere $\rho = 2$.

49. Find the volume enclosed by the surface $\rho = 1 - \cos \phi$.

50. The circle $\rho = 2 \sin \phi$, $\theta = \pi/2$, is the circle of radius 1 in the yz-plane that is tangent to the z-axis and lies to the right of the z-axis. The region bounded by the circle in the yz-plane is revolved around the z-axis to generate a solid. Find the volume of the solid. (The first theorem of Pappus in the Miscellaneous Problems in Chapter 5 gives a quick way to check your result.)

51. Find the average value of the function $F(\rho, \phi, \theta) = \rho$ over the solid ball $\rho \leq 1$.

52. Figure 18.57 shows a solid that was made by drilling a conical hole in a solid hemisphere. The equation of the cone is $\phi = \pi/3$. The radius of the hemisphere is $\rho = 2$. Find the center of mass.

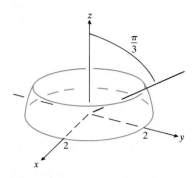

18.57 The solid in Problem 52.

53. Find the centroid of the solid bounded above by the sphere $\rho = a$ and below by the cone $\phi = \pi/3$. (The solid resembles a filled ice cream cone. See Fig. 18.52.)

54. Find the center of mass of a solid hemisphere of radius a and constant density δ.

55. Find the moment of inertia and radius of gyration of a solid sphere of radius a and constant density δ about a diameter of the sphere.

56. A solid ball is bounded by the sphere $\rho = a$. Find the moment of inertia and radius of gyration about the z-axis if the density is
a) $\delta(\rho, \phi, \theta) = \rho^2$
b) $\delta(\rho, \phi, \theta) = r = \rho \sin \phi$
c) $\delta(\rho, \phi, \theta) = r^2 = (\rho \sin \phi)^2$.

Substitutions

57. Evaluate the determinant in Eq. (7) to show that the Jacobian of the transformation from Cartesian $\rho\phi\theta$-space to Cartesian xyz-space is $\rho^2 \sin \phi$.

58. Evaluate the integral in Example 4 by integrating with respect to x, y, and z.

59. Find the volume of the ellipsoid
$$\frac{x^2}{a^2} + \frac{y^2}{b^2} + \frac{z^2}{c^2} = 1.$$

(*Hint:* Let $x = au$, $y = bv$, and $z = cw$. Then find the volume of an appropriate region in uvw-space.)

60. Evaluate
$$\iiint |xyz| \, dx \, dy \, dz$$

over the solid ellipsoid
$$\frac{x^2}{a^2} + \frac{y^2}{b^2} + \frac{z^2}{c^2} \leq 1.$$

(*Hint:* Let $x = au$, $y = bv$, and $z = cw$. Then integrate over an appropriate region in uvw-space.)

61. Let R be the region in xyz-space defined by the inequalities

$$1 \le x \le 2, \quad 0 \le xy \le 2, \quad \text{and} \quad 0 \le z \le 1.$$

Evaluate

$$\iiint\limits_{R} (x^2y + 3xyz) \, dx \, dy \, dz$$

by applying the transformation

$$u = x$$
$$v = xy$$

and

$$w = 3z$$

and integrating over an appropriate region G in uvw-space.

REVIEW QUESTIONS AND EXERCISES

1. Define the double integral of a function of two variables over a bounded region in the coordinate plane.

2. How are double integrals evaluated? How are the limits of integration determined? Give an example.

3. How are double integrals used to calculate areas, masses, moments, centers of mass, and radii of gyration? Give examples.

4. How do you change a double integral from rectangular to polar coordinates? Give an example.

5. Define the triple integral of a function $f(x, y, z)$ over a bounded region in space.

6. How are triple integrals in Cartesian coordinates evaluated? How are the limits of integration determined? Give an example.

7. How are triple integrals in Cartesian coordinates used to calculate volumes, average values, masses, moments, centers of mass, and radii of gyration? Give examples.

8. How are triple integrals defined in cylindrical and spherical coordinates?

9. How are triple integrals in cylindrical coordinates evaluated? How are the limits of integration found? Give examples.

MISCELLANEOUS PROBLEMS

In problems about masses and moments, assume $\delta = 1$ unless otherwise stated.

1. Reverse the order of integration and evaluate

$$\int_0^4 \int_{-\sqrt{4-y}}^{(y-4)/2} dx \, dy.$$

2. Sketch the region over which the integral

$$\int_0^1 \int_{\sqrt{y}}^{2-\sqrt{y}} xy \, dx \, dy$$

is to be evaluated and find its value.

3. Evaluate the integral

$$\int_0^\infty \frac{e^{-ax} - e^{-bx}}{x} \, dx.$$

(*Hint:* Use the relation

$$\frac{e^{-ax} - e^{-bx}}{x} = \int_a^b e^{-xy} \, dy$$

to form a double integral and evaluate the integral by changing the order of integration.)

4. Show that

$$\iint \frac{\partial^2 F(x, y)}{\partial x \, \partial y} \, dx \, dy$$

over the rectangle $x_0 \le x \le x_1, y_0 \le y \le y_1$, is

$$F(x_1, y_1) - F(x_0, y_1) - F(x_1, y_0) + F(x_0, y_0).$$

5. By changing the order of integration, show that the following double integral can be reduced to a single integral:

$$\int_0^x \int_0^u e^{m(x-t)}f(t) \, dt \, du = \int_0^x (x - t)e^{m(x-t)}f(t) \, dt.$$

Similarly, it can be shown that

$$\int_0^x \int_0^v \int_0^u e^{m(x-t)}f(t) \, dt \, du \, dv = \int_0^x \frac{(x - t)^2}{2} e^{m(x-t)}f(t) \, dt.$$

6. Sometimes a multiple integral with variable limits can be changed into one with constant limits. By changing the order of integration, show that

$$\int_0^1 f(x) \left(\int_0^x g(x - y)f(y) \, dy \right) dx$$

$$= \int_0^1 f(y) \left(\int_y^1 g(x - y)f(x) \, dx \right) dy$$

$$= \frac{1}{2} \int_0^1 \int_0^1 g(|x - y|) \, f(x)f(y) \, dx \, dy.$$

7. The integral

$$\int_{-1}^1 \int_{x^2}^1 dy \, dx$$

represents the area of a region of the xy-plane. Sketch the region and express the same area as a double integral with the order of integration reversed.

8. The base of a pile of sand covers the region in the xy-plane that is bounded by the parabola $x^2 + y = 6$ and the line $y = x$. The depth of the sand above the point (x, y) is x^2. Sketch the base of the sand pile and a representative element of volume dV, and find the volume of sand in the pile by double integration.

9. In setting up a double integral for the volume V under the paraboloid $z = x^2 + y^2$ and above a region R of the xy-plane, the following sum of iterated integrals was obtained:

$$V = \int_0^1 \left(\int_0^y (x^2 + y^2)\, dx \right) dy$$
$$+ \int_1^2 \left(\int_0^{2-y} (x^2 + y^2)\, dx \right) dy.$$

Sketch the region R in the xy-plane and express V as an iterated integral in which the order of integration is reversed.

10. Find the centroid of the region between the parabola $x + y^2 - 2y = 0$ and the line $x + 2y = 0$.

11. Find the polar moment of inertia about the origin of a thin triangular plate bounded by the y-axis and the lines $y = 2x$ and $y = 4$.

12. Find the centroid of the infinite region in the second quadrant bounded by the two axes and the curve $y = e^x$.

13. Find the moment about the y-axis of the infinite region under the curve $y = e^{-x^2/2}$ in the first quadrant.

14. Find the moment of inertia of the angle section shown in Fig. 18.58 (a) with respect to the horizontal base, (b) with respect to a horizontal line through the centroid.

18.58   The angle section in Problem 14.

15. *Center of pressure.* The center of pressure on a submerged plane region is the point at which the total hydrostatic force could be applied and still produce the same total moment about any axis in the plane. The hydrostatic pressure at a depth y in a fluid is wy, where w is the weight-density of the fluid. (See the introduction to Article 5.10.) You have immersed a vertical semicircular plate of radius a in a fluid of weight density w as shown in Fig. 18.59. Show that the depth of the center of pressure is

$$\bar{y} = \frac{3\pi a^2 + 32ac + 12\pi c^2}{4(4a + 3\pi c)}.$$

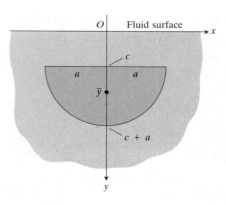

18.59   The plate in Problem 15.

16. You have a thin, flat plate of constant density covering the region in the xy-plane bounded by the ellipse $x^2/a^2 + y^2/b^2 = 1$. Show that the moment of inertia of the plate about a line through the center of the ellipse and making an angle α with the positive x-axis is $M(a^2\sin^2\alpha + b^2\cos^2\alpha)/4$, where M is the plate's mass.

17. Suppose that $f(x, y)$ can be written as a product $f(x, y) = F(x)G(y)$ of a function of x and a function of y. Then the integral of f over the rectangle $R: a \leq x \leq b, c \leq y \leq d$ can be evaluated as a product as well, by the formula

$$\iint_R f(x, y)\, dA = \left(\int_a^b F(x)\, dx \right)\left(\int_c^d G(y)\, dy \right). \quad (1)$$

The argument is that

$$\iint f(x, y)\, dA = \int_c^d \left(\int_a^b F(x)G(y)\, dx \right) dy$$

$$\left(\text{Fubini's theorem}\right)$$

$$= \int_c^d \left(G(y) \int_a^b F(x)\, dx \right) dy$$

$$\left(\begin{array}{c}\text{treating } G(y)\\ \text{as a constant}\end{array}\right)$$

$$= \int_c^d \left(\int_a^b F(x)\, dx \right) G(y)\, dy$$

$$\left(\begin{array}{c}\text{algebraic}\\ \text{rearrangement}\end{array}\right)$$

$$= \left(\int_a^b F(x)\, dx \right) \int_c^d G(y)\, dy.$$

$$\left(\begin{array}{c}\text{treating the integral}\\ \text{of } F(x) \text{ as a constant}\end{array}\right)$$

When it applies, Eq. (1) can be a time-saver. Use it to evaluate the following integrals.

a) $\displaystyle\int_0^{\ln 2} \int_0^{\pi/2} e^x \cos y\, dx\, dy$ b) $\displaystyle\int_1^2 \int_{-1}^1 \frac{x}{y^2}\, dx\, dy$

18. Find the centroid of the region defined by the polar-coordinate inequalities $0 \leq r \leq a$ and $-\alpha \leq \theta \leq \alpha$.

19. Find the centroid of the region in the first quadrant bounded by the rays $\theta = 0$ and $\theta = \pi/4$ and the circles $r = 1$ and $r = 2$.

20. In Article 10.4 we derived the formula

$$A = \int_{\alpha}^{\beta} \frac{1}{2} r^2 \, d\theta$$

for the area of the region R swept out by the radius vector $\overrightarrow{OP}$ as P moves along a curve $r = f(\theta)$ from $\theta = \alpha$ to $\theta = \beta$. Show that this formula is also a consequence of the equation

$$\text{Area of } R = \iint_{R} r \, dr \, d\theta$$

from Article 18.3.

21. Change the following double integral to an equivalent double integral in polar coordinates and sketch the region of integration.

$$\int_{-a}^{a} \int_{0}^{\sqrt{a^2 - y^2}} x \, dx \, dy$$

22. The usual way to evaluate the improper integral $I = \int_{0}^{\infty} e^{-x^2} \, dx$ is to calculate its square:

$$I^2 = \left(\int_{0}^{\infty} e^{-x^2} \, dx \right)\left(\int_{0}^{\infty} e^{-y^2} \, dy \right)$$

$$= \int_{0}^{\infty} \int_{0}^{\infty} e^{-(x^2 + y^2)} \, dx \, dy.$$

Introduce polar coordinates in the last expression and show that

$$I = \int_{0}^{\infty} e^{-x^2} \, dx = \frac{\sqrt{\pi}}{2}.$$

23. Show that if $u = x - y$ and $v = y$, then

$$\int_{0}^{\infty} e^{-sx} \, dx \int_{0}^{x} f(x - y, y) \, dy = \int_{0}^{\infty} \int_{0}^{\infty} e^{-s(u+v)} f(u, v) \, du \, dv.$$

24. A counterweight of a flywheel has the form of the smaller segment cut from a circle of radius a by a chord at a distance b from the center ($b < a$). Find the area of this counterweight and its polar moment of inertia about the center of the circle.

25. Find the radii of gyration about $\theta = 0$ and $\theta = \pi/2$ of a thin plate of density $\delta = 1$ enclosed by one loop of the curve $r^2 = a^2 \cos 2\theta$ ($a > 0$).

26. Find the area of the region enclosed by the lemniscate $r^2 = 2a^2 \cos 2\theta$. Also find the moment of inertia of this region about the y-axis.

27. Show, by transforming to polar coordinates, that

$$K(a) = \int_{0}^{a \sin \beta} \int_{y \cot \beta}^{\sqrt{a^2 - y^2}} \ln(x^2 + y^2) \, dx \, dy$$

$$= a^2 \beta \left(\ln a - \frac{1}{2} \right),$$

where $a > 0$ and $0 < \beta < \pi/2$. What expression do you obtain by changing the order of integration?

28. Find the values of a, b, c for which

$$\int_{-\infty}^{\infty} \int_{-\infty}^{\infty} e^{-(ax^2 + 2bxy + cy^2)} \, dx \, dy = 1$$

(*Hint:* Let $s = \alpha x + \beta y$ and $t = \gamma x + \delta y$, where $(\alpha \delta - \beta \gamma)^2 = ac - b^2$; then $ax^2 + 2bxy + cy^2 = s^2 + t^2$.)

29. Evaluate the integral

$$\iint \frac{dx \, dy}{(1 + x^2 + y^2)^2}$$

over the region enclosed by one loop of the lemniscate $(x^2 + y^2)^2 - (x^2 - y^2) = 0$. (*Hint:* Change to polar coordinates.)

30. Find the volume of the region enclosed by the cylinder $y = \cos x$ and the planes

$$z = y, \quad x = 0, \quad x = \pi/2, \quad \text{and} \quad z = 0.$$

31. Find the center of mass of the homogeneous pyramid whose base is the square region enclosed by the lines $x = 1$, $x = -1$, $y = 1$, $y = -1$, in the plane $z = 0$ and whose vertex is at the point $(0, 0, 1)$.

32. Find the volume of the region bounded above by the sphere $x^2 + y^2 + z^2 = 2a^2$ and below by the paraboloid $az = x^2 + y^2$.

33. Find the volume of the region enclosed by the surfaces

$$z = x^2 + y^2 \quad \text{and} \quad z = \frac{1}{2}(x^2 + y^2 + 1).$$

34. Use a triple integral to find the volume of the region enclosed by the surfaces $x = y^2 + z^2$ and $x = 1 - y^2$.

35. Find the average value of $F(x, y, z) = 15xy\sqrt{x^2 + y^2 + 1}$ over the cube in the first octant bounded by the coordinate planes and the planes $x = 2$, $y = 2$, and $z = 2$.

36. Find the moment of inertia with respect to the z-axis of a solid of constant density δ bounded below by the paraboloid $3az = x^2 + y^2$ and above by the sphere $x^2 + y^2 + z^2 = 4a^2$.

37. The region bounded by the circle $(y - b)^2 + z^2 = a^2$ ($b > a$) in the yz-plane is revolved about the z-axis to generate a solid torus (Fig. 18.60). Find the moment of inertia of the torus about the z-axis.

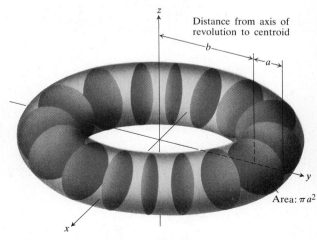

18.60 The torus in Problem 37.

38. A hole is bored through a solid sphere, the axis of the hole being a diameter of the sphere. The volume of the solid remaining is given by the integral

$$V = 2 \int_0^{2\pi} \int_0^{\sqrt{3}} \int_1^{\sqrt{4-z^2}} r \, dr \, dz \, d\theta.$$

a) By inspecting the given integral, find the radius of the hole and the radius of the sphere.

b) Evaluate the integral. (See Example 6, Article 5.4.)

39. Set up an equivalent triple integral in rectangular coordinates. (Arrange the order so that the first integration is with respect to z, the second with respect to y, and the last with respect to x.)

$$\int_0^{\pi/2} \int_1^{\sqrt{3}} \int_1^{\sqrt{4-r^2}} r^3 \sin\theta \cos\theta \, z^2 \, dz \, dr \, d\theta.$$

40. Find the volume of the region bounded by the plane $z = 0$, the cylinder $x^2 + y^2 = a^2$, and the cylinder $az = a^2 - x^2$.

41. Find the volume cut from the solid sphere $r^2 + z^2 \le a^2$ by the cylinder $r = a \sin\theta$.

42. *Spherical vs. cylindrical coordinates.* Triple integrals involving spherical shapes do not always require spherical coordinates for easy evaluation. Some problems may be more easily done with cylindrical coordinates. As a case in point, find the volume of the region bounded above by the sphere $x^2 + y^2 + z^2 = 8$ and below by the plane $z = 2$ by using (a) cylindrical coordinates, (b) spherical coordinates.

43. Find the moment of inertia, about the z-axis, of the region bounded above by the sphere $\rho = a$ and below by the cone $\phi = \pi/3$ (ρ, ϕ, θ are spherical coordinates).

44. Find the volume of the region enclosed by the surface $\rho = a \sin\phi$ in spherical coordinates.

45. Find the moment of inertia of the solid of constant density δ bounded by two concentric spheres of radii a and b ($a < b$) about a diameter.

46. The volume of a solid is given by the triple integral

$$\int_0^2 \left(\int_0^{\sqrt{2x-x^2}} \left(\int_{-\sqrt{4-x^2-y^2}}^{\sqrt{4-x^2-y^2}} dz \right) dy \right) dx.$$

a) Describe the solid by giving equations for the surfaces that form its boundary.

b) Express the volume as a triple integral in cylindrical coordinates. Give the limits of integration but do not evaluate the integral.

47. Find I_x for the region cut from the sphere $x^2 + y^2 + z^2 = 4$ by the cylinder $x^2 + y^2 = 1$.

48. a) The region enclosed by the cardioid $r = 1 - \cos\theta$ in the xy-plane is revolved about the x-axis to generate a solid. Find the volume of the solid. (*Hint:* Use *double* integration. Revolve an area dA about the x-axis to generate a volume dV.)

b) Find the moment of inertia about the x-axis of the solid in part (a).

49. Find the moment of inertia of a solid ball of radius a about a line tangent to the ball.

50. WARNING: Hard problem. Setting up the integral is straightforward, but integrating the result takes hours. (It took MACSYMA 20 minutes.)

A square hole of side $2b$ is cut symmetrically through a sphere of radius a ($a > b\sqrt{2}$). Find the volume removed.

51. Let S be a solid sphere of radius a, constant density δ, and mass $M = (4/3)\pi a^3 \delta$. Let P be a particle of mass m situated at distance b ($b > a$) from the center of S. According to Newton, the force of gravitational attraction of the sphere on P is given by the equation

$$\mathbf{F} = Gm \iiint_S \frac{\mathbf{u} \, \delta \, dV}{r^2},$$

where G is the gravitational constant, $\mathbf{u}$ is a unit vector in the direction from P toward the volume element dV in S, and r^2 is the square of the distance from P to dV. Take the origin at the center of the sphere and P at $(0, 0, b)$ on the z-axis, and show that $\mathbf{F} = -(GMm/b^2)\mathbf{k}$. This equation shows that the force is the same as it would be if all the mass of the sphere were concentrated at its center.

CHAPTER 19

Vector Fields and Integration

OVERVIEW

In this chapter we see the powerful results of combining integral calculus with vectors. We begin with line integrals, which are integrals of functions defined along curves. With these we can calculate the mass of a spring, the amount of work done by a force in moving an object along a path in three dimensions, and the rate at which a fluid circulates around a closed loop. We then come to Green's theorem, one of the finest theorems in calculus, as powerful and as useful as the Fundamental Theorems we studied in Chapter 4. It is named after its discoverer, George Green (1793–1841), a self-taught English mathematician who pioneered the development of the mathematical foundations of electromagnetic theory. Along with Green's theorem, we introduce the notions of the flux density and circulation density of a vector field, known in mathematics as the field's divergence and curl. We then study surface integrals and show how to calculate surface area and how to find the flux of a vector field through a surface in space. After this, we explore the extensions of Green's theorem to three dimensions, in the forms of the Divergence Theorem and Stokes's theorem. These theorems provide the mathematical foundation for much of the theoretical work in electricity, magnetism, fluid mechanics, and orbital mechanics. The applications we discuss will come mostly from the study of fluid flow because fluid flows are relatively easy to visualize. Finally, we explore the remarkable properties of conservative vector fields and show how to find potential functions for these fields.

19.1
Line Integrals

When a curve $\mathbf{R}(t) = x(t)\mathbf{i} + y(t)\mathbf{j} + z(t)\mathbf{k}$, $a \leq t \leq b$, threads its way through the domain of a function $f(x, y, z)$ in space, the values of the function on the curve are given by the composite $f(x(t), y(t), z(t))$. If we integrate this function with respect to arc length from $t = a$ to $t = b$, we calculate what is called the "line integral" of f along the curve. Despite the three-dimensional geometry, the line integral is just an ordinary integral of a real-valued function over an interval of real numbers. We have evaluated integrals like this ever since Chapter 4. The usual techniques apply and nothing new is needed.

The importance of line integrals lies in their application. These are the integrals with which we can calculate the masses and moments of springs and wires that lie along curves in space. We can also combine line integrals with vector products to calculate the work done by a variable force along a curve in space or the rate at which a fluid flows along a curve in space or across a curve in the plane. In the present article we show in detail how line integrals are defined and evaluated and we use line integrals to find the masses and moments of springs, wires, and thin rods. In the next article, we bring line integrals together with vectors to calculate work and measure fluid flow.

Line Integrals

Line integrals are integrals of functions along curves. We define them in the following way. Suppose that $f(x, y, z)$ is a function whose domain contains the curve

$$\mathbf{R}(t) = x(t)\mathbf{i} + y(t)\mathbf{j} + z(t)\mathbf{k}, \qquad a \leq t \leq b.$$

We subdivide the curve from $t = a$ to $t = b$ into a finite number of consecutive subarcs (Fig. 19.1). The typical subarc has length Δs_k and in each subarc we choose a point (x_k, y_k, z_k). We form the sum

$$S_n = \sum_{k=1}^{n} f(x_k, y_k, z_k)\, \Delta s_k. \tag{1}$$

If f is continuous and the functions $x(t)$, $y(t)$, and $z(t)$ have continuous first derivatives, then the sums in (1) approach a limit as n increases and the lengths Δs_k approach zero. We call this limit the **(line) integral of f along the curve from a to b.** When the curve is denoted by a single letter, C for example, the notation for the integral is

$$\int_C f(x, y, z)\, ds. \qquad \text{("The integral of } f \text{ along } C\text{")} \tag{2}$$

To evaluate the integral, we appeal to a remarkable evaluation theorem from advanced calculus. The theorem says that the value of the integral may be calculated from any parameterization of the curve for which

$$\frac{ds}{dt} = \sqrt{\left(\frac{dx}{dt}\right)^2 + \left(\frac{dy}{dt}\right)^2 + \left(\frac{dz}{dt}\right)^2} \tag{3}$$

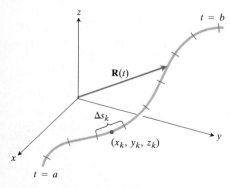

19.1 A curve $\mathbf{R}(t) = x(t)\mathbf{i} + y(t)\mathbf{j} + z(t)\mathbf{k}$ subdivided into small arcs. The length of the typical subarc is Δs_k.

by evaluating the integral

$$\int_{t=a}^{t=b} f(x(t), y(t), z(t)) \sqrt{\left(\frac{dx}{dt}\right)^2 + \left(\frac{dy}{dt}\right)^2 + \left(\frac{dz}{dt}\right)^2}\, dt. \tag{4}$$

Since the square root in Eq. (4) is the length $|\mathbf{v}(t)|$ of the curve's velocity vector,

THEOREM

> **The Evaluation Theorem for Line Integrals**
>
> $$\int_C f(x, y, z)\, ds = \int_{t=a}^{t=b} f(x(t), y(t), z(t))|\mathbf{v}(t)|\, dt. \tag{5}$$

The integral of f along C depends only on the nature of f and the geometry of the curve and may be calculated from almost any convenient parameterization.

Notice that if f is the constant function whose value is 1, the integral of f along C gives the length of C. We calculated the lengths of curves in Article 14.2 and will not repeat the calculations here.

In our first example, we integrate a function over two different paths from the origin to the point $(1, 1, 1)$.

EXAMPLE 1 Figure 19.2 shows two different paths from the origin to the point $(1, 1, 1)$. One path is the union of the line segments C_1 and C_2. The other path is the line segment C_3. Integrate the function

$$f(x, y, z) = x - 3y^2 + z$$

along each path.

Solution We first parameterize the segments that make up the paths. According to the Evaluation Theorem, we can use any parameterization we want as long as $|\mathbf{v}(t)|$ is never zero and we integrate in the direction of increasing t so that $|\mathbf{v}(t)|\, dt$ is positive. We choose the simplest parameterizations we can think of, checking the lengths of the velocity vectors as we go along:

$$\text{For } C_1: \qquad \mathbf{R}_1(t) = t\mathbf{i} + t\mathbf{j}, \qquad 0 \le t \le 1$$
$$|\mathbf{v}_1(t)| = \sqrt{1^2 + 1^2} = \sqrt{2}$$

$$\text{For } C_2: \qquad \mathbf{R}_2(t) = \mathbf{i} + \mathbf{j} + t\mathbf{k}, \qquad 0 \le t \le 1$$
$$|\mathbf{v}_2(t)| = \sqrt{0^2 + 0^2 + 1^2} = 1$$

$$\text{For } C_3: \qquad \mathbf{R}_3(t) = t\mathbf{i} + t\mathbf{j} + t\mathbf{k}, \qquad 0 \le t \le 1$$
$$|\mathbf{v}_3(t)| = |\mathbf{i} + \mathbf{j} + \mathbf{k}| = \sqrt{1^2 + 1^2 + 1^2} = \sqrt{3}.$$

Having decided on the parameterizations, we use Eq. (5) to integrate $f(x, y, z) = x - 3y^2 + z$ along each path in the direction of increasing t:

Path 1:
$$\int_{C_1 \cup C_2} f(x, y, z)\, ds = \int_{C_1} f(x, y, z)\, ds + \int_{C_2} f(x, y, z)\, ds$$

$$= \int_0^1 (t - 3t^2)\sqrt{2}\, dt + \int_0^1 (1 - 3 + t) \cdot 1\, dt$$

$$= \sqrt{2} \int_0^1 (t - 3t^2)\, dt + \int_0^1 (t - 2)\, dt$$

$$= \sqrt{2}\left[\frac{t^2}{2} - t^3\right]_0^1 + \left[\frac{t^2}{2} - 2t\right]_0^1$$

$$= -\frac{\sqrt{2}}{2} - \frac{3}{2} = -\frac{1}{2}(\sqrt{2} + 3),$$

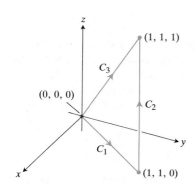

19.2 The paths of integration in Example 1. The arrows show the direction of increasing t in the parameterizations.

Path 2: $\displaystyle\int_{C_3} f(x,\, y,\, z)\, ds = \int_{t=0}^{t=1} f(x(t),\, y(t),\, z(t))|\mathbf{v}(t)|\, dt$

$\displaystyle = \int_0^1 (t - 3t^2 + t)\sqrt{3}\, dt$

$\displaystyle = \sqrt{3}\int_0^1 (2t - 3t^2)\, dt$

$\displaystyle = \sqrt{3}\Big[t^2 - t^3\Big]_0^1 = 0.$ ∎

There are three things to notice about the integrations in Example 1. First, as soon as the components of the appropriate $\mathbf{R}(t)$ are substituted in the formula for f and ds is replaced by the appropriate $|\mathbf{v}(t)|\, dt$ for each segment, the integration becomes a straightforward integration with respect to t. Second, the integral of f over Path 1 is obtained by integrating f over each section of the path and adding the results. Third, the integrals of f over Path 1 and Path 2 have different values. For most functions, the value of the line integral along a path joining two points changes when you change the path. For some functions, however, the value of the integral is the same for all paths joining the two points. We shall explore this amazing phenomenon in Article 19.7.

Additivity

Line integrals have all the algebraic properties we have learned to associate with integrals. The integral of a sum of two functions is the sum of their integrals, the integral of a constant times a function is the constant times the integral of the function, the integral of a function from b to a is the negative of the integral from a to b, and so on. Included in this list is the property that if a curve C is made by joining together a finite number of curves $C_1, C_2, \ldots,$ C_n end to end, then the integral of a function over C is the sum of the integrals over the curves that make it up:

$$\int_C f\, ds = \int_{C_1} f\, ds + \int_{C_2} f\, ds + \cdots + \int_{C_n} f\, ds. \tag{6}$$

This was illustrated in Example 1 when we evaluated the integral of f over the first path by adding the integrals over C_1 and C_2.

Mass and Moment Calculations

Suppose that a wire lies along a curve C in space and that the wire's density at the point (x, y, z) on the curve is $\delta(x, y, z)$. The wire's mass, center of mass, and moments can then be calculated with the formulas given in Table 19.1. The formulas apply to springs and thin rods as well as wires.

The following examples show how the formulas in Table 19.1 are used.

EXAMPLE 2   A coil spring lies along the helix

$$\mathbf{R}(t) = (\cos 4t)\mathbf{i} + (\sin 4t)\mathbf{j} + t\mathbf{k}, \qquad 0 \le t \le 2\pi.$$

The spring's density is a constant, $\delta = 1$. Find the spring's mass, the coordinates of the spring's center of mass, and the spring's moment of inertia and radius of gyration about the z-axis.

TABLE 19.1
Mass and moment formulas for wires, springs, and thin rods

Mass: $M = \int_C \delta(x, y, z)\, ds$

First moments about the coordinate planes:

$$M_{yz} = \int_C x\, \delta\, ds, \qquad M_{xz} = \int_C y\, \delta\, ds, \qquad M_{xy} = \int_C z\, \delta\, ds$$

Coordinates of the center of mass:

$$\bar{x} = M_{yz}/M, \qquad \bar{y} = M_{xz}/M, \qquad \bar{z} = M_{xy}/M$$

Moments of inertia:

$$I_x = \int_C (y^2 + z^2)\, \delta\, ds, \qquad I_y = \int_C (x^2 + z^2)\, \delta\, ds,$$

$$I_z = \int_C (x^2 + y^2)\, \delta\, ds, \qquad I_L = \int_C r^2\, \delta\, ds,$$

$r(x, y, z)$ = distance from point (x, y, z) to line L

Radius of gyration about a line L: $R_L = \sqrt{I_L/M}$.

Solution   We sketch the spring (Fig. 19.3). Because of the symmetries involved, the center of mass lies at the point $(0, 0, \pi)$ on the z-axis.

For the remaining calculations, we first express ds in terms of t:

$$ds = \sqrt{\left(\frac{dx}{dt}\right)^2 + \left(\frac{dy}{dt}\right)^2 + \left(\frac{dz}{dt}\right)^2}\, dt$$

$$= \sqrt{(-4 \sin 4t)^2 + (4 \cos 4t)^2 + 1}\, dt$$

$$= \sqrt{17}\, dt.$$

The formulas in Table 19.1 then give

$$M = \int_{\text{helix}} \delta\, ds = \int_0^{2\pi} (1)\sqrt{17}\, dt = 2\pi\sqrt{17},$$

$$I_z = \int_{\text{helix}} (x^2 + y^2)\, \delta\, ds$$

$$= \int_0^{2\pi} (\cos^2 4t + \sin^2 4t)(1)\sqrt{17}\, dt$$

$$= \int_0^{2\pi} \sqrt{17}\, dt = 2\pi\sqrt{17},$$

and

$$R_z = \sqrt{I_z/M} = \sqrt{2\pi\sqrt{17}/2\pi\sqrt{17}} = 1.$$

Notice that the radius of gyration about the z-axis is the radius of the cylinder around which the helix winds. ■

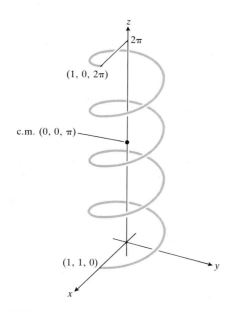

19.3 The helical spring in Example 2.

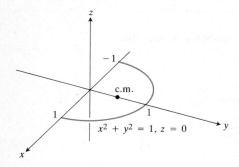

19.4   Example 3 calculates the coordinates of the center of mass of a slender arch lying along this semicircle in the xy-plane.

EXAMPLE 3 A slender metal arch lies along the semicircle $x^2 + y^2 = 1$, $y \geq 0$, in the xy-plane (Fig. 19.4). Find the center of mass if the density at the point (x, y, z) on the arch is $\delta(x, y, z) = 2 - y$.

Solution We know that $\bar{z} = 0$ because the arch lies in the xy-plane and that $\bar{x} = 0$ because the mass in the arch is distributed symmetrically with respect to the y-axis. It remains only to find $\bar{y}$.

To do so, we parameterize the semicircle as

$$C: \quad \mathbf{R}(t) = (\cos t)\mathbf{i} + (\sin t)\mathbf{j}, \quad 0 \leq t \leq \pi,$$

and express ds in terms of dt:

$$ds = \sqrt{\left(\frac{dx}{dt}\right)^2 + \left(\frac{dy}{dt}\right)^2 + \left(\frac{dz}{dt}\right)^2} \, dt$$

$$= \sqrt{(-\sin t)^2 + (\cos t)^2 + (0)^2} \, dt = dt.$$

The formulas in Table 19.1 then give

$$M = \int_C \delta \, ds = \int_C (2 - y) \, ds = \int_0^\pi (2 - \sin t) \, dt$$

$$= \left[2t + \cos t\right]_0^\pi = (2\pi - 1) - (0 + 1)$$

$$= 2\pi - 2,$$

$$M_{xz} = \int_C y \, \delta \, ds = \int_C y(2 - y) \, ds = \int_0^\pi \sin t(2 - \sin t) \, dt$$

$$= \int_0^\pi (2 \sin t - \sin^2 t) \, dt = \int_0^\pi \left(2 \sin t - \frac{1}{2} + \frac{\cos 2t}{2}\right) dt$$

$$= \left[-2 \cos t - \frac{t}{2} + \frac{\sin 2t}{4}\right]_0^\pi$$

$$= \left(-2(-1) - \frac{\pi}{2}\right) - (-2)$$

$$= 4 - \frac{\pi}{2} = \frac{8 - \pi}{2},$$

and

$$\bar{y} = \frac{M_{xz}}{M} = \frac{8 - \pi}{4\pi - 4} \approx 0.57.$$

To the nearest hundredth, the coordinates of the center of mass are $(\bar{x}, \bar{y}, \bar{z}) = (0, 0.57, 0)$. ∎

PROBLEMS

In Problems 1–8, match the vector equation with its graph from the list (a)–(h).

1. $\mathbf{R}(t) = t\mathbf{i} + (1 - t)\mathbf{j}, \quad 0 \leq t \leq 1$

2. $\mathbf{R}(t) = \mathbf{i} + t\mathbf{k}, \quad -1 \leq t \leq 1$

3. $\mathbf{R}(t) = (a \cos t)\mathbf{i} + (a \sin t)\mathbf{j}, \quad 0 \leq t \leq 2\pi$

4. $\mathbf{R}(t) = t\mathbf{i}, \quad 0 \leq t \leq 1$

5. $\mathbf{R}(t) = t\mathbf{i} + t\mathbf{j} + t\mathbf{k}, \quad 0 \leq t \leq 2$

6. $\mathbf{R}(t) = t\mathbf{j} + (2 - 2t)\mathbf{k}, \quad 0 \leq t \leq 1$

7. $\mathbf{R}(t) = (t^2 - 1)\mathbf{j} + 2t\mathbf{k}, \quad 0 \le t \le 1$

8. $\mathbf{R}(t) = (a \cos t)\mathbf{i} + (a \sin t)\mathbf{k}, \quad 0 \le t \le \pi$

a)

b)

c)

d)

e)

f)

g)

h)

In Problems 9–20, evaluate the integral of $f(x, y, z)$ along the given curve.

9. $f(x, y, z) = x + y$
$\mathbf{R}(t) = t\mathbf{i} + (1 - t)\mathbf{j}, \quad 0 \le t \le 1$

10. $f(x, y, z) = x - y$
$\mathbf{R}(t) = t\mathbf{i} + (1 - t)\mathbf{j}, \quad 0 \le t \le 1$

11. $f(x, y, z) = x - y + z - 2$
$\mathbf{R}(t) = t\mathbf{i} + (1 - t)\mathbf{j} + \mathbf{k}, \quad 0 \le t \le 1$

12. $f(x, y, z) = x^2 + y^2$
$\mathbf{R}(t) = t\mathbf{i} + (1 - t)\mathbf{j} + t\mathbf{k}, \quad 0 \le t \le 1$

13. $f(x, y, z) = xy + y + z$
$\mathbf{R}(t) = 2t\mathbf{i} + t\mathbf{j} + (2 - 2t)\mathbf{k}, \quad 0 \le t \le 1$

14. $f(x, y, z) = \sqrt{x^2 + z^2}$
$\mathbf{R}(t) = (a \cos t)\mathbf{j} + (a \sin t)\mathbf{k}, \quad 0 \le t \le 2\pi$

15. $f(x, y, z) = x + y$
$\mathbf{R}(t) = (\cos t)\mathbf{i} + (\sin t)\mathbf{j} + t\mathbf{k}, \quad 0 \le t \le \pi/2$

16. $f(x, y, z) = \sqrt{x^2 + y^2}$
$\mathbf{R}(t) = (a \cos t)\mathbf{i} + (a \sin t)\mathbf{j} + bt\mathbf{k}, \quad -2\pi \le t \le 2\pi$

17. $f(x, y, z) = \sqrt{x^2 + y^2}$
$\mathbf{R}(t) = (t \sin t + \cos t)\mathbf{i} + (t \cos t - \sin t)\mathbf{j}, \quad 0 \le t \le \sqrt{3}$

18. $f(x, y, z) = 3z\sqrt{3x^2 + y^2 + z^2}$
$\mathbf{R}(t) = \mathbf{i} + t\mathbf{k}, \quad -1 \le t \le 1$

19. $f(x, y, z) = \sqrt{3}/(x^2 + y^2 + z^2)$
$\mathbf{R}(t) = t\mathbf{i} + t\mathbf{j} + t\mathbf{k}, \quad 1 \le t \le 2$

20. $f(x, y, z) = \sqrt{3}/(x^2 + y^2 + z^2)$
$\mathbf{R}(t) = t\mathbf{i} + t\mathbf{j} + t\mathbf{k}, \quad 1 \le t \le \infty$

21. Integrate $f(x, y, z) = x + \sqrt{y} - z^2$ over the path from $(0, 0, 0)$ to $(1, 1, 1)$ shown in Fig. 19.5(a). This path is the union of the curves
$$\mathbf{R}_1(t) = t\mathbf{i} + t^2\mathbf{j}, \quad 0 \le t \le 1 \quad \text{and}$$
$$\mathbf{R}_2(t) = \mathbf{i} + \mathbf{j} + t\mathbf{k}, \quad 0 \le t \le 1.$$

22. Integrate $f(x, y, z) = x + \sqrt{y} - z^2$ over the path from $(0, 0, 0)$ to $(1, 1, 1)$ shown in Fig. 19.5(b). This path is the union of the curves
$$\mathbf{R}_1(t) = t\mathbf{k}, \quad 0 \le t \le 1;$$
$$\mathbf{R}_2(t) = t\mathbf{j} + \mathbf{k}, \quad 0 \le t \le 1; \quad \text{and}$$
$$\mathbf{R}_3(t) = t\mathbf{i} + \mathbf{j} + \mathbf{k}, \quad 0 \le t \le 1.$$

(a)

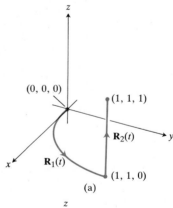

(b)

19.5 The paths of integration in Problems 21 and 22.

23. Find I_z and R_z for the arch in Example 3.

24. Find the mass of a wire that lies along the curve $\mathbf{R}(t) =$ $(t^2 - 1)\mathbf{j} + 2t\mathbf{k}$, $0 \le t \le 1$ if the density is $\delta = \dfrac{3}{2}t$.

25. A circular wire hoop of constant density δ lies along the circle $x^2 + y^2 = a^2$ in the xy-plane. Find the hoop's moment of inertia about the z-axis. Express the moment as a constant times the mass of the hoop. Find the radius of gyration of the hoop about the z-axis.

26. A slim rod of constant density lies along the line segment $\mathbf{R}(t) = t\mathbf{j} + (2 - 2t)\mathbf{k}$, $0 \le t \le 1$ in the yz-plane in space. Find the moments of inertia and radii of gyration of the rod about the three coordinate axes.

27. *The moment of inertia of a rod about an axis normal to one end of the rod.* A slender rod of length L and of constant density δ lies along the interval $0 \le x \le L$ of the x-axis. Find the moment of inertia of the rod about the z-axis. Express the moment as a constant times the mass of the rod.

28. *The moment of inertia of a rod about an axis normal to the rod at its midpoint.* A slender rod of length L and of con-

stant density δ lies along the interval $-L/2 \le x \le L/2$ of the x-axis. Find the moment of inertia of the rod about the z-axis. Express the moment as a constant times the mass of the rod.

29. A wire of constant density δ lies along the semicircle $\mathbf{R}(t) = (a \cos t)\mathbf{i} + (a \sin t)\mathbf{j}$, $0 \le t \le \pi$. Find the center of mass and the moment of inertia and radius of gyration about the x-axis.

30. A spring of constant density δ lies along the helix
$$\mathbf{R}(t) = (a \cos t)\mathbf{i} + (a \sin t)\mathbf{j} + (bt)\mathbf{k}, \quad 0 \le t \le 2\pi.$$
a) Find I_z and R_z.
b) Suppose the length of the spring is doubled to lie along the helix for $0 \le t \le 4\pi$. How do you expect this to affect the values of I_z and R_z? Check your predictions by calculating I_z and R_z for the longer spring.

31. A wire lies along the curve
$$\mathbf{R}(t) = (t^2 - 1)\mathbf{j} + (2t)\mathbf{k}, \quad 0 \le t \le 1.$$
Find the center of mass and the moments of inertia and radii of gyration about the coordinate axes if the wire's density at the point (x, y, z) on the curve is $\delta(x, y, z) = \sqrt{y + 2}$.

19.2
Line Integrals in Vector Fields. Work, Circulation, and Flux

In studies of physical phenomena represented by vectors, we introduce the idea of a vector field on a domain in space, which assigns a vector quantity to each point in the domain. For example, we might be studying the velocity field of a moving fluid or the electric field created by a distribution of electric charges. In situations like these, integrals of real valued functions over closed intervals are replaced by line integrals of functions over paths through the vector field, and the function being integrated is a scalar product of vectors. We can use these line integrals in vector fields to calculate the work done in moving an object along a path against a variable force. (The object might be a vehicle sent into space against Earth's gravitational field.) Or we might want to calculate the work done by a vector field in moving an object along a path through the field (for example, the work done by a particle accelerator in raising the energy of a particle). We can also use line integrals to calculate the rates at which fluids flow along and across curves.

This article shows how the calculations go.

Vector Fields

A **vector field** on a domain in space or in the plane is a function that assigns a vector to each point of the domain. If we attach a projectile's velocity vector to each point of the projectile's trajectory, we have a vector field along the trajectory. If we attach the gradient vector of a function to each point of a level surface of the function, we have a vector field on the surface. If we attach the velocity vector to each point of a three-dimensional fluid flow, we have a vector field on a region in space. These and other vector fields are illustrated in Figures 19.6–19.13. Some illustrations give formulas for the fields as well.

19.6 The field of gradient vectors ∇f on a surface $f(x, y, z) = c$.

19.7 Velocity vectors of a flow around an airfoil in a wind tunnel. The streamlines were made visible by kerosene smoke. (Adapted from *NCFMF Book of Film Notes,* 1974; The MIT Press with Education Development Center, Inc., Newton, Massachusetts.)

19.8 Vectors in the gravitational field
$$\mathbf{F} = -GM(x\,\mathbf{i} + y\,\mathbf{j} + z\,\mathbf{k})/(x^2 + y^2 + z^2)^{3/2}.$$

19.9 Streamlines in a contracting channel. The water speeds up as the channel narrows and the velocity vectors increase in length. (Adapted from *NCFMF Book of Film Notes,* 1974; The MIT Press with Education Development Center, Inc., Newton, Massachusetts.)

19.10 The flow of fluid in a long cylindrical pipe. The vectors $\mathbf{v} = (a^2 - r^2)\mathbf{k}$ inside the cylinder that have their bases in the xy-plane have their tips on the paraboloid $z = a^2 - r^2$.

19.11 The field $\mathbf{F}(x, y) = x\,\mathbf{i} + y\,\mathbf{j}$ of position vectors of points in the plane. Notice the convention that an arrow is drawn with its tail, not its head, at the point where $\mathbf{F}$ is evaluated.

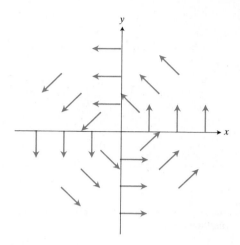

19.12 The two-dimensional field of unit vectors

$$\mathbf{F}(x, y) = (-y\,\mathbf{i} + x\,\mathbf{j})/(x^2 + y^2)^{1/2}.$$

The field is not defined at the origin.

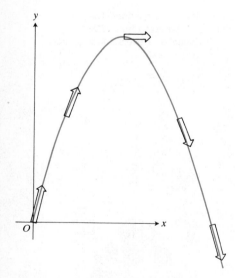

19.13 The velocity vectors $\mathbf{v}(t)$ of a projectile's motion make a vector field along the trajectory.

19.14 The work done by the field $\mathbf{F}$ along the path from (x_k, y_k, z_k) to $(x_{k+1}, y_{k+1}, z_{k+1})$ is approximately $\mathbf{F}(x_k, y_k, z_k) \cdot \Delta\mathbf{R}_k$.

To sketch the fields that had formulas, we picked a representative assortment of domain points and sketched the vectors attached to them. Notice the convention that the arrows representing the vectors are drawn with their tails, not their heads, at the points where the vector function is evaluated. This is different from the way we draw the position vector of a projectile or planet we are studying, with its tail at the origin and its head at the projectile's or planet's location.

Work

To calculate the work done by a vector force field $\mathbf{F}(x, y, z)$ in moving an object from $t = a$ to $t = b$ along a curve

$$\mathbf{R}(t) = x(t)\mathbf{i} + y(t)\mathbf{j} + z(t)\mathbf{k}$$

in the domain of the field, we imagine the curve to be subdivided into a finite number of segments, as shown in Fig. 19.14. Each segment is spanned by a small increment $\Delta\mathbf{R}_k$ in the position vector. The increment runs from the point (x_k, y_k, z_k) to the point $(x_{k+1}, y_{k+1}, z_{k+1})$ in the direction of motion. The amount of work done by $\mathbf{F}$ along the curve from point k to point $k + 1$ is approximately $\mathbf{F}(x_k, y_k, z_k) \cdot \Delta\mathbf{R}_k$. The work done by $\mathbf{F}$ along the entire length of the curve from $t = a$ to $t = b$ is approximately

$$\sum_a^b \mathbf{F}(x_k, y_k, z_k) \cdot \Delta\mathbf{R}_k. \qquad \text{(work approximation)} \qquad (1)$$

If the components of $\mathbf{F}$ are continuous and the functions $x(t)$, $y(t)$, and $z(t)$ have continuous first derivatives, then the sums in (1) approach a limit as the subdivision becomes finer and the magnitudes $|\Delta\mathbf{R}_k|$ approach zero. We call this limit the **work done by F from $t = a$ to $t = b$ along the curve R(t).** The notation for the limit is

$$\text{Work} = \int_{t=a}^{t=b} \mathbf{F} \cdot d\mathbf{R}. \tag{2}$$

To evaluate the work integral, we evaluate $\mathbf{F}$ along the curve, dot with $d\mathbf{R}$, and integrate the resulting scalar function with respect to t:

Formula for Calculating Work

If

$$\mathbf{F} = M\mathbf{i} + N\mathbf{j} + P\mathbf{k}$$

and

$$d\mathbf{R} = dx\mathbf{i} + dy\mathbf{j} + dz\mathbf{k},$$

then

$$\text{Work} = \int_{t=a}^{t=b} \mathbf{F} \cdot d\mathbf{R} = \int_{t=a}^{t=b} M\,dx + N\,dy + P\,dz. \tag{3}$$

The next example shows how this formula is used.

EXAMPLE 1   Find the work done by the force

$$\mathbf{F}(x, y, z) = (x^2 - y)\mathbf{i} + (y^2 - z)\mathbf{j} + (z^2 - x)\mathbf{k}$$

along the following curves from $(0, 0, 0)$ to $(1, 1, 1)$:

$$\mathbf{R}_1(t) = t\mathbf{i} + t\mathbf{j} + t\mathbf{k}, \qquad 0 \le t \le 1,$$
$$\mathbf{R}_2(t) = t\mathbf{i} + t^2\mathbf{j} + t^3\mathbf{k}, \qquad 0 \le t \le 1.$$

Solution   The curves are sketched in Fig. 19.15. In each case, we use Eq. (3) to calculate the work done from $t = 0$ to $t = 1$.

Along the first curve,

$$\mathbf{F}(x, y, z) = (x^2 - y)\mathbf{i} + (y^2 - z)\mathbf{j} + (z^2 - x)\mathbf{k}$$
$$= (t^2 - t)\mathbf{i} + (t^2 - t)\mathbf{j} + (t^2 - t)\mathbf{k},$$
$$d\mathbf{R}_1 = dt\mathbf{i} + dt\mathbf{j} + dt\mathbf{k},$$

and

$$\text{Work} = \int_{t=0}^{t=1} \mathbf{F} \cdot d\mathbf{R}_1 = \int_0^1 (t^2 - t)\,dt + (t^2 - t)\,dt + (t^2 - t)\,dt$$
$$= 3 \int_0^1 (t^2 - t)\,dt = 3\left[\frac{t^3}{3} - \frac{t^2}{2}\right]_0^1 = 1 - \frac{3}{2} = -\frac{1}{2}.$$

The work done by $\mathbf{F}$ along the first curve is $-1/2$.

For the second curve,

$$\mathbf{F}(x, y, z) = (x^2 - y)\mathbf{i} + (y^2 - z)\mathbf{j} + (z^2 - x)\mathbf{k}$$
$$= (t^2 - t^2)\mathbf{i} + (t^4 - t^3)\mathbf{j} + (t^6 - t)\mathbf{k},$$
$$d\mathbf{R}_2 = dt\mathbf{i} + 2t\,dt\mathbf{j} + 3t^2\,dt\mathbf{k},$$

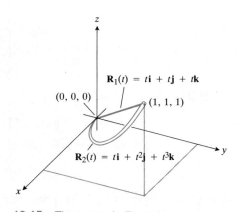

19.15   The curves in Example 1.

and

$$\text{Work} = \int_0^1 \mathbf{F} \cdot d\mathbf{R}_2$$

$$= \int_0^1 (0)\, dt + 2t(t^4 - t^3)\, dt + 3t^2(t^6 - t)\, dt$$

$$= \int_0^1 (2t^5 - 2t^4 + 3t^8 - 3t^3)\, dt$$

$$= \left[\frac{2}{6}t^6 - \frac{2}{5}t^5 + \frac{3}{9}t^9 - \frac{3}{4}t^4 \right]_0^1$$

$$= \frac{1}{3} - \frac{2}{5} + \frac{1}{3} - \frac{3}{4} = -\frac{29}{60}.$$

The work done by $\mathbf{F}$ along the second curve is $-29/60$.   ■

Flow Integrals and Circulation

Suppose that $\mathbf{F}(x, y, z)$ is the velocity field of a fluid flowing through a region in space and that

$$\mathbf{R}(t) = x(t)\mathbf{i} + y(t)\mathbf{j} + z(t)\mathbf{k}, \qquad a \le t \le b,$$

is a curve lying in the region. Then the integral with respect to arc length of $\mathbf{F} \cdot \mathbf{T}$, the scalar component of $\mathbf{F}$ in the direction of the curve's unit tangent vector, is called the **flow integral** of $\mathbf{F}$ along the curve.

Flow Integral and Circulation

Suppose a curve C through the vector field

$$\mathbf{F}(x, y, z) = M(x, y, z)\mathbf{i} + N(x, y, z)\mathbf{j} + P(x, y, z)\mathbf{k}$$

is represented by the vector equation

$$\mathbf{R}(t) = x(t)\mathbf{i} + y(t)\mathbf{j} + z(t)\mathbf{k}, \qquad a \le t \le b.$$

Then the flow integral of $\mathbf{F}$ along C in the direction of increasing t can be calculated from any one of the following three formulas.

$$\text{Flow integral of } \mathbf{F} \text{ along } C = \int_{t=a}^{t=b} \mathbf{F} \cdot \mathbf{T}\, ds$$

$$= \int_{t=a}^{t=b} \mathbf{F} \cdot d\mathbf{R} \qquad \begin{array}{l}\text{(another form,}\\ \text{since } \mathbf{T} = d\mathbf{R}/ds)\end{array} \quad (4)$$

$$= \int_{t=a}^{t=b} M\, dx + N\, dy + P\, dz.$$

(component form)

When the curve is a closed curve, the value of the flow integral is called the **circulation** around the curve.

Notice the way in which the sign of the flow integral in Eq. (4) depends on the direction in which C is traversed as t increases from a to b. If we reverse the direction, we reverse the direction of **T** and thereby change the sign of $\mathbf{F} \cdot \mathbf{T}$ and the sign of its integral with respect to arc length from $t = a$ to $t = b$.

EXAMPLE 2 Evaluate the flow integral of the field

$$\mathbf{F}(x, y, z) = x\mathbf{i} + z\mathbf{j} + y\mathbf{k}$$

along the curve

$$\mathbf{R}(t) = (\cos t)\mathbf{i} + (\sin t)\mathbf{j} + t\mathbf{k}, \qquad 0 \le t \le \pi/2.$$

Solution First express **F** and $d\mathbf{R}$ in terms of t:

$$\mathbf{F}(x, y, z) = x\mathbf{i} + z\mathbf{j} + y\mathbf{k}$$
$$= (\cos t)\mathbf{i} + t\mathbf{j} + (\sin t)\mathbf{k},$$
$$d\mathbf{R} = (-\sin t \, dt)\mathbf{i} + (\cos t \, dt)\mathbf{j} + (dt)\mathbf{k}$$

Then evaluate the flow integral in Eq. (4):

$$\text{Flow integral} = \int_{t=0}^{t=\pi/2} \mathbf{F} \cdot d\mathbf{R}$$

$$= \int_{0}^{\pi/2} (-\sin t \cos t + t \cos t + \sin t) \, dt$$

$$= \left[\frac{\cos^2 t}{2} + t \sin t \right]_{0}^{\pi/2}$$

$$= \left(0 + \frac{\pi}{2} \right) - \left(\frac{1}{2} + 0 \right) = \frac{\pi}{2} - \frac{1}{2}. \qquad \blacksquare$$

EXAMPLE 3 Find the circulation of the field

$$\mathbf{F}(x, y) = (x - y)\mathbf{i} + x\mathbf{j}$$

around the circle

$$\mathbf{R}(t) = (\cos t)\mathbf{i} + (\sin t)\mathbf{j}, \qquad 0 \le t \le 2\pi.$$

Solution We first express **F** and $d\mathbf{R}$ in terms of t:

$$\mathbf{F}(x, y) = (x - y)\mathbf{i} + x\mathbf{j}$$
$$= (\cos t - \sin t)\mathbf{i} + (\cos t)\mathbf{j},$$
$$d\mathbf{R} = (-\sin t \, dt)\mathbf{i} + (\cos t \, dt)\mathbf{j}.$$

Then we evaluate the flow integral:

$$\text{Circulation} = \int_{t=0}^{t=2\pi} \mathbf{F} \cdot d\mathbf{R}$$

$$= \int_{0}^{2\pi} (-\sin t \cos t + \sin^2 t + \cos^2 t) \, dt$$

$$= \int_{0}^{2\pi} (1 - \sin t \cos t) \, dt$$

$$= \left[t - \frac{\sin^2 t}{2} \right]_{0}^{2\pi} = 2\pi. \qquad \blacksquare$$

Flux across a Plane Curve

Suppose we want to know the rate at which a fluid is entering or leaving a region enclosed by a curve C in the xy-plane. To find the rate, we calculate the line integral with respect to arc length around C of $\mathbf{F} \cdot \mathbf{n}$, the scalar component of the fluid's velocity field in the direction of the curve's outward-pointing unit normal vector. The value of this integral is called the **flux** of $\mathbf{F}$ across C. *Flux* is Latin for "flow," but many flux calculations involve no motion at all. If $\mathbf{F}$ were an electric field or a magnetic field, for instance, the integral of $\mathbf{F} \cdot \mathbf{n}$ along C would still be called the flux of $\mathbf{F}$ across C.

DEFINITION

> **Flux across a Closed Plane Curve**
>
> $$\text{Flux of } \mathbf{F} \text{ across } C = \int_C \mathbf{F} \cdot \mathbf{n} \, ds \qquad (5)$$

Notice the difference between flux and circulation. The flux of $\mathbf{F}$ across C is the line integral with respect to arc length of $\mathbf{F} \cdot \mathbf{n}$, the scalar component of $\mathbf{F}$ in the direction of the outward normal. The circulation of $\mathbf{F}$ around C is the line integral with respect to arc length of $\mathbf{F} \cdot \mathbf{T}$, the scalar component of $\mathbf{F}$ in the direction of the unit tangent vector. In short, flux is the integral of the normal component of $\mathbf{F}$ while circulation is the integral of the tangential component of $\mathbf{F}$.

To see how to evaluate the integral in (5), we begin with a parameterization

$$\mathbf{R}(t) = x(t)\mathbf{i} + y(t)\mathbf{j}, \qquad a \le t \le b,$$

that traces the curve C exactly once as t increases from a to b. We can find the outward unit normal vector $\mathbf{n}$ by crossing the curve's unit tangent vector $\mathbf{T}$ with the vector $\mathbf{k}$. But which order do we choose, $\mathbf{T} \times \mathbf{k}$ or $\mathbf{k} \times \mathbf{T}$? Which one points outward? It depends on which way C is traversed as the parameter t increases. If the motion is clockwise, then $\mathbf{k} \times \mathbf{T}$ points outward; if the motion is counterclockwise, then $\mathbf{T} \times \mathbf{k}$ points outward (Fig. 19.16). The usual choice is $\mathbf{n} = \mathbf{T} \times \mathbf{k}$, the choice that assumes counterclockwise motion. Thus, while the value of the arc-length integral in the definition of flux in Eq. (5) does not depend on which way C is traversed, the formulas we are about to derive for evaluating the integral in Eq. (5) will assume counterclockwise motion.

In terms of components,

$$\mathbf{n} = \mathbf{T} \times \mathbf{k} = \left(\frac{dx}{ds}\mathbf{i} + \frac{dy}{ds}\mathbf{j}\right) \times \mathbf{k} = \frac{dy}{ds}\mathbf{i} - \frac{dx}{ds}\mathbf{j}. \qquad (6)$$

If the formula for $\mathbf{F}$ is written in the form

$$\mathbf{F}(x, y) = M(x, y)\mathbf{i} + N(x, y)\mathbf{j},$$

then

$$\mathbf{F} \cdot \mathbf{n} = M(x, y)\frac{dy}{ds} - N(x, y)\frac{dx}{ds}.$$

Hence,

$$\int_C \mathbf{F} \cdot \mathbf{n} \, ds = \int_C \left(M\frac{dy}{ds} - N\frac{dx}{ds}\right) ds = \oint_C M \, dy - N \, dx. \qquad (7)$$

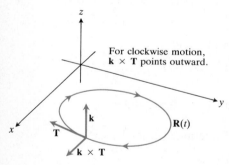

For clockwise motion, $\mathbf{k} \times \mathbf{T}$ points outward.

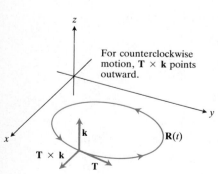

For counterclockwise motion, $\mathbf{T} \times \mathbf{k}$ points outward.

19.16 To find a unit outward normal vector for a curve $\mathbf{R}(t)$ in the xy-plane that is traversed counterclockwise as t increases, we take $\mathbf{n} = \mathbf{T} \times \mathbf{k}$.

We put a directed circle $\circlearrowleft$ on the last integral as a reminder that the integration around the closed curve C is to be in the counterclockwise direction. To evaluate this integral, we express M, dy, N, and dx in terms of t and integrate from $t = a$ to $t = b$. Thus, we do not need to know either **n** or ds to find the flux.

The Formula for Calculating Flux across a Closed Plane Curve

$$\text{Flux of } \mathbf{F} = M\mathbf{i} + N\mathbf{j} \text{ across } C = \oint_C M \, dy - N \, dx \qquad (8)$$

The integral can be evaluated from any parameterization $\mathbf{R}(t)$ that traces C once counterclockwise.

EXAMPLE 4 Find the flux of the field

$$\mathbf{F}(x, y) = (x - y)\mathbf{i} + x\mathbf{j}$$

across the circle

$$C: \qquad \mathbf{R}(t) = (\cos t)\mathbf{i} + (\sin t)\mathbf{j}, \qquad 0 \le t \le 2\pi,$$

in the xy-plane.

Solution The given parameterization traces the circle counterclockwise and covers the circle exactly once as t runs from 0 to 2π. We can therefore use Eq. (8) with this parameterization. In terms of t,

$$M = x - y = \cos t - \sin t,$$
$$N = x = \cos t,$$
$$dy = \cos t \, dt,$$
$$dx = -\sin t \, dt.$$

Hence, the flux is

$$\oint_C M \, dy - N \, dx = \int_0^{2\pi} (\cos t - \sin t)(\cos t \, dt) - (\cos t)(-\sin t \, dt)$$

$$= \int_0^{2\pi} (\cos^2 t - \sin t \cos t + \cos t \sin t) \, dt$$

$$= \int_0^{2\pi} \cos^2 t \, dt$$

$$= \int_0^{2\pi} \frac{1 + \cos 2t}{2} \, dt$$

$$= \left[\frac{t}{2} + \frac{\sin 2t}{4} \right]_0^{2\pi} = \pi.$$

The flux of **F** across the circle is π. Since the answer is positive, the net flow across the curve is outward. A net inward flow would have given a negative flux. ∎

PROBLEMS

In Problems 1–8, find the work done by the force $\mathbf{F}(x, y, z)$ from $(0, 0, 0)$ to $(1, 1, 1)$ along each of the following paths:
   a) the line segment $\mathbf{R}_1(t) = t\mathbf{i} + t\mathbf{j} + t\mathbf{k}, \quad 0 \le t \le 1$,
   b) the curve $\mathbf{R}_2(t) = t\mathbf{i} + t^2\mathbf{j} + t^4\mathbf{k}, \quad 0 \le t \le 1$,
   c) the path consisting of the line segment from $(0, 0, 0)$ to $(1, 1, 0)$ followed by the line segment from $(1, 1, 0)$ to $(1, 1, 1)$.

1. $\mathbf{F}(x, y, z) = 2x\mathbf{i} + 3y\mathbf{j} + 4z\mathbf{k}$

2. $\mathbf{F}(x, y, z) = 3y\mathbf{i} + 2x\mathbf{j} + 4z\mathbf{k}$

3. $\mathbf{F}(x, y, z) = \dfrac{1}{x^2 + 1}\mathbf{j}$

4. $\mathbf{F}(x, y, z) = \sqrt{z}\,\mathbf{i} - 2x\mathbf{j} + \sqrt{y}\,\mathbf{k}$

5. $\mathbf{F}(x, y, z) = xy\mathbf{i} + yz\mathbf{j} + xz\mathbf{k}$

6. $\mathbf{F}(x, y, z) = 3x(x - 1)\mathbf{i} + 3z\mathbf{j} + \mathbf{k}$

7. $\mathbf{F}(x, y, z) = (y + z)\mathbf{i} + (z + x)\mathbf{j} + (x + y)\mathbf{k}$

8. $\mathbf{F}(x, y, z) = (y \sin z)\mathbf{i} + (x \sin z)\mathbf{j} + (xy \cos z)\mathbf{k}$

In Problems 9–12, find the work done by the force $\mathbf{F}(x, y, z)$ along the given path in the direction of increasing t-values.

9. $\mathbf{F}(x, y, z) = xy\mathbf{i} + y\mathbf{j} - yz\mathbf{k}$
   $\mathbf{R}(t) = t\mathbf{i} + t^2\mathbf{j} + t\mathbf{k}, \quad 0 \le t \le 1$

10. $\mathbf{F}(x, y, z) = 2y\mathbf{i} + 3x\mathbf{j} + (x + y)\mathbf{k}$
   $\mathbf{R}(t) = (\cos t)\mathbf{i} + (\sin t)\mathbf{j} + \dfrac{t}{6}\mathbf{k}, \quad 0 \le t \le 2\pi$

11. $\mathbf{F}(x, y, z) = z\mathbf{i} + x\mathbf{j} + y\mathbf{k}$
   $\mathbf{R}(t) = (\sin t)\mathbf{i} + (\cos t)\mathbf{j} + t\mathbf{k}, \quad 0 \le t \le 2\pi$

12. $\mathbf{F}(x, y, z) = 6z\mathbf{i} + y^2\mathbf{j} + 12x\mathbf{k}$
   $\mathbf{R}(t) = (\sin t)\mathbf{i} + (\cos t)\mathbf{j} + \dfrac{t}{6}\mathbf{k}, \quad 0 \le t \le 2\pi$

In Problems 13–16, evaluate the flow integral of the vector field $\mathbf{F}(x, y, z)$ along the given curve.

13. $\mathbf{F}(x, y, z) = -4xy\mathbf{i} + 8y\mathbf{j} + 2\mathbf{k}$
   $\mathbf{R}(t) = t\mathbf{i} + t^2\mathbf{j} + \mathbf{k}, \quad 0 \le t \le 2$

14. $\mathbf{F}(x, y, z) = x^2\mathbf{i} + yz\mathbf{j} + y^2\mathbf{k}$
   $\mathbf{R}(t) = 3t\mathbf{j} + 4t\mathbf{k}, \quad 0 \le t \le 1$

15. $\mathbf{F}(x, y, z) = (x - z)\mathbf{i} + x\mathbf{k}$
   $\mathbf{R}(t) = (\cos t)\mathbf{i} + (\sin t)\mathbf{k}, \quad 0 \le t \le \pi$

16. $\mathbf{F}(x, y, z) = -y\mathbf{i} + x\mathbf{j} + b\mathbf{k}$
   $\mathbf{R}(t) = (a \cos t)\mathbf{i} + (b \sin t)\mathbf{j} + bt\mathbf{k}, \quad 0 \le t \le 2\pi$
   (a and b are positive constants)

17. Evaluate the flow integral of the vector field
   $$\mathbf{F}(x, y) = (x + y)\mathbf{i} - (x^2 + y^2)\mathbf{j}$$

along each of the following paths from $(1, 0)$ to $(-1, 0)$ in the xy-plane.
   a) The upper half of the circle $x^2 + y^2 = 1$
   b) The line segment from $(1, 0)$ to $(-1, 0)$
   c) The line segment from $(1, 0)$ to $(0, -1)$ followed by the line segment from $(0, -1)$ to $(-1, 0)$

18. Find the circulation of $\mathbf{F}(x, y, z) = 2x\mathbf{i} + 2z\mathbf{j} + 2y\mathbf{k}$ along the closed path consisting of the helix
   $$\mathbf{R}_1(t) = (\cos t)\mathbf{i} + (\sin t)\mathbf{j} + t\mathbf{k}, \quad 0 \le t \le \pi/2$$

and the segments
   $$\mathbf{R}_2(t) = \mathbf{j} + \frac{\pi}{2}(1 - t)\mathbf{k}, \quad 0 \le t \le 1$$

and
   $$\mathbf{R}_3(t) = t\mathbf{i} + (1 - t)\mathbf{j}, \quad 0 \le t \le 1.$$

In Problems 19–22, find the circulation and flux of the field $\mathbf{F}(x, y)$ around (across) the closed path that consists of the semicircle
   $$\mathbf{R}_1(t) = (a \cos t)\mathbf{i} + (a \sin t)\mathbf{j}, \quad 0 \le t \le \pi$$

followed by the line segment
   $$\mathbf{R}_2(t) = t\mathbf{i}, \quad -a \le t \le a.$$

19. $\mathbf{F}(x, y) = x\mathbf{i} + y\mathbf{j}$

20. $\mathbf{F}(x, y) = x^2\mathbf{i} + y^2\mathbf{j}$

21. $\mathbf{F}(x, y) = -y\mathbf{i} + x\mathbf{j}$

22. $\mathbf{F}(x, y) = -y^2\mathbf{i} + x^2\mathbf{j}$

23. Find the circulation and flux of the vector fields
   $$\mathbf{F}_1(x, y) = x\mathbf{i} + y\mathbf{j} \quad \text{and} \quad \mathbf{F}_2(x, y) = -y\mathbf{i} + x\mathbf{j}$$
   along (across) each of the following curves.
   a) The ellipse $\mathbf{R}_1(t) = (\cos t)\mathbf{i} + (4 \sin t)\mathbf{j}, \quad 0 \le t \le 2\pi$
   b) The ellipse $\mathbf{R}_2(t) = (4 \cos t)\mathbf{i} + (\sin t)\mathbf{j}, \quad 0 \le t \le 2\pi$

24. Find the flux of the vector fields
   $$\mathbf{F}_1(x, y) = 2x\mathbf{i} - 3y\mathbf{j} \quad \text{and} \quad \mathbf{F}_2(x, y) = 2x\mathbf{i} + (x - y)\mathbf{j}$$
   across the circle
   $$\mathbf{R}(t) = (a \cos t)\mathbf{i} + (a \sin t)\mathbf{j}, \quad 0 \le t \le 2\pi.$$

25. Suppose the function $f(t)$ is positive and continuous for $a \le t \le b$, that $\mathbf{R}(t) = t\mathbf{i} + f(t)\mathbf{j}, \quad a \le t \le b$, and that $\mathbf{F}(x, y) = y\mathbf{i}$. Show that
   $$\int_{t=a}^{t=b} \mathbf{F} \cdot d\mathbf{R}$$

is the area under the graph of f from $t = a$ to $t = b$.

TOOLKIT PROGRAM

Vector Fields

19.3
Green's Theorem in the Plane

We now come to a theorem that describes the relationship between the way a fluid flows along or across the boundary of a plane region and the way the fluid moves around inside the region. The fluid is assumed to be incompressible—like water, for example, and not like a gas. The connection between the fluid's boundary behavior and its internal behavior is made possible by the notions of divergence and curl. The divergence of a fluid's velocity field is a measure of the rate at which fluid is being piped into or out of the region at any given point. The curl of the velocity field is a measure of the fluid's rate of rotation at each point. We'll be more precise about this in a moment.

Green's theorem states that, under conditions usually met in practice, the outward flux of a vector field across the boundary of a plane region equals the double integral of the divergence of the field over the interior of the region. In another form, Green's theorem states that the counterclockwise circulation of a vector field around the boundary of a region equals the double integral of the curl of the vector field over the region.

Green's theorem is one of the great theorems of calculus. It is deep, surprising, and has far-reaching consequences. In pure mathematics, it ranks in importance with the Fundamental Theorems of calculus. In applied mathematics, the generalizations of Green's theorem to three dimensions provide the foundation for important theorems about electricity, magnetism, and fluid flow. We shall explore the three-dimensional forms of Green's theorem in Articles 19.5 and 19.6.

Throughout our discussion of Green's theorem, we talk in terms of velocity fields of fluid flows. We do so because fluid flows are easy to picture and the notions of flux and circulation are easy to interpret. We would like you to be aware, however, that Green's theorem applies to any vector field satisfying certain mathematical conditions. It does not depend for its validity on the field's having a particular physical interpretation.

Flux Density at a Point. Divergence

We need two new ideas for Green's theorem. The first is the idea of the flux density of a vector field at a point, which in mathematics is called the "divergence" of the vector field. We obtain it in the following way.

Suppose that

$$\mathbf{F}(x, y) = M(x, y)\mathbf{i} + N(x, y)\mathbf{j}$$

is the velocity field of a fluid flow in the plane and that the first partial derivatives of M and N are continuous at each point of a region R. Let (x, y) be a point in R and let A be a small rectangle with one corner at (x, y) that, along with its interior, lies entirely in R (Fig. 19.17). The sides of the rectangle, parallel to the coordinate axes, have lengths Δx and Δy. The rate at which fluid leaves the rectangle across the bottom edge is approximately

$$\mathbf{F}(x, y) \cdot (-\mathbf{j}) \, \Delta x = -N(x, y) \, \Delta x. \tag{1}$$

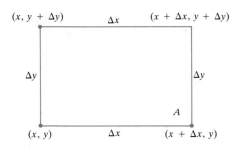

19.17 The rectangle for defining the flux density (divergence) of a vector field at a point (x, y).

■ ■ ■

THE GREEN OF GREEN'S THEOREM

The theorem is due to George Green (1793–1841), a self-taught mathematician who sought to use mathematics in his treatment of static electricity and magnetism. He published his work privately in 1828 as "An Essay on the Application of Analysis to the Theories of Electricity and Magnetism." This profound essay treated line and surface integrals as well, though Gauss's work in 1840 made the explicit connection of line integrals with work. Green's work went unnoticed until the 1850s, when the physicist William Thomson (later Lord Kelvin) published the work after Green's death. Green's theorem was proved simultaneously by the Russian mathematician Mikhail Ostrogradsky in a paper delivered to the St. Petersburg Academy of Sciences in 1828. Green is perhaps better known for this theorem because he inspired later mathematical physicists at Cambridge, such as William Thomson, Gabriel Stokes, Lord Rayleigh, and James Clerk Maxwell. Green was the first great English mathematician to join the mathematical work on the Continent in the eighteenth century, after the split between Newton and Leibniz, and to reintroduce the work into England.

This is the scalar component of the velocity at (x, y) in the direction of the outward normal times the length of the segment. If the velocity is in meters per second, for example, the exit rate will be in meters per second times meters or square meters per second. The rates at which the fluid crosses the other three sides in the directions of their outward normals can be estimated in a similar way. All told, we have

Top: $\mathbf{F}(x, y + \Delta y) \cdot \mathbf{j} \, \Delta x = N(x, y + \Delta y) \, \Delta x,$

Bottom: $\mathbf{F}(x, y) \cdot (-\mathbf{j}) \, \Delta x = -N(x, y) \, \Delta x,$

Right: $\mathbf{F}(x + \Delta x, y) \cdot \mathbf{i} \, \Delta y = M(x + \Delta x, y) \, \Delta y,$ (2)

Left: $\mathbf{F}(x, y) \cdot (-\mathbf{i}) \, \Delta y = -M(x, y) \, \Delta y.$

Combining opposite pairs gives

Top and bottom:

$$(N(x, y + \Delta y) - N(x, y)) \, \Delta x \approx \left(\frac{\partial N}{\partial y} \, \Delta y \right) \Delta x, \qquad (3)$$

Right and left:

$$(M(x + \Delta x, y) - M(x, y)) \, \Delta y \approx \left(\frac{\partial M}{\partial x} \, \Delta x \right) \Delta y. \qquad (4)$$

Adding (3) and (4) gives

$$\text{Flux across rectangle boundary} \approx \left(\frac{\partial M}{\partial x} + \frac{\partial N}{\partial y} \right) \Delta x \, \Delta y. \qquad (5)$$

We now divide by $\Delta x \, \Delta y$ to estimate the total flux per unit area or flux density for the rectangle:

$$\frac{\text{Flux across rectangle boundary}}{\text{Rectangle area}} \approx \left(\frac{\partial M}{\partial x} + \frac{\partial N}{\partial y} \right). \qquad (6)$$

Finally, we let Δx and Δy approach zero to define what we call the *flux density* of $\mathbf{F}$ at the point (x, y).

DEFINITION

Flux Density at a Point

The **flux density** of a vector field $\mathbf{F} = M\mathbf{i} + N\mathbf{j}$ at the point (x, y) is

$$\text{Flux density} = \frac{\partial M}{\partial x} + \frac{\partial N}{\partial y}. \qquad (7)$$

In mathematics, we call the flux density the *divergence* of $\mathbf{F}$. The symbol for it is div $\mathbf{F}$, pronounced "divergence of $\mathbf{F}$" or "div $\mathbf{F}$."

DEFINITION

Divergence of a Vector Field in the Plane

The **divergence** of a vector field $\mathbf{F} = M\mathbf{i} + N\mathbf{j}$ is the scalar function

$$\text{div } \mathbf{F} = \frac{\partial M}{\partial x} + \frac{\partial N}{\partial y}. \qquad (8)$$

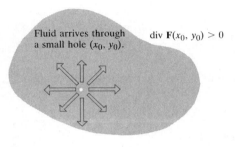

Fluid arrives through a small hole (x_0, y_0).

div $\mathbf{F}(x_0, y_0) > 0$

div $\mathbf{F}(x_0, y_0) < 0$

Fluid leaves through a small hole (x_0, y_0).

19.18 In the flow of an incompressible fluid across a plane region, the divergence is positive at a "source," a point where fluid enters the system, and negative at a "sink," a point where the fluid leaves the system.

If water were flowing into a region through a small hole at the point (x_0, y_0), the lines of flow would diverge there (hence the name) and div $\mathbf{F}(x_0, y_0)$ would be positive. If the water were draining out instead of flowing in, the divergence would be negative. See Fig. 19.18.

EXAMPLE 1 Find the divergence of the vector field

$$\mathbf{F}(x, y) = (x^2 - y)\mathbf{i} + (xy - y^2)\mathbf{j}.$$

Solution We use the formula in Eq. (8):

$$\text{div } \mathbf{F} = \frac{\partial M}{\partial x} + \frac{\partial N}{\partial y}$$

$$= \frac{\partial}{\partial x}(x^2 - y) + \frac{\partial}{\partial y}(xy - y^2)$$

$$= 2x + x - 2y$$

$$= 3x - 2y. \qquad \blacksquare$$

Circulation Density at a Point. The Curl

The second of the two new ideas we need for Green's theorem is the idea of circulation density of a vector field $\mathbf{F}$ at a point, which in mathematics is called the *curl* of $\mathbf{F}$. To obtain it, we return to the velocity field

$$\mathbf{F}(x, y) = M(x, y)\mathbf{i} + N(x, y)\mathbf{j} \tag{9}$$

and the rectangle A. The rectangle is redrawn here as Fig. 19.19.

The counterclockwise circulation of $\mathbf{F}$ around the boundary of A is the sum of flow rates along the sides. For the bottom edge, the flow rate is approximately

$$\mathbf{F}(x, y) \cdot \mathbf{i} \, \Delta x = M(x, y) \, \Delta x.$$

This is the scalar component of the velocity $\mathbf{F}(x, y)$ in the direction of the tangent vector $\mathbf{i}$ times the length of the segment. The rates of flow along the other sides in the counterclockwise direction are expressed in a similar way. In all, we have

Top:	$\mathbf{F}(x, y + \Delta y) \cdot (-\mathbf{i}) \, \Delta x = -M(x, y + \Delta y) \, \Delta x,$
Bottom:	$\mathbf{F}(x, y) \cdot \mathbf{i} \, \Delta x = M(x, y) \, \Delta x,$
Right:	$\mathbf{F}(x + \Delta x, y) \cdot \mathbf{j} \, \Delta y = N(x + \Delta x, y) \, \Delta y,$
Left:	$\mathbf{F}(x, y) \cdot (-\mathbf{j}) \, \Delta y = -N(x, y) \, \Delta y.$

$$(10)$$

We add opposite pairs to get

Top and bottom:

$$-(M(x, y + \Delta y) - M(x, y)) \, \Delta x \approx -\left(\frac{\partial M}{\partial y} \, \Delta y\right) \Delta x, \tag{11}$$

Right and left:

$$(N(x + \Delta x, y) - N(x, y)) \, \Delta y \approx \left(\frac{\partial N}{\partial x} \, \Delta x\right) \Delta y. \tag{12}$$

Adding (11) and (12) and dividing by $\Delta x \, \Delta y$ gives an estimate of the circulation

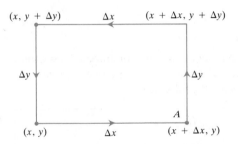

$(x, y + \Delta y)$ Δx $(x + \Delta x, y + \Delta y)$

Δy Δy

A

(x, y) Δx $(x + \Delta x, y)$

19.19 The rectangle for defining the circulation density (curl) of a vector field at a point (x, y).

density for the rectangle:

$$\frac{\text{Circulation around rectangle}}{\text{Rectangle area}} \approx \frac{\partial N}{\partial x} - \frac{\partial M}{\partial y}. \tag{13}$$

Finally, we let Δx and Δy approach zero to define what we call the *circulation density* of $\mathbf{F}$ at the point (x, y).

DEFINITION

> **Circulation Density at a Point in the Plane**
>
> The **circulation density** of a vector field $\mathbf{F} = M\mathbf{i} + N\mathbf{j}$ at the point (x, y) is
>
> $$\text{Circulation density} = \frac{\partial N}{\partial x} - \frac{\partial M}{\partial y}. \tag{14}$$

In mathematics, we call the circulation density the *curl* of $\mathbf{F}$. The symbol for it is curl $\mathbf{F}$, pronounced "curl $\mathbf{F}$," the way it is written.

DEFINITION

> **Curl of a Vector Field in the Plane**
>
> The **curl** of a vector field $\mathbf{F} = M\mathbf{i} + N\mathbf{j}$ is the function
>
> $$\text{curl } \mathbf{F} = \frac{\partial N}{\partial x} - \frac{\partial M}{\partial y}. \tag{15}$$

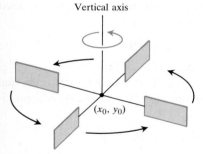

curl $\mathbf{F}(x_0, y_0) > 0$
Counterclockwise circulation

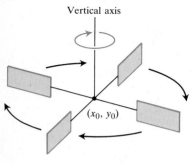

curl $\mathbf{F}(x_0, y_0) < 0$
Clockwise circulation

19.20 In the flow of an incompressible fluid over a plane region, the curl measures the rate of the fluid's rotation at a point. The curl is positive at points where the rotation is counterclockwise and negative where the rotation is clockwise.

If water is moving about a region in the xy-plane in a thin layer, then the circulation, or curl, at a point (x_0, y_0) gives a way to measure how fast and in what direction a small paddle wheel will spin if it is put into the water at (x_0, y_0) with its axis perpendicular to the plane (Fig. 19.20).

EXAMPLE 2   Find the curl of the vector field

$$\mathbf{F}(x, y) = (x^2 - y)\mathbf{i} + (xy - y^2)\mathbf{j}$$

Solution   We use the formula in Eq. (15):

$$\text{curl } \mathbf{F} = \frac{\partial N}{\partial x} - \frac{\partial M}{\partial y}$$

$$= \frac{\partial}{\partial x}(xy - y^2) - \frac{\partial}{\partial y}(x^2 - y) = y + 1. \quad \blacksquare$$

Green's Theorem in the Plane

In one form, Green's theorem states that under suitable conditions the outward flux of a vector field $\mathbf{F} = M\mathbf{i} + N\mathbf{j}$ across a simple closed curve C equals the double integral of div $\mathbf{F}$ over the region R enclosed by C:

$$\underbrace{\oint_C M\, dy - N\, dx}_{\text{Flux}} = \iint_R \underbrace{\left(\frac{\partial M}{\partial x} + \frac{\partial N}{\partial y}\right) dx\, dy}_{\text{Divergence}}. \tag{16}$$

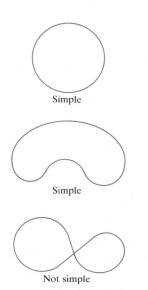

Simple

Simple

Not simple

19.21 In proving Green's theorem, we distinguish between two kinds of closed curves, simple and not simple. Simple curves do not cross themselves.

Simple closed curves are curves that do not cross themselves. A circle is a simple closed curve, but a figure eight is not (Fig. 19.21).

In another form, Green's theorem says that the counterclockwise circulation of **F** around C equals the double integral of curl **F** over R:

$$\oint_C M \, dx + N \, dy = \iint_R \left(\frac{\partial N}{\partial x} - \frac{\partial M}{\partial y} \right) dx \, dy. \qquad (17)$$

Circulation   Curl

The two forms of Green's theorem are equivalent. Applying Eq. (16) to the field $\mathbf{G}_1 = N\mathbf{i} - M\mathbf{j}$ gives Eq. (17), and applying Eq. (17) to $\mathbf{G}_2 = -N\mathbf{i} + M\mathbf{j}$ gives Eq. (16). We do not need to prove them both. We shall prove the curl form shortly.

We need two kinds of assumptions for Green's theorem to hold. First, we need conditions on M and N to ensure the existence of the integrals. The usual assumptions are that M, N, and their first partial derivatives are continuous at every point of some region containing C and R. Second, we need geometric conditions on the curve C. It must be simple, closed, and made up of pieces along which we can integrate M and N. The usual assumptions are that C consists of a finite number of continuously differentiable curves connected end to end. The proof we give for Green's theorem in this article, however, assumes things about the shape of R as well. You can find less restrictive proofs in more advanced texts. (We mention one at the end of this article.)

EXAMPLE 3 Verify both forms of Green's theorem for the field

$$\mathbf{F}(x, y) = (x - y)\mathbf{i} + x\mathbf{j}$$

and the region R bounded by the circle

$$C: \quad \mathbf{R}(t) = (\cos t)\mathbf{i} + (\sin t)\mathbf{j}, \quad 0 \leq t \leq 2\pi.$$

Solution We first express all functions, derivatives, and differentials in terms of t:

$$M = \cos t - \sin t, \qquad dx = d(\cos t) = -\sin t \, dt,$$

$$N = \cos t, \qquad dy = d(\sin t) = \cos t \, dt,$$

$$\frac{\partial M}{\partial x} = 1, \qquad \frac{\partial M}{\partial y} = -1, \qquad \frac{\partial N}{\partial x} = 1, \qquad \frac{\partial N}{\partial y} = 0.$$

The two sides of Eq. (16):

$$\oint_C M \, dy - N \, dx = \int_{t=0}^{t=2\pi} (\cos t - \sin t)(\cos t \, dt) - (\cos t)(-\sin t \, dt)$$

$$= \int_0^{2\pi} \cos^2 t \, dt = \pi,$$

$$\iint_R \left(\frac{\partial M}{\partial x} + \frac{\partial N}{\partial y} \right) dx \, dy = \iint_R (1 + 0) \, dx \, dy = \iint_R dx \, dy = \pi.$$

The two sides of Eq. (17):

$$\oint_C M\,dx + N\,dy = \int_{t=0}^{t=2\pi} (\cos t - \sin t)(-\sin t\,dt) + (\cos t)(\cos t\,dt)$$

$$= \int_0^{2\pi} (-\sin t \cos t + 1)\,dt = 2\pi,$$

$$\iint_R \left(\frac{\partial N}{\partial x} - \frac{\partial M}{\partial y}\right) dx\,dy = \iint_R (1 - (-1))\,dx\,dy$$

$$= 2 \iint_R dx\,dy = 2\pi.\qquad\blacksquare$$

If a closed curve C is constructed by piecing a number of different curves end to end, the process of evaluating a line integral over C can be a lengthy one because there are so many different integrals to evaluate. However, if C bounds a region R to which Green's theorem applies, we can use Green's theorem to change the line integral around C into a single double integral over R.

EXAMPLE 4 Evaluate the integral

$$\oint xy\,dy - y^2\,dx$$

around the square cut from the first quadrant by the lines $x = 1$ and $y = 1$.

Solution We can use either form of Green's theorem to change the line integral into a double integral over the square.

1. *With Eq. (16):* Taking $M = xy$, $N = y^2$, and C and R as the square's boundary and interior gives

$$\oint xy\,dy - y^2\,dx = \iint_R (y + 2y)\,dx\,dy = \int_0^1 \int_0^1 3y\,dx\,dy$$

$$= \int_0^1 \left[3xy\right]_{x=0}^{x=1} dy = \int_0^1 3y\,dy$$

$$= \frac{3}{2}y^2\Big]_0^1 = \frac{3}{2}.$$

2. *With Eq. (17):* Taking $M = -y^2$ and $N = xy$ gives

$$\oint_C -y^2\,dx + xy\,dy = \iint_R (y - (-2y))\,dx\,dy = \frac{3}{2},$$

as before. $\blacksquare$

EXAMPLE 5 Calculate the outward flux of the field $\mathbf{F}(x, y) = x\mathbf{i} + y^2\mathbf{j}$ across the square bounded by the lines $x = \pm 1$ and $y = \pm 1$.

Solution Calculating the flux with a line integral would take four integrations, one for each side of the square. With Green's theorem, we can change the line integral to a simple double integral. With $M = x$, $N = y^2$, C the square, and R

the square's interior, we have

$$\text{Flux} = \oint_C M \, dy - N \, dx$$

$$= \iint_R \left(\frac{\partial M}{\partial x} + \frac{\partial N}{\partial y} \right) dx \, dy$$

$$= \int_{-1}^{1} \int_{-1}^{1} (1 + 2y) \, dx \, dy$$

$$= \int_{-1}^{1} \left[x + 2xy \right]_{x=-1}^{x=1} dy$$

$$= \int_{-1}^{1} (2 + 4y) \, dy = \left[2y + 2y^2 \right]_{-1}^{1} = 4. \qquad \blacksquare$$

A Proof of Green's Theorem (Special Regions)

Let C be a continuously differentiable simple closed curve in the xy-plane with the property that lines parallel to the axes cut it in no more than two points. Let M, N, and their first partial derivatives be continuous at every point of C and let R be the region enclosed by C. Then

$$\oint_C M \, dx + N \, dy = \iint_R \left(\frac{\partial N}{\partial x} - \frac{\partial M}{\partial y} \right) dx \, dy. \qquad (18)$$

Figure 19.22 shows C made up of two parts:

$$C_1: \qquad y_1 = f_1(x), \qquad a \le x \le b,$$

$$C_2: \qquad y_2 = f_2(x), \qquad b \ge x \ge a.$$

For any x between a and b, we can integrate $\partial M / \partial y$ with respect to y from $y_1 = f_1(x)$ to $y_2 = f_2(x)$ and obtain

$$\int_{y_1}^{y_2} \frac{\partial M}{\partial y} \, dy = M(x, y) \Big]_{y=f_1(x)}^{y=f_2(x)} = M(x, f_2(x)) - M(x, f_1(x)). \qquad (19)$$

We can then integrate this with respect to x from a to b:

$$\int_a^b \int_{f_1(x)}^{f_2(x)} \frac{\partial M}{\partial y} \, dy \, dx = \int_a^b \{ M(x, f_2(x)) - M(x, f_1(x)) \} \, dx$$

$$= -\int_b^a M(x, f_2(x)) \, dx - \int_a^b M(x, f_1(x)) \, dx$$

$$= -\int_{C_2} M \, dx - \int_{C_1} M \, dx$$

$$= -\oint_C M \, dx.$$

Therefore

$$\oint_C M \, dx = \iint_R \left(-\frac{\partial M}{\partial y} \right) dx \, dy. \qquad (20)$$

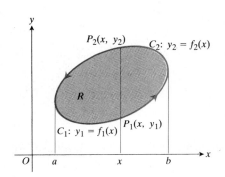

19.22 The boundary curve C is made up of $C_1: y_1 = f_1(x)$ and $C_2: y_2 = f_2(x)$.

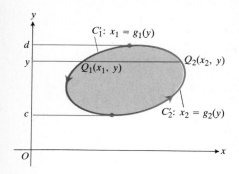

19.23 The boundary curve C' is made up of $C'_1 : x_1 = g_1(y)$ and $C'_2 : x_2 = g_2(y)$.

Equation (20) is half the result we need for Eq. (18). Problem 25 asks you to derive the other half, by integrating $\partial N/\partial x$ first with respect to x and then with respect to y, as suggested by Fig. 19.23. This shows the curve C of Fig. 19.22, decomposed into the two directed parts

$$C'_1 : \qquad x_1 = g_1(y), \qquad d \geq y \geq c,$$
$$C'_2 : \qquad x_2 = g_2(y), \qquad c \leq y \leq d.$$

The result of this double integration is expressed by

$$\oint_C N\, dy = \iint_R \frac{\partial N}{\partial x}\, dx\, dy. \tag{21}$$

Combining Eqs. (20) and (21), we get Eq. (18). This concludes the proof. ∎

Extending the Proof to Other Regions

The argument with which we just derived Green's equation does not apply directly to the rectangular region in Fig. 19.24 because the lines $x = a$, $x = b$, $y = c$, and $y = d$ meet the region's boundary in more than two points. However, if we divide the boundary C into four directed line segments,

$$C_1 : \qquad y = c, \qquad a \leq x \leq b,$$
$$C_2 : \qquad x = b, \qquad c \leq y \leq d,$$
$$C_3 : \qquad y = d, \qquad b \geq x \geq a,$$
$$C_4 : \qquad x = a, \qquad d \geq y \geq c,$$

we can modify the argument in the following way.

Proceeding as in the proof of Eq. (21), we have

$$\int_c^d \int_a^b \frac{\partial N}{\partial x}\, dx\, dy = \int_c^d [N(b, y) - N(a, y)]\, dy$$
$$= \int_c^d N(b, y)\, dy + \int_d^c N(a, y)\, dy \tag{22}$$
$$= \int_{C_2} N\, dy + \int_{C_4} N\, dy.$$

Because y is constant along C_1 and C_3,

$$\int_{C_1} N\, dy = \int_{C_3} N\, dy = 0,$$

so we can add

$$\int_{C_1} N\, dy + \int_{C_3} N\, dy$$

to the right-hand side of Eq. (22) without changing the equality. Doing so, we have

$$\int_c^d \int_a^b \frac{\partial N}{\partial x}\, dx\, dy = \oint_C N\, dy. \tag{23}$$

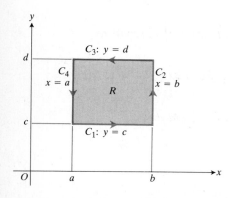

19.24 To prove Green's theorem for the rectangular region shown here, we divide the boundary into four directed line segments.

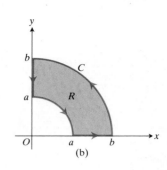

19.25 Other regions to which Green's theorem applies.

Similarly, we could show that

$$\int_a^b \int_c^d \frac{\partial M}{\partial y} \, dy \, dx = -\oint_C M \, dx. \tag{24}$$

Subtracting (24) from (23), we again arrive at

$$\oint_C M \, dx + N \, dy = \iint_R \left(\frac{\partial N}{\partial x} - \frac{\partial M}{\partial y} \right) dx \, dy.$$

Regions like those in Fig. 19.25 can be handled with no greater difficulty. Equation (18) still applies. It also applies to the horseshoe-shaped region R shown in Fig. 19.26, as we see by putting together the regions R_1 and R_2 and their boundaries. Green's theorem applies to C_1, R_1, and to C_2, R_2, yielding

$$\int_{C_1} M \, dx + N \, dy = \iint_{R_1} \left(\frac{\partial N}{\partial x} - \frac{\partial M}{\partial y} \right) dx \, dy,$$

$$\int_{C_2} M \, dx + N \, dy = \iint_{R_2} \left(\frac{\partial N}{\partial x} - \frac{\partial M}{\partial y} \right) dx \, dy.$$

When we add, the line integral along the y-axis from b to a for C_1 cancels the integral over the same segment but in the opposite direction for C_2. Hence

$$\oint_C M \, dx + N \, dy = \iint_R \left(\frac{\partial N}{\partial x} - \frac{\partial M}{\partial y} \right) dx \, dy,$$

where C consists of the two segments of the x-axis from $-b$ to $-a$ and from a to b and of the two semicircles, and where R is the region inside C.

The device of adding line integrals over separate boundaries to build up an integral over a single boundary can be extended to any finite number of subregions. In Fig. 19.27(a), let C_1 be the boundary of the region R_1 in the first quadrant. Similarly for the other three quadrants: C_i is the boundary of the region R_i, $i = 1, 2, 3, 4$. By Green's theorem,

$$\oint_{C_i} M \, dx + N \, dy = \iint_{R_i} \left(\frac{\partial N}{\partial x} - \frac{\partial M}{\partial y} \right) dx \, dy. \tag{25}$$

We add Eqs. (25) for $i = 1, 2, 3, 4$, and get

$$\oint_{r=b} (M \, dx + N \, dy) + \oint_{r=a} (M \, dx + N \, dy) = \iint_{a \leq r \leq b} \left(\frac{\partial N}{\partial x} - \frac{\partial M}{\partial y} \right) dx \, dy. \tag{26}$$

19.26 A region R that combines regions R_1 and R_2.

(a)

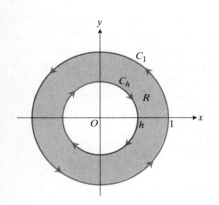

(b)

19.27 The annular region R shown here combines four smaller regions. In polar coordinates, we have $r = a$ for the inner circle, $r = b$ for the outer circle, and $a \leq r \leq b$ for the region itself.

Equation (26) says that the double integral of

$$\left(\frac{\partial N}{\partial x} - \frac{\partial M}{\partial y} \right) dx \, dy$$

over the annular ring R is equal to the line integral of $M \, dx + N \, dy$ over the *entire* boundary of R, in the *direction* along the boundary that keeps the region R on our left as we progress (Fig. 19.27b).

EXAMPLE 6   Verify the circulation form of Green's theorem (Eq. 17) if

$$M = \frac{-y}{x^2 + y^2}, \qquad N = \frac{x}{x^2 + y^2},$$

and the annular region R is defined by the inequalities

$$h^2 \leq x^2 + y^2 \leq 1,$$

where $0 < h < 1$.

Solution   See Fig. 19.28. The boundary of R consists of the circle

$$C_1: \qquad x = \cos \theta, \qquad y = \sin \theta, \qquad 0 \leq \theta \leq 2\pi,$$

around which we shall integrate counterclockwise, and the circle

$$C_h: \qquad x = h \cos \phi, \qquad y = h \sin \phi, \qquad 2\pi \geq \phi \geq 0,$$

around which we shall integrate in the clockwise direction. Notice that the origin is not included in R because h is positive. For all $(x, y) \neq (0, 0)$, the functions M and N and their partial derivatives are continuous. Moreover,

$$\frac{\partial M}{\partial y} = \frac{(x^2 + y^2)(-1) + y(2y)}{(x^2 + y^2)^2} = \frac{y^2 - x^2}{(x^2 + y^2)^2} = \frac{\partial N}{\partial x},$$

so

$$\iint\limits_{R} \left(\frac{\partial N}{\partial x} - \frac{\partial M}{\partial y} \right) dx \, dy = \iint\limits_{R} 0 \, dx \, dy = 0.$$

The line integral is

$$\int_C M \, dx + N \, dy = \oint_{C_1} \frac{x \, dy - y \, dx}{x^2 + y^2} + \oint_{C_h} \frac{x \, dy - y \, dx}{x^2 + y^2}$$

$$= \int_0^{2\pi} (\cos^2\theta + \sin^2\theta) \, d\theta + \int_{2\pi}^{0} \frac{h^2(\cos^2\phi + \sin^2\phi) \, d\phi}{h^2}$$

$$= 2\pi - 2\pi = 0. \qquad \blacksquare$$

In Example 6, the functions M and N are discontinuous at $(0, 0)$, so we cannot apply Green's theorem to $C_1: x^2 + y^2 = 1$ and the region inside it. We must delete the origin, which we did by excluding points inside C_h.

In Example 6, we could replace the outer circle C_1 by an ellipse or any other simple closed curve C that lies outside C_h. The result would be

$$\oint_C (M \, dx + N \, dy) + \oint_{C_h} (M \, dx + N \, dy) = 0,$$

19.28   Green's theorem may be applied to the annulus R by integrating along the boundaries as shown.

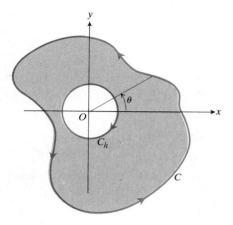

19.29 The region bounded by the circle C_h and the curve C.

which leads to the conclusion

$$\oint_C \frac{x\,dy - y\,dx}{x^2 + y^2} = 2\pi.$$

This result is easily accounted for if we change to polar coordinates for C:

$$x = r\cos\theta, \qquad y = r\sin\theta,$$

$$dx = -r\sin\theta\,d\theta + \cos\theta\,dr,$$

$$dy = r\cos\theta\,d\theta + \sin\theta\,dr.$$

For then

$$\frac{x\,dy - y\,dx}{x^2 + y^2} = \frac{r^2(\cos^2\theta + \sin^2\theta)\,d\theta}{r^2} = d\theta,$$

and θ increases by 2π as we progress once around C counterclockwise (see Fig. 19.29).

The Normal and Tangential Forms of Green's Theorem

As we learned in Article 19.2, the flux of a two-dimensional vector field **F** across a closed curve C in the direction of its outer normal vector **n** is

$$\int_C \mathbf{F} \cdot \mathbf{n}\,ds. \qquad \text{(flux)}$$

Similarly, the circulation around C in the direction of its unit tangent vector **T** is

$$\int_C \mathbf{F} \cdot \mathbf{T}\,ds. \qquad \text{(circulation)}$$

Hence, the flux form of Green's theorem can be written as

$$\int_C \mathbf{F} \cdot \mathbf{n}\,ds = \iint_R \left(\frac{\partial M}{\partial x} + \frac{\partial N}{\partial y}\right)dx\,dy \qquad \text{(normal form)} \qquad (27)$$

and the circulation form of Green's theorem can be written as

$$\oint_C \mathbf{F} \cdot \mathbf{T}\,ds = \iint_R \left(\frac{\partial N}{\partial x} - \frac{\partial M}{\partial y}\right)dx\,dy. \qquad \text{(tangential form)} \qquad (28)$$

The left-hand sides of these equations are now in vector form. How about the right-hand sides? Can they be expressed in vector form too? The answer is yes, and we shall show how to do so with an ingenious vector operator ∇ ("del") in Articles 19.5 and 19.6.

References to Further Reading

For information about the mathematics of Green's theorem, see Tom M. Apostol's *Calculus,* Volume 2, 2nd ed. (New York: John Wiley and Sons, 1969).

For information about the physical interpretations of divergence and curl, see H. M. Schey's *Div, Grad, Curl and All That* (New York: W. W. Norton, 1973).

For historical information about Green's theorem and its development, see Morris Kline's *Mathematical Thought from Ancient to Modern Times* (New York: Oxford University Press, 1972).

PROBLEMS

In Problems 1–4, verify Green's theorem by evaluating both sides of Eqs. (16) and (17) for the vector field $\mathbf{F} = M\mathbf{i} + N\mathbf{j}$. Take the domains of integration to be the disk $R: x^2 + y^2 \leq a^2$ and its bounding circle

$$C: \quad \mathbf{R}(t) = (a \cos t)\mathbf{i} + (a \sin t)\mathbf{j}, \quad 0 \leq t \leq 2\pi.$$

1. $\mathbf{F}(x, y) = -y\mathbf{i} + x\mathbf{j}$
2. $\mathbf{F}(x, y) = y\mathbf{i}$
3. $\mathbf{F}(x, y) = 2x\mathbf{i} - 3y\mathbf{j}$
4. $\mathbf{F}(x, y) = -x^2 y\mathbf{i} + xy^2\mathbf{j}$

In Problems 5–10, apply Green's theorem to find the counterclockwise circulation and outward flux for the field $\mathbf{F}$ and curve C.

5. $\mathbf{F}(x, y) = (x - y)\mathbf{i} + (x + y)\mathbf{j}$
   $C:$ The square bounded by $x = 0$, $x = 1$, $y = 0$, $y = 1$
6. $\mathbf{F}(x, y) = (x^2 - y)\mathbf{i} + (x + y^2)\mathbf{j}$
   $C:$ The square bounded by $x = 0$, $x = 1$, $y = 0$, $y = 1$
7. $\mathbf{F}(x, y) = (xy + x)\mathbf{i} + (xy - y)\mathbf{j}$
   $C:$ The square bounded by $x = 0$, $x = 1$, $y = 0$, $y = 1$
8. $\mathbf{F}(x, y) = (x^2 - y)\mathbf{i} + (x + y^2)\mathbf{j}$
   $C:$ The triangle bounded by $y = 0$, $x = 1$, and $y = x$
9. $\mathbf{F}(x, y) = (x^2 - y^2)\mathbf{i} + (x^2 + y^2)\mathbf{j}$
   $C:$ The triangle bounded by $y = 0$, $x = 1$, and $y = x$
10. $\mathbf{F}(x, y) = (x + y)\mathbf{i} - (x^2 + y^2)\mathbf{j}$
    $C:$ The triangle bounded by $y = 0$, $x = 1$, and $y = x$

11. Find the counterclockwise circulation and outward flux of the field $\mathbf{F}(x, y) = xy\mathbf{i} + y^2\mathbf{j}$ over the boundary of the region enclosed by the parabola $y = x^2$ and the line $y = x$ in the first quadrant.

12. Find the counterclockwise circulation and the outward flux of the field $\mathbf{F}(x, y) = (-\sin y)\mathbf{i} + (x \cos y)\mathbf{j}$ over the square cut from the first quadrant by the lines $x = \pi/2$ and $y = \pi/2$.

Apply Green's theorem in one of its two forms to evaluate the line integrals in Problems 13–18.

13. $\oint_C (y^2 \, dx + x^2 \, dy)$
    C is the triangle bounded by $x = 0$, $x + y = 1$, $y = 0$.

14. $\oint_C (3y \, dx + 2x \, dy)$
    C is the boundary of $0 \leq x \leq \pi$, $0 \leq y \leq \sin x$.

15. $\oint_C (6y + x) \, dx + (y + 2x) \, dy$
    C is the circle $(x - 2)^2 + (y - 3)^2 = 4$.

16. $\oint_C (2x + y^2) \, dx + (2xy + 3y) \, dy$
    C is any simple closed curve in the plane for which Green's theorem holds.

17. $\oint_C 2xy^3 \, dx + 4x^2 y^2 \, dy$
    C is the boundary of the "triangular" region in the first quadrant enclosed by the x-axis, the line $x = 1$, and the curve $y = x^3$.

18. $\oint_C (4x - 2y) \, dx + (2x - 4y) \, dy$
    C is the circle $(x - 2)^2 + (y - 2)^2 = 4$.

19. Let C be the boundary of a region on which Green's theorem holds. Use Green's theorem to calculate

    a) $\oint_C f(x) \, dx + g(y) \, dy$,

    b) $\oint_C k \, y \, dx + h \, x \, dy$ (k and h constants).

20. Show that

    $$\oint_C 4x^3 y \, dx + x^4 \, dy = 0$$

    for any closed curve C to which Green's theorem applies.

21. Show that

    $$\oint_C -y^3 \, dx + x^3 \, dy$$

    is positive for any closed curve C to which Green's theorem applies.

22. Show that the value of

    $$\oint_C xy^2 \, dx + (x^2 y + 2x) \, dy$$

    around any square depends only on the area of the square and not on its location in the plane.

23. *Green's theorem and Laplace's equation.* Assuming that all the necessary derivatives exist and are continuous, show that if $f(x, y)$ satisfies the Laplace equation

    $$\frac{\partial^2 f}{\partial x^2} + \frac{\partial^2 f}{\partial y^2} = 0,$$

    then

    $$\oint_C \frac{\partial f}{\partial y} \, dx - \frac{\partial f}{\partial x} \, dy = 0$$

    for all closed curves C to which Green's theorem applies. (The converse is also true: If the line integral is always zero, then f satisfies the Laplace equation.)

24. Let

    $$\mathbf{F} = \left(\frac{1}{4}x^2 y + \frac{1}{3}y^3\right)\mathbf{i} + x\mathbf{j}.$$

    Among all smooth simple closed curves in the plane, oriented counterclockwise, find the curve around which the work done by $\mathbf{F}$ is the greatest. (*Hint:* Where is curl $\mathbf{F}$ positive?)

25. Establish Eq. (21) to finish the proof of the special case of Green's theorem.

26. Establish Eq. (24) to finish the argument for the extension of Green's theorem.

The Area Corollary of Green's Theorem

If a region R and its simple closed curve boundary C satisfy the hypotheses of Green's theorem, then the area of R is given by the formula

$$\text{Area of } R = \frac{1}{2} \oint_C x \, dy - y \, dx. \qquad (29)$$

The reason is that by Eq. (16), run backward,

$$\text{Area of } R = \iint_R dy \, dx$$

$$= \iint_R \left(\frac{1}{2} + \frac{1}{2} \right) dy \, dx$$

$$= \oint_C \frac{1}{2} x \, dy - \frac{1}{2} y \, dx.$$

Use Eq. (29) to calculate the areas in Problems 27–30.

27. The area of the circle

$$\mathbf{R}(t) = (a \cos t)\mathbf{i} + (a \sin t)\mathbf{j}, \quad 0 \le t \le 2\pi$$

28. The area of the ellipse

$$\mathbf{R}(t) = (a \cos t)\mathbf{i} + (b \sin t)\mathbf{j}, \quad 0 \le t \le 2\pi$$

29. The area enclosed by the hypocycloid

$$\mathbf{R}(t) = (a \cos^3 t)\mathbf{i} + (a \sin^3 t)\mathbf{j}, \quad 0 \le t \le 2\pi$$

(See Fig. 5.35, page 334.)

30. The area enclosed by the curve

$$\mathbf{R}(t) = t^2 \mathbf{i} + ((t^3/3) - t)\mathbf{j}, \quad -\sqrt{3} \le t \le \sqrt{3}$$

(See Fig. 5.36.)

TOOLKIT PROGRAMS

Scalar Fields   Vector Fields
These programs evaluate line integrals.

19.4
Surface Area and Surface Integrals

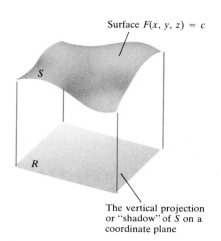

19.30 As we shall soon see, the integral of a function $g(x, y, z)$ over a surface S in space can be calculated by evaluating a closely related double integral over the vertical projection or "shadow" of S on a coordinate plane.

We know how to integrate a function over a flat region in the plane, but what if the function is defined over a curved surface instead? How do we calculate its integral then? The trick to evaluating one of these so-called surface integrals is to rewrite it as a double integral over a region in a coordinate plane beneath the surface (Fig. 19.30). This changes the surface integral into the kind of integral we already know how to evaluate.

Our first step is to find a double-integral formula for calculating the area of a curved surface. We will then see how some of the ideas that arise in the process can be used again to define and evaluate surface integrals. With surface integrals under control, we shall be able to calculate the flux of a three-dimensional vector field through a surface and the masses and moments of thin shells of material. In Articles 19.5 and 19.6 we shall see how surface integrals provide just what we need to generalize the two forms of Green's theorem to three dimensions. One of these generalizations will enable us to express the flux of a vector field through a closed surface as a triple integral over the three-dimensional region enclosed by the surface. The other generalization will enable us to express the circulation of a vector field around a closed curve in space as an integral over a surface bounded by the curve. These results have far-reaching consequences in mathematics as well as in the theories of electromagnetism and fluid flow.

Surface Area

Figure 19.31 shows a piece S of a surface $F(x, y, z) = c$ lying above its "shadow" region R in a plane directly beneath it. If F and its first partial derivatives are continuous, the area of S can be defined and calculated as a

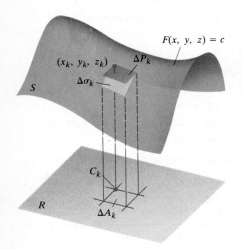

$F(x, y, z) = c$

(x_k, y_k, z_k) ΔP_k

$\Delta \sigma_k$

S

C_k

R ΔA_k

19.31 A surface S and its vertical projection onto a plane beneath it. You can think of R as the shadow of S on the plane. The tangent plate ΔP_k approximates the surface patch $\Delta \sigma_k$ above ΔA_k.

$\nabla F(x_k, y_k, z_k)$ γ $\mathbf{p}$

$\mathbf{u}$ $\mathbf{v}$

ΔP_k

$\Delta \sigma_k$

C_k $\mathbf{p}$

ΔA_k

19.32 Magnified view from Fig. 19.31. The vector $\mathbf{u} \times \mathbf{v}$ (not shown) is parallel to ∇F because both vectors are normal to ΔP_k.

double integral over R. The origin of the integral takes a while to describe, but the integral itself is easy to work with.

The first step in defining the area of S is to divide the region R into small rectangles ΔA_k of the kind we would use if we were defining an integral over R. Directly above each ΔA_k lies a patch of surface $\Delta \sigma_k$ that we may approximate with a portion ΔP_k of the tangent plane. To be specific, we suppose that ΔP_k is a portion of the plane that is tangent to the surface at the point (x_k, y_k, z_k) directly above the back corner C_k of ΔA_k. If the tangent plane is parallel to R, then ΔP_k will be congruent to ΔA_k. Otherwise, it will be a parallelogram whose area is somewhat larger than the area of ΔA_k.

Figure 19.32 gives a magnified view of $\Delta \sigma_k$ and ΔP_k, showing the gradient vector $\nabla F(x_k, y_k, z_k)$ and a unit vector $\mathbf{p}$ that is normal to R. The figure also shows the angle γ between ∇F and $\mathbf{p}$. The other vectors in the picture, $\mathbf{u}$ and $\mathbf{v}$, lie along the edges of the patch ΔP_k in the tangent plane. Thus, both $\mathbf{u} \times \mathbf{v}$ and ∇F are normal to the tangent plane.

We now need a fact we haven't used since Article 13.7, namely that $|(\mathbf{u} \times \mathbf{v}) \cdot \mathbf{p}|$ is the area of the projection of the parallelogram determined by $\mathbf{u}$ and $\mathbf{v}$ onto any plane whose normal is $\mathbf{p}$. In our case, this translates into the statement

$$|(\mathbf{u} \times \mathbf{v}) \cdot \mathbf{p}| = \Delta A_k. \tag{1}$$

Now, $|\mathbf{u} \times \mathbf{v}|$ itself is the area ΔP_k (standard fact about cross products) so Eq. (1) becomes

$$\underbrace{|\mathbf{u} \times \mathbf{v}|}_{\Delta P_k} \underbrace{|\mathbf{p}|}_{1} \underbrace{|\cos (\text{angle between } \mathbf{u} \times \mathbf{v} \text{ and } \mathbf{p})|}_{} = \Delta A_k \tag{2}$$

Same as $|\cos \gamma|$ because ∇F and $\mathbf{u} \times \mathbf{v}$ are both normal to the tangent plane

or

$$\Delta P_k |\cos \gamma| = \Delta A_k$$

or

$$\Delta P_k = \frac{\Delta A_k}{|\cos \gamma|},$$

provided $\cos \gamma \neq 0$. We will have $\cos \gamma \neq 0$ as long as ∇F is not parallel to the ground plane, or as long as $\nabla F \cdot \mathbf{p} \neq 0$.

Since the patches ΔP_k approximate the surface patches $\Delta \sigma_k$ that fit together to make S, the sum

$$\sum \Delta P_k = \sum \frac{\Delta A_k}{|\cos \gamma|} \tag{3}$$

looks like an approximation of what we might like to call the surface area of S. It also looks as if the approximation would improve if we refined the subdivision of R. In fact, the sums on the right-hand side of Eq. (3) are approximating sums for the double integral

$$\iint\limits_{R} \frac{1}{|\cos \gamma|} \, dA. \tag{4}$$

We therefore define the area of S to be the value of this integral whenever the integral exists.

A Practical Formula

For any particular surface $F(x, y, z) = c$,

$$|\nabla F \cdot \mathbf{p}| = |\nabla F| \, |\mathbf{p}| \, |\cos \gamma| = |\nabla F||\cos \gamma|$$

and

$$\frac{1}{|\cos \gamma|} = \frac{|\nabla F|}{|\nabla F \cdot \mathbf{p}|}.$$

This combines with Eq. (4) to give a practical formula for calculating the area of the surface over a region R.

$$\text{Surface area} = \iint\limits_{R} \frac{|\nabla F|}{|\nabla F \cdot \mathbf{p}|} \, dA \tag{5}$$

Thus, the area of the surface $F(x, y, z) = c$ over a plane region R is the double integral over R of the magnitude of the gradient of F divided by the scalar component of the gradient normal to R.

We reached Eq. (5) under the assumptions that $\nabla F \cdot \mathbf{p} \neq 0$ and that F and its first partial derivatives were continuous over R (so that ∇F would be defined and continuous over R). Whenever the integral exists, however, we may define its value to be the area of the surface that lies over R.

The surface area defined by Eq. (5) agrees with our earlier definitions of surface area. We shall not prove this, but see Problem 12.

EXAMPLE 1 Find the area of the surface cut from the bottom of the paraboloid $x^2 + y^2 - z = 0$ by the plane $z = 1$.

Solution We sketch the surface and the region below it in the xy-plane (Fig. 19.33).

The surface S is part of the level surface $F(x, y, z) = x^2 + y^2 - z = 0$ and the region R below it is the disk $x^2 + y^2 \leq 1$. To get a unit vector normal to the plane of R we can take $\mathbf{p} = \mathbf{k}$.

At any point (x, y, z) on the surface,

$$F(x, y, z) = x^2 + y^2 - z,$$

$$\nabla F = 2x\mathbf{i} + 2y\mathbf{j} - \mathbf{k},$$

$$|\nabla F| = \sqrt{(2x)^2 + (2y)^2 + (-1)^2} = \sqrt{4x^2 + 4y^2 + 1},$$

and

$$|\nabla F \cdot \mathbf{p}| = |\nabla F \cdot \mathbf{k}| = |-1| = 1.$$

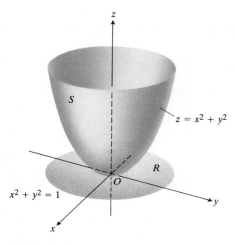

19.33 The area of the parabolic surface above is calculated in Example 1.

The surface area is therefore

$$S = \iint\limits_R \frac{|\nabla F|}{|\nabla F \cdot \mathbf{p}|}\, dA = \iint\limits_{x^2 + y^2 \leq 1} \sqrt{4x^2 + 4y^2 + 1}\, dx\, dy$$

$$= \int_0^{2\pi} \int_0^1 \sqrt{4r^2 + 1}\; r\, dr\, d\theta \qquad \text{(polar coordinates)}$$

$$= \int_0^{2\pi} \left[\frac{1}{12}(4r^2 + 1)^{3/2}\right]_0^1 d\theta$$

$$= \int_0^{2\pi} \frac{1}{12}(5^{3/2} - 1)\, d\theta = \frac{\pi}{6}(5\sqrt{5} - 1).\qquad\blacksquare$$

EXAMPLE 2 Find the area of the cap cut from the hemisphere $x^2 + y^2 + z^2 = 2$, $z \geq 0$, by the cylinder $x^2 + y^2 = 1$ (Fig. 19.34).

Solution The cap is part of the surface $F(x, y, z) = x^2 + y^2 + z^2 = 2$. It projects onto the disk R: $x^2 + y^2 \leq 1$ in the xy-plane. The vector $\mathbf{p} = \mathbf{k}$ is normal to the plane of R.

At any point on the surface,

$$F(x, y, z) = x^2 + y^2 + z^2,$$

$$\nabla F = 2x\mathbf{i} + 2y\mathbf{j} + 2z\mathbf{k},$$

$$|\nabla F| = 2\sqrt{x^2 + y^2 + z^2} = 2\sqrt{2},$$

and

$$|\nabla F \cdot \mathbf{p}| = |\nabla F \cdot \mathbf{k}| = |2z| = 2z.$$

Therefore,

$$\text{Surface area} = \iint\limits_R \frac{|\nabla F|}{|\nabla F \cdot \mathbf{p}|}\, dA = \iint\limits_R \frac{2\sqrt{2}}{2z}\, dA = \sqrt{2} \iint\limits_R \frac{dA}{z}. \qquad (6)$$

What do we do about the z?

Since z is the z-coordinate of a point on the sphere, we may express z in terms of x and y as

$$z = \sqrt{2 - x^2 - y^2}.$$

We continue the work of Eq. (6) with this substitution:

$$\text{Surface area} = \sqrt{2} \iint\limits_R \frac{dA}{z} = \sqrt{2} \iint\limits_{x^2 + y^2 \leq 1} \frac{dA}{\sqrt{2 - x^2 - y^2}}$$

$$= \sqrt{2} \int_0^{2\pi} \int_0^1 \frac{r\, dr\, d\theta}{\sqrt{2 - r^2}} \qquad \text{(polar coordinates)}$$

$$= \sqrt{2} \int_0^{2\pi} \left[-(2 - r^2)^{1/2}\right]_{r=0}^{r=1} d\theta$$

$$= \sqrt{2} \int_0^{2\pi} (\sqrt{2} - 1)\, d\theta$$

$$= 2\pi(2 - \sqrt{2}). \qquad \blacksquare$$

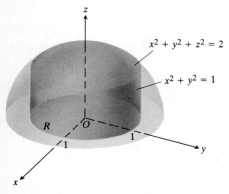

19.34 The cap cut from the hemisphere by the cylinder projects vertically onto the disk R: $x^2 + y^2 \leq 1$.

Surface Integrals

We now show how to integrate a function over a surface, using the ideas we just examined for calculating surface area.

Suppose, for example, that we have an electrical charge distributed over a surface $F(x, y, z) = c$ like the one shown in Fig. 19.31 and that the function $g(x, y, z)$ gives the charge per unit area (charge density) at each point on S. Then we may calculate the total charge on S as an integral in the following way.

We subdivide the shadow region R on the ground plane beneath the surface into small rectangles of the kind we would use if we were defining the surface area of S. Then directly above each ΔA_k lies a patch of surface $\Delta \sigma_k$ that we approximate with a parallelogram-shaped portion of tangent plane, ΔP_k.

Up to this point the construction proceeds as in the definition of surface area, but now we take one additional step: We evaluate g at (x_k, y_k, z_k) and approximate the total charge on the surface patch $\Delta \sigma_k$ by the product

$$g(x_k, y_k, z_k) \, \Delta P_k.$$

The rationale is that when the subdivision of R is sufficiently fine, the value of g throughout $\Delta \sigma_k$ is nearly constant and ΔP_k is nearly the same as $\Delta \sigma_k$. The total charge over S is then approximated by the sum

$$
\begin{aligned}
\text{Total charge} &\approx \sum g(x_k, y_k, z_k) \, \Delta P_k \\
&\approx \sum g(x_k, y_k, z_k) \frac{\Delta A_k}{|\cos \gamma|}.
\end{aligned}
\tag{7}
$$

If F, the function defining the surface S, and its first partial derivatives are continuous, and if g is continuous over S, then the sums on the right-hand side of Eq. (7) approach the limit

$$
\iint\limits_{R} g(x, y, z) \frac{dA}{|\cos \gamma|} = \iint\limits_{R} g(x, y, z) \frac{|\nabla F|}{|\nabla F \cdot \mathbf{p}|} \, dA
\tag{8}
$$

as the rectangular subdivision of R is refined in the usual way. This limit is called the **integral of g over the surface S** and is calculated as a double integral over R. The value of the integral is the total charge on the surface S.

As you might expect, the formula in Eq. (8) defines the integral of *any* function g over the surface S as long as the integral exists. The integral's value takes on different meanings in different applications. If g is the constant function whose value is 1, the integral gives the surface area. If $g(x, y, z)$ is the mass density of a thin shell of material modeled by a surface S, the integral gives the mass of the shell.

We often abbreviate the integral in Eq. (8) as

$$
\iint\limits_{R} g(x, y, z) \frac{|\nabla F|}{|\nabla F \cdot \mathbf{p}|} \, dA = \iint\limits_{S} g \, d\sigma,
\tag{9}
$$

where $d\sigma$ is the **surface area differential**

$$
d\sigma = \frac{|\nabla F|}{|\nabla F \cdot \mathbf{p}|} \, dA = \frac{dA}{|\cos \gamma|}.
\tag{10}
$$

Surface integrals have all the usual algebraic properties of double integrals, the integral of the sum of two functions being the sum of their integrals and so

19.35 To integrate a function over the surface of a cube, we integrate over each face and add the results.

on. The domain additivity property takes the form

$$\iint\limits_{S} g \, d\sigma = \iint\limits_{S_1} g \, d\sigma + \iint\limits_{S_2} g \, d\sigma + \cdots + \iint\limits_{S_n} g \, d\sigma.$$

The idea is that if S is partitioned by smooth curves and line segments into a finite number of nonoverlapping smooth patches, then the integral of a function g over S is the sum of the integrals over the patches. Thus, the integral of a function over the surface of a cube is the sum of the integrals over the faces of the cube. We integrate over a "turtle shell" of welded plates by integrating one plate at a time and adding the results.

EXAMPLE 3   Integrate $g(x, y, z) = xyz$ over the cube cut from the first octant by the planes $x = 1$, $y = 1$, and $z = 1$.

Solution   The cube is shown in Fig. 19.35. We integrate the product xyz over each face and add the results. Since $xyz = 0$ on the faces that lie in the coordinate planes, the integral of xyz over each of these faces is 0 and the integral over the cube reduces to

$$\iint\limits_{cube} xyz \, d\sigma = \iint\limits_{face\ A} xyz \, d\sigma + \iint\limits_{face\ B} xyz \, d\sigma + \iint\limits_{face\ C} xyz \, d\sigma.$$

Face A is the surface $F(x, y, z) = z = 1$ over the square region R_{xy}: $0 \le x \le 1$, $0 \le y \le 1$, in the xy-plane. For this surface and region,

$$\mathbf{p} = \mathbf{k}, \qquad \boldsymbol{\nabla}F = \mathbf{k}, \qquad |\boldsymbol{\nabla}F| = 1, \qquad |\boldsymbol{\nabla}F \cdot \mathbf{p}| = 1,$$

$$d\sigma = \frac{|\boldsymbol{\nabla}F|}{|\boldsymbol{\nabla}F \cdot \mathbf{p}|} \, dx \, dy, \qquad xyz = xy(1) = xy,$$

and

$$\iint\limits_{face\ A} xyz \, d\sigma = \iint\limits_{R_{xy}} xy \, dx \, dy = \int_0^1 \int_0^1 xy \, dx \, dy = \int_0^1 \frac{y}{2} \, dy = \frac{1}{4}.$$

Symmetry tells us that the integrals of xyz over B and C are also 1/4. Hence,

$$\iint\limits_{cube} xyz \, d\sigma = \frac{1}{4} + \frac{1}{4} + \frac{1}{4} = \frac{3}{4}. \qquad \blacksquare$$

The Surface Integral for Flux

If we can choose a unit normal vector $\mathbf{n}$ on a surface S in such a way that $\mathbf{n}$ varies continuously as its initial point moves about the surface, we call the surface **orientable** or **two-sided.** Spheres and other closed surfaces in space (surfaces that enclose solids) are orientable, and by convention we choose $\mathbf{n}$ on a closed surface to point outward. Once $\mathbf{n}$ is chosen we call the surface an **oriented surface.** The direction of $\mathbf{n}$ at any point is called the **positive direction.** Any patch or subportion of an orientable surface is also orientable. The Möbius band shown in Fig. 19.50 in Article 19.6, which is one-sided, is nonorientable.

If **F** is a continuous vector field defined over a surface S, we call the integral over S of $\mathbf{F} \cdot \mathbf{n}$, the scalar component of **F** in the direction of **n,** the flux of **F** across S in the positive direction.

DEFINITION

The **flux** of **F** across S in the direction of **n** is $\displaystyle\iint_S \mathbf{F} \cdot \mathbf{n} \, d\sigma.$ (11)

This definition is a direct generalization of the definition introduced in Article 19.2 for the flux of a two-dimensional vector field **F** across a plane curve C. In the plane, the flux is

$$\int_C \mathbf{F} \cdot \mathbf{n} \, ds,$$

which is the integral of the scalar component of **F** normal to the curve. In space, the flux of a three-dimensional field **F** across a surface is the integral of the scalar component of **F** normal to the surface.

If **F** is the velocity field of a three-dimensional fluid flow, then the flux is the net rate at which the fluid is crossing S in the positive direction. We shall discuss three-dimensional fluid flows in more detail in Article 19.6.

If S is part of a level surface $G(x, y, z) = c$, then **n** may be taken to be one of the two vectors

$$\mathbf{n} = +\frac{\nabla G}{|\nabla G|} \qquad \text{or} \qquad \mathbf{n} = -\frac{\nabla G}{|\nabla G|},$$

depending on which vector gives the preferred direction.

EXAMPLE 4 Find the flux of $\mathbf{F} = yz\mathbf{j} + z^2\mathbf{k}$ outward through the surface S cut from the cylinder $y^2 + z^2 = a^2$, $z \geq 0$, by the planes $x = 0$ and $x = a$ (Fig. 19.36).

Solution The outward unit normal to S may be calculated from the gradient of $G(x, y, z) = y^2 + z^2$ to be

$$\mathbf{n} = +\frac{\nabla G}{|\nabla G|} = \frac{2y\mathbf{j} + 2z\mathbf{k}}{\sqrt{4y^2 + 4z^2}} = \frac{2y\mathbf{j} + 2z\mathbf{k}}{2a} = \frac{y}{a}\mathbf{j} + \frac{z}{a}\mathbf{k}.$$

We also have

$$d\sigma = \frac{|\nabla G|}{|\nabla G \cdot \mathbf{k}|} \, dA = \frac{2a}{|2z|} \, dA = \frac{a}{z} \, dA,$$

where we drop the absolute value signs because $z \geq 0$ on S.

On the surface S, the value $\mathbf{F} \cdot \mathbf{n}$ is given by the formula

$$\mathbf{F} \cdot \mathbf{n} = (yz\mathbf{j} + z^2\mathbf{k}) \cdot \left(\frac{y}{a}\mathbf{j} + \frac{z}{a}\mathbf{k}\right)$$

$$= \frac{y^2z + z^3}{a} = \frac{z}{a}(y^2 + z^2)$$

$$= \frac{z}{a}(a^2) \qquad (y^2 + z^2 = a^2 \text{ on } S)$$

$$= az.$$

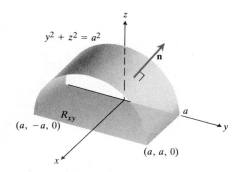

19.36 Example 4 calculates the flux of a vector field outward through this surface. The area of the shadow region R_{xy} is $2a^2$.

Therefore, the flux of **F** outward through S is

$$\iint_S \mathbf{F} \cdot \mathbf{n} \, d\sigma = \iint_S (az)\frac{a}{z} \, dA = \iint_{R_{xy}} a^2 \, dx \, dy = a^2 \cdot \text{area}(R_{xy}) = 2a^4. \quad \blacksquare$$

Masses and Moments of Thin Shells

In engineering and physics, thin shells of material like bowls, metal drums, and domes are modeled with surfaces. Their masses and moments are calculated with surface integrals, using the formulas listed in Table 19.2.

EXAMPLE 5 Find the center of mass of a thin hemispherical shell of radius a and constant density δ.

Solution We model the shell with the hemisphere

$$F(x, y, z) = x^2 + y^2 + z^2 = a^2, \qquad z \geq 0.$$

See Fig. 19.37. The center of mass then lies on the z-axis. Hence, $\bar{x} = \bar{y} = 0$. It remains only to find $\bar{z}$ from the formula $\bar{z} = M_{xy}/M$.

The mass of the shell is

$$M = \iint_S \delta \, d\sigma = \delta \iint_S d\sigma = 2\pi a^2 \, \delta,$$

which is the surface area times the density.

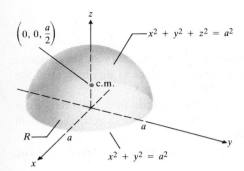

19.37 Example 5 calculates the coordinates of the center of mass of this thin hemispherical shell.

TABLE 19.2
Mass and moment formulas for very thin shells

Mass: $M = \iint_S \delta(x, y, z) \, d\sigma$ ($\delta(x, y, z)$ = density at (x, y, z))

First moments about the coordinate planes:

$$M_{yz} = \iint_S x \, \delta \, d\sigma, \qquad M_{xz} = \iint_S y \, \delta \, d\sigma, \qquad M_{xy} = \iint_S z \, \delta \, d\sigma$$

Coordinates of center of mass:

$$\bar{x} = M_{yz}/M, \qquad \bar{y} = M_{xz}/M, \qquad \bar{z} = M_{xy}/M$$

Moments of inertia:

$$I_x = \iint_S (y^2 + z^2) \, \delta \, d\sigma, \qquad I_y = \iint_S (x^2 + z^2) \, \delta \, d\sigma,$$

$$I_z = \iint_S (x^2 + y^2) \, \delta \, d\sigma, \qquad I_L = \iint_S r^2 \, \delta \, d\sigma$$

$r(x, y, z)$ = distance from point (x, y, z) to line L

Radius of gyration about a line L: $R_L = \sqrt{I_L/M}$

To evaluate the integral for M_{xy}, we take

$$|\nabla F| = |2x\mathbf{i} + 2y\mathbf{j} + 2z\mathbf{k}| = 2\sqrt{x^2 + y^2 + z^2} = 2a,$$

$$|\nabla F \cdot \mathbf{p}| = |\nabla F \cdot \mathbf{k}| = |2z| = 2z,$$

$$d\sigma = \frac{|\nabla F|}{|\nabla F \cdot \mathbf{p}|} dA = \frac{a}{z} dA.$$

This gives

$$M_{xy} = \iint_S z\,\delta\,d\sigma = \delta \iint_R z\,\frac{a}{z}\,dA = \delta\,a \iint_R dA = \delta\,a(\pi a^2) = \pi a^3\,\delta.$$

The z-coordinate of the center of mass is therefore

$$\bar{z} = \frac{M_{xy}}{M} = \frac{\pi a^3\,\delta}{2\pi a^2\,\delta} = \frac{a}{2}.$$

The center of mass is the point $(0, 0, a/2)$. ∎

PROBLEMS

Surface Area

1. Find the area of the surface cut from the paraboloid $x^2 + y^2 - z = 0$ by the plane $z = 2$.

2. Find the area of the band cut from the paraboloid $x^2 + y^2 - z = 0$ by the planes $z = 2$ and $z = 6$.

3. Find the area of the region cut from the plane $x + 2y + 2z = 5$ by the cylinder whose walls are $x = y^2$ and $x = 2 - y^2$.

4. Find the area of the portion of the surface $x^2 - 2z = 0$ that lies above the triangle bounded by the lines $x = \sqrt{3}$, $y = 0$, and $y = x$ in the xy-plane.

5. Find the area of the surface $x^2 - 2y - 2z = 0$ that lies above the triangle bounded by the lines $x = 2$, $y = 0$, and $y = 3x$ in the xy-plane.

6. Find the area of the surface $x^2 - 2\ln x + \sqrt{15}y - z = 0$ above the square $R: 0 \le y \le 1$, $1 \le x \le 2$ in the xy-plane.

7. Find the area of the cap cut from the top of the sphere $x^2 + y^2 + z^2 = 1$ by the plane $z = \sqrt{2}/2$.

8. Find the area of the cap cut from the sphere $x^2 + y^2 + z^2 = 2$ by the cone $z = \sqrt{x^2 + y^2}$.

9. Find the area of the ellipse cut from the plane $z = cx$ by the cylinder $x^2 + y^2 = 1$.

10. Find the area of the portion of the cylinder $x^2 + z^2 = 1$ that lies between the planes $x = \pm 1/2$ and $y = \pm 1/2$.

11. Find the area of the surface cut from the paraboloid $x^2 + y + z^2 = 1$ by the plane $y = 0$. (*Hint:* Project the surface onto the xz-plane.)

12. Whenever we replace an old definition with a new one it is a good idea to try out the new one on familiar objects to see that it gives the answers we expect. For instance, is the surface area of a circular cylinder of base radius r and height h still $2\pi rh$? Find out by using Eq. (5) with $\mathbf{p} = \mathbf{k}$ to calculate the area of the surface cut from the cylinder $y^2 + z^2 = r^2$ by the planes $x = 0$ and $x = h > 0$. (Find the area above the xy-plane, and double.)

Surface Integrals

13. Integrate $g(x, y, z) = x + y + z$ over the surface of the cube cut from the first octant by the planes $x = 1$, $y = 1$, and $z = 1$.

14. Integrate $g(x, y, z) = y + z$ over the surface of the wedge shown in Fig. 19.38.

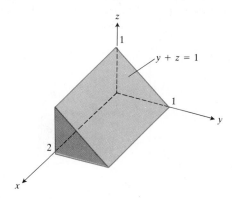

19.38 The wedge in Problem 14.

15. Integrate $g(x, y, z) = xyz$ over the surface of the rectangular solid cut from the first octant by the planes $x = a$, $y = b$, and $z = c$.

16. Integrate $g(x, y, z) = xyz$ over the surface of the rectangular solid bounded by the planes $x = \pm a$, $y = \pm b$, and $z = \pm c$. (*Hint:* You will see what's going on if you integrate over a pair of opposite faces.)

17. Integrate $g(x, y, z) = x + y + z$ over the portion of the plane $2x + 2y + z = 2$ that lies in the first octant.

18. Integrate $g(x, y, z) = yz$ over the portion of the plane $x + y + z = 1$ that lies in the first octant.

19. Integrate $g(x, y, z) = x^2 + y^2 + z^2$ over the surface cut from the cylinder $y^2 + z^2 = a^2$ by the planes $x = 0$ and $x = h$. Take $h > 0$.

20. Integrate $g(x, y, z) = x^2 + y^2$ over the hemisphere $z = \sqrt{a^2 - x^2 - y^2}$.

Flux across a Surface

In Problems 21–26, find the flux of the vector field $\mathbf{F}$ across the portion of the sphere $x^2 + y^2 + z^2 = a^2$ in the first octant in the direction away from the origin.

21. $\mathbf{F}(x, y, z) = z\mathbf{k}$

22. $\mathbf{F}(x, y, z) = -y\mathbf{i} + x\mathbf{j}$

23. $\mathbf{F}(x, y, z) = y\mathbf{i} - x\mathbf{j} + \mathbf{k}$

24. $\mathbf{F}(x, y, z) = zx\mathbf{i} + zy\mathbf{j} + z^2\mathbf{k}$

25. $\mathbf{F}(x, y, z) = x\mathbf{i} + y\mathbf{j} + z\mathbf{k}$

26. $\mathbf{F}(x, y, z) = \dfrac{x\mathbf{i} + y\mathbf{j} + z\mathbf{k}}{\sqrt{x^2 + y^2 + z^2}}$

27. Find the flux of the field $\mathbf{F}(x, y, z) = z^2\mathbf{i} + x\mathbf{j} - 3z\mathbf{k}$ upward through the surface cut from the parabolic cylinder $z = 4 - y^2$ by the planes $x = 0$, $x = 1$, and $z = 0$.

28. Find the flux of the field $\mathbf{F}(x, y, z) = 4x\mathbf{i} + 4y\mathbf{j} + 2\mathbf{k}$ outward through the surface cut from the bottom of the paraboloid $z = x^2 + y^2$ by the plane $z = 1$.

29. Let S be the portion of the cylinder $y = e^x$ in the first octant that projects parallel to the x-axis onto the rectangle R_{yz}: $1 \le y \le 2$, $0 \le z \le 1$ in the yz-plane (Fig. 19.39). Let $\mathbf{n}$ be

the unit vector normal to S that points away from the yz-plane. Find the flux of the field $\mathbf{F}(x, y, z) = -2\mathbf{i} + 2y\mathbf{j} + z\mathbf{k}$ across S in the direction of $\mathbf{n}$.

30. Let S be the portion of the cylinder $y = \ln x$ in the first octant whose projection parallel to the y-axis onto the xz-plane is the rectangle R_{xz}: $1 \le x \le e$, $0 \le z \le 1$. Let $\mathbf{n}$ be the unit vector normal to S that points away from the xz-plane. Find the flux of $\mathbf{F} = 2y\mathbf{j} + z\mathbf{k}$ through S in the direction of $\mathbf{n}$.

31. The sphere $x^2 + y^2 + z^2 = 25$ is cut by the plane $z = 3$, the smaller portion cut off forming a solid that is bounded by a closed surface S made up of a spherical cap S_1 and a flat disk S_2. Find

$$\iint_S \mathbf{F} \cdot \mathbf{n}\, d\sigma = \iint_{S_1} \mathbf{F} \cdot \mathbf{n}\, d\sigma + \iint_{S_2} \mathbf{F} \cdot \mathbf{n}\, d\sigma$$

if $\mathbf{F} = xz\mathbf{i} + yz\mathbf{j} + \mathbf{k}$ and in each integral on the right $\mathbf{n}$ is taken to be the outward-pointing normal to the surface.

32. Find the outward flux of the field $\mathbf{F}(x, y, z) = xy\mathbf{i} + yz\mathbf{j} + xz\mathbf{k}$ across the surface of the cube cut from the first octant by the planes $x = 1$, $y = 1$, and $z = 1$.

Masses and Moments

33. Find the center of mass of the portion of the sphere $x^2 + y^2 + z^2 = 1$ that lies in the first octant if the sphere's density is constant.

34. Find the center of mass of the surface cut from the upper half of the cylinder $y^2 + z^2 = a^2$ by the planes $x = 0$ and $x = a$ if the density is constant. (This is the surface in Example 4.)

35. Find the center of mass, moment of inertia, and radius of gyration about the z-axis of a thin shell of constant density cut from the cone $z = \sqrt{x^2 + y^2}$ by the plane $z = 1$ and $z = 2$.

36. Find the moment of inertia about the z-axis of a thin shell of constant density 1 cut from the cone $z = \sqrt{x^2 + y^2}$ by the cylinder $x^2 + y^2 = 2x$. (*Hint:* The cylindrical coordinate equation for the cylinder is $r = 2\cos\theta$.)

37. Find the mass of a thin spherical shell of radius a if
   a) the density at each point is the distance from the point to a fixed diameter of the sphere,
   b) the density at each point is the square of the distance from the point to a fixed diameter of the sphere.

38. Find the moment of inertia about the z-axis of the surface of the cube cut from the first octant by the planes $x = 1$, $y = 1$, and $z = 1$ if the density is constant.

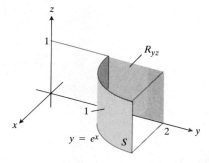

19.39 The surface and region in Problem 29.

19.5
The Divergence Theorem

As we learned in Article 19.3, the divergence form of Green's theorem in the plane states that the net outward flux of a vector field across a simple closed curve in the plane can be calculated by integrating the divergence of the field over the region enclosed by the curve. The corresponding theorem in three dimensions, called the divergence theorem, states that the net outward flux of a vector field across a closed surface in space can be calculated by integrating the divergence of the field over the region enclosed by the surface. In this article, we prove the divergence theorem and show how it simplifies the calculation of flux. We also derive the continuity equation of hydrodynamics (an equation about the conservation of mass in fluid flows with variable densities) and Gauss's law for flux in an electric field. The three-dimensional version of the circulation form of Green's theorem will be treated in the next article.

■ ■ ■

THE DIVERGENCE THEOREM

Mikhail Vassilievich Ostrogradsky (1801–1862) was the first mathematician to publish a proof of the divergence theorem. Having been denied a university diploma from Kharkhov University in 1820 for his atheism and nonattendance of lectures on philosophy and theology, Ostrogradsky left Russia for Paris in 1822. There he kept company with Laplace, Legendre, Fourier, Poisson, and Cauchy. While working on the theory of heat in the mid-1820s, he formulated the divergence theorem as a tool for converting volume integrals to surface integrals.

Carl Friedrich Gauss (1777–1855) had already proved the theorem in his notebooks while working on the theory of gravitation (although his notebooks were not published until many years later), and the theorem is sometimes known as Gauss's theorem. The list of Gauss's accomplishments in science and mathematics is truly astonishing, ranging from the invention of the electric telegraph (with Wilhelm Weber in 1833) to the development of a wonderfully accurate theory of planetary orbits and to work in non-Euclidean geometry that later became fundamental to Einstein's general theory of relativity.

Divergence and del Notation

The **divergence** of a vector field

$$\mathbf{F}(x, y, z) = M(x, y, z)\mathbf{i} + N(x, y, z)\mathbf{j} + P(x, y, z)\mathbf{k}$$

is the scalar function

$$\text{div } \mathbf{F} = \frac{\partial M}{\partial x} + \frac{\partial N}{\partial y} + \frac{\partial P}{\partial z}. \tag{1}$$

The symbol "div **F**" is read "divergence of **F**" or "div **F**." It has the same physical interpretation as the divergence of a two-dimensional vector field. When **F** is the velocity field of a fluid flow, the value of div **F** at a point (x, y, z) is the rate at which fluid is being piped in or drained away at (x, y, z). The divergence is the flux per unit volume or flux density at the point. We shall say more about this at the end of the article.

The divergence of a three-dimensional vector field **F** is usually expressed in terms of the symbolic operation ∇ ("del"), defined by the equation

$$\nabla = \mathbf{i}\frac{\partial}{\partial x} + \mathbf{j}\frac{\partial}{\partial y} + \mathbf{k}\frac{\partial}{\partial z}. \tag{2}$$

This operator can be applied to any differentiable vector field $\mathbf{F} = M\mathbf{i} + N\mathbf{j} + P\mathbf{k}$ to give the divergence of **F** as a symbolic dot product:

$$\nabla \cdot \mathbf{F} = \frac{\partial M}{\partial x} + \frac{\partial N}{\partial y} + \frac{\partial P}{\partial z} = \text{div } \mathbf{F}. \tag{3}$$

The notation $\nabla \cdot \mathbf{F}$ is read "del dot **F**."

When applied to a differentiable scalar function $f(x, y, z)$, the del operator gives the familiar gradient of f:

$$\nabla f = \frac{\partial f}{\partial x}\mathbf{i} + \frac{\partial f}{\partial y}\mathbf{j} + \frac{\partial f}{\partial z}\mathbf{k}. \tag{4}$$

This may now be read as "del f" as well as "grad f."

Since

$$\frac{\partial^2 f}{\partial x^2} + \frac{\partial^2 f}{\partial y^2} + \frac{\partial^2 f}{\partial z^2} = \mathbf{\nabla} \cdot \mathbf{\nabla} f = \mathbf{\nabla}^2 f, \tag{5}$$

Laplace's equation,

$$\frac{\partial^2 f}{\partial x^2} + \frac{\partial^2 f}{\partial y^2} + \frac{\partial^2 f}{\partial z^2} = 0,$$

from Article 16.8, can now be written as

$$\text{div grad } f = 0 \qquad \text{or} \qquad \mathbf{\nabla}^2 f = 0. \tag{6}$$

If we think of a two-dimensional vector field

$$\mathbf{F}\,(x,\,y) = M(x,\,y)\mathbf{i} + N(x,\,y)\mathbf{j}$$

as a three-dimensional field whose **k**-component is zero, then

$$\mathbf{\nabla} \cdot \mathbf{F} = \frac{\partial M}{\partial x} + \frac{\partial N}{\partial y} = \text{div } \mathbf{F},$$

and the normal form of Green's theorem for two-dimensional fields (about flux and divergence; see Eq. 27 in Article 19.3) becomes

$$\int_C \mathbf{F} \cdot \mathbf{n}\, ds = \iint_R \mathbf{\nabla} \cdot \mathbf{F}\, dA. \tag{7}$$

The tangential form of Green's theorem for two-dimensional fields can also be expressed in vector form in del notation, as we shall see in Article 19.6.

As these equations suggest, del notation provides an easy way to express and remember some of the most important relationships in mathematics, engineering, and physics. We shall see more of this as the present chapter continues.

The Divergence Theorem (Special Regions)

In symbols, the divergence theorem says

$$\iint_S \mathbf{F} \cdot \mathbf{n}\, d\sigma = \iiint_D \text{div } \mathbf{F}\, dV \tag{8}$$

or, in del notation,

$$\iint_S \mathbf{F} \cdot \mathbf{n}\, d\sigma = \iiint_D \mathbf{\nabla} \cdot \mathbf{F}\, dV. \tag{9}$$

As you can see, Eq. (9) has much the same form as Eq. (7), which gives the flux-divergence form of Green's theorem for two-dimensional fields. The only difference is that we are now working in three dimensions instead of two.

To prove the divergence theorem, we assume that the components of **F** have continuous partial derivatives and that **n** is the outer unit normal of the surface S that encloses the region D. We also assume that D is a convex region

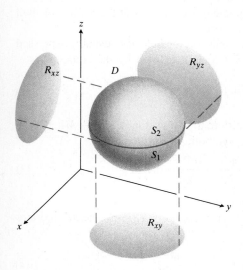

19.40 We first prove the divergence theorem for the kind of three-dimensional region shown here. We then extend the theorem to other regions.

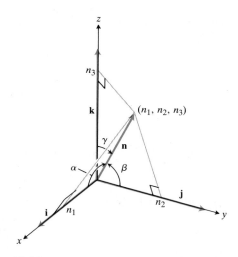

19.41 The scalar components of a unit normal vector **n** are the cosines of the angles α, β, and γ that it makes with **i**, **j**, and **k**.

with no holes or bubbles, such as a sphere, cube, or ellipsoid, and that S is a piecewise smooth surface. In addition, we assume that the projection of D into the xy-plane is a simply connected region R_{xy} and that any line perpendicular to the xy-plane at an interior point of R_{xy} intersects the surface S in exactly two points, producing surfaces S_1 and S_2:

$$S_1: \qquad z = f_1(x, y), \quad (x, y) \text{ in } R_{xy},$$
$$S_2: \qquad z = f_2(x, y), \quad (x, y) \text{ in } R_{xy},$$

with $z_1 \leq z_2$. We make similar assumptions about the projection of D onto the other coordinate planes. See Fig. 19.40.

The components of the unit normal vector $\mathbf{n} = n_1\mathbf{i} + n_2\mathbf{j} + n_3\mathbf{k}$ are the cosines of the angles α, β, and γ that **n** makes with **i**, **j**, and **k** (Fig. 19.41). This is true because all the vectors involved are unit vectors. We have

$$n_1 = \mathbf{n} \cdot \mathbf{i} = |\mathbf{n}| \, |\mathbf{i}| \cos \alpha = \cos \alpha,$$
$$n_2 = \mathbf{n} \cdot \mathbf{j} = |\mathbf{n}| \, |\mathbf{j}| \cos \beta = \cos \beta, \tag{10}$$
$$n_3 = \mathbf{n} \cdot \mathbf{k} = |\mathbf{n}| \, |\mathbf{k}| \cos \gamma = \cos \gamma.$$

Thus,

$$\mathbf{n} = (\cos \alpha)\, \mathbf{i} + (\cos \beta)\, \mathbf{j} + (\cos \gamma)\, \mathbf{k}$$

and

$$\mathbf{F} \cdot \mathbf{n} = M \cos \alpha + N \cos \beta + P \cos \gamma.$$

In component form, the divergence theorem states that

$$\iint_S (M \cos \alpha + N \cos \beta + P \cos \gamma)\, d\sigma = \iiint_D \left(\frac{\partial M}{\partial x} + \frac{\partial N}{\partial y} + \frac{\partial P}{\partial z} \right) dx \, dy \, dz. \tag{11}$$

We prove the theorem by proving three separate equalities that add up to Eq. (11):

$$\iint M \cos \alpha \, d\sigma = \iiint \frac{\partial M}{\partial x} \, dx \, dy \, dz, \tag{12}$$

$$\iint N \cos \beta \, d\sigma = \iiint \frac{\partial N}{\partial y} \, dx \, dy \, dz, \tag{13}$$

$$\iint P \cos \gamma \, d\sigma = \iiint \frac{\partial P}{\partial z} \, dx \, dy \, dz. \tag{14}$$

We prove Eq. (14) by converting the surface integral on the left to a double integral over the projection R_{xy} of D on the xy-plane (Fig. 19.42). The surface S consists of an upper part S_2 whose equation is $z = f_2(x, y)$ and a lower part S_1 whose equation is $z = f_1(x, y)$. On S_2, the outer normal **n** has a positive **k**-component and

$$\cos \gamma \, d\sigma = dx \, dy \qquad \text{because} \qquad d\sigma = \frac{dA}{|\cos \gamma|} = \frac{dx \, dy}{\cos \gamma}. \tag{15}$$

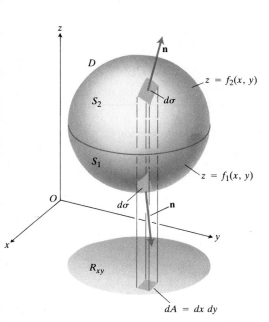

19.42 The three-dimensional region D enclosed by the surfaces S_1 and S_2 shown here projects vertically onto a two-dimensional region R_{xy} in the xy-plane.

See Fig. 19.43. On S_1, the outer normal $\mathbf{n}$ has a negative $\mathbf{k}$-component and

$$\cos\gamma\,d\sigma = -dx\,dy. \qquad (16)$$

Therefore,

$$\iint\limits_{S} P\cos\gamma\,d\sigma = \iint\limits_{S_2} P\cos\gamma\,d\sigma + \iint\limits_{S_1} P\cos\gamma\,d\sigma$$

$$= \iint\limits_{R_{xy}} P(x,\,y,\,z_2)\,dx\,dy - \iint\limits_{R_{xy}} P(x,\,y,\,z_1)\,dx\,dy$$

$$= \iint\limits_{R_{xy}} [P(x,\,y,\,z_2) - P(x,\,y,\,z_1)]\,dx\,dy$$

$$= \iint\limits_{R_{xy}} \left[\int_{z_1}^{z_2} \frac{\partial P}{\partial z}\,dz\right] dx\,dy$$

$$= \iiint\limits_{D} \frac{\partial P}{\partial z}\,dz\,dx\,dy. \qquad (17)$$

This proves Eq. (14).

The proofs for Eqs. (12) and (13) follow the same pattern; or just permute $x,\ y,\ z;\ M,\ N,\ P;\ \alpha,\ \beta,\ \gamma$, in order, and get those results from Eq. (14).

Here γ is acute, so $d\sigma = dx\,dy/\cos\gamma$

Here γ is obtuse, so $d\sigma = -dx\,dy/\cos\gamma$

19.43 An enlarged view of the area patches in Fig. 19.42. The relations $d\sigma = \pm dx\,dy/\cos\gamma$ are derived in Article 19.4.

THEOREM

The Divergence Theorem

$$\iint\limits_{S} \mathbf{F}\cdot\mathbf{n}\,d\sigma = \iiint\limits_{D} \nabla\cdot\mathbf{F}\,dV \qquad (18)$$

Flux integral   Divergence integral

EXAMPLE 1   Evaluate both sides of Eq. (18) for the vector field

$$\mathbf{F}(x,\,y,\,z) = x\mathbf{i} + y\mathbf{j} + z\mathbf{k}$$

over the sphere $x^2 + y^2 + z^2 = a^2$.

Solution   The outer unit normal to S, calculated from the gradient of $f(x,\,y,\,z) = x^2 + y^2 + z^2 - a^2$, is

$$\mathbf{n} = \frac{2(x\mathbf{i} + y\mathbf{j} + z\mathbf{k})}{\sqrt{4(x^2 + y^2 + z^2)}} = \frac{x\mathbf{i} + y\mathbf{j} + z\mathbf{k}}{a}.$$

Hence,

$$\mathbf{F}\cdot\mathbf{n}\,d\sigma = \frac{x^2 + y^2 + z^2}{a}\,d\sigma = \frac{a^2}{a}\,d\sigma = a\,d\sigma$$

because $x^2 + y^2 + z^2 = a^2$ on the surface. Therefore

$$\iint\limits_{S} \mathbf{F}\cdot\mathbf{n}\,d\sigma = \iint\limits_{S} a\,d\sigma = a\iint\limits_{S} d\sigma = a(4\pi a^2) = 4\pi a^3.$$

The divergence of $\mathbf{F}$ is

$$\nabla \cdot \mathbf{F} = \frac{\partial x}{\partial x} + \frac{\partial y}{\partial y} + \frac{\partial z}{\partial z} = 3,$$

so

$$\iiint_D \nabla \cdot \mathbf{F} \, dV = \iiint_D 3 \, dV = 3\left(\frac{4}{3}\pi a^3\right) = 4\pi a^3. \qquad \blacksquare$$

EXAMPLE 2   Calculate the flux of the field

$$\mathbf{F}(x, y, z) = xy\mathbf{i} + yz\mathbf{j} + xz\mathbf{k}$$

over the surface of the cube cut from the first octant by the planes $x = 1$, $y = 1$, and $z = 1$.

Solution   The Divergence Theorem says that instead of calculating the flux as a sum of six separate integrals, one for each face of the cube, we can calculate the flux by integrating the divergence

$$\nabla \cdot \mathbf{F} = \frac{\partial}{\partial x}(xy) + \frac{\partial}{\partial y}(yz) + \frac{\partial}{\partial z}(xz) = y + z + x$$

over the cube's interior:

$$\text{Flux} = \iint_{\substack{\text{cube} \\ \text{surface}}} \mathbf{F} \cdot \mathbf{n} \, d\sigma$$

$$= \iiint_{\substack{\text{cube} \\ \text{interior}}} \nabla \cdot \mathbf{F} \, dV \qquad \text{(the Divergence Theorem)}$$

$$= \int_0^1 \int_0^1 \int_0^1 (x + y + z) \, dx \, dy \, dz$$

$$= \frac{3}{2}. \qquad \text{(after a routine integration)} \qquad \blacksquare$$

The Divergence Theorem for Other Regions

The divergence theorem can be extended to regions that can be split up into a finite number of simple regions of the type just discussed and to regions that can be defined as limits of simpler regions in certain ways. For example, suppose D is the region between two concentric spheres, and $\mathbf{F}$ has continuously differentiable components throughout D and on the bounding surfaces. Split D by an equatorial plane and apply the Divergence Theorem to each half separately. The top half, D_1, is shown in Fig. 19.44. The surface that bounds D_1 consists of an outer hemisphere, a plane washer-shaped base, and an inner hemisphere. The Divergence Theorem says that

$$\iint_{S_1} \mathbf{F} \cdot \mathbf{n}_1 \, d\sigma_1 = \iiint_{D_1} \nabla \cdot \mathbf{F} \, dV_1. \qquad (19)$$

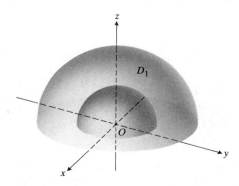

19.44   The upper half of the region between two spheres.

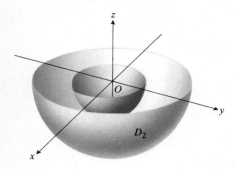

19.45 The lower half of the region between two spheres.

The unit normal $\mathbf{n}_1$ that points outward from D_1 points away from the origin along the outer surface, points down along the flat base, and points toward the origin along the inner surface. Next apply the Divergence Theorem to D_2, as shown in Fig. 19.45:

$$\iint_{S_2} \mathbf{F} \cdot \mathbf{n}_2 \, d\sigma_2 = \iiint_{D_2} \nabla \cdot \mathbf{F} \, dV_2. \tag{20}$$

As we follow $\mathbf{n}_2$ over S_2, pointing outward from D_2, we see that $\mathbf{n}_2$ points upward along the flat surface in the xy-plane, points away from the origin on the outer sphere, and points toward the origin on the inner sphere. When we add (19) and (20), the surface integrals over the flat base cancel because of the opposite signs of $\mathbf{n}_1$ and $\mathbf{n}_2$. We thus arrive at the result

$$\iint_{S} \mathbf{F} \cdot \mathbf{n} \, d\sigma = \iiint_{D} \nabla \cdot \mathbf{F} \, dV. \tag{21}$$

with D the region between the spheres, S the boundary of D consisting of two spheres, and $\mathbf{n}$ the unit normal to S directed outward from D.

EXAMPLE 3 Find the net outward flux of the field

$$\mathbf{F}(x, y, z) = \frac{x\mathbf{i} + y\mathbf{j} + z\mathbf{k}}{\rho^3}, \qquad \rho = \sqrt{x^2 + y^2 + z^2}$$

across the boundary of the region D: $a^2 \leq x^2 + y^2 + z^2 \leq b^2$.

Solution The flux can be calculated by integrating $\nabla \cdot \mathbf{F}$ over D. We have

$$\frac{\partial \rho}{\partial x} = \frac{1}{2}(x^2 + y^2 + z^2)^{-1/2}(2x) = \frac{x}{\rho}$$

and

$$\frac{\partial M}{\partial x} = \frac{\partial}{\partial x}(x\rho^{-3}) = \rho^{-3} - 3x\rho^{-4}\frac{\partial \rho}{\partial x} = \frac{1}{\rho^3} - \frac{3x^2}{\rho^5}.$$

Similarly,

$$\frac{\partial N}{\partial y} = \frac{1}{\rho^3} - \frac{3y^2}{\rho^5} \qquad \text{and} \qquad \frac{\partial P}{\partial z} = \frac{1}{\rho^3} - \frac{3z^2}{\rho^5}.$$

Hence,

$$\text{div } \mathbf{F} = \frac{3}{\rho^3} - \frac{3}{\rho^5}(x^2 + y^2 + z^2) = \frac{3}{\rho^3} - \frac{3\rho^2}{\rho^5} = 0$$

and

$$\iiint_{D} \nabla \cdot \mathbf{F} \, dV = 0.$$

So the integral of $\nabla \cdot \mathbf{F}$ over D is zero and the net outward flux across the boundary of D is zero. But there is more to learn from this example. The flux leaving D across the inner sphere S_a is the negative of the flux leaving D across the outer sphere S_b (because the sum of these fluxes is zero). This means that the flux of $\mathbf{F}$ across S_a in the direction away from the origin equals the flux of $\mathbf{F}$ across S_b in the direction away from the origin. Thus, the flux of $\mathbf{F}$ across a sphere centered at the origin is independent of the radius of the sphere. What is this flux?

To find it, we evaluate the flux integral directly. The outward unit normal on the sphere of radius a is

$$\mathbf{n} = \frac{x\mathbf{i} + y\mathbf{j} + z\mathbf{k}}{\sqrt{x^2 + y^2 + z^2}} = \frac{x\mathbf{i} + y\mathbf{j} + z\mathbf{k}}{a}.$$

Hence, on the sphere,

$$\mathbf{F} \cdot \mathbf{n} = \frac{x\mathbf{i} + y\mathbf{j} + z\mathbf{k}}{a^3} \cdot \frac{x\mathbf{i} + y\mathbf{j} + z\mathbf{k}}{a}$$

$$= \frac{x^2 + y^2 + z^2}{a^4} = \frac{a^2}{a^4} = \frac{1}{a^2}$$

and

$$\iint\limits_{S_a} \mathbf{F} \cdot \mathbf{n} \, d\sigma = \frac{1}{a^2} \iint d\sigma = \frac{1}{a^2}(4\pi a^2) = 4\pi.$$

The outward flux of $\mathbf{F}$ across any sphere centered at the origin is 4π. ◼

Gauss's Law

There is still more to be learned from Example 3. In electromagnetic theory, the electric field created by a point charge q located at the origin is the inverse square field

$$\mathbf{E}(x, y, z) = q \frac{\mathbf{r}}{|\mathbf{r}|^3} = q \frac{x\mathbf{i} + y\mathbf{j} + z\mathbf{k}}{\rho^3},$$

where $\mathbf{r}$ is the position vector of the point (x, y, z) and $\rho = |\mathbf{r}| = \sqrt{x^2 + y^2 + z^2}$. In the notation of Example 3,

$$\mathbf{E} = q\mathbf{F}.$$

The calculations in Example 3 show that the outward flux of $\mathbf{E}$ across any sphere centered at the origin is $4\pi q$. But this result is not confined to spheres. The outward flux of $\mathbf{E}$ across any surface S that encloses the origin (and to which the divergence theorem applies) is also $4\pi q$. To see why, we have only to imagine a large sphere S_a centered at the origin and enclosing the surface S. Since

$$\nabla \cdot \mathbf{E} = \nabla \cdot q\mathbf{F} = q\nabla \cdot \mathbf{F} = 0,$$

the integral of $\nabla \cdot \mathbf{E}$ over the region D between S and S_a is zero. Hence, by the Divergence Theorem,

$$\iint\limits_{\substack{\text{Boundary} \\ \text{of } D}} \mathbf{E} \cdot \mathbf{n} \, d\sigma = 0,$$

and the flux of $\mathbf{E}$ across S in the direction away from the origin must be the same as the flux of $\mathbf{E}$ across S_a in the direction away from the origin, which is

$4\pi q$. This statement is called *Gauss's law,* after Carl Friedrich Gauss (1777–1855), the great German mathematician, physicist, and astronomer who, among many other things, invented the electric telegraph. Gauss's law applies to much more general charge distributions than the one we have assumed here, as you will see if you open nearly any college physics text.

Gauss's Law

$$\iint\limits_{S} \mathbf{E} \cdot \mathbf{n} \, d\sigma = 4\pi q. \tag{22}$$

The Continuity Equation of Hydrodynamics

If $\mathbf{v}(x, y, z)$ is a differentiable vector representing the velocity of a fluid flow through a region D in space, $\delta = \delta(x, y, z, t)$ is the density of the fluid at the point (x, y, z) at time t, and $\mathbf{F} = \delta\mathbf{v}$, then the **continuity equation** is the statement that

$$\nabla \cdot \mathbf{F} + \frac{\partial \delta}{\partial t} = 0. \tag{23}$$

The continuity equation evolves naturally from the Divergence Theorem

$$\iint\limits_{S} \mathbf{F} \cdot \mathbf{n} \, d\sigma = \iiint\limits_{D} \nabla \cdot \mathbf{F} \, dV$$

if the functions involved are continuous, as we shall now show.

First of all, the integral

$$\iint\limits_{S} \mathbf{F} \cdot \mathbf{n} \, d\sigma$$

is the rate at which mass leaves D across S (leaves because $\mathbf{n}$ is the outer normal). To see why, consider a patch of area $\Delta\sigma$ on the surface (Fig. 19.46). In a short time interval Δt, the volume ΔV of fluid that flows across the patch is approximately equal to the volume of a cylinder with base area $\Delta\sigma$ and height $(\mathbf{v}\,\Delta t) \cdot \mathbf{n}$, where $\mathbf{v}$ is a velocity vector rooted at a point of the patch:

$$\Delta V \approx \mathbf{v} \cdot \mathbf{n} \, \Delta\sigma \, \Delta t.$$

The mass of this volume of fluid is about

$$\Delta m \approx \delta\mathbf{v} \cdot \mathbf{n} \, \Delta\sigma \, \Delta t,$$

so the rate at which mass is flowing out of D across the patch is about

$$\frac{\Delta m}{\Delta t} \approx \delta\mathbf{v} \cdot \mathbf{n} \, \Delta\sigma.$$

This leads to the approximation

$$\frac{\sum \Delta m}{\Delta t} \approx \sum \delta\mathbf{v} \cdot \mathbf{n} \, \Delta\sigma \tag{24}$$

as an estimate of the average rate at which mass flows across S. Finally, letting

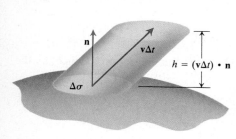

19.46 The fluid that flows upward through the patch $\Delta\sigma$ in a short time Δt fills a "cylinder" whose volume is approximately base × height $= \mathbf{v} \cdot \mathbf{n}\,\Delta t\,\Delta\sigma$.

$\Delta\sigma \to 0$ and $\Delta t \to 0$ gives the instantaneous rate at which mass leaves D across S as

$$\frac{dm}{dt} = \iint_S \delta\mathbf{v} \cdot \mathbf{n} \, d\sigma, \tag{25}$$

which for our particular flow is

$$\frac{dm}{dt} = \iint_S \mathbf{F} \cdot \mathbf{n} \, d\sigma. \tag{26}$$

Now let B be a ball centered at a point Q in the flow. The average value of $\nabla \cdot \mathbf{F}$ over the ball is

$$\frac{1}{\text{volume}} \iiint_B \nabla \cdot \mathbf{F} \, dV. \tag{27}$$

It is a consequence of the continuity of the divergence that $\nabla \cdot \mathbf{F}$ actually takes on this value at some point P in B. Thus,

$$(\nabla \cdot \mathbf{F})_P = \frac{1}{\text{volume}} \iiint_B \nabla \cdot \mathbf{F} \, dV = \frac{\iint_S \mathbf{F} \cdot \mathbf{n} \, d\sigma}{\text{volume}}$$

$$= \frac{\text{rate at which mass leaves } B \text{ across its surface } S}{\text{volume of } B}. \tag{28}$$

The fraction on the right describes decrease in mass per unit volume.

Now let the radius of the ball B approach 0 while the center Q stays fixed. The left-hand side of Eq. (28) converges to $(\nabla \cdot \mathbf{F})_Q$, the right side to $(-\partial\delta/\partial t)_Q$. The equality of these two limits is the continuity equation

$$\nabla \cdot \mathbf{F} = -\frac{\partial\delta}{\partial t}. \tag{29}$$

The continuity equation "explains" $\nabla \cdot \mathbf{F}$: the divergence of $\mathbf{F}$ at a point is the rate at which the density of the fluid is decreasing there.

The Divergence Theorem

$$\iint_S \mathbf{F} \cdot \mathbf{n} \, d\sigma = \iiint_D \nabla \cdot \mathbf{F} \, dV$$

now says that the net decrease in density of the fluid in region D is accounted for by the mass transported across the surface S. In a way, this is a statement about conservation of mass.

PROBLEMS

In Problems 1–12, use the Divergence Theorem to calculate the outward flux of $\mathbf{F}$ across the boundary of the region D.

1. $\mathbf{F}(x, y, z) = (x - y)\mathbf{i} + (y - z)\mathbf{j} + (x - y)\mathbf{k}$
   D: The cube bounded by the planes $x = \pm 1$, $y = \pm 1$, and $z = \pm 1$

2. $\mathbf{F}(x, y, z) = x^2\mathbf{i} + y^2\mathbf{j} + z^2\mathbf{k}$
   D: The cube cut from the first octant by the planes $x = 1$, $y = 1$, and $z = 1$

3. $\mathbf{F}(x, y, z) = x^2\mathbf{i} + y^2\mathbf{j} + z^2\mathbf{k}$
   D: The cube bounded by the planes $x = \pm 1$, $y = \pm 1$, $z = \pm 1$

4. $\mathbf{F}(x, y, z) = x^2\mathbf{i} + y^2\mathbf{j} + z^2\mathbf{k}$
   D: The region cut from the solid cylinder $x^2 + y^2 \le a^2$ by the planes $z = 0$ and $z = 1$

5. $\mathbf{F}(x, y, z) = y\mathbf{i} + xy\mathbf{j} + z\mathbf{k}$
   D: The region inside the solid cylinder $x^2 + y^2 \le 4$ between the plane $z = 0$ and the paraboloid $z = x^2 + y^2$

6. $\mathbf{F}(x, y, z) = x^2\mathbf{i} + xz\mathbf{j} + z\mathbf{k}$
   D: The solid sphere $x^2 + y^2 + z^2 \le 4$

7. $\mathbf{F}(x, y, z) = x^2\mathbf{i} - 2xy\mathbf{j} + 3xz\mathbf{k}$
   D: The region cut from the first octant by the sphere $x^2 + y^2 + z^2 = 4$

8. $\mathbf{F}(x, y, z) = (6x^2 + 2xy)\mathbf{i} + (2y + x^2z)\mathbf{j} + 4x^2y^3\mathbf{k}$
   D: The region cut from the first octant by the cylinder $x^2 + y^2 = 4$ and the plane $z = 4$

9. $\mathbf{F}(x, y, z) = 2xz\mathbf{i} + yj - z^2\mathbf{k}$
   D: The wedge cut from the first octant by the plane $z = y$ and the elliptical cylinder $x^2 + 4y^2 = 16$

10. $\mathbf{F}(x, y, z) = x^3\mathbf{i} + y^3\mathbf{j} + z^3\mathbf{k}$
    D: The solid sphere $x^2 + y^2 + z^2 \le a^2$

11. $\mathbf{F}(x, y, z) = \sqrt{x^2 + y^2 + z^2}\,(x\mathbf{i} + y\mathbf{j} + z\mathbf{k})$
    D: The solid sphere $x^2 + y^2 + z^2 \le a^2$

12. $\mathbf{F}(x, y, z) = \dfrac{x\mathbf{i} + y\mathbf{j} + z\mathbf{k}}{x^2 + y^2 + z^2}$
    D: $1 \le x^2 + y^2 + z^2 \le 4$

13. Let S be the closed cubelike surface shown in Fig. 19.47, with its base the unit square in the xy-plane, its four sides lying in the planes $x = 0$, $x = 1$, $y = 0$, $y = 1$, and its top an arbitrary smooth surface whose identity is unknown. Let $\mathbf{F} = x\mathbf{i} - 2y\mathbf{j} + (z + 3)\mathbf{k}$. Suppose that the outward flux of $\mathbf{F}$ through side A is 1 and through side B is -3. Calculate the outward flux of $\mathbf{F}$ through the top.

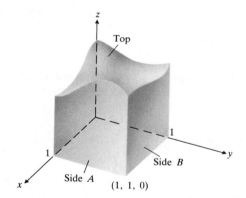

19.47 Problem 13 calculates the flux of a vector field through the top of this solid.

14. Let $\mathbf{F}(x, y, z)$ be a vector field with components that are continuous and differentiable on a portion of space contain-

ing a (finite) region D and its bounding surface S. Suppose that the length of the vector $\mathbf{F}(x, y, z)$ never exceeds 1 when (x, y, z) is a point of S. What bound can be placed on the numerical size of the integral

$$\iiint_D \nabla \cdot \mathbf{F}\, dV?$$

Explain.

15. a) Show that the flux of the position vector $\mathbf{F} = x\mathbf{i} + y\mathbf{j} + z\mathbf{k}$ outward through a piecewise smooth closed surface is three times the volume contained by the surface.

    b) Let $\mathbf{n}$ be the outward unit normal to a piecewise smooth closed surface S. Show that it is impossible for the position vector $\mathbf{F} = x\mathbf{i} + y\mathbf{j} + z\mathbf{k}$ to be orthogonal to $\mathbf{n}$ at every point of S.

Harmonic Functions

16. A function f is said to be *harmonic* in a region D if throughout D it satisfies the Laplace equation

$$\nabla^2 f = 0.$$

Suppose f is harmonic throughout D, S is the boundary of D, $\mathbf{n}$ is the positive unit normal on S, and $\partial f/\partial n$ is the directional derivative of f in the direction of $\mathbf{n}$. This derivative is called the *normal derivative* of f. Prove that

$$\iint_S \frac{\partial f}{\partial n}\, d\sigma = 0.$$

(*Hint:* Let $\mathbf{F} = \nabla f$.)

17. Prove that if f is harmonic in D (see Problem 16), then

$$\iint_S f \frac{\partial f}{\partial n}\, d\sigma = \iiint_D |\nabla f|^2\, dV.$$

(*Hint:* Let $\mathbf{F} = f\nabla f$.)

18. Let S be the eighth of the sphere $x^2 + y^2 + z^2 = a^2$ lying in the first octant, and let $f(x, y, z) = \ln\sqrt{x^2 + y^2 + z^2}$. Calculate

$$\iint_S \frac{\partial f}{\partial n}\, d\sigma.$$

(See Problem 16.)

TOOLKIT PROGRAMS

Vector Fields

19.6
Stokes's Theorem

When we studied vector fields and fluid flows in Article 19.3, we learned that the circulation density or curl of a two-dimensional field $\mathbf{F} = M\mathbf{i} + N\mathbf{j}$ at a point (x, y) is described by the scalar quantity $(\partial N/\partial x - \partial M/\partial y)$. In three

19.48 The circulation vector at a point P in a plane in a three-dimensional fluid flow. Notice its right-hand relation to the circulation line.

dimensions, the circulation around a point P in a plane is described not with a scalar but with a vector. This vector is normal to the plane of the circulation (Fig. 19.48) and points in the direction that gives it a right-hand relation to the circulation line. The length of the vector gives the rate of the fluid's rotation, which usually varies as the circulation plane is tilted about P from one position to another. It turns out that the vector of greatest circulation in a flow with velocity field $\mathbf{F} = M\mathbf{i} + N\mathbf{j} + P\mathbf{k}$ is the vector

$$\text{curl } \mathbf{F} = \left(\frac{\partial P}{\partial y} - \frac{\partial N}{\partial z}\right)\mathbf{i} + \left(\frac{\partial M}{\partial z} - \frac{\partial P}{\partial x}\right)\mathbf{j} + \left(\frac{\partial N}{\partial x} - \frac{\partial M}{\partial y}\right)\mathbf{k}. \qquad (1)$$

We get this information from Stokes's theorem, the generalization of the circulation-curl form of Green's theorem to space.

In this article we present Stokes's theorem and investigate what it says about circulation. We shall also see how Stokes's theorem sometimes enables us to evaluate a line integral around a loop in space by calculating an easier surface integral. This is similar to the way we used Green's theorem to evaluate a line integral around a complicated curve in the plane by calculating an easier double integral over the region enclosed by the curve. In addition, we shall state the circulation-curl or tangential form of Green's theorem in its final vector form. Stokes's theorem has important applications in electromagnetic theory. At the end of this article, we shall mention briefly the connection between Stokes's theorem and Ampère's law for current flow in a magnetic field.

Stokes's Theorem

Stokes's theorem says that, under conditions normally met in practice, the circulation of a vector field $\mathbf{F}$ around the boundary of a surface in space, in a direction counterclockwise with respect to the normal vector $\mathbf{n}$ of the surface, is equal to the double integral of curl $\mathbf{F} \cdot \mathbf{n}$ over the surface.

■ ■ ■

STOKES'S THEOREM AND A MATHEMATICAL TRADITION

George Gabriel Stokes (1819–1903) first published his theorem as a question in an examination for the Smith Prize at Cambridge, England, in 1854. The theorem had been sent to Stokes in a letter in 1850 by a friend, William Thomson (Lord Kelvin). A proponent of the French mathematical physics of Lagrange, Laplace, Fourier, Poisson, and Cauchy, Stokes influenced the Cambridge mathematical tradition by teaching widely and composing pertinent prize examinations. It is from this tradition that a generation of Cambridge mathematicians and scientists emerged, including perhaps the most important mathematical physicist of the nineteenth century, James Clerk Maxwell.

Stokes's Theorem

$$\oint_C \mathbf{F} \cdot d\mathbf{R} = \iint_S (\text{curl } \mathbf{F}) \cdot \mathbf{n} \, d\sigma \qquad (2)$$

Circulation integral   Curl integral

Naturally, we have to place some restrictions on $\mathbf{F}$ and on the nature of the surface for the theorem to hold. We shall discuss these restrictions after looking at our first example.

In del notation, the curl of a three-dimensional vector field $\mathbf{F} = M\mathbf{i} + N\mathbf{j} + P\mathbf{k}$ is written as the symbolic cross product

$$\text{curl } \mathbf{F} = \mathbf{\nabla} \times \mathbf{F} = \begin{vmatrix} \mathbf{i} & \mathbf{j} & \mathbf{k} \\ \dfrac{\partial}{\partial x} & \dfrac{\partial}{\partial y} & \dfrac{\partial}{\partial z} \\ M & N & P \end{vmatrix}$$

$$= \left(\frac{\partial P}{\partial y} - \frac{\partial N}{\partial z}\right)\mathbf{i} + \left(\frac{\partial M}{\partial z} - \frac{\partial P}{\partial x}\right)\mathbf{j} + \left(\frac{\partial N}{\partial x} - \frac{\partial M}{\partial y}\right)\mathbf{k}. \qquad (3)$$

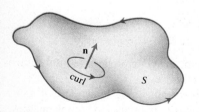

19.49 Green's theorem vs. Stokes's theorem.

Stokes's theorem then takes the following form.

Stokes's Theorem in del Notation

$$\oint_C \mathbf{F} \cdot d\mathbf{R} = \iint_S \boldsymbol{\nabla} \times \mathbf{F} \cdot \mathbf{n} \, d\sigma \qquad (4)$$

If C is a curve in the xy-plane, oriented counterclockwise, and R is the region bounded by C, then $d\sigma = dx \, dy$ and

$$(\boldsymbol{\nabla} \times \mathbf{F}) \cdot \mathbf{n} = (\boldsymbol{\nabla} \times \mathbf{F}) \cdot \mathbf{k} = \left(\frac{\partial N}{\partial x} - \frac{\partial M}{\partial y} \right). \qquad (5)$$

Under these conditions, Stokes's equation becomes

$$\oint_C \mathbf{F} \cdot d\mathbf{R} = \iint_R \left(\frac{\partial N}{\partial x} - \frac{\partial M}{\partial y} \right) dx \, dy,$$

which is the circulation-curl form of the equation in Green's theorem. Conversely, by reversing these steps we can rewrite Green's theorem for two-dimensional fields in del notation as

$$\oint_C \mathbf{F} \cdot d\mathbf{R} = \iint_R \boldsymbol{\nabla} \times \mathbf{F} \cdot \mathbf{k} \, dA. \qquad (6)$$

See Fig. 19.49.

EXAMPLE 1 Evaluate the two integrals in Stokes's theorem for the hemisphere

$$x^2 + y^2 + z^2 = a^2,$$

its bounding circle

$$x^2 + y^2 = a^2, \qquad z = 0,$$

and the field

$$\mathbf{F}(x, y, z) = y\mathbf{i} - x\mathbf{j}.$$

Solution The integrand of the circulation integral is

$$\mathbf{F} \cdot d\mathbf{R} = \mathbf{F} \cdot (dx\mathbf{i} + dy\mathbf{j} + dz\mathbf{k})$$
$$= y \, dx - x \, dy.$$

By the circulation form of Green's theorem for the plane,

$$\oint_C \mathbf{F} \cdot d\mathbf{R} = \oint_C (y \, dx - x \, dy)$$

$$= \iint_R \left(\frac{\partial N}{\partial x} - \frac{\partial M}{\partial y} \right) dx \, dy$$

$$= \iint_{x^2+y^2 \le a^2} -2 \, dx \, dy = -2\pi a^2.$$

For the curl integral of $\mathbf{F} = y\mathbf{i} - x\mathbf{j}$, we have

$$\boldsymbol{\nabla}\times\mathbf{F} = \left(\frac{\partial P}{\partial y} - \frac{\partial N}{\partial z}\right)\mathbf{i} + \left(\frac{\partial M}{\partial z} - \frac{\partial P}{\partial x}\right)\mathbf{j} + \left(\frac{\partial N}{\partial x} - \frac{\partial M}{\partial y}\right)\mathbf{k}$$

$$= (0 - 0)\mathbf{i} + (0 - 0)\mathbf{j} + (-1 - 1)\mathbf{k} = -2\mathbf{k},$$

$$\mathbf{n} = \frac{x\mathbf{i} + y\mathbf{j} + z\mathbf{k}}{\sqrt{x^2 + y^2 + z^2}} = \frac{x\mathbf{i} + y\mathbf{j} + z\mathbf{k}}{a}, \qquad \text{(outer unit normal)}$$

$$d\sigma = \frac{a}{z}\,dA \qquad \text{(obtained as in Article 19.4, Example 5)}$$

$$\boldsymbol{\nabla}\times\mathbf{F} \cdot \mathbf{n}\,d\sigma = -\frac{2z}{a}\frac{a}{z}\,dA = -2\,dA,$$

and

$$\iint\limits_{S} \boldsymbol{\nabla}\times\mathbf{F} \cdot \mathbf{n}\,d\sigma = \iint\limits_{x^2+y^2 \leq a^2} -2\,dA = -2\pi a^2.$$

The circulation around the circle equals the integral of the curl over the hemisphere, as it should. ∎

The Mathematical Requirements for Stokes's Theorem

In Stokes's theorem, we require that the surface be two-sided. This rules out a surface like the Möbius strip shown in Fig. 19.50. This surface is nonorientable because a unit normal vector (think of the shaft of a thumbtack) can be continuously moved around the surface without its touching the boundary of the surface, and in such a way that when it is returned to its initial position it will point in a direction *exactly opposite* to its initial direction.

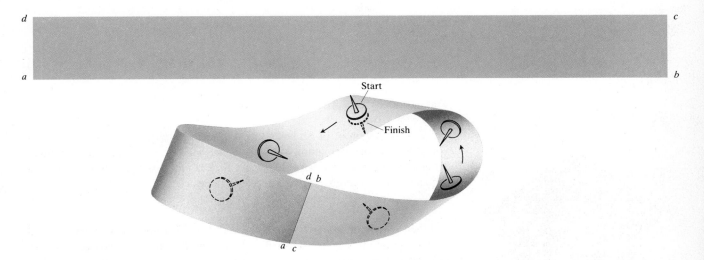

19.50 The Möbius strip can be constructed by taking a rectangular strip of paper *abcd*, giving the end *bc* a single twist to interchange the positions of the vertices *b* and *c*, and then pasting the ends of the strip together so as to bring vertices *a* and *c* together, and also *b* and *d*. The Möbius strip is a nonorientable or one-sided surface and Stokes's theorem does not apply to it.

19.51 The orientation of the boundary C of an oriented surface. Notice the right-hand relation to the positive direction **n**.

We also orient C to have a positive direction that is related by the right-hand rule to the positive direction on S (see Fig. 19.51) and we place restrictions on **F**, C, and S to guarantee the existence of the integrals in Stokes's equation. The usual restrictions are that the functions and derivatives involved be continuous.

Proof of Stokes's Theorem for Polyhedral Surfaces

Let the surface S be a polyhedral surface consisting of a finite number of plane regions. (Think of one of Buckminster Fuller's geodesic domes.) We apply Green's theorem to each separate panel of S. There are two types of panels:

1. those that are surrounded on all sides by other panels, and
2. those that have one or more edges that are not adjacent to other panels.

The boundary Δ of S consists of those edges of the type 2 panels that are not adjacent to other panels. In Fig. 19.52, the triangles ABE, BCE, and CDE represent a part of S, with $ABCD$ part of the boundary Δ. Applying Green's theorem to the three triangles in turn and adding the results, we get

$$\left(\oint_{ABE} + \oint_{BCE} + \oint_{CDE}\right) \mathbf{F} \cdot d\mathbf{R} = \left(\iint_{ABE} + \iint_{BCE} + \iint_{CDE}\right) \text{curl } \mathbf{F} \cdot \mathbf{n} \, d\sigma. \quad (7)$$

The three line integrals on the left-hand side of Eq. (7) combine into a single line integral taken around the periphery $ABCDE$ because the integrals along interior segments cancel in pairs. For example, the integral along the segment BE in triangle ABE is opposite in sign to the integral along the same segment in triangle EBC. Similarly for the segment CE. Hence (7) reduces to

$$\oint_{ABCDE} \mathbf{F} \cdot d\mathbf{R} = \iint_{ABCDE} \text{curl } \mathbf{F} \cdot \mathbf{n} \, d\sigma.$$

In general, when we apply Green's theorem to all the panels and add the results, we get

$$\oint_{\Delta} \mathbf{F} \cdot d\mathbf{R} = \iint_{S} \nabla \times \mathbf{F} \cdot \mathbf{n} \, d\sigma. \quad (8)$$

This is Stokes's theorem for a polyhedral surface S. ∎

A rigorous proof of Stokes's theorem for a more general surface S is beyond the level of a beginning calculus course. However, the following intuitive argument shows why we would expect Stokes's equation to hold. Imagine a sequence of approximating polyhedral surfaces

$$S_1, \quad S_2, \quad \ldots,$$

19.52 Part of a polyhedral surface.

and their corresponding boundaries $\Delta_1, \Delta_2, \ldots$. The surface S_n is constructed so that its boundary Δ_n is inscribed in or tangent to C, the boundary of S, and so that the length of Δ_n approaches the length of C as $n \to \infty$. The faces of S_n might be polygonal regions, approximating pieces of S, and such that the area of S_n approaches the area of S as $n \to \infty$. Assuming that M, N, P, and their partial derivatives are continuous in a region D containing S and C, it is plausible to expect that

$$\oint_{\Delta_n} \mathbf{F} \cdot d\mathbf{R} \qquad \text{approaches} \qquad \oint_C \mathbf{F} \cdot d\mathbf{R}$$

and that

$$\iint_{S_n} \boldsymbol{\nabla} \times \mathbf{F} \cdot \mathbf{n} \, d\sigma_n \qquad \text{approaches} \qquad \iint_S \boldsymbol{\nabla} \times \mathbf{F} \cdot \mathbf{n} \, d\sigma$$

as $n \to \infty$. But if

$$\oint_{\Delta_n} \mathbf{F} \cdot d\mathbf{R} \to \oint_C \mathbf{F} \cdot d\mathbf{R} \tag{9}$$

and

$$\iint_{S_n} \boldsymbol{\nabla} \times \mathbf{F} \cdot \mathbf{n} \, d\sigma \to \iint_S \boldsymbol{\nabla} \times \mathbf{F} \cdot \mathbf{n} \, d\sigma, \tag{10}$$

and if the left-hand sides of (9) and (10) are equal by Stokes's theorem for polyhedra, we then have equality of their limits.

Stokes's Theorem for Surfaces with Holes

Stokes's theorem can also be extended to a surface S that has one or more holes in it (like a curved slice of Swiss cheese), in a way exactly analogous to Green's theorem: The surface integral over S of the normal component of curl $\mathbf{F}$ is equal to the sum of the line integrals around all the boundaries of S (including boundaries of the holes) of the tangential component of $\mathbf{F}$, where the boundary curves are to be traced in the positive direction induced by the positive orientation of S.

Circulation of a Fluid

Stokes's theorem provides the following vector interpretation for curl $\mathbf{F}$. As in the discussion of divergence, let $\mathbf{v}$ be the velocity field of a moving fluid, δ the density, and $\mathbf{F} = \delta \mathbf{v}$. Then

$$\oint_C \mathbf{F} \cdot d\mathbf{R}$$

is the circulation of the fluid around the closed curve C. By Stokes's theorem, this circulation is also equal to the flux of curl $\mathbf{F}$ through a surface S spanning C:

$$\oint_C \mathbf{F} \cdot d\mathbf{R} = \iint_S \boldsymbol{\nabla} \times \mathbf{F} \cdot \mathbf{n} \, d\sigma.$$

Suppose we fix a point Q in the domain of $\mathbf{F}$ and a direction $\mathbf{u}$ at Q. Let C be a circle of radius ρ, with center at Q, whose plane is normal to $\mathbf{u}$. If curl $\mathbf{F}$ is continuous at Q, then the average value of the $\mathbf{u}$-component of curl $\mathbf{F}$ over the

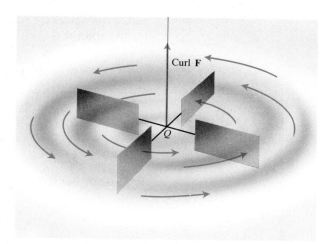

19.53 The paddle-wheel interpretation of curl **F**.

circular disk bounded by C approaches the **u**-component of curl **F** at Q as $\rho \to 0$:

$$(\nabla \times \mathbf{F} \cdot \mathbf{u})_Q = \lim_{\rho \to 0} \frac{1}{\pi \rho^2} \iint\limits_S \nabla \times \mathbf{F} \cdot \mathbf{u} \, d\sigma. \tag{11}$$

If we replace the double integral on the right-hand side of Eq. (11) by the circulation, we get

$$(\nabla \times \mathbf{F} \cdot \mathbf{u})_Q = \lim_{\rho \to 0} \frac{1}{\pi \rho^2} \oint_C \mathbf{F} \cdot d\mathbf{R}. \tag{12}$$

The left-hand side of Eq. (12) is a maximum at Q when **u** has the same direction as curl **F**. When ρ is small, the limit on the right-hand side of Eq. (12) is approximately equal to

$$\frac{1}{\pi \rho^2} \oint_C \mathbf{F} \cdot d\mathbf{R},$$

which is the circulation around C divided by the area of the disk (circulation density). Suppose that a small paddle wheel, of radius ρ, is introduced into the fluid at Q, with its axle directed along **u**. The circulation of the fluid around C will affect the rate of spin of the paddle wheel. The wheel will spin fastest when the circulation integral is maximized; therefore it will spin fastest when the axle of the paddle wheel points in the direction of curl **F**. (See Fig. 19.53.)

EXAMPLE 2   A fluid of constant density δ rotates around the z-axis with velocity $\mathbf{v} = \omega(-y\mathbf{i} + x\mathbf{j})$, where ω is a positive constant called the *angular velocity* of the rotation (Fig. 19.54). If $\mathbf{F} = \delta\mathbf{v}$, find curl **F** and show its relation to the circulation density.

Solution   With $\mathbf{F} = -\delta\omega y\mathbf{i} + \delta\omega x\mathbf{j}$,

$$\nabla \times \mathbf{F} = \left(\frac{\partial P}{\partial y} - \frac{\partial N}{\partial z}\right)\mathbf{i} + \left(\frac{\partial M}{\partial z} - \frac{\partial P}{\partial x}\right)\mathbf{j} + \left(\frac{\partial N}{\partial x} - \frac{\partial M}{\partial y}\right)\mathbf{k}$$

$$= (0 - 0)\mathbf{i} + (0 - 0)\mathbf{j} + (\delta\omega - (-\delta\omega))\mathbf{k}$$

$$= 2\,\delta\omega\mathbf{k}.$$

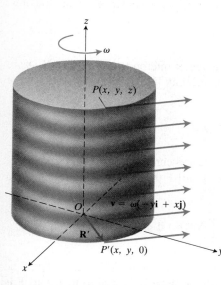

19.54 A steady rotational flow parallel to the *xy*-plane, with constant angular velocity ω in the positive (counterclockwise) direction.

By Stokes's theorem, the circulation of $\mathbf{F}$ around a circle C of radius ρ is

$$\oint_C \mathbf{F} \cdot d\mathbf{R} = \iint_S \boldsymbol{\nabla}\times\mathbf{F} \cdot \mathbf{n} \, d\sigma = \iint 2 \, \delta\omega\mathbf{k} \cdot \mathbf{k} \, dx \, dy = (2 \, \delta\omega)(\pi\rho^2).$$

Thus,

$$(\boldsymbol{\nabla}\times\mathbf{F}) \cdot \mathbf{k} = 2 \, \delta\omega = \frac{1}{\pi\rho^2} \oint_C \mathbf{F} \cdot d\mathbf{R},$$

in agreement with Eq. (12) with $\mathbf{u} = \mathbf{k}$. ∎

An Important Identity

The equation

$$\text{curl grad } f = \mathbf{0} \qquad \text{or} \qquad \boldsymbol{\nabla}\times\boldsymbol{\nabla}f = \mathbf{0} \tag{13}$$

is a vector identity frequently used in connection with vector fields in mathematics and the physical sciences. (We shall use it ourselves in Article 19.7.) The identity holds for any continuous function $f(x, y, z)$ whose first and second partial derivatives are continuous. The proof is straightforward and goes like this: Since

$$\boldsymbol{\nabla}f = \frac{\partial f}{\partial x}\mathbf{i} + \frac{\partial f}{\partial y}\mathbf{j} + \frac{\partial f}{\partial z}\mathbf{k},$$

curl grad f is given by the formula

$$\boldsymbol{\nabla} \times \boldsymbol{\nabla}f = \begin{vmatrix} \mathbf{i} & \mathbf{j} & \mathbf{k} \\ \dfrac{\partial}{\partial x} & \dfrac{\partial}{\partial y} & \dfrac{\partial}{\partial z} \\ \dfrac{\partial f}{\partial x} & \dfrac{\partial f}{\partial y} & \dfrac{\partial f}{\partial z} \end{vmatrix} \tag{14}$$

$$= (f_{zy} - f_{yz})\mathbf{i} - (f_{zx} - f_{xz})\mathbf{j} + (f_{yx} - f_{xy})\mathbf{k}.$$

When f and its first and second partial derivatives are continuous, the mixed second derivatives in parentheses are equal and the vector is zero.

Additional identities appear in Problems 12 and 13.

☐ Ampère's Law

In electromagnetic theory, Stokes's theorem gives Ampère's law

$$\oint_C \mathbf{B} \cdot d\mathbf{R} = \iint_S \boldsymbol{\nabla}\times\mathbf{B} \cdot \mathbf{n} \, d\sigma$$

$$= \frac{4\pi}{c} \iint_S \mathbf{J} \cdot \mathbf{n} \, d\sigma \tag{15}$$

$$= \frac{4\pi}{c}I.$$

In this equation, $\mathbf{B}$ is a magnetic field in space, C is a loop in space enclosing a current I, S is a surface spanning the loop, $\mathbf{J}$ is the current density (current per

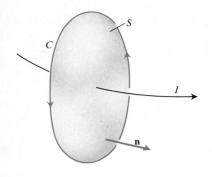

19.55 The directions of C, $\mathbf{n}$, and the current flow in Ampère's law.

unit area), and c is the speed of light. The equation assumes that the direction in which the current flows and the direction of C correspond to the same right-hand rule that relates the direction of C to the normal vector $\mathbf{n}$ on the surface S (Fig. 19.55). Ampère's law is named after its discoverer, André Marie Ampère (1775–1836), French physicist, mathematician, and natural philosopher known for his contributions to electrodynamics. The law tells us that the integral of $\mathbf{B}$ around a closed path is proportional to the current through any surface bounded by the path.

Summary of Integral Theorems in del Notation

We close this article with a table that lists the equations from Green's theorem and its generalizations to three dimensions.

TABLE 19.3
Green's theorem and its generalizations

Green's theorem:	$\displaystyle\oint_C \mathbf{F}\cdot\mathbf{n}\,ds = \iint_R \nabla\cdot\mathbf{F}\,dA$
	$\displaystyle\oint_C \mathbf{F}\cdot d\mathbf{R} = \iint_R \nabla\times\mathbf{F}\cdot\mathbf{k}\,dA$
Divergence theorem:	$\displaystyle\iint_S \mathbf{F}\cdot\mathbf{n}\,d\sigma = \iiint_D \nabla\cdot\mathbf{F}\,dV$
Stokes's theorem:	$\displaystyle\oint_C \mathbf{F}\cdot d\mathbf{R} = \iint_S \nabla\times\mathbf{F}\cdot\mathbf{n}\,d\sigma$

PROBLEMS

In Problems 1–6, use the surface integral in Stokes's theorem to calculate the circulation of the field $\mathbf{F}$ around the curve C in the indicated direction.

1. $\mathbf{F} = x^2\mathbf{i} + 2x\mathbf{j} + z^2\mathbf{k}$
   C: The ellipse $4x^2 + y^2 = 4$, counterclockwise as viewed from above

2. $\mathbf{F} = 2y\mathbf{i} + 3x\mathbf{j} - z^2\mathbf{k}$
   C: The circle $x^2 + y^2 = 9$ in the xy-plane, counterclockwise when viewed from above

3. $\mathbf{F} = y\mathbf{i} + xz\mathbf{j} + x^2\mathbf{k}$
   C: The boundary of the triangle cut from the plane $x + y + z = 1$ by the first octant, counterclockwise when viewed from above

4. $\mathbf{F} = (y^2 + z^2)\mathbf{i} + (x^2 + z^2)\mathbf{j} + (x^2 + y^2)\mathbf{k}$
   C: The boundary of the triangle cut from the plane $x + y + z = 1$ by the first octant, counterclockwise when viewed from above

5. $\mathbf{F} = (y^2 + z^2)\mathbf{i} + (x^2 + y^2)\mathbf{j} + (x^2 + y^2)\mathbf{k}$
   C: The square bounded by the lines $x = \pm 1$ and $y = \pm 1$ in the xy-plane, counterclockwise when viewed from above

6. $\mathbf{F} = x^2y^3\mathbf{i} + \mathbf{j} + z\mathbf{k}$
   C: The boundary of the circle $x^2 + y^2 = a^2$ in the xy-plane, counterclockwise when viewed from above

7. Let S be the region bounded by the ellipse C: $4x^2 + y^2 = 4$ in the plane $z = 1$, let $\mathbf{n} = \mathbf{k}$, and let $\mathbf{F} = x^2\mathbf{i} + 2x\mathbf{j} + z^2\mathbf{k}$. Find the value of

$$\oint_C \mathbf{F}\cdot d\mathbf{R}.$$

8. Let $\mathbf{n}$ be the outer normal of the elliptical shell

$$S:\quad 4x^2 + 9y^2 + 36z^2 = 36, \quad z \geq 0,$$

and let

$$\mathbf{F} = y\mathbf{i} + x^2\mathbf{j} + (x^2 + y^4)^{3/2}\sin e^{\sqrt{xyz}}\mathbf{k}.$$

Use Stokes's theorem to find the value of

$$\iint_S \nabla \times \mathbf{F} \cdot \mathbf{n} \, d\sigma.$$

(*Hint:* One parameterization of the ellipse at the base of the shell is $x = 3 \cos t$, $y = 2 \sin t$, $0 \leq t \leq 2\pi$.)

9. Let S be the cylinder $x^2 + y^2 = a^2$, $0 \leq z \leq h$, together with its top, $x^2 + y^2 \leq a^2$, $z = h$. Let $\mathbf{F} = -y\mathbf{i} + x\mathbf{j} + x^2\mathbf{k}$. Use Stokes's theorem to calculate the flux of curl $\mathbf{F}$ outward through S.

10. Show that

$$\iint_S \nabla \times \mathbf{F} \cdot \mathbf{n} \, d\sigma$$

has the same value for all oriented surfaces S that span C and that induce the same positive direction on C.

11. Show that if $\mathbf{F} = x\mathbf{i} + y\mathbf{j} + z\mathbf{k}$, then $\nabla \cdot \mathbf{F} = 3$ and $\nabla \times \mathbf{F} = \mathbf{0}$.

12. a) Show that if the necessary partial derivatives of the components of

$$\mathbf{F} = M\mathbf{i} + N\mathbf{j} + P\mathbf{k}$$

are continuous, then div(curl $\mathbf{F}$) = 0.

b) Use the result in part (a) to show that

$$\iint_S \nabla \times \mathbf{F} \cdot \mathbf{n} \, d\sigma = 0$$

for any surface to which the Divergence Theorem applies.

13. Let $g(x, y, z)$ be a differentiable scalar function and $\mathbf{F}$, $\mathbf{F}_1$, and $\mathbf{F}_2$ differentiable vector fields. Verify the following identities.

a) div($g\mathbf{F}$) = $\nabla \cdot (g\mathbf{F}) = g\nabla \cdot \mathbf{F} + \mathbf{F} \cdot \nabla g$

b) $\nabla \times (g\mathbf{F}) = g\nabla \times \mathbf{F} + \nabla g \times \mathbf{F}$

c) $\nabla \cdot (\mathbf{F}_1 \times \mathbf{F}_2) = \mathbf{F}_2 \cdot \nabla \times \mathbf{F}_1 - \mathbf{F}_1 \cdot \nabla \times \mathbf{F}_2$

14. Suppose $\mathbf{F} = \text{grad } f$ is the gradient of a continuous scalar function f having continuous first and second order partial derivatives

$$\frac{\partial^2 f}{\partial x^2}, \quad \frac{\partial^2 f}{\partial x \, \partial y}, \quad \cdots$$

throughout a simply connected region D that contains the surface S and its boundary C in the interior of D. What constant value does

$$\oint_C \mathbf{F} \cdot d\mathbf{R}$$

have in such circumstances? Explain.

15. Let $f(x, y, z) = (x^2 + y^2 + z^2)^{-1/2}$, let $\mathbf{F} = \nabla f$, and let C be the circle $x^2 + y^2 = a^2$, $z = 0$, in the xy-plane. Show that

$$\oint_C \mathbf{F} \cdot d\mathbf{R} = 0$$

a) by direct evaluation of the integral, and

b) by applying Stokes's theorem with S the hemisphere $z = \sqrt{a^2 - x^2 - y^2}$, $x^2 + y^2 \leq a^2$.

 TOOLKIT PROGRAMS

Vector Fields

19.7
Path Independence and Conservative Fields

In this article, we describe path independence and discuss the remarkable properties of fields, such as gravitational fields and electric fields, that have path-independent integrals.

Path Independence

If we let A and B be two points in an open region D in space, the work $\int \mathbf{F} \cdot d\mathbf{R}$ done in moving a particle from A to B by a field $\mathbf{F}$ defined on D will usually depend on the path taken. For some fields, however, the value of the integral depends only on the points A and B and will be the same for all paths from A to B. If this is true for all points A and B in D, we say that the integral $\int \mathbf{F} \cdot d\mathbf{R}$ is **independent of the path** in D and that the field $\mathbf{F}$ is **conservative** on D.

Conservative fields are important in science because, under conditions normally met in practice, a field is conservative if and only if it is the gradient field of a scalar function. That is, $\mathbf{F}$ is conservative if and only if $\mathbf{F} = \nabla f$ for some function f. The function f is called a potential function for $\mathbf{F}$. An electric potential is a function whose gradient field is an electric field. A gravitational

Connected and simply connected.

Connected but not simply connected.

Connected and simply connected.

Simply connected but not connected.
No path from A to B lies entirely in the region

19.56   Connectivity and simple connectivity are not the same thing. In fact, neither implies the other, as these pictures illustrate.

potential is a function whose gradient is a gravitational field. As we shall see, once we have found a potential f for a field **F**, we can evaluate all line integrals in **F** by the rule

$$\int_A^B \mathbf{F} \cdot d\mathbf{R} = f(B) - f(A). \tag{1}$$

Conservative fields have other remarkable properties. Saying that **F** is conservative on D is equivalent to saying that the integral of **F** around every closed loop in D is zero. This, in turn, is equivalent in a great many domains to saying that curl **F** = **0**. We explain these relationships by discussing the implications in the following diagram.

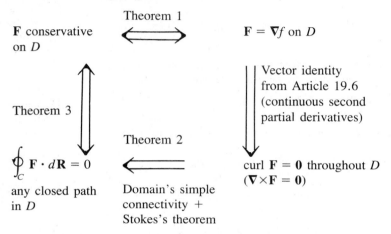

Naturally, we need to impose conditions on the curves, fields, and domains to make these implications hold. We assume that all of the curves are **piecewise smooth,** i.e., made of continuously differentiable pieces connected end to end.

We also assume that the components of **F** are continuous and have continuous first and second partial derivatives. Without this assumption, $\mathbf{F} = \nabla f$ need not imply curl **F** = **0**. As we saw in the previous article, the implication requires the mixed second derivatives of f to be equal. We use the continuity to ensure that.

We assume D to be an *open* region in space. This means that every point in D is the center of a ball that lies entirely in D. We need an added restriction on D if curl **F** = **0** is to imply that the integral of **F** around every closed path in D is zero. For this, we assume D to be **simply connected.** This means that every closed path in D can be contracted to a point without ever leaving D. If D consisted of space with one of the axes removed, for example, D would not be simply connected. There would be no way to contract a loop around the axis to a point without leaving D. (See Problem 27.) On the other hand, space itself *is* simply connected.

Finally, we assume D to be **connected,** which in an open region means that every point can be connected to every other point by a smooth curve that lies in the region. See Fig. 19.56.

The Implications

We now examine the implications in the diagram above, starting with Theorem 1, then Theorem 2, and finally Theorem 3. Theorem 1 has two parts and we shall prove both. A complete proof of Theorem 2 is beyond our reach at this stage but we can describe the main ideas. The proof of Theorem 3 is easy and short.

THEOREM 1

> ### The Fundamental Theorem of Line Integrals
>
> 1. Let $\mathbf{F} = M\mathbf{i} + N\mathbf{j} + P\mathbf{k}$ be a vector field whose components are continuous throughout an open connected region D in space. Then there exists a differentiable function f such that
>
> $$\mathbf{F} = \nabla f = \frac{\partial f}{\partial x}\mathbf{i} + \frac{\partial f}{\partial y}\mathbf{j} + \frac{\partial f}{\partial z}\mathbf{k}$$
>
> if and only if for all points A and B in D the value of the integral
>
> $$\int_A^B \mathbf{F} \cdot d\mathbf{R}$$
>
> is independent of the path joining A to B in D.
>
> 2. If the integral is independent of the path from A to B, its value is
>
> $$\int_A^B \mathbf{F} \cdot d\mathbf{R} = f(B) - f(A).$$

Proof That $\mathbf{F} = \nabla f$ Implies Path Independence of the Integral Suppose that A and B are two points in D and that

$$C: \qquad \mathbf{R}(t) = x(t)\mathbf{i} + y(t)\mathbf{j} + z(t)\mathbf{k}, \qquad a \le t \le b,$$

is a piecewise smooth curve joining A and B. Along the curve, f is a differentiable function of t and

$$\frac{df}{dt} = \frac{\partial f}{\partial x}\frac{dx}{dt} + \frac{\partial f}{\partial y}\frac{dy}{dt} + \frac{\partial f}{\partial z}\frac{dz}{dt} \qquad \text{(Chain Rule)}$$

$$= \nabla f \cdot \left(\frac{dx}{dt}\mathbf{i} + \frac{dy}{dt}\mathbf{j} + \frac{dz}{dt}\mathbf{k}\right)$$

$$= \nabla f \cdot \frac{d\mathbf{R}}{dt}$$

$$= \mathbf{F} \cdot \frac{d\mathbf{R}}{dt}.$$

(2)

Therefore,

$$\int_C \mathbf{F} \cdot d\mathbf{R} = \int_C \mathbf{F} \cdot \frac{d\mathbf{R}}{dt}\, dt$$

$$= \int_C \frac{df}{dt}\, dt$$

$$= \int_{t=a}^{t=b} \frac{df}{dt}\, dt$$

$$= f(x(b), y(b), z(b)) - f(x(a), y(a), z(a))$$

$$= f(B) - f(A).$$

Thus, the value of the line integral is $f(B) - f(A)$, a number independent of the path taken to get from A to B. This proves Part 2 as well as the forward implication in Part 1.

Proof That Path Independence of the Integral Implies F = ∇f   We need to show that if $\mathbf{F}$ is conservative in D, then there exists a differentiable function f, defined at every point of D, for which

$$\frac{\partial f}{\partial x} = M, \qquad \frac{\partial f}{\partial y} = N, \qquad \text{and} \qquad \frac{\partial f}{\partial z} = P. \tag{3}$$

To define f, we choose a point A in D and define f at any point B in D to be the value of the integral of $\mathbf{F} \cdot d\mathbf{R}$ along a smooth path from A to B:

$$f(B) = \int_A^B \mathbf{F} \cdot d\mathbf{R}. \tag{4}$$

All such paths in D give the same value because $\mathbf{F}$ is conservative.

One of the consequences of the definition of f is that for any points B and B_0 in D,

$$\begin{aligned} f(B) - f(B_0) &= \int_A^B \mathbf{F} \cdot d\mathbf{R} - \int_A^{B_0} \mathbf{F} \cdot d\mathbf{R} \\ &= \int_{B_0}^A \mathbf{F} \cdot d\mathbf{R} + \int_A^B \mathbf{F} \cdot d\mathbf{R} \\ &= \int_{B_0}^B \mathbf{F} \cdot d\mathbf{R}. \end{aligned} \tag{5}$$

Let us now calculate the partial derivatives of f at the point $B_0(x_0, y_0, z_0)$. Since B_0 is the center of a ball that lies entirely in D (because D is open), we can find a small positive number h for which the point $B(x_0 + h, y_0, z_0)$ and the line segment C joining B_0 to B also lie in D (Fig. 19.57). The value of the integral

$$\int_{B_0}^B \mathbf{F} \cdot d\mathbf{R}$$

can be calculated by integrating along C with the parameterization

$$\mathbf{R}(t) = (x_0 + th)\mathbf{i} + y_0\mathbf{j} + z_0\mathbf{k}, \qquad 0 \le t \le 1. \tag{6}$$

When we carry out the integration, we find

$$\begin{aligned} \int_{B_0}^B \mathbf{F} \cdot d\mathbf{R} &= \int_{t=0}^{t=1} (M\mathbf{i} + N\mathbf{j} + P\mathbf{k}) \cdot (h\, dt\, \mathbf{i}) \\ &= h \int_0^1 M(x_0 + th, y_0, z_0)\, dt. \end{aligned} \tag{7}$$

Therefore, the difference quotient we use to define $\partial f/\partial x$ at B_0 is

$$\begin{aligned} \frac{f(x_0 + h, y_0, z_0) - f(x_0, y_0, z_0)}{h} &= \frac{1}{h} \int_{B_0}^B \mathbf{F} \cdot d\mathbf{R} \\ &= \int_0^1 M(x_0 + th, y_0, z_0)\, dt. \end{aligned} \tag{8}$$

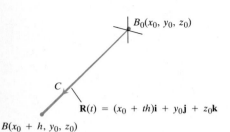

$B_0(x_0, y_0, z_0)$

C

$\mathbf{R}(t) = (x_0 + th)\mathbf{i} + y_0\mathbf{j} + z_0\mathbf{k}$

$B(x_0 + h, y_0, z_0)$

19.57   To calculate the partial derivative with respect to x of the function

$$f(x, y, z) = \int_A^{(x,y,z)} \mathbf{F} \cdot d\mathbf{R}$$

at $B_0(x_0, y_0, z_0)$, we take the limit of $\int_C \mathbf{F} \cdot d\mathbf{R}$ as $h \to 0$.

Since M is continuous, given any $\epsilon > 0$ there is a $\delta > 0$ such that

$$|M(x_0 + th, y_0, z_0) - M(x_0, y_0, t_0)| < \epsilon \qquad \text{when} \qquad |th| < \delta.$$

This implies that when $|h| < \delta$ the integral in Eq. (8) differs from the integral

$$\int_0^1 M(x_0, y_0, z_0)\, dt = M(x_0, y_0, z_0)$$

by less than ϵ. Therefore, as $h \to 0$, the limit of the integral in Eq. (8) is $M(x_0, y_0, z_0)$. The left-hand side of Eq. (8) therefore has the same limit, and we conclude that

$$\frac{\partial f}{\partial x} = M$$

at every point (x_0, y_0, z_0) in D.

The equations $\partial f/\partial y = N$ and $\partial f/\partial z = P$ are derived in a similar way. This concludes the proof of Theorem 1. ∎

THEOREM 2

> If curl $\mathbf{F} = \mathbf{0}$ at every point of a simply connected open region D in space, then
>
> $$\oint_C \mathbf{F} \cdot d\mathbf{R} = 0$$
>
> on any piecewise smooth closed path C in D.

Sketch of a Proof The theorem is usually proved in two steps. The first step is for *simple* closed curves (curves that do not cross themselves). A theorem from topology, a branch of advanced mathematics, states that every differentiable simple closed curve C in a simply connected open region D is the boundary of a smooth two-sided surface S that also lies in D. Hence, by Stokes's theorem,

$$\oint_C \mathbf{F} \cdot d\mathbf{R} = \iint_S (\text{curl } \mathbf{F}) \cdot \mathbf{n}\, d\sigma = 0.$$

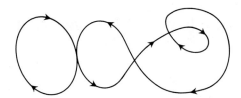

19.58 In a simply-connected region, curves that cross themselves can be divided into loops to which Stokes's theorem applies.

The second step is for curves that cross themselves, like the one in Fig. 19.58. The idea is to break these down into simple loops spanned by orientable surfaces, apply Stokes's theorem one loop at a time, and add the results. ∎

THEOREM 3

> The following statements are equivalent:
>
> 1. $\int \mathbf{F} \cdot d\mathbf{R} = 0$ around every closed loop in D.
> 2. The field $\mathbf{F}$ is conservative on D.

Proof That (1) ⇒ (2) We want to show that the value of the integral

$$\int_A^B \mathbf{F} \cdot d\mathbf{R}$$

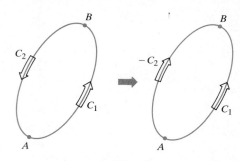

19.59 If we have two paths from A to B, one of them can be reversed to make a loop.

19.60 If A and B lie on a loop, we can reverse part of the loop to make two paths from A to B.

is the same for any two paths C_1 and C_2 from A to B. We reverse the direction on C_2 to make a path $-C_2$ from B to A (Fig. 19.59). Together, C_1 and $-C_2$ make a closed loop C, and

$$\int_{C_1} \mathbf{F} \cdot d\mathbf{R} - \int_{C_2} \mathbf{F} \cdot d\mathbf{R} = \int_{C_1} \mathbf{F} \cdot d\mathbf{R} + \int_{-C_2} \mathbf{F} \cdot d\mathbf{R} = \int_C \mathbf{F} \cdot d\mathbf{R} = 0.$$

Thus the integrals along C_1 and C_2 give the same value.

Proof That (2) $\Rightarrow$ (1) We want to show that

$$\oint_C \mathbf{F} \cdot d\mathbf{R} = 0$$

on any closed loop C. We pick two points A and B on C and use them to break C into two pieces: C_1 from A to B followed by C_2 from B back to A (Fig. 19.60). Then

$$\oint_C \mathbf{F} \cdot d\mathbf{R} = \int_{C_1} \mathbf{F} \cdot d\mathbf{R} + \int_{C_2} \mathbf{F} \cdot d\mathbf{R} = \int_A^B \mathbf{F} \cdot d\mathbf{R} - \int_A^B \mathbf{F} \cdot d\mathbf{R} = 0. \quad \blacksquare$$

Finding Potentials for Conservative Fields

Conservative fields get their name from the fact that bodies acted on by such fields obey the principle of conservation of energy. The test for being conservative, as we have seen, is curl $\mathbf{F} = \mathbf{0}$. In terms of components, $\mathbf{F} = M\mathbf{i} + N\mathbf{j} + P\mathbf{k}$ is conservative if and only if

$$\frac{\partial P}{\partial y} = \frac{\partial N}{\partial z}, \qquad \frac{\partial M}{\partial z} = \frac{\partial P}{\partial x}, \qquad \text{and} \qquad \frac{\partial N}{\partial x} = \frac{\partial M}{\partial y}. \qquad (9)$$

When we know that $\mathbf{F}$ is conservative, we usually want to find a **potential function** or **potential** for $\mathbf{F}$, i.e., a function f with

$$\nabla f = \frac{\partial f}{\partial x}\mathbf{i} + \frac{\partial f}{\partial y}\mathbf{j} + \frac{\partial f}{\partial z}\mathbf{k} = M\mathbf{i} + N\mathbf{j} + P\mathbf{k}. \qquad (10)$$

We do this by integrating the equations

$$\frac{\partial f}{\partial x} = M, \qquad \frac{\partial f}{\partial y} = N, \qquad \frac{\partial f}{\partial z} = P$$

to solve for f, as in the following example.

EXAMPLE 1 Show that

$$\mathbf{F} = (e^x \cos y + yz)\mathbf{i} + (xz - e^x \sin y)\mathbf{j} + (xy + z)\mathbf{k}$$

is conservative and find a potential function for it.

Solution We apply the test of Eq. (9) to

$$M = e^x \cos y + yz, \qquad N = xz - e^x \sin y, \qquad P = xy + z$$

and calculate

$$\frac{\partial M}{\partial z} = y = \frac{\partial P}{\partial x}, \qquad \frac{\partial N}{\partial z} = x = \frac{\partial P}{\partial y}, \qquad \frac{\partial M}{\partial y} = -e^x \sin y + z = \frac{\partial N}{\partial x}.$$

These equalities tell us that there is a function f with $\nabla f = \mathbf{F}$.

We find f by integrating the equations

$$\frac{\partial f}{\partial x} = e^x \cos y + yz, \qquad \frac{\partial f}{\partial y} = xz - e^x \sin y, \qquad \frac{\partial f}{\partial z} = xy + z. \quad (11)$$

We integrate the first equation with respect to x, holding y and z fixed, to get

$$f(x, y, z) = e^x \cos y + xyz + g(y, z).$$

We write the constant of integration as a function of y and z because its value may change if y and z change. We then calculate $\partial f/\partial y$ from this equation and match it with the expression for $\partial f/\partial y$ in (11). This gives

$$-e^x \sin y + xz + \frac{\partial g}{\partial y} = xz - e^x \sin y,$$

or

$$\frac{\partial g}{\partial y} = 0.$$

Therefore, g is a function of z alone, and

$$f(x, y, z) = e^x \cos y + xyz + h(z).$$

We now calculate $\partial f/\partial z$ from this equation and match it to the formula for $\partial f/\partial z$ in (11). This gives

$$xy + \frac{dh}{dz} = xy + z, \qquad \text{or} \qquad \frac{dh}{dz} = z,$$

so

$$h(z) = \frac{z^2}{2} + C.$$

Hence,

$$f(x, y, z) = e^x \cos y + xyz + \frac{z^2}{2} + C.$$

There are infinitely many potential functions for $\mathbf{F}$, one for each value of C. ∎

EXAMPLE 2 Show that

$$\mathbf{F} = (2x - 3)\mathbf{i} - z\mathbf{j} + (\cos z)\mathbf{k}$$

is not conservative.

Solution We apply the test in Eqs. (9) and find right away that

$$\frac{\partial P}{\partial y} = \frac{\partial}{\partial y} (\cos z) = 0$$

while

$$\frac{\partial N}{\partial z} = \frac{\partial}{\partial z} (-z) = -1.$$

Hence $\partial P/\partial y \neq \partial N/\partial z$, so **F** is not conservative. No further testing is required.   ∎

Exact Differential Forms

The forms

$$M(x, y, z)\, dx + N(x, y, z)\, dy + P(x, y, z)\, dz$$

that appear in our line integrals are called **differential forms.** A differential form is said to be **exact** on a region D if it is the differential of a function f at every point of D. That is, the form is exact on D if

$$M\, dx + N\, dy + P\, dz = \frac{\partial f}{\partial x}\, dx + \frac{\partial f}{\partial y}\, dy + \frac{\partial f}{\partial z}\, dz = df \qquad (12)$$

for some function f throughout D. If the form is exact, the field $\mathbf{F} = M\mathbf{i} + N\mathbf{j} + P\mathbf{k}$ is the gradient field of f. Conversely, if $\mathbf{F} = \nabla f$, the differential form is exact. The test for exactness of the differential form is therefore the same as the test for **F** being conservative:

$$\frac{\partial P}{\partial y} = \frac{\partial N}{\partial z}, \qquad \frac{\partial M}{\partial z} = \frac{\partial P}{\partial x}, \qquad \text{and} \qquad \frac{\partial N}{\partial x} = \frac{\partial M}{\partial y}. \qquad (13)$$

This is equivalent to saying that the form is exact if and only if $\nabla \times \mathbf{F} = \mathbf{0}$.

The Test for Exactness of $M\, dx + N\, dy + P\, dz$

The differential form $M\, dx + N\, dy + P\, dz$ is exact if and only if

$$\frac{\partial P}{\partial y} = \frac{\partial N}{\partial z}, \qquad \frac{\partial M}{\partial z} = \frac{\partial P}{\partial x}, \qquad \text{and} \qquad \frac{\partial N}{\partial x} = \frac{\partial M}{\partial y}. \qquad (14)$$

This is equivalent to saying that the field $\mathbf{F} = M\mathbf{i} + N\mathbf{j} + P\mathbf{k}$ is conservative or that

$$\nabla \times \mathbf{F} = \mathbf{0}. \qquad (15)$$

EXAMPLE 3 Show that $y\, dx + x\, dy + 4\, dz$ is exact and evaluate the integral

$$\int_{(1,1,1)}^{(2,3,-1)} y\, dx + x\, dy + 4\, dz$$

along the line segment from $(1, 1, 1)$ to $(2, 3, -1)$.

Solution We let

$$M = y, \qquad N = x, \qquad P = 4$$

and apply the test of Eq. (14):

$$\frac{\partial M}{\partial z} = 0 = \frac{\partial P}{\partial x}, \qquad \frac{\partial N}{\partial z} = 0 = \frac{\partial P}{\partial y}, \qquad \frac{\partial M}{\partial y} = 1 = \frac{\partial N}{\partial x}.$$

These equalities tell us that $y \, dx + x \, dy + 4 \, dz$ is exact.
Because the integrand is exact,

$$y \, dx + x \, dy + 4 \, dz = df$$

for some function f, and the integral's value is $f(2, 3, -1) - f(1, 1, 1)$.
We find f up to a constant by integrating the equations

$$\frac{\partial f}{\partial x} = y, \qquad \frac{\partial f}{\partial y} = x, \qquad \frac{\partial f}{\partial z} = 4. \qquad (16)$$

From the first equation we get

$$f(x, y, z) = xy + g(y, z).$$

The second equation tells us that

$$\frac{\partial f}{\partial y} = x + \frac{\partial g}{\partial y} = x, \qquad \text{or} \qquad \frac{\partial g}{\partial y} = 0.$$

Hence, g is a function of z alone, and

$$f(x, y, z) = xy + h(z).$$

The third equation in (14) tells us that

$$\frac{\partial f}{\partial z} = 0 + \frac{dh}{dz} = 4, \qquad \text{or} \qquad h(z) = 4z + C.$$

Therefore,

$$f(x, y, z) = xy + 4z + C,$$

and the value of the integral is

$$f(2, 3, -1) - f(1, 1, 1) = 2 + C - (5 + C) = -3.$$

Problem 21 suggests another way to evaluate the integral.   ■

In the next chapter, we shall see exact differential forms again when they
arise in our study of differential equations.

PROBLEMS

In Problems 1–6, determine which fields are conservative and
which are not.

1. $\mathbf{F} = yz\mathbf{i} + xz\mathbf{j} + xy\mathbf{k}$

2. $\mathbf{F} = (y \sin z)\mathbf{i} + (x \sin z)\mathbf{j} + (xy \cos z)\mathbf{k}$

3. $\mathbf{F} = y\mathbf{i} + (x + z)\mathbf{j} - y\mathbf{k}$

4. $\mathbf{F} = -y\mathbf{i} + x\mathbf{j}$

5. $\mathbf{F} = (z + y)\mathbf{i} + z\mathbf{j} + (y + x)\mathbf{k}$

6. $\mathbf{F} = (e^x \cos y)\mathbf{i} - (e^x \sin y)\mathbf{j} + z\mathbf{k}$

In Problems 7–10, find a potential f for the field $\mathbf{F}$.

7. $\mathbf{F} = 2x\mathbf{i} + 3y\mathbf{j} + 4z\mathbf{k}$

8. $\mathbf{F} = (y + z)\mathbf{i} + (x + z)\mathbf{j} + (x + y)\mathbf{k}$

9. $\mathbf{F} = e^{y+2z}(\mathbf{i} + x\mathbf{j} + 2x\mathbf{k})$

10. $\mathbf{F} = (y \sin z)\mathbf{i} + (x \sin z)\mathbf{j} + (xy \cos z)\mathbf{k}$

Evaluate the integrals in Problems 11–20.

11. $\displaystyle\int_{(0,0,0)}^{(2,3,-6)} 2x \, dx + 2y \, dy + 2z \, dz$

12. $\displaystyle\int_{(1,1,2)}^{(3,5,0)} yz\,dx + xz\,dy + xy\,dz$

13. $\displaystyle\int_{(0,0)}^{(1,1)} x\,dx + y^2\,dy$

14. $\displaystyle\int_{(1,0)}^{(0,1)} 2x\cos y\,dx - x^2\sin y\,dy$

15. $\displaystyle\int_{(1,0)}^{(1,\pi/2)} 2x\sin y\,dx + x^2\cos y\,dy$

16. $\displaystyle\int_{(1,1)}^{(2,3)} (x^2 + y)\,dx + (y^2 + x)\,dy$

17. $\displaystyle\int_{(1,1)}^{(2,3)} (y^2x + y)\,dx + (x^2y + x)\,dy$

18. $\displaystyle\int_{(0,0,0)}^{(1,2,3)} 2xy\,dx + (x^2 + z^2)\,dy + 2yz\,dz$

19. $\displaystyle\int_{(0,1,1)}^{(2,2,1)} 3x^2\,dx + \frac{z^2}{y}\,dy + 2z\ln y\,dz$

20. $\displaystyle\int_{(1,0,0)}^{(0,1,1)} \sin y\cos x\,dx + \cos y\sin x\,dy + dz$

21. Evaluate the integral

$$\int_{(1,1,1)}^{(2,3,-1)} y\,dx + x\,dy + 4\,dz$$

from Example 3 by finding parametric equations for the line segment from $(1, 1, 1)$ to $(2, 3, -1)$ and evaluating the line integral of $\mathbf{F} = y\mathbf{i} + x\mathbf{j} + 4\mathbf{k}$ along the segment. This is permissible because $\mathbf{F}$ is conservative and the integral is independent of the path.

22. Show that the value of

$$\int_A^B z^2\,dx + 2y\,dy + xz\,dz$$

does not depend on the path taken from A to B.

23. Evaluate

$$\int_C x^2\,dx + yz\,dy + y^2\,dz$$

along the line segment joining $(0, 0, 0)$ to $(0, 3, 4)$.

24. Show that the field

$$\mathbf{F} = \mathbf{i} + z\mathbf{j} - y\mathbf{k}$$

is not conservative.

25. What does Green's theorem say about the circulation of a conservative field around a closed curve in a simply connected open region in the plane?

26. What does Stokes's theorem say about the circulation of a conservative field around a closed curve in a simply connected open region in space?

27. Show that the curl of

$$\mathbf{F} = \frac{-y}{x^2 + y^2}\mathbf{i} + \frac{x}{x^2 + y^2}\mathbf{j} + z\mathbf{k}$$

is zero but that

$$\int_C \mathbf{F}\cdot d\mathbf{R}$$

is not zero if C is the circle $x^2 + y^2 = 1$ in the xy-plane. (Theorem 2 does not apply here because the domain of $\mathbf{F}$ is not simply connected. The field $\mathbf{F}$ is not defined along the z-axis so there is no way to contract C to a point without leaving the domain of $\mathbf{F}$.)

 TOOLKIT PROGRAMS

Scalar Fields   Vector Fields

REVIEW QUESTIONS AND EXERCISES

1. What is a line integral? What are line integrals used for? Give examples.

2. What is a vector field? Give examples.

3. How do you calculate the work done by a force in moving a particle along a curve?

4. What are the flux density and circulation density of a vector field? How are they related to the divergence and curl of the field?

5. What theorem relates flux to the divergence for a vector field defined in the plane? Give an example.

6. What theorem relates circulation to the curl for a vector field defined in the plane? Give an example.

7. How is surface area calculated? Give an example.

8. What is a surface integral? How are surface integrals used to calculate masses and moments? Give examples.

9. What theorem relates flux to the divergence of a vector field in space? Give an example.

10. What theorem relates circulation to the curl of a vector field in space? Give an example.

11. What does it mean for a vector field to be conservative in a region D in space? How are conservative fields characterized?

12. If $f(x, y, z) = 2x - 3y + e^z$ and C is a smooth curve from $A(0, 0, 0)$ to $(1, 2, \ln 3)$, what is the value of

$$\int_C \nabla f\cdot d\mathbf{R}?$$

13. How do you find a potential function for a conservative field? How do you evaluate

$$\int_A^B \mathbf{F} \cdot d\mathbf{R}$$

in a conservative field? Give examples.

14. If $\mathbf{F} = \nabla f$, S is a level surface of f, and C is a curve on S, why is it true (or not true) that

$$\int_C \mathbf{F} \cdot d\mathbf{R} = 0?$$

MISCELLANEOUS PROBLEMS

1. Figure 19.61 shows two line segment paths in space from the origin to the point (1, 1, 1). Integrate $f(x, y, z) = 2x - y^2 - 2z + 1$ over each path.

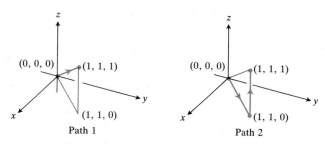

19.61   The paths in Problem 1.

2. Figure 19.62 shows three line segment paths in space from the origin to the point (1, 1, 1). Integrate $f(x, y, z) = x^2 + y - z$ along each path.

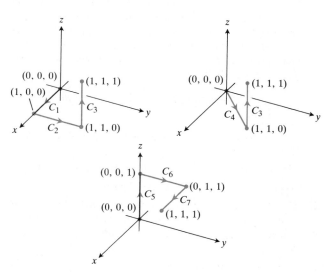

19.62   The paths in Problem 2.

3. A thin wire of constant density lies along the curve

$$\mathbf{R}(t) = (e^t \cos t)\mathbf{i} + (e^t \sin t)\mathbf{j} + e^t\mathbf{k}, \quad 0 \le t \le \ln 2.$$

Find I_z and R_z.

4. A slender metal arch lies along the semicircle $y = \sqrt{a^2 - x^2}$ in the xy-plane. The density at the point (x, y) on the arch is $\delta(x, y) = 2a - y$. Find the center of mass.

5. A wire of constant density δ lies along the curve

$$\mathbf{R}(t) = (t \cos t)\mathbf{i} + (t \sin t)\mathbf{j} + \frac{2\sqrt{2}}{3} t^{3/2}\mathbf{k}, \quad 0 \le t \le 1.$$

Find $\bar{z}$, I_z, and R_z.

6. A wire lies along the curve

$$\mathbf{R}(t) = t\mathbf{i} + t\mathbf{j} + (4 - t^2)\mathbf{k}, \quad 0 \le t \le 1.$$

Find the mass of the wire if a) δ is constant, b) $\delta = t$.

Sketch the vector fields in Problems 7–10.

7. $\mathbf{F} = x\mathbf{i} + y\mathbf{j} + z\mathbf{k}$

8. $\mathbf{F} = -x\mathbf{i} - y\mathbf{j} - z\mathbf{k}$

9. $\mathbf{F} = (x - y)\mathbf{i} + (x + y)\mathbf{j}$

10. $\mathbf{F} = (x\mathbf{i} + y\mathbf{j})/(x^2 + y^2)$

11. Write a formula for a vector field $\mathbf{F}(x, y, z)$ such that $\mathbf{F} = f(\rho)\mathbf{R}$, where $\rho = |\mathbf{R}|$ and $\mathbf{R} = x\mathbf{i} + y\mathbf{j} + z\mathbf{k}$, if
a) $\mathbf{F}$ is directed radially outward from the origin and $|\mathbf{F}| = \rho^{-n}$,
b) $\mathbf{F}$ is directed toward the origin and $|\mathbf{F}| = 2$,
c) $\mathbf{F}$ is a gravitational attraction field for a mass M at the origin in which an inverse-*cube* law applies. (Don't worry about the possible nonexistence of such a field.)

12. Suppose that

$$R = \{(x, y): 0 \le y \le \sqrt{a^2 - x^2}, \quad -a \le x \le a\}$$

and that C is the boundary of R.
a) Sketch R and C.
b) Write out the proof of Green's theorem for this region.

13. Find the area of the surface cut from the end of the paraboloid $y^2 + z^2 = 2x$ by the plane $x = 1$.

14. Find the area of the elliptical region cut from the plane $x + y + z = 1$ by the cylinder $x^2 + y^2 = 1$.

15. Find the area of the triangle cut from the plane $x/a + y/b + z/c = 1$ (a, b, and $c > 0$) by the first octant. Check your answer by vector methods.

16. Find the area of the region above the xy-plane cut from the cone $x^2 + y^2 = z^2$ by the cylinder $x^2 + y^2 = 2ax$.

17. Find a double integral for the area of the surface cut from the cylinder $z = a^2 - y^2$ by the cylinder $x^2 + y^2 = a^2$, and reduce this double integral to a single integral with respect to the variable y.

18. A square hole of side $2\sqrt{2}$ is cut symmetrically through a sphere of radius 2. Show that the area of the surface removed is $16\pi(\sqrt{2} - 1)$.

19. Find the surface area of the torus in Fig. 18.60 in the Miscellaneous Problems at the end of Chapter 18.

20. Find the area of the surface $(x^2 + y^2 + z^2)^2 = x^2 - y^2$. (*Hint:* Use polar coordinates.)

21. Find the area of the spherical part of the boundary of the region

$$x^2 + y^2 + z^2 = 1,$$
$$x^2 + y^2 - x \geq 0,$$
$$x^2 + y^2 + x \geq 0.$$

(*Hint:* Integrate first with respect to x and y.)

22. Find the area of the region cut from the surface $az = y^2 - x^2$ by the cylinder $x^2 + y^2 = a^2$.

23. Find the area of the surface cut from the hemisphere $x^2 + y^2 + z^2 = 4$, $z \geq 0$, by the cylinder $x^2 + y^2 = 2x$ (Fig. 19.63).

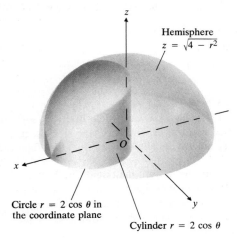

Hemisphere
$z = \sqrt{4 - r^2}$

Circle $r = 2\cos\theta$ in the coordinate plane

Cylinder $r = 2\cos\theta$

19.63 The surface cut from the hemisphere $x^2 + y^2 + z^2 = 4$, $z \geq 0$, by the cylinder $x^2 + y^2 = 2x$ in Problem 23. In cylindrical coordinates, the equations of the hemisphere and cylinder are $z = \sqrt{4 - r^2}$ and $r = 2\cos\theta$.

24. Find the area of the part of the cylinder in Problem 23 that lies inside the hemisphere. (*Hint:* Project onto the xz-plane. Or use single integration, $\int h\,ds$, where h is the altitude of the cylinder and ds is the element of arc length in the xy-plane.)

25. *Special formulas for surface area.* If S is a smooth surface given by an equation $z = f(x, y)$ defined over a region R_{xy} in the xy-plane, we may write

$$F(x, y, z) = f(x, y) - z$$

and regard S as the level surface $F(x, y, z) = 0$ of the func-

tion F. Taking the unit normal to R_{xy} to be $\mathbf{p} = \mathbf{k}$ then gives

$$|\nabla F| = |f_x\mathbf{i} + f_y\mathbf{j} - \mathbf{k}| = \sqrt{f_x^2 + f_y^2 + 1},$$
$$|\nabla F \cdot \mathbf{p}| = |(f_x\mathbf{i} + f_y\mathbf{j} - \mathbf{k}) \cdot \mathbf{k}| = |-1| = 1,$$

and

$$\text{Area of } S = \iint\limits_{R_{xy}} \frac{|\nabla F|}{|\nabla F \cdot \mathbf{p}|}\,dA \qquad (1)$$

$$= \iint\limits_{R_{xy}} \sqrt{f_x^2 + f_y^2 + 1}\,dx\,dy.$$

Similarly, the area of a smooth surface $x = f(y, z)$ over a region R_{yz} in the yz-plane is

$$\text{Area of } S = \iint\limits_{R_{yz}} \sqrt{f_y^2 + f_z^2 + 1}\,dy\,dz, \qquad (2)$$

and the area of a smooth surface $y = f(x, z)$ over a region R_{xz} in the xz-plane is

$$\text{Area of } S = \iint\limits_{R_{xz}} \sqrt{f_x^2 + f_z^2 + 1}\,dx\,dz. \qquad (3)$$

a) Find the area of the surface cut from the bottom of the paraboloid $z = x^2 + y^2$ by the plane $z = 4$.
b) Find the area of the surface cut from the bottom of the paraboloid $x = 1 - y^2 - z^2$ by the yz-plane.
c) Find the area of the surface in the first quadrant cut from the cylinder $2y = z^2$ by the planes $x = 0$, $x = 1$, $y = 0$, and $y = 2$.
d) Derive Eq. (2).
e) Derive Eq. (3).

26. Suppose that the surface of the hemisphere $x^2 + y^2 + z^2 = a^2$, $z \geq 0$, is divided by arcs of great circles on which the spherical coordinate θ remains constant (meridians of longitude) and by circles parallel to the xy-plane on which ϕ remains constant (parallels of latitude). Let the angular spacings be $\Delta\theta$ and $\Delta\phi$, respectively. Express the integral $\iint_S z\,d\sigma$ over the hemisphere in the form

$$\lim_{\substack{\Delta\theta \to 0 \\ \Delta\phi \to 0}} \sum F(\theta, \phi)\,\Delta\theta\,\Delta\phi = \iint F(\theta, \phi)\,d\theta\,d\phi,$$

with appropriate limits of integration, and evaluate. (*Hint:* You should get

$$d\sigma = (r\,d\theta) \cdot (\rho\,d\phi) = \rho^2\sin\phi\,d\theta\,d\phi = a^2\sin\phi\,d\theta\,d\phi,$$

where ρ, ϕ, θ are spherical coordinates.)

27. Use the result $d\sigma = a^2\sin\phi\,d\theta\,d\phi$ from Problem 26 to find the average great-circle distance of a point on the hemisphere $x^2 + y^2 + z^2 = a^2$, $z \geq 0$, from the north pole $N(0, 0, a)$.

28. Heat flows from a hotter to a cooler body. In three-dimensional heat flow, the fundamental equation for the rate at which heat flows out of a region D is

$$\iint\limits_S K\frac{\partial u}{\partial n}\,d\sigma = \iiint\limits_D c\delta\frac{\partial u}{\partial t}\,dV. \qquad (1)$$

The symbolism in this equation is as follows.

$u = u(x, y, z, t)$ the temperature at the point (x, y, z) at time t

K the thermal conductivity coefficient

δ the mass density

c the specific heat coefficient; this is the amount of heat required to raise one unit of mass of the material of the body one degree

S the boundary surface of the region D

$\dfrac{\partial u}{\partial n}$ the directional derivative in the direction of the outward normal to S

How is $\partial u/\partial n$ related to the *gradient* of the temperature? In which direction (described in words) does ∇u point? Why does the left-hand side of Eq. (1) appear to make sense as a measure of the rate of flow? Now look at the right-hand side of Eq. (1): If ΔV is a small volume element in D, what does $\delta\, \Delta V$ represent? If the temperature of this element changes by an amount Δu in time Δt, what is (a) the amount, (b) the average rate of change of heat in the element? In words, what does the right-hand side of Eq. (1) represent physically? Is it reasonable to interpret Eq. (1) as saying that the rate at which heat flows out through the boundary of D is equal to the rate at which heat is being supplied from D?

29. Assuming Eq. (1), Problem 28, and assuming that there is no heat source or sink in D, derive the equation

$$\nabla \cdot (K\, \nabla u) = c\delta\, \frac{\partial u}{\partial t} \qquad (2)$$

as the equation that must be satisfied at each point in D. (*Suggestion:* Apply the divergence theorem to the left-hand side of Eq. 1, and make D be a sphere of radius ϵ; then let $\epsilon \to 0$.)

30. *The Laplace operator.* Assuming the result of Problem 29 and assuming that K, c, and δ are constants, deduce that the condition for steady-state temperature in D is Laplace's equation

$$\nabla^2 u = 0, \qquad \text{or} \qquad \text{div (grad } u) = 0.$$

In higher mathematics, the symbol Δ is used for the *Laplace operator:*

$$\Delta u = \frac{\partial^2 u}{\partial x^2} + \frac{\partial^2 u}{\partial y^2} + \frac{\partial^2 u}{\partial z^2}.$$

Thus, in this notation,

$$\Delta u = \nabla^2 u = \nabla \cdot \nabla u = \text{div (grad } u).$$

Using the Divergence Theorem and assuming that the functions u and v and their first and second order partial derivatives are continuous in the regions considered, verify the formulas in Problems 31–34. Assume that S is the boundary surface of the simply connected region D.

31. $\displaystyle\iint_S u\, \nabla v \cdot d\sigma = \iiint_D [u\, \Delta v + (\nabla u) \cdot (\nabla v)]\, dV$

32. $\displaystyle\iint_S \left(u\, \frac{\partial v}{\partial n} - v\, \frac{\partial u}{\partial n} \right) d\sigma = \iiint_D (u\, \Delta v - v\, \Delta u)\, dV$

(*Suggestion:* Use the result of Problem 31 as given and in the form you get by interchanging u and v.)

33. $\displaystyle\iint_S u\, \frac{\partial u}{\partial n}\, d\sigma = \iiint_D [u\, \Delta u + |\nabla u|^2]\, dV$

(*Suggestion:* Use the result of Problem 31 with $v = u$.)

34. $\displaystyle\iint_S \frac{\partial u}{\partial n}\, d\sigma = \iiint_D \Delta u\, dV$

(*Suggestion:* Use $v = -1$ in the result of Problem 32.)

35. A function u is *harmonic* in a region D if and only if it satisfies *Laplace's equation* $\Delta u = 0$ throughout D. Use the identity in Problem 33 to deduce that if u is harmonic in D and either $u = 0$ or $\partial u/\partial n = 0$ at all points on the surface S that is the boundary of D, then $\nabla u = \mathbf{0}$ throughout D, and therefore u is constant throughout D.

36. The result of Problem 35 can be used to establish the uniqueness of solutions of Laplace's equation in D, provided that either (i) the value of u is prescribed at each point on S or (ii) the value of $\partial u/\partial n$ is prescribed at each point of S. This is done by supposing that u_1 and u_2 are harmonic in D and that both satisfy the same boundary conditions and then letting $u = u_1 - u_2$. Complete this uniqueness proof.

37. Problems 28–30 deal with heat flow. Assume that K, c, and δ are constant and that the temperature $u = u(x, y, z)$ does not vary with time. Use the results of Problems 30 and 34 to conclude that the net rate of outflow of heat through the surface S is zero. (*Note:* This result might apply, for example, to the region D between two concentric spheres if the inner one were maintained at $100°$ and the outer one at $0°$, so that heat would flow into D through the inner surface and out through the outer surface at the same rate.)

38. Let $\rho = (x^2 + y^2 + z^2)^{1/2}$. Show that

$$u = C_1 + \frac{C_2}{\rho}$$

is harmonic where $\rho > 0$, if C_1 and C_2 are constants. Find values of C_1 and C_2 so that the following boundary conditions are satisfied:
a) $u = 100$ when $\rho = 1$, and $u = 0$ when $\rho = 2$;
b) $u = 100$ when $\rho = 1$, and $\partial u/\partial n = 0$ when $\rho = 2$.

(*Note:* Part (a) refers to a steady-state heat flow problem that is like the one discussed at the end of Problem 37; part (b) refers to an insulated boundary on the sphere $\rho = 2$.)

In Problems 39–41, find the circulation of $\mathbf{F}$ around C in the indicated direction.

39. $\mathbf{F} = x\mathbf{i} + y\mathbf{j} + z\mathbf{k}$
   C: The ellipse in which the plane $2x + 3y - z = 0$ meets the cylinder $x^2 + y^2 = 12$, counterclockwise as viewed from above

40. $\mathbf{F} = \nabla(xy^2z^3)$

C: Same as in Problem 39

41. $\mathbf{F} = \nabla \times (x\mathbf{i} + y\mathbf{j} + z\mathbf{k})$

C: The circle $x^2 + y^2 = 1$, $z = 0$, clockwise as viewed from above

42. Evaluate the flow integral of $\mathbf{F} = \nabla \times (x\mathbf{i} + y\mathbf{j} + z\mathbf{k})$ along the line segment from the origin to the point $(1, 2, 3)$.

43. Evaluate the flow integral of $\mathbf{F} = \nabla(xy^2z^3)$ along the line segment from $(1, 1, 1)$ to $(2, 1, -1)$.

44. Let C denote the plane curve whose vector equation is

$$\mathbf{R}(t) = (e^t \cos t)\mathbf{i} + (e^t \sin t)\mathbf{j}.$$

Evaluate the line integral

$$\int \frac{x \, dx + y \, dy}{(x^2 + y^2)^{3/2}}$$

along C from the point $(1, 0)$ to the point $(e^{2\pi}, 0)$.

45. If the density $\rho(x, y, z)$ of a fluid is a function of the pressure $p(x, y, z)$ and

$$f(x, y, z) = \int_{p_0}^{p} \frac{dp}{\rho},$$

where p_0 is a constant, show that $\nabla f = \nabla p / \rho$.

Green's Theorem for Simply Connected Regions

46. Show, by a geometric argument, that Green's theorem holds for any simply connected region R whose boundary

is a simple closed curve C, provided R can be decomposed into a finite number of nonoverlapping regions R_1, $R_2, \ldots, R_n$ with boundaries $C_1, C_2, \ldots, C_n$ of a type for which the theorem holds for each R_i and C_i, $i = 1, \ldots, n$.

47. Suppose R is a region in the xy-plane, C is its boundary, and the area of R is given by

$$A(R) = \oint_C \frac{1}{2}(x \, dy - y \, dx).$$

Suppose the equations $x = f(u, v)$, $y = g(u, v)$ map R and C in a continuous and one-to-one manner onto a region R' and curve C', respectively, in the uv-plane. Use Green's formula to show that

$$\iint_R dx \, dy = \iint_{R'} \begin{vmatrix} f_u & f_v \\ g_u & g_v \end{vmatrix} du \, dv$$

$$= \iint_{R'} \left(\frac{\partial f}{\partial u} \frac{\partial g}{\partial v} - \frac{\partial f}{\partial v} \frac{\partial g}{\partial u} \right) du \, dv.$$

(*Hint:* Note that

$$\iint_R dx \, dy = \frac{1}{2} \int_C (x \, dy - y \, dx)$$

$$= \frac{1}{2} \int_{C'} \left[f(u, v) \left(\frac{\partial g}{\partial u} du + \frac{\partial g}{\partial v} dv \right) \right.$$

$$\left. - g(u, v) \left(\frac{\partial f}{\partial u} du + \frac{\partial f}{\partial v} dv \right) \right],$$

and apply Green's formula to C' and R'.)

CHAPTER 20

Differential Equations

OVERVIEW

The subject of differential equations was introduced in the overviews to Chapters 4 and 6. There we saw that a mathematical model for a physical phenomenon often involves an equation of the form

$$\frac{dy}{dx} = f(x, y) \tag{1a}$$

and an initial condition

$$y = y_0 \quad \text{when} \quad x = x_0. \tag{1b}$$

Equations (1a) and (1b) define what is called an "initial value problem" in differential equations. In addition to the methods we have already studied for solving such problems, we shall learn some new methods, including graphical and numerical methods and the use of infinite series.

The differential equations that describe planetary motion, derived from Newton's second law of motion, involve both acceleration and velocity. We studied these in Chapter 14. For rockets, near the Earth's surface and in space, the differential equations of motion are more complex. Numerical methods are used to solve them, using high-speed and large-scale computers. The computers also direct the rocket engines to fire at times and in ways that will give an appropriate orbit. The requirements for speed and accuracy in such computer applications have been, and are, powerful motivating factors in the research and development of faster and more reliable computer hardware and software.

The purpose of this chapter is to give some idea of the types of equations that can be solved analytically, together with methods for solving them. Sometimes the analytic methods don't work, so that numerical methods are needed, even as Simpson's rule is used to evaluate a definite integral. (Even when an analytic solution can be found, the numerical approach may be more useful, as in space dynamics and control.)

Our treatment is necessarily brief. Scientists and engineers will most likely take courses in advanced calculus and differential equations where they will learn much more about differential equations: how they apply and how to solve them.

20.1
Introduction

A **differential equation** is an equation that involves one or more derivatives, or differentials. Differential equations are classified by

a) type (namely, **ordinary** or **partial**);

b) **order** (that of the highest-order derivative that occurs in the equation); and

c) **degree** (the exponent of the highest power of the highest-order derivative after the equation has been cleared of fractions and radicals in the dependent variable and its derivatives).

For example,

$$\left(\frac{d^3y}{dx^3}\right)^2 + \left(\frac{d^2y}{dx^2}\right)^5 + \frac{y}{x^2 + 1} = e^x \tag{1}$$

is an ordinary differential equation, of order three and degree two.

Only "ordinary" derivatives occur when the dependent variable y is a function of a single independent variable x. On the other hand, if the dependent variable y is a function of two or more independent variables, say

$$y = f(x, t),$$

where x and t are independent variables, then partial derivatives of y may occur. For example, the wave equation

$$\frac{\partial^2 y}{\partial t^2} = c^2 \frac{\partial^2 y}{\partial x^2} \tag{2}$$

(Article 16.8, Problems 32–38) is a partial differential equation, of order two and degree one. (A systematic treatment of partial differential equations lies beyond the scope of this book. For a discussion of partial differential equations, including the wave equation, and solutions of associated physical problems, see Kaplan, *Advanced Calculus,* 3rd ed., Chapter 10, Addison-Wesley, 1984.)

Many physical problems, when formulated in mathematical terms, lead to differential equations. In Article 13.2, for example, we discussed and solved the system of differential equations

$$m \frac{d^2x}{dt^2} = 0, \qquad m \frac{d^2y}{dt^2} = -mg, \tag{3}$$

which describes the motion of a projectile (neglecting air resistance). Indeed, one of the chief sources of differential equations in mechanics is Newton's second law:

$$\mathbf{F} = \frac{d}{dt}(m\mathbf{v}), \tag{4}$$

where $\mathbf{F}$ is the resultant of the forces acting on a body of mass m and $\mathbf{v}$ is its velocity.

An example from the field of chemical kinetics may be represented by a reactant A that undergoes parallel transformations into products B and C at rates that are proportional to the amount of A that is present at time t. (See Fig. 20.1.)

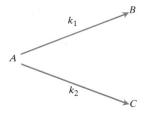

20.1 The reactant A yields products B and C at rates proportional to the amount of A present.

If x, y, and z denote the amounts of A, B, and C present at time t, then the differential equations that describe the process are

$$\frac{dx}{dt} = -(k_1 + k_2)x, \qquad \frac{dy}{dt} = k_1 x, \qquad \frac{dz}{dt} = k_2 x. \tag{5}$$

If, at time $t = 0$, only A is present, the initial conditions are

$$x(0) = x_0, \qquad y(0) = 0, \qquad z(0) = 0. \tag{6}$$

Equations (5) and (6) together provide an example of an **initial value problem.** The first of Eqs. (5) can easily be solved for x to yield

$$x = x_0 e^{-(k_1 + k_2)t}. \tag{7a}$$

When this is substituted into the remaining Eqs. (5), they can be integrated also, to give

$$y = \frac{k_1}{k_1 + k_2}(1 - e^{-(k_1 + k_2)t})\, x_0, \tag{7b}$$

$$z = \frac{k_2}{k_1 + k_2}(1 - e^{-(k_1 + k_2)t})\, x_0. \tag{7c}$$

Together, Eqs. (7a, b, c) give a solution of the initial value problem of Eqs. (5) and (6), as can be verified by direct substitition.

Solutions

A function

$$y = f(x)$$

is said to be a **solution** of a differential equation if the latter is satisfied when y and its derivatives are replaced throughout by $f(x)$ and its corresponding derivatives. For example, if c_1 and c_2 are any constants, then

$$y = c_1 \cos x + c_2 \sin x \tag{8a}$$

is a solution of the differential equation

$$\frac{d^2 y}{dx^2} + y = 0. \tag{8b}$$

A physical problem that translates into a differential equation usually involves additional conditions not expressed by the differential equation itself. In mechanics, for example, the initial position and velocity of the moving body are usually prescribed, as well as the forces. The differential equation, or equations, of motion will usually have solutions in which certain arbitrary constants occur, as in Eq. (8a) above. Specific values are then assigned to these arbitrary constants to meet the prescribed initial conditions. (See the examples worked out in Articles 6.9 and 13.2.)

A differential equation of order n will generally have a solution involving n arbitrary constants. This solution is called the **general solution.** Once this general solution is known, it is only a matter of algebra to determine specific values of the constants if initial conditions are also prescribed. Hence we shall devote our attention to the problem of finding the general solutions of certain types of differential equations.

In the remainder of this chapter, the following topics will be treated. Throughout, only *ordinary* differential equations will be considered.

1. First order equations of first degree
   a) Variables separable
   b) Homogeneous
   c) Linear
   d) Exact differentials
2. Special types of second order equations
3. Linear equations with constant coefficients
   a) Homogeneous
   b) Nonhomogeneous
4. Approximation techniques
   a) Power series
   b) Numerical methods

Several new terms occur in this list, and they will be defined in the appropriate places.

PROBLEMS

Show that each function is a solution of the accompanying differential equation.

1. $xy'' - y' = 0$,
   a) $y = x^2 + 3$
   b) $y = C_1 x^2 + C_2$

2. $x^3y''' + 4x^2y'' + xy' + y = x$, $y = (1/2)x$

3. $yy'' = 2(y')^2 - 2y'$,
   a) $y = C$
   b) $C_1y = \tan(C_1 x + C_2)$

4. $y' + \dfrac{1}{x}y = 1$, $y = \dfrac{C}{x} + \dfrac{x}{2}$

5. $2y' + 3y = e^{-x}$, $y = e^{-x} + Ce^{-(3/2)x}$

6. $(x \sin x)y' + (\sin x + x \cos x)y = e^x/x$,

$$y = \frac{1}{x \sin x} \int_1^x \frac{e^t}{t}\, dt$$

20.2
First Order Differential Equations of First Degree

A differential equation of first order and first degree can always be put into the form

$$M\, dx + N\, dy = 0, \tag{1}$$

where M and N represent functions of x and y. When these functions satisfy certain conditions, the solution is relatively easy to find and can be written as

$$f(x, y) = C. \tag{2}$$

The special types of first order, first degree equations we consider are designated separable, homogeneous, linear, and exact.

Separable Equations

When M is a function of x alone and N is a function of y alone, then Eq. (1) has the form

$$M(x)\,dx + N(y)\,dy = 0.$$

This is the standard form of a first order separable equation. Such equations are very common and are used in many of the applications we discussed in Chapter 6 (compound interest, radioactive decay, Newton's law of cooling, and population growth). Once the equation has been put in standard form, we can find its solution by direct integration of each term, giving us

$$\int M(x)\,dx + \int N(y)\,dy = C.$$

EXAMPLE 1   Solve the differential equation

$$\frac{dy}{dx} = (1 + y^2)\,e^x. \tag{3}$$

Solution   We rewrite the equation in differential form, grouping all the y terms with dy and the x terms with dx:

$$\frac{dy}{1 + y^2} = e^x\,dx \qquad \text{or} \qquad e^x\,dx - \frac{dy}{1 + y^2} = 0.$$

Integration of the first part with respect to x and the second part with respect to y gives the solution

$$e^x - \tan^{-1}(y) = C. \tag{4}$$

If we solve for y, we get

$$y = \tan(e^x - C). \qquad\blacksquare$$

Homogeneous Equations

Occasionally a differential equation whose variables cannot be separated can be transformed by a change of variable into an equation whose variables can be separated. This is the case with any equation that can be put into the form

$$\frac{dy}{dx} = F\!\left(\frac{y}{x}\right). \tag{5}$$

Such an equation is called **homogeneous.**

   To transform Eq. (5) into an equation whose variables may be separated, we introduce the new variable

$$v = \frac{y}{x}. \tag{6}$$

Then

$$y = vx, \qquad \frac{dy}{dx} = v + x\frac{dv}{dx},$$

and (5) becomes

$$v + x\frac{dv}{dx} = F(v). \tag{7}$$

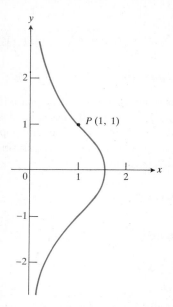

20.2   The solution of $(x^2 + y^2)\, dx + 2xy\, dy = 0$ with $y = 1$ when $x = 1$ is $x(x^2 + 3y^2) = 4$.

Equation (7) can be solved by separation of variables:

$$\frac{dx}{x} + \frac{dv}{v - F(v)} = 0. \tag{8}$$

After (8) is solved, the solution of the original equation is obtained when we replace v by y/x.

EXAMPLE 2   Show that the equation

$$(x^2 + y^2)\, dx + 2xy\, dy = 0$$

is homogeneous and solve it.

Solution   From the given equation, we have

$$\frac{dy}{dx} = -\frac{x^2 + y^2}{2xy} = -\frac{1 + (y/x)^2}{2(y/x)}.$$

This has the form of Eq. (5), with

$$F(v) = -\frac{1 + v^2}{2v}, \qquad \text{where} \quad v = \frac{y}{x}.$$

Then Eq. (8) becomes

$$\frac{dx}{x} + \frac{dv}{v + \dfrac{1 + v^2}{2v}} = 0, \qquad \text{or} \qquad \frac{dx}{x} + \frac{2v\, dv}{1 + 3v^2} = 0.$$

The solution of this equation can be written as

$$\ln |x| + \frac{1}{3} \ln (1 + 3v^2) = C,$$

so that

$$x^3(1 + 3v^2) = \pm e^{3C} = C'.$$

In terms of y and x, the solution is

$$x^3 \left(1 + \frac{3y^2}{x^2} \right) = x(x^2 + 3y^2) = C'. \qquad \blacksquare$$

REMARK   If we wish to satisfy the initial condition $y = 1$ when $x = 1$, we substitute these values into the equation $x(x^2 + 3y^2) = C'$ and get $C' = 4$. The graph of this equation is shown in Fig. 20.2. The graph is symmetric about the x-axis and the equation $x(x^2 + 3y^2) = 4$ does not define y as a function of x. However, there is only one x for each y so we can change our point of view and use the same equation, and graph, to define x as a function of y, satisfying the differential equation $(x^2 + y^2)\, dx + 2xy\, dy = 0$ and the initial condition $x = 1$ when $y = 1$.

Linear Equations

The complexity of a differential equation depends primarily on the way in which the *dependent* variable and its derivatives occur. Of particular importance are those equations that are linear. In a linear differential equation, each term of the equation is of degree one or zero, where, in computing the degree of a term, we add the exponents of the dependent variable and of any of its derivatives that occur. For example, (d^2y/dx^2) is of the first degree, while $y(dy/dx)$ is of the second degree because we must add 1 for the exponent of y and 1 for the exponent of dy/dx.

A linear differential equation of first order can always be put into the standard form

$$\frac{dy}{dx} + Py = Q, \tag{9}$$

where P and Q are functions of x.

Linear differential equations occur in many areas. As a simple example, consider a population that is changing due to births, deaths, immigration, and emigration. These rates may vary with time. If the population at time t is y, $k(t) = $ birth rate minus death rate, and $m(t) = $ immigration rate minus emigration rate, then the rate of change of the population is

$$\frac{dy}{dt} = k(t)y + m(t).$$

This equation is linear in y, and the methods we are now going to study can be used to solve it when $k(t)$ and $m(t)$ are known.

One common method for solving Eq. (9) is to find a function $\rho = \rho(x)$ such that if the equation is multiplied by ρ, the left side becomes the derivative of the product ρy. That is, we multiply (9) by ρ,

$$\rho \frac{dy}{dx} + \rho Py = \rho Q, \tag{9'}$$

and then impose upon ρ the condition that

$$\rho \frac{dy}{dx} + \rho Py = \frac{d}{dx}(\rho y). \tag{10}$$

When we expand the right-hand side of (10) and cancel terms, we obtain, as the condition to be satisfied by ρ,

$$\frac{d\rho}{dx} = \rho P. \tag{11}$$

In Eq. (11), $P = P(x)$ is a known function, so we can separate the variables and solve for ρ:

$$\frac{d\rho}{\rho} = P \, dx, \qquad \ln |\rho| = \int P \, dx + \ln C,$$
$$\rho = \pm C \, e^{\int P \, dx}. \tag{12}$$

Since we do not require the most general function ρ, we may take $\pm C = 1$ in (12) and use

$$\rho = e^{\int P \, dx}. \tag{13}$$

This function is called an **integrating factor** for Eq. (9). With its help, (9') becomes

$$\frac{d}{dx}(\rho y) = \rho Q,$$

whose solution is

$$\rho y = \int \rho Q \ dx. \tag{14}$$

Since ρ is given by (13), while P and Q are known from the given differential equation (9), we have, in Eqs. (13) and (14), a summary of all that is required to solve (9).

EXAMPLE 3 Solve the differential equation

$$\frac{dy}{dx} + y = e^x.$$

Solution This equation is already in standard form with $P = 1$ and $Q = e^x$. We calculate the integrating factor from the equation

$$\rho = e^{\int P \, dx} = e^{\int dx} = e^x.$$

When we multiply both sides of the original equation by this factor, the left side becomes

$$e^x \frac{dy}{dx} + e^x y = \frac{d}{dx}(e^x y)$$

and the right side becomes e^{2x}. We then have

$$\frac{d}{dx}(e^x y) = e^{2x},$$

which we now integrate to get

$$e^x y = \frac{1}{2}e^{2x} + C$$

$$y = \frac{1}{2}e^x + C e^{-x}.$$

■

EXAMPLE 4 Solve the equation

$$x \frac{dy}{dx} - 3y = x^2, \qquad x > 0.$$

Solution We put the equation in standard form,

$$\frac{dy}{dx} - \frac{3}{x}y = x,$$

and then read off

$$P = -\frac{3}{x}, \qquad Q = x.$$

Hence

$$\rho = e^{\int -(3/x) \, dx} = e^{-3 \ln x} = \frac{1}{e^{3 \ln x}} = \frac{1}{x^3},$$

and

$$\frac{1}{x^3} y = \int \frac{x}{x^3} \, dx = -\frac{1}{x} + C,$$

$$y = -x^2 + C \, x^3. \qquad \blacksquare$$

Note that whenever $\int P \, dx$ involves logarithms, as in Example 4, it is profitable to simplify the expression for $e^{\int P \, dx}$ before substituting into Eq. (14). To simplify the expression, we use the properties of the logarithmic and exponential functions:

$$e^{\ln A} = A, \qquad e^{m \ln A} = A^m, \qquad e^{n + m \ln A} = A^m e^n.$$

A differential equation that is linear in y and dy/dx may also be separable, or homogeneous. In such cases, we have a choice of methods of solution. Observe also that an equation that is linear in x and dx/dy can be solved by the integrating factor technique; we need only interchange the roles of x and y in Eqs. (9), (13), and (14).

Exact Equations

An equation that can be written in the form

$$M(x, y) \, dx + N(x, y) \, dy = 0 \qquad (15)$$

and having the property that

$$\frac{\partial M}{\partial y} = \frac{\partial N}{\partial x}, \qquad (16)$$

is said to be **exact,** because its left-hand side is an exact differential. (See Eqs. 12 and 13 in Article 19.7. To apply those equations to our present discussion, take $P = 0$ and notice that $M(x, y)$ and $N(x, y)$ do not depend on z. Then the three conditions in Eq. 13 reduce to the single condition $\partial M/\partial y = \partial N/\partial x$.) The technique of solving an exact equation consists in finding a function $f(x, y)$ such that

$$df = M \, dx + N \, dy. \qquad (17)$$

Then (15) becomes

$$df = 0$$

and the solution is

$$f(x, y) = C, \qquad (18)$$

where C is an arbitrary constant.

To find a function $f(x, y)$ that satisfies Eq. (17), we require that

$$\frac{\partial f}{\partial x} = M(x, y), \qquad \frac{\partial f}{\partial y} = N(x, y).$$

Integrating the first of these with respect to x, holding y constant, we get

$$f(x, y) = \int_x M(x, y) \, dx + g(y), \qquad (19)$$

■ ■ ■

THE SHAPE OF THE EARTH AND THE EXACT DIFFERENTIAL

In the eighteenth century, there were two conflicting ideas about the shape of the earth. One, held by the Cartesians, was that the earth was elongated at the poles. The other, held by the Newtonians, was that the earth was flattened at the poles. Many mathematicians took sides in the controversy, and in 1736 the young French mathematician Alexis Clairaut (1713–1765) joined an expedition to Lapland under the direction of Pierre de Maupertuis to make measurements of longitude to settle the matter. The result confirmed the Newtonian view. Clairaut published his studies on the shape of the earth in 1743 in a paper that also treated the equilibrium of fluids and the gravitational attraction of ellipsoids of revolution. In this work, as well as in earlier papers (in 1739 and 1740), we first find the partial derivative criterion for the exactness of a differential form $M \, dx + N \, dy$.

where $g(y)$ represents an unknown function of y that plays the role of an arbitrary constant of integration. The integral

$$\int_x M(x, y)\, dx$$

is an ordinary indefinite integral with respect to x with y held constant. The function f produced by Eq. (19) satisfies the condition

$$\frac{\partial f}{\partial x} = M(x, y)$$

for any choice of the function $g(y)$, because $g(y)$ is constant under partial differentiation with respect to x.

To see how we may satisfy the second condition, $\partial f/\partial y = N(x, y)$, we differentiate both sides of Eq. (19) with respect to y and set the result equal to $N(x, y)$:

$$\frac{\partial f}{\partial y} = \frac{\partial}{\partial y} \int_x M(x, y)\, dx + \frac{\partial g(y)}{\partial y} = N(x, y).$$

Because $g(y)$ is a function of y alone, we can write dg/dy in place of $\partial g/\partial y$ and get

$$\frac{dg}{dy} = N(x, y) - \frac{\partial}{\partial y} \int_x M(x, y)\, dx. \tag{20}$$

If the condition for exactness is satisfied, the right side of Eq. (20) is a function of y alone, and we can integrate it with respect to y to get $g(y)$. When we substitute this value of $g(y)$ in Eq. (19), we have a function $f(x, y)$ that gives the solution of the equation $M(x, y)\, dx + N(x, y)\, dy = 0$ in the form $f(x, y) = C$.

The next example shows how all this works.

EXAMPLE 5 Show that the equation

$$(x^2 + y^2)\, dx + (2xy + \cos y)\, dy = 0$$

is exact and solve it.

Solution   In this example we have

$$M(x, y) = x^2 + y^2, \qquad \frac{\partial M}{\partial y} = 2y,$$

$$N(x, y) = 2xy + \cos y, \qquad \frac{\partial N}{\partial x} = 2y,$$

so the condition $\partial M/\partial y = \partial N/\partial x$ is satisfied because $2y = 2y$. Next, to find $f(x, y)$ such that

$$\frac{\partial f}{\partial x} = x^2 + y^2, \qquad \frac{\partial f}{\partial y} = 2xy + \cos y,$$

we integrate the first equation with respect to x, holding y constant, and get

$$f(x, y) = \frac{x^3}{3} + xy^2 + g(y).$$

Differentiating both sides of this equation with respect to y and setting the result equal to $N(x, y)$, we get

$$2xy + \frac{dg}{dy} = 2xy + \cos y, \qquad \text{or} \qquad \frac{dg}{dy} = \cos y.$$

When we integrate this last equation with respect to y we get

$$g(y) = \sin y.$$

(We can omit the arbitrary constant of integration here because we set $f(x, y) = C$ for the final solution.) The desired function f is $f(x, y) = x^3/3 + xy^2 + \sin y$ and the solution of the differential equation is

$$\frac{x^3}{3} + xy^2 + \sin y = C. \qquad \blacksquare$$

REMARK The method we used in this example follows the pattern of Eqs. (19) and (20). As you gain more experience, you will probably discover some shortcuts. In this example, the terms $y^2\, dx$ and $2xy\, dy$ can be combined as follows:

$$y^2\, dx + 2xy\, dy = d(xy^2)$$

after the general pattern of

$$d(uv) = u\, dv + v\, du$$

with $u = y^2$ and $v = x$. Then the differential equation is the same as

$$d\!\left(\frac{x^3}{3} + xy^2 + \sin y\right) = 0$$

because

$$d\!\left(\frac{x^3}{3}\right) = x^2\, dx, \qquad d(xy^2) = y^2\, dx + 2xy\, dy, \qquad d(\sin y) = \cos y\, dy.$$

If we had seen this pattern ahead of time, we could immediately have written the solution found by the longer method.

It can be proved that every first order differential equation

$$P(x, y)\, dx + Q(x, y)\, dy = 0$$

can be made exact by multiplication by a suitable **integrating factor** $\rho(x, y)$. Such an integrating factor has the property that

$$\frac{\partial}{\partial y}[\rho(x, y)P(x, y)] = \frac{\partial}{\partial x}[\rho(x, y)Q(x, y)].$$

Unfortunately, it is not easy to determine ρ from this equation. In fact, there is no general technique by which even a single integrating factor can be produced for an arbitrary differential equation, and the search for one can be frustrating. However, we can often recognize certain combinations of differentials that can be made exact by "ingenious devices."

EXAMPLE 6 Solve the equation

$$x\, dy - y\, dx = xy^2\, dx.$$

Solution The combination $x\,dy - y\,dx$ may ring a bell in our memories and cause us to recall the formula

$$d\left(\frac{u}{v}\right) = \frac{v\,du - u\,dv}{v^2} = -\left(\frac{u\,dv - v\,du}{v^2}\right).$$

Therefore we might divide the given equation by x^2 or change signs and divide by y^2. Clearly, the latter approach will be more profitable, so we proceed as follows:

$$x\,dy - y\,dx = xy^2\,dx, \qquad \frac{y\,dx - x\,dy}{y^2} = -x\,dx,$$

$$d\left(\frac{x}{y}\right) + x\,dx = 0, \qquad \frac{x}{y} + \frac{x^2}{2} = C.$$

The same result would be obtained if we wrote our equation in the form

$$(xy^2 + y)\,dx - x\,dy = 0$$

and multiplied by the integrating factor $1/y^2$. This would give

$$\left(x + \frac{1}{y}\right)dx - \left(\frac{x}{y^2}\right)dy = 0,$$

which is exact, since

$$\frac{\partial}{\partial y}\left(x + \frac{1}{y}\right) = \frac{\partial}{\partial x}\left(-\frac{x}{y^2}\right). \qquad ■$$

REMARK When we have a given equation $M\,dx + N\,dy = 0$, there may be more than one way to solve it. For example, we can solve the equation $x\,dy + y\,dx = 0$ by separating the variables after dividing by xy. It is also a homogeneous equation, it is linear (in both x and y), and it is exact. The easiest way to solve it is to recognize that $x\,dy + y\,dx = d(xy)$, so the solution is $xy = C$. In the following set of problems, several are in categories that are labeled by type, and there is another group that is not so labeled. In solving problems from the latter set, you may first want to look for a clever trick (it can simplify things) and then look to see if the equation is exact, separable, homogeneous, or linear. However you get your answer, you can always check it by differentiation. But there are very simple-looking equations that cannot be solved by any of the methods of this article. (One way to deal with such problems is by the use of series, as we illustrate in Article 20.9.) It is also important to realize that this chapter does not cover all of the methods that would be treated in a more advanced textbook. (And new results are being discovered by researchers who specialize in the field of differential equations.)

PROBLEMS

Separable

Separate the variables and solve the following differential equations.

1. $x(2y - 3)\,dx + (x^2 + 1)\,dy = 0$

2. $x^2(y^2 + 1)\,dx + y\sqrt{x^3 + 1}\,dy = 0$

3. $\dfrac{dy}{dx} = e^{x-y}$

4. $\sqrt{2xy}\,\dfrac{dy}{dx} = 1$

5. $\sin x\,\dfrac{dx}{dy} + \cosh 2y = 0$

6. $\ln x\,\dfrac{dx}{dy} = \dfrac{x}{y}$

7. $xe^y \, dy + \dfrac{x^2 + 1}{y} \, dx = 0$

8. $y\sqrt{2x^2 + 3} \, dy + x\sqrt{4 - y^2} \, dx = 0$

9. $\sqrt{1 + x^2} \, dy + \sqrt{y^2 - 1} \, dx = 0$

10. $x^2 y \dfrac{dy}{dx} = (1 + x) \csc y$

11. *Continuous compounding.* An investor has $1000 with which to open an account and plans to add $1000 per year. All funds in the account will earn 10% interest per year, compounded continuously. Assuming that the added deposits are also made continuously, show that $dx/dt = 1000 + 0.10 \, x$, and $x(0) = 1000$, where x is the number of dollars in the account at time t. How many years would it take for the account to reach $100,000?

12. *The snow plow problem.* One morning in January it began to snow and kept on snowing at a constant rate. A snow plow began to plow at noon. It plowed clean and at a constant rate (volume per unit time). From one o'clock until two o'clock it went only half as far as it had gone between noon and one o'clock. When did the snow begin falling?

13. *Newton's law of cooling* assumes that the temperature T of a small hot object placed in a medium of temperature T_a decreases at a rate proportional to $T - T_a$. If an object cools from 100 °C to 80 °C in twenty minutes when the surrounding temperature is 20 °C, how long does it take to cool from 100 °C to 60 °C?

(For additional problems, see Article 6.9.)

Homogeneous

Show that the equations in Problems 14–19 are homogeneous and solve.

14. $(x^2 + y^2) \, dx + xy \, dy = 0$

15. $x^2 \, dy + (y^2 - xy) \, dx = 0$

16. $(xe^{y/x} + y) \, dx - x \, dy = 0$

17. $(x + y) \, dy + (x - y) \, dx = 0$

18. $y' = \dfrac{y}{x} + \cos \dfrac{y - x}{x}$

19. $\left(x \sin \dfrac{y}{x} - y \cos \dfrac{y}{x} \right) dx + x \cos \dfrac{y}{x} \, dy = 0$

20. Solve the equation

$$(x + y + 1) \, dx + (y - x - 3) \, dy = 0$$

by making a change of variable of the form

$$x = r + a, \qquad y = s + b,$$

and choosing the constants a and b so that the resulting equation is

$$(r + s) \, dr + (r - s) \, ds = 0.$$

Then solve this equation and express its solution in terms of x and y.

Orthogonal trajectories. If every member of a family of curves is a solution of the differential equation

$$M(x, y) \, dx + N(x, y) \, dy = 0,$$

while every member of a second family of curves is a solution of the related equation

$$N(x, y) \, dx - M(x, y) \, dy = 0,$$

then each curve of the one family is orthogonal to every curve of the other family. Each family is said to be a family of *orthogonal trajectories* of the other. In Problems 21 and 22, find the family of solutions of the given differential equation and the family of orthogonal trajectories. Sketch both families.

21. $x \, dy - 2y \, dx = 0$

22. $2xy \, dy + (x^2 - y^2) \, dx = 0$

23. Find the orthogonal trajectories of the family of curves $xy = a$, $a \neq 0$ (Fig. 20.3).

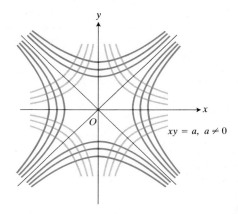

20.3 Each curve is orthogonal to every curve it meets in the other family.

Linear

Solve Problems 24–33.

24. $\dfrac{dy}{dx} + 2y = e^{-x}$

25. $2 \dfrac{dy}{dx} - y = e^{x/2}$

26. $x \dfrac{dy}{dx} + 3y = \dfrac{\sin x}{x^2}$

27. $x \, dy + y \, dx = \sin x \, dx$

28. $x \, dy + y \, dx = y \, dy$

29. $(x - 1)^3 \dfrac{dy}{dx} + 4(x - 1)^2 y = x + 1$

30. $\cosh x \, dy + (y \sinh x + e^x) \, dx = 0$

31. $e^{2y} \, dx + 2(xe^{2y} - y) \, dy = 0$

32. $(x - 2y) \, dy + y \, dx = 0$

33. $(y^2 + 1) \, dx + (2xy + 1) \, dy = 0$

34. *Blood sugar.* If glucose is fed intravenously at a constant rate, the change in the overall concentration $c(t)$ of glucose in the blood with respect to time may be described by the differential equation

$$\frac{dc}{dt} = \frac{G}{100\ V} - kc.$$

In this equation, G, V, and k are positive constants, G being the rate at which glucose is admitted, in milligrams per minute, and V the volume of blood in the body, in liters (around 5 liters for an adult). The concentration $c(t)$ is measured in milligrams per centiliter. The term $-kc$ is included because the glucose is assumed to be changing continually into other molecules at a rate proportional to its concentration.
a) Solve the equation for $c(t)$, using c_0 to denote $c(0)$.
b) Find the steady state concentration, $\lim_{t \to \infty} c(t)$.

35. *An investment program.* An investor has a salary of \$20,000 per year with expected increases of \$1000 per year. If an initial deposit of \$1000 is invested in a program that pays 8% per annum, compounded continuously, and additional deposits are continuously added at the rate of 5% of salary, a model for the investment x at the end of t years is of the form $dx/dt = a + bx + ct$, where a, b, and c are positive constants, and $x(0) = 1000$:

$$\frac{dx}{dt} = 0.08\ x + 0.05(20{,}000 + 1000\ t)$$

$$= 1000 + 0.08\ x + 50\ t.$$

What is x at time t? (If you have a programmable calculator you can easily compute $x(5)$, $x(10)$, $x(20)$, and $x(40)$ to see how the investment grows.)

Exact

In Problems 36–38, use the given integrating factors ρ to make the differential equations exact. Then solve the equations.

36. $(x + 2y)\ dx - x\ dy = 0$, $\rho = 1/x^3$

37. $y\ dx + x\ dy = 0$,

    a) $\rho = \dfrac{1}{xy}$ b) $\rho = \dfrac{1}{(xy)^2}$

38. $y\ dx - x\ dy = 0$,

    a) $\rho = \dfrac{1}{y^2}$ b) $\rho = \dfrac{1}{x^2}$

    c) $\rho = \dfrac{1}{xy}$ d) $\rho = \dfrac{1}{x^2 + y^2}$

Solve Problems 39–48.

39. $(x + y)\ dx + (x + y^2)\ dy = 0$

40. $(2xe^y + e^x)\ dx + (x^2 + 1)e^y\ dy = 0$

41. $(2xy + y^2)\ dx + (x^2 + 2xy - y)\ dy = 0$

42. $(x + \sqrt{y^2 + 1})\ dx - \left(y - \dfrac{xy}{\sqrt{y^2 + 1}}\right)\ dy = 0$

43. $x\ dy - y\ dx + x^3\ dx = 0$

44. $x\ dy - y\ dx = (x^2 + y^2)\ dx$

45. $(x^2 + x - y)\ dx + x\ dy = 0$

46. $\left(e^x + \ln y + \dfrac{y}{x}\right)\ dx + \left(\dfrac{x}{y} + \ln x + \sin y\right)\ dy = 0$

47. $\left(\dfrac{y^2}{1 + x^2} - 2y\right)\ dx + (2y\ \tan^{-1}x - 2x + \sinh y)\ dy = 0$

48. $dy + \dfrac{y - \sin x}{x}\ dx = 0$

49. If a, b, and c are constants and
$$(ax^2 + by^2)\ dx + cxy\ dy = 0$$
is an exact equation, what relation(s) among the constants must hold? Solve the equation subject to such conditions.

50. If a, b, and c are constants and
$$(ax^2 + by^2)\ dx + cxy\ dy = 0$$
is not exact but has the integrating factor $1/x^2$, what relation(s) among the constants must hold? Solve the equation subject to such conditions.

In Problems 51–60, find (a) the general solution and (b) the solution that satisfies the given initial condition.

51. $(\cos x + y \cos x)\ dx + dy = 0$; $y = 0$ when $x = \pi/2$

52. $(x + y^2)\ dx + (2xy + 1)\ dy = 0$; $y = 2$ when $x = 0$

53. $y' + 2y = x$; $y = 1$ when $x = 0$

54. $(x + xy^2)\ dx + (y - x^2y)\ dy = 0$; $y = 1$ when $x = 0$

55. $(1 + y^2)\ dx + e^{-x}\ dy = 0$; $y = 1$ when $x = 0$

56. $(x + e^y)\ dx + (y + xe^y)\ dy = 0$; $y = 0$ when $x = 2$

57. $(x + 2y)\ dx + (y + 2x)\ dy = 0$; $y = 1$ when $x = 1$

58. $(x + 2y)\ dx + (y - 2x)\ dy = 0$; $y = 1$ when $x = 1$

59. $xy' + 2y = x^3$; $y = 1$ when $x = 2$

60. $y\ dx + (2x - y^3)\ dy = 0$; $y = 2$ when $x = 1$

20.3
Second Order Equations Reducible to First Order

Certain types of second order differential equations can be reduced to first order equations by a suitable change of variables. These resulting first order equations can often be solved by the techniques of the previous article.

The general form of a second order differential equation is

$$F\left(x, y, \frac{dy}{dx}, \frac{d^2y}{dx^2}\right) = 0. \tag{1}$$

Two types of equations that can be reduced to first order by substitution arise when either the x or the y variable is missing.

Type 1 *Equations with dependent variable missing.* When Eq. (1) has the special form

$$F\left(x, \frac{dy}{dx}, \frac{d^2y}{dx^2}\right) = 0, \tag{2}$$

we can reduce it to a first order equation by substituting

$$p = \frac{dy}{dx}, \qquad \frac{d^2y}{dx^2} = \frac{dp}{dx}.$$

Then Eq. (2) takes the form

$$F\left(x, p, \frac{dp}{dx}\right) = 0,$$

which is of the first order in p. If this equation can be solved for p as a function of x, say

$$p = \phi\,(x, C_1),$$

then we can find y by one additional integration:

$$y = \int \frac{dy}{dx}\, dx = \int p\, dx = \int \phi\,(x, C_1)\, dx + C_2.$$

The differential equation

$$\frac{d^2y}{dx^2} = \frac{w}{H}\sqrt{1 + \left(\frac{dy}{dx}\right)^2}$$

was solved by this technique in Article 9.3.

EXAMPLE 1 Find the general solution of the second order equation

$$x\,\frac{d^2y}{dx^2} + \frac{dy}{dx} = x^2.$$

Solution In this example, the dependent variable y does not occur (except in the derivatives) and the substitutions

$$\frac{dy}{dx} = p, \qquad \frac{d^2y}{dx^2} = \frac{dp}{dx}$$

reduce the equation to

$$x\,\frac{dp}{dx} + p = x^2.$$

If we multiply through by dx and change to differentials, we get

$$x\,dp + p\,dx = x^2\,dx$$

or $d(xp) = d(x^3/3)$, whose solution is

$$xp = \frac{x^3}{3} + C_1.$$

We now replace p by dy/dx and divide by x to get

$$\frac{dy}{dx} = \frac{x^2}{3} + \frac{C_1}{x}.$$

We can now integrate directly and get

$$y = \frac{x^3}{9} + C_1 \ln |x| + C_2.$$

A typical initial value problem might require that the solution meet pre-scribed values of y and dy/dx at $x = x_0$. For example, if $y = 0$ and $dy/dx = 1$ when $x = 1$, we have the two equations

$$\left(\frac{dy}{dx}\right)_{x=1} = 1 = \frac{1}{3} + C_1, \qquad C_1 = \frac{2}{3}$$

and

$$(y)_{x=1} = 0 = \frac{1}{9} + \frac{2}{3} \ln 1 + C_2, \qquad C_2 = -\frac{1}{9}.$$

The solution of the problem with these initial conditions is

$$y = \frac{x^3}{9} + \frac{2}{3} \ln |x| - \frac{1}{9}.$$ ∎

Type 2 *Equations with independent variable missing.* When Eq. (1) does not contain x explicitly but has the form

$$F\left(y, \frac{dy}{dx}, \frac{d^2y}{dx^2}\right) = 0, \tag{3}$$

the substitutions to use are

$$p = \frac{dy}{dx} \qquad \text{and} \qquad \frac{d^2y}{dx^2} = \frac{dp}{dx} = \frac{dp}{dy} \cdot \frac{dy}{dx} = \frac{dp}{dy} \cdot p.$$

Then Eq. (3) takes the form

$$F\left(y, p, p\frac{dp}{dy}\right) = 0,$$

which is of the first order in p. Its solution gives p in terms of y, and then a further integration gives the solution of Eq. (3).

EXAMPLE 2 Solve the equation

$$\frac{d^2y}{dx^2} + y = 0.$$

Solution Let

$$\frac{dy}{dx} = p, \qquad \frac{d^2y}{dx^2} = \frac{dp}{dy} p.$$

Then we proceed as follows:

$$p \frac{dp}{dy} + y = 0, \qquad p \, dp + y \, dy = 0,$$

$$\frac{p^2}{2} + \frac{y^2}{2} = \frac{C_1^2}{2}, \qquad p = \frac{dy}{dx} = \pm\sqrt{C_1^2 - y^2},$$

$$\frac{dy}{\sqrt{C_1^2 - y^2}} = \pm dx, \qquad \sin^{-1}\frac{y}{C_1} = \pm(x + C_2),$$

$$y = C_1 \sin[\pm(x + C_2)] = \pm C_1 \sin(x + C_2).$$

Since C_1 is arbitrary, there is no need for the $\pm$ sign, and we have

$$y = C_1 \sin(x + C_2)$$

as the general solution.   ■

☐ **Linear Second Order Equations.**   A third type of second order equations can sometimes be reduced to first order equations. These are the second order linear equations for which we can find one solution (possibly by guessing or by having it given to us). By second order linear, we mean an equation of the form

$$y'' + P(x)y' + Q(x)y = F(x).$$

If we can find any solution, say $y = u(x)$, that satisfies the simplified equation

$$u'' + P(x)u' + Q(x)u = 0,$$

then the general solution of the corresponding equation

$$y'' + P(x)y' + Q(x)y = 0$$

can sometimes be found by making the substitution

$$y = uv$$

and solving for the function $v = v(x)$. The next example shows the method.

EXAMPLE 3   Show that $y = e^x$ is a solution of the equation

$$y'' - 2y' + y = 0,$$

and use the result to find the general solution.

Solution   If $y = e^x$, then $y' = y'' = e^x$ and

$$y'' - 2y' + y = e^x - 2e^x + e^x = 0.$$

This shows that $y = e^x$ is one solution. We now substitute $u = e^x$, $y = uv = e^x v$ and calculate the first and second derivatives to get

$$y' = e^x v' + e^x v, \qquad y'' = e^x v'' + 2e^x v' + e^x v.$$

Next we substitute into the equation $y'' - 2y' + y = 0$ and get

$$e^x(v'' + 2v' + v - 2v' - 2v + v) = 0,$$

which simplifies to $v'' = 0$. The solution of this equation is simply $v = C_1 x + C_2$ and this gives

$$y = e^x v = e^x(C_1 x + C_2)$$

as the general solution of $y'' - 2y' + y = 0$.   ■

In Problem 17 you are asked to verify that if $u'' + Pu' + Qu = 0$ and $u \neq 0$, the substitution $y = uv$ transforms $y'' + Py' + Qy = F$ into $v'' + v'[P + 2u'/u] = F/u$. This latter equation then reduces into a first order linear equation for w if we substitute $w = v'$.

PROBLEMS

Solve Problems 1–8.

1. $\dfrac{d^2y}{dx^2} + \dfrac{dy}{dx} = 0$

2. $\dfrac{d^2y}{dx^2} + y\,\dfrac{dy}{dx} = 0$

3. $\dfrac{d^2y}{dx^2} + x\,\dfrac{dy}{dx} = 0$

4. $x\,\dfrac{d^2y}{dx^2} + \dfrac{dy}{dx} = 0$

5. $\dfrac{d^2y}{dx^2} - y = 0$

6. $\dfrac{d^2y}{dx^2} + \omega^2 y = 0$ ($\omega =$ constant $\neq 0$)

7. $xy''' - 2y'' = 0$ (*Hint:* Substitute $y'' = q$.)

8. $2y'' + (y')^2 + 1 = 0$

9. In Example 2 we found the general solution of the second order equation $y'' + y = 0$ in the form $y = C_1 \sin(x + C_2)$. Determine constants C_1 and C_2 if y must also satisfy the conditions
   a) $y(0) = 0$ and $y(\pi/2) = 5$,
   b) $y(0) = 0$ and $y'(0) = 2$.

10. *Hooke's law.* A mass m is suspended from one end of a vertical spring whose other end is attached to a rigid support. The body is allowed to come to rest and is then pulled down an additional slight amount A and released. Find its motion. (*Hint:* Assume Newton's second law of motion and Hooke's law, which says that the tension in the spring is proportional to the amount it is stretched. Let x denote the displacement of the body at time t, measured from the equilibrium position. Then $m(d^2x/dt^2) = -kx$, where k, the

"spring constant," is the proportionality factor in Hooke's law.)

11. *Air resistance proportional to velocity.* A Jeep suspended from a parachute falls through space under the pull of gravity. If air resistance produces a retarding force proportional to the Jeep's velocity and the Jeep starts from rest at time $t = 0$, find the distance it falls in time t.

In Problems 12–16, first show that $y = u(x)$ satisfies the given equation and then find the general solution.

12. $u = \cos x$, $y'' + y = 0$

13. $u = x^2$, $x^2 y'' - 2y = 0$

14. $u = e^{-x}$, $y'' - y' - 2y = 0$

15. $u = x^{1/2}$, $y'' + y/(4x^2) = 0$ ($x > 0$)

16. $u = x$, $x^2 y'' + xy' - y = 0$

17. If P and Q are functions of x, $y = u(x)$ satisfies the linear equation $y'' + Py' + Qy = 0$, and $u \neq 0$, show that the substitution $y = uv$ transforms the equation $y'' + Py' + Qy = F(x)$ into the equation $v'' + v'[P + 2u'/u] = F(x)/u$. The further substitution $w = v'$ then reduces this last equation into a first order linear equation for $w = w(x)$. Show that this can be applied to the equation $y'' - 2y' + y = e^x$ with $u(x) = e^x$ to lead successively to

$$w' = 1, \quad w = x + C_1 = v', \quad v = x^2/2 + C_1 x + C_2,$$

and

$$y = uv = e^x\left(\frac{x^2}{2} + C_1 x + C_2\right).$$

20.4

Linear Second Order Homogeneous Equations with Constant Coefficients

An equation of the form

$$\frac{d^n y}{dx^n} + a_1 \frac{d^{n-1}y}{dx^{n-1}} + a_2 \frac{d^{n-2}y}{dx^{n-2}} + \cdots + a_{n-1}\frac{dy}{dx} + a_n y = F(x),$$

which is linear in y and its derivatives, is called a **linear** equation of order n. If $F(x)$ is identically zero, the equation is said to be **homogeneous;** otherwise it is called **nonhomogeneous.** The equation is called linear even when the coefficients $a_1, a_2, \ldots, a_n$ are functions of x. However, we shall consider only the case where these coefficients are *constants*.

Linear Differential Operators

At this point, it is convenient to introduce the symbol D to represent the operation of differentiation with respect to x. That is, we write $Df(x)$ to mean $(d/dx)f(x)$. Furthermore, we define powers of D to mean taking successive derivatives:

$$D^2f(x) = D\{Df(x)\} = \frac{d^2f(x)}{dx^2},$$

$$D^3f(x) = D\{D^2f(x)\} = \frac{d^3f(x)}{dx^3},$$

and so on. A polynomial in D is to be interpreted as an operator that, when applied to $f(x)$, produces a linear combination of f and its successive derivatives. For example,

$$(D^2 + D - 2)f(x) = D^2f(x) + Df(x) - 2f(x)$$

$$= \frac{d^2f(x)}{dx^2} + \frac{df(x)}{dx} - 2f(x).$$

Such a polynomial in D is called a **linear differential operator** and may be denoted by the single letter L. If L_1 and L_2 are two such linear operators, their sum and product are defined by the equations

$$(L_1 + L_2)f(x) = L_1f(x) + L_2f(x),$$

$$L_1L_2f(x) = L_1(L_2f(x)).$$

Linear differential operators that are polynomials in D with constant coefficients satisfy basic algebraic laws that make it possible to treat them like ordinary polynomials so far as addition, multiplication, and factoring are concerned. Thus,

$$(D^2 + D - 2)f(x) = (D + 2)(D - 1)f(x)$$

$$= (D - 1)(D + 2)f(x).$$

Since this equation holds for any twice-differentiable function f, we also write the equality between operators:

$$D^2 + D - 2 = (D + 2)(D - 1) = (D - 1)(D + 2).$$

The Characteristic Equation

In the remainder of this article, we consider only linear *second order* equations with constant coefficients. Because the solutions of nonhomogeneous equations depend on the solutions of the corresponding homogeneous equations, we focus on equations of the form

$$(D^2 + 2aD + b)y = 0. \tag{1}$$

(Nonhomogeneous equations will be dealt with in the next article, and applications will be treated in Article 20.6.)

The usual method of solving Eq. (1) is to begin by factoring the operator; thus

$$D^2 + 2aD + b = (D - r_1)(D - r_2).$$

We do this by finding the two roots r_1 and r_2 of the equation

$$r^2 + 2ar + b = 0. \tag{1'}$$

The equation (1') is called the **characteristic equation** of the differential equation. Its roots satisfy the equations

$$r_1 + r_2 = -2a, \qquad r_1 r_2 = b,$$

and give us the desired factorization

$$(D - r_1)(D - r_2) = D^2 - (r_1 + r_2)D + r_1 r_2$$
$$= D^2 + 2aD + b.$$

Therefore, Eq. (1) is equivalent to

$$(D - r_1)(D - r_2)y = 0. \tag{2}$$

If we now let

$$(D - r_2)y = u, \tag{3a}$$

we get

$$(D - r_1)u = 0. \tag{3b}$$

From here, we can solve Eq. (2) in two steps. From Eq. (3b), which is first order separable in u and x, we find

$$u = C_1 e^{r_1 x}.$$

We substitute this into (3a), which becomes

$$(D - r_2)y = C_1 e^{r_1 x}$$

or

$$\frac{dy}{dx} - r_2 y = C_1 e^{r_1 x}.$$

This equation is linear in y. Its integrating factor is

$$\rho = e^{-r_2 x},$$

(see Article 20.2), and its solution is

$$e^{-r_2 x} y = C_1 \int e^{(r_1 - r_2)x} \, dx + C_2. \tag{4}$$

How we proceed at this point depends on whether r_1 and r_2 are equal.

CASE 1: If $r_1 \neq r_2$, the evaluation of the integral in Eq. (4) leads to

$$e^{-r_2 x} y = \frac{C_1}{r_1 - r_2} e^{(r_1 - r_2)x} + C_2$$

or

$$y = \frac{C_1}{r_1 - r_2} e^{r_1 x} + C_2 e^{r_2 x}.$$

Since C_1 is an arbitrary constant, so is $C_1/(r_1 - r_2)$, and the solution of Eq. (2)

can be written simply as

$$y = C_1 e^{r_1 x} + C_2 e^{r_2 x}, \qquad \text{if} \quad r_1 \ne r_2. \tag{5}$$

CASE 2: If $r_1 = r_2$, then $e^{(r_1 - r_2)x} = e^0 = 1$, and Eq. (4) reduces to

$$e^{-r_2 x} y = C_1 x + C_2$$

or

$$y = (C_1 x + C_2) e^{r_2 x}, \qquad \text{if} \quad r_1 = r_2. \tag{6}$$

EXAMPLE 1   Solve the equation

$$\frac{d^2 y}{dx^2} + \frac{dy}{dx} - 2y = 0.$$

Solution   The characteristic equation $r^2 + r - 2 = 0$ has roots $r_1 = 1$, $r_2 = -2$. Hence, by Eq. (5), the solution of the differential equation is

$$y = C_1 e^x + C_2 e^{-2x}. \qquad \blacksquare$$

EXAMPLE 2   Solve the equation

$$\frac{d^2 y}{dx^2} + 4 \frac{dy}{dx} + 4y = 0.$$

Solution

$$r^2 + 4r + 4 = (r + 2)^2,$$

$$r_1 = r_2 = -2,$$

$$y = (C_1 x + C_2) e^{-2x}. \qquad \blacksquare$$

Imaginary Roots

If the coefficients a and b in Eq. (1) are real, the roots of the characteristic equation will be either real or a pair of complex conjugate numbers:

$$r_1 = \alpha + i\beta, \qquad r_2 = \alpha - i\beta. \tag{7}$$

If $\beta \ne 0$, Eq. (5) applies, with the result

$$\begin{aligned} y &= c_1 e^{(\alpha + i\beta)x} + c_2 e^{(\alpha - i\beta)x} \\ &= e^{\alpha x}[c_1 e^{i\beta x} + c_2 e^{-i\beta x}], \end{aligned} \tag{8}$$

where c_1 and c_2 are complex constants. By Euler's formula,

$$e^{i\beta x} = \cos \beta x + i \sin \beta x, \qquad e^{-i\beta x} = \cos \beta x - i \sin \beta x.$$

Hence, Eq. (8) may be replaced by

$$y = e^{\alpha x}[(c_1 + c_2) \cos \beta x + i(c_1 - c_2) \sin \beta x]. \tag{9}$$

Finally, if we introduce new arbitrary constants

$$C_1 = c_1 + c_2, \qquad C_2 = i(c_1 - c_2),$$

Eq. (9) takes the form

$$y = e^{\alpha x}(C_1 \cos \beta x + C_2 \sin \beta x),$$

if   (9')

$$r_1 = \alpha + i\beta, \qquad r_2 = \alpha - i\beta.$$

The arbitrary constants C_1 and C_2 in (9') will be real provided the constants c_1 and c_2 in (9) are complex conjugates:

$$c_1 = \frac{1}{2}(C_1 - i\,C_2), \qquad c_2 = \frac{1}{2}(C_1 + i\,C_2).$$

To solve a problem where the roots of the characteristic equation are complex conjugates, we simply write down the appropriate version of Eq. (9').

EXAMPLE 3   Solve the equation

$$\frac{d^2y}{dx^2} + 2\frac{dy}{dx} + 2y = 0.$$

Solution   The characteristic equation $r^2 + 2r + 2 = 0$ has roots $r_1 = -1 + i$, $r_2 = -1 - i$. Hence, in Eq. (9'), we take

$$\alpha = -1, \qquad \beta = 1,$$

and obtain the solution

$$y = e^{-x}[C_1 \cos x + C_2 \sin x].$$

EXAMPLE 4   Solve the equation

$$\frac{d^2y}{dx^2} + \omega^2 y = 0, \qquad \omega \neq 0.$$

Solution   The characteristic equation $r^2 + \omega^2 = 0$ has roots $r_1 = i\omega$, $r_2 = -i\omega$. Hence we take $\alpha = 0$, $\beta = \omega$ in Eq. (9') and obtain the solution

$$y = C_1 \cos \omega x + C_2 \sin \omega x.$$

PROBLEMS

Solve the following equations.

1. $\dfrac{d^2y}{dx^2} + 2\dfrac{dy}{dx} = 0$

2. $\dfrac{d^2y}{dx^2} + 5\dfrac{dy}{dx} + 6y = 0$

5. $\dfrac{d^2y}{dx^2} + \dfrac{dy}{dx} + y = 0$

6. $\dfrac{d^2y}{dx^2} - 4\dfrac{dy}{dx} + 4y = 0$

3. $\dfrac{d^2y}{dx^2} + 6\dfrac{dy}{dx} + 5y = 0$

4. $\dfrac{d^2y}{dx^2} - 2\dfrac{dy}{dx} - 3y = 0$

7. $\dfrac{d^2y}{dx^2} + 6\dfrac{dy}{dx} + 9y = 0$

8. $\dfrac{d^2y}{dx^2} - 6\dfrac{dy}{dx} + 10y = 0$

9. $\dfrac{d^2y}{dx^2} - 2\dfrac{dy}{dx} + 4y = 0$ **10.** $\dfrac{d^2y}{dx^2} - 10\dfrac{dy}{dx} + 16y = 0$

Solve the following equations subject to the given initial conditions ($y'' = d^2y/dx^2$ and $y' = dy/dx$).

11. $y'' - y = 0$, $y(0) = 1$, $y'(0) = -2$

12. $y'' + 2y' + y = 0$, $y(0) = 0$, $y'(0) = 1$

13. $2y'' - y' - y = 0$, $y(0) = -1$, $y'(0) = 0$

14. $y'' - 2y' + 3y = 0$, $y(0) = 2$, $y'(0) = 1$

15. $y'' - 4y = 0$, $y(0) = 0$, $y'(0) = 3$

16. $y'' - 4y = 0$, $y(0) = 0$, $y'(0) = 2$

17. $y'' + 4y = 0$, $y(0) = 0$, $y'(0) = 2$

20.5

Linear Second Order Nonhomogeneous Equations with Constant Coefficients

In Article 20.4, we learned how to solve the homogeneous equation

$$\frac{d^2y}{dx^2} + 2a\frac{dy}{dx} + by = 0. \tag{1}$$

We can now describe a method for solving the nonhomogeneous equation

$$\frac{d^2y}{dx^2} + 2a\frac{dy}{dx} + by = F(x). \tag{2}$$

To solve Eq. (2), we first determine the general solution of the related homogeneous Eq. (1), obtained by replacing $F(x)$ by zero. Let us denote this solution by

$$y_h = C_1\,u_1(x) + C_2\,u_2(x), \tag{3}$$

where C_1 and C_2 are arbitrary constants and $u_1(x)$, $u_2(x)$ are functions of one or more of the following forms:

$$e^{rx}, \qquad xe^{rx}, \qquad e^{\alpha x}\cos\beta x, \qquad e^{\alpha x}\sin\beta x.$$

The procedure outlined in the optional section and in Problem 17 of Article 20.3 could be used to solve Eq. (2), but there are other ways that are usually much faster and easier. These ways involve finding a *particular solution* of Eq. (2) and adding it to the general solution of Eq. (1). We shall discuss three ways of finding a particular solution:

1. inspired guessing;
2. the method of variation of parameters;
3. the method of undetermined coefficients.

But first, let us verify that *all* solutions of the nonhomogeneous equation are of the form

$$y = y_h(x) + y_p(x),$$

where $y_h(x)$, Eq. (3), is the general solution of the homogeneous equation (1) and $y_p(x)$ is any particular solution of the nonhomogeneous equation (2). The crux of the matter is that if L is any linear differential operator (not necessarily one of second order or having constant coefficients), then

$$L(k_1 \, y_1 + k_2 \, y_2) = k_1 \, L(y_1) + k_2 \, L(y_2)$$

for any constants k_1 and k_2 and functions y_1 and y_2 to which the operator L can be applied. In the present context, we can take

$$L = D^2 + 2aD + b$$

and write Eqs. (1) and (2) as $L(y) = 0$ and $L(y) = F(x)$, respectively. The following theorem states the relation between solutions of homogeneous and nonhomogeneous linear equations.

Relation Between Solutions of Homogeneous and Nonhomogeneous Linear Equations

THEOREM

Let $L(y) = F(x)$ be a linear nonhomogeneous differential equation, and $L(y) = 0$ be the related homogeneous equation with the same operator L. If y_h is the general solution of the homogeneous equation and y_p is a particular solution of the nonhomogeneous equation, then the general solution of the nonhomogeneous equation is

$$y = y_h(x) + y_p(x).$$

Proof   Let $L(y_h(x)) = 0$ and $L(y_p(x)) = F(x)$. Then

$$L(y_h(x) + y_p(x)) = L(y_h(x)) + L(y_p(x)) = 0 + F(x) = F(x),$$

which establishes the fact that $y = y_h + y_p$ is a solution of the nonhomogeneous equation. To show that all solutions have this form, let y be any solution of the nonhomogeneous equation $L(y) = F(x)$. Then

$$L(y - y_p) = L(y) - L(y_p) = F(x) - F(x) = 0,$$

so that $y - y_p$ satisfies the homogeneous equation. Therefore, if y_h is the *general solution* of $(D^2 + 2aD + b)y = 0$, there is some choice of the constants C_1 and C_2 such that

$$y - y_p = C_1 \, u_1(x) + C_2 \, u_2(x),$$

which means that

$$y = y_p + C_1 \, u_1(x) + C_2 \, u_2(x). \qquad \blacksquare$$

Inspired Guessing   In the following example, it does not require a genius to guess that there may be a particular solution of the form $y_p(x) = $ Constant. If there is such a solution, the particular constant that works can easily be determined from the equation.

EXAMPLE 1   Solve the equation

$$\frac{d^2y}{dx^2} + 2\frac{dy}{dx} - 3y = 6.$$

Solution The solution y_h satisfies the homogeneous equation

$$\frac{d^2y_h}{dx^2} + 2\frac{dy_h}{dx} - 3y_h = 0.$$

Its characteristic equation is

$$r^2 + 2r - 3 = 0,$$

whose roots are

$$r_1 = -3, \qquad r_2 = 1.$$

Hence

$$y_h = C_1 e^{-3x} + C_2 e^x.$$

Now, to find a particular solution of the original equation, observe that $y =$ Constant would do, provided $-3y = 6$. Hence,

$$y_p = -2$$

is one particular solution. The complete solution is

$$y = y_p + y_h = -2 + C_1 e^{-3x} + C_2 e^x. \qquad \blacksquare$$

Variation of Parameters

Fortunately, there is a general method for finding the solution of the nonhomogeneous Eq. (2) once the general solution of the corresponding homogeneous equation is known. The method is known as the method of **variation of parameters.** It consists of replacing the constants C_1 and C_2 on the right side of Eq. (3) by functions $v_1(x)$ and $v_2(x)$ and then requiring that the new expression

$$y = v_1 u_1 + v_2 u_2$$

be a solution of the nonhomogeneous equation (2). There are two functions v_1 and v_2 to be determined, and requiring that Eq. (2) be satisfied is only one condition. It turns out that a simplification results if we also require that

$$v_1' u_1 + v_2' u_2 = 0. \qquad (4)$$

Then we have

$$y = v_1 u_1 + v_2 u_2,$$

$$\frac{dy}{dx} = v_1 u_1' + v_2 u_2' + v_1' u_1 + v_2' u_2$$

$$= v_1 u_1' + v_2 u_2',$$

$$\frac{d^2y}{dx^2} = v_1 u_1'' + v_2 u_2'' + v_1' u_1' + v_2' u_2'.$$

If we substitute these expressions into the left-hand side of Eq. (2) and rearrange terms, we obtain

$$v_1\left[\frac{d^2u_1}{dx^2} + 2a\frac{du_1}{dx} + bu_1\right] + v_2\left[\frac{d^2u_2}{dx^2} + 2a\frac{du_2}{dx} + bu_2\right]$$

$$+ v_1' u_1' + v_2' u_2' = F(x).$$

■ ■ ■

VARIATION OF PARAMETERS

As early as 1728, Leonhard Euler (1707–1783) focused on second order differential equations, as they came up constantly in his work in mechanics. Such physical problems as pendulum motion in resisting media, the motion of projectiles, and the effect of air resistance captured his attention. Euler used essentially the method of *variation of parameters* to reduce second order equations to first order. Historically, this work begins a systematic study of second order equations. It is here that Euler demonstrated the important role of the exponential function in solving second and higher order equations. Joseph Louis Lagrange (1736–1813), Euler's mathematical heir, continued on to develop the more general method we use today for solving nonhomogeneous differential equations.

The two bracketed terms are zero, since u_1 and u_2 are solutions of the homogeneous equation (1). Hence Eq. (2) is satisfied if, in addition to Eq. (4), we require that

$$v_1'u_1' + v_2'u_2' = F(x). \tag{5}$$

Equations (4) and (5) may be solved together as a pair,

$$v_1'u_1 + v_2'u_2 = 0, \qquad v_1'u_1' + v_2'u_2' = F(x),$$

for the unknown functions v_1' and v_2'. Cramer's rule gives

$$v_1' = \frac{\begin{vmatrix} 0 & u_2 \\ F(x) & u_2' \end{vmatrix}}{\begin{vmatrix} u_1 & u_2 \\ u_1' & u_2' \end{vmatrix}} = \frac{-u_2 F(x)}{D},$$

$$v_2' = \frac{\begin{vmatrix} u_1 & 0 \\ u_1' & F(x) \end{vmatrix}}{\begin{vmatrix} u_1 & u_2 \\ u_1' & u_2' \end{vmatrix}} = \frac{u_1 F(x)}{D}, \tag{6}$$

where

$$D = \begin{vmatrix} u_1 & u_2 \\ u_1' & u_2' \end{vmatrix}.$$

Now v_1 and v_2 can be found by integration.

Method of Variation of Parameters

In applying the method of variation of parameters to solve the equation

$$\frac{d^2y}{dx^2} + 2a\frac{dy}{dx} + by = F(x), \tag{2}$$

we can work directly with the equations in (6). It is not necessary to rederive them. The steps are as follows:

STEP 1: Solve the associated homogeneous equation,

$$\frac{d^2y}{dx^2} + 2a\frac{dy}{dx} + by = 0,$$

to find the functions u_1 and u_2.

STEP 2: Calculate D, v_1', and v_2' from (6).

STEP 3: Integrate v_1' and v_2' to find v_1 and v_2.

STEP 4: Write down the general solution of (2) as

$$y = v_1 u_1 + v_2 u_2.$$

EXAMPLE 2 Solve the equation

$$\frac{d^2y}{dx^2} + 2\frac{dy}{dx} - 3y = 6$$

of Example 1 by variation of parameters.

Solution   We first solve the associated homogeneous equation

$$\frac{d^2y}{dx^2} + 2\frac{dy}{dx} - 3y = 0$$

as in Example 1 to find

$$u_1(x) = e^{-3x}, \qquad u_2(x) = e^x.$$

Then, from the equations in (6), we have

$$D = \begin{vmatrix} e^{-3x} & e^x \\ -3e^{-3x} & e^x \end{vmatrix} = e^{-2x} + 3e^{-2x} = 4e^{-2x},$$

$$v_1' = \frac{-6e^x}{4e^{-2x}} = -\frac{3}{2}e^{3x}, \qquad v_2' = \frac{6e^{-3x}}{4e^{-2x}} = \frac{3}{2}e^{-x}. \tag{7}$$

Hence

$$v_1 = \int -\frac{3}{2}e^{3x}\, dx = -\frac{1}{2}e^{3x} + C_1,$$

$$v_2 = \int \frac{3}{2}e^{-x}\, dx = -\frac{3}{2}e^{-x} + C_2,$$

and

$$y = v_1u_1 + v_2u_2$$

$$= \left(-\frac{1}{2}e^{3x} + C_1\right)e^{-3x} + \left(-\frac{3}{2}e^{-x} + C_2\right)e^x$$

$$= -2 + C_1\, e^{-3x} + C_2\, e^x. \qquad \blacksquare$$

Undetermined Coefficients

The method of variation of parameters is completely general, for it will produce a particular solution of the nonhomogeneous equation (2) for any continuous function $F(x)$. But the calculations involved can be complicated, and in special cases there may be easier methods to use. For instance, we do not need to use variation of parameters to find a particular solution of

$$\frac{d^2y}{dx^2} - \frac{dy}{dx} + 5y = 3, \tag{8}$$

if we can find the particular solution $y_p = 3/5$ by inspection. And even for an equation like

$$\frac{d^2y}{dx^2} + 3y = e^x, \tag{9}$$

we might guess that there is a solution of the form

$$y_p = Ae^x$$

and substitute $y_p = Ae^x$ and its second derivative into Eq. (9) to find A. If we do so, we find that the function $y_p = (1/4)e^x$ is a solution of Eq. (9).

Again, we might guess that the equation

$$\frac{d^2y}{dx^2} + y = 3x^2 + 4 \tag{10}$$

has a particular solution of the form

$$y_p = Cx^2 + Dx + E.$$

When we substitute this polynomial and its second derivative into Eq. (10) to see whether appropriate values for the constants $C, D,$ and E can be found, Eq. (10) becomes

$$2C + (Cx^2 + Dx + E) = 3x^2 + 4,$$

$$Cx^2 + Dx + 2C + E = 3x^2 + 4. \tag{11}$$

This latter equation will hold for all values of x if its two sides are identical as polynomials in x; that is, if

$$C = 3, \qquad D = 0, \qquad \text{and} \qquad 2C + E = 4, \tag{12}$$

or

$$C = 3, \qquad D = 0, \qquad E = -2.$$

We conclude that

$$y_p = 3x^2 + 0\, x - 2 = 3x^2 - 2 \tag{13}$$

is a solution of Eq. (10).

In each of the foregoing examples, the particular solution we found resembled the function $F(x)$ on the right side of the given differential equation. This was no accident, for we guessed the form of the particular solution by looking at $F(x)$ first. The method of first guessing the form of the solution up to certain undetermined constants and then determining the values of these constants by using the differential equation is known as the **method of undetermined coefficients.** It depends for its success on our ability to recognize the form of a particular solution, and for this reason, among others, it lacks the generality of the method of variation of parameters. Nevertheless, its simplicity makes it the method of choice in a number of special cases.

We shall limit our discussion of the method of undetermined coefficients to selected equations in which the function $F(x)$ in Eq. (2) is the sum of one or more terms like

$$e^{rx}, \qquad \cos kx, \qquad \sin kx, \qquad ax^2 + bx + c.$$

Even so, we will find that the particular solutions of Eq. (2) do not always resemble $F(x)$ as closely as the ones we have seen.

EXAMPLE 3   Find a particular solution of

$$\frac{d^2y}{dx^2} - \frac{dy}{dx} = 2 \sin x. \tag{14}$$

Solution   If we try to find a particular solution of the form

$$y_p = A \sin x$$

and substitute the derivatives of y_p in the given equation, we find that A must satisfy the equation

$$-A \sin x - A \cos x = 2 \sin x \tag{15}$$

for all values of x. Since this requires A to be equal to -2 and 0 at the same time, we conclude that Eq. (14) has no solution of the form $A \sin x$.

It turns out that the required form is the sum

$$y_p = A \sin x + B \cos x. \tag{16}$$

The result of substituting the derivatives of this new candidate into Eq. (14) is

$$-A \sin x - B \cos x - (A \cos x - B \sin x) = 2 \sin x,$$
$$(B - A) \sin x - (A + B) \cos x = 2 \sin x. \tag{17}$$

Equation (17) will be an identity if

$$B - A = 2 \quad \text{and} \quad A + B = 0,$$

or

$$A = -1, \quad B = 1.$$

Our particular solution is

$$y_p = \cos x - \sin x. \qquad \blacksquare$$

EXAMPLE 4 Find a particular solution of

$$\frac{d^2y}{dx^2} - 3\frac{dy}{dx} + 2y = 5e^x. \tag{18}$$

Solution If we substitute

$$y_p = Ae^x$$

and its derivatives in Eq. (18), we find that

$$Ae^x - 3Ae^x + 2Ae^x = 5e^x,$$
$$0 = 5e^x.$$

The trouble can be traced to the fact that $y = e^x$ is already a solution of the related homogeneous equation

$$\frac{d^2y}{dx^2} - 3\frac{dy}{dx} + 2y = 0. \tag{19}$$

The characteristic equation of Eq. (19) is

$$r^2 - 3r + 2 = (r - 1)(r - 2) = 0,$$

which has $r = 1$ as a simple root. We may therefore expect Ae^x to "vanish" when substituted into the left-hand side of (18).

The appropriate way to modify the trial solution in this case is to replace Ae^x by Axe^x. When we substitute

$$y_p = Axe^x$$

and its derivatives into (18), we obtain

$$(Axe^x + 2Ae^x) - 3(Axe^x + Ae^x) + 2Axe^x = 5e^x$$
$$-Ae^x = 5e^x$$
$$A = -5.$$

The function

$$y_p = -5xe^x$$

is a particular solution of Eq. (19). ■

EXAMPLE 5 Find a particular solution of

$$\frac{d^2y}{dx^2} - 6\frac{dy}{dx} + 9y = e^{3x}. \tag{20}$$

Solution The characteristic equation,

$$r^2 - 6r + 9 = (r - 3)^2 = 0,$$

has $r = 3$ as a *double* root. The appropriate choice for y_p in this case is neither Ae^{3x} nor Axe^{3x}, but Ax^2e^{3x}. When we substitute

$$y_p = Ax^2e^{3x}$$

and its derivatives in the given differential equation, we get

$$(9Ax^2e^{3x} + 12Axe^{3x} + 2Ae^{3x}) - 6(3Ax^2e^{3x} + 2Axe^{3x}) + 9Ax^2e^{3x} = e^{3x}$$

$$2Ae^{3x} = e^{3x}$$

$$A = \frac{1}{2}.$$

The function

$$y_p = \frac{1}{2}x^2e^{3x}$$

is a solution of Eq. (20). ∎

When we wish to find a particular solution of Eq. (2), and the function $F(x)$ is the sum of two or more terms, we choose a trial function for each term in $F(x)$ and add them.

EXAMPLE 6 Solve the equation

$$\frac{d^2y}{dx^2} - \frac{dy}{dx} = 5e^x - \sin 2x. \tag{21}$$

Solution We first check the characteristic equation, which is

$$r^2 - r = 0.$$

Its roots are

$$r_1 = 1, \qquad r_2 = 0,$$

so the general solution of the related homogeneous equation is

$$y_h = C_1 e^x + C_2.$$

We now seek a particular solution y_p. That is, we seek a function that will produce $5e^x - \sin 2x$ when substituted into the left side of Eq. (21). One part of y_p is to produce $5e^x$, the other $-\sin 2x$.

Since any function of the form C_1e^x is a solution of the related homogeneous equation, we choose our trial y_p to be the sum

$$y_p = Axe^x + B\cos 2x + C\sin 2x,$$

including xe^x where we might otherwise have included e^x. When the derivatives of y_p are substituted in Eq. (21), the resulting equations are

$$(Axe^x + 2Ae^x - 4B \cos 2x - 4C \sin 2x)$$

$$- (Axe^x + Ae^x - 2B \sin 2x + 2C \cos 2x) = 5e^x - \sin 2x,$$

$$Ae^x - (4B + 2C) \cos 2x + (2B - 4C) \sin 2x = 5e^x - \sin 2x.$$

These equations will hold if

$$A = 5, \qquad (4B + 2C) = 0, \qquad (2B - 4C) = -1,$$

or

$$A = 5, \qquad B = -\frac{1}{10}, \qquad C = \frac{1}{5}.$$

Our particular solution is

$$y_p = 5xe^x - \frac{1}{10} \cos 2x + \frac{1}{5} \sin 2x.$$

The complete solution of Eq. (21) is

$$y = y_h + y_p = C_1 e^x + C_2 + 5xe^x - \frac{1}{10} \cos 2x + \frac{1}{5} \sin 2x. \qquad ■$$

You may find Table 20.1 helpful in solving the problems at the end of this article.

TABLE 20.1

The method of undetermined coefficients for selected equations of the form

$$\frac{d^2y}{dx^2} + 2a \frac{dy}{dx} + by = F(x)$$

If $F(x)$ has a term that is a constant multiple of	And if	Then include this expression in the trial function for y_p
e^{rx}	r is not a root of the characteristic equation	Ae^{rx}
	r is a single root of the characteristic equation	Axe^{rx}
	r is a double root of the characteristic equation	Ax^2e^{rx}
$\sin kx, \cos kx$	ki is not a root of the characteristic equation	$B \cos kx + C \sin kx$
$ax^2 + bx + c$	0 is not a root of the characteristic equation	$Dx^2 + Ex + F$ (chosen to match the degree of $ax^2 + bx + c$)
	0 is a single root of the characteristic equation	$Dx^3 + Ex^2 + Fx$ (degree one higher than the degree of $ax^2 + bx + c$)
	0 is a double root of the characteristic equation	$Dx^4 + Ex^3 + Fx^2$ (degree two higher than the degree of $ax^2 + bx + c$)

PROBLEMS

Solve the equations in Problems 1–12 by variation of parameters.

1. $\dfrac{d^2y}{dx^2} + \dfrac{dy}{dx} = x$

2. $\dfrac{d^2y}{dx^2} + y = \tan x, \quad -\dfrac{\pi}{2} < x < \dfrac{\pi}{2}$

3. $\dfrac{d^2y}{dx^2} + y = \sin x$ **4.** $\dfrac{d^2y}{dx^2} + 2\dfrac{dy}{dx} + y = e^x$

5. $\dfrac{d^2y}{dx^2} + 2\dfrac{dy}{dx} + y = e^{-x}$ **6.** $\dfrac{d^2y}{dx^2} - y = x$

7. $\dfrac{d^2y}{dx^2} - y = e^x$ **8.** $\dfrac{d^2y}{dx^2} - y = \sin x$

9. $\dfrac{d^2y}{dx^2} + 4\dfrac{dy}{dx} + 5y = 10$ **10.** $\dfrac{d^2y}{dx^2} - \dfrac{dy}{dx} = 2^x$

11. $\dfrac{d^2y}{dx^2} + y = \sec x, \quad -\dfrac{\pi}{2} < x < \dfrac{\pi}{2}$

12. $\dfrac{d^2y}{dx^2} - \dfrac{dy}{dx} = e^x \cos x, \quad x > 0$

Solve the equations in Problems 13–28 by the method of undetermined coefficients.

13. $\dfrac{d^2y}{dx^2} - 3\dfrac{dy}{dx} - 10y = -3$

14. $\dfrac{d^2y}{dx^2} - 3\dfrac{dy}{dx} - 10y = 2x - 3$

15. $\dfrac{d^2y}{dx^2} - \dfrac{dy}{dx} = \sin x$ **16.** $\dfrac{d^2y}{dx^2} + 2\dfrac{dy}{dx} + y = x^2$

17. $\dfrac{d^2y}{dx^2} + y = \cos 3x$ **18.** $\dfrac{d^2y}{dx^2} + y = e^{2x}$

19. $\dfrac{d^2y}{dx^2} - \dfrac{dy}{dx} - 2y = 20 \cos x$

20. $\dfrac{d^2y}{dx^2} + y = 2x + 3e^x$ **21.** $\dfrac{d^2y}{dx^2} - y = e^x + x^2$

22. $\dfrac{d^2y}{dx^2} + 2\dfrac{dy}{dx} + y = 6 \sin 2x$

23. $\dfrac{d^2y}{dx^2} - \dfrac{dy}{dx} - 6y = e^{-x} - 7 \cos x$

24. $\dfrac{d^2y}{dx^2} + 3\dfrac{dy}{dx} + 2y = e^{-x} + e^{-2x} - x$

25. $\dfrac{d^2y}{dx^2} + 5\dfrac{dy}{dx} = 15x^2$ **26.** $\dfrac{d^2y}{dx^2} - \dfrac{dy}{dx} = -8x + 3$

27. $\dfrac{d^2y}{dx^2} - 3\dfrac{dy}{dx} = e^{3x} - 12x$

28. $\dfrac{d^2y}{dx^2} + 7\dfrac{dy}{dx} = 42x^2 + 5x + 1$

In each of Problems 29–31, the given differential equation has a particular solution y_p of the form given. Determine the coefficients in y_p. Then solve the differential equation.

29. $\dfrac{d^2y}{dx^2} - 5\dfrac{dy}{dx} = xe^{5x}, \quad y_p = A\,x^2e^{5x} + B\,xe^{5x}$

30. $\dfrac{d^2y}{dx^2} - \dfrac{dy}{dx} = \cos x + \sin x, \quad y_p = A \cos x + B \sin x$

31. $\dfrac{d^2y}{dx^2} + y = 2 \cos x + \sin x, \quad y_p = A\,x \cos x + B\,x \sin x$

In Problems 32–35, solve the given differential equations (a) by variation of parameters and (b) by the method of undetermined coefficients.

32. $\dfrac{d^2y}{dx^2} - 4\dfrac{dy}{dx} + 4y = 2e^{2x}$

33. $\dfrac{d^2y}{dx^2} - \dfrac{dy}{dx} = e^x + e^{-x}$ **34.** $\dfrac{d^2y}{dx^2} - 9\dfrac{dy}{dx} = 9e^{9x}$

35. $\dfrac{d^2y}{dx^2} - 4\dfrac{dy}{dx} - 5y = e^x + 4$

Solve the differential equations in Problems 36–45. Some of the equations can be solved by the method of undetermined coefficients, but others cannot.

36. $\dfrac{d^2y}{dx^2} + y = \csc x, \quad 0 < x < \pi$

37. $\dfrac{d^2y}{dx^2} + y = \cot x, \quad 0 < x < \pi$

38. $\dfrac{d^2y}{dx^2} + 4y = \sin x$ **39.** $\dfrac{d^2y}{dx^2} - 8\dfrac{dy}{dx} = e^{8x}$

40. $\dfrac{d^2y}{dx^2} + 4\dfrac{dy}{dx} + 5y = x + 2$

41. $\dfrac{d^2y}{dx^2} - \dfrac{dy}{dx} = x^3$ **42.** $\dfrac{d^2y}{dx^2} + 9y = 9x - \cos x$

43. $\dfrac{d^2y}{dx^2} + 2\dfrac{dy}{dx} = x^2 - e^x$

44. $\dfrac{d^2y}{dx^2} - 3\dfrac{dy}{dx} + 2y = e^x - e^{2x}$

45. $\dfrac{d^2y}{dx^2} + y = \sec x \tan x, \quad -\dfrac{\pi}{2} < x < \dfrac{\pi}{2}$

The method of undetermined coefficients can sometimes be used to solve first order ordinary differential equations. Use the method to solve the equations in Problems 46–49.

46. $\dfrac{dy}{dx} + 4y = x$ **47.** $\dfrac{dy}{dx} - 3y = e^x$

48. $\dfrac{dy}{dx} + y = \sin x$ **49.** $\dfrac{dy}{dx} - 3y = 5e^{3x}$

Solve the differential equations in Problems 50 and 51 subject to the given initial conditions.

50. $\dfrac{d^2y}{dx^2} + y = e^{2x}; \quad y(0) = 0, \quad y'(0) = \dfrac{2}{5}$

51. $\dfrac{d^2y}{dx^2} + y = \sec^2 x, \quad -\dfrac{\pi}{2} < x < \dfrac{\pi}{2}; \quad y(0) = y'(0) = 1$

52. *Bernoulli's equation of order 2.* Solve the equation

$$\frac{dy}{dx} + y = (xy)^2$$

by carrying out the following steps: (1) divide both sides of the equation by y^2; (2) make the change of variable $u = y^{-1}$; (3) solve the resulting equation for u in terms of x; (4) let $y = u^{-1}$.

53. Solve the integral equation

$$y(x) + \int_0^x y(t)\, dt = x.$$

(*Hint:* Differentiate.)

20.6

Vibrations

One of the most important applications of second order linear differential equations with constant coefficients is the study of vibrations. Vibrations, or oscillatory motion, can occur in many practical situations, but the classic example is the motion of a spring. Let us consider a spring of natural length L with its upper end fastened to a rigid support at A (Fig. 20.4). A weight W, of mass m, is suspended from the spring. The weight stretches the spring to a length $L + s$ when the system is allowed to come to rest in a new equilibrium position. By Hooke's law, the tension in the spring is ks, where k is the so-called spring constant. The force of gravity pulling down on the weight is $W = mg$. Equilib-

20.4 A spring of natural length L is stretched a distance s by a weight W. It is then stretched an additional distance x_0 and released to create a vibration of the spring. The weight's position at any subsequent time is described by a different equation.

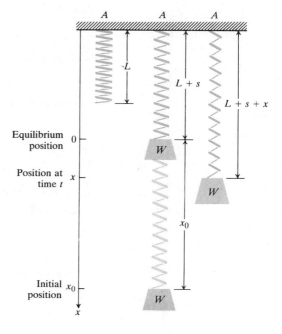

rium requires

$$ks = mg. \tag{1}$$

Suppose now that the weight is pulled down an additional amount x_0 beyond the equilibrium position and released. How can we describe its motion?

Let x, positive direction downward, denote the displacement of the weight away from equilibrium at any time t after the motion has started. Then the forces acting upon the weight are

$$+mg, \quad \text{due to gravity,}$$

$$-k(s + x), \quad \text{due to the spring tension.}$$

The resultant of these forces is, by Newton's second law, also equal to

$$m\frac{d^2x}{dt^2}.$$

Therefore, if there are no other forces (such as friction),

$$m\frac{d^2x}{dt^2} = mg - ks - kx. \tag{2}$$

By Eq. (1), $mg - ks = 0$, so (2) becomes

$$m\frac{d^2x}{dt^2} + kx = 0. \tag{3}$$

In addition to the differential equation (3), the motion satisfies the initial conditions:

$$\text{At } t = 0: \quad x = x_0 \quad \text{and} \quad \frac{dx}{dt} = 0. \tag{4}$$

Let $\omega = \sqrt{k/m}$. Then Eq. (3) becomes

$$\frac{d^2x}{dt^2} + \omega^2 x = 0, \quad \text{or} \quad (D^2 + \omega^2)x = 0.$$

The roots of the characteristic equation

$$r^2 + \omega^2 = 0$$

are complex conjugates

$$r = \pm\omega i.$$

Hence

$$x = c_1 \cos \omega t + c_2 \sin \omega t \tag{5}$$

is the general solution of the differential equation. To fit the initial conditions, we also compute

$$\frac{dx}{dt} = -c_1\omega \sin \omega t + c_2\omega \cos \omega t$$

and then substitute from (4). This yields

$$x_0 = c_1, \quad 0 = c_2\omega.$$

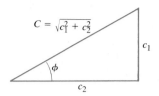

20.5 $c_1 = C \sin \phi$ and $c_2 = C \cos \phi$.

Therefore

$$c_1 = x_0, \qquad c_2 = 0,$$

and

$$x = x_0 \cos \omega t \qquad (6)$$

describes the motion of the weight. Equation (6) represents simple harmonic motion of amplitude a and period $T = 2\pi/\omega$.

The two terms on the right-hand side of Eq. (5) can be combined into a single term by using the trigonometric identity

$$\sin(\omega t + \phi) = \cos \omega t \sin \phi + \sin \omega t \cos \phi.$$

To apply the identity, we take

$$c_1 = C \sin \phi, \qquad c_2 = C \cos \phi, \qquad (7a)$$

where

$$C = \sqrt{c_1^2 + c_2^2}, \qquad \phi = \tan^{-1} \frac{c_1}{c_2}, \qquad (7b)$$

as in Fig. 20.5. Then Eq. (5) can be written in the alternative form

$$x = C \sin(\omega t + \phi). \qquad (8)$$

Here C and ϕ may be taken as two new arbitrary constants, replacing the two constants c_1 and c_2 of Eq. (5). Equation (8) represents simple harmonic motion of amplitude C and period $T = 2\pi/\omega$. The angle ϕ is called the *phase angle*. A graph of Eq. (8) is given in Fig. 20.6.

Equation (3) assumes that there is no friction in the system. Next, consider the case where the motion of the weight is retarded by a friction force $c(dx/dt)$ proportional to the velocity, where c is a positive constant. Then the differential equation is

$$m \frac{d^2x}{dt^2} = -kx - c \frac{dx}{dt},$$

or

$$\frac{d^2x}{dt^2} + 2b \frac{dx}{dt} + \omega^2 x = 0, \qquad (9)$$

20.6   Graph of simple harmonic motion of a frictionless spring, referred to as undamped vibration.

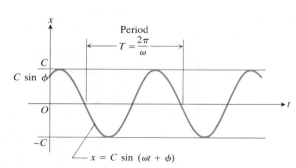

where

$$2b = \frac{c}{m} \quad \text{and} \quad \omega = \sqrt{\frac{k}{m}}.$$

If we introduce the operator $D = d/dt$, Eq. (9) becomes

$$(D^2 + 2bD + \omega^2)\, x = 0.$$

The characteristic equation is

$$r^2 + 2br + \omega^2 = 0$$

with roots

$$r = -b \pm \sqrt{b^2 - \omega^2}. \tag{10}$$

Three cases now present themselves, depending upon the relative sizes of b and ω.

CASE 1: If $b = \omega$, the two roots in Eq. (10) are equal, and the solution of (9) is

$$x = (c_1 + c_2 t)\, e^{-\omega t}. \tag{11}$$

As time goes on, x approaches zero. The motion is not oscillatory.

CASE 2: If $b > \omega$, then the roots (10) are both real but unequal, and

$$x = c_1\, e^{r_1 t} + c_2\, e^{r_2 t}, \tag{12}$$

where

$$r_1 = -b + \sqrt{b^2 - \omega^2} \quad \text{and} \quad r_2 = -b - \sqrt{b^2 - \omega^2}.$$

Here again the motion is not oscillatory. Both r_1 and r_2 are negative, and x approaches zero as time goes on.

CASE 3: If $b < \omega$, let

$$\omega^2 - b^2 = \alpha^2.$$

Then

$$r_1 = -b + \alpha i, \qquad r_2 = -b - \alpha i$$

and

$$x = e^{-bt}[c_1 \cos \alpha t + c_2 \sin \alpha t]. \tag{13a}$$

If we introduce the substitutions (7), we may also write Eq. (13a) in the equivalent form

$$x = Ce^{-bt} \sin(\alpha t + \phi). \tag{13b}$$

This equation represents damped vibratory motion. It is analogous to simple harmonic motion, of period $T = 2\pi/\alpha$, except that the amplitude is not constant but is given by Ce^{-bt}. Since this tends to zero as t increases, the vibrations tend to die out as time goes on. Observe, however, that Eq. (13b) reduces to Eq. (8) in the absence of friction. The effect of friction is twofold:

1. $b = c/(2m)$ appears in the exponential *damping factor* e^{-bt}. The larger b is, the more quickly the vibrations tend to become unnoticeable.

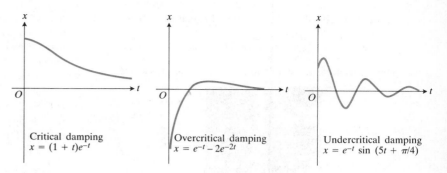

x · O · t

Critical damping
$x = (1 + t)e^{-t}$

x · O · t

Overcritical damping
$x = e^{-t} - 2e^{-2t}$

x · O · t

Undercritical damping
$x = e^{-t} \sin (5t + \pi/4)$

20.7 Three ways to damp vibration.

2. The period $T = 2\pi/\alpha = 2\pi/\sqrt{\omega^2 - b^2}$ is greater than the period $T_0 = 2\pi/\omega$ in the friction-free system.

Curves representing solutions of Eq. (9) in typical cases are shown in Figs. 20.6 and 20.7. The size of b, relative to ω, determines the kind of solution, and b also determines the rate of damping. It is therefore customary to say that there is

a) critical damping if $b = \omega$;

b) overcritical damping if $b > \omega$;

c) undercritical damping if $0 < b < \omega$;

d) no damping if $b = 0$.

PROBLEMS

1. Suppose the motion of the weight in Fig. 20.4 is described by the differential equation (3). Find the motion if $x = x_0$ and $dx/dt = v_0$ at $t = 0$. Express the answer in two equivalent forms (Eqs. 5 and 6).

2. A 5-lb weight is suspended from the lower end of a spring whose upper end is attached to a rigid support. The weight extends the spring by 2 in. If, after the weight has come to rest in its new equilibrium position, it is struck a sharp blow that starts it downward with a velocity of 4 ft/sec, find its subsequent motion, assuming there is no friction.

3. *An R–L–C series circuit.* A simple electrical circuit shown in Fig. 20.8 contains a capacitor of capacitance C farads, a coil of inductance L henrys, a resistance of R ohms, and a generator that produces an electromotive force v volts, in series. If the current intensity at time t in the circuit is i amperes, the differential equation describing the current i is

$$L \frac{d^2i}{dt^2} + R \frac{di}{dt} + \frac{1}{C} i = \frac{dv}{dt}.$$

20.8 An R–L–C series circuit.

Find i as a function of t if
a) $R = 0$, $1/(LC) = \omega^2$, $v = $ constant;
b) $R = 0$, $1/(LC) = \omega^2$, $v = V \sin \alpha t$, $V = $ constant, $\alpha = $ constant $\neq \omega$;
c) $R = 0$, $1/(LC) = \omega^2$, $v = V \sin \omega t$, $V = $ constant;
d) $R = 50$, $L = 5$, $C = 9 \times 10^{-6}$, $v = $ constant.

4. A simple pendulum of length l makes an angle θ with the vertical. As it swings back and forth, its motion, neglecting

friction, is described by the differential equation

$$\frac{d^2\theta}{dt^2} = -\frac{g}{l}\sin\theta,$$

where g (the acceleration due to gravity, $g \approx 32$ ft/sec^2) is

20.9   Motion of a simple pendulum. (Photograph from *PSSC Physics, Second Edition;* 1965; D.C. Heath & Company with Education Development Center, Inc.)

a constant. (See Fig. 20.9.) Solve the differential equation of motion, under the assumption that θ is so small that $\sin\theta$ may be replaced by θ without appreciable error. Assume that $\theta = \theta_0$ and $d\theta/dt = 0$ when $t = 0$.

5. A circular disk of mass m and radius r is suspended by a thin wire attached to the center of one of its flat faces. If the disk is twisted through an angle θ, torsion in the wire tends to turn the disk back in the opposite direction. The differential equation for the motion is

$$\frac{1}{2}mr^2\frac{d^2\theta}{dt^2} = -k\theta,$$

where k is the coefficient of torsion of the wire. Find the motion if $\theta = \theta_0$ and $d\theta/dt = v_0$ at $t = 0$.

6. A cylindrical spar buoy, diameter 1 ft, weight 100 lb, floats partially submerged in an upright position. When it is depressed slightly from its equilibrium position and released, it bobs up and down according to the differential equation

$$\frac{100}{g}\frac{d^2x}{dt^2} = -16\pi x - c\frac{dx}{dt}.$$

Here x (ft) is the vertical displacement of the buoy from its equilibrium position, and $c\,(dx/dt)$ is the frictional resistance of the water. Find c if the period of oscillation is observed to be 1.6 sec. (Take $g = 32$ ft/sec^2.)

7. Suppose the upper end of the spring in Fig. 20.4 is attached not to a rigid support at A but to a member that itself undergoes up and down motion given by a function of time t, say $y = f(t)$. If the positive direction of y is downward, the differential equation of motion is

$$m\frac{d^2x}{dt^2} + kx = kf(t).$$

Let $x = x_0$ and $dx/dt = 0$ when $t = 0$, and solve for x
a) if $f(t) = A\sin\alpha t$ and $\alpha \neq \sqrt{k/m}$,
b) if $f(t) = A\sin\alpha t$ and $\alpha = \sqrt{k/m}$.

    TOOLKIT PROGRAMS

Damped Oscillator   Forced Damped Oscillator

□ **20.7**
Higher Order Linear Equations with Constant Coefficients

The equation-solving methods of Articles 20.4 and 20.5 can be extended to equations of higher order. The characteristic algebraic equation associated with the differential equation

$$(D^n + a_1D^{n-1} + \cdots + a_{n-1}D + a_n)y = F(x) \tag{1}$$

is

$$r^n + a_1 r^{n-1} + \cdots + a_{n-1} r + a_n = 0. \tag{2}$$

If the roots $r_1, r_2, \ldots, r_n$ are all distinct, the solution of the homogeneous equation obtained by replacing $F(x)$ by 0 in Eq. (1) is

$$y_h = c_1 e^{r_1 x} + c_2 e^{r_2 x} + \cdots + c_n e^{r_n x}.$$

Pairs of complex conjugate roots $\alpha \pm i\beta$ can be grouped together, and the corresponding part of y_h can be written in terms of the functions

$$e^{\alpha x}\cos \beta x \qquad \text{and} \qquad e^{\alpha x}\sin \beta x.$$

In case the roots of Eq. (2) are not all distinct, the portion of y_h that corresponds to a root r of multiplicity m is to be replaced by

$$(C_1 x^{m-1} + C_2 x^{m-2} + \cdots + C_m)e^{rx}.$$

Note that the polynomial in parentheses contains m arbitrary constants.

EXAMPLE Solve the equation

$$\frac{d^4 y}{dx^4} - 3 \frac{d^3 y}{dx^3} + 3 \frac{d^2 y}{dx^2} - \frac{dy}{dx} = 0.$$

Solution The characteristic equation is $r^4 - 3r^3 + 3r^2 - r = r(r - 1)^3$. The roots of the characteristic equation are

$$r_1 = 0, \qquad r_2 = r_3 = r_4 = 1.$$

The solution is

$$y = C_1 + (C_2 x^2 + C_3 x + C_4)e^x. \qquad \blacksquare$$

Variation of Parameters If the general solution of the homogeneous equation is

$$y_h = C_1 u_1 + C_2 u_2 + \cdots + C_n u_n,$$

then

$$y = v_1 u_1 + v_2 u_2 + \cdots + v_n u_n$$

will be a solution of the nonhomogeneous equation (1), provided

$$v_1' u_1 \quad + v_2' u_2 \quad + \cdots + v_n' u_n \quad = 0,$$
$$v_1' u_1' \quad + v_2' u_2' \quad + \cdots + v_n' u_n' \quad = 0,$$
$$\vdots$$
$$v_1' u_1^{(n-2)} + v_2' u_2^{(n-2)} + \cdots + v_n' u_n^{(n-2)} = 0,$$
$$v_1' u_1^{(n-1)} + v_2' u_2^{(n-1)} + \cdots + v_n' u_n^{(n-1)} = F(x).$$

These equations may be solved for $v_1', v_2', \ldots, v_n'$ by Cramer's rule, and the results integrated to give $v_1, v_2, \ldots, v_n$.

PROBLEMS

Solve the following equations.

1. $\dfrac{d^3y}{dx^3} - 3\dfrac{d^2y}{dx^2} + 2\dfrac{dy}{dx} = 0$ 2. $\dfrac{d^3y}{dx^3} - y = 0$

3. $\dfrac{d^4y}{dx^4} - 4\dfrac{d^2y}{dx^2} + 4y = 0$ 4. $\dfrac{d^4y}{dx^4} - 16y = 0$

5. $\dfrac{d^4y}{dx^4} + 16y = 0$ 6. $\dfrac{d^3y}{dx^3} - 3\dfrac{dy}{dx} + 2y = e^x$

7. $\dfrac{d^4y}{dx^4} - 4\dfrac{d^3y}{dx^3} + 6\dfrac{d^2y}{dx^2} - 4\dfrac{dy}{dx} + y = 7$

8. $\dfrac{d^4y}{dx^4} + y = x + 1$

20.8
Approximation Methods: Power Series

In the rest of this chapter we introduce some ways of approximating solutions of differential equations. The treatment is necessarily brief and nonrigorous. The purpose is to show some methods for dealing with differential equations that cannot be solved by the earlier methods. However, for simplicity, some of the examples can also be done by those earlier ways, and we can see how the new and old techniques compare.

The following example deals with the equation $u'' + x^2u = 0$. It cannot be solved by our previous methods. But we can solve it by assuming that the solution has a power series representation $\Sigma a_n x^n$. The equation $u'' + x^2u = 0$ is a linear second order equation that belongs to the class of equations of the form $u'' + a_1(x)u' + a_2(x)u = 0$, where the coefficients $a_1(x)$ and $a_2(x)$ are *analytic* in the neighborhood of $x = 0$ (that is, their Maclaurin series converge to the functions in some interval $|x| < h$). In our example, $a_1(x) = 0$ and $a_2(x) = x^2$.

EXAMPLE 1   Find a power series solution for

$$u'' + x^2u = 0. \tag{1}$$

Solution   We assume that there is a solution of the form

$$u = a_0 + a_1 x + a_2 x^2 + \cdots + a_n x^n + \cdots \tag{2}$$

where the coefficients a_k are to be determined to satisfy Eq. (1). If we differentiate the series twice, we get

$$u'' = 2a_2 + 3 \cdot 2a_3x + \cdots + n(n-1) a_n x^{n-2} + \cdots. \tag{3}$$

The series for x^2u is just x^2 times the right-hand side of Eq. (2):

$$x^2u = a_0 x^2 + a_1 x^3 + a_2 x^4 + \cdots + a_n x^{n+2} + \cdots. \tag{4}$$

The series for $u'' + x^2u$ is the sum of the series in Eqs. (3) and (4):

$$u'' + x^2u = 2a_2 + 6a_3x + (12a_4 + a_0) x^2 + (20a_5 + a_1) x^3 + \cdots$$
$$+ [n(n-1) a_n + a_{n-4}] x^{n-2} + \cdots. \tag{5}$$

Notice that the coefficient of x^{n-2} in Eq. (4) is a_{n-4}. To satisfy Eq. (1), the coefficients of the individual powers of x on the right side of Eq. (5) must all be zero:

$$2a_2 = 0, \quad 6a_3 = 0, \quad 12a_4 + a_0 = 0, \quad 20a_5 + a_1 = 0, \tag{6}$$

and for all $n \geq 4$,

$$n(n-1)a_n + a_{n-4} = 0. \tag{7}$$

We can see immediately from Eq. (2) that

$$a_0 = u(0), \qquad a_1 = u'(0). \tag{8}$$

In other words, the first two coefficients are the values of u and of u' at $x = 0$. The recursion formulas (6) and (7) enable us to evaluate all the other coefficients in terms of these two.

The first two of Eqs. (6) give

$$a_2 = 0, \qquad a_3 = 0.$$

Equation (7) shows that if $a_{n-4} = 0$, then $a_n = 0$; so we conclude that

$$a_6 = 0, \qquad a_7 = 0, \qquad a_{10} = 0, \qquad a_{11} = 0,$$

and whenever $n = 4k + 2$ or $4k + 3$, a_n is zero. For the other coefficients we have

$$a_n = \frac{-a_{n-4}}{n(n-1)}$$

so that

$$a_4 = \frac{-a_0}{4 \cdot 3}, \qquad a_8 = \frac{-a_4}{8 \cdot 7} = \frac{a_0}{3 \cdot 4 \cdot 7 \cdot 8},$$

$$a_{12} = \frac{-a_8}{11 \cdot 12} = \frac{-a_0}{3 \cdot 4 \cdot 7 \cdot 8 \cdot 11 \cdot 12};$$

and

$$a_5 = \frac{-a_1}{5 \cdot 4}, \qquad a_9 = \frac{-a_5}{9 \cdot 8} = \frac{a_1}{4 \cdot 5 \cdot 8 \cdot 9},$$

$$a_{13} = \frac{-a_9}{12 \cdot 13} = \frac{-a_1}{4 \cdot 5 \cdot 8 \cdot 9 \cdot 12 \cdot 13}.$$

The answer is best expressed as the sum of two separate series—one multiplied by a_0, the other by a_1:

$$u = a_0\left(1 - \frac{x^4}{3 \cdot 4} + \frac{x^8}{3 \cdot 4 \cdot 7 \cdot 8} - \frac{x^{12}}{3 \cdot 4 \cdot 7 \cdot 8 \cdot 11 \cdot 12} + \cdots\right)$$

$$+ a_1\left(x - \frac{x^5}{4 \cdot 5} + \frac{x^9}{4 \cdot 5 \cdot 8 \cdot 9} - \frac{x^{13}}{4 \cdot 5 \cdot 8 \cdot 9 \cdot 12 \cdot 13} + \cdots\right). \tag{9}$$

Both series converge absolutely for all values of x, as is readily seen by the ratio test.   ■

REMARK   If the differential equation $u'' + x^2u = 0$ were important enough to justify the effort, we could tabulate values of the two series for selected values of x or program the computations for a calculator or computer. In fact, the general solution can be expressed in the form

$$u = C_1\, x^{1/2}J_{1/4}(x^2/2) + C_2\, x^{1/2}J_{-1/4}(x^2/2)$$

where $J_p(x)$ is called the Bessel function of the first kind of order p. Extensive tables of values of Bessel functions have been computed from series. (See

Handbook of Mathematical Functions, AMS 55, edited by M. Abramowitz and L. A. Stegun, Dover Publications, Inc., New York, 10th printing 1972, page 362, Eq. 9.1.51, and page 361, Article 9.1.27.)

Next we show an alternative method for finding a series solution for an initial value problem consisting of a differential equation of the form

$$y' = f(x, y) \tag{10}$$

and the initial condition

$$y = y_0 \quad \text{when} \quad x = x_0. \tag{11}$$

If there is a Taylor series in powers of $x - x_0$ that converges to a solution of the problem, it must have the form

$$y = y_0 + y'(x_0)(x - x_0) + \frac{1}{2!}y''(x_0)(x - x_0)^2 + \cdots.$$

The idea is to use the differential equation itself to calculate $y'(x_0) = f(x_0, y_0)$ and higher order derivatives by differentiating both sides of Eq. (10) repeatedly with respect to x and then substituting previously calculated values, as shown in the next example. In the example, $x_0 = 1$, $y_0 = 0$, and $f(x, y) = x + y$.

EXAMPLE 2   Solve the differential equation

$$y' = x + y \tag{12}$$

subject to the initial condition

$$y = 0 \quad \text{when} \quad x = 1. \tag{13}$$

Solution   The Taylor series for $y(x)$, in powers of $x - 1$, is

$$y(x) = y(1) + y'(1)(x - 1) + \frac{1}{2!}y''(1)(x - 1)^2 + \cdots.$$

Substituting $x = 1$ and $y = 0$ in Eq. (12), we get

$$y'(1) = 1 + 0 = 1. \tag{14}$$

If we differentiate both sides of Eq. (12) with respect to x, we get

$$y'' = 1 + y'. \tag{15}$$

We know that $y'(1) = 1$ and now we can calculate $y''(1)$:

$$y''(1) = 1 + y'(1) = 2. \tag{16}$$

Continue: by differentiating Eq. (15) with respect to x, we get

$$y''' = 0 + y'' = y''. \tag{17}$$

Since we already know that $y''(1) = 2$, we now find

$$y'''(1) = y''(1) = 2.$$

The pattern continues: all higher order derivatives satisfy

$$y^{(n+1)}(x) = y^{(n)}(x) \quad \text{for } n \geq 2. \tag{18}$$

At $x = 1$, all of these derivatives have the value 2:

$$y^{(n)}(1) = 2, \qquad \text{for } n \geq 2. \tag{19}$$

When the values of $y(1)$, $y'(1)$, $y''(1)$, and so on, are substituted into the Taylor series for $y(x)$, we get

$$y(x) = 0 + 1(x - 1) + 2\frac{1}{2!}(x - 1)^2 + 2\frac{1}{3!}(x - 1)^3 + \cdots$$

$$\tag{20}$$

$$= x - 1 + 2 \sum_{n=2}^{\infty} \frac{(x - 1)^n}{n!}.$$

This series brings to mind a comparable series, namely, the series for $2e^{x-1}$, which is

$$2e^{x-1} = 2 + 2(x - 1) + 2 \sum_{n=2}^{\infty} \frac{(x - 1)^n}{n!}. \tag{21}$$

Comparing Eqs. (20) and (21), we see that our series representation for $y(x)$, Eq. (20), is the same as

$$y(x) = 2e^{x-1} - 2 - (x - 1). \tag{22}$$

In Problem 1, you are asked to verify that Eq. (22) does satisfy the differential equation $y' = x + y$ with $y = 0$ when $x = 1$. ∎

REMARK The technique used in solving Example 1 could also have been used in Example 2. To apply the method of Example 1 in solving $y' = x + y$, we could use a series

$$y(x) = b_0 + b_1(x - 1) + b_2(x - 1)^2 + b_3(x - 1)^3 + \cdots$$

together with the series for x as $x = 1 + (x - 1)$. Then

$$y' = b_1 + 2b_2(x - 1) + 3b_3(x - 1)^2 + \cdots$$

and the equation $y' = x + y$ becomes

$$b_1 + 2b_2(x - 1) + 3b_3(x - 1)^2 + \cdots + nb_n(x - 1)^{n-1}$$
$$+ (n + 1)b_{n+1}(x - 1)^n + \cdots$$

$$= 1 + (x - 1) + b_0 + b_1(x - 1) + b_2(x - 1)^2 + \cdots + b_n(x - 1)^n + \cdots$$

$$= (1 + b_0) + (1 + b_1)(x - 1) + b_2(x - 1)^2 + \cdots + b_n(x - 1)^n + \cdots.$$

The coefficients of corresponding powers of $x - 1$ must be the same on both sides of this equation. In other words,

$$b_1 = 1 + b_0,$$

$$2b_2 = 1 + b_1,$$

$$3b_3 = b_2, \quad \ldots, \quad (n + 1)b_{n+1} = b_n \quad \text{for } n \geq 2.$$

These equations can be solved recursively to give all the coefficients in terms of b_0, where $b_0 = y(1)$. In particular, if $y(1) = 0$, then $b_0 = 0$, $b_1 = 1$, $2b_2 = 1 + b_1 = 2$, so $b_2 = 1$; $3b_3 = b_2 = 1$, so $b_3 = 1/3$; and so on.

PROBLEMS

1. Show that $y = 2e^{x-1} - 2 - (x - 1)$ satisfies the differential equation $y' = x + y$ and the initial condition $y = 0$ when $x = 1$.

Find appropriate Maclaurin series or Taylor series solutions for the following initial value problems.

2. $y' = y$; $y = 1$ when $x = 0$

3. $y' + y = 0$; $y = 1$ when $x = 0$

4. $y' = 2y$; $y = 2$ when $x = 0$

5. $y' + 2y = 0$; $y = -1$ when $x = 1$

6. $y' = y$; $y = 0$ and $y' = 1$ when $x = 0$

7. $y'' + y = 0$; $y = 1$ and $y' = 0$ when $x = 0$

8. $y'' + y = x$; $y = 2$ and $y' = 1$ when $x = 0$

9. $y'' - y = x$; $y = -1$ and $y' = 2$ when $x = 0$

10. $y'' + x = y$; $y = 0$ and $y' = -2$ when $x = 2$

11. $y'' - x^2 y = 0$; $y = y_0$ and $y' = y_1$ when $x = 0$

12. $y'' + x^2 y = x$; $y = y_0$ and $y' = y_1$ when $x = 0$

20.9
Slope Fields and Picard's Existence and Uniqueness Theorem

We begin this article by considering the geometric notions of solution curves, slope fields, and isoclines for a differential equation $y' = f(x, y)$.

A **solution curve** is one whose equation satisfies the differential equation. For example, any curve whose equation is $y = Ce^x$ is a solution curve for the equation $y' = y$.

A **slope field** assigns to each point $P(x, y)$ in the domain of f a number $f(x, y)$ which is the *slope* of any solution curve through P.

An **isocline,** associated with a differential equation $y' = f(x, y)$, is a curve with equation $f(x, y) = $ constant. For example, the isoclines associated with the equation $y' = y$ are the lines $y = C$, where C is a constant.

We could represent a slope field for the equation $y' = f(x, y)$ by the surface $z = f(x, y)$, but here we prefer to use a different approach, one that helps us visualize what a solution curve might look like without actually solving the differential equation. Figure 20.10 is a graphical representation for a portion of the slope field associated with the equation

$$y' = x + y. \tag{1}$$

(To simplify the language, we shall say simply that it is the slope field.) This figure was done by a computer, but it can be visualized as a pattern of iron filings that have been sprinkled onto a piece of paper and have arranged themselves so that the bit at the point $P(x, y)$ has slope $x + y$. Since iron filings won't actually do that for us, we resort to other approaches.

One simple approach is to first construct a set of isoclines. For Eq. (1), each isocline is a straight line satisfying the equation

$$x + y = C,$$

where C is a constant. Through some point on the line $x + y = 3$, for example, we draw a short line segment of slope 3. Having drawn one such segment, we then draw others parallel to it. Fig. 20.10 shows 12 such segments on that line. We repeat this process on each of the isoclines, starting with a short segment of slope C at a point on the line $x + y = C$. The resulting pattern is a portion of the slope field for the differential equation $y' = x + y$.

20.10   Slope field for $y' = x + y$, $-3 \le x \le 3$, $-3 \le y \le 3$.

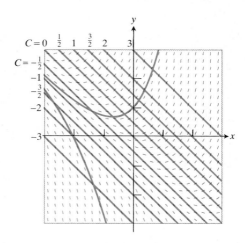

20.11 Slope field, isoclines $x + y = C$, and specific solution curves for the equation $y' = x + y$, $-3 \le x \le 3$, $-3 \le y \le 3$.

Notice one special line—the line $x + y = -1$. This line itself has slope -1 and satisfies the equation $x + y = y'$. At each point of this line, the slope field follows the line itself. Above this line, every solution curve is concave up because $y'' = 1 + y' = 1 + x + y$ is positive if $x + y > -1$. Below the line, any solution curve must be concave down.

Fig. 20.11 is a composite that shows two solution curves as well as a set of isoclines and part of the slope field for $y' = x + y$, in the region $-3 \le x \le 3$, $-3 \le y \le 3$. The upper solution curve is for the initial value problem

$$y' = x + y, \qquad y(0) = 1; \qquad y = -1 - x + 2e^x,$$

and the lower curve is for

$$y' = x + y, \qquad y(-2) = 0; \qquad y = -1 - x - e^{x+2}.$$

In this example it is easy to find the general solution of

$$y' = x + y, \qquad y(x_0) = y_0.$$

Problem 21 asks you to show that the general solution is

$$y = -1 - x + (1 + x_0 + y_0)\, e^{x - x_0}.$$

Picard's Existence and Uniqueness Theorem and Iteration Scheme

We now turn our attention to an important theorem credited to Charles Emile Picard (1856–1941), a French mathematician. The standard proof of the theorem (see reference below) hinges on an iterative process that produces a sequence of functions that converge to a solution of the problem $y' = f(x, y)$ with $y(x_0) = y_0$, provided the hypotheses of the theorem are satisfied. The iteration process, known as **Picard's iteration scheme,** is an ingenious method for finding analytic approximations to a solution. That scheme will be illustrated after we look at the theorem.

Suppose we are given an initial value problem

$$y' = f(x, y), \qquad y(x_0) = y_0, \tag{2}$$

where f is defined and continuous inside a rectangle R in the xy-plane (Fig. 20.12). Does the initial value problem always have at least one solution if (x_0, y_0) is inside R? Might it have more than one solution? The next example shows that the answer to the second question is yes, unless something more is required of f.

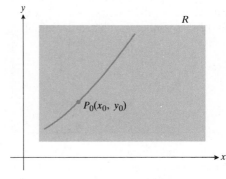

20.12 When is there exactly one solution of $y' = f(x, y)$, with $y(x_0) = y_0$? Answer: When f and f_y are continuous throughout a rectangular region R that contains $P_0(x_0, y_0)$ in its interior.

EXAMPLE 1 The initial value problem

$$y' = y^{4/5}, \qquad y(0) = 0, \tag{3}$$

has the obvious solution $y \equiv 0$. Another solution is found by separating the variables and integrating:

$$y = \left(\frac{x}{5}\right)^5.$$

There are two more solutions (see Fig. 20.13):

$$y = \begin{cases} 0 & \text{for } x \le 0, \\ \left(\dfrac{x}{5}\right)^5 & \text{for } x > 0, \end{cases}$$

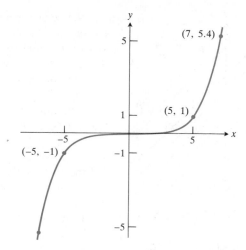

20.13 Part of the graph of $y = (x/5)^5$, one of the solutions of $y' = y^{4/5}$, $y(0) = 0$. Another solution is $y = 0$.

and

$$y = \begin{cases} \left(\dfrac{x}{5}\right)^5 & \text{for } x \le 0, \\ 0 & \text{for } x > 0. \end{cases}$$

In this example, $f(x, y) = y^{4/5}$ is continuous in the entire xy-plane. But its partial derivative $f_y = (4/5)y^{-1/5}$ is not continuous where $y = 0$. ∎

The following theorem gives a sufficient condition for the existence and uniqueness of a solution of the initial value problem of Eqs. (2).

THEOREM

Picard's Uniqueness and Existence Theorem

Let $f(x, y)$ and its first order partial derivative f_y be continuous throughout the interior and on the boundary of a rectangle R. If $P(x_0, y_0)$ is any point in the interior of R, then there exists a positive number r such that the initial value problem

$$y' = f(x, y),$$
$$y(x_0) = y_0 \tag{4}$$

has a unique solution, $y = y(x)$, for $x_0 - r \le x \le x_0 + r$.

We shall not prove this theorem* but introduce Picard's iteration scheme, which is important in the proof. We begin by noting that any solution of the

---

*For a proof of Picard's theorem, see, for example, George F. Simmons, *Differential Equations with Applications and Historical Notes*, McGraw-Hill Book Co., New York, 1972, pages 422–428.

initial value problem of Eqs. (4) must also satisfy the integral equation

$$y(x) = y_0 + \int_{x_0}^{x} f(t, y(t)) \, dt \tag{5}$$

because

$$\int_{x_0}^{x} \frac{dy}{dt} \, dt = y(x) - y(x_0).$$

The converse is also true: If $y(x)$ satisfies Eq. (5), then $y' = f(x, y(x))$ and $y(x_0) = y_0$. So Eqs. (4) may be replaced by Eq. (5). This sets the stage for Picard's iteration method: In the integrand in Eq. (5), replace $y(t)$ by the constant y_0, then integrate and call the resulting right-hand side of Eq. (5) $y_1(x)$:

$$y_1(x) = y_0 + \int_{x_0}^{x} f(t, y_0) \, dt. \tag{6}$$

This starts the process. To keep it going, we use the iterative formulas

$$y_{n+1}(x) = y_0 + \int_{x_0}^{x} f(t, y_n(t)) \, dt. \tag{7}$$

The proof of Picard's theorem consists in showing that this process produces a sequence of functions $\{y_n(x)\}$ that converge to a function $y(x)$ that satisfies Eqs. (4) and (5), for values of x sufficiently near x_0. (And that the solution is unique: that is, no other method will lead to a different solution.)

The following examples illustrate the Picard iteration scheme, but in most practical cases the computations soon become too burdensome to continue.

EXAMPLE 2 Illustrate the Picard iteration scheme for the initial value problem

$$y' = x - y, \qquad y(0) = 1. \tag{8}$$

Solution For the problem at hand, Eq. (6) becomes

$$y_1(x) = 1 + \int_{0}^{x} (t - 1) \, dt \tag{9a}$$

$$= 1 + \frac{x^2}{2} - x.$$

If we now use Eq. (7) with $n = 1$, we get

$$y_2(x) = 1 + \int_{0}^{x} \left(t - 1 - \frac{t^2}{2} + t \right) dt \tag{9b}$$

$$= 1 - x + x^2 - \frac{x^3}{6}.$$

The next iteration, with $n = 2$, gives

$$y_3(x) = 1 + \int_{0}^{x} \left(t - 1 + t - t^2 + \frac{t^3}{6} \right) dt \tag{9c}$$

$$= 1 - x + x^2 - \frac{x^3}{3} + \frac{x^4}{4!}.$$

In this example, it is possible to find the exact solution, because

$$\frac{dy}{dx} + y = x$$

is a first order equation that is linear in y. It has an integrating factor e^x and the general solution is

$$y = x - 1 + C\,e^{-x}.$$

The solution of our particular initial value problem is

$$y = x - 1 + 2e^{-x}. \tag{10}$$

If we substitute the Maclaurin series for e^{-x} in Eq. (10), we get

$$y = x - 1 + 2\left(1 - x + \frac{x^2}{2!} - \frac{x^3}{3!} + \frac{x^4}{4!} - \cdots\right)$$

$$= 1 - x + x^2 - \frac{x^3}{3} + 2\left(\frac{x^4}{4!} - \frac{x^5}{5!} + \cdots\right),$$

and we see that the Picard method has given us the first four terms of this expansion in Eq. (9c) in $y_3(x)$. ∎

In the next example, we cannot find a solution in terms of elementary functions. The Picard scheme is one way we could get an idea of how the solution behaves near the initial point.

EXAMPLE 3   Find $y_n(x)$ for $n = 0, 1, 2,$ and 3 for the initial value problem

$$y' = x^2 + y^2, \qquad y(0) = 0.$$

Solution   By definition, $y_0(x) = y(0) = 0$. The other functions $y_n(x)$ are generated by the integral representation

$$y_{n+1}(x) = 0 + \int_0^x [t^2 + (y_n(t))^2]\,dt$$

$$= \frac{x^3}{3} + \int_0^x (y_n(t))^2\,dt.$$

We successively calculate

$$y_1(x) = \frac{x^3}{3},$$

$$y_2(x) = \frac{x^3}{3} + \frac{x^7}{63},$$

$$y_3(x) = \frac{x^3}{3} + \frac{x^7}{63} + \frac{2x^{11}}{2079} + \frac{x^{15}}{59{,}535}. \qquad ∎$$

REMARK   In the next article we shall introduce numerical methods for solving initial value problems like those in Examples 2 and 3 above. If we program such a numerical solution for a calculator or computer, it is helpful to have an independent check on the program. For x near zero, we would expect $y_2(x)$, or $y_3(x)$, to provide such a check.

PROBLEMS

In Problems 1–6, sketch some of the isoclines and part of the slope field. Using the slope field, sketch a portion of the graph of the solution that includes the point $P(1, -1)$.

1. $y' = x$
2. $y' = y$
3. $y' = 1/x$
4. $y' = 1/y$
5. $y' = xy$
6. $y' = x^2 + y^2$

7. Sketch some of the slope field for the equation $y' = (x + y)^2$. Show isoclines where the slope is 0, 1/4, 1, 4. Sketch (roughly) the solutions of the equation that include (a) $P(0, 0)$, (b) $Q(-1, 1)$, and (c) $R(1, 0)$. Now make the substitution $u = x + y$ and find the general solution of $y' = (x + y)^2$. What is the equation of the solution that passes through the origin? Sketch that solution more accurately.

8. Show that every solution of the equation $y' = (x + y)^2$ has a graph that has a point of inflection but no maximum or minimum. (The graph also has vertical asymptotes, as you may discover by letting $x + y = u$.)

9. Use isoclines to sketch part of the slope field for the equation $y' = x - y$. Include some part of all four quadrants.
  a) Which isocline is also a solution of the differential equation?
  b) Use the differential equation to determine where the solution curves are concave up and concave down.

In Problems 10–13, write an equivalent first order differential equation and initial condition for y.

10. $y = 1 + \int_0^x y(t)\, dt$

11. $y = -1 + \int_1^x (t - y(t))\, dt$

12. $y = \int_1^x \frac{1}{t}\, dt$

13. $y = 2 - \int_0^x (1 + y(t)) \sin t\, dt$

14. What integral equation is equivalent to the initial value problem $y' = f(x)$, $y(x_0) = y_0$?

Use Picard's iteration scheme to find $y_n(x)$ for $n = 0, 1, 2, 3$ in Problems 15–20.

15. $y' = x$, $y(1) = 2$
16. $y' = y$, $y(0) = 1$
17. $y' = xy$, $y(1) = 1$
18. $y' = x + y$, $y(0) = 0$
19. $y' = x + y$, $y(0) = 1$
20. $y' = 2x - y$, $y(-1) = 1$

21. Show that the general solution of

$$y' = x + y, \quad y(x_0) = y_0$$

is

$$y = -1 - x + (1 + x_0 + y_0)\, e^{x - x_0}.$$

22. In Example 3, verify the correctness of the equation for $y_3(x)$. (Find the mistake if there is one.)

TOOLKIT PROGRAMS

Antiderivatives and Direction Fields

20.10
Numerical Methods

We complete our survey of methods for approximating solutions of differential equations with some numerical techniques. We describe how these techniques can be applied to the initial value problem $y' = f(x, y)$ with $y = y_0$ when $x = x_0$. These methods do not produce a general solution of $y' = f(x, y)$; however, they are very important, for example, in solving equations of motion in real situations like launching a rocket or intercepting one in orbit. In such situations, numerical answers are much more useful than algebraic expressions would be. Our examples will be kept simple, but the ideas can be extended to far more complicated equations and systems of equations. Naturally, we would not attempt to solve such complex systems of equations with pencil and paper; instead, we would turn to well-designed computer programs that have been carefully checked for accuracy.

We shall consider three numerical methods: first, a simple method due to Euler, then an improved Euler method, and finally a still more accurate method usually called the Runge-Kutta method.

The Euler Method

The initial value problem

$$y' = f(x, y), \qquad y(x_0) = y_0, \tag{1}$$

provides us with a starting point, $P(x_0, y_0)$, and a slope $f(x_0, y_0)$. We know that the graph of the solution must be a curve through P with that slope. If we use the tangent through P to approximate the actual solution curve, the approximation may be fairly good from $x_0 - h$ to $x_0 + h$, for small values of h. Thus, we might choose $h = 0.1$, say, and move along the tangent line from P to $P'(x_1, y_1)$, where $x_1 = x_0 + h$ and $y_1 = y_0 + h f(x_0, y_0)$. If we think of P' as a new starting point, we can move from P' to $P''(x_2, y_2)$, where $x_2 = x_1 + h$ and $y_2 = y_1 + h f(x_1, y_1)$. If we replace h by $-h$, we move to the left from P instead of to the right. The process can be continued, but the errors are likely to accumulate as we take more steps.

This method of approximating a solution curve by moving along tangent lines in small steps is due to Euler. To see how the process works and to gain some idea of the errors, we illustrate Euler's method for the initial value problem $y' = 1 + y$, $y(0) = 1$.

EXAMPLE 1 Take $h = 0.1$ and investigate the accuracy of the Euler approximation method for the initial value problem

$$y' = 1 + y, \qquad y(0) = 1, \tag{2}$$

by letting

$$x_{n+1} = x_n + h, \qquad y_{n+1} = y_n + h (1 + y_n). \tag{3}$$

Continue to $x = 1$.

Solution The exact solution of Eqs. (2) is

$$y = 2e^x - 1. \tag{4}$$

Table 20.2 shows the results using Eqs. (3) and the exact results rounded to four decimals for comparison. By the time we get to $x = 1$, the error is about 5.6%. ∎

TABLE 20.2

x	y (approx)	y (exact)	Error = y (exact) − y (approx)
0	1	1	0
0.1	1.2	1.2103	0.0103
0.2	1.42	1.4428	0.0228
0.3	1.662	1.6997	0.0377
0.4	1.9282	1.9836	0.0554
0.5	2.2210	2.2974	0.0764
0.6	2.5431	2.6442	0.1011
0.7	2.8974	3.0275	0.1301
0.8	3.2872	3.4511	0.1639
0.9	3.7159	3.9192	0.2033
1.0	4.1875	4.4366	0.2491

The Improved Euler Method

This method first gets an estimate of y_{n+1}, as in the original Euler method, but calls the result z_{n+1}. We then take the average of $f(x_n, y_n)$ and $f(x_{n+1}, z_{n+1})$ in place of $f(x_n, y_n)$ in the next step. Thus

$$z_{n+1} = y_n + h\,f(x_n, y_n), \tag{5a}$$

$$y_{n+1} = y_n + \frac{h}{2}[f(x_n, y_n) + f(x_{n+1}, z_{n+1})]. \tag{5b}$$

If we apply this improved method to Example 1, again with $h = 0.1$, we get the following results at $x = 1$:

$$y \text{ (approx)} = 4.42816\ 1693,$$

$$y \text{ (exact)} = 4.43656\ 3656,$$

$$\text{Error} = y \text{ (exact)} - y \text{ (approx)} = 0.00840\ 1963,$$

and the error is less than $2/10$ of 1%.

The Runge-Kutta Method*

The improved Euler method just described corresponds to approximating an integral by the trapezoidal formula. The Runge-Kutta method now to be described is analogous to using Simpson's approximation to an integral. It requires four intermediate calculations, as given in the following equations:

$$k_1 = h\,f(x_n, y_n)$$

$$k_2 = h\,f\left(x_n + \frac{h}{2}, y_n + \frac{k_1}{2}\right)$$

$$k_3 = h\,f\left(x_n + \frac{h}{2}, y_n + \frac{k_2}{2}\right) \tag{6}$$

$$k_4 = h\,f(x_n + h, y_n + k_3).$$

We then calculate y_{n+1} from the formula

$$y_{n+1} = y_n + \frac{1}{6}(k_1 + 2k_2 + 2k_3 + k_4). \tag{7}$$

When we apply this method to the problem of estimating $y(1)$ for the problem $y' = 1 + y$, $y(0) = 1$, still using $h = 0.1$, we get

$$y \text{ (approx)} = 4.43655\ 9490$$

with an error $0.00000\ 4166$, which is less than $1/10{,}000$ of 1%. This is clearly the most accurate of the three methods and is not difficult to program for a calculator or computer.

The next example shows that the error in the Runge-Kutta approximation did not continue to increase as the process was continued. In fact, with $h = 0.1$,

---

*The method described here is one of many Runge-Kutta methods. It is simple and efficient. For others, see, for example, *Handbook of Mathematical Functions,* edited by Abramowitz and Stegun, Dover Publications, Inc., New York, 1972, page 896.

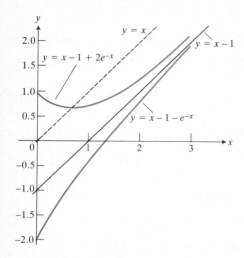

20.14   Two solutions of $y' = x - y$:
a) $y(0) = 1$, $y = x - 1 + 2e^{-x}$, b) $y(0) = -2$,
$y = x - 1 - e^{-x}$

the difference between the exact solutions and the approximations remained less than 10^{-6} for the two initial value problems:

a) $y' = x - y$, $y(0) = 1$,

b) $y' = x - y$, $y(0) = -2$.

REMARK   The fact that the differential equation is linear in y is significant in discussing the accuracy of the Runge-Kutta approximation. Such accuracy is not attained for the initial value problem

$$y' = x^2 + y^2,   \quad y(0) = 0.$$

The Runge-Kutta approximation to $y(2)$, using $h = 0.1$, is 1.47×10^{11}. For this problem, the value of y increases without bound (in the vernacular, we say it "becomes infinite") as x increases beyond $x = 2$. The solution curve has a vertical asymptote at $x = 2.003147$ (approximately). (See Fig. 20.16, later, and Problem 5.) No matter how small we take h to be, we cannot assert any accuracy for our approximations as we approach this asymptote.

EXAMPLE 2   Table 20.3 shows the comparison of $y(x)$ as estimated by the Runge-Kutta method with $h = 0.1$ and the true value, for solutions of $y' = x - y$ (a) with $y(0) = 1$ and (b) with $y(0) = -2$.

More points were actually computed and plotted to give the graphs shown in Fig. 20.14. The upper curve, $y = x - 1 + 2e^{-x}$, is concave up and has a minimum when $x = y = \ln 2$. The lower curve is concave down, is always rising as x increases, and crosses the x-axis at a value of x near 1.3. Both curves approach the line $y = x - 1$ as an asymptote as $x \to \infty$.   ∎

TABLE 20.3

x	y (Runge-Kutta)	y (true value)	Difference
a)			
0	1	1	0
0.5	0.71306 1869	0.71306 1319	5.50×10^{-7}
1.0	0.73575 9549	0.73575 8882	6.67×10^{-7}
1.5	0.94626 0927	0.94626 0320	6.07×10^{-7}
2.0	1.27067 1057	1.27067 0566	4.91×10^{-7}
2.5	1.66417 0370	1.66416 9997	3.73×10^{-7}
3.0	2.09957 4407	2.09957 4137	2.70×10^{-7}
b)			
0	-2	-2	0
0.5	$-1.10653 0935$	$-1.10653 0660$	-2.75×10^{-7}
1.0	$-0.36787 9775$	$-0.36787 9441$	-3.34×10^{-7}
1.5	$+0.27686 9537$	$+0.27686 9840$	-3.03×10^{-7}
2.0	0.86466 4472	0.86466 4717	-2.46×10^{-7}
2.5	1.41791 4816	1.41791 5001	-1.85×10^{-7}
3.0	1.95021 2796	1.95021 2932	-1.36×10^{-7}

TABLE 20.4

x	y (Runge-Kutta)	y (actual)
0	0	0
0.5	0.04179 1	0.04179 1146
1.0	0.35023 4	0.35023 1844
1.5	1.51747 3	1.51744 7544
2.0	71.57899 6	317.22440 0
2.1	$1.47001\ 1 \times 10^{11}$	
2.2	$9.99999\ 9 \times 10^{99}$	(meaning, "you broke the bank!")

EXAMPLE 3 For the initial value problem

$$y' = x^2 + y^2, \qquad y(0) = 0, \tag{8}$$

some of the Runge-Kutta approximations we got with $h = 0.1$ are shown in Table 20.4. Part of the graph is shown in Fig. 20.15.

Although there is no elementary solution of the equation $y' = x^2 + y^2$, it is possible to get an answer in terms of series. In the first place, the Maclaurin series that corresponds to the initial value $y(0) = 0$ begins this way:

$$y = \frac{x^3}{3} + \frac{x^7}{63} + \frac{2x^{11}}{6237} + \cdots. \tag{9}$$

In theory, one can get many more terms of the series, but the coefficients do not follow a simple pattern.

A more productive method involves an ingenious substitution

$$y = -\frac{u'}{u}, \tag{10a}$$

which transforms the equation $y' = x^2 + y^2$ into the second order equation

$$u'' + x^2 u = 0. \tag{10b}$$

In Article 20.8, Eq. (9), we found the solution of Eq. (10b) in the form of two power series:

$$u_1(x) = 1 - \frac{x^4}{3 \cdot 4} + \frac{x^8}{3 \cdot 4 \cdot 7 \cdot 8} - \frac{x^{12}}{3 \cdot 4 \cdot 7 \cdot 8 \cdot 11 \cdot 12} + \cdots \tag{11a}$$

and

$$u_2(x) = x - \frac{x^5}{4 \cdot 5} + \frac{x^9}{4 \cdot 5 \cdot 8 \cdot 9} - \frac{x^{13}}{4 \cdot 5 \cdot 8 \cdot 9 \cdot 12 \cdot 13} + \cdots. \tag{11b}$$

The general solution of Eq. (10b) is

$$u = C_1 u_1 + C_2 u_2.$$

Notice that

$$u_1(0) = 1, \qquad u_2(0) = 0, \qquad u_1'(0) = 0, \qquad u_2'(0) = 1.$$

Therefore, if we wish to solve the general initial value problem

$$y' = x^2 + y^2, \qquad y(0) = y_0, \tag{12a}$$

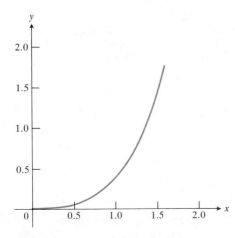

20.15 Approximate solution of $y' = x^2 + y^2$, $y(0) = 0$.

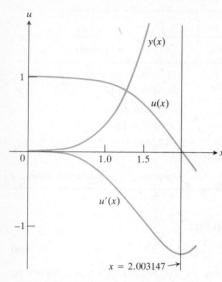

20.16 Graphs of $y(x)$, $u(x)$, and $u'(x)$ for the initial value problems

$$y' = x^2 + y^2, \quad y(0) = 0,$$
$$u'' + x^2 u = 0, \quad u(0) = 1, \quad u'(0) = 0,$$
$$y(x) = -u'/u.$$

The y-curve has an asymptote at $x = 2.003147$, where $u = 0$.

using

$$y(x) = -\frac{u'(x)}{u(x)} = -\frac{C_1\,u_1'(x) + C_2\,u_2'(x)}{C_1\,u_1(x) + C_2\,u_2(x)}, \tag{12b}$$

we need to have

$$y(0) = -\frac{C_2}{C_1} = y_0.$$

We therefore take $C_1 = 1$, $C_2 = -y_0$. For specific values of y_0, it is relatively easy to compute values of the four series for $u_1(x)$, $u_2(x)$, $u_1'(x)$, and $u_2'(x)$ and combine them to obtain

$$y(x) = -\frac{-u_1'(x) - y_0\,u_2'(x)}{-u_1(x) - y_0\,u_2(x)}. \tag{12c}$$

We used this approach with $y_0 = 0$ to get the actual values of y to compare with the values obtained by using the Runge-Kutta method in Table 20.4. (See Fig. 20.16.)

Problems 5 and 6 invite you to consider how you could predict that the solution of Eqs. (12a) becomes infinite near $x = 2$ if $y_0 = 0$ and before $x = 1$ if $y_0 = 1$.

PROBLEMS

1. Use the Euler method with $h = 1/5$ to estimate $y(1)$ if $y' = y$ and $y(0) = 1$. What is the exact value of $y(1)$?

2. Show that the Euler method leads to the estimate $(1 + (1/n))^n$ for $y(1)$ if $h = 1/n$, $y' = y$, and $y(0) = 1$. What is the limit as $n \to \infty$?

3. Use the improved Euler method with $h = 1/5$ to estimate $y(1)$ if $y' = y$ and $y(0) = 1$.

4. Use the Runge-Kutta method with $h = 1/5$ to estimate $y(1)$ if $y' = y$ and $y(0) = 1$.

5. For the initial value problem $y' = x^2 + y^2$, $y(0) = 0$, let $y = f(a)$ be the value obtained at $x = a$, where a is positive. In Example 3 we could take $a = 1.5$, $f(a) = 1.517$.) As x increases beyond $x = a$, $x^2 \geq a^2$ and $y' = x^2 + y^2 \geq a^2 + y^2$. Therefore, $y(x)$ increases faster than the solution of the simpler problem $y' = a^2 + y^2$, $y(a) = f(a)$. Solve this simpler problem and thereby show that the solution of the original problem becomes infinite at a value of x no greater than

$$x^* = a + \frac{1}{a}\left(\frac{\pi}{2} - \tan^{-1}\!\left(\frac{f(a)}{a}\right)\right).$$

If you have a calculator or access to a computer, show that $x^* = 2.0198$ if $a = 1.5$, $f(a) = 1.517$, and calculate x^* for $a = 2.0$, $f(a) = 317.2244$.

6. For the initial value problem $y' = x^2 + y^2$, $y(0) = y_0 = 1$, show that y increases faster than the solution of the simpler problem $y' = y^2$, $y(0) = 1$. Solve this simpler problem and thus show that the solution of the original problem becomes infinite at a value of x not greater than 1. (If you have a calculator or computer, you can use Eq. 12c to find that y becomes infinite as x approaches 0.96981 0654, where $u_1 = u_2 = 0.92743\ 9876$.)

7. Find the first five nonzero terms in the Maclaurin series for the solution of the initial value problem $y' = x^2 + y^2$, $y(0) = 1$. Recall from Chapter 12 that if

$$y = a_0 + a_1\,x + a_2\,x^2 + \cdots + a_n\,x^n + \cdots$$

and

$$y^2 = c_0 + c_1\,x + c_2\,x^2 + \cdots + c_n\,x^n + \cdots,$$

then

$$c_n = \sum_{k=0}^{n} a_k\,a_{n-k},$$

and

$$y' = a_1 + 2a_2\,x + 3a_3\,x^2 + \cdots + na_n\,x^{n-1} + \cdots.$$

To satisfy the differential equation, we must have $a_1 = c_0$, $2a_2 = c_1$, $3a_3 = 1 + c_2$, and $na_n = c_{n-1}$ for $n \geq 4$. You can now determine a_0 (from the initial value), c_0, a_1, c_1, a_2, c_2, a_3, and so on.

8. Solve the problem $y' = 1 + y^2$, $y(0) = 0$, (a) by separating the variables and (b) by using the substitution $y = -u'/u$ and solving the equivalent problem for u. (Note the similarity with the problem $y' = x^2 + y^2$, $y(0) = 0$. In the present problem we can find $u(x)$ without using series.)

TOOLKIT PROGRAMS

First Order Initial   Second Order Initial
Value Problem   Value Problem

REVIEW QUESTIONS AND EXERCISES

1. List some differential equations (having physical interpretations) that you have come across in your courses in chemistry, physics, engineering, or life and social sciences. Or look for some in the articles on differential equations, dynamics, electromagnetic waves, hydromechanics, quantum mechanics, or thermodynamics in a library reference.

2. How are differential equations classified?

3. What is meant by a "solution" of a differential equation?

4. Review methods for solving ordinary, first order, and first degree differential equations
   a) when the variables are separable,
   b) when the equation is homogeneous,
   c) when the equation is linear in one variable,
   d) when the equation is exact.
   Illustrate each with an example.

5. Review methods of solving second order equations
   a) with dependent variable missing,
   b) with independent variable missing.
   Illustrate each with an example.

6. Review methods of solving linear differential equations with constant coefficients
   a) in the homogeneous case,
   b) in the nonhomogeneous case.
   Illustrate each with an example.

7. *Rocket with variable mass.* If an external force F acts on a system whose mass varies with time, Newton's law of motion is

$$\frac{d(mv)}{dt} = F + (v + u)\frac{dm}{dt}.$$

In this equation, m is the mass of the system at time t, v is its velocity, and $v + u$ is the velocity of the mass that is entering (or leaving) the system at the rate dm/dt. Suppose that a rocket of initial mass m_0 starts from rest but is driven upward by firing some of its mass directly backward at the constant rate of $dm/dt = -b$ units per second and at constant speed relative to the rocket $u = -c$. The only external force acting on the rocket is $F = -mg$ due to gravity. Under these assumptions, show that the height of the rocket above the ground at the end of t seconds (t small compared to m_0/b) is

$$y = c\left(t + \frac{m_0 - bt}{b}\ln\frac{m_0 - bt}{m_0}\right) - \frac{1}{2}gt^2.$$

8. Which of the following differential equations *cannot* be solved as a first order, exact differential equation?
   a) $y^2\,dx + 2xy\,dy = 0$

   b) $(2x\sin y + y^3 e^x)\,dx + (x^2\cos y + 3y^2 e^x)\,dy = 0$
   c) $(2x\cos y + 3x^2 y)\,dx + (x^3 - x^2\sin y - y)\,dy = 0$
   d) $(y\ln y - e^{-xy})\,dx + \left(\dfrac{1}{y} + x\ln y\right)dy = 0$

9. The differential equation $(x^2 - y^3)y' = 2xy$ has an integrating factor of the form y^n. Find n and solve the equation.

10. If y is a solution of $(D^2 + 4)y = e^x$, show that it satisfies $(D - 1)(D^2 + 4)y = 0$, because $(D - 1)e^x = 0$, where $D = d/dx$. Find the general solution of $(D - 1)(D^2 + 4)y = 0$ and use the result to solve $(D^2 + 4)y = e^x$.

11. If you were to solve $y' = y^2$, $y(0) = 1$ by finding the Maclaurin series for $y(x)$, for what values of x would you expect the series to converge? Why? (*Suggestion.* Solve the initial value problem without the use of series, then expand your answer in a series.)

12. A student of mathematics and computer science used a computer to graph solutions of

$$y' = (1 + y^2)\cos x,$$

a) for $y(0) = 0$ and
b) for $y(0) = 1$.

One curve was continuous and very well behaved. The other blew up! By solving the equation for an arbitrary initial value $y(0) = y_0$, find the set of values of y_0 for which the solution remains bounded. If a solution does not remain bounded, where are the asymptotes of its graph? See Fig. 20.17.

20.17 Solutions: $y = \tan(\sin x + D)$ for $D = 0$, $\pi/4$.

MISCELLANEOUS PROBLEMS

Solve the differential equations in Problems 1–30.

1. $y \ln y \, dx + (1 + x^2) \, dy = 0$ **2.** $\dfrac{dy}{dx} = \dfrac{y^2 - y - 2}{x^2 + x}$

3. $e^{x+2y} \, dy - e^{y-2} \, dx = 0$

4. $\sqrt{1 + \left(\dfrac{dy}{dx}\right)^2} = ky$

5. $y \, dy = \sqrt{1 + y^4} \, dx$

6. $(2x + y) \, dx + (x - 2y) \, dy = 0$

7. $\dfrac{dy}{dx} = \dfrac{x^2 + y^2}{2xy}$

8. $x \dfrac{dy}{dx} = y + \sqrt{x^2 + y^2}$

9. $x \, dy = \left(y + x \cos^2 \dfrac{y}{x}\right) dx$

10. $x (\ln y - \ln x) \, dy = y (1 + \ln y - \ln x) \, dx$

11. $x \, dy + (2y - x^2 - 1) \, dx = 0$

12. $\cos y \, dx + (x \sin y - \cos^2 y) \, dy = 0$

13. $\cosh x \, dy - (y + \cosh x)\sinh x \, dx = 0$

14. $(x + 1) \, dy + (2y - x) \, dx = 0$

15. $(1 + y^2) \, dx + (2xy + y^2 + 1) \, dy = 0$

16. $(x^2 + y) \, dx + (e^y + x) \, dy = 0$

17. $(x^2 + y^2) \, dx + (2xy + \cosh y) \, dy = 0$

18. $(e^x + \ln y) \, dx + \dfrac{x + y}{y} \, dy = 0$

19. $x (1 + e^y) \, dx + (1/2)(x^2 + y^2)e^y \, dy = 0$

20. $\left(\sin x + \tan^{-1} \dfrac{y}{x}\right) dx - (y - \ln \sqrt{x^2 + y^2}) \, dy = 0$

21. $\dfrac{d^2y}{dx^2} - 2y \dfrac{dy}{dx} = 0$ **22.** $\dfrac{d^2x}{dy^2} + 4x = 0$

23. $\dfrac{d^2y}{dx^2} = 1 + \left(\dfrac{dy}{dx}\right)^2$ **24.** $\dfrac{d^2x}{dy^2} = 1 - \left(\dfrac{dx}{dy}\right)^2$

25. $x^2 \dfrac{d^2y}{dx^2} + x \dfrac{dy}{dx} = 1$ **26.** $\dfrac{d^2y}{dx^2} - 4 \dfrac{dy}{dx} + 3y = 0$

27. $\dfrac{d^3y}{dx^3} - 2 \dfrac{d^2y}{dx^2} + \dfrac{dy}{dx} = 0$ **28.** $\dfrac{d^2y}{dx^2} + 4y = \sec 2x$

29. $\dfrac{d^2y}{dx^2} - \dfrac{dy}{dx} - 2y = e^{2x}$ **30.** $\dfrac{d^2y}{dx^2} - 2 \dfrac{dy}{dx} + 5y = e^{-x}$

31. Find the *general solution* of the differential equation $4 x^2 y'' + 4 xy' - y = 0$, given that there is a solution of the form $y = x^c$ for some constant c.

32. Show that the only plane curves that have constant curvature are circles and straight lines. (Assume the appropriate derivatives exist and are continuous.)

33. Find the orthogonal trajectories of the family of curves $x^2 = Cy^3$. (*Caution:* The differential equation should not contain the arbitrary constant C.)

34. Find the orthogonal trajectories of the family of circles $(x - C)^2 + y^2 = C^2$.

35. Find the orthogonal trajectories of the family of parabolas $y^2 = 4C(C - x)$.

36. The equation $d^2y/dt^2 + 100 \, y = 0$ represents a simple harmonic motion. Find the general solution of the equation and determine the constants of integration if $y = 10$, $dy/dt = 50$, when $t = 0$. Find the period and the amplitude of the motion.

Problems 37–39 refer to the differential equation $dy/dx = x + \sin y$. (See Fig. 20.18.)

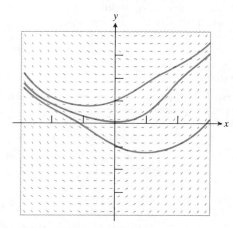

20.18   Solutions to $dy/dx = x + \sin y$ passing through $(0, 0)$, $(0, \pi/2)$, $(0, -\pi/2)$. (See Problems 37–39.)

37. For solutions near $y = 0$, we might use the approximation $\sin y \approx y$ and replace the original problem with $dy/dx = x + y$. Solve this equation with $y(0) = 0$. What is the Maclaurin series for your answer?

38. *Series.* Given the initial value problem $y' = x + \sin y$, $y(0) = 0$. By implicit differentiation we can calculate successive derivatives. For example, $y'' = 1 + (\cos y)y'$, $y''' = -(\sin y)(y')^2 + (\cos y)y''$. These can in turn be evaluated at $x = 0$ by using the given initial value $y(0) = 0$. Using this procedure, find the terms of the Maclaurin series for $y(x)$ through x^4. (Compare the answer with that of Problem 37.)

39. Using the method of the preceding problem, find the terms through x^4 in the Maclaurin series for $y(x)$ if y satisfies $y' = x + \sin y$, and (a) $y(0) = \pi/2$. (b) Repeat for $y(0) = -\pi/2$.

Appendixes

A.1

Proofs of the Limit Theorems from Article 1.9

In this appendix we prove Theorems 1 and 2 of Article 1.9.

> **The Limit Combination Theorem**
>
> If $\lim_{t \to c} F_1(t) = L_1$ and $\lim_{t \to c} F_2(t) = L_2$, then
>
> i) $\lim [F_1(t) + F_2(t)] = L_1 + L_2$,
>
> ii) $\lim [F_1(t) - F_2(t)] = L_1 - L_2$,
>
> iii) $\lim F_1(t)F_2(t) = L_1 L_2$,
>
> iv) $\lim kF_2(t) = kL_2$ (any number k),
>
> v) $\lim \dfrac{F_1(t)}{F_2(t)} = \dfrac{L_1}{L_2}$,   provided $L_2 \neq 0$.
>
> The limits are all to be taken as $t \to c$.

i) $\lim [F_1(t) + F_2(t)] = L_1 + L_2$

To prove that the sum $F_1(t) + F_2(t)$ has the limit $L_1 + L_2$ as $t \to c$ we must show that for any $\epsilon > 0$ there exists a $\delta > 0$ such that for all t,

$$0 < |t - c| < \delta \implies |F_1(t) + F_2(t) - (L_1 + L_2)| < \epsilon. \tag{1a}$$

If ϵ is positive, then so is $\epsilon/2$, and because $\lim_{t \to c} F_1(t) = L_1$ we know that there is a $\delta_1 > 0$ such that for all t,

$$0 < |t - c| < \delta_1 \implies |F_1(t) - L_1| < \epsilon/2. \tag{1b}$$

Likewise, there is a $\delta_2 > 0$ such that for all t,

$$0 < |t - c| < \delta_2 \implies |F_2(t) - L_2| < \epsilon/2. \tag{1c}$$

Now let δ be the minimum of δ_1 and δ_2. Then δ is a positive number, and the ϵ inequalities in (1b) and (1c) both hold when $0 < |t - c| < \delta$. Thus, for all t, the inequality $0 < |t - c| < \delta$ implies

$$
\begin{aligned}
|F_1(t) + F_2(t) - (L_1 + L_2)| &= |F_1(t) - L_1 + F_2(t) - L_2| \\
&\leq |F_1(t) - L_1| + |F_2(t) - L_2| \\
&< \frac{\epsilon}{2} + \frac{\epsilon}{2} = \epsilon.
\end{aligned}
$$

This establishes (1a) and proves part (i) of the theorem.

iii) $\lim [F_1(t) \cdot F_2(t)] = L_1 \cdot L_2$

Let ϵ be an arbitrary positive number, and write

$$
\begin{aligned}
F_1(t) &= L_1 + [F_1(t) - L_1], \\
F_2(t) &= L_2 + [F_2(t) - L_2].
\end{aligned}
$$

When we multiply these expressions and subtract $L_1 L_2$, we get

$$
\begin{aligned}
F_1(t) \cdot F_2(t) - L_1 L_2 &= L_1[F_2(t) - L_2] + L_2[F_1(t) - L_1] \\
&\quad + [F_1(t) - L_1] \cdot [F_2(t) - L_2].
\end{aligned} \tag{3a}
$$

The numbers $\sqrt{\epsilon/3}$, $\epsilon/[3(1 + |L_1|)]$, and $\epsilon/[3(1 + |L_2|)]$ are all positive, and, because $F_1(t)$ has limit L_1 and $F_2(t)$ has limit L_2, there are positive numbers δ_1, δ_2, δ_3, and δ_4 such that

$$
\begin{array}{ll}
|F_1(t) - L_1| < \sqrt{\epsilon/3} & \text{when } 0 < |t - c| < \delta_1, \\
|F_2(t) - L_2| < \sqrt{\epsilon/3} & \text{when } 0 < |t - c| < \delta_2, \\
|F_1(t) - L_1| < \epsilon/[3(1 + |L_2|)] & \text{when } 0 < |t - c| < \delta_3, \\
|F_2(t) - L_2| < \epsilon/[3(1 + |L_1|)] & \text{when } 0 < |t - c| < \delta_4.
\end{array}
$$

We now let δ be the minimum of the four positive numbers δ_1, δ_2, δ_3, δ_4. Then δ is a positive number and all four of the inequalities above are satisfied when $0 < |t - c| < \delta$. By taking absolute values in Eq. (3a) and applying the triangle inequality, we get

$$
\begin{aligned}
|F_1(t)F_2(t) - L_1 L_2| &\leq |L_1| \cdot |F_2(t) - L_2| + |L_2| \cdot |F_1(t) - L_1| \\
&\quad + |F_1(t) - L_1| \cdot |F_2(t) - L_2| \tag{3b} \\
&< \frac{\epsilon}{3} + \frac{\epsilon}{3} + \frac{\epsilon}{3} = \epsilon \qquad \text{when} \quad 0 < |t - c| < \delta.
\end{aligned}
$$

This completes the proof of part (iii).

iv) $\lim kF_2(t) = kL_2$

This is a special case of part (iii) with $F_1(t) = k$, the function whose output value is the constant k for all values of t.

ii) $\lim [F_1(t) - F_2(t)] = L_1 - L_2$

This can be deduced from parts (i) and (iv) in the following way:

$$
\begin{aligned}
\lim [F_1(t) - F_2(t)] &= \lim [F_1(t) + (-1)F_2(t)] \\
&= \lim F_1(t) + \lim (-1)F_2(t) \\
&= \lim F_1(t) + (-1)\lim F_2(t) \\
&= \lim F_1(t) - \lim F_2(t) \\
&= L_1 - L_2.
\end{aligned}
$$

v) $\lim \dfrac{F_1(t)}{F_2(t)} = \dfrac{L_1}{L_2}$ if $L_2 \neq 0$

Since L_2 is not zero, $|L_2|$ is a positive number, and because $F_2(t)$ has L_2 as limit when $t \to c$, we know that there is a positive number δ_1 such that

$$
|F_2(t) - L_2| < \frac{|L_2|}{2} \qquad \text{when} \quad 0 < |t - c| < \delta_1. \tag{5a}
$$

Now, for any numbers A and B, it can be shown that

$$
|A| - |B| \leq |A - B| \qquad \text{and} \qquad |B| - |A| \leq |A - B|,
$$

from which it follows that

$$
\big||A| - |B|\big| \leq |A - B|. \tag{5b}
$$

Taking $A = F_2(t)$ and $B = L_2$ in (5b), we can deduce from (5a) that

$$
-\frac{1}{2}|L_2| < |F_2(t)| - |L_2| < \frac{1}{2}|L_2| \qquad \text{when} \quad 0 < |t - c| < \delta_1.
$$

By adding $|L_2|$ to the three terms of the foregoing inequality, we get

$$
\frac{1}{2}|L_2| < |F_2(t)| < \frac{3}{2}|L_2|,
$$

from which it follows that

$$
\left| \frac{1}{F_2(t)} - \frac{1}{L_2} \right| = \left| \frac{L_2 - F_2(t)}{L_2 F_2(t)} \right| \leq \frac{2}{|L_2|^2}|L_2 - F_2(t)| \tag{5c}
$$

when $0 < |t - c| < \delta_1$.

   All that we have done so far is to show that the difference between the reciprocals of $F_2(t)$ and L_2, at the left-hand side of (5c), is no greater in absolute value than a constant times $|L_2 - F_2(t)|$ when t is close enough to c. The fact that L_2 is the limit of $F_2(t)$ has not yet been used with full force.

   But now let ϵ be an arbitrary positive number. Then $\frac{1}{2}|L_2|^2\epsilon$ is also a positive number and there is a positive number δ_2 such that

$$
|L_2 - F_2(t)| < \frac{\epsilon}{2}|L_2|^2 \qquad \text{when} \quad 0 < |t - c| < \delta_2. \tag{5d}
$$

We now let $\delta = \min \{\delta_1, \delta_2\}$ and have a positive number δ such that the inequalities in (5c) and (5d) combine to produce the result

$$\left| \frac{1}{F_2(t)} - \frac{1}{L_2} \right| < \epsilon \qquad \text{when} \quad 0 < |t - c| < \delta.$$

What we have just shown is that

If $\lim F_2(t) = L_2$ as $t \to c$, and $L_2 \neq 0$, then

$$\lim \frac{1}{F_2(t)} = \frac{1}{L_2}.$$

Having already proved the Product Law, we get the final Quotient Law by applying the Product Law to $F_1(t)$ and $1/F_2(t)$ as follows:

$$\lim \frac{F_1(t)}{F_2(t)} = \lim \left[F_1(t) \cdot \frac{1}{F_2(t)} \right] = [\lim F_1(t)] \cdot \left[\lim \frac{1}{F_2(t)} \right] = L_1 \cdot \frac{1}{L_2}. \quad \blacksquare$$

THEOREM 2

The Sandwich Theorem

Suppose that

$$f(t) \leq g(t) \leq h(t)$$

for all values of $t \neq c$ in some interval about c and that $f(t)$ and $h(t)$ approach the same limit L as t approaches c. Then

$$\lim_{t \to c} g(t) = L.$$

Proof for right-hand limits   If

$$\lim_{t \to c^+} f(t) = \lim_{t \to c^+} h(t) = L,$$

then for any $\epsilon > 0$ there exists a $\delta > 0$ such that for all t,

$$c < t < c + \delta \Rightarrow L - \epsilon < f(t) < L + \epsilon \qquad \text{and} \qquad L - \epsilon < h(t) < L + \epsilon.$$

We combine the ϵ-inequalities on the right with the inequality $f(t) \leq g(t) \leq h(t)$ to obtain

$$L - \epsilon < f(t) \leq g(t) \leq h(t) < L + \epsilon$$

and thereby

$$L - \epsilon < g(t) < L + \epsilon.$$

Therefore, for all t,

$$c < t < c + \delta \Rightarrow |g(t) - L| < \epsilon.$$

This shows that

$$\lim_{t \to c^+} g(t) = L. \qquad \blacksquare$$

Proof for left-hand limits   Given $\epsilon > 0$, there exists a $\delta > 0$ such that for all t,

$$c - \delta < t < c \Rightarrow L - \epsilon < f(t) < L + \epsilon \qquad \text{and} \qquad L - \epsilon < h(t) < L + \epsilon.$$

As before, we conclude that for all t,

$$c - \delta < t < c \Rightarrow L - \epsilon < g(t) < L + \epsilon$$

and therefore that

$$\lim_{t \to c^-} g(t) = L. \qquad \blacksquare$$

Proof for two-sided limits   If $\lim_{t \to c} f(t) = L$ and $\lim_{t \to c} h(t) = L$, then f and h approach L as $t \to c^-$ and as $t \to c^+$. Therefore, as we have shown above, $g(t) \to L$ as $t \to c^-$ and $t \to c^+$. Since the right- and left-hand limits of g as $t \to c$ exist and equal L,

$$\lim_{t \to c} g(t) = L. \qquad \blacksquare$$

PROBLEMS

1. If $F_1(t)$, $F_2(t)$, and $F_3(t)$ have limits L_1, L_2, and L_3, respectively, as t approaches c, show that their sum has limit $L_1 + L_2 + L_3$. Generalize the result to the sum of any finite number of functions.

2. If n is any positive integer greater than 1, and $F_1(t)$, $F_2(t)$, . . . , $F_n(t)$ have the finite limits L_1, L_2, . . . , L_n, respectively, as t approaches c, show that the product of the n functions has limit $L_1 \cdot L_2 \cdot \ldots \cdot L_n$. (Use mathematical induction and part iii of Theorem 1.)

3. Use the results of Article 1.9, Example 2, and Problem 2 above to show that $\lim_{t \to c} t^n = c^n$ for any positive integer n.

4. Use the result of Example 3 in Article 1.9 and the results of Problems 1 and 3 above to show that $\lim_{t \to c} f(t) = f(c)$ for any polynomial function

$$f(t) = a_0 t^n + a_1 t^{n-1} + \cdots + a_n.$$

5. Use Theorem 1 and the result of Problem 4 to show that if $f(t)$ and $g(t)$ are polynomials, and if $g(c) \neq 0$, then

$$\lim_{t \to c} \frac{f(t)}{g(t)} = \frac{f(c)}{g(c)}.$$

6. Figure A.1 gives the diagram for a proof that the composite of two continuous functions is continuous. Reconstruct the proof from the diagram. The statement to be proved is this:

If $f(t)$ is continuous at $t = c$ and $g(x)$ is continuous at $x = f(c)$, then the composite $g \circ f$ is continuous at $t = c$.

Assume also that c is an interior point of the domain of f and that $f(c)$ is an interior point of the domain of g. This will make all the limits involved two-sided. (The arguments for the cases in which one or both of c and $f(c)$ are endpoints of the domains of f and g are similar to the argument that assumes both to be interior points.)

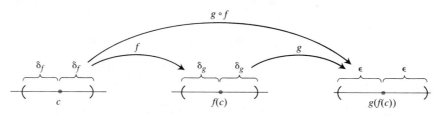

A.1   The diagram for a proof that the composite of two continuous functions is continuous. The continuity of composites holds for any finite number of functions. The only requirement is that each function be continuous where it is applied. In the figure, f is to be continuous at c, and g is to be continuous at $f(c)$.

A.2
Mathematical Induction

Many formulas, like

$$1 + 2 + \cdots + n = \frac{n(n + 1)}{2},$$

can be shown to hold for every positive integer n by applying an axiom called the *mathematical induction principle*. A proof that uses this axiom is called a *proof by mathematical induction* or a *proof by induction*.

The steps in proving a formula by induction are

STEP 1: Check that it holds for $n = 1$.

STEP 2: Prove that if it holds for any positive integer $n = k$, then it also holds for $n = k + 1$.

Once these steps are completed (the axiom says), we know that the formula holds for all positive integers n. By Step 1 it holds for $n = 1$. By Step 2 it holds for $n = 2$, and therefore by Step 2 also for $n = 3$, and by Step 2 again for $n = 4$, and so on. If the first domino falls, and the kth domino always knocks over the $(k + 1)$st when it falls, all the dominoes fall.

From another point of view, suppose we have a sequence of statements

$$S_1, S_2, \ldots, S_n, \ldots,$$

one for each positive integer. Suppose we can show that assuming any one of the statements to be true implies that the next statement in line is true. Suppose that we can also show that S_1 is true. Then we may conclude that the statements are true from S_1 on.

EXAMPLE 1   Show that

$$1 + 2 + \cdots + n = \frac{n(n + 1)}{2}$$

for every positive integer n.

Solution   We accomplish the proof by carrying out the two steps of mathematical induction.

STEP 1: The formula holds for $n = 1$ because

$$1 = \frac{1(1 + 1)}{2}.$$

STEP 2: If the formula holds for $n = k$, does it also hold for $n = k + 1$? The answer is yes, and here's why: If

$$1 + 2 + \cdots + k = \frac{k(k + 1)}{2},$$

then

$$1 + 2 + \cdots + k + (k + 1) = \frac{k(k + 1)}{2} + (k + 1)$$

$$= \frac{k^2 + k + 2k + 2}{2}$$

$$= \frac{(k + 1)(k + 2)}{2}$$

$$= \frac{(k + 1)((k + 1) + 1)}{2}.$$

The last expression in this string of equalities is the expression $n(n + 1)/2$ for $n = (k + 1)$.

The mathematical induction principle now guarantees the original formula for all positive integers n.

Notice that all we have to do here is carry out Steps 1 and 2. The mathematical induction principle does the rest. ■

EXAMPLE 2 Show that

$$\frac{1}{2^1} + \frac{1}{2^2} + \cdots + \frac{1}{2^n} = 1 - \frac{1}{2^n}$$

for all positive integers n.

Solution We accomplish the proof by carrying out the two steps of mathematical induction.

STEP 1: The formula holds for $n = 1$ because

$$\frac{1}{2^1} = 1 - \frac{1}{2^1}.$$

STEP 2: If

$$\frac{1}{2^1} + \frac{1}{2^2} + \cdots + \frac{1}{2^k} = 1 - \frac{1}{2^k},$$

then

$$\frac{1}{2^1} + \frac{1}{2^2} + \cdots + \frac{1}{2^k} + \frac{1}{2^{k+1}} = 1 - \frac{1}{2^k} + \frac{1}{2^{k+1}}$$

$$= 1 - \frac{1 \cdot 2}{2^k \cdot 2} + \frac{1}{2^{k+1}}$$

$$= 1 - \frac{2}{2^{k+1}} + \frac{1}{2^{k+1}}$$

$$= 1 - \frac{1}{2^{k+1}}.$$

Thus, the original formula is seen to hold for $n = k + 1$ whenever it holds for $n = k$.

With these two steps verified, the mathematical induction principle now guarantees the formula for every positive integer n. ∎

Other Starting Integers

Instead of starting at $n = 1$, some induction arguments start at another integer. The steps for such an argument are

STEP 1: Check that the formula holds for $n = n_1$ (whatever the appropriate first integer is).

STEP 2: Prove that if the formula holds for any integer $n = k \geq n_1$, then it also holds for $n = k + 1$.

Once these steps are completed, the mathematical induction principle will guarantee the formula for all $n \geq n_1$.

EXAMPLE 3 Show that $n! > 3^n$ if n is large enough.

Solution How large is large enough? We experiment:

n	1	2	3	4	5	6	7
$n!$	1	2	6	24	120	720	5040
3^n	3	9	27	81	243	729	2187

It looks as if $n! > 3^n$ for $n \geq 7$. To be sure, we apply mathematical induction. We take $n_1 = 7$ in Step 1 and try for Step 2.
Suppose $k! > 3^k$ for some $k \geq 7$. Then

$$(k + 1)! = (k + 1)(k!) > (k + 1)3^k > 7 \cdot 3^k > 3^{k+1}.$$

Thus, for $k \geq 7$,

$$k! > 3^k \quad \Rightarrow \quad (k + 1)! > 3^{k+1}.$$

The mathematical induction principle now guarantees $n! \geq 3^n$ for all $n \geq 7$. ∎

PROBLEMS

1. Assuming that the triangle inequality $|a + b| \leq |a| + |b|$ holds for any two numbers a and b, show that

$$|x_1 + x_2 + \cdots + x_n| \leq |x_1| + |x_2| + \cdots + |x_n|$$

for any n numbers.

2. Show that if $r \neq 1$, then

$$1 + r + r^2 + \cdots + r^n = \frac{1 - r^{n+1}}{1 - r}$$

for all positive integers n.

3. Use the Product Rule

$$\frac{d}{dx}(uv) = u\frac{dv}{dx} + v\frac{du}{dx}$$

and the fact that

$$\frac{d}{dx}(x) = 1$$

to show that

$$\frac{d}{dx}(x^n) = nx^{n-1}$$

for all positive integers n.

4. Suppose that a function $f(x)$ has the property that $f(x_1 x_2) = f(x_1) + f(x_2)$ for any two positive numbers x_1 and x_2. Show that

$$f(x_1 x_2 \cdots x_n) = f(x_1) + f(x_2) + \cdots + f(x_n)$$

for the product of any n positive numbers $x_1, x_2, \ldots, x_n$.

5. Show that

$$\frac{2}{3^1} + \frac{2}{3^2} + \cdots + \frac{2}{3^n} = 1 - \frac{1}{3^n}$$

for all positive integers n.

6. Show that $n! > n^3$ if n is large enough.

7. Show that $2^n > n^2$ if n is large enough.

8. Show that $2^n \geq \frac{1}{8}$ for $n \geq -3$.

A.3
Formulas from Precalculus Mathematics

Algebra

1. Laws of Exponents

$$a^m a^n = a^{m+n}, \quad (ab)^m = a^m b^m, \quad (a^m)^n = a^{mn}, \quad a^{m/n} = \sqrt[n]{a^m}$$

If $a \neq 0$,

$$\frac{a^m}{a^n} = a^{m-n}, \quad a^0 = 1, \quad a^{-m} = \frac{1}{a^m}.$$

2. Zero

$$a \cdot 0 = 0 \cdot a = 0 \text{ for any finite number } a$$

If $a \neq 0$,

$$\frac{0}{a} = 0, \quad 0^a = 0, \quad a^0 = 1.$$

Division by zero is not defined.

3. Fractions

$$\frac{a}{b} + \frac{c}{d} = \frac{ad + bc}{bd}, \quad \frac{a}{b} \cdot \frac{c}{d} = \frac{ac}{bd}, \quad \frac{a/b}{c/d} = \frac{a}{b} \cdot \frac{d}{c}, \quad \frac{-a}{b} = -\frac{a}{b} = \frac{a}{-b},$$

$$\frac{(a/b) + (c/d)}{(e/f) + (g/h)} = \frac{(a/b) + (c/d)}{(e/f) + (g/h)} \cdot \frac{bdfh}{bdfh} = \frac{(ad + bc)fh}{(eh + fg)bd}$$

4. Binomial Theorem, for n = Positive Integer

$$(a + b)^n = a^n + na^{n-1}b + \frac{n(n-1)}{1 \cdot 2} a^{n-2}b^2$$

$$+ \frac{n(n-1)(n-2)}{1 \cdot 2 \cdot 3} a^{n-3}b^3 + \cdots + nab^{n-1} + b^n$$

For instance,

$$(a + b)^1 = a + b,$$

$$(a + b)^2 = a^2 + 2ab + b^2,$$

$$(a + b)^3 = a^3 + 3a^2b + 3ab^2 + b^3,$$

$$(a + b)^4 = a^4 + 4a^3b + 6a^2b^2 + 4ab^3 + b^4$$

5. Difference of Like Integer Powers, $n > 1$

$$a^n - b^n = (a - b)(a^{n-1} + a^{n-2}b + a^{n-3}b^2 + \cdots + ab^{n-2} + b^{n-1})$$

For instance,

$$a^2 - b^2 = (a - b)(a + b),$$
$$a^3 - b^3 = (a - b)(a^2 + ab + b^2),$$
$$a^4 - b^4 = (a - b)(a^3 + a^2b + ab^2 + b^3)$$

6. Proportionality

The statements "y varies directly as x" and "y is directly proportional to x" mean the same thing, namely that

$$y = kx$$

for some constant k. The constant k is called the *proportionality factor* of the equation $y = kx$.

   Similarly, "y varies inversely as x" and "y is inversely proportional to x" both mean that

$$y = k\frac{1}{x}$$

for some constant k. Again, k is the proportionality factor of the equation.

7. Remainders and Factors

If a polynomial

$$f(x) = a_n x^n + a_{n-1} x^{n-1} + \cdots + a_1 x + a_0$$

is divided by $x - r$ until the remainder R is independent of x, then $R = f(r)$. In other words, $x - r$ is a factor of $f(x)$ if and only if r is a solution of the equation $f(x) = 0$. For example, if we divide $f(x) = x^2 + x - 1$ by $x - 2$, we find that $R = 5$:

$$
\begin{array}{r}
x + 3 \\
x - 2\overline{)x^2 + x - 1} \\
\underline{x^2 - 2x} \\
3x - 1 \\
\underline{3x - 6} \\
5
\end{array}
$$

Thus, $f(x) = x^2 + x - 1 = (x - 2)(x + 3) + 5$ and $f(2) = 5$. If we divide $g(x) = x^2 + x - 6$ by $x - 2$, we find that $R = 0$:

$$
\begin{array}{r}
x + 3 \\
x - 2\overline{)x^2 + x - 6} \\
\underline{x^2 - 2x} \\
3x - 6 \\
\underline{3x - 6} \\
0
\end{array}
$$

Therefore, $g(x) = x^2 + x - 6 = (x - 2)(x + 3)$ and $g(2) = 0$.

8. The Quadratic Formula

By completing the square on the first two terms of the equation

$$ax^2 + bx + c = 0$$

and solving the resulting equation for x (details omitted), we obtain the formula

$$x = \frac{-b \pm \sqrt{b^2 - 4ac}}{2a}.$$

This equation is called the **quadratic formula.** For example, the solutions of the equation

$$2x^2 + 3x - 1 = 0$$

are

$$x = \frac{-3 \pm \sqrt{(3)^2 - 4(2)(-1)}}{2(2)} = \frac{-3 \pm \sqrt{9 + 8}}{4},$$

or

$$x = \frac{-3 + \sqrt{17}}{4} \quad \text{and} \quad x = \frac{-3 - \sqrt{17}}{4}.$$

When we apply the quadratic formula to the equation

$$x^2 + 2x + 2 = 0,$$

we find

$$x = \frac{-2 \pm \sqrt{4 - 4(1)(2)}}{2(1)} = \frac{-2 \pm \sqrt{-4}}{2}$$

$$= \frac{-2 \pm 2\sqrt{-1}}{2} = -1 \pm \sqrt{-1} = -1 \pm i.$$

The solutions of the equation are the complex numbers

$$x = -1 + i \quad \text{and} \quad x = -1 - i.$$

For more about complex numbers, see Appendix 11.

9. Horner's Method

The fastest way (usually) to evaluate a polynomial

$$p(x) = a_0 + a_1x + a_2x^2 + \cdots + a_nx^n$$

at $x = c$ is to calculate

$$p(c) = a_0 + c(a_1 + c(a_2 + c(a_3 + \cdots + c(a_{n-1} + ca_n))) \cdots),$$

working from the inside out. Arranged this way, the calculation does not require parenthesis keys. All we do is multiply and add n times, for a total of $2n$ operations (fewer if any of the coefficients are zero).

For example, the value of

$$p(x) = 3x^3 - 6x^2 + 4x + 5$$

at $x = 2$ is

$$p(2) = 5 + 2(4 + 2(-6 + 2 \cdot 3)) = 13.$$

Start here,
work back

Geometry

(A = area, B = area of base, C = circumference, S = lateral area or surface area, V = volume)

1. Triangle

$$A = \tfrac{1}{2}bh$$

2. Similar Triangles

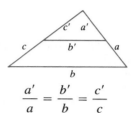

$$\frac{a'}{a} = \frac{b'}{b} = \frac{c'}{c}$$

3. Theorem of Pythagoras

$$a^2 + b^2 = c^2$$

4. Parallelogram

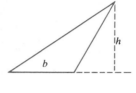

$$A = bh$$

5. Trapezoid

$$A = \tfrac{1}{2}(a + b)h$$

6. Circle

$$A = \pi r^2, \qquad C = 2\pi r$$

7. Any Cylinder or Prism with Parallel Bases

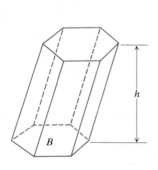

$$V = Bh$$

8. Right Circular Cylinder

$$V = \pi r^2 h, \qquad S = 2\pi rh$$

9. Any Cone or Pyramid

$$V = \tfrac{1}{3}Bh$$

10. Right Circular Cone

$$V = \tfrac{1}{3}\pi r^2 h, \qquad S = \pi rs$$

11. Sphere

$$V = \tfrac{4}{3}\pi r^3, \qquad S = 4\pi r^2$$

Trigonometry

1. Definitions and Fundamental Identities

Sine: $\sin \theta = \dfrac{y}{r} = \dfrac{1}{\csc \theta}$

Cosine: $\cos \theta = \dfrac{x}{r} = \dfrac{1}{\sec \theta}$

Tangent: $\tan \theta = \dfrac{y}{x} = \dfrac{1}{\cot \theta}$

$\sin(-\theta) = -\sin \theta,$

$\cos(-\theta) = \cos \theta$

$\sin^2\theta + \cos^2\theta = 1$

$\sec^2\theta = 1 + \tan^2\theta$

$\csc^2\theta = 1 + \cot^2\theta$

$\sin 2\theta = 2 \sin \theta \cos \theta, \qquad \cos 2\theta = \cos^2\theta - \sin^2\theta$

$$\cos^2\theta = \frac{1 + \cos 2\theta}{2}, \qquad \sin^2\theta = \frac{1 - \cos 2\theta}{2}$$

$\sin(A + B) = \sin A \cos B + \cos A \sin B$

$\sin(A - B) = \sin A \cos B - \cos A \sin B$

$\cos(A + B) = \cos A \cos B - \sin A \sin B$

$\cos(A - B) = \cos A \cos B + \sin A \sin B$

$$\tan(A + B) = \frac{\tan A + \tan B}{1 - \tan A \tan B}$$

$$\tan(A - B) = \frac{\tan A - \tan B}{1 + \tan A \tan B}$$

$\sin\left(A - \dfrac{\pi}{2}\right) = -\cos A, \qquad \cos\left(A - \dfrac{\pi}{2}\right) = \sin A$

$\sin\left(A + \dfrac{\pi}{2}\right) = \cos A, \qquad \cos\left(A + \dfrac{\pi}{2}\right) = -\sin A$

$\sin A \sin B = \frac{1}{2} \cos(A - B) - \frac{1}{2} \cos(A + B)$

$\cos A \cos B = \frac{1}{2} \cos(A - B) + \frac{1}{2} \cos(A + B)$

$\sin A \cos B = \frac{1}{2} \sin(A - B) + \frac{1}{2} \sin(A + B)$

$\sin A + \sin B = 2 \sin \frac{1}{2}(A + B) \cos \frac{1}{2}(A - B)$

$\sin A - \sin B = 2 \cos \frac{1}{2}(A + B) \sin \frac{1}{2}(A - B)$

$\cos A + \cos B = 2 \cos \frac{1}{2}(A + B) \cos \frac{1}{2}(A - B)$

$\cos A - \cos B = -2 \sin \frac{1}{2}(A + B) \sin \frac{1}{2}(A - B)$

2. Common Reference Triangles

3. Angles and Sides of a Triangle

Law of cosines: $c^2 = a^2 + b^2 - 2ab \cos C$

Law of sines: $\dfrac{\sin A}{a} = \dfrac{\sin B}{b} = \dfrac{\sin C}{c}$

Area $= \dfrac{1}{2}bc \sin A = \dfrac{1}{2}ac \sin B = \dfrac{1}{2}ab \sin C$

$= \sqrt{s(s-a)(s-b)(s-c)}$

where $s = \dfrac{1}{2}(a + b + c)$

```
TOOLKIT PROGRAMS

Function Evaluator    Super * Grapher
```

A.4

The Law of Cosines and the Angle Sum Formulas

In Fig. A.2 triangle OAB has been placed with O at the origin and B on the x-axis at $B(b, 0)$. The third vertex A has coordinates

$$x = a \cos \theta, \qquad y = a \sin \theta. \tag{1}$$

The angle BOA has measure θ. By the formula for the distance between two points, the square of the distance c from A to B is

$$c^2 = (a \cos \theta - b)^2 + (a \sin \theta)^2$$
$$= a^2(\cos^2\theta + \sin^2\theta) + b^2 - 2ab \cos \theta,$$

or

$$c^2 = a^2 + b^2 - 2ab \cos \theta. \tag{2}$$

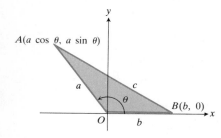

A.2 To derive the law of cosines, we compute the distance between A and B and square.

Equation (2) is the **law of cosines.** In words, it says: "The square of any side of a triangle is equal to the sum of the squares of the other two sides minus

twice the product of those two sides and the cosine of the angle between them.''
When $\theta = \pi/2$, $\cos \theta = 0$ and Eq. (2) reduces to the Pythagorean theorem.
Equation (2) holds for a general angle θ, since it is based solely on the distance
formula and on Eqs. (1). The same equation works with the exterior angle
$(2\pi - \theta)$, or the opposite of $(2\pi - \theta)$, in place of θ, because

$$\cos(2\pi - \theta) = \cos(\theta - 2\pi) = \cos \theta.$$

It is still a valid formula when A is on the x-axis and $\theta = \pi$ or $\theta = 0$, as we can
easily verify if we remember that $\cos 0 = 1$ and $\cos \pi = -1$. In these special
cases, the right-hand side of Eq. (2) becomes $(a - b)^2$ or $(a + b)^2$.

Proofs of the Sum Formulas

We derive the equation $\cos(A - B) = \cos A \cos B + \sin A \sin B$ in Article 2.6
by applying the law of cosines to the triangle OPQ in Fig. A.3. We take
$OP = OQ = r = 1$. Then the coordinates of P are

$$x_P = \cos A, \qquad y_P = \sin A,$$

and of Q,

$$x_Q = \cos B, \qquad y_Q = \sin B.$$

Hence the square of the distance between P and Q is

$$(PQ)^2 = (x_Q - x_P)^2 + (y_Q - y_P)^2$$
$$= (x_Q^2 + y_Q^2) + (x_P^2 + y_P^2) - 2(x_Q x_P + y_Q y_P)$$
$$= 2 - 2 (\cos A \cos B + \sin A \sin B).$$

But angle $QOP = A - B$, and the law of cosines gives

$$(PQ)^2 = (OP)^2 + (OQ)^2 - 2(OP)(OQ) \cos(A - B) = 2 - 2 \cos(A - B).$$

When we equate these two expressions for $(PQ)^2$ we obtain

$$\cos(A - B) = \cos A \cos B + \sin A \sin B, \qquad (11d)$$

which is Eq. (11d) of Article 2.6.

We can now derive Eqs. (11a,b,c) of Article 2.6 from Eq. (11d). We shall
also need the results

$$\sin 0 = 0, \qquad \sin(\pi/2) = 1, \qquad \sin(-\pi/2) = -1,$$
$$\cos 0 = 1, \qquad \cos(\pi/2) = 0, \qquad \cos(-\pi/2) = 0. \qquad (3)$$

1. In Eq. (11d), we put $A = \pi/2$ and use Eqs. (3) to get

$$\cos\left(\frac{\pi}{2} - B\right) = \sin B. \qquad (4)$$

If we replace B by $\pi/2 - B$ and $\pi/2 - B$ by $\pi/2 - (\pi/2 - B)$ in this equation, we get

$$\cos B = \sin\left(\frac{\pi}{2} - B\right). \qquad (5)$$

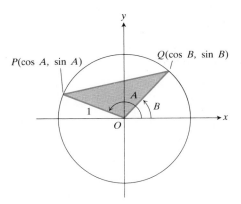

A.3 To derive the formula for cos $(A - B)$,
apply the law of cosines to this triangle.

Equations (4) and (5) tell us that the sine and cosine of an angle are the cosine
and sine, respectively, of the complementary angle.

2. We next put $B = -\pi/2$ in Eq. (11d) and use Eqs. (3) to get

$$\cos\left(A + \frac{\pi}{2}\right) = -\sin A. \qquad (6)$$

3. We can get the formula for $\cos(A + B)$ from Eq. (11d) by substituting $-B$ for B:

$$\cos(A + B) = \cos[A - (-B)]$$
$$= \cos A \cos(-B) + \sin A \sin(-B) \qquad (11b)$$
$$= \cos A \cos B - \sin A \sin B.$$

4. To derive formulas for $\sin(A \pm B)$, we use Eq. (4) with B replaced by $A + B$, and then use Eq. (11d):

$$\sin(A + B) = \cos[\pi/2 - (A + B)]$$
$$= \cos[(\pi/2 - A) - B]$$
$$= \cos(\pi/2 - A) \cos B + \sin(\pi/2 - A) \sin B \qquad (11a)$$
$$= \sin A \cos B + \cos A \sin B.$$

Equation (11c) of Article 2.6 follows from this if we replace B by $-B$:

$$\sin(A - B) = \sin A \cos B - \cos A \sin B. \qquad (11c)$$

A.5
A Proof of the Stronger Form of l'Hôpital's Rule

In this appendix, we prove the theorem in Article 3.8 that we called the stronger form of l'Hôpital's rule.

THEOREM

l'Hôpital's Rule (Stronger Form)

Suppose that

$$f(x_0) = g(x_0) = 0$$

and that the functions f and g are both differentiable on an open interval (a, b) that contains the point x_0. Suppose also that $g' \neq 0$ at every point in (a, b) except possibly x_0. Then

$$\lim_{x \to x_0} \frac{f(x)}{g(x)} = \lim_{x \to x_0} \frac{f'(x)}{g'(x)}, \qquad (1)$$

provided the limit on the right exists.

The proof of the stronger form of l'Hôpital's rule is based on Cauchy's Mean Value Theorem, a mean value theorem that involves two functions instead of one. We prove Cauchy's theorem first and then show how it applies to give l'Hôpital's rule.

THEOREM

> **Cauchy's Mean Value Theorem**
>
> Suppose that the functions $f(x)$ and $g(x)$ are continuous for $a \leq x \leq b$ and differentiable for $a < x < b$, and that $g'(x) \neq 0$ for $a < x < b$. Then there exists a number c between a and b such that
>
> $$\frac{f'(c)}{g'(c)} = \frac{f(b) - f(a)}{g(b) - g(a)}. \tag{2}$$

Proof of Cauchy's Mean Value Theorem We apply the Mean Value Theorem of Article 3.7 twice. First we use it to show that $g(b) \neq g(a)$. For if $g(b)$ did equal $g(a)$, then the Mean Value Theorem would apply to g, to give

$$g'(c) = \frac{g(b) - g(a)}{b - a} = 0$$

for some c between a and b. But this cannot happen because $g'(x) \neq 0$ for $a < x < b$.

We next apply the Mean Value Theorem to the function

$$F(x) = f(x) - f(a) - \frac{f(b) - f(a)}{g(b) - g(a)}[g(x) - g(a)]. \tag{3}$$

This function is continuous and differentiable where f and g are, and $F(b) = F(a) = 0$. Therefore there is a number c between a and b for which $F'(c) = 0$. In terms of f and g this says

$$F'(c) = f'(c) - \frac{f(b) - f(a)}{g(b) - g(a)}[g'(c)] = 0, \tag{4}$$

or

$$\frac{f'(c)}{g'(c)} = \frac{f(b) - f(a)}{g(b) - g(a)},$$

which is Eq. (2). ∎

Proof of the Stronger Form of l'Hôpital's Rule We first establish Eq. (1) for the case $x \rightarrow x_0^+$. The method needs almost no change to apply to $x \rightarrow x_0^-$, and the combination of these two cases establishes the result.

Suppose that x lies to the right of x_0. Then $g'(x) \neq 0$ and we can apply Cauchy's Mean Value Theorem to the closed interval from x_0 to x. This produces a number c between x_0 and x such that

$$\frac{f'(c)}{g'(c)} = \frac{f(x) - f(x_0)}{g(x) - g(x_0)}. \tag{5}$$

But $f(x_0) = g(x_0) = 0$, so

$$\frac{f'(c)}{g'(c)} = \frac{f(x)}{g(x)}. \tag{6}$$

As x approaches x_0, c approaches x_0 because it lies between x and x_0. Therefore,

$$\lim_{x \rightarrow x_0^+} \frac{f(x)}{g(x)} = \lim_{c \rightarrow x_0^+} \frac{f'(c)}{g'(c)} = \lim_{x \rightarrow x_0^+} \frac{f'(x)}{g'(x)}.$$

This establishes l'Hôpital's rule for the case where x approaches x_0 from above. The case where x approaches x_0 from below is proved by applying Cauchy's Mean Value Theorem to the closed interval $[x, x_0]$, $x < x_0$. ∎

PROBLEMS

Although the importance of Cauchy's Mean Value Theorem lies elsewhere, we can sometimes satisfy our curiosity about the identity of the number c in Eq. (2), as in Problems 1–4.

In Problems 1–4, find all values of c that satisfy Eq. (2) in the conclusion of Cauchy's Mean Value Theorem.

1. $f(x) = x$, $\quad g(x) = x^2$, $\quad [a, b] = [-2, 0]$

2. $f(x) = x$, $\quad g(x) = x^2$, $\quad [a, b]$ arbitrary

3. $f(x) = \dfrac{x^3}{3} - 4x$, $\quad g(x) = x^2$, $\quad [a, b] = [0, 3]$

4. $f(x) = \sin x$, $\quad g(x) = \cos x$, $\quad [a, b] = [0, \pi/2]$

A.6
The Distributive Law for Vector Cross Products

In this appendix we prove the distributive law

$$\mathbf{A} \times (\mathbf{B} + \mathbf{C}) = \mathbf{A} \times \mathbf{B} + \mathbf{A} \times \mathbf{C} \tag{1}$$

from Article 13.5.

Proof   To see that Eq. (1) is valid, we construct $\mathbf{A} \times \mathbf{B}$ a new way. We draw $\mathbf{A}$ and $\mathbf{B}$ from the common point O and construct a plane M perpendicular to $\mathbf{A}$ at O (Fig. A.4). We then project $\mathbf{B}$ orthogonally onto M, yielding a vector $\mathbf{B}'$ whose length is $|\mathbf{B}| \sin \theta$. We rotate $\mathbf{B}'$ 90° about $\mathbf{A}$ in the positive sense to produce a vector $\mathbf{B}''$. Finally, we multiply $\mathbf{B}''$ by the length of $\mathbf{A}$. The resulting vector $|\mathbf{A}|\mathbf{B}''$ is equal to $\mathbf{A} \times \mathbf{B}$ since $\mathbf{B}''$ has the same direction as $\mathbf{A} \times \mathbf{B}$ by its construction (Fig. A.4) and

$$|\mathbf{A}||\mathbf{B}''| = |\mathbf{A}||\mathbf{B}'| = |\mathbf{A}||\mathbf{B}| \sin \theta = |\mathbf{A} \times \mathbf{B}|.$$

Now each of these three operations, namely,

1. projection onto M,
2. rotation about $\mathbf{A}$ through 90°,
3. multiplication by the scalar $|\mathbf{A}|$,

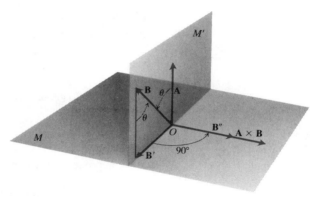

A.4   As explained in the text, $\mathbf{A} \times \mathbf{B} = |\mathbf{A}|\mathbf{B}''$.

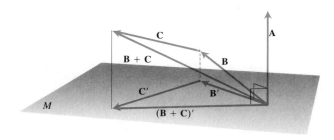

A.5 The vectors, **B**, **C**, **B** + **C**, and their projections onto a plane perpendicular to **A**.

when applied to a triangle whose plane is not parallel to **A,** will produce another triangle. If we start with the triangle whose sides are **B, C,** and **B** + **C** (Fig. A.5) and apply these three steps, we successively obtain

1. a triangle whose sides are **B′, C′,** and (**B** + **C**)′ satisfying the vector equation

$$\mathbf{B'} + \mathbf{C'} = (\mathbf{B} + \mathbf{C})';$$

2. a triangle whose sides are **B″, C″,** and (**B** + **C**)″ satisfying the vector equation

$$\mathbf{B''} + \mathbf{C''} = (\mathbf{B} + \mathbf{C})''$$

(the double-prime on each vector has the same meaning as in Fig. A.4); and, finally,

3. a triangle whose sides are |**A**|**B″**, |**A**|**C″**, and |**A**|(**B** + **C**)″ satisfying the vector equation

$$|\mathbf{A}|\mathbf{B''} + |\mathbf{A}|\mathbf{C''} = |\mathbf{A}|(\mathbf{B} + \mathbf{C})''. \tag{2}$$

When we use the equations |**A**|**B″** = **A** × **B**, |**A**|**C″** = **A** × **C** and |**A**|(**B** + **C**)″ = **A** × (**B** + **C**), which result from our discussion above, Eq. (2) becomes

$$\mathbf{A} \times \mathbf{B} + \mathbf{A} \times \mathbf{C} = \mathbf{A} \times (\mathbf{B} + \mathbf{C}),$$

which is the distributive law (1) that we wanted to establish.   ■

A.7
Determinants and Cramer's Rule

A rectangular array of numbers like

$$A = \begin{bmatrix} 2 & 1 & 3 \\ 1 & 0 & -2 \end{bmatrix}$$

is called a **matrix.** We call A a "2 by 3" matrix because it has two rows and three columns. An "m by n" matrix has m rows and n columns, and the **entry** or **element** (number) in the ith row and jth column is often denoted by a_{ij}:

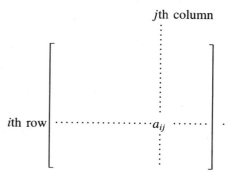

The matrix

$$A = \begin{bmatrix} 2 & 1 & 3 \\ 1 & 0 & -2 \end{bmatrix}$$

has

$$a_{11} = 2, \qquad a_{12} = 1, \qquad a_{13} = 3,$$
$$a_{21} = 1, \qquad a_{22} = 0, \qquad a_{23} = -2.$$

A matrix with the same number of rows as columns is a **square matrix.** It is a **matrix of order** n if the number of rows and columns is n.

With each square matrix A we associate a number det A or $|a_{ij}|$, called the **determinant** of A, calculated from the entries of A in the following way. (The vertical bars in the notation $|a_{ij}|$ do not mean absolute value.) For $n = 1$ and $n = 2$ we define

$$\det [a] = a, \tag{1}$$

$$\det \begin{bmatrix} a_{11} & a_{12} \\ a_{21} & a_{22} \end{bmatrix} = a_{11}a_{22} - a_{21}a_{12}. \tag{2}$$

For a matrix of order 3, we define

$$\det A = \det \begin{bmatrix} a_{11} & a_{12} & a_{13} \\ a_{21} & a_{22} & a_{23} \\ a_{31} & a_{32} & a_{33} \end{bmatrix} = \begin{array}{l} \text{Sum of all signed products} \\ \text{of the form } \pm a_{1i}a_{2j}a_{3k}, \end{array} \tag{3}$$

where i, j, k is a permutation of 1, 2, 3 in some order. There are 3! = 6 such permutations, so there are six terms in the sum. Half of these have plus signs and the other half have minus signs, according to the index of the permutation, where the index is the number we define next. The sign is positive when the index is even and negative when the index is odd.

DEFINITION

> **Index of a Permutation**
>
> Given any permutation of the numbers 1, 2, 3, . . . , n, denote the permutation by $i_1, i_2, i_3, \ldots , i_n$. In this arrangement, some of the numbers following i_1 may be less than i_1, and however many of these there are is called the **number of inversions** in the arrangement pertaining to i_1. Likewise, there is a number of inversions pertaining to each of the other i's; it is the number of indices that come after that particular one in the arrangement and are less than it. The **index** of the permutation is the sum of all of the numbers of inversions pertaining to the separate indices.

EXAMPLE 1 For $n = 5$, the permutation

$$5\ 3\ 1\ 2\ 4$$

has

4 inversions pertaining to the first element, 5,

2 inversions pertaining to the second element, 3,

and no further inversions, so the index is $4 + 2 = 6$. ◼

The following table shows the permutations of 1, 2, 3, the index of each permutation, and the signed product in the determinant of Eq. (3).

Permutation	Index	Signed product
1 2 3	0	$+a_{11}a_{22}a_{33}$
1 3 2	1	$-a_{11}a_{23}a_{32}$
2 1 3	1	$-a_{12}a_{21}a_{33}$
2 3 1	2	$+a_{12}a_{23}a_{31}$
3 1 2	2	$+a_{13}a_{21}a_{32}$
3 2 1	3	$-a_{13}a_{22}a_{31}$

(4)

The sum of the six signed products is

$$a_{11}(a_{22}a_{33} - a_{23}a_{32}) - a_{12}(a_{21}a_{33} - a_{23}a_{31}) + a_{13}(a_{21}a_{32} - a_{22}a_{31})$$

$$= a_{11}\begin{vmatrix} a_{22} & a_{23} \\ a_{32} & a_{33} \end{vmatrix} - a_{12}\begin{vmatrix} a_{21} & a_{23} \\ a_{31} & a_{33} \end{vmatrix} + a_{13}\begin{vmatrix} a_{21} & a_{22} \\ a_{31} & a_{32} \end{vmatrix}$$

$$= \begin{vmatrix} a_{11} & a_{12} & a_{13} \\ a_{21} & a_{22} & a_{23} \\ a_{31} & a_{32} & a_{33} \end{vmatrix}.$$

(5)

The formula

$$\begin{vmatrix} a_{11} & a_{12} & a_{13} \\ a_{21} & a_{22} & a_{23} \\ a_{31} & a_{32} & a_{33} \end{vmatrix} = a_{11}\begin{vmatrix} a_{22} & a_{23} \\ a_{32} & a_{33} \end{vmatrix} - a_{12}\begin{vmatrix} a_{21} & a_{23} \\ a_{31} & a_{33} \end{vmatrix} + a_{13}\begin{vmatrix} a_{21} & a_{22} \\ a_{31} & a_{32} \end{vmatrix}$$ (6)

is often used to calculate 3 by 3 determinants.

Equation (6) reduces the calculation of a 3 by 3 determinant to the calculation of three 2 by 2 determinants.

Many people prefer to remember the following scheme for calculating the six signed products in the determinant of a 3 by 3 matrix:

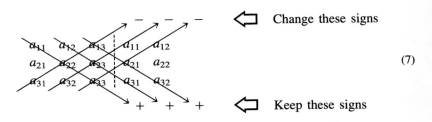

(7)

Change these signs

Keep these signs

Minors and Cofactors

The second order determinants on the right-hand side of Eq. (6) are called the **minors** (short for minor determinant) of the entries they multiply. Thus,

$$\begin{vmatrix} a_{22} & a_{23} \\ a_{32} & a_{33} \end{vmatrix} \text{ is the minor of } a_{11},$$

$$\begin{vmatrix} a_{21} & a_{23} \\ a_{31} & a_{33} \end{vmatrix} \text{ is the minor of } a_{12},$$

and so on. The minor of the element a_{ij} in a matrix A is the determinant of the matrix that remains when the row and column containing a_{ij} are deleted:

$$\begin{vmatrix} a_{11} & a_{12} & a_{13} \\ a_{21} & a_{22} & a_{23} \\ a_{31} & a_{32} & a_{33} \end{vmatrix}. \quad \text{The minor of } a_{22} \text{ is } \begin{vmatrix} a_{11} & a_{13} \\ a_{31} & a_{33} \end{vmatrix}.$$

$$\begin{vmatrix} a_{11} & a_{12} & a_{13} \\ a_{21} & a_{22} & a_{23} \\ a_{31} & a_{32} & a_{33} \end{vmatrix}. \quad \text{The minor of } a_{23} \text{ is } \begin{vmatrix} a_{11} & a_{12} \\ a_{31} & a_{32} \end{vmatrix}.$$

The **cofactor** of a_{ij} is the determinant A_{ij} that is $(-1)^{i+j}$ times the minor of a_{ij}. Thus,

$$A_{22} = (-1)^{2+2} \begin{vmatrix} a_{11} & a_{13} \\ a_{31} & a_{33} \end{vmatrix} = \begin{vmatrix} a_{11} & a_{13} \\ a_{31} & a_{33} \end{vmatrix},$$

$$A_{23} = (-1)^{2+3} \begin{vmatrix} a_{11} & a_{12} \\ a_{31} & a_{32} \end{vmatrix} = - \begin{vmatrix} a_{11} & a_{12} \\ a_{31} & a_{32} \end{vmatrix}.$$

The effect of the factor $(-1)^{i+j}$ is to change the sign of the minor when the sum $i + j$ is odd. There is a checkerboard pattern for remembering these sign changes:

$$\begin{array}{ccc} + & - & + \\ - & + & - \\ + & - & + \end{array}$$

In the upper left corner, $i = 1$, $j = 1$ and $(-1)^{1+1} = +1$. In going from any cell to an adjacent cell in the same row or column, we change i by 1 or j by 1, but not both, so we change the exponent from even to odd or from odd to even, which changes the sign from $+$ to $-$ or from $-$ to $+$.

When we rewrite Eq. (6) in terms of cofactors we get

$$\det A = a_{11}A_{11} + a_{12}A_{12} + a_{13}A_{13}. \tag{8}$$

EXAMPLE 2 Find the determinant of the matrix

$$A = \begin{bmatrix} 2 & 1 & 3 \\ 3 & -1 & -2 \\ 2 & 3 & 1 \end{bmatrix}.$$

Solution 1   The cofactors are

$$A_{11} = (-1)^{1+1} \begin{vmatrix} -1 & -2 \\ 3 & 1 \end{vmatrix}, \qquad A_{12} = (-1)^{1+2} \begin{vmatrix} 3 & -2 \\ 2 & 1 \end{vmatrix},$$

$$A_{13} = (-1)^{1+3} \begin{vmatrix} 3 & -1 \\ 2 & 3 \end{vmatrix}.$$

To find det A, we multiply each element of the first row of A by its cofactor and add:

$$\det A = 2 \begin{vmatrix} -1 & -2 \\ 3 & 1 \end{vmatrix} + (-1) \begin{vmatrix} 3 & -2 \\ 2 & 1 \end{vmatrix} + 3 \begin{vmatrix} 3 & -1 \\ 2 & 3 \end{vmatrix}$$

$$= 2(-1 + 6) - 1(3 + 4) + 3(9 + 2) = 10 - 7 + 33 = 36.$$

Solution 2   From Eq. (7) we find

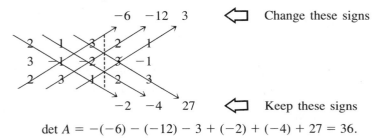

$$\det A = -(-6) - (-12) - 3 + (-2) + (-4) + 27 = 36. \qquad \blacksquare$$

Expanding by Columns or by Other Rows

The determinant of a square matrix can be calculated from the cofactors of any row or any column.

If we were to expand the determinant in Example 2 by cofactors according to elements of its third column, say, we would get

$$+3 \begin{vmatrix} 3 & -1 \\ 2 & 3 \end{vmatrix} - (-2) \begin{vmatrix} 2 & 1 \\ 2 & 3 \end{vmatrix} + 1 \begin{vmatrix} 2 & 1 \\ 3 & -1 \end{vmatrix}$$

$$= 3(9 + 2) + 2(6 - 2) + 1(-2 - 3) = 33 + 8 - 5 = 36.$$

Useful Facts About Determinants

FACT 1: If two rows of a matrix are identical, the determinant is zero.

FACT 2: Interchanging two rows of a matrix changes the sign of its determinant.

FACT 3: The determinant of a matrix is the sum of the products of the elements of the ith row (or column) by their cofactors, for any i.

FACT 4: The determinant of the **transpose** of a matrix is equal to the original determinant. (''Transpose'' means to write the rows as columns.)

FACT 5: If each element of some row (or column) of a matrix is multiplied by a constant c, the determinant is multiplied by c.

FACT 6: If all elements of a matrix above the main diagonal (or all below it) are zero, the determinant of the matrix is the product of the elements on the main diagonal.

EXAMPLE 3

$$\begin{vmatrix} 3 & 4 & 7 \\ 0 & -2 & 5 \\ 0 & 0 & 5 \end{vmatrix} = (3)(-2)(5) = -30.$$ ◼

FACT 7: If the elements of any row of a matrix are multiplied by the cofactors of the corresponding elements of a different row and these products are summed, the sum is zero.

EXAMPLE 4   If A_{11}, A_{12}, A_{13} are the cofactors of the elements of the first row of $A = (a_{ij})$, then the sums

$$a_{21}A_{11} + a_{22}A_{12} + a_{23}A_{13}$$

(elements of second row times cofactors of elements of first row) and

$$a_{31}A_{11} + a_{32}A_{12} + a_{33}A_{13}$$

are both zero. ◼

A similar result holds for columns.

FACT 8: If the elements of any column of a matrix are multiplied by the cofactors of the corresponding elements of a different column and these products are summed, the sum is zero.

FACT 9: If each element of a row of a matrix is multiplied by a constant c and the results added to a different row, the determinant is not changed.

EXAMPLE 5   If we start with

$$A = \begin{bmatrix} 2 & 1 & 3 \\ 3 & -1 & -2 \\ 2 & 3 & 1 \end{bmatrix}$$

and add -2 times row 1 to row 2 (subtract 2 times row 1 from row 2) we get

$$B = \begin{bmatrix} 2 & 1 & 3 \\ -1 & -3 & -8 \\ 2 & 3 & 1 \end{bmatrix}.$$

Since det $A = 36$ (Example 2), we should find that det $B = 36$ as well. Indeed we do, as the following calculation shows:

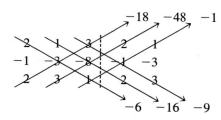

$$\det B = -(-18) - (-48) - (-1) + (-6) + (-16) + (-9)$$

$$= 18 + 48 + 1 - 6 - 16 - 9$$

$$= 67 - 31 = 36.$$ ◼

EXAMPLE 6 Evaluate the fourth order determinant

$$\begin{vmatrix} 1 & -2 & 3 & 1 \\ 2 & 1 & 0 & 2 \\ -1 & 2 & 1 & -2 \\ 0 & 1 & 2 & 1 \end{vmatrix}.$$

Solution We subtract 2 times row 1 from row 2 and add row 1 to row 3 to get

$$\begin{vmatrix} 1 & -2 & 3 & 1 \\ 0 & 5 & -6 & 0 \\ 0 & 0 & 4 & -1 \\ 0 & 1 & 2 & 1 \end{vmatrix}.$$

We then multiply the elements of the first column by their cofactors to get

$$\begin{vmatrix} 5 & -6 & 0 \\ 0 & 4 & -1 \\ 1 & 2 & 1 \end{vmatrix} = 5(4 + 2) - (-6)(0 + 1) + 0 = 36. \qquad ∎$$

Cramer's Rule

When the determinant

$$D = \det A = \begin{vmatrix} a_{11} & a_{12} \\ a_{21} & a_{22} \end{vmatrix}$$

of the coefficient matrix of the system

$$a_{11}x + a_{12}y = b_1$$
$$a_{21}x + a_{22}y = b_2 \tag{9}$$

of linear equations is 0, the system either has infinitely many solutions or no solution at all. The system

$$x + y = 0,$$
$$2x + 2y = 0$$

whose determinant is

$$D = \begin{vmatrix} 1 & 1 \\ 2 & 2 \end{vmatrix} = 2 - 2 = 0$$

has infinitely many solutions. We can find an x to match any given y. The system

$$x + y = 0,$$
$$2x + 2y = 2$$

has no solution. If $x + y = 0$, then $2x + 2y = 2(x + y)$ cannot be 2.

If $D \neq 0$, the system (9) has a unique solution, and Cramer's rule states that it may be found from the formulas

$$x = \frac{\begin{vmatrix} b_1 & a_{12} \\ b_2 & a_{22} \end{vmatrix}}{D}, \qquad y = \frac{\begin{vmatrix} a_{11} & b_1 \\ a_{21} & b_2 \end{vmatrix}}{D}. \tag{10}$$

The numerator in the formula for x comes from replacing the first column in A (the x-column) by the column of constants b_1 and b_2 (the b-column). Replacing the y-column by the b-column gives the numerator of the y-solution.

EXAMPLE 7   Solve the system

$$3x - y = 9,$$
$$x + 2y = -4.$$

Solution   We use Eqs. (10). The determinant of the coefficient matrix is

$$D = \begin{vmatrix} 3 & -1 \\ 1 & 2 \end{vmatrix} = 6 + 1 = 7.$$

Hence,

$$x = \frac{\begin{vmatrix} 9 & -1 \\ -4 & 2 \end{vmatrix}}{D} = \frac{18 - 4}{7} = \frac{14}{7} = 2,$$

$$y = \frac{\begin{vmatrix} 3 & 9 \\ 1 & -4 \end{vmatrix}}{D} = \frac{-12 - 9}{7} = \frac{-21}{7} = -3. \qquad \blacksquare$$

Systems of three equations in three unknowns work the same way. If the determinant

$$D = \det A = \begin{vmatrix} a_{11} & a_{12} & a_{13} \\ a_{21} & a_{22} & a_{23} \\ a_{31} & a_{32} & a_{33} \end{vmatrix}$$

of the system

$$a_{11}x + a_{12}y + a_{13}z = b_1,$$
$$a_{21}x + a_{22}y + a_{23}z = b_2, \qquad (11)$$
$$a_{31}x + a_{32}y + a_{33}z = b_3$$

is zero, the system either has infinitely many solutions or no solution at all. If $D \neq 0$, the system has a unique solution, given by Cramer's rule:

$$x = \frac{1}{D} \begin{vmatrix} b_1 & a_{12} & a_{13} \\ b_2 & a_{22} & a_{23} \\ b_3 & a_{32} & a_{33} \end{vmatrix},$$

$$y = \frac{1}{D} \begin{vmatrix} a_{11} & b_1 & a_{13} \\ a_{21} & b_2 & a_{23} \\ a_{31} & b_3 & a_{33} \end{vmatrix}, \qquad (12)$$

$$z = \frac{1}{D} \begin{vmatrix} a_{11} & a_{12} & b_1 \\ a_{21} & a_{22} & b_2 \\ a_{31} & a_{32} & b_3 \end{vmatrix}.$$

The pattern continues in higher dimensions: If $AX = B$ and det $A = 0$, the system either has infinitely many solutions or no solution at all. If det $A \neq 0$, the system has a unique solution given by

$$x_i = \frac{\det U_i}{\det A},\tag{13}$$

where U_i is the matrix obtained from A by replacing the ith column in A by the b-column.

For a derivation of Cramer's rule, see Appendix 8, Problem 32.

Other Formulas That Use Determinants

1. The equation of a line through two points (x_1, y_1) and (x_2, y_2) in the plane is

$$\begin{vmatrix} x & y & 1 \\ x_1 & y_1 & 1 \\ x_2 & y_2 & 1 \end{vmatrix} = 0.$$

2. The area of a triangle with vertices (x_1, y_1), (x_2, y_2), (x_3, y_3) is

$$\pm \frac{1}{2} \begin{vmatrix} x_1 & y_1 & 1 \\ x_2 & y_2 & 1 \\ x_3 & y_3 & 1 \end{vmatrix}.$$

3. The volume of the parallelepiped spanned by the vectors **A, B,** and **C** is

$$\pm\, \mathbf{A} \cdot (\mathbf{B} \times \mathbf{C}) = \pm \begin{vmatrix} a_1 & a_2 & a_3 \\ b_1 & b_2 & b_3 \\ c_1 & c_2 & c_3 \end{vmatrix}.$$

4. The cross product of $\mathbf{A} = a_1\mathbf{i} + a_2\mathbf{j} + a_3\mathbf{k}$ and $\mathbf{B} = b_1\mathbf{i} + b_2\mathbf{j} + b_3\mathbf{k}$ is

$$\mathbf{A} \times \mathbf{B} = \begin{vmatrix} \mathbf{i} & \mathbf{j} & \mathbf{k} \\ a_1 & a_2 & a_3 \\ b_1 & b_2 & b_3 \end{vmatrix}.$$

5. If $F = M(x, y, z)\mathbf{i} + N(x, y, z)\mathbf{j} + P(x, y, z)\mathbf{k}$, then

$$\operatorname{curl} \mathbf{F} = \boldsymbol{\nabla} \times \mathbf{F} = \begin{vmatrix} \mathbf{i} & \mathbf{j} & \mathbf{k} \\ \dfrac{\partial}{\partial x} & \dfrac{\partial}{\partial y} & \dfrac{\partial}{\partial z} \\ M & N & P \end{vmatrix}.$$

A Reduction Formula for Evaluating Determinants

The following reduction formula is derived in E. Miller's article "Evaluating an nth Order Determinant in n Easy Steps," *MATYC Journal* 12 (1978), 123–128. Evaluating determinants with this formula is relatively fast and readily programmable. For a short FORTRAN IV program that does so, see Alban J. Rogues's article, "Determinants: A Short Program," *Two-Year College Mathematics Journal*, Vol. 10, No. 5 (November 1979), pp. 340–342.

The formula for the determinant of an n by n matrix $A = (a_{ij})$ is

$$\det A = \left(\frac{1}{a_{11}}\right)^{n-2} \begin{vmatrix} \begin{vmatrix} a_{11} & a_{12} \\ a_{21} & a_{22} \end{vmatrix} & \begin{vmatrix} a_{11} & a_{13} \\ a_{21} & a_{23} \end{vmatrix} & \begin{vmatrix} a_{11} & a_{14} \\ a_{21} & a_{24} \end{vmatrix} & \cdots & \begin{vmatrix} a_{11} & a_{1n} \\ a_{21} & a_{2n} \end{vmatrix} \\[2mm] \begin{vmatrix} a_{11} & a_{12} \\ a_{31} & a_{32} \end{vmatrix} & \begin{vmatrix} a_{11} & a_{13} \\ a_{31} & a_{33} \end{vmatrix} & \begin{vmatrix} a_{11} & a_{14} \\ a_{31} & a_{34} \end{vmatrix} & \cdots & \begin{vmatrix} a_{11} & a_{1n} \\ a_{31} & a_{3n} \end{vmatrix} \\[2mm] \begin{vmatrix} a_{11} & a_{12} \\ a_{41} & a_{42} \end{vmatrix} & \begin{vmatrix} a_{11} & a_{13} \\ a_{41} & a_{43} \end{vmatrix} & \begin{vmatrix} a_{11} & a_{14} \\ a_{41} & a_{44} \end{vmatrix} & \cdots & \begin{vmatrix} a_{11} & a_{1n} \\ a_{41} & a_{4n} \end{vmatrix} \\[2mm] \vdots & \vdots & \vdots & & \vdots \\[2mm] \begin{vmatrix} a_{11} & a_{12} \\ a_{n1} & a_{n2} \end{vmatrix} & \begin{vmatrix} a_{11} & a_{13} \\ a_{n1} & a_{n3} \end{vmatrix} & \begin{vmatrix} a_{11} & a_{14} \\ a_{n1} & a_{n4} \end{vmatrix} & \cdots & \begin{vmatrix} a_{11} & a_{1n} \\ a_{n1} & a_{nn} \end{vmatrix} \end{vmatrix}.$$

EXAMPLE 8

$$\begin{vmatrix} 1 & 0 & 2 & -1 \\ 3 & -2 & 6 & 4 \\ 5 & 4 & 3 & 0 \\ 2 & 2 & -5 & 6 \end{vmatrix} = \left(\frac{1}{1}\right)^2 \begin{vmatrix} \begin{vmatrix} 1 & 0 \\ 3 & -2 \end{vmatrix} & \begin{vmatrix} 1 & 2 \\ 3 & 6 \end{vmatrix} & \begin{vmatrix} 1 & -1 \\ 3 & 4 \end{vmatrix} \\[2mm] \begin{vmatrix} 1 & 0 \\ 5 & 4 \end{vmatrix} & \begin{vmatrix} 1 & 2 \\ 5 & 3 \end{vmatrix} & \begin{vmatrix} 1 & -1 \\ 5 & 0 \end{vmatrix} \\[2mm] \begin{vmatrix} 1 & 0 \\ 2 & 2 \end{vmatrix} & \begin{vmatrix} 1 & 2 \\ 2 & -5 \end{vmatrix} & \begin{vmatrix} 1 & -1 \\ 2 & 6 \end{vmatrix} \end{vmatrix}$$

$$= \begin{vmatrix} -2 & 0 & 7 \\ 4 & -7 & 5 \\ 2 & -9 & 8 \end{vmatrix} = \left(\frac{1}{-2}\right)^1 \begin{vmatrix} 14 & -38 \\ 18 & -30 \end{vmatrix} = -132 \quad \blacksquare$$

PROBLEMS

Evaluate the following determinants.

1. $\begin{vmatrix} 2 & 3 & 1 \\ 4 & 5 & 2 \\ 1 & 2 & 3 \end{vmatrix}$

2. $\begin{vmatrix} 2 & -1 & -2 \\ -1 & 2 & 1 \\ 3 & 0 & -3 \end{vmatrix}$

3. $\begin{vmatrix} 1 & 2 & 3 & 4 \\ 0 & 1 & 2 & 3 \\ 0 & 0 & 2 & 1 \\ 0 & 0 & 3 & 2 \end{vmatrix}$

4. $\begin{vmatrix} 1 & -1 & 2 & 3 \\ 2 & 1 & 2 & 6 \\ 1 & 0 & 2 & 3 \\ -2 & 2 & 0 & -5 \end{vmatrix}$

Evaluate the following determinants by expanding according to the cofactors of (a) the third row and (b) the second column.

5. $\begin{vmatrix} 2 & -1 & 2 \\ 1 & 0 & 3 \\ 0 & 2 & 1 \end{vmatrix}$

6. $\begin{vmatrix} 1 & 0 & -1 \\ 0 & 2 & -2 \\ 2 & 0 & 1 \end{vmatrix}$

7. $\begin{vmatrix} 1 & 1 & 0 & 0 \\ 0 & 0 & -2 & 1 \\ 0 & -1 & 0 & 7 \\ 3 & 0 & 2 & 1 \end{vmatrix}$

8. $\begin{vmatrix} 0 & 1 & 0 & 0 \\ 0 & 1 & 1 & 0 \\ 1 & 1 & 1 & 1 \\ 1 & 1 & 0 & 0 \end{vmatrix}$

Solve the following systems of equations by Cramer's rule.

9. $\begin{aligned} x + 8y &= 4 \\ 3x - y &= -13 \end{aligned}$

10. $\begin{aligned} 2x + 3y &= 5 \\ 3x - y &= 2 \end{aligned}$

11. $\begin{aligned} 4x - 3y &= 6 \\ 3x - 2y &= 5 \end{aligned}$

12. $\begin{aligned} x + y + z &= 2 \\ 2x - y + z &= 0 \\ x + 2y - z &= 4 \end{aligned}$

13. $\begin{aligned} 2x + y - z &= 2 \\ x - y + z &= 7 \\ 2x + 2y + z &= 4 \end{aligned}$

14. $\begin{aligned} 2x - 4y &= 6 \\ x + y + z &= 1 \\ 5y + 7z &= 10 \end{aligned}$

15. $\begin{aligned} x \qquad - z &= 3 \\ 2y - 2z &= 2 \\ 2x \quad + z &= 3 \end{aligned}$

16. $\begin{aligned} x_1 + x_2 - x_3 + x_4 &= 2 \\ x_1 - x_2 + x_3 + x_4 &= -1 \\ x_1 + x_2 + x_3 - x_4 &= 2 \\ x_1 \qquad + x_3 + x_4 &= -1 \end{aligned}$

17. Find values of h and k for which the system

$$2x + hy = 8,$$
$$x + 3y = k$$

has (a) infinitely many solutions, (b) no solution at all.

18. For what value of x will

$$\begin{vmatrix} x & x & 1 \\ 2 & 0 & 5 \\ 6 & 7 & 1 \end{vmatrix} = 0?$$

19. Suppose u, v, and w are twice-differentiable functions of x that satisfy the relation $au + bv + cw = 0$, where a, b, and c are constants, not all zero. Show that

$$\begin{vmatrix} u & v & w \\ u' & v' & w' \\ u'' & v'' & w'' \end{vmatrix} = 0.$$

20. Expanding the quotient

$$\frac{ax + b}{(x - r_1)(x - r_2)}$$

by partial fractions calls for finding the values of C and D that make the equation

$$\frac{ax + b}{(x - r_1)(x - r_2)} = \frac{C}{x - r_1} + \frac{D}{x - r_2}$$

hold for all x.

a) Find a system of linear equations that determines C and D.

b) Under what circumstances does the system of equations in part (a) have a unique solution? That is, when is the determinant of the coefficient matrix of the system different from zero?

A.8
Matrices, Linear Equations, and Eigenvalues

A rectangular array of numbers like

$$A = \begin{bmatrix} 2 & 1 & 3 \\ 1 & 0 & -2 \end{bmatrix}$$

is called a **matrix.** We call A a "2 by 3" matrix because it has two rows and three columns. An "m by n" matrix has m rows and n columns, and the **entry** or **element** (number) in the ith row and jth column is often denoted by a_{ij}:

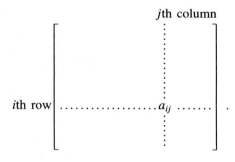

The matrix

$$A = \begin{bmatrix} 2 & 1 & 3 \\ 1 & 0 & -3 \end{bmatrix} \quad \text{has} \quad \begin{matrix} a_{11} = 2, & a_{12} = 1, & a_{13} = 3, \\ a_{21} = 1, & a_{22} = 0, & a_{23} = -2. \end{matrix}$$

Matrix Addition and Multiplication

Two matrices are equal if (and only if) they have the same numbers in the same positions. For example,

$$B = \begin{bmatrix} b_{11} & b_{12} & b_{13} \\ b_{21} & b_{22} & b_{23} \end{bmatrix} = \begin{bmatrix} 2 & 1 & 3 \\ 1 & 0 & -2 \end{bmatrix}$$

means that

$$b_{11} = 2, \quad b_{12} = 1, \quad b_{13} = 3,$$

$$b_{21} = 1, \quad b_{22} = 0, \quad b_{23} = -2.$$

For matrices A and B to be equal they must have the same **shape** (same number of rows and same number of columns) and must have

$$a_{ij} = b_{ij} \qquad \text{for all } i \text{ and } j.$$

Matrices with the same shape can be added by adding corresponding elements. For example,

$$\begin{bmatrix} 2 & 1 & 3 \\ 1 & 0 & -2 \end{bmatrix} + \begin{bmatrix} 1 & -2 & 2 \\ 2 & 3 & -1 \end{bmatrix} = \begin{bmatrix} 3 & -1 & 5 \\ 3 & 3 & -3 \end{bmatrix}.$$

To multiply a matrix by a number c, we multiply each element by c. For example,

$$7\begin{bmatrix} 2 & 1 & 3 \\ 1 & 0 & -2 \end{bmatrix} = \begin{bmatrix} 14 & 7 & 21 \\ 7 & 0 & -14 \end{bmatrix}.$$

A system of simultaneous linear equations

$$a_{11}x + a_{12}y + a_{13}z = b_1,$$
$$a_{21}x + a_{22}y + a_{23}z = b_2 \tag{1}$$

can be written in matrix form as

$$\begin{bmatrix} a_{11} & a_{12} & a_{13} \\ a_{21} & a_{22} & a_{23} \end{bmatrix} \begin{bmatrix} x \\ y \\ z \end{bmatrix} = \begin{bmatrix} b_1 \\ b_2 \end{bmatrix}, \tag{2}$$

or, more compactly, as

$$AX = B, \tag{3}$$

where

$$A = \begin{bmatrix} a_{11} & a_{12} & a_{13} \\ a_{21} & a_{22} & a_{23} \end{bmatrix}, \qquad X = \begin{bmatrix} x \\ y \\ z \end{bmatrix}, \qquad B = \begin{bmatrix} b_1 \\ b_2 \end{bmatrix}. \tag{4}$$

To form the product indicated by AX in Eq. (3), we take the elements of the first row of A in order from left to right, multiply them by the corresponding elements of X from the top down, and add these products to get

$$a_{11}x + a_{12}y + a_{13}z,$$

which we set equal to b_1. The result is the first equation in (1). We then repeat the process with the second row in (2) to obtain the second equation in (1).

An m by n matrix A can multiply an n by p matrix B from the left to give an m by p matrix $C = AB$:

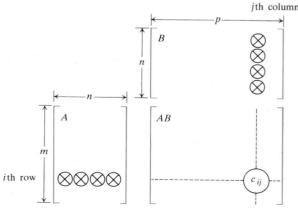

The element in the ith row and jth column of AB is the sum

$$c_{ij} = a_{i1}b_{1j} + a_{i2}b_{2j} + \cdots + a_{in}b_{nj} = \sum_{k=1}^{n} a_{ik}b_{kj},$$

$$i = 1, 2, \ldots, m, \quad \text{and} \quad j = 1, 2, \ldots, p. \tag{6}$$

In words, Eq. (6) says, "To get the element in the ith row and jth column of AB, multiply the entries in the ith row of A from left to right by the corresponding entries in the jth column of B from top to bottom, and add. The sum is c_{ij}." For example,

$$\begin{bmatrix} & a & & b & & c \\ & d & & e & & f \\ & u & & v & & w \end{bmatrix} \tag{7}$$

$$\begin{bmatrix} 2 & -1 & 3 \\ 3 & 2 & 2 \end{bmatrix}\begin{bmatrix} 2a - d + 3u & 2b - e + 3v & 2c - f + 3w \\ 3a + 2d + 2u & 3b + 2e + 2v & 3c + 2f + 2w \end{bmatrix}.$$

In most cases it saves space to write the matrices in a product side by side. Thus we usually write the multiplication in (7) as

$$\begin{bmatrix} 2 & -1 & 3 \\ 3 & 2 & 2 \end{bmatrix}\begin{bmatrix} a & b & c \\ d & e & f \\ u & v & w \end{bmatrix} = \begin{bmatrix} 2a - d + 3u & 2b - e + 3v & 2c - f + 3w \\ 3a + 2d + 2u & 3b + 2e + 2v & 3c + 2f + 2w \end{bmatrix}.$$

Matrix multiplication has the following properties:

$$(AB)C = A(BC) \qquad \text{(Associative Law)}$$

$$A(B + C) = AB + AC \qquad \text{(Left Distributive Law)}$$

$$(A + B)C = AC + BC \qquad \text{(Right Distributive Law)}$$

$$k(AB) = (kA)B = A(kB) \qquad \text{(Scalar Distributive Law, for any number } k\text{)}$$

Matrix multiplication is not commutative because the equation $AB = BA$ does not always hold. For example,

$$\begin{bmatrix} 0 & 1 \\ 2 & 3 \end{bmatrix}\begin{bmatrix} 0 & 1 \\ 0 & 0 \end{bmatrix} = \begin{bmatrix} 0 & 0 \\ 0 & 2 \end{bmatrix} \quad \text{while} \quad \begin{bmatrix} 0 & 1 \\ 0 & 0 \end{bmatrix}\begin{bmatrix} 0 & 1 \\ 2 & 3 \end{bmatrix} = \begin{bmatrix} 2 & 3 \\ 0 & 0 \end{bmatrix}.$$

The matrix

$$I = \begin{bmatrix} 1 & 0 \\ 0 & 1 \end{bmatrix}$$

is a *multiplicative identity* for all 2 by 2 matrices. That is, $IM = MI = M$ for every 2 by 2 matrix M. In terms of entries,

$$\begin{bmatrix} 1 & 0 \\ 0 & 1 \end{bmatrix}\begin{bmatrix} a & b \\ c & d \end{bmatrix} = \begin{bmatrix} a & b \\ c & d \end{bmatrix}\begin{bmatrix} 1 & 0 \\ 0 & 1 \end{bmatrix} = \begin{bmatrix} a & b \\ c & d \end{bmatrix} \tag{8}$$

for any numbers a, b, c, and d. Similarly,

$$I = \begin{bmatrix} 1 & 0 & 0 \\ 0 & 1 & 0 \\ 0 & 0 & 1 \end{bmatrix}$$

is a multiplicative identity for all 3 by 3 matrices.

Elementary Row Operations and Row Reduction

Two systems of linear equations are called **equivalent** if they have the same set of solutions. To solve a system of linear equations it is often possible to transform it step by step into an equivalent system of equations that is so simple it can be solved by inspection. We shall illustrate such a sequence of steps by transforming the system

$$2x + 3y - 4z = -3,$$
$$x + 2y + 3z = 3, \tag{9}$$
$$3x - y - z = 6,$$

into the equivalent system

$$x = 2$$
$$y = -1, \tag{10}$$
$$z = 1.$$

EXAMPLE 1 Solve the system of equations (9).

Solution The system (9) is the same as

$$AX = B, \qquad A = \begin{bmatrix} 2 & 3 & -4 \\ 1 & 2 & 3 \\ 3 & -1 & -1 \end{bmatrix}, \qquad B = \begin{bmatrix} -3 \\ 3 \\ 6 \end{bmatrix}. \tag{11}$$

We start with the 3 by 4 matrix $[A \vdots B]$ whose first three columns are the columns of A and whose fourth column is B. That is,

$$[A \vdots B] = \begin{bmatrix} 2 & 3 & -4 & \vdots & -3 \\ 1 & 2 & 3 & \vdots & 3 \\ 3 & -1 & -1 & \vdots & 6 \end{bmatrix}. \tag{12}$$

This is the *augmented matrix* of the system $AX = B$.

We are going to transform this augmented matrix with a sequence of so-called **elementary row operations.** These operations, which are to be performed on the rows of the matrix, are of three kinds:

1. Multiply any row by a constant different from 0.
2. Add a constant multiple of any row to another row.
3. Interchange two rows.

Our goal is to transform $[A \vdots B]$ into $[I \vdots S]$, where

$$I = \begin{bmatrix} 1 & 0 & 0 \\ 0 & 1 & 0 \\ 0 & 0 & 1 \end{bmatrix} \qquad \text{and} \qquad S = \begin{bmatrix} s_1 \\ s_2 \\ s_3 \end{bmatrix}. \tag{13}$$

The matrix $[I \vdots S]$ is the matrix of the system

$$x = s_1,$$
$$y = s_2, \tag{14}$$
$$z = s_3.$$

The solution of (14) is obvious and, as long as we get from $[A \vdots B]$ to $[I \vdots S]$ by elementary row transformations, it will be the same as the solution of (9). Thus, we solve (9) by producing (14).

Our systematic approach will be to get a 1 in the upper left corner and use Type 2 operations to get zeros elsewhere in the first column. That will make the first column the same as the first column of I. Then we shall use Type 1 or Type 3 operations to get a 1 in the second position in the second row, and follow that by Type 2 operations to get the second column to be what we want: namely, like the second column of I. Then we will work on the third column.

START:

$$[A \vdots B] = \begin{bmatrix} 2 & 3 & -4 & \vdots & -3 \\ 1 & 2 & 3 & \vdots & 3 \\ 3 & -1 & -1 & \vdots & 6 \end{bmatrix}. \tag{15}$$

STEP 1: Interchange rows 1 and 2 and get

$$\begin{bmatrix} 1 & 2 & 3 & \vdots & 3 \\ 2 & 3 & -4 & \vdots & -3 \\ 3 & -1 & -1 & \vdots & 6 \end{bmatrix}. \tag{16}$$

STEP 2: Add -2 times row 1 to row 2.

STEP 3: Add -3 times row 1 to row 3.

The result of steps 2 and 3 is

$$\begin{bmatrix} 1 & 2 & 3 & \vdots & 3 \\ 0 & -1 & -10 & \vdots & -9 \\ 0 & -7 & -10 & \vdots & -3 \end{bmatrix}. \tag{17}$$

STEP 4: Multiply row 2 by -1. Then

STEP 5: Add -2 times row 2 to row 1, and

STEP 6: Add 7 times row 2 to row 3.

The combined result of these steps is

$$\begin{bmatrix} 1 & 0 & -17 & \vdots & -15 \\ 0 & 1 & 10 & \vdots & 9 \\ 0 & 0 & 60 & \vdots & 60 \end{bmatrix}. \tag{18}$$

STEP 7: Multiply row 3 by $1/60$.

STEP 8: Add 17 times row 3 to row 1.

STEP 9: Add -10 times row 3 to row 2.

The final result is

$$[I \vdots S] = \begin{bmatrix} 1 & 0 & 0 & \vdots & 2 \\ 0 & 1 & 0 & \vdots & -1 \\ 0 & 0 & 1 & \vdots & 1 \end{bmatrix}. \tag{19}$$

This represents the system (10). The solution of this system, and therefore of the system (9), is $x = 2$, $y = -1$, $z = 1$. To check the solution, we substitute these values in (9) and find that indeed

$$2(2) + 3(-1) - 4(1) = -3,$$

$$(2) + 2(-1) + 3(1) = \quad 3, \tag{20}$$

$$3(2) - \quad (-1) - \quad (1) = \quad 6. \qquad \blacksquare$$

The method of using elementary row operations to reduce the augmented matrix of a system of linear equations to a simpler form is sometimes called the **method of row reduction.** It works because at each step the system of equations represented by the transformed matrix is equivalent to the original system. Thus, in Example 1, when we finally arrived at the matrix (19), which represented the system (10) whose solution could be found by inspection, we knew that this solution was also the solution of (9). Notice that we checked the solution anyhow. It is always a good idea to do that.

Inverses of Square Matrices

If M and P are $n \times n$ square matrices with the property that $PM = MP = I$, then we call P the *inverse* of M and use the alternative notation

$$P = M^{-1},$$

pronounced "*M* inverse."

The sequence of row operations that we used to find the solution of the system $AX = B$ in Example 1 can also be used to find the inverse of the matrix A. We start with the 3 by 6 matrix $[A \vdots I]$ whose first three columns are the columns of A and whose last three columns are the columns of I, namely,

$$[A \vdots I] = \begin{bmatrix} 2 & 3 & -4 & \vdots & 1 & 0 & 0 \\ 1 & 2 & 3 & \vdots & 0 & 1 & 0 \\ 3 & -1 & -1 & \vdots & 0 & 0 & 1 \end{bmatrix}. \tag{21}$$

We then carry out Steps 1 through 9 of Example 1 on the augmented matrix $[A \vdots I]$. The final result is

$$[I \vdots A^{-1}] = \begin{bmatrix} 1 & 0 & 0 & \vdots & \dfrac{1}{60} & \dfrac{7}{60} & \dfrac{17}{60} \\ 0 & 1 & 0 & \vdots & \dfrac{10}{60} & \dfrac{10}{60} & -\dfrac{10}{60} \\ 0 & 0 & 1 & \vdots & -\dfrac{7}{60} & \dfrac{11}{60} & \dfrac{1}{60} \end{bmatrix}. \tag{22}$$

The 3 by 3 matrix in the last three columns is

$$A^{-1} = \frac{1}{60} \begin{bmatrix} 1 & 7 & 17 \\ 10 & 10 & -10 \\ -7 & 11 & 1 \end{bmatrix}. \tag{23}$$

To check, we multiply A by A^{-1}:

$$A^{-1}A = \frac{1}{60} \begin{bmatrix} 1 & 7 & 17 \\ 10 & 10 & -10 \\ -7 & 11 & 1 \end{bmatrix} \begin{bmatrix} 2 & 3 & -4 \\ 1 & 2 & 3 \\ 3 & -1 & -1 \end{bmatrix}$$

$$= \frac{1}{60} \begin{bmatrix} 2 + 7 + 51 & 3 + 14 - 17 & -4 + 21 - 17 \\ 20 + 10 - 30 & 30 + 20 + 10 & -40 + 30 + 10 \\ -14 + 11 + 3 & -21 + 22 - 1 & 28 + 33 - 1 \end{bmatrix} \tag{24}$$

$$= \frac{1}{60} \begin{bmatrix} 60 & 0 & 0 \\ 0 & 60 & 0 \\ 0 & 0 & 60 \end{bmatrix} = \begin{bmatrix} 1 & 0 & 0 \\ 0 & 1 & 0 \\ 0 & 0 & 1 \end{bmatrix}.$$

Knowing A^{-1} provides a second way to solve the system

$$2x + 3y - 4z = -3,$$
$$x + 2y + 3z = 3, \qquad (25)$$
$$3x - y - z = 6.$$

We write the system in matrix form as $AX = B$ and then multiply B on the left by A^{-1} to find the solution matrix (column vector) X. Thus,

$$X = IX = (A^{-1}A)X = A^{-1}(AX) = A^{-1}B = \frac{1}{60}\begin{bmatrix} 1 & 7 & 17 \\ 10 & 10 & -10 \\ -7 & 11 & 1 \end{bmatrix}\begin{bmatrix} -3 \\ 3 \\ 6 \end{bmatrix}$$

$$= \frac{1}{60}\begin{bmatrix} 120 \\ -60 \\ 60 \end{bmatrix}$$

$$= \begin{bmatrix} 2 \\ -1 \\ 1 \end{bmatrix}.$$

In nonmatrix form, the solution is $x = 2$, $y = -1$, $z = 1$. The general procedure is this:

If the coefficient matrix A of a system

$$AX = B$$

of n equations and n unknowns has an inverse, then the solution of the system is

$$X = A^{-1}B.$$

Only square matrices can have inverses. If an n by n matrix A has an inverse, the method shown above for $n = 3$ will give it: Put the n by n identity matrix I alongside A and use the row operations to get I in place of A. The n by n matrix that is then beside that, where I used to be, is A^{-1}.

Not every n by n matrix has an inverse. For example, the 2 by 2 matrix

$$\begin{bmatrix} 1 & 1 \\ a & a \end{bmatrix}$$

has no inverse. The method we outlined above would reduce this to

$$\begin{bmatrix} 1 & 1 \\ 0 & 0 \end{bmatrix},$$

which cannot be changed by elementary row operations into the 2 by 2 identity matrix. We shall have more to say about inverses in a moment.

A system of m equations in n unknowns may have no solutions, only one solution, or infinitely many solutions. If there are any solutions, they can be found by the method of row reduction described above.

The Determinant Formula for a Matrix Inverse

Another way to find the inverse of a square matrix A (assuming A has one) depends on the fact that A has an inverse if and only if det $A \neq 0$. We describe the method, give an example, and then indicate why the method works. We illustrate with 3 by 3 matrices, but the method works for any square matrix whose determinant is not zero.

To find the inverse of a matrix whose determinant is not zero:

1. Construct the matrix of cofactors of A:

$$\text{cof } A = \begin{bmatrix} A_{11} & A_{12} & A_{13} \\ A_{21} & A_{22} & A_{23} \\ A_{31} & A_{32} & A_{33} \end{bmatrix}.$$

2. Construct the transposed matrix of cofactors (called the **adjoint** of A):

$$\text{adj } A = (\text{cof } A)^T = \begin{bmatrix} A_{11} & A_{21} & A_{31} \\ A_{12} & A_{22} & A_{32} \\ A_{13} & A_{23} & A_{33} \end{bmatrix}.$$

("Transpose" means to write the rows as columns.)

3. Then

$$A^{-1} = \frac{1}{\det A} \text{ adj } A. \tag{25}$$

EXAMPLE 2 Use the determinant formula in (25) to find the inverse of the matrix

$$A = \begin{bmatrix} 2 & 3 & -4 \\ 1 & 2 & 3 \\ 3 & -1 & -1 \end{bmatrix}.$$

Solution We calculate the matrix of minors

$$M = \begin{bmatrix} 1 & -10 & -7 \\ -7 & 10 & -11 \\ 17 & 10 & 1 \end{bmatrix}$$

and apply the sign corrections according to the checkerboard pattern $(-1)^{i+j}$ to get the matrix of cofactors

$$\text{cof } A = \begin{bmatrix} 1 & 10 & -7 \\ 7 & 10 & 11 \\ 17 & -10 & 1 \end{bmatrix}.$$

The adjoint of A is the transposed cofactor matrix

$$\text{adj } A = \begin{bmatrix} 1 & 7 & 17 \\ 10 & 10 & -10 \\ -7 & 11 & 1 \end{bmatrix}.$$

We can find the determinant of A by multiplying the first row of A by the first column of adj A (which is the first row of the matrix of cofactors, so we are multiplying the elements in the first row of A by their own cofactors):

$$\det A = 2(1) + 3(10) + (-4)(-7) = 2 + 30 + 28 = 60.$$

We then divide adj A by det A to obtain

$$A^{-1} = \frac{1}{60} \begin{bmatrix} 1 & 7 & 17 \\ 10 & 10 & -10 \\ -7 & 11 & 1 \end{bmatrix},$$

in agreement with Eq. (23).   ∎

Why does the determinant formula work? Let us take a closer look at the product (adj A)A for $n = 3$. Since adj A is the transposed cofactor matrix,

$$(\text{adj } A)A = \begin{bmatrix} A_{11} & A_{21} & A_{31} \\ A_{12} & A_{22} & A_{32} \\ A_{13} & A_{23} & A_{33} \end{bmatrix} \begin{bmatrix} a_{11} & a_{12} & a_{13} \\ a_{21} & a_{22} & a_{23} \\ a_{31} & a_{32} & a_{33} \end{bmatrix}$$

$$= \begin{bmatrix} \det A & 0 & 0 \\ 0 & \det A & 0 \\ 0 & 0 & \det A \end{bmatrix}.$$

(26)

The entries appear as they do in this last matrix because A_{ij} is the cofactor of a_{ij}. Each entry on the main diagonal is det A because it results from multiplying a column of A by the cofactors of the same column. Each entry off the main diagonal is the result of multiplying a column of A by the cofactors of some other column and is 0 by Fact 8 in Appendix 7. To continue the work begun in Eq. (26), we factor out det A to get

$$(\text{adj } A)A = (\det A)I,$$

$$\left(\frac{1}{\det A} \text{ adj } A \right) A = I.$$

(27)

The matrix in parentheses on the left side of this last equation must be A^{-1} because its product with A is I. (See Problem 31.) It is also true that

$$A \left(\frac{1}{\det A} \text{ adj } A \right) = I$$

Eigenvectors and Eigenvalues

As we have seen, the product of a square matrix A and a column vector X is again a column vector. If $X \neq 0$ (the entries in the column are not all 0) and if AX is a nonzero scalar multiple λ of X, we call X an **eigenvector** of A. (*Eigen* is German for "special.") For instance, the vector

$$X = \begin{bmatrix} 1 \\ -3 \end{bmatrix}$$

is an eigenvector of the matrix

$$A = \begin{bmatrix} 4 & 2 \\ 3 & -1 \end{bmatrix}$$

because

$$\begin{bmatrix} 4 & 2 \\ 3 & -1 \end{bmatrix}\begin{bmatrix} 1 \\ -3 \end{bmatrix} = \begin{bmatrix} 4 - 6 \\ 3 + 3 \end{bmatrix} = \begin{bmatrix} -2 \\ 6 \end{bmatrix} = -2 \cdot \begin{bmatrix} 1 \\ -3 \end{bmatrix}. \qquad (28)$$

The number $\lambda = -2$ is called an **eigenvalue** of A.

Eigenvalues and eigenvectors play important roles in the theory of systems of differential equations. To develop a procedure for finding them, we rewrite the equation $AX = \lambda X$ as a system of linear equations in the following way:

$$AX = \lambda X$$

$$AX - \lambda X = 0 \qquad \left(\begin{array}{l}\text{Here, 0 is the column matrix} \\ \text{whose entries are all zero.}\end{array}\right)$$

$$AX - \lambda IX = 0 \qquad (I \text{ is the identity matrix, so } X = IX.)$$

$$(A - \lambda I)X = 0 \qquad (\text{Matrix multiplication is distributive.})$$

$$\begin{bmatrix} a_{11} - \lambda & a_{12} \\ a_{21} & a_{22} - \lambda \end{bmatrix}\begin{bmatrix} x_1 \\ x_2 \end{bmatrix} = \begin{bmatrix} 0 \\ 0 \end{bmatrix} \qquad (\text{entries displayed}). \qquad (29)$$

Thus, X is an eigenvector of A if and only if this last equation holds for one or more values of λ.

If $\det(A - \lambda I) \neq 0$, the only solution of the system in (29) is the so-called **trivial solution** $x_1 = x_2 = 0$. However, if $\det(A - \lambda I) = 0$, the system will have other solutions as well, and these **nontrivial solutions** are the ones we seek. We therefore look for the values of λ that make this determinant zero. In entry form,

$$\begin{aligned} \det(A - \lambda I) &= (a_{11} - \lambda)(a_{22} - \lambda) - a_{21}a_{12} \\ &= a_{11}a_{22} - a_{22}\lambda - a_{11}\lambda + \lambda^2 - a_{21}a_{12} \\ &= \lambda^2 - (a_{11} + a_{22})\lambda + (a_{11}a_{22} - a_{21}a_{12}) \\ &= \lambda^2 - (a_{11} + a_{22})\lambda + \det A. \end{aligned}$$

Thus $\det(A - \lambda I) = 0$ if and only if

$$\lambda^2 - (a_{11} + a_{22})\lambda + \det A = 0. \qquad (30)$$

This equation is the **characteristic equation** of A, and its solutions are the values of λ for which the equations $(A - \lambda I)X = 0$ and $AX = \lambda X$ have nontrivial solutions. The process of finding the eigenvectors of A therefore reduces to one of accomplishing two steps:

STEP 1: Solve Eq. (30) to find the eigenvalues of A.

STEP 2: For each eigenvalue, solve the equation $(A - \lambda I)X = 0$ to find the corresponding eigenvectors.

The following example shows how to do this when A is a 2 by 2 matrix whose eigenvalues are real and distinct. Information about complex eigenvalues and additional information about real eigenvalues that repeat can be found in most texts in linear algebra and differential equations and in many advanced calculus texts.

EXAMPLE 3 Find the eigenvectors of the matrix

$$A = \begin{bmatrix} 4 & 2 \\ 3 & -1 \end{bmatrix}.$$

Solution The characteristic equation is

$$\lambda^2 - (4 + (-1))\lambda + ((4)(-1) - (3)(2)) = 0,$$

$$\lambda^2 - 3\lambda - 10 = 0,$$

$$(\lambda - 5)(\lambda + 2) = 0,$$

so

$$\lambda = 5 \quad \text{or} \quad \lambda = -2.$$

To find the eigenvectors for $\lambda = 5$, we solve the system $(A - 5I)X = 0$ for X:

$$(A - 5I)X = 0,$$

$$\begin{bmatrix} 4 - 5 & 2 \\ 3 & -1 - 5 \end{bmatrix} \begin{bmatrix} x_1 \\ x_2 \end{bmatrix} = \begin{bmatrix} 0 \\ 0 \end{bmatrix},$$

$$\begin{bmatrix} -1 & 2 \\ 3 & -6 \end{bmatrix} \begin{bmatrix} x_1 \\ x_2 \end{bmatrix} = \begin{bmatrix} 0 \\ 0 \end{bmatrix},$$

$$-x_1 + 2x_2 = 0,$$

$$3x_1 - 6x_2 = 0.$$

The last two equations both tell us that

$$x_1 = 2x_2.$$

In other words, if $x_1 = 2x_2$ then X will be an eigenvector of A with eigenvalue $\lambda = 5$. Thus, any vector of the form

$$X = \begin{bmatrix} 2x_2 \\ x_2 \end{bmatrix} = x_2 \begin{bmatrix} 2 \\ 1 \end{bmatrix}, \ x_2 \neq 0,$$

is an eigenvector of A with eigenvalue $\lambda = 5$. If we wanted just one, we might take it to be

$$X_1 = \begin{bmatrix} 2 \\ 1 \end{bmatrix},$$

the simplest (nonzero) vector of the required form.

We already know from Eq. (28) that

$$X_2 = \begin{bmatrix} 1 \\ -3 \end{bmatrix}$$

is an eigenvector of A for $\lambda = -2$. The other eigenvectors for $\lambda = -2$ are scalar multiples of X_2 (Problem 33). ■

The eigenvalues and eigenvectors of higher-dimensional square matrices are found in an analogous way. For instance, to find the eigenvectors of

$$A = \begin{bmatrix} 3 & 2 & 2 \\ 0 & -4 & -2 \\ 0 & 5 & 3 \end{bmatrix}, \tag{31}$$

we find the values of λ for which $\det(A - \lambda I) = 0$ and, for each value found, find the nontrivial solutions of the system $(A - \lambda I)X = 0$. The characteristic equation of the matrix A in (31) turns out to be

$$\det(A - \lambda I) = (3 - \lambda)(\lambda + 2)(\lambda - 1) = 0,$$

so the eigenvalues are $\lambda = 3$, $\lambda = -2$, and $\lambda = 1$. For $\lambda = 1$, the equation $(A - \lambda I)X = 0$ becomes

$$\begin{bmatrix} 2 & 2 & 2 \\ 0 & -5 & -2 \\ 0 & 5 & 2 \end{bmatrix} \begin{bmatrix} x_1 \\ x_2 \\ x_3 \end{bmatrix} = \begin{bmatrix} 0 \\ 0 \\ 0 \end{bmatrix},$$

or

$$2x_1 + 2x_2 + 2x_3 = 0,$$
$$-5x_2 - 2x_3 = 0,$$
$$5x_2 + 2x_3 = 0.$$

Hence,

$$x_2 = -\frac{2}{5}x_3,$$

$$x_1 = -x_2 - x_3 = \frac{2}{5}x_3 - x_3 = -\frac{3}{5}x_3,$$

and

$$X = \begin{bmatrix} x_1 \\ x_2 \\ x_3 \end{bmatrix} = \begin{bmatrix} -\dfrac{3}{5}x_3 \\ -\dfrac{2}{5}x_3 \\ x_3 \end{bmatrix} = -\frac{x_3}{5} \begin{bmatrix} 3 \\ 2 \\ -5 \end{bmatrix}.$$

The eigenvectors associated with $\lambda = 1$ are the nonzero multiples of

$$X_1 = \begin{bmatrix} 3 \\ 2 \\ -5 \end{bmatrix}.$$

As in the two-dimensional case, the characteristic equations of higher-dimensional matrices may have complex roots and repeated real roots, but we shall not explore these possibilities here.

PROBLEMS

1. a) Write the following system of linear equations in matrix form $AX = B$.

$$2x - 3y + 4z = -19,$$
$$6x + 4y - 2z = 8,$$
$$x + 5y + 4z = 23$$

b) Show that

$$X = \begin{bmatrix} -2 \\ 5 \\ 0 \end{bmatrix}$$

is a solution of the system in part (a).

2. Let A be an arbitrary matrix with 3 rows and 3 columns and let

$$I = \begin{bmatrix} 1 & 0 & 0 \\ 0 & 1 & 0 \\ 0 & 0 & 1 \end{bmatrix}.$$

Show that $IA = A$ and also that $AI = A$. This will show that I is the multiplicative identity matrix for all 3 by 3 matrices.

3. Let A be an arbitrary 3 by 3 matrix and let R_{12} be the matrix obtained from the 3 by 3 identity matrix by interchanging rows 1 and 2:

$$R_{12} = \begin{bmatrix} 0 & 1 & 0 \\ 1 & 0 & 0 \\ 0 & 0 & 1 \end{bmatrix}.$$

Compute $R_{12}A$ and show that you would get the same result by interchanging rows 1 and 2 of A.

4. Let A and R_{12} be as in Problem 3 above. Compute AR_{12} and show that the result is what you would get by interchanging columns 1 and 2 of A. (Note that R_{12} is also the result of interchanging columns 1 and 2 of the 3 by 3 identity matrix I.)

Solve the following systems of equations by row reduction

5. $x + 8y = 4$
$3x - y = -13$

6. $2x + 3y = 5$
$3x - y = 2$

7. $4x - 3y = 6$
$3x - 2y = 5$

8. $x + y + z = 2$
$2x - y + z = 0$
$x + 2y - z = 4$

9. $2x + y - z = 2$
$x - y + z = 7$
$2x + 2y + z = 4$

10. $2x - 4y = 6$
$x + y + z = 1$
$5y + 7z = 10$

11. $x \quad - z = 3$
$2y - 2z = 2$
$2x \quad + z = 3$

12. $x_1 + x_2 - x_3 + x_4 = 2$
$x_1 - x_2 + x_3 + x_4 = -1$
$x_1 + x_2 + x_3 - x_4 = 2$
$x_1 \quad + x_3 + x_4 = -1$

13. Verify that the inverse of

$$A = \begin{bmatrix} a & b \\ c & d \end{bmatrix}$$

is

$$A^{-1} = \frac{1}{ad - bc} \begin{bmatrix} d & -b \\ -c & a \end{bmatrix} \quad \text{if } ad \neq bc.$$

That is, show that

$$AA^{-1} = A^{-1}A = \begin{bmatrix} 1 & 0 \\ 0 & 1 \end{bmatrix}.$$

14. a) Use the result in Problem 13 to write down the inverses of

$$A = \begin{bmatrix} 2 & -1 \\ 3 & 1 \end{bmatrix} \quad \text{and} \quad B = \begin{bmatrix} 2 & 3 \\ -1 & 1 \end{bmatrix}.$$

b) In part (a), B is the transpose of A. Is B^{-1} the transpose of A^{-1}?

15. Use the result in Problem 13 to solve the system of equations in Problem 5.

16. Given

$$A = \begin{bmatrix} 1 & 0 & -1 \\ 0 & 2 & -2 \\ 2 & 0 & 1 \end{bmatrix} \quad \text{and} \quad A^{-1} = \begin{bmatrix} \frac{1}{3} & 0 & \frac{1}{3} \\ -\frac{2}{3} & \frac{1}{2} & \frac{1}{3} \\ -\frac{2}{3} & 0 & \frac{1}{3} \end{bmatrix},$$

solve the system

$$AX = \begin{bmatrix} 3 \\ 2 \\ 3 \end{bmatrix}.$$

17. a) Find the inverse of the matrix

$$A = \begin{bmatrix} 1 & 8 & 9 \\ 0 & 4 & 6 \\ 0 & 0 & 3 \end{bmatrix}.$$

b) Solve the following system of equations.

$$x + 8y + 9z = 10,$$
$$4y + 6z = 10,$$
$$3z = -10$$

18. a) Solve the system

$$2x - y + 2z = 5$$
$$3x + y - 3z = 7 \quad \text{or} \quad \begin{bmatrix} 2 & -1 \\ 3 & 1 \end{bmatrix}\begin{bmatrix} x \\ y \end{bmatrix} = \begin{bmatrix} 5 - 2z \\ 7 + 3z \end{bmatrix}$$

for x and y in terms of z.

b) Each equation on the left in part (a) represents a plane. Use the solution you obtained in (a) to write parametric equations for the line in which the planes intersect.

Find the eigenvalues of the matrices in Problems 19–30. (Some of the characteristic equations have repeated real roots or complex roots.) Then find an eigenvector for each real eigenvalue.

19. $\begin{bmatrix} 2 & 0 \\ 0 & 3 \end{bmatrix}$

20. $\begin{bmatrix} -6 & 4 \\ 3 & 1 \end{bmatrix}$

21. $\begin{bmatrix} 1 & 1 \\ 1 & 1 \end{bmatrix}$

22. $\begin{bmatrix} 1 & -1 \\ 1 & 3 \end{bmatrix}$

23. $\begin{bmatrix} 1 & 1 \\ -1 & -1 \end{bmatrix}$

24. $\begin{bmatrix} 2 & -1 \\ 2 & 3 \end{bmatrix}$

25. $\begin{bmatrix} 4 & 2 \\ 5 & -2 \end{bmatrix}$

26. $\begin{bmatrix} 2 & -5 \\ 5 & 2 \end{bmatrix}$

27. $\begin{bmatrix} 3 & 1 & 0 \\ -1 & 0 & -1 \\ 1 & 2 & 3 \end{bmatrix}$

28. $\begin{bmatrix} 5 & 1 & 2 \\ 0 & 2 & 4 \\ 0 & -1 & -3 \end{bmatrix}$

29. $\begin{bmatrix} 0 & 2 & 1 \\ 1 & 0 & 0 \\ 2 & 0 & 0 \end{bmatrix}$

30. $\begin{bmatrix} 1 & 0 & 0 \\ 0 & 2 & 1 \\ 2 & -1 & 4 \end{bmatrix}$

31. Suppose an n by n matrix A has an inverse A^{-1} and that B is an n by n matrix with the property that $BA = I$. Show that $B = A^{-1}$.

32. *A derivation of Cramer's rule.* We have seen that if the coefficient matrix A of a system $AX = B$ of n linear equations in n unknowns has an inverse, then the solution of the system is $X = A^{-1}B$. We also know that $A^{-1} = (1/\det A)\text{adj } A$. Combine these facts to show that the ith entry in X is

$$x_i = \frac{1}{\det A}(A_{1i}b_1 + \cdots + A_{ki}b_k + \cdots + A_{ni}b_n)$$

and hence that

$$x_i = \frac{\det U_i}{\det A},$$

where U_i is the matrix obtained from A by replacing the ith column of A by B.

33. Show that if X is an eigenvector of a matrix A, then every nonzero scalar multiple of X is an eigenvector of A.

A.9

The Increment Theorem and Euler's Mixed Derivative Theorem

In this appendix we derive Eq. (3) of Article 16.4 (in a slightly different form) and then go on to derive Euler's Mixed Derivative Theorem (Theorem 2 of Article 16.8), which first appeared in 1734 in one of a series of papers Euler wrote on hydrodynamics.

THEOREM

> **Increment Theorem for Functions of Two Variables**
>
> Suppose that $z = f(x, y)$ is continuous and has partial derivatives throughout a region
>
> $$R: \quad |x - x_0| < h, \qquad |y - y_0| < k,$$
>
> in the xy-plane. Suppose also that Δx and Δy are small enough for the point $(x_0 + \Delta x, y_0 + \Delta y)$ to lie in R. If f_x and f_y are continuous at (x_0, y_0), then the increment
>
> $$\Delta z = f(x_0 + \Delta x, y_0 + \Delta y) - f(x_0, y_0) \tag{1}$$
>
> can be written as
>
> $$\Delta z = f_x(x_0, y_0)\, \Delta x + f_y(x_0, y_0)\, \Delta y + \epsilon_1\, \Delta x + \epsilon_2\, \Delta y, \tag{2}$$
>
> where
>
> $$\epsilon_1, \epsilon_2 \to 0 \qquad \text{as} \qquad \Delta x, \Delta y \to 0.$$

Proof The region R (Fig. A.6) is a rectangle centered at $A(x_0, y_0)$ with dimensions $2h$ by $2k$. Since $C(x_0 + \Delta x, y_0 + \Delta y)$ lies in R, the point $B(x_0 + \Delta x, y_0)$ and the line segments AB and BC also lie in R. Thus f is continuous and has partial derivatives f_x and f_y at each point of these segments.

We may think of Δz as the sum

$$\Delta z = \Delta z_1 + \Delta z_2 \tag{3}$$

of two increments, where

$$\Delta z_1 = f(x_0 + \Delta x, y_0) - f(x_0, y_0) \tag{4}$$

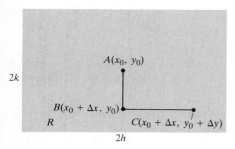

A.6 The rectangular region R in the proof of the Increment Theorem.

is the change from A to B and

$$\Delta z_2 = f(x_0 + \Delta x, y_0 + \Delta y) - f(x_0 + \Delta x, y_0) \tag{5}$$

is the change from B to C (Fig. A.7). Notice that the sum $\Delta z_1 + \Delta z_2$ equals Δz, as it should:

$$\begin{aligned}
\Delta z_1 + \Delta z_2 &= [f(x_0 + \Delta x, y_0) - f(x_0, y_0)] \\
&\quad + [f(x_0 + \Delta x, y_0 + \Delta y) - f(x_0 + \Delta x, y_0)] \\
&= f(x_0 + \Delta x, y_0 + \Delta y) - f(x_0, y_0) \\
&= \Delta z.
\end{aligned} \tag{6}$$

The function

$$F(x) = f(x, y_0)$$

is a continuous and differentiable function of x on the closed interval joining x_0 and $x_0 + \Delta x$, with derivative

$$F'(x) = f_x(x, y_0).$$

Hence, by the Mean Value Theorem of Article 3.7, there is a point c between x and $x + \Delta x$ at which

$$F(x_0 + \Delta x) - F(x_0) = F'(c)\,\Delta x$$

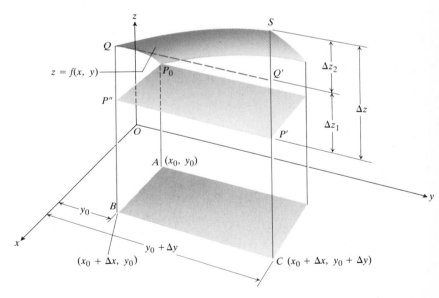

A.7 Part of the surface $z = f(x, y)$ near $P_0(x_0, y_0, f(x_0, y_0))$. The points P_0, P', and P'' have the same height $z_0 = f(x_0, y_0)$ above the xy-plane. The change in z is $\Delta z = P'S$. The change

$$\Delta z_1 = f(x_0 + \Delta x, y_0) - f(x_0, y_0),$$

shown as $P''Q = P'Q'$, is caused by changing x from x_0 to $x_0 + \Delta x$ while holding y equal to y_0. Then, with x held equal to $x_0 + \Delta x$,

$$\Delta z_2 = f(x_0 + \Delta x, y_0 + \Delta y) - f(x_0 + \Delta x, y_0)$$

is the change in z caused by changing y from y_0 to $y_0 + \Delta y$. This is represented by $Q'S$. The total change in z is the sum of Δz_1 and Δz_2.

or

$$f(x_0 + \Delta x, y_0) - f(x_0, y_0) = f_x(c, y_0)\,\Delta x$$

or

$$\Delta z_1 = f_x(c, y_0)\,\Delta x. \tag{7}$$

Similarly, the function

$$G(y) = f(x_0 + \Delta x, y)$$

is a continuous and differentiable function of y on the closed interval joining y_0 and $y_0 + \Delta y$, with derivative

$$G'(y) = f_y(x_0 + \Delta x, y).$$

Hence there is a point d between y_0 and $y_0 + \Delta y$ at which

$$G(y_0 + \Delta y) - G(y_0) = G'(d)\,\Delta y$$

or

$$f(x_0 + \Delta x, y_0 + \Delta y) - f(x_0 + \Delta x, y_0) = f_y(x_0 + \Delta x, d)\,\Delta y$$

or

$$\Delta z_2 = f_y(x_0 + \Delta x, d)\,\Delta y. \tag{8}$$

Now, as Δx and $\Delta y \to 0$, we know $c \to x_0$ and $d \to y_0$. Therefore, since f_x and f_y are continuous at (x_0, y_0), the quantities

$$\begin{aligned}
\epsilon_1 &= f_x(c, y_0) - f_x(x_0, y_0), \\
\epsilon_2 &= f_y(x_0 + \Delta x, d) - f_y(x_0, y_0)
\end{aligned} \tag{9}$$

both approach zero as Δx and $\Delta y \to 0$.

Finally,

$$\begin{aligned}
\Delta z &= \Delta z_1 + \Delta z_2 \\
&= f_x(c, y_0)\,\Delta x + f_y(x_0 + \Delta x, d)\,\Delta y \quad \text{(from (7) and (8))} \\
&= [f_x(x_0, y_0) + \epsilon_1]\,\Delta x + [f_y(x_0, y_0) + \epsilon_2]\,\Delta y \quad \text{(from (9))} \\
&= f_x(x_0, y_0)\,\Delta x + f_y(x_0, y_0)\,\Delta y + \epsilon_1\,\Delta x + \epsilon_2\,\Delta y,
\end{aligned}$$

where ϵ_1 and $\epsilon_2 \to 0$ as Δx and $\Delta y \to 0$. This is what we set out to prove. ∎

Analogous results hold for functions of any finite number of independent variables. For a function of three variables

$$w = f(x, y, z)$$

that is continuous and has partial derivatives f_x, f_y, f_z at and in some neighborhood of the point (x_0, y_0, z_0), and whose derivatives are continuous at the point, we have

$$\begin{aligned}
\Delta w &= f(x_0 + \Delta x, y_0 + \Delta y, z_0 + \Delta z) - f(x_0, y_0, z_0) \\
&= f_x\,\Delta x + f_y\,\Delta y + f_z\,\Delta z + \epsilon_1\,\Delta x + \epsilon_2\,\Delta y + \epsilon_3\,\Delta z,
\end{aligned} \tag{10}$$

where

$$\epsilon_1, \epsilon_2, \epsilon_3 \to 0 \qquad \text{when} \quad \Delta x, \Delta y, \text{ and } \Delta z \to 0.$$

The partial derivatives f_x, f_y, f_z in this formula are to be evaluated at the point (x_0, y_0, z_0).

The result (10) can be proved by treating Δw as the sum of three increments,

$$\Delta w_1 = f(x_0 + \Delta x, y_0, z_0) - f(x_0, y_0, z_0), \tag{11a}$$

$$\Delta w_2 = f(x_0 + \Delta x, y_0 + \Delta y, z_0) - f(x_0 + \Delta x, y_0, z_0), \tag{11b}$$

$$\Delta w_3 = f(x_0 + \Delta x, y_0 + \Delta y, z_0 + \Delta z) - f(x_0 + \Delta x, y_0 + \Delta y, z_0), \tag{11c}$$

and applying the Mean Value Theorem to each of these separately. Two coordinates remain constant and only one varies in each of these partial increments $\Delta w_1, \Delta w_2, \Delta w_3$. In (11b), for example, only y varies, since x is held equal to $x_0 + \Delta x$ and z is held equal to z_0. Since the function $f(x_0 + \Delta x, y, z_0)$ is a continuous function of y with a derivative f_y, it is subject to the Mean Value Theorem, and we have

$$\Delta w_2 = f_y(x_0 + \Delta x, y_1, z_0) \, \Delta y$$

for some y_1 between y_0 and $y_0 + \Delta y$.

THEOREM

> **Euler's Mixed-Derivative Theorem**
>
> If $f(x, y)$ and its partial derivatives f_x, f_y, f_{xy}, and f_{yx} are defined in a region containing a point (a, b) and are all continuous at (a, b), then
>
> $$f_{xy}(a, b) = f_{yx}(a, b). \tag{12}$$

Proof The equality of $f_{xy}(a, b)$ and $f_{yx}(a, b)$ can be established by four applications of the Mean Value Theorem. By hypothesis, the point (a, b) lies in the interior of a rectangle R in the xy-plane on which f, f_x, f_y, f_{xy}, and f_{yx} are all continuous. We let h and k be numbers such that the point $(a + h, b + k)$ also lies in the rectangle R, and we consider the difference

$$\Delta = F(a + h) - F(a), \tag{13}$$

where

$$F(x) = f(x, b + k) - f(x, b). \tag{14}$$

We apply the Mean Value Theorem to F (which is continuous because it is differentiable), and Eq. (13) becomes

$$\Delta = hF'(c_1), \tag{15}$$

where c_1 lies between a and $a + h$. From Eq. (14),

$$F'(x) = f_x(x, b + k) - f_x(x, b),$$

so Eq. (15) becomes

$$\Delta = h[f_x(c_1, b + k) - f_x(c_1, b)]. \tag{16}$$

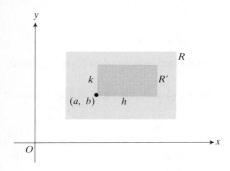

A.8 The key to proving $f_{xy}(a, b) = f_{yx}(a, b)$ is the fact that no matter how small R' is, f_{xy} and f_{yx} take on equal values somewhere inside R' (although not necessarily at the same point of R').

Now we apply the Mean Value Theorem to the function $g(y) = f_x(c_1, y)$ and have

$$g(b + k) - g(b) = kg'(d_1) \tag{17a}$$

or

$$f_x(c_1, b + k) - f_x(c_1, b) = kf_{xy}(c_1, d_1), \tag{17b}$$

for some d_1 between b and $b + k$. By substituting this into Eq. (16), we get

$$\Delta = hkf_{xy}(c_1, d_1), \tag{18}$$

for some point (c_1, d_1) in the rectangle R' whose vertices are the four points (a, b), $(a + h, b)$, $(a + h, b + k)$, and $(a, b + k)$. (See Fig. A.8.)

By substituting from Eq. (14) into Eq. (13), we may also write

$$\Delta = f(a + h, b + k) - f(a + h, b) - f(a, b + k) + f(a, b)$$

$$= [f(a + h, b + k) - f(a, b + k)] - [f(a + h, b) - f(a, b)] \tag{19}$$

$$= \phi(b + k) - \phi(b),$$

where

$$\phi(y) = f(a + h, y) - f(a, y). \tag{20}$$

The Mean Value Theorem applied to Eq. (19) now gives

$$\Delta = k\phi'(d_2), \tag{21}$$

for some d_2 between b and $b + k$. By Eq. (20),

$$\phi'(y) = f_y(a + h, y) - f_y(a, y). \tag{22}$$

Substituting from Eq. (22) into Eq. (21) gives

$$\Delta = k[f_y(a + h, d_2) - f_y(a, d_2)]. \tag{23}$$

Finally, we apply the Mean Value Theorem to the expression in brackets and get

$$\Delta = khf_{yx}(c_2, d_2), \tag{24}$$

for some c_2 between a and $a + h$.

Together, Eqs. (18) and (24) show that

$$f_{xy}(c_1, d_1) = f_{yx}(c_2, d_2), \tag{25}$$

where (c_1, d_1) and (c_2, d_2) both lie in the rectangle R' (Fig. A.8). Equation (25) is not quite the result we want, since it says only that f_{xy} has the same value at (c_1, d_1) that f_{yx} has at (c_2, d_2). But the numbers h and k in our discussion may be made as small as we wish. The hypothesis that f_{xy} and f_{yz} are both continuous at (a, b) means that $f_{xy}(c_1, d_1) = f_{xy}(a, b) + \epsilon_1$ and $f_{yx}(c_2, d_2) = f_{yx}(a, b) + \epsilon_2$, where $\epsilon_1, \epsilon_2 \to 0$ as $h, k \to 0$. Hence, if we let h and $k \to 0$, we have $f_{xy}(a, b) = f_{yx}(a, b)$. ∎

REMARK The equality of $f_{xy}(a, b)$ and $f_{yx}(a, b)$ can be proved with weaker hypotheses than the ones we assumed. For example, it is enough for f, f_x, and f_y to exist in R and for f_{xy} to be continuous at (a, b). Then f_{yx} will exist at (a, b) and be equal to f_{xy} at that point.

A.10
Vector Operator Formulas in Cartesian, Cylindrical, and Spherical Coordinates. Vector Identities

Formulas for Grad, Div, Curl, and the Laplacian

	Cartesian (x, y, z)	Cylindrical (r, θ, z)	Spherical (ρ, ϕ, θ)
	$\mathbf{i}$, $\mathbf{j}$, and $\mathbf{k}$ are unit vectors in the directions of increasing x, y, and z. F_x, F_y, and F_z are the scalar components of $\mathbf{F}(x, y, z)$ in these directions.	$\mathbf{u}_r$, $\mathbf{u}_\theta$, and $\mathbf{k}$ are unit vectors in the directions of increasing r, θ, and z. F_r, F_θ, and F_z are the scalar components of $\mathbf{F}(r, \theta, z)$ in these directions.	$\mathbf{u}_\rho$, $\mathbf{u}_\phi$, and $\mathbf{u}_\theta$ are unit vectors in the directions of increasing ρ, ϕ, and θ. F_ρ, F_ϕ, and F_θ are the scalar components of $\mathbf{F}(\rho, \phi, \theta)$ in these directions.
Gradient	$\nabla f = \dfrac{\partial f}{\partial x}\mathbf{i} + \dfrac{\partial f}{\partial y}\mathbf{j} + \dfrac{\partial f}{\partial z}\mathbf{k}$	$\nabla f = \dfrac{\partial f}{\partial r}\mathbf{u}_r + \dfrac{1}{r}\dfrac{\partial f}{\partial \theta}\mathbf{u}_\theta + \dfrac{\partial f}{\partial z}\mathbf{k}$	$\nabla f = \dfrac{\partial f}{\partial \rho}\mathbf{u}_\rho + \dfrac{1}{\rho}\dfrac{\partial f}{\partial \phi}\mathbf{u}_\phi + \dfrac{1}{\rho \sin \phi}\dfrac{\partial f}{\partial \theta}\mathbf{u}_\theta$
Divergence	$\nabla \cdot \mathbf{F} = \dfrac{\partial F_x}{\partial x} + \dfrac{\partial F_y}{\partial y} + \dfrac{\partial F_z}{\partial z}$	$\nabla \cdot \mathbf{F} = \dfrac{1}{r}\dfrac{\partial}{\partial r}(rF_r) + \dfrac{1}{r}\dfrac{\partial F_\theta}{\partial \theta} + \dfrac{\partial F_z}{\partial z}$	$\nabla \cdot \mathbf{F} = \dfrac{1}{\rho^2}\dfrac{\partial}{\partial \rho}(\rho^2 F_\rho)$ $+ \dfrac{1}{\rho \sin \phi}\dfrac{\partial}{\partial \phi}(F_\phi \sin \phi) + \dfrac{1}{\rho \sin \phi}\dfrac{\partial F_\theta}{\partial \theta}$
Curl	$\nabla \times \mathbf{F} = \begin{vmatrix} \mathbf{i} & \mathbf{j} & \mathbf{k} \\ \dfrac{\partial}{\partial x} & \dfrac{\partial}{\partial y} & \dfrac{\partial}{\partial z} \\ F_x & F_y & F_z \end{vmatrix}$	$\nabla \times \mathbf{F} = \begin{vmatrix} \dfrac{1}{r}\mathbf{u}_r & \mathbf{u}_\theta & \dfrac{1}{r}\mathbf{k} \\ \dfrac{\partial}{\partial r} & \dfrac{\partial}{\partial \theta} & \dfrac{\partial}{\partial z} \\ F_r & rF_\theta & F_z \end{vmatrix}$	$\nabla \times \mathbf{F} = \begin{vmatrix} \dfrac{\mathbf{u}_\rho}{\rho^2 \sin \phi} & \dfrac{\mathbf{u}_\phi}{\rho \sin \phi} & \dfrac{\mathbf{u}_\theta}{\rho} \\ \dfrac{\partial}{\partial \rho} & \dfrac{\partial}{\partial \phi} & \dfrac{\partial}{\partial \theta} \\ F_\rho & \rho F_\phi & \rho \sin \phi\, F_\theta \end{vmatrix}$
Laplacian	$\nabla^2 f = \dfrac{\partial^2 f}{\partial x^2} + \dfrac{\partial^2 f}{\partial y^2} + \dfrac{\partial^2 f}{\partial z^2}$	$\nabla^2 f = \dfrac{1}{r}\dfrac{\partial}{\partial r}\left(r\dfrac{\partial f}{\partial r}\right) + \dfrac{1}{r^2}\dfrac{\partial^2 f}{\partial \theta^2} + \dfrac{\partial^2 f}{\partial z^2}$	$\nabla^2 f = \dfrac{1}{\rho^2}\dfrac{\partial}{\partial \rho}\left(\rho^2 \dfrac{\partial f}{\partial \rho}\right)$ $+ \dfrac{1}{\rho^2 \sin \phi}\dfrac{\partial}{\partial \phi}\left(\sin \phi \dfrac{\partial f}{\partial \phi}\right) + \dfrac{1}{\rho^2 \sin^2\phi}\dfrac{\partial^2 f}{\partial \theta^2}$

Vector Triple Products

$$(\mathbf{A} \times \mathbf{B}) \cdot \mathbf{C} = (\mathbf{B} \times \mathbf{C}) \cdot \mathbf{A} = (\mathbf{C} \times \mathbf{A}) \cdot \mathbf{B}$$

$$\mathbf{A} \times (\mathbf{B} \times \mathbf{C}) = (\mathbf{A} \cdot \mathbf{C})\mathbf{B} - (\mathbf{A} \cdot \mathbf{B})\mathbf{C}$$

Vector Identities for the Cartesian Form of the Operator ∇

In the identities listed here, $f(x, y, z)$ and $g(x, y, z)$ are differentiable scalar functions and $\mathbf{u}(x, y, z)$ and $\mathbf{v}(x, y, z)$ are differentiable vector functions.

$$\nabla \cdot f\mathbf{v} = f\nabla \cdot \mathbf{v} + \mathbf{v} \cdot \nabla f = f\nabla \cdot \mathbf{v} + (\mathbf{v} \cdot \nabla)f$$

$$\nabla \times f\mathbf{v} = f\nabla \times \mathbf{v} + \nabla f \times \mathbf{v}$$

$$\nabla \cdot (\nabla \times \mathbf{v}) = 0$$

$$\nabla \times (\nabla f) = \mathbf{0}$$

$$\nabla(fg) = f\nabla g + g\nabla f$$

$$\nabla(\mathbf{u} \cdot \mathbf{v}) = (\mathbf{u} \cdot \nabla)\mathbf{v} + (\mathbf{v} \cdot \nabla)\mathbf{u} + \mathbf{u} \times (\nabla \times \mathbf{v}) + \mathbf{v} \times (\nabla \times \mathbf{u})$$

$$\nabla \cdot (\mathbf{u} \times \mathbf{v}) = \mathbf{v} \cdot (\nabla \times \mathbf{u}) - \mathbf{u} \cdot (\nabla \times \mathbf{v})$$

$$\nabla \times (\mathbf{u} \times \mathbf{v}) = (\mathbf{v} \cdot \nabla)\mathbf{u} - (\mathbf{u} \cdot \nabla)\mathbf{v} + \mathbf{u}(\nabla \cdot \mathbf{v}) - \mathbf{v}(\nabla \cdot \mathbf{u})$$

$$\nabla \times (\nabla \times \mathbf{v}) = \nabla(\nabla \cdot \mathbf{v}) - (\nabla \cdot \nabla)\mathbf{v} = \nabla(\nabla \cdot \mathbf{v}) - \nabla^2 \mathbf{v}$$

$$(\nabla \times \mathbf{v}) \times \mathbf{v} = (\mathbf{v} \cdot \nabla)\mathbf{v} - \frac{1}{2}\nabla(\mathbf{v} \cdot \mathbf{v})$$

A.11
Complex Numbers

Complex numbers are expressions of the form $a + ib$, where a and b are "real" numbers and i is a symbol for $\sqrt{-1}$. Unfortunately, the words "real" and "imaginary" have connotations that somehow place $\sqrt{-1}$ in a less favorable position than $\sqrt{2}$ in our minds. As a matter of fact, a good deal of imagination, in the sense of *inventiveness,* has been required to construct the *real* number system, which forms the basis of the calculus. In this article we shall review the various stages of this invention. The further invention of a complex number system will then not seem so strange.

The earliest stage of number development was the recognition of the **counting numbers** 1, 2, 3, . . . , which we now call the **natural numbers** or the **positive integers.** Certain simple arithmetical operations can be performed with these numbers without getting outside the system. That is, the system of positive integers is **closed** under the operations of addition and multiplication. By this we mean that if m and n are any positive integers, then

$$m + n = p \quad \text{and} \quad mn = q \tag{1}$$

are also positive integers. Given the two positive integers on the left-hand side of either equation in (1), we can find the corresponding positive integer on the right. More than this, we may sometimes specify the positive integers m and p and find a positive integer n such that $m + n = p$. For instance, $3 + n = 7$ can be solved when the only numbers we know are the positive integers. But the equation $7 + n = 3$ cannot be solved unless the number system is enlarged. The number zero and the negative integers were invented to solve equations like that. In a civilization that recognizes all the **integers**

$$\ldots, -3, -2, -1, 0, 1, 2, 3, \ldots, \tag{2}$$

an educated person may always find the missing integer that solves the equation $m + n = p$ when given the other two integers in the equation.

Suppose our educated people also know how to multiply any two of the integers in (2). If, in Eqs. (1), they are given m and q, they discover that sometimes they can find n and sometimes they can't. If their imagination is still in good working order, they may be inspired to invent still more numbers and introduce fractions, which are just ordered pairs m/n of integers m and n. The number zero has special properties that may bother them for a while, but they ultimately discover that it is handy to have all ratios of integers m/n, excluding

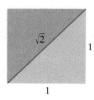

A.9 Segments of irrational length can be constructed with straightedge and compass.

only those having zero in the denominator. This system, called the set of **rational numbers,** is now rich enough for them to perform the so-called **rational operations** of arithmetic:

1. a) addition
   b) subtraction
2. a) multiplication
   b) division

on any two numbers in the system, *except that they cannot divide by zero.*

The geometry of the unit square (Fig. A.9) and the Pythagorean theorem showed that they could construct a geometric line segment that, in terms of some basic unit of length, has length equal to $\sqrt{2}$. Thus they could solve the equation

$$x^2 = 2$$

by a geometric construction. But then they discovered that the line segment representing $\sqrt{2}$ and the line segment representing the unit of length 1 were incommensurable quantities. This means that the ratio $\sqrt{2}/1$ cannot be expressed as the ratio of two *integer* multiples of some other, presumably more fundamental, unit of length. That is, our educated people could not find a rational number solution of the equation $x^2 = 2$.

There *is* no rational number whose square is 2. To see why, suppose that there were such a rational number. Then we could find integers p and q with no common factor other than 1, and such that

$$p^2 = 2q^2. \tag{3}$$

Since p and q are integers, p must be even; otherwise its product with itself would be odd. In symbols, $p = 2p_1$, where p_1 is an integer. This leads to $2p_1^2 = q^2$, which says q must be even, say $q = 2q_1$, where q_1 is an integer. This makes 2 a common factor of p and q, contrary to our choice of p and q as integers with no common factor other than 1. Hence there is no rational number whose square is 2.

Although our educated people could not find a rational solution of the equation $x^2 = 2$, they could get a sequence of rational numbers

$$\frac{1}{1}, \ \frac{7}{5}, \ \frac{41}{29}, \ \frac{239}{169}, \ \cdots, \tag{4}$$

whose squares form a sequence

$$\frac{1}{1}, \ \frac{49}{25}, \ \frac{1681}{841}, \ \frac{57,121}{28,561}, \ \cdots, \tag{5}$$

that converges to 2 as its limit. This time their imagination suggested that they needed the concept of a limit of a sequence of rational numbers. If we accept the fact that an increasing sequence that is bounded from above always approaches a limit and observe that the sequence in (4) has these properties, then we want it to have a limit L. This would also mean, from (5), that $L^2 = 2$, and hence L is *not* one of our rational numbers. If to the rational numbers we further add the limits of all bounded increasing sequences of rational numbers, we arrive at the system of all "real" numbers. The word *real* is placed in quotes because there is nothing that is either "more real" or "less real" about this system than there is about any other mathematical system.

Imagination was called upon at many stages during the development of the real number system. In fact, the art of invention was needed at least three times in constructing the systems we have discussed so far:

1. The *first invented* system: the set of *all integers* as constructed from the counting numbers.

2. The *second invented* system: the set of *rational numbers m/n* as constructed from the integers.

3. The *third invented* system: the set of all *real numbers x* as constructed from the rational numbers.

These invented systems form a hierarchy in which each system contains the previous system. Each system is also richer than its predecessor in that it permits additional operations to be performed without going outside the system:

1. In the system of all integers, we can solve all equations of the form

$$x + a = 0, \tag{6}$$

where a may be any integer.

2. In the system of all rational numbers, we can solve all equations of the form

$$ax + b = 0, \tag{7}$$

provided a and b are rational numbers and $a \neq 0$.

3. In the system of all real numbers, we can solve all of the equations in (6) and (7) and, in addition, all quadratic equations

$$ax^2 + bx + c = 0 \quad \text{having} \quad a \neq 0 \quad \text{and} \quad b^2 - 4ac \geq 0. \tag{8}$$

You are probably familiar with the formula that gives the solutions of (8), namely,

$$x = \frac{-b \pm \sqrt{b^2 - 4ac}}{2a}, \tag{9}$$

and are familiar with the further fact that when the discriminant, $d = b^2 - 4ac$, is negative, the solutions in (9) do *not* belong to any of the systems discussed above. In fact, the very simple quadratic equation

$$x^2 + 1 = 0 \tag{10}$$

is impossible to solve if the only number systems that can be used are the three invented systems mentioned so far.

Thus we come to the *fourth invented* system, the set of all complex numbers $a + ib$. We could, in fact, dispense entirely with the symbol i and use a notation like (a, b). We would then speak simply of a pair of real numbers a and b. Since, under algebraic operations, the numbers a and b are treated somewhat differently, it is essential to keep the *order* straight. We therefore might say that the **complex number system** consists of the set of all ordered pairs of real numbers (a, b), together with the rules by which they are to be equated, added, multiplied, and so on, listed below. We shall use both the (a, b) notation and the notation $a + ib$. We call a the **real part** and b the **imaginary part** of (a, b). We make the following definitions.

Equality

$a + ib = c + id$
if and only if
$a = c$ and $b = d$.

Two complex numbers (a, b) and (c, d) are *equal* if and only if $a = c$ and $b = d$.

Addition

$(a + ib) + (c + id)$
$= (a + c) + i(b + d)$

The sum of the two complex numbers (a, b) and (c, d) is the complex number $(a + c, b + d)$.

Multiplication

$(a + ib)(c + id)$
$= (ac - bd) + i(ad + bc)$

The product of two complex numbers (a, b) and (c, d) is the complex number $(ac - bd, ad + bc)$.

$c(a + ib) = ac + i(bc)$

The product of a real number c and the complex number (a, b) is the complex number (ac, bc).

The set of all complex numbers (a, b) in which the second number is zero has all the properties of the set of real numbers a. For example, addition and multiplication of $(a, 0)$ and $(c, 0)$ give

$$(a, 0) + (c, 0) = (a + c, 0),$$

$$(a, 0) \cdot (c, 0) = (ac, 0),$$

which are numbers of the same type with imaginary part equal to zero. Also, if we multiply a "real number" $(a, 0)$ and the complex number (c, d), we get

$$(a, 0) \cdot (c, d) = (ac, ad) = a(c, d).$$

In particular, the complex number $(0, 0)$ plays the role of zero in the complex number system, and the complex number $(1, 0)$ plays the role of unity.

The number pair $(0, 1)$, which has real part equal to zero and imaginary part equal to one has the property that its square,

$$(0, 1)(0, 1) = (-1, 0),$$

has real part equal to minus one and imaginary part equal to zero. Therefore, in the system of complex numbers (a, b), there is a number $x = (0, 1)$ whose square can be added to unity $= (1, 0)$ to produce zero $= (0, 0)$; that is,

$$(0, 1)^2 + (1, 0) = (0, 0).$$

The equation

$$x^2 + 1 = 0$$

therefore has a solution $x = (0, 1)$ in this new number system.

You are probably more familiar with the $a + ib$ notation than you are with the notation (a, b). And since the laws of algebra for the ordered pairs enable us to write

$$(a, b) = (a, 0) + (0, b) = a(1, 0) + b(0, 1),$$

while $(1, 0)$ behaves like unity and $(0, 1)$ behaves like a square root of minus one, we need not hesitate to write $a + ib$ in place of (a, b). The i associated with b is like a tracer element that tags the imaginary part of $a + ib$. We can pass at will from the realm of ordered pairs (a, b) to the realm of expressions $a + ib$, and conversely. But there is nothing less "real" about the symbol $(0, 1) = i$ than there is about the symbol $(1, 0) = 1$, once we have learned the laws of algebra in the complex number system (a, b).

To reduce any rational combination of complex numbers to a single complex number, we apply the laws of elementary algebra, replacing i^2 wherever it appears by -1. Of course, we cannot divide by the complex number $(0, 0) = 0 + i0$. But if $a + ib \neq 0$, then we may carry out a division as follows:

$$\frac{c + id}{a + ib} = \frac{(c + id)(a - ib)}{(a + ib)(a - ib)} = \frac{(ac + bd) + i(ad - bc)}{a^2 + b^2}.$$

The result is a complex number $x + iy$ with

$$x = \frac{ac + bd}{a^2 + b^2}, \qquad y = \frac{ad - bc}{a^2 + b^2},$$

and $a^2 + b^2 \neq 0$, since $a + ib = (a, b) \neq (0, 0)$.

The number $a - ib$ that is used as multiplier to clear the i out of the denominator is called the **complex conjugate** of $a + ib$. It is customary to use $\bar{z}$ (read "z bar") to denote the complex conjugate of z; thus

$$z = a + ib, \qquad \bar{z} = a - ib.$$

Thus, we multiplied the numerator and denominator of the complex fraction $(c + id)/(a + ib)$ by the complex conjugate of the denominator. This will always replace the denominator by a real number.

Argand Diagrams There are two geometric representations of the complex number $z = x + iy$:

a) as the point $P(x, y)$ in the xy-plane and

b) as the vector $\overrightarrow{OP}$ from the origin to P.

In each representation, the x-axis is called the **real axis** and the y-axis is the **imaginary axis**. Both representations are called **Argand diagrams** for $x + iy$ (Fig. A.10).

In terms of the polar coordinates of x and y, we have

$$x = r \cos \theta, \qquad y = r \sin \theta,$$

and

$$z = x + iy = r(\cos \theta + i \sin \theta). \tag{11}$$

We define the **absolute value** of a complex number $x + iy$ to be the length r of a vector $\overrightarrow{OP}$ from the origin to $P(x, y)$. We denote the absolute value by vertical bars, thus:

$$|x + iy| = \sqrt{x^2 + y^2}. \tag{12a}$$

If we always choose the polar coordinates r and θ so that r is nonnegative, then

$$r = |x + iy|. \tag{12b}$$

The polar angle θ is called the **argument** of z and is written $\theta = \arg z$. Of course, any integer multiple of 2π may be added to θ to produce another appropriate angle.

The following equation gives a useful formula connecting a complex number z, its conjugate $\bar{z}$, and its absolute value $|z|$, namely,

$$z \cdot \bar{z} = |z|^2. \tag{13}$$

The identity

$$e^{i\theta} = \cos \theta + i \sin \theta \tag{14}$$

introduced in Article 12.3 leads to the following rules for calculating products,

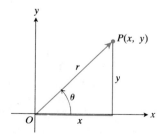

A.10   This Argand diagram represents $z = x + iy$ both as a point $P(x, y)$ and as a vector $\overrightarrow{OP}$.

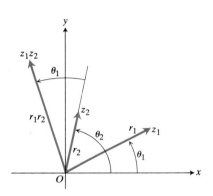

A.11 Argand diagrams for $e^{i\theta} = \cos\theta + i\sin\theta$ (a) as a vector, (b) as a point.

(a)   (b)

quotients, powers, and roots of complex numbers. It also leads to Argand diagrams for $e^{i\theta}$. Since $\cos\theta + i\sin\theta$ is what we get from Eq. (11) by taking $r = 1$, we can say that $e^{i\theta}$ is represented by a unit vector that makes an angle θ with the positive x-axis, as shown in Fig. A.11.

Products To multiply two complex numbers, we multiply their absolute values and add their angles. Let

$$z_1 = r_1 e^{i\theta_1}, \qquad z_2 = r_2 e^{i\theta_2}, \tag{15}$$

so that

$$|z_1| = r_1, \quad \arg z_1 = \theta_1; \qquad |z_2| = r_2, \quad \arg z_2 = \theta_2. \tag{16}$$

Then

$$z_1 z_2 = r_1 e^{i\theta_1} \cdot r_2 e^{i\theta_2} = r_1 r_2 e^{i(\theta_1 + \theta_2)}$$

and hence

$$|z_1 z_2| = r_1 r_2 = |z_1| \cdot |z_2|,$$
$$\arg(z_1 z_2) = \theta_1 + \theta_2 = \arg z_1 + \arg z_2. \tag{17}$$

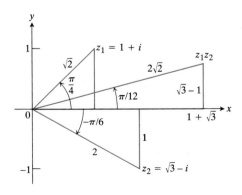

A.12 When z_1 and z_2 are multiplied, $|z_1 z_2| = r_1 \cdot r_2$, and $\arg(z_1 z_2) = \theta_1 + \theta_2$.

Thus the product of two complex numbers is represented by a vector whose length is the product of the lengths of the two factors and whose argument is the sum of their arguments (Fig. A.12). In particular, a vector may be rotated in the counterclockwise direction through an angle θ by simply multiplying it by $e^{i\theta}$. Multiplication by i rotates 90°, by -1 rotates 180°, by $-i$ rotates 270°, etc.

EXAMPLE 1 Let

$$z_1 = 1 + i, \qquad z_2 = \sqrt{3} - i.$$

We plot these complex numbers in an Argand diagram (Fig. A.13) from which we read off the polar representations

$$z_1 = \sqrt{2}\, e^{i\pi/4}, \qquad z_2 = 2 e^{-i\pi/6}.$$

Then

$$z_1 z_2 = 2\sqrt{2} \exp\left(\frac{i\pi}{4} - \frac{i\pi}{6}\right) = 2\sqrt{2} \exp\left(\frac{i\pi}{12}\right)$$

$$= 2\sqrt{2}\left(\cos\frac{\pi}{12} + i\sin\frac{\pi}{12}\right) \approx 2.73 + 0.73i.$$

A.13 To multiply two complex numbers, multiply their absolute values and add their arguments.

Quotients   Suppose $r_2 \neq 0$ in Eq. (15). Then

$$\frac{z_1}{z_2} = \frac{r_1 e^{i\theta_1}}{r_2 e^{i\theta_2}} = \frac{r_1}{r_2} e^{i(\theta_1 - \theta_2)}.$$

Hence

$$\left| \frac{z_1}{z_2} \right| = \frac{r_1}{r_2} = \frac{|z_1|}{|z_2|},$$

$$\arg \left(\frac{z_1}{z_2} \right) = \theta_1 - \theta_2 = \arg z_1 - \arg z_2.$$

That is, we divide lengths and subtract angles.

EXAMPLE 2   Let $z_1 = 1 + i$ and $z_2 = \sqrt{3} - i$, as in Example 1. Then

$$\frac{1 + i}{\sqrt{3} - i} = \frac{\sqrt{2} e^{i\pi/4}}{2 e^{-i\pi/6}} = \frac{\sqrt{2}}{2} e^{5\pi i/12}$$

$$\approx 0.707 \left(\cos \frac{5\pi}{12} + i \sin \frac{5\pi}{12} \right)$$

$$\approx 0.183 + 0.\overline{683}i. \qquad \blacksquare$$

Powers   If n is a positive integer, we may apply the product formulas in (17) to find

$$z^n = z \cdot z \cdot \ldots \cdot z \qquad (n \text{ factors}).$$

With $z = re^{i\theta}$, we obtain

$$z^n = (re^{i\theta})^n = r^n e^{i(\theta + \theta + \cdots + \theta)} \qquad (n \text{ summands})$$

$$= r^n e^{in\theta}. \tag{18}$$

The length $r = |z|$ is raised to the nth power and the angle $\theta = \arg z$ is multiplied by n.

In particular, if we place $r = 1$ in Eq. (18), we obtain De Moivre's theorem.

THEOREM

> **De Moivre's Theorem**
>
> $$(\cos \theta + i \sin \theta)^n = \cos n\theta + i \sin n\theta. \tag{19}$$

If we expand the left-hand side of De Moivre's equation by the binomial theorem and reduce it to the form $a + ib$, we obtain formulas for $\cos n\theta$ and $\sin n\theta$ as polynomials of degree n in $\cos \theta$ and $\sin \theta$.

EXAMPLE 4   If $n = 3$ in Eq. (19), we have

$$(\cos \theta + i \sin \theta)^3 = \cos 3\theta + i \sin 3\theta.$$

The left-hand side of this equation is

$$\cos^3 \theta + 3i \cos^2 \theta \sin \theta - 3 \cos \theta \sin^2 \theta - i \sin^3 \theta.$$

The real part of this must equal $\cos 3\theta$ and the imaginary part must equal $\sin 3\theta$. Therefore,

$$\cos 3\theta = \cos^3\theta - 3 \cos \theta \sin^2\theta,$$

$$\sin 3\theta = 3 \cos^2\theta \sin \theta - \sin^3\theta. \qquad \blacksquare$$

Roots If $z = re^{i\theta}$ is a complex number different from zero and n is a positive integer, then there are precisely n different complex numbers $w_0, w_1, \ldots, w_{n-1}$, that are nth roots of z. To see why, let $w = \rho e^{i\alpha}$ be an nth root of $z = re^{i\theta}$, so that

$$w^n = z$$

or

$$\rho^n e^{in\alpha} = re^{i\theta}. \qquad (20)$$

Then

$$\rho = \sqrt[n]{r} \qquad (21)$$

is the real, positive nth root of r. As regards the angle, although we cannot say that $n\alpha$ and θ must be equal, we can say that they may differ only by an integral multiple of 2π. That is,

$$n\alpha = \theta + 2k\pi, \qquad k = 0, \pm 1, \pm 2, \ldots . \qquad (22)$$

Therefore

$$\alpha = \frac{\theta}{n} + k\frac{2\pi}{n}.$$

Hence all the nth roots of $z = re^{i\theta}$ are given by

$$\sqrt[n]{re^{i\theta}} = \sqrt[n]{r} \exp i\left(\frac{\theta}{n} + k\frac{2\pi}{n}\right), \qquad k = 0, \pm 1, \pm 2, \ldots . \qquad (23)$$

There might appear to be infinitely many different answers corresponding to the infinitely many possible values of k. But one readily sees that $k = n + m$ gives the same answer as $k = m$ in Eq. (23). Thus we need only take n consecutive values for k to obtain all the different nth roots of z. For convenience, we take

$$k = 0, 1, 2, \ldots, n - 1.$$

All the nth roots of $re^{i\theta}$ lie on a circle centered at the origin O and having radius equal to the real, positive nth root of r. One of them has argument $\alpha = \theta/n$. The others are uniformly spaced around the circle, each being separated from its neighbors by an angle equal to $2\pi/n$. Figure A.14 illustrates the placement of the three cube roots, w_0, w_1, w_2, of the complex number $z = re^{i\theta}$.

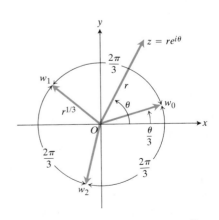

A.14 The three cube roots of $z = re^{i\theta}$.

A.15 The four fourth roots of -16.

EXAMPLE 5 Find the four fourth roots of -16.

Solution As our first step, we plot the number -16 in an Argand diagram (Fig. A.15) and determine its polar representation $re^{i\theta}$. Here,

$$z = -16,$$

$$r = +16,$$

$$\theta = \pi.$$

One of the fourth roots of $16\, e^{i\pi}$ is $2\, e^{i\pi/4}$. We obtain others by successive additions of $2\pi/4 = \pi/2$ to the argument of this first one. Hence

$$\sqrt[4]{16 \exp i\pi} = 2 \exp i\left(\frac{\pi}{4}, \frac{3\pi}{4}, \frac{5\pi}{4}, \frac{7\pi}{4}\right),$$

and the four roots are

$$w_0 = 2\left[\cos\frac{\pi}{4} + i\sin\frac{\pi}{4}\right] = \sqrt{2}(1 + i),$$

$$w_1 = 2\left[\cos\frac{3\pi}{4} + i\sin\frac{3\pi}{4}\right] = \sqrt{2}(-1 + i),$$

$$w_2 = 2\left[\cos\frac{5\pi}{4} + i\sin\frac{5\pi}{4}\right] = \sqrt{2}(-1 - i),$$

$$w_3 = 2\left[\cos\frac{7\pi}{4} + i\sin\frac{7\pi}{4}\right] = \sqrt{2}(1 - i).$$ ∎

The Fundamental Theorem of Algebra One may well say that the invention of $\sqrt{-1}$ is all well and good and leads to a number system that is richer than the real number system alone; but where will this process end? Are we also going to invent still more systems so as to obtain $\sqrt[4]{-1}$, $\sqrt[6]{-1}$, and so on? By now it should be clear that this is not necessary. These numbers are already expressible in terms of the complex number system $a + ib$. In fact, the Fundamental Theorem of Algebra says that with the introduction of the complex numbers we now have enough numbers to factor every polynomial into a product of linear factors and hence enough numbers to solve every possible polynomial equation.

THEOREM

> **The Fundamental Theorem of Algebra**
>
> Every polynomial equation of the form
>
> $$a_0 z^n + a_1 z^{n-1} + a_2 z^{n-2} + \cdots + a_{n-1} z + a_n = 0,$$
>
> in which the coefficients $a_0, a_1, \ldots, a_n$ are any complex numbers, whose degree n is greater than or equal to one, and whose leading coefficient a_0 is not zero, has exactly n roots in the complex number system, provided each multiple root of multiplicity m is counted as m roots.

A proof of this theorem can be found in almost any text on the theory of functions of a complex variable.

PROBLEMS

1. Find $(a, b) \cdot (c, d) = (ac - bd, ad + bc)$.
   a) $(2, 3) \cdot (4, -2)$
   b) $(2, -1) \cdot (-2, 3)$
   c) $(-1, -2) \cdot (2, 1)$

   (This is how complex numbers are multiplied by computers.)

2. Solve the following equations for the real numbers x and y.
   a) $(3 + 4i)^2 - 2(x - iy) = x + iy$
   b) $\left(\dfrac{1 + i}{1 - i}\right)^2 + \dfrac{1}{x + iy} = 1 + i$
   c) $(3 - 2i)(x + iy) = 2(x - 2iy) + 2i - 1$

3. Show with an Argand diagram that the law for adding complex numbers is the same as the parallelogram law for adding vectors.

4. How may the following complex numbers be obtained from $z = x + iy$ geometrically? Sketch.
   a) $\bar{z}$
   b) $\overline{(-z)}$
   c) $-z$
   d) $1/z$

5. Show that the conjugate of the sum (product, or quotient) of two complex numbers z_1 and z_2 is the same as the sum (product, or quotient) of their conjugates.

6. a) Extend the results of Problem 5 to show that
   $$f(\bar{z}) = \overline{f(z)} \quad \text{if}$$
   $$f(z) = a_0 z^n + a_1 z^{n-1} + \cdots + a_{n-1} z + a_n$$
   is a polynomial with real coefficients $a_0, \ldots, a_n$.
   b) If z is a root of the equation $f(z) = 0$, where $f(z)$ is a polynomial with real coefficients as in part (a), show that the conjugate $\bar{z}$ is also a root of the equation. Thus, complex roots of polynomials with real coefficients always come in complex-conjugate pairs. (*Hint:* Let $f(z) = u + iv = 0$; then both u and v are zero. Now use the fact that $f(\bar{z}) = \overline{f(z)} = u - iv$.)

7. Show that $|\bar{z}| = |z|$.

8. If z and $\bar{z}$ are equal, what can you say about the location of the point z in the complex plane?

9. Let $Re(z)$ and $Im(z)$ denote respectively the real and imaginary parts of z. Show that
   a) $z + \bar{z} = 2Re(z)$,
   b) $z - \bar{z} = 2i\,Im(z)$
   c) $|Re(z)| \le |z|$,
   d) $|z_1 + z_2|^2 = |z_1|^2 + |z_2|^2 + 2Re(z_1\bar{z}_2)$,
   e) $|z_1 + z_2| \le |z_1| + |z_2|$.

10. Show that the distance between the two points z_1 and z_2 in an Argand diagram is equal to $|z_1 - z_2|$.

In Problems 11–15, graph the points $z = x + iy$ that satisfy the given conditions.

11. a) $|z| = 2$ b) $|z| < 2$ c) $|z| > 2$

12. $|z - 1| = 2$ 13. $|z + 1| = 1$

14. $|z + 1| = |z - 1|$ 15. $|z + i| = |z - 1|$

Express the answer to each of Problems 16–19 in the form $re^{i\theta}$, with $r \ge 0$ and $-\pi < \theta \le \pi$. Sketch.

16. $(1 + \sqrt{-3})^2$ 17. $\dfrac{1 + i}{1 - i}$

18. $\dfrac{1 + i\sqrt{3}}{1 - i\sqrt{3}}$ 19. $(2 + 3i)(1 - 2i)$

20. Use De Moivre's theorem (Eq. 19) to express $\cos 4\theta$ and $\sin 4\theta$ as polynomials in $\cos \theta$ and $\sin \theta$.

21. Find the three cube roots of 1.

22. Find the two square roots of i.

23. Find the three cube roots of $-8i$.

24. Find the six sixth roots of 64.

25. Find the four solutions of the equation $z^4 - 2z^2 + 4 = 0$.

26. Find the six solutions of the equation $z^6 + 2z^3 + 2 = 0$.

27. Find all solutions of the equation $x^4 + 4x^2 + 16 = 0$.

28. Solve the equation $x^4 + 1 = 0$.

  TOOLKIT PROGRAMS

Complex Number Calculator

A.12
Tables for sin x, cos x,
tan x, e^x, e^{-x}, and ln x

TABLE 1
Natural trigonometric functions

Angle					Angle				
Degree	Radian	Sine	Cosine	Tangent	Degree	Radian	Sine	Cosine	Tangent
0°	0.000	0.000	1.000	0.000					
1°	0.017	0.017	1.000	0.017	46°	0.803	0.719	0.695	1.036
2°	0.035	0.035	0.999	0.035	47°	0.820	0.731	0.682	1.072
3°	0.052	0.052	0.999	0.052	48°	0.838	0.743	0.669	1.111
4°	0.070	0.070	0.998	0.070	49°	0.855	0.755	0.656	1.150
5°	0.087	0.087	0.996	0.087	50°	0.873	0.766	0.643	1.192
6°	0.105	0.105	0.995	0.105	51°	0.890	0.777	0.629	1.235
7°	0.122	0.122	0.993	0.123	52°	0.908	0.788	0.616	1.280
8°	0.140	0.139	0.990	0.141	53°	0.925	0.799	0.602	1.327
9°	0.157	0.156	0.988	0.158	54°	0.942	0.809	0.588	1.376
10°	0.175	0.174	0.985	0.176	55°	0.960	0.819	0.574	1.428
11°	0.192	0.191	0.982	0.194	56°	0.977	0.829	0.559	1.483
12°	0.209	0.208	0.978	0.213	57°	0.995	0.839	0.545	1.540
13°	0.227	0.225	0.974	0.231	58°	1.012	0.848	0.530	1.600
14°	0.244	0.242	0.970	0.249	59°	1.030	0.857	0.515	1.664
15°	0.262	0.259	0.966	0.268	60°	1.047	0.866	0.500	1.732
16°	0.279	0.276	0.961	0.287	61°	1.065	0.875	0.485	1.804
17°	0.297	0.292	0.956	0.306	62°	1.082	0.883	0.469	1.881
18°	0.314	0.309	0.951	0.325	63°	1.100	0.891	0.454	1.963
19°	0.332	0.326	0.946	0.344	64°	1.117	0.899	0.438	2.050
20°	0.349	0.342	0.940	0.364	65°	1.134	0.906	0.423	2.145
21°	0.367	0.358	0.934	0.384	66°	1.152	0.914	0.407	2.246
22°	0.384	0.375	0.927	0.404	67°	1.169	0.921	0.391	2.356
23°	0.401	0.391	0.921	0.424	68°	1.187	0.927	0.375	2.475
24°	0.419	0.407	0.914	0.445	69°	1.204	0.934	0.358	2.605
25°	0.436	0.423	0.906	0.466	70°	1.222	0.940	0.342	2.748
26°	0.454	0.438	0.899	0.488	71°	1.239	0.946	0.326	2.904
27°	0.471	0.454	0.891	0.510	72°	1.257	0.951	0.309	3.078
28°	0.489	0.469	0.883	0.532	73°	1.274	0.956	0.292	3.271
29°	0.506	0.485	0.875	0.554	74°	1.292	0.961	0.276	3.487
30°	0.524	0.500	0.866	0.577	75°	1.309	0.966	0.259	3.732
31°	0.541	0.515	0.857	0.601	76°	1.326	0.970	0.242	4.011
32°	0.559	0.530	0.848	0.625	77°	1.344	0.974	0.225	4.332
33°	0.576	0.545	0.839	0.649	78°	1.361	0.978	0.208	4.705
34°	0.593	0.559	0.829	0.675	79°	1.379	0.982	0.191	5.145
35°	0.611	0.574	0.819	0.700	80°	1.396	0.985	0.174	5.671
36°	0.628	0.588	0.809	0.727	81°	1.414	0.988	0.156	6.314
37°	0.646	0.602	0.799	0.754	82°	1.431	0.990	0.139	7.115
38°	0.663	0.616	0.788	0.781	83°	1.449	0.993	0.122	8.144
39°	0.681	0.629	0.777	0.810	84°	1.466	0.995	0.105	9.514
40°	0.698	0.643	0.766	0.839	85°	1.484	0.996	0.087	11.43
41°	0.716	0.656	0.755	0.869	86°	1.501	0.998	0.070	14.30
42°	0.733	0.669	0.743	0.900	87°	1.518	0.999	0.052	19.08
43°	0.750	0.682	0.731	0.933	88°	1.536	0.999	0.035	28.64
44°	0.768	0.695	0.719	0.966	89°	1.553	1.000	0.017	57.29
45°	0.785	0.707	0.707	1.000	90°	1.571	1.000	0.000	

TABLE 2
Exponential functions

x	e^x	e^{-x}	x	e^x	e^{-x}
0.00	1.0000	1.0000	2.5	12.182	0.0821
0.05	1.0513	0.9512	2.6	13.464	0.0743
0.10	1.1052	0.9048	2.7	14.880	0.0672
0.15	1.1618	0.8607	2.8	16.445	0.0608
0.20	1.2214	0.8187	2.9	18.174	0.0550
0.25	1.2840	0.7788	3.0	20.086	0.0498
0.30	1.3499	0.7408	3.1	22.198	0.0450
0.35	1.4191	0.7047	3.2	24.533	0.0408
0.40	1.4918	0.6703	3.3	27.113	0.0369
0.45	1.5683	0.6376	3.4	29.964	0.0334
0.50	1.6487	0.6065	3.5	33.115	0.0302
0.55	1.7333	0.5769	3.6	36.598	0.0273
0.60	1.8221	0.5488	3.7	40.447	0.0247
0.65	1.9155	0.5220	3.8	44.701	0.0224
0.70	2.0138	0.4966	3.9	49.402	0.0202
0.75	2.1170	0.4724	4.0	54.598	0.0183
0.80	2.2255	0.4493	4.1	60.340	0.0166
0.85	2.3396	0.4274	4.2	66.686	0.0150
0.90	2.4596	0.4066	4.3	73.700	0.0136
0.95	2.5857	0.3867	4.4	81.451	0.0123
1.0	2.7183	0.3679	4.5	90.017	0.0111
1.1	3.0042	0.3329	4.6	99.484	0.0101
1.2	3.3201	0.3012	4.7	109.95	0.0091
1.3	3.6693	0.2725	4.8	121.51	0.0082
1.4	4.0552	0.2466	4.9	134.29	0.0074
1.5	4.4817	0.2231	5	148.41	0.0067
1.6	4.9530	0.2019	6	403.43	0.0025
1.7	5.4739	0.1827	7	1096.6	0.0009
1.8	6.0496	0.1653	8	2981.0	0.0003
1.9	6.6859	0.1496	9	8103.1	0.0001
2.0	7.3891	0.1353	10	22026	0.00005
2.1	8.1662	0.1225			
2.2	9.0250	0.1108			
2.3	9.9742	0.1003			
2.4	11.023	0.0907			

TABLE 3
Natural logarithms

x	$\log_e x$	x	$\log_e x$	x	$\log_e x$	x	$\log_e x$
0.0	*	3.5	1.2528	7.0	1.9459	15	2.7081
0.1	7.6974	3.6	1.2809	7.1	1.9601	16	2.7726
0.2	8.3906	3.7	1.3083	7.2	1.9741	17	2.8332
0.3	8.7960	3.8	1.3350	7.3	1.9879	18	2.8904
0.4	9.0837	3.9	1.3610	7.4	2.0015	19	2.9444
0.5	9.3069	4.0	1.3863	7.5	2.0149	20	2.9957
0.6	9.4892	4.1	1.4110	7.6	2.0281	25	3.2189
0.7	9.6433	4.2	1.4351	7.7	2.0412	30	3.4012
0.8	9.7769	4.3	1.4586	7.8	2.0541	35	3.5553
0.9	9.8946	4.4	1.4816	7.9	2.0669	40	3.6889
1.0	0.0000	4.5	1.5041	8.0	2.0794	45	3.8067
1.1	0.0953	4.6	1.5261	8.1	2.0919	50	3.9120
1.2	0.1823	4.7	1.5476	8.2	2.1041	55	4.0073
1.3	0.2624	4.8	1.5686	8.3	2.1163	60	4.0943
1.4	0.3365	4.9	1.5892	8.4	2.1282	65	4.1744
1.5	0.4055	5.0	1.6094	8.5	2.1401	70	4.2485
1.6	0.4700	5.1	1.6292	8.6	2.1518	75	4.3175
1.7	0.5306	5.2	1.6487	8.7	2.1633	80	4.3820
1.8	0.5878	5.3	1.6677	8.8	2.1748	85	4.4427
1.9	0.6419	5.4	1.6864	8.9	2.1861	90	4.4998
2.0	0.6931	5.5	1.7047	9.0	2.1972	95	4.5539
2.1	0.7419	5.6	1.7228	9.1	2.2083	100	4.6052
2.2	0.7885	5.7	1.7405	9.2	2.2192		
2.3	0.8329	5.8	1.7579	9.3	2.2300		
2.4	0.8755	5.9	1.7750	9.4	2.2407		
2.5	0.9163	6.0	1.7918	9.5	2.2513		
2.6	0.9555	6.1	1.8083	9.6	2.2618		
2.7	0.9933	6.2	1.8245	9.7	2.2721		
2.8	1.0296	6.3	1.8405	9.8	2.2824		
2.9	1.0647	6.4	1.8563	9.9	2.2925		
3.0	1.0986	6.5	1.8718	10	2.3026		
3.1	1.1314	6.6	1.8871	11	2.3979		
3.2	1.1632	6.7	1.9021	12	2.4849		
3.3	1.1939	6.8	1.9169	13	2.5649		
3.4	1.2238	6.9	1.9315	14	2.6391		

*Subtract 10 from $\log_e x$ entries for $x < 1.0$.

Answers

CHAPTER 1

Article 1.1, p. 4

	Q	R	S	T
1.	$(1, 2)$	$(-1, -2)$	$(-1, 2)$	$(-2, 1)$
3.	$(-2, -2)$	$(2, 2)$	$(2, -2)$	$(2, -2)$
5.	$(0, -1)$	$(0, 1)$	$(0, -1)$	$(1, 0)$
7.	$(-2, 0)$	$(2, 0)$	$(2, 0)$	$(0, -2)$
9.	$(-1, 3)$	$(1, -3)$	$(1, 3)$	$(-3, -1)$
11.	$(-\pi, \pi)$	$(\pi, -\pi)$	(π, π)	$(-\pi, -\pi)$
13.	$(x, -y)$	$(-x, y)$	$(-x, -y)$	(y, x)

15. Missing vertices: $(-1, 4)$, $(-1, -2)$, $(5, 2)$

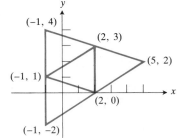

17. $A(-12, 2)$, $B(-12, -5)$, $C(9, -5)$
19. 2

Article 1.2, pp. 9–10

1. $\Delta x = 2$, $\Delta y = 1$ **3.** $\Delta x = 2$, $\Delta y = -4$ **5.** $\Delta x = -5$, $\Delta y = 0$ **7.** a) $\Delta x = 57$, $\Delta y = -10$ b) $\Delta x = -2$, $\Delta y = -12$ c) $\Delta x = 1$, $\Delta y = -14$ **9.** Above, right **11.** Left **13.** Below, right **15.** Below **17.** $m = 3$, $m_\perp = -1/3$ **19.** $m = -1/3$, $m_\perp = 3$
21. $m = -1$, $m_\perp = 1$ **23.** $m = 0$, $m_\perp$ is undefined **25.** $m = -1$, $m_\perp = 1$ **27.** $m = 2$, $m_\perp = -1/2$ **29.** $m = y/x$, $m_\perp = -x/y$
31. m_{AB} is undefined, $m_\perp = 0$ **33.** $17.35° \le \alpha \le 48.37°$ **35.** 0.8391 **37.** Fiber glass is best (-16 deg/in.); gypsum is poorest (-3 deg/in.). **39.** a) Not collinear b) Not collinear c) B, C, D collinear d) Not collinear e) A, B, C, D collinear
41. $(3, -3)$ **43.** $x = -2$, $y = -9$

Article 1.3, pp. 16–17

1. a) $x = 2$ b) $y = 3$ **3.** a) $x = 0$ b) $y = 0$ **5.** a) $x = -4$ b) $y = 0$ **7.** a) $x = 0$ b) $y = b$ **9.** $x - y = 0$ **11.** $x - y = -2$
13. $2x - y = -b$ **15.** $3x - 2y = 0$ **17.** $x = 1$ **19.** $x = -2$ **21.** $F_0 x + Ty = TF_0$ **23.** $x = 0$ **25.** $5x + 15y = 19$ **27.** $y - y_1 = \dfrac{y_1 - y_0}{x_1 - x_0}(x - x_1)$ **29.** $3x - y = 2$ **31.** $x - y = -\sqrt{2}$ **33.** $5x + y = 2.5$ **35.** $m = 3$, $(0, 5)$, $(-5/3, 0)$ **37.** $m = -1$, $(0, 2)$, $(2, 0)$ **39.** $m = 1/2$, $(0, -2)$, $(4, 0)$ **41.** $m = 4/3$, $(0, -4)$, $(3, 0)$ **43.** $m = -4/3$, $(0, 4)$, $(3, 0)$ **45.** $m = 3/2$, $(0, 3)$, $(-2, 0)$ **47.** $m = 3$, $(0, -20)$, $(20/3, 0)$ **49.** $m = -b/a$, $(0, b)$, $(a, 0)$ **51.** $\sqrt{3}x + 3y = 3$ **53.** a) $x - y = 1$ b) $x + y = 3$ c) $3/\sqrt{2}$ **55.** a) $x + \sqrt{3}y = 0$ b) $\sqrt{3}x - y = 0$ c) $3/2$ **57.** a) $2x + y = -2$ b) $x - 2y = -6$ c) $6/\sqrt{5}$ **59.** a) $2x - y = 2$ b) $x + 2y = 1$ c) $4/\sqrt{5}$ **61.** a) $x = 3$ b) $y = 2$ c) 8 **63.** a) $x = a$ b) $y = b$ c) $|a + 1|$ **65.** a) $3y + 4x = 34$ b) $3x - 4y = -12$ c) 22/5 **67.** $45°$ **69.** $150°$ **71.** $116.6°$ **73.** $126.9°$ **75.** $\sqrt{3}x - y = \sqrt{3} - 4$ **77.** $x = -2$ **79.** About 5.97 atm
81. $s = 0.0023t + 34.85$ **83.** 23 ft

Article 1.4, pp. 28–29

1. $x \geq 0$, $y \geq 0$ **3.** $x \geq 0$, $y \leq 0$ **5.** $x \geq -4$, $y \geq 0$ **7.** $x \neq 2$, $y \neq 0$ **9.** All x, $-2 \leq y \leq 2$ **11.** All x, $-3 \leq y \leq 3$
13. All x, $y \geq 1$ **15.** All x, $y \leq 0$ **17.** $x \geq -1$, $y \geq 0$ **19.** $x \geq 0$, $y \geq 1$

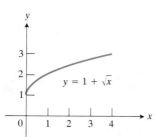

21. $x \geq 0$, $y \geq 0$ **23.** $x \neq 0$, $y \neq 0$ **25.** All x, $-1 \leq y \leq 1$

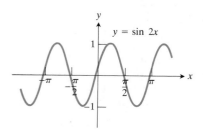

27. All x, $0 \leq y \leq 1$ **29.** All x, $0 \leq y \leq 2$

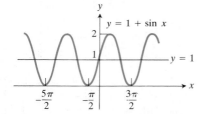

31. a) No b) No c) $x > 0$ **33.** a) No b) No c) No d) $0 < x \leq 1$ **35.** a) $(x/2) \neq \pm(\pi/2)$, $\pm(3\pi/2)$, . . . , $\pm((2n - 1)\pi)/2$
b) $x \neq \pm\pi$, $\pm3\pi$, . . . , $\pm(2n - 1)\pi$ c) All real values d) Domain is answer to part (b), range is answer to part (c)   **37.** Could
not be i, iii, or iv because $f(0) \neq 0$

39.

x	y
0	0
$\frac{1}{2}$	$\frac{1}{2}$
1	1
$\frac{3}{2}$	$\frac{1}{2}$
2	0

41.

43.

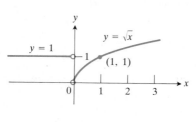

45. D_f: all x, D_g: $x \geq 1$; $D_{f+g} = D_{f-g} = D_{fg} = D_{g/f}$: $x \geq 1$; $D_{f/g}$: $x > 1$ **47.** D_f: $x \geq 0$, D_g: $x \geq -1$; $D_{f+g} = D_{f-g} = D_{fg} = D_{f/g}$:
$x \geq 0$; $D_{g/f}$: $x > 0$ **49.** a) -4 b) 11 c) 2 d) $1 + (1/x)$ e) $1 + (1/2x)$ f) $1 + 5x$, $x \neq 0$ **51.** $f(1 - x) = (x/x - 1)$ **53.** 4

Article 1.5, pp. 34–35

1. 3 **3.** 5 **5.** ±2 **7.** −1/2, −9/2 **9.** 17/3, −1/3 **11.** f **13.** c **15.** b **17.** d **19.** a **21.** −2 < x < 2 **23.** −1 ≤ x ≤ 3
25. −4 < x < 2 **27.** −3/2 < x < −1/2 **29.** 0 ≤ x ≤ 1 **31.** x ≤ −1 or x ≥ 1 **33.** |x| < 8 **35.** |x + 2| < 3 **37.** |y| < a
39. |y − L| < ε **41.** |x − x_0| < 5 **43.** f(x) = 1 − x for x ≤ 1; f(x) = x − 1 for x ≥ 1
45.

47.

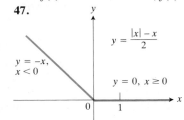

49. Domain of y = $\sqrt{x^2}$ is all x, but domain of y = $(\sqrt{x})^2$ is x ≥ 0. They have the same range, y ≥ 0. **51.** g(x) = $\sqrt{x}$

53. a)

b)

c)

d)

55. If x < 0 is not an integer, then [x] is 1 less than the integer part of the decimal representation of x.

Article 1.6, pp. 41–42

1. Approximately 3.5 flies/day **3.** a) h^2 − 3 b) −3 c) 3x + y = 3 **5.** a) 2x b) 4x − y = 3 c) (0, 1) **7.** a) −2x b) 2x −
y = −5 c) (0, 4) **9.** a) 2x + 3 b) x − y = −1 c) (−3/2, −1/4) **11.** a) 2x + 4 b) y = 0 c) (−2, 0) **13.** a) 2x − 4 b) 2x +
y = 3 c) (2, 0) **15.** a) $3x^2$ b) 3x − y = 2 c) (0, 0) **17.** a) $3x^2$ − 3 b) y = 2 c) (−1, 2), (1, −2) **19.** a) $3x^2$ − 6x b) y +
3x = 5 c) (0, 4), (2, 0)

Article 1.7, pp. 48–49

1. f'(x) = 2x, m = 6, 6x − y = 9 **3.** f'(x) = 2, m = 2, 2x − y = −3 **5.** f'(x) = 1/(2$\sqrt{x}$), m = 1/(2$\sqrt{3}$), x − 2$\sqrt{3}$y =
−(3 + 2$\sqrt{3}$) **7.** f'(x) = −2/(2x + 1)^2, m = −2/49, 2x + 49y = 13 **9.** f'(x) = 4x − 1, m = 11, 11x − y = 13 **11.** f'(x) = $4x^3$,
m = 108, 108x − y = 243 **13.** f'(x) = 1 + (1/x^2), m = 10/9, 10x − 9y = 6 **15.** f'(x) = 1/$\sqrt{2x}$, m = 1/$\sqrt{6}$, x − $\sqrt{6}$y = −3
17. f'(x) = 1/$\sqrt{2x + 3}$, m = 1/3, x − 3y = −6 **19.** f'(x) = −1/(2x + 3)$^{3/2}$, m = −1/27, x + 27y = 12
21. The derivative of f(x) = |x| is f'(x) = |x|/x, x ≠ 0. **23.** a) b) 0, 2, 4, 5 **25.** 0 **27.** Foxes/day

Article 1.8, pp. 53–55

3. a) $\Delta s = 3.2$, $v_{av} = 1.6$ b) $v(t) = 1.6t$ c) $v(2) = 3.2$ **5.** a) $\Delta s = 18$, $v_{av} = 9$ b) $v(t) = 4t + 5$ c) $v(2) = 13$ **7.** a) $\Delta s = -8$, $v_{av} = -4$ b) $v(t) = -2t - 2$ c) $v(2) = -6$ **9.** a) $\Delta s = 8$, $v_{av} = 4$ b) $v(t) = 4$ c) $v(2) = 4$ **11.** a) $s = 490t^2$, $v = 980t$ b) 4/7 sec, $v_{av} = 280$ cm/sec c) About 0.034 sec/flash   **13.** a) 32 ft/sec b) −16 ft/sec c) 0 ft/sec   **15.** 190 ft/sec   **17.** $t = 8$ sec; $v = 0$ **19.** 2.8 sec   **21.** $dQ/dt = 8000$ gal/min, $\Delta Q/\Delta t = 10,000$ gal/min   **23.** a) \$90 per machine b) \$80 per machine c) $f(101) - f(100) = 11,080 - 11,000 = 80$

Article 1.9, pp. 68–70

1. 4 **3.** 4 **5.** 2 **7.** 25 **9.** 1 **11.** $2x$ **13.** 0 **15.** −5 **17.** 18 **19.** 36 **21.** 2 **23.** 4 **25.** 1 **27.** a) −10 b) −20   **29.** a) 4 b) −21 c) −12 d) −7/3 **31.** 5/4 **33.** 0 **35.** Does not exist **37.** −1/2 **39.** −1 **41.** 1/3 **43.** −1/2 **45.** 0   **47.** 3/4 **49.** −3 **51.** 3/4a **53.** $f(x) = 1/x$, $g(x) = -1/x$ **55.** $f(x) = 1/x$, $g(x) = 1/x^2$ **57.** $f(x) = |x - 1|$ **59.** a) All c except 0, 1, 2 b) $c = 2$ c) $c = 0$ **61.** 0 **63.** 0 **65.** 1 **67.** 6 **69.** For all c not an integer **71.** 1 **73.** 1

75. 1 **77.** 0 **79.** 0 **81.** −1 **83.** 2 **85.** 2/3 **89.** a) $L = 5$, any $\delta \le \epsilon/2$ b) $L = -1$, any $\delta \le \epsilon/2$ c) $L = -1$, any $\delta \le \epsilon/3$ d) $L = 7$, any $\delta > 0$ e) $L = 4$, any $\delta \le \epsilon$ f) $L = 6$, any $\delta \le \epsilon$ g) $L = -5$, any $\delta \le 2\epsilon/3$ h) $L = 2$, any $\delta \le$ minimum of $(1, \epsilon/2)$ i) $L = -1/9$, any $\delta \le$ minimum of $(18\epsilon, 1)$ **91.** a) $\delta \le 1/80$ b) $\delta \le 1/800$ c) $\delta \le$ minimum $(\epsilon/8, 1)$

93. $f'(0) = 0$ **95.** a) 11; $1.9975 < x < 2.0025$; $\delta = 0.00025$ b) 2; $10.9601 < x < 11.0401$; $\delta = 0.0399$ c) 21; $\sqrt{24.99} < |x| < \sqrt{25.01}$; $\delta = \sqrt{25.01} - 5$ **97.** $I = (5, 5 + \epsilon^2)$; $\lim_{x \to 5^+} \sqrt{x - 5} = 0$

Article 1.10, pp. 76–77

1. 2/5 **3.** 0 **5.** 0 **7.** 1/2 **9.** 1 **11.** ∞ **13.** 1 **15.** 3/10 **17.** 1 **19.** 2 **21.** 1 **23.** 0 **25.** −2/3 **27.** 1 **29.** 0 **31.** 2   **33.** ∞ **35.** ∞ **37.** 4 **39.** ∞ **41.** $-\infty$ **43.** ∞ **45.** 0 **47.** $-\infty$ **49.** −7 **51.** a) −1/5 b) 0 c) −1/3   **53.**

55.

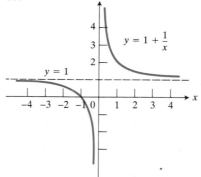

57. 2

Article 1.11, pp. 86–87

1. a) Yes b) Yes c) Yes d) Yes **3.** a) No b) No **5.** a) 0 b) 0 **7.** All x except 0, 1 **9.** All x in domain except 1, 2   **11.** All x except −1, 0, 1 **13.** $x = -2$ **15.** $x = 1, 3$ **17.** $x = -2, 5$ **19.** None **21.** $x = 0$ **23.** 6 **25.** 3/2

27. a)

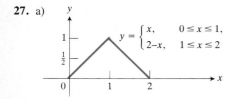

b) Yes c) No

29. 4/3 **31.** 1 **33.** 0 **35.** Maximum is 1, minimum is 0. **37.** No maximum; minimum is 0; because the interval is open.   **39.** $f(0) < 0 < f(1)$ implies existence of x such that $f(x) = 0$, by the Intermediate Value Theorem.

Miscellaneous Problems, Chapter 1, pp. 89–91

1. $A(u - h, v - k)$ **5.** a) $2x + 3y = -7$ b) $\sqrt{13}$ **7.** $x = 0$; $3x + 4y = 0$ **9.** a) $-A/B$ b) C/B c) C/A d) $Bx - Ay = 0$ **13.** 4 circles; $C(-1 - \sqrt{5}, 1)$, $r = (\sqrt{2} + \sqrt{10})/2$; $C(0, -2 - \sqrt{5})$, $r = (3\sqrt{2} + \sqrt{10})/2$; $C(0, -2 + \sqrt{5})$, $r = (3\sqrt{2} - \sqrt{10})/2$; $C(\sqrt{5} - 1, 1)$, $r = (\sqrt{10} - \sqrt{2})/2$. **15.** If $L_1 \not\parallel L_2$, all lines through the point of intersection of L_1, and L_2; if $L_1 \parallel L_2$, all lines parallel to L_1 and L_2. **17.** 2 lines; $x + 5y = 10\sqrt{2}$, $x + 5y = -10\sqrt{2}$ **19.** $A = \pi r^2$; $C = 2\pi r$; $A = C^2/4\pi$ **21.** $x \neq -1$; $y \neq 0$ **23.** $x \geq 0$; $0 < y \leq 1$ **25.** $ad + b = bc + d$ **27.** a) $1/(1 - x)$ b) $x/(x + 1)$ c) x d) $1 - x$
29. **31.** $m(a, b) = (a + b)/2 - |a - b|/2$ **33.** a) b)

35. a) $2/(x + 1)^2$ b) $(3/2)x^{1/2}$ c) $-1/(3x^{2/3})$ **37.** a) 6 b) $(0, 2)$ **39.** $dy/dx = 180 - 32x$; $(45/8, 2025/4)$ **41.** $t = 1$, $s = 16$
43. 0 **45.** 0 **47.** 1/3 **49.** $2a$ **51.** $2x$ **53.** $-1/x^2$ **55.** 1/2 **57.** Does not exist **59.** 1 **61.** -1 **63.** 0 **65.** 0 **67.** a) 0
b) $-1/3$ c) $f(-1/x) = (-x^2 - x)/(5x^2 + 7x + 2)$, $x \neq 0$; $f(0) = -1/5$; $1/f(x) = 2x - 5$, $x \neq 1$, 5/2 **69.** a) 0 b) 1/2 **71.** $M = 1 + (1/\epsilon)$ **75.** No **77.** Every integer and $0 < x < 1$ **79.** Any $\delta \leq \epsilon$ **87.** $\delta = \sqrt{4 + \epsilon} - 2$; (b) $\delta \to 0^+$ **89.** For $\epsilon = 4$, all $x \neq 5$; for $\epsilon \leq 2$, all $x > 5$

CHAPTER 2

Article 2.1, pp. 99–100

1. $y' = 1$, $y'' = 0$ **3.** $y' = 2x$, $y'' = 2$ **5.** $y' = -2x$, $y'' = -2$ **7.** $y' = 2$, $y'' = 0$ **9.** $y' = x^2 + x + 1$, $y'' = 2x + 1$ **11.** $v = 32t$, $a = 32$ **13.** $v = 32t - 60$, $a = 32$ **15.** $v = gt + v_0$, $a = g$ **17.** $y' = 15x^2 - 15x^4$, $y'' = 30x - 60x^3$ **19.** $y' = x^3 - x^2 + x - 1$, $y'' = 3x^2 - 2x + 1$ **21.** $y' = 2x^3 - 3x - 1$, $y'' = 6x^2 - 3$ **23.** $y' = 5x^4 - 2x$, $y'' = 20x^3 - 2$ **25.** $y' = 12x + 13$, $y'' = 12$ **27.** c **29.** $x + 9y = 29$ **31.** $y = 4x - 2$; $y = 4x + 2$; smallest slope $= 1$ at $x = 0$ **33.** $(0, 16)$ and $(-4/3, 0)$ **35.** $a = 1$, $b = 1$, $c = 0$ **37.** Mars: about 4.46 sec; Jupiter: about 0.73 sec **39.** about 29.39 m **41.** $a(3) = 10$, $a(-1/3) = -10$ **43.** $t = x_1(1 - (1/n))$; draw line through the given point and $(t, 0)$

Article 2.2, pp. 109–110

1. $x^2 - x + 1$ **3.** $10x(x^2 + 1)^4$ **5.** $-2(x + 1)(x^2 + 1)^{-4}(2x^2 + 3x - 1)$ **7.** $-19/(3x - 2)^2$ **9.** $(1 + x^2)^{-2}(x^2 - 2x - 1)$
11. $-40/(2x - 3)^5$ **13.** $4x - 14$ **15.** $45(2x - 1)^2(x + 7)^{-4}$ **17.** $-30(2x^3 - 3x^2 + 6x)^{-6}(x^2 - x + 1)$ **19.** $2x^{-2} - 2x^{-3}$
21. $-12x^{-2} + 12x^{-4} - 12x^{-5}$ **23.** $(x + 2)^{-2}$ **25.** $(1 - t^2)/(t^2 + 1)^2$ **27.** $(2 - 4t)/(t^2 - t)^3$ **29.** $(2 - 6t^2)/(3t^2 + 1)^2$
31. $3(t^2 + 3t)^2(2t + 3)$ **33.** a) 13 b) -7 c) 7/25 d) 20 e) -225 f) -30 **35.** $4y - 3x = 4$ **37.** $8(3 - 2x)^{-3}$ **39.** $3x^2 - 1$
41. $-8x + 4x^3$ **43.** $0.09y$ **45.** a) $dP/dt = (a^2 k)/(akt + 1)^2$; b) $t = 0$, $dP/dt = a^2 k$

Article 2.3, pp. 116–117

1. $-x/y$ **3.** $(2x - y)/x$ **5.** $3x^2/2y$ **7.** $-\sqrt{y/x}$ **9.** $(1 - 2xy - 3x^2)/(x^2 + 1)$; also $(y/x) - (x + y)^2$ **11.** $(2x^2 + 1)/\sqrt{x^2 + 1}$
13. $(1 - 2y)/(2x + 2y - 1)$ **15.** $2x/[y(x^2 + 1)^2]$ **17.** $[5(3x + 7)^4]/2y^2$ **19.** $(-y/x)^2$ **21.** $(2x - 1)/2y$
23. $-(x^2 + 9)/[3x^2(x^2 + 3)^{2/3}]$ **25.** $(6y - x^2)/(y^2 - 6x)$ **27.** $-(2/5)(2x + 5)^{-6/5}$ **29.** $-(1/4)(x - x^{3/2})^{-1/2}$ **31.** $(-4, 0)$ and $(0, 1)$ **33.** $(2/3)a^{-1/3}$; $a \neq 0$, in order for db/da to exist. **35.** $-x/y$; $-1/y^3$ **37.** $-(y/x)^{1/3}$; $(1/3)x^{-4/3}y^{-1/3}$ **39.** $(y + 1)^{-1}$; $-(y + 1)^{-3}$ **41.** $(1 + y^{-1/2})^{-1}$; $(1/2)(\sqrt{y} + 1)^{-3}$ **43.** a) $4y - 7x = -2$ b) $7y + 4x = 29$ **45.** a) $y - 3x = 6$ b) $3y + x = 8$
47. a) $y + x = -1$ b) $y - x = 3$ **49.** $y + 2x = 3$, $y + 2x = -3$ **51.** $(-13/4, 17/16)$ **53.** $(\pm\sqrt{7}, 0)$, $m = -2$ **55.** $v = 2/5$ m/sec; $a = -4/125$ m/sec^2

Article 2.4, pp. 126–127

1. $4x - 3$; 1.04 **3.** $-(1/4)x + 1$; 0.475 **5.** $2x - 2$; 0.2 **7.** $10x - 13$; 6 **9.** $(1/4)x + 1$; 2.025 **11.** $-(4/5)x + 9/5$; 5.16
13. $1 + 2x$ **15.** $1 - 5x$ **17.** $2 + 8x$ **19.** $2 + x$ **21.** $1 - x$ **23.** a) 1.02 b) 1.003 c) 1.001 **25.** a) 0.21 b) 0.2 c) 0.01
27. a) 0.231 b) 0.2 c) 0.031 **29.** a) $-0.\overline{3}$ b) -0.4 c) $0.0\overline{6}$ **31.** $4\pi r^2 \, dr$ **33.** $3x^2 \, dx$ **35.** $2\pi hr \, dr$ **37.** $(2/3)\pi hr \, dr$
39. a) 0.08π b) 2% **41.** 3% **43.** Within 1% **45.** Within 1/3% **47.** Within 1% **49.** Within 0.05% **53.** b) $2 + 2x$, $1 + (1/2)x$, $x = 0.04$ c) $f(0.04) \approx 0.003$ **55.** 14% the speed of light

Article 2.5, pp. 133–134

1. $8t - 20$ **3.** $-t^2$ **5.** $-1/9$ **7.** $-t^{-2}(2t^{-1} + 2)^{-1/2}$ **9.** $8t + 10$ **11.** $18x + (1/3x^2)$ **13.** $(16t^2 - 20t + 1)^{-1/2}(16t - 10)$
15. $(1 - 2v)/[v^2(v - 1)^2]$ **17.** $24t$ **19.** $-(1/4t^2)$ **21.** 1 **23.** $\sqrt{2}/4$ **25.** 0 **27.** $dv/dt = (1/2)k^2$ **29.** $kT/2$

Article 2.6, pp. 142–143

1.

3.

5.

7.

9.

11.

13.

15.

17.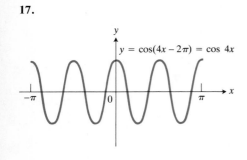

19.

21. a) 37 b) 365 c) 101 d) 25
23. 0 **25.** 0 **27.** $\cos^2 A + \sin^2 A = 1$

Article 2.7, pp. 149–150

1. $\cos(x + 1)$ **3.** $\frac{1}{2}\cos(x/2)$ **5.** $-5 \sin 5x$ **7.** $-2 \sin 2x$ **9.** $3 \cos(3x + 4)$ **11.** $x \cos x + \sin x$ **13.** $x \cos x$ **15.** $\tan x \sec x$
17. $\tan(x - 1)\sec(x - 1)$ **19.** $-\tan(1 - x)\sec(1 - x)$ **21.** $2 \sec^2 2x$ **23.** $2 \sin x \cos x = \sin 2x$ **25.** $-10 \sin 5x \cos 5x =$
$-5 \sin 10x$ **27.** $5 \sec^2(5x - 1)$ **29.** $2 \cos 2x$ **31.** $(-\sin 2x)/\sqrt{2 + \cos 2x} = (-\sin 2x)/y$ **33.** $-(\sin \sqrt{x})/2\sqrt{x}$ **35.** $y =$
$\sqrt{\cos^2 x} = |\cos x|$. Therefore $y' = -\sin x$ when $\cos x > 0$ and $y' = -\sin x$ when $\cos x < 0$. Alternatively, $y' =$
$(-\sin 2x)/\sqrt{2(1 + \cos 2x)}$. Note that in either case, y' does not exist for $x = (2k + 1)(\pi/2)$, k an integer. **37.** $\cos^2 y$
39. $(-\sin 8x)/y$ **41.** $-(1/x)[y + \cos^2(xy)]$ **43.** $y = 2x$ **45.** 1 **47.** 1 **49.** 1 **51.** $-\sin a$

53.

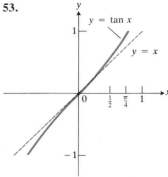

55. $b = 1$; f is continuous at $x = 0$, but $f'(0)$ does not exist.   **57.** $1 + (3/2)x$; their sum

Article 2.8, p. 154

1. $x^2 + y^2 = 1$; begins at $(1, 0)$ and travels in a counterclockwise direction to end at same point.   **3.** Same as Problem 1.
5. $x^2 + y^2 = 9$; begins and ends at $(3, 0)$; travels in a counterclockwise direction.   **7.** $x + y = 1$; from $P(1, 0)$ to $Q(0, 1)$
9. a) $x = 2 \cos t$, $y = -2 \sin t$, $0 \le t \le 2\pi$ b) $x = 2 \cos t$, $y = 2 \sin t$, $0 \le t \le 2\pi$

11.

13.

15.

17.

19.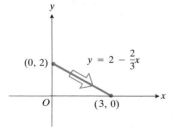

21. $y = 1 + (x/2)$; $dy/dt = 1$, $dx/dt = 2$, $dy/dx = 1/2$ **23.** $x^2 + y^2 = 25$; $dy/dt = 5 \cos t$, $dx/dt = -5 \sin t$, $dy/dx = -x/y = -\cot t$ **25.** $y = \cos x$; $dy/dt = 2t \cos(t^2)$, $dx/dt = 2t$, $dy/dx = \cos[(\pi/2) + x] = -\sin x$ **27.** $x^2 + (y - 1)^2 = 1$; $dy/dt = \cos t$; $dx/dt = -\sin t$; $dy/dx = -x/(y - 1) = -\cos t/\sin t$ **29.** c **31.** a) $5/3$ b) $3y - 5x = -8$ **33.** a) $1/11$ b) $11y - x = 16$
35. a) -2 b) $2y + 4x = 9$ **37.** 2   **39.** -3 **41.** 0   **43.** $-(1/4)t^{-3/2}$ **45.** $2t^{-3}$ **47.** $-1/\sin^3 t$ **49.** $-1/200$ **51.** 3

Article 2.9, pp. 158–159

1. 0.618   **3.** 1.164   **5.** −1.189   **7.** They all equal x_0.   **9.** 1.179509

11. b)

$$y = x^3 - 3x - 1,$$
$$-2 \le x \le 2$$

c) 1.87939   d) −0.34730, −1.53209

13. 0.73909   **15.** 0.630115, 2.573272

Article 2.10, p. 161

1. $(3x^2 - 3)\, dx$ **3.** $(2 - 2x^2)\, dx/(1 + x^2)^2$ **5.** $[(1 - y)/(1 + x)]\, dx$ **7.** $5 \cos 5x\, dx$ **9.** $2 \sec^2(x/2)\, dx$
11. $\csc[1 - (x/3)]\cot[1 - (x/3)]\, dx$ **13.** $dx = dt,\ dy = (1 + t)\, dt,\ dy/dx = 1 + t$
15. $dx = -\sin t\, dt,\ dy = \cos t\, dt,\ dy/dx = -\cot t$

Miscellaneous Problems, Chapter 2, pp. 162–165

1. $-4(x^2 - 4)^{-3/2}$ **3.** $-y/(x + 2y)$ **5.** $-(2xy + y^2)/(2xy + x^2)$ **7.** $2 \sin(1 - 2x)$ **9.** $(x + 1)^{-2}$ **11.** $(3x^3 - 2xa^2)/\sqrt{x^2 - a^2}$
13. $2x/(1 - x^2)^2$ **15.** $10 \sec^2 5x \tan 5x$ **17.** $[(4x + 5)\sqrt{2x^2 + 5x}]/2$ **19.** $(-2y^2\sqrt{xy} - y)/(4xy\sqrt{xy} + x)$ **21.** $-\sqrt[3]{y/x}$
23. $-y/x$ **25.** $-(x + 2y + y^2)/(2x + 4y + 2xy)$ **27.** $1/[2y(x + 1)^2]$ **29.** $-(y + 2)/(x + 3)$ **31.** $(y - 3x^2)/(3y^2 - x)$
33. $(1 + x)^{-1/2}(1 - x)^{-3/2}$ **35.** $x^2(x^3 + 1)^{-2/3}$ **37.** $-2 \csc^2 2x$ **39.** $(\cos^2 x + 2 \sin^2 x)/\cos^3 x$ **41.** $-8x^2\sin 8x + 2x \cos 8x$
43. $(1 + \cos x)^{-1}$ **45.** $-\csc x \cot x$ **47.** $-\sin 2x \sin(\sin^2 x)$ **49.** $2 \sec^2 x \tan x$ **51.** $-3 \sin 6x \sin(\sin^2 3x)$
53. $2(1 + t)/\sqrt{2t + t^2}$; also $(4x + 8)/\sqrt{4x^2 + 16x + 15}$ **55.** $2x + [2x(1 - x^2)]/(1 + x^2)^3$ **57.** $3t^2/(2t + 1)$; also
$[3(y + 1)^{2/3}]/[2(y + 1)^{1/3} + 1]$ **59.** 1, $y = x$ **61.** 2, $y - 2x = -5$ **63.** $\pi(20x - x^2)$ **65.** 56 ft/sec   **67.** $-12 + 18x$, $f(x) =$
$(2 - 3x)^2$ **69.** $-8/5$ **71.** $-3/32$ **73.** $x + y = 3$ **75.** a) $3(2x - 1)^{-5/2}$ b) $-162(3x + 2)^{-4}$ c) $6a$ **79.** $h + 2k = 5$; $h = -4$,
$k = 9/2$, $r = 5\sqrt{5}/2$ **81.** $5y + 8x = 18$; $5y - 3x = -18$ **83.** 3 ft   **85.** $dy = 0.9$, $\Delta y = 0.92$ **87.** 30
89. $2/[3\sqrt{x^2 + x}(2x + 1)(2y + 1)]$ **91.** $[3/(x + 1)^2] \sin[(2x - 1)/(x + 1)]^2$ **93.** $dy/dt = [-x(x - 1)^2]/\sqrt{x^2 + 16}$; $-12/5$
95. $(\sin z + \sin y + xy \cos y \cos z - 2y^2 \cos z)/[\sin y\, (\sin z + \sin y)]$ **99.** Length of a side $= 2r \sin(\pi/n)$; $2\pi r$, yes
101. $-2 \cos 3t$; -2 **103.** 2, -2 **109.** All values of m and b such that $m = -b/\pi$; $m = -1$, $b = \pi$
111. b) $2x \sin(1/x) - \cos(1/x)$ c) No, $\lim_{x\to 0} \cos(1/x)$ does not exist.

CHAPTER 3

Article 3.1, pp. 171–172

1. $y' = 2x - 1$, decreasing on $(-\infty, 1/2)$, increasing on $(1/2, \infty)$; $y(1/2) = 3/4$ is minimum.   **3.** $y' = x^2 - x - 2$, increasing
on $(-\infty, -1) \cup (2, \infty)$, decreasing on $(-1, 2)$; $y(-1) = 3/2$ is local maximum, $y(2) = -3$ is local minimum   **5.** $y' = 3x^2 - 27$,
increasing on $(-\infty, -3) \cup (3, \infty)$, decreasing on $(-3, 3)$; $y(-3) = 90$ is local maximum, $y(3) = -18$ is local minimum.
7. $y' = 6x - 6x^2$, decreasing on $(-\infty, 0) \cup (1, \infty)$, increasing on $(0, 1)$; $y(0) = 0$ is local minimum, $y(1) = 1$ is local maximum.
9. $y' = 4x^3$, decreasing on $(-\infty, 0)$, increasing on $(0, \infty)$; $y(0) = 0$ is minimum.

11. $y' = -3x^{-4}$, decreasing for all $x \neq 0$, no extreme values

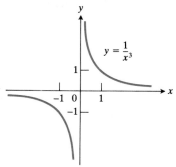

13. $y' = -2(x + 1)^{-3}$, increasing on $(-\infty, -1)$, decreasing on $(-1, \infty)$, no extreme values

15. $y' = -\sin x$, increasing on $(-\pi, 0) \cup (\pi, 3\pi/2]$, decreasing on $[-3\pi/2, -\pi) \cup (0, \pi)$; $y(0) = 1$ is maximum, $y(-\pi) = y(\pi) = -1$ is minimum.

17. $y' = 2x$ for $x > 0$, $-2x$ for $x < 0$, always increasing, no extreme values

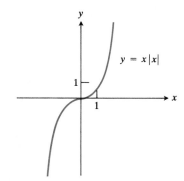

19. Maximum is 1 at $x = 0$, minimum is -1 at $x = -\pi$ or π.

21. $y' = (x + 1)^{-2}$ is always positive for $x \neq -1$.

23. $f(x) = \begin{cases} 1 - x & \text{for } x \leq 0 \\ 1 - 2x & \text{for } x > 0 \end{cases}$ (This answer is not unique.)

25. a) $y' < 0$ if $0 \leq x < \pi/3$, $y'(\pi/3) = 0$, $y' > 0$ if $\pi/3 < x \leq \pi$ b)

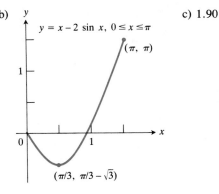

c) 1.90

Article 3.2, pp. 176–177

1. a) $(2, \infty)$ b) $(-\infty, 2)$ c) Always d) Never. Minimum at $(2, -1)$, no inflection points. **3.** a) Always b) Never c) $(0, \infty)$ d) $(-\infty, 0)$. No extreme points, $(0, 0)$ is inflection point. **5.** a) $(-\infty, -1) \cup (1, \infty)$ b) $(-1, 1)$ c) $(0, \infty)$ d) $(-\infty, 0)$. Local maximum at $(-1, 5)$, local minimum at $(1, 1)$, $(0, 3)$ is inflection point. **7.** a) $(-\infty, 1) \cup (3, \infty)$ b) $(1, 3)$ c) $(2, \infty)$ d) $(-\infty, 2)$. Local maximum at $(1, 5)$, local minimum at $(3, 1)$, $(2, 3)$ is inflection point. **9.** a) Always b) Never c) $(2, \infty)$ d) $(-\infty, 2)$. No extreme values, $(2, 1)$ is inflection point. **11.** a) $(-\pi/2, \pi/2)$ b) Never c) $(0, \pi/2)$ d) $(-\pi/2, 0)$. No extreme values, $(0, 0)$ is inflection point. **13.** a) $(-\infty, 0)$ b) $(0, \infty)$ c) Never d) Always. Maximum at $(0, 0)$. **15.** $y = (x - 1)^2$. This answer is not unique. **17.** Maximum at $(-1, 7)$. **19.** Local maximum at $(1, 16)$, local minimum at $(3, 0)$, inflection point at $(2, 8)$ **21.** Local maximum at $(-2, 28)$, local minimum at $(2, -4)$, inflection point at $(0, 12)$ **23.** Inflection point at $(-1, 1)$ **25.** Local maximum at $(-5, 400)$, local minimum at $(9, -972)$, inflection point at $(2, -286)$

27. Local maximum at $(0, 2)$, minimum values at $(-1, 1)$ and $(1, 1)$, inflection points at $(-1/\sqrt{3}, 13/9)$ and $(1/\sqrt{3}, 13/9)$.

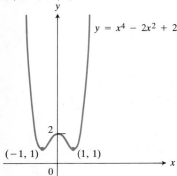

$y = x^4 - 2x^2 + 2$

$(-1, 1)$ $(1, 1)$

29. Minimum at $(1, -1)$, inflection points at $(0, 0)$ and $(2/3, -16/27)$

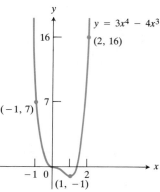

$y = 3x^4 - 4x^3$

$(2, 16)$

$(-1, 7)$

$(1, -1)$

31. Maximum of $\sqrt{2}$ at $x = \pi/4, -7\pi/4, \ldots$; minimum of $-\sqrt{2}$ at $x = 5\pi/4, -3\pi/4, \ldots$; inflection points at all x-intercepts

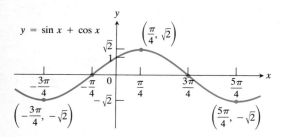

$y = \sin x + \cos x$

$\left(\dfrac{\pi}{4}, \sqrt{2}\right)$

$\left(-\dfrac{3\pi}{4}, -\sqrt{2}\right)$ $\left(\dfrac{5\pi}{4}, -\sqrt{2}\right)$

33. a) T b) P **35.**

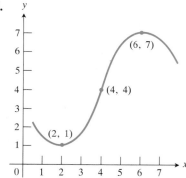

$(6, 7)$

$(4, 4)$

$(2, 1)$

37. Concave down.

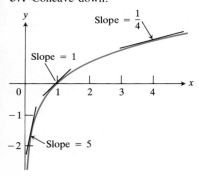

Slope $= \dfrac{1}{4}$

Slope $= 1$

Slope $= 5$

39. a) 3 times, once each in the intervals $(-2, -1)$, $(-1/2, 0)$, and $(1/3, 1)$ b) Once, in the interval $(-2, -1)$ c) Once, in the interval $(1, 2)$

41. Local minimum at $x = -15$, $y = -37,125$, and $x = 9$, $y = -9,477$; local maximum at $x = 0$, $y = 0$; $(-9, -21,141)$ and $(5, -5,125)$ are inflection points.

43. a) $x = 2$ b) $x = 4$

Article 3.3, pp. 184–185

1. Odd **3.** Odd **5.** Neither **7.** Even **9.** Even **11.** Odd
13. a) None b) $(0, -1/3)$ c) $x = 3/2$; $y = 0$ d) $-2/9$ e) Decreases for all $x \neq 3/2$
f) Concave down on $(-\infty, 3/2)$, concave up on $(3/2, \infty)$

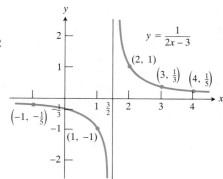

$y = \dfrac{1}{2x - 3}$

$(2, 1)$

$\left(3, \dfrac{1}{3}\right)$ $\left(4, \dfrac{1}{5}\right)$

$\left(-1, -\dfrac{1}{5}\right)$

$(1, -1)$

15. a) Odd function, symmetrical to origin b) $(\pm 1, 0)$
c) $x = 0$, $y = x$ d) 2 e) Rises for all $x \neq 0$ f) Concave up
on $(-\infty, 0)$, concave down on $(0, \infty)$ g) $y \approx x$ for large x,
$y \approx -1/x$ for small x

17. a) None b) $(0, 3/2)$, $(-3, 0)$ c) $x = -2$, $y = 1$
d) -1, $-1/4$ e) Falls for all $x \neq -2$ f) Concave down on
$(-\infty, -2)$, concave up on $(-2, \infty)$ g) $y \approx 1$ for large x,
$y \approx 1/(x + 2)$ near $x = -2$

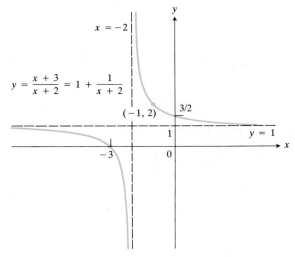

19. a) None b) $(0, -1)$, $(-1, 0)$ c) $x = 1$, $y = 1$
d) -2, $-1/2$ e) Falls for all $x \neq 1$ f) Concave down on
$(-\infty, 1)$, concave up on $(1, \infty)$ g) $y \approx 1$ for large x, $y \approx$
$2/(x - 1)$ near $x = 1$

21. a) y-axis b) $(0, 1)$ c) $y = 0$ d) 0 e) Falling on $(0, \infty)$,
rising on $(-\infty, 0)$ f) Concave down on $(-1/\sqrt{3}, 1/\sqrt{3})$,
concave up on $(-\infty, -1/\sqrt{3}) \cup (1/\sqrt{3}, \infty)$

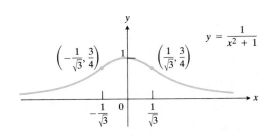

23. a) Origin b) $(0, 0)$ c) $x = \pm 1$, $y = 0$ d) -1 e) Always
falling for $x \neq \pm 1$ f) Concave down on $(-\infty, -1) \cup (0, 1)$,
concave up on $(-1, 0) \cup (1, \infty)$ g) $y \approx 1/(2(x - 1))$ near
$x = 1$, $y \approx 1/(2(x + 1))$ near $x = -1$

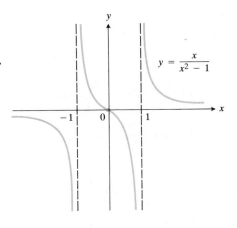

25. a) Even b) (0, 0) c) $x = \pm 1$, $y = 1$ d) 0 e) Rising on $(-\infty, -1) \cup (-1, 0)$, falling on $(0, 1) \cup (1, \infty)$ f) Concave down on $(-1, 1)$, concave up on $(-\infty, -1) \cup (1, \infty)$ g) $y \approx 1$ for x large, $y \approx 1/(2(x - 1))$ near $x = 1$, $y \approx 1/(2(x + 1))$ near $x = -1$

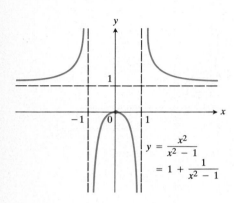

$$y = \frac{x^2}{x^2 - 1} = 1 + \frac{1}{x^2 - 1}$$

27. a) None b) (0, 4), $(\pm 2, 0)$ c) $x = 1$, $y = x + 1$ d) $y'(0) = 4$, $y'(-2) = 4/3$, $y'(2) = 4$ e) Always rising f) Concave up on $(-\infty, 1)$, concave down on $(1, \infty)$ g) $y \approx x$ (or $x + 1$) for x large, $y \approx -3/(x - 1)$ near $x = 1$

$$y = \frac{x^2 - 4}{x - 1} = x + 1 - \frac{3}{x - 1}$$

$$y = x + 1$$

29. a) None b) (0, $-1/4$), $(\pm 1, 0)$ c) $x = -2$, $y = 1/2x - 1$ d) $y'(1) = 1/3$, $y'(-1) = -1$, $y'(0) = 1/8$ e) Rising on $(-\infty, -2 - \sqrt{3}) \cup (-2 + \sqrt{3}, \infty)$, falling on $(-2 - \sqrt{3}, -2 + \sqrt{3})$ f) Concave up on $(-2, \infty)$, concave down on $(-\infty, -2)$ g) $y \approx (x/2) - 1$ (or $x/2$) for x large, $y \approx 3/(2x + 4)$ near $x = -2$

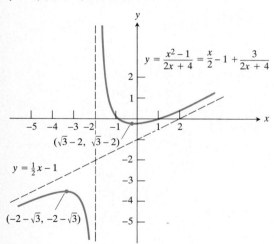

$$y = \frac{x^2 - 1}{2x + 4} = \frac{x}{2} - 1 + \frac{3}{2x + 4}$$

$(\sqrt{3} - 2, \sqrt{3} - 2)$

$y = \frac{1}{2}x - 1$

$(-2 - \sqrt{3}, -2 - \sqrt{3})$

31. a) None b) $(-\sqrt[3]{1/3}, 0)$ c) $x = 0$ d) $-6\sqrt[3]{9}$ e) Falling on $(-\infty, 0) \cup (0, \sqrt[3]{1/6})$, rising on $(\sqrt[3]{1/6}, \infty)$ f) Concave up on $(-\infty, -\sqrt[3]{1/3}) \cup (0, \infty)$, concave down on $(-\sqrt[3]{1/3}, 0)$ g) $y \approx 6x^2$ as $x \to \pm \infty$, $y \approx 2/x$ as $x \to 0$

$$y = \frac{2}{x} + 6x^2$$

$y = 6x^2$

$y = \frac{2}{x}$

$\left(-\frac{1}{\sqrt[3]{3}}, 0\right)$

33. a) To y-axis b) (0, 4), $(\pm 2, 0)$ c) $x = \pm 1$, $y = 1$ d) $y'(0) = 0$, $y'(\pm 2) = \pm 4/3$ e) Falling on $(-\infty, 0)$, rising on $(0, \infty)$ f) Concave down on $(-\infty, -1) \cup (1, \infty)$, concave up on $(-1, 1)$ g) $y \approx 1$ for x large, $y \approx -3/(2(x-1))$ near $x = 1$, $y \approx 3/(2(x + 1))$ near $x = -1$

$$y = \frac{x^2 - 4}{x^2 - 1} = 1 - \frac{3}{x^2 - 1}$$

35. a) None b) $(0, 1/3)$ c) $x = 1$, $x = 3$, $y = 1$ d) $4/9$
e) Rising on $((1 - \sqrt{5})/2, (1 + \sqrt{5})/2)$, falling on
$(-\infty, (1 - \sqrt{5})/2) \cup ((1 + \sqrt{5})/2, \infty)$ f) Concave down
on $(-\infty, -1) \cup (1, 3)$, concave up on $(-1, 1) \cup (3, \infty)$
g) $y \approx 1$ for x large, $y \approx 2/((x-1)(-2)) = 1/(1 - x)$ near
$x = 1$, and $y \approx 10/(2(x - 3)) = 5/(x - 3)$ near $x = 3$

37. a) None b) $(1, 0)$ c) $x = 0$, $x = 2$, $y = 0$ d) $y'(1) =$
-1 e) Rising on $(-\infty, 0)$, falling on $(0, 2) \cup (2, \infty)$
f) Concave up on $(-\infty, 0) \cup (0, 1.22)$ and $(2, \infty)$, concave
down on $(1.22, 2)$ g) $y \approx 1/(2x^2)$ near $x = 0$, $y \approx 1/(4(x - 2))$
near $x = 2$

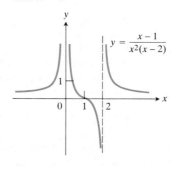

$$y = \frac{x - 1}{x^2(x - 2)}$$

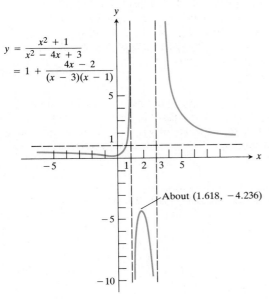

$$y = \frac{x^2 + 1}{x^2 - 4x + 3}$$
$$= 1 + \frac{4x - 2}{(x - 3)(x - 1)}$$

About $(1.618, -4.236)$

41. Odd **43.** Even **45.** Odd **47.** Odd **49.** Even **51.** Odd **53.** Even **55.** Neither **57.** a) Even b) Even **59.** a) Odd
b) Odd c) Even d) Even **61.** a) Increasing b) Decreasing c) Concave down d) Concave down

Article 3.4, p. 191

1. Absolute maximum = $1/4$ at $x = 1/2$; absolute minimum = 0 at $x = 0, 1$ **3.** Absolute maximum = $2\sqrt{3}/9$ at $x = 1/\sqrt{3}$;
absolute minimum = 0 at $x = 0, 1$ **5.** Local maximum = 686 at $x = -7$; local minimum = -686 at $x = 7$, no absolute extrema
7. Absolute minimum = -1 at $x = 2$ **9.** Absolute maximum = $1/4$ at $x = 1/2$ **11.** No critical points; Absolute minimum = 0
at $x = 0$ and absolute maximum = 6 at $x = 3$ **13.** Absolute minimum = 3 at $x = 1$ **15.** No extreme values **17.** Absolute
maximum = $1/4$ at $x = 1/4$; no absolute minimum **19.** Absolute minimum = -3 at $x = 1$; absolute maximum = 8 at $x = 2$
21. Absolute minimum = 0; no maximum **23.** Absolute minimum = 1 at $x = 0, \pi/2$; absolute maximum = $3/2$ at $x = \pi/6$
25. Local maximum = 0 at $x = 0$; local minimum = -9477 at $x = -9$; absolute minimum = $-37,125$ at $x = 15$ **27.** Absolute
minimum = 4 at $x = 1/2$, no maximum value **29.** Local maximum = 1 at $x = 1$; local minimum = 0 at $x = 0$ **31.** No extreme
values **33.** Absolute maximum = 1 at $x = \pm\pi/2$; absolute minimum = -1 at $x = \pm3\pi/2$ **35.** Absolute maximum = 3 at $x =$
2; absolute minimum = 0 at $x = \pm1$
39. Critical points $x = \pm1$; inflection points are $(0, 0)$ and $(\pm\sqrt{3}, \pm\sqrt{3}/4)$; $y = 0$ is asymptote.
41. b **43.** 4

$$y = \frac{x}{x^2 + 1}$$

$(1, \tfrac{1}{2})$

$(-1, -\tfrac{1}{2})$

Article 3.5, pp. 199–203

1. a) 0 and 20 b) 12 and 8 c) $1/4$ and $19\frac{3}{4}$ **3.** 16 **5.** $(1/2, \sqrt{3}/2)$ **7.** 32 **9.** 80,000 m² **11.** 6 ft × 6 ft × 3 ft **13.** $a =$
$b = 10\sqrt{2}$ **15.** a) 4, $8/5$ b) $8/5 < t < 4$ c) $v(11/5) = -2187/125$ **17.** 18 in. × 9 in. **19.** 12 m × 18 m; 72 **21.** $r = h =$
$10/\sqrt[3]{\pi}$ cm **23.** about \$48.28 **27.** Radius = $\sqrt{2/3}H$, height = $H/\sqrt{3}$ **29.** a) 16 b) -54 c) -1 **31.** a) Cut a piece =
$4L/(3\sqrt{3} + 4)$ for the square. b) Maximum = $L^2/16$ occurs if all of the area is in the square. **33.** $32\pi r^3/81$ **37.** $(\sqrt{3}/2)D \times$
$(1/2)D$ (D = diameter) **39.** Ratio of diameter of semicircle to height of rectangle is $8/(4 + \pi)$

41. $r = h = \sqrt{S/5\pi}$ **43.** a) No, only one critical point at $x = 1/2$, which gives absolute minimum of $3/4$. b) No, absolute minimum is 0 at $(2n - 1)\pi$ (n an integer).   **45.** Maximum $= \sqrt{2}$ at $\pi/4 + 2n\pi$; minimum $= -\sqrt{2}$ at $5\pi/4 + 2n\pi$ (n an integer)   **47.** $r = (3V/8\pi)^{1/3}$, $h = (3V/\pi)^{1/3}$ ($r =$ radius of hemisphere, $h =$ height of cylinder, $V =$ total volume)   **53.** b) $d^2y/dx^2 > d^2R/dx^2$ **55.** Maximum of $ka^2/4$ at $x = a/2$

Article 3.6, pp. 207–209

1. $dA/dt = 2\pi r(dr/dt)$ **3.** $dV/dt = 3s^2(ds/dt)$ **5.** a) 1 b) $-1/3$ c) $dR/dt = I^{-1}(dV/dt) - VI^{-2}(dI/dt)$ d) $3/2$ **7.** a) $x =$ distance from second base $= 60$ ft, $y =$ distance from third base $= 30\sqrt{13}$ ft b) $y^2 = x^2 + 90^2$ c) $y(dy/dt) = x(dx/dt)$ d) $-(32/\sqrt{13})$ ft/sec   **9.** $(119/3)$ ft²/sec; $t = (13\sqrt{2}/4)$ sec   **13.** $(25/9\pi)$ ft/min; $(200/3)$ ft²/min   **15.** 33.7 ft/sec   **17.** ± 3 m/sec; $dy/dx = \pm 2$ **19.** 1; $10/10610 \approx 0.00094$ **21.** 8 ft/sec, -3 ft/sec   **23.** 20 ft/sec   **25.** 62.5 mph   **27.** a) 12 knots at noon, 8 knots one hour later b) No, the minimum distance between ships was 6.5 miles.

Article 3.7, pp. 215–216

1. $f(-2) = 11$, $f(-1) = -1$ and $f'(x) = 4x^3 + 3 < 0$ on $(-2, -1)$ **3.** $f(-1) = 1$, $f(0) = -6$, $f'(x) = 6x^2 - 6x - 12 < 0$ on $(-1, 0)$ **5.** i) $y(\pm 2) = 0$, $y'(0) = 0$ ii) $y(-5) = y(-3) = 0$, $y'(-4) = 0$ iii) $y(-1) = y(2) = 0$, $y(2) = 0$, $y'(0) = y'(2) = 0$ iv) $y(0) = y(9) = y(24) = 0$, $y'(4) = y'(18) = 0$. Between every two zeroes of y there lies a zero of y'.   **7.** $1/2$ **9.** 1   **15.** No contradiction; the conditions in the Mean Value Theorem are sufficient but not necessary.   **17.** $y = [x]$ is not continuous on $[0, 1]$.   **23.** There are no others.

Article 3.8, pp. 221–222

1. $1/4$ **3.** $5/7$ **5.** 0   **7.** 5   **9.** 1   **11.** $\sqrt{2}$ **13.** 1   **15.** -1 **17.** $1/2$ **19.** 3   **21.** na **23.** $-1/2$ **25.** $\sqrt{10}$ **27.** b) 0 c)   **29.** a) 0 b) 1 c) ∞

$y = \sec x + \tan x$

(0, 1)

Article 3.9, pp. 228–229

1. x^2 **3.** $(1/2) - (1/2)x$ **5.** $(3/8) + (3/4)x - (1/8)x^2$ **7.** a) 0.00146 b) 0.005   **9.** a) x b) 0.00089   **11.** a) $1 - (1/2)(x - \pi/2)^2$ b) 0.000167   **13.** a) $(\pi/2) - x$ b) 0.000167   **15.** a) $|x| < 0.25$ b) $|x| < 0.06$

Miscellaneous Problems, Chapter 3, pp. 230–235

1. a) $x < 9/2$ b) $x > 9/2$ c) No x d) All x **3.** a) $x < 3$ b) $x > 3$ c) $0 < x < 2$ d) $x < 0$ or $x > 2$ **5.** a) $x > \sqrt[3]{2}$ b) $x < \sqrt[3]{2}$, $x \neq 0$ c) $x < -\sqrt[3]{4}$ or $x > 0$ d) $-\sqrt[3]{4} < x < 0$ **7.** a) $x < 0$ b) $x > 0$ c) $x \neq 0$ d) No x

$(\sqrt[3]{2}, 3\sqrt[3]{4})$

$y = x^2 + 4x^{-1}$

$-\sqrt[3]{4}$ 0

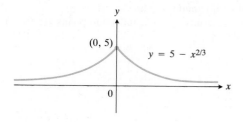

(0, 5)

$y = 5 - x^{2/3}$

9. a) $x \neq 0$ b) No x c) $x < 0$ d) $x > 0$ **11.** a) $x < -2b/a$ or $x > 0$ b) $-2b/a < x < 0$ c) $x > -b/a$ d) $x < -b/a$

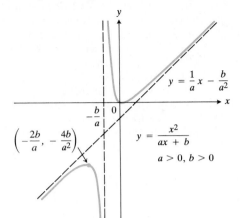

13. a) $x < -1$ or $x > 1/3$ b) $-1 < x < 1/3$ c) $x > -1/3$ d) $x < -1/3$ **15.** a) $x < 1$ or $x > 2$ b) $1 < x < 2$ c) $x > 3/2$
d) $x < 3/2$

17. Symmetric to x-axis, with domain $(0, 1]$. For $y > 0$, the graph decreases and is concave up for $0 < x < 3/4$, concave down for $3/4 < x \leq 1$. For $y < 0$, the graph increases and is concave down for $0 < x < 3/4$, concave up for $3/4 < x \leq 1$. Inflection points are $(3/4, \pm 1)$.

19. a) Domain is all x; there is no symmetry; intercepts are $(0, 1)$ and $(-1, 0)$; $y = 0$ is asymptote; $y' = (1 - 2x - x^2)/(x^2 + 1)^2$; $y'(0) = 1$, $y'(-1) = 1/2$; falling for $x < -1 - \sqrt{2}$ or $x > -1 + \sqrt{2}$, rising for $-1 - \sqrt{2} < x < -1 + \sqrt{2}$; $y'' = (2(x - 1)(x^2 + 4x + 1))/(x^2 + 1)^3$; concave down for $x < -2 - \sqrt{3}$ or $-2 + \sqrt{3} < x < 1$, concave up for $-2 - \sqrt{3} < x < -2 + \sqrt{3}$ or $x > 1$.

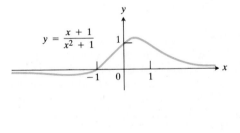

b) Domain is all $x \neq -1$; there is no symmetry; intercept is $(0, 1)$; asymptotes are $x = -1$ and $y = x - 1$; $y' = 1 - 2/(x + 1)^2$; $y'(0) = -1$; rising for $x < -1 - \sqrt{2}$ or $x > -1 + \sqrt{2}$, falling for $-1 - \sqrt{2} < x < -1 + \sqrt{2}$; $y'' = 4/(x + 1)^3$; concave down for $x < -1$, concave up for $x > -1$. Dominant terms: $y \approx x$ (or $x - 1$) for x large, $y \approx 2/(x + 1)$ near $x = -1$

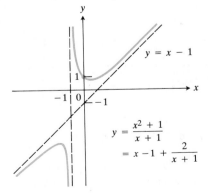

21. Domain is all x; no symmetry; intercepts are $(0, 0)$, $(-1, 0)$ and $(2, 0)$; no asymptotes; $y' = 3x^2 - 2x - 2$; $y'(0) = -2$, $y'(-1) = 3$, $y'(2) = 4$; rising for $x < (1 - \sqrt{7})/3$ or $x > (1 + \sqrt{7})/3$, falling for $(1 - \sqrt{7})/3 < x < (1 + \sqrt{7})/3$; concave up for $x > 1/3$, concave down for $x < 1/3$.   **23.** Domain is all $x \neq \pm 2$; symmetry to y-axis; asymptotes are $x = \pm 2$ and $y = 0$; intercept $(0, 2)$; $y' = 16x/(4 - x^2)^2$; $y'(0) = 0$; falling for $x < 0$, $x \neq -2$; rising for $x > 0$, $x \neq 2$; $y'' = (16(3x^2 + 4))/(4 - x^2)^3$; concave up for $-2 < x < 2$, concave down for $x < -2$ or $x > 2$. Dominant terms: $y \approx 2/(2 - x)$ near $x = 2$, $y \approx 2/(2 + x)$ near $x = -2$
25. Domain is all $x \neq \pm 1$; no symmetry; intercepts are $(0, 8)$ and $(2, 0)$; asymptotes are $y = 0$, $x = \pm 1$; $y' = (-4(x^2 - 4x + 1))/(x^2 - 1)^2$; $y'(0) = -4$, $y'(2) = 4/3$; falling for $x < 2 - \sqrt{3}$ or $x > 2 + \sqrt{3}$, rising for $2 - \sqrt{3} < x < 2 + \sqrt{3}$; $y'' = (8(x^3 - 4x^2 - 3x + 4))/(x^2 - 1)^3$; concave down for $x < -1$ or $1 < x < 4.5$, concave up for $-1 < x < 1$ or $x > 4.5$. Dominant terms: $y \approx -2/(x - 1)$ near $x = 1$, $y \approx 6/(x + 1)$ near $x = -1$

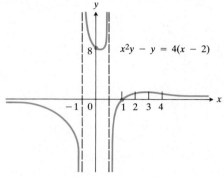

$x^2 y - y = 4(x - 2)$

27. $a = b = c = 1$, $d = e = 0$ **29.** a) $x = 1$ because y' changes from $+$ to $-$ b) $x = 3$ because y' changes from $-$ to $+$
31. 10   **33.** $a = 1$, $b = -3$, $c = -9$, $d = 5$ **35.** 25   **37.** $r = h = 4$ ft   **39.** $4\sqrt{3}$ **41.** a) $(2, \pm\sqrt{3})$ b) $(1, 0)$ c) $(1, 0)$
43. $x = aw/\sqrt{w^2 - l^2}$, $y = b - (al/\sqrt{w^2 - l^2})$ **45.** Let $b = $ fixed base. Then $a = c = s - (b/2)$.   **57.** 6 ft $\times$ 18 ft   **59.** 1 unit
61. $r = \sqrt{A}$, $\theta = 2$ rad   **63.** a) -0.04π cm^2/sec   b) $t = (3a - 5b)/(b^2 - a^2)$ **67.** $x = \dfrac{1}{n}\sum_{j=1}^{n} c_j$ **73.** $(1/4)$ cm/min
75. $dr/dt = -(3/400)$ m/min; $ds/dt = -(6\pi/5)$ m^2/min; $-(3/4000)$ m; $-(3\pi/25)$ m^2 **77.** $\sqrt{3}/2$ **79.** Yes, $dV/dt > 0$ for $0 \leq y \leq 10$ ($c = (6.5\pi/32) + 0.2$)   **83.** $5\sqrt{13}$ mi/hr   **85.** No, y' does not exist at $x = 0$.   **87.** $y' = 0$, so y is a constant function ($y \approx -0.7$).   **89.** 12   **93.** a) 1   b) 2   **97.** b) (i) $1/a$ (iii) $0 < x_0 < 2/a$

CHAPTER 4

Article 4.1, pp. 243–244
1. a) $x^2 + C$ b) $3x + C$ c) $x^2 + 3x + C$ **3.** a) $x^3 + C$ b) $(1/3)x^3 + C$ c) $(1/3)x^3 + x^2 + C$ **5.** a) $x^{-3} + C$ b) $-(1/3x^3) + C$ c) $-(1/3x^3) + x^2 + 3x + C$ **7.** a) $x^{3/2} + C$ b) $(8/3)x^{3/2} + C$ c) $(1/3)x^3 - (8/3)x^{3/2} + C$ **9.** a) $x^2 - x + C$ b) $(1/6)(2x - 1)^3 + C$ c) $(1/8)(2x - 1)^4 + C$ **11.** a) $(x^2 - 3)^2 + C$ b) $(1/4)(x^2 - 3)^2 + C$ **13.** $y = x^2 - 7x + C$
15. $y = (1/3)x^3 + x + C$ **17.** $y = 5x^{-1} + C$ **19.** $y = (x - 2)^5/5 + C$ **21.** $y = (x^2/2) - (1/x) + C$ **23.** $y = \sqrt{x^2 + C}$
25. $y = 1 + \sqrt{(x + 1)^2 + C}$ **27.** $y = ((1/3)x^{3/2} + C)^2$ **29.** $y = -1/(x^2 + C)$ **31.** $y = x^{-2} + C$ **33.** $s = t^3 + 2t^2 - 6t + C$
35. $y = (4t^3/3) + 4t - (1/t) + C$ **37.** 48 m/sec

Article 4.2, pp. 246–248
1. $s = t^3 + 4$ **3.** $s = (1/3)(t + 1)^3 - (1/3)$ **5.** $s = -(t + 1)^{-1} - 4$ **7.** $v = 9.8t + 20$; $s = 4.9t^2 + 20t$ **9.** $v = t^2 + 1$; $s = (1/3)t^3 + t + 1$ **11.** $v = (t + 1)^2$; $s = (1/3)t^3 + t^2 + t$ **13.** $y = x^3 + x^2 + x - 3$ **15.** $y = (x - 7)^4 + 9$ **17.** $y = ((1/4)x^2 + 1)^2$
19. $y = 1/(2 - x^2)$ **21.** $y = -x^3 + x^2 + 4x + 1$ **23.** $y = x^3/16$ **25.** $f(x) = 2x^{3/2} - 50$ **27.** 24.25 m/sec   **29.** $dy = x^2 + C$; $y(1) = 1$ implies that $C = 0$.

Article 4.3, pp. 252–253
1. $(1/244)(x - 1)^{244} + C$ **3.** $-2\sqrt{1 - x} + C$ **5.** $(1/6)(2x^2 - 1)^{3/2} + C$ **7.** $-(3/5)(2 - t)^{5/3} + C$ **9.** $x + (1/2)x^4 + (1/7)x^7 + C$ **11.** $(1/22)(x^2 + 1)^{11} + C$ **13.** $(2/3)(1 + x^3)^{1/2} + C$ **15.** $(-1/3(3x + 2)) + C$ **17.** $-3(1 - r^2)^{1/2} + C$
19. $-(1/20)(7 - x^5)^4 + C$ **21.** $(-1/2(s + 1)^2) + C$ **23.** $-1/(x + 2) + C$ **25.** $(1/3)(x + 1)^{3/2} + C$ **27.** $(1/6)(y + 2)^6 + C$
29. $(-2/(1 + \sqrt{x})) + C$ **31.** $y = (1/3)(1 + x^2)^{3/2} - (1/3)$ **33.** $r = (3z^2 - 1)^4 - 4$ **35.** $2(y^2 + 1)^{1/2} = (x^2 + 1)^{3/2} + 1$
37. a, c

Article 4.4, pp. 258–260
1. $-(1/3)\cos 3x + C$ **3.** $\tan(x + 2) + C$ **5.** $-\csc(x + \pi/2) + C$ **7.** $-(1/4)\cos(2x^2) + C$ **9.** $-(1/2)\cos 2t + C$
11. $(4/3)\sin 3y + C$ **13.** $(\sin^3 x)/3 + C$ **15.** $(1/2)\tan 2\theta + C$ **17.** $2\sec(x/2) + C$ **19.** $-\cot\theta + C$ **21.** $(1/2)y + (1/4)\sin 2y + C$ **23.** $(1/3)\sin 3t - (1/9)\sin^3 3t + C$ **25.** $-\csc x + C$ **27.** $\sqrt{2} - \cos 2t + C$ **29.** $-(1/2)\cos^3(2x/3) + C$
31. $\tan\theta + \sec\theta + C$ **33.** $\tan y - \cot y + C$ **35.** $-(3/2)\cos 2x + (4/3)\sin 3x + C$ **37.** $(1/3)\tan^3 x + C$ **39.** $-(1/4)\cot^4 x + C$

41. $(2/3)\tan^{3/2}x + C$ **43.** $(2/5)\sin^{5/2}x + C$ **45.** $\sin x - (1/3)\sin^3 x + C$ **47.** $\sin x - (2/3)\sin^3 x + (1/5)\sin^5 x + C$
49. $(1/6)\cos^{-3} 2x + C$ **51.** $2\tan\sqrt{x} - 2\sqrt{x} + C$ **53.** $(1/3)\sin\sqrt{3x^2 - 6} + C$ **55.** d   **57.** $y^2 = (5/2)x^2 + 3\cos x - 3$ or
$y = ((5/2)x^2 + 3\cos x - 3)^{1/2}$ **59.** $y = ((3/2)\sin\pi x - (1/2))^{2/3}$ **61.** 6 m   **63.** All 3 are correct; the antiderivatives differ by
a constant.   **65.** $(1/3)[1 + \sin^2(x - 1)]^{3/2} + C$

Article 4.5, pp. 270–271

1. a) 7/4 b) 9/4   **3.** a) $(\sqrt{2}\pi)/4 \approx 1.11$ **5.** a) $1 + \sqrt{2} + \sqrt{3} \approx 4.146$ b) $3 + \sqrt{2} + \sqrt{3} \approx 6.146$
                               b) $[\pi(\sqrt{2} + 2)]/4 \approx 2.68$

7. $\frac{1}{2} + 1 + 2 + 4 + 8$ **9.** 2.5   **11.** 1   **13.** 34   **15.** a, b, c, d   **17.** a) -8 b) 6 c) -5 d) 0

Article 4.6, pp. 276–277

3. $[m(b^2 - a^2)]/2$ **9.** 4/3   **11.** 1/3

Article 4.7, pp. 285–287

1. 12   **3.** 26/3   **5.** 1/15   **7.** 10   **9.** 1   **11.** 1/2   **13.** 2/3   **15.** 4/3   **17.** $\pi/6$ **19.** 5/6   **21.** 8   **23.** 8/3   **25.** 2   **27.** $\sqrt{2} - 1$
29. $\pi/2$ **31.** 2/9   **33.** 38/15   **35.** 0   **37.** 3/2   **39.** $\sqrt{3}/8$ **41.** 16   **45.** $\sqrt{1 + x^2}$ **47.** $-\sqrt{1 - x^2}$ **49.** $2\cos(4x^2)$
51. $-(\cos x)/(2 + \sin x)$ **53.** $\csc x$ **55.** $\cos x$ **57.** a) $2 + 10x$ b) $2 + 10x - 5x^2$ **59.** a) 1 b) 1/4 c) $(12)^{1/3}$

Article 4.8, pp. 289–290

1. a) 14/3 b) 2/3   **3.** a) 1/2 b) $-1/2$ **5.** a) $(1/2)(\sqrt{10} - 3)$ b) $(1/2)(3 - \sqrt{10})$ **7.** a) 45/8 b) $-45/8$ **9.** a) 1/6 b) 1/2
11. a) 0 b) 0   **13.** a) 0 b) $-(1/4)\sin(2\pi^2)$ **15.** $2\sqrt{3}$ **17.** 0   **19.** a)   b) $2\sqrt{3}$

21. $F(6) - F(2)$ **25.** $\displaystyle\int_0^1 x^2\, dx = \int_{-1}^0 (x + 1)^2\, dx = 1/3$ **27.** $\displaystyle\int_4^8 \sqrt{x - 4}\, dx = \int_{-1}^3 \sqrt{x + 1}\, dx = 16/3$

Article 4.9, pp. 299–301

1. a) 2 b) 2 c) 2   **3.** a) 17/4 b) 4 c) 4   **5.** a) 5.146 b) 5.252 c) 5.$\overline{3}$ **7.** a) 1/600 b) $1.\overline{3} \times 10^{-5}$ c) $|E_T| \le 1/96$,
$|E_S| \le 1/1920$ **9.** a) Any n b) Any even n **11.** a) $n \ge 283$ b) Any even n **13.** a) $n \ge 75$ b) $n \ge 12$ **15.** $n > 6$ **17.** 10.6 ft
19. a) 0.00664 b) 2.74 c) 0.24%

Miscellaneous Problems, Chapter 4, pp. 302–304

1. $y = -[2/(C + x^2)]$ **3.** $y^3 + 3y = x^3 - 3x + C$ **5.** $1/(r + 2) = -1/(3 - s) + C$ **7.** a) $y = (1/3)(x^2 - 4)^{3/2} + 3$ b) $y = 1/\sqrt{1 - x^2}$ c) $y = 4/(2 - x^4)$ d) $y = (15/2 - 3/2x)^{2/3} - 1$ **9.** $y = x^3 + 2x - 4$ **11.** a) $v = (2/3)t^{3/2} - 2t^{1/2} + (4/3)$ b) $s = (4/15)t^{5/2} - (4/3)t^{3/2} + (4/3)t - (4/15)$ **13.** -8 **15.** $y = (1/3)(1 + x^2)^{3/2} - (7/3)$ **17.** $y = 2((1/3)x^{3/2} + x^{1/2} - (1/3))^{1/2}$
19. $y = ((1/6)(9 + x^2)^{3/2} + (3/2))^2$ **21.** $s = -16t^2 + 96t$; 144 ft   **23.** a) 4 ft/sec b) 64/3 ft   **25.** 1/6   **27.** $\int_0^1 f(x)\, dx$ **29.** πr^2
31. 0   **33.** 6/5   **35.** 1/3   **37.** -2 **39.** $(3/5)(\sqrt[3]{7} - \sqrt[3]{2})$ **41.** $\sqrt{3}$ **43.** 35/4   **45.** a) $1/(x + \sqrt{x^2 + 1})$ b) $2f(2)$, assuming f
continuous   **49.** No

CHAPTER 5

Article 5.1, p. 308

1. a) $v(t) > 0$ for $0 \le t < 1/2$ or $3/2 < t \le 2$, $v(t) < 0$ for $1/2 < t < 3/2$ b) 20 c) 0 **3.** a) $v(t) > 0$ for $0 \le t < 5$, $v(t) < 0$ for $5 < t \le 10$ b) 245 c) 0 **5.** a) $v(t) > 0$ for $0 \le t < 1$, $v(t) < 0$ for $1 < t < 2$ b) 6 c) 4 **7.** a) $v(t) > 0$ for $0 < t < \pi/3$, $v(t) < 0$ for $\pi/3 < t \le \pi/2$ b) 6 c) 2 **9.** a) $s(t) = \sin t$ b) 4 c) 0 **11.** a) $s(t) = 5 \sin \pi t + 5$ b) 15 c) -5 **13.** 4 **15.** $2g$ **17.** a) Total = 2, net = 2 b) Total = 4, net = 0 c) Total = 4, net = 4 d) Total = 2, net = 2

Article 5.2, pp. 312–313

1. 32/3 **3.** 1/12 **5.** 32/3 **7.** 4/3 **9.** 9/2 **11.** 1/12 **13.** 9/2 **15.** 4 **17.** 1/6 **19.** 104/15 **21.** 2π **23.** $(4/\pi) - 1$ **25.** $\sqrt{2} - 1$ **27.** 5/6 **29.** 32/3 **31.** $2\sqrt{2}$ **33.** 3/4

Article 5.3, pp. 318–319

1. $8\pi/3$ **3.** $\pi/30$ **5.** $16\pi/15$ **7.** $\pi/9$ **9.** 2π **11.** $32\pi/3$ **13.** $16\pi/15$ **15.** $\pi/2$ **17.** π^2 **19.** a) 8π b) $8\pi/3$ **21.** $18\pi/15$ **23.** $3\pi^2$ **25.** $144\pi/15$ **27.** $c = 2/\pi$ **29.** a) $\pi h^2/3(3a - h)$ b) $1/120\pi$ ft/sec **31.** 10,240 **33.** $16a^3/3$ **35.** $8a^3/3$ **37.** 15,990 ft^3

Article 5.4, p. 328

1. $8\pi/3$ **3.** $56\pi/15$ **5.** $256\pi/5$ **7.** $117\pi/5$ **9.** $\pi/3$ **11.** a) $5\pi/3$ b) $4\pi/3$ **13.** a) $11\pi/15$ b) $97\pi/105$ c) $121\pi/210$ d) $23\pi/30$ **15.** a) $512\pi/21$ b) $832\pi/21$ **17.** a) $\pi/6$ b) $\pi/6$ **19.** $\pi/2$ **21.** $32\pi/3$ **23.** $2\pi^2$ **25.** $2\pi^2 a^2 b$

Article 5.5, pp. 333–335

1. 12 **3.** 14/3 **5.** 123/32 **7.** $6a$ **9.** 4 **11.** $a\pi^2/8$ **13.** 21/2 **15.** 134/27 **19.** Approximately \$38,400

Article 5.6, p. 340

1. $(\pi/27) (10\sqrt{10} - 1)$ **3.** $1823\pi/18$ **5.** $253\pi/20$ **7.** $99\pi/2$ **9.** $\pi(r_2 + r_1)l$ **11.** a) $4\pi^2$ b) Curve is circle $x^2 + (y - 1)^2 = 1$. **13.** $12\pi a^2/5$ **15.** $s = 2\pi rh$

Article 5.7, p. 344

1. a) $2/\pi$ b) 0 **3.** 49/12 **5.** a) 1/2 b) 1/2 **9.** Average daily inventory = 300; average daily holding cost is \$6.00. **11.** 25 **13.** b) 32 c) 42.7

Article 5.8, 355–356

1. 4 ft **3.** $M_0 = L^2/2$; $M = L$; $\bar{x} = L/2$ **5.** $M_0 = (17/12)L^2$; $M = (7/3)L$; $\bar{x} = (17/28)L$ **7.** $\bar{y} = 2/3$ **9.** $(16/105, 8/15)$ **11.** $(1, -3/5)$ **13.** $(4a/3\pi, 4a/3\pi)$ **15.** $(2a/[3(4 - \pi)], 2a/[3(4 - \pi)])$ **17.** $(2/5, 1)$ **19.** $(1, -2/5)$ **21.** $(0, 1)$ **23.** $(1, 1)$ **25.** 13/6 **27.** $(3/7)h$ **29.** $(3/5, 1/2)$ **31.** $(0, \pi a/4)$

Article 5.9, pp. 361–363

1. 0.3 newtons **3.** a) 200 b) 400 in.-lb c) 8 in. **5.** 5/2 ft-lb; 12 ft **7.** a) $(1/2)k$ b) $(1/3)k$ **9.** 1944 ft-lb **11.** $4\pi r^2 h(4\pi + h)$ **13.** $20,070,400\pi/3$ newtons **15.** $200\pi/3$ ft-tons **17.** 11,400 ft-tons **19.** 30 hr, 51 min **21.** 85.1 ft-lb

Article 5.10, p. 366

1. 375 lb **3.** $1,666\frac{2}{3}$ lb **5.** $41\frac{2}{3}$ lb **7.** 1033.4 ft^3 **9.** Yes. The end of the tank will have moved only 3.33 ft toward the drain by the time the tank is full.

Miscellaneous Problems, Chapter 5, pp. 367–370

1. 29/2; 27/2 **3.** 9/2 **5.** $(7 - 4\sqrt{2})/2$ **7.** 32/3 **9.** 243/8 **11.** ≈ 0.137 **13.** 8 **15.** 9/8 **17.** $\pi - 2$ **19.** Max at $(0, 0)$; min at $(2, -4)$; 27/4 **21.** $8\pi/3$ **23.** $\pm\sqrt{(2x - a)/\pi}$ **25.** $32\pi/3$ **27.** $(4\pi/3)(4 - 2\sqrt{2})$ **29.** $(112\pi a^3)/15$ **31.** $\pi^2/4$ **33.** $72\pi/35$ **35.** $\pm\sqrt{(2x + 1)/\pi}$ **37.** a) $1022/189$ b) $19/3$ c) $27/20$ **39.** a) $424\pi/15$ b) $153\pi/40$ **43.** $\pi a/2$ **45.** $8a^2/3$ **47.** a) 72 b) $82\frac{2}{3}$ **49.** $(9/10, 9/5)$ **51.** $(8/5, 1)$ **53.** $(3/4, 3/10)$ **55.** a) $(5a/7, 0)$ b) $(2a/3, 0)$ **57.** $(2a/5, 2a/5)$ **59.** Work = 22,500 ft-lb; about 257 sec **61.** $[k(b - a)]/ab$ **63.** 335,153.3 ft-lb **65.** 2200 lb **67.** $2,333\frac{1}{3}$ lb **69.** a) $2h/3$ b) $(6a^2 + 8ah + 3h^2)/[2(3a + 2h)]$ **71.** a) 32π b) $32\sqrt{2}\pi$ **73.** $4\pi^2$ **75.** $(0, 2a/\pi)$ **77.** $(0, 4a/3\pi)$ **79.** $[\pi a^3(4 + 3\pi)]/3\sqrt{2}$

CHAPTER 6

Article 6.1, p. 378

1. a) $g(x) = (x - 3)/2$ b)
c) $f'(x) = 2$,
$g'(x) = 1/2$

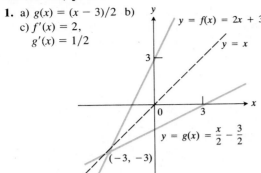

3. a) $g(x) = 5x - 35$ b)
c) $f'(x) = 1/5$,
$g'(x) = 5$

5. a) $g(x) = \sqrt{x - 1}$ b)
c) $f'(5) = 10$,
$g'(26) = 1/10$

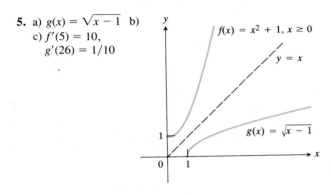

7. a) $g(x) = \sqrt[3]{x + 1}$ b)
c) $f'(2) = 12$,
$g'(7) = 1/12$

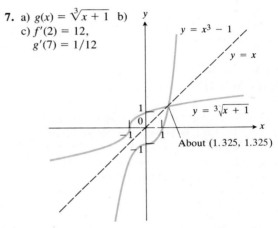

9. $x^{1/5}$ **11.** $x^{3/2}$ **13.** $\sqrt{x} + 1$ **15.** $x^{-1/2}$ **17.** b) $y = \sqrt[3]{x}$ has no derivative at $x = 0$. The slope of $y = x^3$ is 0 at $x = 0$, and the x- and y-axes are the tangent lines at $x = 0$. **19.** $1/6$ **21.** 4.8 min

Article 6.2, pp. 383–384

1. a) $\pi/4$ b) $\pi/3$ c) $\pi/6$ **3.** a) $\pi/6$ b) $\pi/4$ c) $\pi/3$ **5.** a) $\pi/3$ b) $\pi/4$ c) $\pi/6$ **7.** a) $\pi/4$ b) $\pi/6$ c) $\pi/3$ **9.** a) $\pi/4$
b) $\pi/3$ c) $\pi/6$ **11.** a) $\pi/4$ b) $\pi/6$ c) $\pi/3$ **13.** $\cos\alpha = \sqrt{3}/2$, $\tan\alpha = 1/\sqrt{3}$, $\csc\alpha = 2$, $\sec\alpha = 2/\sqrt{3}$ **15.** $\sqrt{2}/2$ **17.** 2
19. $2/\sqrt{3}$ **21.** $\sqrt{2}/2$ **23.** 0 **25.** 0 **27.** π **29.** $-\pi/3$ **31.** 0.6 **33.** $2\pi/3$ **35.** $\pi/2$ **37.** $\pi/2$ **39.** $\pi/2$ **41.** 0 **43.** $42.2°$
45. $\alpha = \tan^{-1}\sqrt{2} \approx 54.7°$ **49.**

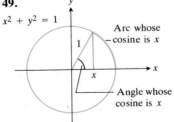

Article 6.3, pp. 389–390

1. $-2x/\sqrt{1 - x^4}$ **3.** $15/(1 + 9x^2)$ **5.** $1/\sqrt{4 - x^2}$ **7.** $1/(|x|\sqrt{25x^2 - 1})$ **9.** $-2x/[(x^2 + 1)\sqrt{x^4 + 2x^2}]$ **11.** 0
13. $-1/2x\sqrt{x - 1}$ **15.** $\sqrt{x^2 - 4}/x$ if $x > 2$; $(x^2 + 4)/(x\sqrt{x^2 - 4})$ if $x < -2$ **17.** $1/(x^2 + 1)$ **19.** $(\sin^{-1}x)^2$ **21.** $\pi/6$
23. $\pi/12$ **25.** π **27.** $\pi/12$ **29.** $\pi/6$ **31.** π **33.** $-\pi/24$ **35.** $\pi/2$ **37.** $\pi/6$ **39.** $-\pi/12$ **41.** 2 **43.** $1/6$ **45.** $\pi/2$

47. (4/5) rad/hr   **49.** Yes; since $\sin^{-1}x = (\pi/2) - \cos^{-1}x$, these functions differ by a constant.   **51.** $-\sqrt{1 - x^2} + C$ **53.** $y = 1/(\tan^{-1}x + 1)$ **55.** $\sin^{-1} y = \sec^{-1}|x| - (\pi/2)$

Article 6.4, pp. 396–397

1. $1/x$ **3.** $1/x$ **5.** $-1/x$ **7.** $(3 \ln^2 x)/x$ **9.** $\sec x$ **11.** $x^2(1 + 3 \ln 2x)$ **13.** $1/[x(1 + \ln^2 x)]$ **15.** $2x(1 + \ln(x^2))$ **17.** $(\tan^{-1}x)/x^2$ **19.** $2 \cos(\ln x)$ **21.** $\ln|x| + C$ **23.** $(1/2)\ln|x| + C$ **25.** $\ln 2$ **27.** $(1/2)\ln 3$ **29.** $(1/8)\ln 5$ **31.** $-(1/3)\ln|\cos 3x| + C$ **33.** $-(1/3)\ln|4 - x^3| + C$ **35.** $\ln|\ln x| + C$ **37.** $(1/3)\ln^3 2$ **39.** $\ln|\sec x + \tan x| + C$ **41.** 0   **43.** -2 **45.** $2 \ln 2 - 1$ **47.** $((2 \ln 2)/\pi, 0)$ **49.** $\ln 2$

Article 6.5, pp. 402–403

1. $4 \ln 2$ **3.** $(3/2)\ln 2$ **5.** $2 \ln 2 - 2 \ln 3$ **7.** $2 \ln 3 - 3 \ln 2$ **9.** $2 \ln 3 - \ln 2$ **11.** $x/(x^2 + 5)$ **13.** $-[1/x + 1/[2(x + 1)]]$ **15.** $(2 \cos 2x + 2 \cos^2 x)/\sin 2x$ or $\cot x + 2 \cot 2x$ **17.** $1/x + 1/[2(x + 2)]$ **19.** $1/x - x^2/(1 + x^3)$ **21.** $10x/(x^2 + 1) + 1/[2(1 - x)]$ **23.** $(y/2)[1/x + 1/(x + 1)]$ **25.** $y[1/[2(x + 3)] + \cot x - \tan x]$ **27.** $(y/3)[1/x + 1/(x - 2) - 2x/(x^2 + 1)]$ **29.** $(y/3)[1/x + 1/(x + 1) + 1/(x - 2) - 2x/(x^2 + 1) - 2/(2x + 3)]$ **31.** $2y[14/(3x) + 2/(3 - 2x) + 1/[(1 + x^2)\tan^{-1}x]]$ **33.** $\ln 2$ **35.** $\ln \sqrt{2}$ **37.** $4 - 5 \ln 2$ **39.** a) $\ln(5/4)$ b) $\sin^{-1}(3/5)$ c) $1/5$ **41.** a)

$y = \ln|x|$

b)

$y = |\ln x|$

43. $\ln 2$ **45.** a) $4 \ln 4 - 3$ b) $27\pi/5$ **47.** 0.005 (linear); $0.000\overline{3}$ (quadratic)

Article 6.6, pp. 409–411

1. x **3.** x^{-2} **5.** x^{-1} **7.** $2x$ **9.** $x - x^2$ **11.** xe^x **13.** $4 \ln^2 x$ **15.** $x^2 + 2x + 1$ **17.** $e^{-x}\sin x + 2$ **19.** $3e^{3x}$ **21.** $-7e^{5-7x}$ **23.** $xe^x(x + 2)$ **25.** $(\cos x)e^{\sin x}$ **27.** $(1 - x)/x$ **29.** $e^{\sin^{-1}x}/\sqrt{1 - x^2}$ **31.** $27x^2 e^{3x}$ **33.** $2xe^{-x^2}(1 - x^2)$ **35.** $e^x/(1 + e^{2x})$ **37.** $y(3/x - 2 - 5 \tan 5x)$ **39.** $y(\sin x + x \cos x)$ **41.** $[2e^{2x} - \cos(x + 3y)]/[3 \cos(x + 3y)]$ **43.** 8   **45.** 0   **47.** a **49.** 2/3   **51.** 7   **53.** $\ln 3$ **55.** $\tan^{-1}2 - (\pi/4)$ **57.** 1/2   **59.** 1   **61.** a) Maximum of e at $x = \pi/2$; minimum of $1/e$ at $x = -(\pi/2)$ or $3\pi/2$ b)

$y = e^{\sin x}$ $\left(\dfrac{\pi}{2}, e\right)$

63. $1/(2e)$ **65.** 0.005 (linear); 0.00017 (quadratic)   **67.** a) 1 b) $\pi/2$ c) π **69.** b) $y = 3e^{-2t}$ **71.** $y = -e^{-x} + e^{-4}$ **73.** $(y + 1)^2 = 2x$

Article 6.7, pp. 417–418

1. $(\ln 2)2^x$ **3.** $(3 \ln 2)2^{3x}$ or $(\ln 8)8^x$ **5.** $(2 \ln 3)3^{2x}$ or $(\ln 9)9^x$ **7.** $(\ln 4)4^x$ or $(2 \ln 2)2^{2x}$ **9.** $y(1 + \sec^2 x \ln(\sin x))$ **11.** $[(2 \ln x)x^{\ln x}]/x$ **13.** $(1 - x)^x[x/(x - 1) + \ln(1 - x)]$ **15.** $2^x[1/x + (\ln 2)(\ln x)]$ **17.** $4/\ln 5$ **19.** $1/(2 \ln 2)$ **21.** $4/\ln 3$ **23.** $(3 \ln 2)/\ln 4$ **25.** $24/\ln 25$ **27.** $1/\ln 2$ **29.** b **31.** 0   **33.** $\ln 3$ **35.** e^2 **37.** a) 1/4 b) $-5/8$ **39.** $x = -0.766664696$, $y = 0.587774756$

Article 6.8, pp. 426–427

1. 2   **3.** -2 **5.** 2   **7.** 4/3   **9.** 12   **11.** $\ln 3/\ln 2$ **13.** 1/2   **15.** $1/(x \ln 4)$ **17.** $1/(\ln 10)$ **19.** $1/x$ **21.** $1/[2 \ln 10(x + 1)]$ **23.** $-1/(x \log_2^2 x \ln 2)$ **25.** $2/[(x + 1)\ln 5]$ **27.** $\cot x/\ln 7$ **29.** $\log_2 x$ is changing faster.   **31.** $(\ln 10)/2$ **33.** $(\ln^2 8)/\ln 4$ **35.** $9 \ln 5$ **37.** $(\ln 10)(\ln 2)$ **39.** a, b, c, e, f, g, h   **41.** a, b, c, h   **43.** $\ln x$ grows faster.   **45.** f grows faster than g if $\deg(f) > \deg(g)$, and at the same rate as g if $\deg(f) = \deg(g)$.   **47.** a) 10^{-7} b) 7 c) 1

Article 6.9, pp. 433–435

1. 2.81×10^{14} **3.** 21.9 years   **5.** $Q = Q_0 e^{kt}$ **7.** 4.5%   **11.** a) $k = (-\ln 2)/5700$ b) 18,935 years c) 6658 years   **13.** 465 days   **15.** $-30°$ **17.** $53.4°$ after 15 min; $23.8°$ after 2 hr; in 3.9 hr   **19.** $(V/R)[1 - (1/e)]A$ **21.** $(mv_0/k)(1 - e^{-kt/m})$ **23.** 585.4 kg

Miscellaneous Problems, Chapter 6, pp. 436–440

1. $1 \pm \sqrt{2}$ **3.** $\pi/2$ **5.** $\pi\sqrt{3}$ **7.** b) $61°$ **11.** $1 + \ln x$ **13.** $(1 - \ln x)/x^2$ **15.** $(\ln x - 1)/\ln^2 x$ **17.** $x^2(1 + 3 \ln x)$ **19.** $2/x$ **21.** $1/(1 - x^2)$ **23.** $1/[x(x^2 + 1)]$ **25.** $\sec^{-1}x$ **27.** $[x(x - 2)/(x^2 + 1)^{1/3}][(2x - 2)/[x(x - 2)] - 2x/[3(x^2 + 1)]]$

29. $-(1/3)\ln 3$ **31.** $\ln \sqrt{3}$ **33.** $-\ln 32$ **35.** $(1/4)\ln(x^4 + 1) + C$ **37.** $(1/3)\ln|\sin(2 + x^3)| + C$ **39.** $\ln(3 + \sin^{-1}x) + C$
41. a) 0 b) 0 **45.** a) x b) $-(3/2) + 2x - (x^2/2)$ **47.** $a \ln|y_1/a|$ **49.** $y = xe^{x-y}$ **51.** $y = (b/a^2)x^2$ **53.** $y^2 = 4x$
55. $-(1/x^2)e^{1/x}$ **57.** $e^{-x}(1 - x)$ **59.** $e^{-x}[(1/x) - \ln x]$ **61.** $(1/x) + 2$ **63.** $(\sin 2x)e^{\sin^2 x}$ **65.** $1/(1 + e^x)$ **67.** a) $2 - e^{2x}$
b) $e^{2x-(1/2)e^{2x}}(2 - e^{2x})$ **69.** $8/(e^{2x} + e^{-2x})^2$ **71.** $2x/\sqrt{1 - x^4} - e^{(x^2)} - 2x^2e^{(x^2)}$ **73.** $5e^{2x}\cos 3x - e^{2x}\sin 3x$ **75.** e **77.** 3/8
79. 0 **81.** 0 **83.** 1 **85.** $1 + 5x$

87. a)

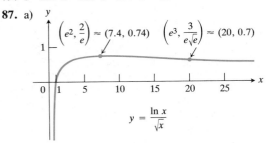

$$y = \frac{\ln x}{\sqrt{x}}$$

b)

$y = e^{-x^2}$

c)

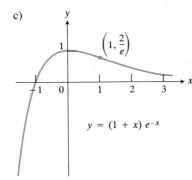

$y = (1 + x)e^{-x}$

89. $y = 1 - (1/2)(e^x + e^{-x})$ **91.** $(\pi/2)((3/4) + \ln 2)$ **93.** $a^2[e - (1/e)]$ **95.** $\sqrt{e}/4$ **99.** $2e^x/(2 - e^x)$ **101.** If m is the
particle's mass, $F = m\omega^2x$. **107.** a) $x^{\tan 3x}[(\tan 3x + 3x \ln x \sec^2 3x)/x]$ b) $-(y \ln y)/(x \ln x)$
c) $(x^2 + 2)^{1-x}(4x - 2x^2 - (x^2 + 2)\ln(x^2 + 2))$ d) $x^{1/x}[(1 - \ln x)/x^2]$ **109.** a) $1/(2 \ln 4)$ b) 1 **113.** 34.7 years **115.** $2\sqrt{2}$

CHAPTER 7

Article 7.1, pp. 447–449

1. $(2/3)(3x^2 + 5)^{3/2} + C$ **3.** 1/2 **5.** 1/4 **7.** $\tan(e^x) + C$ **9.** $(1/6)(3 + 4e^x)^{3/2} + C$ **11.** $\ln|\ln x| + C$ **13.** 2
15. $(1/2)\tan^2x + C$ or $(1/2)\sec^2x + C$ **17.** $\sin x - (1/3)\sin^3x + C$ **19.** $(1/3)\sec^3x - \sec x + C$ **21.** 19/3
23. $-1/[2(2x - 7)] + C$ **25.** $-\ln 3$ **27.** $(2/3)(\sin x)^{3/2} + C$ **29.** $-(1 + \sin x)^{-1} + C$ **31.** 0 **33.** $-(1/4)\sqrt{1 - 4x^2} + C$
35. $\pi/4$ **37.** $(3/2)\ln 2$ **39.** $(\pi - 2)/4$ **41.** $\pi/2$ **43.** $-(1/12)(3x^2 + 4)^{-2} + C$ **45.** $(2/3)\sqrt{x^2 + 5} + C$ **47.** 3/2
49. $-(1/3)e^{-3x} + C$ **51.** $\tan^{-1}(e^x) + C$ **53.** $-(1/3)\cos^3x + C$ **55.** 9/4 **57.** $(1/2)\ln|e^{2x} - e^{-2x}| + C$ **59.** 0
61. $\sec^{-1}|2t| + C$ **63.** $\ln|\sin x| + C$ **65.** $(1/3)\sec^3x + C$ **67.** $(1/3)e^{\tan 3x} + C$ **69.** $(\sqrt{3} - 1)/2$ **71.** $-\sqrt{1 + \cot 2t} + C$
73. $(1/2)(e^{\pi/4} - 1)$ **75.** $2/\ln 2$ **77.** $(1/2)\ln^2x + C$ **79.** $2 \ln 2$ **81.** $(1/2)e^{(x^2)} + C$

Article 7.2, pp. 455–456

1. $-x \cos x + \sin x + C$ **3.** $-x^2\cos x + 2x \sin x + 2 \cos x + C$ **5.** $2 \ln 2 - (3/4)$ **7.** $(x^4/4)\ln x - (x^4/16) + C$
9. $x \tan^{-1}x - (1/2)\ln(1 + x^2) + C$ **11.** $x \sin^{-1}x + \sqrt{1 - x^2} + C$ **13.** $x \tan x + \ln|\cos x| + C$ **15.** $e^x(x^3 - 3x^2 + 6x - 6) + C$
17. $e^x(x^2 - 7x + 7) + C$ **19.** $e^x(x^5 - 5x^4 + 20x^3 - 60x^2 + 120x - 120) + C$ **21.** $(\pi^2/8) - 1/2$ **23.** $x^4\sin x + 4x^3\cos x -$
$12x^2\sin x - 24x \cos x + 24 \sin x + C$ **25.** $(x^2/a) \sin ax + (2x/a^2) \cos ax - (2/a^3)\sin ax + C$ **27.** $(2\pi/3) - (\sqrt{3}/2)$ **29.** 1/2
31. $(x^2/2)\sin^{-1}(1/x) + (1/2)\sqrt{x^2 - 1} + C$ **33.** $(e^x/2)(\sin x - \cos x) + C$ **35.** $(e^{2x}/13)(3 \sin 3x + 2 \cos 3x) + C$
37. $(x/2)[\sin(\ln x) - \cos(\ln x)] + C$ **39.** a) π b) 3π **41.** $2\pi[(e - 2)/e]$ **43.** $\bar{x} = (6 - 2e)/(e - 2); \bar{y} = (e^2 - 3)/[8(e - 2)]$
45. a) $\pi^4/6 - \pi^2/4$ b) 8π **47.** $\pi^2 + \pi - 4$ **49.** $[2\pi\sqrt{2}(e^\pi - 2)]/5$ **51.** $[(x^2 + 1)/2]\tan^{-1}x - (x/2) + C$

Article 7.3, pp. 464–466

1. $-\cos x + (2/3)\cos^3x - (1/5)\cos^5x + C$ **3.** 4/3 **5.** $-\cos x + \cos^3x - (3/5)\cos^5x + (1/7)\cos^7x + C$ **7.** $-(3/5)\cos^{5/3}x +$
$(6/11)\cos^{11/3}x - (3/17)\cos^{17/3}x + C$ **9.** $\ln(2 + \sqrt{3})$ **11.** $\sqrt{3} - (1/2)\ln(2 - \sqrt{3})$ **13.** $(1/2)\sec(e^x)\tan(e^x) +$
$(1/2)\ln|\sec(e^x) + \tan(e^x)| + C$ **15.** 4/3 **17.** 96/35 **19.** $(1/2) + \ln(\sqrt{2}/2)$ **21.** $(\tan^5x)/5 - (\tan^3x)/3 + \tan x - x + C$
23. $-\ln|\csc x + \cot x| + C$ **25.** a) $\ln(\sqrt{2} + 1)$ b) $\ln \sqrt{3}$ **27.** $-6/5$ **29.** π **31.** 0 **33.** $-(1/4)\ln|\csc 4x + \cot 4x| + C$
35. $2\sqrt{2} - 2 \ln(\sqrt{2} + 1)$ **37.** $\ln(\sqrt{2} + 1)$ **39.** a) 8.3 cm b) 10.9 cm

Article 7.4, p. 470

1. π **3.** $(1/2)t - (1/8)\sin 4t + C$ **5.** $\pi/16$ **7.** $3\pi/8a$ **9.** $\tan x - (3/2)x + (1/4)\sin 2x + C$ **11.** $\pi/16$ **13.** $\tan \theta +$
$(1/2)\sin 2\theta - (1/32)\sin 4\theta - (15/8)\theta + C$ **15.** 4 **17.** 2/5 **19.** $4 - \pi$ **21.** $2 \ln(\sqrt{2} + 1)$ **23.** 2 **25.** $2 \ln(\sqrt{2} + 1)$ **27.** 0
29. $-\cot x - (\cot^3x)/3 + C$ **31.** $2 - (\pi/2)$ **33.** $\ln|\sec t + \tan t| - \sin t + C$ **35.** $2\sqrt{2}$ **37.** $\pi^4/6 - \pi^2/4$

Article 7.5, pp. 476–477

1. $\pi/4$ **3.** $(1/2)\tan^{-1}2x + C$ **5.** $\pi/4$ **7.** $\ln|\sqrt{25 + y^2} + y| + C$ **9.** $(1/3)\ln|\sqrt{25 + 9y^2} + 3y| + C$
11. $\ln|3z + \sqrt{9z^2 - 1}| + C$ **13.** $\pi/24$ **15.** $\pi/24$ **17.** $4\sqrt{3} - 2 \ln(2 + \sqrt{3})$ **19.** $-\sqrt{9 - x^2}/9x + C$ **21.** 1/5 **23.** $(25\pi/4)$

25. $\pi/6$ **27.** 2π **29.** $(2 - \sqrt{2})/3$ **31.** $2 \sec^{-1}2x + C$ **33.** $-(\sqrt{1 - x^2}/x) - \sin^{-1}x + C$ **35.** $-(3/4) + (\sqrt{7}/3)$
37. $x/(a^2\sqrt{a^2 - x^2}) + C$ **39.** $-\sqrt{16 - y^2} + C$ **41.** $-(1/\sqrt{x^2 - 1}) + C$ **43.** a) $(1/a)\tan^{-1}(u/a) + C$ b) $\sin^{-1}(u/a) + C$
45. a) $u/\sqrt{a^2 + u^2}$ b) a/u **47.** $(8\pi^2 + 6\pi\sqrt{3})/3\sqrt{3}$ **49.** $y = 1/[\sin^{-1}(x/2) + \sqrt{2} - x]$
51. $(\delta/2)[2\sqrt{3} - \sqrt{2} + \ln(2 + \sqrt{3})/(1 + \sqrt{2})]$

Article 7.6, p. 481

1. $\pi/8$ **3.** $\ln \sqrt{2} + (\pi/8)$ **5.** $\pi/(6\sqrt{3})$ **7.** $(1/3)\ln|\sqrt{9x^2 - 6x + 5} + 3x - 1| + C$ **9.** $(1/9)\sqrt{9x^2 - 6x + 5} +$
$(1/9)\ln|\sqrt{9x^2 - 6x + 5} + 3x - 1| + C$ **11.** $\ln|\sqrt{x^2 + 2x} + x + 1| + C$ **13.** $2 - \ln 9$ **15.** $\ln|\sqrt{x^2 - 2x - 3} + x - 1| + C$
17. $-\sqrt{2x - x^2} + 2 \sin^{-1}(x - 1) + C$ **19.** $-\sqrt{5 + 4x - x^2} + 2 \sin^{-1}[(x - 2)/3] + C$ **21.** $3 - 2\sqrt{2}$ **23.** $\ln \sqrt{2} - (\pi/2)$
25. $100\pi[\tan^{-1}(5/2) - \tan^{-1}(-3/4)]$ **27.** $\pi\sqrt{2}[\sqrt{2} + \ln(1 + \sqrt{2})]$ **29.** $\bar{x} = (4 + \ln 9)/\ln 9,\ \bar{y} = [2 \tan^{-1}(4/3)]/(3 \ln 9)$
31. $\pi/2$

Article 7.7, pp. 489–491

1. $2/(x - 3) + 3/(x - 2)$ **3.** $1/(x + 1) + 3/[(x + 1)^2]$ **5.** $-2/x - 1/x^2 + 2/(x - 1)$ **7.** $1 - 12/(x - 2) + 17/(x - 3)$
9. $1/3x^2 - 1/[3(x^2 + 9)]$ **11.** $\ln \sqrt{3}$ **13.** $4 - \ln 3$ **15.** $2 \ln 3$ **17.** $(2/7)\ln|x + 6| + (5/7)\ln|x - 1| + C$ **19.** $(9/2) - 6 \ln 2$
21. $(5/6)\ln|x + 5| + (1/6)\ln|x - 1| + C$ **23.** $(2/3)\ln|x + 5| + (1/3)\ln|x - 1| + C$ **25.** $\ln(3/2) - 1/4$ **27.** $-(3/8)\ln|x| +$
$(5/16)\ln|x - 2| + (1/16)\ln|x + 2| + C$ **29.** $5(\sqrt{3} - \pi/3)$ **31.** 0 **33.** $4 \ln|x| - (1/2)\ln(x^2 + 1) + \tan^{-1}x + C$ **35.** $x^2/2 +$
$(3/2)\ln|x + 1| - (9/2)\ln|x + 3| + C$ **37.** $\pi/4 + 1/2$ **39.** $1/2 - \pi/4$ **41.** $2 \ln 3 - 3 \ln 2$ **43.** $5/4 - 3\pi/8$ **45.** $4\sqrt{x} - x -$
$4 \ln(\sqrt{x} + 1) + C$ **47.** $2\sqrt{x} - 3\sqrt[3]{x} + 6\sqrt[6]{x} - 6 \ln(\sqrt[6]{x} + 1) + C$ **49.** $\ln 2 - 2 + (\pi/2)$ **51.** $6\pi \ln 5$
53. $(8 + 8 \ln 2 - 3 \ln 3)/(3 \ln 5 - 2 \ln 3)$ **55.** a) $x/(1000 - x) = e^{4t}/499$ b) 1.55 days **57.** $x/(a - x) = [x_0/(a - x_0)]e^{akt}$

Article 7.8, pp. 499–501

1. $\pi/2$ **3.** 6 **5.** 4 **7.** 1000 **9.** $\ln 2$ **11.** Diverges **13.** Converges **15.** Converges **17.** Diverges **19.** Converges
21. Diverges **23.** Diverges **25.** Converges **27.** Converges **29.** Diverges **31.** Diverges **33.** Converges **35.** Diverges
37. Converges **39.** Converges **41.** 0.8862 **45.** 1 **47.** 2π **49.** $(\bar{x}, \bar{y}) = (2/\pi, 0)$

Article 7.9, p. 504

1. $\sqrt{\pi}/2$ **3.** $(2/\sqrt{3})[\tan^{-1}\sqrt{2} - (\pi/4)]$ **5.** $x/[18(9 - x^2)] + (1/108)\ln|(x + 3)/(x - 3)| + C$ **7.** $(44 - 24\sqrt{2})/231$
9. $\sqrt{2}(\pi/4 - 1)$ **11.** $-(1/6)\ln|(5 + 4 \sin 2x + 3 \cos 2x)/(4 + 5 \sin 2x)| + C$ **13.** $[(x + 1)(2x - 3)^{3/2}]/5$ **15.** $10!$ **17.** $\pi/4$
19. $\pi/4 - 1/2$ **21.** $\bar{x} = 4/3,\ \bar{y} = \ln \sqrt{2}$

Article 7.10, pp. 508–509

1. $3\pi/4$ **3.** $(1/6)\cos^5x \sin x + (5/24)\cos^3x \sin x + (5/16)(\cos x \sin x + x) + C$ **5.** $3\pi/8$ **7.** $8/15$ **9.** $(1/4)\tan^22x +$
$(1/2)\ln|\cos 2x| + C$ **11.** $(1/4)\tan^4x - (1/2)\tan^2x + \ln|\cos x| + C$ **13.** $-(1/2)\cot^2x - \ln|\sin x| + C$ **15.** $\pi/2 - 4/3$
17. $4\sqrt{3}$ **19.** $(1/4)\sec^3x \tan x + (3/8)\sec x \tan x + (3/8)\ln|\sec x + \tan x| + C$ **21.** $\sqrt{2} + \ln(\sqrt{2} + 1)$
23. $-(1/3)\csc^2x \cot x - (2/3)\cot x + C$ **25.** $1/\sqrt{2}$ **27.** $(735/1024) + (3/8)\ln 2$ **31.** b) $-(x^n/a)\cos ax +$
$(n/a)\int x^{n-1} \cos ax\, dx$ **33.** $(81(\ln 3)^2 - 2 \ln 3)/4$

Miscellaneous Problems, Chapter 7, pp. 509–513

1. $2\sqrt{1 + \sin x} + C$ **3.** $(\tan^2x)/2 + C$ **5.** 18 **7.** $\ln|\sqrt{x^2 + 2x + 2} + x + 1| + C$ **9.** $e^x(x^2 - 2x + 2) + C$ **11.** $\tan^{-1}(e^t) + C$
13. $2 \tan^{-1}\sqrt{x} + C$ **15.** $(9/25)(t^{5/3} + 1)^{5/3} + C$ **17.** $\ln|(\sqrt{e^t + 1} - 1)/(\sqrt{e^t + 1} + 1)| + C$ **19.** $\pi/4$ **21.** $\pi/4$ **23.** $\ln \sqrt{3}$
25. 2 **27.** $(3/5)(1 + e^x)^{5/3} - (3/2)(1 + e^x)^{2/3} + C$ **29.** $5/12$ **31.** $\ln|2 + \ln x| + C$ **33.** $(4/21)(2\sqrt[4]{27} + \sqrt[4]{8})$ **35.** $3/80$
37. $-2\sqrt{x + 1} \cos \sqrt{x + 1} + 2 \sin \sqrt{x + 1} + C$ **39.** $(1/4)\ln 3$ **41.** $\ln|y| - (1/3)\ln|2y^3 + 1| + 1/[3(2y^3 + 1)] + C$
43. $\ln|x/\sqrt{x^2 + 1}| + 1/[2(x^2 + 1)] + C$ **45.** $\ln|e^x - 1| - x + C$ **47.** $(3/2)\ln|x + 3| - (1/2)\ln|x + 1| + C$
49. $\ln|x/\sqrt{x^2 + 4}| + C$ **51.** $\sqrt{x^2 - 1} - \tan^{-1}\sqrt{x^2 - 1} + C$ **53.** $\ln|x| - 2 \ln|3\sqrt{x} + 1| + C$ **55.** $(1/2)\ln(5/2)$
57. $(3/8)(1 + e^{2t})^{1/3}(e^{2t} - 3) + C$ **59.** $x^2/2 + (4/3)\ln|x + 2| + (2/3)\ln|x - 1| + C$ **61.** $\ln|x - 1| - 1/(x - 1) + C$
63. $\tan^{-1}(2\sqrt{y^2 + y}) + C$ **65.** $(3/8)\sin^{-1}x + (1/2)x\sqrt{1 - x^2} + (1/8)x\sqrt{1 - x^2}(1 - 2x^2) + C$
67. $x \tan x + \ln|\cos x| - (x^2/2) + C$ **69.** $\pi^2 - 4$ **71.** $[(9 - 2\sqrt{3})\pi]/72$ **73.** $(1/2) + \ln 2$
75. $x \sec^{-1}x - \ln|x + \sqrt{x^2 - 1}| + C$ **77.** $(1/16)\ln|(x^2 - 4)/(x^2 + 4)| + C$ **79.** $\ln|\cot x| + C$ **81.** $(2\sqrt{3}\pi - 9)/9$
83. $\sqrt{2} \tan^{-1}(\sqrt{2} \tan t) - t + C$ **85.** $e^t\sin(e^t) + \cos(e^t) + C$ **87.** $(x^2/2)\ln(x^3 + x) + (1/2)\ln(x^2 + 1) - (3/4)x^2 + C$
89. $\ln|\sqrt{3 + \sin^2x} + \sin x| + C$ **91.** $(x^2 - 2)\cos(1 - x) + 2x \sin(1 - x) + C$ **93.** $(\tan^2x)/2 + \ln|\cos x| + C$
95. $\ln\sqrt{x^2 + 1} + 1/[2(x^2 + 1)] + C$ **97.** $298/15$ **99.** $x \ln|x - \sqrt{x^2 - 1}| + \sqrt{x^2 - 1} + C$ **101.** $x \ln|x + \sqrt{x}| - x + \sqrt{x} -$
$\ln(\sqrt{x} + 1) + C$ **103.** $3 \ln 3 - 2$ **105.** 2 **107.** $(1/4)(\sin^{-1}x)^2 + (1/2)x\sqrt{1 - x^2} \sin^{-1}x - (x^2/4) + C$ **109.** $x - \tan x +$
$\sec x + C$ **111.** $(1/2)\ln|\sin x - 1| - (1/18)\ln|\sin x + 1| - (4/9)\ln|\sin x - 2| - 1/[3(\sin x - 2)] + C$
113. $-2 \sin^{-1}[\cos(x/2)/\cos(\alpha/2)] + C$ **115.** $\ln 6 - 2 + 2\sqrt{2} \tan^{-1}(1/\sqrt{2})$ **117.** $\ln|(1 + \sqrt{1 - e^{-t}})/(1 - \sqrt{1 - e^{-t}})| + C$
119. $x(\sin^{-1}x)^2 - 2x + 2\sqrt{1 - x^2} \sin^{-1}x + C$ **121.** $(1/2)x^2\sin^{-1}x - (1/4)\sin^{-1}x + (1/4)x\sqrt{1 - x^2} + C$
123. $(1/4)\ln|\sec 2\theta + \tan 2\theta| + (1/2)\theta + C$ **125.** $(1/4)\ln|1 - 2t^2| - (1/4)\ln|(1 + 2t\sqrt{1 - t^2})/(1 - 2t^2)| - (1/2)\sin^{-1}t + C$
127. $(1/16)\ln|(x^2 + 2x + 2)/(x^2 - 2x + 2)| + (1/8)(\tan^{-1}(x + 1) + \tan^{-1}(x - 1)) + C$ **129.** $\pi/4$ **131.** $\pi/2$ **133.** -1

135. $a^2(\pi - 2)/4$; $\bar{x} = \bar{y} = 2a/[3(\pi - 2)]$ **137.** $2k\pi^2 a^2$ **139.** $\sqrt{1 + e^2} - \ln|(\sqrt{1 + e^2} + 1)/e| - \sqrt{2} + \ln(1 + \sqrt{2})$
141. $\pi[e^2\sqrt{1 + e^4} + \ln(\sqrt{1 + e^4} + e^2) - \sqrt{2} - \ln(1 + \sqrt{2})]$ **143.** 2π **145.** $(2\pi a^3)/3$
147. Converges to 1 **149.** 1/2
151. 2 **153.** $2/[1 - \tan(x/2)] + C$ **155.** $-\cot(x/2) - x + C$
157. $-(\sqrt{2}/2)\ln|[\tan(x/2) + 1 + \sqrt{2}]/[\tan(x/2) + 1 - \sqrt{2}]| + C$

CHAPTER 8

Article 8.1, p. 517

1. $x^2 = 8y + 16$ **3.** $x - y = 1$ **5.** $(x^2 + y^2)^2 + 8(y^2 - x^2) = 0$ **7.** $3x^2 - 20x + 4y^2 + 12 = 0$ **9.** $5x^2 - 4y^2 = 20$ **11.** $x^2 + y^2 - 4x - 6y + 4 = 0$ **13.** $C(2, 2)$, $\sqrt{5}$

Article 8.2, p. 520

1. $x^2 + y^2 - 4y = 0$ **3.** $x^2 + y^2 - 6x + 8y = 0$ **5.** $x^2 + y^2 + 4x + 2y < 1$ **7.** $C(0, 0)$, $r = 4$ **9.** $C(0, 1)$, $r = 2$
11. $C(-2, 0)$, $r = 4$ **13.** $C(-1, -1)$, $r = 1$ **15.** $C(-\frac{1}{4}, -\frac{1}{4})$, $r = \sqrt{2}/4$ **17.** $(-1, 2)$ only **19.** $x^2 + y^2 - 4x - 4y = 5$
21. $5x^2 + 5y^2 + 10x - 10y + 1 = 0$ **23.** $x^2 + y^2 + 2x + 2y - 23 = 0$ **25.** $x^2 + y^2 - 6x - 8y = 0$; $C(3, 4)$, $r = 5$
27. $C(-2, 4)$, $r = 2\sqrt{5}$ **29.** 7/2 by $3\sqrt{7}/2$

Article 8.3, pp. 527–528

1. $x^2 = 8y$, $y = -2$ **3.** $x^2 = -16y$, $y = 4$ **5.** $(x + 2)^2 = 4(y - 3)$, $y = 2$ **7.** $(y - 1)^2 = 12(x + 3)$, $x = -6$ **9.** $y^2 = 8(x - 2)$,
$F(4, 0)$ **11.** $(y - 1)^2 = -16(x + 3)$, $F(-7, 1)$ **13.** $(y - 1)^2 = 4x$, $F(1, 1)$ **15.** $V(0, 0)$, $F(2, 0)$, axis: $y = 0$, directrix: $x = -2$
17. $V(0, 0)$, $F(0, 25)$, axis: $x = 0$, directrix: $y = -25$ **19.** $V(1, 1)$, $F(1, -1)$, axis: $x = 1$, directrix: $y = 3$ **21.** $V(2, 0)$,
$F(1, 0)$, axis: $y = 0$, directrix: $x = 3$ **23.** $V(-1, -1)$, $F(-1, 0)$, axis: $x = -1$, directrix: $y = -2$ **25.** $V(-2, 2)$, $F(0, 2)$, axis:
$y = 2$, directrix: $x = -4$ **27.** All points in the plane except those satisfying $y^2 = x$

Article 8.4, pp. 535–536

1. $x^2/12 + y^2/16 = 1$, $e = 1/2$ **3.** $x^2/5 + (y - 2)^2/9 = 1$, $e = 2/3$ **5.** $C(7, 5)$, V: $(7, 0)$, $(7, 10)$, $F(7, 5 \pm \sqrt{21})$
7. $C(-1, -4)$, V: $(-1, 1)$, $(-1, -9)$, F: $(-1, 0)$, $(-1, -8)$ **9.** $C(3, 1)$, V: $(3, 6)$, $(3, -4)$, $F(3, 1 \pm \sqrt{21})$ **11.** $C(-5, 0)$, V:
$(-10, 0)$, $(0, 0)$, $F(-5 \pm 2\sqrt{6}, 0)$ **13.** $C(2, -1)$, V: $(-1, -1)$, $(5, -1)$, $F(2 \pm 2\sqrt{2}, -1)$ **15.** $C(2, -2)$, V: $(2, 0)$, $(2, -4)$,
$F(2, -2 \pm \sqrt{3})$ **17.** $C(-1, 3)$, V: $(-5, 3)$, $(3, 3)$, $F(-1 \pm \sqrt{7}, 3)$ **21.** $(x - 1)^2/4 + (y - 4)^2/9 = 1$; $F(1, 4 \pm \sqrt{5})$

23. The ellipse and its interior **25.**   **27.**

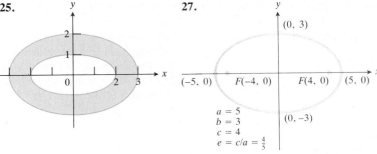

29. a) $(x - 17.5)^2/(18.09)^2 + y^2/(4.56)^2 = 1$ b) 0.584 AU, 5.41×10^7 mi c) 35.60 AU, 3.30×10^9 mi **31.** $(0, 16/3\pi)$
33. $a = 0$, $b = -4$, $c = 0$, or $4x^2 + y^2 - 4y = 0$

Article 8.5, pp. 544–545

1.

3.

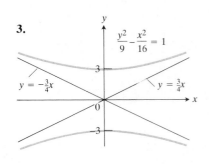

5. $C(2, -3)$, V: $(4, -3)$, $(0, -3)$, $F(2 \pm \sqrt{13}, -3)$, $e = \sqrt{13}/2$, $2(y + 3) = \pm 3(x - 2)$ **7.** $C(2, -3)$, V: $(5, -3, (-1, -3)$,
$F(2 \pm \sqrt{13}, -3)$, $e = \sqrt{13}/3$, $3(y + 3) = \pm 2(x - 2)$ **9.** $C(4, 3)$, V: $(4, 6)$, $(4, 0)$, F: $(4, 8)$, $(4, -2)$, $e = 5/3$, $4(y - 3) =$
$\pm 3(x - 4)$ **11.** $C(-2, 1)$, V: $(0, 1)$, $(-4, 1)$, F: $(-5, 1)$, $(1, 1)$, $e = 3/2$, $2(y - 1) = \pm \sqrt{5}(x + 2)$ **13.** $C(2, 0)$, V: $(0, 0)$,
$(4, 0)$, $F(2 \pm \sqrt{5}, 0)$, $e = \sqrt{5}/2$, $2y = \pm(x - 2)$ **15.** $(y - 2)^2 - (x^2/3) = 1$ **17.** Closest to A on a hyperbola with foci A and
B, and with $a = 129.9$ mi **19.** a) $44\pi/3$ b) $256\pi/3$ **21.** $42\pi\sqrt{7}$ **23.** The circles' radii increase at the same rate, so their
difference is constant.

Article 8.6, p. 550

1. $(x')^2 - (y')^2 = 4$ **3.** $(x')^2 = -4y'$ **5.** $y' = \pm 1$ **7.** $(x')^2 = -4(y' + \frac{1}{2})$ **9.** $4(x')^2 + 2(y')^2 = 19$ **11.** $\sin \theta = 1/\sqrt{5}$,
$\cos \theta = 2/\sqrt{5}$ **15.** $x' = \pm 1/\sqrt{2}$

Article 8.7, p. 552

1. Hyperbola **3.** Parabola **5.** Ellipse **7.** Hyperbola **9.** Ellipse (circle) **11.** Parabola **13.** Ellipse **15.** Hyperbola
17. Parabola

Article 8.8, p. 555

3. Line L and the point D do not exist.

Article 8.9, pp. 561–562

1. $x^2 + y^2 = 1$ **3.** $x^2/16 + y^2/4 = 1$ **5.** $x = 1 - 2y^2$, $|x| \le 1$, $|y| \le 1$ **7.** $x^2 - y^2 = 1$, left branch **9.** $x = \cos^{-1}(1 - y) -$
$\sqrt{2y - y^2}$, one arch of cycloid **11.** $y = x^{2/3}$ **13.** $x = y^2$ **15.** $y = (x - 1)^2 + 4$, $x \ge 1$ **17.** $x = -at/\sqrt{1 + t^2}$, $y = a/\sqrt{1 + t^2}$
19. $x = a(\cos t + t \sin t)$, $y = a(\sin t - t \cos t)$ **21.** $x = (a - b)\cos \theta + b \cos[(a - b)/b]\theta$, $y = (a - b)\sin \theta - b \sin[(a - b)/b]\theta$
23. b) $x = tx_0$, $y = ty_0$ c) $x = -1 + t$, $y = t$ **25.** $(1, 1)$ **27.** $8a$ **29.** $5\pi^2$ **31.** 6723 ft

Miscellaneous Problems, Chapter 8, pp. 563–566

1. $x^2 + xy + y^2 = 3k^2$ **5.** $(-16/5, 53/10)$ **7.** $y^2 = 2(x - (7/2))$ **11.** $x = 2$ P.M. **13.** $(\pm 8\sqrt{3}/9, 4/3)$ **19.** a) $F_1P = 5$;
$F_2P = \sqrt{5}$ b) Outside; $OF_2 + OF_1 = 5 + 3 > 5 + \sqrt{5}$ **21.** $C(3, 2)$, V: $(15, 2)$, $(-9, 2)$, $F(3 \pm 8\sqrt{2}, 2)$, $e = 2\sqrt{2}/3$
23. $(x + 3)^2/36 + (y - 1)^2/20 = 1$ **25.** $x = x' - 2$, $y = y' - 1$, $(2, 1)$ **27.** Circle: $\int_0^a \sqrt{a^2 - x^2}\, dx$;
ellipse: $\int_0^a (b/a)\sqrt{a^2 - x^2}\, dx = \pi ab$ **29.** The plane is at the point $(5, 10\sqrt{10}/3)$ in the chosen coordinate system.
31. $C(1, -2)$, V: $(1, -5)$, $(1, 1)$, $F(1, -2 \pm \sqrt{13})$, $2(y + 2) = \pm 3(x - 1)$ **35.** With $B = (-c, 0)$ and $A = (c, 0)$, the path is
the right branch of hyperbola $3x^2 - y^2 + 2cx - c^2 = 0$. **37.** On one branch of a hyperbola with foci the locations of the rifle
and the target. **39.** $e = \sqrt{2}$ **41.** $3y - 27x - 56 = 0$, $3y - 27x + 104 = 0$ **43.** $x^2 + 4xy + 5y^2 - 1 = 0$ **45.** $C(1/2, -1/2)$,
$V(1/2, -1/2 \pm \sqrt{2})$, $F(1/2, -1/2 \pm 2)$, $e = \sqrt{2}$, $y + 1/2 = \pm(x - (1/2))$ **47.** The curve is part of the parabola $2y'^2 =$
$2a\sqrt{2}x' - a^2$. **49.** Any point (x, y) that satisfies the equation of either the line, the circle, or the parabola will make the product
zero.

51. Any point (x, y) such that either $x + y = 0$ or $x^2 + y^2 - 1 = 0$
53. The points on, or interior to, the ellipse
55. The points inside either ellipse that lie outside the other, since
the product of two values will be negative when either but not both
is negative

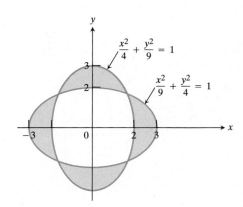

57. The points on either the circle $x^2 + y^2 = 9$ or the hyperbola $y^2 - x^2 = 9$ **59.** $x = a \sin^2 t \tan t$, $y = a \sin^2 t$ **61.** $x = 8t -$
$2 \sin 2t$, $y = 4 - 2 \cos 2t$ **65.** b) 5.868 c) $|E| \le 0.003$

CHAPTER 9

Article 9.1, pp. 571–572

	sinh	cosh	tanh	coth	sech	csch
1.	$-3/4$	$5/4$	$-3/5$	$-5/3$	$4/5$	$-4/3$
3.	$-7/24$	$25/24$	$-7/25$	$-25/7$	$24/25$	$-24/7$
5.	$\pm4/3$	$5/3$	$\pm4/5$	$\pm5/4$	$3/5$	$\pm3/4$

7. $x + (1/x)$ **9.** $(x^2 - 1)/(x^2 + 1)$ **11.** e^{5x} **13.** e^{-3x} **15.** Odd **17.** Even **19.** Even **21.** Odd **23.** Even **25.** ln 2

Article 9.2, pp. 575–577

1. $3 \cosh 3x$ **3.** 0 **5.** $-\text{csch}^2(\tan x)\sec^2 x$ **7.** $-\text{csch}(x/4)\coth(x/4)$ **9.** 0 **11.** sech x **13.** csch x **15.** $2x \cosh 2x$
19. $(2/5)\sinh 5$ **21.** 0 **23.** $\sinh 1 - \cosh 1 + 1 = 1 - e^{-1}$ **25.** $1/(4e)$ **27.** $2(\cosh 1 - \sinh 1) = 2e^{-1}$ **29.** $(1/2)\sinh$
$(2x + 1) + C$ **31.** $-(1/3)\text{sech}^3 x + C$ **33.** $\ln(\cosh x) + C$ **35.** $2 \cosh \sqrt{x} + C$ **37.** $2\sqrt{2} \cosh(x/2) + C$ **39.** $\cosh^2 x + C$
41. $\ln(1 + \cosh x) + C$ **43.** $x^2\sinh x - 2x \cosh x + 2 \sinh x + C$ **45.** $-(1/15)\text{sech}^3 5x + C$ **47.** $(\sinh^3 3x \cosh 3x)/12 -$
$(\sinh 6x)/16 + (3x)/8 + C$ **49.** $e^{5x}/10 + e^x/2 + C$ **51.** $-\ln|\text{csch } x + \coth x| + C$ **53.** 3π **55.** $15\pi/16 + \pi \ln 2$ **57.** $2\pi/3$
59. $(0, 1/\pi)$ **61.** π **63.** $A = 1, B = 1, C = -1, D = -1$

Article 9.3, p. 583

1. $a \sinh(x_1/a)$ **5.** $\bar{x} = 0, \bar{y} = [x_1\text{csch}(x_1/a)]/2 + y_1/2$ **7.** b) $dx/ds = a/\sqrt{a^2 + s^2}, dy/ds = s/\sqrt{a^2 + s^2}$ **9.** b) i) 115 ft
ii) 16 lb

Article 9.4, pp. 590–591

1. 0 **3.** $-\ln 3$ **5.** ln 2 **7.** $(1/2)\ln 3$ **9.** $(1/2)\ln 3$ **11.** $-(1/2)\ln 3$ **13.** ln 3 **15.** ln 5 **17.** $2/\sqrt{1 + 4x^2}$ **19.** sec x
21. $-2 \csc 2x$ **23.** $1/[2\sqrt{x(x - 1)}]$ **25.** $-1/(|x|\sqrt{x^2 + 1})$ **27.** $|\sec x|$ **29.** $x/\sqrt{1 + x^2} + 1/(|x|\sqrt{x^2 + 1})$ **31.** $x^2/\sqrt{x^2 - 4}$
33. $(1/2)\ln[(8 + \sqrt{73})/3]$ **35.** $(1/2)\ln 3$ **37.** $-(1/2)\ln[(1 + \sqrt{2})/(2 + \sqrt{5})]$ **39.** 0 **41.** $(x/2)\sqrt{x^2 + 1} + (1/2)\sinh^{-1}x + C$
43. $24 \ln(3/2) - 3 \ln 3 + 6$ **45.** $\cosh^{-1}(x - 2) + C, x > 3$ **47.** $\ln|u + \sqrt{u^2 - 1}| + C$ **49.** $-\ln|(\sqrt{1 + u^2} + 1)/u| + C$
51. $\sqrt{x^2 - 25} - 5 \sec^{-1}|x/5| + C$ **53.** $(25/2)\cosh^{-1}(x/5) + (x/2)\sqrt{x^2 - 25} + C$ **55.** $(x/2)\sqrt{x^2 - 4} - 2 \cosh^{-1}(x/2) + C$
57. $\sqrt{5}/2 + (1/4)\ln(2 + \sqrt{5})$ **59.** $\sqrt{2} - (5/4) - \ln(1 + \sqrt{2}) + \ln 3$ **61.** $y = x^2/4 - (1/2)\ln|x| - 1/4$

Miscellaneous Problems, Chapter 9, pp. 592–593

3. 0 **5.** $\cosh x = 41/9, \tanh x = -40/41$ **9.** $y = 1$ **13.** $6 \sinh 3x \cosh 3x$ **15.** $-1/[\sqrt{1 - x^2} \sin^{-1}x \sqrt{1 - (\sin^{-1}x)^2}]$
17. $[\sec^2(\tanh^{-1}x)]/(1 - x^2)$ **19.** $e^{\cosh^{-1}x}/\sqrt{x^2 - 1}$ **21.** $|\sec x|$ **23.** 1
25. $(18 - 10\sqrt{3})/27$ **27.** $\ln|\cosh \sqrt{2} - \cosh 1|$ **29.** Diverges **31.** $\ln \sqrt{3}$ **33.** ln 2

CHAPTER 10

Article 10.1, pp. 600–601

1. These label the same points: a, c; b, d; e, j, k; f, i, l; m, o; n, p. **3.** a) $(2, \pi/2 + 2k\pi)$ or $(-2, -\pi/2 + 2k\pi)$, k an integer
b) $(2, 2k\pi)$ or $(-2, (2k + 1)\pi)$, k an integer c) $(-2, \pi/2 + 2k\pi)$ or $(2, 3\pi/2 + 2k\pi)$, k an integer d) $(-2, 2k\pi)$ or
$(2, (2k + 1)\pi)$, k an integer **5.** a) $(1, 1)$ b) $(1, 0)$ c) $(0, 0)$ d) $(-1, -1)$ e) $(3\sqrt{3}/2, -3/2)$ f) $(3, 4)$ g) $(1, 0)$ h) $(-\sqrt{3}, 3)$

23. $x = 2$ **25.** $y = 4$ **27.** $y = 0$ **29.** $x + y = 1$ **31.** $x^2 + y^2 = 1$ **33.** $y = e^x$ **35.** $y - 2x = 5$ **37.** $x^2 + (y - 2)^2 = 4$
39. $x^2 = 4y$ **41.** $r \cos \theta = 7$ **43.** $\theta = \pi/4$ **45.** $r = 2$ **47.** $(r^2\cos^2\theta)/9 + (r^2\sin^2\theta)/4 = 1$ **49.** $r = 4 \cot \theta \csc \theta$
51. $x + \sqrt{3}y = 6$ **53.** $x - y = 2$

Article 10.2, pp. 607–608

1. Symmetric to x-axis **3.** Symmetric to x-axis, y-axis, and origin **5.** Symmetric to y-axis

7. Symmetric to y-axis **9.** a) y **11.** a) y

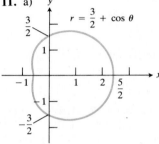

17. $(a(1 - \sqrt{2}/2), (3\pi)/4), (a(1 + \sqrt{2}/2), (7\pi)/4), (0, 0)$ **19.** $((-1 + \sqrt{5})/2, \cos^{-1}[(3 - \sqrt{5})/2]), ((-1 + \sqrt{5})/2,$
$2\pi - \cos^{-1}[(3 - \sqrt{5})/2]$ **21.** $(1, \pm\pi/12), (1, \pm5\pi/12), (1, \pm13\pi/12), (1, \pm17\pi/12)$ **23.** $(1 \pm \sqrt{2}/2,$
$3\pi/2), (1 \pm \sqrt{2}/2, \pi/2), (0, 0)$

Article 10.3, pp. 612–613

1. $(x - 2)^2 + y^2 = 4$ **3.** $(x + 1)^2 + y^2 = 1$ **5.** $(x^2 + y^2)^3 = 4x^2y^2$ **7.** $(x^2 + y^2)^2 = 8(x^2 - y^2)$

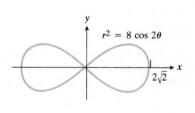

9. $(x^2 + y^2 + 16x)^2 = 64(x^2 + y^2)$ **11.** $(x^2 + y^2 + 4x)^2 = 16(x^2 + y^2)$

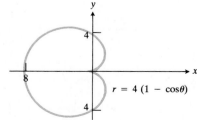

13. $3x^2 - y^2 + 16x + 16 = 0$ **15.** $3x^2 + 4y^2 + 8x = 16$ **17.** $9y^2 = -12x + 4$ **19.** Circle, center $(1, -\pi/4)$, radius $= 1$
21. The line $x + \sqrt{3}y = 10$ **23.**   **25.** d **27.** 1 **29.** k **31.** i

35. Mercury: $r = 0.3707/[1 - (0.2056)\cos \theta]$; Venus: $r = 0.7233/[1 - (0.0068)\cos \theta]$; Earth: $r = 0.9997/[1 - (0.0167)\cos \theta]$;
Mars: $r = 1.5107/[1 - (0.0934)\cos \theta]$; Jupiter: $r = 5.1908/[1 - (0.0484)\cos \theta]$; Saturn: $r = 9.5109/[1 - (0.0543)\cos \theta]$;
Uranus: $r = 19.139/[1 - (0.0460)\cos \theta]$; Neptune: $r = 30.058/[1 - (0.0082)\cos \theta]$ **37.** a) $x^2 + y^2 = 2y$ and $x = 1$
b) $(1, 1)$ or $(\sqrt{2}, \pi/4)$; $(-1, 1)$ or $(\sqrt{2}, 3\pi/4)$ **39.** $r = 4/(1 - \cos \theta)$ **41.** $(50, 0)$, $r = -45/(4 - 5 \cos \theta)$

Article 10.4, p. 620

1. $(\pi + 2)/16$ **3.** 18π **5.** πa^2 **7.** $[(3\sqrt{3} - \pi)/3]a^2$ **9.** $(\pi/2 - 1)a^2$ **11.** $(2\pi + 3\sqrt{3})/6$ **13.** $[(3\pi - 8)/2]a^2$ **15.** 2
17. $4a^2$ **19.** $1/8(2\pi e^{2\pi} - e^{2\pi} + 1)$ **21.** a) $2\pi a$ b) πa c) πa **23.** $2a$ **25.** $(3a\pi)/2$ **27.** $4\pi a^2(2 - \sqrt{2})$
29. $[4\pi(8 - \sqrt{2})]/5$

Miscellaneous Problems, Chapter 10, pp. 621–624

1. Spiral of Archimedes (See graph of Article 10.2, Problem 7.) **3.** a) Vertical line $x = a$ b) Horizontal line $y = a$ c) Curve:
$y = ax/(x - a) = a + a^2/(x - a)$; hyperbola with asymptotes $x = a$, $y = a$ **5.** Circle with center $(-1, -1)$, $r = 3$
7.   **9.** Lemniscate (similar to Fig. 10.19) **11.** a)

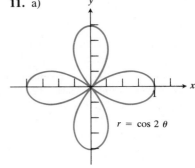

b) Lemniscate (symmetrical to x-axis, y-axis, and origin)

13. a) Parabola b) Parabola **15.** $((a\sqrt{2})/2, \pi/4)$, $(0, 0)$ **17.** (a, π), $(a, 0)$ **19.** $(0, 0)$ **21.** The origin and the points
$(a/2, \pm\pi/3)$, $(a/2, \pm 2\pi/3)$ **23.** a) $y^2 - 2x = 1$ b) $y^2 + 2x = 1$ **25.** a) $3x^2 - y^2 + 4x + 1 = 0$ b) $3x^2 - y^2 - 4x + 1 = 0$
27. $r = 2/(1 + \cos \theta)$ **29.** $r = -2a \cos \theta$ **31.** $r = 8/(3 - \cos \theta)$ **33.** a^2 **35.** $(3a^2\pi)/2$ **37.** $4a^2$ **39.** $(5a^2\pi)/4$
41. $[a^2(3\pi - 8)]/2$ **43.** $(3a\pi)/2$ **45.** When $\tan \psi_2 = -1/\tan \psi_1$ **47.** 0 **51.** Intersect at $(2, \pi/3)$ and $(2, -\pi/3)$; $\beta = \pi/2$ at
both points **55.** $\beta = \pi/2$ at both points of intersection **57.** $\pi/4$ **61.** $(5a/6, 0)$ **63.** $(4a/5, 0)$

CHAPTER 11

Article 11.1, pp. 635–636

1. $2/3$, $3/5$, $4/7$, $5/9$; converges to $1/2$ **3.** $-1/3$, $-3/5$, $-5/7$, $-7/9$; converges to -1 **5.** $1/2$, $1/2$, $1/2$, $1/2$; converges to
$1/2$ **7.** 1, $3/2$, $7/4$, $15/8$, $31/16$, $63/32$ **9.** 2, 1, $1/2$, $1/4$, $1/8$, $1/16$ **11.** 1, 1, 2, 3, 5, 8 **13.** b) $\sqrt{2}$
15. $x = 0.739085133$ **17.** Converges to 0.876726216; $x^2 - \sin x = 0$ **19.** $x = 1.165561185$; yes

Article 11.2, pp. 640–641

1. 0, $-1/4$, $-2/9$, $-3/16$; converges to 0 **3.** $1/3$, $1/9$, $1/27$, $1/81$; converges to 0 **5.** 1, $-1/3$, $1/5$, $-1/7$; converges to 0
7. 0, -1, 0, 1; diverges **9.** 1, $-1/\sqrt{2}$, $1/\sqrt{3}$, $-1/2$; converges to 0 **11.** Converges to 0 **13.** Converges to 1 **15.** Diverges

17. Converges to $-2/3$ **19.** Converges to $\sqrt{2}$ **21.** Converges to 0   **23.** Diverges   **25.** Converges to 1   **27.** Converges to -5
29. Converges to 1   **31.** Converges to 1   **33.** Converges to 5   **35.** Converges to 0   **37.** Converges to 0   **39.** Converges to $\sqrt{2}$
41. Diverges   **43.** Converges to 0   **45.** Diverges   **47.** Converges to 0   **49.** Converges to $\pi/2$ **51.** Converges to 1
53. Converges to $1/2$ **57.** Converges to 1 for any $x > 0$ **61.** 1

Article 11.3, pp. 644–645

1. Converges to 0   **3.** Converges to 0   **5.** Converges to 0   **7.** Converges to e^7 **9.** Converges to 0   **11.** Diverges
13. Converges to 1   **15.** Converges to 0   **17.** Converges to $1/e$ **19.** Converges to 0   **21.** Converges to x **23.** Converges to 1
25. Converges to 1   **27.** Converges to $1/e$ **29.** Converges to 0   **31.** Diverges   **33.** Converges to $1/(p-1)$ **35.** $n \geq 9124$
37. $n > 10^{75}$

Article 11.4, pp. 655–657

1. $S_n = 3[1 - (1/3)^n]$; 3   **3.** $S_n = (1 - e^{-n})/(1 - e^{-1})$; $e/(e-1)$ **5.** $S_n = [1 - (-2)^n]/3$; diverges   **7.** $S_n = -\ln(n+1)$;
diverges   **9.** a) $\sum_{n=-2}^{\infty} [1/(n+4)(n+5)]$ b) $\sum_{n=0}^{\infty} [1/(n+2)(n+3)]$ c) $\sum_{n=5}^{\infty} [1/(n-3)(n-2)]$ **11.** $4/3$ **13.** $7/3$ **15.** $23/2$
17. $5/3$ **19.** 1   **21.** $1/9$ **23.** a) $26/111$ b) Yes, every repeating decimal forms a geometric series with ratio of 10^{-n}, where n
is the number of repeating digits.   **25.** Converges to $2 + \sqrt{2}$ **27.** Converges to 1   **29.** Diverges   **31.** Converges to $e^2/(e^2 - 1)$
33. Diverges   **35.** Converges to $3/2$ **37.** Diverges   **39.** $a = 1$, $r = -x$ **41.** 8   **43.** Take $\sum_{n=1}^{\infty} n$ and $\sum_{n=1}^{\infty} (-n)$ **45.** Take
$\sum_{n=1}^{\infty} a_n = \sum_{n=1}^{\infty} 1/2^n$ and $\sum_{n=1}^{\infty} b_n = \sum_{n=1}^{\infty} 1/3^n$

Article 11.5, pp. 668–669

Note. The tests mentioned in Problems 1–23 may not be the only ones that apply.

1. Converges; geometric series, $r = 1/10 < 1$ **3.** Converges; compare with $\Sigma \, 1/2^n$ **5.** Converges; compare with $\Sigma \, 2/n^2$
7. Diverges; compare with $\Sigma \, 1/n$ **9.** Converges; geometric series with $r = 2/3 < 1$ **11.** Diverges; compare with $\Sigma \, 1/(n+1)$
13. Diverges; $\lim_{n\to\infty} 2^n/(n+1) \neq 0$.   **15.** Converges; compare with $\Sigma \, 1/n^{3/2}$ **17.** Diverges; integral test   **19.** Diverges;
$\lim_{n\to\infty} [1 + (1/n)]^n = e \neq 0$ **21.** Converges; sum of two convergent series, $\Sigma \, 1/n2^n$ and $-\Sigma \, 1/2^n$ **23.** Converges; compare
with $\Sigma \, 1/3^{n-1}$ **27.** $S_n < 41.554$

Article 11.6, p. 677

Note. The reasons given may not be the only appropriate ones.

1. Converges; ratio test   **3.** Converges; root test   **5.** Diverges; $\lim_{n\to\infty} [(n-2)/n]^n = e^{-2} \neq 0$ **7.** Diverges; ratio test
9. Diverges; nth term test   **11.** Diverges; compare with $\Sigma \, 1/n$ **13.** Converges; limit comparison test   **15.** Diverges; limit
comparison test   **17.** Diverges; integral test   **19.** Diverges; limit comparison test   **21.** Converges; ratio test   **23.** Converges; ratio
test   **25.** Converges; ratio test   **27.** Converges for $-2 < x < 2$ **29.** Converges for $-\sqrt{2} < x < \sqrt{2}$
31. Converges for $-1 \leq x \leq 1$ **33.** Converges; ratio test   **35.** Diverges; integral test   **37.** Converges; ratio test   **39.** Converges;
ratio test   **41.** Converges; compare with $\Sigma \, 1/n^4$

Article 11.7, pp. 682–683

Note. The reasons given may not be the only appropriate ones.

1. Converges absolutely; p-series for $p = 2$ **3.** Diverges; $\Sigma \, 1/n$ diverges and $\Sigma \, [(1-n)/n^2] = \Sigma \, 1/n^2 - \Sigma \, 1/n$ **5.** Converges
absolutely; compare with $\Sigma \, 1/n^2$ **7.** Converges absolutely; compare with p-series for $p = 3/2$ **9.** Converges absolutely; ratio
test   **11.** Diverges; nth term test   **13.** Converges absolutely; geometric series with $r = 1/5 < 1$ **15.** Converges absolutely; ratio
test   **17.** Converges absolutely; compare with p-series $\Sigma 1/n^2 - 2\Sigma \, 1/n^3$

Article 11.8, pp. 689–690

1. Converges   **3.** Diverges   **5.** Converges   **7.** Converges   **9.** Converges   **11.** Absolutely convergent   **13.** Absolutely
convergent   **15.** Conditionally convergent   **17.** Diverges   **19.** Conditionally convergent   **21.** Absolutely convergent
23. Absolutely convergent   **25.** Diverges   **27.** Conditionally convergent   **29.** 0.2   **31.** 2×10^{-11} **33.** 0.54030   **35.** a) The a_n
are not decreasing.   b) $-1/2$ **39.** Let $a_n = b_n = (-1)^{n+1}/\sqrt{n}$

Article 11.9, p. 692

1. Converges   **3.** Converges   **5.** Diverges   **7.** Converges   **9.** Converges   **11.** Diverges   **13.** Diverges   **15.** Diverges
17. Diverges   **19.** Converges

Miscellaneous Problems, Chapter 11, pp. 698–699

1. $S_n = \ln[(n+1)/2n]$; converges to $\ln 1/2$ **7.** Diverges; $\lim_{n\to\infty} \tanh n = 1 \neq 0$ **9.** Diverges   **11.** Converges   **13.** Diverges
15. Converges   **17.** Diverges   **19.** Converges

CHAPTER 12

Article 12.2, p. 708

1. $1 - x + x^2/2! - x^3/3!$; $1 - x + x^2/2! - x^3/3! + x^4/4!$ **3.** $1 - x^2/2!$; $1 - x^2/2! + x^4/4!$ **5.** $x + x^3/3!$; $x + x^3/3!$
7. $1 - 2x$; $1 - 2x + x^4$ **9.** $1 - 2x + x^2$; $1 - 2x + x^2$ **11.** x^2 **13.** $1 + (3/2)x + (1/2!)(3/2)(1/2)x^2 +$
$(1/3!)(3/2)(1/2)(-1/2)x^3 + \cdots + (1/n!)[(3)(1)(-1)(-3) \cdots (5 - 2n)/2^n]x^n + \cdots$ **15.** $\Sigma_{n=0}^{\infty} e^{10}(x - 10)^n/n!$
17. $\Sigma_{n=1}^{\infty} (-1)^{n-1}(x - 1)^n/n$ **19.** $-\Sigma_{n=0}^{\infty} (x + 1)^n$ **21.** $1 + 2(x - \pi/4) + 2(x - \pi/4)^2$

Article 12.3, pp. 717–718

1. $\Sigma_{n=0}^{\infty} (x^n/2^n n!)$ **3.** $5 \Sigma_{n=0}^{\infty} (-1)^n[x^{2n}/\pi^{2n}(2n)!]$ **5.** $\Sigma_{n=2}^{\infty} (-1)^n(x^{2n}/2n!)$ **9.** $1 - x + x^2 + R_2(x)$ **11.** $1 + (1/2)x - (1/8)x^2 +$
$R_2(x)$ **13.** $|x| < (0.06)^{1/5} \approx 0.57$ **15.** $|R_2| \le 10^{-9}/3! < 1.67 \times 10^{-10}$; $x < \sin x$ for $-10^{-3} < x < 0$ **17.** $|R_2| \le e^{0.1}/6000 <$
1.85×10^{-4} **19.** $|R_4| \le [(0.5)^5/5!] \cosh 0.5 < 0.0003$ **23.** a) $-1 + 0i$ b) $\sqrt{2}/2 + (\sqrt{2}/2)i$ c) $0 - i$ d) $0 + i$
27. $\int e^{ax} \cos bx \, dx = [e^{ax}/(a^2 + b^2)](a \cos bx + b \sin bx) + C_1$; $\int e^{ax} \sin bx \, dx = [e^{ax}/(a^2 + b^2)](a \sin bx - b \cos bx) + C_2$

Article 12.4, pp. 725–726

1. $31° \approx 0.5411$ radians; use $\cos x \approx 1 - x^2/2! + x^4/4! \approx 0.857$; $|R| \le (0.5411)^6/6! < 3.5 \times 10^{-5}$
3. $\sin 6.3 \approx \sin(2\pi + 0.0168) = \sin(0.0168)$; use $\sin x \approx x = 0.0168$; $|R| \le (0.0168)^3/3! < 7.91 \times 10^{-7}$
5. Use $\ln(1 + x) \approx x - x^2/2 + x^3/3 - x^4/4 \approx 0.223$; $|R| \le (0.25)^5/5 < 2 \times 10^{-4}$ **7.** $\Sigma_{n=1}^{\infty} (-1)^{n-1}[(2^n x^n)/n]$; $-1/2 < x \le 1/2$
9. 0.100   **13.** $\ln(3/2)$ **17.** $C_3 \approx 3.1416127$ **21.** c) 6 d) $1/q$

Article 12.5, pp. 739–740

1. $|x| < 1$; diverges at $x = \pm 1$ **3.** $|x| < 2$; diverges at $x = \pm 2$ **5.** $-\infty < x < \infty$ **7.** $-4 < x < 0$; diverges at $x = -4, 0$
9. $|x| < 1$; converges at $x = \pm 1$ **11.** $-\infty < x < \infty$ **13.** $|x| < 1/e$; diverges at $x = 1/e$; converges at $x = -1/e$ **15.** $x = 0$
17. $1/2 < x < 3/2$; diverges at $x = 1/2, 3/2$ **19.** $|x| < \sqrt{3}$; diverges at $x = \pm\sqrt{3}$ **21.** e^{3x+6} **25.** $\Sigma_{n=1}^{\infty} (2n)x^{2n-1}$ **27.** 0.0027
29. 0.00033   **31.** 0.3636   **33.** 0.100   **35.** a) $\sinh^{-1}x = x - (1/6)x^3 + (3/40)x^5 - (5/112)x^7 + \cdots$ b) $\sinh^{-1}0.25 \approx 0.247$
37. Problem 15 is such an example.   **41.** $x + x^2 + x^3/3 - x^5/30 + \cdots$ **43.** $x^2/2 + x^4/12 + x^6/45 + \cdots$ **45.** c) Yes

Article 12.6, p. 742

1. 1   **3.** $-1/24$ **5.** -2 **7.** 0   **9.** $-1/3$ **11.** $1/120$ **13.** $1/3$ **15.** -1 **17.** 3   **19.** 0   **21.** b) $1/2$ **23.** $\sin x \approx 6x/(6 + x^2)$
is better.

Miscellaneous Problems, Chapter 12, pp. 745–747

1. a) $\Sigma_{n=0}^{\infty} (-1)^n x^{n+2}$ b) No, the series converges for $|x| < 1$.   **3.** $1 + x + x^2/2 - x^4/8$ **5.** a) $-x^2/2 - x^4/12 - x^6/45 - \cdots$
b) -0.00017 **7.** 0.747   **9.** $\Sigma_{n=0}^{\infty} (-1)^{n+1}(x - 2)^n$; $1 < x < 3$ **11.** $\cos x = (1/2) \Sigma_{n=0}^{\infty} (-1)^n(x - \pi/3)^{2n}/2n! +$
$(\sqrt{3}/2) \Sigma_{n=0}^{\infty} (-1)^{n+1}(x - \pi/3)^{2n+1}/(2n + 1)!$ **13.** $f(1) = 1.543$ **15.** $e^{(e^x)} = e[1 + x + x^2 + (5/6)x^3 + \cdots]$ **17.** $|E| \le 0.0011$
19. $\lim_{x \to 0} ((\sin x)/x)^{1/x^2} = e^{-1/6}$ **21.** $-5 \le x < 1$ **23.** $-\infty < x < \infty$ **25.** $1 < x < 5$ **27.** $0 \le x \le 2$ **29.** $-\infty < x < \infty$
31. $x \ge 1/2$ **37.** $x + x^2 + (2/3)x^3 + (2/3)x^4 + (13/15)x^5 + \cdots$

CHAPTER 13

Article 13.1, pp. 755–756

1. a)

d)

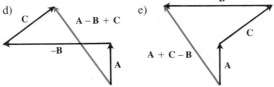

3. $-\mathbf{i} + \mathbf{j}$ 5. 0 7. $(\sqrt{3}/2)\mathbf{i} - (1/2)\mathbf{j}$ 9. $\pm(1/\sqrt{17})\mathbf{i} \pm (4/\sqrt{17})\mathbf{j}$ 11. a) $-(1/\sqrt{17})\mathbf{i} - (4/\sqrt{17})\mathbf{j}$ b) $-(4/\sqrt{17})\mathbf{i} + (1/\sqrt{17})\mathbf{j}$ 13. $\sqrt{2}$, $(1/\sqrt{2})\mathbf{i} + (1/\sqrt{2})\mathbf{j}$, $\pi/4$ 15. 2, $(\sqrt{3}/2)\mathbf{i} + (1/2)\mathbf{j}$, $\pi/6$ 17. 13, $(5/13)\mathbf{i} + (12/13)\mathbf{j}$, $\tan^{-1}(12/5)$ 19. $2\sqrt{5}$

Article 13.2, pp. 761–763

1. 50 sec 3. a) 72.2 sec, 25,510.2 m b) 4020 m c) 6377.6 m 7. 9.9 m/sec; 18.5° or 71.6° 9. 189.6 mph 13. No, $y = 29.9$ ft 15. 149.3 ft/sec; 2.25 sec 17. Yes; marble A will reach point R when $t = R/(v_0\cos \alpha)$. At this moment, the height of both marbles is $R \tan \alpha - \frac{1}{2}[R/(v_0\cos \alpha)]^2$.

Article 13.3, pp. 772–773

1. Line parallel to z-axis through (2, 3, 0) 3. x-axis 5. Circle in xy-plane, center (0, 0, 0), radius = 2 7. Circle in xz-plane, center (0, 0, 0), radius = 2 9. Circle in yz-plane, center (0, 0, 0), radius = 1 11. Circle in xy-plane, center (0, 0, 0), radius = 4 13. a) First quadrant of xy-plane b) Fourth quadrant of xy-plane 17. a) Hemisphere above xy-plane, center at origin, radius = 1 b) Hemisphere with interior 19. a) $x = 3$ b) $y = -1$ c) $z = -2$ 21. a) $z = 1$ b) $x = 3$ c) $y = -1$ 23. a) $x^2 + (y - 2)^2 = 4$, $z = 0$ b) $(y - 2)^2 + z^2 = 4$, $x = 0$ c) $x^2 + z^2 = 4$, $y = 2$ 25. a) $y = 3$, $z = -1$ b) $x = 1$, $z = -1$ c) $x = 1$, $y = 3$ 27. $x^2 + y^2 = 16$, $z = 3$ 29. $0 \le z \le 1$ 31. $z \le 0$ 33. a) $(x - 1)^2 + (y - 1)^2 + (z - 1)^2 < 1$ b) $(x - 1)^2 + (y - 1)^2 + (z - 1)^2 > 1$ 35. 3, $(2/3)\mathbf{i} + (1/3)\mathbf{j} - (2/3)\mathbf{k}$ 37. 9, $(1/9)\mathbf{i} + (4/9)\mathbf{j} - (8/9)\mathbf{k}$ 39. 5, $\mathbf{k}$ 41. 1, $(1/\sqrt{2})\mathbf{i} - (1/\sqrt{2})\mathbf{k}$ 43. 1, $(1/\sqrt{3})\mathbf{i} + (1/\sqrt{3})\mathbf{j} + (1/\sqrt{3})\mathbf{k}$ 45. 6, $\mathbf{i}$ 47. a) $2\mathbf{i}$ b) $-4\mathbf{j}$ c) $\sqrt{3}\mathbf{k}$ d) $(3/10)\mathbf{j} + (4/10)\mathbf{k}$ e) $6\mathbf{i} - 2\mathbf{j} + 3\mathbf{k}$ f) $(au_1/|u|)\mathbf{i} + (au_2/|u|)\mathbf{j} + (au_3/|u|)\mathbf{k}$, $|u| = \sqrt{u_1^2 + u_2^2 + u_3^2}$ 49. $(5/\sqrt{3})\mathbf{i} + (5/\sqrt{3})\mathbf{j} + (5/\sqrt{3})\mathbf{k}$ 51. $C(-2, 0, 2)$, $r = 2\sqrt{2}$ 53. $C(0, 0, a)$, $r = a$ 55. a) $\sqrt{y^2 + z^2}$ b) $\sqrt{x^2 + z^2}$ c) $\sqrt{x^2 + y^2}$ 57. a) $(3/2)\mathbf{i} + (3/2)\mathbf{j} - 3\mathbf{k}$ b) $\mathbf{i} + \mathbf{j} - 2\mathbf{k}$ c) (2, 2, 1)

Article 13.4, pp. 779–780

	$\mathbf{A} \cdot \mathbf{B}$	$|\mathbf{A}|$	$|\mathbf{B}|$	$\cos \theta$	$|\mathbf{B}| \cos \theta$	$\text{Proj}_{\mathbf{A}}\mathbf{B}$
1.	10	$\sqrt{13}$	$\sqrt{26}$	$5\sqrt{2}/13$	$10/\sqrt{13}$	$(30/13)\mathbf{i} + (20/13)\mathbf{j}$
3.	4	$\sqrt{14}$	2	$\sqrt{14}/7$	$2\sqrt{14}/7$	$(6/7)\mathbf{i} - (4/7)\mathbf{j} - (2/7)\mathbf{k}$
5.	2	$\sqrt{34}$	$\sqrt{3}$	$2/\sqrt{102}$	$2/\sqrt{34}$	$(5/17)\mathbf{j} - (3/17)\mathbf{k}$
7.	$-\sqrt{2} + \sqrt{3}$	$\sqrt{2}$	3	$(\sqrt{6} - 2)/6$	$(\sqrt{6} - 2)/2$	$[(\sqrt{3} - \sqrt{2})/2](-\mathbf{i} + \mathbf{j})$
9.	-25	5	5	-1	-5	$-2\mathbf{i} + 4\mathbf{j} - \sqrt{5}\mathbf{k}$
11.	25	15	5	$1/3$	$5/3$	$(10/9)\mathbf{i} + (11/9)\mathbf{j} - (2/9)\mathbf{k}$

13. 37.9°, 71.1°, 71.1° 15. 35.3° 17. (3, 3, 0) 19. $\mathbf{B}_1 = (3/2)\mathbf{i} + (3/2)\mathbf{j}$, $\mathbf{B}_2 = -(3/2)\mathbf{i} + (3/2)\mathbf{j} + 4\mathbf{k}$ 21. a) $2\sqrt{10}$ b) $6/\sqrt{10}$ 23. Verify that $\mathbf{A} \cdot \mathbf{B} = 0$, $\mathbf{A} \cdot \mathbf{C} = 0$, $\mathbf{B} \cdot \mathbf{C} = 0$ 25. -5 N-m 27. $2000\sqrt{3}$ N-m 33. $\pi/4$ 35. $\pi/3$ 37. $\pi/2$, $\pi/4$

Article 13.5, p. 785

1. $\sqrt{26}$, $-(1/\sqrt{26})\mathbf{i} - (3/\sqrt{26})\mathbf{j} + (4/\sqrt{26})\mathbf{k}$, $\sqrt{26}$, $(1/\sqrt{26})\mathbf{i} + (3/\sqrt{26})\mathbf{j} - (4/\sqrt{26})\mathbf{k}$ 3. $\mathbf{A} \times \mathbf{B} = \mathbf{B} \times \mathbf{A} = 0\mathbf{i} + 0\mathbf{j} + 0\mathbf{k}$ 5. 6, $-\mathbf{k}$, 6, $\mathbf{k}$ 7. $6\sqrt{5}$, $(1/\sqrt{5})\mathbf{i} - (2/\sqrt{5})\mathbf{k}$, $6\sqrt{5}$, $-(1/\sqrt{5})\mathbf{i} + (2/\sqrt{5})\mathbf{k}$ 9. $\mathbf{k}$ 11. $\mathbf{i} - \mathbf{j} + \mathbf{k}$ 13. $3\mathbf{i} - \mathbf{j}$ 15. $8\mathbf{i} + 4\mathbf{j} + 4\mathbf{k}$ 17. $\sqrt{3}/2$ 19. 3 21. $2\mathbf{i} + 4\mathbf{j} + 7\mathbf{k}$, 0, 0 23. $\mathbf{A}\|\mathbf{C}$ 27. 2 29. 13 31. 11/2 33. 25/2

Article 13.6, pp. 791–792

1. $x = 3 + t$, $y = -4 + t$, $z = -1 + t$ 3. $x = -2 - 5t$, $y = -5t$, $z = 3 + 5t$ 5. $x = 0$, $y = 2t$, $z = t$ 7. $x = 1$, $y = 1$, $z = 1 + t$ 9. $x = t$, $y = -7 + 2t$, $z = 2t$ 11. $x = t$, $y = 0$, $z = 0$ 13. $x = t$, $y = t$, $z = t$, $0 \le t \le 1$ 15. $x = 1$, $y = t$, $z = 0$, $0 \le t \le 1$ 17. $x = 0$, $y = 1 - 2t$, $z = 1$, $0 \le t \le 1$ 19. $x = 1 + t$, $y = 2$, $z = -2 + 2t$, $0 \le t \le 1$ 21. $3x - 2y - z = -3$ 23. $7x - 5y - 4z = 6$ 25. $x + y + z = 1$ 27. $4x + 4y - 5z = 0$ 29. $x - y + z = 0$ 31. $\sqrt{120}$ 33. 0 35. 2 37. 3 39. 19/5 41. 5/3 43. $(3/2, -3/2, 1/2)$ 45. (1, 1, 0) 47. 78.9° 49. 90° 51. $x = 1 - t$, $y = 1 + t$, $z = -1$ 53. $x = 4$, $y = 1 + 6t$, $z = 3t$ 55. $(1, -1, 0)$, $(-1, 0, -3)$, $(0, -1/2, -3/2)$

Article 13.7, p. 798

1. 245 3. $200\mathbf{i} - 120\mathbf{j} - 90\mathbf{k}$ 5. 4/3 7. 7/6 9. a) $(0, 9, -3)$ b) $\sqrt{963}$ c) 11, 29, 1 11. Parts (a), (e), (i)

Miscellaneous Problems, Chapter 13, pp. 799–802

11. $3/11(3\mathbf{i} - \mathbf{j} + \mathbf{k})$ 13. $\mathbf{C} = (\mathbf{A} \cdot \mathbf{B}/\mathbf{B} \cdot \mathbf{B})\mathbf{B}$, $\mathbf{D} = \mathbf{A} - \mathbf{C}$ 19. $(2/3)\mathbf{i} + (2/3)\mathbf{j} + (4/3)\mathbf{k}$ 23. $2x + 2y + z = 5$ 25. $(4/3, -2/3, -2/3)$ 27. $-3\mathbf{j} + 3\mathbf{k}$ 29. 59.5° 31. $(1, -2, -1)$, $x = 1 - 5t$, $y = -2 + 3t$, $z = -1 + 4t$ 39. $(1/\sqrt{2})\mathbf{j} + (1/\sqrt{2})\mathbf{k}$ 41. 60° 43. To the location $(26, 23, -\frac{1}{3})$ 45. $\sqrt{78}/3$ 47. 3 49. b) $6/\sqrt{14}$ c) $2x - y + 2z = 8$ d) $x - 2y + z = 3 \pm 5\sqrt{6}$ 51. $11/\sqrt{107}$ 53. a) 1 b) $-10\mathbf{i} - 2\mathbf{j} - 12\mathbf{k}$ 59. $\cos \alpha = 2/3$, $\cos \beta = 11/15$, $\cos \gamma = -2/15$ 61. $\cos \alpha = 0$, $\cos \beta = 1/2$, $\cos \gamma = \sqrt{3}/2$ 63. $\cos \alpha = 1$, $\cos \beta = 0$, $\cos \gamma = 0$ 65. $\cos \alpha = \cos \beta = \cos \gamma = 1/\sqrt{3}$ 67. $\cos \alpha = 2/7$, $\cos \beta = -3/7$, $\cos \gamma = 6/7$

CHAPTER 14

Article 14.1, pp. 811–812

1. $\mathbf{v} = (-2\sin t)\mathbf{i} + (3\cos t)\mathbf{j}$; $\mathbf{a} = (-2\cos t)\mathbf{i} - (3\sin t)\mathbf{j}$; at $t = \pi/4$, $|\mathbf{v}| = \sqrt{26}/2$ **3.** $\mathbf{v} = (-2\sin 2t)\mathbf{i} + (2\cos t)\mathbf{j}$; $\mathbf{a} = (-4\cos 2t)\mathbf{i} - (2\sin t)\mathbf{j}$; at $t = 0$, $|\mathbf{v}| = 2$ **5.** $\mathbf{v} = (\sec t \tan t)\mathbf{i} + (\sec^2 t)\mathbf{j}$; $\mathbf{a} = (\sec^3 t + \tan^2 t \sec t)\mathbf{i} + (2\sec^2 t \tan t)\mathbf{j}$; at $t = \pi/6$, $|\mathbf{v}| = 2\sqrt{5}/3$ **7.** $\mathbf{v} = (-a\omega\sin\omega t)\mathbf{i} + (a\omega\cos\omega t)\mathbf{j}$; $\mathbf{a} = (-a\omega^2\cos\omega t)\mathbf{i} - (a\omega^2\sin\omega t)\mathbf{j}$; at $t = \pi/(3\omega)$, $|\mathbf{v}| = a\omega$ **9.** $\mathbf{v} = \mathbf{i} + \sqrt{2}t\mathbf{j} + t^2\mathbf{k}$; $\mathbf{a} = \sqrt{2}\mathbf{j} + 2t\mathbf{k}$; at $t = 1$, $|\mathbf{v}| = 2$ **11.** $\mathbf{v} = (-\sqrt{2}/(t+1)^2)\mathbf{i} + t^2\mathbf{j} + (\sqrt{2}/t)\mathbf{k}$; $\mathbf{a} = (2\sqrt{2}/(t+1)^3)\mathbf{i} + 2t\mathbf{j} - (\sqrt{2}/t^2)\mathbf{k}$; at $t = 1$, $|\mathbf{v}| = 5\sqrt{2}/4$ **13.** Orthogonal for $t = 0, \pi, 2\pi$ **15.** Orthogonal for $t = 0, \pi/2, \pi$ **17.** $\theta = \cos^{-1}(3/\sqrt{15}) \approx 39.2°$ **19.** $\pi/2$ **21.** $\mathbf{v}(t) = -(3/2)t^2\mathbf{i} + 2\mathbf{j}$; $\mathbf{R}(t) = [-(1/2)t^3 + 4]\mathbf{i} + 2t\mathbf{j}$ **23.** $\mathbf{v}(t) = (2\sqrt{1+t} - 3)\mathbf{i} + e^{-t}\mathbf{j}$; $\mathbf{R}(t) = [(4/3)(1+t)^{3/2} - 3t - 1]\mathbf{i} - e^{-t}\mathbf{j} + \mathbf{k}$ **25.** a) $\mathbf{v}(t) = \mathbf{i} + 2t\mathbf{j}$; $\mathbf{a}(t) = 2\mathbf{j}$ b) $\mathbf{v}(2) = \mathbf{i} + 4\mathbf{j}$; $\mathbf{a}(2) = 2\mathbf{j}$

27. $\mathbf{v}(2) = 20\mathbf{i} - 19.6\mathbf{j}$; $|\mathbf{v}(2)| \approx 28$ **29.** Maximum is 3, minimum is 2.

Article 14.2, p. 817

1. $\mathbf{T} = [-(2/3)\sin t]\mathbf{i} + [(2/3)\cos t]\mathbf{j} + (\sqrt{5}/3)\mathbf{k}$; 3π **3.** $\mathbf{T} = (1/\sqrt{1+t})\mathbf{i} + (\sqrt{t}/\sqrt{1+t})\mathbf{j}$; $52/3$ **5.** $\mathbf{T} = (6/7)\mathbf{i} - (2/7)\mathbf{j} - (3/7)\mathbf{k}$; 14 **7.** $\mathbf{T} = [(\cos t - t\sin t)/|1 + t|]\mathbf{i} + [(\sin t + t\cos t)/|1 + t|]\mathbf{j} + (\sqrt{2}t/|1 + t|)\mathbf{k}$; $\pi + (1/2)\pi^2$ **9.** $\mathbf{T} = (-\sin\omega t)\mathbf{i} + (\cos\omega t)\mathbf{j}$; $2\pi a$ **11.** $s(t) = \sqrt{3}(e^t - 1)$; $\sqrt{3}(e^\pi - e^{-\pi})$ **13.** $\mathbf{v}(0) = 6\mathbf{j}$; $|\mathbf{v}(0)| = 6$; $\mathbf{T}(0) = \mathbf{j}$; $\mathbf{a}(0) = 6\mathbf{i}$

Article 14.3, pp. 826–827

1. $\mathbf{T} = (\cos t)\mathbf{i} - (\sin t)\mathbf{j}$; $\mathbf{N} = (-\sin t)\mathbf{i} - (\cos t)\mathbf{j}$; $\kappa = \cos t$ **3.** $\mathbf{T} = (\sin t)\mathbf{i} + (\cos t)\mathbf{j}$; $\mathbf{N} = (\cos t)\mathbf{i} - (\sin t)\mathbf{j}$; $\kappa = \cos t$ **5.** $\mathbf{T} = (t/\sqrt{1+t^2})\mathbf{i} + (1/\sqrt{1+t^2})\mathbf{j}$; $\mathbf{N} = (1/\sqrt{1+t^2})\mathbf{i} - (t/\sqrt{1+t^2})\mathbf{j}$; $\kappa = 1/t(t^2+1)^{3/2}$ **7.** $\mathbf{T} = (-\cos t)\mathbf{i} + (\sin t)\mathbf{j}$; $\mathbf{N} = (\sin t)\mathbf{i} + (\cos t)\mathbf{j}$; $\kappa = -(1/3)\sec t \csc t$ **9.** $\mathbf{T} = [(3/5)\cos t]\mathbf{i} - [(3/5)\sin t]\mathbf{j} + (4/5)\mathbf{k}$; $\mathbf{N} = (-\sin t)\mathbf{i} - (\cos t)\mathbf{j}$; $\kappa = 3/25$; $\mathbf{B} = [(4/5)\cos t]\mathbf{i} - [(4/5)\sin t]\mathbf{j} - (3/5)\mathbf{k}$; $\tau = -4/25$ **11.** $\mathbf{T} = (1/\sqrt{2})(\cos t - \sin t)\mathbf{i} + (1/\sqrt{2})(\sin t + \cos t)\mathbf{j}$; $\mathbf{N} = (1/\sqrt{2})(-\sin t - \cos t)\mathbf{i} + (1/\sqrt{2})(\cos t - \sin t)\mathbf{j}$; $\kappa = (1/\sqrt{2})e^{-t}$; $\mathbf{B} = \mathbf{k}$; $\tau = 0$ **13.** $\mathbf{T} = (2/3)(1+t)^{1/2}\mathbf{i} - (2/3)(1-t)^{1/2}\mathbf{j} + \frac{1}{3}\mathbf{k}$; $\mathbf{N} = \sqrt{(1-t)/2}\,\mathbf{i} + \sqrt{(1+t)/2}\,\mathbf{j}$; $\kappa = \sqrt{2}/(3\sqrt{1-t^2})$ **15.** $\mathbf{a} = (2t/\sqrt{1+t^2})\mathbf{T} + (2/\sqrt{1+t^2})\mathbf{N}$ **17.** $\mathbf{a} = 2\mathbf{N}$ **19.** $\mathbf{a} = a\omega^2\mathbf{N}$ **21.** $\mathbf{a} = (4/3)\mathbf{T} + [(2\sqrt{5})/3]\mathbf{N}$ **23.** $\mathbf{a} = 2\mathbf{N}$ **25.** $\mathbf{a} = 10\mathbf{T} + 6\mathbf{N}$ **27.** Osculating: $z = 0$; normal: $-(\sqrt{2}/2)x + (\sqrt{2}/2)y = 0$; rectifying: $(\sqrt{2}/2)x + (\sqrt{2}/2)y = 1$ **29.** Osculating: $-(1/\sqrt{5})y + (2/\sqrt{5})z = 0$; normal: $(2/\sqrt{5})y + (1/\sqrt{5})z = 0$; rectifying: $x = 2$ **39.** $[x - (\pi/2)]^2 + y^2 = 1$ **45.** $\omega/2b$

Article 14.4, p. 837

1. 93.16 min   **3.** 6763 km   **5.** 1654 min   **7.** 20,426.7 km   **9.** Circle if $v_0 = \sqrt{GM/r_0}$; ellipse if $\sqrt{GM/r_0} < v_0 < \sqrt{2GM/r_0}$; parabola if $v_0 = \sqrt{2GM/r_0}$; hyperbola if $v_0 > \sqrt{2GM/r_0}$ **11.** For planets, 2.9×10^{-19}; for satellites orbiting earth, 9.9×10^{-14}; for satellites orbiting moon, 8.04×10^{-12}.   **15.** $\mathbf{v} = (3a\sin\theta)\mathbf{u}_r + [3a(1 - \cos\theta)]\mathbf{u}_\theta$; $\mathbf{a} = (18a\cos\theta - 9a)\mathbf{u}_r + (18a\sin\theta)\mathbf{u}_\theta$ **17.** $\mathbf{v} = (2ae^{a\theta})\mathbf{u}_r + (2e^{a\theta})\mathbf{u}_\theta$; $\mathbf{a} = (4a^2e^{a\theta} - 4e^{a\theta})\mathbf{u}_r + (8ae^{a\theta})\mathbf{u}_\theta$ **19.** $\mathbf{v} = (-8\sin 4t)\mathbf{u}_r + (4\cos 4t)\mathbf{u}_\theta$; $\mathbf{a} = (-40\cos 4t)\mathbf{u}_r - (32\sin 4t)\mathbf{u}_\theta$

Miscellaneous Problems, Chapter 14, pp. 838–839

1. $s(0) = 1$ **3.** $x = 1 + t$; $y = t$; $z = -t$ **5.** $2\sqrt{3}(\mathbf{i} + \mathbf{j} - 2\mathbf{k})$ **7.** $\mathbf{v} \cdot \mathbf{j} = 12$; $\mathbf{a} \cdot \mathbf{j} = 26$ **9.** c) $\pi/2$ **13.** $(\ln(1/\sqrt{2}), 1/\sqrt{2})$ **17.** $\mathbf{T} = (2/3)\mathbf{i} + (1/3)\mathbf{j} + (2/3)\mathbf{k}$; $\mathbf{N} = (1/\sqrt{5})\mathbf{i} - (2/\sqrt{5})\mathbf{j}$; $\mathbf{B} = (4/3\sqrt{5})\mathbf{i} + (2/3\sqrt{5})\mathbf{j} - (5/3\sqrt{5})\mathbf{k}$; $\kappa = 2\sqrt{5}/9$; $\tau = 4/9$ **19.** $1/a$ **21.** $\kappa = |f^2 + 2(f')^2 - ff''|/[f^2 + (f')^2]^{3/2}$ **23.** a) $\mathbf{v} = -\mathbf{u}_r + 3\mathbf{u}_\theta$; $\mathbf{a} = -9\mathbf{u}_r - 6\mathbf{u}_\theta$ b) $\sqrt{37} + \ln(\sqrt{37} + 6)$ **25.** a) $dx/dt = (\cos\theta)(dr/dt) - (r\sin\theta)(d\theta/dt)$; $dy/dt = (\sin\theta)(dr/dt) + (r\cos\theta)(d\theta/dt)$ b) $dr/dt = (\cos\theta)(dx/dt) + (\sin\theta)(dy/dt)$; $r(d\theta/dt) = (-\sin\theta)(dx/dt) + (\cos\theta)(dy/dt)$ **27.** 3.21

CHAPTER 15

Article 15.1, pp. 844–845

19. True, the generating curve is a line.

21. a) The cylinder $x^2 + y^2 = 1$ with the plane $x + z = 1$ b) $x^2 + y^2 = 1$

c) $s = \displaystyle\int_0^{2\pi} \sqrt{2 - \cos^2 t}\; dt$

Article 15.2, p. 852

1. $9x^2 + y^2 + z^2 = 9$　　　　**5.** $4x^2 + 9y^2 + 4z^2 = 36$　　**7.** $z = x^2 + y^2$　　**11.** $z = 8 - x^2 - y^2$

13. $x = y^2 + 4z^2 + 1$　　**17.** $x^2 + y^2 = z^2$　　**23.** $x^2 + y^2 - z^2 = 1$　　**29.** $z^2 - x^2/4 - y^2 = 1$

33. $y^2 - x^2 = z$

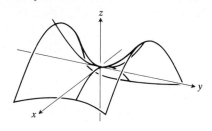

37. $y = x^2 + z$

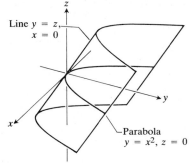

Line $y = z$, $x = 0$

Parabola
$y = x^2, z = 0$

39. The ellipsoid $(x^2/4) + (y^2/9) + z^2 = 1$

Article 15.3, pp. 855–856

	(x, y, z)	(r, θ, z)	(ρ, ϕ, θ)
1.	$(0, 0, 0)$	$(0, 0, 0)$	$(0, 0, 0)$
3.	$(0, 1, 0)$	$(1, \pi/2, 0)$	$(1, \pi/2, \pi/2)$
5.	$(1, 0, 0)$	$(1, 0, 0)$	$(1, \pi/2, 0)$
7.	$(0, 1, 1)$	$(1, \pi/2, 1)$	$(\sqrt{2}, \pi/4, \pi/2)$
9.	$(0, -2\sqrt{2}, 0)$	$(2\sqrt{2}, 3\pi/2, 0)$	$(2\sqrt{2}, \pi/2, 3\pi/2)$

11. Cartesian: $x = y = 0$; spherical: $\rho \sin \phi = 0$; z-axis **13.** $\phi = \pi/2$; $z = 0$; xy-plane **15.** $z = 3$; $z = 3$; plane parallel to xy-plane **17.** $x = 0$; $r \cos \theta = 0$; yz-plane **19.** $\rho = 2$; $r^2 + z^2 = 4$; sphere **21.** A cylinder with xy-section the circle $x^2 + y^2 = 16$ and with z-axis as its axis **23.** A cylinder with xy-section the cardioid $r = 1 - \cos \theta$ and elements parallel to the z-axis

$r = 1 - \cos \theta$

25. A circle in the plane $z = 3$ with center $(0, 0, 3)$ and radius 2 **27.** A helix, passing through the point $(3, 0, 0)$ and spiraling up the surface of the cylinder whose xy-section is the circle $x^2 + y^2 = 9$

The helix $r = 3$, $z = \theta/2$

29. A cone with vertex at the origin, making an angle of $\pi/6$ with the z-axis, and having the z-axis as axis   **31.** A semicircle on the surface of the sphere with center $(0, 0, 0)$ and radius 5. The semicircle passes through the points $(0, 0, 5)$, $(5/\sqrt{2}, 5/\sqrt{2}, 0)$ and $(0, 0, -5)$.

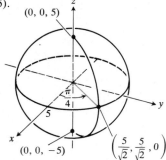

33. A semicircle formed by cutting the hemisphere with center $(0, 0, 2)$, radius $= 2$, $y \geq 0$ by the yz-plane

Miscellaneous Problems, Chapter 15, pp. 856–858

1. Line; plane   **3.** Circle; cylinder   **5.** Parabola; cylinder   **7.** Four-leaf rose; cylinder   **9.** Circle; cylinder   **11.** Four-leaf rose; cylinder   **13.** Region between two concentric spheres   **15.** The wedge-shaped region cut from the cone $\phi = \pi/4$ by the planes $y = 0$ and $y = x$ **17.** Intersection of the sphere $\rho = 1$ and the cylinder $x^2 + y^2 - y = 0$; not a plane curve   **19.** To the z-axis   **21.** $(2, 3, 1)$ **23.** $V = 2\pi R^2 h[1 - (R^2 - r^2)/3R^2]$ **25.** a) $V = (\pi abh/3c^2)(3c^2 + h^2)$ b) $V = (h/3)(2A_0 + A_h)$ **27.** a) $d\theta/dt = \sqrt{2gb\theta/(a^2 + a^2\theta^2 + b^2)}$ b) $s = (a\theta/2)[\sqrt{c^2 + \theta^2} + c^2\ln(\theta + \sqrt{c^2 + \theta^2}) - c^2\ln c]$, $c^2 = (a^2 + b^2)/a^2$
29. $\mathbf{u}_\rho = (\sin\phi\cos\theta)\mathbf{i} + (\sin\phi\sin\theta)\mathbf{j} + (\cos\phi)\mathbf{k}$; $\mathbf{u}_\phi = (\cos\theta\cos\phi)\mathbf{i} + (\sin\theta\cos\phi)\mathbf{j} - (\sin\phi)\mathbf{k}$; $\mathbf{u}_\theta = (-\sin\theta)\mathbf{i} + (\cos\theta)\mathbf{j}$
31. a) $ds^2 = dr^2 + r^2 d\theta^2 + dz^2$ b) $ds^2 = \rho^2 d\phi^2 + d\rho^2 + \rho^2\sin^2\phi d\theta^2$

CHAPTER 16

Article 16.1, pp. 864–865

1. Domain: $(-\infty, \infty)$; range: $(-\infty, \infty)$; level curves: lines $x + y = k$ **3.** Domain: $\{(x, y): y > -x\}$; range: $(-\infty, \infty)$; level curves: lines $x + y = e^k$ **5.** Domain: $\{(x, y): x \neq 0\}$; range: $(-\pi/2, \pi/2)$; level curves: lines $y = mx$, $x \neq 0$ **7.** Domain: $\{(x, y): x \neq 0, y \geq 0\}$; range: $(-\infty, \infty)$; level curves: parabolas $y = kx^2$, $k \geq 0$, $x \neq 0$ **9.** Spheres centered at the origin   **11.** Spheres centered at the origin   **13.** Planes parallel to yz-plane   **15.** Planes $x + y + z = k$ **17.** Cylinders with axis the x-axis
19. a)   b)   **21.** a)   b)

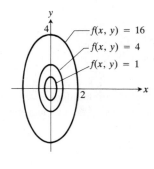

23. The graph of $z = f(x, y)$ is the hemisphere with center $(0, 0, 0)$ and radius 2, above the xy-plane. The level curves are concentric circles with radii ≤ 2.

(a)   (b)

25.

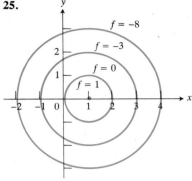

27. $x^2 + y^2 = 10$ **29.** 62.43 km

Article 16.2, p. 872

1. 5/2   **3.** 1/2   **5.** 5   **7.** 1   **9.** 0   **11.** 1   **13.** 0   **15.** $\pi/2$ **23.** $\sqrt{x^2 + y^2} < 0.1$ **25.** $|x| < 0.005$, $|y| < 0.005$
27. $\sqrt{x^2 + y^2 + z^2} < 0.1$ **29.** $|x| < 0.005$, $|y| < 0.005$, $|z| < 0.005$ **31.** Yes   **33.**

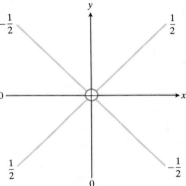

Article 16.3, pp. 876–877

1. $f_x = 2, f_y = 0$ **3.** $f_x = f_y = 0$ **5.** $f_x = y - 1, f_y = x$ **7.** $f_x = 2x - y, f_y = 2y - x$ **9.** $f_x = 5y - 14x + 3, f_y = 5x - 2y - 6$
11. $f_x = x/\sqrt{x^2 + y^2}, f_y = y/\sqrt{x^2 + y^2}$ **13.** $f_x = -(y^2 + 1)/(xy - 1)^2, f_y = -(x^2 + 1)/(xy - 1)^2$ **15.** $f_x = e^x \ln y, f_y = e^x/y$
17. $f_x = e^x \sin(y + 1), f_y = e^x \cos(y + 1)$ **19.** $f_x = 2 \operatorname{sech}^2(2x + 5y), f_y = 5 \operatorname{sech}^2(2x + 5y)$ **21.** $f_x = y + z, f_y = x + z, f_z = y + x$
23. $f_x = 0, f_y = 2y, f_z = 4z$ **25.** $f_P = -1/QRP^2, f_Q = -1/RPQ^2, f_R = -1/PQR^2$ **27.** $f_\rho = \sin \phi \cos \theta, f_\phi = \rho \cos \theta \cos \phi, f_\theta = -\rho \sin \phi \sin \theta$ **29.** -2 **31.** $\partial x/\partial y = 0; \partial x/\partial z = -3$ **33.** $\partial A/\partial a = a/(bc \sin A), \partial A/\partial b = (c \cos A - b)/bc \sin A$
35. $\partial\rho/\partial x = x/\sqrt{x^2 + y^2 + z^2}$ **37.** $\partial\theta/\partial y = x/(x^2 + y^2)$ **39.** $\partial\phi/\partial x = xz/(x^2 + y^2 + z^2)\sqrt{x^2 + y^2}$ **41.** $\partial\mathbf{R}/\partial\rho = \mathbf{u}_\rho,$
$\partial\mathbf{R}/\partial\phi = \rho\mathbf{u}_\phi, \partial\mathbf{R}/\partial\theta = (\rho \sin \phi)\mathbf{u}_\theta$

Article 16.4, pp. 883–884

1. $1/(1 + \pi)$ **3.** 0   **5.** $-3/\sqrt{5}$ **7.** $-1/\sqrt{3}$ **9.** $dz/dt = \partial z/\partial x \cdot dx/dt + \partial z/\partial y \cdot dy/dt$ **11.** $\partial w/\partial y = \partial w/\partial u \cdot \partial u/\partial y + \partial w/\partial v \cdot \partial v/\partial y$ **13.** $\partial y/\partial r = dy/du \cdot \partial u/\partial r$ **15.** $\partial w/\partial p = \partial w/\partial x \cdot \partial x/\partial p + \partial w/\partial y \cdot \partial y/\partial p + \partial w/\partial z \cdot \partial z/\partial p + \partial w/\partial t \cdot \partial t/\partial p$ **17.** a) Yes
b) Yes c) No   **21.** $v + uv + 1$ **23.** $2x + 4y^2$ **25.** 99   **27.** -2π **33.** $t = -2, 1$ **35.** a) Minimum at $\pi/4$ and $5\pi/4$;
maximum at $3\pi/4$ and $7\pi/4$ b) 6 and 2   **37.** $f_x = (\cos \theta)(\partial w/\partial r) - (\sin \theta/r)(\partial w/\partial\theta); f_y = ((\cos \theta)/r)(\partial w/\partial\theta) + (\sin \theta)(\partial w/\partial r)$

Article 16.5, pp. 890–891

1. a) 0 b) $1 + 2z$ c) $1 + 2z$ **3.** a) $\partial U/\partial p + (v/nR)(\partial U/\partial T)$ b) $(nR/v)(\partial U/\partial p) + \partial U/\partial T$ **5.** a) 5 b) 5   **7.** $(\partial x/\partial r)_\theta = \cos \theta$;
$(\partial r/\partial x)_y = x/r = \cos \theta$

Article 16.6, pp. 897–898

1. $2\mathbf{i} + 2\mathbf{j} - 4\mathbf{k}$ **3.** $-2\mathbf{i}$ **5.** $-(1/27)\mathbf{i} - (2/27)\mathbf{j} + (2/27)\mathbf{k}$ **7.** $(1/\sqrt{2})\mathbf{k}$ **9.** $\sqrt{2}$ **11.** $(2 - \pi)/\sqrt{17}$ **13.** $\sqrt{3} - 1$
15. 43/15   **17.** 2/3   **19.** Most rapid increase: $\mathbf{i}$, 2; most rapid decrease: $-\mathbf{i}$, -2 **21.** Most rapid increase: $-(3/\sqrt{10})\mathbf{i} +$
$(1/\sqrt{10})\mathbf{j}, 8\sqrt{10}$; most rapid decrease: $(3/\sqrt{10})\mathbf{i} - (1/\sqrt{10})\mathbf{j}, -8\sqrt{10}$ **23.** Most rapid increase: $(1/\sqrt{216})(10\mathbf{i} + 4\mathbf{j} + 10\mathbf{k})$,
$6\sqrt{6}$; most rapid decrease: $-(1/\sqrt{216})(10\mathbf{i} + 4\mathbf{j} + 10\mathbf{k}), -6\sqrt{6}$ **25.** $3\sqrt{2}/10$ **27.** 1/15   **29.** $-(12/13)\mathbf{i} + (5/13)\mathbf{j}, (12/13)\mathbf{i} - (5/13)\mathbf{j}$ **31.** $-7/\sqrt{5}$ **33.** $\pi/\sqrt{2}$ **37.** $\nabla w = (\partial w/\partial\rho)\mathbf{u}_\rho + (1/\rho)(\partial w/\partial\phi)\mathbf{u}_\phi + (1/\rho \sin \phi)(\partial w/\partial\theta)\mathbf{u}_\theta$

Article 16.7, p. 901

1. a) $x + y + z = 3$ b) $x = 1 + t$, $y = 1 + t$, $z = 1 + t$ **3.** a) $3x + 4y + 5z = 0$ b) $x = 3 - 6t$, $y = 4 - 8t$, $z = -5 - 10t$
5. a) $2x - z = 2$ b) $x = 1 - 2t$, $y = 0$, $z = t$ **7.** a) $2y + 3z = 7$ b) $x = 1$, $y = -1 + 4t$, $z = 3 + 6t$ **9.** a) $x - 3y - z = -1$
b) $x = 1 + t$, $y = 1 - 3t$, $z = -1 - t$ **11.** a) $(\pi/2)x + z = \pi/2$ b) $x = 1 - (\pi/2)t$, $y = 0$, $z = -t$ **13.** a) $9x - 7y - z = 21$
b) $x = 2 + 9t$, $y = -3 - 7t$, $z = 18 - t$ **15.** a) $x + 2z = 2$ b) $x = t$, $y = 0$, $z = 1 + 2t$ **17.** a) $x - y = 0$ b) $x = t$, $y = -t$,
$z = 0$ **19.** $\nabla f = 2\mathbf{i} + 2\mathbf{j} - \mathbf{k}$

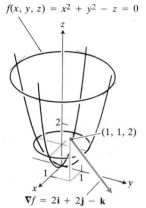

$f(x, y, z) = x^2 + y^2 - z = 0$

$(1, 1, 2)$

$\nabla f = 2\mathbf{i} + 2\mathbf{j} - \mathbf{k}$

21. $\nabla f = -2\mathbf{i} - 2\mathbf{j} + \mathbf{k}$ **23.** $\nabla f = -2\mathbf{i} + 2\sqrt{3}\mathbf{j}$ **25.** $\nabla f = 12\mathbf{j} + 18\sqrt{3}\mathbf{k}$ **27.** $\nabla f = 2\mathbf{i} + 2\mathbf{j} + 2\mathbf{k}$ **29.** $x = 1$, $y = 1 + 2t$, $z = 1 - 2t$ **31.** $x = 1 - 2t$, $y = 1$, $z = 1/2 + 2t$ **33.** $x = 1 + 90t$, $y = 1 - 90t$, $z = 3$
35. Normal: $2\sqrt{2}\mathbf{i} + 2\sqrt{2}\mathbf{j}$; $x + y = 2\sqrt{2}$

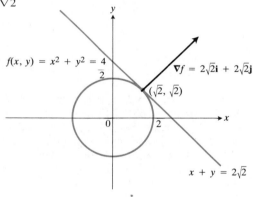

$f(x, y) = x^2 + y^2 = 4$

$\nabla f = 2\sqrt{2}\mathbf{i} + 2\sqrt{2}\mathbf{j}$

$(\sqrt{2}, \sqrt{2})$

$x + y = 2\sqrt{2}$

37. Normal: $16\mathbf{i} + 12\sqrt{5}\mathbf{j}$; $4x + 3\sqrt{5}y = 18$ **39.** Normal: $-2\mathbf{i} + 2\mathbf{j}$; $x - y = 4$ **41.** Two lines: $x = t$, $y = \pm 4 - t$, $z = t$

Article 16.8, pp. 907–908

1. $f_{xx} = 2(y^2 - x^2)/(x^2 + y^2)^2$, $f_{yy} = 2(x^2 - y^2)/(x^2 + y^2)^2$, $f_{yx} = (-4xy)/(x^2 + y^2)^2$ **3.** $f_{xx} = 2y - y \sin x$, $f_{yy} = -\cos y$, $f_{yx} = 2x + \cos x$ **5.** $f_{xx} = 0$, $f_{yy} = 0$, $f_{xy} = z$ **11.** a) x first b) y first c) x first d) x first e) y first f) y first **13.** $f_{uu} + uf_{uv} + vf_{vv} + f_v$ **15.** $2f_x + 4s^2 f_{xx} + 4r^2 f_{yy} + 8rsf_{xy}$ **17.** $(-\sin t)f_x + (\cos^2 t)f_{xx} + 2f_y + 4t^2 f_{yy} + (4t \cos t)f_{xy}$ **19.** 2 **31.** a) $a = -c$, b arbitrary; $f(x, y) = bxy + c^2$ b) $c = -3a$, $b = -3d$, a and d arbitrary; $f(x, y) = ax^3 - 3dx^2 y - 3axy^2 + dy^3$ **39.** $w = e^{-c^2\pi^2 t} \sin \pi x$

Miscellaneous Problems, Chapter 16, pp. 909–913

1. The domain of $z = \sqrt{\sin(\pi xy)}$, $xy \ge 0$ consists of the set of points in the first and third quadrants for which $\sin(\pi xy) \ge 0$. These points make up the closed region between the coordinate axes and the hyperbola $xy = 1$, together with the alternate bands between the hyperbolas $xy = n$, n a positive integer: $2 \leq xy \leq 3$, $4 \leq xy \leq 5$, $\cdots$, $2k \leq xy \leq 2k + 1$, $\cdots$. See the figure.

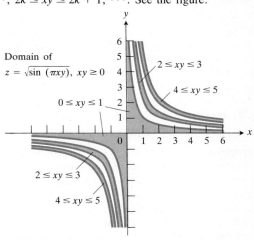

Domain of
$z = \sqrt{\sin (\pi xy)}$, $xy \geq 0$

$0 \leq xy \leq 1$

$2 \leq xy \leq 3$

$4 \leq xy \leq 5$

$2 \leq xy \leq 3$

$4 \leq xy \leq 5$

3. No, $\lim_{x \to 0} f(x, 0) = 1$, $\lim_{x \to 0} f(x, x) = 0$ **7.** a) $f_x = y \sin 2xy$, $f_y = x \sin 2xy$ b) $f_x = 2xy^2 \cos(x, y)^2$, $f_y = 2x^2y \cos(xy^2)$
11. a) 1 b) 0 c) 0 **13.** $\partial w/\partial x = (\cos \theta)(\partial w/\partial r) - ((\sin \theta)/r)(\partial w/\partial \theta)$; $\partial w/\partial y = (\sin \theta)(\partial w/\partial r) + ((\cos \theta)/r)(\partial w/\partial \theta)$
17. $\partial w/\partial r = \partial w/\partial s = 2/(r + s)$ **19.** $\partial F/\partial x = (1/2)(\partial f/\partial u + \partial f/\partial v)$; $\partial F/\partial y = -(1/2)(\partial f/\partial u - \partial f/\partial v)$ **23.** 1
25. $h'(x) = f_x(x, y) - (g_x(x, y)/g_y(x, y))f_y(x, y)$ **29.** $-14/\sqrt{6}$ **31.** 7 **33.** $\sqrt{3}$ **39.** $|\nabla \theta| = \sqrt{1/(x^2 + y^2)} = 1/r$;
direction is $\theta + (\pi/2)$ **43.** $(1/7)(2\mathbf{i} + 6\mathbf{j} - 3\mathbf{k})$ **47.** a) $[1/\sqrt{x^2 + y^2}][(x + 3x^3 + 3xy^2)\mathbf{i} + (y + 3y^3 + 3yx^2)\mathbf{j}] - \mathbf{k}$
b) $-1/\sqrt{(1 + 3(x^2 + y^2))^2 + 1}$; $-1/\sqrt{2}$
49. a)

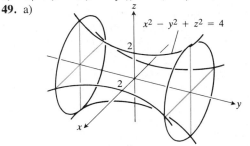

$x^2 - y^2 + z^2 = 4$

b) $4\mathbf{i} + 6\mathbf{j} + 6\mathbf{k}$ c) $2x + 3y + 3z = 4$; $x = 2 + 4t$, $y = -3 + 6t$, $z = 3 + 6t$

51. All points (a, b, c) such that $a^2 + b^2 + c^2 = 2$ **53.** $a = \pm\sqrt{3}$ **59.** $f_{uu} + (y^2 + 2xy)f_{vu} + 2yf_v + 2xy^3f_{vv}$ **61.** 2
65. $[\sin^2(y + z) + \sin(x + y)]/\cos^3(y + z)$ **67.** 0 **69.** $f_{yx}(0, 0) = 1$, $f_{xy}(0, 0) = -1$

CHAPTER 17

Article 17.1, pp. 923–924

1. Minimum of -5 at $(-3, 3)$ **3.** Saddle point at $(6/5, 69/25, -112/25)$ **5.** Saddle point at $(-2, 1, 3)$ **7.** Maximum of
$-28/9$ at $(4/9, 2/9)$ **9.** Minimum of 0 at $(1, -2)$ **11.** Minimum of -6 at $(2, -1)$ **13.** Saddle point at $(1/6, 4/3, -11/12)$
15. Saddle point at $(1, 2, 9)$ **17.** Saddle point at $(0, 0, 0)$ **19.** Minimum of -2 at $(-2, 3)$ **21.** Minimum of -3 at $(1, -2)$
23. Saddle point at $(0, 0, 6)$; maximum of $170/27$ at $(-2/3, 2/3)$. **25.** Saddle point at $(1, -1, 1)$; minimum of 0 at $(0, 0)$.
27. Saddle point at $(0, 0, 15)$; minimum of 14 at $(1, 1)$ **29.** Saddle point at $(0, 0, 0)$; maximum of 2 at $(1, 1)$ and $(-1, 1)$
31. Absolute minimum of -5 at $(1, 2)$, absolute maximum of 1 at $(0, 0)$ **33.** Absolute minimum of 0 at $(0, 0)$, absolute
maximum of 4 at $(0, 2)$ **35.** Absolute maximum of 11 at $(0, -3)$, absolute minimum of -10 at $(4, -2)$ **37.** Absolute
minimum of -4 at $(2, 0)$, absolute maximum of $-3\sqrt{2}/2$ at the corner points $(1, \pm\pi/4)$ and $(3, \pm\pi/4)$ **39.** Hottest points are
$(-1/2, \pm\sqrt{3}/2)$ with temperature of $9/4$, coldest point is $(1/2, 0)$ with temperature of $-1/4$ **41.** Absolute maximum of 4 at
$(1, 1)$, no absolute minimum **43.** Low point at $(0, 0, 0)$; high points at corners $(\pm1, \pm1, \sqrt{2})$; $\partial z/\partial x$ and $\partial y/\partial x$ do not exist at
origin but exist and are nonzero at the corners. **45.** a) (i) 2,0 (ii) 2, -2 (iii) 2, -2 b) (i) $2\sqrt{2}$, 2 (ii) $2\sqrt{2}$, -2
(iii) $2\sqrt{2}$, $-2\sqrt{2}$ c) (i) 8, 4 (ii) 8, 4 (iii) 8, 4 **47.** a) (i) 12, 9 (ii) 12, 9 (iii) 12, 9 b) (i) $6\sqrt{2}$, 6 (ii) $6\sqrt{2}$, -6
(iii) $6\sqrt{2}$, $-6\sqrt{2}$

Article 17.2, pp. 934–935

1. a) $L(x, y) = 1$ b) $L(x, y) = 2x + 2y - 1$ **3.** a) $L(x, y) = 1 + x$ b) $L(x, y) = (\pi/2) - y$ **5.** a) $L(x, y) = 3x - 4y + 5$
b) $L(x, y) = 3x - 4y + 5$ **7.** a) $L(x, y) = x - 6y + 7$, $|E| \le 0.06$ **9.** $L(x, y) = 1 + x + y$, $|E| \le 0.08$ **11.** $L(x, y) = 1 + x$,
$|E| \le 0.02$ **13.** $2h_0 = r_0$ **15.** a) To changes in x b) $dx/dy = -1/2$ **17.** 3% **19.** 0.002 m **21.** 0.1% **23.** a) $L(x, y, z) =$
$2x + 2y + 2z - 3$ b) $L(x, y, z) = y + z$ c) $L(x, y, z) = 0$ **25.** a) $L(x, y, z) = x$ b) $L(x, y, z) = (1/\sqrt{2})(x + y)$ c) $L(x, y, z) =$
$(1/3)(x + 2y + 2z)$ **27.** a) $L(x, y, z) = 2 + x$ b) $L(x, y, z) = 1 + x - [y - (\pi/2)] - z$ c) $L(x, y, z) = 1 + x - [y - (\pi/4)] -$
$[z - (\pi/4)]$ **29.** $L(x, y, z) = 2x - 6y - 2z + 6$; $|E| \le 0.0024$ **31.** $L(x, y, z) = 1 + x + y - z$; $|E| \le 0.00135$ **33.** a) $dS =$
$S_0(dx + dP/100 - 30dh - 5dw)$ b) More sensitive to a change in height **35.** 6% **37.** 338 ft^2 **39.** $dQ = -400dh + 10dK +$
dM; more sensitive to h

Article 17.3, pp. 946–948

1. Maximum at $(\pm1/\sqrt{2}, \pm1/2)$, minimum at $(-1/\sqrt{2}, 1/2)$ and $(1/\sqrt{2}, -1/2)$ **3.** Maximum of 39 at $(1, 3)$ **5.** Local
minimum of 0 at $(0, 3)$, local maximum of 4 at $(2, 1)$ **7.** a) Minimum of 8 at $(4, 4)$ b) Maximum of 64 at $(8, 8)$ **9.** $r =$
2 cm, $h = 4$ cm **11.** $T = 0$ at $(\sqrt{5}, 2\sqrt{5})$ or $(-\sqrt{5}, -2\sqrt{5})$, $T = 125$ at $(2\sqrt{5}, -\sqrt{5})$ or $(-2\sqrt{5}, \sqrt{5})$ **13.** Minimum of
0 at $(0, 0)$, maximum of 20 at $(2, 4)$ **15.** $(3/2, 2, 5/2)$ **17.** Minimum distance is 1 **19.** $(0, 0, \pm2)$ are each closest to the
origin. **21.** 50 **23.** $4096\sqrt{5}/125$ **25.** Maximum of 5 at $(0, 1)$, minimum of $-1/3$ at $(0, -1/3)$ **27.** Hottest points at
$(\pm4/3, -4/3, -4/3)$ **29.** Maximum of 128 at $(8, 14)$ **31.** $\sqrt{a_1^2 + a_2^2 + \cdots + a_n^2}$ **33.** $(1, 2, 5)$ **35.** Maximum of 2000 at
$(10, 10, 20)$ **37.** In substituting $y = x$ into the equation $z = 1 + x + y$, we are treating $y = x$ as if it were one of the
problem's original constraints. It isn't—it's a conclusion we draw about the coordinates of the points on C that lie closest to
and farthest from the origin.

Article 17.4, p. 954

1. $1 + x + (1/2)x^2 - (1/2)y^2$ **3.** $x^2 + y^2$ **5.** $1 + x + y + x^2 + 2xy + y^2$ **7.** $\cos x \cos y \approx 1 - (1/2)(x^2 + y^2)$; $|E| \le 0.0013$

Miscellaneous Problems, Chapter 17, pp. 955–956

1. Absolute maximum is 28; absolute minimum is $-9/4$. **3.** Absolute maximum is 18; absolute minimum is $-17/4$.
5. Absolute maximum is 8; absolute minimum is -1. **7.** Absolute maximum is 4; absolute minimum is -4. **9.** Absolute
maximum is 4; absolute minimum is -4. **11.** $(0, 0)$ is a saddle point; z has a minimum at $(3, 3)$. **13.** $(1, 1, 1)$, $(1, -1, -1)$,
$(-1, -1, 1)$, $(-1, 1, -1)$ **15.** $(\sqrt{3}abc)/2$

CHAPTER 18

Article 18.1, pp. 968–969

1. 16   **3.** 0   **5.** $(\pi^2 + 4)/2$ **7.** $\pi/4$

9. $9 - 9e$ **11.** $(3/2)\ln 2$ **13.** $-1/10$ **15.** $(\ln 2)^2$

17. $\int_0^4 \int_0^{(4-y)/2} dx\, dy = 4$ **19.** $\int_0^1 \int_{x^2}^x dy\, dx = 1/6$ **21.** $\int_1^{e^2} \int_{\ln y}^2 dx\, dy = e^2 - 3$

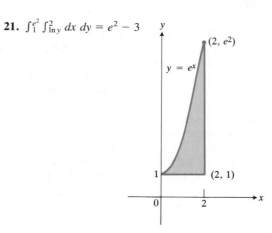

23. $\int_{-2}^{2} \int_{0}^{\sqrt{(4-x^2)/2}} y \, dy \, dx = 8/3$ **25.** $\int_{0}^{1} \int_{y}^{\sqrt{y}} f(x, y) \, dx \, dy$ **27.** $\int_{1}^{e} \int_{\ln y}^{1} dx \, dy$ **29.** $\int_{0}^{8} \int_{\sqrt[3]{y}}^{2} f(x, y) \, dx \, dy$ **31.** 2   **33.** $(e - 2)/2$

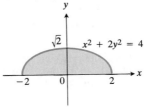

35. $(1/4)\ln 17$ **37.** $625/12$ **39.** 16   **41.** $e - 1$ **43.** $(\pi a^3)/2$

Article 18.2, pp. 974–976

1. $a^2/2$ **3.** 4   **5.** $1/3$ **7.** $2 + (\pi/2)$

9. $1/6$ **11.** $\sqrt{2} - 1$ **13.** $3/2$

        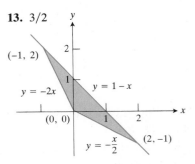

15. $(\bar{x}, \bar{y}) = (5/14, 38/35)$ **17.** $(\bar{x}, \bar{y}) = (64/35, 5/7)$ **19.** $(\bar{x}, \bar{y}) = (-8/5, -1/2); I_x = 189, I_y = 6345/7, I_0 = 7668/7; R_x = \sqrt{7/10}, R_y = \sqrt{47/14}, R_0 = \sqrt{142/35}$ **21.** $I_x = \delta \, bh^3/12, R_x = h/\sqrt{6}$ **23.** $\bar{x} = \bar{y} = 4a/3\pi$ **25.** $(\bar{x}, \bar{y}) = (\pi/2, \pi/8)$
27. $\frac{4}{3}ab(a^2 + b^2)$ **29.** $(\bar{x}, \bar{y}) = (0, 7/10); I_x = 9/10, I_y = 3/10, I_0 = 6/5; R_x = 3\sqrt{6}/10, R_y = 3\sqrt{2}/10, R_0 = 3\sqrt{2}/5$
31. $(\bar{x}, \bar{y}) = (3/8, 17/16)$ **33.** $(\bar{x}, \bar{y}) = (8/5, 8/15); I_x = 1/6$ **35.** $(\bar{x}, \bar{y}) = (11/3, 14/27); I_y = 432, R_y = 4$
37. $(\bar{x}, \bar{y}) = (0, 13/31); I_y = 7/5, R_y = \sqrt{21/31}$ **39.** $(\bar{x}, \bar{y}) = (100/9, 0); I_x = 20, R_x = \sqrt{1/3}$ **41.** $a = 1$ **43.** a) $(u_1 + u_2)/3$

Article 18.3, pp. 984–985

1. $\pi/2$ **3.** $\pi/8$ **5.** πa^2 **7.** $a^3\sqrt{2}/6$ **9.** $3 \ln|2 + \sqrt{3}|$ **11.** $5\pi/6$ **13.** $\pi/2 + 1$ **15.** $(\pi - 1)/2$ **17.** $\pi/12$ **19.** $(3\pi/8) + 1$
21. $\pi \ln 4$ **23.** 4   **25.** $(5/6, 0)$ **27.** $35\pi/16$ **29.** $(32 + 15\pi)/24$ **33.** a) $1/3$ b) $33/4$ **35.** $8 + (52/3)\ln 2$
37. $(\pi ab/4)(a^2 + b^2)$

Article 18.4, pp. 991–992

1. 1   **3.** $\int_{0}^{1} \int_{0}^{2-2x} \int_{0}^{(6-3y-6x)/2} dz \, dy \, dx, \int_{0}^{1} \int_{0}^{3-3x} \int_{0}^{(6-2z-6x)/3} dy \, dz \, dx, \int_{0}^{3} \int_{0}^{(6-2z)/3} \int_{0}^{(6-3y-2z)/6} dx \, dy \, dz,$
$\int_{0}^{3} \int_{0}^{(3-z)/3} \int_{0}^{(6-2z-6x)/3} dy \, dx \, dz, \int_{0}^{2} \int_{0}^{(2-y)/2} \int_{0}^{(6-3y-6x)/3} dz \, dx \, dy, \int_{0}^{2} \int_{0}^{(6-3y)/2} \int_{0}^{(6-3y-2z)/6} dx \, dz \, dy = 1$ **5.** 1   **7.** 1
9. $(\pi^3/2)(1 - \cos 1)$ **11.** $4\pi/3$ **13.** $16/3$ **15.** $\pi/2$ **17.** 0   **19.** $31/3$ **21.** 1   **23.** π **25.** $2/3$ **27.** $2/3$ **29.** $2/7$ **31.** 32π
33. $abc/6$ **35.** $16a^3/3$ **37.** $4\pi abc/3$ **39.** a) $\int_{-1}^{1} \int_{0}^{1-x^2} \int_{x^2}^{1-z} dy \, dz \, dx$ b) $\int_{0}^{1} \int_{-\sqrt{1-z}}^{\sqrt{1-z}} \int_{x^2}^{1-z} dy \, dx \, dz$ c) $\int_{0}^{1} \int_{0}^{1-z} \int_{-\sqrt{y}}^{\sqrt{y}} dx \, dy \, dz$
d) $\int_{0}^{1} \int_{0}^{1-y} \int_{-\sqrt{y}}^{\sqrt{y}} dx \, dz \, dy$ e) $\int_{0}^{1} \int_{-\sqrt{y}}^{\sqrt{y}} \int_{0}^{1-y} dz \, dx \, dy$

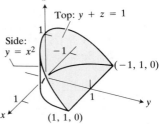

Article 18.5, pp. 994–996

1. $R_x = \sqrt{(b^2 + c^2)/12}, R_y = \sqrt{(a^2 + c^2)/12}, R_z = \sqrt{(a^2 + b^2)/12}$ **3.** $I_x = (M/3)(b^2 + c^2), I_y = (M/3)(a^2 + c^2), I_z = (M/3)(a^2 + b^2)$ **5.** $(\bar{x}, \bar{y}, \bar{z}) = (0, 0, 12/5); I_x = (7904/105)\delta; I_y = (4832/63)\delta; I_z = (256/45)\delta$ **7.** a) $(\bar{x}, \bar{y}, \bar{z}) = (0, 0, 8/3)$
b) $c = 2\sqrt{2}$ **9.** $I_L = 1386; R_L = 77\sqrt{2}/2$ **11.** $I_L = 40/3; R_L = \sqrt{5/3}$ **13.** $(\bar{x}, \bar{y}, \bar{z}) = (4/5, 2/5, 2/5)$ **15.** $(\bar{x}, \bar{y}, \bar{z}) = (8/15, 8/15, 8/15); I_x = I_y = I_z = 11/6; R_x = R_y = R_z = \sqrt{11/15}$

Article 18.6, pp. 1004–1007

1. $2\pi(2\sqrt{2}-2)/3$ **3.** $17\pi/5$ **5.** $2\pi(3\sqrt{2}-4)$ **7.** π^2 **9.** $\pi/3$ **11.** 5π **13.** a) $\int_0^{2\pi}\int_0^{\pi}\int_0^2 \rho^2\sin\phi\,d\rho\,d\phi\,d\theta$
b) $\int_0^{2\pi}\int_0^2\int_{-\sqrt{4-r^2}}^{\sqrt{4-r^2}} r\,dz\,dr\,d\theta$ c) $\int_{-2}^2\int_{-\sqrt{4-x^2}}^{\sqrt{4-x^2}}\int_{-\sqrt{4-x^2-y^2}}^{\sqrt{4-x^2-y^2}} dz\,dy\,dx$ **15.** a) $\int_{-1}^1\int_{-\sqrt{1-x^2}}^{\sqrt{1-x^2}}\int_0^{\sqrt{4-x^2-y^2}}\sqrt{x^2+y^2}\,dz\,dy\,dx$
b) $\int_0^{2\pi}\int_0^{\pi/6}\int_0^2 \rho^3\sin^2\phi\,d\rho\,d\phi\,d\theta + \int_0^{2\pi}\int_{\pi/6}^{\pi/2}\int_0^{\csc\phi}\rho^3\sin^2\phi\,d\rho\,d\phi\,d\theta$ **17.** $\int_{-\pi/2}^{\pi/2}\int_0^{\cos\theta}\int_0^{3r^2} f(r,\theta,z)\,dz\,r\,dr\,d\theta$
19. a) $\int_0^{2\pi}\int_0^1\int_0^{\sqrt{1-r^2}} r^3\,dz\,dr\,d\theta$ b) $\int_0^{2\pi}\int_0^{\pi/2}\int_0^1 \rho^4\sin^3\phi\,d\rho\,d\phi\,d\theta$ **21.** $\pi/2$ **23.** 8π **25.** $5\pi/2$ **27.** $(4\pi a^3/3)(8-3\sqrt{3})$
29. $52\pi/3$ **31.** $2/3$ **33.** $(0, 0, 3/8)$ **35.** $I_z = 30\pi; R_z = \sqrt{10}/2$ **37.** $(0, 0, 7a/2(8\sqrt{2}-7))$ **39.** $\pi/4$ **41.** $(9\sqrt{3}/32, 0, 0)$
43. a) $(0, 0, 4/5); I_z = \pi/12, R_z = \sqrt{1/3}$ b) $(0, 0, 5/6); I_z = \pi/14, R_z = \sqrt{5/14}$ **45.** $(2a^3\pi)/3$ **47.** $5\pi/3$ **49.** $8\pi/3$
51. $3/4$ **53.** $(0, 0, 9a/16)$ **55.** $I_z = (8\pi\delta a^5)/15, R_z = a\sqrt{2/5}$ **59.** $(4/3)\pi abc$ **61.** $2 + 3\ln 2$

Miscellaneous Problems, Chapter 18, pp. 1007–1010

1. $\int_{-2}^2\int_{2x+4}^{4-x^2} dy\,dx = 4/3$ **3.** $\ln(b/a)$ **7.** $\int_0^1\int_{-\sqrt{y}}^{\sqrt{y}} dx\,dy$ **9.** $\int_0^1\int_x^{2-x} (x^2+y^2)\,dy\,dx$ **11.** $104/3$ **13.** 1

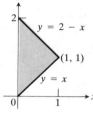

17. a) $\sin(\ln 2)[e^{\pi/2}-1]$ b) 0 **19.** $(28\sqrt{2}/9\pi, 28(2-\sqrt{2})/9\pi)$ **21.** $\int_{-\pi/2}^{\pi/2}\int_0^a (r\cos\theta)r\,dr\,d\theta$ **25.** $R_x = (a/4)\sqrt{(3\pi-8)/3}$,
$R_y = (a/4)\sqrt{(3\pi+8)/3}$ **27.** $K(a) = \int_0^{a\cos\beta}\int_0^{x\tan\beta}\ln(x^2+y^2)\,dy\,dx + \int_{a\cos\beta}^a\int_0^{\sqrt{a^2-x^2}}\ln(x^2+y^2)\,dy\,dx$ **29.** $(\pi-2)/4$
31. $(0, 0, 1/4)$ **33.** $\pi/4$ **35.** $(1/2)(122-25\sqrt{5})$ **37.** $[(\delta\pi^2 a^2 b)/2](3a^2+4b^2)$ **39.** $\int_0^1\int_{\sqrt{1-x^2}}^{\sqrt{3-x^2}}\int_1^{\sqrt{4-x^2-y^2}} xyz^2\,dz\,dy\,dx +$
$\int_1^{\sqrt{3}}\int_0^{\sqrt{3-x^2}}\int_1^{\sqrt{4-x^2-y^2}} xyz^2\,dz\,dy\,dx$ **41.** $(2a^3/9)(3\pi-4)$ **43.** $\pi a^5/12$ **45.** $(8\pi/15)(b^5-a^5)$ **47.** $(2\pi/15)(128-51\sqrt{3})$
49. $(28\pi a^5)/15$

CHAPTER 19

Article 19.1, pp. 1016–1018

1. (c) **3.** (g) **5.** (d) **7.** (f) **9.** $\sqrt{2}$ **11.** $-\sqrt{2}$ **13.** $13/2$ **15.** $2\sqrt{2}$ **17.** $7/3$ **19.** $1/2$ **21.** $(5\sqrt{5}+9)/6$ **23.** $2\pi-2, 1$
25. $I_z = 2\pi\delta a^3 = a^2 M; R_z = a$ **27.** $L^2 M/3$ **29.** $(0, 2a/\pi); I_x = \pi a^3\delta/2; R_x = a/\sqrt{2}$ **31.** Center of mass: $(0, -3/5, 9/8)$;
$I_x = 192/35; I_y = 64/15; I_z = 128/105; R_x = \sqrt{72/35}; R_y = \sqrt{8/5}; R_z = \sqrt{16/35}$

Article 19.2, p. 1026

1. (a) $9/2$ (b) $9/2$ (c) $9/2$ **3.** (a) $\pi/4$ (b) $\ln 2$ (c) $\pi/4$ **5.** (a) 1 (b) $17/18$ (c) $5/6$ **7.** (a) 3 (b) 3 (c) 3 **9.** $1/2$ **11.** $-\pi$
13. 48 **15.** π **17.** (a) $-(\pi/2)$ (b) 0 (c) 1 **19.** Circulation $= 0$; flux $= \pi a^2$ **21.** Circulation $= \pi a^2$; flux $= 0$
23. (a) Circulation of $\mathbf{F}_1 = 0$, flux $= 8\pi$; circulation of $\mathbf{F}_2 = 8\pi$, flux $= 0$ (b) Circulation of $\mathbf{F}_1 = 0$, flux $= 8\pi$; circulation of
$\mathbf{F}_2 = 8\pi$, flux $= 0$

Article 19.3, pp. 1038–1039

1. Flux $= 0$, circulation $= 2\pi a^2$ **3.** Flux $= -\pi a^2$, circulation $= 0$ **5.** Circulation $= 2$; flux $= 2$ **7.** Circulation $= 0$; flux $= 1$
9. Circulation $= 1$; flux $= 1$ **11.** Circulation $= -1/12$; flux $= 1/5$ **13.** 0 **15.** -16π **17.** $2/33$ **19.** (a) 0 (b) $(h-k)A$, where
A is the area of the region bounded by C **27.** πa^2 **29.** $3a^2\pi/8$

Article 19.4, pp. 1047–1048

1. $13\pi/3$ **3.** 4 **5.** $6\sqrt{6}-2\sqrt{2}$ **7.** $(2-\sqrt{2})\pi$ **9.** $\pi\sqrt{c^2+1}$ **11.** $(\pi/6)(5\sqrt{5}-1)$ **13.** 9 **15.** $(abc/4)(ab+bc+ac)$
17. 2 **19.** $2\pi ah(h^2+3a^2)/3$ **21.** $\pi a^3/6$ **23.** $\pi a^2/4$ **25.** $\pi a^3/2$ **27.** -32 **29.** -4 **31.** 128π **33.** $(1/2, 1/2, 1/2)$
35. $(0, 0, 14/9); I_z = 15\sqrt{2}\pi/2; R_z = \sqrt{5/2}$ **37.** (a) $\pi^2 a^3$ (Use $\delta = \sqrt{x^2+y^2} = r$ and Formula 33 from the integral table.)
(b) $8\pi a^4/3$ (Use $\delta = x^2 + y^2 = r^2$.)

Article 19.5, pp. 1057–1058

1. 16 **3.** 0 **5.** 8π **7.** 3π **9.** $16/3$ **11.** $4\pi a^4$ **13.** 5

Article 19.6, pp. 1066–1067

1. 4π **3.** $-5/6$ **5.** 0 **7.** 4π **9.** $2\pi a^2$

Article 19.7, pp. 1075–1076

1. Conservative **3.** Not conservative **5.** Not conservative **7.** $f(x,y,z) = x^2 + \frac{3}{2}y^2 + 2z^2 + C$ **9.** $f(x,y,z) = xe^{y+2z} + C$ **11.** 49
13. 5/6 **15.** 1 **17.** 45/2 **19.** $8 + \ln 2$ **21.** -3 **23.** 24 **25.** $\oint_C \mathbf{F} \cdot d\mathbf{R} = 0$

Miscellaneous Problems, Chapter 19, pp. 1077–1080

1. (a) $2\sqrt{3}/3$ (b) $(5\sqrt{2} + 3)/3$ **3.** $I_z = 7\delta\sqrt{3}/3$; $R_z = \sqrt{7/3}$ **5.** $\bar{z} = 32\sqrt{2}/105$; $I_z = 7/12$; $R_z = \sqrt{7/18}$ **11.** (a) $\mathbf{F} = \rho^{-(n+1)}\mathbf{R}$
(b) $\mathbf{F} = -(2/\rho)\mathbf{R}$ (c) $\mathbf{F} = -GmM\,\mathbf{R}/\rho^4$ **13.** $(2\pi/3)(3\sqrt{3} - 1)$ **15.** $\frac{1}{2}\sqrt{a^2b^2 + a^2c^2 + b^2c^2}$ **17.** $4\int_0^a (1 + 4y^2)^{1/2}\,(a^2 - y^2)^{1/2}\,dy$
19. $4\pi^2 ab$ **21.** 8 **23.** $4 - 8\pi$ **25.** (a) $(\pi/6)(17\sqrt{17} - 1)$ (b) $(\pi/6)(5\sqrt{5} - 1)$ (c) $\sqrt{5} + (1/2)\ln(\sqrt{5} + 2)$ **27.** a **39.** 0
41. 0 **43.** -3

CHAPTER 20

Article 20.2, pp. 1092–1094

1. $|2y - 3| = C/(x^2 + 1)$ **3.** $e^y = e^x + C$ **5.** $\sinh 2y - 2\cos x = C$ **7.** $e^y(y - 1) + x^2/2 + \ln|x| = C$ **9.** $\cosh^{-1}y + \sinh^{-1}x = C$ **11.** $10\ln 10$ years **13.** 48.2 minutes **15.** $y = x/(\ln|x| + C)$ **17.** $(1/2)\ln(x^2 + y^2) + \tan^{-1}(y/x) = C$ **19.** $\sin(y/x) = C/x$
21. $y = Cx^2$, $x^2 + 2y^2 = C$ **23.** $x^2 - y^2 = C$ **25.** $y = (1/2)e^{x/2}(x + C)$ **27.** $y = (1/x)(C - \cos x)$ **29.** $y = (x^3 - 3x + C)/$
$3(x - 1)^4$ **31.** $x = e^{-2y}(y^2 + C)$ **33.** $x = (C - y)/(y^2 + 1)$ **35.** $x(t) = -20312.5 - 625t + 21312.5e^{0.08t}$ **37.** a) $xy = C$
b) $xy = C$ **39.** $x^2/2 + xy + y^3/3 = C$ **41.** $x^2y + xy^2 - y^2/2 = C$ **43.** $y = Cx - x^3/2$ **45.** $y = Cx - x^2 - x\ln|x|$
47. $y^2\tan^{-1}x + \cosh y - 2xy = C$ **49.** $c = 2b$; $ax^3/3 + bxy^2 = C$ **51.** $y = -1 \pm e^{1-\sin x}$ **53.** $y = (1/2)x + (5/4)e^{-2x} - 1/4$
55. $y = \tan(1 - e^x + \pi/4)$ **57.** $x^2 + 4xy + y^2 = 6$ **59.** $y = (1/5)x^3 - 12/5x^2$

Article 20.3, p. 1098

1. $y = C_1 e^{-x} + C_2$ **3.** $y = C_1 \int e^{-x^2/2}\,dx + C_2$ (The integral cannot be calculated explicitly as a finite combination of elementary
functions.) **5.** $y = C_1 \sinh(x + C_2)$ **7.** $y = C_1 x^4 + C_2 x + C_3$ **9.** a) $y = 5\sin x$ b) $y = 2\sin x$ **11.** $x = mgt/k +$
$(m^2g/k^2)(e^{-kt/m} - 1)$ **13.** $y = C_1 x^{-1} + C_2 x^2$ **15.** $y = \sqrt{x}(C_1\ln|x| + C_2)$

Article 20.4, pp. 1103–1104

1. $y = C_1 + C_2 e^{-2x}$ **3.** $y = C_1 e^{-x} + C_2 e^{-5x}$ **5.** $y = e^{-x/2}(C_1\cos(\sqrt{3}/2)x + C_2\sin(\sqrt{3}/2)x)$ **7.** $y = (C_1 + C_2 x)e^{-3x}$ **9.** $y =$
$e^x(C_1\cos\sqrt{3}x + C_2\sin\sqrt{3}x)$ **11.** $y = -(1/2)e^x + (3/2)e^{-x}$ **13.** $y = -(2/3)e^{-x/2} - (1/3)e^x$ **15.** $y = (3/4)(e^{2x} - e^{-2x})$
17. $y = \sin 2x$

Article 20.5, pp. 1112–1113

1. $y = C_1 + C_2 e^{-x} + (1/2)x^2 - x$ **3.** $y = C_1\cos x + C_2\sin x - (1/2)x\cos x$ **5.** $y = C_1 e^{-x} + C_2 xe^{-x} + (1/2)x^2 e^{-x}$ **7.** $y =$
$C_1 e^x + C_2 e^{-x} + xe^x/2$ **9.** $y = e^{-2x}(C_1\cos x + C_2\sin x) + 2$ **11.** $y = C_1\cos x + C_2\sin x + \cos x \ln|\cos x| + x\sin x$ **13.** $y =$
$C_1 e^{5x} + C_2 e^{-2x} + 3/10$ **15.** $y = C_1 + C_2 e^x + (\cos x - \sin x)/2$ **17.** $y = C_1\cos x + C_2\sin x - (1/8)\cos 3x$ **19.** $y = C_1 e^{2x} +$
$C_2 e^{-x} - 6\cos x - 2\sin x$ **21.** $y = C_1 e^x + C_2 e^{-x} + xe^x/2 - x^2 - 2$ **23.** $y = C_1 e^{3x} + C_2 e^{-2x} - e^{-x}/4 + (49\cos x + 7\sin x)/50$
25. $y = C_1 + C_2 e^{-5x} + x^3 - 3x^2/5 + 6x/25$ **27.** $y = C_1 + C_2 e^{3x} + xe^{3x}/3 + 2x^2 + 4x/3$ **29.** $y = C_1 + C_2 e^{5x} + (x^2/10 -$
$x/25)e^{5x}$ **31.** $y = C_1\cos x + C_2\sin x + x\sin x - (x\cos x)/2$ **33.** $y = C_1 + C_2 e^x + xe^x + e^{-x}/2$ **35.** $y = C_1 e^{5x} + C_2 e^{-x} -$
$e^x/8 - 4/5$ **37.** $y = C_1\cos x + C_2\sin x - \sin x \ln|\csc x + \cot x|$ **39.** $y = C_1 + C_2 e^{8x} + xe^{8x}/8$ **41.** $y = C_1 + C_2 e^x - x^4/4 -$
$x^3 - 3x^2 - 6x$ **43.** $y = C_1 + C_2 e^{-2x} + x^3/6 - x^2/4 + x/4 - e^x/3$ **45.** $y = C_1\cos x + C_2\sin x + x\cos x + \sin x \ln|\sec x|$
47. $y = Ce^{3x} - e^x/2$ **49.** $y = Ce^{3x} + 5xe^{3x}$ **51.** $y = 2\cos x + \sin x - 1 + \sin x \ln|\sec x + \tan x|$ **53.** $y = 1 - e^{-x}$

Article 20.6, pp. 1117–1118

1. $x = x_0\cos \omega t + (v_0/\omega)\sin \omega t$; $x = \sqrt{x_0^2 + (v_0^2/\omega^2)}\,\sin(\omega t + \tan^{-1}(x_0\omega/v_0))$ **3.** a) $i = C_1\cos \omega t + C_2\sin \omega t$ b) $i = C_1\cos \omega t +$
$C_2\sin \omega t + \alpha V/L(\omega^2 - \alpha^2)\cos \alpha t$ c) $i = C_1\cos \omega t + C_2\sin \omega t + (V/2L)t\sin \omega t$ d) $i = e^{-5t}(C_1\cos 149t + C_2\sin 149t)$ **5.** $\theta =$
$\theta_0\cos \omega t + (v_0/\omega)\sin \omega t$, $\omega = \sqrt{2k/mr^2}$ **7.** a) $x = C_1\cos \sqrt{k/m}t + C_2\sin \sqrt{k/m}t + [kA/(k - m\alpha^2)]\sin \alpha t$ b) $x =$
$C_1\cos \sqrt{k/m}t + C_2\sin \sqrt{k/m}t - \sqrt{A^2k/4m}t\cos \alpha t$

Article 20.7, p. 1120

1. $y = C_1 e^x + C_2 e^{2x} + C_3$ **3.** $y = (C_1 x + C_2)e^{\sqrt{2}x} + (C_3 x + C_4)e^{-\sqrt{2}x}$ **5.** $y = e^{\sqrt{2}x}(C_1\cos \sqrt{2}x + C_2\sin \sqrt{2}x) +$
$e^{-\sqrt{2}x}(C_3\cos \sqrt{2}x + C_4\sin \sqrt{2}x)$ **7.** $y = (C_1 + C_2 x + C_3 x^2 + C_4 x^3)e^x + 7$

Article 20.8, p. 1124

3. $y = \sum_{n=0}^{\infty} (-1)^n x^n/n!$ **5.** $y = \sum_{n=0}^{\infty} (-1)^{n+1}(2^n/n!)(x - 1)^n$ **7.** $y = \sum_{n=0}^{\infty} (-1)^n x^{2n}/(2n)!$ **9.** $y = 3\sum_{n=0}^{\infty} [x^{2n+1}/(2n + 1)!] -$
$\sum_{n=0}^{\infty} (x^{2n}/(2n)!) - x$ **11.** $y = \sum_{n=0}^{\infty} a_n x^n$, where $a_0 = y_0$, $a_1 = y_1$, $a_2 = 0$, $a_3 = 0$, $a_n = [1/n(n - 1)]a_{n-4}$ for $n \geq 4$

Article 20.9, p. 1129

1. The isoclines are the vertical lines $x = $ constant.   **3.** The isoclines are the vertical lines $x = $ constant.

5. The isoclines are the hyperbolas $xy = $ constant.

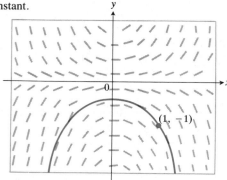

7. The isoclines are the diagonal lines $x + y = $ constant. $y = \tan(x + C) - x$. Passes through $(0, 0)$ when $C = 0$.
9. a) $y = x - 1$ b) Concave up for $y > x - 1$; concave down for $y < x - 1$ **11.** $y' = x - y$; $y(1) = -1$

13. $y' = -(1 + y)\sin x$; $y(0) = 2$
15. $y_0 = 2$; $y_1 = (1/2)x^2 + 3/2$; $y_2 = (1/2)x^2 + 3/2$;
$y_3 = (1/2)x^2 + 3/2$
17. $y_0 = 1$; $y_1 = (1/2)x^2 + 1/2$; $y_2 = 5/8 + (1/4)x^2 + (1/8)x^4$;
$y_3 = 29/48 + (5/16)x^2 + (1/16)x^4 + (1/48)x^6$
19. $y_0 = 1$; $y_1 = 1 + x + (1/2)x^2$; $y_2 = 1 + x + x^2 + (1/6)x^3$;
$y_3 = 1 + x + x^2 + (1/3)x^3 + (1/24)x^4$

Article 20.10, pp. 1134–1135

1. $y(1) \approx 2.488320$; exact value is e.   **3.** $y(1) \approx 2.702708$ **5.** $1 + x + x^2 + (4/3)x^3 + (7/6)x^4$

Miscellaneous Problems, Chapter 20, p. 1136

1. $\ln|\ln y| = C - \tan^{-1}x$ **3.** $y = \ln(C - e^{-x}) - 2$ **5.** $\sinh^{-1}y^2 = 2x + C$ **7.** $y^2 = x^2 - Cx$ **9.** $y = x\tan^{-1}(\ln|x| + C)$
11. $y = 1/2 + x^2/4 + C/x^2$ **13.** $y = \cosh x[\ln(\cosh x) + C]$ **15.** $3x + 3xy^2 + 3y + y^3 = C$ **17.** $x^3 + 3xy^2 + 3\sinh y = C$
19. $x^2(e^y + 1) + e^y(y^2 - 2y + 2) = C$ **21.** $y = -(x + C)^{-1}$ or $y = C_1\tan(C_1x + C_2)$ or $(1/2C_1)\ln|(y - C_1)/(y + C_1)| = x + C_2$
23. $y = -\ln|\cos(x + C_1)| + C_2$ **25.** $y = (1/2)\ln^2|x| + C_1\ln|x| + C_2$ **27.** $y = C_1 + (C_2 + C_3x)e^x$ **29.** $y = [C_1 + (1/3)x]e^{2x} + C_2e^{-x}$ **31.** $y = C_1\sqrt{x} + C_2/\sqrt{x}$ **33.** $3x^2 + 2y^2 = C$ **35.** $y^2 = 2Bx + B^2$ **37.** $y = e^x - x - 1$; $y = \sum_{n=2}^{\infty} (x^n/n!)$
39. a) $\pi/2 + x + x^2/2! - x^3/3! - 3x^4/4!$ b) $-\pi/2 - x + x^2/2! + x^3/3! - 3x^4/4!$

Appendix A.1, p. A-5

1. If $F_1(t) \to L_1$, $F_2(t) \to L_2$, . . . , $F_n(t) \to L_n$, as $t \to c$, then $F_1(t) + F_2(t) + \cdots + F_n(t) \to L_1 + L_2 + \cdots + L_n$ as $t \to c$, for any positive integer n.

Appendix A.5, p. A-18

1. $c = -1$ **3.** $\dfrac{-1 + \sqrt{37}}{3}$

Appendix A.7, p. A-28–A-29

1. -5 **3.** 1 **5.** a) -7 b) -7 **7.** a) 38 b) 38 **9.** $x = -4$, $y = 1$ **11.** $x = 3$, $y = 2$ **13.** $x = 3$, $y = -2$, $z = 2$ **15.** $x = 2$, $y = 0$, $z = -1$ **17.** a) $h = 6$, $k = 4$ b) $h = 6$, $k \neq 4$

Appendix A.8, pp. A-40–A-42

1. a) $\begin{bmatrix} 2 & -3 & 4 \\ 6 & 4 & -2 \\ 1 & 5 & 4 \end{bmatrix} \begin{bmatrix} x \\ y \\ z \end{bmatrix} = \begin{bmatrix} -19 \\ 8 \\ 23 \end{bmatrix}$ **5.** $x = -4$, $y = 1$ **7.** $x = 3$, $y = 2$ **9.** $x = 3$, $y = -2$, $z = 2$

11. $x = 2$, $y = 0$, $z = -1$ **15.** $x = -4$, $y = 1$ **17.** a) $\begin{bmatrix} 1 & -2 & 1 \\ 0 & \frac{1}{4} & -\frac{1}{2} \\ 0 & 0 & \frac{1}{3} \end{bmatrix}$ b) $x = -20$, $y = 15/2$, $z = -10/3$

19. $\lambda = 3$ or $\lambda = 2$; $\begin{bmatrix} 1 \\ 0 \end{bmatrix}, \begin{bmatrix} 0 \\ 1 \end{bmatrix}$ **21.** $\lambda = 0$ or $\lambda = 2$; $\begin{bmatrix} 1 \\ -1 \end{bmatrix}, \begin{bmatrix} 1 \\ 1 \end{bmatrix}$ **23.** $\lambda = 0$; $\begin{bmatrix} 1 \\ -1 \end{bmatrix}$

25. $\lambda = 1 \pm \sqrt{19}$; $\begin{bmatrix} 2 \\ \sqrt{19} - 3 \end{bmatrix}, \begin{bmatrix} 2 \\ -\sqrt{19} - 3 \end{bmatrix}$

27. $\lambda = 2$; $\begin{bmatrix} 1 \\ -1 \\ 1 \end{bmatrix}$

29. $\lambda = 0$ or $\lambda = \pm 2$; $\begin{bmatrix} 0 \\ 1 \\ -2 \end{bmatrix}, \begin{bmatrix} -2 \\ 1 \\ 2 \end{bmatrix}, \begin{bmatrix} 2 \\ 1 \\ 2 \end{bmatrix}$

Appendix A.11, p. A-57

1. a) $(14, 8)$ b) $(-1, 8)$ c) $(0, -5)$ **11.** a) Circle with center at origin, radius $= 2$ b) Interior of circle in (a) c) Exterior of circle in (a)   **13.** Circle with center $(-1, 0)$, radius $= 1$ **15.** Points on the line $y = -x$.   **17.** $e^{(\pi/2)i}$ **19.** $\sqrt{65}\, e^{i\,\tan^{-1}(-0.125)}$

21. $1, -\dfrac{1}{2} \pm \dfrac{\sqrt{3}}{2}i$ **23.** $2i, -\sqrt{3} - i, \sqrt{3} - i$ **25.** $\dfrac{\sqrt{6}}{2} \pm \dfrac{\sqrt{2}}{2}i, -\dfrac{\sqrt{6}}{2} \pm \dfrac{\sqrt{2}}{2}i$ **27.** $1 \pm \sqrt{3}i, -1 \pm \sqrt{3}i$

Index

Note. Numbers in parentheses refer to problems on the pages indicated.

65. $\displaystyle\int \frac{\cos ax}{\sin ax}\,dx = \frac{1}{a}\ln|\sin ax| + C$

66. $\displaystyle\int \cos^n ax \sin ax\,dx = -\frac{\cos^{n+1} ax}{(n+1)a} + C, \qquad n \neq -1$

67. $\displaystyle\int \frac{\sin ax}{\cos ax}\,dx = -\frac{1}{a}\ln|\cos ax| + C$

68. $\displaystyle\int \sin^n ax \cos^m ax\,dx = -\frac{\sin^{n-1} ax \cos^{m+1} ax}{a(m+n)} + \frac{n-1}{m+n}\int \sin^{n-2} ax \cos^m ax\,dx,$

$\qquad\qquad\qquad\qquad n \neq -m \qquad$ (If $n = -m$, use No. 86.)

69. $\displaystyle\int \sin^n ax \cos^m ax\,dx = \frac{\sin^{n+1} ax \cos^{m-1} ax}{a(m+n)} + \frac{m-1}{m+n}\int \sin^n ax \cos^{m-2} ax\,dx,$

$\qquad\qquad\qquad\qquad m \neq -n \qquad$ (If $m = -n$, use No. 87.)

70. $\displaystyle\int \frac{dx}{b + c\sin ax} = \frac{-2}{a\sqrt{b^2 - c^2}}\tan^{-1}\left[\sqrt{\frac{b-c}{b+c}}\tan\left(\frac{\pi}{4} - \frac{ax}{2}\right)\right] + C, \qquad b^2 > c^2$

71. $\displaystyle\int \frac{dx}{b + c\sin ax} = \frac{-1}{a\sqrt{c^2 - b^2}}\ln\left|\frac{c + b\sin ax + \sqrt{c^2 - b^2}\cos ax}{b + c\sin ax}\right| + C, \qquad b^2 < c^2$

72. $\displaystyle\int \frac{dx}{1 + \sin ax} = -\frac{1}{a}\tan\left(\frac{\pi}{4} - \frac{ax}{2}\right) + C$

73. $\displaystyle\int \frac{dx}{1 - \sin ax} = \frac{1}{a}\tan\left(\frac{\pi}{4} + \frac{ax}{2}\right) + C$

74. $\displaystyle\int \frac{dx}{b + c\cos ax} = \frac{2}{a\sqrt{b^2 - c^2}}\tan^{-1}\left[\sqrt{\frac{b-c}{b+c}}\tan\frac{ax}{2}\right] + C, \qquad b^2 > c^2$

75. $\displaystyle\int \frac{dx}{b + c\cos ax} = \frac{1}{a\sqrt{c^2 - b^2}}\ln\left|\frac{c + b\cos ax + \sqrt{c^2 - b^2}\sin ax}{b + c\cos ax}\right| + C, \qquad b^2 < c^2$

76. $\displaystyle\int \frac{dx}{1 + \cos ax} = \frac{1}{a}\tan\frac{ax}{2} + C$ $\qquad$ 77. $\displaystyle\int \frac{dx}{1 - \cos ax} = -\frac{1}{a}\cot\frac{ax}{2} + C$

78. $\displaystyle\int x\sin ax\,dx = \frac{1}{a^2}\sin ax - \frac{x}{a}\cos ax + C$ $\qquad$ 79. $\displaystyle\int x\cos ax\,dx = \frac{1}{a^2}\cos ax + \frac{x}{a}\sin ax + C$

80. $\displaystyle\int x^n \sin ax\,dx = -\frac{x^n}{a}\cos ax + \frac{n}{a}\int x^{n-1}\cos ax\,dx$

81. $\displaystyle\int x^n \cos ax\,dx = \frac{x^n}{a}\sin ax - \frac{n}{a}\int x^{n-1}\sin ax\,dx$

82. $\displaystyle\int \tan ax\,dx = \frac{1}{a}\ln|\sec ax| + C$ $\qquad$ 83. $\displaystyle\int \cot ax\,dx = \frac{1}{a}\ln|\sin ax| + C$

84. $\displaystyle\int \tan^2 ax\,dx = \frac{1}{a}\tan ax - x + C$ $\qquad$ 85. $\displaystyle\int \cot^2 ax\,dx = -\frac{1}{a}\cot ax - x + C$

86. $\displaystyle\int \tan^n ax\,dx = \frac{\tan^{n-1} ax}{a(n-1)} - \int \tan^{n-2} ax\,dx, \qquad n \neq 1$

87. $\displaystyle\int \cot^n ax\,dx = -\frac{\cot^{n-1} ax}{a(n-1)} - \int \cot^{n-2} ax\,dx, \qquad n \neq 1$

88. $\displaystyle\int \sec ax\,dx = \frac{1}{a}\ln|\sec ax + \tan ax| + C$ $\qquad$ 89. $\displaystyle\int \csc ax\,dx = -\frac{1}{a}\ln|\csc ax + \cot ax| + C$

Continued overleaf.

90. $\int \sec^2 ax\,dx = \dfrac{1}{a}\tan ax + C$

91. $\int \csc^2 ax\,dx = -\dfrac{1}{a}\cot ax + C$

92. $\int \sec^n ax\,dx = \dfrac{\sec^{n-2} ax \tan ax}{a(n-1)} + \dfrac{n-2}{n-1}\int \sec^{n-2} ax\,dx, \quad n \neq 1$

93. $\int \csc^n ax\,dx = -\dfrac{\csc^{n-2} ax \cot ax}{a(n-1)} + \dfrac{n-2}{n-1}\int \csc^{n-2} ax\,dx, \quad n \neq 1$

94. $\int \sec^n ax \tan ax\,dx = \dfrac{\sec^n ax}{na} + C, \quad n \neq 0$

95. $\int \csc^n ax \cot ax\,dx = -\dfrac{\csc^n ax}{na} + C, \quad n \neq 0$

96. $\int \sin^{-1} ax\,dx = x\sin^{-1} ax + \dfrac{1}{a}\sqrt{1-a^2x^2} + C$

97. $\int \cos^{-1} ax\,dx = x\cos^{-1} ax - \dfrac{1}{a}\sqrt{1-a^2x^2} + C$

98. $\int \tan^{-1} ax\,dx = x\tan^{-1} ax - \dfrac{1}{2a}\ln(1+a^2x^2) + C$

99. $\int x^n \sin^{-1} ax\,dx = \dfrac{x^{n+1}}{n+1}\sin^{-1} ax - \dfrac{a}{n+1}\int \dfrac{x^{n+1}\,dx}{\sqrt{1-a^2x^2}}, \quad n \neq -1$

100. $\int x^n \cos^{-1} ax\,dx = \dfrac{x^{n+1}}{n+1}\cos^{-1} ax + \dfrac{a}{n+1}\int \dfrac{x^{n+1}\,dx}{\sqrt{1-a^2x^2}}, \quad n \neq -1$

101. $\int x^n \tan^{-1} ax\,dx = \dfrac{x^{n+1}}{n+1}\tan^{-1} ax - \dfrac{a}{n+1}\int \dfrac{x^{n+1}\,dx}{1+a^2x^2}, \quad n \neq -1$

102. $\int e^{ax}\,dx = \dfrac{1}{a}e^{ax} + C$

103. $\int b^{ax}\,dx = \dfrac{1}{a}\dfrac{b^{ax}}{\ln b} + C, \quad b > 0, \ b \neq 1$

104. $\int xe^{ax}\,dx = \dfrac{e^{ax}}{a^2}(ax-1) + C$

105. $\int x^n e^{ax}\,dx = \dfrac{1}{a}x^n e^{ax} - \dfrac{n}{a}\int x^{n-1} e^{ax}\,dx$

106. $\int x^n b^{ax}\,dx = \dfrac{x^n b^{ax}}{a\ln b} - \dfrac{n}{a\ln b}\int x^{n-1} b^{ax}\,dx, \quad b > 0, \ b \neq 1$

107. $\int e^{ax}\sin bx\,dx = \dfrac{e^{ax}}{a^2+b^2}(a\sin bx - b\cos bx) + C$

108. $\int e^{ax}\cos bx\,dx = \dfrac{e^{ax}}{a^2+b^2}(a\cos bx + b\sin bx) + C$

109. $\int \ln ax\,dx = x\ln ax - x + C$

110. $\int x^n \ln ax\,dx = \dfrac{x^{n+1}}{n+1}\ln ax - \dfrac{x^{n+1}}{(n+1)^2} + C, \quad n \neq -1$

111. $\int x^{-1}\ln ax\,dx = \dfrac{1}{2}(\ln ax)^2 + C$

112. $\int \dfrac{dx}{x\ln ax} = \ln|\ln ax| + C$

113. $\int \sinh ax\,dx = \dfrac{1}{a}\cosh ax + C$

114. $\int \cosh ax\,dx = \dfrac{1}{a}\sinh ax + C$

115. $\int \sinh^2 ax\,dx = \dfrac{\sinh 2ax}{4a} - \dfrac{x}{2} + C$

116. $\int \cosh^2 ax\,dx = \dfrac{\sinh 2ax}{4a} + \dfrac{x}{2} + C$

117. $\int \sinh^n ax\,dx = \dfrac{\sinh^{n-1} ax \cosh ax}{na} - \dfrac{n-1}{n}\int \sinh^{n-2} ax\,dx, \quad n \neq 0$